Central Intelligence Agency

The World Factbook 1995

The printed version of the Factbook is published annually in July by the Central Intelligence Agency for the use of US Government officials, and the style, format, coverage, and content are designed to meet their specific requirements. Information was provided by the American Geophysical Union, Bureau of the Census, Central Intelligence Agency, Defense Intelligence Agency, Defense Mapping Agency, Defense Nuclear Agency, Department of State, Foreign Broadcast Information Service, Maritime Administration, National Science Foundation (Polar Information Program), Naval Maritime Intelligence Center, Office of Territorial and International Affairs, US Board on Geographic Names, US Coast Guard, and others.

Comments and queries are welcome and may be addressed to:

Central Intelligence Agency
Attn.: Office of Public and Agency Information
Washington, DC 20505
Telephone: [1] (703) 351-2053

Contents

Notes, Definitions, and Abbreviations

There have been some significant changes in this edition. The Trust Territory of the Pacific Islands became the independent nation of Palau. The gross domestic product (GDP) of all countries is now presented on a purchasing power parity (PPP) basis rather than on the old exchange rate basis. There is a new entry on Age structure and the Airports entry now includes unpaved runways. The Communications category has been restructured and now includes the entries of Telephone system, Radio, and Television. The remainder of the entries in the former Communications category—Railroads, Highways, Inland waterways, Pipelines, Ports, Merchant marine, and Airports—can now be found under a new category called Transportation. There is a new appendix listing estimates of gross domestic product on an exchange rate basis for all nations. A reference map of the Republic of South Africa is included. The electronic files used to produce the Factbook have been restructured into a database. As a result, the formats of some entries in this edition have been changed. Additional changes will occur in the 1996 Factbook.

Abbreviations: (see Appendix B for abbreviations for international organizations and groups and Appendix D for abbreviations for selected international environmental agreements)

avdp.	avoirdupois
c.i.f.	cost, insurance, and freight
CY	calendar year
DWT	deadweight ton
est.	estimate
Ex-Im	Export-Import Bank of the United States
f.o.b.	free on board
FRG	Federal Republic of Germany (West Germany); used for information dated before 3 October 1990 or CY91
FSU	former Soviet Union
FY	fiscal year (FY93/94, for example, began in calendar year 1993 and ended in calendar year 1994)
FYROM	The Former Yugoslav Republic of Macedonia
GDP	gross domestic product
GDR	German Democratic Republic (East Germany); used for information dated before 3 October 1990 or CY91
GNP	gross national product
GRT	gross register ton
GWP	gross world product
km	kilometer
kW	kilowatt
kWh	kilowatt hour
m	meter
NA	not available
NEGL	negligible
nm	nautical mile
NZ	New Zealand
ODA	official development assistance
OOF	other official flows
PDRY	People's Democratic Republic of Yemen [Yemen (Aden) or South Yemen]; used for information dated before 22 May 1990 or CY91
sq km	square kilometer
sq mi	square mile
UAE	United Arab Emirates
UK	United Kingdom
US	United States
USSR	Union of Soviet Socialist Republics (Soviet Union); used for information dated before 25 December 1991
YAR	Yemen Arab Republic [Yemen (Sanaa) or North Yemen]; used for information dated before 22 May 1990 or CY91

Administrative divisions: The numbers, designatory terms, and first-order administrative divisions are generally those approved by the US Board on Geographic Names (BGN). Changes that have been reported but not yet acted on by BGN are noted.

Airports: Only airports with usable runways are included in this listing. For airports with more than one runway, only the longest runway is included. Not all airports have facilities for refueling, maintenance, or air traffic control. Paved runways have concrete or asphalt surfaces; unpaved runways have grass, dirt, sand, or gravel surfaces.

Area: Total area is the sum of all land and water areas delimited by international boundaries and/or coastlines. Land area is the aggregate of all surfaces delimited by international boundaries and/or coastlines, excluding inland water bodies (lakes, reservoirs, rivers). Comparative areas are based on total area equivalents. Most entities are compared with the entire US or one of the 50 states. The smaller entities are compared with Washington, DC (178 sq km, 69 sq mi) or The Mall in Washington, DC (0.59 sq km, 0.23 sq mi, 146 acres).

Birth rate: The average annual number of births during a year per 1,000 population at midyear; also known as crude birth rate.

Dates of information: In general, information available as of 1 January 1995 is used in the preparation of this edition. Population figures are estimates for 1 July 1995, with population growth rates estimated for calendar year 1995. Major political events have been updated through April 1995.

Death rate: The average annual number of deaths during a year per 1,000 population at midyear; also known as crude death rate.

Digraphs: The digraph is a two-letter "country code" that precisely identifies every entity without overlap, duplication, or omission. AF, for example, is the digraph for Afghanistan. It is a standardized geopolitical data element promulgated in the Federal Information Processing Standards Publication (FIPS) 10-3 by the National Bureau of Standards (now called National Institute of Standards and Technology) at the US Department of Commerce and maintained by the Office of the Geographer at the US Department of State. The digraph is used to eliminate confusion and incompatibility in the collection, processing, and dissemination of area-specific data and is particularly useful for interchanging data between databases.

Diplomatic representation: The US Government has diplomatic relations with 184 nations, including 178 of the 185 UN members (excluded UN members are Bhutan, Cuba, Iran, Iraq, North Korea, former Yugoslavia, and the US itself). In addition, the US has diplomatic relations with 6 nations that are not in the UN—Holy See, Kiribati, Nauru, Switzerland, Tonga, and Tuvalu.

Economic aid: This entry refers to bilateral commitments of official development assistance (ODA) and other official flows (OOF). ODA is defined as financial assistance which is concessional in character, has the main objective to promote economic development and welfare of LDCs, and contains a grant element of at least 25%. OOF transactions are also official government assistance, but with a main objective other than development and with a grant element less than 25%. OOF transactions include official export credits (such as Ex-Im Bank credits), official equity and portfolio investment, and debt reorganization by the official sector that does not meet concessional terms. Aid is considered to have been committed when agreements are initialed by the parties involved and constitute a formal declaration of intent.

Entities: Some of the nations, dependent areas, areas of special sovereignty, and governments included in this publication are not independent, and others are not officially recognized by the US Government. "Nation" refers to a people politically organized into a sovereign state with a definite territory. "Dependent area" refers to a broad category of political entities that are associated in some way with a nation. Names used for page headings are usually the short-form names as approved by the US Board on Geographic Names. There are 266 entities in *The World Factbook* that may be categorized as follows:

NATIONS

184 UN members (excluding the former Yugoslavia, which is still counted by the UN)

7 nations that are not members of the UN—Holy See, Kiribati, Nauru, Serbia and Montenegro, Switzerland, Tonga, Tuvalu

OTHER

1 Taiwan

DEPENDENT AREAS

6 Australia—Ashmore and Cartier Islands, Christmas Island, Cocos (Keeling) Islands, Coral Sea Islands, Heard Island and McDonald Islands, Norfolk Island

2 Denmark—Faroe Islands, Greenland

16 France—Bassas da India, Clipperton Island, Europa Island, French Guiana, French Polynesia, French Southern and Antarctic Lands, Glorioso Islands, Guadeloupe, Juan de Nova Island, Martinique, Mayotte, New Caledonia, Reunion, Saint Pierre and Miquelon, Tromelin Island, Wallis and Futuna

2 Netherlands—Aruba, Netherlands Antilles

3 New Zealand—Cook Islands, Niue, Tokelau

3 Norway—Bouvet Island, Jan Mayen, Svalbard

1 Portugal—Macau

16 United Kingdom—Anguilla, Bermuda, British Indian Ocean Territory, British Virgin Islands, Cayman Islands, Falkland Islands, Gibraltar, Guernsey, Hong Kong, Jersey, Isle of Man, Montserrat, Pitcairn Islands, Saint Helena, South Georgia and the South Sandwich Islands, Turks and Caicos Islands

14 United States—American Samoa, Baker Island, Guam, Howland Island, Jarvis Island, Johnston Atoll, Kingman Reef, Midway Islands, Navassa Island, Northern Mariana Islands, Palmyra Atoll, Puerto Rico, Virgin Islands, Wake Island

MISCELLANEOUS

6 Antarctica, Gaza Strip, Paracel Islands, Spratly Islands, West Bank, Western Sahara

OTHER ENTITIES

4 oceans—Arctic Ocean, Atlantic Ocean, Indian Ocean, Pacific Ocean

1 World

266 total

Exchange rate: The official value of a nation's monetary unit at a given date or over a given period of time, as expressed in units of local currency per US dollar and as determined by international market forces or official fiat.

GDP methodology: In the "Economy" section, GDP dollar estimates for all countries are derived from purchasing power parity (PPP) calculations rather than from conversions at official currency exchange rates. The PPP method normally involves the use of international dollar price weights, which are applied to the quantities of goods and services produced in a given economy. In addition to the lack of reliable data from the majority of countries, the statistician faces a major difficulty in specifying, identifying, and allowing for the quality of goods and services. The division of a GDP estimate in local currency by the corresponding PPP estimate in dollars gives the PPP conversion rate. On average, one thousand dollars will buy the same market basket of goods in the US as one thousand dollars—converted to the local currency at the PPP conversion rate—will buy in the other country. Whereas PPP estimates for OECD countries are quite reliable, PPP estimates for developing countries are often rough approximations. Most of the GDP estimates are based on extrapolation of numbers published by the UN International Comparison Program and by Professors Robert Summers and Alan Heston of the University of Pennsylvania and their colleagues. Currency exchange rates depend on a variety of international and domestic financial forces that often have little relation to domestic output. In developing countries with weak currencies the exchange rate estimate of GDP in dollars is typically one-fourth to one-half the PPP estimate. Furthermore, exchange rates may suddenly go up or down by 10% or more because of market forces or official fiat whereas real output has remained unchanged. On 12 January 1994, for example, the 14 countries of the African Financial Community (whose currencies are tied to the French franc) devalued their currencies by 50%. This move, of course, did not cut the real output of these countries by half. One important caution: the proportion of, say, defense expenditures as a percentage of GDP in local currency accounts may differ substantially from the proportion when GDP accounts are expressed in PPP terms, as, for example, when an observer tries to estimate the dollar level of Russian or Japanese military expenditures. Note—the numbers for GDP and other economic data can **not** be chained together from successive volumes of the Factbook because of changes in the US dollar measuring rod, revisions of data by statistical agencies, use of new or different sources of information, and changes in national statistical methods and practices.

Gross domestic product (GDP): The value of all final goods and services produced within a nation in a given year.

Gross national product (GNP): The value of all final goods and services produced within a nation in a given year, plus income earned abroad, minus income earned by foreigners from domestic production.

Gross world product (GWP): The aggregate value of all goods and services produced worldwide in a given year.

Growth rate (population): The annual percent change in the population, resulting from a surplus (or deficit) of births over deaths and the balance of migrants entering and leaving a country. The rate may be positive or negative.

Illicit drugs: There are five categories of illicit drugs—narcotics, stimulants, depressants (sedatives), hallucinogens, and cannabis. These categories include many drugs legally produced and prescribed by doctors as well as those illegally produced and sold outside medical channels.

Cannabis (Cannabis sativa) is the common hemp plant, which provides hallucinogens with some sedative properties, and includes marijuana (pot, Acapulco gold, grass, reefer), tetrahydrocannabinol (THC, Marinol), hashish (hash), and hashish oil (hash oil).

Coca (Erythroxylum coca) is a bush, and the leaves contain the stimulant used to make cocaine. Coca is not to be confused with cocoa, which comes from cacao seeds and is used in making chocolate, cocoa, and cocoa butter.

Cocaine is a stimulant derived from the leaves of the coca bush.

Depressants (sedatives) are drugs that reduce tension and anxiety and include chloral hydrate, barbiturates (Amytal, Nembutal, Seconal, phenobarbital), benzodiazepines (Librium, Valium), methaqualone (Quaalude), glutethimide (Doriden), and others (Equanil, Placidyl, Valmid).

Drugs are any chemical substances that effect a physical, mental, emotional, or behavioral change in an individual.

Drug abuse is the use of any licit or illicit chemical substance that results in physical, mental, emotional, or behavioral impairment in an individual.

Hallucinogens are drugs that affect sensation, thinking, self-awareness, and emotion. Hallucinogens include LSD (acid, microdot), mescaline and peyote (mexc, buttons, cactus), amphetamine variants (PMA, STP, DOB), phencyclidine (PCP, angel dust, hog), phencyclidine analogues (PCE, PCPy, TCP), and others (psilocybin, psilocyn).

Hashish is the resinous exudate of the cannabis or hemp plant (Cannabis sativa).

Heroin is a semisynthetic derivative of morphine.

Mandrax is the Southwest Asian slang term for methaqualone, a pharmaceutical depressant.

Marijuana is the dried leaves of the cannabis or hemp plant (Cannabis sativa).

Methaqualone is a pharmaceutical depressant, in slang referred to as Quaaludes in North America or Mandrax in Southwest Asia

Narcotics are drugs that relieve pain, often induce sleep, and refer to opium, opium derivatives, and synthetic substitutes. Natural narcotics include opium (paregoric, parepectolin), morphine (MS-Contin, Roxanol), codeine (Tylenol with codeine, Empirin with codeine, Robitussan AC), and thebaine. Semisynthetic narcotics include heroin (horse, smack), and hydromorphone (Dilaudid). Synthetic narcotics include meperidine or Pethidine (Demerol, Mepergan), methadone (Dolophine, Methadose), and others (Darvon, Lomotil).

Opium is the milky exudate of the incised, unripe seedpod of the opium poppy.

Opium poppy (Papaver somniferum) is the source for many natural and semisynthetic narcotics.

Poppy straw concentrate is the alkaloid derived from the mature dried opium poppy.

Qat (kat, khat) is a stimulant from the buds or leaves of catha edulis that is chewed or drunk as tea.

Quaaludes is the North American slang term for methaqualone, a pharmaceutical depressant.

Stimulants are drugs that relieve mild depression, increase energy and activity, and include cocaine (coke, snow, crack), amphetamines (Desoxyn, Dexedrine), phenmetrazine (Preludin), methylphenidate (Ritalin), and others (Cylert, Sanorex, Tenuate).

Infant mortality rate: The number of deaths to infants under one year old in a given year per 1,000 live births occurring in the same year.

International disputes: This category includes a wide variety of situations that range from traditional bilateral boundary disputes to unilateral claims of one sort or another. Information regarding disputes over international boundaries and maritime boundaries has been reviewed by the Department of State. References to other situations involving borders or frontiers may also be included, such as resource disputes, geopolitical questions, or irredentist issues. However, inclusion does not necessarily constitute official acceptance or recognition by the US Government.

Irrigated land: The figure refers to the land area that is artificially supplied with water.

Land use: The land surface is categorized as arable land—land cultivated for crops that are replanted after each harvest (wheat, maize, rice); permanent crops—land cultivated for crops that are not replanted after each harvest (citrus, coffee, rubber); meadows and pastures—land permanently used for herbaceous forage crops; forest and woodland—under dense or open stands of trees; and other—any land type not specifically mentioned above (urban areas, roads, desert).

Leaders: The chief of state is the titular leader of the country who represents the state at official and ceremonial functions but is not involved with the day-to-day activities of the government. The head of government is the administrative leader who manages the day-to-day activities of the government. In the UK, the monarch is the chief of state, and the Prime Minister is the head of government. In the US, the President is both the chief of state and the head of government.

Life expectancy at birth: The average number of years to be lived by a group of people all born in the same year, if mortality at each age remains constant in the future.

Literacy: There are no universal definitions and standards of literacy. Unless otherwise noted, all rates are based on the most common definition—the ability to read and write at a specified age. Detailing the standards that individual countries use to assess the ability to read and write is beyond the scope of this publication.

Maritime claims: The proximity of neighboring states may prevent some national claims from being extended the full distance.

Merchant marine: All ships engaged in the carriage of goods. All commercial vessels (as opposed to all nonmilitary ships), which excludes tugs, fishing vessels, offshore oil rigs, etc. Also, a grouping of merchant ships by nationality or register.

Captive register—A register of ships maintained by a territory, possession, or colony primarily or exclusively for the use of ships owned in the parent country; also referred to as an offshore register, the offshore equivalent of an internal register. Ships on a captive register will fly the same flag as the parent country, or a local variant of it, but will be subject to the maritime laws and taxation rules of the offshore territory. Although the nature of a captive register makes it especially desirable for ships owned in the parent country, just as in the internal register, the ships may also be owned abroad. The captive register then acts as a flag of convenience register, except that it is not the register of an independent state.

Flag of convenience register—A national register offering registration to a merchant ship not owned in the flag state. The major flags of convenience (FOC) attract ships to their registers by virtue of low fees, low or nonexistent taxation of profits, and liberal manning requirements. True FOC registers are characterized by having relatively few of the ships registered actually owned in the flag state. Thus, while virtually any flag can be used for ships under a given set of circumstances, an FOC register is one where the majority of the merchant fleet is owned abroad. It is also referred to as an open register.

Flag state—The nation in which a ship is registered and which holds legal jurisdiction over operation of the ship, whether at home or abroad. Flag state maritime legislation determines how a ship is manned and taxed and whether a foreign-owned ship may be placed on the register.

Internal register—A register of ships maintained as a subset of a national register. Ships on the internal register fly the national flag and have that nationality but are subject to a separate set of maritime rules from those on the main national register. These differences usually include lower taxation of profits, manning by foreign nationals, and, usually, ownership outside the flag state (when it functions as an FOC register). The Norwegian International Ship Register and Danish International Ship Register are the most notable examples of an internal register. Both have been instrumental in stemming flight from the national flag to flags of convenience and in attracting foreign owned ships to the Norwegian and Danish flags.

Merchant ship—A vessel that carries goods against payment of freight; commonly used to denote any nonmilitary ship but accurately restricted to commercial vessels only.

Register—The record of a ship's ownership and nationality as listed with the maritime authorities of a country; also, the compendium of such individual ships' registrations. Registration of a ship provides it with a nationality and makes it subject to the laws of the country in which registered (the flag state) regardless of the nationality of the ship's ultimate owner.

Money figures: All money figures are expressed in contemporaneous US dollars unless otherwise indicated.

National product: The total output of goods and services in a country in a given year. See GDP methodology, Gross domestic product (GDP), and Gross national product (GNP).

Net migration rate: The balance between the number of persons entering and leaving a country during the year per 1,000 persons (based on midyear population). An excess of persons entering the country is referred to as net immigration (3.56 migrants/1,000 population); an excess of persons leaving the country as net emigration (-9.26 migrants/1,000 population).

Population: Figures are estimates from the Bureau of the Census based on statistics from population censuses, vital statistics registration systems, or sample surveys pertaining to the recent past, and on assumptions about future trends. Starting with the 1993 Factbook, demographic estimates for some countries (mostly African) have taken into account the effects of the growing incidence of AIDS infections; in 1993 these countries were Burkina, Burundi, Central African Republic, Congo, Cote d'Ivoire, Kenya, Malawi, Rwanda, Tanzania, Uganda, Zaire, Zambia, Zimbabwe, Thailand, Brazil, and Haiti.

Telephone numbers: All telephone numbers presented in the Factbook consist of the country code in brackets, the city or area code (where required) in parentheses, and the local number. The one component that is **not** presented is the international access code which varies from country to country. For example, an international direct dial phone call placed from the United States to Madrid, Spain, would be as follows:

011 [34] (1) 577-xxxx where
011 is the international access code for station-to-station calls
(01 is for calls other than station-to-station calls),
[34] is the country code for Spain,
(1) is the city code for Madrid,
577 is the local exchange,
and xxxx is the local telephone number.

An international direct dial phone call placed from another country to the United States would be as follows:

international access code + [1] (202) 939-xxxx where
[1] is the country code for the United States,
(202) is the area code for Washington, DC,
939 is the local exchange,
and xxxx is the local telephone number.

Total fertility rate: The average number of children that would be born per woman if all women lived to the end of their childbearing years and bore children according to a given fertility rate at each age.

Years: All year references are for the calendar year (CY) unless indicated as fiscal year (FY). FY93/94 refers to the fiscal year that began in calendar year 1993 and ended in calendar year 1994 as defined in the Fiscal Year entry of the Economy section for each nation. FY90-94 refers to the four fiscal years that began in calendar year 1990 and ended in calendar year 1994.

Note: Information for the US and US dependencies was compiled from material in the public domain and does not represent Intelligence Community estimates. The *Handbook of International Economic Statistics*, published annually in September by the Central Intelligence Agency, contains detailed economic information for the Organization for Economic Cooperation and Development (OECD) countries, Eastern Europe, the newly independent republics of the former nations of Yugoslavia and the Soviet Union, and selected other countries. The Handbook can be obtained wherever *The World Factbook* is available.

Afghanistan

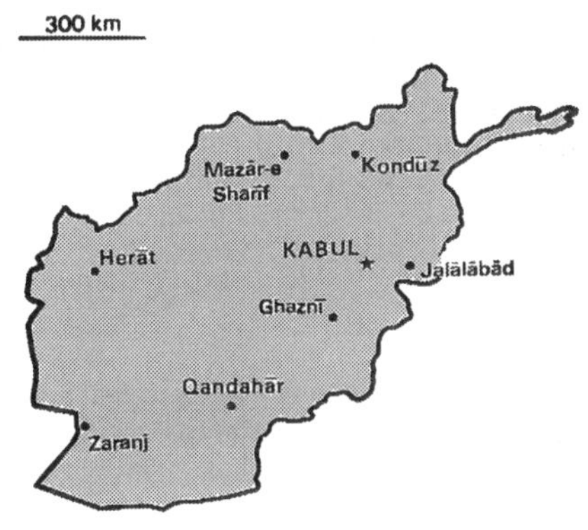

Geography

Location: Southern Asia, north of Pakistan
Map references: Asia
Area:
total area: 647,500 sq km
land area: 647,500 sq km
comparative area: slightly smaller than Texas
Land boundaries: total 5,529 km, China 76 km, Iran 936 km, Pakistan 2,430 km, Tajikistan 1,206 km, Turkmenistan 744 km, Uzbekistan 137 km
Coastline: 0 km (landlocked)
Maritime claims: none; landlocked
International disputes: periodic disputes with Iran over Helmand water rights; Iran supports clients in country, private Pakistani and Saudi sources also are active; power struggles among various groups for control of Kabul, regional rivalries among emerging warlords, traditional tribal disputes continue; support to Islamic fighters in Tajikistan's civil war; border dispute with Pakistan (Durand Line); support to Islamic militants worldwide by some factions
Climate: arid to semiarid; cold winters and hot summers
Terrain: mostly rugged mountains; plains in north and southwest
Natural resources: natural gas, petroleum, coal, copper, talc, barites, sulphur, lead, zinc, iron ore, salt, precious and semiprecious stones
Land use:
arable land: 12%
permanent crops: 0%
meadows and pastures: 46%
forest and woodland: 3%
other: 39%
Irrigated land: 26,600 sq km (1989 est.)
Environment:
current issues: soil degradation; overgrazing; deforestation (much of the remaining forests are being cut down for fuel and building materials); desertification
natural hazards: damaging earthquakes occur in Hindu Kush mountains; flooding
international agreements: party to—Endangered Species, Environmental Modification, Marine Dumping, Nuclear Test Ban; signed, but not ratified—Biodiversity, Climate Change, Hazardous Wastes, Law of the Sea, Marine Life Conservation
Note: landlocked

People

Population: 21,251,821 (July 1995 est.)
Age structure:
0-14 years: 42% (female 4,342,218; male 4,507,141)
15-64 years: 56% (female 5,406,675; male 6,443,734)
65 years and over: 2% (female 256,443; male 295,610) (July 1995 est.)
Population growth rate: 14.47% (1995 est.)
Birth rate: 42.69 births/1,000 population (1995 est.)
Death rate: 18.53 deaths/1,000 population (1995 est.)
Net migration rate: 120.5 migrant(s)/1,000 population (1995 est.)
Infant mortality rate: 152.8 deaths/1,000 live births (1995 est.)
Life expectancy at birth:
total population: 45.37 years
male: 45.98 years
female: 44.72 years (1995 est.)
Total fertility rate: 6.21 children born/woman (1995 est.)
Nationality:
noun: Afghan(s)
adjective: Afghan
Ethnic divisions: Pashtun 38%, Tajik 25%, Uzbek 6%, Hazara 19%, minor ethnic groups (Chahar Aimaks, Turkmen, Baloch, and others)
Religions: Sunni Muslim 84%, Shi'a Muslim 15%, other 1%
Languages: Pashtu 35%, Afghan Persian (Dari) 50%, Turkic languages (primarily Uzbek and Turkmen) 11%, 30 minor languages (primarily Balochi and Pashai) 4%, much bilingualism
Literacy: age 15 and over can read and write (1990 est.)
total population: 29%
male: 44%
female: 14%
Labor force: 4.98 million
by occupation: agriculture and animal husbandry 67.8%, industry 10.2%, construction 6.3%, commerce 5.0%, services and other 10.7% (1980 est.)

Government

Names:
conventional long form: Islamic State of Afghanistan
conventional short form: Afghanistan
local long form: Dowlat-e Eslami-ye Afghanestan
local short form: Afghanestan
former: Republic of Afghanistan
Digraph: AF
Type: transitional government
Capital: Kabul
Administrative divisions: 30 provinces (velayat, singular—velayat); Badakhshan, Badghis, Baghlan, Balkh, Bamian, Farah, Faryab, Ghazni, Ghowr, Helmand, Herat, Jowzjan, Kabol, Kandahar, Kapisa, Konar, Kondoz, Laghman, Lowgar, Nangarhar, Nimruz, Oruzgan, Paktia, Paktika, Parvan, Samangan, Sar-e Pol, Takhar, Vardak, Zabol
note: there may be two new provinces of Nurestan (Nuristan) and Khowst
Independence: 19 August 1919 (from UK)
National holiday: Victory of the Muslim Nation, 28 April; Remembrance Day for Martyrs and Disabled, 4 May; Independence Day, 19 August
Constitution: none
Legal system: a new legal system has not been adopted but the transitional government has declared it will follow Islamic law (Shari'a)
Suffrage: undetermined; previously males 15-50 years of age; universal
Executive branch:
chief of state: President Burhanuddin RABBANI (Interim President July-December 1992; President since 2 January 1993); Vice President Mohammad NABI MOHAMMADI (since NA); election last held 31 December 1992 (next to be held NA); results—Burhanuddin RABBANI was elected to a two-year term by a national shura, later amended by multi-party agreement to 18 months; note—in June 1994 failure to agree on a transfer mechanism resulted in RABBANI's extending the term to 28 December 1994; following the expiration of the term and while negotiations on the formation of a new government go on, RABBANI continues in office
head of government: Prime Minister Gulbuddin HIKMATYAR (since 17 March 1993); note—Prime Minister HIKMATYAR is the nominal head of government and does not have any real authority; First Deputy Prime Minister Qutbuddin HELAL (since 17 March 1993); Deputy Prime Minister Arsala RAHMANI (since 17 March 1993)
cabinet: Council of Ministers
note: term of present government expired 28 December 1994; factional fighting since 1 January 1994 has kept government officers from actually occupying ministries and discharging government responsibilities; the government's authority to remove cabinet members, including the Prime Minister, following the expiration of their term is questionable
Legislative branch: a unicameral parliament consisting of 205 members was chosen by the shura in January 1993; non-functioning as of June 1993
Judicial branch: an interim Chief Justice of the Supreme Court has been appointed, but a new court system has not yet been organized
Political parties and leaders: current political organizations include Jamiat-i-Islami (Islamic Society), Burhanuddin RABBANI,

Ahmad Shah MASOOD; Hizbi Islami-Gulbuddin (Islamic Party), Gulbuddin HIKMATYAR faction; Hizbi Islami-Khalis (Islamic Party), Yunis KHALIS faction; Ittihad-i-Islami Barai Azadi Afghanistan (Islamic Union for the Liberation of Afghanistan), Abdul Rasul SAYYAF; Harakat-Inqilab-i-Islami (Islamic Revolutionary Movement), Mohammad Nabi MOHAMMADI; Jabha-i-Najat-i-Milli Afghanistan (Afghanistan National Liberation Front), Sibghatullah MOJADDEDI; Mahaz-i-Milli-Islami (National Islamic Front), Sayed Ahamad GAILANI; Hizbi Wahdat-Khalili faction (Islamic Unity Party), Abdul Karim KHALILI; Hizbi Wahdat-Akbari faction (Islamic Unity Party), Mohammad Akbar AKBARI; Harakat-i-Islami (Islamic Movement), Mohammed Asif MOHSENI; Jumbesh-i-Milli Islami (National Islamic Movement), Abdul Rashid DOSTAM; Taliban (Religious Students Movement), Mohammad OMAR
note: the former ruling Watan Party has been disbanded
Other political or pressure groups: the former resistance commanders are the major power brokers in the countryside and their shuras (councils) are now administering most cities outside Kabul; tribal elders and religious students are trying to wrest control from them; ulema (religious scholars); tribal elders; religious students (talib)
Member of: AsDB, CP, ECO, ESCAP, FAO, G-77, IAEA, IBRD, ICAO, ICRM, IDA, IDB, IFAD, IFC, IFRCS, ILO, IMF, INTELSAT, IOC, ITU, NAM, OIC, UN, UNCTAD, UNESCO, UNIDO, UPU, WFTU, WHO, WMO, WTO
Diplomatic representation in US:
chief of mission: (vacant); Charge d'Affaires Abdul RAHIM
chancery: 2341 Wyoming Avenue NW, Washington, DC 20008
telephone: [1] (202) 234-3770, 3771
FAX: [1] (202) 328-3516
consulate(s) general: New York
consulate(s): Washington, DC
US diplomatic representation: none; embassy was closed in January 1989
Flag: NA; note—the flag has changed at least twice since 1992

Economy

Overview: Afghanistan is an extremely poor, landlocked country, highly dependent on farming (wheat especially) and livestock raising (sheep and goats). Economic considerations have played second fiddle to political and military upheavals during more than 15 years of war, including the nearly 10-year Soviet military occupation (which ended 15 February 1989). Over the past decade, one-third of the population fled the country, with Pakistan sheltering more than 3 million refugees and Iran about 3 million. About 1.4 million Afghan refugees remain in Pakistan and about 2 million in Iran. Another 1 million probably moved into and around urban areas within Afghanistan. Although reliable data are unavailable, gross domestic product is lower than 13 years ago because of the loss of labor and capital and the disruption of trade and transport.
National product: GDP $NA
National product real growth rate: NA%
National product per capita: $NA
Inflation rate (consumer prices): 56.7% (1991)
Unemployment rate: NA%
Budget:
revenues: $NA
expenditures: $NA, including capital expenditures of $NA
Exports: $188.2 million (f.o.b., 1991)
commodities: fruits and nuts, handwoven carpets, wool, cotton, hides and pelts, precious and semi-precious gems
partners: FSU countries, Pakistan, Iran, Germany, India, UK, Belgium, Luxembourg, Czechoslovakia
Imports: $616.4 million (c.i.f., 1991)
commodities: food and petroleum products; most consumer goods
partners: FSU countries, Pakistan, Iran, Japan, Singapore, India, South Korea, Germany
External debt: $2.3 billion (March 1991 est.)
Industrial production: growth rate 2.3% (FY90/91 est.); accounts for about 25% of GDP
Electricity:
capacity: 480,000 kW
production: 550 million kWh
consumption per capita: 39 kWh (1993)
Industries: small-scale production of textiles, soap, furniture, shoes, fertilizer, and cement; handwoven carpets; natural gas, oil, coal, copper
Agriculture: largely subsistence farming and nomadic animal husbandry; cash products—wheat, fruits, nuts, karakul pelts, wool, mutton
Illicit drugs: an illicit cultivator of opium poppy and cannabis for the international drug trade; world's second-largest opium producer after Burma (950 metric tons in 1994) and a major source of hashish
Economic aid:
recipient: $450 million US assistance provided 1985-1993; the UN provides assistance in the form of food aid, immunization, land mine removal, and a wide range of aid to refugees and displaced persons
Currency: 1 afghani (AF) = 100 puls
Exchange rates: afghanis (Af) per US$1—1,900 (January 1994), 1,019 (March 1993), 850 (1991), 700 (1989-90), 220 (1988-89); note—these rates reflect the free market exchange rates rather than the official exchange rates
Fiscal year: 21 March—20 March

Transportation

Railroads:
total: 24.6 km
broad gauge: 9.6 km 1.524-m gauge from Gushgy (Turkmenistan) to Towraghondi; 15 km 1,524-m gauge from Termiz (Uzbekistan) to Kheyrabad transshipment point on south bank of Amu Darya
Highways:
total: 21,000 km
paved: 2,800 km
unpaved: gravel 1,650 km; earth 16,550 km (1984)
Inland waterways: total navigability 1,200 km; chiefly Amu Darya, which handles vessels up to about 500 metric tons
Pipelines: petroleum products—Uzbekistan to Bagram and Turkmenistan to Shindand; natural gas 180 km
Ports: Keleft, Kheyrabad, Shir Khan
Airports:
total: 48
with paved runways over 3,047 m: 3
with paved runways 2,438 to 3,047 m: 5
with paved runways 1,524 to 2,437 m: 2
with paved runways under 914 m: 15
with unpaved runways 2,438 to 3,047 m: 3
with unpaved runways 1,524 to 2,438 m: 14
with unpaved runways 914 to 1,523 m: 6

Communications

Telephone system: 31,200 telephones; limited telephone, telegraph, and radiobroadcast services; 1 public telephone in Kabul
local: NA
intercity: NA
international: one link between western Afghanistan and Iran (via satellite)
Radio:
broadcast stations: AM 5, FM 0, shortwave 2
radios: NA
Television:
broadcast stations: several television stations run by factions and local councils which provide intermittent service
televisions: NA

Defense Forces

Branches: the military still does not exist on a national scale; some elements of the former Army, Air and Air Defense Forces, National Guard, Border Guard Forces, National Police Force (Sarandoi), and tribal militias still exist but are factionalized among the various mujahedin and former regime leaders
Manpower availability: males age 15-49 5,646,789; males fit for military service 3,011,777; males reach military age (22) annually 200,264 (1995 est.)
Defense expenditures: exchange rate conversion—$450 million, 15% of GDP (1990 est.); the new government has not yet adopted a defense budget

Albania

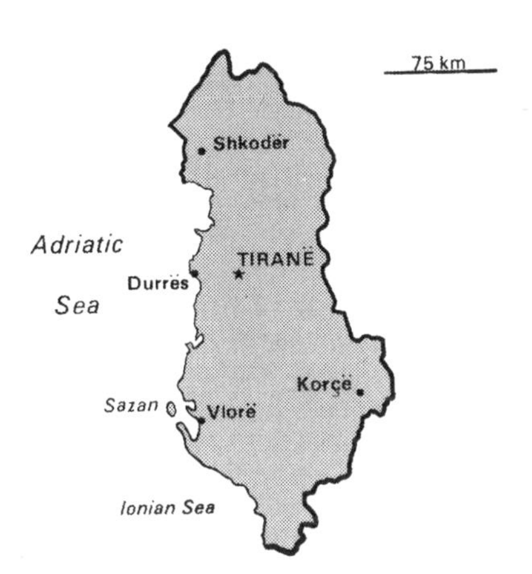

Geography

Location: Southeastern Europe, bordering the Adriatic Sea and Ionian Sea, between Greece and Serbia and Montenegro
Map references: Ethnic Groups in Eastern Europe, Europe
Area:
total area: 28,750 sq km
land area: 27,400 sq km
comparative area: slightly larger than Maryland
Land boundaries: total 720 km, Greece 282 km, The Former Yugoslav Republic of Macedonia 151 km, Serbia and Montenegro 287 km (114 km with Serbia, 173 km with Montenegro)
Coastline: 362 km
Maritime claims:
continental shelf: 200-m depth or to the depth of exploitation
territorial sea: 12 nm
International disputes: the Albanian Government supports protection of the rights of ethnic Albanians outside of its borders; Albanian majority in Kosovo seeks independence from Serbian Republic; Albanians in Macedonia claim discrimination in education, access to public sector jobs and representation in government; Albania is involved in a bilaterlal dispute with Greece over border demarcation, the treatment of Albania's ethnic Greek minority, and migrant Albanian workers in Greece
Climate: mild temperate; cool, cloudy, wet winters; hot, clear, dry summers; interior is cooler and wetter
Terrain: mostly mountains and hills; small plains along coast
Natural resources: petroleum, natural gas, coal, chromium, copper, timber, nickel
Land use:
arable land: 21%
permanent crops: 4%
meadows and pastures: 15%
forest and woodland: 38%
other: 22%
Irrigated land: 4,230 sq km (1989)
Environment:
current issues: deforestation; soil erosion; water pollution from industrial and domestic effluents
natural hazards: destructive earthquakes; tsunami occur along southwestern coast
international agreements: party to—Biodiversity, Climate Change
Note: strategic location along Strait of Otranto (links Adriatic Sea to Ionian Sea and Mediterranean Sea)

People

Population: 3,413,904 (July 1995 est.)
note: IMF, working with Albanian government figures, estimates the population at 3,120,000 in 1993 and that the population has fallen since 1990
Age structure:
0-14 years: 32% (female 520,186; male 563,953)
15-64 years: 62% (female 1,026,321; male 1,104,371)
65 years and over: 6% (female 112,252; male 86,821) (July 1995 est.)
Population growth rate: 1.16% (1995 est.)
Birth rate: 21.7 births/1,000 population (1995 est.)
Death rate: 5.22 deaths/1,000 population (1995 est.)
Net migration rate: -4.88 migrant(s)/1,000 population (1995 est.)
Infant mortality rate: 28.1 deaths/1,000 live births (1995 est.)
Life expectancy at birth:
total population: 73.81 years
male: 70.83 years
female: 77.02 years (1995 est.)
Total fertility rate: 2.71 children born/woman (1995 est.)
Nationality:
noun: Albanian(s)
adjective: Albanian
Ethnic divisions: Albanian 95%, Greeks 3%, other 2% (Vlachs, Gypsies, Serbs, and Bulgarians) (1989 est.)
Religions: Muslim 70%, Albanian Orthodox 20%, Roman Catholic 10%
note: all mosques and churches were closed in 1967 and religious observances prohibited; in November 1990, Albania began allowing private religious practice
Languages: Albanian (Tosk is the official dialect), Greek
Literacy: age 9 and over can read and write (1955)
total population: 72%
male: 80%
female: 63%
Labor force: 1.5 million (1987)
by occupation: agriculture 60%, industry and commerce 40% (1986)

Government

Names:
conventional long form: Republic of Albania
conventional short form: Albania
local long form: Republika e Shqiperise
local short form: Shqiperia
former: People's Socialist Republic of Albania
Digraph: AL
Type: emerging democracy
Capital: Tirane
Administrative divisions: 26 districts (rrethe, singular—rreth); Berat, Dibre, Durres, Elbasan, Fier, Gjirokaster, Gramsh, Kolonje, Korce, Kruje, Kukes, Lezhe, Librazhd, Lushnje, Mat, Mirdite, Permet, Pogradec, Puke, Sarande, Shkoder, Skrapar, Tepelene, Tirane, Tropoje, Vlore
Independence: 28 November 1912 (from Ottoman Empire)
National holiday: Independence Day, 28 November (1912)
Constitution: an interim basic law was approved by the People's Assembly on 29 April 1991; a draft constitution was rejected by popular referendum in the fall of 1994 and a new draft is pending
Legal system: has not accepted compulsory ICJ jurisdiction
Suffrage: 18 years of age; universal and compulsory
Executive branch:
chief of state: President of the Republic Sali BERISHA (since 9 April 1992)
head of government: Prime Minister of the Council of Ministers Aleksander Gabriel MEKSI (since 10 April 1992)
cabinet: Council of Ministers; appointed by the president
Legislative branch: unicameral
People's Assembly (Kuvendi Popullor). elections last held 22 March 1992; results—DP 62.29%, ASP 25.57%, SDP 4.33%, RP 3.15%, UHP 2.92%, other 1.74%; seats—(140 total) DP 92, ASP 38, SDP 7, RP 1, UHP 2
note: 6 members of the Democratic Party defected making the present seating in the Assembly DP 86, ASP 38, SDP 7, DAP 6, RP 1, UHP 2
Judicial branch: Supreme Court
Political parties and leaders: there are at least 28 political parties; most prominent are the Albanian Socialist Party (ASP; formerly the Albania Workers Party), Fatos NANO, first secretary; Democratic Party (DP); Albanian Republican Party (RP), Sabri GODO; Omonia (Greek minority party), Sotir QIRJAZATI, first secretary; Social Democratic Party (SDP), Skender GJINUSHI; Democratic Alliance Party (DAP), Neritan CEKA, chairman; Unity for Human Rights Party (UHP), Vasil MELO, chairman; Ecology Party (EP), Namik HOTI, chairman
Member of: BSEC, CCC, CE (guest), EBRD, ECE, FAO, IAEA, IBRD, ICAO, ICRM, IDA,

IDB, IFAD, IFC, IFRCS, ILO, IMF, IMO, INTELSAT (nonsignatory user), INTERPOL, IOC, IOM, ISO, ITU, NACC, OIC, OSCE, UN, UNCTAD, UNESCO, UNIDO, UPU, WFTU, WHO, WIPO, WMO, WTO
Diplomatic representation in US:
chief of mission: Ambassador Lublin Hasan DILJA
chancery: Suite 1010, 1511 K Street NW, Washington, DC 20005
telephone: [1] (202) 223-4942, 8187
FAX: [1] (202) 628-7342
US diplomatic representation:
chief of mission: Ambassador Joseph E. LAKE
embassy: Rruga E. Elbansanit 103, Tirane
mailing address: PSC 59, Box 100 (A), APO AE 09624
telephone: [355] (42) 328-75, 335-20
FAX: [355] (42) 322-22
Flag: red with a black two-headed eagle in the center

Economy

Overview: An extremely poor country by European standards, Albania is making the difficult transition to a more open-market economy. The economy rebounded in 1993-94 after a severe depression accompanying the collapse of the previous centrally planned system in 1990 and 1991. Stabilization policies—including a strict monetary policy, public sector layoffs, and reduced social services—have improved the government's fiscal situation and reduced inflation. The recovery was spurred by the remittances of some 20% of the population which works abroad, mostly in Greece and Italy. These remittances supplement GDP and help offset the large foreign trade deficit. Foreign assistance and humanitarian aid also supported the recovery. Most agricultural land was privatized in 1992, substantially improving peasant incomes. Albania's limited industrial sector, now less than one-sixth of GDP, continued to decline in 1994. A sharp fall in chromium prices reduced hard currency receipts from the mining sector. Large segments of the population, especially those living in urban areas, continue to depend on humanitarian aid to meet basic food requirements. Unemployment remains a severe problem accounting for approximately one-fifth of the work force. Growth is expected to continue in 1995, but could falter if Albania becomes involved in the conflict in the former Yugoslavia, workers' remittances from Greece are reduced, or foreign assistance declines.
National product: GDP—purchasing power parity—$3.8 billion (1994 est.)
National product real growth rate: 11% (1994 est.)
National product per capita: $1,110 (1994 est.)
Inflation rate (consumer prices): 16% (1994)
Unemployment rate: 18% (1994 est.)
Budget:
revenues: $1.1 billion
expenditures: $1.4 billion, including capital expenditures of $70 million (1991 est.)
Exports: $112 million (f.o.b., 1993)
commodities: asphalt, metals and metallic ores, electricity, crude oil, vegetables, fruits, tobacco
partners: Italy, The Former Yugoslav Republic of Macedonia, Germany, Greece, Czech Republic, Slovakia, Poland, Romania, Bulgaria, Hungary
Imports: $621 million (f.o.b., 1993)
commodities: machinery, consumer goods, grains
partners: Italy, The Former Yugoslav Republic of Macedonia, Germany, Czech Republic, Slovakia, Romania, Poland, Hungary, Bulgaria, Greece
External debt: $920 million (1994 est.)
Industrial production: growth rate -10% (1993 est.); accounts for 16% of GDP (1993 est.)
Electricity:
capacity: 770,000 kW
production: 4 billion kWh
consumption per capita: 1,200 kWh (1994)
Industries: food processing, textiles and clothing, lumber, oil, cement, chemicals, mining, basic metals, hydropower
Agriculture: accounts for 55% of GDP; arable land per capita among lowest in Europe; 80% of arable land now in private hands; 60% of the work force engaged in farming; produces wide range of temperate-zone crops and livestock
Illicit drugs: transshipment point for Southwest Asian heroin transiting the Balkan route and cocaine from South America destined for Western Europe; limited opium production
Economic aid:
recipient: $303 million (1993)
Currency: 1 lek (L) = 100 qintars
Exchange rates: leke (L) per US$1—100 (January 1995), 99 (January 1994), 97 (January 1993), 50 (January 1992), 25 (September 1991)
Fiscal year: calendar year

Transportation

Railroads:
total: 543 km line connecting Podgorica (Serbia and Montenegro) and Shkoder completed August 1986
standard gauge: 509 km 1.435-m gauge
narrow gauge: 34 km 0.950-m gauge (1990)
Highways:
total: 18,450 km
paved: 17,450 km
unpaved: earth 1,000 km (1991)
Inland waterways: 43 km plus Albanian sections of Lake Scutari, Lake Ohrid, and Lake Prespa (1990)
Pipelines: crude oil 145 km; petroleum products 55 km; natural gas 64 km (1991)
Ports: Durres, Sarande, Shergjin, Vlore
Merchant marine:
total: 11 cargo ships (1,000 GRT or over) totaling 52,967 GRT/76,887 DWT
Airports:
total: 11
with paved runways 2,438 to 3,047 m: 3
with paved runways 914 to 1,523 m: 2
with unpaved runways over 3,047 m: 2
with unpaved runways 2,438 to 3,047 m: 1
with unpaved runways 1,524 to 2,438 m: 1
with unpaved runways 914 to 1,523 m: 2

Communications

Telephone system: about 55,000 telephones; about 15 telephones/1,000 persons
local: primitive; about 11,000 telephones in Tirane, the capital city
intercity: obsolete wire system; no longer provides a telephone for every village; in 1992, following the fall of the communist government, peasants cut the wire to about 1,000 villages and used it to build fences
international: inadequate; carried through the Tirane exchange and transmitted through Italy on 240 microwave radio relay circuits and through Greece on 150 microwave radio relay circuits
Radio:
broadcast stations: AM 17, FM 1, shortwave 0
radios: 515,000 (1987 est.)
Television:
broadcast stations: 9
televisions: 255,000 (1987 est.)

Defense Forces

Branches: Army, Navy, Air and Air Defense Forces, Interior Ministry Troops, Border Guards
Manpower availability: males age 15-49 919,085; males fit for military service 755,574; males reach military age (19) annually 33,323 (1995 est.)
Defense expenditures: 330 million leke, NA% of GNP (1993); note—conversion of defense expenditures into US dollars using the current exchange rate could produce misleading results

Algeria

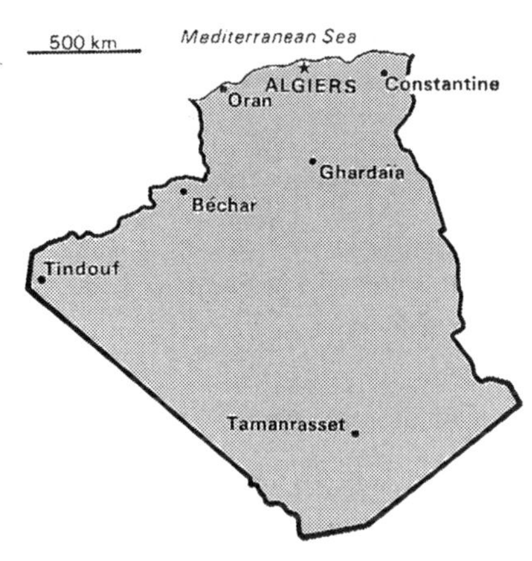

Geography

Location: Northern Africa, bordering the Mediterranean Sea, between Morocco and Tunisia
Map references: Africa
Area:
total area: 2,381,740 sq km
land area: 2,381,740 sq km
comparative area: slightly less than 3.5 times the size of Texas
Land boundaries: total 6,343 km, Libya 982 km, Mali 1,376 km, Mauritania 463 km, Morocco 1,559 km, Niger 956 km, Tunisia 965 km, Western Sahara 42 km
Coastline: 998 km
Maritime claims:
exclusive fishing zone: 32-52 nm
territorial sea: 12 nm
International disputes: Libya claims part of southeastern Algeria; land boundary dispute with Tunisia settled in 1993
Climate: arid to semiarid; mild, wet winters with hot, dry summers along coast; drier with cold winters and hot summers on high plateau; sirocco is a hot, dust/sand-laden wind especially common in summer
Terrain: mostly high plateau and desert; some mountains; narrow, discontinuous coastal plain
Natural resources: petroleum, natural gas, iron ore, phosphates, uranium, lead, zinc
Land use:
arable land: 3%
permanent crops: 0%
meadows and pastures: 13%
forest and woodland: 2%
other: 82%
Irrigated land: 3,360 sq km (1989 est.)
Environment:
current issues: soil erosion from overgrazing and other poor farming practices; desertification; dumping of raw sewage, petroleum refining wastes, and other industrial effluents is leading to the pollution of rivers and coastal waters; Mediterranean Sea, in particular, becoming polluted from oil wastes, soil erosion, and fertilizer runoff; inadequate supplies of potable water
natural hazards: mountainous areas subject to severe earthquakes; mudslides
international agreements: party to—Climate Change, Endangered Species, Environmental Modification, Ozone Layer Protection, Ship Pollution, Wetlands; signed, but not ratified—Biodiversity, Desertification, Law of the Sea, Nuclear Test Ban
Note: second-largest country in Africa (after Sudan)

People

Population: 28,539,321 (July 1995 est.)
Age structure:
0-14 years: 41% (female 5,678,879; male 5,885,246)
15-64 years: 56% (female 7,887,885; male 8,033,508)
65 years and over: 3% (female 557,636; male 496,167) (July 1995 est.)
Population growth rate: 2.25% (1995 est.)
Birth rate: 29.02 births/1,000 population (1995 est.)
Death rate: 6.05 deaths/1,000 population (1995 est.)
Net migration rate: -0.49 migrant(s)/1,000 population (1995 est.)
Infant mortality rate: 50.3 deaths/1,000 live births (1995 est.)
Life expectancy at birth:
total population: 68.01 years
male: 66.94 years
female: 69.13 years (1995 est.)
Total fertility rate: 3.7 children born/woman (1995 est.)
Nationality:
noun: Algerian(s)
adjective: Algerian
Ethnic divisions: Arab-Berber 99%, European less than 1%
Religions: Sunni Muslim (state religion) 99%, Christian and Jewish 1%
Languages: Arabic (official), French, Berber dialects
Literacy: age 15 and over can read and write (1990 est.)
total population: 57%
male: 70%
female: 46%
Labor force: 6.2 million (1992 est.)
by occupation: government 29.5%, agriculture 22%, construction and public works 16.2%, industry 13.6%, commerce and services 13.5%, transportation and communication 5.2% (1989)

Government

Names:
conventional long form: Democratic and Popular Republic of Algeria
conventional short form: Algeria
local long form: Al Jumhuriyah al Jaza'iriyah ad Dimuqratiyah ash Shabiyah
local short form: Al Jaza'ir
Digraph: AG
Type: republic
Capital: Algiers
Administrative divisions: 48 provinces (wilayas, singular—wilaya); Adrar, Ain Defla, Ain Temouchent, Alger, Annaba, Batna, Bechar, Bejaia, Biskra, Blida, Bordj Bou Arreridj, Bouira, Boumerdes, Chlef, Constantine, Djelfa, El Bayadh, El Oued, El Tarf, Ghardaia, Guelma, Illizi, Jijel, Khenchela, Laghouat, Mascara, Medea, Mila, Mostaganem, M'Sila, Naama, Oran, Ouargla, Oum el Bouaghi, Relizane, Saida, Setif, Sidi Bel Abbes, Skikda, Souk Ahras, Tamanghasset, Tebessa, Tiaret, Tindouf, Tipaza, Tissemsilt, Tizi Ouzou, Tlemcen
Independence: 5 July 1962 (from France)
National holiday: Anniversary of the Revolution, 1 November (1954)
Constitution: 19 November 1976, effective 22 November 1976; revised 3 November 1988 and 23 February 1989
Legal system: socialist, based on French and Islamic law; judicial review of legislative acts in ad hoc Constitutional Council composed of various public officials, including several Supreme Court justices; has not accepted compulsory ICJ jurisdiction
Suffrage: 18 years of age; universal
Executive branch:
chief of state: President Lamine ZEROUAL (since 31 January 1994); next election to be held by the end of 1995
head of government: Prime Minister Mokdad SIFI (since 11 April 1994)
cabinet: Council of Ministers; appointed by the prime minister
Legislative branch: unicameral; note—suspended since 1992
National People's Assembly (Al-Majlis Ech-Chaabi Al-Watani): elections first round held on 26 December 1991 (second round canceled by the military after President BENDJEDID resigned 11 January 1992, effectively suspending the Assembly); results—percent of vote by party NA; seats—(281 total); the fundamentalist FIS won 188 of the 231 seats contested in the first round; note—elections (provincial and municipal) were held in June 1990, the first in Algerian history; results—FIS 55%, FLN 27.5%, other 17.5%, with 65% of the voters participating
Judicial branch: Supreme Court (Cour Supreme)
Political parties and leaders: Islamic Salvation Front (FIS, outlawed April 1992), Ali BELHADJ, Dr. Abassi MADANI, Abdelkader HACHANI (all under arrest), Rabeh KEBIR (self-exile in Germany); National Liberation Front (FLN), Abdelhamid MEHRI, Secretary General; Socialist Forces Front (FFS), Hocine Ait AHMED, Secretary General

note: the government established a multiparty system in September 1989 and, as of 31 December 1990, over 50 legal parties existed
Member of: ABEDA, AfDB, AFESD, AL, AMF, AMU, CCC, ECA, FAO, G-15, G-19, G-24, G-77, IAEA, IBRD, ICAO, ICRM, IDA, IDB, IFAD, IFC, IFRCS, ILO, IMF, IMO, INMARSAT, INTELSAT, INTERPOL, IOC, ISO, ITU, NAM, OAPEC, OAS (observer), OAU, OIC, OPEC, UN, UNCTAD, UNESCO, UNHCR, UNIDO, UNMIH, UPU, WCL, WHO, WIPO, WMO, WTO
Diplomatic representation in US:
chief of mission: Ambassador Osmane BENCHERIF
chancery: 2118 Kalorama Road NW, Washington, DC 20008
telephone: [1] (202) 265-2800
US diplomatic representation:
chief of mission: Ambassador Ronald E. NEUMANN
embassy: 4 Chemin Cheikh Bachir El-Ibrahimi, Algiers
mailing address: B. P. Box 549, Alger-Gare, 16000 Algiers
telephone: [213] (2) 69-11-86, 69-18-54, 69-38-75
FAX: [213] (2) 69-39-79
consulate(s): none (Oran closed June 1993)
Flag: two equal vertical bands of green (hoist side) and white with a red five-pointed star within a red crescent; the crescent, star, and color green are traditional symbols of Islam (the state religion)

Economy

Overview: The hydrocarbons sector is the backbone of the economy, accounting for roughly 57% of government revenues, 25% of GDP, and almost all export earnings; Algeria has the fifth largest reserves of natural gas in the world and ranks fourteenth for oil. Algiers' efforts to reform one of the most centrally planned economies in the Arab world began after the 1986 collapse of world oil prices plunged the country into a severe recession. In 1989, the government launched a comprehensive, IMF-supported program to achieve macroeconomic stabilization and to introduce market mechanisms into the economy. Despite substantial progress toward macroeconomic adjustment, in 1992 the reform drive stalled as Algiers became embroiled in political turmoil. In September 1993, a new government was formed, and one priority was the resumption and acceleration of the structural adjustment process. Buffeted by the slump in world oil prices and burdened with a heavy foreign debt, Algiers concluded a one-year standby arrangement with the IMF in April 1994.
National product: GDP—purchasing power parity—$97.1 billion (1994 est.)
National product real growth rate: 0.2% (1994 est.)
National product per capita: $3,480 (1994 est.)
Inflation rate (consumer prices): 30% (1994 est.)
Unemployment rate: 30% (1994 est.)
Budget:
revenues: $14.3 billion
expenditures: $17.9 billion (1995 est.)
Exports: $9.1 billion (f.o.b., 1994)
commodities: petroleum and natural gas 97%
partners: Italy 21%, France 16%, US 14%, Germany 13%, Spain 9%
Imports: $9.2 billion (f.o.b., 1994 est.)
commodities: capital goods 39.7%, food and beverages 21.7%, consumer goods 11.8% (1990)
partners: France 29%, Italy 14%, Spain 9%, US 9%, Germany 7%
External debt: $26 billion (1994)
Industrial production: growth rate NA%; accounts for 35% of GDP (including hydrocarbons)
Electricity:
capacity: 5,370,000 kW
production: 18.3 billion kWh
consumption per capita: 587 kWh (1993)
Industries: petroleum, light industries, natural gas, mining, electrical, petrochemical, food processing
Agriculture: accounts for 12% of GDP (1993) and employs 22% of labor force; products—wheat, barley, oats, grapes, olives, citrus, fruits, sheep, cattle; net importer of food—grain, vegetable oil, sugar
Economic aid:
recipient: US commitments, including Ex-Im (FY70-85), $1.4 billion; Western (non-US) countries, ODA and OOF bilateral commitments (1970-89), $925 million; OPEC bilateral aid (1979-89), $1.8 billion; Communist countries (1970-89), $2.7 billion; net official disbursements (1985-89), $375 million
Currency: 1 Algerian dinar (DA) = 100 centimes
Exchange rates: Algerian dinars (DA) per US$1—42.710 (January 1995), 35.059 (1994), 23.345 (1993), 21.836 (1992), 18.473 (1991), 8.958 (1990)
Fiscal year: calendar year

Transportation

Railroads:
total: 4,733 km
standard gauge: 3,576 km 1.435-m gauge (299 km electrified; 215 km double track)
narrow gauge: 1,157 km 1.055-m gauge
Highways:
total: 95,576 km
paved: concrete, bituminous 57,346 km
unpaved: gravel, crushed stone, earth 38,230 km
Pipelines: crude oil 6,612 km; petroleum products 298 km; natural gas 2,948 km
Ports: Algiers, Annaba, Arzew, Bejaia, Beni Saf, Dellys, Djendjene, Ghazaouet, Jijel, Mostaganem, Oran, Skikda, Tenes
Merchant marine:
total: 75 ships (1,000 GRT or over) totaling 903,179 GRT/1,064,211 DWT
ships by type: bulk 9, cargo 27, chemical tanker 7, liquefied gas tanker 9, oil tanker 5, roll-on/roll-off cargo 12, short-sea passenger 5, specialized tanker 1
Airports:
total: 139
with paved runways over 3,047 m: 9
with paved runways 2,438 to 3,047 m: 23
with paved runways 1,524 to 2,437 m: 14
with paved runways 914 to 1,523 m: 5
with paved runways under 914 m: 20
with unpaved runways 2,438 to 3,047 m: 3
with unpaved runways 1,524 to 2,438 m: 24
with unpaved runways 914 to 1,523 m: 41

Communications

Telephone system: 822,000 telephones; excellent domestic and international service in the north, sparse in the south
local: NA
intercity: 12 domestic satellite links; 20 additional satellite links are planned
international: 5 submarine cables; microwave radio relay to Italy, France, Spain, Morocco, and Tunisia; coaxial cable to Morocco and Tunisia; 2 INTELSAT (1 Atlantic Ocean and 1 Indian Ocean), 1 Intersputnik, 1 ARABSAT earth station
Radio:
broadcast stations: AM 26, FM 0, shortwave 0
radios: 5.2 million
Television:
broadcast stations: 18
televisions: 1.6 million

Defense Forces

Branches: National Popular Army, Navy, Air Force, Territorial Air Defense, National Gendarmerie
Manpower availability: males age 15-49 7,124,894; males fit for military service 4,373,272; males reach military age (19) annually 313,707 (1995 est.)
Defense expenditures: exchange rate conversion—$1.3 billion, 2.7% of GDP (1994)

American Samoa

(territory of the US)

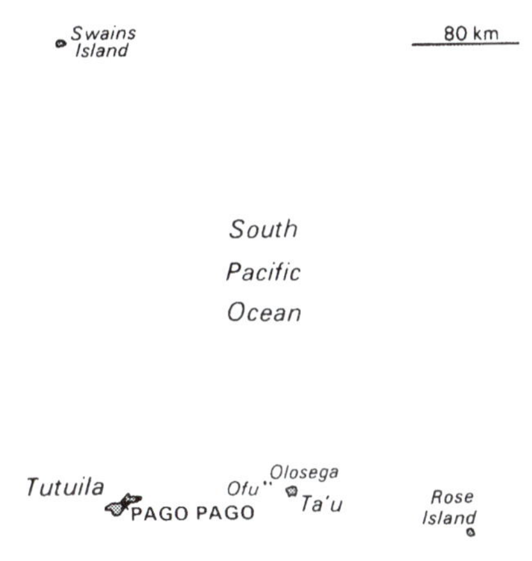

Geography

Location: Oceania, group of islands in the South Pacific Ocean, about one-half of the way from Hawaii to New Zealand
Map references: Oceania
Area:
total area: 199 sq km
land area: 199 sq km
comparative area: slightly larger than Washington, DC
note: includes Rose Island and Swains Island
Land boundaries: 0 km
Coastline: 116 km
Maritime claims:
exclusive economic zone: 200 nm
territorial sea: 12 nm
International disputes: none
Climate: tropical marine, moderated by southeast trade winds; annual rainfall averages 124 inches; rainy season from November to April, dry season from May to October; little seasonal temperature variation
Terrain: five volcanic islands with rugged peaks and limited coastal plains, two coral atolls (Rose Island, Swains Island)
Natural resources: pumice, pumicite
Land use:
arable land: 10%
permanent crops: 5%
meadows and pastures: 0%
forest and woodland: 75%
other: 10%
Irrigated land: NA sq km
Environment:
current issues: limited natural fresh water resources; in many areas of the island water supplies come from roof catchments
natural hazards: typhoons common from December to March
international agreements: NA
Note: Pago Pago has one of the best natural deepwater harbors in the South Pacific Ocean, sheltered by shape from rough seas and protected by peripheral mountains from high winds; strategic location in the South Pacific Ocean

People

Population: 57,366 (July 1995 est.)
Age structure:
0-14 years: NA
15-64 years: NA
65 years and over: NA
Population growth rate: 3.82% (1995 est.)
Birth rate: 36.21 births/1,000 population (1995 est.)
Death rate: 4.01 deaths/1,000 population (1995 est.)
Net migration rate: 6 migrant(s)/1,000 population (1995 est.)
Infant mortality rate: 18.78 deaths/1,000 live births (1995 est.)
Life expectancy at birth:
total population: 72.91 years
male: 71.03 years
female: 74.85 years (1995 est.)
Total fertility rate: 4.3 children born/woman (1995 est.)
Nationality:
noun: American Samoan(s)
adjective: American Samoan
Ethnic divisions: Samoan (Polynesian) 89%, Caucasian 2%, Tongan 4%, other 5%
Religions: Christian Congregationalist 50%, Roman Catholic 20%, Protestant denominations and other 30%
Languages: Samoan (closely related to Hawaiian and other Polynesian languages), English; most people are bilingual
Literacy: age 15 and over can read and write (1980)
total population: 97%
male: 98%
female: 97%
Labor force: 14,400 (1990)
by occupation: government 33%, tuna canneries 34%, other 33% (1990)

Government

Names:
conventional long form: Territory of American Samoa
conventional short form: American Samoa
Abbreviation: AS
Digraph: AQ
Type: unincorporated and unorganized territory of the US; administered by the US Department of Interior, Office of Territorial and International Affairs
Capital: Pago Pago
Administrative divisions: none (territory of the US)
Independence: none (territory of the US)
National holiday: Territorial Flag Day, 17 April (1900)
Constitution: ratified 1966, in effect 1967
Legal system: NA
Suffrage: 18 years of age; universal
Executive branch:
chief of state: President William Jefferson CLINTON (since 20 January 1993); Vice President Albert GORE, Jr. (since 20 January 1993)
head of government: Governor A. P. LUTALI (since 3 January 1993); Lieutenant Governor Tauese P. SUNIA (since 3 January 1993); election last held 3 November 1992 (next to be held NA November 1996); results—A. P. LUTALI (Democrat) 53%, Peter Tali COLEMAN (Republican) 36%
Legislative branch: bicameral Legislative Assembly (Fono)
House of Representatives: elections last held 3 November 1992 (next to be held NA November 1994); results—representatives popularly elected from 17 house districts; seats —(21 total, 20 elected, and 1 nonvoting delegate from Swains Island)
Senate: elections last held 3 November 1992 (next to be held NA November 1996); results—senators elected by village chiefs from 12 senate districts; seats—(18 total) number of seats by party NA
US House of Representatives: elections last held 3 November 1992 (next to be held NA November 1994); results—Eni R. F. H. FALEOMAVAEGA reelected as delegate
Judicial branch: High Court
Political parties and leaders: NA
Member of: ESCAP (associate), INTERPOL (subbureau), IOC, SPC
Diplomatic representation in US: none (territory of the US)
US diplomatic representation: none (territory of the US)
Flag: blue with a white triangle edged in red that is based on the fly side and extends to the hoist side; a brown and white American bald eagle flying toward the hoist side is carrying two traditional Samoan symbols of authority, a staff and a war club

Economy

Overview: Economic activity is strongly linked to the US, with which American Samoa conducts 80%-90% of its foreign trade. Tuna fishing and tuna processing plants are the backbone of the private sector, with canned tuna the primary export. The tuna canneries and the government are by far the two largest employers. Other economic activities include a slowly developing tourist industry. Transfers from the US Government add substantially to American Samoa's economic well-being.
National product: GDP—purchasing power parity—$128 million (1991 est.)
National product real growth rate: NA%
National product per capita: $2,600 (1991)
Inflation rate (consumer prices): 7% (1990)
Unemployment rate: 12% (1991)
Budget:
revenues: $97 million (includes $43,000,000 in local revenue and $54,000,000 in grant revenue)

American Samoa *(continued)*

expenditures: $NA, including capital expenditures of $NA (FY90/91)
Exports: $306 million (f.o.b., 1989)
commodities: canned tuna 93%
partners: US 99.6%
Imports: $360.3 million (c.i.f., 1989)
commodities: materials for canneries 56%, food 8%, petroleum products 7%, machinery and parts 6%
partners: US 62%, Japan 9%, NZ 7%, Australia 11%, Fiji 4%, other 7%
External debt: $NA
Industrial production: growth rate NA%
Electricity:
capacity: 30,000 kW
production: 90 million kWh
consumption per capita: 1,505 kWh (1993)
Industries: tuna canneries (largely dependent on foreign fishing vessels), meat canning, handicrafts
Agriculture: bananas, coconuts, vegetables, taro, breadfruit, yams, copra, pineapples, papayas, dairy farming
Economic aid:
recipient: $21,042,650 in operational funds and $1,227,000 in construction funds for capital improvement projects from the US Department of Interior (1991)
Currency: 1 United States dollar = 100 cents
Exchange rates: US currency is used
Fiscal year: 1 October—30 September

Transportation

Railroads: 0 km
Highways:
total: 350 km
paved: 150 km
unpaved: 200 km
Ports: Aanu'u (new construction), Auasi, Faleosao, Ofu, Pago Pago, Ta'u
Merchant marine: none
Airports:
total: 4
with paved runways 2,438 to 3,047 m: 1
with paved runways under 914 m: 3
note: small airstrips on Fituita and Ofu

Communications

Telephone system: 8,399 telephones; good telex, telegraph, and facsimile services
local: NA
intercity: NA
international: 1 INTELSAT (Pacific Ocean) and 1 COMSAT earth station
Radio:
broadcast stations: AM 1, FM 1, shortwave 0
radios: NA
Television:
broadcast stations: 1
televisions: NA

Defense Forces

Note: defense is the responsibility of the US

Andorra

Geography

Location: Southwestern Europe, between France and Spain
Map references: Europe
Area:
total area: 450 sq km
land area: 450 sq km
comparative area: slightly more than 2.5 times the size of Washington, DC
Land boundaries: total 125 km, France 60 km, Spain 65 km
Coastline: 0 km (landlocked)
Maritime claims: none; landlocked
International disputes: none
Climate: temperate; snowy, cold winters and warm, dry summers
Terrain: rugged mountains dissected by narrow valleys
Natural resources: hydropower, mineral water, timber, iron ore, lead
Land use:
arable land: 2%
permanent crops: 0%
meadows and pastures: 56%
forest and woodland: 22%
other: 20%
Irrigated land: NA sq km
Environment:
current issues: deforestation; overgrazing of mountain meadows contributes to soil erosion
natural hazards: snowslides, avalanches
international agreements: NA
Note: landlocked

People

Population: 65,780 (July 1995 est.)
Age structure:
0-14 years: 18% (female 5,503; male 5,985)
15-64 years: 70% (female 21,873; male 24,334)
65 years and over: 12% (female 4,020; male 4,065) (July 1995 est.)
Population growth rate: 2.72% (1995 est.)
Birth rate: 12.92 births/1,000 population (1995 est.)

Death rate: 7.25 deaths/1,000 population (1995 est.)
Net migration rate: 21.53 migrant(s)/1,000 population (1995 est.)
Infant mortality rate: 7.7 deaths/1,000 live births (1995 est.)
Life expectancy at birth:
total population: 78.52 years
male: 75.65 years
female: 81.66 years (1995 est.)
Total fertility rate: 1.72 children born/woman (1995 est.)
Nationality:
noun: Andorran(s)
adjective: Andorran
Ethnic divisions: Spanish 61%, Andorran 30%, French 6%, other 3%
Religions: Roman Catholic (predominant)
Languages: Catalan (official), French, Castilian
Literacy: NA%
Labor force: NA

Government

Names:
conventional long form: Principality of Andorra
conventional short form: Andorra
local long form: Principat d'Andorra
local short form: Andorra
Digraph: AN
Type: parliamentary democracy (since March 1993) that retains as its heads of state a co-principality; the two princes are the president of France and Spanish bishop of Seo de Urgel, who are represented locally by officials called veguers
Capital: Andorra la Vella
Administrative divisions: 7 parishes (parroquies, singular—parroquia); Andorra, Canillo, Encamp, La Massana, Les Escaldes, Ordino, Sant Julia de Loria
Independence: 1278
National holiday: Mare de Deu de Meritxell, 8 September
Constitution: Andorra's first written constitution was drafted in 1991; adopted 14 March 1993
Legal system: based on French and Spanish civil codes; no judicial review of legislative acts; has not accepted compulsory ICJ jurisdiction
Suffrage: 18 years of age; universal
Executive branch:
chiefs of state: French Co-Prince Francois MITTERRAND (since 21 May 1981), represented by Veguer de Franca Jean Pierre COURTOIS (since NA); note—COURTOIS is to become French ambassador to Libreville and his replacement has not been announced; Spanish Episcopal Co-Prince Mgr. Juan MARTI Alanis (since 31 January 1971), represented by Veguer Episcopal Francesc BADIA Bata (since NA); two permanent delegates (French Prefect Pierre STEINMETZ for the department of Pyrenees-Orientales, since NA, and Spanish Vicar General Nemesi MARQUES Oste for the Seo de Urgel diocese, since NA)
head of government: Executive Council President Marc FORNE (since 21 December 1994) elected by Parliament, following resignation of Oscar RIBAS Reig
cabinet: Executive Council; designated by the executive council president
Legislative branch: unicameral
General Council of the Valleys: (Consell General de las Valls); elections last held 12 December 1993 (next to be held NA); yielded no clear winner; results—percent of vote by party NA; seats—(28 total) number of seats by party NA
Judicial branch: Supreme Court of Andorra at Perpignan (France) for civil cases, the Ecclesiastical Court of the bishop of Seo de Urgel (Spain) for civil cases, Tribunal of the Courts (Tribunal des Cortes) for criminal cases
Political parties and leaders: National Democratic Group (AND), Oscar RIBAS Reig and Jordi FARRAS; Liberal Union (UL), Francesc CERQUEDA; New Democracy (ND), Jaume BARTOMEU; Andorran National Coalition (CNA), Antoni CERQUEDA; National Democratic Initiative (IDN), Vincenc MATEU; Liberal Union (UL), Marc FORNE
note: there are two other small parties
Member of: ECE, IFRCS (associate), INTERPOL, IOC, ITU, UN, UNESCO
Diplomatic representation in US: Andorra has no mission in the US
US diplomatic representation: Andorra is included within the Barcelona (Spain) Consular District, and the US Consul General visits Andorra periodically
Flag: three equal vertical bands of blue (hoist side), yellow, and red with the national coat of arms centered in the yellow band; the coat of arms features a quartered shield; similar to the flags of Chad and Romania that do not have a national coat of arms in the center

Economy

Overview: Tourism, the mainstay of Andorra's economy, accounts for roughly 80% of GDP. An estimated 13 million tourists visit annually, attracted by Andorra's duty-free status and by its summer and winter resorts. The banking sector, with its "tax haven" status, also contributes substantially to the economy. Agricultural production is limited by a scarcity of arable land, and most food has to be imported. The principal livestock activity is sheep raising. Manufacturing consists mainly of cigarettes, cigars, and furniture. Andorra is a member of the EU Customs Union; it is unclear what effect the European Single Market will have on the advantages Andorra obtains from its duty-free status.
National product: GDP—purchasing power parity—$760 million (1992 est.)
National product real growth rate: NA%
National product per capita: $14,000 (1992 est.)
Inflation rate (consumer prices): NA%
Unemployment rate: 0%
Budget:
revenues: $138 million
expenditures: $177 million, including capital expenditures of $NA (1993)
Exports: $30 million (f.o.b., 1993 est.)
commodities: electricity, tobacco products, furniture
partners: France, Spain
Imports: $NA
commodities: consumer goods, food
partners: France, Spain
External debt: $NA
Industrial production: growth rate NA%
Electricity:
capacity: 35,000 kW
production: 140 million kWh
consumption per capita: 2,570 kWh (1992)
Industries: tourism (particularly skiing), sheep, timber, tobacco, banking
Agriculture: sheep raising; small quantities of tobacco, rye, wheat, barley, oats, and some vegetables
Economic aid: none
Currency: 1 French franc (F) = 100 centimes; 1 peseta (Pta) = 100 centimos; the French and Spanish currencies are used
Exchange rates: French francs (F) per US$1—5.2943 (January 1995), 5,5520 (1994), 5.6632 (1993), 5.2938 (1992), 5.6421 (1991), 5.4453 (1990); Spanish pesetas (Ptas) per US$1—132.61 (January 1995), 133.96 (1994), 127.26 (1993), 102.38 (1992), 103.91 (1991), 101.93 (1990)
Fiscal year: calendar year

Transportation

Railroads: 0 km
Highways:
total: 96 km
paved: NA
unpaved: NA
Ports: none
Airports: none

Communications

Telephone system: 17,700 telephones; digital microwave network
local: NA
intercity: NA
international: landline circuits to France and Spain
Radio:
broadcast stations: AM 1, FM 0, shortwave 0
radios: NA

Andorra *(continued)*

Television:
broadcast stations: 0
televisions: NA

Defense Forces

Note: defense is the responsibility of France and Spain

Angola

Geography

Location: Southern Africa, bordering the South Atlantic Ocean, between Namibia and Zaire
Map references: Africa
Area:
total area: 1,246,700 sq km
land area: 1,246,700 sq km
comparative area: slightly less than twice the size of Texas
Land boundaries: total 5,198 km, Congo 201 km, Namibia 1,376 km, Zaire 2,511 km, Zambia 1,110 km
Coastline: 1,600 km
Maritime claims:
exclusive fishing zone: 200 nm
territorial sea: 20 nm
International disputes: none
Climate: semiarid in south and along coast to Luanda; north has cool, dry season (May to October) and hot, rainy season (November to April)
Terrain: narrow coastal plain rises abruptly to vast interior plateau
Natural resources: petroleum, diamonds, iron ore, phosphates, copper, feldspar, gold, bauxite, uranium
Land use:
arable land: 2%
permanent crops: 0%
meadows and pastures: 23%
forest and woodland: 43%
other: 32%
Irrigated land: NA km2
Environment:
current issues: population pressures contributing to overuse of pastures and subsequent soil erosion; desertification; deforestation of tropical rain forest attributable to the international demand for tropical timber and domestic use as a fuel; deforestation contributing to loss of biodiversity; soil erosion contributing to water pollution and siltation of rivers and dams; inadequate supplies of potable water
natural hazards: locally heavy rainfall causes periodic flooding on the plateau
international agreements: party to—Law of the Sea; signed, but not ratified—Biodiversity, Climate Change, Desertification
Note: Cabinda is separated from rest of country by Zaire

People

Population: 10,069,501 (July 1995 est.)
Age structure:
0-14 years: 45% (female 2,208,307; male 2,274,533)
15-64 years: 53% (female 2,641,259; male 2,685,543)
65 years and over: 2% (female 136,573; male 123,286) (July 1995 est.)
Population growth rate: 2.68% (1995 est.)
Birth rate: 45.05 births/1,000 population (1995 est.)
Death rate: 18.1 deaths/1,000 population (1995 est.)
Net migration rate: -0.15 migrant(s)/1,000 population (1995 est.)
Infant mortality rate: 142.1 deaths/1,000 live births (1995 est.)
Life expectancy at birth:
total population: 46.28 years
male: 44.18 years
female: 48.49 years (1995 est.)
Total fertility rate: 6.42 children born/woman (1995 est.)
Nationality:
noun: Angolan(s)
adjective: Angolan
Ethnic divisions: Ovimbundu 37%, Kimbundu 25%, Bakongo 13%, mestico (mixed European and Native African) 2%, European 1%, other 22%
Religions: indigenous beliefs 47%, Roman Catholic 38%, Protestant 15% (est.)
Languages: Portuguese (official), Bantu and other African languages
Literacy: age 15 and over can read and write (1990 est.)
total population: 42%
male: 56%
female: 28%
Labor force: 2.783 million economically active
by occupation: agriculture 85%, industry 15% (1985 est.)

Government

Note: Civil war has been the norm since independence from Portugal on 11 November 1975; a cease-fire lasted from 31 May 1991 until October 1992 when the insurgent National Union for the Total Independence of Angola (UNITA) refused to accept its defeat in internationally monitored elections and fighting resumed throughout much of the countryside. The two sides signed another peace accord on 20 November 1994; the cease-

fire is generally holding but most provisions of the accord remain to be implemented.
Names:
conventional long form: Republic of Angola
conventional short form: Angola
local long form: Republica de Angola
local short form: Angola
former: People's Republic of Angola
Digraph: AO
Type: transitional government nominally a multiparty democracy with a strong presidential system
Capital: Luanda
Administrative divisions: 18 provinces (provincias, singular—provincia); Bengo, Benguela, Bie, Cabinda, Cuando Cubango, Cuanza Norte, Cuanza Sul, Cunene, Huambo, Huila, Luanda, Lunda Norte, Lunda Sul, Malanje, Moxico, Namibe, Uige, Zaire
Independence: 11 November 1975 (from Portugal)
National holiday: Independence Day, 11 November (1975)
Constitution: 11 November 1975; revised 7 January 1978, 11 August 1980, 6 March 1991, and 26 August 1992
Legal system: based on Portuguese civil law system and customary law; recently modified to accommodate political pluralism and increased use of free markets
Suffrage: 18 years of age; universal
Executive branch:
chief of state: President Jose Eduardo DOS SANTOS (since 21 September 1979)
head of government: Prime Minister Marcolino Jose Carlos MOCO (since 2 December 1992)
cabinet: Council of Ministers; appointed by the president
Legislative branch: unicameral
National Assembly (Assembleia Nacional): first nationwide, multiparty elections were held 29-30 September 1992 with disputed results
Judicial branch: Supreme Court (Tribunal da Relacao)
Political parties and leaders: Popular Movement for the Liberation of Angola (MPLA), led by Jose Eduardo DOS SANTOS, is the ruling party and has been in power since 1975; National Union for the Total Independence of Angola (UNITA), led by Jonas SAVIMBI, is a legal party despite its history of armed resistance to the government; five minor parties have small numbers of seats in the National Assembly
Other political or pressure groups: Cabindan State Liberation Front (FLEC), N'ZITA Tiago, leader of largest faction (FLEC-FAC)
note: FLEC is waging a small-scale, highly factionalized, armed struggle for the independence of Cabinda Province
Member of: ACP, AfDB, CCC, CEEAC (observer), ECA, FAO, FLS, G-77, GATT, IBRD, ICAO, ICRM, IDA, IFAD, IFC, IFRCS, ILO, IMF, IMO, INTELSAT, INTERPOL, IOC, IOM, ITU, NAM, OAS (observer), OAU, SADC, UN, UNCTAD, UNESCO, UNIDO, UPU, WCL, WFTU, WHO, WIPO, WMO, WTO
Diplomatic representation in US:
chief of mission: Ambassador Jose Goncalves Martins PATRICIO
embassy: 1819 L Street NW, Washington, DC 20036, Suite 400
telephone: [1] (202) 785-1156
FAX: [1] (202) 785-1258
US diplomatic representation:
chief of mission: Ambassador Edmund T. DE JARNETTE
embassy: 32 Rua Houari Boumedienne, Miramar, Luanda
mailing address: C.P. 6484, Luanda; American Embassy, Luanda, Department of State, Washington, D.C. 20521-2550 (pouch)
telephone: [244] (2) 345-481, 346-418
FAX: [244] (2) 347-884
Flag: two equal horizontal bands of red (top) and black with a centered yellow emblem consisting of a five-pointed star within half a cogwheel crossed by a machete (in the style of a hammer and sickle)

Economy

Overview: Subsistence agriculture provides the main livelihood for 80%-90% of the population but accounts for less than 15% of GDP. Oil production is vital to the economy, contributing about 60% to GDP. Despite the signing of a peace accord in November 1994 between the Angola government and the UNITA insurgents, sporadic fighting continues and many farmers remain reluctant to return to their fields. As a result, much of the country's food requirements must still be imported. Angola has rich natural resources—notably gold, diamonds, and arable land, in addition to large oil deposits—but will need to observe the cease-fire, implement the peace agreement, and reform government policies if it is to achieve its potential.
National product: GDP—purchasing power parity—$6.1 billion (1994 est.)
National product real growth rate: -1% (1994 est.)
National product per capita: $620 (1994 est.)
Inflation rate (consumer prices): 20% average per month (1994 est.)
Unemployment rate: 15% with considerable underemployment (1993 est.)
Budget:
revenues: $928 million
expenditures: $2.5 billion, including capital expenditures of $963 million (1992 est.)
Exports: $3 billion (f.o.b., 1993 est.)
commodities: oil, diamonds, refined petroleum products, gas, coffee, sisal, fish and fish products, timber, cotton
partners: US, France, Germany, Netherlands, Brazil
Imports: $1.6 billion (f.o.b., 1992 est.)
commodities: capital equipment (machinery and electrical equipment), food, vehicles and spare parts, textiles and clothing, medicines, substantial military deliveries
partners: Portugal, Brazil, US, France, Spain
External debt: $11.7 billion (1994 est.)
Industrial production: growth rate NA%; accounts for about 60% of GDP, including petroleum output
Electricity:
capacity: 620,000 kW
production: 1.9 billion kWh
consumption per capita: 189 kWh (1993)
Industries: petroleum; mining—diamonds, iron ore, phosphates, feldspar, bauxite, uranium, and gold; fish processing; food processing; brewing; tobacco; sugar; textiles; cement; basic metal products
Agriculture: cash crops—bananas, sugarcane, coffee, sisal, corn, cotton, cane, manioc, tobacco; food crops—cassava, corn, vegetables, plantains; livestock production accounts for 20%, fishing 4%, forestry 2% of total agricultural output
Illicit drugs: increasingly used as a transshipment point for cocaine destined for Western Europe
Economic aid:
recipient: US commitments, including Ex-Im (FY70-89), $265 million; Western (non-US) countries, ODA and OOF bilateral commitments (1970-89), $1.105 billion; Communist countries (1970-89), $1.3 billion; net official disbursements (1985-89), $750 million
Currency: 1 new kwanza (NKz) = 100 lwei
Exchange rates: new kwanza (NKz) per US$1—900,000 (official rate 25 April 1995), 1,900,000 (black market rate 6 April 1995), 600,000 (official rate 10 January 1995), 90,000 (official rate 1 June 1994), 180,000 (black market rate 1 June 1994); 7,000 (official rate 16 December 1993), 50,000 (black market rate 16 December 1993); 3,884 (July 1993); 550 (April 1992); 90 (November 1991); 60 (October 1990)
Fiscal year: calendar year

Transportation

Railroads:
total: 3,189 km; note—limited trackage in use because of landmines still in place from the civil war; majority of the Benguela Railroad also closed because of civil war
narrow gauge: 2,879 km 1.067-m gauge; 310 km 0.600-m gauge
Highways:
total: 73,828 km
paved: bituminous-surface 8,577 km
unpaved: crushed stone, gravel, improved earth 29,350 km; unimproved earth 35,901 km
Inland waterways: 1,295 km navigable

Angola *(continued)*

Pipelines: crude oil 179 km
Ports: Ambriz, Cabinda, Lobito, Luanda, Malogo, Namibe, Porto Amboim, Soyo
Merchant marine:
total: 12 ships (1,000 GRT or over) totaling 63,776 GRT/99,863 DWT
ships by type: cargo 11, oil tanker 1
Airports:
total: 289
with paved runways over 3,047 m: 4
with paved runways 2,438 to 3,047 m: 9
with paved runways 1,524 to 2,437 m: 12
with paved runways 914 to 1,523 m: 6
with paved runways under 914 m: 93
with unpaved runways over 3,047 m: 1
with unpaved runways 2,438 to 3,047 m: 5
with unpaved runways 1,524 to 2,438 m: 33
with unpaved runways 914 to 1,523 m: 126

Communications

Telephone system: 40,300 telephones; 4.1 telephones/1,000 persons; high frequency radio used extensively for military links; telephone service limited mostly to government and business use
local: NA
intercity: limited system of wire, microwave radio relay, and troposcatter routes
international: 2 INTELSAT (Atlantic Ocean) earth stations
Radio:
broadcast stations: AM 17, FM 13, shortwave 0
radios: NA
Television:
broadcast stations: 6
televisions: NA

Defense Forces

Branches: Army, Navy, Air and Air Defense Forces, National Police Force
Manpower availability: males age 15-49 2,315,717; males fit for military service 1,166,082; males reach military age (18) annually 100,273 (1995 est.)
Defense expenditures: exchange rate conversion—$1.1 billion, 31% of GDP (1993)

Anguilla
(dependent territory of the UK)

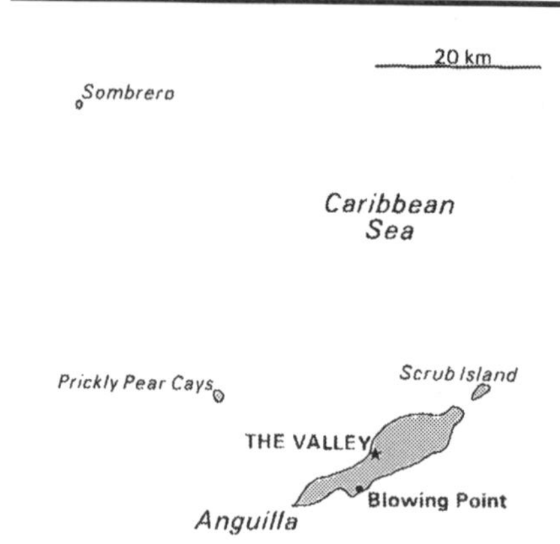

Geography

Location: Caribbean, island in the Caribbean Sea, east of Puerto Rico
Map references: Central America and the Caribbean
Area:
total area: 91 sq km
land area: 91 sq km
comparative area: about half the size of Washington, DC
Land boundaries: 0 km
Coastline: 61 km
Maritime claims:
exclusive fishing zone: 200 nm
territorial sea: 3 nm
International disputes: none
Climate: tropical; moderated by northeast trade winds
Terrain: flat and low-lying island of coral and limestone
Natural resources: negligible; salt, fish, lobster
Land use:
arable land: NA%
permanent crops: NA%
meadows and pastures: NA%
forest and woodland: NA%
other: NA% (mostly rock with sparse scrub oak, few trees, some commercial salt ponds)
Irrigated land: NA sq km
Environment:
current issues: supplies of potable water sometimes cannot meet increasing demand largely because of poor distribution system
natural hazards: frequent hurricanes and other tropical storms (July to October)
international agreements: NA

People

Population: 7,099 (July 1995 est.)
Age structure:
0-14 years: 32% (female 1,129; male 1,115)
15-64 years: 60% (female 2,101; male 2,126)
65 years and over: 8% (female 362; male 266) (July 1995 est.)
Population growth rate: 0.66% (1995 est.)
Birth rate: 24.09 births/1,000 population (1995 est.)
Death rate: 8.03 deaths/1,000 population (1995 est.)
Net migration rate: -9.44 migrant(s)/1,000 population (1995 est.)
Infant mortality rate: 17.3 deaths/1,000 live births (1995 est.)
Life expectancy at birth:
total population: 74.1 years
male: 71.32 years
female: 76.91 years (1995 est.)
Total fertility rate: 3.05 children born/woman (1995 est.)
Nationality:
noun: Anguillan(s)
adjective: Anguillan
Ethnic divisions: black African
Religions: Anglican 40%, Methodist 33%, Seventh-Day Adventist 7%, Baptist 5%, Roman Catholic 3%, other 12%
Languages: English (official)
Literacy: age 12 and over can read and write (1984)
total population: 95%
male: 95%
female: 95%
Labor force: 4,400 (1992)
by occupation: commerce 36%, services 29%, construction 18%, transportation and utilities 10%, manufacturing 3%, agriculture/fishing/forestry/mining 4%

Government

Names:
conventional long form: none
conventional short form: Anguilla
Digraph: AV
Type: dependent territory of the UK
Capital: The Valley
Administrative divisions: none (dependent territory of the UK)
Independence: none (dependent territory of the UK)
National holiday: Anguilla Day, 30 May
Constitution: Anguilla Constitutional Orders 1 April 1982; amended 1990
Legal system: based on English common law
Suffrage: 18 years of age; universal
Executive branch:
chief of state: Queen ELIZABETH II (since 6 February 1952), represented by Governor Alan W. SHAVE (since 14 August 1992)
head of government: Chief Minister Hubert HUGHES (since 16 March 1994)
cabinet: Executive Council; appointed by the governor from the elected members of the House of Assembly
Legislative branch: unicameral
House of Assembly: elections last held 16 March 1994 (next to be held March 1999); results—percent of vote by party NA; seats—(11 total, 7 elected) ANA 2, AUP 2, ADP 2,

independent 1
Judicial branch: High Court
Political parties and leaders: Anguilla National Alliance (ANA); Anguilla United Party (AUP), Hubert HUGHES; Anguilla Democratic Party (ADP), Victor BANKS
Member of: CARICOM (observer), CDB, INTERPOL (subbureau)
Diplomatic representation in US: none (dependent territory of the UK)
US diplomatic representation: none (dependent territory of the UK)
Flag: two horizontal bands of white (top, almost triple width) and light blue with three orange dolphins in an interlocking circular design centered in the white band; a new flag may have been in use since 30 May 1990

Economy

Overview: Anguilla has few natural resources, and the economy depends heavily on lobster fishing, offshore banking, tourism, and remittances from emigrants. In recent years the economy has benefited from a boom in tourism and construction. Development plans center around the improvement of the infrastructure, particularly transport and tourist facilities, and also light industry.
National product: GDP—purchasing power parity—$49 million (1993 est.)
National product real growth rate: 7.5% (1992)
National product per capita: $7,000 (1993 est.)
Inflation rate (consumer prices): 3% (1992 est.)
Unemployment rate: 7% (1992 est.)
Budget:
revenues: $13.8 million
expenditures: $15.2 million, including capital expenditures of $2.4 million (1992 est.)
Exports: $556,000 (f.o.b., 1992)
commodities: lobster and salt
partners: NA
Imports: $33.5 million (f.o.b., 1992)
commodities: NA
partners: NA
External debt: $NA
Industrial production: growth rate NA%
Electricity:
capacity: 2,000 kW
production: 6 million kWh
consumption per capita: 862 kWh (1992)
Industries: tourism, boat building, salt
Agriculture: pigeon peas, corn, sweet potatoes, sheep, goats, pigs, cattle, poultry, fishing (including lobster)
Economic aid:
recipient: Western (non-US) countries, ODA and OOF bilateral commitments (1970-89), $38 million
Currency: 1 EC dollar (EC$) = 100 cents
Exchange rates: East Caribbean dollars (EC$) per US$1—2.70 (fixed rate since 1976)
Fiscal year: NA

Transportation

Railroads: 0 km
Highways:
total: 105 km (1992 est.)
paved: 65 km
unpaved: gravel and earth 40 km
Ports: Blowing Point, Road Bay
Merchant marine: none
Airports:
total: 3
with paved runways 914 to 1,523 m: 1
with paved runways under 914 m: 2

Communications

Telephone system: 890 telephones; modern internal telephone system
local: NA
intercity: NA
international: radio relay microwave link to island of Saint Martin
Radio:
broadcast stations: AM 3, FM 1, shortwave 0
radios: NA
Television:
broadcast stations: 0
televisions: NA

Defense Forces

Note: defense is the responsibility of the UK

Antarctica

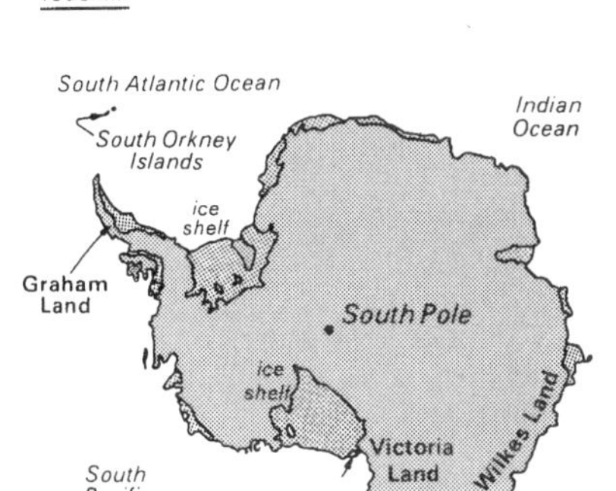

Geography

Location: continent mostly south of the Antarctic Circle
Map references: Antarctic Region
Area:
total area: 14 million sq km (est.)
land area: 14 million sq km (est.)
comparative area: slightly less than 1.5 times the size of the US
note: second-smallest continent (after Australia)
Land boundaries: none, but see entry on International disputes
Coastline: 17,968 km
Maritime claims: none, but see entry on International Disputes
International disputes: Antarctic Treaty defers claims (see Antarctic Treaty Summary below); sections (some overlapping) claimed by Argentina, Australia, Chile, France (Adelie Land), New Zealand (Ross Dependency), Norway (Queen Maud Land), and UK; the US and most other nations do not recognize the territorial claims of other nations and have made no claims themselves (the US reserves the right to do so); no formal claims have been made in the sector between 90 degrees west and 150 degrees west
Climate: severe low temperatures vary with latitude, elevation, and distance from the ocean; East Antarctica is colder than West Antarctica because of its higher elevation; Antarctic Peninsula has the most moderate climate; higher temperatures occur in January along the coast and average slightly below freezing
Terrain: about 98% thick continental ice sheet and 2% barren rock, with average elevations between 2,000 and 4,000 meters; mountain ranges up to 4,897 meters high; ice-free coastal areas include parts of southern Victoria Land, Wilkes Land, the Antarctic Peninsula area, and parts of Ross Island on McMurdo Sound; glaciers form ice shelves along about half of the coastline, and floating ice shelves constitute 11% of the area of the

continent
Natural resources: none presently exploited; iron ore, chromium, copper, gold, nickel, platinum and other minerals, and coal and hydrocarbons have been found in small, uncommercial quantities
Land use:
arable land: 0%
permanent crops: 0%
meadows and pastures: 0%
forest and woodland: 0%
other: 100% (ice 98%, barren rock 2%)
Irrigated land: 0 sq km
Environment:
current issues: in October 1991 it was reported that the ozone shield, which protects the Earth's surface from harmful ultraviolet radiation, had dwindled to the lowest level recorded over Antarctica since 1975 when measurements were first taken
natural hazards: katabatic (gravity-driven) winds blow coastward from the high interior; frequent blizzards form near the foot of the plateau; cyclonic storms form over the ocean and move clockwise along the coast; volcanism on Deception Island and isolated areas of West Antarctica; other seismic activity rare and weak
international agreements: NA
Note: the coldest, windiest, highest, and driest continent; during summer more solar radiation reaches the surface at the South Pole than is received at the Equator in an equivalent period; mostly uninhabitable

---

People

Population: no indigenous inhabitants; note—there are seasonally staffed research stations
Summer (January) population: over 4,115 total; Argentina 207, Australia 268, Belgium 13, Brazil 80, Chile 256, China NA, Ecuador NA, Finland 11, France 78, Germany 32, Greenpeace 12, India 60, Italy 210, Japan 59, South Korea 14, Netherlands 10, NZ 264, Norway 23, Peru 39, Poland NA, South Africa 79, Spain 43, Sweden 10, UK 116, Uruguay NA, US 1,666, former USSR 565 (1989-90)
Winter (July) population: over 1,046 total; Argentina 150, Australia 71, Brazil 12, Chile 73, China NA, France 33, Germany 19, Greenpeace 5, India 1, Japan 38, South Korea 14, NZ 11, Poland NA, South Africa 12, UK 69, Uruguay NA, US 225, former USSR 313 (1989-90)
Year-round stations: 42 total; Argentina 6, Australia 3, Brazil 1, Chile 3, China 2, Finland 1, France 1, Germany 1, India 1, Japan 2, South Korea 1, NZ 1, Poland 1, South Africa 3, UK 5, Uruguay 1, US 3, former USSR 6 (1990-91)
Summer only stations: over 38 total; Argentina 7, Australia 3, Chile 5, Germany 3, India 1, Italy 1, Japan 4, NZ 2, Norway 1, Peru 1, South Africa 1, Spain 1, Sweden 2, UK 1, US numerous, former USSR 5 (1989-90); note—the disintegration of the former USSR has placed the status and future of its Antarctic facilities in doubt; stations may be subject to closings at any time because of ongoing economic difficulties

---

Government

Names:
conventional long form: none
conventional short form: Antarctica
Digraph: AY
Type:
Antarctic Treaty Summary: The Antarctic Treaty, signed on 1 December 1959 and entered into force on 23 June 1961, establishes the legal framework for the management of Antarctica. Administration is carried out through consultative member meetings—the 18th Antarctic Treaty Consultative Meeting was in Japan in April 1993. Currently, there are 42 treaty member nations: 26 consultative and 16 acceding. Consultative (voting) members include the seven nations that claim portions of Antarctica as national territory (some claims overlap) and 19 nonclaimant nations. The US and some other nations that have made no claims have reserved the right to do so. The US does not recognize the claims of others. The year in parentheses indicates when an acceding nation was voted to full consultative (voting) status, while no date indicates the country was an original 1959 treaty signatory. Claimant nations are—Argentina, Australia, Chile, France, New Zealand, Norway, and the UK. Nonclaimant consultative nations are—Belgium, Brazil (1983), China (1985), Ecuador (1990), Finland (1989), Germany (1981), India (1983), Italy (1987), Japan, South Korea (1989), Netherlands (1990), Peru (1989), Poland (1977), South Africa, Spain (1988), Sweden (1988), Uruguay (1985), the US, and Russia. Acceding (nonvoting) members, with year of accession in parentheses, are—Austria (1987), Bulgaria (1978), Canada (1988), Colombia (1988), Cuba (1984), Czech Republic (1993), Denmark (1965), Greece (1987), Guatemala (1991), Hungary (1984), North Korea (1987), Papua New Guinea (1981), Romania (1971), Slovakia (1993), Switzerland (1990), and Ukraine (1992).
Article 1: area to be used for peaceful purposes only; military activity, such as weapons testing, is prohibited, but military personnel and equipment may be used for scientific research or any other peaceful purpose
Article 2: freedom of scientific investigation and cooperation shall continue
Article 3: free exchange of information and personnel in cooperation with the UN and other international agencies
Article 4: does not recognize, dispute, or establish territorial claims and no new claims shall be asserted while the treaty is in force
Article 5: prohibits nuclear explosions or disposal of radioactive wastes
Article 6: includes under the treaty all land and ice shelves south of 60 degrees 00 minutes south
Article 7: treaty-state observers have free access, including aerial observation, to any area and may inspect all stations, installations, and equipment; advance notice of all activities and of the introduction of military personnel must be given
Article 8: allows for jurisdiction over observers and scientists by their own states
Article 9: frequent consultative meetings take place among member nations
Article 10: treaty states will discourage activities by any country in Antarctica that are contrary to the treaty
Article 11: disputes to be settled peacefully by the parties concerned or, ultimately, by the ICJ
Article 12, 13, 14: deal with upholding, interpreting, and amending the treaty among involved nations
Other agreements: more than 170 recommendations adopted at treaty consultative meetings and ratified by governments include—Agreed Measures for the Conservation of Antarctic Fauna and Flora (1964); Convention for the Conservation of Antarctic Seals (1972); Convention on the Conservation of Antarctic Marine Living Resources (1980); a mineral resources agreement was signed in 1988 but was subsequently rejected; in 1991 the Protocol on Environmental Protection to the Antarctic Treaty was signed and awaits ratification; this agreement provides for the protection of the Antarctic environment through five specific annexes on marine pollution, fauna, and flora, environmental impact assessments, waste management, and protected areas; it also prohibits all activities relating to mineral resources except scientific research; 14 parties have ratified Protocol as of April 1995
Legal system: US law, including certain criminal offenses by or against US nationals, such as murder, may apply to areas not under jurisdiction of other countries. Some US laws directly apply to Antarctica. For example, the Antarctic Conservation Act, 16 U.S.C. section 2401 et seq., provides civil and criminal penalties for the following activities, unless authorized by regulation of statute: The taking of native mammals or birds; the introduction of nonindigenous plants and animals; entry into specially protected or scientific areas; the discharge or disposal of pollutants; and the importation into the US of certain items from Antarctica. Violation of the Antarctic Conservation Act carries penalties of up to $10,000 in fines and 1 year in prison. The Departments of Treasury, Commerce, Transportation, and Interior share enforcement responsibilities. Public Law 95-541, the US Antarctic Conservation Act of 1978, requires expeditions from the US to Antarctica to

notify, in advance, the Office of Oceans and Polar Affairs, Room 5801, Department of State, Washington, DC 20520, which reports such plans to other nations as required by the Antarctic Treaty. For more information contact Permit Office, Office of Polar Programs, National Science Foundation, Arlington, Virginia 22230 (703-306-1031).

Economy

Overview: No economic activity at present except for fishing off the coast and small-scale tourism, both based abroad.

Transportation

Ports: none; offshore anchorage
Airports: 42 landing facilities at different locations operated by 15 national governments party to the Treaty; one additional air facility operated by commercial (nongovernmental) tourist organization; helicopter pads at 36 of these locations; runways at 14 locations are gravel, sea ice, glacier ice, or compacted snow surface suitable for wheeled fixed-wing aircraft; no paved runways; 15 locations have snow-surface skiways limited to use by ski-equipped planes—11 runways/skiways 1,000 to 3,000 m, 5 runways/skiways less than 1,000 m, 8 runways/skiways greater than 3,000 m, and 5 of unspecified or variable length; airports generally subject to severe restrictions and limitations resulting from extreme seasonal and geographic conditions; airports do not meet ICAO standards; advance approval from the respective governmental or non-governmental operating organization required for landing

Communications

Telephone system:
local: NA
intercity: NA
international: NA
Radio:
broadcast stations: AM NA, FM NA, shortwave NA
radios: NA
Television:
broadcast stations: NA
televisions: NA

Defense Forces

Note: the Antarctic Treaty prohibits any measures of a military nature, such as the establishment of military bases and fortifications, the carrying out of military maneuvers, or the testing of any type of weapon; it permits the use of military personnel or equipment for scientific research or for any other peaceful purposes

Antigua and Barbuda

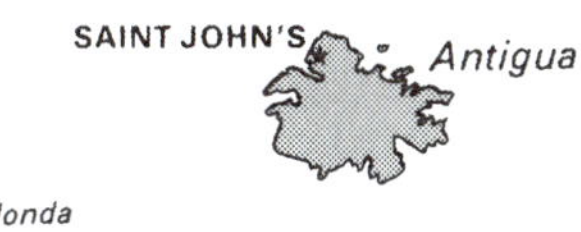

Geography

Location: Caribbean, islands between the Caribbean Sea and the North Atlantic Ocean, east-southeast of Puerto Rico
Map references: Central America and the Caribbean
Area:
total area: 440 sq km
land area: 440 sq km
comparative area: slightly less than 2.5 times the size of Washington, DC
note: includes Redonda
Land boundaries: 0 km
Coastline: 153 km
Maritime claims:
contiguous zone: 24 nm
continental shelf: 200 nm or to the edge of the continental margin
exclusive economic zone: 200 nm
territorial sea: 12 nm
International disputes: none
Climate: tropical marine; little seasonal temperature variation
Terrain: mostly low-lying limestone and coral islands with some higher volcanic areas
Natural resources: negligible; pleasant climate fosters tourism
Land use:
arable land: 18%
permanent crops: 0%
meadows and pastures: 7%
forest and woodland: 16%
other: 59%
Irrigated land: NA sq km
Environment:
current issues: water management—a major concern because of limited natural fresh water resources—is further hampered by the clearing of trees to increase crop production, causing rainfall to run off quickly
natural hazards: hurricanes and tropical storms (July to October); periodic droughts
international agreements: party to—Biodiversity, Climate Change, Environmental Modification, Hazardous Wastes, Law of the Sea, Marine Dumping, Nuclear Test Ban, Ozone Layer Protection, Ship Pollution, Whaling

People

Population: 65,176 (July 1995 est.)
Age structure:
0-14 years: 25% (female 8,062; male 8,390)
15-64 years: 69% (female 22,342; male 22,334)
65 years and over: 6% (female 2,231; male 1,817) (July 1995 est.)
Population growth rate: 0.68% (1995 est.)
Birth rate: 17.08 births/1,000 population (1995 est.)
Death rate: 5.35 deaths/1,000 population (1995 est.)
Net migration rate: -4.91 migrant(s)/1,000 population (1995 est.)
Infant mortality rate: 17.8 deaths/1,000 live births (1995 est.)
Life expectancy at birth:
total population: 73.4 years
male: 71.32 years
female: 75.57 years (1995 est.)
Total fertility rate: 1.68 children born/woman (1995 est.)
Nationality:
noun: Antiguan(s), Barbudan(s)
adjective: Antiguan, Barbudan
Ethnic divisions: black African, British, Portuguese, Lebanese, Syrian
Religions: Anglican (predominant), other Protestant sects, some Roman Catholic
Languages: English (official), local dialects
Literacy: age 15 and over has completed five or more years of schooling (1960)
total population: 89%
male: 90%
female: 88%
Labor force: 30,000
by occupation: commerce and services 82%, agriculture 11%, industry 7% (1983)

Government

Names:
conventional long form: none
conventional short form: Antigua and Barbuda
Digraph: AC
Type: parliamentary democracy
Capital: Saint John's
Administrative divisions: 6 parishes and 2 dependencies*; Barbuda*, Redonda*, Saint George, Saint John, Saint Mary, Saint Paul, Saint Peter, Saint Philip
Independence: 1 November 1981 (from UK)
National holiday: Independence Day, 1 November (1981)
Constitution: 1 November 1981
Legal system: based on English common law
Suffrage: 18 years of age; universal
Executive branch:
chief of state: Queen ELIZABETH II (since 6 February 1952), represented by Governor

General James B. CARLISLE (since NA 1993)
head of government: Prime Minister Lester Bryant BIRD (since 8 March 1994)
cabinet: Council of Ministers; appointed by the governor general on the advice of the prime minister
Legislative branch: bicameral Parliament
Senate: 17 member body appointed by the governor general
House of Representatives: elections last held 8 March 1994 (next to be held NA 1999); results—percent of vote by party NA; seats—(17 total) ALP 11, UPP 5, independent 1
Judicial branch: Eastern Caribbean Supreme Court
Political parties and leaders: Antigua Labor Party (ALP), Lester Bryant BIRD; United Progressive Party (UPP), Baldwin SPENCER
Other political or pressure groups: United Progressive Party (UPP), headed by Baldwin SPENCER, a coalition of three opposition political parties—the United National Democratic Party (UNDP); the Antigua Caribbean Liberation Movement (ACLM); and the Progressive Labor Movement (PLM); Antigua Trades and Labor Union (ATLU), headed by William ROBINSON
Member of: ACP, C, CARICOM, CDB, ECLAC, FAO, G-77, GATT, IBRD, ICAO, ICFTU, ICRM, IFAD, IFC, IFRCS (associate), ILO, IMF, IMO, INTELSAT (nonsignatory user), INTERPOL, IOC, ISO (subscriber), ITU, NAM (observer), OAS, OECS, OPANAL, UN, UNCTAD, UNESCO, UPU, WCL, WFTU, WHO, WMO
Diplomatic representation in US:
chief of mission: Ambassador Patrick Albert LEWIS
chancery: 3216 New Mexico Avenue NW, Washington, DC 20016
telephone: [1] (202) 362-5211, 5166, 5122
FAX: [1] (202) 362-5225
consulate(s) general: Miami
US diplomatic representation: the post was closed 30 June 1994; the US Ambassador to Barbados is accredited to Antigua and Barbuda
Flag: red with an inverted isosceles triangle based on the top edge of the flag; the triangle contains three horizontal bands of black (top), light blue, and white with a yellow rising sun in the black band

Economy

Overview: The economy is primarily service oriented, with tourism the most important determinant of economic performance. In 1993, tourism made a direct contribution to GDP of about 17%, and also spurred growth in other sectors such as construction and transport. While only accounting for roughly 5% of GDP in 1993, agricultural production increased by 4%. Tourist arrivals remained strong in 1994.
National product: GDP—purchasing power parity—$400 million (1993 est.)
National product real growth rate: 3.4% (1993)
National product per capita: $6,000 (1993 est.)
Inflation rate (consumer prices): 7% (1993)
Unemployment rate: 6% (1992 est.)
Budget:
revenues: $105 million
expenditures: $161 million, including capital expenditures of $56 million (1992)
Exports: $54.7 million (f.o.b., 1992)
commodities: petroleum products 48%, manufactures 23%, food and live animals 4%, machinery and transport equipment 17%
partners: OECS 26%, Barbados 15%, Guyana 4%, Trinidad and Tobago 2%, US 0.3%
Imports: $260.9 million (f.o.b., 1992)
commodities: food and live animals, machinery and transport equipment, manufactures, chemicals, oil
partners: US 27%, UK 16%, Canada 4%, OECS 3%, other 50%
External debt: $250 million (1990 est.)
Industrial production: growth rate -4.9% (1993 est.); accounts for 6.5% of GDP
Electricity:
capacity: 52,100 kW
production: 95 million kWh
consumption per capita: 1,242 kWh (1993)
Industries: tourism, construction, light manufacturing (clothing, alcohol, household appliances)
Agriculture: accounts for 5% of GDP; expanding output of cotton, fruits, vegetables, and livestock; other crops—bananas, coconuts, cucumbers, mangoes, sugarcane; not self-sufficient in food
Illicit drugs: a long-time but relatively minor transshipment point for narcotics bound for the US and Europe and recent transshipment point for heroin from Europe to the US; more significant as a drug money laundering center
Economic aid:
recipient: US commitments (1985-88), $10 million; Western (non-US) countries, ODA and OOF bilateral commitments (1970-89), $50 million
Currency: 1 EC dollar (EC$) = 100 cents
Exchange rates: East Caribbean dollars (EC$) per US$1—2.70 (fixed rate since 1976)
Fiscal year: 1 April—31 March

Transportation

Railroads:
total: 77 km
narrow gauge: 64 km 0.760-m gauge; 13 km 0.610-m gauge (used almost exclusively for handling sugar cane)
Highways:
total: 240 km
paved: NA
unpaved: NA
Ports: Saint John's
Merchant marine:
total: 304 ships (1,000 GRT or over) totaling 1,188,113 GRT/1,651,190 DWT
ships by type: bulk 7, cargo 216, chemical tanker 8, container 48, liquefied gas tanker 3, oil tanker 1, refrigerated cargo 10, roll-on/roll-off cargo 11
note: a flag of convenience registry
Airports:
total: 3
with paved runways 2,438 to 3,047 m: 1
with paved runways under 914 m: 2

Communications

Telephone system: 6,700 telephones; good automatic telephone system
local: NA
intercity: NA
international: 1 coaxial submarine cable; 1 INTELSAT (Atlantic Ocean) earth station; tropospheric scatter links with Saba and Guadeloupe
Radio:
broadcast stations: AM 4, FM 2, shortwave 2
radios: NA
Television:
broadcast stations: 2
televisions: NA

Defense Forces

Branches: Royal Antigua and Barbuda Defense Force, Royal Antigua and Barbuda Police Force (includes the Coast Guard)
Defense expenditures: exchange rate conversion—$1.4 million, 1% of GDP (FY90/91)

Arctic Ocean

Geography

Location: body of water mostly north of the Arctic Circle
Map references: Arctic Region
Area:
total area: 14.056 million sq km
comparative area: slightly more than 1.5 times the size of the US; smallest of the world's four oceans (after Pacific Ocean, Atlantic Ocean, and Indian Ocean)
note: includes Baffin Bay, Barents Sea, Beaufort Sea, Chukchi Sea, East Siberian Sea, Greenland Sea, Hudson Bay, Hudson Strait, Kara Sea, Laptev Sea, Northwest Passage, and other tributary water bodies
Coastline: 45,389 km
International disputes: some maritime disputes (see littoral states); Svalbard is the focus of a maritime boundary dispute between Norway and Russia
Climate: polar climate characterized by persistent cold and relatively narrow annual temperature ranges; winters characterized by continuous darkness, cold and stable weather conditions, and clear skies; summers characterized by continuous daylight, damp and foggy weather, and weak cyclones with rain or snow
Terrain: central surface covered by a perennial drifting polar icepack that averages about 3 meters in thickness, although pressure ridges may be three times that size; clockwise drift pattern in the Beaufort Gyral Stream, but nearly straight line movement from the New Siberian Islands (Russia) to Denmark Strait (between Greenland and Iceland); the ice pack is surrounded by open seas during the summer, but more than doubles in size during the winter and extends to the encircling land masses; the ocean floor is about 50% continental shelf (highest percentage of any ocean) with the remainder a central basin interrupted by three submarine ridges (Alpha Cordillera, Nansen Cordillera, and Lomonsov Ridge); maximum depth is 4,665 meters in the Fram Basin
Natural resources: sand and gravel aggregates, placer deposits, polymetallic nodules, oil and gas fields, fish, marine mammals (seals and whales)
Environment:
current issues: endangered marine species include walruses and whales; fragile ecosystem slow to change and slow to recover from disruptions or damage
natural hazards: ice islands occasionally break away from northern Ellesmere Island; icebergs calved from glaciers in western Greenland and extreme northeastern Canada; permafrost in islands; virtually icelocked from October to June; ships subject to superstructure icing from October to May
international agreements: NA
Note: major chokepoint is the southern Chukchi Sea (northern access to the Pacific Ocean via the Bering Strait); strategic location between North America and Russia; shortest marine link between the extremes of eastern and western Russia, floating research stations operated by the US and Russia; maximum snow cover in March or April about 20 to 50 centimeters over the frozen ocean and lasts about 10 months

Government

Digraph: XQ

Economy

Overview: Economic activity is limited to the exploitation of natural resources, including petroleum, natural gas, fish, and seals.

Transportation

Ports: Churchill (Canada), Murmansk (Russia), Prudhoe Bay (US)
Note: sparse network of air, ocean, river, and land routes; the Northwest Passage (North America) and Northern Sea Route (Eurasia) are important seasonal waterways

Communications

Telephone system:
international: no submarine cables

Argentina

Geography

Location: Southern South America, bordering the South Atlantic Ocean, between Chile and Uruguay
Map references: South America
Area:
total area: 2,766,890 sq km
land area: 2,736,690 sq km
comparative area: slightly less than three-tenths the size of the US
Land boundaries: total 9,665 km, Bolivia 832 km, Brazil 1,224 km, Chile 5,150 km, Paraguay 1,880 km, Uruguay 579 km
Coastline: 4,989 km
Maritime claims:
contiguous zone: 24 nm
continental shelf: 200 nm or to the edge of the continental margin
exclusive economic zone: 200 nm
territorial sea: 12 nm
International disputes: short section of the boundary with Uruguay is in dispute; short section of the boundary with Chile is indefinite; claims British-administered Falkland Islands (Islas Malvinas); claims British-administered South Georgia and the South Sandwich Islands; territorial claim in Antarctica
Climate: mostly temperate; arid in southeast; subantarctic in southwest
Terrain: rich plains of the Pampas in northern half, flat to rolling plateau of Patagonia in south, rugged Andes along western border
Natural resources: fertile plains of the pampas, lead, zinc, tin, copper, iron ore, manganese, petroleum, uranium
Land use:
arable land: 9%
permanent crops: 4%
meadows and pastures: 52%
forest and woodland: 22%
other: 13%
Irrigated land: 17,600 sq km (1989 est.)
Environment:
current issues: erosion results from inadequate flood controls and improper land use practices;

irrigated soil degradation; desertification; air pollution in Buenos Aires and other major cites; water pollution in urban areas; rivers becoming polluted due to increased pesticide and fertilizer use
natural hazards: Tucuman and Mendoza areas in the Andes subject to earthquakes; pamperos are violent windstorms that can strike the Pampas and northeast; heavy flooding
international agreements: party to—Antarctic-Environmental Protocol, Antarctic Treaty, Biodiversity, Climate Change, Endangered Species, Environmental Modification, Hazardous Wastes, Marine Dumping, Nuclear Test Ban, Ozone Layer Protection, Ship Pollution, Wetlands, Whaling; signed, but not ratified—Desertification, Law of the Sea, Marine Life Conservation
Note: second-largest country in South America (after Brazil); strategic location relative to sea lanes between South Atlantic and South Pacific Oceans (Strait of Magellan, Beagle Channel, Drake Passage)

People

Population: 34,292,742 (July 1995 est.)
Age structure:
0-14 years: 28% (female 4,706,793; male 4,903,589)
15-64 years: 62% (female 10,680,074; male 10,689,728)
65 years and over: 10% (female 1,922,552; male 1,390,006) (July 1995 est.)
Population growth rate: 1.11% (1995 est.)
Birth rate: 19.51 births/1,000 population (1995 est.)
Death rate: 8.62 deaths/1,000 population (1995 est.)
Net migration rate: 0.19 migrant(s)/1,000 population (1995 est.)
Infant mortality rate: 28.8 deaths/1,000 live births (1995 est.)
Life expectancy at birth:
total population: 71.51 years
male: 68.22 years
female: 74.97 years (1995 est.)
Total fertility rate: 2.65 children born/woman (1995 est.)
Nationality:
noun: Argentine(s)
adjective: Argentine
Ethnic divisions: white 85%, mestizo, Indian, or other nonwhite groups 15%
Religions: nominally Roman Catholic 90% (less than 20% practicing), Protestant 2%, Jewish 2%, other 6%
Languages: Spanish (official), English, Italian, German, French
Literacy: age 15 and over can read and write (1990 est.)
total population: 95%
male: 96%
female: 95%
Labor force: 10.9 million
by occupation: agriculture 12%, industry 31%, services 57% (1985 est.)

Government

Names:
conventional long form: Argentine Republic
conventional short form: Argentina
local long form: Republica Argentina
local short form: Argentina
Digraph: AR
Type: republic
Capital: Buenos Aires
Administrative divisions: 23 provinces (provincias, singular—provincia), and 1 federal district* (distrito federal); Buenos Aires; Catamarca; Chaco; Chubut; Cordoba; Corrientes; Distrito Federal*; Entre Rios; Formosa; Jujuy; La Pampa; La Rioja; Mendoza; Misiones; Neuquen; Rio Negro; Salta; San Juan; San Luis; Santa Cruz; Santa Fe; Santiago del Estero; Tierra del Fuego, Antartida e Islas del Atlantico Sur; Tucuman
note: the US does not recognize any claims to Antarctica or Argentina's claims to the Falkland Islands
Independence: 9 July 1816 (from Spain)
National holiday: Revolution Day, 25 May (1810)
Constitution: 1 May 1853; revised August 1994
Legal system: mixture of US and West European legal systems; has not accepted compulsory ICJ jurisdiction
Suffrage: 18 years of age; universal
Executive branch:
chief of state and head of government: President Carlos Saul MENEM (since 8 July 1989); Vice President (position vacant); election last held 14 May 1995 (next to be held NA May 1999); results—Carlos Saul MENEM was reelected
cabinet: Cabinet; appointed by the president
Legislative branch: bicameral National Congress (Congreso Nacional)
Senate: elections last held May 1989, but provincial elections in late 1991 set the stage for indirect elections by provincial senators for one-third of 48 seats in the national senate in May 1992; seats (48 total)—PJ 29, UCR 11, others 7, vacant 1
Chamber of Deputies: elections last held 3 October 1993 (next to be held October 1995); elections are held every two years and half of the total membership is elected each time for four year terms; seats—(257 total) PJ 122, UCR 83, MODIN 7, UCD 5, other 40
Judicial branch: Supreme Court (Corte Suprema)
Political parties and leaders: Justicialist Party (PJ), Carlos Saul MENEM, Peronist umbrella political organization; Radical Civic Union (UCR),Raul ALFONSIN, moderately left-of-center party; Union of the Democratic Center (UCD), Jorge AGUADO, conservative party; Dignity and Independence Political Party (MODIN), Aldo RICO, right-wing party; Grand Front (Frente Grande), Carlos ALVAREZ, center-left coalition; several provincial parties
Other political or pressure groups: Peronist-dominated labor movement; General Confederation of Labor (CGT; Peronist-leaning umbrella labor organization); Argentine Industrial Union (manufacturers' association); Argentine Rural Society (large landowners' association); business organizations; students; the Roman Catholic Church; the Armed Forces
Member of: AfDB, AG (observer), Australia Group, BCIE, CCC, ECLAC, FAO, G- 6, G-11, G-15, G-19, G-24, G-77, GATT, IADB, IAEA, IBRD, ICAO, ICC, ICFTU, ICRM, IDA, IFAD, IFC, IFRCS, ILO, IMF, IMO, INMARSAT, INTELSAT, INTERPOL, IOC, IOM, ISO, ITU, LAES, LAIA, MERCOSUR, MINURSO, MTCR, NSG (observer), OAS, ONUSAL, OPANAL, PCA, RG, UN, UNAVEM II, UNCTAD, UNESCO, UNFICYP, UNHCR, UNIDO, UNIKOM, UNITAR, UNMIH, UNOMOZ, UNPROFOR, UNTSO, UNU, UPU, WCL, WFTU, WHO, WIPO, WMO, WTO
Diplomatic representation in US:
chief of mission: Ambassador Raul Enrique GRANILLO OCAMPO
chancery: 1600 New Hampshire Avenue NW, Washington, DC 20009
telephone: [1] (202) 939-6400 through 6403
consulate(s) general: Atlanta, Chicago, Houston, Los Angeles, Miami, New Orleans, New York, San Francisco, and San Juan (Puerto Rico)
US diplomatic representation:
chief of mission: Ambassador James R. CHEEK
embassy: 4300 Colombia, 1425 Buenos Aires
mailing address: Unit 4334; APO AA 34034
telephone: [54] (1) 777-4533, 4534
FAX: [54] (1) 777-0197
Flag: three equal horizontal bands of light blue (top), white, and light blue; centered in the white band is a radiant yellow sun with a human face known as the Sun of May

Economy

Overview: Argentina, rich in natural resources, benefits also from a highly literate population, an export-oriented agricultural sector, and a diversified industrial base. Nevertheless, following decades of mismanagement and statist policies, the economy in the late 1980s was plagued with huge external debts and recurring bouts of hyperinflation. Elected in 1989, in the depths of recession, President MENEM has implemented a comprehensive economic restructuring program that shows signs of putting Argentina on a path of stable,

sustainable growth. Argentina's currency has traded at par with the US dollar since April 1991, and inflation has fallen to its lowest level in 20 years. Argentines have responded to the relative price stability by repatriating flight capital and investing in domestic industry. The economy registered an impressive 6% advance in 1994, fueled largely by inflows of foreign capital and strong domestic consumption spending. The government's major short term objective is encouraging exports, e.g., by reducing domestic costs of production. At the start of 1995, the government had to deal with the spillover from international financial movements associated with the devaluation of the Mexican peso. In addition, unemployment had become a serious issue for the government. Despite average annual 7% growth in 1991-94, unemployment surprisingly has doubled—due mostly to layoffs in government bureaus and in privatized industrial firms and utilities and, to a lesser degree, to illegal immigration. Much remains to be done in the 1990s in dismantling the old statist barriers to growth, extending the recent economic gains, and bringing down the rate of unemployment.
National product: GDP—purchasing power parity—$270.8 billion (1994 est.)
National product real growth rate: 6% (1994 est.)
National product per capita: $7,990 (1994 est.)
Inflation rate (consumer prices): 3.9% (1994 est.)
Unemployment rate: 12% (1994 est.)
Budget:
revenues: $48.46 billion
expenditures: $46.5 billion, including capital expenditures of $3.5 billion (1994 est.)
Exports: $15.7 billion (f.o.b., 1994 est.)
commodities: meat, wheat, corn, oilseed, manufactures
partners: US 12%, Brazil, Italy, Japan, Netherlands
Imports: $21.4 billion (c.i.f., 1994 est.)
commodities: machinery and equipment, chemicals, metals, fuels and lubricants, agricultural products
partners: US 22%, Brazil, Germany, Bolivia, Japan, Italy, Netherlands
External debt: $73 billion (April 1994)
Industrial production: growth rate 12.5% accounts for 31% of GDP (1994 est.)
Electricity:
capacity: 17,330,000 kW
production: 54.8 billion kWh
consumption per capita: 1,610 kWh (1993)
Industries: food processing, motor vehicles, consumer durables, textiles, chemicals and petrochemicals, printing, metallurgy, steel
Agriculture: accounts for 8% of GDP (including fishing); produces abundant food for both domestic consumption and exports; among world's top five exporters of grain and beef; principal crops—wheat, corn, sorghum, soybeans, sugar beets
Illicit drugs: increasing use as a transshipment country for cocaine headed for the US and Europe
Economic aid:
recipient: US commitments, including Ex-Im (FY70-89), $1 billion; Western (non-US) countries, ODA and OOF bilateral commitments (1970-89), $4.4 billion; Communist countries (1970-89), $718 million
Currency: 1 nuevo peso argentino = 100 centavos
Exchange rates: pesos per US$1—0.99870 (December 1994), 0.99901 (1994), 0.99895 (1993), 0.99064 (1992), 0.95355 (1991), 0.48759 (1990)
Fiscal year: calendar year

Transportation

Railroads:
total: 34,572 km
broad gauge: NA km 1.676-m gauge
standard gauge: NA km 1.435-m
narrow gauge: 400 km 0.750-m gauge; NA km 1.000-m gauge (209 km electrified)
Highways:
total: 208,350 km
paved: 57,000 km
unpaved: gravel 39,500 km; improved/unimproved earth 111,850 km
Inland waterways: 11,000 km navigable
Pipelines: crude oil 4,090 km; petroleum products 2,900 km; natural gas 9,918 km
Ports: Bahia Blanca, Buenos Aires, Comodoro Rivadavia, Concepcion del Uruguay, La Plata, Mar del Plata, Necochea, Rio Gallegos, Rosario, Santa Fe, Ushuaia
Merchant marine:
total: 44 ships (1,000 GRT or over) totaling 434,525 GRT/667,501 DWT
ships by type: bulk 3, cargo 21, chemical tanker 1, container 4, oil tanker 8, railcar carrier 1, refrigerated cargo 5, roll-on/roll-off cargo 1
Airports:
total: 1,602
with paved runways over 3,047 m: 5
with paved runways 2,438 to 3,047 m: 25
with paved runways 1,524 to 2,437 m: 55
with paved runways 914 to 1,523 m: 48
with paved runways under 914 m: 703
with unpaved runways over 3,047 m: 2
with unpaved runways 2,438 to 3,047 m: 1
with unpaved runways 1,524 to 2,438 m: 70
with unpaved runways 914 to 1,523 m: 693

Communications

Telephone system: 2,650,000 telephones; 12,000 public telephones; 78 telephones/1,000 persons; extensive modern system but many families do not have telephones; microwave widely used; however, during rainstorms, the telephone system frequently grounds out, even in Buenos Aires
local: NA
intercity: microwave radio relay and domestic satellite network with 40 earth stations
international: 2 INTELSAT (Atlantic Ocean) earth stations
Radio:
broadcast stations: AM 171, FM 0, shortwave 13
radios: NA
Television:
broadcast stations: 231
televisions: NA

Defense Forces

Branches: Argentine Army, Navy of the Argentine Republic, Argentine Air Force, National Gendarmerie, Argentine Naval Prefecture (Coast Guard only), National Aeronautical Police Force
Manpower availability: males age 15-49 8,573,780; males fit for military service 6,954,584; males reach military age (20) annually 301,166 (1995 est.)
Defense expenditures: $NA, NA% of GDP

Armenia

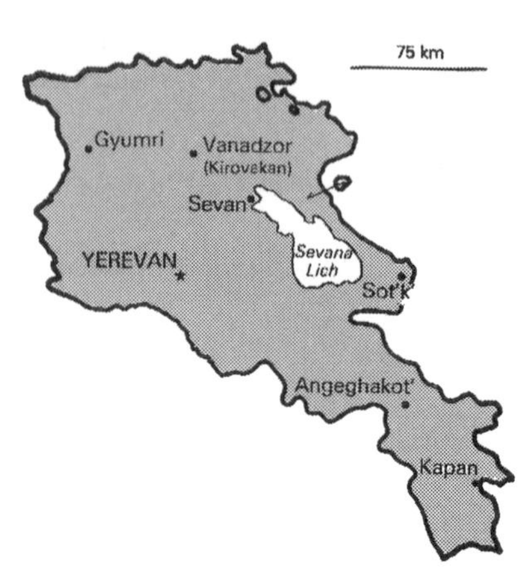

Geography

Location: Southwestern Asia, east of Turkey
Map references: Commonwealth of Independent States—European States
Area:
total area: 29,800 sq km
land area: 28,400 sq km
comparative area: slightly larger than Maryland
Land boundaries: total 1,254 km, Azerbaijan (east) 566 km, Azerbaijan (south) 221 km, Georgia 164 km, Iran 35 km, Turkey 268 km
Coastline: 0 km (landlocked)
Maritime claims: none; landlocked
International disputes: supports ethnic Armenians in Nagorno-Karabakh in their separatist conflict against the Azerbaijani government; traditional demands on former Armenian lands in Turkey have subsided
Climate: highland continental, hot summers, cold winters
Terrain: high Armenian Plateau with mountains; little forest land; fast flowing rivers; good soil in Aras River valley
Natural resources: small deposits of gold, copper, molybdenum, zinc, alumina
Land use:
arable land: 17%
permanent crops: 3%
meadows and pastures: 20%
forest and woodland: 0%
other: 60%
Irrigated land: 3,050 sq km (1990)
Environment:
current issues: soil pollution from toxic chemicals such as DDT; energy blockade, the result of conflict with Azerbaijan, has led to deforestation as citizens scavenge for firewood; pollution of Hrazdan (Razdan) and Aras Rivers; the draining of Sevana Lich, a result of its use as a source for hydropower, threatens drinking water supplies
natural hazards: occasionally severe earthquakes; droughts
international agreements: party to—Biodiversity, Climate Change, Nuclear Test Ban, Wetlands; signed, but not ratified—Desertification
Note: landlocked

People

Population: 3,557,284 (July 1995 est.)
Age structure:
0-14 years: 31% (female 542,664; male 570,998)
15-64 years: 61% (female 1,103,171; male 1,076,226)
65 years and over: 8% (female 154,784; male 109,441) (July 1995 est.)
Population growth rate: 0.94% (1995 est.)
Birth rate: 22.79 births/1,000 population (1995 est.)
Death rate: 6.66 deaths/1,000 population (1995 est.)
Net migration rate: -6.68 migrant(s)/1,000 population (1995 est.)
Infant mortality rate: 26 deaths/1,000 live births (1995 est.)
Life expectancy at birth:
total population: 72.36 years
male: 68.94 years
female: 75.95 years (1995 est.)
Total fertility rate: 3.06 children born/woman (1995 est.)
Nationality:
noun: Armenian(s)
adjective: Armenian
Ethnic divisions: Armenian 93%, Azeri 3%, Russian 2%, other (mostly Yezidi Kurds) 2% (1989)
note: as of the end of 1994, most Azeris had emigrated from Armenia
Religions: Armenian Orthodox 94%
Languages: Armenian 96%, Russian 2%, other 2%
Literacy: age 15 and over can read and write (1989)
total population: 99%
male: 99%
female: 98%
Labor force: 1.578 million
by occupation: industry and construction 34%, agriculture and forestry 31%, other 35% (1992)

Government

Names:
conventional long form: Republic of Armenia
conventional short form: Armenia
local long form: Hayastani Hanrapetut'yun
local short form: Hayastan
former: Armenian Soviet Socialist Republic; Armenian Republic
Digraph: AM
Type: republic
Capital: Yerevan
Administrative divisions: 37 regions (shrjanner, singular—shrjan) and 23 cities* (kaghakner, singular—kaghak); Abovyan*, Akhuryani Shrjan, Alaverdi*, Amasiayi Shrjan, Anii Shrjan, Aparani Shrjan, Aragatsi Shrjan, Ararat*, Ararati Shrjan, Armaviri Shrjan, Artashat*, Artashati Shrjan, Art'ik*, Art'iki Shrjan, Ashots'k'i Shrjan, Ashtarak*, Ashtaraki Shrjan, Baghramyani Shrjan, Ch'arents'avan*, Dilijan*, Ejmiatsin*, Ejmiatsni Shrjan, Goris*, Gorisi Shrjan, Gugark'i Shrjan, Gyumri*, Hoktemberyan*, Hrazdan*, Hrazdani Shrjan, Ijevan*, Ijevani Shrjan, Jermuk*, Kamo*, Kamoyi Shrjan, Kapan*, Kapani Shrjan, Kotayk'i Shrjan, Krasnoselski Shrjan, Martunu Shrjan, Masisi Shrjan, Meghru Shrjan, Metsamor*, Nairii Shrjan, Noyemberyani Shrjan, Sevan*, Sevani Shrjan, Sisiani Shrjan, Spitak*, Spitaki Shrjan, Step'anavan*, Step'anavani Shrjan, T'alini Shrjan, Tashiri Shrjan, Taushi Shrjan, T'umanyani Shrjan, Vanadzor*, Vardenisi Shrjan, Vayk'i Shrjan, Yeghegnadzori Shrjan, Yerevan*
Independence: 28 May 1918 (First Armenian Republic); 23 September 1991 (from Soviet Union)
National holiday: Referendum Day, 21 September
Constitution: adopted NA April 1978; referendum on new constitution to be held 5 July 1995
Legal system: based on civil law system
Suffrage: 18 years of age; universal
Executive branch:
chief of state: President Levon Akopovich TER-PETROSYAN (since October 1991) election last held 16 October 1991 (next to be held NA 1996); results—Levon Akopovich TER-PETROSYAN 86%; radical nationalists about 7%; note—Levon Akopovich TER-PETROSYAN was elected Chairman of the Armenian Supreme Soviet 4 August 1990 before becoming president
head of government: Prime Minister Hrant BAGRATYAN (since 16 February 1993); First Deputy Prime Minister Vigen CHITECHYAN (since 16 February 1993)
cabinet: Council of Ministers; appointed by the president
Legislative branch: unicameral
Supreme Soviet: elections last held 20 May 1990 (next to be held 5 July 1995); results—percent of vote by party NA; seats—(260 total) non-aligned 136, ANM 52, DPA 17, Democratic Liberal Party 17, ARF 12, NDU 9, Christian Democratic Party 1, Constitutional Rights Union 1, ONS 1, Republican Party 1, Nagorno-Karabakh representatives 13
Judicial branch: Supreme Court
Political parties and leaders: Armenian National Movement (ANM), Ter-Husik LAZARYAN, chairman; National Democratic Union (NDU), David VARTANYAN, chairman; Armenian Revolutionary Federation (ARF, Dashnaktsutyun); note—banned until reorganized; Democratic Party of Armenia (DPA; Communist Party), Aram SARKISYAN, chairman; Christian Democratic Party, Azat ARSHAKYAN,

chairman; Greens Party, Hakob SANASARIAN, chairman; Democratic Liberal Party, Rouben MIRZAKHANYAN, chairman; Republican Party, Ashot NAVARSARDYAN, chairman; Union for Self-Determination (ONS), Paruir AIRIKYAN, chairman
Member of: BSEC, CCC, CIS, EBRD, ECE, ESCAP, FAO, IAEA, IBRD, ICAO, IDA, IFAD, ILO, IMF, INTELSAT, INTERPOL, IOC, IOM, ITU, NACC, NAM (observer), OSCE, PFP, UN, UNCTAD, UNESCO, UNIDO, UPU, WHO, WIPO, WMO
Diplomatic representation in US:
chief of mission: Ambassador Ruben SHUGARIAN
chancery: Suite 210, 1660 L Street NW, Washington, DC 20036
telephone: [1] (202) 628-5766
FAX: [1] (202) 628-5769
US diplomatic representation:
chief of mission: Ambassador Harry J. GILMORE
embassy: 18 Gen Bagramian, Yerevan
mailing address: use embassy street address
telephone: [7] (8852) 151-144, 524-661
FAX: [7] (8852) 151-138
Flag: three equal horizontal bands of red (top), blue, and gold

Economy

Overview: Under the old Soviet central planning system, Armenia had developed a more modern industrial sector, supplying machine building equipment, textiles, and other manufactured goods to sister republics in exchange for raw materials and energy resources. Armenia is a large food importer and its mineral deposits (gold, bauxite) are small. The economic decline in recent years (1991-94) has been particularly severe due to the ongoing conflict over the ethnic Armenian-dominated region of Nagorno-Karabakh in Azerbaijan. Azerbaijan and Turkey have blockaded pipeline and railroad traffic to Armenia for its support of the Karabakh Armenians. This has left Armenia with chronic energy shortages because of a lack of capacity and frequent disruptions of natural gas deliveries through unstable Georgia, as well as difficulties in obtaining other types of fuel. In addition, bread is strictly rationed and there are shortages of other goods. In 1994, the economy seemed to bottom out. The government has managed to increase its financial and budgetary discipline, bringing inflation down from around 40% per month in first half 1994 to single digits in second half 1994 and the first quarter of 1995. A full economic recovery cannot be expected until the conflict is settled and the blockade lifted.
National product: GDP—purchasing power parity—$8.1 billion (1994 estimate as extrapolated from World Bank estimate for 1992)
National product real growth rate: -2% (1994 est.)
National product per capita: $2,290 (1994 est.)
Inflation rate (consumer prices): 27% per month average (1994 est.)
Unemployment rate: 6.5% of officially registered unemployed but large numbers of underemployed (1994 est.)
Budget:
revenues: $NA
expenditures: $NA, including capital expenditures of $NA
Exports: $43 million to countries outside the FSU (f.o.b., 1994)
commodities: gold and jewelry, aluminum, transport equipment, electrical equipment
partners: Iran, Russia, Turkmenistan, Georgia
Imports: $120 million from countries outside the FSU (c.i.f., 1994)
commodities: grain, other foods, fuel, other energy
partners: Iran, Russia, Turkmenistan, Georgia, US, EU
External debt: $NA
Industrial production: growth rate 7% (1994 est.); accounts for 41% of GDP
Electricity:
capacity: 4,620,000 kW
production: 5.7 billion kWh
consumption per capita: 1,620 kWh (1994)
Industries: traditionally diverse, including (as a percent of output of former USSR) metalcutting machine tools (5.5%), forging-pressing machines (1.9%), electric motors (9%), tires (1.5%), knitted wear (4.4%), hosiery (3.0%), shoes (2.2%), silk fabric (0.8%), washing machines (2.0%), chemicals, trucks, watches, instruments, and microelectronics (1990); currently, much of industry is shut down
Agriculture: only 17% of land area is arable; employs 31% of labor force as residents increasingly turn to subsistence agriculture; fruits (especially grapes) and vegetable farming, minor livestock sector; vineyards near Yerevan are famous for brandy and other liqueurs
Illicit drugs: illicit cultivator of cannabis mostly for domestic consumption; used as a transshipment point for illicit drugs to Western Europe
Economic aid:
recipient: considerable humanitarian aid, mostly food and energy products, from US and EU; Russia granted 60 billion rubles in technical credits in late 1994 and approved a 110 billion ruble credit almost half of which was to go toward the restart of the Metsamor nuclear power plant
Currency: 1 dram = 100 luma (introduced new currency in November 1993)
Exchange rates: dram per US$1—406 (end December 1994)
Fiscal year: calendar year

Transportation

Railroads:
total: 840 km in common carrier service; does not include industrial lines
broad gauge: 840 km 1.520-m gauge (1990)
Highways:
total: 11,300 km
paved: 10,500 km
unpaved: earth 800 km (1990)
Inland waterways: NA km
Pipelines: natural gas 900 km (1991)
Ports: none
Airports:
total: 11
with paved runways over 3,047 m: 2
with paved runways 1,524 to 2,437 m: 1
with paved runways 914 to 1,523 m: 2
with unpaved runways 1,524 to 2,438 m: 2
with unpaved runways 914 to 1,523 m: 3
with unpaved runways under 914 m: 1

Communications

Telephone system: about 650,000 telephones; 177 telephones/1,000 persons; progress on installation of fiber optic cable and construction of facilities for mobile cellular phone service remains in the negotiation phase for joint venture agreement
local: NA
intercity: NA
international: international connections to other former republics of the USSR are by landline or microwave and to other countries by satellite and by leased connection through the Moscow international gateway switch; 1 INTELSAT earth station
Radio:
broadcast stations: AM NA, FM NA, shortwave NA
radios: NA
Television:
broadcast stations: NA; note—100% of population receives Armenian and Russian TV programs
televisions: NA

Defense Forces

Branches: Army, Air and Air Defense Forces, National Guard, Security Forces (internal and border troops)
Manpower availability: males age 15-49 877,414; males fit for military service 699,167; males reach military age (18) annually 28,634 (1995 est.)
Defense expenditures: 250 million rubles, NA% of GDP (1992 est.); note—conversion of the military budget into US dollars using the current exchange rate could produce misleading results

Aruba

(part of the Dutch realm)

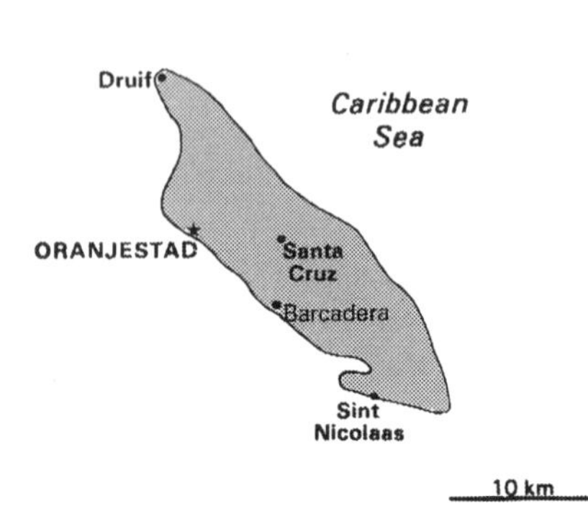

Geography

Location: Caribbean, island in the Caribbean Sea, north of Venezuela
Map references: Central America and the Caribbean
Area:
total area: 193 sq km
land area: 193 sq km
comparative area: slightly larger than Washington, DC
Land boundaries: 0 km
Coastline: 68.5 km
Maritime claims:
territorial sea: 12 nm
International disputes: none
Climate: tropical marine; little seasonal temperature variation
Terrain: flat with a few hills; scant vegetation
Natural resources: negligible; white sandy beaches
Land use:
arable land: 0%
permanent crops: 0%
meadows and pastures: 0%
forest and woodland: 0%
other: 100%
Irrigated land: NA sq km
Environment:
current issues: NA
natural hazards: lies outside the Caribbean hurricane belt
international agreements: NA

People

Population: 65,974 (July 1995 est.)
Age structure:
0-14 years: 23% (female 7,377; male 7,726)
15-64 years: 69% (female 24,269; male 21,141)
65 years and over: 8% (female 3,223; male 2,238) (July 1995 est.)
Population growth rate: 0.65% (1995 est.)
Birth rate: 14.6 births/1,000 population (1995 est.)
Death rate: 6.17 deaths/1,000 population (1995 est.)
Net migration rate: -1.91 migrant(s)/1,000 population (1995 est.)
Infant mortality rate: 8.3 deaths/1,000 live births (1995 est.)
Life expectancy at birth:
total population: 76.56 years
male: 72.89 years
female: 80.42 years (1995 est.)
Total fertility rate: 1.82 children born/woman (1995 est.)
Nationality:
noun: Aruban(s)
adjective: Aruban
Ethnic divisions: mixed European/Caribbean Indian 80%
Religions: Roman Catholic 82%, Protestant 8%, Hindu, Muslim, Confucian, Jewish
Languages: Dutch (official), Papiamento (a Spanish, Portuguese, Dutch, English dialect), English (widely spoken), Spanish
Literacy: NA%
Labor force: NA
by occupation: most employment is in the tourist industry (1995)

Government

Names:
conventional long form: none
conventional short form: Aruba
Digraph: AA
Type: part of the Dutch realm; full autonomy in internal affairs obtained in 1986 upon separation from the Netherlands Antilles
Capital: Oranjestad
Administrative divisions: none (self-governing part of the Netherlands)
Independence: none (part of the Dutch realm; in 1990, Aruba requested and received from the Netherlands cancellation of the agreement to automatically give independence to the island in 1996)
National holiday: Flag Day, 18 March
Constitution: 1 January 1986
Legal system: based on Dutch civil law system, with some English common law influence
Suffrage: 18 years of age; universal
Executive branch:
chief of state: Queen BEATRIX Wilhelmina Armgard (since 30 April 1980), represented by Governor General Olindo KOOLMAN (since 1 January 1992)
head of government: Prime Minister Jan (Henny) H. EMAN (since 29 July 1994)
cabinet: Council of Ministers; appointed with the advice and approval of the legislature
Legislative branch: unicameral
Legislature (Staten): elections last held 29 July 1994 (next to be held by NA July 1998); results—percent of vote by party NA; seats—(21 total) AVP 10, MEP 9, OLA 2
Judicial branch: Joint High Court of Justice
Political parties and leaders: Electoral Movement Party (MEP), Nelson ODUBER; Aruban People's Party (AVP), Jan (Henny) H. EMAN; National Democratic Action (ADN), Pedro Charro KELLY; New Patriotic Party (PPN), Eddy WERLEMEN; Aruban Patriotic Party (PPA), Benny NISBET; Aruban Democratic Party (PDA), Leo BERLINSKI; Democratic Action '86 (AD '86), Arturo ODUBER; Organization for Aruban Liberty (OLA), Glenbert CROES
note: governing coalition includes the MEP, PPA, and ADN
Member of: ECLAC (associate), INTERPOL, IOC, UNESCO (associate), WCL, WTO (associate)
Diplomatic representation in US: none (self-governing part of the Netherlands)
US diplomatic representation: none (self-governing part of the Netherlands)
Flag: blue with two narrow horizontal yellow stripes across the lower portion and a red, four-pointed star outlined in white in the upper hoist-side corner

Economy

Overview: Tourism is the mainstay of the Aruban economy, although offshore banking and oil refining and storage are also important. The rapid growth of the tourism sector over the last decade has resulted in a substantial expansion of other activities. Construction has boomed, with hotel capacity five times the 1985 level. Additionally, the reopening of the country's oil refinery in 1993, a major source of employment and foreign exchange earnings, has further spurred growth. Aruba's small labor force and less than 1% unemployment rate have led to a large number of unfilled job vacancies despite sharp rises in wage rates in recent years.
National product: GDP—purchasing power parity—$1.1 billion (1993 est.)
National product real growth rate: 5% (1993 est.)
National product per capita: $17,000 (1993 est.)
Inflation rate (consumer prices): 7% (1994 est.)
Unemployment rate: 0.6% (1992)
Budget:
revenues: $145 million
expenditures: $185 million, including capital expenditures of $42 million (1988)
Exports: $1.3 billion (including oil re-exports) (f.o.b., 1993 est.)
commodities: mostly refined petroleum products
partners: US 64%, EC
Imports: $1.6 billion (f.o.b., 1993 est.)
commodities: food, consumer goods, manufactures, petroleum products, crude oil for refining and re-export
partners: US 8%, EC
External debt: $81 million (1987)

Industrial production: growth rate NA%
Electricity:
capacity: 90,000 kW
production: 330 million kWh
consumption per capita: 4,761 kWh (1993)
Industries: tourism, transshipment facilities, oil refining
Agriculture: poor quality soils and low rainfall limit agricultural activity to the cultivation of aloes, some livestock, and fishing
Illicit drugs: drug money laundering center and transit point for narcotics bound for the US and Europe
Economic aid:
recipient: Western (non-US) countries ODA and OOF bilateral commitments (1980-89), $220 million
Currency: 1 Aruban florin (Af.) = 100 cents
Exchange rates: Aruban florins (Af.) per US$1—1.7900 (fixed rate since 1986)
Fiscal year: calendar year

Transportation

Railroads: 0 km
Highways:
total: NA
paved: NA
unpaved: NA
Ports: Barcadera, Oranjestad, Sint Nicolaas
Merchant marine: none
Airports:
total: 2
with paved runways 2,438 to 3,047 m: 1
with paved runways 914 to 1,523 m: 1
note: government-owned airport east of Oranjestad accepts transatlantic flights

Communications

Telephone system: 72,168 telephones; 1,100 telephones/1,000 persons; more than adequate
local: NA
intercity: extensive interisland microwave radio relay links
international: 1 submarine cable to Sint Maarten
Radio:
broadcast stations: AM 4, FM 4, shortwave 0
radios: NA
Television:
broadcast stations: 1
televisions: NA

Defense Forces

Note: defense is the responsibility of the Netherlands

Ashmore and Cartier Islands
(territory of Australia)

Geography

Location: Southeastern Asia, islands in the Indian Ocean, northwest of Australia
Map references: Southeast Asia
Area:
total area: 5 sq km
land area: 5 sq km
comparative area: about 8.5 times the size of The Mall in Washington, DC
note: includes Ashmore Reef (West, Middle, and East Islets) and Cartier Island
Land boundaries: 0 km
Coastline: 74.1 km
Maritime claims:
contiguous zone: 12 nm
continental shelf: 200-m depth or to depth of exploitation
exclusive fishing zone: 200 nm
territorial sea: 3 nm
International disputes: none
Climate: tropical
Terrain: low with sand and coral
Natural resources: fish
Land use:
arable land: 0%
permanent crops: 0%
meadows and pastures: 0%
forest and woodland: 0%
other: 100% (all grass and sand)
Irrigated land: 0 sq km
Environment:
current issues: NA
natural hazards: surrounded by shoals and reefs which can pose maritime hazards
international agreements: NA
Note: Ashmore Reef National Nature Reserve established in August 1983

People

Population: no indigenous inhabitants; note—there are only seasonal caretakers

Government

Names:
conventional long form: Territory of Ashmore and Cartier Islands
conventional short form: Ashmore and Cartier Islands
Digraph: AT
Type: territory of Australia administered by the Australian Ministry for the Environment, Sport, and Territories
Capital: none; administered from Canberra, Australia
Administrative divisions: none (territory of Australia)
Independence: none (territory of Australia)
Legal system: relevant laws of the Northern Territory of Australia
Diplomatic representation in US: none (territory of Australia)
US diplomatic representation: none (territory of Australia)

Economy

Overview: no economic activity

Transportation

Ports: none; offshore anchorage only

Defense Forces

Note: defense is the responsibility of Australia; periodic visits by the Royal Australian Navy and Royal Australian Air Force

Atlantic Ocean

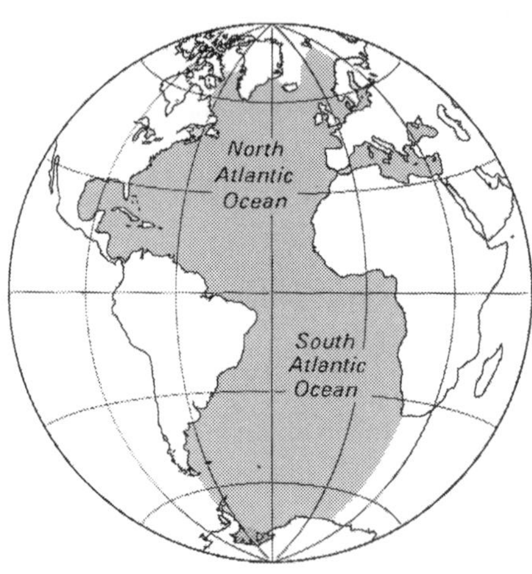

Geography

Location: body of water between Africa, Antarctica, and the Western Hemisphere
Map references: World
Area:
total area: 82.217 million sq km
comparative area: slightly less than nine times the size of the US; second-largest of the world's four oceans (after the Pacific Ocean, but larger than Indian Ocean or Arctic Ocean)
note: includes Baltic Sea, Black Sea, Caribbean Sea, Davis Strait, Denmark Strait, Drake Passage, Gulf of Mexico, Mediterranean Sea, North Sea, Norwegian Sea, Scotia Sea, Weddell Sea, and other tributary water bodies
Coastline: 111,866 km
International disputes: some maritime disputes (see littoral states)
Climate: tropical cyclones (hurricanes) develop off the coast of Africa near Cape Verde and move westward into the Caribbean Sea; hurricanes can occur from May to December, but are most frequent from August to November
Terrain: surface usually covered with sea ice in Labrador Sea, Denmark Strait, and Baltic Sea from October to June; clockwise warm water gyre (broad, circular system of currents) in the northern Atlantic, counterclockwise warm water gyre in the southern Atlantic; the ocean floor is dominated by the Mid-Atlantic Ridge, a rugged north-south centerline for the entire Atlantic basin; maximum depth is 8,605 meters in the Puerto Rico Trench
Natural resources: oil and gas fields, fish, marine mammals (seals and whales), sand and gravel aggregates, placer deposits, polymetallic nodules, precious stones
Environment:
current issues: endangered marine species include the manatee, seals, sea lions, turtles, and whales; driftnet fishing is exacerbating declining fish stocks and contributing to international disputes; municipal sludge pollution off eastern US, southern Brazil, and eastern Argentina; oil pollution in Caribbean Sea, Gulf of Mexico, Lake Maracaibo, Mediterranean Sea, and North Sea; industrial waste and municipal sewage pollution in Baltic Sea, North Sea, and Mediterranean Sea
natural hazards: icebergs common in Davis Strait, Denmark Strait, and the northwestern Atlantic Ocean from February to August and have been spotted as far south as Bermuda and the Madeira Islands; icebergs from Antarctica occur in the extreme southern Atlantic Ocean; ships subject to superstructure icing in extreme northern Atlantic from October to May and extreme southern Atlantic from May to October; persistent fog can be a maritime hazard from May to September
international agreements: NA
Note: major choke points include the Dardanelles, Strait of Gibraltar, access to the Panama and Suez Canals; strategic straits include the Strait of Dover, Straits of Florida, Mona Passage, The Sound (Oresund), and Windward Passage; the Equator divides the Atlantic Ocean into the North Atlantic Ocean and South Atlantic Ocean

Government

Digraph: ZH

Economy

Overview: The Atlantic Ocean provides some of the world's most heavily trafficked sea routes, between and within the Eastern and Western Hemispheres. Other economic activity includes the exploitation of natural resources, e.g., fishing, the dredging of aragonite sands (The Bahamas), and production of crude oil and natural gas (Caribbean Sea, Gulf of Mexico, and North Sea).

Transportation

Ports: Alexandria (Egypt), Algiers (Algeria), Antwerp (Belgium), Barcelona (Spain), Buenos Aires (Argentina), Casablanca (Morocco), Colon (Panama), Copenhagen (Denmark), Dakar (Senegal), Gdansk (Poland), Hamburg (Germany), Helsinki (Finland), Las Palmas (Canary Islands, Spain), Le Havre (France), Lisbon (Portugal), London (UK), Marseille (France), Montevideo (Uruguay), Montreal (Canada), Naples (Italy), New Orleans (US), New York (US), Oran (Algeria), Oslo (Norway), Piraeus (Greece), Rio de Janeiro (Brazil), Rotterdam (Netherlands), Saint Petersburg (Russia), Stockholm (Sweden)
Note: Kiel Canal and Saint Lawrence Seaway are two important waterways

Communications

Telephone system:
international: numerous submarine cables with most between continental Europe and the UK, North America and the UK, and in the Mediterranean; numerous direct links across Atlantic via INTELSAT satellite network

Australia

Geography

Location: Oceania, continent between the Indian Ocean and the South Pacific Ocean
Map references: Oceania
Area:
total area: 7,686,850 sq km
land area: 7,617,930 sq km
comparative area: slightly smaller than the US
note: includes Macquarie Island
Land boundaries: 0 km
Coastline: 25,760 km
Maritime claims:
contiguous zone: 24 nm
continental shelf: 200 nm or to the edge of the continental margin
exclusive economic zone: 200 nm
territorial sea: 12 nm
International disputes: territorial claim in Antarctica (Australian Antarctic Territory)
Climate: generally arid to semiarid; temperate in south and east; tropical in north
Terrain: mostly low plateau with deserts; fertile plain in southeast
Natural resources: bauxite, coal, iron ore, copper, tin, silver, uranium, nickel, tungsten, mineral sands, lead, zinc, diamonds, natural gas, petroleum
Land use:
arable land: 6%
permanent crops: 0%
meadows and pastures: 58%
forest and woodland: 14%
other: 22%
Irrigated land: 18,800 sq km (1989 est.)
Environment:
current issues: soil erosion from overgrazing, industrial development, urbanization, and poor farming practices; soil salinity rising due to the use of poor quality water; desertification; clearing for agricultural purposes threatens the natural habitat of many unique animal and plant species; the Great Barrier Reef off the northeast coast, the largest coral reef in the world, is threatened by increased shipping and its popularity as a tourist site; limited natural fresh water resources
natural hazards: cyclones along the coast; severe droughts
international agreements: party to—Antarctic-Environmental Protocol, Antarctic Treaty, Biodiversity, Climate Change, Endangered Species, Environmental Modification, Hazardous Wastes, Law of the Sea, Marine Dumping, Marine Life Conservation, Nuclear Test Ban, Ozone Layer Protection, Ship Pollution, Tropical Timber 83, Wetlands, Whaling; signed, but not ratified—Desertification
Note: world's smallest continent but sixth-largest country; population concentrated along the eastern and southeastern coasts; regular, tropical, invigorating, sea breeze known as "the Doctor" occurs along the west coast in the summer

People

Population: 18,322,231 (July 1995 est.)
Age structure:
0-14 years: 22% (female 1,929,366; male 2,032,238)
15-64 years: 67% (female 6,017,362; male 6,181,887)
65 years and over: 11% (female 1,227,004; male 934,374) (July 1995 est.)
Population growth rate: 1.31% (1995 est.)
Birth rate: 14.13 births/1,000 population (1995 est.)
Death rate: 7.37 deaths/1,000 population (1995 est.)
Net migration rate: 6.33 migrant(s)/1,000 population (1995 est.)
Infant mortality rate: 7.1 deaths/1,000 live births (1995 est.)
Life expectancy at birth:
total population: 77.78 years
male: 74.67 years
female: 81.04 years (1995 est.)
Total fertility rate: 1.82 children born/woman (1995 est.)
Nationality:
noun: Australian(s)
adjective: Australian
Ethnic divisions: Caucasian 95%, Asian 4%, aboriginal and other 1%
Religions: Anglican 26.1%, Roman Catholic 26%, other Christian 24.3%
Languages: English, native languages
Literacy: age 15 and over can read and write (1980 est.)
total population: 100%
male: 100%
female: 100%
Labor force: 8.63 million (September 1991)
by occupation: finance and services 33.8%, public and community services 22.3%, wholesale and retail trade 20.1%, manufacturing and industry 16.2%, agriculture 6.1% (1987)

Government

Names:
conventional long form: Commonwealth of Australia
conventional short form: Australia
Digraph: AS
Type: federal parliamentary state
Capital: Canberra
Administrative divisions: 6 states and 2 territories*; Australian Capital Territory*, New South Wales, Northern Territory*, Queensland, South Australia, Tasmania, Victoria, Western Australia
Dependent areas: Ashmore and Cartier Islands, Christmas Island, Cocos (Keeling) Islands, Coral Sea Islands, Heard Island and McDonald Islands, Norfolk Island
Independence: 1 January 1901 (federation of UK colonies)
National holiday: Australia Day, 26 January (1788)
Constitution: 9 July 1900, effective 1 January 1901
Legal system: based on English common law; accepts compulsory ICJ jurisdiction, with reservations
Suffrage: 18 years of age; universal and compulsory
Executive branch:
chief of state: Queen ELIZABETH II (since 6 February 1952), represented by Governor General William George HAYDEN (since 16 February 1989)
head of government: Prime Minister Paul John KEATING (since 20 December 1991); Deputy Prime Minister Brian HOWE (since 4 June 1991)
cabinet: Cabinet; prime minister selects his cabinet from members of the House and Senate
Legislative branch: bicameral Federal Parliament
Senate: elections last held 13 March 1993 (next to be held by NA 1996); results—percent of vote by party NA; seats—(76 total) Liberal-National 36, Labor 30, Australian Democrats 7, Greens 2, independents 1
House of Representatives: elections last held 13 March 1993 (next to be held by NA 1996); results—percent of vote by party NA; seats—(147 total) Labor 80, Liberal-National 65, independent 2
Judicial branch: High Court
Political parties and leaders:
government: Australian Labor Party, Paul John KEATING
opposition: Liberal Party, John HOWARD; National Party, Timothy FISCHER; Australian Democratic Party, Cheryl KERNOT; Green Party, leader NA
Other political or pressure groups: Australian Democratic Labor Party (anti-

Communist Labor Party splinter group); Peace and Nuclear Disarmament Action (Nuclear Disarmament Party splinter group)
Member of: AfDB, AG (observer), ANZUS, APEC, AsDB, Australia Group, BIS, C, CCC, CP, EBRD, ESCAP, FAO, G- 8, GATT, IAEA, IBRD, ICAO, ICC, ICFTU, ICRM, IDA, IEA, IFAD, IFC, IFRCS, ILO, IMF, IMO, INMARSAT, INTELSAT, INTERPOL, IOC, IOM, ISO, ITU, MINURSO, MTCR, NAM (guest), NEA, NSG, OECD, PCA, SPARTECA, SPC, SPF, UN, UNCTAD, UNESCO, UNFICYP, UNHCR, UNIDO, UNOSOM, UNTSO, UNU, UPU, WFTU, WHO, WIPO, WMO, ZC
Diplomatic representation in US:
chief of mission: Ambassador Donald Eric RUSSELL
chancery: 1601 Massachusetts Avenue NW, Washington, DC 20036
telephone: [1] (202) 797-3000
FAX: [1] (202) 797-3168
consulate(s) general: Atlanta, Honolulu, Houston, Los Angeles, New York, Pago Pago (American Samoa), and San Francisco
US diplomatic representation:
chief of mission: Ambassador Edward J. PERKINS
embassy: Moonah Place, Yarralumla, Canberra, Australian Capital Territory 2600
mailing address: APO AP 96549
telephone: [61] (6) 270-5000
FAX: [61] (6) 270-5970
consulate(s) general: Melbourne, Perth, and Sydney
consulate(s): Brisbane
Flag: blue with the flag of the UK in the upper hoist-side quadrant and a large seven-pointed star in the lower hoist-side quadrant; the remaining half is a representation of the Southern Cross constellation in white with one small five-pointed star and four, larger, seven-pointed stars

---

Economy

Overview: Australia has a prosperous Western-style capitalist economy, with a per capita GDP comparable to levels in industrialized West European countries. Rich in natural resources, Australia is a major exporter of agricultural products, minerals, metals, and fossil fuels. Primary products account for more than 60% of the value of total exports, so that, as in 1983-84, a downturn in world commodity prices can have a big impact on the economy. The government is pushing for increased exports of manufactured goods, but competition in international markets continues to be severe. Australia has suffered from the low growth and high unemployment characterizing the OECD countries in the early 1990s. In 1992-93 the economy recovered slowly from the prolonged recession of 1990-91, a major restraining factor being weak world demand for Australia's exports. Growth picked up so strongly in 1994 that the government felt the need for fiscal and monetary tightening by yearend. Australia's GDP grew 6.4% in 1994, largely due to increases in industrial output and business investment. A severe drought in 1994 is expected to reduce the value of Australia's net farm production by $825 million in the twelve months through June 1995, but rising world commodity prices are likely to boost rural exports by 7.7% to $14.5 billion in 1995/96, according to government statistics.
National product: GDP—purchasing power parity—$374.6 billion (1994 est.)
National product real growth rate: 6.4% (1994)
National product per capita: $20,720 (1994 est.)
Inflation rate (consumer prices): 2.5% (1994)
Unemployment rate: 8.9% (December 1994)
Budget:
revenues: $83.8 billion
expenditures: $92.3 billion, including capital expenditures of $NA (FY93/94)
Exports: $50.4 billion (1994)
commodities: coal, gold, meat, wool, alumina, wheat, machinery and transport equipment
partners: Japan 25%, US 11%, South Korea 6%, NZ 5.7%, UK, Taiwan, Singapore, Hong Kong (1992)
Imports: $51.1 billion (1994)
commodities: machinery and transport equipment, computers and office machines, crude oil and petroleum products
partners: US 23%, Japan 18%, UK 6%, Germany 5.7%, NZ 4% (1992)
External debt: $147.2 billion (1994)
Industrial production: growth rate 3.9% (FY93/94); accounts for 32% of GDP
Electricity:
capacity: 34,540,000 kW
production: 155 billion kWh
consumption per capita: 8,021 kWh (1993)
Industries: mining, industrial and transportation equipment, food processing, chemicals, steel
Agriculture: accounts for 5% of GDP and over 30% of export revenues; world's largest exporter of beef and wool, second-largest for mutton, and among top wheat exporters; major crops—wheat, barley, sugarcane, fruit; livestock—cattle, sheep, poultry
Illicit drugs: Tasmania is one of the world's major suppliers of licit opiate products; government maintains strict controls over areas of opium poppy cultivation and output of poppy straw concentrate
Economic aid:
donor: ODA and OOF commitments (1970-89), $10.4 billion
Currency: 1 Australian dollar ($A) = 100 cents
Exchange rates: Australian dollars ($A) per US$1—1.3058 (January 1995), 1.3667 (1994), 1.4704 (1993), 1.3600 (1992), 1.2835 (1991), 1.2799 (1990)
Fiscal year: 1 July—30 June

Transportation

Railroads:
total: 40,478 km (1,130 km electrified; 183 km dual gauge)
broad gauge: 7,970 km 1.600-m gauge
standard gauge: 16,201 km 1.435-m gauge
narrow gauge: 16,307 km 1.067-m gauge
Highways:
total: 837,872 km
paved: 243,750 km
unpaved: gravel, crushed stone, stabilized earth 228,396 km; unimproved earth 365,726 km
Inland waterways: 8,368 km; mainly by small, shallow-draft craft
Pipelines: crude oil 2,500 km; petroleum products 500 km; natural gas 5,600 km
Ports: Adelaide, Brisbane, Cairns, Darwin, Devonport, Fremantle, Geelong, Hobart (Tasmania), Launceton (Tasmania), Mackay, Melbourne, Sydney, Townsville
Merchant marine:
total: 81 ships (1,000 GRT or over) totaling 2,620,536 GRT/3,801,970 DWT
ships by type: bulk 30, cargo 7, chemical tanker 3, combination bulk 2, container 7, liquefied gas tanker 6, oil tanker 18, roll-on/roll-off cargo 7, short-sea passenger 1
Airports:
total: 480
with paved runways over 3,047 m: 9
with paved runways 2,438 to 3,047 m: 15
with paved runways 1,524 to 2,437 m: 128
with paved runways 914 to 1,523 m: 125
with paved runways under 914 m: 31
with unpaved runways 1,524 to 2,438 m: 23
with unpaved runways 914 to 1,523 m: 149

Communications

Telephone system: 8,700,000 telephones; good international and domestic service
local: NA
intercity: domestic satellite service
international: submarine cables to New Zealand, Papua New Guinea, and Indonesia; 10 INTELSAT (4 Indian Ocean and 6 Pacific Ocean) earth stations
Radio:
broadcast stations: AM 258, FM 67, shortwave 0
radios: NA
Television:
broadcast stations: 134
televisions: NA

Defense Forces

Branches: Australian Army, Royal Australian Navy, Royal Australian Air Force
Manpower availability: males age 15-49 4,934,175; males fit for military service 4,274,900; males reach military age (17) annually 131,852 (1995 est.)
Defense expenditures: exchange rate conversion—$7.2 billion, 2.2% of GDP (FY94/95)

Geography

Location: Central Europe, north of Italy
Map references: Europe
Area:
total area: 83,850 sq km
land area: 82,730 sq km
comparative area: slightly smaller than Maine
Land boundaries: total 2,496 km, Czech Republic 362 km, Germany 784 km, Hungary 366 km, Italy 430 km, Liechtenstein 37 km, Slovakia 91 km, Slovenia 262 km, Switzerland 164 km
Coastline: 0 km (landlocked)
Maritime claims: none; landlocked
International disputes: none
Climate: temperate; continental, cloudy; cold winters with frequent rain in lowlands and snow in mountains; cool summers with occasional showers
Terrain: in the west and south mostly mountains (Alps); along the eastern and northern margins mostly flat or gently sloping
Natural resources: iron ore, petroleum, timber, magnesite, aluminum, lead, coal, lignite, copper, hydropower
Land use:
arable land: 17%
permanent crops: 1%
meadows and pastures: 24%
forest and woodland: 39%
other: 19%
Irrigated land: 40 sq km (1989)
Environment:
current issues: some forest degradation caused by air and soil pollution; soil pollution results from the use of agricultural chemicals; air pollution results from emissions by coal- and oil-fired power stations and industrial plants and from trucks transiting Austria between northern and southern Europe
natural hazards: NA
international agreements: party to—Air Pollution, Air Pollution-Nitrogen Oxides, Air Pollution-Sulphur 85, Air Pollution-Volatile Organic Compounds, Antarctic Treaty, Biodiversity, Climate Change, Endangered Species, Environmental Modification, Hazardous Wastes, Nuclear Test Ban, Ozone Layer Protection, Ship Pollution, Tropical Timber 83, Wetlands; signed, but not ratified—Air Pollution-Sulpher 94, Antarctic-Environmental Protocol, Law of the Sea, Whaling
Note: landlocked; strategic location at the crossroads of central Europe with many easily traversable Alpine passes and valleys; major river is the Danube; population is concentrated on eastern lowlands because of steep slopes, poor soils, and low temperatures elsewhere

People

Population: 7,986,664 (July 1995 est.)
Age structure:
0-14 years: 17% (female 681,087; male 711,127)
15-64 years: 67% (female 2,672,554; male 2,677,100)
65 years and over: 16% (female 791,762; male 453,034) (July 1995 est.)
Population growth rate: 0.35% (1995 est.)
Birth rate: 11.21 births/1,000 population (1995 est.)
Death rate: 10.27 deaths/1,000 population (1995 est.)
Net migration rate: 2.5 migrant(s)/1,000 population (1995 est.)
Infant mortality rate: 6.9 deaths/1,000 live births (1995 est.)
Life expectancy at birth:
total population: 76.9 years
male: 73.7 years
female: 80.27 years (1995 est.)
Total fertility rate: 1.48 children born/woman (1995 est.)
Nationality:
noun: Austrian(s)
adjective: Austrian
Ethnic divisions: German 99.4%, Croatian 0.3%, Slovene 0.2%, other 0.1%
Religions: Roman Catholic 85%, Protestant 6%, other 9%
Languages: German
Literacy: age 15 and over can read and write (1974 est.)
total population: 99%
Labor force: 3.47 million (1989)
by occupation: services 56.4%, industry and crafts 35.4%, agriculture and forestry 8.1%
note: an estimated 200,000 Austrians are employed in other European countries; foreign laborers in Austria number 177,840, about 5% of labor force (1988)

Government

Names:
conventional long form: Republic of Austria
conventional short form: Austria
local long form: Republik Oesterreich
local short form: Oesterreich

Digraph: AU
Type: federal republic
Capital: Vienna
Administrative divisions: 9 states (bundeslaender, singular—bundesland); Burgenland, Kaernten, Niederoesterreich, Oberoesterreich, Salzburg, Steiermark, Tirol, Vorarlberg, Wien
Independence: 12 November 1918 (from Austro-Hungarian Empire)
National holiday: National Day, 26 October (1955)
Constitution: 1920; revised 1929 (reinstated 1 May 1945)
Legal system: civil law system with Roman law origin; judicial review of legislative acts by a Constitutional Court; separate administrative and civil/penal supreme courts; has not accepted compulsory ICJ jurisdiction
Suffrage: 18 years of age; universal; compulsory for presidential elections
Executive branch:
chief of state: President Thomas KLESTIL (since 8 July 1992); election last held 24 May 1992 (next to be held 1996); results of second ballot—Thomas KLESTIL 57%, Rudolf STREICHER 43%
head of government: Chancellor Franz VRANITZKY (since 16 June 1986); Vice Chancellor Erhard BUSEK (since 2 July 1991)
cabinet: Council of Ministers; chosen by the president on the advice of the chancellor
Legislative branch: bicameral Federal Assembly (Bundesversammlung)
Federal Council (Bundesrat): consists of 63 members representing each of the provinces on the basis of population, but with each province having at least 3 representatives
National Council (Nationalrat): elections last held 9 October 1994 (next to be held October 1998); results—SPOE 34.9%, OEVP 27.7%, FPOE 22.5%, Greens 7.3%, LF 6.0% other 1.6%; seats—(183 total) SPOE 65, OEVP 52, FPOE 42, Greens 13, LF 11
Judicial branch: Supreme Judicial Court (Oberster Gerichtshof) for civil and criminal cases, Administrative Court (Verwaltungsgerichtshof) for bureaucratic cases, Constitutional Court (Verfassungsgerichtshof) for constitutional cases
Political parties and leaders: Social Democratic Party of Austria (SPOE), Franz VRANITZKY, chairman; Austrian People's Party (OEVP), Erhard BUSEK, chairman; Freedom Movement (F) (was the Freedom Party of Austria, FPOE), Joerg HAIDER, chairman; Communist Party (KPOE), Walter SILBERMAYER, chairman; The Greens, Madeleine PETROVIC; Liberal Forum (LF), Heide SCHMIDT
Other political or pressure groups: Federal Chamber of Commerce and Industry; Austrian Trade Union Federation (primarily Socialist); three composite leagues of the Austrian People's Party (OEVP) representing business, labor, and farmers; OEVP-oriented League of Austrian Industrialists; Roman Catholic Church, including its chief lay organization, Catholic Action
Member of: AfDB, AG (observer), AsDB, Australia Group, BIS, CCC, CE, CEI, CERN, EBRD, ECE, EFTA, ESA, EU, FAO, G- 9, GATT, IADB, IAEA, IBRD, ICAO, ICC, ICFTU, ICRM, IDA, IEA, IFAD, IFC, IFRCS, ILO, IMF, IMO, INTELSAT, INTERPOL, IOC, IOM, ISO, ITU, MINURSO, MTCR, NAM (guest), NEA, NSG, OAS (observer), OECD, ONUSAL, OSCE, PCA, UN, UNAMIR, UNCTAD, UNDOF, UNESCO, UNFICYP, UNHCR, UNIDO, UNIKOM, UNMIH, UNOMIL, UNOMOZ, UNTSO, UPU, WCL, WFTU, WHO, WIPO, WMO, WTO, ZC
Diplomatic representation in US:
chief of mission: Ambassador Helmut TUERK
chancery: 3524 International Court NW, Washington, DC 20008-3035
telephone: [1] (202) 895-6700
FAX: [1] (202) 895-6750
consulate(s) general: Chicago, Los Angeles, and New York
US diplomatic representation:
chief of mission: Ambassador Swanee G. HUNT
chancery: Boltzmanngasse 16, A-1091, Vienna
mailing address: use embassy street address
telephone: [43] (1) 313-39
FAX: [43] (1) 310-0682
consulate(s) general: none (Salzburg closed September 1993)
Flag: three equal horizontal bands of red (top), white, and red

Economy

Overview: Austria boasts a prosperous and stable market economy with a sizable but falling proportion of nationalized industry and with extensive welfare benefits. Thanks to its raw material endowment, a technically skilled labor force, and strong links to German industrial firms, Austria occupies specialized niches in European industry and services (tourism, banking) and produces almost enough food to feed itself with only 8% of the labor force in agriculture. After 11 consecutive years of growth, the Austrian economy experienced a mild recession in 1993, but growth resumed in 1994. Unemployment is 4.3% and will likely stay at that level as companies adjust to the competition of EU membership beginning 1 January 1995. To prepare for EU membership, Austria's government has taken measures to open the economy by introducing a major tax reform, privatizing state-owned firms, and liberalizing cross-border capital movements. Problems for the 1990s include an aging population, the high level of industrial subsidies, and the struggle to keep welfare benefits within budgetary capabilities—the deficit climbed to over 4% of GDP in 1994.
National product: GDP—purchasing power parity—$139.3 billion (1994 est.)
National product real growth rate: 2.5% (1994 est.)
National product per capita: $17,500 (1994 est.)
Inflation rate (consumer prices): 3% (1994)
Unemployment rate: 4.3% (1994 est.)
Budget:
revenues: $52.2 billion
expenditures: $60.3 billion, including capital expenditures of $NA (1993 est.)
Exports: $44.1 billion (1994 est.)
commodities: machinery and equipment, iron and steel, lumber, textiles, paper products, chemicals
partners: EC 63.5% (Germany 38.9%), EFTA 9.0%, Eastern Europe/FSU 12.3%, Japan 1.5%, US 3.4% (1993)
Imports: $53.8 billion (1994 est.)
commodities: petroleum, foodstuffs, machinery and equipment, vehicles, chemicals, textiles and clothing, pharmaceuticals
partners: EC 66.8% (Germany 41.3%), EFTA 6.7%, Eastern Europe/FSU 7.5%, Japan 4.4%, US 4.4% (1993)
External debt: $21.5 billion (1994 est.)
Industrial production: growth rate 2.5% (1994 est.)
Electricity:
capacity: 17,230,000 kW
production: 50.2 billion kWh
consumption per capita: 5,824 kWh (1993)
Industries: foods, iron and steel, machines, textiles, chemicals, electrical, paper and pulp, tourism, mining, motor vehicles
Agriculture: accounts for 3.2% of GDP (including forestry); principal crops and animals—grains, fruit, potatoes, sugar beets, sawn wood, cattle, pigs, poultry; 80%-90% self-sufficient in food
Illicit drugs: transshipment point for Southwest Asian heroin transiting the Balkan route and Eastern Europe
Economic aid:
donor: ODA and OOF commitments (1970-89), $2.4 billion
Currency: 1 Austrian schilling (S) = 100 groschen
Exchange rates: Austrian schillings (S) per US$1—10.774 (January 1995), 11.422 (1994), 11.632 (1993), 10.989 (1992), 11.676 (1991), 11.370 (1990)
Fiscal year: calendar year

Transportation

Railroads:
total: 5,624 km
standard gauge: 5,269 km 1.435-m gauge (3,162 km electrified)
narrow gauge: 355 km 1.000-m and 0.760-m

gauge (84 km electrified) (1994)
Highways:
total: 110,000 km
paved: 35,000 km (including 1,554 km of autobahn)
unpaved: mostly gravel and earth 75,000 km (1992)
Inland waterways: 446 km
Pipelines: crude oil 554 km; petroleum products 171 km; natural gas 2,611 km
Ports: Linz, Vienna
Merchant marine:
total: 32 ships (1,000 GRT or over) totaling 152,885 GRT/235,719 DWT
ships by type: bulk 3, cargo 25, oil tanker 1, refrigerated cargo 2, roll-on/roll-off cargo 1
Airports:
total: 55
with paved runways over 3,047 m: 1
with paved runways 2,438 to 3,047 m: 5
with paved runways 1,524 to 2,437 m: 1
with paved runways 914 to 1,523 m: 3
with paved runways under 914 m: 41
with unpaved runways 914 to 1,523 m: 4

Communications

Telephone system: 4,014,000 telephones; highly developed and efficient
local: NA
intercity: NA
international: 2 INTELSAT (1 Atlantic Ocean and 1 Indian Ocean), and EUTELSAT earth stations
Radio:
broadcast stations: AM 6, FM 21 (repeaters 545), shortwave 0
radios: NA
Television:
broadcast stations: 47 (repeaters 870)
televisions: NA

Defense Forces

Branches: Army (includes Flying Division)
Manpower availability: males age 15-49 2,026,567; males fit for military service 1,695,879; males reach military age (19) annually 46,821 (1995 est.)
Defense expenditures: exchange rate conversion—about $1.8 billion, 0.9% of GDP (1994)

Note: Azerbaijan continues to be plagued by an unresolved seven-year-old conflict with Armenian separatists over its Nagorno-Karabakh region. The Karabakh Armenians have declared independence and seized almost 20% of the country's territory, creating almost 1 million Azeri displaced persons in the process. Both sides have generally observed a Russian-mediated cease-fire in place since May 1994, and support the OSCE-mediated peace process, now entering its fourth year. Nevertheless, Baku and Xankandi (Stepanakert) remain far apart on most substantive issues from the placement and composition of a peacekeeping force to the enclave's ultimate political status, and prospects for a negotiated settlement remain dim.

Geography

Location: Southwestern Asia, bordering the Caspian Sea, between Iran and Russia
Map references: Commonwealth of Independent States—European States
Area:
total area: 86,600 sq km
land area: 86,100 sq km
comparative area: slightly larger than Maine
note: includes the exclave of Naxcivan Autonomous Republic and the Nagorno-Karabakh region; the region's autonomy was abolished by Azerbaijani Supreme Soviet on 26 November 1991
Land boundaries: total 2,013 km, Armenia (west) 566 km, Armenia (southwest) 221 km, Georgia 322 km, Iran (south) 432 km, Iran (southwest) 179 km, Russia 284 km, Turkey 9 km
Coastline: 0 km (landlocked)
note: Azerbaijan borders the Caspian Sea (800 km, est.)
Maritime claims: none; landlocked
International disputes: violent and longstanding dispute with ethnic Armenians of Nagorno-Karabakh over its status; Caspian Sea boundaries are not yet determined
Climate: dry, semiarid steppe
Terrain: large, flat Kur-Araz Lowland (much of it below sea level) with Great Caucasus Mountains to the north, Qarabag (Karabakh) Upland in west; Baku lies on Abseron (Apsheron) Peninsula that juts into Caspian Sea
Natural resources: petroleum, natural gas, iron ore, nonferrous metals, alumina
Land use:
arable land: 18%
permanent crops: 4%
meadows and pastures: 25%
forest and woodland: 0%
other: 53%
Irrigated land: 14,010 sq km (1990)
Environment:
current issues: local scientists consider the Abseron (Apsheron) Peninsula (including Baku and Sumqayit) and the Caspian Sea to be the ecologically most devastated area in the world because of severe air, water, and soil pollution; soil pollution results from the use of DDT as a pesticide and also from toxic defoliants used in the production of cotton
natural hazards: droughts; some lowland areas threatened by rising levels of the Caspian Sea
international agreements: signed, but not ratified—Biodiversity, Climate Change
Note: landlocked

People

Population: 7,789,886 (July 1995 est.)
Age structure:
0-14 years: 33% (female 1,241,952; male 1,315,313)
15-64 years: 61% (female 2,437,810; male 2,307,496)
65 years and over: 6% (female 303,926; male 183,389) (July 1995 est.)
Population growth rate: 1.32% (1995 est.)
Birth rate: 22.05 births/1,000 population (1995 est.)
Death rate: 6.56 deaths/1,000 population (1995 est.)
Net migration rate: -2.32 migrant(s)/1,000 population (1995 est.)
Infant mortality rate: 33.9 deaths/1,000 live births (1995 est.)
Life expectancy at birth:
total population: 71.09 years
male: 67.4 years
female: 74.97 years (1995 est.)
Total fertility rate: 2.64 children born/woman (1995 est.)
Nationality:
noun: Azerbaijani(s)
adjective: Azerbaijani
Ethnic divisions: Azeri 90%, Dagestani Peoples 3.2%, Russian 2.5%, Armenian 2.3%, other 2% (1995 est.)
note: almost all Armenians live in the

separatist Nagorno-Karabakh region
Religions: Muslim 93.4%, Russian Orthodox 2.5%, Armenian Orthodox 2.3%, other 1.8% (1995 est.)
note: religious affiliation is still nominal in Azerbaijan; actual practicing adherents are much lower
Languages: Azeri 89%, Russian 3%, Armenian 2%, other 6% (1995 est.)
Literacy: age 15 and over can read and write (1989)
total population: 97%
male: 99%
female: 96%
Labor force: 2.789 million
by occupation: agriculture and forestry 32%, industry and construction 26%, other 42% (1990)

Government

Names:
conventional long form: Azerbaijani Republic
conventional short form: Azerbaijan
local long form: Azarbaycan Respublikasi
local short form: none
former: Azerbaijan Soviet Socialist Republic
Digraph: AJ
Type: republic
Capital: Baku (Baki)
Administrative divisions: 59 rayons (rayonlar; rayon—singular), 11 cities* (saharlar; sahar—singular), 1 autonomous republic** (muxtar respublika); Abscron Rayonu, Agcabadi Rayonu, Agdam Rayonu, Agdas Rayonu, Agstafa Rayonu, Agsu Rayonu, AliBayramli Sahari*, Astara Rayonu, Baki Sahari*, Balakan Rayonu, Barda Rayonu, Beylaqan Rayonu, Bilasuvar Rayonu, Cabrayil Rayonu, Calilabad Rayonu, Daskasan Rayonu, Davaci Rayonu, Fuzuli Rayonu, Gadabay Rayonu, Ganca Sahari*, Goranboy Rayonu, Goycay Rayonu, Haciqabul Rayonu, Imisli Rayonu, Ismayilli Rayonu, Kalbacar Rayonu, Kurdamir Rayonu, Lacin Rayonu, Lankaran Rayonu, Lankaran Sahari*, Lerik Rayonu, Masalli Rayonu, Mingacevir Sahari*, Naftalan Sahari*, Naxcivan Muxtar Respublikasi**, Neftcala Rayonu, Oguz Rayonu, Qabala Rayonu, Qax Rayonu, Qazax Rayonu, Qobustan Rayonu, Quba Rayonu, Qubadli Rayonu, Qusar Rayonu, Saatli Rayonu, Sabirabad Rayonu, Saki Rayonu, Saki Sahari*, Salyan Rayonu, Samaxi Rayonu, Samkir Rayonu, Samux Rayonu, Siyazan Rayonu, Sumqayit Sahari*, Susa Rayonu, Susa Sahari*, Tartar Rayonu, Tovuz Rayonu, Ucar Rayonu, Xacmaz Rayonu, Xankandi Sahari*, Xanlar Rayonu, Xizi Rayonu, Xocali Rayonu, Xocavand Rayonu, Yardimb Rayonu, Yevlax Rayonu, Yevlax Sahari*, Zangilan Rayonu, Zaqatala Rayonu, Zardab Rayonu
Independence: 30 August 1991 (from Soviet Union)
National holiday: Independence Day, 28 May
Constitution: adopted NA April 1978; writing a new constitution
Legal system: based on civil law system
Suffrage: 18 years of age; universal
Executive branch:
chief of state: President Heydar ALIYEV (since 18 June 1993); election last held 3 October 1993 (next to be held NA); results—Heydar ALIYEV won 97% of vote
head of government: Acting Prime Minister Fuad QULIYEV (since 9 October 1994); First Deputy Prime Ministers Abbas ABBASOV, Samed SADYKOV, Vahid AKHMEDOV (since NA)
cabinet: Council of Ministers; appointed by the president and confirmed by the Mejlis
Legislative branch: unicameral
National Assembly (Milli Mejlis): elections last held 30 September and 14 October 1990 for the Supreme Soviet (next expected to be held September 1995 for the National Assembly); seats for Supreme Soviet—(360 total) Communists 280, Democratic Bloc 45 (grouping of opposition parties), other 15, vacant 20; note—on 19 May 1992 the Supreme Soviet was prorogued in favor of a Popular Front-dominated National Council; seats—(50 total) Popular Front 25, opposition elements 25
note: since June 1993 ALIYEV has rotated in several supporters to replace Popular Front adherents
Judicial branch: Supreme Court
Political parties and leaders: Azerbaijan Popular Front (APF), Ebulfez ELCIBEY, chairman; Musavat Party, Isa GAMBAR, chairman; National Independence Party, Etibar MAMEDOV, chairman; Social Democratic Party (SDP), Araz ALIZADE, chairman; Communist Party, Ramiz AKHMEDOV, chairman; People's Freedom Party, Yunus OGUZ, chairman; Independent Social Democratic Party, Arif YUNUSOV and Leila YUNOSOVA, cochairmen; New Azerbaijan Party, Heydar ALIYEV, chairman; Boz Gurd Party, Iskander HAMIDOV, chairman; Azerbaijan Democratic Independence Party, Qabil HUSEYNLI, chairman; Islamic Party of Azerbaijan, Ali Akram, chairman; Ana Veten Party, Fazail AGAMALIYEV; Azerbaijan Democratic Party, Sardar Jalaloglu MAMEDOV; Azerbaijan Democratic Party of Proprietors (DPOP), Makhmud MAMEDOV; Azerbaijan Patriotic Solidarity Party, Sabir RUSTAMHANLI; Azerbaijan Republic Reform Party, Fuad ASADOV; Communist Party of Azerbaijan (unregistered), Sayad SAYADOV; Equality of the Peoples Party, Faukhraddin AYDAYEV; Independent Azerbaijan Party, Nizami SULEYMANOV; Labor Party of Azerbaijan, Sabutai HAJIYEV; Liberal-Democratic Party of Azerbaijan, Lyudmila NIKOLAYEVNA; National Enlightenment Party, Hajy Osman EFENDIYEV; National Liberation Party, Panak SHAKHSEVEV; Peasant Party, Firuz MUSTAFAYEV; Radical Party of Azerbaijan, Malik SHARIFOV; United Azerbaijan Party, Kerrar ABILOV; Vetan Adzhagy Party, Zakir TAGIYEV
Other political or pressure groups: self-proclaimed Armenian Nagorno-Karabakh Republic; Talysh independence movement
Member of: BSEC, CCC, CIS, EBRD, ECE, ECO, ESCAP, IBRD, ICAO, IDB, IFAD, ILO, IMF, INTELSAT, INTERPOL, IOC, ITU, NACC, OIC, OSCE, PFP, UN, UNCTAD, UNESCO, UPU, WHO, WMO
Diplomatic representation in US:
chief of mission: Ambassador Hafiz Mir Jalal PASHAYEV
chancery: (temporary) Suite 700, 927 15th Street NW, Washington, DC 20005
telephone: [1] (202) 842-0001
FAX: [1] (202) 842-0004
US diplomatic representation:
chief of mission: Ambassador Richard D. KAUZLARICH
embassy: Azadliq Prospect 83, Baku
mailing address: use embassy street address
telephone: [9] (9412) 96-00-19, 98-03-37
FAX: [9] (9412) 98-37-55
Flag: three equal horizontal bands of blue (top), red, and green; a crescent and eight-pointed star in white are centered in red band

Economy

Overview: Azerbaijan is less developed industrially than either Armenia or Georgia, the other Transcaucasian states. It resembles the Central Asian states in its majority nominally Muslim population, high structural unemployment, and low standard of living. The economy's most prominent products are oil, cotton, and gas. Production from the Caspian oil and gas field has been in decline for several years, but the November 1994 ratification of the $7.5 billion oil deal with a consortium of Western companies should generate the funds needed to spur future industrial development. Azerbaijan accounted for 1.5% to 2% of the capital stock and output of the former Soviet Union. Azerbaijan shares all the formidable problems of the ex-Soviet republics in making the transition from a command to a market economy, but its considerable energy resources brighten its long-term prospects. Baku has only recently begun making progress on economic reform, and old economic ties and structures have yet to be replaced.
National product: GDP—purchasing power parity—$13.8 billion (1994 estimate as extrapolated from World Bank estimate for 1992)
National product real growth rate: -22% (1994 est.)
National product per capita: $1,790 (1994 est.)
Inflation rate (consumer prices): 28%

monthly average (1994)
Unemployment rate: 0.9% includes officially registered unemployed; also large numbers of other unemployed and underemployed workers (December 1994)
Budget:
revenues: $167.5 million
expenditures: $234.6 million, including capital expenditures of $NA (1994)
Exports: $366 million to non-FSU countries (f.o.b., 1994)
commodities: oil and gas, chemicals, oilfield equipment, textiles, cotton (1991)
partners: mostly CIS and European countries
Imports: $296 million from non-FSU countries (c.i.f., 1994)
commodities: machinery and parts, consumer durables, foodstuffs, textiles (1991)
partners: European countries
External debt: $NA
Industrial production: growth rate -25% (1994)
Electricity:
capacity: 4,900,000 kW
production: 17.5 billion kWh
consumption per capita: 2,270 kWh (1994)
Industries: petroleum and natural gas, petroleum products, oilfield equipment; steel, iron ore, cement; chemicals and petrochemicals; textiles
Agriculture: cotton, grain, rice, grapes, fruit, vegetables, tea, tobacco, cattle, pigs, sheep and goats
Illicit drugs: illicit cultivator of cannabis and opium poppy; mostly for CIS consumption; limited government eradication program; transshipment point for illicit drugs to Western Europe
Economic aid:
recipient: wheat from Turkey
Currency: 1 manat = 100 gopik
Exchange rates: manats per US$1—4500 (April 1995), 4168 (end of December 1994)
Fiscal year: calendar year

Transportation

Railroads:
total: 2,090 km in common carrier service; does not include industrial lines
broad gauge: 2,090 km 1.520-m gauge (1990)
Highways:
total: 36,700 km
paved or graveled: 31,800 km
unpaved: earth 4,900 km (1990)
Pipelines: crude oil 1,130 km; petroleum products 630 km; natural gas 1,240 km
Ports: Baku (Baki)
Airports:
total: 69
with paved runways over 3,047 m: 2
with paved runways 2,438 to 3,047 m: 6
with paved runways 1,524 to 2,437 m: 17
with paved runways 914 to 1,523 m: 3
with paved runways under 914 m: 1
with unpaved runways 914 to 1,523 m: 7
with unpaved runways under 914 m: 33

Communications

Telephone system: 710,000 telephones; 90 telephones/1,000 persons (1991); 202,000 persons waiting for telephone installations (January 1991); domestic telephone service is of poor quality and inadequate
local: a joint venture to establish a cellular telephone system (Bakcel) in the Baku area is supposed to become operational in 1994
intercity: NA
international: connections to other former USSR republics by cable and microwave and to other countries via the Moscow international gateway switch; INTELSAT link installed in late 1992 in Baku with Turkish financial assistance with access to 200 countries through Turkey; since August 1993 an earth station near Baku has provided direct communications with New York through Russia's Stationar-11 satellite
Radio:
broadcast stations: AM NA, FM NA, shortwave NA
radios: NA
Television:
broadcast stations: NA; domestic and Russian TV programs are received locally and Turkish and Iranian TV is received from an INTELSAT satellite through a receive-only earth station
televisions: NA

Defense Forces

Branches: Army, Air Force, Navy, Maritime Border Guard, National Guard, Security Forces (internal and border troops)
Manpower availability: males age 15-49 1,927,955; males fit for military service 1,553,736; males reach military age (18) annually 68,407 (1995 est.)
Defense expenditures: 70.5 billion rubles, 10% of GDP (1993 budget allocation); note—conversion of the military budget into US dollars using the current exchange rate could produce misleading results

The Bahamas

Geography

Location: Caribbean, chain of islands in the North Atlantic Ocean, southeast of Florida
Map references: Central America and the Caribbean
Area:
total area: 13,940 sq km
land area: 10,070 sq km
comparative area: slightly larger than Connecticut
Land boundaries: 0 km
Coastline: 3,542 km
Maritime claims:
continental shelf: 200-m depth or to the depth of exploitation
exclusive fishing zone: 200 nm
territorial sea: 3 nm
International disputes: none
Climate: tropical marine; moderated by warm waters of Gulf Stream
Terrain: long, flat coral formations with some low rounded hills
Natural resources: salt, aragonite, timber
Land use:
arable land: 1%
permanent crops: 0%
meadows and pastures: 0%
forest and woodland: 32%
other: 67%
Irrigated land: NA sq km
Environment:
current issues: coral reef decay
natural hazards: hurricanes and other tropical storms that cause extensive flood and wind damage
international agreements: party to—Biodiversity, Climate Change, Endangered Species, Hazardous Wastes, Law of the Sea, Nuclear Test Ban, Ozone Layer Protection, Ship Pollution
Note: strategic location adjacent to US and Cuba; extensive island chain

People

Population: 256,616 (July 1995 est.)

Age structure:
0-14 years: 28% (female 35,924; male 36,504)
15-64 years: 66% (female 87,868; male 82,780)
65 years and over: 6% (female 8,247; male 5,293) (July 1995 est.)
Population growth rate: 1.09% (1995 est.)
Birth rate: 19.23 births/1,000 population (1995 est.)
Death rate: 5.79 deaths/1,000 population (1995 est.)
Net migration rate: -2.56 migrant(s)/1,000 population (1995 est.)
Infant mortality rate: 24.3 deaths/1,000 live births (1995 est.)
Life expectancy at birth:
total population: 72.12 years
male: 67.37 years
female: 76.97 years (1995 est.)
Total fertility rate: 2.01 children born/woman (1995 est.)
Nationality:
noun: Bahamian(s)
adjective: Bahamian
Ethnic divisions: black 85%, white 15%
Religions: Baptist 32%, Anglican 20%, Roman Catholic 19%, Methodist 6%, Church of God 6%, other Protestant 12%, none or unknown 3%, other 2%
Languages: English, Creole (among Haitian immigrants)
Literacy: age 15 and over can read and write but definition of literacy not available (1963 est.)
total population: 90%
male: 90%
female: 89%
Labor force: 136,900 (1993)
by occupation: government 30%, hotels and restaurants 25%, business services 10%, agriculture 5% (1989)

Government

Names:
conventional long form: Commonwealth of The Bahamas
conventional short form: The Bahamas
Digraph: BF
Type: commonwealth
Capital: Nassau
Administrative divisions: 21 districts; Acklins and Crooked Islands, Bimini, Cat Island, Exuma, Freeport, Fresh Creek, Governor's Harbour, Green Turtle Cay, Harbour Island, High Rock, Inagua, Kemps Bay, Long Island, Marsh Harbour, Mayaguana, New Providence, Nicholls Town and Berry Islands, Ragged Island, Rock Sound, Sandy Point, San Salvador and Rum Cay
Independence: 10 July 1973 (from UK)
National holiday: National Day, 10 July (1973)
Constitution: 10 July 1973
Legal system: based on English common law
Suffrage: 18 years of age; universal
Executive branch:
chief of state: Queen ELIZABETH II (since 6 February 1952), represented by Governor General Sir Clifford DARLING (since 2 January 1992)
head of government: Prime Minister Hubert A. INGRAHAM (since 19 August 1992)
cabinet: Cabinet; appointed by the governor on the prime minister's recommendation
Legislative branch: bicameral Parliament
Senate: a 16-member body appointed by the governor general
House of Assembly: elections last held 19 August 1992 (next to be held by August 1997); results—percent of vote by party NA; seats—(49 total) FNM 32, PLP 17
Judicial branch: Supreme Court
Political parties and leaders: Progressive Liberal Party (PLP), Sir Lynden O. PINDLING; Free National Movement (FNM), Hubert Alexander INGRAHAM;
Member of: ACP, C, CARICOM, CCC, CDB, ECLAC, FAO, G-77, IADB, IBRD, ICAO, ICFTU, ICRM, IFC, IFRCS, ILO, IMF, IMO, INMARSAT, INTELSAT, INTERPOL, IOC, ITU, NAM, OAS, OPANAL, UN, UNCTAD, UNESCO, UNIDO, UPU, WHO, WIPO, WMO
Diplomatic representation in US:
chief of mission: Ambassador Timothy Baswell DONALDSON
chancery: 2220 Massachusetts Avenue NW, Washington, DC 20008
telephone: [1] (202) 319-2660
FAX: [1] (202) 319-2668
consulate(s) general: Miami and New York
US diplomatic representation:
chief of mission: Ambassador Sidney WILLIAMS
embassy: Mosmar Building, Queen Street, Nassau
mailing address: P. O. Box N-8197, Nassau
telephone: [1] (809) 322-1181, 328-2206
FAX: [1] (809) 328-7838
Flag: three equal horizontal bands of aquamarine (top), gold, and aquamarine with a black equilateral triangle based on the hoist side

Economy

Overview: The Bahamas is a stable, developing nation whose economy is based primarily on tourism and offshore banking. Tourism alone provides about 50% of GDP and directly or indirectly employs about 50,000 people or 40% of the local work force. The economy has slackened in recent years, as the annual increase in the number of tourists slowed. Nonetheless, per capita GDP is one of the highest in the region.
National product: GDP—purchasing power parity—$4.4 billion (1994 est.)
National product real growth rate: 3.5% (1994 est.)
National product per capita: $15,900 (1994 est.)
Inflation rate (consumer prices): 2.7% (1994)
Unemployment rate: 13.1% (1993)
Budget:
revenues: $696 million
expenditures: $756 million, including capital expenditures of $NA (FY 94/95)
Exports: $257 million (f.o.b., 1993 est.)
commodities: pharmaceuticals, cement, rum, crawfish, refined petroleum products
partners: US 51%, UK 7%, Norway 7%, France 6%, Italy 5%
Imports: $1.15 billion (f.o.b,,1993 est.)
commodities: foodstuffs, manufactured goods, crude oil, vehicles, electronics
partners: US 55%, Japan 17%, Nigeria 12%, Denmark 7%, Norway 6%
External debt: $455 million (December 1993)
Industrial production: growth rate 3% (1990); accounts for 15% of GDP
Electricity:
capacity: 424,000 kW
production: 929 million kWh
consumption per capita: 3,200 kWh (1993)
Industries: tourism, banking, cement, oil refining and transshipment, salt production, rum, aragonite, pharmaceuticals, spiral welded steel pipe
Agriculture: accounts for 5% of GDP; dominated by small-scale producers; principal products—citrus fruit, vegetables, poultry; large net importer of food
Illicit drugs: transshipment point for cocaine and marijuana bound for US and Europe; also a money-laundering center
Economic aid:
recipient: US commitments, including Ex-Im (FY85-89), $1 million; Western (non-US) countries, ODA and OOF bilateral commitments (1970-89), $345 million
Currency: 1 Bahamian dollar (B$) = 100 cents
Exchange rates: Bahamian dollar (B$) per US$1—1.00 (fixed rate)
Fiscal year: 1 July—30 June

Transportation

Railroads: 0 km
Highways:
total: 2,400 km
paved: 1,350 km
unpaved: gravel 1,050 km
Ports: Freeport, Matthew Town, Nassau
Merchant marine:
total: 936 ships (1,000 GRT or over) totaling 21,815,474 GRT/35,253,416 DWT
ships by type: bulk 162, cargo 181, chemical tanker 39, combination bulk 9, combination ore/oil 19, container 52, liquefied gas tanker 20, oil tanker 182, passenger 55, refrigerated

Bahrain

cargo 146, roll-on/roll-off cargo 43, short-sea passenger 16, vehicle carrier 12
note: a flag of convenience registry; includes 46 countries among which are UK 158 ships, Norway 125, Greece 100, US 94, Denmark 80, Netherlands 53, France 36, Finland 35, Japan 35, Sweden 25
Airports:
total: 60
with paved runways over 3,047 m: 2
with paved runways 2,438 to 3,047 m: 1
with paved runways 1,524 to 2,437 m: 16
with paved runways 914 to 1,523 m: 11
with paved runways under 914 m: 22
with unpaved runways 914 to 1,523 m: 8

Communications

Telephone system: 99,000 telephones; totally automatic system; highly developed
local: NA
intercity: NA
international: tropospheric scatter and submarine cable links to Florida; 3 coaxial submarine cables; 1 INTELSAT (Atlantic Ocean) earth station
Radio:
broadcast stations: AM 3, FM 2, shortwave 0
radios: NA
Television:
broadcast stations: 1
televisions: NA

Defense Forces

Branches: Royal Bahamas Defense Force (Coast Guard only), Royal Bahamas Police Force
Defense expenditures: exchange rate conversion—$65 million, 2.7% of GDP (1990)

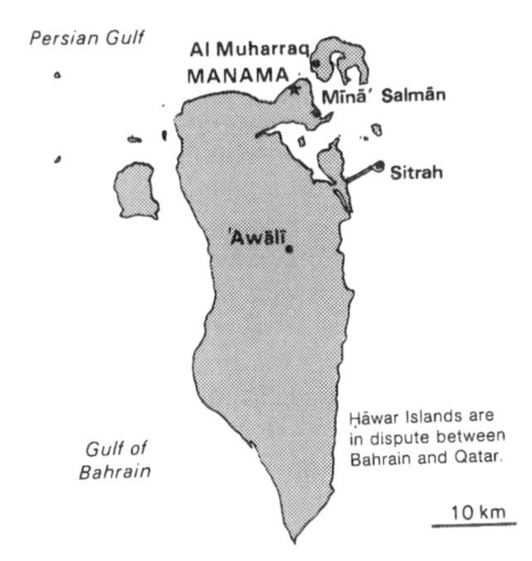

Geography

Location: Middle East, archipelago in the Persian Gulf, east of Saudi Arabia
Map references: Middle East
Area:
total area: 620 sq km
land area: 620 sq km
comparative area: slightly less than 3.5 times the size of Washington, DC
Land boundaries: 0 km
Coastline: 161 km
Maritime claims:
contiguous zone: 24 nm
continental shelf: extending to boundaries to be determined
territorial sea: 12 nm
International disputes: territorial dispute with Qatar over the Hawar Islands; maritime boundary with Qatar
Climate: arid; mild, pleasant winters; very hot, humid summers
Terrain: mostly low desert plain rising gently to low central escarpment
Natural resources: oil, associated and nonassociated natural gas, fish
Land use:
arable land: 2%
permanent crops: 2%
meadows and pastures: 6%
forest and woodland: 0%
other: 90%
Irrigated land: 10 sq km (1989 est.)
Environment:
current issues: desertification resulting from the degradation of limited arable land, periods of drought, and dust storms; coastal degradation (damage to coastlines, coral reefs, and sea vegetation) resulting from oil spills and other discharges from large tankers, oil refineries, and distribution stations; no natural fresh water resources so that groundwater and sea water are the only sources for all water needs
natural hazards: periodic droughts; dust storms
international agreements: party to—Climate Change, Hazardous Wastes, Law of the Sea, Ozone Layer Protection; signed, but not ratified—Biodiversity
Note: close to primary Middle Eastern petroleum sources; strategic location in Persian Gulf through which much of Western world's petroleum must transit to reach open ocean

People

Population: 575,925 (July 1995 est.)
Age structure:
0-14 years: 31% (female 87,398; male 89,976)
15-64 years: 67% (female 152,363; male 231,586)
65 years and over: 2% (female 7,051; male 7,551) (July 1995 est.)
Population growth rate: 2.58% (1995 est.)
Birth rate: 24.12 births/1,000 population (1995 est.)
Death rate: 3.31 deaths/1,000 population (1995 est.)
Net migration rate: 4.95 migrant(s)/1,000 population (1995 est.)
Infant mortality rate: 18 deaths/1,000 live births (1995 est.)
Life expectancy at birth:
total population: 73.94 years
male: 71.46 years
female: 76.49 years (1995 est.)
Total fertility rate: 3.12 children born/woman (1995 est.)
Nationality:
noun: Bahraini(s)
adjective: Bahraini
Ethnic divisions: Bahraini 63%, Asian 13%, other Arab 10%, Iranian 8%, other 6%
Religions: Shi'a Muslim 70%, Sunni Muslim 30%
Languages: Arabic, English, Farsi, Urdu
Literacy: age 15 and over can read and write (1991)
total population: 84%
male: 89%
female: 77%
Labor force: 140,000
by occupation: industry and commerce 85%, agriculture 5%, services 5%, government 3% (1982)
note: 42% of labor force is Bahraini

Government

Names:
conventional long form: State of Bahrain
conventional short form: Bahrain
local long form: Dawlat al Bahrayn
local short form: Al Bahrayn
Digraph: BA
Type: traditional monarchy
Capital: Manama
Administrative divisions: 12 districts (manatiq, singular—mintaqah); Al Hadd, Al Manamah, Al Mintaqah al Gharbiyah, Al Mintaqah al Wusta, Al Mintaqah ash

Shamaliyah, Al Muharraq, Ar Rifa'wa al Mintaqah al Janubiyah, Jidd Hafs, Madinat Hamad, Madinat 'Isa, Mintaqat Juzur Hawar, Sitrah
Independence: 15 August 1971 (from UK)
National holiday: Independence Day, 16 December (1961)
Constitution: 26 May 1973, effective 6 December 1973
Legal system: based on Islamic law and English common law
Suffrage: none
Executive branch:
chief of state: Amir ISA bin Salman Al Khalifa (since 2 November 1961); Heir Apparent HAMAD bin Isa bin Salman Al Khalifa (son of the Amir, born 28 January 1950)
head of government: Prime Minister KHALIFA bin Salman Al Khalifa (since 19 January 1970)
cabinet: Cabinet
Legislative branch: unicameral National Assembly was dissolved 26 August 1975 and legislative powers were assumed by the Cabinet; appointed Advisory Council established 16 December 1992
Judicial branch: High Civil Appeals Court
Political parties and leaders: political parties prohibited; several small, clandestine leftist and Islamic fundamentalist groups are active
Member of: ABEDA, AFESD, AL, AMF, ESCWA, FAO, G-77, GATT, GCC, IBRD, ICAO, ICRM, IDB, IFRCS, ILO, IMF, IMO, INMARSAT, INTELSAT, INTERPOL, IOC, ISO (correspondent), ITU, NAM, OAPEC, OIC, UN, UNCTAD, UNESCO, UNIDO, UPU, WFTU, WHO, WMO
Diplomatic representation in US:
chief of mission: Ambassador Muhammad ABD AL-GHAFFAR al-Abdallah
chancery: 3502 International Drive NW, Washington, DC 20008
telephone: [1] (202) 342-0741, 342-0742
consulate(s) general: New York
US diplomatic representation:
chief of mission: Ambassador David M. RANSOM
embassy: Building No. 979, Road 3119 (next to Ahli Sports Club), Zinj District, Manama
mailing address: FPO AE 09834-5100; P.O. Box 26431, Manama (International Mail)
telephone: [973] 273300; afterhours [973] 275-126
FAX: [973] 272594
Flag: red with a white serrated band (eight white points) on the hoist side

Economy

Overview: Tiny in area, Bahrain is well-to-do in economic resources and per capita income. Petroleum production and processing account for about 80% of export receipts, 60% of government revenues, and 30% of GDP. Economic conditions have fluctuated with the changing fortunes of oil since 1985, for example, during and following the Gulf crisis of 1990-91. With its highly developed communication and transport facilities Bahrain is home to numerous multinational firms with business in the Gulf. A large share of exports consists of petroleum products made from imported crude. Prospects for 1995 are good, with private enterprise the main driving force, e.g., in banking and construction. Unemployment, especially among the young, and the depletion of both oil and underground water resources are major long-term economic problems.
National product: GDP—purchasing power parity—$7.1 billion (1994 est.)
National product real growth rate: 2.2% (1994 est.)
National product per capita: $12,100 (1994 est.)
Inflation rate (consumer prices): 2% (1994 est.)
Unemployment rate: 15% (1991 est.)
Budget:
revenues: $1.2 billion (1989)
expenditures: $1.6 billion, including capital expenditures of $NA (1992)
Exports: $3.69 billion (f.o.b., 1993 est.)
commodities: petroleum and petroleum products 80%, aluminum 7%
partners: Japan 11%, UAE 5%, South Korea 4%, India 4%, Saudi Arabia 3% (1992)
Imports: $3.83 billion (f.o.b., 1993 est.)
commodities: nonoil 59%, crude oil 41%
partners: Saudi Arabia 47%, UK 7%, Japan 7%, US 6%, Germany 5% (1992)
External debt: $2.6 billion (1993)
Industrial production: growth rate 13% (1992); accounts for 38% of GDP, including petroleum
Electricity:
capacity: 1,050,000 kW
production: 3.3 billion kWh
consumption per capita: 5,453 kWh (1993)
Industries: petroleum processing and refining, aluminum smelting, offshore banking, ship repairing
Agriculture: including fishing, accounts for less than 2% of GDP; not self-sufficient in food production; heavily subsidized sector produces fruit, vegetables, poultry, dairy products, shrimp, fish
Economic aid:
recipient: US commitments, including Ex-Im (FY70-79), $24 million; Western (non-US) countries, ODA and OOF bilateral commitments (1970-89), $45 million; OPEC bilateral aid (1979-89), $9.8 billion
Currency: 1 Bahraini dinar (BD) = 1,000 fils
Exchange rates: Bahraini dinars (BD) per US$1—0.3760 (fixed rate)
Fiscal year: calendar year

Transportation

Railroads: 0 km
Highways:
total: 2,670 km
paved: 2,010 km
unpaved: 660 km (1991 est.)
Pipelines: crude oil 56 km; petroleum products 16 km; natural gas 32 km
Ports: Manama, Mina' Salman, Sitrah
Merchant marine:
total: 6 ships (1,000 GRT or over) totaling 79,949 GRT/120,900 DWT
ships by type: bulk 1, cargo 4, chemical tanker 1
Airports:
total: 4
with paved runways over 3,047 m: 2
with paved runways under 914 m: 1
with unpaved runways 1,524 to 2,438 m: 1

Communications

Telephone system: 98,000 telephones; 170 telephones/1,000 people; modern system; good domestic services; excellent international connections
local: NA
intercity: NA
international: 2 INTELSAT (1 Atlantic Ocean and 1 Indian Ocean) and 1 ARABSAT earth station; tropospheric scatter to Qatar, UAE; microwave radio relay to Saudi Arabia; submarine cable to Qatar, UAE, and Saudi Arabia
Radio:
broadcast stations: AM 2, FM 3, shortwave 0
radios: NA
Television:
broadcast stations: 2
televisions: NA

Defense Forces

Branches: Army, Navy, Air Force, Air Defense, Coast Guard, Police Force
Manpower availability: males age 15-49 210,725; males fit for military service 117,414; males reach military age (15) annually 4,346 (1995 est.)
Defense expenditures: exchange rate conversion—$247 million, 5.5% of GDP (1994)

Baker Island

(territory of the US)

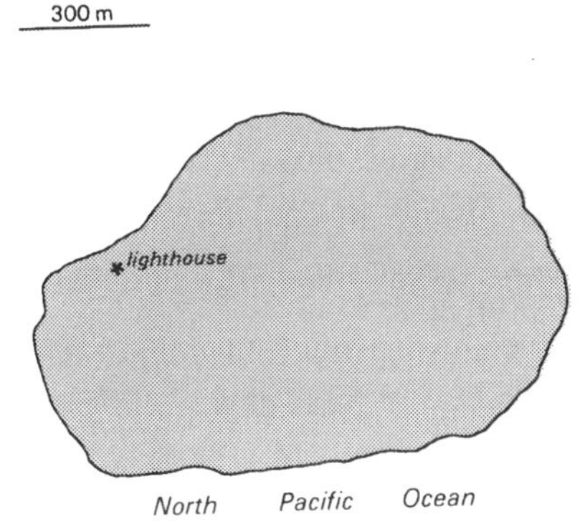

Geography

Location: Oceania, atoll in the North Pacific Ocean, about one-half of the way from Hawaii to Australia
Map references: Oceania
Area:
total area: 1.4 sq km
land area: 1.4 sq km
comparative area: about 2.3 times the size of The Mall in Washington, DC
Land boundaries: 0 km
Coastline: 4.8 km
Maritime claims:
exclusive economic zone: 200 nm
territorial sea: 12 nm
International disputes: none
Climate: equatorial; scant rainfall, constant wind, burning sun
Terrain: low, nearly level coral island surrounded by a narrow fringing reef
Natural resources: guano (deposits worked until 1891)
Land use:
arable land: 0%
permanent crops: 0%
meadows and pastures: 0%
forest and woodland: 0%
other: 100%
Irrigated land: 0 sq km
Environment:
current issues: no natural fresh water resources
natural hazards: the narrow fringing reef surrounding the island can be a maritime hazard
international agreements: NA
Note: treeless, sparse, and scattered vegetation consisting of grasses, prostrate vines, and low growing shrubs; primarily a nesting, roosting, and foraging habitat for seabirds, shorebirds, and marine wildlife

People

Population: uninhabited; note—American civilians evacuated in 1942 after Japanese air and naval attacks during World War II; occupied by US military during World War II, but abandoned after the war; public entry is by special-use permit only and generally restricted to scientists and educators; a cemetery and cemetery ruins are located near the middle of the west coast

Government

Names:
conventional long form: none
conventional short form: Baker Island
Digraph: FQ
Type: unincorporated territory of the US administered by the Fish and Wildlife Service of the US Department of the Interior as part of the National Wildlife Refuge system
Capital: none; administered from Washington, DC

Economy

Overview: no economic activity

Transportation

Ports: none; offshore anchorage only; note—there is one boat landing area along the middle of the west coast
Airports: 1 abandoned World War II runway of 1,665 m
Note: there is a day beacon near the middle of the west coast

Defense Forces

Note: defense is the responsibility of the US; visited annually by the US Coast Guard

Bangladesh

Geography

Location: Southern Asia, bordering the Bay of Bengal, between Burma and India
Map references: Asia
Area:
total area: 144,000 sq km
land area: 133,910 sq km
comparative area: slightly smaller than Wisconsin
Land boundaries: total 4,246 km, Burma 193 km, India 4,053 km
Coastline: 580 km
Maritime claims:
contiguous zone: 18 nm
continental shelf: up to the outer limits of the continental margin
exclusive economic zone: 200 nm
territorial sea: 12 nm
International disputes: a portion of the boundary with India is in dispute; water-sharing problems with upstream riparian India over the Ganges
Climate: tropical; cool, dry winter (October to March); hot, humid summer (March to June); cool, rainy monsoon (June to October)
Terrain: mostly flat alluvial plain; hilly in southeast
Natural resources: natural gas, arable land, timber
Land use:
arable land: 67%
permanent crops: 2%
meadows and pastures: 4%
forest and woodland: 16%
other: 11%
Irrigated land: 27,380 sq km (1989)
Environment:
current issues: many people are landless and forced to live on and cultivate flood-prone land; limited access to potable water; water-borne diseases prevalent; water pollution especially of fishing areas results from the use of commercial pesticides; intermittent water shortages because of falling water tables in the northern and central parts of the country; soil degradation; deforestation; severe overpopulation

natural hazards: droughts, cyclones; much of the country routinely flooded during the summer monsoon season
international agreements: party to—Biodiversity, Climate Change, Endangered Species, Environmental Modification, Hazardous Wastes, Nuclear Test Ban, Ozone Layer Protection, Wetlands; signed, but not ratified—Desertification, Law of the Sea

People

Population: 128,094,948 (July 1995 est.)
Age structure:
0-14 years: 40% (female 25,195,262; male 26,352,299)
15-64 years: 57% (female 34,862,105; male 37,867,705)
65 years and over: 3% (female 1,761,336; male 2,056,241) (July 1995 est.)
Population growth rate: 2.32% (1995 est.)
Birth rate: 34.62 births/1,000 population (1995 est.)
Death rate: 11.43 deaths/1,000 population (1995 est.)
Net migration rate: 0 migrant(s)/1,000 population (1995 est.)
Infant mortality rate: 104.6 deaths/1,000 live births (1995 est.)
Life expectancy at birth:
total population: 55.46 years
male: 55.69 years
female: 55.22 years (1995 est.)
Total fertility rate: 4.39 children born/woman (1995 est.)
Nationality:
noun: Bangladeshi(s)
adjective: Bangladesh
Ethnic divisions: Bengali 98%, Biharis 250,000, tribals less than 1 million
Religions: Muslim 83%, Hindu 16%, Buddhist, Christian, other
Languages: Bangla (official), English
Literacy: age 15 and over can read and write (1990)
total population: 35%
male: 47%
female: 22%
Labor force: 50.1 million
by occupation: agriculture 65%, services 21%, industry and mining 14% (1989)
note: extensive export of labor to Saudi Arabia, UAE, and Oman (1991)

Government

Names:
conventional long form: People's Republic of Bangladesh
conventional short form: Bangladesh
former: East Pakistan
Digraph: BG
Type: republic
Capital: Dhaka
Administrative divisions: 4 divisions; Chittagong, Dhaka, Khulna, Rajshahi
Independence: 16 December 1971 (from Pakistan)
National holiday: Independence Day, 26 March (1971)
Constitution: 4 November 1972, effective 16 December 1972, suspended following coup of 24 March 1982, restored 10 November 1986, amended many times
Legal system: based on English common law
Suffrage: 18 years of age; universal
Executive branch:
chief of state: President Abdur Rahman BISWAS (since 8 October 1991); election last held 8 October 1991 (next to be held by NA October 1996); results—Abdur Rahman BISWAS received 52.1% of parliamentary vote
head of government: Prime Minister Khaleda ZIAur RAHMAN (since 20 March 1991)
cabinet: Council of Ministers; appointed by the president
Legislative branch: unicameral
National Parliament (Jatiya Sangsad): elections last held 27 February 1991 (next to be held by February 1996); results—percent of vote by party NA; seats—(330 total, 300 elected and 30 seats reserved for women) BNP 168, AL 93, JP 35, JI 20, BCP 5, National Awami Party (Muzaffar) 1, Workers Party 1, JSD 1, Ganotantri Party 1, Islami Oikya Jote 1, NDP 1, independents 3
Judicial branch: Supreme Court
Political parties and leaders: Bangladesh Nationalist Party (BNP), Khaleda ZIAur RAHMAN; Awami League (AL), Sheikh Hasina WAJED; Jatiyo Party (JP), Hussain Mohammad ERSHAD (in jail); Jamaat-E-Islami (JI), Ali KHAN; Bangladesh Communist Party (BCP), Saifuddin Ahmed MANIK; National Awami Party (Muzaffar); Workers Party, leader NA; Jatiyo Samajtantik Dal (JSD), Serajul ALAM KHAN; Ganotantri Party, leader NA; Islami Oikya Jote, leader NA; National Democratic Party (NDP), leader NA; Muslim League, Khan A. SABUR; Democratic League, Khondakar MUSHTAQUE Ahmed; Democratic League, Khondakar MUSHTAQUE Ahmed; United People's Party, Kazi ZAFAR Ahmed
Member of: AsDB, C, CCC, CP, ESCAP, FAO, G-77, GATT, IAEA, IBRD, ICAO, ICFTU, ICRM, IDA, IDB, IFAD, IFC, IFRCS, ILO, IMF, IMO, INMARSAT, INTELSAT, INTERPOL, IOC, IOM, ISO, ITU, MINURSO, NAM, OIC, SAARC, UN, UNAMIR, UNCTAD, UNESCO, UNIDO, UNIKOM, UNOMIG, UNOMIL, UNOMOZ, UNOMUR, UNOSOM, UNPROFOR, UNU, UPU, WCL, WFTU, WHO, WIPO, WMO, WTO
Diplomatic representation in US:
chief of mission: Ambassador Humayun KABIR
chancery: 2201 Wisconsin Avenue NW, Washington, DC 20007
telephone: [1] (202) 342-8372 through 8376
consulate(s) general: New York
US diplomatic representation:
chief of mission: Ambassador David N. MERRILL
embassy: Diplomatic Enclave, Madani Avenue, Baridhara, Dhaka
mailing address: G. P. O. Box 323, Dhaka 1212
telephone: [880] (2) 884700 through 884722
FAX: [880] (2) 883-744
Flag: green with a large red disk slightly to the hoist side of center; green is the traditional color of Islam

Economy

Overview: Despite sustained domestic and international efforts to improve economic and demographic prospects, Bangladesh remains one of the world's poorest, most densely populated, and least developed nations. Its economy is overwhelmingly agricultural, with the cultivation of rice the single most important activity in the economy. Major impediments to growth include frequent cyclones and floods, the inefficiency of state-owned enterprises, a rapidly growing labor force that cannot be absorbed by agriculture, delays in exploiting energy resources (natural gas), and inadequate power supplies. Excellent rice crops and expansion of the export garment industry led to real growth of 4% in 1992 and again in 1993. Policy measures intended to reduce government regulation of private industry, to curb population growth, and to expand employment opportunities have had only partial success given the serious nature of Bangladesh's basic problems.
National product: GDP—purchasing power parity—$130.1 billion (1994 est.)
National product real growth rate: 4.5% (1994 est.)
National product per capita: $1,040 (1994 est.)
Inflation rate (consumer prices): 4.3% (1992 est.)
Unemployment rate: NA%
Budget:
revenues: $2.8 billion
expenditures: $4.1 billion, including capital expenditures of $1.8 billion (FY92/93)
Exports: $2.38 billion (1993)
commodities: garments, jute and jute goods, leather, shrimp
partners: US 33%, Western Europe 39% (Germany 8.4%, Italy 6%) (FY91/92 est.)
Imports: $3.99 billion (1993)
commodities: capital goods, petroleum, food, textiles
partners: Hong Kong 7.5%, Singapore 7.4%, China 7.4%, Japan 7.1% (FY91/92 est.)
External debt: $13.5 billion (June 1993)
Industrial production: growth rate 6.9%

(FY92/93 est.); accounts for 9.4% of GDP
Electricity:
capacity: 2,740,000 kW
production: 9.2 billion kWh
consumption per capita: 70 kWh (1993)
Industries: jute manufacturing, cotton textiles, food processing, steel, fertilizer
Agriculture: accounts for 33% of GDP, 65% of employment, and one-fifth of exports; world's largest exporter of jute; commercial products—jute, rice, wheat, tea, sugarcane, potatoes, beef, milk, poultry; shortages include wheat, vegetable oils, cotton
Illicit drugs: transit country for illegal drugs produced in neighboring countries
Economic aid:
recipient: US commitments, including Ex-Im (FY70-89), $3.4 billion; Western (non-US) countries, ODA and OOF bilateral commitments (1980-89), $11.65 million; OPEC bilateral aid (1979-89), $6.52 million; Communist countries (1970-89), $1.5 billion
Currency: 1 taka (Tk) = 100 poiska
Exchange rates: taka (Tk) per US$1—40.250 (January 1995), 40.212 (1994), 39.567 (1993), 38.951 (1992), 36.596 (1991), 34.569 (1990)
Fiscal year: 1 July—30 June

Transportation

Railroads:
total: 2,892 km
broad gauge: 978 km 1.676-m gauge
narrow gauge: 1,914 km 1.000-m gauge (1992)
Highways:
total: 7,240 km
paved: 3,840 km
unpaved: 3,400 km (1985)
Inland waterways: 5,150-8,046 km navigable waterways (includes 2,575-3,058 km main cargo routes)
Pipelines: natural gas 1,220 km
Ports: Barisal, Chandpur, Chittagong, Cox's Bazar, Dacca, Khulna, Mongla (includes Chalna), Narayanganj
Merchant marine:
total: 38 ships (1,000 GRT or over) totaling 293,304 GRT/428,013 DWT
ships by type: bulk 2, cargo 31, oil tanker 2, refrigerated cargo 3
Airports:
total: 16
with paved runways over 3,047 m: 2
with paved runways 2,438 to 3,047 m: 2
with paved runways 1,524 to 2,437 m: 4
with paved runways 914 to 1,523 m: 1
with paved runways under 914 m: 7

Communications

Telephone system: 241,250 telephones; 1 telephone/522 persons; poor domestic telephone service
local: NA
intercity: NA
international: 2 INTELSAT (Indian Ocean) earth stations; adequate international radio communications and landline service
Radio:
broadcast stations: AM 9, FM 6, shortwave 0
radios: NA
Television:
broadcast stations: 11
televisions: NA

Defense Forces

Branches: Army, Navy, Air Force
paramilitary forces: Bangladesh Rifles, Bangladesh Ansars, Armed Police Reserve, Village Defense Parties, National Cadet Corps
Manpower availability: males age 15-49 33,039,035; males fit for military service 19,607,817 (1995 est.)
Defense expenditures: exchange rate conversion—$448 million, 1.7% of GDP (FY93/94)

Barbados

Geography

Location: Caribbean, island between the Caribbean Sea and the North Atlantic Ocean, northeast of Venezuela
Map references: Central America and the Caribbean
Area:
total area: 430 sq km
land area: 430 sq km
comparative area: slightly less than 2.5 times the size of Washington, DC
Land boundaries: 0 km
Coastline: 97 km
Maritime claims:
exclusive economic zone: 200 nm
territorial sea: 12 nm
International disputes: none
Climate: tropical; rainy season (June to October)
Terrain: relatively flat; rises gently to central highland region
Natural resources: petroleum, fishing, natural gas
Land use:
arable land: 77%
permanent crops: 0%
meadows and pastures: 9%
forest and woodland: 0%
other: 14%
Irrigated land: NA sq km
Environment:
current issues: pollution of coastal waters from waste disposal by ships; soil erosion; illegal solid waste disposal threatens contamination of aquifers
natural hazards: hurricanes (especially June to October); periodic landslides
international agreements: party to—Climate Change, Endangered Species, Law of the Sea, Marine Dumping, Ozone Layer Protection, Ship Pollution; signed, but not ratified—Biodiversity
Note: easternmost Caribbean island

People

Population: 256,395 (July 1995 est.)
Age structure:
0-14 years: 24% (female 30,175; male 31,507)
15-64 years: 66% (female 86,103; male 82,727)
65 years and over: 10% (female 15,849; male 10,034) (July 1995 est.)
Population growth rate: 0.24% (1995 est.)
Birth rate: 15.45 births/1,000 population (1995 est.)
Death rate: 8.27 deaths/1,000 population (1995 est.)
Net migration rate: -4.82 migrant(s)/1,000 population (1995 est.)
Infant mortality rate: 19.2 deaths/1,000 live births (1995 est.)
Life expectancy at birth:
total population: 74.16 years
male: 71.47 years
female: 77.06 years (1995 est.)
Total fertility rate: 1.78 children born/woman (1995 est.)
Nationality:
noun: Barbadian(s)
adjective: Barbadian
Ethnic divisions: African 80%, European 4%, other 16%
Religions: Protestant 67% (Anglican 40%, Pentecostal 8%, Methodist 7%, other 12%), Roman Catholic 4%, none 17%, unknown 3%, other 9% (1980)
Languages: English
Literacy: age 15 and over has ever attended school (1970)
total population: 99%
male: 99%
female: 99%
Labor force: 124,800 (1992)
by occupation: services and government 41%, commerce 15%, manufacturing and construction 18%, transportation, storage, communications, and financial institutions 8%, agriculture 6%, utilities 2% (1992 est.)

Government

Names:
conventional long form: none
conventional short form: Barbados
Digraph: BB
Type: parliamentary democracy
Capital: Bridgetown
Administrative divisions: 11 parishes; Christ Church, Saint Andrew, Saint George, Saint James, Saint John, Saint Joseph, Saint Lucy, Saint Michael, Saint Peter, Saint Philip, Saint Thomas
note: the new city of Bridgetown may be given parish status
Independence: 30 November 1966 (from UK)
National holiday: Independence Day, 30 November (1966)
Constitution: 30 November 1966
Legal system: English common law; no judicial review of legislative acts
Suffrage: 18 years of age; universal
Executive branch:
chief of state: Queen ELIZABETH II (since 6 February 1952), represented by Governor General Dame Nita BARROW (since 6 June 1990)
head of government: Prime Minister Owen Seymour ARTHUR (since 6 September 1994); Deputy Prime Minister Billie MILLER (since 6 September 1994)
cabinet: Cabinet; appointed by the governor general on advice of the prime minister
Legislative branch: bicameral Parliament
Senate: consists of a 21-member body appointed by the governor general
House of Assembly: election last held 6 September 1994 (next to be held by January 1999); results—percentage vote by party NA; seats—(28 total) DLP 8, BLP 19, NDP 1
Judicial branch: Supreme Court of Judicature
Political parties and leaders: Democratic Labor Party (DLP),David THOMPSON; Barbados Labor Party (BLP), Owen ARTHUR; National Democratic Party (NDP), Richard HAYNES
Other political or pressure groups: Barbados Workers Union, Leroy TROTMAN; People's Progressive Movement, Eric SEALY; Workers' Party of Barbados, Dr. George BELLE; Clement Payne Labor Union, David COMMISSIONG
Member of: ACP, C, CARICOM, CDB, ECLAC, FAO, G-77, GATT, IADB, IBRD, ICAO, ICFTU, ICRM, IFAD, IFC, IFRCS, ILO, IMF, IMO, INTELSAT, INTERPOL, IOC, ISO (correspondent), ITU, LAES, NAM, OAS, OPANAL, UN, UNCTAD, UNESCO, UNIDO, UPU, WHO, WIPO, WMO
Diplomatic representation in US:
chief of mission: Ambassador Courtney BLACKMAN
chancery: 2144 Wyoming Avenue NW, Washington, DC 20008
telephone: [1] (202) 939-9218, 9219
FAX: [1] (202) 332-7467
consulate(s) general: Miami and New York
consulate(s): Los Angeles
US diplomatic representation:
chief of mission: Ambassador Jeanette W. HYDE
embassy: Canadian Imperial Bank of Commerce Building, Broad Street, Bridgetown
mailing address: P. O. Box 302, Bridgetown; FPO AA 34055
telephone: [1] (809) 436-4950
FAX: [1] (809) 429-5246
Flag: three equal vertical bands of blue (hoist side), yellow, and blue with the head of a black trident centered on the gold band; the trident head represents independence and a break with the past (the colonial coat of arms contained a complete trident)

Economy

Overview: A per capita income of $9,200 gives Barbados one of the highest standards of living of all the small island states of the eastern Caribbean. Historically, the economy was based on the cultivation of sugarcane and related activities. In recent years, however, the economy has diversified into manufacturing and tourism. A moderate recovery that began in late 1993 after 3 years of contraction is mainly due to increased tourism and expansion in the construction sector. Economic prospects for 1995 depend mostly on continued growth in the industrialized countries, especially in Europe, which would spur further expansion in tourism.
National product: GDP—purchasing power parity—$2.4 billion (1994 est.)
National product real growth rate: 3% (1994 est.)
National product per capita: $9,200 (1994 est.)
Inflation rate (consumer prices): 2% (1994 est.)
Unemployment rate: 20.5% (1994 est.)
Budget:
revenues: $509 million
expenditures: $636 million, including capital expenditures of $86 million (FY94/95 est.)
Exports: $161 million (f.o.b., 1993 est.)
commodities: sugar and molasses, rum, other foods and beverages, chemicals, electrical components, clothing
partners: US 13%, UK 10%, Trinidad and Tobago 9%, Windward Islands 8%
Imports: $703 million (c.i.f., 1993 est.)
commodities: consumer goods, machinery, foodstuffs, construction materials, chemicals, fuel, electrical components
partners: US 36%, UK 11%, Trinidad and Tobago 11%, Japan 3%
External debt: $652 million (1991 est.)
Industrial production: growth rate 2% (FY93/94 est.); accounts for about 10% of GDP
Electricity:
capacity: 152,100 kW
production: 510 million kWh
consumption per capita: 1,841 kWh (1993)
Industries: tourism, sugar, light manufacturing, component assembly for export
Agriculture: accounts for 6% of GDP; major cash crop is sugarcane; other crops—vegetables, cotton; not self-sufficient in food
Illicit drugs: one of many Caribbean transshipment points for narcotics bound for the US and Europe
Economic aid:
recipient: US commitments, including Ex-Im (FY70-89), $15 million; Western (non-US) countries, ODA and OOF bilateral commitments (1970-89), $171 million

Currency: 1 Barbadian dollar (Bds$) = 100 cents
Exchange rates: Barbadian dollars (Bds$) per US$1—2.0113 (fixed rate)
Fiscal year: 1 April—31 March

Transportation

Railroads: 0 km
Highways:
total: 1,570 km
paved: 1,475 km
unpaved: gravel, earth 95 km
Ports: Bridgetown
Merchant marine:
total: 12 ships (1,000 GRT or over) totaling 61,563 GRT/103,632 DWT
ships by type: bulk 4, cargo 6, oil tanker 2
Airports:
total: 1
with paved runways over 3,047 m: 1

Communications

Telephone system: 89,000 telephones;
local: island wide automatic telephone system;
intercity: NA
international: 1 INTELSAT (Atlantic Ocean) ground station; tropospheric scatter link to Trinidad and Saint Lucia
Radio:
broadcast stations: AM 3, FM 2, shortwave 0
radios: NA
Television:
broadcast stations: 2 (1 pay)
televisions: NA

Defense Forces

Branches: Royal Barbados Defense Force (includes the Ground Forces and Coast Guard), Royal Barbados Police Force
Manpower availability: males age 15-49 71,153; males fit for military service 49,488 (1995 est.)
Defense expenditures: exchange rate conversion—$NA, NA% of GDP

Bassas da India
(possession of France)

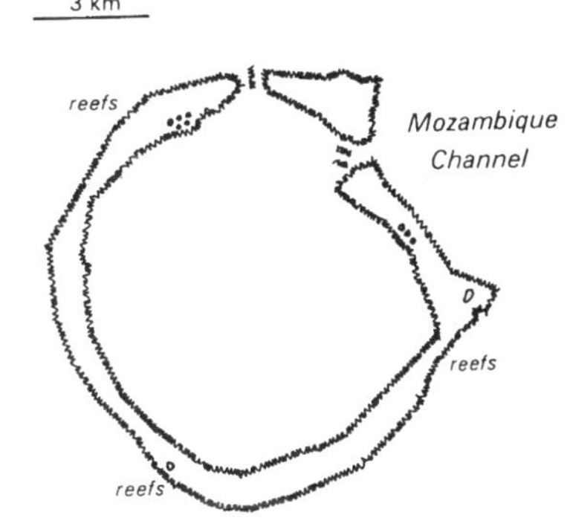

Geography

Location: Southern Africa, islands in the southern Mozambique Channel, about one-half of the way from Mozambique to Madagascar
Map references: Africa
Area:
total area: 0.2 km2
land area: 0.2 km2
comparative area: NA
Land boundaries: 0 km
Coastline: 35.2 km
Maritime claims:
exclusive economic zone: 200 nm
territorial sea: 12 nm
International disputes: claimed by Madagascar
Climate: tropical
Terrain: a volcanic rock 2.4 meters high
Natural resources: none
Land use:
arable land: 0%
permanent crops: 0%
meadows and pastures: 0%
forest and woodland: 0%
other: 100% (all rock)
Irrigated land: 0 sq km
Environment:
current issues: NA
natural hazards: maritime hazard since it is usually under water during high tide and surrounded by reefs; subject to periodic cyclones
international agreements: NA

People

Population: uninhabited

Government

Names:
conventional long form: none
conventional short form: Bassas da India
Digraph: BS
Type: French possession administered by a Commissioner of the Republic, resident in Reunion
Capital: none; administered by France from Reunion
Independence: none (possession of France)

Economy

Overview: no economic activity

Transportation

Ports: none; offshore anchorage only

Defense Forces

Note: defense is the responsibility of France

Belarus

Geography

Location: Eastern Europe, east of Poland
Map references: Commonwealth of Independent States—European States
Area:
total area: 207,600 sq km
land area: 207,600 sq km
comparative area: slightly smaller than Kansas
Land boundaries: total 3,098 km, Latvia 141 km, Lithuania 502 km, Poland 605 km, Russia 959 km, Ukraine 891 km
Coastline: 0 km (landlocked)
Maritime claims: none; landlocked
International disputes: none
Climate: cold winters, cool and moist summers; transitional between continental and maritime
Terrain: generally flat and contains much marshland
Natural resources: forest land, peat deposits, small quantities of oil and natural gas
Land use:
arable land: 29%
permanent crops: 1%
meadows and pastures: 15%
forest and woodland: 0%
other: 55%
Irrigated land: 1,490 sq km (1990)
Environment:
current issues: soil pollution from pesticide use; southern part of the country contaminated with fallout from 1986 nuclear reactor accident at Chornobyl'
natural hazards: NA
international agreements: party to—Air Pollution, Air Pollution-Nitrogen Oxides, Air Pollution-Sulphur 85, Biodiversity, Environmental Modification, Marine Dumping, Nuclear Test Ban, Ozone Layer Protection; signed, but not ratified—Climate Change, Law of the Sea
Note: landlocked

People

Population: 10,437,418 (July 1995 est.)
Age structure:
0-14 years: 22% (female 1,126,062; male 1,166,439)
15-64 years: 65% (female 3,494,891; male 3,293,196)
65 years and over: 13% (female 913,508; male 443,322) (July 1995 est.)
Population growth rate: 0.3% (1995 est.)
Birth rate: 12.98 births/1,000 population (1995 est.)
Death rate: 11.23 deaths/1,000 population (1995 est.)
Net migration rate: 1.27 migrant(s)/1,000 population (1995 est.)
Infant mortality rate: 18.6 deaths/1,000 live births (1995 est.)
Life expectancy at birth:
total population: 71.03 years
male: 66.36 years
female: 75.93 years (1995 est.)
Total fertility rate: 1.87 children born/woman (1995 est.)
Nationality:
noun: Belarusian(s)
adjective: Belarusian
Ethnic divisions: Byelorussian 77.9%, Russian 13.2%, Polish 4.1%, Ukrainian 2.9%, other 1.9%
Religions: Eastern Orthodox, other
Languages: Byelorussian, Russian, other
Literacy: age 15 and over can read and write (1989)
total population: 97%
male: 99%
female: 96%
Labor force: 4.887 million
by occupation: industry and construction 40%, agriculture and forestry 21%, other 39% (1992)

Government

Names:
conventional long form: Republic of Belarus
conventional short form: Belarus
local long form: Respublika Byelarus'
local short form: none
former: Belorussian (Byelorussian) Soviet Socialist Republic
Digraph: BO
Type: republic
Capital: Minsk
Administrative divisions: 6 voblastsi (singular—voblasts') and one municipality* (harady, singular—horad); Brestskaya (Brest), Homyel'skaya (Homyel'), Horad Minsk*, Hrodzyenskaya (Hrodna), Mahilyowskaya (Mahilyow), Minskaya, Vitsyebskaya (Vitsyebsk)
note: the administrative centers of the voblastsi are included in parentheses
Independence: 25 August 1991 (from Soviet Union)
National holiday: Independence Day, 27 July (1990)
Constitution: adopted 15 March 1994; replaces constitution of April 1978
Legal system: based on civil law system
Suffrage: 18 years of age; universal
Executive branch:
chief of state: President Aleksandr LUKASHENKO (since 20 July 1994); election held June 24 and 10 July 1994 (next to be held NA 1999); Aleksandr LUKASHENKO 80%, Vyacheslav KEBICH 14%
head of government: Prime Minister Mikhail CHIGIR (since July 1994); Deputy Prime Ministers Vladimir GARKUN, Viktor GONCHAR, Sergey LING, Mikhail MYASNIKOVICH, Valeriy KOKAREV (since NA)
cabinet: Council of Ministers
note: first presidential elections took place in June-July 1994
Legislative branch: unicameral
Supreme Soviet: elections last held 4 April 1990 (next to be held 14 May 1995); results—Communists 87%; seats—(360 total) number of seats by party NA; note—50 seats are for public bodies; the Communist Party obtained an overwhelming majority
Judicial branch: Supreme Court
Political parties and leaders: Belarusian Popular Front (BPF), Zenon POZNYAK, chairman; Party of Popular Accord, Gennadiy KARPENKO; Union of Belarusian Entreprenuers, V. N. KARYAGIN; Belarusian Party of Communists, Vasiliy NOVIKOV, Viktor CHIKIN, chairmen; Belarus Peasant Party, Yevgeniy LUGIN, chairman; Belarusian Socialist Party, Vyacheslav KUZNETSOV, chairman; Belarusian Social Democrat Party (SDBP), Oleg TRUSOV, Stanislav SHUSHKEVICH, chairmen; Agrarian Party of Belarus, Aleksandr DUBKO; United Democratic Party of Belarus (UDPB), Aleksandr DOBROVOLSKIY, chairman; Independent Trade Unions, Sergey ANTONCHIK, chairman
Member of: CCC, CE (guest), CEI (associate members), CIS, EBRD, ECE, IAEA, IBRD, ICAO, IFC, ILO, IMF, INMARSAT, INTELSAT (nonsignatory user), INTERPOL, IOC, ISO, ITU, NACC, OSCE, PCA, PFP, UN, UNCTAD, UNESCO, UNIDO, UPU, WHO, WIPO, WMO
Diplomatic representation in US:
chief of mission: Ambassador Sergey Nikolayevich MARTYNOV
chancery: 1619 New Hampshire Avenue NW, Washington, DC 20009
telephone: [1] (202) 986-1604
FAX: [1] (202) 986-1805
consulate(s) general: New York
US diplomatic representation:
chief of mission: Ambassador Kenneth Spencer YALOWITZ
embassy: Starovilenskaya #46, Minsk
mailing address: use embassy street address

telephone: [7] (0172) 34-65-37
Flag: three horizontal bands of white (top), red, and white

Economy

Overview: Belarus ranks among the most developed of the former Soviet states, with a relatively modern—by Soviet standards—and diverse machine building sector and a robust agriculture sector. It also serves as a transport link for Russian oil exports to the Baltic states and Eastern and Western Europe. The breakup of the Soviet Union and its command economy has resulted in a sharp economic contraction as traditional trade ties have collapsed. The Belarusian government has lagged behind the governments of most other former Soviet states in economic reform, with privatization almost nonexistent. The system of state orders and distribution persists. In mid-1994, the Belarusian government embarked on an austerity program with IMF support to slash state credits and consumer subsidies in order to bring down the budget deficit and reduce inflation. However, despite its promising start, the regime's drive to reinvigorate the economy has fallen short, and the IMF has criticized its failure to implement the reforms that the Fund had negotiated. As a result, the IMF has suspended talks on introducing a stand-by arrangement. Economic relations with Russia, which will have an important bearing on the future course of the economy, will be strengthened if Minsk adopts the necessary legislation to implement a customs union agreed to in January 1995.
National product: GDP—purchasing power parity—$53.4 billion (1994 estimate as extrapolated from World Bank estimate for 1992)
National product real growth rate: -20% (1994)
National product per capita: $5,130 (1994 est.)
Inflation rate (consumer prices): 29% per month (1994)
Unemployment rate: 1.4% officially registered unemployed (December 1993); large numbers of underemployed workers
Budget:
revenues: $NA
expenditures: $NA, including capital expenditures of $NA
Exports: $968 million to outside of the FSU countries (f.o.b., 1994)
commodities: machinery and transport equipment, chemicals, foodstuffs
partners: Russia, Ukraine, Poland, Bulgaria
Imports: $534 million from outside the FSU countries (c.i.f., 1994)
commodities: fuel, natural gas, industrial raw materials, textiles, sugar
partners: Russia, Ukraine, Poland
External debt: $1.5 billion (July 1994 est.)
Industrial production: growth rate -19% (1994); accounts for about 40% of GDP (1992)
Electricity:
capacity: 7,010,000 kW
production: 31.4 billion kWh
consumption per capita: 3,010 kWh (1994)
Industries: employ about 40% of labor force and produced a wide variety of products including (in percent share of total output of former Soviet Union): tractors (12%); metal-cutting machine tools (11%); off-highway dump trucks up to 110-metric-ton load capacity (100%); wheel-type earthmovers for construction and mining (100%); eight-wheel-drive, high-flotation trucks with cargo capacity of 25 metric tons for use in tundra and roadless areas (100%); equipment for animal husbandry and livestock feeding (25%); motorcycles (21.3%); television sets (11%); chemical fibers (28%); fertilizer (18%); linen fabric (11%); wool fabric (7%); radios; refrigerators; and other consumer goods
Agriculture: accounts for almost 25% of GDP and 5.7% of total agricultural output of former Soviet Union; employs 21% of the labor force; in 1988 produced the following (in percent of total Soviet production): grain (3.6%), potatoes (12.2%), vegetables (3.0%), meat (6.0%), milk (7.0%); net exporter of meat, milk, eggs, flour, potatoes
Illicit drugs: illicit cultivator of opium poppy and cannabis; mostly for the domestic market; transshipment point for illicit drugs to Western Europe
Economic aid: $NA
Currency: Belarusian rubel (BR)
Exchange rates: Belarusian rubels per US$1—10,600 (end December 1994)
Fiscal year: calendar year

Transportation

Railroads:
total: 5,570 km in common carrier service; does not include industrial lines
broad gauge: 5,570 km 1.520-m gauge (1990)
Highways:
total: 98,200 km
paved: 66,100 km
unpaved: earth 32,100 km (1990)
Inland waterways: NA km
Pipelines: crude oil 1,470 km; refined products 1,100 km; natural gas 1,980 km (1992)
Ports: Mazyr
Merchant marine:
note: claims 5% of former Soviet fleet
Airports:
total: 118
with paved runways over 3,047 m: 2
with paved runways 2,438 to 3,047 m: 18
with paved runways 1,524 to 2,437 m: 5
with paved runways under 914 m: 11
with unpaved runways over 3,047 m: 1
with unpaved runways 2,438 to 3,047 m: 6
with unpaved runways 1,524 to 2,438 m: 4
with unpaved runways 914 to 1,523 m: 9
with unpaved runways under 914 m: 62

Communications

Telephone system: 1,849,000 telephones (December 1991); 18 telephones/100 persons; telephone service inadequate for the purposes of either business or the population; about 70% of the telephones are in homes; over 750,000 applications from households for telephones remain unsatisfied (1992); new investment centers on international connections and business needs; the new BelCel NMT 450 cellular system (a joint venture) is now operating in Minsk
local: NA
intercity: NA
international: international traffic is carried by the Moscow international gateway switch and also by 2 satellite earth stations near Minsk—INTELSAT (through Canada) and EUTELSAT (through the UK)
Radio:
broadcast stations: AM NA, FM NA, shortwave 0
radios: 3.14 million (5,615,000 with multiple speaker systems for program diffusion)
Television:
broadcast stations: NA
televisions: 3.538 million

Defense Forces

Branches: Army, Air Force, Air Defense Force, Republic Security Forces (internal and border troops)
Manpower availability: males age 15-49 2,550,500; males fit for military service 1,999,138; males reach military age (18) annually 71,808 (1995 est.)
Defense expenditures: 56.5 billion rubles, NA% of GDP (1993 est.); note—conversion of the military budget into US dollars using the current exchange rate could produce misleading results

Belgium

Geography

Location: Western Europe, bordering the North Sea, between France and the Netherlands
Map references: Europe
Area:
total area: 30,510 sq km
land area: 30,230 sq km
comparative area: slightly larger than Maryland
Land boundaries: total 1,385 km, France 620 km, Germany 167 km, Luxembourg 148 km, Netherlands 450 km
Coastline: 64 km
Maritime claims:
continental shelf: median line with neighbors
exclusive fishing zone: median line with neighbors (extends about 68 km from coast)
territorial sea: 12 nm
International disputes: none
Climate: temperate; mild winters, cool summers; rainy, humid, cloudy
Terrain: flat coastal plains in northwest, central rolling hills, rugged mountains of Ardennes Forest in southeast
Natural resources: coal, natural gas
Land use:
arable land: 24%
permanent crops: 1%
meadows and pastures: 20%
forest and woodland: 21%
other: 34%
Irrigated land: 10 sq km (1989 est.)
Environment:
current issues: Meuse River, a major source of drinking water, polluted from steel production wastes; other rivers polluted by animal wastes and fertilizers; industrial air pollution contributes to acid rain in neighboring countries
natural hazards: flooding is a threat in areas of reclaimed coastal land, protected from the sea by concrete dikes
international agreements: party to—Air Pollution, Air Pollution-Sulphur 85, Antarctic Treaty, Endangered Species, Environmental Modification, Hazardous Wastes, Marine Dumping, Marine Life Conservation, Nuclear Test Ban, Ozone Layer Protection, Ship Pollution, Tropical Timber 83, Wetlands; signed, but not ratified—Air Pollution-Nitrogen Oxides, Air Pollution-Sulphur 94, Air Pollution-Volatile Organic Compounds, Antarctic-Environmental Protocol, Biodiversity, Climate Change, Law of the Sea
Note: crossroads of Western Europe; majority of West European capitals within 1,000 km of Brussels which is the seat of the EU

People

Population: 10,081,880 (July 1995 est.)
Age structure:
0-14 years: 18% (female 875,079; male 919,939)
15-64 years: 66% (female 3,303,219; male 3,363,250)
65 years and over: 16% (female 969,966; male 650,427) (July 1995 est.)
Population growth rate: 0.17% (1995 est.)
Birth rate: 11.46 births/1,000 population (1995 est.)
Death rate: 10.22 deaths/1,000 population (1995 est.)
Net migration rate: 0.5 migrant(s)/1,000 population (1995 est.)
Infant mortality rate: 7 deaths/1,000 live births (1995 est.)
Life expectancy at birth:
total population: 77.21 years
male: 73.94 years
female: 80.67 years (1995 est.)
Total fertility rate: 1.62 children born/woman (1995 est.)
Nationality:
noun: Belgian(s)
adjective: Belgian
Ethnic divisions: Fleming 55%, Walloon 33%, mixed or other 12%
Religions: Roman Catholic 75%, Protestant or other 25%
Languages: Dutch 56%, French 32%, German 1%, legally bilingual 11% divided along ethnic lines
Literacy: age 15 and over can read and write (1980 est.)
total population: 99%
Labor force: 4.126 million
by occupation: services 63.6%, industry 28%, construction 6.1%, agriculture 2.3% (1988)

Government

Names:
conventional long form: Kingdom of Belgium
conventional short form: Belgium
local long form: Royaume de Belgique
local short form: Belgique
Digraph: BE
Type: constitutional monarchy
Capital: Brussels
Administrative divisions: 9 provinces (French: provinces, singular—province; Flemish: provincien, singular—provincie); Antwerpen, Brabant, Hainaut, Liege, Limburg, Luxembourg, Namur, Oost-Vlaanderen, West-Vlaanderen
Independence: 4 October 1830 (from the Netherlands)
National holiday: National Day, 21 July (ascension of King Leopold to the throne in 1831)
Constitution: 7 February 1831, last revised 14 July 1993; parliament approved a constitutional package creating a federal state
Legal system: civil law system influenced by English constitutional theory; judicial review of legislative acts; accepts compulsory ICJ jurisdiction, with reservations
Suffrage: 18 years of age, universal and compulsory
Executive branch:
chief of state: King ALBERT II (since 9 August 1993)
head of government: Prime Minister Jean-Luc DEHAENE (since 6 March 1992)
cabinet: Cabinet; the king appoints the ministers who are approved by the legislature
Legislative branch: bicameral Parliament
Senate: (Flemish—Senaat, French—Senat); elections last held 24 November 1991 (next to be held by the end of 1995); results—percent of vote by party NA; seats—(184 total; of which 106 are directly elected; in the 1995 elections, seats will decrease to 71) CVP 20, SP 14, VLD 13, VU 5, AGALEV 5, VB 5, ROSSEN 1, PS 18, PRL 9, PSC 9, ECOLO 6, FDF 1
Chamber of Deputies: (Flemish—Kamer van Volksvertegenwoordigers, French—Chambre des Representants); elections last held 24 November 1991 (next to be held by 21 May 1995); results—CVP 16.7%, PS 13.6%, SP 12.0%, VLD 11.9%, PRL 8.2%, PSC 7.8%, VB 6.6%, VU 5.9%, ECOLO 5.1%, AGALEV 4.9%, FDF 2.6%, ROSSEM 3.2%, FN 1.5%; seats—(212 total; in 1995 elections, seats will decrease to 150) CVP 39, PS 35, SP 28, VLD 26, PRL 20, PSC 18, VB 12, VU 10, ECOLO 10, AGALEV 7, FDF 3, ROSSEM 3, FN 1
Judicial branch: Supreme Court of Justice (Flemish—Hof van Cassatie, French—Cour de Cassation)
Political parties and leaders: Flemish Christian Democrats (CVP—Christian People's Party), Johan van HECKE, president; Francophone Christian Democrats (PSC—Social Christian Party), Gerard DEPREZ, president; Flemish Socialist Party (SP), Louis TOBBACK, president; Francophone Socialist Party (PS), Philippe BUSQUIN, president; Flemish Liberal Democrats (VLD), Guy VERHOFSTADT, president; Francophone Liberal Reform Party (PRL), Jean GOL, president; Francophone Democratic Front (FDF), Georges CLERFAYT, president; Volksunie (VU), Bert ANCIAUX, president; Vlaams Blok (VB), Karel DILLEN, chairman;

ROSSEM, Jean Pierre VAN ROSSEM; National Front (FN), Daniel FERET, president; AGALEV (Flemish Greens), no president; ECOLO (Francophone Ecologists), no president; other minor parties
Other political or pressure groups: Christian and Socialist Trade Unions; Federation of Belgian Industries; numerous other associations representing bankers, manufacturers, middle-class artisans, and the legal and medical professions; various organizations represent the cultural interests of Flanders and Wallonia; various peace groups such as the Flemish Action Committee Against Nuclear Weapons and Pax Christi
Member of: ACCT, AfDB, AG (observer), AsDB, Australia Group, Benelux, BIS, CCC, CE, CERN, EBRD, EC, ECE, EIB, ESA, FAO, G- 9, G-10, GATT, IADB, IAEA, IBRD, ICAO, ICC, ICFTU, ICRM, IDA, IEA, IFAD, IFC, IFRCS, ILO, IMF, IMO, INMARSAT, INTELSAT, INTERPOL, IOC, IOM, ISO, ITU, MINURSO, MTCR, NACC, NATO, NEA, NSG, OAS (observer), OECD, OSCE, PCA, UN, UNCTAD, UNESCO, UNHCR, UNIDO, UNITAR, UNMOGIP, UNPROFOR, UNRWA, UNTSO, UPU, WCL, WEU, WHO, WIPO, WMO, WTO, ZC
Diplomatic representation in US:
chief of mission: Ambassador Andre ADAM (appointed 3 October 1994)
chancery: 3330 Garfield Street NW, Washington, DC 20008
telephone: [1] (202) 333-6900
FAX: [1] (202) 333-3079
consulate(s) general: Atlanta, Chicago, Los Angeles, and New York
US diplomatic representation:
chief of mission: Ambassador Alan J. BLINKEN
embassy: 27 Boulevard du Regent, B-1000 Brussels
mailing address: APO AE 09724; PSC 82, Box 002, Brussels
telephone: [32] (2) 513 38 30
FAX: [32] (2) 511 27 25
Flag: three equal vertical bands of black (hoist side), yellow, and red; the design was based on the flag of France

Economy

Overview: This small private enterprise economy has capitalized on its central geographic location, highly developed transport network, and diversified industrial and commercial base. Industry is concentrated mainly in the populous Flemish area in the north, although the government is encouraging reinvestment in the southern region of Walloon. With few natural resources Belgium must import substantial quantities of raw materials and export a large volume of manufactures, making its economy unusually dependent on the state of world markets. Three-fourths of its trade is with other EU countries. The economy grew at a strong 4% pace during the period 1988-90, slowed to 1% in 1991-92, dropped by 1.5% in 1993, and recovered with 2.3% growth in 1994. Belgium's public debt has risen to 140% of GDP, and the government is trying to control its expenditures to bring the figure more into line with other industrialized countries.
National product: GDP—purchasing power parity—$181.5 billion (1994 est.)
National product real growth rate: 2.3% (1994 est.)
National product per capita: $18,040 (1994 est.)
Inflation rate (consumer prices): 2.5% (1994)
Unemployment rate: 14.1% (December 1994)
Budget:
revenues: $97.8 billion
expenditures: $109.3 billion, including capital expenditures of $NA (1989)
Exports: $117 billion (f.o.b., 1992) Belgium-Luxembourg Economic Union
commodities: iron and steel, transportation equipment, tractors, diamonds, petroleum products
partners: EC 75.5%, US 3.7%, former Communist countries 1.4% (1991)
Imports: $120 billion (c.i.f., 1992) Belgium-Luxembourg Economic Union
commodities: fuels, grains, chemicals, foodstuffs
partners: EC 73%, US 4.8%, oil-exporting less developed countries 4%, former Communist countries 1.8% (1991)
External debt: $31.3 billion (1992 est.)
Industrial production: growth rate -0.1% (1993 est.); accounts for 25% of GDP
Electricity:
capacity: 14,040,000 kW
production: 66 billion kWh
consumption per capita: 6,334 kWh (1993)
Industries: engineering and metal products, motor vehicle assembly, processed food and beverages, chemicals, basic metals, textiles, glass, petroleum, coal
Agriculture: accounts for 2.0% of GDP; emphasis on livestock production—beef, veal, pork, milk; major crops are sugar beets, fresh vegetables, fruits, grain, tobacco; net importer of farm products
Illicit drugs: source of precursor chemicals for South American cocaine processors; transshipment point for cocaine entering the European market
Economic aid:
donor: ODA and OOF commitments (1970-89), $5.8 billion
Currency: 1 Belgian franc (BF) = 100 centimes
Exchange rates: Belgian francs (BF) per US$1—31.549 (January 1995), 33.456 (1994), 34.597 (1993), 32.150 (1992), 34.148 (1991), 33.418 (1990)
Fiscal year: calendar year

Transportation

Railroads:
total: 3,410 km (2,362 km electrified; 2,563 km double track)
standard gauge: 3,410 km 1.435-m gauge (1994)
Highways:
total: 137,912 km
paved: 129,639 km (including 1,667 km of limited access divided highway)
unpaved: 8,273 km (1992)
Inland waterways: 2,043 km (1,528 km in regular commercial use)
Pipelines: crude oil 161 km; petroleum products 1,167 km; natural gas 3,300 km
Ports: Antwerp, Brugge, Gent, Hasselt, Liege, Mons, Namur, Oostende, Zeebrugge
Merchant marine:
total: 23 ships (1,000 GRT or over) totaling 42,055 GRT/56,842 DWT
ships by type: bulk 1, cargo 9, chemical tanker 6, liquefied gas 2, oil tanker 5
Airports:
total: 43
with paved runways over 3,047 m: 6
with paved runways 2,438 to 3,047 m: 9
with paved runways 1,524 to 2,437 m: 2
with paved runways 914 to 1,523 m: 1
with paved runways under 914 m: 22
with unpaved runways 914 to 1,523 m: 3

Communications

Telephone system: 4,720,000 telephones; highly developed, technologically advanced, and completely automated domestic and international telephone and telegraph facilities
local: NA
intercity: extensive cable network; limited microwave radio relay network; nationwide mobile phone system
international: 5 submarine cables; 2 Atlantic Ocean INTELSAT earth stations and 1 EUTELSAT earth station
Radio:
broadcast stations: AM 3, FM 39, shortwave 0
radios: NA
Television:
broadcast stations: 32
televisions: NA

Defense Forces

Branches: Army, Navy, Air Force, National Gendarmerie
Manpower availability: males age 15-49 2,559,077; males fit for military service 2,126,875; males reach military age (19) annually 61,488 (1995 est.)
Defense expenditures: exchange rate conversion—$3.9 billion, 1.8% of GDP (1994)

Belize

Geography

Location: Middle America, bordering the Caribbean Sea, between Guatemala and Mexico
Map references: Central America and the Caribbean
Area:
total area: 22,960 sq km
land area: 22,800 sq km
comparative area: slightly larger than Massachusetts
Land boundaries: total 516 km, Guatemala 266 km, Mexico 250 km
Coastline: 386 km
Maritime claims:
exclusive economic zone: 200 nm
territorial sea: 12 nm in the north, 3 nm in the south; note—from the mouth of the Sarstoon River to Ranguana Cay, Belize's territorial sea is 3 miles; according to Belize's Maritime Areas Act, 1992, the purpose of this limitation is to provide a framework for the negotiation of a definitive agreement on territorial differences with Guatemala
International disputes: border with Guatemala in dispute; talks to resolve the dispute are stalled
Climate: tropical; very hot and humid; rainy season (May to February)
Terrain: flat, swampy coastal plain; low mountains in south
Natural resources: arable land potential, timber, fish
Land use:
arable land: 2%
permanent crops: 0%
meadows and pastures: 2%
forest and woodland: 44%
other: 52%
Irrigated land: 20 sq km (1989 est.)
Environment:
current issues: deforestation; water pollution from sewage, industrial effluents, agricultural runoff
natural hazards: frequent, devastating hurricanes (September to December) and coastal flooding (especially in south)
international agreements: party to—Biodiversity, Climate Change, Endangered Species, Law of the Sea, Marine Dumping
Note: national capital moved 80 km inland from Belize City to Belmopan because of hurricanes; only country in Central America without a coastline on the North Pacific Ocean

People

Population: 214,061 (July 1995 est.)
Age structure:
0-14 years: 44% (female 45,812; male 47,618)
15-64 years: 53% (female 55,630; male 57,230)
65 years and over: 3% (female 3,970; male 3,801) (July 1995 est.)
Population growth rate: 2.42% (1995 est.)
Birth rate: 33.71 births/1,000 population (1995 est.)
Death rate: 5.86 deaths/1,000 population (1995 est.)
Net migration rate: -3.7 migrant(s)/1,000 population (1995 est.)
Infant mortality rate: 34.7 deaths/1,000 live births (1995 est.)
Life expectancy at birth:
total population: 68.32 years
male: 66.37 years
female: 70.36 years (1995 est.)
Total fertility rate: 4.25 children born/woman (1995 est.)
Nationality:
noun: Belizean(s)
adjective: Belizean
Ethnic divisions: mestizo 44%, Creole 30%, Maya 11%, Garifuna 7%, other 8%
Religions: Roman Catholic 62%, Protestant 30% (Anglican 12%, Methodist 6%, Mennonite 4%, Seventh-Day Adventist 3%, Pentecostal 2%, Jehovah's Witnesses 1%, other 2%), none 2%, other 6% (1980)
Languages: English (official), Spanish, Maya, Garifuna (Carib)
Literacy: age 15 and over has ever attended school (1970)
total population: 91%
male: 91%
female: 91%
Labor force: 51,500
by occupation: agriculture 30%, services 16%, government 15.4%, commerce 11.2%, manufacturing 10.3%
note: shortage of skilled labor and all types of technical personnel (1985)

Government

Names:
conventional long form: none
conventional short form: Belize
former: British Honduras
Digraph: BH
Type: parliamentary democracy
Capital: Belmopan
Administrative divisions: 6 districts; Belize, Cayo, Corozal, Orange Walk, Stann Creek, Toledo
Independence: 21 September 1981 (from UK)
National holiday: Independence Day, 21 September (1981)
Constitution: 21 September 1981
Legal system: English law
Suffrage: 18 years of age; universal
Executive branch:
chief of state: Queen ELIZABETH II (since 6 February 1952), represented by Governor General Sir Colville YOUNG (since 17 November 1993)
head of government: Prime Minister Manuel ESQUIVEL (since July 1993); Deputy Prime Minister Dean BARROW (since NA 1993)
cabinet: Cabinet; appointed by the governor general on advice from the prime minister
Legislative branch: bicameral National Assembly
Senate: consists of an 8-member appointed body; 5 members are appointed on the advice of the prime minister, 2 on the advice of the leader of the opposition, and 1 after consultation with the Belize Advisory Council (this council serves as an independent body to advise the governor-general with respect to difficult decisions such as granting pardons, commutations, stays of execution, the removal of justices of appeal who appear to be incompetent, etc.)
National Assembly: elections last held 30 June 1993 (next to be held June 1998); results—percent of vote by party NA; seats—(28 total) PUP 13 UDP 15
Judicial branch: Supreme Court
Political parties and leaders: People's United Party (PUP), George PRICE, Florencio MARIN, Said MUSA; United Democratic Party (UDP), Manuel ESQUIVEL, Dean LINDO, Dean BARROW; National Alliance for Belizean Rights, Philip GOLDSON
Other political or pressure groups: Society for the Promotion of Education and Research (SPEAR), Assad SHOMAN; United Workers Front, leader NA
Member of: ACP, C, CARICOM, CDB, ECLAC, FAO, G-77, GATT, IADB, IBRD, ICAO, ICFTU, ICRM, IDA, IFAD, IFC, IFRCS, ILO, IMF, IMO, INTELSAT (nonsignatory user), INTERPOL, IOC, IOM (observer), ITU, LAES, NAM, OAS, UN, UNCTAD, UNESCO, UNIDO, UPU, WCL, WHO, WMO
Diplomatic representation in US:
chief of mission: Ambassador Dean R. LINDO
chancery: 2535 Massachusetts Avenue NW, Washington, DC 20008
telephone: [1] (202) 332-9636
FAX: [1] (202) 332-6888
consulate(s) general: Los Angeles
consulate(s): New York
US diplomatic representation:
chief of mission: Ambassador George Charles BRUNO

embassy: Gabourel Lane and Hutson Street, Belize City
mailing address: P. O. Box 286, Belize City; APO: Unit 7401, APO AA 34025
telephone: [501] (2) 77161 through 77163
FAX: [501] (2) 30802
Flag: blue with a narrow red stripe along the top and the bottom edges; centered is a large white disk bearing the coat of arms; the coat of arms features a shield flanked by two workers in front of a mahogany tree with the related motto SUB UMBRA FLOREO (I Flourish in the Shade) on a scroll at the bottom, all encircled by a green garland

Economy

Overview: The small, essentially private enterprise economy is based primarily on agriculture, agro-based industry, and merchandising, with tourism and construction assuming increasing importance. Agriculture accounts for about 30% of GDP and provides 75% of export earnings, while sugar, the chief crop, accounts for almost 40% of hard currency earnings. The US, Belize's main trading partner, is assisting in efforts to reduce dependency on sugar with an agricultural diversification program.
National product: GDP—purchasing power parity—$575 million (1994 est.)
National product real growth rate: 2% (1994 est.)
National product per capita: $2,750 (1994 est.)
Inflation rate (consumer prices): 5.5% (1991)
Unemployment rate: 10% (1993 est.)
Budget:
revenues: $126.8 million
expenditures: $123.1 million, including capital expenditures of $44.8 million (FY90/91 est.)
Exports: $115 million (f.o.b., 1993)
commodities: sugar, citrus fruits, bananas, clothing, fish products, molasses, wood
partners: US 51%, UK, other EC (1992)
Imports: $281 million (c.i.f., 1993)
commodities: machinery and transportation equipment, food, manufactured goods, fuels, chemicals, pharmaceuticals
partners: US 57%, UK 8%, other EC 7%, Mexico (1992)
External debt: $158 million (1992)
Industrial production: growth rate 3.7% (1990); accounts for 12% of GDP
Electricity:
capacity: 34,532 kW
production: 110 million kWh
consumption per capita: 490 kWh (1993)
Industries: garment production, food processing, tourism, construction
Agriculture: commercial crops: bananas, coca, citrus fruits, fish, cultured shrimp, lumber
Illicit drugs: transshipment point for cocaine; an illicit producer of cannabis for the international drug trade; minor money-laundering center
Economic aid:
recipient: US commitments, including Ex-Im (FY70-89), $104 million; Western (non-US) countries, ODA and OOF bilateral commitments (1970-89), $215 million
Currency: 1 Belizean dollar (Bz$) = 100 cents
Exchange rates: Belizean dollars (Bz$) per US$1—2.00 (fixed rate)
Fiscal year: 1 April—31 March

Transportation

Railroads: 0 km
Highways:
total: 2,710 km
paved: 500 km
unpaved: gravel 1,600 km; improved earth 300 km; unimproved earth 310 km
Inland waterways: 825 km river network used by shallow-draft craft; seasonally navigable
Ports: Belize City, Big Creek, Corozol, Punta Gorda
Merchant marine:
total: 41 ships (1,000 GRT or over) totaling 170,002 GRT/270,893 DWT
ships by type: bulk 5, cargo 25, container 4, oil tanker 2, refrigerated cargo 1, roll-on/roll-off cargo 3, vehicle carrier 1
Airports:
total: 46
with paved runways 1,524 to 2,437 m: 1
with paved runways under 914 m: 35
with unpaved runways 2,438 to 3,047 m: 1
with unpaved runways 914 to 1,523 m: 9

Communications

Telephone system: 8,650 telephones; above-average system based on microwave radio relay
local: NA
intercity: microwave radio relay
international: 1 INTELSAT (Atlantic Ocean) earth station
Radio:
broadcast stations: AM 6, FM 5, shortwave 1
radios: NA
Television:
broadcast stations: 1
televisions: NA

Defense Forces

Branches: Belize Defense Force (includes Army, Navy, Air Force, and Volunteer Guard), Belize National Police
Manpower availability: males age 15-49 50,499; males fit for military service 30,040; males reach military age (18) annually 2,285 (1995 est.)
Defense expenditures: exchange rate conversion—$11 million, 2.2% of GDP (FY93/94)

Benin

Geography

Location: Western Africa, bordering the North Atlantic Ocean, between Nigeria and Togo
Map references: Africa
Area:
total area: 112,620 sq km
land area: 110,620 sq km
comparative area: slightly smaller than Pennsylvania
Land boundaries: total 1,989 km, Burkina 306 km, Niger 266 km, Nigeria 773 km, Togo 644 km
Coastline: 121 km
Maritime claims:
territorial sea: 200 nm
International disputes: none
Climate: tropical; hot, humid in south; semiarid in north
Terrain: mostly flat to undulating plain; some hills and low mountains
Natural resources: small offshore oil deposits, limestone, marble, timber
Land use:
arable land: 12%
permanent crops: 4%
meadows and pastures: 4%
forest and woodland: 35%
other: 45%
Irrigated land: 60 sq km (1989 est.)
Environment:
current issues: recent droughts have severely affected marginal agriculture in north; inadequate supplies of potable water; poaching threatens wildlife populations; deforestation; desertification
natural hazards: hot, dry, dusty harmattan wind may affect north in winter
international agreements: party to—Biodiversity, Climate Change, Endangered Species, Environmental Modification, Nuclear Test Ban, Ozone Layer Protection; signed, but not ratified—Desertification, Law of the Sea
Note: no natural harbors

People

Population: 5,522,677 (July 1995 est.)
Age structure:
0-14 years: 48% (female 1,324,553; male 1,333,673)
15-64 years: 49% (female 1,431,630; male 1,299,180)
65 years and over: 3% (female 74,119; male 59,522) (July 1995 est.)
Population growth rate: 3.33% (1995 est.)
Birth rate: 47.25 births/1,000 population (1995 est.)
Death rate: 13.93 deaths/1,000 population (1995 est.)
Net migration rate: 0 migrant(s)/1,000 population (1995 est.)
Infant mortality rate: 107.6 deaths/1,000 live births (1995 est.)
Life expectancy at birth:
total population: 52.24 years
male: 50.34 years
female: 54.2 years (1995 est.)
Total fertility rate: 6.72 children born/woman (1995 est.)
Nationality:
noun: Beninese (singular and plural)
adjective: Beninese
Ethnic divisions: African 99% (42 ethnic groups, most important being Fon, Adja, Yoruba, Bariba), Europeans 5,500
Religions: indigenous beliefs 70%, Muslim 15%, Christian 15%
Languages: French (official), Fon and Yoruba (most common vernaculars in south), tribal languages (at least six major ones in north)
Literacy: age 15 and over can read and write (1990 est.)
total population: 23%
male: 32%
female: 16%
Labor force: 1.9 million (1987)
by occupation: agriculture 60%, transport, commerce, and public services 38%, industry less than 2%

Government

Names:
conventional long form: Republic of Benin
conventional short form: Benin
local long form: Republique du Benin
local short form: Benin
former: Dahomey
Digraph: BN
Type: republic under multiparty democratic rule dropped Marxism-Leninism December 1989; democratic reforms adopted February 1990; transition to multiparty system completed 4 April 1991
Capital: Porto-Novo
Administrative divisions: 6 provinces; Atakora, Atlantique, Borgou, Mono, Oueme, Zou
Independence: 1 August 1960 (from France)
National holiday: National Day, 1 August (1990)
Constitution: 2 December 1990
Legal system: based on French civil law and customary law; has not accepted compulsory ICJ jurisdiction
Suffrage: 18 years of age; universal
Executive branch:
chief of state and head of government: President Nicephore SOGLO (since 4 April 1991); election last held 10 and 24 March 1991 (next election 1996); results—Nicephore SOGLO 68%, Mathieu KEREKOU 32%
cabinet: Executive Council; appointed by the president
Legislative branch: unicameral
National Assembly (Assemblee Nationale): elections last held 28 March 1995; results—percent of vote by party NA; seats—(83 total) Renaissance Party and allies 20, PRD 19, FARD-ALAFIA 10, PSD 7, NCC 3, RDL-VIVOTEN 3, Communist Party 2, Alliance Chameleon 1, RDP 1, ADP 1, other 16
Judicial branch: Supreme Court (Cour Supreme)
Political parties and leaders: as of August 1994, 72 political parties were officially recognized; the following are among the most important: Alliance of the Democratic Union for the Forces of Progress (UDFP), Timothee ADANLIN; Movement for Democracy and Social Progress (MDPS), Jean-Roger AHOYO; Union for Liberty and Development (ULD), Marcellin DEGBE; Alliance of the National Party for Democracy and Development (PNDD) and the Democratic Renewal Party (PRD), Pascal Chabi KAO; Alliance of the Social Democratic Party (PSD) and the National Union for Solidarity and Progress (UNSP), Bruno AMOUSSOU; Our Common Cause (NCC), Albert TEVOEDJRE; National Rally for Democracy (RND), Joseph KEKE; Alliance of the National Movement for Democracy and Development (MNDD), leader NA; Movement for Solidarity, Union, and Progress (MSUP), Adebo ADENIYI; Union for Democracy and National Reconstruction (UDRN), Azaria FAKOREDE; Union for Democracy and National Solidarity (UDS), Mama Amadou N'DIAYE; Assembly of Liberal Democrats for National Reconstruction (RDL), Severin ADJOVI; Alliance for Social Democracy (ASD), Robert DOSSOU; Bloc for Social Democracy (BSD), Michel MAGNIDE; Alliance for Democracy and Progress (ADP), Akindes ADEKPEDJOU and the Democratic Union for Social Renewal (UDRS), Bio Gado Seko N'GOYE; National Union for Democracy and Progress (UNDP), Robert TAGNON; Party for Progress and Democracy, Thiophile NATA; FARD-ALAFIA, Mathieu KEREKOU; The Renaissance Party,

Nicephore SOGLO; The Patriotic Union for the Republic (UPR), Jean-Marie ZAHOUN; Union for the Conservation of Democracy, Bernard HOUEGNON
Member of: ACCT, ACP, AfDB, CEAO, ECA, ECOWAS, Entente, FAO, FZ, G-77, GATT, IBRD, ICAO, ICFTU, ICRM, IDA, IDB, IFAD, IFC, IFRCS, ILO, IMF, IMO, INTELSAT, INTERPOL, IOC, ITU, NAM, OAU, OIC, UN, UNCTAD, UNESCO, UNIDO, UPU, WADB, WCL, WFTU, WHO, WIPO, WMO, WTO
Diplomatic representation in US:
chief of mission: Ambassador Lucien Edgar TONOUKOUIN
chancery: 2737 Cathedral Avenue NW, Washington, DC 20008
telephone: [1] (202) 232-6656, 6657, 6658
FAX: [1] (202) 265-1996
US diplomatic representation:
chief of mission: Ambassador Ruth A. DAVIS
embassy: Rue Caporal Bernard Anani, Cotonou
mailing address: B. P. 2012, Cotonou
telephone: [229] 30-06-50, 30-05-13, 30-17-92
FAX: [229] 41-15-22
Flag: two equal horizontal bands of yellow (top) and red with a vertical green band on the hoist side

Economy

Overview: The economy of Benin remains underdeveloped and dependent on subsistence agriculture, cotton production, and regional trade. Growth in real output has averaged a sound 4% in 1991-94 but this rate barely exceeds the rapid population growth of 3.3%. Inflation jumped to 35% in 1994 (compared to 3% in 1993) following the 50% currency devaluation in January. Commercial and transport activities, which make up almost 36% of GDP, are extremely vulnerable to developments in Nigeria as evidenced by decreased reexport trade in 1994 due to a severe contraction in Nigerian demand. The industrial sector accounts for less than 10% of GDP and mainly produces foods, beverages, cement, and textiles. Support by the Paris Club and official bilateral creditors has eased the external debt situation in recent years. The government, still burdened with money-losing state enterprises and a bloated civil service, is gradually implementing a World Bank supported structural adjustment program.
National product: GDP—purchasing power parity—$6.7 billion (1994 est.)
National product real growth rate: 4% (1994 est.)
National product per capita: $1,260 (1994 est.)
Inflation rate (consumer prices): 35% (1994 est.)
Unemployment rate: NA%
Budget:
revenues: $272 million (1993 est.)
expenditures: $375 million, including capital expenditures of $84 million (1993 est.)
Exports: $332 million (f.o.b., 1993 est.)
commodities: cotton, crude oil, palm products, cocoa
partners: FRG 36%, France 16%, Spain 14%, Italy 8%, UK 4%
Imports: $571 million (f.o.b., 1993 est.)
commodities: foodstuffs, beverages, tobacco, petroleum products, intermediate goods, capital goods, light consumer goods
partners: France 20%, Thailand 8%, Netherlands 7%, US 5%
External debt: $1 billion (December 1990 est.)
Industrial production: growth rate -0.7% (1988); accounts for 10% of GDP
Electricity:
capacity: 30,000 kW
production: 10 million kWh
consumption per capita: 25 kWh (1993)
Industries: textiles, cigarettes, construction materials, beverages, food, petroleum
Agriculture: accounts for 35% of GDP; small farms produce 90% of agricultural output; production is dominated by food crops—corn, sorghum, cassava, yams, beans, rice; cash crops include cotton, palm oil, peanuts; poultry and livestock output has not kept up with consumption
Illicit drugs: transshipment point for narcotics associated with Nigerian trafficking organizations and most commonly destined for Western Europe and the US
Economic aid:
recipient: US commitments, including Ex-Im (FY70-89), $46 million; Western (non-US) countries, ODA and OOF bilateral commitments (1970-89), $1.3 billion; OPEC bilateral aid (1979-89), $19 million, Communist countries (1970-89), $101 million
Currency: 1 CFA franc (CFAF) = 100 centimes
Exchange rates: Communaute Financiere Africaine francs (CFAF) per US$1—529.43 (January 1995), 555.20 (1994), 283.16 (1993), 264.69 (1992), 282.11 (1991), 272.26 (1990)
note: beginning 12 January 1994 the CFA franc was devalued to CFAF 100 per French franc from CFAF 50 at which it had been fixed since 1948
Fiscal year: calendar year

Transportation

Railroads:
total: 578 km (single track)
narrow gauge: 578 km 1.000-m gauge
Highways:
total: 8,435 km
paved: 1,038 km
unpaved: crushed stone 2,600 km; improved earth 1,530 km; unimproved earth 3,267 km
Inland waterways: navigable along small sections, important only locally
Ports: Cotonou, Porto-Novo
Merchant marine: none
Airports:
total: 7
with paved runways 2,438 to 3,047 m: 1
with paved runways 1,524 to 2,437 m: 1
with unpaved runways 1,524 to 2,438 m: 1
with unpaved runways 914 to 1,523 m: 4

Communications

Telephone system: NA telephones; fair system of open wire and microwave radio relay
local: NA
intercity: microwave radio relay and open wire
international: 1 Atlantic Ocean INTELSAT earth station, submarine cable
Radio:
broadcast stations: AM 2, FM 2, shortwave 0
radios: NA
Television:
broadcast stations: 2
televisions: NA

Defense Forces

Branches: Armed Forces (includes Army, Navy, Air Force), National Gendarmerie
Manpower availability: males age 15-49 1,165,463; females age 15-49 1,249,234; males fit for military service 596,956; females fit for military service 631,780; males reach military age (18) annually 60,282 (1995 est.); females reach military age (18) annually 58,770 (1995 est.)
note: both sexes are liable for miltary service
Defense expenditures: exchange rate conversion—$33 million, 3.2% of GDP (1994)

Bermuda
(dependent territory of the UK)

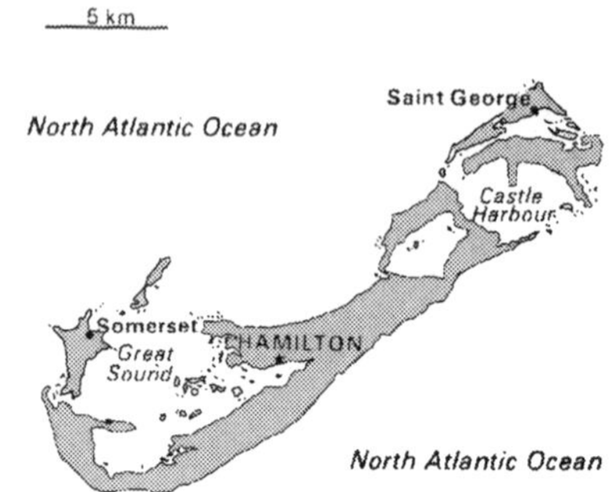

Geography

Location: North America, group of islands in the North Atlantic Ocean, east of North Carolina (US)
Map references: North America
Area:
total area: 50 sq km
land area: 50 sq km
comparative area: about 0.3 times the size of Washington, DC
Land boundaries: 0 km
Coastline: 103 km
Maritime claims:
exclusive fishing zone: 200 nm
territorial sea: 12 nm
International disputes: none
Climate: subtropical; mild, humid; gales, strong winds common in winter
Terrain: low hills separated by fertile depressions
Natural resources: limestone, pleasant climate fostering tourism
Land use:
arable land: 0%
permanent crops: 0%
meadows and pastures: 0%
forest and woodland: 20%
other: 80%
Irrigated land: NA sq km
Environment:
current issues: NA
natural hazards: hurricanes (June to November)
international agreements: NA
Note: consists of about 360 small coral islands with ample rainfall, but no rivers or freshwater lakes; some reclaimed land leased by US Government

People

Population: 61,629 (July 1995 est.)
Age structure:
0-14 years: NA
15-64 years: NA
65 years and over: NA
Population growth rate: 0.76% (1995 est.)
Birth rate: 15.07 births/1,000 population (1995 est.)
Death rate: 7.3 deaths/1,000 population (1995 est.)
Net migration rate: -0.13 migrant(s)/1,000 population (1995 est.)
Infant mortality rate: 13.16 deaths/1,000 live births (1995 est.)
Life expectancy at birth:
total population: 75.03 years
male: 73.36 years
female: 76.97 years (1995 est.)
Total fertility rate: 1.81 children born/woman (1995 est.)
Nationality:
noun: Bermudian(s)
adjective: Bermudian
Ethnic divisions: black 61%, white and other 39%
Religions: Anglican 37%, Roman Catholic 14%, African Methodist Episcopal (Zion) 10%, Methodist 6%, Seventh-Day Adventist 5%, other 28%
Languages: English
Literacy: age 15 and over can read and write (1970)
total population: 98%
male: 98%
female: 99%
Labor force: 32,000
by occupation: clerical 25%, services 22%, laborers 21%, professional and technical 13%, administrative and managerial 10%, sales 7%, agriculture and fishing 2% (1984)

Government

Names:
conventional long form: none
conventional short form: Bermuda
Digraph: BD
Type: dependent territory of the UK
Capital: Hamilton
Administrative divisions: 9 parishes and 2 municipalities*; Devonshire, Hamilton, Hamilton*, Paget, Pembroke, Saint George*, Saint Georges, Sandys, Smiths, Southampton, Warwick
Independence: none (dependent territory of the UK)
National holiday: Bermuda Day, 24 May
Constitution: 8 June 1968
Legal system: English law
Suffrage: 21 years of age; universal
Executive branch:
chief of state: Queen ELIZABETH II (since 6 February 1952), represented by Governor Lord David WADDINGTON (since 25 August 1992)
head of government: Premier John William David SWAN (since NA January 1982); Deputy Premier J. Irving PEARMAN (since 5 October 1993)
cabinet: Cabinet; nominated by the premier, appointed by the governor
Legislative branch: bicameral Parliament
Senate: consists of an 11-member body appointed by the governor
House of Assembly: elections last held 5 October 1993 (next to be held by NA October 1998); results—percent of vote by party UBP 50%, PLP 46%, independents 4%; seats—(40 total) UBP 22, PLP 18
Judicial branch: Supreme Court
Political parties and leaders: United Bermuda Party (UBP), John W. D. SWAN; Progressive Labor Party (PLP), Frederick WADE; National Liberal Party (NLP), Gilbert DARRELL
Other political or pressure groups: Bermuda Industrial Union (BIU), Ottiwell SIMMONS
Member of: CARICOM (observer), CCC, ICFTU, INTERPOL (subbureau), IOC
Diplomatic representation in US: none (dependent territory of the UK)
US diplomatic representation:
chief of mission: Ambassador Robert A. FARMER
consulate(s) general: Crown Hill, 16 Middle Road, Devonshire, Hamilton
mailing address: P. O. Box HM325, Hamilton HMBX; PSC 1002, FPO AE 09727-1002
telephone: [1] (809) 295-1342
FAX: [1] (809) 295-1592
Flag: red with the flag of the UK in the upper hoist-side quadrant and the Bermudian coat of arms (white and blue shield with a red lion holding a scrolled shield showing the sinking of the ship Sea Venture off Bermuda in 1609) centered on the outer half of the flag

Economy

Overview: Bermuda enjoys one of the highest per capita incomes in the world, having successfully exploited its location by providing luxury tourist facilities and financial services. The tourist industry attracts more than 90% of its business from North America. The industrial sector is small, and agriculture is severely limited by a lack of suitable land. About 80% of food needs are imported.
National product: GDP—purchasing power parity—$1.7 billion (1994 est.)
National product real growth rate: 2.5% (1994)
National product per capita: $28,000 (1994 est.)
Inflation rate (consumer prices): 2.5% (1993)
Unemployment rate: 6% (1991)
Budget:
revenues: $327.5 million
expenditures: $308.9 million, including capital expenditures of $35.4 million (FY90/91 est.)
Exports: $60 million (f.o.b., 1991)
commodities: semitropical produce, light manufactures, re-exports of pharmaceuticals

partners: US 62.4%, UK 20%
Imports: $519 million (f.o.b.,1993)
commodities: fuel, foodstuffs, machinery
partners: US 38%, UK 5%, Canada 5%
External debt: $NA
Industrial production: growth rate NA%
Electricity:
capacity: 140,000 kW
production: 504 million kWh
consumption per capita: 7,745 kWh (1993)
Industries: tourism, finance, structural concrete products, paints, pharmaceuticals, ship repairing
Agriculture: accounts for less than 1% of GDP; most basic foods must be imported; produces bananas, vegetables, citrus fruits, flowers, dairy products
Economic aid:
recipient: US commitments, including Ex-Im (FY70-81), $34 million; Western (non-US) countries, ODA and OOF bilateral commitments (1970-89), $277 million
Currency: 1 Bermudian dollar (Bd$) = 100 cents
Exchange rates: Bermudian dollar (Bd$) per US$1 1.0000 (fixed rate)
Fiscal year: 1 April—31 March

Transportation

Railroads: 0 km
Highways:
total: 210 km
paved: 210 km
note: in addition, there are 400 km of paved and unpaved roads that are privately owned
Ports: Hamilton, Saint George
Merchant marine:
total: 65 ships (1,000 GRT or over) totaling 3,144,245 GRT/5,152,030 DWT
ships by type: bulk 14, cargo 4, container 7, liquefied gas tanker 15, oil tanker 16, refrigerated cargo 2, roll-on/roll-off cargo 5, short-sea passenger 1, vehicle carrier 1
note: a flag of convenience registry; includes 12 countries among which are UK 6 ships, Canada 4, US 4, Sweden 3, Hong Kong 2, Mexico 2, Norway 2, Australia 1, Germany 1, NZ 1
Airports:
total: 1
with paved runways 2,438 to 3,047 m: 1

Communications

Telephone system: 52,670 telephones; modern, fully automatic telephone system
local: NA
intercity: NA
international: 3 submarine cables; 2 INTELSAT (Atlantic Ocean) earth stations
Radio:
broadcast stations: AM 5, FM 3, shortwave 0
radios: NA
Television:
broadcast stations: 2
televisions: NA

Defense Forces

Branches: Bermuda Regiment, Bermuda Police Force, Bermuda Reserve Constabulary
Defense expenditures: $NA, NA% of GDP
Note: defense is the responsibility of the UK

Bhutan

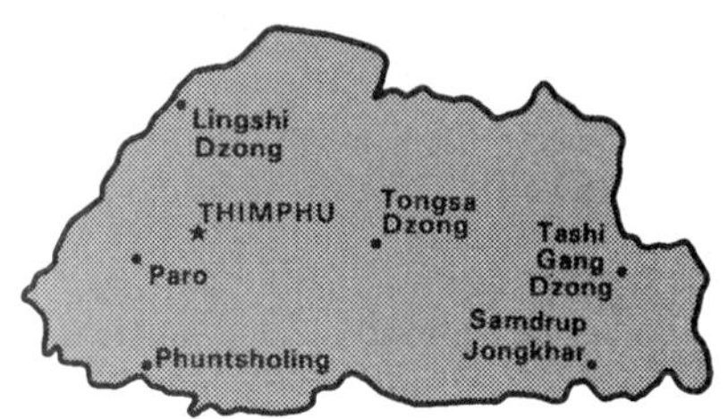

Geography

Location: Southern Asia, between China and India
Map references: Asia
Area:
total area: 47,000 sq km
land area: 47,000 sq km
comparative area: slightly more than half the size of Indiana
Land boundaries: total 1,075 km, China 470 km, India 605 km
Coastline: 0 km (landlocked)
Maritime claims: none; landlocked
International disputes: none
Climate: varies; tropical in southern plains; cool winters and hot summers in central valleys; severe winters and cool summers in Himalayas
Terrain: mostly mountainous with some fertile valleys and savanna
Natural resources: timber, hydropower, gypsum, calcium carbide
Land use:
arable land: 2%
permanent crops: 0%
meadows and pastures: 5%
forest and woodland: 70%
other: 23%
Irrigated land: 340 sq km (1989 est.)
Environment:
current issues: soil erosion; limited access to potable water
natural hazards: violent storms coming down from the Himalayas are the source of the country's name which translates as Land of the Thunder Dragon; frequent landslides during the rainy season
international agreements: party to—Nuclear Test Ban; signed, but not ratified—Biodiversity, Climate Change, Law of the Sea
Note: landlocked; strategic location between China and India; controls several key Himalayan mountain passes

People

Population: 1,780,638 (July 1995 est.)
note: other estimates range as low as 600,000
Age structure:
0-14 years: 40% (female 342,276; male 368,916)
15-64 years: 56% (female 486,258; male 513,560)
65 years and over: 4% (female 34,215; male 35,413) (July 1995 est.)
Population growth rate: 2.34% (1995 est.)
Birth rate: 39.02 births/1,000 population (1995 est.)
Death rate: 15.61 deaths/1,000 population (1995 est.)
Net migration rate: 0 migrant(s)/1,000 population (1995 est.)
Infant mortality rate: 118.6 deaths/1,000 live births (1995 est.)
Life expectancy at birth:
total population: 51.03 years
male: 51.56 years
female: 50.48 years (1995 est.)
Total fertility rate: 5.39 children born/woman (1995 est.)
Nationality:
noun: Bhutanese (singular and plural)
adjective: Bhutanese
Ethnic divisions: Bhote 50%, ethnic Nepalese 35%, indigenous or migrant tribes 15%
Religions: Lamaistic Buddhism 75%, Indian- and Nepalese-influenced Hinduism 25%
Languages: Dzongkha (official), Bhotes speak various Tibetan dialects; Nepalese speak various Nepalese dialects
Literacy: NA%
Labor force: NA
by occupation: agriculture 93%, services 5%, industry and commerce 2%
note: massive lack of skilled labor

Government

Names:
conventional long form: Kingdom of Bhutan
conventional short form: Bhutan
Digraph: BT
Type: monarchy; special treaty relationship with India
Capital: Thimphu
Administrative divisions: 18 districts (dzongkhag, singular and plural); Bumthang, Chhukha, Chirang, Daga, Geylegphug, Ha, Lhuntshi, Mongar, Paro, Pemagatsel, Punakha, Samchi, Samdrup Jongkhar, Shemgang, Tashigang, Thimphu, Tongsa, Wangdi Phodrang
Independence: 8 August 1949 (from India)
National holiday: National Day, 17 December (1907) (Ugyen Wangchuck became first hereditary king)
Constitution: no written constitution or bill of rights
Legal system: based on Indian law and English common law; has not accepted compulsory ICJ jurisdiction
Suffrage: each family has one vote in village-level elections
Executive branch:
Chief of State and Head of Government: King Jigme Singye WANGCHUCK (since 24 July 1972)
Royal Advisory Council (Lodoi Tsokde): nominated by the king
cabinet: Council of Ministers (Lhengye Shungtsog); appointed by the king
Legislative branch: unicameral National Assembly (Tshogdu); no national elections
Judicial branch: High Court
Political parties and leaders: no legal parties
Other political or pressure groups: Buddhist clergy; Indian merchant community; ethnic Nepalese organizations leading militant antigovernment campaign
Member of: AsDB, CP, ESCAP, FAO, G-77, IBRD, ICAO, IDA, IFAD, IMF, INTELSAT, IOC, ITU, NAM, SAARC, UN, UNCTAD, UNESCO, UNIDO, UPU, WHO, WIPO
Diplomatic representation in US: Bhutan has no embassy in the US, but does have a Permanent Mission to the UN, headed by Ugyen TSERING, located at 2 United Nations Plaza, 27th Floor, New York, NY 10017, telephone [1] (212) 826-1919; note—the Bhutanese mission to the UN has consular jurisdiction in the US
consulate(s) general: New York
honorary consulate(s): San Francisco; Washington, DC
US diplomatic representation: no formal diplomatic relations, although informal contact is maintained between the Bhutanese and US Embassy in New Delhi (India)
Flag: divided diagonally from the lower hoist side corner; the upper triangle is orange and the lower triangle is red; centered along the dividing line is a large black and white dragon facing away from the hoist side

Economy

Overview: The economy, one of the world's least developed, is based on agriculture and forestry, which provide the main livelihood for 90% of the population and account for about half of GDP. Agriculture consists largely of subsistence farming and animal husbandry. Rugged mountains dominate the terrain and make the building of roads and other infrastructure difficult and expensive. The economy is closely aligned with India's through strong trade and monetary links. The industrial sector is small and technologically backward, with most production of the cottage industry type. Most development projects, such as road construction, rely on Indian migrant labor. Bhutan's hydropower potential and its attraction for tourists are key resources; however, the government limits the number of tourists to 4,000 per year to minimize foreign influence. Much of the impetus for growth has come from large public-sector companies. Nevertheless, in recent years, Bhutan has shifted toward decentralized development planning and greater private initiative. The government privatized several large public-sector firms, is revamping its trade regime and liberalizing administerial procedures over industrial licensing. The government's industrial contribution to GDP decreased from 13% in 1988 to about 11% in 1993.
National product: GDP—purchasing power parity—$1.2 billion (1994 est.)
National product real growth rate: 5% (1994 est.)
National product per capita: $700 (1994 est.)
Inflation rate (consumer prices): 10% (October 1994)
Unemployment rate: NA%
Budget:
revenues: $52 million
expenditures: $150 million, including capital expenditures of $95 million (FY93/94 est.)
note: the government of India finances nearly three-fifths of Bhutan's budget expenditures
Exports: $66.8 million (f.o.b., FY93/94)
commodities: cardamon, gypsum, timber, handicrafts, cement, fruit, electricity (to India), precious stones, spices
partners: India 87%, Bangladesh
Imports: $97.6 million (c.i.f., FY93/94 est.)
commodities: fuel and lubricants, grain, machinery and parts, vehicles, fabrics, rice
partners: India 79%, Japan, UK, Germany, US
External debt: $141 million (October 1994)
Industrial production: growth rate 7.6% (1992 est.); accounts for 18% of GDP; primarily cottage industry and home based handicrafts
Electricity:
capacity: 360,000 kW
production: 1.7 billion kWh
consumption per capita: 143 kWh (1993)
note: Bhutan exports electricity to India
Industries: cement, wood products, processed fruits, alcoholic beverages, calcium carbide
Agriculture: rice, corn, root crops, citrus fruit, dairy products, foodgrains, eggs
Economic aid:
recipient: Western (non-US) countries, ODA and OOF bilateral commitments (1970-89), $115 million; OPEC bilateral aid (1979-89), $11 million
Currency: 1 ngultrum (Nu) = 100 chetrum; note—Indian currency is also legal tender
Exchange rates: ngultrum (Nu) per US$1—31.374 (January 1995), 31.374 (1994), 30.493 (1993), 25.918 (1992), 22.742 (1991), 17.504 (1990); note—the Bhutanese ngultrum is at par with the Indian rupee
Fiscal year: 1 July—30 June

Transportation

Railroads: 0 km
Highways:
total: 2,165 km
paved: NA
unpaved: gravel 1,703 km
undifferentiated: 462 km
Ports: none
Airports:
total: 2
with paved runways 1,524 to 2,437 m: 1
with unpaved runways 914 to 1,523 m: 1

Communications

Telephone system: NA telephones; domestic telephone service is very poor with very few telephones in use
local: NA
intercity: NA
international: international telephone and telegraph service is by land line through India; an earth station was planned (1990)
Radio:
broadcast stations: AM 1, FM 1, shortwave 0 (1990)
radios: NA
Television:
broadcast stations: 0 (1990)
televisions: NA

Defense Forces

Branches: Royal Bhutan Army, Palace Guard, Militia, Royal Bhutan Police
Manpower availability: males age 15-49 434,586; males fit for military service 232,121; males reach military age (18) annually 17,365 (1995 est.)
Defense expenditures: $NA, NA% of GDP

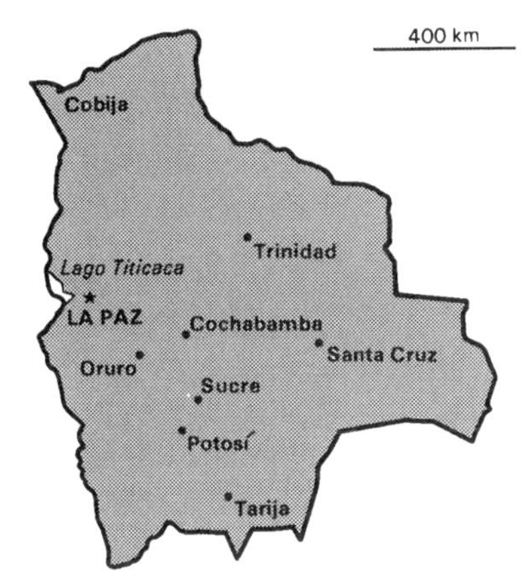

Geography

Location: Central South America, southwest of Brazil
Map references: South America
Area:
total area: 1,098,580 sq km
land area: 1,084,390 sq km
comparative area: slightly less than three times the size of Montana
Land boundaries: total 6,743 km, Argentina 832 km, Brazil 3,400 km, Chile 861 km, Paraguay 750 km, Peru 900 km
Coastline: 0 km (landlocked)
Maritime claims: none; landlocked
International disputes: has wanted a sovereign corridor to the South Pacific Ocean since the Atacama area was lost to Chile in 1884; dispute with Chile over Rio Lauca water rights
Climate: varies with altitude; humid and tropical to cold and semiarid
Terrain: rugged Andes Mountains with a highland plateau (Altiplano), hills, lowland plains of the Amazon Basin
Natural resources: tin, natural gas, petroleum, zinc, tungsten, antimony, silver, iron, lead, gold, timber
Land use:
arable land: 3%
permanent crops: 0%
meadows and pastures: 25%
forest and woodland: 52%
other: 20%
Irrigated land: 1,650 sq km (1989 est.)
Environment:
current issues: the clearing of land for agricultural purposes and the international demand for tropical timber are contributing to deforestation; soil erosion from overgrazing and poor cultivation methods (including slash-and-burn agriculture); desertification; loss of biodiversity; industrial pollution of water supplies used for drinking and irrigation
natural hazards: cold, thin air of high plateau is obstacle to efficient fuel combustion, as well as to physical activity by those unaccustomed to it from birth; flooding in the northeast (March to April)
international agreements: party to—Biodiversity, Climate Change, Endangered Species, Nuclear Test Ban, Tropical Timber 83, Wetlands; signed, but not ratified—Desertification, Environmental Modification, Hazardous Wastes, Law of the Sea, Marine Dumping, Marine Life Conservation, Ozone Layer Protection
Note: landlocked; shares control of Lago Titicaca, world's highest navigable lake (elevation 3,805 m), with Peru

People

Population: 7,896,254 (July 1995 est.)
Age structure:
0-14 years: 39% (female 1,542,931; male 1,565,624)
15-64 years: 57% (female 2,276,308; male 2,188,100)
65 years and over: 4% (female 174,419; male 148,872) (July 1995 est.)
Population growth rate: 2.25% (1995 est.)
Birth rate: 31.61 births/1,000 population (1995 est.)
Death rate: 8.12 deaths/1,000 population (1995 est.)
Net migration rate: -1.01 migrant(s)/1,000 population (1995 est.)
Infant mortality rate: 70.6 deaths/1,000 live births (1995 est.)
Life expectancy at birth:
total population: 63.85 years
male: 61.39 years
female: 66.43 years (1995 est.)
Total fertility rate: 4.1 children born/woman (1995 est.)
Nationality:
noun: Bolivian(s)
adjective: Bolivian
Ethnic divisions: Quechua 30%, Aymara 25%, mestizo (mixed European and Indian ancestry) 25%-30%, European 5%-15%
Religions: Roman Catholic 95%, Protestant (Evangelical Methodist)
Languages: Spanish (official), Quechua (official), Aymara (official)
Literacy: age 15 and over can read and write (1992)
total population: 80%
male: 88%
female: 72%
Labor force: 3.54 million
by occupation: agriculture NA, services and utilities 20%, manufacturing, mining and construction 7% (1993)

Government

Names:
conventional long form: Republic of Bolivia
conventional short form: Bolivia
local long form: Republica de Bolivia

local short form: Bolivia
Digraph: BL
Type: republic
Capital: La Paz (seat of government); Sucre (legal capital and seat of judiciary)
Administrative divisions: 9 departments (departamentos, singular—departamento); Chuquisaca, Cochabamba, Beni, La Paz, Oruro, Pando, Potosi, Santa Cruz, Tarija
Independence: 6 August 1825 (from Spain)
National holiday: Independence Day, 6 August (1825)
Constitution: 2 February 1967
Legal system: based on Spanish law and Code Napoleon; has not accepted compulsory ICJ jurisdiction
Suffrage: 18 years of age, universal and compulsory (married); 21 years of age, universal and compulsory (single)
Executive branch:
chief of state and head of government: President Gonzalo SANCHEZ DE LOZADA Bustamente (since 6 August 1993); Vice President Victor Hugo CARDENAS Conde (since 6 August 1993); election last held 6 June 1993 (next to be held May 1997); results—Gonzalo SANCHEZ DE LOZADA (MNR) 34%, Hugo BANZER Suarez (ADN/MIR alliance) 20%, Carlos PALENQUE Aviles (CONDEPA) 14%, Max FERNANDEZ Rojas (UCS) 13%, Antonio ARANIBAR Quiroga (MBL) 5%; no candidate received a majority of the popular vote; Gonzalo SANCHEZ DE LOZADA won a congressional runoff election on 4 August 1993 after forming a coalition with Max FERNANDEZ and Antonio ARANIBAR; FERNANDEZ left the coalition in 1994
cabinet: Cabinet; appointed by the president from panel proposed by the Senate
Legislative branch: bicameral National Congress (Congreso Nacional)
Chamber of Deputies (Camara de Diputados): elections last held 6 June 1993 (next to be held May 1997); results—percent of vote by party NA; seats—(130 total) MNR 52, UCS 20, ADN 17, MIR 17, CONDEPA 13, MBL 7, ARBOL 1, ASD 1, EJE 1, PCD 1
Chamber of Senators (Camara de Senadores): elections last held 6 June 1993 (next to be held May 1997); results—percent of vote by party NA; seats—(27 total) MNR 17, ADN 4, MIR 4, CONDEPA 1, UCS 1
Judicial branch: Supreme Court (Corte Suprema)
Political parties and leaders:
Left parties: Free Bolivia Movement (MBL), Antonio ARANIBAR; April 9 Revolutionary Vanguard (VR-9), Carlos SERRATE; Alternative of Democratic Socialism (ASD), Jerjes JUSTIANO; Revolutionary Front of the Left (FRI), Oscar ZAMORA; Bolivian Socialist Falange (FSB); Socialist Unzaguista Movement (MAS); Socialist Party One (PS-1); Bolivian Communist Party (PCB)
Center-Left parties: Nationalist Revolutionary Movement (MNR), Gonzalo SANCHEZ DE LOZADA; Movement of the Revolutionary Left (MIR), Jaime PAZ Zamora, Oscar EID; Christian Democrat (PCD), Jorge AGREDA
Center-Right party: Nationalist Democratic Action (ADN), Jorge LANDIVAR, Hugo BANZER
Populist parties: Civic Solidarity Union (UCS), Max FERNANDEZ Rojas; Conscience of the Fatherland (CONDEPA), Carlos PALENQUE Aviles; Popular Patriotic Movement (MPP), Julio MANTILLA; Unity and Progress Movement (MUP), Ivo KULJIS
Evangelical: Bolivian Renovating Alliance (ARBOL), Hugo VILLEGAS
Indigenous: Tupac Katari Revolutionary Liberation Movement (MRTK-L), Victor Hugo CARDENAS Conde; Patriotic Axis of Convergence (EJE-P), Ramiro BARRANCHEA; National Katarista Movement (MKN), Fernando UNTOJA
Member of: AG, ECLAC, FAO, G-11, G-77, GATT, IADB, IAEA, IBRD, ICAO, ICRM, IDA, IFAD, IFC, IFRCS, ILO, IMF, IMO, INTELSAT, INTERPOL, IOC, IOM, ITU, LAES, LAIA, NAM, OAS, OPANAL, PCA, RG, UN, UNCTAD, UNESCO, UNIDO, UPU, WCL, WFTU, WHO, WIPO, WMO, WTO
Diplomatic representation in US:
chief of mission: Ambassador Andres PETRICEVIC Raznatovic
chancery: 3014 Massachusetts Avenue NW, Washington, DC 20008
telephone: [1] (202) 483-4410 through 4412
FAX: [1] (202) 328-3712
consulate(s) general: Miami, New York, and San Francisco
US diplomatic representation:
chief of mission: Ambassador Curt Warren KAMMAN
embassy: Avenida Arce 2780, San Jorge, La Paz
mailing address: P. O. Box 425, La Paz; APO AA 34032
telephone: [591] (2) 350251, 350120
FAX: [591] (2) 359875
Flag: three equal horizontal bands of red (top), yellow, and green with the coat of arms centered on the yellow band; similar to the flag of Ghana, which has a large black five-pointed star centered in the yellow band

Economy

Overview: With its long history of semifeudal social controls, dependence on volatile prices for its mineral exports, and bouts of hyperinflation, Bolivia has remained one of the poorest and least developed Latin American countries. However, Bolivia has experienced generally improving economic conditions since the PAZ Estenssoro administration (1985-89) introduced market-oriented policies which reduced inflation from 11,700% in 1985 to about 20% in 1988. PAZ Estenssoro was followed as President by Jaime PAZ Zamora (1989-93) who continued the free-market policies of his predecessor, despite opposition from his own party and from Bolivia's once powerful labor movement. By maintaining fiscal discipline, PAZ Zamora helped reduce inflation to 9.3% in 1993, while GDP grew by an annual average of 3.25% during his tenure. Inaugurated in August 1993, President SANCHEZ DE LOZADA has vowed to advance the market-oriented economic reforms he helped launch as PAZ Estenssoro's planning minister. His successes so far have included an inflation rate that continues to decrease—the 1994 rate of 8.5% was the lowest in ten years—the signing of a free trade agreement with Mexico, and progress on his unique privatization plan. The main privatization bill was passed by the Bolivian legislature in late March 1994. Related laws—one that establishes SIRESE, the regulatory agency that will oversee the privatizations, and another that outlines the rules for privatization in the electricity sector—were approved later in the year.
National product: GDP—purchasing power parity—$18.3 billion (1994 est.)
National product real growth rate: 4.2% (1994 est.)
National product per capita: $2,370 (1994 est.)
Inflation rate (consumer prices): 8.5% (1994 est.)
Unemployment rate: 6.2% (1994 est.)
Budget:
revenues: $3.75 billion
expenditures: $3.75 billion, including capital expenditures of $556.2 million (1995 est.)
Exports: $1.1 billion (f.o.b., 1994 est.)
commodities: metals 39%, natural gas 9%, soybeans 11%, jewelry 11%, wood 8%
partners: US 26%, Argentina 15% (1993 est.)
Imports: $1.21 billion (c.i.f., 1994 est.)
commodities: capital goods 48%, chemicals 11%, petroleum 5%, food 5% (1993 est.)
partners: US 24%, Argentina 13%, Brazil 11%, Japan 11% (1993 est.)
External debt: $4.2 billion (January 1995)
Industrial production: growth rate 5% (1994 est.)
Electricity:
capacity: 756,200 kW
production: 2.116 billion kWh
consumption per capita: 367 kWh (1994)
Industries: mining, smelting, petroleum, food and beverage, tobacco, handicrafts, clothing; illicit drug industry reportedly produces 15% of its revenues
Agriculture: accounts for about 21% of GDP (including forestry and fisheries); principal commodities—coffee, coca, cotton, corn, sugarcane, rice, potatoes, timber; self-sufficient in food
Illicit drugs: world's second-largest producer of coca (after Peru) with an estimated 48,100

hectares under cultivation in 1994; voluntary and forced eradication programs unable to prevent production from rising to 89,800 metric tons in 1994 from 84,400 tons in 1993; government considers all but 12,000 hectares illicit; intermediate coca products and cocaine exported to or through Colombia and Brazil to the US and other international drug markets; alternative crop program aims to reduce illicit coca cultivation
Economic aid:
recipient: US commitments, including Ex-Im (FY70-89), $990 million; Western (non-US) countries, ODA and OOF bilateral commitments (1970-89), $2.025 billion; Communist countries (1970-89), $340 million
Currency: 1 boliviano ($B) = 100 centavos
Exchange rates: bolivianos ($B) per US$1—4.72 (January 1995), 4.6205 (1994), 4.2651 (1993), 3.9005 (1992), 3.5806 (1991), 3.1727 (1990)
Fiscal year: calendar year

Transportation

Railroads:
total: 3,684 km (single track)
narrow gauge: 3,652 km 1.000-m gauge; 32 km 0.760-m gauge
Highways:
total: 42,815 km
paved: 1,865 km
unpaved: gravel 12,000 km; improved/unimproved earth 28,950 km
Inland waterways: 10,000 km of commercially navigable waterways
Pipelines: crude oil 1,800 km; petroleum products 580 km; natural gas 1,495 km
Ports: none; however, Bolivia has free port privileges in the maritime ports of Argentina, Brazil, Chile, and Paraguay
Merchant marine:
total: 1 cargo ship (1,000 GRT or over) totaling 4,214 GRT/6,390 DWT
Airports:
total: 1,382
with paved runways over 3,047 m: 3
with paved runways 2,438 to 3,047 m: 3
with paved runways 1,524 to 2,437 m: 3
with paved runways under 914 m: 1,016
with unpaved runways 2,438 to 3,047 m: 5
with unpaved runways 1,524 to 2,438 m: 77
with unpaved runways 914 to 1,523 m: 275

Communications

Telephone system: about 150,000 telephones; about 2.0 telephones/100 persons; new subscribers face bureaucratic difficulties; most telephones in La Paz and other cities; microwave radio relay system being expanded; improved international services
local: NA
intercity: microwave radio relay system
international: 1 INTELSAT (Atlantic Ocean) earth station
Radio:
broadcast stations: AM 129, FM 0, shortwave 68
radios: NA
Television:
broadcast stations: 43
televisions: NA

Defense Forces

Branches: Army (Ejercito Boliviano), Navy (Fuerza Naval Boliviana, includes Marines), Air Force (Fuerza Aerea Boliviana), National Police Force (Policia Nacional de Bolivia)
Manpower availability: males age 15-49 1,885,485; males fit for military service 1,226,218; males reach military age (19) annually 81,065 (1995 est.)
Defense expenditures: exchange rate conversion—$134 million; 1.9% of GDP (1994)

Bosnia and Herzegovina

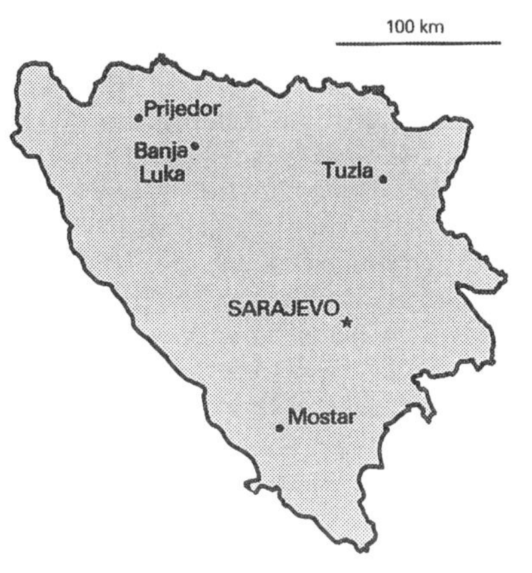

Note: Bosnia and Herzegovina is set to enter its third year of interethnic civil strife which began in the spring of 1992 after the Government of Bosnia and Herzegovina held a referendum on independence. Bosnia's Serbs—supported by neighboring Serbia—responded with armed resistance aimed at partitioning the republic along ethnic lines and joining Serb-held areas to 'greater Serbia'. In March 1994, Bosnia's Muslims and Croats reduced the number of warring factions from three to two by signing an agreement in Washington, DC, creating the Federation of Bosnia and Herzegovina. A group of rebel Muslims, however, continues to battle government forces in the northwest enclave of Bihac. A Contact Group of countries, the US, UK, France, Germany, and Russia, continues to seek a resolution between the Federation and the Bosnian Serbs. In July of 1994 the Contact Group presented a plan to the warring parties that roughly equally divides the country between the two, while maintaining Bosnia in its current internationally recognized borders. The Federation agreed to the plan almost immediately, while the Bosnian Serbs rejected it.

Geography

Location: Southeastern Europe, bordering the Adriatic Sea and Croatia
Map references: Ethnic Groups in Eastern Europe, Europe
Area:
total area: 51,233 sq km
land area: 51,233 sq km
comparative area: slightly larger than Tennessee
Land boundaries: total 1,459 km, Croatia 932 km, Serbia and Montenegro 527 km (312 km with Serbia; 215 km with Montenegro)
Coastline: 20 km
Maritime claims: NA
International disputes: as of January 1995, Bosnian Government and Bosnian Serb leaders

Bosnia and Herzegovina

(continued)

remain far apart on territorial and constitutional solutions for Bosnia; the two sides did, however, sign a four-month cessation of hostilities agreement effective January 1; the Bosnian Serbs continue to reject the Contact Group Plan submitted by the United States, United Kingdom, France, Germany, and Russia, and accepted by the Bosnian Government, which stands firm in its desire to regain lost territory and preserve Bosnia as a multiethnic state within its current borders; Bosnian Serb forces control approximately 70% of Bosnian territory
Climate: hot summers and cold winters; areas of high elevation have short, cool summers and long, severe winters; mild, rainy winters along coast
Terrain: mountains and valleys
Natural resources: coal, iron, bauxite, manganese, timber, wood products, copper, chromium, lead, zinc
Land use:
arable land: 20%
permanent crops: 2%
meadows and pastures: 25%
forest and woodland: 36%
other: 17%
Irrigated land: NA sq km
Environment:
current issues: air pollution from metallurgical plants; sites for disposing of urban waste are limited; widespread casualties, water shortages, and destruction of infrastructure because of civil strife
natural hazards: frequent and destructive earthquakes
international agreements: party to—Air Pollution, Law of the Sea, Marine Dumping, Marine Life Conservation, Nuclear Test Ban, Ozone Layer Protection

People

Population: 3,201,823 (July 1995 est.)
note: all data dealing with population is subject to considerable error because of the dislocations caused by military action and ethnic cleansing
Age structure:
0-14 years: 22% (female 337,787; male 370,966)
15-64 years: 68% (female 1,082,357; male 1,085,610)
65 years and over: 10% (female 190,992; male 134,111) (July 1995 est.)
Population growth rate: 0.65% (1995 est.)
Birth rate: 11.29 births/1,000 population (1995 est.)
Death rate: 7.51 deaths/1,000 population (1995 est.)
Net migration rate: 2.72 migrant(s)/1,000 population (1995 est.)
Infant mortality rate: 11.6 deaths/1,000 live births (1995 est.)
Life expectancy at birth:
total population: 75.47 years
male: 72.75 years
female: 78.37 years (1995 est.)
Total fertility rate: 1.65 children born/woman (1995 est.)
Nationality:
noun: Bosnian(s), Herzegovinian(s)
adjective: Bosnian, Herzegovinian
Ethnic divisions: Muslim 38%, Serb 40%, Croat 22% (est.)
Religions: Muslim 40%, Orthodox 31%, Catholic 15%, Protestant 4%, other 10%
Languages: Serbo-Croatian 99%
Literacy: NA%
Labor force: 1,026,254
by occupation: NA%

Government

Note: The US recognizes the Republic of Bosnia and Herzegovina. The Federation of Bosnia and Herzegovina, formed by the Muslims and Croats in March 1994, remains in the implementation stages.
Names:
conventional long form: Republic of Bosnia and Herzegovina
conventional short form: Bosnia and Herzegovina
local long form: Republika Bosna i Hercegovina
local short form: Bosna i Hercegovina
Digraph: BK
Type: emerging democracy
Capital: Sarajevo
Administrative divisions: 109 districts (opstinas, singular—opstina) Banovici, Banja Luka, Bihac, Bijeljina, Bileca, Bosanska Dubica, Bosanska Gradiska, Bosanska Krupa, Bosanski Brod, Bosanski Novi, Bosanski Petrovac, Bosanski Samac, Bosansko Grahovo, Bratunac, Brcko, Breza, Bugojno, Busovaca, Cazin, Cajnice, Capljina, Celinac, Citluk, Derventa, Doboj, Donji Vakuf, Foca, Fojnica, Gacko, Glamoc, Gorazde, Gornji Vakuf, Gracanica, Gradacac, Grude, Han Pijesak, Jablanica, Jajce, Kakanj, Kalesija, Kalinovik, Kiseljak, Kladanj, Kljuc, Konjic, Kotor Varos, Kresevo, Kupres, Laktasi, Listica, Livno, Lopare, Lukavac, Ljubinje, Ljubuski, Maglaj, Modrica, Mostar, Mrkonjic-Grad, Neum, Nevesinje, Odzak, Olovo, Orasje, Posusje, Prijedor, Prnjavor, Prozor, (Pucarevo) Novi Travnik, Rogatica, Rudo, Sanski Most, Sarajevo-Centar, Sarajevo-Hadzici, Sarajevo-Ilidza, Sarajevo-Ilijas, Sarajevo-Novi Grad, Sarajevo-Novo, Sarajevo-Pale, Sarajevo-Stari Grad, Sarajevo-Trnovo, Sarajevo-Vogosca, Skender Vakuf, Sokolac, Srbac, Srebrenica, Srebrenik, Stolac, Sekovici, Sipovo, Teslic, Tesanj, Drvar, Duvno, Travnik, Trebinje, Tuzla, Ugljevik, Vares, Velika Kladusa, Visoko, Visegrad, Vitez, Vlasenica, Zavidovici, Zenica, Zvornik, Zepce, Zivinice
note: currently under negotiation with the assistance of international mediators
Independence: NA April 1992 (from Yugoslavia)
National holiday: NA
Constitution: promulgated in 1974 (under the Communists), amended 1989, 1990, and 1991; the Assembly planned to draft a new constitution in 1991, before conditions deteriorated; constitution of Federation of Bosnia and Herzegovina (including Muslim and Croatian controlled parts of Republic) ratified April 1994
Legal system: based on civil law system
Suffrage: 16 years of age, if employed; 18 years of age, universal
Executive branch:
chief of state: President Alija IZETBEGOVIC (since 20 December 1990), other members of the collective presidency: Ejup GANIC (since NA November 1990), Nijaz DURAKOVIC (since NA October 1993), Stjepan KLJUJIC (since NA October 1993), Ivo KOMSIC (since NA October 1993), Mirko PEJANOVIC (since NA June 1992), Tatjana LJUJIC-MIJATOVIC (since NA December 1992)
head of government: Prime Minister Haris SILAJDZIC (since NA October 1993)
cabinet: executive body of ministers; members of, and responsible to, the National Assembly
note: the president of the Federation of Bosnia and Herzegovina is Kresimir ZUBAK (since 31 May 1994); Vice President Ejup GANIC (since 31 May 1994)
Legislative branch: bicameral National Assembly
Chamber of Municipalities (Vijece Opeina): elections last held November-December 1990 (next to be held NA); percent of vote by party NA; seats—(110 total) SDA 43, SDS BiH 38, HDZ BiH 23, Party of Democratic Changes 4, DSS 1, SPO 1
Chamber of Citizens (Vijece Gradanstvo): elections last held November-December 1990 (next to be held NA); percent of vote by party NA; seats—(130 total) SDA 43, SDS BiH 34, HDZ BiH 21, Party of Democratic Changes 15, SRSJ BiH 12, LBO 2, DSS 1, DSZ 1, LS 1
note: legislative elections for Federation of Bosnia and Herzegovina are slated for late 1994
Judicial branch: Supreme Court, Constitutional Court
Political parties and leaders: Party of Democratic Action (SDA), Alija IZETBEGOVIC; Croatian Democratic Union of Bosnia and Herzegovina (HDZ BiH), Dario KORDIC; Serbian Democratic Party of Bosnia and Herzegovina (SDS BiH), Radovan KARADZIC, president; Liberal Bosnian Organization (LBO), Adil ZULFIKARPASIC, president; Democratic Party of Socialists (DSS), Nijaz DURAKOVIC, president; Party of Democratic Changes, leader NA; Serbian Movement for Renewal (SPO), Milan TRIVUNCIC; Alliance of Reform Forces of Yugoslavia for Bosnia and Herzegovina (SRSJ

BiH), Dr. Nenad KECMANOVIC, president; Democratic League of Greens (DSZ), Drazen PETROVIC; Liberal Party (LS), Rasim KADIC, president
Other political or pressure groups: NA
Member of: CE (guest), CEI, ECE, FAO, ICAO, IFAD, ILO, IMO, INTELSAT (nonsignatory user), INTERPOL, IOC, IOM (observer), ITU, NAM (guest), OSCE, UN, UNCTAD, UNESCO, UNIDO, UPU, WHO, WIPO, WTO
Diplomatic representation in US:
chief of mission: Ambassador Sven ALKALAJ
chancery: Suite 760, 1707 L Street NW, Washington, DC 20036
telephone: [1] (202) 833-3612, 3613, 3615
FAX: [1] (202) 833-2061
consulate(s) general: New York
US diplomatic representation:
chief of mission: Ambassador Victor JACKOVICH
embassy: address NA
mailing address: American Embassy Bosnia, c/o AmEmbassy Vienna Boltzmangasse 16, A-1091, Vienna, Austria; APO: (Bosnia) Vienna, Department of State, Washington, DC 20521-9900
telephone: [43] (1) 313-39
FAX: [43] (1) 310-0682
Flag: white with a large blue shield; the shield contains white Roman crosses with a white diagonal band running from the upper hoist corner to the lower fly side

Economy

Overview: Bosnia and Herzegovina ranked next to The Former Yugoslav Republic of Macedonia as the poorest republic in the old Yugoslav federation. Although agriculture has been almost all in private hands, farms have been small and inefficient, and the republic traditionally has been a net importer of food. Industry has been greatly overstaffed, one reflection of the rigidities of Communist central planning and management. TITO had pushed the development of military industries in the republic with the result that Bosnia hosted a large share of Yugoslavia's defense plants. As of February 1995, Bosnia and Herzegovina was being torn apart by the continued bitter interethnic warfare that has caused production to plummet, unemployment and inflation to soar, and human misery to multiply. No economic statistics for 1992-94 are available, although output clearly has fallen substantially below the levels of earlier years and almost certainly is well below $1,000 per head. The country receives substantial amounts of humanitarian aid from the international community.
National product: GDP—purchasing power parity—$NA
National product real growth rate: NA%
National product per capita: $NA
Inflation rate (consumer prices): NA%
Unemployment rate: NA%
Budget:
revenues: $NA
expenditures: $NA, including capital expenditures of $NA
Exports: $NA
commodities: NA
partners: NA
Imports: $NA
commodities: NA
partners: NA
External debt: $NA
Industrial production: growth rate NA%; production is sharply down because of interethnic and interrepublic warfare (1991-94)
Electricity:
capacity: 3,800,000 kW
production: NA kWh
consumption per capita: NA kWh (1993)
Industries: steel production, mining (coal, iron ore, lead, zinc, manganese, and bauxite), manufacturing (vehicle assembly, textiles, tobacco products, wooden furniture, 40% of former Yugoslavia's armaments including tank and aircraft assembly, domestic appliances), oil refining (1991)
Agriculture: accounted for 9.0% of GDP in 1989; regularly produces less than 50% of food needs; the foothills of northern Bosnia support orchards, vineyards, livestock, and some wheat and corn; long winters and heavy precipitation leach soil fertility reducing agricultural output in the mountains; farms are mostly privately held, small, and not very productive (1991)
Illicit drugs: NA
Economic aid: $NA
Currency: 1 dinar = 100 para; Croatian dinar used in Croat-held area, presumably to be replaced by new Croatian kuna; old and new Serbian dinars used in Serb-held area; hard currencies probably supplanting local currencies in areas held by Bosnian government
Exchange rates: NA
Fiscal year: calendar year

Transportation

Railroads:
total: 1,021 km (electrified 795 km)
standard gauge: 1,021 km 1.435-m gauge (1994)
Highways:
total: 21,168 km
paved: 11,436 km
unpaved: gravel 8,146 km; earth 1,586 km (1991)
Inland waterways: NA km
Pipelines: crude oil 174 km; natural gas 90 km (1992); note—pipelines now disrupted
Ports: Bosanski Brod
Merchant marine: none
Airports:
total: 27
with paved runways 2,438 to 3,047 m: 4
with paved runways 1,524 to 2,437 m: 3
with paved runways under 914 m: 11
with unpaved runways 1,524 to 2,438 m: 1
with unpaved runways 914 to 1,523 m: 8

Communications

Telephone system: 727,000 telephones; telephone and telegraph network is in need of modernization and expansion; many urban areas are below average when compared with services in other former Yugoslav republics
local: NA
intercity: NA
international: no earth stations
Radio:
broadcast stations: AM 9, FM 2, shortwave 0
radios: 840,000
Television:
broadcast stations: 6
televisions: 1,012,094

Defense Forces

Branches: Army
Manpower availability: males age 15-49 815,055; males fit for military service 657,454; males reach military age (19) annually 38,201 (1995 est.)
Defense expenditures: $NA, NA% of GDP

Botswana

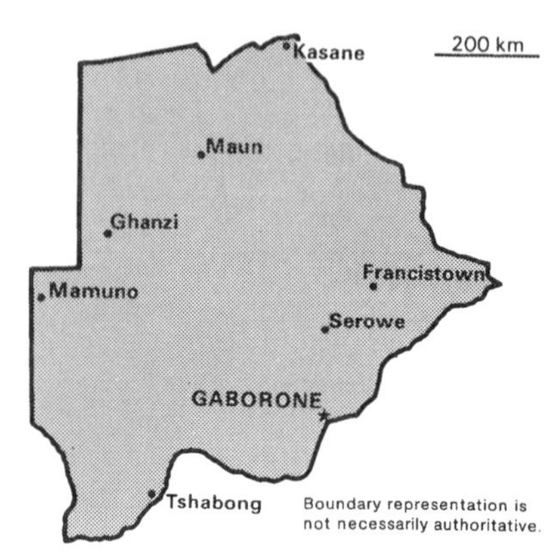

Geography

Location: Southern Africa, north of South Africa
Map references: Africa
Area:
total area: 600,370 sq km
land area: 585,370 sq km
comparative area: slightly smaller than Texas
Land boundaries: total 4,013 km, Namibia 1,360 km, South Africa 1,840 km, Zimbabwe 813 km
Coastline: 0 km (landlocked)
Maritime claims: none; landlocked
International disputes: short section of boundary with Namibia is indefinite; quadripoint with Namibia, Zambia, and Zimbabwe is in disagreement; dispute with Namibia over uninhabited Kasikili (Sidudu) Island in Linyanti (Chobe) River remained unresolved in mid-February 1995 and the parties agreed to refer the matter to the International Court of Justice
Climate: semiarid; warm winters and hot summers
Terrain: predominately flat to gently rolling tableland; Kalahari Desert in southwest
Natural resources: diamonds, copper, nickel, salt, soda ash, potash, coal, iron ore, silver
Land use:
arable land: 2%
permanent crops: 0%
meadows and pastures: 75%
forest and woodland: 2%
other: 21%
Irrigated land: 20 sq km (1989 est.)
Environment:
current issues: overgrazing, primarily as a result of the expansion of the cattle population; desertification; limited natural fresh water resources
natural hazards: periodic droughts; seasonal August winds blow from the west, carrying sand and dust across the country, which can obscure visibility
international agreements: party to—Climate Change, Endangered Species, Law of the Sea, Nuclear Test Ban, Ozone Layer Protection; signed, but not ratified—Biodiversity
Note: landlocked; population concentrated in eastern part of the country

People

Population: 1,392,414 (July 1995 est.)
Age structure:
0-14 years: 43% (female 300,598; male 303,333)
15-64 years: 53% (female 398,347; male 344,838)
65 years and over: 4% (female 25,773; male 19,525) (July 1995 est.)
Population growth rate: 2.36% (1995 est.)
Birth rate: 31.01 births/1,000 population (1995 est.)
Death rate: 7.41 deaths/1,000 population (1995 est.)
Net migration rate: 0 migrant(s)/1,000 population (1995 est.)
Infant mortality rate: 38 deaths/1,000 live births (1995 est.)
Life expectancy at birth:
total population: 63.56 years
male: 60.54 years
female: 66.67 years (1995 est.)
Total fertility rate: 3.86 children born/woman (1995 est.)
Nationality:
noun: Motswana (singular), Batswana (plural)
adjective: Motswana (singular), Batswana (plural)
Ethnic divisions: Batswana 95%, Kalanga, Basarwa, and Kgalagadi 4%, white 1%
Religions: indigenous beliefs 50%, Christian 50%
Languages: English (official), Setswana
Literacy: age 15 and over can read and write (1990 est.)
total population: 23%
male: 32%
female: 16%
Labor force: 428,000 (1992)
by occupation: 220,000 formal sector employees, most others are engaged in cattle raising and subsistence agriculture (1992 est.); 14,300 are employed in various mines in South Africa (March 1992)

Government

Names:
conventional long form: Republic of Botswana
conventional short form: Botswana
former: Bechuanaland
Digraph: BC
Type: parliamentary republic
Capital: Gaborone
Administrative divisions: 10 districts; Central, Chobe, Ghanzi, Kgalagadi, Kgatleng, Kweneng, Ngamiland, North-East, South-East, Southern; in addition, there are 4 town councils—Francistown, Gaborone, Lobatse, Selebi-Phikwe
Independence: 30 September 1966 (from UK)
National holiday: Independence Day, 30 September (1966)
Constitution: March 1965, effective 30 September 1966
Legal system: based on Roman-Dutch law and local customary law; judicial review limited to matters of interpretation; has not accepted compulsory ICJ jurisdiction
Suffrage: 21 years of age; universal
Executive branch:
chief of state and head of government: President Sir Ketumile MASIRE (since 13 July 1980); Vice President Festus MOGAE (since 9 March 1992); election last held 15 October 1994 (next to be held October 1999); results—President Sir Ketumile MASIRE was reelected by the National Assembly
cabinet: Cabinet; appointed by the president
Legislative branch: bicameral Parliament
House of Chiefs: is a largely advisory 15-member body consisting of chiefs of the 8 principal tribes, 4 elected subchiefs, and 3 members selected by the other 12
National Assembly: elections last held 15 October 1994 (next to be held October 1999); results—percent of vote by party NA; seats—(44 total of which 40 are elected and 4 are appointed) BDP 27, BNF 13
Judicial branch: High Court, Court of Appeal
Political parties and leaders: Botswana Democratic Party (BDP), Sir Ketumile MASIRE; Botswana National Front (BNF), Kenneth KOMA; Botswana People's Party (BPP), Knight MARIPE; Botswana Independence Party (BIP), Motsamai MPHO
Member of: ACP, AfDB, C, CCC, ECA, FAO, FLS, G-77, GATT, IBRD, ICAO, ICFTU, ICRM, IDA, IFAD, IFC, IFRCS, ILO, IMF, INTELSAT (nonsignatory user), INTERPOL, IOC, ITU, NAM, OAU, SACU, SADC, UN, UNCTAD, UNESCO, UNIDO, UNOMOZ, UNOMUR, UNOSOM, UPU, WCL, WFTU, WHO, WMO
Diplomatic representation in US:
chief of mission: Ambassador Botsweletse Kingsley SEBELE
chancery: Suite 7M, 3400 International Drive NW, Washington, DC 20008
telephone: [1] (202) 244-4990, 4991
FAX: [1] (202) 244-4164
US diplomatic representation:
chief of mission: Ambassador Howard F. JETER
embassy: address NA, Gaborone
mailing address: P. O. Box 90, Gaborone
telephone: [267] 353982
FAX: [267] 356947
Flag: light blue with a horizontal white-edged black stripe in the center

Economy

Overview: The economy has historically been based on cattle raising and crops. Agriculture today provides a livelihood for more than 80% of the population but supplies only about 50% of food needs and accounts for only 5% of GDP. Subsistence farming and cattle raising predominate. The driving force behind the rapid economic growth of the 1970s and 1980s has been the mining industry. This sector, mostly on the strength of diamonds, has gone from generating 25% of GDP in 1980 to 39% in 1994. No other sector has experienced such growth, especially not agriculture, which is plagued by erratic rainfall and poor soils. The unemployment rate remains a problem at 25%. Hampered by a still sluggish diamond market in 1994, GDP grew by only 1%.
National product: GDP—purchasing power parity—$4.3 billion (1994 est.)
National product real growth rate: 1% (1994 est.)
National product per capita: $3,130 (1994 est.)
Inflation rate (consumer prices): 10% (1994 est.)
Unemployment rate: 25% (1994 est.)
Budget:
revenues: $1.7 billion
expenditures: $1.99 billion, including capital expenditures of $652 million (FY93/94)
Exports: $1.8 billion (f.o.b. 1994)
commodities: diamonds 78%, copper and nickel 6%, meat 5%
partners: Switzerland, UK, SACU (Southern African Customs Union)
Imports: $1.8 billion (c.i.f., 1992)
commodities: foodstuffs, vehicles and transport equipment, textiles, petroleum products
partners: Switzerland, SACU (Southern African Customs Union), UK, US
External debt: $344 million (December 1991)
Industrial production: growth rate 4.6% (FY92/93); accounts for about 43% of GDP, including mining
Electricity:
capacity: 220,000 kW
production: 900 million kWh
consumption per capita: 694 kWh (1993)
Industries: mining of diamonds, copper, nickel, coal, salt, soda ash, potash; livestock processing
Agriculture: sorghum, maize, millet, pulses, groundnuts, beans, cowpeas, sunflower seeds; livestock
Economic aid:
recipient: US aid (1992), $13 million; Norway (1992), $16 million; Sweden (1992), $15.5 million; Germany (1992), $3.6 million; EC/Lome-IV (1992), $3 million-$6 million in grants; $28.7 million in long-term projects (1992)
Currency: 1 pula (P) = 100 thebe
Exchange rates: pula (P) per US$1—1.7086 (January 1995), 2.6976 (November 1994), 2.4190 (1993), 2.1327 (1992), 2.0173 (1991), 1.8601 (1990)
Fiscal year: 1 April—31 March

Transportation

Railroads:
total: 888 km
narrow gauge: 888 km 1.067-m gauge (1992)
Highways:
total: 11,514 km
paved: 1,600 km
unpaved: crushed stone, gravel 1,700 km; improved earth 5,177 km; unimproved earth 3,037 km
Ports: none
Airports:
total: 100
with paved runways over 3,047 m: 1
with paved runways 2,438 to 3,047 m: 1
with paved runways 1,524 to 2,437 m: 6
with paved runways 914 to 1,523 m: 2
with paved runways under 914 m: 23
with unpaved runways 1,524 to 2,438 m: 5
with unpaved runways 914 to 1,523 m: 62

Communications

Telephone system: 26,000 telephones; sparse system; telephone density—18.67 telephones/1,000 persons
local: NA
intercity: small system of open wire lines, microwave radio relay links, and a few radio communication stations
international: 1 INTELSAT (Indian Ocean) earth station
Radio:
broadcast stations: AM 7, FM 13, shortwave 0
radios: NA
Television:
broadcast stations: 0
televisions: NA

Defense Forces

Branches: Botswana Defense Force (includes Army and Air Wing), Botswana National Police
Manpower availability: males age 15-49 306,878; males fit for military service 161,376; males reach military age (18) annually 15,403 (1995 est.)
Defense expenditures: exchange rate conversion—$198 million, 5.2% of GDP (FY93/94)

Bouvet Island
(territory of Norway)

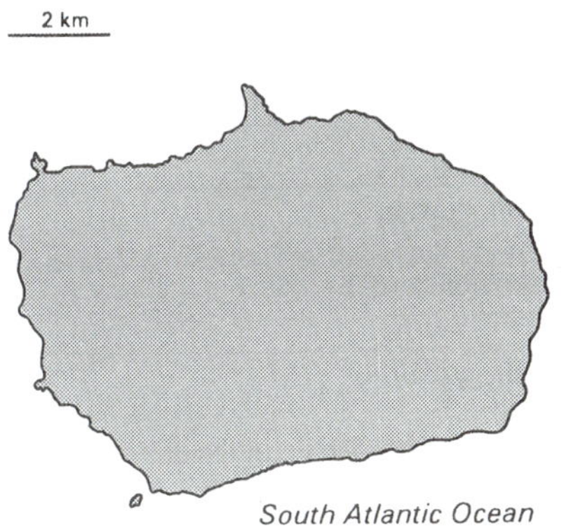

Geography

Location: Southern Africa, island in the South Atlantic Ocean, south-southwest of the Cape of Good Hope (South Africa)
Map references: Antarctic Region
Area:
total area: 58 sq km
land area: 58 sq km
comparative area: about 0.3 times the size of Washington, DC
Land boundaries: 0 km
Coastline: 29.6 km
Maritime claims:
territorial sea: 4 nm
International disputes: none
Climate: antarctic
Terrain: volcanic; maximum elevation about 800 meters; coast is mostly inaccessible
Natural resources: none
Land use:
arable land: 0%
permanent crops: 0%
meadows and pastures: 0%
forest and woodland: 0%
other: 100% (all ice)
Irrigated land: 0 sq km
Environment:
current issues: NA
natural hazards: NA
international agreements: NA
Note: covered by glacial ice

People

Population: uninhabited

Government

Names:
conventional long form: none
conventional short form: Bouvet Island
Digraph: BV
Type: territory of Norway
Capital: none; administered from Oslo, Norway

Independence: none (territory of Norway)

Economy

Overview: no economic activity

Transportation

Ports: none; offshore anchorage only

Communications

Note: automatic meteorological station

Defense Forces

Note: defense is the responsibility of Norway

Brazil

Geography

Location: Eastern South America, bordering the Atlantic Ocean
Map references: South America
Area:
total area: 8,511,965 sq km
land area: 8,456,510 sq km
comparative area: slightly smaller than the US
note: includes Arquipelago de Fernando de Noronha, Atol das Rocas, Ilha da Trindade, Ilhas Martin Vaz, and Penedos de Sao Pedro e Sao Paulo
Land boundaries: total 14,691 km, Argentina 1,224 km, Bolivia 3,400 km, Colombia 1,643 km, French Guiana 673 km, Guyana 1,119 km, Paraguay 1,290 km, Peru 1,560 km, Suriname 597 km, Uruguay 985 km, Venezuela 2,200 km
Coastline: 7,491 km
Maritime claims:
contiguous zone: 24 nm
continental shelf: 200 nm
exclusive economic zone: 200 nm
territorial sea: 12 nm
International disputes: short section of the boundary with Paraguay, just west of Salto das Sete Quedas (Guaira Falls) on the Rio Parana, is in dispute; two short sections of boundary with Uruguay are in dispute—Arroio Invernada (Arroyo de la Invernada) area of the Rio Quarai (Rio Cuareim) and the islands at the confluence of the Rio Quarai and the Uruguay River
Climate: mostly tropical, but temperate in south
Terrain: mostly flat to rolling lowlands in north; some plains, hills, mountains, and narrow coastal belt
Natural resources: bauxite, gold, iron ore, manganese, nickel, phosphates, platinum, tin, uranium, petroleum, hydropower, timber
Land use:
arable land: 7%
permanent crops: 1%
meadows and pastures: 19%
forest and woodland: 67%
other: 6%
Irrigated land: 27,000 sq km (1989 est.)
Environment:
current issues: deforestation in Amazon Basin destroys the habitat and endangers the existence of a multitude of plant and animal species indigenous to the area; air and water pollution in Rio de Janeiro, Sao Paulo, and several other large cities; land degradation and water pollution caused by improper mining activities
natural hazards: recurring droughts in northeast; floods and occasional frost in south
international agreements: party to—Antarctic Treaty, Biodiversity, Climate Change, Endangered Species, Environmental Modification, Hazardous Wastes, Law of the Sea, Marine Dumping, Nuclear Test Ban, Ozone Layer Protection, Ship Pollution, Tropical Timber 83, Wetlands, Whaling; signed, but not ratified—Antarctic-Environmental Protocol, Desertification
Note: largest country in South America; shares common boundaries with every South American country except Chile and Ecuador

People

Population: 160,737,489 (July 1995 est.)
Age structure:
0-14 years: 31% (female 24,641,868; male 25,515,775)
15-64 years: 64% (female 51,966,272; male 51,254,165)
65 years and over: 5% (female 4,393,530; male 2,965,879) (July 1995 est.)
Population growth rate: 1.22% (1995 est.)
Birth rate: 21.16 births/1,000 population (1995 est.)
Death rate: 8.98 deaths/1,000 population (1995 est.)
Net migration rate: 0 migrant(s)/1,000 population (1995 est.)
Infant mortality rate: 57.2 deaths/1,000 live births (1995 est.)
Life expectancy at birth:
total population: 61.82 years
male: 56.57 years
female: 67.32 years (1995 est.)
Total fertility rate: 2.39 children born/woman (1995 est.)
Nationality:
noun: Brazilian(s)
adjective: Brazilian
Ethnic divisions: Caucasion (includes Portuguese, German, Italian, Spanish, Polish) 55%, mixed Caucasion and African 38%, African 6%, other (includes Japanese, Arab, Amerindian) 1%
Religions: Roman Catholic (nominal) 70%
Languages: Portuguese (official), Spanish, English, French
Literacy: age 15 and over can read and write (1991)
total population: 80%

male: 80%
female: 80%
Labor force: 57 million (1989 est.)
by occupation: services 42%, agriculture 31%, industry 27%

Government

Names:
conventional long form: Federative Republic of Brazil
conventional short form: Brazil
local long form: Republica Federativa do Brasil
local short form: Brasil
Digraph: BR
Type: federal republic
Capital: Brasilia
Administrative divisions: 26 states (estados, singular—estado) and 1 federal district* (distrito federal); Acre, Alagoas, Amapa, Amazonas, Bahia, Ceara, Distrito Federal*, Espirito Santo, Goias, Maranhao, Mato Grosso, Mato Grosso do Sul, Minas Gerais, Para, Paraiba, Parana, Pernambuco, Piaui, Rio de Janeiro, Rio Grande do Norte, Rio Grande do Sul, Rondonia, Roraima, Santa Catarina, Sao Paulo, Sergipe, Tocantins
Independence: 7 September 1822 (from Portugal)
National holiday: Independence Day, 7 September (1822)
Constitution: 5 October 1988
Legal system: based on Roman codes; has not accepted compulsory ICJ jurisdiction
Suffrage: voluntary between 16 and 18 years of age and over 70; compulsory over 18 and under 70 years of age
Executive branch:
chief of state and head of government: President Fernando Henrique CARDOSO (since 1 January 1995) election last held 3 October 1994; next to be held October 1998); results—Fernando Henrique CARDOSO 53%, Luis Inacio LULA da Silva 26%, Eneas CARNEIRO 7%, Orestes QUERCIA 4%, Leonel BRIZOLA 3%, Espiridiao AMIN 3%; note—second free, direct presidential election since 1960
cabinet: Cabinet; appointed by the president
Legislative branch: bicameral National Congress (Congresso Nacional)
Federal Senate (Senado Federal): election last held 3 October 1994 for two-thirds of Senate (next to be held October 1996 for one-third of the Senate); results—PMDB 28%, PFL 22%, PSDB 12%, PPR 7%, PDT 7%, PT 6%, PTB 6%, other 12%
Chamber of Deputies (Camara dos Deputados): election last held 3 October 1994 (next to be held October 1998); results—PMDB 21%, PFL 18%, PDT 7%, PSDB 12%, PPR 10%, PTB 6%, PT 10%, other 16%
Judicial branch: Supreme Federal Tribunal
Political parties and leaders: National Reconstruction Party (PRN), Daniel TOURINHO, president; Brazilian Democratic Movement Party (PMDB), Luiz HENRIQUE da Silveira, president; Liberal Front Party (PFL), Jorge BORNHAUSEN, president; Workers' Party (PT), Rui Goethe da Costa FALCAO, president; Brazilian Workers' Party (PTB), Jose Eduardo ANDRADE VIEIRA, president; Democratic Workers' Party (PDT), Anthony GAROTINHO, president; Progressive Renewal Party (PPR), Espiridiao AMIN, president; Brazilian Social Democracy Party (PSDB), Artur DA TAVOLA, president; Popular Socialist Party (PPS), Roberto FREIRE, president; Communist Party of Brazil (PCdoB), Joao AMAZONAS, secretary general; Liberal Party (PL), Alvero VALLE, president
Other political or pressure groups: left wing of the Catholic Church and labor unions allied to leftist Workers' Party are critical of government's social and economic policies
Member of: AfDB, AG (observer), CCC, ECLAC, FAO, G-11, G-15, G-19, G-24, G-77, GATT, IADB, IAEA, IBRD, ICAO, ICC, ICFTU, ICRM, IDA, IFAD, IFC, IFRCS, ILO, IMF, IMO, INMARSAT, INTELSAT, INTERPOL, IOC, IOM (observer), ISO, ITU, LAES, LAIA, MERCOSUR, NAM (observer), OAS, ONUSAL, OPANAL, PCA, RG, UN, UNAVEM II, UNCTAD, UNESCO, UNHCR, UNIDO, UNOMOZ, UNOMUR, UNPROFOR, UNU, UPU, WCL, WFTU, WHO, WIPO, WMO, WTO
Diplomatic representation in US:
chief of mission: Ambassador Paulo Tarso FLECHA de LIMA
chancery: 3006 Massachusetts Avenue NW, Washington, DC 20008
telephone: [1] (202) 745-2700
FAX: [1] (202) 745-2827
consulate(s) general: Boston, Chicago, Los Angeles, Miami, New York, San Juan (Puerto Rico), and San Francisco
consulate(s): Houston
US diplomatic representation:
chief of mission: Ambassador Melvyn LEVITSKY
embassy: Avenida das Nacoes, Lote 3, Brasilia, Distrito Federal
mailing address: Unit 3500; APO AA 34030
telephone: [55] (61) 321-7272
FAX: [55] (61) 225-9136
consulate(s) general: Rio de Janeiro, Sao Paulo
consulate(s): Porto Alegre, Recife
Flag: green with a large yellow diamond in the center bearing a blue celestial globe with 27 white five-pointed stars (one for each state and the Federal District) arranged in the same pattern as the night sky over Brazil; the globe has a white equatorial band with the motto ORDEM E PROGRESSO (Order and Progress)

Economy

Overview: The economy, with large agrarian, mining, and manufacturing sectors, entered the 1990s with declining real growth, runaway inflation, an unserviceable foreign debt of $122 billion, and a lack of policy direction. In addition, the economy remained highly regulated, inward-looking, and protected by substantial trade and investment barriers. Ownership of major industrial and mining facilities is divided among private interests—including several multinationals—and the government. Most large agricultural holdings are private, with the government channeling financing to this sector. Conflicts between large landholders and landless peasants have produced intermittent violence. The COLLOR government, which assumed office in March 1990, launched an ambitious reform program that sought to modernize and reinvigorate the economy by stabilizing prices, deregulating the economy, and opening it to increased foreign competition. Itamar FRANCO, who assumed the presidency following President COLLOR's resignation in December 1992, was out of step with COLLOR's reform agenda; initiatives to redress fiscal problems, privatize state enterprises, and liberalize trade and investment policies lost momentum. Galloping inflation—by June 1994 the monthly rate had risen to nearly 50%—had undermined economic stability. In response, the then finance minister, Fernando Henrique CARDOSO, launched the third phase of his stabilization plan, known as Plano Real, that called for a new currency, the real, which was introduced on 1 July 1994. Inflation subsequently dropped to under 3% per month through the end of 1994. The newly elected President CARDOSO has called for the implementation of sweeping market-oriented reform, including public sector and fiscal reform, privatization, deregulation, and elimination of barriers to increased foreign investment. Brazil's natural resources remain a major, long-term economic strength.
National product: GDP—purchasing power parity—$886.3 billion (1994 est.)
National product real growth rate: 5.3% (1994 est.)
National product per capita: $5,580 (1994 est.)
Inflation rate (consumer prices): 1,094% (1994 est.)
Unemployment rate: 4.9% (1993)
Budget:
revenues: $113 billion
expenditures: $109 billion, including capital expenditures of $23 billion (1992)
Exports: $43.6 billion (f.o.b., 1994 est.)
commodities: iron ore, soybean bran, orange juice, footwear, coffee, motor vehicle parts
partners: EC 27.6%, Latin America 21.8%, US 17.4%, Japan 6.3% (1993)
Imports: $33.2 billion (f.o.b., 1994 est.)

commodities: crude oil, capital goods, chemical products, foodstuffs, coal
partners: US 23.3%, EC 22.5%, Middle East 13.0%, Latin America 11.8%, Japan 6.5% (1993)
External debt: $134 billion (1994)
Industrial production: growth rate 9.5% (1993); accounts for 39% of GDP
Electricity:
capacity: 55,130,000 kW
production: 241.4 billion kWh
consumption per capita: 1,589 kWh (1993)
Industries: textiles, shoes, chemicals, cement, lumber, mining (iron ore, tin), steel making, machine building—including aircraft, motor vehicles, motor vehicle parts and assemblies, and other machinery and equipment
Agriculture: accounts for 11% of GDP; world's largest producer and exporter of coffee and orange juice concentrate and second-largest exporter of soybeans; other products—rice, corn, sugarcane, cocoa, beef; self-sufficient in food, except for wheat
Illicit drugs: illicit producer of cannabis and coca, mostly for domestic consumption; government has a small-scale eradication program to control cannabis and coca cultivation; important transshipment country for Bolivian and Colombian cocaine headed for the US and Europe
Economic aid:
recipient: US commitments, including Ex-Im (FY70-89), $2.5 billion; Western (non-US) countries, ODA and OOF bilateral commitments (1970-89), $10.2 million; OPEC bilateral aid (1979-89), $284 million; former Communist countries (1970-89), $1.3 billion
Currency: 1 real (R$) = 100 centavos
Exchange rates: R$ per US$1—0.85 (January 1995); CR$ per US$1—390.845 (January 1994), 88.449 (1993), 4.513 (1992), 0.407 (1991), 0.068 (1990)
note: on 1 August 1993 the cruzeiro real (CR$), equal to 1,000 cruzeiros, was introduced; another new currency, the real, was introduced on 1 July 1994, equal to 2,750 cruzeiro reals
Fiscal year: calendar year

Transportation

Railroads:
total: 30,612 km (1992)
broad gauge: 5,369 km 1.600-m gauge (1,108 km electrified)
standard gauge: 194 km 1.440-m gauge
narrow gauge: 24,739 km 1.000-m gauge (112 km electrified); 13 km 0.760-m gauge
dual gauge: 310 km 1.600-m/1.000-m gauge (78 km electrified)
Highways:
total: 1,670,148 km
paved: 161,503 km
unpaved: gravel/earth 1,508,645 km (1990)
Inland waterways: 50,000 km navigable
Pipelines: crude oil 2,000 km; petroleum products 3,804 km; natural gas 1,095 km
Ports: Belem, Fortaleza, Ilheus, Imbituba, Manaus, Paranagua, Porto Alegre, Recife, Rio de Janeiro, Rio Grande, Salvador, Santos, Vitoria
Merchant marine:
total: 215 ships (1,000 GRT or over) totaling 5,128,654 GRT/8,664,776 DWT
ships by type: bulk 52, cargo 34, chemical tanker 13, combination ore/oil 12, container 12, liquefied gas tanker 11, oil tanker 64, passenger-cargo 5, refrigerated cargo 1, roll-on/roll-off cargo 11
Airports:
total: 3,467
with paved runways over 3,047 m: 5
with paved runways 2,438 to 3,047 m: 19
with paved runways 1,524 to 2,437 m: 126
with paved runways 914 to 1,523 m: 286
with paved runways under 914 m: 1,652
with unpaved runways 1,524 to 2,438 m: 76
with unpaved runways 914 to 1,523 m: 1,303

Communications

Telephone system: 9.86 million telephones; telephone density—61/1,000 persons; good working system
local: NA
intercity: extensive microwave radio relay systems and 64 domestic satellite earth stations
international: 3 coaxial submarine cables; 3 Atlantic Ocean INTELSAT earth stations
Radio:
broadcast stations: AM 1,223, FM 0, shortwave 151
radios: NA
Television:
broadcast stations: 112 (Brazil has the world's fourth largest television broadcasting system)
televisions: NA

Defense Forces

Branches: Brazilian Army, Brazilian Navy (includes Marines), Brazilian Air Force, Federal Police (paramilitary)
Manpower availability: males age 15-49 44,301,765; males fit for military service 29,815,576; males reach military age (18) annually 1,703,438 (1995 est.)
Defense expenditures: exchange rate conversion—$5.0 billion, 0.9% of GDP (1994)

British Indian Ocean Territory

(dependent territory of the UK)

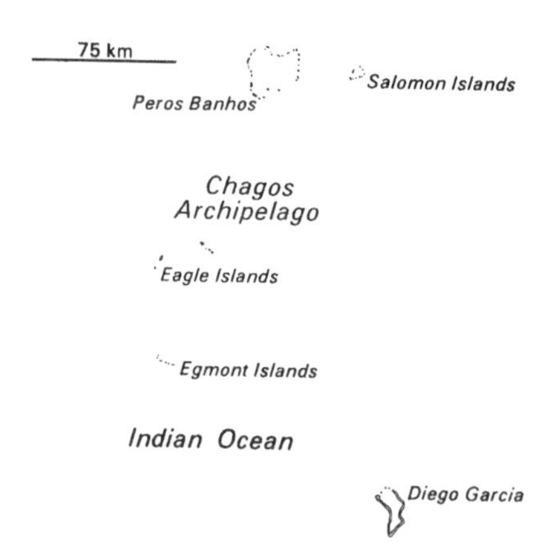

Geography

Location: Southern Asia, archipelago in the Indian Ocean, about one-half of the way from Africa to Indonesia
Map references: World
Area:
total area: 60 sq km
land area: 60 sq km
comparative area: about 0.3 times the size of Washington, DC
note: includes the island of Diego Garcia
Land boundaries: 0 km
Coastline: 698 km
Maritime claims:
exclusive fishing zone: 200 nm
territorial sea: 3 nm
International disputes: the entire Chagos Archipelago is claimed by Mauritius
Climate: tropical marine; hot, humid, moderated by trade winds
Terrain: flat and low (up to 4 meters in elevation)
Natural resources: coconuts, fish
Land use:
arable land: 0%
permanent crops: 0%
meadows and pastures: 0%
forest and woodland: 0%
other: 100%
Irrigated land: 0 sq km
Environment:
current issues: NA
natural hazards: NA
international agreements: NA
Note: archipelago of 2,300 islands; Diego Garcia, largest and southernmost island, occupies strategic location in central Indian Ocean; island is site of joint US-UK military facility

People

Population: no indigenous inhabitants
note: there are UK-US military personnel; civilian inhabitants, known as the Ilois,

evacuated to Mauritius before construction of UK-US military facilities

Government

Names:
conventional long form: British Indian Ocean Territory
conventional short form: none
Abbreviation: BIOT
Digraph: IO
Type: dependent territory of the UK
Capital: none
Independence: none (dependent territory of the UK)
Executive branch:
chief of state: Queen ELIZABETH II (since 6 February 1952)
head of government: Commissioner Mr. D. R. MACLENNAN); Administrator Mr. David Smith; note—both reside in the UK
Diplomatic representation in US: none (dependent territory of the UK)
US diplomatic representation: none (dependent territory of the UK)
Flag: white with the flag of the UK in the upper hoist-side quadrant and six blue wavy horizontal stripes bearing a palm tree and yellow crown centered on the outer half of the flag

Economy

Overview: All economic activity is concentrated on the largest island of Diego Garcia, where joint UK-US defense facilities are located. Construction projects and various services needed to support the military installations are done by military and contract employees from the UK, Mauritius, the Philippines, and the US. There are no industrial or agricultural activities on the islands.
Electricity: provided by the US military

Transportation

Railroads: 0 km
Highways:
total: NA
paved: short stretch of paved road between port and airfield on Diego Garcia
unpaved: NA
Ports: Diego Garcia
Airports:
total: 1
with paved runways over 3,047 m: 1

Communications

Telephone system: NA telephones; minimal facilities
local: NA
intercity: NA
international: 1 INTELSAT (Atlantic Ocean) earth station
Radio:
broadcast stations: AM 1, FM 1, shortwave 0
radios: NA
Television:
broadcast stations: 1
televisions: NA

Defense Forces

Note: defense is the responsibility of the UK

British Virgin Islands

(dependent territory of the UK)

Geography

Location: Caribbean, between the Caribbean Sea and the North Atlantic Ocean, east of Puerto Rico
Map references: Central America and the Caribbean
Area:
total area: 150 sq km
land area: 150 sq km
comparative area: about 0.8 times the size of Washington, DC
note: includes the island of Anegada
Land boundaries: 0 km
Coastline: 80 km
Maritime claims:
exclusive fishing zone: 200 nm
territorial sea: 3 nm
International disputes: none
Climate: subtropical; humid; temperatures moderated by trade winds
Terrain: coral islands relatively flat; volcanic islands steep, hilly
Natural resources: negligible
Land use:
arable land: 20%
permanent crops: 7%
meadows and pastures: 33%
forest and woodland: 7%
other: 33%
Irrigated land: NA sq km
Environment:
current issues: limited natural fresh water resources (except for a few seasonal streams and springs on Tortola, most of the island's water supply comes from wells and rainwater catchment)
natural hazards: hurricanes and tropical storms (July to October)
international agreements: NA
Note: strong ties to nearby US Virgin Islands and Puerto Rico

People

Population: 13,027 (July 1995 est.)

British Virgin Islands *(continued)*

Age structure:
0-14 years: NA
15-64 years: NA
65 years and over: NA
Population growth rate: 1.27% (1995 est.)
Birth rate: 20.25 births/1,000 population (1995 est.)
Death rate: 6.07 deaths/1,000 population (1995 est.)
Net migration rate: -1.5 migrant(s)/1,000 population (1995 est.)
Infant mortality rate: 19.33 deaths/1,000 live births (1995 est.)
Life expectancy at birth:
total population: 72.73 years
male: 70.88 years
female: 74.7 years (1995 est.)
Total fertility rate: 2.27 children born/woman (1995 est.)
Nationality:
noun: British Virgin Islander(s)
adjective: British Virgin Islander
Ethnic divisions: black 90%, white, Asian
Religions: Protestant 86% (Methodist 45%, Anglican 21%, Church of God 7%, Seventh-Day Adventist 5%, Baptist 4%, Jehovah's Witnesses 2%, other 2%), Roman Catholic 6%, none 2%, other 6% (1981)
Languages: English (official)
Literacy: age 15 and over can read and write (1970)
total population: 98%
male: 98%
female: 98%
Labor force: 4,911 (1980)
by occupation: NA

Government

Names:
conventional long form: none
conventional short form: British Virgin Islands
Abbreviation: BVI
Digraph: VI
Type: dependent territory of the UK
Capital: Road Town
Administrative divisions: none (dependent territory of the UK)
Independence: none (dependent territory of the UK)
National holiday: Territory Day, 1 July
Constitution: 1 June 1977
Legal system: English law
Suffrage: 18 years of age; universal
Executive branch:
chief of state: Queen ELIZABETH II (since 6 February 1952), represented by Governor Peter Alfred PENFOLD (since 14 October 1991)
head of government: Chief Minister H. Lavity STOUTT (since NA September 1986)
cabinet: Executive Council; appointed by the governor
Legislative branch: unicameral
Legislative Council: election last held 20 February 1995 (next to be held on NA February 2000); results—percent of vote by party NA; seats—(13 total) VIP 6, CCM 2, UP 2, independents 3
note: legislature was expanded to 13 seats as of election on 20 February 1995
Judicial branch: Eastern Caribbean Supreme Court
Political parties and leaders: United Party (UP), Conrad MADURO; Virgin Islands Party (VIP), H. Lavity STOUTT; Concerned Citizens Movement (CCM), E. Walwyln BREWLEY
Member of: CARICOM (associate), CDB, ECLAC (associate), INTERPOL (subbureau), IOC, OECS (associate), UNESCO (associate)
Diplomatic representation in US: none (dependent territory of the UK)
US diplomatic representation: none (dependent territory of the UK)
Flag: blue with the flag of the UK in the upper hoist-side quadrant and the Virgin Islander coat of arms centered in the outer half of the flag; the coat of arms depicts a woman flanked on either side by a vertical column of six oil lamps above a scroll bearing the Latin word VIGILATE (Be Watchful)

Economy

Overview: The economy, one of the most prosperous in the Caribbean area, is highly dependent on the tourist industry, which generates about 21% of the national income. In 1985 the government offered offshore registration to companies wishing to incorporate in the islands, and, in consequence, incorporation fees generated about $2 million in 1987. The economy slowed in 1991 because of the poor performances of the tourist sector and tight commercial bank credit. Livestock raising is the most significant agricultural activity. The islands' crops, limited by poor soils, are unable to meet food requirements.
National product: GDP—purchasing power parity—$133 million (1991)
National product real growth rate: 2% (1991)
National product per capita: $10,600 (1991)
Inflation rate (consumer prices): 2.5% (1990 est.)
Unemployment rate: NEGL% (1992)
Budget:
revenues: $51 million
expenditures: $88 million, including capital expenditures of $38 million (1991)
Exports: $2.7 million (f.o.b., 1988)
commodities: rum, fresh fish, gravel, sand, fruits, animals
partners: Virgin Islands (US), Puerto Rico, US
Imports: $11.5 million (c.i.f., 1988)
commodities: building materials, automobiles, foodstuffs, machinery
partners: Virgin Islands (US), Puerto Rico, US
External debt: $4.5 million (1985)
Industrial production: growth rate 4% (1985)
Electricity:
capacity: 10,500 kW
production: 50 million kWh
consumption per capita: 3,148 kWh (1993)
Industries: tourism, light industry, construction, rum, concrete block, offshore financial center
Agriculture: livestock (including poultry), fish, fruit, vegetables
Economic aid: $NA
Currency: 1 United States dollar (US$) = 100 cents
Exchange rates: US currency is used
Fiscal year: 1 April—31 March

Transportation

Railroads: 0 km
Highways:
total: 106 km (1983)
paved: NA
unpaved: NA
Ports: Road Town
Merchant marine: none
Airports:
total: 3
with paved runways 914 to 1,523 m: 1
with paved runways under 914 m: 1
with unpaved runways 914 to 1,523 m: 1

Communications

Telephone system: 3,000 telephones; worldwide external telephone service
local: NA
intercity: NA
international: submarine cable communication links to Bermuda
Radio:
broadcast stations: AM 1, FM 0, shortwave 0
radios: NA
Television:
broadcast stations: 1
televisions: NA

Defense Forces

Note: defense is the responsibility of the UK

Brunei

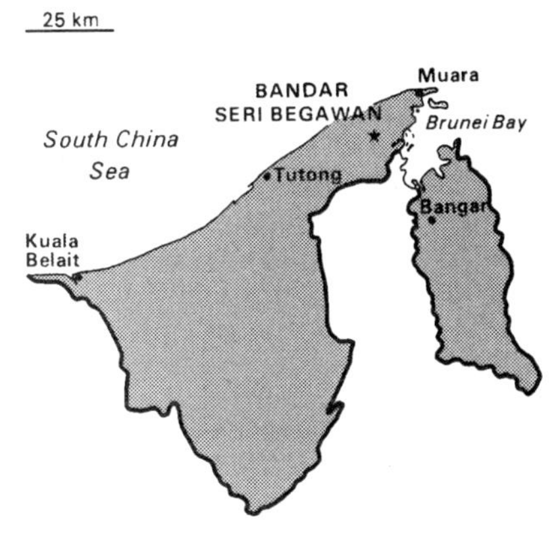

Geography

Location: Southeastern Asia, bordering the South China Sea and Malaysia
Map references: Southeast Asia
Area:
total area: 5,770 sq km
land area: 5,270 sq km
comparative area: slightly larger than Delaware
Land boundaries: total 381 km, Malysia 381 km
Coastline: 161 km
Maritime claims:
exclusive economic zone: 200 nm or to median line
territorial sea: 12 nm
International disputes: may wish to purchase the Malaysian salient that divides the country; all of the Spratly Islands are claimed by China, Taiwan, and Vietnam; parts of them are claimed by Malaysia and the Philippines; in 1984, Brunei established an exclusive fishing zone that encompasses Louisa Reef, but has not publicly claimed the island
Climate: tropical; hot, humid, rainy
Terrain: flat coastal plain rises to mountains in east; hilly lowland in west
Natural resources: petroleum, natural gas, timber
Land use:
arable land: 1%
permanent crops: 1%
meadows and pastures: 1%
forest and woodland: 79%
other: 18%
Irrigated land: 10 sq km (1989 est.)
Environment:
current issues: NA
natural hazards: typhoons, earthquakes, and severe flooding are very rare
international agreements: party to—Endangered Species, Ozone Layer Protection, Ship Pollution; signed, but not ratified—Law of the Sea
Note: close to vital sea lanes through South China Sea linking Indian and Pacific Oceans; two parts physically separated by Malaysia; almost an enclave of Malaysia

People

Population: 292,266 (July 1995 est.)
Age structure:
0-14 years: 34% (female 48,458; male 50,624)
15-64 years: 62% (female 85,581; male 95,955)
65 years and over: 4% (female 5,172; male 6,476) (July 1995 est.)
Population growth rate: 2.63% (1995 est.)
Birth rate: 25.83 births/1,000 population (1995 est.)
Death rate: 5.07 deaths/1,000 population (1995 est.)
Net migration rate: 5.49 migrant(s)/1,000 population (1995 est.)
Infant mortality rate: 24.7 deaths/1,000 live births (1995 est.)
Life expectancy at birth:
total population: 71.24 years
male: 69.65 years
female: 72.91 years (1995 est.)
Total fertility rate: 3.41 children born/woman (1995 est.)
Nationality:
noun: Bruneian(s)
adjective: Bruneian
Ethnic divisions: Malay 64%, Chinese 20%, other 16%
Religions: Muslim (official) 63%, Buddhism 14%, Christian 8%, indigenous beliefs and other 15% (1981)
Languages: Malay (official), English, Chinese
Literacy: age 15 and over can read and write (1991)
total population: 88%
male: 92%
female: 82%
Labor force: 119,000 (1993 est.); note—includes members of the Army
by occupation: government 47.5%, production of oil, natural gas, services, and construction 41.9%, agriculture, forestry, and fishing 3.8% (1986)
note: 33% of labor force is foreign (1988)

Government

Names:
conventional long form: Negara Brunei Darussalam
conventional short form: Brunei
Digraph: BX
Type: constitutional sultanate
Capital: Bandar Seri Begawan
Administrative divisions: 4 districts (daerah-daerah, singular—daerah); Belait, Brunei and Muara, Temburong, Tutong
Independence: 1 January 1984 (from UK)
National holiday: National Day 23 February (1984)
Constitution: 29 September 1959 (some provisions suspended under a State of Emergency since December 1962, others since independence on 1 January 1984)
Legal system: based on Islamic law
Suffrage: none
Executive branch:
chief of state and head of government: Sultan and Prime Minister His Majesty Paduka Seri Baginda Sultan Haji HASSANAL Bolkiah Mu'izzaddin Waddaulah (since 5 October 1967)
cabinet: Council of Cabinet Ministers; composed chiefly of members of the royal family
Legislative branch: unicameral
Legislative Council (Majlis Masyuarat Megeri): elections last held in March 1962; in 1970 the Council was changed to an appointive body by decree of the sultan; an elected legislative Council is being considered as part of constitution reform, but elections are unlikely for several years
Judicial branch: Supreme Court
Political parties and leaders: Brunei United National Party (inactive), Anak HASANUDDIN, chairman; Brunei National Solidarity Party (the first legal political party and now banned), leader NA; Brunei Peoples Party (banned), leader NA
Member of: APEC, ASEAN, C, ESCAP, FAO, G-77, GATT, ICAO, IDB, IMO, INMARSAT, INTELSAT (nonsignatory user), INTERPOL, IOC, ISO (correspondent), ITU, NAM, OIC, UN, UNCTAD, UPU, WHO, WIPO, WMO
Diplomatic representation in US:
chief of mission: Ambassador Haji JAYA bin Abdul Latif
chancery: Watergate, Suite 300, 3rd floor, 2600 Virginia Avenue NW, Washington, DC 20037
telephone: [1] (202) 342-0159
FAX: [1] (202) 342-0158
US diplomatic representation:
chief of mission: Ambassador Theresa A. TULL
embassy: Third Floor, Teck Guan Plaza, Jalan Sultan, Bandar Seri Begawan
mailing address: American Embassy Box B, APO AP 96440
telephone: [673] (2) 229670
FAX: [673] (2) 225293
Flag: yellow with two diagonal bands of white (top, almost double width) and black starting from the upper hoist side; the national emblem in red is superimposed at the center; the emblem includes a swallow-tailed flag on top of a winged column within an upturned crescent above a scroll and flanked by two upraised hands

Economy

Overview: The economy is a mixture of

foreign and domestic entrepreneurship, government regulation and welfare measures, and village tradition. It is almost totally supported by exports of crude oil and natural gas, with revenues from the petroleum sector accounting for more than 40% of GDP. Per capita GDP is among the highest in the Third World, and substantial income from overseas investment supplements domestic production. The government provides for all medical services and subsidizes food and housing.
National product: GDP—purchasing power parity—$4.43 billion (1993 est.)
National product real growth rate: -4% (1993 est.)
National product per capita: $16,000 (1993 est.)
Inflation rate (consumer prices): 2.5% (1993 est.)
Unemployment rate: 5% (1993 est.)
Budget:
revenues: $1.5 billion
expenditures: $1.5 billion, including capital expenditures of $255 million (1990 est.)
Exports: $2.2 billion (f.o.b., 1993 est.)
commodities: crude oil, liquefied natural gas, petroleum products
partners: Japan 52%, South Korea 10%, UK 9%, Thailand 7%, Singapore 6% (1991)
Imports: $1.2 billion (c.i.f., 1993 est.)
commodities: machinery and transport equipment, manufactured goods, food, chemicals
partners: Singapore 34%, UK 23%, US 10%, Japan 8%, Malaysia 7%, Switzerland 4% (1991)
External debt: $0
Industrial production: growth rate 12.9% (1987); accounts for 41.6% of GDP (1990), includes mining, quarrying, and manufacturing
Electricity:
capacity: 380,000 kW
production: 1.2 billion kWh
consumption per capita: 3,971 kWh (1993)
Industries: petroleum, petroleum refining, liquefied natural gas, construction
Agriculture: imports about 80% of its food needs; principal crops and livestock include rice, cassava, bananas, buffaloes, and pigs
Economic aid:
recipient: US commitments, including Ex-Im (FY70-87), $20.6 million; Western (non-US) countries, ODA and OOF bilateral commitments (1970-89), $153 million
Currency: 1 Bruneian dollar (B$) = 100 cents
Exchange rates: Bruneian dollars (B$) per US$1—1.4524 (January 1995), 1.5274 (1994), 1.6158 (1993), 1.6290 (1992), 1.7276 (1991), 1.8125 (1990); note—the Bruneian dollar is at par with the Singapore dollar
Fiscal year: calendar year

Transportation

Railroads:
total: 13 km private line
narrow gauge: 13 km 0.610-m gauge
Highways:
total: 1,090 km
paved: bituminous 370 km (with another 52 km under construction)
unpaved: gravel or earth 720 km
Inland waterways: 209 km; navigable by craft drawing less than 1.2 meters
Pipelines: crude oil 135 km; petroleum products 418 km; natural gas 920 km
Ports: Bandar Seri Begawar, Kuala Belait, Muara, Seria, Tutong
Merchant marine:
total: 7 liquefied gas carriers (1,000 GRT or over) totaling 348,476 GRT/340,635 DWT
Airports:
total: 5
with paved runways over 3,047 m: 1
with paved runways under 914 m: 3
with unpaved runways 914 to 1,523 m: 1

Communications

Telephone system: 33,000 telephones (1987); service throughout country is adequate for present needs; international service good to adjacent Malaysia
local: NA
intercity: NA
international: INTELSAT (NA Indian Ocean and 1 Pacific Ocean) earth stations
Radio:
broadcast stations: AM 4, FM 4, shortwave 0
radios: 74,000 (1987)
note: radiobroadcast coverage good
Television:
broadcast stations: 1
televisions: NA

Defense Forces

Branches: Land Forces, Navy, Air Force, Royal Brunei Police
Manpower availability: males age 15-49 81,560; males fit for military service 47,403; males reach military age (18) annually 2,835 (1995 est.)
Defense expenditures: exchange rate conversion—$312 million, 6.2% of GDP (1994)

Bulgaria

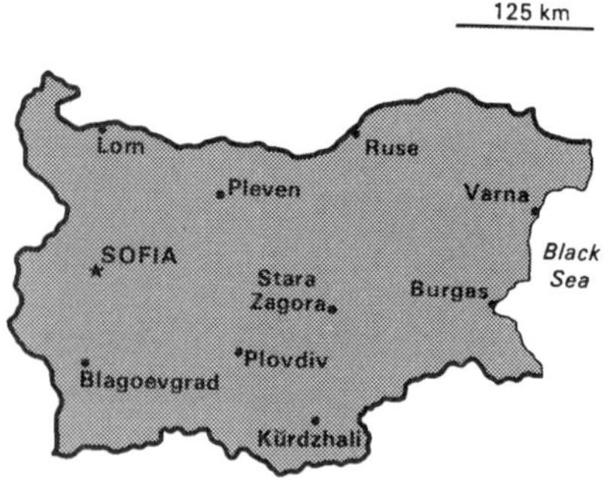

Geography

Location: Southeastern Europe, bordering the Black Sea, between Romania and Turkey
Map references: Ethnic Groups in Eastern Europe, Europe
Area:
total area: 110,910 sq km
land area: 110,550 sq km
comparative area: slightly larger than Tennessee
Land boundaries: total 1,808 km, Greece 494 km, The Former Yugoslav Republic of Macedonia 148 km, Romania 608 km, Serbia and Montenegro 318 km (all with Serbia), Turkey 240 km
Coastline: 354 km
Maritime claims:
contiguous zone: 24 nm
exclusive economic zone: 200 nm
territorial sea: 12 nm
International disputes: none
Climate: temperate; cold, damp winters; hot, dry summers
Terrain: mostly mountains with lowlands in north and southeast
Natural resources: bauxite, copper, lead, zinc, coal, timber, arable land
Land use:
arable land: 34%
permanent crops: 3%
meadows and pastures: 18%
forest and woodland: 35%
other: 10%
Irrigated land: 10 sq km (1989 est.)
Environment:
current issues: air pollution from industrial emissions; rivers polluted from raw sewage, heavy metals, detergents; deforestation; forest damage from air pollution and resulting acid rain; soil contamination from heavy metals from metallurgical plants and industrial wastes
natural hazards: earthquakes, landslides
international agreements: party to—Air Pollution, Air Pollution-Nitrogen Oxides, Air

Pollution-Sulphur 85, Antarctic Treaty, Endangered Species, Environmental Modification, Nuclear Test Ban, Ozone Layer Protection, Ship Pollution, Wetlands; signed, but not ratified—Air Pollution-Sulphur 94, Air Pollution-Volatile Organic Compounds, Antarctic-Environmental Protocol, Biodiversity, Climate Change, Law of the Sea
Note: strategic location near Turkish Straits; controls key land routes from Europe to Middle East and Asia

People

Population: 8,775,198 (July 1995 est.)
Age structure:
0-14 years: 19% (female 800,413; male 841,697)
15-64 years: 66% (female 2,927,880; male 2,910,133)
65 years and over: 15% (female 735,706; male 559,369) (July 1995 est.)
Population growth rate: -0.25% (1995 est.)
Birth rate: 11.75 births/1,000 population (1995 est.)
Death rate: 11.31 deaths/1,000 population (1995 est.)
Net migration rate: -2.91 migrant(s)/1,000 population (1995 est.)
Infant mortality rate: 11.4 deaths/1,000 live births (1995 est.)
Life expectancy at birth:
total population: 73.68 years
male: 70.43 years
female: 77.1 years (1995 est.)
Total fertility rate: 1.71 children born/woman (1995 est.)
Nationality:
noun: Bulgarian(s)
adjective: Bulgarian
Ethnic divisions: Bulgarian 85.3%, Turk 8.5%, Gypsy 2.6%, Macedonian 2.5%, Armenian 0.3%, Russian 0.2%, other 0.6%
Religions: Bulgarian Orthodox 85%, Muslim 13%, Jewish 0.8%, Roman Catholic 0.5%, Uniate Catholic 0.2%, Protestant, Gregorian-Armenian, and other 0.5%
Languages: Bulgarian; secondary languages closely correspond to ethnic breakdown
Literacy: age 15 and over can read and write (1992)
total population: 98%
male: 99%
female: 97%
Labor force: 4.3 million
by occupation: industry 33%, agriculture 20%, other 47% (1987)

Government

Names:
conventional long form: Republic of Bulgaria
conventional short form: Bulgaria
Digraph: BU
Type: emerging democracy
Capital: Sofia
Administrative divisions: 9 provinces (oblasti, singular—oblast); Burgas, Grad Sofiya, Khaskovo, Lovech, Montana, Plovdiv, Ruse, Sofiya, Varna
Independence: 22 September 1908 (from Ottoman Empire)
National holiday: Independence Day 3 March (1878)
Constitution: adopted 12 July 1991
Legal system: based on civil law system, with Soviet law influence; has accepted compulsory ICJ jurisdiction
Suffrage: 18 years of age; universal and compulsory
Executive branch:
chief of state: President Zhelyu Mitev ZHELEV (since 1 August 1990); Vice President (vacant); election last held January 1992; results—Zhelyu ZHELEV was elected by popular vote
head of government: Chairman of the Council of Ministers (Prime Minister) Zhan VIDENOV (since 25 January 1995); Deputy Prime Ministers Doncho KONAKCHIEV, Kiril TSOCHEV, Rumen GECHEV, Svetoslav SHIVAROV (since 25 January 1995)
cabinet: Council of Ministers; elected by the National Assembly
Legislative branch: unicameral
National Assembly (Narodno Sobranie): last held 18 December 1994 (next to be held NA 1997); results—BSP 43.5%, UDF 24.2%, PU 6.5%, MRF 5.4%, BBB 4.7%; seats—(240 total) BSP 125, UDF 69, PU 18, MRF 15, BBB 13
Judicial branch: Supreme Court, Constitutional Court
Political parties and leaders: Bulgarian Socialist Party (BSP), Zhan VIDENOV, chairman; Union of Democratic Forces (UDF), Ivan KOSTOV—an alliance of pro-Democratic parties; People's Union (PU), Stefan SAVOV; Movement for Rights and Freedoms (mainly ethnic Turkish party) (MRF), Ahmed DOGAN; Bulgarian Business Bloc (BBB), George GANCHEV
Other political or pressure groups: Democratic Alliance for the Republic (DAR); New Union for Democracy (NUD); Ecoglasnost; Podkrepa Labor Confederation; Fatherland Union; Bulgarian Communist Party (BCP); Confederation of Independent Trade Unions of Bulgaria (KNSB); Bulgarian Agrarian National Union—United (BZNS); Bulgarian Democratic Center; "Nikola Petkov" Bulgarian Agrarian National Union; Internal Macedonian Revolutionary Organization—Union of Macedonian Societies (IMRO-UMS); numerous regional, ethnic, and national interest groups with various agendas
Member of: ACCT, BIS, BSEC, CCC, CE, CEI (associate members), EBRD, ECE, FAO, G- 9, IAEA, IBRD, ICAO, ICFTU, ICRM, IFC, IFRCS, ILO, IMF, IMO, INMARSAT, INTELSAT (nonsignatory user), INTERPOL, IOC, IOM (observer), ISO, ITU, NACC, NAM (guest), NSG, OSCE, PCA, PFP, UN, UNCTAD, UNESCO, UNIDO, UPU, WEU (associate partner), WFTU, WHO, WIPO, WMO, WTO, ZC
Diplomatic representation in US:
chief of mission: Ambassador Snezhana Damianova BOTUSHAROVA
chancery: 1621 22nd Street NW, Washington, DC 20008
telephone: [1] (202) 387-7969
FAX: [1] (202) 234-7973
US diplomatic representation:
chief of mission: Ambassador William D. MONTGOMERY
embassy: 1 Saborna Street, Sofia
mailing address: Unit 1335, Sofia; APO AE 09213-1335
telephone: [359] (2) 88-48-01 through 05
FAX: [359] (2) 80-19-77
Flag: three equal horizontal bands of white (top), green, and red; the national emblem formerly on the hoist side of the white stripe has been removed—it contained a rampant lion within a wreath of wheat ears below a red five-pointed star and above a ribbon bearing the dates 681 (first Bulgarian state established) and 1944 (liberation from Nazi control)

Economy

Overview: The Bulgarian economy continued its painful adjustment in 1994 from the misdirected development undertaken during four decades of Communist rule. Many aspects of a market economy have been put in place and have begun to function, but much of the economy, especially the industrial sector, has yet to re-establish market links lost with the collapse of the other centrally planned Soviet Bloc economies. The prices of many imported industrial inputs, especially energy products, have risen markedly, and falling real wages have not sufficed to restore competitiveness. The government plans more extensive privatization in 1995 to improve the management of enterprises and to encourage foreign investment. Bulgaria resumed payments on its $10 billion in commercial debt in 1993 following the negotiation of a 50% write-off.
National product: GDP—purchasing power parity—$33.7 billion (1994 est.)
National product real growth rate: 0.2% (1994 est.)
National product per capita: $3,830 (1994 est.)
Inflation rate (consumer prices): 122% (1994)
Unemployment rate: 16% (1994)
Budget:
revenues: $14 billion
expenditures: $17.4 billion, including capital expenditures of $610 million (1993 est.)
Exports: $3.6 billion (f.o.b., 1993)

commodities: machinery and equipment 30.6%; agricultural products 24%; manufactured consumer goods 22.2%; fuels, minerals, raw materials, and metals 10.5%; other 12.7% (1991)
partners: former CEMA countries 57.7% (FSU 48.6%, Poland 2.1%, Czechoslovakia 0.9%); developed countries 26.3% (Germany 4.8%, Greece 2.2%); less developed countries 15.9% (Libya 2.1%, Iran 0.7%) (1991)
Imports: $4.3 billion (c.i.f., 1993)
commodities: fuels, minerals, and raw materials 58.7%; machinery and equipment 15.8%; manufactured consumer goods 4.4%; agricultural products 15.2%; other 5.9%
partners: former CEMA countries 51.0% (FSU 43.2%, Poland 3.7%); developed countries 32.8% (Germany 7.0%, Austria 4.7%); less developed countries 16.2% (Iran 2.8%, Libya 2.5%)
External debt: $12 billion (1994)
Industrial production: growth rate 4% (1994); accounts for about 37% of GDP (1990)
Electricity:
capacity: 11,500,000 kW
production: 35.9 billion kWh
consumption per capita: 3,827 kWh (1993)
Industries: machine building and metal working, food processing, chemicals, textiles, building materials, ferrous and nonferrous metals
Agriculture: climate and soil conditions support livestock raising and the growing of various grain crops, oilseeds, vegetables, fruits, and tobacco; more than one-third of the arable land devoted to grain; world's fourth-largest tobacco exporter; surplus food producer
Illicit drugs: transshipment point for southwest Asian heroin and South American cocaine transiting the Balkan route; limited producer of precursor chemicals
Economic aid:
recipient: $700 million in balance of payments support (1994)
Currency: 1 lev (Lv) = 100 stotinki
Exchange rates: leva (Lv) per US$1—67.04 (January 1995), 32.00 (January 1994), 24.56 (January 1993), 17.18 (January 1992), 16.13 (March 1991), 0.7446 (November 1990); note—floating exchange rate since February 1991
Fiscal year: calendar year

Transportation

Railroads:
total: 4,294 km
standard gauge: 4,049 km 1.435-m gauge (2,650 km electrified; 917 double track)
other: 245 km NA-m gauge (1994)
Highways:
total: 36,932 km
paved: 33,904 km (including 276 km expressways)
unpaved: earth 3,028 km (1992)
Inland waterways: 470 km (1987)
Pipelines: crude oil 193 km; petroleum products 525 km; natural gas 1,400 km (1992)
Ports: Burgas, Lom, Nesebur, Ruse, Varna, Vidin
Merchant marine:
total: 109 ships (1,000 GRT or over) totaling 1,191,231 GRT/1,762,461 DWT
ships by type: bulk 47, cargo 29, chemical carrier 4, container 2, oil tanker 15, passenger-cargo 2, railcar carrier 2, roll-on/roll-off cargo 6, short-sea passenger 1, refrigerated cargo 1
note: Bulgaria owns 2 ships (1,000 GRT or over) totaling 12,960 DWT operating under Liberian registry
Airports:
total: 355
with paved runways over 3,047 m: 1
with paved runways 2,438 to 3,047 m: 17
with paved runways 1,524 to 2,437 m: 10
with paved runways under 914 m: 88
with unpaved runways 2,438 to 3,047 m: 2
with unpaved runways 1,524 to 2,438 m: 1
with unpaved runways 914 to 1,523 m: 10
with unpaved runways under 914 m: 226

Communications

Telephone system: 2,600,000 telephones; 29 telephones/100 persons (1992); extensive but antiquated transmission system of coaxial cable and microwave radio relay; direct dialing to 36 countries; telephone service is available in most villages; almost two-thirds of the lines are residential; 67% of Sofia households have phones (November 1988)
local: NA
intercity: NA
international: 1 earth station using Intersputnik; INTELSAT link used through a Greek earth station
Radio:
broadcast stations: AM 20, FM 15, shortwave 0
radios: NA
Television:
broadcast stations: 29 (Russian repeater in Sofia 1)
televisions: 2.1 million (May 1990)

Defense Forces

Branches: Army, Navy, Air and Air Defense Forces, Border Troops, Internal Troops
Manpower availability: males age 15-49 2,171,414; males fit for military service 1,810,989; males reach military age (19) annually 69,200 (1995 est.)
Defense expenditures: 13 billion leva, NA% of GDP (1994 est.); note—conversion of defense expenditures into US dollars using the current exchange rate could produce misleading results

Burkina

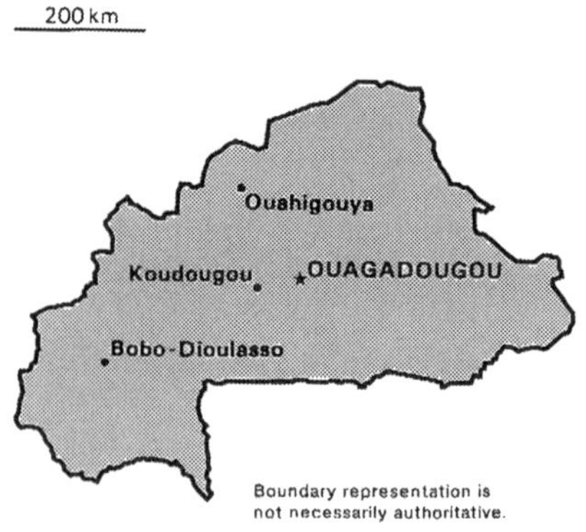

Geography

Location: Western Africa, north of Ghana
Map references: Africa
Area:
total area: 274,200 sq km
land area: 273,800 sq km
comparative area: slightly larger than Colorado
Land boundaries: total 3,192 km, Benin 306 km, Ghana 548 km, Cote d'Ivoire 584 km, Mali 1,000 km, Niger 628 km, Togo 126 km
Coastline: 0 km (landlocked)
Maritime claims: none; landlocked
International disputes: following mutual acceptance of an International Court of Justice (ICJ) ruling in December 1986 on their international boundary dispute, Burkina and Mali are proceeding with boundary demarcation, including the tripoint with Niger
Climate: tropical; warm, dry winters; hot, wet summers
Terrain: mostly flat to dissected, undulating plains; hills in west and southeast
Natural resources: manganese, limestone, marble; small deposits of gold, antimony, copper, nickel, bauxite, lead, phosphates, zinc, silver
Land use:
arable land: 10%
permanent crops: 0%
meadows and pastures: 37%
forest and woodland: 26%
other: 27%
Irrigated land: 160 sq km (1989 est.)
Environment:
current issues: recent droughts and desertification severely affecting agricultural activities, population distribution, and the economy; overgrazing; soil degradation; deforestation
natural hazards: recurring droughts
international agreements: party to—Biodiversity, Climate Change, Endangered Species, Marine Life Conservation, Ozone Layer Protection, Wetlands; signed, but not ratified—Desertification, Law of the Sea,

Nuclear Test Ban
Note: landlocked

People

Population: 10,422,828 (July 1995 est.)
Age structure:
0-14 years: 48% (female 2,488,662; male 2,517,245)
15-64 years: 49% (female 2,707,601; male 2,378,957)
65 years and over: 3% (female 184,578; male 145,785) (July 1995 est.)
Population growth rate: 2.79% (1995 est.)
Birth rate: 48.05 births/1,000 population (1995 est.)
Death rate: 18.22 deaths/1,000 population (1995 est.)
Net migration rate: -1.9 migrant(s)/1,000 population (1995 est.)
Infant mortality rate: 116.9 deaths/1,000 live births (1995 est.)
Life expectancy at birth:
total population: 46.6 years
male: 45.71 years
female: 47.51 years (1995 est.)
Total fertility rate: 6.88 children born/woman (1995 est.)
Nationality:
noun: Burkinabe (singular and plural)
adjective: Burkinabe
Ethnic divisions: Mossi (about 2.5 million), Gurunsi, Senufo, Lobi, Bobo, Mande, Fulani
Religions: indigenous beliefs 40%, Muslim 50%, Christian (mainly Roman Catholic) 10%
Languages: French (official), tribal languages belonging to Sudanic family, spoken by 90% of the population
Literacy: age 15 and over can read and write (1990 est.)
total population: 18%
male: 28%
female: 9%
Labor force: NA (most adults are employed in subsistance agriculture)
by occupation: agriculture 80%, industry 15%, commerce, services, and government 5%
note: 20% of male labor force migrates annually to neighboring countries for seasonal employment (1984)

Government

Names:
conventional long form: Burkina Faso
conventional short form: Burkina
former: Upper Volta
Digraph: UV
Type: parliamentary
Capital: Ouagadougou
Administrative divisions: 30 provinces; Bam, Bazega, Bougouriba, Boulgou, Boulkiemde, Ganzourgou, Gnagna, Gourma, Houet, Kadiogo, Kenedougou, Komoe, Kossi, Kouritenga, Mouhoun, Namentenga, Naouri, Oubritenga, Oudalan, Passore, Poni, Sanguie, Sanmatenga, Seno, Sissili, Soum, Sourou, Tapoa, Yatenga, Zoundweogo
Independence: 5 August 1960 (from France)
National holiday: Anniversary of the Revolution, 4 August (1983)
Constitution: 2 June 1991
Legal system: based on French civil law system and customary law
Suffrage: none
Executive branch:
chief of state: President Captain Blaise COMPAORE (since 15 October 1987); election last held December 1991
head of government: Prime Minister Roch KABORE (since March 1994)
cabinet: Council of Ministers; appointed by the president
Legislative branch: unicameral
Assembly of People's Deputies: elections last held 24 May 1992 (next to be held 1997); results—percent of vote by party NA; seats—(107 total), ODP-MT 78, CNPP-PSD 12, RDA 6, ADF 4, other 7
note: the current law also provides for a second consultative chamber, which has not been formally constituted
Judicial branch: Appeals Court
Political parties and leaders: Organization for People's Democracy—Labor Movement (ODP-MT), ruling party, Simon COMPAORE, Secretary General; National Convention of Progressive Patriots-Social Democratic Party (CNPP-PSD), Moussa BOLY; African Democratic Rally (RDA), Gerard Kango OUEDRAOGO; Alliance for Democracy and Federation (ADF), Amadou Michel NANA
Other political or pressure groups: committees for the defense of the revolution; watchdog/political action groups throughout the country in both organizations and communities
Member of: ACCT, ACP, AfDB, CCC, CEAO, ECA, ECOWAS, Entente, FAO, FZ, G-77, GATT, IBRD, ICAO, ICC, ICFTU, ICRM, IDA, IDB, IFAD, IFC, IFRCS, ILO, IMF, INTELSAT, INTERPOL, IOC, ITU, NAM, OAU, OIC, PCA, UN, UNCTAD, UNESCO, UNIDO, UPU, WADB, WCL, WFTU, WHO, WIPO, WMO, WTO
Diplomatic representation in US:
chief of mission: Ambassador Gaetan R. OUEDRAOGO
chancery: 2340 Massachusetts Avenue NW, Washington, DC 20008
telephone: [1] (202) 332-5577, 6895
US diplomatic representation:
chief of mission: Ambassador Donald J. McCONNELL
embassy: Avenue Raoul Follerau, Ouagadougou
mailing address: 01 B. P. 35, Ouagadougou
telephone: [226] 306723 through 306725
FAX: [226] 312368
Flag: two equal horizontal bands of red (top) and green with a yellow five-pointed star in the center; uses the popular pan-African colors of Ethiopia

Economy

Overview: One of the poorest countries in the world, Burkina has a high population density and a high population growth rate, few natural resources, and a fragile soil. Economic development is hindered by a poor communications network within a landlocked country. Agriculture provides about 40% of GDP and is mainly of a subsistence nature. Industry, dominated by unprofitable government-controlled corporations, accounts for about 15% of GDP. Following the 50% currency devaluation in January 1994, the government updated its development program in conjunction with international agencies. Even with the best of plans, however, the government faces formidable problems on all sides.
National product: GDP—purchasing power parity—$6.5 billion (1993 est.)
National product real growth rate: 0.4% (1993 est.)
National product per capita: $660 (1993 est.)
Inflation rate (consumer prices): -0.6% (1993 est.)
Unemployment rate: NA%
Budget:
revenues: $483 million
expenditures: $548 million, including capital expenditures of $189 million (1992)
Exports: $273 million (f.o.b., 1993)
commodities: cotton, gold, animal products
partners: EC 42%, Cote d'Ivoire 11%, Taiwan 15% (1992)
Imports: $636 million (f.o.b., 1993)
commodities: machinery, food products, petroleum
partners: EC 49%, Africa 24%, Japan 6% (1992)
External debt: $865 million (December 1991 est.)
Industrial production: growth rate 6.7% (1992); accounts for about 15% of GDP
Electricity:
capacity: 60,000 kW
production: 190 million kWh
consumption per capita: 17 kWh (1993)
Industries: cotton lint, beverages, agricultural processing, soap, cigarettes, textiles, gold mining and extraction
Agriculture: accounts for about 40% of GDP; cash crops—peanuts, shea nuts, sesame, cotton; food crops—sorghum, millet, corn, rice; livestock; not self sufficient in food grains
Economic aid:
recipient: US commitments, including Ex-Im (FY70-89), $294 million; Western (non-US) countries, ODA and OOF bilateral commitments (1970-89), $2.9 billion;

Communist countries (1970-89), $113 million
Currency: 1 CFA franc (CFAF) = 100 centimes
Exchange rates: CFA francs (CFAF) per US$1—529.43 (January 1995), 555.20 (1995), 283.16 (1993), 264.69 (1992), 282.11 (1991), 272.26 (1990)
note: beginning 12 January 1994 the CFA franc was devalued to CFAF 100 per French franc from CFAF 50 at which it had been fixed since 1948
Fiscal year: calendar year

Transportation

Railroads:
total: 620 km (520 km Ouagadougou to Cote d'Ivoire border and 100 km Ouagadougou to Kaya; single track)
narrow gauge: 620 km 1.000-m gauge
Highways:
total: 16,500 km
paved: 1,300 km
unpaved: improved earth 7,400 km; unimproved earth 7,800 km (1985)
Ports: none
Airports:
total: 48
with paved runways over 3,047 m: 1
with paved runways 2,438 to 3,047 m: 1
with paved runways under 914 m: 26
with unpaved runways 1,524 to 2,438 m: 4
with unpaved runways 914 to 1,523 m: 16

Communications

Telephone system: NA telephones; all services only fair
local: NA
intercity: microwave radio relay, wire, and radio communication stations
international: 1 Atlantic Ocean INTELSAT earth station
Radio:
broadcast stations: AM 2, FM 1, shortwave 0
radios: NA
Television:
broadcast stations: 2
televisions: NA

Defense Forces

Branches: Army, Air Force, National Gendarmerie, National Police, People's Militia
Manpower availability: males age 15-49 2,081,999; males fit for military service 1,065,605 (1995 est.)
Defense expenditures: exchange rate conversion—$104 million, 6.4% of GDP (1994)

Burma

Geography

Location: Southeastern Asia, bordering the Andaman Sea and the Bay of Bengal, between Bangladesh and Thailand
Map references: Southeast Asia
Area:
total area: 678,500 sq km
land area: 657,740 sq km
comparative area: slightly smaller than Texas
Land boundaries: total 5,876 km, Bangladesh 193 km, China 2,185 km, India 1,463 km, Laos 235 km, Thailand 1,800 km
Coastline: 1,930 km
Maritime claims:
contiguous zone: 24 nm
continental shelf: 200 nm or to the edge of the continental margin
exclusive economic zone: 200 nm
territorial sea: 12 nm
International disputes: none
Climate: tropical monsoon; cloudy, rainy, hot, humid summers (southwest monsoon, June to September); less cloudy, scant rainfall, mild temperatures, lower humidity during winter (northeast monsoon, December to April)
Terrain: central lowlands ringed by steep, rugged highlands
Natural resources: petroleum, timber, tin, antimony, zinc, copper, tungsten, lead, coal, some marble, limestone, precious stones, natural gas
Land use:
arable land: 15%
permanent crops: 1%
meadows and pastures: 1%
forest and woodland: 49%
other: 34%
Irrigated land: 10,180 sq km (1989)
Environment:
current issues: deforestation; industrial pollution of air, soil, and water; inadequate sanitation and water treatment contribute to disease
natural hazards: destructive earthquakes and cyclones; flooding and landslides common during rainy season (June to September); periodic droughts
international agreements: party to—Biodiversity, Climate Change, Nuclear Test Ban, Ozone Layer Protection, Ship Pollution, Tropical Timber 83; signed, but not ratified—Law of the Sea
Note: strategic location near major Indian Ocean shipping lanes

People

Population: 45,103,809 (July 1995 est.)
Age structure:
0-14 years: 36% (female 7,963,544; male 8,285,459)
15-64 years: 60% (female 13,478,211; male 13,404,987)
65 years and over: 4% (female 1,080,922; male 890,686) (July 1995 est.)
Population growth rate: 1.84% (1995 est.)
Birth rate: 28.02 births/1,000 population (1995 est.)
Death rate: 9.63 deaths/1,000 population (1995 est.)
Net migration rate: 0 migrant(s)/1,000 population (1995 est.)
Infant mortality rate: 61.6 deaths/1,000 live births (1995 est.)
Life expectancy at birth:
total population: 60.47 years
male: 58.38 years
female: 62.69 years (1995 est.)
Total fertility rate: 3.58 children born/woman (1995 est.)
Nationality:
noun: Burmese (singular and plural)
adjective: Burmese
Ethnic divisions: Burman 68%, Shan 9%, Karen 7%, Rakhine 4%, Chinese 3%, Mon 2%, Indian 2%, other 5%
Religions: Buddhist 89%, Christian 4% (Baptist 3%, Roman Catholic 1%), Muslim 4%, animist beliefs 1%, other 2%
Languages: Burmese; minority ethnic groups have their own languages
Literacy: age 15 and over can read and write (1990 est.)
total population: 81%
male: 89%
female: 72%
Labor force: 16.007 million (1992)
by occupation: agriculture 65.2%, industry 14.3%, trade 10.1%, government 6.3%, other 4.1% (FY88/89 est.)

Government

Names:
conventional long form: Union of Burma
conventional short form: Burma
local long form: Pyidaungzu Myanma Naingngandaw (translated by the US Government as Union of Myanma and by the Burmese as Union of Myanmar)

local short form: Myanma Naingngandaw
former: Socialist Republic of the Union of Burma
Digraph: BM
Type: military regime
Capital: Rangoon (regime refers to the capital as Yangon)
Administrative divisions: 7 divisions* (yin-mya, singular—yin) and 7 states (pyine-mya, singular—pyine); Chin State, Ayeyarwady*, Bago*, Kachin State, Kayin State, Kayah State, Magway*, Mandalay*, Mon State, Rakhine State, Sagaing*, Shan State, Tanintharyi*, Yangon*
Independence: 4 January 1948 (from UK)
National holiday: Independence Day, 4 January (1948)
Constitution: 3 January 1974 (suspended since 18 September 1988); National Convention started on 9 January 1993 to draft a new constitution; chapter headings and three of 15 sections have been approved
Legal system: has not accepted compulsory ICJ jurisdiction
Suffrage: 18 years of age; universal
Executive branch:
chief of state and head of government: Chairman of the State Law and Order Restoration Council Gen. THAN SHWE (since 23 April 1992)
State Law and Order Restoration Council: military junta which assumed power 18 September 1988
Legislative branch:
People's Assembly (Pyithu Hluttaw): election last held 27 May 1990, but Assembly never convened; results—NLD 80%; seats—(485 total) NLD 396, the regime-favored NUP 10, other 79; was dissolved after the coup of 18 September 1988
Judicial branch: limited; remnants of the British-era legal system in place, but there is no guarantee of a fair public trial; the judiciary is not independent of the executive
Political parties and leaders: Union Solidarity and Development Association (USDA), THAN AUNG, Secretary; National Unity Party (NUP; proregime), THA KYAW; National League for Democracy (NLD), U AUNG SHWE; and eight other minor legal parties
Other political or pressure groups: National Coalition Government of the Union of Burma (NCGUB), headed by the elected prime minister SEIN WIN (consists of individuals legitimately elected to Parliament but not recognized by the military regime; the group fled to a border area and joined with insurgents in December 1990 to form a parallel government; Kachin Independence Army (KIA); United Wa State Army (UWSA); Karen National Union (KNU); several Shan factions, including the Mong Tai Army (MTA); All Burma Student Democratic Front (ABSDF)
Member of: AsDB, CCC, CP, ESCAP, FAO, G-77, GATT, IAEA, IBRD, ICAO, ICRM, IDA, IFAD, IFC, IFRCS, ILO, IMF, IMO, INTELSAT (nonsignatory user), INTERPOL, IOC, ITU, NAM, UN, UNCTAD, UNESCO, UNIDO, UPU, WHO, WMO
Diplomatic representation in US:
chief of mission: Ambassador U THAUNG
chancery: 2300 S Street NW, Washington, DC 20008
telephone: [1] (202) 332-9044, 9045
consulate(s) general: New York
US diplomatic representation:
chief of mission: (vacant); Charge d'Affaires Marilyn A. MEYERS
embassy: 581 Merchant Street, Rangoon (GPO 521)
mailing address: American Embassy, Box B, APO AP 96546
telephone: [95] (1) 82055, 82182 (operator assistance required)
FAX: [95] (1) 80409
Flag: red with a blue rectangle in the upper hoist-side corner bearing, all in white, 14 five-pointed stars encircling a cogwheel containing a stalk of rice; the 14 stars represent the 14 administrative divisions

Economy

Overview: Burma has a mixed economy with about 75% private activity, mainly in agriculture, light industry, and transport, and with about 25% state-controlled activity, mainly in energy, heavy industry, and foreign trade. Government policy in the last six years, 1989-94, has aimed at revitalizing the economy after four decades of tight central planning. Thus, private activity has markedly increased; foreign investment has been encouraged, so far with moderate success; and efforts continue to increase the efficiency of state enterprises. Published estimates of Burma's foreign trade are greatly understated because of the volume of black market trade. A major ongoing problem is the failure to achieve monetary and fiscal stability. Although Burma remains a poor Asian country, its rich resources furnish the potential for substantial long-term increases in income, exports, and living standards.
National product: GDP—purchasing power parity—$41.4 billion (1994 est.)
National product real growth rate: 6.4% (1994)
National product per capita: $930 (1994 est.)
Inflation rate (consumer prices): 38% (1994 est.)
Unemployment rate: NA%
Budget:
revenues: $4.4 billion
expenditures: $6.7 billion, including capital expenditures of $NA (FY93/94 est.)
Exports: $674 million (FY93/94 est.)
commodities: pulses and beans, teak, rice, hardwood
partners: Singapore, China, Thailand, India, Hong Kong
Imports: $1.2 billion (FY93/94 est.)
commodities: machinery, transport equipment, chemicals, food products
partners: Japan, China, Thailand, Singapore, Malaysia
External debt: $5.4 billion (FY93/94 est.)
Industrial production: growth rate 4.9% (FY92/93 est.); accounts for 10% of GDP
Electricity:
capacity: 1,100,000 kW
production: 2.6 billion kWh
consumption per capita: 55 kWh (1993)
Industries: agricultural processing; textiles and footwear; wood and wood products; petroleum refining; mining of copper, tin, tungsten, iron; construction materials; pharmaceuticals; fertilizer
Agriculture: accounts for 65% of GDP and 65% of employment (including fishing, animal husbandry, and forestry); self-sufficient in food; principal crops—paddy rice, corn, oilseed, sugarcane, pulses; world's largest stand of hardwood trees; rice and timber account for 55% of export revenues
Illicit drugs: world's largest illicit producer of opium (2,030 metric tons in 1994—dropped 21% due to regional drought in 1994) and minor producer of cannabis for the international drug trade; opium production continues to be almost double since the collapse of Rangoon's antinarcotic programs; growing role in amphetamine production for regional consumption
Economic aid:
recipient: US commitments, including Ex-Im (FY70-89), $158 million; Western (non-US) countries, ODA and OOF bilateral commitments (1970-89), $3.9 billion; Communist countries (1970-89), $424 million
Currency: 1 kyat (K) = 100 pyas
Exchange rates: kyats (K) per US$1—5.8640 (January 1995), 5.9749 (1994), 6.1570 (1993), 6.1045 (1992), 6.2837 (1991), 6.3386 (1990); unofficial—120
Fiscal year: 1 April—31 March

Transportation

Railroads:
total: 3,991 km (3,878 km common carrier lines, 113 km industrial lines)
standard gauge: 3,878 km 1.435-m gauge
other: 113 km NA-m gauge
Highways:
total: 27,000 km
paved: bituminous 3,200 km
unpaved: gravel, improved earth 17,700 km; unimproved earth 6,100 km
Inland waterways: 12,800 km; 3,200 km navigable by large commercial vessels
Pipelines: crude oil 1,343 km; natural gas 330 km
Ports: Bassein, Bhamo, Chauk, Mandalay,

Moulmein, Myitkyina, Rangoon, Sittwe, Tavoy
Merchant marine:
total: 49 ships (1,000 GRT or over) totaling 638,297 GRT/884,492 DWT
ships by type: bulk 19, cargo 15, chemical tanker 1, container 2, oil tanker 3, passenger-cargo 3, refrigerated cargo 4, vehicle carrier 2
Airports:
total: 80
with paved runways over 3,047 m: 2
with paved runways 2,438 to 3,047 m: 2
with paved runways 1,524 to 2,437 m: 10
with paved runways 914 to 1,523 m: 11
with paved runways under 914 m: 33
with unpaved runways 1,524 to 2,438 m: 5
with unpaved runways 914 to 1,523 m: 17

Communications

Telephone system: 53,000 telephones (1986); meets minimum requirements for local and intercity service for business and government; international service is good
local: NA
intercity: NA
international: 1 INTELSAT (Indian Ocean) earth station
Radio:
broadcast stations: AM 2, FM 1, shortwave 0 (1985)
radios: NA
note: radiobroadcast coverage is limited to the most populous areas
Television:
broadcast stations: 1 (1985)
televisions: NA

Defense Forces

Branches: Army, Navy, Air Force
Manpower availability: males age 15-49 11,553,094; females age 15-49 11,463,189; males fit for military service 6,180,091; females fit for military service 6,116,421; males reach military age (18) annually 457,445 (1995 est.); females reach military age (18) annually 441,628 (1995 est.)
note: both sexes liable for military service
Defense expenditures: $NA, NA% of GDP

Burundi

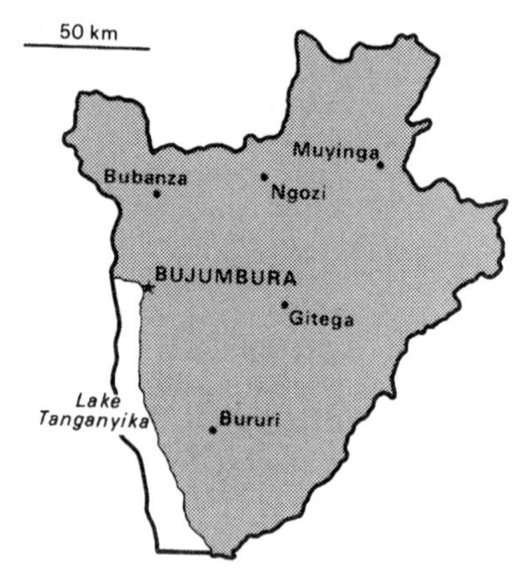

Geography

Location: Central Africa, east of Zaire
Map references: Africa
Area:
total area: 27,830 sq km
land area: 25,650 sq km
comparative area: slightly larger than Maryland
Land boundaries: total 974 km, Rwanda 290 km, Tanzania 451 km, Zaire 233 km
Coastline: 0 km (landlocked)
Maritime claims: none; landlocked
International disputes: none
Climate: temperate; warm; occasional frost in uplands; dry season from June to September
Terrain: hilly and mountainous, dropping to a plateau in east, some plains
Natural resources: nickel, uranium, rare earth oxide, peat, cobalt, copper, platinum (not yet exploited), vanadium
Land use:
arable land: 43%
permanent crops: 8%
meadows and pastures: 35%
forest and woodland: 2%
other: 12%
Irrigated land: 720 sq km (1989 est.)
Environment:
current issues: soil erosion as a result of overgrazing and the expansion of agriculture into marginal lands; deforestation (little forested land remains because of uncontrolled cutting of trees for fuel); habitat loss threatens wildlife populations
natural hazards: flooding, landslides
international agreements: party to—Endangered Species; signed, but not ratified—Biodiversity, Climate Change, Desertification, Law of the Sea, Nuclear Test Ban
Note: landlocked; straddles crest of the Nile-Congo watershed

People

Population: 6,262,429 (July 1995 est.)
Age structure:
0-14 years: 48% (female 1,489,721; male 1,494,730)
15-64 years: 50% (female 1,606,307; male 1,498,021)
65 years and over: 2% (female 105,446; male 68,204) (July 1995 est.)
Population growth rate: 2.18% (1995 est.)
Birth rate: 43.35 births/1,000 population (1995 est.)
Death rate: 21.51 deaths/1,000 population (1995 est.)
Net migration rate: NA migrant(s)/1,000 population (1995 est.)
note: in a number of waves since April 1994, hundreds of thousands of refugees have fled the civil strife between the Hutu and Tutsi factions in Burundi and crossed into Rwanda, Tanzania, and Zaire; the refugee flows are continuing in 1995 as the ethnic violence has persisted
Infant mortality rate: 111.9 deaths/1,000 live births (1995 est.)
Life expectancy at birth:
total population: 39.86 years
male: 37.84 years
female: 41.95 years (1995 est.)
Total fertility rate: 6.63 children born/woman (1995 est.)
Nationality:
noun: Burundian(s)
adjective: Burundi
Ethnic divisions:
Africans: Hutu (Bantu) 85%, Tutsi (Hamitic) 14%, Twa (Pygmy) 1%
non-Africans: Europeans 3,000, South Asians 2,000
Religions: Christian 67% (Roman Catholic 62%, Protestant 5%), indigenous beliefs 32%, Muslim 1%
Languages: Kirundi (official), French (official), Swahili (along Lake Tanganyika and in the Bujumbura area)
Literacy: age 15 and over can read and write (1990 est.)
total population: 50%
male: 61%
female: 40%
Labor force: 1.9 million (1983 est.)
by occupation: agriculture 93.0%, government 4.0%, industry and commerce 1.5%, services 1.5%

Government

Names:
conventional long form: Republic of Burundi
conventional short form: Burundi
local long form: Republika y'u Burundi
local short form: Burundi
Digraph: BY
Type: republic
Capital: Bujumbura
Administrative divisions: 15 provinces; Bubanza, Bujumbura, Bururi, Cankuzo, Cibitoke, Gitega, Karuzi, Kayanza, Kirundo,

Makamba, Muramvya, Muyinga, Ngozi, Rutana, Ruyigi
Independence: 1 July 1962 (from UN trusteeship under Belgian administration)
National holiday: Independence Day, 1 July (1962)
Constitution: 13 March 1992; provides for establishment of a plural political system
Legal system: based on German and Belgian civil codes and customary law; has not accepted compulsory ICJ jurisdiction
Suffrage: universal adult at age NA
Executive branch:
chief of state: President Sylvestre NTIBANTUNGANYA (since September 1994)
note: President Melchior NDADAYE, Burundi's first democratically elected president, died in the military coup of 21 October 1993 and was succeeded on 5 February 1994 by President Cyprien NTARYAMIRA, who was killed in a mysterious airplane explosion on 6 April 1994
head of government: Prime Minister Antoine NDUWAYO (since February 1995); selected by President NTIBANTUNGANYA following the resignation of Anatole KANYENKIKO on 15 February 1995
cabinet: Council of Ministers; appointed by prime minister
Legislative branch: unicameral
National Assembly (Assemblee Nationale): elections last held 29 June 1993 (next to be held NA); results—FRODEBU 71%, UPRONA 21.4%; seats—(81 total) FRODEBU 65, UPRONA 16; other parties won too small shares of the vote to win seats in the assembly
note: The National Unity Charter outlining the principles for constitutional government was adopted by a national referendum on 5 February 1991
Judicial branch: Supreme Court (Cour Supreme)
Political parties and leaders: Unity for National Progress (UPRONA); Burundi Democratic Front (FRODEBU); Organization of the People of Burundi (RBP); Socialist Party of Burundi (PSB); People's Reconciliation Party (PRP); opposition parties, legalized in March 1992, include Burundi African Alliance for the Salvation (ABASA); Rally for Democracy and Economic and Social Development (RADDES); and Party for National Redress (PARENA)
Other political or pressure groups: NA
Member of: ACCT, ACP, AfDB, CCC, CEEAC, CEPGL, ECA, FAO, G-77, GATT, IBRD, ICAO, ICRM, IDA, IFAD, IFC, IFRCS, ILO, IMF, INTELSAT (nonsignatory user), INTERPOL, IOC, ISO (subscriber), ITU, NAM, OAU, UN, UNCTAD, UNESCO, UNIDO, UPU, WHO, WIPO, WMO, WTO
Diplomatic representation in US:
chief of mission: post vacant since recall of Ambassador Jacques BACAMURWANKO in November 1994
chancery: Suite 212, 2233 Wisconsin Avenue NW, Washington, DC 20007
telephone: [1] (202) 342-2574
US diplomatic representation:
chief of mission: Ambassador Robert C. KRUEGER
embassy: Avenue des Etats-Unis, Bujumbura
mailing address: B. P. 1720, Bujumbura
telephone: [257] (2) 23454
FAX: [257] (2) 22926
Flag: divided by a white diagonal cross into red panels (top and bottom) and green panels (hoist side and outer side) with a white disk superimposed at the center bearing three red six-pointed stars outlined in green arranged in a triangular design (one star above, two stars below)

Economy

Overview: A landlocked, resource-poor country in an early stage of economic development, Burundi since October 1993 has suffered from massive ethnic-based violence that has displaced an estimated million people, disrupted production, and set back needed reform programs. Burundi is predominately agricultural with roughly 90% of the population dependent on subsistence agriculture. Its economic health depends on the coffee crop, which accounts for 80% of foreign exchange earnings. The ability to pay for imports therefore continues to rest largely on the vagaries of the climate and the international coffee market. As part of its economic reform agenda, launched in February 1991 with IMF and World Bank support, Burundi is trying to diversify its agricultural exports, attract foreign investment in industry, and modernize government budgetary practices. Although the government remains committed to reforms, it fears new austerity measures would add to ethnic tensions.
National product: GDP—purchasing power parity—$3.7 billion (1994 est.)
National product real growth rate: -13.5% (1994 est.)
National product per capita: $600 (1994 est.)
Inflation rate (consumer prices): 10% (1993 est.)
Unemployment rate: NA%
Budget:
revenues: $318 million
expenditures: $326 million, including capital expenditures of $150 million (1991 est.)
Exports: $68 million (f.o.b., 1993)
commodities: coffee 81%, tea, cotton, hides, and skins
partners: EC 57%, US 19%, Asia 1%
Imports: $203 million (c.i.f., 1993)
commodities: capital goods 31%, petroleum products 15%, foodstuffs, consumer goods
partners: EC 45%, Asia 29%, US 2%
External debt: $1.05 billion (1994 est.)
Industrial production: growth rate 11% (1991 est.); accounts for about 15% of GDP
Electricity:
capacity: 55,000 kW
production: 100 million kWh
consumption per capita: 20 kWh (1993)
Industries: light consumer goods such as blankets, shoes, soap; assembly of imported components; public works construction; food processing
Agriculture: accounts for 50% of GDP; cash crops—coffee, cotton, tea; food crops—corn, sorghum, sweet potatoes, bananas, manioc; livestock—meat, milk, hides and skins
Economic aid:
recipient: US commitments, including Ex-Im (FY70-89), $71 million; Western (non-US) countries, ODA and OOF bilateral commitments (1970-89), $10.2 billion; OPEC bilateral aid (1979-89), $32 million; Communist countries (1970-89), $175 million
Currency: 1 Burundi franc (FBu) = 100 centimes
Exchange rates: Burundi francs (FBu) per US$1—248.51 (December 1994), 252.66 (1994), 242.78 (1993), 208.30 (1992), 181.51 (1991), 171.26 (1990), 158.67 (1989), 140.40 (1988)
Fiscal year: calendar year

Transportation

Railroads: 0 km
Highways:
total: 5,900 km
paved: 640 km
unpaved: gravel, crushed stone 2,260 km; improved, unimproved earth 3,000 km (1990)
Inland waterways: Lake Tanganyika
Ports: Bujumbura
Airports:
total: 4
with paved runways over 3,047 m: 1
with paved runways under 914 m: 1
with unpaved runways 914 to 1,523 m: 2

Communications

Telephone system: 8,000 telephones; primative system; telephone density—1.3 telephones/1,000 persons
local: NA
intercity: sparse system of wire, radiocommunications, and low-capacity microwave radio relay links
international: 1 INTELSAT (Indian Ocean) earth station
Radio:
broadcast stations: AM 2, FM 2, shortwave 0
radios: NA
Television:
broadcast stations: 1
televisions: NA

Defense Forces

Branches: Army (includes naval and air units), paramilitary Gendarmerie
Manpower availability: males age 15-49 1,350,042; males fit for military service 705,864; males reach military age (16) annually 73,308 (1995 est.)
Defense expenditures: exchange rate conversion—$25 million, 2.6% of GDP (1993)

Cambodia

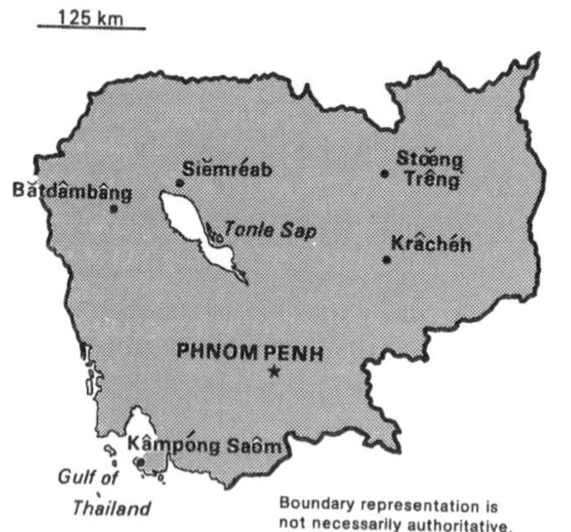

Geography

Location: Southeastern Asia, bordering the Gulf of Thailand, between Thailand and Vietnam
Map references: Southeast Asia
Area:
total area: 181,040 sq km
land area: 176,520 sq km
comparative area: slightly smaller than Oklahoma
Land boundaries: total 2,572 km, Laos 541 km, Thailand 803 km, Vietnam 1,228 km
Coastline: 443 km
Maritime claims:
contiguous zone: 24 nm
continental shelf: 200 nm
exclusive economic zone: 200 nm
territorial sea: 12 nm
International disputes: offshore islands and sections of the boundary with Vietnam are in dispute; maritime boundary with Vietnam not defined; parts of border with Thailand in dispute; maritime boundary with Thailand not clearly defined
Climate: tropical; rainy, monsoon season (May to November); dry season (December to April); little seasonal temperature variation
Terrain: mostly low, flat plains; mountains in southwest and north
Natural resources: timber, gemstones, some iron ore, manganese, phosphates, hydropower potential
Land use:
arable land: 16%
permanent crops: 1%
meadows and pastures: 3%
forest and woodland: 76%
other: 4%
Irrigated land: 920 sq km (1989 est.)
Environment:
current issues: logging activities throughout the country and strip mining for gems in the western region along the border with Thailand are resulting in habitat loss and declining biodiversity (in particular, destruction of mangrove swamps threatens natural fisheries); deforestation; soil erosion; in rural areas, a majority of the population does not have access to potable water
natural hazards: monsoonal rains (June to November); flooding; occasional droughts
international agreements: party to—Marine Life Conservation, Ship Pollution; signed, but not ratified—Desertification, Endangered Species, Law of the Sea, Marine Dumping
Note: a land of paddies and forests dominated by the Mekong River and Tonle Sap

People

Population: 10,561,373 (July 1995 est.)
Age structure:
0-14 years: 46% (female 2,367,414; male 2,438,104)
15-64 years: 51% (female 2,932,788; male 2,494,203)
65 years and over: 3% (female 185,337; male 143,527) (July 1995 est.)
Population growth rate: 2.83% (1995 est.)
Birth rate: 44.42 births/1,000 population (1995 est.)
Death rate: 16.16 deaths/1,000 population (1995 est.)
Net migration rate: 0 migrant(s)/1,000 population (1995 est.)
Infant mortality rate: 109.6 deaths/1,000 live births (1995 est.)
Life expectancy at birth:
total population: 49.46 years
male: 48 years
female: 51 years (1995 est.)
Total fertility rate: 5.81 children born/woman (1995 est.)
Nationality:
noun: Cambodian(s)
adjective: Cambodian
Ethnic divisions: Khmer 90%, Vietnamese 5%, Chinese 1%, other 4%
Religions: Theravada Buddhism 95%, other 5%
Languages: Khmer (official), French
Literacy: age 15 and over can read and write (1990 est.)
total population: 35%
male: 48%
female: 22%
Labor force: 2.5 million to 3 million
by occupation: agriculture 80% (1988 est.)

Government

Names:
conventional long form: Kingdom of Cambodia
conventional short form: Cambodia
local long form: Reacheanachak Kampuchea
local short form: Kampuchea
Digraph: CB
Type: multiparty liberal democracy under a constitutional monarchy established in September 1993

Capital: Phnom Penh
Administrative divisions: 21 provinces (khet, singular and plural); Banteay Meanchey, Batdambang, Kampong Cham, Kampong Chhnang, Kampong Spoe, Kampong Thum, Kampot, Kandal, Kaoh Kong, Kracheh, Mondol Kiri, Phnum Penh, Pouthisat, Preah Vihear, Prey Veng, Rotanokiri, Siemreab-Otdar Meanchey, Sihanoukville, Stoeng Treng, Svay Rieng, Takev
note: Siemreab-Otdar Meanchey may have been divided into two provinces named Siemreab and Otdar Meanchey
Independence: 9 November 1949 (from France)
National holiday: Independence Day, 9 November 1949
Constitution: promulgated September 1993
Legal system: currently being defined
Suffrage: 18 years of age; universal
Executive branch:
chief of state: King Norodom SIHANOUK (reinstated 24 September 1993)
head of government: power shared between First Prime Minister Prince Norodom RANARIDDH and Second Prime Minister HUN SEN
cabinet: Council of Ministers; elected by the National Assembly
Legislative branch: unicameral; a 120-member constituent assembly based on proportional representation within each province was established following the UN supervised election in May 1993; the constituent assembly was transformed into a legislature in September 1993 after delegates promulgated the constitution
Judicial branch: Supreme Court provided for by the constitution has not yet been established and the future judicial system is yet to be defined by law
Political parties and leaders: National United Front for an Independent, Neutral, Peaceful, and Cooperative Cambodia (FUNCINPEC), Prince NORODOM RANARIDDH; Cambodian Prachcachon Party or Cambodian People's Party (CPP), CHEA SIM; Buddhist Liberal Democratic Party, SON SANN; Democratic Kampuchea (DK, also known as the Khmer Rouge), KHIEU SAMPHAN; Molinaka, PROM NEAKAREACH
Member of: ACCT, AsDB, CP, ESCAP, FAO, G-77, IAEA, IBRD, ICAO, ICRM, IDA, IFAD, IFRCS, ILO, IMF, IMO, INTELSAT (nonsignatory user), INTERPOL, ITU, NAM, PCA, UN, UNCTAD, UNESCO, UPU, WFTU, WHO, WMO, WTO
Diplomatic representation in US: Ambassador SISOWATH SIRIRATH represents Cambodia at the United Nations
US diplomatic representation:
chief of mission: Ambassador Charles H. TWINING
embassy: 27 EO Street 240, Phnom Penh
mailing address: Box P, APO AP 96546
telephone: [855] (23) 26436, 26438
FAX: [855] (23) 26437
Flag: horizontal band of red separates two equal horizontal bands of blue with a white three-towered temple representing Angkor Wat in the center

Economy

Overview: The Cambodian economy—virtually destroyed by decades of war—is slowly recovering. Government leaders are moving toward restoring fiscal and monetary discipline and have established good working relations with international financial institutions. Growth, starting from a low base, has been strong in 1991-94. Despite such positive developments, the reconstruction effort faces many tough challenges because of the persistence of internal political divisions and the related lack of confidence of foreign investors. Rural Cambodia, where 90% of about 9.5 million Khmer live, remains mired in poverty. The almost total lack of basic infrastructure in the countryside will hinder development and will contribute to a growing imbalance in growth between urban and rural areas over the near term. Moreover, the government's lack of experience in administering economic and technical assistance programs and rampant corruption among officials will slow the growth of critical public sector investment. Inflation for 1994 as a whole was less than a quarter of the 1992 rate and was declining during the year.
National product: GDP—purchasing power parity—$6.4 billion (1994 est.)
National product real growth rate: 5% (1994 est.)
National product per capita: $630 (1994 est.)
Inflation rate (consumer prices): 26%-30% (1994 est.)
Unemployment rate: NA%
Budget:
revenues: $190 million
expenditures: $365 million, including capital expenditures of $120 million (1994 est.)
Exports: $283.6 million (f.o.b., 1993)
commodities: timber, rubber, soybeans, sesame
partners: Singapore, Japan, Thailand, Hong Kong, Indonesia, Malaysia
Imports: $479.3 million (c.i.f., 1993)
commodities: cigarettes, construction materials, petroleum products, machinery
partners: Singapore, Vietnam, Japan, Australia, Hong Kong, Indonesia
External debt: $383 million to OECD members (1993)
Industrial production: growth rate 7.9% (1993 est.); accounts for 8% of GDP
Electricity:
capacity: 40,000 kW
production: 160 million kWh
consumption per capita: 14 kWh (1993)
Industries: rice milling, fishing, wood and wood products, rubber, cement, gem mining
Agriculture: mainly subsistence farming except for rubber plantations; main crops—rice, rubber, corn; food shortages—rice, meat, vegetables, dairy products, sugar, flour
Illicit drugs: increasingly used as a transshipment country for heroin produced in the Golden Triangle; growing money-laundering center; high-level narcotics-related corruption in government; possible small-scale heroin production; large producer of cannibis
Economic aid:
recipient: US commitments, including Ex-Im (FY70-89), $725 million; Western (non-US countries) (1970-89), $300 million; Communist countries (1970-89), $1.8 billion; donor countries and multilateral institutions pledged $880 million in assistance in 1992; IMF pledged $120 million in aid for 1995-98
Currency: 1 new riel (CR) = 100 sen
Exchange rates: riels (CR) per US$1—2,470 (December 1993), 2,800 (September 1992), 500 (December 1991), 560 (1990), 159.00 (1988), 100.00 (1987)
Fiscal year: calendar year

Transportation

Railroads:
total: 655 km
narrow gauge: 655 km 1.000-m gauge
Highways:
total: 34,100 km (some roads in serious disrepair)
paved: bituminous 3,000 km
unpaved: crushed stone, gravel, or improved earth 3,100 km; unimproved earth 28,000 km
Inland waterways: 3,700 km navigable all year to craft drawing 0.6 meters; 282 km navigable to craft drawing 1.8 meters
Ports: Kampong Saom (Sihanoukville), Kampot, Krong Kaoh Kong, Phnom Penh
Merchant marine: none
Airports:
total: 22
with paved runways 2,438 to 3,047 m: 2
with paved runways 1,524 to 2,437 m: 2
with paved runways 914 to 1,523 m: 3
with paved runways under 914 m: 2
with unpaved runways 1,524 to 2,438 m: 3
with unpaved runways 914 to 1,523 m: 10

Communications

Telephone system: NA telephones; service barely adequate for government requirements and virtually nonexistent for general public
local: NA
intercity: NA
international: international service limited to Vietnam and other adjacent countries
Radio:
broadcast stations: AM 1, FM 0, shortwave 0

Cambodia *(continued)*

radios: NA
Television:
broadcast stations: 1
televisions: NA

Defense Forces

Branches:
Khmer Royal Armed Forces (KRAF): created in 1993 by the merger of the Cambodian People's Armed Forces and the two non-Communist resistance armies; note—the KRAF is also known as the Royal Cambodian Armed Forces (RCAF)
Resistance forces: National Army of Democratic Kampuchea (Khmer Rouge)
Manpower availability: males age 15-49 2,255,050; males fit for military service 1,256,632; males reach military age (18) annually 70,707 (1995 est.)
Defense expenditures: exchange rate conversion—$85 million, 1.4% of GDP (1995 est.)

Cameroon

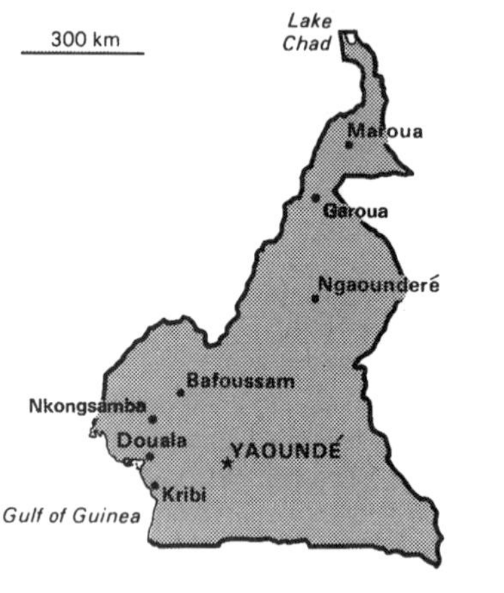

Geography

Location: Western Africa, bordering the North Atlantic Ocean, between Equatorial Guinea and Nigeria
Map references: Africa
Area:
total area: 475,440 sq km
land area: 469,440 sq km
comparative area: slightly larger than California
Land boundaries: total 4,591 km, Central African Republic 797 km, Chad 1,094 km, Congo 523 km, Equatorial Guinea 189 km, Gabon 298 km, Nigeria 1,690 km
Coastline: 402 km
Maritime claims:
territorial sea: 50 nm
International disputes: demarcation of international boundaries in Lake Chad, the lack of which led to border incidents in the past, is completed and awaits ratification by Cameroon, Chad, Niger, and Nigeria; dispute with Nigeria over land and maritime boundaries in the vicinity of the Bakasi Peninsula has been referred to the International Court of Justice
Climate: varies with terrain, from tropical along coast to semiarid and hot in north
Terrain: diverse, with coastal plain in southwest, dissected plateau in center, mountains in west, plains in north
Natural resources: petroleum, bauxite, iron ore, timber, hydropower potential
Land use:
arable land: 13%
permanent crops: 2%
meadows and pastures: 18%
forest and woodland: 54%
other: 13%
Irrigated land: 280 sq km (1989 est.)
Environment:
current issues: water-borne diseases are prevalent; deforestation; overgrazing; desertification; poaching; overfishing
natural hazards: recent volcanic activity with release of poisonous gases
international agreements: party to—Biodiversity, Climate Change, Endangered Species, Law of the Sea, Ozone Layer Protection, Tropical Timber 83; signed, but not ratified—Desertification, Nuclear Test Ban, Tropical Timber 94
Note: sometimes referred to as the hinge of Africa

People

Population: 13.521 million (July 1995 est.)
Age structure:
0-14 years: 44% (female 2,978,216; male 3,001,487)
15-64 years: 52% (female 3,562,247; male 3,523,100)
65 years and over: 4% (female 248,314; male 207,636) (July 1995 est.)
Population growth rate: 2.92% (1995 est.)
Birth rate: 40.42 births/1,000 population (1995 est.)
Death rate: 11.19 deaths/1,000 population (1995 est.)
Net migration rate: 0 migrant(s)/1,000 population (1995 est.)
Infant mortality rate: 75.4 deaths/1,000 live births (1995 est.)
Life expectancy at birth:
total population: 57.48 years
male: 55.41 years
female: 59.6 years (1995 est.)
Total fertility rate: 5.8 children born/woman (1995 est.)
Nationality:
noun: Cameroonian(s)
adjective: Cameroonian
Ethnic divisions: Cameroon Highlanders 31%, Equatorial Bantu 19%, Kirdi 11%, Fulani 10%, Northwestern Bantu 8%, Eastern Nigritic 7%, other African 13%, non-African less than 1%
Religions: indigenous beliefs 51%, Christian 33%, Muslim 16%
Languages: 24 major African language groups, English (official), French (official)
Literacy: age 15 and over can read and write (1987)
total population: 55%
male: 66%
female: 45%
Labor force: NA
by occupation: agriculture 74.4%, industry and transport 11.4%, other services 14.2% (1983)

Government

Names:
conventional long form: Republic of Cameroon
conventional short form: Cameroon
former: French Cameroon
Digraph: CM
Type: unitary republic; multiparty

presidential regime (opposition parties legalized 1990)
Capital: Yaounde
Administrative divisions: 10 provinces; Adamaoua, Centre, Est, Extreme-Nord, Littoral, Nord, Nord-Ouest, Ouest, Sud, Sud-Ouest
Independence: 1 January 1960 (from UN trusteeship under French administration)
National holiday: National Day, 20 May (1972)
Constitution: 20 May 1972
Legal system: based on French civil law system, with common law influence; has not accepted compulsory ICJ jurisdiction
Suffrage: 20 years of age; universal
Executive branch:
chief of state: President Paul BIYA (since 6 November 1982); election last held 11 October 1992; results—President Paul BIYA reelected with about 40% of the vote amid widespread allegations of fraud; SDF candidate John FRU NDI got 36% of the vote; UNDP candidate Bello Bouba MAIGARI got 19% of the vote
head of government: Prime Minister Simon ACHIDI ACHU (since 9 April 1992)
cabinet: Cabinet; appointed by the president
Legislative branch: unicameral
National Assembly (Assemblee Nationale): elections last held 1 March 1992 (next scheduled for March 1997); results—(180 seats) CPDM 88, UNDP 68, UPC 18, MDR 6
Judicial branch: Supreme Court
Political parties and leaders: Cameroon People's Democratic Movement (CPDM), Paul BIYA, president, is government-controlled and was formerly the only party, but opposition parties were legalized in 1990
major opposition parties: National Union for Democracy and Progress (UNDP); Social Democratic Front (SDF); Cameroonian Democratic Union (UDC); Union of Cameroonian Populations (UPC); Movement for the Defense of the Republic (MDR)
Other political or pressure groups: Alliance for Change (FAC), Cameroon Anglophone Movement (CAM)
Member of: ACCT, ACP, AfDB, BDEAC, CCC, CEEAC, ECA, FAO, FZ, G 19, G 77, GATT, IAEA, IBRD, ICAO, ICC, ICFTU, ICRM, IDA, IDB, IFAD, IFC, IFRCS, ILO, IMF, IMO, INMARSAT, INTELSAT, INTERPOL, IOC, ITU, NAM, OAU, OIC, PCA, UDEAC, UN, UNCTAD, UNESCO, UNIDO, UPU, WCL, WFTU, WHO, WIPO, WMO, WTO
Diplomatic representation in US:
chief of mission: Ambassador Jerome MENDOUGA
chancery: 2349 Massachusetts Avenue NW, Washington, DC 20008
telephone: [1] (202) 265-8790 through 8794
US diplomatic representation:
chief of mission: Ambassador Harriet W. ISOM
embassy: Rue Nachtigal, Yaounde
mailing address: B. P. 817, Yaounde
telephone: [237] 23-40-14
FAX: [237] 23-07-53
consulate(s): none (Douala closed September 1993)
Flag: three equal vertical bands of green (hoist side), red, and yellow with a yellow five-pointed star centered in the red band; uses the popular pan-African colors of Ethiopia

Economy

Overview: Because of its offshore oil resources and favorable agricultural conditions, Cameroon has one of the best-endowed, most diversified primary commodity economies in sub-Saharan Africa. Still, it faces many of the serious problems facing other underdeveloped countries, such as political instability, a top-heavy civil service, and a generally unfavorable climate for business enterprise. The development of the oil sector led rapid economic growth between 1970 and 1985. Growth came to an abrupt halt in 1986, precipitated by steep declines in the prices of major exports: coffee, cocoa, and petroleum. Export earnings were cut by almost one-third, and inefficiencies in fiscal management were exposed. In 1990-93, with support from the IMF and World Bank, the government began to introduce reforms designed to spur business investment, increase efficiency in agriculture, and recapitalize the nation's banks. Political instability, following suspect elections in 1992, brought IMF/WB structural adjustment to a halt. Although the 50% devaluation of the currency in January 1994 improved the potential for export growth, mismanagement remains and is the main barrier to economic improvement.
National product: GDP—purchasing power parity—$15.7 billion (1994 est.)
National product real growth rate: -2.9% (1994 est.)
National product per capita: $1,200 (1994 est.)
Inflation rate (consumer prices): -0.8% (FY91/92)
Unemployment rate: 25% (1990 est.)
Budget:
revenues: $1.6 billion
expenditures: $2.3 billion, including capital expenditures of $226 million (FY92/93 est.)
Exports: $1.6 billion (f.o.b., 1993)
commodities: petroleum products, lumber, cocoa beans, aluminum, coffee, cotton
partners: EC (particularly France) about 40%, African countries, US
Imports: $1.96 billion (c.i.f., 1993)
commodities: machines and electrical equipment, food, consumer goods, transport equipment
partners: EC about 60% (France 38%, Germany 9%), African countries, Japan, US 5%
External debt: $6 billion (1991)
Industrial production: growth rate -2.1% (FY90/91); accounts for about 20% of GDP
Electricity:
capacity: 630,000 kW
production: 2.7 billion kWh
consumption per capita: 196 kWh (1993)
Industries: petroleum production and refining, food processing, light consumer goods, textiles, lumber
Agriculture: the agriculture and forestry sectors provide employment for the majority of the population, contributing about 25% to GDP and providing a high degree of self-sufficiency in staple foods; commercial and food crops include coffee, cocoa, timber, cotton, rubber, bananas, oilseed, grains, livestock, root starches
Economic aid:
recipient: US commitments, including Ex-Im (FY70-90), $479 million; Western (non-US) countries, ODA and OOF bilateral commitments (1970-90), $4.75 billion; OPEC bilateral aid (1979-89), $29 million; Communist countries (1970-89), $125 million
Currency: 1 CFA franc (CFAF) = 100 centimes
Exchange rates: Communaute Financiere Africaine francs (CFAF) per US$1—529.43 (January 1995), 555.20 (1994), 283.16 (1993), 264.69 (1992), 282.11 (1991), 272.26 (1990)
note: beginning 12 January 1994, the CFA franc was devalued to CFAF 100 per French franc from CFAF 50 at which it had been fixed since 1948
Fiscal year: 1 July—30 June

Transportation

Railroads:
total: 1,111 km
narrow gauge: 1,111 km 1.000-m gauge
Highways:
total: 65,000 km
paved: 2,682 km
unpaved: gravel, improved earth 32,318 km; unimproved earth 30,000 km
Inland waterways: 2,090 km; of decreasing importance
Ports: Bonaberi, Douala, Garoua, Kribi, Tiko
Merchant marine:
total: 2 cargo ships (1,000 GRT or over) totaling 24,122 GRT/33,509 DWT
Airports:
total: 60
with paved runways over 3,047 m: 2
with paved runways 2,438 to 3,047 m: 4
with paved runways 1,524 to 2,437 m: 3
with paved runways 914 to 1,523 m: 1
with paved runways under 914 m: 20
with unpaved runways 1,524 to 2,438 m: 9
with unpaved runways 914 to 1,523 m: 21

Communications

Telephone system: 26,000 telephones; telephone density—2 telephones/1,000 persons; available only to business and government
local: NA
intercity: cable, microwave radio relay, and troposcatter
international: 2 Atlantic Ocean INTELSAT earth stations
Radio:
broadcast stations: AM 11, FM 11, shortwave 0
radios: NA
Television:
broadcast stations: 1
televisions: NA

Defense Forces

Branches: Army, Navy (includes Naval Infantry), Air Force, National Gendarmerie, Presidential Guard
Manpower availability: males age 15-49 3,038,007; males fit for military service 1,532,303; males reach military age (18) annually 147,293 (1995 est.)
Defense expenditures: exchange rate conversion—$102 million, NA% of GDP (1994)

Canada

Geography

Location: Northern North America, bordering the North Atlantic Ocean and North Pacific Ocean, north of the conterminous US
Map references: North America
Area:
total area: 9,976,140 sq km
land area: 9,220,970 sq km
comparative area: slightly larger than US
Land boundaries: total 8,893 km, US 8,893 km (includes 2,477 km with Alaska)
Coastline: 243,791 km
Maritime claims:
continental shelf: 200 nm or to the edge of the continental margin
exclusive fishing zone: 200 nm
territorial sea: 12 nm
International disputes: maritime boundary disputes with the US; Saint Pierre and Miquelon is focus of maritime boundary dispute between Canada and France
Climate: varies from temperate in south to subarctic and arctic in north
Terrain: mostly plains with mountains in west and lowlands in southeast
Natural resources: nickel, zinc, copper, gold, lead, molybdenum, potash, silver, fish, timber, wildlife, coal, petroleum, natural gas
Land use:
arable land: 5%
permanent crops: 0%
meadows and pastures: 3%
forest and woodland: 35%
other: 57%
Irrigated land: 8,400 sq km (1989 est.)
Environment:
current issues: air pollution and resulting acid rain severely affecting lakes and damaging forests; metal smelting, coal-burning utilities, and vehicle emissions impacting on agricultural and forest productivity; ocean waters becoming contaminated due to agricultural, industrial, mining, and forestry activities
natural hazards: continuous permafrost in north is a serious obstacle to development; cyclonic storms form east of the Rocky Mountains, a result of the mixing of air masses from the Arctic, Pacific, and American interior, and produce most of the country's rain and snow
international agreements: party to—Air Pollution, Air Pollution-Nitrogen Oxides, Air Pollution-Sulphur 85, Antarctic Treaty, Biodiversity, Climate Change, Endangered Species, Environmental Modification, Hazardous Wastes, Marine Dumping, Nuclear Test Ban, Ozone Layer Protection, Ship Pollution, Tropical Timber 83, Wetlands; signed, but not ratified—Air Pollution-Sulphur 94, Air Pollution-Volatile Organic Compounds, Antarctic-Environmental Protocol, Desertification, Law of the Sea
Note: second-largest country in world (after Russia); strategic location between Russia and US via north polar route; nearly 90% of the population is concentrated in the region near the US/Canada border

People

Population: 28,434,545 (July 1995 est.)
Age structure:
0-14 years: 21% (female 2,874,705; male 3,016,050)
15-64 years: 67% (female 9,529,272; male 9,531,107)
65 years and over: 12% (female 2,022,324; male 1,461,087) (July 1995 est.)
Population growth rate: 1.09% (1995 est.)
Birth rate: 13.74 births/1,000 population (1995 est.)
Death rate: 7.43 deaths/1,000 population (1995 est.)
Net migration rate: 4.55 migrant(s)/1,000 population (1995 est.)
Infant mortality rate: 6.8 deaths/1,000 live births (1995 est.)
Life expectancy at birth:
total population: 78.29 years
male: 74.93 years
female: 81.81 years (1995 est.)
Total fertility rate: 1.83 children born/woman (1995 est.)
Nationality:
noun: Canadian(s)
adjective: Canadian
Ethnic divisions: British Isles origin 40%, French origin 27%, other European 20%, indigenous Indian and Eskimo 1.5%
Religions: Roman Catholic 46%, United Church 16%, Anglican 10%, other 28%
Languages: English (official), French (official)
Literacy: age 15 and over can read and write (1986)
total population: 97%
Labor force: 13.38 million
by occupation: services 75%, manufacturing 14%, agriculture 4%, construction 3%, other 4% (1988)

Government

Names:
conventional long form: none
conventional short form: Canada
Digraph: CA
Type: confederation with parliamentary democracy
Capital: Ottawa
Administrative divisions: 10 provinces and 2 territories*; Alberta, British Columbia, Manitoba, New Brunswick, Newfoundland, Northwest Territories*, Nova Scotia, Ontario, Prince Edward Island, Quebec, Saskatchewan, Yukon Territory*
Independence: 1 July 1867 (from UK)
National holiday: Canada Day, 1 July (1867)
Constitution: amended British North America Act 1867 patriated to Canada 17 April 1982; charter of rights and unwritten customs
Legal system: based on English common law, except in Quebec, where civil law system based on French law prevails; accepts compulsory ICJ jurisdiction, with reservations
Suffrage: 18 years of age; universal
Executive branch:
chief of state: Queen ELIZABETH II (since 6 February 1952), represented by Governor General Romeo LeBLANC (since 8 February 1995)
head of government: Prime Minister Jean CHRETIEN (since 4 November 1993) was elected on 25 October 1993, replacing Kim CAMBELL; Deputy Prime Minister Sheila COPPS
cabinet: Federal Ministry; chosen by the prime minister from members of his own party sitting in Parliament
Legislative branch: bicameral Parliament (Parlement)
Senate (Senat): consisting of a body whose members are appointed to serve until 75 years of age by the governor general and selected on the advice of the prime minister; its normal limit 104 senators
House of Commons (Chambre des Communes): elections last held 25 October 1993 (next to be held by NA October 1998); results—percent of votes by party NA; seats—(295 total) Liberal Party 178, Bloc Quebecois 54, Reform Party 52, New Democratic Party 8, Progressive Conservative Party 2, independents 1
Judicial branch: Supreme Court
Political parties and leaders: Liberal Party, Jean CHRETIEN; Bloc Quebecois, Lucien BOUCHARD; Reform Party, Preston MANNING; New Democratic Party, Audrey McLAUGHLIN; Progressive Conservative Party, Jean CHAREST
Member of: ACCT, AfDB, AG (observer), APEC, AsDB, Australia Group, BIS, C, CCC, CDB (non-regional), EBRD, ECE, ECLAC, ESA (cooperating state), FAO, G- 7, G- 8, G-10, GATT, IADB, IAEA, IBRD, ICAO, ICC, ICFTU, ICRM, IDA, IEA, IFAD, IFC, IFRCS, ILO, IMF, IMO, INMARSAT, INTELSAT, INTERPOL, IOC, IOM, ISO, ITU, MINURSO, MTCR, NACC, NAM (guest), NATO, NEA, NSG, OAS, OECD, ONUSAL, OSCE, PCA, UN, UNAMIR, UNCTAD, UNDOF, UNESCO, UNFICYP, UNHCR, UNIDO, UNIKOM, UNITAR, UNOMOZ, UNOSOM, UNPROFOR, UNTSO, UNU, UPU, WCL, WFTU, WHO, WIPO, WMO, WTO, ZC
Diplomatic representation in US:
chief of mission: Ambassador Raymond A.J. CHRETIEN
chancery: 501 Pennsylvania Avenue NW, Washington, DC 20001
telephone: [1] (202) 682-1740
FAX: [1] (202) 682-7726
consulate(s) general: Atlanta, Boston, Buffalo, Chicago, Dallas, Detroit, Los Angeles, Minneapolis, New York, and Seattle
consulate(s): Cincinnati, Cleveland, Miami, Philadelphia, Pittsburgh, Princeton, San Diego, San Francisco, San Jose, and San Juan (Puerto Rico)
US diplomatic representation:
chief of mission: Ambassador James Johnston BLANCHARD
embassy: 100 Wellington Street, K1P 5T1, Ottawa
mailing address: P. O. Box 5000, Ogdensburg, NY 13669-0430
telephone: [1] (613) 238-5335, 4470
FAX: [1] (613) 238-5720
consulate(s) general: Calgary, Halifax, Montreal, Quebec, Toronto, and Vancouver
Flag: three vertical bands of red (hoist side), white (double width, square), and red with a red maple leaf centered in the white band

Economy

Overview: As an affluent, high-tech industrial society, Canada today closely resembles the US in per capita output, market-oriented economic system, and pattern of production. Since World War II the impressive growth of the manufacturing, mining, and service sectors has transformed the nation from a largely rural economy into one primarily industrial and urban. In the 1980s, Canada registered one of the highest rates of real growth among the OECD nations, averaging about 3.2%. With its great natural resources, skilled labor force, and modern capital plant, Canada has excellent economic prospects, although the country still faces high unemployment and a growing debt. Moreover, the continuing constitutional impasse between English- and French-speaking areas has observers discussing a possible split in the confederation; foreign investors have become edgy.
National product: GDP—purchasing power parity—$639.8 billion (1994 est.)
National product real growth rate: 4.5% (1994)
National product per capita: $22,760 (1994)
Inflation rate (consumer prices): 0.2% (1994)
Unemployment rate: 9.6% (December 1994)
Budget:
revenues: $85 billion (Federal)
expenditures: $115.3 billion, including capital expenditures of $NA (FY93/94 est.)
Exports: $164.3 billion (f.o.b., 1994 est.)
commodities: newsprint, wood pulp, timber, crude petroleum, machinery, natural gas, aluminum, motor vehicles and parts; telecommunications equipment
partners: US, Japan, UK, Germany, South Korea, Netherlands, China
Imports: $151.5 billion (c.i.f., 1994 est.)
commodities: crude oil, chemicals, motor vehicles and parts, durable consumer goods, electronic computers; telecommunications equipment and parts
partners: US, Japan, UK, Germany, France, Mexico, Taiwan, South Korea
External debt: $243 billion (1993)
Industrial production: growth rate 4.8% (1993)
Electricity:
capacity: 108,090,000 kW
production: 511 billion kWh
consumption per capita: 16,133 kWh (1993)
Industries: processed and unprocessed minerals, food products, wood and paper products, transportation equipment, chemicals, fish products, petroleum and natural gas
Agriculture: accounts for about 3% of GDP; one of the world's major producers and exporters of grain (wheat and barley); key source of US agricultural imports; large forest resources cover 35% of total land area; commercial fisheries provide annual catch of 1.5 million metric tons, of which 75% is exported
Illicit drugs: illicit producer of cannabis for the domestic drug market; use of hydroponics technology permits growers to plant large quantities of high-quality marijuana indoors; growing role as a transit point for heroin and cocaine entering the US market
Economic aid:
donor: ODA and OOF commitments (1970-89), $7.2 billion
Currency: 1 Canadian dollar (Can$) = 100 cents
Exchange rates: Canadian dollars (Can$) per US$1—1.4129 (January 1995), 1.3656 (1994), 1.2901 (1993), 1.2087 (1992), 1.1457 (1991), 1.1668 (1990)
Fiscal year: 1 April—31 March

Transportation

Railroads:
total: 78,148 km; note—there are two major transcontinental freight railway systems:

Canada *(continued)*

Canadian National (government owned) and Canadian Pacific Railway; passenger service provided by VIA (government operated)
standard gauge: 78,148 km 1.435-m gauge (185 km electrified) (1994)
Highways:
total: 849,404 km
paved: 253,692 km (15,983 km of expressways)
unpaved: gravel 595,712 km (1991)
Inland waterways: 3,000 km, including Saint Lawrence Seaway
Pipelines: crude and refined oil 23,564 km; natural gas 74,980 km
Ports: Becancour, Churchill, Halifax, Montreal, New Westminister, Prince Rupert, Quebec, Saint John (New Brunswick), Saint John's (Newfoundland), Seven Islands, Sydney, Three Rivers, Toronto, Vancouver, Windsor
Merchant marine:
total: 71 ships (1,000 GRT or over) totaling 617,010 GRT/878,819 DWT
ships by type: bulk 17, cargo 10, chemical tanker 5, oil tanker 23, passenger 1, passenger-cargo 1, railcar carrier 2, roll-on/roll-off cargo 7, short-sea passenger 3, specialized tanker 2
note: does not include ships used exclusively in the Great Lakes
Airports:
total: 1,386
with paved runways over 3,047 m: 17
with paved runways 2,438 to 3,047 m: 16
with paved runways 1,524 to 2,437 m: 147
with paved runways 914 to 1,523 m: 234
with paved runways under 914 m: 550
with unpaved runways 1,524 to 2,438 m: 69
with unpaved runways 914 to 1,523 m: 353

Communications

Telephone system: 18,000,000 telephones; excellent service provided by modern media
local: NA
intercity: about 300 earth stations for domestic satellite communications
international: 5 coaxial submarine cables; over 5 INTELSAT earth stations (4 Atlantic Ocean and 1 Pacific Ocean)
Radio:
broadcast stations: AM 900, FM 29, shortwave 0
radios: NA
Television:
broadcast stations: 53 (repeaters 1,400)
televisions: NA

Defense Forces

Branches: Canadian Armed Forces (includes Land Forces Command or LC, Maritime Command or MC, Air Command or AC, Communications Command or CC, Training Command or TC), Royal Canadian Mounted Police (RCMP)
Manpower availability: males age 15-49 7,570,877; males fit for military service 6,522,092; males reach military age (17) annually 151,590 (1995 est.)
Defense expenditures: exchange rate conversion—$9.0 billion, 1.6% of GDP (FY95/96)

Cape Verde

Geography

Location: Western Africa, group of Islands in the North Atlantic Ocean, west of Senegal
Map references: World
Area:
total area: 4,030 sq km
land area: 4,030 sq km
comparative area: slightly larger than Rhode Island
Land boundaries: 0 km
Coastline: 965 km
Maritime claims: measured from claimed archipelagic baselines
exclusive economic zone: 200 nm
territorial sea: 12 nm
International disputes: none
Climate: temperate; warm, dry, summer; precipitation very erratic
Terrain: steep, rugged, rocky, volcanic
Natural resources: salt, basalt rock, pozzolana, limestone, kaolin, fish
Land use:
arable land: 9%
permanent crops: 0%
meadows and pastures: 6%
forest and woodland: 0%
other: 85%
Irrigated land: 20 sq km (1989 est.)
Environment:
current issues: overgrazing of livestock and improper land use such as the cultivation of crops on steep slopes has led to soil erosion; demand for wood used as fuel has resulted in deforestation; desertification; environmental damage has threatened several indigenous species of birds and reptiles; overfishing
natural hazards: prolonged droughts; harmattan wind can obscure visibility; volcanically and seismically active
international agreements: party to—Environmental Modification, Law of the Sea, Marine Dumping, Nuclear Test Ban; signed, but not ratified—Biodiversity, Climate Change, Desertification
Note: strategic location 500 km from west coast of Africa near major north-south sea

routes; important communications station; important sea and air refueling site

People

Population: 435,983 (July 1995 est.)
Age structure:
0-14 years: 50% (female 106,539; male 110,301)
15-64 years: 47% (female 114,931; male 88,029)
65 years and over: 3% (female 9,781; male 6,402) (July 1995 est.)
Population growth rate: 2.98% (1995 est.)
Birth rate: 45.32 births/1,000 population (1995 est.)
Death rate: 8.65 deaths/1,000 population (1995 est.)
Net migration rate: -6.88 migrant(s)/1,000 population (1995 est.)
Infant mortality rate: 55.9 deaths/1,000 live births (1995 est.)
Life expectancy at birth:
total population: 63.01 years
male: 61.1 years
female: 65.01 years (1995 est.)
Total fertility rate: 6.23 children born/woman (1995 est.)
Nationality:
noun: Cape Verdean(s)
adjective: Cape Verdean
Ethnic divisions: Creole (mulatto) 71%, African 28%, European 1%
Religions: Roman Catholicism fused with indigenous beliefs
Languages: Portuguese, Crioulo, a blend of Portuguese and West African words
Literacy: age 15 and over can read and write (1990)
total population: 63%
male: 75%
female: 53%
Labor force: 102,000 (1985 est.)
by occupation: agriculture (mostly subsistence) 57%, services 29%, industry 14% (1981)

Government

Names:
conventional long form: Republic of Cape Verde
conventional short form: Cape Verde
local long form: Republica de Cabo Verde
local short form: Cabo Verde
Digraph: CV
Type: republic
Capital: Praia
Administrative divisions: 14 districts (concelhos, singular—concelho); Boa Vista, Brava, Fogo, Maio, Paul, Praia, Porto Novo, Ribeira Grande, Sal, Santa Catarina, Santa Cruz, Sao Nicolau, Sao Vicente, Tarrafal
Independence: 5 July 1975 (from Portugal)
National holiday: Independence Day, 5 July (1975)
Constitution: new constitution came into force 25 September 1992
Legal system: NA
Suffrage: 18 years of age; universal
Executive branch:
chief of state: President Antonio MASCARENHAS Monteiro (since 22 March 1991; election last held 17 February 1991 (next to be held February 1996); results—Antonio Monteiro MASCARENHAS (independent) received 72.6% of vote
head of government: Prime Minister Carlos Alberto Wahnon de Carvalho VEIGA (since 13 January 1991)
cabinet: Council of Ministers; appointed by prime minister from members of the Assembly
Legislative branch: unicameral
People's National Assembly (Assembleia Nacional Popular): elections last held 13 January 1991 (next to be held January 1996); results—percent of vote by party NA; seats—(79 total) MPD 56, PAICV 23; note—the 1991 multiparty Assembly election ended 15 years of single party rule
Judicial branch: Supreme Tribunal of Justice (Supremo Tribunal de Justia)
Political parties and leaders: Movement for Democracy (MPD), Prime Minister Carlos VEIGA, founder and chairman; African Party for Independence of Cape Verde (PAICV), Pedro Verona Rodrigues PIRES, chairman
Member of: ACP, AfDB, CCC, ECA, ECOWAS, FAO, G-77, IBRD, ICAO, ICFTU, ICRM, IDA, IFAD, IFC, IFRCS, ILO, IMF, IMO, INTELSAT, INTERPOL, IOC, IOM (observer), ITU, NAM, OAU, UN (Cape Verde assumed a nonpermanent seat on the Security Council on 1 January 1992), UNCTAD, UNESCO, UNIDO, UNOMOZ, UPU, WCL, WHO, WMO
Diplomatic representation in US:
chief of mission: (vacant); Charge d'Affaires Jose Eduardo BARBOSA (since 12 February 1994)
chancery: 3415 Massachusetts Avenue NW, Washington, DC 20007
telephone: [1] (202) 965-6820
FAX: [1] (202) 965-1207
consulate(s) general: Boston
US diplomatic representation:
chief of mission: Ambassador Joseph M. SEGARS
embassy: Rua Abilio Macedo 81, Praia
mailing address: C. P. 201, Praia
telephone: [238] 61 56 16
FAX: [238] 61 13 55
Flag: three horozontal bands of light blue (top, double width), white (with a horozontal red stripe in the middle third), and light blue; a circle of 10 yellow five-pointed stars is centered on the hoist end of the red stripe and extends into the upper and lower blue bands

Economy

Overview: Cape Verde's low per capita GDP reflects a poor natural resource base, serious water shortages exacerbated by cycles of long-term drought, and a high birthrate. The economy is service oriented, with commerce, transport, and public services accounting for 60% of GDP. Although nearly 70% of the population lives in rural areas, agriculture's share of GDP is only 20%; the fishing sector accounts for 4%. About 90% of food must be imported. The fishing potential, mostly lobster and tuna, is not fully exploited. Cape Verde annually runs a high trade deficit, financed by remittances from emigrants and foreign aid, which form important supplements to GDP. Economic reforms, launched by the new democratic government in 1991, are aimed at developing the private sector and attracting foreign investment to diversify the economy. Prospects for 1995 depend heavily on the maintenance of aid flows, remittances, and the momentum of the government's development program.
National product: GDP—purchasing power parity—$410 million (1993 est.)
National product real growth rate: 3.5% (1992 est.)
National product per capita: $1,000 (1993 est.)
Inflation rate (consumer prices): 7% (1992)
Unemployment rate: 26% (1990 est.)
Budget:
revenues: $174 million
expenditures: $235 million, including capital expenditures of $165 million (1993 est.)
Exports: $4.4 million (f.o.b., 1992 est.)
commodities: fish, bananas, hides and skins
partners: Netherlands, Portugal, Angola
Imports: $173 million (c.i.f., 1992 est.)
commodities: foodstuffs, consumer goods, industrial products, transport equipment
partners: Portugal, Netherlands, Germany, Spain
External debt: $156 million (1991)
Industrial production: growth rate 3.6% (1990 est.); accounts for 8% of GDP
Electricity:
capacity: 15,000 kW
production: 40 million kWh
consumption per capita: 73 kWh (1993)
Industries: fish processing, salt mining, garment industry, ship repair, construction materials, food and beverage production
Agriculture: accounts for 20% of GDP (including fishing); largely subsistence farming; bananas are the only export crop; other crops—corn, beans, sweet potatoes, coffee; growth potential of agricultural sector limited by poor soils and scanty rainfall; annual food imports required; fish catch provides for both domestic consumption and small exports
Illicit drugs: increasingly used as a transshipment point for illicit drugs moving

Cape Verde *(continued)*

from Latin America and Africa destined for Western Europe
Economic aid:
recipient: US commitments, including Ex-Im (FY75-90), $93 million; Western (non-US) countries, ODA and OOF bilateral commitments (1970-90), $586 million; OPEC bilateral aid (1979-89), $12 million; Communist countries (1970-89), $36 million
Currency: 1 Cape Verdean escudo (CVEsc) = 100 centavos
Exchange rates: Cape Verdean escudos (CVEsc) per US$1—85.537 (1st Quarter 1994), 80.427 (1993), 68.018 (1992), 71.408 (1991), 70.031 (1990)
Fiscal year: calendar year

Transportation

Railroads: 0 km
Highways:
total: 1,100 km (1992)
paved: 680 km
unpaved: 420 km
Ports: Mindelo, Praia, Tarrafal
Merchant marine:
total: 7 (1,000 GRT or over) totaling 11,609 GRT/19,052 DWT
cargo 6, chemical tanker 1
Airports:
total: 6
with paved runways over 3,047 m: 1
with paved runways 914 to 1,523 m: 5

Communications

Telephone system: over 1,700 telephones; telephine density—about 4 telephones/1,000 persons
local: NA
intercity: interisland microwave radio relay system, high frequency radio links to Senegal and Guinea-Bissau
international: 2 coaxial submarine cables; 1 Atlantic Ocean INTELSAT earth station
Radio:
broadcast stations: AM 1, FM 6, shortwave 0
radios: NA
Television:
broadcast stations: 1
televisions: NA

Defense Forces

Branches: People's Revolutionary Armed Forces (FARP; includes Army and Navy), Security Service
Manpower availability: males age 15-49 80,867; males fit for military service 47,225 (1995 est.)
Defense expenditures: exchange rate conversion—$3.4 million, NA% of GDP (1994)

Cayman Islands
(dependent territory of the UK)

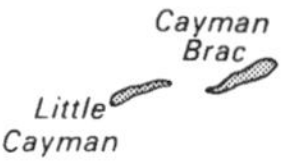

Geography

Location: Caribbean, island group in Caribbean Sea, nearly one-half of the way from Cuba to Honduras
Map references: Central America and the Caribbean
Area:
total area: 260 sq km
land area: 260 sq km
comparative area: slightly less than 1.5 times the size of Washington, DC
Land boundaries: 0 km
Coastline: 160 km
Maritime claims:
exclusive fishing zone: 200 nm
territorial sea: 12 nm
International disputes: none
Climate: tropical marine; warm, rainy summers (May to October) and cool, relatively dry winters (November to April)
Terrain: low-lying limestone base surrounded by coral reefs
Natural resources: fish, climate and beaches that foster tourism
Land use:
arable land: 0%
permanent crops: 0%
meadows and pastures: 8%
forest and woodland: 23%
other: 69%
Irrigated land: NA sq km
Environment:
current issues: no natural fresh water resources, drinking water supplies must be met by rainwater catchment
natural hazards: hurricanes (July to November)
international agreements: NA
Note: important location between Cuba and Central America

People

Population: 33,192 (July 1995 est.)
Age structure:
0-14 years: NA
15-64 years: NA
65 years and over: NA
Population growth rate: 4.3% (1995 est.)
Birth rate: 14.79 births/1,000 population (1995 est.)
Death rate: 4.98 deaths/1,000 population (1995 est.)
Net migration rate: 33.2 migrant(s)/1,000 population (1995 est.)
Infant mortality rate: 8.4 deaths/1,000 live births (1995 est.)
Life expectancy at birth:
total population: 77.1 years
male: 75.37 years
female: 78.81 years (1995 est.)
Total fertility rate: 1.43 children born/woman (1995 est.)
Nationality:
noun: Caymanian(s)
adjective: Caymanian
Ethnic divisions: mixed 40%, white 20%, black 20%, expatriates of various ethnic groups 20%
Religions: United Church (Presbyterian and Congregational), Anglican, Baptist, Roman Catholic, Church of God, other Protestant denominations
Languages: English
Literacy: age 15 and over has ever attended school (1970)
total population: 98%
male: 98%
female: 98%
Labor force: 8,061
by occupation: service workers 18.7%, clerical 18.6%, construction 12.5%, finance and investment 6.7%, directors and business managers 5.9% (1979)

Government

Names:
conventional long form: none
conventional short form: Cayman Islands
Digraph: CJ
Type: dependent territory of the UK
Capital: George Town
Administrative divisions: 8 districts; Creek, Eastern, Midland, South Town, Spot Bay, Stake Bay, West End, Western
Independence: none (dependent territory of the UK)
National holiday: Constitution Day (first Monday in July)
Constitution: 1959, revised 1972 and 1992
Legal system: British common law and local statutes
Suffrage: 18 years of age; universal
Executive branch:
chief of state: Queen ELIZABETH II (since 6 February 1952)
head of government: Governor and President of the Executive Council Michael GORE (since 15 September 1992)
cabinet: Executive Council; 3 members are

appointed by the governor, 4 members elected by the Legislative Assembly
Legislative branch: unicameral
Legislative Assembly: election last held November 1992 (next to be held November 1996); results—percent of vote by party NA; seats—(15 total, 12 elected)
Judicial branch: Grand Court, Cayman Islands Court of Appeal
Political parties and leaders: no formal political parties
Member of: CARICOM (observer), CDB, INTERPOL (subbureau), IOC
Diplomatic representation in US: none (dependent territory of the UK)
US diplomatic representation: none (dependent territory of the UK)
Flag: blue, with the flag of the UK in the upper hoist-side quadrant and the Caymanian coat of arms on a white disk centered on the outer half of the flag; the coat of arms includes a pineapple and turtle above a shield with three stars (representing the three islands) and a scroll at the bottom bearing the motto HE HATH FOUNDED IT UPON THE SEAS

Economy

Overview: The economy depends heavily on tourism (70% of GDP and 75% of foreign currency earnings) and offshore financial services, with the tourist industry aimed at the luxury market and catering mainly to visitors from North America. About 90% of the islands' food and consumer goods must be imported. The Caymanians enjoy one of the highest outputs per capita and one of the highest standards of living in the world.
National product: GDP—purchasing power parity—$700 million (1993 est.)
National product real growth rate: 1.4% (1991)
National product per capita: $23,000 (1993 est.)
Inflation rate (consumer prices): 2.5% (1993 est.)
Unemployment rate: 7% (1992)
Budget:
revenues: $141.5 million
expenditures: $160.7 million, including capital expenditures of $NA (1991)
Exports: $10 million (f.o.b., 1993 est.)
commodities: turtle products, manufactured consumer goods
partners: mostly US
Imports: $312 million (c.i.f., 1993 est.)
commodities: foodstuffs, manufactured goods
partners: US, Trinidad and Tobago, UK, Netherlands Antilles, Japan
External debt: $15 million (1986)
Industrial production: growth rate NA%
Electricity:
capacity: 80,000 kW
production: 230 million kWh
consumption per capita: 6,899 kWh (1993)
Industries: tourism, banking, insurance and finance, construction, building materials, furniture making
Agriculture: minor production of vegetables, fruit, livestock; turtle farming
Illicit drugs: a major money-laundering center for illicit drug profits; transshipment point for narcotics bound for the US and Europe
Economic aid:
recipient: US commitments, including Ex-Im (FY70-89), $26.7 million; Western (non-US) countries, ODA and OOF bilateral commitments (1970-89), $35 million
Currency: 1 Caymanian dollar (CI$) = 100 cents
Exchange rates: Caymanian dollars (CI$) per US$1—0.83 (18 November 1993), 0.85 (22 November 1993)
Fiscal year: 1 April—31 March

Transportation

Railroads: 0 km
Highways:
total: 160 km (main roads)
paved: NA
unpaved: NA
Ports: Cayman Brac, George Town
Merchant marine:
total: 26 ships (1,000 GRT or over) totaling 321,434 GRT/583,348 DWT
ships by type: bulk 7, cargo 6, chemical tanker 2, container 1, oil tanker 3, roll-on/roll-off cargo 7
note: a flag of convenience registry; UK owns 6 ships, India 5, Norway 3, US 3, Greece 1, Sweden 1, UAE 1
Airports:
total: 3
with paved runways 1,524 to 2,437 m: 2
with unpaved runways 914 to 1,523 m: 1

Communications

Telephone system: 35,000 telephones
local: NA
intercity: NA
international: 1 submarine coaxial cable; 1 INTELSAT (Atlantic Ocean) earth station
Radio:
broadcast stations: AM 2, FM 1, shortwave 0
radios: NA
Television:
broadcast stations: 0
televisions: NA

Defense Forces

Branches: Royal Cayman Islands Police Force (RCIPF)
Note: defense is the responsibility of the UK

Central African Republic

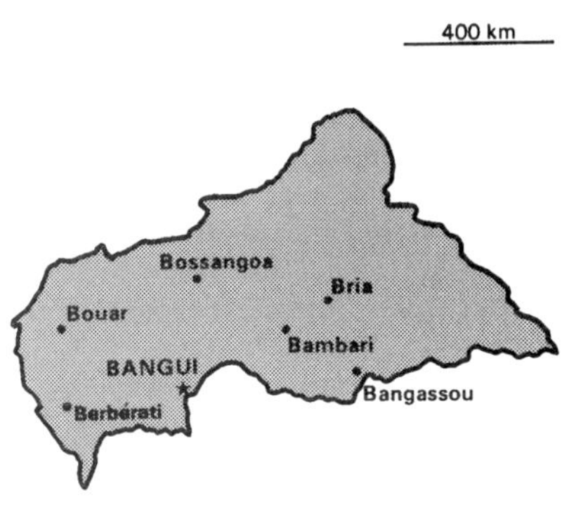

Geography

Location: Central Africa, north of Zaire
Map references: Africa
Area:
total area: 622,980 sq km
land area: 622,980 sq km
comparative area: slightly smaller than Texas
Land boundaries: total 5,203 km, Cameroon 797 km, Chad 1,197 km, Congo 467 km, Sudan 1,165 km, Zaire 1,577 km
Coastline: 0 km (landlocked)
Maritime claims: none; landlocked
International disputes: none
Climate: tropical; hot, dry winters; mild to hot, wet summers
Terrain: vast, flat to rolling, monotonous plateau; scattered hills in northeast and southwest
Natural resources: diamonds, uranium, timber, gold, oil
Land use:
arable land: 3%
permanent crops: 0%
meadows and pastures: 5%
forest and woodland: 64%
other: 28%
Irrigated land: NA sq km
Environment:
current issues: tap water is not potable; poaching has diminished reputation as one of last great wildlife refuges; desertification
natural hazards: hot, dry, dusty harmattan winds affect northern areas; floods are common
international agreements: party to—Endangered Species, Nuclear Test Ban, Ozone Layer Protection; signed, but not ratified—Biodiversity, Climate Change, Desertification, Law of the Sea
Note: landlocked; almost the precise center of Africa

People

Population: 3,209,759 (July 1995 est.)

Central African Republic

(continued)

Age structure:
0-14 years: 43% (female 690,290; male 694,153)
15-64 years: 53% (female 886,421; male 825,268)
65 years and over: 4% (female 64,846; male 48,781) (July 1995 est.)
Population growth rate: 2.1% (1995 est.)
Birth rate: 41.84 births/1,000 population (1995 est.)
Death rate: 20.89 deaths/1,000 population (1995 est.)
Net migration rate: 0 migrant(s)/1,000 population (1995 est.)
Infant mortality rate: 135.6 deaths/1,000 live births (1995 est.)
Life expectancy at birth:
total population: 42.15 years
male: 40.68 years
female: 43.67 years (1995 est.)
Total fertility rate: 5.37 children born/woman (1995 est.)
Nationality:
noun: Central African(s)
adjective: Central African
Ethnic divisions: Baya 34%, Banda 27%, Sara 10%, Mandjia 21%, Mboum 4%, M'Baka 4%, Europeans 6,500 (including 3,600 French)
Religions: indigenous beliefs 24%, Protestant 25%, Roman Catholic 25%, Muslim 15%, other 11%
note: animistic beliefs and practices strongly influence the Christian majority
Languages: French (official), Sangho (lingua franca and national language), Arabic, Hunsa, Swahili
Literacy: age 15 and over can read and write (1990 est.)
total population: 38%
male: 52%
female: 25%
Labor force: 775,413 (1986 est.)
by occupation: agriculture 85%, commerce and services 9%, industry 3%, government 3%
note: about 64,000 salaried workers

Government

Names:
conventional long form: Central African Republic
conventional short form: none
local long form: Republique Centrafricaine
local short form: none
former: Central African Empire
Abbreviation: CAR
Digraph: CT
Type: republic;
Capital: Bangui
Administrative divisions: 14 prefectures (prefectures, singular—prefecture), 2 economic prefectures* (prefectures economiques, singular—prefecture economique), and 1 commune**; Bamingui-Bangoran, Bangui** Basse-Kotto, Gribingui*, Haute-Kotto, Haute-Sangha, Haut-Mbomou, Kemo-Gribingui, Lobaye, Mbomou, Nana-Mambere, Ombella-Mpoko, Ouaka, Ouham, Ouham-Pende, Sangha*, Vakaga
Independence: 13 August 1960 (from France)
National holiday: National Day, 1 December (1958) (proclamation of the republic)
Constitution: 21 November 1986
Legal system: based on French law
Suffrage: 21 years of age; universal
Executive branch:
chief of state: President Ange PATASSE (since 22 October 1993); election last held 19 September 1993 (next scheduled for 1998); PATASSE received 52.45% of the votes and Abel GOUMBA received 45.62%
head of government: Prime Minister (vacant) (Dr. Jean-Luc MANDABA resigned on 11 April 1995)
cabinet: Council of Ministers; appointed by the president
Legislative branch: unicameral
National Assembly (Assemblee Nationale): elections last held 19 September 1993; results—percentage vote by party NA; seats—(85 total) MLPC 33, RDC 14, PLD 7, ADP 6, PSD 3, others 22
note: the National Assembly is advised by the Economic and Regional Council (Conseil Economique et Regional); when they sit together they are called the Congress (Congres)
Judicial branch: Supreme Court (Cour Supreme)
Political parties and leaders: Movement for the Liberation of the Central African People (MLPC), the party of the new president, Ange Felix PATASSE; Movement for Democracy and Development (MDD), David DACKO; Marginal Movement for Democracy, Renaissance and Evolution (MDREC), Joseph BENDOUNGA; Central African Democratic Assembly (RDC), Andre KOLINGBA; Patriotic Front for Progress (FFP), Abel GOUMBA; Civic Forum (FC), Gen. Timothee MALENDOMA
Member of: ACCT, ACP, AfDB, BDEAC, CCC, CEEAC, ECA, FAO, FZ, G-77, GATT, IBRD, ICAO, ICRM, IDA, IFAD, IFC, IFRCS, ILO, IMF, INTELSAT, INTERPOL, IOC, ITU, NAM, OAU, UDEAC, UN, UNCTAD, UNESCO, UNIDO, UPU, WCL, WHO, WIPO, WMO
Diplomatic representation in US:
chief of mission: Ambassador Henri KOBA (appointed 19 September 1994)
chancery: 1618 22nd Street NW, Washington, DC 20008
telephone: [1] (202) 483-7800, 7801
FAX: [1] (202) 332-9893
US diplomatic representation:
chief of mission: Ambassador Robert E. GRIBBIN III
embassy: Avenue David Dacko, Bangui
mailing address: B. P. 924, Bangui
telephone: [236] 61 02 00, 61 25 78, 61 02 10
FAX: [236] 61 44 94
Flag: four equal horizontal bands of blue (top), white, green, and yellow with a vertical red band in center; there is a yellow five-pointed star on the hoist side of the blue band

Economy

Overview: Subsistence agriculture, together with forestry, remains the backbone of the CAR economy, with more than 70% of the population living in outlying areas. The agricultural sector generates about half of GDP. Timber has accounted for about 26% of export earnings and the diamond industry for 54%. Important constraints to economic development include the CAR's landlocked position, a poor transportation system, a largely unskilled work force, and a legacy of misdirected macroeconomic policies. A major plus is the large forest reserves, which the government is moving to protect from overexploitation. The 50% devaluation of the currencies of 14 Francophone African nations on 12 January 1994 had mixed effects on CAR's economy. While diamond, timber, coffee, and cotton exports increased—leading GDP to increase by 5.5%—inflation rose to 40%, fueled by the rising prices of imports on which the economy depends. CAR's poor resource base and primitive infrastructure will keep it dependent on multilateral donors and France for the foreseeable future.
National product: GDP—purchasing power parity—$2.2 billion (1994 est.)
National product real growth rate: 5.5% (1994 est.)
National product per capita: $700 (1994 est.)
Inflation rate (consumer prices): 40% (1994 est.)
Unemployment rate: 30% (1988 est.) in Bangui
Budget:
revenues: $175 million
expenditures: $312 million, including capital expenditures of $122 million (1991 est.)
Exports: $123.5 million (f.o.b., 1992)
commodities: diamonds, timber, cotton, coffee, tobacco
partners: France, Belgium, Italy, Japan, US
Imports: $165.1 million (f.o.b., 1992)
commodities: food, textiles, petroleum products, machinery, electrical equipment, motor vehicles, chemicals, pharmaceuticals, consumer goods, industrial products
partners: France, other EC countries, Japan, Algeria
External debt: $859 million (1991)
Industrial production: growth rate 4% (1990 est.); accounts for 14% of GDP
Electricity:
capacity: 40,000 kW
production: 100 million kWh

consumption per capita: 29 kWh (1993)
Industries: diamond mining, sawmills, breweries, textiles, footwear, assembly of bicycles and motorcycles
Agriculture: self-sufficient in food production except for grain; commercial crops—cotton, coffee, tobacco, timber; food crops—manioc, yams, millet, corn, bananas
Economic aid:
recipient: US commitments, including Ex-Im (FY70-90), $52 million; Western (non-US) countries, ODA and OOF bilateral commitments (1970-90), $1.6 billion; OPEC bilateral aid (1979-89), $6 million; Communist countries (1970-89), $38 million
Currency: 1 CFA franc (CFAF) = 100 centimes
Exchange rates: Communaute Financiere Africaine francs (CFAF) per US$1—529.43 (January 1995), 555.20 (1994), 283.16 (1993), 264.69 (1992), 282.11 (1991), 272.26 (1990)
note: beginning 12 January 1994, the CFA franc was devalued to CFAF 100 per French franc from CFAF 50 at which it had been fixed since 1948
Fiscal year: calendar year

Transportation

Railroads: 0 km
Highways:
total: 22,000 km
paved: bituminous 458 km
unpaved: improved earth 10,542 km; unimproved earth 11,000 km
Inland waterways: 800 km; traditional trade carried on by means of shallow-draft dugouts; Oubangui is the most important river
Ports: Bangui, Nola
Airports:
total: 61
with paved runways 2,438 to 3,047 m: 1
with paved runways 1,524 to 2,437 m: 2
with paved runways under 914 m: 19
with unpaved runways 2,438 to 3,047 m: 1
with unpaved runways 1,524 to 2,438 m: 9
with unpaved runways 914 to 1,523 m: 29

Communications

Telephone system: NA telephones; system is only fair
local: NA
intercity: network consists principally of micowave radio relay and low capacity, low powered radio communication
international: 1 Atlantic Ocean INTELSAT earth station
Radio:
broadcast stations: AM 1, FM 1, shortwave 0
radios: NA
Television:
broadcast stations: 1
televisions: NA

Defense Forces

Branches: Central African Army (includes Republican Guard), Air Force, National Gendarmerie, Police Force
Manpower availability: males age 15-49 718,487; males fit for military service 375,950 (1995 est.)
Defense expenditures: exchange rate conversion—$30 million, 2.3% of GDP (1994)

Chad

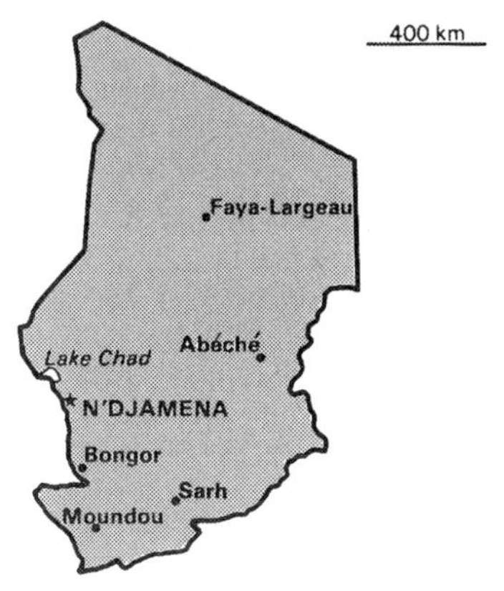

Geography

Location: Central Africa, south of Libya
Map references: Africa
Area:
total area: 1.284 million sq km
land area: 1,259,200 sq km
comparative area: slightly more than three times the size of California
Land boundaries: total 5,968 km, Cameroon 1,094 km, Central African Republic 1,197 km, Libya 1,055 km, Niger 1,175 km, Nigeria 87 km, Sudan 1,360 km
Coastline: 0 km (landlocked)
Maritime claims: none; landlocked
International disputes: the International Court of Justice (ICJ) ruled in February 1994 that the 100,000 sq km Aozou Strip between Chad and Libya belongs to Chad; Libya has withdrawn some of its forces in response to the ICJ ruling, but still maintains an airfield in the disputed area; demarcation of international boundaries in Lake Chad, the lack of which has led to border incidents in the past, is completed and awaiting ratification by Cameroon, Chad, Niger, and Nigeria
Climate: tropical in south, desert in north
Terrain: broad, arid plains in center, desert in north, mountains in northwest, lowlands in south
Natural resources: petroleum (unexploited but exploration under way), uranium, natron, kaolin, fish (Lake Chad)
Land use:
arable land: 2%
permanent crops: 0%
meadows and pastures: 36%
forest and woodland: 11%
other: 51%
Irrigated land: 100 sq km (1989 est.)
Environment:
current issues: inadequate supplies of potable water; improper waste disposal in rural areas contributes to soil and water pollution; desertification
natural hazards: hot, dry, dusty harmattan winds occur in north; periodic droughts; locust plagues

international agreements: party to—Biodiversity, Climate Change, Endangered Species, Nuclear Test Ban, Ozone Layer Protection, Wetlands; signed, but not ratified—Law of the Sea, Marine Dumping
Note: landlocked; Lake Chad is the most significant water body in the Sahel

People

Population: 5,586,505 (July 1995 est.)
Age structure:
0-14 years: 44% (female 1,198,619; male 1,267,470)
15-64 years: 54% (female 1,563,678; male 1,456,481)
65 years and over: 2% (female 71,971; male 28,286) (July 1995 est.)
Population growth rate: 2.18% (1995 est.)
Birth rate: 42.05 births/1,000 population (1995 est.)
Death rate: 20.26 deaths/1,000 population (1995 est.)
Net migration rate: 0 migrant(s)/1,000 population (1995 est.)
Infant mortality rate: 129.7 deaths/1,000 live births (1995 est.)
Life expectancy at birth:
total population: 41.19 years
male: 40.04 years
female: 42.38 years (1995 est.)
Total fertility rate: 5.33 children born/woman (1995 est.)
Nationality:
noun: Chadian(s)
adjective: Chadian
Ethnic divisions:
north and center: Muslims (Arabs, Toubou, Hadjerai, Fulbe, Kotoko, Kanembou, Baguirmi, Boulala, Zaghawa, and Maba)
south: non-Muslims (Sara, Ngambaye, Mbaye, Goulaye, Moundang, Moussei, Massa)
nonindigenous 150,000, of whom 1,000 are French
Religions: Muslim 50%, Christian 25%, indigenous beliefs, animism 25%
Languages: French (official), Arabic (official), Sara (in south), Sango (in south), more than 100 different languages and dialects are spoken
Literacy: age 15 and over has the ability to read and write in French and Arabic (1990 est.)
total population: 30%
male: 42%
female: 18%
Labor force: NA
by occupation: agriculture 85% (engaged in unpaid subsistence farming, herding, and fishing)

Government

Names:
conventional long form: Republic of Chad
conventional short form: Chad
local long form: Republique du Tchad
local short form: Tchad
Digraph: CD
Type: republic
Capital: N'Djamena
Administrative divisions: 14 prefectures (prefectures, singular—prefecture); Batha, Biltine, Borkou-Ennedi-Tibesti, Chari-Baguirmi, Guera, Kanem, Lac, Logone Occidental, Logone Oriental, Mayo-Kebbi, Moyen-Chari, Ouaddai, Salamat, Tandjile
Independence: 11 August 1960 (from France)
National holiday: Independence Day 11 August (1960)
Constitution: 22 December 1989 (suspended 3 December 1990); Provisional National Charter 1 March 1991 is in effect (note—the constitutional commission, which was drafting a new constitution to submit to transitional parliament for ratification in April 1994, failed to do so but expects to submit a new draft to the parliament before the end of April 1995)
Legal system: based on French civil law system and Chadian customary law; has not accepted compulsory ICJ jurisdiction
Suffrage: universal at age NA
Executive branch:
chief of state: President Lt. Gen. Idriss DEBY, since 4 December 1990 (after seizing power on 3 December 1990—transitional government's mandate expires April 1996)
head of government: Prime Minister Djimasta KOIBLA (since 9 April 1995)
cabinet: Council of State; appointed by the president on recommendation of the prime minister
Legislative branch: unicameral
National Consultative Council (Conceil National Consultatif): elections, formerly scheduled for April 1995, were postponed by mutual agreement of the parties concerned until some time prior to April 1996; elections last held 8 July 1990; the National Consultative Council was disbanded 3 December 1990 and replaced by the Provisional Council of the Republic having 30 members appointed by President DEBY on 8 March 1991; this, in turn, was replaced by a 57-member Higher Transitional Council (Conseil Superieur de Transition) elected by a specially convened Sovereign National Conference on 6 April 1993
Judicial branch: Court of Appeal
Political parties and leaders: Patriotic Salvation Movement (MPS), former dissident group, Idriss DEBY, chairman
note: President DEBY, who promised political pluralism, a new constitution, and free elections by April 1994, subsequently twice postponed these initiatives, first until April 1995 and again until sometime before April 1996; there are numerous dissident groups and at least 45 opposition political parties
Other political or pressure groups: NA
Member of: ACCT, ACP, AfDB, BDEAC, CEEAC, ECA, FAO, FZ, G-77, GATT, IBRD, ICAO, ICFTU, ICRM, IDA, IDB, IFAD, IFRCS, ILO, IMF, INTELSAT, INTERPOL, IOC, ITU, NAM, OAU, OIC, UDEAC, UN, UNCTAD, UNESCO, UNIDO, UPU, WCL, WHO, WIPO, WMO, WTO
Diplomatic representation in US:
chief of mission: Ambassador Mahamat Saleh AHMAT
chancery: 2002 R Street NW, Washington, DC 20009
telephone: [1] (202) 462-4009
FAX: [1] (202) 265-1937
US diplomatic representation:
chief of mission: Ambassador Laurence E. POPE II
embassy: Avenue Felix Eboue, N'Djamena
mailing address: B. P. 413, N'Djamena
telephone: [235] (51) 62 18, (51) 40 09, (51) 47 59
FAX: [235] (51) 33 72
Flag: three equal vertical bands of blue (hoist side), yellow, and red; similar to the flag of Romania; also similar to the flag of Andorra, which has a national coat of arms featuring a quartered shield centered in the yellow band; design was based on the flag of France

Economy

Overview: Climate, geographic remoteness, poor resource endowment, and lack of infrastructure make Chad one of the most underdeveloped countries in the world. Its economy is hobbled by political turmoil, conflict with Libya, drought, and food shortages. Consequently the economy has shown little progress in recent years in overcoming a severe setback brought on by civil war in the late 1980s. More than 80% of the work force is involved in subsistence farming and fishing. Cotton is the major cash crop, accounting for at least half of exports. Chad is highly dependent on foreign aid, especially food credits, given chronic shortages in several regions. Of all the Francophone countries in Africa, Chad has benefited the least from the 50% devaluation of their currencies on 12 January 1994. Despite an increase in external financial aid and favorable price increases for cotton—the primary source of foreign exchange—the corrupt and enfeebled government bureaucracy continues to dampen economic enterprise by neglecting payments to domestic suppliers and public sector salaries. Oil production in the Lake Chad area remains a distant prospect and the subsistence-driven economy probably will continue to limp along in the near term.
National product: GDP—purchasing power parity—$2.8 billion (1993 est.)
National product real growth rate: 3.5% (1993 est.)
National product per capita: $530 (1993 est.)

Inflation rate (consumer prices): -4.1% (1992)
Unemployment rate: NA%
Budget:
revenues: $120 million
expenditures: $363 million, including capital expenditures of $104 million (1992 est.)
Exports: $190 million (f.o.b., 1992)
commodities: cotton 48%, cattle 35%, textiles 5%, fish
partners: France, Nigeria, Cameroon
Imports: $261 million (f.o.b., 1992)
commodities: machinery and transportation equipment 39%, industrial goods 20%, petroleum products 13%, foodstuffs 9%; note—excludes military equipment
partners: US, France, Nigeria, Cameroon
External debt: $492 million (December 1990 est.)
Industrial production: growth rate 2.7% (1992 est.); accounts for nearly 15% of GDP
Electricity:
capacity: 40,000 kW
production: 80 million kWh
consumption per capita: 13 kWh (1993)
Industries: cotton textile mills, slaughterhouses, brewery, natron (sodium carbonate), soap, cigarettes
Agriculture: accounts for about 45% of GDP; largely subsistence farming; cotton most important cash crop; food crops include sorghum, millet, peanuts, rice, potatoes, manioc; livestock—cattle, sheep, goats, camels; self-sufficient in food in years of adequate rainfall
Economic aid:
recipient: US commitments, including Ex-Im (FY70-89), $198 million; Western (non-US) countries, ODA and OOF bilateral commitments (1970-89), $1.5 billion; OPEC bilateral aid (1979-89), $28 million; Communist countries (1970-89), $80 million
Currency: 1 CFA franc (CFAF) = 100 centimes
Exchange rates: Communaute Financiere Africaine Francs (CFAF) per US$1—529.43 (January 1995), 555.20 (1994), 283.16 (1993), 264.69 (1992), 282.11 (1991), 272.26 (1990)
note: beginning 12 January 1994 the CFA franc was devalued to CFAF 100 per French franc from CFAF 50 at which it had been fixed since 1948
Fiscal year: calendar year

Transportation

Railroads: 0 km
Highways:
total: 31,322 km
paved: bituminous 263 km
unpaved: gravel, crushed stone 7,069 km; earth 23,990 km
Inland waterways: 2,000 km navigable
Ports: none
Airports:
total: 66
with paved runways 2,438 to 3,047 m: 3
with paved runways 1,524 to 2,437 m: 1
with paved runways under 914 m: 23
with unpaved runways over 3,047 m: 1
with unpaved runways 1,524 to 2,438 m: 17
with unpaved runways 914 to 1,523 m: 21

Communications

Telephone system: NA telephones; primitive system
local: NA
intercity: fair system of radio communication stations for intercity links
international: 1 INTELSAT (Atlantic Ocean) earth station
Radio:
broadcast stations: AM 6, FM 1, shortwave 0
radios: NA
Television:
broadcast stations: NA; note—limited TV service; many facilties are inoperative
televisions: NA

Defense Forces

Branches: Armed Forces (includes Ground Force, Air Force, and Gendarmerie), Republican Guard, Police
Manpower availability: males age 15-49 1,307,210; males fit for military service 679,640; males reach military age (20) annually 54,945 (1995 est.)
Defense expenditures: exchange rate conversion—$74 million, 11.1% of GDP (1994)

Chile

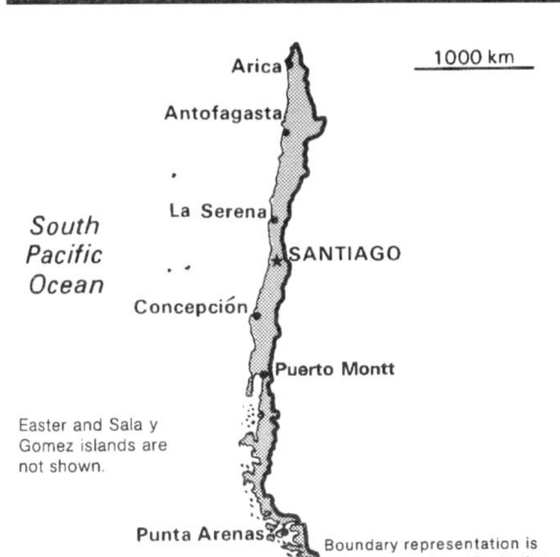

Geography

Location: Southern South America, bordering the South Atlantic Ocean and South Pacific Ocean, between Argentina and Peru
Map references: South America
Area:
total area: 756,950 sq km
land area: 748,800 sq km
comparative area: slightly smaller than twice the size of Montana
note: includes Isla de Pascua (Easter Island) and Isla Sala y Gomez
Land boundaries: total 6,171 km, Argentina 5,150 km, Bolivia 861 km, Peru 160 km
Coastline: 6,435 km
Maritime claims:
contiguous zone: 24 nm
continental shelf: 200 nm
exclusive economic zone: 200 nm
territorial sea: 12 nm
International disputes: short section of the southern boundary with Argentina is indefinite; Bolivia has wanted a sovereign corridor to the South Pacific Ocean since the Atacama area was lost to Chile in 1884; dispute with Bolivia over Rio Lauca water rights; territorial claim in Antarctica (Chilean Antarctic Territory) partially overlaps Argentine and British claims
Climate: temperate; desert in north; cool and damp in south
Terrain: low coastal mountains; fertile central valley; rugged Andes in east
Natural resources: copper, timber, iron ore, nitrates, precious metals, molybdenum
Land use:
arable land: 7%
permanent crops: 0%
meadows and pastures: 16%
forest and woodland: 21%
other: 56%
Irrigated land: 12,650 sq km (1989 est.)
Environment:
current issues: air pollution from industrial and vehicle emissions; water pollution from raw sewage; deforestation contributing to loss

of biodiversity; soil erosion; desertification
natural hazards: severe earthquakes; active volcanism; tsunamis
international agreements: party to—Antarctic-Environmental Protocol, Antarctic Treaty, Biodiversity, Climate Change, Endangered Species, Environmental Modification, Hazardous Wastes, Marine Dumping, Nuclear Test Ban, Ozone Layer Protection, Ship Pollution, Wetlands, Whaling; signed, but not ratified—Law of the Sea
Note: strategic location relative to sea lanes between Atlantic and Pacific Oceans (Strait of Magellan, Beagle Channel, Drake Passage); Atacama Desert one of world's driest regions

People

Population: 14,161,216 (July 1995 est.)
Age structure:
0-14 years: 29% (female 2,014,877; male 2,099,450)
15-64 years: 64% (female 4,574,947; male 4,529,251)
65 years and over: 7% (female 549,385; male 393,306) (July 1995 est.)
Population growth rate: 1.49% (1995 est.)
Birth rate: 20.29 births/1,000 population (1995 est.)
Death rate: 5.42 deaths/1,000 population (1995 est.)
Net migration rate: 0 migrant(s)/1,000 population (1995 est.)
Infant mortality rate: 14.3 deaths/1,000 live births (1995 est.)
Life expectancy at birth:
total population: 74.88 years
male: 71.89 years
female: 78.01 years (1995 est.)
Total fertility rate: 2.49 children born/woman (1995 est.)
Nationality:
noun: Chilean(s)
adjective: Chilean
Ethnic divisions: European and European-Indian 95%, Indian 3%, other 2%
Religions: Roman Catholic 89%, Protestant 11%, Jewish
Languages: Spanish
Literacy: age 15 and over can read and write (1992)
total population: 94%
male: 95%
female: 94%
Labor force: 4.728 million
by occupation: services 38.3% (includes government 12%), industry and commerce 33.8%, agriculture, forestry, and fishing 19.2%, mining 2.3%, construction 6.4% (1990)

Government

Names:
conventional long form: Republic of Chile
conventional short form: Chile
local long form: Republica de Chile
local short form: Chile
Digraph: CI
Type: republic
Capital: Santiago
Administrative divisions: 13 regions (regiones, singular—region); Aisen del General Carlos Ibanez del Campo, Antofagasta, Araucania, Atacama, Bio-Bio, Coquimbo, Libertador General Bernardo O'Higgins, Los Lagos, Magallanes y de la Antartica Chilena, Maule, Region Metropolitana, Tarapaca, Valparaiso
note: the US does not recognize claims to Antarctica
Independence: 18 September 1810 (from Spain)
National holiday: Independence Day, 18 September (1810)
Constitution: 11 September 1980, effective 11 March 1981; amended 30 July 1989
Legal system: based on Code of 1857 derived from Spanish law and subsequent codes influenced by French and Austrian law; judicial review of legislative acts in the Supreme Court; has not accepted compulsory ICJ jurisdiction
Suffrage: 18 years of age; universal and compulsory
Executive branch:
chief of state and head of government: President Eduardo FREI Ruiz-Tagle (since 11 March 1994) election last held 11 December 1993 (next to be held December 1999); results—Eduardo FREI Ruiz-Tagle (PDC) 58%, Arturo ALESSANDRI 24.4%, other 17.6%
cabinet: Cabinet; appointed by the president
Legislative branch: bicameral National Congress (Congreso Nacional)
Senate (Senado): election last held 11 December 1993 (next to be held December 1997); results—percent of vote by party NA; seats—(46 total, 38 elected) Concertation of Parties for Democracy 21 (PDC 13, PS 4, PPD 3, PR 1), Union for the Progress of Chile 15 (RN 11, UDI 3, UCC 1), right-wing independents 10
Chamber of Deputies (Camara de Diputados): election last held 11 December 1993 (next to be held December 1997); results—Concertation of Parties for Democracy 53.95% (PDC 27.16%, PS 12.01%, PPD 11.82%, PR 2.96%,); Union for the Progress of Chile 30.57% (RN 15.25%, UDI 12.13%, UCC 3.19%); seats—(120 total) Concertation of Parties for Democracy 70 (PDC 37, PPD 15, PR 2, PS 15, left-wing independent 1), Union for the Progress of Chile 47 (RN 30, UDI 15, UCC 2), right-wing independents 3
Judicial branch: Supreme Court (Corte Suprema)
Political parties and leaders: Concertation of Parties for Democracy consists mainly of three parties: Christian Democratic Party (PDC), Alejandro FOXLEY; Socialist Party (PS), Camilo ESCALONA; Party for Democracy (PPD), Jorge SCHAULSOHN; Radical Party (PR);
Union for the Progress of Chile consists mainly of three parties: National Renewal (RN), Andres ALLAMAND; Independent Democratic Union (UDI), Jovino NOVOA; Center Center Union (UCC), Francisco Javier ERRAZURIZ
Other political or pressure groups: revitalized university student federations at all major universities; labor—United Labor Central (CUT) includes trade unionists from the country's five largest labor confederations; Roman Catholic Church
Member of: APEC, CCC, ECLAC, FAO, G-11, G-77, GATT, IADB, IAEA, IBRD, ICAO, ICFTU, ICRM, IDA, IFAD, IFC, IFRCS, ILO, IMF, IMO, INMARSAT, INTELSAT, INTERPOL, IOC, IOM, ISO, ITU, LAES, LAIA, NAM, OAS, ONUSAL, OPANAL, PCA, RG, UN, UNCTAD, UNESCO, UNIDO, UNMOGIP, UNTSO, UNU, UPU, WCL, WFTU, WHO, WIPO, WMO, WTO
Diplomatic representation in US:
chief of mission: Ambassador Gabriel GUERRA-MONDRAGON
chancery: 1732 Massachusetts Avenue NW, Washington, DC 20036
telephone: [1] (202) 785-1746
FAX: [1] (202) 887-5579
consulate(s) general: Chicago, Houston, Los Angeles, Miami, New York, Philadelphia, San Francisco, and San Juan (Puerto Rico)
US diplomatic representation:
chief of mission: Ambassador Gabriel GUERRA-MONDRAGON
embassy: Codina Building, 1343 Agustinas, Santiago
mailing address: Unit 4127, Santiago; APO AA 34033
telephone: [56] (2) 232-2600
FAX: [56] (2) 330-3710
Flag: two equal horizontal bands of white (top) and red; there is a blue square the same height as the white band at the hoist-side end of the white band; the square bears a white five-pointed star in the center; design was based on the US flag

Economy

Overview: Chile has a prosperous, essentially free market economy, with the degree of government intervention varying according to the philosophy of the different regimes. Under the center-left government of President AYLWIN, which took power in March 1990, spending on social welfare rose steadily. At the same time business investment, exports, and consumer spending also grew substantially. The new president, FREI, who took office in March 1994, has emphasized social spending even more. Growth in 1991-94 has averaged

6.5% annually, with an estimated one million Chileans having moved out of poverty in the last four years. Copper remains vital to the health of the economy; Chile is the world's largest producer and exporter of copper. Success in meeting the government's goal of sustained annual growth of 5% depends on world copper prices, the level of confidence of foreign investors and creditors, and the government's own ability to maintain a conservative fiscal stance.
National product: GDP—purchasing power parity—$97.7 billion (1994 est.)
National product real growth rate: 4.3% (1994 est.)
National product per capita: $7,010 (1994 est.)
Inflation rate (consumer prices): 8.7% (1994 est.)
Unemployment rate: 6% (1994 est.)
Budget:
revenues: $10.9 billion
expenditures: $10.9 billion, including capital expenditures of $1.2 billion (1993)
Exports: $11.5 billion (f.o.b., 1994)
commodities: copper 41%, other metals and minerals 8.7%, wood products 7.1%, fish and fishmeal 9.8%, fruits 8.4% (1991)
partners: EC 29%, Japan 17%, US 16%, Argentina 5%, Brazil 5% (1992)
Imports: $10.9 billion (f.o.b., 1994)
commodities: capital goods 25.2%, spare parts 24.8%, raw materials 15.4%, petroleum 10%, foodstuffs 5.7%
partners: EC 24%, US 21%, Brazil 10%, Japan 10% (1992)
External debt: $20 billion (1994 est.)
Industrial production: growth rate 4.3% (1993 est.); accounts for 34% of GDP
Electricity:
capacity: 4,810,000 kW
production: 22 billion kWh
consumption per capita: 1,499 kWh (1993)
Industries: copper, other minerals, foodstuffs, fish processing, iron and steel, wood and wood products, transport equipment, cement, textiles
Agriculture: accounts for about 7% of GDP (including fishing and forestry); major exporter of fruit, fish, and timber products; major crops—wheat, corn, grapes, beans, sugar beets, potatoes, deciduous fruit; livestock products—beef, poultry, wool, self-sufficient in most foods; 1991 fish catch of 6.6 million metric tons; net agricultural importer
Illicit drugs: a minor transshipment country for cocaine destined for the US and Europe; booming economy has made it more attractive to traffickers seeking to launder drug profits
Economic aid:
recipient: US commitments, including Ex-Im (FY70-89), $521 million; Western (non-US) countries, ODA and OOF bilateral commitments (1970-89), $1.6 billion; Communist countries (1970-89), $386 million
Currency: 1 Chilean peso (Ch$) = 100 centavos
Exchange rates: Chilean pesos (Ch$) per US$1—408 (January 1995), 420.08 (1994), 404.35 (1993), 362.59 (1992), 349.37 (1991), 305.06 (1990)
Fiscal year: calendar year

Transportation

Railroads:
total: 7,766 km
broad gauge: 3,974 km 1.676-m gauge (1,865 km electrified)
standard gauge: 150 km 1.435-m gauge
narrow gauge: 3,642 km 1.000-m gauge (80 km electrified)
Highways:
total: 79,599 km
paved: 10,984 km
unpaved: gravel or earth 68,615 km (1990)
Inland waterways: 725 km
Pipelines: crude oil 755 km; petroleum products 785 km; natural gas 320 km
Ports: Antofagasta, Arica, Chanarol, Coquimbo, Iquique, Puerto Montt, Punta Arenas, San Antonio, San Vicente, Talcahuano, Valparaiso
Merchant marine:
total: 36 ships (1,000 GRT or over) totaling 510,006 GRT/879,891 DWT
ships by type: bulk 13, cargo 7, chemical tanker 3, combination ore/oil 2, liquefied gas tanker 3, oil tanker 3, roll-on/roll-off cargo 3, vehicle carrier 2
Airports:
total: 390
with paved runways over 3,047 m: 5
with paved runways 2,438 to 3,047 m: 5
with paved runways 1,524 to 2,437 m: 18
with paved runways 914 to 1,523 m: 17
with paved runways under 914 m: 252
with unpaved runways over 3,047 m: 1
with unpaved runways 2,438 to 3,047 m: 3
with unpaved runways 1,524 to 2,438 m: 13
with unpaved runways 914 to 1,523 m: 76

Communications

Telephone system: 768,000 telephones; modern telephone system based on extensive microwave radio relay facilities
local: NA
intercity: extensive microwave radio relay links and 3 domestic satellite stations
international: 2 INTELSAT (Atlantic Ocean) earth stations
Radio:
broadcast stations: AM 159, FM 0, shortwave 11
radios: NA
Television:
broadcast stations: 131
televisions: NA

Defense Forces

Branches: Army of the Nation, National Navy (includes Naval Air, Coast Guard, and Marines), Air Force of the Nation, Carabineros of Chile (National Police), Investigations Police
Manpower availability: males age 15-49 3,758,770; males fit for military service 2,796,740; males reach military age (19) annually 121,831 (1995 est.)
Defense expenditures: exchange rate conversion—$1 billion, 3.4% of GDP (1991 est.)

China

(also see separate Taiwan entry)

Geography

Location: Eastern Asia, bordering the East China Sea, Korea Bay, Yellow Sea, and South China Sea, between North Korea and Vietnam
Map references: Asia
Area:
total area: 9,596,960 sq km
land area: 9,326,410 sq km
comparative area: slightly larger than the US
Land boundaries: total 22,143.34 km, Afghanistan 76 km, Bhutan 470 km, Burma 2,185 km, Hong Kong 30 km, India 3,380 km, Kazakhstan 1,533 km, North Korea 1,416 km, Kyrgyzstan 858 km, Laos 423 km, Macau 0.34 km, Mongolia 4,673 km, Nepal 1,236 km, Pakistan 523 km, Russia (northeast) 3,605 km, Russia (northwest) 40 km, Tajikistan 414 km, Vietnam 1,281 km
Coastline: 14,500 km
Maritime claims:
continental shelf: claim to shallow areas of East China Sea and Yellow Sea
territorial sea: 12 nm
International disputes: boundary with India in dispute; disputed sections of the boundary with Russia remain to be settled; boundary with Tajikistan in dispute; a short section of the boundary with North Korea is indefinite; involved in a complex dispute over the Spratly Islands with Malaysia, Philippines, Taiwan, Vietnam, and possibly Brunei; maritime boundary dispute with Vietnam in the Gulf of Tonkin; Paracel Islands occupied by China, but claimed by Vietnam and Taiwan; claims Japanese-administered Senkaku-shoto (Senkaku Islands/Diaoyu Tai), as does Taiwan
Climate: extremely diverse; tropical in south to subarctic in north
Terrain: mostly mountains, high plateaus, deserts in west; plains, deltas, and hills in east
Natural resources: coal, iron ore, petroleum, mercury, tin, tungsten, antimony, manganese, molybdenum, vanadium, magnetite, aluminum, lead, zinc, uranium, hydropower potential (world's largest)
Land use:
arable land: 10%
permanent crops: 0%
meadows and pastures: 31%
forest and woodland: 14%
other: 45%
Irrigated land: 478,220 sq km (1991—Chinese data)
Environment:
current issues: air pollution from the overwhelming use of high-sulfur coal as a fuel, produces acid rain which is damaging forests; water shortages experienced throughout the country, particularly in urban areas; future growth in water usage threatens to outpace supplies; water pollution from industrial effluents; much of the population does not have access to potable water; less than 10% of sewage receives treatment; deforestation; estimated loss of one-fifth of agricultural land since 1957 to soil erosion and economic development; desertification; trade in endangered species
natural hazards: frequent typhoons (about five per year along southern and eastern coasts); damaging floods; tsunamis; earthquakes; droughts
international agreements: party to—Antarctic-Environmental Protocol, Antarctic Treaty, Biodiversity, Climate Change, Endangered Species, Hazardous Wastes, Marine Dumping, Nuclear Test Ban, Ozone Layer Protection, Ship Pollution, Tropical Timber 83, Wetlands, Whaling; signed, but not ratified—Desertification, Law of the Sea
Note: world's third-largest country (after Russia and Canada)

People

Population: 1,203,097,268 (July 1995 est.)
Age structure:
0-14 years: 26% (female 151,266,866; male 167,234,782)
15-64 years: 67% (female 391,917,572; male 419,103,994)
65 years and over: 7% (female 39,591,692; male 33,982,362) (July 1995 est.)
Population growth rate: 1.04% (1995 est.)
Birth rate: 17.78 births/1,000 population (1995 est.)
Death rate: 7.36 deaths/1,000 population (1995 est.)
Net migration rate: 0 migrant(s)/1,000 population (1995 est.)
Infant mortality rate: 52.1 deaths/1,000 live births (1995 est.)
Life expectancy at birth:
total population: 68.08 years
male: 67.09 years
female: 69.18 years (1995 est.)
Total fertility rate: 1.84 children born/woman (1995 est.)
Nationality:
noun: Chinese (singular and plural)
adjective: Chinese
Ethnic divisions: Han Chinese 91.9%, Zhuang, Uygur, Hui, Yi, Tibetan, Miao, Manchu, Mongol, Buyi, Korean, and other nationalities 8.1%
Religions: Daoism (Taoism), Buddhism, Muslim 2%-3%, Christian 1% (est.)
note: officially atheist, but traditionally pragmatic and eclectic
Languages: Standard Chinese or Mandarin (Putonghua, based on the Beijing dialect), Yue (Cantonese), Wu (Shanghainese), Minbei (Fuzhou), Minnan (Hokkien-Taiwanese), Xiang, Gan, Hakka dialects, minority languages (see Ethnic divisions entry)
Literacy: age 15 and over can read and write (1990)
total population: 78%
male: 87%
female: 68%
Labor force: 583.6 million (1991)
by occupation: agriculture and forestry 60%, industry and commerce 25%, construction and mining 5%, social services 5%, other 5% (1990 est.)

Government

Names:
conventional long form: People's Republic of China
conventional short form: China
local long form: Zhonghua Renmin Gongheguo
local short form: Zhong Guo
Abbreviation: PRC
Digraph: CH
Type: Communist state
Capital: Beijing
Administrative divisions: 23 provinces (sheng, singular and plural), 5 autonomous regions* (zizhiqu, singular and plural), and 3 municipalities** (shi, singular and plural); Anhui, Beijing**, Fujian, Gansu, Guangdong, Guangxi*, Guizhou, Hainan, Hebei, Heilongjiang, Henan, Hubei, Hunan, Jiangsu, Jiangxi, Jilin, Liaoning, Nei Mongol*, Ningxia*, Qinghai, Shaanxi, Shandong, Shanghai**, Shanxi, Sichuan, Tianjin**, Xinjiang*, Xizang* (Tibet), Yunnan, Zhejiang
note: China considers Taiwan its 23rd province
Independence: 221 BC (unification under the Qin or Ch'in Dynasty 221 BC; Qing or Ch'ing Dynasty replaced by the Republic on 12 February 1912; People's Republic established 1 October 1949)
National holiday: National Day, 1 October (1949)
Constitution: most recent promulgated 4 December 1982
Legal system: a complex amalgam of custom and statute, largely criminal law; rudimentary civil code in effect since 1 January 1987; new legal codes in effect since 1 January 1980;

continuing efforts are being made to improve civil, administrative, criminal, and commercial law
Suffrage: 18 years of age; universal
Executive branch:
chief of state: President JIANG Zemin (since 27 March 1993); Vice President RONG Yiren (since 27 March 1993); election last held 27 March 1993 (next to be held 1998); results—JIANG Zemin was nominally elected by the Eighth National People's Congress
head of government: Premier LI Peng (Acting Premier since 24 November 1987, Premier since 9 April 1988) Vice Premier ZHU Rongji (since 8 April 1991); Vice Premier ZOU Jiahua (since 8 April 1991); Vice Premier QIAN Qichen (since 29 March 1993); Vice Premier LI Lanqing (29 March 1993); Vice Premier WU Bangguo (since 17 March 1995); Vice Premier JIANG Chunyun (since 17 March 1995)
cabinet: State Council; appointed by the National People's Congress (NPC)
Legislative branch: unicameral
National People's Congress: (Quanguo Renmin Daibiao Dahui) elections last held March 1993 (next to be held March 1998); results—CCP is the only party but there are also independents; seats—(2,977 total) (elected at county or xian level)
Judicial branch: Supreme People's Court
Political parties and leaders: Chinese Communist Party (CCP), JIANG Zemin, general secretary of the Central Committee (since 24 June 1989); eight registered small parties controlled by CCP
Other political or pressure groups: such meaningful opposition as exists consists of loose coalitions, usually within the party and government organization, that vary by issue
Member of: AfDB, APEC, AsDB, CCC, ESCAP, FAO, IAEA, IBRD, ICAO, ICFTU, ICRM, IDA, IFAD, IFC, IFRCS, ILO, IMF, IMO, INMARSAT, INTELSAT, INTERPOL, IOC, ISO, ITU, MINURSO, NAM (observer), PCA, UN, UN Security Council, UNCTAD, UNESCO, UNHCR, UNIDO, UNIKOM, UNITAR, UNOMIL, UNOMOZ, UNTSO, UNU, UPU, WHO, WIPO, WMO, WTO
Diplomatic representation in US:
chief of mission: Ambassador LI Daoyu
chancery: 2300 Connecticut Avenue NW, Washington, DC 20008
telephone: [1] (202) 328-2500 through 2502
consulate(s) general: Chicago, Houston, Los Angeles, New York, and San Francisco
US diplomatic representation:
chief of mission: Ambassador J. Stapleton ROY
embassy: Xiu Shui Bei Jie 3, 100600 Beijing
mailing address: PSC 461, Box 50, Beijing; FPO AP 96521-0002
telephone: [86] (1) 5323831
FAX: [86] (1) 5323178
consulate(s) general: Chengdu, Guangzhou, Shanghai, Shenyang
Flag: red with a large yellow five-pointed star and four smaller yellow five-pointed stars (arranged in a vertical arc toward the middle of the flag) in the upper hoist-side corner

Economy

Overview: Beginning in late 1978 the Chinese leadership has been trying to move the economy from the sluggish Soviet-style centrally planned economy to a more productive and flexible economy with market elements, but still within the framework of monolithic Communist control. To this end the authorities switched to a system of household responsibility in agriculture in place of the old collectivization, increased the authority of local officials and plant managers in industry, permitted a wide variety of small-scale enterprise in services and light manufacturing, and opened the economy to increased foreign trade and investment. The result has been a strong surge in production, particularly in agriculture in the early 1980s. Industry also has posted major gains, especially in coastal areas near Hong Kong and opposite Taiwan, where foreign investment and modern production methods have helped spur production of both domestic and export goods. Aggregate output has more than doubled since 1978. On the darker side, the leadership has often experienced in its hybrid system the worst results of socialism (bureaucracy, lassitude, corruption) and of capitalism (windfall gains and stepped-up inflation). Beijing thus has periodically backtracked, retightening central controls at intervals. In 1992-94 annual growth of GDP accelerated, particularly in the coastal areas—to more than 10% annually according to official claims. In late 1993 China's leadership approved additional long-term reforms aimed at giving more play to market-oriented institutions and at strengthening the center's control over the financial system. In 1994 strong growth continued in the widening market-oriented areas of the economy. At the same time, the government struggled to (a) collect revenues due from provinces, businesses, and individuals; (b) keep inflation within bounds; (c) reduce extortion and other economic crimes; and (d) keep afloat the large state-owned enterprises, most of which had not participated in the vigorous expansion of the economy. From 60 to 100 million surplus rural workers are adrift between the villages and the cities, many barely subsisting through part-time low-pay jobs. Popular resistance, changes in central policy, and loss of authority by rural cadres have weakened China's population control program, which is essential to the nation's long-term economic viability. One of the most dangerous long-term threats to continued rapid economic growth is the deterioration in the environment, notably air pollution, soil erosion, and the steady fall of the water table especially in the north.
National product: GDP—purchasing power parity—$2.9788 trillion (1994 estimate as extrapolated from World Bank estimate for 1992 by use of official Chinese growth statistics for 1993-94; because of the difficulties with official statistics in this time of rapid change, the result may overstate China's GDP by as much as 25%)
National product real growth rate: 11.8% (1994 est.)
National product per capita: $2,500 (1994 est.)
Inflation rate (consumer prices): 25.5% (December 1994 over December 1993)
Unemployment rate: 2.7% in urban areas (1994); substantial underemployment
Budget: deficit $13.7 billion (1994)
Exports: $121 billion (f.o.b., 1994)
commodities: textiles, garments, footwear, toys, machinery and equipment, weapon systems
partners: Hong Kong, Japan, US, Germany, South Korea, Russia (1993)
Imports: $115.7 billion (c.i.f., 1994)
commodities: rolled steel, motor vehicles, textile machinery, oil products, aircraft
partners: Japan, Taiwan, US, Hong Kong, Germany, South Korea (1993)
External debt: $100 billion (1994 est.)
Industrial production: growth rate 17.5% (1994 est.)
Electricity:
capacity: 162,000,000 kW
production: 746 billion kWh
consumption per capita: 593 kWh (1993)
Industries: iron and steel, coal, machine building, armaments, textiles and apparel, petroleum, cement, chemical fertilizers, consumer durables, food processing, autos, consumer electronics, telecommunications
Agriculture: accounts for almost 30% of GDP; among the world's largest producers of rice, potatoes, sorghum, peanuts, tea, millet, barley, and pork; commercial crops include cotton, other fibers, and oilseeds; produces variety of livestock products; basically self-sufficient in food; fish catch of 13.35 million metric tons (including fresh water and pond raised) (1991)
Illicit drugs: illicit producer of opium; bulk of production is in Yunnan Province (which produced 25 metric tons in 1994); transshipment point for heroin produced in the Golden Triangle
Economic aid:
donor: to less developed countries (1970-89) $7 billion
recipient: US commitments, including Ex-Im (FY70-87), $220.7 million; Western (non-US) countries, ODA and OOF bilateral commitments (1970-87), $13.5 billion
Currency: 1 yuan (¥) = 10 jiao
Exchange rates: yuan (¥) per US$1—8.4413 (January 1995), 8.6187 (1994), 5.7620 (1993), 5.5146 (1992), 5.3234 (1991), 4.7832 (1990)
note: beginning 1 January 1994, the People's

Bank of China quotes the midpoint rate against the US dollar based on the previous day's prevailing rate in the interbank foreign exchange market
Fiscal year: calendar year

Transportation

Railroads:
total: 65,780 km
standard gauge: 55,180 km 1.435-m gauge (7,174 km electrified; more than 11,000 km double track)
narrow gauge: 600 km 1.000-m gauge; 10,000 km 0.762-m to 1.067-m gauge dedicated industrial lines
Highways:
total: 1.029 million km
paved: 170,000 km
unpaved: gravel/improved earth 648,000 km; unimproved earth 211,000 km (1990)
Inland waterways: 138,600 km; about 109,800 km navigable
Pipelines: crude oil 9,700 km; petroleum products 1,100 km; natural gas 6,200 km (1990)
Ports: Aihui, Changsha, Dalian, Fuzhou, Guangzhou, Hangzhou, Harbin, Huangpu, Nanning, Ningbo, Qingdao, Qinhuangdao, Shanghai, Shantou, Tanggu, Xiamen, Xingang, Zhanjiang
Merchant marine:
total: 1,628 ships (1,000 GRT or over) totaling 16,013,532 GRT/24,027,766 DWT
ships by type: barge carrier 3, bulk 298, cargo 849, chemical tanker 14, combination bulk 10, container 98, liquefied gas tanker 4, multifunction large load carrier 1, oil tanker 212, passenger 24, passenger-cargo 25, refrigerated cargo 21, roll-on/roll-off cargo 24, short-sea passenger 44, vehicle carrier 1
note: China beneficially owns an additional 250 ships (1,000 GRT or over) totaling approximately 8,831,462 DWT that operate under Panamanian, Hong Kong, Maltese, Liberian, Vanuatu, Cypriot, Saint Vincent and the Grenadines, Bahamian, and Singaporean registry
Airports:
total: 204
with paved runways over 3,047 m: 17
with paved runways 2,438 to 3,047 m: 69
with paved runways 1,524 to 2,437 m: 89
with paved runways 914 to 1,523 m: 9
with paved runways under 914 m: 7
with unpaved runways 1,524 to 2,438 m: 7
with unpaved runways 914 to 1,523 m: 3
with unpaved runways under 914 m: 3

Communications

Telephone system: 20,000,000 telephones (summer 1994); domestic and international services are increasingly available for private use; unevenly distributed internal system serves principal cities, industrial centers, and most townships; expanding phone lines, interprovincial fiber optic links, satellite communications, cellullar/mobile communications, etc.
local: NA
intercity: fiber optic trunk lines, 55 earth stations for domestic satellites
international: 5 INTELSAT earth stations (4 Pacific Ocean and 1 Indian Ocean), 1 INMARSAT earth station; several international fiber optic links to Japan and Hong Kong
Radio:
broadcast stations: AM 274, FM NA, shortwave 0
radios: 215 million
Television:
broadcast stations: 202 (repeaters 2,050)
televisions: 75 million

Defense Forces

Branches: People's Liberation Army (PLA), which includes the Ground Forces, Navy (includes Marines and Naval Aviation), Air Force, Second Artillery Corps (the strategic missile force), People's Armed Police (internal security troops, nominally subordinate to Ministry of Public Security, but included by the Chinese as part of the "armed forces" and considered to be an adjunct to the PLA in war time)
Manpower availability: males age 15-49 351,330,411; males fit for military service 194,286,619; males reach military age (18) annually 9,841,658 (1995 est.)
Defense expenditures: defense budget—63.09 billion yuan, NA% of GDP (1995 est.); note—conversion of the defense budget into US dollars using the current exchange rate could produce misleading results

Christmas Island
(territory of Australia)

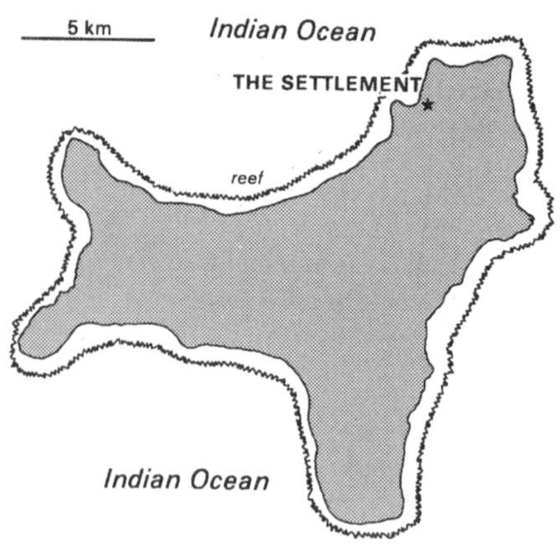

Geography

Location: Southeastern Asia, island in the Indian Ocean, south of Indonesia
Map references: Southeast Asia
Area:
total area: 135 sq km
land area: 135 sq km
comparative area: about 0.8 times the size of Washington, DC
Land boundaries: 0 km
Coastline: 138.9 km
Maritime claims:
contiguous zone: 12 nm
exclusive fishing zone: 200 nm
territorial sea: 3 nm
International disputes: none
Climate: tropical; heat and humidity moderated by trade winds
Terrain: steep cliffs along coast rise abruptly to central plateau
Natural resources: phosphate
Land use:
arable land: 0%
permanent crops: 0%
meadows and pastures: 0%
forest and woodland: 0%
other: 100%
Irrigated land: NA sq km
Environment:
current issues: NA
natural hazards: almost completely surrounded by a reef which can be a maritime hazard
international agreements: NA
Note: located along major sea lanes of Indian Ocean

People

Population: 889 (July 1995 est.)
Age structure:
0-14 years: NA
15-64 years: NA
65 years and over: NA
Population growth rate: -9% (1995 est.)
Birth rate: NA

Death rate: NA
Net migration rate: NA
Infant mortality rate: NA
Life expectancy at birth:
total population: NA
male: NA
female: NA
Total fertility rate: NA
Nationality:
noun: Christmas Islander(s)
adjective: Christmas Island
Ethnic divisions: Chinese 61%, Malay 25%, European 11%, other 3%, no indigenous population
Religions: Buddhist 36.1%, Muslim 25.4%, Christian 17.7% (Roman Catholic 8.2%, Church of England 3.2%, Presbyterian 0.9%, Uniting Church 0.4%, Methodist 0.2%, Baptist 0.1%, and other 4.7%), none 12.7%, unknown 4.6%, other 3.5% (1981)
Languages: English
Labor force: NA
by occupation: all workers are employees of the Phosphate Mining Company of Christmas Island, Ltd.

Government

Names:
conventional long form: Territory of Christmas Island
conventional short form: Christmas Island
Digraph: KT
Type: territory of Australia
Capital: The Settlement
Administrative divisions: none (territory of Australia)
Independence: none (territory of Australia)
National holiday: NA
Constitution: Christmas Island Act of 1958
Legal system: under the authority of the governor general of Australia
Executive branch:
chief of state: Queen ELIZABETH II (since 6 February 1952)
head of government: Administrator M. J. GRIMES (since NA)
cabinet: Advisory Council
Legislative branch: none
Judicial branch: none
Political parties and leaders: none
Member of: none
Diplomatic representation in US: none (territory of Australia)
US diplomatic representation: none (territory of Australia)
Flag: the flag of Australia is used

Economy

Overview: Phosphate mining had been the only significant economic activity, but in December 1987 the Australian Government closed the mine as no longer economically viable. Plans have been under way to reopen the mine and also to build a casino and hotel to develop tourism.
National product: GDP $NA
National product real growth rate: NA%
National product per capita: $NA
Inflation rate (consumer prices): NA%
Unemployment rate: NA%
Budget:
revenues: $NA
expenditures: $NA, including capital expenditures of $NA
Exports: $NA
commodities: phosphate
partners: Australia, NZ
Imports: $NA
commodities: consumer goods
partners: principally Australia
External debt: $NA
Industrial production: growth rate NA%
Electricity:
capacity: 11,000 kW
production: 30 million kWh
consumption per capita: 17,800 kWh (1990)
Industries: phosphate extraction (near depletion)
Agriculture: NA
Economic aid: none
Currency: 1 Australian dollar ($A) = 100 cents
Exchange rates: Australian dollars ($A) per US$1—1.3058 (January 1995), 1.3667 (1994), 1.4704, (1993), 1.3600 (1992), 1.2836 (1991), 1.2799 (1990)
Fiscal year: 1 July—30 June

Transportation

Railroads: 0 km
Highways:
total: NA km
paved: NA km
unpaved: NA km
Ports: Flying Fish Cove
Merchant marine: none
Airports:
total: 1
with paved runways 1,524 to 2,437 m: 1

Communications

Telephone system: NA telephones
local: NA
intercity: NA
international: NA
Radio:
broadcast stations: AM 1, FM 0, shortwave 0
radios: NA
Television:
broadcast stations: 1
televisions: NA

Defense Forces

Note: defense is the responsibility of Australia

Clipperton Island

(possession of France)

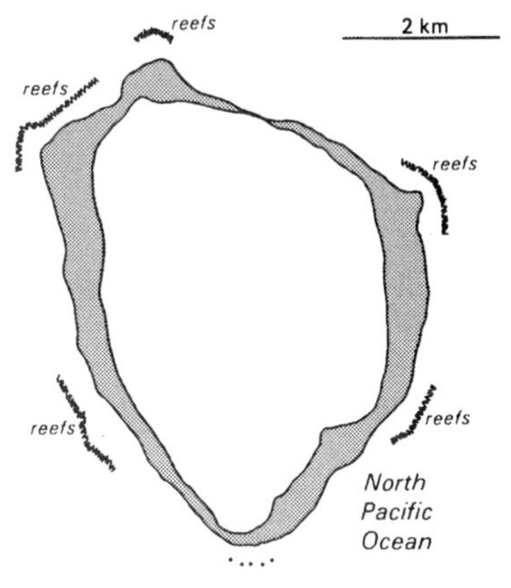

Geography

Location: Middle America, atoll in the North Pacific Ocean, southwest of Mexico
Map references: World
Area:
total area: 7 sq km
land area: 7 sq km
comparative area: about 12 times the size of The Mall in Washington, DC
Land boundaries: 0 km
Coastline: 11.1 km
Maritime claims:
exclusive economic zone: 200 nm
territorial sea: 12 nm
International disputes: claimed by Mexico
Climate: tropical
Terrain: coral atoll
Natural resources: none
Land use:
arable land: 0%
permanent crops: 0%
meadows and pastures: 0%
forest and woodland: 0%
other: 100% (all coral)
Irrigated land: 0 sq km
Environment:
current issues: NA
natural hazards: NA
international agreements: NA
Note: reef about 8 km in circumference

People

Population: uninhabited

Government

Names:
conventional long form: none
conventional short form: Clipperton Island
local long form: none
local short form: Ile Clipperton
former: sometimes called Ile de la Passion
Digraph: IP
Type: French possession administered by

Clipperton Island *(continued)*

France from French Polynesia by High Commissioner of the Republic
Capital: none; administered by France from French Polynesia
Independence: none (possession of France)

Economy

Overview: The only economic activity is a tuna fishing station.

Transportation

Ports: none; offshore anchorage only

Defense Forces

Note: defense is the responsibility of France

Cocos (Keeling) Islands
(territory of Australia)

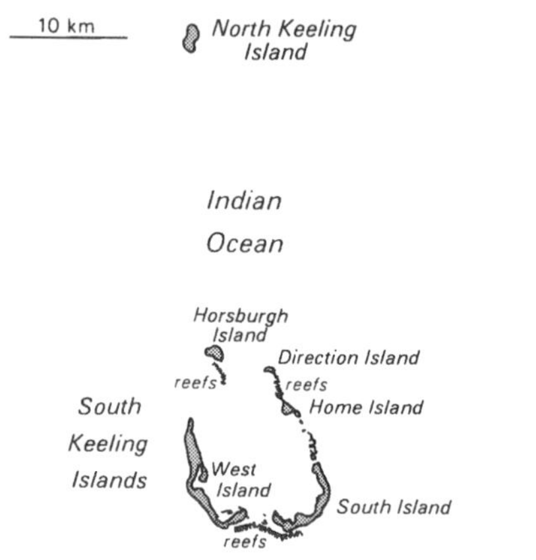

Geography

Location: Southeastern Asia, group of islands in the Indian Ocean, south of Indonesia, about one-half of the way from Australia to Sri Lanka
Map references: Southeast Asia
Area:
total area: 14 sq km
land area: 14 sq km
comparative area: about 24 times the size of The Mall in Washington, DC
note: includes the two main islands of West Island and Home Island
Land boundaries: 0 km
Coastline: 2.6 km
Maritime claims:
exclusive fishing zone: 200 nm
territorial sea: 3 nm
International disputes: none
Climate: pleasant, modified by the southeast trade wind for about nine months of the year; moderate rain fall
Terrain: flat, low-lying coral atolls
Natural resources: fish
Land use:
arable land: NA%
permanent crops: NA%
meadows and pastures: NA%
forest and woodland: NA%
other: NA%
Irrigated land: NA sq km
Environment:
current issues: there are no natural fresh water resources on the island, groundwater does accumulate in natural underground reservoirs
natural hazards: cyclones may occur in the early months of the year
international agreements: NA
Note: two coral atolls thickly covered with coconut palms and other vegetation

People

Population: 604 (July 1995 est.)
Age structure:
0-14 years: NA
15-64 years: NA
65 years and over: NA
Population growth rate: 0.98% (1995 est.)
Birth rate: NA
Death rate: NA
Net migration rate: NA
Infant mortality rate: NA
Life expectancy at birth: NA
Total fertility rate: NA
Nationality:
noun: Cocos Islander(s)
adjective: Cocos Islander
Ethnic divisions:
West Island: Europeans
Home Island: Cocos Malays
Religions: Sunni Muslims
Languages: English
Labor force: NA

Government

Names:
conventional long form: Territory of Cocos (Keeling) Islands
conventional short form: Cocos (Keeling) Islands
Digraph: CK
Type: territory of Australia
Capital: West Island
Administrative divisions: none (territory of Australia)
Independence: none (territory of Australia)
National holiday: NA
Constitution: Cocos (Keeling) Islands Act of 1955
Legal system: based upon the laws of Australia and local laws
Suffrage: NA
Executive branch:
chief of state: Queen ELIZABETH II (since 6 February 1952)
head of government: Administrator B. CUNNINGHAM (since NA)
cabinet: Islands Council; Chairman of the Islands Council Haji WAHIN bin Bynie (since NA)
Legislative branch: unicameral Islands Council
Judicial branch: Supreme Court
Political parties and leaders: NA
Member of: none
Diplomatic representation in US: none (territory of Australia)
US diplomatic representation: none (territory of Australia)
Flag: the flag of Australia is used

Economy

Overview: Grown throughout the islands, coconuts are the sole cash crop. Copra and fresh coconuts are the major export earners. Small local gardens and fishing contribute to the food supply, but additional food and most other necessities must be imported from Australia.

National product: GDP $NA
National product real growth rate: NA%
National product per capita: $NA
Inflation rate (consumer prices): NA%
Budget:
revenues: $NA
expenditures: $NA, including capital expenditures of $NA
Exports: $NA
commodities: copra
partners: Australia
Imports: $NA
commodities: foodstuffs
partners: Australia
External debt: $NA
Industrial production: growth rate NA%
Electricity:
capacity: 1,000 kW
production: 2 million kWh
consumption per capita: 2,980 kWh (1990)
Industries: copra products
Agriculture: gardens provide vegetables, bananas, pawpaws, coconuts
Economic aid: none
Currency: 1 Australian dollar ($A) = 100 cents
Exchange rates: Australian dollars ($A) per US$1—1.3058 (January 1995), 1.3667 (1994), 1.4704 (1993), 1.3600 (1992), 1.2836 (1991), 1.2799 (1990)
Fiscal year: 1 July—30 June

Transportation

Railroads: 0 km
Highways:
total: NA km
paved: NA km
unpaved: NA km
Ports: none; lagoon anchorage only
Merchant marine: none
Airports:
total: 1
with paved runways 1,524 to 2,437 m: 1

Communications

Telephone system: NA telephones
local: NA
intercity: NA
international: linked by telephone, telex, and facsimile communications via satellite with Australia
Radio:
broadcast stations: AM 1, FM 0, shortwave 0
radios: 250 (1985)
Television:
broadcast stations: 0
televisions: NA

Defense Forces

Note: defense is the responsibility of Australia

Colombia

Geography

Location: Northern South America, bordering the Caribbean Sea, between Panama and Venezuela, and bordering the North Pacific Ocean, between Ecuador and Panama
Map references: South America
Area:
total area: 1,138,910 sq km
land area: 1,038,700 sq km
comparative area: slightly less than three times the size of Montana
note: includes Isla de Malpelo, Roncador Cay, Serrana Bank, and Serranilla Bank
Land boundaries: total 7,408 km, Brazil 1,643 km, Ecuador 590 km, Panama 225 km, Peru 2,900 km, Venezuela 2,050 km
Coastline: 3,208 km (Caribbean Sea 1,760 km, North Pacific Ocean 1,448 km)
Maritime claims:
continental shelf: 200-m depth or to the depth of exploitation
exclusive economic zone: 200 nm
territorial sea: 12 nm
International disputes: maritime boundary dispute with Venezuela in the Gulf of Venezuela; territorial dispute with Nicaragua over Archipelago de San Andres y Providencia and Quita Sueno Bank
Climate: tropical along coast and eastern plains; cooler in highlands
Terrain: flat coastal lowlands, central highlands, high Andes Mountains, eastern lowland plains
Natural resources: petroleum, natural gas, coal, iron ore, nickel, gold, copper, emeralds
Land use:
arable land: 4%
permanent crops: 2%
meadows and pastures: 29%
forest and woodland: 49%
other: 16%
Irrigated land: 5,150 sq km (1989 est.)
Environment:
current issues: deforestation; soil damage from overuse of pesticides; air pollution, especially in Bogota, from vehicle emissions
natural hazards: highlands subject to volcanic eruptions; occasional earthquakes; periodic droughts
international agreements: party to—Antarctic Treaty, Biodiversity, Endangered Species, Marine Life Conservation, Nuclear Test Ban, Ozone Layer Protection, Ship Pollution, Tropical Timber 83; signed, but not ratified—Antarctic-Environmental Protocol, Climate Change, Desertification, Hazardous Wastes, Law of the Sea, Marine Dumping
Note: only South American country with coastlines on both North Pacific Ocean and Caribbean Sea

People

Population: 36,200,251 (July 1995 est.)
Age structure:
0-14 years: 32% (female 5,784,010; male 5,925,600)
15-64 years: 63% (female 11,642,870; male 11,245,235)
65 years and over: 5% (female 888,358; male 714,178) (July 1995 est.)
Population growth rate: 1.7% (1995 est.)
Birth rate: 21.89 births/1,000 population (1995 est.)
Death rate: 4.69 deaths/1,000 population (1995 est.)
Net migration rate: -0.17 migrant(s)/1,000 population (1995 est.)
Infant mortality rate: 26.9 deaths/1,000 live births (1995 est.)
Life expectancy at birth:
total population: 72.48 years
male: 69.68 years
female: 75.38 years (1995 est.)
Total fertility rate: 2.4 children born/woman (1995 est.)
Nationality:
noun: Colombian(s)
adjective: Colombian
Ethnic divisions: mestizo 58%, white 20%, mulatto 14%, black 4%, mixed black-Indian 3%, Indian 1%
Religions: Roman Catholic 95%
Languages: Spanish
Literacy: age 15 and over can read and write (1985)
total population: 88%
male: 88%
female: 88%
Labor force: 12 million (1990)
by occupation: services 46%, agriculture 30%, industry 24% (1990)

Government

Names:
conventional long form: Republic of Colombia
conventional short form: Colombia
local long form: Republica de Colombia
local short form: Colombia

Digraph: CO
Type: republic; executive branch dominates government structure
Capital: Bogota
Administrative divisions: 32 departments (departamentos, singular—departamento) and 1 capital district* (distrito capital); Amazonas, Antioquia, Arauca, Atlantico, Bogota*, Bolivar, Boyaca, Caldas, Caqueta, Casanare, Cauca, Cesar, Choco, Cordoba, Cundinamarca, Guainia, Guaviare, Huila, La Guajira, Magdalena, Meta, Narino, Norte de Santander, Putumayo, Quindio, Risaralda, San Andres y Providencia, Santander, Sucre, Tolima, Valle del Cauca, Vaupes, Vichada
Independence: 20 July 1810 (from Spain)
National holiday: Independence Day, 20 July (1810)
Constitution: 5 July 1991
Legal system: based on Spanish law; a new criminal code modeled after US procedures was enacted in 1992-93; judicial review of executive and legislative acts; accepts compulsory ICJ jurisdiction, with reservations
Suffrage: 18 years of age; universal and compulsory
Executive branch:
chief of state and head of government: President Ernesto SAMPER Pizano (since 7 August 1994); election last held 29 May 1994 (next to be held May 1998) and resulted in no candidate receiving more than 50% of the total vote; a run-off election to select a president from the two leading candidates was held on 19 June 1994; results—Ernesto SAMPER Pizano (Liberal Party) 50.4%, Andres PASTRANA Arango (Conservative Party) 48.6%, blank votes 1%; Humberto de la CALLE was elected vice president in a new procedure that replaces the traditional designation of vice presidents by newly elected presidents.
cabinet: Cabinet
Legislative branch: bicameral Congress (Congreso)
Senate (Senado): elections last held 13 March 1994 (next to be held NA March 1998); preliminary results—percent of vote by party NA; seats—(102 total) Liberal Party 59, conservatives (includes PC, MSN, and NDF) 31, other 12
House of Representatives (Camara de Representantes): elections last held 13 March 1994 (next to be held NA March 1998); preliminary results—percent of vote by party NA; seats—(161 total) Liberal Party 89, conservatives (includes PC, MSN, and NDF) 53, AD/M-19 2, other 17
Judicial branch: Supreme Court of Justice (Corte Suprema de Justical), Constitutional Court, Council of State
Political parties and leaders: Liberal Party (PL), Juan Guillermo ANGEL; Conservative Party (PC), Fabio VALENCIA Cossio; National Salvation Movement (MSN), Alvaro GOMEZ Hurtado; New Democratic Force (NDF), Andres PASTRANA Arango; Democratic Alliance M-19 (AD/M-19) is a coalition of small leftist parties and dissident liberals and conservatives; Patriotic Union (UP) is a legal political party formed by Revolutionary Armed Forces of Colombia (FARC) and Colombian Communist Party (PCC), Carlos ROMERO
Other political or pressure groups: three insurgent groups are active in Colombia—Revolutionary Armed Forces of Colombia (FARC), Manuel MARULANDA and Alfonso CANO; National Liberation Army (ELN), Manuel PEREZ; and dissidents of the recently demobilized People's Liberation Army (EPL), Francisco CARABALLO; Francisco CARABALLO was captured by the government in June 1994
Member of: AG, CCC, CDB, CG, ECLAC, FAO, G-11, G-24, G-77, GATT, IADB, IAEA, IBRD, ICAO, ICC, ICFTU, ICRM, IDA, IFAD, IFC, IFRCS, ILO, IMF, IMO, INMARSAT, INTELSAT, INTERPOL, IOC, IOM, ISO, ITU, LAES, LAIA, NAM, OAS, ONUSAL, OPANAL, PCA, RG, UN, UNCTAD, UNESCO, UNHCR, UNIDO, UNPROFOR, UNU, UPU, WCL, WFTU, WHO, WIPO, WMO, WTO
Diplomatic representation in US:
chief of mission: Ambassador Carlos LLERAS de la Fuente
chancery: 2118 Leroy Place NW, Washington, DC 20008
telephone: [1] (202) 387-8338
FAX: [1] (202) 232-8643
consulate(s) general: Boston, Chicago, Houston, Los Angeles, Miami, New Orleans, New York, San Francisco, San Juan (Puerto Rico), and Washington, DC
consulate(s): Atlanta and Tampa
US diplomatic representation:
chief of mission: Ambassador Myles R. R. FRECHETTE
embassy: Calle 38, No. 8-61, Bogota
mailing address: Apartado Aereo 3831, Bogota; APO AA 34038
telephone: [57] (1) 320-1300
FAX: [57] (1) 288-5687
consulate(s): Barranquilla
Flag: three horizontal bands of yellow (top, double-width), blue, and red; similar to the flag of Ecuador, which is longer and bears the Ecuadorian coat of arms superimposed in the center

Economy

Overview: Colombia's economy has grown steadily since 1991, when the government implemented sweeping economic reform measures. President SAMPER, who took office in August 1994, has pledged to maintain those reforms while expanding government assistance for poor Colombians, who continue to make up about 40% of the population. In an effort to bring down inflation, SAMPER has arranged a "social pact" with business and labor to curtail price hikes and trim inflation to 18%. The rapid development of oil, coal, and other nontraditional industries, along with copious inflows of capital and strengthening of prices for coffee, have helped keep growth at 5%-6%. Development of the massive Cusiana oilfield provides the means to sustain this level over the next several years. Exporters say, however, that their sales have been hampered by the appreciation of the Colombian peso, and farmers have sought government help in adjusting to greater foreign competition. Moreover, increased foreign investment and even greater domestic growth have been hindered by an inadequate energy and transportation infrastructure and by violence stemming from drug trafficking and persistent rural insurgency.
National product: GDP—purchasing power parity—$172.4 billion (1994 est.)
National product real growth rate: 5.7% (1994 est.)
National product per capita: $4,850 (1994 est.)
Inflation rate (consumer prices): 22.6% (1994 est.)
Unemployment rate: 7.9% (1994 est.)
Budget:
revenues: $16 billion (1995 est.)
expenditures: $21 billion (1995 est.)
Exports: $8.3 billion (f.o.b., 1994 est.)
commodities: petroleum, coffee, coal, bananas, fresh cut flowers
partners: US 39%, EC 25.7%, Japan 2.9%, Venezuela 8.5% (1992)
Imports: $10.6 billion (c.i.f., 1994 est.)
commodities: industrial equipment, transportation equipment, consumer goods, chemicals, paper products
partners: US 36%, EC 18%, Brazil 4%, Venezuela 6.5%, Japan 8.7% (1992)
External debt: $12.6 billion (1994 est.)
Industrial production: growth rate 5% (1994 est.); accounts for about 20% of GDP
Electricity:
capacity: 10,220,000 kW
production: 33 billion kWh
consumption per capita: 890 kWh (1993)
Industries: textiles, food processing, oil, clothing and footwear, beverages, chemicals, metal products, cement; mining—gold, coal, emeralds, iron, nickel, silver, salt
Agriculture: growth rate 3.8% (1994 est.); accounts for about 15% of GDP; crops make up two-thirds and livestock one-third of agricultural output; climate and soils permit a wide variety of crops, such as coffee, rice, tobacco, corn, sugarcane, cocoa beans, oilseeds, vegetables; forest products and shrimp farming are becoming more important
Illicit drugs: illicit producer of coca, opium poppies, and cannabis; about 45,000 hectares of coca under cultivation; the world's largest processor of coca derivatives into cocaine; supplier of cocaine to the US and other

international drug markets; active eradication program against narcotics crop
Economic aid:
recipient: US commitments, including Ex-Im (FY70-89), $1.6 billion; Western (non-US) countries, ODA and OOF bilateral commitments (1970-89), $3.3 billion; Communist countries (1970-89), $399 million
Currency: 1 Colombian peso (Col$) = 100 centavos
Exchange rates: Colombian pesos (Col$) per US$1—846.67 (January 1995), 844.84 (1994), 863.06 (1993), 759.28 (1992), 633.05 (1991), 502.26 (1990)
Fiscal year: calendar year

Transportation

Railroads:
total: 3,386 km
standard gauge: 150 km 1.435-m gauge
narrow gauge: 3,236 km 0.914-m gauge (2,611 km in use)
Highways:
total: 107,377 km (1991)
paved: 12,778 km
unpaved: gravel/earth 94,599 km
Inland waterways: 14,300 km, navigable by river boats
Pipelines: crude oil 3,585 km; petroleum products 1,350 km; natural gas 830 km; natural gas liquids 125 km
Ports: Barranquilla, Buenaventura, Cartagena, Leticia, Puerto Bolivar, San Andres, Santa Marta, Tumaco, Turbo
Merchant marine:
total: 22 ships (1,000 GRT or over) totaling 104,577 GRT/142,617 DWT
ships by type: bulk 6, cargo 9, container 4, oil tanker 3
Airports:
total: 1,307
with paved runways over 3,047 m: 2
with paved runways 2,438 to 3,047 m: 7
with paved runways 1,524 to 2,437 m: 34
with paved runways 914 to 1,523 m: 31
with paved runways under 914 m: 734
with unpaved runways 1,524 to 2,438 m: 80
with unpaved runways 914 to 1,523 m: 419

Communications

Telephone system: 1,890,000 telephones; modern system in many respects
local: NA
intercity: nationwide microwave radio relay system; 11 domestic satellite earth stations
international: 2 INTELSAT (Atlantic Ocean) earth stations
Radio:
broadcast stations: AM 413, FM 0, shortwave 28
radios: NA
Television:
broadcast stations: 33
televisions: NA

Defense Forces

Branches: Army (Ejercito Nacional), Navy (Armada Nacional, includes Marines and Coast Guard), Air Force (Fuerza Aerea Colombiana), National Police (Policia Nacional)
Manpower availability: males age 15-49 9,851,980; males fit for military service 6,640,348; males reach military age (18) annually 349,599 (1995 est.)
Defense expenditures: exchange rate conversion—$1.2 billion (1992 est.)

Comoros

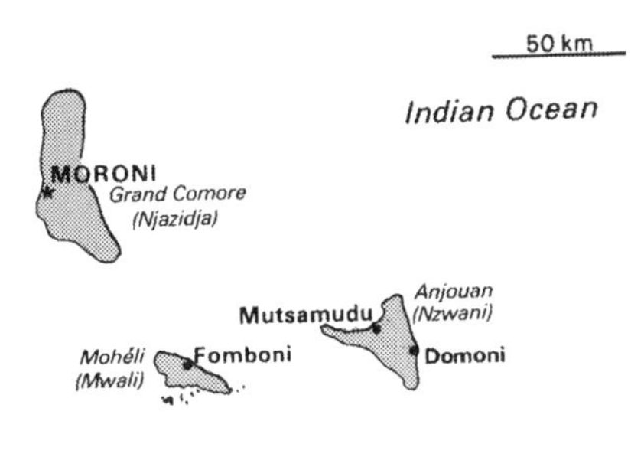

Geography

Location: Southern Africa, group of islands in the Mozambique Channel, about two-thirds of the way between northern Madagascar and northern Mozambique
Map references: Africa
Area:
total area: 2,170 sq km
land area: 2,170 sq km
comparative area: slightly more than 12 times the size of Washington, DC
Land boundaries: 0 km
Coastline: 340 km
Maritime claims:
exclusive economic zone: 200 nm
territorial sea: 12 nm
International disputes: claims French-administered Mayotte
Climate: tropical marine; rainy season (November to May)
Terrain: volcanic islands, interiors vary from steep mountains to low hills
Natural resources: negligible
Land use:
arable land: 35%
permanent crops: 8%
meadows and pastures: 7%
forest and woodland: 16%
other: 34%
Irrigated land: NA sq km
Environment:
current issues: soil degradation and erosion results from crop cultivation on slopes without proper terracing; deforestation
natural hazards: cyclones and tsunamis possible during rainy season (December to April); Mount Kartala on Grand Comore is an active volcano
international agreements: party to—Biodiversity, Climate Change, Endangered Species, Hazardous Wastes, Law of the Sea, Ozone Layer Protection; signed, but not ratified—Desertification
Note: important location at northern end of Mozambique Channel

People

Population: 549,338 (July 1995 est.)
Age structure:
0-14 years: 48% (female 131,334; male 132,327)
15-64 years: 49% (female 137,083; male 133,629)
65 years and over: 3% (female 7,860; male 7,105) (July 1995 est.)
Population growth rate: 3.56% (1995 est.)
Birth rate: 46.22 births/1,000 population (1995 est.)
Death rate: 10.6 deaths/1,000 population (1995 est.)
Net migration rate: 0 migrant(s)/1,000 population (1995 est.)
Infant mortality rate: 77.3 deaths/1,000 live births (1995 est.)
Life expectancy at birth:
total population: 58.27 years
male: 56.04 years
female: 60.57 years (1995 est.)
Total fertility rate: 6.73 children born/woman (1995 est.)
Nationality:
noun: Comoran(s)
adjective: Comoran
Ethnic divisions: Antalote, Cafre, Makoa, Oimatsaha, Sakalava
Religions: Sunni Muslim 86%, Roman Catholic 14%
Languages: Arabic (official), French (official), Comoran (a blend of Swahili and Arabic)
Literacy: age 15 and over can read and write (1980)
total population: 48%
male: 56%
female: 40%
Labor force: 140,000 (1982)
by occupation: agriculture 80%, government 3%

Government

Names:
conventional long form: Federal Islamic Republic of the Comoros
conventional short form: Comoros
local long form: Republique Federale Islamique des Comores
local short form: Comores
Digraph: CN
Type: independent republic
Capital: Moroni
Administrative divisions: three islands; Grand Comore (Njazidja), Anjouan (Nzwani), and Moheli (Mwali)
note: there are also four municipalities named Domoni, Fomboni, Moroni, and Mutsamudu
Independence: 6 July 1975 (from France)
National holiday: Independence Day, 6 July (1975)
Constitution: 7 June 1992
Legal system: French and Muslim law in a new consolidated code
Suffrage: 18 years of age; universal
Executive branch:
chief of state: President Said Mohamed DJOHAR (since 11 March 1990); election last held 11 March 1990 (next to be held March 1996); results—Said Mohamed DJOHAR (UDZIMA) 55%, Mohamed TAKI Abdulkarim (UNDC) 45%
head of government: Prime Minister Halifa HOUMADI (since 13 October 1994); note—HOUMADI is the fifteenth prime minister appointed by President DJOHAR in the last three years
cabinet: Council of Ministers; appointed by the president
Legislative branch: unicameral
Federal Assembly (Assemblee Federale): elections last held 12-20 December 1993 (next to be held by NA January 1998); results—percent of vote by party NA; seats—(42 total) Ruling Coalition: RDR 15, UNDC 5, MWANGAZA 2; Opposition: UDZIMA 8, other smaller parties 10; 2 seats remained unfilled
Judicial branch: Supreme Court (Cour Supreme)
Political parties and leaders: over 20 political parties are currently active, the most important of which are; Comoran Union for Progress (UDZIMA), Omar TAMOU; Islands' Fraternity and Unity Party (CHUMA), Said Ali KEMAL; Comoran Party for Democracy and Progress (PCDP), Ali MROUDJAE; Realizing Freedom's Capability (UWEZO), Mouazair ABDALLAH; Democratic Front of the Comoros (FDR), Moustapha CHELKH; Dialogue Proposition Action (DPA/MWANGAZA), Said MCHAWGAMA; Rally for Change and Democracy (RACHADE), Hassan HACHIM; Union for Democracy and Decentralization (UNDC), Mohamed Taki Halidi IBRAHAM; Rally for Democracy and Renewal (RDR); Comoran Popular Front (FPC), Mohamed HASSANALI, Mohamed El Arif OUKACHA, Abdou MOUSTAKIM (Secretary General)
Member of: ACCT, ACP, AfDB, AL, CCC, ECA, FAO, FZ, G-77, IBRD, ICAO, IDA, IDB, IFAD, IFC, IFRCS (associate), ILO, IMF, INTELSAT (nonsignatory user), IOC, ITU, NAM, OAU, OIC, UN, UNCTAD, UNESCO, UNIDO, UPU, WHO, WMO
Diplomatic representation in US:
chief of mission: Ambassador Mohamed Ahamadu DJIMBANAO (ambassador to the US and Canada)
chancery: (temporary) care of the Permanent Mission of the Federal and Islamic Republic of the Comoros to the United Nations, 336 East 45th Street, 2nd Floor, New York, NY 10017
telephone: [1] (212) 972-8010
FAX: [1] (212) 983-4712
US diplomatic representation: none; ambassador to Port Louis, Mauritius, is accredited to Comoros
Flag: green with a white crescent in the center of the field, its points facing upward; there are four white five-pointed stars placed in a line between the points of the crescent; the crescent, stars, and color green are traditional symbols of Islam; the four stars represent the four main islands of the archipelago—Mwali, Njazidja, Nzwani, and Mayotte (a territorial collectivity of France, but claimed by Comoros); the design, the most recent of several, is described in the constitution approved by referendum on 7 June 1992

Economy

Overview: One of the world's poorest countries, Comoros is made up of several islands that have poor transportation links, a young and rapidly increasing population, and few natural resources. The low educational level of the labor force contributes to a subsistence level of economic activity, high unemployment, and a heavy dependence on foreign grants and technical assistance. Agriculture, including fishing, hunting, and forestry, is the leading sector of the economy. It contributes 40% to GDP, employs 80% of the labor force, and provides most of the exports. The country is not self-sufficient in food production; rice, the main staple, accounts for 90% of imports. The government is struggling to upgrade education and technical training, to privatize commercial and industrial enterprises, to improve health services, to diversify exports, and to reduce the high population growth rate. Continued foreign support is essential if the goal of 4% annual GDP growth is to be reached in the late 1990s.
National product: GDP—purchasing power parity—$370 million (1994 est.)
National product real growth rate: 0.9% (1994 est.)
National product per capita: $700 (1994 est.)
Inflation rate (consumer prices): 15% (1993 est.)
Unemployment rate: 15.8% (1989)
Budget:
revenues: $83 million
expenditures: $92 million, including capital expenditures of $32 million (1992)
Exports: $13.7 million (f.o.b., 1993 est.)
commodities: vanilla, ylang-ylang, cloves, perfume oil, copra
partners: US 44%, France 40%, Germany 6%, Africa 5% (1992)
Imports: $40.9 million (f.o.b., 1993 est.)
commodities: rice and other foodstuffs, petroleum products, cement, consumer goods
partners: France 34%, South Africa 14%, Kenya 8%, Japan 4% (1992)
External debt: $160 million (1992 est.)
Industrial production: growth rate -6.5% (1989 est.); accounts for 6% of GDP

Electricity:
capacity: 16,000 kW
production: 17 million kWh
consumption per capita: 27 kWh (1993)
Industries: perfume distillation, textiles, furniture, jewelry, construction materials, soft drinks
Agriculture: accounts for 40% of GDP; most of population works in subsistence agriculture and fishing; plantations produce cash crops for export—vanilla, cloves, perfume essences, copra; principal food crops—coconuts, bananas, cassava; world's leading producer of essence of ylang-ylang (for perfumes) and second-largest producer of vanilla; large net food importer
Economic aid:
recipient: US commitments, including Ex-Im (FY80-89), $10 million; Western (non-US) countries, ODA and OOF bilateral commitments (1970-89), $435 million; OPEC bilateral aid (1979-89), $22 million; Communist countries (1970-89), $18 million
Currency: 1 Comoran franc (CF) = 100 centimes
Exchange rates: Comoran francs (CF) per US$1—297.07 (January 1995), 416.40 (1994), 254.57 (1993), 264.69 (1992), 282.11 (1991), 272.26 (1990)
note: beginning 12 January 1994, the Comoran franc was devalued to 75 per French franc from 50 per French franc at which it had been fixed since 1948
Fiscal year: calendar year

Transportation

Railroads: 0 km
Highways:
total: 750 km
paved: bituminous 210 km
unpaved: crushed stone, gravel 540 km
Ports: Fomboni, Moroni, Mutsamudo
Merchant marine: none
Airports:
total: 4
with paved runways 2,438 to 3,047 m: 1
with paved runways 914 to 1,523 m: 3

Communications

Telephone system: over 1,800 telephones; sparse system of radio relay and high-frequency radio communication stations for interisland and external communications to Madagascar and Reunion
local: NA
intercity: high frequency radio and microwave radio relay
international: high frequency radio
Radio:
broadcast stations: AM 2, FM 1, shortwave 0
radios: NA
Television:
broadcast stations: 0
televisions: NA

Defense Forces

Branches: Comoran Security Force
Manpower availability: males age 15-49 117,349; males fit for military service 70,178 (1995 est.)
Defense expenditures: $NA, NA% of GDP

Congo

Geography

Location: Western Africa, bordering the South Atlantic Ocean, between Angola and Gabon
Map references: Africa
Area:
total area: 342,000 sq km
land area: 341,500 sq km
comparative area: slightly smaller than Montana
Land boundaries: total 5,504 km, Angola 201 km, Cameroon 523 km, Central African Republic 467 km, Gabon 1,903 km, Zaire 2,410 km
Coastline: 169 km
Maritime claims:
territorial sea: 200 nm
International disputes: long segment of boundary with Zaire along the Congo River is indefinite (no division of the river or its islands has been made)
Climate: tropical; rainy season (March to June); dry season (June to October); constantly high temperatures and humidity; particularly enervating climate astride the Equator
Terrain: coastal plain, southern basin, central plateau, northern basin
Natural resources: petroleum, timber, potash, lead, zinc, uranium, copper, phosphates, natural gas
Land use:
arable land: 2%
permanent crops: 0%
meadows and pastures: 29%
forest and woodland: 62%
other: 7%
Irrigated land: 40 sq km (1989)
Environment:
current issues: air pollution from vehicle emissions; water pollution from the dumping of raw sewage; tap water is not potable; deforestation
natural hazards: seasonal flooding

international agreements: party to—Endangered Species, Ozone Layer Protection, Tropical Timber 83; signed, but not ratified—Biodiversity, Climate Change, Desertification, Law of the Sea, Tropical Timber 94
Note: about 70% of the population lives in Brazzaville, Pointe Noire, or along the railroad between them

People

Population: 2,504,996 (July 1995 est.)
Age structure:
0-14 years: 44% (female 543,324; male 548,840)
15-64 years: 53% (female 682,927; male 645,045)
65 years and over: 3% (female 49,879; male 34,981) (July 1995 est.)
Population growth rate: 2.32% (1995 est.)
Birth rate: 39.86 births/1,000 population (1995 est.)
Death rate: 16.7 deaths/1,000 population (1995 est.)
Net migration rate: 0 migrant(s)/1,000 population (1995 est.)
Infant mortality rate: 109.4 deaths/1,000 live births (1995 est.)
Life expectancy at birth:
total population: 47.09 years
male: 45.23 years
female: 49 years (1995 est.)
Total fertility rate: 5.23 children born/woman (1995 est.)
Nationality:
noun: Congolese (singular and plural)
adjective: Congolese or Congo
Ethnic divisions:
south: Kongo 48%
north: Sangha 20%, M'Bochi 12%
center: Teke 17%, Europeans 8,500 (mostly French)
Religions: Christian 50%, animist 48%, Muslim 2%
Languages: French (official), African languages (Lingala and Kikongo are the most widely used)
Literacy: age 15 and over can read and write (1984)
total population: 60%
male: 71%
female: 49%
Labor force: 79,100 wage earners
by occupation: agriculture 75%, commerce, industry, and government 25%

Government

Names:
conventional long form: Republic of the Congo
conventional short form: Congo
local long form: Republique Populaire du Congo
local short form: Congo
former: Congo/Brazzaville
Digraph: CF
Type: republic
Capital: Brazzaville
Administrative divisions: 9 regions (regions, singular—region) and 1 commune*; Bouenza, Brazzaville*, Cuvette, Kouilou, Lekoumou, Likouala, Niari, Plateaux, Pool, Sangha
Independence: 15 August 1960 (from France)
National holiday: Congolese National Day, 15 August (1960)
Constitution: new constitution approved by referendum March 1992
Legal system: based on French civil law system and customary law
Suffrage: 18 years of age; universal
Executive branch:
chief of state: President Pascal LISSOUBA (since August 1992); election last held August 1992 (next to be held August 1997); results—President Pascal LISSOUBA won with 61% of the vote
head of government: Prime Minister Jacques Joachim YHOMBI-OPANGO (since 23 June 1993)
cabinet: Council of Ministers; named by the president
Legislative branch: bicameral
National Assembly (Assemblee Nationale): election last held 3 October 1993; results—percentage vote by party NA; seats—(125 total) UPADS 64, URD/PCT 58, others 3
Senate: election last held 26 July 1992 (next to be held July 1998); results—percent vote by party NA; seats—(60 total) UPADS 23, MCDDI 14, RDD 8, RDPS 5, PCT 2, others 8
Judicial branch: Supreme Court (Cour Supreme)
Political parties and leaders: Congolese Labor Party (PCT), Denis SASSOU-NGUESSO, president; Pan-African Union for Social Development (UPADS), Pascal LISSOUBA, leader; Association for Democracy and Development (RDD), Joachim Yhombi OPANGO, president; Congolese Movement for Democracy and Integral Development (MCDDI), Bernard KOLELAS, leader; Association for Democracy and Social Progress (RDPS), Jean-Pierre Thystere TCHICAYA, president; Union of Democratic Forces (UFD), David Charles GANAO, leader; Union for Development and Social Progress (UDPS), Jean-Michael BOKAMBA-YANGOUMA, leader
note: Congo has many political parties of which these are among the most important
Other political or pressure groups: Union of Congolese Socialist Youth (UJSC); Congolese Trade Union Congress (CSC); Revolutionary Union of Congolese Women (URFC); General Union of Congolese Pupils and Students (UGEEC)
Member of: ACCT, ACP, AfDB, BDEAC, CCC, CEEAC, ECA, FAO, FZ, G-77, GATT, IBRD, ICAO, ICRM, IDA, IFAD, IFC, IFRCS, ILO, IMF, IMO, INTELSAT, INTERPOL, IOC, ITU, NAM, OAU, UDEAC, UN, UNAMIR, UNAVEM II, UNCTAD, UNESCO, UNIDO, UPU, WFTU, WHO, WIPO, WMO, WTO
Diplomatic representation in US:
chief of mission: Ambassador Pierre Damien BOUSSOUKOU-BOUMBA
chancery: 4891 Colorado Avenue NW, Washington, DC 20011
telephone: [1] (202) 726-0825
FAX: [1] (202) 726-1860
US diplomatic representation:
chief of mission: Ambassador William C. RAMSEY
embassy: Avenue Amilcar Cabral, Brazzaville
mailing address: B. P. 1015, Brazzaville
telephone: [242] 83 20 70
FAX: [242] 83 63 38
Flag: red, divided diagonally from the lower hoist side by a yellow band; the upper triangle (hoist side) is green and the lower triangle is red; uses the popular pan-African colors of Ethiopia

Economy

Overview: Congo's economy is a mixture of village agriculture and handicrafts, an industrial sector based largely on oil, support services, and a government characterized by budget problems and overstaffing. A reform program, supported by the IMF and World Bank, ran into difficulties in 1990-91 because of problems in changing to a democratic political regime and a heavy debt-servicing burden. Oil has supplanted forestry as the mainstay of the economy, providing about two-thirds of government revenues and exports. In the early 1980s rapidly rising oil revenues enabled Congo to finance large-scale development projects with growth averaging 5% annually, one of the highest rates in Africa. Subsequently, growth has slowed to an average of roughly 1.5% annually, only two thirds of the population growth rate. Political turmoil and misguided government investment have derailed economic reform programs sponsored by the IMF and World Bank. Even with these difficulties Congo enjoys one of the highest incomes per capita in sub-Saharan Africa
National product: GDP—purchasing power parity—$6.7 billion (1993 est.)
National product real growth rate: -2.1% (1993 est.)
National product per capita: $2,820 (1994 est.)
Inflation rate (consumer prices): 2.2% (1992 est.)
Unemployment rate: NA%
Budget:
revenues: $765 million
expenditures: $952 million, including capital expenditures of $65 million (1990)
Exports: $1.1 billion (f.o.b., 1993)

commodities: crude oil 83%, lumber, plywood, sugar, cocoa, coffee, diamonds
partners: US, Italy, France, Spain, other EC countries
Imports: $472 million (c.i.f., 1991)
commodities: intermediate manufactures, capital equipment, construction materials, foodstuffs
partners: France, US, Italy, Japan, other EC countries
External debt: $4 billion (1993)
Industrial production: growth rate 8% (1993 est.); accounts for 35% of GDP; includes petroleum
Electricity:
capacity: 120,000 kW
production: 400 million kWh
consumption per capita: 201 kWh (1993)
Industries: petroleum, cement, lumbering, brewing, sugar milling, palm oil, soap, cigarette
Agriculture: accounts for 12% of GDP (including fishing and forestry); cassava accounts for 90% of food output; other crops—rice, corn, peanuts, vegetables; cash crops include coffee and cocoa; forest products important export earner; imports over 90% of food needs
Economic aid:
recipient: US commitments, including Ex-Im (FY70-90), $63 million; Western (non-US) countries, ODA and OOF bilateral commitments (1970-90), $2.5 billion; OPEC bilateral aid (1979-89), $15 million; Communist countries (1970-89), $338 million
Currency: 1 CFA franc (CFAF) = 100 centimes
Exchange rates: Communaute Financiere Africaine francs (CFAF) per US$1—529.43 (January 1994), 555.20 (1994), 283.16 (1993), 264.69 (1992), 282.11 (1991), 272.26 (1990)
note: beginning 12 January 1994, the CFA franc was devalued to CFAF 100 per French franc from CFAF 50 at which it had been fixed since 1948
Fiscal year: calendar year

Transportation

Railroads:
total: 797 km (includes 285 km that are privately owned)
narrow gauge: 797 km 1.067-m gauge
Highways:
total: 11,960 km
paved: 560 km
unpaved: gravel or crushed stone 850 km; improved earth 5,350 km, unimproved earth 5,200 km
Inland waterways: the Congo and Ubangi (Oubangui) Rivers provide 1,120 km of commercially navigable water transport; the rest are used for local traffic only
Pipelines: crude oil 25 km
Ports: Brazzaville, Impfondo, Ouesso, Oyo, Pointe-Noire
Merchant marine: none
Airports:
total: 41
with paved runways over 3,047 m: 1
with paved runways 1,524 to 2,437 m: 3
with paved runways under 914 m: 11
with unpaved runways 1,524 to 2,438 m: 8
with unpaved runways 914 to 1,523 m: 18

Communications

Telephone system: 18,100 telephones; telephone density—7 telephones/1,000 persons; services adequate for government use; key centers are Brazzaville, Pointe-Noire, and Loubomo
local: NA
intercity: primary network consists of microwave radio relay and coaxial cable
international: 1 Atlantic Ocean INTELSAT earth station
Radio:
broadcast stations: AM 4, FM 1, shortwave 0
radios: NA
Television:
broadcast stations: 4
televisions: NA

Defense Forces

Branches: Army, Navy (includes Marines), Air Force, National Police
Manpower availability: males age 15-49 568,663; males fit for military service 289,335; males reach military age (20) annually 24,749 (1995 est.)
Defense expenditures: exchange rate conversion—$110 million, 3.8% of GDP (1993)

Cook Islands
(free association with New Zealand)

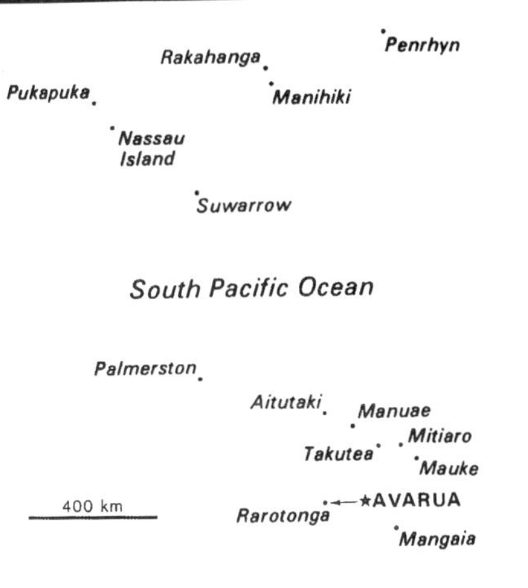

Geography

Location: Oceania, group of islands in the South Pacific Ocean, about one-half of the way from Hawaii to New Zealand
Map references: Oceania
Area:
total area: 240 sq km
land area: 240 sq km
comparative area: slightly less than 1.3 times the size of Washington, DC
Land boundaries: 0 km
Coastline: 120 km
Maritime claims:
continental shelf: 200 nm or to the edge of the continental margin
exclusive economic zone: 200 nm
territorial sea: 12 nm
International disputes: none
Climate: tropical; moderated by trade winds
Terrain: low coral atolls in north; volcanic, hilly islands in south
Natural resources: negligible
Land use:
arable land: 4%
permanent crops: 22%
meadows and pastures: 0%
forest and woodland: 0%
other: 74%
Irrigated land: NA sq km
Environment:
current issues: NA
natural hazards: typhoons (November to March)
international agreements: party to—Biodiversity, Climate Change; signed, but not ratified—Law of the Sea

People

Population: 19,343 (July 1995 est.)
Age structure:
0-14 years: NA
15-64 years: NA
65 years and over: NA
Population growth rate: 1.13% (1995 est.)

Birth rate: 23.05 births/1,000 population (1995 est.)
Death rate: 5.2 deaths/1,000 population (1995 est.)
Net migration rate: -6.53 migrant(s)/1,000 population (1995 est.)
Infant mortality rate: 24.7 deaths/1,000 live births (1995 est.)
Life expectancy at birth:
total population: 71.14 years
male: 69.2 years
female: 73.1 years (1995 est.)
Total fertility rate: 3.27 children born/woman (1995 est.)
Nationality:
noun: Cook Islander(s)
adjective: Cook Islander
Ethnic divisions: Polynesian (full blood) 81.3%, Polynesian and European 7.7%, Polynesian and other 7.7%, European 2.4%, other 0.9%
Religions: Christian (majority of populace members of Cook Islands Christian Church)
Languages: English (official), Maori
Literacy: NA%
Labor force: 5,810
by occupation: agriculture 29%, government 27%, services 25%, industry 15%, other 4% (1981)

Government

Names:
conventional long form: none
conventional short form: Cook Islands
Digraph: CW
Type: self-governing parliamentary government in free association with New Zealand; Cook Islands is fully responsible for internal affairs; New Zealand retains responsibility for external affairs, in consultation with the Cook Islands
Capital: Avarua
Administrative divisions: none
Independence: none (became self-governing in free association with New Zealand on 4 August 1965 and has the right at any time to move to full independence by unilateral action)
National holiday: Constitution Day, 4 August
Constitution: 4 August 1965
Legal system: NA
Suffrage: universal adult at age NA
Executive branch:
chief of state: Queen ELIZABETH II (since 6 February 1952); Representative of the Queen Apenera SHORT (since NA); Representative of New Zealand Adrian SINCOCK (since NA)
head of government: Prime Minister Geoffrey HENRY (since 1 February 1989); Deputy Prime Minister Inatio AKARURU (since 1 February 1989)
cabinet: Cabinet; collectively responsible to the Parliament
Legislative branch: unicameral
Parliament: elections last held 24 March 1994 (next to be held NA); results—percent of vote by party NA; seats—(25 total) Cook Islands Party 20, Democratic Party 3, Alliance Party 2
note: the House of Arikis (chiefs) advises on traditional matters, but has no legislative powers
Judicial branch: High Court
Political parties and leaders: Cook Islands Party, Geoffrey HENRY; Democratic Party, Sir Thomas DAVIS; Cook Islands Labor Party, Rena JONASSEN; Cook Islands People's Party, Sadaraka SADARAKA; Alliance Party, Norman GEORGE
Member of: AsDB, ESCAP (associate), ICAO, ICFTU, IFAD, INTELSAT (nonsignatory user), IOC, SPARTECA, SPC, SPF, UNESCO, WHO
Diplomatic representation in US: none (self-governing in free association with New Zealand)
US diplomatic representation: none (self-governing in free association with New Zealand)
Flag: blue, with the flag of the UK in the upper hoist-side quadrant and a large circle of 15 white five-pointed stars (one for every island) centered in the outer half of the flag

Economy

Overview: Agriculture provides the economic base. The major export earners are fruit, copra, and clothing. Manufacturing activities are limited to a fruit-processing plant and several clothing factories. Economic development is hindered by the isolation of the islands from foreign markets and a lack of natural resources and good transportation links. A large trade deficit is annually made up for by remittances from emigrants and from foreign aid, largely from New Zealand. Current economic development plans call for exploiting the tourism potential and expanding the fishing industry.
National product: GDP—purchasing power parity—$57 million (1993 est.)
National product real growth rate: NA%
National product per capita: $3,000 (1993 est.)
Inflation rate (consumer prices): 6.2% (1990)
Unemployment rate: NA%
Budget:
revenues: $38 million
expenditures: $34.4 million, including capital expenditures of $NA (1993 est.)
Exports: $3.4 million (f.o.b., 1990)
commodities: copra, fresh and canned fruit, clothing
partners: NZ 80%, Japan
Imports: $50 million (c.i.f., 1990)
commodities: foodstuffs, textiles, fuels, timber
partners: NZ 49%, Japan, Australia, US
External debt: $124 million (1994)
Industrial production: growth rate NA%; accounts for 5% of GDP
Electricity:
capacity: 14,000 kW
production: 21 million kWh
consumption per capita: 741 kWh (1993)
Industries: fruit processing, tourism
Agriculture: accounts for 12% of GDP, export crops—copra, citrus fruits, pineapples, tomatoes, bananas; subsistence crops—yams, taro
Economic aid:
recipient: Western (non-US) countries, ODA and OOF bilateral commitments (1970-89), $128 million; in 1994, Cook Islands received $4.3 million in budget support and $2.7 million in project aid from New Zealand, the country's largest source of aid
Currency: 1 New Zealand dollar (NZ$) = 100 cents
Exchange rates: New Zealand dollars (NZ$) per US$1—1.5601 (January 1995), 1.6844 (1994), 1.8495 (1993), 1.8584 (1992), 1.7265 (1991), 1.6750 (1990)
Fiscal year: 1 April—31 March

Transportation

Railroads: 0 km
Highways:
total: 187 km
paved: 35 km
unpaved: gravel 35 km; improved earth 84 km; unimproved earth 33 km (1980)
Ports: Avarua, Avatiu
Merchant marine:
total: 1 cargo ship (1,000 GRT or over) totaling 1,464 GRT/2,181 DWT
Airports:
total: 7
with paved runways 1,524 to 2,437 m: 1
with unpaved runways 1,524 to 2,438 m: 3
with unpaved runways 914 to 1,523 m: 3

Communications

Telephone system: 2,052 telephones
local: NA
intercity: NA
international: 1 INTELSAT (Pacific Ocean) earth station
Radio:
broadcast stations: AM 1, FM 1, shortwave 0
radios: 11,000
Television:
broadcast stations: 1
televisions: 17,000 (1989)

Defense Forces

Note: defense is the responsibility of New Zealand

Coral Sea Islands

(territory of Australia)

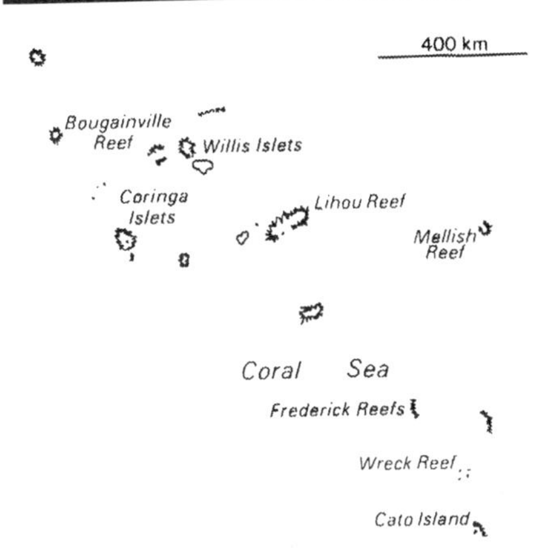

Geography

Location: Oceania, islands in the Coral Sea, northeast of Australia
Map references: Oceania
Area:
total area: less than 3 sq km
land area: less than 3 sq km
comparative area: NA
note: includes numerous small islands and reefs scattered over a sea area of about 1 million sq km, with Willis Islets the most important
Land boundaries: 0 km
Coastline: 3,095 km
Maritime claims:
exclusive fishing zone: 200 nm
territorial sea: 3 nm
International disputes: none
Climate: tropical
Terrain: sand and coral reefs and islands (or cays)
Natural resources: negligible
Land use:
arable land: 0%
permanent crops: 0%
meadows and pastures: 0%
forest and woodland: 0%
other: 100% (mostly grass or scrub cover)
Irrigated land: 0 sq km
Environment:
current issues: no permanent fresh water resources
natural hazards: occasional, tropical cyclones
international agreements: NA
Note: important nesting area for birds and turtles

People

Population: no indigenous inhabitants; note—there are 3 meteorologists

Government

Names:
conventional long form: Coral Sea Islands Territory
conventional short form: Coral Sea Islands
Digraph: CR
Type: territory of Australia administered by the Ministry for Environment, Sport, and Territories
Capital: none; administered from Canberra, Australia
Independence: none (territory of Australia)
Flag: the flag of Australia is used

Economy

Overview: no economic activity

Transportation

Ports: none; offshore anchorage only

Defense Forces

Note: defense is the responsibility of Australia; visited regularly by the Royal Australian Navy; Australia has control over the activities of visitors

Costa Rica

Geography

Location: Middle America, bordering both the Caribbean Sea and the North Pacific Ocean, between Nicaragua and Panama
Map references: Central America and the Caribbean
Area:
total area: 51,100 sq km
land area: 50,660 sq km
comparative area: slightly smaller than West Virginia
note: includes Isla del Coco
Land boundaries: total 639 km, Nicaragua 309 km, Panama 330 km
Coastline: 1,290 km
Maritime claims:
exclusive economic zone: 200 nm
territorial sea: 12 nm
International disputes: none
Climate: tropical; dry season (December to April); rainy season (May to November)
Terrain: coastal plains separated by rugged mountains
Natural resources: hydropower potential
Land use:
arable land: 6%
permanent crops: 7%
meadows and pastures: 45%
forest and woodland: 34%
other: 8%
Irrigated land: 1,180 sq km (1989 est.)
Environment:
current issues: deforestation, largely a result of the clearing of land for cattle ranching; soil erosion
natural hazards: occasional earthquakes, hurricanes along Atlantic coast; frequent flooding of lowlands at onset of rainy season; active volcanoes
international agreements: party to—Biodiversity, Climate Change, Endangered Species, Law of the Sea, Marine Dumping, Nuclear Test Ban, Ozone Layer Protection, Wetlands, Whaling; signed, but not ratified—Desertification, Marine Life Conservation

People

Population: 3,419,114 (July 1995 est.)
Age structure:
0-14 years: 35% (female 585,976; male 617,456)
15-64 years: 60% (female 1,013,491; male 1,036,195)
65 years and over: 5% (female 88,050; male 77,946) (July 1995 est.)
Population growth rate: 2.24% (1995 est.)
Birth rate: 24.88 births/1,000 population (1995 est.)
Death rate: 3.47 deaths/1,000 population (1995 est.)
Net migration rate: 1.02 migrant(s)/1,000 population (1995 est.)
Infant mortality rate: 10.3 deaths/1,000 live births (1995 est.)
Life expectancy at birth:
total population: 78.11 years
male: 76.21 years
female: 80.1 years (1995 est.)
Total fertility rate: 3.01 children born/woman (1995 est.)
Nationality:
noun: Costa Rican(s)
adjective: Costa Rican
Ethnic divisions: white (including mestizo) 96%, black 2%, Indian 1%, Chinese 1%
Religions: Roman Catholic 95%
Languages: Spanish (official), English; spoken around Puerto Limon
Literacy: age 15 and over can read and write (1984)
total population: 93%
male: 93%
female: 93%
Labor force: 868,300
by occupation: industry and commerce 35.1%, government and services 33%, agriculture 27%, other 4.9% (1985 est.)

Government

Names:
conventional long form: Republic of Costa Rica
conventional short form: Costa Rica
local long form: Republica de Costa Rica
local short form: Costa Rica
Digraph: CS
Type: democratic republic
Capital: San Jose
Administrative divisions: 7 provinces (provincias, singular—provincia); Alajuela, Cartago, Guanacaste, Heredia, Limon, Puntarenas, San Jose
Independence: 15 September 1821 (from Spain)
National holiday: Independence Day, 15 September (1821)
Constitution: 9 November 1949
Legal system: based on Spanish civil law system; judicial review of legislative acts in the Supreme Court; has not accepted compulsory ICJ jurisdiction
Suffrage: 18 years of age; universal and compulsory
Executive branch:
chief of state and head of government: President Jose Maria FIGUERES Olsen (since 8 May 1994); First Vice President Rodrigo OREAMUNO Blanco (since 8 May 1994); Second Vice President Rebeca GRYNSPAN Mayufis (since 8 May 1994); election last held 6 February 1994 (next to be held February 1998); results—President FIGUERES (PLN party) 49.7%, Miquel Angel RODRIGUEZ (PUSC party) 47.5%
cabinet: Cabinet; selected by the president
Legislative branch: unicameral
Legislative Assembly (Asamblea Legislativa): elections last held 6 February 1994 (next to be held February 1998); results—percent of vote by party NA; seats—(61 total) PLN 28, PUSC 29, minority parties 4
Judicial branch: Supreme Court (Corte Suprema)
Political parties and leaders: National Liberation Party (PLN), Manuel AGUILAR Bonilla; Social Christian Unity Party (PUSC), Rafael Angel CALDERON Fournier; Marxist Popular Vanguard Party (PVP), Humberto VARGAS Carbonell; New Republic Movement (MNR), Sergio Erick ARDON Ramirez; Progressive Party (PP), Isaac Felipe AZOFEIFA Bolanos; People's Party of Costa Rica (PPC), Lenin CHACON Vargas; Radical Democratic Party (PRD), Juan Jose ECHEVERRIA Brealey
Other political or pressure groups: Costa Rican Confederation of Democratic Workers (CCTD, Liberation Party affiliate); Confederated Union of Workers (CUT, Communist Party affiliate); Authentic Confederation of Democratic Workers (CATD, Communist Party affiliate); Chamber of Coffee Growers; National Association for Economic Development (ANFE); Free Costa Rica Movement (MCRL, rightwing militants); National Association of Educators (ANDE)
Member of: AG (observer), BCIE, CACM, ECLAC, FAO, G-77, GATT, IADB, IAEA, IBRD, ICAO, ICFTU, ICRM, IDA, IFAD, IFC, IFRCS, ILO, IMF, IMO, INTELSAT, INTERPOL, IOC, IOM, ITU, LAES, LAIA (observer), NAM (observer), OAS, OPANAL, UN, UNCTAD, UNESCO, UNIDO, UNU, UPU, WCL, WFTU, WHO, WIPO, WMO
Diplomatic representation in US:
chief of mission: Ambassador Sonia PICADO
chancery: 2114 S Street NW, Washington, DC 20008
telephone: [1] (202) 234-2945
FAX: [1] (202) 265-4795
consulate(s) general: Albuquerque, Atlanta, Chicago, Durham, Houston, Los Angeles, Miami, New Orleans, New York, Orlando, Philadelphia, San Antonio, San Diego, San Francisco, and San Juan (Puerto Rico)
consulate(s): Austin
US diplomatic representation:
chief of mission: US Ambassador to Costa Rica Peter DE VOS
embassy: Pavas Road, San Jose
mailing address: APO AA 34020
telephone: [506] 220-3939
FAX: [506] 220-2305
Flag: five horizontal bands of blue (top), white, red (double width), white, and blue, with the coat of arms in a white disk on the hoist side of the red band

Economy

Overview: Costa Rica's basically stable and progressive economy depends especially on tourism and export of bananas, coffee, and other agricultural products. In 1994 the economy grew at an estimated 4.3%, compared with 6.5% in 1993, 7.7% in 1992, and 2.1% in 1991. Inflation in 1993 dropped to 9% from 17% in 1992 and 25% in 1991, an indication of basic financial stability. Unemployment is officially reported at only 4.0%, but there is much underemployment. Costa Rica signed a free trade agreement with Mexico in 1994.
National product: GDP—purchasing power parity—$16.9 billion (1994 est.)
National product real growth rate: 4.3% (1994 est.)
National product per capita: $5,050 (1994 est.)
Inflation rate (consumer prices): 9% (1993 est.)
Unemployment rate: 4% (1993); much underemployment
Budget:
revenues: $1.1 billion
expenditures: $1.34 billion, including capital expenditures of $110 million (1991 est.)
Exports: $2.1 billion (f.o.b., 1993)
commodities: coffee, bananas, textiles, sugar
partners: US, Germany, Italy, Guatemala, El Salvador, Netherlands, UK, France
Imports: $2.9 billion (c.i.f., 1993)
commodities: raw materials, consumer goods, capital equipment, petroleum
partners: US, Japan, Mexico, Guatemala, Venezuela, Germany
External debt: $3.2 billion (1991)
Industrial production: growth rate 10.5% (1992); accounts for 22% of GDP
Electricity:
capacity: 1,040,000 kW
production: 4.1 billion kWh
consumption per capita: 1,164 kWh (1993)
Industries: food processing, textiles and clothing, construction materials, fertilizer, plastic products
Agriculture: accounts for 19% of GDP and 70% of exports; cash commodities—coffee, beef, bananas, sugar; other food crops include corn, rice, beans, potatoes; normally self-sufficient in food except for grain; depletion of

forest resources resulting in lower timber output
Illicit drugs: transshipment country for cocaine and heroin from South America; illicit production of cannabis on small, scattered plots
Economic aid:
recipient: US commitments, including Ex-Im (FY70-89), $1.4 billion; Western (non-US) countries, ODA and OOF bilateral commitments (1970-89), $935 million; Communist countries (1971-89), $27 million
Currency: 1 Costa Rican colon (C) = 100 centimos
Exchange rates: Costa Rican colones (C) per US$1—164.39 (December 1994), 157.07 (1994), 142.17 (1993), 134.51 (1992), 122.43 (1991), 91.58 (1990)
Fiscal year: calendar year

Transportation

Railroads:
total: 950 km (260 km electrified)
narrow gauge: 950 km 1.067-m gauge
Highways:
total: 35,560 km
paved: 5,600 km
unpaved: gravel and earth 29,960 km (1992)
Inland waterways: about 730 km, seasonally navigable
Pipelines: petroleum products 176 km
Ports: Caldera, Golfito, Moin, Puerto Limon, Puerto Quepos, Puntarenas
Merchant marine: none
Airports:
total: 174
with paved runways 2,438 to 3,047 m: 2
with paved runways 1,524 to 2,437 m: 1
with paved runways 914 to 1,523 m: 17
with paved runways under 914 m: 117
with unpaved runways 1,524 to 2,438 m: 1
with unpaved runways 914 to 1,523 m: 36

Communications

Telephone system: 292,000 telephones; very good domestic telephone service
local: NA
intercity: NA
international: connection into Central American Microwave System; 1 INTELSAT (Atlantic Ocean) earth station
Radio:
broadcast stations: AM 71, FM 0, shortwave 13
radios: NA
Television:
broadcast stations: 18
televisions: NA

Defense Forces

Branches: Civil Guard, Coast Guard, Air Section, Rural Assistance Guard; note—the Constitution prohibits armed forces
Manpower availability: males age 15-49 896,516; males fit for military service 602,785; males reach military age (18) annually 32,815 (1995 est.)
Defense expenditures: exchange rate conversion—$22 million, 0.5% of GDP (1989)

Cote d'Ivoire

(also known as Ivory Coast)

Geography

Location: Western Africa, bordering the North Atlantic Ocean, between Ghana and Liberia
Map references: Africa
Area:
total area: 322,460 sq km
land area: 318,000 sq km
comparative area: slightly larger than New Mexico
Land boundaries: total 3,110 km, Burkina 584 km, Ghana 668 km, Guinea 610 km, Liberia 716 km, Mali 532 km
Coastline: 515 km
Maritime claims:
continental shelf: 200 nm
exclusive economic zone: 200 nm
territorial sea: 12 nm
International disputes: none
Climate: tropical along coast, semiarid in far north; three seasons—warm and dry (November to March), hot and dry (March to May), hot and wet (June to October)
Terrain: mostly flat to undulating plains; mountains in northwest
Natural resources: petroleum, diamonds, manganese, iron ore, cobalt, bauxite, copper
Land use:
arable land: 9%
permanent crops: 4%
meadows and pastures: 9%
forest and woodland: 26%
other: 52%
Irrigated land: 620 sq km (1989 est.)
Environment:
current issues: deforestation (most of the country's forests—once the largest in West Africa—have been cleared by the timber industry); water pollution from sewage and industrial and agricultural effluents
natural hazards: coast has heavy surf and no natural harbors; during the rainy season torrential flooding is possible
international agreements: party to—

Biodiversity, Climate Change, Endangered Species, Hazardous Wastes, Law of the Sea, Marine Dumping, Nuclear Test Ban, Ozone Layer Protection, Ship Pollution, Tropical Timber 83; signed, but not ratified—Desertification

People

Population: 14,791,257 (July 1995 est.)
Age structure:
0-14 years: 48% (female 3,506,147; male 3,534,751)
15-64 years: 50% (female 3,619,759; male 3,820,999)
65 years and over: 2% (female 142,366; male 167,235) (July 1995 est.)
Population growth rate: 3.38% (1995 est.)
Birth rate: 46.17 births/1,000 population (1995 est.)
Death rate: 14.95 deaths/1,000 population (1995 est.)
Net migration rate: NA migrant(s)/1,000 population (1995 est.)
note: since 1989, over 350,000 refugees have fled to Cote d'Ivoire to escape the civil war in Liberia; if a lasting peace is achieved in Liberia in 1995, large numbers of refugees can be expected to return to their homes
Infant mortality rate: 93.1 deaths/1,000 live births (1995 est.)
Life expectancy at birth:
total population: 48.87 years
male: 46.52 years
female: 51.29 years (1995 est.)
Total fertility rate: 6.61 children born/woman (1995 est.)
Nationality:
noun: Ivorian(s)
adjective: Ivorian
Ethnic divisions: Baoule 23%, Bete 18%, Senoufou 15%, Malinke 11%, Agni, foreign Africans (mostly Burkinabe and Malians, about 3 million), non-Africans 130,000 to 330,000 (French 30,000 and Lebanese 100,000 to 300,000)
Religions: indigenous 25%, Muslim 60%, Christian 12%
Languages: French (official), 60 native dialects, Dioula is the most widely spoken
Literacy: age 15 and over can read and write (1988)
total population: 34%
male: 44%
female: 23%
Labor force: 5.718 million
by occupation: over 85% of population engaged in agriculture, forestry, livestock raising; about 11% of labor force are wage earners, nearly half in agriculture and the remainder in government, industry, commerce, and professions

Government

Names:
conventional long form: Republic of Cote d'Ivoire
conventional short form: Cote d'Ivoire
local long form: Republique de Cote d'Ivoire
local short form: Cote d'Ivoire
former: Ivory Coast
Digraph: IV
Type: republic; multiparty presidential regime established 1960
Capital: Yamoussoukro
note: although Yamoussoukro has been the capital since 1983, Abidjan remains the administrative center; foreign governments, including the United States, maintain presence in Abidjan
Administrative divisions: 50 departments (departements, singular—departement); Abengourou, Abidjan, Aboisso, Adzope, Agboville, Agnibilekrou, Bangolo, Beoumi, Biankouma, Bondoukou, Bongouanou, Bouafle, Bouake, Bouna, Boundiali, Dabakala, Daloa, Danane, Daoukro, Dimbokro, Divo, Duekoue, Ferkessedougou, Gagnoa, Grand-Lahou, Guiglo, Issia, Katiola, Korhogo, Lakota, Man, Mankono, Mbahiakro, Odienne, Oume, Sakassou, San-Pedro, Sassandra, Seguela, Sinfra, Soubre, Tabou, Tanda, Tingrela, Tiassale, Touba, Toumodi, Vavoua, Yamoussoukro, Zuenoula
Independence: 7 August 1960 (from France)
National holiday: National Day, 7 December
Constitution: 3 November 1960; has been amended numerous times, last time November 1990
Legal system: based on French civil law system and customary law; judicial review in the Constitutional Chamber of the Supreme Court; has not accepted compulsory ICJ jurisdiction
Suffrage: 21 years of age; universal
Executive branch:
chief of state: President Henri Konan BEDIE (since 7 December 1993) constitutional successor who will serve during the remainder of the term of former President Felix HOUPHOUET-BOIGNY who died in office after continuous service from November 1960 (next election October 1995)
head of government: Prime Minister Daniel Kablan DUNCAN (since 10 December 1993)
cabinet: Council of Ministers; appointed by the prime minister
Legislative branch: unicameral
National Assembly (Assemblee Nationale): elections last held 25 November 1990 (next to be held November 1995); results—percent of vote by party NA; seats—(175 total) PDCI 163, FPI 9, PIT 1, independents 2
Judicial branch: Supreme Court (Court Supreme)
Political parties and leaders: Democratic Party of the Cote d'Ivoire (PDCI), Henri Konan BEDIE; Rally of the Republicans (RDR), Djeny KOBINA; Ivorian Popular Front (FPI), Laurent GBAGBO; Ivorian Worker's Party (PIT), Francis WODIE; Ivorian Socialist Party (PSI), Morifere BAMBA; over 20 smaller parties
Member of: ACCT, ACP, AfDB, CCC, CEAO, ECA, ECOWAS, Entente, FAO, FZ, G-24, G-77, GATT, IAEA, IBRD, ICAO, ICC, ICRM, IDA, IFAD, IFC, IFRCS, ILO, IMF, IMO, INTELSAT, INTERPOL, IOC, ITU, NAM, OAU, UN, UNCTAD, UNESCO, UNIDO, UNITAR, UPU, WADB, WCL, WFTU, WHO, WIPO, WMO, WTO
Diplomatic representation in US:
chief of mission: Ambassador Moise KOUMOUE-KOFFI
chancery: 2424 Massachusetts Avenue NW, Washington, DC 20008
telephone: [1] (202) 797-0300
US diplomatic representation:
chief of mission: Ambassador Hume A. HORAN
embassy: 5 Rue Jesse Owens, Abidjan
mailing address: 01 B. P. 1712, Abidjan
telephone: [225] 21 09 79, 21 46 72
FAX: [225] 22 32 59
Flag: three equal vertical bands of orange (hoist side), white, and green; similar to the flag of Ireland, which is longer and has the colors reversed—green (hoist side), white, and orange; also similar to the flag of Italy, which is green (hoist side), white, and red; design was based on the flag of France

Economy

Overview: Cote d'Ivoire is among the world's largest producers and exporters of coffee, cocoa beans, and palm-kernel oil. Consequently, the economy is highly sensitive to fluctuations in international prices for coffee and cocoa and to weather conditions. Despite attempts by the government to diversify, the economy is still largely dependent on agriculture and related industries. After several years of lagging performance, the Ivorian economy began a comeback in 1994, due to improved prices for cocoa and coffee, growth in non-traditional primary exports such as pineapples and rubber, trade and banking liberalization, offshore oil and gas discoveries, and generous external financing and debt rescheduling by multilateral lenders and France. The 50% devaluation in January 1994 caused a one time jump in the inflation rate. Government adherence to a renewed structural adjustment program has led to a budget surplus for the first time in several years, a smaller personnel budget, and an increase in public

investment. While real growth in 1994 was only 1.5%, the IMF and World Bank expect it will surpass 6% in 1995.
National product: GDP—purchasing power parity—$20.5 billion (1994 est.)
National product real growth rate: 1.5% (1994 est.)
National product per capita: $1,430 (1994 est.)
Inflation rate (consumer prices): NA%
Unemployment rate: 14% (1985)
Budget:
revenues: $1.9 billion
expenditures: $3.4 billion, including capital expenditures of $408 million (1993)
Exports: $2.7 billion (f.o.b., 1993)
commodities: cocoa 30%, coffee 20%, tropical woods 11%, petroleum, cotton, bananas, pineapples, palm oil, cotton
partners: France, Netherlands, Germany, Italy, Burkina, US, Belgium, UK (1992)
Imports: $1.6 billion (f.o.b., 1993)
commodities: food, capital goods, consumer goods, fuel
partners: France, Nigeria, Japan, Netherlands, US (1992)
External debt: $17.3 billion (1993 est.)
Industrial production: growth rate 0% (1993 est.); accounts for 20% of GDP, including petroleum
Electricity:
capacity: 1,170,000 kW
production: 1.8 billion kWh
consumption per capita: 123 kWh (1993)
Industries: foodstuffs, wood processing, oil refining, automobile assembly, textiles, fertilizer, beverages
Agriculture: most important sector, contributing one-third to GDP and 80% to exports; cash crops include coffee, cocoa beans, timber, bananas, palm kernels, rubber; food crops—corn, rice, manioc, sweet potatoes; not self-sufficient in bread grain and dairy products
Illicit drugs: illicit producer of cannabis; mostly for local consumption; some international drug trade; transshipment point for Southwest and Southeast Asian heroin to Europe and occasionally to the US
Economic aid:
recipient: US commitments, including Ex-Im (FY70-89), $356 million; Western (non-US) countries, ODA and OOF bilateral commitments (1970-88), $5.2 billion
Currency: 1 CFA franc (CFAF) = 100 centimes
Exchange rates: Communaute Financiere Africaine francs (CFAF) per US$1—529.43 (January 1995), 555.20 (1994), 283.16 (1993), 264.69 (1992), 282.11 (1991), 272.26 (1990)
note: beginning 12 January 1994, the CFA franc was devalued to CFAF 100 per French franc from CFAF 50 at which it had been fixed since 1948
Fiscal year: calendar year

Transportation

Railroads:
total: 660 km (25 km double track)
narrow gauge: 660 km 1.000-meter gauge
Highways:
total: 46,600 km
paved: 3,600 km
unpaved: gravel, crushed stone, improved earth 32,000 km; unimproved earth 11,000 km
Inland waterways: 980 km navigable rivers, canals, and numerous coastal lagoons
Ports: Abidjan, Aboisso, Dabou, San-Pedro
Merchant marine:
total: 5 ships (1,000 GRT or over) totaling 49,671 GRT/69,216 DWT
ships by type: chemical tanker 1, container 2, oil tanker 1, roll-on/roll-off cargo 1
Airports:
total: 40
with paved runways over 3,047 m: 1
with paved runways 2,438 to 3,047 m: 2
with paved runways 1,524 to 2,437 m: 4
with paved runways under 914 m: 11
with unpaved runways 1,524 to 2,438 m: 6
with unpaved runways 914 to 1,523 m: 16

Communications

Telephone system: 87,700 telephones; well-developed by African standards but operating well below capacity; consists of open-wire lines and radio relay microwave links
local: NA
intercity: microwave radio relay
international: 2 INTELSAT (1 Atlantic Ocean and 1 Indian Ocean) earth stations; 2 coaxial submarine cables
Radio:
broadcast stations: AM 71, FM 0, shortwave 13
radios: NA
Television:
broadcast stations: 18
televisions: NA

Defense Forces

Branches: Army, Navy, Air Force, paramilitary Gendarmerie, Presidential Guard, Military Fire Group
Manpower availability: males age 15-49 3,318,314; males fit for military service 1,724,020; males reach military age (18) annually 154,120 (1995 est.)
Defense expenditures: exchange rate conversion—$140 million, 1.4% of GDP (1993)

Croatia

Geography

Location: Southeastern Europe, bordering the Adriatic Sea, between Bosnia and Herzegovina and Slovenia
Map references: Ethnic Groups in Eastern Europe, Europe
Area:
total area: 56,538 sq km
land area: 56,410 sq km
comparative area: slightly smaller than West Virginia
Land boundaries: total 2,028 km, Bosnia and Herzegovina 932 km, Hungary 329 km, Serbia and Montenegro 266 km (241 km with Serbia; 25 km with Montenego), Slovenia 501 km
Coastline: 5,790 km (mainland 1,778 km, islands 4,012 km)
Maritime claims:
continental shelf: 200-m depth or to the depth of exploitation
International disputes: Ethnic Serbs have occupied UN protected areas in eastern Croatia and along the western Bosnia and Herzegovinian border
Climate: Mediterranean and continental; continental climate predominant with hot summers and cold winters; mild winters, dry summers along coast
Terrain: geographically diverse; flat plains along Hungarian border, low mountains and highlands near Adriatic coast, coastline, and islands
Natural resources: oil, some coal, bauxite, low-grade iron ore, calcium, natural asphalt, silica, mica, clays, salt
Land use:
arable land: 32%
permanent crops: 20%
meadows and pastures: 18%
forest and woodland: 15%
other: 15%
Irrigated land: NA sq km
Environment:
current issues: air pollution (from metallurgical plants) and resulting acid rain is

damaging the forests; coastal pollution from industrial and domestic waste; widespread casualties and destruction of infrastructure in border areas affected by civil strife
natural hazards: frequent and destructive earthquakes
international agreements: party to—Air Pollution, Hazardous Wastes, Marine Dumping, Nuclear Test Ban, Ozone Layer Protection, Ship Pollution, Wetlands; signed, but not ratified—Air Pollution-Sulphur 94, Biodiversity, Climate Change, Desertification
Note: controls most land routes from Western Europe to Aegean Sea and Turkish Straits

People

Population: 4,665,821 (July 1995 est.)
Age structure:
0-14 years: 19% (female 418,272; male 442,064)
15-64 years: 68% (female 1,592,187; male 1,588,455)
65 years and over: 13% (female 394,650; male 230,193) (July 1995 est.)
Population growth rate: 0.13% (1995 est.)
Birth rate: 11.02 births/1,000 population (1995 est.)
Death rate: 10.55 deaths/1,000 population (1995 est.)
Net migration rate: 0.77 migrant(s)/1,000 population (1995 est.)
Infant mortality rate: 8.4 deaths/1,000 live births (1995 est.)
Life expectancy at birth:
total population: 74.02 years
male: 70.59 years
female: 77.65 years (1995 est.)
Total fertility rate: 1.62 children born/woman (1995 est.)
Nationality:
noun: Croat(s)
adjective: Croatian
Ethnic divisions: Croat 78%, Serb 12%, Muslim 0.9%, Hungarian 0.5%, Slovenian 0.5%, others 8.1% (1991)
Religions: Catholic 76.5%, Orthodox 11.1%, Slavic Muslim 1.2%, Protestant 0.4%, others and unknown 10.8%
Languages: Serbo-Croatian 96%, other 4%
Literacy: age 15 and over can read and write (1991)
total population: 97%
male: 99%
female: 95%
Labor force: 1,509,489
by occupation: industry and mining 37%, agriculture 16% (1981 est.), government NA%, other

Government

Names:
conventional long form: Republic of Croatia
conventional short form: Croatia
local long form: Republika Hrvatska
local short form: Hrvatska
Digraph: HR
Type: parliamentary democracy
Capital: Zagreb
Administrative divisions: 21 counties (zupanijas, zupanija—singular): Bjelovar-Bilogora, City of Zagreb, Dubrovnik-Neretva, Istra, Karlovac, Koprivnica-Krizevci, Krapina-Zagorje, Lika-Senj, Medimurje, Osijek-Baranja, Pozega-Slavonija, Primorje-Gorski Kotar, Sibenik, Sisak-Moslavina, Slavonski Brod-Posavina, Split-Dalmatia, Varazdin, Virovitica-Podravina, Vukovar-Srijem, Zadar-Knin, Zagreb
Independence: 25 June 1991 (from Yugoslavia)
National holiday: Statehood Day, 30 May (1990)
Constitution: adopted on 22 December 1990
Legal system: based on civil law system
Suffrage: 18 years of age; universal (16 years of age, if employed)
Executive branch:
chief of state: President Franjo TUDJMAN (since 30 May 1990); election last held 4 August 1992 (next to be held NA 1997); results—Franjo TUDJMAN reelected with about 56% of the vote; his opponent Dobroslav PARAGA got 5% of the vote
head of government: Prime Minister Nikica VALENTIC (since 3 April 1993); Deputy Prime Ministers Mato GRANIC (since 8 September 1992); Ivica KOSTOVIC (since 14 October 1993); Jure RADIC (since NA); Borislav SKEGRO (since 3 April 1993)
cabinet: Council of Ministers; appointed by the president
Legislative branch: bicameral parliament Assembly (Sabor)
House of Districts (Zupanije Dom): elections last held 7 and 21 February 1993 (next to be held NA February 1997); results—percent of vote by party NA; seats—(68 total; 63 elected, 5 presidentially appointed) HDZ 37, HSLS 16, HSS 5, Istrian Democratic Assembly 3, SPH-SDP 1, HNS 1
House of Representatives (Predstavnicke Dom): elections last held 2 August 1992 (next to be held NA August 1996); results—percent of vote by party NA; seats—(138 total) HDZ 85, HSLS 14, SPH-SDP 11, HNS 6, Dalmatian Action/Istrian Democratic Assembly/ Rijeka Democratic Alliance coalition 6, HSP 5, HSS 3, SNS 3, independents 5
Judicial branch: Supreme Court, Constitutional Court
Political parties and leaders: Croatian Democratic Union (HDZ), Zlatko CANJUGA, secretary general; Croatian Democratic Independents (HND), Stjepan MESIC, president; Croatian Social Liberal Party (HSLS), Drazen BUDISA, president; Croatian Democratic Peasant Party (HDSS), Ante BABIC; Croatian Party of Rights (HSP), Ante DAPIC; Croatian Peasants' Party (HSS), Josip PANKRETIC; Croatian People's Party (HNS), Radimir CACIC, president; Dalmatian Action (DA), Mira LJUBIC-LORGER; Serb National Party (SNS), Milan DJUKIC; Social Democratic Action (SDP), Miko TRIPALO; other small parties include the Istrian Democratic Assembly and the Rijeka Democratic Alliance
Other political or pressure groups: NA
Member of: CCC, CE (guest), CEI, EBRD, ECE, FAO, IADB, IAEA, IBRD, ICAO, ICRM, IDA, IFAD, IFC, ILO, IMF, IMO, INMARSAT, INTELSAT, INTERPOL, IOC, IOM, ISO, ITU, NAM (observer), OSCE, UN, UNCTAD, UNESCO, UNIDO, UPU, WHO, WIPO, WMO, WTO
Diplomatic representation in US:
chief of mission: Ambassador Petar A. SARCEVIC
chancery: 2343 Massachusetts Avenue NW, Washington, DC 20008
telephone: [1] (202) 588-5899
FAX: [1] (202) 588-8936
consulate(s) general: New York
US diplomatic representation:
chief of mission: Ambassador Peter W. GALBRAITH
embassy: Andrije Hebranga 2, Zagreb
mailing address: US Embassy, Zagreb, Unit 1345, APO AE 09213-1345
telephone: [385] (41) 456-000
FAX: [385] (41) 440-235
Flag: red, white, and blue horizontal bands with Croatian coat of arms (red and white checkered)

Economy

Overview: Before the dissolution of Yugoslavia, the republic of Croatia, after Slovenia, was the most prosperous and industrialized area, with a per capita output perhaps one-third above the Yugoslav average. At present, Croatian Serb Separatists control approximately one-third of the Croatian territory, and one of the overriding determinants of Croatia's long-term political and economic prospects will be the resolution of this territorial dispute. Croatia faces serious economic problems stemming from: the legacy of longtime Communist mismanagement of the economy; large foreign debt; damage during the fighting to bridges, factories, power lines, buildings, and houses; the large refugee population, both Croatian and Bosnian; and the disruption of economic ties to Serbia and the other former Yugoslav republics, as well as within its own territory. At the minimum, extensive Western aid and investment, especially in the tourist and oil industries, would seem necessary to revive the moribund economy. However, peace and political stability must come first; only then will recent government moves toward a "market-friendly" economy restore old levels of output. As of February 1995, fighting continues among Croats, Serbs, and Muslims, and national

boundaries and final political arrangements are still in doubt.
National product: GDP—purchasing power parity—$12.4 billion (1994 est.)
National product real growth rate: 3.4% (1994 est.)
National product per capita: $2,640 (1994 est.)
Inflation rate (consumer prices): 3% (1994 est.)
Unemployment rate: 17% (December 1994)
Budget:
revenues: $NA
expenditures: $NA, including capital expenditures of $NA
Exports: $3.9 billion (f.o.b., 1993)
commodities: machinery and transport equipment 30%, other manufacturers 37%, chemicals 11%, food and live animals 9%, raw materials 6.5%, fuels and lubricants 5% (1990)
partners: EC countries, Slovenia
Imports: $4.7 billion (c.i.f., 1993)
commodities: machinery and transport equipment 21%, fuels and lubricants 19%, food and live animals 16%, chemicals 14%, manufactured goods 13%, miscellaneous manufactured articles 9%, raw materials 6.5%, beverages and tobacco 1% (1990)
partners: EC countries, Slovenia, FSU countries
External debt: $2.9 billion (September 1994)
Industrial production: growth rate -4% (1994 est.)
Electricity:
capacity: 3,570,000 kW
production: NA kWh
consumption per capita: NA kWh (1993)
Industries: chemicals and plastics, machine tools, fabricated metal, electronics, pig iron and rolled steel products, aluminum reduction, paper, wood products (including furniture), building materials (including cement), textiles, shipbuilding, petroleum and petroleum refining, food processing and beverages
Agriculture: Croatia normally produces a food surplus; most agricultural land in private hands and concentrated in Croat-majority districts in Slavonia and Istria; much of Slavonia's land has been put out of production by fighting; wheat, corn, sugar beets, sunflowers, alfalfa, and clover are main crops in Slavonia; central Croatian highlands are less fertile but support cereal production, orchards, vineyards, livestock breeding, and dairy farming; coastal areas and offshore islands grow olives, citrus fruits, and vegetables
Economic aid:
recipient: IMF, $192 million
Currency: 1 Croatian kuna (HRK) = 100 paras
Exchange rates: Croatian kuna per US $1—5.6144 (November 1994)
Fiscal year: calendar year

Transportation

Railroads:
total: 2,699 km
standard gauge: 2,699 km 1.435-m gauge (963 km electrified)
note: disrupted by territorial dispute (1994)
Highways:
total: 27,368 km
paved: 22,176 km (302 km of expressways)
unpaved: 5,192 km (1991)
Inland waterways: 785 km perennially navigable
Pipelines: crude oil 670 km; petroleum products 20 km; natural gas 310 km (1992); note—now disrupted because of territorial dispute
Ports: Dubrovnik, Omis, Ploce, Pula, Rijeka, Sibenik, Split, Zadar
Merchant marine:
total: 35 ships (1,000 GRT or over) totaling 181,565 GRT/225,533 DWT
ships by type: bulk 1, cargo 20, chemical tanker 1, container 2, oil tanker 2, passenger 2, refrigerated cargo 1, roll-on/roll-off cargo 2, short-sea passenger 4
note: also controlled by Croatian shipowners are 134 ships (1,000 GRT or over) totaling 3,286,231 DWT that operate under Maltese and Saint Vincent and the Grenadines registry
Airports:
total: 76
with paved runways over 3,047 m: 2
with paved runways 2,438 to 3,047 m: 6
with paved runways 1,524 to 2,437 m: 2
with paved runways 914 to 1,523 m: 1
with paved runways under 914 m: 55
with unpaved runways 1,524 to 2,438 m: 2
with unpaved runways 914 to 1,523 m: 8

Communications

Telephone system: 350,000 telephones
local: NA
intercity: NA
international: no satellite links
Radio:
broadcast stations: AM 14, FM 8, shortwave 0
radios: 1.1 million
Television:
broadcast stations: 12 (repeaters 2)
televisions: 1.027 million

Defense Forces

Branches: Ground Forces, Naval Forces, Air and Air Defense Forces, Frontier Guard, Home Guard
Manpower availability: males age 15-49 1,183,184; males fit for military service 943,749; males reach military age (19) annually 32,831 (1995 est.)
Defense expenditures: 337 billion to 393 billion dinars, NA% of GDP (1993 est.); note—conversion of defense expenditures into US dollars using the current exchange rate could produce misleading results

Cuba

Geography

Location: Caribbean, island between the Caribbean Sea and the North Atlantic Ocean, south of Florida
Map references: Central America and the Caribbean
Area:
total area: 110,860 sq km
land area: 110,860 sq km
comparative area: slightly smaller than Pennsylvania
Land boundaries: total 29 km, US Naval Base at Guantanamo Bay 29 km
note: Guantanamo Naval Base is leased by the US and thus remains part of Cuba
Coastline: 3,735 km
Maritime claims:
exclusive economic zone: 200 nm
territorial sea: 12 nm
International disputes: US Naval Base at Guantanamo Bay is leased to US and only mutual agreement or US abandonment of the area can terminate the lease
Climate: tropical; moderated by trade winds; dry season (November to April); rainy season (May to October)
Terrain: mostly flat to rolling plains with rugged hills and mountains in the southeast
Natural resources: cobalt, nickel, iron ore, copper, manganese, salt, timber, silica, petroleum
Land use:
arable land: 23%
permanent crops: 6%
meadows and pastures: 23%
forest and woodland: 17%
other: 31%
Irrigated land: 8,960 sq km (1989)
Environment:
current issues: pollution of Havana Bay; overhunting threatens wildlife populations; deforestation

natural hazards: the east coast is subject to hurricanes from August to October (in general, the country averages about one hurricane every other year); droughts are common
international agreements: party to—Antarctic Treaty, Biodiversity, Climate Change, Endangered Species, Environmental Modification, Hazardous Wastes, Law of the Sea, Marine Dumping, Ozone Layer Protection, Ship Pollution; signed, but not ratified—Antarctic-Environmental Protocol, Desertification, Marine Life Conservation
Note: largest country in Caribbean

People

Population: 10,937,635 (July 1995 est.)
Age structure:
0-14 years: 22% (female 1,191,320; male 1,256,928)
15-64 years: 68% (female 3,732,434; male 3,751,464)
65 years and over: 10% (female 528,104; male 477,385) (July 1995 est.)
Population growth rate: 0.65% (1995 est.)
Birth rate: 14.54 births/1,000 population (1995 est.)
Death rate: 6.53 deaths/1,000 population (1995 est.)
Net migration rate: -1.55 migrant(s)/1,000 population (1995 est.)
Infant mortality rate: 8.1 deaths/1,000 live births (1995 est.)
Life expectancy at birth:
total population: 77.05 years
male: 74.86 years
female: 79.37 years (1995 est.)
Total fertility rate: 1.63 children born/woman (1995 est.)
Nationality:
noun: Cuban(s)
adjective: Cuban
Ethnic divisions: mulatto 51%, white 37%, black 11%, Chinese 1%
Religions: nominally Roman Catholic 85% prior to Castro assuming power
Languages: Spanish
Literacy: age 15-49 and over can read and write (1981)
total population: 98%
Labor force: 4,620,800 economically active population (1988); 3,578,800 in state sector
by occupation: services and government 30%, industry 22%, agriculture 20%, commerce 11%, construction 10%, transportation and communications 7% (June 1990)

Government

Names:
conventional long form: Republic of Cuba
conventional short form: Cuba
local long form: Republica de Cuba
local short form: Cuba
Digraph: CU
Type: Communist state
Capital: Havana
Administrative divisions: 14 provinces (provincias, singular—provincia) and 1 special municipality* (municipio especial); Camaguey, Ciego de Avila, Cienfuegos, Ciudad de La Habana, Granma, Guantanamo, Holguin, Isla de la Juventud*, La Habana, Las Tunas, Matanzas, Pinar del Rio, Sancti Spiritus, Santiago de Cuba, Villa Clara
Independence: 20 May 1902 (from Spain 10 December 1898; administered by the US from 1898 to 1902)
National holiday: Rebellion Day, 26 July (1953)
Constitution: 24 February 1976
Legal system: based on Spanish and American law, with large elements of Communist legal theory; does not accept compulsory ICJ jurisdiction
Suffrage: 16 years of age; universal
Executive branch:
chief of state and head of government: President of the Council of State and President of the Council of Ministers Fidel CASTRO Ruz (Prime Minister from February 1959 until 24 February 1976 when office was abolished; President since 2 December 1976); First Vice President of the Council of State and First Vice President of the Council of Ministers Gen. Raul CASTRO Ruz (since 2 December 1976)
cabinet: Council of Ministers; proposed by the president of the Council of State, appointed by the National Assembly
Legislative branch: unicameral
National Assembly of People's Power: (Asamblea Nacional del Poder Popular) elections last held February 1993 (next to be held NA); seats—589 total, elected directly from slates approved by special candidacy commissions
Judicial branch: People's Supreme Court (Tribunal Supremo Popular)
Political parties and leaders: only party—Cuban Communist Party (PCC), Fidel CASTRO Ruz, first secretary
Member of: CCC, ECLAC, FAO, G-77, GATT, IAEA, ICAO, ICRM, IFAD, IFRCS, ILO, IMO, INMARSAT, INTELSAT (nonsignatory user), INTERPOL, IOC, ISO, ITU, LAES, LAIA (observer), NAM, OAS (excluded from formal participation since 1962), PCA, UN, UNCTAD, UNESCO, UNIDO, UPU, WCL, WFTU, WHO, WIPO, WMO, WTO
Diplomatic representation in US:
chief of mission: Principal Officer Alfonso FRAGA PEREZ (since August 1992) represented by the Cuban Interests Section of the Swiss Embassy in Washington, DC
chancery: 2630 and 2639 16th Street NW, Cuban Interests Section, Swiss Embassy, Washington, DC 20009
telephone: [1] (202) 797-8609, 8610, 8615
US diplomatic representation:
chief of mission: Principal Officer Joseph G. SULLIVAN
US Interests Section: USINT, Swiss Embassy, Calzada Entre L Y M, Vedado Seccion, Havana
mailing address: use street address
telephone: 33-3551 through 3559, 33-3543 through 3547, 33-3700 (operator assistance required)
FAX: Telex 512206
note: protecting power in Cuba is Switzerland—US Interests Section, Swiss Embassy
Flag: five equal horizontal bands of blue (top and bottom) alternating with white; a red equilateral triangle based on the hoist side bears a white five-pointed star in the center

Economy

Overview: Cuba's heavily statist economy remains severely depressed as the result of its own inefficiencies and the loss of massive amounts of economic aid from the former Soviet Bloc. Total output in 1994 was only about half the output of 1989. The fall in output and in imports is reflected in the deterioration of food supplies, shortages of electricity, inability to get spare parts, and the replacement of motor-driven vehicles by bicycles and draft animals. Higher world market prices for sugar and nickel in 1994, however, resulted in a slight increase in export earnings for the first time in six years, despite lower production of both commodities. The growth of tourism slowed in late 1994 as a result of negative publicity surrounding the exodus of Cubans from the island and other international factors. The government continued its aggressive search for foreign investment and announced preliminary agreements to form large joint ventures with Mexican investors in telecommunications and oil refining. In mid-1994, the National Assembly began introducing several new taxes and price increases to stem growing excess liquidity and restore some of the peso's value as a monetary instrument. In October the government attempted to stimulate food production by permitting the sale of any surplus production (over state quotas) at unrestricted prices at designated markets. Similar but much smaller markets were also introduced for the sale of manufactured goods in December. The various government measures have influenced a remarkable appreciation of the black market value of the peso, from more than 100 pesos to

the dollar in September 1994 to 40 pesos to the dollar in early 1995. Policy discussions continue in the bureaucracy over the proper pace and scope of economic reform.
National product: GDP—purchasing power parity—$14 billion (1994 est.)
National product real growth rate: 0.4% (1994 est.)
National product per capita: $1,260 (1994 est.)
Inflation rate (consumer prices): NA%
Unemployment rate: NA%
Budget:
revenues: $9.3 billion
expenditures: $12.5 billion, including capital expenditures of $NA (1994 est.)
Exports: $1.6 billion (f.o.b., 1994 est.)
commodities: sugar, nickel, shellfish, tobacco, medical products, citrus, coffee
partners: Russia 15%, Canada 9%, China 8%, Egypt 6%, Spain 5%, Japan 4%, Morocco 4% (1994 est.)
Imports: $1.7 billion (c.i.f., 1994 est.)
commodities: petroleum, food, machinery, chemicals
partners: Spain 17%, Mexico 10%, France 8%, China 8%, Venezuela 7%, Italy 4%, Canada 3%, (1994 est.)
External debt: $10.8 billion (convertible currency, December 1993)
Industrial production: growth rate NA%
Electricity:
capacity: 3,990,000 kW
production: 12 billion kWh
consumption per capita: 1,022 kWh (1993)
Industries: sugar milling and refining, petroleum refining, food and tobacco processing, textiles, chemicals, paper and wood products, metals (particularly nickel), cement, fertilizers, consumer goods, agricultural machinery
Agriculture: key commercial crops—sugarcane, tobacco, and citrus fruits; other products—coffee, rice, potatoes, meat, beans; world's largest sugar exporter; not self-sufficient in food (excluding sugar); sector hurt by persistent shortages of fuels and parts
Economic aid:
recipient: Western (non-US) countries, ODA and OOF bilateral commitments (1970-89), $710 million; Communist countries (1970-89), $18.5 billion
Currency: 1 Cuban peso (Cu$) = 100 centavos
Exchange rates: Cuban pesos (Cu$) per US$1—1.0000 (non-convertible, official rate, linked to the US dollar)
Fiscal year: calendar year

Transportation

Railroads:
total: 12,623 km
standard gauge: 4,881 km 1.435-m gauge (151.7 km electrified)
other: 7,742 km 0.914- and 1.435-m gauge for sugar plantation lines
Highways:
total: 26,477 km
paved: 14,477 km
unpaved: gravel or earth 12,000 km (1989)
Inland waterways: 240 km
Ports: Cienfuegos, La Habana, Manzanillo, Mariel, Matanzas, Nuevitas, Santiago de Cuba
Merchant marine:
total: 48 ships (1,000 GRT or over) totaling 278,103 GRT/396,138 DWT
ships by type: bulk 1, cargo 22, chemical tanker 1, liquefied gas tanker 4, oil tanker 10, passenger-cargo 1, refrigerated cargo 9
note: Cuba beneficially owns an additional 24 ships (1,000 GRT or over) totaling 215,703 DWT under the registry of Panama, Cyprus, Malta, and Mauritius
Airports:
total: 181
with paved runways over 3,047 m: 7
with paved runways 2,438 to 3,047 m: 8
with paved runways 1,524 to 2,437 m: 13
with paved runways 914 to 1,523 m: 10
with paved runways under 914 m: 106
with unpaved runways 1,524 to 2,438 m: 1
with unpaved runways 914 to 1,523 m: 36

Communications

Telephone system: 229,000 telephones; 20.7 telephones/1,000 persons; among the world's least developed telephone systems
local: NA
intercity: NA
international: 1 INTELSAT (Atlantic Ocean) earth station
Radio:
broadcast stations: AM 150, FM 5, shortwave 0
radios: 2.14 million
Television:
broadcast stations: 58
televisions: 1.53 million

Defense Forces

Branches: Revolutionary Armed Forces (FAR) includes ground forces, Revolutionary Navy (MGR), Air and Air Defense Force (DAAFAR), Territorial Militia Troops (MTT), and Youth Labor Army (EJT); Interior Ministry Border Guards (TGF),
Manpower availability: males age 15-49 3,065,751; females age 15-49 3,023,997; males fit for military service 1,909,901; females fit for military service 1,878,768; males reach military age (17) annually 72,582; females reach military age (17) annually 69,361 (1995 est.)
Defense expenditures: exchange rate conversion—approx. $600 million, 4% of GSP (gross social product) in 1994 was for defense
Note: Moscow, for decades the key military supporter and supplier of Cuba, cut off military aid by 1993

Cyprus

Geography

Location: Middle East, island in the Mediterreanean Sea, south of Turkey
Map references: Middle East
Area:
total area: 9,250 sq km (note—3,355 sq km are in the Turkish area)
land area: 9,240 sq km
comparative area: about 0.7 times the size of Connecticut
Land boundaries: 0 km
Coastline: 648 km
Maritime claims:
continental shelf: 200-m depth or to the depth of exploitation
territorial sea: 12 nm
International disputes: 1974 hostilities divided the island into two de facto autonomous areas, a Greek area controlled by the Cypriot Government (59% of the island's land area) and a Turkish-Cypriot area (37% of the island), that are separated by a UN buffer zone (4% of the island); there are two UK sovereign base areas within the Greek Cypriot portion of the island
Climate: temperate, Mediterranean with hot, dry summers and cool, wet winters
Terrain: central plain with mountains to north and south; scattered but significant plains along southern coast
Natural resources: copper, pyrites, asbestos, gypsum, timber, salt, marble, clay earth pigment
Land use:
arable land: 40%
permanent crops: 7%
meadows and pastures: 10%
forest and woodland: 18%
other: 25%
Irrigated land: 350 sq km (1989)
Environment:
current issues: water resource problems (no natural reservoir catchments, seasonal disparity in rainfall, and most potable resources concentrated in the Turkish Cypriot area); water pollution from sewage and industrial wastes; coastal degradation; loss of wildlife habitats from urbanization
natural hazards: moderate earthquake activity
international agreements: party to—Air Pollution, Endangered Species, Environmental Modification, Hazardous Wastes, Law of the Sea, Marine Dumping, Nuclear Test Ban, Ozone Layer Protection, Ship Pollution; signed, but not ratified—Biodiversity, Climate Change

People

Population:
total: 736,636 (July 1995 est.) (78% Greek, 18% Turk, 4% other)
Greek area: 602,656 (July 1995 est.) (94.9% Greek, 0.3% Turk, 4.8% other)
Turkish area: 133,980 (July 1995 est.) (2.1% Greek, 97.7% Turk, 0.2% other)
Age structure:
0-14 years: 26% (female 92,179; male 97,723)
15-64 years: 64% (female 234,929; male 236,693)
65 years and over: 10% (female 42,190; male 32,922) (July 1995 est.)
Population growth rate: 0.88% (1995 est.)
Birth rate: 16.27 births/1,000 population (1995 est.)
Death rate: 7.48 deaths/1,000 population (1995 est.)
Net migration rate: 0 migrant(s)/1,000 population (1995 est.)
Infant mortality rate: 8.6 deaths/1,000 live births (1995 est.)
Life expectancy at birth:
total population: 76.47 years
male: 74.19 years
female: 78.85 years (1995 est.)
Total fertility rate: 2.3 children born/woman (1995 est.)
Nationality:
noun: Cypriot(s)
adjective: Cypriot
Ethnic divisions:
total: Greek 78% (99.5% of the Greeks live in the Greek area; 0.5% of the Greeks live in the Turkish area), Turkish 18% (1.3% of the Turks live in the Greek area; 98.7% of the Turks live in the Turkish area), other 4% (99.2% of the other ethnic groups live in the Greek area; 0.8% of the other ethnic groups live in the Turkish area)
Religions: Greek Orthodox 78%, Muslim 18%, Maronite, Armenian Apostolic, and other 4%
Languages: Greek, Turkish, English
Literacy: age 15 and over can read and write (1987 est.)
total population: 94%
male: 98%
female: 91%
Labor force:
Greek area: 285,500
by occupation: services 57%, industry 29%, agriculture 14% (1992)
Turkish area: 74,000
by occupation: services 52%, industry 23%, agriculture 25% (1992)

Government

Names:
conventional long form: Republic of Cyprus
conventional short form: Cyprus
note: the Turkish area refers to itself as the "Turkish Republic" or the "Turkish Republic of Northern Cyprus"
Abbreviation: the Turkish area is sometimes referred to as the TRNC which is short for "Turkish Republic of Northern Cyprus"
Digraph: CY
Type: republic
note: a disaggregation of the two ethnic communities inhabiting the island began after the outbreak of communal strife in 1963; this separation was further solidified following the Turkish invasion of the island in July 1974, which gave the Turkish Cypriots de facto control in the north; Greek Cypriots control the only internationally recognized government; on 15 November 1983 Turkish Cypriot President Rauf DENKTASH declared independence and the formation of a "Turkish Republic of Northern Cyprus" (TRNC), which has been recognized only by Turkey; both sides publicly call for the resolution of intercommunal differences and creation of a new federal system of government
Capital: Nicosia
note: the Turkish area's capital is Lefkosa (Nicosia)
Administrative divisions: 6 districts; Famagusta, Kyrenia, Larnaca, Limassol, Nicosia, Paphos; note—Turkish area administrative divisions include Kyrenia, all but a small part of Famagusta, and small parts of Nicosia and Larnaca
Independence: 16 August 1960 (from UK)
note: Turkish area proclaimed self-rule on NA February 1975 from Republic of Cyprus
National holiday: Independence Day, 1 October
note: Turkish area celebrates 15 November as Independence Day
Constitution: 16 August 1960; negotiations to create the basis for a new or revised constitution to govern the island and to better relations between Greek and Turkish Cypriots have been held intermittently; in 1975 Turkish Cypriots created their own Constitution and governing bodies within the "Turkish Federated State of Cyprus," which was renamed the "Turkish Republic of Northern Cyprus" in 1983; a new Constitution for the Turkish area passed by referendum on 5 May 1985
Legal system: based on common law, with civil law modifications
Suffrage: 18 years of age; universal

Executive branch:
chief of state and head of government: President Glafcos CLERIDES (since 28 February 1993); election last held 14 February 1993 (next to be held February 1998); results—Glafkos CLERIDES 50.3%, George VASSILIOU 49.7%
cabinet: Council of Ministers; appointed jointly by the president and vice-president
note: Rauf R. DENKTASH has been president of the Turkish area since 13 February 1975; Hakki ATUN has been prime minister of the Turkish area since 1 January 1994; there is a Council of Ministers (cabinet) in the Turkish area; elections last held 15 and 22 April 1995 (next to be held April 2000); results—Rauf R. DENKTASH 62.5%, Dervis EROGLU 37.5%
Legislative branch: unicameral
Greek area: House of Representatives (Vouli Antiprosopon): elections last held 19 May 1991 (next to be held NA); results—DISY 35.8%, AKEL (Communist) 30.6%, DIKO 19.5%, EDEK 10.9%; others 3.2%; seats—(56 total) DISY 20, AKEL (Communist) 18, DIKO 11, EDEK 7
Turkish area: Assembly of the Republic (Cumhuriyet Meclisi): elections last held 12 December 1993 (next to be held NA); results—UBP 29.9%, DP 29.2%, CTP 24.2% TKP 13.3%, others 3.4%; seats—(50 total) UBP (conservative) 15, DP 16, CTP 13, TKP 5, UDP 1
Judicial branch: Supreme Court; note—there is also a Supreme Court in the Turkish area
Political parties and leaders:
Greek area: Progressive Party of the Working People (AKEL, Communist Party), Dimitrios CHRISTOFIAS; Democratic Rally (DISY), John MATSIS; Democratic Party (DIKO), Spyros KYPRIANOU; United Democratic Union of the Center (EDEK), Vassos LYSSARIDIS; Socialist Democratic Renewal Movement (ADISOK), Mikhalis PAPAPETROU; Liberal Party, Nikos ROLANDIS; Free Democrats, George VASSILIOU
Turkish area: National Unity Party (UBP), Dervis EROGLU; Communal Liberation Party (TKP), Mustafa AKINCI; Republican Turkish Party (CTP), Ozker OZGUR; New Cyprus Party (YKP), Alpay DURDURAN; Free Democratic Party (HDP), Ismet KOTAK; National Justice Party (MAP), Zorlu TORE; Unity and Sovereignty Party (BEP), Arif Salih KIRDAG; Democratic Party (DP), Hakki ATUN; Fatherland Party (VP), Orhan UCOK; National Birth Party (UDP) note—the HDP, MAP, and VP merged under the label National Struggle Unity Party (MMBP) to compete in the 12 December 1993 legislative election
Other political or pressure groups: United Democratic Youth Organization (EDON, Communist controlled); Union of Cyprus Farmers (EKA, Communist controlled); Cyprus Farmers Union (PEK, pro-West); Pan-Cyprian Labor Federation (PEO, Communist controlled); Confederation of Cypriot Workers (SEK, pro-West); Federation of Turkish Cypriot Labor Unions (Turk-Sen); Confederation of Revolutionary Labor Unions (Dev-Is)
Member of: C, CCC, CE, EBRD, ECE, FAO, G-77, GATT, IAEA, IBRD, ICAO, ICC, ICFTU, IDA, IFAD, IFC, IFRCS (associate), ILO, IMF, IMO, INMARSAT, INTELSAT, INTERPOL, IOC, IOM, ISO, ITU, NAM, OAS (observer), OSCE, PCA, UN, UNCTAD, UNESCO, UNIDO, UPU, WCL, WFTU, WHO, WIPO, WMO, WTO
Diplomatic representation in US:
chief of mission: Ambassador Andreas J. JACOVIDES
chancery: 2211 R Street NW, Washington, DC 20008
telephone: [1] (202) 462-5772
consulate(s) general: New York
note: Representative of the Turkish area in the US is Namik KORMAN, office at 1667 K Street NW, Washington, DC, telephone [1] (202) 887-6198
US diplomatic representation:
chief of mission: Ambassador Richard A. BOUCHER
embassy: corner of Metochiou and Ploutarchou Streets, Engomi, Nicosia
mailing address: P. O. Box 4536 APO AE 09836
telephone: [357] (2) 476100
FAX: [357] (2) 465944
Flag: white with a copper-colored silhouette of the island (the name Cyprus is derived from the Greek word for copper) above two green crossed olive branches in the center of the flag; the branches symbolize the hope for peace and reconciliation between the Greek and Turkish communities
note: the Turkish Cypriot flag has a horizontal red stripe at the top and bottom between which is a red crescent and red star on a white field

---

Economy

Overview: The Greek Cypriot economy is small, diversified, and prosperous. Industry contributes 14% to GDP and employs 29% of the labor force, while the service sector contributes 53% to GDP and employs 57% of the labor force. An average 6.8% rise in real GDP between 1986 and 1990 was temporarily checked in 1991, because of the adverse effects of the Gulf war on tourism. After surging 8.5% in 1992, growth slowed to 2.0% in 1993—its lowest level in two decades—because of the decline in tourist arrivals associated with the recession in Western Europe, Cyprus' main trading partner, and the loss in export competitiveness due to a sharp rise in unit labor costs. Real GDP is likely to have picked up in 1994, and inflation is estimated to have risen to between 5% and 6%. The Turkish Cypriot economy has less than one-third the per capita GDP of the south. Because it is recognized only by Turkey, it has had much difficulty arranging foreign financing, and foreign firms have hesitated to invest there. The economy remains heavily dependent on agriculture, which employs one-quarter of the work force. Moreover, because the Turkish lira is legal tender, the Turkish Cypriot economy has suffered the same high inflation as mainland Turkey. The small, vulnerable economy is estimated to have experienced a sharp drop in growth during 1994 because of the severe economic crisis affecting the mainland. To compensate for the economy's weakness, Turkey provides direct and indirect aid to nearly every sector; financial support has risen in value to about one-third of Turkish Cypriot GDP.
National product:
Greek area: GDP—purchasing power parity—$7.3 billion (1994 est.)
Turkish area: GDP—purchasing power parity—$510 million (1994 est.)
National product real growth rate:
Greek area: 5% (1994 est.)
Turkish area: -4% (1994 est.)
National product per capita:
Greek area: $12,500 (1994 est.)
Turkish area: $3,500 (1994 est.)
Inflation rate (consumer prices):
Greek area: 4.8% (1993)
Turkish area: 63.4% (1992)
Unemployment rate:
Greek area: 2.3% (1993)
Turkish area: 1.2% (1992)
Budget:
revenues: Greek area—$1.8 billion
Turkish area—$285 million
expenditures: Greek area—$2.4 billion, including capital expenditures of $400 million
Turkish area—$377 million, including capital expenditures of $80 million (1995 est.)
Exports: $868 million (f.o.b., 1993)
commodities: citrus, potatoes, grapes, wine, cement, clothing and shoes
partners: UK 18%, Greece 9%, Lebanon 14%, Germany 6%
Imports: $2.6 billion (f.o.b., 1993)
commodities: consumer goods, petroleum and lubricants, food and feed grains, machinery
partners: UK 13%, Japan 9%, Italy 10%, Germany 8%, US 8%
External debt: $2.4 billion (1993)
Industrial production: growth rate 0.1% (1993); accounts for 14% of GDP
Electricity:
capacity: 550,000 kW
production: 2.3 billion kWh
consumption per capita: 2,903 kWh (1993)
Industries: food, beverages, textiles, chemicals, metal products, tourism, wood products
Agriculture: contributes 6% to GDP and employs 25% of labor force in the south; major crops—potatoes, vegetables, barley, grapes,

olives, citrus fruits; vegetables and fruit provide 25% of export revenues
Illicit drugs: transit point for heroin via air routes and container traffic to Europe, especially from Lebanon and Turkey
Economic aid:
recipient: US commitments, including Ex-Im (FY70-89), $292 million; Western (non-US) countries, ODA and OOF bilateral commitments (1970-89), $250 million; OPEC bilateral aid (1979-89), $62 million; Communist countries (1970-89), $24 million
Currency: 1 Cypriot pound (£C) = 100 cents; 1 Turkish lira (TL) = 100 kurus
Exchange rates: Cypriot pounds per $US1—0.4725 (January 1995), 0.4915 (1994), 0.4970 (1993), 0.4502 (1992), 0.4615 (1991), 0.4572 (1990); Turkish liras (TL) per US$1—37,444.1 (December 1994), 29,608.7 (1994), 10,984.6 (1993), 6,872.4 (1992), 4,171.8 (1991), 2,608.6 (1990)
Fiscal year: calendar year

Transportation

Railroads: 0 km
Highways:
Greek area:
total: 10,448 km
paved: 5,694 km
unpaved: gravel, crushed stone, earth 4,754 km (1992)
Turkish area:
total: 6,116 km
paved: 5,278 km
unpaved: 838 km
Ports: Famagusta, Kyrenia, Larnaca, Limassol, Vasilikos Bay
Merchant marine:
total: 1,446 ships (1,000 GRT or over) totaling 22,911,818 GRT/39,549,216 DWT
ships by type: bulk 473, cargo 530, chemical tanker 28, combination bulk 55, combination ore/oil 24, container 92, liquefied gas tanker 3, multifunction large-load carrier 5, oil tanker 120, passenger 5, passenger-cargo 1, railcar carrier 1, refrigerated cargo 58, roll-on/roll-off cargo 33, short-sea passenger 14, specialized tanker 2, vehicle carrier 2
note: a flag of convenience registry; includes 48 countries among which are ships of Greece 705, Germany 174, Russia 56, Netherlands 45, Japan 27, Belgium 25, UK 21, Spain 17, Switzerland 14, Hong Kong 13
Airports:
total: 15
with paved runways 2,438 to 3,047 m: 7
with paved runways 914 to 1,523 m: 3
with paved runways under 914 m: 4
with unpaved runways 914 to 1,523 m: 1

Communications

Telephone system: 210,000 telephones; excellent in both the area controlled by the Cypriot Government (Greek area), and in the Turkish-Cypriot administered area; largely open-wire and microwave radio relay
local: NA
intercity: microwave radio/relay
international: international service by tropospheric scatter, 3 submarine cables, and 2 INTELSAT (1 Atlantic Ocean and 1 Indian Ocean) and 1 EUTELSAT earth station
Radio:
Greek sector: NA
broadcast stations: AM 11, FM 8, shortwave 0
radios: NA
Turkish sector: NA
broadcast stations: AM 2, FM 6, shortwave 0
radios: NA
Television:
Greek sector: NA
broadcast stations: 1 (repeaters 34)
televisions: NA
Turkish sector: NA
broadcast stations: 1
televisions: NA

Defense Forces

Branches:
Greek area: Greek Cypriot National Guard (GCNG; includes air and naval elements), Greek Cypriot Police
Turkish area: Turkish Cypriot Security Force
Manpower availability: males age 15-49 188,231; males fit for military service 129,397; males reach military age (18) annually 5,467 (1995 est.)
Defense expenditures: exchange rate conversion—$457 million, 5.6% of GDP (1995)

Czech Republic

Geography

Location: Central Europe, southeast of Germany
Map references: Ethnic Groups in Eastern Europe, Europe
Area:
total area: 78,703 sq km
land area: 78,645 sq km
comparative area: slightly smaller than South Carolina
Land boundaries: total 1,880 km, Austria 362 km, Germany 646 km, Poland 658 km, Slovakia 214 km
Coastline: 0 km (landlocked)
Maritime claims: none; landlocked
International disputes: Liechtenstein claims restitution for 1,600 square kilometers of Czech territory confiscated from its royal family in 1918; Sudeten German claims for restitution of property confiscated in connection with their expulsion after World War II versus the Czech Republic claims that restitution does not preceed before February 1948 when the Communists seized power; unresolved property issues with Slovakia over redistribution of property of the former Czechoslovak federal government
Climate: temperate; cool summers; cold, cloudy, humid winters
Terrain: two main regions: Bohemia in the west, consisting of rolling plains, hills, and plateaus surrounded by low mountains; and Moravia in the east, consisting of very hilly country
Natural resources: hard coal, soft coal, kaolin, clay, graphite
Land use:
arable land: NA%
permanent crops: NA%
meadows and pastures: NA%
forest and woodland: NA%
other: NA%
Irrigated land: NA sq km
Environment:
current issues: air and water pollution in areas of northwest Bohemia centered around Zeplica

and in northern Moravia around Ostrava present health risks; acid rain damaging forests
natural hazards: NA
international agreements: party to—Air Pollution, Air Pollution-Nitrogen Oxides, Air Pollution-Sulphur 85, Antarctic Treaty, Biodiversity, Climate Change, Endangered Species, Environmental Modification, Hazardous Wastes, Nuclear Test Ban, Ozone Layer Protection, Ship Pollution, Wetlands; signed, but not ratified—Air Pollution-Sulphur 94, Antarctic-Environmental Protocol, Law of the Sea
Note: landlocked; strategically located astride some of oldest and most significant land routes in Europe; Moravian Gate is a traditional military corridor between the North European Plain and the Danube in central Europe

People

Population: 10,432,774 (July 1995 est.)
Age structure:
0-14 years: 19% (female 981,918; male 1,030,003)
15-64 years: 68% (female 3,529,411; male 3,530,112)
65 years and over: 13% (female 848,599; male 512,731) (July 1995 est.)
Population growth rate: 0.26% (1995 est.)
Birth rate: 13.46 births/1,000 population (1995 est.)
Death rate: 10.85 deaths/1,000 population (1995 est.)
Net migration rate: 0 migrant(s)/1,000 population (1995 est.)
Infant mortality rate: 8.9 deaths/1,000 live births (1995 est.)
Life expectancy at birth:
total population: 73.54 years
male: 69.87 years
female: 77.41 years (1995 est.)
Total fertility rate: 1.84 children born/ woman (1995 est.)
Nationality:
noun: Czech(s)
adjective: Czech
note: 300,000 Slovaks declared themselves Czech citizens in 1994
Ethnic divisions: Czech 94.4%, Slovak 3%, Polish 0.6%, German 0.5%, Gypsy 0.3%, Hungarian 0.2%, other 1%
Religions: atheist 39.8%, Roman Catholic 39.2%, Protestant 4.6%, Orthodox 3%, other 13.4%
Languages: Czech, Slovak
Literacy: can read and write
total population: 99%
Labor force: 5.389 million
by occupation: industry 37.9%, agriculture 8.1%, construction 8.8%, communications and other 45.2% (1990)

Government

Names:
conventional long form: Czech Republic
conventional short form: Czech Republic
local long form: Ceska Republika
local short form: Cechy
Digraph: EZ
Type: parliamentary democracy
Capital: Prague
Administrative divisions: 8 regions (kraje, kraj—singular); Jihocesky, Jihomoravsky, Praha, Severocesky, Severomoravsky, Stredocesky, Vychodocesky, Zapadocesky
Independence: 1 January 1993 (from Czechoslovakia)
National holiday: National Liberation Day, 9 May; Founding of the Republic, 28 October
Constitution: ratified 16 December 1992; effective 1 January 1993
Legal system: civil law system based on Austro-Hungarian codes; has not accepted compulsory ICJ jurisdiction; legal code modified to bring it in line with Organization on Security and Cooperation in Europe (OSCE) obligations and to expunge Marxist Leninist legal theory
Suffrage: 18 years of age; universal
Executive branch:
chief of state: President Vaclav HAVEL (since 26 January 1993); election last held 26 January 1993 (next to be held NA January 1998); results—Vaclav HAVEL elected by the National Council
head of government: Prime Minister Vaclav KLAUS (since NA June 1992); Deputy Prime Ministers Ivan KOCARNIK, Josef LUX, Jan KALVODA (since NA June 1992)
cabinet: Cabinet; appointed by the president on recommendation of the prime minister
Legislative branch: bicameral National Council (Narodni rada)
Senate: elections not yet held; seats (81 total)
Chamber of Deputies: elections last held 5-6 June 1992 (next to be held NA 1996); results—percent of vote by party NA given breakup and realignment of all parliamentary opposition parties since 1992; seats—(200 total) governing coalition: ODS 65, KDS 10, ODA 16, KDU-CSL 15, opposition: CSSD 18, LB 25, KSCM 10, LSU 9, LSNS 5, CMSS 9, SPR-RSC 6, independents 12
Judicial branch: Supreme Court, Constitutional Court
Political parties and leaders:
governing coalition: Civic Democratic Party (ODS), Vaclav KLAUS, chairman; Christian Democratic Party (KDS), Ivan PILIP, chairman; Civic Democratic Alliance (ODA), Jan KALVODA, chairman; Christian Democratic Union/Czech People's Party (KDU-CSL), Josef LUX, chairman
opposition: Czech Social Democrats (CSSD—left opposition), Milos ZEMAN, chairman; Left Bloc (LB—left opposition), Marie STIBOROVA, chairman; Communist Party (KSCM—left opposition), Miroslav GREBENICEK, chairman; Liberal Social Union (LSU—left opposition), Frantisek TRNKA, chairman; Liberal National Social Party (LSNS—center party), Pavel HIRS, chairman; Bohemian-Moravian Center Party (CMSS—center party), Jan KYCER, chairman; Assembly for the Republic (SPR-RSC—right radical) , Miroslav SLADEK, chairman
Other political or pressure groups: Czech-Moravian Chamber of Trade Unions; Civic Movement
Member of: Australia Group, BIS, CCC, CE (guest), CEI, CERN, EBRD, ECE, FAO, GATT, IAEA, IBRD, ICAO, ICFTU, ICRM, IDA, IFC, IFRCS, ILO, IMF, IMO, INMARSAT, INTELSAT, INTERPOL, IOC, IOM (observer), ISO, ITU, NACC, NSG, OSCE, PCA, PFP, UN, UNCTAD, UNESCO, UNIDO, UNOMIL, UNOMOZ, UNPROFOR, UPU, WEU (associate partner), WFTU, WHO, WIPO, WMO, WTO, ZC
Diplomatic representation in US:
chief of mission: Ambassador Michael ZANTOVSKY
chancery: 3900 Spring of Freedom Street NW, Washington, DC 20008
telephone: [1] (202) 363-6315, 6316
FAX: [1] (202) 966-8540
US diplomatic representation:
chief of mission: Ambassador Adrian A. BASORA
embassy: Trziste 15, 11801 Prague 1
mailing address: Unit 1330; APO AE 09213-1330
telephone: [42] (2) 2451-0847
FAX: [42] (2) 2451-1001
Flag: two equal horizontal bands of white (top) and red with a blue isosceles triangle based on the hoist side (almost identical to the flag of the former Czechoslovakia)

Economy

Overview: The government of the Czech Republic, using successful stabilization policies to bolster its claims to full membership in the western economic community, has reduced inflation to 10%, kept unemployment at 3%, balanced the budget, run trade surpluses, and reoriented exports to the EU since the breakup of the Czechoslovak federation on 1 January 1993. GDP grew 2% in 1994 after stagnating in 1993 and contracting nearly 20% since 1990. Prague's mass privatization program, including its innovative distribution of ownership shares to Czech citizens via 'coupon vouchers,' has made the most rapid progress in Eastern Europe. When

coupon shares are distributed in early 1995, 75%-80% of the economy will be in private hands or partially privatized, according to the Czech government. Privatized companies still face major problems in restructuring; the number of annual bankruptcies quadrupled in 1994. In September 1994, Prague repaid $471 million in IMF loans five years ahead of schedule, making the Czech Republic the first East European country to pay off all IMF debts. Despite these outlays, hard-currency reserves in the banking system totaled more than $8.5 billion in October. Standard & Poor's boosted the Republic's credit rating to BBB+ in mid-1994—up from a BBB rating that was already two steps higher than Hungary's and one step above Greece's rating. Prague forecasts a balanced budget, at least 3% GDP growth, 5% unemployment, and single-digit inflation for 1995. Inflationary pressures—primarily as a result of foreign bank lending to Czech enterprises but perhaps also due to eased currency convertibility controls—are likely to be the most troublesome issues in 1995. Continuing economic recovery in Western Europe should boost Czech exports and production but a substantial increase in prices could erode the Republic's comparative advantage in low wages and exchange rates. Prague already took steps in 1994 to increase control over banking policies to neutralize the impact of foreign inflows on the money supply. Although Czech unemployment is currently the lowest in Central Europe, it will probably increase 1-2 percentage points in 1995 as large state firms go bankrupt or are restructured and service sector growth slows.
National product: GDP—purchasing power parity—$76.5 billion (1994 est.)
National product real growth rate: 2.2% (1994 est.)
National product per capita: $7,350 (1994 est.)
Inflation rate (consumer prices): 10.2% (1994 est.)
Unemployment rate: 3.2% (1994 est.)
Budget:
revenues: $14 billion
expenditures: $13.6 billion, including capital expenditures of $NA (1994 est.)
Exports: $13.4 billion (f.o.b., 1994 est.)
commodities: manufactured goods, machinery and transport equipment, chemicals, fuels, minerals, metals, agricultural products (January-November 1994)
partners: Germany 28.7%, Slovakia 15.5%, Austria 7.9%, Italy 6.4%, France 3.2%, Russia 3.2%, Poland 3.1%, UK 2.9%, Netherlands 2.4%, Hungary 2.2%, US 2.1%, Belgium 1.3% (January-June 1994)
Imports: $13.3 billion (f.o.b., 1994 est.)
commodities: machinery and transport equipment, manufactured goods, chemicals, fuels and lubricants, raw materials, agricultural products (January-November 1994)
partners: Germany 24.1%, Slovakia 15.6%, Russia 9.8%, Austria 7.6%, Italy 4.9%, France 3.6%, US 3.2%, Netherlands 2.9%, UK 2.8%, Poland 2.7%, Switzerland 2.2%, Belgium 2.0% (January-June 1994)
External debt: $8.7 billion (October 1994)
Industrial production: growth rate 4.9% (January-September 1994)
Electricity:
capacity: 14.470,000 kW
production: 56.3 billion kWh
consumption per capita: 4,842 kWh (1993)
Industries: fuels, ferrous metallurgy, machinery and equipment, coal, motor vehicles, glass, armaments
Agriculture: largely self-sufficient in food production; diversified crop and livestock production, including grains, potatoes, sugar beets, hops, fruit, hogs, cattle, and poultry; exporter of forest products
Illicit drugs: transshipment point for Southwest Asian heroin and Latin American cocaine to Western Europe
Economic aid:
donor: 1.4 million annually to IMF beginning in 1994
Currency: 1 koruna (Kc) = 100 haleru
Exchange rates: koruny (Kcs) per US$1—27.762 (January 1995), 28.785 (1994), 29.153 (1993), 28.26 (1992), 29.53 (1991), 17.95 (1990)
note: values before 1993 reflect Czechoslovak exchange rates
Fiscal year: calendar year

Transportation

Railroads:
total: 9,434 km (include 1.520-m broad, 1.435-m standard, and several narrow gauges) (1988)
Highways:
total: 55,890 km (1988)
paved: NA
unpaved: NA
Inland waterways: NA km; the Elbe (Labe) is the principal river
Pipelines: natural gas 5,400 km
Ports: Decin, Prague, Usti nad Labem
Merchant marine:
total: 14 ships (1,000 GRT or over) totaling 181,646 GRT/282,296 DWT
ships by type: bulk 5, cargo 9
Airports:
total: 116
with paved runways over 3,047 m: 2
with paved runways 2,438 to 3,047 m: 9
with paved runways 1,524 to 2,437 m: 13
with paved runways under 914 m: 5
with unpaved runways over 3,047 m: 1
with unpaved runways 2,438 to 3,047 m: 3
with unpaved runways 1,524 to 2,438 m: 10
with unpaved runways 914 to 1,523 m: 32
with unpaved runways under 914 m: 41

Communications

Telephone system: NA telephones
local: NA
intercity: NA
international: NA
Radio:
broadcast stations: AM, FM, shortwave
radios: NA
Television:
broadcast stations: NA
televisions: NA

Defense Forces

Branches: Army, Air and Air Defense Forces, Civil Defense, Railroad Units
Manpower availability: males age 15-49 2,753,301; males fit for military service 2,095,661; males reach military age (18) annually 91,177 (1995 est.)
Defense expenditures: 27 billion koruny, NA% of GNP (1994 est.); note—conversion of defense expenditures into US dollars using the current exchange rate could produce misleading results

Denmark

Geography

Location: Northern Europe, bordering the Baltic Sea and the North Sea, on a peninsula north of Germany
Map references: Europe
Area:
total area: 43,070 sq km
land area: 42,370 sq km
comparative area: slightly more than twice the size of Massachusetts
note: includes the island of Bornholm in the Baltic Sea and the rest of metropolitan Denmark, but excludes the Faroe Islands and Greenland
Land boundaries: total 68 km, Germany 68 km
Coastline: 3,379 km
Maritime claims:
contiguous zone: 4 nm
continental shelf: 200-m depth or to the depth of exploitation
exclusive fishing zone: 200 nm
territorial sea: 3 nm
International disputes: Rockall continental shelf dispute involving Iceland, Ireland, and the UK (Ireland and the UK have signed a boundary agreement in the Rockall area)
Climate: temperate; humid and overcast; mild, windy winters and cool summers
Terrain: low and flat to gently rolling plains
Natural resources: petroleum, natural gas, fish, salt, limestone
Land use:
arable land: 61%
permanent crops: 0%
meadows and pastures: 6%
forest and woodland: 12%
other: 21%
Irrigated land: 4,300 sq km (1989 est.)
Environment:
current issues: air pollution, principally from vehicle emissions; nitrogen and phosphorus pollution of the North Sea; drinking and surface water becoming polluted from animal wastes
natural hazards: flooding is a threat in some areas of the country (e.g., parts of Jutland, along the southern coast of the island of Lolland) that are protected from the sea by a system of dikes
international agreements: party to—Air Pollution, Air Pollution-Nitrogen Oxides, Air Pollution-Sulphur 85, Antarctic Treaty, Biodiversity, Climate Change, Endangered Species, Environmental Modification, Hazardous Wastes, Marine Dumping, Marine Life Conservation, Nuclear Test Ban, Ozone Layer Protection, Ship Pollution, Tropical Timber 83, Wetlands, Whaling; signed, but not ratified—Air Pollution-Sulphur 94, Air Pollution-Volatile Organic Compounds, Antarctic-Environmental Protocol, Desertification, Law of the Sea
Note: controls Danish Straits linking Baltic and North Seas; about one-quarter of the population lives in Copenhagen

People

Population: 5,199,437 (July 1995 est.)
Age structure:
0-14 years: 17% (female 430,598; male 451,993)
15-64 years: 68% (female 1,731,531; male 1,780,083)
65 years and over: 15% (female 473,537; male 331,695) (July 1995 est.)
Population growth rate: 0.22% (1995 est.)
Birth rate: 12.38 births/1,000 population (1995 est.)
Death rate: 11.14 deaths/1,000 population (1995 est.)
Net migration rate: 0.96 migrant(s)/1,000 population (1995 est.)
Infant mortality rate: 6.8 deaths/1,000 live births (1995 est.)
Life expectancy at birth:
total population: 76.11 years
male: 73.23 years
female: 79.16 years (1995 est.)
Total fertility rate: 1.69 children born/woman (1995 est.)
Nationality:
noun: Dane(s)
adjective: Danish
Ethnic divisions: Scandinavian, Eskimo, Faroese, German
Religions: Evangelical Lutheran 91%, other Protestant and Roman Catholic 2%, other 7% (1988)
Languages: Danish, Faroese, Greenlandic (an Eskimo dialect), German (small minority)
Literacy: age 15 and over can read and write (1980 est.)
total population: 99%
Labor force: 2,553,900
by occupation: private services 37.1%, government services 30.4%, manufacturing and mining 20%, construction 6.3%, agriculture, forestry, and fishing 5.6%, electricity/gas/water 0.6% (1991)

Government

Names:
conventional long form: Kingdom of Denmark
conventional short form: Denmark
local long form: Kongeriget Danmark
local short form: Danmark
Digraph: DA
Type: constitutional monarchy
Capital: Copenhagen
Administrative divisions: metropolitan Denmark—14 counties (amter, singular—amt) and 1 city* (stad); Arhus, Bornholm, Frederiksborg, Fyn, Kbenhavn, Nordjylland, Ribe, Ringkbing, Roskilde, Snderjylland, Staden Kbenhavn*, Storstrm, Vejle, Vestsjaelland, Viborg
note: see separate entries for the Faroe Islands and Greenland, which are part of the Danish realm and self-governing administrative divisions
Independence: 1849 (became a constitutional monarchy)
National holiday: Birthday of the Queen, 16 April (1940)
Constitution: 5 June 1953
Legal system: civil law system; judicial review of legislative acts; accepts compulsory ICJ jurisdiction, with reservations
Suffrage: 18 years of age; universal
Executive branch:
chief of state: Queen MARGRETHE II (since NA January 1972); Heir Apparent Crown Prince FREDERIK, elder son of the Queen (born 26 May 1968)
head of government: Prime Minister Poul Nyrup RASMUSSEN (since NA January 1993)
cabinet: Cabinet; appointed by the monarch
Legislative branch: unicameral
Parliament (Folketing): elections last held 21 September 1994 (next to be held by December 1998); results—Social Democrats 34.6%, Liberals 23.3%, Conservatives 15.0%, Social People's Party 7.3%, Progress Party 6.4%, Radical Liberals 4.6%, Unity Party 3.1%, Center Democrats 2.8%, Christian People's Party 1.8%; seats—(179 total) Social Democrats 63, Liberals 44, Conservatives 28, Social People's Party 13, Progress Party 11, Radical Liberals 8, Unity Party 6, Center Democrats 5, independent 1
Judicial branch: Supreme Court
Political parties and leaders: Social Democratic Party, Poul Nyrup RASMUSSEN; Conservative Party, Hans ENGELL; Liberal Party, Uffe ELLEMANN-JENSEN; Socialist People's Party, Holger K. NIELSEN; Progress Party, Group Chairman Kim BEHNKE and Policy Spokesman Jan Kopke CHRISTENSEN; Center Democratic Party, Mimi Stilling JAKOBSEN; Radical Liberal Party, Marianne JELVED; Christian People's Party, Jann SJURSEN; Common Course, Preben Moller HANSEN; Danish Workers'

Party; Unity Party
Member of: AfDB, AG (observer), AsDB, Australia Group, BIS, CBSS, CCC, CE, CERN, EBRD, EC, ECE, EIB, ESA, FAO, G-9, GATT, IADB, IAEA, IBRD, ICAO, ICC, ICFTU, ICRM, IDA, IEA, IFAD, IFC, IFRCS, ILO, IMF, IMO, INMARSAT, INTELSAT, INTERPOL, IOC, IOM, ISO, ITU, MTCR, NACC, NATO, NC, NEA, NIB, NSG, OECD, OSCE, PCA, UN, UNCTAD, UNESCO, UNFICYP, UNHCR, UNIDO, UNIKOM, UNMOGIP, UNOMIG, UNPROFOR, UNTSO, UPU, WEU, WFTU, WHO, WIPO, WMO, ZC
Diplomatic representation in US:
chief of mission: Ambassador Peter Pedersen DYVIG (Knud-Erik TYGESEN is Ambassador Elect for 1995)
chancery: 3200 Whitehaven Street NW, Washington, DC 20008
telephone: [1] (202) 234-4300
FAX: [1] (202) 328-1470
consulate(s) general: Chicago, Los Angeles, and New York
US diplomatic representation:
chief of mission: Ambassador Edward E. ELSON
embassy: Dag Hammarskjolds Alle 24, 2100 Copenhagen 0
mailing address: APO AE 09716
telephone: [45] (31) 42 31 44
FAX: [45] (35) 43 02 23
Flag: red with a white cross that extends to the edges of the flag; the vertical part of the cross is shifted to the hoist side, and that design element of the DANNEBROG (Danish flag) was subsequently adopted by the other Nordic countries of Finland, Iceland, Norway, and Sweden

Economy

Overview: This thoroughly modern economy features high-tech agriculture, up-to-date small-scale and corporate industry, extensive government welfare measures, comfortable living standards, and high dependence on foreign trade. Denmark is self-sufficient in food production. The new center-left coalition government will concentrate on reducing the persistent high unemployment rate and the budget deficit as well as following the previous government's policies of maintaining low inflation and a current account surplus. In the face of recent international market pressure on the Danish krone, the coalition has also vowed to maintain a stable currency. The coalition hopes to lower marginal income taxes while maintaining overall tax revenues; boost industrial competitiveness through labor market and tax reforms and increased research and development funds; and improve welfare services for the neediest while cutting paperwork and delays. Prime Minister RASMUSSEN's reforms will focus on adapting Denmark to the criteria for European integration by 1999; although Copenhagen has won from the European Union (EU) the right to opt out of the European Monetary Union (EMU) if a national referendum rejects it. Denmark is, in fact, one of the few EU countries likely to fit into the EMU on time. Denmark is weathering the current worldwide slump better than many West European countries. After posting 4.5% real GDP growth in 1994, Copenhagen is predicting a continued strong showing in 1995, with real GDP up by 3.2%. The government expects an upswing in business investment in 1995 to drive economic growth. Although unemployment is high, it remains stable compared to most European countries.
National product: GDP—purchasing power parity—$103 billion (1994 est.)
National product real growth rate: 4.5% (1994 est.)
National product per capita: $19,860 (1994 est.)
Inflation rate (consumer prices): 2% (1994 est.)
Unemployment rate: 12.3% (1994 est.)
Budget:
revenues: $56.5 billion
expenditures: $64.4 billion, including capital expenditures of $NA (1994 est.)
Exports: $42.9 billion (f.o.b., 1994)
commodities: meat and meat products, dairy products, transport equipment (shipbuilding), fish, chemicals, industrial machinery
partners: EC 54.3% (Germany 23.6%, UK 10.1%, France 5.7%), Sweden 10.5%, Norway 5.8%, US 4.9%, Japan 3.6% (1992)
Imports: $37.1 billion (c.i.f., 1994 est.)
commodities: petroleum, machinery and equipment, chemicals, grain and foodstuffs, textiles, paper
partners: EC 53.4% (Germany 23.1%, UK 8.2%, France 5.6%), Sweden 10.8%, Norway 5.4%, US 5.7%, Japan 4.1% (1992)
External debt: $40.9 billion (1994 est.)
Industrial production: growth rate -2.5% (1993 est.)
Electricity:
capacity: 10,030,000 kW
production: 32 billion kWh
consumption per capita: 5,835 kWh (1993)
Industries: food processing, machinery and equipment, textiles and clothing, chemical products, electronics, construction, furniture, and other wood products, shipbuilding
Agriculture: accounts for 4% of GDP; principal products—meat, dairy, grain, potatoes, rape, sugar beets, fish
Economic aid:
donor: ODA and OOF commitments (1970-89), $5.9 billion
Currency: 1 Danish krone (DKr) = 100 oere
Exchange rates: Danish kroner (DKr) per US$1—6.034 (January 1995), 6.361 (1994), 6.484 (1993), 6.036 (1992), 6.396 (1991), 6.189 (1990)
Fiscal year: calendar year

Transportation

Railroads:
total: 2,838 km (494 km privately owned and operated)
standard gauge: 2,838 km 1.435-m gauge (440 km electrified; 760 km double track) (1994)
Highways:
total: 71,042 km
paved: concrete, asphalt, stone block 71,042 km (696 km of expressways)
Inland waterways: 417 km
Pipelines: crude oil 110 km; petroleum products 578 km; natural gas 700 km
Ports: Alborg, Arhus, Copenhagen, Esbjerg, Fredericia, Grenaa, Koge, Odense, Struer
Merchant marine:
total: 345 ships (1,000 GRT or over) totaling 5,005,470 GRT/6,974,750 DWT
ships by type: bulk 17, cargo 109, chemical tanker 24, combination bulk 1, container 61, liquefied gas tanker 32, livestock carrier 4, oil tanker 32, railcar carrier 1, refrigerated cargo 18, roll-on/roll-off cargo 35, short-sea passenger 11
note: Denmark has created its own internal register, called the Danish International Ship register (DIS); DIS ships do not have to meet Danish manning regulations, and they amount to a flag of convenience within the Danish register
Airports:
total: 118
with paved runways over 3,047 m: 2
with paved runways 2,438 to 3,047 m: 7
with paved runways 1,524 to 2,437 m: 3
with paved runways 914 to 1,523 m: 13
with paved runways under 914 m: 85
with unpaved runways 1,524 to 2,438 m: 1
with unpaved runways 914 to 1,523 m: 7

Communications

Telephone system: 4,509,000 telephones; excellent telephone and telegraph services; buried and submarine cables and microwave radio relay support trunk network
local: NA
intercity: microwave radio relay
international: 19 submarine coaxial cables; 7 INTELSAT, EUTELSAT, and INMARSAT earth stations
Radio:
broadcast stations: AM 3, FM 2, shortwave 0
radios: NA
Television:
broadcast stations: 50
televisions: NA

Defense Forces

Branches: Royal Danish Army, Royal Danish Navy, Royal Danish Air Force, Home Guard
Manpower availability: males age 15-49 1,347,774; males fit for military service 1,158,223; males reach military age (20) annually 36,191 (1995 est.)
Defense expenditures: exchange rate conversion—$2.7 billion, 1.9% of GDP (1994)

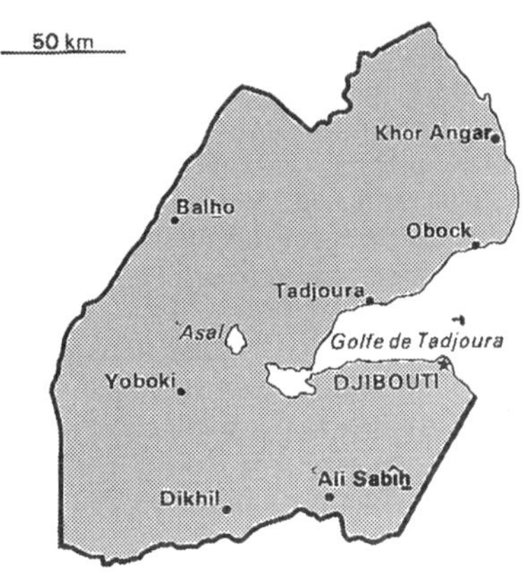

Geography

Location: Eastern Africa, bordering the Gulf of Aden and the Red Sea, between Eritrea and Somalia
Map references: Africa
Area:
total area: 22,000 sq km
land area: 21,980 sq km
comparative area: slightly larger than Massachusetts
Land boundaries: total 508 km, Eritrea 113 km, Ethiopia 337 km, Somalia 58 km
Coastline: 314 km
Maritime claims:
contiguous zone: 24 nm
exclusive economic zone: 200 nm
territorial sea: 12 nm
International disputes: none
Climate: desert; torrid, dry
Terrain: coastal plain and plateau separated by central mountains
Natural resources: geothermal areas
Land use:
arable land: 0%
permanent crops: 0%
meadows and pastures: 9%
forest and woodland: 0%
other: 91%
Irrigated land: NA sq km
Environment:
current issues: inadequate supplies of potable water; desertification
natural hazards: earthquakes; droughts; occasional cyclonic disturbances from the Indian Ocean bring heavy rains and flash floods
international agreements: party to Biodiversity, Endangered Species, Law of the Sea, Ship Pollution; signed, but not ratified—Climate Change, Desertification
Note: strategic location near world's busiest shipping lanes and close to Arabian oilfields; terminus of rail traffic into Ethiopia; a vast wasteland

People

Population: 421,320 (July 1995 est.)
Age structure:
0-14 years: 43% (female 90,070; male 90,631)
15-64 years: 55% (female 108,824; male 121,715)
65 years and over: 2% (female 4,900; male 5,180) (July 1995 est.)
Population growth rate: 1.48% (1995 est.)
Birth rate: 42.79 births/1,000 population (1995 est.)
Death rate: 15.51 deaths/1,000 population (1995 est.)
Net migration rate: -12.46 migrant(s)/1,000 population (1995 est.)
Infant mortality rate: 108.8 deaths/1,000 live births (1995 est.)
Life expectancy at birth:
total population: 49.7 years
male: 47.83 years
female: 51.62 years (1995 est.)
Total fertility rate: 6.15 children born/woman (1995 est.)
Nationality:
noun: Djiboutian(s)
adjective: Djiboutian
Ethnic divisions: Somali 60%, Afar 35%, French, Arab, Ethiopian, and Italian 5%
Religions: Muslim 94%, Christian 6%
Languages: French (official), Arabic (official), Somali, Afar
Literacy: age 15 and over can read and write (1990)
total population: 48%
male: 63%
female: 34%

Government

Names:
conventional long form: Republic of Djibouti
conventional short form: Djibouti
former: French Territory of the Afars and Issas French Somaliland
Digraph: DJ
Type: republic
Capital: Djibouti
Administrative divisions: 5 districts (cercles, singular—cercle); 'Ali Sabih, Dikhil, Djibouti, Obock, Tadjoura
Independence: 27 June 1977 (from France)
National holiday: Independence Day, 27 June (1977)
Constitution: multiparty constitution approved in referendum 4 September 1992
Legal system: based on French civil law system, traditional practices, and Islamic law
Suffrage: universal adult at age NA
Executive branch:
chief of state: President HASSAN GOULED Aptidon (since 24 June 1977); election last held 7 May 1993 (next to be held NA 1999);

results—President Hassan GOULED Aptidon was reelected
head of government: Prime Minister BARKAT Gourad Hamadou (since 30 September 1978)
cabinet: Council of Ministers; responsible to the president
Legislative branch: unicameral
Chamber of Deputies (Chambre des Deputes): elections last held 18 December 1992; results—RPP (the ruling party) dominated; seats—(65 total) RPP 65
Judicial branch: Supreme Court (Cour Supreme)
Political parties and leaders:
ruling party: People's Progress Assembly (RPP), Hassan GOULED Aptidon
other parties: Democratic Renewal Party (PRD), Mohamed Jama ELABE; Democratic National Party (PND), ADEN Robleh Awaleh
Other political or pressure groups: Front for the Restoration of Unity and Democracy (FRUD) and affiliates; Movement for Unity and Democracy (MUD)
Member of: ACCT, ACP, AfDB, AFESD, AL, ECA, FAO, G-77, IBRD, ICAO, ICRM, IDA, IDB, IFAD, IFC, IFRCS, IGADD, ILO, IMF, IMO, INTELSAT (nonsignatory user), INTERPOL, IOC, ITU, NAM, OAU, OIC, UN, UNCTAD, UNESCO, UNIDO, UPU, WFTU, WHO, WMO
Diplomatic representation in US:
chief of mission: Ambassador Roble OLHAYE
chancery: Suite 515, 1156 15th Street NW, Washington, DC 20005
telephone: [1] (202) 331-0270
FAX: [1] (202) 331-0302
US diplomatic representation:
chief of mission: Ambassador Martin L. CHESHES
embassy: Plateau du Serpent, Boulevard Marechal Joffre, Djibouti
mailing address: B. P. 185, Djibouti
telephone: [253] 35 39 95
FAX: [253] 35 39 40
Flag: two equal horizontal bands of light blue (top) and light green with a white isosceles triangle based on the hoist side bearing a red five-pointed star in the center

Economy

Overview: The economy is based on service activities connected with the country's strategic location and status as a free trade zone in northeast Africa. Two-thirds of the inhabitants live in the capital city, the remainder being mostly nomadic herders. Scanty rainfall limits crop production to fruits and vegetables, and most food must be imported. Djibouti provides services as both a transit port for the region and an international transshipment and refueling center. It has few natural resources and little industry. The nation is, therefore, heavily dependent on foreign assistance (an important supplement to GDP) to help support its balance of payments and to finance development projects. An unemployment rate of over 30% continues to be a major problem. Per capita consumption dropped an estimated 35% over the last six years because of recession, civil war, and a high population growth rate (including immigrants and refugees).
National product: GDP—purchasing power parity—$500 million (1994 est.)
National product real growth rate: -3% (1994 est.)
National product per capita: $1,200 (1994 est.)
Inflation rate (consumer prices): 6% (1993 est.)
Unemployment rate: over 30% (1994 est.)
Budget:
revenues: $164 million
expenditures: $201 million, including capital expenditures of $16 million (1993 est.)
Exports: $184 million (f.o.b., 1994 est.)
commodities: hides and skins, coffee (in transit)
partners: Somalia 48%, Yemen 42%
Imports: $384 million (f.o.b., 1994 est.)
commodities: foods, beverages, transport equipment, chemicals, petroleum products
partners: France, UK, Saudi Arabia, Bahrain, South Korea
External debt: $227 million (1993 est.)
Industrial production: growth rate 3% (1991 est.); accounts for 14% of GDP
Electricity:
capacity: 90,000 kW
production: 170 million kWh
consumption per capita: 398 kWh (1993)
Industries: limited to a few small-scale enterprises, such as dairy products and mineral-water bottling
Agriculture: mostly fruit and vegetables; herding of goats, sheep, and camels
Economic aid:
recipient: US commitments, including Ex-Im (FY78-89), $39 million; Western (non-US) countries, including ODA and OOF bilateral commitments (1970-89), $1.1 billion; OPEC bilateral aid (1979-89), $149 million; Communist countries (1970-89), $35 million
Currency: 1 Djiboutian franc (DF) = 100 centimes
Exchange rates: Djiboutian francs (DF) per US$1—177.721 (fixed rate since 1973)
Fiscal year: calendar year

Transportation

Railroads:
total: 97 km (Djibouti segment of the Addis Ababa-Djibouti railroad)
narrow gauge: 97 km 1.000-m gauge
Highways:
total: 2,900 km
paved: 280 km
unpaved: improved, unimproved earth 2,620 km (1982)
Ports: Djibouti
Merchant marine:
total: 1 cargo ship (1,000 GRT or over) totaling 1,369 GRT/3,030 DWT
Airports:
total: 13
with paved runways over 3,047 m: 1
with paved runways 2,438 to 3,047 m: 1
with paved runways under 914 m: 3
with unpaved runways 1,524 to 2,438 m: 2
with unpaved runways 914 to 1,523 m: 6

Communications

Telephone system: NA telephones; telephone facilities in the city of Djibouti are adequate as are the microwave radio relay connections to outlying areas of the country
local: NA
intercity: microwave radio relay network
international: international connections via submarine cable to Saudi Arabia and by satellite link to other countries; 1 INTELSAT (Indian Ocean) and 1 ARABSAT earth station
Radio:
broadcast stations: AM 2, FM 2, shortwave 0
radios: NA
Television:
broadcast stations: 1
televisions: NA

Defense Forces

Branches: Djibouti National Army (includes Navy and Air Force), National Security Force (Force Nationale de Securite), National Police Force
Manpower availability: males age 15-49 101,385; males fit for military service 59,337 (1995 est.)
Defense expenditures: exchange rate conversion—$26 million, NA% of GDP (1989)

Dominica

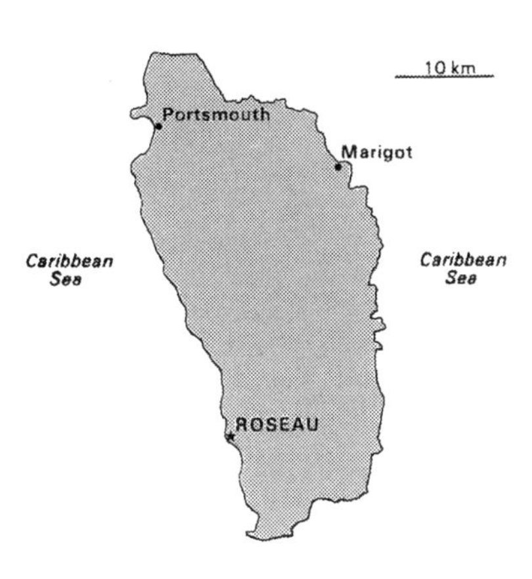

Geography

Location: Caribbean, island between the Caribbean Sea and the North Atlantic Ocean, about one-half of the way from Puerto Rico to Trinidad and Tobago
Map references: Central America and the Caribbean
Area:
total area: 750 sq km
land area: 750 sq km
comparative area: slightly more than four times the size of Washington, DC
Land boundaries: 0 km
Coastline: 148 km
Maritime claims:
contiguous zone: 24 nm
exclusive economic zone: 200 nm
territorial sea: 12 nm
International disputes: none
Climate: tropical; moderated by northeast trade winds; heavy rainfall
Terrain: rugged mountains of volcanic origin
Natural resources: timber
Land use:
arable land: 9%
permanent crops: 13%
meadows and pastures: 3%
forest and woodland: 41%
other: 34%
Irrigated land: NA sq km
Environment:
current issues: NA
natural hazards: flash floods are a constant threat; destructive hurricanes can be expected during the late summer months
international agreements: party to—Biodiversity, Climate Change, Environmental Modification, Law of the Sea, Ozone Layer Protection, Whaling

People

Population: 82,608 (July 1995 est.)
Age structure:
0-14 years: 29% (female 11,665; male 12,130)
15-64 years: 64% (female 25,606; male 26,890)
65 years and over: 7% (female 3,724; male 2,593) (July 1995 est.)
Population growth rate: 0.4% (1995 est.)
Birth rate: 18.63 births/1,000 population (1995 est.)
Death rate: 5.33 deaths/1,000 population (1995 est.)
Net migration rate: -9.36 migrant(s)/1,000 population (1995 est.)
Infant mortality rate: 9.9 deaths/1,000 live births (1995 est.)
Life expectancy at birth:
total population: 77.2 years
male: 74.35 years
female: 80.2 years (1995 est.)
Total fertility rate: 1.95 children born/woman (1995 est.)
Nationality:
noun: Dominican(s)
adjective: Dominican
Ethnic divisions: black, Carib Indians
Religions: Roman Catholic 77%, Protestant 15% (Methodist 5%, Pentecostal 3%, Seventh-Day Adventist 3%, Baptist 2%, other 2%), none 2%, unknown 1%, other 5%
Languages: English (official), French patois
Literacy: age 15 and over has ever attended school (1970)
total population: 94%
male: 94%
female: 94%
Labor force: 25,000
by occupation: agriculture 40%, industry and commerce 32%, services 28% (1984)

Government

Names:
conventional long form: Commonwealth of Dominica
conventional short form: Dominica
Digraph: DO
Type: parliamentary democracy
Capital: Roseau
Administrative divisions: 10 parishes; Saint Andrew, Saint David, Saint George, Saint John, Saint Joseph, Saint Luke, Saint Mark, Saint Patrick, Saint Paul, Saint Peter
Independence: 3 November 1978 (from UK)
National holiday: Independence Day, 3 November (1978)
Constitution: 3 November 1978
Legal system: based on English common law
Suffrage: 18 years of age; universal
Executive branch:
chief of state: President Crispin Anselm SORHAINDO (since 25 October 1993) election last held 4 October 1993 (next to be held NA October 1998); results—President Crispin Anselm SORHAINDO was elected by the House of Assembly to a five-year term
head of government: Prime Minister (Mary) Eugenia CHARLES (since 21 July 1980, elected for a third term 28 May 1990)
cabinet: Cabinet; appointed by the president on the advice of the prime minister
Legislative branch: unicameral
House of Assembly: elections last held 28 May 1990 (next to be held by October 1995); results—percent of vote by party NA; seats—(30 total; 9 appointed senators and 21 elected representatives) DFP 11, UWP 6, DLP 4
Judicial branch: Eastern Caribbean Supreme Court
Political parties and leaders: Dominica Freedom Party (DFP), Brian ALLEYNE; Dominica Labor Party (DLP), Rosie DOUGLAS; United Workers Party (UWP), Edison JAMES
Other political or pressure groups: Dominica Liberation Movement (DLM), a small leftist group
Member of: ACCT, ACP, C, CARICOM, CDB, ECLAC, FAO, G-77, GATT, IBRD, ICFTU, ICRM, IDA, IFAD, IFC, IFRCS, ILO, IMF, IMO, INTERPOL, IOC, NAM (observer), OAS, OECS, OPANAL, UN, UNCTAD, UNESCO, UNIDO, UPU, WCL, WHO, WMO
Diplomatic representation in US: Dominica has no embassy in the US
consulate(s) general: New York
US diplomatic representation: no official presence since the Ambassador resides in Bridgetown (Barbados), but travels frequently to Dominica
Flag: green with a centered cross of three equal bands—the vertical part is yellow (hoist side), black, and white—the horizontal part is yellow (top), black, and white; superimposed in the center of the cross is a red disk bearing a sisserou parrot encircled by 10 green five-pointed stars edged in yellow; the 10 stars represent the 10 administrative divisions (parishes)

Economy

Overview: The economy is dependent on agriculture and thus is highly vulnerable to climatic conditions. Agriculture accounts for about 30% of GDP and employs 40% of the labor force. Principal products include bananas, citrus, mangoes, root crops, and coconuts. Development of the tourist industry remains difficult because of the rugged coastline and the lack of an international airport. In 1994 a tropical storm devastated the banana industry.
National product: GDP—purchasing power parity—$200 million (1994 est.)
National product real growth rate: 1.6% (1994 est.)
National product per capita: $2,260 (1994 est.)
Inflation rate (consumer prices): 1.6% (1993 est.)

Unemployment rate: 15% (1992 est.)
Budget:
revenues: $70 million
expenditures: $84 million, including capital expenditures of $26 million (FY90/91 est.)
Exports: $48.3 million (f.o.b., 1993)
commodities: bananas, soap, bay oil, vegetables, grapefruit, oranges
partners: UK 55%, CARICOM countries, Italy, US
Imports: $98.8 million (f.o.b., 1993)
commodities: manufactured goods, machinery and equipment, food, chemicals
partners: US 25%, CARICOM, UK, Japan, Canada
External debt: $92.8 million (1992)
Industrial production: growth rate -10% (1994 est.); accounts for 7% of GDP
Electricity:
capacity: 7,000 kW
production: 30 million kWh
consumption per capita: 347 kWh (1993)
Industries: soap, coconut oil, tourism, copra, furniture, cement blocks, shoes
Agriculture: accounts for 30% of GDP; principal crops—bananas, citrus, mangoes, root crops, coconuts; bananas provide the bulk of export earnings; forestry and fisheries potential not exploited
Illicit drugs: transshipment point for narcotics bound for the US and Europe; minor cannabis producer
Economic aid:
recipient: Western (non-US) countries, ODA and OOF bilateral commitments (1970-89), $120 million
Currency: 1 EC dollar (EC$) = 100 cents
Exchange rates: East Caribbean dollars (EC$) per US$1—2.70 (fixed rate since 1976)
Fiscal year: 1 July—30 June

Transportation

Railroads: 0 km
Highways:
total: 750 km
paved: 370 km
unpaved: gravel or earth 380 km
Ports: Portsmouth, Roseau
Merchant marine: none
Airports:
total: 2
with paved runways 914 to 1,523 m: 1
with paved runways under 914 m: 1

Communications

Telephone system: 4,600 telephones; fully automatic network
local: NA
intercity: NA
international: VHF and UHF radio telephone communications link to Saint Lucia; new SHF links to Martinique and Guadeloupe
Radio:
broadcast stations: AM 3, FM 2, shortwave 0
radios: NA
Television:
broadcast stations: 1 cable
televisions: NA

Defense Forces

Branches: Commonwealth of Dominica Police Force (includes Special Service Unit, Coast Guard)
Defense expenditures: $NA, NA% of GDP

Dominican Republic

Geography

Location: Caribbean, eastern two-thirds of the island of Hispaniola, between the Caribbean Sea and the North Atlantic Ocean, east of Haiti
Map references: Central America and the Caribbean
Area:
total area: 48,730 sq km
land area: 48,380 sq km
comparative area: slightly more than twice the size of New Hampshire
Land boundaries: total 275 km, Haiti 275 km
Coastline: 1,288 km
Maritime claims:
contiguous zone: 24 nm
continental shelf: 200 nm or to the edge of the continental margin
exclusive economic zone: 200 nm
territorial sea: 6 nm
International disputes: none
Climate: tropical maritime; little seasonal temperature variation; seasonal variation in rainfall
Terrain: rugged highlands and mountains with fertile valleys interspersed
Natural resources: nickel, bauxite, gold, silver
Land use:
arable land: 23%
permanent crops: 7%
meadows and pastures: 43%
forest and woodland: 13%
other: 14%
Irrigated land: 2,250 sq km (1989)
Environment:
current issues: water shortages; soil eroding into the sea damages coral reefs; deforestation
natural hazards: occasional hurricanes (July to October)
international agreements: party to—Endangered Species, Marine Dumping, Marine Life Conservation, Nuclear Test Ban, Ozone Layer Protection; signed, but not ratified—Biodiversity, Climate Change, Law of the Sea
Note: shares island of Hispaniola with Haiti

(eastern two-thirds is the Dominican Republic, western one-third is Haiti)

People

Population: 7,511,263 (July 1995 est.)
Age structure:
0-14 years: 35% (female 1,288,210; male 1,336,162)
15-64 years: 61% (female 2,246,791; male 2,312,555)
65 years and over: 4% (female 178,388; male 149,157) (July 1995 est.)
Population growth rate: 1.17% (1995 est.)
Birth rate: 23.92 births/1,000 population (1995 est.)
Death rate: 6.15 deaths/1,000 population (1995 est.)
Net migration rate: -6.04 migrant(s)/1,000 population (1995 est.)
Infant mortality rate: 49.5 deaths/1,000 live births (1995 est.)
Life expectancy at birth:
total population: 68.73 years
male: 66.57 years
female: 70.99 years (1995 est.)
Total fertility rate: 2.72 children born/woman (1995 est.)
Nationality:
noun: Dominican(s)
adjective: Dominican
Ethnic divisions: white 16%, black 11%, mixed 73%
Religions: Roman Catholic 95%
Languages: Spanish
Literacy: age 15 and over can read and write (1990 est.)
total population: 83%
male: 85%
female: 82%
Labor force: 2.3 million to 2.6 million
by occupation: agriculture 49%, services 33%, industry 18% (1986)

Government

Names:
conventional long form: Dominican Republic
conventional short form: none
local long form: Republica Dominicana
local short form: none
Digraph: DR
Type: republic
Capital: Santo Domingo
Administrative divisions: 29 provinces (provincias, singular—provincia) and 1 district* (distrito); Azua, Baoruco, Barahona, Dajabon, Distrito Nacional*, Duarte, Elias Pina, El Seibo, Espaillat, Hato Mayor, Independencia, La Altagracia, La Romana, La Vega, Maria Trinidad Sanchez, Monsenor Nouel, Monte Cristi, Monte Plata, Pedernales, Peravia, Puerto Plata, Salcedo, Samana, Sanchez Ramirez, San Cristobal, San Juan, San Pedro de Macoris, Santiago, Santiago Rodriguez, Valverde
Independence: 27 February 1844 (from Haiti)
National holiday: Independence Day, 27 February (1844)
Constitution: 28 November 1966
Legal system: based on French civil codes
Suffrage: 18 years of age; universal and compulsory or married persons regardless of age
note: members of the armed forces and police cannot vote
Executive branch:
chief of state and head of government: President Joaquin BALAGUER Ricardo (since 16 August 1986, sixth elected term began 16 August 1994); Vice President Jacinto PEYNADO (since 16 August 1994) election last held 16 May 1994 (next to be held May 1996); results—Joaquin BALAGUER (PRSC) 42.6%, Juan BOSCH Gavino (PLD) 13.2%, Jose Francisco PENA Gomez (PRD) 41.9%, Jacobo MAJLUTA (PRI) 2.3%
cabinet: Cabinet; nominated by the president
Legislative branch: bicameral National Congress (Congreso Nacional)
Senate (Senado): elections last held 16 May 1994 (next to be held May 1998); results—percent of vote by party NA; seats—(30 total) PRSC 15, PLD 1, PRD 14
Chamber of Deputies (Camara de Diputados): elections last held 16 May 1994 (next to be held May 1998); results—percent of vote by party NA; seats—(120 total) PLD 13, PRSC 50, PRD 57
Judicial branch: Supreme Court (Corte Suprema)
Political parties and leaders:
Major parties: Social Christian Reformist Party (PRSC), Joaquin BALAGUER Ricardo; Dominican Liberation Party (PLD), (vacant following retirement of Juan BOSCH Gavino); Dominican Revolutionary Party (PRD), Jose Francisco PENA Gomez; Independent Revolutionary Party (PRI), Jacobo MAJLUTA
Minor parties: National Veterans and Civilian Party (PNVC), Juan Rene BEAUCHAMPS Javier; Liberal Party of the Dominican Republic (PLRD), Andres Van Der HORST; Democratic Quisqueyan Party (PQD), Elias WESSIN Chavez; National Progressive Force (FNP), Marino VINICIO Castillo; Popular Christian Party (PPC), Rogelio DELGADO Bogaert; Dominican Communist Party (PCD), Narciso ISA Conde; Dominican Workers' Party (PTD), Ivan RODRIGUEZ; Anti-Imperialist Patriotic Union (UPA), Ignacio RODRIGUEZ Chiappini; Alliance for Democracy Party (APD), Maximilano Rabelais PUIG Miller, Nelsida MARMOLEJOS, Vicente BENGOA; Democratic Union (UD), Fernando ALVAREZ Bogaert
note: in 1983 several leftist parties, including the PCD, joined to form the Dominican Leftist Front (FID); however, they still retain individual party structures
Other political or pressure groups: Collective of Popular Organzations (COP), leader NA
Member of: ACP, CARICOM (observer), ECLAC, FAO, G-11, G-77, GATT, IADB, IAEA, IBRD, ICAO, ICFTU, ICRM, IDA, IFAD, IFC, IFRCS, ILO, IMF, IMO, INTELSAT, INTERPOL, IOC, IOM, ITU, LAES, LAIA (observer), NAM (guest), OAS, OPANAL, PCA, UN, UNCTAD, UNESCO, UNIDO, UPU, WCL, WFTU, WHO, WMO, WTO
Diplomatic representation in US:
chief of mission: Ambassador Jose del Carmen ARIZA Gomez
chancery: 1715 22nd Street NW, Washington, DC 20008
telephone: [1] (202) 332-6280
FAX: [1] (202) 265-8057
consulate(s) general: Boston, Chicago, Los Angeles, Mayaguez (Puerto Rico), Miami, New Orleans, New York, Philadelphia, San Francisco, and San Juan (Puerto Rico)
consulate(s): Charlotte Amalie (Virgin Islands), Detroit, Houston, Jacksonville, Minneapolis, Mobile, and Ponce (Puerto Rico)
US diplomatic representation:
chief of mission: Ambassador Donna Jean HRINAK
embassy: corner of Calle Cesar Nicolas Penson and Calle Leopoldo Navarro, Santo Domingo
mailing address: Unit 5500, Santo Domingo; APO AA 34041
telephone: [1] (809) 541-2171, 8100
FAX: [1] (809) 686-7437
Flag: a centered white cross that extends to the edges, divides the flag into four rectangles—the top ones are blue (hoist side) and red, the bottom ones are red (hoist side) and blue; a small coat of arms is at the center of the cross

Economy

Overview: The Dominican economy showed some signs of slippage in 1994, although its overall performance in recent years has been relatively strong. After posting an increase of nearly 8% in 1992, GDP growth fell to 3% in 1993 and 1994 as mining output decreased and erosion of real wages caused private consumption to decline. A pre-election boost in government spending in early 1994 led to the first government deficit in four years and bumped inflation up to 14% for the year. Continued dynamism in construction and the services sector, especially tourism, should keep the economy growing in 1995. Tourism, agriculture, and manufacturing for export remain key sectors of the economy. Domestic industry is based on the processing of agricultural products, oil refining, and chemicals.
National product: GDP—purchasing power

parity—$24 billion (1994 est.)
National product real growth rate: 2.9% (1994 est.)
National product per capita: $3,070 (1994 est.)
Inflation rate (consumer prices): 14% (1994)
Unemployment rate: 30% (1994 est.)
Budget:
revenues: $1.8 billion
expenditures: $2.2 billion, including capital expenditures of $NA (1994 est.)
Exports: $585 million (f.o.b., 1994)
commodities: ferronickel, sugar, gold, coffee, cocoa
partners: US 52%, EC 23%, Puerto Rico 9%, Asia 7% (1992)
Imports: $2.5 billion (c.i.f., 1994 est.)
commodities: foodstuffs, petroleum, cotton and fabrics, chemicals and pharmaceuticals
partners: US 60% (1993)
External debt: $4.3 billion (1994 est.)
Industrial production: growth rate 3.4% (1994); accounts for 14% of GDP
Electricity:
capacity: 1,450,000 kW
production: 5.4 billion kWh
consumption per capita: 651 kWh (1993)
Industries: tourism, sugar processing, ferronickel and gold mining, textiles, cement, tobacco
Agriculture: accounts for 15% of GDP and employs 49% of labor force; commercial crops—sugarcane, coffee, cotton, cocoa, and tobacco; food crops—rice, beans, potatoes, corn, bananas; animal output—cattle, hogs, dairy products, meat, eggs; not self-sufficient in food
Illicit drugs: transshipment point for South American drugs destined for the US and Europe
Economic aid:
recipient: US commitments, including Ex-Im (FY85-89), $575 million; Western (non-US) countries, ODA and OOF bilateral commitments (1970-89), $655 million
Currency: 1 Dominican peso (RD$) = 100 centavos
Exchange rates: Dominican pesos (RD$) per US$1—13.258 (January 1995), 13.160 (1994), 12.679 (1993), 12.774 (1992), 12.692 (1991), 8.525 (1990)
Fiscal year: calendar year

Transportation

Railroads:
total: 1,655 km (in numerous segments; includes 4 different gauges from 0.558-m narrow gauge to 1.435-m standard gauge)
Highways:
total: 12,000 km
paved: 5,800 km
unpaved: gravel or improved earth 5,600 km; unimproved earth 600 km
Pipelines: crude oil 96 km; petroleum products 8 km
Ports: Barahona, La Romana, Puerto Plata, San Pedro de Macoris, Santo Domingo
Merchant marine:
total: 1 cargo ship (1,000 GRT or over) totaling 1,587 GRT/1,165 DWT
Airports:
total: 36
with paved runways over 3,047 m: 2
with paved runways 2,438 to 3,047 m: 1
with paved runways 1,524 to 2,437 m: 4
with paved runways 914 to 1,523 m: 5
with paved runways under 914 m: 16
with unpaved runways 2,438 to 3,047 m: 1
with unpaved runways 1,524 to 2,438 m: 1
with unpaved runways 914 to 1,523 m: 6

Communications

Telephone system: 190,000 telephones; relatively efficient domestic system based on islandwide microwave radio relay network
local: NA
intercity: islandwide microwave radio relay network
international: 1 coaxial submarine cable; 1 INTELSAT (Atlantic Ocean) earth station
Radio:
broadcast stations: AM 120, FM 0, shortwave 6
radios: NA
Television:
broadcast stations: 18
televisions: NA

Defense Forces

Branches: Army, Navy, Air Force, National Police
Manpower availability: males age 15-49 2,008,597; males fit for military service 1,266,812; males reach military age (18) annually 79,769 (1995 est.)
Defense expenditures: exchange rate conversion—$116 million, 1.4% of GDP (1994)

Ecuador

Geography

Location: Western South America, bordering the Pacific Ocean at the Equator, between Colombia and Peru
Map references: South America
Area:
total area: 283,560 sq km
land area: 276,840 sq km
comparative area: slightly smaller than Nevada
note: includes Galapagos Islands
Land boundaries: total 2,010 km, Colombia 590 km, Peru 1,420 km
Coastline: 2,237 km
Maritime claims:
continental shelf: claims continental shelf between mainland and Galapagos Islands
territorial sea: 200 nm
International disputes: three sections of the boundary with Peru are in dispute
Climate: tropical along coast becoming cooler inland
Terrain: coastal plain (costa), inter-Andean central highlands (sierra), and flat to rolling eastern jungle (oriente)
Natural resources: petroleum, fish, timber
Land use:
arable land: 6%
permanent crops: 3%
meadows and pastures: 17%
forest and woodland: 51%
other: 23%
Irrigated land: 5,500 sq km (1989 est.)
Environment:
current issues: deforestation; soil erosion; desertification; water pollution
natural hazards: frequent earthquakes, landslides, volcanic activity; periodic droughts
international agreements: party to—Antarctic-Environmental Protocol, Antarctic Treaty, Biodiversity, Climate Change, Endangered Species, Hazardous Wastes, Nuclear Test Ban, Ozone Layer Protection, Ship Pollution, Tropical Timber 83, Wetlands; signed, but not ratified—Tropical Timber 94
Note: Cotopaxi in Andes is highest active volcano in world

People

Population: 10,890,950 (July 1995 est.)
Age structure:
0-14 years: 36% (female 1,928,977; male 1,990,036)
15-64 years: 60% (female 3,281,575; male 3,230,082)
65 years and over: 4% (female 244,862; male 215,418) (July 1995 est.)
Population growth rate: 1.95% (1995 est.)
Birth rate: 25.08 births/1,000 population (1995 est.)
Death rate: 5.55 deaths/1,000 population (1995 est.)
Net migration rate: 0 migrant(s)/1,000 population (1995 est.)
Infant mortality rate: 37.7 deaths/1,000 live births (1995 est.)
Life expectancy at birth:
total population: 70.35 years
male: 67.83 years
female: 72.99 years (1995 est.)
Total fertility rate: 2.97 children born/woman (1995 est.)
Nationality:
noun: Ecuadorian(s)
adjective: Ecuadorian
Ethnic divisions: mestizo (mixed Indian and Spanish) 55%, Indian 25%, Spanish 10%, black 10%
Religions: Roman Catholic 95%
Languages: Spanish (official), Indian languages (especially Quechua)
Literacy: age 15 and over can read and write (1990)
total population: 87%
male: 90%
female: 84%
Labor force: 2.8 million
by occupation: agriculture 35%, manufacturing 21%, commerce 16%, services and other activities 28% (1982)

Government

Names:
conventional long form: Republic of Ecuador
conventional short form: Ecuador
local long form: Republica del Ecuador
local short form: Ecuador
Digraph: EC
Type: republic
Capital: Quito
Administrative divisions: 21 provinces (provincias, singular—provincia); Azuay, Bolivar, Canar, Carchi, Chimborazo, Cotopaxi, El Oro, Esmeraldas, Galapagos, Guayas, Imbabura, Loja, Los Rios, Manabi, Morona-Santiago, Napo, Pastaza, Pichincha, Sucumbios, Tungurahua, Zamora-Chinchipe
Independence: 24 May 1822 (from Spain)
National holiday: Independence Day, 10 August (1809) (independence of Quito)
Constitution: 10 August 1979
Legal system: based on civil law system; has not accepted compulsory ICJ jurisdiction
Suffrage: 18 years of age; universal, compulsory for literate persons ages 18-65, optional for other eligible voters
Executive branch:
chief of state and head of government: President Sixto DURAN-BALLEN Cordovez (since 10 August 1992); Vice President Alberto DAHIK Garzoni (since 10 August 1992); election runoff election held 5 July 1992 (next to be held NA 1996); results—Sixto DURAN-BALLEN elected as president and Alberto DAHIK elected as vice president
cabinet: Cabinet; appointed by the president
Legislative branch: unicameral
National Congress (Congreso Nacional): elections last held 1 May 1994 (next to be held 1 May 1996); results—percent of vote by party NA; seats—(77 total) PSC 25, PRE 11, MPD 8, ID 7, DP 7, PCE 7, PUR 2, CFP 2, APRE 2, PSE 1, FRA 1, PLRE 1, LN 1, independents 2
Judicial branch: Supreme Court (Corte Suprema)
Political parties and leaders:
Center Right parties: Social Christian Party (PSC), Jaime NEBOT Saadi, president; Republican Unity Party (PUR), President Sixto DURAN-BALLEN, leader; Ecuadorian Conservative Party (PCE), Vice President Alberto DAHIK, president
Center-Left parties: Democratic Left (ID), Andres VALLEJO Arcos, Rodrigo BORJA Cevallos, leaders; Popular Democracy (DP), Rodrigo PAZ, leader; Ecuadorian Radical Liberal Party (PLRE), Medardo MORA, leader; Radical Alfarista Front (FRA), Jaime ASPIAZU Seminario, director
Populist parties: Roldista Party (PRE), Abdala BUCARAM Ortiz, director; Concentration of Popular Forces (CFP), Rodolfo BAQUERIZO Nazur, leader; Popular Revolutionary Action (APRE), Frank VARGAS Passos, leader
Far-Left parties: Popular Democratic Movement (MPD), Juan Jose CASTELLO, leader; Ecuadorian Socialist Party (PSE), Leon ROLDOS, leader; Broad Leftist Front (FADI), Rene Mauge MOSQUERA, chairman; Ecuadorian National Liberation (LN), Alfredo CASTILLO, director
Communists: Communist Party of Ecuador (PCE, pro-North Korea), Rene Mauge MOSQUERA, Secretary General; Communist Party of Ecuador/Marxist-Leninist (PCMLE, Maoist)
Member of: AG, ECLAC, FAO, G-11, G-77, IADB, IAEA, IBRD, ICAO, ICC, ICFTU, ICRM, IDA, IFAD, IFC, IFRCS, ILO, IMF, IMO, INTELSAT, INTERPOL, IOC, IOM, ITU, LAES, LAIA, NAM, OAS, OPANAL, PCA, RG, UN, UNCTAD, UNESCO, UNIDO, UPU, WCL, WFTU, WHO, WIPO, WMO, WTO
Diplomatic representation in US:
chief of mission: Ambassador Edgar TERAN Teran
chancery: 2535 15th Street NW, Washington, DC 20009
telephone: [1] (202) 234-7200
consulate(s) general: Chicago, Houston, Los Angeles, Miami, New Orleans, New York, and San Francisco
consulate(s): Newark
US diplomatic representation:
chief of mission: Ambassador Peter F. ROMERO
embassy: Avenida 12 de Octubre y Avenida Patria, Quito
mailing address: APO AA 34039-3420
telephone: [593] (2) 562-890, 561-624, 561-749
FAX: [593] (2) 502-052
consulate(s) general: Guayaquil
Flag: three horizontal bands of yellow (top, double width), blue, and red with the coat of arms superimposed at the center of the flag; similar to the flag of Colombia that is shorter and does not bear a coat of arms

Economy

Overview: Ecuador has substantial oil resources and rich agricultural areas. Growth has been uneven in recent years because of fluctuations in prices for Ecuador's primary exports—oil and bananas—as well as because of government policies designed to curb inflation. President Sixto DURAN-BALLEN launched a series of macroeconomic reforms when he came into office in August 1992 which included raising domestic fuel prices and utility rates, eliminating most subsidies, and bringing the government budget into balance. These measures helped to reduce inflation from 55% in 1992 to 25% in 1994. DURAN-BALLEN has a much more favorable attitude toward foreign investment than his predecessor and has supported several laws designed to encourage foreign investment. Ecuador has implemented free or complementary trade agreements with Bolivia, Chile, Colombia, Peru, and Venezuela, as well as applied for World Trade Organization membership. Ecuador signed a standby agreement with the IMF and rescheduled its $7.6 billion commercial debt in 1994 thereby regaining access to multilateral lending. Growth in 1994 speeded up to 3.9%, based on increased exports of bananas and non-traditional products, while international reserves increased to a record $1.6 billion.
National product: GDP—purchasing power parity—$41.1 billion (1994 est.)
National product real growth rate: 3.9% (1994 est.)
National product per capita: $3,840 (1994 est.)
Inflation rate (consumer prices): 25% (1994)
Unemployment rate: 7.1% (1994)
Budget:
revenues: $2.76 billion (1994)

expenditures: $2.76 billion, including capital expenditures of $NA (1994)
Exports: $3.3 billion (f.o.b., 1994 est.)
commodities: petroleum 39%, bananas 17%, shrimp 16%, cocoa 3%, coffee 6%
partners: US 42%, Latin America 29%, Caribbean, EU countries 17%
Imports: $3 billion (f.o.b., 1994 est.)
commodities: transport equipment, consumer goods, vehicles, machinery, chemicals
partners: US 28%, EU 17%, Latin America 31%, Caribbean, Japan
External debt: $13.2 billion (yearend 1993 est.)
Industrial production: growth rate 6.4% (1993); accounts for almost 35% of GDP, including petroleum
Electricity:
capacity: 2,230,000 kW
production: 6.9 billion kWh
consumption per capita: 612 kWh (1993)
Industries: petroleum, food processing, textiles, metal work, paper products, wood products, chemicals, plastics, fishing, lumber
Agriculture: accounts for 14% of GDP (including fishing and forestry); leading producer and exporter of bananas and balsawood; other agricultural exports—coffee, cocoa, fish, shrimp; other crops—rice, potatoes, manioc, plantains, sugarcane; livestock products—cattle, sheep, hogs, beef, pork, dairy products; net importer of foodgrains, dairy products, and sugar
Illicit drugs: significant transit country for derivatives of coca originating in Colombia, Bolivia, and Peru; minor illicit producer of coca; importer of precursor chemicals used in production of illicit narcotics; important money-laundering hub
Economic aid:
recipient: US commitments, including Ex-Im (FY70-89), $498 million; Western (non-US) countries, ODA and OOF bilateral commitments (1970-91), $2.39 billion; Communist countries (1970-89), $64 million
Currency: 1 sucre (S/) = 100 centavos
Exchange rates: sucres (S/) per US$1—1,198.1 (December 1994), 2,196.7 (1994), 1,919.1 (1993), 1,534.0 (1992), 1,046.25 (1991), 767.8 (1990), 767.78 (1990), 526.35 (1989)
Fiscal year: calendar year

Transportation

Railroads:
total: 965 km (single track)
narrow gauge: 965 km 1.067-m gauge
Highways:
total: 43,709 km
paved: 5,245 km
unpaved: 38,464 km
Inland waterways: 1,500 km
Pipelines: crude oil 800 km; petroleum products 1,358 km
Ports: Esmeraldas, Guayaquil, La Libertad, Manta, Puerto Bolivar, San Lorenzo
Merchant marine:
total: 33 ships (1,000 GRT or over) totaling 222,822 GRT/326,447 DWT
ships by type: bulk 1, cargo 2, container 2, liquefied gas tanker 2, oil tanker 13, passenger 3, refrigerated cargo 10
Airports:
total: 175
with paved runways over 3,047 m: 2
with paved runways 2,438 to 3,047 m: 7
with paved runways 1,524 to 2,437 m: 8
with paved runways 914 to 1,523 m: 15
with paved runways under 914 m: 107
with unpaved runways 1,524 to 2,438 m: 5
with unpaved runways 914 to 1,523 m: 31

Communications

Telephone system: 318,000 telephones; 30 telephones/1,000 persons; domestic facilities generally inadequate and unreliable
local: NA
intercity: NA
international: 1 INTELSAT (Atlantic Ocean) earth station
Radio:
broadcast stations: AM 272, FM 0, shortwave 39
radios: NA
Television:
broadcast stations: 33
televisions: NA

Defense Forces

Branches: Army (Ejercito Ecuatoriano), Navy (Armada Ecuatoriana, includes Marines), Air Force (Fuerza Aerea Ecuatoriana), National Police
Manpower availability: males age 15-49 2,814,867; males fit for military service 1,903,979; males reach military age (20) annually 113,985 (1995 est.)
Defense expenditures: $NA, NA% of GDP

Egypt

Geography

Location: Northern Africa, bordering the Mediterranean Sea, between Libya and the Gaza Strip
Map references: Africa
Area:
total area: 1,001,450 sq km
land area: 995,450 sq km
comparative area: slightly more than three times the size of New Mexico
Land boundaries: total 2,689 km, Gaza Strip 11 km, Israel 255 km, Libya 1,150 km, Sudan 1,273 km
Coastline: 2,450 km
Maritime claims:
contiguous zone: 24 nm
continental shelf: 200-m depth or to the depth of exploitation
exclusive economic zone: 200 nm
territorial sea: 12 nm
International disputes: administrative boundary with Sudan does not coincide with international boundary creating the "Hala'ib Triangle," a barren area of 20,580 sq km, tensions over this disputed area began to escalate in 1992 and remain high
Climate: desert; hot, dry summers with moderate winters
Terrain: vast desert plateau interrupted by Nile valley and delta
Natural resources: petroleum, natural gas, iron ore, phosphates, manganese, limestone, gypsum, talc, asbestos, lead, zinc
Land use:
arable land: 3%
permanent crops: 2%
meadows and pastures: 0%
forest and woodland: 0%
other: 95%
Irrigated land: 25,850 sq km (1989 est.)
Environment:
current issues: agricultural land being lost to urbanization and windblown sands; increasing soil salinization below Aswan High Dam; desertification; oil pollution threatening coral reefs, beaches, and marine habitats; other water

pollution from agricultural pesticides, raw sewage, and industrial effluents; very limited natural fresh water resources away from the Nile which is the only perennial water source; rapid growth in population overstraining natural resources
natural hazards: periodic droughts; frequent earthquakes, flash floods, landslides, volcanic activity; hot, driving windstorm called khamsin occurs in spring; duststorms, sandstorms
international agreements: party to—Biodiversity, Climate Change, Endangered Species, Environmental Modification, Hazardous Wastes, Law of the Sea, Marine Dumping, Nuclear Test Ban, Ozone Layer Protection, Ship Pollution, Tropical Timber 83, Wetlands; signed, but not ratified—Desertification, Tropical Timber 94
Note: controls Sinai Peninsula, only land bridge between Africa and remainder of Eastern Hemisphere; controls Suez Canal, shortest sea link between Indian Ocean and Mediterranean Sea; size, and juxtaposition to Israel, establish its major role in Middle Eastern geopolitics

People

Population: 62,359,623 (July 1995 est.)
Age structure:
0-14 years: 37% (female 11,380,668; male 11,872,728)
15-64 years: 59% (female 18,250,706; male 18,641,830)
65 years and over: 4% (female 1,204,477; male 1,009,214) (July 1995 est.)
Population growth rate: 1.95% (1995 est.)
Birth rate: 28.69 births/1,000 population (1995 est.)
Death rate: 8.86 deaths/1,000 population (1995 est.)
Net migration rate: -0.35 migrant(s)/1,000 population (1995 est.)
Infant mortality rate: 74.5 deaths/1,000 live births (1995 est.)
Life expectancy at birth:
total population: 61.12 years
male: 59.22 years
female: 63.12 years (1995 est.)
Total fertility rate: 3.67 children born/woman (1995 est.)
Nationality:
noun: Egyptian(s)
adjective: Egyptian
Ethnic divisions: Eastern Hamitic stock (Egyptians, Bedouins, and Berbers) 99%, Greek, Nubian, Armenian, other European (primarily Italian and French) 1%
Religions: Muslim (mostly Sunni) 94% (official estimate), Coptic Christian and other 6% (official estimate)
Languages: Arabic (official), English and French widely understood by educated classes
Literacy: age 15 and over can read and write (1990 est.)
total population: 48%
male: 63%
female: 34%
Labor force: 16 million (1994 est.)
by occupation: government, public sector enterprises, and armed forces 36%, agriculture 34%, privately owned service and manufacturing enterprises 20% (1984)
note: shortage of skilled labor; 2,500,000 Egyptians work abroad, mostly in Saudi Arabia and the Gulf Arab states (1993 est.)

Government

Names:
conventional long form: Arab Republic of Egypt
conventional short form: Egypt
local long form: Jumhuriyat Misr al-Arabiyah
local short form: none
former: United Arab Republic (with Syria)
Digraph: EG
Type: republic
Capital: Cairo
Administrative divisions: 26 governorates (muhafazat, singular—muhafazah); Ad Daqahliyah, Al Bahr al Ahmar, Al Buhayrah, Al Fayyum, Al Gharbiyah, Al Iskandariyah, Al Isma'iliyah, Al Jizah, Al Minufiyah, Al Minya, Al Qahirah, Al Qalyubiyah, Al Wadi al Jadid, Ash Sharqiyah, As Suways, Aswan, Asyu't, Bani Suwayf, Bur Sa'id, Dumyat, Janub Sina, Kafr ash Shaykh, Matruh, Qina, Shamal Sina, Suhaj
Independence: 28 February 1922 (from UK)
National holiday: Anniversary of the Revolution, 23 July (1952)
Constitution: 11 September 1971
Legal system: based on English common law, Islamic law, and Napoleonic codes; judicial review by Supreme Court and Council of State (oversees validity of administrative decisions); accepts compulsory ICJ jurisdiction, with reservations
Suffrage: 18 years of age; universal and compulsory
Executive branch:
chief of state: President Mohammed Hosni MUBARAK (sworn in as president on 14 October 1981, eight days after the assassination of President SADAT); national referendum held 4 October 1993 validated Mubarak's nomination by the People's Assembly to a third 6-year presidential term
head of government: Prime Minister Atef Mohammed Najib SEDKY (since 12 November 1986)
cabinet: Cabinet; appointed by the president
Legislative branch: bicameral
People's Assembly (Majlis al-Cha'b): elections last held 29 November 1990 (next to be held NA November 1995); results—NDP 86.3%, NPUG 1.3%, independents 12.4%; seats—(454 total, 444 elected, 10 appointed by the president) NDP 383, NPUG 6, independents 55; note—most opposition parties boycotted; NDP figures include NDP members who ran as independents and other NDP-affiliated independents
Advisory Council (Majlis al-Shura): functions only in a consultative role; elections last held 8 June 1989 (next to be held NA June 1995); results—NDP 100%; seats—(258 total, 172 elected, 86 appointed by the president) NDP 172
Judicial branch: Supreme Constitutional Court
Political parties and leaders: National Democratic Party (NDP), President Mohammed Hosni MUBARAK, leader, is the dominant party; legal opposition parties are; New Wafd Party (NWP), Fu'ad SIRAJ AL-DIN; Socialist Labor Party, Ibrahim SHUKRI; National Progressive Unionist Grouping (NPUG), Khalid MUHYI-AL-DIN; Socialist Liberal Party (SLP), Mustafa Kamal MURAD; Democratic Unionist Party, Mohammed 'Abd-al-Mun'im TURK; Umma Party, Ahmad al-SABAHI; Misr al-Fatah Party (Young Egypt Party), Gamal RABIE; Nasserist Arab Democratic Party, Dia' al din DAWUD; Democratic Peoples' Party, Anwar AFIFI; The Greens Party, Kamal KIRAH; Social Justice Party, Muhammad 'ABD-AL-'AL
note: formation of political parties must be approved by government
Other political or pressure groups: despite a constitutional ban against religious-based parties, the technically illegal Muslim Brotherhood constitutes MUBARAK's potentially most significant political opposition; MUBARAK tolerated limited political activity by the Brotherhood for his first two terms, but has moved more aggressively in the past year to block its influence; trade unions and professional associations are officially sanctioned
Member of: ABEDA, ACC, AFESD, AL, AMF, CAEU, CCC, ESCWA, FAO, G-19, G-77, IAEA, IBRD, ICAO, ICRM, IDA, IDB, IFAD, IFC, IFRCS, ILO, IMF, IMO, INMARSAT, INTELSAT, INTERPOL, IOC, ISO, ITU, NAM, OAPEC, OIC, OPEC, PCA, UN, UNAMIR, UNCTAD, UNESCO, UNIDO, UNOMIL, UNPROFOR, UPU, WFTU, WHO, WIPO, WMO, WTO
Diplomatic representation in US:
chief of mission: Ambassador Ahmed Maher El SAYED
chancery: 3521 International Court NW, Washington, DC 20008
telephone: [1] (202) 895-5400
FAX: [1] (202) 244-4319, 5131
consulate(s) general: Chicago, Houston, New York, and San Francisco
US diplomatic representation:
chief of mission: Ambassador Edward S. WALKER, Jr.
embassy: (North Gate) 8, Kamel El-Din Salah Street, Garden City, Cairo
mailing address: APO AE 09839-4900

telephone: [20] (2) 3557371
FAX: [20] (2) 3573200
Flag: three equal horizontal bands of red (top), white, and black with the national emblem (a shield superimposed on a golden eagle facing the hoist side above a scroll bearing the name of the country in Arabic) centered in the white band; similar to the flag of Yemen, which has a plain white band; also similar to the flag of Syria that has two green stars and to the flag of Iraq, which has three green stars (plus an Arabic inscription) in a horizontal line centered in the white band

Economy

Overview: Half of Egypt's GDP originates in the public sector, most industrial plants being owned by the government. Overregulation holds back technical modernization and foreign investment. Even so, the economy grew rapidly during the late 1970s and early 1980s, but in 1986 the collapse of world oil prices and an increasingly heavy burden of debt servicing led Egypt to begin negotiations with the IMF for balance-of-payments support. Egypt's first IMF standby arrangement concluded in mid-1987 was suspended in early 1988 because of the government's failure to adopt promised reforms. Egypt signed a follow-on program with the IMF and also negotiated a structural adjustment loan with the World Bank in 1991. In 1991-93 the government made solid progress on administrative reforms such as liberalizing exchange and interest rates but resisted implementing major structural reforms like streamlining the public sector. As a result, the economy has not gained momentum and unemployment has become a growing problem. Egypt probably will continue making uneven progress in implementing the successor programs with the IMF and World Bank it signed onto in late 1993. Tourism has plunged since 1992 because of sporadic attacks by Islamic extremists on tourist groups. President MUBARAK has cited population growth as the main cause of the country's economic troubles. The addition of about 1.2 million people a year to the already huge population of 62 million exerts enormous pressure on the 5% of the land area available for agriculture along the Nile.
National product: GDP—purchasing power parity—$151.5 billion (1994 est.)
National product real growth rate: 1.5% (1994 est.)
National product per capita: $2,490 (1994 est.)
Inflation rate (consumer prices): 8% (1994 est.)
Unemployment rate: 20% (1994 est.)
Budget:
revenues: $18 billion
expenditures: $19.4 billion, including capital expenditures of $3.8 billion (FY94/95 est.)
Exports: $3.1 billion (f.o.b., FY93/94 est.)
commodities: crude oil and petroleum products, cotton yarn, raw cotton, textiles, metal products, chemicals
partners: EU, US, Japan
Imports: $11.2 billion (c.i.f., FY93/94 est.)
commodities: machinery and equipment, foods, fertilizers, wood products, durable consumer goods, capital goods
partners: EU, US, Japan
External debt: $31.2 billion (December 1994 est.)
Industrial production: growth rate 2.7% (FY92/93 est.)
Electricity:
capacity: 11,830,000 kW
production: 44.5 billion kWh
consumption per capita: 695 kWh (1993)
Industries: textiles, food processing, tourism, chemicals, petroleum, construction, cement, metals
Agriculture: cotton, rice, corn, wheat, beans, fruit, vegetables; cattle, water buffalo, sheep, goats; annual fish catch about 140,000 metric tons
Illicit drugs: a transit point for Southwest Asian and Southeast Asian heroin and opium moving to Europe and the US; popular transit stop for Nigerian couriers; large domestic consumption of hashish from Lebanon and Syria
Economic aid:
recipient: US commitments, including Ex-Im (FY70-89), $15.7 billion; Western (non-US) countries, ODA and OOF bilateral commitments (1970-88), $10.1 billion; OPEC bilateral aid (1979-89), $2.9 billion; Communist countries (1970-89), $2.4 billion
Currency: 1 Egyptian pound (£E) = 100 piasters
Exchange rates: Egyptian pounds (£E) per US$1—3.4 (November 1994), 3.369 (November 1993), 3.345 (November 1992), 2.7072 (1990); market rate: 3.3920 (January 1995), 3.3920 (1994), 3.3704 (1993), 3.3300 (1992), 2.0000 (1991), 1.1000 (1990)
Fiscal year: 1 July—30 June

Transportation

Railroads:
total: 4,895 km (42 km electrified; 951 km double track)
standard gauge: 4,548 km 1.435-m gauge (42 km electrified; 951 km double track)
narrow gauge: 347 km 0.750-m gauge
Highways:
total: 47,387 km
paved: 34,593 km
unpaved: 12,794 km
Inland waterways: 3,500 km (including the Nile, Lake Nasser, Alexandria-Cairo Waterway, and numerous smaller canals in the delta); Suez Canal, 193.5 km long (including approaches), used by oceangoing vessels drawing up to 16.1 meters of water
Pipelines: crude oil 1,171 km; petroleum products 596 km; natural gas 460 km
Ports: Alexandria, Al Ghurdaqah, Aswan, Asyut, Bur Safajah, Damietta, Marsa Matruh, Port Said, Suez
Merchant marine:
total: 168 ships (1,000 GRT or over) totaling 1,187,442 GRT/1,821,327 DWT
ships by type: bulk 19, cargo 83, container 2, oil tanker 15, passenger 30, refrigerated cargo 1, roll-on/roll-off cargo 14, short-sea passenger 4
Airports:
total: 91
with paved runways over 3,047 m: 11
with paved runways 2,438 to 3,047 m: 35
with paved runways 1,524 to 2,437 m: 17
with paved runways 914 to 1,523 m: 3
with paved runways under 914 m: 14
with unpaved runways 2,438 to 3,047 m: 2
with unpaved runways 1,524 to 2,438 m: 2
with unpaved runways 914 to 1,523 m: 7

Communications

Telephone system: 600,000 telephones; 11 telephones/1,000 persons; large system by Third World standards but inadequate for present requirements and undergoing extensive upgrading
local: NA
intercity: principal centers at Alexandria, Cairo, Al Mansurah, Ismailia Suez, and Tanta are connected by coaxial cable and microwave radio relay
international: 2 INTELSAT Atlantic Ocean and Indian Ocean), 1 ARABSAT, and 1 INMARSAT earth station; 5 coaxial submarine cables, microwave troposcatter (to Sudan), and microwave radio relay (to Libya, Israel, and Jordan)
Radio:
broadcast stations: AM 39, FM 6, shortwave 0
radios: NA
Television:
broadcast stations: 41
televisions: NA

Defense Forces

Branches: Army, Navy, Air Force, Air Defense Command
Manpower availability: males age 15-49 16,113,413; males fit for military service 10,455,955; males reach military age (20) annually 648,724 (1995 est.)
Defense expenditures: exchange rate conversion—$3.5 billion, 8.2% of total government budget (FY94/95)

El Salvador

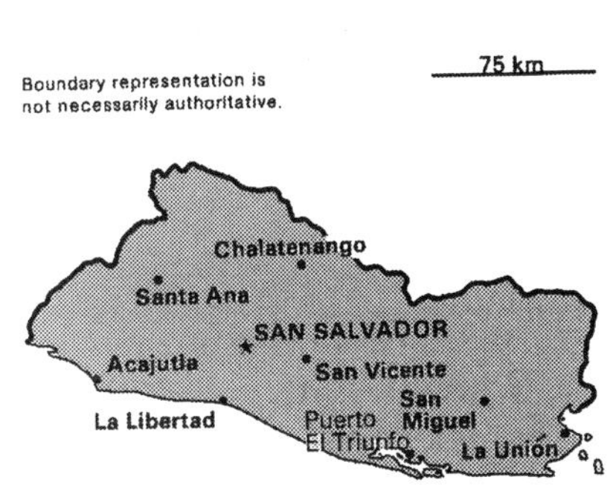

Geography

Location: Middle America, bordering the North Pacific Ocean, between Guatemala and Honduras
Map references: Central America and the Caribbean
Area:
total area: 21,040 sq km
land area: 20,720 sq km
comparative area: slightly smaller than Massachusetts
Land boundaries: total 545 km, Guatemala 203 km, Honduras 342 km
Coastline: 307 km
Maritime claims:
territorial sea: 200 nm
International disputes: land boundary dispute with Honduras mostly resolved by 11 September 1992 International Court of Justice (ICJ) decision; with respect to the maritime boundary in the Golfo de Fonseca, ICJ referred to an earlier agreement in this century and advised that some tripartite resolution among El Salvador, Honduras and Nicaragua likely would be required
Climate: tropical; rainy season (May to October); dry season (November to April)
Terrain: mostly mountains with narrow coastal belt and central plateau
Natural resources: hydropower, geothermal power, petroleum
Land use:
arable land: 27%
permanent crops: 8%
meadows and pastures: 29%
forest and woodland: 6%
other: 30%
Irrigated land: 1,200 sq km (1989)
Environment:
current issues: deforestation; soil erosion; water pollution; contamination of soils from disposal of toxic wastes
natural hazards: known as the Land of Volcanoes; frequent and sometimes very destructive earthquakes and volcanic activity
international agreements: party to—Biodiversity, Endangered Species, Hazardous Wastes, Nuclear Test Ban, Ozone Layer Protection; signed, but not ratified—Climate Change, Law of the Sea
Note: smallest Central American country and only one without a coastline on Caribbean Sea

People

Population: 5,870,481 (July 1995 est.)
Age structure:
0-14 years: 40% (female 1,165,152; male 1,200,759)
15-64 years: 56% (female 1,677,958; male 1,602,230)
65 years and over: 4% (female 122,368; male 102,014) (July 1995 est.)
Population growth rate: 2.02% (1995 est.)
Birth rate: 32.39 births/1,000 population (1995 est.)
Death rate: 6.19 deaths/1,000 population (1995 est.)
Net migration rate: -5.96 migrant(s)/1,000 population (1995 est.)
Infant mortality rate: 38.9 deaths/1,000 live births (1995 est.)
Life expectancy at birth:
total population: 67.5 years
male: 64.89 years
female: 70.23 years (1995 est.)
Total fertility rate: 3.69 children born/woman (1995 est.)
Nationality:
noun: Salvadoran(s)
adjective: Salvadoran
Ethnic divisions: mestizo 94%, Indian 5%, white 1%
Religions: Roman Catholic 75%
note: there is extensive activity by Protestant groups throughout the country; by the end of 1992, there were an estimated 1 million Protestant evangelicals in El Salvador
Languages: Spanish, Nahua (among some Indians)
Literacy: age 15 and over can read and write (1990 est.)
total population: 73%
male: 76%
female: 70%
Labor force: 1.7 million (1982 est.)
by occupation: agriculture 40%, commerce 16%, manufacturing 15%, government 13%, financial services 9%, transportation 6%, other 1%
note: shortage of skilled labor and a large pool of unskilled labor, but training programs improving situation (1984 est.)

Government

Names:
conventional long form: Republic of El Salvador
conventional short form: El Salvador
local long form: Republica de El Salvador
local short form: El Salvador
Digraph: ES
Type: republic
Capital: San Salvador
Administrative divisions: 14 departments (departamentos, singular—departamento); Ahuachapan, Cabanas, Chalatenango, Cuscatlan, La Libertad, La Paz, La Union, Morazan, San Miguel, San Salvador, Santa Ana, San Vicente, Sonsonate, Usulutan
Independence: 15 September 1821 (from Spain)
National holiday: Independence Day, 15 September (1821)
Constitution: 20 December 1983
Legal system: based on civil and Roman law, with traces of common law; judicial review of legislative acts in the Supreme Court; accepts compulsory ICJ jurisdiction, with reservations
Suffrage: 18 years of age; universal
Executive branch:
chief of state and head of government: President Armando CALDERON SOL (since 1 June 1994); Vice President Enrique BORGO Bustamante (since 1 June 1994) election last held 20 March 1994 (next to be held March 1999); results—Armando CALDERON SOL (ARENA) 49.03%, Ruben ZAMORA Rivas (CD/FMLN/MNR) 24.09%, Fidel CHAVEZ Mena (PDC) 16.39%, other 10.49%; because no candidate received a majority, a run off election was held 24 April 1994; results—Armando CALDERON SOL (ARENA) 68.35%, Ruben ZAMORA Rivas (CD/FMLN/MNR) 31.65%
cabinet: Council of Ministers
Legislative branch: unicameral
Legislative Assembly (Asamblea Legislativa): elections last held 20 March 1994 (next to be held March 1997); results—ARENA 46.4%, FMLN 25.0%, PDC 21.4%, PCN 4.8%, other 2.4%; seats—(84 total) ARENA 39, FMLN 21, PDC 18, PCN 4, other 2
Judicial branch: Supreme Court (Corte Suprema)
Political parties and leaders: National Republican Alliance (ARENA), Juan Jose DOMENECH, president; Farabundo Marti National Liberation Front (FMLN), Salvador SANCHEZ Ceren (aka Leonel GONZALEZ), general coordinator; Christian Democratic Party (PDC), Ronal UMANA, secretary general; National Conciliation Party (PCN), Ciro CRUZ Zepeda, secretary general; Democratic Convergence (CD), Juan Jose MARTEL, secretary general; Unity Movement, Jorge MARTINEZ Menendez, president
note: newly formed parties not yet officially recognized by the Supreme Electoral Tribunal. Liberal Democratic Party (PLD), Kirio Waldo SALGADO, founder; Social Democratic Party (breakaway from FMLN), Joaquin VILLALOBOS, founder; Social Christian Renovation Movement (MRSC) (breakaway from PDC), Abraham RODRIGUEZ, founder

Other political or pressure groups:
labor organizations: Salvadoran Communal Union (UCS), peasant association; General Confederation of Workers (CGT), moderate; United Workers Front (FUT)
business organizations: Productive Alliance (AP), conservative; National Federation of Salvadoran Small Businessmen (FENAPES), conservative
Member of: BCIE, CACM, ECLAC, FAO, G-77, GATT, IADB, IAEA, IBRD, ICAO, ICFTU, ICRM, IDA, IFAD, IFC, IFRCS, ILO, IMF, IMO, INTELSAT, INTERPOL, IOC, IOM, ITU, LAES, LAIA (observer), NAM (observer), OAS, OPANAL, PCA, UN, UNCTAD, UNESCO, UNIDO, UPU, WCL, WFTU, WHO, WIPO, WMO, WTO
Diplomatic representation in US:
chief of mission: Ambassador Ana Cristina SOL
chancery: 2308 California Street NW, Washington, DC 20008
telephone: [1] (202) 265-9671, 9672
consulate(s) general: Chicago, Dallas, Houston, Los Angeles, Miami, New Orleans, New York, and San Francisco
US diplomatic representation:
chief of mission: Ambassador Alan H. FLANIGAN
embassy: Final Boulevard, Station Antiguo Cuscatlan, San Salvador
mailing address: Unit 3116, San Salvador; APO AA 34023
telephone: [503] 78-4444
FAX: [503] 78-6011
Flag: three equal horizontal bands of blue (top), white, and blue with the national coat of arms centered in the white band; the coat of arms features a round emblem encircled by the words REPUBLICA DE EL SALVADOR EN LA AMERICA CENTRAL; similar to the flag of Nicaragua, which has a different coat of arms centered in the white band—it features a triangle encircled by the words REPUBLICA DE NICARAGUA on top and AMERICA CENTRAL on the bottom; also similar to the flag of Honduras, which has five blue stars arranged in an X pattern centered in the white band

Economy

Overview: The agricultural sector accounts for 24% of GDP, employs about 40% of the labor force, and contributes about 66% to total exports. Coffee is the major commercial crop, accounting for 45% of export earnings. The manufacturing sector, based largely on food and beverage processing, accounts for 19% of GDP and 15% of employment. In 1992-94 the government made substantial progress toward privatization and deregulation of the economy. Growth in national output in 1991-94 nearly averaged 5%, exceeding growth in population for the first time since 1987; and inflation in 1994 of 10% was down from 19% in 1993.
National product: GDP—purchasing power parity—$9.8 billion (1994 est.)
National product real growth rate: 5% (1994 est.)
National product per capita: $1,710 (1994 est.)
Inflation rate (consumer prices): 10% (1994 est.)
Unemployment rate: 6.7% (1993)
Budget:
revenues: $846 million
expenditures: $890 million, including capital expenditures of $NA (1992 est.)
Exports: $823 million (f.o.b., 1994 est.)
commodities: coffee, sugarcane, shrimp
partners: US, Guatemala, Costa Rica, Germany
Imports: $2.1 billion (c.i.f., 1994 est.)
commodities: raw materials, consumer goods, capital goods
partners: US, Guatemala, Mexico, Venezuela, Germany
External debt: $2.6 billion (December 1992)
Industrial production: growth rate 7.6% (1993)
Electricity:
capacity: 750,000 kW
production: 2.4 billion kWh
consumption per capita: 408 kWh (1993)
Industries: food processing, beverages, petroleum, nonmetallic products, tobacco, chemicals, textiles, furniture
Agriculture: accounts for 24% of GDP and 40% of labor force (including fishing and forestry); coffee most important commercial crop; other products—sugarcane, corn, rice, beans, oilseeds, beef, dairy products, shrimp; not self-sufficient in food
Illicit drugs: transshipment point for cocaine; marijuana produced for local consumption
Economic aid:
recipient: US commitments, including Ex-Im (FY70-90), $2.95 billion (plus $250 million for 1992-96); Western (non-US) countries, ODA and OOF bilateral commitments (1970-89), $525 million
Currency: 1 Salvadoran colon (C) = 100 centavos
Exchange rates: Salvadoran colones (C) per US$1—8.760 (January 1995), 8.750 (1994), 8.670 (1993), 8.4500 (1992), 8.080 (1991), 8.0300 (1990)
Fiscal year: calendar year

Transportation

Railroads:
total: 602 km (single track; note—some sections abandoned, unusable, or operating at reduced capacity)
narrow gauge: 602 km 0.914-m gauge
Highways:
total: 10,000 km
paved: 1,500 km
unpaved: gravel 4,100 km; improved, unimproved earth 4,400 km
Inland waterways: Rio Lempa partially navigable
Ports: Acajutla, Puerto Cutuco, La Libertad, La Union, Puerto El Triunfo
Merchant marine: none
Airports:
total: 106
with paved runways over 3,047 m: 1
with paved runways 1,524 to 2,437 m: 2
with paved runways 914 to 1,523 m: 2
with paved runways under 914 m: 78
with unpaved runways 914 to 1,523 m: 23

Communications

Telephone system: 116,000 telephones; 21 telephones/1,000 persons
local: NA
intercity: nationwide microwave radio relay system
international: 1 INTELSAT (Atlantic Ocean) earth station; connected to Central American Microwave System
Radio:
broadcast stations: AM 77, FM 0, shortwave 2
radios: NA
Television:
broadcast stations: 5
televisions: NA

Defense Forces

Branches: Army, Navy, Air Force
Manpower availability: males age 15-49 1,393,480; males fit for military service 892,958; males reach military age (18) annually 77,562 (1995 est.)
Defense expenditures: exchange rate conversion—$103 million, 0.7% of GDP (1994); $91.9 million, less than 1% of GDP (1995 est.)

Equatorial Guinea

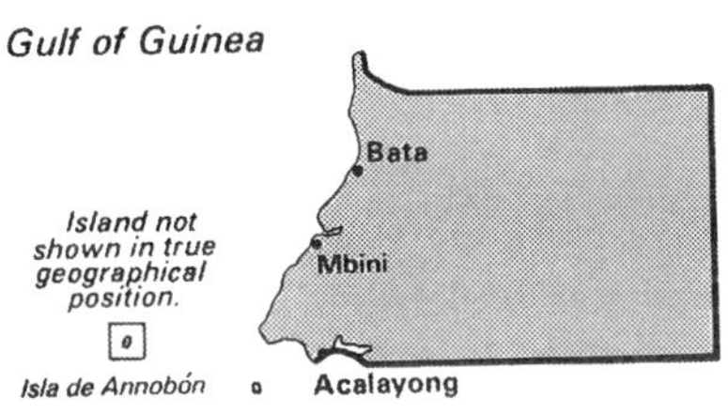

Geography

Location: Western Africa, bordering the North Atlantic Ocean, between Cameroon and Gabon
Map references: Africa
Area:
total area: 28,050 sq km
land area: 28,050 sq km
comparative area: slightly larger than Maryland
Land boundaries: total 539 km, Cameroon 189 km, Gabon 350 km
Coastline: 296 km
Maritime claims:
exclusive economic zone: 200 nm
territorial sea: 12 nm
International disputes: maritime boundary dispute with Gabon because of disputed sovereignty over islands in Corisco Bay
Climate: tropical; always hot, humid
Terrain: coastal plains rise to interior hills; islands are volcanic
Natural resources: timber, petroleum, small unexploited deposits of gold, manganese, uranium
Land use:
arable land: 8%
permanent crops: 4%
meadows and pastures: 4%
forest and woodland: 51%
other: 33%
Irrigated land: NA sq km
Environment:
current issues: tap water is not potable; desertification
natural hazards: violent windstorms
international agreements: party to—Biodiversity, Endangered Species, Nuclear Test Ban; signed, but not ratified—Desertification, Law of the Sea
Note: insular and continental regions rather widely separated

People

Population: 420,293 (July 1995 est.)
Age structure:
0-14 years: 43% (female 90,404; male 90,997)
15-64 years: 53% (female 117,124; male 105,724)
65 years and over: 4% (female 8,969; male 7,075) (July 1995 est.)
Population growth rate: 2.59% (1995 est.)
Birth rate: 40.22 births/1,000 population (1995 est.)
Death rate: 14.36 deaths/1,000 population (1995 est.)
Net migration rate: 0 migrant(s)/1,000 population (1995 est.)
Infant mortality rate: 100.2 deaths/1,000 live births (1995 est.)
Life expectancy at birth:
total population: 52.56 years
male: 50.39 years
female: 54.79 years (1995 est.)
Total fertility rate: 5.23 children born/woman (1995 est.)
Nationality:
noun: Equatorial Guinean(s) or Equatoguinean(s)
adjective: Equatorial Guinean or Equatoguinean
Ethnic divisions: Bioko (primarily Bubi, some Fernandinos), Rio Muni (primarily Fang), Europeans less than 1,000, mostly Spanish
Religions: nominally Christian and predominantly Roman Catholic, pagan practices
Languages: Spanish (official), pidgin English, Fang, Bubi, Ibo
Literacy: age 15 and over can read and write (1983)
total population: 62%
male: 77%
female: 48%
Labor force: 172,000 (1986 est.)
by occupation: agriculture 66%, services 23%, industry 11% (1980)
note: labor shortages on plantations

Government

Names:
conventional long form: Republic of Equatorial Guinea
conventional short form: Equatorial Guinea
local long form: Republica de Guinea Ecuatorial
local short form: Guinea Ecuatorial
former: Spanish Guinea
Digraph: EK
Type: republic in transition to multiparty democracy
Capital: Malabo
Administrative divisions: 7 provinces (provincias, singular—provincia); Annobon, Bioko Norte, Bioko Sur, Centro Sur, Kie-Ntem, Litoral, Wele-Nzas
Independence: 12 October 1968 (from Spain)
National holiday: Independence Day, 12 October (1968)
Constitution: new constitution 17 November 1991
Legal system: partly based on Spanish civil law and tribal custom
Suffrage: universal adult at age NA
Executive branch:
chief of state: President Brig. Gen. (Ret.) Teodoro OBIANG NGUEMA MBASOGO (since 3 August 1979); election last held 25 June 1989 (next to be held 25 June 1996); results—President Brig. Gen. (Ret.) Teodoro OBIANG NGUEMA MBASOGO was reelected without opposition
head of government: Prime Minister Silvestre SIALE BILEKA (since 17 January 1992); Vice Prime Minister Anatolio NDONG MBA (since November 1993)
cabinet: Council of Ministers; appointed by the president
Legislative branch: unicameral
House of People's Representatives: (Camara de Representantes del Pueblo) elections last held 21 November 1993; seats—(82 total) PDGE 72, various opposition parties 10
Judicial branch: Supreme Tribunal
Political parties and leaders:
ruling party: Democratic Party for Equatorial Guinea (PDGE), Brig. Gen. (Ret.) Teodoro OBIANG NGUEMA MBASOGO, party leader
opposition parties: Progressive Democratic Alliance (ADP), Antonio-Ebang Mbele Abang, president; Popular Action of Equatorial Guinea (APGE),Casiano Masi Edu, leader; Liberal Democratic Convention (CLD), Alfonso Nsue MOKUY, president; Convergence for Social Democracy (CPDS),Santiago Obama Ndong, president; Social Democratic and Popular Convergence (CSDP), Secundino Oyono Agueng Ada, general secretary; Party of the Social Democratic Coalition (PCSD), Buenaventura Moswi M'Asumu, general coordinater; Liberal Party (PL), leaders unknown; Party of Progress (PP), Severo MOTO Nsa, president; Social Democratic Party (PSD), Benjamin-Gabriel Balingha Balinga Alene, general secretary, Socialist Party of Equatorial Guinea (PSGE), Tomas MICHEBE Fernandez, general secretary; National Democratic Union (UDENA), Jose MECHEBA Ikaka, president; Democratic Social Union (UDS), Jesus Nze Obama Avomo, general secretary; Popular Union (UP), Juan Bitui, president
Member of: ACCT, ACP, AfDB, BDEAC, CEEAC, ECA, FAO, FZ, G-77, IBRD, ICAO, IDA, IFAD, IFC, IFRCS (associate), ILO, IMF, IMO, INTELSAT (nonsignatory user), INTERPOL, IOC, ITU, NAM, OAS (observer), OAU, UDEAC, UN, UNCTAD, UNESCO, UNIDO, UPU, WHO
Diplomatic representation in US:
chief of mission: (vacant); Charge d'Affaires

ad interim Teodoro Biyogo NSUE
chancery: (temporary) 57 Magnolia Avenue, Mount Vernon, NY 10553
telephone: [1] (914) 738-9584, 667-6913
FAX: [1] (914) 667-6838
US diplomatic representation:
chief of mission: Charge d'Affaires Joseph P. O'NEILL
embassy: Calle de Los Ministros, Malabo
mailing address: P.O. Box 597, Malabo
telephone: [240] (9) 21-85, 24-06, 25-07
FAX: [240] (9) 21-64
Flag: three equal horizontal bands of green (top), white, and red with a blue isosceles triangle based on the hoist side and the coat of arms centered in the white band; the coat of arms has six yellow six-pointed stars (representing the mainland and five offshore islands) above a gray shield bearing a silk-cotton tree and below which is a scroll with the motto UNIDAD, PAZ, JUSTICIA (Unity, Peace, Justice)

Economy

Overview: Agriculture, forestry, and fishing account for about half of GDP and nearly all exports. Subsistence farming predominates. Although pre-independence Equatorial Guinea counted on cocoa production for hard currency earnings, the deterioration of the rural economy under successive brutal regimes has diminished potential for agriculture-led growth. A number of aid programs sponsored by the World Bank and the international donor community have failed to revitalize export agriculture. Businesses for the most part are owned by government officials and their family members. Commerce accounts for about 8% of GDP and the construction, public works, and service sectors for about 38%. Undeveloped natural resources include titanium, iron ore, manganese, uranium, and alluvial gold. Oil exploration, taking place under concessions offered to US, French, and Spanish firms, has been moderately successful. Increased production from recently discovered natural gas fields will provide a greater share of exports in 1995.
National product: GDP—purchasing power parity—$280 million (1993 est.)
National product real growth rate: 7.3% (1993 est.)
National product per capita: $700 (1993 est.)
Inflation rate (consumer prices): 1.6% (1992 est.)
Unemployment rate: NA%
Budget:
revenues: $32.5 million
expenditures: $35.9 million, including capital expenditures of $3 million (1992 est.)
Exports: $56 million (f.o.b., 1993)
commodities: coffee, timber, cocoa beans
partners: Spain 55.2%, Nigeria 11.4%, Cameroon 9.1% (1992)
Imports: $62 million (c.i.f., 1993)
commodities: petroleum, food, beverages, clothing, machinery
partners: Cameroon 23.1%, Spain 21.8%, France 14.1%, US 4.3% (1992)
External debt: $260 million (1992 est)
Industrial production: growth rate 11.3% (1993 est.)
Electricity:
capacity: 23,000 kW
production: 20 million kWh
consumption per capita: 50 kWh (1993)
Industries: fishing, sawmilling
Agriculture: accounts for almost 50% of GDP, cash crops—timber and coffee from Rio Muni, cocoa from Bioko; food crops—rice, yams, cassava, bananas, oil palm nuts, manioc, livestock
Economic aid:
recipient: US commitments, including Ex-Im (FY81-89), $14 million; Western (non-US) countries, ODA and OOF bilateral commitments (1970-89), $130 million; Communist countries (1970-89), $55 million
Currency: 1 CFA franc (CFAF) = 100 centimes
Exchange rates: Communaute Financiere Africaine francs (CFAF) per US$1—529.43 (January 1995), 555.20 (1994), 273,16 (1993), 264.69 (1992), 282.11 (1991), 272.26 (1990)
note: beginning 12 January 1994, the CFA franc was devalued to CFAF 100 per French franc from CFAF 50 at which it had been fixed since 1948
Fiscal year: 1 April—31 March

Transportation

Railroads:
total: 0 km
Highways:
total: 2,760 km (2,460 km on Rio Muni and 300 km on Bioko)
paved: NA
unpaved: NA
Ports: Bata, Luba, Malabo
Merchant marine:
total: 2 ships (1,000 GRT or over) totaling 6,412 GRT/6,699 DWT
ships by type: cargo 1, passenger-cargo 1
Airports:
total: 3
with paved runways 2,438 to 3,047 m: 1
with paved runways 1,524 to 2,437 m: 1
with paved runways under 914 m: 1

Communications

Telephone system: 2,000 telephones; poor system with adequate government services
local: NA
intercity: NA
international: international communications from Bata and Malabo to African and European countries; 1 Indian Ocean INTELSAT earth station
Radio:
broadcast stations: AM 2, FM 0, shortwave 0
radios: NA
Television:
broadcast stations: 1
televisions: NA

Defense Forces

Branches: Army, Navy, Air Force, Rapid Intervention Force, National Police
Manpower availability: males age 15-49 89,752; males fit for military service 45,611 (1995 est.)
Defense expenditures: exchange rate conversion—$2.5 million, NA% of GDP (FY93/94)

Eritrea

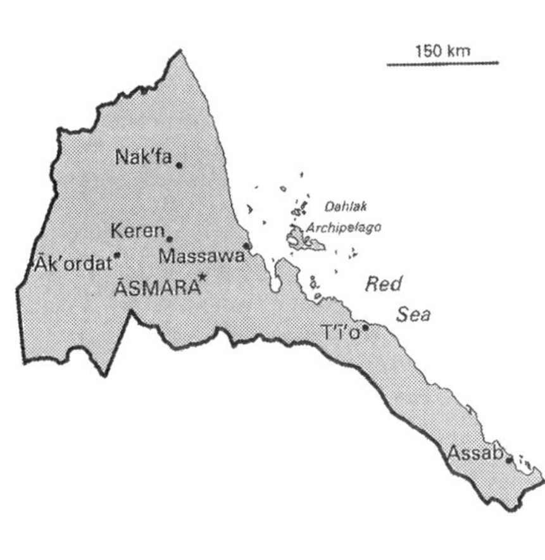

Geography

Location: Eastern Africa, bordering the Red Sea, between Djibouti and Sudan
Map references: Africa
Area:
total area: 121,320 sq km
land area: 121,320 sq km
comparative area: slightly larger than Pennsylvania
Land boundaries: total 1,630 km, Djibouti 113 km, Ethiopia 912 km, Sudan 605 km
Coastline: 1,151 km (land and island coastline is 2,234 km)
Maritime claims: NA
International disputes: none
Climate: hot, dry desert strip along Red Sea coast; cooler and wetter in the central highlands (up to 61 cm of rainfall annually); semiarid in western hills and lowlands; rainfall heaviest during June-September except on coastal desert
Terrain: dominated by extension of Ethiopian north-south trending highlands, descending on the east to a coastal desert plain, on the northwest to hilly terrain and on the southwest to flat-to-rolling plains
Natural resources: gold, potash, zinc, copper, salt, probably oil (petroleum geologists are prospecting for it), fish
Land use:
arable land: 3%
permanent crops: 2% (coffee)
meadows and pastures: 40%
forest and woodland: 5%
other: 50%
Irrigated land: NA sq km
Environment:
current issues: famine; deforestation; desertification; soil erosion; overgrazing; loss of infrastructure from civil warfare
natural hazards: frequent droughts
international agreements: party to—Endangered Species; signed, but not ratified—Desertification
Note: strategic geopolitical position along world's busiest shipping lanes; Eritrea retained the entire coastline of Ethiopia along the Red Sea upon de jure independence from Ethiopia on 27 April 1993

People

Population: 3,578,709 (July 1995 est.)
Age structure:
0-14 years: 43% (female 763,416; male 774,922)
15-64 years: 54% (female 965,124; male 965,435)
65 years and over: 3% (female 52,950; male 56,862) (July 1995 est.)
Population growth rate: 9.04% (1995 est.)
Birth rate: 44.34 births/1,000 population (1995 est.)
Death rate: 15.67 deaths/1,000 population (1995 est.)
Net migration rate: NA migrant(s)/1,000 population (1995 est.)
note: repatriation of up to a half million Eritrean refugees in Sudan is now underway; 100,000 are expected to return during 1995
Infant mortality rate: 120.6 deaths/1,000 live births (1995 est.)
Life expectancy at birth:
total population: 50 years
male: 48.28 years
female: 51.78 years (1995 est.)
Total fertility rate: 6.53 children born/woman (1995 est.)
Nationality:
noun: Eritrean(s)
adjective: Eritrean
Ethnic divisions: ethnic Tigrays 50%, Tigre and Kunama 40%, Afar 4%, Saho (Red Sea coast dwellers) 3%
Religions: Muslim, Coptic Christian, Roman Catholic, Protestant
Languages: Tigre and Kunama, Cushitic dialects, Tigre, Nora Bana, Arabic
Labor force: NA

Government

Names:
conventional long form: State of Eritrea
conventional short form: Eritrea
local long form: none
local short form: none
former: Eritrea Autonomous Region in Ethiopia
Digraph: ER
Type: transitional government
note: on 29 May 1991 ISAIAS Afworke, secretary general of the Peoples' Front for Democracy and Justice (PFDJ), which then served and still serves as the country's legislative body, announced the formation of the Provisional Government in Eritrea (PGE) in preparation for the 23-25 April 1993 referendum on independence for the autonomous region of Eritrea; the result was a landslide vote for independence which was proclaimed on 27 April 1993
Capital: Asmara (formerly Asmera)
Administrative divisions: 9 provinces; Akole Guzay, Baraka, Danakil, Hamasen, Sahil, Samhar, Senhit, Seraye, Sahil
Independence: 27 May 1993 (from Ethiopia; formerly the Eritrea Autonomous Region)
National holiday: National Day (independence from Ethiopia), 24 May (1993)
Constitution: transitional "constitution" decreed 19 May 1993
Legal system: NA
Suffrage: NA
Executive branch:
chief of state and head of government: President ISAIAS Afworke (since 22 May 1993)
cabinet: State Council; the collective executive authority
note: election to be held before 20 May 1997
Legislative branch: unicameral
National Assembly: PFDJ Central Committee serves as the country's legislative body until country-wide elections are held (before 20 May 1997)
Judicial branch: Judiciary
Political parties and leaders: People's Front for Democracy and Justice (PFDJ), ISAIAS Afworke, PETROS Solomon (the only party recognized by the government)
Other political or pressure groups: Eritrean Islamic Jihad (EIJ); Islamic Militant Group; Eritrean Liberation Front (ELF), ABDULLAH Muhammed; Eritrean Liberation Front—United Organization (ELF-UO), Mohammed Said NAWUD; Eritrean Liberation Front—Revolutionary Council (ELF-RC), Ahmed NASSER
Member of: ACP, ECA, FAO, IBRD, ICAO, IDA, IFAD, IGADD, ILO, IMF, IMO, INTELSAT (nonsignatory user), ITU, OAU, UN, UNCTAD, UNESCO, UPU, WFTU
Diplomatic representation in US:
chief of mission: Ambassador AMDEMICHAEL Berhane Khasai
chancery: Suite 400, 910 17th Street NW, Washington, DC 20006
telephone: [1] (202) 429-1991
FAX: [1] (202) 429-9004
US diplomatic representation:
chief of mission: Ambassador Robert G. HOUDEK
embassy: 34 Zera Yacob St., Asmara
mailing address: P.O. Box 211, Asmara
telephone: [291] (1) 120004
FAX: [291] (1) 127584
Flag: red isosceles triangle (based on the hoist side) dividing the flag into two right triangles; the upper triangle is green, the lower one is blue; a gold wreath encircling a gold olive branch is centered on the hoist side of the red triangle

Economy

Overview: With independence from Ethiopia on 27 April 1993, Eritrea faces the bitter economic problems of a small, desperately poor African country. Most of the population will continue to depend on subsistence farming. Domestic output is substantially augmented by worker remittances from abroad. Government revenues come from custom duties and income and sales taxes. Eritrea has inherited the entire coastline of Ethiopia and has long-term prospects for revenues from the development of offshore oil, offshore fishing, and tourism. For the time being, Ethiopia will be largely dependent on Eritrean ports for its foreign trade.
National product: GDP—purchasing power parity—$1.8 billion (1994 est.)
National product real growth rate: 2% (1994 est.)
National product per capita: $500 (1994 est.)
Inflation rate (consumer prices): NA%
Unemployment rate: NA%
Budget:
revenues: $NA
expenditures: $NA, including capital expenditures of $NA
Exports: $NA
commodities: NA
partners: NA
Imports: $NA
commodities: NA
partners: NA
External debt: $NA
Industrial production: growth rate NA%
Electricity:
capacity: NA kW
production: NA kWh
consumption per capita: NA kWh
Industries: food processing, beverages, clothing and textiles
Agriculture: products—sorghum, livestock (including goats), fish, lentils, vegetables, maize, cotton, tobacco, coffee, sisal (for making rope)
Economic aid: $NA
Currency: 1 birr (Br) = 100 cents; at present, Ethiopian currency used
Exchange rates: 1 birr (Br) per US$1—5.9500 (January 1995), 5.9500 (1994), 5.000 (fixed rate 1992-93); note—official rate pegged to US$
Fiscal year: NA

Transportation

Railroads:
total: 307 km; note—nonoperational since 1978; links Ak'ordat and Asmara (formerly Asmera) with the port of Massawa (formerly Mits'iwa)
narrow gauge: 307 km 1.000-m gauge (1993 est.)
Highways:
total: 3,845 km
paved: 807 km
unpaved: gravel 840 km; improved earth 402 km; unimproved earth 1,796 km
Ports: Assab (Aseb), Massawa (Mits'iwa)
Merchant marine: none
Airports:
total: 20
with paved runways over 3,047 m: 1
with paved runways 2,438 to 3,047 m: 1
with paved runways under 914 m: 2
with unpaved runways over 3,047 m: 1
with unpaved runways 2,438 to 3,047 m: 2
with unpaved runways 1,524 to 2,438 m: 6
with unpaved runways 914 to 1,523 m: 7

Communications

Telephone system: NA
local: NA
intercity: NA
international: NA
Radio:
broadcast stations: AM NA, FM NA, shortwave 0
radios: NA
Television:
broadcast stations: NA
televisions: NA

Defense Forces

Branches: Eritrean People's Liberation Front (EPLF)
Defense expenditures: $NA, NA% of GDP

Estonia

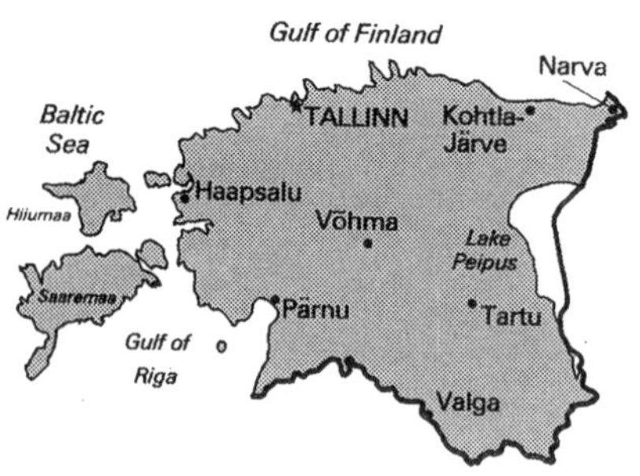

Geography

Location: Eastern Europe, bordering the Baltic Sea and Gulf of Finland, between Latvia and Russia
Map references: Europe
Area:
total area: 45,100 sq km
land area: 43,200 sq km
comparative area: slightly larger than New Hampshire and Vermont combined
note: includes 1,520 islands in the Baltic Sea
Land boundaries: total 557 km, Latvia 267 km, Russia 290 km
Coastline: 1,393 km
Maritime claims:
exclusive economic zone: limits to be fixed in coordination with neighboring states
territorial sea: 12 nm
International disputes: claims over 2,000 sq km of Russian territory in the Narva and Pechora regions—based on boundary established under the 1921 Peace Treaty of Tartu
Climate: maritime, wet, moderate winters, cool summers
Terrain: marshy, lowlands
Natural resources: shale oil, peat, phosphorite, amber
Land use:
arable land: 22%
permanent crops: 0%
meadows and pastures: 11%
forest and woodland: 31%
other: 36%
Irrigated land: 110 sq km (1990)
Environment:
current issues: air heavily polluted with sulfur dioxide from oil-shale burning power plants in northeast; contamination of soil and groundwater with petroleum products, chemicals at military bases
natural hazards: flooding occurs frequently in the spring
international agreements: party to—Biodiversity, Climate Change, Endangered Species, Hazardous Wastes, Ship Pollution, Wetlands

People

Population: 1,625,399 (July 1995 est.)
Age structure:
0-14 years: 22% (female 174,304; male 181,101)
15-64 years: 65% (female 549,473; male 515,426)
65 years and over: 13% (female 139,722; male 65,373) (July 1995 est.)
Population growth rate: 0.53% (1995 est.)
Birth rate: 13.9 births/1,000 population (1995 est.)
Death rate: 11.93 deaths/1,000 population (1995 est.)
Net migration rate: 3.31 migrant(s)/1,000 population (1995 est.)
Infant mortality rate: 18.7 deaths/1,000 live births (1995 est.)
Life expectancy at birth:
total population: 70.17 years
male: 65.2 years
female: 75.39 years (1995 est.)
Total fertility rate: 1.98 children born/woman (1995 est.)
Nationality:
noun: Estonian(s)
adjective: Estonian
Ethnic divisions: Estonian 61.5%, Russian 30.3%, Ukrainian 3.17%, Byelorussian 1.8%, Finn 1.1%, other 2.13% (1989)
Religions: Lutheran
Languages: Estonian (official), Latvian, Lithuanian, Russian, other
Literacy: age 15 and over can read and write (1989)
total population: 100%
male: 100%
female: 100%
Labor force: 750,000 (1992)
by occupation: industry and construction 42%, agriculture and forestry 20%, other 38% (1990)

Government

Names:
conventional long form: Republic of Estonia
conventional short form: Estonia
local long form: Eesti Vabariik
local short form: Eesti
former: Estonian Soviet Socialist Republic
Digraph: EN
Type: republic
Capital: Tallinn
Administrative divisions: 15 counties (maakonnad, singular—maakond): Harju maakond (Tallinn), Hiiu maakond (Kardla), Ida-Viru maakond (Johvi), Jarva maakond (Paide), Jogeva maakond (Jogeva), Laane maakond (Haapsalu), Laane-Viru maakond (Rakvere), Parnu maakond (Parnu), Polva maakond (Polva), Rapla maakond (Rapla), Saare maakond (Kuessaare), Tartu maakond (Tartu), Valga maakond (Valga), Viljandi maakond (Viljandi), Voru maakond (Voru)
note: county centers are in parentheses
Independence: 6 September 1991 (from Soviet Union)
National holiday: Independence Day, 24 February (1918)
Constitution: adopted 28 June 1992
Legal system: based on civil law system; no judicial review of legislative acts
Suffrage: 18 years of age; universal
Executive branch:
chief of state: President Lennart MERI (since 21 October 1992); election last held 20 September 1992; (next to be held fall 1996); results—no candidate received majority; newly elected Parliament elected Lennart MERI (21 October 1992)
head of government: Prime Minister Andres TARAND (since NA October 1994)
cabinet: Council of Ministers; appointed by the prime minister, authorized by the legislature
Legislative branch: unicameral
Parliament (Riigikogu): elections last held 5 March 1995 (next to be held NA 1998); results—KMU 32.22%, RE 16.18%, K 14.17%, Pro Patria and ERSP 7.85%, M 5.98%, Our Home is Estonia and Right-Wingers 5.0%; seats—(101 total) KMU 41, RE 19, K I6, Pro Patria 8, Our Home is Estonia 6, M 6, Right-Wingers 5
Judicial branch: Supreme Court
Political parties and leaders: Coalition Party and Rural Union (KMU) made up of 4 parties: Coalition Party, Country People's Party, Farmer's Assembly, and Pensioners' and Families' League; Coalition Party, Tiit VAHI, chairman; Country People's Party, Arnold RUUTEL, chairman; Farmer's Assembly, Jaak-Hans KUKS, chairman; Pensioners' and Families' League; Reform Party-Liberals (RE), Siim KALLAS, chairman; Center Party (K), Edgar SAVISAAR, chairman; Union of Pro Patria (Isaama of Fatherland), Mart LAAR, chairman; National Independence Party (ERSP), Kelam TUNNE, chairman; Our Home is Estonia made up of 2 parties: United Peoples Party and the Russian Party in Estonia; United Peoples Party, Viktor ANDREJEV, chairman; Russian Party in Estonia, Sergei KUZNETSOV, chairman; Moderates (M) made up of 2 parties: Social Democratic Party and Rural Center Party; Social Democratic Party, Eiki NESTOR, chairman; Rural Center Party, Vambo KAAL, chairman; Right-Wingers, Ulo NUGIS, chairman
Member of: BIS, CBSS, CCC, CE, EBRD, ECE, FAO, IAEA, IBRD, ICAO, ICFTU, ICRM, IFC, ILO, IMF, IMO, INTERPOL, IOC, ISO (correspondent), ITU, NACC, OSCE, PFP, UN, UNCTAD, UNESCO, UPU, WEU (associate partner), WHO, WIPO, WMO
Diplomatic representation in US:
chief of mission: Ambassador Toomas Hendrik ILVES
chancery: 1030 15th Street NW, Washington, DC 20005, Suite 1000
telephone: [1] (202) 789-0320
FAX: [1] (202) 789-0471
consulate(s) general: New York
US diplomatic representation:
chief of mission: (vacant); Charge d'Affaires Keith SMITH
embassy: Kentmanni 20, Tallinn EE 0001
mailing address: use embassy street address
telephone: [372] (2) 312-021 through 024
FAX: [372] (2) 312-025
Flag: pre-1940 flag restored by Supreme Soviet in May 1990—three equal horizontal bands of blue (top), black, and white

Economy

Overview: Bolstered by a widespread national desire to reintegrate into Western Europe, the Estonian government has pursued an ambitious program of market reforms and stabilization measures, which is rapidly transforming the economy. Three years after independence—and two years after the introduction of the kroon—Estonians are beginning to reap tangible benefits; inflation, though still high, was brought down to about 2% per month in second half 1994; production declines have bottomed out with estimated growth of 4% in 1994; and living standards are rising. Economic restructuring has been dramatic. By 1994 the service sector accounted for over 55% of GDP, while the once-dominant heavy industrial sector continues to shrink. The private sector is growing rapidly; the share of the state enterprises in the economy has steadily declined and by late 1994 accounted for only about 40% of GDP. Estonia's foreign trade has shifted rapidly from East to West; the Western industrialized countries now account for two-thirds of foreign trade.
National product: GDP—purchasing power parity—$10.4 billion (1994 estimate as extrapolated from World Bank estimate for 1992)
National product real growth rate: 4% (1994 est.)
National product per capita: $6,460 (1994 est.)
Inflation rate (consumer prices): 3.3% per month (1994 average)
Unemployment rate: about 2% in 1994 (official estimate but large number of underemployed workers)
Budget:
revenues: $643 million
expenditures: $639 million, including capital expenditures of $NA (1993 est.)
Exports: $1.65 billion (f.o.b., 1994)
commodities: textile 14%, food products 11%, vehicles 11%, metals 11% (1993)
partners: Russia, Finland, Sweden, Germany
Imports: $1 billion (c.i.f., 1994)
commodities: machinery 18%, fuels 15%, vehicles 14%, textiles 10% (1993)

partners: Finland, Russia, Germany, Sweden
External debt: $650 million (end of 1991)
Industrial production: growth rate -27% (1993)
Electricity:
capacity: 3,420,000 kW
production: 11.3 billion kWh
consumption per capita: 6,528 kWh (1993)
Industries: oil shale, shipbuilding, phosphates, electric motors, excavators, cement, furniture, clothing, textiles, paper, shoes, apparel
Agriculture: accounts for 10% of GDP; employs 20% of work force; very efficient by Soviet standards; net exports of meat, fish, dairy products, and potatoes; imports of feedgrains for livestock; fruits and vegetables
Illicit drugs: transshipment point for illicit drugs from Central and Southwest Asia and Latin America to Western Europe; very limited illicit opium producer; mostly for domestic consumption
Economic aid:
recipient: US commitments, including Ex-Im (1992), $10 million
Currency: 1 Estonian kroon (EEK) = 100 cents (introduced in August 1992)
Exchange rates: kroons (EEK) per US$1—12.25 (January 1995); note—kroons are tied to the German Deutschmark at a fixed rate of 8 to 1
Fiscal year: calendar year

Transportation

Railroads:
total: 1,030 km common carrier lines only; does not include dedicated industrial lines
broad gauge: 1,030 km 1.520-m gauge (1990)
Highways:
total: 30,300 km
paved or graveled: 29,200 km
unpaved: earth 1,100 km (1990)
Inland waterways: 500 km perennially navigable
Pipelines: natural gas 420 km (1992)
Ports: Haapsalu, Narva, Novotallin, Paldiski, Parnu, Tallinn
Merchant marine:
total: 65 ships (1,000 GRT or over) totaling 415,332 GRT/532,749 DWT
ships by type: bulk 6, cargo 44, container 2, oil tanker 2, roll-on/roll-off cargo 7, short-sea passenger 4
Airports:
total: 22
with paved runways 2,438 to 3,047 m: 7
with paved runways 914 to 1,523 m: 3
with unpaved runways 2,438 to 3,047 m: 1
with unpaved runways 1,524 to 2,438 m: 2
with unpaved runways 914 to 1,523 m: 4
with unpaved runways under 914 m: 5

Communications

Telephone system: about 400,000 telephones; 246 telephones/1,000 persons; telephone system is antiquated; improvements are being made piecemeal, with emphasis on business needs and international connections; there are still about 150,000 unfulfilled requests for telephone service
local: NA
intercity: NA
international: international traffic is carried to the other former USSR republics by land line or microwave and to other countries partly by leased connection to the Moscow international gateway switch, and partly by a new Tallinn-Helsinki fiber optic submarine cable which gives Estonia access to international circuits everywhere; substantial investment has been made in cellular systems which are operational throughout Estonia and also Latvia and which have access to the international packet switched digital network via Helsinki
Radio:
broadcast stations: AM NA, FM NA, shortwave 0
radios: NA
Television:
broadcast stations: 3; note—provide Estonian programs as well as Moscow Ostenkino's first and second programs
televisions: NA

Defense Forces

Branches: Ground Forces, Navy, Air and Air Defense Force (not officially sanctioned), Maritime Border Guard, Volunteer Defense League (Kaitseliit), Security Forces (internal and border troops), Coast Guard
Manpower availability: males age 15-49 396,588; males fit for military service 311,838; males reach military age (18) annually 11,915 (1995 est.)
Defense expenditures: exchange rate conversion—$34.1 million, almost 5% of the overall State budget and 1.5% of GDP (1995)

Ethiopia

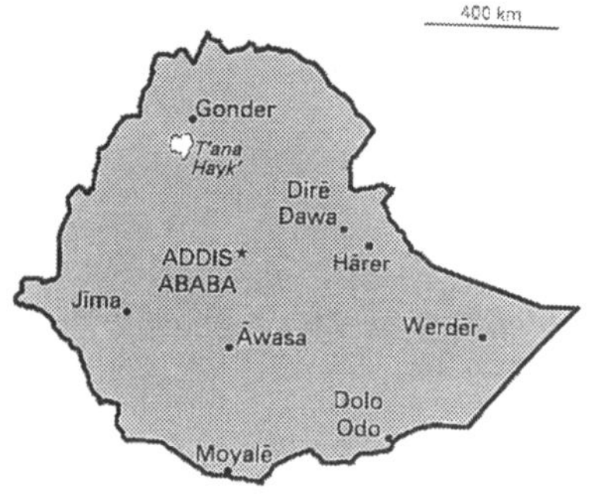

Geography

Location: Eastern Africa, west of Somalia
Map references: Africa
Area:
total area: 1,127,127 sq km
land area: 1,119,683 sq km
comparative area: slightly less than twice the size of Texas
Land boundaries: total 5,311 km, Djibouti 337 km, Eritrea 912 km, Kenya 830 km, Somalia 1,626 km, Sudan 1,606 km
Coastline: none—landlocked
Maritime claims: none; landlocked
International disputes: southern half of the boundary with Somalia is a Provisional Administrative Line; territorial dispute with Somalia over the Ogaden
Climate: tropical monsoon with wide topographic-induced variation
Terrain: high plateau with central mountain range divided by Great Rift Valley
Natural resources: small reserves of gold, platinum, copper, potash
Land use:
arable land: 12%
permanent crops: 1%
meadows and pastures: 41%
forest and woodland: 24%
other: 22%
Irrigated land: 1,620 sq km (1989 est.)
Environment:
current issues: deforestation; overgrazing; soil erosion; desertification; famine
natural hazards: geologically active Great Rift Valley susceptible to earthquakes, volcanic eruptions; frequent droughts
international agreements: party to—Biodiversity, Climate Change, Endangered Species, Ozone Layer Protection; signed, but not ratified—Desertification, Environmental Modification, Law of the Sea, Nuclear Test Ban
Note: landlocked—entire coastline along the Red Sea was lost with the de jure independence of Eritrea on 27 April 1993

People

Population: 55,979,018 (July 1995 est.)
note: Ethiopian demographic data, except population and population growth rate, include Eritrea
Age structure:
0-14 years: 46% (female 12,782,345; male 12,802,187)
15-64 years: 52% (female 14,352,059; male 14,511,342)
65 years and over: 2% (female 815,974; male 715,111) (July 1995 est.)
Population growth rate: 3.09% (1995 est.)
Birth rate: 46.68 births/1,000 population (1995 est.)
Death rate: 15.77 deaths/1,000 population (1995 est.)
Net migration rate: NA migrant(s)/1,000 population (1995 est.)
note: repatriation of Ethiopian refugees from Sudan, Kenya and Somalia, where they had taken refuge from war and famine in earlier years, is expected to continue in 1995; additional influxes of Sudanese and Somalis fleeing fighting in their countries can be expected in 1995
Infant mortality rate: 120.6 deaths/1,000 live births (1995 est.)
Life expectancy at birth:
total population: 50 years
male: 48.28 years
female: 51.78 years (1995 est.)
Total fertility rate: 7.07 children born/woman (1995 est.)
Nationality:
noun: Ethiopian(s)
adjective: Ethiopian
Ethnic divisions: Oromo 40%, Amhara and Tigrean 32%, Sidamo 9%, Shankella 6%, Somali 6%, Afar 4%, Gurage 2%, other 1%
Religions: Muslim 45%-50%, Ethiopian Orthodox 35%-40%, animist 12%, other 5%
Languages: Amharic (official), Tigrinya, Orominga, Guaraginga, Somali, Arabic, English (major foreign language taught in schools)
Literacy: age 10 and over can read and write (1984)
total population: 24%
male: 33%
female: 16%
Labor force: 18 million
by occupation: agriculture and animal husbandry 80%, government and services 12%, industry and construction 8% (1985)

Government

Names:
conventional long form: none
conventional short form: Ethiopia
local long form: none
local short form: Ityop'iya
Digraph: ET
Type: transitional government
note: on 28 May 1991 the Ethiopian People's Revolutionary Democratic Front (EPRDF) toppled the authoritarian government of MENGISTU Haile-Mariam and took control in Addis Ababa; a new constitution was promulgated in December 1994 and national and regional elections are scheduled for May 1995; the administrative regions will elect regional assemblies by popular vote; the National Assembly will have two chambers—one elected by popular vote and the other selected as representatives by the regional assemblies; the lower house of the National Assembly will select or confirm the president, the prime minister and the cabinet officers and judges; the prime minister will be the chief executive officer and the duties of the president will be mostly ceremonial
Capital: Addis Ababa
Administrative divisions: 14 ethnically-based administrative regions (astedader akababiwach, singular—astedader akababi) Addis Ababa, Afar, Amhara, Benishangul, Gambela, Gurage-Hadiya-Kambata, Hareri, Kefa, Omo, Oromo, Sidama, Somali, Tigray, Wolayta
note: the following named four administrative regions may have been abolished and their territories distributed among the remaining ten regions: Kefa, Omo, Sidama, and Wolayta
Independence: oldest independent country in Africa and one of the oldest in the world—at least 2,000 years
National holiday: National Day, 28 May (1991) (defeat of Mengistu regime)
Constitution: new constitution promulgated in December 1994
Legal system: NA
Suffrage: 18 years of age; universal
Executive branch:
chief of state: President MELES Zenawi (since 1 June 1991); appointed by the Council of Representatives following the military defeat of the MENGISTU government; following the elections to the National Assembly scheduled for May 1995 the lower house of the National Assembly will nominate a new president
head of government: Prime Minister TAMIRAT Layne (since 6 June 1991); a new prime minister will be designated by the party in power following the elections to the General Assembly in May 1995
cabinet: Council of Ministers; presently designated by the chairman of the Council of Representatives; under the new constitution and following the elections in May 1995 the cabinet officers will be selected by the prime minister
Legislative branch:
Constituent Assembly: elections were held on 5 June 1994; results—government parties swept almost all seats; in December 1994 the Constituent Assembly ratified the new constitution with few changes; the new constitution prescribes two chambers for the new National Assembly—one which is elected by popular vote and one which represents the ethnic interests of the regional governments
Judicial branch: Supreme Court
Political parties and leaders: Ethiopian People's Revolutionary Democratic Front (EPRDF), MELES Zenawi
Other political or pressure groups: Oromo Liberation Front (OLF); All Amhara People's Organization; Southern Ethiopia People's Democratic Coalition; numerous small, ethnic-based groups have formed since Mengistu's resignation, including several Islamic militant groups
Member of: ACP, AfDB, CCC, ECA, FAO, G-24, G-77, IAEA, IBRD, ICAO, ICRM, IDA, IFAD, IFC, IFRCS, IGADD, ILO, IMF, IMO, INTELSAT, INTERPOL, IOC, ISO, ITU, NAM, OAU, UN, UNCTAD, UNESCO, UNHCR, UNIDO, UNU, UPU, WFTU, WHO, WMO, WTO
Diplomatic representation in US:
chief of mission: Ambassador BERHANE Gebre-Christos
chancery: 2134 Kalorama Road NW, Washington, DC 20008
telephone: [1] (202) 234-2281, 2282
FAX: [1] (202) 328-7950
US diplomatic representation:
chief of mission: Ambassador Irvin HICKS
embassy: Entoto Street, Addis Ababa
mailing address: P. O. Box 1014, Addis Ababa
telephone: [251] (1) 550666
FAX: [251] (1) 552191
Flag: three equal horizontal bands of green (top), yellow, and red; Ethiopia is the oldest independent country in Africa, and the colors of her flag were so often adopted by other African countries upon independence that they became known as the pan-African colors

Economy

Overview: With the independence of Eritrea on 27 April 1993, Ethiopia continues to face difficult economic problems as one of the poorest and least developed countries in Africa. Its economy is based on agriculture, which accounts for about 45% of GDP, 90% of exports, and 80% of total employment; coffee generates 60% of export earnings. The agricultural sector suffers from frequent periods of drought, poor cultivation practices, and deterioration of internal security conditions. The manufacturing sector is heavily dependent on inputs from the agricultural sector. Over 90% of large-scale industry, but less than 10% of agriculture, is state run. The government is considering selling off a portion of state-owned plants, and is implementing reform measures that are gradually liberalizing the economy. A major medium-term problem is the improvement of roads, water supply, and other parts of an infrastructure badly neglected during years of civil strife.

Ethiopia *(continued)*

National product: GDP—purchasing power parity—$20.3 billion (1993 est.)
National product real growth rate: 3% (1994 est.)
National product per capita: $380 (1993 est.)
Inflation rate (consumer prices): 10% (FY93/94)
Unemployment rate: NA%
Budget:
revenues: $1.2 billion
expenditures: $1.7 billion, including capital expenditures of $707 million (FY93/94)
Exports: $219.8 million (f.o.b., 1993 est.)
commodities: coffee, leather products, gold
partners: Germany, Japan, Saudi Arabia, France, Italy
Imports: $1.04 billion (c.i.f., 1993 est.)
commodities: capital goods, consumer goods, fuel
partners: US, Germany, Italy, Saudi Arabia, Japan
External debt: $3.7 billion (1993 est.)
Industrial production: growth rate -3.3% (FY91/92); accounts for 12% of GDP
Electricity:
capacity: 460,000 kW
production: 1.3 billion kWh
consumption per capita: 23 kWh (1993)
Industries: food processing, beverages, textiles, chemicals, metals processing, cement
Agriculture: accounts for 45% of GDP; export crops of coffee and oilseeds are grown partly on state farms; estimated 50% of agricultural production is at subsistence level; principal crops and livestock—cereals, pulses, coffee, oilseeds, sugarcane, potatoes and other vegetables, hides and skins, cattle, sheep, goats
Illicit drugs: transit hub for heroin originating in Southwest and Southeast Asia and destined for Europe and North America as well as cocaine destined for southern African markets; cultivates qat (chat) for local use and regional export
Economic aid:
recipient: US commitments, including Ex-Im (FY70-89), $504 million; Western (non-US) countries, ODA and OOF bilateral commitments (1970-89), $3.4 billion; OPEC bilateral aid (1979-89), $8 million; Communist countries (1970-89), $2 billion
Currency: 1 birr (Br) = 100 cents
Exchange rates: birr (Br) per US$1—5.9500 (January 1995), 5.9500 (1994), 5.0000 (fixed rate 1992-93); fixed at 2.070 before 1992; note—official rate pegged to the US$
Fiscal year: 8 July—7 July

Transportation

Railroads:
total: 681 km (Ethiopian segment of the Addis Ababa-Djibouti railroad)
narrow gauge: 681 km 1.000-m gauge
Highways:
total: 24,127 km
paved: 3,289 km
unpaved: gravel 6,664 km; improved earth 1,652 km; unimproved earth 12,522 km (1993)
Ports: none
Merchant marine:
total: 12 ships (1,000 GRT or over) totaling 62,627 GRT/88,909 DWT
ships by type: cargo 8, livestock carrier 1, oil tanker 2, roll-on/roll-off cargo 1
Airports:
total: 98
with paved runways over 3,047 m: 2
with paved runways 2,438 to 3,047 m: 3
with paved runways 1,524 to 2,437 m: 2
with paved runways 914 to 1,523 m: 1
with paved runways under 914 m: 24
with unpaved runways over 3,047 m: 4
with unpaved runways 2,438 to 3,047 m: 6
with unpaved runways 1,524 to 2,438 m: 14
with unpaved runways 914 to 1,523 m: 42

Communications

Telephone system: NA telephones; open-wire and radio relay system adequate for government use
local: NA
intercity: open wire and microwave radio relay links
international: open-wire to Sudan and Djibouti; microwave radio relay to Kenya and Djibouti; 3 INTELSAT (1 Atlantic Ocean and 2 Pacific Ocean) earth stations
Radio:
broadcast stations: AM 4, FM 0, shortwave 0
radios: 9 million
Television:
broadcast stations: 1
televisions: 100,000

Defense Forces

Branches: Transitional Government of Ethiopia Forces, Air Force, Police
Manpower availability: males age 15-49 12,658,084; males fit for military service 6,569,759; males reach military age (18) annually 565,976 (1995 est.)
Defense expenditures: exchange rate conversion—$140 million, 4.1% of GDP (FY94/95)

Europa Island
(possession of France)

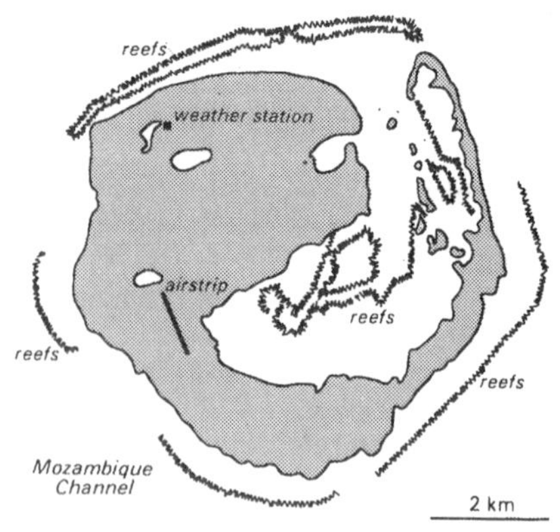

Geography

Location: Southern Africa, island in the Mozambique Channel, about one-half of the way from southern Mozambique to southern Madagascar
Map references: Africa
Area:
total area: 28 sq km
land area: 28 sq km
comparative area: about 0.2 times the size of Washington, DC
Land boundaries: 0 km
Coastline: 22.2 km
Maritime claims:
exclusive economic zone: 200 nm
territorial sea: 12 nm
International disputes: claimed by Madagascar
Climate: tropical
Terrain: NA
Natural resources: negligible
Land use:
arable land: NA%
permanent crops: NA%
meadows and pastures: NA%
forest and woodland: NA% (heavily wooded)
other: NA%
Irrigated land: 0 sq km
Environment:
current issues: NA
natural hazards: NA
international agreements: NA
Note: wildlife sanctuary

People

Population: uninhabited

Government

Names:
conventional long form: none
conventional short form: Europa Island
local long form: none
local short form: Ile Europa

Digraph: EU
Type: French possession administered by Commissioner of the Republic; resident in Reunion
Capital: none; administered by France from Reunion
Independence: none (possession of France)

Economy

Overview: no economic activity

Transportation

Ports: none; offshore anchorage only
Airports:
total: 1
with unpaved runways 914 to 1,523 m: 1

Communications

Note: 1 meteorological station

Defense Forces

Note: defense is the responsibility of France

Falkland Islands (Islas Malvinas)

(dependent territory of the UK)

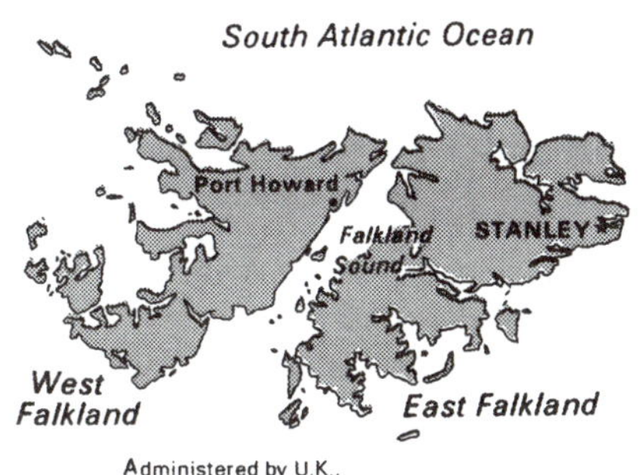

Geography

Location: Southern South America, islands in the South Atlantic Ocean, east of southern Argentina
Map references: South America
Area:
total area: 12,170 sq km
land area: 12,170 sq km
comparative area: slightly smaller than Connecticut
note: includes the two main islands of East and West Falkland and about 200 small islands
Land boundaries: 0 km
Coastline: 1,288 km
Maritime claims:
continental shelf: 200 nm
exclusive fishing zone: 200 nm
territorial sea: 12 nm
International disputes: administered by the UK, claimed by Argentina
Climate: cold marine; strong westerly winds, cloudy, humid; rain occurs on more than half of days in year; occasional snow all year, except in January and February, but does not accumulate
Terrain: rocky, hilly, mountainous with some boggy, undulating plains
Natural resources: fish, wildlife
Land use:
arable land: 0%
permanent crops: 0%
meadows and pastures: 99%
forest and woodland: 0%
other: 1%
Irrigated land: NA sq km
Environment:
current issues: NA
natural hazards: strong winds persist throughout the year
international agreements: NA
Note: deeply indented coast provides good natural harbors; short growing season

People

Population: 2,317 (July 1995 est.)
Age structure:
0-14 years: NA
15-64 years: NA
65 years and over: NA
Population growth rate: 2.43% (1995 est.)
Birth rate: NA
Death rate: NA
Net migration rate: NA
Infant mortality rate: NA
Life expectancy at birth: NA
Total fertility rate: NA
Nationality:
noun: Falkland Islander(s)
adjective: Falkland Island
Ethnic divisions: British
Religions: primarily Anglican, Roman Catholic, United Free Church, Evangelist Church, Jehovah's Witnesses, Lutheran, Seventh-Day Adventist
Languages: English
Labor force: 1,100 (est.)
by occupation: agriculture 95% (mostly sheepherding)

Government

Names:
conventional long form: Colony of the Falkland Islands
conventional short form: Falkland Islands (Islas Malvinas)
Digraph: FA
Type: dependent territory of the UK
Capital: Stanley
Administrative divisions: none (dependent territory of the UK)
Independence: none (dependent territory of the UK)
National holiday: Liberation Day, 14 June (1982)
Constitution: 3 October 1985
Legal system: English common law
Suffrage: 18 years of age; universal
Executive branch:
chief of state: Queen ELIZABETH II (since 6 February 1952)
head of government: Governor David Everard TATHAM (since August 1992)
cabinet: Executive Council; 3 members elected by the Legislative Council, 2 ex-officio members (chief executive and the financial secretary), and the governor
Legislative branch: unicameral
Legislative Council: elections last held 11 October 1989 (next to be held October 1994); results—percent of vote by party NA; seats—(10 total, 8 elected) independents 8
Judicial branch: Supreme Court
Political parties and leaders: NA
Member of: ICFTU
Diplomatic representation in US: none (dependent territory of the UK)
US diplomatic representation: none (dependent territory of the UK)
Flag: blue with the flag of the UK in the upper

Falkland Islands (Islas Malvinas)
(continued)

hoist-side quadrant and the Falkland Island coat of arms in a white disk centered on the outer half of the flag; the coat of arms contains a white ram (sheep raising is the major economic activity) above the sailing ship Desire (whose crew discovered the islands) with a scroll at the bottom bearing the motto DESIRE THE RIGHT

Economy

Overview: The economy was formerly based on agriculture, mainly sheep farming, which directly or indirectly employs most of the work force. Dairy farming supports domestic consumption; crops furnish winter fodder. Exports feature shipments of high-grade wool to the UK and the sale of postage stamps and coins. Rich stocks of fish in the surrounding waters are not presently exploited by the islanders. So far, efforts to establish a domestic fishing industry have been unsuccessful. The economy has diversified since 1987 when the government began selling fishing licenses to foreign trawlers operating within the Falklands exclusive fishing zone. These license fees total more than $40 million per year and support the island's health, education, and welfare system. To encourage tourism, the Falkland Islands Development Corporation has built three lodges for visitors attracted by the abundant wildlife and trout fishing. The islands are now self-financing except for defense. The British Geological Survey announced a 200-mile oil exploration zone around the islands in 1993 and early seismic surveys suggest substantial reserves capable of producing 500,000 barrels per day.
National product: GDP $NA
National product real growth rate: NA%
National product per capita: $NA
Inflation rate (consumer prices): 7.4% (1980-87 average)
Unemployment rate: NA%; labor shortage
Budget:
revenues: $65 million
expenditures: $55.2 million, including capital expenditures of $NA (1992-93)
Exports: at least $14.7 million
commodities: wool, hides and skins, and meat
partners: UK, Netherlands, Japan (1987 est.)
Imports: at least $13.9 million
commodities: food, clothing, timber, and machinery
partners: UK, Netherlands Antilles (Curacao), Japan (1987 est.)
External debt: $NA
Industrial production: growth rate NA%
Electricity:
capacity: 9,200 kW
production: 17 million kWh
consumption per capita: 7,253 kWh (1993)
Industries: wool and fish processing
Agriculture: predominantly sheep farming; small dairy herds; some fodder and vegetable crops
Economic aid:
recipient: Western (non-US) countries, ODA and OOF bilateral commitments (1992-93), $87 million
Currency: 1 Falkland pound (£F) = 100 pence
Exchange rates: Falkland pound (£F) per US$1—0.6350 (January 1995), 0.6529 (1994), 0.6658 (1993), 0.5664 (1992), 0.5652 (1991), 0.5604 (1990); note—the Falkland pound is at par with the British pound
Fiscal year: 1 April—31 March

Transportation

Railroads: 0 km
Highways:
total: 510 km
paved: 30 km
unpaved: gravel 80 km; unimproved earth 400 km
Ports: Stanley
Merchant marine: none
Airports:
total: 5
with paved runways 2,438 to 3,047 m: 1
with paved runways under 914 m: 4

Communications

Telephone system: 590 telephones
local: NA
intercity: government-operated radiotelephone and private VHF/CB radio networks provide effective service to almost all points on both islands
international: 1 INTELSAT (Atlantic Ocean) earth station with links through London to other countries
Radio:
broadcast stations: AM 2, FM 3, shortwave 0
radios: NA
Television:
broadcast stations: 0
televisions: NA

Defense Forces

Branches: British Forces Falkland Islands (includes Army, Royal Air Force, Royal Navy, and Royal Marines), Police Force
Defense expenditures: $NA, NA% of GDP
Note: defense is the responsibility of the UK

Faroe Islands
(part of the Danish realm)

Geography

Location: Northern Europe, island group between the Norwegian Sea and the north Atlantic Ocean, about one-half of the way from Iceland to Norway
Map references: Europe
Area:
total area: 1,400 sq km
land area: 1,400 sq km
comparative area: slightly less than eight times the size of Washington, DC
Land boundaries: 0 km
Coastline: 764 km
Maritime claims:
exclusive fishing zone: 200 nm
territorial sea: 3 nm
International disputes: none
Climate: mild winters, cool summers; usually overcast; foggy, windy
Terrain: rugged, rocky, some low peaks; cliffs along most of coast
Natural resources: fish
Land use:
arable land: 2%
permanent crops: 0%
meadows and pastures: 0%
forest and woodland: 0%
other: 98%
Irrigated land: NA sq km
Environment:
current issues: NA
natural hazards: NA
international agreements: NA
Note: archipelago of 18 inhabited islands and a few uninhabited islets; strategically located along important sea lanes in northeastern Atlantic; precipitous terrain limits habitation to small coastal lowlands

People

Population: 48,871 (July 1995 est.)
Age structure:
0-14 years: 24% (female 5,673; male 6,119)
15-64 years: 63% (female 14,164; male 16,835)

65 years and over: 13% (female 3,335; male 2,745) (July 1995 est.)
Population growth rate: 0.99% (1995 est.)
Birth rate: 17.54 births/1,000 population (1995 est.)
Death rate: 7.59 deaths/1,000 population (1995 est.)
Net migration rate: 0 migrant(s)/1,000 population (1995 est.)
Infant mortality rate: 7.9 deaths/1,000 live births (1995 est.)
Life expectancy at birth:
total population: 78.29 years
male: 74.91 years
female: 81.8 years (1995 est.)
Total fertility rate: 2.42 children born/woman (1995 est.)
Nationality:
noun: Faroese (singular and plural)
adjective: Faroese
Ethnic divisions: Scandinavian
Religions: Evangelical Lutheran
Languages: Faroese (derived from Old Norse), Danish
Literacy: NA%
Labor force: 17,585
by occupation: largely engaged in fishing, manufacturing, transportation, and commerce

Government

Names:
conventional long form: none
conventional short form: Faroe Islands
local long form: none
local short form: Foroyar
Digraph: FO
Type: part of the Danish realm; self-governing overseas administrative division of Denmark
Capital: Torshavn
Administrative divisions: none (self-governing overseas administrative division of Denmark)
Independence: none (part of the Danish realm; self-governing overseas administrative division of Denmark)
National holiday: Birthday of the Queen, 16 April (1940)
Constitution: 5 June 1953 (Danish constitution)
Legal system: Danish
Suffrage: 20 years of age; universal
Executive branch:
chief of state: Queen MARGRETHE II (since 14 January 1972), represented by High Commissioner Bent KLINTE (since NA)
head of government: Prime Minister Edmund JOENSEN (since 15 September 1994)
cabinet: Landsstyri; elected by the local legislature
Legislative branch: unicameral
Faroese Parliament (Logting): elections last held 8 July 1994 (next to be held by July 1998); results—percent of vote by party NA; seats—(32 total) Liberal Party 8, People's Party 6, Social Democrats 5, Republicans 4, Workers' Party 3, Christian Democrats 2, Center Party 2, Home Rule Party 2
Danish Parliament: elections last held on 21 September 1994 (next to be held by September 1998); results—percent of vote by party NA; seats—(2 total) Liberals 2
Judicial branch: none
Political parties and leaders: Social Democratic Party, Marita PETERSEN; Workers Front, Oli JACOBSEN; Home Rule Party, Helena Dam A NEYSTABOE; The 'Coalition Party', Edmund JOENSEN; Republican Party, Finnbogir ESAKSON; Centrist Party, Tordur NICLASEN; Christian People's Party, Niels Pauli DANIELSEN; People's Party, Arnfinn KALLSBERG; Liberal Party; Christian Democratic Party
Member of: none
Diplomatic representation in US: none (self-governing overseas administrative division of Denmark)
US diplomatic representation: none (self-governing overseas administrative division of Denmark)
Flag: white with a red cross outlined in blue that extends to the edges of the flag; the vertical part of the cross is shifted to the hoist side in the style of the DANNEBROG (Danish flag)

Economy

Overview: The Faroese, who have long enjoyed the affluent living standards of the Danes and other Scandinavians, now must cope with the decline of the all-important fishing industry and one of the world's heaviest per capita external debts of about $25,000. When the nations of the world extended their fishing zones to 200 nautical miles in the early 1970s, the Faroese no longer could continue their traditional long-distance fishing and subsequently depleted their own nearby fishing areas. The government's tight controls on fish stocks and its austerity measures have caused a recession, and subsidy cuts will force nationalization in the fishing industry, which has already been plagued with bankruptcies. Copenhagen has threatened to withhold its annual subsidy of $130 million—roughly one-third of the islands' budget revenues—unless the Faroese make significant efforts to balance their budget. To this extent the Faroe government is expected to continue its tough policies, including introducing a 20% value-added tax (VAT) in 1993, and has agreed to an IMF economic-political stabilization plan. In addition to its annual subsidy, the Danish government has bailed out the second largest Faroe bank to the tune of $140 million since October 1992.
National product: GDP—purchasing power parity—$662 million (1989 est.)
National product real growth rate: -10.8% (1993 est.)
National product per capita: $14,000 (1989 est.)
Inflation rate (consumer prices): 6.8% (1993 est.)
Unemployment rate: 23% (1993)
Budget:
revenues: $407.2 million
expenditures: $482.7 million, including capital expenditures of $NA (1993 est.)
Exports: $345.3 million (f.o.b., 1993 est.)
commodities: fish and fish products 88%, animal feedstuffs, transport equipment (ships) (1989)
partners: Denmark 20%, Germany 18.3%, UK 14.2%, France 11.2%, Spain 7.9%, US 4.5%
Imports: $234.4 million (c.i.f., 1993 est.)
commodities: machinery and transport equipment 24.4%, manufactures 24%, food and livestock 19%, fuels 12%, chemicals 6.5%
partners: Denmark 43.8%, Norway 19.8%, Sweden 4.9%, Germany 4.2%, US 1.3%
External debt: $1.2 billion (1993 est.)
Industrial production: growth rate NA%
Electricity:
capacity: 90,000 kW
production: 200 million kWh
consumption per capita: 3,953 kWh (1992)
Industries: fishing, shipbuilding, handicrafts
Agriculture: accounts for 27% of GDP; principal crops—potatoes and vegetables; livestock—sheep; annual fish catch about 360,000 metric tons
Economic aid:
recipient: receives an annual subsidy from Denmark of about $130 million
Currency: 1 Danish krone (DKr) = 100 oere
Exchange rates: Danish kroner (DKr) per US$1—6.034 (January 1995), 6.361 (1994), 6.484 (1993), 6.036 (1992), 6.396 (1991), 6.189 (1990)
Fiscal year: 1 April—31 March

Transportation

Railroads: 0 km
Highways:
total: 200 km
paved: NA
unpaved: NA
Ports: Klaksvick, Torshavn, Tvoroyri
Merchant marine:
total: 7 ships (1,000 GRT or over) totaling 19,879 GRT/18,444 DWT
ships by type: cargo 5, roll-on/roll-off cargo 1, short-sea passenger 1
Airports:
total: 1
with paved runways 914 to 1,523 m: 1

Communications

Telephone system: 27,900 telephones; good international communications; fair domestic facilities
local: NA

Faroe Islands *(continued)*

intercity: NA
international: 3 coaxial submarine cables
Radio:
broadcast stations: AM 1, FM 3 repeaters 10, shortwave 0
radios: NA
Television:
broadcast stations: 3 (repeaters 29)
televisions: NA

Defense Forces

Branches: no organized native military forces; only a small Police Force and Coast Guard are maintained
Defense expenditures: $NA, NA% of GDP
Note: defense is the responsibility of Denmark

Fiji

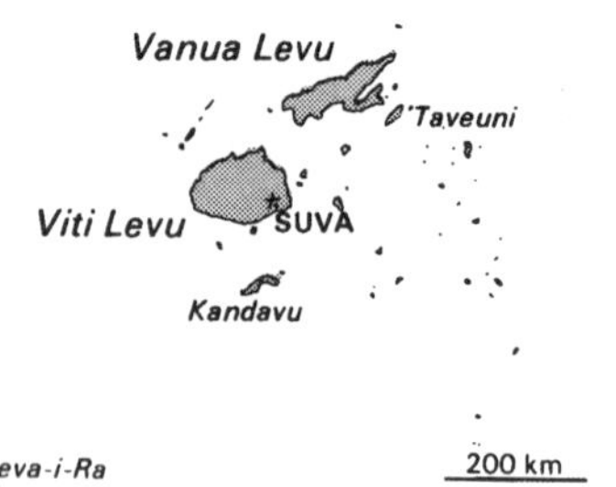

Geography

Location: Oceania, island group in the South Pacific Ocean, about two-thirds of the way from Hawaii to New Zealand
Map references: Oceania
Area:
total area: 18,270 sq km
land area: 18,270 sq km
comparative area: slightly smaller than New Jersey
Land boundaries: 0 km
Coastline: 1,129 km
Maritime claims: measured from claimed archipelagic baselines
continental shelf: 200-m depth or to the depth of exploitation; rectilinear shelf claim added
exclusive economic zone: 200 nm
territorial sea: 12 nm
International disputes: none
Climate: tropical marine; only slight seasonal temperature variation
Terrain: mostly mountains of volcanic origin
Natural resources: timber, fish, gold, copper, offshore oil potential
Land use:
arable land: 8%
permanent crops: 5%
meadows and pastures: 3%
forest and woodland: 65%
other: 19%
Irrigated land: 10 sq km (1989 est.)
Environment:
current issues: deforestation; soil erosion
natural hazards: cyclonic storms can occur from November to January
international agreements: party to—Biodiversity, Climate Change, Law of the Sea, Marine Life Conservation, Nuclear Test Ban, Ozone Layer Protection, Tropical Timber 94
Note: includes 332 islands of which approximately 110 are inhabited

People

Population: 772,891 (July 1995 est.)
Age structure:
0-14 years: 36% (female 136,570; male 142,581)
15-64 years: 61% (female 235,491; male 235,411)
65 years and over: 3% (female 11,943; male 10,895) (July 1995 est.)
Population growth rate: 1.16% (1995 est.)
Birth rate: 23.69 births/1,000 population (1995 est.)
Death rate: 6.42 deaths/1,000 population (1995 est.)
Net migration rate: -5.67 migrant(s)/1,000 population (1995 est.)
Infant mortality rate: 17.7 deaths/1,000 live births (1995 est.)
Life expectancy at birth:
total population: 65.42 years
male: 63.13 years
female: 67.82 years (1995 est.)
Total fertility rate: 2.87 children born/woman (1995 est.)
Nationality:
noun: Fijian(s)
adjective: Fijian
Ethnic divisions: Fijian 49%, Indian 46%, European, other Pacific Islanders, overseas Chinese, and other 5%
Religions: Christian 52% (Methodist 37%, Roman Catholic 9%), Hindu 38%, Muslim 8%, other 2%
note: Fijians are mainly Christian, Indians are Hindu, and there is a Muslim minority (1986)
Languages: English (official), Fijian, Hindustani
Literacy: age 15 and over can read and write (1986)
total population: 87%
male: 90%
female: 84%
Labor force: 235,000
by occupation: subsistence agriculture 67%, wage earners 18%, salary earners 15% (1987)

Government

Names:
conventional long form: Republic of Fiji
conventional short form: Fiji
Digraph: FJ
Type: republic
note: military coup leader Maj. Gen. Sitiveni RABUKA formally declared Fiji a republic on 6 October 1987
Capital: Suva
Administrative divisions: 4 divisions and 1 dependency*; Central, Eastern, Northern, Rotuma*, Western
Independence: 10 October 1970 (from UK)
National holiday: Independence Day, 10 October (1970)
Constitution: 10 October 1970 (suspended 1 October 1987); a new Constitution was proposed on 23 September 1988 and promulgated on 25 July 1990; the 1990

Constitution is under review; the review is scheduled to be complete by 1997
Legal system: based on British system
Suffrage: 21 years of age; universal
Executive branch:
chief of state: President Ratu Sir Kamisese MARA (since 12 January 1994); First Vice President Ratu Sir Josaia TAIVAIQIA (since 12 January 1994); Second Vice President Ratu Inoke TAKIVEIKATA (since 12 January 1994); note—President GANILAU died on 15 December 1993 and Vice President MARA became acting president; MARA was elected president by the Great Council of Chiefs on 12 January 1994
head of government: Prime Minister Sitiveni RABUKA (since 2 June 1992)
Presidential Council: appointed by the governor general
Great Council of Chiefs: highest ranking members of the traditional chiefly system
cabinet: Cabinet; appointed by prime minister from members of Parliament and responsible to Parliament
Legislative branch: the bicameral Parliament was dissolved following the coup of 14 May 1987
Senate: nonelective body containing 34 seats, 24 reserved for ethnic Fijians, 9 for Indians and others, 1 for the island of Rotuma; appointed by President
House of Representatives: elections last held 18-25 February 1994 (next to be held NA 1999); results—percent of vote by party NA; seats—(70 total, with ethnic Fijians allocated 37 seats, ethnic Indians 27 seats, and independents and other 6 seats) number of seats by party SVT 31, NFP 20, FLP 7, FA 5, GVP 4, independents 2, ANC 1
Judicial branch: Supreme Court
Political parties and leaders: Fijian Political Party (SVT—primarily Fijian), leader Maj. Gen. Sitivini RABUKA; National Federation Party (NFP; primarily Indian), Jai Ram REDDY; Fijian Nationalist Party (FNP), Sakeasi BUTADROKA; Fiji Labor Party (FLP), Mahendra CHAUDHRY; General Voters Party (GVP), Bill SORBY; Fiji Conservative Party (FCP), Isireli VUIBAU; Conservative Party of Fiji (CPF), Jolale ULUDOLE and Viliame SAVU; Fiji Indian Liberal Party, Swami MAHARAJ; Fiji Indian Congress Party, Ishwari BAJPAI; Fiji Independent Labor (Muslim), leader NA; Four Corners Party, David TULVANUAVOU; Fijian Association (FA), leader NA; General Electors' Association, leader NA
note: in early 1995, ethnic Fijian members of the All National Congress (ANC) merged with the Fijian Association (FA), the new FA is scheduled to hold its first meeting in April 1995 at which time the leaders of the party will be chosen; it is likely that Josevata KAMIKAMICA, the leader of the FA before the merger, will be elected leader and Adi Kuini Bavadra SPEED, the leader of the ANC before the merger, will be elected deputy leader; the remaining members of the ANC have renamed their party the General Electors' Association
Member of: ACP, AsDB, CP, ESCAP, FAO, G-77, GATT, IBRD, ICAO, ICFTU, ICRM, IDA, IFAD, IFC, IFRCS, ILO, IMF, IMO, INTELSAT, INTERPOL, IOC, ITU, PCA, SPARTECA, SPC, SPF, UN, UNAMIR, UNCTAD, UNESCO, UNIDO, UNIFIL, UNIKOM, UPU, WFTU, WHO, WIPO, WMO
Diplomatic representation in US:
chief of mission: Ambassador Pita Kewa NACUVA
chancery: Suite 240, 2233 Wisconsin Avenue NW, Washington, DC 20007
telephone: [1] (202) 337-8320
FAX: [1] (202) 337-1996
consulate(s): New York
US diplomatic representation:
chief of mission: (vacant); Charge d'Affaires Michael W. MARINE
embassy: 31 Loftus Street, Suva
mailing address: P. O. Box 218, Suva
telephone: [679] 314466
FAX: [679] 300081
Flag: light blue with the flag of the UK in the upper hoist-side quadrant and the Fijian shield centered on the outer half of the flag; the shield depicts a yellow lion above a white field quartered by the cross of Saint George featuring stalks of sugarcane, a palm tree, bananas, and a white dove

Economy

Overview: Fiji's economy is primarily agricultural, with a large subsistence sector. Sugar exports and tourism are the major sources of foreign exchange. Industry contributes 13% to GDP, with sugar processing accounting for one-third of industrial activity. Roughly 250,000 tourists visit each year. Political uncertainty and drought, however, contribute to substantial fluctuations in earnings from tourism and sugar and to the emigration of skilled workers. In 1992, growth was approximately 3%, based on growth in tourism and a lessening of labor-management disputes in the sugar and gold-mining sectors. In 1993, the government's budgeted growth rate of 3% was not achieved because of a decline in non-sugar agricultural output and damage from Cyclone Kina. Growth in 1994 is estimated to be 5%, largely attributed to increased tourism and expansion in domestic production, particularly in the manufacturing sector.
National product: GDP—purchasing power parity—$4.3 billion (1994 est.)
National product real growth rate: 5% (1994 est.)
National product per capita: $5,650 (1994 est.)
Inflation rate (consumer prices): 1.5% (1994)
Unemployment rate: 5.4% (1992)
Budget:
revenues: $485 million
expenditures: $579 million, including capital expenditures of $58 million (1994)
Exports: $405 million (f.o.b., 1993)
commodities: sugar 40%, clothing, gold, processed fish, lumber
partners: EC 26%, Australia 15%, Pacific Islands 11%, Japan 6%
Imports: $634 million (c.i.f., 1993)
commodities: machinery and transport equipment, petroleum products, food, consumer goods, chemicals
partners: Australia 30%, NZ 17%, Japan 13%, EC 6%, US 6%
External debt: $670 million (1994 est.)
Industrial production: growth rate 0% (1993 est.); accounts for 13% of GDP
Electricity:
capacity: 200,000 kW
production: 480 million kWh
consumption per capita: 581 kWh (1993)
Industries: sugar, tourism, copra, gold, silver, clothing, lumber, small cottage industries
Agriculture: accounts for 23% of GDP; principal cash crop is sugarcane; coconuts, cassava, rice, sweet potatoes, bananas; small livestock sector includes cattle, pigs, horses, and goats; fish catch nearly 33,000 tons (1989)
Economic aid:
recipient: Western (non-US) countries, ODA and OOF bilateral commitments (1980-89), $815 million
Currency: 1 Fijian dollar (F$) = 100 cents
Exchange rates: Fijian dollars (F$) per US$1—1.4140 (January 1995), 1.4641 (1994), 1.5418 (1993), 1.5030 (1992), 1.4756 (1991), 1.4809 (1990)
Fiscal year: calendar year

Transportation

Railroads:
total: 644 km; note—belongs to the government owned Fiji Sugar Corporation
narrow gauge: 644 km 0.610-m gauge
Highways:
total: 3,300 km
paved: 1,590 km
unpaved: gravel, crushed stone, stabilized earth 1,290 km; unimproved earth 420 km (1984)
Inland waterways: 203 km; 122 km navigable by motorized craft and 200-metric-ton barges
Ports: Labasa, Lautoka, Levuka, Savusavu, Suva
Merchant marine:
total: 5 ships (1,000 GRT or over) totaling 16,267 GRT/17,884 DWT
ships by type: chemical tanker 2, oil tanker 1, roll-on/roll-off cargo 2
Airports:
total: 23
with paved runways over 3,047 m: 1

with paved runways 1,524 to 2,437 m: 1
with paved runways 914 to 1,523 m: 1
with paved runways under 914 m: 16
with unpaved runways 914 to 1,523 m: 4

Communications

Telephone system: 53,228 telephones; 71 telephones/1,000 persons; modern local, interisland, and international (wire/radio integrated) public and special-purpose telephone, telegraph, and teleprinter facilities; regional radio center
local: NA
intercity: NA
international: important COMPAC cable link between US-Canada and NZ-Australia; 1 INTELSAT (Pacific Ocean) earth station
Radio:
broadcast stations: AM 7, FM 1, shortwave 0
radios: NA
Television:
broadcast stations: 0
televisions: NA

Defense Forces

Branches: Republic of Fiji Military Forces (RFMF; includes army, navy, and air elements)
Manpower availability: males age 15-49 201,441; males fit for military service 111,046; males reach military age (18) annually 8,466 (1995 est.)
Defense expenditures: exchange rate conversion—$22.4 million, about 2% of GDP (FY91/92)

Finland

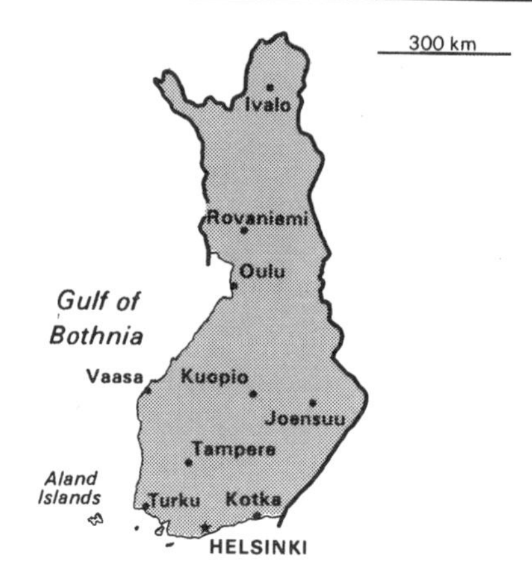

Geography

Location: Northern Europe, bordering the Baltic Sea, Gulf of Bothnia, and Gulf of Finland, between Sweden and Russia
Map references: Europe
Area:
total area: 337,030 sq km
land area: 305,470 sq km
comparative area: slightly smaller than Montana
Land boundaries: total 2,628 km, Norway 729 km, Sweden 586 km, Russia 1,313 km
Coastline: 1,126 km (excludes islands and coastal indentations)
Maritime claims:
contiguous zone: 6 nm
continental shelf: 200-m depth or to the depth of exploitation
exclusive fishing zone: 12 nm
territorial sea: 4 nm
International disputes: none
Climate: cold temperate; potentially subarctic, but comparatively mild because of moderating influence of the North Atlantic Current, Baltic Sea, and more than 60,000 lakes
Terrain: mostly low, flat to rolling plains interspersed with lakes and low hills
Natural resources: timber, copper, zinc, iron ore, silver
Land use:
arable land: 8%
permanent crops: 0%
meadows and pastures: 0%
forest and woodland: 76%
other: 16%
Irrigated land: 620 sq km (1989 est.)
Environment:
current issues: air pollution from manufacturing and power plants contributing to acid rain; water pollution from industrial wastes, agricultural chemicals; habitat loss threatens wildlife populations
natural hazards: NA
international agreements: party to—Air Pollution, Air Pollution-Nitrogen Oxides, Air Pollution-Sulphur 85, Air Pollution-Volatile Organic Compounds, Antarctic Treaty, Biodiversity, Climate Change, Endangered Species, Environmental Modification, Hazardous Wastes, Marine Dumping, Marine Life Conservation, Nuclear Test Ban, Ozone Layer Protection, Ship Pollution, Tropical Timber 83, Wetlands, Whaling; signed, but not ratified—Air Pollution-Sulphur 94, Antarctic-Environmental Protocol, Desertification, Law of the Sea
Note: long boundary with Russia; Helsinki is northernmost national capital on European continent; population concentrated on small southwestern coastal plain

People

Population: 5,085,206 (July 1995 est.)
Age structure:
0-14 years: 19% (female 469,666; male 491,484)
15-64 years: 67% (female 1,683,371; male 1,716,307)
65 years and over: 14% (female 457,061; male 267,317) (July 1995 est.)
Population growth rate: 0.3% (1995 est.)
Birth rate: 12.22 births/1,000 population (1995 est.)
Death rate: 9.77 deaths/1,000 population (1995 est.)
Net migration rate: 0.59 migrant(s)/1,000 population (1995 est.)
Infant mortality rate: 5.2 deaths/1,000 live births (1995 est.)
Life expectancy at birth:
total population: 76.22 years
male: 72.51 years
female: 80.11 years (1995 est.)
Total fertility rate: 1.79 children born/woman (1995 est.)
Nationality:
noun: Finn(s)
adjective: Finnish
Ethnic divisions: Finn, Swede, Lapp, Gypsy, Tatar
Religions: Evangelical Lutheran 89%, Greek Orthodox 1%, none 9%, other 1%
Languages: Finnish 93.5% (official), Swedish 6.3% (official), small Lapp- and Russian-speaking minorities
Literacy: age 15 and over can read and write (1980 est.)
total population: 100%
Labor force: 2.533 million
by occupation: public services 30.4%, industry 20.9%, commerce 15.0%, finance, insurance, and business services 10.2%, agriculture and forestry 8.6%, transport and communications 7.7%, construction 7.2%

Government

Names:
conventional long form: Republic of Finland

conventional short form: Finland
local long form: Suomen Tasavalta
local short form: Suomi
Digraph: FI
Type: republic
Capital: Helsinki
Administrative divisions: 12 provinces (laanit, singular—laani); Ahvenanmaa, Hame, Keski-Suomi, Kuopio, Kymi, Lappi, Mikkeli, Oulu, Pohjois-Karjala, Turku ja Pori, Uusimaa, Vaasa
Independence: 6 December 1917 (from Soviet Union)
National holiday: Independence Day, 6 December (1917)
Constitution: 17 July 1919
Legal system: civil law system based on Swedish law; Supreme Court may request legislation interpreting or modifying laws; accepts compulsory ICJ jurisdiction, with reservations
Suffrage: 18 years of age; universal
Executive branch:
chief of state: President Martti AHTISAARI (since 1 March 1994); election last held 31 January-6 February 1994 (next to be held January 2000); results—Martti AHTISAARI 54%, Elisabeth REHN 46%
head of government: Prime Minister Paavo LIPPONEN (since 13 April 1995); Deputy Prime Minister Sauli NIINISTO (since 13 April 1995)
cabinet: Council of State (Valtioneuvosto); appointed by the president, responsible to Parliament
Legislative branch: unicameral
Parliament (Eduskunta): elections last held 19 March 1995 (next to be held March 1999); results—Social Democratic Party 28.3%, Center Party 19.9%, National Coalition (Conservative) Party 17.9%, Leftist Alliance (Communist) 11.2%, Swedish People's Party 5.1%, Green League 6.5%, Ecology Party 0.3%, Rural 1.3%, Finnish Christian League 3.0%, Liberal People's Party 0.6%, Young Finns 2.8%; seats—(200 total) Social Democratic Party 63, Center Party 44, National Coalition (Conservative) Party 39, Leftist Alliance (Communist) 22, Swedish People's Party 11, Green League 9, Ecology Party 1, Rural 1, Finnish Christian League 7, Young Finns 2, Aaland Islands 1
Judicial branch: Supreme Court (Korkein Oikeus)
Political parties and leaders:
government coalition: Social Democratic Party, Paavo LIPPONEN; National Coalition (conservative) Party, Sauli NIINISTO; Leftist Alliance (Communist) People's Democratic League and Democratic Alternative, Claes ANDERSON; Swedish People's Party, (Johan) Ole NORRBACK; Green League, Pekka HAAVISTO
other: Center Party, Esko AHO; Finnish Christian League, Toimi KANKAANNIEMI; Rural Party, Tina MAKELA; Liberal People's Party, Tuulikki UKKOLA; Greens Ecological Party (EPV); Young Finns
Other political or pressure groups: Finnish Communist Party-Unity, Yrjo HAKANEN; Constitutional Rightist Party; Finnish Pensioners Party; Communist Workers Party, Timo LAHDENMAKI
Member of: AfDB, AG (observer), AsDB, Australia Group, BIS, CBSS, CCC, CE, CERN, EBRD, ECE, EFTA, ESA (associate), EU, FAO, G- 9, GATT, IADB, IAEA, IBRD, ICAO, ICC, ICFTU, ICRM, IDA, IEA, IFAD, IFC, IFRCS, ILO, IMF, IMO, INMARSAT, INTELSAT, INTERPOL, IOC, IOM, ISO, ITU, MTCR, NACC (observer), NAM (guest), NC, NEA, NIB, NSG, OAS (observer), OECD, OSCE, PCA, PFP, UN, UNCTAD, UNDOF, UNESCO, UNFICYP, UNHCR, UNIDO, UNIFIL, UNIKOM, UNMOGIP, UNPROFOR, UNTSO, UPU, WFTU, WHO, WIPO, WMO, WTO, ZC
Diplomatic representation in US:
chief of mission: Ambassador Jukka VALTASAARI
chancery: 3301 Massachusetts Avenue NW, Washington, DC 20008
telephone: [1] (202) 298-5800
FAX: [1] (202) 298-6030
consulate(s) general: Los Angeles and New York
US diplomatic representation:
chief of mission: Ambassador Derek N. SHEARER
embassy: Itainen Puistotie 14A, FIN-00140, Helsinki
mailing address: APO AE 09723
telephone: [358] (0) 171931
FAX: [358] (0) 174681
Flag: white with a blue cross that extends to the edges of the flag; the vertical part of the cross is shifted to the hoist side in the style of the DANNEBROG (Danish flag)

Economy

Overview: Finland has a highly industrialized, largely free market economy, with per capita output two-thirds of the US figure. Its key economic sector is manufacturing—principally the wood, metals, and engineering industries. Trade is important, with the export of goods representing about 30% of GDP. Except for timber and several minerals, Finland depends on imports of raw materials, energy, and some components for manufactured goods. Because of the climate, agricultural development is limited to maintaining self-sufficiency in basic products. Forestry, an important export earner, provides a secondary occupation for the rural population. The economy, which experienced an average of 4.9% annual growth between 1987 and 1989, sank into deep recession in 1991 as GDP contracted by 6.5%. The recession—which continued in 1992 with GDP contracting by 4.1%—has been caused by economic overheating, depressed foreign markets, and the dismantling of the barter system between Finland and the former Soviet Union under which Soviet oil and gas had been exchanged for Finnish manufactured goods. The Finnish Government has proposed efforts to increase industrial competitiveness and efficiency by an increase in exports to Western markets, cuts in public expenditures, partial privatization of state enterprises, and changes in monetary policy. In June 1991 Helsinki had tied the markka to the European Union's (EU) European Currency Unit (ECU) to promote stability. Ongoing speculation resulting from a lack of confidence in the government's policies forced Helsinki to devalue the markka by about 12% in November 1991 and to indefinitely break the link in September 1992. The devaluations have boosted the competitiveness of Finnish exports. The recession bottomed out in 1993, and Finland participated in the general European upturn of 1994. Unemployment probably will remain a serious problem during the next few years; the majority of Finnish firms face a weak domestic market and the troubled German and Swedish export markets. The Finns voted in an October 1994 referendum to enter the EU, and Finland officially joined the Union on 1 January 1995. Increasing integration with Western Europe will dominate the economic picture over the next few years.
National product: GDP—purchasing power parity—$81.8 billion (1994 est.)
National product real growth rate: 3.5% (1994 est.)
National product per capita: $16,140 (1994 est.)
Inflation rate (consumer prices): 2.1% (1992)
Unemployment rate: 22% (1993)
Budget:
revenues: $21.7 billion
expenditures: $31.7 billion, including capital expenditures of $NA (1993 est.)
Exports: $23.4 billion (f.o.b., 1993)
commodities: paper and pulp, machinery, chemicals, metals, timber
partners: EC 53.2% (Germany 15.6%, UK 10.7%), EFTA 19.5% (Sweden 12.8%), US 5.9%, Japan 1.3%, Russia 2.8% (1992)
Imports: $18 billion (c.i.f., 1993)
commodities: foodstuffs, petroleum and petroleum products, chemicals, transport equipment, iron and steel, machinery, textile yarn and fabrics, fodder grains
partners: EC 47.2% (Germany 16.9%, UK 8.7%), EFTA 19.0% (Sweden 11.7%), US 6.1%, Japan 5.5%, Russia 7.1% (1992)
External debt: $30 billion (December 1993)
Industrial production: growth rate 5% (1993 est.); accounts for 28% of GDP
Electricity:
capacity: 13,360,000 kW
production: 58 billion kWh

consumption per capita: 12,196 kWh (1993)
Industries: metal products, shipbuilding, forestry and wood processing (pulp, paper), copper refining, foodstuffs, chemicals, textiles, clothing
Agriculture: accounts for 7% of GDP (including forestry); livestock production, especially dairy cattle, predominates; main crops—cereals, sugar beets, potatoes; 85% self-sufficient, but short of foodgrains and fodder grains; annual fish catch about 160,000 metric tons
Illicit drugs: transshipment point for Latin American cocaine for the West European market
Economic aid:
donor: ODA and OOF commitments (1970-89), $2.7 billion
Currency: 1 markka (FMk) or Finmark = 100 pennia
Exchange rates: markkaa (FMk) per US$1—4.7358 (January 1995), 5.2235 (1994), 5.7123 (1993), 4.4794 (1992), 4.0440 (1991), 3.8235 (1990)
Fiscal year: calendar year

Transportation

Railroads
total: 5,864 km
broad gauge: 5,864 km 1.524-m gauge (1,710 km electrified; 480 km multiple track)
Highways:
total: 76,755 km
paved: bituminous concrete, bituminous treated soil 47,588 km (318 km of expressways)
unpaved: gravel 29,167 km (1992)
Inland waterways: 6,675 km total (including Saimaa Canal); 3,700 km suitable for steamers
Pipelines: natural gas 580 km
Ports: Hamina, Helsinki, Kokkola, Kotka, Loviisa, Oulu, Pori, Rauma, Turku, Uusikaupunki, Varkaus
Merchant marine:
total: 93 ships (1,000 GRT or over) totaling 1,050,270 GRT/1,080,150 DWT
ships by type: bulk 7, cargo 20, chemical tanker 5, liquefied gas tanker 3, oil tanker 12, passenger 3, refrigerated cargo 1, roll-on/roll-off cargo 31, short-sea passenger 10, vehicle carrier 1
Airports:
total: 159
with paved runways over 3,047 m: 3
with paved runways 2,438 to 3,047 m: 23
with paved runways 1,524 to 2,437 m: 13
with paved runways 914 to 1,523 m: 21
with paved runways under 914 m: 94
with unpaved runways 914 to 1,523 m: 5

Communications

Telephone system: 3,140,000 telephones; good service from cable and microwave radio relay network
local: NA
intercity: cable and microwave radio relay
international: 1 submarine cable; INTELSAT satellite transmission service via Swedish earth station and a receive-only INTELSAT earth station near Helsinki for TV programs
Radio:
broadcast stations: AM 6, FM 105, shortwave 0
radios: NA
Television:
broadcast stations: 235
televisions: NA

Defense Forces

Branches: Army, Navy, Air Force, Frontier Guard (includes Sea Guard)
Manpower availability: males age 15-49 1,318,231; males fit for military service 1,083,749; males reach military age (17) annually 33,085 (1995 est.)
Defense expenditures: exchange rate conversion—$1.86 billion, about 1.9% of GDP (1994)

France

Geography

Location: Western Europe, bordering the Bay of Biscay and English Channel, between Belgium and Spain southeast of the UK; bordering the Mediterranean Sea, between Italy and Spain
Map references: Europe
Area:
total area: 547,030 sq km
land area: 545,630 sq km
comparative area: slightly more than twice the size of Colorado
note: includes Corsica and the rest of metropolitan France, but excludes the overseas administrative divisions
Land boundaries: total 2,892.4 km, Andorra 60 km, Belgium 620 km, Germany 451 km, Italy 488 km, Luxembourg 73 km, Monaco 4.4 km, Spain 623 km, Switzerland 573 km
Coastline: 3,427 km (mainland 2,783 km, Corsica 644 km)
Maritime claims:
contiguous zone: 24 nm
continental shelf: 200-m depth or to the depth of exploitation
exclusive economic zone: 200 nm
territorial sea: 12 nm
International disputes: Madagascar claims Bassas da India, Europa Island, Glorioso Islands, Juan de Nova Island, and Tromelin Island; Comoros claims Mayotte; Mauritius claims Tromelin Island; Seychelles claims Tromelin Island; Suriname claims part of French Guiana; Mexico claims Clipperton Island; territorial claim in Antarctica (Adelie Land); Saint Pierre and Miquelon is focus of maritime boundary dispute between Canada and France
Climate: generally cool winters and mild summers, but mild winters and hot summers along the Mediterranean
Terrain: mostly flat plains or gently rolling hills in north and west; remainder is mountainous, especially Pyrenees in south, Alps in east
Natural resources: coal, iron ore, bauxite,

fish, timber, zinc, potash
Land use:
arable land: 32%
permanent crops: 2%
meadows and pastures: 23%
forest and woodland: 27%
other: 16%
Irrigated land: 11,600 sq km (1989 est.)
Environment:
current issues: some forest damage from acid rain; air pollution from industrial and vehicle emissions; water pollution from urban wastes, agricultural runoff
natural hazards: flooding
international agreements: party to—Air Pollution, Air Pollution-Nitrogen Oxides, Air Pollution-Sulphur 85, Antarctic-Environmental Protocol, Antarctic Treaty, Biodiversity, Climate Change, Endangered Species, Hazardous Wastes, Marine Dumping, Marine Life Conservation, Ozone Layer Protection, Ship Pollution, Tropical Timber 83, Wetlands, Whaling; signed, but not ratified—Air Pollution-Sulphur 94, Air Pollution-Volatile Organic Compounds, Desertification, Law of the Sea
Note: largest West European nation; occasional warm tropical wind known as mistral

People

Population: 58,109,160 (July 1995 est.)
Age structure:
0-14 years: 19% (female 5,438,447; male 5,700,143)
15-64 years: 65% (female 18,889,771; male 19,001,536)
65 years and over: 16% (female 5,433,276; male 3,645,987) (July 1995 est.)
Population growth rate: 0.46% (1995 est.)
Birth rate: 13 births/1,000 population (1995 est.)
Death rate: 9.29 deaths/1,000 population (1995 est.)
Net migration rate: 0.86 migrant(s)/1,000 population (1995 est.)
Infant mortality rate: 6.5 deaths/1,000 live births (1995 est.)
Life expectancy at birth:
total population: 78.37 years
male: 74.5 years
female: 82.44 years (1995 est.)
Total fertility rate: 1.8 children born/woman (1995 est.)
Nationality:
noun: Frenchman(men), Frenchwoman(women)
adjective: French
Ethnic divisions: Celtic and Latin with Teutonic, Slavic, North African, Indochinese, Basque minorities
Religions: Roman Catholic 90%, Protestant 2%, Jewish 1%, Muslim (North African workers) 1%, unaffiliated 6%
Languages: French 100%, rapidly declining regional dialects and languages (Provencal, Breton, Alsatian, Corsican, Catalan, Basque, Flemish)
Literacy: age 15 and over can read and write (1991 est.)
total population: 99%
Labor force: 24.17 million
by occupation: services 61.5%, industry 31.3%, agriculture 7.2% (1987)

Government

Names:
conventional long form: French Republic
conventional short form: France
local long form: Republique Francaise
local short form: France
Digraph: FR
Type: republic
Capital: Paris
Administrative divisions: 22 regions (regions, singular—region); Alsace, Aquitaine, Auvergne, Basse-Normandie, Bourgogne, Bretagne, Centre, Champagne-Ardenne, Corse, Franche-Comte, Haute-Normandie, Ile-de-France, Languedoc-Roussillon, Limousin, Lorraine, Midi-Pyrenees, Nord-Pas-de-Calais, Pays de la Loire, Picardie, Poitou-Charentes, Provence-Alpes-Cote d'Azur, Rhone-Alpes
note: the 22 regions are subdivided into 96 departments; see separate entries for the overseas departments (French Guiana, Guadeloupe, Martinique, Reunion) and the territorial collectivities (Mayotte, Saint Pierre and Miquelon)
Dependent areas: Bassas da India, Clipperton Island, Europa Island, French Polynesia, French Southern and Antarctic Lands, Glorioso Islands, Juan de Nova Island, New Caledonia, Tromelin Island, Wallis and Futuna
note: the US does not recognize claims to Antarctica
Independence: 486 (unified by Clovis)
National holiday: National Day, Taking of the Bastille, 14 July (1789)
Constitution: 28 September 1958, amended concerning election of president in 1962, amended to comply with provisions of EC Maastricht Treaty in 1992; amended to tighten immigration laws 1993
Legal system: civil law system with indigenous concepts; review of administrative but not legislative acts
Suffrage: 18 years of age, universal
Executive branch:
chief of state: President Francois MITTERRAND (since 21 May 1981); election last held 8 May 1988 (next to be held by May 1995); results—Second Ballot Francois MITTERRAND 54%, Jacques CHIRAC 46%
head of government: Prime Minister Edouard BALLADUR (since 29 March 1993)
cabinet: Council of Ministers; appointed by the president on the suggestion of the prime minister
Legislative branch: bicameral Parliament (Parlement)
Senate (Senat): elections last held 27 September 1992 (next to be held September 1995; nine-year term, elected by thirds every three years); results—percent of vote by party NA; seats—(321 total; 296 metropolitan France, 13 for overseas departments and territories, and 12 for French nationals abroad) RPR 91, UDF 142, PS 66, PCF 16, independents 2, other 4
National Assembly (Assemblee Nationale): elections last held 21 and 28 March 1993 (next to be held NA 1998); results—percent of vote by party NA; seats—(577 total) RPR 247, UDF 213, PS 67, PCF 24, independents 26
Judicial branch: Constitutional Court (Cour Constitutionnelle)
Political parties and leaders: Rally for the Republic (RPR), Alain JUPPE, interim head; Union for French Democracy (UDF, coalition of PR, CDS, RAD, PSD), Valery Giscard d'ESTAING; Republican Party (PR), Gerard LONGUET; Center for Social Democrats (CDS), Francois BAYROU; Radical (RAD), Yves GALLAND; Socialist Party (PS), Henri EMMANUELLI; Left Radical Movement (MRG), Jean-Francois HORY; Communist Party (PCF), Robert HUE; National Front (FN), Jean-Marie LE PEN; The Greens, Antoine WAECHTER, Jean-Louis VIDAL, Guy CAMBOT; Generation Ecology (GE), Brice LALONDE
Other political or pressure groups: Communist-controlled labor union (Confederation Generale du Travail—CGT) nearly 2.4 million members (claimed); Socialist-leaning labor union (Confederation Francaise Democratique du Travail or CFDT) about 800,000 members (est.); independent labor union (Force Ouvriere) 1 million members (est.); independent white-collar union (Confederation Generale des Cadres) 340,000 members (claimed); National Council of French Employers (Conseil National du Patronat Francais—CNPF or Patronat)
Member of: ACCT, AfDB, AG (observer), AsDB, Australia Group, BDEAC, BIS, CCC, CDB (non-regional), CE, CERN, EBRD, EC, ECA (associate), ECE, ECLAC, EIB, ESA, ESCAP, FAO, FZ, G- 5, G- 7, G-10, GATT, IADB, IAEA, IBRD, ICAO, ICC, ICFTU, ICRM, IDA, IEA, IFAD, IFC, IFRCS, ILO, IMF, IMO, INMARSAT, INTELSAT, INTERPOL, IOC, IOM, ISO, ITU, MINURSO, MTCR, NACC, NATO, NEA, NSG, OAS (observer), OECD, ONUSAL, OSCE, PCA, SPC, UN, UN Security Council, UNCTAD, UNESCO, UNHCR, UNIDO, UNIFIL, UNIKOM, UNITAR, UNMIH, UNPROFOR, UNRWA, UNTSO, UNU, UPU, WCL, WEU, WFTU, WHO, WIPO, WMO, WTO, ZC

Diplomatic representation in US:
chief of mission: Ambassador Jacques ANDREANI
chancery: 4101 Reservoir Road NW, Washington, DC 20007
telephone: [1] (202) 944-6000
consulate(s) general: Atlanta, Boston, Chicago, Honolulu, Houston, Los Angeles, Miami, New Orleans, New York, San Francisco, and San Juan (Puerto Rico)
US diplomatic representation:
chief of mission: Ambassador Pamela C. HARRIMAN
embassy: 2 Avenue Gabriel, 75382 Paris Cedex 08
mailing address: Unit 21551, Paris; APO AE 09777
telephone: [33] (1) 42 96 12 02, 42 61 80 75
FAX: [33] (1) 42 66 97 83
consulate(s) general: Bordeaux, Marseille, Strasbourg
Flag: three equal vertical bands of blue (hoist side), white, and red; known as the French Tricouleur (Tricolor); the design and colors are similar to a number of other flags, including those of Belgium, Chad, Ireland, Cote d'Ivoire, and Luxembourg; the official flag for all French dependent areas

Economy

Overview: One of the world's most highly developed economies, France has substantial agricultural resources and a diversified modern industrial sector. Large tracts of fertile land, the application of modern technology, and subsidies have combined to make it the leading agricultural producer in Western Europe. Largely self-sufficient in agricultural products, France is a major exporter of wheat and dairy products. The industrial sector generates about one-quarter of GDP, and the growing services sector has become crucial to the economy. Following stagnation and recession in 1991-93, French GDP in 1994 expanded 2.4%. Growth in 1995 is expected to be in the 3.0% to 3.5% range. Persistently high unemployment will still pose a major problem for the government. Paris remains committed to maintaining the franc-deutsche mark parity, which has kept French interest rates high despite France's low inflation. Although the pace of economic and financial integration within the European Union has slowed down, integration presumably will remain a major force shaping the fortunes of the various economic sectors over the next few years.
National product: GDP—purchasing power parity—$1.0801 trillion (1994 est.)
National product real growth rate: 2.4% (1994 est.)
National product per capita: $18,670 (1994 est.)
Inflation rate (consumer prices): 1.6% (1994)
Unemployment rate: 12.6% (yearend 1994)
Budget:
revenues: $220.5 billion
expenditures: $249.1 billion, including capital expenditures of $47 billion (1993 budget)
Exports: $249.2 billion (f.o.b., 1994 est.)
commodities: machinery and transportation equipment, chemicals, foodstuffs, agricultural products, iron and steel products, textiles and clothing
partners: Germany 18.6%, Italy 11.0%, Spain 11.0%, Belgium-Luxembourg 9.1%, UK 8.8%, Netherlands 7.9%, US 6.4%, Japan 2.0%, FSU 0.7% (1991 est.)
Imports: $238.1 billion (c.i.f., 1994 est.)
commodities: crude oil, machinery and equipment, agricultural products, chemicals, iron and steel products
partners: Germany 17.8%, Italy 10.9%, US 9.5%, Netherlands 8.9%, Spain 8.8%, Belgium-Luxembourg 8.5%, UK 7.5%, Japan 4.1%, FSU 1.3% (1991 est.)
External debt: $300 billion (1993 est.)
Industrial production: growth rate 2.6% (1994 est.)
Electricity:
capacity: 105,250,000 kW
production: 447 billion kWh
consumption per capita: 6,149 kWh (1993)
Industries: steel, machinery, chemicals, automobiles, metallurgy, aircraft, electronics, mining, textiles, food processing, tourism
Agriculture: accounts for 4% of GDP (including fishing and forestry); one of the world's top five wheat producers; other principal products—beef, dairy products, cereals, sugar beets, potatoes, wine grapes; self-sufficient for most temperate-zone foods; shortages include fats and oils and tropical produce, but overall net exporter of farm products; fish catch of 850,000 metric tons ranks among world's top 20 countries and is all used domestically
Economic aid:
donor: ODA and OOF commitments (1970-89), $75.1 billion
Currency: 1 French franc (F) = 100 centimes
Exchange rates: French francs (F) per US$1—5.9243 (January 1995), 5.5520 (1994), 5.6632 (1993), 5.2938 (1992), 5.6421 (1991), 5.4453 (1990)
Fiscal year: calendar year

Transportation

Railroads:
total: 34,074 km
standard gauge: 33,975 km 1.435-m gauge (5,850 km electrified; 12,132 km double or multiple track)
other: 99 km various gauges including 1.000-m (privately owned and operated) (1994)
Highways:
total: 1,511,200 km
paved: 811,200 km (including 7,700 km of controlled access divided highway)
unpaved: 700,000 km (1992)
Inland waterways: 14,932 km; 6,969 km heavily traveled
Pipelines: crude oil 3,059 km; petroleum products 4,487 km; natural gas 24,746 km
Ports: Bordeaux, Boulogne, Cherbourg, Dijon, Dunkerque, La Pallice, Le Havre, Lyon, Marseille, Mullhouse, Nantes, Paris, Rouen, Saint Nazaire, Saint Malo, Strasbourg
Merchant marine:
total: 78 ships (1,000 GRT or over) totaling 2,186,183 GRT/3,323,068 DWT
ships by type: bulk 6, cargo 7, chemical tanker 6, container 15, liquefied gas tanker 4, oil tanker 21, passenger 1, roll-on/roll-off cargo 11, short-sea passenger 5, specialized tanker 2
note: France also maintains a captive register for French-owned ships in the Kerguelen Islands (French Southern and Antarctic Lands) and French Polynesia
Airports:
total: 476
with paved runways over 3,047 m: 12
with paved runways 2,438 to 3,047 m: 29
with paved runways 1,524 to 2,437 m: 96
with paved runways 914 to 1,523 m: 74
with paved runways under 914 m: 188
with unpaved runways 1,524 to 2,438 m: 3
with unpaved runways 914 to 1,523 m: 74

Communications

Telephone system: 39,200,000 telephones; highly developed; extensive cable and microwave radio relay networks; large-scale introduction of optical-fiber systems; satellite systems for domestic traffic
local: NA
intercity: microwave radio relay, optical fiber cable, and domestic satellites
international: 2 INTELSAT earth stations (with total of 5 antennas—2 Indian Ocean and 3 for Atlantic Ocean); HF radio communications with more than 20 countries; INMARSAT service; EUTELSAT TV service
Radio:
broadcast stations: AM 41, FM 800 (mostly repeaters), shortwave 0
radios: 48 million
Television:
broadcast stations: 846 (mostly repeaters)
televisions: 36 million

Defense Forces

Branches: Army, Navy (includes Naval Air), Air Force and Air Defense, National Gendarmerie
Manpower availability: males age 15-49 14,740,155; males fit for military service 12,258,691; males reach military age (18) annually 378,489 (1995 est.)
Defense expenditures: exchange rate conversion—$47.1 billion, 3.1% of GDP (1995)

French Guiana
(overseas department of France)

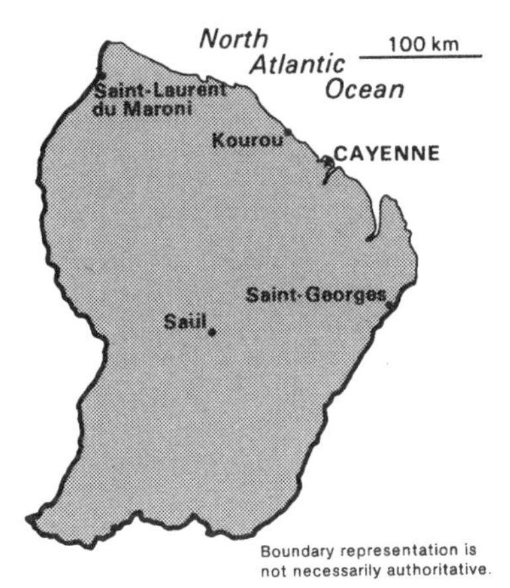

Geography

Location: Northern South America, bordering the North Atlantic Ocean, between Brazil and Suriname
Map references: South America
Area:
total area: 91,000 sq km
land area: 89,150 sq km
comparative area: slightly smaller than Indiana
Land boundaries: total 1,183 km, Brazil 673 km, Suriname 510 km
Coastline: 378 km
Maritime claims:
exclusive economic zone: 200 nm
territorial sea: 12 nm
International disputes: Suriname claims area between Riviere Litani and Riviere Marouini (both headwaters of the Lawa)
Climate: tropical; hot, humid; little seasonal temperature variation
Terrain: low-lying coastal plains rising to hills and small mountains
Natural resources: bauxite, timber, gold (widely scattered), cinnabar, kaolin, fish
Land use:
arable land: 0%
permanent crops: 0%
meadows and pastures: 0%
forest and woodland: 82%
other: 18%
Irrigated land: NA sq km
Environment:
current issues: NA
natural hazards: high frequency of heavy showers and severe thunderstorms; flooding
international agreements: NA
Note: mostly an unsettled wilderness

People

Population: 145,270 (July 1995 est.)
Age structure:
0-14 years: 32% (female 22,511; male 23,535)
15-64 years: 63% (female 41,995; male 50,064)
65 years and over: 5% (female 3,608; male 3,557) (July 1995 est.)
Population growth rate: 4.13% (1995 est.)
Birth rate: 25.23 births/1,000 population (1995 est.)
Death rate: 4.61 deaths/1,000 population (1995 est.)
Net migration rate: 20.65 migrant(s)/1,000 population (1995 est.)
Infant mortality rate: 15.1 deaths/1,000 live births (1995 est.)
Life expectancy at birth:
total population: 75.52 years
male: 72.27 years
female: 78.94 years (1995 est.)
Total fertility rate: 3.46 children born/woman (1995 est.)
Nationality:
noun: French Guianese (singular and plural)
adjective: French Guianese
Ethnic divisions: black or mulatto 66%, Caucasian 12%, East Indian, Chinese, Amerindian 12%, other 10%
Religions: Roman Catholic
Languages: French
Literacy: age 15 and over can read and write (1982)
total population: 83%
male: 84%
female: 82%
Labor force: 23,265
by occupation: services, government, and commerce 60.6%, industry 21.2%, agriculture 18.2% (1980)

Government

Names:
conventional long form: Department of Guiana
conventional short form: French Guiana
local long form: none
local short form: Guyane
Digraph: FG
Type: overseas department of France
Capital: Cayenne
Administrative divisions: none (overseas department of France)
Independence: none (overseas department of France)
National holiday: National Day, Taking of the Bastille, 14 July (1789)
Constitution: 28 September 1958 (French Constitution)
Legal system: French legal system
Suffrage: 18 years of age, universal
Executive branch:
chief of state: President Francois MITTERRAND (since 21 May 1981)
head of government: Prefect Jean-Francois CORDET (since NA 1992); President of the General Council Elie CASTOR (since NA); President of the Regional Council Antoine KARAM (22 March 1993)
cabinet: Council of Ministers
Legislative branch: unicameral General Council and a unicameral Regional Council
General Council: elections last held 25 September and 8 October 1988 (next to be held NA); results—percent of vote by party NA; seats—(19 total) PSG 12, URC 7
Regional Council: elections last held 22 March 1992 (next to be held NA); results—percent of vote by party NA; seats—(31 total) PSG 16, FDG 10, RPR 2, independents 3
French Senate: elections last held 24 September 1989 (next to be held September 1998); results—percent of vote by party NA; seats—(1 total) PSG 1
French National Assembly: elections last held 21 and 28 March 1993 (next to be held NA 1998); results—percent of vote by party NA; seats—(2 total) RPR 1, independent 1
Judicial branch: Court of Appeals (highest local court based in Martinique with jurisdiction over Martinique, Guadeloupe, and French Guiana)
Political parties and leaders: Guianese Socialist Party (PSG), Elie CASTRO; Conservative Union for the Republic (UPR), Leon BERTRAND; Rally for the Center Right (URC); Rally for the Republic (RPR); Guyana Democratic Front (FDG), Georges OTHILY; Walwari Committee, Christine TAUBIRA-DELANON
Member of: FZ, WCL, WFTU
Diplomatic representation in US: none (overseas department of France)
US diplomatic representation: none (overseas department of France)
Flag: the flag of France is used

Economy

Overview: The economy is tied closely to that of France through subsidies and imports. Besides the French space center at Kourou, fishing and forestry are the most important economic activities, with exports of fish and fish products (mostly shrimp) accounting for more than 60% of total revenue in 1992. The large reserves of tropical hardwoods, not fully exploited, support an expanding sawmill industry that provides sawn logs for export. Cultivation of crops—rice, cassava, bananas, and sugarcane—is limited to the coastal area, where the population is largely concentrated. French Guiana is heavily dependent on imports of food and energy. Unemployment is a serious problem, particularly among younger workers.
National product: GDP—purchasing power parity—$800 million (1993 est.)
National product real growth rate: NA%
National product per capita: $6,000 (1993 est.)
Inflation rate (consumer prices): 2.5% (1992)
Unemployment rate: 13% (1990)
Budget:
revenues: $735 million

expenditures: $735 million, including capital expenditures of $NA (1987)
Exports: $59 million (f.o.b., 1992)
commodities: shrimp, timber, rum, rosewood essence
partners: France 52%, Spain 15%, US 5% (1992)
Imports: $1.5 billion (c.i.f., 1992)
commodities: food (grains, processed meat), other consumer goods, producer goods, petroleum
partners: France 77%, Germany 11%, US 5% (1992)
External debt: $1.2 billion (1988)
Industrial production: growth rate NA%
Electricity:
capacity: 180,000 kW
production: 450 million kWh
consumption per capita: 3,149 kWh (1993)
Industries: construction, shrimp processing, forestry products, rum, gold mining
Agriculture: some vegetables for local consumption; rice, corn, manioc, cocoa, bananas, sugar; livestock—cattle, pigs, poultry
Illicit drugs: small amount of marijuana grown for local consumption
Economic aid:
recipient: Western (non-US) countries, ODA and OOF bilateral commitments (1970-89), $1.51 billion
Currency: 1 French franc (F) = 100 centimes
Exchange rates: French francs (F) per US$1—5.9243 (January 1995), 5.5520 (1994), 5.6632 (1993), 5.2938 (1992), 5.6421 (1991), 5.4453 (1990)
Fiscal year: calendar year

Transportation

Railroads:
total: 22 km (est.)
Highways:
total: 1,137 km
paved: 455 km
unpaved: improved, unimproved earth 682 km (1988)
Inland waterways: 460 km, navigable by small oceangoing vessels and river and coastal steamers; 3,300 km navigable by native craft
Ports: Cayenne, Degrad des Cannes, Saint-Laurent du Maroni
Merchant marine: none
Airports:
total: 11
with paved runways over 3,047 m: 1
with paved runways 914 to 1,523 m: 2
with paved runways under 914 m: 5
with unpaved runways 914 to 1,523 m: 3

Communications

Telephone system: 18,100 telephones; fair open-wire and microwave radio relay system
local: NA
intercity: open wire and microwave radio relay
international: 1 INTELSAT (Atlantic Ocean) earth station
Radio:
broadcast stations: AM 5, FM 7, shortwave 0
radios: NA
Television:
broadcast stations: 9
televisions: NA

Defense Forces

Branches: French Forces, Gendarmerie
Manpower availability: males age 15-49 41,986; males fit for military service 27,298 (1995 est.)
Defense expenditures: $NA, NA% of GDP
Note: defense is the responsibility of France

French Polynesia
(overseas territory of France)

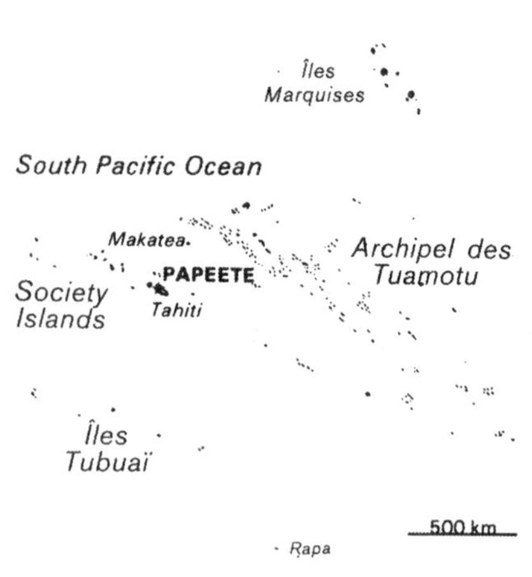

Geography

Location: Oceania, archipelago in the South Pacific Ocean, about one-half of the way from South America to Australia
Map references: Oceania
Area:
total area: 3,941 sq km
land area: 3,660 sq km
comparative area: slightly less than one-third the size of Connecticut
Land boundaries: 0 km
Coastline: 2,525 km
Maritime claims:
exclusive economic zone: 200 nm
territorial sea: 12 nm
International disputes: none
Climate: tropical, but moderate
Terrain: mixture of rugged high islands and low islands with reefs
Natural resources: timber, fish, cobalt
Land use:
arable land: 1%
permanent crops: 19%
meadows and pastures: 5%
forest and woodland: 31%
other: 44%
Irrigated land: NA sq km
Environment:
current issues: NA
natural hazards: occasional cyclonic storms in January
international agreements: NA
Note: includes five archipelagoes; Makatea in French Polynesia is one of the three great phosphate rock islands in the Pacific Ocean—the others are Banaba (Ocean Island) in Kiribati and Nauru

People

Population: 219,999 (July 1995 est.)
Age structure:
0-14 years: 36% (female 38,361; male 39,744)
15-64 years: 60% (female 64,034; male 69,024)
65 years and over: 4% (female 4,437; male

4,399) (July 1995 est.)
Population growth rate: 2.23% (1995 est.)
Birth rate: 27.56 births/1,000 population (1995 est.)
Death rate: 5.27 deaths/1,000 population (1995 est.)
Net migration rate: 0 migrant(s)/1,000 population (1995 est.)
Infant mortality rate: 14.6 deaths/1,000 live births (1995 est.)
Life expectancy at birth:
total population: 70.75 years
male: 68.32 years
female: 73.29 years (1995 est.)
Total fertility rate: 3.3 children born/woman (1995 est.)
Nationality:
noun: French Polynesian(s)
adjective: French Polynesian
Ethnic divisions: Polynesian 78%, Chinese 12%, local French 6%, metropolitan French 4%
Religions: Protestant 54%, Roman Catholic 30%, other 16%
Languages: French (official), Tahitian (official)
Literacy: age 14 and over can read and write but definition of literary not available (1977)
total population: 98%
male: 98%
female: 98%
Labor force: 76,630 employed (1988)

Government

Names:
conventional long form: Territory of French Polynesia
conventional short form: French Polynesia
local long form: Territoire de la Polynesie Francaise
local short form: Polynesie Francaise
Digraph: FP
Type: overseas territory of France since 1946
Capital: Papeete
Administrative divisions: none (overseas territory of France); there are no first-order administrative divisions as defined by the US Government, but there are 5 archipelagic divisions named Archipel des Marquises, Archipel des Tuamotu, Archipel des Tubuai, Iles du Vent, and Iles Sous-le-Vent
note: Clipperton Island is administered by France from French Polynesia
Independence: none (overseas territory of France)
National holiday: National Day, Taking of the Bastille, 14 July (1789)
Constitution: 28 September 1958 (French Constitution)
Legal system: based on French system
Suffrage: 18 years of age; universal
Executive branch:
chief of state: President Francois MITTERRAND (since 21 May 1981); High Commissioner of the Republic Paul RONCIERE (since 8 August 1994)
head of government: President of the Territorial Government of French Polynesia Gaston FLOSSE (since 10 May 1991); Deputy to the French Assembly and President of the Territorial Assembly Jean JUVENTIN (since NA November 1992); Territorial Vice President and Minister of Health Michel BUILLARD (since 12 September 1991)
cabinet: Council of Ministers; president submits a list of members of the Assembly for approval by them to serve as ministers
Legislative branch: unicameral
Territorial Assembly: elections last held 17 March 1991 (next to be held March 1996); results—percent of vote by party NA; seats—(41 total) People's Rally for the Republic (Gaullist) 18, Polynesian Union Party 12, New Fatherland Party 7, other 4
French Senate: elections last held 24 September 1989 (next to be held September 1998); results—percent of vote by party NA; seats—(1 total) party NA
French National Assembly: elections last held 21 and 28 March 1993 (next to be held NA March 1998); results—percent of vote by party NA; seats—(2 total) People's Rally for the Republic (Gaullist) 2
Judicial branch: Court of Appeal, Court of the First Instance, Court of Administrative Law
Political parties and leaders: People's Rally for the Republic (Tahoeraa Huiraatira), Gaston FLOSSE; Polynesian Union Party (includes Te Tiarama), Alexandre LEONTIEFF; Here Ai'a Party, Jean JUVENTIN; New Fatherland Party (Ai'a Api), Emile VERNAUDON; Polynesian Liberation Front (Tavini Hviraatira No Te Ao Maohi), Oscar TEMARU; Independent Party (Ia Mana Te Nunaa), Jacques DROLLET; other small parties
Member of: ESCAP (associate), FZ, ICFTU, SPC, WMO
Diplomatic representation in US: none (overseas territory of France)
US diplomatic representation: none (overseas territory of France)
Flag: the flag of France is used

Economy

Overview: Since 1962, when France stationed military personnel in the region, French Polynesia has changed from a subsistence economy to one in which a high proportion of the work force is either employed by the military or supports the tourist industry. Tourism accounts for about 20% of GDP and is a primary source of hard currency earnings.
National product: GDP—purchasing power parity—$1.5 billion (1993 est.)
National product real growth rate: NA%
National product per capita: $7,000 (1993 est.)
Inflation rate (consumer prices): 1.7% (1991)
Unemployment rate: 10% (1990 est.)
Budget:
revenues: $614 million
expenditures: $957 million, including capital expenditures of $NA (1988)
Exports: $88.9 million (f.o.b., 1989)
commodities: coconut products 79%, mother-of-pearl 14%, vanilla, shark meat
partners: France 54%, US 17%, Japan 17%
Imports: $765 million (c.i.f., 1989)
commodities: fuels, foodstuffs, equipment
partners: France 53%, US 11%, Australia 6%, NZ 5%
External debt: $NA
Industrial production: growth rate NA%; accounts for 15% of GDP
Electricity:
capacity: 75,000 kW
production: 275 million kWh
consumption per capita: 1,189 kWh (1993)
Industries: tourism, pearls, agricultural processing, handicrafts
Agriculture: coconut and vanilla plantations; vegetables and fruit; poultry, beef, dairy products
Economic aid:
recipient: Western (non-US) countries, ODA and OOF bilateral commitments (1970-88), $3.95 billion
Currency: 1 CFP franc (CFPF) = 100 centimes
Exchange rates: Comptoirs Francais du Pacifique francs (CFPF) per US$1—96.25 (January 1995), 100.94 (1994), 102.96 (1993), 96.24 (1992), 102.57 (1991), 99.00 (1990); note—linked at the rate of 18.18 to the French franc
Fiscal year: calendar year

Transportation

Railroads: 0 km
Highways:
total: 600 km (1982)
paved: NA
unpaved: NA
Ports: Mataura, Papeete, Rikitea, Uturoa
Merchant marine:
total: 3 ships (1,000 GRT or over) totaling 4,127 GRT/6,710 DWT
ships by type: passenger-cargo 2, refrigerated cargo 1
note: a subset of the French register allowing French-owned ships to operate under more liberal taxation and manning regulations than permissable under the main French register
Airports:
total: 43
with paved runways over 3,047 m: 2
with paved runways 1,524 to 2,437 m: 5
with paved runways 914 to 1,523 m: 14

with paved runways under 914 m: 18
with unpaved runways 914 to 1,523 m: 4

Communications

Telephone system: 33,200 telephones
local: NA
intercity: NA
international: 1 INTELSAT (Pacific Ocean) earth station
Radio:
broadcast stations: AM 5, FM 2, shortwave 0
radios: 84,000
Television:
broadcast stations: 6
televisions: 26,400

Defense Forces

Branches: French Forces (includes Army, Navy, Air Force), Gendarmerie
Note: defense is responsibility of France

French Southern and Antarctic Lands

(overseas territory of France)

700 km
Indian Ocean
Île Amsterdam
Île Saint-Paul
Îles Crozet
Îles Kerguelen

Geography

Location: Southern Africa, islands in the southern Indian Ocean, about equidistant between Africa, Antarctica, and Australia; note—French Southern and Antarctic Lands includes Ile Amsterdam, Ile Saint-Paul, Iles Crozet, and Iles Kerguelen in the southern Indian Ocean, along with the French-claimed sector of Antartica, "Terre Adelie"; the United States does not recognize the French claim to "Terre Adelie"
Map references: Antarctic Region
Area:
total area: 7,781 sq km
land area: 7,781 sq km
comparative area: slightly less than 1.5 times the size of Delaware
note: includes Ile Amsterdam, Ile Saint-Paul, Iles Crozet and Iles Kerguelen; excludes "Terre Adelie" claim of about 500,000 sq km in Antarctica that is not recognized by the US
Land boundaries: 0 km
Coastline: 1,232 km
Maritime claims:
exclusive economic zone: 200 nm from Iles Kerguelen only
territorial sea: 12 nm
International disputes: "Terre Adelie" claim in Antarctica is not recognized by the US
Climate: antarctic
Terrain: volcanic
Natural resources: fish, crayfish
Land use:
arable land: 0%
permanent crops: 0%
meadows and pastures: 0%
forest and woodland: 0%
other: 100%
Irrigated land: 0 sq km
Environment:
current issues: NA
natural hazards: Ile Amsterdam and Ile Saint-Paul are extinct volcanoes
international agreements: NA
Note: remote location in the southern Indian Ocean

People

Population: no indigenous inhabitants; note—there are researchers whose numbers vary from 150 in winter (July) to 200 in summer (January)

Government

Names:
conventional long form: Territory of the French Southern and Antarctic Lands
conventional short form: French Southern and Antarctic Lands
local long form: Territoire des Terres Australes et Antarctiques Francaises
local short form: Terres Australes et Antarctiques Francaises
Digraph: FS
Type: overseas territory of France since 1955; governed by High Administrator Bernard de GOUTTES (since May 1990), who is assisted by a 7-member Consultative Council and a 12-member Scientific Council
Capital: none; administered from Paris, France
Administrative divisions: none (overseas territory of France); there are no first-order administrative divisions as defined by the US Government, but there are 3 districts named Ile Crozet, Iles Kerguelen, and Iles Saint-Paul et Amsterdam; excludes "Terre Adelie" claim in Antarctica that is not recognized by the US
Independence: none (overseas territory of France)
Flag: the flag of France is used

Economy

Overview: Economic activity is limited to servicing meteorological and geophysical research stations and French and other fishing fleets. The fish catches landed on Iles Kerguelen by foreign ships are exported to France and Reunion.
Budget:
revenues: $17.5 million
expenditures: $NA, including capital expenditures of $NA (1992)

Transportation

Highways:
total: NA
paved: NA
unpaved: NA
Ports: none; offshore anchorage only
Merchant marine:
total: 48 ships (1,000 GRT or over) totaling 1,290,975 GRT/2,403,050 DWT
ships by type: bulk 5, cargo 6, chemical tanker 4, container 1, liquefied gas tanker 3, multifunction large-load carrier 1, oil tanker 15, refrigerated cargo 4, roll-on/roll-off cargo

8, specialized liquefied tanker 1
note: a subset of the French register allowing French-owned ships to operate under more liberal taxation and manning regulations than permissable under the main French register
Airports: none

Communications

Telephone system: NA telephones
local: NA
intercity: NA
international: NA
Radio:
broadcast stations: AM NA, FM NA, shortwave NA
radios: NA
Television:
broadcast stations: NA
televisions: NA

Defense Forces

Note: defense is the responsibility of France

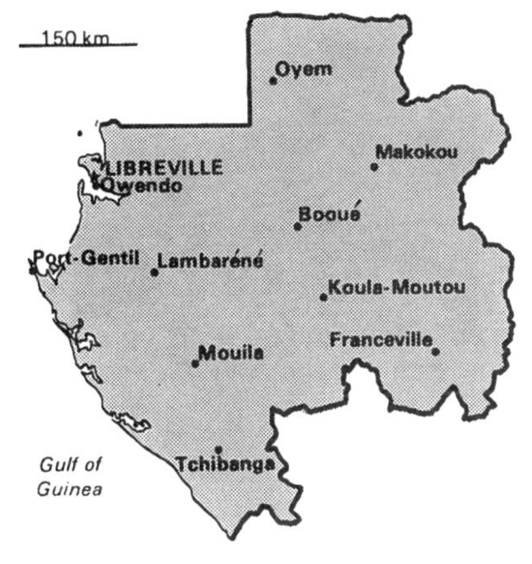

Geography

Location: Western Africa, bordering the Atlantic Ocean at the Equator, between Congo and Equatorial Guinea
Map references: Africa
Area:
total area: 267,670 sq km
land area: 257,670 sq km
comparative area: slightly smaller than Colorado
Land boundaries: total 2,551 km, Cameroon 298 km, Congo 1,903 km, Equatorial Guinea 350 km
Coastline: 885 km
Maritime claims:
contiguous zone: 24 nm
exclusive economic zone: 200 nm
territorial sea: 12 nm
International disputes: maritime boundary dispute with Equatorial Guinea because of disputed sovereignty over islands in Corisco Bay
Climate: tropical; always hot, humid
Terrain: narrow coastal plain; hilly interior; savanna in east and south
Natural resources: petroleum, manganese, uranium, gold, timber, iron ore
Land use:
arable land: 1%
permanent crops: 1%
meadows and pastures: 18%
forest and woodland: 78%
other: 2%
Irrigated land: NA sq km
Environment:
current issues: deforestation; poaching
natural hazards: NA
international agreements: party to—Endangered Species, Marine Dumping, Nuclear Test Ban, Ozone Layer Protection, Ship Pollution, Tropical Timber 83, Wetlands; signed, but not ratified—Biodiversity, Climate Change, Law of the Sea, Tropical Timber 94

People

Population: 1,155,749 (July 1995 est.)
Age structure:
0-14 years: 34% (female 193,859; male 194,761)
15-64 years: 61% (female 347,839; male 359,997)
65 years and over: 5% (female 30,218; male 29,075) (July 1995 est.)
Population growth rate: 1.46% (1995 est.)
Birth rate: 28.34 births/1,000 population (1995 est.)
Death rate: 13.72 deaths/1,000 population (1995 est.)
Net migration rate: 0 migrant(s)/1,000 population (1995 est.)
Infant mortality rate: 92.4 deaths/1,000 live births (1995 est.)
Life expectancy at birth:
total population: 55.14 years
male: 52.31 years
female: 58.06 years (1995 est.)
Total fertility rate: 3.93 children born/woman (1995 est.)
Nationality:
noun: Gabonese (singular and plural)
adjective: Gabonese
Ethnic divisions: Bantu tribes including four major tribal groupings (Fang, Eshira, Bapounou, Bateke), other Africans and Europeans 100,000, including 27,000 French
Religions: Christian 55%-75%, Muslim less than 1%, animist
Languages: French (official), Fang, Myene, Bateke, Bapounou/Eschira, Bandjabi
Literacy: age 15 and over can read and write (1990 est.)
total population: 61%
male: 74%
female: 48%
Labor force: 120,000 salaried
by occupation: agriculture 65.0%, industry and commerce 30.0%, services 2.5%, government 2.5%

Government

Names:
conventional long form: Gabonese Republic
conventional short form: Gabon
local long form: Republique Gabonaise
local short form: Gabon
Digraph: GB
Type: republic; multiparty presidential regime (opposition parties legalized 1990)
Capital: Libreville
Administrative divisions: 9 provinces; Estuaire, Haut-Ogooue, Moyen-Ogooue, Ngounie, Nyanga, Ogooue-Ivindo, Ogooue-Lolo, Ogooue-Maritime, Woleu-Ntem
Independence: 17 August 1960 (from France)

National holiday: Renovation Day, 12 March (1968) (Gabonese Democratic Party established)
Constitution: adopted 14 March 1991
Legal system: based on French civil law system and customary law; judicial review of legislative acts in Constitutional Chamber of the Supreme Court; compulsory ICJ jurisdiction not accepted
Suffrage: 21 years of age; universal
Executive branch:
chief of state: President El Hadj Omar BONGO (since 2 December 1967); election last held on 5 December 1993 (next to be held 1998); results—President Omar BONGO was reelected with 51% of the vote
head of government: Prime Minister Paulin OBAME Nguema (since 9 December 1994)
cabinet: Council of Ministers; appointed by the prime minister in consultation with the president
Legislative branch: unicameral
National Assembly (Assemblee Nationale): elections last held on 5 December 1993 (next to be held by 1998); results—percent of vote by party NA; seats—(120 total) PDG 62, Morena-Bucherons/RNB 19, PGP 18, National Recovery Movement (Morena-Original) 7, APSG 6, USG 4, CRP 1, independents 3
Judicial branch: Supreme Court (Cour Supreme)
Political parties and leaders: Gabonese Democratic Party (PDG, former sole party), Jaques ADIAHENOT, Secretary General; National Recovery Movement—Lumberjacks (Morena-Bucherons/RNB), Fr. Paul M'BA-ABESSOLE, leader; Gabonese Party for Progress (PGP), Pierre-Louis AGONDHO-OKAWE, President; National Recovery Movement (Morena-Original), Pierre ZONGUE-NGUEMA, Chairman; Association for Socialism in Gabon (APSG), leader NA; Gabonese Socialist Union (USG), leader NA; Circle for Renewal and Progress (CRP), leader NA; Union for Democracy and Development (UDD), leader NA; Rally of Democrats (RD), leader NA; Forces of Change for Democratic Union, leader NA
Member of: ACCT, ACP, AfDB, BDEAC, CCC, CEEAC, ECA, FAO, FZ, G-24, G-77, GATT, IAEA, IBRD, ICAO, ICC, ICFTU, IDA, IDB, IFAD, IFC, IFRCS (associate), ILO, IMF, IMO, INMARSAT, INTELSAT, INTERPOL, IOC, ITU, NAM, OAU, OIC, OPEC, UDEAC, UN, UNCTAD, UNESCO, UNIDO, UPU, WCL, WHO, WIPO, WMO, WTO
Diplomatic representation in US:
chief of mission: Ambassador Paul BOUNDOUKOU-LATHA
chancery: 2233 Wisconsin Avenue NW, Washington, DC 20007, Suite 200
telephone: [1] (202) 797-1000
US diplomatic representation:
chief of mission: Ambassador Joseph C. WILSON IV
embassy: Boulevard de la Mer, Libreville
mailing address: B. P. 4000, Libreville
telephone: [241] 76 20 03 through 76 20 04, 74 34 92
FAX: [241] 74 55 07
Flag: three equal horizontal bands of green (top), yellow, and blue

Economy

Overview: Notwithstanding its serious ongoing economic problems, Gabon enjoys a per capita income more than twice that of most nations of sub-Saharan Africa. Gabon depended on timber and manganese until oil was discovered offshore in the early 1970s. The oil sector now accounts for 50% of GDP. Real growth was feeble in 1992 and Gabon continues to face the problem of fluctuating prices for its oil, timber, manganese, and uranium exports. Despite an abundance of natural wealth, and a manageable rate of population growth, the economy is hobbled by poor fiscal management. In 1992, the fiscal deficit widened to 2.4% of GDP, and Gabon failed to settle arrears on its bilateral debt, leading to a cancellation of rescheduling agreements with official and private creditors. Devaluation of its Francophone currency by 50% in January 1994 did not set off an expected inflationary spiral but the government must continue to keep a tight reign on spending and wage increases.
National product: GDP—purchasing power parity—$5.6 billion (1994 est.)
National product real growth rate: 1.9% (1994 est.)
National product per capita: $4,900 (1994 est.)
Inflation rate (consumer prices): 35% (1994 est.)
Unemployment rate: NA%
Budget:
revenues: $1.3 billion
expenditures: $1.6 billion, including capital expenditures of $311 million (1993 est.)
Exports: $2.1 billion (f.o.b., 1993 est)
commodities: crude oil 80%, timber 10%, manganese 6%, uranium 2%
partners: US 38%, France 26%, Japan, Germany
Imports: $832 million (c.i.f., 1993 est.)
commodities: foodstuffs, chemical products, petroleum products, construction materials, manufactures, machinery
partners: France 42%, African countries 23%, US, Japan
External debt: $3.3 billion (1993 est.)
Industrial production: growth rate -3% (1991)
Electricity:
capacity: 315,000 kW
production: 910 million kWh
consumption per capita: 757 kWh (1993)
Industries: food and beverages, lumbering and plywood, textiles, cement, petroleum refining, mining—manganese, uranium, gold, petroleum
Agriculture: cash crops—cocoa, coffee, palm oil; livestock raising not developed; importer of food; small fishing operations provide a catch of about 20,000 metric tons; okoume (a tropical softwood) is the most important timber product
Economic aid:
recipient: US commitments, including Ex-Im (FY70-90), $68 million; Western (non-US) countries, ODA and OOF bilateral commitments (1970-90), $2.342 billion; Communist countries (1970-89), $27 million
Currency: 1 CFA franc (CFAF) = 100 centimes
Exchange rates: Communaute Financiere Africaine francs (CFAF) per US$1—529.43 (January 1995), 555.20 (1994), 283.16 (1993), 264.69 (1992), 282.11 (1991), 272.26 (1990)
note: beginning 12 January 1994, the CFA franc was devalued to CFAF 100 per French franc from CFAF 50 at which it had been fixed since 1948
Fiscal year: calendar year

Transportation

Railroads:
total: 649 km single track (Transgabonese Railroad)
standard gauge: 649 km 1.437-m gauge
Highways:
total: 7,500 km
paved: 560 km
unpaved: crushed stone 960 km; earth 5,980 km
Inland waterways: 1,600 km perennially navigable
Pipelines: crude oil 270 km; petroleum products 14 km
Ports: Cape Lopez, Kango, Lambarene, Libreville, Owendo, Port-Gentil
Merchant marine:
total: 1 cargo ships (1,000 GRT or over) totaling 9,281 GRT/12,665 DWT
Airports:
total: 69
with paved runways over 3,047 m: 1
with paved runways 2,438 to 3,047 m: 1
with paved runways 1,524 to 2,437 m: 7
with paved runways 914 to 1,523 m: 1
with paved runways under 914 m: 28
with unpaved runways 1,524 to 2,438 m: 8
with unpaved runways 914 to 1,523 m: 23

Communications

Telephone system: 15,000 telephones; telephone density—13/1,000 persons
local: NA
intercity: adequate system, comprising cable, microwave radio relay, tropospheric scatter, radiocommunication stations, and 12 domestic

satellite links
international: 3 Atlantic Ocean INTELSAT earth stations
Radio:
broadcast stations: AM 6, FM 6, shortwave 0
radios: NA
Television:
broadcast stations: 3 (repeaters 5)
televisions: NA

Defense Forces

Branches: Army, Navy, Air Force, Presidential Guard, National Gendarmerie, National Police
Manpower availability: males age 15-49 272,025; males fit for military service 138,197; males reach military age (20) annually 10,516 (1995 est.)
Defense expenditures: exchange rate conversion—$154 million, 2.4% of GDP (1993)

Geography

Location: Western Africa, bordering the North Atlantic Ocean and Senegal
Map references: Africa
Area:
total area: 11,300 sq km
land area: 10,000 sq km
comparative area: slightly more than twice the size of Delaware
Land boundaries: total 740 km, Senegal 740 km
Coastline: 80 km
Maritime claims:
contiguous zone: 18 nm
continental shelf: not specified
exclusive fishing zone: 200 nm
territorial sea: 12 nm
International disputes: short section of boundary with Senegal is indefinite
Climate: tropical; hot, rainy season (June to November); cooler, dry season (November to May)
Terrain: flood plain of the Gambia River flanked by some low hills
Natural resources: fish
Land use:
arable land: 16%
permanent crops: 0%
meadows and pastures: 9%
forest and woodland: 20%
other: 55%
Irrigated land: 120 sq km (1989 est.)
Environment:
current issues: deforestation; desertification; water-borne diseases prevalent
natural hazards: rainfall has dropped by 30% in the last thirty years
international agreements: party to—Biodiversity, Climate Change, Endangered Species, Law of the Sea, Nuclear Test Ban, Ozone Layer Protection, Ship Pollution; signed, but not ratified—Desertification
Note: almost an enclave of Senegal; smallest country on the continent of Africa

People

Population: 989,273 (July 1995 est.)
Age structure:
0-14 years: 47% (female 231,636; male 231,053)
15-64 years: 51% (female 257,329; male 244,947)
65 years and over: 2% (female 11,850; male 12,458) (July 1995 est.)
Population growth rate: 3.08% (1995 est.)
Birth rate: 45.97 births/1,000 population (1995 est.)
Death rate: 15.19 deaths/1,000 population (1995 est.)
Net migration rate: 0 migrant(s)/1,000 population (1995 est.)
Infant mortality rate: 120.8 deaths/1,000 live births (1995 est.)
Life expectancy at birth:
total population: 50.55 years
male: 48.25 years
female: 52.92 years (1995 est.)
Total fertility rate: 6.23 children born/woman (1995 est.)
Nationality:
noun: Gambian(s)
adjective: Gambian
Ethnic divisions: African 99% (Mandinka 42%, Fula 18%, Wolof 16%, Jola 10%, Serahuli 9%, other 4%), non-Gambian 1%
Religions: Muslim 90%, Christian 9%, indigenous beliefs 1%
Languages: English (official), Mandinka, Wolof, Fula, other indigenous vernaculars
Literacy: age 15 and over can read and write (1990 est.)
total population: 27%
male: 39%
female: 16%
Labor force: 400,000 (1986 est.)
by occupation: agriculture 75.0%, industry, commerce, and services 18.9%, government 6.1%

Government

Names:
conventional long form: Republic of The Gambia
conventional short form: The Gambia
Digraph: GA
Type: republic under multiparty democratic rule
Capital: Banjul
Administrative divisions: 5 divisions and 1 city*; Banjul*, Lower River, MacCarthy Island, North Bank, Upper River, Western
Independence: 18 February 1965 (from UK; The Gambia and Senegal signed an agreement on 12 December 1981 that called for the creation of a loose confederation to be known as Senegambia, but the agreement was dissolved on 30 September 1989)

National holiday: Independence Day, 18 February (1965)
Constitution: 24 April 1970
Legal system: based on a composite of English common law, Koranic law, and customary law; accepts compulsory ICJ jurisdiction, with reservations
Suffrage: 21 years of age; universal
Executive branch:
chief of state and head of government: Chairman of the Armed Forces Provisional Ruling Council Capt. Yahya A. J. J. JAMMEH (since the military coup of 22 July 1994); Vice Chairman of the Armed Forces Provisional Ruling Council Capt. Edward SINGHATEH (since March 1995); election last held on 29 April 1992; results—Sir Dawda JAWARA (PPP) 58.5%, Sherif Mustapha DIBBA (NCP) 22.2%, Assan Musa CAMARA (GPP) 8.0% (prior to the 22 July 1994 coup, next election was scheduled for April 1997)
cabinet: Cabinet; appointed by the president from members of the House of Representatives (present cabinet appointed by Chairman of the Armed Forces Provisional Ruling Council)
Legislative branch: unicameral
House of Representatives: elections last held on 29 April 1992 (next to be held April 1997); results—PPP 58.1%; seats—(43 total, 36 elected) PPP 30, NCP 6
Judicial branch: Supreme Court
Political parties and leaders: People's Progressive Party (PPP), Dawda K. JAWARA (in exile), secretary general; National Convention Party (NCP), Sheriff DIBBA (in exile); Gambian People's Party (GPP), Hassan Musa CAMARA; United Party (UP), leader NA; People's Democratic Organization of Independence and Socialism (PDOIS), leader NA; People's Democratic Party (PDP), Jabel SALLAH
Member of: ACP, AfDB, C, CCC, ECA, ECOWAS, FAO, G-77, GATT, IBRD, ICAO, ICFTU, ICRM, IDA, IDB, IFAD, IFC, IFRCS, IMF, IMO, INTELSAT (nonsignatory user), INTERPOL, IOC, ITU, NAM, OAU, OIC, UN, UNCTAD, UNESCO, UNIDO, UPU, WCL, WFTU, WHO, WIPO, WMO, WTO
Diplomatic representation in US:
chief of mission: (vacant); Charge d'Affaires Aminatta DIBBA
chancery: Suite 1000, 1155 15th Street NW, Washington, DC 20005
telephone: [1] (202) 785-1399, 1379, 1425
FAX: [1] (202) 785-1430
US diplomatic representation:
chief of mission: Ambassador Andrew J. WINTER
embassy: Fajara, Kairaba Avenue, Banjul
mailing address: P. M. B. No. 19, Banjul
telephone: [220] 392856, 392858, 391970, 391971
FAX: [220] 392475
Flag: three equal horizontal bands of red (top), blue with white edges, and green

Economy

Overview: The Gambia has no important mineral or other natural resources and has a limited agricultural base. About 75% of the population is engaged in crop production and livestock raising, which contribute 30% to GDP. Small-scale manufacturing activity—processing peanuts, fish, and hides—accounts for less than 10% of GDP. A sustained structural adjustment program, including a liberalized trade policy, had fostered a respectable 4% rate of growth in recent years. Reexport trade constitutes one-third of economic activity; however, border closures associated with Senegal's monetary crisis in late 1993 led to a halving of reexport trade, reducing government revenues in turn. The 50% devaluation of the CFA franc in January 1994 has made Senegalese goods more competitive and apparently prompted a relaxation of Senegalese controls, paving the way for a comeback in reexports. But overwhelming these developments were the devastating effects of the military's takeover in July 1994. By October, traffic at the Port of Banjul had fallen precipitously as importers nervously scaled back their activities with the commencement of the anticorruption drive by the new regime. Concerned with the growing potential for serious unrest after a countercoup attempt was bloodily put down by the regime, the United Kingdom and the EU in November issued a travelers advisory for The Gambia, which brought a halt to tourism almost immediately. The Gambia faces additional problems in 1995 if, as is likely, economic sanctions by Western governments remain in effect in response to indications that the military regime intends to stay in power far longer than expected by the donors.
National product: GDP—purchasing power parity—$1 billion (1993 est.)
National product real growth rate: NA%
National product per capita: $1,050 (1993 est.)
Inflation rate (consumer prices): 6.5% (1993)
Unemployment rate: NA%
Budget:
revenues: $94 million
expenditures: $89 million, including capital expenditures of $24 million (FY92/93 est.)
Exports: $81 million (f.o.b., FY92/93 est.)
commodities: peanuts and peanut products, fish, cotton lint, palm kernels
partners: Japan 60%, Europe 29%, Africa 5%, US 1%, other 5% (1989)
Imports: $154 million (f.o.b., FY92/93 est.)
commodities: foodstuffs, manufactures, raw materials, fuel, machinery and transport equipment
partners: Europe 57%, Asia 25%, USSR and Eastern Europe 9%, US 6%, other 3% (1989)
External debt: $286 million (FY92/93 est.)
Industrial production: growth rate 6.7%
Electricity:
capacity: 30,000 kW
production: 70 million kWh
consumption per capita: 64 kWh (1993)
Industries: peanut processing, tourism, beverages, agricultural machinery assembly, woodworking, metalworking, clothing
Agriculture: accounts for 30% of GDP; one-third of food requirements is imported; major export crop is peanuts; other principal crops—millet, sorghum, rice, corn, cassava, palm kernels; livestock—cattle, sheep, goats; forestry and fishing resources not fully exploited
Economic aid:
recipient: US commitments, including Ex-Im (FY70-89), $93 million; Western (non-US) countries, ODA and OOF bilateral commitments (1970-89), $535 million; Communist countries (1970-89), $39 million
Currency: 1 dalasi (D) = 100 butut
Exchange rates: dalasi (D) per US$1—9.565 (January 1995), 9.576 (1994), 9.129 (1993), 8.888 (1992), 8.803 (1991), 7.883 (1990)
Fiscal year: 1 July—30 June

Transportation

Railroads: 0 km
Highways:
total: 3,083 km
paved: 431 km
unpaved: gravel, crushed stone 501 km; unimproved earth 2,151 km
Inland waterways: 400 km
Ports: Banjul
Merchant marine:
total: 1 bulk ship (1,000 GRT or over) totaling 11,194 GRT/19,394 DWT
Airports:
total: 1
with paved runways over 3,047 m: 1

Communications

Telephone system: 3,500 telephones; telephone density—4 telephones/1,000 persons
local: NA
intercity: adequate network of radio relay and wire
international: 1 Atlantic Ocean INTELSAT earth station
Radio:
broadcast stations: AM 3, FM 2, shortwave 0
radios: NA
Television:
broadcast stations: NA
televisions: NA

Defense Forces

Branches: Army, Navy, National Police

Manpower availability: males age 15-49 214,680; males fit for military service 108,659 (1995 est.)
Defense expenditures: exchange rate conversion—$14 million, 3.8% of GDP (FY93/94)

Gaza Strip

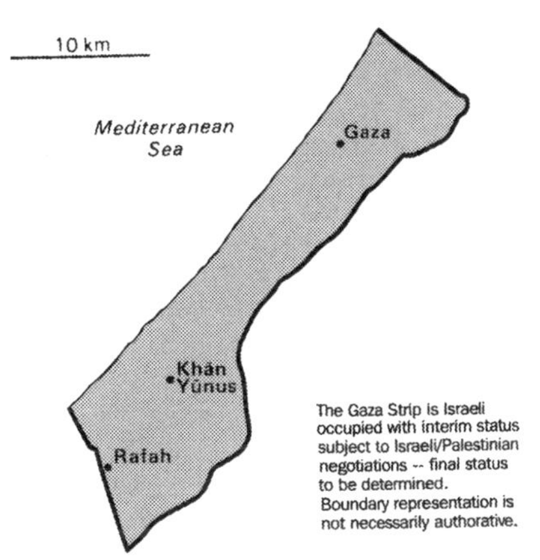

10 km
Mediterranean Sea
Gaza
Khān Yūnus
Rafah
The Gaza Strip is Israeli occupied with interim status subject to Israeli/Palestinian negotiations -- final status to be determined.
Boundary representation is not necessarily authorative.

Note: The Israel-PLO Declaration of Principles on Interim Self-Government Arrangements ("the DOP"), signed in Washington on 13 September 1993, provides for a transitional period not exceeding five years of Palestinian interim self-government in the Gaza Strip and the West Bank. Under the DOP, final status negotiations are to begin no later than the beginning of the third year of the transitional period.

Geography

Location: Middle East, bordering the Mediterranean Sea, between Egypt and Israel
Map references: Middle East
Area:
total area: 360 sq km
land area: 360 sq km
comparative area: slightly more than twice the size of Washington, DC
Land boundaries: total 62 km, Egypt 11 km, Israel 51 km
Coastline: 40 km
Maritime claims: Israeli occupied with interim status subject to Israeli/Palestinian negotiations—final status to be determined
International disputes: West Bank and Gaza Strip are Israeli occupied with interim status subject to Israeli/Palestinian negotiations—final status to be determined
Climate: temperate, mild winters, dry and warm to hot summers
Terrain: flat to rolling, sand- and dune-covered coastal plain
Natural resources: negligible
Land use:
arable land: 13%
permanent crops: 32%
meadows and pastures: 0%
forest and woodland: 0%
other: 55%
Irrigated land: 115 sq km (1992 est.)
Environment:
current issues: desertification
natural hazards: NA
international agreements: NA

Note: there are 24 Jewish settlements and civilian land use sites in the Gaza Strip (August 1994 est.)

People

Population: 813,322 (July 1995 est.)
note: in addition, there are 4,800 Jewish settlers in the Gaza Strip (August 1994 est.)
Age structure:
0-14 years: 52% (female 205,192; male 215,158)
15-64 years: 45% (female 185,748; male 183,886)
65 years and over: 3% (female 13,106; male 10,232) (July 1995 est.)
Population growth rate: 4.55% (1995 est.)
Birth rate: 50.24 births/1,000 population (1995 est.)
Death rate: 4.75 deaths/1,000 population (1995 est.)
Net migration rate: 0 migrant(s)/1,000 population (1995 est.)
Infant mortality rate: 30.6 deaths/1,000 live births (1995 est.)
Life expectancy at birth:
total population: 71.09 years
male: 69.56 years
female: 72.69 years (1995 est.)
Total fertility rate: 7.74 children born/woman (1995 est.)
Nationality:
noun: NA
adjective: NA
Ethnic divisions: Palestinian Arab and other 99.4%, Jewish 0.6%
Religions: Muslim (predominantly Sunni) 98.7%, Christian 0.7%, Jewish 0.6%
Languages: Arabic, Hebrew (spoken by Israeli settlers), English (widely understood)
Literacy: NA%
Labor force: NA
by occupation: construction 33.4%, agriculture 20.0%, commerce, restaurants, and hotels 14.9%, industry 10.0%, other services 21.7% (1991)
note: excluding Jewish settlers

Government

Note: Under the Israeli-PLO Declaration of Principles on Interim Self-Government Arrangements ("the DOP"), Israel agreed to transfer certain powers and responsibilities to the Palestinian Authority, and subsequently to an elected Palestinian Council, as part of interim self-governing arrangements in the West Bank and Gaza Strip. A transfer of powers and responsibilities for the Gaza Strip and Jericho has taken place pursuant to the Israel-PLO 4 May 1994 Cairo Agreement on the Gaza Strip and the Jericho Area. The DOP provides that Israel will retain responsibility during the transitional period for external security and for internal security and public order of settlements and Israelis. Final status is

to be determined through direct negotiations within five years.
Names:
conventional long form: none
conventional short form: Gaza Strip
local long form: none
local short form: Qita Ghazzah
Digraph: GZ

Economy

Overview: In 1991 roughly 40% of Gaza Strip workers were employed across the border by Israeli industrial, construction, and agricultural enterprises, with worker remittances supplementing GDP by roughly 50%. Gaza depends upon Israel for nearly 90% of its external trade. Aggravating the impact of Israeli military administration, unrest in the territory since 1988 (intifadah) has raised unemployment and lowered the standard of living of Gazans. The Persian Gulf crisis and its aftershocks also have dealt blows to Gaza since August 1990. Worker remittances from the Gulf states have dropped, unemployment has increased, and exports have fallen. The withdrawal of Israel from the Gaza Strip in May 1994 brings a new set of adjustment problems.
National product: GDP—purchasing power parity—$1.7 billion (1993 est.)
National product real growth rate: NA%
National product per capita: $2,400 (1993 est.)
Inflation rate (consumer prices): 5.7% (1993)
Unemployment rate: 45% (1994 est.)
Budget:
revenues: $33.6 million
expenditures: $34.5 million, including capital expenditures of $NA (FY89/90)
Exports: $83 million (f.o.b., 1992)
commodities: citrus
partners: Israel, Egypt
Imports: $365 million (c.i.f., 1992)
commodities: food, consumer goods, construction materials
partners: Israel, Egypt
External debt: $NA
Industrial production: growth rate 11% (1991 est.)
Electricity: power supplied by Israel
Industries: generally small family businesses that produce textiles, soap, olive-wood carvings, and mother-of-pearl souvenirs; the Israelis have established some small-scale modern industries in an industrial center
Agriculture: olives, citrus and other fruits; vegetables; beef and dairy products
Economic aid: $240 million disbursed from international aid pledges in 1994
Currency: 1 new Israeli shekel (NIS) = 100 new agorot
Exchange rates: new Israeli shekels (NIS) per US$1—3.0270 (December 1994), 3.0111 (1994), 2.8301 (1993), 2.4591 (1992), 2.2791 (1991), 2.0162 (1990)
Fiscal year: calendar year (since 1 January 1992)

Transportation

Railroads:
total: NA km; note—one line, abandoned and in disrepair, little trackage remains
Highways:
total: NA
paved: NA
unpaved: NA
note: small, poorly developed road network
Ports: Gaza
Airports:
total: 1
with paved runways under 914 m: 1

Communications

Telephone system: NA; note—10% of Palestinian households have telephones (1992 est.)
local: NA
intercity: NA
international: NA
Radio:
broadcast stations: AM 0, FM 0, shortwave 0
radios: NA; note—95% of Palestinian households have radios (1992 est.)
Television:
broadcast stations: 0
televisions: NA; note—59% of Palestinian households have televisions (1992 est.)

Defense Forces

Branches: NA
Defense expenditures: $NA, NA% of GDP

Georgia

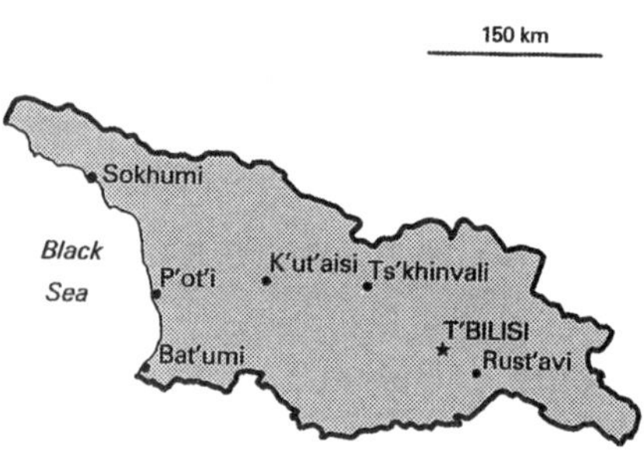

Note: Georgia has been beset by ethnic and civil strife since independence. In late 1991, the country's first elected president, Zviad GAMSAKHURDIA, was ousted in an armed coup. In October 1993, GAMSAKHURDIA and his supporters sponsored a failed attempt to retake power from the current government led by former Soviet Foreign Minister Eduard SHEVARDNADZE. The Georgian government has also faced armed separatist conflicts in the Abkhazia and South Ossetia regions. A cease-fire went into effect in South Ossetia in June 1992 and a joint Georgian-Ossetian-Russian peacekeeping force has been in place since that time. Georgian forces were driven out of the Abkhaz region in September 1993 after a yearlong war with Abkhaz separatists. Nearly 200,000 Georgian refugees have since fled Abkhazia, adding substantially to the estimated 100,000 internally displaced persons already in Georgia. Russian peacekeepers are deployed along the border of Abkhazia and the rest of Georgia.

Geography

Location: Southwestern Asia, bordering the Black Sea, between Turkey and Russia
Map references: Middle East
Area:
total area: 69,700 sq km
land area: 69,700 sq km
comparative area: slightly larger than South Carolina
Land boundaries: total 1,461 km, Armenia 164 km, Azerbaijan 322 km, Russia 723 km, Turkey 252 km
Coastline: 310 km
Maritime claims: NA
International disputes: none
Climate: warm and pleasant; Mediterranean-like on Black Sea coast
Terrain: largely mountainous with Great Caucasus Mountains in the north and Lesser Caucasus Mountains in the south; Kolkhida Lowland opens to the Black Sea in the west; Mtkvari River Basin in the east; good soils in

river valley flood plains, foothills of Kolkhida Lowland
Natural resources: forest lands, hydropower, manganese deposits, iron ores, copper, minor coal and oil deposits; coastal climate and soils allow for important tea and citrus growth
Land use:
arable land: 11%
permanent crops: 4%
meadows and pastures: 29%
forest and woodland: 38%
other: 18%
Irrigated land: 4,660 sq km (1990)
Environment:
current issues: air pollution, particularly in Rust'avi; heavy pollution of Mtkvari River and the Black Sea; inadequate supplies of potable water; soil pollution from toxic chemicals
natural hazards: NA
international agreements: party to—Biodiversity, Climate Change, Ship Pollution; signed, but not ratified—Desertification

People

Population: 5,725,972 (July 1995 est.)
Age structure:
0-14 years: 24% (female 674,331; male 707,355)
15-64 years: 64% (female 1,894,681; male 1,791,847)
65 years and over: 12% (female 410,703; male 247,055) (July 1995 est.)
Population growth rate: 0.77% (1995 est.)
Birth rate: 15.77 births/1,000 population (1995 est.)
Death rate: 8.73 deaths/1,000 population (1995 est.)
Net migration rate: 0.66 migrant(s)/1,000 population (1995 est.)
Infant mortality rate: 22.6 deaths/1,000 live births (1995 est.)
Life expectancy at birth:
total population: 73.1 years
male: 69.43 years
female: 76.95 years (1995 est.)
Total fertility rate: 2.16 children born/woman (1995 est.)
Nationality:
noun: Georgian(s)
adjective: Georgian
Ethnic divisions: Georgian 70.1%, Armenian 8.1%, Russian 6.3%, Azeri 5.7%, Ossetian 3%, Abkhaz 1.8%, other 5%
Religions: Georgian Orthodox 65%, Russian Orthodox 10%, Muslim 11%, Armenian Orthodox 8%, unknown 6%
Languages: Armenian 7%, Azeri 6%, Georgian 71% (official), Russian 9%, other 7%
Literacy: age 15 and over can read and write (1989)
total population: 99%
male: 100%
female: 98%
Labor force: 2.763 million
by occupation: industry and construction 31%, agriculture and forestry 25%, other 44% (1990)

Government

Names:
conventional long form: Republic of Georgia
conventional short form: Georgia
local long form: Sak'art'velos Respublika
local short form: Sak'art'velo
former: Georgian Soviet Socialist Republic
Digraph: GG
Type: republic
Capital: T'bilisi
Administrative divisions: 2 autonomous republics (avtomnoy respubliki, singular—avtom respublika); Abkhazia (Sokhumi), Ajaria (Bat'umi)
note: the administrative centers of the autonomous republics are included in parentheses; there are no oblasts—the rayons around T'bilisi are under direct republic jurisdiction
Independence: 9 April 1991 (from Soviet Union)
National holiday: Independence Day, 26 May (1991)
Constitution: adopted 21 February 1921; currently amending constitution for Parliamentary and popular review by late 1995
Legal system: based on civil law system
Suffrage: 18 years of age; universal
Executive branch:
chief of state: Chairman of Parliament Eduard Amvrosiyevich SHEVARDNADZE (Chairman of the Government Council since 10 March 1992; elected Chairman of Parliament in 11 October 1992; note—the Government Council has since been disbanded); election last held 11 October 1992 (next to be held October 1995); results—Eduard SHEVARDNADZE 95%
head of government: Prime Minister Otar PATSATSIA (since September 1993); Deputy Prime Ministers Avtandil MARGIANI, Zurab KERVALISHVILI (since 25 November 1992), Tamaz NADAREISHVILI (since September 1993), Temur BASILIA (since 17 March 1994), Bakur GULA (since NA)
cabinet: Council of Ministers
Legislative branch: unicameral
Georgian Parliament (Supreme Soviet): elections last held 11 October 1992 (next to be held October 1995); results—percent of vote by party NA; seats—(225 total) number of seats by party NA
Judicial branch: Supreme Court
Political parties and leaders: Citizens Union (CU), Eduard SHEVARDNADZE, Zurab SHVANIA, general secretary; National Democratic Party (NDP), Georgi (Gia) CHANTURIA, Ivane GIORGADZE; United Republican Party, umbrella organization for parties including the GPF and the Charter 1991 Party, cochairmen Bakhtand DZABIRADZE, Nodar NATADZE, and Theodor PAATASHVILI; Georgian Popular Front (GPF), Nodar NATADZE, chairman; Charter 1991 Party, Theodor PAATASHVILI; Georgian Social Democratic Party (GSDP), Guram MUCHAIDZE, secretary general; National Reconstruction and Rebirth of Georgia Union, Valerian ADVADZE; Christian Democratic Union (CDU), Irakli SHENGELAYA; Democratic Georgia Union (DGU), El'dar SHENGELAYA; National Independence Party (NIP), Irakliy TSERETELI, chairman; Georgian Monarchists' Party (GMP), Temur ZHORZHOLIANI; Green Party, Zurab ZHVANIA; Republican Party (RP), Ivliane KHAINDRAVA; Workers' Union of Georgia (WUG), Vakhtang GABUNIA; Agrarian Party of Georgia (APG), Roin LIPARTELIANI; Choice Society (Archevani), Jaba IOSELIANI, chairman; Georgian Workers Communist Party, Panteleimon GIORGADZE, chairman; National Liberation Front, Tengiz SIGULA, chairman
Other political or pressure groups: supporters of ousted President Zviad GAMSAKHURDIA (deceased 1 January 1994) boycotted the October elections and remain a source of opposition
Member of: BSEC, CCC, CIS, EBRD, ECE, IBRD, ICAO, IDA, ILO, IMF, IMO, INMARSAT, INTERPOL, IOC, IOM (observer), ITU, NACC, OSCE, PFP, UN, UNCTAD, UNESCO, UNIDO, UPU, WHO, WIPO, WMO, WTO
Diplomatic representation in US:
chief of mission: Ambassador Tedo JAPARIDZE
chancery: (temporary) Suite 424, 1511 K Street NW, Washington, DC 20005
telephone: [1] (202) 393-6060, 5959
US diplomatic representation:
chief of mission: Ambassador Kent N. BROWN
embassy: #25 Antoneli Street, T'bilisi 380026
mailing address: use embassy street address
telephone: [7] (8832) 98-99-67, 93-38-03
FAX: [7] (8832) 93-37-59
Flag: maroon field with small rectangle in upper hoist side corner; rectangle divided horizontally with black on top, white below

Economy

Overview: Georgia's economy has traditionally revolved around Black Sea tourism; cultivation of citrus fruits, tea, and grapes; mining of manganese and copper; and a small industrial sector producing wine, metals, machinery, chemicals, and textiles. The country imports the bulk of its energy needs, including natural gas and oil products. Its only sizable domestic energy resource is hydropower. Since 1990, widespread conflicts, e.g., in Abkhazia, South Ossetia, and

Mingreliya, have severely aggravated the economic crisis resulting from the disintegration of the Soviet command economy in December 1991. Throughout 1993 and 1994, much of industry was functioning at only 20% of capacity; heavy disruptions in agricultural cultivation were reported; and tourism was shut down. The country is precariously dependent on US and EU humanitarian grain shipments, as most other foods are priced beyond reach of the average citizen. Georgia is also suffering from an acute energy crisis, as it is having problems paying for even minimal imports. Georgia is pinning its hopes for recovery on reestablishing trade ties with Russia and on developing international transportation through the key Black Sea ports of P'ot'i and Bat'umi. The government began a tenuous program in 1994 aiming to stabilize prices and reduce large consumer subsidies.
National product: GDP—purchasing power parity—$6 billion (1994 estimate as extrapolated from World Bank estimate for 1992)
National product real growth rate: -30% (1994 est.)
National product per capita: $1,060 (1994 est.)
Inflation rate (consumer prices): 40.5% per month (2nd half 1993 est.)
Unemployment rate: officially less than 5% but real unemployment may be more than 20%, with even larger numbers of underemployed workers
Budget:
revenues: $NA
expenditures: $NA, including capital expenditures of $NA
Exports: $NA
commodities: citrus fruits, tea, wine, other agricultural products; diverse types of machinery; ferrous and nonferrous metals; textiles; chemicals; fuel re-exports
partners: Russia, Turkey, Armenia, Azerbaijan (1992)
Imports: $NA
commodities: fuel, grain and other foods, machinery and parts, transport equipment
partners: Russia, Azerbaijan, Turkey (1993); note—EU and US sent humanitarian food shipments
External debt: NA (T'bilisi owes about $400 million to Turkmenistan for natural gas as of January 1995)
Industrial production: growth rate -27% (1993); accounts for 36% of GDP
Electricity:
capacity: 4,410,000 kW
production: 9.1 billion kWh
consumption per capita: 1,526 kWh (1993)
Industries: heavy industrial products include raw steel, rolled steel, airplanes; machine tools, foundry equipment, electric locomotives, tower cranes, electric welding equipment, machinery for food preparation and meat packing, electric motors, process control equipment, instruments; trucks, tractors, and other farm machinery; light industrial products, including cloth, hosiery, and shoes; chemicals; wood-working industries; the most important food industry is wine
Agriculture: accounted for 97% of former USSR citrus fruits and 93% of former USSR tea; important producer of grapes; also cultivates vegetables and potatoes; dependent on imports for grain, dairy products, sugar; small livestock sector
Illicit drugs: illicit cultivator of cannabis and opium poppy; mostly for domestic consumption; used as transshipment point for illicit drugs to Western Europe
Economic aid:
recipient: heavily dependent on US and EU for humanitarian grain shipments; EC granted around $70 million in trade credits in 1992 and another $40 million in 1993; Turkey granted $50 million in 1993; smaller scale credits granted by Russia and China
Currency: coupons introduced in April 1993 to be followed by introduction of the lari at undetermined future date; in July 1993 use of the Russian ruble was banned
Exchange rates: coupons per $US1—1,280,000 (end December 1994)
Fiscal year: calendar year

Transportation

Railroads:
total: 1,570 km in common carrier service; does not include industrial lines
broad gauge: 1,570 km 1.520-m gauge (1990)
Highways:
total: 33,900 km
paved and graveled: 29,500 km
unpaved: earth 4,400 km (1990)
Pipelines: crude oil 370 km; refined products 300 km; natural gas 440 km (1992)
Ports: Bat'umi, P'ot'i, Sokhumi
Merchant marine:
total: 32 ships (1,000 GRT or over) totaling 419,416 GRT/640,897 DWT
ships by type: bulk 11, cargo 1, oil tanker 19, short-sea passenger 1
Airports:
total: 28
with paved runways over 3,047 m: 1
with paved runways 2,438 to 3,047 m: 7
with paved runways 1,524 to 2,437 m: 4
with paved runways 914 to 1,523 m: 1
with paved runways under 914 m: 1
with unpaved runways over 3,047 m: 1
with unpaved runways 2,438 to 3,047 m: 1
with unpaved runways 1,524 to 2,438 m: 1
with unpaved runways 914 to 1,523 m: 5
with unpaved runways under 914 m: 6
Note: transportation network is in poor condition and disrupted by ethnic conflict, criminal activities, and fuel shortages; network lacks maintenance and repair

Communications

Telephone system: 672,000 telephones (mid-1993); 117 telephones/1,000 persons; poor telephone service; 339,000 unsatisfied applications for telephones (December 1990)
local: NA
intercity: NA
international: links via landline to CIS members and Turkey; low-capacity satellite link and leased international connections via the Moscow international gateway switch with other countries; international electronic mail and telex service available
Radio:
broadcast stations: AM NA, FM NA, shortwave NA
radios: NA
Television:
broadcast stations: NA
televisions: NA

Defense Forces

Branches: Army, Navy, Air Force, Interior Ministry Troops, Border Guards/National Guard
Manpower availability: males age 15-49 1,385,593; males fit for military service 1,095,835; males reach military age (18) annually 42,207 (1995 est.)
Defense expenditures: exchange rate conversion—$85 million, NA% of GDP (1992)
Note: Georgian forces are poorly organized and not fully under the government's control

Germany

Geography

Location: Central Europe, bordering the Baltic Sea and the North Sea, between the Netherlands and Poland, south of Denmark
Map references: Europe
Area:
total area: 356,910 sq km
land area: 349,520 sq km
comparative area: slightly smaller than Montana
note: includes the formerly separate Federal Republic of Germany, the German Democratic Republic, and Berlin following formal unification on 3 October 1990
Land boundaries: total 3,621 km, Austria 784 km, Belgium 167 km, Czech Republic 646 km, Denmark 68 km, France 451 km, Luxembourg 138 km, Netherlands 577 km, Poland 456 km, Switzerland 334 km
Coastline: 2,389 km
Maritime claims:
continental shelf: 200-m depth or to the depth of exploitation
exclusive economic zone: 200 nm
territorial sea: 12 nm
International disputes: none
Climate: temperate and marine; cool, cloudy, wet winters and summers; occasional warm, tropical foehn wind; high relative humidity
Terrain: lowlands in north, uplands in center, Bavarian Alps in south
Natural resources: iron ore, coal, potash, timber, lignite, uranium, copper, natural gas, salt, nickel
Land use:
arable land: 34%
permanent crops: 1%
meadows and pastures: 16%
forest and woodland: 30%
other: 19%
Irrigated land: 4,800 sq km (1989 est.)
Environment:
current issues: emissions from coal-burning utilities and industries and lead emissions from vehicle exhausts (the result of continued use of leaded fuels) contribute to air pollution; acid rain, resulting from sulfur dioxide emissions, is damaging forests; heavy pollution in the Baltic Sea from raw sewage and industrial effluents from rivers in eastern Germany
natural hazards: NA
international agreements: party to—Air Pollution, Air Pollution-Nitrogen Oxides, Air Pollution-Sulphur 85, Air Pollution-Volatile Organic Compounds, Antarctic-Environmental Protocol, Antarctic Treaty, Biodiversity, Climate Change, Endangered Species, Environmental Modification, Law of the Sea, Marine Dumping, Nuclear Test Ban, Ozone Layer Protection, Ship Pollution, Tropical Timber 83, Wetlands, Whaling; signed, but not ratified—Air Pollution-Sulphur 94, Desertification, Hazardous Wastes
Note: strategic location on North European Plain and along the entrance to the Baltic Sea

People

Population: 81,337,541 (July 1995 est.)
Age structure:
0-14 years: 16% (female 6,518,108; male 6,857,577)
15-64 years: 68% (female 27,167,824; male 28,130,083)
65 years and over: 16% (female 8,127,938; male 4,536,011) (July 1995 est.)
Population growth rate: 0.26% (1995 est.)
Birth rate: 10.98 births/1,000 population (1995 est.)
Death rate: 10.83 deaths/1,000 population (1995 est.)
Net migration rate: 2.46 migrant(s)/1,000 population (1995 est.)
Infant mortality rate: 6.3 deaths/1,000 live births (1995 est.)
Life expectancy at birth:
total population: 76.62 years
male: 73.5 years
female: 79.92 years (1995 est.)
Total fertility rate: 1.5 children born/woman (1995 est.)
Nationality:
noun: German(s)
adjective: German
Ethnic divisions: German 95.1%, Turkish 2.3%, Italians 0.7%, Greeks 0.4%, Poles 0.4%, other 1.1% (made up largely of people fleeing the war in the former Yugoslavia)
Religions: Protestant 45%, Roman Catholic 37%, unaffiliated or other 18%
Languages: German
Literacy: age 15 and over can read and write (1991 est.)
total population: 99%
Labor force: 36.75 million
by occupation: industry 41%, agriculture 6%, other 53% (1987)

Government

Names:
conventional long form: Federal Republic of Germany
conventional short form: Germany
local long form: Bundesrepublik Deutschland
local short form: Deutschland
Digraph: GM
Type: federal republic
Capital: Berlin
note: the shift from Bonn to Berlin will take place over a period of years with Bonn retaining many administrative functions and several ministries
Administrative divisions: 16 states (laender, singular—land); Baden-Wuerttemberg, Bayern, Berlin, Brandenburg, Bremen, Hamburg, Hessen, Mecklenburg-Vorpommern, Niedersachsen, Nordrhein-Westfalen, Rheinland-Pfalz, Saarland, Sachsen, Sachsen-Anhalt, Schleswig-Holstein, Thueringen
Independence: 18 January 1871 (German Empire unification); divided into four zones of occupation (UK, US, USSR, and later, France) in 1945 following World War II; Federal Republic of Germany (FRG or West Germany) proclaimed 23 May 1949 and included the former UK, US, and French zones; German Democratic Republic (GDR or East Germany) proclaimed 7 October 1949 and included the former USSR zone; unification of West Germany and East Germany took place 3 October 1990; all four power rights formally relinquished 15 March 1991
National holiday: German Unity Day (Day of Unity), 3 October (1990)
Constitution: 23 May 1949, known as Basic Law; became constitution of the united German people 3 October 1990
Legal system: civil law system with indigenous concepts; judicial review of legislative acts in the Federal Constitutional Court; has not accepted compulsory ICJ jurisdiction
Suffrage: 18 years of age; universal
Executive branch:
chief of state: President Roman HERZOG (since 1 July 1994)
head of government: Chancellor Dr. Helmut KOHL (since 4 October 1982)
cabinet: Cabinet; appointed by the president upon the proposal of the chancellor
Legislative branch: bicameral chamber (no official name for the two chambers as a whole)
Federal Assembly (Bundestag): last held 16 October 1994 (next to be held by NA 1998); results—CDU 34.2%, SPD 36.4%, Alliance 90/Greens 7.3%, CSU 7.3%, FDP 6.9%, PDS 4.4%, Republicans 1.9% ; seats—(662 total, but number can vary) CDU 244, SPD 252, Alliance 90/Greens 49, CSU 50, FDP 47, PDS

30; elected by direct popular vote under a system combining direct and proportional representation; a party must win 5% of the national vote or 3 direct mandates to gain representation
Federal Council (Bundesrat): State governments are directly represented by votes; each has 3 to 6 votes depending on size and are required to vote as a block; current composition: votes—(68 total) SPD-led states 37, CDU-led states 31
Judicial branch: Federal Constitutional Court (Bundesverfassungsgericht)
Political parties and leaders: Christian Democratic Union (CDU), Helmut KOHL, chairman; Christian Social Union (CSU), Theo WAIGEL, chairman; Free Democratic Party (FDP), Klaus KINKEL, chairman; Social Democratic Party (SPD), Rudolf SCHARPING, chairman; Alliance '90/Greens, Krista SAGER, Juergen TRITTIN, cochairpersons; Party of Democratic Socialism (PDS), Lothar BISKY, chairman; Republikaner, Rolf SCHLIERER, chairman; National Democratic Party (NPD), Guenter DECKERT; Communist Party (DKP), Rolf PRIEMER
Other political or pressure groups: expellee, refugee, and veterans groups
Member of: AfDB, AG (observer), AsDB, Australia Group, BDEAC, BIS, CBSS, CCC, CDB (non-regional), CE, CERN, EBRD, EC, ECE, EIB, ESA, FAO, G- 5, G- 7, G-10, GATT, IADB, IAEA, IBRD, ICAO, ICC, ICFTU, ICRM, IDA, IEA, IFAD, IFC, IFRCS, ILO, IMF, IMO, INMARSAT, INTELSAT, INTERPOL, IOC, IOM, ISO, ITU, MINURSO, MTCR, NACC, NAM (guest), NATO, NEA, NSG, OAS (observer), OECD, OSCE, PCA, UN, UNCTAD, UNESCO, UNHCR, UNIDO, UNITAR, UNOMIG, UPU, WEU, WHO, WIPO, WMO, WTO, ZC
Diplomatic representation in US:
chief of mission: Ambassador Juergen CHROBOG
chancery: 4645 Reservoir Road NW, Washington, DC 20007
telephone: [1] (202) 298-4000
FAX: [1] (202) 298-4249
consulate(s) general: Atlanta, Boston, Chicago, Detroit, Houston, Los Angeles, Miami, New York, San Francisco, Seattle
consulate(s): Manila (Trust Territories of the Pacific Islands) and Wellington (America Samoa)
US diplomatic representation:
chief of mission: Ambassador Charles E. REDMAN
embassy: Deichmanns Aue 29, 53170 Bonn
mailing address: Unit 21701, Bonn; APO AE 09080
telephone: [49] (228) 3391
FAX: [49] (228) 339-2663
branch office: Berlin
consulate(s) general: Frankfurt, Hamburg, Leipzig, Munich, and Stuttgart
Flag: three equal horizontal bands of black (top), red, and yellow

Economy

Overview: Five years after the fall of the Berlin Wall, progress towards economic integration between eastern and western Germany is clearly visible, yet the eastern region almost certainly will remain dependent on subsidies funded by western Germany until well into the next century. The staggering $390 billion in western German assistance that the eastern states have received since 1990—40 times the amount in real terms of US Marshall Fund aid sent to West Germany after World War II—is just beginning to have an impact on the eastern German standard of living, which plummeted after unification. Assistance to the east continues to run at roughly $100 billion annually. Although the growth rate in the east was much greater than in the west in 1993-94, eastern GDP per capita nonetheless remains well below preunification levels; it will take 10-15 years for the eastern states to match western Germany's living standards. The economic recovery in the east is led by the construction industries which account for one-third of industrial output, with growth increasingly supported by the service sectors and light manufacturing industries. Eastern Germany's economy is changing from one anchored on manufacturing to a more service-oriented economy. Western Germany, with three times the per capita output of the eastern states, has an advanced market economy and is a world leader in exports. The strong recovery in 1994 from recession began in the export sector and spread to the investment and consumption sectors in response to falling interest rates. Western Germany has a highly urbanized and skilled population that enjoys excellent living standards, abundant leisure time, and comprehensive social welfare benefits. It is relatively poor in natural resources, coal being the most important mineral. Western Germany's world-class companies manufacture technologically advanced goods. The region's economy is mature: services and manufacturing account for the dominant share of economic activities, and raw materials and semimanufactured goods constitute a large portion of imports.
National product:
Germany: GDP—purchasing power parity—$1.3446 trillion (1994 est.)
western: GDP—purchasing power parity—$1.2363 trillion (1994 est.)
eastern: GDP—purchasing power parity—$108.3 billion (1994 est.)
National product real growth rate:
Germany: 2.9% (1994 est.)
western: 2.3% (1994 est.)
eastern: 9.2% (1994 est.)
National product per capita:
Germany: $16,580 (1994 est.)
western: $19,660 (1994 est.)
eastern: $5,950 (1994 est.)
Inflation rate (consumer prices):
western: 3% (1994)
eastern: 3.2% (1994 est.)
Unemployment rate:
western: 8.2% (December 1994)
eastern: 13.5% (December 1994)
Budget:
revenues: $690 billion
expenditures: $780 billion, including capital expenditures of $96.5 billion (1994)
Exports: $437 billion (f.o.b., 1994)
commodities: manufactures 89.3% (including machines and machine tools, chemicals, motor vehicles, iron and steel products), agricultural products 5.5%, raw materials 2.7%, fuels 1.3% (1993)
partners: EC 47.9% (France 11.7%, Netherlands 7.4%, Italy 7.5%, UK 7.7%, Belgium-Luxembourg 6.6%), EFTA 15.5%, US 7.7%, Eastern Europe 5.2%, OPEC 3.0% (1993)
Imports: $362 billion (f.o.b., 1994)
commodities: manufactures 75.1%, agricultural products 10.0%, fuels 8.3%, raw materials 5.0% (1993)
partners: EC 46.4% (France 11.3%, Netherlands 8.4%, Italy 8.1%, UK 6.0%, Belgium-Luxembourg 5.7%), EFTA 14.3%, US 7.3%, Japan 6.3%, Eastern Europe 5.1%, OPEC 2.6% (1993)
External debt: $NA
Industrial production:
western: growth rate 2.8% (1994)
eastern: growth rate $NA
Electricity:
capacity: 115,430,000 kW
production: 493 billion kWh
consumption per capita: 5,683 kWh (1993)
Industries:
western: among world's largest and technologically advanced producers of iron, steel, coal, cement, chemicals, machinery, vehicles, machine tools, electronics; food and beverages
eastern: metal fabrication, chemicals, brown coal, shipbuilding, machine building, food and beverages, textiles, petroleum refining
Agriculture:
western: accounts for about 1% of GDP (including fishing and forestry); diversified crop and livestock farming; principal crops and livestock include potatoes, wheat, barley, sugar beets, fruit, cabbage, cattle, pigs, poultry; net importer of food
eastern: accounts for about 10% of GDP (including fishing and forestry); principal crops—wheat, rye, barley, potatoes, sugar beets, fruit; livestock products include pork, beef, chicken, milk, hides and skins; net importer of food
Illicit drugs: source of precursor chemicals for South American cocaine processors; transshipment point for Southwest Asian

heroin and Latin American cocaine for West European markets
Economic aid:
western-donor: ODA and OOF commitments (1970-89), $75.5 billion
eastern-donor: bilateral to non-Communist less developed countries (1956-89) $4 billion
Currency: 1 deutsche mark (DM) = 100 pfennige
Exchange rates: deutsche marks (DM) per US$1—1.5313 (January 1995), 1.6228 (1994), 1.6533 (1993), 1.5617 (1992), 1.6595 (1991), 1.6157 (1990)
Fiscal year: calendar year

Transportation

Railroads:
total: 43,457 km
standard gauge: 43,190 km (electrified 16,694 km)
narrow gauge: 267 km (1994)
Highways:
total: 636,282 km
paved: 501,282 km (10,955 km of autobahn)
unpaved: 135,000 km (1991)
Inland waterways:
western: 5,222 km, of which almost 70% are usable by craft of 1,000-metric-ton capacity or larger; major rivers include the Rhine and Elbe; Kiel Canal is an important connection between the Baltic Sea and North Sea
eastern: 2,319 km (1988)
Pipelines: crude oil 3,644 km; petroleum products 3,946 km; natural gas 97,564 km (1988)
Ports: Berlin, Bonn, Brake, Bremen, Bremerhaven, Cologne, Dresden, Duisburg, Emden, Hamburg, Karlsruhe, Kiel, Lubeck, Magdeburg, Mannheim, Rostock, Stuttgart
Merchant marine:
total: 481 ships (1,000 GRT or over) totaling 5,065,074 GRT/6,409,198 DWT
ships by type: barge carrier 6, bulk 8, cargo 224, chemical tanker 16, combination bulk 4, combination ore/oil 5, container 158, liquefied gas tanker 13, oil tanker 10, passenger 3, railcar carrier 4, refrigerated cargo 7, roll-on/roll-off cargo 18, short-sea passenger 5
note: the German register includes ships of the former East and West Germany
Airports:
total: 660
with paved runways over 3,047 m: 13
with paved runways 2,438 to 3,047 m: 64
with paved runways 1,524 to 2,437 m: 68
with paved runways 914 to 1,523 m: 53
with paved runways under 914 m: 381
with unpaved runways over 3,047 m: 2
with unpaved runways 2,438 to 3,047 m: 8
with unpaved runways 1,524 to 2,438 m: 9
with unpaved runways 914 to 1,523 m: 62

Communications

Telephone system:
western: 40,300,000 telephones; highly developed, modern telecommunication service to all parts of the country; fully adequate in all respects; intensively developed, highly redundant cable and microwave radio relay networks, all completely automatic
local: very modern
intercity: domestic satellite, microwave radio relay, and cable systems
international: 12 INTELSAT (Atlantic Ocean), 2 INTELSAT (Indian Ocean), and 1 EUTELSAT earth station; 2 HF radiocommunication centers; tropospheric scatter links
eastern: 3,970,000 telephones; badly needs modernization
local: NA
intercity: NA
international: 1 INTELSAT earth station and 1 Intersputnik system
Radio:
western: NA
broadcast stations: AM 80, FM 470, shortwave 0
radios: NA
eastern: NA
broadcast stations: AM 23, FM 17, shortwave 0
radios: 67 million
Television:
broadcast stations: 246 (repeaters 6,000); note—there are 15 Russian repeaters in eastern Germany
televisions: 25 million in western Germany, 6 million in eastern Germany

Defense Forces

Branches: Army, Navy (includes Naval Air Arm), Air Force, Border Police, Coast Guard
Manpower availability: males 15-49 20,274,127; males fit for military service 17,472,940; males reach military age (18) annually 428,082 (1995 est.)
Defense expenditures: exchange rate conversion—$40 billion, 1.8% of GNP (1994)

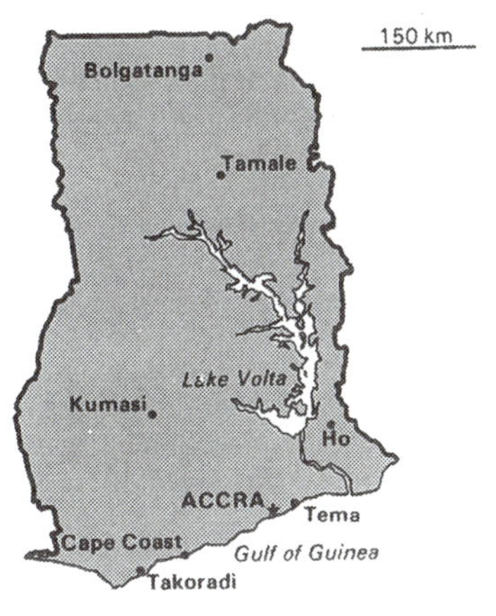

Geography

Location: Western Africa, bordering the North Atlantic Ocean, between Cote d'Ivoire and Togo
Map references: Africa
Area:
total area: 238,540 sq km
land area: 230,020 sq km
comparative area: slightly smaller than Oregon
Land boundaries: total 2,093 km, Burkina 548 km, Cote d'Ivoire 668 km, Togo 877 km
Coastline: 539 km
Maritime claims:
contiguous zone: 24 nm
continental shelf: 200 nm
exclusive economic zone: 200 nm
territorial sea: 12 nm
International disputes: none
Climate: tropical; warm and comparatively dry along southeast coast; hot and humid in southwest; hot and dry in north
Terrain: mostly low plains with dissected plateau in south-central area
Natural resources: gold, timber, industrial diamonds, bauxite, manganese, fish, rubber
Land use:
arable land: 5%
permanent crops: 7%
meadows and pastures: 15%
forest and woodland: 37%
other: 36%
Irrigated land: 80 sq km (1989)
Environment:
current issues: recent drought in north severely affecting agricultural activities; deforestation; overgrazing; soil erosion; poaching and habitat destruction threatens wildlife populations; water pollution; inadequate supplies of potable water
natural hazards: dry, dusty, harmattan winds occur from January to March; droughts
international agreements: party to—Biodiversity, Endangered Species, Environmental Modification, Law of the Sea, Nuclear Test Ban, Ozone Layer Protection,

Ship Pollution, Tropical Timber 83, Wetlands; signed, but not ratified—Climate Change, Desertification, Marine Life Conservation
Note: Lake Volta is the world's largest artificial lake; northeasterly harmattan wind (January to March)

People

Population: 17,763,138 (July 1995 est.)
Age structure:
0-14 years: 46% (female 4,030,154; male 4,069,945)
15-64 years: 51% (female 4,638,451; male 4,494,533)
65 years and over: 3% (female 276,186; male 253,869) (July 1995 est.)
Population growth rate: 3.06% (1995 est.)
Birth rate: 43.57 births/1,000 population (1995 est.)
Death rate: 12.02 deaths/1,000 population (1995 est.)
Net migration rate: -0.94 migrant(s)/1,000 population (1995 est.)
Infant mortality rate: 81.7 deaths/1,000 live births (1995 est.)
Life expectancy at birth:
total population: 55.85 years
male: 53.88 years
female: 57.88 years (1995 est.)
Total fertility rate: 6.09 children born/woman (1995 est.)
Nationality:
noun: Ghanaian(s)
adjective: Ghanaian
Ethnic divisions: black African 99.8% (major tribes—Akan 44%, Moshi-Dagomba 16%, Ewe 13%, Ga 8%), European and other 0.2%
Religions: indigenous beliefs 38%, Muslim 30%, Christian 24%, other 8%
Languages: English (official), African languages (including Akan, Moshi-Dagomba, Ewe, and Ga)
Literacy: age 15 and over can read and write (1990 est.)
total population: 60%
male: 70%
female: 51%
Labor force: 3.7 million
by occupation: agriculture and fishing 54.7%, industry 18.7%, sales and clerical 15.2%, services, transportation, and communications 7.7%, professional 3.7%

Government

Names:
conventional long form: Republic of Ghana
conventional short form: Ghana
former: Gold Coast
Digraph: GH
Type: constitutional democracy
Capital: Accra
Administrative divisions: 10 regions; Ashanti, Brong-Ahafo, Central, Eastern, Greater Accra, Northern, Upper East, Upper West, Volta, Western
Independence: 6 March 1957 (from UK)
National holiday: Independence Day, 6 March (1957)
Constitution: new constitution approved 28 April 1992
Legal system: based on English common law and customary law; has not accepted compulsory ICJ jurisdiction
Suffrage: 18 years of age; universal
Executive branch:
chief of state and head of government: President Jerry John RAWLINGS (since 3 November 1992) election last held 3 November 1992 (next to be held November1996)
cabinet: Cabinet; president nominates members subject to approval by the Parliament
Legislative branch: unicameral
National Assembly: elections last held 29 December 1992 (next to be held December 1996); results—opposition boycotted the election; the National Democratic Congress won 198 of 200 total seats and independents won 2
Judicial branch: Supreme Court
Political parties and leaders: National Democratic Congress, Jerry John RAWLINGS; New Patriotic Party, Albert Adu BOAHEN; People's Heritage Party, Alex ERSKINE; various other smaller parties
Member of: ACP, AfDB, C, CCC, ECA, ECOWAS, FAO, G-24, G-77, GATT, IAEA, IBRD, ICAO, ICFTU, ICRM, IDA, IFAD, IFC, IFRCS, ILO, IMF, IMO, INTELSAT, INTERPOL, IOC, IOM (observer), ITU, MINURSO, NAM, OAU, UN, UNAMIR, UNCTAD, UNESCO, UNIDO, UNIFIL, UNIKOM, UNPROFOR, UNU, UPU, WCL, WFTU, WHO, WIPO, WMO, WTO
Diplomatic representation in US:
chief of mission: Ambassador Ekwow SPIO-GARBRAH
chancery: 3512 International Drive NW, Washington, DC 20008
telephone: [1] (202) 686-4520
FAX: [1] (202) 686-4527
consulate(s) general: New York
US diplomatic representation:
chief of mission: Ambassador Kenneth L. BROWN (scheduled to leave in June 1995)
embassy: Ring Road East, East of Danquah Circle, Accra
mailing address: P. O. Box 194, Accra
telephone: [233] (21) 775348, 775349, 775297, 775298
FAX: [233] (21) 776008
Flag: three equal horizontal bands of red (top), yellow, and green with a large black five-pointed star centered in the gold band; uses the popular pan-African colors of Ethiopia; similar to the flag of Bolivia, which has a coat of arms centered in the yellow band

Economy

Overview: Well endowed with natural resources, Ghana is relatively well off, having twice the per capita output of the poorer countries in West Africa. Heavily reliant on international assistance, Ghana has made steady progress in liberalizing its economy since 1983. Overall growth continued at a rate of approximately 5% in 1994, due largely to increased gold, timber, and cocoa production—major sources of foreign exchange. The economy, however, continues to revolve around subsistence agriculture, which accounts for 45% of GDP and employs 55% of the work force, mainly small landholders. Public sector wage increases, regional peacekeeping commitments, and the containment of internal unrest in the underdeveloped north have placed substantial demands on the government's budget and have led to inflationary deficit financing and a 27% depreciation of the cedi in 1994.
National product: GDP—purchasing power parity—$22.6 billion (1994 est.)
National product real growth rate: 5% (1994 est.)
National product per capita: $1,310 (1994 est.)
Inflation rate (consumer prices): 25% (1993 est.)
Unemployment rate: 10% (1991)
Budget:
revenues: $1.05 billion
expenditures: $1.2 billion, including capital expenditures of $178 million (1993)
Exports: $1 billion (f.o.b., 1993 est.)
commodities: cocoa 40%, gold, timber, tuna, bauxite, and aluminum
partners: Germany 31%, US 12%, UK 11%, Netherlands 6%, Japan 5% (1991)
Imports: $1.7 billion (c.i.f., 1993 est.)
commodities: petroleum 16%, consumer goods, foods, intermediate goods, capital equipment
partners: UK 22%, US 11%, Germany 9%, Japan 6%
External debt: $4.6 billion (December 1993 est.)
Industrial production: growth rate 3.4% in manufacturing (1993); accounts for almost 15% of GDP
Electricity:
capacity: 1,180,000 kW
production: 6.1 billion kWh
consumption per capita: 323 kWh (1993)
Industries: mining, lumbering, light manufacturing, aluminum, food processing
Agriculture: accounts for almost 50% of GDP (including fishing and forestry); the major cash crop is cocoa; other principal crops—rice, coffee, cassava, peanuts, corn, shea nuts, timber; normally self-sufficient in food

Illicit drugs: illicit producer of cannabis for the international drug trade; transit hub for Southwest and Southeast Asian heroin destined for Europe and the US
Economic aid:
recipient: US commitments, including Ex-Im (FY70-89), $455 million; Western (non-US) countries, ODA and OOF bilateral commitments (1970-89), $2.6 billion; OPEC bilateral aid (1979-89), $78 million; Communist countries (1970-89) $106 million
Currency: 1 new cedi (C) = 100 pesewas
Exchange rates: new cedis per US$1—1,046.74 (December 1994), 936.71 (1994), 649.06 (1993), 437.09 (1992), 367.83 (1991), 326.33 (1990)
Fiscal year: calendar year

Transportation

Railroads:
total: 953 km; note—undergoing major renovation
narrow gauge: 953 km 1.067-m gauge (32 km double track)
Highways:
total: 32,250 km
paved: concrete, bituminous 6,084 km
unpaved: gravel, crushed stone, improved earth 26,166 km
Inland waterways: Volta, Ankobra, and Tano Rivers provide 168 km of perennial navigation for launches and lighters; Lake Volta provides 1,125 km of arterial and feeder waterways
Pipelines: none
Ports: Takoradi, Tema
Merchant marine:
total: 3 ships (1,000 GRT or over) totaling 27,427 GRT/35,894 DWT
ships by type: cargo 2, refrigerated cargo 1
Airports:
total: 12
with paved runways 2,438 to 3,047 m: 3
with paved runways 1,524 to 2,437 m: 1
with paved runways 914 to 1,523 m: 2
with paved runways under 914 m: 2
with unpaved runways 1,524 to 2,438 m: 2
with unpaved runways 914 to 1,523 m: 2

Communications

Telephone system: 42,300 telephones; poor to fair system; telephone density—2.4/1,000 persons
local: NA
intercity: primarily microwave radio relay
international: 1 Atlantic Ocean INTELSAT earth station
Radio:
broadcast stations: AM 4, FM 1, shortwave 0
radios: NA
Television:
broadcast stations: 4 (translators 8)
televisions: NA

Defense Forces

Branches: Army, Navy, Air Force, Police Force, Palace Guard, Civil Defense
Manpower availability: males age 15-49 3,975,767; males fit for military service 2,217,032; males reach military age (18) annually 170,723 (1995 est.)
Defense expenditures: exchange rate conversion—$108 million, 1.5% of GDP (1993)

Gibraltar
(dependent territory of the UK)

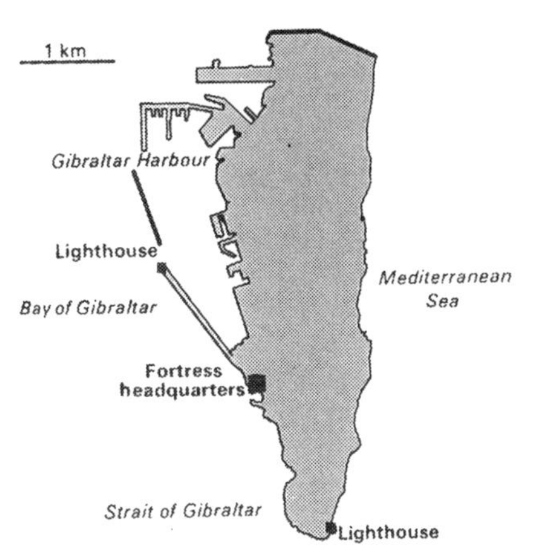

Geography

Location: Southwestern Europe, bordering the Strait of Gibraltar, which links the Mediterranean Sea and the North Atlantic Ocean, on the southern coast of Spain
Map references: Europe
Area:
total area: 6.5 sq km
land area: 6.5 sq km
comparative area: about 11 times the size of The Mall in Washington, DC
Land boundaries: total 1.2 km, Spain 1.2 km
Coastline: 12 km
Maritime claims:
territorial sea: 3 nm
International disputes: source of occasional friction between Spain and the UK
Climate: Mediterranean with mild winters and warm summers
Terrain: a narrow coastal lowland borders The Rock
Natural resources: negligible
Land use:
arable land: 0%
permanent crops: 0%
meadows and pastures: 0%
forest and woodland: 0%
other: 100%
Irrigated land: NA sq km
Environment:
current issues: limited natural freshwater resources, so large concrete or natural rock water catchments collect rain water
natural hazards: NA
international agreements: NA
Note: strategic location on Strait of Gibraltar that links the North Atlantic Ocean and Mediterranean Sea

People

Population: 31,874 (July 1995 est.)
Age structure:
0-14 years: 24% (female 3,757; male 3,835)
15-64 years: 63% (female 9,730; male 10,485)
65 years and over: 13% (female 2,360; male

1,707) (July 1995 est.)
Population growth rate: 0.62% (1995 est.)
Birth rate: 15 births/1,000 population (1995 est.)
Death rate: 8.85 deaths/1,000 population (1995 est.)
Net migration rate: 0 migrant(s)/1,000 population (1995 est.)
Infant mortality rate: 7.9 deaths/1,000 live births (1995 est.)
Life expectancy at birth:
total population: 76.61 years
male: 73.7 years
female: 79.48 years (1995 est.)
Total fertility rate: 2.29 children born/woman (1995 est.)
Nationality:
noun: Gibraltarian(s)
adjective: Gibraltar
Ethnic divisions: Italian, English, Maltese, Portuguese, Spanish
Religions: Roman Catholic 74%, Protestant 11% (Church of England 8%, other 3%), Moslem 8%, Jewish 2%, none or other 5% (1981)
Languages: English (used in schools and for official purposes), Spanish, Italian, Portuguese, Russian
Literacy: NA%
Labor force: 14,800 (including non-Gibraltar laborers)
note: UK military establishments and civil government employ nearly 50% of the labor force

Government

Names:
conventional long form: none
conventional short form: Gibraltar
Digraph: GI
Type: dependent territory of the UK
Capital: Gilbraltar
Administrative divisions: none (dependent territory of the UK)
Independence: none (dependent territory of the UK)
National holiday: Commonwealth Day (second Monday of March)
Constitution: 30 May 1969
Legal system: English law
Suffrage: 18 years of age; universal, plus other UK subjects resident six months or more
Executive branch:
chief of state: Queen ELIZABETH II (since 6 February 1952), represented by Governor and Commander in Chief Gen. Sir John CHAPPLE (since NA March 1993)
head of government: Chief Minister Joe BOSSANO (since 25 March 1988)
Gibraltar Council: advises the governor
cabinet: Council of Ministers; appointed from the elected members of the Assembly by the governor in consultation with the chief minister
Legislative branch: unicameral
House of Assembly: elections last held on 16 January 1992 (next to be held January 1996); results—SL 73.3%; seats—(18 total, 15 elected) number of seats by party NA
Judicial branch: Supreme Court, Court of Appeal
Political parties and leaders: Gibraltar Socialist Labor Party (SL), Joe BOSSANO; Gibraltar Labor Party/Association for the Advancement of Civil Rights (GCL/AACR), leader NA; Gibraltar Social Democrats, Peter CARUANA; Gibraltar National Party, Joe GARCIA
Other political or pressure groups: Housewives Association; Chamber of Commerce; Gibraltar Representatives Organization
Member of: INTERPOL (subbureau)
Diplomatic representation in US: none (dependent territory of the UK)
US diplomatic representation: none (dependent territory of the UK)
Flag: two horizontal bands of white (top, double width) and red with a three-towered red castle in the center of the white band; hanging from the castle gate is a gold key centered in the red band

Economy

Overview: Gibraltar benefits from an extensive shipping trade and offshore banking. The British military presence has been severely reduced and now only contributes about 11% to the local economy. The financial sector accounts for 15% of GDP; tourism, shipping services fees, and duties on consumer goods also generate revenue. Because more than 70% of the economy is in the public sector, changes in government spending have a major impact on the level of employment.
National product: GDP—purchasing power parity—$205 million (1993 est.)
National product real growth rate: NA%
National product per capita: $6,600 (1993 est.)
Inflation rate (consumer prices): 3.6% (1988)
Unemployment rate: NA%
Budget:
revenues: $116 million
expenditures: $124 million, including capital expenditures of $NA (1992-93)
Exports: $57 million (f.o.b., 1992)
commodities: (principally re-exports) petroleum 51%, manufactured goods 41%, other 8%
partners: UK, Morocco, Portugal, Netherlands, Spain, US, FRG
Imports: $420 million (c.i.f., 1992)
commodities: fuels, manufactured goods, and foodstuffs
partners: UK, Spain, Japan, Netherlands
External debt: $318 million (1987)
Industrial production: growth rate NA%
Electricity:
capacity: 47,000 kW
production: 90 million kWh
consumption per capita: 2,539 kWh (1993)
Industries: tourism, banking and finance, construction, commerce; support to large UK naval and air bases; transit trade and supply depot in the port; light manufacturing of tobacco, roasted coffee, ice, mineral waters, candy, beer, and canned fish
Agriculture: none
Economic aid:
recipient: US commitments, including Ex-Im (FY70-88), $800,000; Western (non-US) countries and ODA bilateral commitments (1992-93), $2.5 million
Currency: 1 Gibraltar pound (£G) = 100 pence
Exchange rates: Gibraltar pounds (£G) per US$1—0.6350 (January 1995), 0.6529 (1994), 0.6658 (1993), 0.5664 (1992), 0.5652 (1991), 0.5603 (1990); note—the Gibraltar pound is at par with the British pound
Fiscal year: 1 July—30 June

Transportation

Railroads:
total: NA km; 1.000-m gauge system in dockyard area only
Highways:
total: 50 km
paved: 50 km
Pipelines: none
Ports: Gibraltar
Merchant marine:
total: 23 ships (1,000 GRT or over) totaling 419,707 GRT/721,110 DWT
ships by type: bulk 3, cargo 3, chemical tanker 1, container 2, oil tanker 14
Airports:
total: 1
with paved runways 1,524 to 2,437 m: 1

Communications

Telephone system: 9,400 telephones; adequate, automatic domestic system and adequate international radiocommunication and microwave facilities
local: NA
intercity: NA
international: 1 INTELSAT (Atlantic Ocean) earth station
Radio:
broadcast stations: AM 1, FM 6, shortwave 0
radios: NA
Television:
broadcast stations: 4
televisions: NA

Defense Forces

Branches: British Army, Royal Navy, Royal Air Force
Note: defense is the responsibility of the UK

Glorioso Islands

(possession of France)

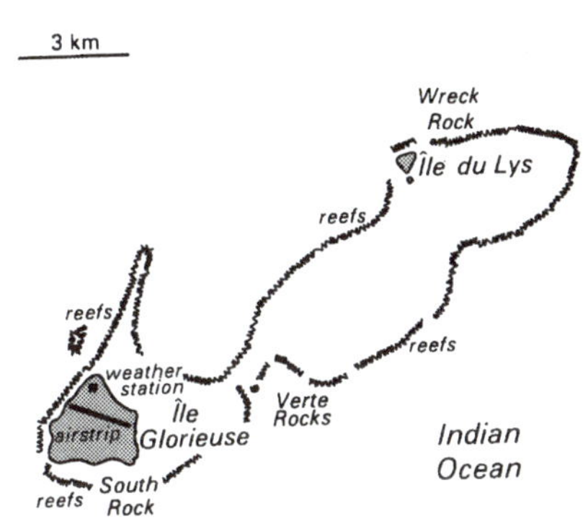

Geography

Location: Southern Africa, group of islands in the Indian Ocean, northwest of Madagascar
Map references: Africa
Area:
total area: 5 sq km
land area: 5 sq km
comparative area: about 8.5 times the size of The Mall in Washington, DC
note: includes Ile Glorieuse, Ile du Lys, Verte Rocks, Wreck Rock, and South Rock
Land boundaries: 0 km
Coastline: 35.2 km
Maritime claims:
exclusive economic zone: 200 nm
territorial sea: 12 nm
International disputes: claimed by Madagascar
Climate: tropical
Terrain: NA
Natural resources: guano, coconuts
Land use:
arable land: 0%
permanent crops: 0%
meadows and pastures: 0%
forest and woodland: 0%
other: 100% (all lush vegetation and coconut palms)
Irrigated land: 0 sq km
Environment:
current issues: NA
natural hazards: periodic cyclones
international agreements: NA

People

Population: uninhabited

Government

Names:
conventional long form: none
conventional short form: Glorioso Islands
local long form: none
local short form: Iles Glorieuses
Digraph: GO
Type: French possession administered by Commissioner of the Republic, resident in Reunion
Capital: none; administered by France from Reunion
Independence: none (possession of France)

Economy

Overview: no economic activity

Transportation

Ports: none; offshore anchorage only
Airports:
total: 1
with unpaved runways 914 to 1,523 m: 1

Defense Forces

Note: defense is the responsibility of France

Greece

Geography

Location: Southern Europe, bordering the Aegean Sea, Ionian Sea, and the Mediterranean Sea, between Albania and Turkey
Map references: Europe
Area:
total area: 131,940 sq km
land area: 130,800 sq km
comparative area: slightly smaller than Alabama
Land boundaries: total 1,210 km, Albania 282 km, Bulgaria 494 km, Turkey 206 km, The Former Yugoslav Republic of Macedonia 228 km
Coastline: 13,676 km
Maritime claims:
continental shelf: 200-m depth or to the depth of exploitation
territorial sea: 6 nm
International disputes: complex maritime, air, and territorial disputes with Turkey in Aegean Sea; Cyprus question; dispute with The Former Yugoslav Republic of Macedonia over name, symbols, and certain constitutional provisions; Greece is involved in a bilateral dispute with Albania over border demarcation, the treatment of Albania's ethnic Greek minority, and migrant Albanian workers in Greece
Climate: temperate; mild, wet winters; hot, dry summers
Terrain: mostly mountains with ranges extending into sea as peninsulas or chains of islands
Natural resources: bauxite, lignite, magnesite, petroleum, marble
Land use:
arable land: 23%
permanent crops: 8%
meadows and pastures: 40%
forest and woodland: 20%
other: 9%
Irrigated land: 11,900 sq km (1989 est.)
Environment:
current issues: air pollution; water pollution
natural hazards: severe earthquakes

international agreements: party to—Air Pollution, Antarctic Treaty, Biodiversity, Climate Change, Endangered Species, Environmental Modification, Hazardous Wastes, Marine Dumping, Nuclear Test Ban, Ozone Layer Protection, Ship Pollution, Tropical Timber 83, Wetlands; signed, but not ratified—Air Pollution-Nitrogen Oxides, Air Pollution-Sulphur 94, Air Pollution-Volatile Organic Compounds, Antarctic-Environmental Protocol, Desertification, Law of the Sea
Note: strategic location dominating the Aegean Sea and southern approach to Turkish Straits; a peninsular country, possessing an archipelago of about 2,000 islands

People

Population: 10,647,511 (July 1995 est.)
Age structure:
0-14 years: 18% (female 904,374; male 947,494)
15-64 years: 67% (female 3,601,029; male 3,565,931)
65 years and over: 15% (female 919,044; male 709,639) (July 1995 est.)
Population growth rate: 0.72% (1995 est.)
Birth rate: 10.56 births/1,000 population (1995 est.)
Death rate: 9.31 deaths/1,000 population (1995 est.)
Net migration rate: 5.99 migrant(s)/1,000 population (1995 est.)
Infant mortality rate: 8.3 deaths/1,000 live births (1995 est.)
Life expectancy at birth:
total population: 77.92 years
male: 75.39 years
female: 80.59 years (1995 est.)
Total fertility rate: 1.46 children born/woman (1995 est.)
Nationality:
noun: Greek(s)
adjective: Greek
Ethnic divisions: Greek 98%, other 2%
note: the Greek Government states there are no ethnic divisions in Greece
Religions: Greek Orthodox 98%, Muslim 1.3%, other 0.7%
Languages: Greek (official), English, French
Literacy: age 15 and over can read and write (1991)
total population: 95%
male: 98%
female: 93%
Labor force: 4.077 million
by occupation: services 52%, agriculture 23%, industry 25% (1994)

Government

Names:
conventional long form: Hellenic Republic
conventional short form: Greece
local long form: Elliniki Dhimokratia
local short form: Ellas
former: Kingdom of Greece
Digraph: GR
Type: presidential parliamentary government; monarchy rejected by referendum 8 December 1974
Capital: Athens
Administrative divisions: 52 prefectures (nomoi, singular—nomos); Aitolia kai Akarnania, Akhaia, Argolis, Arkadhia, Arta, Attiki, Dhodhekanisos, Dhrama, Evritania, Evros, Evvoia, Florina, Fokis, Fthiotis, Grevena, Ilia, Imathia, Ioannina, Iraklion, Kardhitsa, Kastoria, Kavala, Kefallinia, Kerkira, Khalkidhiki, Khania, Khios, Kikladhes, Kilkis, Korinthia, Kozani, Lakonia, Larisa, Lasithi, Lesvos, Levkas, Magnisia, Messinia, Pella, Pieria, Piraievs, Preveza, Rethimni, Rodhopi, Samos, Serrai, Thesprotia, Thessaloniki, Trikala, Voiotia, Xanthi, Zakinthos, autonomous region: Agion Oros (Mt. Athos)
Independence: 1829 (from the Ottoman Empire)
National holiday: Independence Day, 25 March (1821) (proclamation of the war of independence)
Constitution: 11 June 1975
Legal system: based on codified Roman law; judiciary divided into civil, criminal, and administrative courts
Suffrage: 18 years of age; universal and compulsory
Executive branch:
chief of state: President Konstantinos (Kostis) STEPHANOPOULOS (since 10 March 1995) election last held 10 March 1995 (next to be held by NA 2000); results—Konstantinos STEPHANOPOULOS was elected by Parliament
head of government: Prime Minister Andreas PAPANDREOU (since 10 October 1993)
cabinet: Cabinet; appointed by the president on recommendation of the prime minister
Legislative branch: unicameral
Chamber of Deputies (Vouli ton Ellinon): elections last held 10 October 1993 (next to be held by NA October 1997); results—PASOK 46.88%, ND 39.30%, Political Spring 4.87%, KKE 4.54%, and Progressive Left (replaced by Coalition of the Left and Progress) 2.94%; seats—(300 total) PASOK 170, ND 111, Political Spring 10, KKE 9
Judicial branch: Supreme Judicial Court, Special Supreme Tribunal
Political parties and leaders: New Democracy (ND; conservative), Miltiades EVERT; Panhellenic Socialist Movement (PASOK), Andreas PAPANDREOU; Communist Party (KKE), Aleka PAPARIGA; Ecologist-Alternative List, leader rotates; Political Spring, Antonis SAMARAS; Coalition of the Left and Progress (Synaspismos), Nikolaos KONSTANTOPOULOS
Member of: Australia Group, BIS, BSEC, CCC, CE, CERN, EBRD, EC, ECE, EIB, FAO, G- 6, GATT, IAEA, IBRD, ICAO, ICC, ICFTU, ICRM, IDA, IEA, IFAD, IFC, IFRCS, ILO, IMF, IMO, INMARSAT, INTELSAT, INTERPOL, IOC, IOM, ISO, ITU, MINURSO, MTCR, NACC, NAM (guest), NATO, NEA, NSG, OAS (observer), OECD, OSCE, PCA, UN, UNCTAD, UNESCO, UNHCR, UNIDO, UNIKOM, UPU, WEU, WFTU, WHO, WIPO, WMO, WTO, ZC
Diplomatic representation in US:
chief of mission: Ambassador Loucas TSILAS
chancery: 2221 Massachusetts Avenue NW, Washington, DC 20008
telephone: [1] (202) 939-5800
FAX: [1] (202) 939-5824
consulate(s) general: Atlanta, Boston, Chicago, Houston, Los Angeles, New York, and San Francisco
consulate(s): New Orleans
US diplomatic representation:
chief of mission: Ambassador Thomas M.T. NILES
embassy: 91 Vasilissis Sophias Boulevard, 10160 Athens
mailing address: PSC 108, Athens; APO AE 09842
telephone: [30] (1) 721-2951, 8401
FAX: [30] (1) 645-6282
consulate(s) general: Thessaloniki
Flag: nine equal horizontal stripes of blue alternating with white; there is a blue square in the upper hoist-side corner bearing a white cross; the cross symbolizes Greek Orthodoxy, the established religion of the country

Economy

Overview: Greece has a mixed capitalist economy with the basic entrepreneurial system overlaid in 1981-89 by a socialist system that enlarged the public sector from 55% of GDP in 1981 to about 70% in 1989. Since then, the public sector has been reduced to about 60% of GDP. Tourism continues as a major source of foreign exchange, and agriculture is self-sufficient except for meat, dairy products, and animal feedstuffs. Over the last decade, real GDP growth has averaged 1.6% a year, compared with the European Union average of 2.2%. Inflation continues to be well above the EU average, and the national debt has reached 140% of GDP, the highest in the EU. Prime Minister PAPANDREOU will probably make only limited progress correcting the economy's problems of high inflation, large budget deficit, and decaying infrastructure. His economic program suggests that although he will shun his expansionary policies of the 1980s, he will avoid tough measures needed to slow inflation or reduce the state's role in the economy. He has limited the previous government's privatization plans, for example, and has called for generous welfare spending and real wage

increases. Athens continues to rely heavily on EU aid, which recently has amounted to about 6% of GDP. Greece almost certainly will not meet the EU's Maastricht Treaty convergence targets of public deficit held to 3% of GDP and national debt to 60% of GDP by 1999. Per capita GDP has fallen below Portugal's level, the lowest among EU members.
National product: GDP—purchasing power parity—$93.7 billion (1994 est.)
National product real growth rate: 0.4% (1994 est.)
National product per capita: $8,870 (1994 est.)
Inflation rate (consumer prices): 10.9% (1994 est.)
Unemployment rate: 10.1% (1994 est.)
Budget:
revenues: $28.3 billion
expenditures: $37.6 billion, including capital expenditures of $5.2 billion (1994)
Exports: $9 billion (f.o.b., 1993)
commodities: manufactured goods 53%, foodstuffs 34%, fuels 5%
partners: Germany 24%, Italy 14%, France 7%, UK 6%, US 4% (1993)
Imports: $19.2 billion (f.o.b., 1993)
commodities: manufactured goods 72%, foodstuffs 15%, fuels 10%
partners: Germany 16%, Italy 14%, France 7%, Japan 7%, UK 6% (1993)
External debt: $26.9 billion (1993)
Industrial production: growth rate 3.2% (1993 est.); accounts for 18% of GDP
Electricity:
capacity: 8,970,000 kW
production: 35.8 billion kWh
consumption per capita: 3,257 kWh (1993)
Industries: tourism, food and tobacco processing, textiles, chemicals, metal products, mining, petroleum
Agriculture: including fishing and forestry, accounts for 12% of GDP; principal products—wheat, corn, barley, sugar beets, olives, tomatoes, wine, tobacco, potatoes; self-sufficient in food except meat, dairy products, and animal feedstuffs
Illicit drugs: illicit producer of cannabis and limited opium; mostly for domestic production; serves as a gateway to Europe for traffickers smuggling cannabis and heroin from the Middle East and Southwest Asia to the West and precursor chemicals to the East; transshipment point for Southwest Asian heroin transiting the Balkan route
Economic aid:
recipient: US commitments, including Ex-Im (FY70-81), $525 million; Western (non-US) countries, ODA and OOF bilateral commitments (1970-89), $1.39 billion
Currency: 1 drachma (Dr) = 100 lepta
Exchange rates: drachmae (Dr) per US$1—238.20 (January 1995), 242.60 (1994), 229.26 (1993), 190.62 (1992), 182.27 (1991), 158.51 (1990)
Fiscal year: calendar year

Transportation

Railroads:
total: 2,503 km
standard gauge: 1,565 km 1.435-m gauge (36 km electrified; 100 km double track)
narrow gauge: 887 km 1,000-m gauge; 22 km 0.750-m gauge; 29 km 0.600-m gauge
Highways:
total: 130,000 km
paved: 119,210 km (116 km expressways)
unpaved: 10,790 km (1990)
Inland waterways: 80 km; system consists of three coastal canals; including the Corinth Canal (6 km) which crosses the Isthmus of Corinth connecting the Gulf of Corinth with the Saronic Gulf and shortens the sea voyage from the Adriatic to Piraievs (Piraeus) by 325 km; and three unconnected rivers
Pipelines: crude oil 26 km; petroleum products 547 km
Ports: Alexandroupolis, Elevsis, Iraklion (Crete), Kavala, Kerkira, Khalkis, Igoumenitsa, Lavrion, Patrai, Piraievs (Piraeus), Thessaloniki, Volos
Merchant marine:
total: 1,046 ships (1,000 GRT or over) totaling 29,076,911 GRT/53,618,024 DWT
ships by type: bulk 469, cargo 105, chemical tanker 22, combination bulk 21, combination ore/oil 31, container 40, liquefied gas tanker 5, oil tanker 239, passenger 14, passenger-cargo 3, refrigerated cargo 10, roll-on/roll-off cargo 16, short-sea passenger 67, specialized tanker 3, vehicle carrier 1
note: ethnic Greeks also own 125 ships under Liberian registry, 323 under Panamanian, 705 under Cypriot, 351 under Maltese, and 100 under Bahamian
Airports:
total: 79
with paved runways over 3,047 m: 5
with paved runways 2,438 to 3,047 m: 15
with paved runways 1,524 to 2,437 m: 16
with paved runways 914 to 1,523 m: 17
with paved runways under 914 m: 22
with unpaved runways 1,524 to 2,438 m: 1
with unpaved runways 914 to 1,523 m: 3

Communications

Telephone system: 4,080,000 telephones; adequate, modern networks reach all areas; microwave radio relay carries most traffic; extensive open-wire network; submarine cables to off-shore islands
local: NA
intercity: microwave radio relay and open wire
international: tropospheric links, 8 submarine cables; 2 INTELSAT (1 Atlantic Ocean and 1 Indian Ocean) and 1 EUTELSAT earth station
Radio:
broadcast stations: AM 29, FM 17 (repeaters 20), shortwave 0
radios: NA
Television:
broadcast stations: 361
televisions: NA

Defense Forces

Branches: Hellenic Army, Hellenic Navy, Hellenic Air Force, National Guard, Police
Manpower availability: males age 15-49 2,676,152; males fit for military service 2,046,996; males reach military age (21) annually 75,857 (1995 est.)
Defense expenditures: exchange rate conversion—$4.1 billion, 5.4% of GDP (1994)

Greenland

(part of the Danish realm)

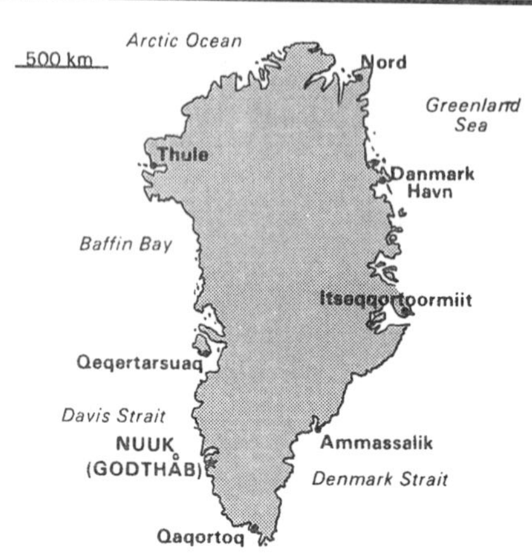

Geography

Location: Northern North America, island between the Arctic Ocean and the North Atlantic Ocean, northeast of Canada
Map references: Arctic Region
Area:
total area: 2,175,600 sq km
land area: 383,600 sq km (ice free)
comparative area: slightly more than three times the size of Texas
Land boundaries: 0 km
Coastline: 44,087 km
Maritime claims:
exclusive fishing zone: 200 nm
territorial sea: 3 nm
International disputes: none
Climate: arctic to subarctic; cool summers, cold winters
Terrain: flat to gradually sloping icecap covers all but a narrow, mountainous, barren, rocky coast
Natural resources: zinc, lead, iron ore, coal, molybdenum, cryolite, uranium, fish
Land use:
arable land: 0%
permanent crops: 0%
meadows and pastures: 1%
forest and woodland: 0%
other: 99%
Irrigated land: 0 sq km
Environment:
current issues: NA
natural hazards: continuous permafrost over northern two-thirds of the island
international agreements: NA
Note: dominates North Atlantic Ocean between North America and Europe; sparse population confined to small settlements along coast

People

Population: 57,611 (July 1995 est.)
Age structure:
0-14 years: 27% (female 7,664; male 7,881)
15-64 years: 68% (female 17,761; male 21,580)
65 years and over: 5% (female 1,500; male 1,225) (July 1995 est.)
Population growth rate: 1.05% (1995 est.)
Birth rate: 17.7 births/1,000 population (1995 est.)
Death rate: 7.2 deaths/1,000 population (1995 est.)
Net migration rate: 0 migrant(s)/1,000 population (1995 est.)
Infant mortality rate: 25.1 deaths/1,000 live births (1995 est.)
Life expectancy at birth:
total population: 67.65 years
male: 63.33 years
female: 71.98 years (1995 est.)
Total fertility rate: 2.25 children born/woman (1995 est.)
Nationality:
noun: Greenlander(s)
adjective: Greenlandic
Ethnic divisions: Greenlander 86% (Eskimos and Greenland-born Caucasians), Danish 14%
Religions: Evangelical Lutheran
Languages: Eskimo dialects, Danish
Literacy: NA%
Labor force: 22,800
by occupation: largely engaged in fishing, hunting, sheep breeding

Government

Names:
conventional long form: none
conventional short form: Greenland
local long form: none
local short form: Kalaallit Nunaat
Digraph: GL
Type: part of the Danish realm; self-governing overseas administrative division
Capital: Nuuk (Godthab)
Administrative divisions: 3 municipalities (kommuner, singular—kommun); Nordgronland, Ostgronland, Vestgronland
Independence: none (part of the Danish realm; self-governing overseas administrative division)
National holiday: Birthday of the Queen, 16 April (1940)
Constitution: 5 June 1953 (Danish constitution)
Legal system: Danish
Suffrage: 18 years of age; universal
Executive branch:
chief of state: Queen MARGRETHE II (since 14 January 1972), represented by High Commissioner Steen SPORE (since NA 1993)
head of government: Home Rule Chairman Lars Emil JOHANSEN (since 15 March 1991)
cabinet: Landsstyre; formed from the Landsting on basis of strength of parties
Legislative branch: unicameral
Parliament (Landsting): elections last held on 4 March 1995 (next to be held 5 March 1999); results—Siumut 38.5%, Inuit Ataqatigiit 20.3%, Atassut Party 29.7%; seats—(31 total) Siumut 12, Atassut Party 10, Inuit Ataqatigiit 6, conservative splinter grouping 2, independent 1
Danish Folketing: last held on 21 September 1994 (next to be held by September 1998); Greenland elects two representatives to the Folketing; results—percent of vote by party NA; seats—(2 total) Liberals 1, Social Democrats 1; note—Greenlandic representatives are affiliated with Danish political parties
Judicial branch: High Court (Landsret)
Political parties and leaders: two-party ruling coalition; Siumut (Forward Party, a moderate socialist party that advocates more distinct Greenlandic identity and greater autonomy from Denmark), Lars Emil JOHANSEN, chairman; Inuit Ataqatigiit (IA) (Eskimo Brotherhood, a Marxist-Leninist party that favors complete independence from Denmark rather than home rule), Josef MOTZFELDT; Atassut Party (Solidarity, a more conservative party that favors continuing close relations with Denmark), Daniel SKIFTE; AKULLIIT, Bjarne KREUTZMANN; Issituup (Polar Party), Nicolai HEINRICH
Diplomatic representation in US: none (self-governing overseas administrative division of Denmark)
US diplomatic representation: none (self-governing overseas administrative division of Denmark)
Flag: two equal horizontal bands of white (top) and red with a large disk slightly to the hoist side of center—the top half of the disk is red, the bottom half is white

Economy

Overview: Greenland's economic situation at present is difficult. Unemployment is increasing, and prospects for economic growth in the immediate future are dim. Following the closing of the Black Angel lead and zinc mine in 1989, Greenland became almost completely dependent on fishing and fish processing, the sector accounting for 95% of exports. Prospects for fisheries are not bright, as the important shrimp catches will at best stabilize and cod catches have dropped. Resumption of mining and hydrocarbon activities is not around the corner, thus leaving only tourism with some potential for the near future. The public sector in Greenland, i.e., the central government and its commercial entities and the municipalities, plays a dominant role in Greenland accounting for about two-thirds of total employment. About half the government's revenues come from grants from the Danish Government.
National product: GDP—purchasing power parity—$NA
National product real growth rate: NA%
National product per capita: $NA

Inflation rate (consumer prices): 1.3% (1993 est.)
Unemployment rate: 6.6% (1993 est.)
Budget:
revenues: $667 million
expenditures: $635 million, including capital expenditures of $103.8 million (1993 est.)
Exports: $330.5 million (f.o.b., 1993 est.)
commodities: fish and fish products 95%
partners: Denmark 79%, Benelux 9%, Germany 5%
Imports: $369.6 million (c.i.f., 1993 est.)
commodities: manufactured goods 28%, machinery and transport equipment 24%, food and live animals 12.4%, petroleum products 12%
partners: Denmark 65%, Norway 8.8%, US 4.6%, Germany 3.8%, Japan 3.8%, Sweden 2.4%
External debt: $297.1 million (1993)
Industrial production: growth rate NA%
Electricity:
capacity: 84,000 kW
production: 210 million kWh
consumption per capita: 3,361 kWh (1993)
Industries: fish processing (mainly shrimp), handicrafts, some small shipyards, potential for platinum and gold mining
Agriculture: sector dominated by fishing and sheep raising; crops limited to forage and small garden vegetables; 1988 fish catch of 133,500 metric tons
Economic aid: none
Currency: 1 Danish krone (DKr) = 100 oere
Exchange rates: Danish kroner (DKr) per US$1—6.034 (January 1995), 6.361 (1994), 6.484 (1993), 6.036 (1992), 6.396 (1991), 6.189 (1990)
Fiscal year: calendar year

Transportation

Railroads: 0 km
Highways:
total: 150 km
paved: 60 km
unpaved: 90 km
Ports: Faeringehavn, Frederikshaab, Holsteinsborg, Nanortalik, Narsaq, Nuuk (Godthaab), Sondrestrom
Merchant marine: none
Airports:
total: 10
with paved runways over 3,047 m: 1
with paved runways 2,438 to 3,047 m: 1
with paved runways 1,524 to 2,437 m: 1
with paved runways 914 to 1,523 m: 1
with paved runways under 914 m: 2
with unpaved runways 1,524 to 2,438 m: 1
with unpaved runways 914 to 1,523 m: 3

Communications

Telephone system: 17,900 telephones; adequate domestic and international service provided by cables and microwave radio relay
local: NA
intercity: microwave radio relay
international: 2 coaxial submarine cables; 1 INTELSAT (Atlantic Ocean) earth station
Radio:
broadcast stations: AM 5, FM 7 (repeaters 35), shortwave 0
radios: NA
Television:
broadcast stations: 4 (repeaters 9)
televisions: NA

Defense Forces

Note: defense is responsibility of Denmark

Grenada

Geography

Location: Caribbean, island in the Caribbean Sea, north of Trinidad and Tobago
Map references: Central America and the Caribbean
Area:
total area: 340 sq km
land area: 340 sq km
comparative area: slightly less than twice the size of Washington, DC
Land boundaries: 0 km
Coastline: 121 km
Maritime claims:
exclusive economic zone: 200 nm
territorial sea: 12 nm
International disputes: none
Climate: tropical; tempered by northeast trade winds
Terrain: volcanic in origin with central mountains
Natural resources: timber, tropical fruit, deepwater harbors
Land use:
arable land: 15%
permanent crops: 26%
meadows and pastures: 3%
forest and woodland: 9%
other: 47%
Irrigated land: NA sq km
Environment:
current issues: NA
natural hazards: lies on edge of hurricane belt; hurricane season lasts from June to November
international agreements: party to—Biodiversity, Climate Change, Law of the Sea, Ozone Layer Protection, Whaling
Note: the administration of the islands of the Grenadines group is divided between Saint Vincent and the Grenadines and Grenada

People

Population: 94,486 (July 1995 est.)
Age structure:
0-14 years: 43% (female 20,076; male 20,824)
15-64 years: 52% (female 23,123; male 25,828)

65 years and over: 5% (female 2,514; male 2,121) (July 1995 est.)
Population growth rate: 0.45% (1995 est.)
Birth rate: 29.69 births/1,000 population (1995 est.)
Death rate: 5.95 deaths/1,000 population (1995 est.)
Net migration rate: -19.24 migrant(s)/1,000 population (1995 est.)
Infant mortality rate: 12.1 deaths/1,000 live births (1995 est.)
Life expectancy at birth:
total population: 70.67 years
male: 68.2 years
female: 73.17 years (1995 est.)
Total fertility rate: 3.85 children born/woman (1995 est.)
Nationality:
noun: Grenadian(s)
adjective: Grenadian
Ethnic divisions: black African
Religions: Roman Catholic, Anglican, other Protestant sects
Languages: English (official), French patois
Literacy: age 15 and over has ever attended school (1970)
total population: 98%
male: 98%
female: 98%
Labor force: 36,000
by occupation: services 31%, agriculture 24%, construction 8%, manufacturing 5%, other 32% (1985)

Government

Names:
conventional long form: none
conventional short form: Grenada
Digraph: GJ
Type: parliamentary democracy
Capital: Saint George's
Administrative divisions: 6 parishes and 1 dependency*; Carriacou and Petit Martinique*, Saint Andrew, Saint David, Saint George, Saint John, Saint Mark, Saint Patrick
Independence: 7 February 1974 (from UK)
National holiday: Independence Day, 7 February (1974)
Constitution: 19 December 1973
Legal system: based on English common law
Suffrage: 18 years of age; universal
Executive branch:
chief of state: Queen ELIZABETH II (since 6 February 1952), represented by Governor General Reginald Oswald PALMER (since 6 August 1992)
head of government: Prime Minister George BRIZAN (since 1 February 1994)
cabinet: Cabinet; appointed by the governor general on advice of the prime minister
Legislative branch: bicameral Parliament
Senate: consists of a 13-member body, 10 appointed by the government and 3 by the Leader of the Opposition
House of Representatives: elections last held on 13 March 1990 (next to be held by NA July 1995); results—percent of vote by party NA; seats—(15 total) NDC 7, GULP 4, TNP 2, NNP 2
Judicial branch: Supreme Court
Political parties and leaders: National Democratic Congress (NDC), George BRIZAN; Grenada United Labor Party (GULP), Sir Eric GAIRY; The National Party (TNP), Ben JONES; New National Party (NNP), Keith MITCHELL; Maurice Bishop Patriotic Movement (MBPM), Terrence MARRYSHOW
Member of: ACP, C, CARICOM, CDB, ECLAC, FAO, G-77, GATT, IBRD, ICAO, ICFTU, ICRM, IDA, IFAD, IFC, IFRCS, ILO, IMF, INTERPOL, IOC, ISO (subscriber), ITU, LAES, NAM, OAS, OECS, OPANAL, UN, UNCTAD, UNESCO, UNIDO, UPU, WCL, WHO, WTO
Diplomatic representation in US:
chief of mission: Ambassador Denneth MODESTE
chancery: 1701 New Hampshire Avenue NW, Washington, DC 20009
telephone: [1] (202) 265-2561
US diplomatic representation:
chief of mission: (vacant); Charge d'Affaires Ollie P. ANDERSON, Jr.
embassy: Point Salines, Saint George's
mailing address: P. O. Box 54, Saint George's, Grenada, W.I.
telephone: [1] (809) 444-1173 through 1178
FAX: [1] (809) 444-4820
Flag: a rectangle divided diagonally into yellow triangles (top and bottom) and green triangles (hoist side and outer side) with a red border around the flag; there are seven yellow five-pointed stars with three centered in the top red border, three centered in the bottom red border, and one on a red disk superimposed at the center of the flag; there is also a symbolic nutmeg pod on the hoist-side triangle (Grenada is the world's second-largest producer of nutmeg, after Indonesia); the seven stars represent the seven administrative divisions

Economy

Overview: The economy is essentially agricultural and centers on the traditional production of spices and tropical plants. Agriculture accounts for about 15% of GDP and 80% of exports and employs 24% of the labor force. Tourism is the leading foreign exchange earner, followed by agricultural exports. Manufacturing remains relatively undeveloped, but is expected to grow, given a more favorable private investment climate since 1983. The economy achieved an impressive average annual growth rate of 5.5% in 1986-91 but has stalled since 1992. Unemployment remains high at about 25%.
National product: GDP—purchasing power parity—$258 million (1993 est.)
National product real growth rate: 0.5% (1993 est.)
National product per capita: $2,750 (1993 est.)
Inflation rate (consumer prices): 2.6% (1993 est.)
Unemployment rate: 25% (1994 est.)
Budget:
revenues: $82.2 million (1993 est.)
expenditures: $74.3 million, including capital expenditures of $11.8 million (1993 est.)
Exports: $18.6 million (f.o.b., 1993 est.)
commodities: bananas, cocoa, nutmeg, fruit and vegetables, clothing, mace
partners: Netherlands, UK, Trinidad and Tobago, United States
Imports: $133.8 million (f.o.b., 1993 est.)
commodities: food 25%, manufactured goods 22%, machinery 20%, chemicals 10%, fuel 6% (1989)
partners: US 29%, UK, Trinidad and Tobago, Japan, Canada (1989)
External debt: $89.9 million (1993)
Industrial production: growth rate 1.8% (1992 est.); accounts for 9% of GDP
Electricity:
capacity: 12,500 kW
production: 60 million kWh
consumption per capita: 639 kWh (1993)
Industries: food and beverage, textile, light assembly operations, tourism, construction
Agriculture: accounts for 14% of GDP and 80% of exports; bananas, cocoa, nutmeg, and mace account for two-thirds of total crop production; world's second-largest producer and fourth-largest exporter of nutmeg and mace; small-sized farms predominate, growing a variety of citrus fruits, avocados, root crops, sugarcane, corn, and vegetables
Economic aid:
recipient: US commitments, including Ex-Im (FY84-89), $60 million; Western (non-US) countries, ODA and OOF bilateral commitments (1970-89), $70 million; Communist countries (1970-89), $32 million
Currency: 1 EC dollar (EC$) = 100 cents
Exchange rates: East Caribbean dollars (EC$) per US$1—2.70 (fixed rate since 1976)
Fiscal year: calendar year

Transportation

Railroads: 0 km
Highways:
total: 1,000 km
paved: 600 km
unpaved: otherwise improved 300 km; unimproved earth 100 km
Ports: Grenville, Saint George's
Merchant marine: none
Airports:
total: 3
with paved runways 2,438 to 3,047 m: 1
with paved runways 1,524 to 2,437 m: 1
with paved runways under 914 m: 1

Communications

Telephone system: 5,650 telephones; automatic, islandwide telephone system; new SHF radio links to the islands of Trinidad, Tobago, and Saint Vincent; VHF and UHF radio links to the islands of Trinidad and Carriacou
local: NA
intercity: NA
international: SHF, VHF, and UHF radio communications
Radio:
broadcast stations: AM 1, FM 0, shortwave 0
radios: NA
Television:
broadcast stations: 1
televisions: NA

Defense Forces

Branches: Royal Grenada Police Force, Coast Guard
Defense expenditures: $NA, NA% of GDP

Guadeloupe

(overseas department of France)

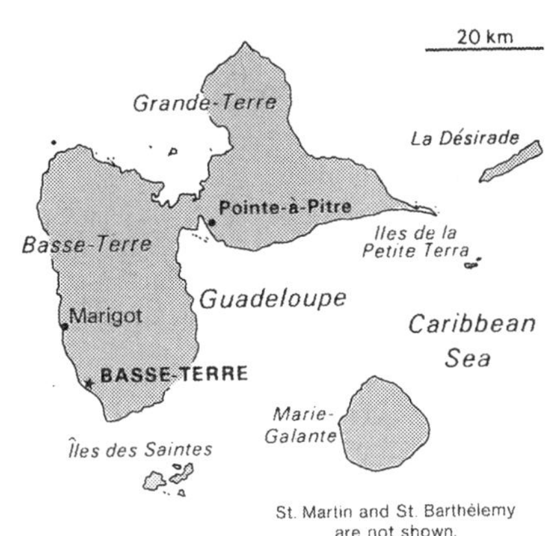

Geography

Location: Caribbean, islands in the eastern Caribbean Sea, southeast of Puerto Rico
Map references: Central America and the Caribbean
Area:
total area: 1,780 sq km
land area: 1,706 sq km
comparative area: 10 times the size of Washington, DC
note: Guadeloupe is an archipelago of nine inhabited islands, of which Basse Terre, Grande Terre, and Marie-Galante are the three largest
Land boundaries: 0 km
Coastline: 306 km
Maritime claims:
exclusive economic zone: 200 nm
territorial sea: 12 nm
International disputes: none
Climate: subtropical tempered by trade winds; relatively high humidity
Terrain: Basse-Terre is volcanic in origin with interior mountains; Grand-Terre is low limestone formation; most of the seven other islands are volcanic in origin
Natural resources: cultivable land, beaches and climate that foster tourism
Land use:
arable land: 18%
permanent crops: 5%
meadows and pastures: 13%
forest and woodland: 40%
other: 24%
Irrigated land: 30 sq km (1989 est.)
Environment:
current issues: NA
natural hazards: hurricanes (June to October); La Soufriere is an active volcano
international agreements: NA

People

Population: 402,815 (July 1995 est.)
Age structure:
0-14 years: 26% (female 51,069; male 52,922)
15-64 years: 66% (female 134,328; male 130,875)
65 years and over: 8% (female 19,318; male 14,303) (July 1995 est.)
Population growth rate: 1.24% (1995 est.)
Birth rate: 18.15 births/1,000 population (1995 est.)
Death rate: 5.58 deaths/1,000 population (1995 est.)
Net migration rate: -0.16 migrant(s)/1,000 population (1995 est.)
Infant mortality rate: 8.5 deaths/1,000 live births (1995 est.)
Life expectancy at birth:
total population: 77.2 years
male: 74.16 years
female: 80.38 years (1995 est.)
Total fertility rate: 1.95 children born/woman (1995 est.)
Nationality:
noun: Guadeloupian(s)
adjective: Guadeloupe
Ethnic divisions: black or mulatto 90%, white 5%, East Indian, Lebanese, Chinese less than 5%
Religions: Roman Catholic 95%, Hindu and pagan African 5%
Languages: French, creole patois
Literacy: age 15 and over can read and write (1982)
total population: 90%
male: 90%
female: 90%
Labor force: 120,000
by occupation: services, government, and commerce 53.0%, industry 25.8%, agriculture 21.2%

Government

Names:
conventional long form: Department of Guadeloupe
conventional short form: Guadeloupe
local long form: Departement de la Guadeloupe
local short form: Guadeloupe
Digraph: GP
Type: overseas department of France
Capital: Basse-Terre
Administrative divisions: none (overseas department of France)
Independence: none (overseas department of France)
National holiday: National Day, Taking of the Bastille, 14 July (1789)
Constitution: 28 September 1958 (French Constitution)
Legal system: French legal system
Suffrage: 18 years of age; universal
Executive branch:
chief of state: President Francois MITTERRAND (since 21 May 1981)
head of government: Prefect Franck PERRIEZ (since NA 1992); President of the General

Council Dominique LARIFLA (since NA); President of the Regional Council Lucette MICHAUX-CHEVRY (since 22 March 1992)
cabinet: Council of Ministers
Legislative branch: unicameral General Council and unicameral Regional Council
General Council: elections last held NA March 1992 (next to be held by NA 1996); results—percent of vote by party NA; seats—(43 total) FRUI.G 13, RPR/DUD 13, PPDG 8, FGPS 3, PCG 3, UPLG 1, PSG 1, independent 1
Regional Council: elections last held on 31 January 1993 (next to be held by 16 March 1998); results—RPR/DUD 48.30%, FGPS 17.09%, FRUI.G 7.44%, PPDG 8.90%, UPLG 7.75% PCG 6.05%; seats—(41 total) seats by party NA
French Senate: elections last held in September 1986 (next to be held September 1995); Guadeloupe elects two representatives; results—percent of vote by party NA; seats—(2 total) PCG 1, FGPS 1
French National Assembly: elections last held on 21 and 28 March 1993 (next to be held March 1998); Guadeloupe elects four representatives; results—percent of vote by party NA; seats—(4 total) FGPS 1, RPR 1, PPDG 1, independent 1
Judicial branch: Court of Appeal (Cour d'Appel) with jurisdiction over Guadeloupe, French Guiana, and Martinique
Political parties and leaders: Rally for the Republic (RPR), Aldo BLAISE; Communist Party of Guadeloupe (PCG), Christian Medard CELESTE; Socialist Party (FGPS), Georges LOUISOR; Popular Union for the Liberation of Guadeloupe (UPLG), Lucien PERATIN; FGPS Dissidents (FRUI.G); Union for French Democracy (UDF), Simon BARLAGNE; Progressive Democratic Party (PPDG), Henri BANGOU
Other political or pressure groups: Popular Union for the Liberation of Guadeloupe (UPLG); Movement for Independent Guadeloupe (MPGI); General Union of Guadeloupe Workers (UGTG); General Federation of Guadeloupe Workers (CGT-G); Christian Movement for the Liberation of Guadeloupe (KLPG)
Member of: FZ, WCL, WFTU
Diplomatic representation in US: none (overseas department of France)
US diplomatic representation: none (overseas department of France)
Flag: the flag of France is used

Economy

Overview: The economy depends on agriculture, tourism, light industry, and services. It is also dependent upon France for large subsidies and imports. Tourism is a key industry, with most tourists from the US. In addition, an increasingly large number of cruise ships visit the islands. The traditionally important sugarcane crop is slowly being replaced by other crops, such as bananas (which now supply about 50% of export earnings), eggplant, and flowers. Other vegetables and root crops are cultivated for local consumption, although Guadeloupe is still dependent on imported food, which comes mainly from France. Light industry consists mostly of sugar and rum production. Most manufactured goods and fuel are imported. Unemployment is especially high among the young.
National product: GDP—purchasing power parity—$3.8 billion (1993 est.)
National product real growth rate: NA%
National product per capita: $9,000 (1993 est.)
Inflation rate (consumer prices): 3.7% (1990)
Unemployment rate: 31.3% (1990)
Budget:
revenues: $400 million
expenditures: $671 million, including capital expenditures of $NA (1989)
Exports: $130 million (f.o.b., 1992)
commodities: bananas, sugar, rum
partners: France 70%, Martinique 17% (1991)
Imports: $1.5 billion (c.i.f., 1992)
commodities: foodstuffs, fuels, vehicles, clothing and other consumer goods, construction materials
partners: France 60%, EC, US, Japan (1991)
External debt: $NA
Industrial production: growth rate NA%
Electricity:
capacity: 320,000 kW
production: 650 million kWh
consumption per capita: 1,421 kWh (1993)
Industries: construction, cement, rum, sugar, tourism
Agriculture: cash crops—bananas, sugarcane; other products include tropical fruits and vegetables; livestock—cattle, pigs, goats; not self-sufficient in food
Economic aid:
recipient: US commitments, including Ex-Im (FY70-88), $4 million; Western (non-US) countries, ODA and OOF bilateral commitments (1970-89), $8.235 billion
Currency: 1 French franc (F) = 100 centimes
Exchange rates: French francs (F) per US$1—5.9243 (January 1995), 5.5520 (1994), 5.6632 (1993), 5.2938 (1992), 5.6421 (1991), 5.4453 (1990)
Fiscal year: calendar year

Transportation

Railroads:
total: NA km; privately owned, narrow-gauge plantation lines
Highways:
total: 1,940 km
paved: 1,600 km
unpaved: gravel, earth 340 km
Ports: Basse-Terre, Gustavia, Marigot, Pointe-a-Pitre
Merchant marine: none
Airports:
total: 9
with paved runways over 3,047 m: 1
with paved runways 914 to 1,523 m: 2
with paved runways under 914 m: 6

Communications

Telephone system: 57,300 telephones; domestic facilities inadequate
local: NA
intercity: NA
international: 1 INTELSAT (Atlantic Ocean) earth station; interisland microwave radio relay to Antigua and Barbuda, Dominica, and Martinique
Radio:
broadcast stations: AM 2, FM 8 (private stations licensed to broadcast FM 30), shortwave 0
radios: NA
Television:
broadcast stations: 9
televisions: NA

Defense Forces

Branches: French Forces, Gendarmerie
Note: defense is responsibility of France

Guam

(territory of the US)

Geography

Location: Oceania, island in the North Pacific Ocean, about three-quarters of the way from Hawaii to the Philippines
Map references: Oceania
Area:
total area: 541.3 sq km
land area: 541.3 sq km
comparative area: slightly more than three times the size of Washington, DC
Land boundaries: 0 km
Coastline: 125.5 km
Maritime claims:
exclusive economic zone: 200 nm
territorial sea: 12 nm
International disputes: none
Climate: tropical marine; generally warm and humid, moderated by northeast trade winds; dry season from January to June, rainy season from July to December; little seasonal temperature variation
Terrain: volcanic origin, surrounded by coral reefs; relatively flat coraline limestone plateau (source of most fresh water) with steep coastal cliffs and narrow coastal plains in north, low-rising hills in center, mountains in south
Natural resources: fishing (largely undeveloped), tourism (especially from Japan)
Land use:
arable land: 11%
permanent crops: 11%
meadows and pastures: 15%
forest and woodland: 18%
other: 45%
Irrigated land: NA sq km
Environment:
current issues: NA
natural hazards: frequent squalls during rainy season; relatively rare, but potentially very destructive typhoons (especially in August)
international agreements: NA
Note: largest and southernmost island in the Mariana Islands archipelago; strategic location in western North Pacific Ocean

People

Population: 153,307 (July 1995 est.)
Age structure:
0-14 years: NA
15-64 years: NA
65 years and over: NA
Population growth rate: 2.42% (1995 est.)
Birth rate: 25.01 births/1,000 population (1995 est.)
Death rate: 3.86 deaths/1,000 population (1995 est.)
Net migration rate: 3 migrant(s)/1,000 population (1995 est.)
Infant mortality rate: 15.17 deaths/1,000 live births (1995 est.)
Life expectancy at birth:
total population: 74.29 years
male: 72.42 years
female: 76.13 years (1995 est.)
Total fertility rate: 2.32 children born/woman (1995 est.)
Nationality:
noun: Guamanian(s)
adjective: Guamanian
Ethnic divisions: Chamorro 47%, Filipino 25%, Caucasian 10%, Chinese, Japanese, Korean, and other 18%
Religions: Roman Catholic 98%, other 2%
Languages: English, Chamorro, Japanese
Literacy: age 15 and over can read and write (1990)
total population: 99%
male: 99%
female: 99%
Labor force: 46,930 (1990)
by occupation: federal and territorial government 40%, private 60% (trade 18%, services 15.6%, construction 13.8%, other 12.6%) (1990)

Government

Names:
conventional long form: Territory of Guam
conventional short form: Guam
Digraph: GQ
Type: organized, unincorporated territory of the US with policy relations between Guam and the US under the jurisdiction of the Office of Territorial and International Affairs, US Department of the Interior
Capital: Agana
Administrative divisions: none (territory of the US)
Independence: none (territory of the US)
National holiday: Guam Discovery Day (first Monday in March) (1521); Liberation Day, 21 July
Constitution: Organic Act of 1 August 1950
Legal system: modeled on US; federal laws apply
Suffrage: 18 years of age; universal; US citizens, but do not vote in US presidential elections
Executive branch:
chief of state: President William Jefferson CLINTON (since 20 January 1993); Vice President Albert GORE, Jr. (since 20 January 1993)
head of government: Governor Carl GUTIERREZ (since 8 November 1994); Lieutenant Governor Madeleine BORDALLO (since 8 November 1994); election last held 8 November 1994 (next to be held NA November 1998); results—Carl GUTIERREZ, a Democrat, was elected Govenor, and Madeleine BORDALLO, a Democrat was elected Lieutenant Governor
cabinet: executive departments; heads appointed by the governor with the consent of the Guam legislature
Legislative branch: unicameral
Legislature: elections last held 8 November 1994 (next to be held NA November 1996); results—percent of vote by party NA; seats—(21 total) number of seats by party NA
US House of Representatives: elections last held 8 November 1994 (next to be held NA November 1996); Guam elects one delegate; results—Robert UNDERWOOD was reelected as delegate; seats—(1 total) Democrat 1
Judicial branch: Federal District Court, Territorial Superior Court
Political parties and leaders: Democratic Party (controls the legislature); Republican Party (party of the Governor)
Member of: ESCAP (associate), INTERPOL (subbureau), IOC, SPC
Diplomatic representation in US: none (territory of the US)
US diplomatic representation: none (territory of the US)
Flag: territorial flag is dark blue with a narrow red border on all four sides; centered is a red-bordered, pointed, vertical ellipse containing a beach scene, outrigger canoe with sail, and a palm tree with the word GUAM superimposed in bold red letters; US flag is the national flag

Economy

Overview: The economy depends mainly on US military spending and on revenues from tourism. Over the past 20 years the tourist industry has grown rapidly, creating a construction boom for new hotels and the expansion of older ones. Visitors numbered about 900,000 in 1992. The slowdown in Japanese economic growth has been reflected in less vigorous growth in the tourism sector. About 60% of the labor force works for the private sector and the rest for government. Most food and industrial goods are imported, with about 75% from the US. Guam faces the problem of building up the civilian economic sector to offset the impact of military downsizing.

National product: GDP—purchasing power parity—$2 billion (1991 est.)
National product real growth rate: NA%
National product per capita: $14,000 (1991 est.)
Inflation rate (consumer prices): 4% (1992 est.)
Unemployment rate: 2% (1992 est.)
Budget:
revenues: $525 million
expenditures: $395 million, including capital expenditures of $NA (1991)
Exports: $34 million (f.o.b., 1984)
commodities: mostly transshipments of refined petroleum products, construction materials, fish, food and beverage products
partners: US 25%, Trust Territory of the Pacific Islands 63%, other 12%
Imports: $493 million (c.i.f., 1984)
commodities: petroleum and petroleum products, food, manufactured goods
partners: US 23%, Japan 19%, other 58%
External debt: $NA
Industrial production: growth rate NA%
Electricity:
capacity: 300,000 kW
production: 750 million kWh
consumption per capita: 4,797 kWh (1993)
Industries: US military, tourism, construction, transshipment services, concrete products, printing and publishing, food processing, textiles
Agriculture: relatively undeveloped with most food imported; fruits, vegetables, eggs, pork, poultry, beef, copra
Economic aid: although Guam receives no foreign aid, it does receive large transfer payments from the general revenues of the US Federal Treasury into which Guamanians pay no income or excise taxes; under the provisions of a special law of Congress, the Guamanian Treasury, rather than the US Treasury, receives federal income taxes paid by military and civilian Federal employees stationed in Guam
Currency: 1 United States dollar (US$) = 100 cents
Exchange rates: US currency is used
Fiscal year: 1 October—30 September

Transportation

Railroads: 0 km
Highways:
total: 674 km (all-weather roads)
paved: NA
unpaved: NA
Ports: Apra Harbor
Merchant marine: none
Airports:
total: 5
with paved runways over 3,047 m: 2
with paved runways 2,438 to 3,047 m: 1
with paved runways 1,524 to 2,437 m: 1
with paved runways under 914 m: 1

Communications

Telephone system: 26,317 telephones (1989)
local: NA
intercity: NA
international: 2 INTELSAT (Pacific Ocean) earth stations
Radio:
broadcast stations: AM 3, FM 3, shortwave 0
radios: NA
Television:
broadcast stations: 3
televisions: NA

Defense Forces

Note: defense is the responsibility of the US

Guatemala

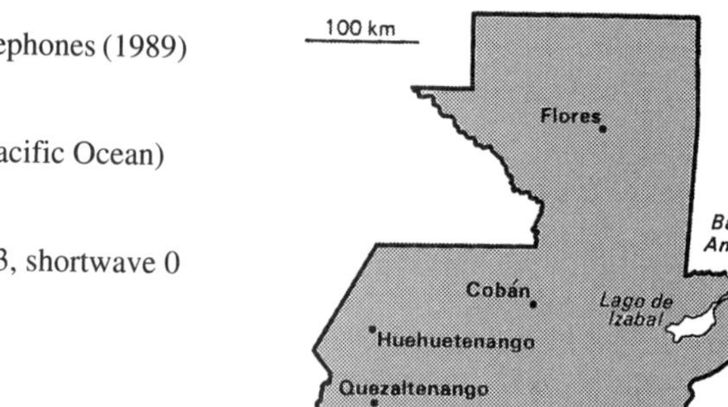

Geography

Location: Middle America, bordering the Caribbean Sea, between Honduras and Belize and bordering the North Pacific Ocean, between El Salvador and Mexico
Map references: Central America and the Caribbean
Area:
total area: 108,890 sq km
land area: 108,430 sq km
comparative area: slightly smaller than Tennessee
Land boundaries: total 1,687 km, Belize 266 km, El Salvador 203 km, Honduras 256 km, Mexico 962 km
Coastline: 400 km
Maritime claims:
continental shelf: 200-m depth or to the depth of exploitation
exclusive economic zone: 200 nm
territorial sea: 12 nm
International disputes: border with Belize in dispute; talks to resolve the dispute are stalled
Climate: tropical; hot, humid in lowlands; cooler in highlands
Terrain: mostly mountains with narrow coastal plains and rolling limestone plateau (Peten)
Natural resources: petroleum, nickel, rare woods, fish, chicle
Land use:
arable land: 12%
permanent crops: 4%
meadows and pastures: 12%
forest and woodland: 40%
other: 32%
Irrigated land: 780 sq km (1989 est.)
Environment:
current issues: deforestation; soil erosion; water pollution
natural hazards: numerous volcanoes in mountains, with frequent violent earthquakes; Caribbean coast subject to hurricanes and other tropical storms
international agreements: party to—Antarctic Treaty, Endangered Species, Environmental

Modification, Marine Dumping, Nuclear Test Ban, Ozone Layer Protection, Wetlands; signed, but not ratified—Antarctic-Environmental Protocol, Biodiversity, Climate Change, Hazardous Wastes, Law of the Sea
Note: no natural harbors on west coast

People

Population: 10,998,602 (July 1995 est.)
Age structure:
0-14 years: 43% (female 2,324,041; male 2,424,686)
15-64 years: 53% (female 2,939,170; male 2,934,334)
65 years and over: 4% (female 198,807; male 177,564) (July 1995 est.)
Population growth rate: 2.53% (1995 est.)
Birth rate: 34.65 births/1,000 population (1995 est.)
Death rate: 7.33 deaths/1,000 population (1995 est.)
Net migration rate: -2.04 migrant(s)/1,000 population (1995 est.)
Infant mortality rate: 52.2 deaths/1,000 live births (1995 est.)
Life expectancy at birth:
total population: 64.85 years
male: 62.27 years
female: 67.56 years (1995 est.)
Total fertility rate: 4.63 children born/woman (1995 est.)
Nationality:
noun: Guatemalan(s)
adjective: Guatemalan
Ethnic divisions: Mestizo—mixed Amerindian-Spanish ancestry (in local Spanish called Ladino) 56%, Amerindian or predominently Amerindian 44%
Religions: Roman Catholic, Protestant, traditional Mayan
Languages: Spanish 60%, Indian language 40% (23 Indian dialects, including Quiche, Cakchiquel, Kekchi)
Literacy: age 15 and over can read and write (1990 est.)
total population: 55%
male: 63%
female: 47%
Labor force: 3.2 million (1994 est.)
by occupation: agriculture 60%, services 13%, manufacturing 12%, commerce 7%, construction 4%, transport 3%, utilities 0.7%, mining 0.3% (1985)

Government

Names:
conventional long form: Republic of Guatemala
conventional short form: Guatemala
local long form: Republica de Guatemala
local short form: Guatemala
Digraph: GT
Type: republic
Capital: Guatemala
Administrative divisions: 22 departments (departamentos, singular—departamento); Alta Verapaz, Baja Verapaz, Chimaltenango, Chiquimula, El Progreso, Escuintla, Guatemala, Huehuetenango, Izabal, Jalapa, Jutiapa, Peten, Quetzaltenango, Quiche, Retalhuleu, Sacatepequez, San Marcos, Santa Rosa, Solola, Suchitepequez, Totonicapan, Zacapa
Independence: 15 September 1821 (from Spain)
National holiday: Independence Day, 15 September (1821)
Constitution: 31 May 1985, effective 14 January 1986
note: suspended 25 May 1993 by President SERRANO; reinstated 5 June 1993 following ouster of president
Legal system: civil law system; judicial review of legislative acts; has not accepted compulsory ICJ jurisdiction
Suffrage: 18 years of age; universal
Executive branch:
chief of state and head of government: President Ramiro DE LEON Carpio (since 6 June 1993); Vice President Arturo HERBRUGER (since 18 June 1993); election runoff held on 11 January 1991 (next to be held November 1995); results—Jorge SERRANO Elias (MAS) 68.1%, Jorge CARPIO Nicolle (UCN) 31.9%
note: President SERRANO resigned on 1 June 1993 shortly after dissolving Congress and the judiciary; on 6 June 1993, Ramiro DE LEON Carpio was chosen as the new president by a vote of Congress; he will finish off the remainder of SERRANO's term which expires 14 January 1996
cabinet: Council of Ministers; named by the president
Legislative branch: unicameral
Congress of the Republic (Congreso de la Republica): by agreement of 11 November 1993, a special election was held on 14 August 1994 to select 80 new congressmen (next election to be held in November 1995 for full four year terms); results—percent of vote by party; FRG 40%, PAN 31.25%, DCG 15%, UCN 10%, MLN 2.5%, UD 1.25%; seats—(80 total) FRG 32, PAN 25, DCG 12, UCN 8, MLN 2, UD 1
note: on 11 November 1993 the congress approved a procedure that would reduce its membership from 116 seats to 80; the procedure provided for a special election in mid 1994 to elect an interim congress of 80 members to serve until replaced in a general election in November 1995; the plan was approved in a general referendum in January 1994 and the special election was held on 14 August 1994
Judicial branch: Supreme Court of Justice (Corte Suprema de Justicia); additionally the Court of Constitutionality is presided over by the President of the Supreme Court
Political parties and leaders: National Centrist Union (UCN), (vacant); Solidarity Action Movement (MAS), Oliverio GARCIA Rodas; Christian Democratic Party (DCG), Alfonso CABRERA Hidalgo; National Advancement Party (PAN), Alvaro ARZU Irigoyen; National Liberation Movement (MLN), Mario SANDOVAL Alarcon; Social Democratic Party (PSD), Mario SOLORZANO Martinez; Revolutionary Party (PR), Carlos CHAVARRIA Perez; Guatemalan Republican Front (FRG), Efrain RIOS Montt; Democratic Union (UD)
Other political or pressure groups: Coordinating Committee of Agricultural, Commercial, Industrial, and Financial Associations (CACIF); Mutual Support Group (GAM); Agrarian Owners Group (UNAGRO); Committee for Campesino Unity (CUC); leftist guerrilla movement known as Guatemalan National Revolutionary Union (URNG) has four main factions—Guerrilla army of the Poor (EGP); Revolutionary Organization of the People in Arms (ORPA); Rebel Armed Forces (FAR); Guatemalan Labor Party (PGT/O)
Member of: BCIE, CACM, CCC, ECLAC, FAO, G-24, G-77, GATT, IADB, IAEA, IBRD, ICAO, ICFTU, ICRM, IDA, IFAD, IFC, IFRCS, ILO, IMF, IMO, INTELSAT, INTERPOL, IOC, IOM, ITU, LAES, LAIA (observer), NAM, OAS, OPANAL, PCA, UN, UNCTAD, UNESCO, UNIDO, UNU, UPU, WCL, WFTU, WHO, WIPO, WMO, WTO
Diplomatic representation in US:
chief of mission: Ambassador Edmond MULET
chancery: 2220 R Street NW, Washington, DC 20008
telephone: [1] (202) 745-4952 through 4954
FAX: [1] (202) 745-1908
consulate(s) general: Chicago, Houston, Los Angeles, Miami, New York, and San Francisco
US diplomatic representation:
chief of mission: Ambassador Marilyn McAFEE
embassy: 7-01 Avenida de la Reforma, Zone 10, Guatemala City
mailing address: APO AA 34024
telephone: [502] (2) 311541
FAX: [502] (2) 318885
Flag: three equal vertical bands of light blue (hoist side), white, and light blue with the coat of arms centered in the white band; the coat of arms includes a green and red quetzal (the national bird) and a scroll bearing the inscription LIBERTAD 15 DE SEPTIEMBRE DE 1821 (the original date of independence from Spain) all superimposed on a pair of crossed rifles and a pair of crossed swords and framed by a wreath

Economy

Overview: The economy is based on family and corporate agriculture, which accounts for 25% of GDP, employs about 60% of the labor

force, and supplies two-thirds of exports. Manufacturing, predominantly in private hands, accounts for about 15% of GDP and 12% of the labor force. In both 1990 and 1991, the economy grew by 3%, the fourth and fifth consecutive years of mild growth. In 1992 growth picked up to almost 5% as government policies favoring competition and foreign trade and investment took stronger hold. In 1993-94, despite political unrest, this momentum continued, foreign investment held up, and annual growth was 4%.
National product: GDP—purchasing power parity—$33 billion (1994 est.)
National product real growth rate: 4% (1994 est.)
National product per capita: $3,080 (1994 est.)
Inflation rate (consumer prices): 12% (1994 est.)
Unemployment rate: 4.9%; underemployment 30%-40% (1994 est.)
Budget:
revenues: $604 million (1990)
expenditures: $808 million, including capital expenditures of $134 million (1990)
Exports: $1.38 billion (f.o.b., 1994 est.)
commodities: coffee, sugar, bananas, cardamon, beef
partners: US 30%, El Salvador, Costa Rica, Germany, Honduras
Imports: $2.6 billion (c.i.f., 1994 est.)
commodities: fuel and petroleum products, machinery, grain, fertilizers, motor vehicles
partners: US 44%, Mexico, Venezuela, Japan, Germany
External debt: $2.2 billion (1992 est.)
Industrial production: growth rate 1.9% (1991 est.); accounts for 18% of GDP
Electricity:
capacity: 700,000 kW
production: 2.3 billion kWh
consumption per capita: 211 kWh (1993)
Industries: sugar, textiles and clothing, furniture, chemicals, petroleum, metals, rubber, tourism
Agriculture: accounts for 25% of GDP; most important sector of economy; contributes two-thirds of export earnings; principal crops—sugarcane, corn, bananas, coffee, beans, cardamom; livestock—cattle, sheep, pigs, chickens; food importer
Illicit drugs: transit country for cocaine shipments; illicit producer of opium poppy and cannabis for the international drug trade; the government has an active eradication program for cannabis and opium poppy
Economic aid:
recipient: US commitments, including Ex-Im (FY70-90), $1.1 billion; Western (non-US) countries, ODA and OOF bilateral commitments (1970-89), $7.92 billion
Currency: 1 quetzal (Q) = 100 centavos
Exchange rates: free market quetzales (Q) per US$1—5.7372 (January 1995), 5.7512 (1994), 5,6354 (1993), 5.1706 (1992), 5.0289 (1991), 4.4858 (1990); note—black-market rate 2.800 (May 1989)
Fiscal year: calendar year

Transportation

Railroads:
total: 1,019 km (102 km privately owned)
narrow gauge: 1,019 km 0.914-m gauge (single track)
Highways:
total: 26,429 km
paved: 2,868 km
unpaved: gravel 11,421 km; unimproved earth 12,140 km
Inland waterways: 260 km navigable year round; additional 730 km navigable during high-water season
Pipelines: crude oil 275 km
Ports: Champerico, Puerto Barrios, Puerto Quetzal, San Jose, Santo Tomas de Castilla
Merchant marine: none
Airports:
total: 528
with paved runways over 3,047 m: 1
with paved runways 2,438 to 3,047 m: 1
with paved runways 1,524 to 2,437 m: 2
with paved runways 914 to 1,523 m: 5
with paved runways under 914 m: 360
with unpaved runways 2,438 to 3,047 m: 1
with unpaved runways 1,524 to 2,438 m: 12
with unpaved runways 914 to 1,523 m: 146

Communications

Telephone system: 97,670 telephones; fairly modern network centered in the city of Guatemala
local: NA
intercity: NA
international: connection into Central American Microwave System; 1 INTELSAT (Atlantic Ocean) earth station
Radio:
broadcast stations: AM 91, FM 0, shortwave 15
radios: NA
Television:
broadcast stations: 25
televisions: NA

Defense Forces

Branches: Army, Navy, Air Force
Manpower availability: males age 15-49 2,574,501; males fit for military service 1,683,028; males reach military age (18) annually 123,715 (1995 est.)
Defense expenditures: exchange rate conversion—$121 million, 1% of GDP (1993)

Guernsey
(British crown dependency)

Geography

Location: Western Europe, islands in the English Channel, northwest of France
Map references: Europe
Area:
total area: 194 sq km
land area: 194 sq km
comparative area: slightly larger than Washington, DC
note: includes Alderney, Guernsey, Herm, Sark, and some other smaller islands
Land boundaries: 0 km
Coastline: 50 km
Maritime claims:
exclusive fishing zone: 200 nm
territorial sea: 3 nm
International disputes: none
Climate: temperate with mild winters and cool summers; about 50% of days are overcast
Terrain: mostly level with low hills in southwest
Natural resources: cropland
Land use:
arable land: NA%
permanent crops: NA%
meadows and pastures: NA%
forest and woodland: NA%
other: NA%
Irrigated land: NA sq km
Environment:
current issues: NA
natural hazards: NA
international agreements: NA
Note: large, deepwater harbor at Saint Peter Port

People

Population: 64,353 (July 1995 est.)
Age structure:
0-14 years: 18% (female 5,664; male 5,892)
15-64 years: 66% (female 21,574; male 21,030)
65 years and over: 16% (female 6,059; male 4,134) (July 1995 est.)
Population growth rate: 0.98% (1995 est.)

Birth rate: 13.29 births/1,000 population (1995 est.)
Death rate: 9.93 deaths/1,000 population (1995 est.)
Net migration rate: 6.4 migrant(s)/1,000 population (1995 est.)
Infant mortality rate: 6.4 deaths/1,000 live births (1995 est.)
Life expectancy at birth:
total population: 78.34 years
male: 75.63 years
female: 81.07 years (1995 est.)
Total fertility rate: 1.7 children born/woman (1995 est.)
Nationality:
noun: Channel Islander(s)
adjective: Channel Islander
Ethnic divisions: UK and Norman-French descent
Religions: Anglican, Roman Catholic, Presbyterian, Baptist, Congregational, Methodist
Languages: English, French; Norman-French dialect spoken in country districts
Literacy: NA%
Labor force: NA

Government

Names:
conventional long form: Bailiwick of Guernsey
conventional short form: Guernsey
Digraph: GK
Type: British crown dependency
Capital: Saint Peter Port
Administrative divisions: none (British crown dependency)
Independence: none (British crown dependency)
National holiday: Liberation Day, 9 May (1945)
Constitution: unwritten; partly statutes, partly common law and practice
Legal system: English law and local statute; justice is administered by the Royal Court
Suffrage: 18 years of age; universal
Executive branch:
chief of state: Queen ELIZABETH II (since 6 February 1952)
head of government: Lieutenant Governor and Commander in Chief Vice-Admiral Sir John COWARD (since NA 1994); Bailiff Mr. Graham Martyn DOREY (since February 1992)
cabinet: Advisory and Finance Committee (other committees); appointed by the States
Legislative branch: unicameral
Assembly of the States: elections last held NA (next to be held NA); results—no percent of vote by party since all are independents; seats—(60 total, 33 elected), all independents
Judicial branch: Royal Court
Political parties and leaders: none; all independents
Member of: none
Diplomatic representation in US: none (British crown dependency)
US diplomatic representation: none (British crown dependency)
Flag: white with the red cross of Saint George (patron saint of England) extending to the edges of the flag

Economy

Overview: Financial services account for more than 50% of total income. Tourism, manufacturing, and horticulture, mainly tomatoes and cut flowers, have been declining. Bank profits (1992) registered a record 26% growth. Fund management and insurance are the two other major income generators. Per capita output and living standards are somewhat lower than the levels of the less affluent EU countries.
National product: GDP $NA
National product real growth rate: 9% (1987)
National product per capita: $NA
Inflation rate (consumer prices): 7% (1988)
Unemployment rate: NA%
Budget:
revenues: $208.9 million
expenditures: $173.9 million, including capital expenditures of $NA (1988)
Exports: $NA
commodities: tomatoes, flowers and ferns, sweet peppers, eggplant, other vegetables
partners: UK (regarded as internal trade)
Imports: $NA
commodities: coal, gasoline, and oil
partners: UK (regarded as internal trade)
External debt: $NA
Industrial production: growth rate NA%
Electricity:
capacity: 173,000 kW
production: 525 million kWh
consumption per capita: 9,060 kWh (1992)
Industries: tourism, banking
Agriculture: tomatoes, flowers (mostly grown in greenhouses), sweet peppers, eggplant, other vegetables, fruit; Guernsey cattle
Economic aid: none
Currency: 1 Guernsey (£G) pound = 100 pence
Exchange rates: Guernsey pounds (£G) per US$1—0.6350 (January 1995), 0.6529 (1994), 0.6658 (1993), 0.5664 (1992), 0.5652 (1991), 0.5603 (1990); note—the Guernsey pound is at par with the British pound
Fiscal year: calendar year

Transportation

Railroads: 0 km
Highways:
total: NA
paved: NA
unpaved: NA
Ports: Saint Peter Port, Saint Sampson
Merchant marine: none
Airports:
total: 2
with paved runways 914 to 1,523 m: 1
with paved runways under 914 m: 1

Communications

Telephone system: 41,900 telephones
local: NA
intercity: NA
international: 1 submarine cable
Radio:
broadcast stations: AM 1, FM 0, shortwave 0
radios: NA
Television:
broadcast stations: 1
televisions: NA

Defense Forces

Note: defense is the responsibility of the UK

Guinea

Geography

Location: Western Africa, bordering the North Atlantic Ocean, between Guinea-Bissau and Sierra Leone
Map references: Africa
Area:
total area: 245,860 sq km
land area: 245,860 sq km
comparative area: slightly smaller than Oregon
Land boundaries: total 3,399 km, Guinea-Bissau 386 km, Cote d'Ivoire 610 km, Liberia 563 km, Mali 858 km, Senegal 330 km, Sierra Leone 652 km
Coastline: 320 km
Maritime claims:
exclusive economic zone: 200 nm
territorial sea: 12 nm
International disputes: none
Climate: generally hot and humid; monsoonal-type rainy season (June to November) with southwesterly winds; dry season (December to May) with northeasterly harmattan winds
Terrain: generally flat coastal plain, hilly to mountainous interior
Natural resources: bauxite, iron ore, diamonds, gold, uranium, hydropower, fish
Land use:
arable land: 6%
permanent crops: 0%
meadows and pastures: 12%
forest and woodland: 42%
other: 40%
Irrigated land: 240 sq km (1989 est.)
Environment:
current issues: deforestation; inadequate supplies of potable water; desertification; soil contamination and erosion; overfishing
natural hazards: hot, dry, dusty harmattan haze may reduce visibility during dry season
international agreements: party to—Biodiversity, Climate Change, Endangered Species, Law of the Sea, Ozone Layer Protection, Wetlands; signed, but not ratified—Desertification

People

Population: 6,549,336 (July 1995 est.)
Age structure:
0-14 years: 44% (female 1,450,501; male 1,448,164)
15-64 years: 53% (female 1,784,420; male 1,691,502)
65 years and over: 3% (female 102,735; male 72,014) (July 1995 est.)
Population growth rate: 2.43% (1995 est.)
Birth rate: 43.43 births/1,000 population (1995 est.)
Death rate: 19.13 deaths/1,000 population (1995 est.)
Net migration rate: NA migrant(s)/1,000 population (1995 est.)
note: Guinea has received about 400,000 refugees from the civil wars in Liberia and Sierra Leone; the continued fighting in Sierra Leone will likely drive more refugees into Guinea in 1995; on the other hand, peace may be achieved in Liberia and permit Liberian refugees to return home
Infant mortality rate: 136.6 deaths/1,000 live births (1995 est.)
Life expectancy at birth:
total population: 44.6 years
male: 42.31 years
female: 46.95 years (1995 est.)
Total fertility rate: 5.79 children born/woman (1995 est.)
Nationality:
noun: Guinean(s)
adjective: Guinean
Ethnic divisions: Peuhl 40%, Malinke 30%, Soussou 20%, smaller tribes 10%
Religions: Muslim 85%, Christian 8%, indigenous beliefs 7%
Languages: French (official); each tribe has its own language
Literacy: age 15 and over can read and write (1990 est.)
total population: 24%
male: 35%
female: 13%
Labor force: 2.4 million (1983)
by occupation: agriculture 80.0%, industry and commerce 11.0%, services 5.4%, civil servants 3.6%

Government

Names:
conventional long form: Republic of Guinea
conventional short form: Guinea
local long form: Republique de Guinee
local short form: Guinee
former: French Guinea
Digraph: GV
Type: republic
Capital: Conakry
Administrative divisions: 33 administrative regions (regions administratives, singular—region administrative); Beyla, Boffa, Boke, Conakry, Coyah, Dabola, Dalaba, Dinguiraye, Faranah, Forecariah, Fria, Gaoual, Gueckedou, Kankan, Kerouane, Kindia, Kissidougou, Koubia, Koundara, Kouroussa, Labe, Lelouma, Lola, Macenta, Mali, Mamou, Mandiana, Nzerekore, Pita, Siguiri, Telimele, Tougue, Yomou
Independence: 2 October 1958 (from France)
National holiday: Anniversary of the Second Republic, 3 April (1984)
Constitution: 23 December 1990 (Loi Fundamentale)
Legal system: based on French civil law system, customary law, and decree; legal codes currently being revised; has not accepted compulsory ICJ jurisdiction
Suffrage: 18 years of age; universal
Executive branch:
chief of state and head of government: President Lansana CONTE, elected in the first multi-party election 19 December 1993; prior to the election he had ruled as head of military government since 5 April 1984
cabinet: Council of Ministers; appointed by the president
Legislative branch: unicameral
People's National Assembly (Assemblee Nationale Populaire): the People's National Assembly was dissolved after the 3 April 1984 coup; framework established in December 1991 for a new National Assembly with 114 seats; legislative elections, tentatively scheduled for 1994, were not held and are now rescheduled for 11 June 1995
Judicial branch: Court of Appeal (Cour d'Appel)
Political parties and leaders: political parties were legalized on 1 April 1992
pro-government: Party for Unity and Progress (PUP)
other: Rally for the Guinean People (RPG), Alpha CONDE; Union for a New Republic (UNR), Mamadou BAH; Party for Renewal and Progress (PRP), Siradiou DIALLO; Movement of Patriotic Democrats (MDP), Ahmed Tidiane CISSE
Member of: ACCT, ACP, AfDB, CCC, CEAO (observer), ECA, ECOWAS, FAO, G-77, IBRD, ICAO, ICRM, IDA, IDB, IFAD, IFC, IFRCS, ILO, IMF, IMO, INTELSAT, INTERPOL, IOC, ITU, MINURSO, NAM, OAU, OIC, UN, UNCTAD, UNESCO, UNIDO, UPU, WCL, WFTU, WHO, WIPO, WMO, WTO
Diplomatic representation in US:
chief of mission: Ambassador Elhadj Boubacar BARRY
chancery: 2112 Leroy Place NW, Washington, DC 20008
telephone: [1] (202) 483-9420
FAX: [1] (202) 483-8688
US diplomatic representation:
chief of mission: Ambassador Joseph A. SALOOM III
embassy: 2nd Boulevard and 9th Avenue, Conakry

mailing address: B. P. 603, Conakry
telephone: [224] 44 15 20 through 44 15 23
FAX: [224] 44 15 22
Flag: three equal vertical bands of red (hoist side), yellow, and green; uses the popular pan-African colors of Ethiopia; similar to the flag of Rwanda, which has a large black letter R centered in the yellow band

Economy

Overview: Although possessing major mineral and hydropower resources and considerable potential for agricultural development, Guinea remains one of the poorest countries in the world. The agricultural sector contributes about 40% to GDP and employs 80% of the work force, while industry accounts for 27% of GDP. Guinea possesses over 25% of the world's bauxite reserves. The mining sector accounted for 85% of exports in 1991. Long-run improvements in literacy, financial institutions, and the legal framework are needed if the country is to move out of poverty. Except in the bauxite industry, foreign investment remains minimal.
National product: GDP—purchasing power parity—$6.3 billion (1994 est.)
National product real growth rate: 0.8% (1994 est.)
National product per capita: $980 (1994 est.)
Inflation rate (consumer prices): 16.6% (1992 est.)
Unemployment rate: NA%
Budget:
revenues: $449 million
expenditures: $708 million, including capital expenditures of $361 million (1990 est.)
Exports: $622 million (f.o.b., 1992 est.)
commodities: bauxite, alumina, diamonds, gold, coffee, pineapples, bananas, palm kernels
partners: US 23%, Belgium 12%, Ireland 12%, Spain 12%
Imports: $768 million (c.i.f., 1992 est.)
commodities: petroleum products, metals, machinery, transport equipment, foodstuffs, textiles, and other grain
partners: France 26%, Cote d'Ivoire 12%, Hong Kong 6%, Germany 6%
External debt: 2.5 billion (1992)
Industrial production: growth rate NA%; accounts for 27% of GDP
Electricity:
capacity: 180,000 kW
production: 520 million kWh
consumption per capita: 77 kWh (1993)
Industries: mining—bauxite, gold, diamonds; alumina refining; light manufacturing and agricultural processing industries
Agriculture: accounts for 40% of GDP (includes fishing and forestry); mostly subsistence farming; principal products—rice, coffee, pineapples, palm kernels, cassava, bananas, sweet potatoes, timber; livestock—cattle, sheep and goats; not self-sufficient in food grains
Economic aid:
recipient: US commitments, including Ex-Im (FY70-89), $227 million; Western (non-US) countries, ODA and OOF bilateral commitments (1970-89), $1.465 billion; OPEC bilateral aid (1979-89), $120 million; Communist countries (1970-89), $446 million
Currency: 1 Guinean franc (FG) = 100 centimes
Exchange rates: Guinean francs (FG) per US$1—810.94 (1 July 1993), 922.9 (30 September 1992), 675 (1990), 618 (1989), 515 (1988), 440 (1987), 383 (1986)
Fiscal year: calendar year

Transportation

Railroads:
total: 1,048 km
standard gauge: 241 km 1.435-m gauge
narrow gauge: 807 km 1.000-m gauge
Highways:
total: 30,100 km
paved: 1,145 km
unpaved: gravel, crushed stone 12,955 km (of which barely 4,500 are currently all-weather roads); unimproved earth 16,000 km (1987)
Inland waterways: 1,295 km navigable by shallow-draft native craft
Ports: Boke, Conakry, Kamsar
Merchant marine: none
Airports:
total: 15
with paved runways over 3,047 m: 1
with paved runways 2,438 to 3,047 m: 2
with paved runways 1,524 to 2,437 m: 1
with paved runways under 914 m: 1
with unpaved runways 1,524 to 2,438 m: 7
with unpaved runways 914 to 1,523 m: 3

Communications

Telephone system: 15,000 telephones; poor to fair system of open-wire lines, small radiocommunication stations, and new radio relay system
local: NA
intercity: microwave radio relay and radio communications stations
international: 1 INTELSAT (Atlantic Ocean) earth station
Radio:
broadcast stations: AM 3, FM 1, shortwave 0
radios: 200,000
Television:
broadcast stations: 1
televisions: 65,000

Defense Forces

Branches: Army, Navy (acts primarily as a coast guard), Air Force, Republican Guard, Presidential Guard, paramilitary National Gendarmerie, National Police Force (Surete National)
Manpower availability: males age 15-49 1,478,653; males fit for military service 745,990 (1995 est.)
Defense expenditures: exchange rate conversion—$50 million, 1.6% of GDP (1994)

Guinea-Bissau

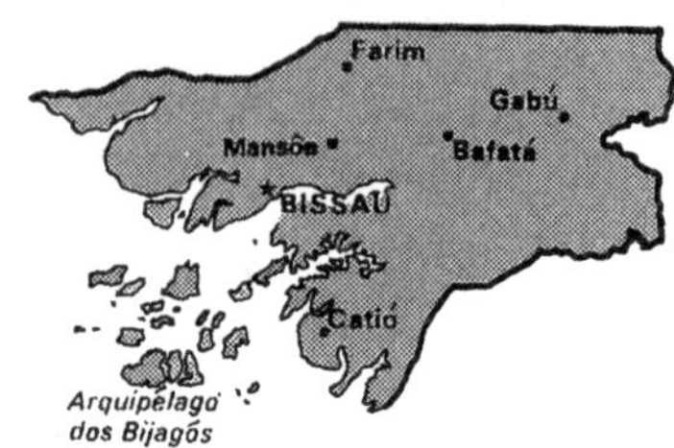

North Atlantic Ocean

Geography

Location: Western Africa, bordering the North Atlantic Ocean, between Guinea and Senegal
Map references: Africa
Area:
total area: 36,120 sq km
land area: 28,000 sq km
comparative area: slightly less than three times the size of Connecticut
Land boundaries: total 724 km, Guinea 386 km, Senegal 338 km
Coastline: 350 km
Maritime claims:
exclusive economic zone: 200 nm
territorial sea: 12 nm
International disputes: none
Climate: tropical; generally hot and humid; monsoonal-type rainy season (June to November) with southwesterly winds; dry season (December to May) with northeasterly harmattan winds
Terrain: mostly low coastal plain rising to savanna in east
Natural resources: unexploited deposits of petroleum, bauxite, phosphates, fish, timber
Land use:
arable land: 11%
permanent crops: 1%
meadows and pastures: 43%
forest and woodland: 38%
other: 7%
Irrigated land: NA sq km
Environment:
current issues: deforestation; soil erosion; overgrazing; overfishing
natural hazards: hot, dry, dusty harmattan haze may reduce visibility during dry season; brush fires
international agreements: party to—Endangered Species, Law of the Sea, Nuclear Test Ban, Wetlands; signed, but not ratified—Biodiversity, Climate Change, Desertification

People

Population: 1,124,537 (July 1995 est.)
Age structure:
0-14 years: 43% (female 242,518; male 243,093)
15-64 years: 54% (female 320,987; male 286,308)
65 years and over: 3% (female 16,129; male 15,502) (July 1995 est.)
Population growth rate: 2.36% (1995 est.)
Birth rate: 40.24 births/1,000 population (1995 est.)
Death rate: 16.62 deaths/1,000 population (1995 est.)
Net migration rate: 0 migrant(s)/1,000 population (1995 est.)
Infant mortality rate: 117.9 deaths/1,000 live births (1995 est.)
Life expectancy at birth:
total population: 47.87 years
male: 46.21 years
female: 49.57 years (1995 est.)
Total fertility rate: 5.43 children born/woman (1995 est.)
Nationality:
noun: Guinea-Bissauan(s)
adjective: Guinea-Bissauan
Ethnic divisions: African 99% (Balanta 30%, Fula 20%, Manjaca 14%, Mandinga 13%, Papel 7%), European and mulatto less than 1%
Religions: indigenous beliefs 65%, Muslim 30%, Christian 5%
Languages: Portuguese (official), Criolo, African languages
Literacy: age 15 and over can read and write (1990 est.)
total population: 36%
male: 50%
female: 24%
Labor force: 403,000 (est.)
by occupation: agriculture 90%, industry, services, and commerce 5%, government 5%

Government

Names:
conventional long form: Republic of Guinea-Bissau
conventional short form: Guinea-Bissau
local long form: Republica de Guine-Bissau
local short form: Guine-Bissau
former: Portuguese Guinea
Digraph: PU
Type: republic, formerly highly centralized, multiparty since mid-1991
Capital: Bissau
Administrative divisions: 9 regions (regioes, singular—regiao); Bafata, Biombo, Bissau, Bolama, Cacheu, Gabu, Oio, Quinara, Tombali
Independence: 10 September 1974 (from Portugal)
National holiday: Independence Day, 10 September (1974)
Constitution: 16 May 1984, amended 4 May 1991 (currently undergoing revision to liberalize popular participation in the government)
Legal system: NA
Suffrage: 15 years of age; universal
Executive branch:
chief of state: President of the Republic of Guinea-Bissau Joao Bernardo VIEIRA (assumed power 14 November 1980); election last held August 1994 (next to be held 1999); results—Joao Bernardo VIEIRA 52%, Kumba YALLA 48%
head of government: Prime Minister Manuel SATURNINO, since 5 November 1994
cabinet: Council of Ministers; appointed by the president
Legislative branch: unicameral
National People's Assembly: (Assembleia Nacional Popular) elections last held 3 July and 7 August 1994 (next to be held 1999); results—percent of vote by party NA; seats—(100 total) PAIGC 62, RGB 19, PRS 12, Union for Change Coalition 6, FLING 1
Judicial branch: none; there is a Ministry of Justice in the Council of Ministers
Political parties and leaders: African Party for the Independence of Guinea-Bissau and Cape Verde (PAIGC), President Joao Bernardo VIEIRA, leader; Democratic Social Front (FDS), Rafael BARBOSA, leader; Bafata Movement, Domingos Fernandes GARNER, leader; Democratic Front (FD), Aristides MENEZES, leader; Social Renovation Party (PRS); Guinea-Bissau Resistance (RGB); Union for Change Coalition; Front for the Liberation and Independence of Guinea (FLING)
Member of: ACCT (associate), ACP, AfDB, ECA, ECOWAS, FAO, G-77, GATT, IBRD, ICAO, ICRM, IDA, IDB, IFAD, IFC, IFRCS, ILO, IMF, IMO, INTELSAT (nonsignatory user), INTERPOL, IOM (observer), ITU, NAM, OAU, OIC, UN, UNAVEM II, UNCTAD, UNESCO, UNIDO, UNOMIL, UNOMOZ, UPU, WFTU, WHO, WIPO, WMO, WTO
Diplomatic representation in US:
chief of mission: Ambassador Alfredo Lopes CABRAL
chancery: 918 16th Street NW, Mezzanine Suite, Washington, DC 20006
telephone: [1] (202) 872-4222
FAX: [1] (202) 872-4226
US diplomatic representation:
chief of mission: Ambassador Roger A. McGUIRE
embassy: Bairro de Penha, Bissau
mailing address: C.P. 297, 1067 Bissau Codex, Bissau, Guinea-Bissau
telephone: [245] 252273, 252274, 252275, 252276
FAX: [245] 252282
Flag: two equal horizontal bands of yellow (top) and green with a vertical red band on the hoist side; there is a black five-pointed star centered in the red band; uses the popular

pan-African colors of Ethiopia; similar to the flag of Cape Verde, which has the black star raised above the center of the red band and is framed by two corn stalks and a yellow clam shell

Economy

Overview: Guinea-Bissau ranks among the poorest countries in the world. Agriculture and fishing are the main economic activities. Cashew nuts, peanuts, and palm kernels are the primary exports. Exploitation of known mineral deposits is unlikely at present because of a weak infrastructure and the high cost of development. With IMF support the country is committed to an economic reform program emphasizing monetary stability and private sector growth. This process will continue at a slow pace because of a heavy foreign debt burden and internal constraints.
National product: GDP—purchasing power parity—$900 million (1993 est.)
National product real growth rate: 2.9% (1993 est.)
National product per capita: $840 (1994 est.)
Inflation rate (consumer prices): 55% (1991 est.)
Unemployment rate: NA%
Budget:
revenues: $33.6 million
expenditures: $44.8 million, including capital expenditures of $570,000 (1991 est.)
Exports: $19 million (f.o.b., 1993)
commodities: cashews, fish, peanuts, palm kernels
partners: Portugal, Spain, Senegal, India, Nigeria
Imports: $56 million (f.o.b., 1993)
commodities: foodstuffs, transport equipment, petroleum products, machinery and equipment
partners: Portugal, Netherlands, China, Germany, Senegal
External debt: $462 million (December 1990 est.)
Industrial production: growth rate NA (1991 est.); accounts for 8% of GDP
Electricity:
capacity: 22,000 kW
production: 40 million kWh
consumption per capita: 37 kWh (1993)
Industries: agricultural processing, beer, soft drinks
Agriculture: accounts for over 45% of GDP, nearly 100% of exports, and 90% of employment; rice is the staple food; other crops include corn, beans, cassava, cashew nuts, peanuts, palm kernels, and cotton; not self-sufficient in food; fishing and forestry potential not fully exploited
Economic aid:
recipient: US commitments, including Ex-Im (FY70-89), $49 million; Western (non-US) countries, ODA and OOF bilateral commitments (1970-89), $615 million; OPEC bilateral aid (1979-89), $41 million; Communist countries (1970-89), $68 million
Currency: 1 Guinea-Bissauan peso (PG) = 100 centavos
Exchange rates: Guinea-Bissauan pesos (PG) per US$1—14,482 (December 1994), 12,892 (1994), 10,082 (1993), 6,934 (1992), 3,659 (1991), 2,185 (1990)
Fiscal year: calendar year

Transportation

Railroads: 0 km
Highways:
total: 3,218 km
paved: bituminous 2,698 km
unpaved: earth 520 km
Inland waterways: scattered stretches are important to coastal commerce
Ports: Bissau
Merchant marine: none
Airports:
total: 32
with paved runways over 3,047 m: 1
with paved runways 1,524 to 2,437 m: 2
with paved runways 914 to 1,523 m: 1
with paved runways under 914 m: 22
with unpaved runways 914 to 1,523 m: 6

Communications

Telephone system: 3,000 telephones; poor system; telephone density—2.7/1,000 persons
local: NA
intercity: combination of microwave radio relay, open wire lines and radiocommunications
international: NA
Radio:
broadcast stations: AM 2, FM 3, shortwave 0
radios: NA
Television:
broadcast stations: 1
televisions: NA

Defense Forces

Branches: People's Revolutionary Armed Force (FARP; includes Army, Navy, and Air Force), paramilitary force
Manpower availability: males age 15-49 251,636; males fit for military service 143,694 (1995 est.)
Defense expenditures: exchange rate conversion—$9 million, 4.5% of GDP (1994)

Guyana

Geography

Location: Northern South America, bordering the North Atlantic Ocean, between Suriname and Venezuela
Map references: South America
Area:
total area: 214,970 sq km
land area: 196,850 sq km
comparative area: slightly smaller than Idaho
Land boundaries: total 2,462 km, Brazil 1,119 km, Suriname 600 km, Venezuela 743 km
Coastline: 459 km
Maritime claims:
continental shelf: 200 nm or to the outer edge of the continental margin
exclusive fishing zone: 200 nm
territorial sea: 12 nm
International disputes: all of the area west of the Essequibo River claimed by Venezuela; Suriname claims area between New (Upper Courantyne) and Courantyne/Kutari Rivers (all headwaters of the Courantyne)
Climate: tropical; hot, humid, moderated by northeast trade winds; two rainy seasons (May to mid-August, mid-November to mid-January)
Terrain: mostly rolling highlands; low coastal plain; savanna in south
Natural resources: bauxite, gold, diamonds, hardwood timber, shrimp, fish
Land use:
arable land: 3%
permanent crops: 0%
meadows and pastures: 6%
forest and woodland: 83%
other: 8%
Irrigated land: 1,300 sq km (1989 est.)
Environment:
current issues: water pollution from sewage and agricultural and industrial chemicals; deforestation
natural hazards: flash floods are a constant threat during rainy seasons
international agreements: party to—Biodiversity, Climate Change, Endangered

Species, Law of the Sea, Ozone Layer Protection, Tropical Timber 83

People

Population: 723,774 (July 1995 est.)
Age structure:
0-14 years: 33% (female 118,515; male 123,048)
15-64 years: 62% (female 224,484; male 225,543)
65 years and over: 5% (female 17,540; male 14,644) (July 1995 est.)
Population growth rate: -0.81% (1995 est.)
Birth rate: 19.41 births/1,000 population (1995 est.)
Death rate: 7.34 deaths/1,000 population (1995 est.)
Net migration rate: -20.19 migrant(s)/1,000 population (1995 est.)
Infant mortality rate: 47.7 deaths/1,000 live births (1995 est.)
Life expectancy at birth:
total population: 65.1 years
male: 61.86 years
female: 68.5 years (1995 est.)
Total fertility rate: 2.23 children born/ woman (1995 est.)
Nationality:
noun: Guyanese (singular and plural)
adjective: Guyanese
Ethnic divisions: East Indian 51%, black and mixed 43%, Amerindian 4%, European and Chinese 2%
Religions: Christian 57%, Hindu 33%, Muslim 9%, other 1%
Languages: English, Amerindian dialects
Literacy: age 15 and over has ever attended school (1990 est.)
total population: 96%
male: 98%
female: 95%
Labor force: 268,000
by occupation: industry and commerce 44.5%, agriculture 33.8%, services 21.7%
note: public-sector employment amounts to 60%-80% of the total labor force (1985)

Government

Names:
conventional long form: Co-operative Republic of Guyana
conventional short form: Guyana
former: British Guiana
Digraph: GY
Type: republic
Capital: Georgetown
Administrative divisions: 10 regions; Barima-Waini, Cuyuni-Mazaruni, Demerara-Mahaica, East Berbice-Corentyne, Essequibo Islands-West Demerara, Mahaica-Berbice, Pomeroon-Supenaam, Potaro-Siparuni, Upper Demerara-Berbice, Upper Takutu-Upper Essequibo
Independence: 26 May 1966 (from UK)
National holiday: Republic Day, 23 February (1970)
Constitution: 6 October 1980
Legal system: based on English common law with certain admixtures of Roman-Dutch law; has not accepted compulsory ICJ jurisdiction
Suffrage: 18 years of age; universal
Executive branch:
chief of state: Executive President Cheddi JAGAN (since 5 October 1992); election last held 5 October 1992; results—Cheddi JAGAN was elected president since he was leader of the party with the most votes in the National Assembly elections
head of government: Prime Minister Sam HINDS (since 5 October 1992)
cabinet: Cabinet of Ministers; appointed by the president, responsible to the legislature
Legislative branch: unicameral
National Assembly: elections last held on 5 October 1992 (next to be held in 1997); results—PPP 53.4%, PNC 42.3%, WPA 2%, TUF 1.2%; seats—(65 total, 53 elected) PPP 36, PNC 26, WPA 2, TUF 1
Judicial branch: Supreme Court of Judicature
Political parties and leaders: People's Progressive Party (PPP), Cheddi JAGAN; People's National Congress (PNC), Hugh Desmond HOYTE; Good and Green Georgetown (GGG), Hamilton GREEN; Working People's Alliance (WPA), Eusi KWAYANA, Rupert ROOPNARINE; Democratic Labor Movement (DLM), Paul TENNASSEE; People's Democratic Movement (PDM), Llewellyn JOHN; National Democratic Front (NDF), Joseph BACCHUS; The United Force (TUF), Manzoor NADIR; United Republican Party (URP), Leslie RAMSAMMY; National Republican Party (NRP), Robert GANGADEEN; Guyana Labor Party (GLP), Nanda GOPAUL
Other political or pressure groups: Trades Union Congress (TUC); Guyana Council of Indian Organizations (GCIO); Civil Liberties Action Committee (CLAC)
note: the latter two organizations are small and active but not well organized
Member of: ACP, C, CARICOM, CCC, CDB, ECLAC, FAO, G-77, GATT, IADB, IBRD, ICAO, ICFTU, ICRM, IDA, IFAD, IFC, IFRCS, ILO, IMF, IMO, INTELSAT (nonsignatory user), INTERPOL, IOC, ITU, LAES, NAM, OAS, ONUSAL, UN, UNCTAD, UNESCO, UNIDO, UPU, WCL, WFTU, WHO, WMO
Diplomatic representation in US:
chief of mission: Ambassador Dr. Ali Odeen ISHMAEL
chancery: 2490 Tracy Place NW, Washington, DC 20008
telephone: [1] (202) 265-6900, 6901
consulate(s) general: New York
US diplomatic representation:
chief of mission: Ambassador George F. JONES
embassy: 99-100 Young and Duke Streets, Kingston, Georgetown
mailing address: P. O. Box 10507, Georgetown
telephone: [592] (2) 54900 through 54909, 57960 through 57969
FAX: [592] (2) 58497
Flag: green with a red isosceles triangle (based on the hoist side) superimposed on a long yellow arrowhead; there is a narrow black border between the red and yellow, and a narrow white border between the yellow and the green

Economy

Overview: Guyana, one of the poorest countries in the Western Hemisphere, has pushed ahead strongly in 1992-94, with an 8% average annual economic growth rate, led by gold mining, and rice, sugar, and forestry products for export. Favorable factors include recovery in the key agricultural and mining sectors, a more favorable atmosphere for business initiative, a more realistic exchange rate, a sharp drop in the inflation rate, and the continued support of international organizations. Serious underlying economic problems will continue. Electric power has been in short supply and constitutes a major barrier to future gains in national output. The government will have to persist in efforts to manage its large $2.2 billion external debt, control inflation, and to extend the privatization program.
National product: GDP—purchasing power parity—$1.4 billion (1994 est.)
National product real growth rate: 8.5% (1994 est.)
National product per capita: $1,950 (1994 est.)
Inflation rate (consumer prices): 15.5% (1994 est.)
Unemployment rate: 12% (1992 est.)
Budget:
revenues: $23.7 million
expenditures: $19.6 million, including capital expenditures of $NA (1994 est.)
Exports: $475 million (f.o.b., 1994)
commodities: sugar, bauxite/alumina, rice, shrimp, molasses
partners: UK 33%, US 31%, Canada 9%, France 5%, Japan 3% (1992)
Imports: $456 million (c.i.f., 1994 est.)
commodities: manufactures, machinery, petroleum, food
partners: US 37%, Trinidad and Tobago 13%, UK 11%, Italy 8%, Japan 5% (1992)
External debt: $2.2 billion (1994 est.)
Industrial production: growth rate 5.6% (1994 est.)
Electricity:
capacity: 110,000 kW
production: 230 million kWh

consumption per capita: 286 kWh (1993)
Industries: bauxite mining, sugar, rice milling, timber, fishing (shrimp), textiles, gold mining
Agriculture: most important sector, accounting for 25% of GDP and about half of exports; sugar and rice are key crops; development potential exists for fishing and forestry; not self-sufficient in food, especially wheat, vegetable oils, and animal products
Illicit drugs: transshipment point for narcotics from South America—primarily Venezuela—to the US and Europe; producer of cannabis
Economic aid:
recipient: US commitments, including Ex-Im (FY70-89), $116 million; Western (non-US) countries, ODA and OOF bilateral commitments (1970-89), $325 million; Communist countries 1970-89, $242 million
Currency: 1 Guyanese dollar (G$) = 100 cents
Exchange rates: Guyanese dollars (G$) per US$1—142.7 (January 1995), 138.3 (1994), 126.7 (1993), 125.0 (1992), 111.8 (1991), 39.533 (1990)
Fiscal year: calendar year

Transportation

Railroads:
total: 100 km NA-m gauge industrial lines for the transport of minerals, including bauxite
Highways:
total: 7,665 km
paved: 550 km
unpaved: gravel 5,000 km; earth 2,115 km
Inland waterways: 6,000 km total of navigable waterways; Berbice, Demerara, and Essequibo Rivers are navigable by oceangoing vessels for 150 km, 100 km, and 80 km, respectively
Ports: Bartica, Georgetown, Linden, New Amsterdam, Parika
Merchant marine:
total: 1 cargo ship (1,000 GRT or over) totaling 1,317 GRT/2,558 DWT
Airports:
total: 54
with paved runways 1,524 to 2,437 m: 3
with paved runways 914 to 1,523 m: 1
with paved runways under 914 m: 34
with unpaved runways 1,524 to 2,438 m: 2
with unpaved runways 914 to 1,523 m: 14

Communications

Telephone system: over 27,000 telephones; fair system for long distance calling
local: NA
intercity: microwave radio relay network for trunk lines
international: tropospheric scatter link to Trinidad; 1 INTELSAT (Atlantic Ocean) earth station
Radio:
broadcast stations: AM 4, FM 3, shortwave 1
radios: NA
Television:
broadcast stations: 0
televisions: NA

Defense Forces

Branches: Guyana Defense Force (GDF; includes Ground Forces, Coast Guard, and Air Corps), Guyana People's Militia (GPM), Guyana National Service (GNS)
Manpower availability: males age 15-49 198,665; males fit for military service 150,573 (1995 est.)
Defense expenditures: $NA, NA% of GDP

Haiti

Geography

Location: Caribbean, western one-third of the island of Hispaniola, between the Caribbean Sea and the North Atlantic Ocean, west of the Dominican Republic
Map references: Central America and the Caribbean
Area:
total area: 27,750 sq km
land area: 27,560 sq km
comparative area: slightly larger than Maryland
Land boundaries: total 275 km, Dominican Republic 275 km
Coastline: 1,771 km
Maritime claims:
contiguous zone: 24 nm
continental shelf: to depth of exploitation
exclusive economic zone: 200 nm
territorial sea: 12 nm
International disputes: claims US-administered Navassa Island
Climate: tropical; semiarid where mountains in east cut off trade winds
Terrain: mostly rough and mountainous
Natural resources: bauxite
Land use:
arable land: 20%
permanent crops: 13%
meadows and pastures: 18%
forest and woodland: 4%
other: 45%
Irrigated land: 750 sq km (1989 est.)
Environment:
current issues: extensive deforestation (much of the remaining forested land is being cleared for agriculture and use as fuel); soil erosion; inadequate supplies of potable water
natural hazards: lies in the middle of the hurricane belt and subject to severe storms from June to October; occasional flooding and earthquakes; periodic droughts
international agreements: party to—Marine Dumping, Marine Life Conservation; signed, but not ratified—Biodiversity, Climate Change, Desertification, Hazardous Wastes,

Law of the Sea, Nuclear Test Ban
Note: shares island of Hispaniola with Dominican Republic (western one-third is Haiti, eastern two-thirds is the Dominican Republic)

People

Population: 6,539,983 (July 1995 est.)
Age structure:
0-14 years: 46% (female 1,490,939; male 1,535,607)
15-64 years: 50% (female 1,692,032; male 1,557,568)
65 years and over: 4% (female 133,291; male 130,546) (July 1995 est.)
Population growth rate: 1.5% (1995 est.)
Birth rate: 38.64 births/1,000 population (1995 est.)
Death rate: 18.65 deaths/1,000 population (1995 est.)
Net migration rate: -4.99 migrant(s)/1,000 population (1995 est.)
Infant mortality rate: 107.5 deaths/1,000 live births (1995 est.)
Life expectancy at birth:
total population: 44.77 years
male: 43.04 years
female: 46.59 years (1995 est.)
Total fertility rate: 5.82 children born/woman (1995 est.)
Nationality:
noun: Haitian(s)
adjective: Haitian
Ethnic divisions: black 95%, mulatto and European 5%
Religions: Roman Catholic 80% (of which an overwhelming majority also practice Voodoo), Protestant 16% (Baptist 10%, Pentecostal 4%, Adventist 1%, other 1%), none 1%, other 3% (1982)
Languages: French (official) 10%, Creole
Literacy: age 15 and over can read and write (1982)
total population: 35%
male: 37%
female: 32%
Labor force: 2.3 million
by occupation: agriculture 66%, services 25%, industry 9%
note: shortage of skilled labor, unskilled labor abundant (1982)

Government

Names:
conventional long form: Republic of Haiti
conventional short form: Haiti
local long form: Republique d'Haiti
local short form: Haiti
Digraph: HA
Type: republic
Capital: Port-au-Prince
Administrative divisions: 9 departments, (departements, singular—departement); Artibonite, Centre, Grand'Anse, Nord, Nord-Est, Nord-Ouest, Ouest, Sud, Sud-Est
Independence: 1 January 1804 (from France)
National holiday: Independence Day, 1 January (1804)
Constitution: approved March 1987, suspended June 1988, most articles reinstated March 1989; October 1991, government claims to be observing the Constitution
Legal system: based on Roman civil law system; accepts compulsory ICJ jurisdiction
Suffrage: 18 years of age; universal
Executive branch:
chief of state: President Jean-Bertrand ARISTIDE (since 7 February 1991), ousted in a coup in September 1991 but, with US military support, returned to power on 15 October 1994; election last held 16 December 1990 (next to be held by December 1995); results—Rev. Jean-Bertrand ARISTIDE 67.5%, Marc BAZIN 14.2%, Louis DEJOIE 4.9%
head of government: Prime Minister Smarck MICHEL (since October 1994)
cabinet: Cabinet; chosen by prime minister in consultation with the president
Legislative branch: bicameral National Assembly (Assemblee Nationale)
Senate: elections last held 18 January 1993, widely condemned as illegitimate (next to be held December 1994); results—percent of vote by party NA; seats—(27 total) FNCD 12, MIDH-PANPRA 8, PAIN 2, MRN 1, RDNP 1, PNT 1, independent 2
Chamber of Deputies: elections last held 16 December 1990, with runoff held 20 January 1991 (next to be held by 25 June 1995); results—percent of vote by party NA; seats—(83 total) FNCD 27, MIDH-PANPRA 17, PDCH 7, PAIN 6, RDNP 6, MDN 5, PNT 3, MKN 2, MODELH 2, MRN 1, independents 5, other 2
Judicial branch: Court of Appeal (Cour de Cassation)
Political parties and leaders: National Front for Change and Democracy (FNCD), Evans PAUL, including National Cooperative Action Movement (MKN), Volvick Remy JOSEPH; National Congress of Democratic Movements (CONACOM), Victor BENOIT; Movement for the Installation of Democracy in Haiti (MIDH), Marc BAZIN; National Progressive Revolutionary Party (PANPRA), Serge GILLES; National Patriotic Movement of November 28 (MNP-28), Dejean BELIZAIRE; National Agricultural and Industrial Party (PAIN), Louis DEJOIE; Movement for National Reconstruction (MRN), Rene THEODORE; Haitian Christian Democratic Party (PDCH), Joseph DOUZE; Assembly of Progressive National Democrats (RDNP), Leslie MANIGAT; National Party of Labor (PNT), Thomas DESULME; Mobilization for National Development (MDN), Hubert DE RONCERAY; Democratic Movement for the Liberation of Haiti (MODELH), Francois LATORTUE; Haitian Social Christian Party (PSCH), Gregoire EUGENE; Movement for the Organization of the Country (MOP), Gesner COMEAU and Jean MOLIERE; Democratic Unity Confederation (KID), Evans PAUL; National Lavalas Political Organization (OPL), Gerard PIERRE/CHARLES
Other political or pressure groups: Roman Catholic Church; Confederation of Haitian Workers (CTH); Federation of Workers Trade Unions (FOS); Autonomous Haitian Workers (CATH); National Popular Assembly (APN); Revolutionary Front for Haitian Advancement and Progress (FRAPH)
Member of: ACCT, ACP, CARICOM (observer), CCC, ECLAC, FAO, G-77, GATT, IADB, IAEA, IBRD, ICAO, ICRM, IDA, IFAD, IFC, IFRCS, ILO, IMF, IMO, INTELSAT, INTERPOL, IOC, ITU, LAES, OAS, OPANAL, PCA, UN, UNCTAD, UNESCO, UNIDO, UPU, WCL, WFTU, WHO, WIPO, WMO, WTO
Diplomatic representation in US:
chief of mission: Ambassador Jean CASIMIR
chancery: 2311 Massachusetts Avenue NW, Washington, DC 20008
telephone: [1] (202) 332-4090 through 4092
FAX: [1] (202) 745-7215
consulate(s) general: Boston, Chicago, Miami, New York, and San Juan (Puerto Rico)
US diplomatic representation:
chief of mission: Ambassador William Lacy SWING
embassy: Harry Truman Boulevard, Port-au-Prince
mailing address: P. O. Box 1761, Port-au-Prince
telephone: [509] 22-0354, 22-0368, 22-0200, 22-0612
FAX: [509] 23-1641
Flag: two equal horizontal bands of blue (top) and red with a centered white rectangle bearing the coat of arms, which contains a palm tree flanked by flags and two cannons above a scroll bearing the motto L'UNION FAIT LA FORCE (Union Makes Strength)

Economy

Overview: About 75% of the population live in abject poverty. Agriculture is mainly small-scale subsistence farming and employs two-thirds of the work force. The majority of the population does not have ready access to safe drinking water, adequate medical care, or sufficient food. The lack of employment opportunities remains one of the most critical problems facing the economy, along with soil erosion and political instability. International trade sanctions in response to the September 1991 coup against President ARISTIDE further damaged the economy. The restoration of President ARISTIDE, the lifting of sanctions in late 1994, and foreign aid will

alleviate some economic problems. Haiti will continue to depend heavily on foreign aid.
National product: GDP—purchasing power parity—$5.6 billion (1994 est.)
National product real growth rate: -15% (1994 est.)
National product per capita: $870 (1994 est.)
Inflation rate (consumer prices): 52% (FY93/94 est.)
Unemployment rate: 50% (1994 est.)
Budget:
revenues: $56 million
expenditures: $131 million, including capital expenditures of $6 million (1994 est.)
Exports: $173.3 million (f.o.b., 1993 est.)
commodities: light manufactures 65%, coffee 19%, other agriculture 8%, other 8%
partners: US 81%, Europe 12% (1993)
Imports: $476.8 million (f.o.b., 1993 est.)
commodities: machines and manufactures 34%, food and beverages 22%, petroleum products 14%, chemicals 10%, fats and oils 9%
partners: US 51%, Europe 16%, Latin America 18% (1993)
External debt: $871 million (September 1994)
Industrial production: growth rate -2% (1991 est.); accounts for 15% of GDP
Electricity:
capacity: 150,000 kW
production: 590 million kWh
consumption per capita: 86 kWh (1993)
Industries: sugar refining, textiles, flour milling, cement manufacturing, tourism, light assembly industries based on imported parts
Agriculture: accounts for 28% of GDP and employs two-thirds of work force; mostly small-scale subsistence farms; commercial crops—coffee, mangoes, sugarcane, wood; staple crops—rice, corn, sorghum; shortage of wheat flour
Illicit drugs: transshipment point for cocaine and marijuana en route to the US and Europe
Economic aid:
recipient: US commitments, including Ex-Im (1970-89), $700 million; Western (non-US) countries, ODA and OOF bilateral commitments (1970-89), $770 million
Currency: 1 gourde (G) = 100 centimes
Exchange rates: gourdes (G) per US$1—14.10 (1 December 1994), 12.00 (1 July 1993), 8.4 (December 1991), fixed rate of 5.000 through second quarter of 1991
Fiscal year: 1 October—30 September

Transportation

Railroads:
total: 40 km (single track; privately owned industrial line)
narrow gauge: 40 km 0.760-m gauge
Highways:
total: 4,000 km
paved: 950 km
unpaved: otherwise improved 900 km; unimproved earth 2,150 km
Inland waterways: negligible; less than 100 km navigable
Ports: Cap-Haitien, Gonaives, Jacmel, Jeremie, Cayes, Miragoane, Port-au-Prince, Port-de-Paix, Saint-Marc
Merchant marine: none
Airports:
total: 14
with paved runways 2,438 to 3,047 m: 2
with paved runways 1,524 to 2,437 m: 1
with paved runways under 914 m: 6
with unpaved runways 914 to 1,523 m: 5

Communications

Telephone system: 36,000 telephones; domestic facilities barely adequate, international facilities slightly better
local: NA
intercity: NA
international: 1 INTELSAT (Atlantic Ocean) earth station
Radio:
broadcast stations: AM 33, FM 0, shortwave 2
radios: NA
Television:
broadcast stations: 4
televisions: NA

Defense Forces

Branches: Army, Navy, Air Force, Police
note: the regular Haitian Army, Navy and Air Force are currently suspended and replaced by the Interim Public Security Force (IPSF)
Manpower availability: males age 15-49 1,323,034; males fit for military service 716,233; males reach military age (18) annually 64,371 (1995 est.)
Defense expenditures: exchange rate conversion—$34 million, 1.5% of GDP (1988 est.)

Heard Island and McDonald Islands

(territory of Australia)

Geography

Location: Southern Africa, islands in the Indian Ocean, about two-thirds of the way from Madagascar to Antarctica
Map references: Antarctic Region
Area:
total area: 412 sq km
land area: 412 sq km
comparative area: slightly less than 2.5 times the size of Washington, DC
Land boundaries: 0 km
Coastline: 101.9 km
Maritime claims:
exclusive fishing zone: 200 nm
territorial sea: 3 nm
International disputes: none
Climate: antarctic
Terrain: Heard Island—bleak and mountainous, with a quiescent volcano; McDonald Islands—small and rocky
Natural resources: none
Land use:
arable land: 0%
permanent crops: 0%
meadows and pastures: 0%
forest and woodland: 0%
other: 100%
Irrigated land: 0 sq km
Environment:
current issues: NA
natural hazards: Heard Island is dominated by a dormant volcano called Big Ben
international agreements: NA
Note: primarily used for research stations

People

Population: uninhabited

Government

Names:
conventional long form: Territory of Heard

Heard Island and McDonald Islands *(continued)*

Island and McDonald Islands
conventional short form: Heard Island and McDonald Islands
Digraph: HM
Type: territory of Australia administered by the Ministry for Environment, Sport, and Territories
Capital: none; administered from Canberra, Australia
Independence: none (territory of Australia)

Economy

Overview: no economic activity

Transportation

Ports: none; offshore anchorage only

Defense Forces

Note: defense is the responsibility of Australia

Holy See (Vatican City)

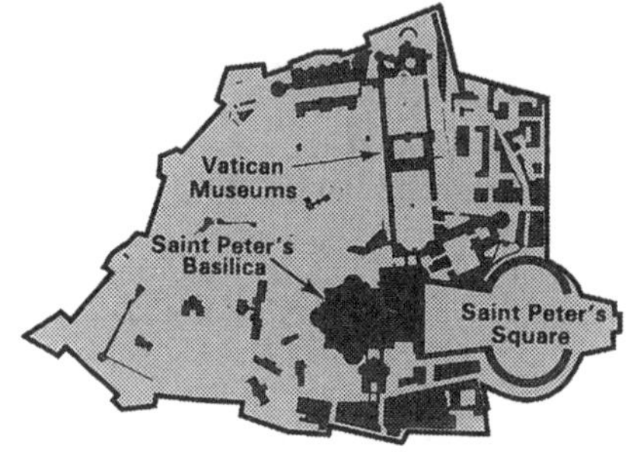

Geography

Location: Southern Europe, an enclave of Rome (Italy)
Map references: Europe
Area:
total area: 0.44 sq km
land area: 0.44 sq km
comparative area: about 0.7 times the size of The Mall in Washington, DC
Land boundaries: total 3.2 km, Italy 3.2 km
Coastline: 0 km (landlocked)
Maritime claims: none; landlocked
International disputes: none
Climate: temperate; mild, rainy winters (September to mid-May) with hot, dry summers (May to September)
Terrain: low hill
Natural resources: none
Land use:
arable land: 0%
permanent crops: 0%
meadows and pastures: 0%
forest and woodland: 0%
other: 100%
Irrigated land: 0 sq km
Environment:
current issues: NA
natural hazards: NA
international agreements: signed, but not ratified—Air Pollution, Environmental Modification
Note: urban; landlocked; enclave of Rome, Italy; world's smallest state; outside the Vatican City, 13 buildings in Rome and Castel Gandolfo (the pope's summer residence) enjoy extraterritorial rights

People

Population: 830 (July 1995 est.)
Population growth rate: 1.15% (1995 est.)
Birth rate: NA
Death rate: NA
Net migration rate: NA
Infant mortality rate: NA
Life expectancy at birth: NA
Total fertility rate: NA
Nationality:
noun: none
adjective: none
Ethnic divisions: Italians, Swiss
Religions: Roman Catholic
Languages: Italian, Latin, various other languages
Labor force: NA
by occupation: dignitaries, priests, nuns, guards, and 3,000 lay workers who live outside the Vatican

Government

Names:
conventional long form: The Holy See (State of the Vatican City)
conventional short form: Holy See (Vatican City)
local long form: Santa Sede (Stato della Citta del Vaticano)
local short form: Santa Sede (Citta del Vaticano)
Digraph: VT
Type: monarchical-sacerdotal state
Capital: Vatican City
Independence: 11 February 1929 (from Italy)
National holiday: Installation Day of the Pope, 22 October (1978) (John Paul II)
note: Pope John Paul II was elected on 16 October 1978
Constitution: Apostolic Constitution of 1967 (effective 1 March 1968)
Legal system: NA
Suffrage: limited to cardinals less than 80 years old
Executive branch:
chief of state: Pope JOHN PAUL II (Karol WOJTYLA; since 16 October 1978); election last held 16 October 1978 (next to be held after the death of the current pope); results—Karol WOJTYLA was elected for life by the College of Cardinals
head of government: Secretary of State Archbishop Angelo Cardinal SODANO (since NA 1991)
cabinet: Pontifical Commission; appointed by Pope
Legislative branch: unicameral Pontifical Commission
Judicial branch: none; normally handled by Italy
Political parties and leaders: none
Other political or pressure groups: none (exclusive of influence exercised by church officers)
Member of: IAEA, ICFTU, INTELSAT, IOM (observer), ITU, OAS (observer), OSCE, UN (observer), UNCTAD, UNHCR, UPU, WIPO, WTO (observer)
Diplomatic representation in US:
chief of mission: Apostolic Pro-Nuncio Archbishop Agostino CACCIAVILLAN
chancery: 3339 Massachusetts Avenue NW,

Washington, DC 20008
telephone: [1] (202) 333-7121
US diplomatic representation:
chief of mission: Ambassador Raymond L. FLYNN
embassy: Via Delle Terme Deciane 26, Rome 00153
mailing address: PSC 59, APO AE 09624
telephone: [39] (6) 46741
FAX: [39] (6) 6380159
Flag: two vertical bands of yellow (hoist side) and white with the crossed keys of Saint Peter and the papal miter centered in the white band

Economy

Overview: This unique, noncommercial economy is supported financially by contributions (known as Peter's Pence) from Roman Catholics throughout the world, the sale of postage stamps and tourist mementos, fees for admission to museums, and the sale of publications. The incomes and living standards of lay workers are comparable to, or somewhat better than, those of counterparts who work in the city of Rome.
Budget:
revenues: $169 million
expenditures: $167.5 million, including capital expenditures of $NA (1993)
Electricity:
capacity: 5,000 kW standby
production: power supplied by Italy
consumption per capita: NA kWh (1992)
Industries: printing and production of a small amount of mosaics and staff uniforms; worldwide banking and financial activities
Currency: 1 Vatican lira (VLit) = 100 centesimi
Exchange rates: Vatican lire (VLit) per US$1—1,609.5 (January 1995), 1,612.4 (1994), 1,573.7 (1993), 1,232.4 (1992), 1,240.6 (1991), 1,198.1 (1990); note—the Vatican lira is at par with the Italian lira which circulates freely
Fiscal year: calendar year

Transportation

Railroads:
total: 862 meters; note—connects to Italy's network at Rome's Saint Peter's station
narrow gauge: 862 meters 1.435-m gauge
Highways: none; all city streets
Ports: none
Airports: none

Communications

Telephone system: 2,000 telephones; automatic exchange
local: NA
intercity: tied into Italian system
international: uses Italian system
Radio:
broadcast stations: AM 3, FM 4, shortwave 0
radios: NA
Television:
broadcast stations: 0
televisions: NA

Defense Forces

Note: defense is the responsibility of Italy; Swiss Papal Guards are posted at entrances to the Vatican City

Honduras

Geography

Location: Middle America, bordering the Caribbean Sea, between Guatemala and Nicaragua and bordering the North Pacific Ocean, between El Salvador and Nicaragua
Map references: Central America and the Caribbean
Area:
total area: 112,090 sq km
land area: 111,890 sq km
comparative area: slightly larger than Tennessee
Land boundaries: total 1,520 km, Guatemala 256 km, El Salvador 342 km, Nicaragua 922 km
Coastline: 820 km
Maritime claims:
contiguous zone: 24 nm
continental shelf: natural extension of territory or to 200 nm
exclusive economic zone: 200 nm
territorial sea: 12 nm
International disputes: land boundary dispute with El Salvador mostly resolved by 11 September 1992 International Court of Justice (ICJ) decision; with respect to the maritime boundary in the Golfo de Fonseca, ICJ referred to an earlier agreement in this century and advised that some tripartite resolution among El Salvador, Honduras and Nicaragua likely would be required
Climate: subtropical in lowlands, temperate in mountains
Terrain: mostly mountains in interior, narrow coastal plains
Natural resources: timber, gold, silver, copper, lead, zinc, iron ore, antimony, coal, fish
Land use:
arable land: 14%
permanent crops: 2%
meadows and pastures: 30%
forest and woodland: 34%
other: 20%
Irrigated land: 900 sq km (1989 est.)
Environment:

current issues: urban population expanding; deforestation results from logging and the clearing of land for agricultural purposes; further land degradation and soil erosion hastened by uncontrolled development and improper land use practices such as farming of marginal lands; mining activities polluting Lago de Yojoa (the country's largest source of freshwater) with heavy metals as well as several rivers and streams
natural hazards: frequent, but generally mild, earthquakes; damaging hurricanes and floods along Caribbean coast
international agreements: party to—Endangered Species, Law of the Sea, Marine Dumping, Nuclear Test Ban, Ozone Layer Protection, Tropical Timber 83, Wetlands; signed, but not ratified—Biodiversity, Climate Change

People

Population: 5,459,743 (July 1995 est.)
Age structure:
0-14 years: 43% (female 1,159,846; male 1,201,927)
15-64 years: 53% (female 1,468,950; male 1,444,959)
65 years and over: 4% (female 95,361; male 88,700) (July 1995 est.)
Population growth rate: 2.66% (1995 est.)
Birth rate: 34.12 births/1,000 population (1995 est.)
Death rate: 6 deaths/1,000 population (1995 est.)
Net migration rate: -1.56 migrant(s)/1,000 population (1995 est.)
Infant mortality rate: 43.4 deaths/1,000 live births (1995 est.)
Life expectancy at birth:
total population: 68.04 years
male: 65.64 years
female: 70.55 years (1995 est.)
Total fertility rate: 4.55 children born/woman (1995 est.)
Nationality:
noun: Honduran(s)
adjective: Honduran
Ethnic divisions: mestizo (mixed Indian and European) 90%, Indian 7%, black 2%, white 1%
Religions: Roman Catholic 97%, Protestant minority
Languages: Spanish, Indian dialects
Literacy: age 15 and over can read and write (1990 est.)
total population: 73%
male: 76%
female: 71%
Labor force: 1.3 million
by occupation: agriculture 62%, services 20%, manufacturing 9%, construction 3%, other 6% (1985)

Government

Names:
conventional long form: Republic of Honduras
conventional short form: Honduras
local long form: Republica de Honduras
local short form: Honduras
Digraph: HO
Type: republic
Capital: Tegucigalpa
Administrative divisions: 18 departments (departamentos, singular—departamento); Atlantida, Choluteca, Colon, Comayagua, Copan, Cortes, El Paraiso, Francisco Morazan, Gracias a Dios, Intibuca, Islas de la Bahia, La Paz, Lempira, Ocotepeque, Olancho, Santa Barbara, Valle, Yoro
Independence: 15 September 1821 (from Spain)
National holiday: Independence Day, 15 September (1821)
Constitution: 11 January 1982, effective 20 January 1982
Legal system: rooted in Roman and Spanish civil law; some influence of English common law; accepts ICJ jurisdiction, with reservations
Suffrage: 18 years of age; universal and compulsory
Executive branch:
chief of state and head of government: President Carlos Roberto REINA Idiaquez (since 27 January 1994); election last held 28 November 1993 (next to be held November 1997); results—Carlos Roberto REINA Idiaquez (PLH) 53%, Oswaldo RAMOS Soto (PNH) 41%, other 6%
cabinet: Cabinet
Legislative branch: unicameral
National Congress (Congreso Nacional): elections last held on 27 November 1993 (next to be held November 1997); results—PNH 53%, PLH 41%, PDCH 1.0%, PINU-SD 2.5%, other 2.5%; seats—(134 total) PNH 55, PLH 77, PINU-SD 2
Judicial branch: Supreme Court of Justice (Corte Suprema de Justica)
Political parties and leaders: Liberal Party (PLH), Rafael PINEDA Ponce, president; National Party of Honduras (PNH), Oswaldo RAMOS Soto, president; National Innovation and Unity Party (PINU), Olban VALLADARES, president; Christian Democratic Party (PDCH), Efrain DIAZ Arrivillaga, president
Other political or pressure groups: National Association of Honduran Campesinos (ANACH); Honduran Council of Private Enterprise (COHEP); Confederation of Honduran Workers (CTH); National Union of Campesinos (UNC); General Workers Confederation (CGT); United Federation of Honduran Workers (FUTH); Committee for the Defense of Human Rights in Honduras (CODEH); Coordinating Committee of Popular Organizations (CCOP)
Member of: BCIE, CACM, ECLAC, FAO, G-77, GATT, IADB, IBRD, ICAO, ICFTU, ICRM, IDA, IFAD, IFC, IFRCS, ILO, IMF, IMO, INTELSAT, INTERPOL, IOC, IOM, ITU, LAES, LAIA (observer), MINURSO, NAM, OAS, OPANAL, PCA, UN, UNCTAD, UNESCO, UNIDO, UPU, WCL, WFTU, WHO, WIPO, WMO
Diplomatic representation in US:
chief of mission: Ambassador Roberto FLORES Bermudez
chancery: 3007 Tilden Street NW, Washington, DC 20008
telephone: [1] (202) 966-7702, 2604, 5008, 4596
FAX: [1] (202) 966-9751
consulate(s) general: Chicago, Houston, Los Angeles, Miami, New Orleans, New York, San Francisco, and San Juan (Puerto Rico)
consulate(s): Boston, Detroit, and Jacksonville
US diplomatic representation:
chief of mission: Ambassador William T. PRYCE
embassy: Avenida La Paz, Apartado Postal No 3453, Tegucigalpa
mailing address: American Embassy, APO AA 34022, Tegucigalpa
telephone: [504] 36-9320, 38-5114
FAX: [504] 36-9037
Flag: three equal horizontal bands of blue (top), white, and blue with five blue five-pointed stars arranged in an X pattern centered in the white band; the stars represent the members of the former Federal Republic of Central America—Costa Rica, El Salvador, Guatemala, Honduras, and Nicaragua; similar to the flag of El Salvador, which features a round emblem encircled by the words REPUBLICA DE EL SALVADOR EN LA AMERICA CENTRAL centered in the white band; also similar to the flag of Nicaragua, which features a triangle encircled by the word REPUBLICA DE NICARAGUA on top and AMERICA CENTRAL on the bottom, centered in the white band

Economy

Overview: Honduras is one of the poorest countries in the Western Hemisphere. Agriculture, the most important sector of the economy, accounts for 28% of GDP, employs 62% of the labor force, and produces two-thirds of exports. Productivity remains low. Manufacturing, still in its early stages, employs 9% of the labor force, accounts for 15% of GDP, and generates 20% of exports. The service sectors, including public administration, account for 50% of GDP and employ 20% of the labor force. Many basic problems face the economy, including rapid population growth, high unemployment, inflation, a lack of basic services, a large and

inefficient public sector, and the dependence of the export sector mostly on coffee and bananas, which are subject to sharp price fluctuations. A far-reaching reform program, initiated by former President CALLEJAS in 1990 and scaled back by President REINA, is beginning to take hold.
National product: GDP—purchasing power parity—$9.7 billion (1994 est.)
National product real growth rate: -1.9% (1994 est.)
National product per capita: $1,820 (1994 est.)
Inflation rate (consumer prices): 30% (1994 est.)
Unemployment rate: 10%; underemployed 30%-40% (1992)
Budget:
revenues: $527 million
expenditures: $668 million, including capital expenditures of $166 million (1993 est.)
Exports: $850 million (f.o.b., 1993 est)
commodities: bananas, coffee, shrimp, lobster, minerals, meat, lumber
partners: US 53%, Germany 11%, Belgium 8%, UK 5%
Imports: $990 million (c.i.f. 1994 est)
commodities: machinery and transport equipment, chemical products, manufactured goods, fuel and oil, foodstuffs
partners: US 50%, Mexico 8%, Guatemala 6%
External debt: $4 billion (1994 est.)
Industrial production: growth rate 10% (1992 est.); accounts for 22% of GDP
Electricity:
capacity: 290,000 kW
production: 2.3 billion kWh
consumption per capita: 445 kWh (1993)
Industries: agricultural processing (sugar and coffee), textiles, clothing, wood products
Agriculture: most important sector, accounting for 28% of GDP, more than 60% of the labor force, and two-thirds of exports; principal products include bananas, coffee, timber, beef, citrus fruit, shrimp; importer of wheat
Illicit drugs: transshipment point for narcotics; illicit producer of cannabis, cultivated on small plots and used principally for local consumption
Economic aid:
recipient: US commitments, including Ex-Im (FY70-89), $1.4 billion; Western (non-US) countries, ODA and OOF bilateral commitments (1970-89), $1.1 billion
Currency: 1 lempira (L) = 100 centavos
Exchange rates: lempiras (L) per US$1—9.1283 (October 1994), 7.2600 (1993), 5.8300 (1992), 5.4000 (1991); 2.0000 (fixed rate until 1991) 5.70 parallel black-market rate (November 1990); the lempira was allowed to float in 1992
Fiscal year: calendar year

Transportation

Railroads:
total: 785 km
narrow gauge: 508 km 1.067-m gauge; 277 km 0.914-m gauge
Highways:
total: 8,950 km
paved: 1,700 km
unpaved: otherwise improved 5,000 km; unimproved earth 2,250 km
Inland waterways: 465 km navigable by small craft
Ports: La Ceiba, Puerto Castilla, Puerto Cortes, San Lorenzo, Tela, Puerto Lempira
Merchant marine:
total: 271 ships (1,000 GRT or over) totaling 802,990 GRT/1,210,553 DWT
ships by type: bulk 31, cargo 171, chemical tanker 1, combination bulk 1, container 6, liquefied gas tanker 2, livestock carrier 3, oil tanker 21, passenger 2, passenger-cargo 3, refrigerated cargo 19, roll-on/roll-off cargo 7, short-sea passenger 2, specialized tanker 1, vehicle carrier 1
note: a flag of convenience registry; Russia owns 14 ships, Vietnam 7, North Korea 4, US 3, Hong Kong 2, South Korea 2, Greece 1
Airports:
total: 159
with paved runways 2,438 to 3,047 m: 3
with paved runways 1,524 to 2,437 m: 2
with paved runways 914 to 1,523 m: 4
with paved runways under 914 m: 118
with unpaved runways 2,438 to 3,047 m: 1
with unpaved runways 1,524 to 2,438 m: 4
with unpaved runways 914 to 1,523 m: 27

Communications

Telephone system: NA telephones; 7 telephones/1,000 persons; inadequate system
local: NA
intercity: NA
international: 2 INTELSAT (Atlantic Ocean) earth stations and the Central American microwave radio relay system
Radio:
broadcast stations: AM 176, FM 0, shortwave 7
radios: NA
Television:
broadcast stations: 28
televisions: NA

Defense Forces

Branches: Army, Navy (includes Marines), Air Force, Public Security Forces (FUSEP)
Manpower availability: males age 15-49 1,275,670; males fit for military service 760,113; males reach military age (18) annually 62,405 (1995 est.)
Defense expenditures: exchange rate conversion—$41 million, about 0.4% of GDP (1994)

Hong Kong
(dependent territory of the UK)

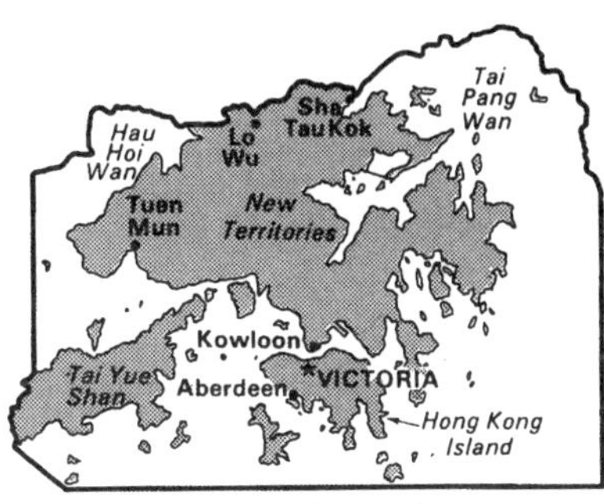

Geography

Location: Eastern Asia, bordering the South China Sea and China
Map references: Southeast Asia
Area:
total area: 1,040 sq km
land area: 990 sq km
comparative area: slightly less than six times the size of Washington, DC
Land boundaries: total 30 km, China 30 km
Coastline: 733 km
Maritime claims:
territorial sea: 3 nm
International disputes: none
Climate: tropical monsoon; cool and humid in winter, hot and rainy from spring through summer, warm and sunny in fall
Terrain: hilly to mountainous with steep slopes; lowlands in north
Natural resources: outstanding deepwater harbor, feldspar
Land use:
arable land: 7%
permanent crops: 1%
meadows and pastures: 1%
forest and woodland: 12%
other: 79%
Irrigated land: 20 sq km (1989)
Environment:
current issues: air and water pollution from rapid urbanization
natural hazards: occasional typhoons
international agreements: NA
Note: more than 200 islands

People

Population: 5,542,869 (July 1995 est.)
Age structure:
0-14 years: 19% (female 499,460; male 549,734)
15-64 years: 70% (female 1,866,540; male 2,016,684)
65 years and over: 11% (female 331,391; male 279,060) (July 1995 est.)
Population growth rate: -0.12% (1995 est.)

Birth rate: 12.02 births/1,000 population (1995 est.)
Death rate: 6.02 deaths/1,000 population (1995 est.)
Net migration rate: -7.22 migrant(s)/1,000 population (1995 est.)
Infant mortality rate: 5.8 deaths/1,000 live births (1995 est.)
Life expectancy at birth:
total population: 80.18 years
male: 76.78 years
female: 83.78 years (1995 est.)
Total fertility rate: 1.39 children born/woman (1995 est.)
Nationality:
noun: Chinese
adjective: Chinese
Ethnic divisions: Chinese 95%, other 5%
Religions: eclectic mixture of local religions 90%, Christian 10%
Languages: Chinese (Cantonese), English
Literacy: age 15 and over has ever attended school (1971)
total population: 77%
male: 90%
female: 64%
Labor force: 2.8 million (1990)
by occupation: manufacturing 28.5%, wholesale and retail trade, restaurants, and hotels 27.9%, services 17.7%, financing, insurance, and real estate 9.2%, transport and communications 4.5%, construction 2.5%, other 9.7% (1989)

Government

Names:
conventional long form: none
conventional short form: Hong Kong
Abbreviation: HK
Digraph: HK
Type: dependent territory of the UK scheduled to revert to China in 1997
Capital: Victoria
Administrative divisions: none (dependent territory of the UK)
Independence: none (dependent territory of the UK; the UK signed an agreement with China on 19 December 1984 to return Hong Kong to China on 1 July 1997; in the joint declaration, China promises to respect Hong Kong's existing social and economic systems and lifestyle)
National holiday: Liberation Day, 29 August (1945)
Constitution: unwritten; partly statutes, partly common law and practice; new Basic Law approved in March 1990 in preparation for 1997
Legal system: based on English common law
Suffrage: direct election 21 years of age; universal for permanent residents living in the territory of Hong Kong for the past seven years; indirect election limited to about 100,000 professionals of electoral college and functional constituencies
Executive branch:
chief of state: Queen ELIZABETH II (since 6 February 1952)
head of government: Governor Chris PATTEN (since 9 July 1992); Chief Secretary Anson CHAN Fang On-Sang (since 29 November 1993)
cabinet: Executive Council; appointed by the governor
Legislative branch: unicameral
Legislative Council: indirect elections last held 12 September 1991 and direct elections were held for the first time 15 September 1991 (next to be held 17 September 1995 when the number of directly-elected seats increases to 50); results—percent of vote by party NA; seats—(60 total; 21 indirectly elected by functional constituencies, 18 directly elected, 18 appointed by governor, 3 ex officio members); indirect elections—number of seats by functional constituency NA; direct elections—UDHK 12, Meeting Point 3, ADPL 1, other 2
Judicial branch: Supreme Court
Political parties and leaders: Democratic Party, Martin LEE, chairman; Democratic Alliance for the Betterment of Hong Kong, TSANG Yuk-shing, chairman; Hong Kong Democratic Foundation, Dr. Patrick SHIU Kin-ying, chairman
note: in April 1994, the United Democrats of Hong Kong (UDHK) and Meeting Point merged to form the Democratic Party; the merger became effective in October 1994
Other political or pressure groups: Liberal Party, Allen LEE, chairman; Association for Democracy and People's Livelihood (ADPL), Frederick FUNG Kin Kee, chairman; Liberal Democratic Federation, HU Fa-kuang, chairman; Federation of Trade Unions (pro-China), LEE Chark-tim, president; Hong Kong and Kowloon Trade Union Council (pro-Taiwan); Confederation of Trade Unions (pro-democracy), LAU Chin-shek, chairman; Hong Kong General Chamber of Commerce; Chinese General Chamber of Commerce (pro-China); Federation of Hong Kong Industries; Chinese Manufacturers' Association of Hong Kong; Hong Kong Professional Teachers' Union, CHEUNG Man-kwong, president; Hong Kong Alliance in Support of the Patriotic Democratic Movement in China, Szeto WAH, chairman
Member of: APEC, AsDB, CCC, ESCAP (associate), GATT, ICFTU, IMO (associate), INTERPOL (subbureau), IOC, ISO (correspondent), WCL, WMO
Diplomatic representation in US: none (dependent territory of the UK)
US diplomatic representation:
chief of mission: Consul General Richard W. MUELLER
consulate(s) general: 26 Garden Road, Hong Kong
mailing address: PSC 464, Box 30, Hong Kong, or FPO AP 96522-0002
telephone: [852] 523-9011
FAX: [852] 845-4845
Flag: blue with the flag of the UK in the upper hoist-side quadrant with the Hong Kong coat of arms on a white disk centered on the outer half of the flag; the coat of arms contains a shield (bearing two junks below a crown) held by a lion (representing the UK) and a dragon (representing China) with another lion above the shield and a banner bearing the words HONG KONG below the shield

Economy

Overview: Hong Kong has a bustling free market economy with few tariffs or nontariff barriers. Natural resources are limited, and food and raw materials must be imported. Manufacturing accounts for about 17% of GDP. Goods and services exports account for about 50% of GDP. Real GDP growth averaged a remarkable 8% in 1987-88, slowed to 3.0% in 1989-90, and picked up to 4.2% in 1991, 5.0% in 1992, 5.2% in 1993, and 5.5% in 1994. Unemployment, which has been declining since the mid-1980s, is now about 2%. A shortage of labor continues to put upward pressure on prices and the cost of living. Prospects for 1995-96 remain bright so long as major trading partners continue to be reasonably prosperous and so long as investors feel China will support free market practices after the takeover in 1997.
National product: GDP—purchasing power parity—$136.1 billion (1994 est.)
National product real growth rate: 5.5% (1994 est.)
National product per capita: $24,530 (1994 est.)
Inflation rate (consumer prices): 8.5% (1994)
Unemployment rate: 1.9% (1994 est.)
Budget:
revenues: $19.2 billion
expenditures: $19.7 billion, including capital expenditures of $NA (FY93/94)
Exports: $168.7 billion (including re-exports of $121.0 billion)(f.o.b., 1994 est.)
commodities: clothing, textiles, yarn and fabric, footwear, electrical appliances, watches and clocks, toys
partners: China 32%, US 23%, Germany 5%, Japan 5%, UK 3% (1993 est.)
Imports: $160 billion (c.i.f., 1994 est.)
commodities: foodstuffs, transport equipment, raw materials, semimanufactures, petroleum; a large share is re-exported
partners: China 36%, Japan 19%, Taiwan 9%, US 7% (1993 est.)
External debt: none (1993)
Industrial production: growth rate 2% (1993 est.)
Electricity:
capacity: 8,930,000 kW

production: 33 billion kWh
consumption per capita: 4,628 kWh (1993)
Industries: textiles, clothing, tourism, electronics, plastics, toys, watches, clocks
Agriculture: minor role in the economy; local farmers produce 26% fresh vegetables, 27% live poultry; 8% of land area suitable for farming
Illicit drugs: a hub for Southeast Asian heroin trade; transshipment and major financial and money-laundering center; increasing indigenous amphetamine and cocaine abuse
Economic aid:
recipient: US commitments, including Ex-Im (FY70-87), $152 million; Western (non-US) countries, ODA and OOF bilateral commitments (1970-89), $923 million
Currency: 1 Hong Kong dollar (HK$) = 100 cents
Exchange rates: Hong Kong dollars (HK$) per US$—7.800 (1994), 7.800 (1993), 7.741 (1992), 7.771 (1991), 7.790 (1990); note—linked to the US dollar at the rate of about 7.8 HK$ per 1 US$ since 1985
Fiscal year: 1 April—31 March

Transportation

Railroads:
total: 35 km
standard gauge: 35 km 1.435-m gauge
Highways:
total: 1,100 km
paved: 794 km
unpaved: gravel, crushed stone, earth 306 km
Ports: Hong Kong
Merchant marine:
total: 217 ships (1,000 GRT or over) totaling 7,657,749 GRT/13,181,496 DWT
ships by type: bulk 116, cargo 29, chemical tanker 2, combination bulk 2, combination ore/oil 6, container 28, liquefied gas tanker 5, oil tanker 18, refrigerated cargo 7, short-sea passenger 1, vehicle carrier 3
note: a flag of convenience registry; includes 15 countries among which are UK with 53 ships, China 15, Bermuda 7, Japan 6, Belgium 3, Germany 3, Greece 3, Canada 2, Netherlands 2, Singapore 2
Airports:
total: 3
with paved runways over 3,047 m: 1
with paved runways under 914 m: 2

Communications

Telephone system: 3,000,000 telephones; modern facilities provide excellent domestic and international services
local: NA
intercity: microwave transmission links and extensive optical fiber transmission network
international: 3 INTELSAT (1 Pacific Ocean and 2 Indian Ocean) earth stations; coaxial cable to Guangzhou, China; links to 5 international submarine cables providing access to ASEAN member nations, Japan, Taiwan, Australia, Middle East, and Western Europe
Radio:
broadcast stations: AM 6, FM 6, shortwave 0
radios: 2.5 million
Television:
broadcast stations: 4 (British Broadcasting Corporation repeater 1; British Forces Broadcasting Service repeater 1)
televisions: 1.312 million (1,224,000 color TV sets)

Defense Forces

Branches: Headquarters of British Forces, Army, Royal Navy, Royal Air Force, Royal Hong Kong Auxiliary Air Force, Royal Hong Kong Police Force
Manpower availability: males age 15-49 1,634,559; males fit for military service 1,245,905; males reach military age (18) annually 40,996 (1995 est.)
Defense expenditures: exchange rate conversion—$207 million, 0.2% of GDP (FY92/93); this represents 65% of the total cost of defending the colony, the remainder being paid by the UK
Note: defense is the responsibility of the UK

Howland Island
(territory of the US)

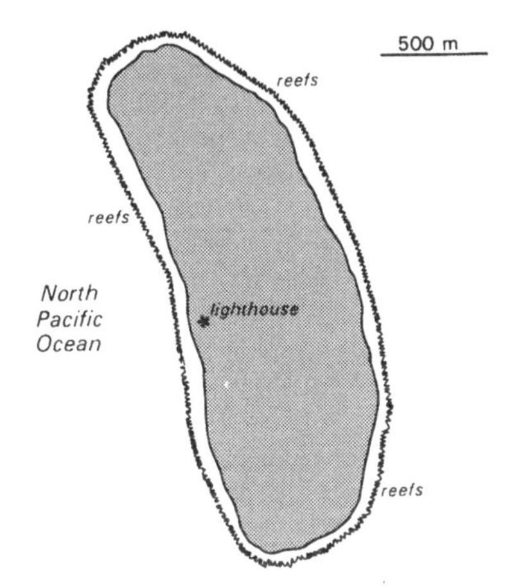

Geography

Location: Oceania, island in the North Pacific Ocean, about one-half of the way from Hawaii to Australia
Map references: Oceania
Area:
total area: 1.6 sq km
land area: 1.6 sq km
comparative area: about 2.7 times the size of The Mall in Washington, DC
Land boundaries: 0 km
Coastline: 6.4 km
Maritime claims:
exclusive economic zone: 200 nm
territorial sea: 12 nm
International disputes: none
Climate: equatorial; scant rainfall, constant wind, burning sun
Terrain: low-lying, nearly level, sandy, coral island surrounded by a narrow fringing reef; depressed central area
Natural resources: guano (deposits worked until late 1800s)
Land use:
arable land: 0%
permanent crops: 0%
meadows and pastures: 0%
forest and woodland: 5%
other: 95%
Irrigated land: 0 sq km
Environment:
current issues: no natural fresh water resources
natural hazards: the narrow fringing reef surrounding the island can be a maritime hazard
international agreements: NA
Note: almost totally covered with grasses, prostrate vines, and low-growing shrubs; small area of trees in the center; primarily a nesting, roosting, and foraging habitat for seabirds, shorebirds, and marine wildlife; feral cats

People

Population: uninhabited; note—American

Howland Island *(continued)*

civilians evacuated in 1942 after Japanese air and naval attacks during World War II; occupied by US military during World War II, but abandoned after the war; public entry is by special-use permit only and generally restricted to scientists and educators

Government

Names:
conventional long form: none
conventional short form: Howland Island
Digraph: HQ
Type: unincorporated territory of the US administered by the Fish and Wildlife Service of the US Department of the Interior as part of the National Wildlife Refuge System
Capital: none; administered from Washington, DC

Economy

Overview: no economic activity

Transportation

Ports: none; offshore anchorage only; note—there is one boat landing area along the middle of the west coast
Airports: airstrip constructed in 1937 for scheduled refueling stop on the round-the-world flight of Amelia Earhart and Fred Noonan—they left Lae, New Guinea, for Howland Island, but were never seen again; the airstrip is no longer serviceable
Note: Earhart Light is a day beacon near the middle of the west coast that was partially destroyed during World War II, but has since been rebuilt in memory of famed aviatrix Amelia Earhart

Defense Forces

Note: defense is the responsibility of the US; visited annually by the US Coast Guard

Hungary

Geography

Location: Central Europe, northwest of Romania
Map references: Ethnic Groups in Eastern Europe, Europe
Area:
total area: 93,030 sq km
land area: 92,340 sq km
comparative area: slightly smaller than Indiana
Land boundaries: total 1,989 km, Austria 366 km, Croatia 329 km, Romania 443 km, Serbia and Montenegro 151 km (all with Serbia), Slovakia 515 km, Slovenia 82 km, Ukraine 103 km
Coastline: 0 km (landlocked)
Maritime claims: none; landlocked
International disputes: Gabcikovo Dam dispute with Slovakia
Climate: temperate; cold, cloudy, humid winters; warm summers
Terrain: mostly flat to rolling plains; hills and low mountains on the Slovakian border
Natural resources: bauxite, coal, natural gas, fertile soils
Land use:
arable land: 50.7%
permanent crops: 6.1%
meadows and pastures: 12.6%
forest and woodland: 18.3%
other: 12.3%
Irrigated land: 1,750 sq km (1989)
Environment:
current issues: air pollution; industrial and municipal pollution of Lake Balaton
natural hazards: levees are common along many streams, but flooding occurs almost every year
international agreements: party to—Air Pollution, Air Pollution-Nitrogen Oxides, Air Pollution-Sulphur 85, Antarctic Treaty, Biodiversity, Climate Change, Endangered Species, Environmental Modification, Hazardous Wastes, Marine Dumping, Nuclear Test Ban, Ozone Layer Protection, Ship Pollution, Wetlands; signed, but not ratified—Air Pollution-Sulphur 94, Air Pollution-Volatile Organic Compounds, Antarctic-Environmental Protocol, Law of the Sea
Note: landlocked; strategic location astride main land routes between Western Europe and Balkan Peninsula as well as between Ukraine and Mediterranean basin

People

Population: 10,318,838 (July 1995 est.)
Age structure:
0-14 years: 18% (female 918,281; male 958,027)
15-64 years: 68% (female 3,534,218; male 3,440,036)
65 years and over: 14% (female 914,221; male 554,055) (July 1995 est.)
Population growth rate: 0.02% (1995 est.)
Birth rate: 12.65 births/1,000 population (1995 est.)
Death rate: 12.44 deaths/1,000 population (1995 est.)
Net migration rate: 0 migrant(s)/1,000 population (1995 est.)
Infant mortality rate: 11.9 deaths/1,000 live births (1995 est.)
Life expectancy at birth:
total population: 71.9 years
male: 67.94 years
female: 76.06 years (1995 est.)
Total fertility rate: 1.82 children born/woman (1995 est.)
Nationality:
noun: Hungarian(s)
adjective: Hungarian
Ethnic divisions: Hungarian 89.9%, Gypsy 4%, German 2.6%, Serb 2%, Slovak 0.8%, Romanian 0.7%
Religions: Roman Catholic 67.5%, Calvinist 20%, Lutheran 5%, atheist and other 7.5%
Languages: Hungarian 98.2%, other 1.8%
Literacy: age 15 and over can read and write (1980)
total population: 99%
male: 99%
female: 98%
Labor force: 5.4 million
by occupation: services, trade, government, and other 44.8%, industry 29.7%, agriculture 16.1%, construction 7.0% (1991)

Government

Names:
conventional long form: Republic of Hungary
conventional short form: Hungary
local long form: Magyar Koztarsasag
local short form: Magyarorszag
Digraph: HU
Type: republic
Capital: Budapest
Administrative divisions: 38 counties (megyek, singular—megye) and 1 capital city* (fovaros); Bacs-Kiskun, Baranya, Bekes,

Bekescsaba, Borsod-Abauj-Zemplen, Budapest*, Csongrad, Debrecen, Dunaujvaros, Eger, Fejer, Gyor, Gyor-Moson-Sopron, Hajdu-Bihar, Heves, Hodmezovasarhely, Jasz-Nagykun-Szolnok, Kaposvar, Kecskemet, Komarom-Esztergom, Miskolc, Nagykanizsa, Nograd, Nyiregyhaza, Pecs, Pest, Somogy, Sopron, Szabolcs-Szatmar-Bereg, Szeged, Szekesfehervar, Szolnok, Szombathely, Tatabanya, Tolna, Vas, Veszprem, Zala, Zalaegerszeg
Independence: 1001 (unification by King Stephen I)
National holiday: St. Stephen's Day (National Day), 20 August (commemorates the founding of Hungarian state circa 1000 A.D.)
Constitution: 18 August 1949, effective 20 August 1949, revised 19 April 1972; 18 October 1989 revision ensured legal rights for individuals and constitutional checks on the authority of the prime minister and also established the principle of parliamentary oversight
Legal system: in process of revision, moving toward rule of law based on Western model
Suffrage: 18 years of age; universal
Executive branch:
chief of state: President Arpad GONCZ (since 3 August 1990; previously interim president from 2 May 1990); election last held 3 August 1990 (next to be held NA 1995); results—President GONCZ elected by parliamentary vote; note—President GONCZ was elected by the National Assembly with a total of 295 votes out of 304 as interim President from 2 May 1990 until elected President
head of government: Prime Minister Gyula HORN (since 15 July 1994)
cabinet: Council of Ministers; elected by the National Assembly on recommendation of the president
Legislative branch: unicameral
National Assembly (Orszaggyules): elections last held on 8 and 29 May 1994 (next to be held spring 1998); results—percent of vote by party NA; seats—(386 total) MSzP 209, SzDSz 70, MDF 37, FKgP 26, KDNP 22, FiDeSz 20, other 2
Judicial branch: Constitutional Court
Political parties and leaders: Hungarian Democratic Forum (MDF), Lajos FUR, chairman; Independent Smallholders (FKgP), Jozsef TORGYAN, president; Hungarian Socialist Party (MSzP), Gyula HORN, president; Christian Democratic People's Party (KDNP), Dr. Lazlo SURJAN, president; Federation of Young Democrats (FiDeSz), Viktor ORBAN, chairman; Alliance of Free Democrats (SzDSz), Ivan PETO, chairman
note: the Hungarian Socialist (Communist) Workers' Party (MSzMP) renounced Communism and became the Hungarian Socialist Party (MSzP) in October 1989; there is still a small MMP
Member of: Australia Group, BIS, CCC, CE, CEI, CERN, EBRD, ECE, FAO, G- 9, GATT, IAEA, IBRD, ICAO, ICRM, IDA, IFC, IFRCS, ILO, IMF, IMO, INTELSAT, INTERPOL, IOC, IOM, ISO, ITU, MTCR, NACC, NAM (guest), NSG, OAS (observer), OSCE, PCA, PFP, UN, UNAVEM II, UNCTAD, UNESCO, UNHCR, UNIDO, UNIKOM, UNOMIG, UNOMOZ, UNOMUR, UNU, UPU, WEU (associate partner), WFTU, WHO, WIPO, WMO, WTO, ZC
Diplomatic representation in US:
chief of mission: Ambassador Gyorgy BANLAKI (since 27 October 1994)
chancery: 3910 Shoemaker Street NW, Washington, DC 20008
telephone: [1] (202) 362-6730
FAX: [1] (202) 966-8135
consulate(s) general: Los Angeles and New York
US diplomatic representation:
chief of mission: Ambassador Donald M. BLINKEN
embassy: V. Szabadsag Ter 12, Budapest
mailing address: Am Embassy, Unit 1320, Budapest; APO AE 09213-1320
telephone: [36] (1) 112-6450
FAX: [36] (1) 132-8934
Flag: three equal horizontal bands of red (top), white, and green

Economy

Overview: Since 1989 Hungary has been a leader in the transition from a socialist command economy to a market economy—thanks in large part to its initial economic reforms during the Communist era. The private sector now accounts for about 55% of GDP. Nonetheless, the transformation is proving difficult, and many citizens say life was better under the old system. On the bright side, the four-year decline in output finally ended in 1994, as real GDP increased an estimated 3%. This growth helped reduce unemployment to just over 10% by yearend, down from a peak of 13%. However, no progress was made against inflation, which remained stuck at about 20%, and the already-large current account deficit in the balance of payments actually got worse, reaching almost $4 billion. Underlying Hungary's other economic problems is the large budget deficit, which probably exceeded 7% of GDP in 1994, despite some late-year budget cutting by the new leftist government. In 1995 the government has pledged to accelerate privatization and lower the budget deficit to 5.5% of GDP. It believes this fiscal tightening will reduce the current account deficit to $2.5 billion but at the cost of holding economic growth to only 1%
National product: GDP—purchasing power parity—$58.8 billion (1994 est.)
National product real growth rate: 3% (1994 est.)
National product per capita: $5,700 (1994 est.)
Inflation rate (consumer prices): 21% (1994)
Unemployment rate: 10.4% (yearend 1994)
Budget:
revenues: $11.3 billion
expenditures: $14.2 billion, including capital expenditures of $NA (1994)
Exports: $10.3 billion (f.o.b., 1994 est.)
commodities: raw materials and semi-finished goods 30.0%, machinery and transport equipment 20.1%, consumer goods 25.2%, food and agriculture 21.4%, fuels and energy 3.4% (1993)
partners: Germany 25.3%, Italy 8.3%, Austria 10.5%, the FSU 14.0%, US 4.3% (1993)
Imports: $14.2 billion (f.o.b., 1994 est.)
commodities: fuels and energy 12.6%, raw materials and semi-finished goods 27.3%, machinery and transport equipment 33.0%, consumer goods 21.2%, food and agriculture 5.9% (1993)
partners: Germany 21.5%, Italy 6.1%, Austria 11.8%, the FSU 20.9%, US 4.3% (1993); note—about one-fourth of the imports from the FSU were MiGs delivered as a debt payment
External debt: $27 billion (September 1994)
Industrial production: growth rate 7% (1994 est.)
Electricity:
capacity: 6,740,000 kW
production: 31 billion kWh
consumption per capita: 3,012 kWh (1993)
Industries: mining, metallurgy, construction materials, processed foods, textiles, chemicals (especially pharmaceuticals), buses, automobiles
Agriculture: including forestry, accounts for 15% of GDP and 16% of employment; highly diversified crop and livestock farming; principal crops—wheat, corn, sunflowers, potatoes, sugar beets; livestock—hogs, cattle, poultry, dairy products; self-sufficient in food output
Illicit drugs: transshipment point for Southeast Asia heroin and South American cocaine destined for Western Europe; limited producer of precursor chemicals
Economic aid:
recipient: assistance pledged by OECD countries since 1989 about $9 billion
Currency: 1 forint (Ft) = 100 filler
Exchange rates: forints per US$1—112 (January 1995), 105.16 (1994), 91.93 (1993), 78.99 (1992), 74.74 (1991), 63.21 (1990), 59.07 (1989)
Fiscal year: calendar year

Transportation

Railroads:
total: 7,785 km
broad gauge: 35 km 1.520-m gauge
standard gauge: 7,574 km 1.435-m gauge (2,277 km electrified; 1,236 km double track)
narrow gauge: 176 km mostly 0.760-m gauge (1994)

Hungary *(continued)*

Highways:
total: 158,711 km
paved: 69,992 km (441 km expressways)
unpaved: 88,719 km (1992)
Inland waterways: 1,622 km (1988)
Pipelines: crude oil 1,204 km; natural gas 4,387 km (1991)
Ports: Budapest, Dunaujvaros
Merchant marine:
total: 10 cargo ships (1,000 GRT or over) totaling 46,121 GRT/61,613 DWT
Airports:
total: 78
with paved runways over 3,047 m: 2
with paved runways 2,438 to 3,047 m: 7
with paved runways 1,524 to 2,437 m: 4
with paved runways under 914 m: 1
with unpaved runways 2,438 to 3,047 m: 7
with unpaved runways 1,524 to 2,438 m: 9
with unpaved runways 914 to 1,523 m: 14
with unpaved runways under 914 m: 34

Communications

Telephone system: 1,520,000 phones; 14.7 telephones/100 inhabitants (1993); 14,213 telex lines; automatic telephone network based on microwave radio relay system; 608,000 telephones on order; 12-15 year wait for a telephone; 49% of all phones are in Budapest (1991)
local: NA
intercity: microwave radio relay
international: 1 INTELSAT and Intersputnik earth stations
Radio:
broadcast stations: AM 32, FM 15, shortwave 0
radios: NA
Television:
broadcast stations: 41 (Russian repeaters 8)
televisions: NA

Defense Forces

Branches: Ground Forces, Air and Air Defense Forces, Border Guard, Territorial Defense
Manpower availability: males age 15-49 2,639,860; males fit for military service 2,105,632; males reach military age (18) annually 86,298 (1995 est.)
Defense expenditures: 66.5 billion forints, NA% of GDP (1994 est.); note—conversion of defense expenditures into US dollars using the prevailing exchange rate could produce misleading results

Iceland

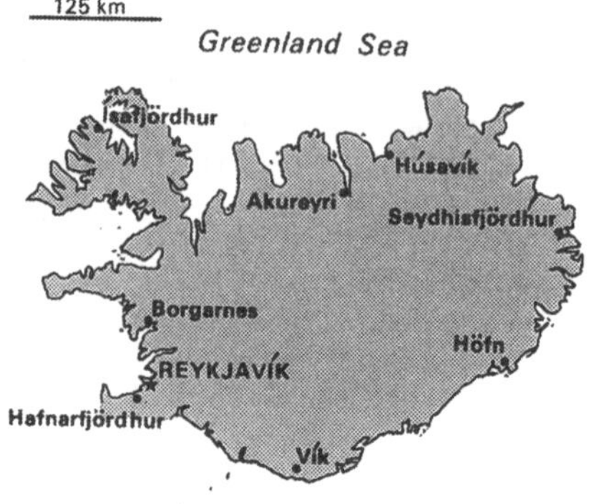

Geography

Location: Northern Europe, island between the Greenland Sea and the North Atlantic Ocean, northwest of the UK
Map references: Arctic Region
Area:
total area: 103,000 sq km
land area: 100,250 sq km
comparative area: slightly smaller than Kentucky
Land boundaries: 0 km
Coastline: 4,988 km
Maritime claims:
continental shelf: 200 nm or to the edge of the continental margin
exclusive economic zone: 200 nm
territorial sea: 12 nm
International disputes: Rockall continental shelf dispute involving Denmark, Ireland, and the UK (Ireland and the UK have signed a boundary agreement in the Rockall area)
Climate: temperate; moderated by North Atlantic Current; mild, windy winters; damp, cool summers
Terrain: mostly plateau interspersed with mountain peaks, icefields; coast deeply indented by bays and fiords
Natural resources: fish, hydropower, geothermal power, diatomite
Land use:
arable land: 1%
permanent crops: 0%
meadows and pastures: 20%
forest and woodland: 1%
other: 78%
Irrigated land: NA sq km
Environment:
current issues: water pollution from fertilizer runoff; inadequate wastewater treatment
natural hazards: earthquakes and volcanic activity
international agreements: party to—Air Pollution, Biodiversity, Climate Change, Law of the Sea, Marine Dumping, Nuclear Test Ban, Ozone Layer Protection, Ship Pollution, Wetlands; signed, but not ratified—Environmental Modification, Marine Life Conservation
Note: strategic location between Greenland and Europe; westernmost European country; more land covered by glaciers than in all of continental Europe

People

Population: 265,998 (July 1995 est.)
note: population data estimates based on average growth rate may differ slightly from official population data because of volatile migration rates
Age structure:
0-14 years: 24% (female 31,482; male 32,912)
15-64 years: 65% (female 84,559; male 87,089)
65 years and over: 11% (female 16,554; male 13,402) (July 1995 est.)
Population growth rate: 0.92% (1995 est.)
Birth rate: 15.85 births/1,000 population (1995 est.)
Death rate: 6.7 deaths/1,000 population (1995 est.)
Net migration rate: 0 migrant(s)/1,000 population (1995 est.)
Infant mortality rate: 4 deaths/1,000 live births (1995 est.)
Life expectancy at birth:
total population: 78.98 years
male: 76.69 years
female: 81.39 years (1995 est.)
Total fertility rate: 2.06 children born/woman (1995 est.)
Nationality:
noun: Icelander(s)
adjective: Icelandic
Ethnic divisions: homogeneous mixture of descendants of Norwegians and Celts
Religions: Evangelical Lutheran 96%, other Protestant and Roman Catholic 3%, none 1% (1988)
Languages: Icelandic
Literacy: age 15 and over can read and write (1976 est.)
total population: 100%
Labor force: 127,900
by occupation: commerce, transportation, and services 60.0%, manufacturing 12.5%, fishing and fish processing 11.8%, construction 10.8%, agriculture 4.0% (1990)

Government

Names:
conventional long form: Republic of Iceland
conventional short form: Iceland
local long form: Lyoveldio Island
local short form: Island
Digraph: IC
Type: republic
Capital: Reykjavik
Administrative divisions: 23 counties (syslar, singular—sysla) and 14 independent

towns* (kaupstadhir, singular—kaupstadhur); Akranes*, Akureyri*, Arnessysla, Austur-Bardhastrandarsysla, Austur-Hunavatnssysla, Austur-Skaftafellssysla, Borgarfjardharsysla, Dalasysla, Eyjafjardharsysla, Gullbringusysla, Hafnarfjordhur*, Husavik*, Isafjordhur*, Keflavik*, Kjosarsysla, Kopavogur*, Myrasysla, Neskaupstadhur*, Nordhur-Isafjardharsysla, Nordhur-Mulasys-la, Nordhur-Thingeyjarsysla, Olafsfjordhur*, Rangarvallasysla, Reykjavik*, Saudharkrokur*, Seydhisfjordhur*, Siglufjordhur*, Skagafjardharsysla, Snaefellsnes-og Hnappadalssysla, Strandasysla, Sudhur-Mulasysla, Sudhur-Thingeyjarsysla, Vesttmannaeyjar*, Vestur-Bardhastrandarsysla, Vestur-Hunavatnssysla, Vestur-Isafjardharsysla, Vestur-Skaftafellssysla
Independence: 17 June 1944 (from Denmark)
National holiday: Anniversary of the Establishment of the Republic, 17 June (1944)
Constitution: 16 June 1944, effective 17 June 1944
Legal system: civil law system based on Danish law; does not accept compulsory ICJ jurisdiction
Suffrage: 18 years of age; universal
Executive branch:
chief of state: President Vigdis FINNBOGADOTTIR (since 1 August 1980); election last held on 29 June 1988 (next scheduled for June 1996); results—there was no election in 1992 as President Vigdis FINNBOGADOTTIR was unopposed
head of government: Prime Minister David ODDSSON (since 30 April 1991)
cabinet: Cabinet; appointed by the president
Legislative branch: unicameral
Parliament (Althing): elections last held on 8 April 1995 (next to be held by April 1999); results—Independence Party 37.1%, Progressive Party 23.3%, Social Democratic Party 11.4%, Socialists 14.3%, People's Movement 7.2%, Women's Party 4.9%; seats (63 total) Independence 25, Progressive 15, Social Democratic 7, Socialists 9, People's Movement 4, Women's Party 3
Judicial branch: Supreme Court (Haestirettur)
Political parties and leaders: Independence Party (conservative), David ODDSSON; Progressive Party, Halldor ASGRIMSSON; Social Democratic Party, Jon Baldvin HANNIBALSSON; People's Alliance (left socialist), Olafur Ragnar GRIMSSON; Women's Party; People's Movement (moderate left); National Awakening, Johanna SIGURDARDOTTIR
Member of: Australia Group, BIS, CCC, CE, EBRD, ECE, EFTA, FAO, GATT, IAEA, IBRD, ICAO, ICC, ICFTU, ICRM, IDA, IFC, IFRCS, ILO, IMF, IMO, INMARSAT, INTELSAT, INTERPOL, IOC, ISO, ITU, MTCR, NACC, NATO, NC, NEA, NIB, OECD, OSCE, PCA, UN, UNCTAD, UNESCO, UNU, UPU, WEU (associate), WHO, WIPO, WMO
Diplomatic representation in US:
chief of mission: Ambassador Einar BENEDIKTSSON
chancery: Suite 1200, 1156 15th Street NW, Washington, DC 20005
telephone: [1] (202) 265-6653 through 6655
FAX: [1] (202) 265-6656
consulate(s) general: New York
US diplomatic representation:
chief of mission: Ambassador Parker W. BORG
embassy: Laufasvegur 21, Box 40, Reykjavik
mailing address: US Embassy, PSC 1003, Box 40, Reykjavik; FPO AE 09728-0340
telephone: [354] (1) 629100
FAX: [354] (1) 629139
Flag: blue with a red cross outlined in white that extends to the edges of the flag; the vertical part of the cross is shifted to the hoist side in the style of the Dannebrog (Danish flag)

Economy

Overview: Iceland's Scandinavian-type economy is basically capitalistic, but with an extensive welfare system, relatively low unemployment, and comparatively even distribution of income. The economy is heavily dependent on the fishing industry, which provides nearly 75% of export earnings and employs 12% of the work force. In the absence of other natural resources—except energy—Iceland's economy is vulnerable to changing world fish prices. The economy, in recession since 1988, began to recover in 1993, posting 0.4% growth, but was still hampered by cutbacks in fish quotas as well as falling world prices for its main exports: fish and fish products, aluminum, and ferrosilicon. Real GDP grew by perhaps 2.4% in 1994. The center-right government plans to continue its policies of reducing the budget and current account deficits, limiting foreign borrowing, containing inflation, revising agricultural and fishing policies, diversifying the economy, and privatizing state-owned industries. The government, however, remains divided on the issue of EU membership, primarily because of Icelanders' concern about losing control over their fishing resources.
National product: GDP—purchasing power parity—$4.5 billion (1994 est.)
National product real growth rate: 2.4% (1994 est.)
National product per capita: $17,250 (1994 est.)
Inflation rate (consumer prices): 1.3% (1994 est.)
Unemployment rate: 7% (1994 est.)
Budget:
revenues: $1.9 billion
expenditures: $2.1 billion, including capital expenditures of $NA (1994 est.)
Exports: $1.4 billion (f.o.b., 1993)
commodities: fish and fish products, animal products, aluminum, ferrosilicon, diatomite
partners: EC 68% (UK 25%, Germany 12%), US 11%, Japan 8% (1992)
Imports: $1.3 billion (c.i.f., 1993)
commodities: machinery and transportation equipment, petroleum products, foodstuffs, textiles
partners: EC 53% (Germany 14%, Denmark 10%, UK 9%), Norway 14%, US 9% (1992)
External debt: $2.5 billion (1993 est.)
Industrial production: growth rate 1.75% (1991 est.)
Electricity:
capacity: 1,070,000 kW
production: 4.7 billion kWh
consumption per capita: 16,458 kWh (1993)
Industries: fish processing, aluminum smelting, ferro-silicon production, geothermal power
Agriculture: accounts for about 15% of GDP; fishing is most important economic activity, contributing nearly 75% to export earnings; principal crops—potatoes, turnips; livestock—cattle, sheep; fish catch of about 1.1 million metric tons in 1992
Economic aid:
recipient: US commitments, including Ex-Im (FY70-81), $19.1 million
Currency: 1 Icelandic krona (IKr) = 100 aurar
Exchange rates: Icelandic kronur (IKr) per US$1—67.760 (January 1995), 69.944 (1994), 67.603 (1993), 57.546 (1992), 58.996 (1991), 58.284 (1990)
Fiscal year: calendar year

Transportation

Railroads: 0 km
Highways:
total: 11,373 km
paved: 2,513 km
unpaved: gravel, earth 8,860 km (1992)
Ports: Akureyri, Hornafjordur, Isafjordur, Keflavik, Raufarhofn, Reykjavik, Seydhisfjordhur, Straumsvik, Vestmannaeyjar
Merchant marine:
total: 6 ships (1,000 GRT or over) totaling 30,025 GRT/40,410 DWT
ships by type: cargo 1, chemical tanker 1, oil tanker 1, refrigerated cargo 1, roll-on/roll-off cargo 2
Airports:
total: 90
with paved runways over 3,047 m: 1
with paved runways 1,524 to 2,437 m: 3
with paved runways 914 to 1,523 m: 6
with paved runways under 914 m: 53
with unpaved runways 1,524-2,437 m: 4
with unpaved runways 914-1,523 m: 23

Communications

Telephone system: 140,000 telephones; adequate domestic service
local: NA
intercity: the trunk network consists of coaxial and fiber-optic cables and microwave radio relay links
international: 2 earth stations carry all international traffic through an Atlantic Ocean INTELSAT satellite
Radio:
broadcast stations: AM 5, FM 147 (transmitters and repeaters), shortwave 0
radios: NA
Television:
broadcast stations: 202 (transmitters and repeaters)
televisions: NA

Defense Forces

Branches: no regular armed forces; Police, Coast Guard; note—Iceland's defense is provided by the US-manned Icelandic Defense Force (IDF) headquartered at Keflavik
Manpower availability: males age 15-49 70,743; males fit for military service 62,698 (1995 est.)
Defense expenditures: none

India

Geography

Location: Southern Asia, bordering the Arabian Sea and the Bay of Bengal, between Bangladesh and Pakistan
Map references: Asia
Area:
total area: 3,287,590 km2
land area: 2,973,190 km2
comparative area: slightly more than one-third the size of the US
Land boundaries: total 14,103 km, Bangladesh 4,053 km, Bhutan 605 km, Burma 1,463 km, China 3,380 km, Nepal 1,690 km, Pakistan 2,912 km
Coastline: 7,000 km
Maritime claims:
contiguous zone: 24 nm
continental shelf: 200 nm or to the edge of the continental margin
exclusive economic zone: 200 nm
territorial sea: 12 nm
International disputes: boundaries with Bangladesh and China; status of Kashmir with Pakistan; water-sharing problems with downstream riparians, Bangladesh over the Ganges and Pakistan over the Indus
Climate: varies from tropical monsoon in south to temperate in north
Terrain: upland plain (Deccan Plateau) in south, flat to rolling plain along the Ganges, deserts in west, Himalayas in north
Natural resources: coal (fourth-largest reserves in the world), iron ore, manganese, mica, bauxite, titanium ore, chromite, natural gas, diamonds, petroleum, limestone
Land use:
arable land: 55%
permanent crops: 1%
meadows and pastures: 4%
forest and woodland: 23%
other: 17%
Irrigated land: 430,390 sq km (1989)
Environment:
current issues: deforestation; soil erosion; overgrazing; desertification; air pollution from industrial effluents and vehicle emissions; water pollution from raw sewage and runoff of agricultural pesticides; tap water is not potable throughout the country; huge and rapidly growing population is overstraining natural resources
natural hazards: droughts, flash floods, severe thunderstorms common; earthquakes
international agreements: party to—Antarctic Treaty, Biodiversity, Climate Change, Endangered Species, Environmental Modification, Hazardous Wastes, Nuclear Test Ban, Ozone Layer Protection, Ship Pollution, Tropical Timber, Wetlands, Whaling; signed, but not ratified—Antarctic-Environmental Protocol, Desertification, Law of the Sea
Note: dominates South Asian subcontinent; near important Indian Ocean trade routes

People

Population: 936,545,814 (July 1995 est.)
Age structure:
0-14 years: 35% (female 159,921,309; male 168,812,255)
15-64 years: 61% (female 274,105,407; male 296,145,798)
65 years and over: 4% (female 18,870,762; male 18,690,283) (July 1995 est.)
Population growth rate: 1.77% (1995 est.)
Birth rate: 27.78 births/1,000 population (1995 est.)
Death rate: 10.07 deaths/1,000 population (1995 est.)
Net migration rate: 0 migrant(s)/1,000 population (1995 est.)
Infant mortality rate: 76.3 deaths/1,000 live births (1995 est.)
Life expectancy at birth:
total population: 59.04 years
male: 58.5 years
female: 59.61 years (1995 est.)
Total fertility rate: 3.4 children born/woman (1995 est.)
Nationality:
noun: Indian(s)
adjective: Indian
Ethnic divisions: Indo-Aryan 72%, Dravidian 25%, Mongoloid and other 3%
Religions: Hindu 80%, Muslim 14%, Christian 2.4%, Sikh 2%, Buddhist 0.7%, Jains 0.5%, other 0.4%
Languages: English enjoys associate status but is the most important language for national, political, and commercial communication, Hindi the national language and primary tongue of 30% of the people, Bengali (official), Telugu (official), Marathi (official), Tamil (official), Urdu (official), Gujarati (official), Malayalam (official), Kannada (official), Oriya (official), Punjabi (official), Assamese (official), Kashmiri (official), Sindhi (official), Sanskrit (official), Hindustani a popular variant of Hindu/Urdu, is spoken widely throughout northern India
note: 24 languages each spoken by a million or

more persons; numerous other languages and dialects, for the most part mutually unintelligible
Literacy: age 7 and over can read and write (1991)
total population: 52%
male: 64%
female: 39%
Labor force: 314.751 million (1990)
by occupation: agriculture 65% (1993 est.)

Government

Names:
conventional long form: Republic of India
conventional short form: India
Digraph: IN
Type: federal republic
Capital: New Delhi
Administrative divisions: 25 states and 7 union territories*; Andaman and Nicobar Islands*, Andhra Pradesh, Arunachal Pradesh, Assam, Bihar, Chandigarh*, Dadra and Nagar Haveli*, Daman and Diu*, Delhi*, Goa, Gujarat, Haryana, Himachal Pradesh, Jammu and Kashmir, Karnataka, Kerala, Lakshadweep*, Madhya Pradesh, Maharashtra, Manipur, Meghalaya, Mizoram, Nagaland, Orissa, Pondicherry*, Punjab, Rajasthan, Sikkim, Tamil Nadu, Tripura, Uttar Pradesh, West Bengal
Independence: 15 August 1947 (from UK)
National holiday: Anniversary of the Proclamation of the Republic, 26 January (1950)
Constitution: 26 January 1950
Legal system: based on English common law; limited judicial review of legislative acts; accepts compulsory ICJ jurisdiction, with reservations
Suffrage: 18 years of age; universal
Executive branch:
chief of state: President Shankar Dayal SHARMA (since 25 July 1992); Vice President Kicheril Raman NARAYANAN (since 21 August 1992)
head of government: Prime Minister P. V. Narasimha RAO (since 21 June 1991)
cabinet: Council of Ministers; appointed by the president on recommendation of the prime minister
Legislative branch: bicameral Parliament (Sansad)
Council of States (Rajya Sabha): body consisting of not more than 250 members, up to 12 appointed by the president, the remainder chosen by the elected members of the state and territorial assemblies
People's Assembly (Lok Sabha): elections last held 21 May, 12 and 15 June 1991 (next to be held by 1996); results—percent of vote by party NA; seats—(545 total, 543 elected, 2 appointed) Congress (I) Party 245, BJP 119, Janata Dal Party 39, Janata Dal (Ajit Singh) 20, CPI/M 35, CPI 14, Telugu Desam 13, AIADMK 11, Samajwadi Janata Party 5, Shiv Sena 4, RSP 4, BSP 1, Congress (S) Party 1, other 23, vacant 9; note—the distribution of seats as of 18 January 1995 is as follows: Congress (I) Party 260, BJP 117, CPI/M 36, Janata Dal Party 24, Samta Party 14, CPI 14, AIADMK 12, Janata Dal (Ajit) 7, Telugu Desam 7, RSP 4, Janata Dal (Ex-Ajit) 3, Samajwadi Party 3, BSP 3, AIFB 3, Shiv Sena 2, Congress (S) Party 1, Kerala Congress (Mani faction) 1, Bihar Peoples Party 1, India National League 1, other 14, vacant 16
Judicial branch: Supreme Court
Political parties and leaders: Congress (I) Party, P. V. Narasimha RAO, president; Bharatiya Janata Party (BJP), L.K. ADVANI; Janata Dal Party, S.R. BOMMAI; Janata Dal (Ajit), Ajit SINGH; Janata Dal (Ex-Ajit), leader NA; Communist Party of India/Marxist (CPI/M), Harkishan Singh SURJEET; Communist Party of India (CPI), Indrajit GUPTA; Telugu Desam (a regional party in Andhra Pradesh), N. T. Rama RAO; All-India Anna Dravida Munnetra Kazagham (AIADMK; a regional party in Tamil Nadu), Jayaram JAYALALITHA; Samajwadi Party (SP), Mulayam Singh YADAV (President), Om Prakash CHAUTALA, Devi LAL; Shiv Sena, Bal THACKERAY; Revolutionary Socialist Party (RSP), Tridip CHOWDHURY; Bahujan Samaj Party (BSP), Kanshi RAM; Congress (S) Party, leader NA; Communist Party of India/Marxist-Leninist (CPI/ML), Vinod MISHRA; Dravida Munnetra Kazagham (a regional party in Tamil Nadu), M. KARUNANIDHI; Akali Dal factions representing Sikh religious community in the Punjab; National Conference (NC; a regional party in Jammu and Kashmir), Farooq ABDULLAH; Bihar Peoples Party, Lovely ANAND; Samta Party (formerly Janata Dal members), Natish KUMAR; Indian National League, Suliaman SAIT; Kerala Congress (Mani faction), K.M. MANI; All India Forward Bloc (AIFB), Prem Dutta PALIWAL (Chairman), Chitta BASU (General Secretary)
Other political or pressure groups: various separatist groups seeking greater communal and/or regional autonomy; numerous religious or militant/chauvinistic organizations, including Adam Sena, Ananda Marg, Vishwa Hindu Parishad, and Rashtriya Swayamsevak Sangh
Member of: AfDB, AG (observer), AsDB, C, CCC, CP, ESCAP, FAO, G- 6, G-15, G-19, G-24, G-77, GATT, IAEA, IBRD, ICAO, ICC, ICFTU, ICRM, IDA, IFAD, IFC, IFRCS, ILO, IMF, IMO, INMARSAT, INTELSAT, INTERPOL, IOC, IOM (observer), ISO, ITU, NAM, OAS (observer), PCA, SAARC, UN, UNAVEM II, UNCTAD, UNESCO, UNIDO, UNIKOM, UNITAR, UNOMIL, UNOMOZ, UNOSOM, UNU, UPU, WFTU, WHO, WIPO, WMO, WTO
Diplomatic representation in US:
chief of mission: Ambassador Siddhartha Shankar RAY
chancery: 2107 Massachusetts Avenue NW, Washington, DC 20008
telephone: [1] (202) 939-7000
consulate(s) general: Chicago, New York, and San Francisco
US diplomatic representation:
chief of mission: Ambassador Frank G. WISNER
embassy: Shanti Path, Chanakyapuri 110021, New Delhi
mailing address: use embassy street address
telephone: [91] (11) 600651
FAX: [91] (11) 6872028
consulate(s) general: Bombay, Calcutta, Madras
Flag: three equal horizontal bands of orange (top), white, and green with a blue chakra (24-spoked wheel) centered in the white band; similar to the flag of Niger, which has a small orange disk centered in the white band

Economy

Overview: India's economy is a mixture of traditional village farming, modern agriculture, handicrafts, a wide range of modern industries, and a multitude of support services. Faster economic growth in the 1980s permitted a significant increase in real per capita private consumption. A large share of the population, perhaps as much as 40%, remains too poor to afford an adequate diet. Financial strains in 1990 and 1991 prompted government austerity measures that slowed industrial growth but permitted India to meet its international payment obligations without rescheduling its debt. Production, trade, and investment reforms since 1991 have provided new opportunities for Indian businessmen and an estimated 100 million to 200 million middle class consumers. New Delhi has always paid its foreign debts on schedule and has stimulated exports, attracted foreign investment, and revived confidence in India's economic prospects. Foreign exchange reserves, precariously low three years ago, now total more than $19 billion. Positive factors for the remainder of the 1990s are India's strong entrepreneurial class and the central government's recognition of the continuing need for market-oriented approaches to economic development, for example in upgrading the wholly inadequate communications facilities. Negative factors include the desperate poverty of hundreds of millions of Indians and the impact of the huge and expanding population on an already overloaded environment.
National product: GDP—purchasing power parity—$1.2539 trillion (1994 est.)
National product real growth rate: 5% (1994 est.)
National product per capita: $1,360 (1994 est.)
Inflation rate (consumer prices): 10% (1994 est.)

India *(continued)*

Unemployment rate: NA%
Budget:
revenues: $30.85 billion
expenditures: $48.35 billion, including capital expenditures of $10.5 billion (FY93/94)
Exports: $24.4 billion (f.o.b., 1994 est.)
commodities: clothing, gems and jewelry, engineering goods, chemicals, leather manufactures, cotton yarn, and fabric
partners: US, Japan, Germany, UK, Hong Kong
Imports: $25.5 billion (c.i.f., 1994 est.)
commodities: crude oil and petroleum products, machinery, gems, fertilizer, chemicals
partners: US, Germany, Saudi Arabia, UK, Belgium, Japan
External debt: $89.2 billion (November 1994)
Industrial production: growth rate 7% (1994 est.); accounts for 28% of GDP
Electricity:
capacity: 81,200,000 kW
production: 314 billion kWh
consumption per capita: 324 kWh (1993)
Industries: textiles, chemicals, food processing, steel, transportation equipment, cement, mining, petroleum, machinery
Agriculture: accounts for 34% of GDP; principal crops—rice, wheat, oilseeds, cotton, jute, tea, sugarcane, potatoes; livestock—cattle, buffaloes, sheep, goats, poultry; fish catch of about 3 million metric tons ranks India among the world's top 10 fishing nations
Illicit drugs: licit producer of opium poppy for the pharmaceutical trade, but an undetermined quantity of opium is diverted to illicit international drug markets; major transit country for illicit narcotics produced in neighboring countries; illicit producer of hashish and methaqualone; produced 82 metric tons of illicit opium in 1994
Economic aid:
recipient: US commitments, including Ex-Im (FY70-89), $4.4 billion; Western (non-US) countries, ODA and OOF bilateral commitments (1980-89), $31.7 billion; OPEC bilateral aid (1979-89), $315 million; USSR (1970-89), $11.6 billion; Eastern Europe (1970-89), $105 million
Currency: 1 Indian rupee (Re) = 100 paise
Exchange rates: Indian rupees (Rs) per US$1—31.374 (January 1995), 31.374 (1994), 30.493 (1993), 25.918 (1992), 22.742 (1991), 17.504 (1990)
Fiscal year: 1 April—31 March

Transportation

Railroads:
total: 62,211 km (6,500 km electrified; 12,617 km double track)
broad gauge: 34,544 km 1.676-m gauge
narrow gauge: 23,599 km 1.000-m gauge; 4,068 km 0.762-m and 0.610-m gauge (1994 est.)
Highways:
total: 1.97 million km
paved: 960,000 km
unpaved: gravel, crushed stone, earth 1.01 million km (1989)
Inland waterways: 16,180 km; 3,631 km navigable by large vessels
Pipelines: crude oil 3,497 km; petroleum products 1,703 km; natural gas 902 km (1989)
Ports: Bombay, Calcutta, Cochin, Haldia, Kandla, Madras, Mormugao, New Mangalore, Pondicherry, Port Blair (Andaman Islands), Tuticorin, Vishakhapatnam
Merchant marine:
total: 299 ships (1,000 GRT or over) totaling 6,288,902 GRT/10,454,178 DWT
ships by type: bulk 114, cargo 78, chemical tanker 9, combination bulk 2, combination ore/oil 5, container 10, liquefied gas tanker 6, oil tanker 68, passenger-cargo 5, roll-on/roll-off cargo 1, short-sea passenger 1
Airports:
total: 352
with paved runways over 3,047 m: 11
with paved runways 2,438 to 3,047 m: 48
with paved runways 1,524 to 2,437 m: 85
with paved runways 914 to 1,523 m: 72
with paved runways under 914 m: 81
with unpaved runways 2,438 to 3,047 m: 2
with unpaved runways 1,524 to 2,438 m: 7
with unpaved runways 914 to 1,523 m: 46

Communications

Telephone system: NA telephones; 5 telephones/1,000 persons; domestic telephone system is poor; long-distance telephoning has been improved by a domestic satellite system which also carries TV
local: NA
intercity: NA
international: 3 INTELSAT (Indian Ocean) earth stations and submarine cables to Malaysia and the United Arab Emirates
Radio:
broadcast stations: AM 96, FM 4, shortwave 0
radios: 60 million
Television:
broadcast stations: 274 (government controlled)
televisions: 21 million

Defense Forces

Branches: Army, Navy, Air Force, various security or paramilitary forces (includes Border Security Force, Assam Rifles, and Coast Guard)
Manpower availability: males age 15-49 253,134,487; males fit for military service 148,814,104; males reach military age (17) annually 9,461,907 (1995 est.)
Defense expenditures: exchange rate conversion—$7.8 billion, 2.8% of GDP (FY94/95)

Indian Ocean

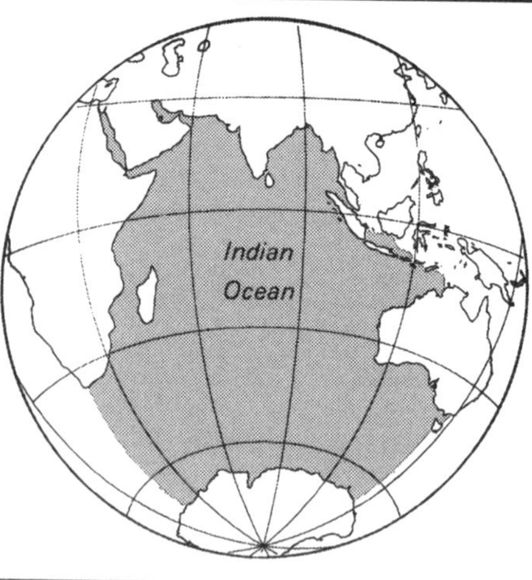

Geography

Location: body of water between Africa, Antarctica, Asia, and Australia
Map references: World
Area:
total area: 73.6 million sq km
comparative area: slightly less than eight times the size of the US; third-largest ocean (after the Pacific Ocean and Atlantic Ocean, but larger than the Arctic Ocean)
note: includes Arabian Sea, Bass Straight, Bay of Bengal, Great Australian Bight, Gulf of Oman, Persian Gulf, Red Sea, Strait of Malacca, and other tributary water bodies
Coastline: 66,526 km
International disputes: some maritime disputes (see littoral states)
Climate: northeast monsoon (December to April), southwest monsoon (June to October); tropical cyclones occur during May/June and October/November in the northern Indian Ocean and January/February in the southern Indian Ocean
Terrain: surface dominated by counterclockwise gyre (broad, circular system of currents) in the southern Indian Ocean; unique reversal of surface currents in the northern Indian Ocean, low atmospheric pressure over southwest Asia from hot, rising, summer air results in the southwest monsoon and southwest-to-northeast winds and currents, while high pressure over northern Asia from cold, falling, winter air results in the northeast monsoon and northeast-to-southwest winds and currents; ocean floor is dominated by the Mid-Indian Ocean Ridge and subdivided by the Southeast Indian Ocean Ridge, Southwest Indian Ocean Ridge, and Ninety East Ridge; maximum depth is 7,258 meters in the Java Trench
Natural resources: oil and gas fields, fish, shrimp, sand and gravel aggregates, placer deposits, polymetallic nodules
Environment:
current issues: endangered marine species include the dugong, seals, turtles, and whales;

Indonesia

oil pollution in the Arabian Sea, Persian Gulf, and Red Sea
natural hazards: ships subject to superstructure icing in extreme south near Antarctica from May to October
international agreements: NA
Note: major chokepoints include Bab el Mandeb, Strait of Hormuz, Strait of Malacca, southern access to the Suez Canal, and the Lombok Strait

Government

Digraph: XO

Economy

Overview: The Indian Ocean provides major sea routes connecting the Middle East, Africa, and East Asia with Europe and the Americas. It carries a particularly heavy traffic of petroleum and petroleum products from the oilfields of the Persian Gulf and Indonesia. Its fish are of great and growing importance to the bordering countries for domestic consumption and export. Fishing fleets from Russia, Japan, Korea, and Taiwan also exploit the Indian Ocean, mainly for shrimp and tuna. Large reserves of hydrocarbons are being tapped in the offshore areas of Saudi Arabia, Iran, India, and western Australia. An estimated 40% of the world's offshore oil production comes from the Indian Ocean. Beach sands rich in heavy minerals and offshore placer deposits are actively exploited by bordering countries, particularly India, South Africa, Indonesia, Sri Lanka, and Thailand.
Industries: based on exploitation of natural resources, particularly fish, minerals, oil and gas, fishing, sand and gravel

Transportation

Ports: Bombay (India), Calcutta (India), Colombo (Sri Lanka), Durban (South Africa), Jakarta (Indonesia), Madras (India), Melbourne (Australia), Richard's Bay (South Africa)

Communications

Telephone system:
international: submarine cables from India to United Arab Emirates and Malaysia, and from Sri Lanka to Djibouti and Indonesia

Geography

Location: Southeastern Asia, archipelago between the Indian Ocean and the Pacific Ocean
Map references: Southeast Asia
Area:
total area: 1,919,440 sq km
land area: 1,826,440 sq km
comparative area: slightly less than three times the size of Texas
Land boundaries: total 2,602 km, Malaysia 1,782 km, Papua New Guinea 820 km
Coastline: 54,716 km
Maritime claims: measured from claimed archipelagic baselines
exclusive economic zone: 200 nm
territorial sea: 12 nm
International disputes: sovereignty over Timor Timur (East Timor Province) disputed with Portugal and not recognized by the UN; two islands in dispute with Malaysia
Climate: tropical; hot, humid; more moderate in highlands
Terrain: mostly coastal lowlands; larger islands have interior mountains
Natural resources: petroleum, tin, natural gas, nickel, timber, bauxite, copper, fertile soils, coal, gold, silver
Land use:
arable land: 8%
permanent crops: 3%
meadows and pastures: 7%
forest and woodland: 67%
other: 15%
Irrigated land: 75,500 sq km (1989 est.)
Environment:
current issues: deforestation; water pollution from industrial wastes, sewage; air pollution in urban areas
natural hazards: occasional floods, severe droughts, and tsunamis
international agreements: party to—Biodiversity, Climate Change, Endangered Species, Hazardous Wastes, Law of the Sea, Nuclear Test Ban, Ozone Layer Protection, Ship Pollution, Tropical Timber 83, Wetlands; signed, but not ratified—Desertification, Marine Life Conservation, Tropical Timber 94
Note: archipelago of 13,500 islands (6,000 inhabited); straddles Equator; strategic location astride or along major sea lanes from Indian Ocean to Pacific Ocean

People

Population: 203,583,886 (July 1995 est.)
Age structure:
0-14 years: 32% (female 32,548,039; male 33,485,810)
15-64 years: 64% (female 65,394,816; male 64,914,362)
65 years and over: 4% (female 4,027,367; male 3,213,492) (July 1995 est.)
Population growth rate: 1.56% (1995 est.)
Birth rate: 24.06 births/1,000 population (1995 est.)
Death rate: 8.48 deaths/1,000 population (1995 est.)
Net migration rate: 0 migrant(s)/1,000 population (1995 est.)
Infant mortality rate: 65 deaths/1,000 live births (1995 est.)
Life expectancy at birth:
total population: 61.22 years
male: 59.13 years
female: 63.42 years (1995 est.)
Total fertility rate: 2.74 children born/woman (1995 est.)
Nationality:
noun: Indonesian(s)
adjective: Indonesian
Ethnic divisions: Javanese 45%, Sundanese 14%, Madurese 7.5%, coastal Malays 7.5%, other 26%
Religions: Muslim 87%, Protestant 6%, Roman Catholic 3%, Hindu 2%, Buddhist 1%, other 1% (1985)
Languages: Bahasa Indonesia (modified form of Malay; official), English, Dutch, local dialects the most widely spoken of which is Javanese
Literacy: age 15 and over can read and write (1990)
total population: 82%
male: 88%
female: 75%
Labor force: 67 million
by occupation: agriculture 55%, manufacturing 10%, construction 4%, transport and communications 3% (1985 est.)

Government

Names:
conventional long form: Republic of Indonesia
conventional short form: Indonesia
local long form: Republik Indonesia
local short form: Indonesia
former: Netherlands East Indies; Dutch East Indies
Digraph: ID

Type: republic
Capital: Jakarta
Administrative divisions: 24 provinces (propinsi-propinsi, singular—propinsi), 2 special regions* (daerah-daerah istimewa, singular—daerah istimewa), and 1 special capital city district** (daerah khusus ibukota); Aceh*, Bali, Bengkulu, Irian Jaya, Jakarta Raya**, Jambi, Jawa Barat, Jawa Tengah, Jawa Timur, Kalimantan Barat, Kalimantan Selatan, Kalimantan Tengah, Kalimantan Timur, Lampung, Maluku, Nusa Tenggara Barat, Nusa Tenggara Timur, Riau, Sulawesi Selatan, Sulawesi Tengah, Sulawesi Tenggara, Sulawesi Utara, Sumatera Barat, Sumatera Selatan, Sumatera Utara, Timor Timur, Yogyakarta*
Independence: 17 August 1945 (proclaimed independence; on 27 December 1949, Indonesia became legally independent from the Netherlands)
National holiday: Independence Day, 17 August (1945)
Constitution: August 1945, abrogated by Federal Constitution of 1949 and Provisional Constitution of 1950, restored 5 July 1959
Legal system: based on Roman-Dutch law, substantially modified by indigenous concepts and by new criminal procedures code; has not accepted compulsory ICJ jurisdiction
Suffrage: 17 years of age; universal and married persons regardless of age
Executive branch:
chief of state and head of government: President Gen. (Ret.) SOEHARTO (since 27 March 1968); Vice President Gen. (Ret.) Try SUTRISNO (since 11 March 1993)
cabinet: Cabinet
Legislative branch: unicameral
House of Representatives (Dewan Perwakilan Rakyat or DPR): elections last held on 8 June 1992 (next to be held NA 1997); results—GOLKAR 68%, PPP 17%, PDI 15%; seats—(500 total, 400 elected, 100 military representatives appointed) GOLKAR 282, PPP 62, PDI 56
note: the People's Consultative Assembly (Majelis Permusyawaratan Rakyat or MPR) includes the DPR plus 500 indirectly elected members who meet every five years to elect the president and vice president and, theoretically, to determine national policy
Judicial branch: Supreme Court (Mahkamah Agung)
Political parties and leaders: GOLKAR (quasi-official party based on functional groups), Lt. Gen. (Ret.) HARMOKO, general chairman; Indonesia Democracy Party (PDI—federation of former Nationalist and Christian Parties), Megawati SUKARNOPUTRI, chairman; Development Unity Party (PPP, federation of former Islamic parties), Ismail Hasan METAREUM, chairman
Member of: APEC, AsDB, ASEAN, CCC, CP, ESCAP, FAO, G-15, G-19, G-77, GATT, IAEA, IBRD, ICAO, ICC, ICFTU, ICRM, IDA, IDB, IFAD, IFC, IFRCS, ILO, IMF, IMO, INMARSAT, INTELSAT, INTERPOL, IOC, IOM (observer), ISO, ITU, NAM, OIC, OPEC, UN, UNCTAD, UNESCO, UNIDO, UNIKOM, UNMIH, UNPROFOR, UPU, WCL, WFTU, WHO, WIPO, WMO, WTO
Diplomatic representation in US:
chief of mission: Ambassador Arifin Mohamad SIREGAR
chancery: 2020 Massachusetts Avenue NW, Washington, DC 20036
telephone: [1] (202) 775-5200
FAX: [1] (202) 775-5365
consulate(s) general: Chicago, Houston, Los Angeles, New York, San Francisco
US diplomatic representation:
chief of mission: Ambassador Robert L. BARRY
embassy: Medan Merdeka Selatan 5, Box 1, Jakarta
mailing address: APO AP 96520
telephone: [62] (21) 360360
FAX: [62] (21) 3862259
consulate(s) general: Medan, Surabaya
Flag: two equal horizontal bands of red (top) and white; similar to the flag of Monaco, which is shorter; also similar to the flag of Poland, which is white (top) and red

Economy

Overview: Indonesia is a mixed economy with some socialist institutions and central planning but with a recent emphasis on deregulation and private enterprise. Indonesia has extensive natural wealth, yet, with a large and rapidly increasing population, it remains a rather poor country. Real GDP growth in 1985-94 averaged about 6%, quite impressive, but not sufficient to both slash underemployment and absorb the 2.3 million workers annually entering the labor force. Agriculture, including forestry and fishing, is an important sector, accounting for 21% of GDP and over 50% of the labor force. The staple crop is rice. Once the world's largest rice importer, Indonesia is now nearly self-sufficient. Plantation crops—rubber and palm oil—and textiles and plywood are being encouraged for both export and job generation. Industrial output now accounts for almost 40% of GDP and is based on a supply of diverse natural resources, including crude oil, natural gas, timber, metals, and coal. Foreign investment has also boosted manufacturing output and exports in recent years. Indeed, the economy's growth is highly dependent on the continuing expansion of nonoil exports. Japan remains Indonesia's most important customer and supplier of aid. Rapid growth in the money supply in 1989-90 prompted Jakarta to implement a tight monetary policy in 1991, forcing the private sector to go to foreign banks for investment financing. Real interest rates remained above 10% and off-shore commercial debt grew. The growth in off-shore debt prompted Jakarta to limit foreign borrowing beginning in late 1991. Despite the continued problems in moving toward a more open financial system and the persistence of a fairly tight credit situation, GDP growth in 1992-94 has matched the government target of 6%-7% annual growth.
National product: GDP—purchasing power parity—$619.4 billion (1994 est.)
National product real growth rate: 6.7% (1994 est.)
National product per capita: $3,090 (1994 est.)
Inflation rate (consumer prices): 9.3% (1994 est.)
Unemployment rate: 3% official rate; underemployment 40% (1994 est.)
Budget:
revenues: $32.8 billion
expenditures: $32.8 billion, including capital expenditures of $12.9 billion (FY94/95)
Exports: $41.3 billion (f.o.b, 1994 est.)
commodities: manufactures 56.7%, fuels 24.8%, foodstuffs 11.1%, raw materials 7.4% (1994 est.)
partners: Japan 30%, US 14%, Singapore 9%, South Korea 6%, Taiwan 4% (1993)
Imports: $31.4 billion (f.o.b., 1994 est.)
commodities: capital equipment 44.2%, intermed and raw materials 37.0%, consumer goods 11.5%, fuels 7.2% (1994 est.)
partners: Japan 22%, US 11%, South Korea 7%, Germany 7%, Singapore 6%, Australia 5%, Taiwan 5% (1993)
External debt: $87 billion (1994)
Industrial production: growth rate 8.4% (1993 est.); accounts for 40% of GDP
Electricity:
capacity: 12,100,000 kW
production: 44 billion kWh
consumption per capita: 207 kWh (1993)
Industries: petroleum and natural gas, textiles, mining, cement, chemical fertilizers, plywood, food, rubber
Agriculture: accounts for 21% of GDP; subsistence food production; small-holder and plantation production for export; main products are rice, cassava, peanuts, rubber, cocoa, coffee, palm oil, copra, other tropical products, poultry, beef, pork, eggs
Illicit drugs: illicit producer of cannabis for the international drug trade, but not a major player; government actively eradicating plantings and prosecuting traffickers; growing role as transshipment point for Golden Triangle heroin; increasing indigenous methamphetamine abuse
Economic aid:
recipient: US commitments, including Ex-Im (FY70-89), $4.4 billion; Western (non-US) countries, ODA and OOF bilateral commitments (1970-89), $25.9 billion; OPEC bilateral aid (1979-89), $213 million; Communist countries (1970-89), $175 million
Currency: 1 Indonesian rupiah (Rp) = 100 sen (sen no longer used)

Exchange rates: Indonesian rupiahs (Rp) per US$1—2,203.6 (January 1995), 2,160.7 (1994), 2,087.1 (1993), 2,029.9 (1992), 1,950.3 (1991), 1,842.8 (1990)
Fiscal year: 1 April—31 March

Transportation

Railroads:
total: 6,964 km
narrow gauge: 6,389 km 1.067-m gauge (101 km electrified; 101 km double track); 497 km 0.750-m gauge; 78 km 0.600-m gauge
Highways:
total: 119,500 km
paved: NA
unpaved: NA
undifferentiated: provincial 34,180 km; district 73,508 km; state 11,812 km
Inland waterways: 21,579 km total; Sumatra 5,471 km, Java and Madura 820 km, Kalimantan 10,460 km, Celebes 241 km, Irian Jaya 4,587 km
Pipelines: crude oil 2,505 km; petroleum products 456 km; natural gas 1,703 km (1989)
Ports: Cilacap, Cirebon, Jakarta, Kupang, Palembang, Semarang, Surabaya, Ujungpandang
Merchant marine:
total: 438 ships (1,000 GRT or over) totaling 1,942,527 GRT/2,818,296 DWT
ships by type: bulk 26, cargo 259, chemical tanker 7, container 11, liquefied gas tanker 6, livestock carrier 1, oil tanker 85, passenger 6, passenger-cargo 12, roll-on/roll-off cargo 7, short-sea passenger 7, specialized tanker 7, vehicle carrier 4
Airports:
total: 450
with paved runways over 3,047 m: 3
with paved runways 2,438 to 3,047 m: 10
with paved runways 1,524 to 2,437 m: 35
with paved runways 914 to 1,523 m: 42
with paved runways under 914 m: 324
with unpaved runways 1,524 to 2,438 m: 4
with unpaved runways 914 to 1,523 m: 32

Communications

Telephone system: 763,000 telephones (1986); domestic service fair, international service good
local: NA
intercity: interisland microwave system and HF police net; 1 earth station for a domestic satellite
international: 2 INTELSAT (1 Indian Ocean and 1 Pacific Ocean) earth stations
Radio:
broadcast stations: AM 618, FM 38, shortwave 0
radios: NA
note: radiobroadcast coverage good
Television:
broadcast stations: 9
televisions: NA

Defense Forces

Branches: Army, Navy, Air Force, National Police
Manpower availability: males age 15-49 55,883,688; males fit for military service 32,952,204; males reach military age (18) annually 2,247,586 (1995 est.)
Defense expenditures: exchange rate conversion—$2.4 billion, 1.5% of GNP (FY94/95)

Iran

Geography

Location: Middle East, bordering the Gulf of Oman and the Persian Gulf, between Iraq and Pakistan
Map references: Middle East
Area:
total area: 1.648 million sq km
land area: 1.636 million sq km
comparative area: slightly larger than Alaska
Land boundaries: total 5,440 km, Afghanistan 936 km, Armenia 35 km, Azerbaijan (north) 432 km, Azerbaijan (northwest) 179 km, Iraq 1,458 km, Pakistan 909 km, Turkey 499 km, Turkmenistan 992 km
Coastline: 2,440 km
note: Iran also borders the Caspian Sea (740 km)
Maritime claims:
contiguous zone: 24 nm
continental shelf: natural prolongation
exclusive economic zone: bilateral agreements, or median lines in the Persian Gulf
territorial sea: 12 nm
International disputes: Iran and Iraq restored diplomatic relations in 1990 but are still trying to work out written agreements settling outstanding disputes from their eight-year war concerning border demarcation, prisoners-of-war, and freedom of navigation and sovereignty over the Shatt al Arab waterway; Iran occupies two islands in the Persian Gulf claimed by the UAE: Tunb as Sughra (Arabic), Jazireh-ye Tonb-e Kuchek (Persian) or Lesser Tunb, and Tunb al Kubra (Arabic), Jazireh-ye Tonb-e Bozorg (Persian) or Greater Tunb; it jointly administers with the UAE an island in the Persian Gulf claimed by the UAE, Abu Musa (Arabic) or Jazireh-ye Abu Musa (Persian); in 1992 the dispute over Abu Musa and the Tunb islands became more acute when Iran unilaterally tried to control the entry of third country nationals into the UAE portion of Abu Musa island, Tehran subsequently backed off in the face of significant diplomatic support for the UAE in the region, but in 1994 it increased its military

presence on the disputed islands; periodic disputes with Afghanistan over Helmand water rights; Caspian Sea boundaries are not yet determined
Climate: mostly arid or semiarid, subtropical along Caspian coast
Terrain: rugged, mountainous rim; high, central basin with deserts, mountains; small, discontinuous plains along both coasts
Natural resources: petroleum, natural gas, coal, chromium, copper, iron ore, lead, manganese, zinc, sulfur
Land use:
arable land: 8%
permanent crops: 0%
meadows and pastures: 27%
forest and woodland: 11%
other: 54%
Irrigated land: 57,500 sq km (1989 est.)
Environment:
current issues: air pollution, especially in urban areas, from vehicle emissions, refinery operations, and industrial effluents; deforestation; overgrazing; desertification; oil pollution in the Persian Gulf; inadequate supplies of potable water
natural hazards: periodic droughts, floods; duststorms, sandstorms; earthquakes along the Western border
international agreements: party to—Endangered Species, Hazardous Wastes, Nuclear Test Ban, Ozone Layer Protection, Wetlands; signed, but not ratified—Biodiversity, Climate Change, Desertification, Environmental Modification, Law of the Sea, Marine Life Conservation

People

Population: 64,625,455 (July 1995 est.)
Age structure:
0-14 years: 45% (female 14,113,933; male 14,995,015)
15-64 years: 51% (female 16,237,810; male 16,803,943)
65 years and over: 4% (female 1,197,869; male 1,276,885) (July 1995 est.)
Population growth rate: 2.29% (1995 est.)
Birth rate: 34.85 births/1,000 population (1995 est.)
Death rate: 6.85 deaths/1,000 population (1995 est.)
Net migration rate: -5.11 migrant(s)/1,000 population (1995 est.)
Infant mortality rate: 54.6 deaths/1,000 live births (1995 est.)
Life expectancy at birth:
total population: 66.97 years
male: 65.77 years
female: 68.22 years (1995 est.)
Total fertility rate: 4.93 children born/woman (1995 est.)
Nationality:
noun: Iranian(s)
adjective: Iranian
Ethnic divisions: Persian 51%, Azerbaijani 24%, Gilaki and Mazandarani 8%, Kurd 7%, Arab 3%, Lur 2%, Baloch 2%, Turkmen 2%, other 1%
Religions: Shi'a Muslim 95%, Sunni Muslim 4%, Zoroastrian, Jewish, Christian, and Baha'i 1%
Languages: Persian and Persian dialects 58%, Turkic and Turkic dialects 26%, Kurdish 9%, Luri 2%, Baloch 1%, Arabic 1%, Turkish 1%, other 2%
Literacy: age 15 and over can read and write (1991)
total population: 66%
male: 74%
female: 56%
Labor force: 15.4 million
by occupation: agriculture 33%, manufacturing 21%
note: shortage of skilled labor (1988 est.)

Government

Names:
conventional long form: Islamic Republic of Iran
conventional short form: Iran
local long form: Jomhuri-ye Eslami-ye Iran
local short form: Iran
Digraph: IR
Type: theocratic republic
Capital: Tehran
Administrative divisions: 24 provinces (ostanha, singular—ostan); Azarbayjan-e Bakhtari (West Azerbaijan), Azarbayjan-e Khavari (East Azerbaijan), Bakhtaran, Bushehr, Chahar Mahall va Bakhtiari, Esfahan, Fars, Gilan, Hamadan, Hormozgan, Ilam, Kerman, Khorasan, Khuzestan, Kohkiluyeh va Buyer Ahmadi, Kordestan, Lorestan, Markazi, Mazandaran, Semnan, Sistan va Baluchestan, Tehran, Yazd, Zanjan
note: there may be a new province named Ardabil formed from a part of Azarbayjan-e Khavari (East Azerbaijan) which may have been renamed Azarbayjan-e Markazi (Central Azerbaijan); the name Bakhtaran may have been changed to Kermanshahan
Independence: 1 April 1979 (Islamic Republic of Iran proclaimed)
National holiday: Islamic Republic Day, 1 April (1979)
Constitution: 2-3 December 1979; revised 1989 to expand powers of the presidency and eliminate the prime ministership
Legal system: the Constitution codifies Islamic principles of government
Suffrage: 15 years of age; universal
Executive branch:
supreme leader (rahbar) and functional chief of state: Leader of the Islamic Revolution Ayatollah Ali Hoseini-KHAMENEI (since 4 June 1989)
head of government: President Ali Akbar Hashemi-RAFSANJANI (since 3 August 1989); election last held June 1993 (next to be held June 1997); results—Ali Akbar Hashemi-RAFSANJANI was elected with 63% of the vote
cabinet: Council of Ministers; selected by the president with legislative approval
Legislative branch: unicameral
Islamic Consultative Assembly (Majles-e-Shura-ye-Eslami): elections last held 8 April 1992 (next to be held April 1996); results—percent of vote by party NA; seats—(270 seats total) number of seats by party NA
Judicial branch: Supreme Court
Political parties and leaders: there are at least 76 licensed parties; the three most important are—Tehran Militant Clergy Association, Mohammad Reza MAHDAVI-KANI; Militant Clerics Association, Mehdi MAHDAVI-KARUBI and Mohammad Asqar MUSAVI-KHOINIHA; Fedaiyin Islam Organization, Sadeq KHALKHALI
Other political or pressure groups: groups that generally support the Islamic Republic include Hizballah, Mojahedin of the Islamic Revolution, Muslim Students Following the Line of the Imam; armed political groups that have been almost completely repressed by the government include Mojahedin-e Khalq Organization (MEK), People's Fedayeen, Kurdish Democratic Party; the Society for the Defense of Freedom
Member of: CCC, CP, ECO, ESCAP, FAO, G-19, G-24, G-77, IAEA, IBRD, ICAO, ICC, ICRM, IDA, IDB, IFAD, IFC, IFRCS, ILO, IMF, IMO, INMARSAT, INTELSAT, INTERPOL, IOC, IOM (observer), ISO, ITU, NAM, OIC, OPEC, PCA, UN, UNCTAD, UNESCO, UNHCR, UNIDO, UPU, WCL, WFTU, WHO, WMO, WTO
Diplomatic representation in US:
chief of mission: Iran has an Interests Section in the Pakistani Embassy in Washington, DC
chancery: Iranian Interests Section, 2209 Wisconsin Avenue NW, Washington, DC 20007
telephone: [1] (202) 965-4990
US diplomatic representation: protecting power in Iran is Switzerland
Flag: three equal horizontal bands of green (top), white, and red; the national emblem (a stylized representation of the word Allah) in red is centered in the white band; Allah Alkbar (God is Great) in white Arabic script is repeated 11 times along the bottom edge of the green band and 11 times along the top edge of the red band

Economy

Overview: Iran's economy is a mixture of central planning, state ownership of oil and other large enterprises, village agriculture, and small-scale private trading and service ventures. Over the past several years, the government has introduced several measures to

liberalize the economy and reduce government intervention, but most of these changes have moved slowly because of political opposition. Iran has faced increasingly severe financial difficulties since mid-1992 due to an import surge that began in 1989 and general financial mismanagement. At yearend 1993 the Iranian Government estimated that it owed foreign creditors about $30 billion; an estimated $8 billion of this debt was in arrears. At yearend 1994, Iran rescheduled $12 billion in debt. Earnings from oil exports—which provide 90% of Iran's export revenues—are providing less relief to Iran than usual because of reduced oil prices.
National product: GDP—purchasing power parity—$310 billion (1994 est.)
National product real growth rate: -2% (1994 est.)
National product per capita: $4,720 (1994 est.)
Inflation rate (consumer prices): 35% (1994)
Unemployment rate: over 30% (1994 est.)
Budget:
revenues: $NA
expenditures: $NA, including capital expenditures of $NA
Exports: $16 billion (f.o.b., FY92/93 est.)
commodities: petroleum 90%, carpets, fruits, nuts, hides
partners: Japan, Italy, France, Netherlands, Belgium/Luxembourg, Spain, and Germany
Imports: $18 billion (c.i.f., FY92/93 est.)
commodities: machinery, military supplies, metal works, foodstuffs, pharmaceuticals, technical services, refined oil products
partners: Germany, Japan, Italy, UK, UAE
External debt: $30 billion (December 1993)
Industrial production: growth rate 4.6% (1993 est.); accounts for almost 30% of GDP, including petroleum
Electricity:
capacity: 19,080,000 kW
production: 50.8 billion kWh
consumption per capita: 745 kWh (1993)
Industries: petroleum, petrochemicals, textiles, cement and other building materials, food processing (particularly sugar refining and vegetable oil production), metal fabricating, armaments and military equipment
Agriculture: accounts for about 20% of GDP; principal products—wheat, rice, other grains, sugar beets, fruits, nuts, cotton, dairy products, wool, caviar; not self-sufficient in food
Illicit drugs: illicit producer of opium poppy for the domestic and international drug trade; produced 35-70 metric tons in 1993; net opiate importer but also a key transshipment point for Southwest Asian heroin to Europe
Economic aid:
recipient: US commitments, including Ex-Im (FY70-80), $1 billion; Western (non-US) countries, ODA and OOF bilateral commitments (1970-89), $1.675 billion; Communist countries (1970-89), $976 million
note: aid fell sharply following the 1979 revolution
Currency: 10 Iranian rials (IR) = 1 toman; note—domestic figures are generally referred to in terms of the toman
Exchange rates: Iranian rials (IR) per US$1—1,749.04 (January 1995), 1,748.75 (1994), 1,267.77 (1993), 65.552 (1992), 67.505 (1991); black market rate: 3,000 rials per US$1 (December 1994)
Fiscal year: 21 March—20 March

Transportation

Railroads:
total: 4,850 km; note—480 km under construction from Bafq to Bandar-e 'Abbas; segment from Bafq to Sirjan has been completed and is operational; section from Sirjan to Bandar-e 'Abbas still under construction
broad gauge: 90 km 1.676-m gauge
narrow gauge: 4,760 km 1.432-m gauge
Highways:
total: 140,200 km
paved: 42,694 km
unpaved: gravel, crushed stone 46,866 km; improved earth 49,440 km; unimproved earth 1,200 km
Inland waterways: 904 km; the Shatt al Arab is usually navigable by maritime traffic for about 130 km; channel has been dredged to 3 meters and is in use
Pipelines: crude oil 5,900 km; petroleum products 3,900 km; natural gas 4,550 km
Ports: Abadan (largely destroyed in fighting during 1980-88 war), Ahvaz, Bandar Beheshti, Bandar-e 'Abbas, Bandar-e Anzali, Bandar-e Bushehr, Bandar-e Khomeyni, Bandar-e Mah Shahr, Bandar-e Torkeman, Jazireh-ye Khark, Jazireh-ye Lavan, Jazireh-ye Sirri, Khorramshahr (limited operation since November 1992), Now Shahr
Merchant marine:
total: 132 ships (1,000 GRT or over) totaling 3,816,820 GRT/6,991,693 DWT
ships by type: bulk 48, cargo 38, chemical tanker 5, combination bulk 2, liquefied gas tanker 1, oil tanker 26, refrigerated cargo 3, roll-on/roll-off cargo 8, short-sea passenger 1
Airports:
total: 261
with paved runways over 3,047 m: 28
with paved runways 2,438 to 3,047 m: 12
with paved runways 1,524 to 2,437 m: 32
with paved runways 914 to 1,523 m: 20
with paved runways under 914 m: 46
with unpaved runways over 3,047 m: 2
with unpaved runways 2,438 to 3,047 m: 2
with unpaved runways 1,524 to 2,438 m: 18
with unpaved runways 914 to 1,523 m: 101

Communications

Telephone system: 2,143,000 telephones; 35 telephones/1,000 persons
local: NA
intercity: microwave radio relay extends throughout country; system centered in Tehran
international: 3 INTELSAT (2 Atlantic Ocean and 1 Indian Ocean) earth stations; HF radio and microwave radio relay to Turkey, Pakistan, Syria, Kuwait, Tajikistan, and Uzbekistan; submarine fiber optic cable to UAE
Radio:
broadcast stations: AM 77, FM 3, shortwave 0
radios: NA
Television:
broadcast stations: 28
televisions: NA

Defense Forces

Branches: Islamic Republic of Iran Ground Forces, Navy, Air and Air Defense Force, Revolutionary Guards (includes Basij militia with its ground, air, and naval forces), Law Enforcement Forces
Manpower availability: males age 15-49 14,639,290; males fit for military service 8,703,732; males reach military age (21) annually 615,096 (1995 est.)
Defense expenditures: according to official Iranian data, Iran spent 1,785 billion rials, including $808 million in hard currency, in 1992 and budgeted 2,507 billion rials, including $850 million in hard currency, for 1993
note: conversion of rial expenditures into US dollars using the current exchange rate could produce misleading results

Iraq

Geography

Location: Middle East, bordering the Persian Gulf, between Iran and Kuwait
Map references: Middle East
Area:
total area: 437,072 sq km
land area: 432,162 sq km
comparative area: slightly more than twice the size of Idaho
Land boundaries: total 3,631 km, Iran 1,458 km, Jordan 181 km, Kuwait 242 km, Saudi Arabia 814 km, Syria 605 km, Turkey 331 km
Coastline: 58 km
Maritime claims:
continental shelf: not specified
territorial sea: 12 nm
International disputes: Iran and Iraq restored diplomatic relations in 1990 but are still trying to work out written agreements settling outstanding disputes from their eight-year war concerning border demarcation, prisoners-of-war, and freedom of navigation and sovereignty over the Shatt al Arab waterway; in November 1994, Iraq formally accepted the UN-demarcated border with Kuwait which had been spelled out in Security Council Resolutions 687 (1991), 773 (1993), and 883 (1993); this formally ends earlier claims to Kuwait and to Bubiyan and Warbah islands; potential dispute over water development plans by Turkey for the Tigris and Euphrates Rivers
Climate: mostly desert; mild to cool winters with dry, hot, cloudless summers; northern mountainous regions along Iranian and Turkish borders experience cold winters with occasionally heavy snows which melt in early spring, sometimes causing extensive flooding in central and southern Iraq
Terrain: mostly broad plains; reedy marshes along Iranian border in south; mountains along borders with Iran and Turkey
Natural resources: petroleum, natural gas, phosphates, sulfur
Land use:
arable land: 12%
permanent crops: 1%
meadows and pastures: 9%
forest and woodland: 3%
other: 75%
Irrigated land: 25,500 sq km (1989 est)
Environment:
current issues: government water control projects have drained most of the inhabited marsh areas west of Al Qurnah by drying up or diverting the feeder streams and rivers; a once sizable population of Shi'a Muslims, who have inhabited these areas for thousands of years, has been displaced; furthermore, the destruction of the natural habitat poses serious threats to the area's wildlife populations; inadequate supplies of potable water; development of Tigris-Euphrates Rivers system contingent upon agreements with upstream riparian Turkey; air and water pollution; soil degradation (salinization) and erosion; desertification
natural hazards: duststorms, sandstorms, floods
international agreements: party to—Law of the Sea, Nuclear Test Ban; signed, but not ratified—Environmental Modification

People

Population: 20,643,769 (July 1995 est.)
Age structure:
0-14 years: 48% (female 4,850,028; male 5,009,513)
15-64 years: 49% (female 5,021,710; male 5,125,191)
65 years and over: 3% (female 338,790; male 298,537) (July 1995 est.)
Population growth rate: 3.72% (1995 est.)
Birth rate: 43.6 births/1,000 population (1995 est.)
Death rate: 6.82 deaths/1,000 population (1995 est.)
Net migration rate: 0.39 migrant(s)/1,000 population (1995 est.)
Infant mortality rate: 62.4 deaths/1,000 live births (1995 est.)
Life expectancy at birth:
total population: 66.52 years
male: 65.54 years
female: 67.56 years (1995 est.)
Total fertility rate: 6.56 children born/woman (1995 est.)
Nationality:
noun: Iraqi(s)
adjective: Iraqi
Ethnic divisions: Arab 75%-80%, Kurdish 15%-20%, Turkoman, Assyrian or other 5%
Religions: Muslim 97% (Shi'a 60%-65%, Sunni 32%-37%), Christian or other 3%
Languages: Arabic, Kurdish (official in Kurdish regions), Assyrian, Armenian
Literacy: age 15-45 can read and write (1985)
total population: 89%
male: 90%
female: 88%
Labor force: 4.4 million (1989)
by occupation: services 48%, agriculture 30%, industry 22%
note: severe labor shortage; expatriate labor force was about 1,600,000 (July 1990); since then, it has declined substantially

Government

Names:
conventional long form: Republic of Iraq
conventional short form: Iraq
local long form: Al Jumhuriyah al Iraqiyah
local short form: Al Iraq
Digraph: IZ
Type: republic
Capital: Baghdad
Administrative divisions: 18 provinces (muhafazat, singular—muhafazah); Al Anbar, Al Basrah, Al Muthanna, Al Qadisiyah, An Najaf, Arbil, As Sulaymaniyah, At Ta'mim, Babil, Baghdad, Dahuk, Dhi Qar, Diyala, Karbala', Maysan, Ninawa, Salah ad Din, Wasit
Independence: 3 October 1932 (from League of Nations mandate under British administration)
National holiday: Anniversary of the Revolution, 17 July (1968)
Constitution: 22 September 1968, effective 16 July 1970 (provisional Constitution); new constitution drafted in 1990 but not adopted
Legal system: based on Islamic law in special religious courts, civil law system elsewhere; has not accepted compulsory ICJ jurisdiction
Suffrage: 18 years of age; universal
Executive branch:
chief of state: President SADDAM Husayn (since 16 July 1979); Vice President Taha Muhyi al-Din MARUF (since 21 April 1974); Vice President Taha Yasin RAMADAN (since 23 March 1991)
head of government: Prime Minister SADDAM Husayn (since NA May 1994); Deputy Prime Minister Tariq Mikhail AZIZ (since NA 1979)
Revolutionary Command Council: Chairman SADDAM Husayn, Vice Chairman Izzat IBRAHIM al-Duri
cabinet: Council of Ministers
Legislative branch: unicameral
National Assembly (Majlis al-Watani): elections last held on 1 April 1989 (next to be held NA); results—Sunni Arabs 53%, Shi'a Arabs 30%, Kurds 15%, Christians 2% (est.); seats—(250 total) number of seats by party NA
note: in northern Iraq, a "Kurdish Assembly" was elected in May 1992 and calls for Kurdish self-determination within a federated Iraq; the assembly is not recognized by the Baghdad government
Judicial branch: Court of Cassation
Political parties and leaders: Ba'th Party
Other political or pressure groups: political parties and activity severely restricted;

opposition to regime from disaffected members of the Ba'th Party, Army officers, and Shi'a religious and ethnic Kurdish dissidents; the Green Party (government-controlled)
Member of: ABEDA, ACC, AFESD, AL, AMF, CAEU, CCC, ESCWA, FAO, G-19, G-77, IAEA, IBRD, ICAO, ICRM, IDA, IDB, IFAD, IFC, IFRCS, ILO, IMF, IMO, INMARSAT, INTELSAT, INTERPOL, IOC, ITU, NAM, OAPEC, OIC, OPEC, PCA, UN, UNCTAD, UNESCO, UNIDO, UPU, WFTU, WHO, WIPO, WMO, WTO
Diplomatic representation in US:
chief of mission: Iraq has an Interest Section in the Algerian Embassy in Washington, DC
chancery: Iraqi Interests Section, 1801 P Street NW, Washington, DC 20036
telephone: [1] (202) 483-7500
FAX: [1] (202) 462-5066
US diplomatic representation:
chief of mission: (vacant); note—operations have been temporarily suspended; a US Interests Section is located in Poland's embassy in Baghdad
embassy: Masbah Quarter (opposite the Foreign Ministry Club), Baghdad
mailing address: P. O. Box 2447 Alwiyah, Baghdad
telephone: [964] (1) 719-6138, 719-6139, 718-1840, 719-3791
FAX: Telex 212287
Flag: three equal horizontal bands of red (top), white, and black with three green five-pointed stars in a horizontal line centered in the white band; the phrase ALLAHU AKBAR (God is Great) in green Arabic script—Allahu to the right of the middle star and Akbar to the left of the middle star—was added in January 1991 during the Persian Gulf crisis; similar to the flag of Syria that has two stars but no script and the flag of Yemen that has a plain white band; also similar to the flag of Egypt that has a symbolic eagle centered in the white band

Economy

Overview: The Ba'thist regime engages in extensive central planning and management of industrial production and foreign trade while leaving some small-scale industry and services and most agriculture to private enterprise. The economy has been dominated by the oil sector, which has traditionally provided about 95% of foreign exchange earnings. In the 1980s, financial problems caused by massive expenditures in the eight year war with Iran and damage to oil export facilities by Iran, led the government to implement austerity measures and to borrow heavily and later reschedule foreign debt payments. After the end of hostilities in 1988, oil exports gradually increased with the construction of new pipelines and restoration of damaged facilities. Agricultural development remained hampered by labor shortages, salinization, and dislocations caused by previous land reform and collectivization programs. The industrial sector, although accorded high priority by the government, also was under financial constraints. Iraq's seizure of Kuwait in August 1990, subsequent international economic embargoes, and military action by an international coalition beginning in January 1991 drastically changed the economic picture. Industrial and transportation facilities, which suffered severe damage, have been partially restored. Oil exports remain at less than 5% of the previous level. Shortages of spare parts continue. Living standards deteriorated even further in 1993 and 1994; consumer prices have more than doubled in both 1993 and 1994. The UN-sponsored economic embargo has reduced exports and imports and has contributed to the sharp rise in prices. The Iraqi government has been unwilling to abide by UN resolutions so that the economic embargo can be removed. The government's policies of supporting large military and internal security forces and of allocating resources to key supporters of the regime have exacerbated shortages. In brief, per capita output in 1993-94 is far below the 1989-90 level, but no precise estimate is available.
National product: GDP—purchasing power parity—$NA
National product real growth rate: NA%
National product per capita: $NA
Inflation rate (consumer prices): NA%
Unemployment rate: NA%
Budget:
revenues: $NA
expenditures: $NA, including capital expenditures of $NA
Exports: $10.4 billion (f.o.b., 1990)
commodities: crude oil and refined products, fertilizer, sulfur
partners: US, Brazil, Turkey, Japan, Netherlands, Spain (1990)
Imports: $6.6 billion (c.i.f., 1990)
commodities: manufactures, food
partners: Germany, US, Turkey, France, UK (1990)
External debt: $50 billion (1989 est.), excluding debt of about $35 billion owed to Gulf Arab states
Industrial production: growth rate NA%; manufacturing accounts for 10% of GNP (1989)
Electricity:
capacity: 7,170,000 kW
production: 25.7 billion kWh
consumption per capita: 1,247 kWh (1993)
Industries: petroleum production and refining, chemicals, textiles, construction materials, food processing
Agriculture: accounted for 11% of GNP and 30% of labor force before the Gulf war; principal products—wheat, barley, rice, vegetables, dates, other fruit, cotton, wool; livestock—cattle, sheep; not self-sufficient in food output
Economic aid:
recipient: US commitments, including Ex-Im (FY70-80), $3 million; Western (non-US) countries, ODA and OOF bilateral commitments (1970-89), $647 million; Communist countries (1970-89), $3.9 billion
Currency: 1 Iraqi dinar (ID) = 1,000 fils
Exchange rates: Iraqi dinars (ID) per US$1—3.2 (fixed official rate since 1982); black-market rate (March 1995) US$1 = 1200 Iraqi dinars; semi-official rate US$1 = 650 Iraqi dinars
Fiscal year: calendar year

Transportation

Railroads:
total: 2,457 km
standard gauge: 2,457 km 1.435-m gauge
Highways:
total: 45,550 km
paved: 38,400 km
unpaved: 7,150 km (1989 est.)
Inland waterways: 1,015 km; Shatt al Arab is usually navigable by maritime traffic for about 130 km; channel has been dredged to 3 meters and is in use; Tigris and Euphrates Rivers have navigable sections for shallow-draft watercraft; Shatt al Basrah canal was navigable by shallow-draft craft before closing in 1991 because of the Persian Gulf war
Pipelines: crude oil 4,350 km; petroleum products 725 km; natural gas 1,360 km
Ports: Umm Qasr, Khawr az Zubayr, and Al Basrah have limited functionality
Merchant marine:
total: 36 ships (1,000 GRT or over) totaling 795,346 GRT/1,431,154 DWT
ships by type: cargo 14, oil tanker 16, passenger 1, passenger-cargo 1, refrigerated cargo 1, roll-on/roll-off cargo 3
Airports:
total: 121
with paved runways over 3,047 m: 21
with paved runways 2,438 to 3,047 m: 34
with paved runways 1,524 to 2,437 m: 8
with paved runways 914 to 1,523 m: 7
with paved runways under 914 m: 22
with unpaved runways over 3,047 m: 3
with unpaved runways 2,438 to 3,047 m: 5
with unpaved runways 1,524 to 2,438 m: 5
with unpaved runways 914 to 1,523 m: 16

Communications

Telephone system: 632,000 telephones; reconstitution of damaged telecommunication facilities began after the Gulf war; most damaged facilities have been rebuilt
local: NA
intercity: the network consists of coaxial cables and microwave radio relay links
international: 2 INTELSAT (1 Atlantic Ocean

and 1 Indian Ocean), 1 GORIZONT (Atlantic Ocean) in the Intersputnik system, and 1 ARABSAT earth station; coaxial cable and microwave radio relay to Jordan, Kuwait, Syria, and Turkey; Kuwait line is probably non-operational
Radio:
broadcast stations: AM 16, FM 1, shortwave 0
radios: NA
Television:
broadcast stations: 13
televisions: NA

Defense Forces

Branches: Army, Republican Guard and Special Republican Guard, Navy, Air Force, Air Defense Force, Border Guard Force, Internal Security Forces
Manpower availability: males age 15-49 4,626,610; males fit for military service 2,597,687; males reach military age (18) annually 229,015 (1995 est.)
Defense expenditures: $NA, NA% of GNP

Ireland

Geography

Location: Western Europe, occupying five-sixths of the island of Ireland in the North Atlantic Ocean, west of Great Britain
Map references: Europe
Area:
total area: 70,280 sq km
land area: 68,890 sq km
comparative area: slightly larger than West Virginia
Land boundaries: total 360 km, UK 360 km
Coastline: 1,448 km
Maritime claims:
continental shelf: not specified
exclusive fishing zone: 200 nm
territorial sea: 12 nm
International disputes: Northern Ireland question with the UK; Rockall continental shelf dispute involving Denmark, Iceland, and the UK (Ireland and the UK have signed a boundary agreement in the Rockall area)
Climate: temperate maritime; modified by North Atlantic Current; mild winters, cool summers; consistently humid; overcast about half the time
Terrain: mostly level to rolling interior plain surrounded by rugged hills and low mountains; sea cliffs on west coast
Natural resources: zinc, lead, natural gas, petroleum, barite, copper, gypsum, limestone, dolomite, peat, silver
Land use:
arable land: 14%
permanent crops: 0%
meadows and pastures: 71%
forest and woodland: 5%
other: 10%
Irrigated land: NA sq km
Environment:
current issues: water pollution, especially of lakes, from agricultural runoff
natural hazards: NA
international agreements: party to—Air Pollution, Air Pollution-Nitrogen Oxides, Climate Change, Environmental Modification, Hazardous Wastes, Marine Dumping, Nuclear Test Ban, Ozone Layer Protection, Tropical Timber 83, Wetlands, Whaling; signed, but not ratified—Air Pollution-Sulphur 94, Biodiversity, Desertification, Endangered Species, Law of the Sea, Marine Life Conservation
Note: strategic location on major air and sea routes between North America and northern Europe; over 40% of the population resides within 60 miles of Dublin

People

Population: 3,550,448 (July 1995 est.)
Age structure:
0-14 years: 24% (female 415,640; male 440,468)
15-64 years: 64% (female 1,125,638; male 1,155,823)
65 years and over: 12% (female 237,098; male 175,781) (July 1995 est.)
Population growth rate: 0.33% (1995 est.)
Birth rate: 14.04 births/1,000 population (1995 est.)
Death rate: 8.48 deaths/1,000 population (1995 est.)
Net migration rate: -2.22 migrant(s)/1,000 population (1995 est.)
Infant mortality rate: 7.2 deaths/1,000 live births (1995 est.)
Life expectancy at birth:
total population: 75.99 years
male: 73.15 years
female: 79 years (1995 est.)
Total fertility rate: 1.95 children born/woman (1995 est.)
Nationality:
noun: Irishman(men), Irishwoman(men), Irish (collective plural)
adjective: Irish
Ethnic divisions: Celtic, English
Religions: Roman Catholic 93%, Anglican 3%, none 1%, unknown 2%, other 1% (1981)
Languages: Irish (Gaelic), spoken mainly in areas located along the western seaboard, English is the language generally used
Literacy: age 15 and over can read and write (1981 est.)
total population: 98%
Labor force: 1.37 million
by occupation: services 57.0%, manufacturing and construction 28%, agriculture, forestry, and fishing 13.5%, energy and mining 1.5% (1992)

Government

Names:
conventional long form: none
conventional short form: Ireland
Digraph: EI
Type: republic
Capital: Dublin
Administrative divisions: 26 counties; Carlow, Cavan, Clare, Cork, Donegal, Dublin,

Galway, Kerry, Kildare, Kilkenny, Laois, Leitrim, Limerick, Longford, Louth, Mayo, Meath, Monaghan, Offaly, Roscommon, Sligo, Tipperary, Waterford, Westmeath, Wexford, Wicklow
Independence: 6 December 1921 (from UK)
National holiday: Saint Patrick's Day, 17 March
Constitution: 29 December 1937; adopted 1 July 1937 by plebescite
Legal system: based on English common law, substantially modified by indigenous concepts; judicial review of legislative acts in Supreme Court; has not accepted compulsory ICJ jurisdiction
Suffrage: 18 years of age; universal
Executive branch:
chief of state: President Mary Bourke ROBINSON (since 9 November 1990); election last held 9 November 1990 (next to be held November 1997); results—Mary Bourke ROBINSON 52.8%, Brian LENIHAN 47.2%
head of government: Prime Minister John BRUTON (since 15 December 1994)
cabinet: Cabinet; appointed by president with previous nomination of the prime minister and approval of the House of Representatives
Legislative branch: bicameral Parliament (Oireachtas)
Senate (Seanad Eireann): elections last held NA February 1992 (next to be held NA February 1997); results—percent of vote by party NA; seats—(60 total, 49 elected) Fianna Fail 26, Fine Gael 16, Labor 9, Progressive Democrats 2, Democratic Left 1, independents 6
House of Representatives (Dail Eireann): elections last held on 25 November 1992 (next to be held by November 1997); results—Fianna Fail 39.1%, Fine Gael 24.5%, Labor Party 19.3%, Progressive Democrats 4.7%, Democratic Left 2.8%, Sinn Fein 1.6%, Workers' Party 0.7%, independents 5.9%; seats—(166 total) Fianna Fail 68, Fine Gael 45, Labor Party 33, Progressive Democrats 10 Democratic Left 4, Greens 1, independents 5
Judicial branch: Supreme Court
Political parties and leaders: Democratic Left, Proinsias DE ROSSA; Fianna Fail, Bertie AHERN; Labor Party, Richard SPRING; Fine Gael, John BRUTON; Communist Party of Ireland, Michael O'RIORDAN; Sinn Fein, Gerry ADAMS; Progressive Democrats, Desmond O'MALLEY; The Workers' Party, Marion DONNELLY; Green Alliance, Bronwen MAHER
note: Prime Minister BRUTON heads a three-party coalition consisting of the Fine Gael, the Labor Party, and the Democratic Left
Member of: Australia Group, BIS, CCC, CE, EBRD, EC, ECE, EIB, ESA, FAO, GATT, IAEA, IBRD, ICAO, ICC, ICRM, IDA, IEA, IFAD, IFC, IFRCS, ILO, IMF, IMO, INTELSAT, INTERPOL, IOC, ISO, ITU, MINURSO, MTCR, NEA, NSG, OECD, ONUSAL, OSCE, UN, UNCTAD, UNESCO, UNFICYP, UNIDO, UNIFIL, UNIKOM, UNOMOZ, UNOSOM, UNPROFOR, UNTSO, UPU, WEU (observer), WHO, WIPO, WMO, ZC
Diplomatic representation in US:
chief of mission: Ambassador Dermot A. GALLAGHER
chancery: 2234 Massachusetts Avenue NW, Washington, DC 20008
telephone: [1] (202) 462-3939
consulate(s) general: Boston, Chicago, New York, and San Francisco
US diplomatic representation:
chief of mission: Ambassador Jean Kennedy SMITH
embassy: 42 Elgin Road, Ballsbridge, Dublin
mailing address: use embassy street address
telephone: [353] (1) 6687122
FAX: [353] (1) 6689946
Flag: three equal vertical bands of green (hoist side), white, and orange; similar to the flag of the Cote d'Ivoire, which is shorter and has the colors reversed—orange (hoist side), white, and green; also similar to the flag of Italy, which is shorter and has colors of green (hoist side), white, and red

Economy

Overview: The economy is small and trade dependent. Agriculture, once the most important sector, is now dwarfed by industry, which accounts for 37% of GDP, about 80% of exports, and employs 28% of the labor force. Although exports remain the primary engine for Ireland's robust growth, the economy is also benefiting from a rise in consumer spending and recovery in both construction and business investment. Ireland has substantially reduced its external debt since 1987, to 40% of GDP in 1994. Over the same period, inflation has fallen sharply and chronic trade deficits have been transformed into annual surpluses. Unemployment remains a serious problem, however, and job creation is the main focus of government policy. To ease unemployment, Dublin aggressively courts foreign investors and recently created a new industrial development agency to aid small indigenous firms. Government assistance is constrained by Dublin's continuing deficit reduction measures.
National product: GDP—purchasing power parity—$49.8 billion (1994 est.)
National product real growth rate: 5.5% (1994 est.)
National product per capita: $14,060 (1994 est.)
Inflation rate (consumer prices): 2.7% (1994 est.)
Unemployment rate: 16% (1994 est.)
Budget:
revenues: $16 billion
expenditures: $16.6 billion, including capital expenditures of $NA (1994)
Exports: $28 billion (f.o.b., 1994 est.)
commodities: chemicals, data processing equipment, industrial machinery, live animals, animal products
partners: EU 75% (UK 32%, Germany 13%, France 10%), US 9%
Imports: $26 billion (c.i.f., 1994 est.)
commodities: food, animal feed, data processing equipment, petroleum and petroleum products, machinery, textiles, clothing
partners: EU 66% (UK 41%, Germany 8%, France 4%), US 15%
External debt: $20 billion (1994 est.)
Industrial production: growth rate 8.5% (1994 est.); accounts for 37% of GDP
Electricity:
capacity: 3,930,000 kW
production: 14.9 billion kWh
consumption per capita: 3,938 kWh (1993)
Industries: food products, brewing, textiles, clothing, chemicals, pharmaceuticals, machinery, transportation equipment, glass and crystal
Agriculture: accounts for 10% of GDP; principal crops—turnips, barley, potatoes, sugar beets, wheat; livestock—meat and dairy products; 85% self-sufficient in food; food shortages include bread grain, fruits, vegetables
Illicit drugs: transshipment point for hashish from North Africa to the UK and Netherlands
Economic aid:
donor: ODA commitments (1980-89), $90 million
Currency: 1 Irish pound (£Ir) = 100 pence
Exchange rates: Irish pounds (£Ir) per US$1—0.6420 (January 1995), 0.6676 (1994), 0.6816 (1993), 0.5864 (1992), 0.6190 (1991), 0.6030 (1990)
Fiscal year: calendar year

Transportation

Railroads:
total: 1,947 km
broad gauge: 1,947 km 1.600-m gauge (36 km electrified; 485 km double track)
Highways:
total: 92,327 km
paved: 86,787 km (32 km of expressways)
unpaved: gravel, crushed stone 5,540 km (1992)
Inland waterways: limited for commercial traffic
Pipelines: natural gas 225 km
Ports: Arklow, Cork, Drogheda, Dublin, Foynes, Galway, Limerick, New Ross, Waterford
Merchant marine:
total: 47 ships (1,000 GRT or over) totaling 129,996 GRT/160,419 DWT
ships by type: bulk 4, cargo 33, chemical tanker 2, container 2, oil tanker 1, short-sea passenger 3, specialized tanker 2

Ireland *(continued)*

Airports:
total: 44
with paved runways over 3,047 m: 1
with paved runways 2,438 to 3,047 m: 1
with paved runways 1,524 to 2,437 m: 4
with paved runways 914 to 1,523 m: 2
with paved runways under 914 m: 32
with unpaved runways 914 to 1,523 m: 4

Communications

Telephone system: 900,000 telephones; modern digital system using cable and microwave radio relay
local: NA
intercity: microwave radio relay
international: 1 INTELSAT (Atlantic Ocean) earth station
Radio:
broadcast stations: AM 9, FM 45, shortwave 0
radios: NA
Television:
broadcast stations: 86
televisions: NA

Defense Forces

Branches: Army (includes Naval Service and Air Corps), National Police (Garda Siochana)
Manpower availability: males age 15-49 926,831; males fit for military service 749,646; males reach military age (17) annually 34,215 (1995 est.)
Defense expenditures: exchange rate conversion—$500 million, 1.3% of GDP (1994)

Israel
(also see separate Gaza Strip and West Bank entries)

Note: The territories occupied by Israel since the 1967 war are not included in the data below. In keeping with the framework established at the Madrid Conference in October 1991, bilateral negotiations are being conducted between Israel and Palestinian representatives, Syria, and Jordan to determine the final status of the occupied territories. On 25 April 1982, Israel withdrew from the Sinai pursuant to the 1979 Israel-Egypt Peace treaty. Outstanding territorial and other disputes with Jordan were resolved in the 26 October 1994 Israel-Jordan Treaty of Peace.

Geography

Location: Middle East, bordering the Mediterranean Sea, between Egypt and Lebanon
Map references: Middle East
Area:
total area: 20,770 sq km
land area: 20,330 sq km
comparative area: slightly larger than New Jersey
Land boundaries: total 1,006 km, Egypt 255 km, Gaza Strip 51 km, Jordan 238 km, Lebanon 79 km, Syria 76 km, West Bank 307 km
Coastline: 273 km
Maritime claims:
continental shelf: to depth of exploitation
territorial sea: 12 nm
International disputes: separated from Lebanon, Syria, and the West Bank by the 1949 Armistice Line; the Gaza Strip and Jericho area, formerly occupied by Israel, are now administered largely by the Palestinian Authority; other areas of the West Bank outside Jericho are administered jointly by Israel and the Palestinian Authority; Golan Heights is Israeli occupied; Israeli troops in southern Lebanon since June 1982
Climate: temperate; hot and dry in southern and eastern desert areas
Terrain: Negev desert in the south; low coastal plain; central mountains; Jordan Rift Valley
Natural resources: copper, phosphates, bromide, potash, clay, sand, sulfur, asphalt, manganese, small amounts of natural gas and crude oil
Land use:
arable land: 17%
permanent crops: 5%
meadows and pastures: 40%
forest and woodland: 6%
other: 32%
Irrigated land: 2,140 sq km (1989)
Environment:
current issues: limited arable land and natural fresh water resources pose serious constraints; desertification; air pollution from industrial and vehicle emissions; groundwater pollution from industrial and domestic waste, chemical fertilizers, and pesticides
natural hazards: sandstorms may occur during spring and summer
international agreements: party to—Biodiversity, Endangered Species, Hazardous Wastes, Nuclear Test Ban, Ozone Layer Protection, Ship Pollution; signed, but not ratified—Climate Change, Desertification, Marine Life Conservation
Note: there are 199 Jewish settlements and civilian land use sites in the West Bank, 42 in the Israeli-occupied Golan Heights, 24 in the Gaza Strip, and 25 in East Jerusalem (August 1994 est.)

People

Population: 5,433,134 (July 1995 est.)
note: includes 122,000 Jewish settlers in the West Bank, 14,500 in the Israeli-occupied Golan Heights, 4,800 in the Gaza Strip, and 149,000 in East Jerusalem (August 1994 est.)
Age structure:
0-14 years: 29%
15-64 years: 61%
65 years and over: 10%
Population growth rate: 1.4% (1995 est.)
Birth rate: 20.39 births/1,000 population (1995 est.)
Death rate: 6.38 deaths/1,000 population (1995 est.)
Net migration rate: 0 migrant(s)/1,000 population (1995 est.)
Infant mortality rate: 8.4 deaths/1,000 live births (1995 est.)
Life expectancy at birth:
total population: 78.14 years
male: 76 years
female: 80.39 years (1995 est.)
Total fertility rate: 2.81 children born/woman (1995 est.)
Nationality:
noun: Israeli(s)
adjective: Israeli
Ethnic divisions: Jewish 82% (Israel born 50%, Europe/Americas/Oceania born 20%,

Africa born 7%, Asia born 5%), non-Jewish 18% (mostly Arab) (1993 est.)
Religions: Judaism 82%, Islam 14% (mostly Sunni Muslim), Christian 2%, Druze and other 2%
Languages: Hebrew (official), Arabic used officially for Arab minority, English most commonly used foreign language
Literacy: age 15 and over can read and write (1992)
total population: 95%
male: 97%
female: 93%
Labor force: 1.9 million (1992)
by occupation: public services 29.3%, industry 22.1%, commerce 13.9%, finance and business 10.4%, personal and other services 7.4%, construction 6.5%, transport, storage, and communications 6.3%, agriculture, forestry, and fishing 3.5%, other 0.6% (1992)

Government

Names:
conventional long form: State of Israel
conventional short form: Israel
local long form: Medinat Yisra'el
local short form: Yisra'el
Digraph: IS
Type: republic
Capital: Jerusalem
note: Israel proclaimed Jerusalem its capital in 1950, but the US, like nearly all other countries, does not recognize this status, and maintains its Embassy in Tel Aviv
Administrative divisions: 6 districts (mehozot, singular—mehoz); Central, Haifa, Jerusalem, Northern, Southern, Tel Aviv
Independence: 14 May 1948 (from League of Nations mandate under British administration)
National holiday: Independence Day, 14 May 1948 (Israel declared independence on 14 May 1948, but the Jewish calendar is lunar and the holiday may occur in April or May)
Constitution: no formal constitution; some of the functions of a constitution are filled by the Declaration of Establishment (1948), the basic laws of the parliament (Knesset), and the Israeli citizenship law
Legal system: mixture of English common law, British Mandate regulations, and, in personal matters, Jewish, Christian, and Muslim legal systems; in December 1985, Israel informed the UN Secretariat that it would no longer accept compulsory ICJ jurisdiction
Suffrage: 18 years of age; universal
Executive branch:
chief of state: President Ezer WEIZMAN (since 13 May 1993) election last held 24 March 1993 (next to be held NA March 1999); results—Ezer WEIZMAN elected by Knesset
head of government: Prime Minister Yitzhak RABIN (since NA July 1992)
cabinet: Cabinet; selected from and approved by the Knesset
Legislative branch: unicameral
parliament (Knesset): elections last held NA June 1992 (next to be held by NA 1996); results—percent of vote by party NA; seats—(120 total) Labor 44, Likud 32, MERETZ 12, Tzomet 8, National Religious Party 6, SHAS 6, United Torah Jewry 4, Democratic Front for Peace and Equality (Hadash) 3, Moledet 3, Arab Democratic Party 2; note—in 1994 four legislators broke party ranks, resulting in the following new distribution of seats—Labor Party 44, Likud bloc 32, MERETZ 12, National Religious Party 6, SHAS 6, Tzomet 5, United Torah Jewry 4, Democratic Front for Peace and Equality (Hadash) 3, Moledet 2, Arab Democratic Party 2, independents 4 (1 in coalition, 3 voting with opposition)
Judicial branch: Supreme Court
Political parties and leaders:
members of the government: Labor Party, Prime Minister Yitzhak RABIN; MERETZ, Minister of Communications Shulamit ALONI; independent, Gonen SEGEV
not in coalition, but voting with the government: Democratic Front for Peace and Equality (Hadash), Hashim MAHAMID; Arab Democratic Party, Abd al Wahab DARAWSHAH
opposition parties: Likud Party, Binyamin NETANYAHU; Tzomet, Rafael EITAN; National Religious Party, Zevulun HAMMER; United Torah Jewry, Avraham SHAPIRA; Moledet, Rehavam ZEEVI; Peace Guard (independent), Shaul GUTMAN; SHAS, Arieh DERI
note: Israel currently has a coalition government comprising 2 parties and an independent that hold 57 seats of the Knesset's 120 seats
Other political or pressure groups: Gush Emunim, Israeli nationalists advocating Jewish settlement on the West Bank and Gaza Strip; Peace Now supports territorial concessions in the West Bank and is critical of government's Lebanon policy
Member of: AG (observer), CCC, CE (observer), CERN (observer), EBRD, ECE, FAO, GATT, IADB, IAEA, IBRD, ICAO, ICC, ICFTU, IDA, IFAD, IFC, ILO, IMF, IMO, INMARSAT, INTELSAT, INTERPOL, IOC, IOM, ISO, ITU, OAS (observer), PCA, UN, UNCTAD, UNESCO, UNHCR, UNIDO, UPU, WHO, WIPO, WMO, WTO
Diplomatic representation in US:
chief of mission: Ambassador Itamar RABINOVICH
chancery: 3514 International Drive NW, Washington, DC 20008
telephone: [1] (202) 364-5500
FAX: [1] (202) 364-5610
consulate(s) general: Atlanta, Boston, Chicago, Houston, Los Angeles, Miami, New York, Philadelphia, and San Francisco
US diplomatic representation:
chief of mission: Ambassador Martin INDYK
embassy: 71 Hayarkon Street, Tel Aviv
mailing address: PSC 98, Box 100, Tel Aviv; APO AE 09830
telephone: [972] (3) 517-4338
FAX: [972] (3) 663-449
consulate(s) general: Jerusalem
Flag: white with a blue hexagram (six-pointed linear star) known as the Magen David (Shield of David) centered between two equal horizontal blue bands near the top and bottom edges of the flag

Economy

Overview: Israel has a market economy with substantial government participation. It depends on imports of crude oil, grains, raw materials, and military equipment. Despite limited natural resources, Israel has intensively developed its agricultural and industrial sectors over the past 20 years. Industry employs about 22% of Israeli workers, construction 6.5%, agriculture, forestry, and fishing 3.5%, and services most of the rest. Israel is largely self-sufficient in food production except for grains. Diamonds, high-technology equipment, and agricultural products (fruits and vegetables) are leading exports. Israel usually posts current account deficits, which are covered by large transfer payments from abroad and by foreign loans. Roughly half of the government's external debt is owed to the United States, which is its major source of economic and military aid. To earn needed foreign exchange, Israel has been targeting high-technology niches in international markets, such as medical scanning equipment. The influx of Jewish immigrants from the former USSR, which topped 450,000 during the period 1990-94, increased unemployment, intensified housing problems, and strained the government budget. At the same time, the immigrants bring to the economy valuable scientific and professional expertise.
National product: GDP—purchasing power parity—$70.1 billion (1994 est.)
National product real growth rate: 6.8% (1994 est.)
National product per capita: $13,880 (1994 est.)
Inflation rate (consumer prices): 14.5% (1994)
Unemployment rate: 7.5% (1994 est.)
Budget:
revenues: $42.3 billion
expenditures: $45.4 billion, including capital expenditures of $11.1 billion (FY92/93)
Exports: $16.2 billion (f.o.b., 1994 est.)
commodities: machinery and equipment, cut diamonds, chemicals, textiles and apparel, agricultural products, metals

partners: US, EU, Japan
Imports: $22.5 billion (c.i.f., 1994 est.)
commodities: military equipment, investment goods, rough diamonds, oil, other productive inputs, consumer goods
partners: EU, US, Japan
External debt: $25.9 billion (November 1994 est.)
Industrial production: growth rate 8% (1994 est.); accounts for about 30% of GDP
Electricity:
capacity: 4,140,000 kW
production: 23 billion kWh
consumption per capita: 4,290 kWh (1993)
Industries: food processing, diamond cutting and polishing, textiles and apparel, chemicals, metal products, military equipment, transport equipment, electrical equipment, miscellaneous machinery, potash mining, high-technology electronics, tourism
Agriculture: citrus and other fruits, vegetables, cotton; beef, poultry, dairy products
Illicit drugs: increasingly concerned about cocaine and heroin abuse and trafficking
Economic aid:
recipient: US commitments, including Ex-Im (FY70-90), $18.2 billion; Western (non-US) countries, ODA and OOF bilateral commitments (1970-89), $2.8 billion
Currency: 1 new Israeli shekel (NIS) = 100 new agorot
Exchange rates: new Israeli shekels (NIS) per US$1—3.070 (December 1994), 3.0111 (1994), 2.8301 (1993), 2.4591 (1992), 2.2791 (1991), 2.0162 (1990), 1.9164 (1989)
Fiscal year: calendar year (since 1 January 1992)

Transportation

Railroads:
total: 520 km (diesel operated; single track)
standard gauge: 520 km 1.435-m gauge
Highways:
total: 13,461 km
paved: 13,461 km
Pipelines: crude oil 708 km; petroleum products 290 km; natural gas 89 km
Ports: Ashdod, Ashqelon, Elat, Hadera, Haifa, Tel Aviv-Yafo
Merchant marine:
total: 32 ships (1,000 GRT or over) totaling 624,861 GRT/720,765 DWT
ships by type: cargo 7, container 22, refrigerated cargo 2, roll-on/roll-off cargo 1
Airports:
total: 57
with paved runways over 3,047 m: 2
with paved runways 2,438 to 3,047 m: 6
with paved runways 1,524 to 2,437 m: 8
with paved runways 914 to 1,523 m: 7
with paved runways under 914 m: 31
with unpaved runways 914 to 1,523 m: 3

Communications

Telephone system: 1,800,000 telephones; most highly developed in the Middle East although not the largest
local: NA
intercity: good system of coaxial cable and microwave radio relay
international: 3 submarine cables; 3 INTELSAT (2 Atlantic Ocean and 1 Indian Ocean) earth stations
Radio:
broadcast stations: AM 9, FM 45, shortwave 0
radios: NA
Television:
broadcast stations: 20
televisions: NA

Defense Forces

Branches: Israel Defense Forces (includes ground, naval, and air components), Pioneer Fighting Youth (Nahal), Frontier Guard, Chen (women); note—historically there have been no separate Israeli military services
Manpower availability: males age 15-49 1,309,502; females age 15-49 1,283,923; males fit for military service 1,072,501; females fit for military service 1,047,575; males reach military age (18) annually 47,950; females reach military age (18) annually 45,839 (1995 est.)
note: military service mandatory for men and women
Defense expenditures: exchange rate conversion—$6.5 billion, about 10% of GDP (1995)

Italy

Geography

Location: Southern Europe, a peninsula extending into the central Mediterranean Sea, northeast of Tunisia
Map references: Europe
Area:
total area: 301,230 sq km
land area: 294,020 sq km
comparative area: slightly larger than Arizona
note: includes Sardinia and Sicily
Land boundaries: total 1,899.2 km, Austria 430 km, France 488 km, Holy See (Vatican City) 3.2 km, San Marino 39 km, Slovenia 199 km, Switzerland 740 km
Coastline: 4,996 km
Maritime claims:
continental shelf: 200-m depth or to the depth of exploitation
territorial sea: 12 nm
International disputes: none
Climate: predominantly Mediterranean; Alpine in far north; hot, dry in south
Terrain: mostly rugged and mountainous; some plains, coastal lowlands
Natural resources: mercury, potash, marble, sulfur, dwindling natural gas and crude oil reserves, fish, coal
Land use:
arable land: 32%
permanent crops: 10%
meadows and pastures: 17%
forest and woodland: 22%
other: 19%
Irrigated land: 31,000 sq km (1989 est.)
Environment:
current issues: air pollution from industrial emissions such as sulfur dioxide; coastal and inland rivers polluted from industrial and agricultural effluents; acid rain damaging lakes; inadequate industrial waste treatment and disposal facilities
natural hazards: regional risks include landslides, mudflows, avalanches, earthquakes, volcanic eruptions, flooding; land subsidence in Venice
international agreements: party to—Air

Pollution, Air Pollution-Nitrogen Oxides, Air Pollution-Sulphur 85, Antarctic Treaty, Biodiversity, Climate Change, Endangered Species, Environmental Modification, Hazardous Wastes, Law of the Sea, Marine Dumping, Nuclear Test Ban, Ozone Layer Protection, Ship Pollution, Tropical Timber 83, Wetlands; signed, but not ratified—Air Pollution-Sulphur 94, Air Pollution-Volatile Organic Compounds, Antarctic-Environmental Protocol, Desertification
Note: strategic location dominating central Mediterranean as well as southern sea and air approaches to Western Europe

People

Population: 58,261,971 (July 1995 est.)
Age structure:
0-14 years: 15% (female 4,352,325; male 4,603,083)
15-64 years: 68% (female 19,969,086; male 19,874,528)
65 years and over: 17% (female 5,630,747; male 3,832,202) (July 1995 est.)
Population growth rate: 0.21% (1995 est.)
Birth rate: 10.89 births/1,000 population (1995 est.)
Death rate: 9.78 deaths/1,000 population (1995 est.)
Net migration rate: 1.03 migrant(s)/1,000 population (1995 est.)
Infant mortality rate: 7.4 deaths/1,000 live births (1995 est.)
Life expectancy at birth:
total population: 77.85 years
male: 74.67 years
female: 81.23 years (1995 est.)
Total fertility rate: 1.41 children born/woman (1995 est.)
Nationality:
noun: Italian(s)
adjective: Italian
Ethnic divisions: Italian (includes small clusters of German-, French-, and Slovene-Italians in the north and Albanian-Italians and Greek-Italians in the south), Sicilians, Sardinians
Religions: Roman Catholic 98%, other 2%
Languages: Italian, German (parts of Trentino-Alto Adige region are predominantly German speaking), French (small French-speaking minority in Valle d'Aosta region), Slovene (Slovene-speaking minority in the Trieste-Gorizia area)
Literacy: age 15 and over can read and write (1990 est.)
total population: 97%
male: 98%
female: 96%
Labor force: 23.988 million
by occupation: services 58%, industry 32.2%, agriculture 9.8% (1988)

Government

Names:
conventional long form: Italian Republic
conventional short form: Italy
local long form: Repubblica Italiana
local short form: Italia
former: Kingdom of Italy
Digraph: IT
Type: republic
Capital: Rome
Administrative divisions: 20 regions (regioni, singular—regione); Abruzzi, Basilicata, Calabria, Campania, Emilia-Romagna, Friuli-Venezia Giulia, Lazio, Liguria, Lombardia, Marche, Molise, Piemonte, Puglia, Sardegna, Sicilia, Toscana, Trentino-Alto Adige, Umbria, Valle d'Aosta, Veneto
Independence: 17 March 1861 (Kingdom of Italy proclaimed)
National holiday: Anniversary of the Republic, 2 June (1946)
Constitution: 1 January 1948
Legal system: based on civil law system, with ecclesiastical law influence; appeals treated as trials de novo; judicial review under certain conditions in Constitutional Court; has not accepted compulsory ICJ jurisdiction
Suffrage: 18 years of age; universal (except in senatorial elections, where minimum age is 25)
Executive branch:
chief of state: President Oscar Luigi SCALFARO (since 28 May 1992)
head of government: Prime Minister (referred to in Italy as the President of the Council of Ministers) Lamberto DINI (since 1 February 1995)
cabinet: Council of Ministers; nominated by the President of the Council (i.e., Prime Minister) and approved by the President of the Republic
Legislative branch: bicameral Parliament (Parlamento)
Senate (Senato della Repubblica): elections last held 27-28 March 1994 (next must be held by spring 1999, but may be held by end of 1995); results—percent of vote by party NA; seats (326 total, 315 elected, 11 appointed senators-for-life) PDS 61, Northern League 60, National Alliance 48, Forza Italia 36, Italian Popular Party 31, Communist Refoundation 18, Greens and The Network 13, Italian Socialists 13, Christian Democratic Center 12, Democratic Alliance 8, Christian Socialists 5, Pact for Italy 4, Radical Party (Pannella List) 1, others 5
Chamber of Deputies (Camera dei Deputati): elections last held 27-28 March 1994 (next must be held by spring 1999, but may be held by end of 1995); results—percent of vote by party NA; seats—(630 total) Northern League 117, PDS 114, Forza Italia 113, National Alliance 109, Communist Refoundation 39, Christian Democratic Center 33, Italian Popular Party 33, Greens and The Network 20, Democratic Alliance 18, Italian Socialists 16, Pact for Italy 13, Christian Socialists 5
Judicial branch: Constitutional Court (Corte Costituzionale)
Political parties and leaders: Forza Italia (FI), Silvio BERLUSCONI; National Alliance, Gianfranco FINI, party secretary; Northern League—Federal Italy (NL), Umberto BOSSI, president; Italian Social Movement, Pino RAUTI; Democratic Party of the Left (PDS, Massimo D'ALEMA, secretary; Communist Refoundation (RC), Fausto BERTINOTTI; Greens, Gianni MATTIOLI; Italian Socialists, Ottaviano DELTURCO; Rete (The Network), Leoluca ORLANDO; Christian Socialists, Ermanno GORRIERI; Pact for Italy, Mario SEGNI; Italian Popular Party (PPI), Rocco BUTTIGLIONE, Gerardo BIANCO; Christian Democratic Center (CCD), Pier Ferdinando CASINI; Union of the Democratic Center (UDC), Raffaele COSTA; Pannella List, Marco PANNELLA
Other political or pressure groups: the Roman Catholic Church; three major trade union confederations (Confederazione Generale Italiana del Lavoro or CGIL which is PDS-dominated, Confederazione Italiana dei Sindacati Lavoratori or CISL which is centrist, and Unione Italiana del Lavoro or UIL which is center-left); Italian manufacturers and merchants associations (Confindustria, Confcommercio); organized farm groups (Confcoltivatori, Confagricoltura)
Member of: AfDB, AG (observer), AsDB, Australia Group, BIS, CCC, CDB (non-regional), CE, CEI, CERN, EBRD, EC, ECE, ECLAC, EIB, ESA, FAO, G- 7, G-10, GATT, IADB, IAEA, IBRD, ICAO, ICC, ICFTU, ICRM, IDA, IEA, IFAD, IFC, IFRCS, ILO, IMF, IMO, INMARSAT, INTELSAT, INTERPOL, IOC, IOM, ISO, ITU, LAIA (observer), MINURSO, MTCR, NACC, NATO, NEA, NSG, OAS (observer), OECD, ONUSAL, OSCE, PCA, UN, UNCTAD, UNESCO, UNHCR, UNIDO, UNIFIL, UNIKOM, UNITAR, UNMOGIP, UNOMOZ, UNTSO, UPU, WCL, WEU, WHO, WIPO, WMO, WTO, ZC
Diplomatic representation in US:
chief of mission: Ambassador Boris BIANCHERI-CHIAPPORI
chancery: 1601 Fuller Street NW, Washington, DC 20009
telephone: [1] (202) 328-5500
consulate(s) general: Boston, Chicago, Houston, Miami, New York, Los Angeles, Philadelphia, San Francisco
consulate(s): Detroit and New Orleans
US diplomatic representation:
chief of mission: Ambassador Reginald BARTHOLOMEW
embassy: Via Veneto 119/A, 00187-Rome
mailing address: PSC 59, Box 100, Rome;

APO AE 09624
telephone: [39] (6) 46741
FAX: [39] (6) 4882672
consulate(s) general: Florence, Milan, Naples
Flag: three equal vertical bands of green (hoist side), white, and red; similar to the flag of Ireland, which is longer and is green (hoist side), white, and orange; also similar to the flag of the Cote d'Ivoire, which has the colors reversed—orange (hoist side), white, and green

Economy

Overview: Since World War II the Italian economy has changed from one based on agriculture into a ranking industrial economy, with approximately the same total and per capita output as France and the UK. The country is still divided into a developed industrial north, dominated by private companies, and an undeveloped agricultural south, dominated by large public enterprises. Services account for 48% of GDP, industry 35%, agriculture 4%, and public administration 13%. Most raw materials needed by industry and over 75% of energy requirements must be imported. After growing at an average annual rate of 3% in 1983-90, growth slowed to about 1% in 1991 and 1992, fell by 0.7% in 1993, and recovered to 2% in 1994. In the second half of 1992, Rome became unsettled by the prospect of not qualifying to participate in EU plans for economic and monetary union later in the decade; thus it finally began to address its huge fiscal imbalances. Subsequently, the government has adopted fairly stringent budgets, abandoned its highly inflationary wage indexation system, and started to scale back its extremely generous social welfare programs, including pension and health care benefits. Monetary officials were forced to withdraw the lira from the European monetary system in September 1992 when it came under extreme pressure in currency markets. For the 1990s, Italy faces the problems of pushing ahead with fiscal reform, refurbishing a tottering communications system, curbing pollution in major industrial centers, and adjusting to the new competitive forces accompanying the ongoing expansion and economic integration of the European Union.
National product: GDP—purchasing power parity—$998.9 billion (1994 est.)
National product real growth rate: 2.2% (1994 est.)
National product per capita: $17,180 (1994 est.)
Inflation rate (consumer prices): 3.9% (1994)
Unemployment rate: 12.2% (January 1995)
Budget:
revenues: $339 billion
expenditures: $431 billion, including capital expenditures of $NA (1994 est.)
Exports: $190.8 billion (f.o.b., 1994)
commodities: metals, textiles and clothing, production machinery, motor vehicles, transportation equipment, chemicals, other
partners: EU 53.4%, US 7.8%, OPEC 3.8% (1994)
Imports: $168.7 billion (c.i.f., 1994)
commodities: industrial machinery, chemicals, transport equipment, petroleum, metals, food, agricultural products
partners: EU 56.3%, OPEC 5.3%, US 4.6% (1994)
External debt: $67 billion (1993 est.)
Industrial production: growth rate 4.3% (1994 est.); accounts for 35% of GDP
Electricity:
capacity: 61,630,000 kW
production: 209 billion kWh
consumption per capita: 4,033 kWh (1993)
Industries: machinery, iron and steel, chemicals, food processing, textiles, motor vehicles, clothing, footwear, ceramics
Agriculture: accounts for about 4% of GDP; self-sufficient in foods other than meat, dairy products, and cereals; principal crops—fruits, vegetables, grapes, potatoes, sugar beets, soybeans, grain, olives; fish catch of 525,000 metric tons in 1990
Illicit drugs: important gateway country for Latin American cocaine and Southwest Asian heroin entering the European market
Economic aid:
donor: ODA and OOF commitments (1970-89), $25.9 billion
Currency: 1 Italian lira (Lit) = 100 centesimi
Exchange rates: Italian lire (Lit) per US$1—1,609.5 (January 1995), 1,612.4 (1994), 1,573.7 (1993), 1,232.4 (1992), 1,240.6 (1991), 1,198.1 (1990)
Fiscal year: calendar year

Transportation

Railroads:
total: 19,503 km
standard gauge: 18,230 km 1.435-m gauge (10,499 km electrified; 2,112 km privately owned)
narrow gauge: 1,273 km 0.950-m to 1.000-m gauge (224 km electrified; 1,273 km privately owned)
Highways:
total: 305,388 km
paved: 277,388 km (6,940 km of expressways)
unpaved: gravel, crushed stone 23,000 km; earth 5,000 km (1992)
Inland waterways: 2,400 km for various types of commercial traffic, although of limited overall value
Pipelines: crude oil 1,703 km; petroleum products 2,148 km; natural gas 19,400 km
Ports: Ancona, Augusta, Bari, Cagliari (Sardinia), Catania, Gaeta, Genoa, La Spezia, Livorno, Naples, Oristano (Sardinia), Palermo (Sicily), Piombino, Porto Torres (Sardinia), Ravenna, Savona, Trieste, Venice
Merchant marine:
total: 441 ships (1,000 GRT or over) totaling 5,767,969 GRT/8,547,221 DWT
ships by type: bulk 40, cargo 62, chemical tanker 34, combination ore/oil 3, container 18, liquefied gas tanker 37, multifunction large-load carrier 1, oil tanker 136, passenger 7, roll-on/roll-off cargo 54, short-sea passenger 30, specialized tanker 11, vehicle carrier 8
Airports:
total: 138
with paved runways over 3,047 m: 5
with paved runways 2,438 to 3,047 m: 34
with paved runways 1,524 to 2,437 m: 15
with paved runways 914 to 1,523 m: 26
with paved runways under 914 m: 34
with unpaved runways 1,524 to 2,438 m: 2
with unpaved runways 914 to 1,523 m: 22

Communications

Telephone system: 25,600,000 telephones; modern, well-developed, fast; fully automated telephone, telex, and data services
local: NA
intercity: high-capacity cable and microwave radio relay trunks
international: international service by 21 submarine cables, 3 satellite earth stations operating in INTELSAT with 3 Atlantic Ocean antennas and 2 Indian Ocean antennas; also participates in INMARSAT and EUTELSAT systems
Radio:
broadcast stations: AM 135, FM 28 (repeaters 1,840), shortwave 0
radios: 16 million
Television:
broadcast stations: 83 (repeaters 1,000)
televisions: 18 million

Defense Forces

Branches: Army, Navy, Air Force, Carabinieri
Manpower availability: males age 15-49 14,934,657; males fit for military service 12,962,594; males reach military age (18) annually 382,142 (1995 est.)
Defense expenditures: exchange rate conversion—$21.5 billion, 2% of GDP (1994)

Jamaica

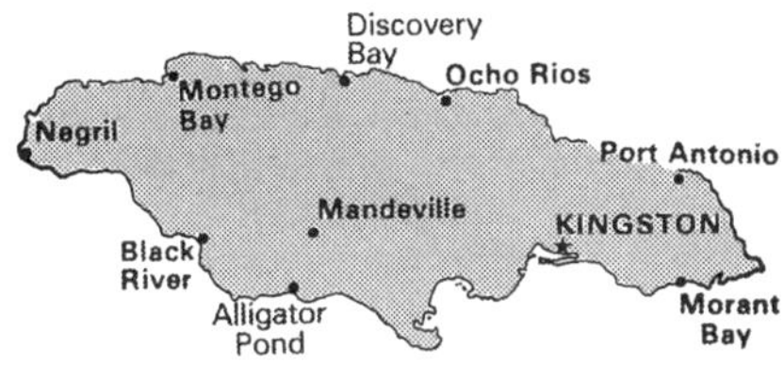

Geography

Location: Caribbean, island in the Caribbean Sea, south of Cuba
Map references: Central America and the Caribbean
Area:
total area: 10,990 sq km
land area: 10,830 sq km
comparative area: slightly smaller than Connecticut
Land boundaries: 0 km
Coastline: 1,022 km
Maritime claims:
continental shelf: 200-m depth or to the depth of exploitation
exclusive economic zone: 200 nm
territorial sea: 12 nm
International disputes: none
Climate: tropical; hot, humid; temperate interior
Terrain: mostly mountains with narrow, discontinuous coastal plain
Natural resources: bauxite, gypsum, limestone
Land use:
arable land: 19%
permanent crops: 6%
meadows and pastures: 18%
forest and woodland: 28%
other: 29%
Irrigated land: 350 sq km (1989 est.)
Environment:
current issues: deforestation; coastal waters polluted by industrial waste, sewage, and oil spills; damage to coral reefs; air pollution in Kingston results from vehicle emissions
natural hazards: hurricanes (especially July to November)
international agreements: party to—Biodiversity, Climate Change, Law of the Sea, Marine Dumping, Marine Life Conservation, Nuclear Test Ban, Ozone Layer Protection, Ship Pollution
Note: strategic location between Cayman Trench and Jamaica Channel, the main sea lanes for Panama Canal

People

Population: 2,574,291 (July 1995 est.)
Age structure:
0-14 years: 33% (female 412,565; male 431,043)
15-64 years: 60% (female 786,700; male 770,681)
65 years and over: 7% (female 96,348; male 76,954) (July 1995 est.)
Population growth rate: 0.78% (1995 est.)
Birth rate: 22.03 births/1,000 population (1995 est.)
Death rate: 5.62 deaths/1,000 population (1995 est.)
Net migration rate: -8.65 migrant(s)/1,000 population (1995 est.)
Infant mortality rate: 16.1 deaths/1,000 live births (1995 est.)
Life expectancy at birth:
total population: 74.65 years
male: 72.39 years
female: 77.01 years (1995 est.)
Total fertility rate: 2.42 children born/woman (1995 est.)
Nationality:
noun: Jamaican(s)
adjective: Jamaican
Ethnic divisions: African 76.3%, Afro-European 15.1%, East Indian and Afro-East Indian 3%, white 3.2%, Chinese and Afro-Chinese 1.2%, other 1.2%
Religions: Protestant 55.9% (Church of God 18.4%, Baptist 10%, Anglican 7.1%, Seventh-Day Adventist 6.9%, Pentecostal 5.2%, Methodist 3.1%, United Church 2.7%, other 2.5%), Roman Catholic 5%, other, including some spiritual cults 39.1% (1982)
Languages: English, Creole
Literacy: age 15 and over has ever attended school (1987)
total population: 82%
male: 77%
female: 86%
Labor force: 1,062,100
by occupation: services 41%, agriculture 22.5%, industry 19%, unemployed 17.5% (1989)

Government

Names:
conventional long form: none
conventional short form: Jamaica
Digraph: JM
Type: parliamentary democracy
Capital: Kingston
Administrative divisions: 14 parishes; Clarendon, Hanover, Kingston, Manchester, Portland, Saint Andrew, Saint Ann, Saint Catherine, Saint Elizabeth, Saint James, Saint Mary, Saint Thomas, Trelawny, Westmoreland
Independence: 6 August 1962 (from UK)
National holiday: Independence Day (first Monday in August) (1962)
Constitution: 6 August 1962
Legal system: based on English common law; has not accepted compulsory ICJ jurisdiction
Suffrage: 18 years of age; universal
Executive branch:
chief of state: Queen ELIZABETH II (since 6 February 1952), represented by Governor General Sir Howard COOKE (since 1 August 1991)
head of government: Prime Minister P. J. PATTERSON (since 30 March 1992); Deputy Prime Minister Seymour MULLINGS (since NA 1993)
cabinet: Cabinet; appointed by the governor general on the advice of the prime minister
Legislative branch: bicameral Parliament
Senate: consists of a 21-member body appointed by the governor general
House of Representatives: elections last held 30 March 1993 (next to be held by March 1998); results—percent of vote by party NA; seats—(60 total) PNP 52, JLP 8
Judicial branch: Supreme Court
Political parties and leaders: People's National Party (PNP) P. J. PATTERSON; Jamaica Labor Party (JLP), Edward SEAGA
Other political or pressure groups: Rastafarians (black religious/racial cultists, pan-Africanists); New Beginnings Movement (NBM)
Member of: ACP, C, CARICOM, CCC, CDB, ECLAC, FAO, G-15, G-19, G-77, GATT, IADB, IAEA, IBRD, ICAO, ICFTU, ICRM, IFAD, IFC, IFRCS, ILO, IMF, IMO, INTELSAT, INTERPOL, IOC, ISO, ITU, LAES, NAM, OAS, OPANAL, UN, UNCTAD, UNESCO, UNIDO, UNITAR, UPU, WCL, WFTU, WHO, WIPO, WMO, WTO
Diplomatic representation in US:
chief of mission: Ambassador Richard Leighton BERNAL
chancery: 1520 New Hampshire Avenue NW, Washington, DC 20036
telephone: [1] (202) 452-0660
FAX: [1] (202) 452-0081
consulate(s) general: Miami and New York
US diplomatic representation:
chief of mission: Ambassador J. Gary COOPER (since October 1994)
embassy: Jamaica Mutual Life Center, 2 Oxford Road, 3rd floor, Kingston
mailing address: use embassy street address
telephone: [1] (809) 929-4850 through 4859
FAX: [1] (809) 926-6743
Flag: diagonal yellow cross divides the flag into four triangles—green (top and bottom) and black (hoist side and fly side)

Economy

Overview: Key sectors in this island economy are bauxite (alumina and bauxite account for more than half of exports) and tourism. The government's tight fiscal and monetary

policies, which have been partially successful in curbing inflation, have held growth to 1.2% in 1993 and 2.0% in 1994.
National product: GDP—purchasing power parity—$7.8 billion (1994 est.)
National product real growth rate: 2% (1994 est.)
National product per capita: $3,050 (1994 est.)
Inflation rate (consumer prices): 26.7% (1994)
Unemployment rate: 15.7% (1992)
Budget:
revenues: $600 million
expenditures: $736 million, including capital expenditures of $NA (FY90/91 est.)
Exports: $1.2 billion (f.o.b., 1994 est.)
commodities: alumina, bauxite, sugar, bananas, rum
partners: US 47%, UK 11%, Canada 9%, Norway 7%; France 4% (1993)
Imports: $2.2 billion (f.o.b., 1994 est.)
commodities: machinery and transport equipment, construction materials, fuel, food, chemicals
partners: US 54%, Japan 4.0%, Mexico 6%, UK 4%, Venezuela 3% (1993)
External debt: $3.6 billion (1994 est.)
Industrial production: growth rate 0.4% (1992); accounts for almost 30% of GDP
Electricity:
capacity: 730,000 kW
production: 2.6 billion kWh
consumption per capita: 988 kWh (1993)
Industries: bauxite mining, tourism, textiles, food processing, light manufactures
Agriculture: accounts for about 7% of GDP, 22% of work force, and 17% of exports; commercial crops—sugarcane, bananas, coffee, citrus, potatoes, vegetables; livestock and livestock products include poultry, goats, milk; not self-sufficient in grain, meat, and dairy products
Illicit drugs: transshipment point for cocaine from Central and South America to North America and Europe; illicit cultivation of cannabis; government has an active cannabis eradication program
Economic aid:
recipient: US commitments, including Ex-Im (FY70-89), $1.2 billion; other countries, ODA and OOF bilateral commitments (1970-89), $1.6 billion
Currency: 1 Jamaican dollar (J$) = 100 cents
Exchange rates: Jamaican dollars (J$) per US$1—33.195 (December 1994), 33.986 (1994), 24.949 (1993), 22.960 (1992), 12.116 (1991), 7.184 (1990)
Fiscal year: 1 April—31 March

Transportation

Railroads:
total: 370 km
standard gauge: 370 km 1.435-m gauge
Highways:
total: 18,200 km
paved: 12,600 km
unpaved: gravel 3,200 km; improved earth 2,400 km
Pipelines: petroleum products 10 km
Ports: Alligator Pond, Discovery Bay, Kingston, Montego Bay, Ocho Rios, Port Antonio, Longs Wharf, Rocky Point
Merchant marine:
total: 3 ships (1,000 GRT or over) totaling 5,931 GRT/10,545 DWT
ships by type: bulk 1, oil tanker 1, roll-on/roll-off cargo 1
Airports:
total: 41
with paved runways 2,438 to 3,047 m: 2
with paved runways 1,524 to 2,437 m: 1
with paved runways 914 to 1,523 m: 3
with paved runways under 914 m: 31
with unpaved runways 914 to 1,523 m: 4

Communications

Telephone system: 127,000 telephones; fully automatic domestic telephone network
local: NA
intercity: NA
international: 2 INTELSAT (Atlantic Ocean) earth stations; 3 coaxial submarine cables
Radio:
broadcast stations: AM 10, FM 17, shortwave 0
radios: NA
Television:
broadcast stations: 8
televisions: NA

Defense Forces

Branches: Jamaica Defense Force (includes Ground Forces, Coast Guard and Air Wing), Jamaica Constabulary Force
Manpower availability: males age 15-49 670,958; males fit for military service 475,235; males reach military age (18) annually 26,244 (1995 est.)
Defense expenditures: exchange rate conversion—$19.3 million, 1% of GDP (FY91/92)

Jan Mayen
(territory of Norway)

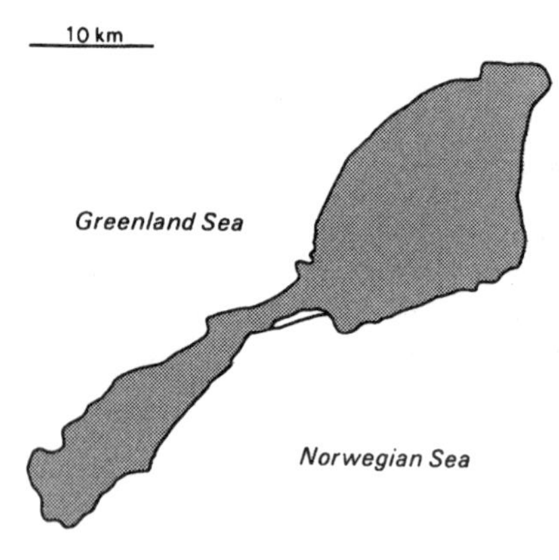

Geography

Location: Northern Europe, island between the Greenland Sea and the Norwegian Sea, northeast of Iceland
Map references: Arctic Region
Area:
total area: 373 sq km
land area: 373 sq km
comparative area: slightly more than twice the size of Washington, DC
Land boundaries: 0 km
Coastline: 124.1 km
Maritime claims:
contiguous zone: 10 nm
continental shelf: 200-m depth or to depth of exploitation
exclusive economic zone: 200 nm
territorial sea: 4 nm
International disputes: none
Climate: arctic maritime with frequent storms and persistent fog
Terrain: volcanic island, partly covered by glaciers; Beerenberg is the highest peak, with an elevation of 2,277 meters
Natural resources: none
Land use:
arable land: 0%
permanent crops: 0%
meadows and pastures: 0%
forest and woodland: 0%
other: 100%
Irrigated land: 0 sq km
Environment:
current issues: NA
natural hazards: dominated by the volcano Beerenberg; volcanic activity resumed in 1970
international agreements: NA
Note: barren volcanic island with some moss and grass

People

Population: no permanent inhabitants; note—there are personnel who man the LORAN C base and the weather and coastal services radio station

Government

Names:
conventional long form: none
conventional short form: Jan Mayen
Digraph: JN
Type: territory of Norway
Capital: none; administered from Oslo, Norway, through a governor (sysselmann) resident in Longyearbyen (Svalbard)
Independence: none (territory of Norway)

Economy

Overview: Jan Mayen is a volcanic island with no exploitable natural resources. Economic activity is limited to providing services for employees of Norway's radio and meteorological stations located on the island.
Electricity:
capacity: 15,000 kW
production: 40 million kWh
consumption per capita: NA kWh (1992)

Transportation

Highways:
total: NA
paved: NA
unpaved: NA
Ports: none; offshore anchorage only
Airports:
total: 1
with unpaved runways 914 to 1,523 m: 1

Communications

Telephone system: NA telephones
local: NA
intercity: NA
international: NA
Radio:
broadcast stations: AM NA, FM NA, shortwave NA
radios: NA
note: radio and meteorological station
Television:
broadcast stations: NA
televisions: NA

Defense Forces

Note: defense is the responsibility of Norway

Japan

Geography

Location: Eastern Asia, island chain between the North Pacific Ocean and the Sea of Japan, east of the Korean peninsula
Map references: Asia
Area:
total area: 377,835 sq km
land area: 374,744 sq km
comparative area: slightly smaller than California
note: includes Bonin Islands (Ogasawara-gunto), Daito-shoto, Minami-jima, Okinotori-shima, Ryukyu Islands (Nansei-shoto), and Volcano Islands (Kazan-retto)
Land boundaries: 0 km
Coastline: 29,751 km
Maritime claims:
exclusive fishing zone: 200 nm
territorial sea: 12 nm; 3 nm in the international straits—La Perouse or Soya, Tsugaru, Osumi, and Eastern and Western Channels of the Korea or Tsushima Strait
International disputes: islands of Etorofu, Kunashiri, Shikotan, and the Habomai group occupied by the Soviet Union in 1945, now administered by Russia, claimed by Japan; Liancourt Rocks disputed with South Korea; Senkaku-shoto (Senkaku Islands) claimed by China and Taiwan
Climate: varies from tropical in south to cool temperate in north
Terrain: mostly rugged and mountainous
Natural resources: negligible mineral resources, fish
Land use:
arable land: 13%
permanent crops: 1%
meadows and pastures: 1%
forest and woodland: 67%
other: 18%
Irrigated land: 28,680 sq km (1989)
Environment:
current issues: air pollution from power plant emissions results in acid rain; acidification of lakes and reservoirs degrading water quality and threatening aquatic life; Japan's appetite for fish and tropical timber is contributing to the depletion of these resources in Asia and elsewhere
natural hazards: many dormant and some active volcanoes; about 1,500 seismic occurrences (mostly tremors) every year; tsunamis
international agreements: party to—Antarctic Treaty, Biodiversity, Climate Change, Endangered Species, Environmental Modification, Hazardous Wastes, Marine Dumping, Nuclear Test Ban, Ozone Layer Protection, Ship Pollution, Tropical Timber 83, Tropical Timber 94, Wetlands, Whaling; signed, but not ratified—Antarctic-Environmental Protocol, Desertification, Law of the Sea
Note: strategic location in northeast Asia

People

Population: 125,506,492 (July 1995 est.)
Age structure:
0-14 years: 16% (female 9,955,603; male 10,542,973)
15-64 years: 69% (female 43,377,425; male 43,843,645)
65 years and over: 15% (female 10,514,017; male 7,272,829) (July 1995 est.)
Population growth rate: 0.32% (1995 est.)
Birth rate: 10.66 births/1,000 population (1995 est.)
Death rate: 7.46 deaths/1,000 population (1995 est.)
Net migration rate: 0 migrant(s)/1,000 population (1995 est.)
Infant mortality rate: 4.3 deaths/1,000 live births (1995 est.)
Life expectancy at birth:
total population: 79.44 years
male: 76.6 years
female: 82.42 years (1995 est.)
Total fertility rate: 1.56 children born/woman (1995 est.)
Nationality:
noun: Japanese (singular and plural)
adjective: Japanese
Ethnic divisions: Japanese 99.4%, other 0.6% (mostly Korean)
Religions: observe both Shinto and Buddhist 84%, other 16% (including 0.7% Christian)
Languages: Japanese
Literacy: age 15 and over can read and write (1970 est.)
total population: 99%
Labor force: 65.87 million (December 1994)
by occupation: trade and services 54%, manufacturing, mining, and construction 33%, agriculture, forestry, and fishing 7%, government 3% (1988)

Government

Names:
conventional long form: none

conventional short form: Japan
Digraph: JA
Type: constitutional monarchy
Capital: Tokyo
Administrative divisions: 47 prefectures; Aichi, Akita, Aomori, Chiba, Ehime, Fukui, Fukuoka, Fukushima, Gifu, Gumma, Hiroshima, Hokkaido, Hyogo, Ibaraki, Ishikawa, Iwate, Kagawa, Kagoshima, Kanagawa, Kochi, Kumamoto, Kyoto, Mie, Miyagi, Miyazaki, Nagano, Nagasaki, Nara, Niigata, Oita, Okayama, Okinawa, Osaka, Saga, Saitama, Shiga, Shimane, Shizuoka, Tochigi, Tokushima, Tokyo, Tottori, Toyama, Wakayama, Yamagata, Yamaguchi, Yamanashi
Independence: 660 BC (traditional founding by Emperor Jimmu)
National holiday: Birthday of the Emperor, 23 December (1933)
Constitution: 3 May 1947
Legal system: modeled after European civil law system with English-American influence; judicial review of legislative acts in the Supreme Court; accepts compulsory ICJ jurisdiction, with reservations
Suffrage: 20 years of age; universal
Executive branch:
chief of state: Emperor AKIHITO (since 7 January 1989)
head of government: Prime Minister Tomiichi MURAYAMA (since 30 June 1994); Deputy Prime Minister Yohei KONO (since 30 June 1994)
cabinet: Cabinet; appointed by the prime minister
Legislative branch: bicameral Diet (Kokkai) consists of an upper house or House of Councillors and a lower house or House of Representatives
House of Councillors (Sangi-in): half of the members elected every three years to six-year terms; elections last held on 26 July 1992 (next set to be held 23 July 1995); results—percent of vote by party NA; seats—(252 total) LDP 106, SDPJ 73, Komeito 24, DSP 12, JCP 11, JNP 4, others 16, independents 6; note—the distribution of seats as of 1 April 1995 is as follows—LDP 94, SDPJ 68, Heisei-kai 47, Shin Ryokufu-kai 16, JCP 11, others 15, vacant 1
House of Representatives (Shugi-in): all members elected every four years to four-year terms; elections last held on 18 July 1993 (next to be held by 1997); results—percent of vote by party NA; seats—(511 total) LDP 223, SDPJ 70, Shinseito 55, Komeito 51, JNP 35, JCP 15, DSP 15, Sakigake 13, others 4, independents 30; note—the distribution of seats as of 1 April 1995 is as follows—LDP 207, Shinshinto 173, SDPJ 70, Sakigake 21, JCP 15, others 19, vacant 6
Judicial branch: Supreme Court
Political parties and leaders: Liberal Democratic Party (LDP), Yohei KONO, president and Yoshiro MORI, secretary general; Social Democratic Party of Japan (SDPJ), Tomiichi MURAYAMA; Japan Communist Party (JCP), Tetsuzo FUWA, Presidium chairman; Sakigake (Harbinger), Masayoshi TAKEMURA, chairman; Shinshinto (New Frontier Party, NFP), Toshiki KAIFU, chairman and Ichiro OZAWA, secretary general
note: Shinshinto was formed in December 1994 by the merger of Shinseito (Japan Renewal Party, JRP), Komeito (Clean Government Party, CGP), Japan New Party (JNP), Democratic Socialist Party (DSP), and several minor groups; Shin Ryokufu-kai is a parliamentary alliance which exists only in the upper house, it includes remnants of Shinseito, JNP, DSP, and a minor labor group; Heisei-kai is a joint bloc of Shinshinto and Komei members; Komei is a group formed from what remains of Komeito in the upper house
Member of: AfDB, AG (observer), APEC, AsDB, Australia Group, BIS, CCC, CP, EBRD, ESCAP, FAO, G- 2, G- 5, G- 7, G- 8, G-10, GATT, IADB, IAEA, IBRD, ICAO, ICC, ICFTU, ICRM, IDA, IEA, IFAD, IFC, IFRCS, ILO, IMF, IMO, INMARSAT, INTELSAT, INTERPOL, IOC, IOM, ISO, ITU, MTCR, NEA, NSG, OAS (observer), OECD, PCA, UN, UNCTAD, UNESCO, UNHCR, UNIDO, UNITAR, UNOMOZ, UNRWA, UNU, UPU, WFTU, WHO, WIPO, WMO, WTO, ZC
Diplomatic representation in US:
chief of mission: Ambassador Takakazu KURIYAMA
chancery: 2520 Massachusetts Avenue NW, Washington, DC 20008
telephone: [1] (202) 939-6700
FAX: [1] (202) 328-2187
consulate(s) general: Agana (Guam), Anchorage, Atlanta, Boston, Chicago, Detroit, Honolulu, Houston, Kansas City (Missouri), Los Angeles, Miami, New Orleans, New York, Portland (Oregon), San Francisco, and Seattle
consulate(s): Saipan (Northern Mariana Islands)
US diplomatic representation:
chief of mission: Ambassador Walter F. MONDALE
embassy: 10-5, Akasaka 1-chome, Minato-ku (107), Tokyo
mailing address: Unit 45004, Box 258, Tokyo; APO AP 96337-0001
telephone: [81] (3) 3224-5000
FAX: [81] (3) 3505-1862
consulate(s) general: Naha (Okinawa), Osaka-Kobe, Sapporo
consulate(s): Fukuoka, Nagoya
Flag: white with a large red disk (representing the sun without rays) in the center

---

Economy

Overview: Government-industry cooperation, a strong work ethic, mastery of high technology, and a comparatively small defense allocation (roughly 1% of GDP) have helped Japan advance with extraordinary rapidity to the rank of second most powerful economy in the world. Industry, the most important sector of the economy, is heavily dependent on imported raw materials and fuels. Usually self-sufficient in rice, Japan must import about 50% of its requirements of other grain and fodder crops. Japan maintains one of the world's largest fishing fleets and accounts for nearly 15% of the global catch. Overall economic growth has been spectacular: a 10% average in the 1960s, a 5% average in the 1970s and 1980s. Economic growth came to a halt in 1992-93 largely because of contractionary domestic policies intended to wring speculative excesses from the stock and real estate markets. Growth resumed at a 0.6% pace in 1994 largely because of consumer demand. As for foreign trade, the stronger yen and slower global growth are containing export growth. Unemployment and inflation remain remarkably low in comparison with the other industrialized nations. Japan continues to run a huge trade surplus—$121 billion in 1994, roughly the same size as in 1993—which supports extensive investment in foreign assets. Prime Minister MURAYAMA has yet to formalize his government's plans for administrative and economic reform, including reduction in the trade surplus. As leader of a coalition government, he has softened his own socialist positions. The crowding of the habitable land area and the aging of the population are two major long-run problems.
National product: GDP—purchasing power parity—$2.5274 trillion (1994 est.)
National product real growth rate: 0.6% (1994 est.)
National product per capita: $20,200 (1994 est.)
Inflation rate (consumer prices): 0.7% (1994)
Unemployment rate: 2.9% (1994)
Budget:
revenues: $569 billion
expenditures: $671 billion, including capital expenditures (public works only) of about $126 billion (1994 est.)
Exports: $395.5 billion (f.o.b., 1994)
commodities: manufactures 97% (including machinery 46%, motor vehicles 20%, consumer electronics 10%)
partners: Southeast Asia 33%, US 29%, Western Europe 18%, China 5%
Imports: $274.3 billion (c.i.f., 1994)
commodities: manufactures 52%, fossil fuels 20%, foodstuffs and raw materials 28%
partners: Southeast Asia 25%, US 23%, Western Europe 15%, China 9%
External debt: $NA
Industrial production: growth rate 1% (1994); accounts for 30% of GDP
Electricity:
capacity: 205,140,000 kW

production: 840 billion kWh
consumption per capita: 6,262 kWh (1993)
Industries: steel and non-ferrous metallurgy, heavy electrical equipment, construction and mining equipment, motor vehicles and parts, electronic and telecommunication equipment and components, machine tools and automated production systems, locomotives and railroad rolling stock, shipbuilding, chemicals, textiles, food processing
Agriculture: accounts for only 2% of GDP; highly subsidized and protected sector, with crop yields among highest in world; principal crops—rice, sugar beets, vegetables, fruit; animal products include pork, poultry, dairy and eggs; about 50% self-sufficient in food production; shortages of wheat, corn, soybeans; world's largest fish catch of 10 million metric tons in 1991
Economic aid:
donor: ODA and OOF commitments (1970-94), $132 billion
note: ODA outlay of $9.9 billion in 1994 (est.)
Currency: yen (¥)
Exchange rates: yen (¥) per US$1—99.75 (January 1995), 102.21 (1994), 111.20 (1993), 126.65 (1992), 134.71 (1991), 144.79 (1990)
Fiscal year: 1 April—31 March

Transportation

Railroads:
total: 27,327 km (5,724 km double track and multitrack sections)
standard gauge: 2,012 km 1.435-m gauge (2,012 km electrified)
narrow gauge: 25,315 km predominantly 1.067-m gauge (9,038 km electrified) (1987)
Highways:
total: 1,111,974 km
paved: 754,102 km (including 4,869 km of national expressways)
unpaved: gravel, crushed stone, or earth 357,872 km (1991)
Inland waterways: about 1,770 km; seagoing craft ply all coastal inland seas
Pipelines: crude oil 84 km; petroleum products 322 km; natural gas 1,800 km
Ports: Akita, Amagasaki, Chiba, Hachinohe, Hakodate, Higashi-Harima, Himeji, Hiroshima, Kawasaki, Kinuura, Kobe, Kushiro, Mizushima, Moji, Nagoya, Osaka, Sakai, Sakaide, Shimizu, Tokyo, Tomakomai
Merchant marine:
total: 851 ships (1,000 GRT or over) totaling 18,195,386 GRT/27,292,044 DWT
ships by type: bulk 210, cargo 63, chemical tanker 7, combination ore/oil 7, container 41, liquefied gas tanker 41, multifunction large-load carrier 1, oil tanker 264, passenger 10, passenger-cargo 5, refrigerated cargo 48, roll-on/roll-off cargo 43, short-sea passenger 30, specialized tanker 2, vehicle carrier 79
note: Japan owns an additional 1,537 ships (1,000 GRT or over) totaling 45,490,202 DWT that operate under Panamanian, Liberian, Vanuatu, Bahamian, Singaporian, Cypriot, Philippines, Hong Kong, and Maltese registry
Airports:
total: 175
with paved runways over 3,047 m: 6
with paved runways 2,438 to 3,047 m: 31
with paved runways 1,524 to 2,437 m: 36
with paved runways 914 to 1,523 m: 30
with paved runways under 914 m: 70
with unpaved runways 914 to 1,523 m: 2

Communications

Telephone system: 64,000,000 telephones; excellent domestic and international service
local: NA
intercity: NA
international: 5 INTELSAT (4 Pacific Ocean and 1 Indian Ocean) earth stations; submarine cables to US (via Guam), Philippines, China, and Russia
Radio:
broadcast stations: AM 318, FM 58, shortwave 0
radios: 95 million
Television:
broadcast stations: 12,350 (1 kW or greater 196)
televisions: 100 million

Defense Forces

Branches: Japan Ground Self-Defense Force (Army), Japan Maritime Self-Defense Force (Navy), Japan Air Self-Defense Force (Air Force)
Manpower availability: males age 15-49 31,947,532; males fit for military service 27,494,758; males reach military age (18) annually 910,970 (1995 est.)
Defense expenditures: exchange rate conversion—$47.2 billion, 1% of GDP (FY95/96)

Jarvis Island

(territory of the US)

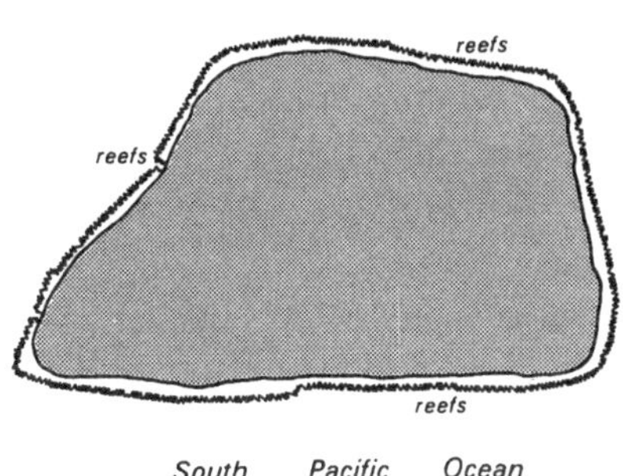

Geography

Location: Oceania, island in the South Pacific Ocean, about one-half of the way from Hawaii to the Cook Islands
Map references: Oceania
Area:
total area: 4.5 sq km
land area: 4.5 sq km
comparative area: about 7.5 times the size of The Mall in Washington, DC
Land boundaries: 0 km
Coastline: 8 km
Maritime claims:
exclusive economic zone: 200 nm
territorial sea: 12 nm
International disputes: none
Climate: tropical; scant rainfall, constant wind, burning sun
Terrain: sandy, coral island surrounded by a narrow fringing reef
Natural resources: guano (deposits worked until late 1800s)
Land use:
arable land: 0%
permanent crops: 0%
meadows and pastures: 0%
forest and woodland: 0%
other: 100%
Irrigated land: 0 sq km
Environment:
current issues: no natural fresh water resources
natural hazards: the narrow fringing reef surrounding the island can be a maritime hazard
international agreements: NA
Note: sparse bunch grass, prostrate vines, and low-growing shrubs; primarily a nesting, roosting, and foraging habitat for seabirds, shorebirds, and marine wildlife; feral cats

People

Population: uninhabited; note—Millersville settlement on western side of island occasionally used as a weather station from

Jarvis Island *(continued)*

1935 until World War II, when it was abandoned; reoccupied in 1957 during the International Geophysical Year by scientists who left in 1958; public entry is by special-use permit only and generally restricted to scientists and educators

Government

Names:
conventional long form: none
conventional short form: Jarvis Island
Digraph: DQ
Type: unincorporated territory of the US administered by the Fish and Wildlife Service of the US Department of the Interior as part of the National Wildlife Refuge System
Capital: none; administered from Washington, DC

Economy

Overview: no economic activity

Transportation

Ports: none; offshore anchorage only; note—there is one boat landing area in the middle of the west coast and another near the southwest corner of the island
Note: there is a day beacon near the middle of the west coast

Defense Forces

Note: defense is the responsibility of the US; visited annually by the US Coast Guard

Jersey
(British crown dependency)

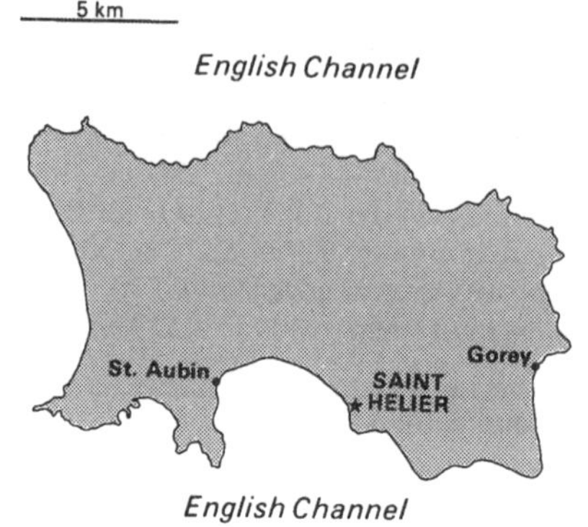

Geography

Location: Western Europe, island in the English Channel, northwest of France
Map references: Europe
Area:
total area: 117 sq km
land area: 117 sq km
comparative area: about 0.7 times the size of Washington, DC
Land boundaries: 0 km
Coastline: 70 km
Maritime claims:
exclusive fishing zone: 200 nm
territorial sea: 3 nm
International disputes: none
Climate: temperate; mild winters and cool summers
Terrain: gently rolling plain with low, rugged hills along north coast
Natural resources: agricultural land
Land use:
arable land: 57%
permanent crops: NA%
meadows and pastures: NA%
forest and woodland: NA%
other: NA%
Irrigated land: NA sq km
Environment:
current issues: NA
natural hazards: NA
international agreements: NA
Note: largest and southernmost of Channel Islands; about 30% of population concentrated in Saint Helier

People

Population: 86,649 (July 1995 est.)
Age structure:
0-14 years: 17% (female 7,029; male 7,450)
15-64 years: 69% (female 30,156; male 29,916)
65 years and over: 14% (female 7,202; male 4,896) (July 1995 est.)
Population growth rate: 0.7% (1995 est.)
Birth rate: 12.83 births/1,000 population (1995 est.)
Death rate: 9.97 deaths/1,000 population (1995 est.)
Net migration rate: 4.11 migrant(s)/1,000 population (1995 est.)
Infant mortality rate: 4.6 deaths/1,000 live births (1995 est.)
Life expectancy at birth:
total population: 76.9 years
male: 73.81 years
female: 80.32 years (1995 est.)
Total fertility rate: 1.44 children born/woman (1995 est.)
Nationality:
noun: Channel Islander(s)
adjective: Channel Islander
Ethnic divisions: UK and Norman-French descent
Religions: Anglican, Roman Catholic, Baptist, Congregational New Church, Methodist, Presbyterian
Languages: English (official), French (official), Norman-French dialect spoken in country districts
Literacy: NA%
Labor force: NA

Government

Names:
conventional long form: Bailiwick of Jersey
conventional short form: Jersey
Digraph: JE
Type: British crown dependency
Capital: Saint Helier
Administrative divisions: none (British crown dependency)
Independence: none (British crown dependency)
National holiday: Liberation Day, 9 May (1945)
Constitution: unwritten; partly statutes, partly common law and practice
Legal system: English law and local statute
Suffrage: NA years of age; universal adult
Executive branch:
Chief of State: Queen ELIZABETH II (since 6 February 1952)
Head of Government: Lieutenant Governor and Commander in Chief Air Marshal Sir John SUTTON (since NA 1990); Bailiff Sir Peter L. CRILL (since NA)
cabinet: committees; appointed by the States
Legislative branch: unicameral
Assembly of the States: elections last held NA (next to be held NA); results—no percent of vote by party since all are independents; seats—(56 total, 52 elected) 52 independents
Judicial branch: Royal Court
Political parties and leaders: none; all independents
Member of: none
Diplomatic representation in US: none (British crown dependency)
US diplomatic representation: none (British

crown dependency)
Flag: white with the diagonal red cross of Saint Patrick (patron saint of Ireland) extending to the corners of the flag

Economy

Overview: The economy is based largely on financial services, agriculture, and tourism. Potatoes, cauliflower, tomatoes, and especially flowers are important export crops, shipped mostly to the UK. The Jersey breed of dairy cattle is known worldwide and represents an important export earner. Milk products go to the UK and other EU countries. In 1986 the finance sector overtook tourism as the main contributor to GDP, accounting for 40% of the island's output. In recent years the government has encouraged light industry to locate in Jersey, with the result that an electronics industry has developed alongside the traditional manufacturing of knitwear. All raw material and energy requirements are imported, as well as a large share of Jersey's food needs.
National product: GDP $NA
National product real growth rate: 8% (1987 est.)
National product per capita: $NA
Inflation rate (consumer prices): 8% (1988 est.)
Unemployment rate: NA%
Budget:
revenues: $308 million
expenditures: $284.4 million, including capital expenditures of $NA (1985)
Exports: $NA
commodities: light industrial and electrical goods, foodstuffs, textiles
partners: UK
Imports: $NA
commodities: machinery and transport equipment, manufactured goods, foodstuffs, mineral fuels, chemicals
partners: UK
External debt: $NA
Industrial production: growth rate NA%
Electricity:
capacity: 50,000 kW standby
production: power supplied by France
consumption per capita: NA kWh (1992)
Industries: tourism, banking and finance, dairy
Agriculture: potatoes, cauliflowers, tomatoes; dairy and cattle farming
Economic aid: none
Currency: 1 Jersey pound (£J) = 100 pence
Exchange rates: Jersey pounds (£J) per US$1—0.6250 (January 1995), 0.6529 (1994), 0.6658 (1993), 0.5664 (1992), 0.5652 (1991), 0.5603 (1990); the Jersey pound is at par with the British pound
Fiscal year: 1 April—31 March

Transportation

Railroads: 0 km
Highways:
total: NA
paved: NA
unpaved: NA
Ports: Gorey, Saint Aubin, Saint Helier
Merchant marine: none
Airports:
total: 1
with paved runways 1,524 to 2,437 m: 1

Communications

Telephone system: 63,700 telephones
local: NA
intercity: NA
international: 3 submarine cables
Radio:
broadcast stations: AM 1, FM 0, shortwave 0
radios: NA
Television:
broadcast stations: 1
televisions: NA

Defense Forces

Note: defense is the responsibility of the UK

Johnston Atoll
(territory of the US)

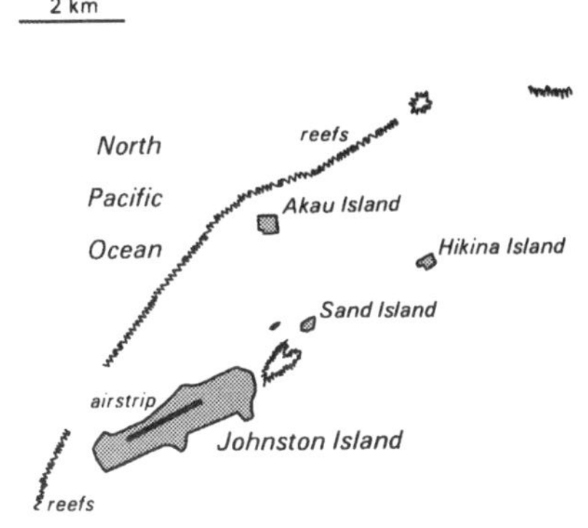

Geography

Location: Oceania, atoll in the North Pacific Ocean, about one-third of the way from Hawaii to the Marshall Islands
Map references: Oceania
Area:
total area: 2.8 sq km
land area: 2.8 sq km
comparative area: about 4.7 times the size of The Mall in Washington, DC
Land boundaries: 0 km
Coastline: 10 km
Maritime claims:
exclusive economic zone: 200 nm
territorial sea: 12 nm
International disputes: none
Climate: tropical, but generally dry; consistent northeast trade winds with little seasonal temperature variation
Terrain: mostly flat with a maximum elevation of 4 meters
Natural resources: guano (deposits worked until about 1890)
Land use:
arable land: 0%
permanent crops: 0%
meadows and pastures: 0%
forest and woodland: 0%
other: 100%
Irrigated land: 0 sq km
Environment:
current issues: no natural fresh water resources
natural hazards: NA
international agreements: NA
Note: strategic location in the North Pacific Ocean; Johnston Island and Sand Island are natural islands; North Island (Akau) and East Island (Hikina) are manmade islands formed from coral dredging; closed to the public; former nuclear weapons test site; site of Johnston Atoll Chemical Agent Disposal System (JACADS); some low-growing vegetation

People

Population: 327 (July 1995 est.)
Population growth rate: 0% (1995 est.)
Birth rate: NA
Death rate: NA
Net migration rate: NA
Infant mortality rate: NA
Life expectancy at birth: NA
Total fertility rate: NA

Government

Names:
conventional long form: none
conventional short form: Johnston Atoll
Digraph: JQ
Type: unincorporated territory of the US administered by the US Defense Nuclear Agency (DNA) and managed cooperatively by DNA and the Fish and Wildlife Service of the US Department of the Interior as part of the National Wildlife Refuge system
Capital: none
Diplomatic representation in US: none (territory of the US)
US diplomatic representation: none (territory of the US)
Flag: the flag of the US is used

Economy

Overview: Economic activity is limited to providing services to US military personnel and contractors located on the island. All food and manufactured goods must be imported.
Electricity: supplied by the management and operations contractor

Transportation

Railroads: 0 km
Highways:
total: NA
paved: NA
unpaved: NA
Ports: Johnston Island
Airports:
total: 1
with paved runways 2,438 to 3,047 m: 1

Communications

Telephone system: NA telephones; excellent system including 60-channel submarine cable, Autodin/SRT terminal, digital telephone switch, Military Affiliated Radio System (MARS station), and UHF/VHF air-ground radio
local: NA
intercity: NA
international: NA
Radio:
broadcast stations: AM NA, FM NA, shortwave NA
radios: NA
Television:
broadcast stations: commercial satellite television system
televisions: NA

Defense Forces

Note: defense is the responsibility of the US

Jordan
(also see separate West Bank entry)

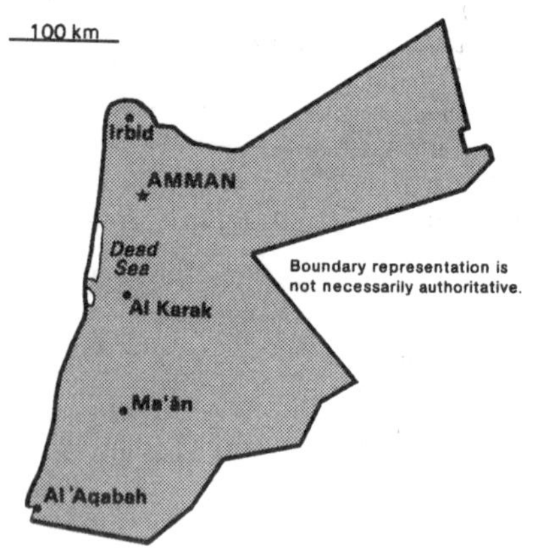

Geography

Location: Middle East, northwest of Saudi Arabia
Map references: Middle East
Area:
total area: 89,213 sq km
land area: 88,884 sq km
comparative area: slightly smaller than Indiana
Land boundaries: total 1,619 km, Iraq 181 km, Israel 238 km, Saudi Arabia 728 km, Syria 375 km, West Bank 97 km
Coastline: 26 km
Maritime claims:
territorial sea: 3 nm
International disputes: none
Climate: mostly arid desert; rainy season in west (November to April)
Terrain: mostly desert plateau in east, highland area in west; Great Rift Valley separates East and West Banks of the Jordan River
Natural resources: phosphates, potash, shale oil
Land use:
arable land: 4%
permanent crops: 0.5%
meadows and pastures: 1%
forest and woodland: 0.5%
other: 94%
Irrigated land: 570 sq km (1989 est.)
Environment:
current issues: limited natural fresh water resources; deforestation; overgrazing; soil erosion; desertification
natural hazards: NA
international agreements: party to—Biodiversity, Climate Change, Endangered Species, Hazardous Wastes, Marine Dumping, Nuclear Test Ban, Ozone Layer Protection, Wetlands

People

Population: 4,100,709 (July 1995 est.)
Age structure:

0-14 years: 44% (female 884,462; male 930,266)
15-64 years: 53% (female 1,058,060; male 1,119,347)
65 years and over: 3% (female 53,709; male 54,865) (July 1995 est.)
Population growth rate: 2.69% (1995 est.)
Birth rate: 37.32 births/1,000 population (1995 est.)
Death rate: 4.02 deaths/1,000 population (1995 est.)
Net migration rate: -6.4 migrant(s)/1,000 population (1995 est.)
Infant mortality rate: 32.3 deaths/1,000 live births (1995 est.)
Life expectancy at birth:
total population: 72.27 years
male: 70.43 years
female: 74.21 years (1995 est.)
Total fertility rate: 5.25 children born/woman (1995 est.)
Nationality:
noun: Jordanian(s)
adjective: Jordanian
Ethnic divisions: Arab 98%, Circassian 1%, Armenian 1%
Religions: Sunni Muslim 92%, Christian 8%
Languages: Arabic (official), English widely understood among upper and middle classes
Literacy: age 15 and over can read and write (1991)
total population: 83%
male: 91%
female: 75%
Labor force: 600,000 (1992)
by occupation: industry 11.4%, commerce, restaurants, and hotels 10.5%, construction 10.0%, transport and communications 8.7%, agriculture 7.4%, other services 52.0% (1992)

Government

Names:
conventional long form: Hashemite Kingdom of Jordan
conventional short form: Jordan
local long form: Al Mamlakah al Urduniyah al Hashimiyah
local short form: Al Urdun
former: Transjordan
Digraph: JO
Type: constitutional monarchy
Capital: Amman
Administrative divisions: 8 governorates (muhafazat, singular—muhafazah); Al Balqa', Al Karak, Al Mafraq, 'Amman, At Tafilah, Az Zarqa', Irbid, Ma'an
Independence: 25 May 1946 (from League of Nations mandate under British administration)
National holiday: Independence Day, 25 May (1946)
Constitution: 8 January 1952
Legal system: based on Islamic law and French codes; judicial review of legislative acts in a specially provided High Tribunal; has not accepted compulsory ICJ jurisdiction
Suffrage: 20 years of age; universal
Executive branch:
chief of state: King HUSSEIN Bin Talal Al Hashimi (since 11 August 1952)
head of government: Prime Minister Zayd BIN SHAKIR (since 8 January 1995)
cabinet: Cabinet appointed by the monarch
Legislative branch: bicameral National Assembly (Majlis al-'Umma)
House of Notables (Majlis al-A'ayan): consists of a 40-member body appointed by the king from designated categories of public figures
House of Representatives: elections last held 8 November 1993 (next to be held NA November 1997); results—percent of vote by party NA; seats—(80 total) Muslim Brotherhood (fundamentalist) 16, Independent Islamic bloc (generally traditionalist) 6, Radical leftist 3, pro-government 55
note: the House of Representatives has been convened and dissolved by the King several times since 1974 and in November 1989 the first parliamentary elections in 22 years were held
Judicial branch: Court of Cassation
Political parties and leaders: Al-'Ahd (Pledge) Party, Sec. Gen. 'Abd al-Hadi al-MAJALI; Al-Ahrar (Liberals) Party, Sec. Gen. Ahmad al-ZU'BI; Al-Hurriyah (Freedom) Party, Sec. Gen. Fawwaz al-ZUBI; Al-Watan (Homeland) Party, leader 'Akif al-FAYIZ; Al-Yaqazah (Awakening) Party, Sec. Gen. 'Abd al-Ra'uf al-RAWABIDAH; Constitutional Jordanian Arab Front Party, leader Milhim al-TALL; Democratic Arab Islamic Movement Party-Du'a', Sec. Gen. Yusuf Abu BAKR; Democratic Arab Unionist Party-Wad, Sec. Gen. Anis al-MU'ASHIR; Islamic Action Front (IAF), Sec. Gen. Ishaq al-FARHAN; Jordanian Arab Democratic Party, Sec. Gen. Mu'nis al-RAZZAZ; Jordanian Arab Masses Party, Sec. Gen. 'Abd al-Khaliq SHATAT; Jordanian Arab Socialist Ba'th Party, Command First Secretary Taysir al-HIMSI; Jordanian Communist Party (JCP), Sec. Gen. Ya'qub ZAYADIN; Jordanian Democratic Popular Unity Party, Sec. Gen. 'Azmi al-KHAWAJA; Jordanian Democratic Progressive Party, Sec. Gen. 'Ali 'AMIR; Jordanian National Alliance Party, Sec. Gen. Mijhim al-KHURAYSHAH; Jordanian People's Democratic Party-Hashd, Sec. Gen. Taysir al-ZIBRI; Jordanian Socialist Democratic Party, Sec. Gen. 'Isa MADANAT; Pan-Arab Action Front Party, Sec. Gen. Muhammad al-ZU'BI; Popular Unity Party-the Unionists, Sec. Gen. Talal al-RAMAHI; Progress and Justice Party, Sec. Gen. 'Ali al-SA'D; Progressive Arab Ba'th Party, Command Secretary Mahmud al-MA'AYITAH; Al-Mustaqbal (Future) Party, Sec. Gen. Sulayman 'ARAR
Member of: ABEDA, ACC, AFESD, AL, AMF, CAEU, CCC, ESCWA, FAO, G-77, IAEA, IBRD, ICAO, ICC, ICRM, IDA, IDB, IFAD, IFC, IFRCS, ILO, IMF, IMO, INTELSAT, INTERPOL, IOC, IOM (observer), ISO (correspondent), ITU, NAM, OIC, PCA, UN, UNAVEM II, UNCTAD, UNESCO, UNIDO, UNOMIL, UNOMOZ, UNPROFOR, UNRWA, UPU, WFTU, WHO, WIPO, WMO, WTO
Diplomatic representation in US:
chief of mission: Ambassador Fayiz A. TARAWNEH
chancery: 3504 International Drive NW, Washington, DC 20008
telephone: [1] (202) 966-2664
FAX: [1] (202) 966-3110
US diplomatic representation:
chief of mission: Ambassador Wesley E. EGAN, Jr.
embassy: Jabel Amman, Amman
mailing address: P. O. Box 354, Amman 11118 Jordan; APO AE 09892-0200
telephone: [962] (6) 820101
FAX: [962] (6) 820159
Flag: three equal horizontal bands of black (top), white, and green with a red isosceles triangle based on the hoist side bearing a small white seven-pointed star; the seven points on the star represent the seven fundamental laws of the Koran

Economy

Overview: Jordan benefited from increased Arab aid during the oil boom of the late 1970s and early 1980s, when its annual real GNP growth averaged more than 10%. In the remainder of the 1980s, however, reductions in both Arab aid and worker remittances slowed real economic growth to an average of roughly 2% per year. Imports—mainly oil, capital goods, consumer durables, and food outstripped exports, with the difference covered by aid, remittances, and borrowing. In mid-1989, the Jordanian Government began debt-rescheduling negotiations and agreed to implement an IMF-supported program designed to gradually reduce the budget deficit and implement badly needed structural reforms. The Persian Gulf crisis that began in August 1990, however, aggravated Jordan's already serious economic problems, forcing the government to shelve the IMF program, stop most debt payments, and suspend rescheduling negotiations. Aid from Gulf Arab states, worker remittances, and trade contracted; and refugees flooded the country, producing serious balance-of-payments problems, stunting GDP growth, and straining government resources. The economy rebounded in 1992, largely due to the influx of capital repatriated by workers returning from the Gulf, but the recovery was uneven throughout 1994. The government is implementing the reform program adopted in 1992 and continues to secure rescheduling and

write-offs of its heavy foreign debt. Debt, poverty, and unemployment remain Jordan's biggest on-going problems.
National product: GDP—purchasing power parity—$17 billion (1994 est.)
National product real growth rate: 5.5% (1994 est.)
National product per capita: $4,280 (1994 est.)
Inflation rate (consumer prices): 6% (1994 est.)
Unemployment rate: 16% (1994 est.)
Budget:
revenues: $2 billion
expenditures: $2.4 billion, including capital expenditures of $630 million (1995 est.)
Exports: $1.4 billion (f.o.b., 1994)
commodities: phosphates, fertilizers, potash, agricultural products, manufactures
partners: India, Iraq, Saudi Arabia, EU, Indonesia, UAE
Imports: $3.5 billion (c.i.f., 1994)
commodities: crude oil, machinery, transport equipment, food, live animals, manufactured goods
partners: EU, US, Iraq, Japan, Turkey
External debt: $6 billion (March 1995 est.)
Industrial production: growth rate 3% (1993 est.); accounts for 20% of GDP
Electricity:
capacity: 1,050,000 kW
production: 4.2 billion kWh
consumption per capita: 1,072 kWh (1993)
Industries: phosphate mining, petroleum refining, cement, potash, light manufacturing
Agriculture: accounts for about 8% of GDP; wheat, barley, citrus fruit, tomatoes, melons, olives; sheep, goats, poultry; large net importer of food
Economic aid:
recipient: US commitments, including Ex-Im (FY70-89), $1.7 billion; Western (non-US) countries, ODA and OOF bilateral commitments (1970-89), $1.5 billion; OPEC bilateral aid (1979-89), $9.5 billion; Communist countries (1970-89), $44 million
Currency: 1 Jordanian dinar (JD) = 1,000 fils
Exchange rates: Jordanian dinars (JD) per US$1—0.6994 (January 1995), 0.5987 (1994), 0.6928 (1993), 0.6797 (1992), 0.6808 (1991), 0.6636 (1990)
Fiscal year: calendar year

Transportation

Railroads:
total: 789 km
narrow gauge: 789 km 1.050-m gauge
Highways:
total: 7,500 km
paved: asphalt 5,500 km
unpaved: gravel, crushed stone 2,000 km
Pipelines: crude oil 209 km
Ports: Al'Aqabah
Merchant marine:
total: 2 ships (1,000 GRT or over) totaling 61,678 GRT/113,080 DWT
ships by type: bulk 1, oil tanker 1
Airports:
total: 17
with paved runways over 3,047 m: 9
with paved runways 2,438 to 3,047 m: 4
with paved runways 914 to 1,523 m: 1
with paved runways under 914 m: 2
with unpaved runways 914 to 1,523 m: 1

Communications

Telephone system: 81,500 telephones; adequate telephone system
local: NA
intercity: microwave, cable, and radio links
international: 2 INTELSAT (1 Atlantic Ocean and 1 Indian Ocean), 1 ARABSAT earth station, coaxial cable and microwave to Iraq, Saudi Arabia, and Syria; microwave link to Lebanon is inactive; participant in MEDARABTEL, a microwave radio relay network linking Syria, Jordan, Egypt, Libya, Tunisia, Algeria, and Morocco
Radio:
broadcast stations: AM 5, FM 7, shortwave 0
radios: NA
Television:
broadcast stations: 8 and 1 TV receive-only satellite link
televisions: NA

Defense Forces

Branches: Jordanian Armed Forces (JAF; includes Royal Jordanian Land Force, Royal Naval Force, and Royal Jordanian Air Force); Ministry of the Interior's Public Security Force (falls under JAF only in wartime or crisis situations)
Manpower availability: males age 15-49 981,004; males fit for military service 699,891; males reach military age (18) annually 45,494 (1995 est.)
Defense expenditures: exchange rate conversion—$564.2 million, 9.1% of GDP (1995 est.)

Juan de Nova Island
(possession of France)

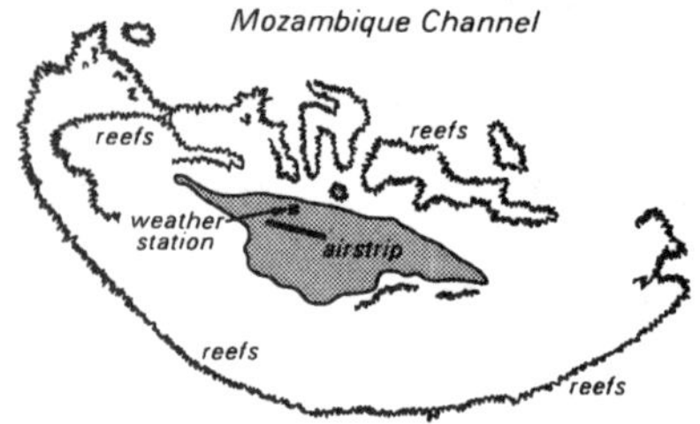

Geography

Location: Southern Africa, island in the Mozambique Channel, about one-third of the way between Madagascar and Mozambique
Map references: Africa
Area:
total area: 4.4 sq km
land area: 4.4 sq km
comparative area: about 7.5 times the size of The Mall in Washington, DC
Land boundaries: 0 km
Coastline: 24.1 km
Maritime claims:
contiguous zone: 12 nm
continental shelf: 200-m depth or to depth of exploitation
exclusive economic zone: 200 nm
territorial sea: 12 nm
International disputes: claimed by Madagascar
Climate: tropical
Terrain: NA
Natural resources: guano deposits and other fertilizers
Land use:
arable land: 0%
permanent crops: 0%
meadows and pastures: 0%
forest and woodland: 90%
other: 10%
Irrigated land: 0 sq km
Environment:
current issues: NA
natural hazards: periodic cyclones
international agreements: NA
Note: wildlife sanctuary

People

Population: uninhabited

Government

Names:
conventional long form: none

conventional short form: Juan de Nova Island
local long form: none
local short form: Ile Juan de Nova
Digraph: JU
Type: French possession administered by Commissioner of the Republic, resident in Reunion
Capital: none; administered by France from Reunion
Independence: none (possession of France)

Economy

Overview: no economic activity

Transportation

Railroads:
total: NA km; short line going to a jetty
Ports: none; offshore anchorage only
Airports:
total: 1
with unpaved runways 914 to 1,523 m: 1

Defense Forces

Note: defense is the responsibility of France

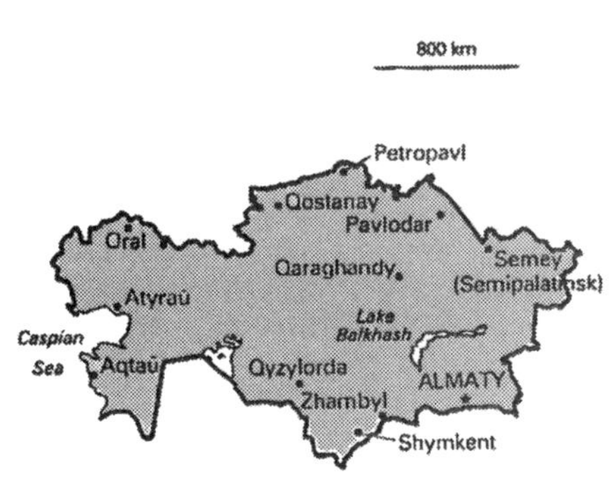

Geography

Location: Central Asia, northwest of China
Map references: Commonwealth of Independent States—Central Asian States
Area:
total area: 2,717,300 sq km
land area: 2,669,800 sq km
comparative area: slightly less than four times the size of Texas
Land boundaries: total 12,012 km, China 1,533 km, Kyrgyzstan 1,051 km, Russia 6,846 km, Turkmenistan 379 km, Uzbekistan 2,203 km
Coastline: 0 km (landlocked)
note: Kazakhstan borders the Aral Sea (1,015 km) and the Caspian Sea (1,894 km)
Maritime claims: none; landlocked
International disputes: Caspian Sea boundaries are not yet determined
Climate: continental, cold winters and hot summers, arid and semiarid
Terrain: extends from the Volga to the Altai Mountains and from the plains in western Siberia to oasis and desert in Central Asia
Natural resources: major deposits of petroleum, coal, iron ore, manganese, chrome ore, nickel, cobalt, copper, molybdenum, lead, zinc, bauxite, gold, uranium
Land use:
arable land: 15%
permanent crops: NEGL%
meadows and pastures: 57%
forest and woodland: 4%
other: 24%
Irrigated land: 23,080 sq km (1990)
Environment:
current issues: radioactive or toxic chemical sites associated with its former defense industries and test ranges are found throughout the country and pose health risks for humans and animals; industrial pollution is severe in some cities; because the two main rivers which flowed into the Aral Sea have been diverted for irrigation, it is drying up and leaving behind a harmful layer of chemical pesticides and natural salts; these substances are then picked up by the wind and blown into noxious dust storms; pollution in the Caspian Sea; soil pollution from overuse of agricultural chemicals and salinization from faulty irrigation practices
natural hazards: NA
international agreements: party to—Biodiversity, Ship Pollution; signed, but not ratified—Climate Change, Desertification
Note: landlocked

People

Population: 17,376,615 (July 1995 est.)
Age structure:
0-14 years: 30% (female 2,589,509; male 2,664,952)
15-64 years: 63% (female 5,531,519; male 5,371,563)
65 years and over: 7% (female 820,900; male 398,172) (July 1995 est.)
Population growth rate: 0.62% (1995 est.)
Birth rate: 19.26 births/1,000 population (1995 est.)
Death rate: 7.93 deaths/1,000 population (1995 est.)
Net migration rate: -5.11 migrant(s)/1,000 population (1995 est.)
Infant mortality rate: 40 deaths/1,000 live births (1995 est.)
Life expectancy at birth:
total population: 68.25 years
male: 63.61 years
female: 73.13 years (1995 est.)
Total fertility rate: 2.43 children born/woman (1995 est.)
Nationality:
noun: Kazakhstani(s)
adjective: Kazakhstani
Ethnic divisions: Kazakh (Qazaq) 41.9%, Russian 37%, Ukrainian 5.2%, German 4.7%, Uzbek 2.1%, Tatar 2%, other 7.1% (1991 official data)
Religions: Muslim 47%, Russian Orthodox 44%, Protestant 2%, other 7%
Languages: Kazakh (Qazaqz) official language spoken by over 40% of population, Russian (language of interethnic communication) spoken by two-thirds of population and used in everyday business
Literacy: age 15 and over can read and write (1989)
total population: 98%
male: 99%
female: 96%
Labor force: 7.356 million
by occupation: industry and construction 31%, agriculture and forestry 26%, other 43% (1992)

Government

Names:
conventional long form: Republic of Kazakhstan
conventional short form: Kazakhstan

local long form: Qazaqstan Respublikasy
local short form: none
former: Kazakh Soviet Socialist Republic
Digraph: KZ
Type: republic
Capital: Almaty
Administrative divisions: 19 oblystar (singular—oblys) and 1 city (qalalar, singular—qala)*; Almaty Qalasy*, Almaty Oblysy, Aqmola Oblysy, Aqtobe Oblysy, Atyrau Oblysy, Batys Qazaqstan Oblysy (Oral), Kokshetau Oblysy, Mangghystau Oblysy (Aqtau), Ongtustik Qazaqstan Oblysy (Shymkent), Qaraghandy Oblysy, Qostanay Oblysy, Qyzylorda Oblysy, Pavlodar Oblysy, Semey Oblysy, Shyghys Qazaqstan Oblysy (Oskemen; formerly Ust'-Kamenogorsk), Soltustik Qazaqstan Oblysy (Petropavl), Taldyqorghan Oblysy, Torghay Oblysy, Zhambyl Oblysy, Zhezqazghan Oblysy
note: names in parentheses are administrative centers when name differs from oblys name
Independence: 16 December 1991 (from the Soviet Union)
National holiday: Independence Day, 16 December (1991)
Constitution: adopted 28 January 1993
Legal system: based on civil law system
Suffrage: 18 years of age; universal
Executive branch:
chief of state: President Nursultan NAZARBAYEV (since NA April 1990); Vice President Yerik ASANBAYEV (since 1 December 1991); election last held 1 December 1991 (next to be held NA 1996); results—Nursultan A. NAZARBAYEV ran unopposed; note—NAZARBAYEV has extended his term to the year 2000 by a nationwide referendum held 30 April 1995
head of government: Prime Minister Akezhan KAZHEGELDIN (since 12 October 1994); First Deputy Prime Ministers Nigmatzhan ISINGARIN (since 12 October 1994) and Vitalia METTE (since March 1995)
cabinet: Council of Ministers; appointed by the prime minister
Legislative branch: unicameral
Supreme Council: elections last held 7 March 1994 (next to be held NA 1999); results—percent of vote by party NA; seats—(177 total) Union Peoples' Unity of Kazakhstan 33, Confederation of Trade Unions of the Republic of Kazakhstan 11, Peoples' Congress of Kazakhstan Party 9, Socialist Party of Kazakhstan 8, Peasant Union of the Republic Kazakhstan 4, Social Movement LAD 4, Organization of Veterans 1, Union of Youth of Kazakhstan 1, Democratic Committee for Human Rights 1, Association of Lawyers of Kazakhstan 1, International Public Committee ''Aral-Asia-Kazakhstan'' 1, Congress of Entrepreneurs of Kazakhstan 1, Deputies of the 12th Supreme Soviet 40, independents 62
note: the Supreme Council disbanded 12 March 1995 following a Constitutional Court ruling that the March 1994 elections were invalid
Judicial branch: Supreme Court
Political parties and leaders: People's Unity Party (PUP; was Union of People's Unity), Kuanysh SULTANOV, chairman; People's Congress of Kazakhstan (PCK), Olzhas SULEYMENOV, chairman; Socialist Party of Kazakhstan (SPK; former Communist Party), Yermukhamet YERTYSHBAYEV, co-chairman; Republican Party (Azat), Kamal ORMANTAYEV, chairman; Democratic Progress (Russian) Party, Alexandra DOKUCHAYEVA, chairman; Confederation of Trade Unions of the Republic of Kazakhstan; Peasant Union of the Republic Kazakhstan (KPU); Social Movement LAD, V. MIKHAYLOV, chairman; Union of Youth of Kazakhstan; Democratic Committee for Human Rights; Association of Lawyers of Kazakhstan; International Public Committee ''Aral-Asia-Kazakhstan''; Congress of Entrepreneurs of Kazakhstan; Deputies of the 12th Supreme Soviet; People's Cooperative Party, Umirzak SARSENOV, chairman; Organization of Veterans
Other political or pressure groups: Independent Trade Union Center (Birlesu; an association of independent trade union and business associations), Leonid SOLOMIN, president
Member of: AsDB, CCC, CIS, EBRD, ECO, ESCAP, IAEA, IBRD, ICAO, IDA, IFC, ILO, IMF, IMO, INTELSAT (nonsignatory user), INTERPOL, IOC, ITU, NACC, OIC (observer), OSCE, PFP, UN, UNCTAD, UNESCO, UPU, WHO, WIPO, WMO, WTO
Diplomatic representation in US:
chief of mission: Ambassador Tuleutai S. SULEYMENOV
chancery: (temporary) 3421 Massachusetts Avenue, NW, Washington, DC 20008
telephone: [1] (202) 333-4504 through 4507
FAX: [1] (202) 333-4509
US diplomatic representation:
chief of mission: Ambassador William H. COURTNEY
embassy: 99/97 Furmanova Street, Almaty, Republic of Kazakhstan 480012
mailing address: use embassy street address
telephone: [7] (3272) 63-24-26
FAX: [7] (3272) 63-38-83
Flag: sky blue background representing the endless sky and a gold sun with 32 rays soaring above a golden steppe eagle in the center; on the hoist side is a ''national ornamentation'' in yellow

Economy

Overview: Kazakhstan, the second largest of the former Soviet states in territory, possesses enormous untapped fossil-fuel reserves as well as plentiful supplies of other minerals and metals. It also has considerable agricultural potential with its vast steppe lands accommodating both livestock and grain production. Kazakhstan's industrial sector rests on the extraction and processing of these natural resources and also on a relatively large machine building sector specializing in construction equipment, tractors, agricultural machinery, and some defense items. The breakup of the USSR and the collapse of demand for Kazakhstan's traditional heavy industry products have resulted in a sharp contraction of the economy since 1991, with the steepest annual decline occurring in 1994. The government has pursued a moderate program of economic reform and privatization which is gradually lifting state controls over economic activity and shifting assets into the private sector. Nevertheless, government control over key sectors of the economy remains strong. Sustained economic hardships and continued pressures from industrial elites will make it difficult for the government to sustain its policies of monetary and fiscal discipline which had brought down inflation by the end of 1994. Continued lack of pipeline transportation for expanded oil exports has closed off a likely source of economic recovery.
National product: GDP—purchasing power parity—$55.2 billion (1994 estimate as extrapolated from World Bank estimate for 1992)
National product real growth rate: -25% (1994 est.)
National product per capita: $3,200 (1994 est.)
Inflation rate (consumer prices): 24% per month (1994 est.)
Unemployment rate: 1.1% includes only officially registered unemployed; also large numbers of underemployed workers (1994)
Budget:
revenues: $NA
expenditures: $NA, including capital expenditures of $NA
Exports: $3.1 billion (1994)
commodities: oil, ferrous and nonferrous metals, chemicals, grain, wool, meat, coal
partners: Russia, Ukraine, Uzbekistan
Imports: $3.5 billion (1994)
commodities: machinery and parts, industrial materials, oil and gas
partners: Russia and other former Soviet republics, China
External debt: less than $1 billion debt to Russia
Industrial production: growth rate -28% (1994)
Electricity:
capacity: 17,380,000 kW
production: 65.1 billion kWh
consumption per capita: 3,750 kWh (1994)
Industries: accounts for 26% of net national product; extractive industries (oil, coal, iron ore, manganese, chromite, lead, zinc, copper, titanium, bauxite, gold, silver, phosphates, sulfur), iron and steel, nonferrous metal,

tractors and other agricultural machinery, electric motors, construction materials
Agriculture: accounts for 20% of GDP; employs about 26% of the labor force; grain, mostly spring wheat; meat, cotton, wool
Illicit drugs: illicit cultivation of cannabis and opium poppy; mostly for CIS consumption; limited government eradication program; used as transshipment point for illicit drugs to Western Europe and North America from Southwest Asia
Economic aid:
recipient: approximately $1 billion in foreign loans and credits allocated in 1994; disbursements projected at $700 billion through 1995
Currency: national currency the tenge introduced on 15 November 1993
Exchange rates: tenges per US$1—54 (yearend 1994)
Fiscal year: calendar year

Transportation

Railroads:
total: 14,460 km in common carrier service; does not include industrial lines
broad gauge: 14,460 km 1.520-m gauge (1990)
Highways:
total: 189,000 km
paved and graveled: 108,100 km
unpaved: earth 80,900 km (1990)
Inland waterways: Syrdariya River, Ertis River
Pipelines: crude oil 2,850 km; refined products 1,500 km; natural gas 3,480 km (1992)
Ports: Aqtau (Shevchenko), Atyrau (Gur'yev), Oskemen (Ust-Kamenogorsk), Pavlodar, Semey (Semipalatinsk)
Airports:
total: 352
with paved runways over 3,047 m: 7
with paved runways 2,438 to 3,047 m: 23
with paved runways 1,524 to 2,437 m: 11
with paved runways 914 to 1,523 m: 5
with paved runways under 914 m: 9
with unpaved runways over 3,047 m: 9
with unpaved runways 2,438 to 3,047 m: 8
with unpaved runways 1,524 to 2,438 m: 25
with unpaved runways 914 to 1,523 m: 65
with unpaved runways under 914 m: 190

Communications

Telephone system: 2.2 million telephones; telephone service is poor; about 17 telephones/100 persons in urban areas and 7.6 telephones/100 persons in rural areas; Almaty has 184,000 telephones
local: NA
intercity: land line and microwave radio relay
international: international traffic with other former USSR republics and China carried by landline and microwave, and with other countries by satellite and through 8 international telecommunications circuits at the Moscow international gateway switch; INTELSAT earth station; new satellite ground station established at Almaty with Turkish financial help (December 1992) with 2500 channel band width
Radio:
broadcast stations: AM NA, FM NA, shortwave NA
radios: 4.088 million (with multiple speakers for program diffusion 6,082,000)
Television:
broadcast stations: Orbita (TV receive only)
televisions: 4.75 million

Defense Forces

Branches: Army, Republic National Guard, Republic Security Forces (internal and border troops)
Manpower availability: males age 15-49 4,513,089; males fit for military service 3,605,584; males reach military age (18) annually 154,280 (1995 est.)
Defense expenditures: 69.3 billion rubles, NA% of GDP (forecast for 1993); note—conversion of the military budget into US dollars using the current exchange rate could produce misleading results

Kenya

Geography

Location: Eastern Africa, bordering the Indian Ocean, between Somalia and Tanzania
Map references: Africa
Area:
total area: 582,650 sq km
land area: 569,250 sq km
comparative area: slightly more than twice the size of Nevada
Land boundaries: total 3,446 km, Ethiopia 830 km, Somalia 682 km, Sudan 232 km, Tanzania 769 km, Uganda 933 km
Coastline: 536 km
Maritime claims:
continental shelf: 200-m depth or to the depth of exploitation
exclusive economic zone: 200 nm
territorial sea: 12 nm
International disputes: administrative boundary with Sudan does not coincide with international boundary; possible claim by Somalia based on unification of ethnic Somalis
Climate: varies from tropical along coast to arid in interior
Terrain: low plains rise to central highlands bisected by Great Rift Valley; fertile plateau in west
Natural resources: gold, limestone, soda ash, salt barytes, rubies, fluorspar, garnets, wildlife
Land use:
arable land: 3%
permanent crops: 1%
meadows and pastures: 7%
forest and woodland: 4%
other: 85%
Irrigated land: 520 sq km (1989)
Environment:
current issues: water pollution from urban and industrial wastes; degradation of water quality from increased use of pesticides and fertilizers; deforestation; soil erosion; desertification; poaching
natural hazards: NA
international agreements: party to—Biodiversity, Climate Change, Endangered Species, Law of the Sea, Marine Dumping,

Marine Life Conservation, Nuclear Test Ban, Ozone Layer Protection, Ship Pollution, Wetlands, Whaling; signed, but not ratified—Desertification
Note: the Kenyan Highlands comprise one of the most successful agricultural production regions in Africa; glaciers on Mt. Kenya; unique physiography supports abundant and varied wildlife of scientific and economic value

People

Population: 28,817,227 (July 1995 est.)
Age structure:
0-14 years: 48% (female 6,841,235; male 6,957,908)
15-64 years: 50% (female 7,277,061; male 7,085,925)
65 years and over: 2% (female 359,659; male 295,439) (July 1995 est.)
Population growth rate: 0.99% (1995 est.)
Birth rate: 41.66 births/1,000 population (1995 est.)
Death rate: 12.04 deaths/1,000 population (1995 est.)
Net migration rate: -19.69 migrant(s)/1,000 population (1995 est.)
Infant mortality rate: 73.5 deaths/1,000 live births (1995 est.)
Life expectancy at birth:
total population: 52.41 years
male: 50.72 years
female: 54.16 years (1995 est.)
Total fertility rate: 5.76 children born/woman (1995 est.)
Nationality:
noun: Kenyan(s)
adjective: Kenyan
Ethnic divisions: Kikuyu 22%, Luhya 14%, Luo 13%, Kalenjin 12%, Kamba 11%, Kisii 6%, Meru 6%, Asian, European, and Arab 1%, other 15%
Religions: Protestant (including Anglican) 38%, Roman Catholic 28%, indigenous beliefs 26%, other 2%
Languages: English (official), Swahili (official), numerous indigenous languages
Literacy: age 15 and over can read and write (1989)
total population: 71%
male: 81%
female: 62%
Labor force:
by occupation: agriculture 75%-80% (1993 est.), non-agriculture 20%-25% (1993 est.)

Government

Names:
conventional long form: Republic of Kenya
conventional short form: Kenya
former: British East Africa
Digraph: KE
Type: republic
Capital: Nairobi
Administrative divisions: 7 provinces and 1 area*; Central, Coast, Eastern, Nairobi Area*, North Eastern, Nyanza, Rift Valley, Western
Independence: 12 December 1963 (from UK)
National holiday: Independence Day, 12 December (1963)
Constitution: 12 December 1963, amended as a republic 1964; reissued with amendments 1979, 1983, 1986, 1988, 1991, and 1992
Legal system: based on English common law, tribal law, and Islamic law; judicial review in High Court; accepts compulsory ICJ jurisdiction, with reservations; constitutional amendment of 1982 making Kenya a de jure one-party state repealed in 1991
Suffrage: 18 years of age; universal
Executive branch:
chief of state and head of government: President Daniel Toroitich arap MOI (since 14 October 1978); Vice President George SAITOTI (since 10 May 1989); election last held on 29 December 1992 (next to be held NA 1997); results—President Daniel T. arap MOI was reelected with 37% of the vote; Kenneth Matiba (FORD-ASILI) 26%; Mwai Kibaki (SP) 19%, Oginga Odinga (FORD-Kenya) 17%
cabinet: Cabinet; appointed by the president
Legislative branch: unicameral
National Assembly (Bunge): elections last held on 29 December 1992 (next to be held NA); results—percent of vote by party NA; seats—(188 total) KANU 100, FORD-Kenya 31, FORD-Asili 31, DP 23, smaller parties 3; president nominates 12 additional members
note: first multiparty election since repeal of one-party state law in 1991
Judicial branch: Court of Appeal, High Court
Political parties and leaders: ruling party is Kenya African National Union (KANU), President Daniel Toroitich arap MOI; opposition parties include Forum for the Restoration of Democracy (FORD-Kenya), Michael WAMALWA; Forum for the Restoration of Democracy (FORD-Asili), Kenneth MATIBA; Democratic Party of Kenya (DP), Mwai KIBAKI
Other political or pressure groups: labor unions; Roman Catholic Church
Member of: ESCAP, FAO, G-77, ICAO, ICRM, IFAD, IFRCS, IMO, INTELSAT (nonsignatory user), IOC, ISO, ITU, NAM, UN, UNCTAD, UNESCO, UNIDO, UNOMIL, UNU, UPU, WFTU, WHO, WIPO, WMO, WTO
Diplomatic representation in US:
chief of mission: Benjamin Edgar KIPKORIR
chancery: 2249 R Street NW, Washington, DC 20008
telephone: [1] (202) 387-6101
FAX: [1] (202) 462-3829
consulate(s) general: Los Angeles and New York
US diplomatic representation:
chief of mission: Ambassador Aurelia BRAZEAL
embassy: corner of Moi Avenue and Haile Selassie Avenue, Nairobi
mailing address: P. O. Box 30137, Unit 64100, Nairobi; APO AE 09831
telephone: [254] (2) 334141
FAX: [254] (2) 340838
Flag: three equal horizontal bands of black (top), red, and green; the red band is edged in white; a large warrior's shield covering crossed spears is superimposed at the center

Economy

Overview: Kenya in recent years has had one of the highest natural rates of growth in population, but the statistics have been complicated by the large-scale movement of nomadic groups and of Somalis back and forth across the border. Population growth has been accompanied by deforestation, deterioration in the road system, the water supply, and other parts of the infrastructure. In industry and services, Nairobi's reluctance to embrace IMF-supported reforms had held back investment and growth in 1991-93. Nairobi's push on economic reform in 1994, however, helped support a 3.3% increase in output.
National product: GDP—purchasing power parity—$33.1 billion (1994 est.)
National product real growth rate: 3.3% (1994 est.)
National product per capita: $1,170 (1994 est.)
Inflation rate (consumer prices): 30% (1994 est.)
Unemployment rate: 35% urban (1994 est.)
Budget:
revenues: $2.4 billion
expenditures: $2.8 billion, including capital expenditures of $740 million (1990 est.)
Exports: $1.45 billion (f.o.b., 1994 est.)
commodities: tea 25%, coffee 18%, petroleum products 11% (1990)
partners: EC 47%, Africa 23%, Asia 11%, US 4%, Middle East 3% (1991)
Imports: $1.85 billion (f.o.b., 1994 est.)
commodities: machinery and transportation equipment 29%, petroleum and petroleum products 15%, iron and steel 7%, raw materials, food and consumer goods (1989)
partners: EC 46%, Asia 23%, Middle East 20%, US 5% (1991)
External debt: $7 billion (1994 est.)
Industrial production: growth rate 3.9% (1991 est.); accounts for 14% of GDP
Electricity:
capacity: 810,000 kW
production: 3.3 billion kWh
consumption per capita: 117 kWh (1993)
Industries: small-scale consumer goods (plastic, furniture, batteries, textiles, soap, cigarettes, flour), processing agricultural

products, oil refining, cement, tourism
Agriculture: most important sector, accounting for 27% of GDP and 65% of exports; cash crops—coffee, tea; food products—corn, wheat, sugarcane, fruit, vegetables, dairy products, beef, pork, poultry, eggs
Illicit drugs: widespread harvesting of small, wild plots of marijuana and qat; most locally consumed; transit country for Southwest Asian heroin moving to West Africa and onward to Europe and North America; Indian methaqualone also transits on way to South Africa
Economic aid:
recipient: US commitments, including Ex-Im (FY70-89), $839 million; Western (non-US) countries, ODA and OOF bilateral commitments (1970-89), $7.49 billion; OPEC bilateral aid (1979-89), $74 million; Communist countries (1970-89), $83 million
Currency: 1 Kenyan shilling (KSh) = 100 cents
Exchange rates: Kenyan shillings (KSh) per US$1—44.478 (January 1995), 56.051 (1994), 58.001 (1993), 32.217 (1992), 27.508 (1991), 22.915 (1990)
Fiscal year: 1 July—30 June

Transportation

Railroads:
total: 2,650 km
narrow gauge: 2,650 km 1.000 m gauge
Highways:
total: 64,590 km
paved: 7,000 km
unpaved: gravel 4,150 km; improved earth 53,440 km
Inland waterways: part of Lake Victoria system is within boundaries of Kenya
Pipelines: petroleum products 483 km
Ports: Kisumu, Lamu, Mombasa
Merchant marine:
total: 2 ships (1,000 GRT or over) totaling 4,883 GRT/6,255 DWT
ships by type: barge carrier 1, oil tanker 1
Airports:
total: 246
with paved runways over 3,047 m: 3
with paved runways 2,438 to 3,047 m: 2
with paved runways 1,524 to 2,437 m: 2
with paved runways 914 to 1,523 m: 22
with paved runways under 914 m: 83
with unpaved runways 2,438 to 3,047 m: 1
with unpaved runways 1,524 to 2,438 m: 14
with unpaved runways 914 to 1,523 m: 119

Communications

Telephone system: over 260,000 telephones; in top group of African systems
local: NA
intercity: consists primarily of microwave radio relay links
international: 2 INTELSAT (1 Atlantic Ocean and 1 Indian Ocean) earth stations
Radio:
broadcast stations: AM 16, FM 4, shortwave 0
radios: NA
Television:
broadcast stations: 6
televisions: NA

Defense Forces

Branches: Army, Navy, Air Force, paramilitary General Service Unit of the Police
Manpower availability: males age 15-49 6,358,344; males fit for military service 3,932,506 (1995 est.)
Defense expenditures: exchange rate conversion—$136 million, 1.9% of GDP (FY93/94)

Kingman Reef
(territory of the US)

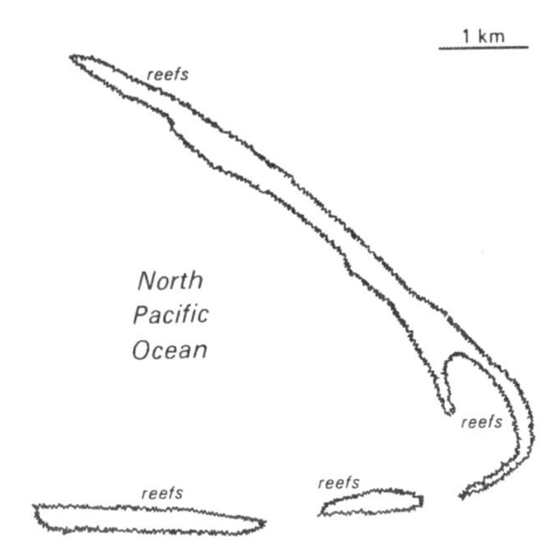

Geography

Location: Oceania, reef in the North Pacific Ocean, about one-half of the way from Hawaii to American Samoa
Map references: Oceania
Area:
total area: 1 sq km
land area: 1 sq km
comparative area: about 1.7 times the size of The Mall in Washington, DC
Land boundaries: 0 km
Coastline: 3 km
Maritime claims:
exclusive economic zone: 200 nm
territorial sea: 12 nm
International disputes: none
Climate: tropical, but moderated by prevailing winds
Terrain: low and nearly level with a maximum elevation of about 1 meter
Natural resources: none
Land use:
arable land: 0%
permanent crops: 0%
meadows and pastures: 0%
forest and woodland: 0%
other: 100%
Irrigated land: 0 sq km
Environment:
current issues: NA
natural hazards: wet or awash most of the time, maximum elevation of about 1 meter makes this a maritime hazard
international agreements: NA
Note: barren coral atoll with deep interior lagoon; closed to the public

People

Population: uninhabited

Government

Names:
conventional long form: none

conventional short form: Kingman Reef
Digraph: KQ
Type: unincorporated territory of the US administered by the US Navy, however it is awash the majority of the time, so it is not usable and is uninhabited
Capital: none; administered from Washington, DC

Economy

Overview: no economic activity

Transportation

Ports: none; offshore anchorage only
Airports: lagoon was used as a halfway station between Hawaii and American Samoa by Pan American Airways for flying boats in 1937 and 1938

Defense Forces

Note: defense is the responsibility of the US

Kiribati

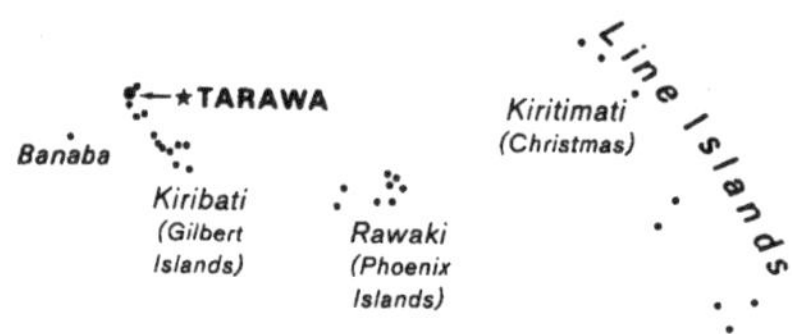

Geography

Location: Oceania, group of islands in the Pacific Ocean, straddling the equator and the International Date Line, about one-half of the way from Hawaii to Australia
Map references: Oceania
Area:
total area: 717 sq km
land area: 717 sq km
comparative area: slightly more than four times the size of Washington, DC
note: includes three island groups—Gilbert Islands, Line Islands, Phoenix Islands
Land boundaries: 0 km
Coastline: 1,143 km
Maritime claims:
exclusive economic zone: 200 nm
territorial sea: 12 nm
International disputes: none
Climate: tropical; marine, hot and humid, moderated by trade winds
Terrain: mostly low-lying coral atolls surrounded by extensive reefs
Natural resources: phosphate (production discontinued in 1979)
Land use:
arable land: 0%
permanent crops: 51%
meadows and pastures: 0%
forest and woodland: 3%
other: 46%
Irrigated land: NA sq km
Environment:
current issues: NA
natural hazards: typhoons can occur any time, but usually November to March; occasional tornadoes
international agreements: party to—Biodiversity, Endangered Species, Marine Dumping, Ozone Layer Protection; signed, but not ratified—Climate Change
Note: 20 of the 33 islands are inhabited; Banaba (Ocean Island) in Kiribati is one of the three great phosphate rock islands in the Pacific Ocean—the others are Makatea in French Polynesia and Nauru

People

Population: 79,386 (July 1995 est.)
Age structure:
0-14 years: NA
15-64 years: NA
65 years and over: NA
Population growth rate: 1.95% (1995 est.)
Birth rate: 31.25 births/1,000 population (1995 est.)
Death rate: 12.31 deaths/1,000 population (1995 est.)
Net migration rate: 0.56 migrant(s)/1,000 population (1995 est.)
Infant mortality rate: 98.4 deaths/1,000 live births (1995 est.)
Life expectancy at birth:
total population: 54.16 years
male: 52.56 years
female: 55.78 years (1995 est.)
Total fertility rate: 3.73 children born/woman (1995 est.)
Nationality:
noun: I-Kiribati (singular and plural)
adjective: I-Kiribati
Ethnic divisions: Micronesian
Religions: Roman Catholic 52.6%, Protestant (Congregational) 40.9%, Seventh-Day Adventist, Baha'i, Church of God, Mormon 6% (1985)
Languages: English (official), Gilbertese
Literacy: NA%
Labor force: 7,870 economically active, not including subsistence farmers (1985 est.)

Government

Names:
conventional long form: Republic of Kiribati
conventional short form: Kiribati
former: Gilbert Islands
Digraph: KR
Type: republic
Capital: Tarawa
Administrative divisions: 3 units; Gilbert Islands, Line Islands, Phoenix Islands
note: in addition, there are 6 districts (Banaba, Central Gilberts, Line Islands, Northern Gilberts, Southern Gilberts, Tarawa) and 21 island councils (Abaiang, Abemama, Aranuka, Arorae, Banaba, Beru, Butaritari, Kanton, Kiritimati, Kuria, Maiana, Makin, Marakei, Nikunau, Nonouti, Onotoa, Tabiteuea, Tabuaeran, Tamana, Tarawa, Teraina; note—one council for each of the inhabited islands)
Independence: 12 July 1979 (from UK)
National holiday: Independence Day, 12 July (1979)
Constitution: 12 July 1979
Legal system: NA
Suffrage: 18 years of age; universal
Executive branch:
chief of state and head of government: President (Beretitenti) Teburoro TITO (since 1 October 1994); Vice President (Kauoman-ni-

Beretitenti) Tewareka TENTOA (since 12 October 1994); election last held on 30 September 1994 (next to be held by NA 1999)
cabinet: Cabinet; appointed by the president from an elected parliament
Legislative branch: unicameral
House of Assembly (Maneaba Ni Maungatabu): elections last held on 22 July 1994 (next to be held by NA 1999); results—percent of vote by party NA; seats—(40 total; 39 elected) Maneaban Te Mauri 13, National Progressive Party 7, independents 19
Judicial branch: Court of Appeal, High Court
Political parties and leaders: National Progressive Party, Teatao TEANNAKI; Christian Democratic Party, Teburoro TITO; New Movement Party, leader NA; Liberal Party, Tewareka TENTOA; Maneaba Party, Roniti TEIWAKI; Maneaban Te Mauri, leader NA
note: there is no tradition of formally organized political parties in Kiribati; they more closely resemble factions or interest groups because they have no party headquarters, formal platforms, or party structures
Member of: ACP, AsDB, C, ESCAP, IBRD, ICAO, ICFTU, IDA, IFC, IFRCS (associate), IMF, INTELSAT (nonsignatory user), INTERPOL, ITU, SPARTECA, SPC, SPF, UNESCO, UPU, WHO
Diplomatic representation in US: Kiribati has no mission in the US
US diplomatic representation: the ambassador to Fiji is accredited to Kiribati
Flag: the upper half is red with a yellow frigate bird flying over a yellow rising sun, and the lower half is blue with three horizontal wavy white stripes to represent the ocean

Economy

Overview: A remote country of 33 scattered coral atolls, Kiribati has few national resources. Commercially viable phosphate deposits were exhausted at the time of independence in 1979. Copra and fish now represent the bulk of production and exports. The economy has fluctuated widely in recent years. Real GDP declined about 5% in 1987, as the fish catch fell sharply to only one-fourth the level of 1986 and copra production was hampered by repeated rains. Output rebounded strongly in 1988, with real GDP growing by 10%. The upturn in economic growth came from an increase in copra production and a good fish catch. GDP then fell by 2.2% in 1989 and by 2.9% in 1990, but has risen by about 3% annually in 1991-93. Foreign financial aid, largely from the UK and Japan, is a critical supplement to GDP, amounting to 25%-50% of GDP in recent years.
National product: GDP—purchasing power parity—$62 million (1993 est.)
National product real growth rate: 2.9% (1993 est.)
National product per capita: $800 (1993 est.)
Inflation rate (consumer prices): 6.5% (1993 est.)
Unemployment rate: 2%; underemployment 70% (1992 est.)
Budget:
revenues: $29.6 million
expenditures: $32.8 million, including capital expenditures of $14 million (1993 est.)
Exports: $4.2 million (f.o.b., 1992 est.)
commodities: copra 50%, seaweed 16%, fish 15%
partners: Denmark, Fiji, US
Imports: $33.1 million (c.i.f., 1992 est.)
commodities: foodstuffs, machinery and equipment, miscellaneous manufactured goods, fuel
partners: Australia 40%, Japan 18%, Fiji 17%, NZ 6%, US 4% (1991)
External debt: $2 million (December 1989 est.)
Industrial production: growth rate 0.7% (1992 est.); accounts for less than 4% of GDP
Electricity:
capacity: 5,000 kW
production: 13 million kWh
consumption per capita: 131 kWh (1993)
Industries: fishing, handicrafts
Agriculture: accounts for 23% of GDP (including fishing); copra and fish contribute about 65% to exports; subsistence farming predominates; food crops—taro, breadfruit, sweet potatoes, vegetables; not self-sufficient in food
Economic aid:
recipient: Western (non-US) countries, ODA and OOF bilateral commitments (1970-89), $273 million
Currency: 1 Australian dollar ($A) = 100 cents
Exchange rates: Australian dollars ($A) per US$1—1.3058 (January 1995), 1.3667 (1994), 1.4704 (1993), 1.3600 (1992), 1.2835 (1991), 1.2799 (1990)
Fiscal year: NA

Transportation

Railroads: 0 km
Highways:
total: 640 km
paved: NA
unpaved: NA
Inland waterways: small network of canals, totaling 5 km, in Line Islands
Ports: Banaba, Betio, English Harbor, Kanton
Merchant marine:
total: 1 passenger-cargo ship (1,000 GRT or over) totaling 1,291 GRT/1,295 DWT
Airports:
total: 21
with paved runways 1,524 to 2,437 m: 4
with paved runways 914 to 1,523 m: 1
with paved runways under 914 m: 5
with unpaved runways 914 to 1,523 m: 11

Communications

Telephone system: 1,400 telephones
local: NA
intercity: NA
international: 1 INTELSAT (Pacific Ocean) earth station
Radio:
broadcast stations: AM 1, FM 0, shortwave 0
radios: NA
Television:
broadcast stations: 0
televisions: NA

Defense Forces

Branches: Police Force (carries out law enforcement functions and paramilitary duties; there are small police posts on all islands); no military force is maintained
Defense expenditures: $NA, NA% of GDP

Korea, North

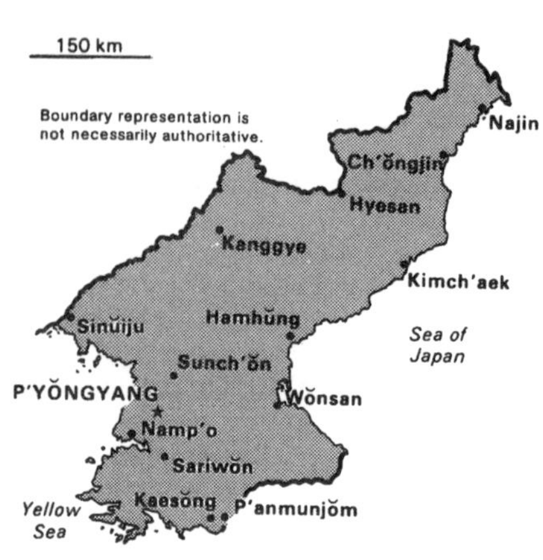

Geography

Location: Eastern Asia, northern half of the Korean peninsula bordering the Korea Bay and the Sea of Japan, between China and Russia
Map references: Asia
Area:
total area: 120,540 sq km
land area: 120,410 sq km
comparative area: slightly smaller than Mississippi
Land boundaries: total 1,673 km, China 1,416 km, South Korea 238 km, Russia 19 km
Coastline: 2,495 km
Maritime claims:
territorial sea: 12 nm
exclusive economic zone: 200 nm
military boundary line: 50 nm in the Sea of Japan and the exclusive economic zone limit in the Yellow Sea where all foreign vessels and aircraft without permission are banned
International disputes: short section of boundary with China is indefinite; Demarcation Line with South Korea
Climate: temperate with rainfall concentrated in summer
Terrain: mostly hills and mountains separated by deep, narrow valleys; coastal plains wide in west, discontinuous in east
Natural resources: coal, lead, tungsten, zinc, graphite, magnesite, iron ore, copper, gold, pyrites, salt, fluorspar, hydropower
Land use:
arable land: 18%
permanent crops: 1%
meadows and pastures: 0%
forest and woodland: 74%
other: 7%
Irrigated land: 14,000 sq km (1989)
Environment:
current issues: localized air pollution attributable to inadequate industrial controls; water pollution; inadequate supplies of potable water
natural hazards: late spring droughts often followed by severe flooding; occasional typhoons during the early fall
international agreements: party to—Antarctic Treaty, Biodiversity, Climate Change, Environmental Modification, Ozone Layer Protection, Ship Pollution; signed, but not ratified—Antarctic-Environmental Protocol, Law of the Sea
Note: strategic location bordering China, South Korea, and Russia; mountainous interior is isolated, nearly inaccessible, and sparsely populated

People

Population: 23,486,550 (July 1995 est.)
Age structure:
0-14 years: 30% (female 3,402,672; male 3,540,313)
15-64 years: 66% (female 7,840,465; male 7,741,155)
65 years and over: 4% (female 622,250; male 339,695) (July 1995 est.)
Population growth rate: 1.78% (1995 est.)
Birth rate: 23.31 births/1,000 population (1995 est.)
Death rate: 5.47 deaths/1,000 population (1995 est.)
Net migration rate: 0 migrant(s)/1,000 population (1995 est.)
Infant mortality rate: 26.8 deaths/1,000 live births (1995 est.)
Life expectancy at birth:
total population: 70.05 years
male: 66.96 years
female: 73.29 years (1995 est.)
Total fertility rate: 2.34 children born/woman (1995 est.)
Nationality:
noun: Korean(s)
adjective: Korean
Ethnic divisions: racially homogeneous
Religions: Buddhism and Confucianism, some Christianity and syncretic Chondogyo
note: autonomous religious activities now almost nonexistent; government-sponsored religious groups exist to provide illusion of religious freedom
Languages: Korean
Literacy: age 15 and over can read and write Korean (1990 est.)
total population: 99%
male: 99%
female: 99%
Labor force: 9.615 million
by occupation: agricultural 36%, nonagricultural 64%
note: shortage of skilled and unskilled labor (mid-1987 est.)

Government

Names:
conventional long form: Democratic People's Republic of Korea
conventional short form: North Korea
local long form: Choson-minjujuui-inmin-konghwaguk
local short form: none
note: the North Koreans generally use the term "Choson" to refer to their country
Abbreviation: DPRK
Digraph: KN
Type: Communist state; Stalinist dictatorship
Capital: P'yongyang
Administrative divisions: 9 provinces (do, singular and plural) and 3 special cities* (jikhalsi, singular and plural); Chagang-do (Chagang Province), Hamgyong-bukto (North Hamgyong Province), Hamgyong-namdo (South Hamgyong Province), Hwanghae-bukto (North Hwanghae Province), Hwanghae-namdo (South Hwanghae Province), Kaesong-si* (Kaesong City), Kangwon-do (Kangwon Province), Namp'o-si* (Namp'o City), P'yongan-bukto (North P'yongan Province), P'yongan-namdo (South P'yongan Province), P'yongyang-si* (P'yongyang City), Yanggang-do (Yanggang Province)
Independence: 9 September 1948
note: 15 August 1945, date of independence from the Japanese and celebrated in North Korea as National Liberation Day
National holiday: DPRK Foundation Day, 9 September (1948)
Constitution: adopted 1948, completely revised 27 December 1972, revised again in April 1992
Legal system: based on German civil law system with Japanese influences and Communist legal theory; no judicial review of legislative acts; has not accepted compulsory ICJ jurisdiction
Suffrage: 17 years of age; universal
Executive branch:
chief of state: KIM Chong-il, is the son of and designated successor to former President KIM Il-song (who died 8 July 1994); formal succession has not yet taken place (January 1995); election last held 24 May 1990 (next to be held by NA); results—President KIM Il-song was reelected without opposition
head of government: Premier KANG Song-san (since December 1992)
cabinet: State Administration Council; appointed by the Supreme People's Assembly
Legislative branch: unicameral
Supreme People's Assembly (Ch'oego Inmin Hoeui): elections last held on 7-9 April 1990 (next to be held NA); results—percent of vote by party NA; seats—(687 total) the KWP approves a single list of candidates who are elected without opposition; minor parties hold a few seats
Judicial branch: Central Court
Political parties and leaders: major party—Korean Workers' Party (KWP), KIM Chong-il, secretary, Central Committee; Korean Social Democratic Party, KIM Pyong-sik, chairman; Chondoist Chongu Party, YU Mi-yong, chairwoman
Member of: ESCAP, FAO, G-77, ICAO,

IFAD, IFRCS, IMO, INTELSAT (nonsignatory user), IOC, ISO, ITU, NAM, UN, UNCTAD, UNESCO, UNIDO, UPU, WFTU, WHO, WIPO, WMO, WTO
Diplomatic representation in US: none
US diplomatic representation: none
Flag: three horizontal bands of blue (top), red (triple width), and blue; the red band is edged in white; on the hoist side of the red band is a white disk with a red five-pointed star

Economy

Overview: More than 90% of this command economy is socialized; agricultural land is collectivized; and state-owned industry produces 95% of manufactured goods. State control of economic affairs is unusually tight even for a Communist country because of the small size and homogeneity of the society and the strict rule of KIM Il-song in the past and now his son, KIM Chong-il. Economic growth during the period 1984-88 averaged 2%-3%, but output declined by 3%-5% annually during 1989-92 because of systemic problems and disruptions in socialist-style economic relations with the former USSR and China. In 1992, output dropped sharply, by perhaps 7%-9%, as the economy felt the cumulative effect of the reduction in outside support. The leadership insisted on maintaining its high level of military outlays from a shrinking economic pie. Moreover, a serious drawdown in inventories and critical shortages in the energy sector have led to increasing interruptions in industrial production. Abundant mineral resources and hydropower have formed the basis of industrial development since World War II. Output of the extractive industries includes coal, iron ore, magnesite, graphite, copper, zinc, lead, and precious metals. Manufacturing is centered on heavy industry, including military industry, with light industry lagging far behind. Despite the use of improved seed varieties, expansion of irrigation, and the heavy use of fertilizers, North Korea has not yet become self-sufficient in food production. Indeed, a shortage of arable lands, several years of poor harvests, and a cumbersome distribution system have resulted in chronic food shortages. The collapse of Communism in the former Soviet Union and Eastern Europe in 1989-91 has disrupted important technological links. North Korea remains far behind South Korea in economic development and living standards. GDP is stagnant
National product: GDP—purchasing power parity—$21.3 billion (1994 est.)
National product real growth rate: 0% (1994 est.)
National product per capita: $920 (1994 est.)
Inflation rate (consumer prices): NA%
Unemployment rate: NA%
Budget:
revenues: $19.3 billion
expenditures: $19.3 billion, including capital expenditures of $NA (1992 est.)
Exports: $1.02 billion (f.o.b., 1993 est.)
commodities: minerals, metallurgical products, agricultural and fishery products, manufactures (including armaments)
partners: China, Japan, Russia, South Korea, Germany, Hong Kong
Imports: $1.64 billion (f.o.b., 1993 est.)
commodities: petroleum, grain, coking coal, machinery and equipment, consumer goods
partners: China, Russia, Japan, Hong Kong, Germany, Singapore
External debt: $8 billion (1992 est.)
Industrial production: growth rate -7% to -9% (1992 est.)
Electricity:
capacity: 9,500,000 kW
production: 50 billion kWh
consumption per capita: 2,053 kWh (1993)
Industries: machine building, military products, electric power, chemicals, mining, metallurgy, textiles, food processing
Agriculture: accounts for about 25% of GDP and 36% of work force; principal crops—rice, corn, potatoes, soybeans, pulses; livestock and livestock products—cattle, hogs, pork, eggs; not self-sufficient in grain
Economic aid:
recipient: Communist countries, $1.4 billion a year in the 1980s, but very little now
Currency: 1 North Korean won (Wn) = 100 chon
Exchange rates: North Korean won (Wn) per US$1—2.15 (May 1994), 2.13 (May 1992), 2.14 (September 1991), 2.1 (January 1990), 2.3 (December 1989)
Fiscal year: calendar year

Transportation

Railroads:
total: 4,915 km
standard gauge: 4,250 km 1.435-m gauge (3,397 km electrified, 159 km double track)
narrow gauge: 665 km 0.762-m gauge (1989)
Highways:
total: 30,000 km
paved: 1,861 km
unpaved: gravel, crushed stone, earth 28,139 km (1992)
Inland waterways: 2,253 km; mostly navigable by small craft only
Pipelines: crude oil 37 km
Ports: Ch'ongjin, Haeju, Hungnam (Hamhung), Kimch'aek, Kosong, Najin, Namp'o, Sinuiju, Songnim, Sonbong (formerly Unggi), Ungsang, Wonsan
Merchant marine:
total: 87 ships (1,000 GRT or over) totaling 727,631 GRT/1,149,291 DWT
ships by type: bulk 9, cargo 70, combination bulk 1, oil tanker 3, passenger 2, passenger-cargo 1, short-sea passenger 1
note: North Korea owns an additional 4 ships (1,000 GRT or over) totaling approximately 32,405 DWT that operate under Honduran registry
Airports:
total: 49
with paved runways over 3,047 m: 2
with paved runways 2,438 to 3,047 m: 15
with paved runways 1,524 to 2,437 m: 2
with paved runways 914 to 1,523 m: 1
with paved runways under 914 m: 2
with unpaved runways 2,438 to 3,047 m: 4
with unpaved runways 1,524 to 2,438 m: 5
with unpaved runways 914 to 1,523 m: 12
with unpaved runways under 914 m: 6

Communications

Telephone system: telephone system is believed to be available only to government officials and not to private individuals
local: NA
intercity: NA
international: 1 earth station near P'yongyang, uses an Indian Ocean INTELSAT satellite; other international connections through Moscow and Beijing
Radio:
broadcast stations: AM 18, FM 0, shortwave 0
radios: 3.5 million
Television:
broadcast stations: 11
televisions: 350,000 (1989)

Defense Forces

Branches: Korean People's Army (includes Army, Navy, Air Force), Civil Security Forces
Manpower availability: males age 15-49 6,753,400; males fit for military service 4,094,854; males reach military age (18) annually 193,480 (1995 est.)
Defense expenditures: exchange rate conversion—about $5 billion, 20%-25% of GDP (1991 est.); note—the officially announced but suspect figure is $2.2 billion (1994), about 12% of total spending

Korea, South

Geography

Location: Eastern Asia, southern half of the Korean peninsula bordering the Sea of Japan and the Yellow Sea, south of North Korea
Map references: Asia
Area:
total area: 98,480 sq km
land area: 98,190 sq km
comparative area: slightly larger than Indiana
Land boundaries: total 238 km, North Korea 238 km
Coastline: 2,413 km
Maritime claims:
continental shelf: not specified
territorial sea: 12 nm; 3 nm in the Korea Strait
International disputes: Demarcation Line with North Korea; Liancourt Rocks claimed by Japan
Climate: temperate, with rainfall heavier in summer than winter
Terrain: mostly hills and mountains; wide coastal plains in west and south
Natural resources: coal, tungsten, graphite, molybdenum, lead, hydropower
Land use:
arable land: 21%
permanent crops: 1%
meadows and pastures: 1%
forest and woodland: 67%
other: 10%
Irrigated land: 13,530 sq km (1989)
Environment:
current issues: air pollution in large cities; water pollution from the discharge of sewage and industrial effluents; driftnet fishing
natural hazards: occasional typhoons bring high winds and floods; earthquakes in southwest
international agreements: party to—Antarctic Treaty, Biodiversity, Climate Change, Endangered Species, Environmental Modification, Hazardous Wastes, Nuclear Test Ban, Ozone Layer Protection, Ship Pollution, Tropical Timber 83, Whaling; signed, but not ratified—Antarctic-Environmental Protocol, Desertification, Law of the Sea

People

Population: 45,553,882 (July 1995 est.)
Age structure:
0-14 years: 24% (female 5,280,998; male 5,640,789)
15-64 years: 71% (female 15,877,182; male 16,291,183)
65 years and over: 5% (female 1,554,512; male 909,218) (July 1995 est.)
Population growth rate: 1.04% (1995 est.)
Birth rate: 15.63 births/1,000 population (1995 est.)
Death rate: 6.18 deaths/1,000 population (1995 est.)
Net migration rate: 0.9 migrant(s)/1,000 population (1995 est.)
Infant mortality rate: 20.9 deaths/1,000 live births (1995 est.)
Life expectancy at birth:
total population: 70.89 years
male: 67.69 years
female: 74.29 years (1995 est.)
Total fertility rate: 1.66 children born/woman (1995 est.)
Nationality:
noun: Korean(s)
adjective: Korean
Ethnic divisions: homogeneous (except for about 20,000 Chinese)
Religions: Christianity 48.6%, Buddhism 47.4%, Confucianism 3%, pervasive folk religion (shamanism), Chondogyo (Religion of the Heavenly Way) 0.2%
Languages: Korean, English widely taught in high school
Literacy: age 15 and over can read and write (1990 est.)
total population: 96%
male: 99%
female: 94%
Labor force: 20 million
by occupation: services and other 52%, mining and manufacturing 27%, agriculture, fishing, forestry 21% (1991)

Government

Names:
conventional long form: Republic of Korea
conventional short form: South Korea
local long form: Taehan-min'guk
local short form: none
note: the South Koreans generally use the term "Hanguk" to refer to their country
Abbreviation: ROK
Digraph: KS
Type: republic
Capital: Seoul
Administrative divisions: 9 provinces (do, singular and plural) and 6 special cities* (jikhalsi, singular and plural); Cheju-do, Cholla-bukto, Cholla-namdo, Ch'ungch'ong-bukto, Ch'ungch'ong-namdo, Inch'on-jikhalsi*, Kangwon-do, Kwangju-jikhalsi*, Kyonggi-do, Kyongsang-bukto, Kyongsang-namdo, Pusan-jikhalsi*, Soul-t'ukpyolsi*, Taegu-jikhalsi*, Taejon-jikhalsi*
Independence: 15 August 1948
National holiday: Independence Day, 15 August (1948)
Constitution: 25 February 1988
Legal system: combines elements of continental European civil law systems, Anglo-American law, and Chinese classical thought
Suffrage: 20 years of age; universal
Executive branch:
chief of state: President KIM Yong-sam (since 25 February 1993); election last held on 18 December 1992 (next to be held NA December 1997); results—KIM Yong-sam (DLP) 41.9%, KIM Tae-chung (DP) 33.8%, CHONG Chu-yong (UPP) 16.3%, other 8%
head of government: Prime Minister YI Hong-ku (since 17 December 1994); Deputy Prime Minister HONG Chae-yong (since 4 October 1994) and Deputy Prime Minister KIM Tok (since 23 December 1994)
cabinet: State Council; appointed by the president on the prime minister's recommendation
Legislative branch: unicameral
National Assembly (Kukhoe): elections last held on 24 March 1992; results—DLP 38.5%, DP 29.2%, Unification National Party (UNP) 17.3% (name later changed to UPP), other 15%; seats—(299 total) DLP 149, DP 97, UNP 31, other 22; the distribution of seats as of January 1994 was DLP 172, DP 96, UPP 11, other 20
note: the change in the distribution of seats reflects the fluidity of the current situation where party members are constantly switching from one party to another
Judicial branch: Supreme Court
Political parties and leaders:
majority party: Democratic Liberal Party (DLP), KIM Yong-sam, president
opposition: Democratic Party (DP), YI Ki-taek, executive chairman; United People's Party (UPP), KIM Tong-kil, chairman; several smaller parties
note: the DLP resulted from a merger of the Democratic Justice Party (DJP), Reunification Democratic Party (RDP), and New Democratic Republican Party (NDRP) on 9 February 1990
Other political or pressure groups: Korean National Council of Churches; National Democratic Alliance of Korea; National Federation of Student Associations; National Federation of Farmers' Associations; National Council of Labor Unions; Federation of Korean Trade Unions; Korean Veterans' Association; Federation of Korean Industries; Korean Traders Association
Member of: AfDB, APEC, AsDB, CCC, CP, EBRD, ESCAP, FAO, G-77, GATT, IAEA, IBRD, ICAO, ICC, ICFTU, ICRM, IDA, IFAD, IFC, IFRCS, ILO, IMF, IMO, INMARSAT, INTELSAT, INTERPOL, IOC,

IOM, ISO, ITU, OAS (observer), UN, UNCTAD, UNESCO, UNIDO, UNU, UPU, WHO, WIPO, WMO, WTO
Diplomatic representation in US:
chief of mission: Ambassador PAK Kun-u
chancery: 2450 Massachusetts Avenue NW, Washington, DC 20008
telephone: [1] (202) 939-5600
consulate(s) general: Agana (Guam), Anchorage, Atlanta, Boston, Chicago, Honolulu, Houston, Los Angeles, Miami, New York, San Francisco, and Seattle
US diplomatic representation:
chief of mission: Ambassador James T. LANEY
embassy: 82 Sejong-Ro, Chongro-ku, Seoul
mailing address: American Embassy, Unit 15550, Seoul; APO AP 96205-0001
telephone: [82] (2) 397-4114
FAX: [82] (2) 738-8845
consulate(s): Pusan
Flag: white with a red (top) and blue yin-yang symbol in the center; there is a different black trigram from the ancient I Ching (Book of Changes) in each corner of the white field

Economy

Overview: The driving force behind the economy's dynamic growth has been the planned development of an export-oriented economy in a vigorously entrepreneurial society. Real GDP increased more than 10% annually between 1986 and 1991. This growth ultimately led to an overheated situation characterized by a tight labor market, strong inflationary pressures, and a rapidly rising current account deficit. As a result, in 1992, economic policy focused on slowing the growth rate of inflation and reducing the deficit. Annual growth slowed to 5%, still above the rate in most other countries of the world, and recovered to 6.3% in 1993. The economy expanded by 8.3% in 1994, driven by booming exports.
National product: GDP—purchasing power parity—$508.3 billion (1994 est.)
National product real growth rate: 8.3% (1994)
National product per capita: $11,270 (1994 est.)
Inflation rate (consumer prices): 5.6% (1994)
Unemployment rate: 2% (November 1994)
Budget:
revenues: $63 billion
expenditures: $63 billion, including capital expenditures of $NA (1995 est.)
Exports: $96.2 billion (f.o.b., 1994)
commodities: electronic and electrical equipment, machinery, steel, automobiles, ships, textiles, clothing, footwear, fish
partners: US 26%, Japan 17%, EU 14%
Imports: $102.3 billion (c.i.f., 1994)
commodities: machinery, electronics and electronic equipment, oil, steel, transport equipment, textiles, organic chemicals, grains
partners: Japan 26%, US 24%, EU 15%
External debt: $44.1 billion (1993)
Industrial production: growth rate 12.1% (1994 est.); accounts for about 45% of GDP
Electricity:
capacity: 26,940,000 kW
production: 137 billion kWh
consumption per capita: 2,847 kWh (1993)
Industries: electronics, automobile production, chemicals, shipbuilding, steel, textiles, clothing, footwear, food processing
Agriculture: accounts for 8% of GDP and employs 21% of work force (including fishing and forestry); principal crops—rice, root crops, barley, vegetables, fruit; livestock and livestock products—cattle, hogs, chickens, milk, eggs; self-sufficient in food, except for wheat; fish catch of 2.9 million metric tons, seventh-largest in world
Economic aid:
recipient: US commitments, including Ex-Im (FY70-89), $3.9 billion; non-US countries (1970-89), $3 billion
Currency: 1 South Korean won (W) = 100 chun (theoretical)
Exchange rates: South Korean won (W) per US$1—790.48 (January 1995), 803.44 (1994), 802.67 (1993), 780.65 (1992), 733.35 (1991), 707.76 (1990)
Fiscal year: calendar year

Transportation

Railroads:
total: 6,763 km
standard gauge: 6,716 km 1.435-meter gauge (525 km electrified; 847 km double track)
narrow gauge: 47 km 0.610-meter gauge
Highways:
total: 63,200 km
paved: expressways 1,550 km
unpaved: NA
undifferentiated: national highway 12,190 km; provincial, local roads 49,460 km (1991)
Inland waterways: 1,609 km; use restricted to small native craft
Pipelines: petroleum products 455 km
Ports: Chinhae, Inch'on, Kunsan, Masan, Mokp'o, Pohang, Pusan, Ulsan, Yosu
Merchant marine:
total: 412 ships (1,000 GRT or over) totaling 6,129,796 GRT/9,985,197 DWT
ships by type: bulk 123, cargo 125, chemical tanker 17, combination bulk 1, combination ore/oil 1, container 61, liquefied gas tanker 13, multifunction large-load carrier 1, oil tanker 51, refrigerated cargo 9, short-sea passenger 1, vehicle carrier 9
Airports:
total: 114
with paved runways over 3,047 m: 1
with paved runways 2,438 to 3,047 m: 22
with paved runways 1,524 to 2,437 m: 10
with paved runways 914 to 1,523 m: 14
with paved runways under 914 m: 63
with unpaved runways 914 to 1,523 m: 4

Communications

Telephone system: 13.3 million telephones; excellent domestic and international services
local: NA
intercity: NA
international: 3 INTELSAT (2 Pacific Ocean and 1 Indian Ocean) earth stations
Radio:
broadcast stations: AM 79, FM 46, shortwave 0
radios: NA
Television:
broadcast stations: 256 (1 kW or greater 57)
televisions: NA

Defense Forces

Branches: Army, Navy, Air Force, Marine Corps, National Maritime Police (Coast Guard)
Manpower availability: males age 15-49 13,580,832; males fit for military service 8,701,742; males reach military age (18) annually 405,290 (1995 est.)
Defense expenditures: exchange rate conversion—$14 billion, 3.3% of GNP (1995 est.)

Kuwait

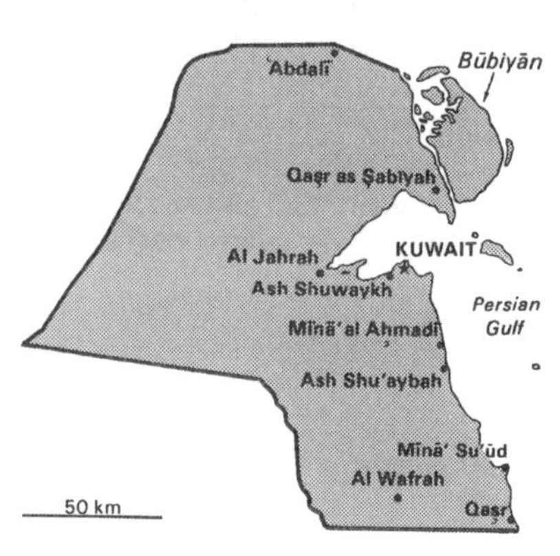

Geography

Location: Middle East, bordering the Persian Gulf, between Iraq and Saudi Arabia
Map references: Middle East
Area:
total area: 17,820 sq km
land area: 17,820 sq km
comparative area: slightly smaller than New Jersey
Land boundaries: total 464 km, Iraq 242 km, Saudi Arabia 222 km
Coastline: 499 km
Maritime claims:
territorial sea: 12 nm
International disputes: in November 1994, Iraq formally accepted the UN-demarcated border with Kuwait which had been spelled out in Security Council Resolutions 687 (1991), 773 (1993), and 883 (1993); this formally ends earlier claims to Kuwait and to Bubiyan and Warbah islands; ownership of Qaruh and Umm al Maradim islands disputed by Saudi Arabia
Climate: dry desert; intensely hot summers; short, cool winters
Terrain: flat to slightly undulating desert plain
Natural resources: petroleum, fish, shrimp, natural gas
Land use:
arable land: 0%
permanent crops: 0%
meadows and pastures: 8%
forest and woodland: 0%
other: 92%
Irrigated land: 20 sq km (1989 est.)
Environment:
current issues: limited natural fresh water resources; some of world's largest and most sophisticated desalination facilities provide much of the water; air and water pollution; desertification
natural hazards: sudden cloudbursts are common from October to April, they bring inordinate amounts of rain which can damage roads and houses; sandstorms and duststorms occur throughout the year, but are most common between March and August
international agreements: party to—Climate Change, Environmental Modification, Hazardous Wastes, Law of the Sea, Nuclear Test Ban, Ozone Layer Protection; signed, but not ratified—Biodiversity, Endangered Species, Marine Dumping
Note: strategic location at head of Persian Gulf

People

Population: 1,817,397 (July 1995 est.)
Age structure:
0-14 years: 34% (female 302,908; male 319,659)
15-64 years: 64% (female 467,163; male 697,849)
65 years and over: 2% (female 13,476; male 16,342) (July 1995 est.)
Population growth rate: 7.46% (1995 est.)
note: this rate reflects the continued post-Gulf crisis return of nationals and expatriates
Birth rate: 21.07 births/1,000 population (1995 est.)
Death rate: 2.2 deaths/1,000 population (1995 est.)
Net migration rate: 55.71 migrant(s)/1,000 population (1995 est.)
Infant mortality rate: 11.5 deaths/1,000 live births (1995 est.)
Life expectancy at birth:
total population: 75.64 years
male: 73.33 years
female: 78.06 years (1995 est.)
Total fertility rate: 2.93 children born/woman (1995 est.)
Nationality:
noun: Kuwaiti(s)
adjective: Kuwaiti
Ethnic divisions: Kuwaiti 45%, other Arab 35%, South Asian 9%, Iranian 4%, other 7%
Religions: Muslim 85% (Shi'a 30%, Sunni 45%, other 10%), Christian, Hindu, Parsi, and other 15%
Languages: Arabic (official), English widely spoken
Literacy: age 15 and over can read and write (1985)
total population: 74%
male: 78%
female: 69%
Labor force: 566,000 (1986)
by occupation: services 45.0%, construction 20.0%, trade 12.0%, manufacturing 8.6%, finance and real estate 2.6%, agriculture 1.9%, power and water 1.7%, mining and quarrying 1.4%
note: 70% of labor force non-Kuwaiti (1986)

Government

Names:
conventional long form: State of Kuwait
conventional short form: Kuwait
local long form: Dawlat al Kuwayt
local short form: Al Kuwayt
Digraph: KU
Type: nominal constitutional monarchy
Capital: Kuwait
Administrative divisions: 5 governorates (muhafazat, singular—muhafazah); Al 'Ahmadi, Al Jahrah, Al Kuwayt, Hawalli, Al Farwaniyah
Independence: 19 June 1961 (from UK)
National holiday: National Day, 25 February (1948)
Constitution: approved and promulgated 11 November 1962
Legal system: civil law system with Islamic law significant in personal matters; has not accepted compulsory ICJ jurisdiction
Suffrage: adult males who resided in Kuwait before 1920 and their male descendants at age 21
note: only 10% of all citizens are eligible to vote; in 1996, naturalized citizens who do not meet the pre-1920 qualification but have been naturalized for thirty years will be eligible to vote
Executive branch:
chief of state: Amir Shaykh JABIR al-Ahmad al-Jabir Al Sabah (since 31 December 1977)
head of government: Prime Minister and Crown Prince SAAD al-Abdallah al-Salim Al Sabah (since 8 February 1978); Deputy Prime Minister SABAH al-Ahmad al-Jabir Al Sabah (since 17 October 1992)
cabinet: Council of Ministers; appointed by the Prime Minister and approved by the Amir
Legislative branch: unicameral
National Assembly (Majlis al-umma): dissolved 3 July 1986; new elections were held on 5 October 1992 with a second election in the 14th and 16th constituencies held February 1993
Judicial branch: High Court of Appeal
Political parties and leaders: none
Other political or pressure groups: small, clandestine leftist and Shi'a fundamentalist groups are active; several groups critical of government policies are publicly active
Member of: ABEDA, AfDB, AFESD, AL, AMF, BDEAC, CAEU, CCC, ESCWA, FAO, G-77, GATT, GCC, IAEA, IBRD, ICAO, ICC, ICRM, IDA, IDB, IFAD, IFC, IFRCS, ILO, IMF, IMO, INMARSAT, INTELSAT, INTERPOL, IOC, ISO (correspondent), ITU, NAM, OAPEC, OIC, OPEC, UN, UNCTAD, UNESCO, UNIDO, UPU, WFTU, WHO, WMO, WTO
Diplomatic representation in US:
chief of mission: Ambassador MUHAMMAD al-Sabah al-Salim Al SABAH
chancery: 2940 Tilden Street NW, Washington, DC 20008
telephone: [1] (202) 966-0702
FAX: [1] (202) 966-0517
US diplomatic representation:
chief of mission: Ambassador Ryan C. CROCKER

embassy: Bneid al-Gar (opposite the Kuwait International Hotel), Kuwait City
mailing address: P.O. Box 77 SAFAT, 13001 SAFAT, Kuwait; Unit 69000, Kuwait; APO AE 09880-9000
telephone: [965] 2424151 through 2424159
FAX: [965] 2442855
Flag: three equal horizontal bands of green (top), white, and red with a black trapezoid based on the hoist side

Economy

Overview: Kuwait is a small and relatively open economy with proved crude oil reserves of about 94 billion barrels—10% of world reserves. Kuwait has rebuilt its war-ravaged petroleum sector; its crude oil production reached at least 2.0 million barrels per day by the end of 1993. The government ran a sizable fiscal deficit in 1993. Petroleum accounts for nearly half of GDP and 90% of export and government revenues. Kuwait lacks water and has practically no arable land, thus preventing development of agriculture. With the exception of fish, it depends almost wholly on food imports. About 75% of potable water must be distilled or imported. Because of its high per capita income, comparable with Western European incomes, Kuwait provides its citizens with extensive health, educational, and retirement benefits. Per capita military expenditures are among the highest in the world. The economy improved moderately in 1994, with the growth in industry and finance, and should see further gains in 1995, especially if oil prices go up. The World Bank has urged Kuwait to push ahead with privatization, including in the oil industry, but the government will move slowly on this front.
National product: GDP—purchasing power parity—$30.7 billion (1994 est.)
National product real growth rate: 9.3% (1994 est.)
National product per capita: $16,900 (1994 est.)
Inflation rate (consumer prices): 3% (1993)
Unemployment rate: NEGL% (1992 est.)
Budget:
revenues: $9 billion
expenditures: $13 billion, including capital expenditures of $NA (FY92/93)
Exports: $10.5 billion (f.o.b., 1993)
commodities: oil
partners: France 16%, Italy 15%, Japan 12%, UK 11%
Imports: $6.6 billion (f.o.b., 1993)
commodities: food, construction materials, vehicles and parts, clothing
partners: US 35%, Japan 12%, UK 9%, Canada 9%
External debt: $7.2 billion (December 1989 est.)
note: external debt has grown substantially in 1991 and 1992 to pay for restoration of war damage
Industrial production: growth rate NA%; accounts for NA% of GDP
Electricity:
capacity: 7,070,000 kW
production: 11 billion kWh
consumption per capita: 6,007 kWh (1993)
Industries: petroleum, petrochemicals, desalination, food processing, building materials, salt, construction
Agriculture: practically none; extensive fishing in territorial waters and Indian Ocean
Economic aid:
donor: pledged bilateral aid to less developed countries (1979-89), $18.3 billion
Currency: 1 Kuwaiti dinar (KD) = 1,000 fils
Exchange rates: Kuwaiti dinars (KD) per US$1—0.2991 (January 1995), 0.2976 (1994), 0.3017 (1993), 0.2934 (1992), 0.2843 (1991), 0.2915 (1990)
Fiscal year: 1 July—30 June

Transportation

Railroads: 0 km
Highways:
total: 4,270 km
paved: bituminous 3,370 km
unpaved: gravel, sand, earth 900 km (est.)
Pipelines: crude oil 877 km; petroleum products 40 km; natural gas 165 km
Ports: Ash Shu'aybah, Ash Shuwaykh, Kuwait, Mina' 'Abd Allah, Mina' al Ahmadi, Mina' Su'ud
Merchant marine:
total: 47 ships (1,000 GRT or over) totaling 2,202,558 GRT/3,618,527 DWT
ships by type: cargo 9, container 3, liquefied gas tanker 7, livestock carrier 4, oil tanker 24
Airports:
total: 8
with paved runways over 3,047 m: 3
with paved runways 2,438 to 3,047 m: 1
with paved runways under 914 m: 2
with unpaved runways 1,524 to 2,438 m: 1
with unpaved runways 914 to 1,523 m: 1

Communications

Telephone system: NA telephones; civil network suffered extensive damage as a result of the Gulf war and reconstruction is still under way with some restored international and domestic capabilities
local: NA
intercity: NA
international: earth stations destroyed during Gulf war and not rebuilt yet; temporary mobile satellite antennae provide international telecommunications; coaxial cable and microwave radio relay to Saudi Arabia; service to Iraq is nonoperational
Radio:
broadcast stations: AM 3, FM 0, shortwave 0
radios: NA
Television:
broadcast stations: 3
televisions: NA

Defense Forces

Branches: Army, Navy, Air Force, National Police Force, National Guard
Manpower availability: males age 15-49 610,205; males fit for military service 363,735; males reach military age (18) annually 16,170 (1995 est.)
Defense expenditures: exchange rate conversion—$3.4 billion, 13.3% of GDP (1995)

Kyrgyzstan

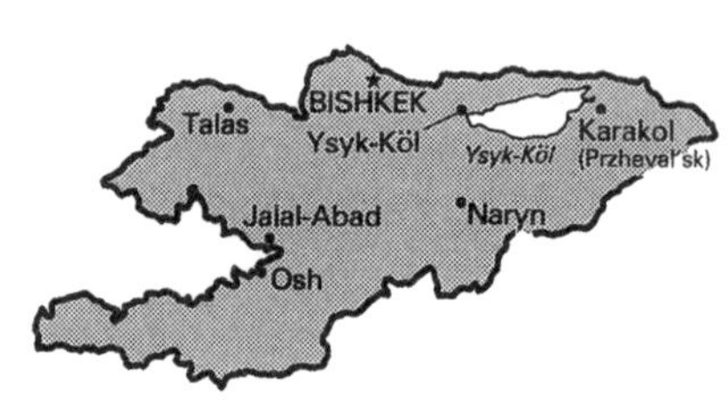

Geography

Location: Central Asia, west of China
Map references: Commonwealth of Independent States—Central Asian States
Area:
total area: 198,500 sq km
land area: 191,300 sq km
comparative area: slightly smaller than South Dakota
Land boundaries: total 3,878 km, China 858 km, Kazakhstan 1,051 km, Tajikistan 870 km, Uzbekistan 1,099 km
Coastline: 0 km (landlocked)
Maritime claims: none; landlocked
International disputes: territorial dispute with Tajikistan on southwestern boundary in Isfara Valley area
Climate: dry continental to polar in high Tien Shan; subtropical in southwest (Fergana Valley); temperate in northern foothill zone
Terrain: peaks of Tien Shan rise to 7,000 meters, and associated valleys and basins encompass entire nation
Natural resources: abundant hydroelectric potential; significant deposits of gold and rare earth metals; locally exploitable coal, oil and natural gas; other deposits of nepheline, mercury, bismuth, lead, and zinc
Land use:
arable land: 7%
permanent crops: NEGL%
meadows and pastures: 42%
forest and woodland: 0%
other: 51%
Irrigated land: 10,320 sq km (1990)
Environment:
current issues: water pollution; many people get their water directly from contaminated streams and wells, as a result, water-borne diseases are prevalent; increasing soil salinity from faulty irrigation practices
natural hazards: NA
international agreements: NA
Note: landlocked

People

Population: 4,769,877 (July 1995 est.)
Age structure:
0-14 years: 37% (female 868,108; male 888,479)
15-64 years: 57% (female 1,377,221; male 1,345,990)
65 years and over: 6% (female 185,807; male 104,272) (July 1995 est.)
Population growth rate: 1.5% (1995 est.)
Birth rate: 25.97 births/1,000 population (1995 est.)
Death rate: 7.32 deaths/1,000 population (1995 est.)
Net migration rate: -3.66 migrant(s)/1,000 population (1995 est.)
Infant mortality rate: 45.8 deaths/1,000 live births (1995 est.)
Life expectancy at birth:
total population: 68.13 years
male: 63.92 years
female: 72.56 years (1995 est.)
Total fertility rate: 3.31 children born/woman (1995 est.)
Nationality:
noun: Kyrgyz(s)
adjective: Kyrgyz
Ethnic divisions: Kirghiz 52.4%, Russian 21.5%, Uzbek 12.9%, Ukrainian 2.5%, German 2.4%, other 8.3%
Religions: Muslim 70%, Russian Orthodox NA%
Languages: Kirghiz (Kyrgyz)—official language, Russian widely used
Literacy: age 15 and over can read and write (1989)
total population: 97%
male: 99%
female: 96%
Labor force: 1.836 million
by occupation: agriculture and forestry 38%, industry and construction 21%, other 41% (1990)

Government

Names:
conventional long form: Kyrgyz Republic
conventional short form: Kyrgyzstan
local long form: Kyrgyz Respublikasy
local short form: none
former: Kirghiz Soviet Socialist Republic
Digraph: KG
Type: republic
Capital: Bishkek
Administrative divisions: 6 oblasttar (singular—oblast) and 1 city* (singular—shaar); Bishkek Shaary*, Chuy Oblasty (Bishkek), Jalal-Abad Oblasty, Naryn Oblasty, Osh Oblasty, Talas Oblasty, Ysyk-Kol Oblasty (Karakol)
note: names in parentheses are administrative centers when name differs from oblast name
Independence: 31 August 1991 (from Soviet Union)
National holiday: National Day, 2 December; Independence Day, 31 August (1991)
Constitution: adopted 5 May 1993
Legal system: based on civil law system
Suffrage: 18 years of age; universal
Executive branch:
chief of state: President Askar AKAYEV (since 28 October 1990); election last held 12 October 1991 (next to be held NA 1996); results—Askar AKAYEV won in uncontested election with 95% of vote and with 90% of electorate voting; note—president elected by Supreme Soviet 28 October 1990, then by popular vote 12 October 1991; AKAYEV won 96% of the vote in a referendum on his status as president on 30 January 1994
head of government: Prime Minister Apas DJUMAGULOV (since NA December 1993)
cabinet: Cabinet of Ministers; subordinate to the president
Legislative branch: bicameral
Assembly of Legislatures: elections last held 5 February 1995 (next to be held no later than NA 1998); 35-member house to which 19 members have been elected so far; next round of runoffs scheduled for 19 April 1995
Assembly of Representatives: elections last held 5 February 1995 (next to be held no later than NA 1998); 70-member house to which 60 members have been elected so far; next round of runoffs scheduled for 19 April 1995
note: the legislature became bicameral for the 5 February 1995 elections
Judicial branch: Supreme Court
Political parties and leaders: Social Democratic Party (SDP), Ishenbai KADYRBEKOV, chairman; Democratic Movement of Kyrgyzstan (DMK), Kazat AKHMATOV, chairman; National Unity, German KUZNETSOV; Communist Party of Kyrgyzstan (PCK), Sherali SYDYKOV, chairman; Democratic Movement of Free Kyrgyzstan (ErK), Topchubek TURGUNALIYEV, chairman; Republican Popular Party of Kyrgyzstan; Agrarian Party of Kyrgyzstan, A. ALIYEV
Other political or pressure groups: National Unity Democratic Movement; Peasant Party; Council of Free Trade Unions; Union of Entrepreneurs; Agrarian Party
Member of: AsDB, CIS, EBRD, ECE, ECO, ESCAP, FAO, IBRD, ICAO, IDA, IDB, IFAD, IFC, ILO, IMF, IOC, IOM (observer), ITU, NACC, OIC, OSCE, PCA, PFP, UN, UNCTAD, UNESCO, UNIDO, UPU, WHO, WIPO, WTO
Diplomatic representation in US:
chief of mission: (vacant); Charge d'Affaires ad interim Almas CHUKIN
chancery: (temporary) Suite 705, 1511 K

Street NW, Washington, DC 20005
telephone: [1] (202) 347-3732, 3733, 3718
FAX: [1] (202) 347-3718
US diplomatic representation:
chief of mission: Ambassador Eileen A. MALLOY
embassy: Erkindik Prospekt #66, Bishkek 720002
mailing address: use embassy street address
telephone: [7] (3312) 22-29-20, 22-27-77, 22-26-31, 22-24-73
FAX: [7] (3312) 22-35-51
Flag: red field with a yellow sun in the center having 40 rays representing the 40 Kirghiz tribes; on the obverse side the rays run counterclockwise, on the reverse, clockwise; in the center of the sun is a red ring crossed by two sets of three lines, a stylized representation of the roof of the traditional Kirghiz yurt

Economy

Overview: Kyrgyzstan is one of the smallest and poorest states of the former Soviet Union. Its economy is heavily agricultural, growing cotton and tobacco on irrigated land in the south and grain in the foothills of the north and raising sheep and goats on mountain pastures. Its small and obsolescent industrial sector, concentrated around Bishkek, has traditionally relied on Russia and other CIS countries for customers and industrial inputs, including most of its fuel. Since 1990, the economy has contracted by almost 50% as subsidies from Moscow vanished and trade links with other former Soviet republics eroded. At the same time, the Kyrgyz government stuck to tight monetary and fiscal policies in 1994 that succeeded in reducing inflation from 23% per month in 1993 to 5.4% per month in 1994. Moreover, Kyrgyzstan has been the most successful of the Central Asian states in reducing state controls over the economy and privatizing state industries. Nevertheless, restructuring proved to be a slow and painful process in 1994 despite relatively large flows of foreign aid and continued progress on economic reform. The decline in output in 1995 may be much smaller, perhaps 5%, compared with an estimated 24% in 1994.
National product: GDP—purchasing power parity—$8.4 billion (1994 estimate as extrapolated from World Bank estimate for 1992)
National product real growth rate: -24% (1994 est.)
National product per capita: $1,790 (1994 est.)
Inflation rate (consumer prices): 5.4% per month (1994 est.)
Unemployment rate: 0.7% includes officially registered unemployed; also large numbers of unregistered unemployed and underemployed workers (1994)
Budget:
revenues: $NA
expenditures: $NA, including capital expenditures of $NA
Exports: $116 million to countries outside the FSU (1994)
commodities: wool, chemicals, cotton, ferrous and nonferrous metals, shoes, machinery, tobacco
partners: Russia 70%, Ukraine, Uzbekistan, Kazakhstan, and others
Imports: $92.4 million from countries outside the FSU (1994)
commodities: grain, lumber, industrial products, ferrous metals, fuel, machinery, textiles, footwear
partners: other CIS republics
External debt: $NA
Industrial production: growth rate -24% (1994 est.)
Electricity:
capacity: 3,660,000 kW
production: 12.7 billion kWh
consumption per capita: 2,700 kWh (1994)
Industries: small machinery, textiles, food-processing industries, cement, shoes, sawn logs, refrigerators, furniture, electric motors, gold, and rare earth metals
Agriculture: wool, tobacco, cotton, livestock (sheep, goats, cattle), vegetables, meat, grapes, fruits and berries, eggs, milk, potatoes
Illicit drugs: illicit cultivator of cannabis and opium poppy; mostly for CIS consumption; limited government eradication program; used as transshipment point for illicit drugs to Western Europe and North America from Southwest Asia
Economic aid:
recipient: IMF aid commitments were $80 million in 1993 and $400 million in 1994
Currency: introduced national currency, the som (10 May 1993)
Exchange rates: soms per US$1—10.6 (yearend 1994)
Fiscal year: calendar year

Transportation

Railroads:
total: 370 km in common carrier service; does not include industrial lines
broad gauge: 370 km 1.520-m gauge (1990)
Highways:
total: 30,300 km
paved and graveled: 22,600 km
unpaved: earth 7,700 km (1990)
Pipelines: natural gas 200 km
Ports: Ysyk-Kol (Rybach'ye)
Airports:
total: 54
with paved runways over 3,047 m: 1
with paved runways 2,438 to 3,047 m: 3
with paved runways 1,524 to 2,437 m: 9
with paved runways under 914 m: 1
with unpaved runways 1,524 to 2,438 m: 4
with unpaved runways 914 to 1,523 m: 4
with unpaved runways under 914 m: 32

Communications

Telephone system: 342,000 telephones (1991); 76 telephones/1,000 persons (December 1991); poorly developed; about 100,000 unsatisfied applications for household telephones
local: NA
intercity: principally by microwave radio relay
international: connections with other CIS countries by landline or microwave and with other countries by leased connections with Moscow international gateway switch and by satellite; 1 GORIZONT and 1 INTELSAT satellite link through Ankara to 200 other countries
Radio:
broadcast stations: AM NA, FM NA, shortwave NA
radios: 825,000 (radio receiver systems with multiple speakers for program diffusion 748,000)
Television:
broadcast stations: NA; note—receives Turkish broadcasts
televisions: 875,000

Defense Forces

Branches: National Guard, Security Forces (internal and border troops), Civil Defense
Manpower availability: males age 15-49 1,154,683; males fit for military service 934,167; males reach military age (18) annually 44,526 (1995 est.)
Defense expenditures: $NA, NA% of GDP

Laos

Geography

Location: Southeastern Asia, northeast of Thailand
Map references: Southeast Asia
Area:
total area: 236,800 sq km
land area: 230,800 sq km
comparative area: slightly larger than Utah
Land boundaries: total 5,083 km, Burma 235 km, Cambodia 541 km, China 423 km, Thailand 1,754 km, Vietnam 2,130 km
Coastline: 0 km (landlocked)
Maritime claims: none; landlocked
International disputes: boundary dispute with Thailand
Climate: tropical monsoon; rainy season (May to November); dry season (December to April)
Terrain: mostly rugged mountains; some plains and plateaus
Natural resources: timber, hydropower, gypsum, tin, gold, gemstones
Land use:
arable land: 4%
permanent crops: 0%
meadows and pastures: 3%
forest and woodland: 58%
other: 35%
Irrigated land: 1,554 sq km (1992 est.)
Environment:
current issues: deforestation; soil erosion; a majority of the population does not have access to potable water
natural hazards: floods, droughts, and blight
international agreements: party to—Climate Change, Environmental Modification, Nuclear Test Ban; signed, but not ratified—Law of the Sea
Note: landlocked

People

Population: 4,837,237 (July 1995 est.)
Age structure:
0-14 years: 45% (female 1,084,615; male 1,111,928)
15-64 years: 51% (female 1,280,142; male 1,199,149)
65 years and over: 4% (female 86,390; male 75,013) (July 1995 est.)
Population growth rate: 2.84% (1995 est.)
Birth rate: 42.64 births/1,000 population (1995 est.)
Death rate: 14.28 deaths/1,000 population (1995 est.)
Net migration rate: 0 migrant(s)/1,000 population (1995 est.)
Infant mortality rate: 99.2 deaths/1,000 live births (1995 est.)
Life expectancy at birth:
total population: 52.2 years
male: 50.66 years
female: 53.81 years (1995 est.)
Total fertility rate: 5.98 children born/woman (1995 est.)
Nationality:
noun: Lao(s) or Laotian(s)
adjective: Lao or Laotian
Ethnic divisions: Lao Loum (lowland) 68%, Lao Theung (upland) 22%, Lao Soung (highland) including the Hmong ("Meo") and the Yao (Mien) 9%, ethnic Vietnamese/Chinese 1%
Religions: Buddhist 60%, animist and other 40%
Languages: Lao (official), French, English, and various ethnic languages
Literacy: age 15 and over can read and write (1992)
total population: 50%
male: 65%
female: 35%
Labor force: 1 million-1.5 million
by occupation: agriculture 80% (1992 est.)

Government

Names:
conventional long form: Lao People's Democratic Republic
conventional short form: Laos
local long form: Sathalanalat Paxathipatai Paxaxon Lao
local short form: none
Digraph: LA
Type: Communist state
Capital: Vientiane
Administrative divisions: 16 provinces (khoueng, singular and plural) and 1 municipality* (kampheng nakhon, singular and plural); Attapu, Bokeo, Bolikhamxai, Champasak, Houaphan, Khammouan, Louangnamtha, Louangphabang, Oudomxai, Phongsali, Salavan, Savannakhet, Viangchan*, Viangchan, Xaignabouli, Xekong, Xiangkhoang
Independence: 19 July 1949 (from France)
National holiday: National Day, 2 December (1975) (proclamation of the Lao People's Democratic Republic)
Constitution: promulgated 14 August 1991
Legal system: based on traditional customs, French legal norms and procedures, and Socialist practice
Suffrage: 18 years of age; universal
Executive branch:
chief of state: President NOUHAK PHOUMSAVAN (since 25 November 1992)
head of government: Prime Minister Gen. KHAMTAI SIPHANDON (since 15 August 1991)
cabinet: Council of Ministers; appointed by the president, approved by the Assembly
Legislative branch: unicameral
National Assembly: elections last held on 20 December 1992 (next to be held NA); results—percent of vote by party NA; seats—(85 total) number of seats by party NA
Judicial branch: Supreme People's Court
Political parties and leaders: Lao People's Revolutionary Party (LPRP), KHAMTAI Siphandon, party president; other parties proscribed
Other political or pressure groups: non-Communist political groups proscribed; most opposition leaders fled the country in 1975
Member of: ACCT, AsDB, ASEAN (observer), CP, ESCAP, FAO, G-77, IBRD, ICAO, ICRM, IDA, IFAD, IFC, IFRCS, ILO, IMF, INTELSAT (nonsignatory user), INTERPOL, IOC, ITU, NAM, PCA, UN, UNCTAD, UNESCO, UNIDO, UPU, WFTU, WHO, WMO, WTO
Diplomatic representation in US:
chief of mission: Ambassador HIEM PHOMMACHANH
chancery: 2222 S Street NW, Washington, DC 20008
telephone: [1] (202) 332-6416, 6417
FAX: [1] (202) 332-4923
US diplomatic representation:
chief of mission: Ambassador Victor L. TOMSETH
embassy: Rue Bartholonie, Vientiane
mailing address: B. P. 114, Vientiane; American Embassy, Box V, APO AP 96546
telephone: [856] (21) 212581, 212582, 212585
FAX: [856] (21) 212584
Flag: three horizontal bands of red (top), blue (double width), and red with a large white disk centered in the blue band

Economy

Overview: The government of Laos—one of the few remaining official Communist states—has been decentralizing control and encouraging private enterprise since 1986. The results, starting from an extremely low base, have been striking—growth has averaged 7.5% annually since 1988. Even so, Laos is a landlocked country with a primitive infrastructure. It has no railroads, a rudimentary road system, and limited external and internal telecommunications. Electricity is available in only a few urban areas.

Subsistence agriculture accounts for half of GDP and provides 80% of total employment. The predominant crop is rice. In non-drought years, Laos is self-sufficient overall in food, but each year flood, pests, and localized drought cause shortages in various parts of the country. For the foreseeable future the economy will continue to depend on aid from the IMF and other international sources; aid from the former USSR and Eastern Europe has been cut sharply. As in many developing countries, deforestation and soil erosion will hamper efforts to maintain the high rate of GDP growth.
National product: GDP—purchasing power parity—$4 billion (1994 est.)
National product real growth rate: 8.4% (1994 est.)
National product per capita: $850 (1994 est.)
Inflation rate (consumer prices): 6.5% (1994 est.)
Unemployment rate: 21% (1992 est.)
Budget:
revenues: $NA
expenditures: $NA
Exports: $277 million (f.o.b., 1994 est.)
commodities: electricity, wood products, coffee, tin, garments
partners: Thailand 57%, Germany 10%, France 10%, Japan 5% (1991)
Imports: $528 million (c.i.f., 1994 est.)
commodities: food, fuel oil, consumer goods, manufactures
partners: Thailand 55%, Japan 16%, China 8%, Italy 4% (1991)
External debt: $NA
Industrial production: growth rate 7.5% (1992 est.); accounts for 18% of GDP (1992 est.)
Electricity:
capacity: 260,000 kW
production: 870 million kWh
consumption per capita: 44 kWh (1993)
Industries: tin and gypsum mining, timber, electric power, agricultural processing, construction
Agriculture: principal crops—rice (80% of cultivated land), sweet potatoes, vegetables, corn, coffee, sugarcane, cotton; livestock—buffaloes, hogs, cattle, poultry
Illicit drugs: illicit producer of cannabis, opium poppy for the international drug trade, fourth largest opium producer (85 metric tons in 1994); heroin producer; increasingly used as transshipment point for heroin produced in Burma
Economic aid:
recipient: US commitments, including Ex-Im (FY70-79), $276 million; Western (non-US) countries, ODA and OOF bilateral commitments (1970-89), $605 million; Communist countries (1970-89), $995 million; international assistance in loans and grant aid (1993/94) $217.7 million

Currency: 1 new kip (NK) = 100 at
Exchange rates: new kips (NK) per US$1—717 (1994 est.), 720 (July 1993), 710 (May 1992), 710 (December 1991), 700 (September 1990), 576 (1989)
Fiscal year: 1 October—30 September

Transportation

Railroads: 0 km
Highways:
total: 14,130 km
paved: 2,260 km
unpaved: 11,870 km (1992 est.)
Inland waterways: about 4,587 km, primarily Mekong and tributaries; 2,897 additional kilometers are sectionally navigable by craft drawing less than 0.5 m
Pipelines: petroleum products 136 km
Ports: none
Merchant marine:
total: 1 cargo ship (1,000 GRT or over) totaling 2,370 GRT/3,000 DWT
Airports:
total: 52
with paved runways over 3,047 m: 1
with paved runways 1,524 to 2,437 m: 5
with paved runways 914 to 1,523 m: 3
with paved runways under 914 m: 25
with unpaved runways 1,524 to 2,438 m: 1
with unpaved runways 914 to 1,523 m: 17

Communications

Telephone system: 7,390 telephones (1986); service to general public very poor; radio communications network provides generally erratic service to government users
local: 16 telephone lines per 1,000 people
intercity: radio communications
international: 1 earth station
Radio:
broadcast stations: AM 10, FM 0, shortwave 0
radios: NA
Television:
broadcast stations: 2
televisions: NA

Defense Forces

Branches: Lao People's Army (LPA; includes riverine naval and militia elements), Air Force, National Police Department
Manpower availability: males age 15-49 1,051,105; males fit for military service 567,017; males reach military age (18) annually 51,437 (1995 est.)
Defense expenditures: exchange rate conversion—$105 million, 8.1% of GDP (FY92/93)

Latvia

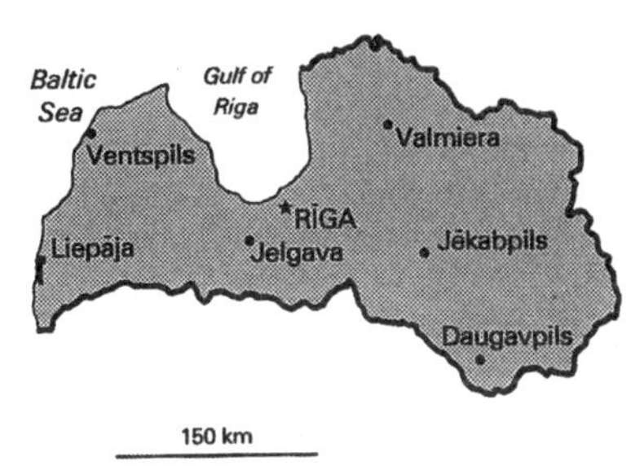

Geography

Location: Eastern Europe, bordering the Baltic Sea, between Estonia and Lithuania
Map references: Europe
Area:
total area: 64,100 sq km
land area: 64,100 sq km
comparative area: slightly larger than West Virginia
Land boundaries: total 1,078 km, Belarus 141 km, Estonia 267 km, Lithuania 453 km, Russia 217 km
Coastline: 531 km
Maritime claims:
exclusive economic zone: 200 nm
territorial sea: 12 nm
continental shelf: 200-m depth or to the depth of exploitation
International disputes: the Abrene section of border ceded by the Latvian Soviet Socialist Republic to Russia in 1944
Climate: maritime; wet, moderate winters
Terrain: low plain
Natural resources: minimal; amber, peat, limestone, dolomite
Land use:
arable land: 27%
permanent crops: 0%
meadows and pastures: 13%
forest and woodland: 39%
other: 21%
Irrigated land: 160 sq km (1990)
Environment:
current issues: air and water pollution because of a lack of waste conversion equipment; Gulf of Riga and Daugava River heavily polluted; contamination of soil and groundwater with chemicals and petroleum products at military bases
natural hazards: NA
international agreements: party to—Air Pollution, Hazardous Wastes, Ship Pollution; signed, but not ratified—Biodiversity, Climate Change

People

Population: 2,762,899 (July 1995 est.)
Age structure:
0-14 years: 22% (female 294,521; male 304,830)
15-64 years: 65% (female 933,003; male 870,128)
65 years and over: 13% (female 247,476; male 112,941) (July 1995 est.)
Population growth rate: 0.5% (1995 est.)
Birth rate: 13.71 births/1,000 population (1995 est.)
Death rate: 12.49 deaths/1,000 population (1995 est.)
Net migration rate: 3.76 migrant(s)/1,000 population (1995 est.)
Infant mortality rate: 21 deaths/1,000 live births (1995 est.)
Life expectancy at birth:
total population: 69.65 years
male: 64.6 years
female: 74.95 years (1995 est.)
Total fertility rate: 1.97 children born/woman (1995 est.)
Nationality:
noun: Latvian(s)
adjective: Latvian
Ethnic divisions: Latvian 51.8%, Russian 33.8%, Byelorussian 4.5%, Ukrainian 3.4%, Polish 2.3%, other 4.2%
Religions: Lutheran, Roman Catholic, Russian Orthodox
Languages: Lettish (official), Lithuanian, Russian, other
Literacy: age 15 and over can read and write (1989)
total population: 100%
male: 100%
female: 99%
Labor force: 1.407 million
by occupation: industry and construction 41%, agriculture and forestry 16%, other 43% (1990)

Government

Names:
conventional long form: Republic of Latvia
conventional short form: Latvia
local long form: Latvijas Republika
local short form: Latvija
former: Latvian Soviet Socialist Republic
Digraph: LG
Type: republic
Capital: Riga
Administrative divisions: 26 counties (singular—rajons) and 7 municipalities*: Aizkraukles Rajons, Aluksnes Rajons, Balvu Rajons, Bauskas Rajons, Cesu Rajons, Daugavpils*, Daugavpils Rajons, Dobeles Rajons, Gulbenes Rajons, Jekabpils Rajons, Jelgava*, Jelgavas Rajons, Jurmala*, Kraslavas Rajons, Kuldigas Rajons, Leipaja*, Liepajas Rajons, Limbazu Rajons, Ludzas Rajons, Madonas Rajons, Ogres Rajons, Preiju Rajons, Rezekne*, Rezeknes Rajons, Riga*, Rigas Rajons, Saldus Rajons, Talsu Rajons, Tukuma Rajons, Valkas Rajons, Valmieras Rajons, Ventspils*, Ventspils Rajons
Independence: 6 September 1991 (from Soviet Union)
National holiday: Independence Day, 18 November (1918)
Constitution: newly elected Parliament in 1993 restored the 1933 constitution
Legal system: based on civil law system
Suffrage: 18 years of age; universal
Executive branch:
chief of state: President Guntis ULMANIS (since 7 July 1993); Parliament (Saeima) elected President ULMANIS in the third round of balloting on 7 July 1993
head of government: Prime Minister Maris GAILIS (since September 1994)
cabinet: Council of Ministers; appointed by the Supreme Council
Legislative branch: unicameral
Parliament (Saeima): elections last held 5-6 June 1993 (next to be held NA October 1995); results—percent of vote by party NA; seats—(100 total) LC 36, LNNK 15, Concord for Latvia 13, LZS 12, Equal Rights 7, LKDS 6, TUB 6, DCP 5
Judicial branch: Supreme Court
Political parties and leaders: Latvian Way Union (LC), Valdis BIRKAVS; Latvian Farmers Union (LZS), Alvars BERKIS; Latvian National Independence Movement (LNNK), Andrejs KRASTINS, Aristids LAMBERGS, cochairmen; Concord for Latvia, Janis JURKANS; Equal Rights, Sergejs DIMANIS; Christian Democrat Union (LKDS), Peteris CIMDINS, Andris SAULITIS, Janis RUSKO; Fatherland and Freedom (TUB), Maris GRINBLATS, Roberts MILBERGS, Oigerts DZENTIS; Democratic Center (DCP), Ints CALITIS; Popular Front of Latvia (LTF), Uldis AUGSTKALNS
Member of: BIS, CBSS, CCC, CE, EBRD, ECE, FAO, IBRD, ICAO, ICRM, IDA, IFC, IFRCS, ILO, IMF, IMO, INTELSAT (nonsignatory user), INTERPOL, IOC, IOM (observer), ITU, NACC, OSCE, PFP, UN, UNCTAD, UNESCO, UNIDO, UPU, WEU (associate partner), WHO, WIPO, WMO
Diplomatic representation in US:
chief of mission: Ambassador Ojars Eriks KALNINS
chancery: 4325 17th Street NW, Washington, DC 20011
telephone: [1] (202) 726-8213, 8214
FAX: [1] (202) 726-6785
US diplomatic representation:
chief of mission: Ambassador Ints M, SILINS
embassy: Raina Boulevard 7, Riga 226050
mailing address: use embassy street address
telephone: [371] (2) 213-962
FAX: [371] 882-0047 (cellular)
Flag: two horizontal bands of maroon (top and bottom), white (middle, narrower than other two bands)

Economy

Overview: Latvia is rapidly becoming a dynamic market economy, rivaled only by Estonia among the former Soviet states in the speed of its transformation. However, the transition has been painful; in 1994 the IMF reported a 2% growth in GDP, following steep declines in 1992-93. The government's tough monetary policies and reform program have kept inflation at less than 2% a month, supported a dynamic private sector now accounting for more than half of GDP, and spurred the growth of trade ties with the West. Much of agriculture is already privatized and the government plans to step up the pace of privatization of state enterprises. Latvia thus is in the midst of recovery, helped by the country's strategic location on the Baltic Sea, its well-educated population, and its diverse—albeit largely obsolete—industrial structure.
National product: GDP—purchasing power parity—$12.3 billion (1994 estimate as extrapolated from World Bank estimate for 1992)
National product real growth rate: 2% (1994 est.)
National product per capita: $4,480 (1994 est.)
Inflation rate (consumer prices): 1.9% (monthly average 1994)
Unemployment rate: 6.5% (December 1994)
Budget:
revenues: $NA
expenditures: $NA, including capital expenditures of $NA
Exports: $1 billion (f.o.b., 1994)
commodities: oil products, timber, ferrous metals, dairy products, furniture, textiles
partners: Russia, Germany, Sweden, Belarus
Imports: $1.2 billion (c.i.f., 1994)
commodities: fuels, cars, ferrous metals, chemicals
partners: Russia, Germany, Sweden, Ukraine
External debt: $NA
Industrial production: growth rate -9.5% (1994 est.); accounts for 27% of GDP
Electricity:
capacity: 2,080,000 kW
production: 5.5 billion kWh
consumption per capita: 1,864 kWh (1993)
Industries: highly diversified; dependent on imports for energy, raw materials, and intermediate products; produces buses, vans, street and railroad cars, synthetic fibers, agricultural machinery, fertilizers, washing machines, radios, electronics, pharmaceuticals, processed foods, textiles
Agriculture: principally dairy farming and livestock feeding; products—meat, milk, eggs, grain, sugar beets, potatoes, vegetables; fishing and fish packing
Illicit drugs: transshipment point for illicit drugs from Central and Southwest Asia and Latin America to Western Europe; limited

producer of illicit opium; mostly for domestic consumption; also produces illicit amphetamines for export
Economic aid: $NA
Currency: 1 lat = 100 cents; introduced NA March 1993
Exchange rates: lats per US$1—0.55 (December 1994), 0.5917 (January 1994), 1.32 (March 1993)
Fiscal year: calendar year

Transportation

Railroads:
total: 2,400 km
broad gauge: 2,400 km 1.520-m gauge (270 km electrified)
Highways:
total: 59,500 km
paved and graveled: 33,000 km
unpaved: earth 26,500 km (1990)
Inland waterways: 300 km perennially navigable
Pipelines: crude oil 750 km; refined products 780 km; natural gas 560 km (1992)
Ports: Daugavpils, Liepaja, Riga, Ventspils
Merchant marine:
total: 85 ships (1,000 GRT or over) totaling 774,182 GRT/1,010,517 DWT
ships by type: cargo 17, oil tanker 37, refrigerated cargo 24, roll-on/roll-off cargo 7
Airports:
total: 50
with paved runways 2,438 to 3,047 m: 6
with paved runways 1,524 to 2,437 m: 2
with paved runways 914 to 1,523 m: 1
with paved runways under 914 m: 27
with unpaved runways 2,438 to 3,047 m: 2
with unpaved runways 914 to 1,523 m: 2
with unpaved runways under 914 m: 10

Communications

Telephone system: 660,000 telephones; 240 telephones/1,000 persons (1993); Latvia is better provided with telephone service than most of the other former Soviet republics; an NMT-450 analog cellular telephone network covers 75% of Latvia's population
local: NA
intercity: NA
international: international traffic carried by leased connection to the Moscow international gateway switch and through the new Ericsson AXE local/transit digital telephone exchange in Riga and through the Finnish cellular net; electronic mail capability by Sprint data network
Radio:
broadcast stations: AM NA, FM NA, shortwave NA
radios: NA
Television:
broadcast stations: NA
televisions: NA

Defense Forces

Branches: Ground Forces, Navy, Air and Air Defense Forces, Security Forces (internal and border troops), Border Guard, Home Guard (Zemessardze)
Manpower availability: males age 15-49 658,193; males fit for military service 517,896; males reach military age (18) annually 18,736 (1995 est.)
Defense expenditures: 176 million rubles, 3% to 5% of GDP (1994); note—conversion of the military budget into US dollars using the prevailing exchange rate could produce misleading results

Lebanon

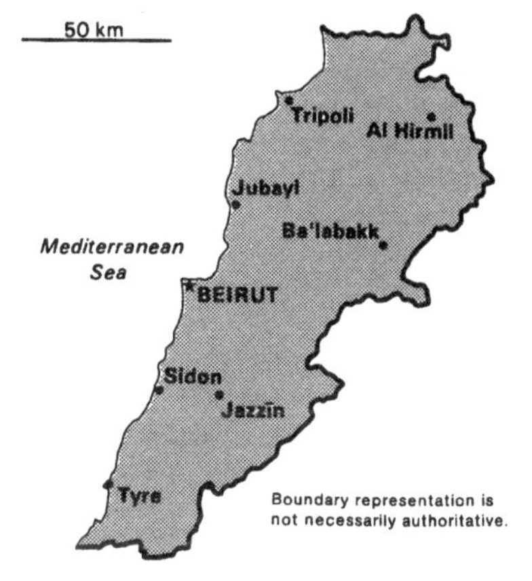

Note: Lebanon has made progress toward rebuilding its political institutions and regaining its national sovereignty since the end of the devastating 16-year civil war which began in 1975. Under the Ta'if accord—the blueprint for national reconciliation—the Lebanese have established a more equitable political system, particularly by giving Muslims a greater say in the political process. Since December 1990, the Lebanese have formed three cabinets and conducted the first legislative election in 20 years. Most of the militias have been weakened or disbanded. The Lebanese Armed Forces (LAF) has seized vast quantities of weapons used by the militias during the war and extended central government authority over about one-half of the country. Hizballah, the radical Sh'ia party, retains most of its weapons. Foreign forces still occupy areas of Lebanon. Israel maintains troops in southern Lebanon and continues to support a proxy militia, The Army of South Lebanon (ASL), along a narrow stretch of territory contiguous to its border. The ASL's enclave encompasses this self-declared security zone and about 20 kilometers north to the strategic town of Jazzine. As of December 1993, Syria maintained about 30,000-35,000 troops in Lebanon. These troops are based mainly in Beirut, North Lebanon, and the Bekaa Valley. Syria's deployment was legitimized by the Arab League early in Lebanon's civil war and in the Ta'if accord. Citing the continued weakness of the LAF, Beirut's requests, and failure of the Lebanese Government to implement all of the constitutional reforms in the Ta'if accord, Damascus has so far refused to withdraw its troops from Beirut.

Geography

Location: Middle East, bordering the Mediterranean Sea, between Israel and Syria
Map references: Middle East
Area:
total area: 10,400 sq km

land area: 10,230 sq km
comparative area: about 0.8 times the size of Connecticut
Land boundaries: total 454 km, Israel 79 km, Syria 375 km
Coastline: 225 km
Maritime claims:
territorial sea: 12 nm
International disputes: separated from Israel by the 1949 Armistice Line; Israeli troops in southern Lebanon since June 1982; Syrian troops in northern, central, and eastern Lebanon since October 1976
Climate: Mediterranean; mild to cool, wet winters with hot, dry summers; Lebanon mountains experience heavy winter snows
Terrain: narrow coastal plain; Al Biqa' (Bekaa Valley) separates Lebanon and Anti-Lebanon Mountains
Natural resources: limestone, iron ore, salt, water-surplus state in a water-deficit region
Land use:
arable land: 21%
permanent crops: 9%
meadows and pastures: 1%
forest and woodland: 8%
other: 61%
Irrigated land: 860 sq km (1989 est.)
Environment:
current issues: deforestation; soil erosion; desertification; air pollution in Beirut from vehicular traffic and the burning of industrial wastes; pollution of coastal waters from raw sewage and oil spills
natural hazards: duststorms, sandstorms
international agreements: party to—Biodiversity, Climate Change, Hazardous Wastes, Law of the Sea, Nuclear Test Ban, Ozone Layer Protection, Ship Pollution; signed, but not ratified—Desertification, Environmental Modification, Marine Dumping, Marine Life Conservation
Note: Nahr al Litani only major river in Near East not crossing an international boundary; rugged terrain historically helped isolate, protect, and develop numerous factional groups based on religion, clan, and ethnicity

People

Population: 3,695,921 (July 1995 est.)
Age structure:
0-14 years: 36% (female 657,403; male 682,757)
15-64 years: 58% (female 1,131,450; male 1,016,859)
65 years and over: 6% (female 111,585; male 95,867) (July 1995 est.)
Population growth rate: 2.15% (1995 est.)
Birth rate: 27.9 births/1,000 population (1995 est.)
Death rate: 6.44 deaths/1,000 population (1995 est.)
Net migration rate: 0 migrant(s)/1,000 population (1995 est.)
Infant mortality rate: 38 deaths/1,000 live births (1995 est.)
Life expectancy at birth:
total population: 69.69 years
male: 67.22 years
female: 72.28 years (1995 est.)
Total fertility rate: 3.31 children born/woman (1995 est.)
Nationality:
noun: Lebanese (singular and plural)
adjective: Lebanese
Ethnic divisions: Arab 95%, Armenian 4%, other 1%
Religions: Islam 70% (5 legally recognized Islamic groups—Alawite or Nusayri, Druze, Isma'ilite, Shi'a, Sunni), Christian 30% (11 legally recognized Christian groups—4 Orthodox Christian, 6 Catholic, 1 Protestant), Judaism NEGL%
Languages: Arabic (official), French (official), Armenian, English
Literacy: age 15 and over can read and write (1990 est.)
total population: 80%
male: 88%
female: 73%
Labor force: 650,000
by occupation: industry, commerce, and services 79%, agriculture 11%, government 10% (1985)

Government

Names:
conventional long form: Republic of Lebanon
conventional short form: Lebanon
local long form: Al Jumhuriyah al Lubnaniyah
local short form: none
Digraph: LE
Type: republic
Capital: Beirut
Administrative divisions: 5 governorates (muhafazat, singular—muhafazah); Al Biqa, 'Al Janub, Ash Shamal, Bayrut, Jabal Lubnan
Independence: 22 November 1943 (from League of Nations mandate under French administration)
National holiday: Independence Day, 22 November (1943)
Constitution: 23 May 1926, amended a number of times
Legal system: mixture of Ottoman law, canon law, Napoleonic code, and civil law; no judicial review of legislative acts; has not accepted compulsory ICJ jurisdiction
Suffrage: 21 years of age; compulsory for all males; authorized for women at age 21 with elementary education
Executive branch:
chief of state: President Ilyas HARAWI (since 24 November 1989); note—by custom, the president is a Maronite Christian, the prime minister is a Sunni Muslim, and the speaker of the legislature is a Shi'a Muslim
head of government: Prime Minister Rafiq HARIRI (since 22 October 1992)
cabinet: Cabinet; chosen by the president in consultation with the members of the National Assembly
Legislative branch: unicameral
National Assembly: (Arabic—Majlis Alnuwab, French—Assemblee Nationale) Lebanon's first legislative election in 20 years was held in the summer of 1992; the National Assembly is composed of 128 deputies, one-half Christian and one-half Muslim; its mandate expires in 1996
Judicial branch: four Courts of Cassation (three courts for civil and commercial cases and one court for criminal cases)
Political parties and leaders: political party activity is organized along largely sectarian lines; numerous political groupings exist, consisting of individual political figures and followers motivated by religious, clan, and economic considerations
Member of: ABEDA, ACCT, AFESD, AL, AMF, CCC, ESCWA, FAO, G-24, G-77, IAEA, IBRD, ICAO, ICC, ICFTU, ICRM, IDA, IDB, IFAD, IFC, IFRCS, ILO, IMF, IMO, INTELSAT, INTERPOL, IOC, ITU, NAM, OIC, PCA, UN, UNCTAD, UNESCO, UNHCR, UNIDO, UNRWA, UPU, WFTU, WHO, WIPO, WMO, WTO
Diplomatic representation in US:
chief of mission: Ambassador Riyad TABBARAH
chancery: 2560 28th Street NW, Washington, DC 20008
telephone: [1] (202) 939-6300
FAX: [1] (202) 939-6324
consulate(s) general: Detroit, New York, and Los Angeles
US diplomatic representation:
chief of mission: (vacant)
embassy: Antelias, Beirut
mailing address: P. O. Box 70-840, Beirut; PSC 815, Box 2, Beirut; FPO AE 09836-0002
telephone: [961] (1) 402200, 403300, 416502, 426183, 417774
FAX: [961] (1) 407112
Flag: three horizontal bands of red (top), white (double width), and red with a green and brown cedar tree centered in the white band

Economy

Overview: The 1975-91 civil war seriously damaged Lebanon's economic infrastructure, cut national output by half, and all but ended Lebanon's position as a Middle Eastern entrepot and banking hub. A tentative peace has enabled the central government to begin restoring control in Beirut, collect taxes, and regain access to key port and government facilities. The battered economy has also been propped up by a financially sound banking system and resilient small- and medium-scale manufacturers. Family remittances, banking transactions, manufactured and farm exports,

the narcotics trade, and international emergency aid are the main sources of foreign exchange. In the relatively settled year of 1991, industrial production, agricultural output, and exports showed substantial gains. The further rebuilding of the war-ravaged country was delayed in 1992 because of an upturn in political wrangling. In October 1992, Rafiq HARIRI was appointed Prime Minister. HARIRI, a wealthy entrepreneur, announced ambitious plans for Lebanon's reconstruction which involve a substantial influx of foreign aid and investment. Progress on restoring basic services is limited. Since Prime Minister HARIRI's appointment, the most significant improvement lies in the stabilization of the Lebanese pound, which had gained over 30% in value by yearend 1993. The years 1993 and 1994 were marked by efforts of the new administration to encourage domestic and foreign investment and to obtain additional international assistance. The construction sector led the 8.5% advance in real GDP in 1994.
National product: GDP—purchasing power parity—$15.8 billion (1994 est.)
National product real growth rate: 8.5% (1994 est.)
National product per capita: $4,360 (1994 est.)
Inflation rate (consumer prices): 12% (1994 est.)
Unemployment rate: 35% (1993 est.)
Budget:
revenues: $1.4 billion
expenditures: $3.2 billion (1994 est.)
Exports: $925 million (f.o.b., 1993 est.)
commodities: agricultural products, chemicals, textiles, precious and semiprecious metals and jewelry, metals and metal products
partners: Saudi Arabia 21%, Switzerland 9.5%, Jordan 6%, Kuwait 12%, US 5%
Imports: $4.1 billion (c.i.f., 1993 est.)
commodities: consumer goods, machinery and transport equipment, petroleum products
partners: Italy 14%, France 12%, US 6%, Turkey 5%, Saudi Arabia 3%
External debt: $765 million (1994 est.)
Industrial production: growth rate 25% (1993 est.)
Electricity:
capacity: 1,220,000 kW
production: 2.5 billion kWh
consumption per capita: 676 kWh (1993)
Industries: banking, food processing, textiles, cement, oil refining, chemicals, jewelry, some metal fabricating
Agriculture: principal products—citrus fruits, vegetables, potatoes, olives, tobacco, hemp (hashish), sheep, goats; not self-sufficient in grain
Illicit drugs: illicit producer of hashish and heroin for the international drug trade; hashish production is shipped to Western Europe, the Middle East, and North and South America; increasingly a key locus of cocaine processing and trafficking; a Lebanese/Syrian 1994 eradication campaign eliminated the opium crop and caused a 50% decrease in the cannabis crop
Economic aid: the government estimates that it has received $1.7 billion in aid and has an additional $725 million in commitments to support its $3 billion National Emergency Recovery Program
Currency: 1 Lebanese pound (£L) = 100 piasters
Exchange rates: Lebanese pounds (£L) per US$1—1,644.6 (January 1995), 1,680.1 (1994), 1,741.4 (1993), 1,712.8 (1992), 928.23 (1991), 695.09 (1990)
Fiscal year: calendar year

Transportation

Railroads:
total: 222 km
standard gauge: 222 km 1.435-m
note: system in disrepair, considered inoperable
Highways:
total: 7,300 km
paved: 6,200 km
unpaved: gravel 450 km; improved earth 650 km
Pipelines: crude oil 72 km (none in operation)
Ports: Al Batrun, Al Mina, An Naqurah, Antilyas, Az Zahrani, Beirut, Jubayl, Juniyah, Shikka Jadidah, Sidon, Tripoli, Tyre
Merchant marine:
total: 64 ships (1,000 GRT or over) totaling 260,383 GRT/381,937 DWT
ships by type: bulk 4, cargo 41, chemical tanker 1, combination bulk 1, combination ore/oil 1, container 2, livestock carrier 6, refrigerated cargo 3, roll-on/roll-off cargo 2, specialized tanker 1, vehicle carrier 2
Airports:
total: 9
with paved runways over 3,047 m: 1
with paved runways 2,438 to 3,047 m: 2
with paved runways 1,524 to 2,437 m: 2
with paved runways 914 to 1,523 m: 1
with paved runways under 914 m: 2
with unpaved runways 914 to 1,523 m: 1

Communications

Telephone system: 325,000 telephones; 95 telephones/1,000 persons; telecommunications system severely damaged by civil war; rebuilding still underway
local: NA
intercity: primarily microwave radio relay and cable
international: 2 INTELSAT (1 Indian Ocean and 1 Atlantic Ocean) satellite links (erratic operations); coaxial cable to Syria; microwave radio relay to Syria but inoperable beyond Syria to Jordan; 3 submarine coaxial cables
Radio:
broadcast stations: AM 5, FM 3, shortwave 0
note: numerous AM and FM stations are operated sporadically by various factions
radios: NA
Television:
broadcast stations: 13
televisions: NA

Defense Forces

Branches: Lebanese Armed Forces (LAF; includes Army, Navy, and Air Force)
Manpower availability: males age 15-49 857,698; males fit for military service 533,640 (1995 est.)
Defense expenditures: exchange rate conversion—$278 million, 5.5% of GDP (1994)

Lesotho

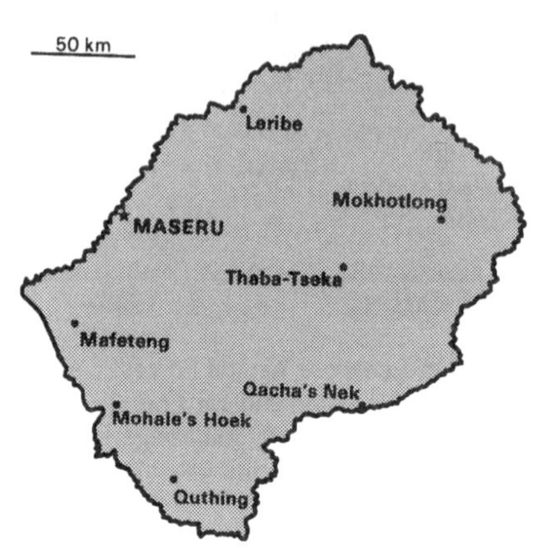

Geography

Location: Southern Africa, an enclave of South Africa
Map references: Africa
Area:
total area: 30,350 sq km
land area: 30,350 sq km
comparative area: slightly larger than Maryland
Land boundaries: total 909 km, South Africa 909 km
Coastline: 0 km (landlocked)
Maritime claims: none; landlocked
International disputes: none
Climate: temperate; cool to cold, dry winters; hot, wet summers
Terrain: mostly highland with plateaus, hills, and mountains
Natural resources: water, agricultural and grazing land, some diamonds and other minerals
Land use:
arable land: 10%
permanent crops: 0%
meadows and pastures: 66%
forest and woodland: 0%
other: 24%
Irrigated land: NA sq km
Environment:
current issues: population pressure forcing settlement in marginal areas results in overgrazing, severe soil erosion, soil exhaustion; desertification; Highlands Water Project will control, store, and redirect water to South Africa
natural hazards: periodic droughts
international agreements: party to—Biodiversity, Marine Life Conservation, Ozone Layer Protection, Wetlands; signed, but not ratified—Climate Change, Desertification, Endangered Species, Law of the Sea, Marine Dumping
Note: landlocked; surrounded by South Africa

People

Population: 1,992,960 (July 1995 est.)
Age structure:
0-14 years: 41% (female 407,213; male 416,709)
15-64 years: 54% (female 558,106; male 520,961)
65 years and over: 5% (female 51,809; male 38,162) (July 1995 est.)
Population growth rate: 2.44% (1995 est.)
Birth rate: 33.39 births/1,000 population (1995 est.)
Death rate: 8.96 deaths/1,000 population (1995 est.)
Net migration rate: 0 migrant(s)/1,000 population (1995 est.)
Infant mortality rate: 67.4 deaths/1,000 live births (1995 est.)
Life expectancy at birth:
total population: 62.56 years
male: 60.74 years
female: 64.43 years (1995 est.)
Total fertility rate: 4.41 children born/woman (1995 est.)
Nationality:
noun: Mosotho (singular), Basotho (plural)
adjective: Basotho
Ethnic divisions: Sotho 99.7%, Europeans 1,600, Asians 800
Religions: Christian 80%, rest indigenous beliefs
Languages: Sesotho (southern Sotho), English (official), Zulu, Xhosa
Literacy: age 15 and over can read and write (1966)
total population: 59%
male: 44%
female: 68%
Labor force: 689,000 economically active
by occupation: 86.2% of resident population engaged in subsistence agriculture; roughly 60% of the active male wage earners work in South Africa

Government

Names:
conventional long form: Kingdom of Lesotho
conventional short form: Lesotho
former: Basutoland
Digraph: LT
Type: constitutional monarchy
Capital: Maseru
Administrative divisions: 10 districts; Berea, Butha-Buthe, Leribe, Mafeteng, Maseru, Mohale's Hoek, Mokhotlong, Qacha's Nek, Quthing, Thaba-Tseka
Independence: 4 October 1966 (from UK)
National holiday: Independence Day, 4 October (1966)
Constitution: 2 April 1993
Legal system: based on English common law and Roman-Dutch law; judicial review of legislative acts in High Court and Court of Appeal; has not accepted compulsory ICJ jurisdiction
Suffrage: 21 years of age; universal
Executive branch:
chief of state: King MOSHOESHOE II (since February 1995)
head of government: Prime Minister Ntsu MOKHEHLE (since 2 April 1993)
cabinet: Cabinet
Legislative branch: bicameral Parliament consisting of the Assembly or lower house whose members are chosen by popular election and the Senate or upper house whose members consist of the 22 principal chiefs and 11 other members appointed by the ruling party; election last held in March 1993 (first since 1971); all 65 seats in the Assembly were won by the BCP
Judicial branch: High Court, Court of Appeal, Magistrate's Court, customary or traditional court
Political parties and leaders: Basotho National Party (BNP), Evaristus SEKHONYANA; Basotho Congress Party (BCP), Ntsu MOKHEHLE; National Independent Party (NIP), A. C. MANYELI; Marematlou Freedom Party (MFP), Vincent MALEBO; United Democratic Party, Charles MOFELI; Communist Party of Lesotho (CPL), Jacob M. KENA
Member of: ACP, AfDB, C, CCC, ECA, FAO, G-77, GATT, IBRD, ICAO, ICFTU, ICRM, IDA, IFAD, IFC, IFRCS, ILO, IMF, INTELSAT (nonsignatory user), INTERPOL, IOC, ITU, NAM, OAU, SACU, SADC, UN, UNCTAD, UNESCO, UNHCR, UNIDO, UPU, WCL, WFTU, WHO, WIPO, WMO, WTO
Diplomatic representation in US:
chief of mission: (vacant); Charge d'Affaires ad interim Mokhali A. LITHEBE (since 2 July 1994)
chancery: 2511 Massachusetts Avenue NW, Washington, DC 20008
telephone: [1] (202) 797-5533 through 5536
FAX: [1] (202) 234-6815
US diplomatic representation:
chief of mission: Ambassador Myrick BISMARCK
embassy: address NA, Maseru
mailing address: P. O. Box 333, Maseru 100, Lesotho
telephone: [266] 312666
FAX: [266] 310116
Flag: divided diagonally from the lower hoist side corner; the upper half is white bearing the brown silhouette of a large shield with crossed spear and club; the lower half is a diagonal blue band with a green triangle in the corner

Economy

Overview: Small, landlocked, and mountainous, Lesotho has no important natural resources other than water. Its economy is based on agriculture, light manufacturing, and remittances from laborers employed in South Africa (these remittances supplement domestic income by as much as 45%). The great majority of households gain their livelihoods from subsistence farming and migrant labor; a large portion of the adult male work force is employed in South African mines. Manufacturing depends largely on farm products to support the milling, canning, leather, and jute industries; other industries include textile, clothing, and construction. Although drought has decreased agricultural activity over the past few years, improvement of a major hydropower facility will permit the sale of water to South Africa and allow Lesotho's economy to continue its moderate growth.
National product: GDP—purchasing power parity—$2.6 billion (1994 est.)
National product real growth rate: 6% (1994 est.)
National product per capita: $1,340 (1994 est.)
Inflation rate (consumer prices): 13.9% (1993)
Unemployment rate: substantial unemployment and underemployment
Budget:
revenues: $438 million
expenditures: $430 million, including capital expenditures of $155 million (FY93/94 est.)
Exports: $109 million (f.o.b., 1992)
commodities: wool, mohair, wheat, cattle, peas, beans, corn, hides, skins, baskets
partners: South Africa 42%, EC 28%, North and South America 25% (1991)
Imports: $964 million (c.i.f., 1992)
commodities: mainly corn, building materials, clothing, vehicles, machinery, medicines, petroleum
partners: South Africa 94%, Asia 3%, EC 1% (1991)
External debt: $512 million (1993)
Industrial production: growth rate 10%; accounts for 17% of GDP (1993 est.)
Electricity: power supplied by South Africa
Industries: food, beverages, textiles, handicrafts, tourism
Agriculture: accounts for 50% of GDP (1993 est.); exceedingly primitive, mostly subsistence farming and livestock; principal crops corn, wheat, pulses, sorghum, barley
Economic aid:
recipient: US commitments, including Ex-Im (FY70-89), $268 million; US (1992), $10.3 million; US (1993 est.), $10.1 million; Western (non-US) countries, ODA and OOF bilateral commitments (1970-89), $819 million; OPEC bilateral aid (1979-89), $4 million; Communist countries (1970-89), $14 million
Currency: 1 loti (L) = 100 lisente
Exchange rates: maloti (M) per US$1—3.5389 (January 1995), 3.5490 (1994), 3.2636 (1993), 2.8497 (1992), 2.7563 (1991), 2.5863 (1990); note—the Basotho loti is at par with the South African rand
Fiscal year: 1 April—31 March

Transportation

Railroads:
total: 2.6 km; note—owned by, operated by, and included in the statistics of South Africa
narrow gauge: 2.6 km 1.067-m gauge
Highways:
total: 7,215 km
paved: 572 km
unpaved: gravel, stabilized earth 2,337 km; improved earth 1,806 km; unimproved earth 2,500 km (1988)
Ports: none
Airports:
total: 29
with paved runways over 3,047 m: 1
with paved runways 914 to 1,523 m: 1
with paved runways under 914 m: 23
with unpaved runways 914 to 1,523 m: 4

Communications

Telephone system: 5,920 telephones; rudimentary system
local: NA
intercity: consists of a few land lines, a small microwave radio relay system, and a minor radio communication system
international: 1 INTELSAT (Atlantic Ocean) earth station
Radio:
broadcast stations: AM 3, FM 4, shortwave 0
radios: NA
Television:
broadcast stations: 1
televisions: NA

Defense Forces

Branches: Lesotho Defense Force (LDF; includes Army and Air Wing), Lesotho Mounted Police
Manpower availability: males age 15-49 453,844; males fit for military service 244,767 (1995 est.)
Defense expenditures: exchange rate conversion—$25 million, NA% of GDP (1994)

Geography

Location: Western Africa, bordering the North Atlantic Ocean, between Cote d'Ivoire and Sierra Leone
Map references: Africa
Area:
total area: 111,370 sq km
land area: 96,320 sq km
comparative area: slightly larger than Tennessee
Land boundaries: total 1,585 km, Guinea 563 km, Cote d'Ivoire 716 km, Sierra Leone 306 km
Coastline: 579 km
Maritime claims:
territorial sea: 200 nm
International disputes: none
Climate: tropical; hot, humid; dry winters with hot days and cool to cold nights; wet, cloudy summers with frequent heavy showers
Terrain: mostly flat to rolling coastal plains rising to rolling plateau and low mountains in northeast
Natural resources: iron ore, timber, diamonds, gold
Land use:
arable land: 1%
permanent crops: 3%
meadows and pastures: 2%
forest and woodland: 39%
other: 55%
Irrigated land: 20 sq km (1989 est.)
Environment:
current issues: tropical rain forest subject to deforestation; soil erosion; loss of biodiversity; pollution of rivers from the dumping of iron ore tailings and of coastal waters from oil residue and raw sewage
natural hazards: dust-laden harmattan winds blow from the Sahara (December to March)
international agreements: party to—Endangered Species, Nuclear Test Ban, Ship Pollution, Tropical Timber 83, Tropical Timber 94; signed, but not ratified—Biodiversity, Climate Change, Environmental

Modification, Law of the Sea, Marine Dumping, Marine Life Conservation

People

Population: 3,073,245 (July 1995 est.)
Age structure:
0-14 years: 44% (female 674,155; male 680,952)
15-64 years: 52% (female 768,147; male 844,326)
65 years and over: 4% (female 55,575; male 50,090) (July 1995 est.)
Population growth rate: 3.32% (1995 est.)
Birth rate: 43.08 births/1,000 population (1995 est.)
Death rate: 12.05 deaths/1,000 population (1995 est.)
Net migration rate: NA migrant(s)/1,000 population (1995 est.)
note: if the Ghanaian-led peace negotiations, under way in 1995, are successful, many Liberian refugees may return from exile
Infant mortality rate: 110.6 deaths/1,000 live births (1995 est.)
Life expectancy at birth:
total population: 58.17 years
male: 55.67 years
female: 60.75 years (1995 est.)
Total fertility rate: 6.3 children born/woman (1995 est.)
Nationality:
noun: Liberian(s)
adjective: Liberian
Ethnic divisions: indigenous African tribes 95% (including Kpelle, Bassa, Gio, Kru, Grebo, Mano, Krahn, Gola, Gbandi, Loma, Kissi, Vai, and Bella), Americo-Liberians 5% (descendants of former slaves)
Religions: traditional 70%, Muslim 20%, Christian 10%
Languages: English 20% (official), Niger-Congo language group about 20 local languages come from this group
Literacy: age 15 and over can read and write (1990 est.)
total population: 40%
male: 50%
female: 29%
Labor force: 510,000 including 220,000 in the monetary economy
by occupation: agriculture 70.5%, services 10.8%, industry and commerce 4.5%, other 14.2%
note: non-African foreigners hold about 95% of the top-level management and engineering jobs

Government

Names:
conventional long form: Republic of Liberia
conventional short form: Liberia
Digraph: LI
Type: republic
Capital: Monrovia
Administrative divisions: 13 counties; Bomi, Bong, Grand Bassa, Grand Cape Mount, Grand Gedeh, Grand Kru, Lofa, Margibi, Maryland, Montserrado, Nimba, River Cess, Sinoe
Independence: 26 July 1847
National holiday: Independence Day, 26 July (1847)
Constitution: 6 January 1986
Legal system: dual system of statutory law based on Anglo-American common law for the modern sector and customary law based on unwritten tribal practices for indigenous sector
Suffrage: 18 years of age; universal
Executive branch:
chief of state and head of government: Chairman of the Council of State David KPOMAKPOR (since March 1994); election last held on 15 October 1985; results—Gen. Dr. Samuel Kanyon DOE (NDPL) 50.9%, Jackson DOE (LAP) 26.4%, other 22.7%
note: constitutional government ended in September 1990 when President Samuel Kanyon DOE was killed by rebel forces; civil war ensued and in July 1993 the Cotonou Peace Treaty was negotiated by the major warring factions under UN auspices; a transitional coalition government under David KROMAKPOR was formed in March 1994 but has been largely ineffective and unable to implement the provisions of the peace treaty; Ghanaian-led negotiations are now underway to seat a new interim government that would oversee elections proposed for late 1995
cabinet: Cabinet; selected by the leaders of the major factions in the civil war
Legislative branch: unicameral Transitional Legislative Assembly, the members of which are appointed by the leaders of the major factions in the civil war
note: the former bicameral legislature no longer exists and there is no assurance that it will be reconstituted very soon
Judicial branch: Supreme Court
Political parties and leaders: National Democratic Party of Liberia (NDPL), Augustus CAINE, chairman; Liberian Action Party (LAP), Emmanuel KOROMAH, chairman; Unity Party (UP), Joseph KOFA, chairman; United People's Party (UPP), Gabriel Baccus MATTHEWS, chairman; National Patriotic Party (NPP), Charles TAYLOR, chairman; Liberian Peoples Party (LPP), Dusty WOLOKOLLIE, chairman
Member of: ACP, AfDB, CCC, ECA, ECOWAS, FAO, G-77, IAEA, IBRD, ICAO, ICFTU, ICRM, IDA, IFAD, IFC, IFRCS, ILO, IMF, IMO, INMARSAT, INTELSAT (nonsignatory user), INTERPOL, IOC, ITU, NAM, OAU, UN, UNCTAD, UNESCO, UNIDO, UPU, WCL, WFTU, WHO, WIPO, WMO
Diplomatic representation in US:
chief of mission: (vacant); Charge d'Affaires Konah K. BLACKETT
chancery: 5201 16th Street NW, Washington, DC 20011
telephone: [1] (202) 723-0437
consulate(s) general: New York
US diplomatic representation:
chief of mission: (vacant); Charge d'Affaires William P. TWADDELL
embassy: 111 United Nations Drive, Monrovia
mailing address: P. O. Box 100098, Mamba Point, Monrovia
telephone: [231] 222991 through 222994
FAX: [231] 223710
Flag: 11 equal horizontal stripes of red (top and bottom) alternating with white; there is a white five-pointed star on a blue square in the upper hoist-side corner; the design was based on the US flag

Economy

Overview: Civil war since 1990 has destroyed much of Liberia's economy, especially the infrastructure in and around Monrovia. Businessmen have fled the country, taking capital and expertise with them. Many will not return. Richly endowed with water, mineral resources, forests, and a climate favorable to agriculture, Liberia had been a producer and exporter of basic products, while local manufacturing, mainly foreign owned, had been small in scope. Political instability threatens prospects for economic reconstruction and repatriation of some 750,000 Liberian refugees who have fled to neighboring countries. The political impasse between the interim government and rebel leader Charles TAYLOR has prevented restoration of normal economic life, including the re-establishment of a strong central government with effective economic development programs. The economy deteriorated further in 1994.
National product: GDP—purchasing power parity—$2.3 billion (1994 est.)
National product real growth rate: NA%
National product per capita: $770 (1994 est.)
Inflation rate (consumer prices): NA%
Unemployment rate: NA%
Budget:
revenues: $242.1 million
expenditures: $435.4 million, including capital expenditures of $29.5 million (1989 est.)
Exports: $505 million (f.o.b., 1989 est.)
commodities: iron ore 61%, rubber 20%, timber 11%, coffee
partners: US, EC, Netherlands
Imports: $394 million (c.i.f., 1989 est.)
commodities: mineral fuels, chemicals, machinery, transportation equipment, rice and other foodstuffs
partners: US, EC, Japan, China, Netherlands, ECOWAS
External debt: $2.1 billion (September 1993 est.)

Industrial production: growth rate NA% (1993-94); much industrial damage caused by factional warfare
Electricity:
capacity: 330,000 kW
production: 440 million kWh
consumption per capita: 143 kWh (1993)
Industries: rubber processing, food processing, construction materials, furniture, palm oil processing, mining (iron ore, diamonds)
Agriculture: accounts for about 40% of GDP (including fishing and forestry); principal products—rubber, timber, coffee, cocoa, rice, cassava, palm oil, sugarcane, bananas, sheep, goats; not self-sufficient in food, imports 25% of rice consumption
Illicit drugs: increasingly a transshipment point for heroin and cocaine
Economic aid:
recipient: US commitments, including Ex-Im (FY70-89), $665 million; Western (non-US) countries, ODA and OOF bilateral commitments (1970-89), $870 million; OPEC bilateral aid (1979-89), $25 million; Communist countries (1970-89), $77 million
Currency: 1 Liberian dollar (L$) = 100 cents
Exchange rates: Liberian dollars (L$) per US$1—1.00 (officially fixed rate since 1940); unofficial parallel exchange rate of US$1—7 (January 1992), unofficial rate floats against the US dollar
Fiscal year: calendar year

Transportation

Railroads:
total: 490 km (single track); note—three rail systems owned and operated by foreign steel and financial interests in conjunction with Liberian Government; one of these, the Lamco Railroad, closed in 1989 after iron ore production ceased; the other two have been shut down by the civil war
standard gauge: 345 km 1.435-m gauge
narrow gauge: 145 km 1.067-m gauge
Highways:
total: 10,087 km
paved: 603 km
unpaved: gravel 5,171 km (includes 2,323 km of private roads of rubber and timber firms, open to the public); earth 4,313 km
Ports: Buchanan, Greenville, Harper, Monrovia
Merchant marine:
total: 1,549 ships (1,000 GRT or over) totaling 56,709,634 GRT/97,038,680 DWT
ships by type: barge carrier 3, bulk 392, cargo 121, chemical tanker 114, combination bulk 33, combination ore/oil 57, container 124, liquefied gas tanker 75, oil tanker 459, passenger 32, passenger-cargo 1, refrigerated cargo 58, roll-on/roll-off cargo 18, short-sea passenger 1, specialized tanker 7, vehicle carrier 54
note: a flag of convenience registry; includes 53 countries; the 10 major fleet flags are: United States 232 ships, Japan 190, Norway 166, Greece 125, Germany 125, United Kingdom 102, Hong Kong 95, China 45, Russia 41, and the Netherlands 34
Airports:
total: 59
with paved runways over 3,047 m: 1
with paved runways 1,524 to 2,437 m: 1
with paved runways under 914 m: 43
with unpaved runways 1,524 to 2,438 m: 3
with unpaved runways 914 to 1,523 m: 11

Communications

Telephone system: NA telephones; telephone and telegraph service via radio relay network; main center is Monrovia; most telecommunications services inoperable due to insurgency movement
local: NA
intercity: NA
international: 1 INTELSAT (Atlantic Ocean) earth station
Radio:
broadcast stations: AM 3, FM 4, shortwave 0
radios: NA
Television:
broadcast stations: 5
televisions: NA

Defense Forces

Branches: NA; the ultimate structure of the Liberian military force will depend on who is the victor in the ongoing civil war
Manpower availability: males age 15-49 732,063; males fit for military service 390,849 (1995 est.)
Defense expenditures: exchange rate conversion—$30 million, 2% of GDP (1994)

Libya

Geography

Location: Northern Africa, bordering the Mediterranean Sea, between Egypt and Tunisia
Map references: Africa
Area:
total area: 1,759,540 sq km
land area: 1,759,540 sq km
comparative area: slightly larger than Alaska
Land boundaries: total 4,383 km, Algeria 982 km, Chad 1,055 km, Egypt 1,150 km, Niger 354 km, Sudan 383 km, Tunisia 459 km
Coastline: 1,770 km
Maritime claims:
territorial sea: 12 nm
Gulf of Sidra closing line: 32 degrees 30 minutes north
International disputes: the International Court of Justice (ICJ) ruled in February 1994 that the 100,000 sq km Aozou Strip between Chad and Libya belongs to Chad, and that Libya must withdraw from it by 31 May 1994; Libya has withdrawn some its forces in response to the ICJ ruling, but still maintains an airfield in the disputed area; maritime boundary dispute with Tunisia; claims part of northern Niger and part of southeastern Algeria
Climate: Mediterranean along coast; dry, extreme desert interior
Terrain: mostly barren, flat to undulating plains, plateaus, depressions
Natural resources: petroleum, natural gas, gypsum
Land use:
arable land: 2%
permanent crops: 0%
meadows and pastures: 8%
forest and woodland: 0%
other: 90%
Irrigated land: 2,420 sq km (1989 est.)
Environment:
current issues: desertification; very limited natural fresh water resources; the Great Manmade River Project, the largest water development scheme in the world, is being built to bring water from large aquifers under the Sahara to coastal cities

natural hazards: hot, dry, dust-laden ghibli is a southern wind lasting one to four days in spring and fall; duststorms, sandstorms
international agreements: party to—Marine Dumping, Nuclear Test Ban, Ozone Layer Protection; signed, but not ratified—Biodiversity, Climate Change, Desertification, Law of the Sea

People

Population: 5,248,401 (July 1995 est.)
Age structure:
0-14 years: 48% (female 1,226,851; male 1,269,813)
15-64 years: 49% (female 1,261,424; male 1,331,093)
65 years and over: 3% (female 76,017; male 83,203) (July 1995 est.)
Population growth rate: 3.7% (1995 est.)
Birth rate: 44.89 births/1,000 population (1995 est.)
Death rate: 7.91 deaths/1,000 population (1995 est.)
Net migration rate: 0 migrant(s)/1,000 population (1995 est.)
Infant mortality rate: 61.4 deaths/1,000 live births (1995 est.)
Life expectancy at birth:
total population: 64.29 years
male: 62.12 years
female: 66.57 years (1995 est.)
Total fertility rate: 6.32 children born/woman (1995 est.)
Nationality:
noun: Libyan(s)
adjective: Libyan
Ethnic divisions: Berber and Arab 97%, Greeks, Maltese, Italians, Egyptians, Pakistanis, Turks, Indians, Tunisians
Religions: Sunni Muslim 97%
Languages: Arabic, Italian, English, all are widely understood in the major cities
Literacy: age 15 and over can read and write (1984)
total population: 60%
male: 77%
female: 42%
Labor force: 1 million (includes about 280,000 resident foreigners)
by occupation: industry 31%, services 27%, government 24%, agriculture 18%

Government

Names:
conventional long form: Socialist People's Libyan Arab Jamahiriya
conventional short form: Libya
local long form: Al Jumahiriyah al Arabiyah al Libiyah ash Shabiyah al Ishirakiyah
local short form: none
Digraph: LY
Type: Jamahiriya (a state of the masses) in theory, governed by the populace through local councils; in fact, a military dictatorship
Capital: Tripoli
Administrative divisions: 25 municipalities (baladiyah, singular—baladiyat); Ajdabiya, Al 'Aziziyah, Al Fatih, Al Jabal al Akhdar, Al Jufrah, Al Khums, Al Kufrah, An Nuqat al Khams, Ash Shati', Awbari, Az Zawiyah, Banghazi, Darnah, Ghadamis, Gharyan, Misratah, Murzuq, Sabha, Sawfajjin, Surt, Tarabulus, Tarhunah, Tubruq, Yafran, Zlitan
Independence: 24 December 1951 (from Italy)
National holiday: Revolution Day, 1 September (1969)
Constitution: 11 December 1969, amended 2 March 1977
Legal system: based on Italian civil law system and Islamic law; separate religious courts; no constitutional provision for judicial review of legislative acts; has not accepted compulsory ICJ jurisdiction
Suffrage: 18 years of age; universal and compulsory
Executive branch:
chief of state: Revolutionary Leader Col. Mu'ammar Abu Minyar al-QADHAFI (since 1 September 1969)
head of government: Chairman of the General People's Committee (Premier) Abd al Majid al-Qa'ud (since 29 January 1994)
cabinet: General People's Committee; established by the General People's Congress
note: national elections are indirect through a hierarchy of peoples' committees
Legislative branch: unicameral
General People's Congress: national elections are indirect through a hierarchy of peoples' committees
Judicial branch: Supreme Court
Political parties and leaders: none
Other political or pressure groups: various Arab nationalist movements with almost negligible memberships may be functioning clandestinely, as well as some Islamic elements
Member of: ABEDA, AfDB, AFESD, AL, AMF, AMU, CAEU, CCC, ECA, FAO, G-77, IAEA, IBRD, ICAO, ICRM, IDA, IDB, IFAD, IFC, IFRCS, ILO, IMF, IMO, INTELSAT, INTERPOL, IOC, ISO, ITU, NAM, OAPEC, OAU, OIC, OPEC, UN, UNCTAD, UNESCO, UNIDO, UNITAR, UPU, WFTU, WHO, WIPO, WMO, WTO
Diplomatic representation in US: none
US diplomatic representation: none
Flag: plain green; green is the traditional color of Islam (the state religion)

Economy

Overview: The socialist-oriented economy depends primarily upon revenues from the oil sector, which contributes practically all export earnings and about one-third of GDP. In 1990 per capita GDP was the highest in Africa at $5,410, but GDP growth rates have slowed and fluctuated sharply in response to changes in the world oil market. Import restrictions and inefficient resource allocations have led to periodic shortages of basic goods and foodstuffs. Windfall revenues from the hike in world oil prices in late 1990 improved the foreign payments position and resulted in a current account surplus through 1992. The nonoil manufacturing and construction sectors, which account for about 20% of GDP, have expanded from processing mostly agricultural products to include petrochemicals, iron, steel, and aluminum. Although agriculture accounts for only 5% of GDP, it employs 18% of the labor force. Climatic conditions and poor soils severely limit farm output, and Libya imports about 75% of its food requirements. The UN sanctions imposed in April 1992 have not yet had a major impact on the economy because Libya's oil revenues generate sufficient foreign exchange which sustains imports of food, consumer goods, and equipment for the oil industry and ongoing development projects.
National product: GDP—purchasing power parity—$32.9 billion (1994 est.)
National product real growth rate: -0.9% (1994 est.)
National product per capita: $6,510 (1994 est.)
Inflation rate (consumer prices): 25% (1993 est.)
Unemployment rate: NA%
Budget:
revenues: $8.1 billion
expenditures: $9.8 billion, including capital expenditures of $3.1 billion (1989 est.)
Exports: $7.2 billion (f.o.b., 1994 est.)
commodities: crude oil, refined petroleum products, natural gas
partners: Italy, Germany, Spain, France, UK, Turkey, Greece, Egypt
Imports: $6.9 billion (f.o.b., 1994 est.)
commodities: machinery, transport equipment, food, manufactured goods
partners: Italy, Germany, UK, France, Spain, Turkey, Tunisia, Eastern Europe
External debt: $3.5 billion excluding military debt (1991 est.)
Industrial production: growth rate 10.5% (1990)
Electricity:
capacity: 4,600,000 kW
production: 16.1 billion kWh
consumption per capita: 3,078 kWh (1993)
Industries: petroleum, food processing, textiles, handicrafts, cement
Agriculture: 5% of GDP; cash crops—wheat, barley, olives, dates, citrus fruits, peanuts; 75% of food is imported
Economic aid:
recipient: Western (non-US) countries, ODA and OOF bilateral commitments (1970-87), $242 million
note: no longer a recipient
Currency: 1 Libyan dinar (LD) = 1,000 dirhams

Exchange rates: Libyan dinars (LD) per US$1—0.3555 (January 1995), 0.3596 (1994), 0.3250 (1993), 0.3013 (1992), 0.2684 (1991), 0.2699 (1990)
Fiscal year: calendar year

Transportation

Railroads:
note: Libya has had no railroad in operation since 1965, all previous systems having been dismantled; current plans are to construct a 1.435-m standard gauge line from the Tunisian frontier to Tripoli and Misratah, then inland to Sabha, center of a mineral-rich area, but there has been no progress; other plans made jointly with Egypt would establish a rail line from As Sallum, Egypt, to Tobruk with completion set for mid-1994; no progress has been reported
Highways:
total: 19,300 km
paved: bituminous 10,800 km
unpaved: gravel, earth 8,500 km
Inland waterways: none
Pipelines: crude oil 4,383 km; petroleum products 443 km (includes liquified petroleum gas 256 km); natural gas 1,947 km
Ports: Al Khums, Banghazi, Darnah, Marsa al Burayqah, Misratah, Ra's Lanuf, Tobruk, Tripoli, Zuwarah
Merchant marine:
total: 30 ships (1,000 GRT or over) totaling 686,136 GRT/1,208,194 DWT
ships by type: cargo 10, chemical tanker 1, liquefied gas tanker 2, oil tanker 10, roll-on/roll-off cargo 3, short-sea passenger 4
Airports:
total: 146
with paved runways over 3,047 m: 24
with paved runways 2,438 to 3,047 m: 5
with paved runways 1,524 to 2,437 m: 22
with paved runways 914 to 1,523 m: 6
with paved runways under 914 m: 21
with unpaved runways over 3,047 m: 4
with unpaved runways 2,438 to 3,047 m: 3
with unpaved runways 1,524 to 2,438 m: 17
with unpaved runways 914 to 1,523 m: 44

Communications

Telephone system: 370,000 telephones; modern telecommunications system
local: NA
intercity: microwave radio relay, coaxial cable, tropospheric scatter, and 14 domestic satellites
international: 2 INTELSAT (1 Atlantic Ocean and 1 Indian Ocean) earth stations; submarine cables to France and Italy; microwave radio relay to Tunisia and Egypt; tropospheric scatter to Greece; planned ARABSAT and Intersputnik satellite earth stations
Radio:
broadcast stations: AM 17, FM 3, shortwave 0
radios: NA
Television:
broadcast stations: 12
televisions: NA

Defense Forces

Branches: Armed Peoples of the Libyan Arab Jamahiriyah (includes Army, Navy, and Air and Air Defense Command), Police
Manpower availability: males age 15-49 1,131,175; males fit for military service 672,571; males reach military age (17) annually 54,676 (1995 est.)
Defense expenditures: exchange rate conversion—$1.4 billion, 6.1% of GDP (1994 est.)

Liechtenstein

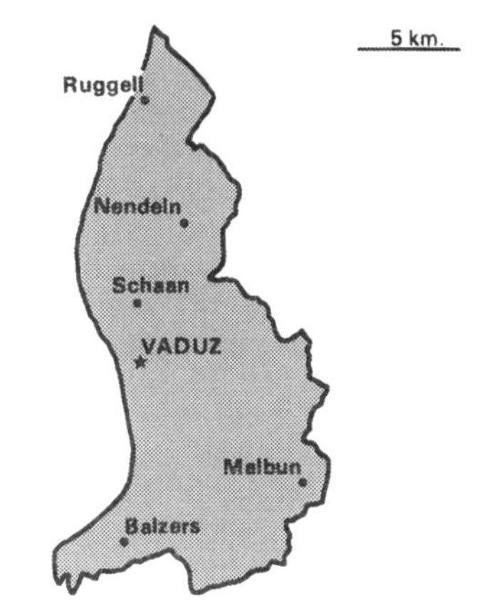

Geography

Location: Central Europe, between Austria and Switzerland
Map references: Europe
Area:
total area: 160 sq km
land area: 160 sq km
comparative area: about 0.9 times the size of Washington, DC
Land boundaries: total 78 km, Austria 37 km, Switzerland 41 km
Coastline: 0 km (landlocked)
Maritime claims: none; landlocked
International disputes: claims 1,600 square kilometers of Czech territory confiscated from its royal family in 1918; the Czech Republic insists that restitution does not go back before February 1948, when the Communists seized power
Climate: continental; cold, cloudy winters with frequent snow or rain; cool to moderately warm, cloudy, humid summers
Terrain: mostly mountainous (Alps) with Rhine Valley in western third
Natural resources: hydroelectric potential
Land use:
arable land: 25%
permanent crops: 0%
meadows and pastures: 38%
forest and woodland: 19%
other: 18%
Irrigated land: NA sq km
Environment:
current issues: NA
natural hazards: NA
international agreements: party to—Air Pollution, Air Pollution-Nitrogen Oxides, Air Pollution-Sulphur 85, Air Pollution-Volatile Organic Compounds, Climate Change, Endangered Species, Hazardous Wastes, Ozone Layer Protection, Wetlands; signed, but not ratified—Air Pollution-Sulphur 94, Biodiversity, Law of the Sea
Note: landlocked; variety of microclimatic variations based on elevation

People

Population: 30,654 (July 1995 est.)
Age structure:
0-14 years: 19% (female 2,897; male 2,974)
15-64 years: 71% (female 10,853; male 10,777)
65 years and over: 10% (female 1,930; male 1,223) (July 1995 est.)
Population growth rate: 1.2% (1995 est.)
Birth rate: 12.95 births/1,000 population (1995 est.)
Death rate: 6.56 deaths/1,000 population (1995 est.)
Net migration rate: 5.58 migrant(s)/1,000 population (1995 est.)
Infant mortality rate: 5.3 deaths/1,000 live births (1995 est.)
Life expectancy at birth:
total population: 77.52 years
male: 73.86 years
female: 81.17 years (1995 est.)
Total fertility rate: 1.47 children born/woman (1995 est.)
Nationality:
noun: Liechtensteiner(s)
adjective: Liechtenstein
Ethnic divisions: Alemannic 95%, Italian and other 5%
Religions: Roman Catholic 87.3%, Protestant 8.3%, unknown 1.6%, other 2.8% (1988)
Languages: German (official), Alemannic dialect
Literacy: age 10 and over can read and write (1981)
total population: 100%
male: 100%
female: 100%
Labor force: 19,905 of which 11,933 are foreigners; 6,885 commute from Austria and Switzerland to work each day
by occupation: industry, trade, and building 53.2%, services 45%, agriculture, fishing, forestry, and horticulture 1.8% (1990)

Government

Names:
conventional long form: Principality of Liechtenstein
conventional short form: Liechtenstein
local long form: Furstentum Liechtenstein
local short form: Liechtenstein
Digraph: LS
Type: hereditary constitutional monarchy
Capital: Vaduz
Administrative divisions: 11 communes (gemeinden, singular—gemeinde); Balzers, Eschen, Gamprin, Mauren, Planken, Ruggell, Schaan, Schellenberg, Triesen, Triesenberg, Vaduz
Independence: 23 January 1719 (Imperial Principality of Liechtenstein established)
National holiday: Assumption Day, 15 August
Constitution: 5 October 1921
Legal system: local civil and penal codes; accepts compulsory ICJ jurisdiction, with reservations
Suffrage: 18 years of age; universal
Executive branch:
chief of state: Prince Hans ADAM II (since 13 November 1989; assumed executive powers 26 August 1984); Heir Apparent Prince ALOIS von und zu Liechtenstein (born 11 June 1968)
head of government: Mario FRICK (since 15 December 1993); Deputy Head of Government Dr. Thomas BUECHEL (since 15 December 1993)
cabinet: Cabinet; elected by the Diet; confirmed by the sovereign
Legislative branch: unicameral
Diet (Landtag): elections last held on 24 October 1993 (next to be held by March 1997); results—VU 50.1%, FBP 41.3%, FL 8.5%; seats—(25 total) VU 13, FBP 11, FL 1
Judicial branch: Supreme Court (Oberster Gerichtshof) for criminal cases, Superior Court (Obergericht) for civil cases
Political parties and leaders: Fatherland Union (VU), Dr. Oswald KRANTZ; Progressive Citizens' Party (FBP), Otmar HASLER; The Free List (FL)
Member of: CE, EBRD, ECE, EFTA, GATT, IAEA, ICRM, IFRCS, INTELSAT, INTERPOL, IOC, ITU, OSCE, UN, UNCTAD, UPU, WCL, WIPO
Diplomatic representation in US: in routine diplomatic matters, Liechtenstein is represented in the US by the Swiss Embassy
US diplomatic representation: the US has no diplomatic or consular mission in Liechtenstein, but the US Consul General at Zurich (Switzerland) has consular accreditation at Vaduz
Flag: two equal horizontal bands of blue (top) and red with a gold crown on the hoist side of the blue band

Economy

Overview: Despite its small size and limited natural resources, Liechtenstein has developed into a prosperous, highly industrialized, free-enterprise economy with a vital service sector and living standards on par with its large European neighbors. Low business taxes—the maximum tax rate is 20%—and easy incorporation rules have induced about 25,000 holding or so-called letter box companies to establish nominal offices in Liechtenstein, providing 30% of state revenues. The country participates in a customs union with Switzerland and uses the Swiss franc as its national currency. Liechtenstein plans to join the European Economic Area (an organization serving as a bridge between EFTA and EU) in 1995.
National product: GDP—purchasing power parity—$630 million (1990 est.)
National product real growth rate: NA%
National product per capita: $22,300 (1990 est.)
Inflation rate (consumer prices): 5.4% (1990)
Unemployment rate: 1.5% (1994)
Budget:
revenues: $259 million
expenditures: $292 million, including capital expenditures of $NA (1990 est.)
Exports: $NA
commodities: small specialty machinery, dental products, stamps, hardware, pottery
partners: EC countries 42.7%, EFTA countries 20.9% (Switzerland 15.4%), other 36.4% (1990)
Imports: $NA
commodities: machinery, metal goods, textiles, foodstuffs, motor vehicles
partners: NA
External debt: $NA
Industrial production: growth rate NA%
Electricity:
capacity: 23,000 kW
production: 150 million kWh
consumption per capita: 5,230 kWh (1992)
Industries: electronics, metal manufacturing, textiles, ceramics, pharmaceuticals, food products, precision instruments, tourism
Agriculture: livestock, vegetables, corn, wheat, potatoes, grapes
Economic aid: none
Currency: 1 Swiss franc, franken, or franco (SwF) = 100 centimes, rappen, or centesimi
Exchange rates: Swiss francs, franken, or franchi (SwF) per US$1—1.2880 (January 1995), 1.3677 (1994), 1.4776 (1993), 1.4062 (1992), 1.4340 (1991), 1.3892 (1990)
Fiscal year: calendar year

Transportation

Railroads:
total: 18.5 km; note—owned, operated, and included in statistics of Austrian Federal Railways
standard gauge: 18.5 km 1.435-m gauge (electrified)
Highways:
total: 322.93 km
paved: 322.93 km
Ports: none
Airports: none

Communications

Telephone system: 25,400 telephones; limited, but sufficient automatic telephone system
local: NA
intercity: NA
international: linked to Swiss networks by cable and radio relay
Radio:
broadcast stations: AM NA, FM NA,

shortwave NA
radios: NA
note: linked to Swiss networks
Television:
broadcast stations: NA
televisions: NA
note: linked to Swiss networks

Defense Forces

Note: defense is responsibility of Switzerland

Lithuania

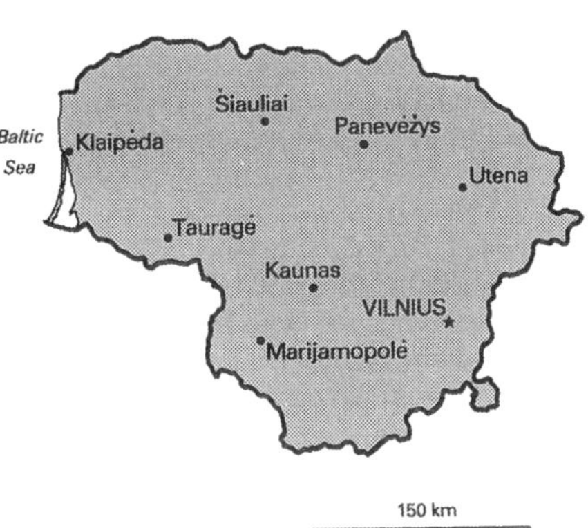

Geography

Location: Eastern Europe, bordering the Baltic Sea, between Latvia and Russia
Map references: Europe
Area:
total area: 65,200 sq km
land area: 65,200 sq km
comparative area: slightly larger than West Virginia
Land boundaries: total 1,273 km, Belarus 502 km, Latvia 453 km, Poland 91 km, Russia (Kaliningrad) 227 km
Coastline: 108 km
Maritime claims:
territorial sea: 12 nm
International disputes: dispute with Russia (Kaliningrad Oblast) over the position of the Nemunas (Nemen) River border presently located on the Lithuanian bank and not in midriver as by international standards
Climate: maritime; wet, moderate winters and summers
Terrain: lowland, many scattered small lakes, fertile soil
Natural resources: peat
Land use:
arable land: 49.1%
permanent crops: 0%
meadows and pastures: 22.2%
forest and woodland: 16.3%
other: 12.4%
Irrigated land: 430 sq km (1990)
Environment:
current issues: contamination of soil and groundwater with petroleum products and chemicals at military bases
natural hazards: NA
international agreements: party to—Ozone Layer Protection, Ship Pollution, Wetlands; signed, but not ratified—Biodiversity, Climate Change

People

Population: 3,876,396 (July 1995 est.)
Age structure:
0-14 years: 23% (female 426,616; male 444,556)
15-64 years: 65% (female 1,299,052; male 1,227,420)
65 years and over: 12% (female 313,217; male 165,535) (July 1995 est.)
Population growth rate: 0.71% (1995 est.)
Birth rate: 14.46 births/1,000 population (1995 est.)
Death rate: 10.95 deaths/1,000 population (1995 est.)
Net migration rate: 3.62 migrant(s)/1,000 population (1995 est.)
Infant mortality rate: 16.5 deaths/1,000 live births (1995 est.)
Life expectancy at birth:
total population: 71.37 years
male: 66.68 years
female: 76.3 years (1995 est.)
Total fertility rate: 2 children born/woman (1995 est.)
Nationality:
noun: Lithuanian(s)
adjective: Lithuanian
Ethnic divisions: Lithuanian 80.1%, Russian 8.6%, Polish 7.7%, Byelorussian 1.5%, other 2.1%
Religions: Roman Catholic, Lutheran, other
Languages: Lithuanian (official), Polish, Russian
Literacy: age 15 and over can read and write (1989)
total population: 98%
male: 99%
female: 98%
Labor force: 1.836 million
by occupation: industry and construction 42%, agriculture and forestry 18%, other 40% (1990)

Government

Names:
conventional long form: Republic of Lithuania
conventional short form: Lithuania
local long form: Lietuvos Respublika
local short form: Lietuva
former: Lithuanian Soviet Socialist Republic
Digraph: LH
Type: republic
Capital: Vilnius
Administrative divisions: 44 regions (rajonai, singular—rajonas) and 11 municipalities*: Akmenes Rajonas, Alytaus Rajonas, Alytus*, Anyksciu Rajonas, Birsionas*, Birzu Rajonas, Druskininkai*, Ignalinos Rajonas, Jonavos Rajonas, Joniskio Rajonas, Jurbarko Rajonas, Kaisiadoriu Rajonas, Marijampoles Rajonas, Kaunas*, Kauno Rajonas, Kedainiu Rajonas, Kelmes Rajonas, Klaipeda*, Klaipedos Rajonas, Kretingos Rajonas, Kupiskio Rajonas, Lazdiju Rajonas, Marijampole*, Mazeikiu Rajonas, Moletu Rajonas, Neringa* Pakruojo Rajonas, Palanga*, Panevezio Rajonas, Panevezys*, Pasvalio Rajonas, Plunges Rajonas, Prienu

Rajonas, Radviliskio Rajonas, Raseiniu Rajonas, Rokiskio Rajonas, Sakiu Rajonas, Salcininky Rajonas, Siauliai*, Siauliu Rajonas, Silales Rajonas, Siltues Rajonas, Sirvinty Rajonas, Skuodo Rajonas, Svencioniu Rajonas, Taurages Rajonas, Telsiu Rajonas, Traky Rajonas, Ukmerges Rajonas, Utenos Rajonas, Varenos Rajonas, Vilkaviskio Rajonas, Vilniaus Rajonas, Vilnius*, Zarasu Rajonas
Independence: 6 September 1991 (from Soviet Union)
National holiday: Independence Day, 16 February (1918)
Constitution: adopted 25 October 1992
Legal system: based on civil law system; no judicial review of legislative acts
Suffrage: 18 years of age; universal
Executive branch:
chief of state: President Algirdas Mykolas BRAZAUSKAS (since 25 November 1992; elected acting president by Parliament 25 November 1992 and elected by direct vote 15 February 1993); election last held 14 February 1993 (next to be held NA 1997); results—Algirdas BRAZAUSKAS was elected; note—on 25 November 1992 BRAZAUSKAS was elected chairman of Parliament and, as such, acting president of the Republic; he was confirmed in office by direct balloting 15 February 1993
head of government: Premier Adolfas SLEZEVICIUS (since 10 March 1993)
cabinet: Council of Ministers; appointed by the president on the nomination of the prime minister
Legislative branch: unicameral
Seimas (parliament): elections last held 26 October and 25 November 1992 (next to be held NA 1996); results—LDDP 51%; seats—(141 total) LDDP 73, Conservative Party 30, LKDP 17, LTS 8, Farmers' Union 4, LLS 4, Center Union 2, others 3
Judicial branch: Supreme Court, Court of Appeals
Political parties and leaders: Christian Democratic Party (LKDP), Povilas KATILIUS, chairman; Democratic Labor Party of Lithuania (LDDP), Adolfas SLEZEVICIUS, chairman; Lithuanian Nationalist Union (LTS), Rimantas SMETONA, chairman; Lithuanian Social Democratic Party (LSDP), Aloyzas SAKALAS, chairman; Farmers' Union, Jonas CIULEVICIUS, chairman; Center Union, Romualdas OZOLAS, chairman; Conservative Party, Vytautas LANDSBERGIS, chairman; Lithuanian Polish Union (LLS), Rytardas MACIKIANEC, chairman
Other political or pressure groups: Homeland Union; Lithuanian Future Forum; Farmers Union
Member of: BIS, CBSS, CCC, CE, EBRD, ECE, FAO, IBRD, ICAO, ICRM, IFC, IFRCS, ILO, IMF, INTELSAT (nonsignatory user), INTERPOL, IOC, ISO (correspondent), ITU, NACC, OSCE, PFP, UN, UNCTAD, UNESCO, UNIDO, UPU, WEU (associate partner), WHO, WIPO, WMO
Diplomatic representation in US:
chief of mission: Ambassador Alfonsas EIDINTAS
chancery: 2622 16th Street NW, Washington, DC 20009
telephone: [1] (202) 234-5860, 2639
FAX: [1] (202) 328-0466
consulate(s) general: New York
US diplomatic representation:
chief of mission: Ambassador James W. SWIHART, Jr.
embassy: Akmenu 6, Vilnius 2600
mailing address: APO AE 09723
telephone: [370] (2) 223-031
FAX: [370] (2) 222-779
Flag: three equal horizontal bands of yellow (top), green, and red

Economy

Overview: Since independence in September 1991, Lithuania has made steady progress in developing a market economy. Almost 50% of state property has been privatized and trade is diversifying with a gradual shift away from the former Soviet Union to Western markets. In addition, the Lithuanian government has adhered to a disciplined budgetary and financial policy which has brought inflation down from a monthly average of around 14% in first half 1993 to an average of 3.1% in 1994. Nevertheless, the process has been painful with industrial output in 1993 less than half the 1991 level. The economy appeared to have bottomed out in 1994, and Vilnius's policies have laid the groundwork for a vigorous recovery over the next few years. Recovery will build on Lithuanian's strategic location with its ice-free port at Klaipeda and its rail and highway hub in Vilnius connecting it with Eastern Europe, Belarus, Russia, and Ukraine, and on its agriculture potential, highly skilled labor force, and diversified industrial sector. Lacking important natural resources, it will remain dependent on imports of fuels and raw materials.
National product: GDP—purchasing power parity—$13.5 billion (1994 estimate as extrapolated from World Bank estimate for 1992)
National product real growth rate: -0.5% (1994 est.)
National product per capita: $3,500 (1994 est.)
Inflation rate (consumer prices): 3.1% (monthly average 1994)
Unemployment rate: 4.5% (January 1995)
Budget:
revenues: $258.5 million
expenditures: $270.2 million, including capital expenditures of $NA (1992 est.)
Exports: $2.2 billion (1994)
commodities: electronics 18%, petroleum products 5%, food 10%, chemicals 6% (1989)
partners: Russia, Ukraine, Germany
Imports: $2.7 billion (1994)
commodities: oil 24%, machinery 14%, chemicals 8%, grain NA% (1989)
partners: Russia, Germany, Belarus
External debt: $NA
Industrial production: growth rate -52% (1992); accounts for 35% of GDP
Electricity:
capacity: 6,190,000 kW
production: 18.9 billion kWh
consumption per capita: 4,608 kWh (1993)
Industries: industry's share in the economy has been declining substantially over the past year, due to the economic crisis and the growth of services in the economy; among branches which are still important: metal-cutting machine tools 6.6%, electric motors 4.6%, television sets 6.2%, refrigerators and freezers 5.4%; other branches: petroleum refining, shipbuilding (small ships), furniture making, textiles, food processing, fertilizers, agricultural machinery, optical equipment, electronic components, computers, and amber
Agriculture: employs around 18% of labor force; sugar, grain, potatoes, sugar beets, vegetables, meat, milk, dairy products, eggs, fish; most developed are the livestock and dairy branches, which depend on imported grain; net exporter of meat, milk, and eggs
Illicit drugs: transshipment point for illicit drugs from Central and Southwest Asia and Latin America to Western Europe; limited producer of illicit opium; mostly for domestic consumption
Economic aid:
recipient: US commitments, including Ex-Im (1992), $10 million; Western (non-US) countries, ODA and OOF bilateral commitments (1970-86), $NA million; Communist countries (1971-86), $NA million
Currency: introduced the convertible litas in June 1993
Exchange rates: litai per US$1—4 (fixed rate 1 May 1994)
Fiscal year: calendar year

Transportation

Railroads:
total: 2,010 km
broad gauge: 2,010 km 1.524-m gauge (120 km electrified) (1990)
Highways:
total: 44,200 km
paved: 35,500 km
unpaved: earth 8,700 km (1990)
Inland waterways: 600 km perennially navigable
Pipelines: crude oil, 105 km; natural gas 760 km (1992)
Ports: Kaunas, Klaipeda
Merchant marine:

total: 44 ships (1,000 GRT or over) totaling 275,893 GRT/321,440 DWT
ships by type: bulk 1, cargo 28, combination bulk 11, railcar carrier 3, roll-on/roll-off cargo 1
Airports:
total: 96
with paved runways over 3,047 m: 3
with paved runways 2,438 to 3,047 m: 2
with paved runways 1,524 to 2,437 m: 4
with paved runways 914 to 1,523 m: 2
with paved runways under 914 m: 14
with unpaved runways 2,438 to 3,047 m: 1
with unpaved runways 1,524 to 2,438 m: 1
with unpaved runways 914 to 1,523 m: 6
with unpaved runways under 914 m: 63

Communications

Telephone system: 900,000 telephones; 240 telephones/1,000 persons; telecommunications system ranks among the most modern of the former Soviet republics
local: NA
intercity: land lines and microwave radio relay
international: international connections no longer depend on the Moscow gateway switch, but are established by satellite through Oslo from Vilnius and through Copenhagen from Kaunas; 1 EUTELSAT and 1 INTELSAT earth station; an NMT-450 analog cellular network operates in Vilnius and other cities and is linked internationally through Copenhagen by EUTELSAT; international electronic mail is available; land lines or microwave to former USSR republics
Radio:
broadcast stations: AM 13, FM 26, shortwave 1, longwave 1
radios: NA
Television:
broadcast stations: 3
televisions: NA

Defense Forces

Branches: Ground Forces, Navy, Air and Air Defense Force, Security Forces (internal and border troops), National Guard (Skat)
Manpower availability: males age 15-49 949,663; males fit for military service 750,386; males reach military age (18) annually 27,630 (1995 est.)
Defense expenditures: exchange rate conversion—$30 million, 2% of GDP (1994); note—for 1995 defense expenditures were $54 million at exchange rate conversion

Luxembourg

Geography

Location: Western Europe, between France and Germany
Map references: Europe
Area:
total area: 2,586 sq km
land area: 2,586 sq km
comparative area: slightly smaller than Rhode Island
Land boundaries: total 359 km, Belgium 148 km, France 73 km, Germany 138 km
Coastline: 0 km (landlocked)
Maritime claims: none; landlocked
International disputes: none
Climate: modified continental with mild winters, cool summers
Terrain: mostly gently rolling uplands with broad, shallow valleys; uplands to slightly mountainous in the north; steep slope down to Moselle floodplain in the southeast
Natural resources: iron ore (no longer exploited)
Land use:
arable land: 24%
permanent crops: 1%
meadows and pastures: 20%
forest and woodland: 21%
other: 34%
Irrigated land: NA sq km
Environment:
current issues: deforestation; air and water pollution in urban areas
natural hazards: NA
international agreements: party to—Air Pollution, Air Pollution-Nitrogen Oxides, Air Pollution-Sulphur 85, Air Pollution-Volatile Organic Compounds, Biodiversity, Climate Change, Endangered Species, Hazardous Wastes, Marine Dumping, Nuclear Test Ban, Ozone Layer Protection, Ship Pollution, Tropical Timber 83; signed, but not ratified—Air Pollution-Sulphur 94, Desertification, Environmental Modification, Law of the Sea
Note: landlocked

People

Population: 404,660 (July 1995 est.)
Age structure:
0-14 years: 18% (female 35,372; male 36,645)
15-64 years: 68% (female 136,960; male 137,792)
65 years and over: 14% (female 35,774; male 22,117) (July 1995 est.)
Population growth rate: 0.57% (1995 est.)
Birth rate: 12.61 births/1,000 population (1995 est.)
Death rate: 9.42 deaths/1,000 population (1995 est.)
Net migration rate: 2.47 migrant(s)/1,000 population (1995 est.)
Infant mortality rate: 6.6 deaths/1,000 live births (1995 est.)
Life expectancy at birth:
total population: 76.95 years
male: 73.31 years
female: 80.75 years (1995 est.)
Total fertility rate: 1.65 children born/woman (1995 est.)
Nationality:
noun: Luxembourger(s)
adjective: Luxembourg
Ethnic divisions: Celtic base (with French and German blend), Portuguese, Italian, and European (guest and worker residents)
Religions: Roman Catholic 97%, Protestant and Jewish 3%
Languages: Luxembourgisch, German, French, English
Literacy: age 15 and over can read and write (1980 est.)
total population: 100%
male: 100%
female: 100%
Labor force: 177,300 (one-third of labor force is foreign workers, mostly from Portugal, Italy, France, Belgium, and Germany)
by occupation: services 65%, industry 31.6%, agriculture 3.4% (1988)

Government

Names:
conventional long form: Grand Duchy of Luxembourg
conventional short form: Luxembourg
local long form: Grand-Duche de Luxembourg
local short form: Luxembourg
Digraph: LU
Type: constitutional monarchy
Capital: Luxembourg
Administrative divisions: 3 districts; Diekirch, Grevenmacher, Luxembourg
Independence: 1839
National holiday: National Day, 23 June (1921) (public celebration of the Grand Duke's birthday)

Constitution: 17 October 1868, occasional revisions
Legal system: based on civil law system; accepts compulsory ICJ jurisdiction
Suffrage: 18 years of age; universal and compulsory
Executive branch:
chief of state: Grand Duke JEAN (since 12 November 1964); Heir Apparent Prince HENRI (son of Grand Duke JEAN, born 16 April 1955)
head of government: Prime Minister Jean-Claude JUNKER (since 1 January 1994); Vice Prime Minister Jacques F. POOS (since 21 July 1984)
cabinet: Council of Ministers; appointed by the sovereign
Legislative branch: unicameral
Chamber of Deputies (Chambre des Deputes): elections last held on 12 June 1994 (next to be held by June 1999); results—percent of vote by party NA; seats—(60 total) CSV 21, LSAP 17, DP 12, Action Committee for Democracy and Pension Rights 5, Greens 5
note: the Council of State (Conseil d'Etat) is an advisory body whose views are considered by the Chamber of Deputies
Judicial branch: Superior Court of Justice (Cour Superieure de Justice)
Political parties and leaders: Christian Social People's Party (CSV), Erna HENNICOT-SCHOEPGES; Socialist Workers Party (LSAP), Ben FAYOT; Democratic Party (DP), Henri GRETHEN; Action Committee for Democracy and Pension Rights, Roby MEHLEN; other minor parties
Other political or pressure groups: group of steel companies representing iron and steel industry; Centrale Paysanne representing agricultural producers; Christian and Socialist labor unions; Federation of Industrialists; Artisans and Shopkeepers Federation
Member of: ACCT, Australia Group, Benelux, CCC, CE, EBRD, EC, ECE, EIB, FAO, GATT, IAEA, IBRD, ICAO, ICC, ICFTU, ICRM, IDA, IEA, IFAD, IFC, IFRCS, ILO, IMF, IMO, INTELSAT, INTERPOL, IOC, IOM, ITU, MTCR, NACC, NATO, NEA, NSG, OECD, OSCE, PCA, UN, UNCTAD, UNESCO, UNIDO, UPU, WCL, WEU, WHO, WIPO, WMO, ZC
Diplomatic representation in US:
chief of mission: Ambassador Alphonse BERNS
chancery: 2200 Massachusetts Avenue NW, Washington, DC 20008
telephone: [1] (202) 265-4171
FAX: [1] (202) 328-8270
consulate(s) general: New York and San Francisco
US diplomatic representation:
chief of mission: Ambassador Clay CONSTANTINOU
embassy: 22 Boulevard Emmanuel-Servais, 2535 Luxembourg City
mailing address: PSC 11, Luxembourg City; APO AE 09132-5380
telephone: [352] 46 01 23
FAX: [352] 46 14 01
Flag: three equal horizontal bands of red (top), white, and light blue; similar to the flag of the Netherlands, which uses a darker blue and is shorter; design was based on the flag of France

Economy

Overview: The stable, prosperous economy features moderate growth, low inflation, and negligible unemployment. Agriculture is based on small but highly productive family-owned farms. The industrial sector, until recently dominated by steel, has become increasingly more diversified, particularly toward high-technology firms. During the past decade, growth in the financial sector has more than compensated for the decline in steel. Services, especially banking, account for a growing proportion of the economy. Luxembourg participates in an economic union with Belgium on trade and most financial matters, is also closely connected economically to the Netherlands, and as a member of the 15-member European Union enjoys the advantages of the open European market.
National product: GDP—purchasing power parity—$9.2 billion (1994 est.)
National product real growth rate: 2.6% (1994 est.)
National product per capita: $22,830 (1994 est.)
Inflation rate (consumer prices): 3.6% (1992)
Unemployment rate: 2.4% (1994)
Budget:
revenues: $4 billion
expenditures: $4.05 billion, including capital expenditures of $NA (1994 est.)
Exports: $6.4 billion (f.o.b., 1991 est.)
commodities: finished steel products, chemicals, rubber products, glass, aluminum, other industrial products
partners: EC 76%, US 5%
Imports: $8.3 billion (c.i.f., 1991 est.)
commodities: minerals, metals, foodstuffs, quality consumer goods
partners: Belgium 37%, Germany 31%, France 12%, US 2%
External debt: $800 million (1994 est.)
Industrial production: growth rate -0.5% (1990); accounts for 25% of GDP
Electricity:
capacity: 1,238,750 kW
production: 1,374 million kWh
consumption per capita: 3,395 kWh (1993)
Industries: banking, iron and steel, food processing, chemicals, metal products, engineering, tires, glass, aluminum
Agriculture: accounts for less than 3% of GDP (including forestry); principal products—barley, oats, potatoes, wheat, fruits, wine grapes; cattle raising widespread
Economic aid: none
Currency: 1 Luxembourg franc (LuxF) = 100 centimes
Exchange rates: Luxembourg francs (LuxF) per US$1—31.549 (January 1995), 33,456 (1994), 34.597 (1993), 32.150 (1992), 34.148 (1991), 33.418 (1990); note—the Luxembourg franc is at par with the Belgian franc, which circulates freely in Luxembourg
Fiscal year: calendar year

Transportation

Railroads:
total: 271 km
standard gauge: 271 km 1.435-m gauge (243 km electrified; 178 km double track) (1994)
Highways:
total: 5,108 km
paved: 5,062 km (95 km of limited access divided highway)
unpaved: 46 km (1992)
Inland waterways: 37 km; Moselle River
Pipelines: petroleum products 48 km
Ports: Mertert
Merchant marine:
total: 45 ships (1,000 GRT or over) totaling 1,129,466 GRT/1,790,988 DWT
ships by type: bulk 6, cargo 2, chemical tanker 4, combination bulk 6, container 2, liquefied gas tanker 8, oil tanker 7, passenger 2, refrigerated cargo 6, roll-on/roll-off cargo 2
Airports:
total: 2
with paved runways over 3,047 m: 1
with paved runways under 914 m: 1

Communications

Telephone system: 230,000 telephones; highly developed, completely automated and efficient system, mainly buried cables; nationwide mobile phone system
local: NA
intercity: buried cable
international: 3 channels leased on TAT-6 coaxial submarine cable
Radio:
broadcast stations: AM 2, FM 3, shortwave 0
radios: NA
Television:
broadcast stations: 3 and 1 direct-broadcast satellite link
televisions: NA

Defense Forces

Branches: Army, National Gendarmerie
Manpower availability: males age 15-49 103,990; males fit for military service 85,912; males reach military age (19) annually 2,190 (1995 est.)
Defense expenditures: exchange rate conversion—$129 million, 1.2% of GDP (1994)

Macau

(overseas territory of Portugal)

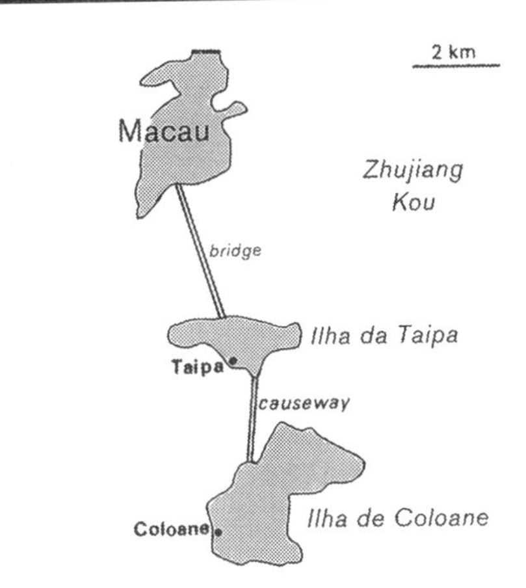

Geography

Location: Eastern Asia, bordering the South China Sea and China
Map references: Southeast Asia
Area:
total area: 16 sq km
land area: 16 sq km
comparative area: about 0.1 times the size of Washington, DC
Land boundaries: total 0.34 km, China 0.34 km
Coastline: 40 km
Maritime claims: not specified
International disputes: none
Climate: subtropical; marine with cool winters, warm summers
Terrain: generally flat
Natural resources: negligible
Land use:
arable land: 0%
permanent crops: 0%
meadows and pastures: 0%
forest and woodland: 0%
other: 100%
Irrigated land: NA sq km
Environment:
current issues: NA
natural hazards: NA
international agreements: party to—Ozone Layer Protection (extended from Portugal)
Note: essentially urban; one causeway and one bridge connect the two islands to the peninsula on mainland

People

Population: 490,901 (July 1995 est.)
Age structure:
0-14 years: 24% (female 56,991; male 60,944)
15-64 years: 68% (female 167,366; male 165,168)
65 years and over: 8% (female 23,537; male 16,895) (July 1995 est.)
Population growth rate: 1.25% (1995 est.)
Birth rate: 14.5 births/1,000 population (1995 est.)
Death rate: 4.21 deaths/1,000 population (1995 est.)
Net migration rate: 2.24 migrant(s)/1,000 population (1995 est.)
Infant mortality rate: 5.4 deaths/1,000 live births (1995 est.)
Life expectancy at birth:
total population: 79.86 years
male: 77.41 years
female: 82.43 years (1995 est.)
Total fertility rate: 1.49 children born/woman (1995 est.)
Nationality:
noun: Macanese (singular and plural)
adjective: Macau
Ethnic divisions: Chinese 95%, Portuguese 3%, other 2%
Religions: Buddhist 45%, Roman Catholic 7%, Protestant 1%, none 45.8%, other 1.2% (1981)
Languages: Portuguese (official), Cantonese is the language of commerce
Literacy: age 15 and over can read and write (1981)
total population: 90%
male: 93%
female: 86%
Labor force: 180,000 (1986)
by occupation: NA

Government

Names:
conventional long form: none
conventional short form: Macau
local long form: none
local short form: Ilha de Macau
Digraph: MC
Type: overseas territory of Portugal scheduled to revert to China in 1999
Capital: Macau
Administrative divisions: 2 districts (concelhos, singular—concelho); Ilhas, Macau
Independence: none (territory of Portugal; Portugal signed an agreement with China on 13 April 1987 to return Macau to China on 20 December 1999; in the joint declaration, China promises to respect Macau's existing social and economic systems and lifestyle for 50 years after transition)
National holiday: Day of Portugal, 10 June (1580)
Constitution: 17 February 1976, Organic Law of Macau; basic law drafted primarily by Beijing awaiting final approval
Legal system: Portuguese civil law system
Suffrage: 18 years of age; universal
Executive branch:
chief of state: President (of Portugal) Mario Alberto SOARES (since 9 March 1986)
head of government: Governor Gen. Vasco Joachim Rocha VIEIRA (since 20 March 1991)
cabinet: Consultative Council; consists of five members appointed by the governor, two nominated by the governor, five members elected for a four-year term (2 represent administrative bodies, 1 represents moral, cultural, and welfare interests, and 2 economic interests), and three statuatory members
Legislative branch: unicameral
Legislative Assembly: elections last held on 10 March 1991 (next to be held NA); results—percent of vote by party NA; seats—(23 total, 8 elected by universal suffrage, 8 by indirect suffrage, and 7 appointed by the governor) number of seats by party NA
Judicial branch: Supreme Court
Political parties and leaders: Association to Defend the Interests of Macau; Macau Democratic Center; Group to Study the Development of Macau; Macau Independent Group
Other political or pressure groups: wealthy Macanese and Chinese representing local interests, wealthy pro-Communist merchants representing China's interests; in January 1967 the Macau Government acceded to Chinese demands that gave China veto power over administration
Member of: CCC, ESCAP (associate), GATT, IMO (associate), INTERPOL (subbureau), WTO (associate)
Diplomatic representation in US: none (Chinese territory under Portuguese administration)
US diplomatic representation: the US has no offices in Macau, and US interests are monitored by the US Consulate General in Hong Kong
Flag: the flag of Portugal is used

Economy

Overview: The economy is based largely on tourism (including gambling) and textile and fireworks manufacturing. Efforts to diversify have spawned other small industries—toys, artificial flowers, and electronics. The tourist sector has accounted for roughly 25% of GDP, and the clothing industry has provided about two-thirds of export earnings; the gambling industry represented well over 40% of GDP in 1992. Macau depends on China for most of its food, fresh water, and energy imports. Japan and Hong Kong are the main suppliers of raw materials and capital goods.
National product: GDP—purchasing power parity—$4.8 billion (1993 est.)
National product real growth rate: NA%
National product per capita: $10,000 (1993 est.)
Inflation rate (consumer prices): 7.7% (1992 est.)
Unemployment rate: 2% (1992 est.)
Budget:
revenues: $305 million
expenditures: $298 million, including capital expenditures of $NA (1989 est.)
Exports: $1.8 billion (1992 est.)
commodities: textiles, clothing, toys
partners: US 35%, Hong Kong 12.5%,

Macau *(continued)*

Germany 12%, China 9.9%, France 8% (1992 est.)
Imports: $2 billion (1992 est.)
commodities: raw materials, foodstuffs, capital goods
partners: Hong Kong 33%, China 20%, Japan 18% (1992 est.)
External debt: $91 million (1985)
Industrial production: growth rate NA%
Electricity:
capacity: 258,000 kW
production: 950 million kWh
consumption per capita: 2,093 kWh (1993)
Industries: clothing, textiles, toys, plastic products, furniture, tourism
Agriculture: rice, vegetables; food shortages—rice, vegetables, meat; depends mostly on imports for food requirements
Economic aid: none
Currency: 1 pataca (P) = 100 avos
Exchange rates: patacas (P) per US$1—8.034 (1991-94), 8.024 (1990), 8.030 (1989); note—linked to the Hong Kong dollar at the rate of 1.03 patacas per Hong Kong dollar
Fiscal year: calendar year

Transportation

Railroads: 0 km
Highways:
total: 42 km
paved: 42 km
Ports: Macau
Merchant marine: none
Airports: none usable, 1 under construction; 1 seaplane station

Communications

Telephone system: 52,000 telephones; fairly modern communication facilities maintained for domestic and international services
local: NA
intercity: NA
international: high-frequency radio communication facility; access to international communications carriers provided via Hong Kong and China; 1 INTELSAT (Indian Ocean) earth station
Radio:
broadcast stations: AM 4, FM 3, shortwave 0
radios: 115,000
Television:
broadcast stations: 0; note—TV programs received from Hong Kong
televisions: NA

Defense Forces

Branches: NA
Manpower availability: males age 15-49 141,160; males fit for military service 78,578 (1995 est.)
Note: defense is responsibility of Portugal

Macedonia, The Former Yugoslav Republic of

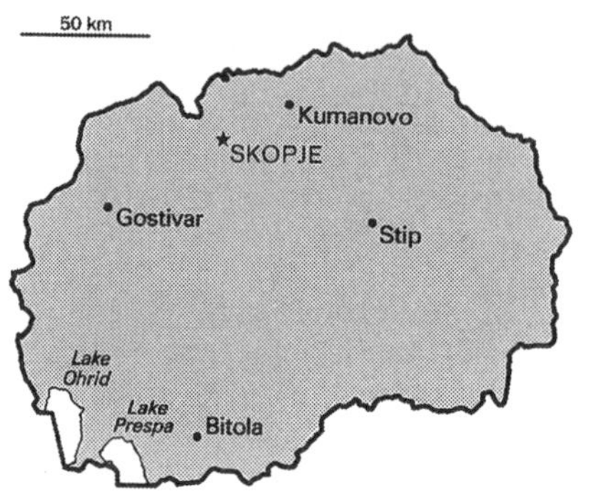

Geography

Location: Southeastern Europe, north of Greece
Map references: Ethnic Groups in Eastern Europe, Europe
Area:
total area: 25,333 sq km
land area: 24,856 sq km
comparative area: slightly larger than Vermont
Land boundaries: total 748 km, Albania 151 km, Bulgaria 148 km, Greece 228 km, Serbia and Montenegro 221 km (all with Serbia)
Coastline: 0 km (landlocked)
Maritime claims: none; landlocked
International disputes: dispute with Greece over name, symbols, and certain constitutional provisions
Climate: hot, dry summers and autumns and relatively cold winters with heavy snowfall
Terrain: mountainous territory covered with deep basins and valleys; there are three large lakes, each divided by a frontier line; country bisected by the Vardar River
Natural resources: chromium, lead, zinc, manganese, tungsten, nickel, low-grade iron ore, asbestos, sulphur, timber
Land use:
arable land: 5%
permanent crops: 5%
meadows and pastures: 20%
forest and woodland: 30%
other: 40%
Irrigated land: NA sq km
Environment:
current issues: air pollution from metallurgical plants
natural hazards: high seismic risks
international agreements: party to—Law of the Sea, Ozone Layer Protection
Note: landlocked; major transportation corridor from Western and Central Europe to Aegean Sea and Southern Europe to Western Europe

People

Population: 2,159,503 (July 1995 est.)
note: the Macedonian government census of July 1994 put the population at 1.94 million, but ethnic allocations were likely undercounted
Age structure:
0-14 years: 25% (female 257,876; male 277,314)
15-64 years: 67% (female 711,810; male 733,903)
65 years and over: 8% (female 97,475; male 81,125) (July 1995 est.)
Population growth rate: 0.9% (1995 est.)
Birth rate: 15.82 births/1,000 population (1995 est.)
Death rate: 6.7 deaths/1,000 population (1995 est.)
Net migration rate: -0.14 migrant(s)/1,000 population (1995 est.)
Infant mortality rate: 24.2 deaths/1,000 live births (1995 est.)
Life expectancy at birth:
total population: 74 years
male: 71.87 years
female: 76.3 years (1995 est.)
Total fertility rate: 2.02 children born/woman (1995 est.)
Nationality:
noun: Macedonian(s)
adjective: Macedonian
Ethnic divisions: Macedonian 65%, Albanian 22%, Turkish 4%, Serb 2%, Gypsies 3%, other 4%
Religions: Eastern Orthodox 67%, Muslim 30%, other 3%
Languages: Macedonian 70%, Albanian 21%, Turkish 3%, Serbo-Croatian 3%, other 3%
Literacy: NA%
Labor force: 591,773 (June 1994)
by occupation: manufacturing and mining 40% (1992)

Government

Names:
conventional long form: The Former Yugoslav Republic of Macedonia
conventional short form: none
local long form: Republika Makedonija
local short form: Makedonija
Abbreviation: F.Y.R.O.M.
Digraph: MK
Type: emerging democracy
Capital: Skopje
Administrative divisions: 34 counties (opstinas, singular—opstina) Berovo, Bitola, Brod, Debar, Delcevo, Gevgelija, Gostivar, Kavadarci, Kicevo, Kocani, Kratovo, Kriva Palanka, Krusevo, Kumanovo, Murgasevo, Negotino, Ohrid, Prilep, Probistip, Radovis, Resen, Skopje-Centar, Skopje-Cair, Skopje-

Karpos, Skopje-Kisela Voda, Skopje-Gazi Baba, Stip, Struga, Strumica, Sveti Nikole, Tetovo, Titov Veles, Valandovo, Vinica
Independence: 17 September 1991 (from Yugoslavia)
National holiday: 8 September
Constitution: adopted 17 November 1991, effective 20 November 1991
Legal system: based on civil law system; judicial review of legislative acts
Suffrage: 18 years of age; universal
Executive branch:
chief of state: President Kiro GLIGOROV (since 27 January 1991); election last held 16 October 1994 (next to be held NA 1997); results—Kiro GLIGOROV was elected by the Assembly in 1991; reelected by popular vote in 1994
head of government: Prime Minister Branko CRVENKOVSKI (since 4 September 1992)
cabinet: Council of Ministers; elected by the majority vote of all the deputies in the Sobranje
Legislative branch: unicameral
Assembly (Sobranje): elections last held 16 and 30 October 1994 (next to be held November 1998); results—percent of vote by party NA; seats—(120 total) seats by party NA
Judicial branch: Constitutional Court, Judicial Court of the Republic
Political parties and leaders: Social-Democratic Alliance of Macedonia (SDSM; former Communist Party), Branko CRVENKOVSKI, president; Party for Democratic Prosperity (PDP); note—two factions competing for party name; one faction is led by Abdurahman HALITI and the other faction is led by Arber XHAFFERI; National Democratic Party (NDP), Ilijas HALINI, president; Alliance of Reform Forces of Macedonia—Liberal Party (SRSM-LP), Stojan ANDOV, president; Socialist Party of Macedonia (SPM), Kiro POPOVSKI, president; Internal Macedonian Revolutionary Organization—Democratic Party for Macedonian National Unity (VMRO-DPMNE), Ljupco GEORGIEVSKI, president; Party of Yugoslavs in Macedonia (SJM), Milan DURCINOV, president; Democratic Party (DP), Petar GOSEV, president
Other political or pressure groups: Movement for All Macedonian Action (MAAK); Democratic Party of Serbs; Democratic Party of Turks; Party for Democratic Action (Slavic Muslim)
Member of: CCC, CE (guest), CEI, EBRD, ECE, IAEA, IBRD, ICAO, IDA, IFAD, IFC, ILO, IMF, IMO, INTELSAT (nonsignatory user), INTERPOL, IOC, ITU, OSCE (observer), UN, UNCTAD, UNESCO, UNIDO, UPU, WHO, WIPO, WMO
Diplomatic representation in US: the US recognized The Former Yugoslav Republic of Macedonia on 8 February 1994
US diplomatic representation:
chief of mission: Victor D. COMRAS
liaison office: ul. 27 Mart No. 5, 9100 Skopje
mailing address: USLO Skopje, Department of State, Washington, DC 20521-7120 (pouch)
telephone: [389] (91) 116-180
FAX: [389] (91) 117-103
Flag: 16-point gold sun (Vergina, Sun) centered on a red field

Economy

Overview: The Former Yugoslav Republic of Macedonia, although the poorest republic in the former Yugoslav federation, can meet basic food and energy needs through its own agricultural and coal resources. Its economic decline will continue unless ties are reforged or enlarged with its neighbors Serbia and Montenegro, Albania, Greece, and Bulgaria. The economy depends on outside sources for all of its oil and gas and most of its modern machinery and parts. An important supplement of GDP is the remittances from thousands of Macedonians working in Germany and other West European nations. Continued political turmoil, both internally and in the region as a whole, prevents any swift readjustments of trade patterns and economic programs. The country's industrial output and GDP are expected to decline further in 1995. The Former Yugoslav Republic of Macedonia's geographical isolation, technological backwardness, and potential political instability place it far down the list of countries of interest to Western investors. Resolution of the dispute with Greece and an internal commitment to economic reform would encourage foreign investment over the long run. In the immediate future, the worst scenario for the economy would be the spread of fighting across its borders.
National product: GDP—purchasing power parity—$1.9 billion (1994 est.)
National product real growth rate: -15% (1994 est.)
National product per capita: $900 (1994 est.)
Inflation rate (consumer prices): 54% (1994)
Unemployment rate: 30% (1993 est.)
Budget:
revenues: $NA
expenditures: $NA, including capital expenditures of $NA
Exports: $1.06 billion (1993)
commodities: manufactured goods 40%, machinery and transport equipment 14%, miscellaneous manufactured articles 23%, raw materials 7.6%, food (rice) and live animals 5.7%, beverages and tobacco 4.5%, chemicals 4.7% (1990)
partners: principally Serbia and Montenegro and the other former Yugoslav republics, Germany, Greece, Albania
Imports: $1.2 billion (1993)
commodities: fuels and lubricants 19%, manufactured goods 18%, machinery and transport equipment 15%, food and live animals 14%, chemicals 11.4%, raw materials 10%, miscellaneous manufactured articles 8.0%, beverages and tobacco 3.5% (1990)
partners: other former Yugoslav republics, Greece, Albania, Germany, Bulgaria
External debt: $840 million (1992)
Industrial production: growth rate -14% (1993)
Electricity:
capacity: 1,600,000 kW
production: NA kWh
consumption per capita: NA kWh (1993)
Industries: low levels of technology predominate, such as, oil refining by distillation only; produces basic liquid fuels, coal, metallic chromium, lead, zinc, and ferronickel; light industry produces basic textiles, wood products, and tobacco
Agriculture: meets the basic needs for food; principal crops are rice, tobacco, wheat, corn, and millet; also grown are cotton, sesame, mulberry leaves, citrus fruit, and vegetables; agricultural production is highly labor intensive
Illicit drugs: limited illicit opium cultivation; transshipment point for Southwest Asian heroin
Economic aid:
recipient: US $10 million (for humanitarian and technical assistance)
EC promised a 100 ECU million economic aid package (1993)
Currency: the denar, which was adopted by the Macedonian legislature 26 April 1992, was initially issued in the form of a coupon pegged to the German mark; subsequently repegged to a basket of seven currencies
Exchange rates: denar per US$1—39 (November 1994), 865 (October 1992)
Fiscal year: calendar year

Transportation

Railroads:
total: 922 km
standard gauge: 922 km 1.435-m gauge (1994)
Highways:
total: 10,591 km
paved: 5,091 km
unpaved: gravel 1,404 km; earth 4,096 km (1991)
Inland waterways: none, lake transport only
Pipelines: none
Ports: none
Airports:
total: 16
with paved runways 2,438 to 3,047 m: 2
with paved runways under 914 m: 11
with unpaved runways 1,524 to 2,438 m: 1
with unpaved runways 914 to 1,523 m: 2

Communications

Telephone system: 125,000 telephones
local: NA

Macedonia, The Former Yugoslav Republic of *(continued)*

intercity: NA
international: no satellite links
Radio:
broadcast stations: AM 6, FM 2, shortwave 0
radios: 370,000
Television:
broadcast stations: 5 (relays 2)
televisions: 325,000

Defense Forces

Branches: Army, Police Force
Manpower availability: males age 15-49 585,403; males fit for military service 474,467; males reach military age (19) annually 19,693 (1995 est.)
Defense expenditures: 7 billion denars, NA% of GNP (1993 est.); note—conversion of the military budget into US dollars using the prevailing exchange rate could produce misleading results

Madagascar

Geography

Location: Southern Africa, island in the Indian Ocean, east of Mozambique
Map references: Africa
Area:
total area: 587,040 sq km
land area: 581,540 sq km
comparative area: slightly less than twice the size of Arizona
Land boundaries: 0 km
Coastline: 4,828 km
Maritime claims:
contiguous zone: 24 nm
continental shelf: 200 nm or 100 nm from the 2,500-m isobath
exclusive economic zone: 200 nm
territorial sea: 12 nm
International disputes: claims Bassas da India, Europa Island, Glorioso Islands, Juan de Nova Island, and Tromelin Island (all administered by France)
Climate: tropical along coast, temperate inland, arid in south
Terrain: narrow coastal plain, high plateau and mountains in center
Natural resources: graphite, chromite, coal, bauxite, salt, quartz, tar sands, semiprecious stones, mica, fish
Land use:
arable land: 4%
permanent crops: 1%
meadows and pastures: 58%
forest and woodland: 26%
other: 11%
Irrigated land: 9,000 sq km (1989 est.)
Environment:
current issues: soil erosion results from deforestation and overgrazing; desertification; surface water contaminated with raw sewage and other organic wastes; several species of flora and fauna unique to the island are endangered
natural hazards: periodic cyclones
international agreements: party to—Endangered Species, Marine Life Conservation, Nuclear Test Ban; signed, but not ratified—Biodiversity, Climate Change, Desertification, Law of the Sea
Note: world's fourth-largest island; strategic location along Mozambique Channel

People

Population: 13,862,325 (July 1995 est.)
Age structure:
0-14 years: 47% (female 3,231,647; male 3,265,715)
15-64 years: 50% (female 3,511,699; male 3,413,564)
65 years and over: 3% (female 225,205; male 214,495) (July 1995 est.)
Population growth rate: 3.18% (1995 est.)
Birth rate: 44.82 births/1,000 population (1995 est.)
Death rate: 12.99 deaths/1,000 population (1995 est.)
Net migration rate: 0 migrant(s)/1,000 population (1995 est.)
Infant mortality rate: 86.9 deaths/1,000 live births (1995 est.)
Life expectancy at birth:
total population: 54.45 years
male: 52.47 years
female: 56.48 years (1995 est.)
Total fertility rate: 6.62 children born/woman (1995 est.)
Nationality:
noun: Malagasy (singular and plural)
adjective: Malagasy
Ethnic divisions: Malayo-Indonesian (Merina and related Betsileo), Cotiers (mixed African, Malayo-Indonesian, and Arab ancestry—Betsimisaraka, Tsimihety, Antaisaka, Sakalava), French, Indian, Creole, Comoran
Religions: indigenous beliefs 52%, Christian 41%, Muslim 7%
Languages: French (official), Malagasy (official)
Literacy: age 15 and over can read and write (1990 est.)
total population: 80%
male: 88%
female: 73%
Labor force:
total workers: 4.9 million
workers not receiving money wages: 4.7 million (96% of total labor force); note—4.3 million workers are in subsistence agriculture
wage earners: 175,000 (3.6% of total work force)
wage earners by occupation: agriculture 45,500, domestic service 29,750, industry 26,250, commerce 24,500, construction 19,250, service 15,750, transportation 10,500, other 3,500 (1985 est.)

Government

Names:
conventional long form: Republic of

Madagascar
conventional short form: Madagascar
local long form: Republique de Madagascar
local short form: Madagascar
former: Malagasy Republic
Digraph: MA
Type: republic
Capital: Antananarivo
Administrative divisions: 6 provinces; Antananarivo, Antsiranana, Fianarantsoa, Mahajanga, Toamasina, Toliary
Independence: 26 June 1960 (from France)
National holiday: Independence Day, 26 June (1960)
Constitution: 19 August 1992 by national referendum
Legal system: based on French civil law system and traditional Malagasy law; has not accepted compulsory ICJ jurisdiction
Suffrage: 18 years of age; universal
Executive branch:
chief of state: President Albert ZAFY (since 9 March 1993); election last held on 10 February 1993 (next to be held 1998); results—Albert ZAFY (UNDD), 67%; Didier RATSIRAKA (AREMA), 33%
head of government: Prime Minister Francisque RAVONY (since 9 August 1993)
cabinet: Council of Ministers; appointed by the prime minister
Legislative branch: bicameral Parliament
Senate (Senat): two-thirds of upper house seats are to be filled from popularly elected regional assemblies; the remaining third is to be filled by presidential appointment; decentralization and formation of regional assemblies is not expected before 1997
National Assembly (Assemblee Nationale): elections last held on 16 June 1993 (next to be held June 1997); results—percent of vote by party NA; seats—(138 total) CFV coalition 76, PMDM/MFM 16, CSCD 11, Famima 10, RPSD 7, various pro-Ratsiraka groups 10, others 8
Judicial branch: Supreme Court (Cour Supreme), High Constitutional Court (Haute Cour Constitutionnelle)
Political parties and leaders: Committee of Living Forces (CFV), an alliance of National Union for Development and Democracy (UNDD), Support Group for Democracy and Development in Madagascar (CSDDM), Action and Reflection Group for the Development of Madagascar (GRAD), Congress Party for Madagascar Independence—Renewal (AKFM Fanavaozana), and some 12 other parties, trade unions, and religious groups; Militant Party for the Development of Madagascar (PMDM/MFM), formerly the Movement for Proletarian Power, Manandafy RAKOTONIRINA; Confederation of Civil Societies for Development (CSCD), Guy Willy RAZANAMASY; Association of United Malagasys (Famima); Rally for Social Democracy (RPSD), Pierre TSIRANANA
Other political or pressure groups: National Council of Christian Churches (FFKM); Federalist Movement
Member of: ACCT, ACP, AfDB, CCC, ECA, FAO, G-77, GATT, IAEA, IBRD, ICAO, ICC, ICFTU, ICRM, IDA, IFAD, IFC, IFRCS, ILO, IMF, IMO, INTELSAT, INTERPOL, IOC, ITU, NAM, OAU, UN, UNCTAD, UNESCO, UNHCR, UNIDO, UNMIH, UPU, WCL, WFTU, WHO, WIPO, WMO, WTO
Diplomatic representation in US:
chief of mission: Ambassador Pierrot Jocelyn RAJAONARIVELO
chancery: 2374 Massachusetts Avenue NW, Washington, DC 20008
telephone: [1] (202) 265-5525, 5526
consulate(s) general: New York
US diplomatic representation:
chief of mission: Ambassador Dennis P. BARRETT
embassy: 14-16 Rue Rainitovo, Antsahavola, Antananarivo
mailing address: B. P. 620, Antananarivo
telephone: [261] (2) 212-57, 200-89, 207-18
FAX: [261] (2) 345-39
Flag: two equal horizontal bands of red (top) and green with a vertical white band of the same width on hoist side

Economy

Overview: Madagascar is one of the poorest countries in the world, suffering from chronic malnutrition, underfunded health and education facilities, a 3% annual population growth rate, and severe loss of forest cover, accompanied by erosion. Agriculture, including fishing and forestry, is the mainstay of the economy, accounting for over 30% of GDP and contributing more than 70% of total export earnings. Industry is largely confined to the processing of agricultural products and textile manufacturing; in 1991 it accounted for only 13% of GDP. In 1986 the government introduced a five-year development plan that stressed self-sufficiency in food (mainly rice) by 1990, increased production for exports, and reduced energy imports. Subsequently, growth in output has been held back because of protracted antigovernment strikes and demonstrations for political reform. Since 1993, corruption and political instability have caused the economy and infrastructure to decay further. Since April 1994, the government commitment to economic reforms has been erratic. Enormous obstacles stand in the way of Madagascar's realizing its considerable growth potential.
National product: GDP—purchasing power parity—$10.6 billion (1994 est.)
National product real growth rate: 2.8% (1994 est.)
National product per capita: $790 (1994 est.)
Inflation rate (consumer prices): 35% (1994 est.)
Unemployment rate: NA%
Budget:
revenues: $250 million
expenditures: $265 million, including capital expenditures of $180 million (1991 est.)
Exports: $240 million (f.o.b., 1993 est.)
commodities: coffee 45%, vanilla 20%, cloves 11%, shellfish, sugar, petroleum products
partners: France, US, Germany, Japan, Russia
Imports: $510 million (f.o.b., 1993 est.)
commodities: intermediate manufactures 30%, capital goods 28%, petroleum 15%, consumer goods 14%, food 13%
partners: France, Germany, Japan, UK, Italy, Netherlands
External debt: $4.3 billion (1993 est.)
Industrial production: growth rate 3.8% (1993 est.); accounts for 13% of GDP
Electricity:
capacity: 220,000 kW
production: 560 million kWh
consumption per capita: 40 kWh (1993)
Industries: agricultural processing (meat canneries, soap factories, breweries, tanneries, sugar refining plants), light consumer goods industries (textiles, glassware), cement, automobile assembly plant, paper, petroleum
Agriculture: accounts for 31% of GDP; cash crops—coffee, vanilla, sugarcane, cloves, cocoa; food crops—rice, cassava, beans, bananas, peanuts; cattle raising widespread; almost self-sufficient in rice
Illicit drugs: illicit producer of cannabis (cultivated and wild varieties) used mostly for domestic consumption
Economic aid:
recipient: US commitments, including Ex-Im (FY70-89), $136 million; Western (non-US) countries, ODA and OOF bilateral commitments (1970-89), $3.125 billion; Communist countries (1970-89), $491 million
Currency: 1 Malagasy franc (FMG) = 100 centimes
Exchange rates: Malagasy francs (FMG) per US$1—3,718.0 (November 1994), 1,913.8 (1993), 1,864.0 (1992), 1,835.4 (1991), 1,454.6 (December 1990)
Fiscal year: calendar year

Transportation

Railroads:
total: 1,020 km
narrow gauge: 1,020 km 1.000-m gauge
Highways:
total: 40,000 km
paved: 4,694 km
unpaved: gravel, crushed stone, stabilized earth 811 km; other earth 34,495 km (est.)
Inland waterways: of local importance only; isolated streams and small portions of Canal des Pangalanes
Ports: Antsiranana, Mahajanga, Port Saint-Louis, Toamasina, Toliaria

Merchant marine:
total: 10 ships (1,000 GRT or over) totaling 20,261 GRT/28,193 DWT
ships by type: cargo 5, chemical tanker 1, liquefied gas tanker 1, oil tanker 1, roll-on/roll-off cargo 2
Airports:
total: 138
with paved runways over 3,047 m: 1
with paved runways 2,438 to 3,047 m: 2
with paved runways 1,524 to 2,437 m: 3
with paved runways 914 to 1,523 m: 21
with paved runways under 914 m: 42
with unpaved runways 1,524 to 2,438 m: 5
with unpaved runways 914 to 1,523 m: 64

Communications

Telephone system: NA telephones; above average system
local: NA
intercity: open-wire lines, coaxial cables, microwave radio relay, and tropospheric scatter links
international: submarine cable to Bahrain; 1 earth station for Indian Ocean INTELSAT
Radio:
broadcast stations: AM 17, FM 3, shortwave 0
radios: NA
Television:
broadcast stations: 1 (repeaters 36)
televisions: NA

Defense Forces

Branches: Popular Armed Forces (includes Intervention Forces, Development Forces, Aeronaval Forces—includes Navy and Air Force), Gendarmerie, Presidential Security Regiment
Manpower availability: males age 15-49 3,027,156; males fit for military service 1,800,127; males reach military age (20) annually 130,071 (1995 est.)
Defense expenditures: exchange rate conversion—$35 million, 1.3% of GDP (1991)

Malawi

Geography

Location: Southern Africa, east of Zambia
Map references: Africa
Area:
total area: 118,480 sq km
land area: 94,080 sq km
comparative area: slightly larger than Pennsylvania
Land boundaries: total 2,881 km, Mozambique 1,569 km, Tanzania 475 km, Zambia 837 km
Coastline: 0 km (landlocked)
Maritime claims: none; landlocked
International disputes: dispute with Tanzania over the boundary in Lake Nyasa (Lake Malawi)
Climate: tropical; rainy season (November to May); dry season (May to November)
Terrain: narrow elongated plateau with rolling plains, rounded hills, some mountains
Natural resources: limestone, unexploited deposits of uranium, coal, and bauxite
Land use:
arable land: 25%
permanent crops: 0%
meadows and pastures: 20%
forest and woodland: 50%
other: 5%
Irrigated land: 200 sq km (1989 est.)
Environment:
current issues: deforestation; land degradation; water pollution from agricultural runoff, sewage, industrial wastes; siltation of spawning grounds endangers fish population
natural hazards: NA
international agreements: party to—Biodiversity, Climate Change, Endangered Species, Environmental Modification, Hazardous Wastes, Marine Life Conservation, Nuclear Test Ban, Ozone Layer Protection; signed, but not ratified—Law of the Sea
Note: landlocked

People

Population: 9,808,384 (July 1995 est.)
Age structure:
0-14 years: 48% (female 2,361,309; male 2,384,679)
15-64 years: 49% (female 2,479,108; male 2,335,729)
65 years and over: 3% (female 139,632; male 107,927) (July 1995 est.)
Population growth rate: 2.63% (1995 est.)
Birth rate: 49.81 births/1,000 population (1995 est.)
Death rate: 23.53 deaths/1,000 population (1995 est.)
Net migration rate: NA migrant(s)/1,000 population (1995 est.)
note: the return of refugees to Mozambique is much reduced compared with 1994
Infant mortality rate: 140.2 deaths/1,000 live births (1995 est.)
Life expectancy at birth:
total population: 39.01 years
male: 38.28 years
female: 39.76 years (1995 est.)
Total fertility rate: 7.36 children born/woman (1995 est.)
Nationality:
noun: Malawian(s)
adjective: Malawian
Ethnic divisions: Chewa, Nyanja, Tumbuko, Yao, Lomwe, Sena, Tonga, Ngoni, Ngonde, Asian, European
Religions: Protestant 55%, Roman Catholic 20%, Muslim 20%, traditional indigenous beliefs
Languages: English (official), Chichewa (official), other languages important regionally
Literacy: age 15 and over can read and write (1987)
total population: 48%
male: 65%
female: 34%
Labor force: 428,000 wage earners
by occupation: agriculture 43%, manufacturing 16%, personal services 15%, commerce 9%, construction 7%, miscellaneous services 4%, other permanently employed 6% (1986)

Government

Names:
conventional long form: Republic of Malawi
conventional short form: Malawi
former: Nyasaland
Digraph: MI
Type: multiparty democracy following a referendum on 14 June 1993; formerly a one-party republic
Capital: Lilongwe

Administrative divisions: 24 districts; Blantyre, Chikwawa, Chiradzulu, Chitipa, Dedza, Dowa, Karonga, Kasungu, Lilongwe, Machinga (Kasupe), Mangochi, Mchinji, Mulanje, Mwanza, Mzimba, Ntcheu, Nkhata Bay, Nkhotakota, Nsanje, Ntchisi, Rumphi, Salima, Thyolo, Zomba
Independence: 6 July 1964 (from UK)
National holiday: Independence Day, 6 July (1964)
Constitution: 6 July 1966; republished as amended January 1974
Legal system: based on English common law and customary law; judicial review of legislative acts in the Supreme Court of Appeal; has not accepted compulsory ICJ jurisdiction
Suffrage: 21 years of age; universal
Executive branch:
chief of state and head of government: President Bakili MULUZI (since 21 May 1994), leader of the United Democratic Front
cabinet: Cabinet; named by the president
Legislative branch: unicameral
National Assembly: elections last held 17 May 1994 (next to be held 1999); results—percent of vote by party NA; seats—(177 total) UDF 84, AFORD 33, MCP 55, others 5
Judicial branch: High Court, Supreme Court of Appeal
Political parties and leaders:
ruling party: United Democratic Front (UDF), Bakili MULUZI
opposition groups: Malawi Congress Party (MCP), Gwanda CHAKUAMBA Phiri, secretary general (top party position); Alliance for Democracy (AFORD), Chakufwa CHIHANA; Socialist League of Malawi (Lesoma), Kapote MWAKUSULA, secretary general; Malawi Democratic Union (MDU), Harry BWANAUSI; Congress for the Second Republic (CSR), Kanyama CHIUME; Malawi Socialist Labor Party (MSLP), Stanford SAMBANEMANJA
Member of: ACP, AfDB, C, CCC, ECA, FAO, G-77, GATT, IBRD, ICAO, ICFTU, ICRM, IDA, IFAD, IFC, IFRCS, ILO, IMF, IMO, INTELSAT, INTERPOL, IOC, ISO (correspondent), ITU, NAM, OAU, SADC, UN, UNAMIR, UNCTAD, UNESCO, UNIDO, UPU, WFTU, WHO, WIPO, WMO, WTO
Diplomatic representation in US:
chief of mission: (vacant); Charge d'Affaires ad interim Patrick NYASULU (since 14 October 1994)
chancery: 2408 Massachusetts Avenue NW, Washington, DC 20008
telephone: [1] (202) 797-1007
US diplomatic representation:
chief of mission: Ambassador Peter R. CHAVEAS
embassy: address NA, in new capital city development area in Lilongwe
mailing address: P. O. Box 30016, Lilongwe 3, Malawi
telephone: [265] 783 166
FAX: [265] 780 471
Flag: three equal horizontal bands of black (top), red, and green with a radiant, rising, red sun centered in the black band; similar to the flag of Afghanistan, which is longer and has the national coat of arms superimposed on the hoist side of the black and red bands

Economy

Overview: Landlocked Malawi ranks among the world's least developed countries. The economy is predominately agricultural, with about 90% of the population living in rural areas. Agriculture accounts for 40% of GDP and 90% of export revenues. After two years of weak performance, economic growth improved significantly in 1988-91 as a result of good weather and a broadly based economic adjustment effort by the government. Drought cut overall output sharply in 1992, but the lost ground was recovered in 1993. The economy depends on substantial inflows of economic assistance from the IMF, the World Bank, and individual donor nations. The new government faces strong challenges, e.g., to spur exports, to improve educational and health facilities, and to deal with environmental problems of deforestation and erosion.
National product: GDP—purchasing power parity—$7.3 billion (1994 est.)
National product real growth rate: 9.3% (1994 est.)
National product per capita: $750 (1994 est.)
Inflation rate (consumer prices): 30% (1994 est.)
Unemployment rate: NA%
Budget:
revenues: $416 million
expenditures: $498 million, including capital expenditures of $NA (1992 est.)
Exports: $311 million (f.o.b., 1993 est.)
commodities: tobacco, tea, sugar, coffee, peanuts, wood products
partners: US, UK, Zambia, South Africa, Germany
Imports: $308 million (c.i.f., 1993 est.)
commodities: food, petroleum products, semimanufactures, consumer goods, transportation equipment
partners: South Africa, Japan, US, UK, Zimbabwe
External debt: $1.8 billion (December 1993 est.)
Industrial production: growth rate 3.5% accounts for about 15% of GDP (1992 est.)
Electricity:
capacity: 190,000 kW
production: 820 million kWh
consumption per capita: 77 kWh (1993)
Industries: agricultural processing (tea, tobacco, sugar), sawmilling, cement, consumer goods
Agriculture: accounts for 40% of GDP; cash crops—tobacco, sugarcane, cotton, tea, and corn; subsistence crops—potatoes, cassava, sorghum, pulses; livestock—cattle, goats
Economic aid:
recipient: US commitments, including Ex-Im (FY70-89), $215 million; Western (non-US) countries, ODA and OOF bilateral commitments (1970-89), $2.15 billion
Currency: 1 Malawian kwacha (MK) = 100 tambala
Exchange rates: Malawian kwacha (MK) per US$1—7.8358 (August 1994), 4.4028 (1993), 3.6033 (1992), 2.8033 (1991), 2.7289 (1990), 2.7595 (1989)
Fiscal year: 1 April—31 March

Transportation

Railroads:
total: 789 km
narrow gauge: 789 km 1.067-m gauge
Highways:
total: 13,135 km
paved: 2,364 km
unpaved: gravel, crushed stone, stabilized earth 251 km; earth, improved earth 10,520 km
Inland waterways: Lake Nyasa (Lake Malawi); Shire River, 144 km
Ports: Chipoka, Monkey Bay, Nkhata Bay, Nkotakota
Airports:
total: 47
with paved runways over 3,047 m: 1
with paved runways 1,524 to 2,437 m: 1
with paved runways 914 to 1,523 m: 4
with paved runways under 914 m: 25
with unpaved runways 1,524 to 2,438 m: 1
with unpaved runways 914 to 1,523 m: 15

Communications

Telephone system: 42,250 telephones
local: NA
intercity: fair system of open-wire lines, radio relay links, and radio communications stations
international: 2 INTELSAT (1 Indian Ocean and 1 Atlantic Ocean) earth stations
Radio:
broadcast stations: AM 10, FM 17, shortwave 0
radios: NA
Television:
broadcast stations: 0
televisions: NA

Defense Forces

Branches: Army (includes Air Wing and Naval Detachment), Police (includes paramilitary Mobile Force Unit), paramilitary Malawi Young Pioneers
Manpower availability: males age 15-49 2,069,302; males fit for military service 1,056,372 (1995 est.)
Defense expenditures: exchange rate conversion—$13 million, 0.7% of GDP (FY93/94)

Malaysia

Geography

Location: Southeastern Asia, peninsula and northern one-third of the island of Borneo bordering the Java Sea and the South China Sea, south of Vietnam
Map references: Southeast Asia
Area:
total area: 329,750 sq km
land area: 328,550 sq km
comparative area: slightly larger than New Mexico
Land boundaries: total 2,669 km, Brunei 381 km, Indonesia 1,782 km, Thailand 506 km
Coastline: 4,675 km (Peninsular Malaysia 2,068 km, East Malaysia 2,607 km)
Maritime claims:
continental shelf: 200-m depth or to depth of exploitation; specified boundary in the South China Sea
exclusive fishing zone: 200 nm
exclusive economic zone: 200 nm
territorial sea: 12 nm
International disputes: involved in a complex dispute over the Spratly Islands with China, Philippines, Taiwan, Vietnam, and possibly Brunei; State of Sabah claimed by the Philippines; Brunei may wish to purchase the Malaysian salient that divides Brunei into two parts; two islands in dispute with Singapore; two islands in dispute with Indonesia
Climate: tropical; annual southwest (April to October) and northeast (October to February) monsoons
Terrain: coastal plains rising to hills and mountains
Natural resources: tin, petroleum, timber, copper, iron ore, natural gas, bauxite
Land use:
arable land: 3%
permanent crops: 10%
meadows and pastures: 0%
forest and woodland: 63%
other: 24%
Irrigated land: 3,420 sq km (1989 est.)
Environment:
current issues: air pollution from industrial and vehicular emissions; water pollution from raw sewage; deforestation
natural hazards: flooding
international agreements: party to—Biodiversity, Climate Change, Endangered Species, Hazardous Wastes, Marine Life Conservation, Nuclear Test Ban, Ozone Layer Protection, Tropical Timber 83; signed, but not ratified—Law of the Sea
Note: strategic location along Strait of Malacca and southern South China Sea

People

Population: 19,723,587 (July 1995 est.)
Age structure:
0-14 years: 37% (female 3,559,434; male 3,690,310)
15-64 years: 59% (female 5,871,131; male 5,844,568)
65 years and over: 4% (female 423,539; male 334,605) (July 1995 est.)
Population growth rate: 2.24% (1995 est.)
Birth rate: 27.95 births/1,000 population (1995 est.)
Death rate: 5.56 deaths/1,000 population (1995 est.)
Net migration rate: 0 migrant(s)/1,000 population (1995 est.)
Infant mortality rate: 24.7 deaths/1,000 live births (1995 est.)
Life expectancy at birth:
total population: 69.48 years
male: 66.55 years
female: 72.56 years (1995 est.)
Total fertility rate: 3.47 children born/woman (1995 est.)
Nationality:
noun: Malaysian(s)
adjective: Malaysian
Ethnic divisions: Malay and other indigenous 59%, Chinese 32%, Indian 9%
Religions:
Peninsular Malaysia: Muslim (Malays), Buddhist (Chinese), Hindu (Indians)
Sabah: Muslim 38%, Christian 17%, other 45%
Sarawak: tribal religion 35%, Buddhist and Confucianist 24%, Muslim 20%, Christian 16%, other 5%
Languages:
Peninsular Malaysia: Malay (official), English, Chinese dialects, Tamil
Sabah: English, Malay, numerous tribal dialects, Chinese (Mandarin and Hakka dialects predominate)
Sarawak: English, Malay, Mandarin, numerous tribal languages
Literacy: age 15 and over can read and write (1990 est.)
total population: 78%
male: 86%
female: 70%
Labor force: 7.627 million (1993)

Government

Names:
conventional long form: none
conventional short form: Malaysia
former: Malayan Union
Digraph: MY
Type: constitutional monarchy
note: Federation of Malaysia formed 9 July 1963; nominally headed by the paramount ruler (king) and a bicameral Parliament; Peninsular Malaysian states—hereditary rulers in all but Melaka, where governors are appointed by Malaysian Pulau Pinang Government; powers of state governments are limited by federal Constitution; Sabah—self-governing state, holds 20 seats in House of Representatives, with foreign affairs, defense, internal security, and other powers delegated to federal government; Sarawak—self-governing state, holds 27 seats in House of Representatives, with foreign affairs, defense, internal security, and other powers delegated to federal government
Capital: Kuala Lumpur
Administrative divisions: 13 states (negeri-negeri, singular—negeri) and 2 federal territories* (wilayah-wilayah persekutuan, singular—wilayah persekutuan); Johor, Kedah, Kelantan, Labuan*, Melaka, Negeri Sembilan, Pahang, Perak, Perlis, Pulau Pinang, Sabah, Sarawak, Selangor, Terengganu, Wilayah Persekutuan*
Independence: 31 August 1957 (from UK)
National holiday: National Day, 31 August (1957)
Constitution: 31 August 1957, amended 16 September 1963
Legal system: based on English common law; judicial review of legislative acts in the Supreme Court at request of supreme head of the federation; has not accepted compulsory ICJ jurisdiction
Suffrage: 21 years of age; universal
Executive branch:
chief of state: Paramount Ruler JA'AFAR ibni Abdul Rahman (since 26 April 1994); Deputy Paramount Ruler SALAHUDDIN ibni Hisammuddin Alam Shah (since 26 April 1994)
head of government: Prime Minister Dr. MAHATHIR bin Mohamad (since 16 July 1981); Deputy Prime Minister ANWAR bin Ibrahim (since 1 December 1993)
cabinet: Cabinet; appointed by the Paramount Ruler from members of parliament
Legislative branch: bicameral Parliament (Parlimen)
Senate (Dewan Negara): consists of 58 members, 32 appointed by the paramount ruler and 26 elected by the state legislatures (2 from each state) for six-year terms; elections last held NA (next to be held NA); results—NA
House of Representatives (Dewan Rakyat): consists of 180 members, elected for five-year

terms; elections last held 21 October 1990 (next to be held by December 1995); results—National Front 52%, other 48%; seats—(180 total) National Front 127, DAP 20, PAS 7, independents 4, other 22; note—within the National Front, UMNO won 71 seats and MCA won 18 seats
Judicial branch: Supreme Court
Political parties and leaders:
Peninsular Malaysia: National Front, a confederation of 13 political parties dominated by United Malays National Organization Baru (UMNO Baru), MAHATHIR bin Mohamad; Malaysian Chinese Association (MCA), LING Liong Sik; Gerakan Rakyat Malaysia, LIM Keng Yaik; Malaysian Indian Congress (MIC), S. Samy VELLU
Sabah: National Front, SALLEH Said Keruak, Sabah Chief Minister, Sakaran DANDAI, head of Sabah State; United Sabah National Organizaton (USNO), leader NA
Sarawak: coalition Sarawak National Front composed of the Party Pesaka Bumiputra Bersatu (PBB), Datuk Patinggi Amar Haji Abdul TAIB Mahmud; Sarawak United People's Party (SUPP), Datuk Amar James WONG Soon Kai; Sarawak National Party (SNAP), Datuk Amar James WONG; Parti Bansa Dayak Sarawak (PBDS), Datuk Leo MOGGIE; major opposition parties are Democratic Action Party (DAP), LIM Kit Siang and Pan-Malaysian Islamic Party (PAS), Fadzil NOOR
Member of: APEC, AsDB, ASEAN, C, CCC, CP, ESCAP, FAO, G-15, G-77, GATT, IAEA, IBRD, ICAO, ICFTU, ICRM, IDA, IDB, IFAD, IFC, IFRCS, ILO, IMF, IMO, INMARSAT, INTELSAT, INTERPOL, IOC, ISO, ITU, MINURSO, NAM, OIC, UN, UNAVEM II, UNCTAD, UNESCO, UNIDO, UNIKOM, UNOMIL, UNOMOZ, UNOSOM, UNPROFOR, UPU, WCL, WFTU, WHO, WIPO, WMO, WTO
Diplomatic representation in US:
chief of mission: Ambassador Abdul MAJID bin Mohamed
chancery: 2401 Massachusetts Avenue NW, Washington, DC 20008
telephone: [1] (202) 328-2700
FAX: [1] (202) 483-7661
consulate(s) general: Los Angeles and New York
US diplomatic representation:
chief of mission: Ambassador John S. WOLF
embassy: 376 Jalan Tun Razak, 50400 Kuala Lumpur
mailing address: P. O. Box No. 10035, 50700 Kuala Lumpur; APO AP 96535-8152
telephone: [60] (3) 2489011
FAX: [60] (3) 2422207
Flag: fourteen equal horizontal stripes of red (top) alternating with white (bottom); there is a blue rectangle in the upper hoist-side corner bearing a yellow crescent and a yellow fourteen-pointed star; the crescent and the star are traditional symbols of Islam; the design was based on the flag of the US

Economy

Overview: The Malaysian economy, a mixture of private enterprise and a soundly managed public sector, has posted a remarkable record of 9% average annual growth in 1988-94. The official growth target for 1995 is 8.5%. This growth has resulted in a substantial reduction in poverty and a marked rise in real wages. Manufactured goods exports expanded rapidly, and foreign investors continued to commit large sums in the economy. The government is aware of the inflationary potential of this rapid development and is closely monitoring fiscal and monetary policies.
National product: GDP—purchasing power parity—$166.8 billion (1994 est.)
National product real growth rate: 8.7% (1994)
National product per capita: $8,650 (1994 est.)
Inflation rate (consumer prices): 3.7% (1994)
Unemployment rate: 2.9% (1994)
Budget:
revenues: $18.7 billion
expenditures: $19.1 billion, including capital expenditures of $4.8 billion (1994)
Exports: $56.6 billion (f.o.b., 1994)
commodities: electronic equipment, petroleum and petroleum products, palm oil, wood and wood products, rubber, textiles
partners: Singapore 22%, US 20%, Japan 13%, UK 4%, Germany 4%, Thailand 4% (1993)
Imports: $55.2 billion (c.i.f., 1994)
commodities: machinery and equipment, chemicals, food, petroleum products
partners: Japan 27%, US 17%, Singapore 15%, Taiwan 5%, Germany 4%, UK 3%, South Korea 3% (1993)
External debt: $35.5 billion (1994 est.)
Industrial production: growth rate 12% (1994); accounts for 38% of GDP (1993 est.)
Electricity:
capacity: 6,700,000 kW
production: 31 billion kWh
consumption per capita: 1,528 kWh (1993)
Industries:
Peninsular Malaysia: rubber and oil palm processing and manufacturing, light manufacturing industry, electronics, tin mining and smelting, logging and processing timber
Sabah: logging, petroleum production
Sarawak: agriculture processing, petroleum production and refining, logging
Agriculture: accounts for 16% of GDP (1993 est.)
Peninsular Malaysia: natural rubber, palm oil, rice
Sabah: mainly subsistence, but also rubber, timber, coconut, rice
Sarawak: rubber, timber, pepper; deficit of rice in all areas
Illicit drugs: transit point for Golden Triangle heroin going to the US, Western Europe, and the Third World despite severe penalties for drug trafficking; increasing indigenous abuse of methamphetamine
Economic aid:
recipient: US commitments, including Ex-Im (FY70-84), $170 million; Western (non-US) countries, ODA and OOF bilateral commitments (1970-89), $4.7 million; OPEC bilateral aid (1979-89), $42 million
Currency: 1 ringgit (M$) = 100 sen
Exchange rates: ringgits (M$) per US$1—2.5542 (January 1995), 2.6242 (1994), 2.5741 (1993), 2.5474 (1992), 2.7501 (1991), 1.7048 (1990)
Fiscal year: calendar year

Transportation

Railroads:
total: 1,801 km (Peninsular Malaysia 1,665 km; Sabah 136 km; Sarawak 0 km)
narrow gauge: 1,801 km 1.000-m gauge (Peninsular Malaysia 1,665 km; Sabah 136 km)
Highways:
total: 29,028 km (Peninsular Malaysia 23,602 km, Sabah 3,782 km, Sarawak 1,644 km)
paved: NA (Peninsular Malaysia 19,354 km mostly bituminous treated)
unpaved: NA (Peninsular Malaysia 4,248 km)
Inland waterways:
Peninsular Malaysia: 3,209 km
Sabah: 1,569 km
Sarawak: 2,518 km
Pipelines: crude oil 1,307 km; natural gas 379 km
Ports: Kota Kinabalu, Kuantan, Kuching, Kudat, Lahad Datu, Labuan, Lumut, Miri, Pasir Gudang, Penang, Port Dickson, Port Kelang, Sandakan, Sibu, Tanjong Berhala, Tanjong Kidurong, Tawau
Merchant marine:
total: 213 ships (1,000 GRT or over) totaling 2,410,823 GRT/3,635,966 DWT
ships by type: bulk 34, cargo 73, chemical tanker 11, container 27, liquefied gas tanker 9, livestock carrier 1, oil tanker 50, roll-on/roll-off cargo 4, short-sea passenger 1, vehicle carrier 3
Airports:
total: 115
with paved runways over 3,047 m: 3
with paved runways 2,438 to 3,047 m: 5
with paved runways 1,524 to 2,437 m: 11
with paved runways 914 to 1,523 m: 6
with paved runways under 914 m: 82
with unpaved runways 1,524 to 2,438 m: 1
with unpaved runways 914 to 1,523 m: 7

Communications

Telephone system: 994,860 telephones (1984); international service good
local: NA
intercity: good intercity service provided on Peninsular Malaysia mainly by microwave radio relay; adequate intercity microwave radio relay network between Sabah and Sarawak via Brunei; 2 domestic satellite links
international: submarine cables extend to India and Sarawak; SEACOM submarine cable links to Hong Kong and Singapore; satellite earth stations—2 INTELSAT (1 Indian Ocean and 1 Pacific Ocean)
Radio:
broadcast stations: AM 28, FM 3, shortwave 0
radios: NA
Television:
broadcast stations: 33
televisions: NA

Defense Forces

Branches: Malaysian Army, Royal Malaysian Navy, Royal Malaysian Air Force, Royal Malaysian Police Force, Marine Police, Sarawak Border Scouts
Manpower availability: males age 15-49 5,041,003; males fit for military service 3,058,445; males reach military age (21) annually 183,760 (1995 est.)
Defense expenditures: exchange rate conversion—$2.1 billion, 2.9% of GDP (1994)

Maldives

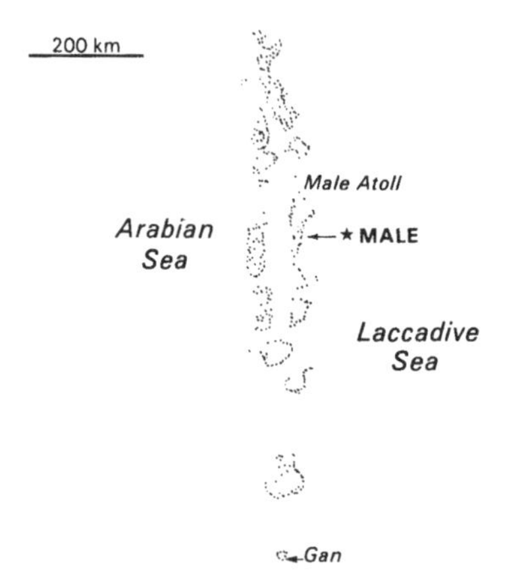

Geography

Location: Southern Asia, group of atolls in the Indian Ocean, south-southwest of India
Map references: Asia
Area:
total area: 300 sq km
land area: 300 sq km
comparative area: slightly more than 1.5 times the size of Washington, DC
Land boundaries: 0 km
Coastline: 644 km
Maritime claims:
exclusive economic zone: 35-310 nm as defined by geographic coordinates; segment of zone coincides with maritime boundary with India
territorial sea: 12 nm
International disputes: none
Climate: tropical; hot, humid; dry, northeast monsoon (November to March); rainy, southwest monsoon (June to August)
Terrain: flat with elevations only as high as 2.5 meters
Natural resources: fish
Land use:
arable land: 10%
permanent crops: 0%
meadows and pastures: 3%
forest and woodland: 3%
other: 84%
Irrigated land: NA sq km
Environment:
current issues: depletion of freshwater aquifers threatens water supplies
natural hazards: low level of islands makes them very sensitive to sea level rise
international agreements: party to—Biodiversity, Climate Change, Hazardous Wastes, Ozone Layer Protection; signed, but not ratified—Law of the Sea
Note: 1,190 coral islands grouped into 26 atolls; archipelago of strategic location astride and along major sea lanes in Indian Ocean

People

Population: 261,310 (July 1995 est.)
Age structure:
0-14 years: 47% (female 60,038; male 63,042)
15-64 years: 50% (female 63,526; male 67,020)
65 years and over: 3% (female 3,537; male 4,147) (July 1995 est.)
Population growth rate: 3.58% (1995 est.)
Birth rate: 42.8 births/1,000 population (1995 est.)
Death rate: 7 deaths/1,000 population (1995 est.)
Net migration rate: 0 migrant(s)/1,000 population (1995 est.)
Infant mortality rate: 50 deaths/1,000 live births (1995 est.)
Life expectancy at birth:
total population: 65.49 years
male: 63.99 years
female: 67.07 years (1995 est.)
Total fertility rate: 6.17 children born/woman (1995 est.)
Nationality:
noun: Maldivian(s)
adjective: Maldivian
Ethnic divisions: Sinhalese, Dravidian, Arab, African
Religions: Sunni Muslim
Languages: Divehi (dialect of Sinhala; script derived from Arabic), English spoken by most government officials
Literacy: age 15 and over can read and write (1985)
total population: 91%
male: 91%
female: 92%
Labor force: 66,000 (est.)
by occupation: fishing industry 25%

Government

Names:
conventional long form: Republic of Maldives
conventional short form: Maldives
Digraph: MV
Type: republic
Capital: Male
Administrative divisions: 19 districts (atolls); Aliff, Baa, Daalu, Faafu, Gaafu Aliff, Gaafu Daalu, Haa Aliff, Haa Daalu, Kaafu, Laamu, Laviyani, Meemu, Naviyani, Noonu, Raa, Seenu, Shaviyani, Thaa, Waavu
Independence: 26 July 1965 (from UK)
National holiday: Independence Day, 26 July (1965)
Constitution: 4 June 1968
Legal system: based on Islamic law with admixtures of English common law primarily in commercial matters; has not accepted compulsory ICJ jurisdiction
Suffrage: 21 years of age; universal

Executive branch:
chief of state and head of government: President Maumoon Abdul GAYOOM (since 11 November 1978); election last held 1 October 1993 (next to be held 1998); results—President Maumoon Abdul GAYOOM was reelected with 92.76% of the vote
cabinet: Ministry of Atolls; appointed by the president
Legislative branch: unicameral
Citizens' Council (Majlis): elections last held 2 December 1994 (next to be held NA December 1999); results—percent of vote NA; seats—(48 total, 40 elected, 8 appointed by the president) independents 40
Judicial branch: High Court
Political parties and leaders: although political parties are not banned, none exist; country governed by the Didi clan for the past eight centuries
Member of: AsDB, C, CP, ESCAP, FAO, G-77, GATT, IBRD, ICAO, IDA, IDB, IFAD, IFC, IMF, IMO, INTELSAT (nonsignatory user), INTERPOL, IOC, ITU, NAM, OIC, SAARC, UN, UNCTAD, UNESCO, UNIDO, UPU, WHO, WMO, WTO
Diplomatic representation in US: Maldives has no embassy in the US, but does have a UN mission in New York; Permanent Representative to the UN Ahmed ZAKI
US diplomatic representation:
chief of mission: the US Ambassador to Sri Lanka is accredited to Maldives and makes periodic visits there
consular agency: Midhath Hilmy, Male
telephone: 322581
Flag: red with a large green rectangle in the center bearing a vertical white crescent; the closed side of the crescent is on the hoist side of the flag

Economy

Overview: Fishing is the largest industry, employing 25% of the work force and accounting for over 60% of exports. Over 90% of government tax revenue comes from import duties and tourism-related taxes. During the 1980s tourism became one of the most important and highest growth sectors of the economy. In 1993, tourism accounted for 17% of GDP and more than 60% of the Maldives' foreign exchange receipts. The Maldivian government initiated an economic reform program in 1989 initially by lifting import quotas and opening some exports to the private sector. Subsequently, it has liberalized regulations to allow more foreign investment. Agriculture and manufacturing continue to play a minor role in the economy, constrained by the limited availability of cultivatable land and the shortage of domestic labor. Most staple foods must be imported. In 1993, industry which consisted mainly of garment production, boat building, and handicrafts accounted for about 6% of GDP.
National product: GDP—purchasing power parity—$360 million (1993 est.)
National product real growth rate: 5.4% (1993 est.)
National product per capita: $1,500 (1993 est.)
Inflation rate (consumer prices): 20% (1993)
Unemployment rate: NEGL%
Budget:
revenues: $95 million (excluding foreign transfers)
expenditures: $143 million, including capital expenditures of $71 million (1993 est.)
Exports: $38.5 million (f.o.b., 1993 est.)
commodities: fish, clothing
partners: US, UK, Sri Lanka, Singapore, Germany
Imports: $177.8 million (c.i.f., 1993)
commodities: consumer goods, intermediate and capital goods, petroleum products
partners: Singapore, Germany, Sri Lanka, India, Japan
External debt: $130 million (1993 est.)
Industrial production: growth rate 24% (1990); accounts for 6% of GDP
Electricity:
capacity: 5,000 kW
production: 30 million kWh
consumption per capita: 123 kWh (1993)
Industries: fishing and fish processing, tourism, shipping, boat building, some coconut processing, garments, woven mats, coir (rope), handicrafts
Agriculture: fishing, coconuts, corn, sweet potatoes
Economic aid:
recipient: US commitments, including Ex-Im (FY70-88), $28 million; Western (non-US) countries, ODA and OOF bilateral commitments (1970-89), $125 million; OPEC bilateral aid (1979-89), $14 million
Currency: 1 rufiyaa (Rf) = 100 laari
Exchange rates: rufiyaa (Rf) per US$1—11.770 (January 1995), 11.586 (1994), 10.957 (1993), 10.569 (1992), 10.253 (1991), 9.509 (1990)
Fiscal year: calendar year

Transportation

Railroads: 0 km
Highways:
total: NA
paved: NA
unpaved: NA (Male has 9.6 km of coral highways within the city)
Ports: Gan, Male
Merchant marine:
total: 16 ships (1,000 GRT or over) totaling 50,384 GRT/77,771 DWT
ships by type: cargo 14, container 1, oil tanker 1
Airports:
total: 2
with paved runways over 3,047 m: 1
with paved runways 2,438 to 3,047 m: 1

Communications

Telephone system: 2,804 telephones; minimal domestic and international facilities
local: NA
intercity: NA
international: 1 INTELSAT (Indian Ocean) earth station
Radio:
broadcast stations: AM 2, FM 1, shortwave 0
radios: NA
Television:
broadcast stations: 1
televisions: NA

Defense Forces

Branches: National Security Service (paramilitary police force)
Manpower availability: males age 15-49 57,172; males fit for military service 31,911 (1995 est.)
Defense expenditures: $NA, NA% of GDP

Mali

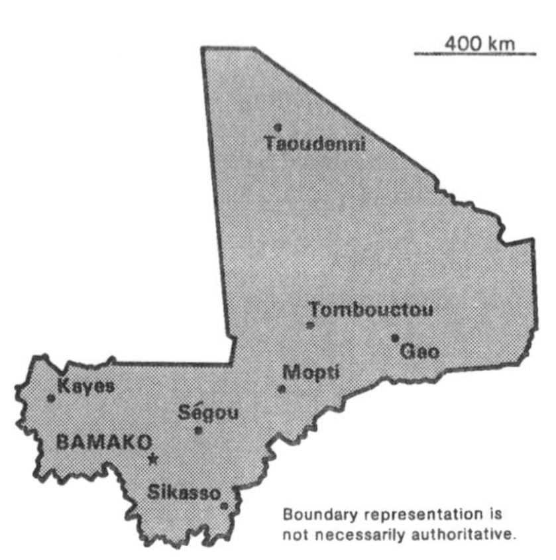

Geography

Location: Western Africa, southwest of Algeria
Map references: Africa
Area:
total area: 1.24 million sq km
land area: 1.22 million sq km
comparative area: slightly less than twice the size of Texas
Land boundaries: total 7,243 km, Algeria 1,376 km, Burkina 1,000 km, Guinea 858 km, Cote d'Ivoire 532 km, Mauritania 2,237 km, Niger 821 km, Senegal 419 km
Coastline: 0 km (landlocked)
Maritime claims: none; landlocked
International disputes: the disputed international boundary between Burkina and Mali was submitted to the International Court of Justice (ICJ) in October 1983 and the ICJ issued its final ruling in December 1986, which both sides agreed to accept; Burkina and Mali are proceeding with boundary demarcation, including the tripoint with Niger
Climate: subtropical to arid; hot and dry February to June; rainy, humid, and mild June to November; cool and dry November to February
Terrain: mostly flat to rolling northern plains covered by sand; savanna in south, rugged hills in northeast
Natural resources: gold, phosphates, kaolin, salt, limestone, uranium, bauxite, iron ore, manganese, tin, and copper deposits are known but not exploited
Land use:
arable land: 2%
permanent crops: 0%
meadows and pastures: 25%
forest and woodland: 7%
other: 66%
Irrigated land: 50 sq km (1989 est.)
Environment:
current issues: deforestation; soil erosion; desertification; inadequate supplies of potable water; poaching
natural hazards: hot, dust-laden harmattan haze common during dry seasons; recurring droughts
international agreements: party to—Climate Change, Desertification, Endangered Species, Law of the Sea, Ozone Layer Protection, Wetlands; signed, but not ratified—Biodiversity, Nuclear Test Ban
Note: landlocked

People

Population: 9,375,132 (July 1995 est.)
Age structure:
0-14 years: 48% (female 2,240,565; male 2,242,373)
15-64 years: 49% (female 2,416,952; male 2,165,043)
65 years and over: 3% (female 162,234; male 147,965) (July 1995 est.)
Population growth rate: 2.89% (1995 est.)
Birth rate: 51.88 births/1,000 population (1995 est.)
Death rate: 19.93 deaths/1,000 population (1995 est.)
Net migration rate: -3 migrant(s)/1,000 population (1995 est.)
Infant mortality rate: 104.5 deaths/1,000 live births (1995 est.)
Life expectancy at birth:
total population: 46.37 years
male: 44.7 years
female: 48.09 years (1995 est.)
Total fertility rate: 7.33 children born/woman (1995 est.)
Nationality:
noun: Malian(s)
adjective: Malian
Ethnic divisions: Mande 50% (Bambara, Malinke, Sarakole), Peul 17%, Voltaic 12%, Songhai 6%, Tuareg and Moor 10%, other 5%
Religions: Muslim 90%, indigenous beliefs 9%, Christian 1%
Languages: French (official), Bambara 80%, numerous African languages
Literacy: age 6 and over can read and write (1988)
total population: 19%
male: 27%
female: 12%
Labor force: 2.666 million (1986 est.)
by occupation: agriculture 80%, services 19%, industry and commerce 1% (1981)

Government

Names:
conventional long form: Republic of Mali
conventional short form: Mali
local long form: Republique de Mali
local short form: Mali
former: French Sudan
Digraph: ML
Type: republic
Capital: Bamako
Administrative divisions: 8 regions (regions, singular—region); Gao, Kayes, Kidal, Koulikoro, Mopti, Segou, Sikasso, Tombouctou
Independence: 22 September 1960 (from France)
National holiday: Anniversary of the Proclamation of the Republic, 22 September (1960)
Constitution: adopted 12 January 1992
Legal system: based on French civil law system and customary law; judicial review of legislative acts in Constitutional Court (which was formally established on 9 March 1994); has not accepted compulsory ICJ jurisdiction
Suffrage: 21 years of age; universal
Executive branch:
chief of state: President Alpha Oumar KONARE (since 8 June 1992); election last held in April 1992 (next to be held April 1997); Alpha KONARE was elected in runoff race against Montaga TALL
head of government: Prime Minister Ibrahima Boubacar KEITA (since March 1994)
cabinet: Council of Ministers; appointed by the prime minister
Legislative branch: unicameral
National Assembly: elections last held on 8 March 1992 (next to be held February 1997); results—percent of vote by party NA; seats—(116 total) Adema 76, CNID 9, US/RAD 8, Popular Movement for the Development of the Republic of West Africa 6, RDP 4, UDD 4, RDT 3, UFDP 3, PDP 2, UMDD 1
Judicial branch: Supreme Court (Cour Supreme)
Political parties and leaders: Association for Democracy (Adema), Ibrahim Baubacar KEITA; National Congress for Democratic Initiative (CNID), Mountaga TALL; Sudanese Union/African Democratic Rally (US/RDA), Mamadou Madeira KEITA; Popular Movement for the Development of the Republic of West Africa; Rally for Democracy and Progress (RDP), Almamy SYLLA; Union for Democracy and Development (UDD), Moussa Balla COULIBALY; Rally for Democracy and Labor (RDT); Union of Democratic Forces for Progress (UFDP), Dembo DIALLO; Party for Democracy and Progress (PDP), Idrissa TRAORE; Malian Union for Democracy and Development (UMDD)
Member of: ACCT, ACP, AfDB, CCC, CEAO, ECA, ECOWAS, FAO, FZ, G-77, GATT, IAEA, IBRD, ICAO, ICFTU, ICRM, IDA, IDB, IFAD, IFC, IFRCS, ILO, IMF, INTELSAT, INTERPOL, IOC, ITU, NAM, OAU, OIC, UN, UNAMIR, UNCTAD, UNESCO, UNIDO, UPU, WADB, WCL, WFTU, WHO, WIPO, WMO, WTO
Diplomatic representation in US:
chief of mission: Ambassador Ibrahim Siragatou CISSE
chancery: 2130 R Street NW, Washington, DC 20008
telephone: [1] (202) 332-2249, 939-8950

US diplomatic representation:
chief of mission: (vacant) (Ambassador William H. DAMERON III retired March 1995)
embassy: Rue Rochester NY and Rue Mohamed V, Bamako
mailing address: B. P. 34, Bamako
telephone: [223] 22 54 70
FAX: [223] 22 37 12
Flag: three equal vertical bands of green (hoist side), yellow, and red; uses the popular pan-African colors of Ethiopia

Economy

Overview: Mali is among the poorest countries in the world, with 65% of its land area desert or semidesert. Economic activity is largely confined to the riverine area irrigated by the Niger. About 10% of the population is nomadic and some 80% of the labor force is engaged in agriculture and fishing. Industrial activity is concentrated on processing farm commodities. The economy is beginning to turn around after contracting through 1992-93, largely because of enhanced exports and import substitute production in the wake of the 50% devaluation of January 1994. Post-devaluation inflation appears to have peaked at 35% in 1994 and the government appears to be keeping on track with its IMF structural adjustment program.
National product: GDP—purchasing power parity—$5.4 billion (1994 est.)
National product real growth rate: 2.4% (1994 est.)
National product per capita: $600 (1994 est.)
Inflation rate (consumer prices): 35% (1994 est.)
Unemployment rate: NA%
Budget:
revenues: $376 million
expenditures: $697 million, including capital expenditures of $NA (1992 est.)
Exports: $415 million (f.o.b., 1993)
commodities: cotton, livestock, gold
partners: mostly franc zone and Western Europe
Imports: $842 million (f.o.b., 1993)
commodities: machinery and equipment, foodstuffs, construction materials, petroleum, textiles
partners: mostly franc zone and Western Europe
External debt: $2.6 billion (1991 est.)
Industrial production: growth rate -1.4% (1992 est.); accounts for 13.0% of GDP
Electricity:
capacity: 90,000 kW
production: 310 million kWh
consumption per capita: 33 kWh (1993)
Industries: minor local consumer goods production and food processing, construction, phosphate and gold mining
Agriculture: accounts for 50% of GDP; mostly subsistence farming; cotton and livestock products account for over 70% of exports; other crops—millet, rice, corn, vegetables, peanuts; livestock—cattle, sheep, goats
Economic aid:
recipient: US commitments, including Ex-Im (FY70-89), $349 million; Western (non-US) countries, ODA and OOF bilateral commitments (1970-89), $3.02 billion; OPEC bilateral aid (1979-89), $92 million; Communist countries (1970-89), $190 million
Currency: 1 CFA franc (CFAF) = 100 centimes
Exchange rates: Communaute Financiere Africaine francs (CFAF) per US$1—529.43 (January 1995), 555.20 (1994), 283.16 (1993), 264.69 (1992), 282.11 (1991), 272.26 (1990)
note: beginning 12 January 1994, the CFA franc was devalued to CFAF 100 per French franc from CFAF 50 at which it had been fixed since 1948
Fiscal year: calendar year

Transportation

Railroads:
total: 642 km; note—linked to Senegal's rail system through Kayes
narrow gauge: 642 km 1.000-m gauge
Highways:
total: 15,700 km
paved: 1,670 km
unpaved: gravel, improved earth 3,670 km; unimproved earth 10,360 km
Inland waterways: 1,815 km navigable
Ports: Koulikoro
Airports:
total: 33
with paved runways 2,438 to 3,047 m: 4
with paved runways 1,524 to 2,437 m: 1
with paved runways 914 to 1,523 m: 2
with paved runways under 914 m: 10
with unpaved runways 2,438 to 3,047 m: 1
with unpaved runways 1,524 to 2,438 m: 3
with unpaved runways 914 to 1,523 m: 12

Communications

Telephone system: 11,000 telephones; domestic system poor but improving; provides only minimal service
local: NA
intercity: microwave radio relay, wire, and radio communications stations; expansion of microwave radio relay in progress
international: 2 INTELSAT (1 Atlantic Ocean and 1 Indian Ocean) earth stations
Radio:
broadcast stations: AM 2, FM 2, shortwave 0
radios: NA
Television:
broadcast stations: 2
televisions: NA

Defense Forces

Branches: Army, Air Force, Gendarmerie, Republican Guard, National Guard, National Police (Surete Nationale)
Manpower availability: males age 15-49 1,861,977; males fit for military service 1,062,916 (1995 est.)
Defense expenditures: exchange rate conversion—$66 million, 2.2% of GDP (1994)

Malta

Geography

Location: Southern Europe, islands in the Mediterranean Sea, south of Sicily (Italy)
Map references: Europe
Area:
total area: 320 sq km
land area: 320 sq km
comparative area: slightly less than twice the size of Washington, DC
Land boundaries: 0 km
Coastline: 140 km
Maritime claims:
contiguous zone: 24 nm
continental shelf: 200-m depth or to the depth of exploitation
exclusive fishing zone: 25 nm
territorial sea: 12 nm
International disputes: Malta and Tunisia are discussing the commercial exploitation of the continental shelf between their countries, particularly for oil exploration
Climate: Mediterranean with mild, rainy winters and hot, dry summers
Terrain: mostly low, rocky, flat to dissected plains; many coastal cliffs
Natural resources: limestone, salt
Land use:
arable land: 38%
permanent crops: 3%
meadows and pastures: 0%
forest and woodland: 0%
other: 59%
Irrigated land: 10 sq km (1989)
Environment:
current issues: very limited natural fresh water resources; increasing reliance on desalination
natural hazards: NA
international agreements: party to—Climate Change, Endangered Species, Law of the Sea, Marine Dumping, Nuclear Test Ban, Ozone Layer Protection, Ship Pollution, Wetlands; signed, but not ratified—Biodiversity, Desertification
Note: the country comprises an archipelago, with only the 3 largest islands (Malta, Gozo, and Comino) being inhabited; numerous bays provide good harbors

People

Population: 369,609 (July 1995 est.)
Age structure:
0-14 years: 22% (female 39,199; male 41,581)
15-64 years: 67% (female 123,665; male 124,167)
65 years and over: 11% (female 23,597; male 17,400) (July 1995 est.)
Population growth rate: 0.75% (1995 est.)
Birth rate: 13.22 births/1,000 population (1995 est.)
Death rate: 7.43 deaths/1,000 population (1995 est.)
Net migration rate: 1.7 migrant(s)/1,000 population (1995 est.)
Infant mortality rate: 7.7 deaths/1,000 live births (1995 est.)
Life expectancy at birth:
total population: 77.02 years
male: 74.75 years
female: 79.48 years (1995 est.)
Total fertility rate: 1.92 children born/woman (1995 est.)
Nationality:
noun: Maltese (singular and plural)
adjective: Maltese
Ethnic divisions: Arab, Sicilian, Norman, Spanish, Italian, English
Religions: Roman Catholic 98%
Languages: Maltese (official), English (official)
Literacy: age 15 and over can read and write (1985)
total population: 84%
male: 86%
female: 82%
Labor force: 127,200
by occupation: government (excluding job corps) 37%, services 26%, manufacturing 22%, training programs 9%, construction 4%, agriculture 2% (1990)

Government

Names:
conventional long form: Republic of Malta
conventional short form: Malta
Digraph: MT
Type: parliamentary democracy
Capital: Valletta
Administrative divisions: none (administration directly from Valletta)
Independence: 21 September 1964 (from UK)
National holiday: Independence Day, 21 September (1964)
Constitution: 1964 constitution substantially amended on 13 December 1974
Legal system: based on English common law and Roman civil law; has accepted compulsory ICJ jurisdiction, with reservations
Suffrage: 18 years of age; universal
Executive branch:
chief of state: President Ugo MIFSUD BONNICI (since 4 April 1994)
head of government: Prime Minister and Foreign Minister Dr. Edward (Eddie) FENECH ADAMI (since 12 May 1987); Deputy Prime Minister Dr. Guido DE MARCO (since 14 May 1987)
cabinet: Cabinet; appointed by the president on advice of the prime minister
Legislative branch: unicameral
House of Representatives: elections last held 22 February 1992 (next to be held by February 1997); results—NP 51.8%, MLP 46.5%; seats—(usually 65 total) MLP 36, NP 29; note—additional seats are given to the party with the largest popular vote to ensure a legislative majority; current total: 69 (MLP 33, NP 36 after adjustment)
Judicial branch: Constitutional Court, Court of Appeal
Political parties and leaders: Nationalist Party (NP), Edward FENECH ADAMI; Malta Labor Party (MLP), Alfred SANT
Member of: C, CCC, CE, EBRD, ECE, FAO, G-77, GATT, IBRD, ICAO, ICFTU, ICRM, IFAD, ILO, IMF, IMO, INMARSAT, INTELSAT (nonsignatory user), INTERPOL, IOC, ISO (correspondent), ITU, NAM, OSCE, PCA, UN, UNCTAD, UNESCO, UNIDO, UPU, WCL, WHO, WIPO, WMO, WTO
Diplomatic representation in US:
chief of mission: Ambassador Albert Borg Olivier DE PUGET
chancery: 2017 Connecticut Avenue NW, Washington, DC 20008
telephone: [1] (202) 462-3611, 3612
FAX: [1] (202) 387-5470
consulate(s): New York
US diplomatic representation:
chief of mission: Ambassador Joseph R. PAOLINO, Jr.
embassy: 2nd Floor, Development House, Saint Anne Street, Floriana, Malta
mailing address: P. O. Box 535, Valletta
telephone: [356] 235960
FAX: [356] 243229
Flag: two equal vertical bands of white (hoist side) and red; in the upper hoist-side corner is a representation of the George Cross, edged in red

Economy

Overview: Significant resources are limestone, a favorable geographic location, and a productive labor force. Malta produces only about 20% of its food needs, has limited freshwater supplies, and has no domestic energy sources. Consequently, the economy is highly dependent on foreign trade and services. Manufacturing and tourism are the largest contributors to the economy. Manufacturing accounts for about 24% of GDP, with the electronics and textile industries major

contributors and with the state-owned Malta drydocks employing about 4,300 people. In 1994, over 1,000,000 tourists visited the island. Per capita GDP of $10,760 places Malta in the range of the less affluent EU countries.
National product: GDP—purchasing power parity—$3.9 billion (1994 est.)
National product real growth rate: 4.4% (1994 est.)
National product per capita: $10,760 (1994 est.)
Inflation rate (consumer prices): 5% (1994 est.)
Unemployment rate: 4.5% (March 1994)
Budget:
revenues: $1.4 billion
expenditures: $1.4 billion, including capital expenditures of $215 million (FY94/95 est.)
Exports: $1.3 billion (f.o.b., 1993)
commodities: machinery and transport equipment, clothing and footware, printed matter
partners: Italy 32%, Germany 16%, UK 8%
Imports: $2.1 billion (c.i.f., 1993)
commodities: food, petroleum, machinery and semimanufactured goods
partners: Italy 27%, Germany 14%, UK 13%, US 9%
External debt: $603 million (1992)
Industrial production: growth rate 5.4% (1992); accounts for 27% of GDP
Electricity:
capacity: 250,000 kW
production: 1.1 billion kWh
consumption per capita: 2,749 kWh (1993)
Industries: tourism, electronics, ship repairyard, construction, food manufacturing, textiles, footwear, clothing, beverages, tobacco
Agriculture: accounts for 3% of GDP and 2% of the work force (1992); overall, 20% self-sufficient; main products—potatoes, cauliflower, grapes, wheat, barley, tomatoes, citrus, cut flowers, green peppers, hogs, poultry, eggs; generally adequate supplies of vegetables, poultry, milk, pork products; seasonal or periodic shortages in grain, animal fodder, fruits, other basic foodstuffs
Illicit drugs: transshipment point for hashish from North Africa to Western Europe
Economic aid:
recipient: US commitments, including Ex-Im (FY70-81), $172 million; Western (non-US) countries, ODA and OOF bilateral commitments (1970-89), $336 million; OPEC bilateral aid (1979-89), $76 million; Communist countries (1970-88), $48 million
Currency: 1 Maltese lira (LM) = 100 cents
Exchange rates: Maltese liri (LM) per US$1—0.3656 (January 1995), 0.3776 (1994), 0.3821 (1993), 0.3178 (1992), 0.3226 (1991), 0.3172 (1990)
Fiscal year: 1 April—31 March

Transportation

Railroads: 0 km
Highways:
total: 1,291 km
paved: asphalt 1,179 km
unpaved: gravel, crushed stone 77 km; earth 35 km
Ports: Marsaxlokk, Valletta
Merchant marine:
total: 964 ships (1,000 GRT or over) totaling 15,518,359 GRT/26,604,739 DWT
ships by type: barge carrier 3, bulk 272, cargo 300, chemical tanker 30, combination bulk 26, combination ore/oil 16, container 33, liquefied gas tanker 3, multifunction large-load carrier 3, oil tanker 191, passenger 7, passenger-cargo 3, railcar carrier 1, refrigerated cargo 14, roll-on/roll-off cargo 26, short-sea passenger 20, specialized tanker 5, vehicle carrier 11
note: a flag of convenience registry; includes 49 countries; the 10 major fleet flags are: Greece 351 ships, Russia 66, Croatia 63, Switzerland 31, Montenegro 29, Italy 27, Germany 23, Monaco 20, UK 20, and Georgia 10
Airports:
total: 1
with paved runways over 3,047 m: 1

Communications

Telephone system: 153,000 telephones; automatic system satisfies normal requirements
local: NA
intercity: submarine cable and microwave radio relay between islands
international: 1 submarine cable and 1 INTELSAT (Atlantic Ocean) earth station
Radio:
broadcast stations: AM 8, FM 4, shortwave 0
radios: NA
Television:
broadcast stations: 2
televisions: NA

Defense Forces

Branches: Armed Forces, Maltese Police Force
Manpower availability: males age 15-49 98,525; males fit for military service 78,305 (1995 est.)
Defense expenditures: exchange rate conversion—$21.4 million, about 0.9% of GDP (FY92/93)

Man, Isle of
(British crown dependency)

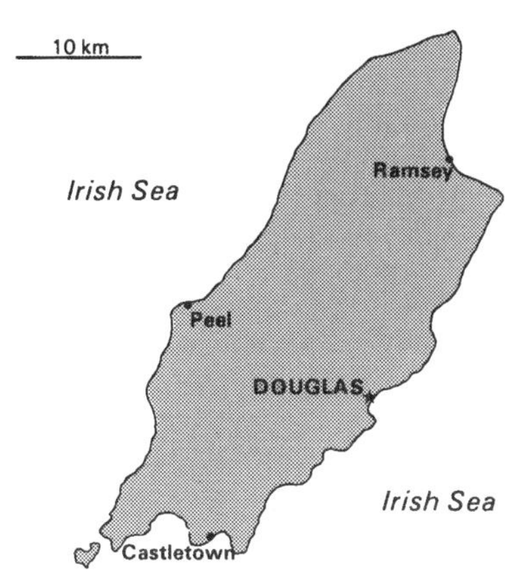

Geography

Location: Western Europe, island in the Irish Sea, between Great Britain and Ireland
Map references: Europe
Area:
total area: 588 sq km
land area: 588 sq km
comparative area: nearly 3.5 times the size of Washington, DC
Land boundaries: 0 km
Coastline: 113 km
Maritime claims:
exclusive fishing zone: 200 nm
territorial sea: 3 nm
International disputes: none
Climate: cool summers and mild winters; humid; overcast about half the time
Terrain: hills in north and south bisected by central valley
Natural resources: lead, iron ore
Land use:
arable land: NA%
permanent crops: NA%
meadows and pastures: NA%
forest and woodland: NA%
other: NA% (extensive arable land and forests)
Irrigated land: NA sq km
Environment:
current issues: NA
natural hazards: NA
international agreements: NA
Note: one small islet, the Calf of Man, lies to the southwest, and is a bird sanctuary

People

Population: 72,751 (July 1995 est.)
Age structure:
0-14 years: 18% (female 6,462; male 6,833)
15-64 years: 64% (female 23,219; male 23,348)
65 years and over: 18% (female 7,759; male 5,130) (July 1995 est.)
Population growth rate: 0.99% (1995 est.)
Birth rate: 13.73 births/1,000 population (1995 est.)

Death rate: 12.36 deaths/1,000 population (1995 est.)
Net migration rate: 8.55 migrant(s)/1,000 population (1995 est.)
Infant mortality rate: 8 deaths/1,000 live births (1995 est.)
Life expectancy at birth:
total population: 76.53 years
male: 73.78 years
female: 79.48 years (1995 est.)
Total fertility rate: 1.8 children born/woman (1995 est.)
Nationality:
noun: Manxman, Manxwoman
adjective: Manx
Ethnic divisions: Manx (Norse-Celtic descent), Briton
Religions: Anglican, Roman Catholic, Methodist, Baptist, Presbyterian, Society of Friends
Languages: English, Manx Gaelic
Literacy: NA%
Labor force: 25,864 (1981)
by occupation: NA

Government

Names:
conventional long form: none
conventional short form: Isle of Man
Digraph: IM
Type: British crown dependency
Capital: Douglas
Administrative divisions: none (British crown dependency)
Independence: none (British crown dependency)
National holiday: Tynwald Day, 5 July
Constitution: 1961, Isle of Man Constitution Act
Legal system: English law and local statute
Suffrage: 21 years of age; universal
Executive branch:
chief of state: Lord of Mann Queen ELIZABETH II (since 6 February 1952), represented by Lieutenant Governor Air Marshal Sir Laurence JONES (since NA 1990)
head of government: President of the Legislative Council Sir Charles KERRUISH (since NA 1990)
cabinet: Council of Ministers
Legislative branch: bicameral Tynwald
Legislative Council: consists of a 10-member body composed of the Lord Bishop of Sodor and Man, a nonvoting attorney general, and 8 others named by the House of Keys
House of Keys: elections last held NA 1991 (next to be held NA 1996); results—percent of vote NA; seats—(24 total) independents 24
Judicial branch: Court of Tynwald
Political parties and leaders: there is no party system and members sit as independents
Member of: none
Diplomatic representation in US: none (British crown dependency)
US diplomatic representation: none (British crown dependency)
Flag: red with the Three Legs of Man emblem (Trinacria), in the center; the three legs are joined at the thigh and bent at the knee; in order to have the toes pointing clockwise on both sides of the flag, a two-sided emblem is used

Economy

Overview: Offshore banking, manufacturing, and tourism are key sectors of the economy. The government's policy of offering incentives to high-technology companies and financial institutions to locate on the island has paid off in expanding employment opportunities in high-income industries. As a result, agriculture and fishing, once the mainstays of the economy, have declined in their shares of GDP. Banking now contributes about 45% to GDP. Trade is mostly with the UK. The Isle of Man enjoys free access to European Union markets.
National product: GDP—purchasing power parity—$780 million (1994 est.)
National product real growth rate: NA%
National product per capita: $10,800 (1994 est.)
Inflation rate (consumer prices): 7% (1992 est.)
Unemployment rate: 1% (1992 est.)
Budget:
revenues: $130.4 million
expenditures: $114.4 million, including capital expenditures of $18.1 million (1985 est.)
Exports: $NA
commodities: tweeds, herring, processed shellfish, beef, lamb
partners: UK
Imports: $NA
commodities: timber, fertilizers, fish
partners: UK
External debt: $NA
Industrial production: growth rate NA%
Electricity:
capacity: 61,000 kW
production: 190 million kWh
consumption per capita: 2,965 kWh (1992)
Industries: financial services, light manufacturing, tourism
Agriculture: cereals and vegetables; cattle, sheep, pigs, poultry
Economic aid: $NA
Currency: 1 Manx pound (£M) = 100 pence
Exchange rates: Manx pounds (£M) per US$1—0.6350 (January 1995), 0.6529 (1994), 0.6658 (1993), 0.5664 (1992), 0.5652 (1991), 0.5603 (1990); the Manx pound is at par with the British pound
Fiscal year: 1 April—31 March

Transportation

Railroads:
total: 60 km (36 km electrified)
Highways:
total: 640 km
paved: NA
unpaved: NA
Ports: Castletown, Douglas, Peel, Ramsey
Merchant marine:
total: 68 ships (1,000 GRT or over) totaling 1,810,355 GRT/3,183,773 DWT
ships by type: bulk 11, cargo 10, chemical tanker 4, container 9, liquefied gas tanker 8, oil tanker 15, roll-on/roll-off cargo 9, vehicle carrier 2
note: a flag of convenience registry; UK owns 9 ships, Switzerland 2, Denmark 1, Netherlands 1
Airports:
total: 1
with paved runways 1,524 to 2,437 m: 1

Communications

Telephone system: 24,435 telephones
local: NA
intercity: NA
international: NA
Radio:
broadcast stations: AM 1, FM 4, shortwave 0
radios: NA
Television:
broadcast stations: 4
televisions: NA

Defense Forces

Note: defense is the responsibility of the UK

Marshall Islands

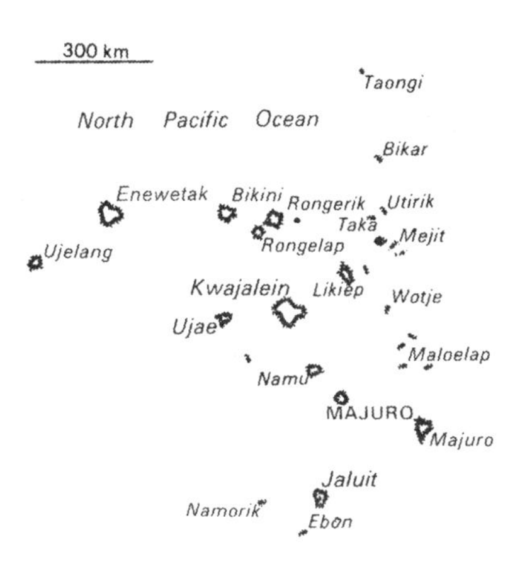

Geography

Location: Oceania, group of atolls and reefs in the North Pacific Ocean, about one-half of the way from Hawaii to Papua New Guinea
Map references: Oceania
Area:
total area: 181.3 sq km
land area: 181.3 sq km
comparative area: slightly larger than Washington, DC
note: includes the atolls of Bikini, Eniwetak, and Kwajalein
Land boundaries: 0 km
Coastline: 370.4 km
Maritime claims:
contiguous zone: 24 nm
exclusive economic zone: 200 nm
territorial sea: 12 nm
International disputes: claims US territory of Wake Island
Climate: wet season May to November; hot and humid; islands border typhoon belt
Terrain: low coral limestone and sand islands
Natural resources: phosphate deposits, marine products, deep seabed minerals
Land use:
arable land: 0%
permanent crops: 60%
meadows and pastures: 0%
forest and woodland: 0%
other: 40%
Irrigated land: NA sq km
Environment:
current issues: inadequate supplies of potable water
natural hazards: occasional typhoons
international agreements: party to Biodiversity, Climate Change, Law of the Sea, Ozone Layer Protection, Ship Pollution
Note: two archipelagic island chains of 30 atolls and 1,152 islands; Bikini and Eniwetak are former US nuclear test sites; Kwajalein, the famous World War II battleground, is now used as a US missile test range

People

Population: 56,157 (July 1995 est.)
Age structure:
0-14 years: 51% (female 13,950; male 14,547)
15-64 years: 47% (female 12,801; male 13,470)
65 years and over: 2% (female 740; male 649) (July 1995 est.)
Population growth rate: 3.86% (1995 est.)
Birth rate: 46.03 births/1,000 population (1995 est.)
Death rate: 7.48 deaths/1,000 population (1995 est.)
Net migration rate: 0 migrant(s)/1,000 population (1995 est.)
Infant mortality rate: 48 deaths/1,000 live births (1995 est.)
Life expectancy at birth:
total population: 63.49 years
male: 61.94 years
female: 65.11 years (1995 est.)
Total fertility rate: 6.89 children born/woman (1995 est.)
Nationality:
noun: Marshallese (singular and plural)
adjective: Marshallese
Ethnic divisions: Micronesian
Religions: Christian (mostly Protestant)
Languages: English (universally spoken and is the official language), two major Marshallese dialects from the Malayo-Polynesian family, Japanese
Literacy: age 15 and over can read and write (1980)
total population: 93%
male: 100%
female: 88%
Labor force: 4,800 (1986)
by occupation: NA

Government

Names:
conventional long form: Republic of the Marshall Islands
conventional short form: Marshall Islands
former: Marshall Islands District (Trust Territory of the Pacific Islands)
Digraph: RM
Type: constitutional government in free association with the US; the Compact of Free Association entered into force 21 October 1986
Capital: Majuro
Administrative divisions: none
Independence: 21 October 1986 (from the US-administered UN trusteeship)
National holiday: Proclamation of the Republic of the Marshall Islands, 1 May (1979)
Constitution: 1 May 1979
Legal system: based on adapted Trust Territory laws, acts of the legislature, municipal, common, and customary laws
Suffrage: 18 years of age; universal
Executive branch:
chief of state and head of government: President Amata KABUA (since 1979); election last held 6 January 1992 (next to be held NA); results—President Amata KABUA was reelected
cabinet: Cabinet; president selects from the parliament
Legislative branch: unicameral
Parliament (Nitijela): elections last held 18 November 1991 (next to be held November 1995); results—percent of vote NA; seats—(33 total) independents 33
Judicial branch: Supreme Court
Political parties and leaders: no formal parties; President KABUA is chief political (and traditional) leader
Member of: AsDB, ESCAP, IAEA, IBRD, ICAO, IDA, IFC, IMF, INTELSAT (nonsignatory user), INTERPOL, SPARTECA, SPC, SPF, UN, UNCTAD, WHO
Diplomatic representation in US:
chief of mission: Ambassador Wilfred I. KENDALL
chancery: 2433 Massachusetts Avenue NW, Washington, DC 20008
telephone: [1] (202) 234-5414
FAX: [1] (202) 232-3236
consulate(s) general: Honolulu and Los Angeles
US diplomatic representation:
chief of mission: Ambassador David C. FIELDS
embassy: address NA, Majuro
mailing address: P. O. Box 1379, Majuro, Republic of the Marshall Islands 96960-1379
telephone: [692] 247-4011
FAX: [692] 247-4012
Flag: blue with two stripes radiating from the lower hoist-side corner orange (top) and white; there is a white star with four large rays and 20 small rays on the hoist side above the two stripes

Economy

Overview: Agriculture and tourism are the mainstays of the economy. Agricultural production is concentrated on small farms, and the most important commercial crops are coconuts, tomatoes, melons, and breadfruit. A few cattle ranches supply the domestic meat market. Small-scale industry is limited to handicrafts, fish processing, and copra. The tourist industry is the primary source of foreign exchange and employs about 10% of the labor force. The islands have few natural resources, and imports far exceed exports. The US Government provides about 70% of the budget.
National product: GDP—purchasing power parity—$75 million (1992 est.)

Marshall Islands *(continued)*

National product real growth rate: 6% (1992)
National product per capita: $1,500 (1992 est.)
Inflation rate (consumer prices): 7% (1992 est.)
Unemployment rate: 16% (1991 est.)
Budget:
revenues: $106 million
expenditures: $128.7 million, including capital expenditures of $NA (1993)
Exports: $3.9 million (f.o.b., 1992 est.)
commodities: coconut oil, fish, live animals, trichus shells
partners: US, Japan, Australia
Imports: $62.9 million (c.i.f., 1992 est.)
commodities: foodstuffs, machinery and equipment, beverages and tobacco, fuels
partners: US, Japan, Australia
External debt: $NA
Industrial production: growth rate NA%
Electricity:
capacity: 42,000 kW
production: 80 million kWh
consumption per capita: 1,840 kWh (1990)
Industries: copra, fish, tourism; craft items from shell, wood, and pearls; offshore banking (embryonic)
Agriculture: coconuts, cacao, taro, breadfruit, fruits, pigs, chickens
Economic aid:
recipient: under the terms of the Compact of Free Association, the US is to provide approximately $40 million in aid annually
Currency: 1 United States dollar (US$) = 100 cents
Exchange rates: US currency is used
Fiscal year: 1 October—30 September

Transportation

Railroads: 0 km
Highways:
total: NA
note: paved roads on major islands (Majuro, Kwajalein), otherwise stone-, coral-, or laterite-surfaced roads and tracks
Ports: Majuro
Merchant marine:
total: 37 ships (1,000 GRT or over) totaling 2,205,275 GRT/4,263,247 DWT
ships by type: bulk carrier 23, cargo 1, combination ore/oil 1, oil tanker 12
Airports:
total: 16
with paved runways 1,524 to 2,437 m: 3
with paved runways 914 to 1,523 m: 1
with paved runways under 914 m: 5
with unpaved runways 1,524 to 2,438 m: 1
with unpaved runways 914 to 1,523 m: 6

Communications

Telephone system: 570 telephones (Majuro) and 186 telephones (Ebeye); telex services
local: NA
intercity: islands interconnected by shortwave radio (used mostly for government purposes)
international: 2 INTELSAT (Pacific Ocean) earth stations; US Government satellite communications system on Kwajalein
Radio:
broadcast stations: AM 1, FM 2, shortwave 1
radios: NA
Television:
broadcast stations: 1
televisions: NA

Defense Forces

Branches: no regular military forces; Police
Note: defense is the responsibility of the US

Martinique
(overseas department of France)

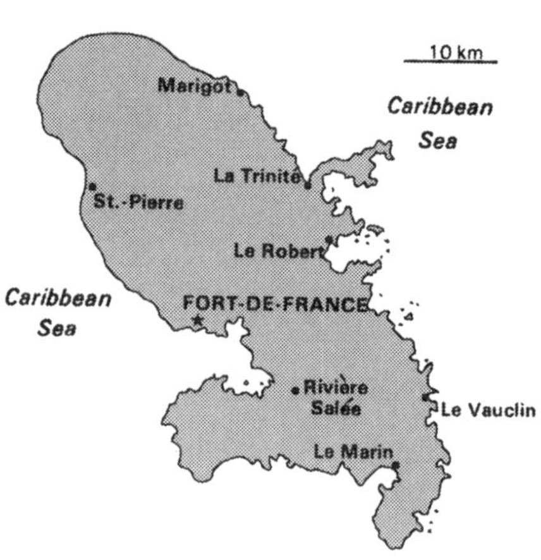

Geography

Location: Caribbean, island in the Caribbean Sea, north of Trinidad and Tobago
Map references: Central America and the Caribbean
Area:
total area: 1,100 sq km
land area: 1,060 sq km
comparative area: slightly more than six times the size of Washington, DC
Land boundaries: 0 km
Coastline: 290 km
Maritime claims:
exclusive economic zone: 200 nm
territorial sea: 12 nm
International disputes: none
Climate: tropical; moderated by trade winds; rainy season (June to October)
Terrain: mountainous with indented coastline; dormant volcano
Natural resources: coastal scenery and beaches, cultivable land
Land use:
arable land: 10%
permanent crops: 8%
meadows and pastures: 30%
forest and woodland: 26%
other: 26%
Irrigated land: 60 sq km (1989 est.)
Environment:
current issues: NA
natural hazards: hurricanes, flooding, and volcanic activity (an average of one major natural disaster every five years)
international agreements: NA

People

Population: 394,787 (July 1995 est.)
Age structure:
0-14 years: 23% (female 44,960; male 46,512)
15-64 years: 67% (female 134,439; male 130,642)
65 years and over: 10% (female 22,058; male 16,176) (July 1995 est.)
Population growth rate: 1.1% (1995 est.)

Birth rate: 16.92 births/1,000 population (1995 est.)
Death rate: 5.82 deaths/1,000 population (1995 est.)
Net migration rate: -0.1 migrant(s)/1,000 population (1995 est.)
Infant mortality rate: 7.3 deaths/1,000 live births (1995 est.)
Life expectancy at birth:
total population: 78.67 years
male: 75.94 years
female: 81.53 years (1995 est.)
Total fertility rate: 1.81 children born/woman (1995 est.)
Nationality:
noun: Martiniquais (singular and plural)
adjective: Martiniquais
Ethnic divisions: African and African-Caucasian-Indian mixture 90%, Caucasian 5%, East Indian, Lebanese, Chinese less than 5%
Religions: Roman Catholic 95%, Hindu and pagan African 5%
Languages: French, Creole patois
Literacy: age 15 and over can read and write (1982)
total population: 93%
male: 92%
female: 93%
Labor force: 100,000
by occupation: service industry 31.7%, construction and public works 29.4%, agriculture 13.1%, industry 7.3%, fisheries 2.2%, other 16.3%

Government

Names:
conventional long form: Department of Martinique
conventional short form: Martinique
local long form: Departement de la Martinique
local short form: Martinique
Digraph: MB
Type: overseas department of France
Capital: Fort-de-France
Administrative divisions: none (overseas department of France)
Independence: none (overseas department of France)
National holiday: National Day, Taking of the Bastille, 14 July (1789)
Constitution: 28 September 1958 (French Constitution)
Legal system: French legal system
Suffrage: 18 years of age; universal
Executive branch:
chief of state: President Francois MITTERRAND (since 21 May 1981)
head of government: Prefect Michel MORIN (since NA); President of the General Council Claude LISE (since 22 March 1992); President of the Regional Council Emile CAPGRAS (since 22 March 1992)
cabinet: Council of Ministers
Legislative branch: unicameral General Council and a unicameral Regional Assembly
General Council: elections last held 25 September and 8 October 1988 (next to be held NA); results—percent of vote by party NA; seats—(44 total) number of seats by party NA; note—a leftist coalition obtained a one-seat margin
Regional Assembly: elections last held on 22 March 1992 (next to be held by March 1998); results—percent of vote by party NA; seats—(41 total) RPR-UDF 16, MIM 9, PPM 9, PCM 5, independents 2
French Senate: elections last held 24 September 1989 (next to be held NA); results—percent of vote by party NA; seats—(2 total) UDF 1, PPM 1
French National Assembly: elections last held NA June 1993 (next to be held NA June 1998); results—percent of vote by party NA; seats—(4 total) RPR 3, FSM 1
Judicial branch: Supreme Court
Political parties and leaders: Rally for the Republic (RPR), Stephen BAGOE; Union for a Martinique of Progress (UMP); Martinique Progressive Party (PPM), Aime CESAIRE; Socialist Federation of Martinique (FSM), Michel YOYO; Martinique Communist Party (PCM); Martinique Patriots (PM); Union for French Democracy (UDF), Jean MARAN; Martinique Independence Movement (MIM), Alfred MARIE-JEANNE; Republican Party (PR), Jean BAILLY
Other political or pressure groups: Proletarian Action Group (GAP); Alhed Marie-Jeanne Socialist Revolution Group (GRS); Caribbean Revolutionary Alliance (ARC); Central Union for Martinique Workers (CSTM), Marc PULVAR; Frantz Fanon Circle; League of Workers and Peasants; Parti Martiniquais Socialiste (PMS); Association for the Protection of Martinique's Heritage (ecologist)
Member of: FZ, WCL, WFTU
Diplomatic representation in US: none (overseas department of France)
US diplomatic representation: the post closed in August 1993 (overseas department of France)
Flag: the flag of France is used

Economy

Overview: The economy is based on sugarcane, bananas, tourism, and light industry. Agriculture accounts for about 10% of GDP and the small industrial sector for 10%. Sugar production has declined, with most of the sugarcane now used for the production of rum. Banana exports are increasing, going mostly to France. The bulk of meat, vegetable, and grain requirements must be imported, contributing to a chronic trade deficit that requires large annual transfers of aid from France. Tourism has become more important than agricultural exports as a source of foreign exchange. The majority of the work force is employed in the service sector and in administration. Banana workers launched protests late in 1992 because of falling banana prices and fears of greater competition in the European market from other producers.
National product: GDP—purchasing power parity—$3.9 billion (1993 est.)
National product real growth rate: NA%
National product per capita: $10,000 (1993 est.)
Inflation rate (consumer prices): 3.9% (1990)
Unemployment rate: 32.1% (1990)
Budget:
revenues: $610 million
expenditures: $1.3 billion, including capital expenditures of $NA (1991)
Exports: $247 million (f.o.b., 1992)
commodities: refined petroleum products, bananas, rum, pineapples
partners: France 57%, Guadeloupe 31%, French Guiana (1991)
Imports: $1.75 billion (c.i.f., 1992)
commodities: petroleum products, crude oil, foodstuffs, construction materials, vehicles, clothing and other consumer goods
partners: France 62%, UK, Italy, Germany, Japan, US (1991)
External debt: $NA
Industrial production: growth rate NA%
Electricity:
capacity: 113,100 kW
production: 700 million kWh
consumption per capita: 1,677 kWh (1993)
Industries: construction, rum, cement, oil refining, sugar, tourism
Agriculture: including fishing and forestry, accounts for about 10% of GDP; principal crops—pineapples, avocados, bananas, flowers, vegetables, sugarcane for rum; dependent on imported food, particularly meat and vegetables
Illicit drugs: transshipment point for cocaine and marijuana bound for the US and Europe
Economic aid:
recipient: Western (non-US) countries, ODA and OOF bilateral commitments (1970-89), $10.1 billion
Currency: 1 French franc (F) = 100 centimes
Exchange rates: French francs (F) per US$1—5.2943 (January 1995), 5.5520 (1994), 5.6632 (1993), 5.2938 (1992), 5.6421 (1991), 5.4453 (1990)
Fiscal year: calendar year

Transportation

Railroads: 0 km
Highways:
total: 1,680 km
paved: 1,300 km
unpaved: gravel, earth 380 km

Ports: Fort-de-France, La Trinite
Merchant marine: none
Airports:
total: 2
with paved runways over 3,047 m: 1
with unpaved runways 914 to 1,523 m: 1

Communications

Telephone system: 68,900 telephones; domestic facilities are adequate
local: NA
intercity: NA
international: interisland microwave radio relay links to Guadeloupe, Dominica, and Saint Lucia; 2 INTELSAT (Atlantic Ocean) earth stations
Radio:
broadcast stations: AM 1, FM 6, shortwave 0
radios: NA
Television:
broadcast stations: 10
televisions: NA

Defense Forces

Branches: French forces (Army, Navy, Air Force), Gendarmerie
Note: defense is the responsibility of France

Mauritania

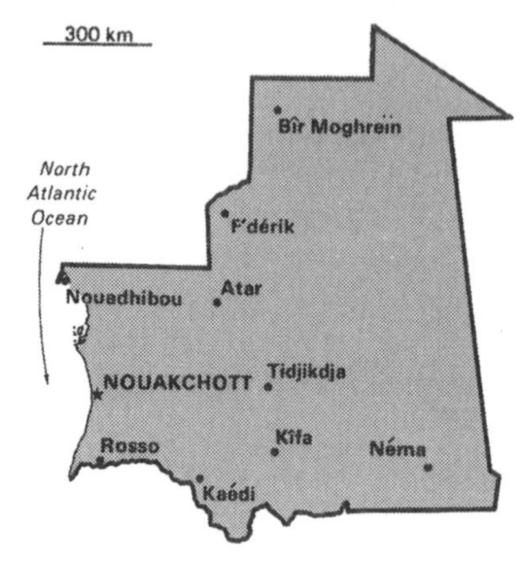

Geography

Location: Northern Africa, bordering the North Atlantic Ocean, between Senegal and Western Sahara
Map references: Africa
Area:
total area: 1,030,700 sq km
land area: 1,030,400 sq km
comparative area: slightly larger than three times the size of New Mexico
Land boundaries: total 5,074 km, Algeria 463 km, Mali 2,237 km, Senegal 813 km, Western Sahara 1,561 km
Coastline: 754 km
Maritime claims:
contiguous zone: 24 nm
continental shelf: 200 nm or to the edge of the continental margin
exclusive economic zone: 200 nm
territorial sea: 12 nm
International disputes: boundary with Senegal in dispute
Climate: desert; constantly hot, dry, dusty
Terrain: mostly barren, flat plains of the Sahara; some central hills
Natural resources: iron ore, gypsum, fish, copper, phosphate
Land use:
arable land: 1%
permanent crops: 0%
meadows and pastures: 38%
forest and woodland: 5%
other: 56%
Irrigated land: 120 sq km (1989 est.)
Environment:
current issues: overgrazing, deforestation, and soil erosion aggravated by drought are contributing to desertification; very limited natural fresh water resources away from the Senegal which is the only perennial river
natural hazards: hot, dry, dust/sand-laden sirocco wind blows primarily in March and April; periodic droughts
international agreements: party to—Climate Change, Nuclear Test Ban, Ozone Layer Protection, Wetlands; signed, but not ratified—Biodiversity, Desertification, Law of the Sea
Note: most of the population concentrated along the Senegal River in the southern part of the country

People

Population: 2,263,202 (July 1995 est.)
Age structure:
0-14 years: 48% (female 544,674; male 551,099)
15-64 years: 49% (female 574,282; male 542,762)
65 years and over: 3% (female 28,955; male 21,430) (July 1995 est.)
Population growth rate: 3.17% (1995 est.)
Birth rate: 47.32 births/1,000 population (1995 est.)
Death rate: 15.66 deaths/1,000 population (1995 est.)
Net migration rate: 0 migrant(s)/1,000 population (1995 est.)
Infant mortality rate: 83.5 deaths/1,000 live births (1995 est.)
Life expectancy at birth:
total population: 48.54 years
male: 45.66 years
female: 51.54 years (1995 est.)
Total fertility rate: 6.92 children born/woman (1995 est.)
Nationality:
noun: Mauritanian(s)
adjective: Mauritanian
Ethnic divisions: mixed Maur/black 40%, Maur 30%, black 30%
Religions: Muslim 100%
Languages: Hasaniya Arabic (official), Pular, Soninke, Wolof (official)
Literacy: age 15 and over can read and write (1988)
total population: 35%
male: 46%
female: 25%
Labor force: 465,000 (1981 est.); 45,000 wage earners (1980)
by occupation: agriculture 47%, services 29%, industry and commerce 14%, government 10%

Government

Names:
conventional long form: Islamic Republic of Mauritania
conventional short form: Mauritania
local long form: Al Jumhuriyah al Islamiyah al Muritaniyah
local short form: Muritaniyah
Digraph: MR
Type: republic
Capital: Nouakchott
Administrative divisions: 12 regions (regions, singular—region); Adrar, Assaba, Brakna, Dakhlet Nouadhibou, Gorgol, Guidimaka, Hodh ech Chargui, Hodh el Gharbi, Inchiri, Tagant, Tiris Zemmour, Trarza

note: there may be a new capital district of Nouakchott
Independence: 28 November 1960 (from France)
National holiday: Independence Day, 28 November (1960)
Constitution: 12 July 1991
Legal system: three-tier system: Islamic (Shari'a) courts, special courts, state security courts (in the process of being eliminated)
Suffrage: 18 years of age; universal
Executive branch:
chief of state and head of government: President Col. Maaouya Ould Sid'Ahmed TAYA (since 12 December 1984); election last held NA January 1992 (next to be held NA January 1998); results—President Col. Maaouya Ould Sid 'Ahmed TAYA elected
cabinet: Council of Ministers
Legislative branch: bicameral legislature
Senate (Majlis al-Shuyukh): elections last held 15 April 1994 (next to be held NA 1996); results—percent of vote by party NA; seats—(56 total, with 17 seats up for election) PRDS 16, UFD/NE 1
National Assembly (Majlis al-Watani): elections last held 6 and 13 March 1992 (next to be held March 1997) ; results—percent of votes by party NA; seats—(79 total) UFD/NE 67, PMR 1, RDU 1, independents 10
Judicial branch: Supreme Court (Cour Supreme)
Political parties and leaders: legalized by constitution passed 12 July 1991, however, politics continue to be tribally based; emerging parties include Democratic and Social Republican Party (PRDS), led by President Col. Maaouya Ould Sid'Ahmed TAYA; Union of Democratic Forces-New Era (UFD/NE), headed by Ahmed Ould DADDAH; Assembly for Democracy and Unity (RDU), Ahmed Ould SIDI BABA; Popular Social and Democratic Union (UPSD), Mohamed Mahmoud Ould MAH; Mauritanian Party for Renewal (PMR), Hameida BOUCHRAYA; National Avant-Garde Party (PAN), Khattry Ould JIDDOU; Mauritanian Party of the Democratic Center (PCDM), Bamba Ould SIDI BADI
Other political or pressure groups: Mauritanian Workers Union (UTM)
Member of: ABEDA, ACCT (associate), ACP, AfDB, AFESD, AL, AMF, AMU, CAEU, CCC, CEAO, ECA, ECOWAS, FAO, G-77, GATT, IBRD, ICAO, ICRM, IDA, IDB, IFAD, IFC, IFRCS, ILO, IMF, IMO, INTELSAT, INTERPOL, IOC, ITU, NAM, OAU, OIC, UN, UNCTAD, UNESCO, UNIDO, UPU, WHO, WIPO, WMO, WTO
Diplomatic representation in US:
chief of mission: Ambassador Ismail Ould IYAHI (since 22 September 1994)
chancery: 2129 Leroy Place NW, Washington, DC 20008
telephone: [1] (202) 232-5700
US diplomatic representation:
chief of mission: Ambassador Dorothy Myers SAMPAS
embassy: address NA, Nouakchott
mailing address: B. P. 222, Nouakchott
telephone: [222] (2) 526-60, 526-63
FAX: [222] (2) 515-92
Flag: green with a yellow five-pointed star above a yellow, horizontal crescent; the closed side of the crescent is down; the crescent, star, and color green are traditional symbols of Islam

Economy

Overview: A majority of the population still depends on agriculture and livestock for a livelihood, even though most of the nomads and many subsistence farmers were forced into the cities by recurrent droughts in the 1970s and 1980s. Mauritania has extensive deposits of iron ore, which account for almost 50% of total exports. The decline in world demand for this ore, however, has led to cutbacks in production. The nation's coastal waters are among the richest fishing areas in the world, but overexploitation by foreigners threatens this key source of revenue. The country's first deepwater port opened near Nouakchott in 1986. In recent years, drought and economic mismanagement have resulted in a substantial buildup of foreign debt. The government has begun the second stage of an economic reform program in consultation with the World Bank, the IMF, and major donor countries. Short-term growth prospects are gloomy because of the heavy debt service burden, rapid population growth, and vulnerability to climatic conditions.
National product: GDP—purchasing power parity—$2.4 billion (1993 est.)
National product real growth rate: 5% (1993 est.)
National product per capita: $1,110 (1994 est.)
Inflation rate (consumer prices): 10% (1993)
Unemployment rate: 20% (1991 est.)
Budget:
revenues: $280 million
expenditures: $346 million, including capital expenditures of $61 million (1989 est.)
Exports: $401 million (f.o.b., 1993 est.)
commodities: iron ore, fish and fish products
partners: Japan 27%, Italy, Belgium, Luxembourg
Imports: $378 million (c.i.f., 1993 est.)
commodities: foodstuffs, consumer goods, petroleum products, capital goods
partners: Algeria 15%, China 6%, US 3%, France, Germany, Spain, Italy
External debt: $1.9 billion (1992 est.)
Industrial production: growth rate NA%; accounts for almost 30% of GDP
Electricity:
capacity: 110,000 kW
production: 135 million kWh
consumption per capita: 61 kWh (1993)
Industries: fish processing, mining of iron ore and gypsum
Agriculture: accounts for 25% of GDP (including fishing); largely subsistence farming and nomadic cattle and sheep herding except in Senegal river valley; crops—dates, millet, sorghum, root crops; fish products number-one export; large food deficit in years of drought
Economic aid:
recipient: US commitments, including Ex-Im (FY70-89), $168 million; Western (non-US) countries, ODA and OOF bilateral commitments (1970-89), $1.3 billion; OPEC bilateral aid (1979-89), $490 million; Communist countries (1970-89), $277 million; Arab Development Bank (1991), $20 million
Currency: 1 ouguiya (UM) = 5 khoums
Exchange rates: ouguiyas (UM) per US$1—125.910 (January 1995), 123.575 (1994), 120.806 (1993),87.027 (1992), 81.946 (1991), 80.609 (1990)
Fiscal year: calendar year

Transportation

Railroads:
total: 690 km (single track); note—owned and operated by government mining company
standard gauge: 690 km 1.435 m gauge
Highways:
total: 7,525 km
paved: 1,685 km
unpaved: gravel, crushed stone, otherwise improved 1,040 km; unimproved earth 4,800 km (roads, trails, tracks)
Inland waterways: mostly ferry traffic on the Senegal River
Ports: Bogue, Kaedi, Nouadhibou, Nouakchott, Rosso
Merchant marine: none
Airports:
total: 28
with paved runways 2,438 to 3,047 m: 3
with paved runways 1,524 to 2,437 m: 4
with paved runways 914 to 1,523 m: 1
with paved runways under 914 m: 2
with unpaved runways 2,438 to 3,047 m: 2
with unpaved runways 1,524 to 2,438 m: 6
with unpaved runways 914 to 1,523 m: 10

Communications

Telephone system: NA telephones; poor system of cable and open-wire lines, minor microwave radio relay links, and radio communications stations (improvements being made)
local: NA
intercity: mostly cable and open wire lines
international: 1 INTELSAT (Atlantic Ocean) and 2 ARABSAT earth stations, with six planned

Mauritania *(continued)*

Radio:
broadcast stations: AM 2, FM 0, shortwave 0
radios: NA
Television:
broadcast stations: 1
televisions: NA

Defense Forces

Branches: Army, Navy, Air Force, National Gendarmerie, National Guard, National Police, Presidential Guard
Manpower availability: males age 15-49 483,916; males fit for military service 236,323 (1995 est.)
Defense expenditures: exchange rate conversion—$36 million, 2.7% of GDP (1994)

Mauritius

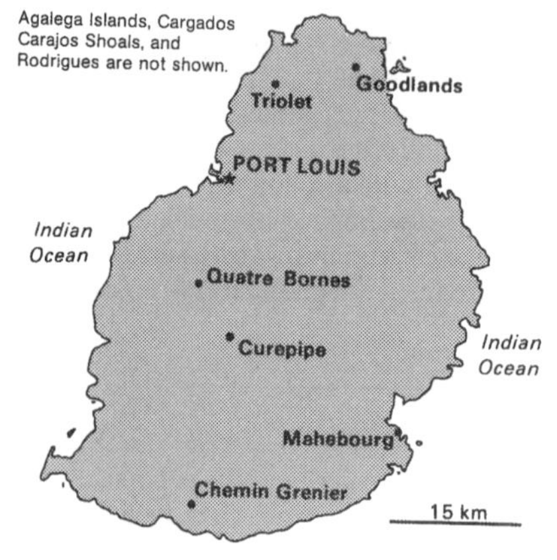

Geography

Location: Southern Africa, island in the Indian Ocean, east of Madagascar
Map references: World
Area:
total area: 1,860 sq km
land area: 1,850 sq km
comparative area: slightly less than 10.5 times the size of Washington, DC
note: includes Agalega Islands, Cargados Carajos Shoals (Saint Brandon), and Rodrigues
Land boundaries: 0 km
Coastline: 177 km
Maritime claims:
continental shelf: 200 nm or to the edge of the continental margin
exclusive economic zone: 200 nm
territorial sea: 12 nm
International disputes: claims UK-administered Chagos Archipelago, which includes the island of Diego Garcia in UK-administered British Indian Ocean Territory; claims French-administered Tromelin Island
Climate: tropical, modified by southeast trade winds; warm, dry winter (May to November); hot, wet, humid summer (November to May)
Terrain: small coastal plain rising to discontinuous mountains encircling central plateau
Natural resources: arable land, fish
Land use:
arable land: 54%
permanent crops: 4%
meadows and pastures: 4%
forest and woodland: 31%
other: 7%
Irrigated land: 170 sq km (1989 est.)
Environment:
current issues: water pollution
natural hazards: cyclones (November to April); almost completely surrounded by reefs that may pose maritime hazards
international agreements: party to—Biodiversity, Climate Change, Endangered Species, Environmental Modification, Hazardous Wastes, Law of the Sea, Marine Life Conservation, Nuclear Test Ban, Ozone Layer Protection

People

Population: 1,127,068 (July 1995 est.)
Age structure:
0-14 years: 28% (female 152,892; male 158,891)
15-64 years: 66% (female 376,049; male 372,910)
65 years and over: 6% (female 39,088; male 27,238) (July 1995 est.)
Population growth rate: 0.89% (1995 est.)
Birth rate: 18.91 births/1,000 population (1995 est.)
Death rate: 6.38 deaths/1,000 population (1995 est.)
Net migration rate: -3.64 migrant(s)/1,000 population (1995 est.)
Infant mortality rate: 17.8 deaths/1,000 live births (1995 est.)
Life expectancy at birth:
total population: 70.84 years
male: 66.9 years
female: 74.95 years (1995 est.)
Total fertility rate: 2.2 children born/woman (1995 est.)
Nationality:
noun: Mauritian(s)
adjective: Mauritian
Ethnic divisions: Indo-Mauritian 68%, Creole 27%, Sino-Mauritian 3%, Franco-Mauritian 2%
Religions: Hindu 52%, Christian 28.3% (Roman Catholic 26%, Protestant 2.3%), Muslim 16.6%, other 3.1%
Languages: English (official), Creole, French, Hindi, Urdu, Hakka, Bojpoori
Literacy: age 15 and over can read and write (1990)
total population: 80%
male: 85%
female: 75%
Labor force: 335,000
by occupation: government services 29%, agriculture and fishing 27%, manufacturing 22%, other 22%

Government

Names:
conventional long form: Republic of Mauritius
conventional short form: Mauritius
Digraph: MP
Type: parliamentary democracy
Capital: Port Louis
Administrative divisions: 9 districts and 3 dependencies*; Agalega Islands*, Black River, Cargados Carajos*, Flacq, Grand Port, Moka, Pamplemousses, Plaines Wilhems, Port Louis, Riviere du Rempart, Rodrigues*, Savanne
Independence: 12 March 1968 (from UK)
National holiday: Independence Day, 12 March (1968)

Constitution: 12 March 1968; amended 12 March 1992
Legal system: based on French civil law system with elements of English common law in certain areas
Suffrage: 18 years of age; universal
Executive branch:
chief of state: President Cassam UTEEM (since 1 July 1992); Vice President Rabindranath GHURBURRON (since 1 July 1992)
head of government: Prime Minister Sir Anerood JUGNAUTH (since 12 June 1982); Deputy Prime Minister Prem NABABSING (since 26 September 1990)
cabinet: Council of Ministers; appointed by the president on recommendation of the prime minister
Legislative branch: unicameral
Legislative Assembly: elections last held on 15 September 1991 (next to be held by 15 September 1996); results—MSM/MMM 53%, MLP/PMSD 38%; seats—(66 total) MSM/MMM alliance 59 (MSM 29, MMM 26, OPR 2, MTD 2); MLP/PMSD 4 (MLP 3, PMSD 1); note—Supreme Court denied the assignment of 3 seats to the MSM
Judicial branch: Supreme Court
Political parties and leaders:
government coalition: Militant Socialist Movement (MSM), A. JUGNAUTH; Mauritian Militant Resurgence (RMM), Prem NABABSING (less 10 legislators under the leadership of Paul BERENGER, now voting with the opposition); Mauritian Social Democratic Party (PMSD), X. DUVAL; Organization of the People of Rodrigues (OPR), Louis Serge CLAIR; Democratic Labor Movement (MTD), Anil BAICHOO
opposition: Mauritian Labor Party (MLP), Navin RAMGOOLMAN; MMM-Berenger Faction, Paul BERENGER; Socialist Workers Front, Sylvio MICHEL
Other political or pressure groups: various labor unions
Member of: ACCT, ACP, AfDB, C, CCC, ECA, FAO, G-77, GATT, IAEA, IBRD, ICAO, ICFTU, ICRM, IDA, IFAD, IFC, IFRCS, ILO, IMF, IMO, INMARSAT, INTELSAT, INTERPOL, IOC, ISO (correspondent), ITU, NAM, OAU, PCA, UN, UNCTAD, UNESCO, UNIDO, UPU, WCL, WFTU, WHO, WIPO, WMO, WTO
Diplomatic representation in US:
chief of mission: Ambassador Anund Priyay NEEWOOR
chancery: Suite 441, 4301 Connecticut Avenue NW, Washington, DC 20008
telephone: [1] (202) 244-1491, 1492
FAX: [1] (202) 966-0983
US diplomatic representation:
chief of mission: Ambassador Leslie M. ALEXANDER
embassy: 4th Floor, Rogers House, John Kennedy Street, Port Louis
mailing address: use embassy street address
telephone: [230] 208-9763 through 9767
FAX: [230] 208-9534
Flag: four equal horizontal bands of red (top), blue, yellow, and green

Economy

Overview: Since independence in 1968, Mauritius has developed from a low income, agriculturally based economy to middle income diversified economy with growing industrial and tourist sectors. For most of the period annual growth has been of the order of 5% to 6%. This remarkable achievement has been reflected in increased life expectancy, lowered infant mortality, and a much improved infrastructure. Sugarcane is grown on about 90% of the cultivated land area and accounts for 40% of export earnings. The government's development strategy centers on industrialization (with a view to modernization and to exports), agricultural diversification, and tourism. Economic performance in 1991-93 continued strong with solid real growth and low unemployment.
National product: GDP—purchasing power parity—$9.3 billion (1993 est.)
National product real growth rate: 4.7% (1993 est.)
National product per capita: $8,600 (1994 est.)
Inflation rate (consumer prices): 9.4% (1993 est.)
Unemployment rate: 2.4% (1991 est.)
Budget:
revenues: $653 million
expenditures: $567 million, including capital expenditures of $143 million (FY92/93 est.)
Exports: $1.32 billion (f.o.b., 1993 est.)
commodities: textiles 44%, sugar 40%, light manufactures 10%
partners: EC and US have preferential treatment, EC 77%, US 15%
Imports: $1.7 billion (f.o.b., 1993 est.)
commodities: manufactured goods 50%, capital equipment 17%, foodstuffs 13%, petroleum products 8%, chemicals 7%
partners: EC, US, South Africa, Japan
External debt: $996.8 million (1993 est.)
Industrial production: growth rate 5.8% (1992); accounts for 25% of GDP
Electricity:
capacity: 340,000 kW
production: 920 million kWh
consumption per capita: 777 kWh (1993)
Industries: food processing (largely sugar milling), textiles, wearing apparel, chemicals, metal products, transport equipment, nonelectrical machinery, tourism
Agriculture: accounts for 10% of GDP; about 90% of cultivated land in sugarcane; other products—tea, corn, potatoes, bananas, pulses, cattle, goats, fish; net food importer, especially rice and fish
Illicit drugs: illicit producer of cannabis for the international drug trade; heroin consumption and transshipment are growing problems
Economic aid:
recipient: US commitments, including Ex-Im (FY70-89), $76 million; Western (non-US) countries (1970-89), $709 million; Communist countries (1970-89), $54 million
Currency: 1 Mauritian rupee (MauR) = 100 cents
Exchange rates: Mauritian rupees (MauRs) per US$1—17.755 (January 1995), 17.960 (1994), 17.648 (1993), 15.563 (1992), 15.652 (1991), 14.839 (1990)
Fiscal year: 1 July—30 June

Transportation

Railroads: 0 km
Highways:
total: 1,800 km
paved: 1,640 km
unpaved: earth 160 km
Ports: Port Louis
Merchant marine:
total: 16 ships (1,000 GRT or over) totaling 191,703 GRT/297,347 DWT
ships by type: bulk 5, cargo 8, liquefied gas tanker 1, oil tanker 1, passenger-cargo 1
Airports:
total: 5
with paved runways 2,438 to 3,047 m: 1
with paved runways 914 to 1,523 m: 1
with paved runways under 914 m: 2
with unpaved runways 914 to 1,523 m: 1

Communications

Telephone system: over 48,000 telephones; small system with good service
local: NA
intercity: utilizes primarily microwave radio relay
international: 1 INTELSAT (Indian Ocean) earth station; new microwave link to Reunion; high-frequency radio links to several countries
Radio:
broadcast stations: AM 2, FM 0, shortwave 0
radios: NA
Television:
broadcast stations: 4
televisions: NA

Defense Forces

Branches: National Police Force (includes the paramilitary Special Mobile Force or SMF, Special Support Units or SSU, and National Coast Guard)
Manpower availability: males age 15-49 321,947; males fit for military service 163,904 (1995 est.)
Defense expenditures: exchange rate conversion—$11.2 million, 0.4% of GDP (FY92/93)

Mayotte

(territorial collectivity of France)

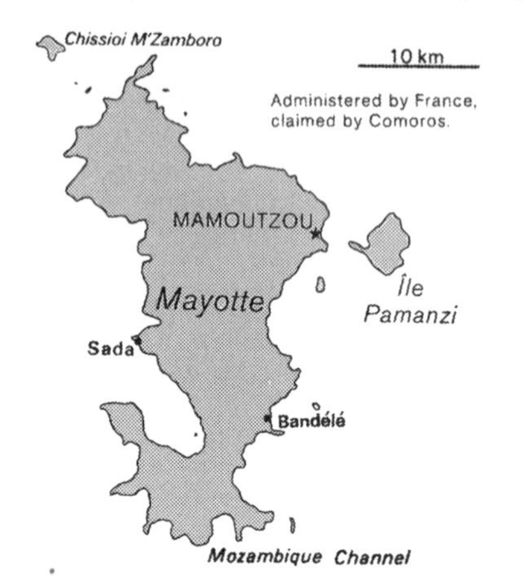

Geography

Location: Southern Africa, island in the Mozambique Channel, about one-half of the way from northern Mozambique to northern Madagascar
Map references: Africa
Area:
total area: 375 sq km
land area: 375 sq km
comparative area: slightly more than twice the size of Washington, DC
Land boundaries: 0 km
Coastline: 185.2 km
Maritime claims:
exclusive economic zone: 200 nm
territorial sea: 12 nm
International disputes: claimed by Comoros
Climate: tropical; marine; hot, humid, rainy season during northeastern monsoon (November to May); dry season is cooler (May to November)
Terrain: generally undulating with ancient volcanic peaks, deep ravines
Natural resources: negligible
Land use:
arable land: NA%
permanent crops: NA%
meadows and pastures: NA%
forest and woodland: NA%
other: NA%
Irrigated land: NA sq km
Environment:
current issues: NA
natural hazards: cyclones during rainy season
international agreements: NA
Note: part of Comoro Archipelago

People

Population: 97,088 (July 1995 est.)
Age structure:
0-14 years: 49% (female 23,910; male 24,120)
15-64 years: 48% (female 22,824; male 23,935)
65 years and over: 3% (female 1,165; male 1,134) (July 1995 est.)
Population growth rate: 3.8% (1995 est.)
Birth rate: 48.44 births/1,000 population (1995 est.)
Death rate: 10.46 deaths/1,000 population (1995 est.)
Net migration rate: 0 migrant(s)/1,000 population (1995 est.)
Infant mortality rate: 77.3 deaths/1,000 live births (1995 est.)
Life expectancy at birth:
total population: 58.27 years
male: 56.04 years
female: 60.57 years (1995 est.)
Total fertility rate: 6.71 children born/woman (1995 est.)
Nationality:
noun: Mahorais (singular and plural)
adjective: Mahoran
Ethnic divisions: NA
Religions: Muslim 99%, Christian (mostly Roman Catholic)
Languages: Mahorian (a Swahili dialect), French
Literacy: NA%
Labor force: NA

Government

Names:
conventional long form: Territorial Collectivity of Mayotte
conventional short form: Mayotte
Digraph: MF
Type: territorial collectivity of France
Capital: Mamoutzou
Administrative divisions: none (territorial collectivity of France)
Independence: none (territorial collectivity of France)
National holiday: National Day, Taking of the Bastille, 14 July (1789)
Constitution: 28 September 1958 (French Constitution)
Legal system: French law
Suffrage: 18 years of age; universal
Executive branch:
chief of state: President Francois MITTERRAND (since 21 May 1981)
head of government: Prefect Jean-Jacques DERACQ (since NA); President of the General Council Younoussa BAMANA (since NA 1976)
Legislative branch: unicameral
General Council (Conseil General): elections last held NA March 1994 (next to be held NA); results—percent of vote by party NA; seats—(19 total) MPM 12, RPR 4, independents 3
French Senate: elections last held on 24 September 1989 (next to be held NA September 1995); results—percent of vote by party NA; seats—(1 total) MPM 1
French National Assembly: elections last held 21 and 28 March 1993 (next to be held NA 1998); results—UDF-CDS 54.3%, RPR 44.3%; seats—(1 total) UDF-CDS 1
Judicial branch: Supreme Court (Tribunal Superieur d'Appel)
Political parties and leaders: Mahoran Popular Movement (MPM), Younoussa BAMANA; Party for the Mahoran Democratic Rally (PRDM), Daroueche MAOULIDA; Mahoran Rally for the Republic (RPR), Mansour KAMARDINE; Union for French Democracy (UDF), Maoulida AHMED; Center of Social Democrats (CDS)
Member of: FZ
Diplomatic representation in US: none (territorial collectivity of France)
US diplomatic representation: none (territorial collectivity of France)
Flag: the flag of France is used

Economy

Overview: Economic activity is based primarily on the agricultural sector, including fishing and livestock raising. Mayotte is not self-sufficient and must import a large portion of its food requirements, mainly from France. The economy and future development of the island are heavily dependent on French financial assistance. Mayotte's remote location is an obstacle to the development of tourism.
National product: GDP—purchasing power parity—$54 million (1993 est.)
National product real growth rate: NA%
National product per capita: $600 (1993 est.)
Inflation rate (consumer prices): NA%
Unemployment rate: NA%
Budget:
revenues: $NA
expenditures: $37.3 million, including capital expenditures of $NA (1985 est.)
Exports: $4 million (f.o.b., 1984)
commodities: ylang-ylang, vanilla
partners: France 79%, Comoros 10%, Reunion 9%
Imports: $21.8 million (f.o.b., 1984)
commodities: building materials, transportation equipment, rice, clothing, flour
partners: France 57%, Kenya 16%, South Africa 11%, Pakistan 8%
External debt: $NA
Industrial production: growth rate NA%
Electricity:
capacity: NA kW
production: NA kWh
consumption per capita: NA kWh
Industries: newly created lobster and shrimp industry
Agriculture: most important sector; provides all export earnings; crops—vanilla, ylang-ylang, coffee, copra; imports major share of food needs
Economic aid:
recipient: Western (non-US) countries, ODA and OOF bilateral commitments (1970-89), $402 million
Currency: 1 French franc (F) = 100 centimes

Exchange rates: French francs (F) per US$1—5.2943 (January 1995), 5.5520 (1994), 5.6632 (1993), 5.2938 (1992), 5.6421 (1991), 5.4453 (1990)
Fiscal year: calendar year

Transportation

Railroads: 0 km
Highways:
total: 42 km
paved: bituminous 18 km
unpaved: 24 km
Ports: Dzaoudzi
Merchant marine: none
Airports:
total: 1
with paved runways 914 to 1,523 m: 1

Communications

Telephone system: 450 telephones; small system administered by French Department of Posts and Telecommunications
local: NA
intercity: NA
international: radio relay and high-frequency radio communications for links to Comoros and international communications
Radio:
broadcast stations: AM 1, FM 0, shortwave 0
radios: NA
Television:
broadcast stations: 0
televisions: NA

Defense Forces

Note: defense is the responsibility of France

Mexico

Geography

Location: Middle America, bordering the Caribbean Sea and the Gulf of Mexico, between Belize and the US and bordering the North Pacific Ocean, between Guatamala and the US
Map references: North America
Area:
total area: 1,972,550 sq km
land area: 1,923,040 sq km
comparative area: slightly less than three times the size of Texas
Land boundaries: total 4,538 km, Belize 250 km, Guatemala 962 km, US 3,326 km
Coastline: 9,330 km
Maritime claims:
contiguous zone: 24 nm
continental shelf: 200 nm or to the edge of the continental margin
exclusive economic zone: 200 nm
territorial sea: 12 nm
International disputes: claims Clipperton Island (French possession)
Climate: varies from tropical to desert
Terrain: high, rugged mountains, low coastal plains, high plateaus, and desert
Natural resources: petroleum, silver, copper, gold, lead, zinc, natural gas, timber
Land use:
arable land: 12%
permanent crops: 1%
meadows and pastures: 39%
forest and woodland: 24%
other: 24%
Irrigated land: 51,500 sq km (1989 est.)
Environment:
current issues: natural fresh water resources scarce and polluted in north, inaccessible and poor quality in center and extreme southeast; raw sewage and industrial effluents polluting rivers in urban areas; deforestation; widespread erosion; desertification; serious air pollution in the national capital and urban centers along US-Mexico border
natural hazards: tsunamis along the Pacific coast, destructive earthquakes in the center and south, and hurricanes on the Gulf and Caribbean coasts
international agreements: party to—Biodiversity, Climate Change, Endangered Species, Hazardous Wastes, Law of the Sea, Marine Dumping, Marine Life Conservation, Nuclear Test Ban, Ozone Layer Protection, Ship Pollution, Wetlands, Whaling; signed, but not ratified—Desertification
Note: strategic location on southern border of US

People

Population: 93,985,848 (July 1995 est.)
Age structure:
0-14 years: 37% (female 17,028,091; male 17,631,110)
15-64 years: 59% (female 28,429,663; male 26,866,886)
65 years and over: 4% (female 2,184,998; male 1,845,100) (July 1995 est.)
Population growth rate: 1.9% (1995 est.)
Birth rate: 26.64 births/1,000 population (1995 est.)
Death rate: 4.64 deaths/1,000 population (1995 est.)
Net migration rate: -3.03 migrant(s)/1,000 population (1995 est.)
Infant mortality rate: 26 deaths/1,000 live births (1995 est.)
Life expectancy at birth:
total population: 73.34 years
male: 69.74 years
female: 77.11 years (1995 est.)
Total fertility rate: 3.09 children born/woman (1995 est.)
Nationality:
noun: Mexican(s)
adjective: Mexican
Ethnic divisions: mestizo (Indian-Spanish) 60%, Amerindian or predominantly Amerindian 30%, Caucasian or predominantly Caucasian 9%, other 1%
Religions: nominally Roman Catholic 89%, Protestant 6%
Languages: Spanish, various Mayan dialects
Literacy: age 15 and over can read and write (1990)
total population: 88%
male: 90%
female: 85%
Labor force: 26.2 million (1990)
by occupation: services 31.7%, agriculture, forestry, hunting, and fishing 28%, commerce 14.6%, manufacturing 11.1%, construction 8.4%, transportation 4.7%, mining and quarrying 1.5%

Government

Names:
conventional long form: United Mexican States
conventional short form: Mexico

local long form: Estados Unidos Mexicanos
local short form: Mexico
Digraph: MX
Type: federal republic operating under a centralized government
Capital: Mexico
Administrative divisions: 31 states (estados, singular—estado) and 1 federal district* (distrito federal); Aguascalientes, Baja California, Baja California Sur, Campeche, Chiapas, Chihuahua, Coahuila de Zaragoza, Colima, Distrito Federal*, Durango, Guanajuato, Guerrero, Hidalgo, Jalisco, Mexico, Michoacan de Ocampo, Morelos, Nayarit, Nuevo Leon, Oaxaca, Puebla, Queretaro de Arteaga, Quintana Roo, San Luis Potosi, Sinaloa, Sonora, Tabasco, Tamaulipas, Tlaxcala, Veracruz-Llave, Yucatan, Zacatecas
Independence: 16 September 1810 (from Spain)
National holiday: Independence Day, 16 September (1810)
Constitution: 5 February 1917
Legal system: mixture of US constitutional theory and civil law system; judicial review of legislative acts; accepts compulsory ICJ jurisdiction, with reservations
Suffrage: 18 years of age; universal and compulsory (but not enforced)
Executive branch:
chief of state and head of government: President Ernesto ZEDILLO Ponce de Leon (since 1 December 1994); election last held on 21 August 1994 (next to be held NA); results—Ernesto ZEDILLO Ponce de Leon (PRI) 50.18%, Cuauhtemoc CARDENAS Solorzano (PRD) 17.08%, Diego FERNANDEZ de Cevallos (PAN) 26.69%; other 6.049%
cabinet: Cabinet; appointed by the president
Legislative branch: bicameral National Congress (Congreso de la Union)
Senate (Camara de Senadores): elections last held on 21 August 1994 (next to be held NA); results—percent of vote by party NA; seats in full Senate—(128 total; Senate expanded from 64 seats at the last election) PRI 93, PRD 25, PAN 10
Chamber of Deputies (Camara de Diputados): elections last held on 24 August 1994 (next to be held NA); results—percent of vote by party NA; seats—(500 total) PRI 300, PAN 119, PRD 71, PFCRN 10
Judicial branch: Supreme Court of Justice (Corte Suprema de Justicia)
Political parties and leaders: (recognized parties) Institutional Revolutionary Party (PRI), Maria de los Angeles MORENO; National Action Party (PAN), Carlos CASTILLO; Popular Socialist Party (PPS), Indalecio SAYAGO Herrera; Democratic Revolutionary Party (PRD), Porfirio MUNOZ Ledo; Cardenist Front for the National Reconstruction Party (PFCRN), Rafael AGUILAR Talamantes; Authentic Party of the Mexican Revolution (PARM), Rosa Maria MARTINEZ Denagri; Democratic Forum Party (PFD), Pablo Emilio MADERO; Mexican Green Ecologist Party (PVEM), Jorge GONZALEZ Torres
Other political or pressure groups: Roman Catholic Church; Confederation of Mexican Workers (CTM); Confederation of Industrial Chambers (CONCAMIN); Confederation of National Chambers of Commerce (CONCANACO); National Peasant Confederation (CNC); Revolutionary Workers Party (PRT); Revolutionary Confederation of Workers and Peasants (CROC); Regional Confederation of Mexican Workers (CROM); Confederation of Employers of the Mexican Republic (COPARMEX); National Chamber of Transformation Industries (CANACINTRA); Coordinator for Foreign Trade Business Organizations (COECE); Federation of Unions Providing Goods and Services (FESEBES)
Member of: AG (observer), APEC, BCIE, CARICOM (observer), CCC, CDB, CG, EBRD, ECLAC, FAO, G- 6, G-11, G-15, G-19, G-24, GATT, IADB, IAEA, IBRD, ICAO, ICC, ICFTU, ICRM, IDA, IFAD, IFC, IFRCS, ILO, IMF, IMO, INMARSAT, INTELSAT, INTERPOL, IOC, IOM (observer), ISO, ITU, LAES, LAIA, NAM (observer), OAS, OECD, ONUSAL, OPANAL, PCA, RG, UN, UNCTAD, UNESCO, UNIDO, UNITAR, UNU, UPU, WCL, WFTU, WHO, WIPO, WMO, WTO
Diplomatic representation in US:
chief of mission: Ambassador Jesus SILVA HERZOG Flores
chancery: 1911 Pennsylvania Avenue NW, Washington, DC 20006
telephone: [1] (202) 728-1600
consulate(s) general: Atlanta, Chicago, Dallas, Denver, El Paso, Houston, Los Angeles, Miami, New Orleans, New York, San Antonio, San Diego, San Francisco, San Juan (Puerto Rico)
consulate(s): Albuquerque, Austin, Boston, Brownsville (Texas), Calexico (California), Corpus Christi, Del Rio (Texas), Detroit, Eagle Pass (Texas), Fresno (California), Loredo, Mc Allen (Texas), Midland (Texas), Nogales (Arizona), Oxnard (California), Philadelphia, Phoenix, Sacramento, St. Louis, Salt Lake City, San Bernardino, San Jose, Santa Ana, Seattle
US diplomatic representation:
chief of mission: Ambassador James R. JONES
embassy: Paseo de la Reforma 305, Colonia Cuauhtemoc, 06500 Mexico, Distrito Federal
mailing address: P. O. Box 3087, Laredo, TX 78044-3087
telephone: [52] (5) 211-0042
FAX: [52] (5) 511-9980, 208-3373
consulate(s) general: Ciudad Juarez, Guadalajara, Monterrey, Tijuana
consulate(s): Hermosillo, Matamoros, Merida, Nuevo Laredo
Flag: three equal vertical bands of green (hoist side), white, and red; the coat of arms (an eagle perched on a cactus with a snake in its beak) is centered in the white band

Economy

Overview: Mexico, under the guidance of new President Ernesto ZEDILLO, entered 1995 in the midst of a severe financial crisis. Mexico's membership in the North American Free Trade Agreement (NAFTA) with the United States and Canada, its solid record of economic reforms, and its strong growth in the second and third quarters of 1994—at an annual rate of 3.8% and 4.5% respectively—seemed to augur bright prospects for 1995. However, an overvalued exchange rate and widening current account deficits created an imbalance that ultimately proved unsustainable. To finance the trade gap, Mexico City had become increasingly reliant on volatile portfolio investment. A series of political shocks in 1994—an uprising in the southern state of Chiapas, the assassination of a presidential candidate, several high profile kidnappings, the killing of a second high-level political figure, and renewed threats from the Chiapas rebels—combined with rising international interest rates and concerns of a devaluation to undermine investor confidence and prompt massive outflows of capital. The dwindling of foreign exchange reserves, which the central bank had been using to defend the currency, forced the new administration to change the exchange rate policy and allow the currency to float freely in the last days of 1994. The adjustment roiled Mexican financial markets, leading to a 30% to 40% weakening of the peso relative to the dollar. ZEDILLO announced an emergency economic program that included federal budget cuts and plans for more privatizations, but it failed to restore investor confidence quickly. While the devaluation is likely to help Mexican exporters, whose products are now cheaper, it also raises the specter of an inflationary spiral if domestic producers increase their prices and workers demand wage hikes. Although strong economic fundamentals bode well for Mexico's longer-term outlook, prospects for solid growth and low inflation have deteriorated considerably, at least through 1995.
National product: GDP—purchasing power parity—$728.7 billion (1994 est.)
National product real growth rate: 3.5% (1994 est.)
National product per capita: $7,900 (1994 est.)
Inflation rate (consumer prices): 7.1% (1994 est.)
Unemployment rate: 9.8% (1994 est.)
Budget:
revenues: $96.99 billion (1994 est.)
expenditures: $96.51 billion (1994 est.), including capital expenditures of $NA (1994 est.)

Exports: $60.8 billion (f.o.b., 1994 est.), includes in-bond industries
commodities: crude oil, oil products, coffee, silver, engines, motor vehicles, cotton, consumer electronics
partners: US 82%, Japan 1.4%, EC 5% (1993 est.)
Imports: $79.4 billion (f.o.b., 1994 est.), includes in-bond industries
commodities: metal-working machines, steel mill products, agricultural machinery, electrical equipment, car parts for assembly, repair parts for motor vehicles, aircraft, and aircraft parts
partners: US 74%, Japan 4.7%, EC 11% (1993 est.)
External debt: $128 billion (1994 est.)
Industrial production: growth rate 4.5% (1994 est.)
Electricity:
capacity: 28,780,000 kW
production: 122 billion kWh
consumption per capita: 1,239 kWh (1993)
Industries: food and beverages, tobacco, chemicals, iron and steel, petroleum, mining, textiles, clothing, motor vehicles, consumer durables, tourism
Agriculture: accounts for 7% of GDP; large number of small farms at subsistence level; major food crops—corn, wheat, rice, beans; cash crops—cotton, coffee, fruit, tomatoes
Illicit drugs: illicit cultivation of opium poppy and cannabis continues in spite of government eradication program; major supplier of heroin and marijuana to the US market; continues as the primary transshipment country for US-bound cocaine and marijuana from South America; increasingly involved in the production and distribution of methamphetamine
Economic aid:
recipient: US commitments, including Ex-Im (FY70-89), $3.1 billion; Western (non-US) countries, ODA and OOF bilateral commitments (1970-89), $7.7 billion; Communist countries (1970-89), $110 million
Currency: 1 New Mexican peso (Mex$) = 100 centavos
Exchange rates: market rate of Mexican pesos (Mex$) per US$1—6.736 (average in March 1995), 5.5133 (January 1995), 3.3751 (1994), 3.1156 (1993), 3,094.9 (1992), 3,018.4 (1991), 2,812.6 (1990)
note: the new peso replaced the old peso on 1 January 1993; 1 new peso = 1,000 old pesos
Fiscal year: calendar year

Transportation

Railroads:
total: 24,500 km
standard gauge: 24,410 km 1.435-m gauge
narrow gauge: 93 km 0.914-m gauge
Highways:
total: 242,300 km
paved: 84,800 km (including 3,166 km of expressways)
unpaved: gravel and earth 157,500 km
Inland waterways: 2,900 km navigable rivers and coastal canals
Pipelines: crude oil 28,200 km; petroleum products 10,150 km; natural gas 13,254 km; petrochemical 1,400 km
Ports: Acapulco, Altamira, Coatzacoalcos, Ensenada, Guaymas, La Paz, Lazaro Cardenas, Manzanillo, Mazatlan, Progreso, Salina Cruz, Tampico, Topolobampo, Tuxpan, Veracruz
Merchant marine:
total: 59 ships (1,000 GRT or over) totaling 949,271 GRT/1,340,595 DWT
ships by type: bulk 1, cargo 2, chemical tanker 4, container 7, liquefied gas tanker 7, oil tanker 30, refrigerated cargo 2, roll-on/roll-off cargo 2, short-sea passenger 4
Airports:
total: 2,055
with paved runways over 3,047 m: 9
with paved runways 2,438 to 3,047 m: 25
with paved runways 1,524 to 2,437 m: 82
with paved runways 914 to 1,523 m: 75
with paved runways under 914 m: 1,262
with unpaved runways over 3,047 m: 1
with unpaved runways 2,438 to 3,047 m: 2
with unpaved runways 1,524 to 2,438 m: 60
with unpaved runways 914 to 1,523 m: 539

Communications

Telephone system: 6,410,000 telephones; highly developed system with extensive microwave radio relay links; privatized in December 1990
local: adequate phone service for business and government, but, at a density of less than 7 telephones/100 persons, the population is poorly served
intercity: includes 120 domestic satellite terminals and an extensive network of microwave radio relay links
international: 5 INTELSAT (4 Atlantic Ocean and 1 Pacific Ocean) earth stations; connected into Central America Microwave System; launched Solidarity I satellite in November 1993
Radio:
broadcast stations: AM 679, FM 0, shortwave 22
radios: NA
Television:
broadcast stations: 238
televisions: NA

Defense Forces

Branches: National Defense (includes Army and Air Force), Navy (includes Marines)
Manpower availability: males age 15-49 23,354,445; males fit for military service 17,029,788; males reach military age (18) annually 1,054,513 (1995 est.)
Defense expenditures: $NA, NA% of GDP

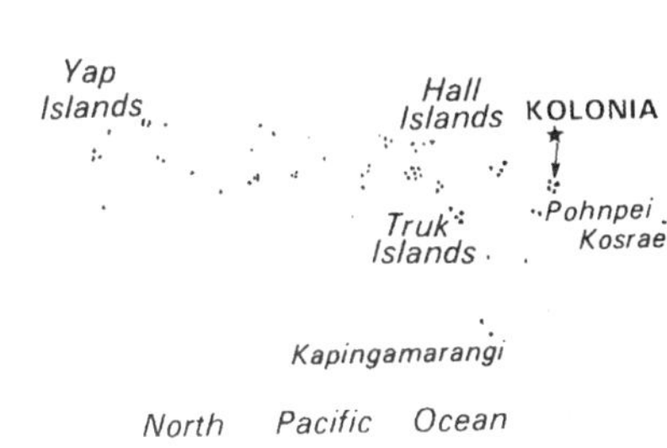

Geography

Location: Oceania, island group in the North Pacific Ocean, about three-quarters of the way from Hawaii to Indonesia
Map references: Oceania
Area:
total area: 702 sq km
land area: 702 sq km
comparative area: slightly less than four times the size of Washington, DC
note: includes Pohnpei (Ponape), Truk (Chuuk), Yap, and Kosrae
Land boundaries: 0 km
Coastline: 6,112 km
Maritime claims:
exclusive economic zone: 200 nm
territorial sea: 12 nm
International disputes: none
Climate: tropical; heavy year-round rainfall, especially in the eastern islands; located on southern edge of the typhoon belt with occasional severe damage
Terrain: islands vary geologically from high mountainous islands to low, coral atolls; volcanic outcroppings on Pohnpei, Kosrae, and Truk
Natural resources: forests, marine products, deep-seabed minerals
Land use:
arable land: NA%
permanent crops: NA%
meadows and pastures: NA%
forest and woodland: NA%
other: NA%
Irrigated land: NA sq km
Environment:
current issues: NA
natural hazards: typhoons (June to December)
international agreements: party to—Biodiversity, Climate Change, Law of the Sea
Note: four major island groups totaling 607 islands

People

Population: 122,950 (July 1995 est.)

Age structure:
0-14 years: NA
15-64 years: NA
65 years and over: NA
Population growth rate: 3.35% (1995 est.)
Birth rate: 28.12 births/1,000 population (1995 est.)
Death rate: 6.3 deaths/1,000 population (1995 est.)
Net migration rate: 11.65 migrant(s)/1,000 population (1995 est.)
Infant mortality rate: 36.52 deaths/1,000 live births (1995 est.)
Life expectancy at birth:
total population: 67.81 years
male: 65.84 years
female: 69.81 years (1995 est.)
Total fertility rate: 3.98 children born/woman (1995 est.)
Nationality:
noun: Micronesian(s)
adjective: Micronesian; Kosrae(s), Pohnpeian(s), Trukese, Yapese
Ethnic divisions: nine ethnic Micronesian and Polynesian groups
Religions: Roman Catholic 50%, Protestant 47%, other and none 3%
Languages: English (official and common language), Trukese, Pohnpeian, Yapese, Kosrean
Literacy: age 15 and over can read and write (1980)
total population: 89%
male: 91%
female: 88%
Labor force: NA
by occupation: two-thirds are government employees
note: 45,000 people are between the ages of 15 and 65

Government

Names:
conventional long form: Federated States of Micronesia
conventional short form: none
former: Kosrae, Ponape, Truk, and Yap Districts (Trust Territory of the Pacific Islands)
Abbreviation: FSM
Digraph: FM
Type: constitutional government in free association with the US; the Compact of Free Association entered into force 3 November 1986
Capital: Kolonia (on the island of Pohnpei)
note: a new capital is being built about 10 km southwest in the Palikir valley
Administrative divisions: 4 states; Kosrae, Pohnpei, Chuuk (Truk), Yap
Independence: 3 November 1986 (from the US-administered UN Trusteeship)
National holiday: Proclamation of the Federated States of Micronesia, 10 May (1979)
Constitution: 10 May 1979
Legal system: based on adapted Trust Territory laws, acts of the legislature, municipal, common, and customary laws
Suffrage: 18 years of age; universal
Executive branch:
chief of state and head of government: President Bailey OLTER (since 21 May 1991); Vice President Jacob NENA (since 21 May 1991); election last held 11 May 1991 (next to be held 7 March 1995); results— Bailey OLTER elected president and Jacob NENA elected vice president
cabinet: Cabinet
Legislative branch: unicameral
Congress: elections last held 5 March 1991 (next to be held 7 March 1995); results— percent of vote NA; seats—(14 total) independents 14
Judicial branch: Supreme Court
Political parties and leaders: no formal parties
Member of: AsDB, ESCAP, IBRD, ICAO, IDA, IFC, IMF, ITU, SPARTECA, SPC, SPF, UN, UNCTAD, WHO
Diplomatic representation in US:
chief of mission: Ambassador Jesse B. MAREHALAU
chancery: 1725 N Street NW, Washington, DC 20036
telephone: [1] (202) 223-4383
FAX: [1] (202) 223-4391
consulate(s) general: Honolulu and Tamuning (Guam)
US diplomatic representation:
chief of mission: Ambassador March Fong EU
embassy: address NA, Kolonia
mailing address: P. O. Box 1286, Pohnpei, Federated States of Micronesia 96941
telephone: [691] 320-2187
FAX: [691] 320-2186
Flag: light blue with four white five-pointed stars centered; the stars are arranged in a diamond pattern

Economy

Overview: Economic activity consists primarily of subsistence farming and fishing. The islands have few mineral deposits worth exploiting, except for high-grade phosphate. The potential for a tourist industry exists, but the remoteness of the location and a lack of adequate facilities hinder development. Financial assistance from the US is the primary source of revenue, with the US pledged to spend $1 billion in the islands in the 1990s. Geographical isolation and a poorly developed infrastructure are major impediments to long-term growth.
National product: GDP—purchasing power parity—$160 million (1990 est.)
note: GDP was supplemented by approximately $100 million in grant aid in 1990
National product real growth rate: 4% (1994)
National product per capita: $1,500 (1990 est.)
Inflation rate (consumer prices): NA%
Unemployment rate: 27% (1989)
Budget:
revenues: $45 million
expenditures: $31 million, including capital expenditures of $NA (FY94/95 est.)
Exports: $3.2 million (f.o.b., 1990)
commodities: fish, copra, bananas, black pepper
partners: Japan, US
Imports: $91.2 million (c.i.f., 1990)
commodities: food, manufactured goods, machinery and equipment, beverages
partners: US, Japan, Australia
External debt: $NA
Industrial production: growth rate NA%
Electricity:
capacity: 18,000 kW
production: 40 million kWh
consumption per capita: 380 kWh (1990)
Industries: tourism, construction, fish processing, craft items from shell, wood, and pearls
Agriculture: mainly a subsistence economy; black pepper; tropical fruits and vegetables, coconuts, cassava, sweet potatoes, pigs, chickens
Economic aid:
recipient: under terms of the Compact of Free Association, the US will provide $1.3 billion in grant aid during the period 1986-2001
Currency: 1 United States dollar (US$) = 100 cents
Exchange rates: US currency is used
Fiscal year: 1 October—30 September

Transportation

Railroads: 0 km
Highways:
total: 226 km
paved: 39 km (on major islands)
unpaved: stone, coral, laterite 187 km
Ports: Colonia (Yap), Kolonia (Pohnpei), Lele, Moen
Merchant marine: none
Airports:
total: 6
with paved runways 1,524 to 2,437 m: 4
with paved runways 914 to 1,523 m: 1
with paved runways under 914 m: 1

Communications

Telephone system: 960 telephones in Kolonia and Truk
local: NA
intercity: islands interconnected by shortwave radio (used mostly for government purposes)
international: 4 INTELSAT (Pacific Ocean) earth stations

Radio:
broadcast stations: AM 5, FM 1, shortwave 1
radios: 16,000
Television:
broadcast stations: 6
televisions: 1,125 (1987 est.)

Defense Forces

Note: defense is the responsibility of the US

Midway Islands
(territory of the US)

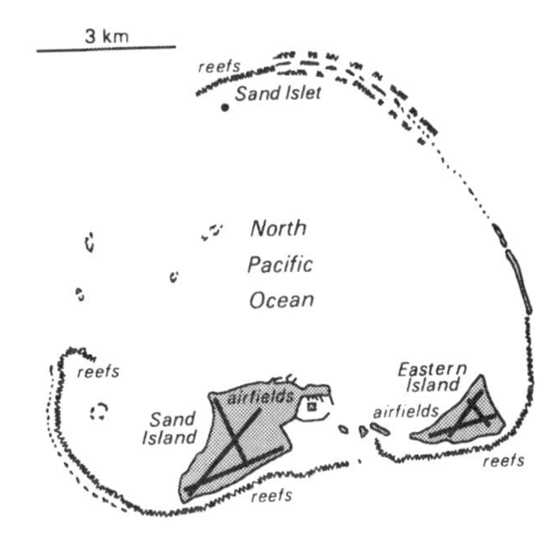

Geography

Location: Oceania, atoll in the North Pacific Ocean, about one-third of the way from Honolulu to Tokyo
Map references: Oceania
Area:
total area: 5.2 sq km
land area: 5.2 sq km
comparative area: about 9 times the size of The Mall in Washington, DC
note: includes Eastern Island and Sand Island
Land boundaries: 0 km
Coastline: 15 km
Maritime claims:
exclusive economic zone: 200 nm
territorial sea: 12 nm
International disputes: none
Climate: tropical, but moderated by prevailing easterly winds
Terrain: low, nearly level
Natural resources: fish, wildlife
Land use:
arable land: 0%
permanent crops: 0%
meadows and pastures: 0%
forest and woodland: 0%
other: 100%
Irrigated land: 0 sq km
Environment:
current issues: NA
natural hazards: NA
international agreements: NA
Note: a coral atoll; closed to the public

People

Population: no indigenous inhabitants; note—there are 453 US military personnel (July 1995 est.)

Government

Names:
conventional long form: none
conventional short form: Midway Islands
Digraph: MQ
Type: unincorporated territory of the US administered by the US Navy, under Naval Facilities Engineering Command, Pacific Division; this facility has been operationally closed since 10 September 1993 and is currently being transferred from Pacific Fleet to Naval Facilities Engineering Command via a Memorandum of Understanding
Capital: none; administered from Washington, DC
Flag: the US flag is used

Economy

Overview: The economy is based on providing support services for US naval operations located on the islands. All food and manufactured goods must be imported.
Electricity: supplied by US Military

Transportation

Railroads: 0 km
Highways:
total: 32 km
paved: NA
Pipelines: 7.8 km
Ports: Sand Island
Airports:
total: 3
with paved runways 1,524 to 2,437 m: 2
with unpaved runways 914 to 1,523 m: 1

Communications

Telephone system:
local: NA
intercity: NA
international: NA
Radio:
broadcast stations: AM NA, FM NA, shortwave NA
radios: NA
Television:
broadcast stations: NA
televisions: NA

Defense Forces

Note: defense is the responsibility of the US

Moldova

Geography

Location: Eastern Europe, northeast of Romania
Map references: Commonwealth of Independent States—European States
Area:
total area: 33,700 sq km
land area: 33,700 sq km
comparative area: slightly more than twice the size of Hawaii
Land boundaries: total 1,389 km, Romania 450 km, Ukraine 939 km
Coastline: 0 km (landlocked)
Maritime claims: none; landlocked
International disputes: certain territory of Moldova and Ukraine—including Bessarabia and Northern Bukovina—are considered by Bucharest as historically a part of Romania; this territory was incorporated into the former Soviet Union following the Molotov-Ribbentrop Pact in 1940
Climate: moderate winters, warm summers
Terrain: rolling steppe, gradual slope south to Black Sea
Natural resources: lignite, phosphorites, gypsum
Land use:
arable land: 50%
permanent crops: 13%
meadows and pastures: 9%
forest and woodland: 0%
other: 28%
Irrigated land: 2,920 sq km (1990)
Environment:
current issues: heavy use of agricultural chemicals, including banned pesticides such as DDT, has contaminated soil and groundwater; extensive soil erosion from poor farming methods
natural hazards: NA
international agreements: signed, but not ratified—Biodiversity, Climate Change
Note: landlocked

People

Population: 4,489,657 (July 1995 est.)
Age structure:
0-14 years: 27% (female 588,155; male 609,372)
15-64 years: 64% (female 1,487,170; male 1,386,293)
65 years and over: 9% (female 258,958; male 159,709) (July 1995 est.)
Population growth rate: 0.36% (1995 est.)
Birth rate: 15.93 births/1,000 population (1995 est.)
Death rate: 10.05 deaths/1,000 population (1995 est.)
Net migration rate: -2.25 migrant(s)/1,000 population (1995 est.)
Infant mortality rate: 29.8 deaths/1,000 live births (1995 est.)
Life expectancy at birth:
total population: 68.22 years
male: 64.81 years
female: 71.8 years (1995 est.)
Total fertility rate: 2.16 children born/woman (1995 est.)
Nationality:
noun: Moldovan(s)
adjective: Moldovan
Ethnic divisions: Moldavian/Romanian 64.5%, Ukrainian 13.8%, Russian 13%, Gagauz 3.5%, Jewish 1.5%, Bulgarian 2%, other 1.7% (1989 figures)
note: internal disputes with ethnic Russians and Ukrainians in the Dniester region and Gagauz Turks in the south
Religions: Eastern Orthodox 98.5%, Jewish 1.5%, Baptist (only about 1,000 members) (1991)
note: the large majority of churchgoers are ethnic Moldavian
Languages: Moldovan (official; virtually the same as the Romanian language), Russian, Gagauz (a Turkish dialect)
Literacy: age 15 and over can read and write (1989)
total population: 96%
male: 99%
female: 94%
Labor force: 2.03 million (January 1994)
by occupation: agriculture 34.4%, industry 20.1%, other 45.5% (1985 figures)

Government

Names:
conventional long form: Republic of Moldova
conventional short form: Moldova
local long form: Republica Moldova
local short form: none
former: Soviet Socialist Republic of Moldova; Moldavia
Digraph: MD
Type: republic
Capital: Chisinau
Administrative divisions: previously divided into 40 rayons; new districts possible under new constitution in 1994
Independence: 27 August 1991 (from Soviet Union)
National holiday: Independence Day, 27 August 1991
Constitution: new constitution adopted NA July 1994; replaces old Soviet constitution of 1979
Legal system: based on civil law system; no judicial review of legislative acts; does not accept compulsory ICJ jurisdiction but accepts many UN and OSCE documents
Suffrage: 18 years of age; universal
Executive branch:
chief of state: President Mircea SNEGUR (since 3 September 1990); election last held 8 December 1991 (next to be held NA 1996); results—Mircea SNEGUR ran unopposed and won 98.17% of vote; note—President SNEGUR was named executive president by the Supreme Soviet on 3 September 1990 and was confirmed by popular election on 8 December 1991
head of government: Prime Minister Andrei SANGHELI (since 1 July 1992; reappointed 5 April 1994 after elections for new legislature); First Deputy Prime Minister Ion GUTU (since NA)
cabinet: Council of Ministers; appointed by the president on recommendation of the prime minister
Legislative branch: unicameral
Parliament: elections last held 27 February 1994 (next to be held NA 1999); results—percent by party NA; seats—(104 total) Agrarian-Democratic Party 56, Socialist/Yedinstvo Bloc 28, Peasants and Intellectual Bloc 11, Christian Democratic Popular Front 9
Judicial branch: Supreme Court
Political parties and leaders: Christian Democratic Popular Front (formerly Moldovan Popular Front), Iurie ROSCA, chairman; Yedinstvo Intermovement, Vladimir SOLONARI, chairman; Social Democratic Party, Oazu NANTOI, chairman, two other chairmen; Agrarian-Democratic Party, Dumitru MOTPAN, chairman; Democratic Party, Gheorghe GHIMPU, chairman; Democratic Labor Party, Alexandru ARSENI, chairman; Reform Party, Anatol SELARU; Republican Party, Victor PUSCAS; Socialist Party, Valeriu SENIC, cochairman; Communist Party, Vladimir VORONIN, cochairman; Peasants and Intellectuals Bloc
Other political or pressure groups: United Council of Labor Collectives (UCLC), Igor SMIRNOV, chairman; Congress of Intellectuals, Alexandru MOSANU; The Ecology Movement of Moldova (EMM), G. MALARCHUK, chairman; The Christian Democratic League of Women of Moldova (CDLWM), L. LARI, chairman; National Christian Party of Moldova (NCPM), D. TODIKE, M. BARAGA, V. NIKU, leaders;

The Peoples Movement Gagauz Khalky (GKh), S. GULGAR, leader; The Democratic Party of Gagauzia (DPG), G. SAVOSTIN, chairman; The Alliance of Working People of Moldova (AWPM), G. POLOGOV, president; Christian Alliance for Greater Romania; Stefan the Great Movement; Liberal Convention of Moldova; Association of Victims of Repression; Christian Democratic Youth League
Member of: BSEC, CE (guest), CIS, EBRD, ECE, IBRD, ICAO, IDA, ILO, IMF, INTELSAT (nonsignatory user), INTERPOL, IOC, IOM (observer), ITU, NACC, OSCE, PFP, UN, UNCTAD, UNESCO, UNIDO, UPU, WHO, WIPO, WTO
Diplomatic representation in US:
chief of mission: Ambassador Nicolae TAU
chancery: Suites 329, 333, 1511 K Street NW, Washington, DC 20005
telephone: [1] (202) 783-3012
FAX: [1] (202) 783-3342
US diplomatic representation:
chief of mission: Ambassador Mary C. PENDLETON
embassy: Strada Alexei Mateevich #103, Chisinau
mailing address: use embassy street address
telephone: [373] (2) 23-37-72
FAX: [373] (2) 23-30-44
Flag: same color scheme as Romania—3 equal vertical bands of blue (hoist side), yellow, and red; emblem in center of flag is of a Roman eagle of gold outlined in black with a red beak and talons carrying a yellow cross in its beak and a green olive branch in its right talons and a yellow scepter in its left talons; on its breast is a shield divided horizontally red over blue with a stylized ox head, star, rose, and crescent all in black-outlined yellow

Economy

Overview: Moldova enjoys a favorable climate and good farmland but has no major mineral deposits. As a result, Moldova's economy is primarily based on agriculture, featuring fruits, vegetables, wine, and tobacco. Moldova must import all of its supplies of oil, coal, and natural gas, and energy shortages have contributed to sharp production declines since the breakup of the Soviet Union in 1991. The Moldovan government is making steady progress on an ambitious economic reform agenda, and the IMF has called Moldova a model for the region. As part of its reform efforts, Chisinau has introduced a stable currency, freed all prices, stopped issuing preferential credits to state enterprises and backed their steady privatization, removed export controls, and freed interest rates. Chisinau appears strongly committed to continuing these reforms in 1995. Meanwhile, privatization of medium and large enterprises got underway in mid-1994 and is expected to pick up speed in 1995. To improve its precarious energy situation, Chisinau reached an agreement with Moscow in December 1994 on gas deliveries for 1995. Gazprom, Russia's national gas company, has agreed to reduce prices for natural gas deliveries to Moldova from the world market price of $80/thousand cubic meters (tcm) to $58/tcm in return for part ownership of the Moldovan pipeline system.
National product: GDP—purchasing power parity—$11.9 billion (1994 estimate as extrapolated from World Bank estimate for 1992)
National product real growth rate: -30% (1994 est.)
National product per capita: $2,670 (1994 est.)
Inflation rate (consumer prices): 7.6% per month (1994)
Unemployment rate: 1% (includes only officially registered unemployed; large numbers of underemployed workers)
Budget:
revenues: $NA
expenditures: $NA, including capital expenditures of $NA
note: budget deficit for 1993 approximately 6% of GDP
Exports: $144 million to outside the FSU countries (1994); over 70% of exports go to FSU countries
commodities: foodstuffs, wine, tobacco, textiles and footwear, machinery, chemicals (1991)
partners: Russia, Kazakhstan, Ukraine, Romania, Germany
Imports: $174 million from outside the FSU countries (1994); over 70% of imports are from FSU countries
commodities: oil, gas, coal, steel, machinery, foodstuffs, automobiles, and other consumer durables
partners: Russia, Ukraine, Uzbekistan, Romania, Germany
External debt: $300 million (as of 11 December 1994)
Industrial production: growth rate -30% (1994 est.)
Electricity:
capacity: 3,000,000 kW
production: 8.2 billion kWh
consumption per capita: 1,830 kWh (1994)
Industries: key products are canned food, agricultural machinery, foundry equipment, refrigerators and freezers, washing machines, hosiery, refined sugar, vegetable oil, shoes, textiles
Agriculture: accounts for about 40% of GDP; Moldova's principal economic activity; products are vegetables, fruits, wine, grain, sugar beets, sunflower seed, meat, milk, tobacco
Illicit drugs: illicit cultivator of opium poppy and cannabis; mostly for CIS consumption; transshipment point for illicit drugs to Western Europe
Economic aid:
recipient: joint EC-US loan (1993), $127 million; IMF STF credit (1993), $64 million; IMF stand-by loan (1993), $72 million; US commitments (1992-93), $61 million in humanitarian aid, $11 million in technical assistance; World Bank loan (1993), $60 million; Russia (1993), 50 billion ruble credit; Romania (1993), 20 billion lei credit
Currency: the leu (plural lei) was introduced in late 1993
Exchange rates: lei per US$1—4.277 (22 December 1994)
Fiscal year: calendar year

Transportation

Railroads:
total: 1,150 km in common carrier service; does not include industrial lines
broad gauge: 1,150 km 1.520-m gauge (1990)
Highways:
total: 20,000 km
paved or graveled: 13,900 km
unpaved: earth 6,100 km (1990)
Pipelines: natural gas 310 km (1992)
Ports: none
Airports:
total: 26
with paved runways over 3,047 m: 1
with paved runways 2,438 to 3,047 m: 2
with paved runways 1,524 to 2,437 m: 2
with paved runways under 914 m: 3
with unpaved runways 2,438 to 3,047 m: 3
with unpaved runways 1,524 to 2,438 m: 2
with unpaved runways 914 to 1,523 m: 5
with unpaved runways under 914 m: 8

Communications

Telephone system: 577,000 telephones; 134 telephones/1,000 persons; telecommunication system not well developed; 215,000 unsatisfied requests for telephone service (1991)
local: NA
intercity: NA
international: international connections to the other former Soviet republics by land line and microwave radio relay through Ukraine, and to other countries by leased connections to the Moscow international gateway switch; 1 EUTELSAT and 1 INTELSAT earth station
Radio:
broadcast stations: AM NA, FM NA, shortwave NA
radios: NA
Television:
broadcast stations: NA
televisions: NA

Defense Forces

Branches: Ground Forces, Air and Air Defense Forces, Republic Security Forces (internal and border troops)
Manpower availability: males age 15-49 1,116,912; males fit for military service 881,642; males reach military age (18) annually 35,447 (1995 est.)
Defense expenditures: $NA, 2% of GDP (1994)

Monaco

Geography

Location: Western Europe, bordering the Mediterranean Sea, on the southern coast of France, near the border with Italy
Map references: Europe
Area:
total area: 1.9 sq km
land area: 1.9 sq km
comparative area: about three times the size of The Mall in Washington, DC
Land boundaries: total 4.4 km, France 4.4 km
Coastline: 4.1 km
Maritime claims:
territorial sea: 12 nm
International disputes: none
Climate: Mediterranean with mild, wet winters and hot, dry summers
Terrain: hilly, rugged, rocky
Natural resources: none
Land use:
arable land: 0%
permanent crops: 0%
meadows and pastures: 0%
forest and woodland: 0%
other: 100%
Irrigated land: NA sq km
Environment:
current issues: NA
natural hazards: NA
international agreements: party to—Biodiversity, Climate Change, Endangered Species, Hazardous Wastes, Marine Dumping, Ozone Layer Protection, Ship Pollution, Whaling; signed, but not ratified—Law of the Sea
Note: second smallest independent state in world (after Holy See); almost entirely urban

People

Population: 31,515 (July 1995 est.)
Age structure:
0-14 years: 17% (female 2,691; male 2,740)
15-64 years: 63% (female 10,233; male 9,645)
65 years and over: 20% (female 3,939; male 2,267) (July 1995 est.)
Population growth rate: 0.7% (1995 est.)
Birth rate: 10.66 births/1,000 population (1995 est.)
Death rate: 12.12 deaths/1,000 population (1995 est.)
Net migration rate: 8.44 migrant(s)/1,000 population (1995 est.)
Infant mortality rate: 7 deaths/1,000 live births (1995 est.)
Life expectancy at birth:
total population: 77.9 years
male: 74.18 years
female: 81.8 years (1995 est.)
Total fertility rate: 1.7 children born/woman (1995 est.)
Nationality:
noun: Monacan(s) or Monegasque(s)
adjective: Monacan or Monegasque
Ethnic divisions: French 47%, Monegasque 16%, Italian 16%, other 21%
Religions: Roman Catholic 95%
Languages: French (official), English, Italian, Monegasque
Literacy: NA%
Labor force: NA

Government

Names:
conventional long form: Principality of Monaco
conventional short form: Monaco
local long form: Principaute de Monaco
local short form: Monaco
Digraph: MN
Type: constitutional monarchy
Capital: Monaco
Administrative divisions: 4 quarters (quartiers, singular—quartier); Fontvieille, La Condamine, Monaco-Ville, Monte-Carlo
Independence: 1419 (rule by the House of Grimaldi)
National holiday: National Day, 19 November
Constitution: 17 December 1962
Legal system: based on French law; has not accepted compulsory ICJ jurisdiction
Suffrage: 25 years of age; universal
Executive branch:
chief of state: Prince RAINIER III (since NA November 1949); Heir Apparent Prince ALBERT Alexandre Louis Pierre (born 14 March 1958)
head of government: Minister of State Paul DIJOUD (since NA)
cabinet: Council of Government; under the authority of the Prince
Legislative branch: unicameral
National Council (Conseil National): elections last held 24 and 31 January 1993 (next to be held NA); results—percent of vote by party NA; seats—(18 total) Campora List 15, Medecin List 2, independent 1
Judicial branch: Supreme Tribunal (Tribunal Supreme)
Political parties and leaders: National and Democratic Union (UND); Campora List, Anne-Marie CAMPORA; Medecin List, Jean-Louis MEDECIN
Member of: ACCT, ECE, IAEA, ICAO, ICRM, IFRCS, IMO, INMARSAT, INTELSAT, INTERPOL, IOC, ITU, OSCE, UN, UNCTAD, UNESCO, UPU, WHO, WIPO
Diplomatic representation in US:
honorary consulate(s) general: Boston, Chicago, Los Angeles, New Orleans, New York, San Francisco, San Juan (Puerto Rico)
honorary consulate(s): Dallas, Palm Beach, Philadelphia, and Washington, DC
US diplomatic representation: no mission in Monaco, but the US Consul General in Marseille, France, is accredited to Monaco
Flag: two equal horizontal bands of red (top) and white; similar to the flag of Indonesia which is longer and the flag of Poland which is white (top) and red

Economy

Overview: Monaco, situated on the French Mediterranean coast, is a popular resort, attracting tourists to its casino and pleasant climate. The Principality has successfully sought to diversify into services and small, high-value-added, nonpolluting industries. The state has no income tax and low business taxes and thrives as a tax haven both for individuals who have established residence and for foreign companies that have set up businesses and offices. About 50% of Monaco's annual revenue comes from value-added taxes on hotels, banks, and the industrial sector; about 25% of revenue comes from tourism. Living standards are high, that is, roughly comparable to those in prosperous French metropolitan suburbs.
National product: GDP—purchasing power parity—$558 million (1993 est.)
National product real growth rate: NA%
National product per capita: $18,000 (1993 est.)
Inflation rate (consumer prices): NA%
Unemployment rate: NEGL%
Budget:
revenues: $424 million
expenditures: $376 million, including capital expenditures of $NA (1991 est.)
Exports: $NA; full customs integration with France, which collects and rebates Monacan trade duties; also participates in EU market system through customs union with France
Imports: $NA; full customs integration with France, which collects and rebates Monacan trade duties; also participates in EU market system through customs union with France
External debt: $NA
Industrial production: growth rate NA%

Electricity:
capacity: 10,000 kW standby; power imported from France
production: NA kWh
consumption per capita: NA kWh (1993)
Agriculture: none
Economic aid: $NA
Currency: 1 French franc (F) = 100 centimes
Exchange rates: French francs (F) per US$1—5.9243 (January 1995), 5.520 (1994), 5.6632 (1993), 5.2938 (1992), 5.6421 (1991), 5.4453 (1990)
Fiscal year: calendar year

Transportation

Railroads:
total: 1.7 km
standard gauge: 1.7 km 1.435-m gauge
Highways: none; city streets
Ports: Monaco
Merchant marine: none
Airports: linked to airport in Nice, France, by helicopter service

Communications

Telephone system: 38,200 telephones; automatic telephone system
local: NA
intercity: NA
international: no satellite links; served by cable into the French communications system
Radio:
broadcast stations: AM 3, FM 4, shortwave 0
radios: NA
Television:
broadcast stations: 5
televisions: NA

Defense Forces

Note: defense is the responsibility of France

Mongolia

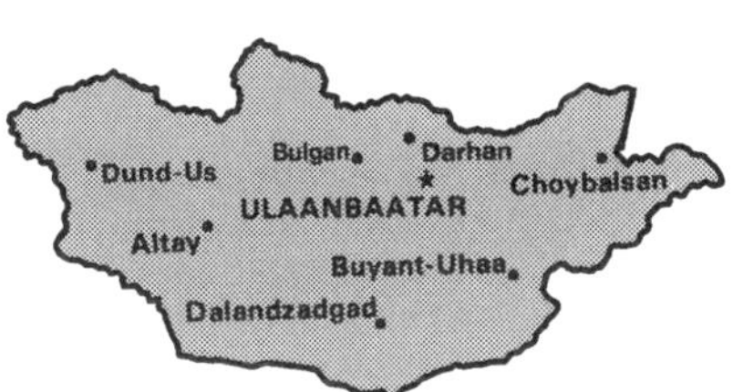

Geography

Location: Northern Asia, north of China
Map references: Asia
Area:
total area: 1.565 million sq km
land area: 1.565 million sq km
comparative area: slightly larger than Alaska
Land boundaries: total 8,114 km, China 4,673 km, Russia 3,441 km
Coastline: 0 km (landlocked)
Maritime claims: none; landlocked
International disputes: none
Climate: desert; continental (large daily and seasonal temperature ranges)
Terrain: vast semidesert and desert plains; mountains in west and southwest; Gobi Desert in southeast
Natural resources: oil, coal, copper, molybdenum, tungsten, phosphates, tin, nickel, zinc, wolfram, fluorspar, gold
Land use:
arable land: 1%
permanent crops: 0%
meadows and pastures: 79%
forest and woodland: 10%
other: 10%
Irrigated land: 770 sq km (1989)
Environment:
current issues: limited natural fresh water resources; policies of the former communist regime promoting rapid urbanization and industrial growth have raised concerns about their negative effects on the environment; the burning of soft coal and the concentration of factories in Ulaanbaatar have severely polluted the air; deforestation, overgrazing, the converting of virgin land to agricultural production have increased soil erosion from wind and rain; desertification
natural hazards: duststorms can occur in the spring
international agreements: party to—Biodiversity, Climate Change, Environmental Modification, Nuclear Test Ban; signed, but not ratified—Desertification, Law of the Sea
Note: landlocked; strategic location between China and Russia

People

Population: 2,493,615 (July 1995 est.)
Age structure:
0-14 years: 40% (female 495,919; male 511,464)
15-64 years: 56% (female 693,037; male 693,776)
65 years and over: 4% (female 54,991; male 44,428) (July 1995 est.)
Population growth rate: 2.58% (1995 est.)
Birth rate: 32.65 births/1,000 population (1995 est.)
Death rate: 6.82 deaths/1,000 population (1995 est.)
Net migration rate: 0 migrant(s)/1,000 population (1995 est.)
Infant mortality rate: 41.8 deaths/1,000 live births (1995 est.)
Life expectancy at birth:
total population: 66.54 years
male: 64.28 years
female: 68.92 years (1995 est.)
Total fertility rate: 4.26 children born/woman (1995 est.)
Nationality:
noun: Mongolian(s)
adjective: Mongolian
Ethnic divisions: Mongol 90%, Kazakh 4%, Chinese 2%, Russian 2%, other 2%
Religions: predominantly Tibetan Buddhist, Muslim 4%
note: previously limited religious activity because of Communist regime
Languages: Khalkha Mongol 90%, Turkic, Russian, Chinese
Literacy: NA%
Labor force: NA
by occupation: primarily herding/agricultural
note: over half the adult population is in the labor force, including a large percentage of women; shortage of skilled labor

Government

Names:
conventional long form: none
conventional short form: Mongolia
local long form: none
local short form: Mongol Uls
former: Outer Mongolia
Digraph: MG
Type: republic
Capital: Ulaanbaatar
Administrative divisions: 18 provinces (aymguud, singular—aymag) and 3 municipalities* (hotuud, singular—hot); Arhangay, Bayanhongor, Bayan-Olgiy, Bulgan, Darhan*, Dornod, Dornogovi, Dundgovi, Dzavhan, Erdenet*, Govi-Altay, Hentiy, Hovd, Hovsgol, Omnogovi, Ovorhangay, Selenge, Suhbaatar, Tov, Ulaanbaatar*, Uvs
Independence: 13 March 1921 (from China)
National holiday: National Day, 11 July (1921)

Constitution: adopted 13 January 1992
Legal system: blend of Russian, Chinese, and Turkish systems of law; no constitutional provision for judicial review of legislative acts; has not accepted compulsory ICJ jurisdiction
Suffrage: 18 years of age; universal
Executive branch:
chief of state: President Punsalmaagiyn OCHIRBAT (since 3 September 1990); election last held 6 June 1993 (next to be held NA 1997); results—Punsalmaagiyn OCHIRBAT (MNDP and MSDP) elected directly with 57.8% of the vote; other candidate Lodongiyn TUDEV (MPRP)
head of government: Prime Minister Putsagiyn JASRAY (since 3 August 1992); Deputy Prime Ministers Lhamsuren ENEBISH and Choijilsurengiyn PUREVDORJ (since NA)
cabinet: Cabinet; appointed by the Great Hural
Legislative branch: unicameral
State Great Hural: elections held for the first time 28 June 1992 (next to be held NA); results—percent of vote by party NA; seats—(76 total) MPRP 71, United Party of Mongolia 4, MSDP 1
note: the People's Small Hural no longer exists
Judicial branch: Supreme Court serves as appeals court for people's and provincial courts, but to date rarely overturns verdicts of lower courts
Political parties and leaders: Mongolian People's Revolutionary Party (MPRP), Budragchagiin DASH-YONDON, secretary general; Mongolian National Democratic Party (MNDP), D. GANBOLD, chairman; Mongolian Social Democratic Party (MSDP), B. BATBAYAR, chairman; United Party of Mongolia, leader NA
note: opposition parties were legalized in May 1990
Member of: AsDB, CCC, ESCAP, FAO, G-77, IAEA, IBRD, ICAO, ICRM, IDA, IFAD, IFC, IFRCS, ILO, IMF, INTELSAT (nonsignatory user), INTERPOL, IOC, ISO, ITU, NAM (observer), UN, UNCTAD, UNESCO, UNIDO, UPU, WFTU, WHO, WIPO, WMO, WTO
Diplomatic representation in US:
chief of mission: Ambassador Luvsandorj DAWAAGIW
chancery: 2833 M Street NW, Washington, DC 20007
telephone: [1] (202) 333-7117
FAX: [1] (202) 298-9227
consulate(s) general: New York
US diplomatic representation:
chief of mission: Ambassador Donald C. JOHNSON
embassy: address NA, Ulaanbaatar
mailing address: c/o American Embassy Beijing, Micro Region 11, Big Ring Road; PSC 461, Box 300, FPO AP 96521-0002
telephone: [976] (1) 329095, 329606
FAX: [976] (1) 320776
Flag: three equal, vertical bands of red (hoist side), blue, and red, centered on the hoist-side red band in yellow is the national emblem (''soyombo''—a columnar arrangement of abstract and geometric representation for fire, sun, moon, earth, water, and the yin-yang symbol)

Economy

Overview: Mongolia's severe climate, scattered population, and wide expanses of unproductive land have constrained economic development. Economic activity traditionally has been based on agriculture and the breeding of livestock. In past years extensive mineral resources had been developed with Soviet support; total Soviet assistance at its height amounted to 30% of GDP. The mining and processing of coal, copper, molybdenum, tin, tungsten, and gold account for a large part of industrial production. Timber and fishing are also important sectors. The Mongolian leadership has been gradually making the transition from Soviet-style central planning to a market economy through privatization and price reform, and is soliciting support from international financial agencies and foreign investors. The economy, however, has still not recovered from the loss of Soviet aid, and the country continues to suffer substantial economic hardships, with one-fourth of the population below the poverty line.
National product: GDP—purchasing power parity—$4.4 billion (1994 est.)
National product real growth rate: 2.5% (1994 est.)
National product per capita: $1,800 (1994 est.)
Inflation rate (consumer prices): 70% (1994 est.)
Unemployment rate: 15% (1991 est.)
Budget:
revenues: $NA
expenditures: $NA, including capital expenditures of $NA (1991 est.)
note: deficit of $67 million
Exports: $360 million (f.o.b., 1993 est.)
commodities: copper, livestock, animal products, cashmere, wool, hides, fluorspar, other nonferrous metals
partners: former CMEA countries 62%, China 17%, EC 8% (1992)
Imports: $361 million (f.o.b., 1993 est.)
commodities: machinery and equipment, fuels, food products, industrial consumer goods, chemicals, building materials, sugar, tea
partners: USSR 75%, Austria 5%, China 5% (1991)
External debt: $NA
Industrial production: growth rate -15% (1992 est.); accounts for about 42% of GDP
Electricity:
capacity: 900,000 kW
production: 3.1 billion kWh
consumption per capita: 1,267 kWh (1993)
Industries: copper, processing of animal products, building materials, food and beverage, mining (particularly coal)
Agriculture: accounts for about 35% of GDP and provides livelihood for about 50% of the population; livestock raising predominates (primarily sheep and goats, but also cattle, camels, and horses); crops—wheat, barley, potatoes, forage
Economic aid: NA
Currency: 1 tughrik (Tug) = 100 mongos
Exchange rates: tughriks (Tug) per US$1—415.34 (January 1995), 412.72 (1994), 42.56 (1992), 9.52 (1991), 5.63 (1990)
note: the exchange rate 40 tughriks = 1US$ was introduced June 1991 and was in force to the end of 1992; beginning 27 May 1993 the exchange rate is the midpoint of the average buying and selling rates that are freely determined on the basis of market transactions between commercial banks and the nonbank public
Fiscal year: calendar year

Transportation

Railroads:
total: 1,750 km
broad gauge: 1,750 km 1.524-m gauge (1988)
Highways:
total: 46,700 km
paved: 1,000 km
unpaved: 45,700 km (1988)
Inland waterways: 397 km of principal routes (1988)
Ports: none
Airports:
total: 34
with paved runways 2,438 to 3,047 m: 7
with paved runways under 914 m: 1
with unpaved runways over 3,047 m: 3
with unpaved runways 2,438 to 3,047 m: 5
with unpaved runways 1,524 to 2,438 m: 10
with unpaved runways 914 to 1,523 m: 3
with unpaved runways under 914 m: 5

Communications

Telephone system: 63,000 telephones (1989)
local: NA
intercity: NA
international: at least 1 satellite earth station
Radio:
broadcast stations: AM 12, FM 1, shortwave 0
radios: 220,000
Television:
broadcast stations: 1 (provincial repeaters—18)
televisions: 120,000

Defense Forces

Branches: Mongolian People's Army (includes Internal Security Forces and Frontier Guards), Air Force

Manpower availability: males age 15-49 605,633; males fit for military service 394,433; males reach military age (18) annually 25,862 (1995 est.)
Defense expenditures: exchange rate conversion—$22.8 million, 1% of GDP (1992)

Montserrat
(dependent territory of the UK)

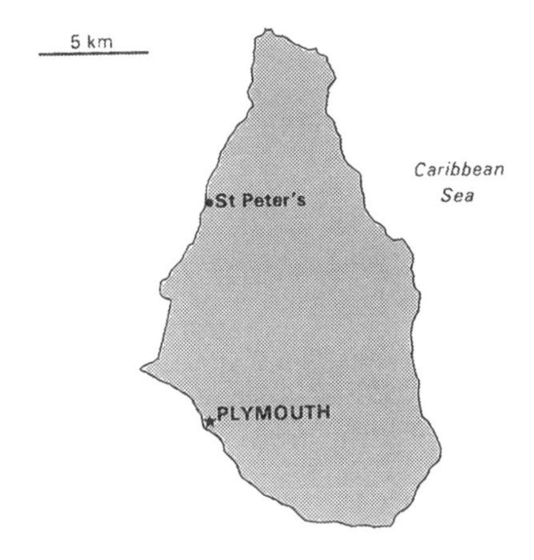

Geography

Location: Caribbean, island in the Caribbean Sea, southeast of Puerto Rico
Map references: Central America and the Caribbean
Area:
total area: 100 sq km
land area: 100 sq km
comparative area: about 0.6 times the size of Washington, DC
Land boundaries: 0 km
Coastline: 40 km
Maritime claims:
exclusive fishing zone: 200 nm
territorial sea: 3 nm
International disputes: none
Climate: tropical; little daily or seasonal temperature variation
Terrain: volcanic islands, mostly mountainous, with small coastal lowland
Natural resources: negligible
Land use:
arable land: 20%
permanent crops: 0%
meadows and pastures: 10%
forest and woodland: 40%
other: 30%
Irrigated land: NA sq km
Environment:
current issues: land erosion occurs on slopes that have been cleared for cultivation
natural hazards: severe hurricanes (June to November); volcanic eruptions (there are seven active volcanoes on the island)
international agreements: NA

People

Population: 12,738 (July 1995 est.)
Age structure:
0-14 years: NA
15-64 years: NA
65 years and over: NA
Population growth rate: 0.3% (1995 est.)
Birth rate: 15.5 births/1,000 population (1995 est.)
Death rate: 9.81 deaths/1,000 population (1995 est.)
Net migration rate: -2.65 migrant(s)/1,000 population (1995 est.)
Infant mortality rate: 11.69 deaths/1,000 live births (1995 est.)
Life expectancy at birth:
total population: 75.69 years
male: 73.93 years
female: 77.49 years (1995 est.)
Total fertility rate: 1.99 children born/woman (1995 est.)
Nationality:
noun: Montserratian(s)
adjective: Montserratian
Ethnic divisions: black, Europeans
Religions: Anglican, Methodist, Roman Catholic, Pentecostal, Seventh-Day Adventist, other Christian denominations
Languages: English
Literacy: age 15 and over has ever attended school (1970)
total population: 97%
male: 97%
female: 97%
Labor force: 5,100
by occupation: community, social, and personal services 40.5%, construction 13.5%, trade, restaurants, and hotels 12.3%, manufacturing 10.5%, agriculture, forestry, and fishing 8.8%, other 14.4% (1983 est.)

Government

Names:
conventional long form: none
conventional short form: Montserrat
Digraph: MH
Type: dependent territory of the UK
Capital: Plymouth
Administrative divisions: 3 parishes; Saint Anthony, Saint Georges, Saint Peter's
Independence: none (dependent territory of the UK)
National holiday: Celebration of the Birthday of the Queen (second Saturday of June)
Constitution: present constitution came into force 19 December 1989
Legal system: English common law and statute law
Suffrage: 18 years of age; universal
Executive branch:
chief of state: Queen ELIZABETH II (since 6 February 1952), represented by Governor Frank SAVAGE (since NA February 1993)
head of government: Chief Minister Reuben T. MEADE (since NA October 1991)
cabinet: Executive Council; consists of the governor, the chief minister, three other ministries, the attorney-general, and the finance secretary
Legislative branch: unicameral

Legislative Council: elections last held 8 October 1991; results—percent of vote by party NA; seats—(11 total, 7 elected) NPP 4, NDP 1, PLM 1, independent 1
Judicial branch: Supreme Court
Political parties and leaders: National Progressive Party (NPP) Reuben T. MEADE; People's Liberation Movement (PLM), Noel TUITT; National Development Party (NDP), Bertrand OSBORNE
Member of: CARICOM, CDB, ECLAC (associate), ICFTU, INTERPOL (subbureau), OECS, WCL
Diplomatic representation in US: none (dependent territory of the UK)
US diplomatic representation: none (dependent territory of the UK)
Flag: blue with the flag of the UK in the upper hoist-side quadrant and the Montserratian coat of arms centered in the outer half of the flag; the coat of arms features a woman standing beside a yellow harp with her arm around a black cross

Economy

Overview: The economy is small and open with economic activity centered on tourism and construction. Tourism is the most important sector and accounts for roughly one-fifth of GDP. Agriculture accounts for about 4% of GDP and industry 10%. The economy is heavily dependent on imports, making it vulnerable to fluctuations in world prices. Exports consist mainly of electronic parts sold to the US.
National product: GDP—purchasing power parity—$55.6 million (1993 est.)
National product real growth rate: 1% (1993 est.)
National product per capita: $4,380 (1993 est.)
Inflation rate (consumer prices): 2.8% (1992)
Unemployment rate: NA
Budget:
revenues: $12.1 million
expenditures: $14.3 million, including capital expenditures of $3.2 million (1988 est.)
Exports: $2.8 million (f.o.b., 1992)
commodities: electronic parts, plastic bags, apparel, hot peppers, live plants, cattle
partners: NA
Imports: $80.6 million (f.o.b., 1992)
commodities: machinery and transportation equipment, foodstuffs, manufactured goods, fuels, lubricants, and related materials
partners: NA
External debt: $2.05 million (1987)
Industrial production: growth rate 8.1% (1986); accounts for 10% of GDP
Electricity:
capacity: 5,271 kW
production: 17 million kWh
consumption per capita: 1,106 kWh (1993)
Industries: tourism; light manufacturing—rum, textiles, electronic appliances
Agriculture: accounts for 4% of GDP; small-scale farming; food crops—tomatoes, onions, peppers; not self-sufficient in food, especially livestock products
Economic aid:
recipient: Western (non-US) countries, ODA and OOF bilateral commitments (1970-89), $90 million
Currency: 1 EC dollar (EC$) = 100 cents
Exchange rates: East Caribbean dollars (EC$) per US$1—2.70 (fixed rate since 1976)
Fiscal year: 1 April—31 March

Transportation

Railroads: 0 km
Highways:
total: 280 km
paved: 200 km
unpaved: gravel, earth 80 km
Ports: Plymouth
Merchant marine: none
Airports:
total: 1
with paved runways 914 to 1,523 m: 1

Communications

Telephone system: 3,000 telephones
local: NA
intercity: NA
international: NA
Radio:
broadcast stations: AM 8, FM 4, shortwave 0
radios: NA
Television:
broadcast stations: 1
televisions: NA

Defense Forces

Branches: Police Force
Note: defense is the responsibility of the UK

Morocco

Geography

Location: Northern Africa, bordering the North Atlantic Ocean and the Mediterranean Sea, between Algeria and Western Sahara
Map references: Africa
Area:
total area: 446,550 sq km
land area: 446,300 sq km
comparative area: slightly larger than California
Land boundaries: total 2,002 km, Algeria 1,559 km, Western Sahara 443 km
Coastline: 1,835 km
Maritime claims:
contiguous zone: 24 nm
continental shelf: 200-m depth or to the depth of exploitation
exclusive economic zone: 200 nm
territorial sea: 12 nm
International disputes: claims and administers Western Sahara, but sovereignty is unresolved; the UN is attempting to hold a referendum; the UN-administered cease-fire has been currently in effect since September 1991; Spain controls five places of sovereignty (plazas de soberania) on and off the coast of Morocco—the coastal enclaves of Ceuta and Melilla which Morocco contests as well as the islands of Penon de Alhucemas, Penon de Velez de la Gomera, and Islas Chafarinas
Climate: Mediterranean, becoming more extreme in the interior
Terrain: northern coast and interior are mountainous with large areas of bordering plateaus, intermontane valleys, and rich coastal plains
Natural resources: phosphates, iron ore, manganese, lead, zinc, fish, salt
Land use:
arable land: 18%
permanent crops: 1%
meadows and pastures: 28%
forest and woodland: 12%
other: 41%
Irrigated land: 12,650 sq km (1989 est.)

Environment:
current issues: land degradation/desertification (soil erosion resulting from farming of marginal areas, overgrazing, destruction of vegetation); water supplies contaminated by raw sewage; siltation of reservoirs; oil pollution of coastal waters
natural hazards: northern mountains geologically unstable and subject to earthquakes; periodic droughts
international agreements: party to—Endangered Species, Marine Dumping, Nuclear Test Ban, Ship Pollution, Wetlands; signed, but not ratified—Biodiversity, Climate Change, Desertification, Environmental Modification, Law of the Sea, Ozone Layer Protection
Note: strategic location along Strait of Gibraltar

People

Population: 29,168,848 (July 1995 est.)
Age structure:
0-14 years: 38% (female 5,486,176; male 5,659,410)
15-64 years: 58% (female 8,456,525; male 8,327,560)
65 years and over: 4% (female 641,236; male 597,941) (July 1995 est.)
Population growth rate: 2.09% (1995 est.)
Birth rate: 27.93 births/1,000 population (1995 est.)
Death rate: 5.97 deaths/1,000 population (1995 est.)
Net migration rate: -1.08 migrant(s)/1,000 population (1995 est.)
Infant mortality rate: 45.8 deaths/1,000 live births (1995 est.)
Life expectancy at birth:
total population: 68.98 years
male: 67.03 years
female: 71.02 years (1995 est.)
Total fertility rate: 3.69 children born/woman (1995 est.)
Nationality:
noun: Moroccan(s)
adjective: Moroccan
Ethnic divisions: Arab-Berber 99.1%, other 0.7%, Jewish 0.2%
Religions: Muslim 98.7%, Christian 1.1%, Jewish 0.2%
Languages: Arabic (official), Berber dialects, French often the language of business, government, and diplomacy
Literacy: age 15 and over can read and write (1990)
total population: 50%
male: 61%
female: 38%
Labor force: 7.4 million
by occupation: agriculture 50%, services 26%, industry 15%, other 9% (1985)

Government

Names:
conventional long form: Kingdom of Morocco
conventional short form: Morocco
local long form: Al Mamlakah al Maghribiyah
local short form: Al Maghrib
Digraph: MO
Type: constitutional monarchy
Capital: Rabat
Administrative divisions: 36 provinces and 5 wilayas*; Agadir, Al Hoceima, Assa-Zag, Azilal, Beni Mellal, Ben Slimane, Boulemane, Casablanca*, Chaouen, El Jadida, El Kelaa des Sraghna, Er Rachidia, Essaouira, Es Smara, Fes*, Figuig, Guelmim, Ifrane, Kenitra, Khemisset, Khenifra, Khouribga, Laayoune, Larache, Marrakech*, Meknes*, Nador, Ouarzazate, Oujda, Rabat-Sale*, Safi, Settat, Sidi Kacem, Tanger, Tan-Tan, Taounate, Taroudannt, Tata, Taza, Tetouan, Tiznit
Independence: 2 March 1956 (from France)
National holiday: National Day, 3 March (1961) (anniversary of King Hassan II's accession to the throne)
Constitution: 10 March 1972, revised 4 September 1992
Legal system: based on Islamic law and French and Spanish civil law system; judicial review of legislative acts in Constitutional Chamber of Supreme Court
Suffrage: 21 years of age; universal
Executive branch:
chief of state: King HASSAN II (since 3 March 1961)
head of government: Prime Minister Abdellatif FILALI (since 29 May 1994)
cabinet: Council of Ministers; appointed by the King
Legislative branch: unicameral
Chamber of Representatives (Majlis Nawab): two-thirds elected by direct universal suffrage and one-third by an electoral college of government, professional, and labor representatives; direct, popular elections last held 15 June 1993 (next to be held NA 1999); results—percent of vote by party NA; seats—(333 total, 222 directly elected) USFP 48, IP 43, MP 33, RNI 28, UC 27, PND 14, MNP 14, PPS 6, PDI 3, SAP 2, PA 2, OADP 2; indirect, special interest elections last held 17 September 1993 (next to be held NA 1999); results—percent of vote by party NA; seats—(333 total, 111 indirectly elected) UC 27, MP 18, RNI 13, MNP 11, PND 10, IP 7, Party of Shura and Istiqlal 6, USFP 4, PPS 4, CDT 4, UTM 3, UGTM 2, SAP 2
Judicial branch: Supreme Court
Political parties and leaders:
opposition: Socialist Union of Popular Forces (USFP), Mohammad al-YAZGHI; Istiqlal Party (IP), M'Hamed BOUCETTA; Party of Progress and Socialism (PPS), Ali YATA; Organization of Democratic and Popular Action (OADP), leader NA
pro-government: Constitutional Union (UC), Maati BOUABID; Popular Movement (MP), Mohamed LAENSER; National Democratic Party (PND), Mohamed Arsalane EL-JADIDI; National Popular Movement (MNP), Mahjoubi AHARDANE
independents: National Rally of Independents (RNI), Ahmed OSMAN; Democracy and Istiqlal Party (PDI), leader NA; Action Party (PA), Abdullah SENHAJI; Non-Obedience Candidates (SAP), leader NA
labor unions and community organizations (indirect elections): Democratic Confederation of Labor (CDT), Nabir AMAOUI; General Union of Moroccan Workers (UGTM), Abderrazzak AFILAL; Moroccan Union of Workers (UTM), leader NA; Party of Shura and Istiqlal, leader NA
Member of: ABEDA, ACCT (associate), AfDB, AFESD, AL, AMF, AMU, CCC, EBRD, ECA, FAO, G-77, GATT, IAEA, IBRD, ICAO, ICC, ICFTU, ICRM, IDA, IDB, IFAD, IFC, IFRCS, ILO, IMF, IMO, INTELSAT, INTERPOL, IOC, IOM (observer), ISO, ITU, NAM, OAS (observer), OIC, UN, UNAVEM II, UNCTAD, UNESCO, UNHCR, UNIDO, UPU, WHO, WIPO, WMO, WTO
Diplomatic representation in US:
chief of mission: Ambassador Mohamed BENAISSA
chancery: 1601 21st Street NW, Washington, DC 20009
telephone: [1] (202) 462-7979 through 7982
FAX: [1] (202) 265-0161
consulate(s) general: New York
US diplomatic representation:
chief of mission: Ambassador Marc C. GINSBERG
embassy: 2 Avenue de Marrakech, Rabat
mailing address: PSC 74, Box 003, APO AE 09718
telephone: [212] (7) 76 22 65
FAX: [212] (7) 76 56 61
consulate(s) general: Casablanca
Flag: red with a green pentacle (five-pointed, linear star) known as Solomon's seal in the center of the flag; green is the traditional color of Islam

Economy

Overview: Morocco faces the typical problems of developing countries—restraining government spending, reducing constraints on private activity and foreign trade, and keeping inflation within bounds. Since the early 1980s the government has pursued an economic program toward these objectives with the support of the IMF, the World Bank, and the Paris Club of creditors. The economy has substantial assets to draw on: the world's largest phosphate reserves, diverse agricultural

and fishing resources, a sizable tourist industry, a growing manufacturing sector, and remittances from Moroccans working abroad. A severe drought in 1992-93 depressed economic activity and held down exports. Real GDP contracted by 4.4% in 1992 and 1.1% in 1993. Despite these setbacks, initiatives to relax capital controls, strengthen the banking sector, and privatize state enterprises went forward in 1993-94. Favorable rainfall in 1994 boosted agricultural production by 40%. Servicing the large debt, high unemployment, and vulnerability to external economic forces remain long-term problems for Morocco.
National product: GDP—purchasing power parity—$87.5 billion (1994 est.)
National product real growth rate: 8% (1994 est.)
National product per capita: $3,060 (1994 est.)
Inflation rate (consumer prices): 5.4% (1994)
Unemployment rate: 16% (1994 est.)
Budget:
revenues: $8.1 billion
expenditures: $8.9 billion (1994 est.)
Exports: $4.1 billion (f.o.b., 1994 est.)
commodities: food and beverages 30%, semiprocessed goods 23%, consumer goods 21%, phosphates 17%
partners: EU 70%, Japan 5%, US 4%, Libya 3%, India 2% (1993)
Imports: $7.5 billion (c.i.f., 1994 est.)
commodities: capital goods 24%, semiprocessed goods 22%, raw materials 16%, fuel and lubricants 16%, food and beverages 13%, consumer goods 9%
partners: EC 59%, US 8%, Saudi Arabia 5%, UAE 3%, Russia 2% (1993)
External debt: $20.5 billion (1994 est.)
Industrial production: growth rate 0.1% accounts for 28% of GDP
Electricity:
capacity: 2,620,000 kW
production: 9.9 billion kWh
consumption per capita: 361 kWh (1993)
Industries: phosphate rock mining and processing, food processing, leather goods, textiles, construction, tourism
Agriculture: accounts for 15% of GDP, 50% of employment, and 30% of export value; not self-sufficient in food; cereal farming and livestock raising predominate; barley, wheat, citrus fruit, wine, vegetables, olives
Illicit drugs: illicit producer of hashish; trafficking on the increase for both domestic and international drug markets; shipments of hashish mostly directed to Western Europe; transit point for cocaine from South America destined for Western Europe
Economic aid:
recipient: US commitments, including Ex-Im (FY70-89), $1.3 billion; US commitments, including Ex-Im (1992), $123.6 million; Western (non-US) countries, ODA and OOF bilateral commitments (1970-89), $7.5 billion; OPEC bilateral aid (1979-89), $4.8 billion; Communist countries (1970-89), $2.5 billion
note: $2.8 billion debt canceled by Saudi Arabia (1991); IMF standby agreement worth $13 million; World Bank, $450 million (1991)
Currency: 1 Moroccan dirham (DH) = 100 centimes
Exchange rates: Moroccan dirhams (DH) per US$1—2.892 (January 1995), 9.203 (1994), 9.299 (1993), 8.538 (1992), 8.707 (1991), 8.242 (1990)
Fiscal year: alendar year

Transportation

Railroads:
total: 1,893 km
standard gauge: 1,893 km 1.435-m gauge (974 km electrified; 246 km double track)
Highways:
total: 59,474 km
paved: 29,440 km
unpaved: gravel, crushed stone, improved earth, unimproved earth 30,034 km
Pipelines: crude oil 362 km; petroleum products (abandoned) 491 km; natural gas 241 km
Ports: Agadir, Al Jadida, Casablanca, El Jorf Lasfar, Kenitra, Mohammedia, Nador, Rabat, Safi, Tangier; also Spanish-controlled Ceuta and Melilla
Merchant marine:
total: 38 ships (1,000 GRT or over) totaling 183,951 GRT/273,057 DWT
ships by type: cargo 6, chemical tanker 9, container 2, oil tanker 4, refrigerated cargo 10, roll-on/roll-off cargo 6, short-sea passenger 1
Airports:
total: 74
with paved runways over 3,047 m: 11
with paved runways 2,438 to 3,047 m: 4
with paved runways 1,524 to 2,437 m: 8
with paved runways 914 to 1,523 m: 3
with paved runways under 914 m: 13
with unpaved runways 2,438 to 3,047 m: 1
with unpaved runways 1,524 to 2,438 m: 10
with unpaved runways 914 to 1,523 m: 24

Communications

Telephone system: 280,000 telephones; 10.5 telephones/1,000 persons
local: NA
intercity: good system composed of wire lines, cables, and microwave radio relay links; principal centers are Casablanca and Rabat; secondary centers are Fes, Marrakech, Oujda, Tangier, and Tetouan
international: 5 submarine cables; 2 INTELSAT (Atlantic Ocean) and 1 ARABSAT earth stations; microwave radio relay to Gibraltar, Spain, and Western Sahara; coaxial cable and microwave radio relay to Algeria; microwave radio relay network linking Syria, Jordan, Egypt, Libya, Tunisia, Algeria, and Morocco
Radio:
broadcast stations: AM 20, FM 7, shortwave 0
radios: NA
Television:
broadcast stations: 26 (repeaters 26)
televisions: NA

Defense Forces

Branches: Royal Moroccan Army, Royal Moroccan Navy, Royal Moroccan Air Force, Royal Gendarmerie, Auxiliary Forces
Manpower availability: males age 15-49 7,307,076; males fit for military service 4,637,453; males reach military age (18) annually 323,921 (1995 est.)
Defense expenditures: exchange rate conversion—$1.3 billion, 3.8% of GDP (1994)

Mozambique

Geography

Location: Southern Africa, bordering the Mozambique Channel, between South Africa and Tanzania
Map references: Africa
Area:
total area: 801,590 sq km
land area: 784,090 sq km
comparative area: slightly less than twice the size of California
Land boundaries: total 4,571 km, Malawi 1,569 km, South Africa 491 km, Swaziland 105 km, Tanzania 756 km, Zambia 419 km, Zimbabwe 1,231 km
Coastline: 2,470 km
Maritime claims:
exclusive economic zone: 200 nm
territorial sea: 12 nm
International disputes: none
Climate: tropical to subtropical
Terrain: mostly coastal lowlands, uplands in center, high plateaus in northwest, mountains in west
Natural resources: coal, titanium
Land use:
arable land: 4%
permanent crops: 0%
meadows and pastures: 56%
forest and woodland: 20%
other: 20%
Irrigated land: 1,150 sq km (1989 est.)
Environment:
current issues: civil strife and recurrent drought in the hinterlands have resulted in increased migration to urban and coastal areas with adverse environmental consequences; desertification; pollution of surface and coastal waters
natural hazards: severe droughts and floods occur in central and southern provinces; devastating cyclones
international agreements: party to—Endangered Species, Ozone Layer Protection; signed, but not ratified—Biodiversity, Climate Change, Law of the Sea

People

Population: 18,115,250 (July 1995 est.)
Age structure:
0-14 years: 45% (female 4,069,117; male 4,078,429)
15-64 years: 53% (female 4,882,292; male 4,630,193)
65 years and over: 2% (female 260,057; male 195,162) (July 1995 est.)
Population growth rate: 2.87% (1995 est.)
Birth rate: 44.6 births/1,000 population (1995 est.)
Death rate: 15.94 deaths/1,000 population (1995 est.)
Net migration rate: NA migrant(s)/1,000 population (1995 est.)
note: by the end of 1994, an estimated 1.6 million Mozambican refugees, who fled to Malawi, Zimbabwa, and South Africa in earlier years from the civil war, had returned; an estimated 100,000 refugees remain to be repatriated from those countries
Infant mortality rate: 126 deaths/1,000 live births (1995 est.)
Life expectancy at birth:
total population: 48.95 years
male: 47.04 years
female: 50.92 years (1995 est.)
Total fertility rate: 6.19 children born/woman (1995 est.)
Nationality:
noun: Mozambican(s)
adjective: Mozambican
Ethnic divisions: indigenous tribal groups, Europeans about 10,000, Euro-Africans 35,000, Indians 15,000
Religions: indigenous beliefs 60%, Christian 30%, Muslim 10%
Languages: Portuguese (official), indigenous dialects
Literacy: age 15 and over can read and write (1990)
total population: 33%
male: 45%
female: 21%
Labor force: NA
by occupation: 90% engaged in agriculture

Government

Names:
conventional long form: Republic of Mozambique
conventional short form: Mozambique
local long form: Republica Popular de Mocambique
local short form: Mocambique
Digraph: MZ
Type: republic
Capital: Maputo
Administrative divisions: 10 provinces (provincias, singular—provincia); Cabo Delgado, Gaza, Inhambane, Manica, Maputo, Nampula, Niassa, Sofala, Tete, Zambezia
Independence: 25 June 1975 (from Portugal)
National holiday: Independence Day, 25 June (1975)
Constitution: 30 November 1990
Legal system: based on Portuguese civil law system and customary law
Suffrage: 18 years of age; universal
Executive branch:
chief of state: President Joaquim Alberto CHISSANO (since 6 November 1986)
head of government: Prime Minister Pascoal MOCUMBI (since December 1994)
cabinet: Cabinet
Legislative branch: unicameral
Assembly of the Republic (Assembleia da Republica): draft electoral law provides for periodic, direct presidential and Assembly elections
note: as called for in the 1992 peace accords, presidential and legislative elections took place during 27-29 October 1994; fourteen parties, including the Mozambique National Resistance (RENAMO) participated; Joaquim Alberto CHISSANO was elected president and his FRELIMO party gathered a slim majority in the 250 seat legislature
Judicial branch: Supreme Court
Political parties and leaders: Front for the Liberation of Mozambique (FRELIMO), Joaquim Alberto CHISSANO, chairman; the ruling party since independence, FRELIMO was the only legal party before 30 November 1990 when the new Constitution went into effect establishing a multiparty system
Member of: ACP, AfDB, CCC, ECA, FAO, FLS, G-77, GATT, IBRD, ICAO, ICRM, IDA, IFAD, IFC, IFRCS, ILO, IMF, IMO, INMARSAT, INTELSAT, INTERPOL, IOC, IOM (observer), ITU, NAM, OAU, OIC, SADC, UN, UNCTAD, UNESCO, UNIDO, UPU, WFTU, WHO, WMO
Diplomatic representation in US:
chief of mission: Ambassador Hipolito Pereira Zozimo PATRICIO
chancery: Suite 570, 1990 M Street NW, Washington, DC 20036
telephone: [1] (202) 293-7146
FAX: [1] (202) 835-0245
US diplomatic representation:
chief of mission: Ambassador Dennis Coleman JETT
embassy: Avenida Kenneth Kuanda, 193 Maputo
mailing address: P. O. Box 783, Maputo
telephone: [258] (1) 492797
FAX: [258] (1) 490114
Flag: three equal horizontal bands of green (top), black, and yellow with a red isosceles triangle based on the hoist side; the black band is edged in white; centered in the triangle is a yellow five-pointed star bearing a crossed rifle and hoe in black superimposed on an open white book

Economy

Overview: One of Africa's poorest countries, Mozambique has failed to exploit the economic potential of its sizable agricultural, hydropower, and transportation resources. Indeed, national output, consumption, and investment declined throughout the first half of the 1980s because of internal disorders, lack of government administrative control, and a growing foreign debt. A sharp increase in foreign aid, attracted by an economic reform policy, resulted in successive years of economic growth in the late 1980s, but aid has declined steadily since 1989. Agricultural output is at only 75% of its 1981 level, and grain has to be imported. Industry operates at only 20%-40% of capacity. The economy depends heavily on foreign assistance to keep afloat. Peace accords signed in October 1992 improved chances of foreign investment, aided IMF-supported economic reforms, and supported continued economic recovery. Elections held in 1994 diverted government attention from the economy, resulting in slippage and delays in the economic reform program. Nonetheless, growth in 1994 was solid and can continue into the late 1990s given continued foreign help in meeting debt obligations.
National product: GDP—purchasing power parity—$10.6 billion (1994 est.)
National product real growth rate: 5.8% (1994 est.)
National product per capita: $610 (1994 est.)
Inflation rate (consumer prices): 50% (1994 est.)
Unemployment rate: 50% (1989 est.)
Budget:
revenues: $252 million
expenditures: $607 million, including capital expenditures of $NA (1992 est.)
Exports: $150 million (f.o.b., 1994 est.)
commodities: shrimp 40%, cashews, cotton, sugar, copra, citrus
partners: Spain, South Africa, US, Portugal, Japan
Imports: $1.14 billion (c.i.f., 1994 est.)
commodities: food, clothing, farm equipment, petroleum
partners: South Africa, UK, France, Japan, Portugal
External debt: $5 billion (1992 est.)
Industrial production: growth rate 5% (1989 est.)
Electricity:
capacity: 2,360,000 kW
production: 1.7 billion kWh
consumption per capita: 58 kWh (1993)
Industries: food, beverages, chemicals (fertilizer, soap, paints), petroleum products, textiles, nonmetallic mineral products (cement, glass, asbestos), tobacco
Agriculture: accounts for 50% of GDP and about 90% of exports; cash crops—cotton, cashew nuts, sugarcane, tea, shrimp; other crops—cassava, corn, rice, tropical fruits; not self-sufficient in food
Economic aid:
recipient: US commitments, including Ex-Im (FY70-89), $350 million; Western (non-US) countries, ODA and OOF bilateral commitments (1970-89), $4.4 billion; OPEC bilateral aid (1979-89), $37 million; Communist countries (1970-89), $890 million
Currency: 1 metical (Mt) = 100 centavos
Exchange rates: meticais (Mt) per US$1—5,220.63 (1st quarter 1994), 3,874.24 (1993), 2,550.40 (1992), 1,763.99 (1991), 1,053.09 (1990)
Fiscal year: calendar year

Transportation

Railroads: *total:* 3,288 km
narrow gauge: 3,140 km 1.067-m gauge; 148 km 0.762-m gauge
Highways:
total: 26,498 km
paved: 4,593 km
unpaved: gravel, crushed stone, stabilized earth 829 km; unimproved earth 21,076 km
Inland waterways: about 3,750 km of navigable routes
Pipelines: crude oil (not operating) 306 km; petroleum products 289 km
Ports: Beira, Inhambane, Maputo, Nacala, Pemba
Merchant marine:
total: 3 cargo ships (1,000 GRT or over) totaling 4,533 GRT/8,024 DWT
Airports:
total: 192
with paved runways over 3,047 m: 1
with paved runways 2,438 to 3,047 m: 4
with paved runways 1,524 to 2,437 m: 11
with paved runways 914 to 1,523 m: 5
wiıh paved runways under 914 m: 112
with unpaved runways 1,524 to 2,438 m: 15
with unpaved runways 914 to 1,523 m: 44
Note:
note: highway traffic impeded by land mines not removed at end of civil war

Communications

Telephone system: NA telephone density; fair system of troposcatter, open-wire lines, and radio relay
local: NA
intercity: microwave radio relay and tropospheric scatter
international: 5 INTELSAT (2 Atlantic Ocean and 3 Indian Ocean) earth stations
Radio:
broadcast stations: AM 29, FM 4, shortwave 0
radios: NA
Television:
broadcast stations: 1
televisions: NA

Defense Forces

Branches: Army, Naval Command, Air and Air Defense Forces, Militia; note—by late 1994, the army and former RENAMO rebels had demobilized; under UN supervision and training, recruits from both the army and rebel forces joined an integrated force that is still forming
Manpower availability: males age 15-49 4,061,109; males fit for military service 2,331,793 (1995 est.)
Defense expenditures: exchange rate conversion—$110 million, 7.3% of GDP (1993)

Namibia

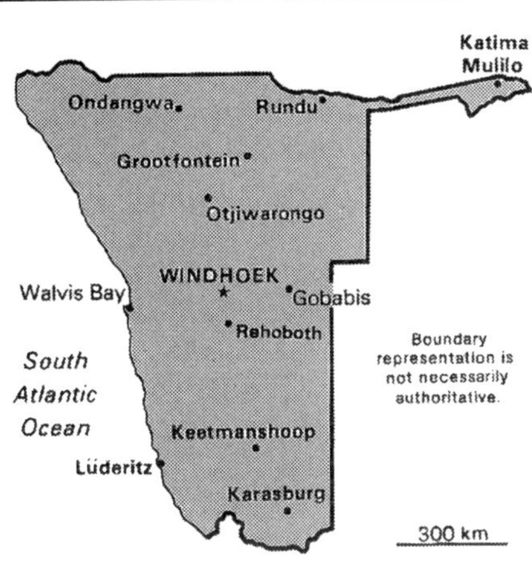

Geography

Location: Southern Africa, bordering the South Atlantic Ocean, between Angola and South Africa
Map references: Africa
Area:
total area: 825,418 sq km
land area: 825,418 sq km
comparative area: slightly more than half the size of Alaska
Land boundaries: total 3,824 km, Angola 1,376 km, Botswana 1,360 km, South Africa 855 km, Zambia 233 km
Coastline: 1,572 km
Maritime claims:
contiguous zone: 24 nm
exclusive economic zone: 200 nm
territorial sea: 12 nm
International disputes: short section of boundary with Botswana is indefinite; quadripoint with Botswana, Zambia, and Zimbabwe is in disagreement; dispute with Botswana over uninhabited Kasikili (Sidudu) Island in Linyanti (Chobe) River remained unresolved in mid-February 1995 and the parties agreed to refer the matter to the International Court of Justice;
Climate: desert; hot, dry; rainfall sparse and erratic
Terrain: mostly high plateau; Namib Desert along coast; Kalahari Desert in east
Natural resources: diamonds, copper, uranium, gold, lead, tin, lithium, cadmium, zinc, salt, vanadium, natural gas, fish; suspected deposits of oil, natural gas, coal, iron ore
Land use:
arable land: 1%
permanent crops: 0%
meadows and pastures: 64%
forest and woodland: 22%
other: 13%
Irrigated land: 40 sq km (1989 est.)
Environment:
current issues: very limited natural fresh water resources; desertification
natural hazards: prolonged periods of drought
international agreements: party to—Endangered Species, Law of the Sea, Ozone Layer Protection; signed, but not ratified—Biodiversity, Climate Change

People

Population: 1,651,545 (July 1995 est.)
Age structure:
0-14 years: 47% (female 384,885; male 394,216)
15-64 years: 50% (female 414,283; male 405,938)
65 years and over: 3% (female 26,783; male 25,440) (July 1995 est.)
Population growth rate: 3.44% (1995 est.)
Birth rate: 43.04 births/1,000 population (1995 est.)
Death rate: 8.61 deaths/1,000 population (1995 est.)
Net migration rate: 0 migrant(s)/1,000 population (1995 est.)
Infant mortality rate: 59.8 deaths/1,000 live births (1995 est.)
Life expectancy at birth:
total population: 62.1 years
male: 59.37 years
female: 64.9 years (1995 est.)
Total fertility rate: 6.34 children born/woman (1995 est.)
Nationality:
noun: Namibian(s)
adjective: Namibian
Ethnic divisions: black 86%, white 6.6%, mixed 7.4%
note: about 50% of the population belong to the Ovambo tribe and 9% to the Kavangos tribe; other ethnic groups include (with approximate share of total population): Herero 7%, Damara 7%, Nama 5%, Caprivian 4%, Bushmen 3%, Baster 2%, Tswana 0.5%
Religions: 80%-90% Christian (50% Lutheran; at least 30% other Christian denominations)
Languages: English 7% (official), Afrikaans common language of most of the population and about 60% of the white population, German 32%, indigenous languages: Oshivambo, Herero, Nama
Literacy: age 15 and over can read and write (1960)
total population: 38%
male: 45%
female: 31%
Labor force: 500,000
by occupation: agriculture 60%, industry and commerce 19%, services 8%, government 7%, mining 6% (1981 est.)

Government

Names:
conventional long form: Republic of Namibia
conventional short form: Namibia
Digraph: WA
Type: republic
Capital: Windhoek
Administrative divisions: 13 districts; Erongo, Hardap, Karas, Khomas, Kunene, Caprivi (Liambezi), Ohangwena, Okavango, Omaheke, Omusati, Oshana, Oshikoto, Otjozondjupa
Independence: 21 March 1990 (from South African mandate)
National holiday: Independence Day, 21 March (1990)
Constitution: ratified 9 February 1990; effective 12 March 1990
Legal system: based on Roman-Dutch law and 1990 constitution
Suffrage: 18 years of age; universal
Executive branch:
chief of state and head of government: President Sam NUJOMA (since 21 March 1990); election last held 7-8 December 1994 (next to be held NA); results—Sam NUJOMA elected president by popular vote
cabinet: Cabinet; appointed by the president from the National Assembly
Legislative branch: bicameral legislature
National Council: elections last held 30 November-3 December 1992 (next to be held by December 1998); results—percent of vote by party NA; seats—(26 total) SWAPO 19, DTA 6, UDF 1
National Assembly: elections last held on 7-8 December 1994 (next to be held NA); results—percent of vote by party NA; seats—(72 total) SWAPO 53, DTA 15, UDF 2, MAG 1, DCN 1
Judicial branch: Supreme Court
Political parties and leaders: South West Africa People's Organization (SWAPO), Sam NUJOMA; DTA of Namibia (formerly Democratic Turnhalle Alliance) (DTA), Mishake MUYONGO; United Democratic Front (UDF), Justus GAROEB; Federal Convention of Namibia (FCN), Kephics CONRUDIE; Monitor Action Group (MAG), Kosie PRETORIUS; Workers Revolutionary Party (WRP); Southwest African National Union (SWANU), Hitjevi VEII; Democratic Coalition of Namibia (DCN), Moses KATJIUONGA
Other political or pressure groups: NA
Member of: ACP, AfDB, C, CCC, ECA, FAO, FLS, G-77, GATT, IAEA, IBRD, ICAO, ICRM, IFAD, IFC, IFRCS (associate), ILO, IMF, INTELSAT (nonsignatory user), INTERPOL, IOC, IOM (observer), ITU, NAM, OAU, SACU, SADC, UN, UNCTAD, UNESCO, UNHCR, UNIDO, UPU, WCL, WHO, WIPO, WMO
Diplomatic representation in US:
chief of mission: Ambassador Tuliameni KALOMOH
chancery: 1605 New Hampshire Avenue NW, Washington, DC 20009
telephone: [1] (202) 986-0540
FAX: [1] (202) 986-0443

US diplomatic representation:
chief of mission: Ambassador Marshall F. McCALLIE
embassy: Ausplan Building, 14 Lossen St., Windhoek
mailing address: Private Bag 12029 Ausspannplatz, Windhoek
telephone: [264] (61) 221601
FAX: [264] (61) 229792
Flag: a large blue triangle with a yellow sunburst fills the upper left section, and an equal green triangle (solid) fills the lower right section; the triangles are separated by a red stripe that is contrasted by two narrow white-edge borders

Economy

Overview: The economy is heavily dependent on the mining industry to extract and process minerals for export. Mining accounts for almost 25% of GDP. Namibia is the fourth-largest exporter of nonfuel minerals in Africa and the world's fifth-largest producer of uranium. Alluvial diamond deposits are among the richest in the world, making Namibia a primary source for gem-quality diamonds. Namibia also produces large quantities of lead, zinc, tin, silver, and tungsten. More than half the population depends on agriculture (largely subsistence agriculture) for its livelihood. Namibia must import some of its food.
National product: GDP—purchasing power parity—$5.8 billion (1994 est.)
National product real growth rate: 5.8% (1994 est.)
National product per capita: $3,600 (1994 est.)
Inflation rate (consumer prices): 11% (1994)
Unemployment rate: 35% in urban areas (1993 est.)
Budget:
revenues: $941 million
expenditures: $1.05 billion, including capital expenditures of $157 million (FY93/94)
Exports: $1.3 billion (f.o.b., 1993 est.)
commodities: diamonds, copper, gold, zinc, lead, uranium, cattle, processed fish, karakul skins
partners: Switzerland, South Africa, Germany, Japan
Imports: $1.1 billion (f.o.b., 1993 est.)
commodities: foodstuffs, petroleum products and fuel, machinery and equipment
partners: South Africa, Germany, US, Switzerland
External debt: about $385 million (1994 est.)
Industrial production: growth rate -14% (1993); accounts for 30% of GDP, including mining
Electricity:
capacity: 406,000 kW
production: 1.29 billion kWh
consumption per capita: 658 kWh (1991)
Industries: meat packing, fish processing, dairy products, mining (copper, lead, zinc, diamond, uranium)
Agriculture: accounts for 10% of GDP; livestock raising major source of cash income; crops—millet, sorghum, peanuts; fish catch potential of over 1 million metric tons not being fulfilled
Economic aid:
recipient: Western (non-US) countries, ODA and OOF bilateral commitments (1970-87), $47.2 million
Currency: 1 South African rand (R) = 100 cents
Exchange rates: South African rand (R) per US$1—3.539 (January 1995), 3.5489 (1994), 3.2678 (1993), 2.8497 (1992), 2.7653 (1991), 2.5863 (1990)
Fiscal year: 1 April—31 March

Transportation

Railroads: *total:* 2,341 km (single track)
narrow gauge: 2,341 km 1.067-m gauge
Highways:
total: 54,500 km
paved: 4,080 km
unpaved: gravel 2,540 km; earth 47,880 km (roads and tracks)
Ports: Luderitz, Walvis Bay
Merchant marine: none
Airports:
total: 135
with paved runways over 3,047 m: 2
with paved runways 2,438 to 3,047 m: 2
with paved runways 1,524 to 2,437 m: 14
with paved runways 914 to 1,523 m: 3
with paved runways under 914 m: 20
with unpaved runways 2,438 to 3,047 m: 1
with unpaved runways 1,524 to 2,438 m: 23
with unpaved runways 914 to 1,523 m: 70

Communications

Telephone system: 62,800 telephones; telephone density—38/1,000 persons
local: good urban services
intercity: fair rural service; microwave radio relay links major towns; connections to other populated places are by open wire
international: NA
Radio:
broadcast stations: AM 4, FM 40, shortwave 0
radios: NA
Television:
broadcast stations: 3
televisions: NA

Defense Forces

Branches: National Defense Force (Army), Police
Manpower availability: males age 15-49 348,380; males fit for military service 206,684 (1995 est.)
Defense expenditures: exchange rate conversion—$54 million, 2% of GDP (FY93/94)

Nauru

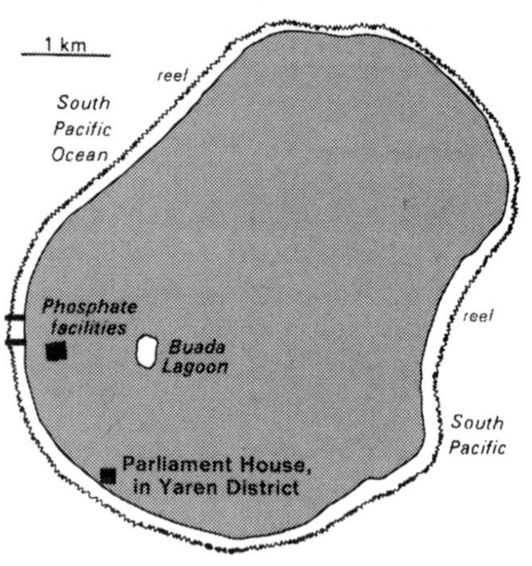

Geography

Location: Oceania, island in the South Pacific Ocean, south of the Marshall Islands
Map references: Oceania
Area:
total area: 21 sq km
land area: 21 sq km
comparative area: about one-tenth the size of Washington, DC
Land boundaries: 0 km
Coastline: 30 km
Maritime claims:
exclusive fishing zone: 200 nm
territorial sea: 12 nm
International disputes: none
Climate: tropical; monsoonal; rainy season (November to February)
Terrain: sandy beach rises to fertile ring around raised coral reefs with phosphate plateau in center
Natural resources: phosphates
Land use:
arable land: 0%
permanent crops: 0%
meadows and pastures: 0%
forest and woodland: 0%
other: 100%
Irrigated land: NA sq km
Environment:
current issues: limited natural fresh water resources, roof storage tanks collect rainwater; phosphate mining threatens limited remaining land resources
natural hazards: periodic droughts
international agreements: party to—Biodiversity, Climate Change, Marine Dumping; signed, but not ratified—Law of the Sea
Note: Nauru is one of the three great phosphate rock islands in the Pacific Ocean—the others are Banaba (Ocean Island) in Kiribati and Makatea in French Polynesia; only 53 km south of Equator

People

Population: 10,149 (July 1995 est.)
Age structure:
0-14 years: NA
15-64 years: NA
65 years and over: NA
Population growth rate: 1.33% (1995 est.)
Birth rate: 18.03 births/1,000 population (1995 est.)
Death rate: 5.1 deaths/1,000 population (1995 est.)
Net migration rate: 0.4 migrant(s)/1,000 population (1995 est.)
Infant mortality rate: 40.6 deaths/1,000 live births (1995 est.)
Life expectancy at birth:
total population: 66.68 years
male: 64.3 years
female: 69.18 years (1995 est.)
Total fertility rate: 2.08 children born/woman (1995 est.)
Nationality:
noun: Nauruan(s)
adjective: Nauruan
Ethnic divisions: Nauruan 58%, other Pacific Islander 26%, Chinese 8%, European 8%
Religions: Christian (two-thirds Protestant, one-third Roman Catholic)
Languages: Nauruan (official; a distinct Pacific Island language), English widely understood, spoken, and used for most government and commercial purposes
Literacy: NA%
Labor force:
by occupation: NA

Government

Names:
conventional long form: Republic of Nauru
conventional short form: Nauru
former: Pleasant Island
Digraph: NR
Type: republic
Capital: no official capital; government offices in Yaren District
Administrative divisions: 14 districts; Aiwo, Anabar, Anetan, Anibare, Baiti, Boe, Buada, Denigomodu, Ewa, Ijuw, Meneng, Nibok, Uaboe, Yaren
Independence: 31 January 1968 (from Australia, New Zealand, and UK—administered UN trusteeship)
National holiday: Independence Day, 31 January (1968)
Constitution: 29 January 1968
Legal system: own Acts of Parliament and British common law
Suffrage: 20 years of age; universal and compulsory
Executive branch:
chief of state and head of government: President Bernard DOWIYOGO (since 12 December 1989); election last held 19 November 1992 (next to be held NA November 1995); results—Bernard DOWIYOGO elected by Parliament
cabinet: Cabinet; appointed by the president from the parliament
Legislative branch: unicameral
Parliament: elections last held on 14 November 1992 (next to be held NA November 1995); results—percent of vote NA; seats—(18 total) independents 18
Judicial branch: Supreme Court
Political parties and leaders: none
Member of: AsDB, C (special), ESCAP, ICAO, INTELSAT (nonsignatory user), INTERPOL, ITU, SPARTECA, SPC, SPF, UPU
Diplomatic representation in US:
consulate(s): Agana (Guam)
US diplomatic representation: the US Ambassador to Fiji is accredited to Nauru
Flag: blue with a narrow, horizontal, yellow stripe across the center and a large white 12-pointed star below the stripe on the hoist side; the star indicates the country's location in relation to the Equator (the yellow stripe) and the 12 points symbolize the 12 original tribes of Nauru

Economy

Overview: Revenues come from the export of phosphates, the reserves of which are expected to be exhausted by the year 2000. Phosphates have given Nauruans one of the highest per capita incomes in the Third World. Few other resources exist, so most necessities must be imported, including fresh water from Australia. The rehabilitation of mined land and the replacement of income from phosphates are serious long-term problems. Substantial amounts of phosphate income are invested in trust funds to help cushion the transition.
National product: GDP—purchasing power parity—$100 million (1993 est.)
National product real growth rate: NA%
National product per capita: $10,000 (1993 est.)
Inflation rate (consumer prices): NA%
Unemployment rate: 0%
Budget:
revenues: $69.7 million
expenditures: $51.5 million, including capital expenditures of $NA (1986 est.)
Exports: $93 million (f.o.b., 1984)
commodities: phosphates
partners: Australia, NZ
Imports: $73 million (c.i.f., 1984)
commodities: food, fuel, manufactures, building materials, machinery
partners: Australia, UK, NZ, Japan
External debt: $33.3 million
Industrial production: growth rate NA%
Electricity:
capacity: 14,000 kW
production: 30 million kWh
consumption per capita: 3,036 kWh (1993)
Industries: phosphate mining, financial services, coconut products
Agriculture: coconuts; other agricultural activity negligible; almost completely dependent on imports for food and water
Economic aid:
recipient: Western (non-US) countries (1970-89), $2 million
Currency: 1 Australian dollar ($A) = 100 cents
Exchange rates: Australian dollars ($A) per US$1—1.3058 (January 1995), 1.3667 (1994), 1.4704 (1993), 1.3600 (1992), 1.2834 (1991), 1.2799 (1990)
Fiscal year: 1 July—30 June

Transportation

Railroads:
total: 3.9 km; note—used to haul phosphates from the center of the island to processing facilities on the southwest coast
Highways:
total: 27 km
paved: 21 km
unpaved: improved earth 6 km
Ports: Nauru
Merchant marine: none
Airports:
total: 1
with paved runways 1,524 to 2,437 m: 1

Communications

Telephone system: 1,600 telephones; adequate local and international radio communications provided via Australian facilities
local: NA
intercity: NA
international: 1 INTELSAT (Pacific Ocean) earth station
Radio:
broadcast stations: AM 1, FM 0, shortwave 0
radios: 4,000
Television:
broadcast stations: 0
televisions: NA

Defense Forces

Branches: no regular armed forces; Directorate of the Nauru Police Force
Defense expenditures: $NA; note—no formal defense structure

Navassa Island
(territory of the US)

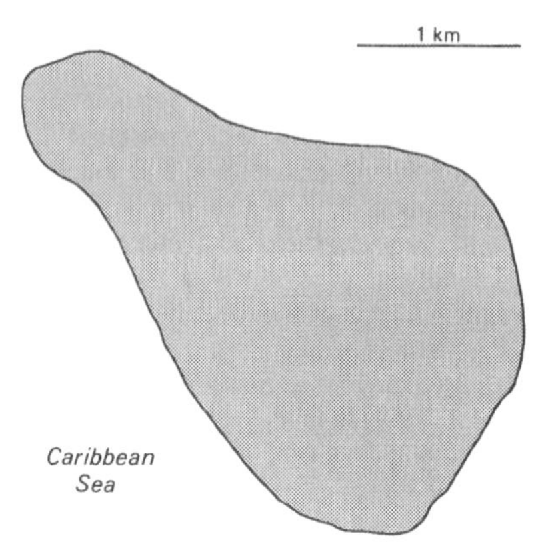

Geography

Location: Caribbean, island in the Caribbean Sea, about one-fourth of the way from Haiti to Jamaica
Map references: Central America and the Caribbean
Area:
total area: 5.2 sq km
land area: 5.2 sq km
comparative area: about nine times the size of The Mall in Washington, DC
Land boundaries: 0 km
Coastline: 8 km
Maritime claims:
exclusive economic zone: 200 nm
territorial sea: 12 nm
International disputes: claimed by Haiti
Climate: marine, tropical
Terrain: raised coral and limestone plateau, flat to undulating; ringed by vertical white cliffs (9 to 15 meters high)
Natural resources: guano
Land use:
arable land: 0%
permanent crops: 0%
meadows and pastures: 10%
forest and woodland: 0%
other: 90%
Irrigated land: 0 sq km
Environment:
current issues: NA
natural hazards: NA
international agreements: NA
Note: strategic location 160 km south of the US Naval Base at Guantanamo Bay, Cuba; mostly exposed rock, but enough grassland to support goat herds; dense stands of fig-like trees, scattered cactus

People

Population: uninhabited; note—transient Haitian fishermen and others camp on the island

Government

Names:
conventional long form: none
conventional short form: Navassa Island
Digraph: BQ
Type: unincorporated territory of the US administered by the US Coast Guard
Capital: none; administered from Washington, DC

Economy

Overview: no economic activity

Transportation

Ports: none; offshore anchorage only

Defense Forces

Note: defense is the responsibility of the US

Nepal

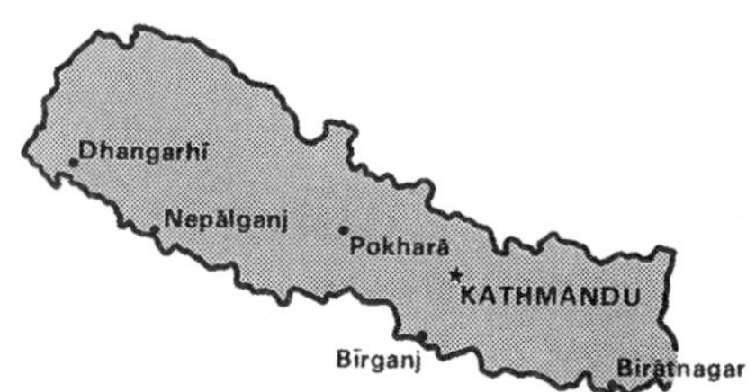

Geography

Location: Southern Asia, between China and India
Map references: Asia
Area:
total area: 140,800 sq km
land area: 136,800 sq km
comparative area: slightly larger than Arkansas
Land boundaries: total 2,926 km, China 1,236 km, India 1,690 km
Coastline: 0 km (landlocked)
Maritime claims: none; landlocked
International disputes: none
Climate: varies from cool summers and severe winters in north to subtropical summers and mild winters in south
Terrain: Terai or flat river plain of the Ganges in south, central hill region, rugged Himalayas in north
Natural resources: quartz, water, timber, hydroelectric potential, scenic beauty, small deposits of lignite, copper, cobalt, iron ore
Land use:
arable land: 17%
permanent crops: 0%
meadows and pastures: 13%
forest and woodland: 33%
other: 37%
Irrigated land: 9,430 sq km (1989)
Environment:
current issues: the almost total dependence on wood for fuel and cutting down trees to expand agricultural land without replanting has resulted in widespread deforestation; soil erosion; water pollution (use of contaminated water presents human health risks)
natural hazards: severe thunderstorms, flooding, landslides, drought, and famine depending on the timing, intensity, and duration of the summer monsoons
international agreements: party to—Biodiversity, Climate Change, Endangered Species, Nuclear Test Ban, Ozone Layer Protection, Tropical Timber 83, Wetlands; signed, but not ratified—Law of the Sea,

Marine Dumping, Marine Life Conservation
Note: landlocked; strategic location between China and India; contains eight of world's 10 highest peaks

People

Population: 21,560,869 (July 1995 est.)
Age structure:
0-14 years: 43% (female 4,479,950; male 4,692,575)
15-64 years: 55% (female 5,778,107; male 5,994,147)
65 years and over: 2% (female 305,502; male 310,588) (July 1995 est.)
Population growth rate: 2.44% (1995 est.)
Birth rate: 37.31 births/1,000 population (1995 est.)
Death rate: 12.9 deaths/1,000 population (1995 est.)
Net migration rate: 0 migrant(s)/1,000 population (1995 est.)
Infant mortality rate: 81.2 deaths/1,000 live births (1995 est.)
Life expectancy at birth:
total population: 53.09 years
male: 52.86 years
female: 53.34 years (1995 est.)
Total fertility rate: 5.15 children born/woman (1995 est.)
Nationality:
noun: Nepalese (singular and plural)
adjective: Nepalese
Ethnic divisions: Newars, Indians, Tibetans, Gurungs, Magars, Tamangs, Bhotias, Rais, Limbus, Sherpas
Religions: Hindu 90%, Buddhist 5%, Muslim 3%, other 2% (1981)
note: only official Hindu state in world, although no sharp distinction between many Hindu and Buddhist groups
Languages: Nepali (official), 20 languages divided into numerous dialects
Literacy: age 15 and over can read and write (1990)
total population: 26%
male: 38%
female: 13%
Labor force: 8.5 million (1991 est.)
by occupation: agriculture 93%, services 5%, industry 2%
note: severe lack of skilled labor

Government

Names:
conventional long form: Kingdom of Nepal
conventional short form: Nepal
Digraph: NP
Type: parliamentary democracy as of 12 May 1991
Capital: Kathmandu
Administrative divisions: 14 zones (anchal, singular and plural); Bagmati, Bheri, Dhawalagiri, Gandaki, Janakpur, Karnali, Kosi, Lumbini, Mahakali, Mechi, Narayani, Rapti, Sagarmatha, Seti
Independence: 1768 (unified by Prithvi Narayan Shah)
National holiday: Birthday of His Majesty the King, 28 December (1945)
Constitution: 9 November 1990
Legal system: based on Hindu legal concepts and English common law; has not accepted compulsory ICJ jurisdiction
Suffrage: 18 years of age; universal
Executive branch:
head of government: Prime Minister Man Mohan ADHIKARI (since 30 November 1994)
chief of state: King BIRENDRA Bir Bikram Shah Dev (since 31 January 1972, crowned King 24 February 1985); Heir Apparent Crown Prince DIPENDRA Bir Bikram Shah Dev, son of the King (born 21 June 1971)
cabinet: Cabinet; appointed by the king on recommendation of the prime minister
Legislative branch: bicameral Parliament
National Council: consists of a 60-member body, 50 appointed by House of Representatives and 10 by the King
House of Representatives: elections last held on 15 November 1994 (next to be held NA); results—NCP 33%, CPN/UML 31%, NDP 18%, Terai Rights Sadbhavana Party 3%, NWPP 1%; seats—(205 total) CPN/UML 88, NCP 83, NDP 20, NWPP 4, Terai Rights Sadbhavana Party 3, independents 7; note—the new Constitution of 9 November 1990 gave Nepal a multiparty democracy system for the first time in 32 years
Judicial branch: Supreme Court (Sarbochha Adalat)
Political parties and leaders: Communist Party of Nepal/United Marxist and Leninist (CPN/UML), Prime Minister Man Mohan ADHIKARI, Deputy Prime Minister Madhav Kumar NEPAL; Nepali Congress Party (NCP), president Krishna Prasad BHATTARAI, former Prime Minister Girija Prasad KOIRALA, Leader of the Opposition Sher Bahadur DEUBA; National Democratic Party (NDP), Surya Bahadur THAPA; Terai Rights Sadbhavana (Goodwill) Party, Gajendra Narayan SINGH; United People's Front (UPF), Niranjan Govinda BAIDYA; Nepal Workers and Peasants Party (NWPP), Narayan Man BIJUKCHHE; Communist Party of Nepal (Democratic-Manandhar), B. B. MANANDHAR
Other political or pressure groups: numerous small, left-leaning student groups in the capital; several small, radical Nepalese antimonarchist groups
Member of: AsDB, CCC, CP, ESCAP, FAO, G-77, IBRD, ICAO, ICRM, IDA, IFAD, IFC, IFRCS, ILO, IMF, IMO, INTELSAT, INTERPOL, IOC, ISO (correspondent), ITU, NAM, SAARC, UN, UNCTAD, UNESCO, UNIDO, UNIFIL, UNOSOM, UNPROFOR, UPU, WFTU, WHO, WMO, WTO
Diplomatic representation in US:
chief of mission: (vacant); Charge d'Affaires ad interim Pradeep KHATIWADA
chancery: 2131 Leroy Place NW, Washington, DC 20008
telephone: [1] (202) 667-4550
consulate(s) general: New York
US diplomatic representation:
chief of mission: Ambassador Sandra L. VOGELGESANG
embassy: Pani Pokhari, Kathmandu
mailing address: use embassy street address
telephone: [977] (1) 411179
FAX: [977] (1) 419963
Flag: red with a blue border around the unique shape of two overlapping right triangles; the smaller, upper triangle bears a white stylized moon and the larger, lower triangle bears a white 12-pointed sun

Economy

Overview: Nepal is among the poorest and least developed countries in the world. Agriculture is the mainstay of the economy, providing a livelihood for over 90% of the population and accounting for half of GDP. Industrial activity is limited, mainly involving the processing of agricultural produce (jute, sugarcane, tobacco, and grain). Production of textiles and carpets has expanded recently and accounted for 85% of foreign exchange earnings in FY93/94. Apart from agricultural land and forests, exploitable natural resources are mica, hydropower, and tourism. Agricultural production in the late 1980s grew by about 5%, as compared with annual population growth of 2.6%. More than 40% of the population is undernourished. Since May 1991, the government has been encouraging trade and foreign investment, e.g., by eliminating business licenses and registration requirements in order to simplify domestic and foreign investment. The government also has been cutting public expenditures by reducing subsidies, privatizing state industries, and laying off civil servants. Prospects for foreign trade and investment in the 1990s remain poor, however, because of the small size of the economy, its technological backwardness, its remoteness, and susceptibility to natural disaster. The international community provides funding for 70% of Nepal's development budget and for 30% of total budgetary expenditures. The government, realizing that attempts to reverse three years of liberalization would jeopardize this vital support, almost certainly will move ahead with its reform program in 1995-96.
National product: GDP—purchasing power parity—$22.4 billion (1994 est.)
National product real growth rate: 5% (1994 est.)
National product per capita: $1,060 (1994 est.)
Inflation rate (consumer prices): 9.6% (June 1994)
Unemployment rate: NA%; note—there is substantial underemployment (1994)

Budget:
revenues: $455 million
expenditures: $854 million, including capital expenditures of $427 million (FY93/94 est.)
Exports: $593 million (f.o.b., 1993) but does not include unrecorded border trade with India
commodities: carpets, clothing, leather goods, jute goods, grain
partners: India, US, Germany, UK
Imports: $899 million (c.i.f., 1993)
commodities: petroleum products 20%, fertilizer 11%, machinery 10%
partners: India, Singapore, Japan, Germany
External debt: $2 billion (1993 est.)
Industrial production: NA
Electricity:
capacity: 280,000 kW
production: 920 million kWh
consumption per capita: 41 kWh (1993)
Industries: small rice, jute, sugar, and oilseed mills; cigarette, textile, carpet, cement, and brick production; tourism
Agriculture: rice, corn, wheat, sugarcane, root crops, milk, buffalo meat; not self-sufficient in food, particularly in drought years
Illicit drugs: illicit producer of cannabis for the domestic and international drug markets; transit point for heroin from Southeast Asia to the West
Economic aid:
recipient: US commitments, including Ex-Im (FY70-89), $304 million; Western (non-US) countries, ODA and OOF bilateral commitments (1980-89), $2.23 billion; OPEC bilateral aid (1979-89), $30 million; Communist countries (1970-89), $286 million
Currency: 1 Nepalese rupee (NR) = 100 paisa
Exchange rates: Nepalese rupees (NRs) per US$1—49.884 (January 1995), 49.398 (1994), 48.607 (1993), 42.742 (1992), 37.255 (1991), 29.370 (1990)
Fiscal year: 16 July—15 July

Transportation

Railroads:
total: 101 km; note—all in Terai close to Indian border
narrow gauge: 101 km 0.762-m gauge
Highways:
total: 7,400 km
paved: 3,000 km
unpaved: 4,400 km
Ports: none
Airports:
total: 44
with paved runways over 3,047 m: 1
with paved runways 1,524 to 2,437 m: 3
with paved runways 914 to 1,523 m: 1
with paved runways under 914 m: 28
with unpaved runways 1,524 to 2,438 m: 1
with unpaved runways 914 to 1,523 m: 10

Communications

Telephone system: 50,000 telephones (1990); poor telephone and telegraph service; fair radio communication
local: NA
intercity: NA
international: international radio communication service is fair; 1 INTELSAT (Indian Ocean) earth station
Radio:
broadcast stations: AM 88, FM 0, shortwave 0
radios: NA
Television:
broadcast stations: 1
televisions: NA

Defense Forces

Branches: Royal Nepalese Army, Royal Nepalese Army Air Service, Nepalese Police Force
Manpower availability: males age 15-49 5,163,703; males fit for military service 2,682,284; males reach military age (17) annually 247,978 (1995 est.)
Defense expenditures: exchange rate conversion—$36 million, 1.2% of GDP (FY92/93)

Netherlands

Geography

Location: Western Europe, bordering the North Sea, between Belgium and Germany
Map references: Europe
Area:
total area: 37,330 sq km
land area: 33,920 sq km
comparative area: slightly less than twice the size of New Jersey
Land boundaries: total 1,027 km, Belgium 450 km, Germany 577 km
Coastline: 451 km
Maritime claims:
exclusive fishing zone: 200 nm
territorial sea: 12 nm
International disputes: none
Climate: temperate; marine; cool summers and mild winters
Terrain: mostly coastal lowland and reclaimed land (polders); some hills in southeast
Natural resources: natural gas, petroleum, fertile soil
Land use:
arable land: 26%
permanent crops: 1%
meadows and pastures: 32%
forest and woodland: 9%
other: 32%
Irrigated land: 5,500 sq km (1989 est.)
Environment:
current issues: water pollution in the form of heavy metals, organic compounds, and nutrients such as nitrates and phosphates; air pollution from vehicles and refining activities; acid rain
natural hazards: the extensive system of dikes and dams, protects nearly one-half of the total area from being flooded
international agreements: party to—Air Pollution, Air Pollution-Nitrogen Oxides, Air Pollution-Sulphur 85, Air Pollution-Volatile Organic Compounds, Antarctic-Environmental Protocol, Antarctic Treaty, Biodiversity, Climate Change, Endangered Species, Environmental Modification,

Hazardous Wastes, Marine Dumping, Marine Life Conservation, Nuclear Test Ban, Ozone Layer Protection, Ship Pollution, Tropical Timber 83, Wetlands, Whaling; signed, but not ratified—Air Pollution-Sulphur 94, Biodiversity, Desertification, Law of the Sea
Note: located at mouths of three major European rivers (Rhine, Maas or Meuse, and Schelde)

People

Population: 15,452,903 (July 1995 est.)
Age structure:
0-14 years: 18% (female 1,382,057; male 1,445,451)
15-64 years: 68% (female 5,184,224; male 5,369,018)
65 years and over: 14% (female 1,238,336; male 833,817) (July 1995 est.)
Population growth rate: 0.52% (1995 est.)
Birth rate: 12.42 births/1,000 population (1995 est.)
Death rate: 8.48 deaths/1,000 population (1995 est.)
Net migration rate: 1.29 migrant(s)/1,000 population (1995 est.)
Infant mortality rate: 6 deaths/1,000 live births (1995 est.)
Life expectancy at birth:
total population: 77.95 years
male: 74.9 years
female: 81.17 years (1995 est.)
Total fertility rate: 1.56 children born/woman (1995 est.)
Nationality:
noun: Dutchman(men), Dutchwoman(women)
adjective: Dutch
Ethnic divisions: Dutch 96%, Moroccans, Turks, and other 4% (1988)
Religions: Roman Catholic 34%, Protestant 25%, Muslim 3%, other 2%, unaffiliated 36% (1991)
Languages: Dutch
Literacy: age 15 and over can read and write (1979 est.)
total population: 99%
Labor force: 6.4 million (1993)
by occupation: services 71.4%, manufacturing and construction 24.6%, agriculture 4.0% (1992)

Government

Names:
conventional long form: Kingdom of the Netherlands
conventional short form: Netherlands
local long form: Koninkrijk der Nederlanden
local short form: Nederland
Digraph: NL
Type: constitutional monarchy
Capital: Amsterdam; The Hague is the seat of government
Administrative divisions: 12 provinces (provincien, singular—provincie); Drenthe, Flevoland, Friesland, Gelderland, Groningen, Limburg, Noord-Brabant, Noord-Holland, Overijssel, Utrecht, Zeeland, Zuid-Holland
Dependent areas: Aruba, Netherlands Antilles
Independence: 1579 (from Spain)
National holiday: Queen's Day, 30 April (1938)
Constitution: 17 February 1983
Legal system: civil law system incorporating French penal theory; judicial review in the Supreme Court of legislation of lower order rather than Acts of the States General; accepts compulsory ICJ jurisdiction, with reservations
Suffrage: 18 years of age; universal
Executive branch:
chief of state: Queen BEATRIX Wilhelmina Armgard (since 30 April 1980); Heir Apparent WILLEM-ALEXANDER, Prince of Orange, son of Queen Beatrix (born 27 April 1967)
head of government: Prime Minister Willem (Wim) KOK (since 22 August 1994); Vice Prime Minister Hans DIJKSTAL and Hans VAN MIERLO (since 22 August 1994)
cabinet: Cabinet; appointed by the prime minister
Legislative branch: bicameral legislature (Staten Generaal)
First Chamber (Eerste Kamer): members indirectly elected by the country's 12 provincial councils for four-year terms; elections last held 9 June 1991 (next to be held 9 June 1995); results—percent of vote by party NA; seats—(75 total) number of seats by party NA
Second Chamber (Tweede Kamer): members directly elected for four-year terms; elections last held on 3 May 1994 (next to be held in May 1999); results—PvdA 24.3%, CDA 22.3%, VVD 20.4%, D'66 16.5%, other 16.5%; seats—(150 total) PvdA 37, CDA 34, VVD 31, D'66 24, other 24
Judicial branch: Supreme Court (De Hoge Raad)
Political parties and leaders: Christian Democratic Appeal (CDA), Hans HELGERS; Labor (PvdA), Wim KOK; Liberal (VVD People's Party for Freedom and Democracy), Frits BOLKESTEIN; Democrats '66 (D'66), Hans van MIERLO; a host of minor parties
Other political or pressure groups: large multinational firms; Federation of Netherlands Trade Union Movement (comprising Socialist and Catholic trade unions) and a Protestant trade union; Federation of Catholic and Protestant Employers Associations; the nondenominational Federation of Netherlands Enterprises, and Interchurch Peace Council (IKV)
Member of: AfDB, AG (observer), AsDB, Australia Group, Benelux, BIS, CCC, CE, CERN, EBRD, EC, ECE, ECLAC, EIB, ESA, ESCAP, FAO, G-10, GATT, IADB, IAEA, IBRD, ICAO, ICC, ICFTU, ICRM, IDA, IEA, IFAD, IFC, IFRCS, ILO, IMF, IMO, INMARSAT, INTELSAT, INTERPOL, IOC, IOM, ISO, ITU, MTCR, NACC, NAM (guest), NATO, NEA, NSG, OAS (observer), OECD, OSCE, PCA, UN, UNAVEM II, UNCTAD, UNESCO, UNHCR, UNIDO, UNITAR, UNOMOZ, UNOMUR, UNPROFOR, UNTSO, UNU, UPU, WCL, WEU, WHO, WIPO, WMO, WTO, ZC
Diplomatic representation in US:
chief of mission: Ambassador Adriaan JACOBOVITS DE SZEGED
chancery: 4200 Linnean Avenue NW, Washington, DC 20008
telephone: [1] (202) 244-5300
FAX: [1] (202) 362-3430
consulate(s) general: Chicago, Houston, Los Angeles, New York
US diplomatic representation:
chief of mission: Ambassador Kirk Terry DORNBUSH
embassy: Lange Voorhout 102, 2514 EJ The Hague
mailing address: PSC 71, Box 1000, the Hague; APO AE 09715
telephone: [31] (70) 310-9209
FAX: [31] (70) 361-4688
consulate(s) general: Amsterdam
Flag: three equal horizontal bands of red (top), white, and blue; similar to the flag of Luxembourg, which uses a lighter blue and is longer

Economy

Overview: This highly developed and affluent economy is based on private enterprise. The government makes its presence felt, however, through many regulations, permit requirements, and welfare programs affecting most aspects of economic activity. The trade and financial services sector contributes over 50% of GDP. Industrial activity provides about 25% of GDP and is led by the food-processing, oil refining, and metalworking industries. The highly mechanized agricultural sector employs only 4% of the labor force, but provides large surpluses for export and the domestic food-processing industry. Indeed the Netherlands ranks third worldwide in value of agricultural exports, behind the US and France. High unemployment and a sizable budget deficit are currently the most serious economic problems. Many of the economic issues of the 1990s will reflect the course of European economic integration.
National product: GDP—purchasing power parity—$275.8 billion (1994 est.)
National product real growth rate: 2% (1994 est.)
National product per capita: $17,940 (1994 est.)

Inflation rate (consumer prices): 2.5% (December 1994)
Unemployment rate: 8.8% (December 1994)
Budget:
revenues: $109.9 billion
expenditures: $122.1 billion, including capital expenditures of $NA (1992 est.)
Exports: $153 billion (f.o.b., 1994 est.)
commodities: metal products, chemicals, processed food and tobacco, agricultural products
partners: EC 77% (Germany 27%, Belgium-Luxembourg 15%, UK 10%), Central and Eastern Europe 10%, US 4% (1991)
Imports: $137 billion (f.o.b., 1994 est.)
commodities: raw materials and semifinished products, consumer goods, transportation equipment, crude oil, food products
partners: EC 64% (Germany 26%, Belgium-Luxembourg 14%, UK 8%), US 8% (1991)
External debt: $0
Industrial production: growth rate -1.5% (1993 est.); accounts for 25% of GDP
Electricity:
capacity: 17,520,000 kW
production: 72.4 billion kWh
consumption per capita: 5,100 kWh (1993)
Industries: agroindustries, metal and engineering products, electrical machinery and equipment, chemicals, petroleum, fishing, construction, microelectronics
Agriculture: accounts for 4.6% of GDP; animal production predominates; crops—grains, potatoes, sugar beets, fruits, vegetables; shortages of grain, fats, and oils
Illicit drugs: important gateway for cocaine, heroin, and hashish entering Europe; European producer of illicit amphetamines and other synthetic drugs
Economic aid:
donor: ODA and OOF commitments (1970-89), $19.4 billion
Currency: 1 Netherlands guilder, gulden, or florin (f.) = 100 cents
Exchange rates: Netherlands guilders, gulden, or florins (f.) per US$1—1.7178 (January 1995), 1.8200 (1994), 1.8573 (1993), 1.7585 (1992), 1.8697 (1991), 1.8209 (1990)
Fiscal year: calendar year

Transportation

Railroads:
total: 2,757 km
standard gauge: 2,757 km km 1.435-m gauge (1,991 km electrified; 1,800 km double track) (1994)
Highways:
total: 104,831 km
paved: 92,251 km (2,118 km of expressway)
unpaved: gravel, crushed stone 12,580 km (1992)
Inland waterways: 6,340 km, of which 35% is usable by craft of 1,000 metric ton capacity or larger
Pipelines: crude oil 418 km; petroleum products 965 km; natural gas 10,230 km
Ports: Amsterdam, Delfzijl, Dordrecht, Eemshaven, Groningen, Haarlem, Ijmuiden, Maastricht, Rotterdam, Terneuzen, Utrecht
Merchant marine:
total: 343 ships (1,000 GRT or over) totaling 2,629,578 GRT/3,337,307 DWT
ships by type: bulk 2, cargo 195, chemical tanker 21, combination bulk 3, container 33, liquefied gas tanker 12, livestock carrier 1, multifunction large-load carrier 1, oil tanker 37, railcar carrier 1, refrigerated cargo 18, roll-on/roll-off cargo 14, short-sea passenger 3, specialized tanker 2
note: many Dutch-owned ships are also registered on the Netherlands Antilles register
Airports:
total: 29
with paved runways over 3,047 m: 1
with paved runways 2,438 to 3,047 m: 9
with paved runways 1,524 to 2,437 m: 5
with paved runways 914 to 1,523 m: 3
with paved runways under 914 m: 8
with unpaved runways 914 to 1,523 m: 3

Communications

Telephone system: 9,418,000 telephones; highly developed, well maintained, and integrated; extensive redundant system of multiconductor cables, supplemented by microwave radio relay links
local: nationwide mobile phone system
intercity: microwave radio relay
international: 5 submarine cables; 3 INTELSAT (1 Indian Ocean and 2 Atlantic Ocean) and 1 EUTELSAT earth station
Radio:
broadcast stations: AM 3 (relays 3), FM 12 (repeaters 39), shortwave 0
radios: NA
Television:
broadcast stations: 8 (repeaters 7)
televisions: NA

Defense Forces

Branches: Royal Netherlands Army, Royal Netherlands Navy (includes Naval Air Service and Marine Corps), Royal Netherlands Air Force, Royal Constabulary
Manpower availability: males age 15-49 4,177,555; males fit for military service 3,656,529; males reach military age (20) annually 94,771 (1995 est.)
Defense expenditures: exchange rate conversion—$7.1 billion, 2.2% of GDP (1994)

Netherlands Antilles

(part of the Dutch realm)

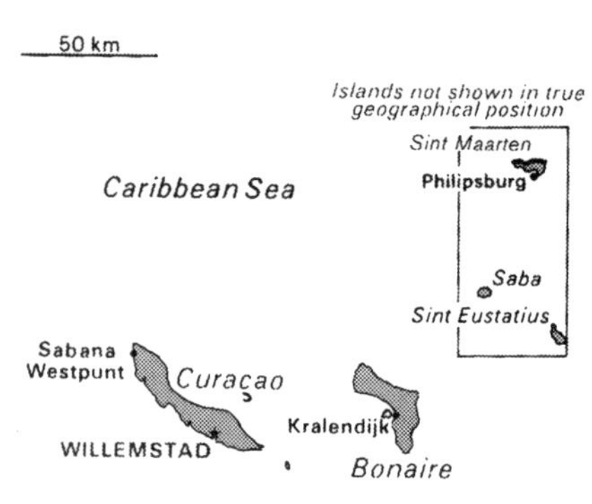

Geography

Location: Caribbean, two island groups in the Caribbean Sea—one includes Curacao and Bonaire north of Venezuela and the other is east of the Virgin Islands
Map references: Central America and the Caribbean
Area:
total area: 960 sq km
land area: 960 sq km
comparative area: slightly less than 5.5 times the size of Washington, DC
note: includes Bonaire, Curacao, Saba, Sint Eustatius, and Sint Maarten (Dutch part of the island of Saint Martin)
Land boundaries: 0 km
Coastline: 364 km
Maritime claims:
exclusive fishing zone: 12 nm
territorial sea: 12 nm
International disputes: none
Climate: tropical; ameliorated by northeast trade winds
Terrain: generally hilly, volcanic interiors
Natural resources: phosphates (Curacao only), salt (Bonaire only)
Land use:
arable land: 8%
permanent crops: 0%
meadows and pastures: 0%
forest and woodland: 0%
other: 92%
Irrigated land: NA sq km
Environment:
current issues: NA
natural hazards: Curacao and Bonaire are south of Caribbean hurricane belt, so rarely threatened; Sint Maarten, Saba, and Sint Eustatius are subject to hurricanes from July to October
international agreements: party to—Whaling (extended from Netherlands)

People

Population: 203,505 (July 1995 est.)

Age structure:
0-14 years: 26% (female 25,349; male 26,577)
15-64 years: 67% (female 69,273; male 67,485)
65 years and over: 7% (female 8,599; male 6,222) (July 1995 est.)
Population growth rate: 1.06% (1995 est.)
Birth rate: 16.23 births/1,000 population (1995 est.)
Death rate: 5.26 deaths/1,000 population (1995 est.)
Net migration rate: -0.38 migrant(s)/1,000 population (1995 est.)
Infant mortality rate: 9 deaths/1,000 live births (1995 est.)
Life expectancy at birth:
total population: 76.94 years
male: 74.67 years
female: 79.33 years (1995 est.)
Total fertility rate: 1.9 children born/woman (1995 est.)
Nationality:
noun: Netherlands Antillean(s)
adjective: Netherlands Antillean
Ethnic divisions: mixed African 85%, Carib Indian, European, Latin, Oriental
Religions: Roman Catholic, Protestant, Jewish, Seventh-Day Adventist
Languages: Dutch (official), Papiamento a Spanish-Portuguese-Dutch-English dialect predominates, English widely spoken, Spanish
Literacy: age 15 and over can read and write (1981)
total population: 98%
male: 98%
female: 99%
Labor force: 89,000
by occupation: government 65%, industry and commerce 28% (1983)

Government

Names:
conventional long form: none
conventional short form: Netherlands Antilles
local long form: none
local short form: Nederlandse Antillen
Digraph: NT
Type: part of the Dutch realm; full autonomy in internal affairs granted in 1954
Capital: Willemstad
Administrative divisions: none (part of the Dutch realm)
Independence: none (part of the Dutch realm)
National holiday: Queen's Day, 30 April (1938)
Constitution: 29 December 1954, Statute of the Realm of the Netherlands, as amended
Legal system: based on Dutch civil law system, with some English common law influence
Suffrage: 18 years of age; universal
Executive branch:
chief of state: Queen BEATRIX Wilhelmina Armgard (since 30 April 1980), represented by Governor General Jaime SALEH (since NA October 1989)
head of government: Prime Minister Miguel POURIER (since 25 February 1994)
cabinet: Council of Ministers; appointed with the advice and approval of the unicameral legislature
Legislative branch: unicameral
Staten: elections last held on 25 February 1994 (next to be held March 1998); results—percent of vote by party NA; seats—(23 total) PAR 8, PNP 3, SPA 2, PDB 2, UPB 1, MAN 2, DP 1, WIPM 1, DP-St.E 1, DP-St.M 1, Nos Patria 1
note: the government of Prime Minister Miguel POURIER is a coalition of several parties
Judicial branch: Joint High Court of Justice
Political parties and leaders: political parties are indigenous to each island
Bonaire: Patriotic Union of Bonaire (UPB), Rudy ELLIS; Democratic Party of Bonaire (PDB), Franklin CRESTIAN
Curacao: Antillean Restructuring Party (PAR), Miguel POURIER; National People's Party (PNP), Maria LIBERIA-PETERS; New Antilles Movement (MAN), Domenico Felip Don MARTINA; Workers' Liberation Front (FOL), Wilson (Papa) GODETT; Socialist Independent (SI), George HUECK and Nelson MONTE; Democratic Party of Curacao (DP), Augustin DIAZ; Nos Patria, Chin BEHILIA
Saba: Windward Islands People's Movement (WIPM Saba), Will JOHNSON; Saba Democratic Labor Movement, Vernon HASSELL; Saba Unity Party, Carmen SIMMONDS
Sint Eustatius: Democratic Party of Sint Eustatius (DP-St.E), K. Van PUTTEN; Windward Islands People's Movement (WIPM); St. Eustatius Alliance (SEA), Ralph BERKEL
Sint Maarten: Democratic Party of Sint Maarten (DP-St.M), Claude WATHEY; Patriotic Movement of Sint Maarten (SPA), Vance JAMES
Member of: CARICOM (observer), ECLAC (associate), ICFTU, INTERPOL, IOC, UNESCO (associate), UPU, WMO, WTO (associate)
Diplomatic representation in US: none (self-governing part of the Netherlands)
US diplomatic representation:
chief of mission: Consul General Bernard J. WOERZ
consulate(s) general: Saint Anna Boulevard 19, Willemstad, Curacao
mailing address: P. O. Box 158, Willemstad, Curacao
telephone: [599] (9) 61-3066
FAX: [599] (9) 61-6489
Flag: white with a horizontal blue stripe in the center superimposed on a vertical red band also centered; five white five-pointed stars are arranged in an oval pattern in the center of the blue band; the five stars represent the five main islands of Bonaire, Curacao, Saba, Sint Eustatius, and Sint Maarten

Economy

Overview: Tourism and offshore finance are the mainstays of the economy. The islands enjoy a high per capita income and a well-developed infrastructure as compared with other countries in the region. Almost all consumer and capital goods are imported, with Venezuela and the US being the major suppliers. Poor soils and inadequate water supplies hamper the development of agriculture.
National product: GDP—purchasing power parity—$1.85 billion (1993 est.)
National product real growth rate: 1.8% (1994 est.)
National product per capita: $10,000 (1993 est.)
Inflation rate (consumer prices): 1.5% (1994 est.)
Unemployment rate: 13.4% (1993 est.)
Budget:
revenues: $209 million
expenditures: $232 million, including capital expenditures of $8 million (1992 est.)
Exports: $240 million (f.o.b., 1993)
commodities: petroleum products 98%
partners: US 39%, Brazil 9%, Colombia 6%
Imports: $1.2 billion (f.o.b., 1993)
commodities: crude petroleum 64%, food, manufactures
partners: Venezuela 26%, US 18%, Colombia 6%, Netherlands 6%, Japan 5%
External debt: $672 million (December 1991)
Industrial production: growth rate NA%
Electricity:
capacity: 200,000 kW
production: 810 million kWh
consumption per capita: 4,054 kWh (1993)
Industries: tourism (Curacao and Sint Maarten), petroleum refining (Curacao), petroleum transshipment facilities (Curacao and Bonaire), light manufacturing (Curacao)
Agriculture: chief products—aloes, sorghum, peanuts, vegetables, tropical fruit
Illicit drugs: money-laundering center; transshipment point for South American cocaine and marijuana bound for the US and Europe
Economic aid:
recipient: Western (non-US) countries, ODA and OOF bilateral commitments (1970-89), $513 million
Currency: 1 Netherlands Antillean guilder, gulden, or florin (NAf.) = 100 cents
Exchange rates: Netherlands Antillean guilders, gulden, or florins (NAf.) per US$1—1.79 (fixed rate since 1989; 1.80 fixed rate 1971-88)
Fiscal year: calendar year

Netherlands Antilles *(continued)*

Transportation

Railroads: 0 km
Highways:
total: 950 km
paved: 300 km
unpaved: gravel, earth 650 km
Ports: Kralendijk, Philipsburg, Willemstad
Merchant marine:
total: 110 ships (1,000 GRT or over) totaling 1,044,553 GRT/1,343,842 DWT
ships by type: barge carrier 1, bulk 2, cargo 36, chemical tanker 6, combination ore/oil 1, liquefied gas tanker 4, multifunction large-load carrier 20, oil tanker 2, passenger 4, refrigerated cargo 27, roll-on/roll-off cargo 7
Airports:
total: 5
with paved runways over 3,047 m: 1
with paved runways 1,524 to 2,437 m: 2
with paved runways 914 to 1,523 m: 1
with paved runways under 914 m: 1

Communications

Telephone system: NA telephones; generally adequate facilities
local: NA
intercity: extensive interisland microwave radio relay links
international: 2 submarine cables; 2 INTELSAT (Atlantic Ocean) earth stations
Radio:
broadcast stations: AM 9, FM 4, shortwave 0
radios: NA
Television:
broadcast stations: 1
televisions: NA

Defense Forces

Branches: Royal Netherlands Navy, Marine Corps, Royal Netherlands Air Force, National Guard, Police Force
Manpower availability: males age 15-49 56,058; males fit for military service 31,558; males reach military age (20) annually 1,734 (1995 est.)
Note: defense is responsibility of the Netherlands

New Caledonia
(overseas territory of France)

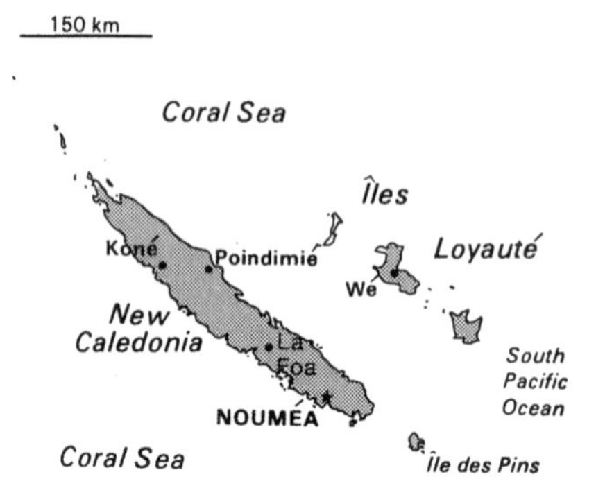

Geography

Location: Oceania, islands in the South Pacific Ocean, east of Australia
Map references: Oceania
Area:
total area: 19,060 sq km
land area: 18,760 sq km
comparative area: slightly smaller than New Jersey
Land boundaries: 0 km
Coastline: 2,254 km
Maritime claims:
exclusive economic zone: 200 nm
territorial sea: 12 nm
International disputes: none
Climate: tropical; modified by southeast trade winds; hot, humid
Terrain: coastal plains with interior mountains
Natural resources: nickel, chrome, iron, cobalt, manganese, silver, gold, lead, copper
Land use:
arable land: 0%
permanent crops: 0%
meadows and pastures: 14%
forest and woodland: 51%
other: 35%
Irrigated land: NA sq km
Environment:
current issues: NA
natural hazards: typhoons most frequent from November to March
international agreements: NA

People

Population: 184,552 (July 1995 est.)
Age structure:
0-14 years: 31% (female 27,712; male 28,677)
15-64 years: 64% (female 58,462; male 60,169)
65 years and over: 5% (female 4,997; male 4,535) (July 1995 est.)
Population growth rate: 1.75% (1995 est.)
Birth rate: 22.04 births/1,000 population (1995 est.)
Death rate: 4.9 deaths/1,000 population (1995 est.)
Net migration rate: 0.4 migrant(s)/1,000 population (1995 est.)
Infant mortality rate: 14.4 deaths/1,000 live births (1995 est.)
Life expectancy at birth:
total population: 74.02 years
male: 70.73 years
female: 77.48 years (1995 est.)
Total fertility rate: 2.57 children born/woman (1995 est.)
Nationality:
noun: New Caledonian(s)
adjective: New Caledonian
Ethnic divisions: Melanesian 42.5%, European 37.1%, Wallisian 8.4%, Polynesian 3.8%, Indonesian 3.6%, Vietnamese 1.6%, other 3%
Religions: Roman Catholic 60%, Protestant 30%, other 10%
Languages: French, 28 Melanesian-Polynesian dialects
Literacy: age 15 and over can read and write (1976)
total population: 91%
male: 92%
female: 90%
Labor force: 50,469 foreign workers for plantations and mines from Wallis and Futuna, Vanuatu, and French Polynesia (1980 est.)
by occupation: NA

Government

Names:
conventional long form: Territory of New Caledonia and Dependencies
conventional short form: New Caledonia
local long form: Territoire des Nouvelle-Caledonie et Dependances
local short form: Nouvelle-Caledonie
Digraph: NC
Type: overseas territory of France since 1956
Capital: Noumea
Administrative divisions: none (overseas territory of France); there are no first-order administrative divisions as defined by the US Government, but there are 3 provinces named Iles Loyaute, Nord, and Sud
Independence: none (overseas territory of France; a referendum on independence will be held in 1998)
National holiday: National Day, Taking of the Bastille, 14 July (1789)
Constitution: 28 September 1958 (French Constitution)
Legal system: the 1988 Matignon Accords grant substantial autonomy to the islands; formerly under French law
Suffrage: 18 years of age; universal
Executive branch:
chief of state: President Francois MITTERRAND (since 21 May 1981)
head of government: High Commissioner and

President of the Council of Government Didier CULTIAUX (since NA July 1994; appointed by the French Ministry of the Interior); President of the Territorial Congress Simon LOUECKHOTE (since 26 June 1989)
cabinet: Consultative Committee
Legislative branch: unicameral
Territorial Assembly: elections last held 11 June 1989 (next to be held July 1995); results—RPCR 44.5%, FLNKS 28.5%, FN 7%, CD 5%, UO 4%, other 11%; seats—(54 total) RPCR 27, FLNKS 19, FN 3, other 5; note—election boycotted by FULK
French Senate: elections last held 27 September 1992 (next to be held September 2001); results—percent of vote by party NA; seats—(1 total) RPCR 1
French National Assembly: elections last held 21 March 1993 (next to be held 21 and 28 March 1998); results—percent of vote by party NA; seats—(2 total) RPCR 2
Judicial branch: Court of Appeal
Political parties and leaders: white-dominated Rassemblement pour la Caledonie dans la Republique (RPCR), conservative, Jacques LAFLEUR, president—affiliated to France's Rassemblement pour la Republique (RPR; also called South Province Party); Melanesian proindependence Kanaka Socialist National Liberation Front (FLNKS), Paul NEAOUTYINE; Melanesian moderate Kanak Socialist Liberation (LKS), Nidoish NAISSELINE; National Front (FN), extreme right, Guy GEORGE; Caledonie Demain (CD), right-wing, Bernard MARANT; Union Oceanienne (UO), conservative, Michel HEMA; Front Uni de Liberation Kanak (FULK), proindependence, Clarence UREGEI; Union Caledonian (UC), Francois BURCK, president; "1999" (new party calling for an autonomous state), Philippe PENTECOST
Member of: ESCAP (associate), FZ, ICFTU, SPC, WFTU, WMO
Diplomatic representation in US: none (overseas territory of France)
US diplomatic representation: none (overseas territory of France)
Flag: the flag of France is used

Economy

Overview: New Caledonia has more than 25% of the world's known nickel resources. In recent years the economy has suffered because of depressed international demand for nickel, the principal source of export earnings. Only a negligible amount of the land is suitable for cultivation, and food accounts for about 25% of imports.
National product: GDP—purchasing power parity—$1 billion (1991 est.)
National product real growth rate: 2.4% (1988)
National product per capita: $6,000 (1991 est.)
Inflation rate (consumer prices): 1.4% (1990)
Unemployment rate: 16% (1989)
Budget:
revenues: $224 million
expenditures: $211 million, including capital expenditures of $NA (1985 est.)
Exports: $671 million (f.o.b., 1989)
commodities: nickel metal 87%, nickel ore
partners: France 32%, Japan 23.5%, US 3.6%
Imports: $764 million (c.i.f., 1989)
commodities: foods, fuels, minerals, machines, electrical equipment
partners: France 44.0%, US 10%, Australia 9%
External debt: $NA
Industrial production: growth rate NA%
Electricity:
capacity: 250,000 kW
production: 1.2 billion kWh
consumption per capita: 6,178 kWh (1993)
Industries: nickel mining and smelting
Agriculture: large areas devoted to cattle grazing; coffee, corn, wheat, vegetables; 60% self-sufficient in beef
Illicit drugs: illicit cannabis cultivation is becoming a principal source of income for some families
Economic aid:
recipient: Western (non-US) countries, ODA and OOF bilateral commitments (1970-89), $4.185 billion
Currency: 1 CFP franc (CFPF) = 100 centimes
Exchange rates: Comptoirs Francais du Pacifique francs (CFPF) per US$1—96.25 (January 1995), 100.93 (1994), 102.96 (1993), 96.24 (1992), 102.57 (1991), 99.00 (1990); note—linked at the rate of 18.18 to the French franc
Fiscal year: calendar year

Transportation

Railroads: 0 km
Highways:
total: 6,340 km
paved: 634 km
unpaved: 5,706 km (1987)
Ports: Muco, Noumea, Thio
Merchant marine:
total: 1 roll-on/roll-off ship (1,000 GRT or over) totaling 3,079 GRT/724 DWT
Airports:
total: 36
with paved runways over 3,047 m: 1
with paved runways 914 to 1,523 m: 3
with paved runways under 914 m: 19
with unpaved runways 914 to 1,523 m: 13

Communications

Telephone system: 32,578 telephones (1987)
local: NA
intercity: NA
international: 1 INTELSAT (Pacific Ocean) satellite link
Radio:
broadcast stations: AM 5, FM 3, shortwave 0
radios: NA
Television:
broadcast stations: 7
televisions: NA

Defense Forces

Branches: French Armed Forces (Army, Navy, Air Force, Gendarmerie); Police Force
Defense expenditures: $NA, NA% of GDP
Note: defense is the responsibility of France

New Zealand

Geography

Location: Oceania, islands in the South Pacific Ocean, southeast of Australia
Map references: Oceania
Area:
total area: 268,680 sq km
land area: 268,670 sq km
comparative area: about the size of Colorado
note: includes Antipodes Islands, Auckland Islands, Bounty Islands, Campbell Island, Chatham Islands, and Kermadec Islands
Land boundaries: 0 km
Coastline: 15,134 km
Maritime claims:
continental shelf: 200 nm or to the edge of the continental margin
exclusive economic zone: 200 nm
territorial sea: 12 nm
International disputes: territorial claim in Antarctica (Ross Dependency)
Climate: temperate with sharp regional contrasts
Terrain: predominately mountainous with some large coastal plains
Natural resources: natural gas, iron ore, sand, coal, timber, hydropower, gold, limestone
Land use:
arable land: 2%
permanent crops: 0%
meadows and pastures: 53%
forest and woodland: 38%
other: 7%
Irrigated land: 2,800 sq km (1989 est.)
Environment:
current issues: deforestation; soil erosion; native flora and fauna hard-hit by species introduced from outside
natural hazards: earthquakes are common, though usually not severe
international agreements: party to—Antarctic-Environmental Protocol, Antarctic Treaty, Biodiversity, Climate Change, Endangered Species, Environmental Modification, Hazardous Wastes, Marine Dumping, Nuclear Test Ban, Ozone Layer Protection, Tropical Timber 83, Wetlands, Whaling; signed, but not ratified—Law of the Sea, Marine Life Conservation
Note: about 80% of the population lives in cities

People

Population: 3,407,277 (July 1995 est.)
Age structure:
0-14 years: 23% (female 381,027; male 401,285)
15-64 years: 65% (female 1,109,402; male 1,111,079)
65 years and over: 12% (female 234,339; male 170,145) (July 1995 est.)
Population growth rate: 0.52% (1995 est.)
Birth rate: 15.14 births/1,000 population (1995 est.)
Death rate: 8.03 deaths/1,000 population (1995 est.)
Net migration rate: -1.87 migrant(s)/1,000 population (1995 est.)
Infant mortality rate: 8.6 deaths/1,000 live births (1995 est.)
Life expectancy at birth:
total population: 76.65 years
male: 73.08 years
female: 80.42 years (1995 est.)
Total fertility rate: 1.99 children born/woman (1995 est.)
Nationality:
noun: New Zealander(s)
adjective: New Zealand
Ethnic divisions: European 88%, Maori 8.9%, Pacific Islander 2.9%, other 0.2%
Religions: Anglican 24%, Presbyterian 18%, Roman Catholic 15%, Methodist 5%, Baptist 2%, other Protestant 3%, unspecified or none 33% (1986)
Languages: English (official), Maori
Literacy: age 15 and over can read and write (1980 est.)
total population: 99%
Labor force: 1,603,500 (June 1991)
by occupation: services 66.6%, industry 22.6, agriculture 10.8% (1992)

Government

Names:
conventional long form: none
conventional short form: New Zealand
Abbreviation: NZ
Digraph: NZ
Type: parliamentary democracy
Capital: Wellington
Administrative divisions: 93 counties, 9 districts*, and 3 town districts**; Akaroa, Amuri, Ashburton, Bay of Islands, Bruce, Buller, Chatham Islands, Cheviot, Clifton, Clutha, Cook, Dannevirke, Egmont, Eketahuna, Ellesmere, Eltham, Eyre, Featherston, Franklin, Golden Bay, Great Barrier Island, Grey, Hauraki Plains, Hawera*, Hawke's Bay, Heathcote, Hikurangi**, Hobson, Hokianga, Horowhenua, Hurunui, Hutt, Inangahua, Inglewood, Kaikoura, Kairanga, Kiwitea, Lake, Mackenzie, Malvern, Manaia**, Manawatu, Mangonui, Maniototo, Marlborough, Masterton, Matamata, Mount Herbert, Ohinemuri, Opotiki, Oroua, Otamatea, Otorohanga*, Oxford, Pahiatua, Paparua, Patea, Piako, Pohangina, Raglan, Rangiora*, Rangitikei, Rodney, Rotorua*, Runanga, Saint Kilda, Silverpeaks, Southland, Stewart Island, Stratford, Strathallan, Taranaki, Taumarunui, Taupo, Tauranga, Thames-Coromandel*, Tuapeka, Vincent, Waiapu, Waiheke, Waihemo, Waikato, Waikohu, Waimairi, Waimarino, Waimate, Waimate West, Waimea, Waipa, Waipawa*, Waipukurau*, Wairarapa South, Wairewa, Wairoa, Waitaki, Waitomo*, Waitotara, Wallace, Wanganui, Waverley**, Westland, Whakatane*, Whangarei, Whangaroa, Woodville
note: there may be a new administrative structure of 16 regions (Auckland, Bay of Plenty, Canterbury, Gisborne, Hawke's Bay, Marlborough, Nelson, Northland, Otago, Southland, Taranaki, Tasman, Waikato, Wanganui-Manawatu, Wellington, West Coast) that are subdivided into 57 districts and 16 cities* (Ashburton, Auckland*, Banks Peninsula, Buller, Carterton, Central Hawke's Bay, Central Otago, Christchurch*, Clutha, Dunedin*, Far North, Franklin, Gisborne, Gore, Grey, Hamilton*, Hastings, Hauraki, Horowhenua, Hurunui, Hutt*, Invercargill*, Kaikoura, Kaipara, Kapiti Coast, Kawerau, Mackenzie, Manawatu, Manukau*, Marlborough, Masterton, Matamata Piako, Napier*, Nelson*, New Plymouth, North Shore*, Opotiki, Otorohanga, Palmerston North*, Papakura*, Porirua*, Queenstown Lakes, Rangitikei, Rodney, Rotorua, Ruapehu, Selwyn, Southland, South Taranaki, South Waikato, South Wairarapa, Stratford, Tararua, Tasman, Taupo, Tauranga, Thames Coromandel, Timaru, Upper Hutt*, Waikato, Waimakariri, Waimate, Waipa, Wairoa, Waitakere*, Waitaki, Waitomo, Wanganui, Wellington*, Western Bay of Plenty, Westland, Whakatane, Whangarei)
Dependent areas: Cook Islands, Niue, Tokelau
Independence: 26 September 1907 (from UK)
National holiday: Waitangi Day, 6 February (1840) (Treaty of Waitangi established British sovereignty)
Constitution: no formal, written constitution; consists of various documents, including certain acts of the UK and New Zealand Parliaments; Constitution Act 1986 was to have come into force 1 January 1987, but has not been enacted
Legal system: based on English law, with special land legislation and land courts for Maoris; accepts compulsory ICJ jurisdiction, with reservations

Suffrage: 18 years of age; universal
Executive branch:
chief of state: Queen ELIZABETH II (since 6 February 1952), represented by Governor General Dame Catherine TIZARD (since 12 December 1990)
head of government: Prime Minister James BOLGER (since 29 October 1990); Deputy Prime Minister Donald McKINNON (since 2 November 1990)
cabinet: Executive Council; appointed by the governor general on recommendation of the prime minister
Legislative branch: unicameral
House of Representatives: (commonly called Parliament) elections last held 6 November 1993 (next to be held NA November 1996); results—NP 35.2%, NZLP 34.7%, Alliance 18.3%, New Zealand First 8.3%; seats—(99 total) NP 50, NZLP 45, Alliance 2, New Zealand First Party 2
Judicial branch: High Court, Court of Appeal
Political parties and leaders: National Party (NP, government), James BOLGER; New Zealand Labor Party (NZLP, opposition), Helen CLARK; Alliance, Sandra LEE; Democratic Party, Dick RYAN; New Zealand Liberal Party, Hanmish MACINTYRE and Gilbert MYLES; Green Party, no official leader; Mana Motuhake, Martin RATA; Socialist Unity Party (SUP, pro-Soviet), Kenneth DOUGLAS; New Zealand First, Winston PETERS
note: the New Labor, Democratic, and Mana Motuhake parties formed a coalition called the Alliance Party, Sandra LEE, president, in September 1991; the Green Party joined the coalition in May 1992
Member of: ANZUS (US suspended security obligations to NZ on 11 August 1986), APEC, AsDB, Australia Group, C, CCC, CP, EBRD, ESCAP, FAO, GATT, IAEA, IBRD, ICAO, ICFTU, ICRM, IDA, IEA, IFAD, IFC, IFRCS, ILO, IMF, IMO, INMARSAT, INTELSAT, INTERPOL, IOC, IOM (observer), ISO, ITU, MTCR, NAM (guest), OECD, PCA, SPARTECA, SPC, SPF, UN, UNAVEM II, UNCTAD, UNESCO, UNIDO, UNOSOM, UNPROFOR, UNTSO, UPU, WFTU, WHO, WIPO, WMO
Diplomatic representation in US:
chief of mission: Ambassador Lionel John WOOD
chancery: 37 Observatory Circle NW, Washington, DC 20008
telephone: [1] (202) 328-4800
consulate(s) general: Apia (Western Samoa), Los Angeles
US diplomatic representation:
chief of mission: Ambassador Josiah Horton BEEMAN
embassy: 29 Fitzherbert Terrace, Thorndon, Wellington
mailing address: P. O. Box 1190, Wellington; PSC 467, Box 1, FPO AP 96531-1001
telephone: [64] (4) 472-2068
FAX: [64] (4) 472-3537
consulate(s) general: Auckland
Flag: blue with the flag of the UK in the upper hoist-side quadrant with four red five-pointed stars edged in white centered in the outer half of the flag; the stars represent the Southern Cross constellation

Economy

Overview: Since 1984 the government has been reorienting an agrarian economy dependent on a guaranteed British market to a more industrialized, open free market economy that can compete on the global scene. The government has hoped that dynamic growth would boost real incomes, broaden and deepen the technological capabilities of the industrial sector, reduce inflationary pressures, and permit the expansion of welfare benefits. The initial results were mixed: inflation is down from double-digit levels, but growth was sluggish in 1988-91. In 1992-93, growth picked up to 3% annually, a sign that the new economic approach was beginning to pay off. Business confidence strengthened in 1994, and export demand picked up in the Asia-Pacific region, resulting in 6.2% growth. Inflation remains among the lowest in the industrial world.
National product: GDP—purchasing power parity—$56.4 billion (1994 est.)
National product real growth rate: 6.2% (1994)
National product per capita: $16,640 (1994 est.)
Inflation rate (consumer prices): 1.6% (FY93/94)
Unemployment rate: 7.5% (December 1994)
Budget:
revenues: $18.94 billion
expenditures: $18.82 billion, including capital expenditures of $NA (FY94/95)
note: surplus $120 million (FY94/95)
Exports: $11.2 billion (1994)
commodities: wool, lamb, mutton, beef, fish, cheese, chemicals, forestry products, fruits and vegetables, manufactures
partners: Australia 20%, Japan 15%, US 12%, UK 6%
Imports: $10.4 billion (1994)
commodities: machinery and equipment, vehicles and aircraft, petroleum, consumer goods
partners: Australia 21%, US 18%, Japan 16%, UK 6%
External debt: $38.5 billion (September 1994)
Industrial production: growth rate 1.9% (1990); accounts for about 20% of GDP
Electricity:
capacity: 7,520,000 kW
production: 30.5 billion kWh
consumption per capita: 8,401 kWh (1993)
Industries: food processing, wood and paper products, textiles, machinery, transportation equipment, banking and insurance, tourism, mining
Agriculture: accounts for about 9% of GDP and 11% of the work force; livestock predominates—wool, meat, dairy products all export earners; crops—wheat, barley, potatoes, pulses, fruits, vegetables; surplus producer of farm products; fish catch reached a record 503,000 metric tons in 1988
Economic aid:
donor: ODA and OOF commitments (1970-89), $526 million
Currency: 1 New Zealand dollar (NZ$) = 100 cents
Exchange rates: New Zealand dollars (NZ$) per US$1—1.5601 (January 1995), 1.6844 (1994), 1.8495 (1993), 1.8584 (1992), 1.7265 (1991), 1.6750 (1990)
Fiscal year: 1 July—30 June

Transportation

Railroads:
total: 4,716 km
narrow gauge: 4,716 km 1.067-m gauge (113 km electrified; 274 km double track)
Highways:
total: 92,648 km
paved: 49,547 km
unpaved: gravel, crushed stone 43,101 km
Inland waterways: 1,609 km; of little importance to transportation
Pipelines: petroleum products 160 km; natural gas 1,000 km; condensate (liquified petroleum gas—LPG) 150 km
Ports: Auckland, Christchurch, Dunedin, Tauranga, Wellington
Merchant marine:
total: 18 ships (1,000 GRT or over) totaling 165,504 GRT/218,699 DWT
ships by type: bulk 6, cargo 2, liquefied gas tanker 1, oil tanker 3, railcar carrier 1, roll-on/roll-off cargo 5
Airports:
total: 102
with paved runways over 3,047 m: 2
with paved runways 1,524 to 2,437 m: 8
with paved runways 914 to 1,523 m: 28
with paved runways under 914 m: 41
with unpaved runways 1,524 to 2,438 m: 2
with unpaved runways 914 to 1,523 m: 21

Communications

Telephone system: 2,110,000 telephones; excellent international and domestic systems
local: NA
intercity: NA
international: submarine cables extend to Australia and Fiji; 2 INTELSAT (Pacific Ocean) earth stations

New Zealand *(continued)*

Radio:
broadcast stations: AM 64, FM 2, shortwave 0
radios: NA
Television:
broadcast stations: 14
televisions: NA

Defense Forces

Branches: New Zealand Army, Royal New Zealand Navy, Royal New Zealand Air Force
Manpower availability: males age 15-49 883,668; males fit for military service 742,871; males reach military age (20) annually 27,162 (1995 est.)
Defense expenditures: exchange rate conversion—$792 million, 2% of GDP (FY90/91)

Nicaragua

Geography

Location: Middle America, bordering both the Caribbean Sea and the North Pacific Ocean, between Costa Rica and Honduras
Map references: Central America and the Caribbean
Area:
total area: 129,494 sq km
land area: 120,254 sq km
comparative area: slightly larger than New York State
Land boundaries: total 1,231 km, Costa Rica 309 km, Honduras 922 km
Coastline: 910 km
Maritime claims:
contiguous zone: 25-nm security zone
continental shelf: natural prolongation
territorial sea: 200 nm
International disputes: territorial disputes with Colombia over the Archipelago de San Andres y Providencia and Quita Sueno Bank; with respect to the maritime boundary question in the Golfo de Fonseca, the International Court of Justice (ICJ) referred the disputants to an earlier agreement in this century and advised that some tripartite resolution among El Salvador, Honduras, and Nicaragua likely would be required
Climate: tropical in lowlands, cooler in highlands
Terrain: extensive Atlantic coastal plains rising to central interior mountains; narrow Pacific coastal plain interrupted by volcanoes
Natural resources: gold, silver, copper, tungsten, lead, zinc, timber, fish
Land use:
arable land: 9%
permanent crops: 1%
meadows and pastures: 43%
forest and woodland: 35%
other: 12%
Irrigated land: 850 sq km (1989 est.)
Environment:
current issues: deforestation; soil erosion; water pollution
natural hazards: destructive earthquakes, volcanoes, landslides, and occasionally severe hurricanes
international agreements: party to—Endangered Species, Nuclear Test Ban, Ozone Layer Protection; signed, but not ratified—Biodiversity, Climate Change, Environmental Modification, Law of the Sea

People

Population: 4,206,353 (July 1995 est.)
Age structure:
0-14 years: 44% (female 921,356; male 930,594)
15-64 years: 53% (female 1,146,485; male 1,097,811)
65 years and over: 3% (female 62,607; male 47,500) (July 1995 est.)
Population growth rate: 2.61% (1995 est.)
Birth rate: 33.73 births/1,000 population (1995 est.)
Death rate: 6.45 deaths/1,000 population (1995 est.)
Net migration rate: -1.19 migrant(s)/1,000 population (1995 est.)
Infant mortality rate: 50.3 deaths/1,000 live births (1995 est.)
Life expectancy at birth:
total population: 64.54 years
male: 61.67 years
female: 67.53 years (1995 est.)
Total fertility rate: 4.17 children born/woman (1995 est.)
Nationality:
noun: Nicaraguan(s)
adjective: Nicaraguan
Ethnic divisions: mestizo (mixed Amerindian and Caucasian) 69%, white 17%, black 9%, Indian 5%
Religions: Roman Catholic 95%, Protestant 5%
Languages: Spanish (official)
note: English- and Indian-speaking minorities on Atlantic coast
Literacy: age 15 and over can read and write (1971)
total population: 57%
male: 57%
female: 57%
Labor force: 1.086 million
by occupation: services 43%, agriculture 44%, industry 13% (1986)

Government

Names:
conventional long form: Republic of Nicaragua
conventional short form: Nicaragua
local long form: Republica de Nicaragua
local short form: Nicaragua
Digraph: NU
Type: republic
Capital: Managua
Administrative divisions: 16 departments

(departamentos, singular—departamento); Boaco, Carazo, Chinandega, Chontales, Esteli, Granada, Jinotega, Leon, Madriz, Managua, Masaya, Matagalpa, Nueva Segovia, Rio San Juan, Rivas, Zelaya
Independence: 15 September 1821 (from Spain)
National holiday: Independence Day, 15 September (1821)
Constitution: 9 January 1987
Legal system: civil law system; Supreme Court may review administrative acts
Suffrage: 16 years of age; universal
Executive branch:
chief of state and head of government: President Violeta Barrios de CHAMORRO (since 25 April 1990); Vice President Virgilio GODOY Reyes (since 25 April 1990); election last held 25 February 1990 (next to be held November 1996); results—Violeta Barrios de CHAMORRO (UNO) 54.7%, Daniel ORTEGA Saavedra (FSLN) 40.8%, other 4.5%
cabinet: Cabinet
Legislative branch: unicameral
National Assembly (Asamblea Nacional): elections last held 25 February 1990 (next to be held November 1996); results—UNO 53.9%, FSLN 40.8%, PSC 1.6%, MUR 1.0%; seats—(92 total) UNO 41, FSLN 39, "Centrist" (Dissident UNO) 12
Judicial branch: Supreme Court (Corte Suprema)
Political parties and leaders:
far right: Liberal Constitutionalist Party* (PLC), Arnold ALEMAN; Conservative Popular Alliance Party (APC), Myriam ARGUELLO; Central American Unionist Party (PUCA), Blanca ROJAS Echaverry; Independent Liberal Party for National Unity (PLUIN), Alfonso MOCADO Guillen; Conservative Party of Nicaragua (PCN), Fernando AGUERO (PCN was formed in 1992 by the merger of the Conservative Social Party (PSC) with the Democratic Conservative Party (PCD) and PCL, the Conservative Party of Labor); National Justice Party (PJN), Jorge DIAZ Cruz; National Conservative Party* (PNC), Adolfo CALERO
center right: Neoliberal Party* (PALI), Adolfo GARCIA Esquivel; National Action Party* (PAN), Delvis MONTIEL; Independent Liberal Party* (PLI), Wilfredo NAVARRO
center left: Christian Democratic Union (UDC), Luis Humberto GUZMAN; Nicaraguan Democratic Movement (MDN), Roberto URROZ; Social Democratic Party (PSD), Adolfo JARQUIN; Movement of Revolutionary Unity (MUR), Pablo HERNANDEZ; Sandinista Renovation Movement (MRS), Sergio RAMIREZ; Democratic Action Movement (MAD), Eden PASTORA; Communist Party of Nicaragua* (PCdeN), Eli ALTIMIRANO Perez
far left: Sandinista National Liberation Front (FSLN), Daniel ORTEGA; Revolutionary Workers' Party (PRT), Bonifacio MIRANDA; Popular Action Movement-Marxist-Leninist (MAP-ML), Isidro TELLEZ; Nicaraguan Socialist Party (PSN), Gustavo TABLADA; Unidad Nicaraguense de Obreros, Campesinos, y Profesionales (UNOCP), Rosalio GONZALEZ Urbina
note: parties marked with an asterisk belong to the National Opposition Union (UNO), an alliance of moderate parties, which, however, does not always follow a unified political agenda
Other political or pressure groups: National Workers Front (FNT) is a Sandinista umbrella group of eight labor unions: Sandinista Workers' Central (CST); Farm Workers Association (ATC); Health Workers Federation (FETASALUD); National Union of Employees (UNE); National Association of Educators of Nicaragua (ANDEN); Union of Journalists of Nicaragua (UPN); Heroes and Martyrs Confederation of Professional Associations (CONAPRO); and the National Union of Farmers and Ranchers (UNAG); Permanent Congress of Workers (CPT) is an umbrella group of four non-Sandinista labor unions: Confederation of Labor Unification (CUS); Autonomous Nicaraguan Workers' Central (CTN-A); Independent General Confederation of Labor (CGT-I); and Labor Action and Unity Central (CAUS); Nicaraguan Workers' Central (CTN) is an independent labor union; Superior Council of Private Enterprise (COSEP) is a confederation of business groups
Member of: BCIE, CACM, ECLAC, FAO, G-77, GATT, IADB, IAEA, IBRD, ICAO, ICFTU, ICRM, IDA, IFAD, IFC, IFRCS, ILO, IMF, IMO, INTELSAT, INTERPOL, IOC, IOM, ITU, LAES, LAIA (observer), NAM, OAS, OPANAL, PCA, UN, UNCTAD, UNESCO, UNHCR, UNIDO, UPU, WCL, WFTU, WHO, WIPO, WMO, WTO
Diplomatic representation in US:
chief of mission: Ambassador Roberto Genaro MAYORGA Cortes
chancery: 1627 New Hampshire Avenue NW, Washington, DC 20009
telephone: [1] (202) 939-6570
consulate(s) general: Houston, Los Angeles, Miami, New Orleans, New York, San Francisco
US diplomatic representation:
chief of mission: Ambassador John F. MAISTO
embassy: Kilometer 4.5 Carretera Sur., Managua
mailing address: APO AA 34021
telephone: [505] (2) 666010, 666013, 666015 through 18, 666026, 666027, 666032 through 34
FAX: [505] (2) 666046
Flag: three equal horizontal bands of blue (top), white, and blue with the national coat of arms centered in the white band; the coat of arms features a triangle encircled by the words REPUBLICA DE NICARAGUA on the top and AMERICA CENTRAL on the bottom; similar to the flag of El Salvador, which features a round emblem encircled by the words REPUBLICA DE EL SALVADOR EN LA AMERICA CENTRAL centered in the white band; also similar to the flag of Honduras, which has five blue stars arranged in an X pattern centered in the white band

Economy

Overview: Since March 1991, when President CHAMORRO began an ambitious economic stabilization program, Nicaragua has had considerable success in reducing inflation and obtaining substantial economic aid from abroad. Annual inflation fell from more than 750% in 1991 to less than 5% in 1992. Inflation rose again to an estimated 20% in 1993, although this increase was due almost entirely to a large currency devaluation in January. As of early 1994, the government was close to finalizing an enhanced structural adjustment facility with the IMF, after the previous standby facility expired in early 1993. Despite these successes, achieving overall economic growth in an economy scarred by misguided economic values and civil war during the 1980s has proved elusive. Economic growth was flat in 1992 and slightly negative in 1993. Nicaragua's per capita foreign debt is one of the highest in the world; nonetheless, as of late 1993, Nicaragua was current on its post-1988 debt as well as on payments to the international financial institutions. Definition of property rights remains a problem; ownership disputes over large tracts of land, businesses, and homes confiscated by the previous government have yet to be resolved. A rise in exports of coffee and other products led growth in 1994.
National product: GDP—purchasing power parity—$6.4 billion (1994 est.)
National product real growth rate: 3.2% (1994 est.)
National product per capita: $1,570 (1994 est.)
Inflation rate (consumer prices): 19.5% (1994 est.)
Unemployment rate: 21.8%; underemployment 50% (1993)
Budget:
revenues: $375 million (1992)
expenditures: $410 million (1992), including capital expenditures of $115 million (1991 est.)
Exports: $329 million (f.o.b., 1994 est.)
commodities: meat, coffee, cotton, sugar, seafood, gold, bananas
partners: US, Central America, Canada, Germany
Imports: $786 million (c.i.f., 1994 est.)
commodities: consumer goods, machinery and equipment, petroleum products
partners: Central America, US, Venezuela, Japan

Nicaragua *(continued)*

External debt: $11 billion (1993)
Industrial production: growth rate -0.8% (1993 est.); accounts for 26% of GDP
Electricity:
capacity: 460,000 kW
production: 1.6 billion kWh
consumption per capita: 376 kWh (1993)
Industries: food processing, chemicals, metal products, textiles, clothing, petroleum refining and distribution, beverages, footwear
Agriculture: crops account for about 15% of GDP; export crops—coffee, bananas, sugarcane, cotton; food crops—rice, corn, cassava, citrus fruit, beans; also produces a variety of animal products—beef, veal, pork, poultry, dairy products; normally self-sufficient in food
Illicit drugs: transshipment point for cocaine destined for the US
Economic aid:
recipient: US commitments, including Ex-Im (FY70-92), $620 million; Western (non-US) countries, ODA and OOF bilateral commitments (1970-89), $1.381 billion
Currency: 1 gold cordoba (C$) = 100 centavos
Exchange rates: gold cordobas (C$) per US$1—7.08 (December 1994), 6.72 (1994), 5.62 (1993), 5.00 (1992); note—gold cordoba replaced cordoba as Nicaragua's currency in 1991 (exchange rate of old cordoba had reached per US$1—25,000,000 by March 1992)
Fiscal year: calendar year

Transportation

Railroads:
total: 376 km; note—majority of system is nonoperational
standard gauge: 3 km 1.435-m gauge line at Puerto Cabezas; note—does not connect with mainline
narrow gauge: 373 km 1.067-m gauge
Highways:
total: 15,286 km
paved: 1,598 km
unpaved: 13,688 km
note: there is a 368.5 km portion of the Pan-American highway which is not in the total
Inland waterways: 2,220 km, including 2 large lakes
Pipelines: crude oil 56 km
Ports: Bluefields, Corinto, El Bluff, Puerto Cabezas, Puerto Sandino, Rama, San Juan del Sur
Merchant marine: none
Airports:
total: 198
with paved runways over 3,047 m: 1
with paved runways 2,438 to 3,047 m: 1
with paved runways 1,524 to 2,437 m: 3
with paved runways 914 to 1,523 m: 3
with paved runways under 914 m: 149
with unpaved runways 1,524 to 2,438 m: 2
with unpaved runways 914 to 1,523 m: 39

Communications

Telephone system: 60,000 telephones; low-capacity radio relay and wire system being expanded; connection into Central American Microwave System
local: NA
intercity: wire and radio relay
international: 1 Intersputnik and 1 INTELSAT (Atlantic Ocean) earth station
Radio:
broadcast stations: AM 45, FM 0, shortwave 3
radios: NA
Television:
broadcast stations: 7
televisions: NA

Defense Forces

Branches: Ground Forces, Navy, Air Force
note: total strength of all branches—14,500
Manpower availability: males age 15-49 982,345; males fit for military service 604,721; males reach military age (18) annually 47,064 (1995 est.)
Defense expenditures: exchange rate conversion—$32 million, 1.7% of GDP (1994), 8.1% of government budget

Niger

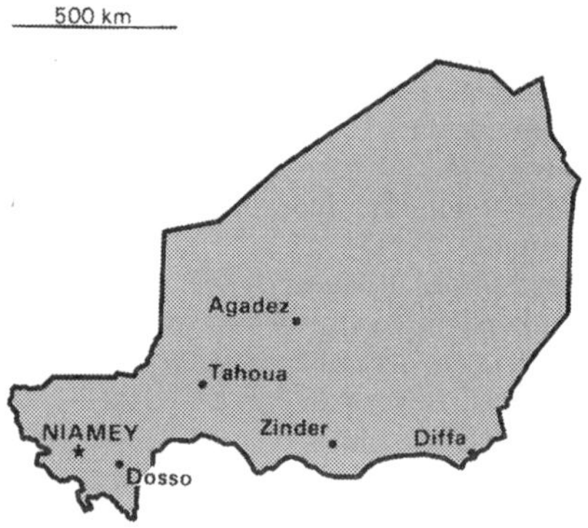

Geography

Location: Western Africa, southeast of Algeria
Map references: Africa
Area:
total area: 1.267 million sq km
land area: 1,266,700 sq km
comparative area: slightly less than twice the size of Texas
Land boundaries: total 5,697 km, Algeria 956 km, Benin 266 km, Burkina 628 km, Chad 1,175 km, Libya 354 km, Mali 821 km, Nigeria 1,497 km
Coastline: 0 km (landlocked)
Maritime claims: none; landlocked
International disputes: Libya claims about 19,400 sq km in northern Niger; demarcation of international boundaries in Lake Chad, the lack of which has led to border incidents in the past, is completed and awaiting ratification by Cameroon, Chad, Niger, and Nigeria; Burkina and Mali are proceeding with boundary demarcation, including the tripoint with Niger
Climate: desert; mostly hot, dry, dusty; tropical in extreme south
Terrain: predominately desert plains and sand dunes; flat to rolling plains in south; hills in north
Natural resources: uranium, coal, iron ore, tin, phosphates
Land use:
arable land: 3%
permanent crops: 0%
meadows and pastures: 7%
forest and woodland: 2%
other: 88%
Irrigated land: 320 sq km (1989 est.)
Environment:
current issues: overgrazing; soil erosion; deforestation; desertification; wildlife populations (such as elephant, hippopotamus, and lion) threatened because of poaching and habitat destruction
natural hazards: recurring droughts
international agreements: party to—Endangered Species, Environmental

Modification, Nuclear Test Ban, Ozone Layer Protection, Wetlands; signed, but not ratified—Biodiversity, Climate Change, Desertification, Law of the Sea
Note: landlocked

People

Population: 9,280,208 (July 1995 est.)
Age structure:
0-14 years: 49% (female 2,275,338; male 2,275,999)
15-64 years: 49% (female 2,314,857; male 2,188,938)
65 years and over: 2% (female 107,432; male 117,644) (July 1995 est.)
Population growth rate: 3.4% (1995 est.)
Birth rate: 54.8 births/1,000 population (1995 est.)
Death rate: 20.8 deaths/1,000 population (1995 est.)
Net migration rate: 0 migrant(s)/1,000 population (1995 est.)
Infant mortality rate: 109.3 deaths/1,000 live births (1995 est.)
Life expectancy at birth:
total population: 45.07 years
male: 43.42 years
female: 46.77 years (1995 est.)
Total fertility rate: 7.35 children born/woman (1995 est.)
Nationality:
noun: Nigerien(s)
adjective: Nigerien
Ethnic divisions: Hausa 56%, Djerma 22%, Fula 8.5%, Tuareg 8%, Beri Beri (Kanouri) 4.3%, Arab, Toubou, and Gourmantche 1.2%, about 4,000 French expatriates
Religions: Muslim 80%, remainder indigenous beliefs and Christians
Languages: French (official), Hausa, Djerma
Literacy: age 15 and over can read and write (1988)
total population: 11%
male: 17%
female: 5%
Labor force: 2.5 million wage earners (1982)
by occupation: agriculture 90%, industry and commerce 6%, government 4%

Government

Names:
conventional long form: Republic of Niger
conventional short form: Niger
local long form: Republique du Niger
local short form: Niger
Digraph: NG
Type: republic
Capital: Niamey
Administrative divisions: 7 departments (departements, singular—departement); Agadez, Diffa, Dosso, Maradi, Niamey, Tahoua, Zinder
Independence: 3 August 1960 (from France)
National holiday: Republic Day, 18 December (1958)
Constitution: approved by national referendum 16 December 1992; promulgated January 1993
Legal system: based on French civil law system and customary law; has not accepted compulsory ICJ jurisdiction
Suffrage: 18 years of age; universal
Executive branch:
chief of state: President Mahamane OUSMANE (since 16 April 1993); election last held 27 March 1993 (next to be held NA February 1998)
head of government: Prime Minister Hama AMADOU (since 21 February 1995)
cabinet: Cabinet; appointed by the president on recommendation of the prime minister
Legislative branch: unicameral
National Assembly: elected by proportional representation for 5 year terms; elections last held 12 January 1995 (next to be held NA); results—percent of vote by party NA; seats—(83 total) MNSD-NASSERA 29, CDS 24, PNDS 12, ANDP-Z 9, UDFP 3, UDPS 2, PADN 2, PPN-RDA 1, UPDP 1
Judicial branch: State Court (Cour d'Etat), Court of Appeal (Cour d'Apel)
Political parties and leaders: National Movement of the Development Society (MNSD-NASSARA), Mamadou TANDJA, chairman; Democratic and Social Convention (CDS), Jacoub SANOUSSI; Nigerien Party for Democracy and Socialism (PNDS), Mahamadou ISSOUFOU; Nigerien Alliance for Democracy and Progress-Zamanlahia (ANDP-Z), Moumouni Adamou DJERMAKOYE; Union of Popular Forces for Democracy and Progress-Sawaba (UDFP), Djibo BAKARY, chairman; Union for Democracy and Social Progress (UDPS), Akoli DAOUEL; Niger Social Democrat Party (PADN), Malam Adji WAZIRI; Niger Progressive Party-African Democratic Rally (PPN-RDA), Dori ABDOULAI, chairman; Union of Patriots, Democrats, and Progressives (UPDP), Professor Andre SALIFOU, chairman
Member of: ACCT, ACP, AfDB, CCC, CEAO, ECA, ECOWAS, Entente, FAO, FZ, G-77, GATT, IAEA, IBRD, ICAO, ICRM, IDA, IDB, IFAD, IFC, IFRCS, ILO, IMF, INTELSAT, INTERPOL, IOC, ITU, NAM, OAU, OIC, UN, UNCTAD, UNESCO, UNIDO, UPU, WADB, WCL, WFTU, WHO, WIPO, WMO, WTO
Diplomatic representation in US:
chief of mission: Ambassador Adamou SEYDOU
chancery: 2204 R Street NW, Washington, DC 20008
telephone: [1] (202) 483-4224 through 4227
US diplomatic representation:
chief of mission: Ambassador John S. DAVISON
embassy: Rue Des Ambassades, Niamey
mailing address: B. P. 11201, Niamey
telephone: [227] 72 26 61 through 72 26 64
FAX: [227] 73 31 67
Flag: three equal horizontal bands of orange (top), white, and green with a small orange disk (representing the sun) centered in the white band; similar to the flag of India, which has a blue spoked wheel centered in the white band

Economy

Overview: Niger is one of the world's poorest countries, with GDP growth lagging behind the rapid growth of population. The economy is centered on subsistence agriculture, animal husbandry, and reexport trade, and increasingly less on uranium, its major export throughout the 1970s and 1980s. Uranium revenues dropped by almost 50% between 1983 and 1990 with the end of the uranium boom. Terms of trade with Nigeria, Niger's largest regional trade partner, have improved dramatically since the 50% devaluation of the African franc in January 1994; this devaluation boosted exports of livestock, peas, onions, and the products of Niger's small cotton industry. The government relies on bilateral and multilateral aid for operating expenses and public investment and is strongly induced to adhere to structural adjustment programs designed by the IMF and the World Bank.
National product: GDP—purchasing power parity—$4.6 billion (1993 est.)
National product real growth rate: 1.4% (1993 est.)
National product per capita: $550 (1994 est.)
Inflation rate (consumer prices): NA%
Unemployment rate: NA%
Budget:
revenues: $188 million
expenditures: $400 million, including capital expenditures of $125 million (1993 est.)
Exports: $246 million (f.o.b., 1993 est.)
commodities: uranium ore 67%, livestock products 20%, cowpeas, onions
partners: France 77%, Nigeria 8%, Cote d'Ivoire, Italy
Imports: $286 million (c.i.f., 1993 est.)
commodities: consumer goods, primary materials, machinery, vehicles and parts, petroleum, cereals
partners: France 23%, Cote d'Ivoire, Germany, Italy, Japan
External debt: $1.2 billion (December 1991 est.)
Industrial production: growth rate -2.7% (1992 est.); accounts for 15% of GDP
Electricity:
capacity: 60,000 kW
production: 200 million kWh
consumption per capita: 42 kWh (1992)
Industries: cement, brick, textiles, food processing, chemicals, slaughterhouses, and a

few other small light industries; uranium mining began in 1971
Agriculture: accounts for roughly 40% of GDP and 90% of labor force; cash crops—cowpeas, cotton, peanuts; food crops—millet, sorghum, cassava, rice; livestock—cattle, sheep, goats; self-sufficient in food except in drought years
Economic aid:
recipient: US commitments, including Ex-Im (FY70-89), $380 million; Western (non-US) countries, ODA and OOF bilateral commitments (1970-89), $3.165 billion; OPEC bilateral aid (1979-89), $504 million; Communist countries (1970-89), $61 million
Currency: 1 CFA franc (CFAF) = 100 centimes
Exchange rates: Communaute Financiere Africaine francs (CFAF) per US$1—529.43 (January 1995), 555.20 (1994), 283.16 (1993), 264.69 (1992), 282.11 (1991), 272.26 (1990)
note: the official rate is pegged to the French franc, and beginning 12 January 1994, the CFA franc was devalued to CFAF 100 per French franc from CFAF 50 at which it had been fixed since 1948
Fiscal year: 1 October—30 September

Transportation

Railroads: 0 km
Highways:
total: 39,970 km
paved: bituminous 3,170 km
unpaved: gravel, laterite 10,330 km; earth 3,470 km; tracks 23,000 km
Inland waterways: Niger River is navigable 300 km from Niamey to Gaya on the Benin frontier from mid-December through March
Ports: none
Airports:
total: 29
with paved runways 2,438 to 3,047 m: 2
with paved runways 1,524 to 2,437 m: 6
with paved runways 914 to 1,523 m: 1
with paved runways under 914 m: 3
with unpaved runways 1,524 to 2,438 m: 1
with unpaved runways 914 to 1,523 m: 16

Communications

Telephone system: 14,260 telephones; small system of wire, radiocommunications, and radio relay links concentrated in southwestern area
local: NA
intercity: wire, radiocommunications, and radio relay; 3 domestic satellite links, with 1 planned
international: 2 INTELSAT (1 Atlantic Ocean and 1 Indian Ocean) earth stations
Radio:
broadcast stations: AM 15, FM 5, shortwave 0
radios: NA
Television:
broadcast stations: 18
televisions: NA

Defense Forces

Branches: Army, Air Force, National Gendarmerie, Republican Guard, National Police
Manpower availability: males age 15-49 1,908,767; males fit for military service 1,029,384; males reach military age (18) annually 94,506 (1995 est.)
Defense expenditures: exchange rate conversion—$32 million, 1.3% of GDP (FY92/93)

Nigeria

Geography

Location: Western Africa, bordering the North Atlantic Ocean, between Benin and Cameroon
Map references: Africa
Area:
total area: 923,770 sq km
land area: 910,770 sq km
comparative area: slightly more than twice the size of California
Land boundaries: total 4,047 km, Benin 773 km, Cameroon 1,690 km, Chad 87 km, Niger 1,497 km
Coastline: 853 km
Maritime claims:
continental shelf: 200-m depth or to the depth of exploitation
exclusive economic zone: 200 nm
territorial sea: 30 nm
International disputes: demarcation of international boundaries in Lake Chad, the lack of which led to border incidents in the past, is completed and awaits ratification by Cameroon, Chad, Niger, and Nigeria; dispute with Cameroon over land and maritime boundaries in the vicinity of the Bakasi Peninsula has been referred to the International Court of Justice
Climate: varies; equatorial in south, tropical in center, arid in north
Terrain: southern lowlands merge into central hills and plateaus; mountains in southeast, plains in north
Natural resources: petroleum, tin, columbite, iron ore, coal, limestone, lead, zinc, natural gas
Land use:
arable land: 31%
permanent crops: 3%
meadows and pastures: 23%
forest and woodland: 15%
other: 28%
Irrigated land: 8,650 sq km (1989 est.)
Environment:
current issues: soil degradation; rapid deforestation; desertification; recent droughts

in north severely affecting marginal agricultural activities
natural hazards: periodic droughts
international agreements: party to—Biodiversity, Climate Change, Endangered Species, Hazardous Wastes, Law of the Sea, Marine Dumping, Marine Life Conservation, Nuclear Test Ban, Ozone Layer Protection

People

Population: 101,232,251 (July 1995 est.)
Age structure:
0-14 years: 45% (female 22,643,026; male 22,850,322)
15-64 years: 52% (female 25,842,286; male 26,978,906)
65 years and over: 3% (female 1,438,392; male 1,479,319) (July 1995 est.)
Population growth rate: 3.16% (1995 est.)
Birth rate: 43.26 births/1,000 population (1995 est.)
Death rate: 12.01 deaths/1,000 population (1995 est.)
Net migration rate: 0.35 migrant(s)/1,000 population (1995 est.)
Infant mortality rate: 72.6 deaths/1,000 live births (1995 est.)
Life expectancy at birth:
total population: 55.98 years
male: 54.69 years
female: 57.3 years (1995 est.)
Total fertility rate: 6.31 children born/woman (1995 est.)
Nationality:
noun: Nigerian(s)
adjective: Nigerian
Ethnic divisions:
north: Hausa and Fulani
southwest: Yoruba
southeast: Ibos
non-Africans 27,000
note: Hausa and Fulani, Yoruba, and Ibos together make up 65% of population
Religions: Muslim 50%, Christian 40%, indigenous beliefs 10%
Languages: English (official), Hausa, Yoruba, Ibo, Fulani
Literacy: age 15 and over can read and write (1990 est.)
total population: 51%
male: 62%
female: 40%
Labor force: 42.844 million
by occupation: agriculture 54%, industry, commerce, and services 19%, government 15%

Government

Names:
conventional long form: Federal Republic of Nigeria
conventional short form: Nigeria
Digraph: NI
Type: military government since 31 December 1983; plans to institute a constitutional conference to prepare for a new transition to civilian rule after plans for a transition in 1993 were negated by General BABANGIDA
Capital: Abuja
note: on 12 December 1991 the capital was officially moved from Lagos to Abuja; many government offices remain in Lagos pending completion of facilities in Abuja
Administrative divisions: 30 states and 1 territory*; Abia, Abuja Capital Territory*, Adamawa, Akwa Ibom, Anambra, Bauchi, Benue, Borno, Cross River, Delta, Edo, Enugu, Imo, Jigawa, Kaduna, Kano, Katsina, Kebbi, Kogi, Kwara, Lagos, Niger, Ogun, Ondo, Osun, Oyo, Plateau, Rivers, Sokoto, Taraba, Yobe
Independence: 1 October 1960 (from UK)
National holiday: Independence Day, 1 October (1960)
Constitution: 1979 constitution still in force; plan for 1989 constitution to take effect in 1993 was not implemented
Legal system: based on English common law, Islamic law, and tribal law
Suffrage: 21 years of age; universal
Executive branch:
chief of state and head of government: Chairman of the Provisional Ruling Council and Commander in Chief of Armed Forces and Defense Minister Gen. Sani ABACHA (since 17 November 1993), Vice-Chairman of the Provisional Ruling Council Oladipo DIYA (since 17 November 1993)
cabinet: Federal Executive Council
Legislative branch: bicameral National Assembly
Senate: suspended after coup of 17 November 1993
House of Representatives: suspended after coup of 17 November 1993
Judicial branch: Supreme Court, Federal Court of Appeal
Political parties and leaders:
note: two political party system suspended after the coup of 17 November 1993
Member of: ACP, AfDB, C, CCC, ECA, ECOWAS, FAO, G-15, G-19, G-24, G-77, GATT, IAEA, IBRD, ICAO, ICC, ICRM, IDA, IFAD, IFC, IFRCS, ILO, IMF, IMO, INMARSAT, INTELSAT, INTERPOL, IOC, ITU, MINURSO, NAM, OAU, OPEC, PCA, UN, UNAMIR, UNAVEM II, UNCTAD, UNESCO, UNHCR, UNIDO, UNIKOM, UNPROFOR, UNU, UPU, WCL, WFTU, WHO, WMO, WTO
Diplomatic representation in US:
chief of mission: Ambassador Zubair Mahmud KAZAURE
chancery: 1333 16th Street NW, Washington, DC 20036
telephone: [1] (202) 986-8400
consulate(s) general: New York
US diplomatic representation:
chief of mission: Ambassador Walter C. CARRINGTON
embassy: 2 Eleke Crescent, Lagos
mailing address: P. O. Box 554, Lagos
telephone: [234] (1) 261-0097
FAX: [234] (1) 261-0257
branch office: Abuja
consulate(s) general: Kaduna
Flag: three equal vertical bands of green (hoist side), white, and green

Economy

Overview: The oil-rich Nigerian economy continues to be hobbled by political instability and poor macroeconomic management. Nigeria's unpopular military rulers show no sign of wanting to restore democratic civilian rule in the near future and appear divided on how to redress fundamental economic imbalances that cause troublesome inflation and the steady depreciation of the naira. The government's domestic and international arrears continue to limit economic growth—even in the oil sector—and prevent an agreement with the IMF and bilateral creditors on debt relief. The inefficient (largely subsistence) agricultural sector has failed to keep up with rapid population growth, and Nigeria, once a large net exporter of food, now must import food.
National product: GDP—purchasing power parity—$122.6 billion (1994 est.)
National product real growth rate: -0.8% (1994 est.)
National product per capita: $1,250 (1994 est.)
Inflation rate (consumer prices): 53% (1993 est.)
Unemployment rate: 28% (1992 est.)
Budget:
revenues: $9 billion
expenditures: $10.8 billion, including capital expenditures of $NA (1992 est.)
Exports: $11.9 billion (f.o.b., 1992)
commodities: oil 95%, cocoa, rubber
partners: US 54%, EC 23%
Imports: $8.3 billion (c.i.f., 1992)
commodities: machinery and equipment, manufactured goods, food and animals
partners: EC 64%, US 10%, Japan 7%
External debt: $29.5 billion (1992)
Industrial production: growth rate 7.7% (1991); accounts for 43% of GDP, including petroleum
Electricity:
capacity: 4,570,000 kW
production: 11.3 billion kWh
consumption per capita: 109 kWh (1993)
Industries: crude oil and mining—coal, tin, columbite; primary processing industries—palm oil, peanut, cotton, rubber, wood, hides and skins; manufacturing industries—textiles, cement, building materials, food products, footwear, chemical, printing, ceramics, steel

Nigeria *(continued)*

Agriculture: accounts for 35% of GDP and half of labor force; cash crops—cocoa, peanuts, palm oil, rubber; food crops—corn, rice, sorghum, millet, cassava, yams; livestock—cattle, sheep, goats, pigs; fishing and forestry resources extensively exploited
Illicit drugs: passenger and cargo air hub for West Africa; facilitates movement of heroin en route from Southeast and Southwest Asia to Western Europe and North America; increasingly a transit route for cocaine from South America intended for West European, East Asian, and North American markets
Economic aid:
recipient: US commitments, including Ex-Im (FY70-89), $705 million; Western (non-US) countries, ODA and OOF bilateral commitments (1970-89), $3 billion; Communist countries (1970-89), $2.2 billion
Currency: 1 naira (N) = 100 kobo
Exchange rates: naira (N) per US$1—21.996 (January 1995), 21.996 (1994), 22.065 (1993), 17.298 (1992), 9.909 (1991), 8.038 (1990)
Fiscal year: calendar year

Transportation

Railroads:
total: 3,567 km
narrow gauge: 3,505 km 1.067-m gauge
standard gauge: 62 km 1.435-m gauge
Highways:
total: 107,990 km
paved: mostly bituminous-surface treatment 30,019 km
unpaved: gravel, crushed stone, improved earth 25,411 km; unimproved earth 52,560 km
Inland waterways: 8,575 km consisting of Niger and Benue Rivers and smaller rivers and creeks
Pipelines: crude oil 2,042 km; petroleum products 3,000 km; natural gas 500 km
Ports: Calabar, Lagos, Onne, Port Harcourt, Sapele, Warri
Merchant marine:
total: 32 ships (1,000 GRT or over) totaling 404,064 GRT/661,850 DWT
ships by type: bulk 1, cargo 14, chemical tanker 3, liquefied gas tanker 1, oil tanker 12, roll-on/roll-off cargo 1
Airports:
total: 80
with paved runways over 3,047 m: 6
with paved runways 2,438 to 3,047 m: 10
with paved runways 1,524 to 2,437 m: 10
with paved runways 914 to 1,523 m: 7
with paved runways under 914 m: 25
with unpaved runways 1,524 to 2,438 m: 1
with unpaved runways 914 to 1,523 m: 21

Communications

Telephone system: NA telephones; above-average system limited by poor maintenance; major expansion in progress
local: NA
intercity: microwave radio relay, coaxial cable, and 20 domestic satellite earth stations carry intercity traffic
international: 3 INTELSAT earth stations (2 Atlantic Ocean and 1 Indian Ocean) and 1 coaxial submarine cable carry international traffic
Radio:
broadcast stations: AM 35, FM 17, shortwave 0
radios: NA
Television:
broadcast stations: 28
televisions: NA

Defense Forces

Branches: Army, Navy, Air Force, paramilitary Police Force
Manpower availability: males age 15-49 23,167,009; males fit for military service 13,246,223; males reach military age (18) annually 1,024,059 (1995 est.)
Defense expenditures: exchange rate conversion—$172 million, about 1% of GDP (1992)

Niue
(free association with New Zealand)

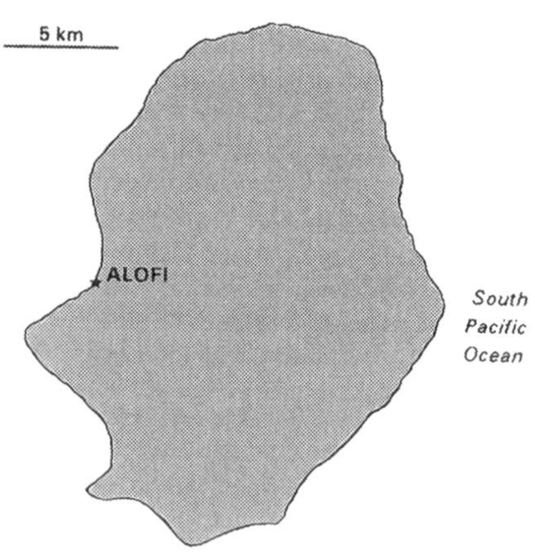

Geography

Location: Oceania, island in the South Pacific Ocean, east of Tonga
Map references: Oceania
Area:
total area: 260 sq km
land area: 260 sq km
comparative area: slightly less than 1.5 times the size of Washington, DC
Land boundaries: 0 km
Coastline: 64 km
Maritime claims:
exclusive economic zone: 200 nm
territorial sea: 12 nm
International disputes: none
Climate: tropical; modified by southeast trade winds
Terrain: steep limestone cliffs along coast, central plateau
Natural resources: fish, arable land
Land use:
arable land: 61%
permanent crops: 4%
meadows and pastures: 4%
forest and woodland: 19%
other: 12%
Irrigated land: NA sq km
Environment:
current issues: traditional methods of burning brush and trees to clear land for agriculture have threatened soil supplies which naturally are not very abundant
natural hazards: typhoons
international agreements: signed, but not ratified—Law of the Sea
Note: one of world's largest coral islands

People

Population: 1,837 (July 1995 est.)
Age structure:
0-14 years: NA
15-64 years: NA
65 years and over: NA
Population growth rate: -3.66% (1995 est.)
Birth rate: NA

Death rate: NA
Net migration rate: NA
Infant mortality rate: NA
Life expectancy at birth: NA
Total fertility rate: NA
Nationality:
noun: Niuean(s)
adjective: Niuean
Ethnic divisions: Polynesian (with some 200 Europeans, Samoans, and Tongans)
Religions: Ekalesia Nieue (Niuean Church) 75%—a Protestant church closely related to the London Missionary Society, Morman 10%, other 15% (mostly Roman Catholic, Jehovah's Witnesses, Seventh-Day Adventist)
Languages: Polynesian closely related to Tongan and Samoan, English
Labor force: 1,000 (1981 est.)
by occupation: most work on family plantations; paid work exists only in government service, small industry, and the Niue Development Board

Government

Names:
conventional long form: none
conventional short form: Niue
Digraph: NE
Type: self-governing territory in free association with New Zealand; Niue fully responsible for internal affairs; New Zealand retains responsibility for external affairs
Capital: Alofi
Administrative divisions: none
Independence: 19 October 1974 (became a self-governing territory in free association with New Zealand on 19 October 1974)
National holiday: Waitangi Day, 6 February (1840) (Treaty of Waitangi established British sovereignty)
Constitution: 19 October 1974 (Niue Constitution Act)
Legal system: English common law
Suffrage: 18 years of age; universal
Executive branch:
chief of state: Queen ELIZABETH II (since 6 February 1952), represented by New Zealand Representative Kurt MEYER (since NA)
head of government: Premier Frank F. LUI (since 12 March 1993; Acting Premier since December 1992)
cabinet: Cabinet; consists of the premier and three other ministers
Legislative branch: unicameral
Legislative Assembly: elections last held 6 March 1993 (next to be held NA 1996); results—percent of vote NA; seats—(20 total, 6 elected)
Judicial branch: Appeal Court of New Zealand, High Court
Political parties and leaders: Niue Peoples Party (NPP), Young VIVIAN
Member of: ESCAP (associate), INTELSAT (nonsignatory user), SPARTECA, SPC, SPF, UNESCO, WHO
Diplomatic representation in US: none (self-governing territory in free association with New Zealand)
US diplomatic representation: none (self-governing territory in free association with New Zealand)
Flag: yellow with the flag of the UK in the upper hoist-side quadrant; the flag of the UK bears five yellow five-pointed stars—a large one on a blue disk in the center and a smaller one on each arm of the bold red cross

Economy

Overview: The economy is heavily dependent on aid from New Zealand. Government expenditures regularly exceed revenues, with the shortfall made up by grants from New Zealand—the grants are used to pay wages to public employees. The agricultural sector consists mainly of subsistence gardening, although some cash crops are grown for export. Industry consists primarily of small factories to process passion fruit, lime oil, honey, and coconut cream. The sale of postage stamps to foreign collectors is an important source of revenue. The island in recent years has suffered a serious loss of population because of migration of Niueans to New Zealand.
National product: GDP—purchasing power parity—$2.4 million (1993 est.)
National product real growth rate: NA%
National product per capita: $1,200 (1993 est.)
Inflation rate (consumer prices): 5% (1992)
Unemployment rate: NA%
Budget:
revenues: $5.5 million
expenditures: $6.3 million, including capital expenditures of $NA (1985 est.)
Exports: $117,500 (f.o.b., 1989)
commodities: canned coconut cream, copra, honey, passion fruit products, pawpaw, root crops, limes, footballs, stamps, handicrafts
partners: NZ 89%, Fiji, Cook Islands, Australia
Imports: $4.1 million (c.i.f., 1989)
commodities: food, live animals, manufactured goods, machinery, fuels, lubricants, chemicals, drugs
partners: NZ 59%, Fiji 20%, Japan 13%, Western Samoa, Australia, US
External debt: $NA
Industrial production: growth rate NA%
Electricity:
capacity: 1,500 kW
production: 2.7 million kWh
consumption per capita: 1,490 kWh (1992)
Industries: tourism, handicrafts, food processing
Agriculture: coconuts, passion fruit, honey, limes; subsistence crops—taro, yams, cassava (tapioca), sweet potatoes; pigs, poultry, beef cattle
Economic aid:
recipient: Western (non-US) countries, ODA and OOF bilateral commitments (1970-89), $62 million
Currency: 1 New Zealand dollar (NZ$) = 100 cents
Exchange rates: New Zealand dollars (NZ$) per US$1—1.5601 (January 1995), 1.6844 (1994), 1.8495 (1993), 1.8584 (1992), 1.7265 (1991), 1.6750 (1990)
Fiscal year: 1 April—31 March

Transportation

Railroads: 0 km
Highways:
total: 229 km
unpaved: all-weather 123 km; plantation access 106 km
Ports: none; offshore anchorage only
Merchant marine: none
Airports:
total: 1
with paved runways 1,524 to 2,437 m: 1

Communications

Telephone system: 383 telephones
local: NA
intercity: single-line telephone system connects all villages on island
international: NA
Radio:
broadcast stations: AM 1,000, FM 1, shortwave 0 (1987 est.)
radios: NA
Television:
broadcast stations: 0
televisions: NA

Defense Forces

Branches: Police Force
Note: defense is the responsibility of New Zealand

Norfolk Island

(territory of Australia)

Geography

Location: Oceania, island in the South Pacific Ocean, east of Australia
Map references: Oceania
Area:
total area: 34.6 sq km
land area: 34.6 sq km
comparative area: about 0.2 times the size of Washington, DC
Land boundaries: 0 km
Coastline: 32 km
Maritime claims:
exclusive fishing zone: 200 nm
territorial sea: 3 nm
International disputes: none
Climate: subtropical, mild, little seasonal temperature variation
Terrain: volcanic formation with mostly rolling plains
Natural resources: fish
Land use:
arable land: 0%
permanent crops: 0%
meadows and pastures: 25%
forest and woodland: 0%
other: 75%
Irrigated land: NA sq km
Environment:
current issues: NA
natural hazards: typhoons (especially May to July)
international agreements: NA

People

Population: 2,756 (July 1995 est.)
Age structure:
0-14 years: NA
15-64 years: NA
65 years and over: NA
Population growth rate: 1.69% (1995 est.)
Birth rate: NA
Death rate: NA
Net migration rate: NA
Infant mortality rate: NA
Life expectancy at birth: NA
Total fertility rate: NA
Nationality:
noun: Norfolk Islander(s)
adjective: Norfolk Islander(s)
Ethnic divisions: descendants of the Bounty mutineers, Australian, New Zealander
Religions: Anglican 39%, Roman Catholic 11.7%, Uniting Church in Australia 16.4%, Seventh-Day Adventist 4.4%, none 9.2%, unknown 16.9%, other 2.4% (1986)
Languages: English (official), Norfolk a mixture of 18th century English and ancient Tahitian
Labor force: NA

Government

Names:
conventional long form: Territory of Norfolk Island
conventional short form: Norfolk Island
Digraph: NF
Type: territory of Australia
Capital: Kingston (administrative center); Burnt Pine (commercial center)
Administrative divisions: none (territory of Australia)
Independence: none (territory of Australia)
National holiday: Pitcairners Arrival Day Anniversary, 8 June (1856)
Constitution: Norfolk Island Act of 1979
Legal system: wide legislative and executive responsibility under the Norfolk Island Act of 1979; Supreme Court
Suffrage: 18 years of age; universal
Executive branch:
chief of state: Queen ELIZABETH II (since 6 February 1952), represented by Administrator Alan Gardner KERR (since NA April 1992), who is appointed by the Governor General of Australia
head of government: Assembly President David Ernest BUFFETT (since NA May 1992)
cabinet: Executive Council
Legislative branch: unicameral
Legislative Assembly: elections last held 20 May 1992 (next to be held NA May 1995); results—percent of vote by party NA; seats—(9 total) independents 9
Judicial branch: Supreme Court
Political parties and leaders: none
Member of: none
Diplomatic representation in US: none (territory of Australia)
US diplomatic representation: none (territory of Australia)
Flag: three vertical bands of green (hoist side), white, and green with a large green Norfolk Island pine tree centered in the slightly wider white band

Economy

Overview: The primary economic activity is tourism, which has brought a level of prosperity unusual among inhabitants of the Pacific islands. The number of visitors has increased steadily over the years and reached 29,000 in FY88/89. Revenues from tourism have given the island a favorable balance of trade and helped the agricultural sector to become self-sufficient in the production of beef, poultry, and eggs.
National product: GDP $NA
National product real growth rate: NA%
National product per capita: $NA
Inflation rate (consumer prices): NA%
Unemployment rate: NA%
Budget:
revenues: $NA
expenditures: $4.2 million, including capital expenditures of $400,000 (1989 est.)
Exports: $1.7 million (f.o.b., FY85/86)
commodities: postage stamps, seeds of the Norfolk Island pine and Kentia palm, small quantities of avocados
partners: Australia, Pacific Islands, NZ, Asia, Europe
Imports: $15.6 million (c.i.f., FY85/86)
commodities: NA
partners: Australia, Pacific Islands, NZ, Asia, Europe
External debt: $NA
Industrial production: growth rate NA%
Electricity:
capacity: 7,000 kW
production: 8 million kWh
consumption per capita: 3,160 kWh (1990)
Industries: tourism
Agriculture: Norfolk Island pine seed, Kentia palm seed, cereals, vegetables, fruit, cattle, poultry
Economic aid: none
Currency: 1 Australian dollar ($A) = 100 cents
Exchange rates: Australian dollars ($A) per US$1—1.3058 (January 1995), 1.3667 (1994), 1.4704 (1993), 1.3600 (1992), 1.2835 (1991), 1.2799 (1990)
Fiscal year: 1 July—30 June

Transportation

Railroads: 0 km
Highways:
total: 80 km
paved: 53 km
unpaved: earth, coral 27 km
Ports: none; loading jetties at Kingston and Cascade
Merchant marine: none
Airports:
total: 1
with paved runways 1,524 to 2,437 m: 1

Communications

Telephone system: 987 telephones (1983)
local: NA
intercity: NA
international: radio link service with Sydney

Radio:
broadcast stations: AM 1, FM 0, shortwave 0
radios: 1,000 (1987 est.)
Television:
broadcast stations: 0
televisions: NA

Defense Forces

Note: defense is the responsibility of Australia

Northern Mariana Islands

(commonwealth in political union with the US)

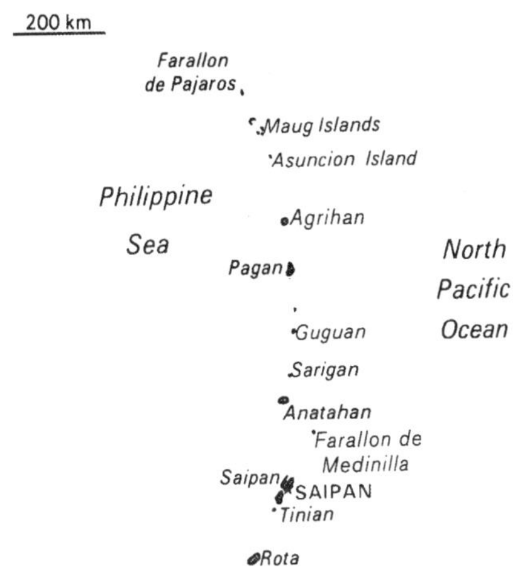

Geography

Location: Oceania, islands in the North Pacific Ocean, about three-quarters of the way from Hawaii to the Philippines
Map references: Oceania
Area:
total area: 477 sq km
land area: 477 sq km
comparative area: slightly more than 2.5 times the size of Washington, DC
note: includes 14 islands including Saipan, Rota, and Tinian
Land boundaries: 0 km
Coastline: 1,482 km
Maritime claims:
exclusive economic zone: 200 nm
territorial sea: 12 nm
International disputes: none
Climate: tropical marine; moderated by northeast trade winds, little seasonal temperature variation; dry season December to June, rainy season July to October
Terrain: southern islands are limestone with level terraces and fringing coral reefs; northern islands are volcanic; highest elevation is 471 m (Mt. Okso' Takpochao on Saipan)
Natural resources: arable land, fish
Land use:
arable land: 5% on Saipan
permanent crops: NA%
meadows and pastures: 19%
forest and woodland: NA%
other: NA%
Irrigated land: NA sq km
Environment:
current issues: contamination of groundwater on Saipan by raw sewage contributes to disease
natural hazards: active volcanoes on Pagan and Agrihan; typhoons (especially August to November)
international agreements: NA
Note: strategic location in the North Pacific Ocean

People

Population: 51,033 (July 1995 est.)
Age structure:
0-14 years: NA
15-64 years: NA
65 years and over: NA
Population growth rate: 3.04% (1995 est.)
Birth rate: 33.05 births/1,000 population (1995 est.)
Death rate: 4.61 deaths/1,000 population (1995 est.)
Net migration rate: 0 migrant(s)/1,000 population (1995 est.)
Infant mortality rate: 37.96 deaths/1,000 live births (1995 est.)
Life expectancy at birth:
total population: 67.43 years
male: 65.53 years
female: 69.48 years (1995 est.)
Total fertility rate: 2.69 children born/woman (1995 est.)
Nationality:
noun: NA
adjective: NA
Ethnic divisions: Chamorro, Carolinians and other Micronesians, Caucasian, Japanese, Chinese, Korean
Religions: Christian (Roman Catholic majority, although traditional beliefs and taboos may still be found)
Languages: English, Chamorro, Carolinian
note: 86% of population speaks a language other than English at home
Literacy: age 15 and over can read and write (1980)
total population: 97%
male: 97%
female: 96%
Labor force: 7,476 total indigenous labor force, 2,699 unemployed; 21,188 foreign workers (1990)
by occupation: NA

Government

Names:
conventional long form: Commonwealth of the Northern Mariana Islands
conventional short form: Northern Mariana Islands
Digraph: CQ
Type: commonwealth in political union with the US; self-governing with locally elected governor, lieutenant governor, and legislature; federal funds to the Commonwealth administered by the US Department of the Interior, Office of Territorial and International Affairs
Capital: Saipan
Administrative divisions: none
Independence: none (commonwealth in political union with the US)

Northern Mariana Islands

(continued)

National holiday: Commonwealth Day, 8 January (1978)
Constitution: Covenant Agreement effective 3 November 1986 and the Constitution of the Commonwealth of the Northern Mariana Islands
Legal system: based on US system except for customs, wages, immigration laws, and taxation
Suffrage: 18 years of age; universal; indigenous inhabitants are US citizens but do not vote in US presidential elections
Executive branch:
chief of state: President William Jefferson CLINTON (since 20 January 1993); Vice President Albert GORE, Jr. (since 20 January 1993)
head of government: Governor Froilan C. TENORIO (since January 1994); Lieutenant Governor Jesus C. BORJA (since January 1994); election last held NA November 1993 (next to be held NA November 1997); results—Froilan C. TENORIO (Democratic Party) was elected governor
Legislative branch: bicameral Legislature
Senate: elections last held NA November 1993 (next to be held NA November 1995); results—percent of vote by party NA; seats—(9 total) Republicans retained a majority
House of Representatives: elections last held NA November 1993 (next to be held NA November 1995); results—percent of vote by party NA; seats—(18 total) Republicans retained a majority
US House of Representatives: the Commonwealth does not have a nonvoting delegate in Congress; instead, it has an elected official ''resident representative'' located in Washington, DC; seats—(1 total) Juan N. BABAUTA, Republican
Judicial branch: Commonwealth Supreme Court, Superior Court, Federal District Court
Political parties and leaders: Republican Party, Benigno R. FITIAL, leader; Democratic Party, Dr. Carlos S. CAMACHO, chairman
Member of: ESCAP (associate), INTERPOL (subbureau), SPC
Flag: blue with a white five-pointed star superimposed on the gray silhouette of a latte stone (a traditional foundation stone used in building) in the center

Economy

Overview: The economy benefits substantially from financial assistance from the US. The rate of funding has declined as locally generated government revenues have grown. An agreement for the years 1986 to 1992 entitled the islands to $228 million for capital development, government operations, and special programs. A rapidly growing major source of income is the tourist industry, which now employs about 50% of the work force. Japanese tourists predominate. The agricultural sector is of minor importance and is made up of cattle ranches and small farms producing coconuts, breadfruit, tomatoes, and melons. Industry is small scale, mostly handicrafts, light manufacturing, and garment production
National product: GDP—purchasing power parity—$524 million (1994 est.)
note: GDP numbers reflect US spending
National product real growth rate: NA%
National product per capita: $10,500 (1994 est.)
Inflation rate (consumer prices): 6.5% (1994 est.)
Unemployment rate: NA%
Budget:
revenues: $190.4 million
expenditures: $190.4 million, including capital expenditures of $19.1 million (FY94/95)
Exports: $263.4 million (f.o.b. 1991 est.)
commodities: garments
partners: NA
Imports: $392.4 million (c.i.f. 1991 est.)
commodities: food, construction equipment and materials, petroleum products
partners: US, Japan
External debt: $NA
Industrial production: growth rate NA%
Electricity:
capacity: 105,000 kW
production: NA
consumption per capita: NA
Industries: tourism, construction, light industry, handicrafts
Agriculture: coconuts, fruits, cattle, vegetables; food is a major import
Economic aid: none
Currency: 1 United States dollar (US$) = 100 cents
Exchange rates: US currency is used
Fiscal year: 1 October—30 September

Transportation

Railroads: 0 km
Highways:
total: 381.5 km
paved: NA
unpaved: NA
undifferentiated: primary 134.5 km; secondary 55 km; local 192 km (1991)
Inland waterways: none
Ports: Saipan, Tinian
Merchant marine: none
Airports:
total: 8
with paved runways 2,438 to 3,047 m: 1
with paved runways 1,524 to 2,437 m: 3
with paved runways under 914 m: 3
with unpaved runways 2,438 to 3,047 m: 1

Communications

Telephone system: NA telephones
local: NA
intercity: NA
international: 2 INTELSAT (Pacific Ocean) earth stations
Radio:
broadcast stations: AM 2, FM 1, shortwave 0 (1984)
radios: NA
Television:
broadcast stations: 1; note—there are 2 cable TV stations
televisions: NA

Defense Forces

Note: defense is the responsibility of the US

Norway

Geography

Location: Northern Europe, bordering the North Sea and the North Atlantic Ocean, west of Sweden
Map references: Europe
Area:
total area: 324,220 sq km
land area: 307,860 sq km
comparative area: slightly larger than New Mexico
Land boundaries: total 2,515 km, Finland 729 km, Sweden 1,619 km, Russia 167 km
Coastline: 21,925 km (includes mainland 3,419 km, large islands 2,413 km, long fjords, numerous small islands, and minor indentations 16,093 km)
Maritime claims:
contiguous zone: 10 nm
continental shelf: 200 nm
exclusive economic zone: 200 nm
territorial sea: 4 nm
International disputes: territorial claim in Antarctica (Queen Maud Land); maritime boundary dispute with Russia over portion of Barents Sea
Climate: temperate along coast, modified by North Atlantic Current; colder interior; rainy year-round on west coast
Terrain: glaciated; mostly high plateaus and rugged mountains broken by fertile valleys; small, scattered plains; coastline deeply indented by fjords; arctic tundra in north
Natural resources: petroleum, copper, natural gas, pyrites, nickel, iron ore, zinc, lead, fish, timber, hydropower
Land use:
arable land: 3%
permanent crops: 0%
meadows and pastures: 0%
forest and woodland: 27%
other: 70%
Irrigated land: 950 sq km (1989)
Environment:
current issues: water pollution; acid rain damaging forests and adversely affecting lakes, threatening fish stocks; air pollution from vehicle emissions
natural hazards: NA
international agreements: party to—Air Pollution, Air Pollution-Nitrogen Oxides, Air Pollution-Sulphur 85, Air Pollution-Volatile Organic Compounds, Antarctic-Environmental Protocol, Antarctic Treaty, Biodiversity, Climate Change, Endangered Species, Environmental Modification, Hazardous Wastes, Marine Dumping, Nuclear Test Ban, Ozone Layer Protection, Ship Pollution, Tropical Timber 83, Wetlands, Whaling; signed, but not ratified—Air Pollution-Sulphur 94, Desertification, Law of the Sea, Tropical Timber 94
Note: about two-thirds mountains; some 50,000 islands off its much indented coastline; strategic location adjacent to sea lanes and air routes in North Atlantic; one of most rugged and longest coastlines in world; Norway and Turkey only NATO members having a land boundary with Russia

People

Population: 4,330,951 (July 1995 est.)
Age structure:
0-14 years: 19% (female 390,344; male 444,570)
15-64 years: 65% (female 1,375,493; male 1,424,027)
65 years and over: 16% (female 408,675; male 287,842) (July 1995 est.)
Population growth rate: 0.37% (1995 est.)
Birth rate: 12.86 births/1,000 population (1995 est.)
Death rate: 10.35 deaths/1,000 population (1995 est.)
Net migration rate: 1.15 migrant(s)/1,000 population (1995 est.)
Infant mortality rate: 6.1 deaths/1,000 live births (1995 est.)
Life expectancy at birth:
total population: 77.61 years
male: 74.26 years
female: 81.15 years (1995 est.)
Total fertility rate: 1.76 children born/woman (1995 est.)
Nationality:
noun: Norwegian(s)
adjective: Norwegian
Ethnic divisions: Germanic (Nordic, Alpine, Baltic), Lapps (Sami) 20,000
Religions: Evangelical Lutheran 87.8% (state church), other Protestant and Roman Catholic 3.8%, none 3.2%, unknown 5.2% (1980)
Languages: Norwegian (official)
note: small Lapp- and Finnish-speaking minorities
Literacy: age 15 and over can read and write (1976 est.)
total population: 99%
Labor force: 2.13 million (1992)
by occupation: services 71.0%, industry 23.5%, agriculture, forestry, and fishing 5.6% (1992)

Government

Names:
conventional long form: Kingdom of Norway
conventional short form: Norway
local long form: Kongeriket Norge
local short form: Norge
Digraph: NO
Type: constitutional monarchy
Capital: Oslo
Administrative divisions: 19 provinces (fylker, singular—fylke); Akershus, Aust-Agder, Buskerud, Finnmark, Hedmark, Hordaland, More og Romsdal, Nordland, Nord-Trondelag, Oppland, Oslo, Ostfold, Rogaland, Sogn og Fjordane, Sor-Trondelag, Telemark, Troms, Vest-Agder, Vestfold
Dependent areas: Bouvet Island, Jan Mayen, Svalbard
Independence: 26 October 1905 (from Sweden)
National holiday: Constitution Day, 17 May (1814)
Constitution: 17 May 1814, modified in 1884
Legal system: mixture of customary law, civil law system, and common law traditions; Supreme Court renders advisory opinions to legislature when asked; accepts compulsory ICJ jurisdiction, with reservations
Suffrage: 18 years of age; universal
Executive branch:
chief of state: King HARALD V (since 17 January 1991); Heir Apparent Crown Prince HAAKON MAGNUS (born 20 July 1973)
head of government: Prime Minister Gro Harlem BRUNDTLAND (since 3 November 1990)
cabinet: State Council; appointed by the king in accordance with the will of the Storting
Legislative branch: modified unicameral Parliament (Storting) which, for certain purposes, divides itself into two chambers
Storting: elections last held 13 September 1993 (next to be held September 1997); results—Labor 37.1%, Center Party 18.5%, Conservatives 15.6%, Christian People's 8.4%, Socialist Left 7.9%, Progress 6%, Left Party 3.6%, Red Electoral Alliance 1.2%; seats—(165 total) Labor 67, Center Party 32, Consevatives 18, Christian People's 13, Socialist Left 13, Progress 10, Left Party 1, Red Electoral Alliance 1, unawarded 10
note: for certain purposes, the Storting divides itself into two chambers and elects one-fourth of its membership to an upper house or Lagting
Judicial branch: Supreme Court (Hoyesterett)
Political parties and leaders: Labor Party, Thorbjorn JAGLUND; Conservative Party, Jan PETERSEN; Center Party, Anne ENGER LAHNSTEIN; Christian People's Party, Kjell Magne BONDEVIK; Socialist Left, Kjellbjorg LUNDE; Norwegian Communist, Kare Andre NILSEN; Progress Party, Carl I. HAGEN; Liberal, Odd Einar DORUM; Left Party; Red Electoral Alliance, Erling FOLKVORD

Member of: AfDB, AsDB, Australia Group, BIS, CBSS, CCC, CE, CERN, EBRD, ECE, EFTA, ESA, FAO, GATT, IADB, IAEA, IBRD, ICAO, ICC, ICFTU, ICRM, IDA, IEA, IFAD, IFC, IFRCS, ILO, IMF, IMO, INMARSAT, INTELSAT, INTERPOL, IOC, IOM, ISO, ITU, MTCR, NACC, NAM (guest), NATO, NC, NEA, NIB, NSG, OECD, OSCE, PCA, UN, UNAVEM II, UNCTAD, UNESCO, UNHCR, UNIDO, UNIFIL, UNIKOM, UNITAR, UNMOGIP, UNOMOZ, UNPROFOR, UNTSO, UPU, WEU (associate), WHO, WIPO, WMO, ZC
Diplomatic representation in US:
chief of mission: Ambassador Kjeld VIBE
chancery: 2720 34th Street NW, Washington, DC 20008
telephone: [1] (202) 333-6000
FAX: [1] (202) 337-0870
consulate(s) general: Houston, Los Angeles, Minneapolis, New York, and San Francisco
consulate(s): Miami
US diplomatic representation:
chief of mission: Ambassador Thomas A. LOFTUS
embassy: Drammensveien 18, 0244 Oslo
mailing address: PSC 69, Box 1000, APO AE 09707
telephone: [47] 22 44 85 50
FAX: [47] 22 44 33 63
Flag: red with a blue cross outlined in white that extends to the edges of the flag; the vertical part of the cross is shifted to the hoist side in the style of the Dannebrog (Danish flag)

Economy

Overview: Norway has a mixed economy involving a combination of free market activity and government intervention. The government controls key areas, such as the vital petroleum sector (through large-scale state enterprises) and extensively subsidizes agriculture, fishing, and areas with sparse resources. Norway also maintains an extensive welfare system that helps propel public sector expenditures to slightly more than 50% of the GDP and results in one of the highest average tax burdens in the world (54%). A small country with a high dependence on international trade, Norway is basically an exporter of raw materials and semiprocessed goods, with an abundance of small- and medium-sized firms, and is ranked among the major shipping nations. The country is richly endowed with natural resources—petroleum, hydropower, fish, forests, and minerals—and is highly dependent on its oil sector to keep its economy afloat. Norway imports more than half its food needs. Although one of the government's main priorities is to reduce this dependency, this situation is not likely to improve for years to come. The government also hopes to reduce unemployment and strengthen and diversify the economy through tax reform and a series of expansionary budgets. The budget deficit is expected to hit a record 8% of GDP because of welfare spending and bail-outs of the banking system. Unemployment is currently running at 8.4%—including those in job programs—because of the weakness of the economy outside the oil sector. Economic growth, only 1.6% in 1993, moved up to 5.5% in 1994. Oslo opted to stay out of the EU during a referendum in November 1994.
National product: GDP—purchasing power parity—$95.7 billion (1994 est.)
National product real growth rate: 5.5% (1994 est.)
National product per capita: $22,170 (1994 est.)
Inflation rate (consumer prices): 1.3% (1994 est.)
Unemployment rate: 8.4% (including people in job-training programs; 1994 est.)
Budget:
revenues: $50.9 billion
expenditures: $55.5 billion, including capital expenditures of $NA (1994 est.)
Exports: $36.6 billion (f.o.b., 1994)
commodities: petroleum and petroleum products 40%, metals and products 10.6%, fish and fish products 6.9%, chemicals 6.4%, natural gas 6.0%, ships 5.4%
partners: EC 66.3%, Nordic countries 16.3%, developing countries 8.4%, US 6.0%, Japan 1.8% (1993)
Imports: $29.3 billion (c.i.f., 1994)
commodities: machinery and equipment 38.9%, chemicals and other industrial inputs 26.6%, manufactured consumer goods 17.8%, foodstuffs 6.4%
partners: EC 48.6%, Nordic countries 25.1%, developing countries 9.6%, US 8.1%, Japan 8.0% (1993)
External debt: $NA
Industrial production: growth rate 4.6% (1994); accounts for 14% of GDP
Electricity:
capacity: 27,280,000 kW
production: 118 billion kWh
consumption per capita: 23,735 kWh (1993)
Industries: petroleum and gas, food processing, shipbuilding, pulp and paper products, metals, chemicals, timber, mining, textiles, fishing
Agriculture: accounts for 3% of GDP and about 6% of labor force; among world's top 10 fishing nations; livestock output exceeds value of crops; fish catch of 1.76 million metric tons in 1989
Illicit drugs: transshipment point for drugs shipped via the CIS and Baltic states for the European market
Economic aid:
donor: ODA and OOF commitments (1970-89), $4.4 billion
Currency: 1 Norwegian krone (NKr) = 100 oere
Exchange rates: Norwegian kroner (NKr) per US$1—6.7014 (January 1995), 7.0469 (1994), 7.0941 (1993), 6.2145 (1992), 6.4829 (1991), 6.2597 (1990)
Fiscal year: calendar year

Transportation

Railroads:
total: 4,026 km
standard gauge: 4,026 km 1.435-m gauge (2,422 km electrified; 96 km double track) (1994)
Highways:
total: 88,922 km
paved: 61,356 km (75 km of expressway)
unpaved: gravel, crushed stone, earth 27,566 km (1990)
Inland waterways: 1,577 km along west coast; 2.4 m draft vessels maximum
Pipelines: refined products 53 km
Ports: Bergen, Drammen, Flora, Hammerfest, Harstad, Haugesund, Kristiansand, Larvik, Narvik, Oslo, Porsgrunn, Stavanger, Tromso, Trondheim
Merchant marine:
total: 764 ships (1,000 GRT or over) totaling 20,793,968 GRT/35,409,472 DWT
ships by type: bulk 159, cargo 92, chemical tanker 85, combination bulk 8, combination ore/oil 28, container 17, liquefied gas tanker 81, oil tanker 162, passenger 13, passenger-cargo 2, railcar carrier 1, refrigerated cargo 13, roll-on/roll-off cargo 54, short-sea passenger 21, vehicle carrier 28
note: the government has created a captive register, the Norwegian International Ship Register (NIS), as a subset of the Norwegian register; ships on the NIS enjoy many benefits of flags of convenience and do not have to be crewed by Norwegians
Airports:
total: 104
with paved runways over 3,047 m: 1
with paved runways 2,438 to 3,047 m: 12
with paved runways 1,524 to 2,437 m: 13
with paved runways 914 to 1,523 m: 10
with paved runways under 914 m: 62
with unpaved runways 914 to 1,523 m: 6

Communications

Telephone system: 3,102,000 telephones; high-quality domestic and international telephone, telegraph, and telex services
local: NA
intercity: domestic earth stations
international: 2 buried coaxial cable systems; 4 coaxial submarine cables; EUTELSAT, INTELSAT (Atlantic Ocean), and MARISAT earth stations
Radio:
broadcast stations: AM 46, FM 493 (350 private and 143 government), shortwave 0
radios: 3.3 million
Television:
broadcast stations: 54 (repeaters 2,100)
televisions: 1.5 million

Oman

Defense Forces

Branches: Norwegian Army, Royal Norwegian Navy, Royal Norwegian Air Force, Home Guard
Manpower availability: males age 15-49 1,116,130; males fit for military service 928,774; males reach military age (20) annually 29,123 (1995 est.)
Defense expenditures: exchange rate conversion—$3.4 billion, 3.2% of GDP (1994)

Geography

Location: Middle East, bordering the Arabian Sea, Gulf of Oman, and Persian Gulf, between Yemen and the United Arab Emirates
Map references: Middle East
Area:
total area: 212,460 sq km
land area: 212,460 sq km
comparative area: slightly smaller than Kansas
Land boundaries: total 1,374 km, Saudi Arabia 676 km, UAE 410 km, Yemen 288 km
Coastline: 2,092 km
Maritime claims:
contiguous zone: 24 nm
exclusive economic zone: 200 nm
territorial sea: 12 nm
International disputes: no defined boundary with most of UAE; Administrative Line with UAE in far north
Climate: dry desert; hot, humid along coast; hot, dry interior; strong southwest summer monsoon (May to September) in far south
Terrain: vast central desert plain, rugged mountains in north and south
Natural resources: petroleum, copper, asbestos, some marble, limestone, chromium, gypsum, natural gas
Land use:
arable land: less than 2%
permanent crops: 0%
meadows and pastures: 5%
forest and woodland: 0%
other: 93%
Irrigated land: 410 sq km (1989 est.)
Environment:
current issues: rising soil salinity, beach pollution from oil spills; very limited natural fresh water resources
natural hazards: summer winds often raise large sandstorms and duststorms in interior; periodic droughts
international agreements: party to—Law of the Sea, Marine Dumping, Ship Pollution, Whaling; signed, but not ratified—Biodiversity, Climate Change

Note: strategic location with small foothold on Musandam Peninsula controlling Strait of Hormuz, a vital transit point for world crude oil

People

Population: 2,125,089 (July 1995 est.)
Age structure:
0-14 years: 46% (female 480,974; male 498,619)
15-64 years: 51% (female 493,685; male 593,740)
65 years and over: 3% (female 31,826; male 26,245) (July 1995 est.)
Population growth rate: 3.71% (1995 est.)
Birth rate: 38.05 births/1,000 population (1995 est.)
Death rate: 5 deaths/1,000 population (1995 est.)
Net migration rate: 4.09 migrant(s)/1,000 population (1995 est.)
Infant mortality rate: 34.3 deaths/1,000 live births (1995 est.)
Life expectancy at birth:
total population: 70.25 years
male: 68.31 years
female: 72.29 years (1995 est.)
Total fertility rate: 6.16 children born/woman (1995 est.)
Nationality:
noun: Omani(s)
adjective: Omani
Ethnic divisions: Arab, Baluchi, South Asian (Indian, Pakistani, Sri Lankan, Bangladeshi)
Religions: Ibadhi Muslim 75%, Sunni Muslim, Shi'a Muslim, Hindu
Languages: Arabic (official), English, Baluchi, Urdu, Indian dialects
Literacy: NA%
Labor force: 430,000 (est.)
by occupation: agriculture 40% (est.)

Government

Names:
conventional long form: Sultanate of Oman
conventional short form: Oman
local long form: Saltanat Uman
local short form: Uman
Digraph: MU
Type: monarchy
Capital: Muscat
Administrative divisions: 6 regions (mintaqah, singular—mintaqat) and 2 governorates* (muhafazah, singular—muhafazat) Ad Dakhiliyah, Al Batinah, Al Wusta, Ash Sharqiyah, Az Zahirah, Masqat, Musandam*, Zufar*
Independence: 1650 (expulsion of the Portuguese)
National holiday: National Day, 18 November (1940)
Constitution: none
Legal system: based on English common law and Islamic law; ultimate appeal to the sultan;

has not accepted compulsory ICJ jurisdiction
Suffrage: none
Executive branch:
chief of state and head of government: Sultan and Prime Minister QABOOS bin Said Al Said (since 23 July 1970)
cabinet: Cabinet
Legislative branch: unicameral Consultative Council
Judicial branch: none; traditional Islamic judges and a nascent civil court system
Political parties and leaders: none
Other political or pressure groups: NA
Member of: ABEDA, AFESD, AL, AMF, ESCWA, FAO, G-77, GCC, IBRD, ICAO, IDA, IDB, IFAD, IFC, ILO, IMF, IMO, INMARSAT, INTELSAT, INTERPOL, IOC, ISO (correspondent), ITU, NAM, OIC, UN, UNCTAD, UNESCO, UNIDO, UPU, WFTU, WHO, WMO
Diplomatic representation in US:
chief of mission: Ambassador Abdallah bin Muhammad bin Aqil al-DHAHAB
chancery: 2535 Belmont Rd. NW, Washington, DC 20008
telephone: [1] (202) 387-1980 through 1982
FAX: [1] (202) 745-4933
US diplomatic representation:
chief of mission: Ambassador David J. DUNFORD
embassy: address NA, Muscat
mailing address: P. O. Box 202, Code No. 115, Muscat
telephone: [968] 698989
FAX: [968] 699779
Flag: three horizontal bands of white (top, double width), red, and green (double width) with a broad, vertical, red band on the hoist side; the national emblem (a khanjar dagger in its sheath superimposed on two crossed swords in scabbards) in white is centered at the top of the vertical band

Economy

Overview: Economic performance is closely tied to the fortunes of the oil industry, including trends in international oil prices and the ability of OPEC producers to agree on output quotas. Petroleum accounts for more than 85% of export earnings, about 80% of government revenues, and roughly 40% of GDP. Oman has proved oil reserves of 4 billion barrels, equivalent to about 20 years' supply at the current rate of extraction. Agriculture is carried on at a subsistence level and the general population depends on imported food. The government is encouraging private investment, both domestic and foreign, as a prime force for further economic development.
National product: GDP—purchasing power parity—$17 billion (1994 est.)
National product real growth rate: 0.5% (1994 est.)
National product per capita: $10,020 (1994 est.)
Inflation rate (consumer prices): 1.2% (1994 est.)
Unemployment rate: NA%
Budget:
revenues: $4.4 billion
expenditures: $5.2 billion, including capital expenditures of $1 billion (1994 est.)
Exports: $4.8 billion (f.o.b., 1994 est.)
commodities: petroleum 87%, re-exports, fish, processed copper, textiles
partners: UAE 33%, Japan 20%, South Korea 14%, China 7% (1993)
Imports: $4.1 billion (c.i.f., 1994 est.)
commodities: machinery, transportation equipment, manufactured goods, food, livestock, lubricants
partners: UAE 24% (largely re-exports), Japan 21%, UK 12%, US 7%, France 6% (1993)
External debt: $3 billion (1993)
Industrial production: growth rate 8.6% (1991); accounts for almost 60% of GDP, including petroleum
Electricity:
capacity: 1,540,000 kW
production: 6 billion kWh
consumption per capita: 3,407 kWh (1993)
Industries: crude oil production and refining, natural gas production, construction, cement, copper
Agriculture: accounts for 4% of GDP and 40% of the labor force (including fishing); less than 2% of land cultivated; largely subsistence farming (dates, limes, bananas, alfalfa, vegetables, camels, cattle); not self-sufficient in food; annual fish catch averages 100,000 metric tons
Economic aid:
recipient: US commitments, including Ex-Im (FY70-89), $137 million; Western (non-US) countries, ODA and OOF bilateral commitments (1970-89), $148 million; OPEC bilateral aid (1979-89), $797 million
Currency: 1 Omani rial (RO) = 1,000 baiza
Exchange rates: Omani rials (RO) per US$1—0.3845 (fixed rate since 1986)
Fiscal year: calendar year

Transportation

Railroads: 0 km
Highways:
total: 26,000 km
paved: 5,000 km
unpaved: 21,000 km (1992 est.)
Pipelines: crude oil 1,300 km; natural gas 1,030 km
Ports: Mina' al Fahl, Mina' Qabus, Mina' Raysut
Merchant marine:
total: 1 passenger ship (1,000 GRT or over) totaling 4,442 GRT/1,320 DWT
Airports:
total: 140
with paved runways over 3,047 m: 4
with paved runways 1,524 to 2,437 m: 1
with paved runways 914 to 1,523 m: 1
with paved runways under 914 m: 36
with unpaved runways over 3,047 m: 3
with unpaved runways 2,438 to 3,047 m: 3
with unpaved runways 1,524 to 2,438 m: 61
with unpaved runways 914 to 1,523 m: 31

Communications

Telephone system: 50,000 telephones; modern system consisting of open-wire, microwave, and radio communications stations; limited coaxial cable
local: NA
intercity: open-wire, microwave, radio communications, and 8 domestic satellite links
international: 2 INTELSAT (Indian Ocean) and 1 ARABSAT earth station
Radio:
broadcast stations: AM 2, FM 3, shortwave 0
radios: NA
Television:
broadcast stations: 7
televisions: NA

Defense Forces

Branches: Army, Navy, Air Force, Royal Oman Police
Manpower availability: males age 15-49 520,428; males fit for military service 294,993; males reach military age (14) annually 26,065 (1995 est.)
Defense expenditures: exchange rate conversion—$1.7 billion, 14.2% of GDP (1995 est.)

Pacific Ocean

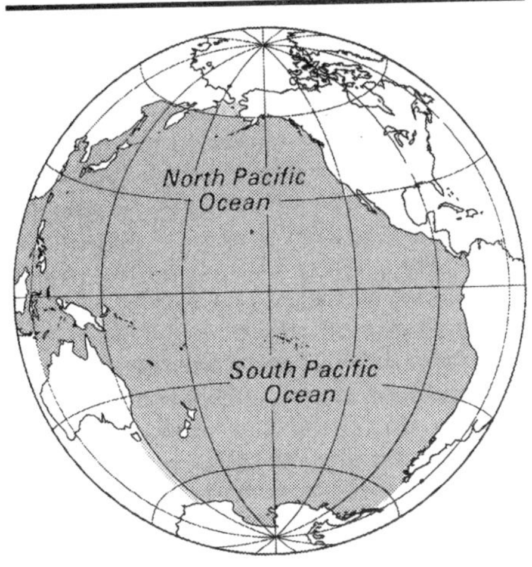

Geography

Location: body of water between Antarctica, Asia, Australia, and the Western Hemisphere
Map references: World
Area:
total area: 165.384 million sq km
comparative area: about 18 times the size of the US; the largest ocean (followed by the Atlantic Ocean, the Indian Ocean, and the Arctic Ocean); covers about one-third of the global surface; larger than the total land area of the world
note: includes Bali Sea, Bellingshausen Sea, Bering Sea, Bering Strait, Coral Sea, East China Sea, Flores Sea, Gulf of Alaska, Gulf of Tonkin, Java Sea, Philippine Sea, Ross Sea, Savu Sea, Sea of Japan, Sea of Okhotsk, South China Sea, Tasman Sea, Timor Sea, and other tributary water bodies
Coastline: 135,663 km
International disputes: some maritime disputes (see littoral states)
Climate: the western Pacific is monsoonal—a rainy season occurs during the summer months, when moisture-laden winds blow from the ocean over the land, and a dry season during the winter months, when dry winds blow from the Asian land mass back to the ocean
Terrain: surface currents in the northern Pacific are dominated by a clockwise, warm-water gyre (broad circular system of currents) and in the southern Pacific by a counterclockwise, cool-water gyre; in the northern Pacific sea ice forms in the Bering Sea and Sea of Okhotsk in winter; in the southern Pacific sea ice from Antarctica reaches its northernmost extent in October; the ocean floor in the eastern Pacific is dominated by the East Pacific Rise, while the western Pacific is dissected by deep trenches, including the world's deepest, the 10,924 meter Marianas Trench
Natural resources: oil and gas fields, polymetallic nodules, sand and gravel aggregates, placer deposits, fish
Environment:
current issues: endangered marine species include the dugong, sea lion, sea otter, seals, turtles, and whales; oil pollution in Philippine Sea and South China Sea
natural hazards: surrounded by a zone of violent volcanic and earthquake activity sometimes referred to as the Pacific Ring of Fire; subject to tropical cyclones (typhoons) in southeast and east Asia from May to December (most frequent from July to October); tropical cyclones (hurricanes) may form south of Mexico and strike Central America and Mexico from June to October (most common in August and September); southern shipping lanes subject to icebergs from Antarctica; occasional El Nino phenomenon occurs off the coast of Peru when the trade winds slacken and the warm Equatorial Countercurrent moves south, killing the plankton that is the primary food source for anchovies; consequently, the anchovies move to better feeding grounds, causing resident marine birds to starve by the thousands because of their lost food source; ships subject to superstructure icing in extreme north from October to May and in extreme south from May to October; persistent fog in the northern Pacific can be a maritime hazard from June to December
international agreements: NA
Note: the major choke points are the Bering Strait, Panama Canal, Luzon Strait, and the Singapore Strait; the Equator divides the Pacific Ocean into the North Pacific Ocean and the South Pacific Ocean; dotted with low coral islands and rugged volcanic islands in the southwestern Pacific Ocean

Government

Digraph: ZN

Economy

Overview: The Pacific Ocean is a major contributor to the world economy and particularly to those nations its waters directly touch. It provides low-cost sea transportation between East and West, extensive fishing grounds, offshore oil and gas fields, minerals, and sand and gravel for the construction industry. In 1985 over half (54%) of the world's fish catch came from the Pacific Ocean, which is the only ocean where the fish catch has increased every year since 1978. Exploitation of offshore oil and gas reserves is playing an ever-increasing role in the energy supplies of Australia, NZ, China, US, and Peru. The high cost of recovering offshore oil and gas, combined with the wide swings in world prices for oil since 1985, has slowed but not stopped new drillings.
Industries: fishing, oil and gas production

Transportation

Ports: Bangkok (Thailand), Hong Kong, Los Angeles (US), Manila (Philippines), Pusan (South Korea), San Francisco (US), Seattle (US), Shanghai (China), Singapore, Sydney (Australia), Vladivostok (Russia), Wellington (NZ), Yokohama (Japan)

Communications

Telephone system:
international: several submarine cables with network nodal points on Guam and Hawaii

Pakistan

Geography

Location: Southern Asia, bordering the Arabian Sea, between India and Iran
Map references: Asia
Area:
total area: 803,940 sq km
land area: 778,720 sq km
comparative area: slightly less than twice the size of California
Land boundaries: total 6,774 km, Afghanistan 2,430 km, China 523 km, India 2,912 km, Iran 909 km
Coastline: 1,046 km
Maritime claims:
contiguous zone: 24 nm
continental shelf: 200 nm or to the edge of the continental margin
exclusive economic zone: 200 nm
territorial sea: 12 nm
International disputes: status of Kashmir with India; border question with Afghanistan (Durand Line); water-sharing problems (Wular Barrage) over the Indus with upstream riparian India
Climate: mostly hot, dry desert; temperate in northwest; arctic in north
Terrain: flat Indus plain in east; mountains in north and northwest; Balochistan plateau in west
Natural resources: land, extensive natural gas reserves, limited petroleum, poor quality coal, iron ore, copper, salt, limestone
Land use:
arable land: 23%
permanent crops: 0%
meadows and pastures: 6%
forest and woodland: 4%
other: 67% (1993)
Irrigated land: 170,000 sq km (1992)
Environment:
current issues: water pollution from raw sewage, industrial wastes, and agricultural runoff; limited natural fresh water resources; a majority of the population does not have access to potable water; deforestation; soil erosion; desertification
natural hazards: frequent earthquakes, occasionally severe especially in north and west; flooding along the Indus after heavy rains (July and August)
international agreements: party to—Biodiversity, Climate Change, Endangered Species, Environmental Modification, Hazardous Wastes, Nuclear Test Ban, Ozone Layer Protection, Ship Pollution, Wetlands; signed, but not ratified—Desertification, Law of the Sea, Marine Life Conservation
Note: controls Khyber Pass and Bolan Pass, traditional invasion routes between Central Asia and the Indian Subcontinent

People

Population: 131,541,920 (July 1995 est.)
Age structure:
0-14 years: 44% (female 28,033,354; male 29,777,818)
15-64 years: 52% (female 33,456,410; male 35,109,482)
65 years and over: 4% (female 2,556,846; male 2,608,010) (July 1995 est.)
Population growth rate: 1.28% (1995 est.)
Birth rate: 41.8 births/1,000 population (1995 est.)
Death rate: 12.07 deaths/1,000 population (1995 est.)
Net migration rate: -16.93 migrant(s)/1,000 population (1995 est.)
Infant mortality rate: 99.5 deaths/1,000 live births (1995 est.)
Life expectancy at birth:
total population: 57.86 years
male: 57.18 years
female: 58.56 years (1995 est.)
Total fertility rate: 6.35 children born/woman (1995 est.)
Nationality:
noun: Pakistani(s)
adjective: Pakistani
Ethnic divisions: Punjabi, Sindhi, Pashtun (Pathan), Baloch, Muhajir (immigrants from India and their descendents)
Religions: Muslim 97% (Sunni 77%, Shi'a 20%), Christian, Hindu, and other 3%
Languages: Urdu (official), English (official; lingua franca of Pakistani elite and most government ministries), Punjabi 64%, Sindhi 12%, Pashtu 8%, Urdu 7%, Balochi and other 9%
Literacy: age 15 and over can read and write (1990 est.)
total population: 35%
male: 47%
female: 21%
Labor force: 36 million
by occupation: agriculture 46%, mining and manufacturing 18%, services 17%, other 19%
note: extensive export of labor

Government

Names:
conventional long form: Islamic Republic of Pakistan
conventional short form: Pakistan
former: West Pakistan
Digraph: PK
Type: republic
Capital: Islamabad
Administrative divisions: 4 provinces, 1 territory*, and 1 capital territory**; Balochistan, Federally Administered Tribal Areas*, Islamabad Capital Territory**, North-West Frontier, Punjab, Sindh
note: the Pakistani-administered portion of the disputed Jammu and Kashmir region includes Azad Kashmir and the Northern Areas
Independence: 14 August 1947 (from UK)
National holiday: Pakistan Day, 23 March (1956) (proclamation of the republic)
Constitution: 10 April 1973, suspended 5 July 1977, restored with amendments 30 December 1985
Legal system: based on English common law with provisions to accommodate Pakistan's stature as an Islamic state; accepts compulsory ICJ jurisdiction, with reservations
Suffrage: 21 years of age; universal; separate electorates and reserved parliamentary seats for non-Muslims
Executive branch:
chief of state: President Sardar Farooq LEGHARI; election last held 13 November 1993 (next to be held no later than 14 October 1998); results—LEGHARI was elected by Parliament and the four provincial assemblies
head of government: Prime Minister Benazir BHUTTO
cabinet: Cabinet
Legislative branch: bicameral Parliament (Majlis-e-Shoora)
Senate: elections last held NA March 1994 (next to be held NA March 1997); results—percent of vote by party NA; seats—(87 total) PPP 22, PML/N 17, Tribal Area Representatives (nonparty) 8, ANP 6, PML/J 5, JWP 5, MQM/A 5, JUI/F 2, PKMAP 2, JI 2, NPP 2, BNM/H 1, BNM/M 1, JUP/NI 1, JUP/NO 1, JAH 1, JUI/S 1, PML/F 1, PNP 1, independents 2, vacant 1
National Assembly: elections last held 6 October 1993 (next to be held by October 1998); results—percent of vote by party NA; seats—(217 total) PPP 92, PML/N 75, PML/J 6, IJM-Islamic Democratic Front 4, ANP 3, PKMAP 4, PIF 3, JWP 2, MDM 2, BNM/H 1, BNM/M 1, NDA 1, NPP 1, PKQP 1, religious minorities 10 reserved seats, independents 9, results pending 2
Judicial branch: Supreme Court, Federal Islamic (Shari'at) Court

Political parties and leaders:
government: Pakistan People's Party (PPP), Benazir BHUTTO; Pakistan Muslim League, Junejo faction (PML/J), Hamid Nasir CHATTHA; National People's Party (NPP), Ghulam Mustapha JATOI; Pakhtun Khwa Milli Awami Party (PKMAP), Mahmood Khan ACHAKZAI; Balochistan National Movement, Hayee Group (BNM/H), Dr. HAYEE Baluch; National Democratic Alliance (NDA), Maulana Kausar NIAZI; Pakhtun Quami Party (PKQP), Mohammed AFZAL Khan; Jamhoori Watan Party (JWP), Akbar Khan BUGTI
opposition: Pakistan Muslim League, Nawaz Sharif faction (PML/N), Nawaz SHARIF; Awami National Party (ANP), Khan Abdul WALI KHAN; Pakistan Islamic Front (PIF), Qazi Hussain AHMED; Balochistan National Movement, Mengal Group (BNM/M), Sardar Akhtar MENGAL; Mohajir Quami Movement, Altaf faction (MQM/A), Altaf HUSSAIN; Jamaat-i-Islami (JI), Qazi Hussain AHMED; Jamiat-al-Hadith (JAH)
frequently shifting: Mutaheda Deeni Mahaz (MDM), Maulana Sami-ul-HAQ, the MDM includes Jamiat Ulema-i-Pakistan, Niazi faction (JUP/NI) and Anjuman Sepah-i-Sahaba Pakistan (ASSP); Islami-Jamhoori-Mahaz (IJM-Islamic Democratic Party), the IJM includes Jamiat Ulema-i-Islami, Fazlur Rehman group (JUI/F); Jamiat Ulema-i-Pakistan, Noorani faction (JUP/NO); Jamiat Ulema-i-Islam, Sami-ul-Haq faction (JUI/S); Pakistan Muslim League, Functional Group (PML/F); Pakistan National Party (PNP)
note: political alliances in Pakistan can shift frequently
Other political or pressure groups: military remains important political force; ulema (clergy), landowners, industrialists, and small merchants also influential
Member of: AsDB, C, CCC, CP, ECO, ESCAP, FAO, G-19, G-24, G-77, GATT, IAEA, IBRD, ICAO, ICC, ICFTU, ICRM, IDA, IDB, IFAD, IFC, IFRCS, ILO, IMF, IMO, INMARSAT, INTELSAT, INTERPOL, IOC, IOM, ISO, ITU, MINURSO, NAM, OAS (observer), OIC, PCA, SAARC, UN, UNCTAD, UNESCO, UNHCR, UNIDO, UNIKOM, UNITAR, UNOMIL, UNOSOM, UNPROFOR, UPU, WCL, WFTU, WHO, WIPO, WMO, WTO
Diplomatic representation in US:
chief of mission: Ambassador Maleeha LODHI
chancery: 2315 Massachusetts Avenue NW, Washington, DC 20008
telephone: [1] (202) 939-6200
FAX: [1] (202) 387-0484
consulate(s) general: Los Angeles and New York
US diplomatic representation:
chief of mission: Ambassador John C. MONJO
embassy: Diplomatic Enclave, Ramna 5, Islamabad
mailing address: P. O. Box 1048, PSC 1212, Box 2000, Unit 6220, Islamabad; APO AE 09812-2000
telephone: [92] (51) 826161 through 826179
FAX: [92] (51) 214222
consulate(s) general: Karachi, Lahore
consulate(s): Peshawar
Flag: green with a vertical white band (symbolizing the role of religious minorities) on the hoist side; a large white crescent and star are centered in the green field; the crescent, star, and color green are traditional symbols of Islam

Economy

Overview: The Pakistani economy has made progress in several key areas since Benazir BHUTTO became Prime Minister in October 1993. She has been under pressure from international donors and the IMF—which gave Pakistan a $1.3 billion structural adjustment credit in February 1994—to continue the economic reforms and austerity measures begun by her predecessor, caretaker Prime Minister Moeen QURESHI (July-October 1993). Foreign exchange reserves climbed to more than $3 billion in 1994, and the budget deficit was substantially reduced. Real GDP growth was 4% in FY93/94, up from 2.3% in FY92/93. Foreign direct and portfolio investment also have increased. Privatization of large public sector utilities began in 1994 with the sale of 12% of the Pakistan Telecommunications Corporation (PTC) and the Water and Power Development Authority (WAPDA); the sale of state-owned banks and other large units are planned for 1995. Still, the government must cope with long-standing economic vulnerabilities—high levels of debt service and defense spending, a small tax base, a huge population, and dependence on cotton-based exports—which hamper its ability to create a stable economic environment. In addition, Pakistan's infrastructure is inadequate and deteriorating, low levels of literacy constrain industrial growth, and increasing sectarian, ethnic, and tribal violence disrupt production.
National product: GDP—purchasing power parity—$248.5 billion (1994 est.)
National product real growth rate: 4% (1994 est.)
National product per capita: $1,930 (1994 est.)
Inflation rate (consumer prices): 12% (FY93/94)
Unemployment rate: 10% (FY90/91 est.)
Budget:
revenues: $10.5 billion
expenditures: $11.2 billion, including capital expenditures of $3.1 billion (FY93/94)
Exports: $6.7 billion (1993)
commodities: cotton, textiles, clothing, rice, leather, carpets
partners: US, Japan, Hong Kong, Germany, UK, UAE, France
Imports: $9.5 billion (1993)
commodities: petroleum, petroleum products, machinery, transportation equipment, vegetable oils, animal fats, chemicals
partners: Japan, US, Germany, UK, Saudi Arabia, Malaysia, South Korea
External debt: $24 billion (1993 est.)
Industrial production: growth rate 5.6% (FY93/94); accounts for 18% of GDP
Electricity:
capacity: 10,800,000 kW (1994)
production: 52.4 billion kWh
consumption per capita: 389 kWh (1993)
Industries: textiles, food processing, beverages, construction materials, clothing, paper products, shrimp
Agriculture: 24% of GDP; world's largest contiguous irrigation system; major crops—cotton, wheat, rice, sugarcane, fruits, vegetables; livestock products—milk, beef, mutton, eggs
Illicit drugs: major illicit producer of opium and hashish for the international drug trade; remains world's third largest opium producer (160 metric tons in 1994); major center for processing Afghan heroin and key transit area for Southwest Asian heroin moving to Western market
Economic aid:
recipient: $2.5 billion (FY91/92); $2.5 billion (FY92/93); $2.5 billion (FY93/94); no US commitments, includes bi- and multilateral aid
Currency: 1 Pakistani rupee (PRe) = 100 paisa
Exchange rates: Pakistani rupees (PRs) per US$1—30.860 (January 1995), 30.570 (1994), 28.107 (1993), 25.083 (1992), 23.801 (1991), 21.707 (1990)
Fiscal year: 1 July—30 June

Transportation

Railroads:
total: 8,773 km
broad gauge: 7,718 km 1.676-m gauge (286 km electrified; 1,037 double track)
narrow gauge: 445 km 1.000-m gauge; 610 km less than 1.000-m gauge (1985)
Highways:
total: 177,410 km
paved: 94,027 km
unpaved: 83,383 km (1991 est.)
Pipelines: crude oil 250 km; petroleum products 885 km; natural gas 4,044 km (1987)
Ports: Gwadar, Karachi, Ormaro (under construction), Port Muhammad bin Qasim
Merchant marine:
total: 30 ships (1,000 GRT or over) totaling 352,189 GRT/532,782 DWT
ships by type: bulk 1, cargo 25, oil tanker 1, passenger-cargo 3

Airports:
total: 119
with paved runways over 3,047 m: 12
with paved runways 2,438 to 3,047 m: 21
with paved runways 1,524 to 2,437 m: 33
with paved runways 914 to 1,523 m: 14
with paved runways under 914 m: 24
with unpaved runways 1,524 to 2,438 m: 7
with unpaved runways 914 to 1,523 m: 8

Communications

Telephone system: NA telephones; about 7 telephones/1,000 persons; the domestic telephone system is poor, adequate only for government and business use; the system for international traffic is better
local: NA
intercity: microwave radio relay
international: 3 INTELSAT (1 Atlantic Ocean and 2 Indian Ocean) earth stations; microwave radio relay
Radio:
broadcast stations: AM 19, FM 8, shortwave 0
radios: NA
Television:
broadcast stations: 29
televisions: NA

Defense Forces

Branches: Army, Navy, Air Force, Civil Armed Forces, National Guard, paramilitary/security forces
Manpower availability: males age 15-49 30,219,551; males fit for military service 18,544,008; males reach military age (17) annually 1,429,719 (1995 est.)
Defense expenditures: exchange rate conversion—$3.2 billion, 5.6% of GDP (FY94/95)

Palau

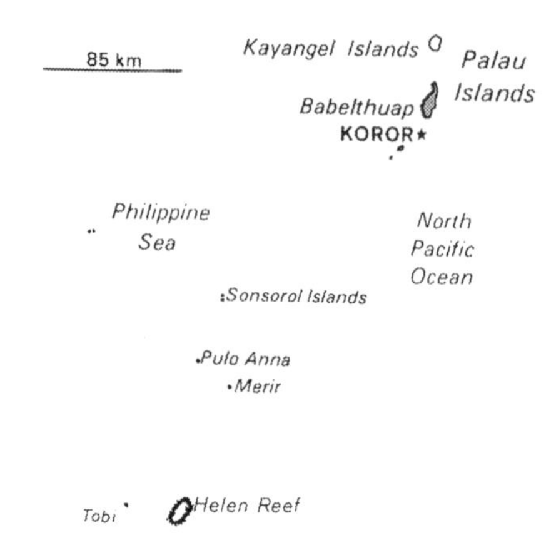

Geography

Location: Oceania, group of islands in the North Pacific Ocean, southeast of the Philippines
Map references: Oceania
Area:
total area: 458 sq km
land area: 458 sq km
comparative area: slightly more than 2.5 times the size of Washington, DC
Land boundaries: 0 km
Coastline: 1,519 km
Maritime claims:
continental shelf: 200-m depth or to depth of exploitation
exclusive fishing zone: 200 nm
territorial sea: 3 nm
International disputes: none
Climate: wet season May to November; hot and humid
Terrain: about 200 islands varying geologically from the high, mountainous main island of Babelthuap to low, coral islands usually fringed by large barrier reefs
Natural resources: forests, minerals (especially gold), marine products, deep-seabed minerals
Land use:
arable land: NA%
permanent crops: NA%
meadows and pastures: NA%
forest and woodland: NA%
other: NA%
Irrigated land: NA sq km
Environment:
current issues: inadequate facilities for disposal of solid waste; threats to the marine ecosystem from sand and coral dredging and illegal fishing practices that involve the use of dynamite
natural hazards: typhoons (June to December)
international agreements: NA
Note: includes World War II battleground of Beliliou (Peleliu) and world-famous rock islands; archipelago of six island groups totaling over 200 islands in the Caroline chain

People

Population: 16,661 (July 1995 est.)
Age structure:
0-14 years: NA
15-64 years: NA
65 years and over: NA
Population growth rate: 1.76% (1995 est.)
Birth rate: 22.11 births/1,000 population (1995 est.)
Death rate: 6.61 deaths/1,000 population (1995 est.)
Net migration rate: 2.12 migrant(s)/1,000 population (1995 est.)
Infant mortality rate: 25.07 deaths/1,000 live births (1995 est.)
Life expectancy at birth:
total population: 71.01 years
male: 69.14 years
female: 73.02 years (1995 est.)
Total fertility rate: 2.85 children born/woman (1995 est.)
Nationality:
noun: Palauan(s)
adjective: Palauan
Ethnic divisions: Palauans are a composite of Polynesian, Malayan, and Melanesian races
Religions: Christian (Catholics, Seventh-Day Adventists, Jehovah's Witnesses, the Assembly of God, the Liebenzell Mission, and Latter-Day Saints), Modekngei religion (one-third of the population observes this religion which is indigenous to Palau)
Languages: English (official in all of Palau's 16 states), Sonsorolese (official in the state of Sonsoral), Angaur and Japanese (in the state of Anguar), Tobi (in the state of Tobi), Palauan (in the other 13 states)
Literacy: age 15 and over can read and write (1980)
total population: 92%
male: 93%
female: 90%
Labor force: NA
by occupation: NA

Government

Names:
conventional long form: Republic of Palau
conventional short form: Palau
former: Trust Territory of the Pacific Islands
Digraph: PS
Type: self-governing territory in free association with the US pursuant to Compact of Free Association which entered into force 1 October 1994; Palau is fully responsible for internal affairs; US retains responsibility for external affairs
Capital: Koror
note: a new capital is being built about 20 km northeast in eastern Babelthuap
Administrative divisions: there are no first-order administrative divisions as defined by the

US Government, but there are 16 states: Aimeliik, Airai, Angaur, Kayangel, Koror, Melekeok, Ngaraard, Ngardmau, Ngaremlengui, Ngatpang, Ngchesar, Ngerchelong, Ngiwal, Peleliu, Sonsorol, Tobi
Independence: 1 October 1994 (from the US-administered UN Trusteeship)
National holiday: Constitution Day, 9 July (1979)
Constitution: 1 January 1981
Legal system: based on Trust Territory laws, acts of the legislature, municipal, common, and customary laws
Suffrage: 18 years of age; universal
Executive branch:
chief of state and head of government: President Kuniwo NAKAMURA (since 1 January 1993), Vice-President Tommy E. REMENGESAU Jr. (since 1 January 1993); election last held 4 November 1992 (next to be held NA November 1996); results—Kuniwo NAKAMURA 50.7%, Johnson TORIBIONG 49.3%
Legislative branch: bicameral Parliament (Olbiil Era Kelulau or OEK)
Senate: elections last held 4 November 1992 (next to be held NA November 1996); results—percent of vote by party NA; seats—(14 total) number of seats by party NA
House of Delegates: elections last held 4 November 1992 (next to be held NA November 1996); results—percent of vote by party NA; seats—(16 total) number of seats by party NA
Judicial branch: Supreme Court, National Court, Court of Common Pleas
Member of: ESCAP (associate), SPC, SPF (observer), UN
Diplomatic representation in US:
chief of mission: Liaison Officer NA
liaison office: 444 North Capital Street NW, Washington, DC 20036
telephone: (202) 624-7793
FAX: NA
note: relationship of free association with the US pursuant to compact of free association which entered into force 1 October 1994
US diplomatic representation:
chief of mission: Liaison Officer Lloyd W. MOSS
liaison office: Erenguul Street, Koror, Republic of Palau
mailing address: P.O. Box 6028, Republic of Palau 96940
telephone: [680] 488-2920
FAX: [680] 488-2911
note: relationship of free association with the US pursuant to compact of free association which entered into force 1 October 1994
Flag: light blue with a large yellow disk (representing the moon) shifted slightly to the hoist side

Economy

Overview: The economy consists primarily of subsistence agriculture and fishing. The government is the major employer of the work force, relying heavily on financial assistance from the US. The compact of "free association" with the United States, entered into after the end of the UN trusteeship on 1 October 1994, provides Palau with $500 million in US aid over 15 years in return for furnishing some military facilities. The population, in effect, enjoys a per capita income of $5,000, twice that of the Philippines and much of Micronesia. Long-run prospects for the tourist sector have been greatly bolstered by the expansion of air travel in the Pacific and the rapidly rising prosperity of leading East Asian countries.
National product: GDP—purchasing power parity—$81.8 million (1994 est.)
note: GDP numbers reflect US spending
National product real growth rate: NA%
National product per capita: $5,000 (1994 est.)
Inflation rate (consumer prices): NA%
Unemployment rate: 20% (1986)
Budget:
revenues: $6 million
expenditures: $NA, including capital expenditures of $NA (1986 est.)
Exports: $600,000 (f.o.b., 1989)
commodities: trochus (type of shellfish), tuna, copra, handicrafts
partners: US, Japan
Imports: $24.6 million (c.i.f., 1989)
commodities: NA
partners: US
External debt: about $100 million (1989)
Industrial production: growth rate NA%
Electricity:
capacity: 16,000 kW
production: 22 million kWh
consumption per capita: 1,540 kWh (1990)
Industries: tourism, craft items (shell, wood, pearl), some commercial fishing and agriculture
Agriculture: subsistence-level production of coconut, copra, cassava, sweet potatoes
Economic aid:
recipient: US commitments, including Ex-Im (FY70-89), $2.56 billion; Western (non-US) countries, ODA and OOF bilateral commitments (1970-89), $92 million
Currency: 1 United States dollar (US$) = 100 cents
Exchange rates: US currency is used
Fiscal year: 1 October—30 September

Transportation

Railroads: 0 km
Highways:
total: 61 km
paved: 36 km
unpaved: gravel 25 km
Ports: Koror
Merchant marine: none
Airports:
total: 3
with paved runways 1,524 to 2,437 m: 1
with unpaved runways 1,524 to 2,438 m: 2

Communications

Telephone system: NA telephones
local: NA
intercity: NA
international: 1 INTELSAT (Pacific Ocean) earth station
Radio:
broadcast stations: AM 1, FM 1, shortwave 0
radios: NA
Television:
broadcast stations: 2
televisions: NA

Defense Forces

Branches: NA
Defense expenditures: $NA, NA% of GDP
Note: defense is the responsibility of the US pursuant to Compact of Free Association which entered into force 1 October 1994

Palmyra Atoll
(territory of the US)

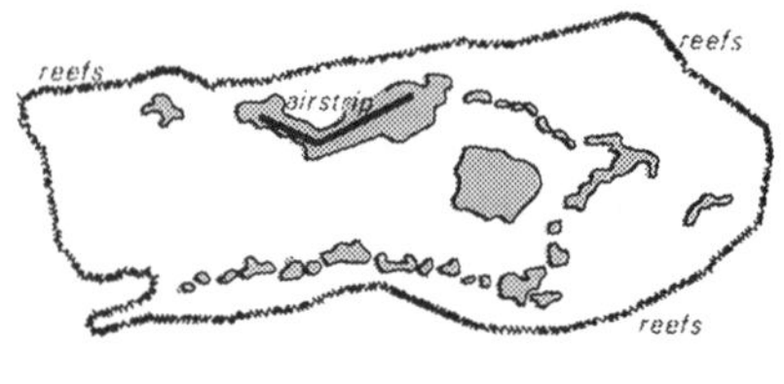

Geography

Location: Oceania, atoll in the North Pacific Ocean, about one-half of the way from Hawaii to American Samoa
Map references: Oceania
Area:
total area: 11.9 sq km
land area: 11.9 sq km
comparative area: about 20 times the size of The Mall in Washington, DC
Land boundaries: 0 km
Coastline: 14.5 km
Maritime claims:
exclusive economic zone: 200 nm
territorial sea: 12 nm
International disputes: none
Climate: equatorial, hot, and very rainy
Terrain: low, with maximum elevations of about 2 meters
Natural resources: none
Land use:
arable land: 0%
permanent crops: 0%
meadows and pastures: 0%
forest and woodland: 100%
other: 0%
Irrigated land: 0 sq km
Environment:
current issues: NA
natural hazards: NA
international agreements: NA
Note: about 50 islets covered with dense vegetation, coconut trees, and balsa-like trees up to 30 meters tall

People

Population: uninhabited

Government

Names:
conventional long form: none
conventional short form: Palmyra Atoll
Digraph: LQ
Type: incorporated territory of the US; privately owned, but administered by the Office of Territorial and International Affairs, US Department of the Interior
Capital: none; administered from Washington, DC

Economy

Overview: no economic activity

Transportation

Highways: much of the road and many causeways built during the war are unserviceable and overgrown
Ports: West Lagoon
Airports:
total: 1
with unpaved runways 1,524 to 2,438 m: 1

Defense Forces

Note: defense is the responsibility of the US

Panama

Geography

Location: Middle America, bordering both the Caribbean Sea and the North Pacific Ocean, between Colombia and Costa Rica
Map references: Central America and the Caribbean
Area:
total area: 78,200 sq km
land area: 75,990 sq km
comparative area: slightly smaller than South Carolina
Land boundaries: total 555 km, Colombia 225 km, Costa Rica 330 km
Coastline: 2,490 km
Maritime claims:
territorial sea: 200 nm
International disputes: none
Climate: tropical; hot, humid, cloudy; prolonged rainy season (May to January), short dry season (January to May)
Terrain: interior mostly steep, rugged mountains and dissected, upland plains; coastal areas largely plains and rolling hills
Natural resources: copper, mahogany forests, shrimp
Land use:
arable land: 6%
permanent crops: 2%
meadows and pastures: 15%
forest and woodland: 54%
other: 23%
Irrigated land: 320 sq km (1989 est.)
Environment:
current issues: water pollution from agricultural runoff threatens fishery resources; deforestation of tropical rain forest; land degradation
natural hazards: NA
international agreements: party to—Biodiversity, Endangered Species, Hazardous Wastes, Marine Dumping, Nuclear Test Ban, Ozone Layer Protection, Ship Pollution, Tropical Timber 83, Wetlands; signed, but not ratified—Climate Change, Law of the Sea, Marine Life Conservation, Tropical Timber 94
Note: strategic location on eastern end of

isthmus forming land bridge connecting North and South America; controls Panama Canal that links North Atlantic Ocean via Caribbean Sea with North Pacific Ocean

People

Population: 2,680,903 (July 1995 est.)
Age structure:
0-14 years: 34% (female 439,491; male 458,817)
15-64 years: 61% (female 812,876; male 823,124)
65 years and over: 5% (female 74,672; male 71,923) (July 1995 est.)
Population growth rate: 1.9% (1995 est.)
Birth rate: 24.12 births/1,000 population (1995 est.)
Death rate: 4.79 deaths/1,000 population (1995 est.)
Net migration rate: -0.36 migrant(s)/1,000 population (1995 est.)
Infant mortality rate: 15.8 deaths/1,000 live births (1995 est.)
Life expectancy at birth:
total population: 75.2 years
male: 72.57 years
female: 77.97 years (1995 est.)
Total fertility rate: 2.8 children born/woman (1995 est.)
Nationality:
noun: Panamanian(s)
adjective: Panamanian
Ethnic divisions: mestizo (mixed Indian and European ancestry) 70%, West Indian 14%, white 10%, Indian 6%
Religions: Roman Catholic 85%, Protestant 15%
Languages: Spanish (official), English 14%
note: many Panamanians bilingual
Literacy: age 15 and over can read and write (1990)
total population: 89%
male: 89%
female: 88%
Labor force: 979,000 (1994 est.)
by occupation: government and community services 31.8%, agriculture, hunting, and fishing 26.8%, commerce, restaurants, and hotels 16.4%, manufacturing and mining 9.4%, construction 3.2%, transportation and communications 6.2%, finance, insurance, and real estate 4.3%
note: shortage of skilled labor, but an oversupply of unskilled labor

Government

Names:
conventional long form: Republic of Panama
conventional short form: Panama
local long form: Republica de Panama
local short form: Panama
Digraph: PM
Type: constitutional republic
Capital: Panama
Administrative divisions: 9 provinces (provincias, singular—provincia) and 1 territory* (comarca); Bocas del Toro, Chiriqui, Cocle, Colon, Darien, Herrera, Los Santos, Panama, San Blas*, Veraguas
Independence: 3 November 1903 (from Colombia; became independent from Spain 28 November 1821)
National holiday: Independence Day, 3 November (1903)
Constitution: 11 October 1972; major reforms adopted April 1983
Legal system: based on civil law system; judicial review of legislative acts in the Supreme Court of Justice; accepts compulsory ICJ jurisdiction, with reservations
Suffrage: 18 years of age; universal and compulsory
Executive branch:
chief of state and head of government: President Ernesto PEREZ BALLADARES Gonzalez Revilla (since 1 September 1994, elected 8 May 1994); First Vice President Tomas Gabriel ALTAMIRANO DUQUE (since 1 September 1994); Second Vice President Felipe Alejandro VIRZI Lopez (since 1 September 1994); election last held 8 May 1994 (next to be held 9 May 1999); results—Ernesto PEREZ BALLADARES (PRD) 33%, Mireya MOSCOSO DE GRUBER (PA) 29%, Ruben BLADES (MPE) 17%, Ruben Dario CARLES (MOLIRENA) 16%
cabinet: Cabinet; appointed by the president
Legislative branch: unicameral
Legislative Assembly (Asamblea Legislativa): legislators from outlying rural districts chosen on a plurality basis while districts located in more populous towns and cities elect multiple legislators by means of a proportion-based formula; elections last held 8 May 1994 (next to be held 9 May 1999); results—percent of vote by party NA); seats—(72 total) PRD 32, PS 4, PALA 1, PA 14, MPE 6, MOLIRENA 4, PLA 3, PRC 3, PL 2, PDC 1, UDI 1, MORENA 1
Judicial branch: Supreme Court of Justice (Corte Suprema de Justicia), 5 superior courts, 3 courts of appeal
Political parties and leaders:
governing coalition: Democratic Revolutionary Party (PRD), Gerardo GONZALEZ; Solidarity Party (PS), Samuel LEWIS GALINDO; Liberal Republican Party (PLR), Rodolfo CHIARI; Labor Party (PALA), Carlos Lopez GUEVARA
other parties: Nationalist Republican Liberal Movement (MOLIRENA), Alfredo RAMIREZ; Authentic Liberal Party (PLA), Arnulfo ESCALONA; Arnulfista Party (PA), Mireya MOSCOSO DE GRUBER; Christian Democratic Party (PDC), Raul OSSA; Liberal Party (PL), Roberto ALEMAN Zubieta; Papa Egoro Movement (MPE), Ruben BLADES; Civic Renewal Party (PRC), Tomas HERRERA; National Unity Mission Party (MUN), Jose Manuel PAREDES; Independent Democratic Union (UDI), Jacinto CARDENAS; National Renovation Movement (MORENA), Pedro VALLERINO
Other political or pressure groups: National Council of Organized Workers (CONATO); National Council of Private Enterprise (CONEP); Panamanian Association of Business Executives (APEDE); National Civic Crusade; Chamber of Commerce; Panamanian Industrialists Society (SIP); Workers Confederation of the Republic of Panama (CTRP)
Member of: AG (associate), CG, ECLAC, FAO, G-77, IADB, IAEA, IBRD, ICAO, ICFTU, ICRM, IDA, IFAD, IFC, IFRCS, ILO, IMF, IMO, INMARSAT, INTELSAT, INTERPOL, IOC, IOM, ITU, LAES, LAIA (observer), NAM, OAS, OPANAL, PCA, UN, UNCTAD, UNESCO, UNIDO, UPU, WCL, WFTU, WHO, WIPO, WMO, WTO
Diplomatic representation in US:
chief of mission: Ambassador Ricardo Alberto ARIAS
chancery: 2862 McGill Terrace NW, Washington, DC 20008
telephone: [1] (202) 483-1407
consulate(s) general: Atlanta, Houston, Miami, New Orleans, New York, San Francisco, San Juan (Puerto Rico), Tampa
US diplomatic representation:
chief of mission: (vacant); Charge d'Affaires Oliver P. GARZA
embassy: Avenida Balboa and Calle 38, Apartado 6959, Panama City 5
mailing address: American Embassy Panama, Unit 0945; APO AA 34002
telephone: [507] 27-1777
FAX: [507] 27-1964
Flag: divided into four, equal rectangles; the top quadrants are white (hoist side) with a blue five-pointed star in the center and plain red, the bottom quadrants are plain blue (hoist side) and white with a red five-pointed star in the center

Economy

Overview: Because of its key geographic location, Panama's economy is service-based, heavily weighted toward banking, commerce, and tourism. Trade and financial ties with the US are especially close. GDP grew at 3.6% in 1994, a respectable rate, yet below the 7.1% average of the early 1990s. Banking and financial services and trade through the Colon Free Zone continued to expand rapidly, with the industrial and agricultural sectors experiencing little growth. The new administration, inaugurated 1 September 1994, has launched an economic plan designed to reverse rising unemployment, attract foreign investment, cut back the size of government, and modernize the economy. The success of the plan in meeting its goals for 1995 and

beyond depends largely on the success of the administration in reforming the labor code and instituting the reforms necessary to join the GATT.
National product: GDP—purchasing power parity—$12.3 billion (1994 est.)
National product real growth rate: 3.6% (1994 est.)
National product per capita: $4,670 (1994 est.)
Inflation rate (consumer prices): 1.8% (1994 est.)
Unemployment rate: 12.9% (1994 est.)
Budget:
revenues: $1.93 billion
expenditures: $1.93 billion, including captial expenditures of $NA (1994)
Exports: $520 million (f.o.b., 1994 est.)
commodities: bananas 43%, shrimp 11%, sugar 4%, clothing 5%, coffee 2%
partners: US 45%, EU, Central America and Caribbean
Imports: $2.205 billion (c.i.f., 1994 est.)
commodities: capital goods 21%, crude oil 11%, foodstuffs 9%, consumer goods, chemicals
partners: US 40%, EU, Central America and Caribbean, Japan
External debt: $6.7 billion (yearend 1993 est.)
Industrial production: growth rate 1.8% (1994 est.); accounts for about 9% of GDP
Electricity:
capacity: 960,000 kW
production: 2.8 billion kWh
consumption per capita: 1,047 kWh (1993)
Industries: manufacturing and construction, petroleum refining, brewing, cement and other construction materials, sugar milling
Agriculture: accounts for 10% of GDP (1992 est.); crops—bananas, rice, corn, coffee, sugarcane; livestock; fishing; importer of food grain, vegetables
Illicit drugs: major cocaine transshipment point and drug money laundering center
Economic aid:
recipient: US commitments, including Ex-Im (FY70-89), $516 million; Western (non-US) countries, ODA and OOF bilateral commitments (1970-89), $582 million; Communist countries (1970-89), $4 million
Currency: 1 balboa (B) = 100 centesimos
Exchange rates: balboas (B) per US$1—1.000 (fixed rate)
Fiscal year: calendar year

Transportation

Railroads:
total: 238 km
broad gauge: 78 km 1.524-m gauge
narrow gauge: 160 km 0.914-m gauge
Highways:
total: 8,530 km
paved: 2,745 km
unpaved: gravel, crushed stone 3,270 km; improved, unimproved earth 2,515 km
Inland waterways: 800 km navigable by shallow draft vessels; 82 km Panama Canal
Pipelines: crude oil 130 km
Ports: Bahia de las Minas, Balboa, Colon, Cristobal, Panama
Merchant marine:
total: 3,526 ships (1,000 GRT or over) totaling 60,748,525 GRT/95,102,552 DWT
ships by type: barge carrier 1, bulk 787, cargo 1,070, chemical tanker 175, combination bulk 33, combination ore/oil 25, container 259, liquefied gas tanker 125, livestock carrier 8, multifunction large-load carrier 6, oil tanker 465, passenger 24, passenger-cargo 3, refrigerated cargo 284, roll-on/roll-off cargo 81, short-sea passenger 34, specialized tanker 9, vehicle carrier 137
note: a flag of convenience registry; includes 93 countries; the 10 major fleet flags are: Japan 1,171 ships, Greece 323, Hong Kong 276, US 212, Taiwan 184, Singapore 181, South Korea 172, China 145 ships, UK 102, and Norway 70
Airports:
total: 115
with paved runways over 3,047 m: 1
with paved runways 2,438 to 3,047 m: 1
with paved runways 1,524 to 2,437 m: 5
with paved runways 914 to 1,523 m: 14
with paved runways under 914 m: 74
with unpaved runways 914 to 1,523 m: 20

Communications

Telephone system: 220,000 telephones; domestic and international facilities well developed
local: NA
intercity: NA
international: 1 coaxial submarine cable; 2 INTELSAT (Atlantic Ocean) earth stations; connected to the Central American Microwave System
Radio:
broadcast stations: AM 91, FM 0, shortwave 0
radios: NA
Television:
broadcast stations: 23
televisions: NA

Defense Forces

Branches: Panamanian Public Forces (PPF; includes the National Police or PNP, Maritime Service, National Air Service, and Institutional Protective Service); Judicial Branch Technical Police
Manpower availability: males age 15-49 701,691; males fit for military service 481,927 (1995 est.)
Defense expenditures: expenditures for the Panamanian security forces amounted to $105 million, 1.0% of GDP (1993 est.)

Papua New Guinea

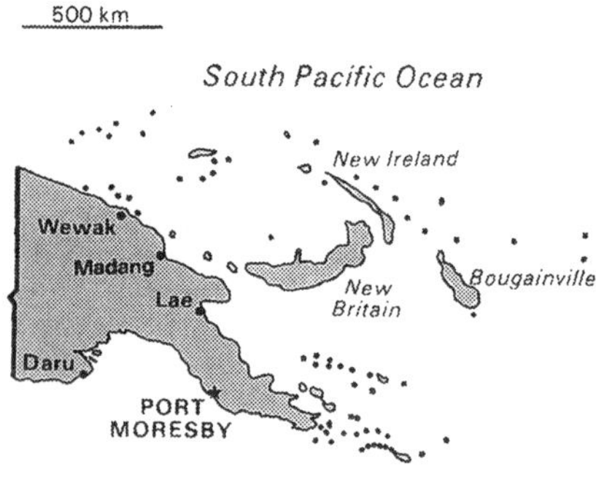

Geography

Location: Southeastern Asia, group of islands including the eastern half of the island of New Guinea between the Coral Sea and the South Pacific Ocean, east of Indonesia
Map references: Oceania
Area:
total area: 461,690 sq km
land area: 451,710 sq km
comparative area: slightly larger than California
Land boundaries: total 820 km, Indonesia 820 km
Coastline: 5,152 km
Maritime claims: measured from claimed archipelagic baselines
continental shelf: 200-m depth or to the depth of exploitation
exclusive fishing zone: 200 nm
territorial sea: 12 nm
International disputes: none
Climate: tropical; northwest monsoon (December to March), southeast monsoon (May to October); slight seasonal temperature variation
Terrain: mostly mountains with coastal lowlands and rolling foothills
Natural resources: gold, copper, silver, natural gas, timber, oil potential
Land use:
arable land: 0%
permanent crops: 1%
meadows and pastures: 0%
forest and woodland: 71%
other: 28%
Irrigated land: NA sq km
Environment:
current issues: rain forest subject to deforestation as a result of growing commercial demand for tropical timber; pollution from mining projects
natural hazards: active volcanism; situated along the Pacific "Rim of Fire"; the country is subject to frequent and sometimes severe earthquakes; mudslides
international agreements: party to—Antarctic

Treaty, Biodiversity, Climate Change, Endangered Species, Environmental Modification, Marine Dumping, Nuclear Test Ban, Ozone Layer Protection, Ship Pollution, Tropical Timber 83, Wetlands; signed, but not ratified—Antarctic-Environmental Protocol, Law of the Sea
Note: shares island of New Guinea with Indonesia; one of world's largest swamps along southwest coast

People

Population: 4,294,750 (July 1995 est.)
Age structure:
0-14 years: 41% (female 847,208; male 892,718)
15-64 years: 57% (female 1,161,961; male 1,268,266)
65 years and over: 2% (female 66,759; male 57,838) (July 1995 est.)
Population growth rate: 2.3% (1995 est.)
Birth rate: 33.2 births/1,000 population (1995 est.)
Death rate: 10.18 deaths/1,000 population (1995 est.)
Net migration rate: 0 migrant(s)/1,000 population (1995 est.)
Infant mortality rate: 61.6 deaths/1,000 live births (1995 est.)
Life expectancy at birth:
total population: 56.85 years
male: 56.01 years
female: 57.74 years (1995 est.)
Total fertility rate: 4.55 children born/woman (1995 est.)
Nationality:
noun: Papua New Guinean(s)
adjective: Papua New Guinean
Ethnic divisions: Melanesian, Papuan, Negrito, Micronesian, Polynesian
Religions: Roman Catholic 22%, Lutheran 16%, Presbyterian/Methodist/London Missionary Society 8%, Anglican 5%, Evangelical Alliance 4%, Seventh-Day Adventist 1%, other Protestant sects 10%, indigenous beliefs 34%
Languages: English spoken by 1%-2%, pidgin English widespread, Motu spoken in Papua region
note: 715 indigenous languages
Literacy: age 15 and over can read and write (1990 est.)
total population: 52%
male: 65%
female: 38%
Labor force: NA

Government

Names:
conventional long form: Independent State of Papua New Guinea
conventional short form: Papua New Guinea
Digraph: PP
Type: parliamentary democracy
Capital: Port Moresby
Administrative divisions: 20 provinces; Central, Chimbu, Eastern Highlands, East New Britain, East Sepik, Enga, Gulf, Madang, Manus, Milne Bay, Morobe, National Capital, New Ireland, Northern, North Solomons, Sandaun, Southern Highlands, Western, Western Highlands, West New Britain
Independence: 16 September 1975 (from the Australian-administered UN trusteeship)
National Holiday: Independence Day, 16 September (1975)
Constitution: 16 September 1975
Legal system: based on English common law
Suffrage: 18 years of age; universal
Executive branch:
chief of state: Queen ELIZABETH II (since 6 February 1952), represented by Governor General Wiwa KOROWI (since NA November 1991)
head of government: Prime Minister Sir Julius CHAN (since 30 August 1994); Deputy Prime Minister Chris HAIVETA (since 7 September 1994)
cabinet: National Executive Council; appointed by the governor on recommendation of the prime minister
Legislative branch: unicameral
National Parliament: (sometimes referred to as the House of Assembly) elections last held 13-26 June 1992 (next to be held NA 1997); results—percent of vote by party NA; seats—(109 total) Pangu Party 24, PDM 17, PPP 10, PAP 10, independents 30, others 18; note—association with political parties is fluid
Judicial branch: Supreme Court
Political parties and leaders: Papua New Guinea United Party (Pangu Party), Jack GENIA; People's Democratic Movement (PDM), Paias WINGTI; People's Action Party (PAP), Akoka DOI; People's Progress Party (PPP), Sir Julius CHAN; United Party (UP), Paul TORATO; Papua Party (PP), Galeva KWARARA; National Party (NP), Paul PORA; Melanesian Alliance (MA), Fr. John MOMIS
Member of: ACP, APEC, AsDB, ASEAN (observer), C, CP, ESCAP, FAO, G-77, IBRD, ICAO, ICFTU, ICRM, IDA, IFAD, IFC, IFRCS, ILO, IMF, IMO, INTELSAT, INTERPOL, IOC, ISO (correspondent), ITU, NAM, NAM (observer), SPARTECA, SPC, SPF, UN, UNCTAD, UNESCO, UNIDO, UPU, WFTU, WHO, WMO
Diplomatic representation in US:
chief of mission: Ambassador Kepas Isimel WATANGIA
chancery: 3rd floor, 1615 New Hampshire Avenue NW, Washington, DC 20009
telephone: [1] (202) 745-3680
FAX: [1] (202) 745-3679
US diplomatic representation:
chief of mission: Ambassador Richard W. TEARE
embassy: Armit Street, Port Moresby
mailing address: P. O. Box 1492, Port Moresby, or APO AE 96553
telephone: [675] 211455, 211594, 211654
FAX: [675] 213423
Flag: divided diagonally from upper hoist-side corner; the upper triangle is red with a soaring yellow bird of paradise centered; the lower triangle is black with five white five-pointed stars of the Southern Cross constellation centered

Economy

Overview: Papua New Guinea is richly endowed with natural resources, but exploitation has been hampered by the rugged terrain and the high cost of developing an infrastructure. Agriculture provides a subsistence livelihood for 85% of the population. Mining of numerous deposits, including copper and gold, accounts for about 60% of export earnings. Budgetary support from Australia and development aid under World Bank auspices have helped sustain the economy. Robust growth in 1991-92 was led by the mining sector; the opening of a large new gold mine helped the advance. At the start of 1995, Port Moresby is looking primarily to the exploitation of mineral and petroleum resources to drive economic development but new prospecting in Papua New Guinea has slumped as other mineral-rich countries have stepped up their competition for international investment. Output from current projects will probably begin to taper off in 1996, but no new large ventures are being developed to succeed them.
National product: GDP—purchasing power parity—$9.2 billion (1994 est.)
National product real growth rate: 6.1% (1994 est.)
National product per capita: $2,200 (1994 est.)
Inflation rate (consumer prices): 1.6% (1994)
Unemployment rate: NA%
Budget:
revenues: $1.33 billion
expenditures: $1.36 billion, including capital expenditures of $NA (1995 est.)
Exports: $2.4 billion (f.o.b., 1993 est.)
commodities: gold, copper ore, oil, logs, palm oil, coffee, cocoa, lobster
partners: Australia, Japan, US, Singapore, New Zealand
Imports: $1.2 billion (c.i.f., 1993 est.)
commodities: machinery and transport equipment, manufactured goods, food, fuels, chemicals
partners: Australia, Japan, UK, New Zealand, Netherlands
External debt: $3.2 billion (1992)
Industrial production: accounts for 32% of GDP
Electricity:
capacity: 490,000 kW

production: 1.8 billion kWh
consumption per capita: 390 kWh (1993)
Industries: copra crushing, palm oil processing, plywood production, wood chip production, mining of gold, silver, and copper, construction, tourism
Agriculture: Accounts for 25% of GDP; livelihood for 85% of population; fertile soils and favorable climate permits cultivating a wide variety of crops; cash crops—coffee, cocoa, coconuts, palm kernels; other products—tea, rubber, sweet potatoes, fruit, vegetables, poultry, pork; net importer of food for urban centers
Economic aid:
recipient: US commitments, including Ex-Im (FY70-89), $40.6 million; Western (non-US) countries, ODA and OOF bilateral commitments (1970-89), $6.5 billion; OPEC bilateral aid (1979-89), $17 million
Currency: 1 kina (K) = 100 toea
Exchange rates: kina (K) per US$1—0.8565 (December 1994), 0.9950 (1994), 1.0221 (1993), 1.0367 (1992), 1.0504 (1991), 1.0467 (1990); note—the government floated the kina on 10 October 1994
Fiscal year: calendar year

Transportation

Railroads: 0 km
Highways:
total: 19,200 km
paved: 640 km
unpaved: gravel, crushed stone, stabilized earth 10,960 km; unimproved earth 7,600 km
Inland waterways: 10,940 km
Ports: Kieta, Lae, Madang, Port Moresby, Rabaul
Merchant marine:
total: 12 ships (1,000 GRT or over) totaling 22,565 GRT/27,071 DWT
ships by type: bulk 2, cargo 3, combination ore/oil 5, container 1, roll-on/roll-off 1
Airports:
total: 505
with paved runways 2,438 to 3,047 m: 1
with paved runways 1,524 to 2,437 m: 13
with paved runways 914 to 1,523 m: 5
with paved runways under 914 m: 411
with unpaved runways 1,524 to 2,438 m: 12
with unpaved runways 914 to 1,523 m: 63

Communications

Telephone system: more than 70,000 telephones (1987); services are adequate and being improved; facilities provide radiobroadcast, radiotelephone and telegraph, coastal radio, aeronautical radio, and international radiocommunication services
local: NA
intercity: mostly radio telephone
international: submarine cables extend to Australia and Guam; 1 INTELSAT (Pacific Ocean) earth station; international radio communication service
Radio:
broadcast stations: AM 31, FM 2, shortwave 0
radios: NA
Television:
broadcast stations: 2 (1987)
televisions: NA

Defense Forces

Branches: Papua New Guinea Defense Force (includes Army, Navy, and Air Force)
Manpower availability: males age 15-49 1,111,661; males fit for military service 618,696 (1995 est.)
Defense expenditures: exchange rate conversion—$55 million, 1.8% of GDP (1993 est.)

Paracel Islands

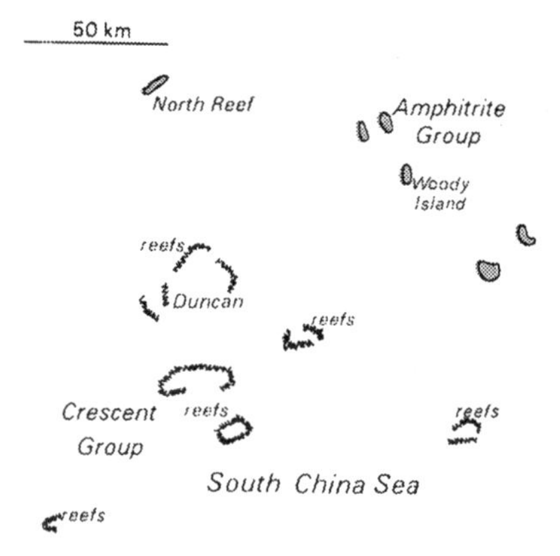

Geography

Location: Southeastern Asia, group of small islands and reefs in the South China Sea, about one-third of the way from central Vietnam to the northern Philippines
Map references: Southeast Asia
Area:
total area: NA sq km
land area: NA sq km
comparative area: NA
Land boundaries: 0 km
Coastline: 518 km
Maritime claims: NA
International disputes: occupied by China, but claimed by Taiwan and Vietnam
Climate: tropical
Terrain: NA
Natural resources: none
Land use:
arable land: 0%
permanent crops: 0%
meadows and pastures: 0%
forest and woodland: 0%
other: 100%
Irrigated land: 0 sq km
Environment:
current issues: NA
natural hazards: typhoons
international agreements: NA

People

Population: no indigenous inhabitants; note—there are scattered Chinese garrisons

Government

Names:
conventional long form: none
conventional short form: Paracel Islands
Digraph: PF

Economy

Overview: no economic activity

Transportation

Ports: small Chinese port facilities on Woody Island and Duncan Island being expanded
Airports:
total: 1
with paved runways 1,524 to 2,437 m: 1 (on Woody Island)

Communications

Telephone system:
local: NA
intercity: NA
international: NA
Radio:
broadcast stations: AM, FM, shortwave
radios: NA
Television:
broadcast stations: NA
televisions: NA

Defense Forces

Note: occupied by China

Paraguay

Geography

Location: Central South America, northeast of Argentina
Map references: South America
Area:
total area: 406,750 sq km
land area: 397,300 sq km
comparative area: slightly smaller than California
Land boundaries: total 3,920 km, Argentina 1,880 km, Bolivia 750 km, Brazil 1,290 km
Coastline: 0 km (landlocked)
Maritime claims: none; landlocked
International disputes: short section of the boundary with Brazil, just west of Salto del Guaira (Guaira Falls) on the Rio Parana, has not been determined
Climate: varies from temperate in east to semiarid in far west
Terrain: grassy plains and wooded hills east of Rio Paraguay; Gran Chaco region west of Rio Paraguay mostly low, marshy plain near the river, and dry forest and thorny scrub elsewhere
Natural resources: hydropower, timber, iron ore, manganese, limestone
Land use:
arable land: 20%
permanent crops: 1%
meadows and pastures: 39%
forest and woodland: 35%
other: 5%
Irrigated land: 670 sq km (1989 est.)
Environment:
current issues: deforestation (an estimated 2 million hectares of forest land have been lost from 1958-1985); water pollution; inadequate means for waste disposal present health risks for many urban residents
natural hazards: local flooding in southeast (early September to June); poorly drained plains may become boggy (early October to June)
international agreements: party to—Biodiversity, Climate Change, Endangered Species, Law of the Sea, Ozone Layer Protection; signed, but not ratified—Nuclear Test Ban
Note: landlocked; buffer between Argentina and Brazil

People

Population: 5,358,198 (July 1995 est.)
Age structure:
0-14 years: 41% (female 1,077,284; male 1,123,776)
15-64 years: 55% (female 1,465,147; male 1,468,642)
65 years and over: 4% (female 120,776; male 102,573) (July 1995 est.)
Population growth rate: 2.71% (1995 est.)
Birth rate: 31.48 births/1,000 population (1995 est.)
Death rate: 4.38 deaths/1,000 population (1995 est.)
Net migration rate: 0 migrant(s)/1,000 population (1995 est.)
Infant mortality rate: 24.1 deaths/1,000 live births (1995 est.)
Life expectancy at birth:
total population: 73.58 years
male: 72.06 years
female: 75.18 years (1995 est.)
Total fertility rate: 4.22 children born/woman (1995 est.)
Nationality:
noun: Paraguayan(s)
adjective: Paraguayan
Ethnic divisions: mestizo (mixed Spanish and Indian) 95%, Caucasians plus Amerindians 5%
Religions: Roman Catholic 90%, Mennonite and other Protestant denominations
Languages: Spanish (official), Guarani
Literacy: age 15 and over can read and write (1990 est.)
total population: 90%
male: 92%
female: 88%
Labor force: 1.692 million (1993 est.)
by occupation: agriculture 45%

Government

Names:
conventional long form: Republic of Paraguay
conventional short form: Paraguay
local long form: Republica del Paraguay
local short form: Paraguay
Digraph: PA
Type: republic
Capital: Asuncion
Administrative divisions: 19 departments (departamentos, singular—departamento); Alto Paraguay, Alto Parana, Amambay, Boqueron, Caaguazu, Caazapa, Canindeyu, Central, Chaco, Concepcion, Cordillera, Guaira, Itapua, Misiones, Neembucu, Nueva Asuncion, Paraguari, Presidente Hayes, San Pedro

Independence: 14 May 1811 (from Spain)
National holiday: Independence Days, 14-15 May (1811)
Constitution: promulgated 20 June 1992
Legal system: based on Argentine codes, Roman law, and French codes; judicial review of legislative acts in Supreme Court of Justice; does not accept compulsory ICJ jurisdiction
Suffrage: 18 years of age; universal and compulsory up to age 60
Executive branch:
chief of state and head of government: President Juan Carlos WASMOSY (since 15 August 1993); Vice President Roberto Angel SEIFART (since 15 August 1993); election last held 9 May 1993 (next to be held May 1998); results—Juan Carlos WASMOSY 40.09%, Domingo LAINO 32.06%, Guillermo CABALLERO VARGAS 23.04%
cabinet: Council of Ministers; nominated by the president
Legislative branch: bicameral Congress (Congreso)
Chamber of Senators (Camara de Senadores): elections last held 9 May 1993 (next to be held May 1998); results—percent of vote by party NA; seats—(45 total) Colorado Party 20, PLRA 17, EN 8
Chamber of Deputies (Camara de Diputados): elections last held on 9 May 1993 (next to be held by May 1998); results—percent of vote by party NA; seats—(80 total) Colorado Party 38, PLRA 33, EN 9
Judicial branch: Supreme Court of Justice (Corte Suprema de Justicia)
Political parties and leaders: Colorado Party, Eugenio SANABRIA CANTERO, president; Authentic Radical Liberal Party (PLRA), Domingo LAINO; National Encounter (EN), Guillermo CABALLERO VARGAS (the EN party includes the following minor parties: Christian Democratic Party (PDC), Jose Angel BURRO; Febrerista Revolutionary Party (PRF), Euclides ACEVEDO; Popular Democratic Party (PDP), Hugo RICHER)
Other political or pressure groups: Confederation of Workers (CUT); Roman Catholic Church
Member of: AG (observer), CCC, ECLAC, FAO, G-77, GATT, IADB, IAEA, IBRD, ICAO, ICFTU, ICRM, IDA, IFAD, IFC, IFRCS, ILO, IMF, IMO, INTELSAT, INTERPOL, IOC, IOM, ITU, LAES, LAIA, MERCOSUR, OAS, OPANAL, PCA, RG, UN, UNCTAD, UNESCO, UNIDO, UPU, WCL, WHO, WIPO, WMO, WTO
Diplomatic representation in US:
chief of mission: Ambassador Jorge Genaro Andres PRIETO CONTI
chancery: 2400 Massachusetts Avenue NW, Washington, DC 20008
telephone: [1] (202) 483-6960 through 6962
FAX: [1] (202) 234-4508
consulate(s) general: Miami, New Orleans, and New York
US diplomatic representation:
chief of mission: Ambassador Robert SERVICE
embassy: 1776 Avenida Mariscal Lopez, Asuncion
mailing address: C. P. 402, Asuncion; Unit 4711, APO AA 34036-0001
telephone: [595] (21) 213-715
FAX: [595] (21) 213-728
Flag: three equal, horizontal bands of red (top), white, and blue with an emblem centered in the white band; unusual flag in that the emblem is different on each side; the obverse (hoist side at the left) bears the national coat of arms (a yellow five-pointed star within a green wreath capped by the words REPUBLICA DEL PARAGUAY, all within two circles); the reverse (hoist side at the right) bears the seal of the treasury (a yellow lion below a red Cap of Liberty and the words Paz y Justicia (Peace and Justice) capped by the words REPUBLICA DEL PARAGUAY, all within two circles)

Economy

Overview: Agriculture, including forestry, accounts for about 25% of GDP, employs about 45% of the labor force, and provides the bulk of exports, in which soybeans and cotton are the most important. Paraguay lacks substantial mineral or petroleum resources but possesses a large hydropower potential. In a major step to increase its economic activity in the region, Paraguay in March 1991 joined the Southern Cone Common Market (MERCOSUR), which includes Brazil, Argentina, and Uruguay. In 1992, the government, through an unorthodox approach, reduced external debt with both commercial and official creditors by purchasing a sizable amount of the delinquent commercial debt in the secondary market at a substantial discount. The government had paid 100% of remaining official debt arrears to the US, Germany, France, and Spain. All commercial debt arrears have been rescheduled. For the long run, the government must press forward with general, market-oriented economic reforms. Growth of 3.5% in 1993 was spurred by higher-than-expected agricultural output and rising international commodity prices. Inflation picked up steam in fourth quarter 1993 because of rises in public sector salaries and utility rates. GDP growth continued in 1994 at 3.5%. Although inflation declined a bit over 1993, increases in food prices, and crop and infrastructure damage from heavy rains at the end of the year, forced inflation to 18%, above the government's target of 15%. Paraguay reaffirmed its commitment to MERCOSUR on 1 January 1995 by implementing the organization's common external tariff.
National product: GDP—purchasing power parity—$15.4 billion (1994 est.)
National product real growth rate: 3.5% (1994 est.)
National product per capita: $2,950 (1994 est.)
Inflation rate (consumer prices): 18% (1994 est.)
Unemployment rate: 11.2% (1994 est.)
Budget:
revenues: $1.2 billion
expenditures: $1.4 billion, including capital expenditures of $487 million (1992 est.)
Exports: $728 million (f.o.b., 1993 est.)
commodities: cotton, soybeans, timber, vegetable oils, meat products, coffee, tung oil
partners: EC 37%, Brazil 25%, Argentina 10%, Chile 6%, US 6%
Imports: $1.38 billion (c.i.f., 1993 est.)
commodities: capital goods, foodstuffs, consumer goods, raw materials, fuels
partners: Brazil 30%, EC 20%, US 18%, Argentina 8%, Japan 7%
External debt: $1.4 billion (yearend 1993 est.)
Industrial production: growth rate 3.6% (1993 est.); accounts for 20% of GDP
Electricity:
capacity: 6,530,000 kW
production: 26.5 billion kWh (1992)
consumption per capita: NA
note: much of the electricity produced in Paraguay is exported to Brazil and domestic consumption cannot be determined
Industries: meat packing, oilseed crushing, milling, brewing, textiles, other light consumer goods, cement, construction
Agriculture: accounts for 26% of GDP; cash crops—cotton, sugarcane, soybeans; other crops—corn, wheat, tobacco, cassava, fruits, vegetables; animal products—beef, pork, eggs, milk; surplus producer of timber; self-sufficient in most foods
Illicit drugs: illicit producer of cannabis for the international drug trade; important transshipment point for Bolivian cocaine headed for the US and Europe
Economic aid:
recipient: US commitments, including Ex-Im (FY70-89), $172 million; Western (non-US) countries, ODA and OOF bilateral commitments (1970-89), $1.1 billion
Currency: 1 guarani (G) = 100 centimos
Exchange rates: guaranies (G) per US$—1,949.6 (January 1995), 1,911.5 (1994), 1,744.3 (1993), 1,500.3 (1992), 1,325.2 (1991), 1,229.8 (1990)
Fiscal year: calendar year

Transportation

Railroads:
total: 970 km
standard gauge: 440 km 1.435-m gauge
narrow gauge: 60 km 1.000-m gauge
other: 470 km various gauges (privately owned)

Highways:
total: 28,300 km
paved: 2,600 km
unpaved: gravel 500 km; earth 25,200 km
Inland waterways: 3,100 km
Ports: Asuncion, Villeta, San Antonio, Encarnacion
Merchant marine:
total: 13 ships (1,000 GRT or over) totaling 16,747 GRT/19,513 DWT
ships by type: cargo 11, oil tanker 2
note: in addition, 1 naval cargo ship is sometimes used commercially
Airports:
total: 929
with paved runways over 3,047 m: 2
with paved runways 1,524 to 2,437 m: 2
with paved runways 914 to 1,523 m: 3
with paved runways under 914 m: 578
with unpaved runways over 3,047 m: 2
with unpaved runways 2,438 to 3,047 m: 1
with unpaved runways 1,524 to 2,438 m: 27
with unpaved runways 914 to 1,523 m: 314

Communications

Telephone system: 78,300 telephones; 16 telephones/1,000 persons; meager telephone service; principal switching center in Asuncion
local: NA
intercity: fair microwave radio relay network
international: 1 INTELSAT (Atlantic Ocean) earth station
Radio:
broadcast stations: AM 40, FM 0, shortwave 7
radios: NA
Television:
broadcast stations: 5
televisions: NA

Defense Forces

Branches: Army, Navy (includes Naval Air and Marines), Air Force
Manpower availability: males age 15-49 1,290,894; males fit for military service 937,054; males reach military age (17) annually 55,551 (1995 est.)
Defense expenditures: exchange rate conversion—$100 million, 1.6% of GDP (1994 est.)

Peru

Geography

Location: Western South America, bordering the South Pacific Ocean, between Chile and Ecuador
Map references: South America
Area:
total area: 1,285,220 sq km
land area: 1.28 million sq km
comparative area: slightly smaller than Alaska
Land boundaries: total 6,940 km, Bolivia 900 km, Brazil 1,560 km, Chile 160 km, Colombia 2,900 km, Ecuador 1,420 km
Coastline: 2,414 km
Maritime claims:
continental shelf: 200 nm
territorial sea: 200 nm
International disputes: three sections of the boundary with Ecuador are in dispute
Climate: varies from tropical in east to dry desert in west
Terrain: western coastal plain (costa), high and rugged Andes in center (sierra), eastern lowland jungle of Amazon Basin (selva)
Natural resources: copper, silver, gold, petroleum, timber, fish, iron ore, coal, phosphate, potash
Land use:
arable land: 3%
permanent crops: 0%
meadows and pastures: 21%
forest and woodland: 55%
other: 21%
Irrigated land: 12,500 sq km (1989 est.)
Environment:
current issues: deforestation; overgrazing of the slopes of the costa and sierra leading to soil erosion; desertification; air pollution in Lima; pollution of rivers and coastal waters from municipal and mining wastes
natural hazards: earthquakes, tsunamis, flooding, landslides, mild volcanic activity
international agreements: party to—Antarctic-Environmental Protocol, Antarctic Treaty, Biodiversity, Climate Change, Endangered Species, Hazardous Wastes, Nuclear Test Ban, Ozone Layer Protection, Ship Pollution, Tropical Timber 83, Wetlands, Whaling; signed, but not ratified—Desertification, Tropical Timber 94
Note: shares control of Lago Titicaca, world's highest navigable lake, with Bolivia

People

Population: 24,087,372 (July 1995 est.)
Age structure:
0-14 years: 35% (female 4,152,520; male 4,296,293)
15-64 years: 61% (female 7,280,287; male 7,378,227)
65 years and over: 4% (female 535,156; male 444,889) (July 1995 est.)
Population growth rate: 1.8% (1995 est.)
Birth rate: 24.88 births/1,000 population (1995 est.)
Death rate: 6.84 deaths/1,000 population (1995 est.)
Net migration rate: 0 migrant(s)/1,000 population (1995 est.)
Infant mortality rate: 52.1 deaths/1,000 live births (1995 est.)
Life expectancy at birth:
total population: 66.07 years
male: 63.86 years
female: 68.38 years (1995 est.)
Total fertility rate: 3 children born/woman (1995 est.)
Nationality:
noun: Peruvian(s)
adjective: Peruvian
Ethnic divisions: Indian 45%, mestizo (mixed Indian and European ancestry) 37%, white 15%, black, Japanese, Chinese, and other 3%
Religions: Roman Catholic
Languages: Spanish (official), Quechua (official), Aymara
Literacy: age 15 and over can read and write (1990 est.)
total population: 82%
male: 92%
female: 74%
Labor force: 8 million (1992)
by occupation: government and other services 44%, agriculture 37%, industry 19% (1988 est.)

Government

Names:
conventional long form: Republic of Peru
conventional short form: Peru
local long form: Republica del Peru
local short form: Peru
Digraph: PE
Type: republic
Capital: Lima
Administrative divisions: 24 departments (departamentos, singular—departamento) and 1 constitutional province* (provincia

constitucional); Amazonas, Ancash, Apurimac, Arequipa, Ayacucho, Cajamarca, Callao*, Cusco, Huancavelica, Huanuco, Ica, Junin, La Libertad, Lambayeque, Lima, Loreto, Madre de Dios, Moquegua, Pasco, Piura, Puno, San Martin, Tacna, Tumbes, Ucayali
note: the 1979 Constitution mandated the creation of regions (regiones, singular—region) to function eventually as autonomous economic and administrative entities; so far, 12 regions have been constituted from 23 of the 24 departments—Amazonas (from Loreto), Andres Avelino Caceres (from Huanuco, Pasco, Junin), Arequipa (from Arequipa), Chavin (from Ancash), Grau (from Tumbes, Piura), Inca (from Cusco, Madre de Dios, Apurimac), La Libertad (from La Libertad), Los Libertadores-Huari (from Ica, Ayacucho, Huancavelica), Mariategui (from Moquegua, Tacna, Puno), Nor Oriental del Maranon (from Lambayeque, Cajamarca, Amazonas), San Martin (from San Martin), Ucayali (from Ucayali); formation of another region has been delayed by the reluctance of the constitutional province of Callao to merge with the department of Lima; because of inadequate funding from the central government and organizational and political difficulties, the regions have yet to assume major responsibilities; the 1993 Constitution retains the regions but limits their authority; the 1993 Constitution also reaffirms the roles of departmental and municipal governments
Independence: 28 July 1821 (from Spain)
National holiday: Independence Day, 28 July (1821)
Constitution: 31 December 1993
Legal system: based on civil law system; has not accepted compulsory ICJ jurisdiction
Suffrage: 18 years of age; universal
Executive branch:
chief of state and head of government: President Alberto Kenyo FUJIMORI Fujimori (since 28 July 1990); election last held 9 April 1995 (next to be held NA 2000); results—Alberto FUJIMORI 64.42%, Javier PEREZ de CUELLAR 21.80%, Mercedes CABANILLAS 4.11%, other 9.67%
cabinet: Council of Ministers; appointed by the president
note: Prime Minister Efrain GOLDENBERG Schreiber (since NA February 1994) does not exercise executive power; this power is in the hands of the president
Legislative branch: unicameral
Congress: elections last held 9 April 1995 (next to be held NA 2000); results—C90/NM 52.1% of the total vote, UPP 14%, eleven other parties 33.9%; seats—(120 total, when installed on 28 July 1995) C90/NM 67, UPP 17, APRA 8, FIM 6, (CODE)-Pais Posible 5, AP 4, PPC 3, Renovacion 3, IU 2, OBRAS 2, MIA 1, FRENATRACA 1, (FREPAP) 1
Judicial branch: Supreme Court of Justice (Corte Suprema de Justicia)
Political parties and leaders: Change 90-New Majority (C90/NM), Alberto FUJIMORI; Union for Peru (UPP), Javier PEREZ de CUELLAR; American Popular Revolutionary Alliance (APRA), Agustin MANTILLA Campos; Independent Moralizing Front (FIM), Fernando OLIVERA Vega; Democratic Coordinator (CODE)-Pais Posible, Jose BARBA Caballero and Alejandro TOLEDO; Popular Action Party (AP), Raul DIEZ CANSECO; Popular Christian Party (PPC), Luis BEDOYA Reyes; Renovacion, Rafael REY Rey; Civic Works Movement (OBRAS), Ricardo BELMONT; United Left (IU), Agustin HAYA de la TORRE; Independent Agrarian Movement (MIA), Rolando SALVATERRIE; Peru 2000-National Front of Workers and Peasants (FRENATRACA), Roger CACARES; Popular Agricultural Front (FREPAP), Ezequiel ATAUCUSI
Other political or pressure groups: leftist guerrilla groups include Shining Path, Abimael GUZMAN Reynoso (imprisoned); Tupac Amaru Revolutionary Movement, Nestor SERPA and Victor POLAY (imprisoned)
Member of: AG, CCC, ECLAC, FAO, G-11, G-15, G-19, G-24, G-77, GATT, IADB, IAEA, IBRD, ICAO, ICFTU, ICRM, IDA, IFAD, IFC, IFRCS, ILO, IMF, IMO, INMARSAT, INTELSAT, INTERPOL, IOC, IOM, ISO (correspondent), ITU, LAES, LAIA, NAM, OAS, OPANAL, PCA, RG (suspended), UN, UNCTAD, UNESCO, UNIDO, UPU, WCL, WFTU, WHO, WIPO, WMO, WTO
Diplomatic representation in US:
chief of mission: Ambassador Ricardo V. LUNA Mendoza
chancery: 1700 Massachusetts Avenue NW, Washington, DC 20036
telephone: [1] (202) 833-9860 through 9869
FAX: [1] (202) 659-8124
consulate(s) general: Chicago, Houston, Los Angeles, Miami, New York, Paterson (New Jersey), and San Francisco
US diplomatic representation:
chief of mission: Ambassador Alvin P. ADAMS, Jr.
embassy: corner of Avenida Inca Garcilaso de la Vega and Avenida Espana, Lima
mailing address: P. O. Box 1995, Lima 1; American Embassy (Lima), APO AA 34031
telephone: [51] (14) 338000
FAX: [51] (14) 316682
Flag: three equal, vertical bands of red (hoist side), white, and red with the coat of arms centered in the white band; the coat of arms features a shield bearing a llama, cinchona tree (the source of quinine), and a yellow cornucopia spilling out gold coins, all framed by a green wreath

Economy

Overview: The Peruvian economy has become increasingly market-oriented, with major privatizations completed in 1994 in the mining and telecommunications industries. In the 1980s the economy suffered from hyperinflation, declining per capita output, and mounting external debt. Peru was shut off from IMF and World Bank support in the mid-1980s because of its huge debt arrears. An austerity program implemented shortly after the FUJIMORI government took office in July 1990 contributed to a third consecutive yearly contraction of economic activity, but the slide came to a halt late that year, and in 1991 output rose 2.4%. After a burst of inflation as the austerity program eliminated government price subsidies, monthly price increases eased to the single-digit level and by December 1991 dropped to the lowest increase since mid-1987. Lima obtained a financial rescue package from multilateral lenders in September 1991, although it faced $14 billion in arrears on its external debt. By working with the IMF and World Bank on new financial conditions and arrangements, the government succeeded in ending its arrears by March 1993. In 1992, GDP had fallen by 2.8%, in part because a warmer-than-usual El Nino current resulted in a 30% drop in the fish catch, but the economy rebounded as strong foreign investment helped push growth to 6% in 1993 and 8.6% in 1994.
National product: GDP—purchasing power parity—$73.6 billion (1994 est.)
National product real growth rate: 8.6% (1994 est.)
National product per capita: $3,110 (1994 est.)
Inflation rate (consumer prices): 15% (1994 est.)
Unemployment rate: 15%; extensive underemployment (1992 est.)
Budget:
revenues: $2 billion
expenditures: $1.7 billion, including capital expenditures of $300 million (1992 est.)
Exports: $4.1 billion (f.o.b., 1994 est.)
commodities: copper, zinc, fishmeal, crude petroleum and byproducts, lead, refined silver, coffee, cotton
partners: US 19%, Japan 9%, Italy, Germany
Imports: $5.1 billion (f.o.b., 1994 est.)
commodities: machinery, transport equipment, foodstuffs, petroleum, iron and steel, chemicals, pharmaceuticals
partners: US 21%, Colombia, Argentina, Japan, Germany, Brazil
External debt: $22.4 billion (1994 est.)
Industrial production: NA
Electricity:
capacity: 4,190,000 kW
production: 11.2 billion kWh
consumption per capita: 448 kWh (1993)
Industries: mining of metals, petroleum, fishing, textiles, clothing, food processing, cement, auto assembly, steel, shipbuilding, metal fabrication
Agriculture: accounts for 12% of GDP, about 35% of labor force; commercial crops—coffee,

cotton, sugarcane; other crops—rice, wheat, potatoes, plantains, coca; animal products—poultry, red meats, dairy, wool; not self-sufficient in grain or vegetable oil; fish catch of 6.9 million metric tons (1990)
Illicit drugs: world's largest coca leaf producer with about 108,600 hectares under cultivation in 1994; source of supply for most of the world's coca paste and cocaine base; at least 85% of coca cultivation is for illicit production; most of cocaine base is shipped to Colombian drug dealers for processing into cocaine for the international drug market, but exports of finished cocaine are increasing
Economic aid:
recipient: US commitments, including Ex-Im (FY70-89), $1.7 billion; Western (non-US) countries, ODA and OOF bilateral commitments (1970-89), $4.3 billion; Communist countries (1970-89), $577 million
Currency: 1 nuevo sol (S/.) = 100 centimos
Exchange rates: nuevo sol (S/.) per US$1—2.20 (February 1995), 2.195 (1994),1.988 (1993), 1.245 (1992), 0.772 (1991), 0.187 (1990)
Fiscal year: calendar year

Transportation

Railroads:
total: 1,801 km
standard gauge: 1,501 km 1.435-m gauge
narrow gauge: 300 km 0.914-m gauge
Highways:
total: 69,942 km
paved: 7,459 km
unpaved: improved earth 13,538 km; unimproved earth 48,945 km
Inland waterways: 8,600 km of navigable tributaries of Amazon system and 208 km of Lago Titicaca
Pipelines: crude oil 800 km; natural gas and natural gas liquids 64 km
Ports: Callao, Chimbote, Ilo, Iquitos, Matarani, Paita, Pucallpa, Salaverry, San Martin, Talara, Yurimaguas
note: Iquitos, Pucallpa, and Yurimaguas are all on the upper reaches of the Amazon and its tributaries
Merchant marine:
total: 10 ships (1,000 GRT or over) totaling 90,501 GRT/144,913 DWT
ships by type: bulk 3, cargo 6, refrigerated cargo 1
note: in addition, 4 naval tankers and 1 naval cargo are sometimes used commercially
Airports:
total: 236
with paved runways over 3,047 m: 6
with paved runways 2,438 to 3,047 m: 16
with paved runways 1,524 to 2,437 m: 11
with paved runways 914 to 1,523 m: 5
with paved runways under 914 m: 97
with unpaved runways over 3,047 m: 1
with unpaved runways 2,438 to 3,047 m: 2
with unpaved runways 1,524 to 2,438 m: 21
with unpaved runways 914 to 1,523 m: 77

Communications

Telephone system: 544,000 telephones; fairly adequate for most requirements
local: NA
intercity: nationwide microwave radio relay system and 12 domestic satellite links
international: 2 Atlantic Ocean INTELSAT earth stations
Radio:
broadcast stations: AM 273, FM 0, shortwave 144
radios: NA
Television:
broadcast stations: 140
televisions: NA

Defense Forces

Branches: Army (Ejercito Peruano), Navy (Marina de Guerra del Peru), Air Force (Fuerza Aerea del Peru), National Police
Manpower availability: males age 15-49 6,369,157; males fit for military service 4,300,772; males reach military age (20) annually 251,798 (1995 est.)
Defense expenditures: exchange rate conversion—$810 million, about 2.7% of GDP (1994)

Philippines

Geography

Location: Southeastern Asia, archipelago between the Philippine Sea and the South China Sea, east of Vietnam
Map references: Southeast Asia
Area:
total area: 300,000 sq km
land area: 298,170 sq km
comparative area: slightly larger than Arizona
Land boundaries: 0 km
Coastline: 36,289 km
Maritime claims: measured from claimed archipelagic baselines
continental shelf: to depth of exploitation
exclusive economic zone: 200 nm
territorial sea: irregular polygon extending up to 100 nm from coastline as defined by 1898 treaty; since late 1970s has also claimed polygonal-shaped area in South China Sea up to 285 nm in breadth
International disputes: involved in a complex dispute over the Spratly Islands with China, Malaysia, Taiwan, Vietnam, and possibly Brunei; claims Malaysian state of Sabah
Climate: tropical marine; northeast monsoon (November to April); southwest monsoon (May to October)
Terrain: mostly mountains with narrow to extensive coastal lowlands
Natural resources: timber, petroleum, nickel, cobalt, silver, gold, salt, copper
Land use:
arable land: 26%
permanent crops: 11%
meadows and pastures: 4%
forest and woodland: 40%
other: 19%
Irrigated land: 16,200 sq km (1989 est.)
Environment:
current issues: uncontrolled deforestation in watershed areas; soil erosion; air and water pollution in Manila; increasing pollution of coastal mangrove swamps which are important fish breeding grounds
natural hazards: astride typhoon belt, usually

affected by 15 and struck by five to six cyclonic storms per year; landslides, active volcanoes, destructive earthquakes, tsunamis
international agreements: party to—Biodiversity, Climate Change, Endangered Species, Hazardous Wastes, Law of the Sea, Marine Dumping, Nuclear Test Ban, Ozone Layer Protection, Tropical Timber 83, Wetlands

People

Population: 73,265,584 (July 1995 est.)
Age structure:
0-14 years: 38% (female 13,841,552; male 14,214,234)
15-64 years: 58% (female 21,603,818; male 20,923,307)
65 years and over: 4% (female 1,425,706; male 1,256,967) (July 1995 est.)
Population growth rate: 2.23% (1995 est.)
Birth rate: 30.42 births/1,000 population (1995 est.)
Death rate: 6.97 deaths/1,000 population (1995 est.)
Net migration rate: -1.14 migrant(s)/1,000 population (1995 est.)
Infant mortality rate: 49.6 deaths/1,000 live births (1995 est.)
Life expectancy at birth:
total population: 65.65 years
male: 63.16 years
female: 68.25 years (1995 est.)
Total fertility rate: 3.81 children born/woman (1995 est.)
Nationality:
noun: Filipino(s)
adjective: Philippine
Ethnic divisions: Christian Malay 91.5%, Muslim Malay 4%, Chinese 1.5%, other 3%
Religions: Roman Catholic 83%, Protestant 9%, Muslim 5%, Buddhist and other 3%
Languages: Pilipino (official; based on Tagalog), English (official)
Literacy: age 15 and over can read and write (1990)
total population: 94%
male: 94%
female: 93%
Labor force: 24.12 million
by occupation: agriculture 46%, industry and commerce 16%, services 18.5%, government 10%, other 9.5% (1989)

Government

Names:
conventional long form: Republic of the Philippines
conventional short form: Philippines
local long form: Republika ng Pilipinas
local short form: Pilipinas
Digraph: RP
Type: republic
Capital: Manila
Administrative divisions: 72 provinces and 61 chartered cities*; Abra, Agusan del Norte, Agusan del Sur, Aklan, Albay, Angeles*, Antique, Aurora, Bacolod*, Bago*, Baguio*, Bais*, Basilan, Basilan City*, Bataan, Batanes, Batangas, Batangas City*, Benguet, Bohol, Bukidnon, Bulacan, Butuan*, Cabanatuan*, Cadiz*, Cagayan, Cagayan de Oro*, Calbayog*, Caloocan*, Camarines Norte, Camarines Sur, Camiguin, Canlaon*, Capiz, Catanduanes, Cavite, Cavite City*, Cebu, Cebu City*, Cotabato*, Dagupan*, Danao*, Dapitan*, Davao City* Davao, Davao del Sur, Davao Oriental, Dipolog*, Dumaguete*, Eastern Samar, General Santos*, Gingoog*, Ifugao, Iligan*, Ilocos Norte, Ilocos Sur, Iloilo, Iloilo City*, Iriga*, Isabela, Kalinga-Apayao, La Carlota*, Laguna, Lanao del Norte, Lanao del Sur, Laoag*, Lapu-Lapu*, La Union, Legaspi*, Leyte, Lipa*, Lucena*, Maguindanao, Mandaue*, Manila*, Marawi*, Marinduque, Masbate, Mindoro Occidental, Mindoro Oriental, Misamis Occidental, Misamis Oriental, Mountain, Naga*, Negros Occidental, Negros Oriental, North Cotabato, Northern Samar, Nueva Ecija, Nueva Vizcaya, Olongapo*, Ormoc*, Oroquieta*, Ozamis*, Pagadian*, Palawan, Palayan*, Pampanga, Pangasinan, Pasay*, Puerto Princesa*, Quezon, Quezon City*, Quirino, Rizal, Romblon, Roxas*, Samar, San Carlos* (in Negros Occidental), San Carlos* (in Pangasinan), San Jose*, San Pablo*, Silay*, Siquijor, Sorsogon, South Cotabato, Southern Leyte, Sultan Kudarat, Sulu, Surigao*, Surigao del Norte, Surigao del Sur, Tacloban*, Tagaytay*, Tagbilaran*, Tangub*, Tarlac, Tawitawi, Toledo*, Trece Martires*, Zambales, Zamboanga*, Zamboanga del Norte, Zamboanga del Sur
Independence: 4 July 1946 (from US)
National holiday: Independence Day, 12 June (1898) (from Spain)
Constitution: 2 February 1987, effective 11 February 1987
Legal system: based on Spanish and Anglo-American law; accepts compulsory ICJ jurisdiction, with reservations
Suffrage: 15 years of age; universal
Executive branch:
chief of state and head of government: President Fidel Valdes RAMOS (since 30 June 1992); Vice President Joseph Ejercito ESTRADA (since 30 June 1992); election last held 11 May 1992 (next to be held NA May 1998); results—Fidel Valdes RAMOS won 23.6% of the vote, a narrow plurality
cabinet: Executive Secretary; appointed by the president with the consent of the Commission of Appointments
Legislative branch: bicameral Congress (Kongreso)
Senate (Senado): elections last held 11 May 1992 (next to be held NA May 1995); results—LDP 66%, NPC 20%, Lakas/NUCD 8%, Liberal 6%; seats—(24 total) LDP 15, NPC 5, Lakas/NUCD 2, Liberal 1, independent 1
House of Representatives (Kapulungan Ng Mga Kinatawan): elections last held 11 May 1992 (next election to be held NA May 1995); results—LDP 43.5%; Lakas/NUCD 25%, NPC 23.5%, Liberal 5%, KBL 3%; seats—(200 total) LDP 87, NPC 45, Lakas/NUCD 41, Liberal 15, NP 6, KBL 3, independents 3
Judicial branch: Supreme Court
Political parties and leaders: Democratic Filipino Struggle (Laban ng Demokratikong Pilipinas, LDP), Edgardo ESPIRITU; People Power-National Union of Christian Democrats (Lakas ng Edsa, NUCD and Partido Lakas Tao, Lakas/NUCD); Fidel V. RAMOS, President of the Republic, Raul MANGLAPUS, Jose DE VENECIA, secretary general; Nationalist People's Coalition (NPC), Eduardo COJUANGCO; Liberal Party, Jovito SALONGA; People's Reform Party (PRP), Miriam DEFENSOR-SANTIAGO; New Society Movement (Kilusan Bagong Lipunan; KBL), Imelda MARCOS; Nacionalista Party (NP), Salvador H. LAUREL, president
Member of: APEC, AsDB, ASEAN, CCC, CP, ESCAP, FAO, G-24, G-77, GATT, IAEA, IBRD, ICAO, ICFTU, ICRM, IDA, IFAD, IFC, IFRCS, ILO, IMF, IMO, INMARSAT, INTELSAT, INTERPOL, IOC, IOM, ISO, ITU, NAM, UN, UNCTAD, UNESCO, UNHCR, UNIDO, UNU, UPU, WCL, WFTU, WHO, WIPO, WMO, WTO
Diplomatic representation in US:
chief of mission: Ambassador Raul Chaves RABE
chancery: 1600 Massachusetts Avenue NW, Washington, DC 20036
telephone: [1] (202) 467-9300
FAX: [1] (202) 328-7614
consulate(s) general: Agana (Guam), Chicago, Honolulu, Houston, Los Angeles, New York, San Francisco, and Seattle
consulate(s): San Diego and San Jose (Saipan)
US diplomatic representation:
chief of mission: Ambassador John D. NEGROPONTE
embassy: 1201 Roxas Boulevard, Ermita Manila 1000
mailing address: APO AP 96440
telephone: [63] (2) 521-71-16
FAX: [63] (2) 522-43-61
consulate(s): Cebu
Flag: two equal horizontal bands of blue (top) and red with a white equilateral triangle based on the hoist side; in the center of the triangle is a yellow sun with eight primary rays (each containing three individual rays) and in each corner of the triangle is a small yellow five-pointed star

Economy

Overview: Domestic output in this primarily agricultural economy failed to grow in 1992 and rose only slightly in 1993. Drought and

power supply problems hampered production, while inadequate revenues prevented government pump priming. Worker remittances helped to supplement GDP. A marked increase in capital goods imports, particularly power generating equipment, telecommunications equipment, and electronic data processors, contributed to 20% annual import growth in 1992-94. Provided the government can cope with the substantial trade deficit and meet the fiscal targets agreed to with the IMF, the Philippines should duplicate the strong growth performance of 1994 in 1995-96.
National product: GDP—purchasing power parity—$161.4 billion (1994 est.)
National product real growth rate: 4.3% (1994 est.)
National product per capita: $2,310 (1994 est.)
Inflation rate (consumer prices): 7.1% (1994 est.)
Unemployment rate: 9% (1994)
Budget:
revenues: $14 billion
expenditures: $15.4 billion, including capital expenditures of $NA (FY94/95 est.)
Exports: $13.4 billion (f.o.b., 1994)
commodities: electronics, textiles, coconut products, copper, fish
partners: US 39%, Japan 16%, Germany 5%, Hong Kong 5%, UK 4% (1993)
Imports: $21.3 billion (f.o.b., 1994)
commodities: raw materials 40%, capital goods 25%, petroleum products 10%
partners: Japan 23%, US 20%, Taiwan 6%, Singapore 5%, South Korea 5% (1993)
External debt: $40 billion (1994 est.)
Industrial production: growth rate 1.4% (1993); accounts for 28% of GDP
Electricity:
capacity: 6,770,000 kW
production: 20.4 billion kWh
consumption per capita: 278 kWh (1993)
Industries: textiles, pharmaceuticals, chemicals, wood products, food processing, electronics assembly, petroleum refining, fishing
Agriculture: accounts for 22% of GDP and about 45% of labor force; major crops—rice, coconuts, corn, sugarcane, bananas, pineapples, mangos; animal products—pork, eggs, beef; net exporter of farm products; fish catch of 2 million metric tons annually
Illicit drugs: illicit producer of cannabis for the international drug trade; growers are producing more and better quality cannabis despite government eradication efforts; transit point for Southwest Asian heroin bound for the US
Economic aid:
recipient: US commitments, including Ex-Im (FY70-89), $3.6 billion; Western (non-US) countries, ODA and OOF bilateral commitments (1970-88), $7.9 billion; OPEC bilateral aid (1979-89), $5 million; Communist countries (1975-89), $123 million
Currency: 1 Philippine peso (P) = 100 centavos
Exchange rates: Philippine pesos (P) per US$1—24.622 (January 1995), 26.417 (1994), 22.120 (1993), 25.512 (1992), 27.479 (1991), 24.311 (1990)
Fiscal year: calendar year

Transportation

Railroads:
total: 800 km (est.); note—including about 390 km in Luzon
narrow gauge: 800 km 1.067-m gauge
Highways:
total: 160,700 km
paved: 29,000 km
unpaved: 131,700 km
Inland waterways: 3,219 km; limited to shallow-draft (less than 1.5 m) vessels
Pipelines: petroleum products 357 km
Ports: Batangas, Cagayan de Oro, Cebu, Davao, Guimaras, Iligan, Iloilo, Jolo, Legaspi, Manila, Masao, Puerto Princesa, San Fernando, Subic Bay, Zamboanga
Merchant marine:
total: 552 ships (1,000 GRT or over) totaling 8,748,083 GRT/14,373,730 DWT
ships by type: bulk 237, cargo 134, chemical tanker 4, combination bulk 10, combination ore/oil 1, container 10, liquefied gas tanker 6, livestock carrier 9, oil tanker 46, passenger 1, passenger-cargo 11, refrigerated cargo 24, roll-on/roll-off cargo 13, short-sea passenger 17, vehicle carrier 29
note: a flag of convenience registry; Japan owns 13 ships, Norway 2, Switzerland 1, Taiwan 1, and South Korea 1
Airports:
total: 269
with paved runways over 3,047 m: 2
with paved runways 2,438 to 3,047 m: 7
with paved runways 1,524 to 2,437 m: 24
with paved runways 914 to 1,523 m: 32
with paved runways under 914 m: 133
with unpaved runways 1,524 to 2,438 m: 4
with unpaved runways 914 to 1,523 m: 67

Communications

Telephone system: 872,900 telephones; good international radio and submarine cable services; domestic and interisland service adequate
local: NA
intercity: 11 domestic satellite links
international: submarine cables extended to Hong Kong, Guam, Singapore, Taiwan, and Japan; 3 INTELSAT (1 Indian Ocean and 2 Pacific Ocean) earth stations
Radio:
broadcast stations: AM 267 (including 6 US), FM 55, shortwave 0
radios: NA
Television:
broadcast stations: 33 (including 4 US)
televisions: NA

Defense Forces

Branches: Army, Navy (includes Coast Guard and Marine Corps), Air Force
Manpower availability: males age 15-49 18,238,568; males fit for military service 12,876,771; males reach military age (20) annually 752,622 (1995 est.)
Defense expenditures: exchange rate conversion—$731 million, 1.4% of GNP (1992)

Pitcairn Islands

(dependent territory of the UK)

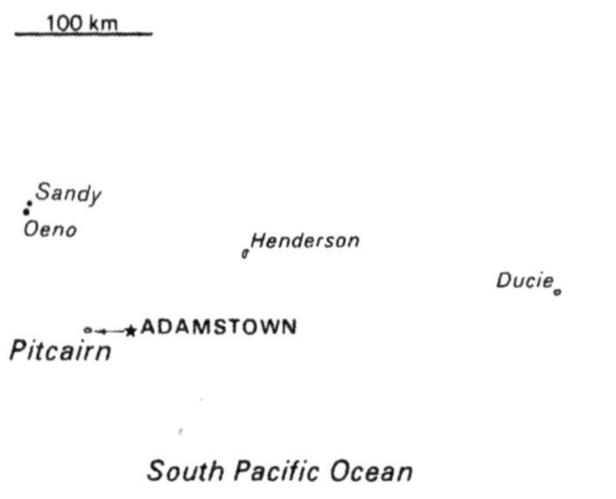

Geography

Location: Oceania, islands in the South Pacific Ocean, about one-half of the way from Peru to New Zealand
Map references: Oceania
Area:
total area: 47 sq km
land area: 47 sq km
comparative area: about 0.3 times the size of Washington, DC
Land boundaries: 0 km
Coastline: 51 km
Maritime claims:
exclusive fishing zone: 200 nm
territorial sea: 3 nm
International disputes: none
Climate: tropical, hot, humid, modified by southeast trade winds; rainy season (November to March)
Terrain: rugged volcanic formation; rocky coastline with cliffs
Natural resources: miro trees (used for handicrafts), fish
Land use:
arable land: NA%
permanent crops: NA%
meadows and pastures: NA%
forest and woodland: NA%
other: NA%
Irrigated land: NA sq km
Environment:
current issues: deforestation (only a small portion of the original forest remains because of burning and clearing for settlement)
natural hazards: typhoons (especially November to March)
international agreements: NA

People

Population: 73 (July 1995 est.)
Age structure:
0-14 years: NA
15-64 years: NA
65 years and over: NA
Population growth rate: 2.8% (1995 est.)
Birth rate: NA births/1,000 population
Death rate: NA deaths/1,000 population
Net migration rate: NA migrant(s)/1,000 population
Infant mortality rate: NA deaths/1,000 live births
Life expectancy at birth:
total population: NA years
male: NA years
female: NA years
Total fertility rate: NA children born/women
Nationality:
noun: Pitcairn Islander(s)
adjective: Pitcairn Islander
Ethnic divisions: descendants of the Bounty mutineers
Religions: Seventh-Day Adventist 100%
Languages: English (official), Tahitian/English dialect
Labor force: NA
by occupation: no business community in the usual sense; some public works; subsistence farming and fishing

Government

Names:
conventional long form: Pitcairn, Henderson, Ducie, and Oeno Islands
conventional short form: Pitcairn Islands
Digraph: PC
Type: dependent territory of the UK
Capital: Adamstown
Administrative divisions: none (dependent territory of the UK)
Independence: none (dependent territory of the UK)
National holiday: Celebration of the Birthday of the Queen (second Saturday in June)
Constitution: Local Government Ordinance of 1964
Legal system: local island by-laws
Suffrage: 18 years of age; universal with three years residency
Executive branch:
chief of state: Queen ELIZABETH II (since 6 February 1952), represented by UK High Commissioner to New Zealand and Governor (non-resident) of the Pitcairn Islands Robert John ALSTON (since NA); Commissioner (non-resident) G.D. HARRAWAY (since NA; is the liaison person between the governor and the Island Council)
head of government: Island Magistrate and Chairman of the Island Council Jay WARREN (since NA)
Legislative branch: unicameral
Island Council: elections last held NA December 1994 (next to be held NA December 1995); results—percent of vote NA; seats—(11 total, 5 elected) all independents
Judicial branch: Island Court
Political parties and leaders: none
Other political or pressure groups: NA
Member of: SPC
Diplomatic representation in US: none (dependent territory of the UK)
US diplomatic representation: none (dependent territory of the UK)
Flag: blue with the flag of the UK in the upper hoist-side quadrant and the Pitcairn Islander coat of arms centered on the outer half of the flag; the coat of arms is yellow, green, and light blue with a shield featuring a yellow anchor

Economy

Overview: The inhabitants exist on fishing and subsistence farming. The fertile soil of the valleys produces a wide variety of fruits and vegetables, including citrus, sugarcane, watermelons, bananas, yams, and beans. Bartering is an important part of the economy. The major sources of revenue are the sale of postage stamps to collectors and the sale of handicrafts to passing ships.
National product: GDP $NA
National product real growth rate: NA%
National product per capita: $NA
Inflation rate (consumer prices): NA%
Unemployment rate: NA%
Budget:
revenues: $430,000
expenditures: $429,000, including capital expenditures of $NA (1987 est.)
Exports: $NA
commodities: fruits, vegetables, curios
partners: NA
Imports: $NA
commodities: fuel oil, machinery, building materials, flour, sugar, other foodstuffs
partners: NA
External debt: $NA
Industrial production: growth rate NA%
Electricity:
capacity: 110 kW
production: 300,000 kWh
consumption per capita: 5,360 kWh (1990)
Industries: postage stamps, handicrafts
Agriculture: based on subsistence fishing and farming; wide variety of fruits and vegetables; must import grain products
Economic aid:
recipient: ODA bilateral commitments (1992-93), $84,000
Currency: 1 New Zealand dollar (NZ$) = 100 cents
Exchange rates: New Zealand dollars (NZ$) per US$1—1.5601 (January 1995), 1.6844 (1994), 1.8495 (1993), 1.8584 (1992), 1.7265 (1991), 1.6750 (1990)
Fiscal year: 1 April—31 March

Transportation

Railroads: 0 km
Highways:
total: 6.4 km
unpaved: earth 6.4 km

Ports: Bounty Bay
Merchant marine: none
Airports: none

Communications

Telephone system: 24 telephones; party line telephone service on the island
local: NA
intercity: NA
international: NA
Radio:
broadcast stations: AM 1, FM 0, shortwave 0
radios: NA
Television:
broadcast stations: 0
televisions: NA

Defense Forces

Note: defense is the responsibility of the UK

Poland

Geography

Location: Central Europe, east of Germany
Map references: Ethnic Groups in Eastern Europe, Europe
Area:
total area: 312,680 sq km
land area: 304,510 sq km
comparative area: slightly smaller than New Mexico
Land boundaries: total 3,114 km, Belarus 605 km, Czech Republic 658 km, Germany 456 km, Lithuania 91 km, Russia (Kaliningrad Oblast) 432 km, Slovakia 444 km, Ukraine 428 km
Coastline: 491 km
Maritime claims:
exclusive economic zone: defined by international treaties
territorial sea: 12 nm
International disputes: none
Climate: temperate with cold, cloudy, moderately severe winters with frequent precipitation; mild summers with frequent showers and thundershowers
Terrain: mostly flat plain; mountains along southern border
Natural resources: coal, sulfur, copper, natural gas, silver, lead, salt
Land use:
arable land: 46%
permanent crops: 1%
meadows and pastures: 13%
forest and woodland: 28%
other: 12%
Irrigated land: 1,000 sq km (1989 est.)
Environment:
current issues: forest damage due to air pollution and resulting acid rain; improper means for disposal of large amounts of hazardous and industrial waste; severe water pollution from industrial and municipal sources; severe air pollution results from emissions of sulfur dioxide from coal-fired power plants, which also drifts into Germany and the Netherlands
natural hazards: NA
international agreements: party to—Air Pollution, Antarctic Treaty, Climate Change, Endangered Species, Environmental Modification, Hazardous Wastes, Marine Dumping, Nuclear Test Ban, Ozone Layer Protection, Ship Pollution, Wetlands; signed, but not ratified—Air Pollution-Nitrogen Oxides, Air Pollution-Sulphur 94, Antarctic-Environmental Protocol, Biodiversity, Law of the Sea
Note: historically, an area of conflict because of flat terrain and the lack of natural barriers on the North European Plain

People

Population: 38,792,442 (July 1995 est.)
Age structure:
0-14 years: 23% (female 4,349,467; male 4,559,536)
15-64 years: 66% (female 12,849,300; male 12,698,179)
65 years and over: 11% (female 2,693,407; male 1,642,553) (July 1995 est.)
Population growth rate: 0.36% (1995 est.)
Birth rate: 13.34 births/1,000 population (1995 est.)
Death rate: 9.23 deaths/1,000 population (1995 est.)
Net migration rate: -0.52 migrant(s)/1,000 population (1995 est.)
Infant mortality rate: 12.4 deaths/1,000 live births (1995 est.)
Life expectancy at birth:
total population: 73.13 years
male: 69.15 years
female: 77.33 years (1995 est.)
Total fertility rate: 1.92 children born/woman (1995 est.)
Nationality:
noun: Pole(s)
adjective: Polish
Ethnic divisions: Polish 97.6%, German 1.3%, Ukrainian 0.6%, Byelorussian 0.5% (1990 est.)
Religions: Roman Catholic 95% (about 75% practicing), Eastern Orthodox, Protestant, and other 5%
Languages: Polish
Literacy: age 15 and over can read and write (1978)
total population: 99%
male: 99%
female: 98%
Labor force: 17.321 million (1993 annual average)
by occupation: industry and construction 32.0%, agriculture 27.6%, trade, transport, and communications 14.7%, government and other 25.7% (1992)

Government

Names:
conventional long form: Republic of Poland

conventional short form: Poland
local long form: Rzeczpospolita Polska
local short form: Polska
Digraph: PL
Type: democratic state
Capital: Warsaw
Administrative divisions: 49 provinces (wojewodztwa, singular—wojewodztwo); Biala Podlaska, Bialystok, Bielsko Biala, Bydgoszcz, Chelm, Ciechanow, Czestochowa, Elblag, Gdansk, Gorzow, Jelenia Gora, Kalisz, Katowice, Kielce, Konin, Koszalin, Krakow, Krosno, Legnica, Leszno, Lodz, Lomza, Lublin, Nowy Sacz, Olsztyn, Opole, Ostroleka, Pila, Piotrkow, Plock, Poznan, Przemysl, Radom, Rzeszow, Siedlce, Sieradz, Skierniewice, Slupsk, Suwalki, Szczecin, Tarnobrzeg, Tarnow, Torun, Walbrzych, Warszawa, Wloclawek, Wroclaw, Zamosc, Zielona Gora
Independence: 11 November 1918 (independent republic proclaimed)
National holiday: Constitution Day, 3 May (1791)
Constitution: interim "small constitution" came into effect in December 1992 replacing the Communist-imposed constitution of 22 July 1952; new democratic constitution being drafted
Legal system: mixture of Continental (Napoleonic) civil law and holdover Communist legal theory; changes being gradually introduced as part of broader democratization process; limited judicial review of legislative acts; has not accepted compulsory ICJ jurisdiction
Suffrage: 18 years of age; universal
Executive branch:
chief of state: President Lech WALESA (since 22 December 1990); election first round held 25 November 1990, second round held 9 December 1990 (next to be held NA November 1995); results—second round Lech WALESA 74.7%, Stanislaw TYMINSKI 25.3%
head of government: Prime Minister Jozef OLEKSY (since 6 March 1995); Deputy Prime Ministers Roman JAGIELINSKI, Grzegorz KOLODKO, and Aleksander LUCZAK (since NA)
cabinet: Council of Ministers; responsible to the president and the Sejm
Legislative branch: bicameral National Assembly (Zgromadzenie Narodowe)
Senate (Senat): elections last held 19 September 1993 (next to be held no later than October 1997); results—percent of vote by party NA; seats—(100 total) Communist origin or linked 71 (PSL 34, SLD 37), post-Solidarity parties 20 (NSZZ 12, UW 6, BBWR 2), non-Communist and non-Solidarity 9 (independants 7, unaffiliated 1, vacant 1)
Diet (Sejm): elections last held 19 September 1993 (next to be held no later than October 1997); results—percent of vote by party NA; seats—(460 total) Communist origin or linked 303 (SLD 171, PSL 132), post-Solidarity parties 131 (UW 74, UP 41, BBWR 16), non-Communist and non-Solidarity 22 (KPN 22); note—four seats are constitutionally assigned to ethnic German parties
Judicial branch: Supreme Court
Political parties and leaders:
post-Solidarity parties: Freedom Union (UW; Democratic Union and Liberal Democratic Congress merged to form Freedom Union), Leszek BALCEROWICZ; Christian-National Union (ZCHN), Ryszard CZARNECKI; Centrum (PC), Jaroslaw KACZYNSKI; Peasant Alliance (PL), Gabriel JANOWSKI; Solidarity Trade Union (NSZZ), Marian KRZAKLEWSKI; Union of Labor (UP), Ryszard BUGAJ; Christian-Democratic Party (PCHD), Pawel LACZKOWSKI; Conservative Party, Alexander HALL; Nonparty Bloc for the Support of the Reforms (BBWR)
non-Communist, non-Solidarity: Confederation for an Independent Poland (KPN), Leszek MOCZULSKI; Polish Economic Program (PPG), Janusz REWINSKI; Christian Democrats (CHD), Andrzej OWSINSKI; German Minority (MN), Henryk KROL; Union of Real Politics (UPR), Janusz KORWIN-MIKKE; Democratic Party (SD), Antoni MACKIEWICZ
Communist origin: Polish Peasant Party (PSL), Waldemar PAWLAK; Democratic Left Alliance (SLD), Aleksander KWASNIEWSKI
Other political or pressure groups: powerful Roman Catholic Church; Solidarity (trade union); All Poland Trade Union Alliance (OPZZ), populist program
Member of: Australia Group, BIS, BSEC (observer), CBSS, CCC, CE, CEI, CERN, EBRD, ECE, FAO, GATT, IAEA, IBRD, ICAO, ICFTU, ICRM, IDA, IFC, IFRCS, ILO, IMF, IMO, INMARSAT, INTELSAT (nonsignatory user), INTERPOL, IOC, IOM, ISO, ITU, MINURSO, NACC, NAM (guest), NSG, OAS (observer), OSCE, PCA, PFP, UN, UNAMIR, UNCTAD, UNDOF, UNESCO, UNIDO, UNIFIL, UNIKOM, UNOMIG, UNPROFOR, UPU, WCL, WEU (associate partner), WFTU, WHO, WIPO, WMO, WTO, ZC
Diplomatic representation in US:
chief of mission: Ambassador Jerzy KOZMINSKI
chancery: 2640 16th Street NW, Washington, DC 20009
telephone: [1] (202) 234-3800 through 3802
FAX: [1] (202) 328-6271
consulate(s) general: Chicago, Los Angeles, and New York
US diplomatic representation:
chief of mission: Ambassador Nicholas Andrew REY
embassy: Aleje Ujazdowskie 29/31, Warsaw
mailing address: American Embassy Warsaw, Box 5010, Unit 1340, APO AE 09213-1340
telephone: [48] (2) 628-30-41
FAX: [48] (2) 628-82-98
consulate(s) general: Krakow, Poznan
Flag: two equal horizontal bands of white (top) and red; similar to the flags of Indonesia and Monaco which are red (top) and white

Economy

Overview: Poland continues to make good progress in the difficult transition to a market economy that began on 1 January 1990, when the new democratic government instituted "shock therapy" by decontrolling prices, slashing subsidies, and drastically reducing import barriers. Real GDP fell sharply in 1990 and 1991, but in 1992 Poland became the first country in the region to resume economic growth with a 2.6% increase. Growth increased to 3.8% in 1993 and 5.5% in 1994—the highest rate in Europe except for Albania. All of the growth since 1991 has come from the booming private sector, which now accounts for at least 55% of GDP, even though privatization of the state-owned enterprises is proceeding slowly and most industry remains in state hands. Industrial production increased 12% in 1994—led by 50% jumps in the output of motor vehicles, radios and televisions, and pulp and paper—and is now well above the 1990 level. Inflation, which had approached 1,200% annually in early 1990, was down to about 30% in 1994, as the government held the budget deficit to 1.5% of GDP. After five years of steady increases, unemployment has leveled off at about 16% nationwide, although it approaches 30% in some regions. The trade deficit was sharply reduced in 1994, due mainly to increased exports to Western Europe, Poland's main customer. The leftist government elected in September 1993 gets generally good marks from foreign observers for its management of the budget but is often criticized for not moving faster on privatization.
National product: GDP—purchasing power parity—$191.1 billion (1994 est.)
National product real growth rate: 5.5% (1994 est.)
National product per capita: $4,920 (1994 est.)
Inflation rate (consumer prices): 30% (1994)
Unemployment rate: 16.1% (November 1994)
Budget:
revenues: $27.1 billion
expenditures: $30 billion, including capital expenditures of $NA (1994 est.)
Exports: $16.3 billion (f.o.b., 1994 est.)
commodities: intermediate goods 26.5%, machinery and transport equipment 18.1%, miscellaneous manufactures 16.7%, foodstuffs 9.4%, fuels 8.4% (1993)
partners: Germany 33.4%, Russia 10.2%, Italy 5.3%, UK 4.3% (1993)

Imports: $18.1 billion (f.o.b., 1994 est.)
commodities: machinery and transport equipment 29.6%, intermediate goods 18.5%, chemicals 13.3%, fuels 12.5%, miscellaneous manufactures 10.1%
partners: Germany 35.8%, Italy 9.2%, Russia 8.5%, UK 6.6% (1993)
External debt: $47 billion (1993); note—Poland's Western government creditors promised in 1991 to forgive 30% of Warsaw's $35 billion official debt immediately and to forgive another 20% in 1994; foreign banks agreed in early 1994 to forgive 45% of their $12 billion debt claim
Industrial production: growth rate 12% (1994 est.)
Electricity:
capacity: 31,120,000 kW
production: 124 billion kWh
consumption per capita: 2,908 kWh (1993)
Industries: machine building, iron and steel, extractive industries, chemicals, shipbuilding, food processing, glass, beverages, textiles
Agriculture: accounts for 7% of GDP; 75% of output from private farms, 25% from state farms; productivity remains low by European standards; leading European producer of rye, rapeseed, and potatoes; wide variety of other crops and livestock; major exporter of pork products; normally self-sufficient in food
Illicit drugs: illicit producer of opium for domestic consumption and amphetamines for the international market; transshipment point for Asian and Latin American illicit drugs to Western Europe; producer of precursor chemicals
Economic aid:
donor: bilateral aid to non-Communist less developed countries (1954-89), $2.2 billion
recipient: Western governments and institutions have pledged $8 billion in grants and loans since 1989, but most of the money has not been disbursed
Currency: 1 zloty (Zl) = 100 groszy
Exchange rates: zlotych (Zl) per US$1—2.45 (January 1995; a currency reform on 1 January 1995 replaced 10,000 old zlotys with 1 new zloty), 22,723 (1994), 18,115 (1993), 13,626 (1992), 10,576 (1991), 9,500 (1990)
Fiscal year: calendar year

Transportation

Railroads:
total: 25,528 km
broad gauge: 659 km 1.520-m gauge
standard gauge: 23,014 km 1.435-m gauge (11,496 km electrified; 8,978 km double track)
narrow gauge: 1,855 km various gauges including 1.000-m, 0.785-m, 0.750-m, and 0.600-m (1994)
Highways:
total: 367,000 km (excluding farm, factory and forest roads)
paved: 235,247 km (257 km of which are limited access expressways)
unpaved: 131,753 km (1992)
Inland waterways: 3,997 km navigable rivers and canals (1991)
Pipelines: crude oil 1,986 km; petroleum products 360 km; natural gas 4,600 km (1992)
Ports: Gdansk, Gdynia, Gliwice, Kolobrzeg, Szczecin, Swinoujscie, Ustka, Warsaw, Wrocaw
Merchant marine:
total: 152 ships (1,000 GRT or over) totaling 2,186,405 GRT/3,270,914 DWT
ships by type: bulk 89, cargo 38, chemical tanker 4, container 7, oil tanker 1, passenger 1, roll-on/roll-off cargo 8, short-sea passenger 4
note: in addition, Poland owns 9 ships (1,000 GRT or over) totaling 76,501 DWT that operate under Bahamian, Liberian, Saint Vincent and the Grenadines, Vanuatu, Panamanian, and Cypriot registry
Airports:
total: 134
with paved runways over 3,047 m: 2
with paved runways 2,438 to 3,047 m: 30
with paved runways 1,524 to 2,437 m: 27
with paved runways 914 to 1,523 m: 3
with paved runways under 914 m: 7
with unpaved runways 2,438 to 3,047 m: 5
with unpaved runways 1,524 to 2,438 m: 10
with unpaved runways 914 to 1,523 m: 32
with unpaved runways under 914 m: 18

Communications

Telephone system: 4.9 million telephones; 12.7 phones/100 residents (1994); severely underdeveloped and outmoded system; exchanges are 86% automatic (1991)
local: NA
intercity: cable, open wire and microwave
international: INTELSAT, EUTELSAT, INMARSAT and Intersputnik earth stations
Radio:
broadcast stations: AM 27, FM 27, shortwave 0
radios: NA
Television:
broadcast stations: 40 (Russian repeaters 5)
televisions: 9.6 million

Defense Forces

Branches: Army, Navy, Air and Air Defense Force
Manpower availability: males age 15-49 10,181,069; males fit for military service 7,940,634; males reach military age (19) annually 323,133 (1995 est.)
Defense expenditures: 50.7 billion zlotych, NA% of GNP (1994 est.); note—conversion of defense expenditures into US dollars using the current exchange rate could produce misleading results

Portugal

Geography

Location: Southwestern Europe, bordering the North Atlantic Ocean, west of Spain
Map references: Europe
Area:
total area: 92,080 sq km
land area: 91,640 sq km
comparative area: slightly smaller than Indiana
note: includes Azores and Madeira Islands
Land boundaries: total 1,214 km, Spain 1,214 km
Coastline: 1,793 km
Maritime claims:
continental shelf: 200-m depth or to the depth of exploitation
exclusive economic zone: 200 nm
territorial sea: 12 nm
International disputes: sovereignty over Timor Timur (East Timor Province) disputed with Indonesia
Climate: maritime temperate; cool and rainy in north, warmer and drier in south
Terrain: mountainous north of the Tagus, rolling plains in south
Natural resources: fish, forests (cork), tungsten, iron ore, uranium ore, marble
Land use:
arable land: 32%
permanent crops: 6%
meadows and pastures: 6%
forest and woodland: 40%
other: 16%
Irrigated land: 6,340 sq km (1989 est.)
Environment:
current issues: soil erosion; air pollution caused by industrial and vehicle emissions; water pollution, especially in coastal areas
natural hazards: Azores subject to severe earthquakes
international agreements: party to—Air Pollution, Biodiversity, Climate Change, Endangered Species, Hazardous Wastes, Marine Dumping, Marine Life Conservation, Ozone Layer Protection, Ship Pollution, Tropical Timber 83, Wetlands; signed, but not

ratified—Air Pollution-Volatile Organic Compounds, Desertification, Environmental Modification, Law of the Sea, Nuclear Test Ban
Note: Azores and Madeira Islands occupy strategic locations along western sea approaches to Strait of Gibraltar

People

Population: 10,562,388 (July 1995 est.)
Age structure:
0-14 years: 18% (female 943,412; male 1,000,971)
15-64 years: 68% (female 3,625,086; male 3,499,176)
65 years and over: 14% (female 889,142; male 604,601) (July 1995 est.)
Population growth rate: 0.36% (1995 est.)
Birth rate: 11.72 births/1,000 population (1995 est.)
Death rate: 9.65 deaths/1,000 population (1995 est.)
Net migration rate: 1.55 migrant(s)/1,000 population (1995 est.)
Infant mortality rate: 9.1 deaths/1,000 live births (1995 est.)
Life expectancy at birth:
total population: 75.53 years
male: 72.11 years
female: 79.16 years (1995 est.)
Total fertility rate: 1.47 children born/woman (1995 est.)
Nationality:
noun: Portuguese (singular and plural)
adjective: Portuguese
Ethnic divisions: homogeneous Mediterranean stock in mainland, Azores, Madeira Islands; citizens of black African descent who immigrated to mainland during decolonization number less than 100,000
Religions: Roman Catholic 97%, Protestant denominations 1%, other 2%
Languages: Portuguese
Literacy: age 15 and over can read and write (1990)
total population: 85%
male: 89%
female: 82%
Labor force: 4.24 million (1994 est.)
by occupation: services 54.5%, manufacturing 24.4%, agriculture, forestry, fisheries 11.2%, construction 8.3%, utilites 1.0%, mining 0.5% (1992)

Government

Names:
conventional long form: Portuguese Republic
conventional short form: Portugal
local long form: Republica Portuguesa
local short form: Portugal
Digraph: PO
Type: republic
Capital: Lisbon
Administrative divisions: 18 districts (distritos, singular—distrito) and 2 autonomous regions* (regioes autonomas, singular—regiao autonoma); Aveiro, Acores (Azores)*, Beja, Braga, Braganca, Castelo Branco, Coimbra, Evora, Faro, Guarda, Leiria, Lisboa, Madeira*, Portalegre, Porto, Santarem, Setubal, Viana do Castelo, Vila Real, Viseu
Dependent areas: Macau (scheduled to become a Special Administrative Region of China on 20 December 1999)
Independence: 1140 (independent republic proclaimed 5 October 1910)
National holiday: Day of Portugal, 10 June (1580)
Constitution: 25 April 1976, revised 30 October 1982 and 1 June 1989
Legal system: civil law system; the Constitutional Tribunal reviews the constitutionality of legislation; accepts compulsory ICJ jurisdiction, with reservations
Suffrage: 18 years of age; universal
Executive branch:
chief of state: President Dr. Mario Alberto Nobre Lopes SOARES (since 9 March 1986); election last held 13 February 1991 (next to be held NA February 1996); results—Dr. Mario Lopes SOARES 70%, Basilio HORTA 14%, Carlos CARVALHAS 13%, Carlos MARQUES 3%; note—SOARES is finishing his second term and by law cannot run for a third consecutive term
head of government: Prime Minister Anibal CAVACO SILVA (since 6 November 1985); note—will be replaced in the October 1995 elections
Council of State: acts as a consultative body to the president
cabinet: Council of Ministers; appointed by the president on recommendation of the prime minister
Legislative branch: unicameral
Assembly of the Republic (Assembleia da Republica): elections last held 6 October 1991 (next to be held NA October 1995); results—PSD 50.4%, PS 29.3%, CDU 8.8%, CDS 4.4%, PSN 1.7%, PRD 0.6%, other 4.8%; seats—(230 total) PSD 136, PS 71, CDU 17, CDS 5, PSN 1
Judicial branch: Supreme Tribunal of Justice (Supremo Tribunal de Justica)
Political parties and leaders: Social Democratic Party (PSD), Fernando NOGUEIRA; Portuguese Socialist Party (PS), Antonio GUTERRES; Party of Democratic Renewal (PRD), Pedro CANAVARRO; Portuguese Communist Party (PCP), Carlos CARVALHAS; Social Democratic Center (CDS), Manuel MONTEIRO; National Solidarity Party (PSN), Manuel SERGIO; Center Democratic Party (CDS); United Democratic Coalition (CDU; Communists)
Member of: AfDB, Australia Group, BIS, CCC, CE, CERN, EBRD, EC, ECE, ECLAC, EIB, FAO, GATT, IADB, IAEA, IBRD, ICAO, ICC, ICFTU, ICRM, IDA, IEA, IFAD, IFC, IFRCS, ILO, IMF, IMO, INMARSAT, INTELSAT, INTERPOL, IOC, IOM, ISO, ITU, LAIA (observer), MTCR, NACC, NAM (guest), NATO, NEA, NSG, OAS (observer), OECD, OSCE, PCA, UN, UNCTAD, UNESCO, UNIDO, UNOMOZ, UNPROFOR, UPU, WCL, WEU, WFTU, WHO, WIPO, WMO, WTO, ZC
Diplomatic representation in US:
chief of mission: Ambassador Francisco Jose Laco Treichler KNOPFLI
chancery: 2125 Kalorama Road NW, Washington, DC 20008
telephone: [1] (202) 328-8610
FAX: [1] (202) 462-3726
consulate(s) general: Boston, New York, Newark (New Jersey), and San Francisco
consulate(s): Los Angeles, New Bedford (Massachusetts), Providence (Rhode Island), and Washington, DC
US diplomatic representation:
chief of mission: Ambassador Elizabeth Frawley BAGLEY
embassy: Avenida das Forcas Armadas, 1600 Lisbon
mailing address: PSC 83, Lisbon; APO AE 09726
telephone: [351] (1) 7266600, 7266659, 7268670, 7268880
FAX: [351] (1) 7269109
consulate(s): Ponta Delgada (Azores)
Flag: two vertical bands of green (hoist side, two-fifths) and red (three-fifths) with the Portuguese coat of arms centered on the dividing line

Economy

Overview: Portugal's economy contracted 0.4% in 1993 but registered a 1.4% growth in 1994, with 3% growth expected in 1995 and 1996. This comeback rests on high levels of public investment, continuing strong export growth, and a gradual recovery in consumer spending. The government's long-run economic goal is the modernization of Portuguese markets, industry, infrastructure, and work force in order to catch up with productivity and income levels of the more advanced EU countries. Per capita income now equals only 55% of the EU average. Economic policy in 1994 focused on reducing inflationary pressures by lowering the fiscal deficit, maintaining a stable escudo, moderating wage increases, and encouraging increased competition. The government's medium-term objective is to be in the first tier of the EU countries eligible to join the economic and monetary union (EMU) as early as 1997. To this end, the 1995 budget posits a cut in total deficit to 5.8% of GDP.
National product: GDP—purchasing power parity—$107.3 billion (1994 est.)
National product real growth rate: 1.4% (1994 est.)

National product per capita: $10,190 (1994 est.)
Inflation rate (consumer prices): 6.1% (May 1994)
Unemployment rate: 6.7% (May 1994)
Budget:
revenues: $31 billion
expenditures: $41 billion, including capital expenditures of $NA (1994)
Exports: $15.4 billion (f.o.b., 1993)
commodities: clothing and footwear, machinery, cork and paper products, hides and skins
partners: EU 75.5%, other developed countries 12.4%, US 4.3% (1994)
Imports: $24.3 billion (c.i.f., 1993)
commodities: machinery and transport equipment, agricultural products, chemicals, petroleum, textiles
partners: EC 72%, other developed countries 10.9%, less developed countries 12.9%, US 3.4%
External debt: $20 billion (1993 est.)
Industrial production: growth rate 1.5% (1994 est.); accounts for 30.6% of GDP
Electricity:
capacity: 8,220,000 kW
production: 29.5 billion kWh
consumption per capita: 2,642 kWh (1993)
Industries: textiles and footwear; wood pulp, paper, and cork; metalworking; oil refining; chemicals; fish canning; wine; tourism
Agriculture: accounts for 5% of GDP; small, inefficient farms; imports more than half of food needs; major crops—grain, potatoes, olives, grapes; livestock sector—sheep, cattle, goats, poultry, meat, dairy products
Illicit drugs: increasingly important gateway country for Latin American cocaine entering the European market; transshipment point for hashish from North Africa to Europe
Economic aid:
recipient: US commitments, including Ex-Im (FY70-89), $1.8 billion; Western (non-US) countries, ODA and OOF bilateral commitments (1970-89), $1.2 billion
Currency: 1 Portuguese escudo (Esc) = 100 centavos
Exchange rates: Portuguese escudos (Esc) per US$1—158.02 (January 1995), 165.99 (1994), 160.80 (1993), 135.00 (1992), 144.48 (1991), 142.55 (1990)
Fiscal year: calendar year

Transportation

Railroads:
total: 3,068 km
broad gauge: 2,761 km 1.668-m gauge (439 km electrified; 426 km double track)
narrow gauge: 307 km 1.000-m gauge
Highways:
total: 70,176 km
paved and graveled: 60,351 km (519 km of expressways)
unpaved: earth 9,825 km
Inland waterways: 820 km navigable; relatively unimportant to national economy, used by shallow-draft craft limited to 300 metric-ton cargo capacity
Pipelines: crude oil 22 km; petroleum products 58 km
Ports: Aveiro, Funchal (Madeira Islands), Horta (Azores), Leixoes, Lisbon, Porto, Ponta Delgada (Azores), Praia da Vitoria (Azores), Setubal, Viana do Castelo
Merchant marine:
total: 65 ships (1,000 GRT or over) totaling 852,785 GRT/1,545,804 DWT
ships by type: bulk 5, cargo 28, chemical tanker 5, container 4, liquefied gas tanker 2, oil tanker 17, refrigerated cargo 2, roll-on/roll-off cargo 1, short-sea passenger 1
note: Portugal has created a captive register on Madeira for Portuguese-owned ships; ships on the Madeira Register (MAR) will have taxation and crewing benefits of a flag of convenience; in addition, Portugal owns 25 ships (1,000 GRT or over) totaling 155,776 DWT that operate under Panamanian and Maltese registry
Airports:
total: 65
with paved runways over 3,047 m: 5
with paved runways 2,438 to 3,047 m: 8
with paved runways 1,524 to 2,437 m: 3
with paved runways 914 to 1,523 m: 18
with paved runways under 914 m: 29
with unpaved runways 914 to 1,523 m: 2

Communications

Telephone system: 2,690,000 telephones
local: NA
intercity: generally adequate integrated network of coaxial cables, open wire and microwave radio relay, domestic satellite earth stations
international: 6 submarine cables; 3 INTELSAT (2 Atlantic Ocean and 1 Indian Ocean), EUTELSAT earth stations; tropospheric link to Azores
Radio:
broadcast stations: AM 57, FM 66 (repeaters 22), shortwave 0
radios: NA
Television:
broadcast stations: 66 (repeaters 23)
televisions: NA

Defense Forces

Branches: Army, Navy (includes Marines), Air Force, National Republican Guard, Fiscal Guard, Public Security Police
Manpower availability: males age 15-49 2,747,357; males fit for military service 2,223,299; males reach military age (20) annually 90,402 (1995 est.)
Defense expenditures: exchange rate conversion—$2.4 billion, 2.9% of GDP (1994)

Puerto Rico
(commonwealth associated with the US)

Geography

Location: Caribbean, island between the Caribbean Sea and the North Atlantic Ocean, east of the Dominican Republic
Map references: Central America and the Caribbean
Area:
total area: 9,104 sq km
land area: 8,959 sq km
comparative area: slightly less than three times the size of Rhode Island
Land boundaries: 0 km
Coastline: 501 km
Maritime claims:
exclusive economic zone: 200 nm
territorial sea: 12 nm
International disputes: none
Climate: tropical marine, mild, little seasonal temperature variation
Terrain: mostly mountains with coastal plain belt in north; mountains precipitous to sea on west coast; sandy beaches along most coastal areas
Natural resources: some copper and nickel, potential for onshore and offshore crude oil
Land use:
arable land: 8%
permanent crops: 9%
meadows and pastures: 41%
forest and woodland: 20%
other: 22%
Irrigated land: 390 sq km (1989 est.)
Environment:
current issues: the recent drought has caused water levels in reservoirs to drop and prompted water rationing for more than one-half of the population
natural hazards: periodic droughts
international agreements: NA
Note: important location along the Mona Passage—a key shipping lane to the Panama Canal; San Juan is one of the biggest and best natural harbors in the Caribbean; many small rivers and high central mountains ensure land is well watered; south coast relatively dry; fertile coastal plain belt in north

People

Population: 3,812,569 (July 1995 est.)
Age structure:
0-14 years: 25% (female 466,596; male 489,127)
15-64 years: 65% (female 1,274,765; male 1,195,785)
65 years and over: 10% (female 213,716; male 172,580) (July 1995 est.)
Population growth rate: 0.16% (1995 est.)
Birth rate: 15.92 births/1,000 population (1995 est.)
Death rate: 7.47 deaths/1,000 population (1995 est.)
Net migration rate: -6.81 migrant(s)/1,000 population (1995 est.)
Infant mortality rate: 12.8 deaths/1,000 live births (1995 est.)
Life expectancy at birth:
total population: 75.1 years
male: 70.78 years
female: 79.66 years (1995 est.)
Total fertility rate: 1.98 children born/woman (1995 est.)
Nationality:
noun: Puerto Rican(s) (US citizens)
adjective: Puerto Rican
Ethnic divisions: Hispanic
Religions: Roman Catholic 85%, Protestant denominations and other 15%
Languages: Spanish, English
Literacy: age 15 and over can read and write (1980)
total population: 89%
male: 90%
female: 88%
Labor force: 1.2 million (1993)
by occupation: government 22%, manufacturing 17%, trade 20%, construction 6%, communications and transportation 5%, other 30% (1993)

Government

Names:
conventional long form: Commonwealth of Puerto Rico
conventional short form: Puerto Rico
Digraph: RQ
Type: commonwealth associated with the US
Capital: San Juan
Administrative divisions: none (commonwealth associated with the US); note—there are 78 municipalities
Independence: none (commonwealth associated with the US)
National holiday: US Independence Day, 4 July (1776)
Constitution: ratified 3 March 1952; approved by US Congress 3 July 1952; effective 25 July 1952
Legal system: based on Spanish civil code
Suffrage: 18 years of age; universal; indigenous inhabitants are US citizens but do not vote in US presidential elections
Executive branch:
chief of state: President William Jefferson CLINTON (since 20 January 1993); Vice President Albert GORE, Jr. (since 20 January 1993)
head of government: Governor Pedro ROSSELLO (since 2 January 1993); election last held 3 November 1992 (next to be held 5 November 1996); results—Pedro ROSSELLO (PNP) 50%, Victoria MUNOZ (PPD) 46%, Fernando MARTIN (PIP) 4%
Legislative branch: bicameral Legislative Assembly
Senate: elections last held 3 November 1992 (next to be held 5 November 1996); results—percent of vote by party NA; seats—(29 total) PNP 20, PPD 8, PIP 1
House of Representatives: elections last held 3 November 1992 (next to be held NA November 1996); results—percent of vote by party NA; seats—(53 total) PNP 36, PPD 16, PIP 1
US House of Representatives: elections last held 3 November 1992 (next to be held 5 November 1996); results—percent of vote by party NA; seats—(1 total) PNP 1 (Carlos Romero BARCELO); note—Puerto Rico elects one representative to the US House of Representatives
Judicial branch: Supreme Court, Superior Courts, Municipal Courts
Political parties and leaders: National Republican Party of Puerto Rico, Luis FERRE; Popular Democratic Party (PPD), Hector ACEVEDO; New Progressive Party (PNP), Pedro ROSSELLO; Puerto Rican Socialist Party (PSP) has been disbanded (1994); Puerto Rican Independence Party (PIP), Ruben BERRIOS Martinez; Puerto Rican Communist Party (PCP), leader(s) unknown
Other political or pressure groups: Armed Forces for National Liberation (FALN); Volunteers of the Puerto Rican Revolution; Boricua Popular Army (also known as the Macheteros); Armed Forces of Popular Resistance
Member of: CARICOM (observer), ECLAC (associate), FAO (associate), ICFTU, INTERPOL (subbureau), IOC, WCL, WFTU, WHO (associate), WTO (associate)
Diplomatic representation in US: none (commonwealth associated with the US)
US diplomatic representation: none (commonwealth associated with the US)
Flag: five equal horizontal bands of red (top and bottom) alternating with white; a blue isosceles triangle based on the hoist side bears a large white five-pointed star in the center; design based on the US flag

Economy

Overview: Puerto Rico has one of the most dynamic economies in the Caribbean region. Industry has surpassed agriculture as the primary sector of economic activity and income. Encouraged by duty free access to the US and by tax incentives, US firms have invested heavily in Puerto Rico since the 1950s. US minimum wage laws apply. Important industries include pharmaceuticals, electronics, textiles, petrochemicals, and processed foods. Sugar production has lost out to dairy production and other livestock products as the main source of income in the agricultural sector. Tourism has traditionally been an important source of income for the island, with estimated arrivals of nearly 3.9 million tourists in 1993.
National product: GDP—purchasing power parity—$26.8 billion (1994 est.)
National product real growth rate: 2.6% (1994 est.)
National product per capita: $7,050 (1994 est.)
Inflation rate (consumer prices): 2.9% (1994)
Unemployment rate: 16% (1994)
Budget:
revenues: $5.1 billion
expenditures: $5.1 billion, including capital expenditures of $NA (FY94/95)
Exports: $21.8 billion (1994)
commodities: pharmaceuticals, electronics, apparel, canned tuna, rum, beverage concentrates, medical equipment, instruments
partners: US 86.2% (1993)
Imports: $16.7 billion (1994)
commodities: chemicals, clothing, food, fish, petroleum products
partners: US 69.2% (1993)
External debt: $NA
Industrial production: growth rate 5% (1994 est.)
Electricity:
capacity: 4,230,000 kW
production: 15.6 billion kWh
consumption per capita: 3,819 kWh (1993)
Industries: manufacturing accounts for 39.4% of GDP; manufacturing of pharmaceuticals, electronics, apparel, food products, instruments; tourism
Agriculture: accounts for only 3% of labor force and just over 1% of GDP; crops—sugarcane, coffee, pineapples, plantains, bananas; livestock—cattle, chickens; imports a large share of food needs (1993)
Economic aid: none
Currency: 1 United States dollar (US$) = 100 cents
Exchange rates: US currency is used
Fiscal year: 1 July—30 June

Transportation

Railroads:
total: 96 km rural narrow-gauge system for hauling sugarcane; note—no passenger railroads

Highways:
total: 13,762 km
paved: 13,762 km (1982)
Ports: Guanica, Guayanilla, Guayama, Playa de Ponce, San Juan
Merchant marine: none
Airports:
total: 31
with paved runways over 3,047 m: 3
with paved runways 1,524 to 2,437 m: 3
with paved runways 914 to 1,523 m: 9
with paved runways under 914 m: 14
with unpaved runways 914 to 1,523 m: 2

Communications

Telephone system: NA telephones; modern system, integrated with that of the US by high capacity submarine cable and INTELSAT with high-speed data capability; digital telephone system with about 1 million lines; cellular telephone service (1990)
local: NA
intercity: NA
international: 1 INTELSAT earth station and submarine cable
Radio:
broadcast stations: AM 50, FM 63, shortwave 0
radios: NA
Television:
broadcast stations: 9; note—cable television available with US programs (1990)
televisions: NA

Defense Forces

Branches: paramilitary National Guard, Police Force
Note: defense is the responsibility of the US

Qatar

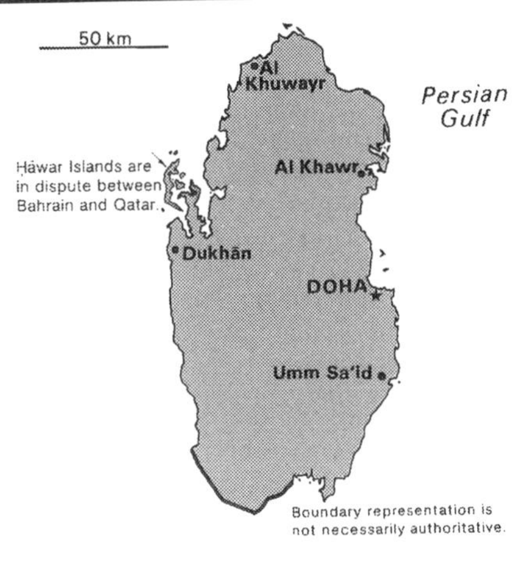

Geography

Location: Middle East, peninsula bordering the Persian Gulf and Saudi Arabia
Map references: Middle East
Area:
total area: 11,000 sq km
land area: 11,000 sq km
comparative area: slightly smaller than Connecticut
Land boundaries: total 60 km, Saudi Arabia 60 km
Coastline: 563 km
Maritime claims:
contiguous zone: 24 nm
exclusive economic zone: 200 nm
territorial sea: 12 nm
International disputes: territorial dispute with Bahrain over the Hawar Islands; maritime boundary with Bahrain
Climate: desert; hot, dry; humid and sultry in summer
Terrain: mostly flat and barren desert covered with loose sand and gravel
Natural resources: petroleum, natural gas, fish
Land use:
arable land: 0%
permanent crops: 0%
meadows and pastures: 5%
forest and woodland: 0%
other: 95%
Irrigated land: NA sq km
Environment:
current issues: limited natural fresh water resources are increasing dependence on large-scale desalination facilities
natural hazards: haze, dust storms, sandstorms common
international agreements: signed, but not ratified—Biodiversity, Law of the Sea
Note: strategic location in central Persian Gulf near major petroleum deposits

People

Population: 533,916 (July 1995 est.)
Age structure:
0-14 years: 30% (female 81,443; male 80,591)
15-64 years: 68% (female 104,921; male 258,135)
65 years and over: 2% (female 2,941; male 5,885) (July 1995 est.)
Population growth rate: 2.74% (1995 est.)
Birth rate: 22.72 births/1,000 population (1995 est.)
Death rate: 3.59 deaths/1,000 population (1995 est.)
Net migration rate: 8.25 migrant(s)/1,000 population (1995 est.)
Infant mortality rate: 20.4 deaths/1,000 live births (1995 est.)
Life expectancy at birth:
total population: 73.03 years
male: 70.45 years
female: 75.5 years (1995 est.)
Total fertility rate: 4.63 children born/woman (1995 est.)
Nationality:
noun: Qatari(s)
adjective: Qatari
Ethnic divisions: Arab 40%, Pakistani 18%, Indian 18%, Iranian 10%, other 14%
Religions: Muslim 95%
Languages: Arabic (official), English commonly used as a second language
Literacy: age 15 and over can read and write (1986)
total population: 76%
male: 77%
female: 72%
Labor force: NA

Government

Names:
conventional long form: State of Qatar
conventional short form: Qatar
local long form: Dawlat Qatar
local short form: Qatar
Digraph: QA
Type: traditional monarchy
Capital: Doha
Administrative divisions: 9 municipalities (baladiyat, singular—baladiyah); Ad Dawhah, Al Ghuwayriyah, Al Jumayliyah, Al Khawr, Al Wakrah, Ar Rayyan, Jarayan al Batinah, Madinat ash Shamal, Umm Salal
Independence: 3 September 1971 (from UK)
National holiday: Independence Day, 3 September (1971)
Constitution: provisional constitution enacted 2 April 1970
Legal system: discretionary system of law controlled by the amir, although civil codes are being implemented; Islamic law is significant in personal matters
Suffrage: none
Executive branch:
chief of state and head of government: Amir and Prime Minister KHALIFA bin Hamad Al Thani (since 22 February 1972); Crown Prince

HAMAD bin Khalifa Al Thani (appointed 31 May 1977; son of Amir and Minister of Defense)
cabinet: Council of Ministers; appointed by the amir
Legislative branch: unicameral
Advisory Council (Majlis al-Shura): constitution calls for elections for part of this consultative body, but no elections have been held; seats—(30 total)
Judicial branch: Court of Appeal
Political parties and leaders: none
Member of: ABEDA, AFESD, AL, AMF, CCC, ESCWA, FAO, G-77, GATT, GCC, IAEA, IBRD, ICAO, ICRM, IDB, IFAD, IFRCS, ILO, IMF, IMO, INMARSAT, INTELSAT, INTERPOL, IOC, ISO (correspondent), ITU, NAM, OAPEC, OIC, OPEC, UN, UNCTAD, UNESCO, UNIDO, UPU, WHO, WIPO, WMO
Diplomatic representation in US:
chief of mission: Ambassador ABD AL-RAHMAN bin Saud bin Fahd Al Thani
chancery: Suite 1180, 600 New Hampshire Avenue NW, Washington, DC 20037
telephone: [1] (202) 338-0111
US diplomatic representation:
chief of mission: Ambassador Kenton W. KEITH
embassy: 149 Ali Bin Ahmed St., Farig Bin Omran (opposite the television station), Doha
mailing address: P. O. Box 2399, Doha
telephone: [974] 864701 through 864703
FAX: [974] 861669
Flag: maroon with a broad white serrated band (nine white points) on the hoist side

Economy

Overview: Oil is the backbone of the economy and accounts for more than 30% of GDP, roughly 75% of export earnings, and 70% of government revenues. Proved oil reserves of 3.3 billion barrels should ensure continued output at current levels for about 25 years. Oil has given Qatar a per capita GDP comparable to the leading West European industrial countries. Production and export of natural gas are becoming increasingly important. Long-term goals feature the development of off-shore oil and the diversification of the economy.
National product: GDP—purchasing power parity—$10.7 billion (1994 est.)
National product real growth rate: -1% (1994 est.)
National product per capita: $20,820 (1994 est.)
Inflation rate (consumer prices): 3% (1993 est.)
Unemployment rate: NA%
Budget:
revenues: $2.5 billion
expenditures: $3 billion, including capital expenditures of $440 million (1992 est.)
Exports: $3.13 billion (f.o.b., 1993 est.)
commodities: petroleum products 75%, steel, fertilizers
partners: Japan 57%, South Korea 9%, Brazil 4%, UAE 4%, Singapore 3% (1992)
Imports: $1.75 billion (f.o.b., 1993 est.)
commodities: machinery and equipment, consumer goods, food, chemicals
partners: Japan 16%, UK 11%, US 11%, Germany 7%, France 5% (1992)
External debt: $1.5 billion (1993 est.)
Industrial production: accounts for 50% of GDP, including oil
Electricity:
capacity: 1,520,000 kW
production: 4.5 billion kWh
consumption per capita: 8,415 kWh (1993)
Industries: crude oil production and refining, fertilizers, petrochemicals, steel (rolls reinforcing bars for concrete construction), cement
Agriculture: farming and grazing on small scale, less than 2% of GDP; agricultural area is small and government-owned; commercial fishing increasing in importance; most food imported
Economic aid:
donor: pledged in ODA to less developed countries (1979-88), $2.7 billion
Currency: 1 Qatari riyal (QR) = 100 dirhams
Exchange rates: Qatari riyals (QR) per US$1—3.6400 riyals (fixed rate)
Fiscal year: 1 April—31 March

Transportation

Railroads: 0 km
Highways:
total: 1,190 km
paved: 1,030 km
unpaved: 160 km (1988 est.)
Pipelines: crude oil 235 km; natural gas 400 km
Ports: Doha, Halul Island, Umm Sa'id
Merchant marine:
total: 19 ships (1,000 GRT or over) totaling 463,227 GRT/763,507 DWT
ships by type: combination ore/oil 1, container 3, cargo 11, oil tanker 3, refrigerated cargo 1
Airports:
total: 6
with paved runways over 3,047 m: 1
with paved runways under 914 m: 2
with unpaved runways 914 to 1,523 m: 3

Communications

Telephone system: 110,000 telephones; modern system centered in Doha
local: NA
intercity: NA
international: tropospheric scatter to Bahrain; microwave radio relay to Saudi Arabia and UAE; submarine cable to Bahrain and UAE; 2 INTELSAT (1 Atlantic Ocean and 1 Indian Ocean) and 1 ARABSAT earth station
Radio:
broadcast stations: AM 2, FM 3, shortwave 0
radios: NA
Television:
broadcast stations: 3
televisions: NA

Defense Forces

Branches: Army, Navy, Air Force, Public Security
Manpower availability: males age 15-49 219,442; males fit for military service 115,103; males reach military age (18) annually 3,915 (1995 est.)
Defense expenditures: $NA, NA%, of GDP

Reunion

(overseas department of France)

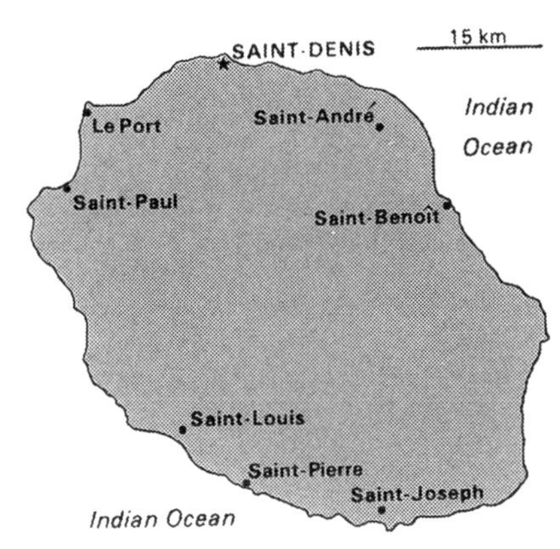

Geography

Location: Southern Africa, island in the Indian Ocean, east of Madagascar
Map references: World
Area:
total area: 2,510 sq km
land area: 2,500 sq km
comparative area: slightly smaller than Rhode Island
Land boundaries: 0 km
Coastline: 201 km
Maritime claims:
exclusive economic zone: 200 nm
territorial sea: 12 nm
International disputes: none
Climate: tropical, but moderates with elevation; cool and dry from May to November, hot and rainy from November to April
Terrain: mostly rugged and mountainous; fertile lowlands along coast
Natural resources: fish, arable land
Land use:
arable land: 20%
permanent crops: 2%
meadows and pastures: 4%
forest and woodland: 35%
other: 39%
Irrigated land: 60 sq km (1989 est.)
Environment:
current issues: NA
natural hazards: periodic, devastating cyclones (December to April); Piton de la Fournaise on the southeastern coast is an active volcano
international agreements: NA

People

Population: 666,067 (July 1995 est.)
Age structure:
0-14 years: 32% (female 104,924; male 109,972)
15-64 years: 62% (female 210,762; male 203,774)
65 years and over: 6% (female 21,606; male 15,029) (July 1995 est.)
Population growth rate: 1.98% (1995 est.)
Birth rate: 24.59 births/1,000 population (1995 est.)
Death rate: 4.79 deaths/1,000 population (1995 est.)
Net migration rate: 0 migrant(s)/1,000 population (1995 est.)
Infant mortality rate: 7.7 deaths/1,000 live births (1995 est.)
Life expectancy at birth:
total population: 74.46 years
male: 71.39 years
female: 77.67 years (1995 est.)
Total fertility rate: 2.75 children born/woman (1995 est.)
Nationality:
noun: Reunionese (singular and plural)
adjective: Reunionese
Ethnic divisions: French, African, Malagasy, Chinese, Pakistani, Indian
Religions: Roman Catholic 94%
Languages: French (official), Creole widely used
Literacy: age 15 and over can read and write (1982)
total population: 79%
male: 76%
female: 80%
Labor force: NA
by occupation: agriculture 30%, industry 21%, services 49% (1981)

Government

Names:
conventional long form: Department of Reunion
conventional short form: Reunion
local long form: none
local short form: Ile de la Reunion
Digraph: RE
Type: overseas department of France
Capital: Saint-Denis
Administrative divisions: none (overseas department of France)
Independence: none (overseas department of France)
National holiday: National Day, Taking of the Bastille, 14 July (1789)
Constitution: 28 September 1958 (French Constitution)
Legal system: French law
Suffrage: 18 years of age; universal
Executive branch:
chief of state: President Francois MITTERRAND (since 21 May 1981)
head of government: Prefect of Reunion Island Hubert FOURNIER (since NA)
cabinet: Council of Ministers
Legislative branch: unicameral General Council and unicameral Regional Council
General Council: elections last held March 1994 (next to be held NA); results—percent of vote by party NA; seats—(47 total) PCR 12, PS 12, UDF 11, RPR 5, others 7
Regional Council: elections last held 25 June 1993 (next to be held NA); results—percent of vote by party NA; seats—(45 total) UPF 17, Free-Dom Movement 13, PCR 9, PS 6
French Senate: elections last held 24 September 1992 (next to be held NA); results—percent of vote by party NA; seats—(3 total) RPR 1, FRA 1, independent 1
French National Assembly: elections last held 21 and 28 March 1993 (next to be held NA 1998); results—percent of vote by party NA; seats—(5 total) PS 1, PCR 1, UPF 1, RPR 1, UDF-CDS 1
Judicial branch: Court of Appeals (Cour d'Appel)
Political parties and leaders: Rally for the Republic (RPR), Alain DEFAUD; Union for French Democracy (UDF), Gilbert GERARD; Communist Party of Reunion (PCR), Elie HOARAU; France-Reunion Future (FRA), Andre THIEN AH KOON; Socialist Party (PS), Jean-Claude FRUTEAU; Social Democrats (CDS), leader NA; Union for France (UPF), included RPR and UDF; Free-Dom Movement, Marguerite SUDRE
Member of: FZ, WFTU
Diplomatic representation in US: none (overseas department of France)
US diplomatic representation: none (overseas department of France)
Flag: the flag of France is used

Economy

Overview: The economy has traditionally been based on agriculture. Sugarcane has been the primary crop for more than a century, and in some years it accounts for 85% of exports. The government has been pushing the development of a tourist industry to relieve high unemployment, which recently amounted to one-third of the labor force. The gap in Reunion between the well-off and the poor is extraordinary and accounts for the persistent social tensions. The white and Indian communities are substantially better off than other segments of the population, often approaching European standards, whereas indigenous groups suffer the poverty and unemployment typical of the poorer nations of the African continent. The outbreak of severe rioting in February 1991 illustrates the seriousness of socioeconomic tensions. The economic well-being of Reunion depends heavily on continued financial assistance from France.
National product: GDP—purchasing power parity—$2.5 billion (1993 est.)
National product real growth rate: NA%
National product per capita: $3,900 (1993 est.)
Inflation rate (consumer prices): NA%
Unemployment rate: 35% (February 1991)

Budget:
revenues: $358 million
expenditures: $914 million, including capital expenditures of $NA (1986 est.)
Exports: $166 million (f.o.b., 1988)
commodities: sugar 75%, rum and molasses 4%, perfume essences 4%, lobster 3%, vanilla and tea 1%
partners: France, Mauritius, Bahrain, South Africa, Italy
Imports: $1.7 billion (c.i.f., 1988)
commodities: manufactured goods, food, beverages, tobacco, machinery and transportation equipment, raw materials, and petroleum products
partners: France, Mauritius, Bahrain, South Africa, Italy
External debt: $NA
Industrial production: growth rate NA%; about 25% of GDP
Electricity:
capacity: 180,000 kW
production: 1 billion kWh
consumption per capita: 1,454 kWh (1993)
Industries: sugar, rum, cigarettes, several small shops producing handicraft items
Agriculture: accounts for 30% of labor force; dominant sector of economy; cash crops—sugarcane, vanilla, tobacco; food crops—tropical fruits, vegetables, corn; imports large share of food needs
Economic aid:
recipient: Western (non-US) countries, ODA and OOF bilateral commitments (1970-89), $14.8 billion
Currency: 1 French franc (F) = 100 centimes
Exchange rates: French francs (F) per US$1—5.2943 (January 1995), 5.5520 (1994), 5.6632 (1993), 5.2938 (1992), 5.6421 (1991), 5.4453 (1990)
Fiscal year: calendar year

Transportation

Railroads: 0 km
Highways:
total: 2,800 km
paved: 2,200 km
unpaved: gravel, crushed stone, stabilized earth 600 km
Ports: Le Port, Pointe des Galets
Merchant marine: none
Airports:
total: 2
with paved runways 2,438 to 3,047 m: 1
with paved runways 914 to 1,523 m: 1

Communications

Telephone system: 85,900 telephones; adequate system; principal center Saint-Denis
local: NA
intercity: modern open-wire and microwave network
international: radiocommunication to Comoros, France, Madagascar; new microwave route to Mauritius; 1 INTELSAT (Indian Ocean) earth station
Radio:
broadcast stations: AM 3, FM 13, shortwave 0
radios: NA
Television:
broadcast stations: 1 (repeaters 18)
televisions: NA

Defense Forces

Branches: French forces (Army, Navy, Air Force, and Gendarmerie)
Manpower availability: males age 15-49 173,693; males fit for military service 89,438; males reach military age (18) annually 5,781 (1995 est.)
Note: defense is the responsibility of France

Romania

Geography

Location: Southeastern Europe, bordering the Black Sea, between Bulgaria and Ukraine
Map references: Ethnic Groups in Eastern Europe, Europe
Area:
total area: 237,500 sq km
land area: 230,340 sq km
comparative area: slightly smaller than Oregon
Land boundaries: total 2,508 km, Bulgaria 608 km, Hungary 443 km, Moldova 450 km, Serbia and Montenegro 476 km (all with Serbia), Ukraine (north) 362 km, Ukraine (south) 169 km
Coastline: 225 km
Maritime claims:
contiguous zone: 24 nm
continental shelf: 200-m depth or to the depth of exploitation
exclusive economic zone: 200 nm
territorial sea: 12 nm
International disputes: certain territory of Moldova and Ukraine—including Bessarabia and Northern Bukovina—are considered by Bucharest as historically a part of Romania; this territory was incorporated into the former Soviet Union following the Molotov-Ribbentrop Pact in 1940
Climate: temperate; cold, cloudy winters with frequent snow and fog; sunny summers with frequent showers and thunderstorms
Terrain: central Transylvanian Basin is separated from the Plain of Moldavia on the east by the Carpathian Mountains and separated from the Walachian Plain on the south by the Transylvanian Alps
Natural resources: petroleum (reserves declining), timber, natural gas, coal, iron ore, salt
Land use:
arable land: 43%
permanent crops: 3%
meadows and pastures: 19%
forest and woodland: 28%
other: 7%

Irrigated land: 34,500 sq km (1989 est.)
Environment:
current issues: soil erosion and degradation; water pollution; air pollution in south from industrial effluents; contamination of Danube delta wetlands
natural hazards: earthquakes most severe in south and southwest; geologic structure and climate promote landslides
international agreements: party to—Air Pollution, Antarctic Treaty, Biodiversity, Climate Change, Endangered Species, Environmental Modification, Hazardous Wastes, Nuclear Test Ban, Ozone Layer Protection, Ship Pollution, Wetlands; signed, but not ratified—Antarctic-Environmental Protocol, Law of the Sea
Note: controls most easily traversable land route between the Balkans, Moldova, and Ukraine

---

People

Population: 23,198,330 (July 1995 est.)
note: the Romanian census of January 1992 gives the population for that date as 22.749 million; the government estimates that population declined in 1993 by 0.3%
Age structure:
0-14 years: 21% (female 2,413,933; male 2,534,019)
15-64 years: 67% (female 7,737,531; male 7,732,038)
65 years and over: 12% (female 1,604,210; male 1,176,599) (July 1995 est.)
Population growth rate: 0.09% (1995 est.)
Birth rate: 13.71 births/1,000 population (1995 est.)
Death rate: 9.93 deaths/1,000 population (1995 est.)
Net migration rate: -2.88 migrant(s)/1,000 population (1995 est.)
Infant mortality rate: 18.7 deaths/1,000 live births (1995 est.)
Life expectancy at birth:
total population: 72.24 years
male: 69.31 years
female: 75.35 years (1995 est.)
Total fertility rate: 1.82 children born/woman (1995 est.)
Nationality:
noun: Romanian(s)
adjective: Romanian
Ethnic divisions: Romanian 89.1%, Hungarian 8.9%, German 0.4%, Ukrainian, Serb, Croat, Russian, Turk, and Gypsy 1.6%
Religions: Romanian Orthodox 70%, Roman Catholic 6% (of which 3% are Uniate), Protestant 6%, unaffiliated 18%
Languages: Romanian, Hungarian, German
Literacy: age 15 and over can read and write (1992)
total population: 97%
male: 98%
female: 95%
Labor force: 11.3 million (1992)
by occupation: industry 38%, agriculture 28%, other 34% (1989)

Government

Names:
conventional long form: none
conventional short form: Romania
local long form: none
local short form: Romania
Digraph: RO
Type: republic
Capital: Bucharest
Administrative divisions: 40 counties (judete, singular—judet) and 1 municipality* (municipiu); Alba, Arad, Arges, Bacau, Bihor, Bistrita-Nasaud, Botosani, Braila, Brasov, Bucuresti*, Buzau, Calarasi, Caras-Severin, Cluj, Constanta, Covasna, Dimbovita, Dolj, Galati, Gorj, Giurgiu, Harghita, Hunedoara, Ialomita, Iasi, Maramures, Mehedinti, Mures, Neamt, Olt, Prahova, Salaj, Satu Mare, Sibiu, Suceava, Teleorman, Timis, Tulcea, Vaslui, Vilcea, Vrancea
Independence: 1881 (from Turkey; republic proclaimed 30 December 1947)
National holiday: National Day of Romania, 1 December (1990)
Constitution: 8 December 1991
Legal system: former mixture of civil law system and Communist legal theory; is now based on the Constitution of France's Fifth Republic
Suffrage: 18 years of age; universal
Executive branch:
chief of state: President Ion ILIESCU (since 20 June 1990, previously President of Provisional Council of National Unity since 23 December 1989); election last held 27 September 1992, with runoff between top two candidates on 11 October 1992 (next to be held NA 1996); results—Ion ILIESCU 61.4%, Emil CONSTANTINESCU 38.6%
head of government: Prime Minister Nicolae VACAROIU (since November 1992)
cabinet: Council of Ministers; appointed by the prime minister
Legislative branch: bicameral Parliament
Senate (Senat): elections last held 27 September 1992 (next to be held NA 1996); results—PSDR 34.3%, CDR 18.2%, DP-FSN 12.6%, others 34.9%; seats—(143 total) PSDR 49, CDR 26, DP-FSN 18, PUNR 13, UDMR 12, PRM 6, PAC 6, PDAR 5, PSM 5, PL-93 2 other 1
House of Deputies (Adunarea Deputatilor): elections last held 27 September 1992 (next to be held NA 1996); results—PSDR 34.0%, CDR 16,4%, DP-FSN 12.3%, others 37.3%; seats—(341 total) PSDR 116, CDR 56, DP-FSN 42, PUNR 29, UDMR 27, PL-93 19, PRM 15, PSM 13, PAC 5, other 19
Judicial branch: Supreme Court of Justice, Constitutional Court
Political parties and leaders: Democratic Party (DP-FSN), Petre ROMAN; Social Democratic Party of Romania (PSDR), Adrian NASTASE; Democratic Union of Hungarians in Romania (UDMR), Bela MARKO; National Liberal Party (PNL), Mircea IONESCU-QUINTUS; National Peasants' Christian and Democratic Party (PNTCD), Corneliu COPOSU; Romanian National Unity Party (PUNR), Gheorghe FUNAR; Socialist Labor Party (PSM), Ilie VERDET; Agrarian Democratic Party of Romania (PDAR), Victor SURDU; The Democratic Convention (CDR), Emil CONSTANTINESCU; Romania Mare Party (PRM), Corneliu Vadim TUDOR; Civic Alliance Party (PAC), Nicolae MANOLESCU, chairman
note: numerous other small parties exist but almost all failed to gain representation in the most recent election
Other political or pressure groups: various human rights and professional associations
Member of: ACCT, BIS, BSEC, CCC, CE, CEI (associate members), EBRD, ECE, FAO, G- 9, G-77, GATT, IAEA, IBRD, ICAO, ICFTU, ICRM, IFAD, IFC, IFRCS, ILO, IMF, IMO, INMARSAT, INTELSAT, INTERPOL, IOC, IOM (observer), ISO, ITU, NACC, NAM (guest), NSG, OAS (observer), OSCE, PCA, PFP, UN, UNCTAD, UNESCO, UNIDO, UNIKOM, UNOSOM, UPU, WCL, WEU (associate partner), WFTU, WHO, WIPO, WMO, WTO, ZC
Diplomatic representation in US:
chief of mission: Ambassador Mihai Horia BOTEZ
chancery: 1607 23rd Street NW, Washington, DC 20008
telephone: [1] (202) 332-4846, 4848, 4851
FAX: [1] (202) 232-4748
consulate(s) general: Los Angeles and New York
US diplomatic representation:
chief of mission: Ambassador Alfred H. MOSES
embassy: Strada Tudor Arghezi 7-9, Bucharest
mailing address: American Consulate General (Bucharest), Unit 1315, Bucharest; APO AE 09213-1315
telephone: [40] (1) 210 01 49, 210 40 42
FAX: [40] (1) 210 03 95
branch office: Cluj-Napoca
Flag: three equal vertical bands of blue (hoist side), yellow, and red; the national coat of arms that used to be centered in the yellow band has been removed; now similar to the flags of Andorra and Chad

---

Economy

Overview: Despite the continuing difficulties in moving away from the former command system, the Romanian economy seems to have bottomed out in 1993-94. Market oriented

reforms have been introduced fitfully since the downfall of CEAUSESCU in December 1989, with the result a growing private sector, especially in services. The slow pace of structural reform, however, has exacerbated Romania's high inflation rate and eroded real wages. Agricultural production rebounded in 1993 from the drought-reduced harvest of 1992. The economy continued its recovery in 1994, further gains being realized in agriculture, construction, services, and trade. Food supplies are adequate but expensive. Romania's infrastructure had deteriorated over the last five years due to reduced levels of public investment. Residents of the capital reported frequent disruptions of heating and water services. The slow and painful process of conversion to a more open economy will continue in 1995.
National product: GDP—purchasing power parity—$64.7 billion (1994 est.)
National product real growth rate: 3.4% (1994 est.)
National product per capita: $2,790 (1994 est.)
Inflation rate (consumer prices): 62% (1994)
Unemployment rate: 10.9% (December 1994)
Budget:
revenues: $8.3 billion
expenditures: $9.4 billion, including capital expenditures of $NA (1995 est.)
Exports: $6 billion (f.o.b., 1994)
commodities: metals and metal products 17.6%, mineral products 11.9%, textiles 18.5%, electric machines and equipment 8.4%, transport materials 6.5% (1994)
partners: EC 36.1%, developing countries 27.4%, East and Central Europe 14.9%, EFTA 5.1%, Russia 5%, Japan 1.4%, US 1.3% (1993)
Imports: $6.3 billion (f.o.b., 1994)
commodities: minerals 21.1%, machinery and equipment 19.7%, textiles 11.5%, agricultural goods 9.2% (1994)
partners: EC 45.8%, East and Central Europe 8.6%, developing countries 22.6%, Russia 11%, EFTA 6.2%, US 5.0%, Japan 0.8% (1993)
External debt: $4.4 billion (1994)
Industrial production: growth rate -1% (1993 est.); accounts for 45% of GDP
Electricity:
capacity: 22,180,000 kW
production: 50.8 billion kWh
consumption per capita: 2,076 kWh (1993)
Industries: mining, timber, construction materials, metallurgy, chemicals, machine building, food processing, petroleum production and refining
Agriculture: accounts for 18% of GDP and 28% of labor force; major wheat and corn producer; other products—sugar beets, sunflower seed, potatoes, milk, eggs, meat, grapes
Illicit drugs: transshipment point for southwest Asian heroin and Latin American cocaine transiting the Balkan route
Economic aid: $NA
Currency: 1 leu (L) = 100 bani
Exchange rates: lei (L) per US$1—1,776.00 (January 1995), 1,655.09 (1994), 760.05 (1993), 307.95 (1992), 76.39 (1991), 22.432 (1990)
Fiscal year: calendar year

Transportation

Railroads:
total: 11,365 km
broad gauge: 45 km 1.524-m gauge
standard gauge: 10,893 km 1.435-m gauge (3,723 km electrified; 3,060 km double track)
narrow gauge: 427 km 0.760-m gauge (1994)
Highways:
total: 461,880 km
paved: 235,559 km (113 km of expressways)
unpaved: 226,321 km (1992)
Inland waterways: 1,724 km (1984)
Pipelines: crude oil 2,800 km; petroleum products 1,429 km; natural gas 6,400 km (1992)
Ports: Braila, Constanta, Galatz, Mangalia, Sulina, Tulcea
Merchant marine:
total: 238 ships (1,000 GRT or over) totaling 2,528,971 GRT/3,849,943 DWT
ships by type: bulk 46, cargo 167, container 2, oil tanker 14, passenger-cargo 1, railcar carrier 1, roll-on/roll-off cargo 7
note: in addition, Romania owns 20 ships (1,000 GRT or over) totaling 1,207,388 DWT that operate under Liberian, Maltese, Cypriot, and Bahamian registry
Airports:
total: 156
with paved runways over 3,047 m: 4
with paved runways 2,438 to 3,047 m: 9
with paved runways 1,524 to 2,437 m: 14
with unpaved runways 2,438 to 3,047 m: 3
with unpaved runways 1,524 to 2,438 m: 1
with unpaved runways 914 to 1,523 m: 17
with unpaved runways under 914 m: 108

Communications

Telephone system: about 2.3 million telephones; 99 telephones/1,000 persons; 89% of phone network is automatic; poor service; cable and open wire
local: NA
intercity: trunk network is microwave; roughly 3,300 villages with no service (February 1990)
international: 1 INTELSAT earth station; new digital international direct dial exchanges are in Bucharest (1993)
Radio:
broadcast stations: AM 12, FM 5, shortwave 0
radios: NA
Television:
broadcast stations: 13 (1990)
televisions: NA

Defense Forces

Branches: Army, Navy, Air and Air Defense Forces, Paramilitary Forces, Civil Defense
Manpower availability: males age 15-49 5,934,524; males fit for military service 5,002,287; males reach military age (20) annually 196,587 (1995 est.)
Defense expenditures: 1,260 billion lei, 3% of GDP (1994); note—conversion of defense expenditures into US dollars using the current exchange rate could produce misleading results

Russia

Boundary representation is not necessarily authoritative.

Geography

Location: Northern Asia (that part west of the Urals is sometimes included with Europe), bordering the Arctic Ocean, between Europe and the North Pacific Ocean
Map references: Asia
Area:
total area: 17,075,200 sq km
land area: 16,995,800 sq km
comparative area: slightly more than 1.8 times the size of the US
Land boundaries: total 20,139 km, Azerbaijan 284 km, Belarus 959 km, China (southeast) 3,605 km, China (south) 40 km, Estonia 290 km, Finland 1,313 km, Georgia 723 km, Kazakhstan 6,846 km, North Korea 19 km, Latvia 217 km, Lithuania (Kaliningrad Oblast) 227 km, Mongolia 3,441 km, Norway 167 km, Poland (Kaliningrad Oblast) 432 km, Ukraine 1,576 km
Coastline: 37,653 km
Maritime claims:
continental shelf: 200-m depth or to the depth of exploitation
exclusive economic zone: 200 nm
territorial sea: 12 nm
International disputes: inherited disputes from former USSR including: sections of the boundary with China; islands of Etorofu, Kunashiri, and Shikotan and the Habomai group occupied by the Soviet Union in 1945, administered by Russia, claimed by Japan; maritime dispute with Norway over portion of the Barents Sea; Caspian Sea boundaries are not yet determined; potential dispute with Ukraine over Crimea; Estonia claims over 2,000 sq km of Russian territory in the Narva and Pechora regions; the Abrene section of the border ceded by the Latvian Soviet Socialist Republic to Russia in 1944; has made no territorial claim in Antarctica (but has reserved the right to do so) and does not recognize the claims of any other nation
Climate: ranges from steppes in the south through humid continental in much of European Russia; subarctic in Siberia to tundra climate in the polar north; winters vary from cool along Black Sea coast to frigid in Siberia; summers vary from warm in the steppes to cool along Arctic coast
Terrain: broad plain with low hills west of Urals; vast coniferous forest and tundra in Siberia; uplands and mountains along southern border regions
Natural resources: wide natural resource base including major deposits of oil, natural gas, coal, and many strategic minerals, timber
note: formidable obstacles of climate, terrain, and distance hinder exploitation of natural resources
Land use:
arable land: 8%
permanent crops: NEGL%
meadows and pastures: 5%
forest and woodland: 45%
other: 42%
Irrigated land: 56,000 sq km (1992)
Environment:
current issues: air pollution from heavy industry, emissions of coal-fired electric plants, and transportation in major cities; industrial and agricultural pollution of inland waterways and sea coasts; deforestation; soil erosion; soil contamination from improper application of agricultural chemicals; scattered areas of sometimes intense radioactive contamination
natural hazards: permafrost over much of Siberia is a major impediment to development; volcanic activity in the Kuril Islands; volcanoes and earthquakes on the Kamchatka Peninsula
international agreements: party to—Air Pollution, Air Pollution-Nitrogen Oxides, Air Pollution-Sulphur 85, Antarctic Treaty, Climate Change, Endangered Species, Environmental Modification, Hazardous Wastes, Marine Dumping, Nuclear Test Ban, Ozone Layer Protection, Ship Pollution, Tropical Timber 83, Wetlands, Whaling; signed, but not ratified—Air Pollution-Sulphur 94, Antarctic-Environmental Protocol, Biodiversity, Law of the Sea
Note: largest country in the world in terms of area but unfavorably located in relation to major sea lanes of the world; despite its size, much of the country lacks proper soils and climates (either too cold or too dry) for agriculture

People

Population: 149,909,089 (July 1995 est.)
note: official Russian statistics put the population at 148,200,000 for 1994
Age structure:
0-14 years: 22% (female 16,208,640; male 16,784,017)
15-64 years: 66% (female 50,711,209; male 48,247,101)
65 years and over: 12% (female 12,557,447; male 5,400,675) (July 1995 est.)
Population growth rate: 0.2% (1995 est.)
note: official Russian statistics put the population growth rate at -6.0% for 1994
Birth rate: 12.64 births/1,000 population (1995 est.)
note: official Russian statistics put the birth rate at 9.5 births per 1,000 population for 1994
Death rate: 11.36 deaths/1,000 population (1995 est.)
note: official Russian statistics put the death rate at 15.5 deaths per 1,000 population in 1994
Net migration rate: 0.7 migrant(s)/1,000 population (1995 est.)
Infant mortality rate: 26.4 deaths/1,000 live births (1995 est.)
note: official Russian statistics put the infant mortality rate at 19.9 deaths per 1,000 live births in 1994
Life expectancy at birth:
total population: 69.1 years
male: 64.1 years
female: 74.35 years (1995 est.)
note: official Russian statistics put life expectancy at birth as 64 years for total population in 1994
Total fertility rate: 1.82 children born/woman (1995 est.)
Nationality:
noun: Russian(s)
adjective: Russian
Ethnic divisions: Russian 81.5%, Tatar 3.8%, Ukrainian 3%, Chuvash 1.2%, Bashkir 0.9%, Byelorussian 0.8%, Moldavian 0.7%, other 8.1%
Religions: Russian Orthodox, Muslim, other
Languages: Russian, other
Literacy: age 15 and over can read and write (1989)
total population: 98%
male: 100%
female: 97%
Labor force: 85.0 million (1993)
by occupation: production and economic services 83.9%, government 16.1%

Government

Names:
conventional long form: Russian Federation
conventional short form: Russia
local long form: Rossiyskaya Federatsiya
local short form: Rossiya
former: Russian Soviet Federative Socialist Republic
Digraph: RS
Type: federation
Capital: Moscow
Administrative divisions: 21 autonomous republics (avtomnykh respublik, singular—avtomnaya respublika); Adygea (Maykop), Bashkortostan (Ufa), Buryatia (Ulan-Ude), Chechnya (Groznyy), Chuvashia (Cheboksary), Dagestan (Makhachkala), Gorno-Altay (Gorno-Altaysk), Ingushetia (Nazran'), Kabardino-Balkaria (Nal'chik),

Kalmykia (Elista), Karachay-Cherkessia (Cherkessk), Karelia (Petrozavodsk), Khakassia (Abakan), Komi (Syktyvkar), Mari El (Yoshkar-Ola), Mordovia (Saransk), North Ossetia (Vladikavkaz), Tatarstan (Kazan'), Tuva (Kyzyl), Udmurtia (Izhevsk), Yakutia—also known as Sakha (Yakutsk); 49 oblasts (oblastey, singular—oblast'); Amur (Blagoveshchensk), Arkhangel'sk, Astrakhan', Belgorod, Bryansk, Chelyabinsk, Chita, Irkutsk, Ivanovo, Kaliningrad, Kaluga, Kamchatka (Petropavlovsk-Kamchatskiy), Kemerovo, Kirov, Kostroma, Kurgan, Kursk, Leningrad (St. Petersburg), Lipetsk, Magadan, Moscow, Murmansk, Nizhniy Novgorod, Novgorod, Novosibirsk, Omsk, Orel, Orenburg, Penza, Perm', Pskov, Rostov, Ryazan', Sakhalin (Yuzhno-Sakhalinsk), Samara, Saratov, Smolensk, Sverdlovsk (Yekaterinburg), Tambov, Tomsk, Tula, Tver' Tyumen', Ul'yanovsk, Vladimir, Volgograd, Vologda, Voronezh, Yaroslavl'; 6 krays (krayev, singular—kray); Altay (Barnaul), Khabarovsk, Krasnodar, Krasnoyarsk, Primorskiy (Vladivostok), Stavropol'; 10 autonomous okrugs; Aga (Aginskoye), Chukotka (Anadyr'), Evenkia (Tura), Khantia-Mansia (Khanty-Mansiysk), Koryakia (Palana), Nenetsia (Nar'yan-Mar), Permyakia (Kudymkar), Taymyria (Dudinka), Ust'-Onda (Ust'-Ordynskiy), Yamalia (Salekhard); 1 autonomous oblast (avtomnykh oblast'); Birobijan
note: the autonomous republics of Chechnya and Ingushetia were formerly the autonomous republic of Checheno-Ingushetia (the boundary between Chechenia and Ingushetia has yet to be determined); the cities of Moscow and St. Petersburg are federal cities; an administrative division has the same name as its administrative center (exceptions have the administrative center name following in parentheses)
Independence: 24 August 1991 (from Soviet Union)
National holiday: Independence Day, June 12 (1990)
Constitution: adopted 12 December 1993
Legal system: based on civil law system; judicial review of legislative acts
Suffrage: 18 years of age; universal
Executive branch:
chief of state: President Boris Nikolayevich YEL'TSIN (since 12 June 1991); election last held 12 June 1991 (next to be held NA 1996); results—percent of vote by party NA; note—no vice president; if the president dies in office, cannot exercise his powers because of ill health, is impeached, or resigns, the premier succeeds him; the premier serves as acting president until a new presidential election is held, which must be within three months
head of government: Premier and Chairman of the Council of Ministers Viktor Stepanovich CHERNOMYRDIN (since 14 December 1992); First Deputy Chairmen of the Council of Ministers Oleg SOSKOVETS (since 30 April 1993) and Anatoliy CHUBAYS (since 5 November 1994)
Security Council: originally established as a presidential advisory body in June 1991, but restructured in March 1992 with responsibility for managing individual and state security
Presidential Administration: drafts presidential edicts and provides staff and policy support to the entire executive branch
cabinet: Council of Ministers; appointed by the president
Group of Assistants: schedules president's appointments, processes presidential edicts and other official documents, and houses the president's press service and primary speechwriters
Council of Heads of Republics: includes the leaders of the 21 ethnic-based Republics
Council of Heads of Administrations: includes the leaders of the 66 autonomous territories and regions, and the mayors of Moscow and St. Petersburg
Presidential Council: prepares policy papers for the president
Legislative branch: bicameral Federal Assembly
Federation Council: elections last held 12 December 1993 (next to be held NA); results—two members elected from each of Russia's 89 territorial units for a total of 176 deputies; 2 seats unfilled as of 15 May 1994 (Chechnya did not participate in the election); Speaker Vladimir SHUMEYKO (Russia's Democratic Choice)
State Duma: elections last held 12 December 1993 (next to be held NA December 1995); results—percent of vote by party NA; seats—(450 total) Russia's Democratic Choice 78, New Regional Policy 66, Liberal Democrats 63, Agrarian Party 55, Communist Party of the Russian Federation 45, Unity and Accord 30, Yavlinskiy-Boldyrev-Lukin Bloc (Yabloko) 27, Women of Russia 23, Democratic Party of Russia 15, Russia's Path 12, other parties 23, affiliation unknown 12, unfilled (as of 13 March 1994; Chechnya did not participate in the election) 1; Speaker Ivan RYBKIN (Agrarian Party); note—as of 11 April 1995, seats were as follows: Russia's Democratic Choice 54, New Regional Policy 32, Liberal Democrats 54, Agrarian Party 51, Communist Party of the Russian Federation 45, Unity and Accord 25, Yavlinskiy-Boldyrev-Lukin Bloc (Yabloko) 28, Liberal Democratic Union of 12 December 9, Women of Russia 22, Democratic Party of Russia 10, Russia's Path 12, Duma 96 23, Russia 35, Stability 36, affiliation unknown 14
Judicial branch: Constitutional Court, Supreme Court (highest court for criminal, civil, and administrative cases), Superior Court of Arbitration (highest court that resolves economic disputes)
Political parties and leaders:
pro-market democrats: Party of Russian Unity and Accord, Sergey SHAKHRAY; Russia's Democratic Choice Party, Yegor GAYDAR; Russian Movement for Democratic Reforms, Anatoliy SOBCHAK; Yavlinskiy-Boldyrev-Lukin Bloc (Yabloko), Grigoriy YAVLINSKIY; Liberal Democratic Union of 12 December, Boris FEDOROV
centrists/special interest parties: Civic Union for Stability, Justice, and Progress, Arkadiy VOL'SKIY; Democratic Party of Russia, Sergey GLAZ'YEV; Women of Russia, Alevtina FEDULOVA; Social Democratic Peoples' Party, Vasiliy LIPITSKIY; New Regional Policy (NPR), Vladimir MEDVEDEV
anti-market and/or ultranationalist parties: Agrarian Party, Mikhail LAPSHIN; Communist Party of the Russian Federation, Gennadiy ZYUGANOV; Liberal Democratic Party of Russia, Vladimir ZHIRINOVSKIY; Derzhava, Aleksandr RUTSKOY
note: more than 20 political parties and associations tried to gather enough signatures to run slates of candidates in the 12 December 1993 legislative elections, but only 13 succeeded
Other political or pressure groups: NA
Member of: BSEC, CBSS, CCC, CE (guest), CERN (observer), CIS, EBRD, ECE, ESCAP, IAEA, IBRD, ICAO, ICRM, IDA, IFC, IFRCS, ILO, IMF, IMO, INMARSAT, INTELSAT, INTERPOL, IOC, IOM (observer), ISO, ITU, MINURSO, NACC, NSG, OAS (observer), OSCE, PCA, PFP, UN, UN Security Council, UNAMIR, UNCTAD, UNESCO, UNIDO, UNIKOM, UNITAR, UNMIH, UNOMOZ, UNPROFOR, UNTSO, UPU, WHO, WIPO, WMO, WTO, ZC
Diplomatic representation in US:
chief of mission: Ambassador Sergey LAVROV
chancery: 2650 Wisconsin Avenue NW, Washington, DC 20007
telephone: [1] (202) 298-5700 through 5704
FAX: [1] (202) 298-5735
consulate(s) general: New York, San Francisco, and Seattle
US diplomatic representation:
chief of mission: Ambassador Thomas R. PICKERING
embassy: Novinskiy Bul'var 19/23, Moscow
mailing address: APO AE 09721
telephone: [7] (095) 252-24-51 through 59
FAX: [7] (095) 956-42-61
consulate(s) general: St. Petersburg, Vladivostok, Yekaterinburg
Flag: three equal horizontal bands of white (top), blue, and red

---

Economy

Overview: Russia, a vast country with a wealth of natural resources, a well-educated population, and a diverse industrial base, continues to experience formidable difficulties in moving from its old centrally planned economy to a modern market economy. President YEL'TSIN's government has made substantial strides in converting to a market economy since launching its economic reform program in January 1992 by freeing nearly all prices, slashing defense spending, eliminating the old centralized distribution system, completing an ambitious voucher privatization program, establishing private financial institutions, and decentralizing foreign trade. Russia, however, has made little progress in a number of key areas that are needed to provide a solid foundation for the transition to a market economy. Financial stabilization has remained elusive, with wide swings in monthly inflation rates. Only limited restructuring of industry has occurred so far because of a scarcity of investment funds and the failure of enterprise managers to make hard cost-cutting decisions. In addition, Moscow has yet to develop a social safety net that would allow faster restructuring by relieving enterprises of the burden of providing social benefits for their workers and has been slow to develop the legal framework necessary to fully support a market economy and to encourage foreign investment. As a result, output has continued to fall. According to Russian official data, which probably overstate the fall, GDP declined by 15% in 1994 compared with a 12% decline in 1993. Industrial output in 1994 fell 21% with all major sectors taking a hit. Agricultural production in 1994 was down 9%. The grain harvest totaled 81 million tons, some 15 million tons less than in 1993. Unemployment climbed to an estimated 6.6 million or about 7% of the work force by yearend 1994. Floundering Russian firms have already had to put another 4.8 million workers on involuntary, unpaid leave or shortened workweeks. Government fears of large-scale unemployment continued to hamper industrial restructuring efforts. According to official Russian data, real per capita income was up nearly 18% in 1994 compared with 1993, in part because many Russians are working second jobs. Most Russians perceive that they are worse off now because of growing crime and health problems and mounting wage arrears. Russia has made significant headway in privatizing state assets, completing its voucher privatization program at midyear 1994. At least a portion of about 110,000 state enterprises were transferred to private hands by the end of 1994. Including partially privatized firms, the private sector accounted for roughly half of GDP in 1994. Financial stabilization continued to remain a challenge for the government. Moscow tightened financial policies in late 1993 and early 1994, including postponing planned budget spending, and succeeded in reducing monthly inflation from 18% in January to about 5% in July and August. At midyear, however, the government relaxed austerity measures in the face of mounting pressure from industry and agriculture, sparking a new round of inflation; the monthly inflation rate jumped to roughly 15% per month during the fourth quarter. In response, Moscow announced a fairly tight government budget for 1995 designed to bring monthly inflation down to around 1% by the end of 1995. According to official statistics, Russia's 1994 trade with nations outside the former Soviet Union produced a $12.3 billion surplus, up from $11.3 billion in 1993. Foreign sales—comprised largely of oil, natural gas, and other raw materials—grew more than 8%. Imports also were up 8% as demand for food and other consumer goods surged. Russian trade with other former Soviet republics continued to decline. At the same time, Russia paid only a fraction of the roughly $20 billion in debt that came due in 1994, and by the end of the year, Russia's hard currency foreign debt had risen to nearly $100 billion. Moscow reached agreement to restructure debts with Paris Club official creditors in mid-1994 and concluded a preliminary deal with its commercial bank creditors late in the year to reschedule debts owed them in early 1995. Capital flight continued to be a serious problem in 1994, with billions of additional dollars in assets being moved abroad, primarily to bank accounts in Europe. Russia's physical plant continues to deteriorate because of insufficient maintenance and new construction. Plant and equipment on average are twice the age of the West's. Many years will pass before Russia can take full advantage of its natural resources and its human assets.

National product: GDP—purchasing power parity—$721.2 billion (1994 estimate as extrapolated from World Bank estimate for 1992)
National product real growth rate: -15% (1994 est.)
National product per capita: $4,820 (1994 est.)
Inflation rate (consumer prices): 10% per month (average 1994)
Unemployment rate: 7.1% (December 1994) with considerable additional underemployment
Budget:
revenues: $NA
expenditures: $NA, including capital expenditures of $NA
Exports: $48 billion (f.o.b., 1994)
commodities: petroleum and petroleum products, natural gas, wood and wood products, metals, chemicals, and a wide variety of civilian and military manufactures
partners: Europe, North America, Japan, Third World countries, Cuba
Imports: $35.7 billion (f.o.b., 1994)
commodities: machinery and equipment, consumer goods, medicines, meat, grain, sugar, semifinished metal products
partners: Europe, North America, Japan, Third World countries, Cuba
External debt: $95 billion-$100 billion (yearend 1994)
Industrial production: growth rate -21% (1994)
Electricity:
capacity: 213,100,000 KW
production: 876 billion kWh
consumption per capita: 5,800 kWh (1994)
Industries: complete range of mining and extractive industries producing coal, oil, gas, chemicals, and metals; all forms of machine building from rolling mills to high-performance aircraft and space vehicles; ship-building; road and rail transportation equipment; communications equipment; agricultural machinery, tractors, and construction equipment; electric power generating and transmitting equipment; medical and scientific instruments; consumer durables
Agriculture: grain, sugar beets, sunflower seeds, meat, milk, vegetables, fruits; because of its northern location does not grow citrus, cotton, tea, and other warm climate products
Illicit drugs: illicit cultivator of cannabis and opium poppy; mostly for domestic consumption; government has active eradication program; used as transshipment point for Asian and Latin American illicit drugs to Western Europe and Latin America
Economic aid:
recipient: US commitments, including Ex-Im (1990-94), $15 billion; other countries, ODA and OOF bilateral commitments (1990-93), $120 billion
Currency: 1 ruble (R) = 100 kopeks
Exchange rates: rubles per US$1—3,550 (29 December 1994), 1,247 (27 December 1993); nominal exchange rate still deteriorating but real exchange rate holding steady
Fiscal year: calendar year

Transportation

Railroads:
total: 154,000 km; note—87,000 km in common carrier service (49,000 km diesel; and 38,000 km electrified); 67,000 km serve specific industries and are not available for common carrier use
broad gauge: 154,000 km 1.520-m gauge (1 January 1994)
Highways:
total: 934,000 km (445,000 km serve specific industries or farms and are not available for common carrier use)
paved and graveled: 725,000 km
unpaved: 209,000 km (1 January 1994)
Inland waterways: total navigable routes in general use 101,000 km; routes with navigation guides serving the Russian River Fleet 95,900 km; of which routes with night navigational aids 60,400 km; man-made navigable routes 16,900 km (1 January 1994)
Pipelines: crude oil 48,000 km; petroleum

products 15,000 km; natural gas 140,000 km (30 June 1993)
Ports: Arkhangel'sk, Astrakhan', Kaliningrad, Kazan', Khabarovsk, Kholmsk, Krasnoyarsk, Moscow, Murmansk, Nakhodka, Nevel'sk, Novorossiysk, Petropavlovsk, St. Petersburg, Rostov, Sochi, Tuapse, Vladivostok, Volgograd, Vostochnyy, Vyborg
Merchant marine:
total: 800 ships (1,000 GRT or over) totaling 7,295,109 GRT/10,128,579 DWT
ships by type: barge carrier 2, bulk cargo 26, cargo 424, chemical tanker 7, combination bulk 22, combination ore/oil 16, container 81, multifunction large-load carrier 3, oil tanker 111, passenger 4, passenger-cargo 5, refrigerated cargo 19, roll-on/roll-off cargo 62, short-sea passenger 16, specialized tanker 2
note: in addition, Russia owns 235 ships (1,000 GRT or over) totaling 5,084,439 DWT that operate under Maltese, Cypriot, Liberian, Panamanian, Saint Vincent and the Grenadines, Honduran, Marshall Islands, Bahamian, and Vanuatu registry
Airports:
total: 2,517
with paved runways over 3,047 m: 54
with paved runways 2,438 to 3,047 m: 202
with paved runways 1,524 to 2,437 m: 108
with paved runways 914 to 1,523 m: 115
with paved runways under 914 m: 151
with unpaved runways over 3,047 m: 25
with unpaved runways 2,438 to 3,047 m: 45
with unpaved runways 1,524 to 2,438 m: 134
with unpaved runways 914 to 1,523 m: 291
with unpaved runways under 914 m: 1,392

Communications

Telephone system: 24,400,000 telephones; 20,900,000 telephones in urban areas and 3,500,000 telephones in rural areas; of these, total installed in homes 15,400,000; total pay phones for long distant calls 34,100; about 164 telephones/1,000 persons; Russia is enlisting foreign help, by means of joint ventures, to speed up the modernization of its telecommunications system; in 1992, only 661,000 new telephones were installed compared with 855,000 in 1991, and in 1992 the number of unsatisfied applications for telephones reached 11,000,000; expanded access to international E-mail service available via Sprint network; the inadequacy of Russian telecommunications is a severe handicap to the economy, especially with respect to international connections
local: NMT-450 analog cellular telephone networks are operational and growing in Moscow and St. Petersburg
intercity: intercity fiberoptic cable installation remains limited
international: international traffic is handled by an inadequate system of satellites, land lines, microwave radio relay and outdated submarine cables; this traffic passes through the international gateway switch in Moscow which carries most of the international traffic for the other countries of the Commonwealth of Independent States; a new Russian Raduga satellite will link Moscow and St. Petersburg with Rome from whence calls will be relayed to destinations in Europe and overseas; satellite earth stations—INTELSAT, Intersputnik, Eutelsat (Moscow), INMARSAT, Orbita
Radio:
broadcast stations: AM 1,050, FM 1,050, shortwave 1,050
radios: 48.8 million (radio receivers with multiple speaker systems for program diffusion 74,300,000)
Television:
broadcast stations: 7,183
televisions: 54.2 million

Defense Forces

Branches: Ground Forces, Navy, Air Forces, Air Defense Forces, Strategic Rocket Forces
Manpower availability: males age 15-49 38,264,699; males fit for military service 29,951,977; males reach military age (18) annually 1,106,176 (1995 est.)
Defense expenditures: $NA, NA% of GDP
note: the Intelligence Community estimates that defense spending in Russia fell about 15% in real terms in 1994, reducing Russian defense outlays to about one-fourth of peak Soviet levels in the late 1980s; although Russia may still spend as much as 10% of its GDP on defense, this is significantly below the 15% to 17% burden the former USSR carried during much of the 1980s; conversion of military expenditures into US dollars using the current exchange rate could produce misleading results

Rwanda

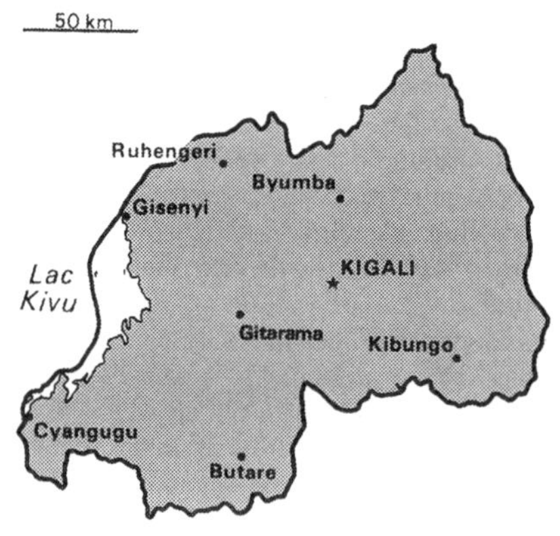

Geography

Location: Central Africa, east of Zaire
Map references: Africa
Area:
total area: 26,340 sq km
land area: 24,950 sq km
comparative area: slightly smaller than Maryland
Land boundaries: total 893 km, Burundi 290 km, Tanzania 217 km, Uganda 169 km, Zaire 217 km
Coastline: 0 km (landlocked)
Maritime claims: none; landlocked
International disputes: none
Climate: temperate; two rainy seasons (February to April, November to January); mild in mountains with frost and snow possible
Terrain: mostly grassy uplands and hills; relief is mountainous with altitude declining from west to east
Natural resources: gold, cassiterite (tin ore), wolframite (tungsten ore), natural gas, hydropower
Land use:
arable land: 29%
permanent crops: 11%
meadows and pastures: 18%
forest and woodland: 10%
other: 32%
Irrigated land: 40 sq km (1989 est.)
Environment:
current issues: deforestation results from uncontrolled cutting of trees for fuel; overgrazing; soil exhaustion; soil erosion
natural hazards: periodic droughts; the volcanic Virunga mountains are in the northwest along the border with Zaire
international agreements: party to—Endangered Species, Nuclear Test Ban; signed, but not ratified—Biodiversity, Climate Change, Law of the Sea
Note: landlocked; predominantly rural population

People

Population: 8,605,307 (July 1995 est.)
note: the demographic estimates were prepared before civil strife, starting in April 1994, set in motion substantial and continuing population changes
Age structure:
0-14 years: 51% (female 2,184,549; male 2,201,049)
15-64 years: 47% (female 2,034,278; male 1,968,298)
65 years and over: 2% (female 126,255; male 90,878) (July 1995 est.)
Population growth rate: 2.67% (1995 est.)
Birth rate: 48.52 births/1,000 population (1995 est.)
Death rate: 21.82 deaths/1,000 population (1995 est.)
Net migration rate: NA migrant(s)/1,000 population (1995 est.)
note: since April 1994, more than one million refugees have fled the civil strife between the Hutu and Tutsi factions in Rwanda and crossed into Zaire, Burundi, and Tanzania; close to 350,000 Rwandan Tutsis who fled civil strife in earlier years are returning to Rwanda and a few of the recent Hutu refugees are going home despite the danger of doing so; the ethnic violence continues and in 1995 could produce further refugee flows as well as deter returns
Infant mortality rate: 118.1 deaths/1,000 live births (1995 est.)
Life expectancy at birth:
total population: 39.33 years
male: 38.5 years
female: 40.19 years (1995 est.)
Total fertility rate: 8.12 children born/woman (1995 est.)
Nationality:
noun: Rwandan(s)
adjective: Rwandan
Ethnic divisions: Hutu 90%, Tutsi 9%, Twa (Pygmoid) 1%
Religions: Roman Catholic 65%, Protestant 9%, Muslim 1%, indigenous beliefs and other 25%
Languages: Kinyarwanda (official), French (official), Kiswahili used in commercial centers
Literacy: age 15 and over can read and write (1990 est.)
total population: 50%
male: 64%
female: 37%
Labor force: 3.6 million
by occupation: agriculture 93%, government and services 5%, industry and commerce 2%

Government

Names:
conventional long form: Republic of Rwanda
conventional short form: Rwanda
local long form: Republika y'u Rwanda
local short form: Rwanda
Digraph: RW
Type: republic; presidential system
note: after genocide and civil war in April 1994, the Tutsi Rwandan Patriotic Front, in July 1994, took power and formed a new government
Capital: Kigali
Administrative divisions: 10 prefectures (prefectures, singular—prefecture in French; plural—NA, singular—prefegitura in Kinyarwanda); Butare, Byumba, Cyangugu, Gikongoro, Gisenyi, Gitarama, Kibungo, Kibuye, Kigali, Ruhengeri
Independence: 1 July 1962 (from Belgium-administered UN trusteeship)
National holiday: Independence Day, 1 July (1962)
Constitution: 18 June 1991
Legal system: based on German and Belgian civil law systems and customary law, judicial review of legislative acts in the Supreme Court; has not accepted compulsory ICJ jurisdiction
Suffrage: NA years of age; universal adult
Executive branch:
chief of state: President Pasteur BIZIMUNGU (since 19 July 1994); took office following the siezure of the government by the Tutsi Rwandan Patriotic Front and the exiling of interim President Dr. Theodore SINDIKUBWABO; no future election dates have been set
head of government: Prime Minister Faustin TWAGIRAMUNGU (since the siezure of power by the Tutsi Rwandan Patriotic Front in July 1994)
cabinet: Council of Ministers; appointed by the president
Legislative branch: unicameral
National Development Council: (Conseil National de Developpement) elections last held 19 December 1988 (next to be held NA 1995); results—MRND was the only party; seats—(70 total) MRND 70
Judicial branch: Constitutional Court consists of the Court of Cassation and the Council of State in joint session
Political parties and leaders: Rwandan Patriotic Front (RPF), Alexis KANYARENGWE, Chairman; National Revolutionary Movement for Democracy and Development (MRND); significant independent parties include: Democratic Republican Movement (MDR); Liberal Party (PL); Democratic and Socialist Party (PSD); Coalition for the Defense of the Republic (CDR); Party for Democracy in Rwanda (PADER); Christian Democratic Party (PDL)
note: formerly a one-party state, Rwanda legalized independent parties in mid-1991
Other political or pressure groups: Rwanda Patriotic Army (RPA), the RPF military wing, Maj. Gen. Paul KAGAME, commander
Member of: ACCT, ACP, AfDB, CCC, CEEAC, CEPGL, ECA, FAO, G-77, GATT, IBRD, ICAO, ICFTU, ICRM, IDA, IFAD, IFC, IFRCS, ILO, IMF, INTELSAT, INTERPOL, IOC, ITU, NAM, OAU, UN, UNCTAD, UNESCO, UNIDO, UPU, WCL, WHO, WIPO, WMO, WTO
Diplomatic representation in US:
chief of mission: (vacant); Charge d'Affaires ad interim Joseph W. MUTABOBA
chancery: 1714 New Hampshire Avenue NW, Washington, DC 20009
telephone: [1] (202) 232-2882
FAX: [1] (202) 232-4544
US diplomatic representation:
note: US Embassy closed indefinitely
chief of mission: Ambassador David P. RAWSON
embassy: Boulevard de la Revolution, Kigali
mailing address: B. P. 28, Kigali
telephone: [250] 756 01 through 03
FAX: [250] 721 28
Flag: three equal vertical bands of red (hoist side), yellow, and green with a large black letter R centered in the yellow band; uses the popular pan-African colors of Ethiopia; similar to the flag of Guinea, which has a plain yellow band

Economy

Overview: Rwanda is a poor African nation suffering bitterly from ethnic-based civil war. Almost 50% of GDP comes from the agricultural sector; coffee and tea make up 80%-90% of total exports. The amount of fertile land is limited, however, and deforestation and soil erosion continue to create problems. The industrial sector in Rwanda is small, contributing only 17% to GDP. Manufacturing focuses mainly on the processing of agricultural products. The Rwandan economy remains dependent on coffee/tea exports and foreign aid. Weak international prices since 1986 have caused the economy to contract and per capita GDP to decline. A structural adjustment program with the World Bank began in October 1990. Ethnic-based insurgency since 1990 has devastated wide areas, especially in the north, and displaced hundreds of thousands of people. A peace accord in mid-1993 temporarily ended most of the fighting, but massive resumption of civil warfare in April 1994 in the capital city Kigali and elsewhere has been taking thousands of lives and severely affecting short-term economic prospects. The economy suffers massively from failure to maintain the infrastructure, looting, neglect of important cash crops, and lack of health care facilities.
National product: GDP—purchasing power parity—$7.9 billion (1993 est.)
National product real growth rate: -8% (1993 est.)
National product per capita: $950 (1993 est.)
Inflation rate (consumer prices): NA%

Unemployment rate: NA%
Budget:
revenues: $350 million
expenditures: $NA, including capital expenditures of $NA (1992 est.)
Exports: $44 million (f.o.b., 1993 est.)
commodities: coffee 63%, tea, cassiterite, wolframite, pyrethrum
partners: Germany, Belgium, Italy, Uganda, UK, France, US
Imports: $250 million (f.o.b., 1993 est.)
commodities: textiles, foodstuffs, machines and equipment, capital goods, steel, petroleum products, cement and construction material
partners: US, Belgium, Germany, Kenya, Japan
External debt: $873 million (1993 est.)
Industrial production: growth rate -2.2% (1991); accounts for 17% of GDP
Electricity:
capacity: 60,000 kW
production: 190 million kWh
consumption per capita: 23 kWh (1993)
Industries: mining of cassiterite (tin ore) and wolframite (tungsten ore), tin, cement, agricultural processing, small-scale beverage production, soap, furniture, shoes, plastic goods, textiles, cigarettes
Agriculture: cash crops—coffee, tea, pyrethrum (insecticide made from chrysanthemums); main food crops—bananas, beans, sorghum, potatoes; stock raising
Economic aid:
recipient: US commitments, including Ex-Im (FY70-89), $128 million; Western (non-US) countries, ODA and OOF bilateral commitments (1970-89), $2 billion; OPEC bilateral aid (1979-89), $45 million; Communist countries (1970-89), $58 million
note: in October 1990 Rwanda launched a Structural Adjustment Program with the IMF; since September 1991, the EC has given $46 million and the US $25 million in support of this program (1993)
Currency: 1 Rwandan franc (RF) = 100 centimes
Exchange rates: Rwandan francs (RF) per US$1—144.3 (3rd quarter 1994), 144.25 (1993), 133.35 (1992), 125.14 (1991), 82.60 (1990)
Fiscal year: calendar year

Transportation

Railroads: 0 km
Highways:
total: 4,885 km
paved: 880 km
unpaved: gravel, sand and gravel 1,305 km; unimproved earth 2,700 km
Inland waterways: Lac Kivu navigable by shallow-draft barges and native craft
Ports: Cyangugu, Gisenyi, Kibuye
Airports:
total: 7
with paved runways over 3,047 m: 1
with paved runways 914 to 1,523 m: 2
with paved runways under 914 m: 3
with unpaved runways 914 to 1,523 m: 1

Communications

Telephone system: NA telephones; telephone system does not provide service to the general public but is intended for business and government use
local: NA
intercity: the capital, Kigali, is connected to the centers of the prefectures by microwave radio relay; the remainder of the network depends on wire and high frequency radio
international: international connections employ microwave radio relay to neighboring countries and satellite communications to more distant countries; 1 INTELSAT (Indian Ocean) and 1 SYMPHONIE earth station in Kigali (includes telex and telefax service)
Radio:
broadcast stations: AM 1, FM 1, shortwave 0
radios: NA
Television:
broadcast stations: 1
televisions: NA

Defense Forces

Branches: Army, Gendarmerie
Manpower availability: males age 15-49 1,792,326; males fit for military service 913,711 (1995 est.)
Defense expenditures: exchange rate conversion—$112.5 million, 7% of GDP (1992)

Saint Helena
(dependent territory of the UK)

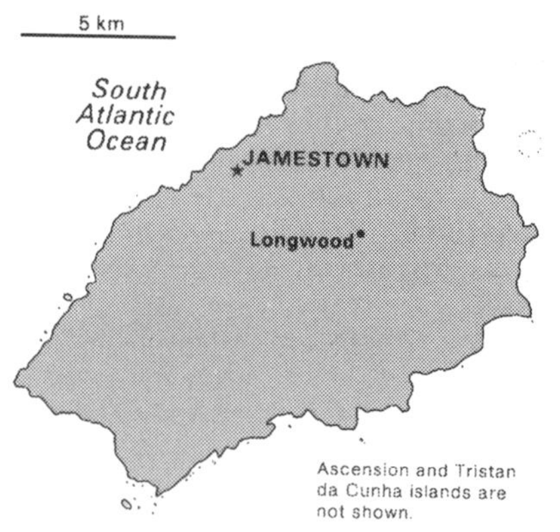

Geography

Location: Southern Africa, island in the South Atlantic Ocean, west of Angola, about two-thirds of the way from South America to Africa
Map references: Africa
Area:
total area: 410 sq km
land area: 410 sq km
comparative area: slightly more than 2.3 times the size of Washington, DC
note: includes Ascension, Gough Island, Inaccessible Island, Nightingale Island, and Tristan da Cunha
Land boundaries: 0 km
Coastline: 60 km
Maritime claims:
exclusive fishing zone: 200 nm
territorial sea: 12 nm
International disputes: none
Climate: tropical; marine; mild, tempered by trade winds
Terrain: rugged, volcanic; small scattered plateaus and plains
Natural resources: fish; Ascension is a breeding ground for sea turtles and sooty terns, no minerals
Land use:
arable land: 7%
permanent crops: 0%
meadows and pastures: 7%
forest and woodland: 3%
other: 83%
Irrigated land: NA sq km
Environment:
current issues: NA
natural hazards: active volcanism on Tristan da Cunha
international agreements: NA
Note: Napoleon Bonaparte's place of exile and burial (the remains were taken to Paris in 1840); harbors at least 40 species of plants unknown anywhere else in the world

People

Population: 6,762 (July 1995 est.)
Age structure:
0-14 years: NA
15-64 years: NA
65 years and over: NA
Population growth rate: 0.31% (1995 est.)
Birth rate: 9.5 births/1,000 population (1995 est.)
Death rate: 6.43 deaths/1,000 population (1995 est.)
Net migration rate: 0 migrant(s)/1,000 population (1995 est.)
Infant mortality rate: 36.1 deaths/1,000 live births (1995 est.)
Life expectancy at birth:
total population: 75.07 years
male: 73.01 years
female: 76.89 years (1995 est.)
Total fertility rate: 1.13 children born/woman (1995 est.)
Nationality:
noun: Saint Helenian(s)
adjective: Saint Helenian
Ethnic divisions: NA
Religions: Anglican (majority), Baptist, Seventh-Day Adventist, Roman Catholic
Languages: English
Literacy: age 20 and over can read and write (1987)
total population: 97%
male: 97%
female: 98%
Labor force: 2,516
by occupation: professional, technical, and related workers 8.7%, managerial, administrative, and clerical 12.8%, sales people 8.1%, farmer, fishermen, etc. 5.4%, craftspersons, production process workers 14.7%, others 50.3% (1987)

Government

Names:
conventional long form: none
conventional short form: Saint Helena
Digraph: SH
Type: dependent territory of the UK
Capital: Jamestown
Administrative divisions: 1 administrative area and 2 dependencies*; Ascension*, Saint Helena, Tristan da Cunha*
Independence: none (dependent territory of the UK)
National holiday: Celebration of the Birthday of the Queen, 10 June 1989 (second Saturday in June)
Constitution: 1 January 1989
Legal system: NA
Suffrage: NA
Executive branch:
chief of state: Queen ELIZABETH II (since 6 February 1952)
head of government: Governor A. N. HOOLE (since NA 1991)
cabinet: Executive Council
Legislative branch: unicameral
Legislative Council: elections last held July 1993 (next to be held NA); results—percent of vote NA; seats—(15 total, 12 elected) independents 15
Judicial branch: Supreme Court
Political parties and leaders: none
Member of: ICFTU
Diplomatic representation in US: none (dependent territory of the UK)
US diplomatic representation: none (dependent territory of the UK)
Flag: blue with the flag of the UK in the upper hoist-side quadrant and the Saint Helenian shield centered on the outer half of the flag; the shield features a rocky coastline and three-masted sailing ship

Economy

Overview: The economy depends primarily on financial assistance from the UK. The local population earns some income from fishing, the raising of livestock, and sales of handicrafts. Because there are few jobs, a large proportion of the work force has left to seek employment overseas.
National product: GDP $NA
National product real growth rate: NA%
National product per capita: $NA
Inflation rate (consumer prices): -1.1% (1986)
Unemployment rate: NA%
Budget:
revenues: $11.2 million
expenditures: $11 million, including capital expenditures of $NA (FY92/93)
Exports: $27,400 (f.o.b., FY92/93)
commodities: fish (frozen and salt-dried skipjack, tuna), handicrafts
partners: South Africa, UK
Imports: $9.8 million (c.i.f., FY92/93)
commodities: food, beverages, tobacco, fuel oils, animal feed, building materials, motor vehicles and parts, machinery and parts
partners: UK, South Africa
External debt: $NA
Industrial production: growth rate NA%
Electricity:
capacity: 9,800 kW
production: 10 million kWh
consumption per capita: NA kWh (1993)
Industries: crafts (furniture, lacework, fancy woodwork), fishing
Agriculture: maize, potatoes, vegetables; timber production being developed; crawfishing on Tristan da Cunha
Economic aid:
recipient: Western (non-US) countries, ODA and OOF bilateral commitments (1992-93), $13.5 million
Currency: 1 Saint Helenian pound (£S) = 100 pence
Exchange rates: Saint Helenian pounds (£S) per US$1—0.6350 (January 1995), 0.6529 (1994), 0.6033 (1993), 0.5664 (1992), 0.5652 (1991), 0.5603 (1990); note—the Saint Helenian pound is at par with the British pound
Fiscal year: 1 April—31 March

Transportation

Railroads: 0 km
Highways:
total: NA (mainland 107 km, Ascension NA, Tristan da Cunha NA)
paved: 169.7 km (mainland 87 km, Ascension 80 km, Tristan da Cunha 2.70 km)
unpaved: NA (mainland 20 km earth roads, Ascension NA, Tristan da Cunha NA)
Ports: Georgetown, Jamestown
Merchant marine: none
Airports:
total: 1
with paved runways over 3,047 m: 1

Communications

Telephone system: 550 telephones; automatic network
local: NA
intercity: HF radio links to Ascension, then into worldwide submarine cable and satellite networks
international: major coaxial submarine cable relay point between South Africa, Portugal, and UK at Ascension; 2 INTELSAT (Atlantic Ocean) earth stations
Radio:
broadcast stations: AM 1, FM 0, shortwave 0
radios: 1,500
Television:
broadcast stations: 0
televisions: NA

Defense Forces

Note: defense is the responsibility of the UK

Saint Kitts and Nevis

Geography

Location: Caribbean, islands in the Caribbean Sea, about one-third of the way from Puerto Rico to Trinidad and Tobago
Map references: Central America and the Caribbean
Area:
total area: 269 sq km
land area: 269 sq km
comparative area: slightly more than 1.5 times the size of Washington, DC
Land boundaries: 0 km
Coastline: 135 km
Maritime claims:
contiguous zone: 24 nm
exclusive economic zone: 200 nm or to the edge of the continental margin
territorial sea: 12 nm
International disputes: none
Climate: subtropical tempered by constant sea breezes; little seasonal temperature variation; rainy season (May to November)
Terrain: volcanic with mountainous interiors
Natural resources: negligible
Land use:
arable land: 22%
permanent crops: 17%
meadows and pastures: 3%
forest and woodland: 17%
other: 41%
Irrigated land: NA sq km
Environment:
current issues: NA
natural hazards: hurricanes (July to October)
international agreements: party to—Biodiversity, Climate Change, Endangered Species, Hazardous Wastes, Law of the Sea, Ozone Layer Protection, Whaling

People

Population: 40,992 (July 1995 est.)
Age structure:
0-14 years: 35% (female 7,072; male 7,430)
15-64 years: 57% (female 11,784; male 11,756)
65 years and over: 8% (female 1,729; male 1,221) (July 1995 est.)
Population growth rate: 0.85% (1995 est.)
Birth rate: 23.49 births/1,000 population (1995 est.)
Death rate: 9.56 deaths/1,000 population (1995 est.)
Net migration rate: -5.39 migrant(s)/1,000 population (1995 est.)
Infant mortality rate: 19.4 deaths/1,000 live births (1995 est.)
Life expectancy at birth:
total population: 66.51 years
male: 63.51 years
female: 69.69 years (1995 est.)
Total fertility rate: 2.56 children born/woman (1995 est.)
Nationality:
noun: Kittsian(s), Nevisian(s)
adjective: Kittsian, Nevisian
Ethnic divisions: black African
Religions: Anglican, other Protestant sects, Roman Catholic
Languages: English
Literacy: age 15 and over has ever attended school (1980)
total population: 97%
male: 97%
female: 98%
Labor force: 20,000 (1981)

Government

Names:
conventional long form: Federation of Saint Kitts and Nevis
conventional short form: Saint Kitts and Nevis
former: Federation of Saint Christopher and Nevis
Digraph: SC
Type: constitutional monarchy
Capital: Basseterre
Administrative divisions: 14 parishes; Christ Church Nichola Town, Saint Anne Sandy Point, Saint George Basseterre, Saint George Gingerland, Saint James Windward, Saint John Capisterre, Saint John Figtree, Saint Mary Cayon, Saint Paul Capisterre, Saint Paul Charlestown, Saint Peter Basseterre, Saint Thomas Lowland, Saint Thomas Middle Island, Trinity Palmetto Point
Independence: 19 September 1983 (from UK)
National holiday: Independence Day, 19 September (1983)
Constitution: 19 September 1983
Legal system: based on English common law
Suffrage: NA years of age; universal adult
Executive branch:
chief of state: Queen ELIZABETH II (since 6 February 1952), represented by Governor General Sir Clement Athelston ARRINDELL (since 19 September 1983, previously Governor General of the West Indies Associated States since NA November 1981)
head of government: Prime Minister Dr. Kennedy Alphonse SIMMONDS (since 19 September 1983, previously Premier of the West Indies Associated States since NA February 1980); Deputy Prime Minister Hugh HEYLIGER (since November 1994)
cabinet: Cabinet; appointed by the governor general in consultation with the prime minister
Legislative branch: unicameral
House of Assembly: elections last held 29 November 1993 (next to be held by 15 November 1995); results—percent of vote by party NA; seats—(14 total, 11 elected) PAM 4, SKNLP 4, NRP 1, CCM 2
Judicial branch: Eastern Caribbean Supreme Court (based on Saint Lucia)
Political parties and leaders: People's Action Movement (PAM), Dr. Kennedy SIMMONDS; Saint Kitts and Nevis Labor Party (SKNLP), Dr. Denzil DOUGLAS; Nevis Reformation Party (NRP), Simeon DANIEL; Concerned Citizens Movement (CCM), Vance AMORY
Member of: ACP, C, CARICOM, CDB, ECLAC, FAO, G-77, GATT, IBRD, ICFTU, ICRM, IDA, IFAD, IFRCS (associate), IMF, INTERPOL, IOC, OAS, OECS, UN, UNCTAD, UNESCO, UNIDO, UPU, WCL, WHO
Diplomatic representation in US:
chief of mission: Ambassador Erstein Mallet EDWARDS
chancery: Suite 608, 2100 M Street NW, Washington, DC 20037
telephone: [1] (202) 833-3550
FAX: [1] (202) 833-3553
US diplomatic representation: no official presence; covered by embassy in Bridgetown, Barbados
Flag: divided diagonally from the lower hoist side by a broad black band bearing two white five-pointed stars; the black band is edged in yellow; the upper triangle is green, the lower triangle is red

Economy

Overview: The economy has traditionally depended on the growing and processing of sugarcane; decreasing world prices have hurt the industry in recent years. Tourism and export-oriented manufacturing have begun to assume larger roles, although they still only account for 7% and 4% of GDP respectively. Growth in the construction and tourism sectors spurred the economic expansion in 1994. Most food is imported.
National product: GDP—purchasing power parity—$210 million (1994 est.)
National product real growth rate: 4.5% (1994 est.)
National product per capita: $5,300 (1994 est.)
Inflation rate (consumer prices): 1.6% (1993)

Unemployment rate: 12.2% (1990)
Budget:
revenues: $103.2 million
expenditures: $102.6 million, including capital expenditures of $50.1 million (1995 est.)
Exports: $32.4 million (f.o.b., 1992)
commodities: machinery, food, electronics, beverages and tobacco
partners: US 50%, UK 30%, CARICOM nations 11% (1992)
Imports: $100 million (f.o.b., 1992)
commodities: machinery, manufactures, food, fuels
partners: US 43%, CARICOM nations 18%, UK 12%, Canada 4%, Japan 4%, OECS 4% (1992)
External debt: $43.3 million (1992)
Industrial production: growth rate 5.9% (1992 est.)
Electricity:
capacity: 15,800 kW
production: 45 million kWh
consumption per capita: 990 kWh (1993)
Industries: sugar processing, tourism, cotton, salt, copra, clothing, footwear, beverages
Agriculture: accounts for 17% of GDP; cash crop—sugarcane; subsistence crops—rice, yams, vegetables, bananas; fishing potential not fully exploited
Illicit drugs: transshipment point for South American drugs destined for the US
Economic aid:
recipient: US commitments, including Ex-Im (FY85-88), $10.7 million; Western (non-US) countries, ODA and OOF bilateral commitments (1970-89), $67 million
Currency: 1 EC dollar (EC$) = 100 cents
Exchange rates: East Caribbean dollars (EC$) per US$1—2.70 (fixed rate since 1976)
Fiscal year: calendar year

Transportation

Railroads:
total: 58 km on Saint Kitts for sugarcane
narrow gauge: 58 km 0.760-m gauge
Highways:
total: 300 km
paved: 125 km
unpaved: otherwise improved 125 km; unimproved earth 50 km
Ports: Basseterre, Charlestown
Merchant marine: none
Airports:
total: 2
with paved runways 1,524 to 2,437 m: 1
with paved runways under 914 m: 1

Communications

Telephone system: 2,400 telephones; good interisland VHF/UHF/SHF radio connections and international link via Antigua and Barbuda and Saint Martin
local: NA
intercity: interisland links are handled by VHF/UHF/SHF radio; within the islands all calls are local
international: international calls are carried by radio to Antigua and Barbuda and there switched to submarine cable or to INTELSAT, or carried to Saint Martin by radio and switched to INTELSAT
Radio:
broadcast stations: AM 2, FM 0, shortwave 0
radios: NA
Television:
broadcast stations: 4
televisions: NA

Defense Forces

Branches: Royal Saint Kitts and Nevis Police Force, Coast Guard
Defense expenditures: $NA, NA% of GDP

Saint Lucia

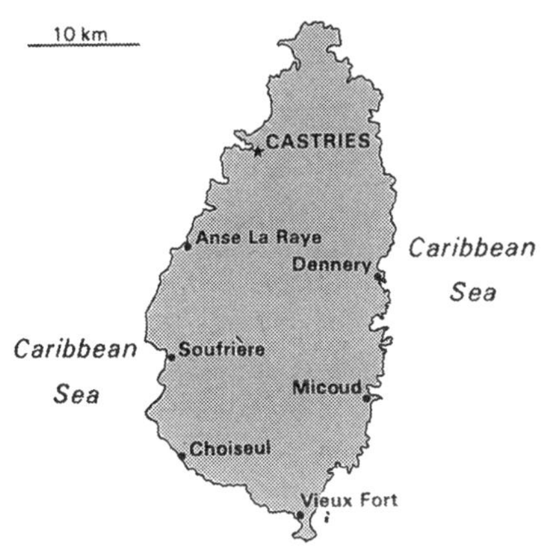

Geography

Location: Caribbean, island in the Caribbean Sea, north of Trinidad and Tobago
Map references: Central America and the Caribbean
Area:
total area: 620 sq km
land area: 610 sq km
comparative area: slightly less than 3.5 times the size of Washington, DC
Land boundaries: 0 km
Coastline: 158 km
Maritime claims:
contiguous zone: 24 nm
exclusive economic zone: 200 nm or to the edge of the continental margin
territorial sea: 12 nm
International disputes: none
Climate: tropical, moderated by northeast trade winds; dry season from January to April, rainy season from May to August
Terrain: volcanic and mountainous with some broad, fertile valleys
Natural resources: forests, sandy beaches, minerals (pumice), mineral springs, geothermal potential
Land use:
arable land: 8%
permanent crops: 20%
meadows and pastures: 5%
forest and woodland: 13%
other: 54%
Irrigated land: 10 sq km (1989 est.)
Environment:
current issues: deforestation; soil erosion, particularly in the northern region
natural hazards: hurricanes and volcanic activity
international agreements: party to—Biodiversity, Climate Change, Endangered Species, Environmental Modification, Hazardous Wastes, Law of the Sea, Marine Dumping, Ozone Layer Protection, Whaling

People

Population: 156,050 (July 1995 est.)
Age structure:
0-14 years: 35% (female 26,710; male 27,255)
15-64 years: 60% (female 47,584; male 46,326)
65 years and over: 5% (female 5,040; male 3,135) (July 1995 est.)
Population growth rate: 1.17% (1995 est.)
Birth rate: 22.48 births/1,000 population (1995 est.)
Death rate: 6.1 deaths/1,000 population (1995 est.)
Net migration rate: -4.67 migrant(s)/1,000 population (1995 est.)
Infant mortality rate: 20.5 deaths/1,000 live births (1995 est.)
Life expectancy at birth:
total population: 69.88 years
male: 66.33 years
female: 73.67 years (1995 est.)
Total fertility rate: 2.37 children born/woman (1995 est.)
Nationality:
noun: Saint Lucian(s)
adjective: Saint Lucian
Ethnic divisions: African descent 90.3%, mixed 5.5%, East Indian 3.2%, Caucasian 0.8%
Religions: Roman Catholic 90%, Protestant 7%, Anglican 3%
Languages: English (official), French patois
Literacy: age 15 and over has ever attended school (1980)
total population: 67%
male: 65%
female: 69%
Labor force: 43,800
by occupation: agriculture 43.4%, services 38.9%, industry and commerce 17.7% (1983 est.)

Government

Names:
conventional long form: none
conventional short form: Saint Lucia
Digraph: ST
Type: parliamentary democracy
Capital: Castries
Administrative divisions: 11 quarters; Anse La Raye, Castries, Choiseul, Dauphin, Dennery, Gros Islet, Laborie, Micoud, Praslin, Soufriere, Vieux Fort
Independence: 22 February 1979 (from UK)
National holiday: Independence Day, 22 February (1979)
Constitution: 22 February 1979
Legal system: based on English common law
Suffrage: 18 years of age; universal
Executive branch:
chief of state: Queen ELIZABETH II (since 6 February 1952), represented by Governor General Sir Stanislaus Anthony JAMES (since 10 October 1988)
head of government: Prime Minister John George Melvin COMPTON (since 3 May 1982)
cabinet: Cabinet; appointed by the governor general on advice of the prime minister
Legislative branch: bicameral Parliament
Senate: consists of an 11-member body, 6 appointed on the advice of the prime minister, 3 on the advice of the leader of the opposition, and 2 after consultation with religious, economic, and social groups
House of Assembly: elections last held 27 April 1992 (next to be held by April 1997); results—percent of vote by party NA; seats—(17 total) UWP 11, SLP 6
Judicial branch: Eastern Caribbean Supreme Court
Political parties and leaders: United Workers' Party (UWP), John COMPTON; Saint Lucia Labor Party (SLP), Julian HUNTE; Progressive Labor Party (PLP), Jon ODLUM
Member of: ACCT (associate), ACP, C, CARICOM, CDB, ECLAC, FAO, G-77, GATT, IBRD, ICAO, ICFTU, ICRM, IDA, IFAD, IFC, IFRCS, ILO, IMF, IMO, INTELSAT (nonsignatory user), INTERPOL, IOC, ISO (subscriber), NAM, OAS, OECS, UN, UNCTAD, UNESCO, UNIDO, UPU, WCL, WFTU, WHO, WIPO, WMO
Diplomatic representation in US:
chief of mission: Ambassador Dr. Joseph Edsel EDMUNDS
chancery: 3216 New Mexico Avenue NW, Washington, DC 20016
telephone: [1] (202) 364-6792 through 6795
FAX: [1] (202) 364-6728
consulate(s) general: New York
US diplomatic representation: no official presence since the Ambassador resides in Bridgetown (Barbados)
Flag: blue with a gold isosceles triangle below a black arrowhead; the upper edges of the arrowhead have a white border

Economy

Overview: Though foreign investment in manufacturing and information processing in recent years has increased Saint Lucia's industrial base, the economy remains vulnerable due to its heavy dependence on banana production, which is subject to periodic droughts and tropical storms. Indeed, the destructive effect of Tropical Storm Debbie in mid-1994 caused the loss of 60% of the year's banana crop. Increased competition from Latin American bananas will probably further reduce market prices, exacerbating Saint Lucia's need to diversify its economy in coming years, e.g., by expanding tourism, manufacturing, and construction
National product: GDP—purchasing power parity—$610 million (1994 est.)
National product real growth rate: 2% (1994 est.)
National product per capita: $4,200 (1994 est.)
Inflation rate (consumer prices): 0.8% (1993)
Unemployment rate: 25% (1993 est.)
Budget:
revenues: $121 million
expenditures: $127 million, including capital expenditures of $104 million (1992 est.)
Exports: $122.8 million (f.o.b., 1992)
commodities: bananas 60%, clothing, cocoa, vegetables, fruits, coconut oil
partners: UK 56%, US 22%, CARICOM 19% (1991)
Imports: $276 million (f.o.b., 1992)
commodities: manufactured goods 21%, machinery and transportation equipment 21%, food and live animals, chemicals, fuels
partners: US 34%, CARICOM 17%, UK 14%, Japan 7%, Canada 4% (1991)
External debt: $96.4 million (1992 est.)
Industrial production: growth rate 3.5% (1990 est.); accounts for 12% of GDP
Electricity:
capacity: 20,000 kW
production: 112 million kWh
consumption per capita: 693 kWh (1993)
Industries: clothing, assembly of electronic components, beverages, corrugated cardboard boxes, tourism, lime processing, coconut processing
Agriculture: accounts for 14% of GDP and 43% of labor force; crops—bananas, coconuts, vegetables, citrus fruit, root crops, cocoa; imports food for the tourist industry
Illicit drugs: transit country for South American drugs destined for the US and Europe
Economic aid:
recipient: Western (non-US) countries, ODA and OOF bilateral commitments (1970-89), $120 million
Currency: 1 EC dollar (EC$) = 100 cents
Exchange rates: East Caribbean dollars (EC$) per US$1—2.70 (fixed rate since 1976)
Fiscal year: 1 April— 31 March

Transportation

Railroads: 0 km
Highways:
total: 760 km
paved: 500 km
unpaved: otherwise improved 260 km
Ports: Castries, Vieux Fort
Merchant marine: none
Airports:
total: 3
with paved runways 2,438 to 3,047 m: 1
with paved runways 1,524 to 2,437 m: 1
with paved runways under 914 m: 1

Communications

Telephone system: 9,500 telephones
local: low density (6 telephones/100 persons) but the system is automatically switched
intercity: no intercity traffic
international: direct microwave link with Martinique and Saint Vincent and the Grenadines; interisland troposcatter link to Barbados
Radio:
broadcast stations: AM 4, FM 1, shortwave 0
radios: NA
Television:
broadcast stations: 1 cable
televisions: NA

Defense Forces

Branches: Royal Saint Lucia Police Force, Coast Guard
Defense expenditures: $NA, NA% of GDP

Saint Pierre and Miquelon

(territorial collectivity of France)

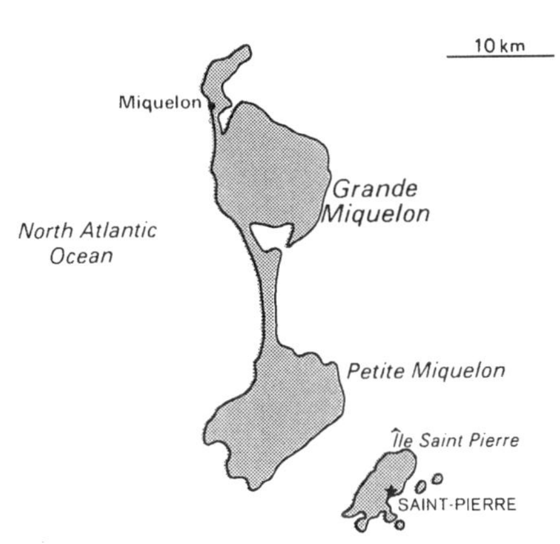

Geography

Location: Northern North America, islands in the North Atlantic Ocean, south of Newfoundland (Canada)
Map references: North America
Area:
total area: 242 sq km
land area: 242 sq km
comparative area: slightly less than 1.5 times the size of Washington, DC
note: includes eight small islands in the Saint Pierre and the Miquelon groups
Land boundaries: 0 km
Coastline: 120 km
Maritime claims:
exclusive economic zone: 200 nm
territorial sea: 12 nm
International disputes: focus of maritime boundary dispute between Canada and France
Climate: cold and wet, with much mist and fog; spring and autumn are windy
Terrain: mostly barren rock
Natural resources: fish, deepwater ports
Land use:
arable land: 13%
permanent crops: 0%
meadows and pastures: 0%
forest and woodland: 4%
other: 83%
Irrigated land: NA sq km
Environment:
current issues: NA
natural hazards: persistent fog throughout the year can be a maritime hazard
international agreements: NA
Note: vegetation scanty

People

Population: 6,757 (July 1995 est.)
Age structure:
0-14 years: NA
15-64 years: NA
65 years and over: NA
Population growth rate: 0.78% (1995 est.)
Birth rate: 13.02 births/1,000 population (1995 est.)
Death rate: 5.83 deaths/1,000 population (1995 est.)
Net migration rate: 0.59 migrant(s)/1,000 population (1995 est.)
Infant mortality rate: 10.77 deaths/1,000 live births (1995 est.)
Life expectancy at birth:
total population: 76 years
male: 74.4 years
female: 77.92 years (1995 est.)
Total fertility rate: 1.67 children born/woman (1995 est.)
Nationality:
noun: Frenchman(men), Frenchwoman(women)
adjective: French
Ethnic divisions: Basques and Bretons (French fishermen)
Religions: Roman Catholic 98%
Languages: French
Literacy: age 15 and over can read and write (1982)
total population: 99%
male: 99%
female: 99%
Labor force: 2,850 (1988)
by occupation: NA

Government

Names:
conventional long form: Territorial Collectivity of Saint Pierre and Miquelon
conventional short form: Saint Pierre and Miquelon
local long form: Departement de Saint-Pierre et Miquelon
local short form: Saint-Pierre et Miquelon
Digraph: SB
Type: territorial collectivity of France
Capital: Saint-Pierre
Administrative divisions: none (territorial collectivity of France)
Independence: none (territorial collectivity of France; has been under French control since 1763)
National holiday: National Day, Taking of the Bastille, 14 July
Constitution: 28 September 1958 (French Constitution)
Legal system: French law
Suffrage: 18 years of age; universal
Executive branch:
chief of state: President Francois MITTERRAND (since 21 May 1981)
head of government: Commissioner of the Republic Yves HENRY (since NA December 1993); President of the General Council Gerard GRIGNON (since NA April 1994)
cabinet: Council of Ministers
Legislative branch: unicameral
General Council: elections last held NA April 1994 (next to be held NA April 2000);

Saint Pierre and Miquelon

(continued)

results—percent of vote by party NA; seats—(19 total) seats by party NA
French Senate: elections last held NA September 1986 (next to be held NA September 1995); results—percent of vote by party NA; seats—(1 total) PS 1
French National Assembly: elections last held 21 and 28 March 1993 (next to be held NA June 1998); results—percent of vote by party NA; seats—(1 total) UDF 1
Judicial branch: Superior Tribunal of Appeals (Tribunal Superieur d'Appel)
Political parties and leaders: Socialist Party (PS), Albert PEN; Union for French Democracy (UDF/CDS), Gerard GRIGNON
Member of: FZ, WFTU
Diplomatic representation in US: none (territorial collectivity of France)
US diplomatic representation: none (territorial collectivity of France)
Flag: the flag of France is used

Economy

Overview: The inhabitants have traditionally earned their livelihood by fishing and by servicing fishing fleets operating off the coast of Newfoundland. The economy has been declining, however, because the number of ships stopping at Saint Pierre has dropped steadily over the years. In March 1989, an agreement between France and Canada set fish quotas for Saint Pierre's trawlers fishing in Canadian and Canadian-claimed waters for three years. The agreement settles a longstanding dispute that had virtually brought fish exports to a halt. The islands are heavily subsidized by France. Imports come primarily from Canada and France.
National product: GDP—purchasing power parity—$66 million (1993 est.)
National product real growth rate: NA%
National product per capita: $10,000 (1993 est.)
Inflation rate (consumer prices): NA%
Unemployment rate: 9.6% (1990)
Budget:
revenues: $18.3 million
expenditures: $18.3 million, including capital expenditures of $5.5 million (1989 est.)
Exports: $30 million (f.o.b., 1991 est.)
commodities: fish and fish products, fox and mink pelts
partners: US 58%, France 17%, UK 11%, Canada, Portugal (1990)
Imports: $82 million (c.i.f., 1991 est.)
commodities: meat, clothing, fuel, electrical equipment, machinery, building materials
partners: Canada, France, US, Netherlands, UK
External debt: $NA
Industrial production: growth rate NA%
Electricity:
capacity: 10,000 kW
production: 50 million kWh
consumption per capita: 6,013 kWh (1993)
Industries: fish processing and supply base for fishing fleets; tourism
Agriculture: vegetables, cattle, sheep, pigs for local consumption; fish catch of 20,500 metric tons (1989)
Economic aid:
recipient: Western (non-US) countries, ODA and OOF bilateral commitments (1970-89), $500 million
Currency: 1 French franc (F) = 100 centimes
Exchange rates: French francs (F) per US$1—5.2943 (January 1995), 5.520 (1994), 5.6632 (1993), 5.2938 (1992), 5.6421 (1991), 5.4453 (1990)
Fiscal year: calendar year

Transportation

Railroads: 0 km
Highways:
total: 120 km
paved: 60 km
unpaved: earth 60 km (1985)
Ports: Saint Pierre
Merchant marine: none
Airports:
total: 2
with paved runways 914 to 1,523 m: 2

Communications

Telephone system: 3,601 telephones
local: NA
intercity: NA
international: radio communication with most countries in the world; 1 satellite link in French domestic satellite system
Radio:
broadcast stations: AM 1, FM 3, shortwave 0
radios: NA
Television:
broadcast stations: 0
televisions: NA

Defense Forces

Note: defense is the responsibility of France

Saint Vincent and the Grenadines

Geography

Location: Caribbean, islands in the Caribbean Sea, north of Trinidad and Tobago
Map references: Central America and the Caribbean
Area:
total area: 340 sq km
land area: 340 sq km
comparative area: slightly less than twice the size of Washington, DC
Land boundaries: 0 km
Coastline: 84 km
Maritime claims:
contiguous zone: 24 nm
continental shelf: 200 nm
exclusive economic zone: 200 nm
territorial sea: 12 nm
International disputes: none
Climate: tropical; little seasonal temperature variation; rainy season (May to November)
Terrain: volcanic, mountainous; Soufriere volcano on the island of Saint Vincent
Natural resources: negligible
Land use:
arable land: 38%
permanent crops: 12%
meadows and pastures: 6%
forest and woodland: 41%
other: 3%
Irrigated land: 10 sq km (1989 est.)
Environment:
current issues: pollution of coastal waters and shorelines from discharges by pleasure yachts and other effluents; in some areas pollution is severe enough to make swimming prohibitive
natural hazards: hurricanes; Soufriere volcano is a constant threat
international agreements: party to—Endangered Species, Law of the Sea, Ship Pollution, Whaling; signed, but not ratified—Desertification
Note: the administration of the islands of the Grenadines group is divided between Saint Vincent and the Grenadines and Grenada

People

Population: 117,344 (July 1995 est.)
Age structure:
0-14 years: 34% (female 19,551; male 20,185)
15-64 years: 61% (female 35,565; male 35,573)
65 years and over: 5% (female 3,793; male 2,677) (July 1995 est.)
Population growth rate: 0.65% (1995 est.)
Birth rate: 19.62 births/1,000 population (1995 est.)
Death rate: 5.46 deaths/1,000 population (1995 est.)
Net migration rate: -7.63 migrant(s)/1,000 population (1995 est.)
Infant mortality rate: 17.2 deaths/1,000 live births (1995 est.)
Life expectancy at birth:
total population: 72.66 years
male: 71.15 years
female: 74.21 years (1995 est.)
Total fertility rate: 2.08 children born/woman (1995 est.)
Nationality:
noun: Saint Vincentian(s) or Vincentian(s)
adjective: Saint Vincentian or Vincentian
Ethnic divisions: African descent, Caucasian, East Indian, Carib Indian
Religions: Anglican, Methodist, Roman Catholic, Seventh-Day Adventist
Languages: English, French patois
Literacy: age 15 and over has ever attended school (1970)
total population: 96%
male: 96%
female: 96%
Labor force: 67,000 (1984 est.)
by occupation: NA

Government

Names:
conventional long form: none
conventional short form: Saint Vincent and the Grenadines
Digraph: VC
Type: constitutional monarchy
Capital: Kingstown
Administrative divisions: 6 parishes; Charlotte, Grenadines, Saint Andrew, Saint David, Saint George, Saint Patrick
Independence: 27 October 1979 (from UK)
National holiday: Independence Day, 27 October (1979)
Constitution: 27 October 1979
Legal system: based on English common law
Suffrage: 18 years of age; universal
Executive branch:
chief of state: Queen ELIZABETH II (since 6 February 1952), represented by Governor General David JACK (since 29 September 1989)
head of government: Prime Minister James F. MITCHELL (since 30 July 1984); Deputy Prime Minister Parnel CAMPBELL (since NA February 1994); note—governor general appoints leader of the majority party to position of prime minister
cabinet: Cabinet; appointed by the governor general on the advice of the prime minister
Legislative branch: unicameral
House of Assembly: elections last held 21 February 1994 (next to be held NA July 1999); results—percent of vote by party NA; seats—(21 total, 15 elected representatives and 6 appointed senators) NDP 12, ULP 3
Judicial branch: Eastern Caribbean Supreme Court (based on Saint Lucia)
Political parties and leaders: New Democratic Party (NDP), James MITCHELL, son of Prime Minister James F. MITCHELL; United People's Movement (UPM), Adrian SAUNDERS; National Reform Party (NRP), Joel MIGUEL; Unity Labor Party (ULP)—formed by the coalition of Saint Vincent Labor Party (SVLP) and the Movement for National Unity (MNU), Vincent BEACHE
Member of: ACP, C, CARICOM, CDB, ECLAC, FAO, G-77, GATT, IBRD, ICAO, ICFTU, ICRM, IDA, IFAD, IFRCS, IMF, IMO, INTELSAT (nonsignatory user), INTERPOL, IOC, ITU, OAS, OECS, OPANAL, UN, UNCTAD, UNESCO, UNIDO, UPU, WCL, WFTU, WHO
Diplomatic representation in US:
chief of mission: Ambassador Kingsley C.A. LAYNE
chancery: 1717 Massachusetts Avenue NW, Suite 102, Washington, DC 20036
telephone: [1] (202) 462-7806, 7846
FAX: [1] (202) 462-7807
US diplomatic representation: no official presence since the Ambassador resides in Bridgetown (Barbados)
Flag: three vertical bands of blue (hoist side), gold (double width), and green; the gold band bears three green diamonds arranged in a V pattern

Economy

Overview: Agriculture, dominated by banana production, is the most important sector of the economy. The services sector, based mostly on a growing tourist industry, is also important. In 1993, economic growth slowed to 1.4%, reflecting a sharp decline in agricultural production caused by drought. The government has been relatively unsuccessful at introducing new industries, and high unemployment rates of 35%-40% continue.
National product: GDP—purchasing power parity—$235 million (1994 est.)
National product real growth rate: 2% (1994 est.)
National product per capita: $2,000 (1994 est.)
Inflation rate (consumer prices): 4% (1993 est.)
Unemployment rate: 35%-40% (1994 est.)
Budget:
revenues: $66.2
expenditures: $77.3 million, including capital expenditures of $23 million (1993 est.)
Exports: $57.1 million (f.o.b., 1993)
commodities: bananas, eddoes and dasheen (taro), arrowroot starch, tennis racquets
partners: UK 54%, CARICOM 34%, US 10%
Imports: $134.6 million (f.o.b., 1993)
commodities: foodstuffs, machinery and equipment, chemicals and fertilizers, minerals and fuels
partners: US 36%, CARICOM 21%, UK 18%, Trinidad and Tobago 13%
External debt: $74.9 million (1993)
Industrial production: NA
Electricity:
capacity: 16,600 kW
production: 50 million kWh
consumption per capita: 436 kWh (1993)
Industries: food processing, cement, furniture, clothing, starch
Agriculture: accounts for 14% of GDP and 60% of labor force; provides bulk of exports; products—bananas, coconuts, sweet potatoes, spices; small numbers of cattle, sheep, hogs, goats; small fish catch used locally
Illicit drugs: transshipment point for South American drugs destined for the US and Europe
Economic aid:
recipient: US commitments, including Ex-Im (FY70-87), $11 million; Western (non-US) countries, ODA and OOF bilateral commitments (1970-89), $81 million
Currency: 1 EC dollar (EC$) = 100 cents
Exchange rates: East Caribbean dollars (EC$) per US$1—2.70 (fixed rate since 1976)
Fiscal year: calendar year

Transportation

Railroads: 0 km
Highways:
total: 1,000 km
paved: 300 km
unpaved: improved earth 400 km; unimproved earth 300 km
Ports: Kingstown
Merchant marine:
total: 580 ships (1,000 GRT or over) totaling 5,212,812 GRT/8,530,725 DWT
ships by type: bulk 106, cargo 289, chemical tanker 15, combination bulk 10, combination ore/oil 3, container 36, liquefied gas tanker 5, livestock carrier 2, oil tanker 53, passenger 2, passenger-cargo 1, refrigerated cargo 30, roll-on/roll-off cargo 25, short-sea passenger 1, specialized tanker 1, vehicle carrier 1
note: a flag of convenience registry; includes 16 countries among which are Croatia 49 ships,

Saint Vincent and the Grenadines

(continued)

Russia 23, Slovenia 11, China 8, Germany 3, Serbia 2, Latvia 1, Montenegro 1, Georgia 1, UAR 1
Airports:
total: 6
with paved runways 914 to 1,523 m: 2
with paved runways under 914 m: 4

Communications

Telephone system: 6,500 telephones; islandwide fully automatic telephone system
local: NA
intercity: VHF/UHF interisland links from Saint Vincent to the other islands of the Grenadines
international: VHF/UHF interisland links from Saint Vincent to Barbados; new SHF links to Grenada and to Saint Lucia
Radio:
broadcast stations: AM 2, FM 0, shortwave 0
radios: NA
Television:
broadcast stations: 1 cable
televisions: NA

Defense Forces

Branches: Royal Saint Vincent and the Grenadines Police Force, Coast Guard
Defense expenditures: $NA, NA% of GDP

San Marino

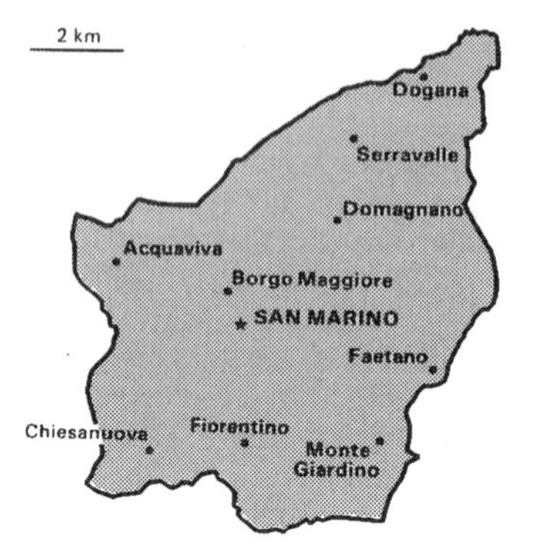

Geography

Location: Southern Europe, an enclave in central Italy
Map references: Europe
Area:
total area: 60 sq km
land area: 60 sq km
comparative area: about 0.3 times the size of Washington, DC
Land boundaries: total 39 km, Italy 39 km
Coastline: 0 km (landlocked)
Maritime claims: none; landlocked
International disputes: none
Climate: Mediterranean; mild to cool winters; warm, sunny summers
Terrain: rugged mountains
Natural resources: building stone
Land use:
arable land: 17%
permanent crops: 0%
meadows and pastures: 0%
forest and woodland: 0%
other: 83%
Irrigated land: NA sq km
Environment:
international agreements: NA
current issues: NA
natural hazards: party to—Biodiversity, Climate Change, Nuclear Test Ban; signed, but not ratified—Air Pollution
Note: landlocked; smallest independent state in Europe after the Holy See and Monaco; dominated by the Apennines

People

Population: 24,313 (July 1995 est.)
Age structure:
0-14 years: 16% (female 1,944; male 1,962)
15-64 years: 68% (female 8,243; male 8,354)
65 years and over: 16% (female 2,198; male 1,612) (July 1995 est.)
Population growth rate: 0.88% (1995 est.)
Birth rate: 10.98 births/1,000 population (1995 est.)
Death rate: 7.61 deaths/1,000 population (1995 est.)
Net migration rate: 5.47 migrant(s)/1,000 population (1995 est.)
Infant mortality rate: 5.6 deaths/1,000 live births (1995 est.)
Life expectancy at birth:
total population: 81.27 years
male: 77.26 years
female: 85.29 years (1995 est.)
Total fertility rate: 1.52 children born/woman (1995 est.)
Nationality:
noun: Sammarinese (singular and plural)
adjective: Sammarinese
Ethnic divisions: Sammarinese, Italian
Religions: Roman Catholic
Languages: Italian
Literacy: age 10 and over can read and write (1976)
total population: 96%
male: 97%
female: 95%
Labor force: 4,300 (est.)
by occupation: industry 42%, agriculture 3%

Government

Names:
conventional long form: Republic of San Marino
conventional short form: San Marino
local long form: Repubblica di San Marino
local short form: San Marino
Digraph: SM
Type: republic
Capital: San Marino
Administrative divisions: 9 municipalities (castelli, singular—castello); Acquaviva, Borgo Maggiore, Chiesanuova, Domagnano, Faetano, Fiorentino, Monte Giardino, San Marino, Serravalle
Independence: 301 AD (by tradition)
National holiday: Anniversary of the Foundation of the Republic, 3 September
Constitution: 8 October 1600; electoral law of 1926 serves some of the functions of a constitution
Legal system: based on civil law system with Italian law influences; has not accepted compulsory ICJ jurisdiction
Suffrage: 18 years of age; universal
Executive branch:
co-chiefs of state: Captain Regent Marino BOLLINI and Captain Regent Settimio LONFERNINI (for the period 1 April 1995-30 September 1995)
head of government: Secretary of State Gabriele GATTI (since July 1986)
cabinet: Congress of State
note: the popularly elected parliament (Great and General Council) selects two of its members to serve as the Captains Regent (Co-Chiefs of State) for a six-month period; they preside over meetings of the Great and General Council and its cabinet (Congress of State)

which has ten other members, all selected by the Great and General Council; assisting the Captains Regent are three Secretaries of State—Foreign Affairs, Internal Affairs, and Finance—and several additional secretaries; the Secretary of State for Foreign Affairs has come to assume many of the prerogatives of a prime minister
Legislative branch: unicameral
Great and General Council: (Consiglio Grande e Generale) elections last held 30 May 1993 (next to be held by May 1998); results—PDCS 41.4%, PSS 23.7%, PDP 18.6%, ADP 7.7%, MD 5.3%, RC 3.3%; seats—(60 total) PDCS 26, PSS 14, PDP 11, ADP 4, MD 3, RC 2
Judicial branch: Council of Twelve (Consiglio dei XII)
Political parties and leaders: Christian Democratic Party (PDCS), Cesare GASPERONI, secretary general; Democratic Progressive Party (PDP—formerly San Marino Communist Party (PSS)), Stefano MACINA, secretary general; San Marino Socialist Party (PSS), Maurizio RATTINI, secretary general; Democratic Movement (MD), Emilio Della BALDA; Popular Democratic Alliance (ADP); Communist Refoundation (RC), Guiseppe AMICHI, Renato FABBRI; Moderate Group, Alvaro SELVA; Social Democratic Party
Member of: CE, ECE, ICAO, ICFTU, ICRM, IFRCS, ILO, IMF, IOC, IOM (observer), ITU, NAM (guest), OSCE, UN, UNCTAD, UNESCO, UPU, WHO, WIPO, WTO
Diplomatic representation in US:
honorary consulate(s) general: Washington and New York
honorary consulate(s): Detroit
US diplomatic representation: no mission in San Marino, but the Consul General in Florence (Italy) is accredited to San Marino
Flag: two equal horizontal bands of white (top) and light blue with the national coat of arms superimposed in the center; the coat of arms has a shield (featuring three towers on three peaks) flanked by a wreath, below a crown and above a scroll bearing the word LIBERTAS (Liberty)

Economy

Overview: The tourist sector contributes over 50% of GDP. In 1993 more than 3 million tourists visited San Marino. The key industries are banking, wearing apparel, electronics, and ceramics. Main agricultural products are wine and cheeses. The per capita level of output and standard of living are comparable to those of Italy, which supplies much of its food.
National product: GDP—purchasing power parity—$380 million (1993 est.)
National product real growth rate: 2.4% (1993 est.)
National product per capita: $15,800 (1993 est.)
Inflation rate (consumer prices): 5.5% (1993)
Unemployment rate: 4.9% (December 1993)
Budget:
revenues: $275 million
expenditures: $275 million, including capital expenditures of $NA (1992 est.)
Exports: trade data are included with the statistics for Italy; commodities: building stone, lime, wood, chestnuts, wheat, wine, baked goods, hides, and ceramics
Imports: wide variety of consumer manufactures, food
External debt: $NA
Industrial production: growth rate NA%; accounts for 42% of labor force
Electricity: supplied by Italy
Industries: tourism, textiles, electronics, ceramics, cement, wine
Agriculture: employs 3% of labor force; products—wheat, grapes, maize, olives, meat, cheese, hides; small numbers of cattle, pigs, horses
Economic aid: $NA
Currency: 1 Italian lire (Lit) = 100 centesimi; note—also mints its own coins
Exchange rates: Italian lire (Lit) per US$1—1,609.5 (January 1995), 1,612.4 (1994), 1,573.7 (1993), 1,232.4 (1992), 1,240.6 (1991), 1,198.1 (1990)
Fiscal year: calendar year

Transportation

Railroads: 0 km
Highways:
total: 104 km
paved: NA
unpaved: NA
Ports: none
Airports: none

Communications

Telephone system: 11,700 telephones; automatic telephone system completely integrated into Italian system
local: NA
intercity: NA
international: microwave and cable links into Italian networks; no communication satellite facilities
Radio:
broadcast stations: AM NA, FM NA, shortwave NA
radios: NA
Television:
broadcast stations: NA; note—receives broadcasts from Italy
televisions: NA

Defense Forces

Branches: public security or police force
Defense expenditures: $3.7 million (1992 est.), 1% of GDP

Sao Tome and Principe

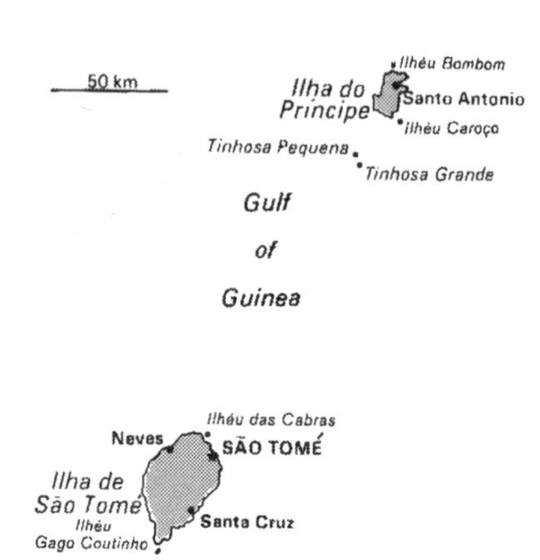

Geography

Location: Western Africa, island in the Atlantic Ocean, straddling the equator, west of Gabon
Map references: Africa
Area:
total area: 960 sq km
land area: 960 sq km
comparative area: slightly less than 5.5 times the size of Washington, DC
Land boundaries: 0 km
Coastline: 209 km
Maritime claims: measured from claimed archipelagic baselines
exclusive economic zone: 200 nm
territorial sea: 12 nm
International disputes: none
Climate: tropical; hot, humid; one rainy season (October to May)
Terrain: volcanic, mountainous
Natural resources: fish
Land use:
arable land: 1%
permanent crops: 20%
meadows and pastures: 1%
forest and woodland: 75%
other: 3%
Irrigated land: NA sq km
Environment:
current issues: deforestation; soil erosion and exhaustion
natural hazards: NA
international agreements: party to—Environmental Modification, Law of the Sea; signed, but not ratified—Biodiversity, Climate Change

People

Population: 140,423 (July 1995 est.)
Age structure:
0-14 years: 40% (female 27,995; male 28,452)
15-64 years: 55% (female 38,846; male 38,619)
65 years and over: 5% (female 3,615; male 2,896) (July 1995 est.)
Population growth rate: 2.62% (1995 est.)
Birth rate: 34.94 births/1,000 population (1995 est.)
Death rate: 8.7 deaths/1,000 population (1995 est.)
Net migration rate: 0 migrant(s)/1,000 population (1995 est.)
Infant mortality rate: 62.1 deaths/1,000 live births (1995 est.)
Life expectancy at birth:
total population: 63.65 years
male: 61.76 years
female: 65.59 years (1995 est.)
Total fertility rate: 4.44 children born/woman (1995 est.)
Nationality:
noun: Sao Tomean(s)
adjective: Sao Tomean
Ethnic divisions: mestico, angolares (descendents of Angolan slaves), forros (descendents of freed slaves), servicais (contract laborers from Angola, Mozambique, and Cape Verde), tongas (children of servicais born on the islands), Europeans (primarily Portuguese)
Religions: Roman Catholic, Evangelical Protestant, Seventh-Day Adventist
Languages: Portuguese (official)
Literacy: age 15 and over can read and write (1991)
total population: 73%
male: 85%
female: 62%
Labor force: most of population mainly engaged in subsistence agriculture and fishing; labor shortages on plantations and of skilled workers

Government

Names:
conventional long form: Democratic Republic of Sao Tome and Principe
conventional short form: Sao Tome and Principe
local long form: Republica Democratica de Sao Tome e Principe
local short form: Sao Tome e Principe
Digraph: TP
Type: republic
Capital: Sao Tome
Administrative divisions: 2 districts (concelhos, singular—concelho); Principe, Sao Tome
Independence: 12 July 1975 (from Portugal)
National holiday: Independence Day, 12 July (1975)
Constitution: approved March 1990; effective 10 September 1990
Legal system: based on Portuguese law system and customary law; has not accepted compulsory ICJ jurisdiction
Suffrage: 18 years of age; universal
Executive branch:
chief of state: President Miguel TROVOADA (since 4 April 1991); election last held 3 March 1991 (next to be held NA March 1996); results—Miguel TROVOADA was elected without opposition in Sao Tome's first multiparty presidential election
head of government: Prime Minister Carlos da GRACA (since 25 October 1994)
cabinet: Council of Ministers; appointed by the president on the proposal of the prime minister
Legislative branch: unicameral
National People's Assembly: (Assembleia Popular Nacional) parliament dissolved by President TROVOADA in July 1994; early elections held 2 October 1994; results—MLSTP 27%, PCD-GR 25.5%, ADI 25.5%; seats—(55 total) MLSTP 27, PCD-GR 14, ADI 14
Judicial branch: Supreme Court
Political parties and leaders: Party for Democratic Convergence-Reflection Group (PCD-GR), Daniel Lima Dos Santos DAIO, secretary general; Movement for the Liberation of Sao Tome and Principe (MLSTP), Carlos da GRACA; Christian Democratic Front (FDC), Alphonse Dos SANTOS; Democratic Opposition Coalition (CODO), leader NA; Independent Democratic Action (ADI), Gabriel COSTA; other small parties
Member of: ACP, AfDB, CEEAC, ECA, FAO, G-77, IBRD, ICAO, ICRM, IDA, IFAD, IFRCS, ILO, IMF, IMO, INTELSAT (nonsignatory user), INTERPOL, IOC, IOM (observer), ITU, NAM, OAU, UN, UNCTAD, UNESCO, UNIDO, UPU, WHO, WMO, WTO
Diplomatic representation in US: Sao Tome and Principe has no embassy in the US, but does have a Permanent Mission to the UN, headed by First Secretary Domingos AUGUSTO Ferreira, located at 122 East 42nd Street, Suite 1604, New York, NY 10168, telephone [1] (212) 697-4211
US diplomatic representation: ambassador to Gabon is accredited to Sao Tome and Principe on a nonresident basis and makes periodic visits to the islands
Flag: three horizontal bands of green (top), yellow (double width), and green with two black five-pointed stars placed side by side in the center of the yellow band and a red isosceles triangle based on the hoist side; uses the popular pan-African colors of Ethiopia

Economy

Overview: This small poor island economy has remained dependent on cocoa since independence 20 years ago. Since then, however, cocoa production has gradually declined because of drought and mismanagement, so that by 1987 annual output had fallen from 10,000 tons to 3,900 tons. As a result, a shortage of cocoa for export has created a serious balance-of-payments problem. Production of less important crops,

such as coffee, copra, and palm kernels, has also declined. The value of imports generally exceeds that of exports by a ratio of 4:1 or more. The emphasis on cocoa production at the expense of other food crops has meant that Sao Tome has to import 90% of food needs. It also has to import all fuels and most manufactured goods. Over the years, Sao Tome has been unable to service its external debt and has had to depend on concessional aid and debt rescheduling. Considerable potential exists for development of a tourist industry, and the government has taken steps to expand facilities in recent years. The government also has attempted to reduce price controls and subsidies and to encourage market-based mechanisms, e. g., to facilitate the distribution of imported food. Annual GDP growth is estimated in the 3%-4% range for 1994-96.
National product: GDP—purchasing power parity—$133 million (1993 est.)
National product real growth rate: NA%
National product per capita: $1,000 (1993 est.)
Inflation rate (consumer prices): 27% (1992 est.)
Unemployment rate: NA%
Budget:
revenues: $10.2 million
expenditures: $36.8 million, including capital expenditures of $22.5 million (1989 est.)
Exports: $5.5 million (f.o.b., 1993 est.)
commodities: cocoa 78%, copra, coffee, palm oil (1992)
partners: Netherlands, Germany, China, Portugal
Imports: $31.5 million (f.o.b., 1993 est.)
commodities: machinery and electrical equipment 44%, food products 18%, petroleum 11% (1992)
partners: Portugal, Japan, Spain, France, Angola
External debt: $237 million (1993)
Industrial production: growth rate 1% (1991); accounts for 7% of GDP
Electricity:
capacity: 5,000 kW
production: 17 million kWh
consumption per capita: 105 kWh (1993)
Industries: light construction, shirts, soap, beer, fisheries, shrimp processing
Agriculture: accounts for 25% of GDP; dominant sector of economy, primary source of exports; cash crops—cocoa, coconuts, palm kernels, coffee; food products—bananas, papaya, beans, poultry, fish; not self-sufficient in food grain and meat
Economic aid:
recipient: US commitments, including Ex-Im (FY70-89), $8 million; Western (non-US) countries, ODA and OOF bilateral commitments (1970-89), $89 million
Currency: 1 dobra (Db) = 100 centimos
Exchange rates: dobras (Db) per US$1—129.59 (1 July 1993), 230 (1992), 260.0 (November 1991), 122.48 (December 1988), 72.827 (1987)
Fiscal year: calendar year

Transportation

Railroads: 0 km
Highways:
total: 300 km
paved: 200 km
unpaved: 100 km
note: roads on Principe are mostly unpaved and in need of repair
Ports: Santo Antonio, Sao Tome
Merchant marine:
total: 1 cargo ship (1,000 GRT or over) totaling 1,096 GRT/1,105 DWT
Airports:
total: 2
with paved runways 1,524 to 2,437 m: 1
with paved runways 914 to 1,523 m: 1

Communications

Telephone system: NA; minimal system
local: NA
intercity: NA
international: 1 Atlantic Ocean INTELSAT earth station
Radio:
broadcast stations: AM 1, FM 2, shortwave 0
radios: NA
Television:
broadcast stations: 0
televisions: NA

Defense Forces

Branches: Army, Navy, National Police
Manpower availability: males age 15-49 33,789; males fit for military service 17,752 (1995 est.)
Defense expenditures: $NA, NA% of GDP

Saudi Arabia

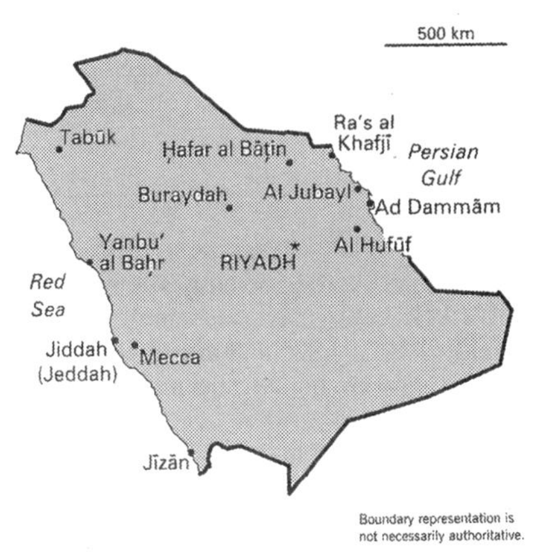

Geography

Location: Middle East, bordering the Persian Gulf and the Red Sea, north of Yemen
Map references: Middle East
Area:
total area: 1,960,582 sq km
land area: 1,960,582 sq km
comparative area: slightly less than one-fourth the size of the US
Land boundaries: total 4,415 km, Iraq 814 km, Jordan 728 km, Kuwait 222 km, Oman 676 km, Qatar 60 km, UAE 457 km, Yemen 1,458 km
Coastline: 2,640 km
Maritime claims:
contiguous zone: 18 nm
continental shelf: not specified
territorial sea: 12 nm
International disputes: large section of boundary with Yemen not defined; status of boundary with UAE not final; Kuwaiti ownership of Qaruh and Umm al Maradim islands is disputed by Saudi Arabia
Climate: harsh, dry desert with great extremes of temperature
Terrain: mostly uninhabited, sandy desert
Natural resources: petroleum, natural gas, iron ore, gold, copper
Land use:
arable land: 1%
permanent crops: 0%
meadows and pastures: 39%
forest and woodland: 1%
other: 59%
Irrigated land: 4,350 sq km (1989 est.)
Environment:
current issues: desertification, depletion of underground water resources; the lack of perennial rivers or permanent water bodies has prompted the development of extensive seawater desalination facilities; coastal pollution from oil spills
natural hazards: frequent sand and dust storms
international agreements: party to—Climate Change, Hazardous Wastes, Ozone Layer

Protection; signed, but not ratified—Law of the Sea
Note: extensive coastlines on Persian Gulf and Red Sea provide great leverage on shipping (especially crude oil) through Persian Gulf and Suez Canal

People

Population: 18,729,576 (July 1995 est.)
note: a 1992 census gives the number of Saudi citizens as 12,304,835 and the number of residents who are not citizens as 4,624,459
Age structure:
0-14 years: 43% (female 3,952,573; male 4,065,224)
15-64 years: 55% (female 4,078,001; male 6,219,737)
65 years and over: 2% (female 203,372; male 210,669) (July 1995 est.)
Population growth rate: 3.68% (1995 est.)
Birth rate: 38.78 births/1,000 population (1995 est.)
Death rate: 5.54 deaths/1,000 population (1995 est.)
Net migration rate: 3.56 migrant(s)/1,000 population (1995 est.)
Infant mortality rate: 48.9 deaths/1,000 live births (1995 est.)
Life expectancy at birth:
total population: 68.5 years
male: 66.79 years
female: 70.3 years (1995 est.)
Total fertility rate: 6.48 children born/woman (1995 est.)
Nationality:
noun: Saudi(s)
adjective: Saudi or Saudi Arabian
Ethnic divisions: Arab 90%, Afro-Asian 10%
Religions: Muslim 100%
Languages: Arabic
Literacy: age 15 and over can read and write (1990)
total population: 62%
male: 73%
female: 48%
Labor force: 5 million-6 million
by occupation: government 34%, industry and oil 28%, services 22%, agriculture 16%

Government

Names:
conventional long form: Kingdom of Saudi Arabia
conventional short form: Saudi Arabia
local long form: Al Mamlakah al Arabiyah as Suudiyah
local short form: Al Arabiyah as Suudiyah
Digraph: SA
Type: monarchy
Capital: Riyadh
Administrative divisions: 13 provinces (mintaqah, singular—mintaqat); Al Bahah, Al Hudud ash Shamaliyah, Al Jawf, Al Madinah, Al Qasim, Ar Riyad, Ash Sharqiyah, Asir, Hail, Jizan, Makkah, Najran, Tabuk
Independence: 23 September 1932 (unification)
National holiday: Unification of the Kingdom, 23 September (1932)
Constitution: none; governed according to Shari'a (Islamic law)
Legal system: based on Islamic law, several secular codes have been introduced; commercial disputes handled by special committees; has not accepted compulsory ICJ jurisdiction
Suffrage: none
Executive branch:
chief of state and head of government: King and Prime Minister FAHD bin Abd al-Aziz Al Saud (since 13 June 1982); Crown Prince and First Deputy Prime Minister ABDALLAH bin Abd al-Aziz Al Saud (half-brother to the King, appointed heir to the throne 13 June 1982)
cabinet: Council of Ministers; dominated by royal family members appointed by the king
Legislative branch: a consultative council composed of 60 members and a chairman who are appointed by the King for a term of four years
Judicial branch: Supreme Council of Justice
Political parties and leaders: none allowed
Member of: ABEDA, AfDB, AFESD, AL, AMF, CCC, ESCWA, FAO, G-19, G-77, GCC, IAEA, IBRD, ICAO, ICC, ICRM, IDA, IDB, IFAD, IFC, IFRCS, ILO, IMF, IMO, INMARSAT, INTELSAT, INTERPOL, IOC, ISO, ITU, NAM, OAPEC, OAS (observer), OIC, OPEC, UN, UNCTAD, UNESCO, UNIDO, UPU, WFTU, WHO, WIPO, WMO
Diplomatic representation in US:
chief of mission: Ambassador BANDAR bin Sultan Abd al-Aziz Al Saud
chancery: 601 New Hampshire Avenue NW, Washington, DC 20037
telephone: [1] (202) 342-3800
consulate(s) general: Houston, Los Angeles, and New York
US diplomatic representation:
chief of mission: Ambassador Raymond E. MABUS, Jr.
embassy: Collector Road M, Diplomatic Quarter, Riyadh
mailing address: American Embassy, Unit 61307, Riyadh; International Mail: P. O. Box 94309, Riyadh 11693; APO AE 09803-1307
telephone: [966] (1) 488-3800
FAX: [966] (1) 482-4364
consulate(s) general: Dhahran, Jiddah (Jeddah)
Flag: green with large white Arabic script (that may be translated as There is no God but God; Muhammad is the Messenger of God) above a white horizontal saber (the tip points to the hoist side); green is the traditional color of Islam

Economy

Overview: This is a well-to-do oil-based economy with strong government controls over major economic activities. About 46% of GDP comes from the private sector. Economic (as well as political) ties with the US are especially strong. The petroleum sector accounts for roughly 75% of budget revenues, 35% of GDP, and almost all export earnings. Saudi Arabia has the largest reserves of petroleum in the world (26% of the proved total), ranks as the largest exporter of petroleum, and plays a leading role in OPEC. For the 1990s the government intends to bring its budget, which has been in deficit since 1983, back into balance, and to encourage private economic activity. Roughly four million foreign workers play an important role in the Saudi economy, for example, in the oil and banking sectors. For about a decade, Saudi Arabia's domestic and international outlays have outstripped its income, and the government has cut its foreign assistance and is beginning to rein in domestic programs. For 1995, the country looks for improvement in oil prices and will continue its policies of restraining public spending and encouraging non-oil exports.
National product: GDP—purchasing power parity—$173.1 billion (1994 est.)
National product real growth rate: -3% (1994 est.)
National product per capita: $9,510 (1994 est.)
Inflation rate (consumer prices): 1% (1993 est.)
Unemployment rate: 6.5% (1992 est.)
Budget:
revenues: $39 billion
expenditures: $50 billion, including capital expenditures of $7.5 billion (1993 est.)
Exports: $39.4 billion (f.o.b., 1993 est.)
commodities: petroleum and petroleum products 92%
partners: US 20%, Japan 18%, Singapore 5%, France 5%, South Korea 5% (1992)
Imports: $28.9 billion (f.o.b., 1993 est.)
commodities: machinery and equipment, chemicals, foodstuffs, motor vehicles, textiles
partners: US 21%, Japan 14%, UK 11%, Germany 8%, Italy 6%, France 5% (1992)
External debt: $18.9 billion (December 1989 est., includes short-term trade credits)
Industrial production: growth rate 20% (1991 est.); accounts for 35% of GDP, including petroleum
Electricity:
capacity: 17,550,000 kW
production: 46 billion kWh
consumption per capita: 2,430 kWh (1993)
Industries: crude oil production, petroleum refining, basic petrochemicals, cement, two small steel-rolling mills, construction, fertilizer, plastics

Agriculture: accounts for about 10% of GDP, 16% of labor force; subsidized by government; products—wheat, barley, tomatoes, melons, dates, citrus fruit, mutton, chickens, eggs, milk; approaching self-sufficiency in food
Illicit drugs: death penalty for traffickers; increasing consumption of heroin and cocaine
Economic aid:
donor: pledged bilateral aid (1979-89), $64.7 billion; pledged $100 million in 1993 to fund reconstruction of Lebanon
Currency: 1 Saudi riyal (SR) = 100 halalah
Exchange rates: Saudi riyals (SR) per US$1—3.7450 (fixed rate since late 1986), 3.7033 (1986)
Fiscal year: calendar year

Transportation

Railroads:
total: 1,390 km
standard gauge: 1,390 km 1.435-m gauge (448 km double track)
Highways:
total: 151,530 km
paved: 60,610 km
unpaved: 90,920 km (1992 est.)
Pipelines: crude oil 6,400 km; petroleum products 150 km; natural gas 2,200 km (includes natural gas liquids 1,600 km)
Ports: Ad Dammam, Al Jubayl, Duba, Jiddah, Jizan, Rabigh, Ras al Khafji, Ras al Mishab, Ras Tanura, Yanbu' al Bahr, Yanbu' al Sinaiyah
Merchant marine:
total: 71 ships (1,000 GRT or over) totaling 855,452 GRT/1,233,477 DWT
ships by type: bulk 1, cargo 12, chemical tanker 5, container 3, liquefied gas tanker 1, livestock carrier 4, oil tanker 22, passenger 1, refrigerated cargo 4, roll-on/roll-off cargo 11, short-sea passenger 7
Airports:
total: 211
with paved runways over 3,047 m: 30
with paved runways 2,438 to 3,047 m: 12
with paved runways 1,524 to 2,437 m: 22
with paved runways 914 to 1,523 m: 4
with paved runways under 914 m: 21
with unpaved runways 2,438 to 3,047 m: 6
with unpaved runways 1,524 to 2,438 m: 73
with unpaved runways 914 to 1,523 m: 43

Communications

Telephone system: 1,624,000 telephones; modern system
local: NA
intercity: extensive microwave and coaxial and fiber optic cable systems
international: microwave radio relay to Bahrain, Jordan, Kuwait, Qatar, UAE, Yemen, and Sudan; coaxial cable to Kuwait and Jordan; submarine cable to Djibouti, Egypt and Bahrain; earth stations—5 INTELSAT (3 Atlantic Ocean and 2 Indian Ocean), 1 ARABSAT, and 1 INMARSAT
Radio:
broadcast stations: AM 43, FM 13, shortwave 0
radios: NA
Television:
broadcast stations: 80
televisions: NA

Defense Forces

Branches: Land Force (Army), Navy, Air Force, Air Defense Force, National Guard, Coast Guard, Frontier Forces, Special Security Force, Public Security Force
Manpower availability: males age 15-49 5,303,679; males fit for military service 2,949,842; males reach military age (17) annually 164,220 (1995 est.)
Defense expenditures: exchange rate conversion—$17.2 billion, 13.8% of GDP (1994)

Senegal

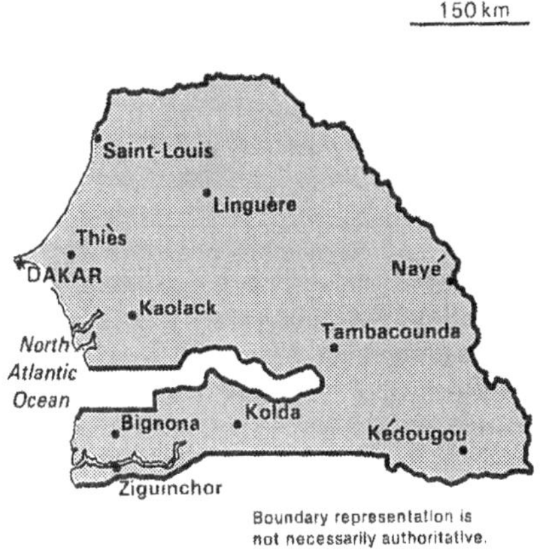

Geography

Location: Western Africa, bordering the North Atlantic Ocean, between Guinea-Bissau and Mauritania
Map references: Africa
Area:
total area: 196,190 sq km
land area: 192,000 sq km
comparative area: slightly smaller than South Dakota
Land boundaries: total 2,640 km, The Gambia 740 km, Guinea 330 km, Guinea-Bissau 338 km, Mali 419 km, Mauritania 813 km
Coastline: 531 km
Maritime claims:
contiguous zone: 24 nm
continental shelf: 200 nm or to the edge of the continental margin
exclusive economic zone: 200 nm
territorial sea: 12 nm
International disputes: short section of the boundary with The Gambia is indefinite; boundary with Mauritania in dispute;
Climate: tropical; hot, humid; rainy season (December to April) has strong southeast winds; dry season (May to November) dominated by hot, dry harmattan wind
Terrain: generally low, rolling, plains rising to foothills in southeast
Natural resources: fish, phosphates, iron ore
Land use:
arable land: 27%
permanent crops: 0%
meadows and pastures: 30%
forest and woodland: 31%
other: 12%
Irrigated land: 1,800 sq km (1989 est.)
Environment:
current issues: wildlife populations threatened by poaching; deforestation; overgrazing; soil erosion; desertification; overfishing
natural hazards: lowlands seasonally flooded; periodic droughts
international agreements: party to—Biodiversity, Climate Change, Endangered

Species, Hazardous Wastes, Law of the Sea, Marine Life Conservation, Nuclear Test Ban, Ozone Layer Protection, Wetlands, Whaling; signed, but not ratified—Desertification, Marine Dumping
Note: The Gambia is almost an enclave of Senegal

People

Population: 9,007,080 (July 1995 est.)
Age structure:
0-14 years: 45% (female 2,004,514; male 2,021,251)
15-64 years: 52% (female 2,398,609; male 2,301,236)
65 years and over: 3% (female 140,128; male 141,342) (July 1995 est.)
Population growth rate: 3.12% (1995 est.)
Birth rate: 42.87 births/1,000 population (1995 est.)
Death rate: 11.64 deaths/1,000 population (1995 est.)
Net migration rate: NA migrant(s)/1,000 population (1995 est.)
Infant mortality rate: 73.6 deaths/1,000 live births (1995 est.)
Life expectancy at birth:
total population: 57.16 years
male: 55.65 years
female: 58.71 years (1995 est.)
Total fertility rate: 6.03 children born/woman (1995 est.)
Nationality:
noun: Senegalese (singular and plural)
adjective: Senegalese
Ethnic divisions: Wolof 36%, Fulani 17%, Serer 17%, Toucouleur 9%, Diola 9%, Mandingo 9%, European and Lebanese 1%, other 2%
Religions: Muslim 92%, indigenous beliefs 6%, Christian 2% (mostly Roman Catholic)
Languages: French (official), Wolof, Pulaar, Diola, Mandingo
Literacy: age 15 and over can read and write (1988)
total population: 27%
male: 37%
female: 18%
Labor force: 2.509 million (77% are engaged in subsistence farming; 175,000 wage earners)
by occupation: private sector 40%, government and parapublic 60%

Government

Names:
conventional long form: Republic of Senegal
conventional short form: Senegal
local long form: Republique du Senegal
local short form: Senegal
Digraph: SG
Type: republic under multiparty democratic rule
Capital: Dakar
Administrative divisions: 10 regions (regions, singular—region); Dakar, Diourbel, Fatick, Kaolack, Kolda, Louga, Saint-Louis, Tambacounda, Thies, Ziguinchor
Independence: 20 August 1960 (from France; The Gambia and Senegal signed an agreement on 12 December 1981 that called for the creation of a loose confederation to be known as Senegambia, but the agreement was dissolved on 30 September 1989)
National holiday: Independence Day, 4 April (1960)
Constitution: 3 March 1963, revised 1991
Legal system: based on French civil law system; judicial review of legislative acts in Supreme Court, which also audits the government's accounting office; has not accepted compulsory ICJ jurisdiction
Suffrage: 18 years of age; universal
Executive branch:
chief of state: President Abdou DIOUF (since 1 January 1981); election last held 21 February 1993 (next to be held NA February 2000); results—Abdou DIOUF (PS) 58.4%, Abdoulaye WADE (PDS) 32.03%, other 9.57%
head of government: Prime Minister Habib THIAM (since 7 April 1991)
cabinet: Council of Ministers; appointed by the prime minister in consultation with the president
Legislative branch: unicameral
National Assembly (Assemblee Nationale): elections last held 9 May 1993 (next to be held NA May 1998); results—PS 70%, PDS 23%, other 7%; seats—(120 total) PS 84, PDS 27, LD-MPT 3, Let Us Unite Senegal 3, PIT 2, UDS-R 1
Judicial branch: Supreme Court (Cour Supreme)
Political parties and leaders: Socialist Party (PS), President Abdou DIOUF; Senegalese Democratic Party (PDS), Abdoulaye WADE; Democratic League-Labor Party Movement (LD-MPT), Dr. Abdoulaye BATHILY; Independent Labor Party (PIT), Amath DANSOKHO; Senegalese Democratic Union-Renewal (UDS-R), Mamadou Puritain FALL; Let Us Unite Senegal (Coalition of African Party for Democracy and Socialism and National Democratic Rally); other small uninfluential parties
Other political or pressure groups: students; teachers; labor; Muslim Brotherhoods
Member of: ACCT, ACP, AfDB, CCC, CEAO, ECA, ECOWAS, FAO, FZ, G-15, G-77, GATT, IAEA, IBRD, ICAO, ICC, ICFTU, ICRM, IDA, IDB, IFAD, IFC, IFRCS, ILO, IMF, IMO, INMARSAT, INTELSAT, INTERPOL, IOC, IOM (observer), ITU, NAM, OAU, OIC, PCA, UN, UNAMIR, UNCTAD, UNESCO, UNIDO, UNIKOM, UNMIH, UNOMUR, UPU, WADB, WCL, WFTU, WHO, WIPO, WMO, WTO
Diplomatic representation in US:
chief of mission: Ambassador Mamadou Mansour SECK
chancery: 2112 Wyoming Avenue NW, Washington, DC 20008
telephone: [1] (202) 234-0540, 0541
US diplomatic representation:
chief of mission: Ambassador Mark JOHNSON
embassy: Avenue Jean XXIII at the corner of Avenue Kleber, Dakar
mailing address: B. P. 49, Dakar
telephone: [221] 23 42 96, 23 34 24
FAX: [221] 22 29 91
Flag: three equal vertical bands of green (hoist side), yellow, and red with a small green five-pointed star centered in the yellow band; uses the popular pan-African colors of Ethiopia

Economy

Overview: In 1994 Senegal embarked on its most concerted structural adjustment effort yet to exploit the 50% devaluation of the currencies of the 14 Francophone African nations on 12 January. After years of foot-dragging, the government finally passed a liberalized labor code which should significantly help lower the cost of labor and improve the manufacturing sector's competitiveness. Inroads also have been made in closing tax loopholes and eliminating monopoly power in several sectors. At the same time the government is holding the line on current fiscal expenditure under the watchful eyes of international organizations on which it depends for substantial support. A bumper peanut crop—Senegal's main source of foreign exchange—coincided with an improvement of international prices and probably resulted in 1994 in a doubling of earnings over 1993. The country's narrow resource base, environmental degradation, and untamed population growth will continue to hold back growth in living standards over the medium term.
National product: GDP—purchasing power parity—$12.3 billion (1993 est.)
National product real growth rate: -2% (1993 est.)
National product per capita: $1,450 (1993 est.)
Inflation rate (consumer prices): -1.8% (1991 est.)
Unemployment rate: NA%
Budget:
revenues: $1.2 billion
expenditures: $1.2 billion, including capital expenditures of $269 million (1992 est.)
Exports: $904 million (f.o.b., 1991 est.)
commodities: fish, ground nuts (peanuts), petroleum products, phosphates, cotton
partners: France, other EC countries, Cote d'Ivoire, Mali
Imports: $1.2 billion (c.i.f., 1991 est.)

commodities: foods and beverages, consumer goods, capital goods, petroleum
partners: France, other EC countries, Nigeria, Cote d'Ivoire, Algeria, China, Japan
External debt: $2.9 billion (1990)
Industrial production: growth rate 1.9% (1991); accounts for 15% of GDP
Electricity:
capacity: 230,000 kW
production: 720 million kWh
consumption per capita: 79 kWh (1993)
Industries: agricultural and fish processing, phosphate mining, petroleum refining, building materials
Agriculture: accounts for 20% of GDP; major products—peanuts (cash crop), millet, corn, sorghum, rice, cotton, tomatoes, green vegetables; estimated two-thirds self-sufficient in food; fish catch of 354,000 metric tons in 1990
Illicit drugs: transshipment point for Southwest and Southeast Asian heroin moving to Europe and North America
Economic aid:
recipient: US commitments, including Ex-Im (FY70-89), $551 million; Western (non-US) countries, ODA and OOF bilateral commitments (1970-89), $5.23 billion; OPEC bilateral aid (1979-89), $589 million; Communist countries (1970-89), $295 million
Currency: 1 CFA franc (CFAF) = 100 centimes
Exchange rates: Communaute Financiere Africaine francs (CFAF) per US$1—529.43 (January 1995), 555.20 (1994), 283.16 (1993), 264.69 (1992), 282.11 (1991), 272.26 (1990)
note: the official rate is pegged to the French franc, and beginning 12 January 1994, the CFA franc was devalued to CFAF 100 per French franc from CFAF 50 at which it had been fixed since 1948
Fiscal year: calendar year

Transportation

Railroads:
total: 905 km
narrow gauge: 905 km 1.000-meter gauge (70 km double track)
Highways:
total: 14,007 km
paved: 3,777 km
unpaved: crushed stone, improved earth 10,230 km
Inland waterways: 897 km total; 785 km on the Senegal, 112 km on the Saloum
Ports: Dakar, Kaolack, Matam, Podor, Richard-Toll, Saint-Louis, Ziguinchor
Merchant marine:
total: 1 bulk ship (1,000 GRT or over) totaling 1,995 GRT/3,775 DWT
Airports:
total: 24
with paved runways over 3,047 m: 1
with paved runways 1,524 to 2,437 m: 9
with paved runways 914 to 1,523 m: 1
with paved runways under 914 m: 2
with unpaved runways 1,524 to 2,438 m: 4
with unpaved runways 914 to 1,523 m: 7

Communications

Telephone system: NA telephones; above-average urban system
local: NA
intercity: microwave and cable
international: 3 submarine cables; 1 INTELSAT (Atlantic Ocean) earth station
Radio:
broadcast stations: AM 8, FM 0, shortwave 0
radios: NA
Television:
broadcast stations: 1
televisions: NA

Defense Forces

Branches: Army, Navy, Air Force, National Gendarmerie, National Police (Surete Nationale)
Manpower availability: males age 15-49 2,021,019; males fit for military service 1,054,855; males reach military age (18) annually 96,589 (1995 est.)
Defense expenditures: exchange rate conversion—$134 million, 2.1% of GDP (1993)

Serbia and Montenegro

Subotica
Novi Sad
BELGRADE
Serbia
Niš
Montenegro
Priština
Podgorica
Adriatic Sea
100 km
Serbia and Montenegro have asserted the formation of a joint independent state, but this entity has not been formally recognized as a state by the United States.

Note: Serbia and Montenegro have asserted the formation of a joint independent state, but this entity has not been formally recognized as a state by the US; the US view is that the Socialist Federal Republic of Yugoslavia (SFRY) has dissolved and that none of the successor republics represents its continuation.

Geography

Location: Southeastern Europe, bordering the Adriatic Sea, between Albania and Bosnia and Herzegovina
Map references: Ethnic Groups in Eastern Europe, Europe
Area:
total area: 102,350 sq km
land area: 102,136 sq km
comparative area: slightly larger than Kentucky
note: Serbia has a total area and a land area of 88,412 sq km making it slightly larger than Maine; Montenegro has a total area of 13,938 sq km and a land area of 13,724 sq km making it slightly larger than Connecticut
Land boundaries: total 2,246 km, Albania 287 km (114 km with Serbia; 173 km with Montenegro), Bosnia and Herzegovina 527 km (312 km with Serbia; 215 km with Montenegro), Bulgaria 318 km, Croatia (north) 241 km, Croatia (south) 25 km, Hungary 151 km, The Former Yugoslav Republic of Macedonia 221 km, Romania 476 km
note: the internal boundary between Montenegro and Serbia is 211 km
Coastline: 199 km (Montenegro 199 km, Serbia 0 km)
Maritime claims: NA
International disputes: Sandzak region bordering northern Montenegro and southeastern Serbia—Muslims seeking autonomy; disputes with Bosnia and Herzegovina and Croatia over Serbian populated areas; Albanian majority in Kosovo seeks independence from Serbian Republic
Climate: in the north, continental climate (cold winter and hot, humid summers with well

distributed rainfall); central portion, continental and Mediterranean climate; to the south, Adriatic climate along the coast, hot, dry summers and autumns and relatively cold winters with heavy snowfall inland
Terrain: extremely varied; to the north, rich fertile plains; to the east, limestone ranges and basins; to the southeast, ancient mountain and hills; to the southwest, extremely high shoreline with no islands off the coast
Natural resources: oil, gas, coal, antimony, copper, lead, zinc, nickel, gold, pyrite, chrome
Land use:
arable land: 30%
permanent crops: 5%
meadows and pastures: 20%
forest and woodland: 25%
other: 20%
Irrigated land: NA sq km
Environment:
current issues: pollution of coastal waters from sewage outlets, especially in tourist-related areas such as Kotor; air pollution around Belgrade and other industrial cities; water pollution from industrial wastes dumped into the Sava which flows into the Danube
natural hazards: destructive earthquakes
international agreements: NA
Note: controls one of the major land routes from Western Europe to Turkey and the Near East; strategic location along the Adriatic coast

People

Population:
total population: 11,101,833 (July 1995 est.)
Montenegro: 708,248 (July 1995 est.)
Serbia: 10,393,585 (July 1995 est.)
Age structure:
Montenegro:
0-14 years: 22% (female 77,498; male 82,005)
15-64 years: 68% (female 236,987; male 241,397)
65 years and over: 10% (female 41,625; male 28,736) (July 1995 est.)
Serbia:
0-14 years: 22% (female 1,095,121; male 1,173,224)
15-64 years: 66% (female 3,431,823; male 3,483,066)
65 years and over: 12% (female 699,488; male 510,863) (July 1995 est.)
Population growth rate:
Montenegro: 0.79% (1995 est.)
Serbia: 0.51% (1995 est.)
Birth rate:
Montenegro: 14.39 births/1,000 population (1995 est.)
Serbia: 14.15 births/1,000 population (1995 est.)
Death rate:
Montenegro: 5.7 deaths/1,000 population (1995 est.)
Serbia: 8.72 deaths/1,000 population (1995 est.)
Net migration rate:
Montenegro: -0.78 migrant(s)/1,000 population (1995 est.)
Serbia: -0.36 migrant(s)/1,000 population (1995 est.)
Infant mortality rate:
Montenegro: 9.8 deaths/1,000 live births (1995 est.)
Serbia: 18.6 deaths/1,000 live births (1995 est.)
Life expectancy at birth:
Montenegro:
total population: 79.56 years
male: 76.69 years
female: 82.61 years (1995 est.)
Serbia:
total population: 73.94 years
male: 71.4 years
female: 76.68 years (1995 est.)
Total fertility rate:
Montenegro: 1.79 children born/woman (1995 est.)
Serbia: 2 children born/woman (1995 est.)
Nationality:
noun: Serb(s) and Montenegrin(s)
adjective: Serbian and Montenegrin
Ethnic divisions: Serbs 63%, Albanians 14%, Montenegrins 6%, Hungarians 4%, other 13%
Religions: Orthodox 65%, Muslim 19%, Roman Catholic 4%, Protestant 1%, other 11%
Languages: Serbo-Croatian 95%, Albanian 5%
Literacy: NA%
Labor force: 2,640,909
by occupation: industry, mining 40% (1990)

Government

Names:
conventional long form: none
conventional short form: Serbia and Montenegro
local long form: none
local short form: Srbija-Crna Gora
Digraph:
Serbia: SR
Montenegro: MW
Type: republic
Capital: Belgrade
Administrative divisions: 2 republics (pokajine, singular—pokajina); and 2 nominally autonomous provinces*; Kosovo*, Montenegro, Serbia, Vojvodina*
Independence: 11 April 1992 (Federal Republic of Yugoslavia formed as self-proclaimed successor to the Socialist Federal Republic of Yugoslavia—SFRY)
National holiday: NA
Constitution: 27 April 1992
Legal system: based on civil law system
Suffrage: 16 years of age, if employed; 18 years of age, universal
Executive branch:
chief of state: President Zoran LILIC (since 25 June 1993); note—Slobodan MILOSEVIC is president of Serbia (since 9 December 1990); Momir BULATOVIC is president of Montenegro (since 23 December 1990); Federal Assembly elected Zoran LILIC on 25 June 1993
head of government: Prime Minister Radoje KONTIC (since 29 December 1992); Deputy Prime Ministers Jovan ZEBIC (since NA March 1993), Uros KLIKOVAC (since 15 September 1994), Nikola SAINOVIC (since 15 September 1995)
cabinet: Federal Executive Council
Legislative branch: bicameral Federal Assembly
Chamber of Republics: elections last held 20 December 1992 (next to be held NA 1996); results—percent of vote by party NA; seats—(40 total, 20 Serbian, 20 Montenegrin)
Chamber of Citizens: elections last held 20 December 1992 (next to be held NA 1996); results—percent of votes by party NA; seats—(138 total, 108 Serbian, 30 Montenegrin) SPS 47, SRS 34, Depos 20, DPSCG 17, DS 5, SP 5, NS 4, DZVM 3, other 3
Judicial branch: Savezni Sud (Federal Court), Constitutional Court
Political parties and leaders: Serbian Socialist Party (SPS, former Communist Party), Slobodan MILOSEVIC; Serbian Radical Party (SRS), Vojislav SESELJ; Serbian Renewal Movement (SPO), Vuk DRASKOVIC, president; Democratic Party (DS), Zoran DJINDJIC; Democratic Party of Serbia (Depos), Vojlslav KOSTUNICA; Democratic Party of Socialists of Montenegro (DPSCG), Momir BULATOVIC, president; People's Party of Montenegro (NS), Milan PAROSKI; Liberal Alliance of Montenegro, Slavko PEROVIC; Democratic Community of Vojvodina Hungarians (DZVM), Andras AGOSTON; League of Communists-Movement for Yugoslavia (SK-PJ), Dragan ATANASOVSKI; Democratic Alliance of Kosovo (LDK), Dr. Ibrahim RUGOVA, president; Party of Democratic Action (SDA), Sulejman UGLJANIN; Civic Alliance of Serbia (GSS), Vesna PESIC, chairman; Socialist Party of Montenegro (SP), leader NA
Other political or pressure group: NA
Diplomatic representation in US: US and Serbia and Montenegro do not maintain full diplomatic relations; the Embassy of the former Socialist Federal Republic of Yugoslavia continues to function in the US
US diplomatic representation:
chief of mission: (vacant); Charge d'Affaires Rudolf V. PERINA
embassy: address NA, Belgrade
mailing address: Box 5070, Unit 1310, APO AE 09213-1310
telephone: [381] (11) 645655
FAX: [381] (11) 645221
Flag: three equal horizontal bands of blue (top), white, and red

Economy

Overview: The swift collapse of the Yugoslav federation in 1991 has been followed by bloody ethnic warfare, the destabilization of republic boundaries, and the breakup of important interrepublic trade flows. Serbia and Montenegro faces major economic problems; output has dropped sharply, particularly in 1993. First, like the other former Yugoslav republics, it depended on its sister republics for large amounts of foodstuffs, energy supplies, and manufactures. Wide differences in climate, mineral resources, and levels of technology among the republics accentuated this interdependence, as did the communist practice of concentrating much industrial output in a small number of giant plants. The breakup of many of the trade links, the sharp drop in output as industrial plants lost suppliers and markets, and the destruction of physical assets in the fighting all have contributed to the economic difficulties of the republics. One singular factor in the economic situation of Serbia and Montenegro is the continuation in office of a communist government that is primarily interested in political and military mastery, not economic reform. A further complication is the imposition of economic sanctions by the UN in 1992. Hyperinflation ended with the establishment of a new currency unit in June 1993; prices were relatively stable in 1994. Reliable statistics are hard to come by; the GDP estimate of $1,000 per capita in 1994 is extremely rough. Output in 1994 seems to have leveled off after the plunge in 1993.
National product: GDP—purchasing power parity—$10 billion (1994 est.)
National product real growth rate: NA%
National product per capita: $1,000 (1994 est.)
Inflation rate (consumer prices): 20% (January-November 1994 est.)
Unemployment rate: more than 40% (1994 est.)
Budget:
revenues: $NA
expenditures: $NA, including capital expenditures of $NA
Exports: $NA
commodities: prior to the breakup of the federation, Yugoslavia exported machinery and transport equipment, manufactured goods, chemicals, food and live animals, raw materials
partners: prior to the imposition of UN sanctions trade partners were the other former Yugoslav republics, Italy, Germany, other EC, the FSU countries, East European countries, US
Imports: $NA
commodities: prior to the breakup of the federation, Yugoslavia imported machinery and transport equipment, fuels and lubricants, manufactured goods, chemicals, food and live animals, raw materials including coking coal for the steel industry
partners: prior to the imposition of UN sanctions trade partners were the other former Yugoslav republics, the FSU countries, EC countries (mainly Italy and Germany), East European countries, US
External debt: $4.2 billion (1993 est.)
Industrial production: NA%
Electricity:
capacity: 10,400,000 kW
production: 34 billion kWh
consumption per capita: 2,400 kWh (1994 est.)
Industries: machine building (aircraft, trucks, and automobiles; armored vehicles and weapons; electrical equipment; agricultural machinery), metallurgy (steel, aluminum, copper, lead, zinc, chromium, antimony, bismuth, cadmium), mining (coal, bauxite, nonferrous ore, iron ore, limestone), consumer goods (textiles, footwear, foodstuffs, appliances), electronics, petroleum products, chemicals, and pharmaceuticals
Agriculture: the fertile plains of Vojvodina produce 80% of the cereal production of the former Yugoslavia and most of the cotton, oilseeds, and chicory; Vojvodina also produces fodder crops to support intensive beef and dairy production; Serbia proper, although hilly, has a well-distributed rainfall and a long growing season; produces fruit, grapes, and cereals; in this area, livestock production (sheep and cattle) and dairy farming prosper; Kosovo produces fruits, vegetables, tobacco, and a small amount of cereals; the mountainous pastures of Kosovo and Montenegro support sheep and goat husbandry; Montenegro has only a small agriculture sector, mostly near the coast where a Mediterranean climate permits the culture of olives, citrus, grapes, and rice
Illicit drugs: NA
Economic aid: $NA
Currency: 1 Yugoslav New Dinar (YD) = 100 paras
Exchange rates: Yugoslav New Dinars (YD) per US $1—102.6 (February 1995 black market rate)
Fiscal year: calendar year

Transportation

Railroads:
total: 3,960 km
standard gauge: 3,960 km 1.435-m gauge (partially electrified) (1992)
Highways:
total: 46,019 km
paved: 26,949 km
unpaved: gravel 10,373 km; earth 8,697 km (1990)
Inland waterways: NA km
Pipelines: crude oil 415 km; petroleum products 130 km; natural gas 2,110 km
Ports: Bar, Belgrade, Kotor, Novi Sad, Pancevo, Tivat
Merchant marine:
Montenegro: total 35 ships (1,000 GRT or over) totaling 543,511 GRT/891,664 DWT (controlled by Montenegrin beneficial owners)
ships by type: bulk 15, cargo 14, container 5, short-sea passenger ferry 1
note: under Maltese and Saint Vincent and the Grenadines flags; no ships remain under Yugoslav flag
Serbia: total 2 (1,000 GRT or over) totaling 113,471 GRT/212,742 DWT (controlled by Serbian beneficial owners)
ships by type: bulk 2
note: all under the flag of Saint Vincent and the Grenadines; no ships remain under Yugoslav flag
Airports:
total: 54
with paved runways over 3,047 m: 2
with paved runways 2,438 to 3,047 m: 5
with paved runways 1,524 to 2,437 m: 5
with paved runways 914 to 1,523 m: 2
with paved runways under 914 m: 24
with unpaved runways 1,524 to 2,438 m: 2
with unpaved runways 914 to 1,523 m: 14

Communications

Telephone system: 700,000 telephones
local: NA
intercity: NA
international: 1 INTELSAT (Atlantic Ocean) earth station
Radio:
broadcast stations: AM 26, FM 9, shortwave 0
radios: 2.015 million
Television:
broadcast stations: 18
televisions: 1 million

Defense Forces

Branches: People's Army (includes Ground Forces with internal and border troops, Naval Forces, and Air and Air Defense Forces), Civil Defense
Manpower availability:
Montenegro: males age 15-49 194,154; males fit for military service 157,611; males reach military age (19) annually 5,498 (1995 est.)
Serbia: males age 15-49 2,652,224; males fit for military service 2,131,894 (1995 est.)
Defense expenditures: 245 billion dinars, 4% to 6% of GDP (1992 est.); note—conversion of defense expenditures into US dollars using the current exchange rate could produce misleading results

Seychelles

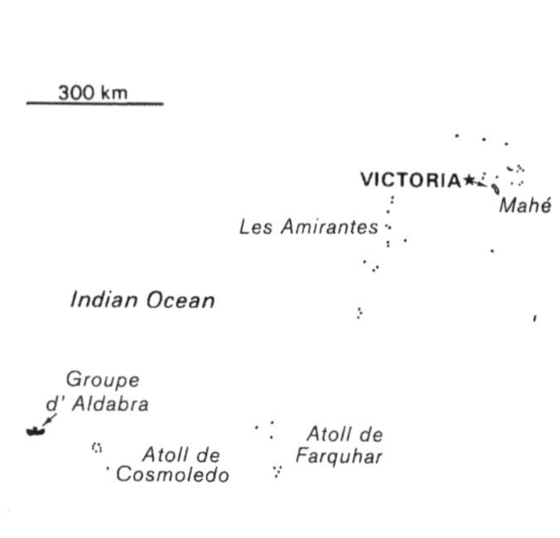

Geography

Location: Eastern Africa, group of islands in the Indian Ocean, northeast of Madagascar
Map references: Africa
Area:
total area: 455 sq km
land area: 455 sq km
comparative area: slightly more than 2.5 times the size of Washington, DC
Land boundaries: 0 km
Coastline: 491 km
Maritime claims:
continental shelf: 200 nm or to the edge of the continental margin
exclusive economic zone: 200 nm
territorial sea: 12 nm
International disputes: claims Tromelin Island
Climate: tropical marine; humid; cooler season during southeast monsoon (late May to September); warmer season during northwest monsoon (March to May)
Terrain: Mahe Group is granitic, narrow coastal strip, rocky, hilly; others are coral, flat, elevated reefs
Natural resources: fish, copra, cinnamon trees
Land use:
arable land: 4%
permanent crops: 18%
meadows and pastures: 0%
forest and woodland: 18%
other: 60%
Irrigated land: NA sq km
Environment:
current issues: no natural fresh water resources, catchments collect rain water
natural hazards: lies outside the cyclone belt, so severe storms are rare; short droughts possible
international agreements: party to—Biodiversity, Climate Change, Endangered Species, Hazardous Wastes, Law of the Sea, Marine Dumping, Nuclear Test Ban, Ozone Layer Protection, Ship Pollution, Whaling; signed, but not ratified—Desertification
Note: 40 granitic and about 50 coralline islands

People

Population: 72,709 (July 1995 est.)
Age structure:
0-14 years: 32% (female 11,630; male 11,811)
15-64 years: 62% (female 23,229; male 21,679)
65 years and over: 6% (female 2,875; male 1,485) (July 1995 est.)
Population growth rate: 0.81% (1995 est.)
Birth rate: 21.35 births/1,000 population (1995 est.)
Death rate: 6.7 deaths/1,000 population (1995 est.)
Net migration rate: -6.6 migrant(s)/1,000 population (1995 est.)
Infant mortality rate: 11.4 deaths/1,000 live births (1995 est.)
Life expectancy at birth:
total population: 70.08 years
male: 66.54 years
female: 73.73 years (1995 est.)
Total fertility rate: 2.16 children born/woman (1995 est.)
Nationality:
noun: Seychellois (singular and plural)
adjective: Seychelles
Ethnic divisions: Seychellois (mixture of Asians, Africans, Europeans)
Religions: Roman Catholic 90%, Anglican 8%, other 2%
Languages: English (official), French (official), Creole
Literacy: age 15 and over can read and write (1971)
total population: 58%
male: 56%
female: 60%
Labor force: 27,700 (1985)
by occupation: industry and commerce 31%, services 21%, government 20%, agriculture, forestry, and fishing 12%, other 16% (1985)

Government

Names:
conventional long form: Republic of Seychelles
conventional short form: Seychelles
Digraph: SE
Type: republic
Capital: Victoria
Administrative divisions: 23 administrative districts; Anse aux Pins, Anse Boileau, Anse Etoile, Anse Louis, Anse Royale, Baie Lazare, Baie Sainte Anne, Beau Vallon, Bel Air, Bel Ombre, Cascade, Glacis, Grand' Anse (on Mahe Island), Grand' Anse (on Praslin Island), La Digue, La Riviere Anglaise, Mont Buxton, Mont Fleuri, Plaisance, Pointe Larue, Port Glaud, Saint Louis, Takamaka
Independence: 29 June 1976 (from UK)
National holiday: National Day, 18 June (1993) (adoption of new constitution)
Constitution: 18 June 1993
Legal system: based on English common law, French civil law, and customary law
Suffrage: 17 years of age; universal
Executive branch:
chief of state and head of government: President France Albert RENE (since 5 June 1977); election last held 20-23 July 1993 (next to be held NA); results—President France Albert RENE (SPPF) reelected with 59.5% of the vote, Sir James MANCHAM (DP) 36.72%
cabinet: Council of Ministers; appointed by the president
Legislative branch: unicameral
People's Assembly (Assemblee du Peuple): elections last held 20-23 July 1993 (next to be held NA); results—SPPF 82%, DP 15%, UO 3%; seats—(33 total, 22 elected, 11 awarded) seats elected—SPPF 21, DP 1; seats awarded—SPPF 6, DP 4, UO 1; total seats by party—SPPF 27, DP 5, UO 1
note: the 11 awarded seats are apportioned according to the share of each party in the total vote
Judicial branch: Court of Appeal, Supreme Court
Political parties and leaders: ruling party—Seychelles People's Progressive Front (SPPF), France Albert RENE; Democratic Party (DP), Sir James MANCHAM; United Opposition (UO), Annette GEORGES—a coalition of the following parties: Seychelles Party (PS), Wavel RAMKALAWAN; Seychelles Democratic Movement (MSPD), Jacques HONDOUL; Seychelles Liberal Party (SLP), Ogilvie BERLOUIS
Other political or pressure groups: trade unions; Roman Catholic Church
Member of: ACCT, ACP, AfDB, C, ECA, FAO, G-77, IBRD, ICAO, ICFTU, ICRM, IFAD, IFC, IFRCS (associate), ILO, IMF, IMO, INTELSAT (nonsignatory user), INTERPOL, IOC, NAM, OAU, UN, UNCTAD, UNESCO, UNIDO, UPU, WCL, WHO, WMO, WTO
Diplomatic representation in US:
chief of mission: Ambassador Marc R. MARENGO
chancery: (temporary) 820 Second Avenue, Suite 900F, New York, NY 10017
telephone: [1] (212) 687-9766, 9767
FAX: [1] (212) 922-9177
US diplomatic representation:
chief of mission: Ambassador Carl Burton STOKES
embassy: 4th Floor, Victoria House, Box 251, Victoria, Mahe
mailing address: Box 148, Unit 62501, Victoria, Seychelles; APO AE 09815-2501
telephone: [248] 225256
FAX: [248] 225189
Flag: three horizontal bands of red (top), white (wavy), and green; the white band is the thinnest, the red band is the thickest

Sierra Leone

Economy

Overview: Since independence in 1976, per capita output has grown to roughly seven times the old near-subsistence level, led by the tourist sector, which employs about 30% of the labor force and provides more than 70% of hard currency earnings. In recent years the government has encouraged foreign investment in order to upgrade hotels and other services. At the same time, the government has moved to reduce the high dependence on tourism by promoting the development of farming, fishing, and small-scale manufacturing. The vulnerability of the tourist sector was illustrated by the sharp drop in 1991-92 due largely to the Gulf war. Although the industry has rebounded, the government recognizes the continuing need for upgrading the sector in the face of stiff international competition.
National product: GDP—purchasing power parity—$430 million (1993 est.)
National product real growth rate: -2% (1993 est.)
National product per capita: $6,000 (1993 est.)
Inflation rate (consumer prices): 3.9% (1993 est.)
Unemployment rate: 9% (1987)
Budget:
revenues: $227.4 million
expenditures: $263 million, including capital expenditures of $54 million (1993 est.)
Exports: $50 million (f.o.b., 1993 est.)
commodities: fish, cinnamon bark, copra, petroleum products (re-exports)
partners: France 43%, UK 22%, Reunion 11%, (1992)
Imports: $261 million (f.o.b., 1993 est.)
commodities: manufactured goods, food, petroleum products, tobacco, beverages, machinery and transportation equipment
partners: Singapore 16%, Bahrain 16%, South Africa, 14%, UK 13% (1992)
External debt: $181 million (1993 est.)
Industrial production: growth rate 4% (1992); accounts for 12% of GDP
Electricity:
capacity: 30,000 kW
production: 110 million kWh
consumption per capita: 1,399 kWh (1993)
Industries: tourism, processing of coconut and vanilla, fishing, coir rope factory, boat building, printing, furniture, beverage
Agriculture: accounts for 5% of GDP, mostly subsistence farming; cash crops—coconuts, cinnamon, vanilla; other products—sweet potatoes, cassava, bananas; broiler chickens; large share of food needs imported; expansion of tuna fishing under way
Economic aid:
recipient: US commitments, including Ex-Im (FY78-89), $26 million; Western (non-US) countries, ODA and OOF bilateral commitments (1978-89), $315 million; OPEC bilateral aid (1979-89), $5 million; Communist countries (1970-89), $60 million
Currency: 1 Seychelles rupee (SRe) = 100 cents
Exchange rates: Seychelles rupees (SRe) per US$1—4.9371 (January 1995), 5.0559 (1994), 5.1815 (1993), 5.1220 (1992), 5.2893 (1991), 5.3369 (1990)
Fiscal year: calendar year

Transportation

Railroads: 0 km
Highways:
total: 260 km
paved: 160 km
unpaved: crushed stone, earth 100 km
Ports: Victoria
Merchant marine: none
Airports:
total: 14
with paved runways 2,438 to 3,047 m: 1
with paved runways 914 to 1,523 m: 5
with paved runways under 914 m: 6
with unpaved runways 914 to 1,523 m: 2

Communications

Telephone system: 13,000 telephones; direct radio communications with adjacent islands and African coastal countries
local: NA
intercity: radio communciations
international: 1 INTELSAT (Indian Ocean) earth station
Radio:
broadcast stations: AM 2, FM 0, shortwave 0
radios: NA
Television:
broadcast stations: 2
*televisions*NA:

Defense Forces

Branches: Army, Coast Guard, Marines, National Guard, Presidential Protection Unit, Police Force
Manpower availability: males age 15-49 19,829; males fit for military service 10,099 (1995 est.)
Defense expenditures: exchange rate conversion—$12 million, 4% of GDP (1990 est.)

Geography

Location: Western Africa, bordering the North Atlantic Ocean, between Guinea and Liberia
Map references: Africa
Area:
total area: 71,740 sq km
land area: 71,620 sq km
comparative area: slightly smaller than South Carolina
Land boundaries: total 958 km, Guinea 652 km, Liberia 306 km
Coastline: 402 km
Maritime claims:
territorial sea: 200 nm
continental shelf: 200-m depth or to the depth of exploitation
International disputes: none
Climate: tropical; hot, humid; summer rainy season (May to December); winter dry season (December to April)
Terrain: coastal belt of mangrove swamps, wooded hill country, upland plateau, mountains in east
Natural resources: diamonds, titanium ore, bauxite, iron ore, gold, chromite
Land use:
arable land: 25%
permanent crops: 2%
meadows and pastures: 31%
forest and woodland: 29%
other: 13%
Irrigated land: 340 sq km (1989 est.)
Environment:
current issues: rapid population growth pressuring the environment; overharvesting of timber, expansion of cattle grazing, and slash-and-burn agriculture have resulted in deforestation and soil exhaustion; civil war depleting natural resources; overfishing
natural hazards: dry, sand-laden harmattan winds blow from the Sahara (November to May); sandstorms, dust storms
international agreements: party to—Biodiversity, Endangered Species, Law of the Sea, Marine Life Conservation, Nuclear Test

Ban; signed, but not ratified—Climate Change, Environmental Modification

People

Population: 4,753,120 (July 1995 est.)
Age structure:
0-14 years: 44% (female 1,054,826; male 1,020,943)
15-64 years: 53% (female 1,310,506; male 1,216,510)
65 years and over: 3% (female 72,982; male 77,353) (July 1995 est.)
Population growth rate: 2.63% (1995 est.)
Birth rate: 44.65 births/1,000 population (1995 est.)
Death rate: 18.38 deaths/1,000 population (1995 est.)
Net migration rate: NA migrant(s)/1,000 population (1995 est.)
note: thousands of refugees, fleeing the civil strife in Sierra Leone, are taking refuge in Guinea
Infant mortality rate: 138.8 deaths/1,000 live births (1995 est.)
Life expectancy at birth:
total population: 46.94 years
male: 44.07 years
female: 49.89 years (1995 est.)
Total fertility rate: 5.9 children born/woman (1995 est.)
Nationality:
noun: Sierra Leonean(s)
adjective: Sierra Leonean
Ethnic divisions: 13 native African tribes 99% (Temne 30%, Mende 30%, other 39%), Creole, European, Lebanese, and Asian 1%
Religions: Muslim 60%, indigenous beliefs 30%, Christian 10%
Languages: English (official; regular use limited to literate minority), Mende (principal vernacular in the south), Temne (principal vernacular in the north), Krio (the language of the re-settled ex-slave population of the Freetown area and is lingua franca)
Literacy: age 15 and over can read and write English, Mende, Temne, or Arabic (1990 est.)
total population: 21%
male: 31%
female: 11%
Labor force: 1.369 million (1981 est.)
by occupation: agriculture 65%, industry 19%, services 16% (1981 est.)
note: only about 65,000 wage earners (1985)

Government

Names:
conventional long form: Republic of Sierra Leone
conventional short form: Sierra Leone
Digraph: SL
Type: military government
Capital: Freetown
Administrative divisions: 3 provinces and 1 area*; Eastern, Northern, Southern, Western*
Independence: 27 April 1961 (from UK)
National holiday: Republic Day, 27 April (1961)
Constitution: 1 October 1991; suspended following 19 April 1992 coup
Legal system: based on English law and customary laws indigenous to local tribes; has not accepted compulsory ICJ jurisdiction
Suffrage: 18 years of age; universal
Executive branch:
chief of state and head of government: Chairman of the Supreme Council of State Capt. Valentine E. M. STRASSER (since 29 April 1992)
cabinet: Council of Secretaries; responsible to the Supreme Council of State (SCS)
Legislative branch: unicameral House of Representatives (suspended after coup of 29 April 1992); Chairman STRASSER promises multi-party elections sometime in 1995
Judicial branch: Supreme Court (suspended after coup of 29 April 1992)
Political parties and leaders: status of existing political parties is unknown following 29 April 1992 coup
Member of: ACP, AfDB, C, CCC, ECA, ECOWAS, FAO, G-77, GATT, IAEA, IBRD, ICAO, ICFTU, ICRM, IDA, IDB, IFAD, IFC, IFRCS, ILO, IMF, IMO, INTELSAT (nonsignatory user), INTERPOL, IOC, ITU, NAM, OAU, OIC, UN, UNCTAD, UNESCO, UNIDO, UPU, WCL, WFTU, WHO, WIPO, WMO, WTO
Diplomatic representation in US:
chief of mission: Ambassador Thomas Kahota KARGBO
chancery: 1701 19th Street NW, Washington, DC 20009
telephone: [1] (202) 939-9261
US diplomatic representation:
chief of mission: Ambassador Lauralee M. PETERS
embassy: Corner of Walpole and Siaka Stevens Streets, Freetown
mailing address: use embassy street address
telephone: [232] (22) 226481 trough 226485
FAX: [232] (22) 225471
Flag: three equal horizontal bands of light green (top), white, and light blue

Economy

Overview: Sierra Leone has substantial mineral, agricultural, and fishery resources, but the economic and social infrastructure is not well developed. Agriculture generates about 40% of GDP and employs about two-thirds of the working population, with subsistence agriculture dominating the sector. Manufacturing, which accounts for roughly 10% of GDP, consists mainly of the processing of raw materials and of light manufacturing for the domestic market. Diamond mining provides an important source of hard currency. Since 1990, the government has been able to meet its IMF- and World Bank-mandated stabilization targets, holding down fiscal deficits, increasing foreign exchange reserves, and retiring much of its domestic debt—but at a steep cost in terms of capital investments and social spending. Moreover, the economic infrastructure has nearly collapsed due to neglect and war-related disruptions in the mining and agricultural export sectors. The continuing civil war in Liberia has led to a large influx of refugees, who place additional burdens on Sierra Leon's fragile economy.
National product: GDP—purchasing power parity—$4.5 billion (1993 est.)
National product real growth rate: 0.7% (1993 est.)
National product per capita: $1,000 (1993 est.)
Inflation rate (consumer prices): 22% (1993 est.)
Unemployment rate: NA%
Budget:
revenues: $68 million
expenditures: $118 million, including capital expenditures of $28 million (1992 est.)
Exports: $149 million (f.o.b., 1993)
commodities: rutile 48%, bauxite 25%, diamonds 16%, coffee, cocoa, fish
partners: US, UK, Belgium, Germany, other Western Europe
Imports: $149 million (c.i.f., 1993)
commodities: foodstuffs 48%, machinery and equipment 32%, fuels 9%
partners: US, EC countries, Japan, China, Nigeria
External debt: $1.15 billion (yearend 1993)
Industrial production: growth rate -1.5% (FY91/92); accounts for 11% of GDP
Electricity:
capacity: 130,000 kW
production: 220 million kWh
consumption per capita: 44 kWh (1993)
Industries: mining (diamonds, bauxite, rutile), small-scale manufacturing (beverages, textiles, cigarettes, footwear), petroleum refinery
Agriculture: largely subsistence farming; cash crops—coffee, cocoa, palm kernels; harvests of food staple rice meets 80% of domestic needs; annual fish catch averages 53,000 metric tons
Economic aid:
recipient: US commitments, including Ex-Im (FY70-89), $161 million; Western (non-US) countries, ODA and OOF bilateral commitments (1970-89), $848 million; OPEC bilateral aid (1979-89), $18 million; Communist countries (1970-89), $101 million
Currency: 1 leone (Le) = 100 cents
Exchange rates: leones (Le) per US$1—617.67 (January 1995), 586.74 (1994), 567.46 (1993), 499.44 (1992), 295.34 (1991), 144.9275 (1990)
Fiscal year: 1 July—30 June

Singapore

Transportation

Railroads:
total: 84 km mineral line is used on a limited basis because the mine at Marampa is closed
narrow gauge: 84 km 1.067-m gauge
Highways:
total: 7,400 km
paved: 1,150 km
unpaved: crushed stone, gravel 490 km; improved earth 5,760 km
Inland waterways: 800 km; 600 km navigable year round
Ports: Bonthe, Freetown, Pepel
Merchant marine: none
Airports:
total: 11
with paved runways over 3,047 m: 1
with paved runways 914 to 1,523 m: 3
with paved runways under 914 m: 3
with unpaved runways 914 to 1,523 m: 4

Communications

Telephone system: 23,650 telephones; telephone density—5 telephones/1,000 persons; marginal telephone and telegraph service
local: NA
intercity: national microwave radio relay system made unserviceable by military activities
international: 1 INTELSAT (Atlantic Ocean) earth station
Radio:
broadcast stations: AM 1, FM 1, shortwave 0
radios: NA
Television:
broadcast stations: 1
televisions: NA

Defense Forces

Branches: Army, Navy, Police, Security Forces
Manpower availability: males age 15-49 1,030,332; males fit for military service 498,945 (1995 est.)
Defense expenditures: exchange rate conversion—$14 million, 2.6% of GDP (FY92/93)

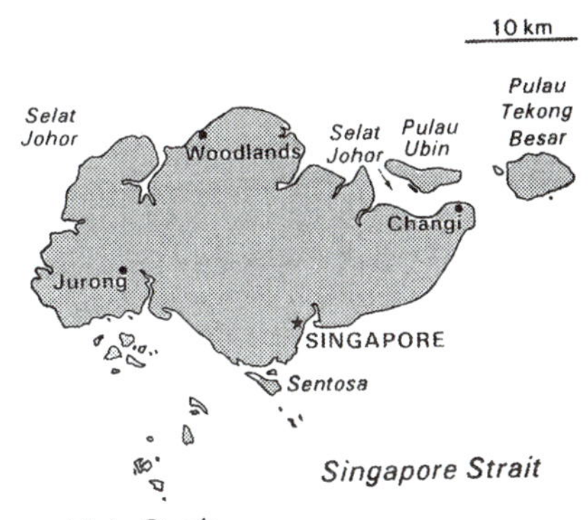

Geography

Location: Southeastern Asia, islands between Malaysia and Indonesia
Map references: Southeast Asia
Area:
total area: 632.6 sq km
land area: 622.6 sq km
comparative area: slightly less than 3.5 times the size of Washington, DC
Land boundaries: 0 km
Coastline: 193 km
Maritime claims:
exclusive fishing zone: within and beyond territorial sea, as defined in treaties and practice
territorial sea: 3 nm
International disputes: two islands in dispute with Malaysia
Climate: tropical; hot, humid, rainy; no pronounced rainy or dry seasons; thunderstorms occur on 40% of all days (67% of days in April)
Terrain: lowland; gently undulating central plateau contains water catchment area and nature preserve
Natural resources: fish, deepwater ports
Land use:
arable land: 4%
permanent crops: 7%
meadows and pastures: 0%
forest and woodland: 5%
other: 84%
Irrigated land: NA sq km
Environment:
current issues: industrial pollution; limited natural fresh water resources; limited land availability presents waste disposal problems
natural hazards: NA
international agreements: party to—Endangered Species, Law of the Sea, Nuclear Test Ban, Ozone Layer Protection, Ship Pollution; signed, but not ratified—Biodiversity, Climate Change
Note: focal point for Southeast Asian sea routes

People

Population: 2,890,468 (July 1995 est.)
Age structure:
0-14 years: 23% (female 327,417; male 348,345)
15-64 years: 70% (female 991,015; male 1,030,668)
65 years and over: 7% (female 105,081; male 87,942) (July 1995 est.)
Population growth rate: 1.06% (1995 est.)
Birth rate: 15.93 births/1,000 population (1995 est.)
Death rate: 5.35 deaths/1,000 population (1995 est.)
Net migration rate: 0 migrant(s)/1,000 population (1995 est.)
Infant mortality rate: 5.7 deaths/1,000 live births (1995 est.)
Life expectancy at birth:
total population: 76.16 years
male: 73.28 years
female: 79.25 years (1995 est.)
Total fertility rate: 1.87 children born/woman (1995 est.)
Nationality:
noun: Singaporean(s)
adjective: Singapore
Ethnic divisions: Chinese 76.4%, Malay 14.9%, Indian 6.4%, other 2.3%
Religions: Buddhist (Chinese), Muslim (Malays), Christian, Hindu, Sikh, Taoist, Confucianist
Languages: Chinese (official), Malay (official and national), Tamil (official), English (official)
Literacy: age 15 and over can read and write (1990)
total population: 89%
male: 95%
female: 83%
Labor force: 1.649 million (1994)
by occupation: financial, business, and other services 33.5%, manufacturing 25.6%, commerce 22.9%, construction 6.6%, other 11.4% (1994)

Government

Names:
conventional long form: Republic of Singapore
conventional short form: Singapore
Digraph: SN
Type: republic within Commonwealth
Capital: Singapore
Administrative divisions: none
Independence: 9 August 1965 (from Malaysia)
National holiday: National Day, 9 August (1965)
Constitution: 3 June 1959, amended 1965 (based on preindependence State of Singapore Constitution)

Legal system: based on English common law; has not accepted compulsory ICJ jurisdiction
Suffrage: 20 years of age; universal and compulsory
Executive branch:
chief of state: President ONG Teng Cheong (since 1 September 1993); election last held 28 August 1993 (next to be held NA August 1997); results—President ONG Teng Cheong was elected with 59% of the vote in the country's first popular election for president
head of government: Prime Minister GOH Chok Tong (since 28 November 1990); Deputy Prime Minister LEE Hsien Loong (since 28 November 1990)
cabinet: Cabinet; appointed by the president, responsible to parliament
Legislative branch: unicameral
Parliament: elections last held 31 August 1991 (next to be held by 31 August 1996); results—percent of vote by party NA; seats—(81 total) PAP 77, SDP 3, WP 1
Judicial branch: Supreme Court
Political parties and leaders:
government: People's Action Party (PAP), GOH Chok Tong, secretary general
opposition: Workers' Party (WP), J. B. JEYARETNAM; Singapore Democratic Party (SDP), CHEE Soon Juan; National Solidarity Party (NSP), leader NA; Barisan Sosialis (BS, Socialist Front), leader NA; Singapore People's Party (SPP), SIN Kek Tong
Member of: APEC, AsDB, ASEAN, C, CCC, CP, ESCAP, G-77, GATT, IAEA, IBRD, ICAO, ICC, ICFTU, ICRM, IFC, IFRCS, ILO, IMF, IMO, INMARSAT, INTELSAT, INTERPOL, IOC, ISO, ITU, NAM, PCA, UN, UNCTAD, UNIKOM, UPU, WHO, WIPO, WMO
Diplomatic representation in US:
chief of mission: Ambassador Sellapan Rama NATHAN
chancery: 3501 International Place NW, Washington, DC 20008
telephone: [1] (202) 537-3100
FAX: [1] (202) 537-0876
US diplomatic representation:
chief of mission: Ambassador Timothy A. CHORBA
embassy: 30 Hill Street, Singapore 0617
mailing address: FPO AP 96534
telephone: [65] 3380251
FAX: [65] 3384550
Flag: two equal horizontal bands of red (top) and white; near the hoist side of the red band, there is a vertical, white crescent (closed portion is toward the hoist side) partially enclosing five white five-pointed stars arranged in a circle

Economy

Overview: Singapore has an open entrepreneurial economy with strong service and manufacturing sectors and excellent international trading links derived from its entrepot history. The economy registered 10.1% growth in 1994, with prospects for 7.5%-8.5% growth in 1995. In 1994, the manufacturing and financial and business services sectors have led economic growth. Exports boomed, led by the electronics sector, particularly US demand for disk drives. Rising labor costs continue to be a threat to Singapore's competitiveness, but there are indications that productivity is keeping up. In applied technology, per capita output, investment, and labor discipline, Singapore has key attributes of a developed country.
National product: GDP—purchasing power parity—$57 billion (1994 est.)
National product real growth rate: 10.1% (1994)
National product per capita: $19,940 (1994 est.)
Inflation rate (consumer prices): 3.6% (1994)
Unemployment rate: 2.6% (1994)
Budget:
revenues: $11.9 billion
expenditures: $10.5 billion, including capital expenditures of $3.9 billion (FY93/94 est.)
Exports: $96.4 billion (f.o.b., 1994)
commodities: computer equipment, rubber and rubber products, petroleum products, telecommunications equipment
partners: Malaysia 20%, US 19%, Hong Kong 9%, Japan 7%, Thailand 6% (1994)
Imports: $102.4 billion (c.i.f., 1994)
commodities: aircraft, petroleum, chemicals, foodstuffs
partners: Japan 22%, Malaysia 16%, US 15%, Taiwan 4%, Saudi Arabia 4% (1994)
External debt: $20 million (1993 est.)
Industrial production: growth rate 13% (1994 est.); accounts for 28% of GDP (1993)
Electricity:
capacity: 4,510,000 kW
production: 17 billion kWh
consumption per capita: 5,590 kWh (1993)
Industries: petroleum refining, electronics, oil drilling equipment, rubber processing and rubber products, processed food and beverages, ship repair, entrepot trade, financial services, biotechnology
Agriculture: minor importance in the economy; self-sufficient in poultry and eggs; must import much of other food; major crops—rubber, copra, fruit, vegetables
Illicit drugs: transit point for Golden Triangle heroin going to the US, Western Europe, and the Third World; also a major money-laundering center
Economic aid:
recipient: US commitments, including Ex-Im (FY70-83), $590 million; Western (non-US) countries, ODA and OOF bilateral commitments (1970-89), $1 billion
Currency: 1 Singapore dollar (S$) = 100 cents
Exchange rates: Singapore dollars (S$) per US$1—1.4524 (January 1995), 1.5275 (1994), 1.6158 (1993), 1.6290 (1992), 1.7276 (1991), 1.8125 (1990)
Fiscal year: 1 April—31 March

Transportation

Railroads:
total: 38.6 km
narrow gauge: 38.6 km 1.000-m gauge
Highways:
total: 2,883 km
paved: 2,796 km
unpaved: 87 km (1991 est.)
Ports: Singapore
Merchant marine:
total: 563 ships (1,000 GRT or over) totaling 11,167,596 GRT/17,845,687 DWT
ships by type: bulk 96, cargo 121, chemical tanker 16, combination bulk 2, combination ore/oil 7, container 78, liquefied gas tanker 6, oil tanker 198, refrigerated cargo 1, roll-on/roll-off cargo 12, short-sea passenger 1, specialized tanker 3, vehicle carrier 22
note: a flag of convenience registry; includes 20 countries among which are Japan 35 ships, Denmark 21, Germany 21, Hong Kong 18, Belgium 14, Thailand 11, Sweden 8, US 7, Indonesia 6, and Norway 5; Singapore owns 1 ship under Malaysia registry
Airports:
total: 10
with paved runways over 3,047 m: 3
with paved runways 2,438 to 3,047 m: 2
with paved runways 1,524 to 2,437 m: 4
with paved runways 914 to 1,523 m: 1

Communications

Telephone system: 1,110,000 telephones; good domestic facilities; good international service; good radio and television broadcast coverage
local: NA
intercity: NA
international: submarine cables extend to Malaysia (Sabah and Peninsular Malaysia), Indonesia, and the Philippines; 2 INTELSAT (1 Indian Ocean and 1 Pacific Ocean) earth stations
Radio:
broadcast stations: AM 13, FM 4, shortwave 0
radios: NA
Television:
broadcast stations: 2
televisions: NA

Defense Forces

Branches: Army, Navy, Air Force, People's Defense Force, Police Force
Manpower availability: males age 15-49 860,437; males fit for military service 629,973 (1995 est.)
Defense expenditures: exchange rate conversion—$2.7 billion, 6% of GDP (1993 est.)

Slovakia

Geography

Location: Central Europe, south of Poland
Map references: Ethnic Groups in Eastern Europe, Europe
Area:
total area: 48,845 sq km
land area: 48,800 sq km
comparative area: about twice the size of New Hampshire
Land boundaries: total 1,355 km, Austria 91 km, Czech Republic 215 km, Hungary 515 km, Poland 444 km, Ukraine 90 km
Coastline: 0 km (landlocked)
Maritime claims: none; landlocked
International disputes: Gabcikovo Dam dispute with Hungary; unresolved property issues with Czech Republic over redistribution of former Czechoslovak federal property
Climate: temperate; cool summers; cold, cloudy, humid winters
Terrain: rugged mountains in the central and northern part and lowlands in the south
Natural resources: brown coal and lignite; small amounts of iron ore, copper and manganese ore; salt
Land use:
arable land: NA%
permanent crops: NA%
meadows and pastures: NA%
forest and woodland: NA%
other: NA%
Irrigated land: NA sq km
Environment:
current issues: air pollution from metallurgical plants presents human health risks; acid rain damaging forests
natural hazards: NA
international agreements: party to—Air Pollution, Air Pollution-Nitrogen Oxides, Air Pollution-Sulphur 85, Antarctic Treaty, Biodiversity, Climate Change, Endangered Species, Environmental Modification, Hazardous Wastes, Nuclear Test Ban, Ozone Layer Protection, Ship Pollution, Wetlands; signed, but not ratified—Air Pollution-Sulphur 94, Antarctic-Environmental Protocol, Law of the Sea
Note: landlocked

People

Population: 5,432,383 (July 1995 est.)
Age structure:
0-14 years: 23% (female 609,795; male 638,346)
15-64 years: 66% (female 1,807,312; male 1,778,712)
65 years and over: 11% (female 364,610; male 233,608) (July 1995 est.)
Population growth rate: 0.54% (1995 est.)
Birth rate: 14.51 births/1,000 population (1995 est.)
Death rate: 9.12 deaths/1,000 population (1995 est.)
Net migration rate: 0 migrant(s)/1,000 population (1995 est.)
Infant mortality rate: 10 deaths/1,000 live births (1995 est.)
Life expectancy at birth:
total population: 73.24 years
male: 69.15 years
female: 77.57 years (1995 est.)
Total fertility rate: 1.93 children born/woman (1995 est.)
Nationality:
noun: Slovak(s)
adjective: Slovak
Ethnic divisions: Slovak 85.7%, Hungarian 10.7%, Gypsy 1.5% (the 1992 census figures underreport the Gypsy/Romany community, which could reach 500,000 or more), Czech 1%, Ruthenian 0.3%, Ukrainian 0.3%, German 0.1%, Polish 0.1%, other 0.3%
Religions: Roman Catholic 60.3%, atheist 9.7%, Protestant 8.4%, Orthodox 4.1%, other 17.5%
Languages: Slovak (official), Hungarian
Literacy: NA%
Labor force: 2.484 million
by occupation: industry 33.2%, agriculture 12.2%, construction 10.3%, communication and other 44.3% (1990)

Government

Names:
conventional long form: Slovak Republic
conventional short form: Slovakia
local long form: Slovenska Republika
local short form: Slovensko
Digraph: LO
Type: parliamentary democracy
Capital: Bratislava
Administrative divisions: 4 departments (kraje, singular—kraj) Bratislava, Zapadoslovensky, Stredoslovensky, Vychodoslovensky
Independence: 1 January 1993 (from Czechoslovakia)
National holiday: Anniversary of Slovak National Uprising, August 29 (1944)
Constitution: ratified 1 September 1992, fully effective 1 January 1993
Legal system: civil law system based on Austro-Hungarian codes; has not accepted compulsory ICJ jurisdiction; legal code modified to comply with the obligations of Organization on Security and Cooperation in Europe (OSCE) and to expunge Marxist-Leninist legal theory
Suffrage: 18 years of age; universal
Executive branch:
chief of state: President Michal KOVAC (since 8 February 1993); election last held 8 February 1993 (next to be held NA 1998); results—Michal KOVAC elected by the National Council
head of government: Prime Minister Vladimir MECIAR (since 12 December 1994)
cabinet: Cabinet; appointed by the president on recommendation of the prime minister
Legislative branch: unicameral
National Council (Narodni Rada): elections last held 30 September-1 October 1994 (next to be held by October 1998); results—HZDS 35%, SDL 10.4%, Hungarian coalition (Hungarian Christian Democrats, Hungarian Civic Party, Coexistence) 10.2%, KDH 10.1%, DU 8.6%, ZRS 7.3%, SNS 5.4%; seats—(150 total) governing coalition 83 (HZDS 61, ZRS 13, SNS 9), opposition 67 (SDL 18, Hungarian coalition 17, KDH 17, DU 15)
Judicial branch: Supreme Court
Political parties and leaders: Movement for a Democratic Slovakia (HZDS), Vladimir MECIAR, chairman; Common Choice/Party of the Democratic Left (SDL), Peter WEISS, chairman; Hungarian Christian Democrats, Vojtech BUGAR; Hungarian Civic Party; Coexistence, Miklos DURAY, chairman; Christian Democratic Movement (KDH), Jan CARNOGURSKY; Democratic Union (DU), Jozef MORAVCIK, chairman; Association of Slovak Workers (ZRS), Jan LUPTAK, chairman; Slovak National Party (SNS), Jan SLOTA, chairman
Other political or pressure groups: Green Party; Social Democratic Party of Slovakia; Slovak Christian Union
Member of: Australia Group, BIS, CCC, CE (guest), CEI, CERN, EBRD, ECE, FAO, GATT, IAEA, IBRD, ICAO, ICFTU, ICRM, IDA, IFC, IFRCS, ILO, IMF, IMO, INMARSAT, INTELSAT (nonsignatory user), INTERPOL, IOC, IOM (observer), ISO, ITU, NACC, NSG, OSCE, PCA, PFP, UN, UNAVEM II, UNCTAD, UNESCO, UNIDO, UNOMIL, UNOMUR, UNPROFOR, UPU, WEU (associate partner), WHO, WIPO, WMO, WTO, ZC
Diplomatic representation in US:
chief of mission: Ambassador Branislav LICHARDUS
chancery: (temporary) Suite 380, 2201 Wisconsin Avenue NW, Washington, DC 20007
telephone: [1] (202) 965-5161
FAX: [1] (202) 965-5166
US diplomatic representation:
chief of mission: Ambassador Theodore E. RUSSELL

embassy: Hviezdoslavovo Namestie 4, 81102 Bratislava
mailing address: use embassy street address
telephone: [42] (7) 330-861, 333-338
FAX: [42] (7) 330-096
Flag: three equal horizontal bands of white (top), blue, and red superimposed with the Slovak cross in a shield centered on the hoist side; the cross is white centered on a background of red and blue

Economy

Overview: In 1994 macroeconomic performance improved steadily but privatization progressed only in fits and starts. Most of Slovakia's IMF-approved targets were met by an interim government that lasted 9 months. Annual inflation fell from 23% in 1993 to 12%; unemployment at 14.6% was still well below forecasts of 17%; and the budget deficit was around half that in 1993. Slovakia's nearly $200 million trade surplus also compares favorably with a more than $800 million deficit in 1993. Furthermore, after contracting almost 25% in the three years following 1990, GDP grew 4.3% in 1994, according to official statistics. Bratislava in June qualified for a $254 million IMF stand-by loan and the second $90 million tranche of its Systemic Transformation Facility and, in December, received approval for a European Union loan worth about $160 million. By the end of September 1994, the Central Bank's foreign currency reserves had tripled since the end of 1993. Slovakia continued to have difficulty attracting foreign investment, however, because of perceived political instability and halting progress in privatization. The interim government prepared property worth nearly $2 billion for the second wave of coupon privatization and sold participation in the program to over 80% of Slovakia's eligible citizens. Parties controlling the new Parliament in November 1994, however, put the second wave of coupon privatization on hold and suspended sales of 38 firms until the new government could evaluate the interim government's decisions in early 1995. The new government's targets for 1995 include GDP growth of 3%, inflation of 8%-10%, unemployment of 15%, and a budget deficit under 3% of GDP. Continuing economic recovery in western Europe should boost Slovak exports and production, but Slovakia's image with foreign creditors and investors could suffer setbacks in 1995 if progress on privatization stalls or budget deficits mount beyond IMF-recommended levels.
National product: GDP—purchasing power parity—$32.8 billion (1994 est.)
National product real growth rate: 4.3% (1994 est.)
National product per capita: $6,070 (1994 est.)
Inflation rate (consumer prices): 12% (1994 est.)
Unemployment rate: 14.6% (1994 est.)
Budget:
revenues: $4.4 billion
expenditures: $4.8 billion, including capital expenditures of $350 million (1994 est.)
Exports: $6.3 billion (f.o.b., January-November 1994)
commodities: machinery and transport equipment; chemicals; fuels, minerals, and metals; agricultural products
partners: Czech Republic 37.7%, Germany 17.1%, Hungary 5.3%, Austria 5.3%, Italy 4.6%, Russia 4.0%, Poland 2.6%, Ukraine 1.8%, US 1.6% (January-September 1994)
Imports: $6.1 billion (f.o.b., January-November 1994)
commodities: machinery and transport equipment; fuels and lubricants; manufactured goods; raw materials; chemicals; agricultural products
partners: Czech Republic 29.9%, Russia 19.0%, Germany 13.2%, Austria 5.8%, Italy 4.3%, US 2.6%, Poland 2.4%, Ukraine 1.9%, Hungary 1.6% (January-September 1994)
External debt: $4.2 billion hard currency indebtedness (1994 est.)
Industrial production: growth rate NA%
Electricity:
capacity: 6,300,000 kW
production: 20.9 billion kWh
consumption per capita: 3,609 kWh (1993)
Industries: metal and metal products; food and beverages; electricity, gas, and water; coking, oil production, and nuclear fuel production; chemicals and manmade fibers; machinery; paper and printing; earthenware and ceramics; transport vehicles; textiles; electrical and optical apparatus; rubber products
Agriculture: largely self-sufficient in food production; diversified crop and livestock production, including grains, potatoes, sugar beets, hops, fruit, hogs, cattle, and poultry; exporter of forest products
Illicit drugs: transshipment point for Southwest Asian heroin bound for Western Europe
Economic aid:
donor: the former Czechoslovakia was a donor—$4.2 billion in bilateral aid to non-Communist less developed countries (1954-89)
Currency: 1 koruna (Sk) = 100 halierov
Exchange rates: koruny (Sk) per US$1—31.14 (September 1994), 32.9 (December 1993), 28.59 (December 1992), 28.26 (1992), 29.53 (1991), 17.95 (1990), 15.05 (1989); note—values before 1993 reflect Czechoslovak exchange rate
Fiscal year: calendar year

Transportation

Railroads:
total: 3,660 km (electrified 635 km)
broad gauge: 102 km 1.520-m gauge
standard gauge: 3,511 km 1.435-m gauge
narrow gauge: 47 km (35 km 1.000-m gauge; 12 km 0.750-m gauge) (1994)
Highways:
total: 17,650 km (1990)
paved: NA
unpaved: NA
Inland waterways: NA km
Pipelines: petroleum products NA km; natural gas 2,700 km
Ports: Bratislava, Komarno
Merchant marine:
total: 2 cargo ships (1,000 GRT or over) totaling 4,160 GRT/6,163 DWT
Airports:
total: 37
with paved runways over 3,047 m: 1
with paved runways 2,438 to 3,047 m: 3
with paved runways 1,524 to 2,437 m: 2
with paved runways 914 to 1,523 m: 2
with paved runways under 914 m: 4
with unpaved runways 2,438 to 3,047 m: 2
with unpaved runways 1,524 to 2,438 m: 2
with unpaved runways 914 to 1,523 m: 10
with unpaved runways under 914 m: 11

Communications

Telephone system: NA telephones
local: NA
intercity: NA
international: NA
Radio:
broadcast stations: AM NA, FM NA, shortwave NA
radios: NA
Television:
broadcast stations: NA
televisions: NA

Defense Forces

Branches: Army, Air and Air Defense Forces, Civil Defense, Railroad Units
Manpower availability: males age 15-49 1,443,719; males fit for military service 1,107,453; males reach military age (18) annually 49,045 (1995 est.)
Defense expenditures: 9.59 billion koruny, 3.1% of GDP (1994 est.); note—conversion of defense expenditures into US dollars using the current exchange rate could produce misleading results

Slovenia

Geography

Location: Southeastern Europe, bordering the Adriatic Sea, between Croatia and Italy
Map references: Ethnic Groups in Eastern Europe, Europe
Area:
total area: 20,296 sq km
land area: 20,296 sq km
comparative area: slightly larger than New Jersey
Land boundaries: total 1,045 km, Austria 262 km, Croatia 501 km, Italy 199 km, Hungary 83 km
Coastline: 32 km
Maritime claims: NA
International disputes: dispute with Croatia over fishing rights in the Adriatic and over some border areas; the border issue is currently under negotiation
Climate: Mediterranean climate on the coast, continental climate with mild to hot summers and cold winters in the plateaus and valleys to the east
Terrain: a short coastal strip on the Adriatic, an alpine mountain region adjacent to Italy, mixed mountain and valleys with numerous rivers to the east
Natural resources: lignite coal, lead, zinc, mercury, uranium, silver
Land use:
arable land: 10%
permanent crops: 2%
meadows and pastures: 20%
forest and woodland: 45%
other: 23%
Irrigated land: NA sq km
Environment:
current issues: Sava River polluted with domestic and industrial waste; pollution of coastal waters with heavy metals and toxic chemicals; forest damage near Koper from air pollution (originating at metallurgical and chemical plants) and resulting acid rain
natural hazards: flooding and earthquakes
international agreements: party to—Air Pollution, Hazardous Wastes, Marine Dumping, Nuclear Test Ban, Ozone Layer Protection, Ship Pollution, Wetlands; signed, but not ratified—Air Pollution-Sulphur 94, Biodiversity, Climate Change

People

Population: 2,051,522 (July 1995 est.)
Age structure:
0-14 years: 19% (female 191,318; male 200,957)
15-64 years: 69% (female 701,082; male 708,482)
65 years and over: 12% (female 160,662; male 89,021) (July 1995 est.)
Population growth rate: 0.24% (1995 est.)
Birth rate: 11.85 births/1,000 population (1995 est.)
Death rate: 9.27 deaths/1,000 population (1995 est.)
Net migration rate: -0.19 migrant(s)/1,000 population (1995 est.)
Infant mortality rate: 7.9 deaths/1,000 live births (1995 est.)
Life expectancy at birth:
total population: 74.73 years
male: 70.91 years
female: 78.76 years (1995 est.)
Total fertility rate: 1.64 children born/woman (1995 est.)
Nationality:
noun: Slovene(s)
adjective: Slovenian
Ethnic divisions: Slovene 91%, Croat 3%, Serb 2%, Muslim 1%, other 3%
Religions: Roman Catholic 96% (including 2% Uniate), Muslim 1%, other 3%
Languages: Slovenian 91%, Serbo-Croatian 7%, other 2%
Literacy: NA%
Labor force: 786,036
by occupation: agriculture 2%, manufacturing and mining 46%

Government

Names:
conventional long form: Republic of Slovenia
conventional short form: Slovenia
local long form: Republika Slovenije
local short form: Slovenija
Digraph: SI
Type: emerging democracy
Capital: Ljubljana
Administrative divisions: 60 provinces (pokajine, singular—pokajina) Ajdovscina, Brezice, Celje, Cerknica, Crnomelj, Dravograd, Gornja Radgona, Grosuplje, Hrastnik Lasko, Idrija, Ilirska Bistrica, Izola, Jesenice, Kamnik, Kocevje, Koper, Kranj, Krsko, Lenart, Lendava, Litija, Ljubljana-Bezigrad, Ljubljana-Center, Ljubljana-Moste-Polje, Ljubljana-Siska, Ljubljana-Vic-Rudnik, Ljutomer, Logatec, Maribor, Metlika, Mozirje, Murska Sobota, Nova Gorica, Novo Mesto, Ormoz, Pesnica, Piran, Postojna, Ptuj, Radlje Ob Dravi, Radovljica, Ravne Na Koroskem, Ribnica, Ruse, Sentjur Pri Celju, Sevnica, Sezana, Skofja Loka, Slovenj Gradec, Slovenska Bistrica, Slovenske Konjice, Smarje Pri Jelsah, Tolmin, Trbovlje, Trebnje, Trzic, Velenje, Vrhnika, Zagorje Ob Savi, Zalec
Independence: 25 June 1991 (from Yugoslavia)
National holiday: Statehood Day, 25 June (1991)
Constitution: adopted 23 December 1991, effective 23 December 1991
Legal system: based on civil law system
Suffrage: 18 years of age; universal (16 years of age, if employed)
Executive branch:
chief of state: President Milan KUCAN (since 22 April 1990); election last held 6 December 1992 (next to be held NA 1996); results—Milan KUCAN reelected by direct popular vote
head of government: Prime Minister Janez DRNOVSEK (since 14 May 1992)
cabinet: Council of Ministers
Legislative branch: bicameral National Assembly
State Assembly: elections last held 6 December 1992 (next to be held NA 1996); results—percent of vote by party NA; seats—(90 total) LDS 22, SKD 15, United List (former Communists and allies) 14, Slovene National Party 12, SLS 10, Democratic Party 6, ZS 5, SDSS 4, Hungarian minority 1, Italian minority 1
State Council: will become operational after next election in 1996; in the election of 6 December 1992, 40 members were elected to represent local and socioeconomic interests
Judicial branch: Supreme Court, Constitutional Court
Political parties and leaders: Liberal Democratic (LDS), Janez DRNOVSEK, chairman; Slovene Christian Democrats (SKD), Lozje PETERLE, chairman; Social Democratic Party of Slovenia (SDSS), Janez JANSA, chairman; Slovene People's National Party, Marjan PODOBNIK, chairman; United List (former Communists and allies), Janez KOCJANCIC, chairman; Slovene People's Party (SLS), Ivan OMAN, chairman; Democratic Party, Igor BAVCAR, chairman; Greens of Slovenia (ZS), Dusan PLUT, chairman
note: parties have changed as of the December 1992 elections
Other political or pressure groups: none
Member of: CCC, CE, CEI, EBRD, ECE, FAO, IADB, IAEA, IBRD, ICAO, ICRM, IDA, IFC, ILO, IMF, IMO, INTELSAT (nonsignatory user), INTERPOL, IOC, IOM (observer), ISO, ITU, NAM (guest), OSCE, PFP, UN, UNCTAD, UNESCO, UNIDO, UPU, WHO, WIPO, WMO, WTO
Diplomatic representation in US:
chief of mission: Ambassador Ernest PETRIC

chancery: 1525 New Hampshire Avenue NW, Washington, DC 20036
telephone: [1] (202) 667-5363
FAX: [1] (202) 667-4563
consulate(s) general: New York
US diplomatic representation:
chief of mission: Ambassador E. Allan WENDT
embassy: P.O. Box 254, Prazakova 4, 61000 Ljubljana
mailing address: American Embassy, Ljubljana, Department of State, Washington, DC 20521-7140
telephone: [386] (61) 301-427, 472, 485
FAX: [386] (61) 301-401
Flag: three equal horizontal bands of white (top), blue, and red with the Slovenian seal (a shield with the image of Triglav in white against a blue background at the center, beneath it are two wavy blue lines depicting seas and rivers, and around it, there are three six-sided stars arranged in an inverted triangle); the seal is located in the upper hoist side of the flag centered in the white and blue bands

Economy

Overview: Slovenia appears to be making a solid economic recovery, fulfilling the promise it showed at the time of Yugoslavia's breakup. It was by far the most prosperous of the former Yugoslav republics, with a per capita income more than twice the national average. It also benefited from strong ties to Western Europe and suffered comparatively small physical damage in the dismemberment process. The beginning was difficult, however. Real GDP fell 15% during 1991-92, while inflation jumped to 247% in 1991 and unemployment topped 8%—nearly three times the 1989 level. The turning point came in 1993 when real GDP grew 1%, unemployment leveled off at about 9%, and inflation slowed dramatically to 23%. In 1994, the rate of growth of GDP rose to 4%, unemployment remained stable, and inflation dropped to 20%. This was accomplished, moreover, without balance-of-payments problems. The government gets generally good economic marks from foreign observers, particularly with regard to fiscal policy—the budget deficit in 1994 was only about 1% of GDP, following several years of small surpluses. Prospects for 1995 appear good, with economic growth expected to remain strong while unemployment and inflation may decline slightly. Privatization, sluggish to date, is expected to pick up in 1995.
National product: GDP—purchasing power parity—$16 billion (1994 est.)
National product real growth rate: 4% (1994 est.)
National product per capita: $8,110 (1994 est.)
Inflation rate (consumer prices): 20% (1994)
Unemployment rate: 9% (1994 est.)
Budget:
revenues: $9.9 billion
expenditures: $9.8 billion, including capital expenditures of $NA (1993)
Exports: $6.5 billion (f.o.b., 1994 est.)
commodities: machinery and transport equipment 27%, intermediate manufactured goods 26%, chemicals 9%, food 4.8%, raw materials 3%, consumer goods 26% (1993)
partners: Germany 29.5%, former Yugoslavia 15.8%, Italy 12.4%, France 8.7%, Austria 5.0% (1993)
Imports: $6.5 billion (f.o.b., 1994 est.)
commodities: machinery and transport equipment 30%, intermediate manufactured goods 17.6%, chemicals 11.5%, raw materials 5.3%, fuels and lubricants 10.8%, food 8.4% (1993)
partners: Germany 25.0%, Italy 16.1%, former Yugoslavia 10.7%, France 8.0%, Austria 8.5% (1993)
External debt: $2.1 billion (1994)
Industrial production: growth rate 6% (1994 est.); accounts for 37% of GDP (1993)
Electricity:
capacity: 2,700,000 kW
production: 8.9 billion kWh
consumption per capita: 4,470 kWh (1993)
Industries: ferrous metallurgy and rolling mill products, aluminum reduction and rolled products, lead and zinc smelting, electronics (including military electronics), trucks, electric power equipment, wood products, textiles, chemicals, machine tools
Agriculture: accounts for 4.8% of GDP (1993); dominated by stock breeding (sheep and cattle) and dairy farming; main crops—potatoes, hops, hemp, flax; an export surplus in these commodities; Slovenia must import many other agricultural products and has a negative overall trade balance in this sector
Illicit drugs: NA
Economic aid: $NA
Currency: 1 tolar (SIT) = 100 stotins
Exchange rates: tolars (SIT) per US$1—127 (January 1995), 112 (June 1993), 28 (January 1992)
Fiscal year: calendar year

Transportation

Railroads:
total: 1,201 km
standard gauge: 1,201 km 1.435-m gauge (electrified 499 km) (1994)
Highways:
total: 14,726 km
paved: 11,046 km (187 km expressways)
unpaved: gravel 3,680 km (1992)
Inland waterways: NA
Pipelines: crude oil 290 km; natural gas 305 km
Ports: Izola, Koper, Piran
Merchant marine:
total: 17 ships (1,000 GRT or over) totaling 265,937 GRT/449,205 DWT (controlled by Slovenian owners)
ships by type: bulk 11, cargo 6
note: ships under the flag of Saint Vincent and the Grenadines, Singapore, Liberia; no ships remain under the Slovenian flag
Airports:
total: 14
with paved runways over 3,047 m: 1
with paved runways 2,438 to 3,047 m: 1
with paved runways 1,524 to 2,437 m: 1
with paved runways 914 to 1,523 m: 2
with paved runways under 914 m: 5
with unpaved runways 1,524 to 2,438 m: 2
with unpaved runways 914 to 1,523 m: 2

Communications

Telephone system: 130,000 telephones
local: NA
intercity: NA
international: NA
Radio:
broadcast stations: AM 6, FM 5, shortwave 0
radios: 370,000
Television:
broadcast stations: 7
televisions: 330,000

Defense Forces

Branches: Slovene Defense Forces
Manpower availability: males age 15-49 542,815; males fit for military service 434,302; males reach military age (19) annually 15,350 (1995 est.)
Defense expenditures: 13.5 billion tolars, 4.5% of GDP (1993 est.); note—conversion of the military budget into US dollars using the current exchange rate could produce misleading results

Solomon Islands

Geography

Location: Oceania, group of islands in the South Pacific Ocean, east of Papua New Guinea
Map references: Oceania
Area:
total area: 28,450 sq km
land area: 27,540 sq km
comparative area: slightly larger than Maryland
Land boundaries: 0 km
Coastline: 5,313 km
Maritime claims: measured from claimed archipelagic baselines
continental shelf: 200 nm
exclusive economic zone: 200 nm
territorial sea: 12 nm
International disputes: none
Climate: tropical monsoon; few extremes of temperature and weather
Terrain: mostly rugged mountains with some low coral atolls
Natural resources: fish, forests, gold, bauxite, phosphates, lead, zinc, nickel
Land use:
arable land: 1%
permanent crops: 1%
meadows and pastures: 1%
forest and woodland: 93%
other: 4%
Irrigated land: NA sq km
Environment:
current issues: deforestation; soil erosion; much of the surrounding coral reefs are dead or dying
natural hazards: typhoons, but they are rarely destructive; geologically active region with frequent earth tremors; volcanic activity
international agreements: party to—Climate Change, Environmental Modification, Marine Dumping, Marine Life Conservation, Ozone Layer Protection, Whaling; signed, but not ratified—Biodiversity, Law of the Sea

People

Population: 399,206 (July 1995 est.)
Age structure:
0-14 years: 46% (female 90,293; male 93,695)
15-64 years: 51% (female 100,183; male 103,374)
65 years and over: 3% (female 5,738; male 5,923) (July 1995 est.)
Population growth rate: 3.4% (1995 est.)
Birth rate: 38.48 births/1,000 population (1995 est.)
Death rate: 4.51 deaths/1,000 population (1995 est.)
Net migration rate: 0 migrant(s)/1,000 population (1995 est.)
Infant mortality rate: 26.7 deaths/1,000 live births (1995 est.)
Life expectancy at birth:
total population: 70.84 years
male: 68.38 years
female: 73.41 years (1995 est.)
Total fertility rate: 5.59 children born/woman (1995 est.)
Nationality:
noun: Solomon Islander(s)
adjective: Solomon Islander
Ethnic divisions: Melanesian 93%, Polynesian 4%, Micronesian 1.5%, European 0.8%, Chinese 0.3%, other 0.4%
Religions: Anglican 34%, Roman Catholic 19%, Baptist 17%, United (Methodist/Presbyterian) 11%, Seventh-Day Adventist 10%, other Protestant 5%, traditional beliefs 4%
Languages: Melanesian pidgin in much of the country is lingua franca, English spoken by 1%-2% of population
note: 120 indigenous languages
Literacy: NA%
Labor force: NA
by occupation: agriculture, forestry, and fishing 32.4%, services 25%, construction, manufacturing, and mining 7.0%, commerce, transport, and finance 4.7% (1984)

Government

Names:
conventional long form: none
conventional short form: Solomon Islands
former: British Solomon Islands
Digraph: BP
Type: parliamentary democracy
Capital: Honiara
Administrative divisions: 7 provinces and 1 town*; Central, Guadalcanal, Honiara*, Isabel, Makira, Malaita, Temotu, Western
Independence: 7 July 1978 (from UK)
National holiday: Independence Day, 7 July (1978)
Constitution: 7 July 1978
Legal system: common law
Suffrage: 21 years of age; universal
Executive branch:
chief of state: Queen ELIZABETH II (since 6 February 1952), represented by Governor General Moses PITAKAKA (since 10 June 1994)
head of government: Prime Minister Solomon MAMALONI (since 7 November 1994); Deputy Prime Minister Dennis LULEI (since 10 November 1994)
cabinet: Cabinet; appointed by the governor general on advice of the prime minister from members of parliament
Legislative branch: unicameral
National Parliament: elections last held NA November 1994 (next to be held NA 1997); results—percent of vote by party NA; seats—(47 total) number of seats by party NA
Judicial branch: High Court
Political parties and leaders: People's Alliance Party (PAP); United Party (UP), leader NA; Solomon Islands Liberal Party (SILP), Bartholemew ULUFA'ALU; Nationalist Front for Progress (NFP), Andrew NORI; Labor Party (LP), Joses TUHANUKU; National Action Party, leader NA; Christian Fellowship, leader NA; National Unity Group, Solomon MAMALONI
Member of: ACP, AsDB, C, ESCAP, FAO, G-77, IBRD, ICAO, ICRM, IDA, IFAD, IFC, IFRCS, ILO, IMF, IMO, INTELSAT (nonsignatory user), IOC, ITU, SPARTECA, SPC, SPF, UN, UNCTAD, UNESCO, UPU, WFTU, WHO, WMO
Diplomatic representation in US: ambassador traditionally resides in Honiara (Solomon Islands)
US diplomatic representation: embassy closed July 1993; the ambassador to Papua New Guinea is accredited to the Solomon Islands
Flag: divided diagonally by a thin yellow stripe from the lower hoist-side corner; the upper triangle (hoist side) is blue with five white five-pointed stars arranged in an X pattern; the lower triangle is green

Economy

Overview: The bulk of the population depend on subsistence agriculture, fishing, and forestry for at least part of their livelihood. Most manufactured goods and petroleum products must be imported. The islands are rich in undeveloped mineral resources such as lead, zinc, nickel, and gold. The economy suffered from a severe cyclone in mid-1986 that caused widespread damage to the infrastructure. In 1993, the government was working with the IMF to develop a structural adjustment program to address the country's fiscal deficit.
National product: GDP—purchasing power parity—$1 billion (1992 est.)
National product real growth rate: 8% (1992)
National product per capita: $2,590 (1992 est.)

Inflation rate (consumer prices): 13% (1992 est.)
Unemployment rate: NA%
Budget:
revenues: $48 million
expenditures: $107 million, including capital expenditures of $45 million (1991 est.)
Exports: $84 million (f.o.b., 1991)
commodities: fish 46%, timber 31%, palm oil 5%, cocoa, copra
partners: Japan 39%, UK 23%, Thailand 9%, Australia 5%, US 2% (1991)
Imports: $110 million (c.i.f., 1991)
commodities: plant and machinery, manufactured goods, food and live animals, fuel
partners: Australia 34%, Japan 16%, Singapore 14%, NZ 9%
External debt: $128 million (1988 est.)
Industrial production: growth rate -3.8% (1991 est.); accounts for 5% of GDP
Electricity:
capacity: 21,000 kW
production: 30 million kWh
consumption per capita: 80 kWh (1993)
Industries: copra, fish (tuna)
Agriculture: including fishing and forestry, accounts for 31% of GDP; mostly subsistence farming; cash crops—cocoa, beans, coconuts, palm kernels, timber; other products—rice, potatoes, vegetables, fruit, cattle, pigs; not self-sufficient in food grains; 90% of the total fish catch of 44,500 metric tons was exported (1988)
Economic aid:
recipient: Western (non-US) countries, ODA and OOF bilateral commitments (1980-89), $250 million
Currency: 1 Solomon Islands dollar (SI$) = 100 cents
Exchange rates: Solomon Islands dollars (SI$) per US$1—3.3113 (September 1994), 3.1877 (1993), 2.9281 (1992), 2.7148 (1991), 2.5288 (1990)
Fiscal year: calendar year

Transportation

Railroads: 0 km
Highways:
total: 1,300 km
paved: 30 km
unpaved: gravel 290 km; earth 980 km
note: in addition, there are 800 km of private logging and plantation roads of varied construction (1982)
Ports: Aola Bay, Honiara, Lofung, Noro, Viru Harbor, Yandina
Merchant marine: none
Airports:
total: 31
with paved runways 1,524 to 2,437 m: 1
with paved runways 914 to 1,523 m: 1
with paved runways under 914 m: 19
with unpaved runways 1,524 to 2,438 m: 1
with unpaved runways 914 to 1,523 m: 9

Communications

Telephone system: 3,000 telephones
local: NA
intercity: NA
international: 1 INTELSAT (Pacific Ocean) earth station
Radio:
broadcast stations: AM 4, FM 0, shortwave 0
radios: NA
Television:
broadcast stations: 0
televisions: NA

Defense Forces

Branches: no military forces; Royal Solomon Islands Police (RSIP)
Defense expenditures: $NA, NA% of GDP

Somalia

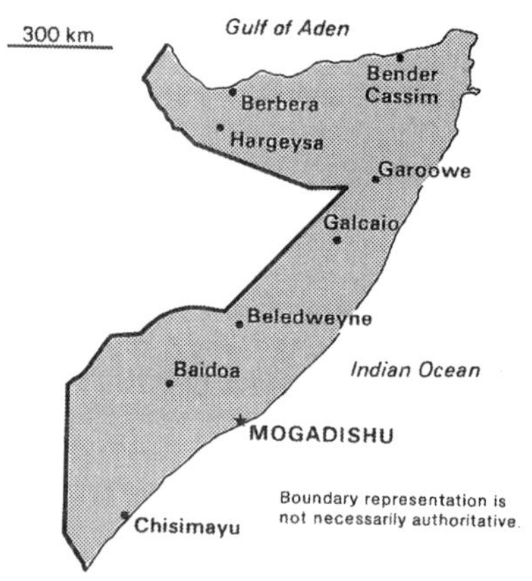

Geography

Location: Eastern Africa, bordering the Gulf of Aden and the Indian Ocean, east of Ethiopia
Map references: Africa
Area:
total area: 637,660 sq km
land area: 627,340 sq km
comparative area: slightly smaller than Texas
Land boundaries: total 2,366 km, Djibouti 58 km, Ethiopia 1,626 km, Kenya 682 km
Coastline: 3,025 km
Maritime claims:
territorial sea: 200 nm
International disputes: southern half of boundary with Ethiopia is a Provisional Administrative Line; territorial dispute with Ethiopia over the Ogaden
Climate: principally desert; December to February—northeast monsoon, moderate temperatures in north and very hot in south; May to October—southwest monsoon, torrid in the north and hot in the south, irregular rainfall, hot and humid periods (tangambili) between monsoons
Terrain: mostly flat to undulating plateau rising to hills in north
Natural resources: uranium and largely unexploited reserves of iron ore, tin, gypsum, bauxite, copper, salt
Land use:
arable land: 2%
permanent crops: 0%
meadows and pastures: 46%
forest and woodland: 14%
other: 38%
Irrigated land: 1,600 sq km (1989 est.)
Environment:
current issues: famine; use of contaminated water contributes to human health problems; deforestation; overgrazing; soil erosion; desertification
natural hazards: recurring droughts; frequent dust storms over eastern plains in summer
international agreements: party to—Endangered Species, Law of the Sea; signed, but not ratified—Marine Dumping, Nuclear Test Ban

Note: strategic location on Horn of Africa along southern approaches to Bab el Mandeb and route through Red Sea and Suez Canal

People

Population: 7,347,554 (July 1995 est.)
Age structure:
0-14 years: 45% (female 1,653,175; male 1,650,377)
15-64 years: 51% (female 1,845,886; male 1,932,012)
65 years and over: 4% (female 138,264; male 127,840) (July 1995 est.)
Population growth rate: 15.58% (1995 est.)
Birth rate: 45.53 births/1,000 population (1995 est.)
Death rate: 13.3 deaths/1,000 population (1995 est.)
Net migration rate: 123.62 migrant(s)/1,000 population (1995 est.)
Infant mortality rate: 119.5 deaths/1,000 live births (1995 est.)
Life expectancy at birth:
total population: 55.74 years
male: 55.48 years
female: 56 years (1995 est.)
Total fertility rate: 7.13 children born/woman (1995 est.)
Nationality:
noun: Somali(s)
adjective: Somali
Ethnic divisions: Somali 85%, Bantu, Arabs 30,000
Religions: Sunni Muslim
Languages: Somali (official), Arabic, Italian, English
Literacy: age 15 and over can read and write (1990 est.)
total population: 24%
male: 36%
female: 14%
Labor force: 2.2 million (very few are skilled laborers)
by occupation: pastoral nomad 70%, agriculture, government, trading, fishing, handicrafts, and other 30%

Government

Names:
conventional long form: none
conventional short form: Somalia
former: Somali Republic
Digraph: SO
Type: none
Capital: Mogadishu
Administrative divisions: 18 regions (plural—NA, singular—gobolka); Awdal, Bakool, Banaadir, Bari, Bay, Galguduud, Gedo, Hiiraan, Jubbada Dhexe, Jubbada Hoose, Mudug, Nugaal, Sanaag, Shabeellaha Dhexe, Shabeellaha Hoose, Sool, Togdheer, Woqooyi Galbeed
Independence: 1 July 1960 (from a merger of British Somaliland, which became independent from the UK on 26 June 1960, and Italian Somaliland, which became independent from the Italian-administered UN trusteeship on 1 July 1960, to form the Somali Republic)
National holiday: NA
Constitution: 25 August 1979, presidential approval 23 September 1979
Legal system: NA
Suffrage: 18 years of age; universal
Executive branch: Somalia has no functioning government; the United Somali Congress (USC) ousted the regime of Maj. Gen. Mohamed SIAD Barre on 27 January 1991; the present political situation is one of anarchy, marked by inter-clan fighting and random banditry
Legislative branch: unicameral People's Assembly
People's Assembly (Golaha Shacbiga): not functioning
Judicial branch: Supreme Court (not functioning)
Political parties and leaders: the United Somali Congress (USC) ousted the former regime on 27 January 1991; formerly the only party was the Somali Revolutionary Socialist Party (SRSP), headed by former President and Commander in Chief of the Army Maj. Gen. Mohamed SIAD Barre
Other political or pressure groups: numerous clan and subclan factions are currently vying for power
Member of: ACP, AfDB, AFESD, AL, AMF, CAEU, ECA, FAO, G-77, IBRD, ICAO, ICRM, IDA, IDB, IFAD, IFC, IFRCS, IGADD, ILO, IMF, IMO, INTELSAT, INTERPOL, IOC, IOM (observer), ITU, NAM, OAU, OIC, UN, UNCTAD, UNESCO, UNHCR, UNIDO, UPU, WFTU, WHO, WIPO, WMO
Diplomatic representation in US: Somalian Embassy ceased operations on 8 May 1991
US diplomatic representation:
note: the US Embassy in Mogadishu was evacuated and closed indefinitely in January 1991; Ambassador Daniel SIMPSON, ambassador to Kenya, represents US interests in Somalia
liaison office: US Embassy, Nairobi, Kenya
address: corner of Moi Avenue and Haile Selassie Avenue, Nairobi
mailing address: P.O. Box 30137, Unit 64100, Nairobi or APO AE 09831
telephone: [254] (2) 334141
FAX: [254] (2) 340838
Flag: light blue with a large white five-pointed star in the center; design based on the flag of the UN (Italian Somaliland was a UN trust territory)

Economy

Overview: One of the world's poorest and least developed countries, Somalia has few resources. Moreover, much of the economy has been devastated by the civil war. Agriculture is the most important sector, with livestock accounting for about 40% of GDP and about 65% of export earnings. Nomads and seminomads who are dependent upon livestock for their livelihood make up about 70% of the population. Crop production generates only 10% of GDP and employs about 20% of the work force. The main export crop is bananas; sugar, sorghum, and corn are grown for the domestic market. The small industrial sector is based on the processing of agricultural products and accounts for less than 10% of GDP; most facilities have been shut down because of the civil strife. The greatly increased political turmoil of 1991-93 has resulted in a substantial drop in output, with widespread famine. In 1994 agricultural output and economic conditions stabilized in the countryside but may turn worse in 1995 if civil strife intensifies after the UN withdrawal.
National product: GDP—purchasing power parity—$3.3 billion (1994 est.)
National product real growth rate: NA%
National product per capita: $500 (1994 est.)
Inflation rate (consumer prices): NA% (1994)
Unemployment rate: NA%
Budget:
revenues: $NA
expenditures: $NA, including capital expenditures of $NA
Exports: $58 million (1990 est.)
commodities: bananas, live animals, fish, hides
partners: Saudi Arabia, Italy, FRG (1986)
Imports: $249 million (1990 est.)
commodities: petroleum products, foodstuffs, construction materials
partners: US 13%, Italy, FRG, Kenya, UK, Saudi Arabia (1986)
External debt: $1.9 billion (1989)
Industrial production: growth rate NA%
Electricity:
capacity: prior to the civil war 75,000 kW, but now almost completely shut down due to war damage; note—UN and relief organizations use their own portable power systems
production: NA kWh
consumption per capita: NA kWh
Industries: a few small industries, including sugar refining, textiles, petroleum refining (mostly shut down) (1994)
Agriculture: dominant sector, led by livestock raising (cattle, sheep, goats); crops—bananas, sorghum, corn, mangoes, sugarcane; not self-sufficient in food; distribution of food disrupted by civil strife; fishing potential largely unexploited

Economic aid:
recipient: US commitments, including Ex-Im (FY70-89), $639 million; Western (non-US) countries, ODA and OOF bilateral commitments (1970-89), $3.8 billion; OPEC bilateral aid (1979-89), $1.1 billion; Communist countries (1970-89), $336 million
Currency: 1 Somali shilling (So. Sh.) = 100 cents
Exchange rates: Somali shillings (So. Sh.) per US$1—approximately 5,000 (1 January 1995), 2,616 (1 July 1993), 4,200 (December 1992), 3,800.00 (December 1990), 490.7 (1989),
Fiscal year: calendar year

Transportation

Railroads: 0 km
Highways:
total: 22,500 km
paved: 2,700 km
unpaved: gravel 3,000 km; improved, stabilized earth 16,800 km (1992)
Pipelines: crude oil 15 km
Ports: Bender Cassim (Boosaaso), Berbera, Chisimayu (Kismaayo), Merca, Mogadishu
Merchant marine:
total: 2 ships (1,000 GRT or over) totaling 5,529 GRT/6,892 DWT
ships by type: cargo 1, refrigerated cargo 1
Airports:
total: 76
with paved runways over 3,047 m: 3
with paved runways 2,438 to 3,047 m: 1
with paved runways 1,524 to 2,437 m: 3
with paved runways 914 to 1,523 m: 1
with paved runways under 914 m: 14
with unpaved runways 2,438 to 3,047 m: 5
with unpaved runways 1,524 to 2,438 m: 16
with unpaved runways 914 to 1,523 m: 33

Communications

Telephone system: NA telephones; the public telecommunications system was completely destroyed or dismantled by the civil war factions; all relief organizations depend on their own private systems (1993)
local: NA
intercity: NA
international: NA
Radio:
broadcast stations: AM NA, FM NA, shortwave NA
radios: NA
Television:
broadcast stations: NA
televisions: NA

Defense Forces

Branches: no functioning central government military forces; clan militias continue to battle for control of key economic or political prizes
Manpower availability: males age 15-49 1,736,673; males fit for military service 972,203 (1995 est.)
Defense expenditures: $NA, NA% of GDP

South Africa

Geography

Location: Southern Africa, at the southern tip of the continent of Africa
Map references: Africa
Area:
total area: 1,219,912 sq km
land area: 1,219,912 sq km
comparative area: slightly less than twice the size of Texas
note: includes Prince Edward Islands (Marion Island and Prince Edward Island)
Land boundaries: total 4,750 km, Botswana 1,840 km, Lesotho 909 km, Mozambique 491 km, Namibia 855 km, Swaziland 430 km, Zimbabwe 225 km
Coastline: 2,798 km
Maritime claims:
continental shelf: 200-m depth or to the depth of exploitation
exclusive fishing zone: 200 nm
territorial sea: 12 nm
International disputes: Swaziland has asked South Africa to open negotiations on reincorporating some nearby South African territories that are populated by ethnic Swazis or that were long ago part of the Swazi Kingdom
Climate: mostly semiarid; subtropical along east coast; sunny days, cool nights
Terrain: vast interior plateau rimmed by rugged hills and narrow coastal plain
Natural resources: gold, chromium, antimony, coal, iron ore, manganese, nickel, phosphates, tin, uranium, gem diamonds, platinum, copper, vanadium, salt, natural gas
Land use:
arable land: 10%
permanent crops: 1%
meadows and pastures: 65%
forest and woodland: 3%
other: 21%
Irrigated land: 11,280 sq km (1989 est.)
Environment:
current issues: lack of important arterial rivers or lakes requires extensive water conservation and control measures; growth in water usage

threatens to outpace supply; pollution of rivers from agricultural runoff and urban discharge; air pollution resulting in acid rain; soil erosion; desertification
natural hazards: prolonged droughts
international agreements: party to—Antarctic Treaty, Endangered Species, Hazardous Wastes, Marine Dumping, Marine Life Conservation, Nuclear Test Ban, Ozone Layer Protection, Ship Pollution, Wetlands, Whaling; signed, but not ratified—Antarctic-Environmental Protocol, Biodiversity, Climate Change, Law of the Sea
Note: South Africa completely surrounds Lesotho and almost completely surrounds Swaziland

People

Population:
total: 45,095,459 (July 1995 est.)
Age structure:
0-14 years: 40% (female 8,842,764; male 9,091,722)
15-64 years: 56% (female 12,825,617; male 12,508,039)
65 years and over: 4% (female 1,047,285; male 780,032) (July 1995 est.)
Population growth rate:
total: 2.61% (1995 est.)
Birth rate: 33.39 births/1,000 population (1995 est.)
Death rate: 7.42 deaths/1,000 population (1995 est.)
Net migration rate: 0.17 migrant(s)/1,000 population (1995 est.)
Infant mortality rate: 45.8 deaths/1,000 live births (1995 est.)
Life expectancy at birth:
total population: 65.42 years
male: 62.68 years
female: 68.25 years (1995 est.)
Total fertility rate: 4.35 children born/woman (1995 est.)
Nationality:
noun: South African(s)
adjective: South African
Ethnic divisions: black 75.2%, white 13.6%, Colored 8.6%, Indian 2.6%
Religions: Christian (most whites and Coloreds and about 60% of blacks), Hindu (60% of Indians), Muslim 2%
Languages: eleven official languages, including Afrikaans, English, Ndebele, Pedi, Sotho, Swazi, Tsonga, Tswana, Venda, Xhosa, Zulu
Literacy: age 15 and over can read and write (1980)
total population: 76%
male: 78%
female: 75%
Labor force: 13.4 million economically active (1990)
by occupation: services 35%, agriculture 30%, industry 20%, mining 9%, other 6%

Government

Names:
conventional long form: Republic of South Africa
conventional short form: South Africa
Abbreviation: RSA
Digraph: SF
Type: republic
Capital: Pretoria (administrative); Cape Town (legislative); Bloemfontein (judicial)
Administrative divisions: 9 provinces; Eastern Cape, Eastern Transvaal, KwaZulu/Natal, Northern Cape, Northern Transvaal, Northwest, Orange Free State, Gauteng, Western Cape
Independence: 31 May 1910 (from UK)
National holiday: Freedom Day, 27 April (1994)
Constitution: 27 April 1994 (interim constitution, replacing the constitution of 3 September 1984)
Legal system: based on Roman-Dutch law and English common law; accepts compulsory ICJ jurisdiction, with reservations
Suffrage: 18 years of age; universal
Executive branch:
chief of state and head of government: Executive President Nelson MANDELA (since 10 May 1994); Deputy Executive President Thabo MBEKI (since 10 May 1994); Deputy Executive President Frederik W. DE KLERK (since 10 May 1994)
note: any political party that wins 20% or more of the National Assembly votes in a general election is entitled to name a Deputy Executive President
cabinet: Cabinet appointed by the Executive President
Legislative branch: bicameral
National Assembly: elections last held 26-29 April 1994 (next to be held NA); results—ANC 62.6%, NP 20.4%, IFP 10.5%, FF 2.2%, DP 1.7%, PAC 1.2%, ACDP 0.5%, other 0.9%; seats—(400 total) ANC 252, NP 82, IFP 43, FF 9, DP 7, PAC 5, ACDP 2
Senate: the Senate is composed of members who are nominated by the nine provincial parliaments (which are elected in parallel with the National Assembly) and has special powers to protect regional interests, including the right to limited self-determination for ethnic minorities; seats—(90 total) ANC 61, NP 17, FF 4, IFP 5, DP 3
note: when the National Assembly meets in joint session with the Senate to consider the provisions of the constitution, the combined group is referred to as the Constitutional Assembly
Judicial branch: Supreme Court
Political parties and leaders: African National Congress (ANC), Nelson MANDELA, president; National Party (NP), Frederik W. DE KLERK, president; Inkatha Freedom Party (IFP), Mangosuthu BUTHELEZI, president; Freedom Front (FF), Constand VILJOEN, president; Democratic Party (DP); Pan Africanist Congress (PAC), Clarence MAKWETU, president; African Christian Democratic Party (ACDP), leader NA
note: in addition to these seven parties which won seats in the National Assembly, twelve other parties received votes in the national elections in April 1994
Other political or pressure groups: NA
Member of: BIS, C, CCC, ECA, FAO, GATT, IAEA, IBRD, ICAO, ICC, ICRM, IDA, IFC, IFRCS, ILO, IMF, INMARSAT, INTELSAT, INTERPOL, IOC, ISO, ITU, NAM, OAU, SACU, SADC, UN, UNCTAD, UPU, WFTU, WHO, WIPO, WMO, ZC
Diplomatic representation in US:
chief of mission: Ambassador Franklin SONN
chancery: 3051 Massachusetts Avenue NW, Washington, DC 20008
telephone: [1] (202) 232-4400
consulate(s) general: Beverly Hills (California), Chicago, and New York
US diplomatic representation:
chief of mission: Ambassador Princeton N. LYMAN
embassy: 877 Pretorius St., Arcadia 0083
mailing address: P.O. Box 9536, Pretoria 0001
telephone: [27] (12) 342-1048
FAX: [27] (12) 342-2244
consulate(s) general: Cape Town, Durban, Johannesburg
Flag: two equal width horizontal bands of red (top) and blue separated by a central green band which splits into a horozontal Y, the arms of which end at the corners of the hoist side, embracing a black isoceles triangle from which the arms are separated by narrow yellow bands; the red and blue bands are separated from the green band and its arms by narrow white stripes
note: prior to 26 April 1994, the flag was actually four flags in one—three miniature flags reproduced in the center of the white band of the former flag of the Netherlands, which has three equal horizontal bands of orange (top), white, and blue; the miniature flags are a vertically hanging flag of the old Orange Free State with a horizontal flag of the UK adjoining on the hoist side and a horizontal flag of the old Transvaal Republic adjoining on the other side

Economy

Overview: Many of the white one-seventh of the South African population enjoy incomes, material comforts, and health and educational standards equal to those of Western Europe. In contrast, most of the remaining population suffers from the poverty patterns of the Third World, including unemployment and lack of job skills. The main strength of the economy lies in its rich mineral resources, which provide two-thirds of exports. Economic developments for the remainder of the 1990s will be driven

largely by the new government's attempts to improve black living conditions, to set the country on an aggressive export-led growth path, and to cut back the enormous numbers of unemployed. The economy in recent years has absorbed less than 5% of the more than 300,000 workers entering the labor force annually. Local economists estimate that the economy must grow between 5% and 6% in real terms annually to absorb all of the new entrants, much less reduce the accumulated total.
National product: GDP—purchasing power parity—$194.3 billion (1994 est.)
National product real growth rate: 2% (1994 est.)
National product per capita: $4,420 (1994 est.)
Inflation rate (consumer prices): 9% (1994 est.)
Unemployment rate: 32.6% (1994 est.); an additional 11% underemployment
Budget:
revenues: $26.3 billion
expenditures: $34 billion, including capital expenditures of $2.5 billion (FY93/94 est.)
Exports: $25.3 billion (f.o.b., 1994)
commodities: gold 27%, other minerals and metals 20%-25%, food 5%, chemicals 3%
partners: Italy, Japan, US, Germany, UK, other EU countries, Hong Kong
Imports: $21.4 billion (f.o.b., 1994)
commodities: machinery 32%, transport equipment 15%, chemicals 11%, oil, textiles, scientific instruments
partners: Germany, US, Japan, UK, Italy
External debt: $18 billion (1994 est.)
Industrial production: growth rate NA%; accounts for about 40% of GDP
Electricity:
capacity: 39,750,000 kW
production: 163 billion kWh
consumption per capita: 3,482 kWh (1993)
Industries: mining (world's largest producer of platinum, gold, chromium), automobile assembly, metalworking, machinery, textile, iron and steel, chemical, fertilizer, foodstuffs
Agriculture: accounts for about 5% of GDP and 30% of labor force; diversified agriculture, with emphasis on livestock; products—cattle, poultry, sheep, wool, milk, beef, corn, wheat, sugarcane, fruits, vegetables; self-sufficient in food
Illicit drugs: transshipment center for heroin and cocaine; cocaine consumption on the rise; world's largest market for illicit methaqualone, usually imported illegally from India through various east African countries
Economic aid: many aid packages for the new government are still being prepared; current aid pledges include US $600 million over 3 years; UK $150 million over 3 years; Australia $21 million over 3 years; Japan $1.3 billion over 2 years
Currency: 1 rand (R) = 100 cents
Exchange rates: rand (R) per US$1—3.5389 (January 1995), 3.5490 (1994), 3.2636 (1993), 2.8497 (1992), 2.7563 (1991), 2.5863 (1990)
Fiscal year: 1 April—31 March

Transportation

Railroads:
total: 20,638 km
narrow gauge: 20,324 km 1.067-m gauge (substantial electrification); 314 km 0.610-m gauge
Highways:
total: 188,309 km
paved: 54,013 km
unpaved: crushed stone, gravel, improved earth 134,296 km
Pipelines: crude oil 931 km; petroleum products 1,748 km; natural gas 322 km
Ports: Cape Town, Durban, East London, Mosselbaai, Port Elizabeth, Richards Bay, Saldanha
Merchant marine:
total: 4 container ships (1,000 GRT or over) totaling 211,276 GRT/198,602 DWT
Airports:
total: 853
with paved runways over 3,047 m: 9
with paved runways 2,438 to 3,047 m: 5
with paved runways 1,524 to 2,437 m: 47
with paved runways 914 to 1,523 m: 72
with paved runways under 914 m: 327
with unpaved runways 1,524 to 2,438 m: 39
with unpaved runways 914 to 1,523 m: 354

Communications

Telephone system: over 4,500,000 telephones; the system is the best developed, most modern, and has the highest capacity in Africa
local: NA
intercity: consists of carrier-equipped open-wire lines, coaxial cables, microwave radio relay links, fiber optic cable, and radiocommunication stations; key centers are Bloemfontein, Cape Town, Durban, Johannesburg, Port Elizabeth, and Pretoria
international: 1 submarine cable; 3 INTELSAT (1 Indian Ocean and 2 Atlantic Ocean) earth stations
Radio:
broadcast stations: AM 14, FM 286, shortwave 0
radios: NA
Television:
broadcast stations: 67
televisions: NA

Defense Forces

Branches: South African National Defence Force (SANDF; includes Army, Navy, Air Force, and Medical Services), South African Police Service (SAPS)
Manpower availability: males age 15-49 10,830,079; males fit for military service 6,601,323; males reach military age (18) annually 439,793 (1995 est.)
Defense expenditures: exchange rate conversion—$3.2 billion, 2.8% of GDP (FY93/94)

South Georgia and the South Sandwich Islands
(dependent territory of the UK)

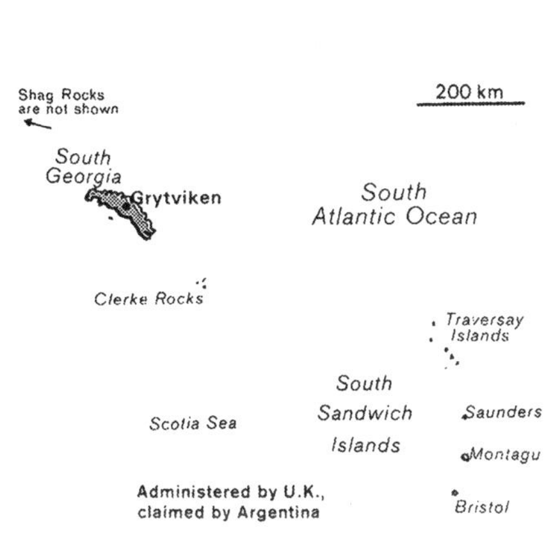

Geography

Location: Southern South America, islands in the South Atlantic Ocean, east of the tip of South America
Map references: Antarctic Region
Area:
total area: 4,066 sq km
land area: 4,066 sq km
comparative area: slightly larger than Rhode Island
note: includes Shag Rocks, Clerke Rocks, Bird Island
Land boundaries: 0 km
Coastline: NA km
Maritime claims:
exclusive fishing zone: 200 nm
territorial sea: 12 nm
International disputes: administered by the UK, claimed by Argentina
Climate: variable, with mostly westerly winds throughout the year, interspersed with periods of calm; nearly all precipitation falls as snow
Terrain: most of the islands, rising steeply from the sea, are rugged and mountainous; South Georgia is largely barren and has steep, glacier-covered mountains; the South Sandwich Islands are of volcanic origin with some active volcanoes
Natural resources: fish
Land use:
arable land: 0%
permanent crops: 0%
meadows and pastures: 0%
forest and woodland: 0%
other: 100% (largely covered by permanent ice and snow with some sparse vegetation consisting of grass, moss, and lichen)
Irrigated land: 0 sq km
Environment:
current issues: NA
natural hazards: the South Sandwich Islands have prevailing weather conditions that generally make them difficult to approach by ship; they are also subject to active volcanism
international agreements: NA
Note: the north coast of South Georgia has several large bays, which provide good anchorage; reindeer, introduced early in this century, live on South Georgia

People

Population: no indigenous population; there is a small military garrison on South Georgia, and the British Antarctic Survey has a biological station on Bird Island; the South Sandwich Islands are uninhabited

Government

Names:
conventional long form: South Georgia and the South Sandwich Islands
conventional short form: none
Digraph: SX
Type: dependent territory of the UK
Capital: none; Grytviken on South Georgia is the garrison town
Administrative divisions: none (dependent territory of the UK)
Independence: none (dependent territory of the UK)
National holiday: Liberation Day, 14 June (1982)
Constitution: 3 October 1985
Legal system: English common law
Executive branch:
chief of state: Queen ELIZABETH II (since 6 February 1952), represented by Commissioner David Everard TATHAM (since August 1992; resident at Stanley, Falkland Islands)
Legislative branch: no elections
Judicial branch: none

Economy

Overview: Some fishing takes place in adjacent waters. There is a potential source of income from harvesting fin fish and krill. The islands receive income from postage stamps produced in the UK.
Budget:
revenues: $291,777
expenditures: $451,000, including capital expenditures of $NA (1988 est.)
Electricity:
capacity: 900 kW
production: 2 million kWh
consumption per capita: NA kWh (1992)

Transportation

Highways:
total: NA
paved: NA
unpaved: NA
Ports: Grytviken
Airports: none

Communications

Telephone system: NA telephones; coastal radio station at Grytviken
local: NA
intercity: NA
international: NA
Radio:
broadcast stations: AM 0, FM 0, shortwave 0
radios: NA
Television:
broadcast stations: 0
televisions: NA

Defense Forces

Note: defense is the responsibility of the UK

Spain

Geography

Location: Southwestern Europe, bordering the Bay of Biscay, Mediterranean Sea, and North Atlantic Ocean, southwest of France
Map references: Europe
Area:
total area: 504,750 sq km
land area: 499,400 sq km
comparative area: slightly more than twice the size of Oregon
note: includes Balearic Islands, Canary Islands, and five places of sovereignty (plazas de soberania) on and off the coast of Morocco—Ceuta, Mellila, Islas Chafarinas, Penon de Alhucemas, and Penon de Velez de la Gomera
Land boundaries: total 1,903.2 km, Andorra 65 km, France 623 km, Gibraltar 1.2 km, Portugal 1,214 km
Coastline: 4,964 km
Maritime claims:
exclusive economic zone: 200 nm
territorial sea: 12 nm
International disputes: Gibraltar question with UK; Spain controls five places of sovereignty (plazas de soberania) on and off the coast of Morocco—the coastal enclaves of Ceuta and Melilla, which Morocco contests, as well as the islands of Penon de Alhucemas, Penon de Velez de la Gomera, and Islas Chafarinas
Climate: temperate; clear, hot summers in interior, more moderate and cloudy along coast; cloudy, cold winters in interior, partly cloudy and cool along coast
Terrain: large, flat to dissected plateau surrounded by rugged hills; Pyrenees in north
Natural resources: coal, lignite, iron ore, uranium, mercury, pyrites, fluorspar, gypsum, zinc, lead, tungsten, copper, kaolin, potash, hydropower
Land use:
arable land: 31%
permanent crops: 10%
meadows and pastures: 21%
forest and woodland: 31%
other: 7%
Irrigated land: 33,600 sq km (1989 est.)
Environment:
current issues: pollution of the Mediterranean Sea from raw sewage and effluents from the offshore production of oil and gas; air pollution; deforestation; desertification
natural hazards: periodic droughts
international agreements: party to—Air Pollution, Air Pollution-Nitrogen Oxides, Air Pollution-Volatile Organic Compounds, Antarctic-Environmental Protocol, Antarctic Treaty, Biodiversity, Climate Change, Endangered Species, Environmental Modification, Hazardous Wastes, Marine Dumping, Marine Life Conservation, Nuclear Test Ban, Ozone Layer Protection, Ship Pollution, Tropical Timber 83, Wetlands, Whaling; signed, but not ratified—Air Pollution-Sulphur 94, Desertification, Law of the Sea
Note: strategic location along approaches to Strait of Gibraltar

People

Population: 39,404,348 (July 1995 est.)
Age structure:
0-14 years: 17% (female 3,214,606; male 3,446,643)
15-64 years: 68% (female 13,377,839; male 13,457,683)
65 years and over: 15% (female 3,461,367; male 2,446,210) (July 1995 est.)
Population growth rate: 0.27% (1995 est.)
Birth rate: 11.21 births/1,000 population (1995 est.)
Death rate: 8.86 deaths/1,000 population (1995 est.)
Net migration rate: 0.31 migrant(s)/1,000 population (1995 est.)
Infant mortality rate: 6.7 deaths/1,000 live births (1995 est.)
Life expectancy at birth:
total population: 77.91 years
male: 74.67 years
female: 81.39 years (1995 est.)
Total fertility rate: 1.41 children born/woman (1995 est.)
Nationality:
noun: Spaniard(s)
adjective: Spanish
Ethnic divisions: composite of Mediterranean and Nordic types
Religions: Roman Catholic 99%, other sects 1%
Languages: Castilian Spanish, Catalan 17%, Galician 7%, Basque 2%
Literacy: age 15 and over can read and write (1986)
total population: 96%
male: 98%
female: 94%
Labor force: 14.621 million
by occupation: services 53%, industry 24%, agriculture 14%, construction 9% (1988)

Government

Names:
conventional long form: Kingdom of Spain
conventional short form: Spain
local short form: Espana
Digraph: SP
Type: parliamentary monarchy
Capital: Madrid
Administrative divisions: 17 autonomous communities (comunidades autonomas, singular—comunidad autonoma); Andalucia, Aragon, Asturias, Canarias, Cantabria, Castilla-La Mancha, Castilla y Leon, Cataluna, Communidad Valencia, Extremadura, Galicia, Islas Baleares, La Rioja, Madrid, Murcia, Navarra, Pais Vasco
note: there are five places of sovereignty on and off the coast of Morocco (Ceuta, Mellila, Islas Chafarinas, Penon de Alhucemas, and Penon de Velez de la Gomera) with administrative status unknown
Independence: 1492 (expulsion of the Moors and unification)
National holiday: National Day, 12 October
Constitution: 6 December 1978, effective 29 December 1978
Legal system: civil law system, with regional applications; does not accept compulsory ICJ jurisdiction
Suffrage: 18 years of age; universal
Executive branch:
chief of state: King JUAN CARLOS I (since 22 November 1975)
head of government: Prime Minister Felipe GONZALEZ Marquez (since 2 December 1982); Deputy Prime Minister Narcis SERRA y Serra (since 13 March 1991)
cabinet: Council of Ministers; designated by the prime minister
Council of State: is the supreme consultative organ of the government
Legislative branch: bicameral The General Courts or National Assembly (Las Cortes Generales)
Senate (Senado): elections last held 6 June 1993 (next to be held by June 1997); results—percent of vote by party NA; seats—(255 total) PSOE 117, PP 107, CiU 15, PNV 5, IU 2, other 9
Congress of Deputies (Congreso de los Diputados): elections last held 6 June 1993 (next to be held by June 1997); results—percent of vote by party NA; seats—(350 total) PSOE 159, PP 141, IU 18, CiU 17, PNV 5, CC 4, HB 2, other 4
Judicial branch: Supreme Court (Tribunal Supremo)
Political parties and leaders:
principal national parties, from right to left: Popular Party (PP), Jose Maria AZNAR Lopez; Democratic Social Center (CDS),

Rafael CALVO Ortega; Spanish Socialist Workers Party (PSOE), Felipe GONZALEZ Marquez, secretary general; Socialist Democracy Party (DS), Ricardo GARCIA Damborenea; Spanish Communist Party (PCE), Julio ANGUITA Gonzalez; United Left (IU—a coalition of parties including the PCE, a branch of the PSOE, and other small parties), Julio ANGUITA Gonzalez
chief regional parties: Convergence and Union (CiU), Miquel ROCA i Junyent, secretary general; Basque Nationalist Party (PNV), Xabier ARZALLUS Antia and Jose Antonio ARDANZA; Basque United People (HB), Jon IDIGORAS Guerricabeitia and Inaki ESNAOLA; Canarian Coalition (CC), a coalition of five parties
Other political or pressure groups: on the extreme left, the Basque Fatherland and Liberty (ETA) and the First of October Antifascist Resistance Group (GRAPO) use terrorism to oppose the government; free labor unions (authorized in April 1977) include the Communist-dominated Workers Commissions (CCOO); the Socialist General Union of Workers (UGT), and the smaller independent Workers Syndical Union (USO); business and landowning interests; the Catholic Church; Opus Dei; university students
Member of: AfDB, AG (observer), AsDB, Australia Group, BIS, CCC, CE, CERN, EBRD, EC, ECE, ECLAC, EIB, ESA, FAO, G-8, GATT, IADB, IAEA, IBRD, ICAO, ICC, ICFTU, ICRM, IDA, IEA, IFAD, IFC, IFRCS, ILO, IMF, IMO, INMARSAT, INTELSAT, INTERPOL, IOC, IOM (observer), ISO, ITU, LAIA (observer), MTCR, NACC, NAM (guest), NATO, NEA, NSG, OAS (observer), OECD, ONUSAL, OSCE, PCA, UN, UNCTAD, UNESCO, UNHCR, UNIDO, UNMIH, UNOMOZ, UNPROFOR, UNU, UPU, WCL, WEU, WFTU, WHO, WIPO, WMO, WTO, ZC
Diplomatic representation in US:
chief of mission: Ambassador Jaime De OJEDA Eiseley
chancery: 2375 Pennsylvania Avenue NW, Washington, DC 20037
telephone: [1] (202) 452-0100, 728-2340
FAX: [1] (202) 833-5670
consulate(s) general: Boston, Chicago, Houston, Los Angeles, Miami, New Orleans, New York, San Francisco, and San Juan (Puerto Rico)
US diplomatic representation:
chief of mission: Ambassador Richard N. GARDNER
embassy: Serrano 75, 28006 Madrid
mailing address: APO AE 09642
telephone: [34] (1) 577-4000
FAX: [34] (1) 577-5735
consulate(s) general: Barcelona
consulate(s): Bilbao
Flag: three horizontal bands of red (top), yellow (double width), and red with the national coat of arms on the hoist side of the yellow band; the coat of arms includes the royal seal framed by the Pillars of Hercules, which are the two promontories (Gibraltar and Ceuta) on either side of the eastern end of the Strait of Gibraltar

Economy

Overview: Spain, with a per capita output approximately two-thirds that of the four leading economies of Western Europe, has shared with these countries the recession of the early 1990s and the upturn of their economic fortunes in 1994. But whereas unemployment in these countries has hovered just above 10%, Spain has been forced to cope with a 25% unemployment rate. Continued political turmoil has complicated the establishment of stable government policies toward budgetary restraint, interest rates, labor law reform, and Spain's role in the evolving economic integration of Western Europe. Because the recession has been so deep, the growth in industrial output, tourism, and other sectors in 1994, while welcome, falls far short of the growth required to bring unemployment down to, say, 10%. The recovery in the economies of major trade partners, the comparatively low inflation rate, lower interest rates, and prospects in the tourist sector suggest that Spain can make substantial progress in 1995.
National product: GDP—purchasing power parity—$515.8 billion (1994 est.)
National product real growth rate: 1.8% (1994 est.)
National product per capita: $13,120 (1994 est.)
Inflation rate (consumer prices): 4.9% (1994)
Unemployment rate: 24.5% (yearend 1994)
Budget:
revenues: $97.7 billion
expenditures: $128 billion, including capital expenditures of $NA (1993 est.)
Exports: $72.8 billion (f.o.b., 1993)
commodities: cars and trucks, semifinished manufactured goods, foodstuffs, machinery
partners: EC 71.2%, US 4.8%, other developed countries 7.9% (1992)
Imports: $92.5 billion (c.i.f., 1993)
commodities: machinery, transport equipment, fuels, semifinished goods, foodstuffs, consumer goods, chemicals
partners: EC 60.7%, US 7.4%, other developed countries 11.5%, Middle East 5.9% (1992)
External debt: $90 billion (1993 est.)
Industrial production: growth rate 4% (1994 est.)
Electricity:
capacity: 43,800,000 kW
production: 148 billion kWh
consumption per capita: 3,545 kWh (1993)
Industries: textiles and apparel (including footwear), food and beverages, metals and metal manufactures, chemicals, shipbuilding, automobiles, machine tools, tourism
Agriculture: accounts for about 5% of GDP and 14% of labor force; major products—grain, vegetables, olives, wine grapes, sugar beets, citrus fruit, beef, pork, poultry, dairy; largely self-sufficient in food; fish catch of 1.4 million metric tons is among top 20 nations
Illicit drugs: key European gateway country for Latin American cocaine and North African hashish entering the European market; transshipment point for Southwest Asian heroin
Economic aid:
recipient: US commitments, including Ex-Im (FY70-87), $1.9 billion; Western (non-US) countries, ODA and OOF bilateral commitments (1970-79), $545 million
note: not currently a recipient
Currency: 1 peseta (Pta) = 100 centimos
Exchange rates: pesetas (Ptas) per US$1—132.61 (January 1995), 133.96 (1994), 127.26 (1993), 102.38 (1992), 103.91 (1991), 101.93 (1990)
Fiscal year: calendar year

Transportation

Railroads:
total: 14,400 km
broad gauge: 12,111 km 1.668-m gauge (6,404 km electrified; 2,295 km double track)
standard gauge: 515 km 1.435-m gauge (515 km electrified)
narrow gauge: 1,746 km 1.000-m gauge (privately owned: 1,727 km, 560 km electrified; government owned: 19 km, all electrified); 28 km 0.914-m gauge (privately owned, 28 km electrified) (1994)
Highways:
total: 331,961 km
paved: 328,641 km (2,700 km of expressways)
unpaved: 3,320 km (1991)
Inland waterways: 1,045 km, but of minor economic importance
Pipelines: crude oil 265 km; petroleum products 1,794 km; natural gas 1,666 km
Ports: Aviles, Barcelona, Bilbao, Cadiz, Cartagena, Castellon de la Plana, Ceuta, Huelva, La Coruna, Las Palmas (Canary Islands), Malaga, Melilla, Pasajes, Puerto de Gijon, Santa Cruz de Tenerife (Canary Islands), Santander, Tarragona, Valencia, Vigo
Merchant marine:
total: 157 ships (1,000 GRT or over) totaling 868,326 GRT/1,382,335 DWT
ships by type: bulk 12, cargo 41, chemical tanker 11, container 9, liquefied gas tanker 4, oil tanker 25, passenger 2, refrigerated cargo 12, roll-on/roll-off cargo 34, short-sea passenger 5, specialized tanker 2
Airports:
total: 106
with paved runways over 3,047 m: 15
with paved runways 2,438 to 3,047 m: 11

with paved runways 1,524 to 2,437 m: 16
with paved runways 914 to 1,523 m: 12
with paved runways under 914 m: 34
with unpaved runways 2,438 to 3,047 m: 1
with unpaved runways 1,524 to 2,438 m: 1
with unpaved runways 914 to 1,523 m: 16

Communications

Telephone system: 15,350,464 telephones; generally adequate, modern facilities
local: NA
intercity: NA
international: 22 coaxial submarine cables; 2 earth stations for INTELSAT (Atlantic Ocean and Indian Ocean); earth stations for working the EUTELSAT, INMARSAT, and MARECS satellite communications systems; microwave tropospheric scatter links to adjacent countries
Radio:
broadcast stations: AM 190, FM 406 (repeaters 134), shortwave 0
radios: NA
Television:
broadcast stations: 100 (repeaters 1,297)
televisions: NA

Defense Forces

Branches: Army, Navy, Air Force, Marines, Civil Guard, National Police, Coastal Civil Guard
Manpower availability: males age 15-49 10,435,970; males fit for military service 8,434,460; males reach military age (20) annually 335,967 (1995 est.)
Defense expenditures: exchange rate conversion—$8 billion, 1.6% of GDP (1994)

Spratly Islands

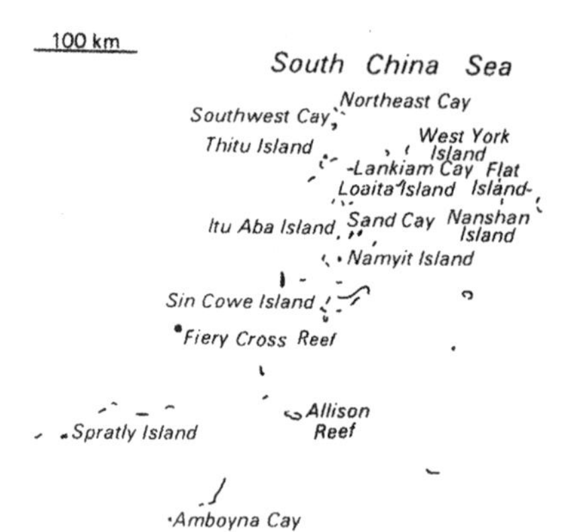

Geography

Location: Southeastern Asia, group of reefs in the South China Sea, about two-thirds of the way from southern Vietnam to the southern Philippines
Map references: Southeast Asia
Area:
total area: NA sq km but less than 5 km2
land area: less than 5 sq km
comparative area: NA
note: includes 100 or so islets, coral reefs, and sea mounts scattered over the South China Sea
Land boundaries: 0 km
Coastline: 926 km
Maritime claims: NA
International disputes: all of the Spratly Islands are claimed by China, Taiwan, and Vietnam; parts of them are claimed by Malaysia and the Philippines; in 1984, Brunei established an exclusive economic zone, which encompasses Louisa Reef, but has not publicly claimed the island
Climate: tropical
Terrain: flat
Natural resources: fish, guano, undetermined oil and natural gas potential
Land use:
arable land: 0%
permanent crops: 0%
meadows and pastures: 0%
forest and woodland: 0%
other: 100%
Irrigated land: 0 sq km
Environment:
current issues: NA
natural hazards: typhoons; serious maritime hazard because of numerous reefs and shoals
international agreements: NA
Note: strategically located near several primary shipping lanes in the central South China Sea; includes numerous small islands, atolls, shoals, and coral reefs

People

Population: no indigenous inhabitants; note—there are scattered garrisons

Government

Names:
conventional long form: none
conventional short form: Spratly Islands
Digraph: PG

Economy

Overview: Economic activity is limited to commercial fishing. The proximity to nearby oil- and gas-producing sedimentary basins suggests the potential for oil and gas deposits, but the region is largely unexplored, and there are no reliable estimates of potential reserves; commercial exploitation has yet to be developed.
Industries: none

Transportation

Ports: none
Airports:
total: 4
with paved runways under 914 m: 3
with unpaved runways 914 to 1,523 m: 1

Communications

Telephone system:
local: NA
intercity: NA
international: NA
Radio:
broadcast stations: AM NA, FM NA, shortwave NA
radios: NA
Television:
broadcast stations: NA
televisions: NA

Defense Forces

Note: about 50 small islands or reefs are occupied by China, Malaysia, the Philippines, Taiwan, and Vietnam

Sri Lanka

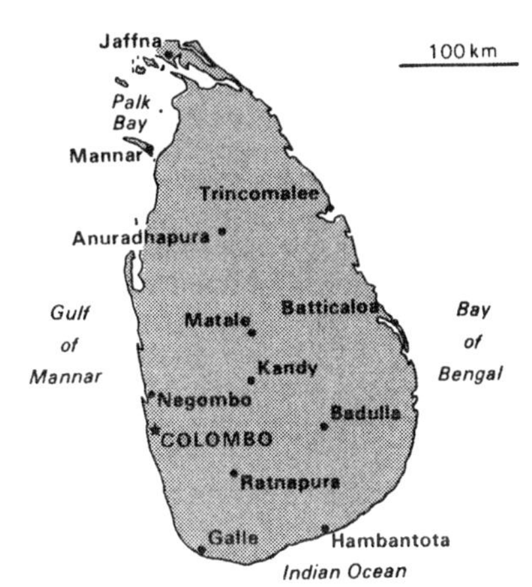

Geography

Location: Southern Asia, island in the Indian Ocean, south of India
Map references: Asia
Area:
total area: 65,610 sq km
land area: 64,740 sq km
comparative area: slightly larger than West Virginia
Land boundaries: 0 km
Coastline: 1,340 km
Maritime claims:
contiguous zone: 24 nm
continental shelf: 200 nm or to the edge of the continental margin
exclusive economic zone: 200 nm
territorial sea: 12 nm
International disputes: none
Climate: tropical monsoon; northeast monsoon (December to March); southwest monsoon (June to October)
Terrain: mostly low, flat to rolling plain, mountains in south-central interior
Natural resources: limestone, graphite, mineral sands, gems, phosphates, clay
Land use:
arable land: 16%
permanent crops: 17%
meadows and pastures: 7%
forest and woodland: 37%
other: 23%
Irrigated land: 5,600 sq km (1989 est.)
Environment:
current issues: deforestation; soil erosion; wildlife populations threatened by poaching; coastal degradation from mining activities and increased pollution; freshwater resources being polluted by industrial wastes and sewage runoff
natural hazards: occasional cyclones and tornadoes
international agreements: party to—Biodiversity, Climate Change, Endangered Species, Environmental Modification, Hazardous Wastes, Law of the Sea, Nuclear Test Ban, Ozone Layer Protection, Wetlands; signed, but not ratified—Marine Life Conservation
Note: strategic location near major Indian Ocean sea lanes

People

Population: 18,342,660 (July 1995 est.)
note: since the outbreak of hostilities between the government and armed Tamil separatists in the mid-1980s, several hundred thousand Tamil civilians have fled the island; as of late 1992, nearly 115,000 were housed in refugee camps in south India, another 95,000 lived outside the Indian camps, and more than 200,000 Tamils have sought political asylum in the West
Age structure:
0-14 years: 29% (female 2,597,969; male 2,713,696)
15-64 years: 65% (female 6,042,228; male 5,902,343)
65 years and over: 6% (female 547,715; male 538,709) (July 1995 est.)
Population growth rate: 1.15% (1995 est.)
Birth rate: 18.13 births/1,000 population (1995 est.)
Death rate: 5.78 deaths/1,000 population (1995 est.)
Net migration rate: -0.84 migrant(s)/1,000 population (1995 est.)
Infant mortality rate: 21.3 deaths/1,000 live births (1995 est.)
Life expectancy at birth:
total population: 72.14 years
male: 69.58 years
female: 74.82 years (1995 est.)
Total fertility rate: 2.08 children born/woman (1995 est.)
Nationality:
noun: Sri Lankan(s)
adjective: Sri Lankan
Ethnic divisions: Sinhalese 74%, Tamil 18%, Moor 7%, Burgher, Malay, and Vedda 1%
Religions: Buddhist 69%, Hindu 15%, Christian 8%, Muslim 8%
Languages: Sinhala (official and national language) 74%, Tamil (national language) 18%
note: English is commonly used in government and is spoken by about 10% of the population
Literacy: age 15 and over can read and write (1990 est.)
total population: 88%
male: 93%
female: 84%
Labor force: 6.6 million
by occupation: agriculture 45.9%, mining and manufacturing 13.3%, trade and transport 12.4%, services and other 28.4% (1985 est.)

Government

Names:
conventional long form: Democratic Socialist Republic of Sri Lanka
conventional short form: Sri Lanka
former: Ceylon
Digraph: CE
Type: republic
Capital: Colombo
Administrative divisions: 8 provinces; Central, North Central, North Eastern, North Western, Sabaragamuwa, Southern, Uva, Western
Independence: 4 February 1948 (from UK)
National holiday: Independence and National Day, 4 February (1948)
Constitution: adopted 16 August 1978
Legal system: a highly complex mixture of English common law, Roman-Dutch, Muslim, Sinhalese, and customary law; has not accepted compulsory ICJ jurisdiction
Suffrage: 18 years of age; universal
Executive branch:
chief of state and head of government: President Chandrika Bandaranaike KUMARATUNGA (since 12 November 1994); note—Sirimavo BANDARANAIKE is the Prime Minister; in Sri Lanka the president is considered to be both the chief of state and the head of the government, this is in contrast to the more common practice of dividing the roles between the president and the prime minister when both offices exist; election last held 9 November 1994 (next to be held NA November 2000); results—Chandrika Bandaranaike KUMARATUNGA (People's Alliance) 62%, Srima DISSANAYAKE (United National Party) 37%, other 1%
cabinet: Cabinet; appointed by the president in consultation with the prime minister
Legislative branch: unicameral
Parliament: elections last held 16 August 1994 (next to be held by August 2000); results—PA 49.0%, UNP 44.0%, SLMC 1.8%, TULF 1.7%, SLPF 1.1%, EPDP 0.3%, UPF 0.3%, PLOTE 0.1%, other 1.7%; seats—(225 total) PA 105, UNP 94, EPDP 9, SLMC 7, TULF 5, PLOTE 3, SLPF 1, UPF 1
Judicial branch: Supreme Court
Political parties and leaders: All Ceylon Tamil Congress (ACTC), C. G. Kumar PONNAMBALAM; Ceylon Workers Congress (CLDC), S. THONDAMAN; Communist Party, K. P. SILVA; Communist Party/Beijing (CP/B), N. SHANMUGATHASAN; Democratic People's Liberation Front (DPLF), leader NA; Democratic United National Front (DUNF), G. M. PREMACHANDRA; Eelam People's Democratic Party (EPDP), Douglas DEVANANDA; Eelam People's Revolutionary Liberation Front (EPRL), Suresh PREMACHANDRAN; Eelam Revolutionary Organization of Students (EROS), Shankar RAJI; Lanka Socialist Party/Trotskyite (LSSP, or Lanka Sama Samaja Party), Colin R. DE SILVA; Liberal Party (LP), Chanaka AMARATUNGA; New Socialist Party (NSSP, or Nava Sama Samaja

Party), Vasudeva NANAYAKKARA; People's Alliance (PA), Chandrika Bandaranaike KUMARATUNGA; People's Liberation Organization of Tamil Eelam (PLOTE), Dharmalingam SIDARTHAN; People's United Front (MEP, or Mahajana Eksath Peramuna), Dinesh GUNAWARDENE; Sri Lanka Freedom Party (SLFP), Sirimavo BANDARANAIKE; Sri Lanka Muslim Congress (SLMC), M. H. M. ASHRAFF; Sri Lanka People's Party (SLMP, or Sri Lanka Mahajana Party), Ossie ABEYGUNASEKERA; Sri Lanka Progressive Front (SLPF), leader NA; Tamil Eelam Liberation Organization (TELO), leader NA; Tamil United Liberation Front (TULF), M. SIVASITHAMBARAM; United National Party (UNP), Ranil WICKREMANSINGHE; Upcountry People's Front (UPF), leader NA; several ethnic Tamil and Muslim parties, represented in either parliament or provincial councils
note: the United Socialist Alliance (USA), which was formed in 1987 and included the NSSP, LSSP, SLMP, CP/M, and CP/B, was defunct as of 1993, following the formation of the People's Alliance Party (PA)
Other political or pressure groups: Liberation Tigers of Tamil Eelam (LTTE) and other smaller Tamil separatist groups; other radical chauvinist Sinhalese groups; Buddhist clergy; Sinhalese Buddhist lay groups; labor unions
Member of: AsDB, C, CCC, CP, ESCAP, FAO, G-24, G-77, GATT, IAEA, IBRD, ICAO, ICC, ICFTU, ICRM, IDA, IFAD, IFC, IFRCS, ILO, IMF, IMO, INMARSAT, INTELSAT, INTERPOL, IOC, IOM, ISO, ITU, NAM, PCA, SAARC, UN, UNCTAD, UNESCO, UNIDO, UNU, UPU, WCL, WFTU, WHO, WIPO, WMO, WTO
Diplomatic representation in US:
chief of mission: Ambassador Jayantha DHANAPALA
chancery: 2148 Wyoming Avenue NW, Washington, DC 20008
telephone: [1] (202) 483-4025 through 4028
FAX: [1] (202) 232-7181
consulate(s): New York
US diplomatic representation:
chief of mission: Ambassador Teresita C. SCHAFFER
embassy: 210 Galle Road, Colombo 3
mailing address: P. O. Box 106, Colombo
telephone: [94] (1) 448007
FAX: [94] (1) 437345
Flag: yellow with two panels; the smaller hoist-side panel has two equal vertical bands of green (hoist side) and orange; the other panel is a large dark red rectangle with a yellow lion holding a sword, and there is a yellow bo leaf in each corner; the yellow field appears as a border that goes around the entire flag and extends between the two panels

Economy

Overview: Industry—dominated by the fast-growing apparel industry—has surpassed agriculture as the main source of export earnings and accounts for over 16% of GDP. The economy has been plagued by high rates of unemployment since the late 1970s. Economic growth, which has been depressed by ethnic unrest, accelerated in 1991-94 as domestic conditions began to improve and conditions for foreign investment brightened. Currently, however, the new government's emphasis on populist measures has clouded Sri Lanka's economic prospects.
National product: GDP—purchasing power parity—$57.6 billion (1994 est.)
National product real growth rate: 5% (1994 est.)
National product per capita: $3,190 (1994 est.)
Inflation rate (consumer prices): 12% (1994 est.)
Unemployment rate: 13.6% (1993 est.)
Budget:
revenues: $2.3 billion
expenditures: $3.6 billion, including capital expenditures of $1.5 billion (1993)
Exports: $2.9 billion (f.o.b., 1993)
commodities: garments and textiles, teas, diamonds, other gems, petroleum products, rubber products, other agricultural products, marine products, graphite
partners: US 35.2%, Germany, UK, Belgium-Luxembourg, Japan, Netherlands, France (1993)
Imports: $4 billion (c.i.f., 1993)
commodities: textiles and textile materials, machinery and equipment, transport equipment, petroleum, building materials
partners: Japan, India, Hong Kong, South Korea, Taiwan, Singapore, China (1993)
External debt: $7.2 billion (1993 est.)
Industrial production: growth rate 9% (1993 est.); accounts for 16% of GDP
Electricity:
capacity: 1,410,000 kW
production: 3.2 billion kWh
consumption per capita: 168 kWh (1993)
Industries: processing of rubber, tea, coconuts, and other agricultural commodities; clothing, cement, petroleum refining, textiles, tobacco
Agriculture: accounts for one-fourth of GDP; field crops—rice, sugarcane, grains, pulses, oilseeds, roots, spices; cash crops—tea, rubber, coconuts; animal products—milk, eggs, hides, meat; not self-sufficient in rice production
Economic aid:
recipient: US commitments, including Ex-Im (FY70-89), $1 billion; Western (non-US) countries, ODA and OOF bilateral commitments (1980-89), $5.1 billion; OPEC bilateral aid (1979-89), $169 million; Communist countries (1970-89), $369 million
Currency: 1 Sri Lankan rupee (SLRe) = 100 cents
Exchange rates: Sri Lankan rupees (SLRes) per US$1—50.115 (January 1995), 49.415 (1994), 48.322 (1993), 43.830 (1992), 41.372 (1991), 40.063 (1990)
Fiscal year: calendar year

Transportation

Railroads:
total: 1,948 km
broad gauge: 1,948 km 1.868-m gauge (102 km double track) (1990)
Highways:
total: 75,263 km
paved: mostly bituminous treated 27,637 km
unpaved: crushed stone, gravel 32,887 km; improved, unimproved earth 14,739 km
Inland waterways: 430 km; navigable by shallow-draft craft
Pipelines: crude oil and petroleum products 62 km (1987)
Ports: Colombo, Galle, Jaffna, Trincomalee
Merchant marine:
total: 26 ships (1,000 GRT or over) totaling 289,115 GRT/453,609 DWT
ships by type: bulk 2, cargo 12, container 1, oil tanker 3, refrigerated cargo 8
Airports:
total: 14
with paved runways over 3,047 m: 1
with paved runways 1,524 to 2,437 m: 5
with paved runways 914 to 1,523 m: 7
with unpaved runways 1,524 to 2,438 m: 1

Communications

Telephone system: 114,000 telephones (1982); very inadequate domestic service, good international service
local: NA
intercity: NA
international: submarine cables extend to Indonesia and Djibouti; 2 INTELSAT (Indian Ocean) earth stations
Radio:
broadcast stations: AM 12, FM 5, shortwave 0
radios: NA
Television:
broadcast stations: 5
televisions: NA

Defense Forces

Branches: Army, Navy, Air Force, Police Force
Manpower availability: males age 15-49 4,990,661; males fit for military service 3,888,372; males reach military age (18) annually 178,926 (1995 est.)
Defense expenditures: exchange rate conversion—$412 million, 3.6% of GDP (1994)

Sudan

Geography

Location: Northern Africa, bordering the Red Sea, between Egypt and Eritrea
Map references: Africa
Area:
total area: 2,505,810 sq km
land area: 2.376 million sq km
comparative area: slightly more than one-quarter the size of the US
Land boundaries: total 7,687 km, Central African Republic 1,165 km, Chad 1,360 km, Egypt 1,273 km, Eritrea 605 km, Ethiopia 1,606 km, Kenya 232 km, Libya 383 km, Uganda 435 km, Zaire 628 km
Coastline: 853 km
Maritime claims:
contiguous zone: 18 nm
continental shelf: 200-m depth or to the depth of exploitation
territorial sea: 12 nm
International disputes: administrative boundary with Kenya does not coincide with international boundary; administrative boundary with Egypt does not coincide with international boundary creating the "Hala'ib Triangle," a barren area of 20,580 sq km, tensions over this disputed area began to escalate in 1992 and remain high
Climate: tropical in south; arid desert in north; rainy season (April to October)
Terrain: generally flat, featureless plain; mountains in east and west
Natural resources: small reserves of petroleum, iron ore, copper, chromium ore, zinc, tungsten, mica, silver, gold
Land use:
arable land: 5%
permanent crops: 0%
meadows and pastures: 24%
forest and woodland: 20%
other: 51%
Irrigated land: 18,900 sq km (1989 est.)
Environment:
current issues: inadequate supplies of potable water; wildlife populations threatened by excessive hunting; soil erosion; desertification
natural hazards: dust storms
international agreements: party to—Climate Change, Endangered Species, Law of the Sea, Nuclear Test Ban, Ozone Layer Protection; signed, but not ratified—Biodiversity, Desertification
Note: largest country in Africa; dominated by the Nile and its tributaries

People

Population: 30,120,420 (July 1995 est.)
Age structure:
0-14 years: 46% (female 6,801,001; male 7,124,892)
15-64 years: 52% (female 7,706,864; male 7,830,980)
65 years and over: 2% (female 280,297; male 376,386) (July 1995 est.)
Population growth rate: 2.35% (1995 est.)
Birth rate: 41.29 births/1,000 population (1995 est.)
Death rate: 11.74 deaths/1,000 population (1995 est.)
Net migration rate: NA migrant(s)/1,000 population (1995 est.)
note: the flow of refugees from the civil war in Sudan into neighboring countries continues, often at the rate of tens of thousands annually; Uganda was the main recipient of Sudanese refugees in the past year; repatriation of Eritrean and Ethiopean refugees in Sudan continues
Infant mortality rate: 77.7 deaths/1,000 live births (1995 est.)
Life expectancy at birth:
total population: 54.71 years
male: 53.81 years
female: 55.65 years (1995 est.)
Total fertility rate: 6 children born/woman (1995 est.)
Nationality:
noun: Sudanese (singular and plural)
adjective: Sudanese
Ethnic divisions: black 52%, Arab 39%, Beja 6%, foreigners 2%, other 1%
Religions: Sunni Muslim 70% (in north), indigenous beliefs 25%, Christian 5% (mostly in south and Khartoum)
Languages: Arabic (official), Nubian, Ta Bedawie, diverse dialects of Nilotic, Nilo-Hamitic, Sudanic languages, English
note: program of Arabization in process
Literacy: age 15 and over can read and write (1983)
total population: 32%
male: 44%
female: 21%
Labor force: 6.5 million
by occupation: agriculture 80%, industry and commerce 10%, government 6%
note: labor shortages for almost all categories of skilled employment (1983 est.)

Government

Names:
conventional long form: Republic of the Sudan
conventional short form: Sudan
local long form: Jumhuriyat as-Sudan
local short form: As-Sudan
former: Anglo-Egyptian Sudan
Digraph: SU
Type: ruling military junta—Revolutionary Command Council (RCC)—dissolved on 16 October 1993 and government civilianized
Capital: Khartoum
Administrative divisions: 9 states (wilayat, singular—wilayat or wilayah*); A'ali an Nil, Al Wusta*, Al Istiwa'iyah*, Al Khartum, Ash Shamaliyah*, Ash Sharqiyah*, Bahr al Ghazal, Darfur, Kurdufan
note: on 14 February 1994, the 9 states comprising Sudan were divided into 26 new states; the new state boundary alignments are undetermined
Independence: 1 January 1956 (from Egypt and UK)
National holiday: Independence Day, 1 January (1956)
Constitution: 12 April 1973, suspended following coup of 6 April 1985; interim constitution of 10 October 1985 suspended following coup of 30 June 1989
Legal system: based on English common law and Islamic law; as of 20 January 1991, the now defunct Revolutionary Command Council imposed Islamic law in the northern states; the council is still studying criminal provisions under Islamic law; Islamic law applies to all residents of the northern states regardless of their religion; some separate religious courts; accepts compulsory ICJ jurisdiction, with reservations
Suffrage: none
Executive branch:
Chief of State and Head of Government: President Lt. General Umar Hasan Ahmad al-BASHIR (since 16 October 1993); prior to 16 October 1993, BASHIR served concurrently as Chief of State, Chairman of the RCC, Prime Minister, and Minister of Defence (since 30 June 1989); First Vice President Major General al-Zubayr Muhammad SALIH (since 19 October 1993); Second Vice President (Police) Maj. General George KONGOR (since NA February 1994), note—upon its dissolution on 16 October 1993, the RCC's executive and legislative powers were devolved to the President and the Transitional National Assembly (TNA), Sudan's appointed legislative body
cabinet: Cabinet; appointed by the president; note—on 30 October 1993, President BASHIR announced a new, predominantly civilian cabinet, consisting of 20 federal ministers, most of whom retained their previous cabinet positions; on 9 February 1995, he abolished

three ministries and redivided their portfolios to create several new ministries; these changes increased National Islamic Front presence at the ministerial level and consolidated its control over the Ministry of Foreign Affairs; President BASHIR's government is dominated by members of Sudan's National Islamic Front, a fundamentalist political organization formed from the Muslim Brotherhood in 1986; front leader Hasan al-TURABI controls Khartoum's overall domestic and foreign policies
Legislative branch: appointed 300-member Transitional National Assembly; officially assumes all legislative authority for Sudan until the proposed 1995 resumption of national elections
Judicial branch: Supreme Court, Special Revolutionary Courts
Political parties and leaders: none; banned following 30 June 1989 coup
Other political or pressure groups: National Islamic Front, Hasan al-TURABI
Member of: ABEDA, ACP, AfDB, AFESD, AL, AMF, CAEU, CCC, ECA, FAO, G-77, IAEA, IBRD, ICAO, ICRM, IDA, IDB, IFAD, IFC, IGADD, ILO, IMF, IMO, INTELSAT, INTERPOL, IOC, ITU, NAM, OAU, OIC, PCA, UN, UNCTAD, UNESCO, UNHCR, UNIDO, UNU, UPU, WFTU, WHO, WIPO, WMO, WTO
Diplomatic representation in US:
chief of mission: Ambassador Ahmad SULAYMAN
chancery: 2210 Massachusetts Avenue NW, Washington, DC 20008
telephone: [1] (202) 338-8565 through 8570
FAX: [1] (202) 667-2406
US diplomatic representation:
chief of mission: Ambassador Donald K. PETTERSON
embassy: Shar'ia Ali Abdul Latif, Khartoum
mailing address: P. O. Box 699, Khartoum; APO AE 09829
telephone: 74700, 74611 (operator assistance required)
FAX: Telex 22619 AMEMSD
Flag: three equal horizontal bands of red (top), white, and black with a green isosceles triangle based on the hoist side

Economy

Overview: Sudan is buffeted by civil war, chronic political instability, adverse weather, high inflation, a drop in remittances from abroad, and counterproductive economic policies. Governmental entities account for more than 70% of new investment. The private sector's main areas of activity are agriculture and trading, with most private industrial investment predating 1980. Agriculture employs 80% of the work force. Industry mainly processes agricultural items. Sluggish economic performance over the past decade, attributable largely to declining annual rainfall, has reduced levels of per capita income and consumption. A large foreign debt and huge arrearages continue to cause difficulties. In 1990 the International Monetary Fund took the unusual step of declaring Sudan noncooperative because of its nonpayment of arrearages to the Fund. After Sudan backtracked on promised reforms in 1992-93, the IMF threatened to expel Sudan from the Fund. To avoid expulsion, Khartoum agreed to make payments on its arrears to the Fund, liberalize exchange rates, and reduce subsidies. These measures have been partially implemented. The government's continued prosecution of the civil war and its growing international isolation led to a further deterioration of the nonagricultural sectors of the economy during 1994. Agriculture, on the other hand, after several disappointing years, enjoyed a bumper fall harvest in 1994; its strong performance produced an overall growth rate in GDP of perhaps 7%.
National product: GDP—purchasing power parity—$23.7 billion (1994 est.)
National product real growth rate: 7% (1994 est.)
National product per capita: $870 (1994 est.)
Inflation rate (consumer prices): 112% (FY93/94 est.)
Unemployment rate: 30% (FY92/93 est.)
Budget:
revenues: $493 million
expenditures: $1.1 billion, including capital expenditures of $225 million (1994 est.)
Exports: $419 million (f.o.b., FY93/94)
commodities: gum arabic 29%, livestock/meat 24%, cotton 13%, sesame, peanuts
partners: Western Europe 46%, Saudi Arabia 14%, Eastern Europe 9%, Japan 9%, US 3% (FY87/88)
Imports: $1.7 billion (c.i.f., FY93/94)
commodities: foodstuffs, petroleum products, manufactured goods, machinery and equipment, medicines and chemicals, textiles
partners: Western Europe 32%, Africa and Asia 15%, US 13%, Eastern Europe 3% (FY87/88)
External debt: $17 billion (June 1993 est.)
Industrial production: growth rate 6.8% (FY92/93 est.); accounts for 11% of GDP
Electricity:
capacity: 500,000 kW
production: 1.3 billion kWh
consumption per capita: 42 kWh (1993)
Industries: cotton ginning, textiles, cement, edible oils, sugar, soap distilling, shoes, petroleum refining
Agriculture: accounts for 35% of GDP; major products—cotton, oilseeds, sorghum, millet, wheat, gum arabic, sheep; marginally self-sufficient in most foods
Economic aid:
recipient: US commitments, including Ex-Im (FY70-89), $1.5 billion; Western (non-US) countries, ODA and OOF bilateral commitments (1970-89), $5.1 billion; OPEC bilateral aid (1979-89), $3.1 billion; Communist countries (1970-89), $588 million
Currency: 1 Sudanese pound (£Sd) = 100 piastres
Exchange rates: official rate—Sudanese pounds (£Sd) per US$1—434.8 (January 1995), 277.8 (1994), 153.8 (1993), 69.4 (1992), 5.4288 (1991), 4.5004 (1990); note—the commercial rate is 300 Sudanese pounds per US $1
Fiscal year: 1 July—30 June

Transportation

Railroads:
total: 5,516 km
narrow gauge: 4,800 km 1.067-m gauge; 716 km 1.6096-m gauge plantation line
Highways:
total: 20,703 km
paved: bituminous treated 2,000 km
unpaved: gravel 4,000 km; improved earth 2,304 km; unimproved earth 12,399 km
Inland waterways: 5,310 km navigable
Pipelines: refined products 815 km
Ports: Juba, Khartoum, Kusti, Malakal, Nimule, Port Sudan, Sawakin
Merchant marine:
total: 5 ships (1,000 GRT or over) totaling 43,024 GRT/122,379 DWT
ships by type: cargo 3, roll-on/roll-off cargo 2
Airports:
total: 70
with paved runways over 3,047 m: 1
with paved runways 2,438 to 3,047 m: 5
with paved runways 1,524 to 2,437 m: 3
with paved runways under 914 m: 13
with unpaved runways 2,438 to 3,047 m: 1
with unpaved runways 1,524 to 2,438 m: 14
with unpaved runways 914 to 1,523 m: 33

Communications

Telephone system: NA telephones; large, well-equipped system by African standards, but barely adequate and poorly maintained by modern standards
local: NA
intercity: consists of microwave radio relay, cable, radio communications, troposcatter, and a domestic satellite system with 14 stations
international: 1 INTELSAT (Atlantic Ocean) and 1 ARABSAT earth station
Radio:
broadcast stations: AM 11, FM 0, shortwave 0
radios: NA
Television:
broadcast stations: 3
televisions: NA

Defense Forces

Branches: Army, Navy, Air Force, Popular Defense Force Militia
Manpower availability: males age 15-49

6,806,588; males fit for military service 4,185,206; males reach military age (18) annually 313,958 (1995 est.)
Defense expenditures: exchange rate conversion—$600 million, 7.3% of GDP (FY93/94 est.)

Suriname

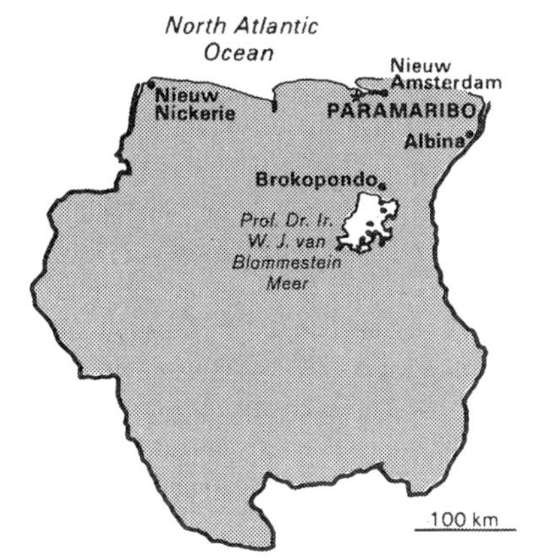

Geography

Location: Northern South America, bordering the North Atlantic Ocean, between French Guiana and Guyana
Map references: South America
Area:
total area: 163,270 sq km
land area: 161,470 sq km
comparative area: slightly larger than Georgia
Land boundaries: total 1,707 km, Brazil 597 km, French Guiana 510 km, Guyana 600 km
Coastline: 386 km
Maritime claims:
exclusive economic zone: 200 nm
territorial sea: 12 nm
International disputes: claims area in French Guiana between Litani Rivier and Riviere Marouini (both headwaters of the Lawa Rivier); claims area in Guyana between New (Upper Courantyne) and Courantyne/Koetari Rivers (all headwaters of the Courantyne)
Climate: tropical; moderated by trade winds
Terrain: mostly rolling hills; narrow coastal plain with swamps
Natural resources: timber, hydropower potential, fish, shrimp, bauxite, iron ore, and small amounts of nickel, copper, platinum, gold
Land use:
arable land: 0%
permanent crops: 0%
meadows and pastures: 0%
forest and woodland: 97%
other: 3%
Irrigated land: 590 sq km (1989 est.)
Environment:
current issues: deforestation as foreign producers obtain timber concessions
natural hazards: NA
international agreements: party to—Endangered Species, Marine Dumping, Nuclear Test Ban, Ship Pollution, Wetlands; signed, but not ratified—Biodiversity, Climate Change, Law of the Sea
Note: mostly tropical rain forest; great diversity of flora and fauna which for the most part is not threatened because of the lack of development; relatively small population most of which lives along the coast

People

Population: 429,544 (July 1995 est.)
Age structure:
0-14 years: 34% (female 70,845; male 74,330)
15-64 years: 61% (female 130,153; male 133,693)
65 years and over: 5% (female 10,897; male 9,626) (July 1995 est.)
Population growth rate: 1.58% (1995 est.)
Birth rate: 24.72 births/1,000 population (1995 est.)
Death rate: 5.91 deaths/1,000 population (1995 est.)
Net migration rate: -3 migrant(s)/1,000 population (1995 est.)
Infant mortality rate: 30.2 deaths/1,000 live births (1995 est.)
Life expectancy at birth:
total population: 69.76 years
male: 67.24 years
female: 72.41 years (1995 est.)
Total fertility rate: 2.73 children born/woman (1995 est.)
Nationality:
noun: Surinamer(s)
adjective: Surinamese
Ethnic divisions: Hindustani (also known locally as "East" Indians; their ancestors emigrated from northern India in the latter part of the 19th century) 37%, Creole (mixed European and African ancestry) 31%, Javanese 15.3%, "Bush Black" (also known as "Bush Creole" whose ancestors were brought to the country in the 17th and 18th centuries as slaves) 10.3%, Amerindian 2.6%, Chinese 1.7%, Europeans 1%, other 1.1%
Religions: Hindu 27.4%, Muslim 19.6%, Roman Catholic 22.8%, Protestant 25.2% (predominantly Moravian), indigenous beliefs 5%
Languages: Dutch (official), English (widely spoken), Sranang Tongo (Surinamese, sometimes called Taki-Taki, is native language of Creoles and much of the younger population and is lingua franca among others), Hindustani (a dialect of Hindi), Javanese
Literacy: age 15 and over can read and write (1990 est.)
total population: 95%
male: 95%
female: 95%
Labor force: NA

Government

Names:
conventional long form: Republic of Suriname
conventional short form: Suriname
local long form: Republiek Suriname
local short form: Suriname

former: Netherlands Guiana, Dutch Guiana
Digraph: NS
Type: republic
Capital: Paramaribo
Administrative divisions: 10 districts (distrikten, singular—distrikt); Brokopondo, Commewijne, Coronie, Marowijne, Nickerie, Para, Paramaribo, Saramacca, Sipaliwini, Wanica
Independence: 25 November 1975 (from Netherlands)
National holiday: Independence Day, 25 November (1975)
Constitution: ratified 30 September 1987
Legal system: NA
Suffrage: 18 years of age; universal
Executive branch:
chief of state and head of government: President Ronald R. VENETIAAN (since 16 September 1991); Prime Minister Jules R. AJODHIA (since 16 September 1991); election last held 6 September 1991 (next to be held NA May 1996); results—elected by the National Assembly—Ronald VENETIAAN (NF) 80% (645 votes), Jules WIJDENBOSCH (NDP) 14% (115 votes), Hans PRADE (DA '91) 6% (49 votes)
cabinet: Cabinet of Ministers; appointed by the president from members of the National Assembly
note: Commander in Chief of the National Army maintains significant power
Legislative branch: unicameral
National Assembly (Assemblee Nationale): elections last held 25 May 1991 (next to be held NA May 1996); results—percent of vote by party NA; seats—(51 total) NF 30, NDP 10, DA '91 9, independents 2
Judicial branch: Supreme Court
Political parties and leaders: The New Front (NF), a coalition of four parties (NPS, VHP, KTPI, SPA), leader Ronald R. VENETIAAN; Progressive Reform Party (VHP), Jaggernath LACHMON; National Party of Suriname (NPS), Ronald VENETIAAN; Party of National Unity and Solidarity (KTPI), Willy SOEMITA; Suriname Labor Party (SPA), Fred DERBY; Democratic Alternative '91 (DA '91), Winston JESSURUN, a coalition of four parties (AF, HPP, Pendawa Lima, BEP) formed in January 1991; Alternative Forum (AF), Gerard BRUNINGS, Winston JESSURUN; Reformed Progressive Party (HPP), Panalal PARMESSAR; Party for Brotherhood and Unity in Politics (BEP), Caprino ALLENDY; Pendawa Lima, Marsha JAMIN; National Democratic Party (NDP), Desire BOUTERSE; Progressive Workers' and Farm Laborers' Union (PALU), Ir Iwan KROLIS, chairman;
Other political or pressure groups: Surinamese Liberation Army (SLA), Ronnie BRUNSWIJK, Johan "Castro" WALLY; Union for Liberation and Democracy, Kofi AFONGPONG; Mandela Bushnegro Liberation Movement, Leendert ADAMS; Tucayana Amazonica, Alex JUBITANA, Thomas SABAJO
Member of: ACP, CARICOM, ECLAC, FAO, G-77, GATT, IADB, IBRD, ICAO, ICFTU, ICRM, IFAD, IFRCS (associate), ILO, IMF, IMO, INTELSAT (nonsignatory user), INTERPOL, IOC, ITU, LAES, NAM, OAS, OPANAL, PCA, UN, UNCTAD, UNESCO, UNIDO, UPU, WCL, WHO, WIPO, WMO
Diplomatic representation in US:
chief of mission: Ambassador Willem A. UDENHOUT
chancery: Suite 108, 4301 Connecticut Avenue NW, Washington, DC 20008
telephone: [1] (202) 244-7488, 7490 through 7492
FAX: [1] (202) 244-5878
consulate(s) general: Miami
US diplomatic representation:
chief of mission: Ambassador Roger R. GAMBLE
embassy: Dr. Sophie Redmondstraat 129, Paramaribo
mailing address: P. O. Box 1821, Paramaribo
telephone: [597] 472900, 477881, 476459
FAX: [597] 410025
Flag: five horizontal bands of green (top, double width), white, red (quadruple width), white, and green (double width); there is a large yellow five-pointed star centered in the red band

Economy

Overview: The economy is dominated by the bauxite industry, which accounts for 15% of GDP and about 70% of export earnings. Paramaribo has failed to initiate the economic reforms necessary to stabilize the economy or win renewed Dutch aid disbursements. The government continues to finance deficit spending with monetary emissions. As a result, high inflation, high unemployment, widespread black market activity, and hard currency shortfalls continue to mark the economy.
National product: GDP—purchasing power parity—$1.2 billion (1994 est.)
National product real growth rate: -0.8% (1994 est.)
National product per capita: $2,800 (1994 est.)
Inflation rate (consumer prices): 225% (1994 est.)
Unemployment rate: NA
Budget:
revenues: $300 million
expenditures: $700 million, including capital expenditures of $70 million (1994 est.)
Exports: $443.3 million (f.o.b., 1993 est.)
commodities: alumina, aluminum, shrimp and fish, rice, bananas
partners: Norway 33%, Netherlands 26%, US 13%, Japan 6%, Brazil 6%, UK 3% (1992)
Imports: $520.5 million (f.o.b., 1993 est.)
commodities: capital equipment, petroleum, foodstuffs, cotton, consumer goods
partners: US 42%, Netherlands 22%, Trinidad and Tobago 10%, Brazil 5% (1992)
External debt: $180 million (March 1993 est.)
Industrial production: growth rate 3.5% (1992 est.); accounts for 18% of GDP
Electricity:
capacity: 420,000 kW
production: 1.4 billion kWh
consumption per capita: 3,123 kWh (1993)
Industries: bauxite mining, alumina and aluminum production, lumbering, food processing, fishing
Agriculture: accounts for 15% of GDP and 25% of export earnings; paddy rice planted on 85% of arable land and represents 60% of total farm output; other products—bananas, palm kernels, coconuts, plantains, peanuts, beef, chicken; shrimp and forestry products of increasing importance; self-sufficient in most foods
Illicit drugs: transshipment point for South American drugs destined for the US and Europe
Economic aid:
recipient: US commitments, including Ex-Im (FY70-83), $2.5 billion; Western (non-US) countries, ODA and OOF bilateral commitments (1970-89), $1.5 billion
Currency: 1 Surinamese guilder, gulden, or florin (Sf.) = 100 cents
Exchange rates: Surinamese guilders, gulden, or florins (Sf.) per US$1—1.7850 (fixed rate); parallel rate 510 (December 1994), 109 (January 1994)
Fiscal year: calendar year

Transportation

Railroads:
total: 166 km (single track)
standard gauge: 80 km 1.435-m gauge
narrow gauge: 86 km 1.000-m gauge
Highways:
total: 8,800 km
paved: 500 km
unpaved: bauxite, gravel, crushed stone 5,400 km; improved and unimproved earth 2,900 km
Inland waterways: 1,200 km; most important means of transport; oceangoing vessels with drafts ranging up to 7 m can navigate many of the principal waterways
Ports: Albina, Moengo, Nieuw Nickerie, Paramaribo, Paranam, Wageningen
Merchant marine:
total: 2 ships (1,000 GRT or over) totaling 2,421 GRT/2,990 DWT
ships by type: cargo 1, container 1
Airports:
total: 46
with paved runways over 3,047 m: 1
with paved runways under 914 m: 38
with unpaved runways 914 to 1,523 m: 7

Communications

Telephone system: 27,500 telephones; international facilities good; domestic microwave system
local: NA
intercity: microwave radio relay network
international: 2 INTELSAT (Atlantic Ocean) earth stations
Radio:
broadcast stations: AM 5, FM 14, shortwave 1
radios: NA
Television:
broadcast stations: 6
televisions: NA

Defense Forces

Branches: National Army (includes small Navy and Air Force elements), Civil Police
Manpower availability: males age 15-49 116,456; males fit for military service 69,011 (1995 est.)
Defense expenditures: $NA, NA% of GDP

Svalbard

(territory of Norway)

Geography

Location: Northern Europe, islands between the Arctic Ocean, Barents Sea, Greenland Sea, and Norwegian Sea, north of Norway
Map references: Arctic Region
Area:
total area: 62,049 sq km
land area: 62,049 sq km
comparative area: slightly smaller than West Virginia
note: includes Spitsbergen and Bjornoya (Bear Island)
Land boundaries: 0 km
Coastline: 3,587 km
Maritime claims:
exclusive fishing zone: 200 nm unilaterally claimed by Norway but not recognized by Russia
territorial sea: 4 nm
International disputes: focus of maritime boundary dispute in the Barents Sea between Norway and Russia
Climate: arctic, tempered by warm North Atlantic Current; cool summers, cold winters; North Atlantic Current flows along west and north coasts of Spitsbergen, keeping water open and navigable most of the year
Terrain: wild, rugged mountains; much of high land ice covered; west coast clear of ice about half the year; fjords along west and north coasts
Natural resources: coal, copper, iron ore, phosphate, zinc, wildlife, fish
Land use:
arable land: 0%
permanent crops: 0%
meadows and pastures: 0%
forest and woodland: 0%
other: 100% (no trees and the only bushes are crowberry and cloudberry)
Irrigated land: NA sq km
Environment:
current issues: NA
natural hazards: ice floes often block up the entrance to Bellsund (a transit point for coal export) on the west coast and occasionally make parts of the northeastern coast inaccessible to maritime traffic
international agreements: NA
Note: northernmost part of the Kingdom of Norway; consists of nine main islands; glaciers and snowfields cover 60% of the total area

People

Population: 2,914 (July 1995 est.)
Age structure:
0-14 years: NA
15-64 years: NA
65 years and over: NA
Population growth rate: -3.5% (1995 est.)
Birth rate: NA births/1,000 population
Death rate: NA deaths/1,000 population
Net migration rate: NA migrant(s)/1,000 population
Infant mortality rate: NA deaths/1,000 live births
Life expectancy at birth:
total population: NA years
male: NA years
female: NA years
Total fertility rate: NA children born/woman
Ethnic divisions: Russian 64%, Norwegian 35%, other 1% (1981)
Languages: Russian, Norwegian
Labor force: NA

Government

Names:
conventional long form: none
conventional short form: Svalbard
Digraph: SV
Type: territory of Norway administered by the Ministry of Industry, Oslo, through a governor (sysselmann) residing in Longyearbyen, Spitsbergen; by treaty (9 February 1920) sovereignty was given to Norway
Capital: Longyearbyen
Independence: none (territory of Norway)
National holiday: NA
Legal system: NA
Executive branch:
Chief of State: King HARALD V (since 17 January 1991)
Head of Government: Governor Odd BLOMDAL (since NA), Assistant Governor Jan-Atle HANSEN (since NA September 1993)
Member of: none
Flag: the flag of Norway is used

Economy

Overview: Coal mining is the major economic activity on Svalbard. By treaty (9 February 1920), the nationals of the treaty powers have equal rights to exploit mineral deposits, subject to Norwegian regulation.

Svalbard *(continued)*

Although US, UK, Dutch, and Swedish coal companies have mined in the past, the only companies still mining are Norwegian and Russian. The settlements on Svalbard are essentially company towns. The Norwegian state-owned coal company employs nearly 60% of the Norwegian population on the island, runs many of the local services, and provides most of the local infrastructure. There is also some trapping of seal, polar bear, fox, and walrus.
Budget:
revenues: $13.3 million
expenditures: $13.3 million, including capital expenditures of $NA (1990 est.)
Electricity:
capacity: 21,000 kW
production: 45 million kWh
consumption per capita: 13,860 kWh (1992)
Currency: 1 Norwegian krone (NKr) = 100 oere
Exchange rates: Norwegian kroner (NKr) per US$1—6.7014 (January 1995), 7.0469 (1994), 7.0941 (1993), 6.2145 (1992), 6.4829 (1991), 6.2597 (1990)

Transportation

Railroads: 0 km
Highways:
total: NA
paved: NA
unpaved: NA
Ports: Barentsburg, Longyearbyen, Ny-Alesund, Pyramiden
Merchant marine: none
Airports:
total: 4
with paved runways 1,524 to 2,437 m: 1
with paved runways under 914 m: 3

Communications

Telephone system: NA telephones; local telephone service
local: NA
intercity: NA
international: satellite communication with Norwegian mainland
Radio:
broadcast stations: AM 1, FM 1 (repeaters 2), shortwave 0
radios: NA
Television:
broadcast stations: 1
televisions: NA
Note: there are 5 meteorological/radio stations

Defense Forces

Note: demilitarized by treaty (9 February 1920)

Swaziland

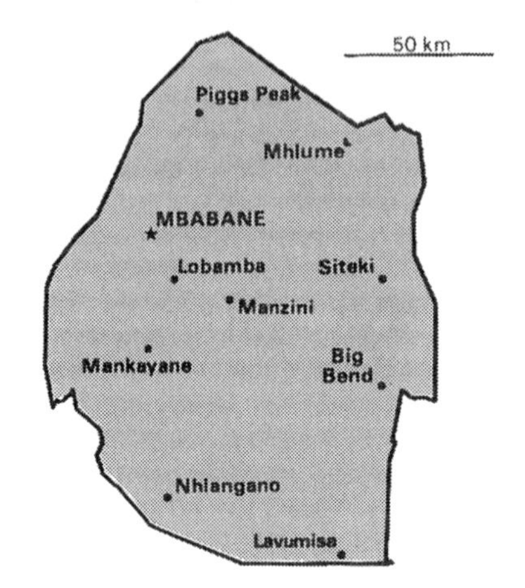

Geography

Location: Southern Africa, between Mozambique and South Africa
Map references: Africa
Area:
total area: 17,360 sq km
land area: 17,200 sq km
comparative area: slightly smaller than New Jersey
Land boundaries: total 535 km, Mozambique 105 km, South Africa 430 km
Coastline: 0 km (landlocked)
Maritime claims: none; landlocked
International disputes: Swaziland has asked South Africa to open negotiations on reincorporating some nearby South African territories that are populated by ethnic Swazis or that were long ago part of the Swazi Kingdom
Climate: varies from tropical to near temperate
Terrain: mostly mountains and hills; some moderately sloping plains
Natural resources: asbestos, coal, clay, cassiterite, hydropower, forests, small gold and diamond deposits, quarry stone, and talc
Land use:
arable land: 10.9%
permanent crops: 0.2%
meadows and pastures: 62.2%
forest and woodland: 6.9%
other: 19.8%
Irrigated land: 640 sq km (1993 est.)
Environment:
current issues: limited access to potable water; wildlife populations being depleted because of excessive hunting; overgrazing; soil degradation; soil erosion
natural hazards: NA
international agreements: party to—Biodiversity, Nuclear Test Ban, Ozone Layer Protection; signed, but not ratified—Climate Change, Law of the Sea
Note: landlocked; almost completely surrounded by South Africa

People

Population: 966,977 (July 1995 est.)
Age structure:
0-14 years: 46% (female 222,544; male 221,003)
15-64 years: 52% (female 261,973; male 238,726)
65 years and over: 2% (female 13,291; male 9,440) (July 1995 est.)
Population growth rate: 3.23% (1995 est.)
Birth rate: 43.06 births/1,000 population (1995 est.)
Death rate: 10.8 deaths/1,000 population (1995 est.)
Net migration rate: 0 migrant(s)/1,000 population (1995 est.)
Infant mortality rate: 90.7 deaths/1,000 live births (1995 est.)
Life expectancy at birth:
total population: 56.84 years
male: 52.83 years
female: 60.96 years (1995 est.)
Total fertility rate: 6.1 children born/woman (1995 est.)
Nationality:
noun: Swazi(s)
adjective: Swazi
Ethnic divisions: African 97%, European 3%
Religions: Christian 60%, indigenous beliefs 40%
Languages: English (official; government business conducted in English), siSwati (official)
Literacy: age 15 and over can read and write (1986)
total population: 67%
male: 70%
female: 65%
Labor force: NA
by occupation: private sector about 65%, public sector 35%

Government

Names:
conventional long form: Kingdom of Swaziland
conventional short form: Swaziland
Digraph: WZ
Type: monarchy; independent member of Commonwealth
Capital: Mbabane (administrative); Lobamba (legislative)
Administrative divisions: 4 districts; Hhohho, Lubombo, Manzini, Shiselweni
Independence: 6 September 1968 (from UK)
National holiday: Somhlolo (Independence) Day, 6 September (1968)
Constitution: none; constitution of 6 September 1968 was suspended 12 April 1973; a new constitution was promulgated 13 October 1978, but has not been formally presented to the people

Legal system: based on South African Roman-Dutch law in statutory courts, Swazi traditional law and custom in traditional courts; has not accepted compulsory ICJ jurisdiction
Suffrage: none
Executive branch:
chief of state: King MSWATI III (since 25 April 1986)
head of government: Prime Minister Prince Jameson Mbilini DLAMINI (since 12 November 1993)
cabinet: Cabinet; designated by the monarch
Legislative branch: bicameral Parliament is advisory
Senate: consists of 30 members (10 appointed by the House of Assembly and 20 by the king)
House of Assembly: consists of 65 members (55 directly elected and 10 appointed by the king); elections last held NA October 1993 (next to be held NA); results—balloting held on a non-party basis
Judicial branch: High Court, Court of Appeal
Political parties and leaders:
note: political parties are banned by the Constitution promulgated on 13 October 1978; illegal parties are prohibited from holding large public gatherings
illegal parties: Peoples' United Democratic Movement (PUDEMO), Kilson SHONOWE; Swaziland Youth Congress (SWAYCO), Benedict TSABEDZE; Swaziland Communist Party (SWACOPA), Mphandlana SHONGWE
Member of: ACP, AfDB, C, CCC, ECA, FAO, G-77, GATT, IBRD, ICAO, ICFTU, ICRM, IDA, IFAD, IFC, IFRCS, ILO, IMF, INTELSAT, INTERPOL, IOC, ITU, NAM, OAU, PCA, SACU, SADC, UN, UNCTAD, UNESCO, UNIDO, UPU, WHO, WIPO, WMO
Diplomatic representation in US:
chief of mission: Ambassador Madzandza Mary KHANYA
chancery: 3400 International Drive NW, Washington, DC 20008
telephone: [1] (202) 362-6683, 6685
FAX: [1] (202) 244-8059
US diplomatic representation:
chief of mission: Ambassador John T. SPROTT
embassy: Central Bank Building, Warner Street, Mbabane
mailing address: P. O. Box 199, Mbabane
telephone: [268] 46441 through 46445
FAX: [268] 45959
Flag: three horizontal bands of blue (top), red (triple width), and blue; the red band is edged in yellow; centered in the red band is a large black and white shield covering two spears and a staff decorated with feather tassels, all placed horizontally

Economy

Overview: The economy is based on subsistence agriculture, which occupies more than 60% of the population and contributes nearly 25% to GDP. Manufacturing, which includes a number of agroprocessing factories, accounts for another quarter of GDP. Mining has declined in importance in recent years; high-grade iron ore deposits were depleted by 1978, and health concerns cut world demand for asbestos. Exports of sugar and forestry products are the main earners of hard currency. Surrounded by South Africa, except for a short border with Mozambique, Swaziland is heavily dependent on South Africa, from which it receives 90% of its imports and to which it sends about half of its exports. Remittances from Swazi workers in South African mines may supplement domestically produced income by as much as 20%.
National product: GDP—purchasing power parity—$3.3 billion (1994 est.)
National product real growth rate: 4.5% (1994 est.)
National product per capita: $3,490 (1994 est.)
Inflation rate (consumer prices): 11.3% (1993 est.)
Unemployment rate: 15% (1992 est.)
Budget:
revenues: $342 million
expenditures: $410 million, including capital expenditures of $130 million (1994 est.)
Exports: $632 million (f.o.b., 1993 est.)
commodities: sugar, edible concentrates, wood pulp, cotton yarn, asbestos
partners: South Africa 50% (est.), EC countries, Canada
Imports: $734 million (f.o.b., 1993 est.)
commodities: motor vehicles, machinery, transport equipment, petroleum products, foodstuffs, chemicals
partners: South Africa 90% (est.), Switzerland, UK
External debt: $240 million (1992)
Industrial production: growth rate 4.2% (1993 est.)
Electricity:
capacity: 120,000 kW
production: 410 million kWh
consumption per capita: 1,003 kWh (1993)
Industries: mining (coal and asbestos), wood pulp, sugar
Agriculture: accounts for over 60% of labor force; mostly subsistence agriculture; cash crops—sugarcane, cotton, maize, tobacco, rice, citrus fruit, pineapples; other crops and livestock—corn, sorghum, peanuts, cattle, goats, sheep; not self-sufficient in grain
Economic aid:
recipient: bilateral aid (1991) $35 million of which US disbursements $12 million, UK disbursements $6 million, and Denmark $2 million; multilateral aid (1991) $24 million of which EC disbursements $8 million
Currency: 1 lilangeni (E) = 100 cents
Exchange rates: emalangeni (E) per US$1—3.5389 (January 1995), 3.5490 (1994), 3.2636 (1993), 2.8497 (1992), 2.7563 (1991), 2.5863 (1990); note—the Swazi emalangeni is at par with the South African rand
Fiscal year: 1 April—31 March

Transportation

Railroads:
total: 297 km; note—includes 71 km which are not in use
narrow gauge: 297 km 1.067-m gauge (single track)
Highways:
total: 2,853 km
paved: 510 km
unpaved: crushed stone, gravel, stabilized earth 1,230 km; improved earth 1,113 km
Ports: none
Airports:
total: 18
with paved runways 2,438 to 3,047 m: 1
with paved runways under 914 m: 9
with unpaved runways 914 to 1,523 m: 8

Communications

Telephone system: 17,000 telephones; telephone density is only 17.6/1,000 persons
local: NA
intercity: system consists of carrier-equipped open-wire lines and low-capacity radio relay microwave links
international: 1 INTELSAT (Atlantic Ocean) earth station
Radio:
broadcast stations: AM 7, FM 6, shortwave 0
radios: NA
Television:
broadcast stations: 10
televisions: NA

Defense Forces

Branches: Umbutfo Swaziland Defense Force (Army), Royal Swaziland Police Force
Manpower availability: males age 15-49 212,239; males fit for military service 122,782 (1995 est.)
Defense expenditures: exchange rate conversion—$22 million, NA% of GDP (FY93/94)

Sweden

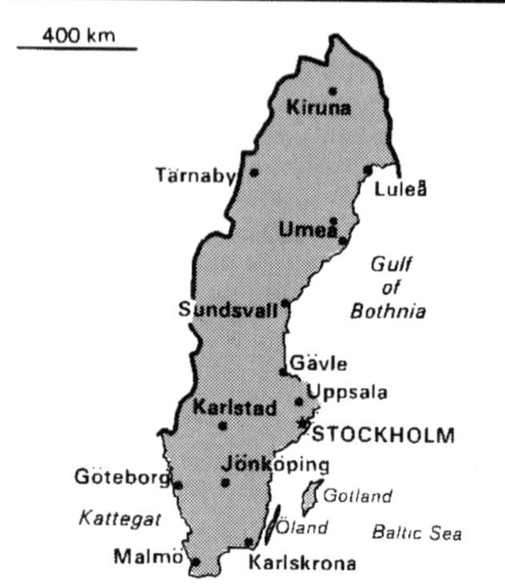

Geography

Location: Northern Europe, bordering the Baltic Sea, Gulf of Bothnia, and Skagerrak, between Finland and Norway
Map references: Europe
Area:
total area: 449,964 sq km
land area: 410,928 sq km
comparative area: slightly smaller than California
Land boundaries: total 2,205 km, Finland 586 km, Norway 1,619 km
Coastline: 3,218 km
Maritime claims:
continental shelf: 200-m depth or to the depth of exploitation
exclusive economic zone: agreed boundaries or midlines
territorial sea: 12 nm
International disputes: none
Climate: temperate in south with cold, cloudy winters and cool, partly cloudy summers; subarctic in north
Terrain: mostly flat or gently rolling lowlands; mountains in west
Natural resources: zinc, iron ore, lead, copper, silver, timber, uranium, hydropower potential
Land use:
arable land: 7%
permanent crops: 0%
meadows and pastures: 2%
forest and woodland: 64%
other: 27%
Irrigated land: 1,120 sq km (1989 est.)
Environment:
current issues: acid rain damaging soils and lakes; pollution of the North Sea and the Baltic Sea
natural hazards: ice floes in the surrounding waters, especially in the Gulf of Bothnia, can interfere with maritime traffic
international agreements: party to—Air Pollution, Air Pollution-Nitrogen Oxides, Air Pollution-Sulphur 85, Air Pollution-Volatile Organic Compounds, Antarctic-Environmental Protocol, Antarctic Treaty, Biodiversity, Climate Change, Endangered Species, Environmental Modification, Hazardous Wastes, Marine Dumping, Nuclear Test Ban, Ozone Layer Protection, Ship Pollution, Tropical Timber 83, Wetlands, Whaling; signed, but not ratified—Air Pollution-Sulphur 94, Desertification, Law of the Sea
Note: strategic location along Danish Straits linking Baltic and North Seas

People

Population: 8,821,759 (July 1995 est.)
Age structure:
0-14 years: 19% (female 810,859; male 854,553)
15-64 years: 64% (female 2,761,060; male 2,856,012)
65 years and over: 17% (female 887,597; male 651,678) (July 1995 est.)
Population growth rate: 0.46% (1995 est.)
Birth rate: 13.19 births/1,000 population (1995 est.)
Death rate: 10.84 deaths/1,000 population (1995 est.)
Net migration rate: 2.27 migrant(s)/1,000 population (1995 est.)
Infant mortality rate: 5.6 deaths/1,000 live births (1995 est.)
Life expectancy at birth:
total population: 78.43 years
male: 75.64 years
female: 81.39 years (1995 est.)
Total fertility rate: 1.97 children born/woman (1995 est.)
Nationality:
noun: Swede(s)
adjective: Swedish
Ethnic divisions: white, Lapp (Sami), foreign born or first-generation immigrants 12% (Finns, Yugoslavs, Danes, Norwegians, Greeks, Turks)
Religions: Evangelical Lutheran 94%, Roman Catholic 1.5%, Pentecostal 1%, other 3.5% (1987)
Languages: Swedish
note: small Lapp- and Finnish-speaking minorities; immigrants speak native languages
Literacy: age 15 and over can read and write (1991 est.)
total population: 99%
Labor force: 4.552 million (84% unionized,1992)
by occupation: community, social and personal services 38.3%, mining and manufacturing 21.2%, commerce, hotels, and restaurants 14.1%, banking, insurance 9.0%, communications 7.2%, construction 7.0%, agriculture, fishing, and forestry 3.2% (1991)

Government

Names:
conventional long form: Kingdom of Sweden
conventional short form: Sweden
local long form: Konungariket Sverige
local short form: Sverige
Digraph: SW
Type: constitutional monarchy
Capital: Stockholm
Administrative divisions: 24 provinces (lan, singular and plural); Alvsborgs Lan, Blekinge Lan, Gavleborgs Lan, Goteborgs och Bohus Lan, Gotlands Lan, Hallands Lan, Jamtlands Lan, Jonkopings Lan, Kalmar Lan, Kopparbergs Lan, Kristianstads Lan, Kronobergs Lan, Malmohus Lan, Norrbottens Lan, Orebro Lan, Ostergotlands Lan, Skaraborgs Lan, Sodermanlands Lan, Stockholms Lan, Uppsala Lan, Varmlands Lan, Vasterbottens Lan, Vasternorrlands Lan, Vastmanlands Lan
Independence: 6 June 1809 (constitutional monarchy established)
National holiday: Day of the Swedish Flag, 6 June
Constitution: 1 January 1975
Legal system: civil law system influenced by customary law; accepts compulsory ICJ jurisdiction, with reservations
Suffrage: 18 years of age; universal
Executive branch:
chief of state: King CARL XVI GUSTAF (since 19 September 1973); Heir Apparent Princess VICTORIA Ingrid Alice Desiree, daughter of the King (born 14 July 1977)
head of government: Prime Minister Ingvar CARLSSON (since 6 October 1994); Deputy Prime Minister Mona SAHLIN (since 6 October 1994)
cabinet: Cabinet; appointed by the prime minister
Legislative branch: unicameral
Parliament (Riksdag): elections last held 18 September 1994 (next to be held NA September 1998); results—Social Democrats 45.4%, Moderate Party (Conservatives) 22.3%, Center Party 7.7%, Liberals 7.2%, Left Party 6.2%, Greens 5.8%, Christian Democrats 4.1%, New Democracy Party 1.2%; seats—(349 total) Social Democrats 162, Moderate Party (Conservatives) 80, Center Party 27, Liberals 26, Left Party 22, Greens 18, Christian Democrats 14; note—the New Democracy Party did not receive a seat because parties require a minimum of 4.8% of votes for a seat in parliament
Judicial branch: Supreme Court (Hogsta Domstolen)
Political parties and leaders: Social Democratic Party, Ingvar CARLSSON; Moderate Party (conservative), Carl BILDT; Liberal People's Party, Maria LEISSNER; Center Party, Olof JOHANSSON; Christian

Democratic Party, Alf SVENSSON; New Democracy Party, Vivianne FRANZEN; Left Party (VP; Communist), Gudrun SCHYMAN; Communist Workers' Party, Rolf HAGEL; Green Party, no formal leader but party spokesperson is Birger SHLAUG
Member of: AfDB, AG (observer), AsDB, Australia Group, BIS, CBSS, CCC, CE, CERN, EBRD, ECE, EFTA, ESA, EU, FAO, G- 6, G- 8, G- 9, G-10, GATT, IADB, IAEA, IBRD, ICAO, ICC, ICFTU, ICRM, IDA, IEA, IFAD, IFC, IFRCS, ILO, IMF, IMO, INMARSAT, INTELSAT, INTERPOL, IOC, IOM, ISO, ITU, MTCR, NAM (guest), NC, NEA, NIB, NSG, OECD, ONUSAL, OSCE, PCA, PFP, UN, UNAVEM II, UNCTAD, UNESCO, UNHCR, UNIDO, UNIKOM, UNITAR, UNMOGIP, UNOMIG, UNOMOZ, UNPROFOR, UNTSO, UPU, WFTU, WHO, WIPO, WMO, ZC
Diplomatic representation in US:
chief of mission: Ambassador Carl Henrik Sihver LILJEGREN
chancery: 1501 M Street NW, Washington, DC 20005
telephone: [1] (202) 467-2600
FAX: [1] (202) 467-2699
consulate(s) general: Los Angeles and New York
US diplomatic representation:
chief of mission: Ambassador Thomas L. SIEBERT
embassy: Strandvagen 101, S-115 89 Stockholm
mailing address: use embassy street address
telephone: [46] (8) 783 53 00
FAX: [46] (8) 661 19 64
Flag: blue with a yellow cross that extends to the edges of the flag; the vertical part of the cross is shifted to the hoist side in the style of the Dannebrog (Danish flag)

Economy

Overview: Aided by a long period of peace and neutrality during World War I through World War II, Sweden has achieved an enviable standard of living under a mixed system of high-tech capitalism and extensive welfare benefits. It has a modern distribution system, excellent internal and external communications, and a skilled labor force. Timber, hydropower, and iron ore constitute the resource base of an economy that is heavily oriented toward foreign trade. Privately owned firms account for about 90% of industrial output, of which the engineering sector accounts for 50% of output and exports. In 1990, agriculture accounted for only 1.2% of GDP and 1.9% of the jobs, Sweden being about 50% sufficient in most products. In the last few years, however, this extraordinarily favorable picture has been clouded by inflation, growing unemployment, and a gradual loss of competitiveness in international markets. Although Prime Minister BILDT's center-right minority coalition had hoped to charge ahead with free-market-oriented reforms, a skyrocketing budget deficit—about 14% of GDP in FY93/94 projections—and record unemployment have forestalled many of the plans. Unemployment in 1994 is estimated at around 9% with another 5% in job training. Continued heavy foreign exchange speculation forced the government to cooperate in late 1992 with the opposition Social Democrats on two crisis packages—one a severe austerity pact and the other a program to spur industrial competitiveness—which basically set economic policy through 1997. In November 1992, Sweden broke its tie to the EC's ECU, and the krona has since depreciated about 25% against the dollar. The boost in export competitiveness from the depreciation helped lift Sweden out of its 3-year recession. To curb the budget deficit and bolster confidence in the economy, the new Social Democratic government is proposing cuts in welfare benefits, subsidies, defense, and foreign aid. Sweden has harmonized its economic policies with those of the EU, which it joined at the start of 1995.
National product: GDP—purchasing power parity—$163.1 billion (1994 est.)
National product real growth rate: 2.4% (1994 est.)
National product per capita: $18,580 (1994 est.)
Inflation rate (consumer prices): 2.5% (1994 est.)
Unemployment rate: 8.8% (1994 est.)
Budget:
revenues: $47.9 billion
expenditures: $70.9 billion, including capital expenditures of $NA (FY93/94)
Exports: $59.9 billion (f.o.b., 1994)
commodities: machinery, motor vehicles, paper products, pulp and wood, iron and steel products, chemicals, petroleum and petroleum products
partners: EC 55.8% (Germany 15%, UK 9.7%, Denmark 7.2%, France 5.8%), EFTA 17.4% (Norway 8.4%, Finland 5.1%), US 8.2%, Central and Eastern Europe 2.5% (1992)
Imports: $49.6 billion (c.i.f., 1994)
commodities: machinery, petroleum and petroleum products, chemicals, motor vehicles, foodstuffs, iron and steel, clothing
partners: EC 53.6% (Germany 17.9%, UK 6.3%, Denmark 7.5%, France 4.9%), EFTA (Norway 6.6%, Finland 6%), US 8.4%, Central and Eastern Europe 3% (1992)
External debt: $NA
Industrial production: growth rate 9% (1994)
Electricity:
capacity: 34,560,000 kW
production: 141 billion kWh
consumption per capita: 14,891 kWh (1993)
Industries: iron and steel, precision equipment (bearings, radio and telephone parts, armaments), wood pulp and paper products, processed foods, motor vehicles
Agriculture: animal husbandry predominates, with milk and dairy products accounting for 37% of farm income; main crops—grains, sugar beets, potatoes; 100% self-sufficient in grains and potatoes; Sweden is about 50% self-sufficient in most products
Illicit drugs: transshipment point for narcotics shipped via the CIS and Baltic states for the European market
Economic aid:
donor: ODA and OOF commitments (1970-89), $10.3 billion
Currency: 1 Swedish krona (SKr) = 100 oere
Exchange rates: Swedish kronor (SKr) per US$1—7.4675 (January 1995), 7.7160 (1994), 7.7834 (1993), 5.8238 (1992), 6.0475 (1991) 5.9188 (1990)
Fiscal year: 1 July—30 June

Transportation

Railroads:
total: 12,000 km (includes 953 km of privately owned railways)
standard gauge: 10,742 km 1.435-m gauge (7,502 km electrified and 1,152 km double track); 8 km 1.435-m gauge (electrified; privately owned)
narrow gauge: 61 km 0.891-m gauge (electrified; privately owned)
other: 1,189 km NA-m gauge (1994)
Highways:
total: 135,859 km
paved: 97,818 km (including 936 km of expressways)
unpaved: gravel 38,041 km (1991)
Inland waterways: 2,052 km navigable for small steamers and barges
Pipelines: natural gas 84 km
Ports: Gavle, Goteborg, Halmstad, Helsingborg, Hudiksvall, Kalmar, Karlshamn, Malmo, Solvesborg, Stockholm, Sundsvall
Merchant marine:
total: 157 ships (1,000 GRT or over) totaling 1,872,350 GRT/2,075,722 DWT
ships by type: bulk 10, cargo 24, chemical tanker 25, combination ore/oil 1, container 2, oil tanker 31, railcar carrier 2, refrigerated cargo 1, roll-on/roll-off cargo 37, short-sea passenger 8, specialized tanker 4, vehicle carrier 12
Airports:
total: 253
with paved runways over 3,047 m: 2
with paved runways 2,438 to 3,047 m: 8
with paved runways 1,524 to 2,437 m: 84
with paved runways 914 to 1,523 m: 26
with paved runways under 914 m: 129
with unpaved runways 914 to 1,523 m: 4

Communications

Telephone system: 8,200,000 telephones; excellent domestic and international facilities; automatic system
local: NA
intercity: coaxial and multiconductor cable carry most voice traffic; parallel microwave network carries TV, radio, and some additional telephone channels
international: 5 submarine coaxial cables; 1 INTELSAT (Atlantic Ocean) and 1 EUTELSAT earth station
Radio:
broadcast stations: AM 5, FM 360 (mostly repeaters), shortwave 0
radios: 7 million
Television:
broadcast stations: 880 (mostly repeaters)
televisions: 3.5 million

Defense Forces

Branches: Swedish Army, Royal Swedish Navy, Swedish Air Force
Manpower availability: males age 15-49 2,133,420; males fit for military service 1,864,258; males reach military age (19) annually 52,937 (1995 est.)
Defense expenditures: exchange rate conversion—$5.4 billion, 2.4% of GDP (FY94/95)

Switzerland

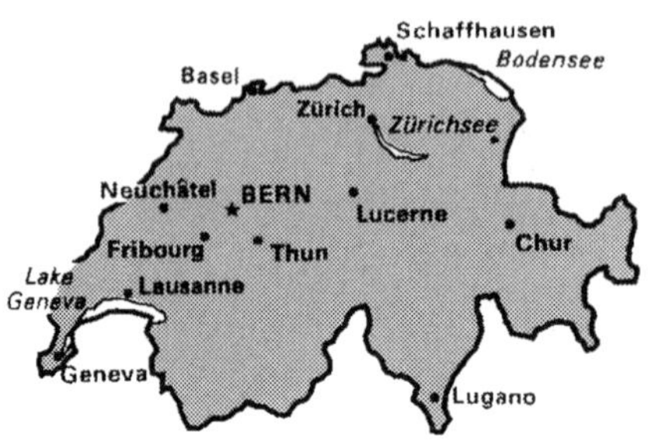

Geography

Location: Central Europe, east of France
Map references: Europe
Area:
total area: 41,290 sq km
land area: 39,770 sq km
comparative area: slightly more than twice the size of New Jersey
Land boundaries: total 1,852 km, Austria 164 km, France 573 km, Italy 740 km, Liechtenstein 41 km, Germany 334 km
Coastline: 0 km (landlocked)
Maritime claims: none; landlocked
International disputes: none
Climate: temperate, but varies with altitude; cold, cloudy, rainy/snowy winters; cool to warm, cloudy, humid summers with occasional showers
Terrain: mostly mountains (Alps in south, Jura in northwest) with a central plateau of rolling hills, plains, and large lakes
Natural resources: hydropower potential, timber, salt
Land use:
arable land: 10%
permanent crops: 1%
meadows and pastures: 40%
forest and woodland: 26%
other: 23%
Irrigated land: 250 sq km (1989)
Environment:
current issues: air pollution from vehicle emissions and open air burning; acid rain; water pollution from increased use of agricultural fertilizers; loss of biodiversity
natural hazards: avalanches, landslides, flash floods
international agreements: party to—Air Pollution, Air Pollution-Nitrogen Oxides, Air Pollution-Sulphur 85, Air Pollution-Volatile Organic Compounds, Antarctic Treaty, Biodiversity, Climate Change, Endangered Species, Environmental Modification, Hazardous Wastes, Marine Dumping, Marine Life Conservation, Nuclear Test Ban, Ozone Layer Protection, Ship Pollution, Tropical Timber 83, Wetlands, Whaling; signed, but not ratified—Air Pollution-Sulphur 94, Antarctic-Environmental Protocol, Desertification, Law of the Sea
Note: landlocked; crossroads of northern and southern Europe; along with southeastern France and northern Italy, contains the highest elevations in Europe

People

Population: 7,084,984 (July 1995 est.)
Age structure:
0-14 years: 17% (female 594,565; male 622,436)
15-64 years: 68% (female 2,375,792; male 2,448,213)
65 years and over: 15% (female 623,136; male 420,842) (July 1995 est.)
Population growth rate: 0.57% (1995 est.)
Birth rate: 12.04 births/1,000 population (1995 est.)
Death rate: 9.16 deaths/1,000 population (1995 est.)
Net migration rate: 2.82 migrant(s)/1,000 population (1995 est.)
Infant mortality rate: 6.3 deaths/1,000 live births (1995 est.)
Life expectancy at birth:
total population: 78.36 years
male: 74.99 years
female: 81.88 years (1995 est.)
Total fertility rate: 1.6 children born/woman (1995 est.)
Nationality:
noun: Swiss (singular and plural)
adjective: Swiss
Ethnic divisions:
total population: German 65%, French 18%, Italian 10%, Romansch 1%, other 6%
Swiss nationals: German 74%, French 20%, Italian 4%, Romansch 1%, other 1%
Religions: Roman Catholic 47.6%, Protestant 44.3%, other 8.1% (1980)
Languages: German 65%, French 18%, Italian 12%, Romansch 1%, other 4%
note: figures for Swiss nationals only—German 74%, French 20%, Italian 4%, Romansch 1%, other 1%
Literacy: age 15 and over can read and write (1980 est.)
total population: 99%
Labor force: 3.48 million (900,000 foreign workers, mostly Italian)
by occupation: services 50%, industry and crafts 34%, government 10%, agriculture and forestry 6% (1992)

Government

Names:
conventional long form: Swiss Confederation
conventional short form: Switzerland
local long form: Schweizerische Eidgenossenschaft (German) Confederation

Suisse (French) Confederazione Svizzera (Italian)
local short form: Schweiz (German) Suisse (French) Svizzera (Italian)
Digraph: SZ
Type: federal republic
Capital: Bern
Administrative divisions: 26 cantons (cantons, singular—canton in French; cantoni, singular—cantone in Italian; kantone, singular—kanton in German); Aargau, Ausser-Rhoden, Basel-Landschaft, Basel-Stadt, Bern, Fribourg, Geneve, Glarus, Graubunden, Inner-Rhoden, Jura, Luzern, Neuchatel, Nidwalden, Obwalden, Sankt Gallen, Schaffhausen, Schwyz, Solothurn, Thurgau, Ticino, Uri, Valais, Vaud, Zug, Zurich
Independence: 1 August 1291
National holiday: Anniversary of the Founding of the Swiss Confederation, 1 August (1291)
Constitution: 29 May 1874
Legal system: civil law system influenced by customary law; judicial review of legislative acts, except with respect to federal decrees of general obligatory character; accepts compulsory ICJ jurisdiction, with reservations
Suffrage: 18 years of age; universal
Executive branch:
chief of state and head of government: President Kaspar VILLIGER (1995 calendar year; presidency rotates annually); Vice President Jean-Pascal DELAMURAZ (term runs concurrently with that of president)
cabinet: Federal Council (German—Bundesrat, French—Censeil Federal, Italian—Consiglio Federale); elected by the Federal Assembly from own members
Legislative branch: bicameral Federal Assembly (German—Bundesversammlung, French—Assemblee Federale, Italian—Assemblea Federale)
Council of States: German—Standerat, French—Conseil des Etats, Italian—Consiglio degli Stati; elections last held throughout 1991 (next to be held NA 1995); results—percent of vote by party NA; seats—(46 total) FDP 18, CVP 16, SVP 4, SPS 3, LPS 3, LdU 1, Ticino League 1
National Council: German—Nationalrat, French—Conseil National, Italian—Consiglio Nazionale; elections last held 20 October 1991 (next to be held NA October 1995); results—percent of vote by party NA; seats—(200 total) FDP 44, SPS 42, CVP 37, SVP 25, GPS 14, LPS 10, AP 8, LdU 6, SD 5, EVP 3, PdA 2, Ticino League 2, other 2
Judicial branch: Federal Supreme Court
Political parties and leaders: Free Democratic Party (FDP), Franz STEINEGGER, president; Social Democratic Party (SPS), Peter BODENMANN, president; Christian Democratic People's Party (CVP), Anton COTTIER, president; Swiss People's Party (SVP), Hans UHLMANN, president; Green Party (GPS), Verena DIENER, president; Freedom Party (FPS), Roland BORER, president; Liberal Party (LPS), Christoph EYMANN, president; Alliance of Independents' Party (LdU), Monica WEBER, president; Ticino League, Giuliano BIGNASCA, president; and other minor parties including the Automobile Party (AP), Swiss Democratic Party (SD), Workers' Party (PdA), and the Evangelical People's Party (EVP); note—see elections
Member of: AfDB, AG (observer), AsDB, Australia Group, BIS, CCC, CE, CERN, EBRD, ECE, EFTA, ESA, FAO, G- 8, G-10, GATT, IADB, IAEA, IBRD, ICAO, ICC, ICFTU, ICRM, IDA, IEA, IFAD, IFC, IFRCS, ILO, IMF, IMO, INMARSAT, INTELSAT, INTERPOL, IOC, IOM, ISO, ITU, MINURSO, MTCR, NAM (guest), NEA, NSG, OAS (observer), OECD, OSCE, PCA, UN (observer), UNCTAD, UNESCO, UNHCR, UNIDO, UNITAR, UNMIH, UNOMIG, UNPROFOR, UNTSO, UNU, UPU, WCL, WHO, WIPO, WMO, WTO, ZC
Diplomatic representation in US:
chief of mission: Ambassador Carlo JAGMETTI
chancery: 2900 Cathedral Avenue NW, Washington, DC 20008
telephone: [1] (202) 745-7900
FAX: [1] (202) 387-2564
consulate(s) general: Atlanta, Chicago, Houston, Los Angeles, New York, Pago Pago (American Samoa), and San Francisco
US diplomatic representation:
chief of mission: Ambassador M. Larry LAWRENCE
embassy: Jubilaeumstrasse 93, 3005 Bern
mailing address: use embassy street address
telephone: [41] (31) 357 70 11
FAX: [41] (31) 357 73 44
branch office: Geneva
consulate(s) general: Zurich
Flag: red square with a bold, equilateral white cross in the center that does not extend to the edges of the flag

Economy

Overview: Switzerland's economy—one of the most prosperous and stable in the world—is nonetheless undergoing a stressful adjustment after both the inflationary boom of the late 1980s and the electorate's rejection of membership in the European Economic Area (EEA) in 1992. So far the decision to remain outside the European single market structure does not appear to have harmed Swiss interests. In December 1994, the Swiss began bilateral negotiations with the EU aimed at establishing closer ties in areas of mutual interest and progressing toward the free circulation of persons, goods, capital, and services between the two parties. The Swiss emerged from a three-year recession in mid-1993 and posted 1.8% GDP growth in 1994. The Swiss central bank's tight monetary policies brought inflation down from about 4% in 1992 to just under 1% in 1994. Unemployment has fallen slightly from 5.1% in 1993 to 4.7% in 1994. Swiss per capita output, living standards, education, and health care remain unsurpassed in Europe. The country has few mineral resources, but its spectacular natural beauty sustains a substantial tourism industry.
National product: GDP—purchasing power parity—$148.4 billion (1994 est.)
National product real growth rate: 1.8% (1994 est.)
National product per capita: $22,080 (1994 est.)
Inflation rate (consumer prices): 0.9% (1994 est.)
Unemployment rate: 4.7% (1994 est.)
Budget:
revenues: $26.7 billion
expenditures: $32 billion, including capital expenditures of $NA (1994 est.)
Exports: $69.6 billion (f.o.b., 1994 est.)
commodities: machinery and equipment, precision instruments, metal products, foodstuffs, textiles and clothing
partners: Western Europe 63.1% (EU countries 56%, other 7.1%), US 8.8%, Japan 3.4%
Imports: $68.2 billion (c.i.f., 1994 est.)
commodities: agricultural products, machinery and transportation equipment, chemicals, textiles, construction materials
partners: Western Europe 79.2% (EU countries 72.3%, other 6.9%), US 6.4%
External debt: $NA
Industrial production: growth rate 0% (1993 est.)
Electricity:
capacity: 15,430,000 kW
production: 58 billion kWh
consumption per capita: 6,699 kWh (1993)
Industries: machinery, chemicals, watches, textiles, precision instruments
Agriculture: dairy farming predominates; less than 50% self sufficient in food; must import fish, refined sugar, fats and oils (other than butter), grains, eggs, fruits, vegetables, meat
Illicit drugs: money-laundering center
Economic aid:
donor: ODA and OOF commitments (1970-89), $3.5 billion
Currency: 1 Swiss franc, franken, or franco (SwF) = 100 centimes, rappen, or centesimi
Exchange rates: Swiss francs, franken, or franchi (SwF) per US$1—1.2880 (January 1995), 1.3677 (1994), 1.4776 (1993), 1.4062 (1992), 1.4340 (1991), 1.3892 (1990)
Fiscal year: calendar year

Switzerland *(continued)*

Transportation

Railroads:
total: 5,763 km (1,432 km double track)
standard gauge: 3,533 km 1.435-m gauge (99% electrified; 560 km nongovernment owned)
narrow gauge: 1,094 km 1.000-m gauge (99% electrified; 1,020 km nongovernment owned)
other: 1,136 km NA-m gauge (1994)
Highways:
total: 71,118 km
paved: 71,118 km (including 1,514 km of expressways)
Inland waterways: 65 km; Rhine (Basel to Rheinfelden, Schaffhausen to Bodensee); 12 navigable lakes
Pipelines: crude oil 314 km; natural gas 1,506 km
Ports: Basel
Merchant marine:
total: 22 ships (1,000 GRT or over) totaling 374,935 GRT/669,353 DWT
ships by type: bulk 12, cargo 2, chemical tanker 4, oil tanker 2, roll-on/roll-off cargo 1, specialized tanker 1
Airports:
total: 69
with paved runways over 3,047 m: 4
with paved runways 2,438 to 3,047 m: 3
with paved runways 1,524 to 2,437 m: 14
with paved runways 914 to 1,523 m: 5
with paved runways under 914 m: 42
with unpaved runways 914 to 1,523 m: 1

Communications

Telephone system: 5,890,000 telephones; excellent domestic, international, and broadcast services
local: NA
intercity: extensive cable and microwave networks
international: 2 INTELSAT (Atlantic Ocean and Indian Ocean) earth stations
Radio:
broadcast stations: AM 7, FM 265, shortwave 0
radios: NA
Television:
broadcast stations: 18 (repeaters 1,322)
televisions: NA

Defense Forces

Branches: Army, Air Force and Antiaircraft Command
Manpower availability: males age 15-49 1,847,639; males fit for military service 1,582,335; males reach military age (20) annually 41,831 (1995 est.)
Defense expenditures: exchange rate conversion—$4.1 billion, 1.4% of GDP (1995)

Syria

Geography

Location: Middle East, bordering the Mediterranean Sea, between Lebanon and Turkey
Map references: Middle East
Area:
total area: 185,180 sq km
land area: 184,050 sq km
comparative area: slightly larger than North Dakota
note: includes 1,295 sq km of Israeli-occupied territory
Land boundaries: total 2,253 km, Iraq 605 km, Israel 76 km, Jordan 375 km, Lebanon 375 km, Turkey 822 km
Coastline: 193 km
Maritime claims:
contiguous zone: 41 nm
territorial sea: 35 nm
International disputes: separated from Israel by the 1949 Armistice Line; Golan Heights is Israeli occupied; Hatay question with Turkey; ongoing dispute over water development plans by Turkey for the Tigris and Euphrates Rivers; Syrian troops in northern Lebanon since October 1976
Climate: mostly desert; hot, dry, sunny summers (June to August) and mild, rainy winters (December to February) along coast; cold weather with snow or sleet periodically hits Damascus
Terrain: primarily semiarid and desert plateau; narrow coastal plain; mountains in west
Natural resources: petroleum, phosphates, chrome and manganese ores, asphalt, iron ore, rock salt, marble, gypsum
Land use:
arable land: 28%
permanent crops: 3%
meadows and pastures: 46%
forest and woodland: 3%
other: 20%
Irrigated land: 10,000 sq km (1992)
Environment:
current issues: deforestation; overgrazing; soil erosion; desertification; water pollution from dumping of raw sewage and wastes from petroleum refining; inadequate supplies of potable water
natural hazards: dust storms, sandstorms
international agreements: party to—Hazardous Wastes, Nuclear Test Ban, Ozone Layer Protection, Ship Pollution; signed, but not ratified—Biodiversity, Desertification, Environmental Modification
Note: there are 42 Jewish settlements and civilian land use sites in the Israeli-occupied Golan Heights (August 1994 est.)

People

Population: 15,451,917 (July 1995 est.)
note: in addition, there are 31,000 people living in the Israeli-occupied Golan Heights—16,500 Arabs (15,000 Druze and 1,500 Alawites) and 14,500 Jewish settlers (August 1994 est.)
Age structure:
0-14 years: 48% (female 3,639,776; male 3,826,154)
15-64 years: 49% (female 3,691,862; male 3,854,989)
65 years and over: 3% (female 219,251; male 219,885) (July 1995 est.)
Population growth rate: 3.71% (1995 est.)
Birth rate: 43.21 births/1,000 population (1995 est.)
Death rate: 6.07 deaths/1,000 population (1995 est.)
Net migration rate: 0 migrant(s)/1,000 population (1995 est.)
Infant mortality rate: 41.1 deaths/1,000 live births (1995 est.)
Life expectancy at birth:
total population: 66.81 years
male: 65.67 years
female: 68.01 years (1995 est.)
Total fertility rate: 6.55 children born/woman (1995 est.)
Nationality:
noun: Syrian(s)
adjective: Syrian
Ethnic divisions: Arab 90.3%, Kurds, Armenians, and other 9.7%
Religions: Sunni Muslim 74%, Alawite, Druze, and other Muslim sects 16%, Christian (various sects) 10%, Jewish (tiny communities in Damascus, Al Qamishli, and Aleppo)
Languages: Arabic (official), Kurdish, Armenian, Aramaic, Circassian, French widely understood
Literacy: age 15 and over can read and write (1990 est.)
total population: 64%
male: 78%
female: 51%
Labor force: 4.3 million (1994 est.)
by occupation: miscellaneous and government services 36%, agriculture 32%, industry and construction 32%; note—shortage of skilled labor (1984)

Government

Names:
conventional long form: Syrian Arab Republic
conventional short form: Syria
local long form: Al Jumhuriyah al Arabiyah as Suriyah
local short form: Suriyah
former: United Arab Republic (with Egypt)
Digraph: SY
Type: republic under leftwing military regime since March 1963
Capital: Damascus
Administrative divisions: 14 provinces (muhafazat, singular—muhafazah); Al Hasakah, Al Ladhiqiyah, Al Qunaytirah, Ar Raqqah, As Suwayda', Dar'a, Dayr az Zawr, Dimashq, Halab, Hamah, Hims, Idlib, Rif Dimashq, Tartus
Independence: 17 April 1946 (from League of Nations mandate under French administration)
National holiday: National Day, 17 April (1946)
Constitution: 13 March 1973
Legal system: based on Islamic law and civil law system; special religious courts; has not accepted compulsory ICJ jurisdiction
Suffrage: 18 years of age; universal
Executive branch:
chief of state: President Hafiz al-ASAD (since 22 February 1971 see note); Vice Presidents 'Abd al-Halim ibn Said KHADDAM, Rif'at al-ASAD, and Muhammad Zuhayr MASHARIQA (since 11 March 1984); election last held 2 December 1991 (next to be held NA December 1998); results—President Hafiz al-ASAD was reelected for a fourth seven-year term with 99.98% of the vote; note—President ASAD seized power in the November 1970 coup, assumed presidential powers 22 February 1971, and was confirmed as president in the 12 March 1971 national elections
head of government: Prime Minister Mahmud ZU'BI (since 1 November 1987); Deputy Prime Minister Lt. Gen. Mustafa TALAS (since 11 March 1984); Deputy Prime Minister Salim YASIN (since NA December 1981); Deputy Prime Minister Rashid AKHTARINI (since 4 July 1992)
cabinet: Council of Ministers; appointed by the president
Legislative branch: unicameral
People's Council (Majlis al-Chaab): elections last held 24-25 August 1994 (next to be held NA); results—percent of vote by party NA; seats—(250 total) National Progressive Front 167, independents 83
Judicial branch: Supreme Constitutional Court, High Judicial Council, Court of Cassation, State Security Courts
Political parties and leaders:
National Progressive Front includes: the ruling Arab Socialist Resurrectionist (Ba'th) Party, Hafiz al-ASAD, President of the Republic, Secretary General of the party, and Chairman of the National Progressive Front; Syrian Arab Socialist Party (ASP), 'Abd al-Ghani KANNUT; Arab Socialist Union (ASU), Jamal ATASSI; Syrian Communist Party (SCP), Khalid BAKDASH; Arab Socialist Unionist Movement, Sami SOUFAN; and Democratic Socialist Union Party, leader NA
Other political or pressure groups: non-Ba'th parties have little effective political influence; Communist party ineffective; conservative religious leaders; Muslim Brotherhood
Member of: ABEDA, AFESD, AL, AMF, CAEU, CCC, ESCWA, FAO, G-24, G-77, IAEA, IBRD, ICAO, ICC, ICRM, IDA, IDB, IFAD, IFC, IFRCS, ILO, IMF, IMO, INTELSAT, INTERPOL, IOC, ISO, ITU, NAM, OAPEC, OIC, UN, UNCTAD, UNESCO, UNIDO, UNRWA, UPU, WFTU, WHO, WMO, WTO
Diplomatic representation in US:
chief of mission: Ambassador Walid MUALEM
chancery: 2215 Wyoming Avenue NW, Washington, DC 20008
telephone: [1] (202) 232-6313
FAX: [1] (202) 234-9548
US diplomatic representation:
chief of mission: Ambassador Christopher W. S. ROSS
embassy: Abou Roumaneh, Al-Mansur Street No. 2, Damascus
mailing address: P. O. Box 29, Damascus
telephone: [963] (11) 333-2814, 714-108, 333-3788
FAX: [963] (11) 224-7938
Flag: three equal horizontal bands of red (top), white, and black with two small green five-pointed stars in a horizontal line centered in the white band; similar to the flag of Yemen, which has a plain white band and of Iraq, which has three green stars (plus an Arabic inscription) in a horizontal line centered in the white band; also similar to the flag of Egypt, which has a symbolic eagle centered in the white band

Economy

Overview: In 1990-93 Syria's state-dominated Ba'thist economy benefited from the Gulf war, increased oil production, good weather, and economic deregulation. Economic growth averaged roughly 10%. The Gulf war provided Syria an aid windfall of nearly $5 billion dollars from Arab, European, and Japanese donors. However, the benefits of the 1990-93 boom were not evenly distributed and the gap between rich and poor is widening. A nationwide financial scandal and increasing inflation were accompanied by a decline in GDP growth to 4% in 1994. For the long run, Syria's economy is still saddled with a large number of poorly performing public sector firms, and industrial productivity remains to be improved. Oil production is likely to fall off dramatically by the end of the decade. Unemployment will become a problem for the government when the more than 60% of the population under the age of 20 enter the labor force.
National product: GDP—purchasing power parity—$74.4 billion (1994 est.)
National product real growth rate: 4% (1994 est.)
National product per capita: $5,000 (1994 est.)
Inflation rate (consumer prices): 16.3% (1993 est.)
Unemployment rate: 7.5% (1993 est.)
Budget: NA
Exports: $3.6 billion (f.o.b., 1994 est.)
commodities: petroleum 53%, textiles 22%, cotton, fruits and vegetables, wheat, barley, chickens
partners: EC 48%, former CEMA countries 24%, Arab countries 18% (1991)
Imports: $4 billion (c.i.f., 1994 est.)
commodities: foodstuffs 21%, metal products 17%, machinery 15%
partners: EC 37%, former CEMA countries 15%, US and Canada 10% (1991)
External debt: $19.4 billion (1993 est.)
Industrial production: growth rate NA%
Electricity:
capacity: 4,160,000 kW
production: 13.2 billion kWh
consumption per capita: 865 kWh (1993)
Industries: textiles, food processing, beverages, tobacco, phosphate rock mining, petroleum
Agriculture: accounts for 30% of GDP and one-third of labor force; all major crops (wheat, barley, cotton, lentils, chickpeas) grown mainly on rain-watered land causing wide swings in production; animal products—beef, lamb, eggs, poultry, milk; not self-sufficient in grain or livestock products
Illicit drugs: a transit country for Lebanese and Turkish refined cocaine going to Europe and heroin and hashish bound for regional and Western markets
Economic aid:
recipient: no US aid; about $4.2 billion in loans and grants from Arab and Western donors 1990-92 as a result of Gulf war stance
Currency: 1 Syrian pound (£S) = 100 piasters
Exchange rates: Syrian pounds (£S) per US$1—11.2 (official fixed rate), 26.6 (blended rate used by the UN and diplomatic missions), 42.0 (neighboring country rate—applies to most state enterprise imports), 46.0—53.0 (offshore rate) (yearend 1993)
Fiscal year: calendar year

Syria *(continued)*

Transportation

Railroads:
total: 1,998 km
broad gauge: 1,766 km 1.435-m gauge
narrow gauge: 232 km 1.050-m gauge
Highways:
total: 31,569 km
paved: 24,308 km (including 670 km of expressways)
unpaved: 7,261 km
Inland waterways: 870 km; minimal economic importance
Pipelines: crude oil 1,304 km; petroleum products 515 km
Ports: Baniyas, Jablah, Latakia, Tartus
Merchant marine:
total: 80 ships (1,000 GRT or over) totaling 233,701 GRT/364,714 DWT
ships by type: bulk 10, cargo 68, vehicle carrier 2
Airports:
total: 107
with paved runways over 3,047 m: 5
with paved runways 2,438 to 3,047 m: 16
with paved runways 914 to 1,523 m: 1
with paved runways under 914 m: 67
with unpaved runways 1,524 to 2,438 m: 3
with unpaved runways 914 to 1,523 m: 15

Communications

Telephone system: 512,600 telephones; 37 telephones/1,000 persons; fair system currently undergoing significant improvement and digital upgrades, including fiber optic technology
local: NA
intercity: coaxial cable and microwave radio relay network
international: 1 INTELSAT (Indian Ocean) and 1 Intersputnik earth station; 1 submarine cable; coaxial cable and microwave radio relay to Iraq, Jordan, Lebanon, and Turkey
Radio:
broadcast stations: AM 9, FM 1, shortwave 0
radios: NA
Television:
broadcast stations: 17
televisions: NA

Defense Forces

Branches: Syrian Arab Army, Syrian Arab Navy, Syrian Arab Air Force, Syrian Arab Air Defense Forces, Police and Security Force
Manpower availability: males age 15-49 3,440,030; males fit for military service 1,927,930; males reach military age (19) annually 159,942 (1995 est.)
Defense expenditures: exchange rate conversion—$2.2 billion, 6% of GDP (1992)

Taiwan

Entry

follows

Zimbabwe

Tajikistan

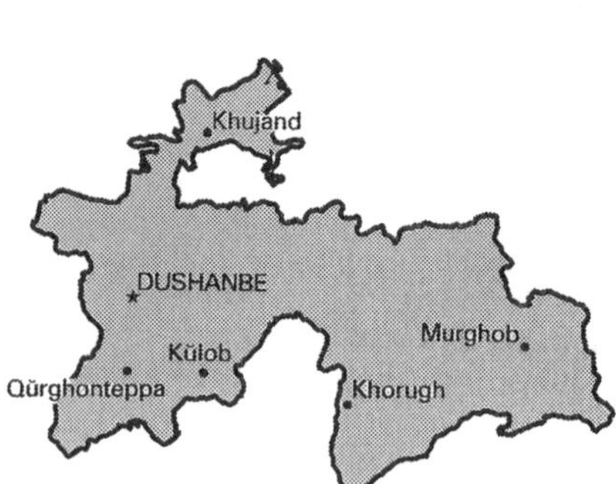

Note: Tajikistan has experienced three changes of government since it gained independence in September 1991. The current president, Emomali RAKHMONOV, was elected to the presidency in November 1994, yet has been in power since 1992. The country is suffering through its third year of a civil war, with no clear end in sight. Underlying the conflict are deeply-rooted regional and clan-based animosities that pit a government consisting of people primarily from the Kulob (Kulyab), Khujand (Leninabad), and Hisor (Hissar) regions against a secular and Islamic-led opposition from the Gharm, Gorno-Badakhshan, and Qurghonteppa (Kurgan-Tyube) regions. Government and opposition representatives have held periodic rounds of UN-mediated peace talks and agreed in September 1994 to a cease-fire. Russian-led peacekeeping troops are deployed throughout the country, and Russian border guards are stationed along the Tajik-Afghan border.

Geography

Location: Central Asia, west of China
Map references: Commonwealth of Independent States—Central Asian States
Area:
total area: 143,100 sq km
land area: 142,700 sq km
comparative area: slightly smaller than Wisconsin
Land boundaries: total 3,651 km, Afghanistan 1,206 km, China 414 km, Kyrgyzstan 870 km, Uzbekistan 1,161 km
Coastline: 0 km (landlocked)
Maritime claims: none; landlocked
International disputes: boundary with China in dispute; territorial dispute with Kyrgyzstan on northern boundary in Isfara Valley area; Afghanistan's and other foreign support to Tajik rebels based in northern Afghanistan
Climate: midlatitude continental, hot summers, mild winters; semiarid to polar in Pamir Mountains

Terrain: Pamir and Altay Mountains dominate landscape; western Fergana Valley in north, Kofarnihon and Vakhsh Valleys in southwest
Natural resources: significant hydropower potential, some petroleum, uranium, mercury, brown coal, lead, zinc, antimony, tungsten
Land use:
arable land: 6%
permanent crops: 0%
meadows and pastures: 23%
forest and woodland: 0%
other: 71%
Irrigated land: 6,940 sq km (1990)
Environment:
current issues: inadequate sanitation facilities; increasing levels of soil salinity; industrial pollution; excessive pesticides; part of the basin of the shrinking Aral Sea which suffers from severe overutilization of available water for irrigation and associated pollution
natural hazards: NA
international agreements: NA
Note: landlocked

People

Population: 6,155,474 (July 1995 est.)
Age structure:
0-14 years: 43% (female 1,303,627; male 1,340,086)
15-64 years: 53% (female 1,612,429; male 1,624,379)
65 years and over: 4% (female 157,841; male 117,112) (July 1995 est.)
Population growth rate: 2.6% (1995 est.)
Birth rate: 34.06 births/1,000 population (1995 est.)
Death rate: 6.58 deaths/1,000 population (1995 est.)
Net migration rate: -1.44 migrant(s)/1,000 population (1995 est.)
Infant mortality rate: 60.4 deaths/1,000 live births (1995 est.)
Life expectancy at birth:
total population: 69.03 years
male: 66.11 years
female: 72.1 years (1995 est.)
Total fertility rate: 4.55 children born/woman (1995 est.)
Nationality:
noun: Tajik(s)
adjective: Tajik
Ethnic divisions: Tajik 64.9%, Uzbek 25%, Russian 3.5% (declining because of emigration), other 6.6%
Religions: Sunni Muslim 80%, Shi'a Muslim 5%
Languages: Tajik (official), Russian widely used in government and business
Literacy: age 15 and over can read and write (1989)
total population: 98%
male: 99%
female: 97%
Labor force: 1.95 million (1992)
by occupation: agriculture and forestry 43%, government and services 24%, industry 14%, trade and communications 11%, construction 8% (1990)

Government

Names:
conventional long form: Republic of Tajikistan
conventional short form: Tajikistan
local long form: Jumhurii Tojikistan
local short form: none
former: Tajik Soviet Socialist Republic
Digraph: TI
Type: republic
Capital: Dushanbe
Administrative divisions: 2 oblasts (viloyatho, singular—viloyat) and one autonomous oblast* (viloyati avtonomii); Viloyati Avtonomii Badakhshoni Kuni* (Khorugh—formerly Khorog), Viloyati Khatlon (Qurghonteppa—formerly Kurgan-Tyube), Viloyati Leninobad (Khujand—formerly Leninabad)
note: the administrative center names are in parentheses
Independence: 9 September 1991 (from Soviet Union)
National holiday: National Day, 9 September (1991)
Constitution: new constitution adopted 6 November 1994
Legal system: based on civil law system; no judicial review of legislative acts
Suffrage: 18 years of age; universal
Executive branch:
chief of state: President Emomili RAKHMONOV (since 6 November 1994; was Head of State and Assembly Chairman since NA November 1992); election last held 6 November 1994 (next to be held NA 1998); results—Emomili RAKHMONOV 58%, Abdumalik ABDULLAJANOV 40%
head of government: Prime Minister Jamshed KARIMOV (since 2 December 1994)
cabinet: Council of Ministers
Legislative branch: unicameral
Supreme Soviet: elections last held 26 February 1994 (next to be held NA); results—percent of vote by party NA; estimated seats—(181 total) Communist Party and affiliates 100, Popular Party 10, Party of Political and Economic Progress 1, Party of Popular Unity 6, other 64
Judicial branch: Prosecutor General
Political parties and leaders: Communist Party (People's Party of Tajikistan—PPT), Abdumalik ABDULAJANOV; Party of Economic Freedom (PEF), Abdumalik ABDULAJANOV; Tajik Socialist Party (TSP), Shodi SHABDOLOV; Tajik Democratic Party (TDP), Abdu-Nabi SATARZADE, chairman; note—suspended for six months; Islamic Renaissance Party (IRP), Sayed Abdullo NURI, chairman; Rebirth (Rastokhez), Takhir ABDUZHABOROV; Lali Badakhshan Society, Atobek AMIRBEK; People's Democratic Party (PDP), Abdujalil HAMIDOV, chairman; Tajikistan Party of Economic and Political Renewal (TPEPR), Mukhtor BOBOYEV
note: all the above-listed parties except the Communist Party, the Party of National Unity, and the People's Party were banned in June 1993
Other political or pressure groups: Tajikistan Opposition Movement based in northern Afghanistan
Member of: CIS, EBRD, ECO, ESCAP, IBRD, ICAO, IDA, IDB, IFAD, ILO, IMF, INTELSAT (nonsignatory user), IOC, IOM (observer), ITU, NACC, OIC, OSCE, UN, UNCTAD, UNESCO, UNIDO, UPU, WHO, WIPO, WMO
Diplomatic representation in US:
chief of mission: NA
chancery: NA
telephone: NA
US diplomatic representation:
chief of mission: Ambassador Stanley T. ESCUDERO
embassy: Interim Chancery, #39 Ainii Street, Oktyabrskaya Hotel, Dushanbe
mailing address: use embassy street address
telephone: [7] (3772) 21-03-56
Flag: three horizontal stripes of red (top), a wider stripe of white, and green; a crown surmounted by seven five-pointed stars is located in the center of the white stripe

Economy

Overview: Tajikistan had the next-to-lowest per capita GDP in the former USSR, the highest rate of population growth, and an extremely low standard of living. Agriculture dominates the economy, cotton being the most important crop. Mineral resources, varied but limited in amount, include silver, gold, uranium, and tungsten. Industry is limited to a large aluminum plant, hydropower facilities, and small obsolete factories mostly in light industry and food processing. The Tajik economy has been gravely weakened by three years of civil war and by the loss of subsidies and markets for its products, which has left Tajikistan dependent on Russia and Uzbekistan and on international humanitarian assistance for much of its basic subsistence needs. Moreover, constant political turmoil and the continued dominance by former Communist officials have impeded the introduction of meaningful economic reforms. In the meantime, Tajikistan's efforts to adopt the Russian ruble as its domestic currency despite Russia's unwillingness to supply sufficient rubles left the country in a severe monetary

crisis throughout 1994, keeping inflation low but leaving workers and pensioners unpaid for months at a time. The government has announced plans to introduce its own currency in 1995 to help resolve the problem.
National product: GDP—purchasing power parity—$8.5 billion (1994 estimate as extrapolated from World Bank estimate for 1992)
National product real growth rate: -12% (1994 est.)
National product per capita: $1,415 (1994 est.)
Inflation rate (consumer prices): NA%
Unemployment rate: 1.5% includes only officially registered unemployed; also large numbers of underemployed workers and unregistered unemployed people (September 1994)
Budget:
revenues: $NA
expenditures: $NA, including capital expenditures of $NA
Exports: $320 million to outside the FSU countries (1994)
commodities: cotton, aluminum, fruits, vegetable oil, textiles
partners: Russia, Kazakhstan, Ukraine, Uzbekistan, Turkmenistan
Imports: $318 million from outside the FSU countries (1994)
commodities: fuel, chemicals, machinery and transport equipment, textiles, foodstuffs
partners: Russia, Uzbekistan, Kazakhstan
External debt: $NA
Industrial production: growth rate -31% (1994)
Electricity:
capacity: 3,800,000 kW
production: 17 billion kWh
consumption per capita: 2,800 kWh (1994)
Industries: aluminum, zinc, lead, chemicals and fertilizers, cement, vegetable oil, metal-cutting machine tools, refrigerators and freezers
Agriculture: cotton, grain, fruits, grapes, vegetables; cattle, sheep and goats
Illicit drugs: illicit cultivation of cannabis and opium poppy; mostly for CIS consumption; used as transshipment points for illicit drugs from Southwest Asia to Western Europe and North America
Economic aid:
recipient: Russia and Uzbekistan reportedly provided substantial general assistance throughout 1993 and 1994; Western aid and credits promised through the end of 1993 were $700 million but disbursements were only $104 million; large scale development loans await IMF approval of a reform and stabilization plan
Currency: 1 ruble (R) = 100 kopeks; Tajikistan uses the Russian ruble as its currency by agreement with Russia; government has plans to introduce its own currency, the Tajik ruble, in 1995
Exchange rates: NA
Fiscal year: calendar year

Transportation

Railroads:
total: 480 km in common carrier service; does not include industrial lines (1990)
Highways:
total: 29,900 km
paved: 21,400 km
unpaved: earth 8,500 km (1990)
Pipelines: natural gas 400 km (1992)
Ports: none
Airports:
total: 59
with paved runways over 3,047 m: 1
with paved runways 2,438 to 3,047 m: 5
with paved runways 1,524 to 2,437 m: 7
with paved runways 914 to 1,523 m: 1
with unpaved runways 914 to 1,523 m: 9
with unpaved runways under 914 m: 36

Communications

Telephone system: 303,000 telephones (December 1991); about 55 telephones/1,000 persons (1991); poorly developed and not well maintained; many towns are not reached by the national network
local: NA
intercity: cable and microwave radio relay
international: linked by cable and microwave to other CIS republics, and by leased connections to the Moscow international gateway switch; Dushanbe linked by INTELSAT to international gateway switch in Ankara; 1 Orbita and 2 INTELSAT earth stations
Radio:
broadcast stations: AM NA, FM NA, shortwave NA
radios: NA
Television:
broadcast stations: NA
televisions: NA
note: 1 INTELSAT earth station provides TV receive-only service from Turkey

Defense Forces

Branches: Army (being formed), National Guard, Security Forces (internal and border troops)
Manpower availability: males age 15-49 1,410,229; males fit for military service 1,153,638; males reach military age (18) annually 57,942 (1995 est.)
Defense expenditures: $NA, NA% of GDP

Tanzania

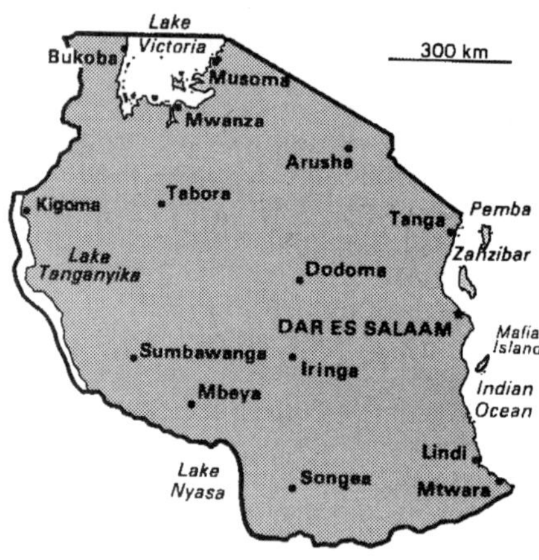

Geography

Location: Eastern Africa, bordering the Indian Ocean, between Kenya and Mozambique
Map references: Africa
Area:
total area: 945,090 sq km
land area: 886,040 sq km
comparative area: slightly larger than twice the size of California
note: includes the islands of Mafia, Pemba, and Zanzibar
Land boundaries: total 3,402 km, Burundi 451 km, Kenya 769 km, Malawi 475 km, Mozambique 756 km, Rwanda 217 km, Uganda 396 km, Zambia 338 km
Coastline: 1,424 km
Maritime claims:
exclusive economic zone: 200 nm
territorial sea: 12 nm
International disputes: boundary dispute with Malawi in Lake Nyasa; Tanzania-Zaire-Zambia tripoint in Lake Tanganyika may no longer be indefinite since it is reported that the indefinite section of the Zaire-Zambia boundary has been settled
Climate: varies from tropical along coast to temperate in highlands
Terrain: plains along coast; central plateau; highlands in north, south
Natural resources: hydropower potential, tin, phosphates, iron ore, coal, diamonds, gemstones, gold, natural gas, nickel
Land use:
arable land: 5%
permanent crops: 1%
meadows and pastures: 40%
forest and woodland: 47%
other: 7%
Irrigated land: 1,530 sq km (1989 est.)
Environment:
current issues: soil degradation; deforestation; desertification; destruction of coral reefs threatens marine habitats; recent droughts affected marginal agriculture
natural hazards: the tsetse fly and lack of

water limit agriculture; flooding on the central plateau during the rainy season
international agreements: party to—Endangered Species, Hazardous Wastes, Law of the Sea, Nuclear Test Ban, Ozone Layer Protection; signed, but not ratified—Biodiversity, Climate Change, Desertification
Note: Mount Kilimanjaro is highest point in Africa

People

Population: 28,701,077 (July 1995 est.)
Age structure:
0-14 years: 47% (female 6,724,575; male 6,676,652)
15-64 years: 50% (female 7,462,615; male 7,027,551)
65 years and over: 3% (female 425,211; male 384,473) (July 1995 est.)
Population growth rate: 2.55% (1995 est.)
Birth rate: 45.25 births/1,000 population (1995 est.)
Death rate: 19.81 deaths/1,000 population (1995 est.)
Net migration rate: NA migrant(s)/1,000 population (1995 est.)
note: in February 1995, a fresh influx of refugees from civil strife in Burundi brought the total number of Burundian refugees in Tanzania to about 60,000; in addition, since April 1994 more than a half million refugees from Rwanda have taken refuge in Tanzania to escape civil strife in Rwanda
Infant mortality rate: 109 deaths/1,000 live births (1995 est.)
Life expectancy at birth:
total population: 42.53 years
male: 40.88 years
female: 44.22 years (1995 est.)
Total fertility rate: 6.15 children born/woman (1995 est.)
Nationality:
noun: Tanzanian(s)
adjective: Tanzanian
Ethnic divisions:
mainland: native African 99% (consisting of well over 100 tribes), Asian, European, and Arab 1%
Zanzibar: NA
Religions:
mainland: Christian 45%, Muslim 35%, indigenous beliefs 20%
Zanzibar: Muslim 99% plus
Languages: Swahili (official; widely understood and generally used for communication between ethnic groups and is used in primary education), English (official; primary language of commerce, administration, and higher education)
note: first language of most people is one of the local languages
Literacy: age 15 and over has ability to read and write a letter or message in Kisahili (1988)
total population: 59%
male: 71%
female: 48%
Labor force: 732,200 wage earners
by occupation: agriculture 90%, industry and commerce 10% (1986 est.)

Government

Names:
conventional long form: United Republic of Tanzania
conventional short form: Tanzania
former: United Republic of Tanganyika and Zanzibar
Digraph: TZ
Type: republic
Capital: Dar es Salaam
note: some government offices have been transferred to Dodoma, which is planned as the new national capital by the end of the 1990s
Administrative divisions: 25 regions; Arusha, Dar es Salaam, Dodoma, Iringa, Kigoma, Kilimanjaro, Lindi, Mara, Mbeya, Morogoro, Mtwara, Mwanza, Pemba North, Pemba South, Pwani, Rukwa, Ruvuma, Shinyanga, Singida, Tabora, Tanga, Zanzibar Central/South, Zanzibar North, Zanzibar Urban/West, Ziwa Magharibi
Independence: 26 April 1964; Tanganyika became independent 9 December 1961 (from UN trusteeship under British administration); Zanzibar became independent 19 December 1963 (from UK); Tanganyika united with Zanzibar 26 April 1964 to form the United Republic of Tanganyika and Zanzibar; renamed United Republic of Tanzania 29 October 1964
National holiday: Union Day, 26 April (1964)
Constitution: 25 April 1977; major revisions October 1984
Legal system: based on English common law; judicial review of legislative acts limited to matters of interpretation; has not accepted compulsory ICJ jurisdiction
Suffrage: 18 years of age; universal
Executive branch:
chief of state: President Ali Hassan MWINYI (since 5 November 1985); First Vice President Cleopa MSUYA (since 5 December 1994); Second Vice President and President of Zanzibar Salmin AMOUR (since 9 November 1990) election last held 28 October 1990 (next to be held 29 October 1995); results—Ali Hassan MWINYI was elected without opposition
head of government: Prime Minister Cleopa David MSUYA (since 7 December 1994)
cabinet: Cabinet; appointed by the president from the National Assembly
Legislative branch: unicameral
National Assembly (Bunge): elections last held 28 October 1990 (next to be held 29 October 1995); results—CCM was the only party; seats—(241 total, 168 elected) CCM 168
Judicial branch: Court of Appeal, High Court
Political parties and leaders: Chama Cha Mapinduzi (CCM or Revolutionary Party), Ali Hassan MWINYI; Civic United Front (CUF), James MAPALALA; National Convention for Construction and Reform (NCCR), Lyatonga (Augustine) MREMA; Union for Multiparty Democracy (UMD), Abdullah FUNDIKIRA; Chama Cha Demokrasia na Maendeleo (CHADEMA), Edwin I. M. MTEI, chairman; Democratic Party (unregistered), Reverend MTIKLA
Member of: ACP, AfDB, C, CCC, EADB, ECA, FAO, FLS, G- 6, G-77, GATT, IAEA, IBRD, ICAO, ICRM, IDA, IFAD, IFC, IFRCS, ILO, IMF, IMO, INTELSAT, INTERPOL, IOC, ISO, ITU, NAM, OAU, SADC, UN, UNCTAD, UNESCO, UNHCR, UNIDO, UPU, WCL, WFTU, WHO, WIPO, WMO, WTO
Diplomatic representation in US:
chief of mission: Ambassador Charles Musama NYIRABU
chancery: 2139 R Street NW, Washington, DC 20008
telephone: [1] (202) 939-6125
FAX: [1] (202) 797-7408
US diplomatic representation:
chief of mission: Ambassador Brady ANDERSON
embassy: 36 Laibon Road (off Bagamoyo Road), Dar es Salaam
mailing address: P. O. Box 9123, Dar es Salaam
telephone: [255] (51) 66010 through 66015
FAX: [255] (51) 66701
Flag: divided diagonally by a yellow-edged black band from the lower hoist-side corner; the upper triangle (hoist side) is green and the lower triangle is blue

Economy

Overview: Tanzania is one of the poorest countries in the world. The economy is heavily dependent on agriculture, which accounts for about 58% of GDP, provides 85% of exports, and employs 90% of the work force. Topography and climatic conditions, however, limit cultivated crops to only 5% of the land area. Industry accounts for 8% of GDP and is mainly limited to processing agricultural products and light consumer goods. The economic recovery program announced in mid-1986 has generated notable increases in agricultural production and financial support for the program by bilateral donors. The World Bank, the International Monetary Fund, and bilateral donors have provided funds to rehabilitate Tanzania's deteriorated economic infrastructure. Growth in 1991-94 has featured a pickup in industrial production and a substantial increase in output of minerals, led by gold. Recent banking reforms have helped increase private sector growth and investment.

Tanzania *(continued)*

National product: GDP—purchasing power parity—$21 billion (1994 est.)
National product real growth rate: 3% (1994 est.)
National product per capita: $750 (1994 est.)
Inflation rate (consumer prices): 25% (1994 est.)
Unemployment rate: NA%
Budget:
revenues: $495 million
expenditures: $631 million, including capital expenditures of $118 million (1990 est.)
Exports: $462 million (f.o.b., 1994)
commodities: coffee, cotton, tobacco, tea, cashew nuts, sisal
partners: Germany, UK, Japan, Netherlands, Kenya, Hong Kong, US
Imports: $1.4 billion (c.i.f., 1994)
commodities: manufactured goods, machinery and transportation equipment, cotton piece goods, crude oil, foodstuffs
partners: Germany, UK, US, Japan, Italy, Denmark
External debt: $6.7 billion (1993)
Industrial production: growth rate 9.3% (1990); accounts for 8% of GDP
Electricity:
capacity: 440,000 kW
production: 880 million kWh
consumption per capita: 30 kWh (1993)
Industries: primarily agricultural processing (sugar, beer, cigarettes, sisal twine), diamond and gold mining, oil refining, shoes, cement, textiles, wood products, fertilizer
Agriculture: accounts for about 58% of GDP; cash crops—coffee, sisal, tea, cotton, pyrethrum (insecticide made from chrysanthemums), cashews, tobacco, cloves (Zanzibar); food crops—corn, wheat, cassava, bananas, fruits, vegetables; small numbers of cattle, sheep, and goats; not self-sufficient in food grain production
Illicit drugs: growing role in transshipment of Southwest Asian heroin destined for European and US markets
Economic aid:
recipient: US commitments, including Ex-Im (FY70-89), $400 million; Western (non-US) countries, ODA and OOF bilateral commitments (1970-89), $9.8 billion; OPEC bilateral aid (1979-89), $44 million; Communist countries (1970-89), $614 million
Currency: 1 Tanzanian shilling (TSh) = 100 cents
Exchange rates: Tanzanian shillings (TSh) per US$1—523.40 (December 1994), 509.63 (1994), 405.27 (1993), 297.71 (1992), 219.16 (1991), 195.06 (1990)
Fiscal year: 1 July—30 June

Transportation

Railroads:
total: 2,600 km; note—not a part of Tanzania Railways Corporation is the Tanzania-Zambia Railway Authority (TAZARA), which operates 1,860 km of 1.067-m narrow gauge track between Dar es Salaam and New Kapiri M'poshi in Zambia; 969 km are in Tanzania and 891 km are in Zambia; because of the difference in gauge, this system does not connect to Tanzania Railways
narrow gauge: 2,600 km 1.000-m gauge
Highways:
total: 81,900 km
paved: 3,600 km
unpaved: gravel, crushed stone 5,600 km; improved, unimproved earth 72,700 km
Inland waterways: Lake Tanganyika, Lake Victoria, Lake Nyasa
Pipelines: crude oil 982 km
Ports: Bukoba, Dar es Salaam, Kigoma, Lindi, Mkoani, Mtwara, Musoma, Mwanza, Tanga, Wete, Zanzibar
Merchant marine:
total: 7 ships (1,000 GRT or over) totaling 29,145 GRT/39,186 DWT
ships by type: cargo 3, oil tanker 1, passenger-cargo 2, roll-on/roll-off cargo 1
Airports:
total: 108
with paved runways over 3,047 m: 2
with paved runways 2,438 to 3,047 m: 2
with paved runways 1,524 to 2,437 m: 6
with paved runways 914 to 1,523 m: 1
with paved runways under 914 m: 30
with unpaved runways 1,524 to 2,438 m: 16
with unpaved runways 914 to 1,523 m: 51

Communications

Telephone system: 103,800 telephones; fair system operating below capacity
local: NA
intercity: open wire, microwave radio relay, troposcatter
international: 2 satellite earth stations—1 Indian Ocean INTELSAT and 1 Atlantic Ocean INTELSAT
Radio:
broadcast stations: AM 12, FM 4, shortwave 0
radios: NA
Television:
broadcast stations: 2
televisions: NA

Defense Forces

Branches: Tanzanian People's Defense Force (TPDF; includes Army, Navy, and Air Force), paramilitary Police Field Force Unit, Militia
Manpower availability: males age 15-49 6,188,455; males fit for military service 3,584,912 (1995 est.)
Defense expenditures: exchange rate conversion—$69 million, NA% of GDP (FY94/95)

Thailand

Geography

Location: Southeastern Asia, bordering the Andaman Sea and the Gulf of Thailand, southeast of Burma
Map references: Southeast Asia
Area:
total area: 514,000 sq km
land area: 511,770 sq km
comparative area: slightly more than twice the size of Wyoming
Land boundaries: total 4,863 km, Burma 1,800 km, Cambodia 803 km, Laos 1,754 km, Malaysia 506 km
Coastline: 3,219 km
Maritime claims:
continental shelf: 200-m depth or to the depth of exploitation
exclusive economic zone: 200 nm
territorial sea: 12 nm
International disputes: boundary dispute with Laos; unresolved maritime boundary with Vietnam; parts of border with Thailand in dispute; maritime boundary with Thailand not clearly defined
Climate: tropical; rainy, warm, cloudy southwest monsoon (mid-May to September); dry, cool northeast monsoon (November to mid-March); southern isthmus always hot and humid
Terrain: central plain; Khorat plateau in the east; mountains elsewhere
Natural resources: tin, rubber, natural gas, tungsten, tantalum, timber, lead, fish, gypsum, lignite, fluorite
Land use:
arable land: 34%
permanent crops: 4%
meadows and pastures: 1%
forest and woodland: 30%
other: 31%
Irrigated land: 42,300 sq km (1989 est.)
Environment:
current issues: air pollution from vehicle emissions; water pollution from organic and factory wastes; deforestation; soil erosion;

wildlife populations threatened by illegal hunting
natural hazards: land subsidence in Bangkok area resulting from the depletion of the water table; droughts
international agreements: party to—Climate Change, Endangered Species, Marine Life Conservation, Nuclear Test Ban, Ozone Layer Protection, Tropical Timber 83; signed, but not ratified—Biodiversity, Hazardous Wastes, Law of the Sea
Note: controls only land route from Asia to Malaysia and Singapore

People

Population: 60,271,300 (July 1995 est.)
Age structure:
0-14 years: 29% (female 8,545,362; male 8,866,271)
15-64 years: 66% (female 19,733,773; male 20,185,392)
65 years and over: 5% (female 1,636,426; male 1,304,076) (July 1995 est.)
Population growth rate: 1.24% (1995 est.)
Birth rate: 18.87 births/1,000 population (1995 est.)
Death rate: 6.48 deaths/1,000 population (1995 est.)
Net migration rate: 0 migrant(s)/1,000 population (1995 est.)
Infant mortality rate: 35.7 deaths/1,000 live births (1995 est.)
Life expectancy at birth:
total population: 68.42 years
male: 64.94 years
female: 72.08 years (1995 est.)
Total fertility rate: 2.04 children born/woman (1995 est.)
Nationality:
noun: Thai (singular and plural)
adjective: Thai
Ethnic divisions: Thai 75%, Chinese 14%, other 11%
Religions: Buddhism 95%, Muslim 3.8%, Christianity 0.5%, Hinduism 0.1%, other 0.6% (1991)
Languages: Thai, English the secondary language of the elite, ethnic and regional dialects
Literacy: age 15 and over can read and write (1990)
total population: 93%
male: 96%
female: 91%
Labor force: 30.87 million
by occupation: agriculture 62%, industry 13%, commerce 11%, services (including government) 14% (1989 est.)

Government

Names:
conventional long form: Kingdom of Thailand
conventional short form: Thailand
Digraph: TH
Type: constitutional monarchy
Capital: Bangkok
Administrative divisions: 76 provinces (changwat, singular and plural); Amnat Charoen, Ang Thong, Buriram, Chachoengsao, Chai Nat, Chaiyaphum, Chanthaburi, Chiang Mai, Chiang Rai, Chon Buri, Chumphon, Kalasin, Kamphaeng Phet, Kanchanaburi, Khon Kaen, Krabi, Krung Thep Mahanakhon, Lampang, Lamphun, Loei, Lop Buri, Mae Hong Son, Maha Sarakham, Mukdahan, Nakhon Nayok, Nakhon Pathom, Nakhon Phanom, Nakhon Ratchasima, Nakhon Sawan, Nakhon Si Thammarat, Nan, Narathiwat, Nong Bua Lamphu, Nong Khai, Nonthaburi, Pathum Thani, Pattani, Phangnga, Phatthalung, Phayao, Phetchabun, Phetchaburi, Phichit, Phitsanulok, Phra Nakhon Si Ayutthaya, Phrae, Phuket, Prachin Buri, Prachuap Khiri Khan, Ranong, Ratchaburi, Rayong, Roi Et, Sa Kaeo, Sakon Nakhon, Samut Prakan, Samut Sakhon, Samut Songkhram, Sara Buri, Satun, Sing Buri, Sisaket, Songkhla, Sukhothai, Suphan Buri, Surat Thani, Surin, Tak, Trang, Trat, Ubon Ratchathani, Udon Thani, Uthai Thani, Uttaradit, Yala, Yasothon
Independence: 1238 (traditional founding date; never colonized)
National holiday: Birthday of His Majesty the King, 5 December (1927)
Constitution: new constitution approved 7 December 1991; amended 10 June 1992
Legal system: based on civil law system, with influences of common law; has not accepted compulsory ICJ jurisdiction; martial law in effect since 23 February 1991 military coup
Suffrage: 21 years of age; universal
Executive branch:
chief of state: King PHUMIPHON Adunyadet (since 9 June 1946); Heir Apparent Crown Prince WACHIRALONGKON (born 28 July 1952)
head of government: Prime Minister CHUAN Likphai (since 23 September 1992)
cabinet: Council of Ministers
Privy Council: NA
Legislative branch: bicameral National Assembly (Rathasatha)
Senate (Vuthisatha): consists of a 270-member appointed body
House of Representatives (Saphaphoothan-Rajsadhorn): elections last held 13 September 1992 (next to be held NA); results—percent of vote by party NA; seats—(360 total) DP 79, TNP 77, NDP 60, NAP 51, Phalang Tham 47, SAP 22, LDP 8, SP 8, Mass Party 4, Thai Citizen's Party 3, People's Party 1, People's Force Party 0
Judicial branch: Supreme Court (Sarndika)
Political parties and leaders: Democrat Party (DP), CHUAN Likphai; Thai Nation Party (TNP or Chat Thai Party), Banhan SINLAPA-ACHA; National Development Party (NDP or Chat Phattana), Chatchai CHUNHAWAN; New Aspiration Party (NAP), Gen. Chawalit YONGCHAIYUT; Phalang Tham (Palang Dharma), CHAMLONG Simuang; Social Action Party (SAP), Montri PHONGPHANIT; Liberal Democratic Party (LDP or Seri Tham), Athit URAIRAT; Solidarity Party (SP), Uthai PHIMCHAICHON; Mass Party (Muanchon), Pol. Cpt. Choem YUBAMRUNG; Thai Citizen's Party (Prachakon Thai), Samak SUNTHONWET; People's Party (Ratsadon), Chaiphak SIRIWAT; People's Force Party (Phalang Prachachon), Col. Sophon HANCHAREON
Member of: APEC, AsDB, ASEAN, CCC, CP, ESCAP, FAO, G-77, GATT, IAEA, IBRD, ICAO, ICFTU, ICRM, IDA, IFAD, IFC, IFRCS, ILO, IMF, IMO, INTELSAT, INTERPOL, IOC, IOM, ISO, ITU, NAM, PCA, UN, UNCTAD, UNESCO, UNHCR, UNIDO, UNIKOM, UNU, UPU, WCL, WFTU, WHO, WIPO, WMO
Diplomatic representation in US:
chief of mission: Ambassador MANATPHAT Chuto
chancery: 1024 Wisconsin Avenue NW, Washington, DC 20007
telephone: [1] (202) 944-3600
FAX: [1] (202) 944-3611
consulate(s) general: Chicago, Los Angeles, and New York
US diplomatic representation:
chief of mission: Ambassador David F. LAMBERTSON
embassy: 95 Wireless Road, Bangkok
mailing address: APO AP 96546
telephone: [66] (2) 252-5040
FAX: [66] (2) 254-2990
consulate(s) general: Chiang Mai
consulate(s): Udorn (Udon Thani)
Flag: five horizontal bands of red (top), white, blue (double width), white, and red

Economy

Overview: Thailand's economy recovered rapidly from the political unrest in May 1992 to post an impressive 7.5% growth rate for the year, 7.8% in 1993, and 8% in 1994. One of the more advanced developing countries in Asia, Thailand depends on exports of manufactures and the development of the service sector to fuel the country's rapid growth. Much of Thailand's recent imports have been for capital equipment, suggesting that the export sector is poised for further growth. With foreign investment slowing, Bangkok is working to increase the generation of domestic capital. Prime Minister CHUAN's government—Thailand's fifth government in less than three years—is pledged to continue Bangkok's probusiness policies, and the return of a

democratically elected government has improved business confidence. Even so, CHUAN must overcome divisions within his ruling coalition to complete much needed infrastructure development programs if Thailand is to remain an attractive place for business investment. Over the longer-term, Bangkok must produce more college graduates with technical training and upgrade workers' skills to continue its rapid economic development.
National product: GDP—purchasing power parity—$355.2 billion (1994 est.)
National product real growth rate: 8% (1994 est.)
National product per capita: $5,970 (1994 est.)
Inflation rate (consumer prices): 5% (1994 est.)
Unemployment rate: 3.2% (1993 est.)
Budget:
revenues: $28.4 billion
expenditures: $28.4 billion, including capital expenditures of $9.6 billion (FY94/95 est.)
Exports: $46 billion (f.o.b., 1994 est.)
commodities: machinery and manufactures 83%, agricultural products and fisheries 16%, others 1% (1994 est.)
partners: US 22%, Japan 17%, Singapore 12%, Hong Kong 5%, Germany 4% (1993)
Imports: $52.6 billion (c.i.f., 1994 est.)
commodities: capital goods 44%, intermediate goods and raw materials 37%, consumer goods 16%, other 3% (1994 est.)
partners: Japan 30%, US 12%, Singapore 6%, Germany 5%, Taiwan 5% (1993)
External debt: $64.3 billion (1994 est.)
Industrial production: growth rate 11.5% (1993 est.); accounts for about 26% of GDP
Electricity:
capacity: 12,810,000 kW
production: 56.8 billion kWh
consumption per capita: 909 kWh (1993)
Industries: tourism is the largest source of foreign exchange; textiles and garments, agricultural processing, beverages, tobacco, cement, light manufacturing, such as jewelry; electric appliances and components, integrated circuits, furniture, plastics; world's second-largest tungsten producer and third-largest tin producer
Agriculture: accounts for 11% of GDP and 62% of labor force; leading producer and exporter of rice and cassava (tapioca); other crops—rubber, corn, sugarcane, coconuts, soybeans; except for wheat, self-sufficient in food
Illicit drugs: a minor producer of opium and marijuana; major illicit transit point for heroin, particularly from Burma and Laos, for the international drug market; eradication efforts have reduced the area of cannabis cultivation and shifted some production to neighboring countries; opium poppy cultivation has been reduced by eradication efforts; also a major drug money laundering center; rapidly growing role in amphetamine production for regional consumption; increasing indigenous abuse of heroin and cocaine
Economic aid:
recipient: US commitments, including Ex-Im (FY70-89), $870 million; Western (non-US) countries, ODA and OOF bilateral commitments (1970-89), $8.6 billion; OPEC bilateral aid (1979-89), $19 million
Currency: 1 baht (B) = 100 satang
Exchange rates: baht (B) per US$1—25.074 (January 1995), 25.150 (1994), 25.319 (1993), 25.400 (1992), 25.517 (1991), 25.585 (1990)
Fiscal year: 1 October—30 September

Transportation

Railroads:
total: 3,940 km
narrow gauge: 3,940 km 1.000-m gauge (99 km double track)
Highways:
total: 77,697 km
paved: 35,855 km (including 88 km of expressways)
unpaved: gravel, other stabilization 14,092 km; earth 27,750 km (1988)
Inland waterways: 3,999 km principal waterways; 3,701 km with navigable depths of 0.9 m or more throughout the year; numerous minor waterways navigable by shallow-draft native craft
Pipelines: petroleum products 67 km; natural gas 350 km
Ports: Bangkok, Laem Chabang, Pattani, Phuket, Sattahip, Si Racha, Songkhla
Merchant marine:
total: 229 ships (1,000 GRT or over) totaling 1,231,172 GRT/1,931,117 DWT
ships by type: bulk 22, cargo 122, chemical tanker 3, combination bulk 1, container 15, liquefied gas tanker 9, oil tanker 45, passenger 1, refrigerated cargo 7, roll-on/roll-off cargo 2, short-sea passenger 1, specialized tanker 1
Airports:
total: 105
with paved runways over 3,047 m: 6
with paved runways 2,438 to 3,047 m: 9
with paved runways 1,524 to 2,437 m: 10
with paved runways 914 to 1,523 m: 23
with paved runways under 914 m: 42
with unpaved runways 1,524 to 2,438 m: 1
with unpaved runways 914 to 1,523 m: 14

Communications

Telephone system: 739,500 telephones (1987); service to general public inadequate; bulk of service to government activities provided by multichannel cable and microwave radio relay network
local: NA
intercity: microwave radio relay and multichannel cable; domestic satellite system being developed
international: 2 INTELSAT (1 Indian Ocean and 1 Pacific Ocean) earth stations
Radio:
broadcast stations: AM 200 (in government-controlled network), FM 100 (in government-controlled network), shortwave 0
radios: NA
Television:
broadcast stations: 11 (in government-controlled network)
televisions: NA

Defense Forces

Branches: Royal Thai Army, Royal Thai Navy (includes Royal Thai Marine Corps), Royal Thai Air Force, Paramilitary Forces
Manpower availability: males age 15-49 17,297,854; males fit for military service 10,489,564; males reach military age (18) annually 585,009 (1995 est.)
Defense expenditures: exchange rate conversion—$4.0 billion, 2.5% of GNP (FY94/95)

Togo

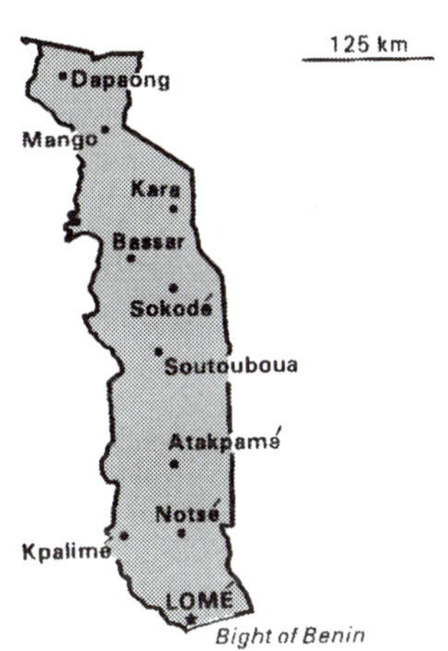

Geography

Location: Western Africa, bordering the North Atlantic Ocean, between Benin and Ghana
Map references: Africa
Area:
total area: 56,790 sq km
land area: 54,390 sq km
comparative area: slightly smaller than West Virginia
Land boundaries: total 1,647 km, Benin 644 km, Burkina 126 km, Ghana 877 km
Coastline: 56 km
Maritime claims:
exclusive economic zone: 200 nm
territorial sea: 30 nm
International disputes: none
Climate: tropical; hot, humid in south; semiarid in north
Terrain: gently rolling savanna in north; central hills; southern plateau; low coastal plain with extensive lagoons and marshes
Natural resources: phosphates, limestone, marble
Land use:
arable land: 25%
permanent crops: 1%
meadows and pastures: 4%
forest and woodland: 28%
other: 42%
Irrigated land: 70 sq km (1989 est.)
Environment:
current issues: deforestation attributable to slash-and-burn agriculture and the use of wood for fuel; recent droughts affecting agriculture
natural hazards: hot, dry harmattan wind can reduce visibility in north during winter; periodic droughts
international agreements: party to—Endangered Species, Law of the Sea, Nuclear Test Ban, Ozone Layer Protection, Ship Pollution, Tropical Timber 83; signed, but not ratified—Biodiversity, Climate Change, Desertification, Tropical Timber 94

People

Population: 4,410,370 (July 1995 est.)
Age structure:
0-14 years: 49% (female 1,069,171; male 1,079,999)
15-64 years: 49% (female 1,121,685; male 1,043,000)
65 years and over: 2% (female 51,392; male 45,123) (July 1995 est.)
Population growth rate: 3.58% (1995 est.)
Birth rate: 46.78 births/1,000 population (1995 est.)
Death rate: 11.01 deaths/1,000 population (1995 est.)
Net migration rate: 0 migrant(s)/1,000 population (1995 est.)
Infant mortality rate: 86.5 deaths/1,000 live births (1995 est.)
Life expectancy at birth:
total population: 57.42 years
male: 55.29 years
female: 59.6 years (1995 est.)
Total fertility rate: 6.83 children born/woman (1995 est.)
Nationality:
noun: Togolese (singular and plural)
adjective: Togolese
Ethnic divisions: 37 tribes; largest and most important are Ewe, Mina, and Kabye, European and Syrian-Lebanese under 1%
Religions: indigenous beliefs 70%, Christian 20%, Muslim 10%
Languages: French (official and the language of commerce), Ewe and Mina (the two major African languages in the south), Dagomba and Kabye (the two major African languages in the north)
Literacy: age 15 and over can read and write (1990 est.)
total population: 43%
male: 56%
female: 31%
Labor force: NA
by occupation: agriculture 80%
note: about 88,600 wage earners, evenly divided between public and private sectors

Government

Names:
conventional long form: Republic of Togo
conventional short form: Togo
local long form: Republique Togolaise
local short form: none
former: French Togo
Digraph: TO
Type: republic under transition to multiparty democratic rule
Capital: Lome
Administrative divisions: 23 circumscriptions (circonscriptions, singular—circonscription); Amlame (Amou), Aneho (Lacs), Atakpame (Ogou), Badou (Wawa), Bafilo (Assoli), Bassar (Bassari), Dapango (Tone), Kande (Keran), Klouto (Kloto), Pagouda (Binah), Lama-Kara (Kozah), Lome (Golfe), Mango (Oti), Niamtougou (Doufelgou), Notse (Haho), Pagouda, Sotouboua, Tabligbo (Yoto), Tchamba, Nyala, Tchaoudjo, Tsevie (Zio), Vogan (Vo)
note: the 23 units may now be called prefectures (singular—prefecture) and reported name changes for individual units are included in parentheses
Independence: 27 April 1960 (from French-administered UN trusteeship)
National holiday: Independence Day, 27 April (1960)
Constitution: multiparty draft constitution approved by High Council of the Republic 1 July 1992; adopted by public referendum 27 September 1992
Legal system: French-based court system
Suffrage: NA years of age; universal adult
Executive branch:
chief of state: President Gen. Gnassingbe EYADEMA (since 14 April 1967); election last held 25 August 1993 (next election to be held NA 1998); all major opposition parties boycotted the election; Gen. EYADEMA won 96.5% of the vote
head of government: Prime Minister Edem KODJO (since April 1994)
cabinet: Council of Ministers; appointed by the president and the prime minister
Legislative branch: unicameral
National Assembly: elections last held 6 and 20 February 1994 (next to be held NA); results—percent of vote by party NA; seats—(81 total) CAR 36, RPT 35, UTD 7, UJD 2, CFN 1
note: the Supreme Court ordered new elections for 3 seats of the Action Committee for Renewal (CAR) and the Togolese Union for Democracy (UTD), lowering their totals to 34 and 6 seats, respectively; the remaining 3 seats have not been filled
Judicial branch: Court of Appeal (Cour d'Appel), Supreme Court (Cour Supreme)
Political parties and leaders: Rally of the Togolese People (RPT), President Gen. Gnassingbe EYADEMA; Coordination des Forces Nouvelles (CFN), Joseph KOFFIGOH; The Togolese Union for Democracy (UTD), Edem KODJO; The Action Committee for Renewal (CAR), Yao AGBOYIBOR; The Union for Democracy and Solidarity (UDS), Antoine FOLLY; The Pan-African Sociodemocrats Group (GSP), an alliance of three radical parties: The Democratic Convention of African Peoples (CDPA), Leopold GNININVI; The Party for Democracy and Renewal (PDR), Zarifou AYEVA; The Pan-African Social Party (PSP), Francis AGBAGLI; The Union of Forces for Change (UFC), Gilchrist OLYMPIO (in exile); Union of Justice and Democracy (UJD), Lal TAXPANDJAN
note: Rally of the Togolese People (RPT) led by President EYADEMA was the only party

until the formation of multiple parties was legalized 12 April 1991
Member of: ACCT, ACP, AfDB, CCC, CEAO (observer), ECA, ECOWAS, Entente, FAO, FZ, G-77, GATT, IBRD, ICAO, ICC, ICFTU, ICRM, IDA, IFAD, IFC, IFRCS, ILO, IMF, IMO, INTELSAT, INTERPOL, IOC, ITU, MINURSO, NAM, OAU, UN, UNAMIR, UNCTAD, UNESCO, UNIDO, UPU, WADB, WCL, WFTU, WHO, WIPO, WMO, WTO
Diplomatic representation in US:
chief of mission: Charge d'Affaires Edem Frederic HEGBE
chancery: 2208 Massachusetts Avenue NW, Washington, DC 20008
telephone: [1] (202) 234-4212
FAX: [1] (202) 232-3190
US diplomatic representation:
chief of mission: Ambassador Johnny YOUNG (since September 1994)
embassy: Rue Pelletier Caventou and Rue Vauban, Lome
mailing address: B. P. 852, Lome
telephone: [228] 21 77 17, 21 29 91 through 21 29 94
FAX: [228] 21 79 52
Flag: five equal horizontal bands of green (top and bottom) alternating with yellow; there is a white five-pointed star on a red square in the upper hoist-side corner; uses the popular pan-African colors of Ethiopia

Economy

Overview: The economy is heavily dependent on subsistence agriculture, which accounts for about half of GDP and provides employment for 80% of the labor force. Primary agricultural exports are cocoa, coffee, and cotton, which together generate about 30% of total export earnings. Togo is self-sufficient in basic foodstuffs when harvests are normal. In the industrial sector phosphate mining is by far the most important activity, although it has suffered from the collapse of world phosphate prices and increased foreign competition. Togo serves as a regional commercial and trade center. The government's decade-long IMF and World Bank supported effort to implement economic reform measures to encourage foreign investment and bring revenues in line with expenditures has stalled. Political unrest, including private and public sector strikes throughout 1992 and 1993, has jeopardized the reform program, shrunk the tax base, and disrupted vital economic activity. Although strikes had ended in 1994, political unrest and lack of funds prevented the government from taking advantage of the 50% currency devaluation of January 1994. Resumption of World Bank and IMF flows will depend on implementation of several controversial moves toward privatization and on downsizing the military, on which the regime depends to stay in power.
National product: GDP—purchasing power parity—$3.3 billion (1993 est.)
National product real growth rate: NA%
National product per capita: $800 (1993 est.)
Inflation rate (consumer prices): 0.5% (1991 est.)
Unemployment rate: NA%
Budget:
revenues: $284 million
expenditures: $407 million, including capital expenditures of $NA (1991 est.)
Exports: $221 million (f.o.b., 1993)
commodities: phosphates, cotton, cocoa, coffee
partners: EC 40%, Africa 16%, US 1% (1990)
Imports: $292 million (c.i.f., 1993)
commodities: machinery and equipment, consumer goods, food, chemical products
partners: EC 57%, Africa 17%, US 5%, Japan 4% (1990)
External debt: $1.3 billion (1991)
Industrial production: growth rate 9% (1991 est.); accounts for 20% of GDP
Electricity:
capacity: 30,000 kW
production: 60 million kWh
consumption per capita: 83 kWh (1993)
Industries: phosphate mining, agricultural processing, cement, handicrafts, textiles, beverages
Agriculture: accounts for 49% of GDP; cash crops—coffee, cocoa, cotton; food crops—yams, cassava, corn, beans, rice, millet, sorghum; livestock production not significant; annual fish catch of 10,000-14,000 tons
Illicit drugs: increasingly used as transit hub by heroin traffickers
Economic aid:
recipient: US commitments, including Ex-Im (FY70-90), $142 million; Western (non-US) countries, ODA and OOF bilateral commitments (1970-90), $2 billion; OPEC bilateral aid (1979-89), $35 million; Communist countries (1970-89), $51 million
Currency: 1 CFA franc (CFAF) = 100 centimes
Exchange rates: Communaute Financiere Africaine francs (CFAF) per US$1—529.43 (January 1995), 555.20 (1994), 283.16 (1993), 264.69 (1992), 282.11 (1991), 272.26 (1990)
note: the official rate is pegged to the French franc, and beginning 12 January 1994, the CFA franc was devalued to CFAF 100 per French franc from CFAF 50 at which it had been fixed since 1948
Fiscal year: calendar year

Transportation

Railroads:
total: 532 km
narrow gauge: 532 km 1.000-m gauge
Highways:
total: 6,462 km
paved: 1,762 km
unpaved: unimproved earth 4,700 km
Inland waterways: 50 km Mono River
Ports: Kpeme, Lome
Merchant marine: none
Airports:
total: 9
with paved runways 2,438 to 3,047 m: 2
with paved runways under 914 m: 2
with unpaved runways 914 to 1,523 m: 5

Communications

Telephone system: NA telephones; fair system based on network of radio relay routes supplemented by open wire lines
local: NA
intercity: microwave radio relay and open wire lines
international: 1 Atlantic Ocean INTELSAT and 1 SYMPHONIE earth station
Radio:
broadcast stations: AM 2, FM 0, shortwave 0
radios: NA
Television:
broadcast stations: 3 (relays 2)
televisions: NA

Defense Forces

Branches: Army, Navy, Air Force, Gendarmerie
Manpower availability: males age 15-49 936,270; males fit for military service 491,578 (1995 est.)
Defense expenditures: exchange rate conversion—$48 million, 2.9% of GDP (1993)

Tokelau

(territory of New Zealand)

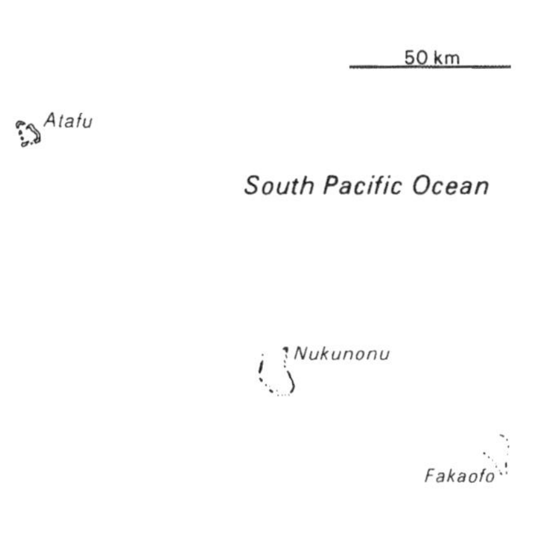

Geography

Location: Oceania, group of islands in the South Pacific Ocean, about one-half of the way from Hawaii to New Zealand
Map references: Oceania
Area:
total area: 10 sq km
land area: 10 sq km
comparative area: about 17 times the size of The Mall in Washington, DC
Land boundaries: 0 km
Coastline: 101 km
Maritime claims:
exclusive economic zone: 200 nm
territorial sea: 12 nm
International disputes: none
Climate: tropical; moderated by trade winds (April to November)
Terrain: coral atolls enclosing large lagoons
Natural resources: negligible
Land use:
arable land: 0%
permanent crops: 0%
meadows and pastures: 0%
forest and woodland: 0%
other: 100%
Irrigated land: NA sq km
Environment:
current issues: very limited natural resources and overcrowding are contributing to emigration to New Zealand
natural hazards: lies in Pacific typhoon belt
international agreements: NA

People

Population: 1,503 (July 1995 est.)
Age structure:
0-14 years: NA
15-64 years: NA
65 years and over: NA
Population growth rate: -1.3% (1995 est.)
Birth rate: NA births/1,000 population
Death rate: NA deaths/1,000 population
Net migration rate: NA migrant(s)/1,000 population
Infant mortality rate: NA deaths/1,000 live births
Life expectancy at birth:
total population: NA years
male: NA years
female: NA years
Total fertility rate: NA children born/woman
Nationality:
noun: Tokelauan(s)
adjective: Tokelauan
Ethnic divisions: Polynesian
Religions: Congregational Christian Church 70%, Roman Catholic 28%, other 2%
note: on Atafu, all Congregational Christian Church of Samoa; on Nukunonu, all Roman Catholic; on Fakaofo, both denominations, with the Congregational Christian Church predominant
Languages: Tokelauan (a Polynesian language), English
Labor force: NA

Government

Names:
conventional long form: none
conventional short form: Tokelau
Digraph: TL
Type: territory of New Zealand
Capital: none; each atoll has its own administrative center
Administrative divisions: none (territory of New Zealand)
Independence: none (territory of New Zealand)
National holiday: Waitangi Day, 6 February (1840) (Treaty of Waitangi established British sovereignty over New Zealand)
Constitution: administered under the Tokelau Islands Act of 1948, as amended in 1970
Legal system: British and local statutes
Suffrage: NA
Executive branch:
Chief of State: Queen ELIZABETH II (since 6 February 1952)
Head of Government: Administrator Graham ANSELL (since NA 1990, appointed by the Minister of Foreign Affairs in New Zealand); Official Secretary Casimilo J. PEREZ (since NA), Office of Tokelau Affairs; Tokelau's governing Council will elect its first head of government
Legislative branch: unicameral Council of Elders (Taupulega) on each atoll
Judicial branch: High Court in Niue, Supreme Court in New Zealand
Political parties and leaders: NA
Member of: SPC, WHO (associate)
Diplomatic representation in US: none (territory of New Zealand)
US diplomatic representation: none (territory of New Zealand)
Flag: the flag of New Zealand is used

Economy

Overview: Tokelau's small size, isolation, and lack of resources greatly restrain economic development and confine agriculture to the subsistence level. The people must rely on aid from New Zealand to maintain public services, annual aid being substantially greater than GDP. The principal sources of revenue come from sales of copra, postage stamps, souvenir coins, and handicrafts. Money is also remitted to families from relatives in New Zealand.
National product: GDP—purchasing power parity—$1.5 million (1993 est.)
National product real growth rate: NA%
National product per capita: $1,000 (1993 est.)
Inflation rate (consumer prices): NA%
Unemployment rate: NA%
Budget:
revenues: $430,830
expenditures: $2.8 billion, including capital expenditures of $37,300 (1987 est.)
Exports: $98,000 (f.o.b., 1983)
commodities: stamps, copra, handicrafts
partners: NZ
Imports: $323,400 (c.i.f., 1983)
commodities: foodstuffs, building materials, fuel
partners: NZ
External debt: $0
Industrial production: growth rate NA%
Electricity:
capacity: 200 kW
production: 300,000 kWh
consumption per capita: 180 kWh (1990)
Industries: small-scale enterprises for copra production, wood work, plaited craft goods; stamps, coins; fishing
Agriculture: coconuts, copra; basic subsistence crops—breadfruit, papaya, bananas; pigs, poultry, goats
Economic aid:
recipient: Western (non-US) countries, ODA and OOF bilateral commitments (1970-89), $24 million
Currency: 1 New Zealand dollar (NZ$) = 100 cents
Exchange rates: New Zealand dollars (NZ$) per US$1—1.5601 (January 1995), 1.6844 (1994), 1.8495 (1993), 1.8584 (1992), 1.7265 (1991), 1.6750 (1990)
Fiscal year: 1 April—31 March

Transportation

Railroads: 0 km
Highways:
total: NA
paved: NA
unpaved: NA
Ports: none; offshore anchorage only
Merchant marine: none
Airports: none; lagoon landings by amphibious aircraft from Western Samoa

Communications

Telephone system: NA telephones
local: NA
intercity: radiotelephone service between islands
international: radiotelephone service to Western Samoa
Radio:
broadcast stations: AM NA, FM NA, shortwave NA
radios: NA
Television:
broadcast stations: NA
televisions: NA

Defense Forces

Note: defense is the responsibility of New Zealand

Tonga

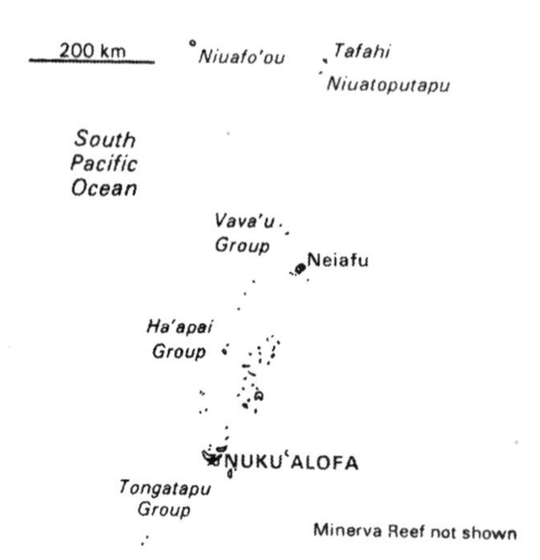

Geography

Location: Oceania, archipelago in the South Pacific Ocean, about two-thirds of the way from Hawaii to New Zealand
Map references: Oceania
Area:
total area: 748 sq km
land area: 718 sq km
comparative area: slightly more than four times the size of Washington, DC
Land boundaries: 0 km
Coastline: 419 km
Maritime claims:
continental shelf: 200-m depth or to the depth of exploitation
exclusive economic zone: 200 nm
territorial sea: 12 nm
International disputes: none
Climate: tropical; modified by trade winds; warm season (December to May), cool season (May to December)
Terrain: most islands have limestone base formed from uplifted coral formation; others have limestone overlying volcanic base
Natural resources: fish, fertile soil
Land use:
arable land: 25%
permanent crops: 55%
meadows and pastures: 6%
forest and woodland: 12%
other: 2%
Irrigated land: NA sq km
Environment:
current issues: deforestation results as more and more land is being cleared for agriculture and settlement; some damage to coral reefs from starfish and indiscriminate coral and shell collectors; overhunting threatens native sea turtle populations
natural hazards: cyclones (October to April); earthquakes and volcanic activity on Fonuafo'ou
international agreements: party to—Marine Life Conservation, Nuclear Test Ban
Note: archipelago of 170 islands (36 inhabited)

People

Population: 105,600 (July 1995 est.)
Age structure:
0-14 years: NA
15-64 years: NA
65 years and over: NA
Population growth rate: 0.78% (1995 est.)
Birth rate: 24.37 births/1,000 population (1995 est.)
Death rate: 6.75 deaths/1,000 population (1995 est.)
Net migration rate: -9.87 migrant(s)/1,000 population (1995 est.)
Infant mortality rate: 20.2 deaths/1,000 live births (1995 est.)
Life expectancy at birth:
total population: 68.16 years
male: 65.8 years
female: 70.62 years (1995 est.)
Total fertility rate: 3.56 children born/woman (1995 est.)
Nationality:
noun: Tongan(s)
adjective: Tongan
Ethnic divisions: Polynesian, Europeans about 300
Religions: Christian (Free Wesleyan Church claims over 30,000 adherents)
Languages: Tongan, English
Literacy: age 15 and over can read and write simple message in Tongan or English (1976)
total population: 100%
male: 100%
female: 100%
Labor force: NA
by occupation: agriculture 70%, mining (600 engaged in mining)

Government

Names:
conventional long form: Kingdom of Tonga
conventional short form: Tonga
former: Friendly Islands
Digraph: TN
Type: hereditary constitutional monarchy
Capital: Nuku'alofa
Administrative divisions: three island groups; Ha'apai, Tongatapu, Vava'u
Independence: 4 June 1970 (emancipation from UK protectorate)
National holiday: Emancipation Day, 4 June (1970)
Constitution: 4 November 1875, revised 1 January 1967
Legal system: based on English law
Suffrage: 21 years of age; universal
Executive branch:
chief of state: King Taufa'ahau TUPOU IV (since 16 December 1965)
head of government: Prime Minister Baron VAEA (since 22 August 1991); Deputy Prime Minister S. Langi KAVALIKU (since 22 August 1991)

cabinet: Cabinet; appointed by the king
Privy Council: consists of the king and the cabinet
Legislative branch: unicameral; consists of twelve cabinet ministers sitting ex-officio, nine nobles selected by the country's thirty-three nobles, and nine people's representatives elected by the populace
Legislative Assembly (Fale Alea): elections last held 3-4 February 1993 (next to be held NA February 1996); results—percent of vote NA; seats—(30 total, 9 elected) 6 proreform, 3 traditionalist
Judicial branch: Supreme Court
Political parties and leaders: Tonga People's Party, Viliami FUKOFUKA
Member of: ACP, AsDB, C, ESCAP, FAO, G-77, IBRD, ICAO, ICFTU, ICRM, IDA, IFAD, IFC, IFRCS, IMF, INTELSAT (nonsignatory user), INTERPOL, IOC, ITU, SPARTECA, SPC, SPF, UNCTAD, UNESCO, UNIDO, UPU, WHO
Diplomatic representation in US: Ambassador Sione KITE, resides in London
consulate(s) general: San Francisco
US diplomatic representation: the US has no offices in Tonga; the ambassador to Fiji is accredited to Tonga
Flag: red with a bold red cross on a white rectangle in the upper hoist-side corner

Economy

Overview: The economy's base is agriculture, which employs about 70% of the labor force and contributes 40% to GDP. Squash, coconuts, bananas, and vanilla beans are the main crops, and agricultural exports make up two-thirds of total exports. The country must import a high proportion of its food, mainly from New Zealand. The manufacturing sector accounts for only 11% of GDP. Tourism is the primary source of hard currency earnings, but the country also remains dependent on sizable external aid and remittances to offset its trade deficit. The economy continued to grow in 1993-94 largely because of a rise in squash exports, increased aid flows, and several large construction projects. The government is now turning its attention to further development of the private sector and the reduction of the budget deficit.
National product: GDP—purchasing power parity—$214 million (1994 est.)
National product real growth rate: 5% (1994 est.)
National product per capita: $2,050 (1994 est.)
Inflation rate (consumer prices): 3% (1993)
Unemployment rate: NA%
Budget:
revenues: $36.4 million
expenditures: $68.1 million, including capital expenditures of $33.2 million (1991 est.)
Exports: $11.3 million (f.o.b., FY92/93)
commodities: squash, vanilla, fish, root crops, coconut oil
partners: Japan 34%, US 17%, Australia 13%, NZ 13% (FY90/91)
Imports: $56 million (c.i.f., FY92/93)
commodities: food products, machinery and transport equipment, manufactures, fuels, chemicals
partners: NZ 33%, Australia 22%, US 8%, Japan 8% (FY90/91)
External debt: $47.5 million (FY90/91)
Industrial production: growth rate 1.5% (FY91/92); accounts for 11% of GDP
Electricity:
capacity: 6,000 kW
production: 30 million kWh
consumption per capita: 231 kWh (1993)
Industries: tourism, fishing
Agriculture: accounts for 40% of GDP; dominated by coconut, copra, and banana production; vanilla beans, cocoa, coffee, ginger, black pepper
Economic aid:
recipient: US commitments, including Ex-Im (FY70-89), $16 million; Western (non-US) countries, ODA and OOF bilateral commitments (1970-89), $258 million
Currency: 1 pa'anga (T$) = 100 seniti
Exchange rates: pa'anga (T$) per US$1—1.2653 (January 1995), 1.3202 (1994), 1.3841 (1993), 1.3471 (1992), 1.2961 (1991), 1.2800 (1990)
Fiscal year: 1 July—30 June

Transportation

Railroads: 0 km
Highways:
total: 366 km
paved: 272 km (198 km on Tongatapu; 74 km on Vava'u)
unpaved: 94 km (usable only in dry weather)
Ports: Neiafu, Nuku'alofa, Pangai
Merchant marine:
total: 2 ships (1,000 GRT or over) totaling 5,440 GRT/8,984 DWT
ships by type: cargo 1, roll-on/roll-off cargo 1
Airports:
total: 6
with paved runways 2,438 to 3,047 m: 1
with paved runways under 914 m: 2
with unpaved runways 1,524 to 2,438 m: 1
with unpaved runways 914 to 1,523 m: 2

Communications

Telephone system: 3,529 telephones
local: NA
intercity: NA
international: 1 INTELSAT (Pacific Ocean) earth station
Radio:
broadcast stations: AM 1, FM 0, shortwave 0
radios: 66,000
Television:
broadcast stations: 0
televisions: NA

Defense Forces

Branches: Tonga Defense Services, Maritime Division, Royal Tongan Marines, Tongan Royal Guards, Police
Defense expenditures: $NA, NA% of GDP

Trinidad and Tobago

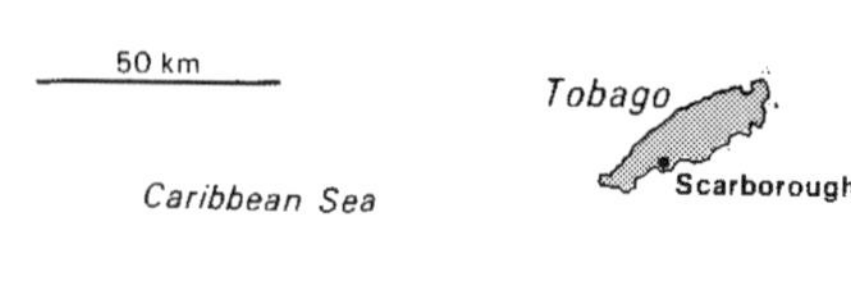

Geography

Location: Caribbean, islands between the Caribbean Sea and the North Atlantic Ocean, northeast of Venezuela
Map references: Central America and the Caribbean
Area:
total area: 5,130 sq km
land area: 5,130 sq km
comparative area: slightly smaller than Delaware
Land boundaries: 0 km
Coastline: 362 km
Maritime claims:
contiguous zone: 24 nm
continental shelf: 200 nm or to the outer edge of the continental margin
exclusive economic zone: 200 nm
territorial sea: 12 nm
International disputes: none
Climate: tropical; rainy season (June to December)
Terrain: mostly plains with some hills and low mountains
Natural resources: petroleum, natural gas, asphalt
Land use:
arable land: 14%
permanent crops: 17%
meadows and pastures: 2%
forest and woodland: 44%
other: 23%
Irrigated land: 220 sq km (1989 est.)
Environment:
current issues: water pollution from agricultural chemicals, industrial wastes, and raw sewage; oil pollution of beaches; deforestation; soil erosion
natural hazards: outside usual path of hurricanes and other tropical storms
international agreements: party to—Climate Change, Endangered Species, Hazardous Wastes, Law of the Sea, Marine Life Conservation, Nuclear Test Ban, Ozone Layer Protection, Tropical Timber 83, Wetlands; signed, but not ratified—Biodiversity

People

Population: 1,271,159 (July 1995 est.)
Age structure:
0-14 years: 31% (female 191,627; male 198,225)
15-64 years: 64% (female 399,726; male 407,495)
65 years and over: 5% (female 40,577; male 33,509) (July 1995 est.)
Population growth rate: 0.12% (1995 est.)
Birth rate: 16.62 births/1,000 population (1995 est.)
Death rate: 6.88 deaths/1,000 population (1995 est.)
Net migration rate: -8.59 migrant(s)/1,000 population (1995 est.)
Infant mortality rate: 18.5 deaths/1,000 live births (1995 est.)
Life expectancy at birth:
total population: 70.14 years
male: 67.75 years
female: 72.6 years (1995 est.)
Total fertility rate: 2.01 children born/woman (1995 est.)
Nationality:
noun: Trinidadian(s), Tobagonian(s)
adjective: Trinidadian, Tobagonian
Ethnic divisions: black 43%, East Indian (a local term—primarily immigrants from northern India) 40%, mixed 14%, white 1%, Chinese 1%, other 1%
Religions: Roman Catholic 32.2%, Hindu 24.3%, Anglican 14.4%, other Protestant 14%, Muslim 6%, none or unknown 9.1%
Languages: English (official), Hindi, French, Spanish
Literacy: age 15 and over can read and write (1990)
total population: 97%
male: 98%
female: 96%
Labor force: 463,900
by occupation: construction and utilities 18.1%, manufacturing, mining, and quarrying 14.8%, agriculture 10.9%, other 56.2% (1985 est.)

Government

Names:
conventional long form: Republic of Trinidad and Tobago
conventional short form: Trinidad and Tobago
Digraph: TD
Type: parliamentary democracy
Capital: Port-of-Spain
Administrative divisions: 8 counties, 3 municipalities*, and 1 ward**; Arima*, Caroni, Mayaro, Nariva, Port-of-Spain*, Saint Andrew, Saint David, Saint George, Saint Patrick, San Fernando*, Tobago**, Victoria
Independence: 31 August 1962 (from UK)
National holiday: Independence Day, 31 August (1962)
Constitution: 1 August 1976
Legal system: based on English common law; judicial review of legislative acts in the Supreme Court; has not accepted compulsory ICJ jurisdiction
Suffrage: 18 years of age; universal
Executive branch:
chief of state: President Noor Mohammed HASSANALI (since 18 March 1987)
head of government: Prime Minister Patrick Augustus Mervyn MANNING (since 17 December 1991)
cabinet: Cabinet; responsible to parliament
Legislative branch: bicameral Parliament
Senate: consists of a 31-member body appointed by the president
House of Representatives: elections last held 16 December 1991 (next to be held by December 1996); results—PNM 32%, UNC 13%, NAR 2%; seats—(36 total) PNM 21, UNC 13, NAR 2
Judicial branch: Court of Appeal, Supreme Court
Political parties and leaders: People's National Movement (PNM), Patrick MANNING; United National Congress (UNC), Basdeo PANDAY; National Alliance for Reconstruction (NAR), Selby WILSON; Movement for Social Transformation (MOTION), David ABDULLAH; National Joint Action Committee (NJAC), Makandal DAAGA; Republican Party, Nello MITCHELL; National Development Party (NDP), Carson CHARLES; Movement for Unity and Progress (MUP), Hulsie BHAGGAN
Member of: ACP, C, CARICOM, CCC, CDB, ECLAC, FAO, G-24, G-77, GATT, IADB, IBRD, ICAO, ICFTU, ICRM, IDA, IFAD, IFC, IFRCS, ILO, IMF, IMO, INTELSAT, INTERPOL, IOC, ISO, ITU, LAES, NAM, OAS, OPANAL, UN, UNCTAD, UNESCO, UNIDO, UNU, UPU, WFTU, WHO, WIPO, WMO
Diplomatic representation in US:
chief of mission: Ambassador Corinne Averille McKNIGHT
chancery: 1708 Massachusetts Avenue NW, Washington, DC 20036
telephone: [1] (202) 467-6490
FAX: [1] (202) 785-3130
consulate(s) general: New York
US diplomatic representation:
chief of mission: Ambassador Brian DONNELLY (since September 1994)
embassy: 15 Queen's Park West, Port-of-Spain
mailing address: P. O. Box 752, Port-of-Spain
telephone: [1] (809) 622-6372 through 6376, 6176
FAX: [1] (809) 628-5462
Flag: red with a white-edged black diagonal band from the upper hoist side

Economy

Overview: Trinidad and Tobago's petroleum-based economy still enjoys a high per capita income by Latin American standards, even though output and living standards are substantially below the boom years of 1973-82. The country suffers from widespread unemployment, large foreign-debt payments, and periods of low international oil prices. The government has begun to make progress in its efforts to diversify exports and to liberalize its trade regime, making 1994 the first year of substantial growth since the early 1980s.
National product: GDP—purchasing power parity—$15 billion (1994 est.)
National product real growth rate: 3% (1994 est.)
National product per capita: $11,280 (1994 est.)
Inflation rate (consumer prices): 10.1% (1994 est.)
Unemployment rate: 18.1% (1994)
Budget:
revenues: $1.6 billion
expenditures: $1.6 billion, including capital expenditures of $158 million (1993 est.)
Exports: $1.9 billion (f.o.b., 1994)
commodities: petroleum and petroleum products, chemicals, steel products, fertilizer, sugar, cocoa, coffee, citrus, flowers
partners: US 44%, CARICOM 15%, Latin America 9%, EC 5% (1993)
Imports: $996 million (c.i.f., 1994)
commodities: machinery, transportation equipment, manufactured goods, food, live animals
partners: US 43%, Venezuela 10%, UK 8%, other EC 8% (1993)
External debt: $2 billion (1994)
Industrial production: growth rate 1% (1994 est.); accounts for 39% of GDP, including petroleum
Electricity:
capacity: 1,150,000 kW
production: 3.9 billion kWh
consumption per capita: 2,740 kWh (1993)
Industries: petroleum, chemicals, tourism, food processing, cement, beverage, cotton textiles
Agriculture: accounts for 3% of GDP; major crops—cocoa, sugarcane; sugarcane acreage is being shifted into rice, citrus, coffee, vegetables; poultry sector most important source of animal protein; must import large share of food needs
Illicit drugs: transshipment point for South American drugs destined for the US and Europe and producer of cannabis
Economic aid:
recipient: US commitments, including Ex-Im (FY70-89), $373 million; Western (non-US) countries, ODA and OOF bilateral commitments (1970-89), $518 million
Currency: 1 Trinidad and Tobago dollar (TT$) = 100 cents
Exchange rates: Trinidad and Tobago dollars (TT$) per US$1—5.8758 (January 1995), 5.9160 (1994), 5.3511 (1993), 4.2500 (fixed rate 1989-1992); note—effective 13 April 1993, the exchange rate of the TT dollar is market-determined as opposed to the prior fixed relationship to the US dollar
Fiscal year: calendar year

Transportation

Railroads:
note: minimal agricultural railroad system near San Fernando
Highways:
total: 8,000 km
paved: 4,000 km
unpaved: improved earth 1,000 km; unimproved earth 3,000 km
Pipelines: crude oil 1,032 km; petroleum products 19 km; natural gas 904 km
Ports: Pointe-a-Pierre, Point Fortin, Point Lisas, Port-of-Spain, Scarborough, Tembladora
Merchant marine:
total: 2 cargo ships (1,000 GRT or over) totaling 12,507 GRT/21,923 DWT
Airports:
total: 6
with paved runways over 3,047 m: 1
with paved runways 2,438 to 3,047 m: 1
with paved runways 1,524 to 2,437 m: 1
with paved runways under 914 m: 2
with unpaved runways 914 to 1,523 m: 1

Communications

Telephone system: 109,000 telephones; excellent international service via tropospheric scatter links to Barbados and Guyana; good local service
local: NA
intercity: NA
international: 1 INTELSAT (Atlantic Ocean) earth station; linked to Barbados and Guyana by tropospheric scatter system
Radio:
broadcast stations: AM 2, FM 4, shortwave 0
radios: NA
Television:
broadcast stations: 5
televisions: NA

Defense Forces

Branches: Trinidad and Tobago Defense Force (includes Ground Forces, Coast Guard, and Air Wing), Trinidad and Tobago Police Service
Manpower availability: males age 15-49 347,841; males fit for military service 249,904 (1995 est.)
Defense expenditures: exchange rate conversion—$83 million, 1.5% of GDP (1994)

Tromelin Island
(possession of France)

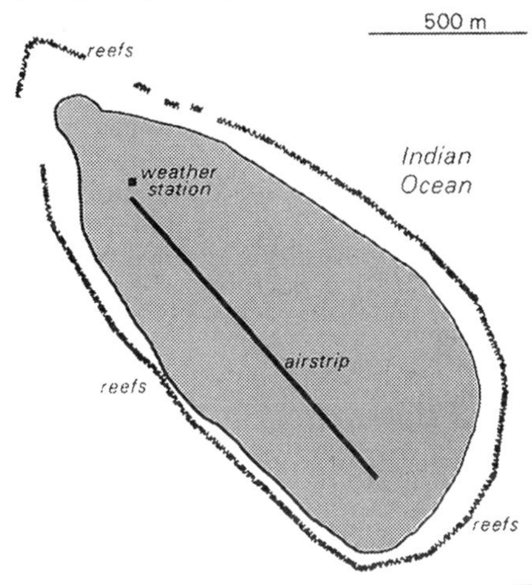

Geography

Location: Southern Africa, island in the Indian Ocean, east of Madagascar
Map references: Africa
Area:
total area: 1 sq km
land area: 1 sq km
comparative area: about 1.7 times the size of The Mall in Washington, DC
Land boundaries: 0 km
Coastline: 3.7 km
Maritime claims:
contiguous zone: 12 nm
continental shelf: 200-m depth or to depth of exploitation
exclusive economic zone: 200 nm
territorial sea: 12 nm
International disputes: claimed by Madagascar, Mauritius, and Seychelles
Climate: tropical
Terrain: sandy
Natural resources: fish
Land use:
arable land: 0%
permanent crops: 0%
meadows and pastures: 0%
forest and woodland: 0%
other: 100% (scattered bushes)
Irrigated land: 0 sq km
Environment:
current issues: NA
natural hazards: NA
international agreements: NA
Note: climatologically important location for forecasting cyclones; wildlife sanctuary

People

Population: uninhabited

Government

Names:
conventional long form: none
conventional short form: Tromelin Island

Tromelin Island *(continued)*

local long form: none
local short form: Ile Tromelin
Digraph: TE
Type: French possession administered by Commissioner of the Republic, resident in Reunion
Capital: none; administered by France from Reunion
Independence: none (possession of France)

Economy

Overview: no economic activity

Transportation

Ports: none; offshore anchorage only
Airports:
total: 1
with paved runways under 914 m: 1

Communications

Note: important meteorological station

Defense Forces

Note: defense is the responsibility of France

Tunisia

Geography

Location: Northern Africa, bordering the Mediterranean Sea, between Algeria and Libya
Map references: Africa
Area:
total area: 163,610 sq km
land area: 155,360 sq km
comparative area: slightly larger than Georgia
Land boundaries: total 1,424 km, Algeria 965 km, Libya 459 km
Coastline: 1,148 km
Maritime claims:
contiguous zone: 24 nm
territorial sea: 12 nm
International disputes: maritime boundary dispute with Libya; land boundary dispute with Algeria settled in 1993; Malta and Tunisia are discussing the commercial exploitation of the continental shelf between their countries, particularly for oil exploration
Climate: temperate in north with mild, rainy winters and hot, dry summers; desert in south
Terrain: mountains in north; hot, dry central plain; semiarid south merges into the Sahara
Natural resources: petroleum, phosphates, iron ore, lead, zinc, salt
Land use:
arable land: 20%
permanent crops: 10%
meadows and pastures: 19%
forest and woodland: 4%
other: 47%
Irrigated land: 2,750 sq km (1989)
Environment:
current issues: toxic and hazardous waste disposal is ineffective and presents human health risks; water pollution from raw sewage; limited natural fresh water resources; deforestation; overgrazing; soil erosion; desertification
natural hazards: NA
international agreements: party to—Biodiversity, Climate Change, Endangered Species, Environmental Modification, Law of the Sea, Marine Dumping, Nuclear Test Ban, Ozone Layer Protection, Ship Pollution, Wetlands; signed, but not ratified—Desertification, Marine Life Conservation
Note: strategic location in central Mediterranean

People

Population: 8,879,845 (July 1995 est.)
Age structure:
0-14 years: 35% (female 1,507,866; male 1,563,411)
15-64 years: 60% (female 2,665,586; male 2,672,712)
65 years and over: 5% (female 226,201; male 244,069) (July 1995 est.)
Population growth rate: 1.69% (1995 est.)
Birth rate: 22.52 births/1,000 population (1995 est.)
Death rate: 4.86 deaths/1,000 population (1995 est.)
Net migration rate: -0.74 migrant(s)/1,000 population (1995 est.)
Infant mortality rate: 32.3 deaths/1,000 live births (1995 est.)
Life expectancy at birth:
total population: 73.25 years
male: 71.16 years
female: 75.44 years (1995 est.)
Total fertility rate: 2.73 children born/woman (1995 est.)
Nationality:
noun: Tunisian(s)
adjective: Tunisian
Ethnic divisions: Arab-Berber 98%, European 1%, Jewish less than 1%
Religions: Muslim 98%, Christian 1%, Jewish 1%
Languages: Arabic (official and one of the languages of commerce), French (commerce)
Literacy: age 15 and over can read and write (1989)
total population: 57%
male: 69%
female: 45%
Labor force: 2.25 million
by occupation: agriculture 32%
note: shortage of skilled labor

Government

Names:
conventional long form: Republic of Tunisia
conventional short form: Tunisia
local long form: Al Jumhuriyah at Tunisiyah
local short form: Tunis
Digraph: TS
Type: republic
Capital: Tunis
Administrative divisions: 23 governorates; Beja, Ben Arous, Bizerte, Gabes, Gafsa, Jendouba, Kairouan, Kasserine, Kebili, L'Ariana, Le Kef, Mahdia, Medenine, Monastir, Nabeul, Sfax, Sidi Bou Zid, Siliana, Sousse, Tataouine, Tozeur, Tunis, Zaghouan
Independence: 20 March 1956 (from France)

National holiday: National Day, 20 March (1956)
Constitution: 1 June 1959; amended 12 July 1988
Legal system: based on French civil law system and Islamic law; some judicial review of legislative acts in the Supreme Court in joint session
Suffrage: 20 years of age; universal
Executive branch:
chief of state: President Zine el Abidine BEN ALI (since 7 November 1987); election last held 20 March 1994 (next to be held NA 1999); results—President Zine el Abidine BEN ALI was reelected without opposition
head of government: Prime Minister Hamed KAROUI (since 26 September 1989)
cabinet: Council of Ministers; appointed by the president
Legislative branch: unicameral
Chamber of Deputies (Majlis al-Nuwaab): elections last held 20 March 1994 (next to be held NA 1999); results—RCD 97.7%, MDS 1.0%, others 1.3%; seats—(163 total) RCD 144, MDS 10, others 9; note—the government changed the electoral code to guarantee that the opposition won seats
Judicial branch: Court of Cassation (Cour de Cassation)
Political parties and leaders: Constitutional Democratic Rally Party (RCD), President BEN ALI (official ruling party); Movement of Democratic Socialists (MDS), Mohammed MOUAADA; five other political parties are legal, including the Communist Party
Other political or pressure groups: the Islamic fundamentalist party, An Nahda (Rebirth), is outlawed
Member of: ABEDA, ACCT, AfDB, AFESD, AL, AMF, AMU, CCC, ECA, FAO, G-77, GATT, IAEA, IBRD, ICAO, ICC, ICFTU, ICRM, IDA, IDB, IFAD, IFC, IFRCS, ILO, IMF, IMO, INMARSAT, INTELSAT, INTERPOL, IOC, ISO, ITU, MINURSO, NAM, OAPEC (withdrew from active membership in 1986), OAS (observer), OAU, OIC, UN, UNAMIR, UNCTAD, UNESCO, UNHCR, UNIDO, UNITAR, UNMIH, UNPROFOR, UPU, WHO, WIPO, WMO, WTO
Diplomatic representation in US:
chief of mission: Ambassador Mohamed Azzouz ENNAIFER
chancery: 1515 Massachusetts Avenue NW, Washington, DC 20005
telephone: [1] (202) 862-1850
US diplomatic representation:
chief of mission: Ambassador Mary Ann CASEY
embassy: 144 Avenue de la Liberte, 1002 Tunis-Belvedere
mailing address: use embassy street address
telephone: [216] (1) 782-566
FAX: [216] (1) 789-719
Flag: red with a white disk in the center bearing a red crescent nearly encircling a red five-pointed star; the crescent and star are traditional symbols of Islam

Economy

Overview: Tunisia has a diverse economy, with important agricultural, mining, energy, tourism, and manufacturing sectors. Detailed governmental control of economic affairs has gradually lessened over the past decade, including increasing privatization of trade and commerce, simplification of the tax structure, and a cautious approach to debt. Real growth has averaged roughly 5% in 1991-94, and inflation has been moderate. Growth in tourism and IMF support have been key elements in this solid record. Further privatization and further improvements in government administrative efficiency are among the challenges for the future.
National product: GDP purchasing power parity—$37.1 billion (1994 est.)
National product real growth rate: 4.4% (1994 est.)
National product per capita: $4,250 (1994 est.)
Inflation rate (consumer prices): 4.5% (1993 est.)
Unemployment rate: 16.2% (1993 est.)
Budget:
revenues: $4.3 billion
expenditures: $5.5 billion, including capital expenditures to $NA (1993 est.)
Exports: $4.6 billion (f.o.b., 1993)
commodities: hydrocarbons, agricultural products, phosphates and chemicals
partners: EC countries 75%, Middle East 10%, Algeria 2%, India 2%, US 1%
Imports: $6.5 billion (c.i.f., 1993)
commodities: industrial goods and equipment 57%, hydrocarbons 13%, food 12%, consumer goods
partners: EC countries 70%, US 5%, Middle East 2%, Japan 2%, Switzerland 1%, Algeria 1%
External debt: $7.7 billion (1993 est.)
Industrial production: growth rate 5% (1989); accounts for 22% of GDP, including petroleum
Electricity:
capacity: 1,410,000 kW
production: 5.4 billion kWh
consumption per capita: 595 kWh (1993)
Industries: petroleum, mining (particularly phosphate and iron ore), tourism, textiles, footwear, food, beverages
Agriculture: accounts for 16% of GDP and one-third of labor force; output subject to severe fluctuations because of frequent droughts; export crops—olives, dates, oranges, almonds; other products—grain, sugar beets, wine grapes, poultry, beef, dairy; not self-sufficient in food
Economic aid:
recipient: US commitments, including Ex-Im (FY70-89), $730 million; Western (non-US) countries, ODA and OOF bilateral commitments (1970-89) $52 million; OPEC bilateral aid (1979-89), $684 million; Communist countries (1970-89), $410 million
Currency: 1 Tunisian dinar (TD) = 1,000 millimes
Exchange rates: Tunisian dinars (TD) per US$1—0.9849 (January 1995), 1.0116 (1994), 1.0037 (1993), 0.8844 (1992), 0.9246 (1991), 0.8783 (1990)
Fiscal year: calendar year

Transportation

Railroads:
total: 2,260 km
standard gauge: 492 km 1.435-m gauge
narrow gauge: 1,758 km 1.000-m gauge
dual gauge: 10 km 1.000-m and 1.435-m gauges
Highways:
total: 29,183 km
paved: bituminous 17,510 km
unpaved: improved, unimproved earth 11,673 km
Pipelines: crude oil 797 km; petroleum products 86 km; natural gas 742 km
Ports: Bizerte, Gabes, La Goulette, Sfax, Sousse, Tunis, Zarzis
Merchant marine:
total: 19 ships (1,000 GRT or over) totaling 129,035 GRT/168,032 DWT
ships by type: bulk 6, cargo 5, chemical tanker 4, oil tanker 1, roll-on/roll-off cargo 2, short-sea passenger 1
Airports:
total: 31
with paved runways over 3,047 m: 3
with paved runways 2,438 to 3,047 m: 6
with paved runways 1,524 to 2,437 m: 2
with paved runways 914 to 1,523 m: 3
with paved runways under 914 m: 8
with unpaved runways 1,524 to 2,438 m: 2
with unpaved runways 914 to 1,523 m: 7

Communications

Telephone system: 233,000 telephones; 28 telephones/1,000 persons; the system is above the African average; key centers are Sfax, Sousse, Bizerte, and Tunis
local: NA
intercity: facilities consist of open-wire lines, coaxial cable, and microwave radio relay
international: 5 submarine cables; 1 INTELSAT (Atlantic Ocean) and 1 ARABSAT earth station with back-up control station; coaxial cable and microwave radio relay to Algeria and Libya
Radio:
broadcast stations: AM 7, FM 8, shortwave 0
radios: NA
Television:
broadcast stations: 19
televisions: NA

Defense Forces

Branches: Army, Navy, Air Force, paramilitary forces, National Guard
Manpower availability: males age 15-49 2,294,912; males fit for military service 1,317,642; males reach military age (20) annually 93,601 (1995 est.)
Defense expenditures: exchange rate conversion—$549 million, 3% of GDP (1994)

Turkey

Geography

Location: Southwestern Asia (that part west of the Bosporus is sometimes included with Europe), bordering the Black Sea, between Bulgaria and Georgia, and bordering the Aegean Sea and the Mediterranean Sea, between Greece and Syria
Map references: Middle East
Area:
total area: 780,580 sq km
land area: 770,760 sq km
comparative area: slightly larger than Texas
Land boundaries: total 2,627 km, Armenia 268 km, Azerbaijan 9 km, Bulgaria 240 km, Georgia 252 km, Greece 206 km, Iran 499 km, Iraq 331 km, Syria 822 km
Coastline: 7,200 km
Maritime claims:
exclusive economic zone: in Black Sea only—to the maritime boundary agreed upon with the former USSR
territorial sea: 6 nm in the Aegean Sea, 12 nm in the Black Sea and in the Mediterranean Sea
International disputes: complex maritime, air and territorial disputes with Greece in Aegean Sea; Cyprus question; Hatay question with Syria; ongoing dispute with downstream riparians (Syria and Iraq) over water development plans for the Tigris and Euphrates Rivers
Climate: temperate; hot, dry summers with mild, wet winters; harsher in interior
Terrain: mostly mountains; narrow coastal plain; high central plateau (Anatolia)
Natural resources: antimony, coal, chromium, mercury, copper, borate, sulphur, iron ore
Land use:
arable land: 30%
permanent crops: 4%
meadows and pastures: 12%
forest and woodland: 26%
other: 28%
Irrigated land: 22,200 sq km (1989 est.)
Environment:
current issues: water pollution from dumping of chemicals and detergents; air pollution, particularly in urban areas; deforestation
natural hazards: very severe earthquakes, especially in northern Turkey, along an arc extending from the Sea of Marmara to Lake Van
international agreements: party to—Air Pollution, Hazardous Wastes, Nuclear Test Ban, Ozone Layer Protection, Ship Pollution, Wetlands; signed, but not ratified—Biodiversity, Desertification, Environmental Modification
Note: strategic location controlling the Turkish Straits (Bosporus, Sea of Marmara, Dardanelles) that link Black and Aegean Seas

People

Population: 63,405,526 (July 1995 est.)
Age structure:
0-14 years: 35% (female 10,815,288; male 11,203,723)
15-64 years: 60% (female 18,723,772; male 19,391,037)
65 years and over: 5% (female 1,764,363; male 1,507,343) (July 1995 est.)
Population growth rate: 1.97% (1995 est.)
Birth rate: 25.33 births/1,000 population (1995 est.)
Death rate: 5.64 deaths/1,000 population (1995 est.)
Net migration rate: 0 migrant(s)/1,000 population (1995 est.)
Infant mortality rate: 45.6 deaths/1,000 live births (1995 est.)
Life expectancy at birth:
total population: 71.48 years
male: 69.11 years
female: 73.96 years (1995 est.)
Total fertility rate: 3.12 children born/woman (1995 est.)
Nationality:
noun: Turk(s)
adjective: Turkish
Ethnic divisions: Turkish 80%, Kurdish 20%
Religions: Muslim 99.8% (mostly Sunni), other 0.2% (Christian and Jews)
Languages: Turkish (official), Kurdish, Arabic
Literacy: age 15 and over can read and write (1990)
total population: 79%
male: 90%
female: 68%
Labor force: 20.4 million
by occupation: agriculture 44%, services 41%, industry 15%
note: between 1.5 million and 1.8 million Turks work abroad (1994)

Government

Names:
conventional long form: Republic of Turkey

conventional short form: Turkey
local long form: Turkiye Cumhuriyeti
local short form: Turkiye
Digraph: TU
Type: republican parliamentary democracy
Capital: Ankara
Administrative divisions: 73 provinces (iller, singular—il); Adana, Adiyaman, Afyon, Agri, Aksaray, Amasya, Ankara, Antalya, Artvin, Aydin, Balikesir, Batman, Bayburt, Bilecik, Bingol, Bitlis, Bolu, Burdur, Bursa, Canakkale, Cankiri, Corum, Denizli, Diyarbakir, Edirne, Elazig, Erzincan, Erzurum, Eskisehir, Gazi Antep, Giresun, Gumushane, Hakkari, Hatay, Icel, Isparta, Istanbul, Izmir, Kahraman Maras, Karaman, Kars, Kastamonu, Kayseri, Kirikkale, Kirklareli, Kirsehir, Kocaeli, Konya, Kutahya, Malatya, Manisa, Mardin, Mugla, Mus, Nevsehir, Nigde, Ordu, Rize, Sakarya, Samsun, Sanli Urfa, Siirt, Sinop, Sirnak, Sivas, Tekirdag, Tokat, Trabzon, Tunceli, Usak, Van, Yozgat, Zonguldak
Independence: 29 October 1923 (successor state to the Ottoman Empire)
National holiday: Anniversary of the Declaration of the Republic, 29 October (1923)
Constitution: 7 November 1982
Legal system: derived from various continental legal systems; accepts compulsory ICJ jurisdiction, with reservations
Suffrage: 21 years of age; universal
Executive branch:
chief of state: President Suleyman DEMIREL (since 16 May 1993)
head of government: Prime Minister Tansu CILLER (since 5 July 1993); Deputy Prime Minister Hikmet CETIN (since 27 March 1995)
National Security Council: advisory body to the President and the Cabinet
cabinet: Council of Ministers; appointed by the president on nomination of the prime minister
Legislative branch: unicameral
Grand National Assembly of Turkey: (Turkiye Buyuk Millet Meclisi) elections last held 20 October 1991 (next to be held NA October 1996); results DYP 27.03%, ANAP 24.01%, SHP 20.75%, RP 16.88%, DSP 10.75%, SBP 0.44%, independent 0.14%; seats—(450 total) DYP 178, ANAP 115, SHP 86, RP 40, MCP 19, DSP 7, other 5
note: seats held by various parties are subject to change due to defections, creation of new parties, and ouster or death of sitting deputies; present seats by party are as follows: DYP 183, ANAP 97, RP 38, CHP 65, MHP 17, BBP 7, DSP 10, YP 3, MP 2, independents 6, vacant 22
Judicial branch: Court of Cassation
Political parties and leaders: True Path Party (DYP), Tansu CILLER; Motherland Party (ANAP), Mesut YILMAZ; Welfare Party (RP), Necmettin ERBAKAN; Democratic Left Party (DSP), Bulent ECEVIT; Nationalist Action Party (MHP; note—members also regroup under the name of National Labor Party or MCP), Alparslan TURKES; Socialist Unity Party (SBP), Sadun AREN; New Party (YP), Yusuf Bozkurt OZAL; Republican People's Party (CHP), Hikmet CETIN; note—Social Democrat Populist Party (SHP) has merged with CHP; Workers Party (IP), Dogu PERINCEK; Nation Party (MP), Aykut EDIBALI; Democrat Party (DP), Aydin MENDERES; Grand Unity Party (BBP), Muhsin YAZICIOGLU; Rebirth Party (YDP), Hasan Celal GUZEL; People's Democracy Party (HADEP), Murat BOZLAK; Main Path Party (ANAYOL), Gurcan BASER; Democratic Target Party (DHP), Abdulkadir Yasar TURK; Liberal Party (LP), Besim TIBUK; New Democracy Movement (YDH), Cem BOYNER; Democracy and Change Party (DDP), Ibrahim AKSOY
Other political or pressure groups: Turkish Confederation of Labor (TURK-IS), Bayram MERAL; Confederation of Revolutionary Workers Unions (DISK), Ridvan BUDAK; Moral Rights Workers Union (HAK-IS), Negati CECIK; Turkish Industrialists' and Businessmen's Association (TUSIAD), Halis KOMILI; Turkish Union of Chambers of Commerce and Commodity Exchanges (TOBB), Yalim EREZ; Turkish Confederation of Employers' Unions (TISK), Refik BAYDUR
Member of: AsDB, BIS, BSEC, CCC, CE, CERN (observer), EBRD, ECE, ECO, FAO, GATT, IAEA, IBRD, ICAO, ICC, ICFTU, ICRM, IDA, IDB, IEA, IFAD, IFC, IFRCS, ILO, IMF, IMO, INMARSAT, INTELSAT, INTERPOL, IOC, IOM (observer), ISO, ITU, NACC, NATO, NEA, OECD, OIC, OSCE, PCA, UN, UNCTAD, UNESCO, UNHCR, UNIDO, UNIKOM, UNRWA, UPU, WEU (associate), WFTU, WHO, WIPO, WMO, WTO
Diplomatic representation in US:
chief of mission: Ambassador Nuzhet KANDEMIR
chancery: 1714 Massachusetts Avenue NW, Washington, DC 20036
telephone: [1] (202) 659-8200
consulate(s) general: Chicago, Houston, Los Angeles, and New York
US diplomatic representation:
chief of mission: Ambassador Marc GROSSMAN
embassy: 110 Ataturk Boulevard, Ankara
mailing address: PSC 93, Box 5000, Ankara; APO AE 09823
telephone: [90] (312) 468-6110 through 6128
FAX: [90] (312) 467-0019
consulate(s) general: Istanbul
consulate(s): Adana
Flag: red with a vertical white crescent (the closed portion is toward the hoist side) and white five-pointed star centered just outside the crescent opening

Economy

Overview: In early 1995, after an impressive economic performance through most of the 1980s, Turkey continues to suffer through its most damaging economic crisis in the last 15 years. Sparked by the downgrading in January 1994 of Turkey's international credit rating by two US credit rating agencies, the crisis stems from years of loose fiscal and monetary policies that had exacerbated inflation and allowed the public debt, money supply, and current account deficit to explode. In April 1994, Prime Minister CILLER introduced an austerity package aimed at restoring domestic and international confidence in her fragile coalition government. Three months later the IMF endorsed the program, paving the way for a $740 million IMF standby loan. Although the economy showed signs of improvement following the stabilization measures, CILLER has been unable to overcome the political obstacles to tough structural reforms necessary for sustained, longer-term growth. As a consequence, the economy is suffering the worst of both worlds: at the end of 1994, inflation hit a record 126% (annual rate), and real GDP dropped an estimated 5% for the year as a whole, the worst decline in Turkey's post-war history. At the same time, the government missed key 1994 targets stipulated in the IMF agreement: the budget deficit is estimated to have overshot the government's goal by 47%; the total public sector borrowing requirement likely reached 10%-12% of GDP, rather than 8.5% called for in the program; and the Turkish lira's value fell 5% to 7% more than expected. The unprecedented effort by the Kurdistan Workers' Party (PKK) to raise the economic costs of its insurgency against the Turkish state is adding to Turkey's economic problems. Attacks against tourists have jeopardized tourist revenues, which account for about 3% of GDP, while economic activity in southeastern Turkey, where most of the violence occurs, has dropped considerably. Turkish officials are now negotiating a new letter of intent with the IMF that will stipulate more realistic macroeconomic goals for 1995 and allow the release of remaining funds of the standby agreement.
National product: GDP—purchasing power parity—$305.2 billion (1994 est.)
National product real growth rate: -5% (1994 est.)
National product per capita: $4,910 (1994 est.)
Inflation rate (consumer prices): 106% (1994)
Unemployment rate: 12.6% (1994)

Budget:
revenues: $28.3 billion
expenditures: $33.3 billion, including capital expenditures of $3.2 billion (1995)
Exports: $15.3 billion (f.o.b., 1993)
commodities: manufactured products 72%, foodstuffs 23%, mining products 4% (1993)
partners: Germany 24%, Russia 7%, US 7%, UK 6% (1993)
Imports: $27.6 billion (f.o.b., 1993)
commodities: manufactured products 71%, fuels 14%, foodstuffs 6% (1993)
partners: Germany 15%, US 11%, Italy 9%, Russia 8% (1993)
External debt: $66.6 billion (1994)
Industrial production: growth rate 6.7% (1993); accounts for 26% of GDP
Electricity:
capacity: 18,710,000 kW
production: 71 billion kWh
consumption per capita: 1,079 kWh (1993)
Industries: textiles, food processing, mining (coal, chromite, copper, boron), steel, petroleum, construction, lumber, paper
Agriculture: accounts for 16% of GDP; products—tobacco, cotton, grain, olives, sugar beets, pulses, citrus fruit, variety of animal products; self-sufficient in food most years
Illicit drugs: major transit route for Southwest Asian heroin and hashish to Western Europe and the US via air, land, and sea routes; major Turkish, Iranian, and other international trafficking organizations operate out of Istanbul; laboratories to convert imported morphine base into heroin are in remote regions of Turkey as well as near Istanbul; government maintains strict controls over areas of legal opium poppy cultivation and output of poppy straw concentrate
Economic aid:
recipient: US commitments, including Ex-Im (FY70-89), $2.3 billion; Western (non-US) countries, ODA and OOF bilateral commitments (1970-89), $10.1 billion; OPEC bilateral aid (1979-89), $665 million; Communist countries (1970-89), $4.5 billion
note: aid for Persian Gulf war efforts from coalition allies (1991), $4.1 billion; aid pledged for Turkish Defense Fund, $2.5 billion
Currency: 1 Turkish lira (TL) = 100 kurus
Exchange rates: Turkish liras (TL) per US$1—37,444.1 (December 1994), 29,608.7 (1994), 10,984.6 (1993), 6,872.4 (1992), 4,171.8 (1991), 2,608.6 (1990)
Fiscal year: calendar year

Transportation

Railroads:
total: 10,413 km
standard gauge: 10,413 km 1.435-m gauge (1,033 km electrified)
Highways:
total: 320,611 km
paved: 29,915 km (including 862 km of expressways)
unpaved: 290,696 km (1992)
Inland waterways: about 1,200 km
Pipelines: crude oil 1,738 km; petroleum products 2,321 km; natural gas 708 km
Ports: Gemlik, Hopa, Iskenderun, Istanbul, Izmir, Izmit, Mersin, Samsun, Trabzon
Merchant marine:
total: 423 ships (1,000 GRT or over) totaling 5,014,004 GRT/8,695,636 DWT
ships by type: bulk 113, cargo 203, chemical tanker 14, combination bulk 7, combination ore/oil 12, container 2, liquefied gas tanker 4, livestock carrier 1, oil tanker 46, passenger-cargo 1, refrigerated cargo 2, roll-on/roll-off cargo 9, short-sea passenger 7, specialized tanker 2
Airports:
total: 116
with paved runways over 3,047 m: 16
with paved runways 2,438 to 3,047 m: 20
with paved runways 1,524 to 2,437 m: 12
with paved runways 914 to 1,523 m: 21
with paved runways under 914 m: 34
with unpaved runways 1,524 to 2,438 m: 2
with unpaved runways 914 to 1,523 m: 11

Communications

Telephone system: 3,400,000 telephones; fair domestic and international systems
local: NA
intercity: trunk radio relay microwave network; limited open wire network
international: 2 INTELSAT (Atlantic Ocean) and 1 EUTELSAT earth station; 1 submarine cable
Radio:
broadcast stations: AM 15, FM 94, shortwave 0
radios: NA
Television:
broadcast stations: 357
televisions: NA

Defense Forces

Branches: Land Forces, Navy (includes Naval Air and Naval Infantry), Air Force, Coast Guard, Gendarmerie
Manpower availability: males age 15-49 16,519,152; males fit for military service 10,067,089; males reach military age (20) annually 625,476 (1995 est.)
Defense expenditures: exchange rate conversion—$6.9 billion, 4.1% of GDP (1993); note—figures do not include about $7 billion for the government's counterinsurgency efforts against the separatist Kurdistan Workers' Party (PKK)

Turkmenistan

Geography

Location: Central Asia, bordering the Caspian Sea, between Iran and Kazakhstan
Map references: Commonwealth of Independent States—Central Asian States
Area:
total area: 488,100 sq km
land area: 488,100 sq km
comparative area: slightly larger than California
Land boundaries: total 3,736 km, Afghanistan 744 km, Iran 992 km, Kazakhstan 379 km, Uzbekistan 1,621 km
Coastline: 0 km
note: Turkmenistan borders the Caspian Sea (1,768 km)
Maritime claims: none; landlocked
International disputes: Caspian Sea boundaries are not yet determined
Climate: subtropical desert
Terrain: flat-to-rolling sandy desert with dunes rising to mountains in the south; low mountains along border with Iran; borders Caspian Sea in west
Natural resources: petroleum, natural gas, coal, sulphur, salt
Land use:
arable land: 2%
permanent crops: 0%
meadows and pastures: 69%
forest and woodland: 0%
other: 29%
Irrigated land: 12,450 sq km (1990)
Environment:
current issues: contamination of soil and groundwater with agricultural chemicals, pesticides; salinization, water-logging of soil due to poor irrigation methods; Caspian Sea pollution; diversion of a large share of the flow of the Amu Darya into irrigation contributes to that river's inability to replenish the Aral Sea; desertification
natural hazards: NA
international agreements: party to—Ozone Layer Protection
Note: landlocked

People

Population: 4,075,316 (July 1995 est.)
Age structure:
0-14 years: 40% (female 798,620; male 821,550)
15-64 years: 56% (female 1,155,392; male 1,128,844)
65 years and over: 4% (female 105,424; male 65,486) (July 1995 est.)
Population growth rate: 1.97% (1995 est.)
Birth rate: 29.93 births/1,000 population (1995 est.)
Death rate: 7.34 deaths/1,000 population (1995 est.)
Net migration rate: -2.92 migrant(s)/1,000 population (1995 est.)
Infant mortality rate: 68.5 deaths/1,000 live births (1995 est.)
Life expectancy at birth:
total population: 65.35 years
male: 61.85 years
female: 69.02 years (1995 est.)
Total fertility rate: 3.72 children born/woman (1995 est.)
Nationality:
noun: Turkmen(s)
adjective: Turkmen
Ethnic divisions: Turkmen 73.3%, Russian 9.8%, Uzbek 9%, Kazakh 2%, other 5.9%
Religions: Muslim 87%, Eastern Orthodox 11%, unknown 2%
Languages: Turkmen 72%, Russian 12%, Uzbek 9%, other 7%
Literacy: age 15 and over can read and write (1989)
total population: 98%
male: 99%
female: 97%
Labor force: 1.642 million (January 1994)
by occupation: agriculture and forestry 44%, industry and construction 20%, other 36% (1992)

Government

Names:
conventional long form: none
conventional short form: Turkmenistan
local long form: none
local short form: Turkmenistan
former: Turkmen Soviet Socialist Republic
Digraph: TX
Type: republic
Capital: Ashgabat
Administrative divisions: 5 welayatlar (singular—welayat): Ahal Welayaty (Ashgabat), Balkan Welayaty (Nebitdag), Dashhowuz Welayaty (formerly Tashauz), Lebap Welayaty (Charjew), Mary Welayaty
note: names in parentheses are administrative centers when name differs from welayat name
Independence: 27 October 1991 (from the Soviet Union)
National holiday: Independence Day, 27 October (1991)
Constitution: adopted 18 May 1992
Legal system: based on civil law system
Suffrage: 18 years of age; universal
Executive branch:
chief of state: President Saparmurad NIYAZOV (since NA October 1990); election last held 21 June 1992 (next to be held NA 2002); results—Saparmurad NIYAZOV 99.5% (ran unopposed); note—a 15 January 1994 referendum extended NIYAZOV's term an additional five years until 2002 (99.99% approval)
head of government: Prime Minister (vacant); Deputy Prime Ministers Orazgeldi AYDOGDIYEV (since NA), Babamurad BAZAROV (since NA), Khekim ISHANOV (since NA), Valeriy OTCHERTSOV (since NA), Yagmur OVEZOV (since NA), Matkarim RAJAPOV (since NA), Abad RIZAYEVA (since NA), Rejep SAPAROV (since NA), Boris SHIKHMURADOV (since NA), Batyr SARJAYEV (since NA)
cabinet: Council of Ministers
Legislative branch: under 1992 constitution there are two parliamentary bodies, a unicameral People's Council (Halk Maslahaty—having more than 100 members and meeting infrequently) and a 50-member unicameral Assembly (Majlis)
Assembly (Majlis): elections last held 11 December 1994 (next to be held NA); results—percent of vote by party NA; seats (50 total) Democratic Party 45, others 5; note—all 50 preapproved by President NIYAZOV
Judicial branch: Supreme Court
Political parties and leaders: Democratic Party of Turkmenistan, Saparmurad NIYAZOV; Party for Democratic Development, Durdymurat HOJA-MUKHAMMED, chairman; Agzybirlik, Nurberdy NURMAMEDOV, cochairman, Hubayberdi HALLIYEV, cochairman
note: formal opposition parties are outlawed; unofficial, small opposition movements exist underground or in foreign countries
Member of: CCC, CIS, EBRD, ECE, ECO, ESCAP, IBRD, ICAO, IDB, ILO, IMF, IMO, INTELSAT (nonsignatory user), IOC, ISO (correspondent), ITU, NACC, OIC, OSCE, PFP, UN, UNCTAD, UNESCO, UPU, WHO, WMO, WTO
Diplomatic representation in US:
chief of mission: Ambassador Khalil UGUR
chancery: 1511 K Street NW, Suite 412, Washington, DC 20005
telephone: [1] (202) 737-4800
FAX: [1] (202) 737-1152
US diplomatic representation:
chief of mission: Ambassador Joseph S. HULINGS III
embassy: 6 Teheran Street, Yubilenaya Hotel, Ashgabat
mailing address: use embassy street address
telephone: [7] (3632) 24-49-25, 24-49-22
FAX: [7] (3632) 25-53-79
Flag: green field, including a vertical stripe on the hoist side, with a claret vertical stripe in between containing five white, black, and orange carpet guls (an assymetrical design used in producing rugs) associated with five different tribes; a white crescent and five white stars in the upper left corner to the right of the carpet guls

Economy

Overview: Turkmenistan is largely desert country with nomadic cattle raising, intensive agriculture in irrigated oases, and huge gas and oil resources. Half its irrigated land is planted in cotton making it the world's tenth largest producer. It also has the world's fifth largest reserves of natural gas and significant oil resources. Until the end of 1993, Turkmenistan had experienced less economic disruption than other former Soviet states because its economy received a boost from higher prices for oil and gas and a sharp increase in hard currency earnings. In 1994, Russia's refusal to export Turkmen gas to hard currency markets and mounting debts of its major customers in the former USSR for gas deliveries contributed to a sharp fall in industrial production and caused the budget to shift from a surplus to a slight deficit. Furthermore, with an authoritarian ex-Communist regime in power and a tribally-based social structure, Turkmenistan has taken a cautious approach to economic reform, hoping to use gas and cotton sales to sustain its inefficient economy. With the onset of economic hard times, even cautious moves toward economic restructuring and privatization have slowed down. For 1995, Turkmenistan will face continuing constraints on its earnings because of its customers' inability to pay for their gas and a low average cotton crop in 1994. Turkmenistan is working hard to open new gas export channels through Iran and Turkey, but these may take many years to realize.
National product: GDP—purchasing power parity—$13.1 billion (1994 estimate as extrapolated from World Bank estimate for 1992)
National product real growth rate: -24% (1994 est.)
National product per capita: $3,280 (1994 est.)
Inflation rate (consumer prices): 25% per month (1994)
Unemployment rate: NA
Budget:
revenues: $NA
expenditures: $NA, including capital expenditures of $NA
Exports: $382 million to states outside the FSU (1994)
commodities: natural gas, cotton, petroleum

products, electricity, textiles, carpets
partners: Ukraine, Russia, Kazakhstan, Uzbekistan, Georgia, Azerbaijan, Armenia, Eastern Europe, Turkey, Argentina
Imports: $304 million from states outside the FSU (1994)
commodities: machinery and parts, grain and food, plastics and rubber, consumer durables, textiles
partners: Russia, Azerbaijan, Uzbekistan, Kazakhstan, Turkey
External debt: NEGL
Industrial production: growth rate -25% (1994)
Electricity:
capacity: 2,480,000 kW
production: 10.5 billion kWh
consumption per capita: 2,600 kWh (1994)
Industries: natural gas, oil, petroleum products, textiles, food processing
Agriculture: cotton, grain, animal husbandry
Illicit drugs: illicit cultivator of cannabis and opium poppy; mostly for CIS consumption; limited government eradication program; used as transshipment point for illicit drugs from Southwest Asia to Western Europe
Economic aid:
recipient: Turkmenistan has received about $200 million in bilateral aid credits
Currency: Turkmenistan introduced its national currency, the manat, on 1 November 1993
Exchange rates: manats per US$1—multiple rate system: 10 (official) and 230 (permitted in transactions between the government and individuals)
Fiscal year: calendar year

Transportation

Railroads:
total: 2,120 km in common carrier service; does not include industrial lines
broad gauge: 2,120 km 1.520-m gauge (1990)
Highways:
total: 23,000 km
paved and graveled: 18,300 km
unpaved: earth 4,700 km (1990)
Pipelines: crude oil 250 km; natural gas 4,400 km
Ports: Turkmenbashi (formerly Krasnowodsk)
Airports:
total: 64
with paved runways 2,438 to 3,047 m: 13
with paved runways 1,524 to 2,437 m: 8
with paved runways 914 to 1,523 m: 1
with unpaved runways 914 to 1,523 m: 7
with unpaved runways under 914 m: 35

Communications

Telephone system: NA telephones; only 7.5 telephones/100 persons (1991)
local: NA
intercity: NA
international: poorly developed; linked by cable and microwave to other CIS republics and to other countries by leased connections to the Moscow international gateway switch; a new telephone link from Ashgabat to Iran has been established; a new exchange in Ashgabat switches international traffic through Turkey via INTELSAT; 1 Orbita and 1 INTELSAT earth station
Radio:
broadcast stations: AM NA, FM NA, shortwave NA
radios: NA
Television:
broadcast stations: NA
televisions: NA

Defense Forces

Branches: National Guard, Republic Security Forces (internal and border troops), Joint Command Turkmenistan/Russia (Ground, Air, and Air Defense)
Manpower availability: males age 15-49 993,321; males fit for military service 810,392; males reach military age (18) annually 40,430 (1995 est.)
Defense expenditures: $NA, NA% of GDP

Turks and Caicos Islands

(dependent territory of the UK)

Geography

Location: Caribbean, two island groups in the North Atlantic Ocean, southeast of The Bahamas
Map references: Central America and the Caribbean
Area:
total area: 430 sq km
land area: 430 sq km
comparative area: slightly less than 2.5 times the size of Washington, DC
Land boundaries: 0 km
Coastline: 389 km
Maritime claims:
exclusive fishing zone: 200 nm
territorial sea: 12 nm
International disputes: none
Climate: tropical; marine; moderated by trade winds; sunny and relatively dry
Terrain: low, flat limestone; extensive marshes and mangrove swamps
Natural resources: spiny lobster, conch
Land use:
arable land: 2%
permanent crops: 0%
meadows and pastures: 0%
forest and woodland: 0%
other: 98%
Irrigated land: NA sq km
Environment:
current issues: limited natural fresh water resources, private cisterns collect rainwater
natural hazards: frequent hurricanes
international agreements: NA
Note: 30 islands (eight inhabited)

People

Population: 13,941 (July 1995 est.)
Age structure:
0-14 years: NA
15-64 years: NA
65 years and over: NA
Population growth rate: 2.41% (1995 est.)
Birth rate: 13.46 births/1,000 population (1995 est.)

Death rate: 5.16 deaths/1,000 population (1995 est.)
Net migration rate: 15.83 migrant(s)/1,000 population (1995 est.)
Infant mortality rate: 12.63 deaths/1,000 live births (1995 est.)
Life expectancy at birth:
total population: 75.37 years
male: 73.44 years
female: 77.04 years (1995 est.)
Total fertility rate: 2.3 children born/woman (1995 est.)
Nationality:
noun: none
adjective: none
Ethnic divisions: African
Religions: Baptist 41.2%, Methodist 18.9%, Anglican 18.3%, Seventh-Day Adventist 1.7%, other 19.9% (1980)
Languages: English (official)
Literacy: age 15 and over has ever attended school (1970)
total population: 98%
male: 99%
female: 98%
Labor force: NA
by occupation: majority engaged in fishing and tourist industries; some subsistence agriculture

Government

Names:
conventional long form: none
conventional short form: Turks and Caicos Islands
Digraph: TK
Type: dependent territory of the UK
Capital: Grand Turk
Administrative divisions: none (dependent territory of the UK)
Independence: none (dependent territory of the UK)
National holiday: Constitution Day, 30 August (1976)
Constitution: introduced 30 August 1976, suspended in 1986, restored and revised 5 March 1988
Legal system: based on laws of England and Wales with a small number adopted from Jamaica and The Bahamas
Suffrage: 18 years of age; universal
Executive branch:
chief of state: Queen ELIZABETH II (since 6 February 1953), represented by Governor Martin BOURKE (since NA February 1993)
head of government: Chief Minister Derek H. TAYLOR (since 31 January 1995)
cabinet: Executive Council; consists of three ex-officio members and five appointed by the governor from the Legislative Council
Legislative branch: unicameral
Legislative Council: elections last held 31 January 1995 (next to be held by NA 2000); results—percent of vote by party NA; seats—(20 total, 13 elected) PDM 8, PNP 4, independent (Norman SAUNDERS) 1
Judicial branch: Supreme Court
Political parties and leaders: Progressive National Party (PNP), Washington MISSICK; People's Democratic Movement (PDM), Derek H. TAYLOR; National Democratic Alliance (NDA), Ariel MISSICK
Member of: CARICOM (associate), CDB, INTERPOL (subbureau)
Diplomatic representation in US: none (dependent territory of the UK)
US diplomatic representation: none (dependent territory of the UK)
Flag: blue with the flag of the UK in the upper hoist-side quadrant and the colonial shield centered on the outer half of the flag; the shield is yellow and contains a conch shell, lobster, and cactus

Economy

Overview: The economy is based on fishing, tourism, and offshore banking. Only subsistence farming—corn, cassava, citrus, and beans—exists on the Caicos Islands, so that most foods, as well as nonfood products, must be imported.
National product: GDP—purchasing power parity—$80.8 million (1992 est.)
National product real growth rate: -1.5% (1992)
National product per capita: $6,000 (1992 est.)
Inflation rate (consumer prices): NA%
Unemployment rate: 12% (1992)
Budget:
revenues: $20.3 million
expenditures: $44 million, including capital expenditures of $23.9 million (1989 est.)
Exports: $6.8 million (f.o.b., 1993)
commodities: lobster, dried and fresh conch, conch shells
partners: US, UK
Imports: $42.8 million (1993)
commodities: food and beverages, tobacco, clothing, manufactures, construction materials
partners: US, UK
External debt: $NA
Industrial production: growth rate NA%
Electricity:
capacity: 9,050 kW
production: 11.1 million kWh
consumption per capita: 860 kWh (1992)
Industries: fishing, tourism, offshore financial services
Agriculture: subsistence farming prevails, based on corn and beans; fishing more important than farming; not self-sufficient in food
Illicit drugs: transshipment point for South American narcotics destined for the US
Economic aid:
recipient: Western (non-US) countries, ODA and OOF bilateral commitments (1970-89), $110 million
Currency: 1 United States dollar (US$) = 100 cents
Exchange rates: US currency is used
Fiscal year: calendar year

Transportation

Railroads: 0 km
Highways:
total: 121 km (including 24 km tarmac)
paved: NA
unpaved: NA
Ports: Cockburn Harbour, Grand Turk, Providenciales, Salt Cay
Merchant marine: none
Airports:
total: 7
with paved runways 1,524 to 2,437 m: 3
with paved runways 914 to 1,523 m: 1
with paved runways under 914 m: 1
with unpaved runways 914 to 1,523 m: 2

Communications

Telephone system: 1,446 telephones; fair cable and radio services
local: NA
intercity: NA
international: 2 submarine cables; 1 INTELSAT (Atlantic Ocean) earth station
Radio:
broadcast stations: AM 3, FM 0, shortwave 0
radios: NA
Television:
broadcast stations: NA
televisions: NA

Defense Forces

Note: defense is the responsibility of the UK

Tuvalu

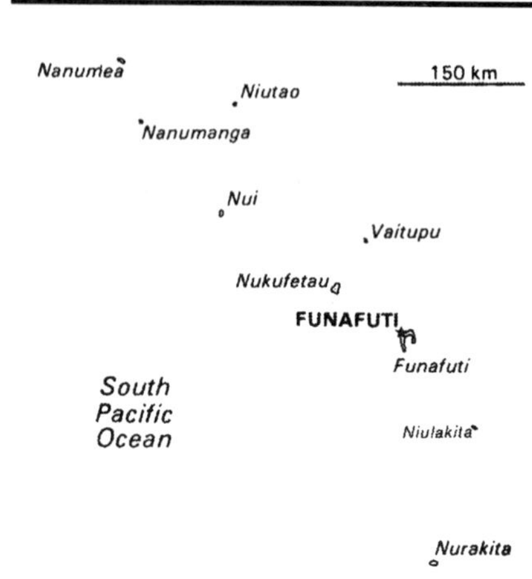

Geography

Location: Oceania, island group consisting of nine coral atolls in the South Pacific Ocean, about one-half of the way from Hawaii to Australia
Map references: Oceania
Area:
total area: 26 sq km
land area: 26 sq km
comparative area: about 0.1 times the size of Washington, DC
Land boundaries: 0 km
Coastline: 24 km
Maritime claims:
contiguous zone: 24 nm
exclusive economic zone: 200 nm
territorial sea: 12 nm
International disputes: none
Climate: tropical; moderated by easterly trade winds (March to November); westerly gales and heavy rain (November to March)
Terrain: very low-lying and narrow coral atolls
Natural resources: fish
Land use:
arable land: 0%
permanent crops: 0%
meadows and pastures: 0%
forest and woodland: 0%
other: 100%
note: Tuvalu's nine coral atolls have enough soil to grow coconuts and support subsistence agriculture
Irrigated land: NA sq km
Environment:
current issues: since there are no streams or rivers and groundwater is not potable, all water needs must be met by catchment systems with storage facilities; beachhead erosion because of the use of sand for building materials; excessive clearance of forest undergrowth for use as fuel; damage to coral reefs from the spread of the crown of thorns starfish
natural hazards: severe tropical storms are rare
international agreements: party to—Climate Change, Endangered Species, Marine Dumping, Ozone Layer Protection, Ship Pollution; signed, but not ratified—Biodiversity, Law of the Sea

People

Population: 9,991 (July 1995 est.)
Age structure:
0-14 years: 36% (female 1,787; male 1,852)
15-64 years: 59% (female 3,105; male 2,764)
65 years and over: 5% (female 258; male 225) (July 1995 est.)
Population growth rate: 1.58% (1995 est.)
Birth rate: 24.82 births/1,000 population (1995 est.)
Death rate: 9.01 deaths/1,000 population (1995 est.)
Net migration rate: 0 migrant(s)/1,000 population (1995 est.)
Infant mortality rate: 27.9 deaths/1,000 live births (1995 est.)
Life expectancy at birth:
total population: 63.15 years
male: 61.87 years
female: 64.34 years (1995 est.)
Total fertility rate: 3.11 children born/woman (1995 est.)
Nationality:
noun: Tuvaluans(s)
adjective: Tuvaluan
Ethnic divisions: Polynesian 96%
Religions: Church of Tuvalu (Congregationalist) 97%, Seventh-Day Adventist 1.4%, Baha'i 1%, other 0.6%
Languages: Tuvaluan, English
Literacy: NA%
Labor force: NA
by occupation: NA

Government

Names:
conventional long form: none
conventional short form: Tuvalu
former: Ellice Islands
Digraph: TV
Type: democracy; began debating republic status in 1992
Capital: Funafuti
Administrative divisions: none
Independence: 1 October 1978 (from UK)
National holiday: Independence Day, 1 October (1978)
Constitution: 1 October 1978
Legal system: NA
Suffrage: 18 years of age; universal
Executive branch:
chief of state: Queen ELIZABETH II (since 6 February 1952), represented by Governor General Tulaga MANUELLA (since NA June 1994)
head of government: Prime Minister Kamuta LATASI (since 10 December 1993); Deputy Prime Minister Otinielu TAUSI (since 10 December 1993)
cabinet: Cabinet; appointed by the governor general on recommendation of the prime minister
Legislative branch: unicameral
Parliament (Palamene): elections last held 25 November 1993 (next to be held by NA 1997); results—percent of vote NA; seats—(12 total)
Judicial branch: High Court
Political parties and leaders: none
Member of: ACP, AsDB, C (special), ESCAP, IFRCS (associate), INTELSAT (nonsignatory user), SPARTECA, SPC, SPF, UNESCO, UPU, WHO
Diplomatic representation in US: Tuvalu has no mission in the US
US diplomatic representation: none
Flag: light blue with the flag of the UK in the upper hoist-side quadrant; the outer half of the flag represents a map of the country with nine yellow five-pointed stars symbolizing the nine islands

Economy

Overview: Tuvalu consists of a scattered group of nine coral atolls with poor soil. The country has no known mineral resources and few exports. Subsistence farming and fishing are the primary economic activities. The islands are too small and too remote for development of a tourist industry. Government revenues largely come from the sale of stamps and coins and worker remittances. Substantial income is received annually from an international trust fund established in 1987 by Australia, NZ, and the UK and supported also by Japan and South Korea.
National product: GDP—purchasing power parity—$7.8 million (1993 est.)
National product real growth rate: NA%
National product per capita: $800 (1993 est.)
Inflation rate (consumer prices): 2.9% (1989)
Unemployment rate: NA%
Budget:
revenues: $4.3 million
expenditures: $4.3 million, including capital expenditures of $NA (1989 est.)
Exports: $165,000 (f.o.b., 1989)
commodities: copra
partners: Fiji, Australia, NZ
Imports: $4.4 million (c.i.f., 1989)
commodities: food, animals, mineral fuels, machinery, manufactured goods
partners: Fiji, Australia, NZ
External debt: $NA
Industrial production: growth rate NA%
Electricity:
capacity: 2,600 kW
production: 3 million kWh
consumption per capita: 330 kWh (1990)
Industries: fishing, tourism, copra

Agriculture: coconuts and fish
Economic aid:
recipient: US commitments, including Ex-Im (FY70-87), $1 million; Western (non-US) countries, ODA and OOF bilateral commitments (1970-89), $101 million
Currency: 1 Tuvaluan dollar ($T) or 1 Australian dollar ($A) = 100 cents
Exchange rates: Tuvaluan dollars ($T) or Australian dollars ($A) per US$1—1.3058 (January 1995), 1.3667 (1994), 1.4704 (1993), 1.3600 (1992), 1.2835 (1991), 1.2799 (1990)
Fiscal year: NA

Transportation

Railroads: 0 km
Highways:
total: 8 km
unpaved: gravel 8 km
Ports: Funafuti, Nukufetau
Merchant marine:
total: 8 ships (1,000 GRT or over) totaling 44,473 GRT/73,652 DWT
ships by type: cargo 1, chemical tanker 4, oil tanker 1, passenger-cargo 1, refrigerated cargo 1
Airports:
total: 1
with unpaved runways 1,524 to 2,438 m: 1

Communications

Telephone system: 108 telephones; 300 radiotelephones
local: NA
intercity: NA
international: NA
Radio:
broadcast stations: AM 1, FM 0, shortwave 0
radios: 4,000
Television:
broadcast stations: 0
televisions: NA

Defense Forces

Branches: no military forces, Police Force
Defense expenditures: $NA, NA% of GDP

Uganda

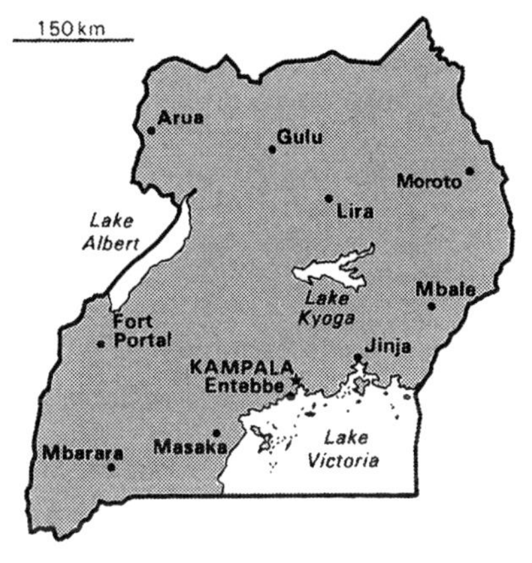

Geography

Location: Eastern Africa, west of Kenya
Map references: Africa
Area:
total area: 236,040 sq km
land area: 199,710 sq km
comparative area: slightly smaller than Oregon
Land boundaries: total 2,698 km, Kenya 933 km, Rwanda 169 km, Sudan 435 km, Tanzania 396 km, Zaire 765 km
Coastline: 0 km (landlocked)
Maritime claims: none; landlocked
International disputes: none
Climate: tropical; generally rainy with two dry seasons (December to February, June to August); semiarid in northeast
Terrain: mostly plateau with rim of mountains
Natural resources: copper, cobalt, limestone, salt
Land use:
arable land: 23%
permanent crops: 9%
meadows and pastures: 25%
forest and woodland: 30%
other: 13%
Irrigated land: 90 sq km (1989 est.)
Environment:
current issues: draining of wetlands for agricultural use; deforestation; overgrazing; soil erosion; poaching is widespread
natural hazards: NA
international agreements: party to—Biodiversity, Climate Change, Endangered Species, Law of the Sea, Marine Life Conservation, Nuclear Test Ban, Ozone Layer Protection, Wetlands; signed, but not ratified—Environmental Modification
Note: landlocked

People

Population: 19,573,262 (July 1995 est.)
Age structure:
0-14 years: 49% (female 4,792,164; male 4,834,757)
15-64 years: 49% (female 4,802,650; male 4,704,159)
65 years and over: 2% (female 215,648; male 223,884) (July 1995 est.)
Population growth rate: 2.25% (1995 est.)
Birth rate: 48.03 births/1,000 population (1995 est.)
Death rate: 24.35 deaths/1,000 population (1995 est.)
Net migration rate: NA migrant(s)/1,000 population (1995 est.)
note: Uganda is host to refugees from a number of neighboring countries, including Zaire, Sudan, and Rwanda; probably in excess of 100,000 southern Sudanese fled to Uganda during the past year; many of the 8,000 Rwandans who took refuge in Uganda have returned home
Infant mortality rate: 112.2 deaths/1,000 live births (1995 est.)
Life expectancy at birth:
total population: 36.58 years
male: 36.26 years
female: 36.91 years (1995 est.)
Total fertility rate: 6.7 children born/woman (1995 est.)
Nationality:
noun: Ugandan(s)
adjective: Ugandan
Ethnic divisions: Baganda 17%, Karamojong 12%, Basogo 8%, Iteso 8%, Langi 6%, Rwanda 6%, Bagisu 5%, Acholi 4%, Lugbara 4%, Bunyoro 3%, Batobo 3%, European, Asian, Arab 1%, other 23%
Religions: Roman Catholic 33%, Protestant 33%, Muslim 16%, indigenous beliefs 18%
Languages: English (official), Luganda, Swahili, Bantu languages, Nilotic languages
Literacy: age 15 and over can read and write (1991)
total population: 56%
male: 68%
female: 45%
Labor force: 4.5 million (est.)
by occupation: agriculture over 80%

Government

Names:
conventional long form: Republic of Uganda
conventional short form: Uganda
Digraph: UG
Type: republic
Capital: Kampala
Administrative divisions: 39 districts; Apac, Arua, Bundibugyo, Bushenyi, Gulu, Hoima, Iganga, Jinja, Kabale, Kabarole, Kalangala, Kampala, Kamuli, Kapchorwa, Kasese, Kibale, Kiboga, Kisoro, Kitgum, Kotido, Kumi, Lira, Luwero, Masaka, Masindi, Mbale, Mbarara, Moroto, Moyo, Mpigi, Mubende, Mukono, Nebbi, Ntungamo, Pallisa, Rakai, Rukungiri, Sototi, Tororo
Independence: 9 October 1962 (from UK)

National holiday: Independence Day, 9 October (1962)
Constitution: 8 September 1967, in process of constitutional revision
Legal system: government plans to restore system based on English common law and customary law and reinstitute a normal judicial system; accepts compulsory ICJ jurisdiction, with reservations
Suffrage: 18 years of age; universal
Executive branch:
chief of state: President Lt. Gen. Yoweri Kaguta MUSEVENI (since 29 January 1986); Vice President Dr. Specioza Wandira KAZIBWE (since 18 November 1994)
head of government: Prime Minister Kintu MUSOKE (since 18 November 1994)
cabinet: Cabinet; appointed by the president
Legislative branch: unicameral
Constituent Assembly: elections last held 28 March 1994 (next to be held end of 1995); results—284 non-partisan delegates elected to an interim Constituent Assembly with the principal task of writing a final draft of a new constitution for Uganda on the basis of which a regular Constituent Assembly will be elected
note: first free and fair election in 30 years is to be held by end of 1995
Judicial branch: Court of Appeal, High Court
Political parties and leaders: only party—National Resistance Movement (NRM), Yoweri MUSEVENI
note: Ugandan People's Congress (UPC), Milton OBOTE; Democratic Party (DP), Paul SSEMOGEERE; and Conservative Party (CP), Joshua S. MAYANJA-NKANGI continue to exist but are all proscribed from conducting public political activities
Other political or pressure groups: Lord's Resistance Army (LRA); Ruwenzori Movement
Member of: ACP, AfDB, C, CCC, EADB, ECA, FAO, G-77, GATT, IAEA, IBRD, ICAO, ICFTU, ICRM, IDA, IDB, IFAD, IFC, IFRCS, IGADD, ILO, IMF, INTELSAT, INTERPOL, IOC, IOM, ISO (correspondent), ITU, NAM, OAU, OIC, PCA, UN, UNCTAD, UNESCO, UNHCR, UNIDO, UNITAR, UPU, WFTU, WHO, WIPO, WMO, WTO
Diplomatic representation in US:
chief of mission: Ambassador Stephen Kapimpina KATENTA-APULI
chancery: 5911 16th Street NW, Washington, DC 20011
telephone: [1] (202) 726-7100 through 7102, 0416
FAX: [1] (202) 726-1727
US diplomatic representation:
chief of mission: Ambassador E. Michael SOUTHWICK
embassy: Parliament Avenue, Kampala
mailing address: P. O. Box 7007, Kampala
telephone: [256] (41) 259792, 259793, 259795
FAX: [256] (41) 259794
Flag: six equal horizontal bands of black (top), yellow, red, black, yellow, and red; a white disk is superimposed at the center and depicts a red-crested crane (the national symbol) facing the staff side

Economy

Overview: Uganda has substantial natural resources, including fertile soils, regular rainfall, and sizable mineral deposits of copper and cobalt. Agriculture is the most important sector of the economy, employing over 80% of the work force. Coffee is the major export crop and accounts for the bulk of export revenues. Since 1986 the government—with the support of foreign countries and international agencies—has acted to rehabilitate and stabilize the economy by undertaking currency reform, raising producer prices on export crops, increasing prices of petroleum products, and improving civil service wages. The policy changes are especially aimed at dampening inflation and boosting production and export earnings. In 1990-94, the economy turned in a solid performance based on continued investment in the rehabilitation of infrastructure, improved incentives for production and exports, and gradually improving domestic security. The economy again prospered in 1994 with rapid growth, low inflation, growing foreign investment, a trimmed bureaucracy, and the continued return of exiled Indian-Ugandan entrepreneurs.
National product: GDP—purchasing power parity—$16.2 billion (1994 est.)
National product real growth rate: 6% (1994 est.)
National product per capita: $850 (1994 est.)
Inflation rate (consumer prices): 5% (1994 est.)
Unemployment rate: NA%
Budget:
revenues: $365 million
expenditures: $545 million, including capital expenditures of $165 million (1989 est.)
Exports: $237 million (f.o.b., 1993 est.)
commodities: coffee 97%, cotton, tea
partners: US 25%, UK 18%, France 11%, Spain 10%
Imports: $696 million (c.i.f., 1993 est.)
commodities: petroleum products, machinery, cotton piece goods, metals, transportation equipment, food
partners: Kenya 25%, UK 14%, Italy 13%
External debt: $2.9 billion (1993 est.)
Industrial production: growth rate 1.5% (1992); accounts for 5% of GDP
Electricity:
capacity: 160,000 kW
production: 780 million kWh
consumption per capita: 32 kWh (1993)
Industries: sugar, brewing, tobacco, cotton textiles, cement
Agriculture: mainly subsistence; accounts for 57% of GDP and over 80% of labor force; cash crops—coffee, tea, cotton, tobacco; food crops—cassava, potatoes, corn, millet, pulses; livestock products—beef, goat meat, milk, poultry; self-sufficient in food
Economic aid:
recipient: US commitments, including Ex-Im (1970-89), $145 million; Western (non-US) countries, ODA and OOF bilateral commitments (1970-89), $1.4 billion; OPEC bilateral aid (1979-89), $60 million; Communist countries (1970-89), $169 million
Currency: 1 Ugandan shilling (USh) = 100 cents
Exchange rates: Ugandan shillings (USh) per US$1—1,195 (December 1994), 1,195.0 (1993), 1.133.8 (1992), 734.0 (1991), 428.85 (1990), 223.1 (1989)
Fiscal year: 1 July—30 June

Transportation

Railroads:
total: 1,300 km single track
narrow gauge: 1,300 km 1.000-m-gauge
Highways:
total: 26,200 km
paved: 1,970 km
unpaved: gravel, crushed stone 5,849 km; earth, tracks 18,381 km
Inland waterways: Lake Victoria, Lake Albert, Lake Kyoga, Lake George, Lake Edward; Victoria Nile, Albert Nile; principal inland water ports are at Jinja and Port Bell, both on Lake Victoria
Ports: Entebbe, Jinja, Port Bell
Merchant marine:
total: 3 roll-on/roll-off cargo ships (1,000 GRT or over) totaling 5,091 GRT/NA DWT
Airports:
total: 29
with paved runways over 3,047 m: 3
with paved runways 1,524 to 2,437 m: 1
with paved runways under 914 m: 9
with unpaved runways 2,438 to 3,047 m: 1
with unpaved runways 1,524 to 2,438 m: 6
with unpaved runways 914 to 1,523 m: 9

Communications

Telephone system: NA telephones; fair system
local: NA
intercity: microwave and radio communications stations
international: 1 INTELSAT (Atlantic Ocean) earth station
Radio:
broadcast stations: AM 10, FM 0, shortwave 0
radios: NA
Television:
broadcast stations: 9
televisions: NA

Defense Forces

Branches: Army, Navy, Air Wing
Manpower availability: males age 15-49 4,231,019; males fit for military service 2,298,654 (1995 est.)
Defense expenditures: exchange rate conversion—$55 million, 1.7% of budget (FY93/94)

Ukraine

Geography

Location: Eastern Europe, bordering the Black Sea, between Poland and Russia
Map references: Commonwealth of Independent States—European States
Area:
total area: 603,700 sq km
land area: 603,700 sq km
comparative area: slightly smaller than Texas
Land boundaries: total 4,558 km, Belarus 891 km, Hungary 103 km, Moldova 939 km, Poland 428 km, Romania (southwest) 169 km, Romania (west) 362 km, Russia 1,576 km, Slovakia 90 km
Coastline: 2,782 km
Maritime claims:
continental shelf: 200-m or to the depth of exploitation
exclusive economic zone: undefined
territorial sea: 12 nm
International disputes: certain territory of Moldova and Ukraine—including Bessarabia and Northern Bukovina—are considered by Bucharest as historically a part of Romania; this territory was incorporated into the former Soviet Union following the Molotov-Ribbentrop Pact in 1940; potential dispute with Russia over Crimea; has made no territorial claim in Antarctica (but has reserved the right to do so) and does not recognize the claims of any other nation
Climate: temperate continental; Mediterranean only on the southern Crimean coast; precipitation disproportionately distributed, highest in west and north, lesser in east and southeast; winters vary from cool along the Black Sea to cold farther inland; summers are warm across the greater part of the country, hot in the south
Terrain: most of Ukraine consists of fertile plains (steppes) and plateaux, mountains being found only in the west (the Carpathians), and in the Crimean Peninsula in the extreme south
Natural resources: iron ore, coal, manganese, natural gas, oil, salt, sulphur, graphite, titanium, magnesium, kaolin, nickel, mercury, timber
Land use:
arable land: 56%
permanent crops: 2%
meadows and pastures: 12%
forest and woodland: 0%
other: 30%
Irrigated land: 26,000 sq km (1990)
Environment:
current issues: inadequate supplies of potable water; air and water pollution; deforestation; radiation contamination in the northeast from 1986 accident at Chornobyl' Nuclear Power Plant
natural hazards: NA
international agreements: party to—Air Pollution, Air Pollution-Nitrogen Oxides, Air Pollution-Sulphur 85, Antarctic Treaty, Environmental Modification, Marine Dumping, Nuclear Test Ban, Ozone Layer Protection, Ship Pollution; signed, but not ratified—Air Pollution-Sulphur 94, Air Pollution-Volatile Organic Compounds, Biodiversity, Climate Change, Law of the Sea
Note: strategic position at the crossroads between Europe and Asia; second largest country in Europe

People

Population: 51,867,828 (July 1995 est.)
Age structure:
0-14 years: 21% (female 5,217,850; male 5,407,450)
15-64 years: 65% (female 17,563,924; male 16,334,299)
65 years and over: 14% (female 4,976,893; male 2,367,412) (July 1995 est.)
Population growth rate: 0.04% (1995 est.)
Birth rate: 12.31 births/1,000 population (1995 est.)
Death rate: 12.67 deaths/1,000 population (1995 est.)
Net migration rate: 0.71 migrant(s)/1,000 population (1995 est.)
Infant mortality rate: 20.5 deaths/1,000 live births (1995 est.)
Life expectancy at birth:
total population: 70.11 years
male: 65.59 years
female: 74.87 years (1995 est.)
Total fertility rate: 1.81 children born/woman (1995 est.)
Nationality:
noun: Ukrainian(s)
adjective: Ukrainian
Ethnic divisions: Ukrainian 73%, Russian 22%, Jewish 1%, other 4%
Religions: Ukrainian Orthodox—Moscow Patriarchate, Ukrainian Orthodox—Kiev Patriarchate, Ukrainian Autocephalous Orthodox, Ukrainian Catholic (Uniate), Protestant, Jewish
Languages: Ukrainian, Russian, Romanian, Polish, Hungarian

Literacy: age 15 and over can read and write (1989)
total population: 98%
male: 100%
female: 97%
Labor force: 23.55 million (January 1994)
by occupation: industry and construction 33%, agriculture and forestry 21%, health, education, and culture 16%, trade and distribution 7%, transport and communication 7%, other 16% (1992)

Government

Names:
conventional long form: none
conventional short form: Ukraine
local long form: none
local short form: Ukrayina
former: Ukrainian Soviet Socialist Republic
Digraph: UP
Type: republic
Capital: Kiev (Kyyiv)
Administrative divisions: 24 oblasti (singular—oblast'), 1 autonomous republic* (avtomnaya respublika), and 2 municipalites (mista, singular—misto) with oblast status**; Cherkas'ka (Cherkasy), Chernihivs'ka (Chernihiv), Chernivets'ka (Chernivtsi), Dnipropetrovs'ka (Dnipropetrovs'k), Donets'ka (Donets'k), Ivano-Frankivs'ka (Ivano-Frankivs'k), Kharkivs'ka (Kharkiv), Khersons'ka (Kherson), Khmel'nyts'ka (Khmel'nyts'kyy), Kirovohrads'ka (Kirovohrad), Kyyiv**, Kyyivs'ka (Kiev), Luhans'ka (Luhans'k), L'vivs'ka (L'viv), Mykolayivs'ka (Mykolayiv), Odes'ka (Odesa), Poltavs'ka (Poltava), Respublika Krym* (Simferopol'), Rivnens'ka (Rivne), Sevastopol'**, Sums'ka (Sevastopol'), Ternopil's'ka (Ternopil'), Vinnyts'ka (Vinnytsya), Volyns'ka (Luts'k), Zakarpats'ka (Uzhhorod), Zaporiz'ka (Zaporizhzhya), Zhytomyrs'ka (Zhytomyr)
note: names in parentheses are administrative centers when name differs from oblast' name
Independence: 1 December 1991 (from Soviet Union)
National holiday: Independence Day, 24 August (1991)
Constitution: using 1978 pre-independence constitution; new constitution currently being drafted
Legal system: based on civil law system; no judicial review of legislative acts
Suffrage: 18 years of age; universal
Executive branch:
chief of state: President Leonid D. KUCHMA (since 19 July 1994); election last held 26 June and 10 July 1994 (next to be held NA 1999); results—Leonid KUCHMA 52.15%, Leonid KRAVCHUK 45.06%
head of government: Acting Prime Minister Yeuben MARCHUK (since 3 March 1995); First Deputy Prime Ministers Yevhen MARCHUK and Viktor PYNZENYK (since 31 October 1994) and six deputy prime ministers
cabinet: Council of Ministers; appointed by the president and approved by the Supreme Council
National Security Council: originally created in 1992, but signficantly revamped and strengthened under President KUCHMA; members include the president, prime minister, Ministers of Finance, Environment, Justice, Internal Affairs, Foreign Economic Relations, Economic and Foreign Affairs; the NSC staff is tasked with developing national security policy on domestic and international matters and advising the president
Presidential Administration: helps draft presidential edicts and provides policy support to the president
Council of Regions: advisory body created by President KUCHMA in September 1994; includes the Chairmen of Oblast and Kiev and Sevastopol City Supreme Councils
Legislative branch: unicameral
Supreme Council: elections last held 27 March 1994 with repeat elections continuing through December 1998 to fill empty seats (next to be held NA); results—percent of vote by party NA; seats—(450 total) Communists 91, Rukh 22, Agrarians 18, Socialists 15, Republicans 11, Congress of Ukrainian Nationalists 5, Labor 5, Party of Democratic Revival 4, Democrats 2, Social Democrats 2, Civil Congress 2, Conservative Republicans 1, Party of Economic Revival of Crimea 1, Christian Democrats 1, independents 225; note—405 deputies have been elected; run-off elections for the remaining 45 seats to be held by December 1998
Judicial branch: joint commission formed in April 1995 to define a program of judicial reform by year-end
Political parties and leaders: Green Party of Ukraine, Vitaliy KONONOV, leader; Liberal Party of Ukraine; Liberal Democratic Party of Ukraine, Volodymyr KLYMCHUK, chairman; Democratic Party of Ukraine, Volodymyr Oleksandrovych YAVORIVSKIY, chairman; People's Party of Ukraine, Leopol'd TABURYANSKYY, chairman; Peasants' Party of Ukraine, Serhiy DOVHRAN', chairman; Party of Democratic Rebirth (Revival) of Ukraine, Volodymyr FILENKO, chairman; Social Democratic Party of Ukraine, Yuriy VUZDUHAN, chairman; Socialist Party of Ukraine, Oleksandr MOROZ, chairman; Ukrainian Christian Democratic Party, Vitaliy ZHURAVSKYY, chairman; Ukrainian Conservative Republican Party, Stepan KHMARA, chairman; Ukrainian Labor Party, Valentyn LANDYK, chairman; Ukrainian Party of Justice, Mykhaylo HRECHKO, chairman; Ukrainian Peasants' Democratic Party, Serhiy PLACHINDA, chairman; Ukrainian Republican Party, Mykhaylo HORYN', chairman; Ukrainian National Conservative Party, Viktor RADIONOV, chairman; Ukrainian People's Movement for Restructuring (Rukh), Vyacheslav CHORNOVIL, chairman; Ukrainian Communist Party, Petr SYMONENKO; Agrarian Party; Congress of Ukrainian Nationalists, S. STESTKO; Civil Congress, O. BAZYLUK; Party of Economic Revival of Crimea; Democratic Party Of Ukraine, Serhiy DOVMAN', chairman
Other political or pressure groups: New Ukraine (Nova Ukrayina); Congress of National Democratic Forces
Member of: BSEC, CCC, CE (guest), CEI (associate members), CIS, EBRD, ECE, IAEA, IBRD, ICAO, ICRM, IFC, ILO, IMF, IMO, INMARSAT, INTELSAT (nonsignatory user), INTERPOL, IOC, IOM (observer), ISO, ITU, NACC, OSCE, PCA, PFP, UN, UNCTAD, UNESCO, UNIDO, UNPROFOR, UPU, WHO, WIPO, WMO
Diplomatic representation in US:
chief of mission: Ambassador Yuriy SHCHERBAK
chancery: 3350 M Street NW, Washington, DC 20007
telephone: [1] (202) 333-0606
FAX: [1] (202) 333-0817
consulate(s) general: Chicago and New York
US diplomatic representation:
chief of mission: Ambassador William Green MILLER
embassy: 10 Yuria Kotsyubinskovo, 252053 Kiev 53
mailing address: use embassy street address
telephone: [7] (044) 244-73-49, 244-37-45
FAX: [7] (044) 244-73-50
Flag: two equal horizontal bands of azure (top) and golden yellow represent grainfields under a blue sky

Economy

Overview: After Russia, the Ukrainian republic was far and away the most important economic component of the former Soviet Union, producing more than three times the output of the next-ranking republic. Its fertile black soil generated more than one-fourth of Soviet agricultural output, and its farms provided substantial quantities of meat, milk, grain, and vegetables to other republics. Likewise, its diversified heavy industry supplied equipment and raw materials to industrial and mining sites in other regions of the former USSR. In early 1992, the Ukrainian government liberalized most prices and erected a legal framework for privatization, but widespread resistance to reform within the government and the legislature soon stalled reform efforts and led to some backtracking. Loose monetary and fiscal policies pushed inflation to hyperinflationary levels in late 1993. Greater monetary and fiscal restraint lowered inflation in 1994, but also contributed

to an accelerated decline in industrial output. Since his election in July 1994, President KUCHMA has developed—and parliament has approved—a comprehensive economic reform program, maintained financial discipline, and reduced state controls over prices, the exchange rate, and foreign trade. Implementation of KUCHMA's economic agenda will encounter considerable resistance from parliament, entrenched bureaucrats, and industrial interests and will contribute to further declines in output and rising unemployment which will sorely test the government's ability to stay the course on reform in 1995.
National product: GDP—purchasing power parity—$189.2 billion (1994 estimate as extrapolated from World Bank estimate for 1992)
National product real growth rate: -19% (1994 est.)
National product per capita: $3,650 (1994 est.)
Inflation rate (consumer prices): 14% per month (1994)
Unemployment rate: 0.4% officially registered; large number of unregistered or underemployed workers
Budget:
revenues: $NA
expenditures: $NA, including capital expenditures of $NA
Exports: $11.8 billion (1994)
commodities: coal, electric power, ferrous and nonferrous metals, chemicals, machinery and transport equipment, grain, meat
partners: FSU countries, China, Italy, Switzerland
Imports: $14.2 billion (1994)
commodities: energy, machinery and parts, transportation equipment, chemicals, textiles
partners: FSU countries, Germany, Poland, Czech Republic
External debt: $7.5 billion (yearend 1994)
Industrial production: growth rate -28% (1994 est.); accounts for 50% of GDP
Electricity:
capacity: 54,380,000 kW
production: 182 billion kWh
consumption per capita: 3,200 kWh (1994)
Industries: coal, electric power, ferrous and nonferrous metals, machinery and transport equipment, chemicals, food-processing (especially sugar)
Agriculture: accounts for about 25% of GDP; grain, vegetables, meat, milk, sugar beets
Illicit drugs: illicit cultivator of cannabis and opium poppy; mostly for CIS consumption; limited government eradication program; used as transshipment point for illicit drugs to Western Europe
Economic aid: $550 million economic aid and $350 million to help disassemble the atomic weapons from the US in 1994
Currency: Ukraine withdrew the Russian ruble from circulation on 12 November 1992 and declared the karbovanets (plural karbovantsi) sole legal tender in Ukrainian markets; Ukrainian officials claim this is an interim move toward introducing a new currency—the hryvnya—possibly in mid-1995
Exchange rates: karbovantsi per 1$US—107,900 (end December 1994), 130,000 (April 1994)
Fiscal year: calendar year

Transportation

Railroads:
total: 23,350 km
broad gauge: 23,350 km 1.524-m gauge (8,600 km electrified)
Highways:
total: 273,700 km
paved and graveled: 236,400 km
unpaved: earth 37,300 km
Inland waterways: 1,672 km perennially navigable (Pryp''yat' and Dnipro Rivers)
Pipelines: crude oil 2,010 km; petroleum products 1,920 km; natural gas 7,800 km (1992)
Ports: Berdyans'k, Illichivs'k, Izmayil, Kerch, Kherson, Kiev (Kyyiv), Mariupol', Mykolayiv, Odesa, Pivdenne, Reni
Merchant marine:
total: 379 ships (1,000 GRT or over) totaling 3,799,253 GRT/5,071,175 DWT
ships by type: barge carrier 7, bulk 55, cargo 221, chemical tanker 2, container 20, multifunction large-load carrier 1, oil tanker 10, passenger 12, passenger-cargo 5, railcar carrier 2, refrigerated cargo 5, roll-on/roll-off cargo 32, short-sea passenger 7
Airports:
total: 706
with paved runways over 3,047 m: 14
with paved runways 2,438 to 3,047 m: 55
with paved runways 1,524 to 2,437 m: 34
with paved runways 914 to 1,523 m: 3
with paved runways under 914 m: 57
with unpaved runways over 3,047 m: 7
with unpaved runways 2,438 to 3,047 m: 7
with unpaved runways 1,524 to 2,438 m: 16
with unpaved runways 914 to 1,523 m: 37
with unpaved runways under 914 m: 476

Communications

Telephone system: 7,886,000 telephone circuits; about 151.4 telephone circuits/1,000 persons (1991); the telephone system is inadequate both for business and for personal use; 3.56 million applications for telephones had not been satisfied as of January 1991; electronic mail services have been established in Kiev, Odesa, and Luhans'k by Sprint
local: an NMT-450 analog cellular telephone network operates in Kiev (Kyyiv) and allows direct dialing of international calls through Kiev's EWSD digital exchange
intercity: NA
international: calls to other CIS countries are carried by land line or microwave; other international calls to 167 countries are carried by satellite or by the 150 leased lines through the Moscow gateway switch; INTELSAT, INMARSAT, and Intersputnik earth stations
Radio:
broadcast stations: AM NA, FM NA, shortwave NA
radios: 15 million
Television:
broadcast stations: NA
televisions: 20 million

Defense Forces

Branches: Army, Navy, Air and Air Defense Forces, Republic Security Forces (internal and border troops), National Guard
Manpower availability: males age 15-49 12,324,832; males fit for military service 9,667,642; males reach military age (18) annually 359,546 (1995 est.)
Defense expenditures: 544.3 billion karbovantsi, less than 4% of GDP (forecast for 1993); note—conversion of defense expenditures into US dollars using the current exchange rate could produce misleading results

United Arab Emirates

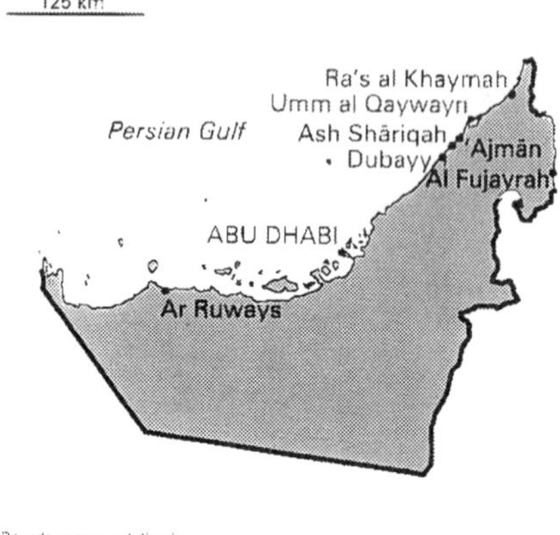

Geography

Location: Middle East, bordering the Gulf of Oman and the Persian Gulf, between Oman and Saudi Arabia
Map references: Middle East
Area:
total area: 75,581 sq km
land area: 75,581 sq km
comparative area: slightly smaller than Maine
Land boundaries: total 867 km, Oman 410 km, Saudi Arabia 457 km
Coastline: 1,318 km
Maritime claims:
contiguous zone: 24 nm
continental shelf: 200 nm or to the edge of the continental margin
exclusive economic zone: 200 nm
territorial sea: 12 nm
International disputes: location and status of boundary with Saudi Arabia is not final; no defined boundary with most of Oman, but Administrative Line in far north; claims two islands in the Persian Gulf occupied by Iran (Jazireh-ye Tonb-e Bozorg or Greater Tunb, and Jazireh-ye Tonb-e Kuchek or Lesser Tunb); claims island in the Persian Gulf jointly administered with Iran (Jazireh-ye Abu Musa or Abu Musa); in 1992, the dispute over Abu Musa and the Tunb islands became more acute when Iran unilaterally tried to control the entry of third country nationals into the UAE portion of Abu Musa island, Tehran subsequently backed off in the face of significant diplomatic support for the UAE in the region
Climate: desert; cooler in eastern mountains
Terrain: flat, barren coastal plain merging into rolling sand dunes of vast desert wasteland; mountains in east
Natural resources: petroleum, natural gas
Land use:
arable land: 0%
permanent crops: 0%
meadows and pastures: 2%
forest and woodland: 0%
other: 98%
Irrigated land: 50 sq km (1989 est.)
Environment:
current issues: lack of natural freshwater resources being overcome by desalination plants; desertification; beach pollution from oil spills
natural hazards: frequent sand and dust storms
international agreements: party to—Endangered Species, Hazardous Wastes, Marine Dumping, Ozone Layer Protection; signed, but not ratified—Biodiversity, Law of the Sea
Note: strategic location along southern approaches to Strait of Hormuz, a vital transit point for world crude oil

People

Population: 2,924,594 (July 1995 est.)
Age structure:
0-14 years: 35% (female 499,559; male 521,415)
15-64 years: 64% (female 643,819; male 1,229,730)
65 years and over: 1% (female 10,296; male 19,775) (July 1995 est.)
Population growth rate: 4.55% (1995 est.)
Birth rate: 27.02 births/1,000 population (1995 est.)
Death rate: 3.03 deaths/1,000 population (1995 est.)
Net migration rate: 21.53 migrant(s)/1,000 population (1995 est.)
Infant mortality rate: 21 deaths/1,000 live births (1995 est.)
Life expectancy at birth:
total population: 72.51 years
male: 70.42 years
female: 74.71 years (1995 est.)
Total fertility rate: 4.53 children born/woman (1995 est.)
Nationality:
noun: Emirian(s)
adjective: Emirian
Ethnic divisions: Emirian 19%, other Arab 23%, South Asian 50%, other expatriates (includes Westerners and East Asians) 8% (1982)
note: less than 20% are UAE citizens (1982)
Religions: Muslim 96% (Shi'a 16%), Christian, Hindu, and other 4%
Languages: Arabic (official), Persian, English, Hindi, Urdu
Literacy: age 15 and over can read and write but definition of literary not available (1985)
total population: 71%
male: 72%
female: 69%
Labor force: 580,000 (1986 est.)
by occupation: industry and commerce 85%, agriculture 5%, services 5%, government 5%
note: 80% of labor force is foreign (est.)

Government

Names:
conventional long form: United Arab Emirates
conventional short form: none
local long form: Al Imarata al Arabiyah al Muttahidah
local short form: none
former: Trucial States
Abbreviation: UAE
Digraph: TC
Type: federation with specified powers delegated to the UAE central government and other powers reserved to member emirates
Capital: Abu Dhabi
Administrative divisions: 7 emirates (imarat, singular—imarah); Abu Zaby (Abu Dhabi), 'Ajman, Al Fujayrah, Dubai, Ra's al Khaymah, Sharjah, Umm al Qaywayn
Independence: 2 December 1971 (from UK)
National holiday: National Day, 2 December (1971)
Constitution: 2 December 1971 (provisional)
Legal system: secular codes are being introduced by the UAE Government and in several member emirates; Islamic law remains influential
Suffrage: none
Executive branch:
chief of state: President ZAYID bin Sultan Al Nuhayyan (since 2 December 1971), ruler of Abu Dhabi; Vice President Shaykh MAKTUM bin Rashid al-Maktum (since 8 October 1990), ruler of Dubayy
head of government: Prime Minister Shaykh MAKTUM bin Rashid al-Maktum (since 8 October 1990), ruler of Dubayy; Deputy Prime Minister SULTAN bin Zayid Al Nuhayyan (since 20 November 1990)
Supreme Council of Rulers: composed of the seven emirate rulers, the council is the highest constitutional authority in the UAE; establishes general policies and sanctions federal legislation, Abu Dhabi and Dubayy rulers have veto power; council meets four times a year
cabinet: Council of Ministers; appointed by the president
Legislative branch: unicameral Federal National Council (Majlis Watani Itihad); no elections
Judicial branch: Union Supreme Court
Political parties and leaders: none
Other political or pressure groups: NA
Member of: ABEDA, AFESD, AL, AMF, CAEU, CCC, ESCWA, FAO, G-77, GATT, GCC, IAEA, IBRD, ICAO, ICRM, IDA, IDB, IFAD, IFC, IFRCS, ILO, IMF, IMO, INMARSAT, INTELSAT, INTERPOL, IOC, ISO (correspondent), ITU, NAM, OAPEC, OIC, OPEC, UN, UNCTAD, UNESCO, UNIDO, UPU, WHO, WIPO, WMO, WTO
Diplomatic representation in US:
chief of mission: Ambassador Muhammad bin Husayn al-SHAALI
chancery: Suite 600, 3000 K Street NW,

Washington, DC 20007
telephone: [1] (202) 338-6500
US diplomatic representation:
chief of mission: Ambassador William A. RUGH
embassy: Al-Sudan Street, Abu Dhabi
mailing address: P. O. Box 4009, Abu Dhabi; American Embassy Abu Dhabi, Department of State, Washington, DC 20521-6010 (pouch)
telephone: [971] (2) 436691, 436692
FAX: [971] (2) 434771
consulate(s) general: Dubayy (Dubai)
Flag: three equal horizontal bands of green (top), white, and black with a thicker vertical red band on the hoist side

---

Economy

Overview: The UAE has an open economy with one of the world's highest incomes per capita and with a sizable annual trade surplus. Its wealth is based on oil and gas output (about 40% of GDP), and the fortunes of the economy fluctuate with the prices of those commodities. Since 1973, the UAE has undergone a profound transformation from an impoverished region of small desert principalities to a modern state with a high standard of living. At present levels of production, crude oil reserves should last for over 100 years. Although much stronger economically than most Gulf states, the UAE faces similar problems with weak international oil prices and the pressures for cuts in OPEC oil production quotas. The UAE government is encouraging increased privatization within the economy.
National product: GDP—purchasing power parity—$62.7 billion (1994 est.)
National product real growth rate: -0.5% (1994 est.)
National product per capita: $22,480 (1994 est.)
Inflation rate (consumer prices): 5.1% (1994 est.)
Unemployment rate: NEGL% (1988)
Budget:
revenues: $4.3 billion
expenditures: $4.8 billion, including capital expenditures of $NA (1993 est)
Exports: $24 billion (f.o.b., 1994 est.)
commodities: crude oil 66%, natural gas, re-exports, dried fish, dates
partners: Japan 35%, South Korea 5%, Iran 4%, Oman 4%, Singapore 4% (1993)
Imports: $20 billion (f.o.b., 1994)
commodities: manufactured goods, machinery and transport equipment, food
partners: Japan 12%, UK 10%, US 9%, Germany 7%, South Korea 5% (1993)
External debt: $11.6 billion (1994 est.)
Industrial production: growth rate 1.7% (1992 est.); accounts for 50% of GDP, including petroleum
Electricity:
capacity: 4,760,000 kW
production: 16.5 billion kWh
consumption per capita: 5,796 kWh (1993)
Industries: petroleum, fishing, petrochemicals, construction materials, some boat building, handicrafts, pearling
Agriculture: accounts for 2% of GDP and 5% of labor force; cash crop—dates; food products—vegetables, watermelons, poultry, eggs, dairy, fish; only 25% self-sufficient in food
Illicit drugs: growing role as heroin transshipment and money-laundering center
Economic aid:
donor: pledged in bilateral aid to less developed countries (1979-89) $9.1 billion
Currency: 1 Emirian dirham (Dh) = 100 fils
Exchange rates: Emirian dirhams (Dh) per US$1—3.6710 (fixed rate)
Fiscal year: calendar year

Transportation

Railroads: 0 km
Highways:
total: 2,000 km
paved: 1,800 km
unpaved: gravel, graded earth 200 km
Pipelines: crude oil 830 km; natural gas, including natural gas liquids, 870 km
Ports: Ajman, Al Fujayrah, Das Island, Khawr Fakkan, Mina' Jabal' Ali, Mina' Khalid, Mina' Rashid, Mina' Saqr, Mina' Zayid, Umm al Qiwain
Merchant marine:
total: 57 ships (1,000 GRT or over) totaling 1,128,253 GRT/1,938,770 DWT
ships by type: bulk 1, cargo 18, chemical tanker 1, container 10, liquefied gas tanker 1, livestock carrier 1, oil tanker 21, refrigerated cargo 1, roll-on/roll-off cargo 3
Airports:
total: 41
with paved runways over 3,047 m: 9
with paved runways 2,438 to 3,047 m: 3
with paved runways 1,524 to 2,437 m: 2
with paved runways 914 to 1,523 m: 3
with paved runways under 914 m: 12
with unpaved runways 2,438 to 3,047 m: 1
with unpaved runways 1,524 to 2,438 m: 3
with unpaved runways 914 to 1,523 m: 8

Communications

Telephone system: 386,600 telephones; modern system consisting of microwave and coaxial cable; key centers are Abu Dhabi and Dubayy
local: NA
intercity: microwave and coaxial cable
international: 3 INTELSAT (1 Atlantic Ocean and 2 Indian Ocean) and 1 ARABSAT earth station; submarine cables to Qatar, Bahrain, India, and Pakistan; tropospheric scatter to Bahrain; microwave radio relay to Saudi Arabia
Radio:
broadcast stations: AM 8, FM 3, shortwave 0
radios: NA
Television:
broadcast stations: 12
televisions: NA

Defense Forces

Branches: Army, Navy, Air Force, paramilitary (includes Federal Police Force)
Manpower availability: males age 15-49 1,072,261; males fit for military service 583,967; males reach military age (18) annually 19,266 (1995 est.)
Defense expenditures: exchange rate conversion—$1.59 billion, 4.3% of GDP (1994)

United Kingdom

Geography

Location: Western Europe, islands including the northern one-sixth of the island of Ireland between the North Atlantic Ocean and the North Sea, northwest of France
Map references: Europe
Area:
total area: 244,820 sq km
land area: 241,590 sq km
comparative area: slightly smaller than Oregon
note: includes Rockall and Shetland Islands
Land boundaries: total 360 km, Ireland 360 km
Coastline: 12,429 km
Maritime claims:
continental shelf: as defined in continental shelf orders or in accordance with agreed upon boundaries
exclusive fishing zone: 200 nm
territorial sea: 12 nm
International disputes: Northern Ireland question with Ireland; Gibraltar question with Spain; Argentina claims Falkland Islands (Islas Malvinas); Argentina claims South Georgia and the South Sandwich Islands; Mauritius claims island of Diego Garcia in British Indian Ocean Territory; Rockall continental shelf dispute involving Denmark, Iceland, and Ireland (Ireland and the UK have signed a boundary agreement in the Rockall area); territorial claim in Antarctica (British Antarctic Territory)
Climate: temperate; moderated by prevailing southwest winds over the North Atlantic Current; more than half of the days are overcast
Terrain: mostly rugged hills and low mountains; level to rolling plains in east and southeast
Natural resources: coal, petroleum, natural gas, tin, limestone, iron ore, salt, clay, chalk, gypsum, lead, silica
Land use:
arable land: 29%
permanent crops: 0%
meadows and pastures: 48%
forest and woodland: 9%
other: 14%
Irrigated land: 1,570 sq km (1989)
Environment:
current issues: sulfur dioxide emissions from power plants contribute to air pollution; some rivers polluted by agricultural wastes and coastal waters polluted because of large-scale disposal of sewage at sea
natural hazards: NA
international agreements: party to—Air Pollution, Air Pollution-Nitrogen Oxides, Air Pollution-Volatile Organic Compounds, Antarctic Treaty, Biodiversity, Climate Change, Endangered Species, Environmental Modification, Hazardous Wastes, Marine Dumping, Marine Life Conservation, Nuclear Test Ban, Ozone Layer Protection, Ship Pollution, Tropical Timber 83, Wetlands, Whaling; signed, but not ratified—Air Pollution-Sulphur 94, Antarctic-Environmental Protocol, Desertification
Note: lies near vital North Atlantic sea lanes; only 35 km from France and now linked by tunnel under the English Channel; because of heavily indented coastline, no location is more than 125 km from tidal waters

People

Population: 58,295,119 (July 1995 est.)
Age structure:
0-14 years: 19% (female 5,572,189; male 5,843,192)
15-64 years: 65% (female 18,723,583; male 18,935,931)
65 years and over: 16% (female 5,471,383; male 3,748,841) (July 1995 est.)
Population growth rate: 0.27% (1995 est.)
Birth rate: 13.18 births/1,000 population (1995 est.)
Death rate: 10.66 deaths/1,000 population (1995 est.)
Net migration rate: 0.17 migrant(s)/1,000 population (1995 est.)
Infant mortality rate: 7 deaths/1,000 live births (1995 est.)
Life expectancy at birth:
total population: 77 years
male: 74.18 years
female: 79.95 years (1995 est.)
Total fertility rate: 1.82 children born/woman (1995 est.)
Nationality:
noun: Briton(s), British (collective plural)
adjective: British
Ethnic divisions: English 81.5%, Scottish 9.6%, Irish 2.4%, Welsh 1.9%, Ulster 1.8%, West Indian, Indian, Pakistani, and other 2.8%
Religions: Anglican 27 million, Roman Catholic 9 million, Muslim 1 million, Presbyterian 800,000, Methodist 760,000, Sikh 400,000, Hindu 350,000, Jewish 300,000 (1991 est.)
note: the UK does not include a question on religion in its census
Languages: English, Welsh (about 26% of the population of Wales), Scottish form of Gaelic (about 60,000 in Scotland)
Literacy: age 15 and over can read and write (1991 est.)
total population: 99%
Labor force: 28.048 million
by occupation: services 62.8%, manufacturing and construction 25.0%, government 9.1%, energy 1.9%, agriculture 1.2% (June 1992)

Government

Names:
conventional long form: United Kingdom of Great Britain and Northern Ireland
conventional short form: United Kingdom
Abbreviation: UK
Digraph: UK
Type: constitutional monarchy
Capital: London
Administrative divisions: 47 counties, 7 metropolitan counties, 26 districts, 9 regions, and 3 islands areas
England: 39 counties, 7 metropolitan counties*; Avon, Bedford, Berkshire, Buckingham, Cambridge, Cheshire, Cleveland, Cornwall, Cumbria, Derby, Devon, Dorset, Durham, East Sussex, Essex, Gloucester, Greater London*, Greater Manchester*, Hampshire, Hereford and Worcester, Hertford, Humberside, Isle of Wight, Kent, Lancashire, Leicester, Lincoln, Merseyside*, Norfolk, Northampton, Northumberland, North Yorkshire, Nottingham, Oxford, Shropshire, Somerset, South Yorkshire*, Stafford, Suffolk, Surrey, Tyne and Wear*, Warwick, West Midlands*, West Sussex, West Yorkshire*, Wiltshire
Northern Ireland: 26 districts; Antrim, Ards, Armagh, Ballymena, Ballymoney, Banbridge, Belfast, Carrickfergus, Castlereagh, Coleraine, Cookstown, Craigavon, Down, Dungannon, Fermanagh, Larne, Limavady, Lisburn, Londonderry, Magherafelt, Moyle, Newry and Mourne, Newtownabbey, North Down, Omagh, Strabane
Scotland: 9 regions, 3 islands areas*; Borders, Central, Dumfries and Galloway, Fife, Grampian, Highland, Lothian, Orkney*, Shetland*, Strathclyde, Tayside, Western Isles*
Wales: 8 counties; Clwyd, Dyfed, Gwent, Gwynedd, Mid Glamorgan, Powys, South Glamorgan, West Glamorgan
Dependent areas: Anguilla, Bermuda, British Indian Ocean Territory, British Virgin Islands, Cayman Islands, Falkland Islands, Gibraltar, Guernsey, Hong Kong (scheduled to become a Special Administrative Region of China on 1 July 1997), Jersey, Isle of Man, Montserrat, Pitcairn Islands, Saint Helena, South Georgia and the South Sandwich Islands, Turks and Caicos Islands

Independence: 1 January 1801 (United Kingdom established)
National holiday: Celebration of the Birthday of the Queen (second Saturday in June)
Constitution: unwritten; partly statutes, partly common law and practice
Legal system: common law tradition with early Roman and modern continental influences; no judicial review of Acts of Parliament; accepts compulsory ICJ jurisdiction, with reservations
Suffrage: 18 years of age; universal
Executive branch:
chief of state: Queen ELIZABETH II (since 6 February 1952); Heir Apparent Prince CHARLES (son of the Queen, born 14 November 1948)
head of government: Prime Minister John MAJOR (since 28 November 1990)
cabinet: Cabinet of Ministers
Legislative branch: bicameral Parliament
House of Lords: consists of a 1,200-member body, four-fifths are hereditary peers, 2 archbishops, 24 other senior bishops, serving and retired Lords of Appeal in Ordinary, other life peers, Scottish peers
House of Commons: elections last held 9 April 1992 (next to be held by NA April 1997); results—Conservative 41.9%, Labor 34.5%, Liberal Democratic 17.9%, other 5.7%; seats—(651 total) Conservative 336, Labor 271, Liberal Democratic 20, other 24
Judicial branch: House of Lords
Political parties and leaders: Conservative and Unionist Party, John MAJOR; Labor Party, Anthony (Tony) Blair; Liberal Democrats (LD), Jeremy (Paddy) ASHDOWN; Scottish National Party, Alex SALMOND; Welsh National Party (Plaid Cymru), Dafydd Iwan WIGLEY; Ulster Unionist Party (Northern Ireland), James MOLYNEAUX; Democratic Unionist Party (Northern Ireland), Rev. Ian PAISLEY; Ulster Popular Unionist Party (Northern Ireland); Social Democratic and Labor Party (SDLP, Northern Ireland), John HUME; Sinn Fein (Northern Ireland), Gerry ADAMS
Other political or pressure groups: Trades Union Congress; Confederation of British Industry; National Farmers' Union; Campaign for Nuclear Disarmament
Member of: AfDB, AG (observer), AsDB, Australia Group, BIS, C, CCC, CDB (non-regional), CE, CERN, EBRD, EC, ECA (associate), ECE, ECLAC, EIB, ESA, ESCAP, FAO, G- 5, G- 7, G-10, GATT, IADB, IAEA, IBRD, ICAO, ICC, ICFTU, ICRM, IDA, IEA, IFAD, IFC, IFRCS, ILO, IMF, IMO, INMARSAT, INTELSAT, INTERPOL, IOC, IOM (observer), ISO, ITU, MTCR, NACC, NATO, NEA, NSG, OECD, OSCE, PCA, SPC, UN, UN Security Council, UNCTAD, UNFICYP, UNHCR, UNIDO, UNIKOM, UNITAR, UNPROFOR, UNRWA, UNU, UPU, WCL, WEU, WHO, WIPO, WMO, ZC
Diplomatic representation in US:
chief of mission: Ambassador Sir Robin William RENWICK
chancery: 3100 Massachusetts Avenue NW, Washington, DC 20008
telephone: [1] (202) 462-1340
FAX: [1] (202) 898-4255
consulate(s) general: Atlanta, Boston, Chicago, Cleveland, Houston, Los Angeles, New York, and San Francisco,
consulate(s): Dallas, Miami, and Seattle
US diplomatic representation:
chief of mission: Ambassador Adm. William W. CROWE
embassy: 24/31 Grosvenor Square, London, W. 1A1AE
mailing address: PSC 801, Box 40, FPO AE 09498-4040
telephone: [44] (71) 499-9000
FAX: [44] (71) 409-1637
consulate(s) general: Belfast, Edinburgh
Flag: blue with the red cross of Saint George (patron saint of England) edged in white superimposed on the diagonal red cross of Saint Patrick (patron saint of Ireland) which is superimposed on the diagonal white cross of Saint Andrew (patron saint of Scotland); known as the Union Flag or Union Jack; the design and colors (especially the Blue Ensign) have been the basis for a number of other flags including dependencies, Commonwealth countries, and others

Economy

Overview: The UK is one of the world's great trading powers and financial centers, and its economy ranks among the four largest in Western Europe. The economy is essentially capitalistic; over the past 13 years the ruling Tories have greatly reduced public ownership and contained the growth of social welfare programs. Agriculture is intensive, highly mechanized, and efficient by European standards, producing about 60% of food needs with only 1% of the labor force. The UK has large coal, natural gas, and oil reserves, and primary energy production accounts for 12% of GDP, one of the highest shares of any industrial nation. Services, particularly banking, insurance, and business services, account by far for the largest proportion of GDP while industry continues to decline in importance, now employing only 25% of the work force and generating only 21% of GDP. The economy registered 4.2% GDP growth in 1994, its fastest annual rate for six years. Exports and manufacturing output are the primary engines of growth. Unemployment is gradually falling. Inflation is at the lowest level in 27 years, but British monetary authorities raised interest rates to 6.25% in 1994 in a preemptive strike on emerging inflationary pressures such as higher taxes and rising manufacturing costs. The combination of a buoyant economy and fiscal tightening is projected to trim the FY94/95 budget shortfall to about $50 billion—down from about $75 billion in FY93/94. The major economic policy question for Britain in the 1990s is the terms on which it participates in the financial and economic integration of Europe.
National product: GDP—purchasing power parity—$1.0452 trillion (1994 est.)
National product real growth rate: 4.2% (1994 est.)
National product per capita: $17,980 (1994 est.)
Inflation rate (consumer prices): 2.4% (1994)
Unemployment rate: 9.3% (1994)
Budget:
revenues: $325.5 billion
expenditures: $400.9 billion, including capital expenditures of $33 billion (FY93/94 est.)
Exports: $200 billion (f.o.b., 1994 est.)
commodities: manufactured goods, machinery, fuels, chemicals, semifinished goods, transport equipment
partners: EU countries 56.7% (Germany 14.0%, France 11.1%, Netherlands 7.9%), US 10.9%
Imports: $215 billion (c.i.f., 1994 est.)
commodities: manufactured goods, machinery, semifinished goods, foodstuffs, consumer goods
partners: EU countries 51.7% (Germany 14.9%, France 9.3%, Netherlands 8.4%), US 11.6%
External debt: $16.2 billion (June 1992)
Industrial production: growth rate 5.6% (1994)
Electricity:
capacity: 65,360,000 kW
production: 303 billion kWh
consumption per capita: 5,123 kWh (1993)
Industries: production machinery including machine tools, electric power equipment, automation equipment, railroad equipment, shipbuilding, aircraft, motor vehicles and parts, electronics and communications equipment, metals, chemicals, coal, petroleum, paper and paper products, food processing, textiles, clothing, and other consumer goods
Agriculture: accounts for only 1.5% of GDP; wide variety of crops and livestock products
Illicit drugs: gateway country for Latin American cocaine entering the European market; producer of synthetic drugs; transshipment point for Southwest Asian heroin; money-laundering center
Economic aid:
donor: ODA and OOF commitments (1992-93), $3.2 billion
Currency: 1 British pound (£) = 100 pence
Exchange rates: British pounds (£) per US$1—0.6350 (January 1995), 0.6529 (1994), 0.6033 (1993), 0.5664 (1992), 0.5652 (1991), 0.5603 (1990)
Fiscal year: 1 April—31 March

United Kingdom *(continued)*

Transportation

Railroads:
total: 16,888 km; note—several additional small standard-gauge and narrow-gauge lines are privately owned and operated
broad gauge: 330 km 1.600-m gauge (190 km double track)
standard gauge: 16,558 km 1.435-m gauge (4,950 km electrified; 12,591 km double or multiple track)
Highways:
total: 360,047 km (includes Northern Ireland)
paved: 360,047 km (includes Northern Ireland; Great Britain has 3,100 km limited access divided highway)
Inland waterways: 2,291 total; British Waterways Board, 606 km; Port Authorities, 706 km; other, 979 km
Pipelines: crude oil (almost all insignificant) 933 km; petroleum products 2,993 km; natural gas 12,800 km
Ports: Aberdeen, Belfast, Bristol, Cardiff, Grangemouth, Hull, Leith, Liverpool, London, Manchester, Medway, Sullom Voe, Tees, Tyne
Merchant marine:
total: 155 ships (1,000 GRT or over) totaling 3,249,823 GRT/3,978,336 DWT
ships by type: bulk 11, cargo 24, chemical tanker 2, container 23, liquefied gas tanker 3, oil tanker 56, passenger 7, passenger-cargo 1, refrigerated cargo 1, roll-on/roll-off cargo 13, short-sea passenger 13, specialized tanker 1
Airports:
total: 505
with paved runways over 3,047 m: 10
with paved runways 2,438 to 3,047 m: 30
with paved runways 1,524 to 2,437 m: 174
with paved runways 914 to 1,523 m: 91
with paved runways under 914 m: 172
with unpaved runways 1,524 to 2,438 m: 1
with unpaved runways 914 to 1,523 m: 27

Communications

Telephone system: 30,200,000 telephones; technologically advanced domestic and international system
local: NA
intercity: equal mix of buried cables, microwave and optical-fiber systems
international: 40 coaxial submarine cables; 10 INTELSAT (7 Atlantic Ocean and 3 Indian Ocean), 1 INMARSAT, and 1 EUTELSAT earth station; at least 8 large international switching centers
Radio:
broadcast stations: AM 225, FM 525 (mostly repeaters), shortwave 0
radios: 70 million
Television:
broadcast stations: 207 (repeaters 3,210)
televisions: 20 million

Defense Forces

Branches: Army, Royal Navy (includes Royal Marines), Royal Air Force
Manpower availability: males age 15-49 14,429,485; males fit for military service 12,041,935 (1995 est.)
Defense expenditures: exchange rate conversion—$35.1 billion, 3.1% of GDP (FY95/96)

United States

Geography

Location: North America, bordering both the North Atlantic Ocean and the North Pacific Ocean, between Canada and Mexico
Map references: North America
Area:
total area: 9,372,610 sq km
land area: 9,166,600 sq km
comparative area: about half the size of Russia; about three-tenths the size of Africa; about one-half the size of South America (or slightly larger than Brazil); slightly smaller than China; about two and one-half times the size of Western Europe
note: includes only the 50 states and District of Columbia
Land boundaries: total 12,248 km, Canada 8,893 km (including 2,477 km with Alaska), Cuba 29 km (US Naval Base at Guantanamo Bay), Mexico 3,326 km
Coastline: 19,924 km
Maritime claims:
contiguous zone: 12 nm
continental shelf: not specified
exclusive economic zone: 200 nm
territorial sea: 12 nm
International disputes: maritime boundary disputes with Canada (Dixon Entrance, Beaufort Sea, Strait of Juan de Fuca, Machias Seal Island); US Naval Base at Guantanamo Bay is leased from Cuba and only mutual agreement or US abandonment of the area can terminate the lease; Haiti claims Navassa Island; US has made no territorial claim in Antarctica (but has reserved the right to do so) and does not recognize the claims of any other nation; Republic of Marshall Islands claims Wake Island
Climate: mostly temperate, but tropical in Hawaii and Florida and arctic in Alaska, semiarid in the great plains west of the Mississippi River and arid in the Great Basin of the southwest; low winter temperatures in the northwest are ameliorated occasionally in January and February by warm chinook winds from the eastern slopes of the Rocky Mountains

Terrain: vast central plain, mountains in west, hills and low mountains in east; rugged mountains and broad river valleys in Alaska; rugged, volcanic topography in Hawaii
Natural resources: coal, copper, lead, molybdenum, phosphates, uranium, bauxite, gold, iron, mercury, nickel, potash, silver, tungsten, zinc, petroleum, natural gas, timber
Land use:
arable land: 20%
permanent crops: 0%
meadows and pastures: 26%
forest and woodland: 29%
other: 25%
Irrigated land: 181,020 sq km (1989 est.)
Environment:
current issues: air pollution resulting in acid rain in both the US and Canada; the US is the largest single emitter of carbon dioxide from the burning of fossil fuels; water pollution from runoff of pesticides and fertilizers; very limited natural fresh water resources in much of the western part of the country require careful management; desertification
natural hazards: tsunamis, volcanoes, and earthquake activity around Pacific Basin; hurricanes along the Atlantic coast; tornadoes in the midwest; mudslides in California; forest fires in the west; flooding; permafrost in northern Alaska is a major impediment to development
international agreements: party to—Air Pollution, Air Pollution-Nitrogen Oxides, Antarctic Treaty, Climate Change, Endangered Species, Environmental Modification, Marine Dumping, Marine Life Conservation, Nuclear Test Ban, Ozone Layer Protection, Ship Pollution, Tropical Timber 83, Wetlands, Whaling; signed, but not ratified—Air Pollution Volatile Organic Compounds, Antarctic-Environmental Protocol, Biodiversity, Desertification, Hazardous Wastes, Tropical Timber 94
Note: world's fourth-largest country (after Russia, Canada, and China)

People

Population: 263,814,032 (July 1995 est.)
Age structure:
0-14 years: 22% (female 28,391,451; male 29,845,630)
15-64 years: 65% (female 86,454,415; male 85,474,002)
65 years and over: 13% (female 19,949,978; male 13,698,559) (July 1995 est.)
Population growth rate: 1.02% (1995 est.)
Birth rate: 15.25 births/1,000 population (1995 est.)
Death rate: 8.38 deaths/1,000 population (1995 est.)
Net migration rate: 3.34 migrant(s)/1,000 population (1995 est.)
Infant mortality rate: 7.88 deaths/1,000 live births (1995 est.)
Life expectancy at birth:
total population: 75.99 years
male: 72.8 years
female: 79.7 years (1995 est.)
Total fertility rate: 2.08 children born/woman (1995 est.)
Nationality:
noun: American(s)
adjective: American
Ethnic divisions: white 83.4%, black 12.4%, Asian 3.3%, Native American 0.8% (1992)
Religions: Protestant 56%, Roman Catholic 28%, Jewish 2%, other 4%, none 10% (1989)
Languages: English, Spanish (spoken by a sizable minority)
Literacy: age 15 and over has completed five or more years of schooling (1979)
total population: 97%
male: 97%
female: 97%
Labor force: 131.056 million (includes unemployed) (1994)
by occupation: managerial and professional 27.5%, technical, sales and administrative support 30.3%, services 13.7%, manufacturing, mining, transportation, and crafts 25.5%, farming, forestry, and fishing 2.9%

Government

Names:
conventional long form: United States of America
conventional short form: United States
Abbreviation: US or USA
Digraph: US
Type: federal republic; strong democratic tradition
Capital: Washington, DC
Administrative divisions: 50 states and 1 district*; Alabama, Alaska, Arizona, Arkansas, California, Colorado, Connecticut, Delaware, District of Columbia*, Florida, Georgia, Hawaii, Idaho, Illinois, Indiana, Iowa, Kansas, Kentucky, Louisiana, Maine, Maryland, Massachusetts, Michigan, Minnesota, Mississippi, Missouri, Montana, Nebraska, Nevada, New Hampshire, New Jersey, New Mexico, New York, North Carolina, North Dakota, Ohio, Oklahoma, Oregon, Pennsylvania, Rhode Island, South Carolina, South Dakota, Tennessee, Texas, Utah, Vermont, Virginia, Washington, West Virginia, Wisconsin, Wyoming
Dependent areas: American Samoa, Baker Island, Guam, Howland Island, Jarvis Island, Johnston Atoll, Kingman Reef, Midway Islands, Navassa Island, Northern Mariana Islands, Palmyra Atoll, Puerto Rico, Virgin Islands, Wake Island
note: from 18 July 1947 until 1 October 1994, the US has administered the Trust Territory of the Pacific Islands, but recently entered into a new political relationship with all four political units: the Northern Mariana Islands is a Commonwealth in political union with the US (effective 3 November 1986); Palau concluded a Compact of Free Association with the US (effective 1 October 1994); the Federated States of Micronesia signed a Compact of Free Association with the US (effective 3 November 1986); the Republic of the Marshall Islands signed a Compact of Free Association with the US (effective 21 October 1986)
Independence: 4 July 1776 (from England)
National holiday: Independence Day, 4 July (1776)
Constitution: 17 September 1787, effective 4 March 1789
Legal system: based on English common law; judicial review of legislative acts; accepts compulsory ICJ jurisdiction, with reservations
Suffrage: 18 years of age; universal
Executive branch:
chief of state and head of government: President William Jefferson CLINTON (since 20 January 1993); Vice President Albert GORE, Jr. (since 20 January 1993); election last held 3 November 1992 (next to be held 5 November 1996); results—William Jefferson CLINTON (Democratic Party) 43.2%, George BUSH (Republican Party) 37.7%, Ross PEROT (Independent) 19.0%, other 0.1%
cabinet: Cabinet; appointed by the president with Senate approval
Legislative branch: bicameral Congress
Senate: elections last held 8 November 1994 (next to be held 5 November 1996); results—percent of vote by party NA; seats—(100 total) Republican Party 54, Democratic Party 46
House of Representatives: elections last held 8 November 1994 (next to be held 5 November 1996); results—percent of vote by party NA; seats—(435 total) Republican Party 231, Democratic Party 203, independent 1
Judicial branch: Supreme Court
Political parties and leaders: Republican Party, Haley BARBOUR, national committee chairman; Jeanie AUSTIN, co-chairman; Democratic Party, David C. WILHELM, national committee chairman; several other groups or parties of minor political significance
Member of: AfDB, AG (observer), ANZUS, APEC, AsDB, Australia Group, BIS, CCC, CP, EBRD, ECE, ECLAC, ESCAP, FAO, G- 2, G- 5, G- 7, G- 8, G 10, GATT, IADB, IAEA, IBRD, ICAO, ICC, ICFTU, ICRM, IDA, IEA, IFAD, IFC, IFRCS, ILO, IMF, IMO, INMARSAT, INTELSAT, INTERPOL, IOC, IOM, ISO, ITU, MINURSO, MTCR, NACC, NATO, NEA, NSG, OAS, OECD, OSCE, PCA, SPC, UN, UN Security Council, UNCTAD, UNHCR, UNIDO, UNIKOM, UNITAR, UNMIH, UNOMOZ, UNPROFOR, UNRWA, UNTSO, UNU, UPU, WCL, WHO, WIPO, WMO, WTO, ZC
Flag: thirteen equal horizontal stripes of red (top and bottom) alternating with white; there is a blue rectangle in the upper hoist-side

corner bearing 50 small white five-pointed stars arranged in nine offset horizontal rows of six stars (top and bottom) alternating with rows of five stars; the 50 stars represent the 50 states, the 13 stripes represent the 13 original colonies; known as Old Glory; the design and colors have been the basis for a number of other flags including Chile, Liberia, Malaysia, and Puerto Rico

Economy

Overview: The US has the most powerful, diverse, and technologically advanced economy in the world, with a per capita GDP of $25,850, the largest among major industrial nations. The economy is market oriented with most decisions made by private individuals and business firms and with government purchases of goods and services made predominantly in the marketplace. In 1989 the economy enjoyed its seventh successive year of substantial growth, the longest in peacetime history. The expansion featured moderation in wage and consumer price increases and a steady reduction in unemployment to 5.2% of the labor force. In 1990, however, growth slowed to 1% because of a combination of factors, such as the worldwide increase in interest rates, Iraq's invasion of Kuwait in August, the subsequent spurt in oil prices, and a general decline in business and consumer confidence. In 1991 output fell by 0.6%, unemployment grew, and signs of recovery proved premature. Growth picked up to 2.3% in 1992 and to 3.1% in 1993. Unemployment, however, declined only gradually, the increase in GDP being mainly attributable to gains in output per worker. The year 1994 witnessed a solid 4% gain in real output, a low inflation rate of 2.6%, and a drop in unemployment below 6%. The capture of both houses of Congress by the Republicans in the elections of 8 November 1994 means substantial changes are likely in US economic policy, including changes in the ways the US will address its major economic problems in 1995-96. These problems include inadequate investment in economic infrastructure, rapidly rising medical costs of an aging population, and sizable budget and trade deficits.
National product: GDP—purchasing power parity—$6.7384 trillion (1994)
National product real growth rate: 4.1% (1994)
National product per capita: $25,850 (1994)
Inflation rate (consumer prices): 2.6% (1994)
Unemployment rate: 5.5% (March 1995)
Budget:
revenues: $1.258 trillion
expenditures: $1.461 trillion, including capital expenditures of $NA (1994)
Exports: $513 billion (f.o.b., 1994)
commodities: capital goods, automobiles, industrial supplies and raw materials, consumer goods, agricultural products
partners: Western Europe 24.3%, Canada 22.1%, Japan 10.5% (1993)
Imports: $664 billion (c.i.f., 1994)
commodities: crude oil and refined petroleum products, machinery, automobiles, consumer goods, industrial raw materials, food and beverages
partners: Canada, 19.3%, Western Europe 18.1%, Japan 18.1% (1993)
External debt: $NA
Industrial production: growth rate 5.4% (1994 est.)
Electricity:
capacity: 695,120,000 kW
production: 3.1 trillion kWh
consumption per capita: 11,236 kWh (1993)
Industries: leading industrial power in the world, highly diversified and technologically advanced; petroleum, steel, motor vehicles, aerospace, telecommunications, chemicals, electronics, food processing, consumer goods, lumber, mining
Agriculture: accounts for 2% of GDP and 2.9% of labor force; favorable climate and soils support a wide variety of crops and livestock production; world's second largest producer and number one exporter of grain; surplus food producer; fish catch of 4.4 million metric tons (1990)
Illicit drugs: illicit producer of cannabis for domestic consumption with 1987 production estimated at 3,500 metric tons or about 25% of the available marijuana; ongoing eradication program aimed at small plots and greenhouses has not reduced production
Economic aid:
donor: commitments, including ODA and OOF, (FY80-89), $115.7 billion
Currency: 1 United States dollar (US$) = 100 cents
Exchange rates:
British pounds: (£) per US$—0.6350 (January 1995), 0.6529 (1994), 0.6033 (1993), 0.5664 (1992), 0.5652 (1991), 0.5603 (1990)
Canadian dollars: (Can$) per US$—1.4129 (January 1995), 1.3656 (1994), 1.2901 (1993), 1.2087 (1992), 1.1457 (1991), 1.1668 (1990)
French francs: (F) per US$—5.2943 (January 1995), 5.5520 (1994), 5.6632 (1993), 5.2938 (1992), 5.6421 (1991), 5.4453 (1990)
Italian lire: (Lit) per US$—1,609.5 (January 1995), 1,612.4 (1994), 1,573.7 (1993), 1,232.4 (1992), 1,240.6 (1991), 1,198.1 (1990)
Japanese yen: (¥) per US$—99.75 (January 1995), 102.21 (1994), 111.20 (1993), 126.65 (1992), 134.71 (1991), 144.79 (1990)
German deutsche marks: (DM) per US$—1.5313 (January 1995), 1.6228 (1994), 1.6533 (1993), 1.5617 (1992), 1.6595 (1991), 1.6157 (1990)
Fiscal year: 1 October—30 September

Transportation

Railroads:
total: 240,000 km mainline routes (nongovernment owned)
standard gauge: 240,000 km 1.435-m gauge (1989)
Highways:
total: 6,243,163 km
paved: 3,633,520 km (including 84,865 km of expressways)
unpaved: 2,609,643 km (1990)
Inland waterways: 41,009 km of navigable inland channels, exclusive of the Great Lakes (est.)
Pipelines: petroleum 276,000 km; natural gas 331,000 km (1991)
Ports: Anchorage, Baltimore, Boston, Charleston, Chicago, Duluth, Hampton Roads, Honolulu, Houston, Jacksonville, Los Angeles, New Orleans, New York, Philadelphia, Port Canaveral, Portland (Oregon), Prudhoe Bay, San Francisco, Savannah, Seattle, Tampa, Toledo
Merchant marine:
total: 354 ships (1,000 GRT or over) totaling 11,462,000 GRT/16,477,000 DWT
ships by type: bulk 22, cargo 28, chemical tanker 16, intermodal 130, liquefied gas tanker 13, passenger-cargo 2, tanker 130, tanker tug-barge 13
note: in addition, there are 189 government-owned vessels
Airports:
total: 15,032
with paved runways over 3,047 m: 181
with paved runways 2,438 to 3,047 m: 208
with paved runways 1,524 to 2,437 m: 1,242
with paved runways 914 to 1,523 m: 2,489
with paved runways under 914 m: 8,994
with unpaved runways over 3,047 m: 1
with unpaved runways 2,438 to 3,047 m: 7
with unpaved runways 1,524 to 2,438 m: 180
with unpaved runways 914 to 1,523 m: 1,730

Communications

Telephone system: 126,000,000 telephones; 7,557,000 cellular telephones
local: NA
intercity: large system of fiber optic cable, microwave radio relay, coaxial cable, and domestic satellites
international: 16 satellites and 24 ocean cable systems in use; 61 INTELSAT (45 Atlantic Ocean and 16 Pacific Ocean) earth stations (1990)
Radio:
broadcast stations: AM 4,987, FM 4,932, shortwave 0
radios: 530 million
Television:
broadcast stations: 1,092 (about 9,000 cable TV systems)
televisions: 193 million

Uruguay

Defense Forces

Branches: Department of the Army, Department of the Navy (includes Marine Corps), Department of the Air Force
Defense expenditures: $284.4 billion, 4.2% of GDP (1994 est.)

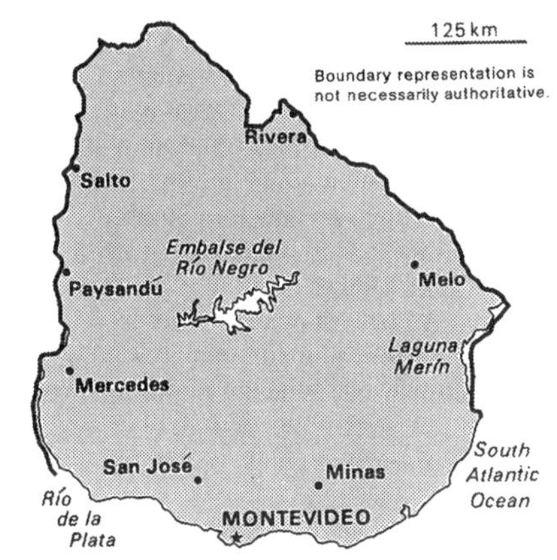

Geography

Location: Southern South America, bordering the South Atlantic Ocean, between Argentina and Brazil
Map references: South America
Area:
total area: 176,220 sq km
land area: 173,620 sq km
comparative area: slightly smaller than Washington State
Land boundaries: total 1,564 km, Argentina 579 km, Brazil 985 km
Coastline: 660 km
Maritime claims:
continental shelf: 200-m depth or to the depth of exploitation
territorial sea: 200 nm; overflight and navigation guaranteed beyond 12 nm
International disputes: short section of boundary with Argentina is in dispute; two short sections of the boundary with Brazil are in dispute—Arroyo de la Invernada (Arroio Invernada) area of the Rio Cuareim (Rio Quarai) and the islands at the confluence of the Rio Cuareim (Rio Quarai) and the Uruguay River
Climate: warm temperate; freezing temperatures almost unknown
Terrain: mostly rolling plains and low hills; fertile coastal lowland
Natural resources: soil, hydropower potential, minor minerals
Land use:
arable land: 8%
permanent crops: 0%
meadows and pastures: 78%
forest and woodland: 4%
other: 10%
Irrigated land: 1,100 sq km (1989 est.)
Environment:
current issues: substantial pollution from Brazilian industry along border; one-fifth of country affected by acid rain generated by Brazil; water pollution from meat packing/tannery industry; inadequate solid/hazardous waste disposal
natural hazards: seasonally high winds (the pampero is a chilly and occasional violent wind which blows north from the Argentine pampas), droughts, floods; because of the absence of mountains, which act as weather barriers, all locations are particularly vulnerable to rapid changes in weather fronts
international agreements: party to—Antarctic-Environmental Protocol, Antarctic Treaty, Biodiversity, Climate Change, Endangered Species, Environmental Modification, Hazardous Wastes, Law of the Sea, Nuclear Test Ban, Ozone Layer Protection, Ship Pollution, Wetlands; signed, but not ratified—Marine Dumping, Marine Life Conservation

People

Population: 3,222,716 (July 1995 est.)
Age structure:
0-14 years: 25% (female 392,262; male 409,580)
15-64 years: 63% (female 1,026,314; male 995,492)
65 years and over: 12% (female 233,377; male 165,691) (July 1995 est.)
Population growth rate: 0.74% (1995 est.)
Birth rate: 17.57 births/1,000 population (1995 est.)
Death rate: 9.27 deaths/1,000 population (1995 est.)
Net migration rate: -0.93 migrant(s)/1,000 population (1995 est.)
Infant mortality rate: 16.3 deaths/1,000 live births (1995 est.)
Life expectancy at birth:
total population: 74.46 years
male: 71.24 years
female: 77.83 years (1995 est.)
Total fertility rate: 2.41 children born/woman (1995 est.)
Nationality:
noun: Uruguayan(s)
adjective: Uruguayan
Ethnic divisions: white 88%, mestizo 8%, black 4%
Religions: Roman Catholic 66% (less than half adult population attends church regularly), Protestant 2%, Jewish 2%, nonprofessing or other 30%
Languages: Spanish, Brazilero (Portuguese-Spanish mix on the Brazilian frontier)
Literacy: age 15 and over can read and write (1990 est.)
total population: 96%
male: 97%
female: 96%
Labor force: 1.355 million (1991 est.)
by occupation: government 25%, manufacturing 19%, agriculture 11%, commerce 12%, utilities, construction, transport, and communications 12%, other services 21% (1988 est.)

Government

Names:
conventional long form: Oriental Republic of Uruguay
conventional short form: Uruguay
local long form: Republica Oriental del Uruguay
local short form: Uruguay
Digraph: UY
Type: republic
Capital: Montevideo
Administrative divisions: 19 departments (departamentos, singular—departamento); Artigas, Canelones, Cerro Largo, Colonia, Durazno, Flores, Florida, Lavalleja, Maldonado, Montevideo, Paysandu, Rio Negro, Rivera, Rocha, Salto, San Jose, Soriano, Tacuarembo, Treinta y Tres
Independence: 25 August 1828 (from Brazil)
National holiday: Independence Day, 25 August (1828)
Constitution: 27 November 1966, effective February 1967, suspended 27 June 1973, new constitution rejected by referendum 30 November 1980
Legal system: based on Spanish civil law system; accepts compulsory ICJ jurisdiction
Suffrage: 18 years of age; universal and compulsory
Executive branch:
chief of state and head of government: President Julio Maria SANGUINETTI (since 1 March 1995); Vice President Hugo BATALLA (since 1 March 1995); election last held 27 November 1994 (next to be held NA November 1999)
cabinet: Council of Ministers; appointed by the president
Legislative branch: bicameral General Assembly (Asamblea General)
Chamber of Senators (Camara de Senadores): elections last held 27 November 1994 (next to be held NA November 1999); results—Colorado 36%, Blanco 34 %, Encuentro Progresista 27%, New Sector 3%; seats—(30 total) Colorado 11, Blanco 10, Encuentro Progresista 8, New Sector 1
Chamber of Representatives (Camera de Representantes): elections last held 27 November 1994 (next to be held NA November 1999); results—Colorado 32%, Blanco 31%, Encuentro Progresista 31%, New Sector 5%; seats—(99 total) Colorado 32, Blanco 31, Encuentro Progresista 31, New Sector 5
Judicial branch: Supreme Court
Political parties and leaders: National (Blanco) Party; Colorado Party, Jorge BATLLE; Broad Front Coalition, Gen. Liber SEREGNI Mosquera; New sector Coalition, Hugo BATALLA; Encuentro Progresista
Member of: AG (observer), CCC, ECLAC, FAO, G-11, G-77, GATT, IADB, IAEA, IBRD, ICAO, ICC, ICRM, IFAD, IFC, IFRCS, ILO, IMF, IMO, INTELSAT, INTERPOL, IOC, IOM, ISO, ITU, LAES, LAIA, MERCOSUR, NAM (observer), OAS, OPANAL, PCA, RG, UN, UNAMIR, UNCTAD, UNESCO, UNIDO, UNIKOM, UNMOGIP, UNOMIL, UNOMOZ, UPU, WCL, WFTU, WHO, WIPO, WMO, WTO
Diplomatic representation in US:
chief of mission: Ambassador Eduardo MACGILLYCUDDY
chancery: 1918 F Street NW, Washington, DC 20006
telephone: [1] (202) 331-1313 through 1316
consulate(s) general: Los Angeles, Miami, and New York
consulate(s): New Orleans
US diplomatic representation:
chief of mission: Ambassador Thomas J. DODD
embassy: Lauro Muller 1776, Montevideo
mailing address: APO AA 34035
telephone: [598] (2) 23 60 61, 48 77 77
FAX: [598] (2) 48 86 11
Flag: nine equal horizontal stripes of white (top and bottom) alternating with blue; there is a white square in the upper hoist-side corner with a yellow sun bearing a human face known as the Sun of May and 16 rays alternately triangular and wavy

Economy

Overview: Uruguay's economy is a small one with favorable climate, good soils, and substantial hydropower potential. Economic development has been restrained in recent years by excessive government regulation of economic detail and 40% to 130% inflation. Although the GDP growth rate slowed in 1993 to 1.7%, following a healthy expansion to 7.5% in 1992, it rebounded in 1994 to an estimated 4%, spurred mostly by increasing agricultural and other exports and a surprise reversal of the downward trend in industrial production. In a major step toward regional economic cooperation, Uruguay confirmed its commitment to the Southern Cone Common Market (MERCOSUR) customs union by implementing MERCOSUR's common external tariff on most tradables on 1 January 1995. Inflation in 1994 declined for the third consecutive year, yet, at 44%, it remains the highest in the region; analysts predict that the expanding fiscal deficit and wage indexation will force the inflation rate back toward the 50% mark in 1995.
National product: GDP—purchasing power parity—$23 billion (1994 est.)
National product real growth rate: 4% (1994 est.)
National product per capita: $7,200 (1994 est.)
Inflation rate (consumer prices): 44% (1994 est.)
Unemployment rate: 9% (1994 est.)
Budget:
revenues: $2.9 billion
expenditures: $3 billion, including capital expenditures of $388 million (1991 est.)
Exports: $1.78 billion (f.o.b., 1994 est.)
commodities: wool and textile manufactures, beef and other animal products, leather, rice
partners: Brazil, Argentina, US, China, Italy
Imports: $2.461 billion (c.i.f., 1994 est.)
commodities: machinery and equipment, vehicles, chemicals, minerals, plastics
partners: Brazil, Argentina, US, Nigeria
External debt: $4.2 billion (1993)
Industrial production: growth rate 3.9% (1992); accounts for 28% of GDP
Electricity:
capacity: 2,070,000 kW
production: 9 billion kWh
consumption per capita: 1,575 kWh (1993)
Industries: meat processing, wool and hides, sugar, textiles, footwear, leather apparel, tires, cement, petroleum refining, wine
Agriculture: accounts for 12% of GDP; large areas devoted to livestock grazing; wheat, rice, corn, sorghum; fishing; self-sufficient in most basic foodstuffs
Economic aid:
recipient: US commitments, including Ex-Im (FY70-88), $105 million; Western (non-US) countries, ODA and OOF bilateral commitments (1970-89), $420 million; Communist countries (1970-89), $69 million
Currency: 1 Uruguayan peso ($Ur) = 100 centesimos
Exchange rates: Uruguayan pesos ($Ur) per US$1—5.6 (January 1995), 4.4710 (January 1994), 3.9484 (1993), 3.0270 (1992), 2.0188 (1991), 1.1710 (1990)
note: on 1 March 1993 the former New Peso (N$Ur) was replaced as Uruguay's unit of currency by the Peso which is equal to 1,000 of the New Pesos
Fiscal year: calendar year

Transportation

Railroads:
total: 3,000 km
standard gauge: 3,000 km 1.435-m gauge
Highways:
total: 49,900 km
paved: 6,700 km
unpaved: gravel 3,000 km; earth 40,200 km
Inland waterways: 1,600 km; used by coastal and shallow-draft river craft
Ports: Fray Bentos, Montevideo, Nueva Palmira, Paysandu, Punta del Este
Merchant marine:
total: 3 ships (1,000 GRT or over) totaling 71,405 GRT/110,939 DWT
ships by type: cargo 1, container 1, oil tanker 1
Airports:
total: 85
with paved runways 2,438 to 3,047 m: 2
with paved runways 1,524 to 2,437 m: 5

with paved runways 914 to 1,523 m: 8
with paved runways under 914 m: 54
with unpaved runways 1,524 to 2,438 m: 2
with unpaved runways 914 to 1,523 m: 14

Communications

Telephone system: 337,000 telephones; telephone density 10/100 persons; some modern facilities
local: most modern facilities concentrated in Montevideo
intercity: new nationwide microwave network
international: 2 INTELSAT (Atlantic Ocean) earth stations
Radio:
broadcast stations: AM 99, FM 0, shortwave 9
radios: NA
Television:
broadcast stations: 26
televisions: NA

Defense Forces

Branches: Army, Navy (includes Naval Air Arm, Coast Guard, Marines), Air Force, Grenadier Guards, Coracero Guard, Police
Manpower availability: males age 15-49 775,060; males fit for military service 629,385 (1995 est.)
Defense expenditures: exchange rate conversion—$216 million, 2.3% of GDP (1991 est.)

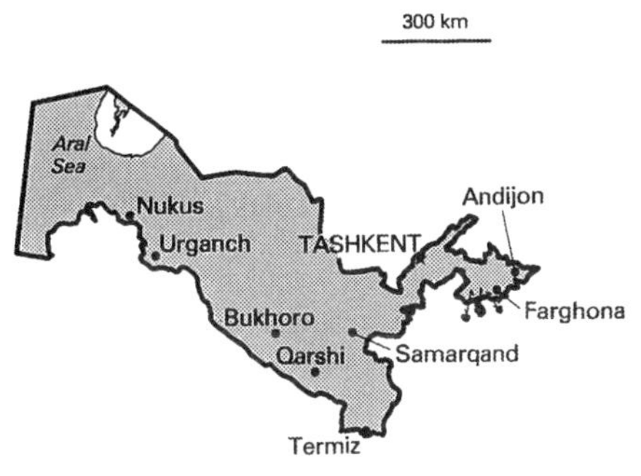

Geography

Location: Central Asia, north of Afghanistan
Map references: Commonwealth of Independent States—Central Asian States
Area:
total area: 447,400 sq km
land area: 425,400 sq km
comparative area: slightly larger than California
Land boundaries: total 6,221 km, Afghanistan 137 km, Kazakhstan 2,203 km, Kyrgyzstan 1,099 km, Tajikistan 1,161 km, Turkmenistan 1,621 km
Coastline: 0 km
note: Uzbekistan borders the Aral Sea (420 km)
Maritime claims: none; landlocked
International disputes: none
Climate: mostly midlatitude desert, long, hot summers, mild winters; semiarid grassland in east
Terrain: mostly flat-to-rolling sandy desert with dunes; broad, flat intensely irrigated river valleys along course of Amu Darya and Sirdaryo Rivers; Fergana Valley in east surrounded by mountainous Tajikistan and Kyrgyzstan; shrinking Aral Sea in west
Natural resources: natural gas, petroleum, coal, gold, uranium, silver, copper, lead and zinc, tungsten, molybdenum
Land use:
arable land: 10%
permanent crops: 1%
meadows and pastures: 47%
forest and woodland: 0%
other: 42%
Irrigated land: 41,550 sq km (1990)
Environment:
current issues: drying up of the Aral Sea is resulting in growing concentrations of chemical pesticides and natural salts; these substances are then blown from the increasingly exposed lake bed and contribute to desertification; water pollution from industrial wastes and the heavy use of fertilizers and pesticides is the cause of many human health disorders; increasing soil salinization; soil contamination from agricultural chemicals, including DDT
natural hazards: NA
international agreements: party to—Climate Change, Environmental Modification, Ozone Layer Protection
Note: landlocked

People

Population: 23,089,261 (July 1995 est.)
Age structure:
0-14 years: 40% (female 4,553,432; male 4,670,496)
15-64 years: 55% (female 6,400,578; male 6,384,862)
65 years and over: 5% (female 656,933; male 422,960) (July 1995 est.)
Population growth rate: 2.08% (1995 est.)
Birth rate: 29.45 births/1,000 population (1995 est.)
Death rate: 6.44 deaths/1,000 population (1995 est.)
Net migration rate: -2.23 migrant(s)/1,000 population (1995 est.)
Infant mortality rate: 52 deaths/1,000 live births (1995 est.)
Life expectancy at birth:
total population: 68.79 years
male: 65.5 years
female: 72.24 years (1995 est.)
Total fertility rate: 3.67 children born/woman (1995 est.)
Nationality:
noun: Uzbek(s)
adjective: Uzbek
Ethnic divisions: Uzbek 71.4%, Russian 8.3%, Tajik 4.7%, Kazakh 4.1%, Tatar 2.4%, Karakalpak 2.1%, other 7%
Religions: Muslim 88% (mostly Sunnis), Eastern Orthodox 9%, other 3%
Languages: Uzbek 74.3%, Russian 14.2%, Tajik 4.4%, other 7.1%
Literacy: age 15 and over can read and write (1989)
total population: 97%
male: 98%
female: 96%
Labor force: 8.234 million
by occupation: agriculture and forestry 43%, industry and construction 22%, other 35% (1992)

Government

Names:
conventional long form: Republic of Uzbekistan
conventional short form: Uzbekistan
local long form: Uzbekiston Respublikasi
local short form: none
former: Uzbek Soviet Socialist Republic
Digraph: UZ
Type: republic

Capital: Tashkent (Toshkent)
Administrative divisions: 12 wiloyatlar (singular—wiloyat), 1 autonomous republic* (respublikasi), and 1 city** (shahri); Andijon Wiloyati, Bukhoro Wiloyati, Jizzakh Wiloyati, Farghona Wiloyati, Qoraqalpoghiston* (Nukus), Qashqadaryo Wiloyati (Qarshi), Khorazm Wiloyati (Urganch), Namangan Wiloyati, Nawoiy Wiloyati, Samarqand Wiloyati, Sirdaryo Wiloyati (Guliston), Surkhondaryo Wiloyati (Termiz), Toshkent Shahri**, Toshkent Wiloyati
note: an administrative division has the same name as its administrative center (exceptions have the administrative center name following in parentheses)
Independence: 31 August 1991 (from Soviet Union)
National holiday: Independence Day, 1 September (1991)
Constitution: new constitution adopted 8 December 1992
Legal system: evolution of Soviet civil law; still lacks independent judicial system
Suffrage: 18 years of age; universal
Executive branch:
chief of state: President Islam KARIMOV (since NA March 1990); election last held 29 December 1991 (next to be held NA); results—Islam KARIMOV 86%, Mukhammad SOLIKH 12%, other 2%; note—a 26 March 1995 referendum extended KARIMOV's term until 2000 (99.6% approval)
head of government: Prime Minister Abdulhashim MUTALOV (since 13 January 1992), First Deputy Prime Minister Ismail DJURABEKOV (since NA); Deputy Prime Ministers Viktor CHIZHEN, Bakhtiyar HAMIDOV, Kayim KHAKKULOV, Yuriy PAYGIN, Saidmukhtar SAIDKASYMOV, Utkur SULTANOV, Mirabror USMANOV, Murat SHARIFKHOJAYEV (since NA)
cabinet: Cabinet of Ministers; appointed by the president with approval of the Supreme Assembly
Legislative branch: unicameral
Supreme Council: elections last held 25 December 1994 (next to be held NA); results—percent of vote by party NA; seats—(250 total) People's Democratic Party 207, Fatherland Progress Party 12, other 31; note—final runoffs were held 22 January 1995; seating was as follows: People's Democratic Party 69, Fatherland Progress Party 14, Social Democratic Party 47, local government 120
Judicial branch: Supreme Court
Political parties and leaders: People's Democratic Party (PDP; formerly Communist Party), Islam A. KARIMOV, chairman; Fatherland Progress Party (FPP), Anwar YULDASHEV, chairman; Social Democratic Party, Anvar JORABAYEV, chairman; Erk (Freedom) Democratic Party (EDP), Muhammad SOLIKH, chairman (in exile); note—EDP was banned 9 December 1992
Other political or pressure groups: Birlik (Unity) People's Movement (BPM), Abdul Rakhim PULATOV, chairman (in exile); Islamic Rebirth Party (IRP), Abdullah UTAYEV, chairman; Adolat-94 (formed by former Vice President Shukhrat MIRSAIDOV and Ibragim BURIEV
note: PULATOV (BPM) is in exile in the West; UTAYEV (IRP) is either in prison or in exile
Member of: AsDB, CCC, CIS, EBRD, ECE, ECO, ESCAP, IAEA, IBRD, ICAO, IDA, IFC, ILO, IMF, INTERPOL, IOC, ISO, ITU, NACC, NAM, OSCE, PFP, UN, UNCTAD, UNESCO, UNIDO, UPU, WHO, WIPO, WMO, WTO
Diplomatic representation in US:
chief of mission: Ambassador Fatikh TESHABAYEV
chancery: (temporary) Suites 619 and 623, 1511 K Street NW, Washington, DC 20005
telephone: [1] (202) 638-4266, 4267
FAX: [1] (202) 638-4268
consulate(s) general: New York
US diplomatic representation:
chief of mission: Ambassador Henry L. CLARKE
embassy: 82 Chilanzarskaya, Tashkent
mailing address: use embassy street address
telephone: [7] (3712) 77-14-07, 77-10-81
FAX: [7] (3712) 77-69-53
Flag: three equal horizontal bands of blue (top), white, and green separated by red fimbriations with a crescent moon and 12 stars in the upper hoist-side quadrant

Economy

Overview: Uzbekistan is a dry, landlocked country of which 10% consists of intensely cultivated, irrigated river valleys. It is one of the poorest states of the former USSR with 60% of its population living in overpopulated rural communities. Nevertheless, Uzbekistan is the world's third largest cotton exporter, a major producer of gold and natural gas, and a regionally significant producer of chemicals and machinery. Since independence, the government has sought to prop up the Soviet-style command economy with subsidies and tight controls on prices and production. Such policies have buffered the economy from the sharp declines in output and high inflation experienced by many other former Soviet republics. They had become increasingly unsustainable, however, as inflation moves along at 14% per month and as Russia has forced the Uzbek government to introduce its own currency. Faced with mounting economic problems, the government has begun to move on a reform agenda and cooperate with international financial institutions, announced an acceleration of privatization, and stepped up efforts to attract foreign investors. Nevertheless, the regime is likely to find it difficult to sustain its drive for economic reform.
National product: GDP—purchasing power parity—$54.5 billion (1994 estimate as extrapolated from World Bank estimate for 1992)
National product real growth rate: -4% (1994 est.)
National product per capita: $2,400 (1994 est.)
Inflation rate (consumer prices): 14% per month (1994 est.)
Unemployment rate: 0.3% includes only officially registered unemployed; large numbers of underemployed workers (December 1994)
Budget:
revenues: $NA
expenditures: $NA, including capital expenditures of $NA
Exports: $943.7 million to outside the FSU countries (1994)
commodities: cotton, gold, natural gas, mineral fertilizers, ferrous metals, textiles, food products
partners: Russia, Ukraine, Eastern Europe, US
Imports: $1.15 billion from outside the FSU countries (1994)
commodities: grain, machinery and parts, consumer durables, other foods
partners: principally other FSU countries, Czech Republic
External debt: $NA
Industrial production: growth rate 1% (1994 est.)
Electricity:
capacity: 11,690,000 kW
production: 47.5 billion kWh
consumption per capita: 2,130 kWh (1994)
Industries: textiles, food processing, machine building, metallurgy, natural gas
Agriculture: cotton, vegetables, fruits, grain, livestock
Illicit drugs: illicit cultivator of cannabis and opium poppy; mostly for CIS consumption; limited government eradication programs; used as transshipment point for illicit drugs to Western Europe
Economic aid:
recipient: the IMF has established a Systemic Transformation Facility of $74 million and the World Bank has made a rehabilitation loan of $160 million with other project loans pending; estimated annual external financing requirements for 1995-96 of $600 million to $700 million
Currency: introduced provisional som-coupons 10 November 1993 which circulated parallel to the Russian rubles; became the sole legal currency 31 January 1994; was replaced in July 1994 by the som currency
Exchange rates: soms per US$1—25 (yearend 1994)
Fiscal year: calendar year

Transportation

Railroads:
total: 3,460 km in common carrier service; does not include industrial lines
broad gauge: 3,460 km 1.520-m gauge (1990)
Highways:
total: 78,400 km
paved and graveled: 67,000 km
unpaved: earth 11,400 km (1990)
Pipelines: crude oil 250 km; petroleum products 40 km; natural gas 810 km (1992)
Ports: Termiz
Airports:
total: 261
with paved runways over 3,047 m: 6
with paved runways 2,438 to 3,047 m: 14
with paved runways 1,524 to 2,437 m: 2
with paved runways 914 to 1,523 m: 8
with paved runways under 914 m: 5
with unpaved runways 2,438 to 3,047 m: 2
with unpaved runways 1,524 to 2,438 m: 1
with unpaved runways 914 to 1,523 m: 7
with unpaved runways under 914 m: 216

Communications

Telephone system: 1,458,000 telephones; 63 telephones/1,000 persons (1995)
local: NMT-450 analog cellular network established in Tashkent
intercity: NA
international: poorly developed; linked by landline or microwave with CIS member states and by leased connection via the Moscow international gateway switch to other countries; new INTELSAT links to Tokyo and Ankara give Uzbekistan international access independent of Russian facilities; Orbita and INTELSAT earth stations
Radio:
broadcast stations: AM NA, FM NA, shortwave NA
radios: NA
Television:
broadcast stations: NA
televisions: NA

Defense Forces

Branches: Army, Air and Air Defense, Republic Security Forces (internal and border troops), National Guard
Manpower availability: males age 15-49 5,567,580; males fit for military service 4,537,455; males reach military age (18) annually 222,506 (1995 est.)
Defense expenditures: $NA, NA% of GDP

Vanuatu

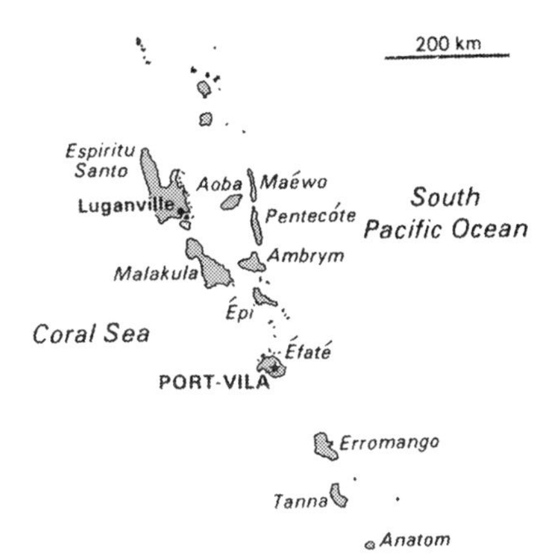

Geography

Location: Oceania, group of islands in the South Pacific Ocean, about three-quarters of the way from Hawaii to Australia
Map references: Oceania
Area:
total area: 14,760 sq km
land area: 14,760 sq km
comparative area: slightly larger than Connecticut
note: includes more than 80 islands
Land boundaries: 0 km
Coastline: 2,528 km
Maritime claims: measured from claimed archipelagic baselines
contiguous zone: 24 nm
continental shelf: 200 nm or to the edge of the continental margin
exclusive economic zone: 200 nm
territorial sea: 12 nm
International disputes: none
Climate: tropical; moderated by southeast trade winds
Terrain: mostly mountains of volcanic origin; narrow coastal plains
Natural resources: manganese, hardwood forests, fish
Land use:
arable land: 1%
permanent crops: 5%
meadows and pastures: 2%
forest and woodland: 1%
other: 91%
Irrigated land: NA sq km
Environment:
current issues: a majority of the population does not have access to a potable and reliable supply of water
natural hazards: tropical cyclones or typhoons (January to April); volcanism causes minor earthquakes
international agreements: party to—Biodiversity, Climate Change, Endangered Species, Marine Dumping, Ozone Layer Protection, Ship Pollution; signed, but not ratified—Law of the Sea

People

Population: 173,648 (July 1995 est.)
Age structure:
0-14 years: 41% (female 34,819; male 36,128)
15-64 years: 56% (female 47,320; male 50,456)
65 years and over: 3% (female 2,217; male 2,708) (July 1995 est.)
Population growth rate: 2.22% (1995 est.)
Birth rate: 31.26 births/1,000 population (1995 est.)
Death rate: 9.06 deaths/1,000 population (1995 est.)
Net migration rate: 0 migrant(s)/1,000 population (1995 est.)
Infant mortality rate: 66.3 deaths/1,000 live births (1995 est.)
Life expectancy at birth:
total population: 59.71 years
male: 57.9 years
female: 61.61 years (1995 est.)
Total fertility rate: 4.14 children born/woman (1995 est.)
Nationality:
noun: Ni-Vanuatu (singular and plural)
adjective: Ni-Vanuatu
Ethnic divisions: indigenous Melanesian 94%, French 4%, Vietnamese, Chinese, Pacific Islanders
Religions: Presbyterian 36.7%, Anglican 15%, Catholic 15%, indigenous beliefs 7.6%, Seventh-Day Adventist 6.2%, Church of Christ 3.8%, other 15.7%
Languages: English (official), French (official), pidgin (known as Bislama or Bichelama)
Literacy: age 15 and over can read and write (1979)
total population: 53%
male: 57%
female: 48%
Labor force: NA
by occupation: NA

Government

Names:
conventional long form: Republic of Vanuatu
conventional short form: Vanuatu
former: New Hebrides
Digraph: NH
Type: republic
Capital: Port-Vila
Administrative divisions: 6 provinces; Malampa, Penama, Sanma, Shefa, Tafea, Torba
Independence: 30 July 1980 (from France and UK)
National holiday: Independence Day, 30 July (1980)
Constitution: 30 July 1980
Legal system: unified system being created from former dual French and British systems
Suffrage: 18 years of age; universal

Executive branch:
chief of state: President Jean Marie LEYE (since 2 March 1994)
head of government: Prime Minister Maxime CARLOT Korman (since 16 December 1991); Deputy Prime Minister Sethy REGENVANU (since 17 December 1991)
cabinet: Council of Ministers; appointed by the prime minister, responsible to parliament
Legislative branch: unicameral
Parliament: elections last held 2 December 1991 (next to be held NA November 1995); note—after election, a coalition was formed by the Union of Moderate Parties and the National United Party to form a new government on 16 December 1991, but political party associations are fluid; seats—(46 total) UMP 19, NUP 10, VP 10, MPP 4, TUP 1, Nagriamel 1, Friend 1
note: the National Council of Chiefs advises on matters of custom and land
Judicial branch: Supreme Court
Political parties and leaders: Vanuatu Party (VP), Donald KALPOKAS; Union of Moderate Parties (UMP), Maxime CARLOT Korman; Melanesian Progressive Party (MPP), Barak SOPE; National United Party (NUP), Walter LINI; Tan Union Party (TUP), Vincent BOULEKONE; Nagriamel Party, Jimmy STEVENS; Friend Melanesian Party, leader NA; People's Democratic Party (PDP), Sethy REGENVANU
note: the VP, MPP, TUP, and Nagriamel Party have formed a coalition called the United Front (UF) heading into the November 1995 elections
Member of: ACCT, ACP, AsDB, C, ESCAP, FAO, G-77, IBRD, ICAO, ICRM, IDA, IFC, IFRCS (associate), IMF, IMO, INTELSAT (nonsignatory user), IOC, ITU, NAM, SPARTECA, SPC, SPF, UN, UNCTAD, UNESCO, UNIDO, UPU, WHO, WMO
Diplomatic representation in US: Vanuatu does not have a mission in the US
US diplomatic representation: the ambassador to Papua New Guinea is accredited to Vanuatu
Flag: two equal horizontal bands of red (top) and green with a black isosceles triangle (based on the hoist side) all separated by a black-edged yellow stripe in the shape of a horizontal Y (the two points of the Y face the hoist side and enclose the triangle); centered in the triangle is a boar's tusk encircling two crossed namele leaves, all in yellow

Economy

Overview: The economy is based primarily on subsistence farming which provides a living for about 80% of the population. Fishing and tourism are the other mainstays of the economy, with 43,000 visitors in 1992. Mineral deposits are negligible; the country has no known petroleum deposits. A small light industry sector caters to the local market. Tax revenues come mainly from import duties.
National product: GDP—purchasing power parity—$200 million (1993 est.)
National product real growth rate: NA%
National product per capita: $1,200 (1993 est.)
Inflation rate (consumer prices): 2.3% (1992 est.)
Unemployment rate: NA%
Budget:
revenues: $90 million
expenditures: $103 million, including capital expenditures of $45 million (1989 est.)
Exports: $14.9 million (f.o.b., 1991)
commodities: copra, beef, cocoa, timber, coffee
partners: Netherlands, Japan, France, New Caledonia, Belgium
Imports: $74 million (f.o.b., 1991)
commodities: machines and vehicles, food and beverages, basic manufactures, raw materials and fuels, chemicals
partners: Australia 36%, Japan 13%, NZ 10%, France 8%, Fiji 8%
External debt: $40 million (yearend 1992)
Industrial production: growth rate 8.1% (1990); accounts for about 10% of GDP
Electricity:
capacity: 17,000 kW
production: 30 million kWh
consumption per capita: 181 kWh (1993)
Industries: food and fish freezing, wood processing, meat canning
Agriculture: export crops—coconuts, cocoa, coffee, fish; subsistence crops—taro, yams, coconuts, fruits, vegetables
Economic aid:
recipient: Western (non-US) countries, ODA and OOF bilateral commitments (1970-89), $606 million
Currency: 1 vatu (VT) = 100 centimes
Exchange rates: vatu (VT) per US$1—112.42 (December 1994), 116.41 (1994), 121.58 (1993), 113.39 (1992), 111.68 (1991), 116.57 (1990)
Fiscal year: calendar year

Transportation

Railroads: 0 km
Highways:
total: 1,027 km
paved: 240 km
unpaved: 787 km
Ports: Forari, Port-Vila, Santo (Espiritu Santo)
Merchant marine:
total: 116 ships (1,000 GRT or over) totaling 1,874,698 GRT/2,758,783 DWT
ships by type: bulk 52, cargo 18, chemical tanker 3, combination bulk 1, container 4, liquefied gas tanker 5, livestock carrier 1, oil tanker 5, refrigerated cargo 17, vehicle carrier 10
note: a flag of convenience registry; includes 21 countries among which are ships of the US 117, Japan 39, Netherlands 12, China 11, UAE 6, Greece 6, Canada 6, Hong Kong 4, Russia 2, Australia 2
Airports:
total: 31
with paved runways 2,438 to 3,047 m: 1
with paved runways 1,524 to 2,437 m: 1
with paved runways under 914 m: 17
with unpaved runways 1,524 to 2,438 m: 1
with unpaved runways 914 to 1,523 m: 11

Communications

Telephone system: 3,000 telephones
local: NA
intercity: NA
international: 1 INTELSAT (Pacific Ocean) earth station
Radio:
broadcast stations: AM 2, FM 0, shortwave 0
radios: NA
Television:
broadcast stations: 0
televisions: NA

Defense Forces

Branches: no regular military forces; Vanuatu Police Force (VPF; includes the paramilitary Vanuatu Mobile Force or VMF)
Defense expenditures: $NA, NA% of GDP

Venezuela

Geography

Location: Northern South America, bordering the Caribbean Sea and the North Atlantic Ocean, between Colombia and Guyana
Map references: South America
Area:
total area: 912,050 sq km
land area: 882,050 sq km
comparative area: slightly more than twice the size of California
Land boundaries: total 4,993 km, Brazil 2,200 km, Colombia 2,050 km, Guyana 743 km
Coastline: 2,800 km
Maritime claims:
contiguous zone: 15 nm
continental shelf: 200-m depth or to the depth of exploitation
exclusive economic zone: 200 nm
territorial sea: 12 nm
International disputes: claims all of Guyana west of the Essequibo River; maritime boundary dispute with Colombia in the Gulf of Venezuela
Climate: tropical; hot, humid; more moderate in highlands
Terrain: Andes Mountains and Maracaibo Lowlands in northwest; central plains (llanos); Guiana Highlands in southeast
Natural resources: petroleum, natural gas, iron ore, gold, bauxite, other minerals, hydropower, diamonds
Land use:
arable land: 3%
permanent crops: 1%
meadows and pastures: 20%
forest and woodland: 39%
other: 37%
Irrigated land: 2,640 sq km (1989 est.)
Environment:
current issues: sewage pollution of Lago de Valencia; oil and urban pollution of Lago de Maracaibo; deforestation; soil degradation; urban and industrial pollution, especially along the Caribbean coast
natural hazards: subject to floods, rockslides, mudslides; periodic droughts
international agreements: party to—Biodiversity, Climate Change, Endangered Species, Marine Life Conservation, Nuclear Test Ban, Ozone Layer Protection, Ship Pollution, Wetlands, Whaling; signed, but not ratified—Hazardous Wastes, Marine Dumping
Note: on major sea and air routes linking North and South America

People

Population: 21,004,773 (July 1995 est.)
Age structure:
0-14 years: 35% (female 3,650,705; male 3,795,032)
15-64 years: 60% (female 6,350,466; male 6,313,887)
65 years and over: 5% (female 486,020; male 408,663) (July 1995 est.)
Population growth rate: 2.1% (1995 est.)
Birth rate: 25.11 births/1,000 population (1995 est.)
Death rate: 4.57 deaths/1,000 population (1995 est.)
Net migration rate: 0.46 migrant(s)/1,000 population (1995 est.)
Infant mortality rate: 26.5 deaths/1,000 live births (1995 est.)
Life expectancy at birth:
total population: 73.31 years
male: 70.48 years
female: 76.29 years (1995 est.)
Total fertility rate: 2.97 children born/woman (1995 est.)
Nationality:
noun: Venezuelan(s)
adjective: Venezuelan
Ethnic divisions: mestizo 67%, white 21%, black 10%, Amerindian 2%
Religions: nominally Roman Catholic 96%, Protestant 2%
Languages: Spanish (official), native dialects spoken by about 200,000 Amerindians in the remote interior
Literacy: age 15 and over can read and write (1990)
total population: 90%
male: 91%
female: 89%
Labor force: 7.6 million
by occupation: services 63%, industry 25%, agriculture 12% (1993)

Government

Names:
conventional long form: Republic of Venezuela
conventional short form: Venezuela
local long form: Republica de Venezuela
local short form: Venezuela
Digraph: VE
Type: republic
Capital: Caracas
Administrative divisions: 21 states (estados, singular—estado), 1 territory* (territorio), 1 federal district** (distrito federal), and 1 federal dependency*** (dependencia federal); Amazonas*, Anzoategui, Apure, Aragua, Barinas, Bolivar, Carabobo, Cojedes, Delta Amacuro, Dependencias Federales***, Distrito Federal**, Falcon, Guarico, Lara, Merida, Miranda, Monagas, Nueva Esparta, Portuguesa, Sucre, Tachira, Trujillo, Yaracuy, Zulia
note: the federal dependency consists of 11 federally controlled island groups with a total of 72 individual islands
Independence: 5 July 1811 (from Spain)
National holiday: Independence Day, 5 July (1811)
Constitution: 23 January 1961
Legal system: based on Napoleonic code; judicial review of legislative acts in Cassation Court only; has not accepted compulsory ICJ jurisdiction
Suffrage: 18 years of age; universal
Executive branch:
chief of state and head of government: President Rafael CALDERA Rodriguez (since 2 February 1994); election last held 5 December 1993 (next to be held NA December 1998); results—Rafael CALDERA (National Convergence) 30.45%, Claudio FERMIN (AD) 23.59%, Oswaldo ALVAREZ PAZ (COPEI) 22.72%, Andres VELASQUEZ (Causa R) 21.94%, other 1.3%
cabinet: Council of Ministers; appointed by the president
Legislative branch: bicameral Congress of the Republic (Congreso de la Republica)
Senate (Senado): elections last held 5 December 1993 (next to be held NA December 1998); results—percent of vote by party NA; seats—(53 total) AD 18, COPEI 15, Causa R 9, MAS 5, National Convergence 6; note—3 former presidents (2 from AD, 1 from COPEI) hold lifetime senate seats
Chamber of Deputies (Camara de Diputados): elections last held 5 December 1993 (next to be held NA December 1998); results—AD 27.9%, COPEI 26.9%, MAS 12.4%, National Convergence 12.9%, Causa R 19.9%; seats—(203 total) AD 55, COPEI 53, MAS 24, National Convergence 26, Causa R 40, other 5
Judicial branch: Supreme Court of Justice (Corte Suprema de Justicia) Roberto YEPES, President
Political parties and leaders: National Convergence (Convergencia), Jose Miguel UZCATEGUI, president, Juan Jose CALDERA, national coordinator; Social Christian Party (COPEI), Luis HERRERA Campins, president, and Donald RAMIREZ, secretary general; Democratic Action (AD), Pedro PARIS Montesinos, president, and Luis ALFARO Ucero, secretary general; Movement Toward Socialism (MAS), Gustavo MARQUEZ, president, and Enrique OCHOA

Other political or pressure groups: FEDECAMARAS, a conservative business group; Venezuelan Confederation of Workers (CTV, labor organization dominated by the Democratic Action); VECINOS groups
Member of: AG, BCIE, CARICOM (observer), CDB, CG, ECLAC, FAO, G-11, G-15, G-19, G-24, G-77, GATT, IADB, IAEA, IBRD, ICAO, ICC, ICFTU, ICRM, IFAD, IFC, IFRCS, ILO, IMF, IMO, INTELSAT, INTERPOL, IOC, IOM, ISO, ITU, LAES, LAIA, MINURSO, NAM, OAS, ONUSAL, OPANAL, OPEC, PCA, RG, UN, UNCTAD, UNESCO, UNHCR, UNIDO, UNIKOM, UNMIH, UNPROFOR, UNU, UPU, WCL, WFTU, WHO, WIPO, WMO, WTO
Diplomatic representation in US:
chief of mission: Ambassador Pedro Luis ECHEVERRIA
chancery: 1099 30th Street NW, Washington, DC 20007
telephone: [1] (202) 342-2214
consulate(s) general: Boston, Chicago, Houston, Miami, New Orleans, New York, San Francisco, and San Juan (Puerto Rico)
US diplomatic representation:
chief of mission: Ambassador Jeffrey DAVIDOW
embassy: Avenida Francisco de Miranda and Avenida Principal de la Floresta, Caracas
mailing address: P. O. Box 62291, Caracas 1060-A; APO AA 34037
telephone: [58] (2) 285-2222, 3111
FAX: [58] (2) 285-0366
Flag: three equal horizontal bands of yellow (top), blue, and red with the coat of arms on the hoist side of the yellow band and an arc of seven white five-pointed stars centered in the blue band

Economy

Overview: Despite efforts to broaden the base of the economy, petroleum continues to play a dominant role. In 1994, as GDP declined 3.3%, the oil sector—which accounts for 24% of the total—enjoyed a 6% expansion, provided 45% of the budget revenues, and generated 70% of the export earnings. President CALDERA, who assumed office in February 1994, has used an interventionist, reactive approach to managing the economy, instituting price and foreign exchange controls in mid-year to slow inflation and stop the loss of foreign exchange reserves. The government claims it will remove these controls once inflationary pressures abate, but the $8 billion bailout of the banking sector in 1994 has made it difficult for the government to make good on its promise. Economic controls, coupled with political uncertainty driven by recurrent coup rumors, continue to deter foreign and domestic investment; private forecasters see the recession persisting for a third year in 1995.
National product: GDP—purchasing power parity—$178.3 billion (1994 est.)
National product real growth rate: -3.3% (1994 est.)
National product per capita: $8,670 (1994 est.)
Inflation rate (consumer prices): 71% (1994 est.)
Unemployment rate: 9% (1994 est.)
Budget:
revenues: $10.3 billion
expenditures: $14.6 billion, including capital expenditures of $103 million (1994 est.)
Exports: $15.2 billion (f.o.b., 1994 est.)
commodities: petroleum 72%, bauxite and aluminum, steel, chemicals, agricultural products, basic manufactures
partners: US and Puerto Rico 55%, Japan, Netherlands, Italy
Imports: $7.6 billion (f.o.b., 1994 est.)
commodities: raw materials, machinery and equipment, transport equipment, construction materials
partners: US 40%, Germany, Japan, Netherlands, Canada
External debt: $40.1 billion (1994)
Industrial production: growth rate -1.4% (1993 est.); accounts for 41% of GDP
Electricity:
capacity: 18,740,000 kW
production: 72 billion kWh
consumption per capita: 3,311 kWh (1993)
Industries: petroleum, iron-ore mining, construction materials, food processing, textiles, steel, aluminum, motor vehicle assembly
Agriculture: accounts for 6% of GDP; products—corn, sorghum, sugarcane, rice, bananas, vegetables, coffee, beef, pork, milk, eggs, fish; not self-sufficient in food other than meat
Illicit drugs: illicit producer of cannabis, opium, and coca leaf for the international drug trade on a small scale; however, large quantities of cocaine and heroin transit the country from Colombia; important money-laundering hub
Economic aid:
recipient: US commitments, including Ex-Im (FY70-86), $488 million; Communist countries (1970-89), $10 million
Currency: 1 bolivar (Bs) = 100 centimos
Exchange rates: bolivares (Bs) per US$1—169.570 (January 1995), 148.503 (1994), 90.826 (1993), 68.38 (1992), 56.82 (1991), 46.90 (1990)
Fiscal year: calendar year

Transportation

Railroads:
total: 542 km (363 km single track; 179 km privately owned)
standard gauge: 542 km 1.435-m gauge
Highways:
total: 81,000 km
paved: 31,200 km
unpaved: gravel 24,800 km; earth and unimproved earth 25,000 km
Inland waterways: 7,100 km; Rio Orinoco and Lago de Maracaibo accept oceangoing vessels
Pipelines: crude oil 6,370 km; petroleum products 480 km; natural gas 4,010 km
Ports: Amuay, Bajo Grande, El Tablazo, La Guaira, La Salina, Maracaibo, Matanzas, Palua, Puerto Cabello, Puerto la Cruz, Puerto Ordaz, Puerto Sucre, Punta Cardon
Merchant marine:
total: 39 ships (1,000 GRT or over) totaling 686,811 GRT/1,110,829 DWT
ships by type: bulk 4, cargo 11, combination bulk 1, liquefied gas tanker 2, oil tanker 15, passenger-cargo 1, roll-on/roll-off cargo 4, short-sea passenger 1
Airports:
total: 431
with paved runways over 3,047 m: 4
with paved runways 2,438 to 3,047 m: 11
with paved runways 1,524 to 2,437 m: 34
with paved runways 914 to 1,523 m: 65
with paved runways under 914 m: 191
with unpaved runways 1,524 to 2,438 m: 12
with unpaved runways 914 to 1,523 m: 114

Communications

Telephone system: 1,440,000 telephones; modern and expanding
local: NA
intercity: 3 domestic satellite earth stations
international: 3 submarine coaxial cables; 1 INTELSAT (Atlantic Ocean) earth station
Radio:
broadcast stations: AM 181, FM 0, shortwave 26
radios: NA
Television:
broadcast stations: 59
televisions: NA

Defense Forces

Branches: National Armed Forces (Fuerzas Armadas Nacionales or FAN) includes Ground Forces or Army (Fuerzas Terrestres or Ejercito), Naval Forces (Fuerzas Navales or Armada), Air Force (Fuerzas Aereas or Aviacion), Armed Forces of Cooperation or National Guard (Fuerzas Armadas de Cooperation or Guardia Nacional)
Manpower availability: males age 15-49 5,491,524; males fit for military service 3,981,190; males reach military age (18) annually 227,292 (1995 est.)
Defense expenditures: exchange rate conversion—$1.95 billion, 4% of GDP (1991)

Vietnam

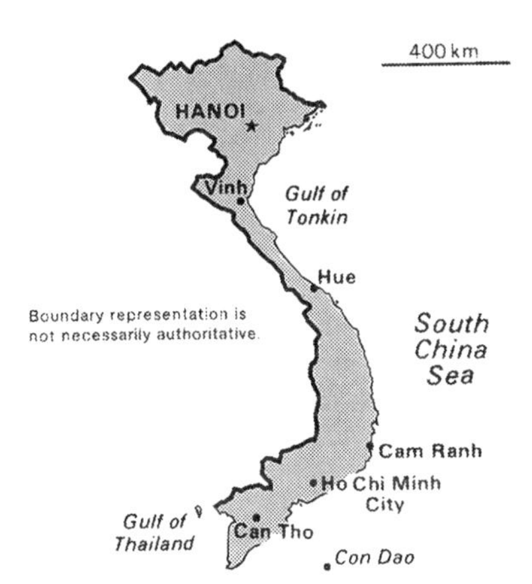

Geography

Location: Southeastern Asia, bordering the Gulf of Thailand, Gulf of Tonkin, and South China Sea, between China and Cambodia
Map references: Southeast Asia
Area:
total area: 329,560 sq km
land area: 325,360 sq km
comparative area: slightly larger than New Mexico
Land boundaries: total 3,818 km, Cambodia 982 km, China 1,281 km, Laos 1,555 km
Coastline: 3,444 km (excludes islands)
Maritime claims:
contiguous zone: 24 nm
continental shelf: 200 nm or to the edge of the continental margin
exclusive economic zone: 200 nm
territorial sea: 12 nm
International disputes: maritime boundary with Cambodia not defined; involved in a complex dispute over the Spratly Islands with China, Malaysia, Philippines, Taiwan, and possibly Brunei; unresolved maritime boundary with Thailand; maritime boundary dispute with China in the Gulf of Tonkin; Paracel Islands occupied by China but claimed by Vietnam and Taiwan
Climate: tropical in south; monsoonal in north with hot, rainy season (mid-May to mid-September) and warm, dry season (mid-October to mid-March)
Terrain: low, flat delta in south and north; central highlands; hilly, mountainous in far north and northwest
Natural resources: phosphates, coal, manganese, bauxite, chromate, offshore oil deposits, forests
Land use:
arable land: 22%
permanent crops: 2%
meadows and pastures: 1%
forest and woodland: 40%
other: 35%
Irrigated land: 18,300 sq km (1989 est.)
Environment:
current issues: logging and slash-and-burn agricultural practices are contributing to deforestation; soil degradation; water pollution and overfishing threatening marine life populations; inadequate supplies of potable water because of groundwater contamination
natural hazards: occasional typhoons (May to January) with extensive flooding
international agreements: party to—Biodiversity, Climate Change, Endangered Species, Environmental Modification, Law of the Sea, Ozone Layer Protection, Ship Pollution, Wetlands; signed, but not ratified—Nuclear Test Ban

People

Population: 74,393,324 (July 1995 est.)
Age structure:
0-14 years: 36% (female 13,225,916; male 13,918,321)
15-64 years: 59% (female 22,353,710; male 21,223,739)
65 years and over: 5% (female 2,236,453; male 1,435,185) (July 1995 est.)
Population growth rate: 1.71% (1995 est.)
Birth rate: 26.25 births/1,000 population (1995 est.)
Death rate: 7.6 deaths/1,000 population (1995 est.)
Net migration rate: -1.51 migrant(s)/1,000 population (1995 est.)
Infant mortality rate: 44.6 deaths/1,000 live births (1995 est.)
Life expectancy at birth:
total population: 65.72 years
male: 63.66 years
female: 67.91 years (1995 est.)
Total fertility rate: 3.21 children born/woman (1995 est.)
Nationality:
noun: Vietnamese (singular and plural)
adjective: Vietnamese
Ethnic divisions: Vietnamese 85%-90%, Chinese 3%, Muong, Thai, Meo, Khmer, Man, Cham
Religions: Buddhist, Taoist, Roman Catholic, indigenous beliefs, Islam, Protestant
Languages: Vietnamese (official), French, Chinese, English, Khmer, tribal languages (Mon-Khmer and Malayo-Polynesian)
Literacy: age 15 and over can read and write (1989)
total population: 88%
male: 93%
female: 83%
Labor force: 32.7 million
by occupation: agricultural 65%, industrial and service 35% (1990 est.)

Government

Names:
conventional long form: Socialist Republic of Vietnam
conventional short form: Vietnam
local long form: Cong Hoa Chu Nghia Viet Nam
local short form: Viet Nam
Abbreviation: SRV
Digraph: VM
Type: Communist state
Capital: Hanoi
Administrative divisions: 50 provinces (tinh, singular and plural), 3 municipalities* (thu do, singular and plural); An Giang, Ba Ria-Vung Tau, Bac Thai, Ben Tre, Binh Dinh, Binh Thuan, Can Tho, Cao Bang, Dac Lac, Dong Nai, Dong Thap, Gia Lai, Ha Bac, Ha Giang, Ha Noi*, Ha Tay, Ha Tinh, Hai Hung, Hai Phong*, Ho Chi Minh*, Hoa Binh, Khanh Hoa, Kien Giang, Kon Tum, Lai Chau, Lam Dong, Lang Son, Lao Cai, Long An, Minh Hai, Nam Ha, Nghe An, Ninh Binh, Ninh Thuan, Phu Yen, Quang Binh, Quang Nam-Da Nang, Quang Ngai, Quang Ninh, Quang Tri, Soc Trang, Son La, Song Be, Tay Ninh, Thai Binh, Thanh Hoa, Thua Thien-Hue, Tien Giang, Tra Vinh, Tuyen Quang, Vinh Long, Vinh Phu, Yen Bai
Independence: 2 September 1945 (from France)
National holiday: Independence Day, 2 September (1945)
Constitution: 15 April 1992
Legal system: based on Communist legal theory and French civil law system
Suffrage: 18 years of age; universal
Executive branch:
chief of state: President Le Duc ANH (since 23 September 1992)
head of government: Prime Minister Vo Van KIET (since 9 August 1991); First Deputy Prime Minister Phan Van KHAI (since 10 August 1991); Deputy Prime Minister Nguyen KHANH (since NA February 1987); Deputy Prime Minister Tran Duc LUONG (since NA February 1987)
cabinet: Cabinet; appointed by the president on proposal of the prime minister and ratification of the Assembly
Legislative branch: unicameral
National Assembly (Quoc-Hoi): elections last held 19 July 1992 (next to be held NA July 1997); results—VCP is the only party; seats—(395 total) VCP or VCP-approved 395
Judicial branch: Supreme People's Court
Political parties and leaders: only party—Vietnam Communist Party (VCP), DO MUOI, general secretary
Member of: ACCT, AsDB, ASEAN (observer), CCC, ESCAP, FAO, G-77, IAEA, IBRD, ICAO, ICRM, IDA, IFAD, IFC,

NAM, UN, UNCTAD, UNESCO, UNIDO, UPU, WCL, WFTU, WHO, WIPO, WMO, WTO
Diplomatic representation in US:
chief of mission: Liaison Officer Le Van BANG
liaison office: address NA, Washington, DC
mailing address: NA
telephone: NA
FAX: NA
note: negotiations between representatives of the US and Vietnam concluded 28 January 1995 with the signing of an agreement to establish liaison offices in Hanoi and Washington
US diplomatic representation:
chief of mission: Liaison Officer James HALL
liaison office: address NA, Hanoi
mailing address: NA
telephone: NA
FAX: NA
note: negotiations between representatives of the US and Vietnam concluded 28 January 1995 with the signing of an agreement to establish liaison offices in Hanoi and Washington
Flag: red with a large yellow five-pointed star in the center

Economy

Overview: Vietnam has made significant progress in recent years moving away from the planned economic model toward a more effective market-based economic system. Most prices are now fully decontrolled, and the Vietnamese currency has been effectively devalued and floated at world market rates. In addition, the scope for private sector activity has been expanded, primarily through decollectivization of the agricultural sector and introduction of laws giving legal recognition to private business. Nearly three-quarters of export earnings are generated by only two commodities, rice and crude oil. Led by industry and construction, the economy did well in 1993 and 1994 with output rising 7% and 9% respectively. However, the industrial sector remains burdened by noncompetitive state-owned enterprises the government is unwilling or unable to privatize. Unemployment looms as a serious problem with roughly 20% of the work force without jobs and with population growth swelling the ranks of the labor force yearly.
National product: GDP—purchasing power parity—$83.5 billion (1994 est.)
National product real growth rate: 8.8% (1994 est.)
National product per capita: $1,140 (1994 est.)
Inflation rate (consumer prices): 14.4% (1994)
Unemployment rate: 20% (1994 est.)
Budget:
revenues: $3.6 billion
expenditures: $4.5 billion, including capital expenditures of $NA (1994 est.)
Exports: $3.6 billion (f.o.b., 1994 est.)
commodities: petroleum, rice, agricultural products, marine products, coffee
partners: Japan, Singapore, Hong Kong, France, South Korea
Imports: $4.2 billion (f.o.b., 1994 est.)
commodities: petroleum products, machinery and equipment, steel products, fertilizer, raw cotton, grain
partners: Singapore, Japan, South Korea, France, Hong Kong, Taiwan
External debt: $4 billion Western countries; $4.5 billion CEMA debts primarily to Russia;
Industrial production: growth rate 13% (1994 est.); accounts for 21% of GDP
Electricity:
capacity: 2,200,000 kW
production: 9.7 billion kWh
consumption per capita: 125 kWh (1993)
Industries: food processing, textiles, machine building, mining, cement, chemical fertilizer, glass, tires, oil
Agriculture: accounts for 36% of GDP; paddy rice, corn, potatoes make up 50% of farm output; commercial crops (rubber, soybeans, coffee, tea, bananas) and animal products 50%; since 1989 self-sufficient in food staple rice; fish catch of 943,100 metric tons (1989 est.); note—the third largest exporter of rice in the World, behind the US and Thailand
Illicit drugs: opium producer and increasingly important transit point for Southeast Asian heroin destined for the US and Europe; growing opium addiction; small-scale heroin producer
Economic aid:
recipient: $2 billion in credits and grants pledged by international donors for 1995, Japan largest contributor with $650 million pledged for 1995
Currency: 1 new dong (D) = 100 xu
Exchange rates: new dong (D) per US$1—11,000 (October 1994), 10,800 (November 1993), 8,100 (July 1991), 7,280 (December 1990), 3,996 (March 1990)
Fiscal year: calendar year

Transportation

Railroads:
total: 3,059 km (including 224 km not restored to service after war damage)
standard gauge: 151 km 1.435-m gauge
narrow gauge: 2,454 km 1.000-m gauge
other gauge: 230 km NA-m dual gauge (three rails)
Highways:
total: 85,000 km
paved: 9,400 km
unpaved: gravel, improved earth 48,700 km; unimproved earth 26,900 km
Inland waterways: 17,702 km navigable; more than 5,149 km navigable at all times by vessels up to 1.8 meter draft
Pipelines: petroleum products 150 km
Ports: Da Nang, Haiphong, Ho Chi Minh City, Hon Gai, Qui Nhon, Nha Trang
Merchant marine:
total: 109 ships (1,000 GRT or over) totaling 449,963 GRT/932,837 DWT
ships by type: bulk 3, cargo 92, oil tanker 10, refrigerated cargo 3, roll-on/roll-off cargo 1
Airports:
total: 48
with paved runways over 3,047 m: 8
with paved runways 2,438 to 3,047 m: 3
with paved runways 1,524 to 2,437 m: 5
with paved runways 914 to 1,523 m: 13
with paved runways under 914 m: 7
with unpaved runways 1,524 to 2,438 m: 2
with unpaved runways 914 to 1,523 m: 5
with unpaved runways under 914 m: 5

Communications

Telephone system: NA telephones; 2 telephones/1,000 persons; the inadequacies of the obsolete switching equipment and cable system are a serious constraint on the business sector and on economic growth, and restrict access to the international links that Vietnam has established with most major countries; the telephone system is not generally available for private use
local: NA
intercity: NA
international: 3 satellite earth stations
Radio:
broadcast stations: AM NA, FM 228, shortwave 0
radios: 7 million (1991)
Television:
broadcast stations: 36 (repeaters 77)
televisions: 2.5 million (1991)

Defense Forces

Branches: People's Army of Vietnam (PAVN; includes Ground forces, Navy (includes Naval Infantry), and Air Force
Manpower availability: males age 15-49 18,799,370; males fit for military service 11,913,116; males reach military age (17) annually 742,394 (1995 est.)
Defense expenditures: exchange rate conversion—$435 million, 2.5% of GDP (1994)

Virgin Islands
(territory of the US)

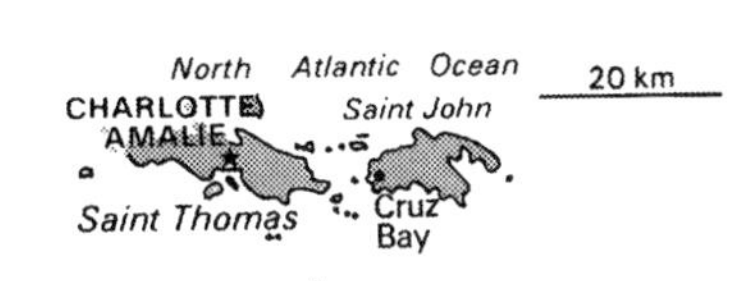

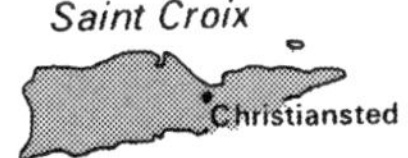

Geography

Location: Caribbean, islands between the Caribbean Sea and the North Atlantic Ocean, east of Puerto Rico
Map references: Central America and the Caribbean
Area:
total area: 352 sq km
land area: 349 sq km
comparative area: slightly less than twice the size of Washington, DC
Land boundaries: 0 km
Coastline: 188 km
Maritime claims:
exclusive economic zone: 200 nm
territorial sea: 12 nm
International disputes: none
Climate: subtropical, tempered by easterly tradewinds, relatively low humidity, little seasonal temperature variation; rainy season May to November
Terrain: mostly hilly to rugged and mountainous with little level land
Natural resources: sun, sand, sea, surf
Land use:
arable land: 15%
permanent crops: 6%
meadows and pastures: 26%
forest and woodland: 6%
other: 47%
Irrigated land: NA sq km
Environment:
current issues: lack of natural freshwater resources
natural hazards: rarely affected by hurricanes; frequent and severe droughts, floods, and earthquakes
international agreements: NA
Note: important location along the Anegada Passage—a key shipping lane for the Panama Canal; Saint Thomas has one of the best natural, deepwater harbors in the Caribbean

People

Population: 97,229 (July 1995 est.)
note: West Indian (45% born in the Virgin Islands and 29% born elsewhere in the West Indies) 74%, US mainland 13%, Puerto Rican 5%, other 8%
Age structure:
0-14 years: NA
15-64 years: NA
65 years and over: NA
Population growth rate: -0.29% (1995 est.)
Birth rate: 18.49 births/1,000 population (1995 est.)
Death rate: 5.2 deaths/1,000 population (1995 est.)
Net migration rate: -16.17 migrant(s)/1,000 population (1995 est.)
Infant mortality rate: 12.54 deaths/1,000 live births (1995 est.)
Life expectancy at birth:
total population: 75.29 years
male: 73.6 years
female: 77.2 years (1995 est.)
Total fertility rate: 2.41 children born/woman (1995 est.)
Nationality:
noun: Virgin Islander(s)
adjective: Virgin Islander
Ethnic divisions: black 80%, white 15%, other 5%
Religions: Baptist 42%, Roman Catholic 34%, Episcopalian 17%, other 7%
Languages: English (official), Spanish, Creole
Literacy: NA%
Labor force: 45,500 (1988)
by occupation: tourism 70%

Government

Names:
conventional long form: Virgin Islands of the United States
conventional short form: Virgin Islands
Digraph: VQ
Type: organized, unincorporated territory of the US administered by the Office of Territorial and International Affairs, US Department of the Interior
Capital: Charlotte Amalie
Administrative divisions: none (territory of the US)
National holiday: Transfer Day, 31 March (1917) (from Denmark to US)
Constitution: Revised Organic Act of 22 July 1954
Legal system: based on US
Suffrage: 18 years of age; universal; note—indigenous inhabitants are US citizens but do not vote in US presidential elections
Executive branch:
chief of state: President William Jefferson CLINTON (since 20 January 1993); Vice President Albert GORE, Jr. (since 20 January 1993)
head of government: Governor Dr. Roy L. SCHNEIDER (since 5 January 1995); Lieutenant Governor Kenneth E. MAPP (since 5 January 1995); election last held 22 November 1994 (next to be held NA November 1998); results—Dr. Roy L. SCHNEIDER (Independent Party) 54.7%, former Lt. Governor Derek HODGE 42.6%
Legislative branch: unicameral
Senate: elections last held 8 November 1994 (next to be held 5 November 1996); results—percent of vote by party NA; seats—(15 total) Democrats 7, Independents 7, Republican 1
US House of Representatives: elections last held 8 November 1994 (next to be held 5 November 1996); results—Victor O. FRAZER (Independent) 54.5%, Eileen R. PETERSON (Democrat) 45.5%; seats—(1 total); seat by party NA; note—the Virgin Islands elects one representative to the US House of Representatives
Judicial branch:
US District Court: handles civil matters over $50,000, felonies (persons 15 years of age and over), and federal cases
Territorial Court: handles civil matters up to $50,000, small claims, juvenile, domestic, misdemeanors, and traffic cases
Political parties and leaders: Democratic Party, Marilyn STAPLETON; Independent Citizens' Movement (ICM), Virdin C. BROWN; Republican Party, Charlotte-Poole DAVIS
Member of: ECLAC (associate), IOC
Diplomatic representation in US: none (territory of the US)
US diplomatic representation: none (territory of the US)
Flag: white with a modified US coat of arms in the center between the large blue initials V and I; the coat of arms shows an eagle holding an olive branch in one talon and three arrows in the other with a superimposed shield of vertical red and white stripes below a blue panel

Economy

Overview: Tourism is the primary economic activity, accounting for more than 70% of GDP and 70% of employment. The manufacturing sector consists of textile, electronics, pharmaceutical, and watch assembly plants. The agricultural sector is small, most food being imported. International business and financial services are a small but growing component of the economy. One of the world's largest petroleum refineries is at Saint Croix.
National product: GDP—purchasing power parity—$1.2 billion (1987 est.)
National product real growth rate: NA%
National product per capita: $11,000 (1987)
Inflation rate (consumer prices): NA%
Unemployment rate: 3.7% (1992)
Budget:
revenues: $364.4 million
expenditures: $364.4 million, including capital expenditures of $NA (1990 est.)
Exports: $2.8 billion (f.o.b., 1990)
commodities: refined petroleum products
partners: US, Puerto Rico

Virgin Islands *(continued)*

Imports: $3.3 billion (c.i.f., 1990)
commodities: crude oil, foodstuffs, consumer goods, building materials
partners: US, Puerto Rico
External debt: $NA
Industrial production: growth rate 12% (year NA); accounts for NA% of GDP
Electricity:
capacity: 320,000 kW
production: 970 million kWh
consumption per capita: 9,172 kWh (1993)
Industries: tourism, petroleum refining, watch assembly, rum distilling, construction, pharmaceuticals, textiles, electronics
Agriculture: truck gardens, food crops (small scale), fruit, sorghum, Senepol cattle
Economic aid:
recipient: Western (non-US) countries, ODA and OOF bilateral commitments (1970-89), $42 million
Currency: 1 United States dollar (US$) = 100 cents
Exchange rates: US currency is used
Fiscal year: 1 October—30 September

Transportation

Railroads: 0 km
Highways:
total: 856 km
paved: NA
unpaved: NA
Ports: Charlotte Amalie, Christiansted, Cruz Bay, Port Alucroix
Merchant marine: none
Airports:
total: 2
with paved runways 1,524 to 2,437 m: 2
note: international airports on Saint Thomas and Saint Croix

Communications

Telephone system: 58,931 telephones; modern telephone system using fiber-optic cable, submarine cable, microwave radio, and satellite facilities
local: NA
intercity: NA
international: NA
Radio:
broadcast stations: AM 4, FM 8, shortwave 0 (1988)
radios: 98,000
Television:
broadcast stations: 4 (1988)
televisions: 63,000

Defense Forces

Note: defense is the responsibility of the US

Wake Island
(territory of the US)

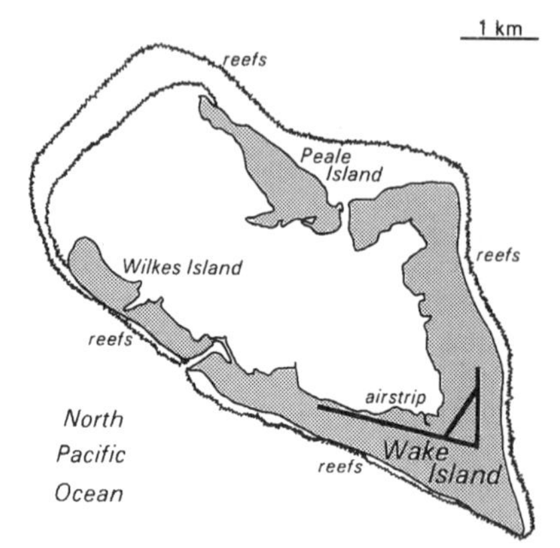

Geography

Location: Oceania, island in the North Pacific Ocean, about two-thirds of the way from Hawaii to the Northern Mariana Islands
Map references: Oceania
Area:
total area: 6.5 sq km
land area: 6.5 sq km
comparative area: about 11 times the size of The Mall in Washington, DC
Land boundaries: 0 km
Coastline: 19.3 km
Maritime claims:
exclusive economic zone: 200 nm
territorial sea: 12 nm
International disputes: claimed by the Republic of the Marshall Islands
Climate: tropical
Terrain: atoll of three coral islands built up on an underwater volcano; central lagoon is former crater, islands are part of the rim; average elevation less than 4 meters
Natural resources: none
Land use:
arable land: 0%
permanent crops: 0%
meadows and pastures: 0%
forest and woodland: 0%
other: 100%
Irrigated land: 0 sq km
Environment:
current issues: NA
natural hazards: occasional typhoons
international agreements: NA
Note: strategic location in the North Pacific Ocean; emergency landing location for transpacific flights

People

Population: 302 (July 1995 est.)
Population growth rate: 0% (1995 est.)
Birth rate: NA births/1,000 population
Death rate: NA deaths/1,000 population
Net migration rate: NA migrant(s)/1,000 population
Infant mortality rate: NA deaths/1,000 live births
Life expectancy at birth:
total population: NA years
male: NA years
female: NA years
Total fertility rate: NA children born/woman

Government

Names:
conventional long form: none
conventional short form: Wake Island
Digraph: WQ
Type: unincorporated territory of the US administered by the US Army and Strategic Defense Command since 1 October 1994
Capital: none; administered from Washington, DC
Independence: none (territory of the US)
Flag: the US flag is used

Economy

Overview: Economic activity is limited to providing services to US military personnel and contractors located on the island. All food and manufactured goods must be imported.
Electricity: supplied by US military

Transportation

Railroads: 0 km
Ports: none; two offshore anchorages for large ships
Merchant marine: none
Airports:
total: 1
with paved runways 2,438 to 3,047 m: 1
Note: formerly an important commercial aviation base, now used by US military, some commercial cargo planes, as well as the US Army Space and Strategic Defense Command for missile launches

Communications

Telephone system: NA telephones; satellite communications; 1 Autovon circuit off the Overseas Telephone System (OTS)
local: NA
intercity: NA
international: NA
Radio:
broadcast stations: AM 0, FM NA, shortwave NA
radios: NA
note: Armed Forces Radio/Television Service (AFRTS) radio service provided by satellite
Television:
broadcast stations: NA
televisions: NA
note: Armed Forces Radio/Television Service (AFRTS) television service provided by satellite
Note: formerly an important commercial aviation base, now used by US military, as well as the US Army Space and Strategic Defense Command for missile launches

Defense Forces

Note: defense is the responsibility of the US

Wallis and Futuna

(overseas territory of France)

50 km

MATA-UTU

Île Uvéa

South Pacific Ocean

Île Futuna

Leava

Île Alofi

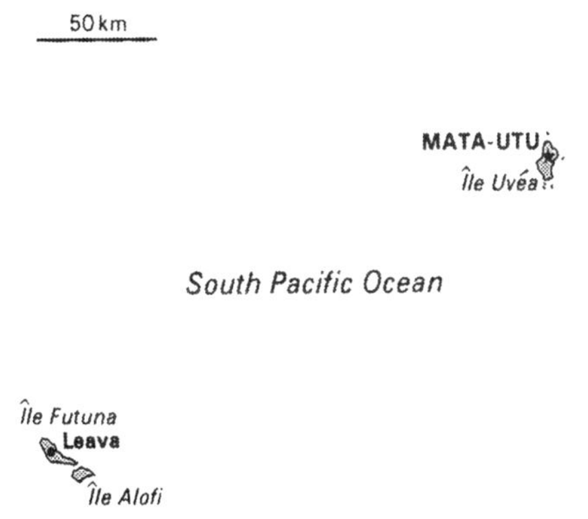

Geography

Location: Oceania, islands in the South Pacific Ocean, about two-thirds of the way from Hawaii to New Zealand
Map references: Oceania
Area:
total area: 274 sq km
land area: 274 sq km
comparative area: slightly larger than Washington, DC
note: includes Ile Uvea (Wallis Island), Ile Futuna (Futuna Island), Ile Alofi, and 20 islets
Land boundaries: 0 km
Coastline: 129 km
Maritime claims:
exclusive economic zone: 200 nm
territorial sea: 12 nm
International disputes: none
Climate: tropical; hot, rainy season (November to April); cool, dry season (May to October)
Terrain: volcanic origin; low hills
Natural resources: negligible
Land use:
arable land: 5%
permanent crops: 20%
meadows and pastures: 0%
forest and woodland: 0%
other: 75%
Irrigated land: NA sq km
Environment:
current issues: deforestation (only small portions of the original forests remain) largely as a result of the continued use of wood as the main fuel source; as a consequence of cutting down the forests, the mountainous terrain of Futuna is particularly prone to erosion; there are no permanent settlements on Alofi because of the lack of natural fresh water resources
natural hazards: NA
international agreements: NA
Note: both island groups have fringing reefs

People

Population: 14,499 (July 1995 est.)
Age structure:
0-14 years: NA
15-64 years: NA
65 years and over: NA
Population growth rate: 1.11% (1995 est.)
Birth rate: 25.06 births/1,000 population (1995 est.)
Death rate: 5.14 deaths/1,000 population (1995 est.)
Net migration rate: -8.85 migrant(s)/1,000 population (1995 est.)
Infant mortality rate: 24.92 deaths/1,000 live births (1995 est.)
Life expectancy at birth:
total population: 72.24 years
male: 71.62 years
female: 72.9 years (1995 est.)
Total fertility rate: 3.11 children born/woman (1995 est.)
Nationality:
noun: Wallisian(s), Futunan(s), or Wallis and Futuna Islanders
adjective: Wallisian, Futunan, or Wallis and Futuna Islander
Ethnic divisions: Polynesian
Religions: Roman Catholic
Languages: French, Wallisian (indigenous Polynesian language)
Literacy: age 15 and over can read and write (1969)
total population: 50%
male: 50%
female: 51%
Labor force: NA
by occupation: agriculture, livestock, and fishing 80%, government 4% (est.)

Government

Names:
conventional long form: Territory of the Wallis and Futuna Islands
conventional short form: Wallis and Futuna
local long form: Territoire des Iles Wallis et Futuna
local short form: Wallis et Futuna
Digraph: WF
Type: overseas territory of France
Capital: Mata-Utu (on Ile Uvea)
Administrative divisions: none (overseas territory of France)
Independence: none (overseas territory of France)
Constitution: 28 September 1958 (French Constitution)
Legal system: French legal system
Suffrage: 18 years of age; universal
Executive branch:
chief of state: President Francois MITTERRAND (since 21 May 1981)
head of government: High Administrator Philippe LEGRIX (since NA); President of the Territorial Assembly Soane Mani UHILA (since NA March 1992)
cabinet: Council of the Territory consists of 3 kings and 3 members appointed by the high administrator on advice of the Territorial Assembly
note: there are three traditional kings with limited powers
Legislative branch: unicameral
Territorial Assembly (Assemblee Territoriale): elections last held 15 March 1987 (next to be held NA); results—percent of vote by party NA; seats—(20 total) RPR 7, UPL 5, UDF 4, UNF 4
French Senate: elections last held 24 September 1989 (next to be held by NA September 1998); results—percent of vote by party NA; seats—(1 total) RPR 1
French National Assembly: elections last held 21 and 28 March 1992 (next to be held by NA September 1996); results—percent of vote by party NA; seats—(1 total) MRG 1; note—Wallis and Futuna elect one deputy
Judicial branch: none; justice generally administered under French law by the chief administrator, but the three traditional kings administer customary law and there is a magistrate in Mata-Utu
Political parties and leaders: Rally for the Republic (RPR); Union Populaire Locale (UPL); Union Pour la Democratie Francaise (UDF); Lua kae tahi (Giscardians); Mouvement des Radicaux de Gauche (MRG)
Member of: FZ, SPC
Diplomatic representation in US: none (overseas territory of France)
US diplomatic representation: none (overseas territory of France)
Flag: the flag of France is used

Economy

Overview: The economy is limited to traditional subsistence agriculture, with about 80% of the labor force earning its livelihood from agriculture (coconuts and vegetables), livestock (mostly pigs), and fishing. About 4% of the population is employed in government. Revenues come from French Government subsidies, licensing of fishing rights to Japan and South Korea, import taxes, and remittances from expatriate workers in New Caledonia. Wallis and Futuna imports food—particularly sugar and beef—fuel, clothing, machinery, and transport equipment, but its exports are negligible, consisting of copra and handicrafts.
National product: GDP—purchasing power parity—$28.7 million (1994 est.)
National product real growth rate: NA%
National product per capita: $2,000 (1994 est.)
Inflation rate (consumer prices): NA%
Unemployment rate: NA%
Budget:
revenues: $2.7 million
expenditures: $2.7 million, including capital expenditures of $NA (1983 est.)
Exports: $6.6 million (f.o.b., 1986)

Wallis and Futuna *(continued)*

commodities: copra, handicrafts
partners: NA
Imports: $13.3 million (c.i.f., 1984)
commodities: foodstuffs, manufactured goods, transportation equipment, fuel, clothing
partners: France, Australia, New Zealand
External debt: $NA
Industrial production: growth rate NA%
Electricity:
capacity: 1,200 kW
production: 1 million kWh
consumption per capita: 70 kWh (1990)
Industries: copra, handicrafts, fishing, lumber
Agriculture: dominated by coconut production, with subsistence crops of yams, taro, bananas, and herds of pigs and goats
Economic aid:
recipient: Western (non-US) countries, ODA and OOF bilateral commitments (1970-89), $118 million
Currency: 1 CFP franc (CFPF) = 100 centimes
Exchange rates: Comptoirs Francais du Pacifique francs (CFPF) per US$1—96.25 (January 1995), 100.94 (1994), 102.96 (1993), 96.24 (1992), 102.57 (1991), 99.0 (1990); note—linked at the rate of 18.18 to the French franc
Fiscal year: NA

Transportation

Railroads: 0 km
Highways:
total: 120 km (Ile Uvea 100 km, Ile Futuna 20km)
paved: 16 km (on Il Uvea)
unpaved: 104 km (Ile Uvea 84 km, Ile Futuna 20 km)
Inland waterways: none
Ports: Leava, Mata-Utu
Merchant marine:
total: 1 oil tanker (1,000 GRT or over) totaling 26,000 GRT/40,000 DWT
Airports:
total: 2
with paved runways 1,524 to 2,437 m: 1
with unpaved runways 914 to 1,523 m: 1

Communications

Telephone system: 225 telephones
local: NA
intercity: NA
international: NA
Radio:
broadcast stations: AM 1, FM 0, shortwave 0
radios: NA
Television:
broadcast stations: 0
televisions: NA

Defense Forces

Note: defense is the responsibility of France

West Bank

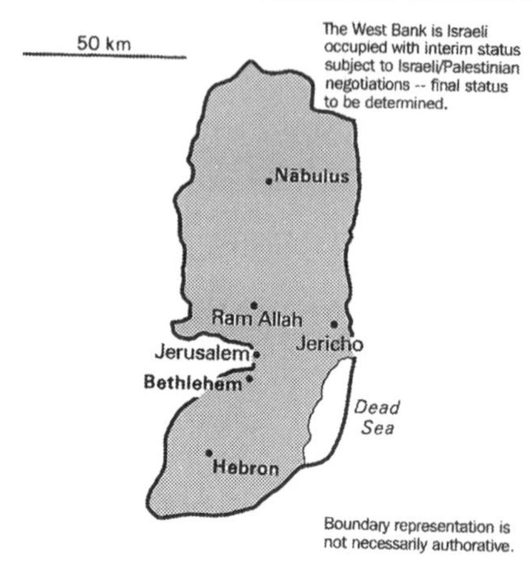

Note: The Israel-PLO Declaration of Principles on Interim Self-Government Arrangements ("the DOP"), signed in Washington on 13 September 1993, provides for a transitional period not exceeding five years of Palestinian interim self-government in the Gaza Strip and the West Bank. Under the DOP, final status negotiations are to begin no later than the beginning of the third year of the transitional period.

Geography

Location
Middle East, west of Jordan
Map references: Middle East
Area:
total area: 5,860 sq km
land area: 5,640 sq km
comparative area: slightly larger than Delaware
note: includes West Bank, Latrun Salient, and the northwest quarter of the Dead Sea, but excludes Mt. Scopus; East Jerusalem and Jerusalem No Man's Land are also included only as a means of depicting the entire area occupied by Israel in 1967
Land boundaries: total 404 km, Israel 307 km, Jordan 97 km
Coastline: 0 km (landlocked)
Maritime claims: none; landlocked
International disputes: West Bank and Gaza Strip are Israeli occupied with interim status subject to Israeli/Palestinian negotiations—final status to be determined
Climate: temperate, temperature and precipitation vary with altitude, warm to hot summers, cool to mild winters
Terrain: mostly rugged dissected upland, some vegetation in west, but barren in east
Natural resources: negligible
Land use:
arable land: 27%
permanent crops: 0%
meadows and pastures: 32%
forest and woodland: 1%
other: 40%
Irrigated land: NA sq km
Environment:
current issues: NA
natural hazards: NA
international agreements: NA
Note: landlocked; highlands are main recharge area for Israel's coastal aquifers; there are 199 Jewish settlements and civilian land use sites in the West Bank and 25 in East Jerusalem (August 1994 est.)

People

Population: 1,319,991 (July 1995 est.)
note: in addition, there are 122,000 Jewish settlers in the West Bank and 149,000 in East Jerusalem (August 1994 est.)
Age structure:
0-14 years: 46% (female 293,269; male 308,775)
15-64 years: 51% (female 335,193; male 337,722)
65 years and over: 3% (female 25,759; male 19,273) (July 1995 est.)
Population growth rate: 3.499% (1995 est.)
Birth rate: 39.83 births/1,000 population (1995 est.)
Death rate: 4.84 deaths/1,000 population (1995 est.)
Net migration rate: 0 migrant(s)/1,000 population (1995 est.)
Infant mortality rate: 29.7 deaths/1,000 live births (1995 est.)
Life expectancy at birth:
total population: 71.42 years
male: 69.91 years
female: 73.00 years (1995 est.)
Total fertility rate: 5.34 children born/woman (1995 est.)
Nationality:
noun: NA
adjective: NA
Ethnic divisions: Palestinian Arab and other 83%, Jewish 17%
Religions: Muslim 75% (predominantly Sunni), Jewish 17%, Christian and other 8%
Languages: Arabic, Hebrew spoken by Israeli settlers, English widely understood
Literacy: NA%
Labor force: NA
by occupation: construction 28.2%, agriculture 21.8%, industry 14.5%, commerce, restaurants, and hotels 12.6%, other services 22.9% (1991)
note: excluding Jewish settlers

Government

Note: Under the Israeli-PLO Declaration of Principles on Interim Self-Government Arragements ("the DOP"), Israel agreed to transfer certain powers and responsibilities to the Palestinian Authority, and subsequently to an elected Palestinian Council, as part of interim self-governing arrangements in the

West Bank and Gaza Strip. A transfer of powers and responsibilities for the Gaza Strip and Jericho has taken place pursuant to the Israel-PLO 4 May 1994 Cairo Agreement on the Gaza Strip and the Jericho Area. A transfer of powers and responsibilities in certain spheres for the rest of the West Bank has taken place pursuant to the Israel-PLO 29 August 1994 Agreement on Preparatory Transfer of Powers and Responsibilities. The DOP provides that Israel will retain responsibility during the transitional period for external security and for internal security and public order of settlements and Israelis. Final status is to be determined through direct negotiations within five years.
Names:
conventional long form: none
conventional short form: West Bank
Digraph: WE

Economy

Overview: Economic progress in the West Bank has been hampered by Israeli military administration and the effects of the Palestinian uprising (intifadah). Industries using advanced technology or requiring sizable investment have been discouraged by a lack of local capital and restrictive Israeli policies. Capital investment consists largely of residential housing, not productive assets that would enable local Palestinian firms to compete with Israeli industry. GDP has been substantially supplemented by remittances of workers employed in Israel and Persian Gulf states. Such transfers from the Gulf dropped after Iraq invaded Kuwait in August 1990. In the wake of the Persian Gulf crisis, many Palestinians have returned to the West Bank, increasing unemployment, and export revenues have dropped because of the decline of markets in Jordan and the Gulf states. Israeli measures to curtail the intifadah also have added to unemployment and lowered living standards. The area's economic situation has worsened since Israel's partial closure of the territories in 1993.
National product: GDP—purchasing power parity—$4 billion (1994 est.)
National product real growth rate: NA%
National product per capita: $2,800 (1994 est.)
Inflation rate (consumer prices): 6.8% (1993)
Unemployment rate: 35% (1994 est.)
Budget:
revenues: $43.4 million
expenditures: $43.7 million, including capital expenditures of $NA (FY89/90)
Exports: $217 million (f.o.b., 1992)
commodities: olives, fruit, vegetables
partners: Jordan, Israel
Imports: $867 million (c.i.f., 1992)
commodities: food, consumer goods, construction materials
partners: Jordan, Israel
External debt: $NA
Industrial production: growth rate NA%
Electricity:
capacity: NA kW
production: NA kWh
consumption per capita: NA kWh
note: most electricity imported from Israel; East Jerusalem Electric Company buys and distributes electricity to Palestinians in East Jerusalem and its concession in the West Bank; the Israel Electric Company directly supplies electricity to most Jewish residents and military facilities; at the same time, some Palestinian municipalities, such as Nabulus and Janin, generate their own electricity from small power plants
Industries: generally small family businesses that produce cement, textiles, soap, olive-wood carvings, and mother-of-pearl souvenirs; the Israelis have established some small-scale modern industries in the settlements and industrial centers
Agriculture: olives, citrus and other fruits, vegetables, beef, and dairy products
Economic aid: $NA
Currency: 1 new Israeli shekel (NIS) = 100 new agorot; 1 Jordanian dinar (JD) = 1,000 fils
Exchange rates: new Israeli shekels (NIS) per US$1—3.0270 (December 1994), 3.0111 (1994), 2.8301 (1993), 2.4591 (1992), 2.2791 (1991), 2.0162 (1990); Jordanian dinars (JD) per US$1—0.6995 (January 1995), 0.6987 (1994), 0.6928 (1993), 0.6797 (1992), 0.6808 (1991), 0.6636 (1990)
Fiscal year: calendar year (since 1 January 1992)

Transportation

Railroads: 0 km
Highways:
total: NA
paved: NA
unpaved: NA
note: small road network; Israelis have developed many highways to service Jewish settlements
Ports: none
Airports:
total: 2
with paved runways 1,524 to 2,437 m: 1
with paved runways under 914 m: 1

Communications

Telephone system: NA telephones; note—8% of Palestinian households have telephones (1992 est.)
local: NA
intercity: NA
international: NA
note: Israeli company BEZEK is responsible for communication services in the West Bank
Radio:
broadcast stations: AM 1, FM 0, shortwave 0
radios: NA; note—82% of Palestinian households have radios (1992 est.)
Television:
broadcast stations: 0; note—1 planned for Jericho
televisions: NA; note—54% of Palestinian households have televisions (1992 est.)

Defense Forces

Branches: NA
Defense expenditures: $NA, NA% of GDP

Western Sahara

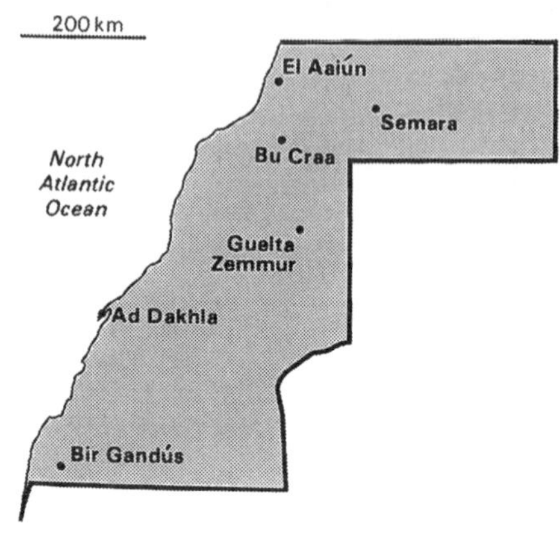

Geography

Location: Northern Africa, bordering the North Atlantic Ocean, between Mauritania and Morocco
Map references: Africa
Area:
total area: 266,000 sq km
land area: 266,000 sq km
comparative area: slightly smaller than Colorado
Land boundaries: total 2,046 km, Algeria 42 km, Mauritania 1,561 km, Morocco 443 km
Coastline: 1,110 km
Maritime claims: contingent upon resolution of sovereignty issue
International disputes: claimed and administered by Morocco, but sovereignty is unresolved and the UN is attempting to hold a referendum on the issue; the UN-administered cease-fire has been currently in effect since September 1991
Climate: hot, dry desert; rain is rare; cold offshore air currents produce fog and heavy dew
Terrain: mostly low, flat desert with large areas of rocky or sandy surfaces rising to small mountains in south and northeast
Natural resources: phosphates, iron ore
Land use:
arable land: 0%
permanent crops: 0%
meadows and pastures: 19%
forest and woodland: 0%
other: 81%
Irrigated land: NA sq km
Environment:
current issues: sparse water and arable land
natural hazards: hot, dry, dust/sand-laden sirocco wind can occur during winter and spring; widespread harmattan haze exists 60% of time, often severely restricting visibility
international agreements: NA

People

Population: 217,211 (July 1995 est.)
Age structure:
0-14 years: NA
15-64 years: NA
65 years and over: NA
Population growth rate: 2.48% (1995 est.)
Birth rate: 46.9 births/1,000 population (1995 est.)
Death rate: 18.52 deaths/1,000 population (1995 est.)
Net migration rate: -3.62 migrant(s)/1,000 population (1995 est.)
Infant mortality rate: 148.95 deaths/1,000 live births (1995 est.)
Life expectancy at birth:
total population: 46.31 years
male: 45.34 years
female: 47.59 years (1995 est.)
Total fertility rate: 6.91 children born/woman (1995 est.)
Nationality:
noun: Sahrawi(s), Sahraoui(s)
adjective: Sahrawian, Sahraouian
Ethnic divisions: Arab, Berber
Religions: Muslim
Languages: Hassaniya Arabic, Moroccan Arabic
Literacy: NA%
Labor force: 12,000
by occupation: animal husbandry and subsistence farming 50%

Government

Names:
conventional long form: none
conventional short form: Western Sahara
Digraph: WI
Type: legal status of territory and question of sovereignty unresolved; territory contested by Morocco and Polisario Front (Popular Front for the Liberation of the Saguia el Hamra and Rio de Oro), which in February 1976 formally proclaimed a government in exile of the Sahrawi Arab Democratic Republic (SADR); territory partitioned between Morocco and Mauritania in April 1976, with Morocco acquiring northern two-thirds; Mauritania, under pressure from Polisario guerrillas, abandoned all claims to its portion in August 1979; Morocco moved to occupy that sector shortly thereafter and has since asserted administrative control; the Polisario's government in exile was seated as an OAU member in 1984; guerrilla activities continued sporadically, until a UN-monitored cease-fire was implemented 6 September 1991
Capital: none
Administrative divisions: none (under de facto control of Morocco)
Executive branch: none
Member of: none
Diplomatic representation in US: none
US diplomatic representation: none

Economy

Overview: Western Sahara, a territory poor in natural resources and having little rainfall, depends on pastoral nomadism, fishing, and phosphate mining as the principal sources of income for the population. Most of the food for the urban population must be imported. All trade and other economic activities are controlled by the Moroccan Government. Incomes and standards of living are substantially below the Moroccan level.
National product: GDP $NA
National product real growth rate: NA%
National product per capita: $NA
Inflation rate (consumer prices): NA%
Unemployment rate: NA%
Budget:
revenues: $NA
expenditures: $NA, including capital expenditures of $NA
Exports: $8 million (f.o.b., 1982 est.)
commodities: phosphates 62%
partners: Morocco claims and administers Western Sahara, so trade partners are included in overall Moroccan accounts
Imports: $30 million (c.i.f., 1982 est.)
commodities: fuel for fishing fleet, foodstuffs
partners: Morocco claims and administers Western Sahara, so trade partners are included in overall Moroccan accounts
External debt: $NA
Industrial production: growth rate NA%
Electricity:
capacity: 60,000 kW
production: 79 million kWh
consumption per capita: 339 kWh (1993)
Industries: phosphate mining, handicrafts
Agriculture: limited largely to subsistence agriculture and fishing; some barley is grown in nondrought years; fruit and vegetables are grown in the few oases; food imports are essential; camels, sheep, and goats are kept by the nomadic natives; cash economy exists largely for the garrison forces
Economic aid: $NA
Currency: 1 Moroccan dirham (DH) = 100 centimes
Exchange rates: Moroccan dirhams (DH) per US$1—8.892 (January 1995), 9.203 (1994), 9.299 (1993), 8.538 (1992), 8.707 (1991), 8.242 (1990)
Fiscal year: NA

Transportation

Railroads: 0 km
Highways:
total: 6,200 km
unpaved: gravel 1,450 km; improved, unimproved earth, tracks 4,750 km
Ports: Ad Dakhla, Cabo Bojador, El Aaiun

Western Samoa

Airports:
total: 14
with paved runways 2,438 to 3,047 m: 3
with paved runways under 914 m: 3
with unpaved runways 1,524 to 2,438 m: 1
with unpaved runways 914 to 1,523 m: 7

Communications

Telephone system: 2,000 telephones; sparse and limited system
local: NA
intercity: NA
international: tied into Morocco's system by microwave radio relay, troposcatter, and 2 INTELSAT (Atlantic Ocean) earth stations linked to Rabat, Morocco
Radio:
broadcast stations: AM 2, FM 0, shortwave 0
radios: NA
Television:
broadcast stations: 2
televisions: NA

Defense Forces

Branches: NA
Defense expenditures: $NA, NA% of GDP

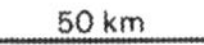

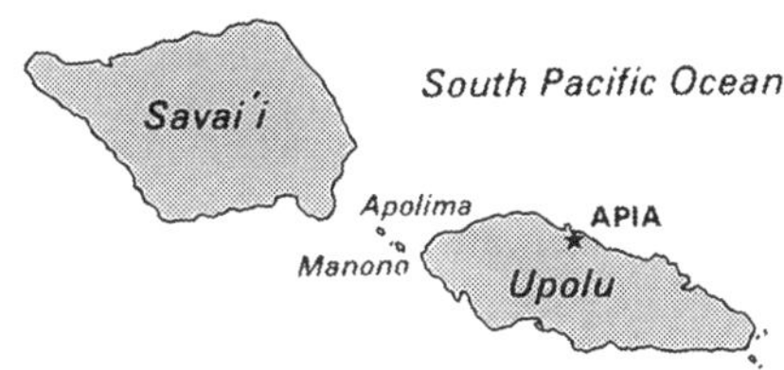

Geography

Location: Oceania, group of islands in the South Pacific Ocean, about one-half of the way from Hawaii to New Zealand
Map references: Oceania
Area:
total area: 2,860 sq km
land area: 2,850 sq km
comparative area: slightly smaller than Rhode Island
Land boundaries: 0 km
Coastline: 403 km
Maritime claims:
exclusive economic zone: 200 nm
territorial sea: 12 nm
International disputes: none
Climate: tropical; rainy season (October to March), dry season (May to October)
Terrain: narrow coastal plain with volcanic, rocky, rugged mountains in interior
Natural resources: hardwood forests, fish
Land use:
arable land: 19%
permanent crops: 24%
meadows and pastures: 0%
forest and woodland: 47%
other: 10%
Irrigated land: NA sq km
Environment:
current issues: soil erosion
natural hazards: occasional typhoons; active volcanism
international agreements: party to—Biodiversity, Climate Change, Nuclear Test Ban, Ozone Layer Protection; signed, but not ratified—Law of the Sea

People

Population: 209,360 (July 1995 est.)
Age structure:
0-14 years: 40% (female 41,503; male 42,844)
15-64 years: 56% (female 55,683; male 61,065)
65 years and over: 4% (female 4,323; male 3,942) (July 1995 est.)
Population growth rate: 2.37% (1995 est.)
Birth rate: 31.74 births/1,000 population (1995 est.)
Death rate: 5.88 deaths/1,000 population (1995 est.)
Net migration rate: -2.14 migrant(s)/1,000 population (1995 est.)
Infant mortality rate: 35.5 deaths/1,000 live births (1995 est.)
Life expectancy at birth:
total population: 68.38 years
male: 65.99 years
female: 70.88 years (1995 est.)
Total fertility rate: 4.04 children born/woman (1995 est.)
Nationality:
noun: Western Samoan(s)
adjective: Western Samoan
Ethnic divisions: Samoan 92.6%, Euronesians 7% (persons of European and Polynesian blood), Europeans 0.4%
Religions: Christian 99.7% (about one-half of population associated with the London Missionary Society; includes Congregational, Roman Catholic, Methodist, Latter Day Saints, Seventh-Day Adventist)
Languages: Samoan (Polynesian), English
Literacy: age 15 and over can read and write (1971)
total population: 97%
male: 97%
female: 97%
Labor force: NA
by occupation: agriculture 60%

Government

Names:
conventional long form: Independent State of Western Samoa
conventional short form: Western Samoa
Digraph: WS
Type: constitutional monarchy under native chief
Capital: Apia
Administrative divisions: 11 districts; A'ana, Aiga-i-le-Tai, Atua, Fa'asaleleaga, Gaga'emauga, Gagaifomauga, Palauli, Satupa'itea, Tuamasaga, Va'a-o-Fonoti, Vaisigano
Independence: 1 January 1962 (from UN trusteeship administered by New Zealand)
National holiday: National Day, 1 June (1962)
Constitution: 1 January 1962
Legal system: based on English common law and local customs; judicial review of legislative acts with respect to fundamental rights of the citizen; has not accepted compulsory ICJ jurisdiction
Suffrage: 21 years of age; universal, but only matai (head of family) are able to run for the Legislative Assembly

Executive branch:
chief of state: Chief Susuga Malietoa TANUMAFILI II (Co-Chief of State from 1 January 1962 until becoming sole Chief of State on 5 April 1963)
head of government: Prime Minister TOFILAU Eti Alesana (since 7 April 1988)
cabinet: Cabinet; appointed by the head of state with the prime minister's advice
Legislative branch: unicameral
Legislative Assembly (Fono): elections last held 5 April 1991 (next to be held by NA 1996); results—percent of vote by party NA; seats—(47 total) HRPP 28, SNDP 18, independents 1
note: only matai (head of family) are able to run for the Legislative Assembly
Judicial branch: Supreme Court, Court of Appeal
Political parties and leaders: Human Rights Protection Party (HRPP), TOFILAU Eti Alesana, chairman; Samoan National Development Party (SNDP), TAPUA Tamasese Efi, chairman
Member of: ACP, AsDB, C, ESCAP, FAO, G-77, IBRD, ICFTU, ICRM, IDA, IFAD, IFC, IFRCS, IMF, INTELSAT (nonsignatory user), IOC, ITU, SPARTECA, SPC, SPF, UN, UNCTAD, UNESCO, UPU, WHO
Diplomatic representation in US:
chief of mission: Ambassador Tuiloma Neroni SLADE
chancery: 820 Second Avenue, Suite 800, New York, NY 10017
telephone: [1] (212) 599-6196, 6197
FAX: [1] (212) 599-0797
US diplomatic representation:
chief of mission: the ambassador to New Zealand is accredited to Western Samoa
embassy: 5th floor, Beach Road, Apia
mailing address: P.O. Box 3430, Apia
telephone: [685] 21631
FAX: [685] 22030
Flag: red with a blue rectangle in the upper hoist-side quadrant bearing five white five-pointed stars representing the Southern Cross constellation

Economy

Overview: Agriculture employs more than half of the labor force, contributes 50% to GDP, and furnishes 90% of exports. The bulk of export earnings comes from the sale of coconut oil and copra. The economy depends on emigrant remittances and foreign aid to support a level of imports much greater than export earnings. Tourism has become the most important growth industry. The economy continued to falter in 1994, as remittances and tourist earnings remained low. Production of taro, the primary food export crop, has dropped 97% since a fungal disease struck the crop in 1993. The rapid growth in 1994 of the giant African snail population in Western Samoa is also threatening the country's basic food crops, such as bananas and coconuts.
National product: GDP—purchasing power parity—$400 million (1992 est.)
National product real growth rate: -4.3% (1992 est.)
National product per capita: $2,000 (1992 est.)
Inflation rate (consumer prices): 14% (1994 est.)
Unemployment rate: NA%
Budget:
revenues: $95.3 million
expenditures: $76.7 million, including capital expenditures of $NA (1994 est.)
Exports: $6.4 million (f.o.b., 1993)
commodities: coconut oil and cream, taro, copra, cocoa
partners: New Zealand 34%, American Samoa 21%, Germany 18%, Australia 11%
Imports: $11.5 million (c.i.f., 1992 est.)
commodities: intermediate goods 58%, food 17%, capital goods 12%
partners: New Zealand 37%, Australia 25%, Japan 11%, Fiji 9%
External debt: $141 million (June 1993)
Industrial production: growth rate -0.3% (1992 est.); accounts for 16% of GDP
Electricity:
capacity: 29,000 kW
production: 50 million kWh
consumption per capita: 200 kWh (1993)
Industries: timber, tourism, food processing, fishing
Agriculture: accounts for about 50% of GDP; coconuts, fruit (including bananas, taro, yams)
Economic aid:
recipient: US commitments, including Ex-Im (FY70-89), $18 million; Western (non-US) countries, ODA and OOF bilateral commitments (1970-89), $306 million; OPEC bilateral aid (1979-89), $4 million
Currency: 1 tala (WS$) = 100 sene
Exchange rates: tala (WS$) per US$1—2.4600 (January 1995), 2.5349 (1994), 2.5681 (1993), 2.4655 (1992), 2.3975 (1991), 2.3095 (1990)
Fiscal year: calendar year

Transportation

Railroads: 0 km
Highways:
total: 2,042 km
paved: 375 km
unpaved: gravel, crushed stone, earth 1,667 km
Ports: Apia, Asau, Mulifanua, Salelologa
Merchant marine:
total: 1 roll-on/roll-off cargo ship (1,000 GRT or over) totaling 3,838 GRT/5,536 DWT
Airports:
total: 3
with paved runways 2,438 to 3,047 m: 1
with paved runways under 914 m: 2

Communications

Telephone system: 7,500 telephones
local: NA
intercity: NA
international: 1 INTELSAT (Pacific Ocean) earth station
Radio:
broadcast stations: AM 1, FM 0, shortwave 0
radios: 70,000
Television:
broadcast stations: 0
televisions: NA

Defense Forces

Branches: no regular armed services; Western Samoa Police Force
Defense expenditures: $NA, NA% of GDP

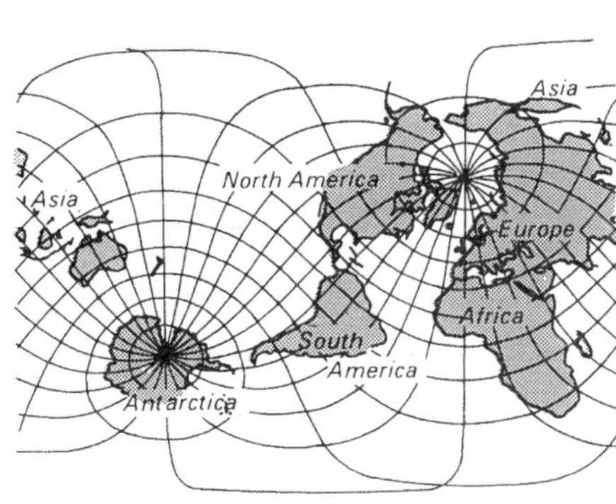

Geography

Map references: World, Time Zones
Area:
total area: 510.072 million sq km
land area: 148.94 million sq km
water area: 361.132 million sq km
comparative area: land area about 16 times the size of the US
note: 70.8% of the world is water, 29.2% is land
Land boundaries: the land boundaries in the world total 250,883.64 km (not counting shared boundaries twice)
Coastline: 356,000 km
Maritime claims:
contiguous zone: 24 nm claimed by most but can vary
continental shelf: 200-m depth claimed by most or to depth of exploitation, others claim 200 nm or to the edge of the continental margin
exclusive fishing zone: 200 nm claimed by most but can vary
exclusive economic zone: 200 nm claimed by most but can vary
territorial sea: 12 nm claimed by most but can vary
note: boundary situations with neighboring states prevent many countries from extending their fishing or economic zones to a full 200 nm; 43 nations and other areas that are landlocked include Afghanistan, Andorra, Armenia, Austria, Azerbaijan, Belarus, Bhutan, Bolivia, Botswana, Burkina, Burundi, Central African Republic, Chad, Czech Republic, Ethiopia, Holy See (Vatican City), Hungary, Kazakhstan, Kyrgyzstan, Laos, Lesotho, Liechtenstein, Luxembourg, Malawi, Mali, Moldova, Mongolia, Nepal, Niger, Paraguay, Rwanda, San Marino, Slovakia, Swaziland, Switzerland, Tajikistan, The Former Yugoslav Republic of Macedonia, Turkmenistan, Uganda, Uzbekistan, West Bank, Zambia, Zimbabwe
Climate: two large areas of polar climates separated by two rather narrow temperate zones from a wide equatorial band of tropical to subtropical climates
Terrain: highest elevation is Mt. Everest at 8,848 meters and lowest depression is the Dead Sea at 392 meters below sea level; greatest ocean depth is the Marianas Trench at 10,924 meters
Natural resources: the rapid using up of nonrenewable mineral resources, the depletion of forest areas and wetlands, the extinction of animal and plant species, and the deterioration in air and water quality (especially in Eastern Europe and the former USSR) pose serious long-term problems that governments and peoples are only beginning to address
Land use:
arable land: 10%
permanent crops: 1%
meadows and pastures: 24%
forest and woodland: 31%
other: 34%
Irrigated land: NA sq km
Environment:
current issues: large areas subject to overpopulation, industrial disasters, pollution (air, water, acid rain, toxic substances), loss of vegetation (overgrazing, deforestation, desertification), loss of wildlife, soil degradation, soil depletion, erosion
natural hazards: large areas subject to severe weather (tropical cyclones), natural disasters (earthquakes, landslides, tsunamis, volcanic eruptions)
international agreements: 23 selected international environmental agreements included under the Environment entry for each country and in Appendix E: Selected International Environmental Agreements

People

Population: 5,733,687,096 (July 1995 est.)
Age structure:
0-14 years: 31.6% (female 882,809,689; male 928,121,801)
15-64 years: 62% (female 1,752,393,539; male 1,802,004,124)
65 years and over: 6.4% (female 209,437,234; male 158,246,581) (July 1995 est.)
Population growth rate: 1.5% (1995 est.)
Birth rate: 24 births/1,000 population (1995 est.)
Death rate: 9 deaths/1,000 population (1995 est.)
Infant mortality rate: 64 deaths/1,000 live births (1995 est.)
Life expectancy at birth:
total population: 62 years
male: 61 years
female: 64 years (1995 est.)
Total fertility rate: 3.1 children born/woman (1995 est.)
Labor force: 2.24 billion (1992)
by occupation: NA

Government

Digraph: XX
Administrative divisions: 265 nations, dependent areas, other, and miscellaneous entries
Legal system: varies by individual country; 186 (not including Yugoslavia) are parties to the United Nations International Court of Justice (ICJ or World Court)

Economy

Overview: Led by recovery in Western Europe and strong performances by the US, Canada, and key Third World countries, real global output—gross world product (GWP)—rose 3% in 1994 compared with 2% in 1993. Results varied widely among regions and countries. Average growth of 3% in the GDP of industrialized countries (60% of GWP in 1994) and average growth of 6% in the GDP of less developed countries (34% of GWP) were partly offset by a further 11% drop in the GDP of the former USSR/Eastern Europe area (now only 6% of GWP). With the notable exception of Japan at 2.9%, unemployment was typically 5%-12% in the industrial world. The US accounted for 22% of GWP in 1994; Western Europe accounted for another 22%; and Japan accounted for 8%. These are the three "economic superpowers" which are presumably destined to compete for mastery in international markets on into the 21st century. As for the less developed countries, China, India, and the Four Dragons—South Korea, Taiwan, Hong Kong, and Singapore—once again posted records of 5% growth or better; however, many other countries, especially in Africa, continued to suffer from drought, rapid population growth, inflation, and civil strife. Central Europe made considerable progress in moving toward "market-friendly" economies, whereas the 15 ex-Soviet countries (with the notable exceptions of the three Baltic states) typically experienced further declines in output, sometimes as high as 30%. Externally, the nation-state, as a bedrock economic-political institution, is steadily losing control over international flows of people, goods, funds, and technology. Internally, the central government in a number of cases is losing control over resources as separatist regional movements—typically based on ethnicity—gain momentum, e.g., in the successor states of the former Soviet Union, in the former Yugoslavia, and in India. In Western Europe, governments face the difficult political problem of channeling resources away from welfare programs in order to increase investment and strengthen incentives to seek employment. The addition of nearly 100 million people each year to an already overcrowded globe is exacerbating the problems of pollution, desertification, underemployment, epidemics,

and famine. Because of their own internal problems, the industrialized countries have inadequate resources to deal effectively with the poorer areas of the world, which, at least from the economic point of view, are becoming further marginalized. (For the specific economic problems of each country, see the individual country entries in this volume.)
National product: GWP (gross world product)—purchasing power parity—$30.7 trillion (1994 est.)
National product real growth rate: 3.2% (1994 est.)
National product per capita: $5,400 (1994 est.)
Inflation rate (consumer prices):
all countries: 25%
developed countries: 5%
developing countries: 50% (1994 est.)
note: national inflation rates vary widely in individual cases, from stable prices to hyperinflation
Unemployment rate: 30% combined unemployment and underemployment in many non-industrialized countries; developed countries typically 5%-12% unemployment
Exports: $4 trillion (f.o.b., 1994 est.)
commodities: the whole range of industrial and agricultural goods and services
partners: in value, about 75% of exports from the developed countries
Imports: $4.1 trillion (c.i.f., 1994 est.)
commodities: the whole range of industrial and agricultural goods and services
partners: in value, about 75% of imports by the developed countries
External debt: $1 trillion for less developed countries (1993 est.)
Industrial production: growth rate 5% (1994 est.)
Electricity:
capacity: 2,773,000,000 kW
production: 11.601 trillion kWh
consumption per capita: 1,937 kWh (1993)
Industries: industry worldwide is dominated by the onrush of technology, especially in computers, robotics, telecommunications, and medicines and medical equipment; most of these advances take place in OECD nations; only a small portion of non-OECD countries have succeeded in rapidly adjusting to these technological forces, and the technological gap between the industrial nations and the less-developed countries continues to widen; the rapid development of new industrial (and agricultural) technology is complicating already grim environmental problems
Agriculture: the production of major food crops has increased substantially in the last 20 years; the annual production of cereals, for instance, has risen by 50%, from about 1.2 billion metric tons to about 1.8 billion metric tons; production increases have resulted mainly from increased yields rather than increases in planted areas; while global production is sufficient for aggregate demand, about one-fifth of the world's population remains malnourished, primarily because local production cannot adequately provide for large and rapidly growing populations, which are too poor to pay for food imports; conditions are especially bad in Africa where drought in recent years has intensified the consequences of overpopulation
Economic aid: $NA

Transportation

Railroads:
total: 1,201,337 km includes about 190,000 to 195,000 km of electrified routes of which 147,760 km are in Europe, 24,509 km in the Far East, 11,050 km in Africa, 4,223 km in South America, and 4,160 km in North America; note—fastest speed in daily service is 300 km/hr attained by France's SNCF TGV-Atlantique line
broad gauge: 251,153 km
standard gauge: 710,754 km
narrow gauge: 239,430 km
Highways:
total: NA
paved: NA
unpaved: NA
Ports: Chiba, Houston, Kawasaki, Kobe, Marseille, Mina' al Ahmadi (Kuwait), New Orleans, New York, Rotterdam, Yokohama
Merchant marine:
total: 25,364 ships (1,000 GRT or over) totaling 435,458,296 GRT/697,171,651 DWT
ships by type: barge carrier 39, bulk 5,202, cargo 8,121, chemical tanker 911, combination bulk 293, combination ore/oil 290, container 1,903, liquefied gas 675, livestock carrier 48, multifunction large-load carrier 53, oil tanker 4,332, passenger 287, passenger-cargo 114, railcar carrier 24, refrigerated cargo 1,023, roll-on/roll-off cargo 1,047, short-sea passenger 465, specialized tanker 77, vehicle carrier 460 (April 1995)

Communications

Telephone system:
local: NA
intercity: NA
international: NA
Radio:
broadcast stations: AM NA, FM NA, shortwave NA
radios: NA
Television:
broadcast stations: NA
televisions: NA

Defense Forces

Branches: ground, maritime, and air forces at all levels of technology
Defense expenditures: a further decline in 1994, by perhaps 5%-10%, to roughly three-quarters of a trillion dollars, or 2.5% of gross world product (1994 est.)

Yemen

Geography

Location: Middle East, bordering the Arabian Sea, Gulf of Aden, and Red Sea, between Oman and Saudi Arabia
Map references: Middle East
Area:
total area: 527,970 sq km
land area: 527,970 sq km
comparative area: slightly larger than twice the size of Wyoming
note: includes Perim, Socotra, the former Yemen Arab Republic (YAR or North Yemen), and the former People's Democratic Republic of Yemen (PDRY or South Yemen)
Land boundaries: total 1,746 km, Oman 288 km, Saudi Arabia 1,458 km
Coastline: 1,906 km
Maritime claims:
contiguous zone: 18 nm in the North; 24 nm in the South
continental shelf: 200 nm or to the edge of the continental margin
exclusive economic zone: 200 nm
territorial sea: 12 nm
International disputes: undefined section of boundary with Saudi Arabia; a treaty with Oman defining the Yemeni-Omani boundary was ratified in December 1992
Climate: mostly desert; hot and humid along west coast; temperate in western mountains affected by seasonal monsoon; extraordinarily hot, dry, harsh desert in east
Terrain: narrow coastal plain backed by flat-topped hills and rugged mountains; dissected upland desert plains in center slope into the desert interior of the Arabian Peninsula
Natural resources: petroleum, fish, rock salt, marble, small deposits of coal, gold, lead, nickel, and copper, fertile soil in west
Land use:
arable land: 6%
permanent crops: 0%
meadows and pastures: 30%
forest and woodland: 7%
other: 57%
Irrigated land: 3,100 sq km (1989 est.)

Environment:
current issues: very limited natural fresh water resources; inadequate supplies of potable water; overgrazing; soil erosion; desertification
natural hazards: sandstorms and dust storms in summer
international agreements: party to—Environmental Modification, Law of the Sea, Nuclear Test Ban; signed, but not ratified—Biodiversity, Climate Change
Note: controls Bab el Mandeb, the strait linking the Red Sea and the Gulf of Aden, one of world's most active shipping lanes

People

Population: 14,728,474 (July 1995 est.)
Age structure:
0-14 years: 50% (female 3,551,953; male 3,776,338)
15-64 years: 48% (female 3,505,735; male 3,508,229)
65 years and over: 2% (female 216,210; male 169,989) (July 1995 est.)
Population growth rate: 4.02% (1995 est.)
Birth rate: 44.85 births/1,000 population (1995 est.)
Death rate: 8.01 deaths/1,000 population (1995 est.)
Net migration rate: 3.39 migrant(s)/1,000 population (1995 est.)
Infant mortality rate: 58.2 deaths/1,000 live births (1995 est.)
Life expectancy at birth:
total population: 62.51 years
male: 61.57 years
female: 63.5 years (1995 est.)
Total fertility rate: 7.15 children born/woman (1995 est.)
Nationality:
noun: Yemeni(s)
adjective: Yemeni
Ethnic divisions: predominantly Arab; Afro-Arab concentrations in western coastal locations; South Asians in southern regions; small European communities in major metropolitan areas
Religions: Muslim including Sha'fi (Sunni) and Zaydi (Shi'a), small numbers of Jewish, Christian, and Hindu
Languages: Arabic
Literacy: age 15 and over can read and write (1990 est.)
total population: 38%
male: 53%
female: 26%
Labor force: no reliable estimates exist, most people are employed in agriculture and herding or as expatriate laborers; services, construction, industry, and commerce account for less than half of the labor force

Government

Names:
conventional long form: Republic of Yemen
conventional short form: Yemen
local long form: Al Jumhuriyah al Yamaniyah
local short form: Al Yaman
Digraph: YM
Type: republic
Capital: Sanaa
Administrative divisions: 17 governorates (muhafazat, singular—muhafazah); Abyan, Adan, Al Bayda, Al Hudaydah, Al Jawf, Al Mahrah, Al Mahwit, Dhamar, Hadramaut, Hajjah, Ibb, Lahij, Marib, Sadah, Sana, Shabwah, Taizz
note: there may be a new governorate for the capital city of Sanaa
Independence: 22 May 1990 Republic of Yemen was established on 22 May 1990 with the merger of the Yemen Arab Republic {Yemen (Sanaa) or North Yemen} and the Marxist-dominated People's Democratic Republic of Yemen {Yemen (Aden) or South Yemen}; previously North Yemen had become independent on NA November 1918 (from the Ottoman Empire) and South Yemen had become independent on 30 November 1967 (from the UK)
National holiday: Proclamation of the Republic, 22 May (1990)
Constitution: 16 May 1991
Legal system: based on Islamic law, Turkish law, English common law, and local tribal customary law; does not accept compulsory ICJ jurisdiction
Suffrage: 18 years of age; universal
Executive branch:
chief of state: President Ali Abdallah SALIH (since 22 May 1990, the former president of North Yemen); Vice President Abd al-Rab Mansur al-HADI (since NA October 1994)
head of government: Prime Minister Abd al-Aziz ABD AL-GHANI (since NA October 1994)
cabinet: Council of Ministers
Legislative branch: unicameral
House of Representatives: elections last held 27 April 1993 (next to be held NA 1997); results—percent of vote by party NA; seats—(301 total) GPC 124, Islaah 61, YSP 55, others 13, independents 47, election nullified 1
Judicial branch: Supreme Court
Political parties and leaders: over 40 political parties are active in Yemen, but only three project significant influence; since the May-July 1994 civil war, President SALIH's General People's Congress (GPC) and Shaykh Abdallah bin Husayn al-AHMAR's Yemeni Grouping for Reform, or Islaah, have joined to form a coalition government; the Yemeni Socialist Party (YSP), headed by Ali Salih UBAYD, has regrouped as a loyal opposition
Other political or pressure groups: NA
Member of: ACC, AFESD, AL, AMF, CAEU, CCC, ESCWA, FAO, G-77, IBRD, ICAO, ICRM, IDA, IDB, IFAD, IFC, IFRCS, ILO, IMF, IMO, INTELSAT, INTERPOL, IOC, ITU, NAM, OIC, UN, UNCTAD, UNESCO, UNIDO, UPU, WFTU, WHO, WIPO, WMO, WTO
Diplomatic representation in US:
chief of mission: Ambassador Muhsin Ahmad al-AYNI
chancery: Suite 705, 2600 Virginia Avenue NW, Washington, DC 20037
telephone: [1] (202) 965-4760, 4761
FAX: [1] (202) 337-2017
US diplomatic representation:
chief of mission: Ambassador David NEWTON
embassy: Dhahr Himyar Zone, Sheraton Hotel District, Sanaa
mailing address: P. O. Box 22347 Sanaa; Sanaa, Department of State, Washington, DC 20521-6330
telephone: [967] (1) 238843 through 238852
FAX: [967] (1) 251563
Flag: three equal horizontal bands of red (top), white, and black; similar to the flag of Syria which has two green stars and of Iraq which has three green stars (plus an Arabic inscription) in a horizontal line centered in the white band; also similar to the flag of Egypt which has a symbolic eagle centered in the white band

Economy

Overview: Whereas the northern city Sanaa is the political capital of a united Yemen, the southern city Aden, with its refinery and port facilities, is the economic and commercial capital. Future economic development depends heavily on Western-assisted development of the country's moderate oil resources. Former South Yemen's willingness to merge stemmed partly from the steady decline in Soviet economic support. The low level of domestic industry and agriculture has made northern Yemen dependent on imports for practically all of its essential needs. Once self-sufficient in food production, northern Yemen has become a major importer. Land once used for export crops—cotton, fruit, and vegetables—has been turned over to growing a shrub called qat, whose leaves are chewed for their stimulant effect by Yemenis and which has no significant export market. Economic growth in former South Yemen has been constrained by a lack of incentives, partly stemming from centralized control over production decisions, investment allocation, and import choices. Yemen's large trade deficits have been compensated for by remittances from Yemenis working abroad and by foreign aid. Since the Gulf crisis, remittances have dropped substantially. Growth in 1994-95 is constrained by low oil prices, rapid inflation, and political deadlock that are causing a lack of economic cooperation

and leadership. However, a peace agreement with Saudi Arabia in February 1995 and the expectation of a rise in oil prices brighten Yemen's economic prospects.
National product: GDP—purchasing power parity—$23.4 billion (1994 est.)
National product real growth rate: -1.4% (1994 est.)
National product per capita: $1,955 (1994 est.)
Inflation rate (consumer prices): 145% (1994 est.)
Unemployment rate: 30% (December 1994)
Budget:
revenues: $NA
expenditures: $NA, including capital expenditures of $NA
Exports: $1.75 billion (f.o.b., 1994 est.)
commodities: crude oil, cotton, coffee, hides, vegetables, dried and salted fish
partners: Germany 28%, Japan 15%, UK 9%, Austria 7%, China 7% (1992)
Imports: $2.65 billion (f.o.b., 1994 est.)
commodities: textiles and other manufactured consumer goods, petroleum products, sugar, grain, flour, other foodstuffs, cement, machinery, chemicals
partners: US 16%, UK 7%, Japan 6%, France 6%, Italy 6% (1992)
External debt: $7 billion (1993)
Industrial production: growth rate NA%, accounts for 18% of GDP
Electricity:
capacity: 810,000 kW
production: 1.8 billion kWh
consumption per capita: 149 kWh (1993)
Industries: crude oil production and petroleum refining; small-scale production of cotton textiles and leather goods; food processing; handicrafts; small aluminum products factory; cement
Agriculture: accounts for 26% of GDP; products—grain, fruits, vegetables, qat (mildly narcotic shrub), coffee, cotton, dairy, poultry, meat, fish; not self-sufficient in grain
Economic aid:
recipient: US commitments, including Ex-Im (FY70-89), $389 million; Western (non-US) countries, ODA and OOF bilateral commitments (1970-89), $2 billion; OPEC bilateral aid (1979-89), $3.2 billion; Communist countries (1970-89), $2.4 billion
Currency: Yemeni rial (new currency); 1 North Yemeni riyal (YR) = 100 fils; 1 South Yemeni dinar (YD) = 1,000 fils
note: following the establishment of the Republic of Yemen on 22 May 1990, the North Yemeni riyal and the South Yemeni dinar are to be replaced with a new Yemeni rial
Exchange rates: Yemeni rials per US$1—12.0 (official); 90 (market rate, December 1994)
Fiscal year: calendar year

Transportation

Railroads: 0 km
Highways:
total: 51,390 km
paved: 4,830 km
unpaved: 46,560 km (1992 est.)
Pipelines: crude oil 644 km; petroleum products 32 km
Ports: Aden, Al Hudaydah, Al Mukalla, Mocha, Nishtun
Merchant marine:
total: 3 ships (1,000 GRT or over) totaling 12,059 GRT/18,563 DWT
ships by type: cargo 1, oil tanker 2
Airports:
total: 46
with paved runways over 3,047 m: 2
with paved runways 2,438 to 3,047 m: 6
with paved runways 1,524 to 2,437 m: 1
with paved runways 914 to 1,523 m: 1
with paved runways under 914 m: 4
with unpaved runways over 3,047 m: 2
with unpaved runways 2,438 to 3,047 m: 8
with unpaved runways 1,524 to 2,438 m: 10
with unpaved runways 914 to 1,523 m: 12

Communications

Telephone system: 65,000 telephones; since unification in 1990, efforts are still being made to create a national domestic civil telecommunications network
local: NA
intercity: the network consists of microwave radio relay, cable, and troposcatter
international: 3 INTELSAT (2 Indian Ocean and 1 Atlantic Ocean), 1 Intersputnik, and 2 ARABSAT earth stations; microwave radio relay to Saudi Arabia and Djibouti
Radio:
broadcast stations: AM 4, FM 1, shortwave 0
radios: NA
Television:
broadcast stations: 10
televisions: NA

Defense Forces

Branches: Army, Navy, Air Force, paramilitary (includes Police)
Manpower availability: males age 15-49 3,135,649; males fit for military service 1,771,226; males reach military age (14) annually 181,057 (1995 est.)
Defense expenditures: exchange rate conversion—$1.65 billion, 7.1% of GDP (1993)

Zaire

Geography

Location: Central Africa, northeast of Angola
Map references: Africa
Area:
total area: 2,345,410 sq km
land area: 2,267,600 sq km
comparative area: slightly more than one-quarter the size of US
Land boundaries: total 10,271 km, Angola 2,511 km, Burundi 233 km, Central African Republic 1,577 km, Congo 2,410 km, Rwanda 217 km, Sudan 628 km, Uganda 765 km, Zambia 1,930 km
Coastline: 37 km
Maritime claims:
exclusive economic zone: boundaries with neighbors
territorial sea: 12 nm
International disputes: Tanzania-Zaire-Zambia tripoint in Lake Tanganyika may no longer be indefinite since it is reported that the indefinite section of the Zaire-Zambia boundary has been settled; long section with Congo along the Congo River is indefinite (no division of the river or its islands has been made)
Climate: tropical; hot and humid in equatorial river basin; cooler and drier in southern highlands; cooler and wetter in eastern highlands; north of Equator—wet season April to October, dry season December to February; south of Equator—wet season November to March, dry season April to October
Terrain: vast central basin is a low-lying plateau; mountains in east
Natural resources: cobalt, copper, cadmium, petroleum, industrial and gem diamonds, gold, silver, zinc, manganese, tin, germanium, uranium, radium, bauxite, iron ore, coal, hydropower potential
Land use:
arable land: 3%
permanent crops: 0%
meadows and pastures: 4%
forest and woodland: 78%
other: 15%

Irrigated land: 100 sq km (1989 est.)
Environment:
current issues: poaching threatens wildlife populations; water pollution; deforestation; 1.2 million Rwandan refugees are responsible for significant deforestation, soil erosion, and wildlife poaching in eastern Zaire
natural hazards: periodic droughts in south; volcanic activity
international agreements: party to—Biodiversity, Climate Change, Endangered Species, Hazardous Wastes, Law of the Sea, Marine Dumping, Nuclear Test Ban, Ozone Layer Protection, Tropical Timber 83; signed, but not ratified—Desertification, Environmental Modification
Note: straddles Equator; very narrow strip of land that controls the lower Congo River and is only outlet to South Atlantic Ocean; dense tropical rain forest in central river basin and eastern highlands

People

Population: 44,060,636 (July 1995 est.)
Age structure:
0-14 years: 48% (female 10,522,368; male 10,527,451)
15-64 years: 50% (female 11,211,353; male 10,630,118)
65 years and over: 2% (female 647,307; male 522,039) (July 1995 est.)
Population growth rate: 3.18% (1995 est.)
Birth rate: 48.33 births/1,000 population (1995 est.)
Death rate: 16.57 deaths/1,000 population (1995 est.)
Net migration rate: NA migrant(s)/1,000 population (1995 est.)
note: in 1994, more than one million refugees fled into Zaire to escape the fighting between the Hutus and the Tutsis in Rwanda and Burundi; a small number of these are returning to their homes in 1995 despite fear of the ongoing violence; additionally, Zaire is host to 105,000 Angolan, more than 250,000 Burundian and 100,000 Sudanese refugees; repatriation of Angolan refugees was suspended in May 1994 because of the recurrence of fighting in Angola; if present peace accords hold, repatriation of Angolans may recommence
Infant mortality rate: 108.7 deaths/1,000 live births (1995 est.)
Life expectancy at birth:
total population: 47.54 years
male: 45.68 years
female: 49.46 years (1995 est.)
Total fertility rate: 6.7 children born/woman (1995 est.)
Nationality:
noun: Zairian(s)
adjective: Zairian
Ethnic divisions: over 200 African ethnic groups, the majority are Bantu; four largest tribes—Mongo, Luba, Kongo (all Bantu), and the Mangbetu-Azande (Hamitic) make up about 45% of the population
Religions: Roman Catholic 50%, Protestant 20%, Kimbanguist 10%, Muslim 10%, other syncretic sects and traditional beliefs 10%
Languages: French, Lingala, Swahili, Kingwana, Kikongo, Tshiluba
Literacy: age 15 and over can read and write (1990 est.)
total population: 72%
male: 84%
female: 61%
Labor force: 15 million (25% of the labor force comprises wage earners)
by occupation: agriculture 75%, industry 13%, services 12% (1985)

Government

Names:
conventional long form: Republic of Zaire
conventional short form: Zaire
local long form: Republique du Zaire
local short form: Zaire
former: Belgian Congo Congo/Leopoldville Congo/Kinshasa
Digraph: CG
Type: republic with a strong presidential system
Capital: Kinshasa
Administrative divisions: 10 regions (regions, singular—region) and 1 town* (ville); Bandundu, Bas-Zaire, Equateur, Haut-Zaire, Kasai-Occidental, Kasai-Oriental, Kinshasa*, Maniema, Nord-Kivu, Shaba, Sud-Kivu
Independence: 30 June 1960 (from Belgium)
National holiday: Anniversary of the Regime (Second Republic), 24 November (1965)
Constitution: 24 June 1967, amended August 1974, revised 15 February 1978; amended April 1990; new transitional constitution promulgated in April 1994
Legal system: based on Belgian civil law system and tribal law; has not accepted compulsory ICJ jurisdiction
Suffrage: 18 years of age; universal and compulsory
Executive branch:
chief of state: President Marshal MOBUTU Sese Seko Kuku Ngbendu wa Za Banga (since 24 November 1965) election last held 29 July 1984 (next to be held by 9 July 1995); results—President MOBUTU was reelected without opposition
head of government: Prime Minister Leon KENGO wa Dondo (since 14 June 1994)
cabinet: National Executive Council; appointed by mutual agreement of the president and the prime minister
Legislative branch: unicameral
parliament: a single body consisting of the High Council of the Republic and the Parliament of the Transition with membership equally divided between presidential supporters and opponents
Judicial branch: Supreme Court (Cour Supreme)
Political parties and leaders: sole legal party until January 1991—Popular Movement of the Revolution (MPR); other parties include Union for Democracy and Social Progress (UDPS), Etienne TSHISEKEDI wa Mulumba; Democratic Social Christian Party (PDSC); Union of Federalists and Independent Republicans (UFERI); Unified Lumumbast Party (PALU), Antoine GIZENGA; Union of Independent Democrats (UDI), Leon KENGO wa Dondo
Member of: ACCT, ACP, AfDB, CCC, CEEAC, CEPGL, ECA, FAO, G-19, G-24, G-77, GATT, IAEA, IBRD, ICAO, ICC, ICRM, IDA, IFAD, IFC, IFRCS, ILO, IMF, IMO, INTELSAT, INTERPOL, IOC, ITU, NAM, OAU, PCA, UN, UNCTAD, UNESCO, UNHCR, UNIDO, UPU, WCL, WFTU, WHO, WIPO, WMO, WTO
Diplomatic representation in US:
chief of mission: Ambassador TATANENE Manata
chancery: 1800 New Hampshire Avenue NW, Washington, DC 20009
telephone: [1] (202) 234-7690, 7691
US diplomatic representation:
chief of mission: (vacant); Charge d'Affaires John M. YATES
embassy: 310 Avenue des Aviateurs, Kinshasa
mailing address: Unit 31550, Kinshasha; APO AE 09828
telephone: [243] (12) 21532, 21628
FAX: [243] (12) 21534 ext. 2308, 21535 ext. 2308; (88) 43805, 43467
Flag: light green with a yellow disk in the center bearing a black arm holding a red flaming torch; the flames of the torch are blowing away from the hoist side; uses the popular pan-African colors of Ethiopia

Economy

Overview: Zaire's economy has continued to disintegrate although Prime Minister KENGO has had some success in slowing the rate of economic decline. While meaningful economic figures are difficult to come by, Zaire's hyperinflation, chronic large government deficits, and plunging mineral production have made the country one of the world's poorest. Most formal transactions are conducted in hard currency as indigenous bank notes have lost almost all value, and a barter economy now flourishes in all but the largest cities. Most individuals and families hang on grimly through subsistence farming and petty trade. The government has not been able to meet its financial obligations to the International Monetary Fund or put in place the financial measures advocated by the IMF. Although short-term prospects for improvement are dim,

improved political stability would boost Zaire's long-term potential to effectively exploit its vast wealth of mineral and agricultural resources.
National product: GDP—purchasing power parity—$18.8 billion (1994 est.)
National product real growth rate: 4% (1994 est.)
National product per capita: $440 (1994 est.)
Inflation rate (consumer prices): 40% per month (1993 est.)
Unemployment rate: NA%
Budget:
revenues: $NA
expenditures: $NA, including capital expenditures of $NA
Exports: $362 million (f.o.b., 1993 est.)
commodities: copper, coffee, diamonds, cobalt, crude oil
partners: US, Belgium, France, Germany, Italy, UK, Japan, South Africa
Imports: $356 million (f.o.b., 1993 est.)
commodities: consumer goods, foodstuffs, mining and other machinery, transport equipment, fuels
partners: South Africa, US, Belgium, France, Germany, Italy, Japan, UK
External debt: $9.2 billion (May 1992 est.)
Industrial production: growth rate -20% (1993); accounts for 16% of GDP
Electricity:
capacity: 2,830,000 kW
production: 6.2 billion kWh
consumption per capita: 133 kWh (1993)
Industries: mining, mineral processing, consumer products (including textiles, footwear, cigarettes, processed foods and beverages), cement, diamonds
Agriculture: cash crops—coffee, palm oil, rubber, quinine; food crops—cassava, bananas, root crops, corn
Illicit drugs: illicit producer of cannabis, mostly for domestic consumption
Economic aid:
recipient: US commitments, including Ex-Im (FY70-89), $1.1 billion; Western (non-US) countries, ODA and OOF bilateral commitments (1970-89), $6.9 billion; OPEC bilateral aid (1979-89), $35 million; Communist countries (1970-89), $263 million
note: except for humanitarian aid to private organizations, no US assistance has been given to Zaire since 1992
Currency: 1 zaire (Z) = 100 makuta
Exchange rates: new zaires (Z) per US$1—3,275.71 (December 1994), 1,194.12 (1994), 2.51 (1993); zaire (Z) per US$1—645,549 (1992), 15,587 (1991), 719 (1990)
note: on 22 October 1993 the new zaire, equal to 3,000,000 old zaires, was introduced
Fiscal year: calendar year

Transportation

Railroads:
total: 5,138 km; note—severely reduced trackage in use because of civil strife
narrow gauge: 3,987 km 1.067-m gauge (858 km electrified); 125 km 1.000-m gauge; 1,026 km 0.600-m gauge
Highways:
total: 146,500 km
paved: 2,800 km
unpaved: gravel, improved earth 46,200 km; unimproved earth 97,500 km
Inland waterways: 15,000 km including the Congo, its tributaries, and unconnected lakes
Pipelines: petroleum products 390 km
Ports: Banana, Boma, Bukavu, Bumba, Goma, Kalemie, Kindu, Kinshasa, Kisangani, Matadi, Mbandaka
Merchant marine: none
Airports:
total: 270
with paved runways over 3,047 m: 4
with paved runways 2,438 to 3,047 m: 3
with paved runways 1,524 to 2,437 m: 15
with paved runways 914 to 1,523 m: 2
with paved runways under 914 m: 97
with unpaved runways 1,524 to 2,438 m: 22
with unpaved runways 914 to 1,523 m: 127

Communications

Telephone system: NA telephones
local: NA
intercity: barely adequate wire and microwave service in and between urban areas; 14 domestic earth stations
international: 1 INTELSAT (Atlantic Ocean earth station)
Radio:
broadcast stations: AM 10, FM 4, shortwave 0
radios: NA
Television:
broadcast stations: 18
televisions: NA

Defense Forces

Branches: Army, Navy, Air Force, National Gendarmerie, paramilitary Civil Guard, Special Presidential Division
Manpower availability: males age 15-49 9,479,245; males fit for military service 4,828,367 (1995 est.)
Defense expenditures: exchange rate conversion—$46 million, 1.5% of GDP (1990)

Zambia

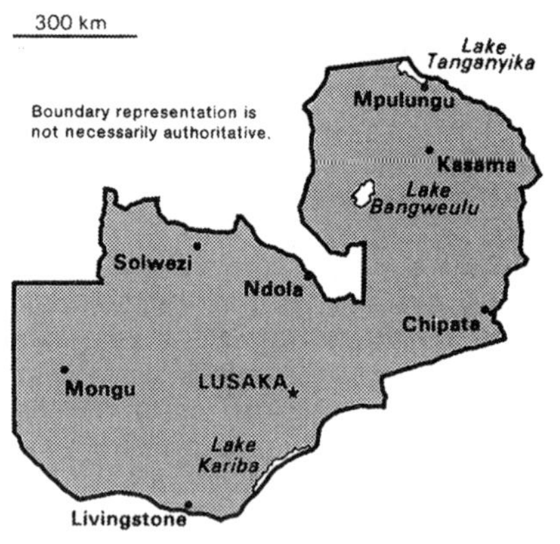

Geography

Location: Southern Africa, east of Angola
Map references: Africa
Area:
total area: 752,610 sq km
land area: 740,720 sq km
comparative area: slightly larger than Texas
Land boundaries: total 5,664 km, Angola 1,110 km, Malawi 837 km, Mozambique 419 km, Namibia 233 km, Tanzania 338 km, Zaire 1,930 km, Zimbabwe 797 km
Coastline: 0 km (landlocked)
Maritime claims: none; landlocked
International disputes: quadripoint with Botswana, Namibia, and Zimbabwe is in disagreement; Tanzania-Zaire-Zambia tripoint in Lake Tanganyika may no longer be indefinite since it is reported that the indefinite section of the Zaire-Zambia boundary has been settled
Climate: tropical; modified by altitude; rainy season (October to April)
Terrain: mostly high plateau with some hills and mountains
Natural resources: copper, cobalt, zinc, lead, coal, emeralds, gold, silver, uranium, hydropower potential
Land use:
arable land: 7%
permanent crops: 0%
meadows and pastures: 47%
forest and woodland: 27%
other: 19%
Irrigated land: 320 sq km (1989 est.)
Environment:
current issues: air pollution and resulting acid rain in the mineral extraction and refining region; poaching seriously threatens rhinoceros and elephant populations; deforestation; soil erosion; desertification; lack of adequate water treatment presents human health risks
natural hazards: tropical storms (November to April)
international agreements: party to—Biodiversity, Climate Change, Endangered

Species, Hazardous Wastes, Law of the Sea, Nuclear Test Ban, Ozone Layer Protection, Wetlands; signed, but not ratified—Desertification
Note: landlocked

People

Population: 9,445,723 (July 1995 est.)
Age structure:
0-14 years: 50% (female 2,331,820; male 2,363,319)
15-64 years: 48% (female 2,332,798; male 2,193,363)
65 years and over: 2% (female 112,484; male 111,939) (July 1995 est.)
Population growth rate: 2.7% (1995 est.)
Birth rate: 45.47 births/1,000 population (1995 est.)
Death rate: 18.42 deaths/1,000 population (1995 est.)
Net migration rate: -0.04 migrant(s)/1,000 population (1995 est.)
Infant mortality rate: 86 deaths/1,000 live births (1995 est.)
Life expectancy at birth:
total population: 42.88 years
male: 42.74 years
female: 43.03 years (1995 est.)
Total fertility rate: 6.62 children born/woman (1995 est.)
Nationality:
noun: Zambian(s)
adjective: Zambian
Ethnic divisions: African 98.7%, European 1.1%, other 0.2%
Religions: Christian 50%-75%, Muslim and Hindu 24%-49%, indigenous beliefs 1%
Languages: English (official)
note: about 70 indigenous languages
Literacy: age 15 and over can read and write (1990 est.)
total population: 73%
male: 81%
female: 65%
Labor force: 3.4 million
by occupation: agriculture 85%, mining, manufacturing, and construction 6%, transport and services 9%

Government

Names:
conventional long form: Republic of Zambia
conventional short form: Zambia
former: Northern Rhodesia
Digraph: ZA
Type: republic
Capital: Lusaka
Administrative divisions: 9 provinces; Central, Copperbelt, Eastern, Luapula, Lusaka, Northern, North-Western, Southern, Western
Independence: 24 October 1964 (from UK)
National holiday: Independence Day, 24 October (1964)
Constitution: 2 August 1991
Legal system: based on English common law and customary law; judicial review of legislative acts in an ad hoc constitutional council; has not accepted compulsory ICJ jurisdiction
Suffrage: 18 years of age; universal
Executive branch:
chief of state and head of government: President Frederick CHILUBA (since 31 October 1991); Vice President General Godfrey MIYANDA (since NA August 1994; he replaced Levy MWANAWASA who was elected 31 October 1991 and resigned NA August 1994) election last held 31 October 1991 (next to be held NA 1996); results—Frederick CHILUBA 84%, Kenneth KAUNDA 16%
cabinet: Cabinet; appointed by the president from members of the National Assembly
Legislative branch: unicameral
National Assembly: elections last held 31 October 1991 (next to be held NA 1996); results—percent of vote by party NA; seats—(150 total) MMD 125, UNIP 25; note the MMD's majority was weakened by the defection of 13 of its parliamentary members during 1993 and the defeat of its candidates in 4 of the resulting by-elections
Judicial branch: Supreme Court
Political parties and leaders: Movement for Multiparty Democracy (MMD), Frederick CHILUBA; United National Independence Party (UNIP), Kebby MUSOKATWANE; National Party (NP), Inonge MBIKUSITA-LEWANIKA;
Member of: ACP, AfDB, C, CCC, ECA, FAO, FLS, G-19, G-77, GATT, IAEA, IBRD, ICAO, ICFTU, ICRM, IDA, IFAD, IFC, IFRCS, ILO, IMF, INTELSAT, INTERPOL, IOC, IOM, ITU, NAM, OAU, SADC, UN, UNCTAD, UNESCO, UNIDO, UNOMOZ, UPU, WCL, WHO, WIPO, WMO, WTO
Diplomatic representation in US:
chief of mission: Ambassador Dunstan Weston KAMANA
chancery: 2419 Massachusetts Avenue NW, Washington, DC 20008
telephone: [1] (202) 265-9717 through 9719
FAX: [1] (202) 332-0826
US diplomatic representation:
chief of mission: Ambassador Roland K. KUCHEL
embassy: corner of Independence Avenue and United Nations Avenue, Lusaka
mailing address: P. O. Box 31617, Lusaka
telephone: [260] (1) 228595, 228601, 228602, 228603
FAX. [260] (1) 261538
Flag: green with a panel of three vertical bands of red (hoist side), black, and orange below a soaring orange eagle, on the outer edge of the flag

Economy

Overview: Prior to 1993 the economy had been in decline for more than a decade with falling imports and growing foreign debt. Economic difficulties stemmed largely from a chronically depressed level of copper production and weak copper prices, generally ineffective economic policies, and high inflation. An annual population growth of 3% brought a decline in per capita GDP of 50% over the decade. However, economic reforms enacted since 1992 have helped reduce inflation, have begun to strengthen the social safety net, and have been accompanied by GDP growth at an estimated 6.8% in 1993 and 4% in 1994. The huge external debt remains a key problem.
National product: GDP—purchasing power parity—$7.9 billion (1994 est.)
National product real growth rate: 4% (1994 est.)
National product per capita: $860 (1994 est.)
Inflation rate (consumer prices): 89% (1994 est.)
Unemployment rate: NA%
Budget:
revenues: $665 million
expenditures: $767 million, including capital expenditures of $300 million (1991 est.)
Exports: $1.01 billion (f.o.b., 1993 est.)
commodities: copper, zinc, cobalt, lead, tobacco
partners: EC countries, Japan, South Africa, US, India
Imports: $1.13 billion (c.i.f., 1993 est.)
commodities: machinery, transportation equipment, foodstuffs, fuels, manufactures
partners: EC countries, Japan, Saudi Arabia, South Africa, US
External debt: $7.3 billion (1993)
Industrial production: growth rate -1% (1992); accounts for 42% of GDP
Electricity:
capacity: 2,440,000 kW
production: 7.8 billion kWh
consumption per capita: 650 kWh (1993)
Industries: copper mining and processing, construction, foodstuffs, beverages, chemicals, textiles, and fertilizer
Agriculture: accounts for 12% of GDP and 85% of labor force; crops—corn (food staple), sorghum, rice, peanuts, sunflower, tobacco, cotton, sugarcane, cassava; cattle, goats, beef, eggs
Illicit drugs: increasingly a regional transshipment center for methaqualone and heroin
Economic aid:
recipient: US commitments, including Ex-Im (1970-89), $4.8 billion; Western (non-US) countries, ODA and OOF bilateral commitments (1970-89), $4.8 billion; OPEC bilateral aid (1979-89), $60 million; Communist countries (1970-89), $533 million
Currency: 1 Zambian kwacha (ZK) = 100 ngwee
Exchange rates: Zambian kwacha (ZK) per US$1—672.8 (September 1994), 434.78 (1993), 156.25 (1992), 61.7284 (1991), 28.9855 (1990)
Fiscal year: calendar year

Zambia *(continued)*

Transportation

Railroads:
total: 1,273 km
narrow gauge: 1,273 km 1.067-m gauge (13 km double track)
note: not a part of Zambia Railways is the Tanzania-Zambia Railway Authority (TAZARA), which operates 1,860 km of 1.067-m narrow gauge track between Dar es Salaam and New Kapiri M'poshi where it connects to the Zambia Railways system; 891 km of the TAZARA line transit Zambia
Highways:
total: 36,370 km
paved: 6,500 km
unpaved: crushed stone, gravel, stabilized earth 7,000 km; improved, unimproved earth 22,870 km
Inland waterways: 2,250 km, including Zambezi and Luapula Rivers, Lake Tanganyika
Pipelines: crude oil 1,724 km
Ports: Mpulungu
Airports:
total: 113
with paved runways over 3,047 m: 1
with paved runways 2,438 to 3,047 m: 4
with paved runways 1,524 to 2,437 m: 4
with paved runways 914 to 1,523 m: 4
with paved runways under 914 m: 39
with unpaved runways 1,524 to 2,438 m: 4
with unpaved runways 914 to 1,523 m: 57

Communications

Telephone system: NA telephones; facilities are among the best in Sub-Saharan Africa
local: NA
intercity: high capacity micrwave radio relay connects most larger towns and cities
international: 2 INTELSAT earth stations (1 Indian Ocean and 1 Atlantic Ocean)
Radio:
broadcast stations: AM 11, FM 5, shortwave 0
radios: NA
Television:
broadcast stations: 9
televisions: NA

Defense Forces

Branches: Army, Air Force, Police
Manpower availability: males age 15-49 1,953,967; males fit for military service 1,028,113 (1995 est.)
Defense expenditures: exchange rate conversion—$45 million, 1.4% of GDP (1994)

Zimbabwe

Geography

Location: Southern Africa, northeast of Botswana
Map references: Africa
Area:
total area: 390,580 sq km
land area: 386,670 sq km
comparative area: slightly larger than Montana
Land boundaries: total 3,066 km, Botswana 813 km, Mozambique 1,231 km, South Africa 225 km, Zambia 797 km
Coastline: 0 km (landlocked)
Maritime claims: none; landlocked
International disputes: quadripoint with Botswana, Namibia, and Zambia is in disagreement
Climate: tropical; moderated by altitude; rainy season (November to March)
Terrain: mostly high plateau with higher central plateau (high veld); mountains in east
Natural resources: coal, chromium ore, asbestos, gold, nickel, copper, iron ore, vanadium, lithium, tin, platinum group metals
Land use:
arable land: 7.25%
permanent crops: 0.25% (coffee is a permanent crop)
meadows and pastures: 12.5%
forest and woodland: 49%
other: 31%
Irrigated land: 2,250 sq km (1993 est.)
Environment:
current issues: deforestation; soil erosion; land degradation; air and water pollution; the black rhinoceros herd—once the largest concentration of the species in the world—has been significantly reduced by poaching
natural hazards: recurring droughts; floods and severe storms are rare
international agreements: party to—Biodiversity, Climate Change, Endangered Species, Law of the Sea, Ozone Layer Protection; signed, but not ratified—Desertification
Note: landlocked

People

Population: 11,139,961 (July 1995 est.)
Age structure:
0-14 years: 47% (female 2,588,193; male 2,617,485)
15-64 years: 51% (female 2,915,697; male 2,723,511)
65 years and over: 2% (female 151,635; male 143,440) (July 1995 est.)
Population growth rate: 1.78% (1995 est.)
Birth rate: 36.35 births/1,000 population (1995 est.)
Death rate: 18.54 deaths/1,000 population (1995 est.)
Net migration rate: NA migrant(s)/1,000 population (1995 est.)
note: following the settlement of hostilities in Mozambique in 1992, refugees from the fighting there began to return to their homes; this process continues at a lesser rate in 1995; there is a small but steady flow of Zimbabweans into South Africa in search of better paid employment
Infant mortality rate: 72.7 deaths/1,000 live births (1995 est.)
Life expectancy at birth:
total population: 41.35 years
male: 39.73 years
female: 43.01 years (1995 est.)
Total fertility rate: 4.93 children born/woman (1995 est.)
Nationality:
noun: Zimbabwean(s)
adjective: Zimbabwean
Ethnic divisions: African 98% (Shona 71%, Ndebele 16%, other 11%), white 1%, mixed and Asian 1%
Religions: syncretic (part Christian, part indigenous beliefs) 50%, Christian 25%, indigenous beliefs 24%, Muslim and other 1%
Languages: English (official), Shona, Sindebele
Literacy: age 15 and over can read and write (1982)
total population: 78%
male: 84%
female: 72%
Labor force: 3.1 million
by occupation: agriculture 74%, transport and services 16%, mining, manufacturing, construction 10% (1987)

Government

Names:
conventional long form: Republic of Zimbabwe
conventional short form: Zimbabwe
former: Southern Rhodesia
Digraph: ZI
Type: parliamentary democracy
Capital: Harare
Administrative divisions: 8 provinces;

Manicaland, Mashonaland Central, Mashonaland East, Mashonaland West, Masvingo (Victoria), Matabeleland North, Matabeleland South, Midlands
Independence: 18 April 1980 (from UK)
National holiday: Independence Day, 18 April (1980)
Constitution: 21 December 1979
Legal system: mixture of Roman-Dutch and English common law
Suffrage: 18 years of age; universal
Executive branch:
chief of state and head of government: Executive President Robert Gabriel MUGABE (since 31 December 1987); Co-Vice President Simon Vengai MUZENDA (since 31 December 1987); Co-Vice President Joshua M. NKOMO (since 6 August 1990); election last held 28-30 March 1990 (next to be held NA March 1996); results—Robert MUGABE 78.3%, Edgar TEKERE 21.7%
cabinet: Cabinet; appointed by the president; responsible to Parliament
Legislative branch: unicameral
Parliament: elections last held 8-9 April 1995 (next to be held NA March 2000); results—percent of vote by party NA; seats—(150 total, 120 elected) ZANU-PF 118, ZANU-S 2
Judicial branch: Supreme Court
Political parties and leaders: Zimbabwe African National Union-Patriotic Front (ZANU-PF), Robert MUGABE; Zimbabwe African National Union-Sithole (ZANU-S), Ndabaningi SITHOLE; Zimbabwe Unity Movement (ZUM), Edgar TEKERE; Democratic Party (DP), Emmanuel MAGOCHE; Forum Party of Zimbabwe, Enock DUMBUTSHENA; United Parties, Abel MUZOREWA
Member of: ACP, AfDB, C, CCC, ECA, FAO, FLS, G-15, G-77, GATT, IAEA, IBRD, ICAO, ICFTU, ICRM, IDA, IFAD, IFC, IFRCS, ILO, IMF, INTELSAT, INTERPOL, IOC, IOM (observer), ISO, ITU, NAM, OAU, PCA, SADC, UN, UNAMIR, UNAVEM II, UNCTAD, UNESCO, UNIDO, UNOMUR, UNOSOM, UPU, WCL, WHO, WIPO, WMO, WTO
Diplomatic representation in US:
chief of mission: Ambassador Amos Bernard Muvengwa MIDZI
chancery: 1608 New Hampshire Avenue NW, Washington, DC 20009
telephone: [1] (202) 332-7100
FAX: [1] (202) 483-9326
US diplomatic representation:
chief of mission: Ambassador Johnny CARSON
embassy: 172 Herbert Chitepo Avenue, Harare
mailing address: P. O. Box 3340, Harare
telephone: [263] (4) 794521
FAX: [263] (4) 796488
Flag: seven equal horizontal bands of green, yellow, red, black, red, yellow, and green with a white equilateral triangle edged in black based on the hoist side; a yellow Zimbabwe bird is superimposed on a red five-pointed star in the center of the triangle

Economy

Overview: Agriculture employs three-fourths of the labor force and supplies almost 40% of exports. The manufacturing sector, based on agriculture and mining, produces a variety of goods and contributes 35% to GDP. Mining accounts for only 5% of both GDP and employment, but minerals and metals account for about 40% of exports. Severe drought caused GDP to drop 8% in 1992, with growth rebounding to 2% in 1993 and 3.5% in 1994. Despite the lingering effects of the drought on economic and social conditions, the government is continuing to push its IMF/World Bank structural adjustment program aimed at encouraging exports and foreign investment.
National product: GDP—purchasing power parity—$17.4 billion (1994 est.)
National product real growth rate: 3.5% (1994 est.)
National product per capita: $1,580 (1994 est.)
Inflation rate (consumer prices): 22% (December 1994 est.)
Unemployment rate: at least 45% (1994 est.)
Budget:
revenues: $1.7 billion
expenditures: $2.2 billion, including capital expenditures of $253 million (FY92/93)
Exports: $1.8 billion (f.o.b., 1994 est.)
commodities: agricultural 35% (tobacco 30%, other 5%), manufactures 25%, gold 12%, ferrochrome 10%, textiles 8% (1992)
partners: UK 14%, Germany 11%, South Africa 10%, Japan 7%, US 5% (1991)
Imports: $1.8 billion (c.i.f., 1992 est.)
commodities: machinery and transportation equipment 41%, other manufactures 23%, chemicals 16%, fuels 12% (1991)
partners: South Africa 25%, UK 15%, Germany 9%, US 6%, Japan 5% (1991)
External debt: $3.5 billion (December 1992 est.)
Industrial production: growth rate 2.3% (1992); accounts for 35% of GDP
Electricity:
capacity: 2,040,000 kW
production: 9 billion kWh
consumption per capita: 913 kWh (1993)
Industries: mining, steel, clothing and footwear, chemicals, foodstuffs, fertilizer, beverage, transportation equipment, wood products
Agriculture: accounts for 20% of GDP; 40% of land area divided into 4,500 large commercial farms and 42% in communal lands; crops—corn (food staple), cotton, tobacco, wheat, coffee, sugarcane, peanuts; livestock—cattle, sheep, goats, pigs; self-sufficient in food
Economic aid: NA
Currency: 1 Zimbabwean dollar (Z$) = 100 cents
Exchange rates: Zimbabwean dollars (Z$) per US$1—8.3752 (January 1995), 8.1500 (1994), 6.4725 (1993), 5.1046 (1992), 3.4282 (1991), 2.4480 (1990)
Fiscal year: 1 July—30 June

Transportation

Railroads:
total: 2,745 km
narrow gauge: 2,745 km 1.067-m gauge (355 km electrified; 42 km double track)
Highways:
total: 85,237 km
paved: 15,800 km
unpaved: crushed stone, gravel, stabilized earth 39,090 km; improved earth 23,097 km; unimproved earth 7,250 km
Inland waterways: Lake Kariba is a potential line of communication
Pipelines: petroleum products 212 km
Ports: Binga, Kariba
Airports:
total: 471
with paved runways over 3,047 m: 3
with paved runways 2,438 to 3,047 m: 2
with paved runways 1,524 to 2,437 m: 6
with paved runways 914 to 1,523 m: 13
with paved runways under 914 m: 222
with unpaved runways 1,524 to 2,438 m: 2
with unpaved runways 914 to 1,523 m: 223

Communications

Telephone system: 247,000 telephones; system was once one of the best in Africa, but now suffers from poor maintenance
local: NA
intercity: consists of microwave links, open-wire lines, and radio communications stations
international: 1 INTELSAT (Atlantic Ocean) earth station
Radio:
broadcast stations: AM 8, FM 18, shortwave 0
radios: NA
Television:
broadcast stations: 8
televisions: NA

Defense Forces

Branches: Zimbabwe National Army, Air Force of Zimbabwe, Zimbabwe Republic Police (includes Police Support Unit, Paramilitary Police)
Manpower availability: males age 15-49 2,435,931; males fit for military service 1,514,068 (1995 est.)
Defense expenditures: exchange rate conversion—$175 million, 3.1% of GDP (FY94/95)

Taiwan

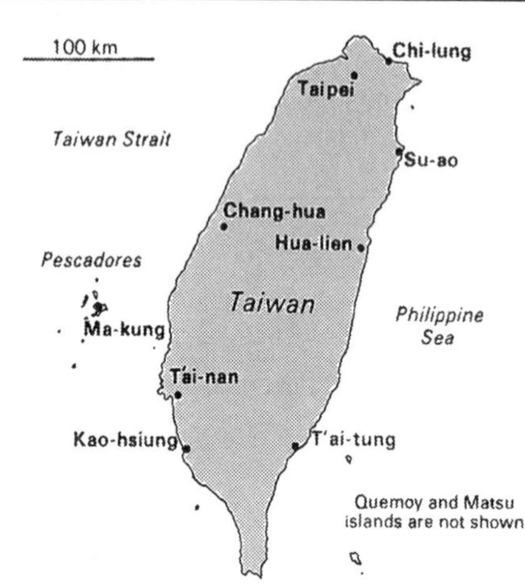

Geography

Location: Eastern Asia, islands bordering the East China Sea, Philippine Sea, South China Sea, and Taiwan Strait, north of the Philippines, off the southeastern coast of China
Map references: Southeast Asia
Area:
total area: 35,980 sq km
land area: 32,260 sq km
comparative area: slightly larger than Maryland and Delaware combined
note: includes the Pescadores, Matsu, and Quemoy
Land boundaries: 0 km
Coastline: 1,448 km
Maritime claims:
exclusive economic zone: 200 nm
territorial sea: 12 nm
International disputes: involved in complex dispute over the Spratly Islands with China, Malaysia, Philippines, Vietnam, and possibly Brunei; Paracel Islands occupied by China, but claimed by Vietnam and Taiwan; Japanese-administered Senkaku-shoto (Senkaku Islands/ Diaoyu Tai) claimed by China and Taiwan
Climate: tropical; marine; rainy season during southwest monsoon (June to August); cloudiness is persistent and extensive all year
Terrain: eastern two-thirds mostly rugged mountains; flat to gently rolling plains in west
Natural resources: small deposits of coal, natural gas, limestone, marble, and asbestos
Land use:
arable land: 24%
permanent crops: 1%
meadows and pastures: 5%
forest and woodland: 55%
other: 15%
Irrigated land: NA sq km
Environment:
current issues: water pollution from industrial emissions, raw sewage; air pollution; contamination of drinking water supplies; trade in endangered species
natural hazards: earthquakes and typhoons
international agreements: signed, but not ratified—Marine Life Conservation

People

Population: 21,500,583 (July 1995 est.)
Age structure:
0-14 years: 24% (female 2,543,134; male 2,665,878)
15-64 years: 68% (female 7,191,964; male 7,482,814)
65 years and over: 8% (female 734,535; male 882,258) (July 1995 est.)
Population growth rate: 0.93% (1995 est.)
Birth rate: 15.33 births/1,000 population (1995 est.)
Death rate: 5.71 deaths/1,000 population (1995 est.)
Net migration rate: -0.37 migrant(s)/1,000 population (1995 est.)
Infant mortality rate: 5.6 deaths/1,000 live births (1995 est.)
Life expectancy at birth:
total population: 75.47 years
male: 72.17 years
female: 78.93 years (1995 est.)
Total fertility rate: 1.81 children born/ woman (1995 est.)
Nationality:
noun: Chinese (singular and plural)
adjective: Chinese
Ethnic divisions: Taiwanese 84%, mainland Chinese 14%, aborigine 2%
Religions: mixture of Buddhist, Confucian, and Taoist 93%, Christian 4.5%, other 2.5%
Languages: Mandarin Chinese (official), Taiwanese (Min), Hakka dialects
Literacy: age 15 and over can read and write (1980)
total population: 86%
male: 93%
female: 79%
Labor force: 7.9 million
by occupation: industry and commerce 53%, services 22%, agriculture 15.6%, civil administration 7% (1989)

Administration

Names:
conventional long form: none
conventional short form: Taiwan
local long form: none
local short form: T'ai-wan
Digraph: TW
Type: multiparty democratic regime; opposition political parties legalized in March, 1989
Capital: Taipei
Administrative divisions: some of the ruling party in Taipei claim to be the government of all China; in keeping with that claim, the central administrative divisions include 2 provinces (sheng, singular and plural) and 2 municipalities* (shih, singular and plural)—Fu-chien (some 20 offshore islands of Fujian Province including Quemoy and Matsu), Kao-hsiung*, T'ai-pei*, and Taiwan (the island of Taiwan and the Pescadores islands); the more commonly referenced administrative divisions are those of Taiwan Province—16 counties (hsien, singular and plural), 5 municipalities* (shih, singular and plural), and 2 special municipalities** (chuan-shih, singular and plural); Chang-hua, Chia-i, Chia-i*, Chi-lung*, Hsin-chu, Hsin-chu*, Hua-lien, I-lan, Kao-hsiung, Kao-hsiung**, Miao-li, Nan-t'ou, P'eng-hu, P'ing-tung, T'ai-chung, T'ai-chung*, T'ai-nan, T'ai-nan*, T'ai-pei, T'ai-pei**, T'ai-tung, T'ao-yuan, and Yun-lin; the provincial capital is at Chung-hsing-hsin-ts'un
note: Taiwan uses the Wade-Giles system for romanization
National holiday: National Day, 10 October (1911) (Anniversary of the Revolution)
Constitution: 1 January 1947, amended in 1992, presently undergoing revision
Legal system: based on civil law system; accepts compulsory ICJ jurisdiction, with reservations
Suffrage: 20 years of age; universal
Executive branch:
chief of state: President LI Teng-hui (since 13 January 1988); Vice President LI Yuan-zu (since 20 May 1990)
head of government: Premier (President of the Executive Yuan) LIEN Chan (since 23 February 1993); Vice Premier (Vice President of the Executive Yuan) HSU Li-teh (since 23 February 1993); presidential election last held 21 March 1990 (next election will probably be a direct popular election and will be held NA March 1996); results—President LI Teng-hui was reelected by the National Assembly; vice presidential election last held 21 March 1990; results—LI Yuan-zu was elected by the National Assembly
cabinet: Executive Yuan; appointed by the president
Legislative branch: unicameral Legislative Yuan and unicameral National Assembly
Legislative Yuan: elections last held 19 December 1992 (next to be held NA December 1995); results—KMT 60%, DPP 31%, independents 9%; seats—(304 total, 161 elected) KMT 96, DPP 50, independents 15
National Assembly: first National Assembly elected in November 1946 with a supplementary election in December 1986; second and present National Assembly elected in December 1991; seats—(403 total) KMT 318, DPP 75, other 10; (next election to be held probably in 1996 and will be a direct popular election)
Judicial branch: Judicial Yuan
Political parties and leaders: Kuomintang (KMT, Nationalist Party), LI Teng-hui, chairman; Democratic Progressive Party (DPP), SHIH Ming-teh, chairman; Chinese New Party (CNP); Labor Party (LP)
Other political or pressure groups: Taiwan independence movement, various environmental groups
note: debate on Taiwan independence has

become acceptable within the mainstream of domestic politics on Taiwan; political liberalization and the increased representation of the opposition Democratic Progressive Party in Taiwan's legislature have opened public debate on the island's national identity; advocates of Taiwan independence, both within the DPP and the ruling Kuomintang, oppose the ruling party's traditional stand that the island will eventually unify with mainland China; the aims of the Taiwan independence movement include establishing a sovereign nation on Taiwan and entering the UN; other organizations supporting Taiwan independence include the World United Formosans for Independence and the Organization for Taiwan Nation Building
Member of: expelled from UN General Assembly and Security Council on 25 October 1971 and withdrew on same date from other charter-designated subsidiary organs; expelled from IMF/World Bank group April/May 1980; seeking to join GATT; attempting to retain membership in INTELSAT; suspended from IAEA in 1972, but still allows IAEA controls over extensive atomic development, APEC, AsDB, BCIE, ICC, IOC, WCL
Diplomatic representation in US: none; unofficial commercial and cultural relations with the people of the US are maintained through a private instrumentality, the Taipei Economic and Cultural Representative Office (TECRO) with headquarters in Taipei and field offices in Washington and 10 other US cities
US diplomatic representation: unofficial commercial and cultural relations with the people of Taiwan are maintained through a private institution, the American Institute in Taiwan (AIT), which has offices in Taipei at #7, Lane 134, Hsin Yi Road, Section 3, telephone [886] (2) 709-2000, and in Kao-hsiung at #2 Chung Cheng 3d Road, telephone [886] (7) 224-0154 through 0157, and the American Trade Center at Room 3207 International Trade Building, Taipei World Trade Center, 333 Keelung Road Section 1, Taipei 10548, telephone [886] (2) 720-1550
Flag: red with a dark blue rectangle in the upper hoist-side corner bearing a white sun with 12 triangular rays

Economy

Overview: Taiwan has a dynamic capitalist economy with considerable government guidance of investment and foreign trade and partial government ownership of some large banks and industrial firms. Real growth in GNP has averaged about 9% a year during the past three decades. Export growth has been even faster and has provided the impetus for industrialization. Inflation and unemployment are remarkably low. Agriculture contributes about 4% to GDP, down from 35% in 1952. Taiwan currently ranks as number 13 among major trading countries. Traditional labor-intensive industries are steadily being replaced with more capital- and technology-intensive industries. Taiwan has become a major investor in China, Thailand, Indonesia, the Philippines, Malaysia, and Vietnam. The tightening of labor markets has led to an influx of foreign workers, both legal and illegal.
National product: GDP—purchasing power parity—$257 billion (1994 est.)
National product real growth rate: 6% (1994 est.)
National product per capita: $12,070 (1994 est.)
Inflation rate (consumer prices): 5.2% (1994 est.)
Unemployment rate: 1.6% (1994)
Budget:
revenues: $30.3 billion
expenditures: $30.1 billion, including capital expenditures of $NA (1991 est.)
Exports: $93 billion (f.o.b., 1994)
commodities: electrical machinery 19.7%, electronic products 19.6%, textiles 10.9%, footwear 3.3%, foodstuffs 1.0%, plywood and wood products 0.9% (1993 est.)
partners: US 27.6%, Hong Kong 21.7%, EC countries 15.2%, Japan 10.5% (1994 est.)
Imports: $85.1 billion (c.i.f., 1994)
commodities: machinery and equipment 15.7%, electronic products 15.6%, chemicals 9.8%, iron and steel 8.5%, crude oil 3.9%, foodstuffs 2.1% (1993 est.)
partners: Japan 30.1%, US 21.7%, EC countries 17.6% (1993 est.)
External debt: $620 million (1992 est.)
Industrial production: growth rate 4.5% (1994 est.); accounts for more than 40% of GDP
Electricity:
capacity: 21,460,000 kW
production: 108 billion kWh
consumption per capita: 4,789 kWh (1993)
Industries: electronics, textiles, chemicals, clothing, food processing, plywood, sugar milling, cement, shipbuilding, petroleum refining
Agriculture: accounts for 4% of GDP and 16% of labor force (includes part-time farmers); heavily subsidized sector; major crops—vegetables, rice, fruit, tea; livestock—hogs, poultry, beef, milk; not self-sufficient in wheat, soybeans, corn; fish catch increasing, reached 1.4 million metric tons in 1988
Illicit drugs: an important heroin transit point; also a major drug money laundering center
Economic aid:
recipient: US, including Ex-Im (FY46-82), $4.6 billion; Western (non-US) countries, ODA and OOF bilateral commitments (1970-89), $500 million
Currency: 1 New Taiwan dollar (NT$) = 100 cents
Exchange rates: New Taiwan dollars per US$1—26.2 (1994), 26.6 (1993), 25.4 (1992), 25.748 (1991), 27.108 (1990), 26.407 (1989)
Fiscal year: 1 July—30 June

Transportation

Railroads:
total: 4,600 km; note—1,075 km in common carrier service and about 3,525 km is dedicated to industrial use
narrow gauge: 4,600 km 1.067-m
Highways:
total: 20,041 km
paved: bituminous, concrete pavement 17,095 km
unpaved: crushed stone, gravel 2,371 km; graded earth 575 km
Pipelines: petroleum products 615 km; natural gas 97 km
Ports: Chi-lung (Keelung), Hua-lien, Kao-hsiung, Su-ao, T'ai-chung
Merchant marine:
total: 198 ships (1,000 GRT or over) totaling 5,635,682 GRT/8,652,111 DWT
ships by type: bulk 55, cargo 30, chemical tanker 1, combination bulk 2, combination ore/oil 1, container 78, oil tanker 17, passenger-cargo 1, refrigerated cargo 12, roll-on/roll-off cargo 1
Airports:
total: 41
with paved runways over 3,047 m: 8
with paved runways 2,438 to 3,047 m: 11
with paved runways 1,524 to 2,437 m: 6
with paved runways 914 to 1,523 m: 6
with paved runways under 914 m: 8
with unpaved runways 1,524 to 2,438 m: 2

Communications

Telephone system: 7,800,000 telephones; best developed system in Asia outside of Japan
local: NA
intercity: extensive microwave radio relay links on east and west coasts
international: 2 INTELSAT (1 Pacific Ocean and 1 Indian Ocean) earth stations, submarine cable links to Japan (Okinawa), Philippines, Guam, Singapore, Hong Kong, Indonesia, Australia, Middle East, and Western Europe
Radio:
broadcast stations: AM 91, FM 23, shortwave 0
radios: 8.62 million
Television:
broadcast stations: 15 (repeaters 13)
televisions: 6.386 million (color 5,680,000, monochrome 706,000)

Defense Forces

Branches: Army, Navy (includes Marines), Air Force, Coastal Patrol and Defense Command, Armed Forces Reserve Command, Military Police Command

Manpower availability: males age 15-49 6,293,884; males fit for military service 4,863,014; males reach military age (19) annually 201,191 (1995 est.)
Defense expenditures: exchange rate conversion—$9.8 billion, 3.4% of GDP (FY94/95); $9.77 billion proposed for FY95/96 budget

Appendix A:

The United Nations System

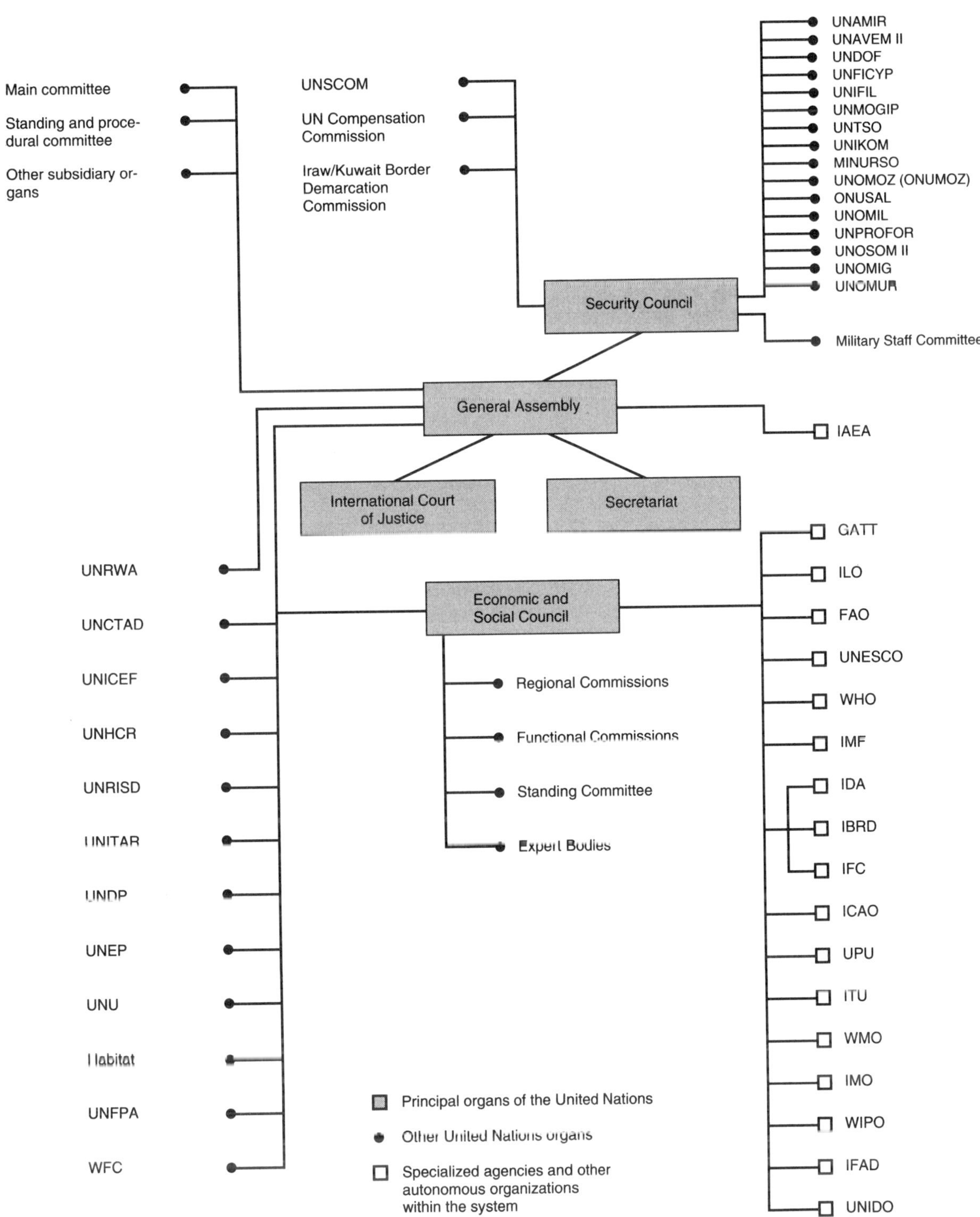

Main committee
Standing and procedural committee
Other subsidiary organs
UNSCOM
UN Compensation Commission
Iraw/Kuwait Border Demarcation Commission
UNAMIR
UNAVEM II
UNDOF
UNFICYP
UNIFIL
UNMOGIP
UNTSO
UNIKOM
MINURSO
UNOMOZ (ONUMOZ)
ONUSAL
UNOMIL
UNPROFOR
UNOSOM II
UNOMIG
UNOMUR
Security Council
Military Staff Committee
General Assembly
IAEA
International Court of Justice
Secretariat
UNRWA
UNCTAD
UNICEF
UNHCR
UNRISD
UNITAR
UNDP
UNEP
UNU
Habitat
UNFPA
WFC
Economic and Social Council
Regional Commissions
Functional Commissions
Standing Committee
Expert Bodies
GATT
ILO
FAO
UNESCO
WHO
IMF
IDA
IBRD
IFC
ICAO
UPU
ITU
WMO
IMO
WIPO
IFAD
UNIDO
Principal organs of the United Nations
Other United Nations organs
Specialized agencies and other autonomous organizations within the system

Based on chart from the *UN Chronicle*

Appendix B:

Abbreviations for International Organizations and Groups

A	ABEDA	Arab Bank for Economic Development in Africa
	ACC	Arab Cooperation Council
	ACCT	Agence de Cooperation Culturelle et Technique; see Agency for Cultural and Technical Cooperation
	ACP	African, Caribbean, and Pacific Countries
	AfDB	African Development Bank
	AFESD	Arab Fund for Economic and Social Development
	AG	Andean Group
	AL	Arab League
	ALADI	Asociacion Latinoamericana de Integracion; see Latin American Integration Association (LAIA)
	AMF	Arab Monetary Fund
	AMU	Arab Maghreb Union
	ANZUS	Australia-New Zealand-United States Security Treaty
	APEC	Asia Pacific Economic Cooperation
	AsDB	Asian Development Bank
	ASEAN	Association of Southeast Asian Nations
B	BAD	Banque Africaine de Developpement; see African Development Bank (AfDB)
	BADEA	Banque Arabe de Developpement Economique en Afrique; see Arab Bank for Economic Development in Africa (ABEDA)
	BCIE	Banco Centroamericano de Integracion Economico; see Central American Bank for Economic Integration (BCIE)
	BDEAC	Banque de Developpment des Etats de l'Afrique Centrale; see Central African States Development Bank (BDEAC)
	Benelux	Benelux Economic Union
	BID	Banco Interamericano de Desarrollo; see Inter-American Development Bank (IADB)
	BIS	Bank for International Settlements
	BOAD	Banque Ouest-Africaine de Developpement; see West African Development Bank (WADB)
	BSEC	Black Sea Economic Cooperation Zone
C	C	Commonwealth
	CACM	Central American Common Market
	CAEU	Council of Arab Economic Unity
	CARICOM	Caribbean Community and Common Market
	CBSS	Council of the Baltic Sea States
	CCC	Customs Cooperation Council
	CDB	Caribbean Development Bank
	CE	Council of Europe
	CEAO	Communaute Economique de l'Afrique de l'Ouest; see West African Economic Community (CEAO)
	CEEAC	Communaute Economique des Etats de l'Afrique Centrale; see Economic Community of Central African States (CEEAC)

	CEI	Central European Initiative
	CEMA	Council for Mutual Economic Assistance; also known as CMEA or Comecon; abolished 1 January 1991
	CEPGL	Communaute Economique des Pays des Grands Lacs; see Economic Community of the Great Lakes Countries (CEPGL)
	CERN	Conseil Europeen pour la Recherche Nucleaire; see European Organization for Nuclear Research (CERN)
	CG	Contadora Group
	CIS	Commonwealth of Independent States
	CMEA	Council for Mutual Economic Assistance (CEMA); also known as Comecon; abolished 1 January 1991
	COCOM	Coordinating Committee on Export Controls
	Comecon	Council for Mutual Economic Assistance (CEMA); also known as CMEA; abolished 1 January 1991
	CP	Colombo Plan
	CSCE	Conference on Security and Cooperation in Europe
D	DC	developed country
E	EADB	East African Development Bank
	EBRD	European Bank for Reconstruction and Development
	EC	European Community; see European Union (EU)
	ECA	Economic Commission for Africa
	ECAFE	Economic Commission for Asia and the Far East; see Economic and Social Commission for Asia and the Pacific (ESCAP)
	ECE	Economic Commission for Europe
	ECLA	Economic Commission for Latin America; see Economic Commission for Latin America and the Caribbean (ECLAC)
	ECLAC	Economic Commission for Latin America and the Caribbean
	ECO	Economic Cooperation Organization
	ECOSOC	Economic and Social Council
	ECOWAS	Economic Community of West African States
	ECSC	European Coal and Steel Community
	ECWA	Economic Commission for Western Asia; see Economic and Social Commission for Western Asia (ESCWA)
	EEC	European Economic Community
	EFTA	European Free Trade Association
	EIB	European Investment Bank
	Entente	Council of the Entente
	ESA	European Space Agency
	ESCAP	Economic and Social Commission for Asia and the Pacific
	ESCWA	Economic and Social Commission for Western Asia
	EU	European Union
	Euratom	European Atomic Energy Community
F	FAO	Food and Agriculture Organization
	FLS	Front Line States
	FZ	Franc Zone
G	G-2	Group of 2
	G-3	Group of 3
	G-5	Group of 5

	G-6	Group of 6 (not to be confused with the Big Six)
	G-7	Group of 7
	G-8	Group of 8
	G-9	Group of 9
	G-10	Group of 10
	G-11	Group of 11
	G-15	Group of 15
	G-19	Group of 19
	G-24	Group of 24
	G-30	Group of 30
	G-33	Group of 33
	G-77	Group of 77
	GATT	General Agreement on Tariffs and Trade
	GCC	Gulf Cooperation Council
H	Habitat	Commission on Human Settlements
I	IADB	Inter-American Development Bank
	IAEA	International Atomic Energy Agency
	IBEC	International Bank for Economic Cooperation
	IBRD	International Bank for Reconstruction and Development
	ICAO	International Civil Aviation Organization
	ICC	International Chamber of Commerce
	ICEM	Intergovernmental Committee for European Migration; see International Organization for Migration (IOM)
	ICFTU	International Confederation of Free Trade Unions
	ICJ	International Court of Justice
	ICM	Intergovernmental Committee for Migration; see International Organization for Migration (IOM)
	ICRC	International Committee of the Red Cross
	ICRM	International Red Cross and Red Crescent Movement
	IDA	International Development Association
	IDB	Islamic Development Bank
	IEA	International Energy Agency
	IFAD	International Fund for Agricultural Development
	IFC	International Finance Corporation
	IFCTU	International Federation of Christian Trade Unions
	IFRCS	International Federation of Red Cross and Red Crescent Societies
	IGADD	Inter-Governmental Authority on Drought and Development
	IIB	International Investment Bank
	ILO	International Labor Organization
	IMCO	Intergovernmental Maritime Consultative Organization; see International Maritime Organization (IMO)
	IMF	International Monetary Fund
	IMO	International Maritime Organization
	INMARSAT	International Maritime Satellite Organization
	INTELSAT	International Telecommunications Satellite Organization
	INTERPOL	International Criminal Police Organization
	IOC	International Olympic Committee

	IOM	International Organization for Migration
	ISO	International Organization for Standardization
	ITU	International Telecommunication Union
L	LAES	Latin American Economic System
	LAIA	Latin American Integration Association
	LAS	League of Arab States; see Arab League (AL)
	LDC	less developed country
	LLDC	least developed country
	LORCS	League of Red Cross and Red Crescent Societies
M	MERCOSUR	Mercado Comun del Cono Sur; see Southern Cone Common Market
	MINURSO	United Nations Mission for the Referendum in Western Sahara
	MTCR	Missile Technology Control Regime
N	NACC	North Atlantic Cooperation Council
	NAM	Nonaligned Movement
	NATO	North Atlantic Treaty Organization
	NC	Nordic Council
	NEA	Nuclear Energy Agency
	NIB	Nordic Investment Bank
	NIC	newly industrializing country; see newly industrializing economy (NIE)
	NIE	newly industrializing economy
	NSG	Nuclear Suppliers Group
O	OAPEC	Organization of Arab Petroleum Exporting Countries
	OAS	Organization of American States
	OAU	Organization of African Unity
	OECD	Organization for Economic Cooperation and Development
	OECS	Organization of Eastern Caribbean States
	OIC	Organization of the Islamic Conference
	ONUMOZ	See UNOMOZ
	ONUSAL	United Nations Observer Mission in El Salvador
	OPANAL	Organismo para la Proscripcion de las Armas Nucleares en la America Latina y el Caribe; see Agency for the Prohibition of Nuclear Weapons in Latin America and the Caribbean
	OPEC	Organization of Petroleum Exporting Countries
	OSCE	Organization on Security and Cooperation in Europe
P	PCA	Permanent Court of Arbitration
	PFP	Partnership for Peace
R	RG	Rio Group
S	SAARC	South Asian Association for Regional Cooperation
	SACU	Southern African Customs Union
	SADC	Southern African Development Community
	SADCC	Southern African Development Coordination Conference
	SELA	Sistema Economico Latinoamericana; see Latin American Economic System (LAES)
	SPARTECA	South Pacific Regional Trade and Economic Cooperation Agreement
	SPC	South Pacific Commission
	SPF	South Pacific Forum
U	UDEAC	Union Douaniere et Economique de l'Afrique Centrale; see Central African Customs and Economic Union (UDEAC)

	UN	United Nations
	UNAVEM II	United Nations Angola Verification Mission
	UNAMIR	United Nations Assistance Mission for Rwanda
	UNCTAD	United Nations Conference on Trade and Development
	UNDOF	United Nations Disengagement Observer Force
	UNDP	United Nations Development Program
	UNEP	United Nations Environment Program
	UNESCO	United Nations Educational, Scientific, and Cultural Organization
	UNFICYP	United Nations Force in Cyprus
	UNFPA	United Nations Fund for Population Activities; see UN Population Fund (UNFPA)
	UNHCR	United Nations Office of the High Commissioner for Refugees
	UNICEF	United Nations Children's Fund
	UNIDO	United Nations Industrial Development Organization
	UNIFIL	United Nations Interim Force in Lebanon
	UNIKOM	United Nations Iraq-Kuwait Observation Mission
	UNITAR	United Nations Institute for Training and Research
	UNMIH	United Nations Mission in Haiti
	UNMOGIP	United Nations Military Observer Group in India and Pakistan
	UNOMIG	United Nations Observer Mission in Georgia
	UNOMIL	United Nations Observer Mission in Liberia
	UNOMOZ (ONUMOZ)	United Nations Operation in Mozambique
	UNOMUR	United Nations Observer Mission Uganda-Rwanda
	UNOSOM	United Nations Operation in Somalia
	UNPROFOR	United Nations Protection Force
	UNRISD	United Nations Research Institute for Social Development
	UNRWA	United Nations Relief and Works Agency for Palestine Refugees in the Near East
	UNTAC	United Nations Transitional Authority in Cambodia
	UNTSO	United Nations Truce Supervision Organization
	UNU	United Nations University
	UPU	Universal Postal Union
	USSR/EE	USSR/Eastern Europe
W	WADB	West African Development Bank
	WCL	World Confederation of Labor
	WEU	Western European Union
	WFC	World Food Council
	WFP	World Food Program
	WFTU	World Federation of Trade Unions
	WHO	World Health Organization
	WIPO	World Intellectual Property Organization
	WMO	World Meteorological Organization
	WP	Warsaw Pact (members met 1 July 1991 to dissolve the alliance)
	WTO	see WToO
	WToO	World Tourism Organization
	WTrO	World Trade Organization (will be added in *The World Factbook 1996*)
Z	ZC	Zangger Committee

Note: Not all international organizations and groups have abbreviations.

Appendix C:

International Organizations and Groups

advanced developing countries	another term for those less developed countries (LDCs) with particularly rapid industrial development; see newly industrializing economies (NIEs)
African, Caribbean, and Pacific Countries (ACP) *address*—Avenue Georges Henri 451, B-1200 Brussels, Belgium *telephone*—[32] (2) 733 96 00 *FAX*—[32] (2) 735 55 73 *established*—1 April 1976 *aim*—to manage their preferential economic and aid relationship with the EU	*members*—(70) Angola, Antigua and Barbuda, The Bahamas, Barbados, Belize, Benin, Botswana, Burkina, Burundi, Cameroon, Cape Verde, Central African Republic, Chad, Comoros, Congo, Cote d'Ivoire, Djibouti, Dominica, Dominican Republic, Equatorial Guinea, Eritrea, Ethiopia, Fiji, Gabon, The Gambia, Ghana, Grenada, Guinea, Guinea-Bissau, Guyana, Haiti, Jamaica, Kenya, Kiribati, Lesotho, Liberia, Madagascar, Malawi, Mali, Mauritania, Mauritius, Mozambique, Namibia, Niger, Nigeria, Papua New Guinea, Rwanda, Saint Kitts and Nevis, Saint Lucia, Saint Vincent and the Grenadines, Sao Tome and Principe, Senegal, Seychelles, Sierra Leone, Solomon Islands, Somalia, Sudan, Suriname, Swaziland, Tanzania, Togo, Tonga, Trinidad and Tobago, Tuvalu, Uganda, Vanuatu, Western Samoa, Zaire, Zambia, Zimbabwe
African Development Bank (AfDB) *note:* also known as Banque Africaine de Developpement (BAD) *address*—01 BP 1387, Abidjan 01, Cote d'Ivoire *telephone*—[225] 20 44 44 *FAX*—[225] 21 77 53, 20 49 01, 20 49 09 *established*—4 August 1963 *aim*—to promote economic and social development	*regional members*—(51) Algeria, Angola, Benin, Botswana, Burkina, Burundi, Cameroon, Cape Verde, Central African Republic, Chad, Comoros, Congo, Cote d'Ivoire, Djibouti, Egypt, Equatorial Guinea, Ethiopia, Gabon, The Gambia, Ghana, Guinea, Guinea-Bissau, Kenya, Lesotho, Liberia, Libya, Madagascar, Malawi, Mali, Mauritania, Mauritius, Morocco, Mozambique, Namibia, Niger, Nigeria, Rwanda, Sao Tome and Principe, Senegal, Seychelles, Sierra Leone, Somalia, Sudan, Swaziland, Tanzania, Togo, Tunisia, Uganda, Zaire, Zambia, Zimbabwe *nonregional members*—(26) Argentina, Austria, Belgium, Brazil, Canada, China, Denmark, Finland, France, Germany, India, Italy, Japan, South Korea, Kuwait, Netherlands, Norway, Portugal, Saudi Arabia, Spain, Sweden, Switzerland, Turkey, UK, US, Yugoslavia
Agence de Cooperation Culturelle et Technique (ACCT)	see Agency for Cultural and Technical Cooperation (ACCT)
Agency for Cultural and Technical Cooperation (ACCT) *note*—acronym from Agence de Cooperation Culturelle et Technique *address*—13 quai Andre-Citroen, F-75015 Paris, France *telephone*—[33] (1) 44 37 33 00 *FAX*—[33] (1) 45 79 14 98 *established*—21 March 1970 *aim*—to promote cultural and technical cooperation among French-speaking countries	*members*—(37) Belgium, Benin, Bulgaria, Burkina, Burundi, Cambodia, Cameroon, Canada, Central African Republic, Chad, Comoros, Congo, Cote d'Ivoire, Djibouti, Dominica, Equatorial Guinea, France, Gabon, Guinea, Haiti, Laos, Lebanon, Luxembourg, Madagascar, Mali, Mauritius, Monaco, Niger, Romania, Rwanda, Senegal, Seychelles, Togo, Tunisia, Vanuatu, Vietnam, Zaire *associate members*—(5) Egypt, Guinea-Bissau, Mauritania, Morocco, Saint Lucia *participating governments*—(2) New Brunswick (Canada), Quebec (Canada)

Note: The Socialist Federal Republic of Yugoslavia (SFRY) has dissolved and ceases to exist. None of the successor states of the former Yugoslavia, including Serbia and Montenegro, have been permitted to participate solely on the basis of the membership of the former Yugoslavia in the United Nations General Assembly and Economic and Social Council and their subsidiary bodies and in various United Nations specialized agencies. The United Nations, however, permits the seat and nameplate of the SFRY to remain, permits the SFRY mission to continue to function, and continues to fly the flag of the former Yugoslavia. For a variety of reasons, a number of other organizations have not yet taken action with regard to the membership of the former Yugoslavia. *The World Factbook* therefore continues to list Yugoslavia under international organizations where the SFRY seat remains or where no action has yet been taken.

Agency for the Prohibition of Nuclear Weapons in Latin America and the Caribbean (OPANAL)

note—acronym from Organismo para la Proscripcion de las Armas Nucleares en la America Latina y el Caribe (OPANAL)
address—Temistocles 78, Col Polanco, CP 011560, Mexico City 5 DF, Mexico
telephone—[52] (5) 280 4923, 280 5064
FAX—[52] (5) 280 2965
established—14 February 1967
aim—to encourage the peaceful uses of atomic energy and prohibit nuclear weapons

members—(28) Antigua and Barbuda, Argentina, The Bahamas, Barbados, Bolivia, Brazil, Chile, Colombia, Costa Rica, Dominica, Dominican Republic, Ecuador, El Salvador, Grenada, Guatemala, Haiti, Honduras, Jamaica, Mexico, Nicaragua, Panama, Paraguay, Peru, Saint Vincent and the Grenadines, Suriname, Trinidad and Tobago, Uruguay, Venezuela

---

Andean Group (AG)

address—c\o JUNAC, Paseo de la Republica 3895, Casilla 18-1177, Lima 27, Peru
telephone—[51] (14) 414212
FAX—[51] (14) 420911
established—26 May 1969
effective—16 October 1969
aim—to promote harmonious development through economic integration

members—(5) Bolivia, Colombia, Ecuador, Peru, Venezuela
associate member—(1) Panama
observers—(26) Argentina, Australia, Austria, Belgium, Brazil, Canada, Costa Rica, Denmark, Egypt, Finland, France, Germany, India, Israel, Italy, Japan, Mexico, Netherlands, Paraguay, Spain, Sweden, Switzerland, UK, US, Uruguay, Yugoslavia

---

Arab Bank for Economic Development in Africa (ABEDA)

note—also known as Banque Arabe de Developpement Economique en Afrique (BADEA)
address—Sayed Abdel Rahman El Mahdi Avenue, P.O. Box 2640, Khartoum, Sudan
telephone—[249] (11) 73646, 73498, 73709
FAX—[249] (11) 70600
established—18 February 1974
effective—16 September 1974
aim—to promote economic development

members—(17 plus the Palestine Liberation Organization) Algeria, Bahrain, Egypt, Iraq, Jordan, Kuwait, Lebanon, Libya, Mauritania, Morocco, Oman, Qatar, Saudi Arabia, Sudan, Syria, Tunisia, UAE, Palestine Liberation Organization; *note*—these are all the members of the Arab League except for Comoros, Djibouti, Somalia, and Yemen

---

Arab Cooperation Council (ACC)

established—16 February 1989
aim—to promote economic cooperation and integration, possibly leading to an Arab Common Market

members—(4) Egypt, Iraq, Jordan, Yemen

Arab Fund for Economic and Social Development (AFESD)

address—P.O. Box 21923, Safat 13080, Kuwait
telephone—[965] 2451580, 2451588
FAX—[965] 2416758
established—16 May 1968
aim—to promote economic and social development

members—(20 plus the Palestine Liberation Organization) Algeria, Bahrain, Djibouti, Egypt (suspended from 1979 to 1988), Iraq, Jordan, Kuwait, Lebanon, Libya, Mauritania, Morocco, Oman, Qatar, Saudi Arabia, Somalia, Sudan, Syria, Tunisia, UAE, Yemen, Palestine Liberation Organization

---

Arab League (AL)

note—also known as League of Arab States (LAS)
address—Midan Attahrir, Tahrir Square, P.O. Box 11642, Cairo, Egypt
telephone—[20] (2) 750 511
FAX—[20] (2) 740 331
established—22 March 1945
aim—to promote economic, social, political, and military cooperation

members—(21 plus the Palestine Liberation Organization) Algeria, Bahrain, Comoros, Djibouti, Egypt, Iraq, Jordan, Kuwait, Lebanon, Libya, Mauritania, Morocco, Oman, Qatar, Saudi Arabia, Somalia, Sudan, Syria, Tunisia, UAE, Yemen, Palestine Liberation Organization

Arab Maghreb Union (AMU) *address*—27 avenue Okba Agdal, Rabat, Morocco *established*—17 February 1989 *aim*—to promote cooperation and integration among the Arab states of northern Africa	*members*—(5) Algeria, Libya, Mauritania, Morocco, Tunisia
Arab Monetary Fund (AMF) *address*—P.O. Box 2818, Abu Dhabi, United Arab Emirates *telephone*—[971] (2) 215000 *FAX*—[971] (2) 326454 *established*—27 April 1976 *effective*—2 February 1977 *aim*—to promote Arab cooperation, development, and integration in monetary and economic affairs	*members*—(19 plus the Palestine Liberation Organization) Algeria, Bahrain, Egypt, Iraq, Jordan, Kuwait, Lebanon, Libya, Mauritania, Morocco, Oman, Qatar, Saudi Arabia, Somalia, Sudan, Syria, Tunisia, UAE, Yemen, Palestine Liberation Organization
Asia Pacific Economic Cooperation (APEC) *address*—Ministry of Trade and Industry, Public Relations, 8 Shenton Way No 48-01, Treasury Building, Singapore, Singapore *established*—7 November 1989 *aim*—to promote trade and investment in the Pacific basin	*members*—(18) all ASEAN members (Brunei, Indonesia, Malaysia, Philippines, Singapore, Thailand) plus Australia, Canada, Chile, China, Hong Kong, Japan, South Korea, Mexico, NZ, Papua New Guinea, Taiwan, US *observers*—(3) Association of Southeast Asian Nations, Pacific Economic Cooperation Conference, South Pacific Forum
Asian Development Bank (AsDB) *address*—6 ADB Avenue, Mandaluyong, METRO Manila, Philippines *telephone*—[63] (2) 711 3851 *FAX*—[63] (2) 741 7961, 631 6816 *established*—19 December 1966 *aim*—to promote regional economic cooperation	*regional members*—(40) Afghanistan, Australia, Bangladesh, Bhutan, Burma, Cambodia, China, Cook Islands, Fiji, Hong Kong, India, Indonesia, Japan, Kazakhstan, Kiribati, South Korea, Kyrgyzstan, Laos, Malaysia, Maldives, Marshall Islands, Federated States of Micronesia, Mongolia, Nauru, Nepal, NZ, Pakistan, Papua New Guinea, Philippines, Singapore, Solomon Islands, Sri Lanka, Taiwan, Thailand, Tonga, Tuvalu, Uzbekistan, Vanuatu, Vietnam, Western Samoa *nonregional members*—(16) Austria, Belgium, Canada, Denmark, Finland, France, Germany, Italy, Netherlands, Norway, Spain, Sweden, Switzerland, Turkey, UK, US
Asociacion Latinoamericana de Integracion (ALADI)	see Latin American Integration Association (LAIA)
Association of Southeast Asian Nations (ASEAN) *address*—Jalan Sisingamangaraja 70A, Kebayoran Baru, P.O. Box 2072, Jakarta 12110, Indonesia *telephone*—[62] (21) 71 22 72, 71 19 88 *FAX*—[62] (21) 739 82 34 *established*—9 August 1967 *aim*—to encourage regional economic, social, and cultural cooperation among the non-Communist countries of Southeast Asia	*members*—(6) Brunei, Indonesia, Malaysia, Philippines, Singapore, Thailand *observers*—(3) Laos, Papua New Guinea, Vietnam
Australia Group *established*—1984 *aim*—to consult on and coordinate export controls related to chemical and biological weapons	*members*—(28) Argentina, Australia, Austria, Belgium, Canada, Czech Republic, Denmark, Finland, France, Germany, Greece, Hungary, Iceland, Ireland, Italy, Japan, Luxembourg, Netherlands, NZ, Norway, Poland, Portugal, Slovakia, Spain, Sweden, Switzerland, UK, US *observer*—(1) Singapore

Australia-New Zealand-United States Security Treaty (ANZUS) *address*—c/o Department of Foreign Affairs and Trade, Bag 8, Queen Victoria Terrace, Canberra ACT 2600, Australia *telephone*—[61] (62) 61 91 11 *FAX*—[61] (62) 61 21 51 *established*—1 September 1951 *effective*—29 April 1952 *aim*—to implement a trilateral mutual security agreement, although the US suspended security obligations to NZ on 11 August 1986	*members*—(3) Australia, NZ, US
Banco Centroamericano de Integracion Economico (BCIE)	see Central American Bank for Economic Integration (BCIE)
Banco Interamericano de Desarrollo (BID)	see Inter-American Development Bank (IADB)
Bank for International Settlements (BIS) *address*—Centralbahnplatz 2, CH-4002 Basel, Switzerland *telephone*—[41] (61) 280 80 80 *FAX*—[41] (61) 280 91 00 *established*—20 January 1930 *effective*—17 March 1930 *aim*—to promote cooperation among central banks in international financial settlements	*members*—(33) Australia, Austria, Belgium, Bulgaria, Canada, Czech Republic, Denmark, Estonia, Finland, France, Germany, Greece, Hungary, Iceland, Ireland, Italy, Japan, Latvia, Lithuania, Netherlands, Norway, Poland, Portugal, Romania, Slovakia, South Africa, Spain, Sweden, Switzerland, Turkey, UK, US, Yugoslavia
Banque Africaine de Developpement (BAD)	see African Development Bank (AfDB)
Banque Arabe de Developpement Economique en Afrique (BADEA)	see Arab Bank for Economic Development in Africa (ABEDA)
Banque de Developpement des Etats de l'Afrique Centrale (BDEAC)	see Central African States Development Bank (BDEAC)
Banque Ouest-Africaine de Developpement (BOAD)	see West African Development Bank (WADB)
Benelux Economic Union (Benelux) *note*—acronym from Belgium, Netherlands, and Luxembourg *address*—Rue de la Regence 39, B-1000 Brussels, Belgium *telephone*—[32] (2) 519 38 11 *FAX*—[32] (2) 513 42 06 *established*—3 February 1958 *effective*—1 November 1960 *aim*—to develop closer economic cooperation and integration	*members*—(3) Belgium, Luxembourg, Netherlands
Big Seven *note*—membership is the same as the Group of 7 *established*—NA 1975 *aim*—to discuss and coordinate major economic policies	*members*—(7) Big Six (Canada, France, Germany, Italy, Japan, UK) plus the US
Big Six *note*—not to be confused with the Group of 6 *established*—NA 1967 *aim*—to foster economic cooperation	*members*—(6) Canada, France, Germany, Italy, Japan, UK

Black Sea Economic Cooperation Zone (BSEC) *established*—25 June 1992 *aim*—to enhance regional stability through economic cooperation	*members*—(11) Albania, Armenia, Azerbaijan, Bulgaria, Georgia, Greece, Moldova, Romania, Russia, Turkey, Ukraine *observer*—(1) Poland
Caribbean Community and Common Market (CARICOM) *address*—CARICOM, P.O. Box 10827, Bank of Guyana Building, 3rd floor, Avenue of the Republic, Georgetown, Guyana *telephone*—[592] (2) 69281 through 69289 FAX—[592] (2) 66091, 67816, 57341 *established*—4 July 1973 *effective*—1 August 1973 *aim*—to promote economic integration and development, especially among the less developed countries	*members*—(14) Antigua and Barbuda, The Bahamas, Barbados, Belize, Dominica, Grenada, Guyana, Jamaica, Montserrat, Saint Kitts and Nevis, Saint Lucia, Saint Vincent and the Grenadines, Suriname, Trinidad and Tobago *associate members*—(2) British Virgin Islands, Turks and Caicos Islands *observers*—(9) Anguilla, Bermuda, Cayman Islands, Dominican Republic, Haiti, Mexico, Netherlands Antilles, Puerto Rico, Venezuela
Caribbean Development Bank (CDB) *address*—P.O. Box 408, Wildey, St. Michael, Barbados *telephone*—[1] (809) 431 1600 *FAX*—[1] (809) 426 7269 *established*—18 October 1969 *effective*—26 January 1970 *aim*—to promote economic development and cooperation	*regional members*—(20) Anguilla, Antigua and Barbuda, The Bahamas, Barbados, Belize, British Virgin Islands, Cayman Islands, Colombia, Dominica, Grenada, Guyana, Jamaica, Mexico, Montserrat, Saint Kitts and Nevis, Saint Lucia, Saint Vincent and the Grenadines, Trinidad and Tobago, Turks and Caicos Islands, Venezuela *nonregional members*—(5) Canada, France, Germany, Italy, UK
Cartagena Group	see Group of 11
Central African Customs and Economic Union (UDEAC) *note*—acronym from Union Douaniere et Economique de l'Afrique Centrale *address*—BP 969, Bangui, Central African Republic *telephone*—[236] 61 09 22, 61 45 77 *FAX*—[236] 61 21 35 *established*—8 December 1964 *effective*—1 January 1966 *aim*—to promote the establishment of a Central African Common Market	*members*—(6) Cameroon, Central African Republic, Chad, Congo, Equatorial Guinea, Gabon
Central African States Development Bank (BDEAC) *note*—acronym from Banque de Developpement des Etats de l'Afrique Centrale *address*—BDEAC, Place du Gouvernement, BP 1177, Brazzaville, Congo *telephone*—[242] 83 01 26, 83 01 49, 81 02 12, 81 02 21 *FAX*—[242] 83 02 66 *established*—3 December 1975 *aim*—to provide loans for economic development	*members*—(9) Cameroon, Central African Republic, Chad, Congo, Equatorial Guinea, France, Gabon, Germany, Kuwait

Central American Bank for Economic Integration (BCIE) *note*—acronym from Banco Centroamericano de Integracion Economico *address*—Apartado Postal 772, Tegucigalpa DC, Honduras *telephone*—[504] 372230 through 372239, 371184 through 371188 *FAX*—[504] 370793 *established*—13 December 1960 *aim*—to promote economic integration and development	*members*—(5) Costa Rica, El Salvador, Guatemala, Honduras, Nicaragua *nonregional members*—(4) Argentina, Mexico, Taiwan, Venezuela
Central American Common Market (CACM) *address*—4A Avda 10-25, Zona 14, Apdo Postal 1237, 01901 Guatemala City, Guatemala *telephone*—[502] (2) 682151 *FAX*—[502] (2) 681071 *established*—13 December 1960 *effective*—3 June 1961 *aim*—to promote establishment of a Central American Common Market	*members*—(5) Costa Rica, El Salvador, Guatemala, Honduras, Nicaragua
Central European Initiative (CEI) *note*—evolved from the Hexagonal Group *address*—Chairman of the National Coordinators, Ministry for Foreign Affairs, Bem rakpart 47, Budapest II, Hungary *established*—27 July 1991 *aim*—to form an economic and political cooperation group for the region between the Adriatic and the Baltic Seas	*members*—(10) Austria, Bosnia and Herzegovina, Croatia, Czech Republic, Hungary, Italy, The Former Yugoslav Republic of Macedonia, Poland, Slovakia, Slovenia *associate members*—(4) Bulgaria, Belarus, Romania, Ukraine
centrally planned economies	a term applied mainly to the traditionally Communist states that looked to the former USSR for leadership; most are now evolving toward more democratic and market-oriented systems; also known formerly as the Second World or as the Communist countries; through the 1980s, this group included Albania, Bulgaria, Cambodia, China, Cuba, Czechoslovakia, GDR, Hungary, North Korea, Laos, Mongolia, Poland, Romania, USSR, Vietnam, Yugoslavia
Colombo Plan (CP) *address*—Colombo Plan Bureau, P.O. Box 596, 12 Melbourne Avenue, Colombo 4, Sri Lanka *telephone*—[94] (1) 581813, 581853, 581754 *FAX*—[94] (1) 580721 *established*—1 July 1951 *aim*—to promote economic and social development in Asia and the Pacific	*members*—(24) Afghanistan, Australia, Bangladesh, Bhutan, Burma, Cambodia, Fiji, India, Indonesia, Iran, Japan, South Korea, Laos, Malaysia, Maldives, Nepal, NZ, Pakistan, Papua New Guinea, Philippines, Singapore, Sri Lanka, Thailand, US
Commission for Social Development *address*—c/o ECOSOC/DPCSD, United Nations, New York, NY 10017, USA *telephone*—[1] (212) 963 2320 *FAX*—[1] (212) 963 5935 *established*—21 June 1946 as the Social Commission, renamed 29 July 1966 *aim*—to deal, as part of the Economic and Social Council, with social development programs of UN	*members*—(32) selected on a rotating basis from all regions

Commission on Crime Prevention and Criminal Justice *established*—6 February 1992 *aim*—to provide guidance, as part of the Economic and Social Council, on crime prevention and criminal justice	*members*—(40) selected on a rotating basis from all regions
Commission on Human Rights *address*—c/o United Nations Office, Centre for Human Rights, Palais des Nations, CH-1211 Geneva 10, Switzerland *telephone*—[41] (22) 917 12 34, 907 12 34 *FAX*—[41] (22) 733 32 46 *established*—18 February 1946 *aim*—to assist, as part of the Economic and Social Council, with human rights programs of UN	*members*—(53) selected on a rotating basis from all regions
Commission on Human Settlements (Habitat) *address*—c/o HABITAT, P.O. Box 30030, Nairobi, Kenya *telephone*—[254] (2) 621234 *FAX*—[254] (2) 226473, 226479 *established*—12 October 1978 *aim*—to assist, as part of the Economic and Social Council, in solving human settlement problems of UN	*members*—(58) selected on a rotating basis from all regions
Commission on Narcotic Drugs *address*—c/o International Drug Control Programme, Treaty Implementation and Legal Affairs Division, P.O. Box 500, A-1400 Vienna, Austria *telephone*—[43] (1) 211 310 *FAX*—[43] (1) 230 7002 *established*—16 February 1946 *aim*—Economic and Social Council organization dealing with illicit drugs programs of UN	*members*—(53) selected on a rotating basis from all regions with emphasis on producing and processing countries
Commission on Science and Technology for Development *established*—30 April 1992 *aim*—to promote international cooperation, as part of the Economic and Social Council, in the field of science and technology	*members*—(53) selected on a rotating basis from all regions
Commission on the Status of Women *address*—c/o Economic and Social Council, Affairs Division, Department for Policy Coordination and Sustainable Development, Room S-2963, United Nations, New York, NY 10017, USA *established*—21 June 1946 *aim*—to deal, as part of the Economic and Social Council, with women's rights goals of UN	*members*—(45) selected on a rotating basis from all regions

Commission on Sustainable Development *established*—12 February 1993 *aim*—to monitor, as part of the Economic and Social Council, implementation of agreements reached at the UN Conference on Environment and Development	*members*—(53) selected on a rotating basis from all regions
Commonwealth (C) *address*—c/o Commonwealth Secretariat, Marlborough House, Pall Mall, London SW1Y5HX, UK *telephone*—[44] (71) 839 3411 *FAX*—[44] (71) 930 0827 *established*—31 December 1931 *aim*—to foster multinational cooperation and assistance, as a voluntary association that evolved from the British Empire	*members*—(49) Antigua and Barbuda, Australia, The Bahamas, Bangladesh, Barbados, Belize, Botswana, Brunei, Canada, Cyprus, Dominica, The Gambia, Ghana, Grenada, Guyana, India, Jamaica, Kenya, Kiribati, Lesotho, Malawi, Malaysia, Maldives, Malta, Mauritius, Namibia, NZ, Nigeria, Pakistan, Papua New Guinea, Saint Kitts and Nevis, Saint Lucia, Saint Vincent and the Grenadines, Seychelles, Sierra Leone, Singapore, Solomon Islands, South Africa, Sri Lanka, Swaziland, Tanzania, Tonga, Trinidad and Tobago, Uganda, UK, Vanuatu, Western Samoa, Zambia, Zimbabwe *special members*—(2) Nauru, Tuvalu
Commonwealth of Independent States (CIS) *address*—Kirov Street 17, 220000 Minsk, Belarus *telephone*—[7] (172) 293434, 293517 *FAX*—[7] (172) 261894, 261944 *established*—8 December 1991 *effective*—21 December 1991 *aim*—to coordinate intercommonwealth relations and to provide a mechanism for the orderly dissolution of the USSR	*members*—(12) Armenia, Azerbaijan, Belarus, Georgia, Kazakhstan, Kyrgyzstan, Moldova, Russia, Tajikistan, Turkmenistan, Ukraine, Uzbekistan
Communaute Economique de l'Afrique de l'Ouest (CEAO)	see West African Economic Community (CEAO)
Communaute Economique des Etats de l'Afrique Centrale (CEEAC)	see Economic Community of Central African States (CEEAC)
Communaute Economique des Pays des Grands Lacs (CEPGL)	see Economic Community of the Great Lakes Countries (CEPGL)
Communist countries	traditionally the Marxist-Leninist states with authoritarian governments and command economies based on the Soviet model; most of the original and the successor states are no longer Communist; see centrally planned economies
Conference on Security and Cooperation in Europe (CSCE)	see Organization on Security and Cooperation in Europe (OSCE)
Conseil Europeen pour la Recherche Nucleaire (CERN)	see European Organization for Nuclear Research (CERN)
Contadora Group (CG)	was established 5 January 1983 (on the Panamanian island of Contadora) to reduce tensions and conflicts in Central America; has evolved into the Rio Group (RG); members included Colombia, Mexico, Panama, Venezuela
Cooperation Council for the Arab States of the Gulf	see Gulf Cooperation Council (GCC)
Coordinating Committee on Export Controls (COCOM)	established in 1949 to control the export of strategic products and technical data from member countries to proscribed destinations; members were Australia, Belgium, Canada, Denmark, France, Germany, Greece, Italy, Japan, Luxembourg, Netherlands, Norway, Portugal, Spain, Turkey, UK, US; abolished 31 March 1994; COCOM members are working on a new organization with expanded membership which focuses on nonproliferation export controls as opposed to East-West control of advanced technology
Council for Mutual Economic Assistance (CEMA) *note*—also known as CMEA or Comecon	established 25 January 1949 to promote the development of socialist economies and was abolished 1 January 1991; members included Afghanistan (observer), Albania (had not participated since 1961 break with USSR), Angola (observer), Bulgaria, Cuba, Czechoslovakia, Ethiopia (observer), GDR, Hungary, Laos (observer), Mongolia, Mozambique (observer), Nicaragua (observer), Poland, Romania, USSR, Vietnam, Yemen (observer), Yugoslavia (associate)

Council of Arab Economic Unity (CAEU) *address*—BP 925100, Amman, Jordan *telephone*—[962] (6) 66 43 26, 66 43 27, 66 43 28 *FAX*—[962] (6) 66 33 43 *established*—3 June 1957 *effective*—30 May 1964 *aim*—to promote economic integration among Arab nations	*members*—(11 plus the Palestine Liberation Organization) Egypt, Iraq, Jordan, Kuwait, Libya, Mauritania, Somalia, Sudan, Syria, UAE, Yemen, Palestine Liberation Organization
Council of Europe (CE) *address*—Palais de l'Europe, F-67075 Strasbourg CEDEX, France *telephone*—[33] 88 41 20 00 *FAX*—[33] 88 41 27 81, 88 41 27 82 *established*—5 May 1949 *effective*—3 August 1949 *aim*—to promote increased unity and quality of life in Europe	*members*—(32) Austria, Belgium, Bulgaria, Cyprus, Czech Republic, Denmark, Estonia, Finland, France, Germany, Greece, Hungary, Iceland, Ireland, Italy, Liechtenstein, Lithuania, Luxembourg, Malta, Netherlands, Norway, Poland, Portugal, Romania, San Marino, Slovakia, Slovenia, Spain, Sweden, Switzerland, Turkey, UK *guests*—(9) Albania, Belarus, Bosnia and Herzegovina, Croatia, Latvia, The Former Yugoslav Republic of Macedonia, Moldova, Russia, Ukraine *observer*—(1) Israel
Council of the Baltic Sea States (CBSS) *established*—5 March 1992 *aim*—to promote cooperation among the Baltic Sea states in the areas of aid to new democratic institutions, economic development, humanitarian aid, energy and the environment, cultural programs and education, and transportation and communication	*members*—(10) Denmark, Estonia, Finland, Germany, Latvia, Lithuania, Norway, Poland, Russia, Sweden
Council of the Entente (Entente) *address*—BP 3734, Abidjan 01, Cote d'Ivoire *telephone*—[225] 33 10 01, 33 28 35 *FAX*—[225] 33 11 49 *established*—29 May 1959 *aim*—to promote economic, social, and political coordination	*members*—(5) Benin, Burkina, Cote d'Ivoire, Niger, Togo
Customs Cooperation Council (CCC) *address*—Rue de l'Industrie 26-38, B-1040 Brussels, Belgium *telephone*—[32] (2) 508 42 11 *FAX*—[32] (2) 508 42 40 *established*—15 December 1950 *aim*—to promote international cooperation in customs matters	*members*—(136) Albania, Algeria, Angola, Argentina, Armenia, Australia, Austria, Azerbaijan, The Bahamas, Bangladesh, Belarus, Belgium, Bermuda, Botswana, Brazil, Bulgaria, Burkina, Burma, Burundi, Cameroon, Canada, Cape Verde, Central African Republic, Chile, China, Colombia, Comoros, Congo, Cote d'Ivoire, Croatia, Cuba, Cyprus, Czech Republic, Denmark, Egypt, Estonia, Ethiopia, Finland, France, Gabon, The Gambia, Georgia, Germany, Ghana, Greece, Guatemala, Guinea, Guyana, Haiti, Hong Kong, Hungary, Iceland, India, Indonesia, Iran, Iraq, Ireland, Israel, Italy, Jamaica, Japan, Jordan, Kazakhstan, Kenya, South Korea, Kuwait, Latvia, Lebanon, Lesotho, Liberia, Libya, Lithuania, Luxembourg, Macau, The Former Yugoslav Republic of Macedonia, Madagascar, Malawi, Malaysia, Mali, Malta, Mauritania, Mauritius, Mexico, Mongolia, Morocco, Mozambique, Namibia, Nepal, Netherlands, NZ, Niger, Nigeria, Norway, Pakistan, Paraguay, Peru, Philippines, Poland, Portugal, Qatar, Romania, Russia, Rwanda, Saudi Arabia, Senegal, Sierra Leone, Singapore, Slovakia, Slovenia, South Africa, Spain, Sri Lanka, Sudan, Swaziland, Sweden, Switzerland, Syria, Tanzania, Thailand, Togo, Trinidad and Tobago, Tunisia, Turkey, Turkmenistan, Uganda, Ukraine, UAE, UK, US, Uruguay, Uzbekistan, Vietnam, Yemen, Zaire, Zambia, Zimbabwe
developed countries (DCs)	the top group in the hierarchy of developed countries (DCs), former USSR/Eastern Europe (former USSR/EE), and less developed countries (LDCs); includes the market-oriented economies of the mainly democratic nations in the Organization for Economic Cooperation and Development (OECD), Bermuda, Israel, South Africa, and the European ministates; also known as the First World, high-income countries, the North, industrial countries; generally have a per capita GDP in excess of $10,000 although four OECD countries and South Africa have figures well under $10,000 and two of the excluded OPEC countries have figures of more than $10,000; the 35 DCs are: Andorra, Australia, Austria, Belgium, Bermuda, Canada, Denmark, Faroe Islands, Finland, France, Germany, Greece, Holy See, Iceland, Ireland, Israel, Italy, Japan, Liechtenstein, Luxembourg, Malta, Mexico, Monaco, Netherlands, NZ, Norway, Portugal, San Marino, South Africa, Spain, Sweden, Switzerland, Turkey, UK, US

developing countries	an imprecise term for the less developed countries with growing economies; see less developed countries (LDCs)
East African Development Bank (EADB) *address*—4 Nile Avenue, P.O. Box 7128, Kampala, Uganda *telephone*—[256] (41) 230021, 230825 *FAX*—[256] (41) 259763 *established*—6 June 1967 *effective*—1 December 1967 *aim*—to promote economic development	*members*—(3) Kenya, Tanzania, Uganda
Economic and Social Commission for Asia and the Pacific (ESCAP) *address*—United Nations Building, Rajadamnern Avenue, Bangkok 10200, Thailand *telephone*—[66] (2) 2829161 through 2829200, 2829381 through 2829389 *FAX*—[66] (2) 2811743 *established*—28 March 1947 as Economic Commission for Asia and the Far East (ECAFE) *aim*—to carryout the commitment of the Economic and Social Council of the UN to promote economic development	*members*—(49) Afghanistan, Armenia, Australia, Azerbaijan, Bangladesh, Bhutan, Brunei, Burma, Cambodia, China, Fiji, France, India, Indonesia, Iran, Japan, Kazakhstan, Kiribati, North Korea, South Korea, Kyrgyzstan, Laos, Malaysia, Maldives, Marshall Islands, Federated States of Micronesia, Mongolia, Nauru, Nepal, Netherlands, NZ, Pakistan, Papua New Guinea, Philippines, Russia, Singapore, Solomon Islands, Sri Lanka, Tajikistan, Thailand, Tonga, Turkmenistan, Tuvalu, UK, US, Uzbekistan, Vanuatu, Vietnam, Western Samoa *associate members*—(10) American Samoa, Cook Islands, French Polynesia, Guam, Hong Kong, Macau, New Caledonia, Niue, Northern Mariana Islands, Palau
Economic and Social Commission for Western Asia (ESCWA) *address*—(temporary) P.O. Box 927115, Amman, Jordan *telephone*—[962] (6) 694351 *FAX*—[962] (6) 694981, 694982 *established*—9 August 1973 as Economic Commission for Western Asia (ECWA) *aim*—to promote economic development as a regional commission for the UN's Economic and Social Council	*members*—(12 plus the Palestine Liberation Organization) Bahrain, Egypt, Iraq, Jordan, Kuwait, Lebanon, Oman, Qatar, Saudi Arabia, Syria, UAE, Yemen, Palestine Liberation Organization
Economic and Social Council (ECOSOC) *address*—United Nations, New York, NY 10017, USA *telephone*—[1] (212) 963 1234 *FAX*—[1] (212) 758 2718 *established*—26 June 1945 *effective*—24 October 1945 *aim*—to coordinate the economic and social work of the UN; includes five regional commissions (see Economic Commission for Africa, Economic Commission for Europe, Economic Commission for Latin America and the Caribbean, Economic and Social Commission for Asia and the Pacific, Economic and Social Commission for Western Asia) and 10 functional commissions (see Commission for Social Development, Commission on Human Rights, Commission on Narcotic Drugs, Commission on the Status of Women, Population Commission, Statistical Commission, Commission on Science and Technology for Development, Commission on Sustainable Development, Commission on Crime Prevention and Criminal Justice, and Commission on Transnational Corporations)	*members*—(54) selected on a rotating basis from all regions

Economic Commission for Africa (ECA) *address*—P.O. Box 3001-3005, Addis Ababa, Ethiopia *telephone*—[251] (1) 51 72 00 *FAX*—[251] (1) 51 44 16 *established*—29 April 1958 *aim*—to promote economic development as a regional commission of the UN's Economic and Social Council	*members*—(53) Algeria, Angola, Benin, Botswana, Burkina, Burundi, Cameroon, Cape Verde, Central African Republic, Chad, Comoros, Congo, Cote d'Ivoire, Djibouti, Egypt, Equatorial Guinea, Eritrea, Ethiopia, Gabon, The Gambia, Ghana, Guinea, Guinea-Bissau, Kenya, Lesotho, Liberia, Libya, Madagascar, Malawi, Mali, Mauritania, Mauritius, Morocco, Mozambique, Namibia, Niger, Nigeria, Rwanda, Sao Tome and Principe, Senegal, Seychelles, Sierra Leone, Somalia, South Africa, Sudan, Swaziland, Tanzania, Togo, Tunisia, Uganda, Zaire, Zambia, Zimbabwe *associate members*—(2) France, UK
Economic Commission for Asia and the Far East (ECAFE)	see Economic and Social Commission for Asia and the Pacific (ESCAP)
Economic Commission for Europe (ECE) *address*—Palais des Nations, CH-1211 Geneva 10, Switzerland *telephone*—[41] (22) 917 1234, 907 2893 *FAX*—[41] (22) 917 0036 *established*—28 March 1947 *aim*—to promote economic development as a regional commission of the UN's Economic and Social Council	*members*—(54) Albania, Andorra, Armenia, Austria, Azerbaijan, Belarus, Belgium, Bosnia and Herzegovina, Bulgaria, Canada, Croatia, Cyprus, Czech Republic, Denmark, Estonia, Finland, France, Georgia, Germany, Greece, Hungary, Iceland, Ireland, Israel, Italy, Kazakhstan, Kyrgyzstan, Latvia, Liechtenstein, Lithuania, Luxembourg, The Former Yugoslav Republic of Macedonia, Malta, Moldova, Monaco, Netherlands, Norway, Poland, Portugal, Romania, Russia, San Marino, Slovakia, Slovenia, Spain, Sweden, Switzerland, Turkey, Turkmenistan, Ukraine, UK, US, Uzbekistan, Yugoslavia
Economic Commission for Latin America (ECLA)	see Economic Commission for Latin America and the Caribbean (ECLAC)
Economic Commission for Latin America and the Caribbean (ECLAC) *address*—Edificio Naciones Unidas, Avenida Dag Hammarskjold, Casilla 179 D, Santiago, Chile *telephone*—[56] (2) 2102000 *FAX* [56] (2) 2080252, 2081946 *established*—25 February 1948 as Economic Commission for Latin America (ECLA) *aim*—to promote economic development as a regional commission of the UN's Economic and Social Council	*members*—(41) Antigua and Barbuda, Argentina, The Bahamas, Barbados, Belize, Bolivia, Brazil, Canada, Chile, Colombia, Costa Rica, Cuba, Dominica, Dominican Republic, Ecuador, El Salvador, France, Grenada, Guatemala, Guyana, Haiti, Honduras, Italy, Jamaica, Mexico, Netherlands, Nicaragua, Panama, Paraguay, Peru, Portugal, Saint Kitts and Nevis, Saint Lucia, Saint Vincent and the Grenadines, Spain, Suriname, Trinidad and Tobago, UK, US, Uruguay, Venezuela *associate members*—(6) Aruba, British Virgin Islands, Montserrat, Netherlands Antilles, Puerto Rico, Virgin Islands
Economic Commission for Western Asia (ECWA)	see Economic and Social Commission for Western Asia (ESCWA)
Economic Community of Central African States (CEEAC) *note*—acronym from Communaute Economique des Etats de l'Afrique Centrale *address*—CEEAC, BP 2112, Libreville, Gabon *telephone*—[241] 73 35 47, 73 35 48, 73 36 77 *established*—18 October 1983 *aim*—to promote regional economic cooperation and establish a Central African Common Market	*members*—(10) Burundi, Cameroon, Central African Republic, Chad, Congo, Equatorial Guinea, Gabon, Rwanda, Sao Tome and Principe, Zaire *observer*—(1) Angola
Economic Community of the Great Lakes Countries (CEPGL) *note*—acronym from Communaute Economique des Pays des Grands Lacs *address*—B.O. Box 58, Gisenyi, Rwanda *telephone*—[250] 40228 *FAX*—[250] 40785 *established*—26 September 1976 *aim*—to promote regional economic cooperation and integration	*members*—(3) Burundi, Rwanda, Zaire

Economic Community of West African States (ECOWAS) *address*—6 King George V Road, PMB 12745, Lagos, Nigeria *telephone*—[234] (1) 636839, 636841, 636064, 630398 *FAX*—[234] (1) 636822 *established*—28 May 1975 *aim*—to promote regional economic cooperation	*members*—(16) Benin, Burkina, Cape Verde, Cote d'Ivoire, The Gambia, Ghana, Guinea, Guinea-Bissau, Liberia, Mali, Mauritania, Niger, Nigeria, Senegal, Sierra Leone, Togo
Economic Cooperation Organization (ECO) *address*—5 Hejab Avenue, Bd Keshavarz, P.O. Box 14155-6176, Teheran, Iran Islamic Republic *telephone*—[98] (21) 658614, 656152, 658045 *FAX*—[98] (21) 658046 *established*—NA 1985 *aim*—to promote regional cooperation in trade, transportation, communications, tourism, cultural affairs, and economic development	*members*—(10) Afghanistan, Azerbaijan, Iran, Kazakhstan, Kyrgyzstan, Pakistan, Tajikistan, Turkey, Turkmenistan, Uzbekistan *associate member*—(1) "Turkish Republic of Northern Cyprus"
European Bank for Reconstruction and Development (EBRD) *address*—One Exchange Square, London EC2A 2EH, UK *telephone*—[44] (71) 338 6000 *FAX*—[44] (71) 338 6100 *established*—15 April 1991 *aim*—to facilitate the transition of seven centrally planned economies in Europe (Bulgaria, former Czechoslovakia, Hungary, Poland, Romania, former USSR, and former Yugoslavia) to market economies by committing 60% of its loans to privatization	*members*—(59) Albania, Armenia, Australia, Austria, Azerbaijan, Belarus, Belgium, Bulgaria, Canada, Croatia, Cyprus, Czech Republic, Denmark, Egypt, European Union (EU), European Investment Bank (EIB), Estonia, Finland, France, Georgia, Germany, Greece, Hungary, Iceland, Ireland, Israel, Italy, Japan, Kazakhstan, South Korea, Kyrgyzstan, Latvia, Liechtenstein, Lithuania, Luxembourg, The Former Yugoslav Republic of Macedonia, Malta, Mexico, Moldova, Morocco, Netherlands, NZ, Norway, Poland, Portugal, Russia, Romania, Slovakia, Slovenia, Spain, Sweden, Switzerland, Tajikistan, Turkey, Turkmenistan, Ukraine, UK, US, Uzbekistan; *note*—includes all 25 members of the OECD; also includes the EU as a single entity
European Community (or European Communities, EC)	was established 8 April 1965 to integrate the European Atomic Energy Community (Euratom), the European Coal and Steel Community (ESC), the European Economic Community (EEC or Common Market), and to establish a completely integrated common market and an eventual federation of Europe; merged into the European Union (EU) on 7 February 1992; member states at the time of merger were Belgium, Denmark, France, Germany, Greece, Ireland, Italy, Luxembourg, Netherlands, Portugal, Spain, UK
European Free Trade Association (EFTA) *address*—9-11 rue de Varembe, CH-1211 Geneva 20, Switzerland *telephone*—[41] (22) 749 11 11 *FAX*—[41] (22) 733 92 91 *established*—4 January 1960 *effective*—3 May 1960 *aim*—to promote expansion of free trade	*members*—(7) Austria, Finland, Iceland, Liechtenstein, Norway, Sweden, Switzerland
European Investment Bank (EIB) *address*—Bd Konrad Adenauer 100, L-2950 Luxembourg, Luxembourg *telephone*—[352] 43791 *FAX*—[352] 437704 *established*—25 March 1957 *effective*—1 January 1958 *aim*—to promote economic development of the EU	*members*—(12) Belgium, Denmark, France, Germany, Greece, Ireland, Italy, Luxembourg, Netherlands, Portugal, Spain, UK

European Organization for Nuclear Research (CERN)

note—acronym retained from the predecessor organization Conseil Europeen pour la Recherche Nucleaire
address—CH-1211 Geneva 23, Switzerland
telephone—[41] (22) 767 61 11
FAX—[41] (22) 767 65 55
established—1 July 1953
effective—29 September 1954
aim—to foster nuclear research for peaceful purposes only

members—(19) Austria, Belgium, Czech Republic, Denmark, Finland, France, Germany, Greece, Hungary, Italy, Netherlands, Norway, Poland, Portugal, Slovakia, Spain, Sweden, Switzerland, UK
observers—(6) EC, Israel, Russia, Turkey, United Nations Educational, Scientific, and Cultural Organization (UNESCO), Yugoslavia

---

European Space Agency (ESA)

address—8-10 rue Mario Nikis, F-75738 Paris CEDEX 15, France
telephone—[33] (1) 42 73 76 54
FAX—[33] (1) 42 73 75 60
established—31 July 1973
effective—1 May 1975
aim—to promote peaceful cooperation in space research and technology

members—(13) Austria, Belgium, Denmark, France, Germany, Ireland, Italy, Netherlands, Norway, Spain, Sweden, Switzerland, UK
associate member—(1) Finland
cooperating state—(1) Canada

European Union (EU)

note—evolved from the European Community (EC)
address—c/o European Commission, Rue de la Loi 200, B-1049 Brussels, Belgium
telephone—[32] (2) 299 11 11
FAX [32] (2) 295 01 38 through 295 01 40
established—7 February 1992
effective—1 November 1993
aim—to coordinate policy among the 15 members in three fields: economics, building on the European Economic Community's (EEC) efforts to establish a common market and eventually a common currency; defense, within the concept of a Common Foreign and Security Policy (CFSP); and justice and home affairs, including immigration, drugs, terrorism, and improved living and working conditions

members—(15) Austria, Belgium, Denmark, Finland, France, Germany, Greece, Ireland, Italy, Luxembourg, Netherlands, Portugal, Spain, Sweden, UK

First World

another term for countries with advanced, industrialized economies; this term is fading from use; see developed countries (DCs)

Food and Agriculture Organization (FAO)

address—Viale delle Terme di Caracalla, I-00100 Rome, Italy
telephone—[39] (6) 52251
FAX [39] (6) 5225 3152, 5225 5155, 578 2610
established—16 October 1945
aim—to raise living standards and increase availability of agricultural products, as a UN specialized agency

members—(170) Afghanistan, Albania, Algeria, Angola, Antigua and Barbuda, Argentina, Armenia, Australia, Austria, The Bahamas, Bahrain, Bangladesh, Barbados, Belgium, Belize, Benin, Bhutan, Bolivia, Bosnia and Herzegovina, Botswana, Brazil, Brunei, Bulgaria, Burkina, Burma, Burundi, Cambodia, Cameroon, Canada, Cape Verde, Central African Republic, Chad, Chile, China, Colombia, Comoros, Congo, Costa Rica, Cote d'Ivoire, Croatia, Cuba, Cyprus, Czech Republic, Denmark, Djibouti, Dominica, Dominican Republic, Ecuador, Egypt, El Salvador, Equatorial Guinea, Eritrea, Estonia, Ethiopia, EU, Fiji, Finland, France, Gabon, The Gambia, Germany, Ghana, Greece, Grenada, Guatemala, Guinea, Guinea-Bissau, Guyana, Haiti, Honduras, Hungary, Iceland, India, Indonesia, Iran, Iraq, Ireland, Israel, Italy, Jamaica, Japan, Jordan, Kenya, North Korea, South Korea, Kuwait, Kyrgyzstan, Laos, Latvia, Lebanon, Lesotho, Liberia, Libya, Lithuania, Luxembourg, The Former Yugoslav Republic of Macedonia, Madagascar, Malawi, Malaysia, Maldives, Mali, Malta, Mauritania, Mauritius, Mexico, Mongolia, Morocco, Mozambique, Namibia, Nepal, Netherlands, NZ, Nicaragua, Niger, Nigeria, Norway, Oman, Pakistan, Panama, Papua New Guinea, Paraguay, Peru, Philippines, Poland, Portugal, Qatar, Romania, Rwanda, Saint Kitts and Nevis, Saint Lucia, Saint Vincent and the Grenadines, Sao Tome and Principe, Saudi Arabia, Senegal, Seychelles, Sierra Leone, Slovakia, Slovenia, Solomon Islands, Somalia, South Africa, Spain, Sri Lanka, Sudan, Suriname, Swaziland, Sweden, Switzerland, Syria, Tanzania, Thailand, Togo, Tonga, Trinidad and Tobago, Tunisia, Turkey, Uganda, UAE, UK, US, Uruguay, Vanuatu, Venezuela, Vietnam, Western Samoa, Yemen, Yugoslavia (suspended), Zaire, Zambia, Zimbabwe
associate member—(1) Puerto Rico

Former USSR/Eastern Europe (former USSR/EE)	the middle group in the hierarchy of developed countries (DCs), former USSR/Eastern Europe (former USSR/EE), and less developed countries (LDCs); these countries are in political and economic transition and may well be grouped differently in the near future; this group of 27 countries consists of Albania, Armenia, Azerbaijan, Belarus, Bosnia and Herzegovina, Bulgaria, Croatia, Czech Republic, Estonia, Georgia, Hungary, Kazakhstan, Kyrgyzstan, Latvia, Lithuania, The Former Yugoslav Republic of Macedonia, Moldova, Poland, Romania, Russia, Serbia and Montenegro, Slovakia, Slovenia, Tajikistan, Turkmenistan, Ukraine, Uzbekistan
Four Dragons	the four small Asian less developed countries (LDCs) that have experienced unusually rapid economic growth; also known as the Four Tigers; this group includes Hong Kong, South Korea, Singapore, Taiwan
Four Tigers	another term for the Four Dragons; see Four Dragons
Franc Zone (FZ) *address*—Direction Generale des Service Etrangers (Service de la Zone Franc), Banque de France, 39 rue Crois-des-Petits-Champs, BP 140-01, Paris Cedex 01, France *telephone*—[33] (1) 42 92 31 26 *FAX*—[33] (1) 42 92 39 88 *established*—20 December 1945 *aim*—to form a monetary union among countries whose currencies are linked to the French franc	*members*—(15) Benin, Burkina, Cameroon, Central African Republic, Chad, Comoros, Congo, Cote d'Ivoire, Equatorial Guinea, France, Gabon, Mali, Niger, Senegal, Togo; *note*—France includes metropolitan France, the four overseas departments of France (French Guiana, Guadeloupe, Martinique, Reunion), the two territorial collectivities of France (Mayotte, Saint Pierre and Miquelon), and the three overseas territories of France (French Polynesia, New Caledonia, Wallis and Futuna)
Front Line States (FLS)	established to achieve black majority rule in South Africa; members included Angola, Botswana, Mozambique, Namibia, Tanzania, Zambia, Zimbabwe
General Agreement on Tariffs and Trade (GATT) *note*—was subsumed by the World Trade Organization (WTrO) on 1 January 1995 *address*—rue de Lausanne 154, CH-1211 Geneva 21, Switzerland *telephone*—[41] (22) 739 51 11 *FAX*—[41] (22) 731 42 06 *established*—30 October 1947 *effective*—1 January 1948 *aim*—to promote the expansion of international trade on a nondiscriminatory basis	*members*—(123) Angola, Antigua and Barbuda, Argentina, Australia, Austria, Bahrain, Bangladesh, Barbados, Belgium, Belize, Benin, Bolivia, Botswana, Brazil, Brunei, Burkina, Burma, Burundi, Cameroon, Canada, Central African Republic, Chad, Chile, Colombia, Congo, Costa Rica, Cote d'Ivoire, Cuba, Cyprus, Czech Republic, Denmark, Dominica, Dominican Republic, Egypt, El Salvador, Fiji, Finland, France, Gabon, The Gambia, Germany, Ghana, Greece, Grenada, Guatemala, Guinea-Bissau, Guyana, Haiti, Honduras, Hong Kong, Hungary, Iceland, India, Indonesia, Ireland, Israel, Italy, Jamaica, Japan, Kenya, South Korea, Kuwait, Lesotho, Liechtenstein, Luxembourg, Macau, Madagascar, Malawi, Malaysia, Maldives, Mali, Malta, Mauritania, Mauritius, Mexico, Morocco, Mozambique, Namibia, Netherlands, NZ, Nicaragua, Niger, Nigeria, Norway, Pakistan, Paraguay, Peru, Philippines, Poland, Portugal, Qatar, Romania, Rwanda, Saint Kitts and Nevis, Saint Lucia, Saint Vincent and the Grenadines, Senegal, Sierra Leone, Singapore, Slovakia, South Africa, Spain, Sri Lanka, Suriname, Swaziland, Sweden, Switzerland, Tanzania, Thailand, Togo, Trinidad and Tobago, Tunisia, Turkey, Uganda, UAE, UK, US, Uruguay, Venezuela, Yugoslavia (suspended), Zaire, Zambia, Zimbabwe
Group of 2 (G-2) *established*—informal term that came into use about 1986 *aim*—to facilitate bilateral economic cooperation between the two most powerful economic giants	*members*—(2) Japan, US
Group of 3 (G-3) *established*—NA October 1990 *aim*—mechanism for policy coordination	*members*—(3) Colombia, Mexico, Venezuela
Group of 5 (G-5) *established*—22 September 1985 *aim*—to coordinate the economic policies of the five major non-Communist economic powers	*members*—(5) France, Germany, Japan, UK, US

Group of 6 (G-6)

note—also known as Groupe des Six Sur le Desarmement not to be confused with the Big Six
established—22 May 1984
aim—to achieve nuclear disarmament

members—(6) Argentina, Greece, India, Mexico, Sweden, Tanzania

Group of 7 (G-7)

note—membership is the same as the Big Seven
established—22 September 1985
aim—to facilitate economic cooperation among the seven major non-Communist economic powers

members—(7) Group of 5 (France, Germany, Japan, UK, US) plus Canada and Italy

Group of 8 (G-8)

established—NA October 1975
aim—to facilitate economic cooperation among the developed countries (DCs) that participated in the Conference on International Economic Cooperation (CIEC), held in several sessions between NA December 1975 and 3 June 1977

members—(8) Australia, Canada, EU (as one member), Japan, Spain, Sweden, Switzerland, US

Group of 9 (G-9)

established—NA
aim—to discuss matters of mutual interest on an informal basis

members—(9) Austria, Belgium, Bulgaria, Denmark, Finland, Hungary, Romania, Sweden, Yugoslavia

Group of 10 (G-10)

note—also known as the Paris Club; includes the wealthiest members of the IMF who provide most of the money to be loaned and act as the informal steering committee; name persists in spite of the addition of Switzerland on NA April 1984
address—c/o IMF Office in Europe, 64-66 ave d'Iena, F-75116 Paris, France
telephone—[33] (1) 40 69 30 80
FAX—[33] (1) 47 23 40 89
established—NA October 1962
aim—to coordinate credit policy

members—(11) Belgium, Canada, France, Germany, Italy, Japan, Netherlands, Sweden, Switzerland, UK, US
nonstate participants—(4) BIS, EU, IMF, OECD

Group of 11 (G-11)

note—also known as the Cartagena Group
established—22 June 1984, in Cartagena, Colombia
aim—to provide a forum for largest debtor nations in Latin America

members—(11) Argentina, Bolivia, Brazil, Chile, Colombia, Dominican Republic, Ecuador, Mexico, Peru, Uruguay, Venezuela

Group of 15 (G-15)

note—byproduct of the Non-Aligned Movement
address—Technical Support Facility, Ch du Champ d'Ancier 17, Case postale 326, CH-1211 Geneva 19, Switzerland
telephone—[41] (22) 798 42 10
FAX—[41] (22) 798 38 49
established—September 1989
aim—to promote economic cooperation among developing nations; to act as the main political organ for the Non-Aligned Movement

members—(15) Algeria, Argentina, Brazil, Egypt, India, Indonesia, Jamaica, Malaysia, Mexico, Nigeria, Peru, Senegal, Venezuela, Yugoslavia, Zimbabwe

Group of 19 (G-19) *established*—NA October 1975 *aim*—to represent the interests of the less developed countries (LDCs) that participated in the Conference on International Economic Cooperation (CIEC) held in several sessions between NA December 1975 and 3 June 1977	*members*—(19) Algeria, Argentina, Brazil, Cameroon, Egypt, India, Indonesia, Iran, Iraq, Jamaica, Mexico, Nigeria, Pakistan, Peru, Saudi Arabia, Venezuela, Yugoslavia, Zaire, Zambia
Group of 24 (G-24) *address*—c/o European Commission, DGI, G-24 Coordination Unit, Rue de la Science 29, B-1049 Brussels, Belgium *telephone*—[32] (2) 299 22 44 *FAX*—[32] (2) 299 06 02 *established*—NA January 1972 *aim*—to promote the interests of developing countries in Africa, Asia, and Latin America within the IMF	*members*—(24) Algeria, Argentina, Brazil, Colombia, Cote d'Ivoire, Egypt, Ethiopia, Gabon, Ghana, Guatemala, India, Iran, Lebanon, Mexico, Nigeria, Pakistan, Peru, Philippines, Sri Lanka, Syria, Trinidad and Tobago, Venezuela, Yugoslavia, Zaire
Group of 30 (G-30) *address*—1990 M Street NW, Suite 450, Washington, DC 20036, USA *telephone*—[1] (202) 331 2472 *established*—NA 1979 *aim*—to discuss and propose solutions to the world's economic problems	*members*—(30) informal group of 30 leading international bankers, economists, financial experts, and businessmen organized by Johannes Witteveen (former managing director of the IMF)
Group of 33 (G-33) *established*—NA 1987 *aim*—to promote solutions to international economic problems	*members*—(33) leading economists from 13 countries
Group of 77 (G-77) *established*—NA October 1967 *aim*—to promote economic cooperation among developing countries; name persists in spite of increased membership	*members*—(127 plus the Palestine Liberation Organization) Afghanistan, Algeria, Angola, Antigua and Barbuda, Argentina, The Bahamas, Bahrain, Bangladesh, Barbados, Belize, Benin, Bhutan, Bolivia, Botswana, Brazil, Brunei, Burkina, Burma, Burundi, Cambodia, Cameroon, Cape Verde, Central African Republic, Chad, Chile, Colombia, Comoros, Congo, Costa Rica, Cote d'Ivoire, Cuba, Cyprus, Djibouti, Dominica, Dominican Republic, Ecuador, Egypt, El Salvador, Equatorial Guinea, Ethiopia, Fiji, Gabon, The Gambia, Ghana, Grenada, Guatemala, Guinea, Guinea-Bissau, Guyana, Haiti, Honduras, India, Indonesia, Iran, Iraq, Jamaica, Jordan, Kenya, North Korea, South Korea, Kuwait, Laos, Lebanon, Lesotho, Liberia, Libya, Madagascar, Malawi, Malaysia, Maldives, Mali, Malta, Mauritania, Mauritius, Mexico, Mongolia, Morocco, Mozambique, Namibia, Nepal, Nicaragua, Niger, Nigeria, Oman, Pakistan, Panama, Papua New Guinea, Paraguay, Peru, Philippines, Qatar, Romania, Rwanda, Saint Kitts and Nevis, Saint Lucia, Saint Vincent and the Grenadines, Sao Tome and Principe, Saudi Arabia, Senegal, Seychelles, Sierra Leone, Singapore, Solomon Islands, Somalia, Sri Lanka, Sudan, Suriname, Swaziland, Syria, Tanzania, Thailand, Togo, Tonga, Trinidad and Tobago, Tunisia, Uganda, UAE, Uruguay, Vanuatu, Venezuela, Vietnam, Western Samoa, Yemen, Yugoslavia, Zaire, Zambia, Zimbabwe, Palestine Liberation Organization
Gulf Cooperation Council (GCC) *note*—also known as the Cooperation Council for the Arab States of the Gulf *address*—P.O. Box 7431, Riyadh 11462 Saudi Arabia *telephone*—[966] (1) 4827777 *FAX*—[966] (1) 4829089 *established*—25 May 1981 *aim*—to promote regional cooperation in economic, social, political, and military affairs	*members*—(6) Bahrain, Kuwait, Oman, Qatar, Saudi Arabia, UAE
Habitat	see Commission on Human Settlements
Hexagonal Group	see Central European Initiative (CEI)
high-income countries	another term for the industrialized countries with high per capita GDPs; see developed countries (DCs)
industrial countries	another term for the developed countries; see developed countries (DCs)

Inter-American Development Bank (IADB) *note*—also known as Banco Interamericano de Desarrollo (BID) *address*—1300 New York Avenue NW, Washington, DC 10577, USA *telephone*—[1] (202) 632 1000 *FAX*—[1] (202) 789 2835 *established*—8 April 1959 *effective*—30 December 1959 *aim*—to promote economic and social development in Latin America	*members*—(46) Argentina, Austria, The Bahamas, Barbados, Belgium, Belize, Bolivia, Brazil, Canada, Chile, Colombia, Costa Rica, Croatia, Denmark, Dominican Republic, Ecuador, El Salvador, Finland, France, Germany, Guatemala, Guyana, Haiti, Honduras, Israel, Italy, Jamaica, Japan, Mexico, Netherlands, Nicaragua, Norway, Panama, Paraguay, Peru, Portugal, Slovenia, Spain, Suriname, Sweden, Switzerland, Trinidad and Tobago, UK, US, Uruguay, Venezuela
Inter-Governmental Authority on Drought and Development (IGADD) *address*—BP 2653, Djibouti, Djibouti *telephone*—[253] 354050, 352880 *FAX*—[253] 356994 *established*—15-16 January 1986 *aim*—to promote cooperation on drought-related matters	*members*—(7) Djibouti, Eritrea, Ethiopia, Kenya, Somalia, Sudan, Uganda
International Atomic Energy Agency (IAEA) *address*—Wagramerstrasse 5, P.O. Box 100, A-1400 Vienna, Austria *telephone*—[43] (1) 2360 2045 *FAX* [43] (1) 234564 *established*—26 October 1956 *effective*—29 July 1957 *aim*—to promote peaceful uses of atomic energy	*members*—(121) Afghanistan, Albania, Algeria, Argentina, Armenia, Australia, Austria, Bangladesh, Belarus, Belgium, Bolivia, Brazil, Bulgaria, Burma, Cambodia, Cameroon, Canada, Chile, China, Colombia, Costa Rica, Cote d'Ivoire, Croatia, Cuba, Cyprus, Czech Republic, Denmark, Dominican Republic, Ecuador, Egypt, El Salvador, Estonia, Ethiopia, Finland, France, Gabon, Germany, Ghana, Greece, Guatemala, Haiti, Holy See, Hungary, Iceland, India, Indonesia, Iran, Iraq, Ireland, Israel, Italy, Jamaica, Japan, Jordan, Kazakhstan, Kenya, South Korea, Kuwait, Lebanon, Liberia, Libya, Liechtenstein, Lithuania, Luxembourg, The Former Yugoslav Republic of Macedonia, Madagascar, Malaysia, Mali, Marshall Islands, Mauritius, Mexico, Monaco, Mongolia, Morocco, Namibia, Netherlands, NZ, Nicaragua, Niger, Nigeria, Norway, Pakistan, Panama, Paraguay, Peru, Philippines, Poland, Portugal, Qatar, Romania, Russia, Saudi Arabia, Senegal, Sierra Leone, Singapore, Slovakia, Slovenia, South Africa, Spain, Sri Lanka, Sudan, Sweden, Switzerland, Syria, Tanzania, Thailand, Tunisia, Turkey, Uganda, Ukraine, UAE, UK, US, Uruguay, Uzbekistan, Venezuela, Vietnam, Yugoslavia (suspended), Zaire, Zambia, Zimbabwe
International Bank for Economic Cooperation (IBEC)	was established on 22 October 1963 to promote economic cooperation and development; members were Bulgaria, Cuba, Czechoslovakia, East Germany, Hungary, Mongolia, Poland, Romania, USSR, Vietnam; now it is a Russian bank with a new charter
International Bank for Reconstruction and Development (IBRD) *note*—also known as the World Bank *address*—1818 H Street NW, Washington, DC 20433, USA *telephone*—[1] (202) 477 1234 *FAX*—[1] (202) 477 6391 *established*—22 July 1944 *effective*—27 December 1945 *aim*—to provide economic development loans; a UN specialized agency	*members*—(178) Afghanistan, Albania, Algeria, Angola, Antigua and Barbuda, Argentina, Armenia, Australia, Austria, Azerbaijan, The Bahamas, Bahrain, Bangladesh, Barbados, Belarus, Belgium, Belize, Benin, Bhutan, Bolivia, Botswana, Brazil, Bulgaria, Burkina, Burma, Burundi, Cambodia, Cameroon, Canada, Cape Verde, Central African Republic, Chad, Chile, China, Colombia, Comoros, Congo, Costa Rica, Cote d'Ivoire, Croatia, Cyprus, Czech Republic, Denmark, Djibouti, Dominica, Dominican Republic, Ecuador, Egypt, El Salvador, Equatorial Guinea, Eritrea, Estonia, Ethiopia, Fiji, Finland, France, Gabon, The Gambia, Georgia, Germany, Ghana, Greece, Grenada, Guatemala, Guinea, Guinea-Bissau, Guyana, Haiti, Honduras, Hungary, Iceland, India, Indonesia, Iran, Iraq, Ireland, Israel, Italy, Jamaica, Japan, Jordan, Kazakhstan, Kenya, Kiribati, South Korea, Kuwait, Kyrgyzstan, Laos, Latvia, Lebanon, Lesotho, Liberia, Libya, Lithuania, Luxembourg, The Former Yugoslav Republic of Macedonia, Madagascar, Malawi, Malaysia, Maldives, Mali, Malta, Marshall Islands, Mauritania, Mauritius, Mexico, Federated States of Micronesia, Moldova, Mongolia, Morocco, Mozambique, Namibia, Nepal, Netherlands, NZ, Nicaragua, Niger, Nigeria, Norway, Oman, Pakistan, Panama, Papua New Guinea, Paraguay, Peru, Philippines, Poland, Portugal, Qatar, Romania, Russia, Rwanda, Saint Kitts and Nevis, Saint Lucia, Saint Vincent and the Grenadines, Sao Tome and Principe, Saudi Arabia, Senegal, Seychelles, Sierra Leone, Singapore, Slovakia, Slovenia, Solomon Islands, Somalia, South Africa, Spain, Sri Lanka, Sudan, Suriname, Swaziland, Sweden, Switzerland, Syria, Tajikistan, Tanzania, Thailand, Togo, Tonga, Trinidad and Tobago, Tunisia, Turkey, Turkmenistan, Uganda, Ukraine, UAE, UK, US, Uruguay, Uzbekistan, Vanuatu, Venezuela, Vietnam, Western Samoa, Yemen, Zaire, Zambia, Zimbabwe

International Chamber of Commerce (ICC) *address*—38 Cours Albert 1st, F-75008 Paris, France *telephone*—[33] (1) 49 53 28 75 *FAX*—[33] (1) 49 53 29 42 *established*—NA 1919 *aim*—to promote free trade and private enterprise and to represent business interests at national and international levels	*members*—(59 national councils) Argentina, Australia, Austria, Belgium, Brazil, Burkina, Cameroon, Canada, Colombia, Cote d'Ivoire, Cyprus, Denmark, Ecuador, Egypt, Finland, France, Gabon, Germany, Greece, Iceland, India, Indonesia, Iran, Ireland, Israel, Italy, Japan, Jordan, South Korea, Kuwait, Lebanon, Luxembourg, Madagascar, Mexico, Morocco, Netherlands, Nigeria, Norway, Pakistan, Portugal, Saudi Arabia, Senegal, Singapore, South Africa, Spain, Sri Lanka, Sweden, Switzerland, Syria, Taiwan, Togo, Tunisia, Turkey, UK, US, Uruguay, Venezuela, Yugoslavia, Zaire
International Civil Aviation Organization (ICAO) *address*—1000 Sherbrooke Street West, Suite 327, Montreal PQ H3A 2R2, Canada *telephone*—[1] (514) 285 8219 *FAX*—[1] (514) 288 4772 *established*—7 December 1944 *effective*—4 April 1947 *aim*—to promote international cooperation in civil aviation; a UN specialized agency	*members*—(183) Afghanistan, Albania, Algeria, Angola, Antigua and Barbuda, Argentina, Armenia, Australia, Austria, Azerbaijan, The Bahamas, Bahrain, Bangladesh, Barbados, Belarus, Belgium, Belize, Benin, Bhutan, Bolivia, Bosnia and Herzegovina, Botswana, Brazil, Brunei, Bulgaria, Burkina, Burma, Burundi, Cambodia, Cameroon, Canada, Cape Verde, Central African Republic, Chad, Chile, China, Colombia, Comoros, Congo, Cook Islands, Costa Rica, Cote d'Ivoire, Croatia, Cuba, Cyprus, Czech Republic, Denmark, Djibouti, Dominican Republic, Ecuador, Egypt, El Salvador, Equatorial Guinea, Eritrea, Estonia, Ethiopia, Fiji, Finland, France, Gabon, The Gambia, Georgia, Germany, Ghana, Greece, Grenada, Guatemala, Guinea, Guinea-Bissau, Guyana, Haiti, Honduras, Hungary, Iceland, India, Indonesia, Iran, Iraq, Ireland, Israel, Italy, Jamaica, Japan, Jordan, Kazakhstan, Kenya, Kiribati, North Korea, South Korea, Kuwait, Kyrgyzstan, Laos, Latvia, Lebanon, Lesotho, Liberia, Libya, Lithuania, Luxembourg, The Former Yugoslav Republic of Macedonia, Madagascar, Malawi, Malaysia, Maldives, Mali, Malta, Marshall Islands, Mauritania, Mauritius, Mexico, Federated States of Micronesia, Moldova, Monaco, Mongolia, Morocco, Mozambique, Namibia, Nauru, Nepal, Netherlands, NZ, Nicaragua, Niger, Nigeria, Norway, Oman, Pakistan, Panama, Papua New Guinea, Paraguay, Peru, Philippines, Poland, Portugal, Qatar, Romania, Russia, Rwanda, Saint Lucia, Saint Vincent and the Grenadines, San Marino, Sao Tome and Principe, Saudi Arabia, Senegal, Seychelles, Sierra Leone, Singapore, Slovakia, Slovenia, Solomon Islands, Somalia, South Africa, Spain, Sri Lanka, Sudan, Suriname, Swaziland, Sweden, Switzerland, Syria, Tajikistan, Tanzania, Thailand, Togo, Tonga, Trinidad and Tobago, Tunisia, Turkey, Turkmenistan, Uganda, Ukraine, UAE, UK, US, Uruguay, Uzbekistan, Vanuatu, Venezuela, Vietnam, Yemen, Zaire, Zambia, Zimbabwe
International Committee of the Red Cross (ICRC) *address*—ICRC, 19 av de la Paix, CH-1202 Geneva, Switzerland *telephone*—[41] (22) 734 60 01 *FAX*—[41] (22) 733 82 80 *established*—NA 1863 *aim*—to provide humanitarian aid in wartime	*members*—(25 individuals) all Swiss nationals
International Confederation of Free Trade Unions (ICFTU) *address*—International Trade Union House, Bd Emile Jacqmain 155, B-1210 Brussels, Belgium *telephone*—[32] (2) 224 02 11 *FAX*—[32] (2) 218 84 15, 219 75 03 *established*—NA December 1949 *aim*—to promote the trade union movement	*members*—(164 national organizations in the following 118 areas) Antigua and Barbuda, Argentina, Australia, Austria, The Bahamas, Bangladesh, Barbados, Basque Country, Belgium, Belize, Benin, Bermuda, Botswana, Brazil, Bulgaria, Burkina, Cameroon, Canada, Cape Verde, Chad, Chile, China, Colombia, Cook Islands, Costa Rica, Curacao, Cyprus, Czech Republic, Denmark, Dominica, Dominican Republic, Ecuador, El Salvador, Estonia, Falkland Islands, Fiji, Finland, France, French Polynesia, Gabon, The Gambia, Germany, Ghana, Greece, Grenada, Guatemala, Guyana, Holy See, Honduras, Hong Kong, Iceland, India, Indonesia, Israel, Italy, Jamaica, Japan, Kenya, Kiribati, South Korea, Lebanon, Lesotho, Liberia, Luxembourg, Madagascar, Malawi, Malaysia, Mali, Malta, Mauritius, Mexico, Montserrat, Morocco, Netherlands, New Caledonia, NZ, Nicaragua, Norway, Pakistan, Panama, Papua New Guinea, Paraguay, Peru, Philippines, Poland, Portugal, Puerto Rico, Romania, Rwanda, Saint Helena, Saint Kitts and Nevis, Saint Lucia, Saint Vincent and the Grenadines, San Marino, Senegal, Seychelles, Sierra Leone, Singapore, Slovakia, Spain, Sri Lanka, Suriname, Swaziland, Sweden, Switzerland, Thailand, Togo, Tonga, Trinidad and Tobago, Tunisia, Turkey, Uganda, UK, US, Venezuela, Western Samoa, Zambia, Zimbabwe
International Court of Justice (ICJ) *note*—also known as the World Court *address*—Peace Palace, NL-2517 KJ The Hague, Netherlands *telephone*—[31] (70) 302 23 23 *FAX*—[31] (70) 364 99 28 *established*—26 June 1945 *effective*—24 October 1945 *aim*—primary judicial organ of the UN	*members*—(15 judges) elected by the UN General Assembly and Security Council to represent all principal legal systems

International Criminal Police Organization (INTERPOL)

address—BP 6041, F-69411 Lyon CEDEX 06, France
telephone—[33] 71 44 70 00
FAX—[33] 72 44 71 63
established—13 June 1956
aim—to promote international cooperation among police authorities in fighting crime

members—(176) Albania, Algeria, Andorra, Angola, Antigua and Barbuda, Argentina, Armenia, Aruba, Australia, Austria, Azerbaijan, The Bahamas, Bahrain, Bangladesh, Barbados, Belarus, Belgium, Belize, Benin, Bolivia, Bosnia and Herzegovina, Botswana, Brazil, Brunei, Bulgaria, Burkina, Burma, Burundi, Cambodia, Cameroon, Canada, Cape Verde, Central African Republic, Chad, Chile, China, Colombia, Congo, Costa Rica, Cote d'Ivoire, Croatia, Cuba, Cyprus, Czech Republic, Denmark, Djibouti, Dominica, Dominican Republic, Ecuador, Egypt, El Salvador, Equatorial Guinea, Estonia, Ethiopia, Fiji, Finland, France, Gabon, The Gambia, Georgia, Germany, Ghana, Greece, Grenada, Guatemala, Guinea, Guinea-Bissau, Guyana, Haiti, Honduras, Hungary, Iceland, India, Indonesia, Iran, Iraq, Ireland, Israel, Italy, Jamaica, Japan, Jordan, Kazakhstan, Kenya, Kiribati, South Korea, Kuwait, Laos, Latvia, Lebanon, Lesotho, Liberia, Libya, Liechtenstein, Lithuania, Luxembourg, The Former Yugoslav Republic of Macedonia, Madagascar, Malawi, Malaysia, Maldives, Mali, Malta, Marshall Islands, Mauritania, Mauritius, Mexico, Moldova, Monaco, Mongolia, Morocco, Mozambique, Namibia, Nauru, Nepal, Netherlands, Netherlands Antilles, NZ, Nicaragua, Niger, Nigeria, Norway, Oman, Pakistan, Panama, Papua New Guinea, Paraguay, Peru, Philippines, Poland, Portugal, Qatar, Romania, Russia, Rwanda, Saint Kitts and Nevis, Saint Lucia, Saint Vincent and the Grenadines, Sao Tome and Principe, Saudi Arabia, Senegal, Seychelles, Sierra Leone, Singapore, Slovakia, Slovenia, Somalia, South Africa, Spain, Sri Lanka, Sudan, Suriname, Swaziland, Sweden, Switzerland, Syria, Tanzania, Thailand, Togo, Tonga, Trinidad and Tobago, Tunisia, Turkey, Uganda, Ukraine, UAE, UK, US, Uruguay, Uzbekistan, Venezuela, Vietnam, Yemen, Zaire, Zambia, Zimbabwe
subbureaus—(13) American Samoa, Anguilla, Bermuda, British Virgin Islands, Cayman Islands, Gibraltar, Guam, Hong Kong, Macau, Montserrat, Northern Mariana Islands, Puerto Rico, Turks and Caicos Islands

International Development Association (IDA)

address—1818 H Street NW, Washington, DC 20433, USA
telephone—[1] (202) 477 12 34
established—26 January 1960
effective—24 September 1960
aim—UN specialized agency and IBRD affiliate that provides economic loans for low income countries

members—(157)
Part I—(24 more economically advanced countries) Australia, Austria, Belgium, Canada, Denmark, Finland, France, Germany, Iceland, Ireland, Italy, Japan, Kuwait, Luxembourg, Netherlands, NZ, Norway, Russia, South Africa, Sweden, Switzerland, UAE, UK, US
Part II—(133 less developed nations) Afghanistan, Albania, Algeria, Angola, Argentina, Armenia, Bangladesh, Belize, Benin, Bhutan, Bolivia, Botswana, Brazil, Burkina, Burma, Burundi, Cambodia, Cameroon, Cape Verde, Central African Republic, Chad, Chile, China, Colombia, Comoros, Congo, Costa Rica, Cote d'Ivoire, Croatia, Cyprus, Czech Republic, Djibouti, Dominica, Dominican Republic, Ecuador, Egypt, El Salvador, Equatorial Guinea, Eritrea, Ethiopia, Fiji, Gabon, The Gambia, Georgia, Ghana, Greece, Grenada, Guatemala, Guinea, Guinea-Bissau, Guyana, Haiti, Honduras, Hungary, India, Indonesia, Iran, Iraq, Israel, Jordan, Kazakhstan, Kenya, Kiribati, South Korea, Kyrgyzstan, Laos, Latvia, Lebanon, Lesotho, Liberia, Libya, The Former Yugoslav Republic of Macedonia, Madagascar, Malawi, Malaysia, Maldives, Mali, Marshall Islands, Mauritania, Mauritius, Mexico, Federated States of Micronesia, Moldova, Mongolia, Morocco, Mozambique, Nepal, Nicaragua, Niger, Nigeria, Oman, Pakistan, Panama, Papua New Guinea, Paraguay, Peru, Philippines, Poland, Portugal, Rwanda, Saint Kitts and Nevis, Saint Lucia, Saint Vincent and the Grenadines, Sao Tome and Principe, Saudi Arabia, Senegal, Sierra Leone, Slovakia, Slovenia, Solomon Islands, Somalia, Spain, Sri Lanka, Sudan, Swaziland, Syria, Tajikistan, Tanzania, Thailand, Togo, Tonga, Trinidad and Tobago, Tunisia, Turkey, Uganda, Uzbekistan, Vanuatu, Vietnam, Western Samoa, Yemen, Zaire, Zambia, Zimbabwe

International Energy Agency (IEA)

address—2 rue Andre Pascal, F-75775 Paris CEDEX 16, France
telephone—[33] (1) 45 24 82 00
FAX—[33] (1) 45 24 99 88
established—15 November 1974
aim—to promote cooperation on energy matters, especially emergency oil sharing and relations between oil consumers and oil producers; established by the OECD

members—(23) Australia, Austria, Belgium, Canada, Denmark, Finland, France, Germany, Greece, Ireland, Italy, Japan, Luxembourg, Netherlands, NZ, Norway, Portugal, Spain, Sweden, Switzerland, Turkey, UK, US

International Federation of Red Cross and Red Crescent Societies (IFRCS) *note*—formerly known as League of Red Cross and Red Crescent Societies (LORCS) *established*—5 May 1919 *aim*—to provide humanitarian aid in peacetime	*members*—(151) Afghanistan, Albania, Algeria, Angola, Argentina, Australia, Austria, The Bahamas, Bahrain, Bangladesh, Barbados, Belgium, Belize, Benin, Bolivia, Botswana, Brazil, Bulgaria, Burkina, Burma, Burundi, Cambodia, Cameroon, Canada, Cape Verde, Central African Republic, Chad, Chile, China, Colombia, Congo, Costa Rica, Cote d'Ivoire, Cuba, Czech Republic, Denmark, Djibouti, Dominica, Dominican Republic, Ecuador, Egypt, El Salvador, Ethiopia, Fiji, Finland, France, The Gambia, Germany, Ghana, Greece, Grenada, Guatemala, Guinea, Guinea-Bissau, Guyana, Haiti, Honduras, Hungary, Iceland, India, Indonesia, Iran, Iraq, Ireland, Italy, Jamaica, Japan, Jordan, Kenya, North Korea, South Korea, Kuwait, Laos, Latvia, Lebanon, Lesotho, Liberia, Libya, Liechtenstein, Lithuania, Luxembourg, Madagascar, Malawi, Malaysia, Mali, Mauritania, Mauritius, Mexico, Monaco, Mongolia, Morocco, Mozambique, Nepal, Netherlands, NZ, Nicaragua, Niger, Nigeria, Norway, Pakistan, Panama, Papua New Guinea, Paraguay, Peru, Philippines, Poland, Portugal, Qatar, Romania, Russia, Rwanda, Saint Lucia, Saint Vincent and the Grenadines, San Marino, Sao Tome and Principe, Saudi Arabia, Senegal, Sierra Leone, Singapore, Slovakia, Solomon Islands, Somalia, South Africa, Spain, Sri Lanka, Sudan, Suriname, Swaziland, Sweden, Switzerland, Syria, Tanzania, Thailand, Togo, Tonga, Trinidad and Tobago, Tunisia, Turkey, Uganda, UAE, UK, US, Uruguay, Venezuela, Vietnam, Western Samoa, Yemen, Yugoslavia, Zaire, Zambia, Zimbabwe *associate members*—(13) Andorra, Antigua and Barbuda, Comoros, Cyprus, Equatorial Guinea, Gabon, Kiribati, Namibia, Saint Kitts and Nevis, Seychelles, Suriname, Tuvalu, Vanuatu
International Finance Corporation (IFC) *address*—1818 H Street NW, Washington, DC 20433, USA *telephone*—[1] (202) 477 1234 *FAX*—[1] (202) 477 6391 *established*—25 May 1955 *effective*—20 July 1956 *aim*—to support private enterprise in international economic development; a UN specialized agency and IBRD affiliate	*members*—(161) Afghanistan, Albania, Algeria, Angola, Antigua and Barbuda, Argentina, Australia, Austria, The Bahamas, Bangladesh, Barbados, Belarus, Belgium, Belize, Benin, Bolivia, Botswana, Brazil, Bulgaria, Burkina, Burma, Burundi, Cameroon, Canada, Cape Verde, Central African Republic, Chile, China, Colombia, Comoros, Congo, Costa Rica, Cote d'Ivoire, Croatia, Cyprus, Czech Republic, Denmark, Djibouti, Dominica, Dominican Republic, Ecuador, Egypt, El Salvador, Equatorial Guinea, Estonia, Ethiopia, Fiji, Finland, France, Gabon, The Gambia, Germany, Ghana, Greece, Grenada, Guatemala, Guinea, Guinea-Bissau, Guyana, Haiti, Honduras, Hungary, Iceland, India, Indonesia, Iran, Iraq, Ireland, Israel, Italy, Jamaica, Japan, Jordan, Kazakhstan, Kenya, Kiribati, South Korea, Kuwait, Kyrgyzstan, Laos, Latvia, Lebanon, Lesotho, Liberia, Libya, Lithuania, Luxembourg, The Former Yugoslav Republic of Macedonia, Madagascar, Malawi, Malaysia, Maldives, Mali, Marshall Islands, Mauritania, Mauritius, Mexico, Federated States of Micronesia, Mongolia, Morocco, Mozambique, Namibia, Nepal, Netherlands, NZ, Nicaragua, Niger, Nigeria, Norway, Oman, Pakistan, Panama, Papua New Guinea, Paraguay, Peru, Philippines, Poland, Portugal, Romania, Russia, Rwanda, Saint Lucia, Saudi Arabia, Senegal, Seychelles, Sierra Leone, Singapore, Slovakia, Slovenia, Solomon Islands, Somalia, South Africa, Spain, Sri Lanka, Sudan, Swaziland, Sweden, Switzerland, Syria, Tanzania, Thailand, Togo, Tonga, Trinidad and Tobago, Tunisia, Turkey, Uganda, Ukraine, UAE, UK, US, Uruguay, Uzbekistan, Vanuatu, Venezuela, Vietnam, Western Samoa, Yemen, Zaire, Zambia, Zimbabwe
International Fund for Agricultural Development (IFAD) *address*—Via del Serafico 107, I-00142 Rome, Italy *telephone*—[39] (6) 54591 *FAX*—[39] (6) 5043463 *established*—NA November 1974 *aim*—to promote agricultural development; a UN specialized agency	*members*—(157) *Category I*—(21 industrialized aid contributors) Australia, Austria, Belgium, Canada, Denmark, Finland, France, Germany, Greece, Ireland, Italy, Japan, Luxembourg, Netherlands, NZ, Norway, Spain, Sweden, Switzerland, UK, US *Category II*—(12 petroleum-exporting aid contributors) Algeria, Gabon, Indonesia, Iran, Iraq, Kuwait, Libya, Nigeria, Qatar, Saudi Arabia, UAE, Venezuela *Category III*—(124 aid recipients) Afghanistan, Albania, Angola, Antigua and Barbuda, Argentina, Armenia, Azerbaijan, Bangladesh, Barbados, Belize, Benin, Bhutan, Bolivia, Bosnia and Herzegovina, Botswana, Brazil, Burkina, Burma, Burundi, Cambodia, Cameroon, Cape Verde, Central African Republic, Chad, Chile, China, Colombia, Comoros, Congo, Cook Islands, Costa Rica, Cote d'Ivoire, Croatia, Cuba, Cyprus, Djibouti, Dominica, Dominican Republic, Ecuador, Egypt, El Salvador, Equatorial Guinea, Eritrea, Ethiopia, Fiji, The Gambia, Ghana, Grenada, Guatemala, Guinea, Guinea-Bissau, Guyana, Haiti, Honduras, India, Israel, Jamaica, Jordan, Kenya, North Korea, South Korea, Kyrgyzstan, Laos, Lebanon, Lesotho, Liberia, The Former Yugoslav Republic of Macedonia, Madagascar, Malawi, Malaysia, Maldives, Mali, Malta, Mauritania, Mauritius, Mexico, Mongolia, Morocco, Mozambique, Namibia, Nepal, Nicaragua, Niger, Oman, Pakistan, Panama, Papua New Guinea, Paraguay, Peru, Philippines, Portugal, Romania, Rwanda, Saint Kitts and Nevis, Saint Lucia, Saint Vincent and the Grenadines, Sao Tome and Principe, Senegal, Seychelles, Sierra Leone, Solomon Islands, Somalia, Sri Lanka, Sudan, Suriname, Swaziland, Syria, Tajikistan, Tanzania, Thailand, Togo, Tonga, Trinidad and Tobago, Tunisia, Turkey, Uganda, Uruguay, Vietnam, Western Samoa, Yemen, Yugoslavia (suspended), Zaire, Zambia, Zimbabwe
International Investment Bank (IIB)	established on 7 July 1970; to promote economic development; members were Bulgaria, Cuba, Czechoslovakia, East Germany, Hungary, Mongolia, Poland, Romania, USSR, Vietnam; now it is a Russian bank with a new charter

International Labor Organization (ILO) *address*—International Labor Office, 4 route des Morillons, CH-1211 Geneva 22, Switzerland *telephone*—[41] (22) 799 61 11 *FAX*—[41] (22) 798 86 85 *established*—11 April 1919 (affiliated with the UN 14 December 1946) *aim*—UN specialized agency concerned with world labor issues	*members*—(171) Afghanistan, Albania, Algeria, Angola, Antigua and Barbuda, Argentina, Armenia, Australia, Austria, Azerbaijan, The Bahamas, Bahrain, Bangladesh, Barbados, Belarus, Belgium, Belize, Benin, Bolivia, Bosnia and Herzegovina, Botswana, Brazil, Bulgaria, Burkina, Burma, Burundi, Cambodia, Cameroon, Canada, Cape Verde, Central African Republic, Chad, Chile, China, Colombia, Comoros, Congo, Costa Rica, Cote d'Ivoire, Croatia, Cuba, Cyprus, Czech Republic, Denmark, Djibouti, Dominica, Dominican Republic, Ecuador, Egypt, El Salvador, Equatorial Guinea, Eritrea, Estonia, Ethiopia, Fiji, Finland, France, Gabon, Georgia, Germany, Ghana, Greece, Grenada, Guatemala, Guinea, Guinea-Bissau, Guyana, Haiti, Honduras, Hungary, Iceland, India, Indonesia, Iran, Iraq, Ireland, Israel, Italy, Jamaica, Japan, Jordan, Kazakhstan, Kenya, South Korea, Kuwait, Kyrgyzstan, Laos, Latvia, Lebanon, Lesotho, Liberia, Libya, Lithuania, Luxembourg, The Former Yugoslav Republic of Macedonia, Madagascar, Malawi, Malaysia, Mali, Malta, Mauritania, Mauritius, Mexico, Moldova, Mongolia, Morocco, Mozambique, Namibia, Nepal, Netherlands, NZ, Nicaragua, Niger, Nigeria, Norway, Oman, Pakistan, Panama, Papua New Guinea, Paraguay, Peru, Philippines, Poland, Portugal, Qatar, Romania, Russia, Rwanda, Saint Lucia, San Marino, Sao Tome and Principe, Saudi Arabia, Senegal, Seychelles, Sierra Leone, Singapore, Slovakia, Slovenia, Solomon Islands, Somalia, South Africa, Spain, Sri Lanka, Sudan, Suriname, Swaziland, Sweden, Switzerland, Syria, Tajikistan, Tanzania, Thailand, Togo, Trinidad and Tobago, Tunisia, Turkey, Turkmenistan, Uganda, Ukraine, UAE, UK, US, Uruguay, Uzbekistan, Venezuela, Vietnam, Yemen, Yugoslavia (suspended), Zaire, Zambia, Zimbabwe
International Maritime Organization (IMO) *note*—name changed from Intergovernmental Maritime Consultative Organization (IMCO) on 22 May 1982 *address*—4 Albert Embankment, London SE1 7SR, UK *telephone*—[44] (71) 735 7611 *FAX* [44] (71) 587 3210 *established*—17 March 1958 *aim* to deal with international maritime affairs; a UN specialized agency	*members*—(149) Albania, Algeria, Angola, Antiqua and Barbuda, Argentina, Australia, Austria, The Bahamas, Bahrain, Bangladesh, Barbados, Belgium, Belize, Benin, Bolivia, Bosnia and Herzegovina, Brazil, Brunei, Bulgaria, Burma, Cambodia, Cameroon, Canada, Cape Verde, Chile, China, Colombia, Congo, Costa Rica, Cote d'Ivoire, Croatia, Cuba, Cyprus, Czech Republic, Denmark, Djibouti, Dominica, Dominican Republic, Ecuador, Egypt, El Salvador, Equatorial Guinea, Eritrea, Estonia, Ethiopia, Fiji, Finland, France, Gabon, The Gambia, Georgia, Germany, Chana, Greece, Guatemala, Guinea, Guinea-Bissau, Guyana, Haiti, Honduras, Hungary, Iceland, India, Indonesia, Iran, Iraq, Ireland, Israel, Italy, Jamaica, Japan, Jordan, Kazakhstan, Kenya, North Korea, South Korea, Kuwait, Latvia, Lebanon, Liberia, Libya, Luxembourg, The Former Yugoslav Republic of Macedonia, Madagascar, Malawi, Malaysia, Maldives, Malta, Mauritania, Mauritius, Mexico, Monaco, Morocco, Mozambique, Nepal, Netherlands, NZ, Nicaragua, Nigeria, Norway, Oman, Pakistan, Panama, Papua New Guinea, Paraguay, Peru, Philippines, Poland, Portugal, Qatar, Romania, Russia, Saint Lucia, Saint Vincent and the Grenadines, Sao Tome and Principe, Saudi Arabia, Senegal, Seychelles, Sierra Leone, Singapore, Slovakia, Slovenia, Solomon Islands, Somalia, Spain, Sri Lanka, Sudan, Suriname, Sweden, Switzerland, Syria, Tanzania, Thailand, Togo, Trinidad and Tobago, Tunisia, Turkey, Turkmenistan, Ukraine, UAE, UK, US, Uruguay, Vanuatu, Venezuela, Vietnam, Yemen, Yugoslavia (suspended), Zaire *associate members*—(2) Hong Kong, Macau
International Maritime Satellite Organization (INMARSAT) *address*—99 City Road, London EC1Y 1AX, UK *telephone*—[44] (71) 728 1000 *FAX*—[44] (71) 728 1044 *established*—3 September 1976 *effective*—26 July 1979 *aim* to provide worldwide communications for maritime shipping and other applications	*members*—(75) Algeria, Argentina, Australia, The Bahamas, Bahrain, Bangladesh, Belarus, Belgium, Brazil, Brunei, Bulgaria, Cameroon, Canada, Chile, China, Colombia, Croatia, Cuba, Cyprus, Czech Republic, Denmark, Egypt, Finland, France, Gabon, Georgia, Germany, Greece, Iceland, India, Indonesia, Iran, Iraq, Israel, Italy, Japan, South Korea, Kuwait, Liberia, Malaysia, Malta, Mauritius, Mexico, Monaco, Mozambique, Netherlands, NZ, Nigeria, Norway, Oman, Pakistan, Panama, Peru, Philippines, Poland, Portugal, Qatar, Romania, Russia, Saudi Arabia, Senegal, Singapore, Slovakia, South Africa, Spain, Sri Lanka, Sweden, Switzerland, Tunisia, Turkey, Ukraine, UAE, UK, US, Yugoslavia

International Monetary Fund (IMF) *address*—700 19th Street NW, Washington, DC 20431, USA *telephone*—[1] (202) 623 7000 *FAX*—[1] (202) 623 4661, 623 7491, 623 4662 *established*—22 July 1944 *effective*—27 December 1945 *aim*—to promote world monetary stability and economic development; a UN specialized agency	*members*—(179) Afghanistan, Albania, Algeria, Angola, Antigua and Barbuda, Argentina, Armenia, Australia, Austria, Azerbaijan, The Bahamas, Bahrain, Bangladesh, Barbados, Belarus, Belgium, Belize, Benin, Bhutan, Bolivia, Botswana, Brazil, Bulgaria, Burkina, Burma, Burundi, Cambodia, Cameroon, Canada, Cape Verde, Central African Republic, Chad, Chile, China, Colombia, Comoros, Congo, Costa Rica, Cote d'Ivoire, Croatia, Cyprus, Czech Republic, Denmark, Djibouti, Dominica, Dominican Republic, Ecuador, Egypt, El Salvador, Equatorial Guinea, Eritrea, Estonia, Ethiopia, Fiji, Finland, France, Gabon, The Gambia, Georgia, Germany, Ghana, Greece, Grenada, Guatemala, Guinea, Guinea-Bissau, Guyana, Haiti, Honduras, Hungary, Iceland, India, Indonesia, Iran, Iraq, Ireland, Israel, Italy, Jamaica, Japan, Jordan, Kazakhstan, Kenya, Kiribati, South Korea, Kuwait, Kyrgyzstan, Laos, Latvia, Lebanon, Lesotho, Liberia, Libya, Lithuania, Luxembourg, The Former Yugoslav Republic of Macedonia, Madagascar, Malawi, Malaysia, Maldives, Mali, Malta, Marshall Islands, Mauritania, Mauritius, Mexico, Federated States of Micronesia, Moldova, Mongolia, Morocco, Mozambique, Namibia, Nepal, Netherlands, NZ, Nicaragua, Niger, Nigeria, Norway, Oman, Pakistan, Panama, Papua New Guinea, Paraguay, Peru, Philippines, Poland, Portugal, Qatar, Romania, Russia, Rwanda, Saint Kitts and Nevis, Saint Lucia, Saint Vincent and the Grenadines, San Marino, Sao Tome and Principe, Saudi Arabia, Senegal, Seychelles, Sierra Leone, Singapore, Slovakia, Slovenia, Solomon Islands, Somalia, South Africa, Spain, Sri Lanka, Sudan, Suriname, Swaziland, Sweden, Switzerland, Syria, Tajikistan, Tanzania, Thailand, Togo, Tonga, Trinidad and Tobago, Tunisia, Turkey, Turkmenistan, Uganda, Ukraine, UAE, UK, US, Uruguay, Uzbekistan, Vanuatu, Venezuela, Vietnam, Western Samoa, Yemen, Zaire, Zambia, Zimbabwe
International Olympic Committee (IOC) *note*—there are 194 National Olympic Committees of which 185 are recognized by the International Olympic Committee *address*—Chateau de Vidy, CH-1007 Lausanne, Switzerland *telephone*—[41] (21) 621 61 11 *FAX*—[41] (21) 621 62 16 *established*—23 June 1894 *aim*—to promote the Olympic ideals and administer the Olympic games: 1996 Summer Olympics in Atlanta, United States (20 July-4 August); 1998 Winter Olympics in Nagano, Japan (date NA); 2000 Summer Olympics in Sydney, Australia (date NA)	*National Olympic Committees*—(193 and the Palestine Liberation Organization) Afghanistan, Albania, Algeria, American Samoa, Andorra, Angola, Antigua and Barbuda, Argentina, Armenia, Aruba, Australia, Austria, Azerbaijan, The Bahamas, Bahrain, Bangladesh, Barbados, Belarus, Belgium, Belize, Benin, Bermuda, Bhutan, Bolivia, Bosnia and Herzegovina, Botswana, Brazil, British Virgin Islands, Brunei, Bulgaria, Burkina, Burma, Burundi, Cameroon, Canada, Cape Verde, Cayman Islands, Central African Republic, Chad, Chile, China, Colombia, Comoros, Congo, Cook Islands, Costa Rica, Cote d'Ivoire, Croatia, Cuba, Cyprus, Czech Republic, Denmark, Djibouti, Dominica, Dominican Republic, Ecuador, Egypt, El Salvador, Equatorial Guinea, Estonia, Ethiopia, Fiji, Finland, France, Gabon, The Gambia, Georgia, Germany, Ghana, Greece, Grenada, Guam, Guatemala, Guinea, Guyana, Haiti, Honduras, Hong Kong, Hungary, Iceland, India, Indonesia, Iran, Iraq, Ireland, Israel, Italy, Jamaica, Japan, Jordan, Kazahkstan, Kenya, North Korea, South Korea, Kuwait, Kyrgyzstan, Laos, Latvia, Lebanon, Lesotho, Liberia, Libya, Liechtenstein, Lithuania, Luxembourg, The Former Yugoslav Republic of Macedonia, Madagascar, Malawi, Malaysia, Maldives, Mali, Malta, Mauritania, Mauritius, Mexico, Moldova, Monaco, Mongolia, Morocco, Mozambique, Namibia, Nepal, Netherlands, Netherlands Antilles, NZ, Nicaragua, Niger, Nigeria, Norway, Oman, Pakistan, Panama, Papua New Guinea, Paraguay, Peru, Philippines, Poland, Portugal, Puerto Rico, Qatar, Romania, Russia, Rwanda, Saint Kitts and Nevis, Saint Lucia, Saint Vincent and the Grenadines, San Marino, Sao Tome and Principe, Saudi Arabia, Senegal, Seychelles, Sierra Leone, Singapore, Slovakia, Slovenia, Solomon Islands, Somalia, South Africa, Spain, Sri Lanka, Sudan, Suriname, Swaziland, Sweden, Switzerland, Syria, Taiwan, Tajikistan, Tanzania, Thailand, Togo, Tonga, Trinidad and Tobago, Tunisia, Turkey, Turkmenistan, Uganda, Ukraine, UAE, UK, US, Uruguay, Uzbekistan, Vanuatu, Venezuela, Vietnam, Virgin Islands, Western Samoa, Yemen, Yugoslavia (Serbia and Montenegro), Zaire, Zambia, Zimbabwe, Palestine Liberation Organization
International Organization for Migration (IOM) *note*—established as Provisional Intergovernmental Committee for the Movement of Migrants from Europe; renamed Intergovernmental Committee for European Migration (ICEM) on 15 November 1952; renamed Intergovernmental Committee for Migration (ICM) in November 1980; current name adopted 14 November 1989 *address*—17 route des Morillons, CP 71, CH-1211 Geneva 19, Switzerland *telephone*—[41] (22) 717 91 11 *FAX*—[41] (22) 798 61 50 *established*—5 December 1951 *aim*—to facilitate orderly international emigration and immigration	*members*—(52) Albania, Angola, Argentina, Armenia, Australia, Austria, Bangladesh, Belgium, Bolivia, Canada, Chile, Colombia, Costa Rica, Croatia, Cyprus, Denmark, Dominican Republic, Ecuador, Egypt, El Salvador, Finland, France, Germany, Greece, Guatemala, Honduras, Hungary, Israel, Italy, Japan, Kenya, South Korea, Luxembourg, Netherlands, Nicaragua, Norway, Pakistan, Panama, Paraguay, Peru, Philippines, Poland, Portugal, Sri Lanka, Sweden, Switzerland, Thailand, Uganda, US, Uruguay, Venezuela, Zambia *observers*—(44) Belize, Bosnia and Herzegovina, Brazil, Bulgaria, Cape Verde, Czech Republic, Federation of Ethnic Communities' Council of Australia Inc., Georgia, Ghana, Guinea-Bissau, Holy See, India, Indonesia, Iran, Japan International Friendship and Welfare Foundation, Jordan, Kyrgyzstan, Latvia, Mexico, Moldova, Morocco, Mozambique, Namibia, NZ, Niwano Peace Foundation, Partnership with the Children of the Third World, Presiding Bishop's Fund for World Relief/Episcopal Church, Refugee Council of Australia, Romania, Russia, San Marino, Sao Tome and Principe, Senegal, Slovakia, Slovenia, Somalia, Spain, Tajikistan, Turkey, Ukraine, UK, Vietnam, Yugoslavia, Zimbabwe

International Organization for Standardization (ISO) *address*—CP 56, 1 rue de Varembe, CH-1211 Geneva 20, Switzerland *telephone*—[41] (22) 749 01 11 *FAX*—[41] (22) 733 34 30 *established*—NA February 1947 *aim*—to promote the development of international standards	*members*—(76 national standards organizations) Albania, Algeria, Argentina, Australia, Austria, Bangladesh, Belarus, Belgium, Brazil, Bulgaria, Canada, Chile, China, Colombia, Croatia, Cuba, Cyprus, Czech Republic, Denmark, Egypt, Ethiopia, Finland, France, Germany, Greece, Hungary, Iceland, India, Indonesia, Iran, Ireland, Israel, Italy, Jamaica, Japan, Kenya, North Korea, South Korea, Libya, Malaysia, Mexico, Mongolia, Morocco, Netherlands, NZ, Norway, Pakistan, Philippines, Poland, Portugal, Romania, Russia, Saudi Arabia, Singapore, Slovakia, Slovenia, South Africa, Spain, Sri Lanka, Sweden, Switzerland, Syria, Tanzania, Thailand, Trinidad and Tobago, Tunisia, Turkey, Ukraine, UK, US, Uruguay, Uzbekistan, Venezuela, Vietnam, Yugoslavia, Zimbabwe *correspondent members*—(19) Bahrain, Barbados, Brunei, Estonia, Hong Kong, Jordan, Kuwait, Lithuania, Malawi, Malta, Mauritius, Nepal, Oman, Papua New Guinea, Peru, Qatar, Turkmenistan, Uganda, UAE *subscriber members*—(4) Antigua and Barbuda, Burundi, Grenada, Saint Lucia
International Red Cross and Red Crescent Movement (ICRM) *address*—CICR, 19 Av de la Paix, CH-1202 Geneva, Switzerland *telephone*—[41] (22) 734 60 01 *FAX*—[41] (22) 733 20 57 *established*—NA 1928 *aim*—to promote worldwide humanitarian aid through the International Committee of the Red Cross (ICRC) in wartime, and International Federation of Red Cross and Red Crescent Societies (IFRCS; formerly League of Red Cross and Red Crescent Societies or LORCS) in peacetime	National Societies—(161 countries) Afghanistan, Albania, Algeria, Angola, Antigua and Barbuda, Argentina, Australia, Austria, The Bahamas, Bahrain, Bangladesh, Barbados, Belgium, Belize, Benin, Bolivia, Botswana, Brazil, Bulgaria, Burkina, Burma, Burundi, Cambodia, Cameroon, Canada, Cape Verde, Central African Republic, Chad, Chile, China, Colombia, Congo, Costa Rica, Cote d'Ivoire, Croatia, Cuba, Czech Republic, Denmark, Djibouti, Dominica, Dominican Republic, Ecuador, Egypt, El Salvador, Estonia, Ethiopia, Fiji, Finland, France, Gambia, Germany, Ghana, Greece, Grenada, Guatemala, Guinea, Guinea-Bissau, Guyana, Haiti, Honduras, Hungary, Iceland, India, Indonesia, Iran, Iraq, Ireland, Italy, Jamaica, Japan, Jordan, Kenya, North Korea, South Korea, Kuwait, Laos, Latvia, Lebanon, Lesotho, Liberia, Libya, Liechtenstein, Lithuania, Luxembourg, Madagascar, Malawi, Malaysia, Mali, Malta, Mauritania, Mauritius, Mexico, Monaco, Mongolia, Morocco, Mozambique, Namibia, Nepal, Netherlands, NZ, Nicaragua, Niger, Nigeria, Norway, Pakistan, Panama, Papua New Guinea, Paraguay, Peru, Philippines, Poland, Portugal, Qatar, Romania, Russia, Rwanda, Saint Kitts and Nevis, Saint Lucia, Saint Vincent and the Grenadines, San Marino, Sao Tome and Principe, Saudi Arabia, Senegal, Seychelles, Sierra Leone, Singapore, Slovakia, Slovenia, Solomon Islands, Somalia, South Africa, Spain, Sri Lanka, Sudan, Suriname, Swaziland, Sweden, Switzerland, Syria, Tanzania, Thailand, Togo, Tonga, Trinidad and Tobago, Tunisia, Turkey, Uganda, Ukraine, UAR, UK, US, Uruguay, Vanuatu, Venezuela, Vietnam, Western Samoa, Yemen, Yugoslavia, Zaire, Zambia, Zimbabwe
International Telecommunication Union (ITU) *address*—Place des Nations, 1211 Geneva 20, Switzerland *telephone*—[41] (22) 730 51 11 *FAX*—[41] (22) 733 72 56 *established*—9 December 1932 *effective*—1 January 1934 *affiliated with the UN*—15 November 1947 *aim*—to deal with world telecommunications issues; UN specialized agency	*members*—(184) Afghanistan, Albania, Algeria, Andorra, Angola, Antigua and Barbuda, Argentina, Armenia, Australia, Austria, Azerbaijan, The Bahamas, Bahrain, Bangladesh, Barbados, Belarus, Belgium, Belize, Benin, Bhutan, Bolivia, Bosnia and Herzegovina, Botswana, Brazil, Brunei, Bulgaria, Burkina, Burma, Burundi, Cambodia, Cameroon, Canada, Cape Verde, Central African Republic, Chad, Chile, China, Colombia, Comoros, Congo, Costa Rica, Cote d'Ivoire, Croatia, Cuba, Cyprus, Czech Republic, Denmark, Djibouti, Dominican Republic, Ecuador, Egypt, El Salvador, Equatorial Guinea, Eritrea, Estonia, Ethiopia, Fiji, Finland, France, Gabon, The Gambia, Georgia, Germany, Ghana, Greece, Grenada, Guatemala, Guinea, Guinea-Bissau, Guyana, Haiti, Holy See, Honduras, Hungary, Iceland, India, Indonesia, Iran, Iraq, Ireland, Israel, Italy, Jamaica, Japan, Jordan, Kazakhstan, Kenya, Kiribati, North Korea, South Korea, Kuwait, Kyrgyzstan, Laos, Latvia, Lebanon, Lesotho, Liberia, Libya, Liechtenstein, Lithuania, Luxembourg, The Former Yugoslav Republic of Macedonia, Madagascar, Malawi, Malaysia, Maldives, Mali, Malta, Mauritania, Mauritius, Mexico, Federated States of Micronesia, Moldova, Monaco, Mongolia, Morocco, Mozambique, Namibia, Nauru, Nepal, Netherlands, NZ, Nicaragua, Niger, Nigeria, Norway, Oman, Pakistan, Panama, Papua New Guinea, Paraguay, Peru, Philippines, Poland, Portugal, Qatar, Romania, Russia, Rwanda, Saint Vincent and the Grenadines, San Marino, Sao Tome and Principe, Saudi Arabia, Senegal, Sierra Leone, Singapore, Slovakia, Slovenia, Solomon Islands, Somalia, South Africa, Spain, Sri Lanka, Sudan, Suriname, Swaziland, Sweden, Switzerland, Syria, Tajikistan, Tanzania, Thailand, Togo, Tonga, Trinidad and Tobago, Tunisia, Turkey, Turkmenistan, Uganda, Ukraine, UAE, UK, US, Uruguay, Uzbekistan, Vanuatu, Venezuela, Vietnam, Western Samoa, Yemen, Yugoslavia (suspended), Zaire, Zambia, Zimbabwe

International Telecommunications Satellite Organization (INTELSAT) *address*—INTELSAT, 3400 International Drive NW, Washington, DC 20008-3098, USA *telephone*—[1] (202) 944 6800 *FAX*—[1] (202) 944 7860 *established*—20 August 1971 *effective*—12 February 1973 *aim*—to develop and operate a global commercial telecommunications satellite system	*members*—(134) Afghanistan, Algeria, Angola, Argentina, Armenia, Australia, Austria, Azerbaijan, The Bahamas, Bahrain, Bangladesh, Barbados, Belgium, Benin, Bhutan, Bolivia, Brazil, Brunei, Burkina, Cameroon, Canada, Cape Verde, Central African Republic, Chad, Chile, China, Colombia, Congo, Costa Rica, Cote d'Ivoire, Croatia, Cyprus, Czech Republic, Denmark, Dominican Republic, Ecuador, Egypt, El Salvador, Ethiopia, Fiji, Finland, France, Gabon, Germany, Ghana, Greece, Guatemala, Guinea, Haiti, Holy See, Honduras, Hungary, Iceland, India, Indonesia, Iran, Iraq, Ireland, Israel, Italy, Jamaica, Japan, Jordan, Kazakhstan, Kenya, South Korea, Kuwait, Kyrgyzstan, Lebanon, Libya, Liechtenstein, Luxembourg, Madagascar, Malawi, Malaysia, Mali, Mauritania, Mauritius, Mexico, Federated States of Micronesia, Monaco, Morocco, Mozambique, Namibia, Nepal, Netherlands, NZ, Nicaragua, Niger, Nigeria, Norway, Oman, Pakistan, Panama, Papua New Guinea, Paraguay, Peru, Philippines, Poland, Portugal, Qatar, Romania, Russia, Rwanda, Saudi Arabia, Senegal, Singapore, Somalia, South Africa, Spain, Sri Lanka, Sudan, Swaziland, Sweden, Switzerland, Syria, Tanzania, Thailand, Togo, Trinidad and Tobago, Tunisia, Turkey, Uganda, UAE, UK, US, Uruguay, Venezuela, Vietnam, Yemen, Yugoslavia, Zaire, Zambia, Zimbabwe *nonsignatory users*—(50) Albania, Antigua and Barbuda, Belarus, Belize, Bosnia and Herzegovina, Botswana, Bulgaria, Burma, Burundi, Cambodia, Comoros, Cook Islands, Cuba, Djibouti, Equatorial Guinea, Eritrea, Gambia, Guinea-Bissau, Guyana, Kiribati, North Korea, Laos, Latvia, Lesotho, Liberia, Lithuania, The Former Yugoslav Republic of Macedonia, Maldives, Malta, Marshall Islands, Moldova, Mongolia, Nauru, Niue, Sao Tome and Principe, Seychelles, Sierra Leone, Slovakia, Slovenia, Solomon Islands, Saint Lucia, Saint Vincent and the Grenadines, Suriname, Tajikistan, Tonga, Turkmenistan, Tuvalu, Ukraine, Vanuatu, Western Samoa
Islamic Development Bank (IDB) *address*—P.O. Box 5925, Jeddah 21432, Saudi Arabia *telephone*—[966] (2) 6361400 *FAX*—[966] (2) 6366871 *established*—15 December 1973 *aim*—to promote Islamic economic aid and social development	*members*—(48 plus the Palestine Liberation Organization) Afghanistan, Albania, Algeria, Azerbaijan, Bahrain, Bangladesh, Benin, Brunei, Burkina, Cameroon, Chad, Comoros, Djibouti, Egypt, Gabon, The Gambia, Guinea, Guinea-Bissau, Indonesia, Iran, Iraq, Jordan, Kuwait, Kyrgyzstan, Lebanon, Libya, Malaysia, Maldives, Mali, Mauritania, Morocco, Niger, Oman, Pakistan, Qatar, Saudi Arabia, Senegal, Sierra Leone, Somalia, Sudan, Syria, Tajikistan, Tunisia, Turkey, Turkmenistan, Uganda, UAE, Yemen, Palestine Liberation Organization
Latin American Economic System (LAES) *note*—also known as Sistema Economico Latinoamericana (SELA) *address*—SELA, Avda Francisco de Miranda, Torre Europa, piso 4, Chacaito, Apartado de Correos 17035, Caracas 1010-A, Venezuela *telephone*—[58] (2) 905 5111 *FAX*—[58] (2) 951 6953, 951 7246 *established*—17 October 1975 *aim*—to promote economic and social development through regional cooperation	*members*—(27) Argentina, Barbados, Belize, Bolivia, Brazil, Chile, Colombia, Costa Rica, Cuba, Dominican Republic, Ecuador, El Salvador, Grenada, Guatemala, Guyana, Haiti, Honduras, Jamaica, Mexico, Nicaragua, Panama, Paraguay, Peru, Suriname, Trinidad and Tobago, Uruguay, Venezuela
Latin American Integration Association (LAIA) *note*—also known as Asociacion Latinoamericana de Integracion (ALADI) *address*—Calle Cebollati 1461, Casilla de Correo 577, 11000 Montevideo, Uruguay *telephone*—[598] (2) 40 11 21, 49 59 15 *FAX*—[598] (2) 49 06 49 *established*—12 August 1980 *effective*—18 March 1981 *aim*—to promote freer regional trade	*members*—(11) Argentina, Bolivia, Brazil, Chile, Colombia, Ecuador, Mexico, Paraguay, Peru, Uruguay, Venezuela *observers*—(16) Commission of the European Communities, Costa Rica, Cuba, Dominican Republic, El Salvador, Guatemala, Honduras, Inter-American Development Bank, Italy, Nicaragua, Organization of American States, Panama, Portugal, Spain, United Nations Development Program, United Nations Economic Commission for Latin America and the Caribbean
League of Arab States (LAS)	see Arab League (AL)
League of Red Cross and Red Crescent Societies (LORCS)	see International Federation of Red Cross and Red Crescent Societies (IFRCS)

least developed countries (LLDCs)	that subgroup of the less developed countries (LDCs) initially identified by the UN General Assembly in 1971 as having no significant economic growth, per capita GDPs normally less than $1,000, and low literacy rates; also known as the undeveloped countries. The 42 LLDCs are: Afghanistan, Bangladesh, Benin, Bhutan, Botswana, Burkina, Burma, Burundi, Cape Verde, Central African Republic, Chad, Comoros, Djibouti, Equatorial Guinea, Eritrea, Ethiopia, The Gambia, Guinea, Guinea-Bissau, Haiti, Kiribati, Laos, Lesotho, Malawi, Maldives, Mali, Mauritania, Mozambique, Nepal, Niger, Rwanda, Sao Tome and Principe, Sierra Leone, Somalia, Sudan, Tanzania, Togo, Tuvalu, Uganda, Vanuatu, Western Samoa, Yemen
less developed countries (LDCs)	the bottom group in the hierarchy of developed countries (DCs), former USSR/Eastern Europe (former USSR/EE), and less developed countries (LDCs); mainly countries and dependent areas with low levels of output, living standards, and technology; per capita GDPs are generally below $5,000 and often less than $1,500; however, the group also includes a number of countries with high per capita incomes, areas of advanced technology, and rapid rates of growth; includes the advanced developing countries, developing countries, Four Dragons (Four Tigers), least developed countries (LLDCs), low-income countries, middle-income countries, newly industrializing economies (NIEs), the South, Third World, underdeveloped countries, undeveloped countries; the 172 LDCs are: Afghanistan, Algeria, American Samoa, Angola, Anguilla, Antigua and Barbuda, Argentina, Aruba, The Bahamas, Bahrain, Bangladesh, Barbados, Belize, Benin, Bhutan, Bolivia, Botswana, Brazil, British Virgin Islands, Brunei, Burkina, Burma, Burundi, Cambodia, Cameroon, Cape Verde, Cayman Islands, Central African Republic, Chad, Chile, China, Christmas Island, Cocos Islands, Colombia, Comoros, Congo, Cook Islands, Costa Rica, Cote d'Ivoire, Cuba, Cyprus, Djibouti, Dominica, Dominican Republic, Ecuador, Egypt, El Salvador, Equatorial Guinea, Eritrea, Ethiopia, Falkland Islands, Fiji, French Guiana, French Polynesia, Gabon, The Gambia, Gaza Strip, Ghana, Gibraltar, Greenland, Grenada, Guadeloupe, Guam, Guatemala, Guernsey, Guinea, Guinea-Bissau, Guyana, Haiti, Honduras, Hong Kong, India, Indonesia, Iran, Iraq, Jamaica, Jersey, Jordan, Kenya, Kiribati, North Korea, South Korea, Kuwait, Laos, Lebanon, Lesotho, Liberia, Libya, Macau, Madagascar, Malawi, Malaysia, Maldives, Mali, Isle of Man, Marshall Islands, Martinique, Mauritania, Mauritius, Mayotte, Federated States of Micronesia, Mongolia, Montserrat, Morocco, Mozambique, Namibia, Nauru, Nepal, Netherlands Antilles, New Caledonia, Nicaragua, Niger, Nigeria, Niue, Norfolk Island, Northern Mariana Islands, Oman, Palau, Pakistan, Panama, Papua New Guinea, Paraguay, Peru, Philippines, Pitcairn Islands, Puerto Rico, Qatar, Reunion, Rwanda, Saint Helena, Saint Kitts and Nevis, Saint Lucia, Saint Pierre and Miquelon, Saint Vincent and the Grenadines, Sao Tome and Principe, Saudi Arabia, Senegal, Seychelles, Sierra Leone, Singapore, Solomon Islands, Somalia, Sri Lanka, Sudan, Suriname, Swaziland, Syria, Taiwan, Tanzania, Thailand, Togo, Tokelau, Tonga, Trinidad and Tobago, Tunisia, Turks and Caicos Islands, Tuvalu, UAE, Uganda, Uruguay, Vanuatu, Venezuela, Vietnam, Virgin Islands, Wallis and Futuna, West Bank, Western Sahara, Western Samoa, Yemen, Zaire, Zambia, Zimbabwe
low-income countries	another term for those less developed countries with below-average per capita GDPs; see less developed countries (LDCs)
London Suppliers Group	see Nuclear Suppliers Group (NSG)
Mercado Comun del Cono Sur (MERCOSUR)	see Southern Cone Common Market
middle-income countries	another term for those less developed countries with above-average per capita GDPs; see less developed countries (LDCs)
Missile Technology Control Regime (MTCR) *established*—April 1987 *aim*—to arrest missile proliferation by controlling the export of key missile technologies and equipment	*members*—(25) Argentina, Australia, Austria, Belgium, Canada, Denmark, Finland, France, Germany, Greece, Hungary, Iceland, Ireland, Italy, Japan, Luxembourg, Netherlands, NZ, Norway, Portugal, Spain, Sweden, Switzerland, UK, US
Near Abroad	the 14 non-Russian successor states of the USSR, in which 25 million ethnic Russians live and in which Moscow has expressed a strong national security interest
newly industrializing countries (NICs)	former term for the newly industrializing economies; see newly industrializing economies (NIEs)
newly industrializing economies (NIEs)	that subgroup of the less developed countries (LDCs) that has experienced particularly rapid industrialization of their economies; formerly known as the newly industrializing countries (NICs); also known as advanced developing countries; usually includes the Four Dragons (Hong Kong, South Korea, Singapore, Taiwan), and Brazil

Nonaligned Movement (NAM) *address*—c/o Ministry of Foreign Affairs, Jalan Taman Pejambon 6, Jakarta PUSAT, Indonesia *established*—1-6 September 1961 *aim*—to establish political and military cooperation apart from the traditional East or West blocs	*members*—(110 plus the Palestine Liberation Organization) Afghanistan, Algeria, Angola, The Bahamas, Bahrain, Bangladesh, Barbados, Belize, Benin, Bhutan, Bolivia, Botswana, Brunei, Burkina, Burma, Burundi, Cambodia, Cameroon, Cape Verde, Central African Republic, Chad, Chile, Colombia, Comoros, Congo, Cote d'Ivoire, Cuba, Cyprus, Djibouti, Ecuador, Egypt, Equatorial Guinea, Ethiopia, Gabon, The Gambia, Ghana, Grenada, Guatemala, Guinea, Guinea-Bissau, Guyana, Honduras, India, Indonesia, Iran, Iraq, Jamaica, Jordan, Kenya, North Korea, Kuwait, Laos, Lebanon, Lesotho, Liberia, Libya, Madagascar, Malawi, Malaysia, Maldives, Mali, Malta, Mauritania, Mauritius, Mongolia, Morocco, Mozambique, Namibia, Nepal, Nicaragua, Niger, Nigeria, Oman, Pakistan, Panama, Papua New Guinea, Peru, Philippines, Qatar, Rwanda, Saint Lucia, Sao Tome and Principe, Saudi Arabia, Senegal, Seychelles, Sierra Leone, Singapore, Somalia, South Africa, Sri Lanka, Sudan, Suriname, Swaziland, Syria, Tanzania, Thailand, Togo, Trinidad and Tobago, Tunisia, Uganda, UAE, Uzbekistan, Vanuatu, Venezuela, Vietnam, Yemen, Yugoslavia, Zaire, Zambia, Zimbabwe, Palestine Liberation Organization *observers*—(19) Afro-Asian Solidarity Organization, Antigua and Barbuda, Arab League, Armenia, Brazil, China, Costa Rica, Croatia, Dominica, El Salvador, Kanaka Socialist National Liberation Front (New Caledonia), Mexico, Mongolia, Organization of African Unity, Organization of the Islamic Conference, Papua New Guinea, Socialist Party of Puerto Rico, UN, Uruguay *guests*—(21) Australia, Austria, Bosnia and Herzegovina, Bulgaria, Canada, Dominican Republic, Finland, Germany, Greece, Hungary, Netherlands, NZ, Norway, Poland, Portugal, Romania, San Marino, Slovenia, Spain, Sweden, Switzerland
Nordic Council (NC) *address*—Tyrgatan 7, Box 19506, S-104 32 Stockholm, Sweden *telephone*—[46] (8) 453 47 00 *FAX*—[46] (8) 411 75 36 *established*—16 March 1952 *effective*—12 February 1953 *aim*—to promote regional economic, cultural, and environmental cooperation	*members*—(5) Denmark (including Faroe Islands and Greenland), Finland (including Aland Islands), Iceland, Norway, Sweden *observers*—the Sami (Lapp) local parliaments of Finland, Norway, and Sweden
Nordic Investment Bank (NIB) *address*—Fabiansgatan 34, PB 249 SF-00171 Helsinki, Finland *telephone*—[358] (0) 18001 *FAX*—[358] (0) 1800309 *established*—4 December 1975 *effective*—1 June 1976 *aim*—to promote economic cooperation and development	*members*—(5) Denmark (including Faroe Islands and Greenland), Finland (including Aland Islands), Iceland, Norway, Sweden
North	a popular term for the rich industrialized countries generally located in the northern portion of the Northern Hemisphere; the counterpart of the South; see developed countries (DCs)
North Atlantic Cooperation Council (NACC) *note*—an extension of NATO *address*—c/o NATO, B-1110 Brussels, Belgium *telephone*—[32] (2) 728 41 11 *FAX*—[32] (2) 728 45 79 *established*—8 November 1991 *effective*—20 December 1991 *aim*—to discuss cooperation on mutual political and security issues	*members*—(38) Albania, Armenia, Azerbaijan, Belarus, Belgium, Bulgaria, Canada, Czech Republic, Denmark, Estonia, France, Georgia, Germany, Greece, Hungary, Iceland, Italy, Kazakhstan, Kyrgyzstan, Latvia, Lithuania, Luxembourg, Moldova, Netherlands, Norway, Poland, Portugal, Romania, Russia, Slovakia, Spain, Tajikistan, Turkey, Turkmenistan, Ukraine, UK, US, Uzbekistan *observer*—(1) Finland
North Atlantic Treaty Organization (NATO) *address*—B-110 Brussels, Belgium *telephone*—[32] (2) 728 41 11 *FAX*—[32] (2) 728 45 79 *established*—17 September 1949 *aim*—to promote mutual defense and cooperation	*members*—(16) Belgium, Canada, Denmark, France, Germany, Greece, Iceland, Italy, Luxembourg, Netherlands, Norway, Portugal, Spain, Turkey, UK, US

Nuclear Energy Agency (NEA) *address*—Le Seine St. Germain, 12 bd des Iles, F-92130 Issy-les-Moulineaux, France *telephone*—[33] (1) 45 24 10 10 *FAX*—[33] (1) 45 24 11 10 *established*—NA 1958 *aim*—to promote the peaceful uses of nuclear energy; associated with OECD	*members*—(23) Australia, Austria, Belgium, Canada, Denmark, Finland, France, Germany, Greece, Iceland, Ireland, Italy, Japan, Luxembourg, Netherlands, Norway, Portugal, Spain, Sweden, Switzerland, Turkey, UK, US
Nuclear Suppliers Group (NSG) *note*—also known as the London Suppliers Group *address*—c/o IAEA, Wagramerstrasse 5, P.O. Box 100, A-1400 Vienna, Austria *telephone*—[43] (1) 2360 2045 *FAX*—[43] (1) 234564 *established*—1974 *aim*—to establish guidelines for exports of technical information, processing equipment for uranium enrichment and nuclear materials to countries of proliferation concern and regions of conflict and instability	*members*—(28) Australia, Austria, Belgium, Bulgaria, Canada, Czech Republic, Denmark, Finland, France, Germany, Greece, Hungary, Ireland, Italy, Japan, Luxembourg, Netherlands, Norway, Poland, Portugal, Romania, Russia, Slovakia, Spain, Sweden, Switzerland, UK, US *observers*—(2) Argentina, European Commission
Organismo para la Proscripcion de las Armas Nucleares en la America Latina y el Caribe (OPANAL)	see Agency for the Prohibition of Nuclear Weapons in Latin America and the Caribbean (OPANAL)
Organization for Economic Cooperation and Development (OECD) *address*—2 rue Andre Pascal, F-75775 Paris CEDEX 16, France *telephone*—[33] (1) 45 24 82 00 *FAX*—[33] (1) 45 24 85 00, 45 24 81 76 *established*—14 December 1960 *effective*—30 September 1961 *aim*—to promote economic cooperation and development	*members*—(25) Australia, Austria, Belgium, Canada, Denmark, Finland, France, Germany, Greece, Iceland, Ireland, Italy, Japan, Luxembourg, Mexico, Netherlands, NZ, Norway, Portugal, Spain, Sweden, Switzerland, Turkey, UK, US *special member*—(1) EU
Organization of African Unity (OAU) *address*—P. O. Box 3243, Addis Ababa, Ethiopia *telephone*—[251] (1) 517700 *FAX*—[251] (1) 512622 *established*—25 May 1963 *aim*—to promote unity and cooperation among African states	*members*—(53) Algeria, Angola, Benin, Botswana, Burkina, Burundi, Cameroon, Cape Verde, Central African Republic, Chad, Comoros, Congo, Cote d'Ivoire, Djibouti, Egypt, Equatorial Guinea, Eritrea, Ethiopia, Gabon, The Gambia, Ghana, Guinea, Guinea-Bissau, Kenya, Lesotho, Liberia, Libya, Madagascar, Malawi, Mali, Mauritania, Mauritius, Mozambique, Namibia, Niger, Nigeria, Rwanda, Sahrawi Arab Democratic Republic, Sao Tome and Principe, Senegal, Seychelles, Sierra Leone, Somalia, South Africa, Sudan, Swaziland, Tanzania, Togo, Tunisia, Uganda, Zaire, Zambia, Zimbabwe
Organization of American States (OAS) *address*—corner of 17th Street and Constitution Avenue NW, Washington, DC 20006, USA *telephone*—[1] (202) 458 3000 *FAX* [1] (202) 458 3967 *established*—30 April 1948 *effective*—13 December 1951 *aim*—to promote regional peace and security as well as economic and social development	*members*—(35) Antigua and Barbuda, Argentina, The Bahamas, Barbados, Belize, Bolivia, Brazil, Canada, Chile, Colombia, Costa Rica, Cuba (excluded from formal participation since 1962), Dominica, Dominican Republic, Ecuador, El Salvador, Grenada, Guatemala, Guyana, Haiti, Honduras, Jamaica, Mexico, Nicaragua, Panama, Paraguay, Peru, Saint Kitts and Nevis, Saint Lucia, Saint Vincent and the Grenadines, Suriname, Trinidad and Tobago, US, Uruguay, Venezuela *observers*—(31) Algeria, Angola, Austria, Belgium, Central American Parliament, Commission of the European Communities, Cyprus, Egypt, Equatorial Guinea, Finland, France, Germany, Greece, Holy See, Hungary, India, Israel, Italy, Japan, South Korea, Morocco, Netherlands, Pakistan, Poland, Portugal, Romania, Russia, Saudi Arabia, Spain, Switzerland, Tunisia

Organization of Arab Petroleum Exporting Countries (OAPEC) *address*—POB 20501, Safat 13066, Kuwait *telephone*—[965] 5340713 *FAX*—[965] 5340694 *established*—9 January 1968 *aim*—to promote cooperation in the petroleum industry	*members*—(10) Algeria, Bahrain, Egypt, Iraq, Kuwait, Libya, Qatar, Saudi Arabia, Syria, UAE
Organization of Eastern Caribbean States (OECS) *address*—P.O. Box 179, The Morne, Castries, St. Lucia *telephone*—[1] (809) 452 2537 *FAX*—[1] (809) 453 1628 *established*—18 June 1981 *effective*—4 July 1981 *aim*—to promote political, economic, and defense cooperation	*members*—(7) Antigua and Barbuda, Dominica, Grenada, Montserrat, Saint Kitts and Nevis, Saint Lucia, Saint Vincent and the Grenadines *associate member*—(1) British Virgin Islands
Organization of Petroleum Exporting Countries (OPEC) *address*—Obere Donaustrasse 93, A-1020 Vienna, Austria *telephone*—[43] (1) 21 11 20 *FAX*—[43] (1) 26 43 20 *established*—14 September 1960 *aim*—to coordinate petroleum policies	*members*—(12) Algeria, Gabon, Indonesia, Iran, Iraq, Kuwait, Libya, Nigeria, Qatar, Saudi Arabia, UAE, Venezuela
Organization of the Islamic Conference (OIC) *address*—Kilo 6, Mecca Road, P.O. Box 178, Jeddah 21411, Saudi Arabia *telephone*—[966] (2) 680-0800 *FAX*—[966] (2) 687-3568 *established*—22-25 September 1969 *aim*—to promote Islamic solidarity in economic, social, cultural, and political affairs	*members*—(48 plus the Palestine Liberation Organization) Afghanistan, Albania, Algeria, Azerbaijan, Bahrain, Bangladesh, Benin, Brunei, Burkina, Cameroon, Chad, Comoros, Djibouti, Egypt, Gabon, The Gambia, Guinea, Guinea-Bissau, Indonesia, Iran, Iraq, Jordan, Kuwait, Kyrgyzstan, Lebanon, Libya, Malaysia, Maldives, Mali, Mauritania, Morocco, Niger, Oman, Pakistan, Qatar, Saudi Arabia, Senegal, Sierra Leone, Somalia, Sudan, Syria, Tajikistan, Tunisia, Turkey, Turkmenistan, Uganda, UAE, Yemen, Palestine Liberation Organization *observers*—(3) Kazakhstan, Mozambique, "Turkish Republic of Northern Cyprus"
Organization on Security and Cooperation in Europe (OSCE) *note*—formerly the Conference on Security and Cooperation in Europe (CSCE) *address*—Thunovska 12, Mala Strana, 110 00 Prague 1, Czech Republic *telephone*—[422] (2) 24311069 *FAX*—[422] (2) 24310629 *established*—1 January 1995 *aim*—to discuss issues of mutual concern and to review implementation of the Helsinki Agreement	*members*—(53) Albania, Armenia, Austria, Azerbaijan, Belarus, Belgium, Bosnia and Herzegovina, Bulgaria, Canada, Croatia, Cyprus, Czech Republic, Denmark, Estonia, Finland, France, Georgia, Germany, Greece, Holy See, Hungary, Iceland, Ireland, Italy, Kazakhstan, Kyrgyzstan, Latvia, Liechtenstein, Lithuania, Luxembourg, Malta, Moldova, Monaco, Netherlands, Norway, Poland, Portugal, Romania, Russia, San Marino, Slovakia, Slovenia, Spain, Sweden, Switzerland, Tajikistan, Turkey, Turkmenistan, Ukraine, UK, US, Uzbekistan, Yugoslavia (suspended) *observer*—(1) The Former Yugoslav Republic of Macedonia
Paris Club	see Group of 10
Partnership for Peace (PFP) *established*—10-11 January 1994 *aim*—to expand and intensify political and military cooperation throughout Europe, increase stability, diminish threats to peace, and build relationships by promoting the spirit of practical cooperation and commitment to democratic principles that underpin NATO; program under the auspices of NATO	*members*—(24) Albania, Armenia, Azerbaijan, Belarus, Bulgaria, Czech Republic, Estonia, Finland, Georgia, Hungary, Kazakhstan, Kyrgyzstan, Latvia, Lithuania, Moldova, Poland, Romania, Russia, Slovakia, Slovenia, Sweden, Turkmenistan, Ukraine, Uzbekistan

Permanent Court of Arbitration (PCA) *address*—Peace Palace, Carnegieplein 2, NL-2517 KJ The Hague, Netherlands *telephone*—[31] (70) 346 96 80 *FAX*—[31] (70) 356 13 38 *established*—29 July 1899 *aim*—to facilitate the settlement of international disputes	*members*—(80) Argentina, Australia, Austria, Belarus, Belgium, Bolivia, Brazil, Bulgaria, Burkina, Cambodia, Cameroon, Canada, Chile, China, Colombia, Cuba, Cyprus, Czech Republic, Denmark, Dominican Republic, Ecuador, Egypt, El Salvador, Fiji, Finland, France, Germany, Greece, Guatemala, Haiti, Honduras, Hungary, Iceland, India, Iran, Iraq, Israel, Italy, Japan, Jordan, Kyrgyzstan, Laos, Lebanon, Luxembourg, Malta, Mauritius, Mexico, Netherlands, NZ, Nicaragua, Nigeria, Norway, Pakistan, Panama, Paraguay, Peru, Poland, Portugal, Romania, Russia, Senegal, Singapore, Slovakia, Spain, Sri Lanka, Sudan, Suriname, Swaziland, Sweden, Switzerland, Thailand, Turkey, Uganda, Ukraine, UK, US, Uruguay, Venezuela, Zaire, Zimbabwe
Population Commission *address*—c/o ECOSOC, United Nations, New York, NY 10017, USA *telephone*—[1] (212) 754 1234 *established*—3 October 1946 *aim*—to deal with population matters of importance to the UN, as part of Economic and Social Council organization	*members*—(27) selected on a rotating basis from all regions
Rio Group (RG) *note*—formerly known as Grupo de los Ocho, established in December 1986 *established*—NA 1988 *aim*—to consult on regional Latin American issues	*members*—(11) Argentina, Bolivia, Brazil, Chile, Colombia, Ecuador, Mexico, Paraguay, Peru, Uruguay, Venezuela; *note*—Panama was expelled in 1988
Second World	another term for the traditionally Marxist-Leninist states of the USSR and Eastern Europe, with authoritarian governments and command economies based on the Soviet model; the term is fading from use; see centrally planned economies
Social Commission	see Commission for Social Development
socialist countries	in general, countries in which the government owns and plans the use of the major factors of production; *note*—the term is sometimes used incorrectly as a synonym for Communist countries
South	a popular term for the poorer, less industrialized countries generally located south of the developed countries; the counterpart of the North; see less developed countries (LDCs)
South Asian Association for Regional Cooperation (SAARC) *address*—P.O. Box 4222, Kathmandu, Nepal *telephone*—[977] (1) 221785, 221787, 221794 *FAX*—[977] (1) 227033 *established*—8 December 1985 *aim*—to promote economic, social, and cultural cooperation	*members*—(7) Bangladesh, Bhutan, India, Maldives, Nepal, Pakistan, Sri Lanka
South Pacific Commission (SPC) *address*—Anse Vata, BP D5 Noumea CEDEX, New Caledonia *telephone*—[687] 26 20 00 *FAX* [687] 26 38 18 *established*—6 February 1947 *effective*—29 July 1948 *aim*—to promote regional cooperation in economic and social matters	*members*—(27) American Samoa, Australia, Cook Islands, Fiji, France, French Polynesia, Guam, Kiribati, Marshall Islands, Federated States of Micronesia, Nauru, New Caledonia, NZ, Niue, Northern Mariana Islands, Palau, Papua New Guinea, Pitcairn Islands, Solomon Islands, Tokelau, Tonga, Tuvalu, UK, US, Vanuatu, Wallis and Futuna, Western Samoa
South Pacific Forum (SPF) *address*—c/o forum Secretariat, Ratu Sukuna Road GPO Box 856, Suva, Fiji *telephone*—[679] 312 600, 303 106 *FAX*—[679] 302 204 *established*—5 August 1971 *aim*—to promote regional cooperation in political matters	*members*—(15) Australia, Cook Islands, Fiji, Kiribati, Marshall Islands, Federated States of Micronesia, Nauru, NZ, Niue, Papua New Guinea, Solomon Islands, Tonga, Tuvalu, Vanuatu, Western Samoa *observer*—(1) Palau

South Pacific Regional Trade and Economic Cooperation Agreement (SPARTECA) *address*—(see South Pacific Forum) *established*—NA 1981 *aim*—to redress unequal trade relationship of Australia and New Zealand with small island economies in Pacific region	*members*—(15) Australia, Cook Islands, Fiji, Kiribati, Marshall Islands, Federated States of Micronesia, Nauru, NZ, Niue, Papua New Guinea, Solomon Islands, Tonga, Tuvalu, Vanuatu, Western Samoa
Southern African Customs Union (SACU) *address*—Director General, Trade and Industry, Private Bag X84, Pretoria 0001, South Africa *established*—11 December 1969 *aim*—to promote free trade and cooperation in customs matters	*members*—(9) Bophuthatswana, Botswana, Ciskei, Lesotho, Namibia, South Africa, Swaziland, Transkei, Venda
Southern African Development Community (SADC) *note*—evolved from the Southern African Development Coordination Conference (SADCC) *address*—Private Bag 008, Gaborone, Botswana *telephone*—[267] (31) 51863, 51864, 51865 *FAX*—[267] (31) 372848 *established*—17 August 1992 *aim*—to promote regional economic development and integration	*members*—(11) Angola, Botswana, Lesotho, Malawi, Mozambique, Namibia, South Africa, Swaziland, Tanzania, Zambia, Zimbabwe
Southern Cone Common Market (MERCOSUR) *address*—c/o Cancilleria de la Republica de Argentina, Buenos Aires, Argentina *established*—26 March 1991 *aim*—to increase regional economic cooperation	*members*—(4) Argentina, Brazil, Paraguay, Uruguay
Statistical Commission *address*—c/o ECOSOC, United Nations, New York, NY 10017, USA *telephone*—[1] (212) 963 1234 *FAX*—[1] (212) 758 2718 *established*—21 June 1946 *aim*—to deal with development and standardization of national statistics of interest to the UN, as part of the Economic and Social Council organization	*members*—(24) selected on a rotating basis from all regions
Third World	another term for the less developed countries; the term is fading from use; see less developed countries (LDCs)
underdeveloped countries	refers to those less developed countries with the potential for above-average economic growth; see less developed countries (LDCs)
undeveloped countries	refers to those extremely poor less developed countries (LDCs) with little prospect for economic growth; see least developed countries (LLDCs)
Union Douaniere et Economique de l'Afrique Centrale (UDEAC)	see Central African Customs and Economic Union (UDEAC)

United Nations (UN) *address*—United Nations, New York, NY 10017, USA *telephone*—[1] (212) 963 1234 *FAX*—[1] (212) 758 2718 *established*—26 June 1945 *effective*—24 October 1945 *aim*—to maintain international peace and security and to promote cooperation involving economic, social, cultural and humanitarian problems	*members*—(184 excluding Yugoslavia) Afghanistan, Albania, Algeria, Andorra, Angola, Antigua and Barbuda, Argentina, Armenia, Australia, Austria, Azerbaijan, The Bahamas, Bahrain, Bangladesh, Barbados, Belarus, Belgium, Belize, Benin, Bhutan, Bolivia, Bosnia and Herzegovina, Botswana, Brazil, Brunei, Bulgaria, Burkina, Burma, Burundi, Cambodia, Cameroon, Canada, Cape Verde, Central African Republic, Chad, Chile, China, Colombia, Comoros, Congo, Costa Rica, Cote d'Ivoire, Croatia, Cuba, Cyprus, Czech Republic, Denmark, Djibouti, Dominica, Dominican Republic, Ecuador, Egypt, El Salvador, Equatorial Guinea, Eritrea, Estonia, Ethiopia, Fiji, Finland, France, Gabon, The Gambia, Georgia, Germany, Ghana, Greece, Grenada, Guatemala, Guinea, Guinea-Bissau, Guyana, Haiti, Honduras, Hungary, Iceland, India, Indonesia, Iran, Iraq, Ireland, Israel, Italy, Jamaica, Japan, Jordan, Kazakhstan, Kenya, North Korea, South Korea, Kuwait, Kyrgyzstan, Laos, Latvia, Lebanon, Lesotho, Liberia, Libya, Liechtenstein, Lithuania, Luxembourg, The Former Yugoslav Republic of Macedonia, Madagascar, Malawi, Malaysia, Maldives, Mali, Malta, Marshall Islands, Mauritania, Mauritius, Mexico, Federated States of Micronesia, Moldova, Monaco, Mongolia, Morocco, Mozambique, Namibia, Nepal, Netherlands, NZ, Nicaragua, Niger, Nigeria, Norway, Oman, Pakistan, Palau, Panama, Papua New Guinea, Paraguay, Peru, Philippines, Poland, Portugal, Qatar, Romania, Russia, Rwanda, Saint Kitts and Nevis, Saint Lucia, Saint Vincent and the Grenadines, San Marino, Sao Tome and Principe, Saudi Arabia, Senegal, Seychelles, Sierra Leone, Singapore, Slovakia, Slovenia, Solomon Islands, Somalia, South Africa, Spain, Sri Lanka, Sudan, Suriname, Swaziland, Sweden, Syria, Tajikistan, Tanzania, Thailand, Togo, Trinidad and Tobago, Tunisia, Turkey, Turkmenistan, Uganda, Ukraine, UAE, UK, US, Uruguay, Uzbekistan, Vanuatu, Venezuela, Vietnam, Western Samoa, Yemen, Yugoslavia (suspended), Zaire, Zambia, Zimbabwe; *note*—all UN members are represented in the General Assembly *observers*—(2 plus the Palestine Liberation Organization) Holy See, Switzerland, Palestine Liberation Organization
United Nations Angola Verification Mission (UNAVEM II) *note*—successor to original UNAVEM *address*—c/o United Nations, UNAVEM II, New York, NY 10017, USA *telephone*—[1] (212) 963 1234 *FAX*—[1] (212) 758 2718 *established*—20 December 1988 *aim*—to verify the withdrawal of Cuban troops from Angola; established by the UN Security Council	*members*—(16) Argentina, Brazil, Congo, Guinea-Bissau, Hungary, India, Jordan, Malaysia, Morocco, Netherlands, NZ, Nigeria, Norway, Slovakia, Sweden, Zimbabwe
United Nations Assistance Mission for Rwanda (UNAMIR) *established*—5 October 1993 *aim*—to monitor ceasefire agreement, to support and provide safe conditions for displaced persons; established by the UN Security Council	*members*—(17) Austria, Bangladesh, Canada, Congo, Egypt, Fiji, Ghana, Malawi, Mali, Nigeria, Poland, Russia, Senegal, Togo, Tunisia, Uruguay, Zimbabwe
United Nations Children's Fund (UNICEF) *note*—acronym retained from the predecessor organization UN International Children's Emergency Fund *address*—UNICEF House, Three United Nations Plaza, New York, NY 10017, USA *telephone*—[1] (212) 326 7000 *established*—11 December 1946 *aim*—to help establish child health and welfare services	*members*—(41) selected on a rotating basis from all regions
United Nations Conference on Trade and Development (UNCTAD) *address*—Palais des Nations, CH-1211 Geneva 10, Switzerland *telephone*—[41] (22) 917 12 34, 907 12 34 *FAX*—[41] (22) 907 00 57 *established*—30 December 1964 *aim*—to promote international trade	*members*—(187) all UN members plus Holy See, Switzerland, Tonga

United Nations Development Program (UNDP)

address—One United National Plaza, New York, NY 10017, USA
telephone—[1] (212) 906 5788, 906 5000
FAX [1] (212) 906 5365
established—22 November 1965
aim—to provide technical assistance to stimulate economic and social development

members—(36) selected on a rotating basis from all regions

United Nations Disengagement Observer Force (UNDOF)

address—c/o UNDOF, P.O. Box 5368, Damascus, Syrian AR
established—31 May 1974
aim—to observe the 1973 Arab-Israeli ceasefire; established by the UN Security Council

members—(4) Austria, Canada, Finland, Poland

United Nations Educational, Scientific, and Cultural Organization (UNESCO)

address—7 place de Fontenoy, F-75700 Paris, France
telephone—[33] (1) 45 68 10 00
FAX—[33] (1) 45 67 16 90
established—16 November 1945
effective—4 November 1946
aim—to promote cooperation in education, science, and culture

members—(182) Afghanistan, Albania, Algeria, Andorra, Angola, Antigua and Barbuda, Argentina, Armenia, Australia, Austria, Azerbaijan, The Bahamas, Bahrain, Bangladesh, Barbados, Belarus, Belgium, Belize, Benin, Bhutan, Bolivia, Bosnia and Herzegovina, Botswana, Brazil, Bulgaria, Burkina, Burma, Burundi, Cambodia, Cameroon, Canada, Cape Verde, Central African Republic, Chad, Chile, China, Colombia, Comoros, Congo, Cook Islands, Costa Rica, Cote d'Ivoire, Croatia, Cuba, Cyprus, Czech Republic, Denmark, Djibouti, Dominica, Dominican Republic, Ecuador, Egypt, El Salvador, Equatorial Guinea, Eritrea, Estonia, Ethiopia, Fiji, Finland, France, Gabon, The Gambia, Georgia, Germany, Ghana, Greece, Grenada, Guatemala, Guinea, Guinea-Bissau, Guyana, Haiti, Honduras, Hungary, Iceland, India, Indonesia, Iran, Iraq, Ireland, Israel, Italy, Jamaica, Japan, Jordan, Kazakhstan, Kenya, Kiribati, North Korea, South Korea, Kuwait, Kyrgyzstan, Laos, Latvia, Lebanon, Lesotho, Liberia, Libya, Lithuania, Luxembourg, The Former Yugoslav Republic of Macedonia, Madagascar, Malawi, Malaysia, Maldives, Mali, Malta, Mauritania, Mauritius, Mexico, Moldova, Monaco, Mongolia, Morocco, Mozambique, Namibia, Nepal, Netherlands, NZ, Nicaragua, Niger, Nigeria, Niue, Norway, Oman, Pakistan, Panama, Papua New Guinea, Paraguay, Peru, Philippines, Poland, Portugal, Qatar, Romania, Russia, Rwanda, Saint Kitts and Nevis, Saint Lucia, Saint Vincent and the Grenadines, San Marino, Sao Tome and Principe, Saudi Arabia, Senegal, Seychelles, Sierra Leone, Slovakia, Slovenia, Solomon Islands, Somalia, Spain, Sri Lanka, Sudan, Suriname, Swaziland, Sweden, Switzerland, Syria, Tajikistan, Tanzania, Thailand, Togo, Tonga, Trinidad and Tobago, Tunisia, Turkey, Turkmenistan, Tuvalu, Uganda, Ukraine, UAE, Uruguay, Uzbekistan, Vanuatu, Venezuela, Vietnam, Western Samoa, Yemen, Yugoslavia (suspended), Zaire, Zambia, Zimbabwe
associate members—(3) Aruba, British Virgin Islands, Netherlands Antilles

United Nations Environment Program (UNEP)

address—One United Nations Plaza, New York, NY 10017, USA
telephone—[1] (212) 906 5000
FAX—[1] (212) 826 2057
established—15 December 1972
aim—to promote international cooperation on all environmental matters

members—(58) selected on a rotating basis from all regions

United Nations Force in Cyprus (UNFICYP)

address—c/o UN Peace Keeping Missions, Office for Special Political Affairs, United Nations, New York, NY 10017, USA
telephone—[1] (212) 963 1234
FAX—[1] (212) 963 4879
established—4 March 1964
aim—to serve as a peacekeeping force between Greek Cypriots and Turkish Cypriots in Cyprus; established by the UN Security Council

members—(8) Argentina, Australia, Austria, Canada, Denmark, Finland, Ireland, UK

United Nations General Assembly

address—see United Nations
established—26 June 1945
effective—24 October 1945
aim—to function as the primary deliberative organ of the UN

members—(185) all UN members are represented in the General Assembly

United Nations Industrial Development Organization (UNIDO)

address—Vienna International Center, P.O. Box 300, A-1400 Vienna, Austria
telephone—[43] (1) 211 310 *FAX*—[43] (1) 23 21 56
established—17 November 1966
effective—1 January 1967
aim—UN specialized agency that promotes industrial development especially among the members

members—(166) Afghanistan, Albania, Algeria, Angola, Argentina, Armenia, Australia, Austria, The Bahamas, Bahrain, Bangladesh, Barbados, Belarus, Belgium, Belize, Benin, Bhutan, Bolivia, Bosnia and Herzegovina, Botswana, Brazil, Bulgaria, Burkina, Burma, Burundi, Cameroon, Canada, Cape Verde, Central African Republic, Chad, Chile, China, Colombia, Comoros, Congo, Costa Rica, Cote d'Ivoire, Croatia, Cuba, Cyprus, Czech Republic, Denmark, Djibouti, Dominica, Dominican Republic, Ecuador, Egypt, El Salvador, Equatorial Guinea, Ethiopia, Fiji, Finland, France, Gabon, The Gambia, Georgia, Germany, Ghana, Greece, Grenada, Guatemala, Guinea, Guinea-Bissau, Guyana, Haiti, Honduras, Hungary, India, Indonesia, Iran, Iraq, Ireland, Israel, Italy, Jamaica, Japan, Jordan, Kenya, North Korea, South Korea, Kuwait, Kyrgyzstan, Laos, Latvia, Lebanon, Lesotho, Liberia, Libya, Luxembourg, The Former Yugoslav Republic of Macedonia, Madagascar, Malawi, Malaysia, Maldives, Mali, Malta, Mauritania, Mauritius, Mexico, Moldova, Mongolia, Morocco, Mozambique, Namibia, Nepal, Netherlands, NZ, Nicaragua, Niger, Nigeria, Norway, Oman, Pakistan, Panama, Papua New Guinea, Paraguay, Peru, Philippines, Poland, Portugal, Qatar, Romania, Russia, Rwanda, Saint Kitts and Nevis, Saint Lucia, Saint Vincent and the Grenadines, Sao Tome and Principe, Saudi Arabia, Senegal, Seychelles, Sierra Leone, Slovakia, Slovenia, Somalia, Spain, Sri Lanka, Sudan, Suriname, Swaziland, Sweden, Switzerland, Syria, Tajikistan, Tanzania, Thailand, Togo, Tonga, Trinidad and Tobago, Tunisia, Turkey, Uganda, Ukraine, UAE, UK, US, Uruguay, Uzbekistan, Vanuatu, Venezuela, Vietnam, Yemen, Yugoslavia (suspended), Zaire, Zambia, Zimbabwe

United Nations Institute for Training and Research (UNITAR)

established—11 December 1963
aim—to help the UN become more effective through training and research

members (Board of Trustees)—(24) Argentina, Belgium, Canada, China, Cote d'Ivoire, France, Germany, India, Italy, Jamaica, Japan, Libya, Mexico, Netherlands, Norway, Pakistan, Russia, Sweden, Switzerland, Tunisia, Uganda, UK, US, Yugoslavia

United Nations Interim Force in Lebanon (UNIFIL)

address—c/o UN Peace Keeping Missions, Office for Special Political Affairs, United Nations, New York, NY 10017, USA
telephone—[1] (212) 963 1234
FAX—[1] (212) 963 4879
established—19 March 1978
aim—to confirm the withdrawal of Israeli forces, restore peace, and reestablish Lebanese authority in southern Lebanon; established by the UN Security Council

members—(9) Fiji, Finland, France, Ghana, Ireland, Italy, Nepal, Norway, Poland

United Nations Iraq-Kuwait Observation Mission (UNIKOM)

address—c/o UN Peace Keeping Mission, Office for Special Political Affairs, United Nations, New York, NY 10017, USA
telephone [1] (212) 963 1234
FAX—[1] (212) 963 4879
established—9 April 1991
aim—to observe and monitor the demilitarized zone established between Iraq and Kuwait; established by the UN Security Council

members—(33) Argentina, Austria, Bangladesh, Canada, China, Denmark, Fiji, Finland, France, Ghana, Greece, Hungary, India, Indonesia, Ireland, Italy, Kenya, Malaysia, Nigeria, Norway, Pakistan, Poland, Romania, Russia, Senegal, Singapore, Sweden, Thailand, Turkey, UK, US, Uruguay, Venezuela

United Nations Military Observer Group in India and Pakistan (UNMOGIP)

address—c/o OUSGSPA, Room 3853, United Nations, New York, NY 10017, USA
telephone—[1] (212) 963 4457
FAX—[1] (212) 758 2718
established—13 August 1948
aim—to observe the 1949 India-Pakistan ceasefire; established by the UN Security Council

members—(8) Belgium, Chile, Denmark, Finland, Italy, Norway, Sweden, Uruguay

United Nations Mission in Haiti (UNMIH)

established—23 September 1993
aim—to assist in implementing the agreement to transfer power back into the civilian government; established by the UN Security Council

members—(14) Algeria, Argentina, Austria, Canada, France, Indonesia, Madagascar, Russia, Senegal, Spain, Switzerland, Tunisia, US, Venezuela

United Nations Mission in Liberia (UNOMIL)

established—22 September 1993
aim—to assist in the implementation of the peace agreement; established by the UN Security Council

members—(13) Austria, Bangladesh, China, Czech Republic, Egypt, Guinea-Bissau, India, Jordan, Kenya, Malaysia, Pakistan, Slovakia, Uruguay

United Nations Mission for the Referendum in Western Sahara (MINURSO)

address—c/o UN Peace Keeping Missions, Office for Special Political Affairs, United Nations, New York, NY 10017, USA
telephone—[1] (212) 963 1234
FAX—[1] (212) 963 4879
established—29 April 1991
aim—to supervise the referendum in Western Sahara; established by the UN Security Council

members—(27) Argentina, Australia, Austria, Bangladesh, Belgium, Canada, China, Egypt, France, Germany, Ghana, Greece, Guinea, Honduras, Ireland, Italy, Kenya, Malaysia, Nigeria, Pakistan, Poland, Russia, Switzerland, Togo, Tunisia, US, Venezuela

United Nations Observer Mission in El Salvador (ONUSAL)

address—c/o UN Peace Keeping Missions, Office for Special Political Affairs, United Nations, New York, NY 10017, USA
telephone—[1] (212) 963 1234
FAX—[1] (212) 963 4879
established—20 May 1991
aim—to verify ceasefire arrangements and to monitor the maintenance of public order pending the organization of a new National Civil Police; established by the UN Security Council

members—(14) Argentina, Austria, Brazil, Canada, Chile, Colombia, France, Guyana, Ireland, Italy, Mexico, Spain, Sweden, Venezuela

United Nations Observer Mission in Georgia (UNOMIG)

address—c/o UN Peace Keeping Missions, Office for Special Political Affairs, United Nations, New York, NY 10017, USA
telephone—[1] (212) 963 1234
FAX—[1] (212) 963 4879
established—1993 for a period of six months
aim—to verify compliance with the cease-fire agreement reached 27 July 1993 and investigate reports of violations of that agreement; established by the UN Security Council

members—(7) Bangladesh, Denmark, Germany, Hungary, Poland, Sweden, Switzerland

United Nations Observer Mission Uganda-Rwanda (UNOMUR)

address—c/o UN Peace Keeping Missions, Office for Special Political Affairs, United Nations, New York, NY 10017, USA
telephone—[1] (212) 963 1234
FAX—[1] (212) 963 4879
established—1993 for six months
aim—to monitor the Uganda/Rwanda border to verify that no military assistance reaches Rwanda across the border; established by the UN Security Council

members—(8) Bangladesh, Botswana, Brazil, Hungary, Netherlands, Senegal, Slovakia, Zimbabwe

United Nations Office of the High Commissioner for Refugees (UNHCR)

address—Case postale 2500, Depot, CH-1211 Geneva, Switzerland
telephone—[41] (22) 739 81 11
FAX—[41] (22) 731 95 46
established—3 December 1949
effective—1 January 1951
aim—to ensure the humanitarian treatment of refugees and find permanent solutions to refugee problems

members—(47) Algeria, Argentina, Australia, Austria, Belgium, Brazil, Canada, China, Colombia, Denmark, Ethiopia, Finland, France, Germany, Greece, Holy See, Hungary, Iran, Israel, Italy, Japan, Lebanon, Lesotho, Madagascar, Morocco, Namibia, Netherlands, Nicaragua, Nigeria, Norway, Pakistan, Philippines, Somalia, Spain, Sudan, Sweden, Switzerland, Tanzania, Thailand, Tunisia, Turkey, Uganda, UK, US, Venezuela, Yugoslavia, Zaire

United Nations Operation in Mozambique (UNOMOZ)

note—supposed to shut down 31 January 1995
address—c/o UN Peace Keeping Missions, Office for Special Political Affairs, United Nations, New York, NY 10017, USA
telephone—[1] (212) 963 1234
FAX—[1] (212) 963 4879
established—16 December 1992
aim—to supervise the ceasefire; established by the UN Security Council

members—(27) Argentina, Austria, Bangladesh, Botswana, Brazil, Canada, Cape Verde, China, Czech Republic, Egypt, Guinea-Bissau, Hungary, India, Ireland, Italy, Japan, Jordan, Malaysia, Netherlands, Norway, Portugal, Russia, Spain, Sweden, US, Uruguay, Zambia

United Nations Operation in Somalia (UNOSOM II) *note*—UN peacekeepers left Somalia on 2 March 1995; some personnel remain in Somalia engaged in humanitarian work *address*—c/o UN Peace Keeping Missions, Office for Special Political Affairs, United Nations, New York, NY 10017, USA *telephone*—[1] (212) 963 1234 *FAX*—[1] (212) 963 4879 *established*—24 April 1992 *aim*—to facilitate an immediate cessation of hostilities, to maintain a ceasefire in order to promote a political settlement, and to provide urgent humanitarian assistance; established by the UN Security Council	*members*—(14) Australia, Bangladesh, Botswana, Canada, Egypt, India, Ireland, Malaysia, Nepal, NZ, Nigeria, Pakistan, Romania, Zimbabwe
United Nations Population Fund (UNFPA) *note*—acronym retained from predecessor organization UN Fund for Population Activities *address*—220 E. 42nd Street, 19th Floor, Room DN-1901, New York, NY 10017, USA *telephone*—[1] (212) 297 5000 *FAX*—[1] (212) 557 6416 *established*—NA July 1967 *aim*—to assist both developed and developing countries to deal with their population problems	*members*—(52) selected on a rotating basis from all regions
United Nations Protection Force (UNPROFOR) *address*—c/o UN Peace Keeping Mission, Office for Special Political Affairs, United Nations, New York, NY 10017, USA *telephone*—[1] (212) 963 1234 *FAX*—[1] (212) 963 4879 *established*—28 February 1992 *aim*—to create conditions for peace and security required for the negotiation of an overall settlement of the "Yugoslav" crisis; established by the UN Security Council	*members*—(35) Argentina, Bangladesh, Belgium, Brazil, Canada, Colombia, Czech Republic, Denmark, Egypt, Finland, France, Ghana, Indonesia, Ireland, Jordan, Kenya, Malaysia, Nepal, Netherlands, NZ, Nigeria, Norway, Pakistan, Poland, Portugal, Russia, Slovakia, Spain, Sweden, Switzerland, Tunisia, Ukraine, UK, US, Venezuela
United Nations Relief and Works Agency for Palestine Refugees in the Near East (UNRWA) *address*—Vienna International Center, P. O. Box 700, A-1400 Vienna, Austria *telephone*—[43] (1) 211 31, ext. 4530 *FAX*—[43] (1) 230 7487 *established*—8 December 1949 *aim*—to provide assistance to Palestinian refugees	*members*—(10) Belgium, Egypt, France, Japan, Jordan, Lebanon, Syria, Turkey, UK, US
United Nations Research Institute for Social Development (UNRISD) *established*—1 July 1964 *aim*—to conduct research into the problems of economic development during different phases of economic growth	*members*—no country members, but a Board of Directors consisting of a chairman appointed by the UN secretary general and 10 individual members

United Nations Secretariat *address*—see United Nations *established*—26 June 1945 *effective*—24 October 1945 *aim*—to serve as the primary administrative organ of the UN; a Secretary General is appointed for a five-year term by the General Assembly on the recommendation of the Security Council	*members*—the UN secretary general and staff
United Nations Security Council *address*—c/o United Nations, New York, NY 10017, USA *telephone*—[1] (212) 963 1234 *FAX*—[1] (212) 758 2718 *established*—26 June 1945 *effective*—24 October 1945 *aim*—to maintain international peace and security	*permanent members*—(5) China, France, Russia, UK, US *nonpermanent members*—(10) elected for two-year terms by the UN General Assembly; Argentina (1994-95), Brazil (1993-94), Czech Republic (1994-95), Djibouti (1993-94), NZ (1993-94), Nigeria (1994-95), Oman (1994-95), Pakistan (1993-94), Rwanda (1994-95), Spain (1993-94)
United Nations Transitional Authority in Cambodia (UNTAC)	established by the UN Security Council on 28 February 1992 to contribute to the restoration and maintenance of peace and to the holding of free elections; disbanded sometime after the UN-supervised election in May 1993; members were Algeria, Argentina, Australia, Austria, Bangladesh, Belgium, Brunei, Bulgaria, Cameroon, Canada, Chile, China, Colombia, Egypt, Fiji, France, Germany, Ghana, Hungary, India, Indonesia, Ireland, Italy, Japan, Jordan, Kenya, Malaysia, Morocco, Nepal, Netherlands, NZ, Nigeria, Norway, Pakistan, Philippines, Poland, Russia, Senegal, Singapore, Sweden, Thailand, Tunisia, UK, US, Uruguay
United Nations Truce Supervision Organization (UNTSO) *address*—Government House, P.O. Box 490, Jerusalem, Israel *telephone*—[972] (2) 734 223 *established*—NA May 1948 *aim*—to supervise the 1948 Arab-Israeli ceasefire and subsequently extended to work in the Sinai, Lebanon, Jordan, Afghanistan, and Pakistan; initially established by the UN Security Council	*members*—(19) Argentina, Australia, Austria, Belgium, Canada, Chile, China, Denmark, Finland, France, Ireland, Italy, Netherlands, NZ, Norway, Russia, Sweden, Switzerland, US
United Nations Trusteeship Council	established on 26 June 1945, effective on 24 October 1945, to supervise the administration of the 11 UN trust territories; members were China, France, Russia, UK, US; its mandate ended on 1 October 1994 when the Trust Territory of the Pacific Islands (Palau) became the Republic of Palau, a self-governing territory in free association with the US
United Nations University (UNU) *established*—6 December 1973 *aim*—to conduct research in development, welfare, and human survival and to train scholars	*members (associated institutes)*—(32) Argentina, Australia, Bangladesh, Brazil, Canada, Chile, China, Colombia, Costa Rica, Ethiopia, France, Ghana, Guatemala, Hungary, Iceland, India, Japan, Kenya, South Korea, Mexico, Netherlands, Nigeria, Philippines, Spain, Sri Lanka, Sudan, Switzerland, Thailand, Trinidad and Tobago, UK, US, Venezuela

Universal Postal Union (UPU) *address*—Bureau International de l'UPU, Weltpoststrasse 4, CH-3000 Berne 15, Switzerland *telephone*—[41] (31) 350 31 11 *FAX*—[41] (31) 350 31 10 *established*—9 October 1874, affiliated with the UN 15 November 1947 *effective*—1 July 1948 *aim*—to promote international postal cooperation; a UN specialized agency	*members*—(189) Afghanistan, Albania, Algeria, Angola, Antigua and Barbuda, Argentina, Armenia, Australia, Austria, Azerbaijan, The Bahamas, Bahrain, Bangladesh, Barbados, Belarus, Belgium, Belize, Benin, Bhutan, Bolivia, Bosnia and Herzegovina, Botswana, Brazil, Brunei, Bulgaria, Burkina, Burma, Burundi, Cambodia, Cameroon, Canada, Cape Verde, Central African Republic, Chad, Chile, China, Colombia, Comoros, Congo, Costa Rica, Cote d'Ivoire, Croatia, Cuba, Cyprus, Czech Republic, Denmark, Djibouti, Dominica, Dominican Republic, Ecuador, Egypt, El Salvador, Equatorial Guinea, Eritrea, Estonia, Ethiopia, Fiji, Finland, France, Gabon, The Gambia, Georgia, Germany, Ghana, Greece, Grenada, Guatemala, Guinea, Guinea-Bissau, Guyana, Haiti, Holy See, Honduras, Hungary, Iceland, India, Indonesia, Iran, Iraq, Ireland, Israel, Italy, Jamaica, Japan, Jordan, Kazakhstan, Kenya, Kiribati, North Korea, South Korea, Kuwait, Kyrgyzstan, Laos, Latvia, Lebanon, Lesotho, Liberia, Libya, Liechtenstein, Lithuania, Luxembourg, The Former Yugoslav Republic of Macedonia, Madagascar, Malawi, Malaysia, Maldives, Mali, Malta, Mauritania, Mauritius, Mexico, Moldova, Monaco, Mongolia, Morocco, Mozambique, Namibia, Nauru, Nepal, Netherlands, Netherlands Antilles, NZ, Nicaragua, Niger, Nigeria, Norway, Oman, Overseas Territories of the UK, Pakistan, Panama, Papua New Guinea, Paraguay, Peru, Philippines, Poland, Portugal, Qatar, Romania, Russia, Rwanda, Saint Kitts and Nevis, Saint Lucia, Saint Vincent and the Grenadines, San Marino, Sao Tome and Principe, Saudi Arabia, Senegal, Seychelles, Sierra Leone, Singapore, Slovakia, Slovenia, Solomon Islands, Somalia, South Africa, Spain, Sri Lanka, Sudan, Suriname, Swaziland, Sweden, Switzerland, Syria, Tajikistan, Tanzania, Thailand, Togo, Tonga, Trinidad and Tobago, Tunisia, Turkey, Turkmenistan, Tuvalu, Uganda, Ukraine, UAE, UK, US, Uruguay, Uzbekistan, Vanuatu, Venezuela, Vietnam, Western Samoa, Yemen, Yugoslavia (suspended), Zaire, Zambia, Zimbabwe
Warsaw Pact (WP)	established 14 May 1955 to promote mutual defense; members met 1 July 1991 to dissolve the alliance; member states at the time of dissolution were Bulgaria, Czechoslovakia, Hungary, Poland, Romania, and the USSR; earlier members included East Germany and Albania
West African Development Bank (WADB) *note*—also known as Banque Ouest-Africaine de Developpement (BOAD) *address*—BOAD, BP 1172, 68 av de la liberation, Lome, Togo *telephone*—[228] 21 59 06, 21 42 44, 21 01 13 *FAX*—[228] 21 52 67, 21 72 69 *established*—14 November 1973 *aim*—to promote regional economic development and integration	*members*—(7) Benin, Burkina, Cote d'Ivoire, Mali, Niger, Senegal, Togo
West African Economic Community (CEAO) *note*—acronym from Communaute Economique l'Afrique de l'Ouest	established on 3 June 1972 to promote regional economic development; its members were Benin, Burkina, Cote d'Ivoire, Mali, Mauritania, Niger, Senegal; it was disbanded in 1994
Western European Union (WEU) *address*—Rue de la Regence 4, B-1000 Brussels, Belgium *telephone*—[32] (2) 500 44 11 *FAX*—[32] (2) 511 35 19 *established*—23 October 1954 *effective*—6 May 1955 *aim*—to provide mutual defense and to move toward political unification	*members*—(10) Belgium, France, Germany, Greece, Italy, Luxembourg, Netherlands, Portugal, Spain, UK *associate members*—(3) Iceland, Norway, Turkey *associate partners*—(9) Bulgaria, Czech Republic, Estonia, Hungary, Latvia, Lithuania, Poland, Romania, Slovakia *observers*—(2) Denmark, Ireland
World Bank	see International Bank for Reconstruction and Development (IBRD)
World Bank Group	includes International Bank for Reconstruction and Development (IBRD), International Development Association (IDA), and International Finance Corporation (IFC)

World Confederation of Labor (WCL) *address*—Rue de Treves 33, B-1040 Brussels, Belgium *telephone*—[32] (2) 230 62 95 *FAX*—[32] (2) 230 87 22 *established*—19 June 1920 as the International Federation of Christian Trade Unions (IFCTU), renamed 4 October 1968 *aim*—to promote the trade union movement	*members*—(99 national organizations) Algeria, Angola, Antigua and Barbuda, Argentina, Aruba, Austria, Bangladesh, Belgium, Belize, Benin, Bolivia, Bonaire Island, Botswana, Brazil, Burkina, Cameroon, Canada, Cape Verde, Central African Republic, Chad, Chile, Colombia, Costa Rica, Cote d'Ivoire, Cuba, Curacao, Cyprus, Dominica, Dominican Republic, Ecuador, El Salvador, France, French Guiana, Gabon, The Gambia, Ghana, Grenada, Guadeloupe, Guatemala, Guinea, Guyana, Haiti, Honduras, Hong Kong, Indonesia, Iran, Italy, Jamaica, Kenya, Lesotho, Liberia, Liechtenstein, Luxembourg, Madagascar, Malaysia, Mali, Malta, Martinique, Mauritius, Mexico, Montserrat, Namibia, Netherlands, Nicaragua, Niger, Nigeria, Pakistan, Panama, Paraguay, Peru, Philippines, Poland, Portugal, Puerto Rico, Romania, Rwanda, Saint Kitts and Nevis, Saint Lucia, Saint Martin, Saint Vincent and the Grenadines, Senegal, Seychelles, Sierra Leone, Spain, Sri Lanka, Suriname, Switzerland, Taiwan, Tanzania, Thailand, Togo, UK, US, Uruguay, Venezuela, Vietnam, Zaire, Zambia, Zimbabwe
World Court	see International Court of Justice (ICJ)
World Federation of Trade Unions (WFTU) *address*—Branicka 112, Branik, CS-14700 Prague 4, Czech Republic *telephone*—[42] (2) 46 21 40 *FAX*—[42] (2) 46 13 78 *established*—3 October 1945 *aim*—to promote the trade union movement	*members*—(116) Afghanistan, Albania, Angola, Antigua and Barbuda, Argentina, Australia, Austria, Bahrain, Bangladesh, Benin, Bolivia, Botswana, Brazil, Bulgaria, Burkina, Cambodia, Cameroon, Canada, Chile, Colombia, Congo, Costa Rica, Cote d'Ivoire, Cuba, Cyprus, Czech Republic, Denmark, Djibouti, Dominican Republic, Ecuador, Egypt, El Salvador, Eritrea, Ethiopia, Fiji, Finland, France, French Guiana, The Gambia, Ghana, Greece, Guadeloupe, Guatemala, Guinea, Guinea-Bissau, Guyana, Haiti, Honduras, Hungary, India, Indonesia, Iran, Iraq, Jamaica, Japan, Jordan, North Korea, Kuwait, Laos, Lebanon, Lesotho, Liberia, Libya, Madagascar, Malawi, Malaysia, Mali, Martinique, Mauritius, Mexico, Mongolia, Mozambique, Nepal, New Caledonia, NZ, Nicaragua, Niger, Nigeria, Oman, Pakistan, Panama, Papua New Guinea, Peru, Philippines, Poland, Portugal, Puerto Rico, Reunion, Romania, Saint Lucia, Saint Pierre and Miquelon, Saint Vincent and the Grenadines, Saudi Arabia, Senegal, Sierra Leone, Solomon Islands, Somalia, South Africa, Spain, Sri Lanka, Sudan, Sweden, Syria, Tanzania, Thailand, Togo, Trinidad and Tobago, Tunisia, Turkey, Uganda, Uruguay, Vanuatu, Venezuela, Vietnam, Yemen, Zaire, Zimbabwe
World Food Council (WFC) *address*—c/o FAO, Via Terme di Caracalla, I-00100 Rome, Italy *telephone*—[39] (6) 522821 *FAX*—[39] (6) 574 5091 *established*—17 December 1974 *aim*—to study world food problems and to recommend solutions; ECOSOC organization	*members*—(36) selected on a rotating basis from all regions
World Food Program (WFP) *address*—Via Cristoforo Colombo 426, I-00145 Rome, Italy *telephone*—[39] (6) 522821 *FAX*—[39] (6) 5123700, 5133537, 52282840 *established*—24 November 1961 *aim*—to provide food aid in support of economic development or disaster relief; an ECOSOC organization	*members*—(41) selected on a rotating basis from all regions

World Health Organization (WHO)

address—CH-1211 Geneva 27, Switzerland
telephone—[41] (22) 791 21 11, 791 32 23
FAX—[41] (22) 791 07 46
established—22 July 1946
effective—7 April 1948
aim—UN specialized agency concerned with health matters

members—(189) Afghanistan, Albania, Algeria, Angola, Antigua and Barbuda, Argentina, Armenia, Australia, Austria, Azerbaijan, The Bahamas, Bahrain, Bangladesh, Barbados, Belarus, Belgium, Belize, Benin, Bhutan, Bolivia, Bosnia and Herzegovina, Botswana, Brazil, Brunei, Bulgaria, Burkina, Burma, Burundi, Cambodia, Cameroon, Canada, Cape Verde, Central African Republic, Chad, Chile, China, Colombia, Comoros, Congo, Cook Islands, Costa Rica, Cote d'Ivoire, Croatia, Cuba, Cyprus, Czech Republic, Denmark, Djibouti, Dominica, Dominican Republic, Ecuador, Egypt, El Salvador, Equatorial Guinea, Estonia, Ethiopia, Fiji, Finland, France, Gabon, The Gambia, Georgia, Germany, Ghana, Greece, Grenada, Guatemala, Guinea, Guinea-Bissau, Guyana, Haiti, Honduras, Hungary, Iceland, India, Indonesia, Iran, Iraq, Ireland, Israel, Italy, Jamaica, Japan, Jordan, Kazakhstan, Kenya, Kiribati, North Korea, South Korea, Kuwait, Kyrgyzstan, Laos, Latvia, Lebanon, Lesotho, Liberia, Libya, Lithuania, Luxembourg, The Former Yugoslav Republic of Macedonia, Madagascar, Malawi, Malaysia, Maldives, Mali, Malta, Marshall Islands, Mauritania, Mauritius, Mexico, Federated States of Micronesia, Moldova, Monaco, Mongolia, Morocco, Mozambique, Namibia, Nepal, Netherlands, NZ, Nicaragua, Niue, Niger, Nigeria, Norway, Oman, Pakistan, Panama, Papua New Guinea, Paraguay, Peru, Philippines, Poland, Portugal, Qatar, Romania, Russia, Rwanda, Saint Kitts and Nevis, Saint Lucia, Saint Vincent and the Grenadines, San Marino, Sao Tome and Principe, Saudi Arabia, Senegal, Seychelles, Sierra Leone, Singapore, Slovakia, Slovenia, Solomon Islands, Somalia, South Africa, Spain, Sri Lanka, Sudan, Suriname, Swaziland, Sweden, Switzerland, Syria, Tajikistan, Tanzania, Thailand, Togo, Tonga, Trinidad and Tobago, Tunisia, Turkey, Turkmenistan, Tuvalu, Uganda, Ukraine, UAE, UK, US, Uruguay, Uzbekistan, Vanuatu, Venezuela, Vietnam, Western Samoa, Yemen, Yugoslavia (suspended), Zaire, Zambia, Zimbabwe
associate members—(2) Puerto Rico, Tokelau

World Intellectual Property Organization (WIPO)

address—34 chemin des Colombettes, Case Postale 18, CH-1211 Geneva 20, Switzerland
telephone—[41] (22) 730 9111
FAX—[41] (22) 733 5428
established—14 July 1967
effective—26 April 1970
aim—to furnish protection for literary, artistic, and scientific works; a UN specialized agency

members—(147) Albania, Algeria, Angola, Argentina, Armenia, Australia, Austria, The Bahamas, Bangladesh, Barbados, Belarus, Belgium, Benin, Bhutan, Bolivia, Bosnia and Herzegovina, Brazil, Brunei, Bulgaria, Burkina, Burundi, Cameroon, Canada, Central African Republic, Chad, Chile, China, Colombia, Congo, Costa Rica, Cote d'Ivoire, Croatia, Cuba, Cyprus, Czech Republic, Denmark, Ecuador, Egypt, El Salvador, Estonia, Fiji, Finland, France, Gabon, The Gambia, Georgia, Germany, Ghana, Greece, Guatemala, Guinea, Guinea-Bissau, Haiti, Holy See, Honduras, Hungary, Iceland, India, Indonesia, Iraq, Ireland, Israel, Italy, Jamaica, Japan, Jordan, Kazakhstan, Kenya, North Korea, South Korea, Kyrgyzstan, Latvia, Lebanon, Lesotho, Liberia, Libya, Liechtenstein, Lithuania, Luxembourg, The Former Yugoslav Republic of Macedonia, Madagascar, Malawi, Malaysia, Maldives, Mali, Malta, Mauritania, Mauritius, Mexico, Moldova, Monaco, Mongolia, Morocco, Namibia, Netherlands, NZ, Nicaragua, Niger, Norway, Pakistan, Panama, Paraguay, Peru, Philippines, Poland, Portugal, Qatar, Romania, Russia, Rwanda, Saint Lucia, San Marino, Saudi Arabia, Senegal, Sierra Leone, Singapore, Slovakia, Slovenia, Somalia, South Africa, Spain, Sri Lanka, Sudan, Suriname, Swaziland, Sweden, Switzerland, Tajikistan, Tanzania, Thailand, Togo, Trinidad and Tobago, Tunisia, Turkey, Uganda, Ukraine, UAE, UK, US, Uruguay, Uzbekistan, Venezuela, Vietnam, Yemen, Yugoslavia (suspended), Zaire, Zambia, Zimbabwe

World Meteorological Organization (WMO)

address—Case Postale 2300, 41 Av Giuseppe-Motta, CH-1211 Geneva 2, Switzerland
telephone—[41] (22) 730 81 11
FAX—[41] (22) 734 23 26
established—11 October 1947
effective—4 April 1951
aim—to sponsor meteorological cooperation; a specialized UN agency

members—(175) Afghanistan, Albania, Algeria, Angola, Antigua and Barbuda, Argentina, Armenia, Australia, Austria, Azerbaijan, The Bahamas, Bahrain, Bangladesh, Barbados, Belarus, Belgium, Belize, Benin, Bolivia, Botswana, Brazil, British Caribbean Territories, Brunei, Bulgaria, Burkina, Burma, Burundi, Cambodia, Cameroon, Canada, Cape Verde, Central African Republic, Chad, Chile, China, Colombia, Comoros, Congo, Costa Rica, Cote d'Ivoire, Croatia, Cuba, Cyprus, Czech Republic, Denmark, Djibouti, Dominica, Dominican Republic, Ecuador, Egypt, El Salvador, Eritrea, Estonia, Ethiopia, Fiji, Finland, France, French Polynesia, Gabon, The Gambia, Georgia, Germany, Ghana, Greece, Guatemala, Guinea, Guinea-Bissau, Guyana, Haiti, Honduras, Hong Kong, Hungary, Iceland, India, Indonesia, Iran, Iraq, Ireland, Israel, Italy, Jamaica, Japan, Jordan, Kazakhstan, Kenya, North Korea, South Korea, Kuwait, Laos, Latvia, Lebanon, Lesotho, Liberia, Libya, Lithuania, Luxembourg, The Former Yugoslav Republic of Macedonia, Madagascar, Malawi, Malaysia, Maldives, Mali, Malta, Mauritania, Mauritius, Mexico, Mongolia, Morocco, Mozambique, Namibia, Nepal, Netherlands, Netherlands Antilles, New Caledonia, NZ, Nicaragua, Niger, Nigeria, Norway, Oman, Pakistan, Panama, Papua New Guinea, Paraguay, Peru, Philippines, Poland, Portugal, Qatar, Romania, Russia, Rwanda, Saint Lucia, Sao Tome and Principe, Saudi Arabia, Senegal, Seychelles, Sierra Leone, Singapore, Slovakia, Slovenia, Solomon Islands, Somalia, South Africa, Spain, Sri Lanka, Sudan, Suriname, Swaziland, Sweden, Switzerland, Syria, Tajikistan, Tanzania, Thailand, Togo, Trinidad and Tobago, Tunisia, Turkey, Turkmenistan, Uganda, Ukraine, UAE, UK, US, Uruguay, Uzbekistan, Vanuatu, Venezuela, Vietnam, Yemen, Yugoslavia (suspended), Zaire, Zambia, Zimbabwe

World Tourism Organization (WToO) *address*—Calle Capitan Haya 41, 28020 Madrid, Spain *telephone*—[34] (1) 571 06 28 *FAX*—[34] (1) 571 37 33 *established*—2 January 1975 *aim*—to promote tourism as a means of contributing to economic development, international understanding, and peace	*members*—(121) Afghanistan, Albania, Algeria, Angola, Argentina, Austria, Bangladesh, Belgium, Benin, Bolivia, Bosnia and Herzegovina, Brazil, Bulgaria, Burkina, Burundi, Cambodia, Cameroon, Canada, Chad, Chile, China, Colombia, Congo, Cote d'Ivoire, Croatia, Cuba, Cyprus, Czech Republic, Dominican Republic, Ecuador, Egypt, El Salvador, Ethiopia, Finland, France, Gabon, The Gambia, Georgia, Germany, Ghana, Greece, Grenada, Guatemala, Guinea, Guinea-Bissau, Haiti, Hungary, India, Indonesia, Iran, Iraq, Israel, Italy, Jamaica, Japan, Jordan, Kazakhstan, Kenya, North Korea, South Korea, Kuwait, Kyrgyzstan, Laos, Lebanon, Lesotho, Libya, Madagascar, Malawi, Malaysia, Maldives, Mali, Malta, Mauritania, Mauritius, Mexico, Moldova, Mongolia, Morocco, Nepal, Netherlands, Nicaragua, Niger, Nigeria, Pakistan, Panama, Paraguay, Peru, Poland, Portugal, Romania, Russia, Rwanda, San Marino, Sao Tome and Principe, Senegal, Seychelles, Sierra Leone, Slovakia, Slovenia, Spain, Sri Lanka, Sudan, Switzerland, Syria, Tanzania, Togo, Tunisia, Turkey, Turkmenistan, Uganda, UAE, US, Uruguay, Uzbekistan, Venezuela, Vietnam, Yemen, Yugoslavia (suspended), Zaire, Zambia, Zimbabwe *associate members*—(4) Aruba, Macau, Netherlands Antilles, Puerto Rico *observer*—(1) Holy See
World Trade Organization (WTrO)	will be added in *The World Factbook 1996*
Zangger Committee (ZC) *established*—early 1970s *aim*—to establish guidelines for the export control provisions of the nuclear Non-Proliferation Treaty	*members*—(29) Australia, Austria, Belgium, Bulgaria, Canada, Czech Republic, Denmark, Finland, France, Germany, Greece, Hungary, Ireland, Italy, Japan, Luxembourg, Netherlands, Norway, Poland, Portugal, Romania, Russia, Slovakia, South Africa, Spain, Sweden, Switzerland, UK, US

Appendix D:

Abbreviations for Selected International Environmental Agreements

A	Air Pollution	Convention on Long-Range Transboundary Air Pollution
	Air Pollution-Nitrogen Oxides	Protocol to the 1979 Convention on Long-Range Transboundary Air Pollution concerning the Control of Emissions of Nitrogen Oxides or Their Transboundary Fluxes
	Air Pollution-Sulphur 85	Protocol to the 1979 Convention on Long-Range Transboundary Air Pollution on the Reduction of Sulphur Emissions or their Transboundary Fluxes by at least 30%
	Air Pollution-Sulphur 94	Protocol to the 1979 Convention on Long-Range Transboundary Air Pollution on Further Reduction of Sulphur Emissions
	Air Pollution-Volatile Organic Compounds	Protocol to the 1979 Convention on Long-Range Transboundary Air Pollution concerning the Control of Emissions of Volatile Organic Compounds or Their Transboundary Fluxes
	Antarctic-Environmental Protocol	Protocol on Environmental Protection to the Antarctic Treaty
B	Biodiversity	Convention on Biological Diversity
C	Climate Change	United Nations Framework Convention on Climate Change
D	Desertification	United Nations Convention to Combat Desertification in those Countries Experiencing Serious Drought and/or Desertification, Particularly in Africa
E	Endangered Species	Convention on the International Trade in Endangered Species of Wild Flora and Fauna (CITES)
	Environmental Modification	Convention on the Prohibition of Military or Any Other Hostile Use of Environmental Modification Techniques
H	Hazardous Wastes	Basel Convention on the Control of Transboundary Movements of Hazardous Wastes and Their Disposal
L	Law of the Sea	United Nations Convention on the Law of the Sea (LOS)
M	Marine Dumping	Convention on the Prevention of Marine Pollution by Dumping Wastes and Other Matter; note—also known as the London Convention
	Marine Life Conservation	Convention on Fishing and Conservation of Living Resources of the High Seas
N	Nuclear Test Ban	Treaty Banning Nuclear Weapons Tests in the Atmosphere, in Outer Space, and Under Water
O	Ozone Layer Protection	Montreal Protocol on Substances That Deplete the Ozone Layer
S	Ship Pollution	Protocol of 1978 Relating to the International Convention for the Prevention of Pollution From Ships, 1973 (MARPOL)
T	Tropical Timber 83	International Tropical Timber Agreement, 1983
	Tropical Timber 94	International Tropical Timber Agreement, 1994
W	Wetlands	Convention on Wetlands of International Importance Especially As Waterfowl Habitat; note—also known as Ramsar
	Whaling	International Convention for the Regulation of Whaling

Note: Not all of the selected international environmental agreements have abbreviations.

Appendix E:

Selected International Environmental Agreements

Air Pollution	see Convention on Long-Range Transboundary Air Pollution
Air Pollution-Nitrogen Oxides	see Protocol to the 1979 Convention on Long-Range Transboundary Air Pollution concerning the Control of Emissions of Nitrogen Oxides or Their Transboundary Fluxes
Air Pollution-Sulphur 85	see Protocol to the 1979 Convention on Long-Range Transboundary Air Pollution on the Reduction of Sulphur Emissions or their Transboundary Fluxes by at least 30%
Air Pollution-Sulphur 94	see Protocol to the 1979 Convention on Long-Range Transboundary Air Pollution on Further Reduction of Sulphur Emissions
Air Pollution-Volatile Organic Compounds	see Protocol to the 1979 Convention on Long-Range Transboundary Air Pollution concerning the Control of Emissions of Volatile Organic Compounds or Their Transboundary Fluxes
Antarctic-Environmental Protocol	see Protocol on Environmental Protection to the Antarctic Treaty
Antarctic Treaty *opened for signature*—1 December 1959 *entered into force*—23 June 1961 *objective*—to ensure that Antarctica is used for peaceful purposes, such as, for international cooperation in scientific research, and that it does not become the scene or object of international discord	*parties*—(42) Argentina, Australia, Austria, Belgium, Brazil, Bulgaria, Canada, Chile, China, Colombia, Cuba, Czech Republic, Denmark, Ecuador, Finland, France, Germany, Greece, Guatemala, Hungary, India, Italy, Japan, North Korea, South Korea, Netherlands, New Zealand, Norway, Papua New Guinea, Peru, Poland, Romania, Russia, Slovakia, South Africa, Spain, Sweden, Switzerland, Ukraine, United Kingdom, United States, Uruguay
Basel Convention on the Control of Transboundary Movements of Hazardous Wastes and Their Disposal *note*—abbreviated as Hazardous Wastes *opened for signature*—22 March 1989 *entered into force*—5 May 1992 *objective*—to reduce transboundary movements of wastes subject to the Convention to a minimum consistent with the environmentally sound and efficient management of such wastes; to minimize the amount and toxicity of wastes generated and ensure their environmentally sound management as closely as possible to the source of generation; and to assist LDCs in environmentally sound management of the hazardous and other wastes they generate	*parties*—(81) Antigua and Barbuda, Argentina, Australia, Austria, The Bahamas, Bahrain, Bangladesh, Belgium, Brazil, Canada, Chile, China, Comoros, Cote d'Ivoire, Croatia, Cuba, Cyprus, Czech Republic, Denmark, Ecuador, Egypt, El Salvador, Estonia, European Union, Finland, France, Greece, Hungary, India, Indonesia, Iran, Ireland, Israel, Italy, Japan, Jordan, South Korea, Kuwait, Latvia, Lebanon, Liechtenstein, Luxembourg, Malawi, Malaysia, Maldives, Mauritius, Mexico, Monaco, Netherlands, New Zealand, Nigeria, Norway, Pakistan, Panama, Peru, Philippines, Poland, Portugal, Romania, Russia, Saint Kitts and Nevis, Saint Lucia, Saudi Arabia, Senegal, Seychelles, Slovakia, Slovenia, South Africa, Spain, Sri Lanka, Sweden, Switzerland, Syria, Tanzania, Trinidad and Tobago, Turkey, United Arab Emirates, United Kingdom, Uruguay, Zaire, Zambia *countries that have signed, but not yet ratified*—(9) Afghanistan, Bolivia, Colombia, Germany, Guatemala, Haiti, Thailand, United States, Venezuela
Biodiversity	see Convention on Biological Diversity

Convention on Biological Diversity *note*—abbreviated as Biodiversity *opened for signature*—5 June 1992 *entered into force*—29 December 1993 *objective*—to develop national strategies for the conservation and sustainable use of biological diversity	*parties*—(111) Albania, Antigua and Barbuda, Argentina, Armenia, Australia, Austria, The Bahamas, Bangladesh, Barbados, Belarus, Belize, Benin, Bolivia, Brazil, Burkina, Burma, Cameroon, Canada, Chad, Chile, China, Colombia, Comoros, Cook Islands, Costa Rica, Cote d'Ivoire, Cuba, Czech Republic, Denmark, Djibouti, Dominica, Ecuador, Egypt, El Salvador, Equatorial Guinea, Estonia, Ethiopia, European Union, Fiji, Finland, France, Gambia, Georgia, Germany, Ghana, Greece, Grenada, Guinea, Guyana, Hungary, Iceland, India, Indonesia, Italy, Jamaica, Japan, Jordan, Kazakhstan, Kenya, Kiribati, North Korea, South Korea, Lebanon, Lesotho, Luxembourg, Malawi, Malaysia, Maldives, Marshall Islands, Mauritius, Mexico, Federated States of Micronesia, Monaco, Mongolia, Nauru, Nepal, Netherlands, New Zealand, Nigeria, Norway, Pakistan, Panama, Papua New Guinea, Paraguay, Peru, Philippines, Portugal, Romania, Saint Kitts and Nevis, Saint Lucia, San Marino, Senegal, Seychelles, Sierra Leone, Slovakia, Spain, Sri Lanka, Swaziland, Sweden, Switzerland, Tunisia, Uganda, United Kingdom, Uruguay, Vanuatu, Venezuela, Vietnam, Western Samoa, Zaire, Zambia, Zimbabwe *countries that have signed, but not yet ratified*—(64) Afghanistan, Algeria, Angola, Azerbaijan, Bahrain, Belgium, Bhutan, Botswana, Bulgaria, Burundi, Cape Verde, Central African Republic, Congo, Croatia, Cyprus, Dominican Republic, Gabon, Guatemala, Guinea-Bissau, Haiti, Honduras, Iran, Ireland, Israel, Kuwait, Latvia, Liberia, Libya, Liechtenstein, Lithuania, Madagascar, Mali, Malta, Mauritania, Moldova, Morocco, Mozambique, Namibia, Nicaragua, Niger, Oman, Poland, Qatar, Russia, Rwanda, Sao Tome and Principe, Singapore, Slovenia, Solomon Islands, South Africa, Sudan, Suriname, Syria, Tanzania, Thailand, Togo, Trinidad and Tobago, Turkey, Tuvalu, Ukraine, United Arab Emirates, United States, Yemen, former Yugoslavia
Climate Change	see United Nations Framework Convention on Climate Change
Convention on Fishing and Conservation of Living Resources of the High Seas *note*—abbreviated as Marine Life Conservation *opened for signature*—29 April 1958 *entered into force*—20 March 1966 *objective*—to solve through international cooperation the problems involved in the conservation of living resources of the high seas, considering that because of the development of modern technology some of these resources are in danger of being overexploited	*parties*—(37) Australia, Belgium, Bosnia and Herzegovina, Burkina, Cambodia, Colombia, Denmark, Dominican Republic, Fiji, Finland, France, Haiti, Jamaica, Kenya, Lesotho, Madagascar, Malawi, Malaysia, Mauritius, Mexico, Netherlands, Nigeria, Portugal, Senegal, Sierra Leone, Solomon Islands, South Africa, Spain, Switzerland, Thailand, Tonga, Trinidad and Tobago, Uganda, United Kingdom, United States, Venezuela, former Yugoslavia *countries that have signed, but not yet ratified*—(21) Afghanistan, Argentina, Bolivia, Costa Rica, Cuba, Ghana, Iceland, Indonesia, Iran, Ireland, Israel, Lebanon, Liberia, Nepal, New Zealand, Pakistan, Panama, Sri Lanka, Taiwan (Canada signed on behalf of Taiwan), Tunisia, Uruguay
Convention on Long-Range Transboundary Air Pollution *note*—abbreviated as Air Pollution *opened for signature*—13 November 1979 *entered into force*—16 March 1983 *objective*—to protect the human environment against air pollution and to gradually reduce and prevent air pollution, including long-range transboundary air pollution	*parties*—(39) Austria, Belarus, Belgium, Bosnia and Herzegovina, Bulgaria, Canada, Croatia, Cyprus, Czech Republic, Denmark, European Union, Finland, France, Germany, Greece, Hungary, Iceland, Ireland, Italy, Latvia, Liechtenstein, Lithuania, Luxembourg, Netherlands, Norway, Poland, Portugal, Romania, Russia, Slovakia, Slovenia, Spain, Sweden, Switzerland, Turkey, Ukraine, United Kingdom, United States, former Yugoslavia *countries that have signed, but not yet ratified*—(2) Holy See, San Marino
Convention on the International Trade in Endangered Species of Wild Flora and Fauna (CITES) *note*—abbreviated as Endangered Species *opened for signature*—3 March 1973 *entered into force*—1 July 1975 *objective*—to protect certain endangered species from overexploitation by means of a system of import/export permits	*parties*—(130) Afghanistan, Algeria, Argentina, Australia, Austria, The Bahamas, Bangladesh, Barbados, Belgium, Belize, Benin, Bolivia, Botswana, Brazil, Brunei, Bulgaria, Burkina, Burundi, Cameroon, Canada, Central African Republic, Chad, Chile, China, Colombia, Comoros, Congo, Costa Rica, Cote d'Ivoire, Cuba, Cyprus, Czech Republic, Denmark, Djibouti, Dominican Republic, Ecuador, Egypt, El Salvador, Equatorial Guinea, Eritrea, Estonia, Ethiopia, Finland, France, Gabon, The Gambia, Germany, Ghana, Greece, Guatemala, Guinea, Guinea-Bissau, Guyana, Honduras, Hungary, India, Indonesia, Iran, Israel, Italy, Japan, Jordan, Kenya, Kiribati, South Korea, Liechtenstein, Liberia, Luxembourg, Madagascar, Malawi, Malaysia, Mali, Malta, Mauritius, Mexico, Monaco, Morocco, Mozambique, Namibia, Nepal, Netherlands, New Zealand, Nicaragua, Niger, Nigeria, Norway, Pakistan, Panama, Papua New Guinea, Paraguay, Peru, Philippines, Poland, Portugal, Romania, Russia, Rwanda, Saint Kitts and Nevis, Saint Lucia, Saint Vincent and the Grenadines, Senegal, Seychelles, Sierra Leone, Singapore, Slovakia, Somalia, South Africa, Spain, Sri Lanka, Sudan, Suriname, Sweden, Switzerland, Tanzania, Thailand, Togo, Trinidad and Tobago, Tunisia, Tuvalu, Uganda, United Arab Emirates, United Kingdom, United States, Uruguay, Vanuatu, Venezuela, Vietnam, Zaire, Zambia, Zimbabwe *countries that have signed, but not yet ratified*—(5) Cambodia, Ireland, Kuwait, Lesotho, Vietnam

Convention on the Prevention of Marine Pollution by Dumping Wastes and Other Matter (London Convention)

note—abbreviated as Marine Dumping
opened for signature—29 December 1972
entered into force—30 August 1975
objective—to control pollution of the sea by dumping and to encourage regional agreements supplementary to the Convention

parties—(76) Afghanistan, Antigua and Barbuda, Argentina, Australia, Barbados, Belarus, Belgium, Belize, Bosnia and Herzegovina, Brazil, Canada, Cape Verde, Chile, China, Costa Rica, Cote d'Ivoire, Croatia, Cuba, Cyprus, Denmark, Dominican Republic, Egypt, European Union, Finland, France, Gabon, Germany, Greece, Guatemala, Haiti, Honduras, Hungary, Iceland, Ireland, Italy, Jamaica, Japan, Jordan, Kenya, Kiribati, Libya, Luxembourg, Malta, Mexico, Monaco, Morocco, Nauru, Netherlands, New Zealand, Nigeria, Norway, Oman, Panama, Papua New Guinea, Philippines, Poland, Portugal, Russia, Saint Lucia, Seychelles, Slovenia, Solomon Islands, South Africa, Spain, Suriname, Sweden, Switzerland, Tunisia, Tuvalu, Ukraine, United Arab Emirates, United Kingdom, United States, Vanuatu, former Yugoslavia, Zaire

Convention on the Prohibition of Military or Any Other Hostile Use of Environmental Modification Techniques

note—abbreviated as Environmental Modification
opened for signature—10 December 1976
entered into force—5 October 1978
objective—to prohibit the military or other hostile use of environmental modification techniques in order to further world peace and trust among nations

parties—(63) Afghanistan, Algeria, Antigua and Barbuda, Argentina, Australia, Austria, Bangladesh, Belarus, Belgium, Benin, Brazil, Bulgaria, Canada, Cape Verde, Chile, Cuba, Cyprus, Czech Republic, Denmark, Dominica, Egypt, Finland, Germany, Ghana, Greece, Guatemala, Hungary, India, Ireland, Italy, Japan, North Korea, South Korea, Kuwait, Laos, Malawi, Mauritius, Mongolia, Netherlands, New Zealand, Niger, Norway, Pakistan, Papua New Guinea, Poland, Romania, Russia, Saint Lucia, Sao Tome and Principe, Slovakia, Solomon Islands, Spain, Sri Lanka, Sweden, Switzerland, Tunisia, Ukraine, United Kingdom, United States, Uruguay, Uzbekistan, Vietnam, Yemen
countries that have signed, but not yet ratified—(17) Bolivia, Ethiopia, Holy See, Iceland, Iran, Iraq, Lebanon, Liberia, Luxembourg, Morocco, Nicaragua, Portugal, Sierra Leone, Syria, Turkey, Uganda, Zaire

Convention on Wetlands of International Importance Especially As Waterfowl Habitat (Ramsar)

note—abbreviated as Wetlands
opened for signature—2 February 1971
entered into force—21 December 1975
objective—to stem the progressive encroachment on and loss of wetlands now and in the future, recognizing the fundamental ecological functions of wetlands and their economic, cultural, scientific, and recreational value

parties—(83) Algeria, Argentina, Armenia, Australia, Austria, Bangladesh, Belgium, Bolivia, Brazil, Bulgaria, Burkina, Canada, Chad, Chile, China, Costa Rica, Croatia, Czech Republic, Denmark, Ecuador, Egypt, Estonia, Finland, France, Gabon, Germany, Ghana, Greece, Guatemala, Guinea, Guinea-Bissau, Honduras, Hungary, Iceland, India, Indonesia, Iran, Ireland, Italy, Japan, Jordan, Kenya, Lesotho, Liechtenstein, Lithuania, Mali, Malta, Mauritania, Mexico, Morocco, Netherlands, New Zealand, Niger, Norway, Pakistan, Panama, Papua New Guinea, Peru, Philippines, Poland, Portugal, Romania, Russia, Senegal, Slovakia, Slovenia, South Africa, Spain, Sri Lanka, Suriname, Sweden, Switzerland, Trinadad and Tobago, Tunisia, Turkey, Uganda, United Kingdom, United States, Uruguay, Venezuela, Vietnam, former Yugoslavia, Zambia

Desertification

see United Nations Convention to Combat Desertification in those Countries Experiencing Serious Drought and/or Desertification, Particularly in Africa

Endangered Species

see Convention on the International Trade in Endangered Species of Wild Flora and Fauna (CITES)

Environmental Modification

see Convention on the Prohibition of Military or Any Other Hostile Use of Environmental Modification Techniques

Hazardous Wastes

see Basel Convention on the Control of Transboundary Movements of Hazardous Wastes and Their Disposal

International Convention for the Regulation of Whaling

note—abbreviated as Whaling
opened for signature—2 December 1946
entered into force—10 November 1948
objective—to protect all species of whales from overfishing; to establish a system of international regulation for the whale fisheries to ensure proper conservation and development of whale stocks; and to safeguard for future generations the great natural resources represented by whale stocks

parties—(39) Antigua and Barbuda, Argentina, Australia, Brazil, Chile, China, Costa Rica, Denmark, Dominica, Finland, France, Germany, Grenada, India, Ireland, Japan, Kenya, South Korea, Mexico, Monaco, Netherlands (Netherlands also extended the convention to Netherlands Antilles), New Zealand, Norway, Oman, Peru, Russia, Saint Kitts and Nevis, Saint Lucia, Saint Vincent and the Grenadines, Senegal, Seychelles (withdrawing effective 30 June 1995), Solomon Islands, South Africa, Spain, Sweden, Switzerland, United Kingdom, United States, Venezuela
countries that have signed, but not yet ratified—(1) Austria
former parties—(10) Belize, Canada, Ecuador, Egypt, Iceland, Jamaica, Mauritius, Panama, Philippines, Uruguay

International Tropical Timber Agreement, 1983 *note*—abbreviated as Tropical Timber 83 *opened for signature*—18 November 1983 *entered into force*—1 April 1985; this agreement will expire when the International Tropical Timber Ageement, 1994 goes into force *objective*—to provide an effective framework for cooperation between tropical timber producers and consumers and to encourage the development of national policies aimed at sustainable utilization and conservation of tropical forests and their genetic resources	*parties*—(52) Australia, Austria, Belgium, Bolivia, Brazil, Burma, Cameroon, Canada, China, Colombia, Congo, Cote d'Ivoire, Denmark, Ecuador, Egypt, European Union, Finland, France, Gabon, Germany, Ghana, Greece, Guyana, Honduras, India, Indonesia, Ireland, Italy, Japan, South Korea, Liberia, Luxembourg, Malaysia, Nepal, Netherlands, New Zealand, Norway, Panama, Papua New Guinea, Peru, Philippines, Portugal, Russia, Spain, Sweden, Switzerland, Thailand, Togo, Trinidad and Tobago, United Kingdom, United States, Zaire
International Tropical Timber Agreement, 1994 *note*—abbreviated as Tropical Timber 94 *opened for signature*—26 January 1994, but not yet in force *objective*—to ensure that by the year 2000 exports of tropical timber originate from sustainably managed sources; to establish a fund to assist tropical timber producers in obtaining the resources necessary to reach this objective	*parties*—(3) Fiji, Japan, Liberia *countries that have signed, but not yet ratified*—(11) Cameroon, Congo, Ecuador, Egypt, Gabon, Indonesia, Norway, Panama, Peru, Togo, United States
Law of the Sea	see United Nations Convention on the Law of the Sea (LOS)
Marine Dumping	see Convention on the Prevention of Marine Pollution by Dumping Wastes and Other Matter (London Convention)
Marine Life Conservation	see Convention on Fishing and Conservation of Living Resources of the High Seas
Montreal Protocol on Substances That Deplete the Ozone Layer *note*—abbreviated as Ozone Layer Protection *opened for signature*—16 September 1987 *entered into force*—1 January 1989 *objective*—to protect the ozone layer by controling emissions of substances that deplete it	*parties*—(148) Algeria, Antigua and Barbuda, Argentina, Australia, Austria, The Bahamas, Bahrain, Bangladesh, Barbados, Belarus, Belgium, Benin, Bolivia, Bosnia and Herzegovina, Botswana, Brazil, Brunei, Bulgaria, Burkina, Burma, Cameroon, Canada, Central African Republic, Chad, Chile, China, Colombia, Comoros, Congo, Costa Rica, Cote d'Ivoire, Croatia, Cuba, Cyprus, Czech Republic, Denmark, Dominica, Dominican Republic, Ecuador, Egypt, El Salvador, Ethiopia, European Union, Fiji, Finland, France, Gabon, The Gambia, Germany, Ghana, Greece, Grenada, Guatemala, Guinea, Guyana, Honduras, Hungary, Iceland, India, Indonesia, Iran, Ireland, Israel, Italy, Jamaica, Japan, Jordan, Kenya, Kiribati, North Korea, South Korea, Kuwait, Lebanon, Lesotho, Libya, Liechtenstein, Lithuania, Luxembourg, The Former Yugoslav Republic of Macedonia, Malawi, Malaysia, Maldives, Mali, Malta, Marshall Islands, Mauritania, Mauritius, Mexico, Monaco, Mozambique, Namibia, Nepal, Netherlands, New Zealand, Nicaragua, Niger, Nigeria, Norway, Pakistan, Panama, Papua New Guinea, Paraguay, Peru, Philippines, Poland, Portugal (Portugal has also extended the protocol to Macau), Romania, Russia, Saint Kitts and Nevis, Saint Lucia, Saudi Arabia, Senegal, Seychelles, Singapore, Slovakia, Slovenia, Solomon Islands, South Africa, Spain, Sri Lanka, Sudan, Swaziland, Sweden, Switzerland, Syria, Tanzania, Thailand, Togo, Trinidad and Tobago, Tunisia, Turkey, Turkmenistan, Tuvalu, Uganda, Ukraine, United Arab Emirates, United Kingdom, United States, Uruguay, Uzbekistan, Vanuatu, Venezuela, Vietnam, Western Samoa, former Yugoslavia, Zaire, Zambia, Zimbabwe *countries that have signed, but not yet ratified*—(1) Morocco
Nuclear Test Ban	see Treaty Banning Nuclear Weapons Tests in the Atmosphere, in Outer Space, and Under Water
Ozone Layer Protection	see Montreal Protocol on Substances that Deplete the Ozone Layer
Protocol of 1978 Relating to the International Convention for the Prevention of Pollution From Ships, 1973 (MARPOL) *note*—abbreviated as Ship Pollution *opened for signature*—17 February 1978 *entered into force*—2 October 1983 *objective*—to preserve the marine environment through the complete elimination of pollution by oil and other harmful substances and the minimization of accidental discharge of such substances	*parties*—(91) Algeria, Antigua and Barbuda, Argentina, Australia, Austria, The Bahamas, Barbados, Belgium, Brazil, Brunei, Bulgaria, Burma, Cambodia, Canada, Chile, China, Colombia, Cote d'Ivoire, Croatia, Cuba, Cyprus, Czech Republic, Denmark, Djibouti, Ecuador, Egypt, Estonia, Finland, France, Gabon, Gambia, Georgia, Germany, Ghana, Greece, Hungary, Iceland, India, Indonesia, Israel, Italy, Jamaica, Japan, Kazakhstan, Kenya, North Korea, South Korea, Latvia, Lebanon, Liberia, Lithuania, Luxembourg, Malta, Marshall Islands, Mexico, Monaco, Morocco, Netherlands, Norway, Oman, Pakistan, Panama, Papua New Guinea, Peru, Poland, Portugal, Romania, Russia, Saint Vincent and the Grenadines, Seychelles, Singapore, Slovakia, Slovenia, South Africa, Spain, Suriname, Sweden, Switzerland, Syria, Togo, Tunisia, Turkey, Tuvalu, Ukraine, United Kingdom, United States, Uruguay, Vanuatu, Venezuela, Vietnam, former Yugoslavia

Protocol on Environmental Protection to the Antarctic Treaty *note*—abbreviated as Antarctic-Environmental Protocol *opened for signature*—4 October 1991, but not yet in force *objective*—to enhance the protection of the Antarctic environment and dependent and associated ecosystems	*parties*—(14) Argentina, Australia, Chile, China, Ecuador, France, Germany, Netherlands, New Zealand, Norway, Peru, Spain, Sweden, Uruguay *countries that have signed, but not yet ratified*—(27) Austria, Belgium, Brazil, Bulgaria, Canada, Colombia, Cuba, Czech Republic, Denmark, Finland, Greece, Guatemala, Hungary, India, Italy, Japan, North Korea, South Korea, Papua New Guinea, Poland, Romania, Russia, Slovakia, South Africa, Switzerland, United Kingdom, United States
Protocol to the 1979 Convention on Long-Range Transboundary Air Pollution concerning the Control of Emissions of Nitrogen Oxides or Their Transboundary Fluxes *note*—abbreviated as Air Pollution-Nitrogen Oxides *opened for signature*—31 October 1988 *entered into force*—14 February 1991 *objective*—to provide for the control or reduction of nitrogen oxides and their transboundary fluxes	*parties*—(25) Austria, Belarus, Bulgaria, Canada, Czech Republic, Denmark, European Union, Finland, France, Germany, Hungary, Ireland, Italy, Liechtenstein, Luxembourg, Netherlands, Norway, Russia, Slovakia, Spain, Sweden, Switzerland, Ukraine, United Kingdom, United States *countries that have signed, but not yet ratified*—(3) Belgium, Greece, Poland
Protocol to the 1979 Convention on Long-Range Transboundary Air Pollution concerning the Control of Emissions of Volatile Organic Compounds or Their Transboundary Fluxes *note*—abbreviated as Air Pollution-Volatile Organic Compounds *opened for signature*—18 November 1991, but not yet in force *objective*—to provide for the control and reduction of emissions of volatile organic compounds in order to reduce their transboundary fluxes so as to protect human health and the environment from adverse effects	*parties*—(11) Austria, Finland, Germany, Liechtenstein, Luxembourg, Netherlands, Norway, Spain, Sweden, Switzerland, United Kingdom *countries that have signed, but not yet ratified*—(12) Belgium, Bulgaria, Canada, Denmark, European Union, France, Greece, Hungary, Italy, Portugal, Ukraine, United States
Protocol to the 1979 Convention on Long-Range Transboundary Air Pollution on Further Reduction of Sulphur Emissions *note*—abbreviated as Air Pollution-Sulphur 94 *opened for signature*—14 June 1994, but not yet in force *objective*—to provide for a further reduction in sulfur emissions or transboundary fluxes	*parties*—(0) *countries that have signed, but not yet ratified*—(28) Austria, Belgium, Bulgaria, Canada, Croatia, Czech Republic, Denmark, European Union, Finland, France, Germany, Greece, Hungary, Ireland, Italy, Liechtenstein, Luxembourg, Netherlands, Norway, Poland, Russia, Slovakia, Slovenia, Spain, Sweden, Switzerland, Ukraine, United Kingdom
Protocol to the 1979 Convention on Long-Range Transboundary Air Pollution on the Reduction of Sulphur Emissions or their Transboundary Fluxes by at least 30% *note*—abbreviated as Air Pollution-Sulphur 85 *opened for signature*—8 July 1985 *entered into force*—2 September 1987 *objective*—to provide for a 30% reduction in sulfur emissions or transboundary fluxes by 1993	*parties*—(21) Austria, Belarus, Belgium, Bulgaria, Canada, Czech Republic, Denmark, Finland, France, Germany, Hungary, Italy, Liechtenstein, Luxembourg, Netherlands, Norway, Russia, Slovakia, Sweden, Switzerland, Ukraine
Ship Pollution	see Protocol of 1978 Relating to the International Convention for the Prevention of Pollution From Ships, 1973 (MARPOL)

Treaty Banning Nuclear Weapon Tests in the Atmosphere, in Outer Space, and Under Water *note*—abbreviated as Nuclear Test Ban *opened for signature*—5 August 1963 *entered into force*—10 October 1963 *objective*—to obtain an agreement on general and complete disarmament under strict international control in accordance with the objectives of the United Nations; to put an end to the armaments race and eliminate incentives for the production and testing of all kinds of weapons, including nuclear weapons	*parties*—(125) Afghanistan, Antigua and Barbuda, Argentina, Armenia, Australia, Austria, Bahamas, Bangladesh, Belarus, Belgium, Benin, Bhutan, Bolivia, Bosnia and Herzegovina, Botswana, Brazil, Bulgaria, Burma, Canada, Cape Verde, Central African Republic, Chad, Chile, China, Colombia, Costa Rica, Cote d'Ivoire, Croatia, Cyprus, Czech Republic, Denmark, Dominican Republic, Ecuador, Egypt, El Salvador, Equatorial Guinea, Fiji, Finland, Gabon, The Gambia, Germany, Ghana, Greece, Guatemala, Guinea-Bissau, Honduras, Hungary, Iceland, India, Indonesia, Iran, Iraq, Ireland, Israel, Italy, Jamaica, Japan, Jordan, Kenya, South Korea, Kuwait, Laos, Lebanon, Liberia, Libya, Luxembourg, Madagascar, Malawi, Malaysia, Malta, Mauritania, Mauritius, Mexico, Mongolia, Morocco, Nepal, Netherlands, New Zealand, Nicaragua, Niger, Nigeria, Norway, Pakistan, Panama, Papua New Guinea, Peru, Philippines, Poland, Romania, Russia, Rwanda, San Marino, Senegal, Seychelles, Sierra Leone, Singapore, Slovakia, Slovenia, South Africa, Spain, Sri Lanka, Sudan, Suriname, Swaziland, Sweden, Switzerland, Syria, Tanzania, Thailand, Togo, Tonga, Trinidad and Tobago, Tunisia, Turkey, Uganda, United Kingdom, United States, Ukraine, Uruguay, Venezuela, Western Samoa, Yemen, former Yugoslavia, Zaire, Zambia *countries that have signed, but not yet ratified*—(11) Algeria, Burkina, Burundi, Cameroon, Ethiopia, Haiti, Mali, Paraguay, Portugal, Somalia, Vietnam
Tropical Timber 83	see International Tropical Timber Agreement, 1983
Tropical Timber 94	see International Tropical Timber Agreement, 1994
United Nations Convention on the Law of the Sea (LOS) *note*—abbreviated as Law of the Sea *opened for signature*—10 December 1982 *entered into force*—16 November 1994 *objective*—to set up a comprehensive new legal regime for the sea and oceans; to include rules concerning environmental standards as well as enforcement provisions dealing with pollution of the marine environment	*parties*—(72) Angola, Antigua and Barbuda, Australia, The Bahamas, Bahrain, Barbados, Belize, Bosnia and Herzegovina, Botswana, Brazil, Cameroon, Cape Verde, Comoros, Costa Rica, Cote d'Ivoire, Cuba, Cyprus, Djibouti, Dominica, Egypt, Fiji, The Gambia, Germany, Ghana, Grenada, Guinea, Guinea-Bissau, Guyana, Honduras, Iceland, Indonesia, Iraq, Italy, Jamaica, Kenya, Kuwait, Lebanon, The Former Yugoslav Republic of Macedonia, Mali, Malta, Marshall Islands, Mauritius, Mexico, Federated States of Micronesia, Namibia, Nigeria, Oman, Paraguay, Philippines, Saint Kitts and Nevis, Saint Lucia, Saint Vincent and the Grenadines, Sao Tome and Principe, Senegal, Seychelles, Sierra Leone, Singapore, Somalia, Sri Lanka, Sudan, Tanzania, Togo, Trinidad and Tobago, Tunisia, Uganda, Uruguay, Vietnam, Yemen, former Yugoslavia, Zaire, Zambia, Zimbabwe *countries that have signed, but not yet ratified*—(91) Afghanistan, Algeria, Argentina, Austria, Bangladesh, Belarus, Belgium, Benin, Bhutan, Bolivia, Brunei, Bulgaria, Burkina, Burma, Burundi, Cambodia, Canada, Central African Republic, Chad, Chile, China, Colombia, Congo, Cook Islands, Czech Republic, Denmark, Dominican Republic, El Salvador, Equatorial Guinea, Ethiopia, European Union, Finland, France, Gabon, Greece, Guatemala, Haiti, Hungary, India, Iran, Ireland, Japan, North Korea, South Korea, Laos, Lesotho, Liberia, Libya, Liechtenstein, Luxembourg, Madagascar, Malawi, Malaysia, Maldives, Mauritania, Monaco, Mongolia, Morocco, Mozambique, Nauru, Nepal, Netherlands, New Zealand, Nicaragua, Niger, Niue, Norway, Pakistan, Panama, Papua New Guinea, Poland, Portugal, Qatar, Romania, Russia, Rwanda, Saudi Arabia, Slovakia, Solomon Islands, South Africa, Spain, Suriname, Swaziland, Sweden, Switzerland, Thailand, Tuvalu, Ukraine, United Arab Emirates, Vanuatu, Western Samoa
United Nations Convention to Combat Desertification in those Countries Experiencing Serious Drought and/or Desertification, Particularly in Africa *note*—abbreviated as Desertification *opened for signature*—14 October 1994, but not yet in force *objective*—to combat desertification and mitigate the effects of drought through national action programs that incorporate long-term strategies supported by international cooperation and partnership arrangements	*parties*—(1) Mexico *countries that have signed, but not yet ratified*—(104) Algeria, Angola, Antigua and Barbuda, Argentina, Armenia, Australia, Bangladesh, Benin, Bolivia, Brazil, Burkina, Burundi, Cambodia, Cameroon, Canada, Cape Verde, Central African Republic, Chad, Chile, China, Colombia, Comoros, Congo, Costa Rica, Cote d'Ivoire, Croatia, Cuba, Denmark, Djibouti, Ecuador, Egypt, Equatorial Guinea, Eritrea, Ethiopia, European Union, Finland, France, Gambia, Georgia, Germany, Ghana, Greece, Guinea, Guinea-Bissau, Haiti, Honduras, India, Indonesia, Iran, Ireland, Israel, Italy, Japan, Jordan, Kazakhstan, Kenya, South Korea, Lebanon, Lesotho, Libya, Luxembourg, Madagascar, Malawi, Mali, Malta, Mauritania, Mauritius, Micronesia, Mongolia, Morocco, Namibia, Netherlands, Nicaragua, Niger, Nigeria, Norway, Pakistan, Panama, Paraguay, Peru, Philippines, Portugal, Saint Vincent and the Grenadines, Senegal, Seychelles, Sierra Leone, South Africa, Spain, Sudan, Sweden, Switzerland, Syria, Tanzania, Togo, Tunisia, Turkey, Turkmenistan, Uganda, United Kingdom, United States, Uzbekistan, Zaire, Zambia, Zimbabwe; *note*—some late changes not included under country entries

United Nations Framework Convention on Climate Change *note*—abbreviated as Climate Change *opened for signature*—9 May 1992 *entered into force*—21 March 1994 *objective*—to achieve stabilization of greenhouse gas concentrations in the atmosphere at a low enough level to prevent dangerous anthropogenic interference with the climate system	*parties*—(119) Albania, Algeria, Antigua and Barbuda, Argentina, Armenia, Australia, Austria, The Bahamas, Bahrain, Bangladesh, Barbados, Belize, Benin, Bolivia, Botswana, Brazil, Burkina, Burma, Cameroon, Canada, Chad, Chile, China, Comoros, Cook Islands, Costa Rica, Cote d'Ivoire, Cuba, Czech Republic, Denmark, Dominica, Ecuador, Egypt, Estonia, Ethiopia, European Union, Fiji, Finland, France, Gambia, Georgia, Germany, Greece, Grenada, Guinea, Guyana, Hungary, Iceland, India, Indonesia, Ireland, Italy, Jamaica, Japan, Jordan, Kenya, North Korea, South Korea, Kuwait, Laos, Lebanon, Liechtenstein, Luxembourg, Malawi, Malaysia, Maldives, Mali, Malta, Marshall Islands, Mauritania, Mauritius, Mexico, Federated States of Micronesia, Monaco, Mongolia, Nauru, Nepal, Netherlands, New Zealand, Nigeria, Norway, Pakistan, Papua New Guinea, Paraguay, Peru, Philippines, Poland, Portugal, Romania, Russia, Saint Kitts and Nevis, Saint Lucia, San Marino, Saudi Arabia, Senegal, Seychelles, Slovakia, Solomon Islands, Spain, Sri Lanka, Sudan, Sweden, Switzerland, Thailand, Trinidad and Tobago, Tunisia, Tuvulu, Uganda, United Kingdom, United States, Uruguay, Uzbekistan, Vanuatu, Venezuela, Vietnam, Western Samoa, Zaire, Zambia, Zimbabwe *countries that have signed, but not yet ratified*—(54) Afghanistan, Angola, Azerbaijan, Belarus, Belgium, Bhutan, Bulgaria, Burundi, Cape Verde, Central African Republic, Colombia, Congo, Croatia, Cyprus, Djibouti, Dominican Republic, El Salvador, Gabon, Ghana, Guatemala, Guinea-Bissau, Haiti, Honduras, Iran, Israel, Kazakhstan, Kiribati, Latvia, Lesotho, Liberia, Libya, Lithuania, Madagascar, Moldova, Morocco, Mozambique, Namibia, Nicaragua, Niger, Oman, Panama, Rwanda, Sao Tome and Principe, Sierra Leone, Singapore, Slovenia, South Africa, Suriname, Swaziland, Tanzania, Togo, Ukraine, Yemen, former Yugoslavia
Wetlands	see Convention on Wetlands of International Importance Especially As Waterfowl Habitat (Ramsar)
Whaling	see International Convention for the Regulation of Whaling

Appendix F:

Weights and Measures

Mathmatical Notation

Mathmatical Power	Name
10^{18} or 1,000,000,000,000,000,000	one quintillion
10^{15} or 1,000,000,000,000,000	one quadrillion
10^{12} or 1,000,000,000,000	one trillion
10^{9} or 1,000,000,000	one billion
10^{6} or 1,000,000	one million
10^{3} or 1,000	one thousand
10^{2} or 100	one hundred
10^{1} or 10	ten
10^{1} or 1	one
10^{-1} or 0.1	one tenth
10^{-2} or 0.01	one hundredth
10^{-3} or 0.001	one thousandth
10^{-6} or 0.000 001	one millionth
10^{-9} or 0.000 000 001	one billionth
10^{-12} or 0.000 000 000 001	one trillionth
10^{-15} or 0.000 000 000 000 001	one quadrillionth
10^{-18} or 0.000 000 000 000 000 001	one quintillionth

Metric Interrelationships

Conversions from a multiple or submultiple to the basic units of meters, liters, or grams can be done using the table. For example, to convert from kilometers to meters, multiply by 1,000 (9.26 kilometers equals 9,260 meters) or to convert from meters to kilometers, multiply by 0.001 (9,260 meters equals 9.26 kilometers)

Prefix	Symbol	Length, weight, or capacity	Area	Volume
exa	E	10^{18}	10^{36}	10^{54}
peta	P	10^{15}	10^{30}	10^{45}
tera	T	10^{12}	10^{24}	10^{36}
giga	G	10^{9}	10^{18}	10^{27}
mega	M	10^{6}	10^{12}	10^{18}
hectokilo	hk	10^{5}	10^{10}	10^{15}
myria	mk	10^{4}	10^{8}	10^{12}
kilo	k	10^{3}	10^{6}	10^{9}
hecto	h	10^{2}	10^{4}	10^{6}
basic unit	—	1 meter,	1 meter^{2}	1 meter^{3}
		1 gram,		
		1 liter		
deci	d	10^{-1}	10^{-2}	10^{-3}
centi	c	10^{-2}	10^{-4}	10^{-6}
miili	m	10^{-3}	10^{-6}	10^{-9}
decimilli	dm	10^{-4}	10^{-8}	10^{-12}
centimilli	cm	10^{-5}	10^{-10}	10^{-15}
micro	u	10^{-6}	10^{-12}	10^{-18}
nano	n	10^{-9}	10^{-18}	10^{-27}
pico	p	10^{-12}	10^{-24}	10^{-36}
femto	f	10^{-15}	10^{-30}	10^{-45}
atto	a	10^{-18}	10^{-36}	10^{-54}

Equivalents

Unit	Metric Equivalent	US Equivalent
acre	0.404 685 64 hectares	43,560 feet^2
acre	4,046,856 4 meters^2	4,840 yards^2
acre	0.004 046 856 4 kilometers^2	0.001 562 5 miles^2, statute
are	100 meters^2	119.599 yards^2
are	100 meters^2	119.599 yards^2
barrel (petroleum, US)	158.987 29 liters	42 gallons
(proof spirits, US)	151.416 47 liters	40 gallons
(beer, US)	117.347 77 liters	31 gallons
bushel	35.239 07 liters	4 pecks
cable	219.456 meters	120 fathoms
chain (surveyor's)	20.116 8 meters	66 feet
cord (wood)	3.624 556 meters^3	128 feet^3
cup	0.236 588 2 liters	8 ounces, liquid (US)
degrees, celsius	(water boils at 100 degrees C, freezes at 0 degrees C)	Multiply by 1.8 and add 32 to obtain degrees F
degrees, fahrenheit	subtract 32 and divide by 1.8 to obtain degrees C	(water boils at 212 degrees F, freezes at 32 degrees F)
dram, avoirdupois	1.771 845 2 grams	0.0625 5 ounces, avoirdupois
dram, troy	3.887 934 6 grams	0.125 ounces, troy
dram, liquid (US)	3.696 69 milliliters	0.125 ounces, liquid
fathom	1.828 8 meters	6 feet
foot	30.48 centimeters	12 inches
foot	0.304 8 meters	0.333 333 3 yards
foot	0.000 304 8 kilometers	0.000 189 39 miles, statute
foot^2	929.030 4 centimeters^2	144 inches^2
foot^2	0.092 903 04 meters^2	0.111 111 1 yards^2
foot^3	28.316 846 592 liters	7.480 519 gallons
foot^3	0.028 316 847 meters^3	1,728 inches^3
furlong	201.168 meters	220 yards
gallon, liquid (US)	3.785 411 784 liters	4 quarts, liquid
gill (US)	118.294 118 milliliters	4 ounces, liquid
grain	64.798 91 milligrams	0.002 285 71 ounces, avdp.
gram	1,000 milligrams	0.035 273 96 ounces, avdp.
hand (height of horse)	10.16 centimeters	4 inches
hectare	10,000 meters^2	2.471 053 8 acres
hundredweight, long	50.802 345 kilograms	112 pounds, avoirdupois
hundredweight, short	45.359 237 kilograms	100 pounds, avoirdupois
inch	2.54 centimeters	0.083 333 33 feet
inch^2	6.451 6 centimeters^2	0.006 944 44 feet^2
inch^3	16.387 064 centimeters^3	0.000 578 7 feet^3
inch^3	16.387 064 milliliters	0.029 761 6 pints, dry
inch^3	16.387 064 milliliters	0.034 632 0 pints, liquid
kilogram	0.001 tons, metric	2.204 623 pounds, avoirdupois
kilometer	1,000 meters	0.621 371 19 miles, statute
kilometer^2	100 hectares	247.105 38 acres
kilometer^2	1,000,000 meters^2	0.386 102 16 miles^2, statute
knot (1 nautical mi/hr)	1.852 kilometers/hour	1.151 statute miles/hour
league, nautical	5.556 kilometers	3 miles, nautical
league, statute	4.828.032 kilometers	3 miles, statute
link (surveyor's)	20.116 8 centimeters	7.92 inches

Equivalents

Unit	Metric Equivalent	US Equivalent
liter	0.001 meters^3	61.023 74 inches^3
liter	0.1 dekaliter	0.908 083 quarts, dry
liter	1,000 milliliters	1.056 688 quarts, liquid
meter	100 centimeters	1.093 613 yards
meter^2	10,000 centimeters^2	1.195 990 yards^2
meter^3	1,000 liters	1.307 951 yards^3
micron	0.000 001 meter	0.000 039 4 inches
mil	0.025 4 millimeters	0.001 inch
mile, nautical	1.852 kilometers	1.150 779 4 miles, statute
mile^2, nautical	3.429 904 kilometers^2	1.325 miles^2, statute
mile, statute	1.609 344 kilometers	5,280 feet or 8 furlongs
mile^2, statute	258.998 811 hectares	640 acres or 1 section
mile^2, statute	2.589 988 11 kilometers^2	0.755 miles^2, nautical
minim (US)	0.061 611 52 milliliters	0.002 083 33 ounces, liquid
ounce, avoirdupois	28.349 523 125 grams	437.5 grains
ounce, liquid (US)	29.573 53 milliliters	0.062 5 pints, liquid
ounce, troy	31.103 476 8 grams	480 grains
pace	76.2 centimeters	30 inches
peck	8.809 767 5 liters	8 quarts, dry
pennyweight	1.555 173 84 grams	24 grains
pint, dry (US)	0.550 610 47 liters	0.5 quarts, dry
pint, liquid (US)	0.473 176 473 liters	0.5 quarts, liquid
point (typographical)	0.351 459 8 millimeters	0.013 837 inches
pound, avoirdupois	453.592 37 grams	16 ounces, avourdupois
pound, troy	373.241 721 6 grams	12 ounces, troy
quart, dry (US)	1.101 221 liters	2 pints, dry
quart, liquid (US)	0.946 352 946 liters	2 pints, liquid
quintal	100 kilograms	220.462 26 pounds, avdp.
rod	5.029 2 meters	5.5 yards
scruple	1.295 978 2 grams	20 grains
section (US)	2.589 988 1 kilometers^2	1 mile^2, statute or 640 acres
span	22.86 centimeters	9 inches
stere	1 meter^3	1.307 95 yards^3
tablespoon	14.786 76 milliliters	3 teaspoons
teaspoon	4.928 922 milliliters	0.333 333 tablespoons
ton, long or deadweight	1,016.046 909 kilograms	2,240 pounds, avoirdupois
ton, metric	1,000 kilograms	2,204.623 pounds, avoirdupois
ton, metric	1,000 kilograms	32,150.75 ounces, troy
ton, register	2.831 684 7 meters^3	100 feet^3
ton, short	907.184 74 kilograms	2,000 pounds, avoirdupois
township (US)	93.239 572 kilometers^2	36 miles^2, statute
yard	0.914 4 meters	3 feet
yard^2	0.836 127 36 meters^2	9 feet^2
yard^3	0.764 554 86 meters^3	27 feet^3
yard^3	764.554 857 984 liters	201.974 gallons

Appendix G:

Estimates of Gross Domestic Product on an Exchange Rate Basis

These estimates of gross domestic product on an exchange rate basis are based on official data from national statistical offices.

	Country	Million US$	Year
A	Afghanistan	-	
	Albania	-	
	Algeria	46,823	1993
	American Samoa	-	
	Andorra	-	
	Angola	-	
	Anguilla	-	
	Antigua and Barbuda	374	1992
	Argentina	-	
	Armenia	-	
	Aruba	-	
	Australia	284,293	1993
	Austria	181,367	1993
	Azerbaijan	-	
B	The Bahamas	-	
	Bahrain	3,903	1990
	Bangladesh	23,957	1993
	Barbados	1,574	1992
	Belarus	-	
	Belgium	207,500	1993
	Belize	524	1993
	Benin	1,898	1991
	Bermuda	-	
	Bhutan	245	1992
	Bolivia	6,058	1991
	Bosnia and Herzegovina	-	
	Botswana	3,702	1992
	Brazil	-	
	British Virgin Islands	-	
	Brunei	-	
	Bulgaria	-	
	Burkina	-	
	Burma	55,073	1993
	Burundi	923	1993
C	Cambodia	-	
	Cameroon	10,918	1992
	Canada	551,645	1993
	Cape Verde	286	1988

	Country	Million US$	Year
	Cayman Islands	-	
	Central African Republic	1,339	1992
	Chad	1,383	1992
	Chile	43,684	1993
	China	544,603	1993
	Christmas Island	-	
	Cocos (Keeling) Islands	-	
	Colombia	-	
	Comoros	43,546	1992
	Congo	-	
	Cook Islands	-	
	Costa Rica	6,722	1992
	Cote d'Ivoire	10,492	1992
	Croatia	-	
	Cuba	-	
	Cyprus	6,700	1992
	Czech Republic	31,664	1993
D	Denmark	135,998	1993
	Djibouti	494	1990
	Dominica	189	1992
	Dominican Republic	8,796	1992
E	Ecuador	14,304	1993
	Egypt	41,855	1992
	El Salvador	7,625	1993
	Equatorial Guinea	181	1992
	Eritrea	-	
	Estonia	-	
	Ethiopia	3,362	1993
F	Falkland Islands (Islas Malvinas)	-	
	Faroe Islands	-	
	Fiji	1,490	1991
	Finland	83,795	1993
	France	1,252,560	1993
	French Guiana	-	
	French Polynesia	-	
G	Gabon	5,913	1992
	The Gambia	332	1992
	Gaza Strip	-	
	Georgia	-	
	Germany	1,880,000	1993
	Ghana	6,884	1992
	Gibraltar	-	
	Greece	73,100	1993
	Greenland	-	
	Grenada	214	1992

	Country	Million US$	Year
	Guadeloupe	-	
	Guam	-	
	Guatemala	11,279	1993
	Guernsey	-	
	Guinea	-	
	Guinea-Bissau	221	1992
	Guyana	447	1993
H	Haiti	2,502	1990
	Honduras	3,343	1993
	Hong Kong	-	
	Hungary	36,113	1993
I	Iceland	6,076	1993
	India	272,231	1992
	Indonesia	142,794	1993
	Iran	1,013,890	1992
	Iraq	-	
	Ireland	47,678	1993
	Israel	65,043	1993
	Italy	999,700	1993
J	Jamaica	3,839	1993
	Japan	4,215,546	1993
	Jersey	-	
	Jordan	5,190	1993
K	Kazakhstan	-	
	Kenya	5,569	1993
	Kiribati	-	
	Korea, North	-	
	Korea, South	-	
	Kuwait	22,416	1993
	Kyrgyzstan	-	
L	Laos	-	
	Latvia	-	
	Lebanon	-	
	Lesotho	710	1992
	Liberia	1,183	1989
	Libya	21,864	1986
	Liechtenstein	-	
	Lithuania		
	Luxembourg	10,600	1993
M	Macau		
	Macedonia, The Former Yugoslav Republic of	-	
	Madagascar	3,371	1993
	Malawi	2,017	1993
	Malaysia	64,434	1993

	Country	Million US$	Year
	Maldives	-	
	Mali	2,451	1991
	Malta	2,743	1992
	Man, Isle of	-	
	Marshall Islands	-	
	Martinique	-	
	Mauritania	1,136	1993
	Mauritius	3,112	1993
	Mayotte	-	
	Mexico	286,631	1991
	Micronesia, Federated States of	-	
	Moldova	-	
	Monaco	-	
	Mongolia	1,111	1992
	Montserrat	-	
	Morocco	28,762	1992
	Mozambique	1,410	1993
N	Namibia	2,508	1993
	Nauru	-	
	Nepal	3,387	1993
	Netherlands	308,995	1993
	Netherlands Antilles	-	
	New Caledonia	-	
	New Zealand	43,698	1993
	Nicaragua	2,214	1990
	Niger	2,506	1990
	Nigeria	37,250	1993
	Niue	-	
	Norfolk Island	-	
	Northern Mariana Islands	-	
	Norway	103,418	1993
O	Oman	11,489	1992
P	Pakistan	48,363	1993
	Palau	-	
	Panama	6,565	1993
	Papua New Guinea	4,292	1992
	Paraguay	6,446	1992
	Peru	39,760	1989
	Philippines	54,068	1992
	Pitcairn Islands	-	
	Poland	85,898	1993
	Portugal	75,100	1993
	Puerto Rico	-	
Q	Qatar	7,473	1992
R	Reunion	-	

	Country	Million US$	Year
	Romania	24,781	1993
	Russia	-	
	Rwanda	1,630	1992
S	Saint Helena	-	
	Saint Kitts and Nevis	158	1990
	Saint Lucia	393	1992
	Saint Pierre and Miquelon	-	
	Saint Vincent and the Grenadines	192	1992
	San Marino	-	
	Sao Tome and Principe	-	
	Saudi Arabia	121,530	1992
	Senegal	4,627	1989
	Serbia and Montenegro	-	
	Seychelles	434	1992
	Sierra Leone	-	
	Singapore	55,086	1993
	Slovakia	-	
	Slovenia	-	
	Solomon Islands	-	
	Somalia	-	
	South Africa	117,442	1993
	Spain	478,391	1993
	Sri Lanka	10,274	1993
	Sudan	27,697	1991
	Suriname	1,872	1991
	Svalbard	-	
	Swaziland	874	1991
	Sweden	186,224	1993
	Switzerland	232,133	1993
	Syria	33,050	1992
T	Tajikistan	-	
	Tanzania	2,086	1993
	Thailand	110,429	1992
	Togo	1,237	1987
	Tokelau	-	
	Tonga	145	1993
	Trinidad and Tobago	4,538	1993
	Tunisia	14,634	1993
	Turkey	138,400	1993
	Turkmenistan	-	
	Turks and Caicos Islands	-	
	Tuvalu	-	
U	Uganda	5,608	1988
	Ukraine	-	
	United Arab Emirates	34,977	1992

	Country	Million US$	Year
	United Kingdom	944,902	1993
	United States	6,738,400	1994
	Uruguay	13,144	1993
	Uzbekistan	-	
V	Vanuatu	153	1990
	Venezuela	59,183	1993
	Vietnam	-	
	Virgin Islands	-	
W	Wallis and Futuna	-	
	West Bank	-	
	Western Sahara	-	
	Western Samoa	-	
	World	-	
Y	Yemen	-	
Z	Zaire	9,078	1991
	Zambia	3,302	1992
	Zimbabwe	6,189	1990
	Taiwan	-	

Appendix H:

Cross-Reference List of Geographic Names

This list indicates where various geographic items—including the location of all United States Foreign Service Posts, alternate names of countries, former names, and political or geographical portions of larger entities—can be found in *The World Factbook*. Spellings are normally, but not always, those approved by the United States Board on Geographic Names (BGN). Alternate names are included in parentheses; additional information is included in brackets.

	Name	Entry in *The World Factbook*
A	Abidjan [US Embassy]	Cote d'Ivoire
	Abu Dhabi [US Embassy]	United Arab Emirates
	Abuja [US Embassy Branch Office]	Nigeria
	Acapulco [US Consular Agency]	Mexico
	Accra [US Embassy]	Ghana
	Adamstown	Pitcairn Islands
	Adana [US Consulate]	Turkey
	Addis Ababa [US Embassy]	Ethiopia
	Adelie Land (Terre Adelie) [claimed by France]	Antarctica
	Aden	Yemen
	Aden, Gulf of	Indian Ocean
	Admiralty Islands	Papua New Guinea
	Adriatic Sea	Atlantic Ocean
	Aegean Islands	Greece
	Aegean Sea	Atlantic Ocean
	Afars and Issas, French Territory of the (F.T.A.I.)	Djibouti
	Agalega Islands	Mauritius
	Agana	Guam
	Aland Islands	Finland
	Alaska	United States
	Alaska, Gulf of	Pacific Ocean
	Aldabra Islands	Seychelles
	Alderney	Guernsey
	Aleutian Islands	United States
	Alexander Island	Antarctica
	Alexandria	Egypt
	Algiers [US Embassy]	Algeria
	Alhucemas, Penon de	Spain
	Alma-Ata (see Almaty)	Kazakhstan
	Almaty (Alma-Ata) [US Embassy]	Kazakhstan
	Alofi	Niue
	Alphonse Island	Seychelles
	Amami Strait	Pacific Ocean
	Amindivi Islands	India
	Amirante Isles	Seychelles
	Amman [US Embassy]	Jordan
	Amsterdam [US Consulate General]	Netherlands
	Amsterdam Island (Ile Amsterdam)	French Southern and Antarctic Lands
	Amundsen Sea	Pacific Ocean
	Amur	China, Russia
	Andaman Islands	India
	Andaman Sea	Indian Ocean
	Andorra la Vella	Andorra
	Anegada Passage	Atlantic Ocean
	Anglo-Egyptian Sudan	Sudan
	Anjouan	Comoros
	Ankara [US Embassy]	Turkey

	Name	Entry in *The World Factbook*
	Annobon	Equatorial Guinea
	Antananarivo [US Embassy]	Madagascar
	Antipodes Islands	New Zealand
	Antwerp [European Logistical Support Office]	Belgium
	Aozou Strip	Chad
	Apia [US Embassy]	Western Samoa
	Aqaba, Gulf of	Indian Ocean
	Arabian Sea	Indian Ocean
	Arafura Sea	Pacific Ocean
	Argun	China; Russia
	Ascension Island	Saint Helena
	Ashgabat [US Embassy]	Turkmenistan
	Ashkhabad (see Ashgabat)	Turkmenistan
	Asmara [US Embassy]	Eritrea
	Asmera (see Asmara)	Eritrea
	Assumption Island	Seychelles
	Asuncion [US Embassy]	Paraguay
	Asuncion Island	Northern Mariana Islands
	Atacama	Chile
	Athens [US Embassy]	Greece
	Attu	United States
	Auckland [US Consulate General]	New Zealand
	Auckland Islands	New Zealand
	Australes Iles (Iles Tubuai)	French Polynesia
	Avarua	Cook Islands
	Axel Heiberg Island	Canada
	Azores	Portugal
	Azov, Sea of	Atlantic Ocean
B	Bab el Mandeb	Indian Ocean
	Babuyan Channel	Pacific Ocean
	Babuyan Islands	Philippines
	Baffin Bay	Arctic Ocean
	Baffin Island	Canada
	Baghdad [US Embassy temporarily suspended; US Interests Section located in Poland's embassy in Baghdad]	Iraq
	Baki (Baku)	Azerbaijan
	Baku [US Embassy]	Azerbaijan
	Baky (Baku)	Azerbaijan
	Balabac Strait	Pacific Ocean
	Balearic Islands	Spain
	Balearic Sea (Iberian Sea)	Atlantic Ocean
	Bali Sea	Indian Ocean
	Balintang Channel	Pacific Ocean
	Balintang Islands	Philippines
	Balleny Islands	Antarctica
	Balochistan	Pakistan
	Baltic Sea	Atlantic Ocean
	Bamako [US Embassy]	Mali
	Banaba (Ocean Island)	Kiribati
	Bandar Seri Begawan [US Embassy]	Brunei
	Banda Sea	Pacific Ocean
	Bangkok [US Embassy]	Thailand
	Bangui [US Embassy]	Central African Republic
	Banjul [US Embassy]	Gambia, The
	Banks Island	Canada
	Banks Islands (Iles Banks)	Vanuatu
	Barcelona [US Consulate General]	Spain

Name	Entry in *The World Factbook*
Barents Sea	Arctic Ocean
Barranquilla [US Consulate]	Colombia
Bashi Channel	Pacific Ocean
Basilan Strait	Pacific Ocean
Bass Strait	Pacific Ocean
Basse-Terre	Guadeloupe
Basseterre	Saint Kitts and Nevis
Basutoland	Lesotho
Batan Islands	Philippines
Bavaria (Bayern)	Germany
Beagle Channel	Atlantic Ocean
Bear Island (Bjornoya)	Svalbard
Beaufort Sea	Arctic Ocean
Bechuanaland	Botswana
Beijing [US Embassy]	China
Beirut [US Embassy]	Lebanon
Belau	Palau
Belem [US Consular Agency]	Brazil
Belep Islands (Iles Belep)	New Caledonia
Belfast [US Consulate General]	United Kingdom
Belgian Congo	Zaire
Belgrade [US Embassy; US does not maintain full diplomatic relations with Serbia and Montenegro]	Serbia and Montenegro
Belize City [US Embassy]	Belize
Belle Isle, Strait of	Atlantic Ocean
Bellingshausen Sea	Pacific Ocean
Belmopan	Belize
Belorussia	Belarus
Bengal, Bay of	Indian Ocean
Bering Sea	Pacific Ocean
Bering Strait	Pacific Ocean
Berkner Island	Antarctica
Berlin [US Branch Office]	Germany
Berlin, East	Germany
Berlin, West	Germany
Bern [US Embassy]	Switzerland
Bessarabia	Romania; Moldova
Bijagos, Arquipelago dos	Guinea-Bissau
Bikini Atoll	Marshall Islands
Bilbao [US Consulate]	Spain
Bioko	Equatorial Guinea
Biscay, Bay of	Atlantic Ocean
Bishkek [US Embassy]	Kyrgyzstan
Bishop Rock	United Kingdom
Bismarck Archipelago	Papua New Guinea
Bismarck Sea	Pacific Ocean
Bissau [US Embassy]	Guinea-Bissau
Bjornoya (Bear Island)	Svalbard
Black Rock	Falkland Islands (Islas Malvinas)
Black Sea	Atlantic Ocean
Bloemfontein	South Africa
Boa Vista	Cape Verde
Bogota [US Embassy]	Colombia
Bombay [US Consulate General]	India
Bonaire	Netherlands Antilles
Bonifacio, Strait of	Atlantic Ocean
Bonin Islands	Japan
Bonn [US Embassy]	Germany

	Name	Entry in *The World Factbook*
	Bophuthatswana	South Africa
	Bora-Bora	French Polynesia
	Bordeaux [US Consulate General]	France
	Borneo	Brunei; Indonesia; Malaysia
	Bornholm	Denmark
	Bosporus	Atlantic Ocean
	Bothnia, Gulf of	Atlantic Ocean
	Bougainville Island	Papua New Guinea
	Bougainville Strait	Pacific Ocean
	Bounty Islands	New Zealand
	Brasilia [US Embassy]	Brazil
	Bratislava [US Embassy]	Slovakia
	Brazzaville [US Embassy]	Congo
	Bridgetown [US Embassy]	Barbados
	Brisbane [US Consulate]	Australia
	British East Africa	Kenya
	British Guiana	Guyana
	British Honduras	Belize
	British Solomon Islands	Solomon Islands
	British Somaliland	Somalia
	Brussels [US Embassy, US Mission to European Union (USEU), US Mission to the North Atlantic Treaty Organization (USNATO)]	Belgium
	Bucharest [US Embassy]	Romania
	Budapest [US Embassy]	Hungary
	Buenos Aires [US Embassy]	Argentina
	Bujumbura [US Embassy]	Burundi
	Burnt Pine	Norfolk Island
	Byelorussia	Belarus
C	Cabinda	Angola
	Cabot Strait	Atlantic Ocean
	Caicos Islands	Turks and Caicos Islands
	Cairo [US Embassy]	Egypt
	Calcutta [US Consulate General]	India
	Calgary [US Consulate General]	Canada
	California, Gulf of	Pacific Ocean
	Campbell Island	New Zealand
	Canal Zone	Panama
	Canary Islands	Spain
	Canberra [US Embassy]	Australia
	Canton (Guangzhou)	China
	Canton Island (Kanton Island)	Kiribati
	Cape Town [US Consulate General]	South Africa
	Caracas [US Embassy]	Venezuela
	Cargados Carajos Shoals	Mauritius
	Caroline Islands	Micronesia, Federated States of; Palau
	Caribbean Sea	Atlantic Ocean
	Carpentaria, Gulf of	Pacific Ocean
	Casablanca [US Consulate General]	Morocco
	Castries	Saint Lucia
	Cato Island	Australia
	Cayenne	French Guiana
	Cebu [US Consulate]	Philippines
	Celebes	Indonesia
	Celebes Sea	Pacific Ocean
	Celtic Sea	Atlantic Ocean
	Central African Empire	Central African Republic
	Ceuta	Spain

	Name	Entry in *The World Factbook*
	Ceylon	Sri Lanka
	Chafarinas, Islas	Spain
	Chagos Archipelago (Oil Islands)	British Indian Ocean Territory
	Channel Islands	Guernsey; Jersey
	Charlotte Amalie	Virgin Islands
	Chatham Islands	New Zealand
	Cheju-do	Korea, South
	Cheju Strait	Pacific Ocean
	Chengdu [US Consulate General]	China
	Chesterfield Islands (Iles Chesterfield)	New Caledonia
	Chiang Mai [US Consulate General]	Thailand
	Chihli, Gulf of (see Bo Hai)	Pacific Ocean
	China, People's Republic of	China
	China, Republic of	Taiwan
	Chisinau [US Embassy]	Moldova
	Choiseul	Solomon Islands
	Christmas Island [Indian Ocean]	Australia
	Christmas Island [Pacific Ocean] (Kiritimati)	Kiribati
	Chukchi Sea	Arctic Ocean
	Ciskei	South Africa
	Ciudad Juarez [US Consulate General]	Mexico
	Cluj-Napoca [US Branch Office]	Romania
	Coco, Isla del	Costa Rica
	Cocos Islands	Cocos (Keeling) Islands
	Colombo [US Embassy]	Sri Lanka
	Colon, Archipielago de (Galapagos Islands)	Ecuador
	Commander Islands (Komandorskiye Ostrova)	Russia
	Conakry [US Embassy]	Guinea
	Congo (Brazzaville)	Congo
	Congo (Kinshasa)	Zaire
	Congo (Leopoldville)	Zaire
	Con Son Islands	Vietnam
	Cook Strait	Pacific Ocean
	Copenhagen [US Embassy]	Denmark
	Coral Sea	Pacific Ocean
	Corn Islands (Islas del Maiz)	Nicaragua
	Corsica	France
	Cosmoledo Group	Seychelles
	Cotonou [US Embassy]	Benin
	Crete	Greece
	Crooked Island Passage	Atlantic Ocean
	Crozet Islands (Iles Crozet)	French Southern and Antarctic Lands
	Curacao [US Consulate General]	Netherlands Antilles
	Czechoslovakia	Czech Republic; Slovakia
D	Dahomey	Benin
	Daito Islands	Japan
	Dakar [US Embassy]	Senegal
	Daman (Damao)	India
	Damascus [US Embassy]	Syria
	Danger Atoll	Cook Islands
	Danish Straits	Atlantic Ocean
	Danzig (Gdansk)	Poland
	Dao Bach Long Vi	Vietnam
	Dardanelles	Atlantic Ocean
	Dar es Salaam [US Embassy]	Tanzania
	Davis Strait	Atlantic Ocean
	Deception Island	Antarctica
	Denmark Strait	Atlantic Ocean

	Name	Entry in *The World Factbook*
	D'Entrecasteaux Islands	Papua New Guinea
	Devon Island	Canada
	Dhahran [US Consulate General]	Saudi Arabia
	Dhaka [US Embassy]	Bangladesh
	Diego Garcia	British Indian Ocean Territory
	Diego Ramirez	Chile
	Diomede Islands	Russia [Big Diomede]; United States [Little Diomede]
	Diu	India
	Djibouti [US Embassy]	Djibouti
	Dodecanese	Greece
	Dodoma	Tanzania
	Doha [US Embassy]	Qatar
	Douala	Cameroon
	Douglas	Man, Isle of
	Dover, Strait of	Atlantic Ocean
	Drake Passage	Atlantic Ocean
	Dubai (see Dubayy)	United Arab Emirates
	Dubayy [US Consulate General]	United Arab Emirates
	Dublin [US Embassy]	Ireland
	Durban [US Consulate General]	South Africa
	Dushanbe [US Embassy]	Tajikistan
	Dutch East Indies	Indonesia
	Dutch Guiana	Suriname
E	East China Sea	Pacific Ocean
	Easter Island (Isla de Pascua)	Chile
	Eastern Channel (East Korea Strait or Tsushima Strait)	Pacific Ocean
	East Germany (German Democratic Republic)	Germany
	East Korea Strait (Eastern Channel or Tsushima Strait)	Pacific Ocean
	East Pakistan	Bangladesh
	East Siberian Sea	Arctic Ocean
	East Timor (Portuguese Timor)	Indonesia
	Edinburgh [US Consulate General]	United Kingdom
	Elba	Italy
	Ellef Ringnes Island	Canada
	Ellesmere Island	Canada
	Ellice Islands	Tuvalu
	Elobey, Islas de	Equatorial Guinea
	Enderbury Island	Kiribati
	Enewetak Atoll (Eniwetok Atoll)	Marshall Islands
	England	United Kingdom
	English Channel	Atlantic Ocean
	Eniwetok Atoll	Marshall Islands
	Epirus, Northern	Albania; Greece
	Essequibo [claimed by Venezuela]	Guyana
	Etorofu	Russia [de facto]
F	Farquhar Group	Seychelles
	Fernando de Noronha	Brazil
	Fernando Po (Bioko)	Equatorial Guinea
	Finland, Gulf of	Atlantic Ocean
	Florence [US Consulate General]	Italy
	Florida, Straits of	Atlantic Ocean
	Formosa	Taiwan
	Formosa Strait (Taiwan Strait)	Pacific Ocean
	Fortaleza [US Consular Agency]	Brazil
	Fort-de-France	Martinique
	Frankfurt am Main [US Consulate General]	Germany
	Franz Josef Land	Russia
	Freetown [US Embassy]	Sierra Leone

	Name	Entry in *The World Factbook*
	French Cameroon	Cameroon
	French Indochina	Cambodia; Laos; Vietnam
	French Guinea	Guinea
	French Sudan	Mali
	French Territory of the Afars and Issas (F.T.A.I.)	Djibouti
	French Togo	Togo
	Friendly Islands	Tonga
	Frunze (Bishkek)	Kyrgyzstan
	Fukuoka [US Consulate]	Japan
	Funafuti	Tuvalu
	Fundy, Bay of	Atlantic Ocean
	Futuna Islands (Hoorn Islands)	Wallis and Futuna
G	Gaborone [US Embassy]	Botswana
	Galapagos Islands (Archipielago de Colon)	Ecuador
	Galleons Passage	Atlantic Ocean
	Gambier Islands (Iles Gambier)	French Polynesia
	Gaspar Strait	Pacific Ocean
	Geneva [Branch Office of the US Embassy, US Mission to European Office of the UN and Other International Organizations]	Switzerland
	Genoa	Italy
	George Town [US Consular Agency]	Cayman Islands
	Georgetown [US Embassy]	Guyana
	German Democratic Republic (East Germany)	Germany
	Gibraltar	Gibraltar
	Gibraltar, Strait of	Atlantic Ocean
	Gilbert Islands	Kiribati
	Goa	India
	Gold Coast	Ghana
	Golan Heights	Syria
	Good Hope, Cape of	South Africa
	Goteborg	Sweden
	Gotland	Sweden
	Gough Island	Saint Helena
	Grand Banks	Atlantic Ocean
	Grand Cayman	Cayman Islands
	Grand Turk	Turks and Caicos Islands
	Great Australian Bight	Indian Ocean
	Great Belt (Store Baelt)	Atlantic Ocean
	Great Britain	United Kingdom
	Great Channel	Indian Ocean
	Greater Sunda Islands	Brunei; Indonesia; Malaysia
	Green Islands	Papua New Guinea
	Greenland Sea	Arctic Ocean
	Grenadines, Northern	Saint Vincent and the Grenadines
	Grenadines, Southern	Grenada
	Grytviken	Georgia
	Guadalajara [US Consulate General]	Mexico
	Guadalcanal	Solomon Islands
	Guadalupe, Isla de	Mexico
	Guangzhou [US Consulate General]	China
	Guantanamo Bay [US Naval Base]	Cuba
	Guatemala [US Embassy]	Guatemala
	Gubal, Strait of	Indian Ocean
	Guinea, Gulf of	Atlantic Ocean
	Guayaquil [US Consulate General]	Ecuador
H	Ha'apai Group	Tonga

	Name	Entry in *The World Factbook*
	Habomai Islands	Russia [de facto]
	Hague, The [US Embassy]	Netherlands
	Hainan Dao	China
	Halifax [US Consulate General]	Canada
	Halmahera	Indonesia
	Hamburg [US Consulate General]	Germany
	Hamilton [US Consulate General]	Bermuda
	Hanoi [US Liaison Office]	Vietnam
	Harare [US Embassy]	Zimbabwe
	Hatay	Turkey
	Havana [US post not maintained; representation by US Interests Section (USINT) of the Swiss Embassy]	Cuba
	Hawaii	United States
	Heard Island	Heard Island and McDonald Islands
	Helsinki [US Embassy]	Finland
	Hermosillo [US Consulate]	Mexico
	Hispaniola	Dominican Republic; Haiti
	Hokkaido	Japan
	Hong Kong [US Consulate General]	Hong Kong
	Honiara	Solomon Islands
	Honshu	Japan
	Hormuz, Strait of	Indian Ocean
	Horn, Cape (Cabo de Hornos)	Chile
	Horne, Iles de	Wallis and Futuna
	Horn of Africa	Ethiopia; Somalia
	Hudson Bay	Arctic Ocean
	Hudson Strait	Arctic Ocean
I	Inaccessible Island	Saint Helena
	Indochina	Cambodia; Laos; Vietnam
	Inner Mongolia (Nei Mongol)	China
	Ionian Islands	Greece
	Ionian Sea	Atlantic Ocean
	Irian Jaya	Indonesia
	Irish Sea	Atlantic Ocean
	Islamabad [US Embassy]	Pakistan
	Islas Malvinas	Falkland Islands (Islas Malvinas)
	Istanbul [US Consulate General]	Turkey
	Italian Somaliland	Somalia
	Ivory Coast	Cote d'Ivoire
	Iwo Jima	Japan
J	Jakarta [US Embassy]	Indonesia
	Jamestown	Saint Helena
	Japan, Sea of	Pacific Ocean
	Java	Indonesia
	Java Sea	Pacific Ocean
	Jeddah (see Jiddah)	Saudi Arabia
	Jerusalem [US Consulate General]	Israel; West Bank
	Jiddah [US Consulate General]	Saudi Arabia
	Johannesburg [US Consulate General]	South Africa
	Juan de Fuca, Strait of	Pacific Ocean
	Juan Fernandez, Isla de	Chile
	Juventud, Isla de la (Isle of Youth)	Cuba
K	Kabul [US Embassy now closed]	Afghanistan
	Kaduna [US Consulate General]	Nigeria
	Kalimantan	Indonesia
	Kamchatka Peninsula (Poluostrov Kamchatka)	Russia
	Kampala [US Embassy]	Uganda

Name	Entry in *The World Factbook*
Kampuchea	Cambodia
Kanton Island	Kiribati
Karachi [US Consulate General]	Pakistan
Kara Sea	Arctic Ocean
Karimata Strait	Pacific Ocean
Kathmandu [US Embassy]	Nepal
Kattegat	Atlantic Ocean
Kauai Channel	Pacific Ocean
Keeling Islands	Cocos (Keeling) Islands
Kerguelen, Iles	French Southern and Antarctic Lands
Kermadec Islands	New Zealand
Khabarovsk	Russia
Khartoum [US Embassy]	Sudan
Khmer Republic	Cambodia
Khuriya Muriya Islands (Kuria Muria Islands)	Oman
Khyber Pass	Pakistan
Kiel Canal (Nord-Ostsee Kanal)	Atlantic Ocean
Kiev [US Embassy]	Ukraine
Kigali [US Embassy]	Rwanda
Kingston [US Embassy]	Jamaica
Kingston	Norfolk Island
Kingstown	Saint Vincent and the Grenadines
Kinshasa [US Embassy]	Zaire
Kirghiziya	Kyrgyzstan
Kiritimati (Christmas Island)	Kiribati
Kishinev (Chisinau)	Moldova
Kithira Strait	Atlantic Ocean
Kodiak Island	United States
Kola Peninsula (Kol'skiy Poluostrov)	Russia
Kolonia [US Embassy]	Micronesia, Federated States of
Korea Bay	Pacific Ocean
Korea, Democratic People's Republic of	Korea, North
Korea, Republic of	Korea, South
Korea Strait	Pacific Ocean
Koror [US Liaison Office]	Palau
Kosovo	Serbia and Montenegro
Kowloon	Hong Kong
Krakow [US Consulate General]	Poland
Kuala Lumpur [US Embassy]	Malaysia
Kunashiri (Kunashir)	Russia [de facto]
Kuril Islands	Russia [de facto]
Kuwait [US Embassy]	Kuwait
Kwajalein Atoll	Marshall Islands
Kyushu	Japan
Kyyiv (Kiev)	Ukraine

L

Name	Entry in *The World Factbook*
Labrador	Canada
Laccadive Islands	India
Laccadive Sea	Indian Ocean
Lagos [US Embassy]	Nigeria
Lahore [US Consulate General]	Pakistan
Lakshadweep	India
La Paz [US Embassy]	Bolivia
La Perouse Strait	Pacific Ocean
Laptev Sea	Arctic Ocean
Las Palmas	Spain
Lau Group	Fiji
Lefkosa (Nicosia)	Cyprus
Leipzig [US Consulate General]	Germany

	Name	Entry in *The World Factbook*
	Leningrad (see Saint Petersburg)	Russia
	Lesser Sunda Islands	Indonesia
	Leyte	Philippines
	Liancourt Rocks [claimed by Japan]	Korea, South
	Libreville [US Embassy]	Gabon
	Ligurian Sea	Atlantic Ocean
	Lilongwe [US Embassy]	Malawi
	Lima [US Embassy]	Peru
	Lincoln Sea	Arctic Ocean
	Line Islands	Kiribati; Palmyra Atoll
	Lisbon [US Embassy]	Portugal
	Ljubljana [US Embassy]	Slovenia
	Lobamba	Swaziland
	Lombok Strait	Indian Ocean
	Lome [US Embassy]	Togo
	London [US Embassy]	United Kingdom
	Longyearbyen	Svalbard
	Lord Howe Island	Australia
	Louisiade Archipelago	Papua New Guinea
	Loyalty Islands (Iles Loyaute)	New Caledonia
	Luanda [US Embassy]	Angola
	Lubumbashi	Zaire
	Lusaka [US Embassy]	Zambia
	Luxembourg [US Embassy]	Luxembourg
	Luzon	Philippines
	Luzon Strait	Pacific Ocean
M	Macao	Macau
	Macau	Macau
	Macedonia	Macedonia, The Former Yugoslav Republic of
	Macquarie Island	Australia
	Madeira Islands	Portugal
	Madras [US Consulate General]	India
	Madrid [US Embassy]	Spain
	Magellan, Strait of	Atlantic Ocean
	Maghreb	Algeria, Libya, Mauritania, Morocco, Tunisia
	Mahe Island	Seychelles
	Maiz, Islas del (Corn Islands)	Nicaragua
	Majorca (Mallorca)	Spain
	Majuro [US Embassy]	Marshall Islands
	Makassar Strait	Pacific Ocean
	Malabo [US Embassy]	Equatorial Guinea
	Malacca, Strait of	Indian Ocean
	Malagasy Republic	Madagascar
	Male [US Consular Agency]	Maldives
	Mallorca (Majorca)	Spain
	Malpelo, Isla de	Colombia
	Malta Channel	Atlantic Ocean
	Malvinas, Islas	Falkland Islands (Islas Malvinas)
	Mamoutzou	Mayotte
	Managua [US Embassy]	Nicaragua
	Manama [US Embassy]	Bahrain
	Manaus [US Consular Agency]	Brazil
	Manchukuo	China
	Manchuria	China
	Manila [US Embassy]	Philippines
	Manipa Strait	Pacific Ocean
	Mannar, Gulf of	Indian Ocean
	Manua Islands	American Samoa

	Name	Entry in *The World Factbook*
	Maputo [US Embassy]	Mozambique
	Marcus Island (Minami-tori-shima)	Japan
	Mariana Islands	Guam; Northern Mariana Islands
	Marion Island	South Africa
	Marmara, Sea of	Atlantic Ocean
	Marquesas Islands (Iles Marquises)	French Polynesia
	Marseille [US Consulate General]	France
	Martin Vaz, Ilhas	Brazil
	Mas a Tierra (Robinson Crusoe Island)	Chile
	Mascarene Islands	Mauritius; Reunion
	Maseru [US Embassy]	Lesotho
	Matamoros [US Consulate]	Mexico
	Mata-Utu	Wallis and Futuna
	Mazatlan	Mexico
	Mbabane [US Embassy]	Swaziland
	McDonald Islands	Heard Island and McDonald Islands
	Medan [US Consulate General]	Indonesia
	Mediterranean Sea	Atlantic Ocean
	Melbourne [US Consulate General]	Australia
	Melilla	Spain
	Merida [US Consulate]	Mexico
	Messina, Strait of	Atlantic Ocean
	Mexico [US Embassy]	Mexico
	Mexico, Gulf of	Atlantic Ocean
	Milan [US Consulate General]	Italy
	Minami-tori-shima	Japan
	Mindanao	Philippines
	Mindoro Strait	Pacific Ocean
	Minicoy Island	India
	Minsk [US Embassy]	Belarus
	Mogadishu	Somalia
	Moldavia	Moldova
	Mombasa	Kenya
	Mona Passage	Atlantic Ocean
	Monaco	Monaco
	Monrovia [US Embassy]	Liberia
	Montenegro	Serbia and Montenegro
	Monterrey [US Consulate General]	Mexico
	Montevideo [US Embassy]	Uruguay
	Montreal [US Consulate General, US Mission to the International Civil Aviation Organization (ICAO)]	Canada
	Moravian Gate	Czech Republic
	Moroni	Comoros
	Mortlock Islands	Micronesia, Federated States of
	Moscow [US Embassy]	Russia
	Mozambique Channel	Indian Ocean
	Munich [US Consulate General]	Germany
	Musandam Peninsula	Oman; United Arab Emirates
	Muscat [US Embassy]	Oman
	Muscat and Oman	Oman
	Myanma, Myanmar	Burma
N	Nagoya [US Consulate]	Japan
	Naha [US Consulate General]	Japan
	Nairobi [US Embassy]	Kenya
	Nampo-shoto	Japan
	Naples [US Consulate General]	Italy
	Nassau [US Embassy]	Bahamas, The
	Natuna Besar Islands	Indonesia

	Name	Entry in *The World Factbook*
	N'Djamena [US Embassy]	Chad
	Netherlands East Indies	Indonesia
	Netherlands Guiana	Suriname
	Nevis	Saint Kitts and Nevis
	New Delhi [US Embassy]	Indi
	Newfoundland	Canada
	New Guinea	Indonesia; Papua New Guinea
	New Hebrides	Vanuatu
	New Siberian Islands	Russia
	New Territories	Hong Kong
	New York, New York [US Mission to the United Nations (USUN)]	United States
	Niamey [US Embassy]	Niger
	Nicobar Islands	India
	Nicosia [US Embassy]	Cyprus
	Nightingale Island	Saint Helena
	North Atlantic Ocean	Atlantic Ocean
	North Channel	Atlantic Ocean
	Northeast Providence Channel	Atlantic Ocean
	Northern Epirus	Albania; Greece
	Northern Grenadines	Saint Vincent and the Grenadines
	Northern Ireland	United Kingdom
	Northern Rhodesia	Zambia
	North Island	New Zealand
	North Korea	Korea, North
	North Pacific Ocean	Pacific Ocean
	North Sea	Atlantic Ocean
	North Vietnam	Vietnam
	Northwest Passages	Arctic Ocean
	North Yemen (Yemen Arab Republic)	Yemen
	Norwegian Sea	Atlantic Ocean
	Nouakchott [US Embassy]	Mauritania
	Noumea	New Caledonia
	Novaya Zemlya	Russia
	Nuku'alofa	Tonga
	Nuevo Laredo [US Consulate]	Mexico
	Nuuk (Godthab)	Greenland
	Nyasaland	Malawi
O	Oahu	United States
	Ocean Island (Banaba)	Kiribati
	Ocean Island (Kure Island)	United States
	Ogaden	Ethiopia; Somalia
	Oil Islands (Chagos Archipelago)	British Indian Ocean Territory
	Okhotsk, Sea of	Pacific Ocean
	Okinawa	Japan
	Oman, Gulf of	Indian Ocean
	Ombai Strait	Pacific Ocean
	Oran	Algeria
	Oranjestad	Aruba
	Oresund (The Sound)	Atlantic Ocean
	Orkney Islands	United Kingdom
	Osaka-Kobe [US Consulate General]	Japan
	Oslo [US Embassy]	Norway
	Otranto, Strait of	Atlantic Ocean
	Ottawa [US Embassy]	Canada
	Ouagadougou [US Embassy]	Burkina
	Outer Mongolia	Mongolia
P	Pacific Islands, Trust Territory of the	Palau

Name	Entry in *The World Factbook*
Pagan	Northern Mariana Islands
Pago Pago	American Samoa
Palawan	Philippines
Palermo	Italy
Palk Strait	Indian Ocean
Pamirs	China; Tajikistan
Panama [US Embassy]	Panama
Panama Canal	Panama
Panama, Gulf of	Pacific Ocean
Papeete	French Polynesia
Paramaribo [US Embassy]	Suriname
Parece Vela	Japan
Paris [US Embassy, US Mission to the Organization for Economic Cooperation and Development (OECD), US Observer Mission at the UN Educational, Scientific, and Cultural Organization (UNESCO)]	France
Pascua, Isla de (Easter Island)	Chile
Passion, Ile de la	Clipperton Island
Pashtunistan	Afghanistan; Pakistan
Peking (Beijing)	China
Peleliu	Palau
Pemba Island	Tanzania
Pentland Firth	Atlantic Ocean
Perim	Yemen
Perouse Strait, La	Pacific Ocean
Persian Gulf	Indian Ocean
Perth [US Consulate General]	Australia
Pescadores	Taiwan
Peshawar [US Consulate]	Pakistan
Peter I Island	Antarctica
Philip Island	Norfolk Island
Philippine Sea	Pacific Ocean
Phnom Penh [US Embassy]	Cambodia
Phoenix Islands	Kiribati
Pines, Isle of (Isla de la Juventud)	Cuba
Pleasant Island	Nauru
Plymouth	Montserrat
Ponape (Pohnpei)	Micronesia
Ponta Delgada [US Consulate]	Portugal
Port-au-Prince [US Embassy]	Haiti
Port Louis [US Embassy]	Mauritius
Port Moresby [US Embassy]	Papua New Guinea
Porto Alegre [US Consulate]	Brazil
Port-of-Spain [US Embassy]	Trinidad and Tobago
Porto-Novo	Benin
Portuguese Guinea	Guinea-Bissau
Portuguese Timor (East Timor)	Indonesia
Port-Vila	Vanuatu
Poznan [US Consulate General]	Poland
Prague [US Embassy]	Czech Republic
Praia [US Embassy]	Cape Verde
Pretoria [US Embassy]	South Africa
Pribilof Islands	United States
Prince Edward Island	Canada
Prince Edward Islands	South Africa
Prince Patrick Island	Canada
Principe	Sao Tome and Principe
Pusan [US Consulate]	Korea, South

	Name	Entry in *The World Factbook*
	P'yongyang	Korea, North
Q	Quebec [US Consulate General]	Canada
	Queen Charlotte Islands	Canada
	Queen Elizabeth Islands	Canada
	Queen Maud Land [claimed by Norway]	Antarctica
	Quito [US Embassy]	Ecuador
R	Rabat [US Embassy]	Morocco
	Ralik Chain	Marshall Islands
	Rangoon [US Embassy]	Burma
	Ratak Chain	Marshall Islands
	Recife [US Consulate]	Brazil
	Redonda	Antigua and Barbuda
	Red Sea	Indian Ocean
	Revillagigedo Island	United States
	Revillagigedo Islands	Mexico
	Reykjavik [US Embassy]	Iceland
	Rhodes	Greece
	Rhodesia	Zimbabwe
	Rhodesia, Northern	Zambia
	Rhodesia, Southern	Zimbabwe
	Riga [US Embassy]	Latvia
	Rio de Janeiro [US Consulate General]	Brazil
	Rio de Oro	Western Sahara
	Rio Muni	Equatorial Guinea
	Riyadh [US Embassy]	Saudi Arabia
	Road Town	British Virgin Islands
	Robinson Crusoe Island (Mas a Tierra)	Chile
	Rocas, Atol das	Brazil
	Rockall [disputed]	United Kingdom
	Rodrigues	Mauritius
	Rome [US Embassy, US Mission to the UN Agencies for Food and Agriculture (FODAG)]	Italy
	Roncador Cay	Colombia
	Roosevelt Island	Antarctica
	Roseau	Dominica
	Ross Dependency [claimed by New Zealand]	Antarctica
	Ross Island	Antarctica
	Ross Sea	Antarctica
	Rota	Northern Mariana Islands
	Rotuma	Fiji
	Ryukyu Islands	Japan
S	Saba	Netherlands Antilles
	Sabah	Malaysia
	Sable Island	Canada
	Sahel	Burkina, Cape Verde, Chad, The Gambia, Guinea-Bissau, Mali, Mauritania, Niger, Senegal
	Saigon (Ho Chi Minh City)	Vietnam
	Saint Brandon	Mauritius
	Saint Christopher and Nevis	Saint Kitts and Nevis
	Saint-Denis	Reunion
	Saint George's [US Embassy]	Grenada
	Saint George's Channel	Atlantic Ocean
	Saint Helier	Jersey
	Saint John's	Antigua and Barbuda
	Saint Lawrence, Gulf of	Atlantic Ocean
	Saint Lawrence Island	United States
	Saint Lawrence Seaway	Atlantic Ocean

Name	Entry in *The World Factbook*
Saint Martin	Guadeloupe
Saint Martin (Sint Maarten)	Netherlands Antilles
Saint Paul Island	Canada
Saint Paul Island	United States
Saint Paul Island (Ile Saint-Paul)	French Southern and Antarctic Lands
Saint Peter and Saint Paul Rocks (Penedos de Sao Pedro e Sao Paulo)	Brazil
Saint Peter Port	Guernsey
Saint Petersburg [US Consulate General]	Russia
Saint-Pierre	Saint Pierre and Miquelon
Saint Vincent Passage	Atlantic Ocean
Saipan	Northern Mariana Islands
Sakhalin Island (Ostrov Sakhalin)	Russia
Sala y Gomez, Isla	Chile
Salisbury (Harare)	Zimbabwe
Salvador de Bahia [US Consular Agency]	Brazil
Salzburg	Austria
Sanaa [US Embassy]	Yemen
San Ambrosio	Chile
San Andres y Providencia, Archipielago	Colombia
San Bernardino Strait	Pacific Ocean
San Felix, Isla	Chile
San Jose [US Embassy]	Costa Rica
San Juan	Puerto Rico
San Luis Potosi	Mexico
San Marino	San Marino
San Salvador [US Embassy]	El Salvador
Santa Cruz	Bolivia
Santa Cruz Islands	Solomon Islands
Santiago [US Embassy]	Chile
Santo Domingo [US Embassy]	Dominican Republic
Sao Paulo [US Consulate General]	Brazil
Sao Pedro e Sao Paulo, Penedos de	Brazil
Sao Tome	Sao Tome and Principe
Sapporo [US Consulate General]	Japan
Sapudi Strait	Pacific Ocean
Sarajevo [US Embassy]	Bosnia and Herzegovina
Sarawak	Malaysia
Sardinia	Italy
Sargasso Sea	Atlantic Ocean
Sark	Guernsey
Scotia Sea	Atlantic Ocean
Scotland	United Kingdom
Scott Island	Antarctica
Senyavin Islands	Micronesia, Federated States of
Seoul [US Embassy]	Korea, South
Serbia	Serbia and Montenegro
Serrana Bank	Colombia
Serranilla Bank	Colombia
Settlement, The	Christmas Island
Severnaya Zemlya (Northland)	Russia
Shag Island	Heard Island and McDonald Islands
Shag Rocks	Falkland Islands (Islas Malvinas)
Shanghai [US Consulate General]	China
Shenyang [US Consulate General]	China
Shetland Islands	United Kingdom
Shikoku	Japan
Shikotan (Shikotan-to)	Japan

Name	Entry in *The World Factbook*
Siam	Thailand
Sibutu Passage	Pacific Ocean
Sicily	Italy
Sicily, Strait of	Atlantic Ocean
Sikkim	India
Sinai	Egypt
Singapore [US Embassy]	Singapore
Singapore Strait	Pacific Ocean
Sinkiang (Xinjiang)	China
Sint Eustatius	Netherlands Antilles
Sint Maarten (Saint Martin)	Netherlands Antilles
Skagerrak	Atlantic Ocean
Skopje [US Liaison Office]	Macedonia, The Former Yugoslav Republic of
Society Islands (Iles de la Societe)	French Polynesia
Socotra	Yemen
Sofia [US Embassy]	Bulgaria
Solomon Islands, northern	Papua New Guinea
Solomon Islands, southern	Solomon Islands
Solomon Sea	Pacific Ocean
Songkhla [US Consulate]	Thailand
Sound, The (Oresund)	Atlantic Ocean
South Atlantic Ocean	Atlantic Ocean
South China Sea	Pacific Ocean
Southern Grenadines	Grenada
Southern Rhodesia	Zimbabwe
South Georgia	South Georgia and the South Sandwich Islands
South Island	New Zealand
South Korea	Korea, South
South Orkney Islands	Antarctica
South Pacific Ocean	Pacific Ocean
South Sandwich Islands	South Georgia and the South Sandwich Islands
South Shetland Islands	Antarctica
South Tyrol	Italy
South Vietnam	Vietnam
South-West Africa	Namibia
South Yemen (People's Democratic Republic of Yemen)	Yemen
Soviet Union [the former]	Armenia, Azerbaijan, Belarus, Estonia, Georgia, Kazakhstan, Kyrgyzstan, Latvia, Lithuania, Moldova, Russia, Tajikistan, Turkmenistan, Ukraine, Uzbekistan
Spanish Guinea	Equatorial Guinea
Spanish Sahara	Western Sahara
Spitsbergen	Svalbard
Stanley	Falkland Islands (Islas Malvinas)
Stockholm [US Embassy]	Sweden
Strasbourg [US Consulate General]	France
Stuttgart [US Consulate General]	Germany
Sucre	Bolivia
Suez, Gulf of	Indian Ocean
Sulu Archipelago	Philippines
Sulu Sea	Pacific Ocean
Sumatra	Indonesia
Sumba	Indonesia
Sunda Islands (Soenda Isles)	Indonesia; Malaysia
Sunda Strait	Indian Ocean
Surabaya [US Consulate General]	Indonesia
Surigao Strait	Pacific Ocean
Surinam	Suriname
Suva [US Embassy]	Fiji

	Name	Entry in *The World Factbook*
	Swains Island	American Samoa
	Swan Islands	Honduras
	Sydney [US Consulate General]	Australia
T	Tahiti	French Polynesia
	Taipei	Taiwan
	Taiwan Strait	Pacific Ocean
	Tallinn [US Embassy]	Estonia
	Tanganyika	Tanzania
	Tangier	Morocco
	Tarawa	Kiribati
	Tartar Strait	Pacific Ocean
	Tashkent [US Embassy]	Uzbekistan
	Tasmania	Australia
	Tasman Sea	Pacific Ocean
	Taymyr Peninsula (Poluostrov Taymyra)	Russia
	T'bilisi [US Embassy]	Georgia
	Tegucigalpa [US Embassy]	Honduras
	Tehran [US post not maintained; representation by Swiss Embassy]	Iran
	Tel Aviv [US Embassy]	Israel
	Terre Adelie (Adelie Land) [claimed by France]	Antarctica
	Thailand, Gulf of	Pacific Ocean
	Thessaloniki [US Consulate General]	Greece
	Thimphu	Bhutan
	Thurston Island	Antarctica
	Tibet (Xizang)	China
	Tibilisi (see T'bilisi)	Georgia
	Tierra del Fuego	Argentina; Chile
	Tijuana [US Consulate General]	Mexico
	Timor	Indonesia
	Timor Sea	Pacific Ocean
	Tinian	Northern Mariana Islands
	Tiran, Strait of	Indian Ocean
	Tirane [US Embassy]	Albania
	Tobago	Trinidad and Tobago
	Tokyo [US Embassy]	Japan
	Tonkin, Gulf of	Pacific Ocean
	Toronto [US Consulate General]	Canada
	Torres Strait	Pacific Ocean
	Torshavn	Faroe Islands
	Toshkent (Tashkent)	Uzbekistan
	Transjordan	Jordan
	Transkei	South Africa
	Transylvania	Romania
	Trindade, Ilha de	Brazil
	Tripoli [US post not maintained; representation by Belgian Embassy]	Libya
	Tristan da Cunha Group	Saint Helena
	Trobriand Islands	Papua New Guinea
	Trucial States	United Arab Emirates
	Truk Islands	Micronesia
	Tsugaru Strait	Pacific Ocean
	Tuamotu Islands (Iles Tuamotu)	French Polynesia
	Tubuai Islands (Iles Tubuai)	French Polynesia
	Tunis [US Embassy]	Tunisia
	Turin	Italy
	Turkish Straits	Atlantic Ocean
	Turkmeniya	Turkmenistan

	Name	Entry in *The World Factbook*
	Turks Island Passage	Atlantic Ocean
	Tyrol, South	Italy
	Tyrrhenian Sea	Atlantic Ocean
U	Udorn (Udon Thani) [US Consulate]	Thailand
	Ulaanbaatar [US Embassy]	Mongolia
	Ullung-do	Korea, South
	Unimak Pass [strait]	Pacific Ocean
	Union of Soviet Socialist Republics [the former USSR]	Armenia, Azerbaijan, Belarus, Estonia, Georgia, Kazakhstan, Kyrgyzstan, Latvia, Lithuania, Moldova, Russia, Tajikistan, Turkmenistan, Ukraine, Uzbekistan
	United Arab Republic	Egypt; Syria
	Upper Volta	Burkina
	USSR [the former]	Armenia, Azerbaijan, Belarus, Estonia, Georgia, Kazakhstan, Kyrgyzstan, Latvia, Lithuania, Moldova, Russia, Tajikistan, Turkmenistan, Ukraine, Uzbekistan
V	Vaduz [US post not maintained; representation from Zurich, Switzerland]	Liechtenstein
	Vakhan (Wakhan Corridor)	Afghanistan
	Valletta [US Embassy]	Malta
	Valley, The	Anguilla
	Vancouver [US Consulate General]	Canada
	Vancouver Island	Canada
	Van Diemen Strait	Pacific Ocean
	Vatican City [US Embassy]	Holy See
	Velez de la Gomera, Penon de	Spain
	Venda	South Africa
	Verde Island Passage	Pacific Ocean
	Victoria	Hong Kong
	Victoria [US Embassy]	Seychelles
	Vienna [US Embassy, US Mission to International Organizations in Vienna (UNVIE)]	Austria
	Vientiane [US Embassy]	Laos
	Vilnius [US Embassy]	Lithuania
	Vladivostok [US Consulate General]	Russia
	Volcano Islands	Japan
	Vostok Island	Kiribati
	Vrangelya, Ostrov (Wrangel Island)	Russia
W	Wakhan Corridor (now Vakhan)	Afghanistan
	Wales	United Kingdom
	Walvis Bay	Namibia
	Warsaw [US Embassy]	Poland
	Washington, DC [The Permanent Mission of the US to the Organization of American States (OAS)]	United States
	Weddell Sea	Atlantic Ocean
	Wellington [US Embassy]	New Zealand
	Western Channel (West Korea Strait)	Pacific Ocean
	West Germany (Federal Republic of Germany)	Germany
	West Island	Cocos (Keeling) Islands
	West Korea Strait (Western Channel)	Pacific Ocean
	West Pakistan	Pakistan
	Wetar Strait	Pacific Ocean
	White Sea	Arctic Ocean
	Willemstad	Netherlands Antilles
	Windhoek [US Embassy]	Namibia
	Windward Passage	Atlantic Ocean
	Wrangel Island (Ostrov Vrangelya)	Russia [de facto]
Y	Yamoussoukro	Cote d'Ivoire

	Name	Entry in *The World Factbook*
	Yangon (Rangoon)	Burma
	Yaounde [US Embassy]	Cameroon
	Yap Islands	Micronesia
	Yaren	Nauru
	Yekaterinburg [US Consulate General]	Russia
	Yellow Sea	Pacific Ocean
	Yemen (Aden) [People's Democratic Republic of Yemen]	Yemen
	Yemen Arab Republic	Yemen
	Yemen, North [Yemen Arab Republic]	Yemen
	Yemen (Sanaa) [Yemen Arab Republic]	Yemen
	Yemen, People's Democratic Republic of	Yemen
	Yemen, South [People's Democratic Republic of Yemen]	Yemen
	Yerevan [US Embassy]	Armenia
	Youth, Isle of (Isla de la Juventud)	Cuba
	Yucatan Channel	Atlantic Ocean
	Yugoslavia [the former]	Bosnia and Herzegovina; Croatia; Serbia and Montenegro; Slovenia; and Macedonia, The Former Yugoslav Republic of
Z	Zagreb [US Embassy]	Croatia
	Zanzibar	Tanzania
	Zurich [US Consulate General]	Switzerland

ASTHEET
CIRCLE'S END
THE EVER STONES
SPACE
LISTAAN
OPTA
NEW GANNATH ALLIANCE
EDGE OF LIGHT
NATH
SOUL'S GRASP
ADOS
GOYA/DANN
WAYWARD
GRASSKAN FREEHOLD
VINNSHASHA
WIRUM
NATHOS
SLAUGHTER
VINN-INFECTED SPACE
TELFAAR
TIN
SASEET
NICTO
CARINA-SAGITTARIUS ARM
LORCA
PAA'Q
Y0-CMQ-858
DRIVE CORPS
N9V6-M4-38AA-CD1
WARNING: THIS MAP, THOUGH INTENTIONALLY GENERALIZED, IS NONETHELESS CODED TO YOUR DNA, AND FALLS UNDER THE CAPTAIN'S DICTATE

ACT TWO

TO MY DEAR FRIEND FRED SCHROEDER,
The only person who's been told the ending of the DRIVE story.
(Even though I think he promptly forgot it.)

SPECIAL THANKS:

To the awesome readers who directly
support DRIVE at Patreon.com/Drive.
Your kindness and support has made it
possible for me to tell this story
in the best possible way.
You make DRIVE a delight to create.

And to David Malki !, who kindly helped
usher this book to production.

small
FISH

DRIVE: ACT TWO

First Printing, 2018. Published by Small Fish Studios, Inc., a California Arts Corporation.

DRIVECOMIC.COM

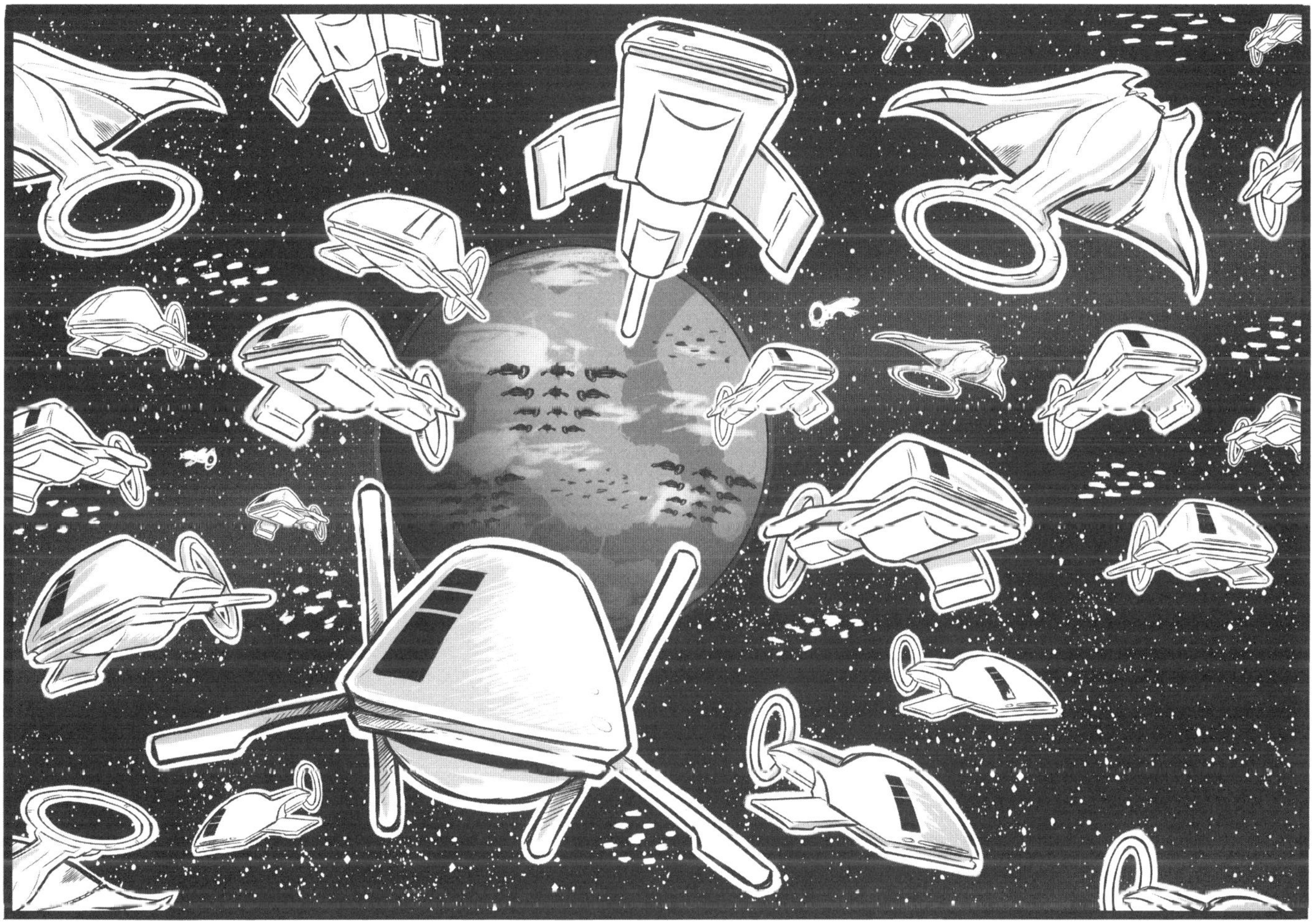

DRIVE is written and drawn by: **DAVE KELLETT** | With colors and zhuzhing by: **BETH MORRELL**

THE STORY SO FAR

The year is 2401, and a Second Spanish Empire rules the Earth.

The Empire was built on the back of an amazing interstellar drive invented by Conrado Cruz. He and his descendants have gone on to rule all of humanity, as they maintain sole control of the drive. Whoever controls space flight, controls the ultimate levers of power. As such, "La Familia" control the banks, the courts, the military, and politics. Under their rule, humanity's domain has come to include other species – such as Veetans, Fillipods, Tesskans, and more.

"Tierra," the homeworld of humanity, reigns supreme.

But a new threat is gnawing at the Empire: The Continuum of Makers. These Makers claim that humanity has stolen their drive – and want it back with an almost religious fervor. To Makers, the "spirits" you "birth" into the world are sacred, personal, and prized. In their complex animistic religion, humanity's theft of such a spirit is a profound offense.

It's war between the Empire and the Continuum.

Humanity will lose, and lose badly, unless it can find some advantage in battle. That hope arrives in the form of a tiny, mysterious creature who can drive a starship like no one's ever seen. The Emperor tasks the scout ship Machito to find these squirrel-sized creatures and get them flying for humanity. The crew of the Machito – Captain Taneel, La Familia engineer Fernando Cruz, the gentle Veetan Nosh, the creative Fillipod Cuddow, and the Jinyiwei spy Orla O'Malley – have searched Earth and Veeta without luck. So they go to the greatest nexus of alien species: The mob trading post, Slaughter.

On Slaughter, the crew accidentally offends the local mob boss, which lands Orla in a coliseum death match. She proves herself surprisingly resilient, until a four-story tall Fekk Dragon almost kills her. Nosh enters the ring to save her, only to become infected by the Vinn – a "species" no one has seen in 70 years. The crew escapes with Nosh and flies to a secret Jinyiwei facility on Nuevo Chile to try and save him.

But a third galactic power has now raised its head: The Vinn are once again on the march.

And so, already losing against one great power, humanity is now staring down a tripartite war against the Continuum **and** the Vinn....

BEGIN  ACT TWO

IN ITS EARLY HEYDAY, *DU FU* WAS THE TECHNOLOGICAL CENTER OF HUMANITY. ENDLESS INNOVATIONS POURED FROM THE PLANET. HENCE THE OLD SLANG, "WHAT'S NEW FROM DU FU?"
BUT BY 2290, THE TECH MANDARINS OF DU FU FOUND THEMSELVES OUTPACED BY A NEW COLONY: ***NUEVO CHILE***

LED BY THE CHARISMATIC FAMILIA GOVERNOR, ***CÉSAR CRUZ DE TALCA,*** NUEVO CHILE GREW FASTER THAN ANY HUMAN COLONY EVER HAD
EMPIEZA tu NUEVA VIDA

THE "CESARISTAS," AS HIS LOYAL COLONISTS CAME TO BE KNOWN, PERFORMED ALMOST SUPER-HUMAN FEATS OF ENGINEERING.
ROADS, PORTS, UNIVERSITIES, MANUFACTURING CENTERS -- THEY ALL APPEARED OVERNIGHT.

AND AS THE TRADE FLOWED IN, POWER AND INFLUENCE FLOWED WITH IT. THE BEST MINDS OF BEIJING UNIVERSITY NO LONGER LEAPED TO DU FU: THEY LEAPED TO NUEVO CHILE.

WITHIN A GENERATION, THE CESARISTAS GREW CONFIDENT IN THEIR POWER, AND SPOKE OPENLY OF DEMOCRACY *POST-FAMILIA*, AND OF REVERSE-ENGINEERING THE DRIVE.
¡Libertad!

EMPEROR PABLO TOLERATED THEIR "MOVEMENT," BUT HIS CRUEL SON OSVALDO DID NOT. OSVALDO CRUSHED THE CESARISTAS, KILLED HIS AGING COUSIN CÉSAR, AND SET OUT TO MAKE AN EXAMPLE OF NUEVO CHILE, WITH ***THE CRUELEST IMPERIAL PUNISHMENT EVER:***

HE CUT THEM OFF. FROM TRADE, FROM TRAVEL ...FROM ***HUMANITY***. ALL RING TRAVEL BANNED.
THEY WERE TO BE ALONE IN THE UNIVERSE ...FOREVER.

BANNING NUEVO CHILE FROM RING TRAVEL SUCCEEDED BEYOND OSVALDO'S WILDEST DREAMS.
IT HUSHED EVERY VOICE CLAMORING FOR DEMOCRACY.
disperse!
¡Libertad!

AND ESTABLISHED THE JINYIWEI AS THE ALL-POWERFUL SWORD-HAND OF THE EMPEROR

MOST OF ALL: IT SOLIDIFIED THE EMPEROR'S GRIP ON THE COLONIES. NO GOVERNOR WOULD EVER CROSS HIM AGAIN.

ALL FLIGHTS TO NUEVO CHILE CANCELLED
BUT AS TERRIFYING AS OSVALDO'S BAN WAS... THE MOST FIENDISH PART WAS THIS: NUEVO CHILE WAS NEVER TRULY BANNED.

OH, IN THE EYES OF THE PUBLIC, LA FAMILIA, AND THE FLEET, IT WAS ABSOLUTELY BANNED FOREVER.
LUNA
MARS
las HERMANITAS
DU FU
DAWN
NUEVO CHILE
PAA'Q

BUT QUIETLY, SECRETLY, POWERFULLY...NUEVO CHILE BECAME HIS AND HIS ALONE.
THE PLANET-WIDE BASE OF THE JINYIWEI

WHERE ARE THESE COORDINATES TAKING US? THIS LOOKS LIKE...THE FAR SIDE OF THE ORION SPUR.
IT IS. THE JINYIWEI ARE TAKING US TO AN UNLUCKY PLANET CALLED "NUEVO CHILE"
YOU KNOW THOSE COORDINATES?? HOW DO YOU KNOW THOSE COORDINATES?
EN BOCA CERRADA, NO ENTRAN MOSCAS.
SUFFICE IT TO SAY, I AM A WOMAN VERY MUCH AWARE OF THE WORLD SHE INHABITS, AND HAS ARMED HERSELF ACCORDINGLY.
YOU WILL NOT CROSS ME AGAIN, ORLA. THE EMPEROR HAS MADE IT CLEAR THAT YOU'RE STAYING HERE, AND I CAN'T OVERRULE THAT...
CAPTAIN, I AM NOT YOUR ENEMY HERE
BUT WHATEVER THE EMPEROR SAYS, THIS SHIP IS MY EMPIRE, AND YOU WILL ANSWER TO ME, WHILE ONBOARD.
Y..YES, MA'AM
CREW! LISTEN UP! WE ARE FLYING INTO A LASHING DEN OF VIPERS...
YOU WILL TOUCH NOTHING, YOU WILL LOOK AT NOTHING, AND YOU WILL TALK TO NO ONE. WE ARE NOT GETTING PULLED INTO THEIR NONSENSE.
WE HAVE ONE JOB ON THAT PLANET: HEAL NOSH. AND WE WILL DO NOTHING ELSE WHILE WE'RE THERE. DO YOU HEAR ME? NOT ONE THING.
C... ...CAN I PEE?
Yes Cuddow.
FOR GOD'S SAKE

CAPTAIN, IS THAT TESSKIL TO STARBOARD?
IT IS. I PROMISED CUDDOW WE'D QUICKLY FLY PAST ON OUR WAY BACK
OH GREATEST JOY! TO CATCH A GLIMPSE OF HOME!
TREMENDOUS THINGS ARE HAPPENING ON TESSKIL!
DEFINE "TREMENDOUS". BECAUSE I'M SEEING SOME FREAK-FACTOR-5000 STUFF HAPPENING WITH THAT PLANET'S GRAVITY
OH, I BET! THE FILLIPOD SCIENCE AUTHORITY IS ATTEMPTING TO RECONSTITUTE THE SOUTHERN HEMISPHERE. NOTHING LIKE IT HAS EVER BEEN ATTEMPTED BEFORE!
IT'S ALL VERY EXCITING..AND TERRIFYING! I'VE BEEN GETTING CONSTANT UPDATES FROM MY COUSIN PUDDY TATT.
Whoa whoa whoa... DID YOU SAY PUDDY TATT?
sigh
YES. THAT IS HIS NAME. AND YES, I KNOW IT MEANS SOMETHING ELSE IN EARTH HYBRID. BUT IT'S A COMPLETE PHONETIC COINCIDENCE.
PUDDY TATT IS AN ANCIENT AND NOBLE FILLIPOD NAME

ANYWAY... THE TASK IS IMMENSE!
THERE ARE CONTINENT-SIZED PIECES OF MANTLE IN UNSTABLE ORBITS AROUND THE PLANET, AND THEY'RE ATTEMPTING TO SAFELY BRING THEM DOWN.
THE REUNION OF THE FIRST PIECE CAUSED MASSIVE EARTHQUAKES... SO THEY'VE HAD TO INVENT ENTIRELY NEW GRAVITY DAMPENING SYSTEMS
AND THE CLOCK IS TICKING! THOSE ORBITS ARE DECAYING RAPIDLY, SO EVERY FILLIPOD IS WORKING THE PROBLEM.
BUT! THE HEAD OF THE FILLIPOD SCIENCE AUTHORITY, ***FIRST MINISTER HUGGSTABLE***, IS CONFIDENT IT CAN BE MADE TO WORK.

NOW YOU'RE MESSING WITH ME
"FIRST MINISTER HUGGSTABLE"

THE SECOND PANOPTICON
OF THE JINYIWEI:
TALCA, NUEVO CHILE

AHHH!
LOOKS LIKE
SHE'LL SEE YOU
NOW...

THIS FIST MOVES MOUNTAINS
THIS FIST MOVES STARS.
ALTHOUGH IT SEEMS LIKE THIS FIST MOVES NEITHER, UNDER YOUR COMMAND
PUÑO, THE OPERATIONAL GOALS WERE --
STOP. YOU RETURNED WITH A THIRD OF THE SHIPS I GAVE YOU.
AS WE FEARED, THE DAISY CUTTERS HAD NO EFFECT. BUT WE WERE ABLE TO CAPTURE THE INFECTED VEETAN.
I SHOULD PUT YOU IN THE PUBLIC STOCKS. KNOW THAT.
BUT I NEED YOU TO FINISH THIS PROJECT. SO! YOU WILL RESTRICT YOURSELF TO 900 CALORIES A DAY FOR TWO MONTHS, AS PUNISHMENT.
I WANT YOU TO FEEL EACH OF THOSE JINYIWEI DEATHS, IN DEPRIVATION.
Y--YES, MA'AM
WE WILL RUN THE SMALLER VINN THROUGH STYX, FIRST, TO TEST ITS EFFECTIVENESS. THEN, THE VEETAN.
AFTER THE MACHITO CREW CONFIRMS HIS PERSONALITY IS INTACT, YOU WILL SEPARATE THEIR PILOT FROM THE REST OF THE SHIP.
I'VE DECIDED WE WILL ASSUME THE MACHITO'S MISSION
WE WILL? AND WHAT OF THE REST OF THAT CREW?
WHY, YOU'LL KILL THEM, OF COURSE.
CAN'T HAVE THE EMPEROR ASKING QUESTIONS

WELCOME, CAPTAIN TANEEL.
WHILE THE PROJECT STYX TEAM PREPS THEIR EQUIPMENT, I'M TO INVITE YOU TO RELAX IN THE LOUNGE.
SSSSSS
UNDERSTOOD. AND WILL THE DIRECTOR BE THERE?
SHORTLY. HE'S STILL SPEAKING WITH EL PUÑO.
SHE'S HERE? ON CHILE? I DIDN'T KNOW SHE EVER LEFT THE PANOPTICON.
EL PUÑO GOES WHEREVER THE EMPEROR COMMANDS, MA'AM. AND YOU'RE STANDING IN THE SECOND PANOPTICON
THE SECOND PANOPTICON?? WELL MY GOODNESS. WHY BUILD ONE WHEN TWO GETS YOU TWICE THE SECRETS.
--FESSA?
CUDDOW?
OF ALL THE PLACES! ...ARE YOU WORKING ON STYX?
I...AM. BUT THE WALLS HAVE EARS, HERE. WE SHOULDN'T SPEAK.
WHY NOT? WHAT'S WRONG WITH TWO OLD FRIENDS SAYING HELLO?
TAKA TAKA TAKA

OH SNAP!
YOU GO, JINYIWEI!
SELF-SERVE ESPRESSO IN THE LOUNNNNNNNGE
YES. HISTORY'S GREATEST HEROES. THAT'S THEM.
ALSO, PLEASE MAKE ME A TRIPLE.
CAP'N! CAP'N! CAP'N!
LORD, CUDDOW. SLOW YOUR ROLL. WHAT'S THE MATTER?
I NEED TO... TELL YOU... SOMETHING IMPORTANT. BUT...UM...
DO YOU SPEAK FILLIPOD?
YOU KNOW I DON'T.
DANG. I WAS HOPING YOU COULD.
NO HUMAN CAN SPEAK IT. OUR EARS CAN'T PICK UP THOSE SING-SONG PITCHES.
AND OUR MOUTHS CAN'T MAKE THOSE WILD UNDULATIONS THAT YOURS CAN.
HA HA HA! THAT'S WHY WE'RE SO DANG GOOD AT SMOOCHIN'! AMIRITE?!?
gross

CON-SARNIT! I CAN'T WRITE DOWN WHAT I NEED TO TELL YOU. I CAN'T SPEAK IT. I CAN'T SING IT.
AND I DON'T HAVE A SUITABLY VARIED SET OF OIL PAINTS TO PAINT IT.
OIL SET
248 COLORS!
THAT ONLY LEAVES ME HISTORY'S MOST USELESS ARTFORM
Interpretive Dance!
SSSAAAH!
NO, CUDDOW, I DON'T NEE—
FEEL
...THE PASSION OF MY BODY'S MOVEMENT
SEE
MY BODY AS A VESSEL OF THOUGHT MADE ACTIVE
UNDERSTAND
MY WIGGLES AS A WEIRD ATTEMPT TO COMMUNICATE SOME SORT OF MESSAGE
CUDDOW.

...SO TO SUMMARIZE: IF THE PROCESS WORKS, THE VIRUS SHOULD BURN OFF COMPLETELY
HEY! MAYBE I COULD TRY WEIRD PAPER CUT-OUTS! LIKE MATISSE! ...DO YOU HAVE SCISSORS?
shhhhh!
HIS BIOLOGICAL SYSTEM WILL BE DRAMATICALLY DIFFERENT, OF COURSE. THE VINN VIRUS IS HUGELY TRANSFORMATIVE
...SHOULD HE WISH, WE CAN TRANSPLANT ORGANS FROM VEETAN CADAVERS, ...BUT THE SIMPLE TRUTH IS, EVERY ORGAN IN HIS BODY SHOULD WORK MARKEDLY BETTER FROM THE INFECTION
WAIT: WHY DID YOU SAY "IF THE PROCESS WORKS"?
WE'VE ONLY TRIED THE DANG THING ONCE... AND NOT ON A VEETAN. AT BEST WE HAVE A FIFTY-FIFTY CHANCE TO---
NO. STEPHEN, NO. THAT IS NOT GOOD ENOUGH.
TANEEL...HEY! YOU CAN'T TAKE THAT KIND OF TONE, HERE
NOT GOOD ENOUGH. DO YOU HEAR ME?? YOU WILL SAVE THAT BOY, BY GOD.
TANEEL, STOP THIS! THIS IS NOT WISDOM: DON'T REPEAT YOUR BROTHER'S MISTAKES...
YOU WILL SAVE THAT BOY
GUARDS! GUARDS!
TAK
YOU.
WILL SAVE.
THAT BOY.

IN LIGHT OF THE CAPTAIN'S UM, TEMPORARY NAP-NAPS, YOU MAY DEAL WITH ME, DIRECTOR. I WILL ASSUME COMMAND OF THE CREW.
YOU??
AND WHY NOT ME?
I AM THE SENIOR FILLIPOD ADVISOR ON TESSKIL. WHICH, BY THE TREATY OF KOCHI, AFFORDS ME THE RANK OF AMBASSADOR.
AND AS YOU KNOW, BY VIRTUE OF SUBSECTION 10, PARAGRAPH 22, LINE 8 OF THE IMPERIAL MILITARY CODE -- AMBASSADORS MAY ASSUME COMMAND OF AN ARMADA SHIP WHEN CAPTAINS ARE INCAPACITATED DURING TIMES OF WAR.
WHATEVER.
...AS I WAS TRYING TO TELL TANEEL BEFORE SHE FLIPPED OUT, WE'VE ALREADY RUN THE SMALLER VINN THROUGH THE PROCEDURE.
WINTER'S KISSES! ALREADY??

IT APPEARS TO HAVE WORKED. THE VIRUS IS COMPLETELY OUT OF ITS SYSTEM.
SO! *MISTER AMBASSADOR.* WHILE I CHECK IN ON THE PREP WORK FOR THE VEETAN, I'LL JUST LEAVE THIS WITH YOU, SINCE YOU'RE IN "COMMAND"
I LIED.
...IT MIGHT'VE BEEN SUBSECTION 6

THE GUARDS SAID THEY'RE STARTING NOSH'S PROCEDURE. SHOULD BE 3-4 HOURS.
WHAT DO WE DO WITH THIS DUDE, IN THE MEANTIME?
HEY, DO YOU TALK, LIL' BUDDY? YOU SPEAK HYBRID?
"HY-BRID"? ...NO?
OK...WELL, I SPEAK FIVE OTHER LANGUAGES. MAYBE YOU SPEAK ONE OF THOSE?
¿O POSIBLEMENTE ESPAÑOL ANTIGUO?
OH!
...YOU'VE PICKED UP A DECORATIVE PLANT. MAYBE YOU'RE A GESTURAL SPECIES? OK, UM...WHAT DOES THE PLANT REPRESENT?
SCHHHLORP
"HUNGER"?
OULD-WAY OU-YAY IKE-LAY ANOTHER ANT-PLAY?

...WE'VE KEPT HIM IN AN INDUCED COMA OVERNIGHT, WHILE WE BROUGHT HIS FEVER DOWN, BUT WE'RE WAKING YOUR VEETAN NOW
THANK YOU FOR LETTING US SEE HIM FIRST, STEPHEN. ESPECIALLY AFTER...I...
DON'T THANK ME. THANK EL PUÑO. SHE INSISTED YOU SEE HIM BEFORE ANYONE ELSE
CAN HE SPEAK?
WE DON'T KNOW YET.
WE'RE ACTUALLY HOPING YOU MIGHT HELP US DETERMINE IF HE'S STILL...HIM.
OF COURSE. OF COURSE.
...READY?
READY.
psshhht
C...
CUH...
NOSH?
HE'S TRYING TO SAY "CAPTAIN"
Y... MUSS HE'P ME
ISZ KILIN ME
OH DEAR BOY! ARE YOU IN PAIN???
ISZ KILIN ME. I CAN SMELL IT.
WHAT'S HURTING YOU, LAD? WE CAN FIX IT.
SOMEWHERE... IN THIS ROOM... THERE'S A MUFFIN. I CAN SMELL IT.
oh my boy!

I'M GOING TO ASK YOU A SERIES OF QUESTIONS. ANSWER AS BEST YOU CAN, OK?
OK
WHAT IS YOUR FULL NAME?
NOSH, IN THE LINE OF TEFFTA
AND WHAT SHIP DO YOU SERVE ON?
MMPH
I AM SENIOR SCIENCE OFFICER ON MASSIVE BATTLE CRUISER SEVILLANO
NO, THAT'S NOT RIGHT. THERE'S A NEW SHIP. DO YOU REMEMBER A NEW SHIP?
OH. HA HA...YES! MACHITO. WHOOPSIE-PIDDLES!
AND WHO AM I? HOW DID WE MEET?
YOU ARE MY MO-- MY CAPTAIN. AND WE MET...
I do not remember
YOU DO REMEMBER. IT WAS AT THE ANNOUNCEMENT.
OH YES! AT THE SCIENCE ACADEMY! ANNOUNCING MCBRIDE'S WORMHOLE.
DID YOU ...SAVE MY LIFE?
WHEN THE FIRST BOMB WENT OFF, YES. THEN YOU SAVED MINE, WITH THE SECOND.
AND YOU COMMISSIONED ME THAT NIGHT, OVER BOWLS OF BABA GANOUSH
THE BABA GANOUSH YOU REMEMBER?
I NEVER FORGET BABA GANOUSH

AHH. CAPTAIN. I'M TOLD YOU'VE CONFIRMED OUR PROCEDURE WORKED.
EL PUÑO. WHAT A -THING- TO SEE YOU.
HA HA HA. "THING" WAS NOT YOUR FIRST CHOICE OF WORDS. DELIGHTFUL.
DIRECTOR, KINDLY ESCORT EVERYONE OUT. I'D LIKE TO SPEAK TO THE PILOT ALONE.
CUDDOW, YOU CAN STOP DANCING: I KNOW WHAT SHE'S DOING.
PUÑO, WE WON'T BE SEPARATED.
WE ARE UNDER DIRECT ORDERS FROM THE EMPEROR.
OH, I KNOW.
HEY!
GRAB
TAK!
Z
BUT THE JINYIWEI ARE FAR BETTER SUITED TO CARRY OUT THOSE ORDERS
...THAN THIS LOW-BUDGET CIRCUS TROUPE YOU'VE ASSEMBLED
KLIK.
VATT
GAAHH!
ORLA. WHA...
PUÑO
I WILL THANK YOU...
TO STEP AWAY FROM MY CIRCUS TROUPE

ORLA! WHAT ARE YOU DOING?

I SERVE THE EMPEROR, PUÑO. NOT YOUR WHIMS.
IN FACT: LET'S CALL HIM NOW, TO CLARIFY

OH, I THINK WE WON'T.
THANK YOU, DIRECTOR.

MA'AM.

NOW, IF YOU'LL KINDLY FIRE AN M-WAVE THROUGH HER BRAINPAN, WE CAN CONTINUE.
NO!

NO SIR!
PLEASE! YOU CAN NOT DO--

GZZZZT!

...that.

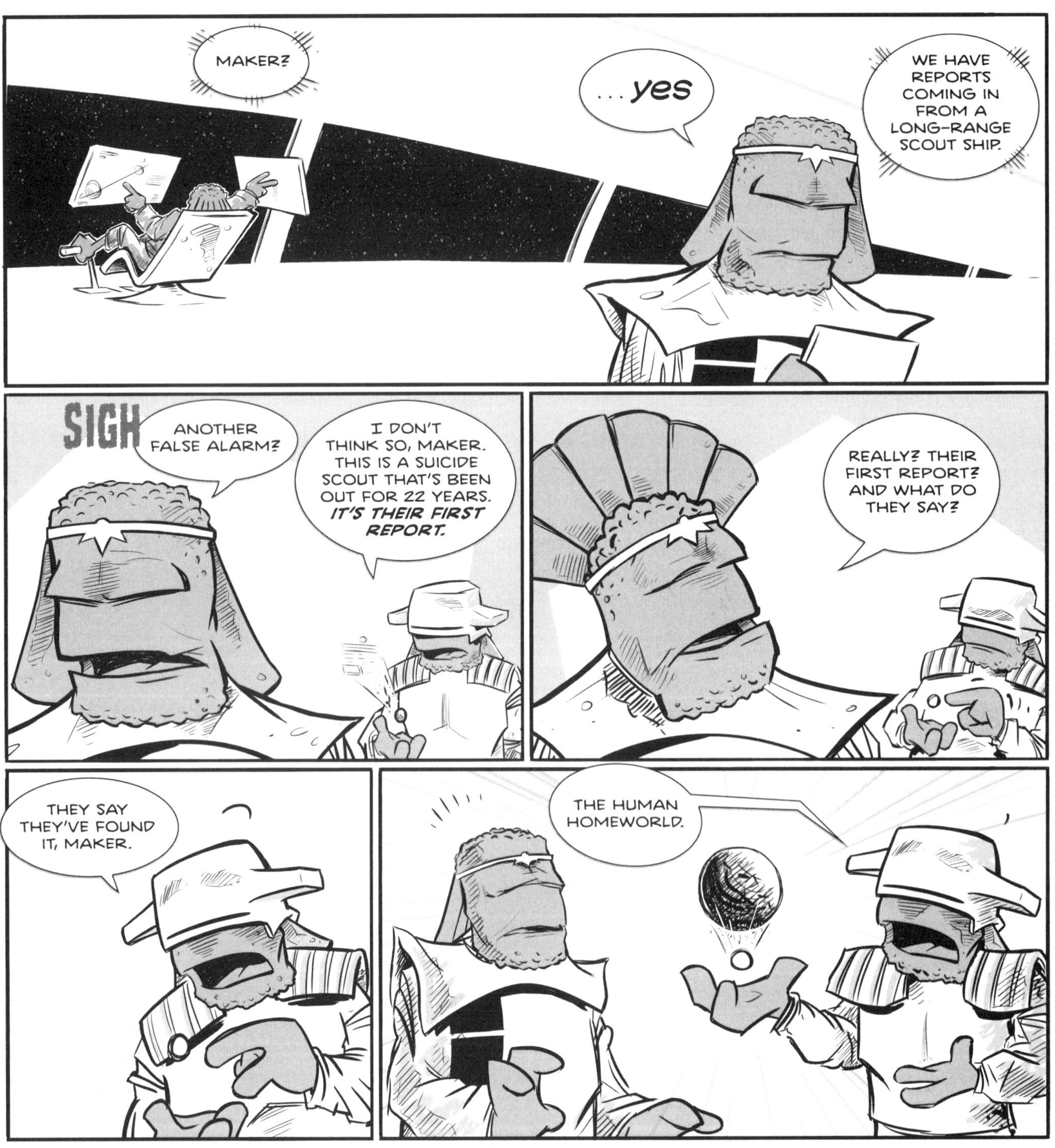
MAKER?
...yes
WE HAVE REPORTS COMING IN FROM A LONG-RANGE SCOUT SHIP.
SIGH
ANOTHER FALSE ALARM?
I DON'T THINK SO, MAKER. THIS IS A SUICIDE SCOUT THAT'S BEEN OUT FOR 22 YEARS. *IT'S THEIR FIRST REPORT.*
REALLY? THEIR FIRST REPORT? AND WHAT DO THEY SAY?
THEY SAY THEY'VE FOUND IT, MAKER.
THE HUMAN HOMEWORLD.

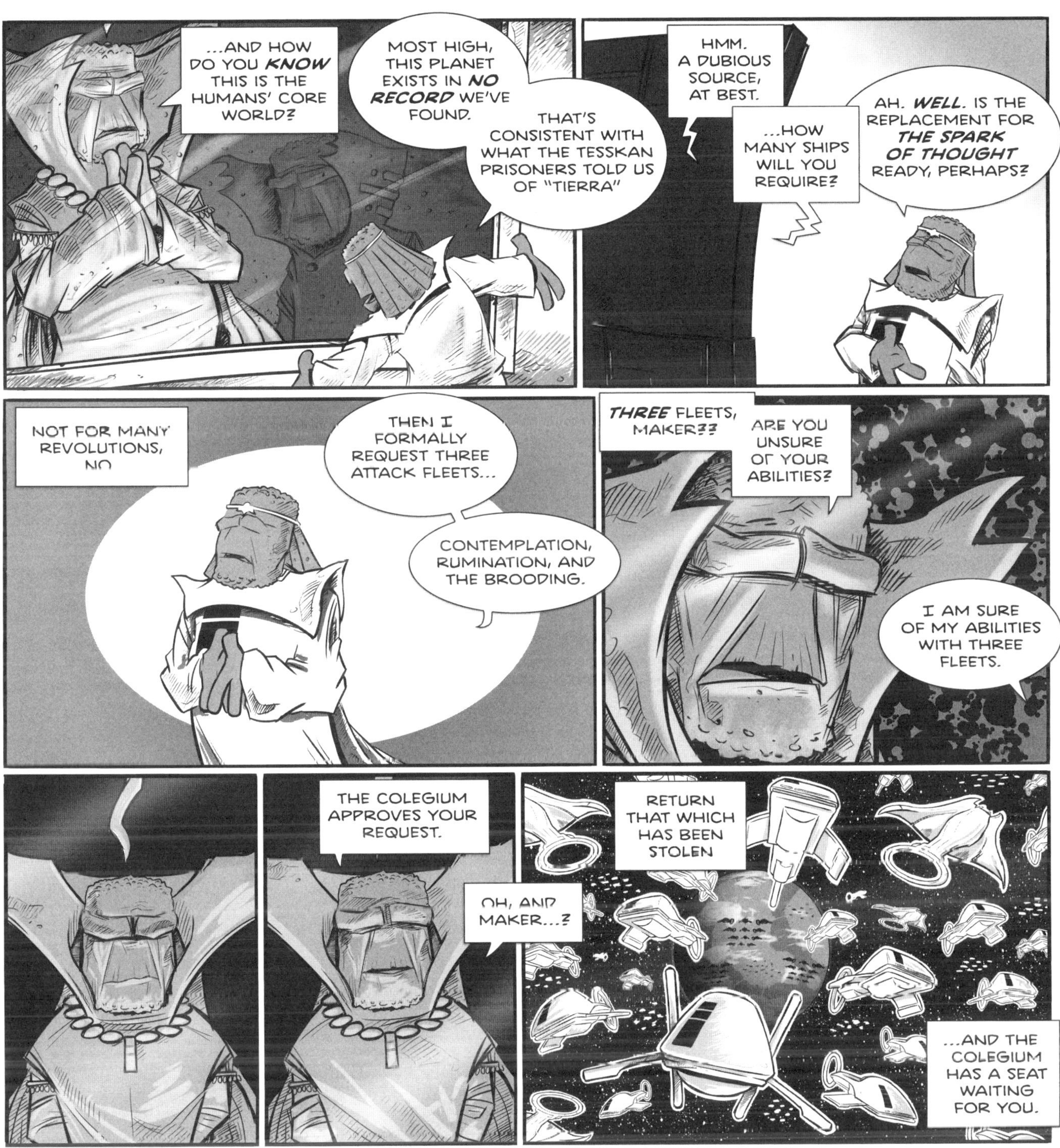
...AND HOW DO YOU KNOW THIS IS THE HUMANS' CORE WORLD?
MOST HIGH, THIS PLANET EXISTS IN NO RECORD WE'VE FOUND.
THAT'S CONSISTENT WITH WHAT THE TESSKAN PRISONERS TOLD US OF "TIERRA"
HMM. A DUBIOUS SOURCE, AT BEST.
...HOW MANY SHIPS WILL YOU REQUIRE?
AH. WELL. IS THE REPLACEMENT FOR THE SPARK OF THOUGHT READY, PERHAPS?
NOT FOR MANY REVOLUTIONS, NO
THEN I FORMALLY REQUEST THREE ATTACK FLEETS...
CONTEMPLATION, RUMINATION, AND THE BROODING.
THREE FLEETS, MAKER??
ARE YOU UNSURE OF YOUR ABILITIES?
I AM SURE OF MY ABILITIES WITH THREE FLEETS.
THE COLEGIUM APPROVES YOUR REQUEST.
OH, AND MAKER...?
RETURN THAT WHICH HAS BEEN STOLEN
...AND THE COLEGIUM HAS A SEAT WAITING FOR YOU.

KRAKK
THANK YOU, ORLA, FOR BEING DISTRACTED BY THE LIGHT SHOW
...I'LL TAKE THAT
AS I WAS SAYING, CAPTAIN: THE JINYIWEI ARE FAR BETTER SUITED FOR THE TASK AT HAND.
THE RISK OF FAILURE FOR THE HUMAN RACE IS TOO LARGE.
NOSH! WHATEVER THAT WAS, DO IT AGAIN!
I DO NOT KNOW, CAPTAIN!
I AM WIGGLING FINGERS, BUT THEY ARE NOT ELECTRICKING!
FOR WHAT IT'S WORTH, TANEEL, I AM SORRY TO SEE YOU GO.
I ALWAYS HAD GREAT RESPECT FOR YOUR CAREER.
VRRRR
NOSH! NOW!
CRACKLE!
ZZZZZZT!
GAHHH

THANK YOU, NOSH
IT WASN'T...
I DIDN'T...
ZZT
ZZT
Daaaaang

NNNNG.
OH, HUSH. YOU'LL LIVE: YOUR WETWARE LEG SEEMS TO HAVE PARTIALLY GROUNDED YOU.
BUT THE DIRECTOR...
DAMMIT, STEPHEN.
Que Dios te perdone
SKITTER! NOSH! GET EVERYONE TO THE MACHITO: I'LL BIND UP EL PUÑO AS YOU PREP THE SHIP.
BOOM
YOU'RE JUST GOING TO...LEAVE HER?
WE'RE NOT KILLING THE HEAD OF THE JINYIWEI
BOOM
I'LL JUST--
BOOM BOOM
WHAT IS THAT?
...PUÑO, WHAT IS THAT?
DON'T KNOW. NOT MY DOING.
GUNNERS, TARGET THE PORTS FIRST.
NOTHING LEAVES THIS PLANET,

BOOM BOOM
TANEEL: PLEASE. SOMETHING IS SERIOUSLY WRONG. AT LEAST LET ME CHECK THE TERMINAL.
FINE. QUICKLY.
L --LORD. IT'S A CONTINUUM ATTACK FLEET.
HERE??
THERE'S OVER 10,000 CRUISERS IN ORBIT. THAT'S...SO MUCH FIREPOWER. I...I CAN'T FIGHT NUMBERS THAT BIG.
GET A HOLD OF YOURSELF. ORDER THE DEFENSES, AND A CIVILIAN EVACUATION.
BOOM BOOM BOOM
"THE DEFENSES"? DAMMIT TANEEL, THERE'S NO DEFENSE AGAINST 10,000 SHIPS.
BOOM
SO WE RUN WITH THE CIVILIANS! PUNCH A HOLE THROUGH THE CONTINUUM PHALANX. LEAP ANYWHERE THAT'S NOT HERE. SAVE AS MANY CITIZENS AS WE CAN!
AT BEST I HAVE 100 CRUISERS IN ORBIT. WHAT WASN'T SENT TO THE REVOLT ON LAS HERMANITAS WAS LOST TO THE VINN.
THEN WE STOP TALKING, AND DO WHAT MUST BE DONE!
DO WHAT MUST BE DONE...
DO WHAT MUST BE DONE...
OK, I'M BROADCASTING PLANET-WIDE.
IN THE NAME OF THE EMPEROR, I'VE ORDERED THE CAPTAIN'S DICTATE TO BE IMMEDIATELY EXECUTED.
THE DICTATE?? BUT THERE'S STILL HOPE!

OFFICE OF THE FLEET ADMIRAL • GUANGDONG, CHINA.

Congratulations on making it to the last week of Captain Candidate School.

By your loyalty, hard work, and dedication, you have demonstrated the highest level of leadership, and a commitment of service to the Empire. As part of your final training, the attached manual will outline the most difficult burden the Empire must now place on you: **The Captain's Dictate.**

It is a unique mantle of responsibility, and one you should hold with pride. Your sacrifice, should it be called upon, ensures the continued invisibility of "Fortress Earth" to her enemies -- present and future. Simply put: **Your swift death means the continued survival of the human race.** So do not see this as a sacrifice to be grudgingly accepted, but an honor to be cherished. You are the best of us, the brightest of us. And you have been trusted with knowledge of inestimable value. You hold the birthplace of humanity in your hands. And with your death, you guard it.

As you turn the pages of the attached manual, you'll find it breaks down into four main sections:
- The historical necessity of the Dictate.
- Common events that would trigger the Dictate, and how to plan for them with 5 simple preparedness steps.
- Famous examples of the Dictate's use, and a few historical close calls that we can all learn from.
- Detailed outlines of fifteen ways to quickly end your life -- five of which require nothing more than your bare hands.

Read this manual, and learn it. You will of course be tested on it during Gauntlet Week, but I want you to commit this to a far deeper part of your memory. You never know when the Dictate will be called for, and I -- *we* -- need you to commit it to muscle memory.

Sincerely yours,

J. Castellano

Juan Castellano
Fleet Admiral, La Grande y Felicísima Armada

PS: If you'll forgive an old man his softer side, allow me to share a few lines of verse with you. I found these to be some small comfort, when I first took on the mantle of the Dictate. They're from a much longer poem by the Nigerian writer Abedi Dangote...where she talks about the child-like trust all space travelers must place in their Captain. I find it a helpful sentiment.

We sat, we two, upon a beach,
with sand unending, within reach.
And plucked you, there, a single grain
And held it safe with quiet pain.

I ran, I flew, I kissed the sun.
I leapt for days! Such joy to run!
And still you kept it safe, that pearl:
You held it safe for all the world.

And when I tired, like a child,
Having run the vastness wild,
I looked up, smiling, at your face
And said, "Where now is that small place?"

And you, with burden, held it high
Most precious grain to which we fly:
Kept there, kept safe, lest it be lost.
Or wrested from us at great cost.

This tiny spark, this prick of light,
from which all joys pour forth so bright.
I take your hand, and you take mine.
...And home for bed, by half-past nine.

2945A-32
5B-5B

TANEEL, THIS MOMENT IS THE REASON THE DICTATE EXISTS. KEEPING EARTH HIDDEN IS MORE IMPORTANT THAN ANY OF OUR LIVES.
RIIIIIP
BUT WE DON'T HAVE TO LOSE OUR LIVES.
STOP. THE ORDER HAS ALREADY GONE OUT. CHILE IS LOST. OUR EMPIRE ASKS ONE LAST TASK OF US, BEFORE THE PLANET FALLS.
DEET DEET DEET
CLICK!
NO! WE CAN RUN! RUN WITH US!
EVEN YOUR MIRACLE PILOT CAN'T GET OUT OF THIS.
THIS IS AN HONORABLE END TO A LIFE OF SERVICE.
THINK ABOUT THIS. MARIA, PLEASE.
I KNOW YOU'VE DESPISED ME SINCE THE ACADEMY.
BUT EVERYTHING I'VE EVER DONE, I DID TO KEEP HUMANITY SAFE.
BELIEVE IT OR NOT, I DO HAVE THE CAPACITY TO LOVE.
JUST GIVE ME TEN MINUTES TO COME UP WITH A PLAN. THAT'S ALL I ASK. TEN MINUTES.
THIS FIST MOVES MOUNTAINS...
NO! NO!
KRAK

REPORTS ARE IN FROM CHILE, TÍO. IT'S A MASSIVE CONTINUUM INVASION.
THE LAST TRANSMISSION SAID DROP SHIPS WERE FALLING LIKE RAIN.
GAHHH
EVERY POTENTIAL WEAPON I TURN TO IS SNATCHED AWAY! MY TESSKAN ARMY IS GONE. MY JINYIWEI NAVY IS IN TATTERS. EVEN MY OVERTURES TO THE GRASSKANS ARE USELESS, NOW.
MY LAST HOPE WAS WITH A TINY SCOUT SHIP, LOOKING BLINDLY FOR PILOTS.
BUT IN A FIT OF IDIOCY, I ORDERED IT TO CHILE AS WELL: TRYING TO CONFIRM OUR CURE FOR THE VINN VIRUS.
...AND STUPIDLY PUT ALL MY EGGS IN ONE BASKET, DAMN IT ALL.
...
HMPH.
ON A BRIGHT NOTE FOR YOU: YOU'LL BE HAPPY TO KNOW PUÑO'S DEAD.
BIT OF COSMIC REVENGE ON YOUR PART, I SUPPOSE?
HER CYBERNETICS TOLD ME SHE DIED QUICKLY, THOUGH.. PERHAPS THE CONTINUUM GRANTED HER ONE LAST MERCY WHEN THEY---
WHAT THE..??
HAVE SOME PRIDE, MAN.
YOU'RE AN EMPEROR, FOR GOD'S SAKE!
AND YOU HAVE PIGEON POOP ON YOUR SHOE. KEEP IT TOGETHER.
SQUINK SQUINK SQUINK SQUINK SQUINK
MANUEL JOSE CRUZ
2339-2401
EMPERADOR

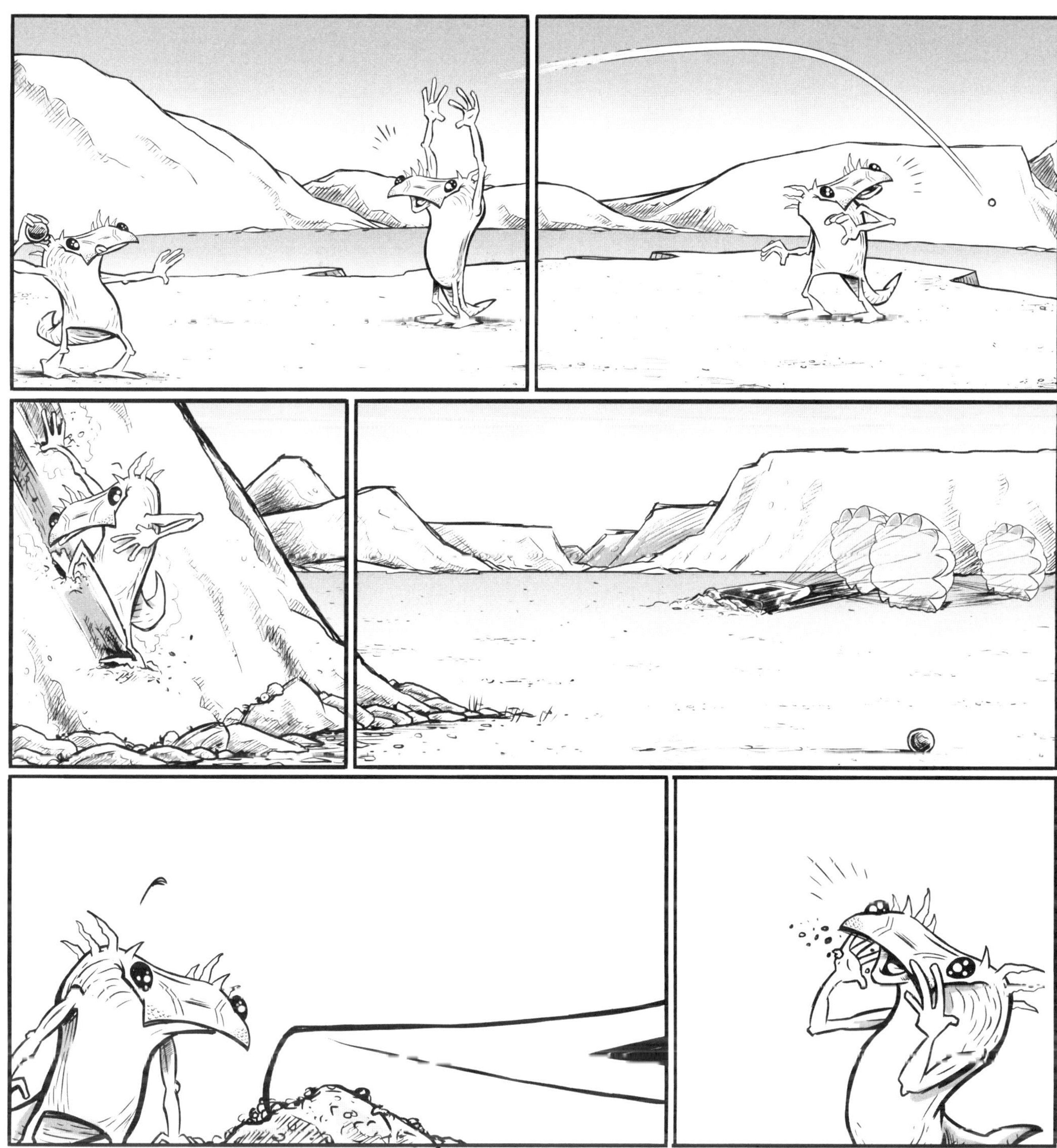

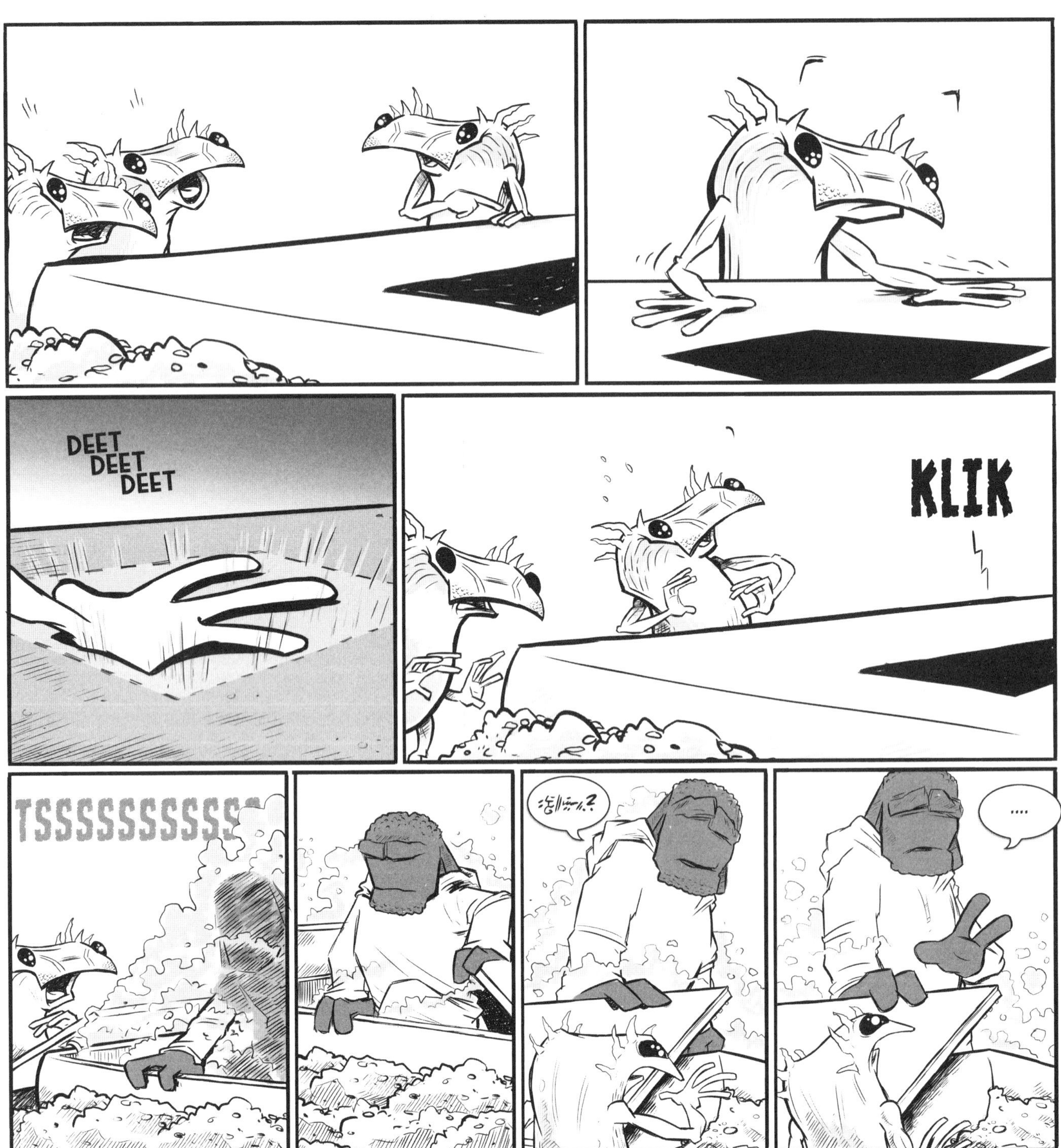
DEET
DEET
DEET
KLIK
TSSSSSSSSSSS
....

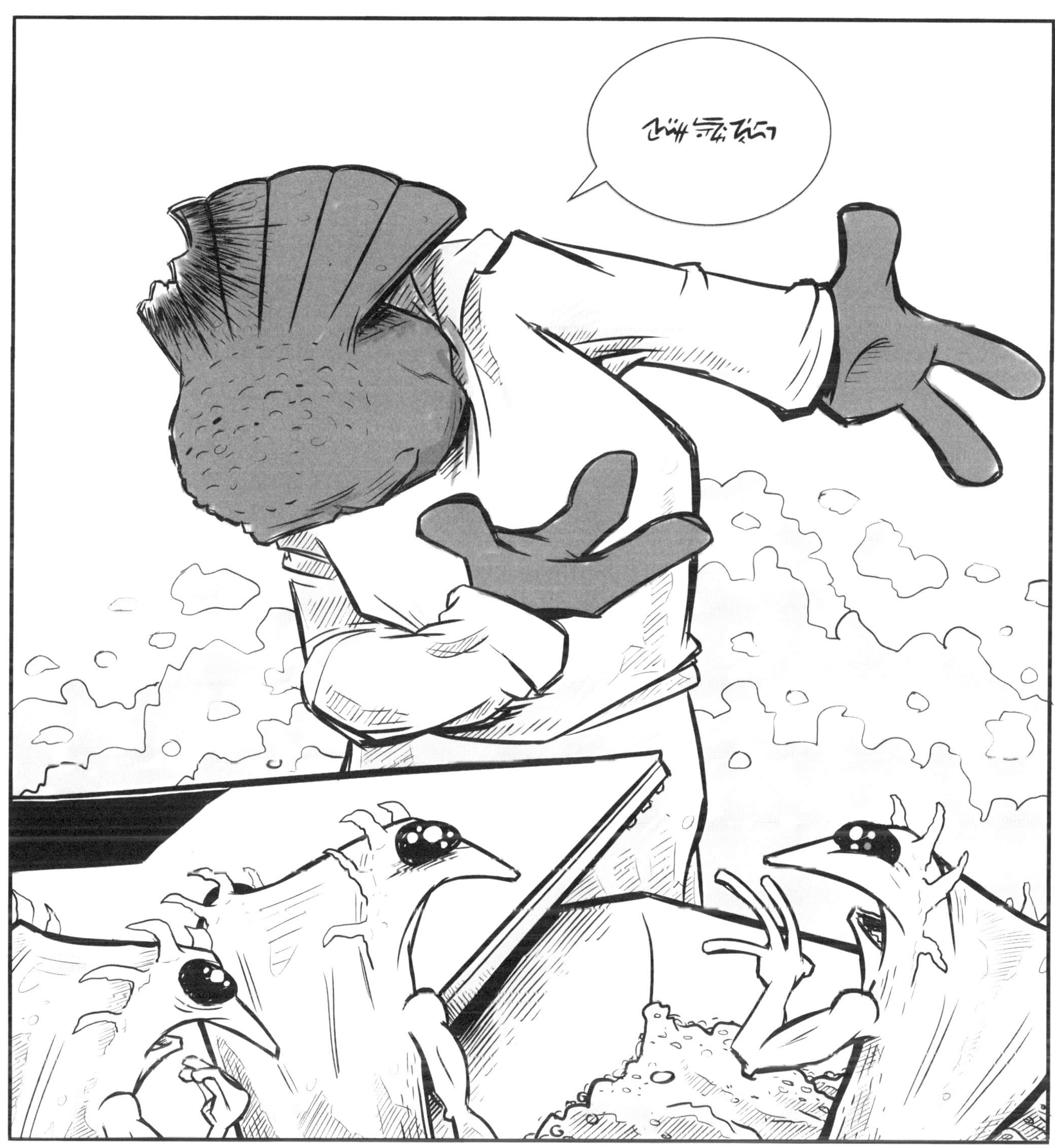

WELL HANG IT ALL: HOW LONG HE BEEN THERE?
FIVE HOURS OR SO. WE BEEN LOOKIN' FOR YOU IN THE SOUTHERN FIELDS EVER SINCE.
IS HE BANGED UP? WHAT'D HE SAY?
HE DIDN'T SAY NUTHIN', PA! HE DON'T SPEAK ANY TI'TAL OR MOQ.
OR GRASSKAN! I TRIED GRASSKAN ON HIM, SIR. MY UNCLE TAUGHT ME GRASSK--
WE **GET IT**, LARR, YOU SPEAK GRASSKAN.
DID YOU FEED HIM?
HE WEREN'T HUNGRY. HE JUST KEPT REACHING FOR OUR **BOOKPADS**. SO WE GAVE 'IM A BIG OL' STACK A THOSE.
...AND THEY HAVE BEEN MOST HELPFUL.
I BELIEVE I HAVE THE HANG OF IT, BUT DO LET ME KNOW IF I MAKE ANY **MISTAKENS** OR MIS-PRONUNCIATIONS. I'VE ONLY HAD A FEW HOURS WITH IT.

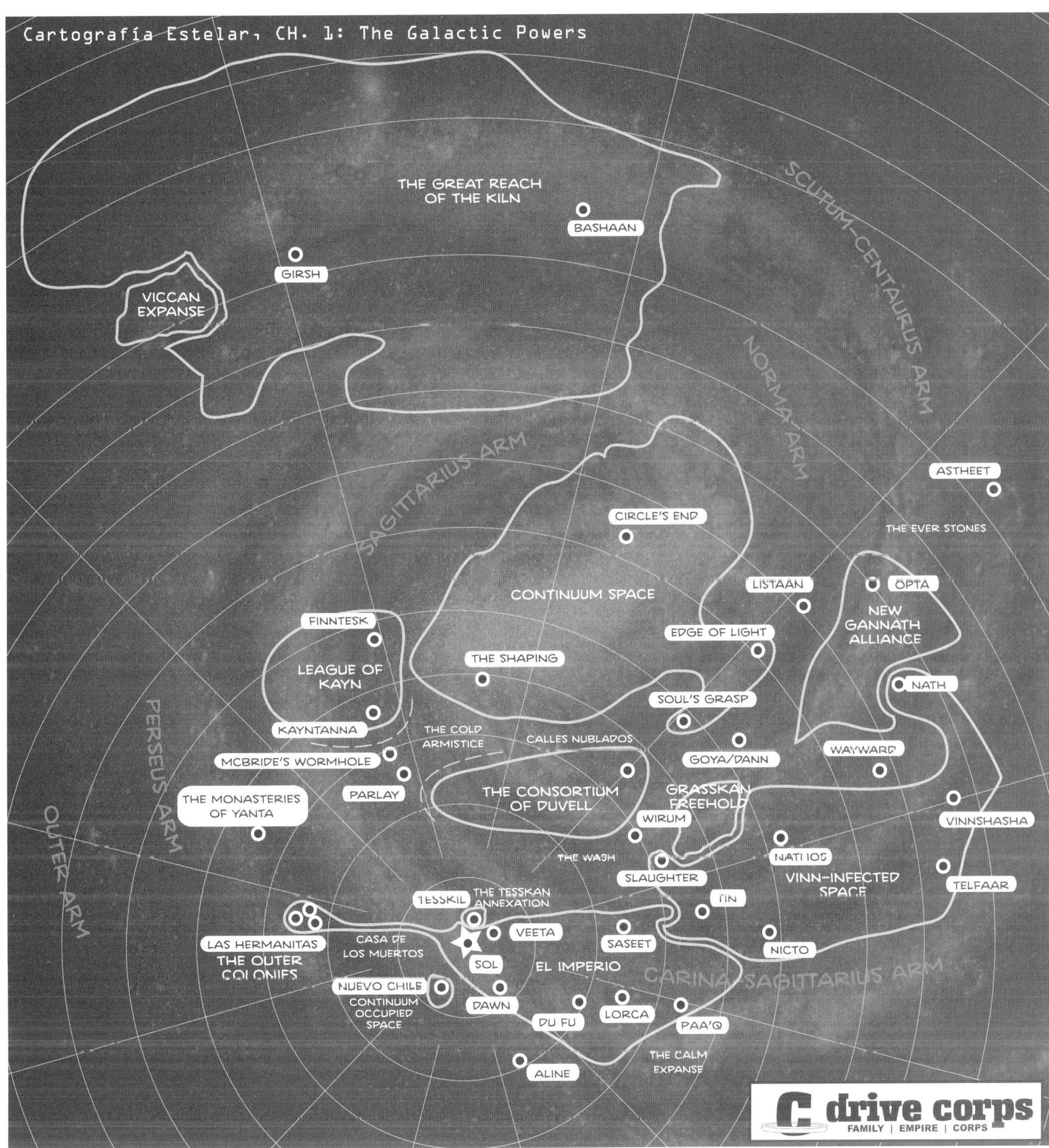

Cartografía Estelar, CH. 1: The Galactic Powers
THE GREAT REACH OF THE KILN
BASHAAN
GIRSH
VICCAN EXPANSE
SCUTUM-CENTAURUS ARM
NORMA ARM
SAGITTARIUS ARM
ASTHEET
CIRCLE'S END
THE EVER STONES
CONTINUUM SPACE
LISTAAN
OPTA
NEW GANNATH ALLIANCE
FINNTESK
EDGE OF LIGHT
LEAGUE OF KAYN
THE SHAPING
NATH
SOUL'S GRASP
KAYNTANNA
THE COLD ARMISTICE
CALLES NUBLADOS
MCBRIDE'S WORMHOLE
GOYA/DANN
WAYWARD
PERSEUS ARM
PARLAY
THE CONSORTIUM OF DUVELL
GRASSKAN FREEHOLD
THE MONASTERIES OF YANTA
VINNSHASHA
WIRUM
OUTER ARM
THE WASH
NATHOS
SLAUGHTER
VINN-INFECTED SPACE
TELFAAR
TESSKIL
THE TESSKAN ANNEXATION
TIN
VEETA
LAS HERMANITAS
CASA DE LOS MUERTOS
SASEET
NICTO
THE OUTER COLONIES
SOL
EL IMPERIO
CARINA-SAGITTARIUS ARM
NUEVO CHILE
CONTINUUM OCCUPIED SPACE
DAWN
DU FU
LORCA
PAA'Q
THE CALM EXPANSE
ALINE
C drive corps
FAMILY | EMPIRE | CORPS

AND YOUR PEOPLE, HAVE THEY EVER LEFT THIS PLANET...?
WIRUM, SIR, AND YESSIR.
GRASSKANS CAME, FEW HUNNER YEARS AGO. TAUGHT US ALL SORTS A' STUFF. GAVE US TECH, TOO, ON THE AGREEMENT WE GROW FOOD FOR 'EM. WHICH WE DONE HAPPILY.
AH YES. GRASSKANS. MY PEOPLE ARE ...AWARE... OF THEM.
AND WHERE MUST I GO TO FIND A STARSHIP?
HA! WELL, YOU COULDN'T BE FARTHER. NEAREST CITY'S A 3-DAY HAUL WITH MY TRACTOR, COME HARVEST.
AND COULD I IMPOSE ON YOU TO DRIVE ME IN THE MORNING?
'FRAID HARVEST AIN'T FOR ANOTHER CYCLE. I'D LOSE MY CROPS, IF I WAS TO GO NOW.
BUT! IF YOU CAN GET HER RUNNIN', YOU CAN TAKE MY OLD TRACTOR. THING HASN'T MOVED IN 20 YEARS.
AHHH! THIS I CAN WORK WITH.
YOU, SIR, HAVE JUST RELIEVED YOURSELF OF A GUEST: I WILL BE GONE BY MORNING.
AND TO THANK YOU, I WILL BUILD YOU A GIFT. A LITTLE SOMETHING TO INCREASE YOUR HARVEST YIELD SIX-FOLD.
DANG-SANG IT! YOU CAN BUILD SUM'IN LIKE THAT OVERNIGHT?
MY FRIEND, I WILL BUILD WORLDS WITH THESE HANDS.
WATCH.

2318.3320.8891.4839.2358.2908.4787.4697.9032.3348.8878.1304.3989.7878.3454.2209.1121.9944.4735.8883.2921

Enciclopedia Xenobiológica

Species: Prahsitt

Homeworld: Wirum

Size: 1.3-1.7m

Color & Markings: Prahsitt children are born with either aquamarine, green, or bright yellow hides; even in the same litter. Extensive testing has shown no difference in aptitude, personality or intelligence between differently colored children. However, the highly superstitious people of Wirum believe a green child (the rarest of the three) portends a truly unique life. This, in turn, begins a self-reinforcing series of parenting choices that result in "Green-Born" children living more interesting lives.

Reproduction & Development: Prahsitts reproduce sexually, producing a litter of 2-3 children after a 5-month gestation period. Conception is only considered after a long checklist of superstitious conditions are met: These can include weather, time, foods prepared, number of red birds seen day-of, ill words spoken or heard, and on and on to a ridiculous degree. Offworlders would think it's a miracle that conception ever happens at all...but the rules are bent all the time, if grain-wine is present.

Prahsitt children are raised in communities of extended family and tight-knit farming communities. When a Prahsitt arrives at their "Time of Choosing," that community parenting flows into a self-selected choice of a new home or guild. That teen Prahsitt, having decided what path they want to pursue in life, moves in with their new family/host to learn a trade or gain knowledge. For most of Prahsitt history, the "Time of Choosing" meant one's offspring moved within a walkable distance, often within the same town. But since the arrival of the Grasskans, it now often means painful, years-long family separations when a teen travels into space. Though a simple, salt-of-the-earth folk by inclination, Prahsitts have adapted well to their spacing age, and can now be found in every port in The Wash, excelling in all manner of jobs.

Like Slaughter, Wirum sits just outside the formal Grasskan Empire, but enjoys most of the trade, protections, and services the Empire has to offer. In the case of Wirum, this is due to an almost impossible series of events wherein a Prahsitt farmer saved the lives of the Grasskan Emperor and her two daughters from an assasination attempt by the Five Familes of Slaughter. The Grasskans, in turn, extended to Wirum their highest life-debt in thanks: The "10 Lifetimes Pledge," during which the Empire would protect and help Wirum. Related: All five Setts were subsequently found murdered in their beds, and replaced by Brood-Setts of the Empire's choosing.

Diet: Omnivores.

Lifespan: 100-120 years.

Language: Moq and Ti'tal are the most commonly spoken languages on Wirum. But the ever-increasing amounts of trade that the Prahsitt conduct in The Wash – especially with the Grasskan Empire – has meant more and more Grasskan and Astina words filtering into daily use.

Social Structure: As a people, Prahsitts value both hard work, and the time it gives them with family. Though advanced technology is peppered throughout their society, it hasn't influenced traditional Prahsitt thinking much. For example, you could find an advanced Grasskan seed recombinator right next to a 100-year old pickax in a Prahsitt toolshed, but it's given no deference: It's just another tool to get things done.

Intelligence Rating: Though most commonly testing in the Juliett/Kilo range of the Chatterjee Scale, Prahsitts have been known to test as high as November/Oscar.

Interaction with Humans: Prahsitts rarely find themselves traveling as far as the Empire, but they are always well-mannered in their human interactions, and get along peaceably with the Empire. Notably, they love drinking and sharing songs with Veetans.

SIR! BEFORE YOU GO!
MY BOY KIK MADE YOU *A GIFT.*
...A PRAHSITT LUCK NECKLACE, TO KEEP ILL WINDS FROM YOUR PATH.

OH! THIS IS... VERY WELL MADE. I AM HONORED BY THIS SPIRITED GIFT.
WELL, UM... I DON'T KNOW IF THIS'LL MEAN MUCH TO YOU, BUT MY BOY IS GREEN-BORN.
MEANS HE GOT A SPECIAL PATH TO FIND IN LIFE. GREAT THINGS TO DO.
DOES IT NOW.
AND WELLSIR, SUM'IN TELLS ME YOU'RE THE ONE TO SHOW HIM THOSE THINGS.
SIGNS IS CAUSED, SIR, AND SIGNS GOT MEANING.
AND WHEN SOMEONE FALLS OUTTA THE SKY, FIXES WHAT CAN'T BE FIXED, PUTS MORE FOOD ON MY FAMILY'S TABLE, WELL....
WE FIGGER I'LL NEVER FIND A BETTER MENTOR, SIR.
EXACTLY.
SO WE WANTED TO ASK IF YOU'D CONSIDER TAKIN' KIK WITH YOU ...AS YER APPRENTICE?
A GREEN-BORN WILL SOAK UP EVERYTHING YOU GOT TO TEACH HIM, AND BE LOYAL AS LOYAL CAN BE.
AND IT'LL BE A TWO-WAY STREET, SIR! I CAN HELP YOU WITH WHAT NEEDS HELPIN'.
I'M AFRAID THE CONTINUUM IS A CLOSED LOOP. IT DOES NOT PERMIT OUTSIDERS.

I AM NO LONGER OF THE CONTINUUM.
BUT THEN...
MY MADE-NAME... IS AHMIS. I WAS ONCE A HIGH MAKER OF THE COLEGIUM, SERVING IN THE GREAT HALL OF SPIRITS ON CIRCLE'S END.
AND NOW... I WILL TEACH YOU WHAT CAN BE TAUGHT.
ALSO PLEASE TAKE HIS DUMB BIRD.

CIRCLE'S END
THE ADOPTED HOME WORLD OF THE CONTINUUM OF MAKERS.
TWO CRUISERS APPROACH FROM OPPOSITE DIRECTIONS: BOTH BEARING BAD NEWS, BOTH DESPERATE TO SEE THE MAKER PRIME IMMEDIATELY.
ONE SHIP IS HEADED UP BY GENERAL MARD, THE FAMED BLIND WARRIOR OF THE GRASSKAN WARS.
THE OTHER, BY THE TEMPORARY COMMANDER OF CONTEMPLATION, RUMINATION & THE BROODING
...WHO TOOK OVER AFTER THE SUDDEN DEATH OF HIS LEADER ON NUEVO CHILE.
BOTH ARRIVE AT THE HALL OF SPIRITS TO FIND THE SAME NEWS: THE MAKER PRIME IS VERY ILL, AND WILL SEE NO ONE

MARD IS INSISTENT:
THEN GET ME FILIA. SOMEONE NEEDS TO TAKE MY NEWS TO THE COLEGIUM IMMEDIATELY.
MINE AS WELL
FILIA IS WITH THE MAKER PRIME. AND ALL NEWS CAN WAIT.
THIS CAN'T. THE GRASSKAN EMPIRE HAS FALLEN! TO A NEW POWER FROM THE OUTER RIM.
NOR CAN MY NEWS WAIT! OUR INTELLIGENCE ON THE HUMAN HOME-WORLD FAILED US! THEY WERE ABLE TO--
STOP. BOTH OF YOU.
FILIA! I...
WE BRING NEWS, FILIA...
AFTER 463 YEARS OF LIFE, AND 122 AS MAKER PRIME... I'M SORRY TO SAY HE IS NO MORE.
"MAY HIS SPIRITS ILLUMINATE THE UNIVERSE."
TING
"MAY HIS SPIRITS ILLUMINATE THE UNIVERSE."

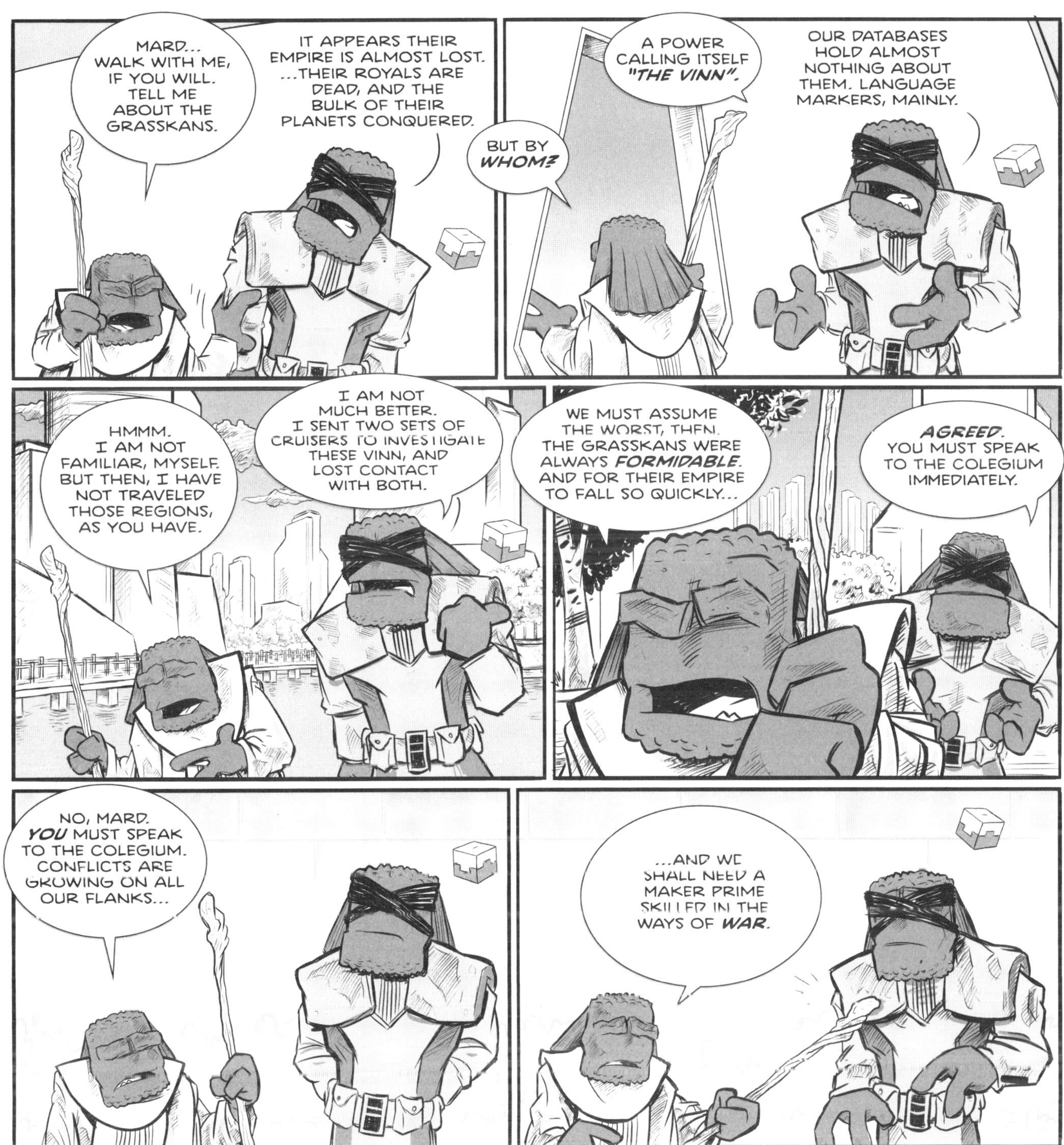
MARD...
WALK WITH ME, IF YOU WILL. TELL ME ABOUT THE GRASSKANS.
IT APPEARS THEIR EMPIRE IS ALMOST LOST. ...THEIR ROYALS ARE DEAD, AND THE BULK OF THEIR PLANETS CONQUERED.
BUT BY WHOM?
A POWER CALLING ITSELF "THE VINN".
OUR DATABASES HOLD ALMOST NOTHING ABOUT THEM. LANGUAGE MARKERS, MAINLY.
HMMM. I AM NOT FAMILIAR, MYSELF. BUT THEN, I HAVE NOT TRAVELED THOSE REGIONS, AS YOU HAVE.
I AM NOT MUCH BETTER. I SENT TWO SETS OF CRUISERS TO INVESTIGATE THESE VINN, AND LOST CONTACT WITH BOTH.
WE MUST ASSUME THE WORST, THEN. THE GRASSKANS WERE ALWAYS FORMIDABLE. AND FOR THEIR EMPIRE TO FALL SO QUICKLY...
AGREED. YOU MUST SPEAK TO THE COLEGIUM IMMEDIATELY.
NO, MARD. YOU MUST SPEAK TO THE COLEGIUM. CONFLICTS ARE GROWING ON ALL OUR FLANKS...
...AND WE SHALL NEED A MAKER PRIME SKILLED IN THE WAYS OF WAR.

CAPTAIN! THANK GOODNESS YOU'RE BACK. ***YOUR PEOPLE ARE CRAZY.***
MY "PEOPLE"?
HUMANS! THEY'RE BROADCASTING FINAL BATTLE CRIES AS THEY CRASH SHIPS INTO THE CONTINUUM.
...?!?
...WELL, I GUESS GOING KAMIKAZE IS ***ONE*** WAY TO DO IT.
DO ***WHAT?*** DIE IN A FIRE??
PUÑO BROADCASTED A SUICIDE ORDER TO ALL CAPTAINS.
...I GUESS THEY'RE TAKING THEIR SHIPS WITH THEM.
AAAAND, *HA HA HA* -- IF I CAN JUST ***JUMP IN*** ON THIS CONVERSATION -- HOW ARE WE ***FEELING*** ABOUT THIS ORDER? *HA HA HA.*
BECAUSE I'M PERSONALLY ***LOVING*** BEING ALIVE. LOTTA ENJOYMENT, HERE. ***BIG*** FAN. *HA HA HA*
WE AIN'T DYIN' TODAY, CUDDOW.
SKITTER, TAKE US UP.
UP? UP WHERE?
GET US BEHIND ANY SUICIDAL CRUISER YOU CAN FIND. WE'RE GONNA DRAFT THROUGH THAT WALL OF SHIPS.
THAT SOUNDS SUSPICIOUSLY LIKE DYING

OK, I'VE PUT US BEHIND A CRUISER: NOW WHAT?
NOW YOU KEEP YOUR MOHAWK ON A SWIVEL... LOOK FOR ANY OPENING WE CAN DUCK THROUGH.
WAIT. I HAVE A BETTER IDEA TO GET US PAST THEM AND DO SOME DAMAGE.
OOO! SEE, NOSH? THAT'S WHY WE PAY CUDDOW THE BIG BUCKS.
ACK, THAT IS NO GOOD! WE SHOULD ALL BE PAID SAME AMOUNTS! WE ARE TEAM OF PALS!
NEVERMINNND
...WHAT'S YOUR IDEA, CUDDOW?
WE ALL KNOW THE EFFECT THAT A DRIVE HAS IN PROXIMITY TO A PLANET'S MASS, YES?
YES
SO HERE'S MY THOUGHT: WE FLY INSIDE THE RINGS OF THAT MOEBIUS HEAVY CRUISER, AND ENGAGE OUR DRIVE RIGHT AS WE PASS THROUGH.
ENGAGING A DRIVE INSIDE A DRIVE? IS THAT SAFE?
N...
NO. NOT FOR BOTH SHIPS INVOLVED. BUT... I THOUGHT THAT WAS THE POINT?
LOOK, I'M NEW AT "WAR"

'NANDO, YOU'RE THE ONLY ONE ONBOARD WHO UNDERSTANDS DRIVE DYNAMICS. WILL CUDDOW'S IDEA WORK?
I...GENUINELY DON'T KNOW. AN ACCIDENT LIKE THIS OCCURRED ONCE NEAR DU FU, BUT IT DESTROYED BOTH SHIPS.
THAT'S BECAUSE BOTH THOSE DRIVES WERE ENGAGING AT THE SAME TIME. IN THIS CASE, THEIR DRIVE WILL BE IN STANDBY: FULLY POWERED UP AND READY, BUT NOT PINCHING SPACE.
OH... RIGHT... THAT'S TRUE.
IN WHICH CASE... HOLD ON...LEMME DO THE MATH.
THERE'S NO TIME.
SKITTER: THREAD THE NEEDLE ON THOSE RINGS, AND PUNCH IT WHEN WE'RE DEAD-CENTER
CUDDOW, YOU'RE EITHER A GENIUS, OR A DASTARDLY MURDERER.
I VOTE GENIUS! HE IS NICE FELLOW!
"...ALSO, I DON'T KNOW WHAT A DASTARD IS..."
VOO VOO VOO

WHAM
WHAM
WHAM
WHAM

WE'RE CLEAR!
WE'RE CLEAR?? ORLA, REAR SCOPES! WHAT DO YOU SEE?
JEEZ LOUISE: THAT TOOK OUT A THIRD OF THEIR FLEET!
AH. HRMMM. SO THAT DOES CREATE A SIGNIFICANT DRIVE EVENT.
MY FRIENDS, I BELIEVE WE'VE JUST WITNESSED THE CUDDOW MANEUVER.
OOO LA LAA! I BET THE LAST TIME WORLD SAW A CUDDOW MANEUVER WAS AT HIGH SCHOOL DANCE!
NO
NO...
THAT MANEUVER WAS JUST ME SIPPING PUNCH IN A CORNER UNTIL MY MOM PICKED ME UP.

AHHHHH, IS GOOD TO BE HOME. IS GOOD TO PUT BACK ON FAVORITE SHIRTINGS.
SHIRT.
"SH-URRRT"
SHIRT
INGS
IS FAVORITE SHIRTINGS. IS SOFT, IS SNUGLY, BUT IS ALSO COOL ON HOT DAYS. IS FEELING LIKE WHOLE WORLD GIVING YOU HUG.
ALSO! HAS SECRET POCKET INSIDE! IS WHERE I KEEP TO HIDE KIT-KATS.
WHAAAT? DON'T THEY GET ALL MELTY AND SOFT IN THERE?
NO NO NO, SKITTER.
SKITTER, LISTEN:
...IS BESTEST PART

WHAT ARE YOU THREE WORKING ON IN HERE?
ON OUR LITTLE VINN! WE'VE BEEN TRYING TO ESTABLISH SOME FORM OF COMMUNICATION.
IT'S FASCINATING! HE SEEMS TO UNDERSTAND US ON SOME RUDIMENTARY LEVEL...
...BUT HAS NO CAPACITY FOR SPEECH.
SO CUDDOW TRIED DR. CHATTERJEE'S PICTOGRAPH SYSTEM ON HIM. AND HE RESPONDED!
AND THEN ORLA HAD THE BRIGHT IDEA OF GIVING HIM ONE OF NOSH'S CRAYONS... AND FOR THE LAST HOUR HE'S BEEN ANSWERING US WITH DRAWINGS.
AND HERE'S WHERE IT GETS INTERESTING: WE WERE ASKING HIM ABOUT THE LEADERSHIP STRUCTURE OF THE VINN, WHEN CUDDOW CHANGED UP THE QUESTIONING...
I BEGAN ASKING HIM ABOUT THE VINN'S MISSING GODS...
SPECIFICALLY, I ASKED HIM TO DRAW THE ICON THAT THE VINN USE TO REPRESENT THEIR GODS.
...AND HERE'S WHAT HE DREW.
IT'S...A ROCKET

creeeak
MYAAAAH!
N...NOSH? IS THAT YOU? ¡DIOS! IT'S FOUR IN THE MORNING!
CAPTAIN...IS REAL? THIS? AM I ALIVE??
YES. OH MY BOY, YES. YOU'RE ABSOLUTELY ALIVE. YOU'RE HOME. YOU'RE SAFE.
I AM LOOKING IN MIRRORS AND NOT SEEING MY FACE. MY BRAIN IS...SCARED.
OH SWEET LAD. I CAN ONLY IMAGINE. WHAT CAN I DO TO MAKE YOU NOT SCARED?

THANKS FOR COMING IN, ORLA. TAKE A SEAT.
CAPTAIN, I...
LISTEN: CHILE OPENED MY EYES TO THE CONSTRAINTS YOU'VE BEEN WORKING UNDER.
WHEN PUÑO BROUGHT MY HUSBAND INTO THE JINYIWEI, I SAW FIRST-HAND HOW MANIPULATIVE & MACHIAVELLIAN SHE COULD BE. I CAN ONLY IMAGINE WHAT YOU'VE BEEN DEALING WITH.
I'VE DONE MY BEST.
I KNOW THAT. YOU DID RIGHT BY YOUR CREW IN AN IMPOSSIBLE SITUATION
BUT WE'RE STILL FACED WITH AN ODD SETUP, HERE. YOU ARE AN AGENT OF THE EMPEROR'S JINYIWEI, YET STILL ANSWERABLE TO ME.
THE CHAIN OF COMMAND'S GONNA GET... WEIRD
OOF.
LIKE...WHICH OF US CALLS THE EMPEROR TO TELL HIM ABOUT THE LOSS OF CHILE?
GURROW.

YES, AMBASSADOR, I CAN *SEE* YOUR PAINTING.
OH, BUT MY WISE LIEGE! IT'S NOT JUST A PAINTING: IT'S A TROMPE-L'ŒIL FAUX HALLWAY! SO AS TO BETTER FRAME MY VISAGE AS I APPEAR BEFORE YOU!
AND THE MUSIC, YOU ASK?
WHY, I COMPOSED IT LAST NIGHT ON A VEETAN HARP. I CALL IT "DARE I TOUCH THE FACE OF GOD??"
WHICH, MY MOST GLORIOUS AND ***FASTIDIOUSLY CLEAN*** KING, IS HOW I FEEL IN OUR TÊTE-À-TÊTE THIS MORNING! SUCH GRAND DELICACIES, TO HEAR THE DICTION OF MY MAGNANIMOUS RULER!
OH! OH! SPEAKING OF WORDS... ***I'VE WRITTEN A SHORT POEM*** FOR THE OCCASION OF THIS CALL!
HOW'S YOUR SCHEDULE?
IT'S VERY SHORT!
I PROMISE IT'S VERY SHORT.
ARE WE PRESSED FOR TIME?
I'LL JUST START, SHALL I?
GET TO WHY YOU'RE CALLING ME
AH!
HA HA HA! YES ***OF COURSE***, O LORD OF LORDS!
IT APPEARS THAT YOU'RE NOW
UMM
"ONE QUESADILLA SHORT, IN THE GREAT COMBO-PLATE OF LIFE."
WHAT?
YOU LOST A PLANET.

The pen is mightier than the sword
...but not when
you work for the empire.
Fillipods!
GET BACK
TO WORK

FILLIPOD PROPAGANDA POSTER

Of all the species in the Empire, Fillipods have proven the most responsive to official propaganda. Fillipods view imperial propaganda as schmaltzy, poorly executed, and yet somehow delicious, delicious eye candy. Ham-fisted in conception, and usually off-key in tone and tenor...imperial Fillipod propaganda nevertheless generated huge fandoms on Tesskil, with Fillipods passing them around or modifying them into memes. The reaction was very similar to humans' reaction to "reality" TV on the Vid-Pack Network. Not able to help themselves, Fillipods would binge and binge on the terrible content, with the result that the campaigns ended up being...effective.

I DON'T UNDERSTAND. WHY IS THIS A BAD DAY TO BOOK PASSAGE OFF-PLANET?
'CAUSE TH' WHOLE DANG GALAXY COME RUNNIN' HERE, THIS MORNIN'
LOOK AROUND YOU! AIN'T NOBODY EVER COMES TO WIRUM'S BACKWATER SPACEPORT! BUT NOW WE GOT FLEETS A' SHIPS PARKIN' EVER-WHERE.
I STILL DON'T UNDERSTAND
THEY ALL REFUGEES! FLEEING WHAT'S LEFT OF TH' GRASSKAN EMPIRE BEFORE THEY GET THEIR SOULS SUCKED OUT.
WHAT'S HAPPENED TO THE GRASSKANS, SIR?
A HORDE OF VAMPIRE CREATURES DONE CONQUERED 'EM, WE HEARS! TOOK THEIR SOULS. THEM VINN MAKES YOU NOT YOU, THEN THEY PAINTS YOUR FACE.
"PAINTS YOUR FACE"?
THE GRASSKANS ARE A MASSIVE POWER. WHO ARE THESE VINN THAT THEY COULD DO THIS?
APPARENTLY THEY DONE SWOOPED IN FROM THE CARINA-SAGITTARIUS. AND AIN'T NUTHIN' STOPPING 'EM.
WELL, KIK. IT SEEMS OUR TRAVEL PLANS HAVE JUST CHANGED.
TO DO MUCH IN THIS LIFE, YOU MUST LEARN MUCH.
AND WE MUST LEARN ABOUT THESE VINN.

IF WE COME ACROSS ANY GRASSKANS IN PORT, KIK, IT'S BEST IF *YOU* DO THE TALKING WITHOUT ME AROUND.
WITHOUT YOU? WHY WOULD WE NEED T--?
BUMP!
MAKER!
WHOA WHOA *WHOA*. EASY NOW: DON'T SHOOT. I GOT NO WEAPON.
I'M NOT HERE FOR TROUBLE. I'M JUST A TRADER, LOOKING TO SETTLE MY FAMILY ON WIRUM.
I NEVER FOUGHT FOR HER MAJESTY...AND SHE'S DEAD NOW, ANYWAY. SO I GOT NO BEEF WITH YOU.
AND I HAVE NO QUARREL WITH YOU, GRASSKAN.
MAY YOUR SPIRITS ILLUMINATE THE--
WAIT
"LOOKING TO SETTLE"? YOU *JUST* ARRIVED?
Y-YES.
AND WHAT OF YOUR FAMILY'S SHIP?
THAT HEAP BARELY MADE IT. WE HAD TO GET TOWED DOWN TO THE PLANET.
I'M ACTUALLY LOOKING TO SELL IT FOR SCRAP, IF YOU KNOW OF ANYO--
MIGHT YOU BE IN THE MARKET FOR A SPEEDER?
ONE THAT CAN ALSO PERFORM TRACTOR-BASED DUTIES?

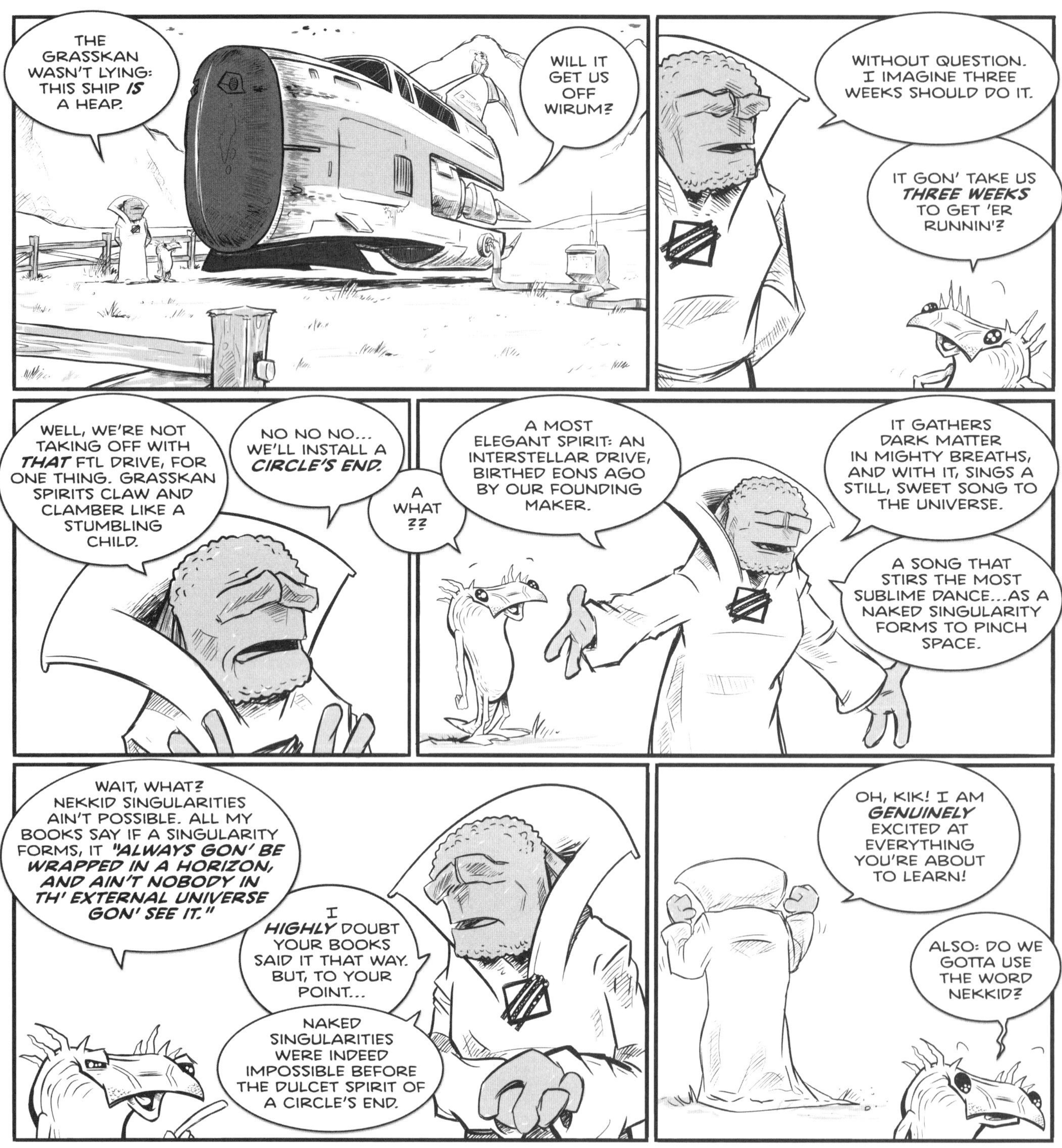
THE GRASSKAN WASN'T LYING: THIS SHIP *IS* A HEAP.
WILL IT GET US OFF WIRUM?
WITHOUT QUESTION. I IMAGINE THREE WEEKS SHOULD DO IT.
IT GON' TAKE US ***THREE WEEKS*** TO GET 'ER RUNNIN'?
WELL, WE'RE NOT TAKING OFF WITH ***THAT*** FTL DRIVE, FOR ONE THING. GRASSKAN SPIRITS CLAW AND CLAMBER LIKE A STUMBLING CHILD.
NO NO NO... WE'LL INSTALL A ***CIRCLE'S END.***
A WHAT ??
A MOST ELEGANT SPIRIT: AN INTERSTELLAR DRIVE, BIRTHED EONS AGO BY OUR FOUNDING MAKER.
IT GATHERS DARK MATTER IN MIGHTY BREATHS, AND WITH IT, SINGS A STILL, SWEET SONG TO THE UNIVERSE.
A SONG THAT STIRS THE MOST SUBLIME DANCE...AS A NAKED SINGULARITY FORMS TO PINCH SPACE.
WAIT, WHAT? NEKKID SINGULARITIES AIN'T POSSIBLE. ALL MY BOOKS SAY IF A SINGULARITY FORMS, IT ***"ALWAYS GON' BE WRAPPED IN A HORIZON, AND AIN'T NOBODY IN TH' EXTERNAL UNIVERSE GON' SEE IT."***
I ***HIGHLY*** DOUBT YOUR BOOKS SAID IT THAT WAY. BUT, TO YOUR POINT...
NAKED SINGULARITIES WERE INDEED IMPOSSIBLE BEFORE THE DULCET SPIRIT OF A CIRCLE'S END.
OH, KIK! I AM ***GENUINELY*** EXCITED AT EVERYTHING YOU'RE ABOUT TO LEARN!
ALSO: DO WE GOTTA USE THE WORD NEKKID?

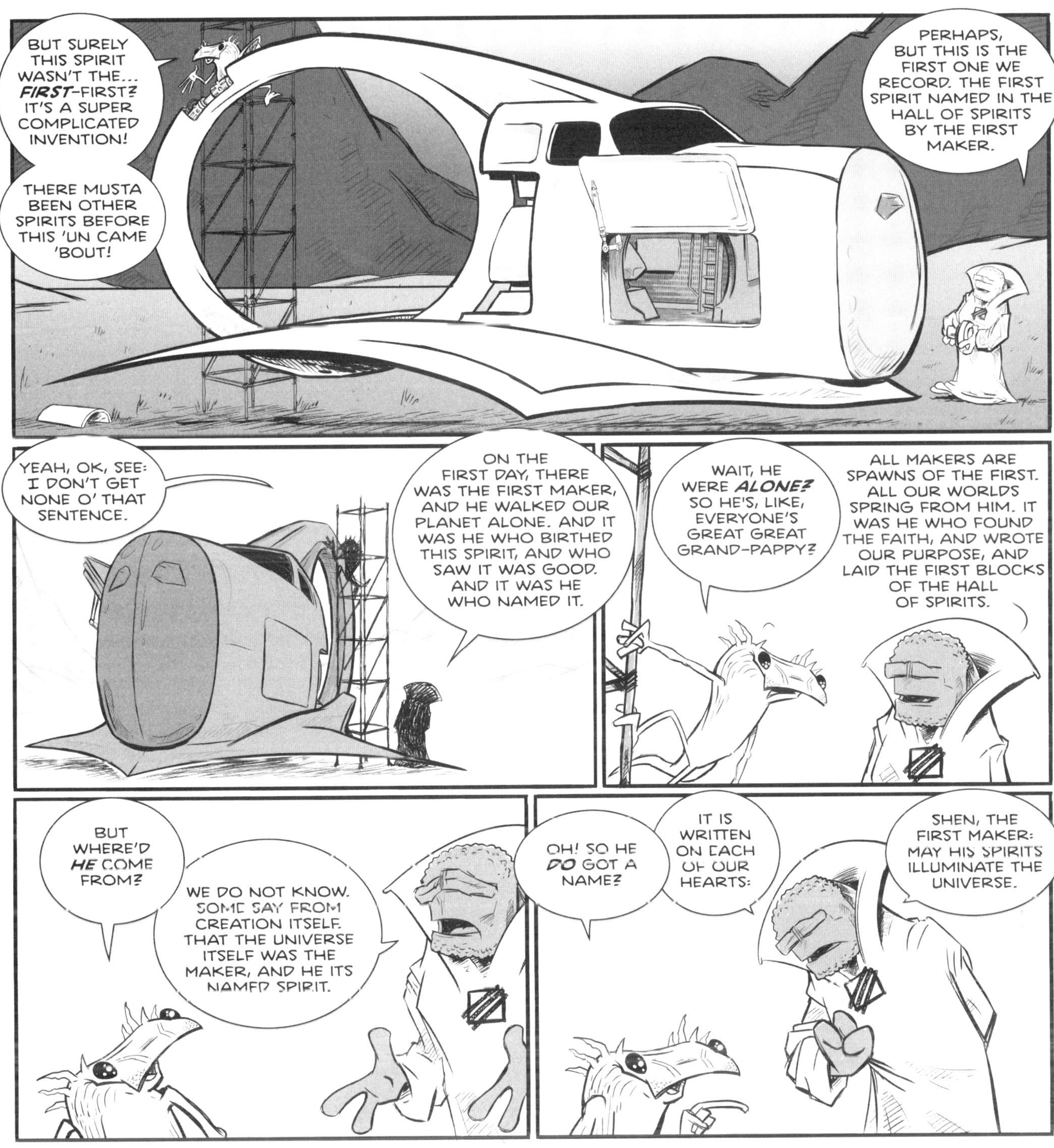
BUT SURELY THIS SPIRIT WASN'T THE... FIRST-FIRST? IT'S A SUPER COMPLICATED INVENTION!
THERE MUSTA BEEN OTHER SPIRITS BEFORE THIS 'UN CAME 'BOUT!
PERHAPS, BUT THIS IS THE FIRST ONE WE RECORD. THE FIRST SPIRIT NAMED IN THE HALL OF SPIRITS BY THE FIRST MAKER.
YEAH, OK, SEE: I DON'T GET NONE O' THAT SENTENCE.
ON THE FIRST DAY, THERE WAS THE FIRST MAKER, AND HE WALKED OUR PLANET ALONE. AND IT WAS HE WHO BIRTHED THIS SPIRIT, AND WHO SAW IT WAS GOOD. AND IT WAS HE WHO NAMED IT.
WAIT, HE WERE ALONE? SO HE'S, LIKE, EVERYONE'S GREAT GREAT GRAND-PAPPY?
ALL MAKERS ARE SPAWNS OF THE FIRST. ALL OUR WORLDS SPRING FROM HIM. IT WAS HE WHO FOUND THE FAITH, AND WROTE OUR PURPOSE, AND LAID THE FIRST BLOCKS OF THE HALL OF SPIRITS.
BUT WHERE'D HE COME FROM?
WE DO NOT KNOW. SOME SAY FROM CREATION ITSELF. THAT THE UNIVERSE ITSELF WAS THE MAKER, AND HE ITS NAMED SPIRIT.
OH! SO HE DO GOT A NAME?
IT IS WRITTEN ON EACH OF OUR HEARTS:
SHEN, THE FIRST MAKER: MAY HIS SPIRITS ILLUMINATE THE UNIVERSE.

SO! ORLA TELLS ME YOU'RE REALLY STARTING TO GRASP HYBRID.
SHE SAYS SCANS OF YOUR BRAIN ARE LOOKING GREAT! AND THE SYNAPSES ARE REWIRING SUPER QUICKLY, AS YOU PICK UP OUR LANGUAGE PHONEMES...?
SO...ARE YOU FEELING MORE CONFIDENT WITH THE LANGUAGE?
WAIT. ARE YOU JUST THUMBS-UPPING ME?

CAPTAIN, THE EMPEROR ASKED IF WE'D "EXTRACTED" ANY FURTHER "RECON" FROM "THE SQUIRREL THING" ABOUT HIS HOMEWORLD.
EVER THE WARM CONVERSATIONALIST, THAT MAN.
LOOK, LIKE I KEEP SAYING: MY ***EARLIEST MEMORY*** IS WAKING UP AND LOOKIN' AT A BIG, SMILIN' NOSH.
OH YES! I REMEMBER! IS GOOD MEMORY! ***IS FRIEND TIMES!***
IS FRIENDS, LOOKING DOWN AT FRIENDS, AND THINKING TO ***BE*** FRIENDS.
AND FRIENDS KNOWING, ***RIGHT THEN,*** THAT THEY WILL HAVE MANY FRIEND ADVENTURES IN THE ***FUTURES***...YET STILL ARE PRESENT IN ***PRESENT FRIEND TIMES.***
reluctant fist bump
IS NOSH HAVING A STROKE?
NO! IS FRIENDS! TALKING FRIEND TIMES!
...VERY SIMILAR RESULT.

WE'VE RECALLED ALL FORCES FROM LAS HERMANITAS, MAJESTY. THEY'LL BEGIN REINFORCING EARTH'S DEFENSES TOMORROW.
GOOD. THOSE BORICUA SETTLERS CAN *HAVE* THEIR IDIOTIC UPRISING. I'VE GOT BIGGER FISH TO FRY WITH THE CONTINUUM NIPPING AT OUR HEELS.

WE'RE STILL WELL SHORT OF THE FLEET WE NEED. EVERY SIMULATION I'VE RUN REQUIRES...
...THOUSANDS MORE SHIPS, *YES*.
HAVE THE JINYIWEI BEGUN MILITARIZING INDÚSTRIAGLOBO'S SHIPYARDS, AS I INSTRUCTED?
WE HAVE. BUT, YOUR COUSIN ARTURO IS...*SLOWING*... OUR WORK. AS CHAIRMAN OF THE I.G. BOARD, HE'S MAKING DIRECT APPEALS TO LA FAMILIA.
ARTURO.
YOU CAN COUNT ON HIM TO SQUEAK WHEN HIS PROFITS START DRYING UP.
YOU HAVE THE FULL AUTHORITY TO OVERRIDE HIM UNDER THE GRAND COUNCIL, OF COURSE...
OVERRIDING DOESN'T SOLVE THE PROBLEM. HE'S NEXT IN LINE FOR THE THRONE. THERE ARE COUSINS WHO WILL *LISTEN* TO HIM.
SAY NO MORE, MAJESTY. I WILL DEAL WITH HIM PERSONALLY.
HA HA HA, ***OF COURSE*** YOU WILL! DIRECTOR, YOU CONTINUE TO DELIGHT.
I'VE FOUND YOUR SERVICE INVALUABLE SINCE PUÑO'S DEATH. ***I'VE BEEN STARING DOWN AN ABYSS*** FOR MONTHS NOW...AND YOU'RE THE FIRST PERSON WHO'S GIVEN ME *HOPE* AGAIN.
THANK YOU, SIR.
WHO COULD'VE GUESSED! MY BEST KILLER TURNS OUT TO BE MY BEST ADMINISTRATOR!
MAKE NO MISTAKE: MARIA WAS A WOMAN OF INCREDIBLE CUNNING. BUT SOMEHOW YOU SURPASS EVEN HER. YOU THINK ***...ORTHOGONALLY.***
FOR THAT, AND SO MUCH MORE, I'VE DECIDED TO NAME ***YOU*** THE NEW FIST OF THE JINYIWEI.
I... I...
THIS FIST MOVES MOUNTAINS.
HMM. THIS FIST DAMN WELL BETTER, ***PUÑO.***

NUEVA SEVILLA, THE OUTER COLONIES
IS IT TRUE?
I'M PATCHED IN TO THE ARMADA'S DEEPEST CRYPTO: IT'S DEFINITELY TRUE. THEY'RE PULLING ALL FORCES BACK TO EARTH.
SO...WE WON? THE EMPIRE IS LEAVING LAS HERMANITAS?!?
GOVERNOR, IT'S MORE THAN JUST THE OUTER COLONIES
...IMPERIAL FORCES ARE PULLING BACK FROM VEETA, TESSKIL, DAWN, DU FU...!
THIS CALLS FOR A CELEBRATION! A FEAST IN YOUR HONOR, PRIMOS!
NO, MA'AM. NO CELEBRATIONS. NOW THE REAL WORK MUST BEGIN.
OH, BUT TAKE A SECOND, FRIENDS! YOU HAVE FOUGHT WITH US, COUNSELED US, EVEN FED US! LET US HONOR THE WORK OF LOS TRES PRIMOS!
IF YOU WISH TO HONOR US, REMEMBER YOUR PROMISE... AND GATHER THE COLONIAL GOVERNORS.
THE COLONIES MUST BEGIN BUILDING A DRIVE FLEET WITH OUR HELP.
BUT...WE'VE ONLY JUST WON OUR FREEDOM.
GOVERNOR, THERE ARE STILL TENS OF BILLIONS OF HUMANS UNDER THE SHACKLES OF THE EMPIRE.
AND FOR THE FIRST TIME, OUR COUSIN THE EMPEROR IS DISTRACTED.
HIS SECRET POLICE ARE LAID LOW. HIS ARMADA IS SHAKEN.
NOW IS OUR CHANCE TO BRING DOWN LA FAMILIA.
ONCE AND FOR ALL.

TESSKIL
FIRST MINISTER....?
I'M TYPING! I'M WORKING!
SIR, THE HUMAN BATTLE CRUISERS STATIONED HERE HAVE...LEFT.
WHAT DO YOU MEAN, "LEFT"? LIKE, LEFT-LEFT? OR LEFT-TO-GET-SOME-FROYO-LEFT?
IT APPEARS LEFT-LEFT.
EVEN THE IMPERIAL GOVERNOR WAS RECALLED FOR CONSULTATIONS.
SO, WE'RE... ALONE, THEN?
IT APPEARS SO.
BUT...BUT THAT WOULD MAKE THIS THE FIRST TIME WE'RE ON OUR OWN IN... A THOUSAND YEARS...
WHAT DO WE DO, SIR?
WELL, WITH NO TESSKANS OR HUMANS OVERSEEING US, THIS IS A CRITICAL MOMENT FOR OUR PEOPLE. IT'S CLEAR WHAT OUR FIRST ORDER OF BUSINESS MUST BE.
SLAM POETRY CONTEST! 7:30!
OHBOYOHBOYOHBOY!
4,000 WORD MINIMUM, PEOPLE!

CONGRATULATIONS ON YOUR WIN!
THANK YOU, SIR!
WINNER WINNER
CHICKEN DINNER
ARE YOU THE MINISTER OF TRANSPORT...? BECAUSE YOUR POETRY MOVED ME.
A HA HA HA
WINNER WINNER
I FEEL TERRIBLE FOR SAYING IT, SIR... BUT LIFE FEELS SO MUCH BETTER WITHOUT THE HUMANS AROUND.
THEY ARE HUGE BUMMERS, GRANTED.
SPEAKING OF HUMANS: DO YOU REMEMBER THAT, UM, TRICKY CHALLENGE YOU WORKED ON WITH SENIOR ADVISOR CUDDOW?
THE ONE TO BAKE EIGHT LAYERS IN A SEVEN-LAYER CAKE?
NO. NO... THE BIG CHALLENGE.
OH! THE ONE TO SNEAK THE WORD "METATEXTUAL" INTO SENTENCES WITHOUT PEOPLE CALLING US ON IT?
NO NO NO. THE SECRET CHALLENGE.
OH!
THE PROJECT TO BUILD THE HUMAN'S DRIVE?
YES.
NOW MIGHT BE A GOOD TIME TO START THAT BACK UP AGAIN.
HEY, YOU'RE RIGHT: WITH NO HUMANS AROUND, IT'S THE PERFECT TIME.
META-TEXTUALLY SPEAKING.

VEETA
LEAST OF US, APOLOGIES FOR DISTURBING YOUR MEDITATION.
NO NEED! I AM TOO MUCH WITH MYSELF. YOU BRING COMPANIONSHIP WITH YOU.
MA'AM, THE HUMAN MILITARY SEEMS TO HAVE...LEFT VEETA.
ALL OF THEM!?!
BUT...
I'D MADE THEIR ADMIRAL A SHEET CAKE.
AND I'M THE BEST AT CAKE.
WAIT.
YOU'VE BEEN HOLDING IT OUT ON THE LAKE?
NOT ANY MORE.
WAN' 'FUM?

RECALLED IMPERIAL PROPAGANDA POSTER, 2401

This digital propaganda poster was pulled almost immediately from the Imperial Vid-Pack System, after it had been made the default lock screen for all users. It was widely derided for a number of clueless design choices, and when Familia themselves were seen to be joking about it, comedy writers and talk shows had a field day with it. Common criticisms included:

- That a Fillipod was taking the lead. And – as would be true of all Fillipods – could in no way be read as "tough looking".
- That an angry-looking, elder-crested Veetan was depicted. And was nonsensically given a cape.
- That Tesskans were included at all, when 98% of the imperial population of Tesskans had just been decimated.
- That a Nyx of Du Fu was seen to be placed too far back in the pack, in what was commonly seen as a "Hey guys, wait up" pose.

DID YOU FIGURE OUT WHY THE DRIVE KEEPS STALLING OUT?
BRRRRRR. ...I *THINK* I DID? I'LL BE HONEST, I WASN'T THE BEST IN DRIVE CORPS.
I PATCHED IT UP, BUT IT LOOKS LIKE CUDDOW'S MANEUVER REALLY BROKE IT GOOD.
THAT SOUNDS LIKE MY TRACK RECORD, YES.
OUR MAIN PROBLEM IS, I DON'T HAVE THE RIGHT PARTS TO FIX IT. SO WE'LL HAVE TO MAKE A PORT CALL SOMEWHERE.
I THOUGHT THAT'D BE THE CASE, SO I CALLED FLEET REPAIR AT GUANG-ZHOU.
UGGGH. WE'RE HEADED BACK TO EARTH *AGAIN*?
NO. APPARENTLY EVERY EARTH FACILITY IS RAMPING UP FOR SHIP PRODUCTION. THEY'VE DIRECTED US TO DU FU FOR REPAIRS.
OOO! I NEVER BEEN TO DU FU!
DU FU! IMMA BUY ME A HOLOSTATION 8 FROM A NYXXXXXX!
IS HOLOSTATION THE ONE HAVE TETRIS ON IT?
NOSH, *EVERY* PLATFORM HAS TETRIS. TETRIS IS A 400-YEAR OLD GAME.
400 YEARS OF BEING BESTEST.
WHAT?
I SAY "400 YEARS OF BEING BESTEST!!"

2318.3320.8891.4839.2358.2908.4787.4697.9032.3348.8878.1304.3989.7878.3454.2209.1121.9944.4735.8883.2921

30421

Enciclopedia Xenobiológica

Página 30421 | Imprimátur del Imperio

Species: Nyx

Homeworld: Du Fu (Adopted Mandarin name. Original Nyxian name untranslatable in Hybrid.)

Size: 1.4-1.8m

Color & Markings: The Nyx are easily identified by their pronounced head crest and series of deep-set, fat-storing rolls across their bodies.

Reproduction & Development: The Nyx reproduce sexually, generating one offspring. Nyxian mating is complicated by a form of haplodiploidism similar to Earth bees, where some of the 16 defined Nyx genders are infertile, or change fertility, based on environment or social status. Offspring are commonly raised by the birthing female and birthing female's mother. Adult Nyx consume a massive 50-70kg of food a day, and most of their waking hours are spent eating. When not eating, the Nyx immediately turn to their second insatiable habit: Tinkering. They are unparalleled inventors, and love to modify and experiment with machines of all types. Before humans arrived, the Nyx had already developed a technologically complex civilization.

Diet: Monophagous. The Nyx stand alone, among intelligent life, in having a diet of only one food: The leaf of the fast-growing Nestr Tree.

Lifespan: 60-70 years.

Language: The Nyx communicate through a series of low-frequency rumblings, which human cochlear implants are incapable of translating. Helpfully, the Nyx have invented a painless throat implant that instantly translates and speaks in Hybrid. The effect is odd, but workable.

Nyxian language is perhaps the most inscrutable ever encountered by humans. The language seems to operate under an endless series of quantum-state future-perfect tenses, shifting under impossible-to-understand rules. Common linguistic dualities such as light/dark, on/off, and alive/dead are not understood in a binary sense, but rather in a spectrum of possibility that's weighted by context, time, location, etc. Meanings operate on multiple levels, and change even while they're being uttered. To the Nyx, this language flux works as effortlessly as breathing. But to other species, it feels closer to madness. (In fact, the Fillipod linguist Sella, who wrote the best-known study of Nyxian language, herself went insane toward the end of her career.) Most surprising to cognitive scientists, though, is that the Nyx *seem* to operate under a constant sense of a "Socratic Ideal". Though their language is constantly in flux, each sentence, each intent, is spoken and received as the purest thought possible *in that moment*. It is the only thing that could be spoken. The only thought that could be expressed. And yet, when asked the same question moments later, they will give a completely different answer. And will remain firm that it, too, is the only thing that could be spoken in *that* moment. Or at least, that is the best guess of linguists. Most have chosen to stop researching Nyxian language altogether, since Sella was committed to a sanitarium. Most linguists now choose to focus on Veetan food poetry.

Social Structure: Imperial anthropologists never had the chance to study, in depth, the traditional social structures of the Nyx before human settlement began to change everything. And sadly, Nyxian written histories, like their spoken language, are completely inscrutable.

Intelligence Rating: Testing of Nyxian subjects has been extensive, but continually inconclusive. Dr. Chatterjee herself was the first to test the intelligence of the Nyx, but found her own methods lacking, due to the impossible language barrier. With unique methods of testing, it was determined the Nyx *perhaps* operate in the Papa/Quebec range. But it could be higher. (Helpfully, the Nyx themselves continue to design and try out self-administered tests that work within the Chatterjee system.)

Interaction with Humans: Though humans were already beginning to settle on Veeta, Du Fu can properly be considered the first human "colony" outside our solar system. In a way that never happened on Veeta, humans completely overtook the administration of Du Fu. And the Nyx, somewhat passively, seemed to be content with that happening. The first human colony on Du Fu was famously started by a cadre of Chinese professors from Beijing University, as a potential "New Eden" of philosophy and erudite governance. But as soon as La Familia learned of the Nyx' abilities with technology, the planet was forced to become the manufacturing engine of the Empire. Inventions and new products now constantly flow from the laboratories of Du Fu, enriching the local human populace, and keeping the Nyx happily tinkering.

"OVERHEAD I CAN COUNT
THE SEVEN STARS OF THE DIPPER,
AND IMAGINE I HEAR THE NOISE
OF THE MILKY WAY AS IT FLOWS...
THE SUN'S BEEN PUSHED BELOW
THE WESTERNMOST HORIZON
A CLEAR AUTUMN MOON
IS TRYING TO RISE IN THE EAST
BECAUSE OF CLOUDS, WE SEE
ONLY SHREDS OF THE GREAT MOUNTAIN.
THE CLEAR RIVER AND THE MUDDY RIVER
MIX TOGETHER AND ARE LOST.

DOWN BELOW YOU CAN'T SEE MUCH,
BECAUSE OF ALL THE MIST.
HOW CAN WE EVEN BE SURE...

...THAT THIS IS OUR GREAT EMPIRE?"
-THE CHINESE POET DU FU,
WRITING IN 753
18
19
20

TSSSSSSH!
TSSSSSSH!
TSSSSSSH!
杜甫
NO ENTRY WITHOUT
NI HAO, SHIP RIDERS.
AND WELCOME TO DU FU.

ORDERS ARE R&R FOR SHIP CREW...
WHILE SHIP CAPTAIN AND SHIP FAMILIA TALK WITH DU FU FAMILIA.
DU FU FAMILIA? WHO?
BONG! BONG!
AH! THAT'D BE ME! AND JUST IN TIME. YES, YES, OFF YOU GO. OFF YOU GO. THANK YOU BOTH.
WHA....WHERE ARE THE NYX OFF TO?
TO EAT, CAPTAIN. EVERY 45 MINUTES. THEY'LL BE BACK IN A FEW.
ALLOW ME TO INTRODUCE MYSELF. I AM SEBASTIÁN CRUZ, PROCONSUL OF DU FU. MY COUSIN THE EMPEROR BID ME WELCOME YOU.
AND TO SPEAK, COUSIN?
INDEED. SEPARATELY! SEPARATELY! MUSTN'T TALK IN PUBLIC! HA HA HA! MUSTN'T-MUSTN'T!
COULD WE EAT BEFORE CHATTING, PROCONSUL? WE'RE ALL A BIT HUNGRY.
OOOOH! LET US GO JOIN THE NYX!
OH NO NO NO. NOT UNLESS YOU LIKE EATING NESTR LEAVES.
THEY TASTE LIKE KALE HAD A LOVE-CHILD WITH A DESICCATED TURTLE SHELL.
THAT IS... VERY SPECIFICS.
I'VE LIVED HERE 8 YEARS. I'VE HAD TIME TO THINK ABOUT IT.

COME-COME, CAPTAIN! COUSIN! TODAY WE DINE AT MY MANSE. AH, SUCH A MERRY BAND OF DINERS WE SHALL BE!
RIGHT BEHIND YOU, PROCONSUL.
ORLA, I'D LIKE YOU TO SPLIT OFF, PLEASE.
FIND THE LOCAL JINYIWEI. TRADE THEM INFO FROM NUEVO CHILE FOR ...ANYTHING YOU CAN GET THAT HELPS US.
ROGER THAT, MA'AM.
TAKE THE REST OF THE CREW AND CHAT UP THE NYX. FIND OUT IF THEY'VE EVER ENCOUNTERED SKITTER'S PEOPLE.
THE NYX AREN'T REALLY A SPACEFARING PEOPLE, BUT MAYBE WE'LL GET LUCKY.
NO NYXES DON'T GO TO SPACINGS?
NOT OFTEN. AS INVENTIVE AS THEY ARE, SPACE TRAVEL IS KIND OF A LOGISTICAL NIGHTMARE FOR THEM.
EACH ONE OF THEM EATS 70 kg A DAY, SO THERE'S A BIG PROBLEM WITH HANDLING ALL THEIR
LEAVES?
LEAVINGS.

HULLO-HULOO! NYXONS! MAY WE JOIN YOU?
I AM NOSH. SEÑOR SCIENCE OFFICER OF THE GOOD SHIP MACHITO!
"SENIOR".
AND THIS IS MY FRIENDZO, CUDDOW. HE WAS SUPER BIG PANTS ON TESSKIL. RAN WHOLE PLANET! BUT NOW HE JUST KIND OF...HANGS OUT? I DUNNO.
I'VE TAKEN UP BAKING!
WE ARE HONORED TO MEET THE FAMED NYX OF DU FU, AND WANTED TO SAY HE---
NOSH! DON'T EAT THEIR LEAVES!
PTUI! PTUI! UGH! TASTES LIKE FOOT!
LIKE MOIST TOE.
I APOLOGIZE FOR MY FRIEND.
I AM SORRY, NYXOLS. APOLOGIZEMENTS. SOMETIME HEAD NOT KNOWING WHAT STOMACH IS DOING.
ANYWAY! WE WANTED TO ASK YOU ABOUT OUR SMALL FRIEND, HERE.
HIS HOMEWORLD IS UNKNOWN TO US, BUT WE GREATLY NEED TO--NOSH!
SORRY! IS GENUINE GROSS! BUT ALSO RIGHT HERE.

IF VEETAN SHIP RIDER REQUIRES EATING
IT WILL ALWAYS HAVE BEEN INVITED TO EAT.
NO NO I SHOULDN'T. IS YOUR FOOD.
ALTHO'... MAYBE IS RUDE TO REFUSE? MAYBE I TAKE TINY SMIDGE?
YOU ARE MOST MAGNANIMOUS IN YOUR WELCOME, KIND SIR.
LADY?
...NYX. BUT IT IS NOT FOOD WE SEEK. IT IS KNOWLEDGE.
YES. FILLIPOD SHIP RIDER WILL HAVE HAD INQUIRED ABOUT SMALLEST SHIP RIDER.
ABOUT UNKNOWNS.
BUT! IS THIS CREATURE UNKNOWN KNOWN TO SHIP RIDERS, OR KNOWN UNKNOWN? AND IN THE UNKNOWING, KNOWS IT IS UNKNOWN?
OR KNOWS NOT WHAT IS UNKNOWN, AND IN SO DOING, KNOWS WHAT IS KNOWN?
UM.
I TAKE SMIDGE MORE.

THE EMPIRE DESPERATELY NEEDS TO KNOW WHERE OUR FRIEND IS HOME IS.
AND AS FELLOW CITIZENS OF THE EMPIRE, WE'RE HOPING YOU CAN HELP.
THE EMPIRE.
THE EMPIRE WILL HAVE ONLY BEEN A CONSTRUCT OF THE MIND.
WILL HAVE BEEN IMPERMANENT.
UM, HA HA! YES! TRUE, I SUPPOSE. EVERYTHING FADES.
BUT FOR NOW, THE EMPIRE STILL VERY MUCH EXISTS, AND NEEDS OUR HELP.
AGREEMENT.
WE WILL HAVE BEEN EXCITED TO HELP.
WAIT. SO YOU WILL HELP??? (...I WASN'T SURE WHERE YOU WERE GOING WITH THE "MIND-CONSTRUCT" THING, THERE.)
DISCOVERY WILL HAVE BEEN AS HAVE WILL TO BE: GREATEST PLEASURE.
THUS! SMALL SHIP RIDER WILL HAVE BEEN MACHINE SAMPLED.
"MACHINE SAMPLED?" WHAT DOES THAT MEAN?
THESE YOUNG NYX NEVER WILL HAVE SEEN A CREATURE AS THIS.
BUT MACHINE SAMPLE WILL HAVE EXCHANGED WITH OLDER MACHINE SAMPLES.
OH! THEY JUST WANT TO COMPARE YOU TO THEIR DATABASE.
HUMMMMMMMMMMMMMMMMMMMMMMM
BUT WHAT ARE YOU GONNA SAMPLE? I AM NOT GETTING BITS OF ME HACKED OFF.
THAT NYX WILL HAVE SAMPLED YOUR ELECTROMAGNETIC FIELD.
THIS NYX WILL HAVE...FOUND A MATCH.
YOU ISOLATE SPECIES WITH THEIR E-M FIELD?? ..DAAAAANG.
THESE CUCKOO KIDS GOT IT GOIN' ONNNN
FROM A SCIENCE PERSPECTIVVVVE
THAT FUTURE-PERFECT TENSE IS KILLING ME. WAS I ALREADY SAMPLED OR NOT???

THIS NYXIAN WINE IS DELIGHTFUL, IS IT NOT? *HIC* GETS THE HEART PUMPING AND THE TALK FLOWING!
THIS IS NYXIAN WINE? I THOUGHT THE NYX DIDN'T DRINK.
OH, THEY DON'T. NO NO NO. BUT WHEN I DESCRIBED WINE FOR THEM, THEY TOOK GREAT JOY IN INVENTING THEIR OWN. DISCOVERY DRIVES THEM LIKE NOTHING ELSE.
BUT BUT BUT! TO DARKER MATTERS. YES, YES. QUICKLY-QUICKLY.
FERNANDO, IT'S YOU I WANTED TO SEE.
HIC BECAUSE YOU'RE IN DANGER, COUSIN. AND THAT'S NOT THE WINE TALKING.
YOUR... YOUR FATHER PROTECTED ME, YOU SEE. AT A TIME IN MY LIFE WHEN I...COULDN'T PROTECT MYSELF.
IN DRIVE CORPS, THE COUSINS CAN BE SUCH...BRUTES. SO CRUEL, AS THEY CLAW FOR THE BEST POSTINGS.
BUT NOT YOUR DEAR FATHER, FERNANDO. HE WAS KIND. AND EVERY TIME I EXCELLED AT EXAMS...HE PROTECTED ME FROM THEIR FISTS. AND THERE WERE SO MANY FISTS.
AND SINCE HE IS NOT HERE TO HELP YOU...IT'S UP TO ME TO PAY IT FORWARD. WITH INFORMATION. YES-YES-YES.
THE HEAD OF INDÚSTRIAGLOBO IS DEAD, FERNANDO. AS IS MARTA CRÚZ, THE CHAIRWOMAN OF BANCO IMPERIAL.
BOTH WERE "NATURAL CAUSES," BUT I HAVE MY SOURCES. THE NYX ARE SUCH CLEVER INVENTORS, YOU SEE. SO SO SO CLEVER.
THEY WERE BOTH KILLED, MY BOY. KILLED. WHICH MEANS...
POR EL AMOR DE DIOS
THE LINE OF SUCCESSION.

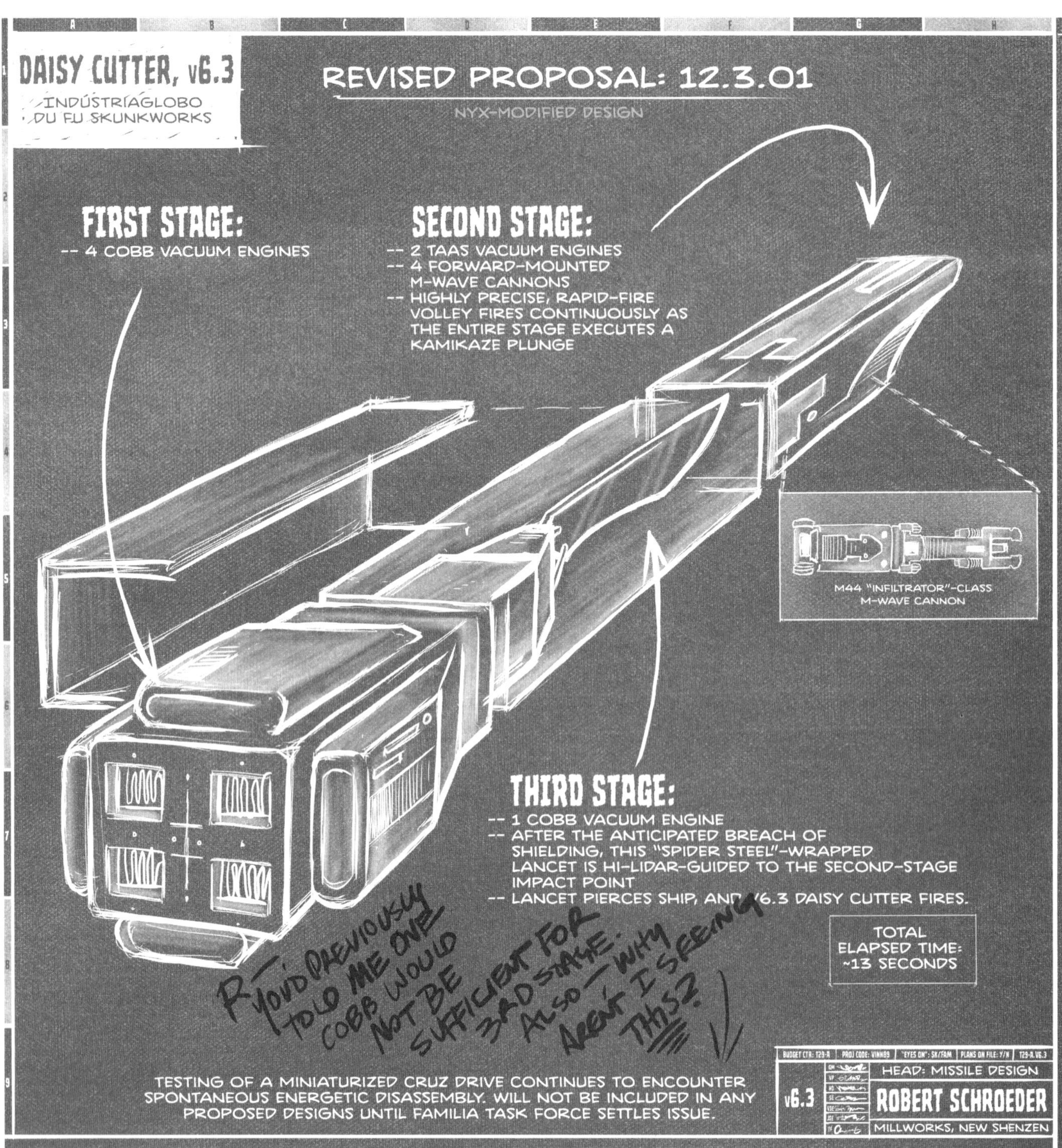

DAISY CUTTER, v6.3
INDÚSTRIAGLOBO
DU FU SKUNKWORKS
REVISED PROPOSAL: 12.3.01
NYX-MODIFIED DESIGN
FIRST STAGE:
-- 4 COBB VACUUM ENGINES
SECOND STAGE:
-- 2 TAAS VACUUM ENGINES
-- 4 FORWARD-MOUNTED M-WAVE CANNONS
-- HIGHLY PRECISE, RAPID-FIRE VOLLEY FIRES CONTINUOUSLY AS THE ENTIRE STAGE EXECUTES A KAMIKAZE PLUNGE
M44 "INFILTRATOR"-CLASS M-WAVE CANNON
THIRD STAGE:
-- 1 COBB VACUUM ENGINE
-- AFTER THE ANTICIPATED BREACH OF SHIELDING, THIS "SPIDER STEEL"-WRAPPED LANCET IS HI-LIDAR-GUIDED TO THE SECOND-STAGE IMPACT POINT
-- LANCET PIERCES SHIP, AND V6.3 DAISY CUTTER FIRES.
TOTAL ELAPSED TIME: ~13 SECONDS
R— YOU'D PREVIOUSLY TOLD ME ONE COBB WOULD NOT BE SUFFICIENT FOR 3RD STAGE. ALSO— WHY AREN'T I SEEING THIS?
TESTING OF A MINIATURIZED CRUZ DRIVE CONTINUES TO ENCOUNTER SPONTANEOUS ENERGETIC DISASSEMBLY. WILL NOT BE INCLUDED IN ANY PROPOSED DESIGNS UNTIL FAMILIA TASK FORCE SETTLES ISSUE.
v6.3
HEAD: MISSILE DESIGN
ROBERT SCHROEDER
MILLWORKS, NEW SHENZEN

ANY SIGNS OF LIFE, KIK?
WELLSIR, IF'N I'M READING YER SCANNERS RIGHT...NOT MUCH.
YOU GOTCHER BASIC SCRAGGLIN' OF FLORA, AND YOU GOTCHER BASIC PROKARYOTIC MICROORGANISMS.
BUT NUTHIN'...BIG. NO CRITTERS, NO PEOPLES.
LOTSA RUINS, THOUGH. AND A LOTTA RADIATION.
MAKER, I DON'T UNNERSTAN WHAT HAPPENED HERE.
...IF THIS IS WHERE ALL THEM VINNS CAME FROM, SHOULDN'T THERE BE A FEW STRAGGLERS THAT SURVIVED HERE ON VINNSHASHA?
PERHAPS... NOT.

MAKER? MAKER?
...MAKER?
AHMIS.
YES, KIK?
YOU BEEN SITTIN' STILL FOR 8 HOURS, NOW. I BEEN CALLIN' YER NAME! YOU DIDNT EVEN LOOK LIKE YOU WAS BREATHIN' NO MORES.
I DONE GOT SKURR'D!
APOLOGIES. I WAS WALKING THE INWARD ROAD. GAMING OUT WHAT HAPPENED HERE.
AND WITH DEDUCTION, I NOW UNDERSTAND THE VINN'S CENTRAL PROBLEM...
ON EVERY PLANET WE'VE BEEN, WE HEAR OF THE VINN'S "QUEST" FOR THEIR MISSING GODS, YES?
BUT NOW I SEE THE TRUE DESPERATION BEHIND THEIR SEARCH: THEY ARE FLEEING THEIR COLLECTIVE DEATH.
THEIR... DEATH?
WHEREVER THEY GO, THEY CONSUME AND EXPAND, AGAIN AND AGAIN.
THEY ARE THE RIPPLES OF A STONE DROPPED INTO A GALAXY-WIDE POND. RIPPLING OUT, OUT, OUT.
BUT LOOK BEHIND THE WAKE OF THOSE RIPPLES, AND I BELIEVE YOU WILL FIND NO LIVING BEINGS.
OH... LIKE WE DONE FOUND HERE?
INDEED. THE VIRUS NEUTERS THEM, KIK. ONCE INFECTED, THE VINN CAN NO LONGER REPRODUCE NATURALLY.
AND SO THEY WILL CONSUME THE GALAXY IN THEIR HOLY QUEST...AND LEAVE ONLY DESOLATION.

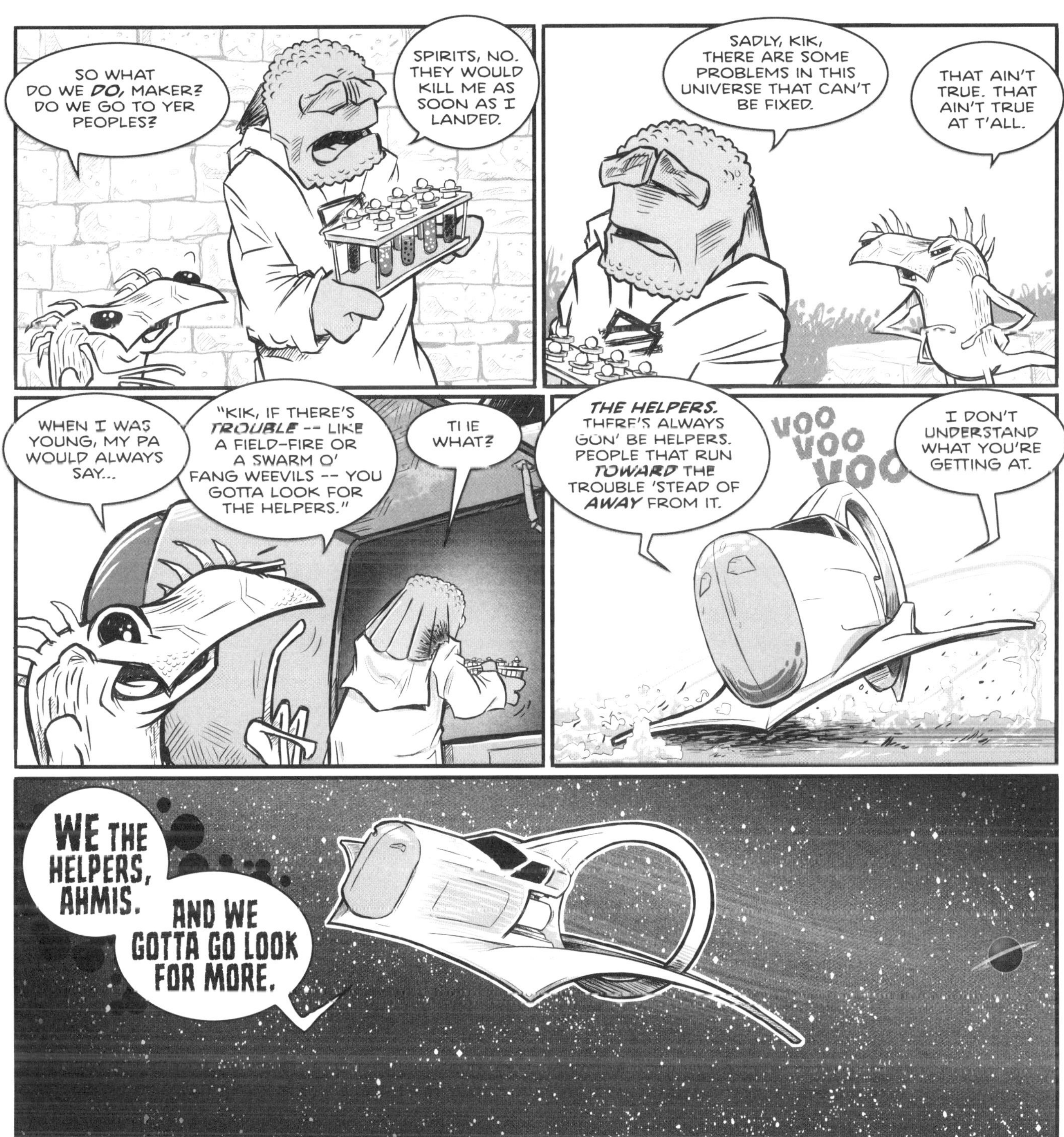
SO WHAT DO WE DO, MAKER? DO WE GO TO YER PEOPLES?
SPIRITS, NO. THEY WOULD KILL ME AS SOON AS I LANDED.
SADLY, KIK, THERE ARE SOME PROBLEMS IN THIS UNIVERSE THAT CAN'T BE FIXED.
THAT AIN'T TRUE. THAT AIN'T TRUE AT T'ALL.
WHEN I WAS YOUNG, MY PA WOULD ALWAYS SAY...
"KIK, IF THERE'S TROUBLE -- LIKE A FIELD-FIRE OR A SWARM O' FANG WEEVILS -- YOU GOTTA LOOK FOR THE HELPERS."
THE WHAT?
THE HELPERS. THERE'S ALWAYS GON' BE HELPERS. PEOPLE THAT RUN TOWARD THE TROUBLE 'STEAD OF AWAY FROM IT.
VOO VOO VOO
I DON'T UNDERSTAND WHAT YOU'RE GETTING AT.
WE THE HELPERS, AHMIS.
AND WE GOTTA GO LOOK FOR MORE.

YOU FOUND A MATCH?
YOU KNOW WHERE I'M FROM???
THIS NYX WILL HAVE SAID...NO.
BUT YOU GOT A MATCH! YOU JUST SAID YOU GOT A MATCH!
THAT NYX WILL HAVE SPOKEN CORRECTLY. WE WILL HAVE FOUND A MATCH.
BUT THE MATCH WILL HAVE BEEN HERE, ON DU FU.
AM I... FROM HERE?
...A SINGLE MATCH. 256 YEARS AGO. TRAVELER TO DU FU.
SHE WAS TO HAVE BEEN VISITOR HERE. WITH A PACK OF YANTAMONKS.
"YANTAMONKS"? WHAT'S A YANTAMONK?
MACHINE SAMPLE...WILL HAVE HAD NO FURTHER INFORMATION.
"YANTAMONKS" SOUNDS LIKE A BUGS BUNNY INSULT.
OR SOME KINDA WHEATGRASS-INFUSED YOGURT DRINK.
IF SO, IS GROSSEST YOGURT FLAVOR.
I WILL HAVE UNDERSTOOD NONE OF THAT.

WAIT, I GET IT! THEY'RE NOT SAYING "YANTAMONKS." THEY'RE SAYING "YANTA **MONKS**."
MONKS⁇
FROM THE MONASTARIES OF YANTA!
YES, YANTAMONKS.
Y-YANTAMONKS.
YANTAMONKS⁇
"YANTAMONKS"
Deet deet Deet Deet deet deet
YANTA.
MONKS.
WHAT A WIN FOR MACHINE TRANSLATION.
YANTA.
MONKS.

YOU'RE REALLY GONNA BUILD DRIVE SHIPS?
WE'RE GONNA TRY.
YOU KNOW WHAT THAT MEANS, DON'T YOU?
MAAAAAH.
DON'T DO THAT!
I HATE WHEN YOU DO THAT!
WHAT, WIND-TALKING?
YES. YOU KNOW I DON'T LIKE IT.
OUR BRAINS LINKED YEARS AGO! IT'S WAY EASIER!
I DON'T CARE! IT'S WEIRD FOR HUMANS.
JUST SAY WHAT YOU GOTTA SAY, OUT LOUD.
FINE.
...WHAT IT MEANS IS, THE TRES PRIMOS AREN'T JUST RUNNIN' AND GUNNIN' ANYMORE.
IT MEANS WE'RE ALL GONNA DIE IN A WAR WE CAN'T WIN.
SHHH! YOU'LL SCARE PEOPLE.
ALSO: WE CAN WIN!
...MAYBE

PORTHOS TELLS ME YOU FEAR THE DAYS AHEAD.
...MAY I SHARE SOMETHING WITH YOU?
SURE. YEAH.
tink tink
AS YOU KNOW, MY COUSINS AND I GREW UP IN THE LUNAR COLONIES.
AND BETWEEN THE .9 ARTIFICIAL GRAVITY, AND THE FAMILIA RICHESSE, IT WAS A CHILDHOOD OF ***EXTREME COMFORT.***
AND WHEN WE ANSWERED OUR DRAFT INTO DRIVE CORPS, WE FOUND WE ***EXCELLED*** THERE.
GOD, IT SEEMED, HAD GRANTED US A QUICK WIT WHEN IT CAME TO STARSHIPS.
HA! NOT LIKE YOU, CERTAINLY, BUT GOOD BY HUMAN STANDARDS.
SO, FOR US, IT WOULD'VE BEEN EASY JUST TO SERVE OUR TIME, AND THEN RETURN TO A RICH POSTING AT INDÚSTRIAGLOBO OR ONE OF THE FAMILY BANKS.
BUT WE'D...
...SEEN TOO MUCH.
WHENEVER WE HAD LEAVE, WE'D JOURNEY TO QUITO, OR ADDIS ABABA...OR EVEN THE RUINS OF NEW YORK.
ANYWHERE WE COULD SEE THE ***REAL*** EMPIRE.

THE POVERTY WAS ALMOST EASY TO COMPREHEND.
WHAT WAS HARDER FOR US TO UNDERSTAND WAS THE OPPRESSIVE SENSE OF ...STAGNATION. OF SEEING ENTIRE VILLAGES, ENTIRE CITIES, WHO KNEW THAT THEIR CHILDREN COULD NEVER ASPIRE TO MORE, AS THEY WERE NOT FAMILIA.
HOPE...IS THE ENGINE OF THE HEART.
AND WHEN YOU LOSE HOPE --NOT ONLY FOR YOURSELF, BUT FOR YOUR CHILDREN AND YOUR CHILDREN'S CHILDREN--WHAT DO YOU HAVE? WHAT FIRES YOUR SOUL?
THE FIRE HAS GONE OUT IN THE SOUL OF OUR EMPIRE.
SO RATHER THAN FEAR THE FIGHTS TO COME, FEAR OUR COMPLACENCY IN THE PRESENT.
FEAR THE PRIVILEGE THAT ALLOWS US TO SAY "OH, EVERY-THING IS FINE," WHEN WE KNOW BILLIONS OF VOICES ARE CRYING OUT IN MUFFLED DESPERATION.
FEAR OUR TENDENCY, EVEN NOW, TO WAVE AWAY THE SINS OF THE EMPIRE BECAUSE WE HAVE FULL BELLIES AND GLAD HEARTS.
THIS UNIVERSE CRIES OUT FOR THOSE WHO CAN HELP, TO HELP.
YOU HAVE GIFTS, MY FRIEND, AS DO WE. AND IN THE FINITE TIME WE'RE GIVEN IN THIS LIFE, OUR JOB IS TO MAKE THIS WORLD BETTER THAN WE FOUND IT.
DON'T FEAR THE DAYS AHEAD.
FEAR NOT WALKING TOWARD THEM.

FAMILIA PROPAGANDA POSTER

Following the assassination of IndústriaGlobo Chairman Arturo Cruz, official surveys of the mood in La Familia showed record lows. As a result, the Central Information Council began issuing new Familia-focused propaganda campaigns, focusing on hope, the future, and service. But the tone of the campaign struck most Familia the wrong way, and the campaign was soon replaced by a campaign showing how they might profit off the war with conflict-focused investments.

THE SPARK OF THOUGHT
GAHHHH. WHAT SMELLS IN HERE?
WELL...I S'POSE IT COULD BE THIS: THERE'S A DEAD MAKER THAT DONE GOT STUCK BEHIND THIS HERE CONSOLE.
NO. AIN'T THAT. SMELLS LIKE SOMEBODY WENT AND GOT SICK ON THEMSELF.
SNIF SNIF
WELL, I DID. BUT THAT WAS A MONTH 'ER SO AGO.
WHAT?? AND YOU AIN'T USED ONE A' THEM MAKER SHOWERS SINCE???
NO. I....
GO. RIGHT NOW. THASSA ORDER.
STOMP
STOMP
STOMP
FASHUNK...
CLICK!
DEET DEET DEET

TSSSH!
AAAAGH! NO! I AIN'T WANTIN' PERFUME! I'M WANTIN' WATER!
TURN OFF! I SAID TURN OFF!
DEET DEET DEET
GAHHH!
WAM WAM WAM!
GLLGLLGLGG! NO! THIS' WORSE!
HELP! HELP! IMMA DROWN IN PERFUME!
SUM'BODY!
YOUSE A MORON.
WELL I THINK HE DONE SMELL LIKE SUNN-SHINE

THERE ARE ENTIRE FACTIONS OF THE FAMILY WATCHING OUT FOR YOU, FERNANDO.
YES-YES-YES.
IN WAYS YOU'LL NEVER KNOW.
WHEN YOU WERE CALLED UP FOR DRIVE CORPS, DID YOU NOTICE YOUR LETTER CAME... EARLIER THAN EXPECTED?
WE DID THAT. ALTERED YOUR RECORDS TO GET YOU POSTED IMMEDIATELY.
YOU WERE SAFER IN DRIVE CORPS LOCK-DOWN THAN AT HOME.
AND YOUR ASSIGNMENT ORDERS, POSTING YOU UNDER THE CAPTAIN? THAT WAS NOT CHANCE.
...A COUSIN YOU'LL NEVER MEET MADE THAT HAPPEN, AT OUR REQUEST.
WHAT??
WE SPRANG INTO ACTION. WE HAD TO! HAD-TO-HAD-TO HAD-TO!
MY NYX ARE SO CLEVER. SO GOOD WITH NETWORKS. AND WHEN THEY DISCOVERED THAT EL PUÑO HAD SABOTAGED YOUR FATHER! AND KILLED OUR DEAR EMPEROR! ...THEY BROUGHT IT RIGHT TO ME. THEY KNEW WHO WAS BEHIND IT ALL.
NYX ARE NOT POLITICAL IN THE SLIGHTEST. NO NO NO. BUT THEY SEE PATTERNS...AND THEY KNEW YOU WERE TARGETED.
OH MY BOY.... I KNOW YOU WANT NOTHING MORE THAN TO LIVE A NORMAL LIFE, BUT THAT CAN'T HAPPEN.
IT'S NOT FAIR, BUT PRECIOUS LITTLE IN THIS FAMILY IS FAIR.
FATE IS HUNTING YOU.
YOU MUST EITHER TAKE UP THE THRONE... OR A GRAVE.

BUT WHY DOES IT HAVE TO BE ME?
BECAUSE IT WAS ALWAYS MEANT TO BE YOU. AS IT WAS YOUR FATHER.
I CAN'T DO IT. I CAN'T...KILL THE EMPEROR!
NO ONE IS ASKING YOU. ALL WE'RE ASKING YOU TO DO IS SURVIVE. AND BE READY.
CAN'T YOU TAKE THE LEAD? YOU'RE A GOVERNOR! YOU HAVE THE NYX BEHIND YOU! AND ALL THOSE...MYSTERIOUS FAMILY MEMBERS, OR WHATEVER.
I AM NOT A... BRAVE MAN, 'NANDO -- OR A LEADER. YOU DON'T GET POSTED TO DU FU IF YOU'RE A RALLYING POINT FOR HEROES.
BUT I HAVE MY SKILLS. AND WE ALL MUST FIGHT WITH THE WEAPONS WE'RE GIVEN, YES?
MY NYX AND I, SOMOS UÑA Y CARNE. THE EMPIRE'S LITTLE OUTCASTS.
THEY THINK WE ARE JUST NERDY TINKERERS ON DU FU, QUIETLY MAKING THEIR GADGETS.
BUT WE PULL MANY HIDDEN LEVERS IN THE EMPIRE THAT THE FAMILY NEVER SEES. YES-YES-YES. A QUIET STRENGTH.
AND THAT STRENGTH IS NOW IN YOUR SERVICE.
I WILL SEND YOU WHAT AID I CAN, 'NANDO: OPENLY AND SECRETLY.
AND MANY OTHER COUSINS STAND READY TO SACRIFICE EVERYTHING FOR YOU. TO SEE YOU STAND UP...
...AND LEAD US ALL.
BUT FIRST... EAT YOUR SOUP.

AH, ORLA! HOW'D IT GO WITH THE JINYIWEI?
I...COULD FIND NO LOCAL STATION OFFICE, MA'AM.
OH YOU WOULDN'T, MY DEAR, NO. NO-NO-*NO.*
ALL THE POSTED OFFICERS CAME DOWN WITH A RARE, EXCRUCIATING FORM OF TINNITUS.
SOMETHING TO DO WITH THEIR JINYIWEI COCHLEAR IMPLANTS. *VERY STRANGE.*
THANKFULLY, THEY SEEM TO HAVE RECOVERED BACK ON EARTH.
SO WE HEAR.
BUT--
FERNANDO!
I HAVE TWO GIFTS FOR YOU BEFORE YOU GO... AND A GIFT FOR THE CAPTAIN!
FIRST, A *TEXTING DEVICE.*
OH. UM. *THANKS,* BUT MY IMPLANT CAN ALREADY TEXT.
NOT LIKE THIS. THE NYX HAVE BEEN WORKING ON THIS LITTLE TRICK FOR DECADES.
IT WILL HAVE HAD A SINGLE, SUSPENDED PHOTON, QUANTUM ENTANGLED WITH ANOTHER HERE ON DU FU.
THEY WILL HAVE SUNG *TOGETHER,* WHEN ONE SINGS ALONE.
YES-YES-YES. VERY HANDY. *YES-YES-YES.*
WHEN THE CHIPS ARE DOWN, YOU MAY SEND ONE *VERY SHORT* CRY FOR HELP... AND WE'LL MAKE SURE THE CAVALRY COMES RUNNING.

ENTANGLEMENT?? HOW DID THEY SOLVE THE---
SECOND! I OFFER YOU ONE OF THE EMPIRE'S BEST-KEPT SECRETS: THE COORDINATES TO McBRIDE'S WORMHOLE.
IF OUR COLLECTIVE PLANS FAIL SPECTACULARLY, AND EVERYONE WE TRUST IS DEAD...THEN YOU WILL FIND NO SAFE HARBOR.
YOU WILL BE A MARKED MAN THROUGHOUT THE WESTERN SPIRAL ARMS OF THE GALAXY.
BUT! PERHAPS YOU CAN FIND PEACEFUL LAST DAYS THROUGH THAT MAGNIFICENT WORMHOLE.
...JUST RUN, RUN, RUN AWAY. YES-YES-YES.
WELL! ON THAT INSPIRING NOTE: WHAT WAS MY GIFT?
WHY, SOME TOP-OF-THE-LINE M-WAVE CANNONS!
....SO THAT HE DOES NOT NEED THOSE FIRST TWO GIFTS.

ALL RIGHT, CREW: LET'S BLOW THIS POPSICLE STAND.
SKITTER, FIRE UP THE SHIP.
CAPTAIN! WAIT!
I KNOW THE MACHITO IS ALL SET FOR FOODS AND REQUISITIONINGS-MENTS, BUT PERMISSION TO BRING ABOARD *A TINY TREAT?*
IT WILL TAKE NO TIMES! I WILL GET IT ON BOARD *RICKETY-SPLITS!* IS TEENY-TINY.
BUHHH. *FINE.* YOU HAVE FIVE MINUTES.
NESTR LEAVES

APOLOGIES FOR THE MIDNIGHT CALL, PROCONSUL. BUT VISITING SHIP WILL HAVE CLEARED THE SYSTEM.
THANK YOU FRIENDS: TRACK THEM AS FAR AS WE CAN. MUST SEE THEM SAFELY OFF. YES-YES-YES.

SCANNERS WILL HAVE HAD TROUBLES, TONIGHT ...TINY BLIPS WILL HAVE...FLICKERED. WE WILL HAVE FIXED IT.
YES. FINE. KEEP ME UPDATED.
AND MORE OF THIS SUMMER WINE, PLEASE!
YOU HAVE OUTDONE YOURSELVES WITH THIS BATCH!

YES-YES-YES.

PROCONSUL.
IT IS AN HONOR TO MEET YOU.
PFFT

HEY CUDDOW, DO YOU HAVE THE-- WAIT. WHAT'S ALL THIS?
--CAPTAIN!
BOOP
AH! HA HA! CAPTAIN! CAPTAIN OF THE SHIP! LA JEFA!
QUIT FUMFERING, MR SNEAKY PANTS: WHAT WAS THAT PROJECTION I JUST SAW?
OH. UM. I WAS. THEORETICALLY. LOOKING. AT HOW SOMEONE COULD BUILD A MISSILE. WITH. A DRIVE. EMBEDDED. INSIDE ANOTHER. DRIVE.
TO TAKE ADVANTAGE OF THE. UM. CUDDOW EFFECT.
WHOA WHOA WHOA. YOU KNOW HOW TO BUILD A DRIVE?
MMMMMMMMMMMMMMMMMMAYBE asitrytoreadyourfacewhileisaythis YES. YES I DO.
WHAT KIND OF AN ANSWER WAS THAT?
LOOK, WHEN YOU GROW UP AROUND TESSKANS, EVERY ANSWER IS FRAMED BY "WILL I GET PUNCHED FOR SAYING THIS?"
SMEK

I CAME IN TO ASK YOU ABOUT YOUR REPORT FROM DU FU: THE MONASTERIES OF YANTA.
YES! IT'S A LEAD! OUR FIRST RECORD OF SKITTER'S PEOPLE!
IT COULD BE OUR CHANCE TO FINALLY COMPLETE OUR MISSION, GO HOME, AND FINISH THE VARIOUS PAINTINGS WE WERE ALL WORKING ON.
...ODDLY SPECIFIC.
...BUT WHAT DO YOU KNOW ABOUT YANTA? ABOUT THE SILL?
MODERN DAY? NOT MUCH. BUT THEY LOOM LARGE IN FILLIPOD PRE-HISTORIES.
THEY CAME FROM THE SKIES, AND TAUGHT US ABOUT ART, AND LANGUAGE, AND INTERNAL RHYME SCHEMES IN POETRY.
AGAIN SPECIFIC.
BUT WERE THEY A... GOOD PEOPLE? SO MUCH OF THEIR HISTORY SEEMS TO HAVE BEEN ABOUT CONQUEST AND EXPANSION.
WE DON'T KNOW WHY, BUT THEY NEVER CONQUERED TESSKIL. THEY SWOOPED IN, SHARED THEIR INTENSE LOVE OF ART AND LIVING WITH PASSION ...AND THEN OFF THEY WENT.
ODD.
...THEY WERE LIKE THE DREAMY SUMMER BOYFRIEND YOU ALWAYS WANTED FOR NANCY DREW.
NO ONE WANTED THAT FOR NANCY.
JUST ME, THEN?

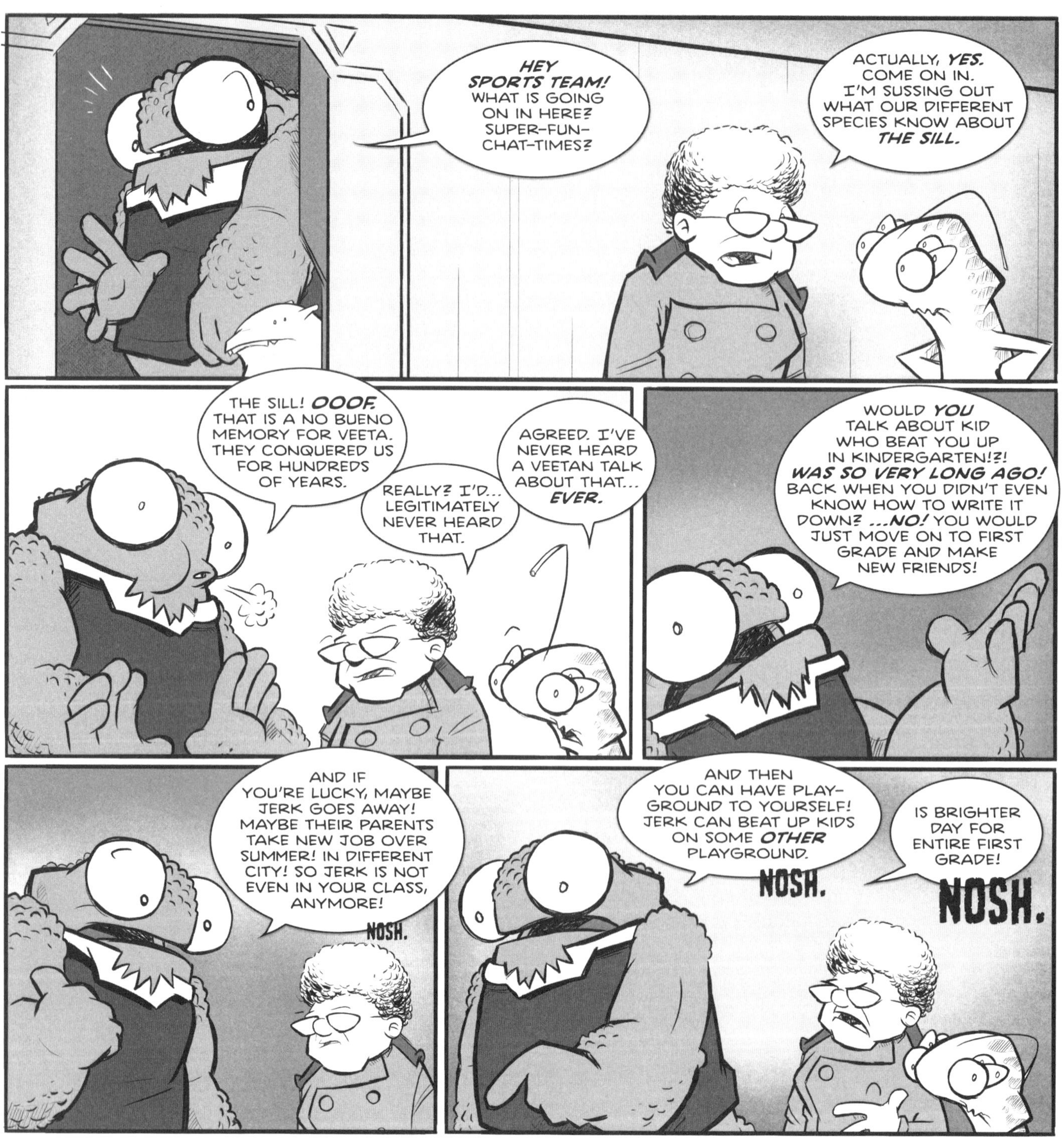
HEY SPORTS TEAM! WHAT IS GOING ON IN HERE? SUPER-FUN-CHAT-TIMES?
ACTUALLY, YES. COME ON IN. I'M SUSSING OUT WHAT OUR DIFFERENT SPECIES KNOW ABOUT THE SILL.
THE SILL! OOOF. THAT IS A NO BUENO MEMORY FOR VEETA. THEY CONQUERED US FOR HUNDREDS OF YEARS.
REALLY? I'D... LEGITIMATELY NEVER HEARD THAT.
AGREED. I'VE NEVER HEARD A VEETAN TALK ABOUT THAT... EVER.
WOULD YOU TALK ABOUT KID WHO BEAT YOU UP IN KINDERGARTEN!?! WAS SO VERY LONG AGO! BACK WHEN YOU DIDN'T EVEN KNOW HOW TO WRITE IT DOWN? ...NO! YOU WOULD JUST MOVE ON TO FIRST GRADE AND MAKE NEW FRIENDS!
AND IF YOU'RE LUCKY, MAYBE JERK GOES AWAY! MAYBE THEIR PARENTS TAKE NEW JOB OVER SUMMER! IN DIFFERENT CITY! SO JERK IS NOT EVEN IN YOUR CLASS, ANYMORE!
NOSH.
AND THEN YOU CAN HAVE PLAY-GROUND TO YOURSELF! JERK CAN BEAT UP KIDS ON SOME OTHER PLAYGROUND.
NOSH.
IS BRIGHTER DAY FOR ENTIRE FIRST GRADE!
NOSH.

THE MACHITO ENTERS THE YANTA SYSTEM, AND IS GREETED BY...AN EERIE STILLNESS. MASSIVE, ANCIENT SPACE STATIONS; HULKING SHIP PLATFORMS...ALL VACANT AND STILL.

A THOUSAND YEARS AGO, THE MACHITO WOULDN'T EVEN BE ABLE TO APPROACH THE SYSTEM. A ONCE-MIRACULOUS GRID, NOW CRASHED AND BROKEN ON THE SHORES OF TIME, WOULD'VE STOPPED THE SHIP COLD.

AND HAD THE MACHITO MADE IT PLANET-SIDE, IT WOULD'VE BEEN REDUCED TO ATOMS BY A DEFENSIVE TECH NO ***CURRENT*** GALACTIC POWER COULD EVEN GUESS AT. ***THIS WAS A TYPE-2 KARDASHEV CIVILIZATION*** -- THE ONLY ONE IN HISTORY -- AND IT WAS CAPABLE OF MIRACLES.

BUT NO TECHNOLOGICAL MIRACLES GREET THE MACHITO NOW...FOR A ***GREATER MIRACLE*** HAS FROZEN THE GEARS OF YANTA. ALL IS QUIET ON THIS CASKET OF A PLANET.

THE CREW STEPS OUT TO HEAR THE EMPTY CREAKS AND GROANS OF MEMORY. AND HERE, AMONG MILE-HIGH SKYSCRAPERS, THEY FIND THEY ARE...***ALONE.***

EXCEPT FOR SKELETONS AND NEAR-SKELETONS: THE REMNANT MONKS OF YANTA, LOST IN EXQUISITE RELIGIOUS ECSTASY...AND DYING, DYING, DYING, DYING.

6325.2765.8891.4839.2358.2908.4787.4697.9032.3348.8878.1304.3989.7878.3454.2209.1121.9944.4735.8883.1189

Enciclopedia Xenobiológica

Página 32591 | Imprimátur del Imperio

Species: The Sill

Homeworld: Yanta

Size: 2m

Color & Markings: Once known for their elaborately carved, exquisitely maintained battle armor, the "warrior reptile" Sill are now most commonly known for their monastic robes.

Reproduction & Development: The Sill reproduce sexually, generating a nested brood of 1-3 offspring. Once hatched, these offspring would have traditionally been raised within an extended family, and schooled for free at the local house of education, run by The Hold. Schooling would have been followed by military training, then by a decade of service, and finally by a caste-appropriate occupation.

Diet: Omnivorous, but with a deathly allergy to pollinating fruits.

Lifespan: 120-140 years.

Language: The Sill originally had three languages: A common tongue used for friendship and familial matters, a logistics tongue for military and tradecraft, and an imperial tongue, used by or with the royal family of The Hold. Today, of course, only the imperial tongue survives with the Monks of Yanta.

Social Structure: For tens of thousands of years, The Hold of the Sill followed a fixed caste system. But early in its expansionist years, The Hold's caste system "stretched," and began allowing for upward mobility within one's caste. This mobility was made possible by the constantly expanding territories of The Hold. Any young Sill could now find military and financial success out among the colonies, and return to the core worlds to bask in wealth.

The Hold lasted between 20,000 and 25,000 years, in various incarnations, and is generally considered the greatest empire the galaxy has ever produced. By technologies now lost, The Hold stood astride many lightyears of the Milky Way, and was resplendent with riches, art, and culture. Even now, when the grand architecture of The Hold is but a crumbling memory, hundreds of species still feel a lingering impact from its art, culture, and literature.

After eons of rule, the complete fall of The Hold was quick – taking only 10-15 years. And it wasn't brought down by armies, or plague, or a massive gamma-ray burst. It was brought down by a slim, 15-page book: The religious teachings of the Prophet Ekka. Ekka taught a simple, step-by-step way for The Sill to align their mind to what she called "The Great Quiet." And the simple fact is, it worked. The technique was transformative for The Sill. Ekka's teaching brought an overwhelming sense of peace, well-being, and interconnectedness with the universe. Many Sill spoke of having reached a new level of existence. But with this mind-expanding sense of oneness came the great bane of The Sill: They lost all desire to reproduce. As was true with other bodily functions like sleeping and eating, the desire to reproduce could only be generated with great mental effort and concentration. As such, within two generations, an empire of 250 billion was reduced to 3 million: The remnant Monks of Yanta.

Intelligence Rating: Oscar-Romeo range

Interaction with Humans: For many species, The Sill were their first contact with alien life. As such, you can find repeated references in the histories of ancient Veeta, Kayntanna, Finntesk, the Fillipods of Tesskil, and hundreds of others. But Earth remained blissfully unaware of The Sill until the early 2300's. As such, humanity's picture of The Sill is colored very differently than that of most other species. For others, The Sill were a massive, conquering, trading empire that straddled space lanes in all directions...only to suddenly turn inward on a religious path. But for humans, The Sill are nothing more than a boring monastic community, with not much to offer. And who frankly need to wash more.

OUR SCANNERS WERE RIGHT: THIS IS THE NEAREST LIVING SILL.
SO SHE'S *ALIVE?* SHE HASN'T MOVED SINCE WE CAME IN.
HER HEARTS JUUUUUST BEAT. ...FIRST BEAT I'VE SEEN IN TWO MINUTES.
IN A TRANCE? OR MAYBE JUST *SUPER TIE-TIE* FROM MORNING JAZZERCISE?
WAIT... *WAIT!* HER HEARTS ARE BEATING FASTER, NOW. I THINK SHE'S--
STIRRING!
MAAAH!
FORGIVE ME. THE GREAT QUIET IS SUCH ENFOLDING JOY! I DID NOT FEEL YOUR MINDS APPROACH.
YOU'RE YELLING! INTO OUR MINDS!
AM I? PERHAPS I FORGET HOW TO SOFTEN. *...AH!*
GREETINGS, O PILGRIMS! YOU SEEK THE GREAT QUIET, AND MY MONKS AND I HAVE STAYED IN THE NOISE FOR JUST SUCH A MOMENT! WE WILL ATTEMPT TO TEACH YOU.
OH. UH. MA'AM? *YOUR MONKS.*
AH. THEY HAVE WRAPPED THEMSELVES IN THE QUIET.
DEAD.
DEAD?
...YES.

WAIT! WHAT ARE THESE THOUGHTS I SEE? YOU HAVE A *TELEPATH* AMONG YOU?
NO MA'AM. WE'RE ALL *VOCAL* SPECIES.

NO! THIS ONE! *WHAT THINGS YOU SEE!* SO LOVELY!
OH. UM. YEAH.

...AND YOU COME BY TELEPATHY IN BIRTH? EXTRAORDINARY!
OH NO NO NO I'M NOT A TELEPATH.

BUT YOU ARE! A STRONG ONE! MY SPECIES *ENGINEERED* THIS GIFT, JUST BEFORE THE QUIET. BUT YOURS IS...*BEAUTIFUL.*
I'M SORRY TO CORRECT YOU, BUT I'M *NOT* A TELEPATH.

HAVE YOU *TRIED?*
TRIED?! NO, WHY WOULD I?

SKITTER, *EVERYONE* TRIES TELEPATHIES, AS A KID!

THE CLOSING OF THE EYES, THE SQUEEZING OF THE HANDS, AND THE THINKING SO HARD: "GRANDPA, YOU *WILL* TAKE ME TO SUPER-FUN-LAND... YOU *WILL* TAKE ME TO SUPER-FUN-LAND...."
I A BILLION PERCENT BELIEVE THAT HAPPENED.

MAKE A TINY CHANGE IN THE COURSE OF A RIVER, AND IT WILL CARVE YOU NEW CANYONS.
SO IT IS WITH TELEPATHY. STEER YOUR WORDS TO DIRECTED THOUGHT, AND YOUR BRAIN WILL FIND NEW PATHS FOR YOU TO WALK.
YEAH, NO, ...THAT DESCRIPTION DOESN'T HELP ME AT ALL.
TAKE THE MENTAL ENERGY TO WITHHOLD A BLINK. GOOD. AND HOLD BACK ONE BREATH. YES. YES. NOW...DIRECT A WORD TO ME BEFORE YOU SAY IT.
...BEFORE YOU SAY IT.
YES!
YES?
THIS COMES EASY TO YOU! YOU HAVE DONE IT BEFORE: I SEE IT IN YOUR THOUGHTS.
I'M DOING IT RIGHT?
INDEED! AND THAT HEADDRESS OF YOURS IS EXTRAORDINARY! IT PAINTS OUR WORLD IN COLORS UNSEEN!
MY MOHAWK? YEAH, IT KINDA SEES...GRAVITY.
IS TRUE! HE CAN SEE IF CREW BATHROOM IS OCCUPIED WITHOUT EVEN KNOCKING!
BUHHHH. HE THINKS THAT'S THE COOLEST THING MY MOHAWK CAN DO.
I HAVE TO KNOCK SO MUCH.

BUT ENOUGH OF THAT: YOU HAVE ALL COME TO DRINK IN THE GREAT QUIET!
AH. NO MA'AM, WE ARE NOT PILGRIMS.
THEN...WHY COME? WHY SEEK OUT THE MONKS?
WE FEW STAY HERE IN THE NOISE ONLY TO HELP OTHERS FIND THE QUIET.
"YOU FEW"? MA'AM, YOU ARE THE ONLY LIVING SILL FOR HUNDREDS OF MILES AROUND.
IS THAT TRUE?
I IMAGINE THAT COULD BE TRUE.
THE JOY OF THE GREAT QUIET IS HARD TO ROUSE FROM. IT TAKES AN IMMENSE ACT OF WILL TO STAY IN THE NOISE. AND EVEN MORE TO CARE FOR THIS PHYSICAL FORM...OR PROCREATE.
...SO THERE ARE LESS AND LESS OF US WITH EACH GENERATION. SOON WE WILL WALK IN THE NOISE NO LONGER.
BUT...HOW DID THIS HAPPEN? AN ENTIRE ADVANCED SPECIES FINDS NIRVANA, AND CHOOSES TO DIE OFF?
CHOOSES? NO NO NO. JUST AS ONE DOES NOT CHOOSE TO WEEP, AT THE DEATH OF A LOVED ONE, ONE DOES NOT CHOOSE TO FEEL JOY, UPON DRINKING IN THE GREAT QUIET.
BUT SURELY THERE WERE SOME AMONG YOU WHO WANTED NOTHING TO DO WITH IT.
IN A DROUGHT, A GRASSFIRE WILL SPREAD TO EVERY BLADE. THUS IT WAS WITH US. WE HAD ENGINEERED TELEPATHY ONLY A GENERATION BEFORE THE PROPHET EKKA CAME ALONG. SO WHEN SHE DISCOVERED HOW TO REACH THE GREAT QUIET, IT SPREAD LIKE WILDFIRE. A WAVE OF JOY, SPREADING FROM PLANET TO PLANET ACROSS THE HOLD...BRINGING ALL INTO ITS EMBRACE!
BUT... EVERYONE DIED!
SO? YOU WILL DIE, AS WILL I! NONE ESCAPE DEATH. SO THE QUESTION BECOMES, FOR WHAT PURPOSE WAS YOUR LIFE AND WORK?
THE GREAT QUIET IS THE ULTIMATE EMBRACE OF PURPOSE! OF TRUTH! OF FULFILLMENT IN ALL CREATION! THE NOISE OF THIS FORM FALLS AWAY, WHEN YOU WALK IN THE GREAT QUIET.
I....IT'S.... PERHAPS MY WORDS FAIL ME.
NO NO NO NO NO NO.
PERHAPS I SHOULD GIVE YOUR TELEPATH A GLIMPSE OF THE GREAT QUIET, AND HE CAN EXPLAIN IT BETTER.
I SENSE HESITATION.
NO NO NO NO NO NO

BUT I DON'T UNDERSTAND. IF YOU CAN REACH THE JOY OF HEAVEN WITH YOUR MIND, WHY WOULD YOU EVER WANNA... STAY HERE?
OUR PROPHET FOUND AN EFFORTLESS DOOR TO THE GREAT QUIET, FOR THE SILL. BUT WE FOUND THAT OTHERS IN OUR EMPIRE WERE...NOT SO BLESSED. SO SOME OF US CHOSE TO STAY IN THE NOISE, AND HELP THEM.
AH. LIKE A BUDDHIST BODHISATTVA, ON EARTH.
AS OUR GREAT EMPIRE COLLAPSED, THOUSANDS OF MONKS TOOK TO THE STARS IN HOLD SHIPS ...SEEKING OUT SPECIES WE MIGHT TEACH!
THOUGH SADLY, AS OUR NUMBERS SHRUNK WITH EACH GENERATION, WE...LOST...THE ABILITY TO PILOT THOSE SHIPS.
...SO WE REMAIN HERE, WAITING FOR PILGRIMS TO COME TO US.
AND DO THEY COME?
THEY DO COME, YES! MY MONKS AND I HAVE TAUGHT MANY. THOUGH...NONE HAVE STEPPED AWAY FROM THE NOISE.
PERHAPS IT ONLY WORKS FOR SILL.
NO! NO! I AM SURE I COULD TEACH YOUR TELEPATH. THIS MIND OF HIS!
IF ANYTHING, HE COULD SHOW ME BEAUTIFUL NEW VISTAS IN THE QUIET, IF I BUT TAUGHT HIM THE PATH.
IT'S ACTUALLY HE THAT BROUGHT US HERE, MA'AM. WE ARE SEEKING HIS PEOPLE, AND HAVE WORD THAT THEY'VE TRAVELED WITH YOUR MONKS, CENTURIES AGO.
POSSIBLY! WITH THEIR GIFTS, THEY WOULD BE MASTER TEACHERS.
GOOD! YES! SO YOU'VE SEEN THEM BEFORE???
HIS MIND IS LIKE NOTHING I HAVE EVER FELT. SO I SAY THIS WITH SURETY...
I HAVE NEVER SEEN HIS KIND BEFORE.

SO YOU'RE AT PEACE WITH YOUR RACE... DYING OFF?
OH FRIEND. DEATH IS NOT THE END IN THE GREAT QUIET.
WAIT. IT'S NOT?

...OR ARE YOU ASKING IF I'LL GRIEVE WHEN THE LAST SILL LEAVES THIS PHYSICAL FORM?
IN EITHER CASE, THE ANSWER IS NO.

FOR OUR RACE LIVES ON, ELSEWHERE. THERE ARE OTHERS.
MA'AM, THAT IS... NOT SO.
THE HOLD WAS THE MIGHTIEST EMPIRE THE GALAXY EVER PRODUCED, BUT IT'S BEEN GONE FOR LIFETIMES. YOU ARE ALL THAT'S LEFT.

NO, THEY LIVE. THERE ARE VAST PLANETS OF SILL WHO DID NOT MODIFY THEMSELVES FOR TELEPATHY. THEY LIVE ON IN THE NOISE.

WE TRAVEL VAST STRETCHES OF SPACE, MA'AM, AND SPEAK TO DOZENS OF SPECIES. ALL REPORT THAT THE ONCE-MIGHTY SILL ARE...GONE.
WE SEE THEM STILL, FRIENDS..FROM THE GREAT QUIET. CONQUERING OTHERS WHILE FAILING TO CONQUER THEMSELVES.
THEY ARE WHO WE WERE, ONCE. PERHAPS WORSE.

WE SEE THEM, STILL.
WE SEE THEM.

...BUT, YOUR TELEPATH HAS WALLS BUILT UP IN HIS MIND. YOU KNOW THIS, YES?
WE DO. HE'S LOST HIS MEMORIES.
NOT LOST. THEY ARE THERE. FULL AND BRIGHT AND DANCING...
...BEHIND PANES OF OPAQUE GLASS.
I CAN ONLY SEE GLIMPSES, BUT THEY ARE ALL OF FAIRER DAYS:
TALL TREES WITH HIGH-BUILT HOMES ...FILLED WITH FOOD AND FAMILY. ALL SET IN ENDLESS JUNGLES OF BLUE.
BLUE JUNGLES?
PLEASE... I HAVE TO KNOW MORE! CAN YOU SHAKE MORE MEMORIES LOOSE?
THIS IS A CHOICE FOR YOU TO MAKE, FRIEND. I CAN NOT UNLOCK A DOOR YOU HOLD CLOSED.
I'M NOT HOLDING IT CLOSED! PLEASE DON'T PLAY WORD GAMES WITH ME: IF YOU CAN SEE MY MEMORIES, YOU CAN FIX THEM! YOU HAVE TO!
FRIEND, I CAN NOT. BUT YOU CAN. IN THE GREAT QUIET.
HERE, I WILL SHOW YOU...
WAIT! NO! SKITTER!

MAKER PRIME: OUR ENEMIES HAVE LANDED ON CIRCLE'S END, AND ARE ASKING TO SPEAK WITH YOU. THE EMISSARY IS FROM *THE ROYAL FAMILY ITSELF.*
THEY SENT A MEMBER OF *THE FAMILY??* AND HAVE THEY BROUGHT THE SPIRITS THEY STOLE FROM US?
THEY HAVE. IN FACT, THEY WILL NOT STOP APOLOGIZING.
THEY SAY IT'S CRITICAL WE MAKE PEACE BETWEEN US.
AMAZING. *...NOW* THEY SEEK PEACE.
THEY'RE CLAIMING A LARGER ENEMY THREATENS US BOTH. THEY SEEK COMMON CAUSE.
FINE. LET'S HEAR THEM OUT. SEND THEM IN.
MAKER PRIME, MAY I PRESENT, *FROM THE ROYAL FAMILY...*
THE MAREETH BARON OF THE GRASSKAN EMPIRE...

OUR EMPIRE IS DEAD, MARD. AS IS OUR EMPRESS. WE ARE DOWN TO OUR LAST STRONGHOLDS.
AND YOU NOW LEAD YOUR PEOPLE?
NO NO NO. BUT AS THE ONLY REMAINING ROYAL, I WAS MADE EMISSARY AS A SIGN OF RESPECT TO YOU.
A SPIRITED GESTURE.
YOU WERE ALWAYS HONORABLE, MARD. EVEN IN WAR.
BUT THESE VINN KNOW NO HONOR. THEY ARE A FIRE THAT CONSUMES THE WORLD.
MY FRIENDS, MY FAMILY, HAVE ALL BEEN TAKEN FROM ME.
NOT KILLED, MIND YOU. TAKEN. RUINED. THEY EXIST NOW AS STRANGERS IN A STRANGE LAND.
MARD, THIS IS WHAT WILL HAPPEN TO ALL OF US ALL IF WE DO NOT STAND TOGETHER.
BARON, YOU KNOW THE CONTINUUM MAKES NO COMMON CAUSE.
WE ARE A CLOSED LOOP.
YOU HAVE NOT SEEN WHAT WE HAVE SEEN! YOU WILL LOSE AS WE HAVE LOST! BUT PERHAPS TOGETHER....
I WILL CONSIDER YOUR WORDS.
AND AS WE WARRED FOR 40 YEARS, YOU SHALL HEAR AN ANSWER IN 40 DAYS.
40 DAYS?? MARD! THEY ARE WIPING US OUT!
THEY....
AS A BEGGAR AT YOUR DOOR, I CAN NOT RUSH YOUR THINKING, BUT YOU MUST KNOW THIS:
OUR END IS CLOSE. AND IF WE ARE TO BE EXTERMINATED, WE HAVE DECIDED TO GO OUT WITH A KILLING BLOW.
WE WILL NOT LEAVE OUR FAMILIES IN THAT HELL. WE WILL SMOTHER THE VINN IN DARKNESS.
IF YOU WILL NOT JOIN US -- OR IF WE CAN NOT BEAT THEM BACK -- WE WILL KILL THEIR STARS.
ALL OF THEM.

THE RETURN OF THE SPIRITS WAS A DEED WELL DONE, BUT DONE UNDER DURESS. THE GRASSKANS HAD DECADES, BUT ONLY NOW BRING THEM HOME TO US?
A FAIR POINT.
BUT THEY WERE ALWAYS HONORABLE IN WAR, AND TREATED OUR PRISONERS WITH RESPECT. THEY JUST NEVER UNDERSTOOD OUR WAYS.
WELL, IT SEEMS EXTERMINATION HAD A WAY OF FOCUSING THEIR UNDERSTANDING.
MARD, WE CAN NOT WAVE AWAY OUR LONG HISTORY WITH THE GRASSKANS, AND SUDDENLY FIGHT SIDE-BY-SIDE.
WE'VE NOW LOST 40 SCOUT SHIPS TO THESE VINN, AND THEIR TERRITORIES GROW EVER CLOSER.
BUT! IMAGINE A SHARED DEFENSE AGAINST THAT THREAT!
THE CONTINUUM IS A CIRCLE UNTO ITSELF. WE STAND ALONE, AS WE ALWAYS HAVE.
BUT ISN'T IT SPIRITED TO TRY? TO EXPERIMENT WITH A NEW WAY OF DEFENDING OURSELVES?
WE WILL WATCH AND LEARN, MARD. IF THE GRASSKANS WIN, THEN NO ACTION NEEDS TAKING. AND IF THEY LOSE, WE SHALL LEARN FROM THEIR LOSS.
MAKERS, HOW CAN YOU NOT SEE THIS AS A CRISIS?
YOU ARE MAKER PRIME, BUT IN THIS MOMENT, YOU ARE OVER-RULED.
THE ENTIRE COLEGIUM HAS VETOED THIS PATH.
THE GRASSKANS WILL FIGHT ALONE.

SIR? YOUR ROYAL ASTRONOMER IS HERE TO SEE YOU.
THAT ONE FILLIPOD GUY? NO NO NO NO
HE SAYS IT'S IMPORTANT. SAYS "THE SKY IS FALLING."
WHAT DOES
WHAT DOES THAT EVEN MEAN?
UGGGH.
SEND HIM IN.
OH, MY MOST BELOVED EMPEROR!
MY ILLUSTRIOUS LIEGE!
MY SUMPTUOUS KING, UPON WHOM ALL EYES FEAST!
EYEROLL
scamper scamper scamper
MOST MAGNIFICENT OF RULERS!
MOST MAGNANIMOUS OF MEN!
I STAND BEFORE YOU, OH MY SUN, MOON, AND STARS...
...SADDENED ONLY THAT I FORGOT TO BRING GLOVES WITH ME!
...BECAUSE DANNNNNG YOU ARE TOO HOT TO HANDLE.
YOU'VE GOT TO LEARN HOW HUMAN PRAISE WORKS

THERE ARE NEW DEVELOPMENTS IN THE GRASSKAN EMPIRE, SIRE! THOUGH...I DON'T EVEN KNOW THE WORDS TO DESCRIBE IT.
HMPH. I THOUGHT YOU PEOPLE WERE ALL "AMAZING POETS".
GIGGLE.
OH MY KING, YOU DO KNOW HOW TO FLATTER!
I MEAN, I DABBLE.
GIGGLE.
MOTHER ALWAYS SAID I WAS "NOT AVERSE TO VERSE..."
GET ON WITH IT.
YES. YES, OF COURSE.
THE STARS, MY EMPEROR. THE GRASSKAN STARS HAVE ALL GONE...DARK.
LIKE... NOVAS?
NO, MY EMINENCE. DARK. ALMOST LIKE THEIR FUSION REACTIONS WERE... SNUFFED OUT?
IT LOOKS LIKE OVER FIFTY STARS, SO FAR...AND ALL FROM HABITABLE SYSTEMS.
I DON'T UNDERSTAND HOW THAT COULD...BE?
AND HOW YOU COULD KNOW THAT IN REAL TIME?
I'M CONFUSED BY THE SCIENCE OF THIS.
SOMETIMES, MY KING, WHEN SCIENCE FAILS US, ART MUST STEP IN.
PERHAPS I SHOULD COMPOSE A FEW DOZEN SONNETS ABOUT IT? oooOOOOOooo
I WILL PUNCH YOU RIGHT IN THE MOUTH.

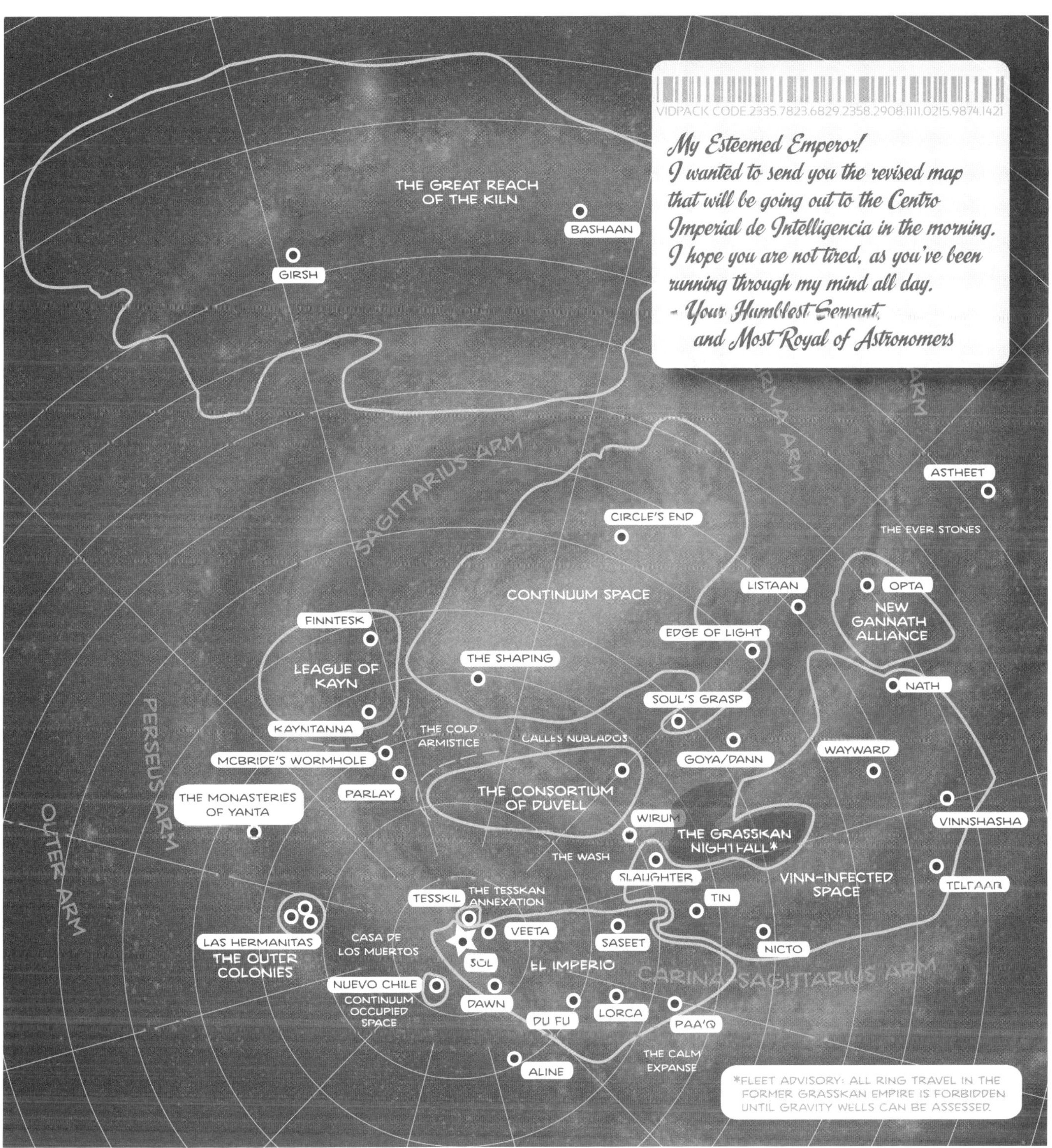

VIDPACK CODE.2335.7823.6829.2358.2908.1111.0215.9874.1421
My Esteemed Emperor!
I wanted to send you the revised map
that will be going out to the Centro
Imperial de Intelligencia in the morning.
I hope you are not tired, as you've been
running through my mind all day.
- Your Humblest Servant,
and Most Royal of Astronomers
THE GREAT REACH OF THE KILN
BASHAAN
GIRSH
SAGITTARIUS ARM
CARINA ARM
ARM
ASTHEET
CIRCLE'S END
THE EVER STONES
LISTAAN
OPTA
NEW GANNATH ALLIANCE
CONTINUUM SPACE
FINNTESK
EDGE OF LIGHT
THE SHAPING
LEAGUE OF KAYN
NATH
SOUL'S GRASP
KAYNTANNA
THE COLD ARMISTICE
CALLES NUBLADOS
GOYA/DANN
WAYWARD
MCBRIDE'S WORMHOLE
PERSEUS ARM
PARLAY
THE CONSORTIUM OF DUVELL
THE MONASTERIES OF YANTA
VINNSHASHA
WIRUM
THE GRASSKAN NIGHTFALL*
OUTER ARM
THE WASH
SLAUGHTER
VINN-INFECTED SPACE
TELFAAR
TESSKIL
THE TESSKAN ANNEXATION
TIN
VEETA
CASA DE LOS MUERTOS
SASEET
NICTO
LAS HERMANITAS
THE OUTER COLONIES
SOL
EL IMPERIO
CARINA-SAGITTARIUS ARM
NUEVO CHILE
CONTINUUM OCCUPIED SPACE
DAWN
DU FU
LORCA
PAA'Q
THE CALM EXPANSE
ALINE
*FLEET ADVISORY: ALL RING TRAVEL IN THE FORMER GRASSKAN EMPIRE IS FORBIDDEN UNTIL GRAVITY WELLS CAN BE ASSESSED.

I WAS 19 THE FIRST DAY I SAW BATTLE AGAINST THE CONTINUUM.
A TERRIBLE, BLOODY DAY.
BUT AT THE END OF IT, I CAME ACROSS A ***MAKER GENERAL*** ON THE BATTLEFIELD... BLEEDING OUT & DYING.
I WAS YOUNG, AND STILL WIDE-EYED, AND DECIDED TO OFFER HIM WATER, AND MY COAT FOR A PILLOW.
I STAYED WITH HIM AS HE DIED, AND HE THANKED ME FOR THE KINDNESS. ***"MAY YOUR SPIRITS,"*** HE WHISPERED, ***"ILLUMINATE THE UNIVERSE."***
ZUU ZUU ZUU
THOOM
WAM
I HAVE THOUGHT ABOUT THAT GENERAL EVERY TIME I CLOSE MY EYES. THERE WAS NO EVIL IN THAT MOMENT. JUST AN HONORABLE END WITH DIGNITY.
AND SO, MARD, AS MY PEOPLE AND I CLIMB INTO OUR GRAVES, I WILL STRIVE FOR MY OWN HONORABLE END... BY OFFERING UP ***A GIFT.***
IN THE CUSTOM OF YOUR PEOPLE, I SHOULD GIVE IT WHILE SAYING "MAY YOUR SPIRITS ***ILLUMINATE*** THE UNIVERSE." BUT THAT WOULD BE A LIE.
FOR THE GIFT I OFFER BRINGS ONLY ***NIGHTFALL.***
AND I PRAY YOU NEVER NEED TO USE IT.

SKITTER! SKITTER!
CUDDOW, WHAT'S THE MEDKIT SAY?
HIS HEART RATE HAS DROPPED TO NOTHING.
BUT HIS BRAIN IS LIT UP.
GOOD GOD, SHE DID IT. SHE TOOK HIM INTO THE GREAT QUIET.
CAPTAIN, WE HAVE TO MOVE. PUT DISTANCE BETWEEN HIM AND SILL.
YES! YES! ...WE LEAP THE MACHITO! MAYBE SHIFTING TO PINCHED SPACE CAN BREAK THE BOND.
THE ONLY OTHER OPTION IS TO SHOOT THIS MONK DEAD...AND GOD KNOWS WHAT THAT WOULD DO. SO 'NANDO, PREP THE DRIVE FOR A LEAP.
WHY WOULD A LEAP WORK?? SHE SAID SHE COULD SENSE OTHERS LIGHTYEARS AWAY.
I DON'T KNOW, ORLA.
I'M JUST GUESSING, HERE. I DIDN'T MAJOR IN "RELIGIOUS TELEPATHY" AT MY COMMUNITY COLLEGE.
WAIT.
IS TRUE? YOU REALLY WENT TO COMMUNITY COLLEGE?
OH, ABSOLUTELY.
THE VALUE-FOR-MONEY PROPOSITION OF A COMMUNITY COLLEGE IS
BAHHHHHHHH NOT THE TIME

'NANDO, LEAP THE SHIP. MAKE IT FAR. MAKE IT LONG.
YES, CAPTAIN. BUT...WITHOUT SKITTER'S MOHAWK TO GUIDE US, WE'LL BE BACK TO DOING IT THE OLD WAY.
THE OLD WAY?
THE CRASHY McCRASHERTON WAY.
AIGHHHH! NO!
NOSH.
NOSH
ARE WE BLOCKING OUR FACE FROM A FASTER-THAN-LIGHT CRASH?
NO
NO, WE AREN'T.

HOW YOU HOLDING UP, 'NANDO?
I AM *ACTIVELY SCHVITZING*, CAPTAIN.
I FORGET HOW TERRIFYING IT IS TO PILOT AT LENGTH.
YOU'VE BEEN FLYING FOR SIX HOURS.
BRING US TO FULL STOP, AND I'LL HAVE ORLA DRIVE FOR A WHILE...
BANNNNNNNNK
WE CAN'T STOP THE SHIP, CAPTAIN. NOT UNTIL SKITTER WAKES UP.
...AND DOES ORLA EVEN HAVE THE FLIGHT HOURS?
WELL, EMERGENCY OR NO, I CAN'T HAVE YOU FLY FOREVER. YOUR REACTION TIMES ARE VISIBLY SLOWING.
I CAN DO THIS, CAPTAIN. LET ME KEEP GOING FOR FIFTEEN MORE MIN--
WAMMMMMMM

HARBOR MASTERS: KNOW YOUR SHIP INSIGNIAS!

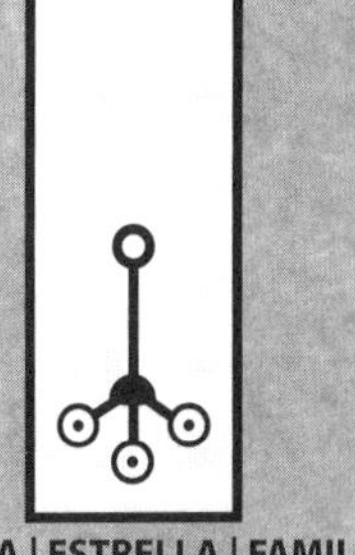

THE SECOND SPANISH EMPIRE
First sketched out by Emperor Conrado Cruz, the imperial insignia represents "Tierra, Estrella y Familia" (the three lower rings) coming together as one, then journeying outward using the Cruz Drive (the upper ring).

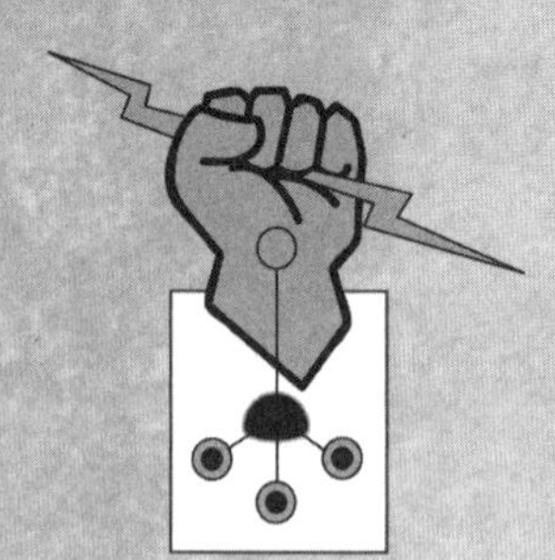

JINYIWEI
Sometimes known as "The Grey Fist," the Jinyiwei are the personal agents and secret police of the Emperor. Ships with this insignia are to be berthed wherever they request.

LA BIBLIOTECA PRIVADA DE LA FAMILIA
Serving as the private library of la Familia, la Biblioteca is the largest data repository in the Empire. Physically located in Sevilla, la Biblioteca also remotely backs up 14 smaller libraries across the Empire. You will sometimes see these massive data-backup ships in port.

DRIVE CORPS
The Corps trains Familia members for their mandatory service as engineers of the Drive. Without Drive Corps knowledge, maintenance and repairs, humanity's intersellar fleet would seize up. As Harbor Master, you will see multiple Corps training vessels operating in port.

INDÚSTRIAGLOBO
In some parts of the Empire, the corporate logo of IndústriaGlobo is as omnipresent as the government's. The largest corporation to ever exist, IG is Familia owned and run, and serves as an extrajudicial arm of Familia control. The Lorca space port, for example, is owned and operated by IG.

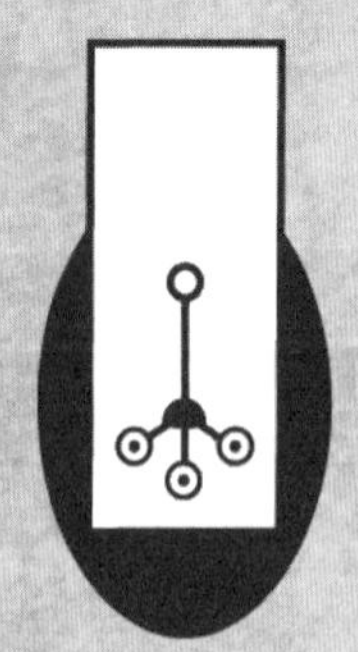

CENTRO IMPERIAL DE INTELLIGENCIA
The official police and intelligence service of the Empire, the Centro Imperial de Intelligencia (C.I.I.) handles security on the core worlds, and works hand-in-hand with the security services operated by colonial governors. Should you need them, their offices are located downstairs.

LA GRANDE Y FELICISSIMA ARMADA
The royal navy of the empire, in which 320 million people serve under Familia and civilian officers. One armada battlecruiser will always be berthed here on Lorca, with occassional corvettes making port calls.

NUEVO CHILE
The flag of the now-defunct human colony is frequently used as a rallying symbol for protesters on Earth or Las Hermanitas. No ship is allowed to dock if they display this symbol, and their ship numbers should be passed to the C.I.I.

HARBOR MASTERS: KNOW YOUR SHIP INSIGNIAS!

THE VEETAN PROTECTORATE
In the complex theological philosophy of Veetans, this symbol indicates the fate-sharing of all life. It has come to be the shorthand symbol of camaraderie, of the greater good, of joy, and of the Veetan people. Though you will not see Veetan ships here on Lorca, most food imports will be marked with this image.

THE CONTINUUM
The symbol of the Continuum is one of triune unity. The Pre-Makers (lowest section), Makers (middle) and Colegium (highest), tied together as a single unit. Immediately inform C.I.I. if scanners pick up this insignia on any cargo, dress, or vessels.

TESSKANS
In the claw-scratched, rough language of the Tesskans, this symbol is a literal "marking of territory." A Tesskan would traditionally scratch this symbol on a tree or rock, along with their name, to claim that area as theirs. There are 14 Tesskans serving as security for the Harbor Master's office. You are advised to steer clear.

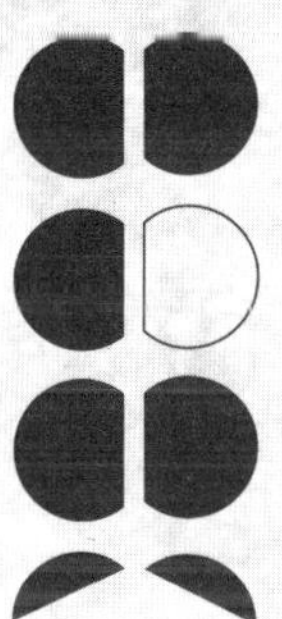

THE GRASSKAN EMPIRE
The symbol of the Grasskan people is actually a logoform meaning "All Will Serve." All species must serve the Grasskans, and all Grasskans must serve the Empress. Check with both C.I.I. and Jinyiwei officers before admitting Grasskan ships into port.

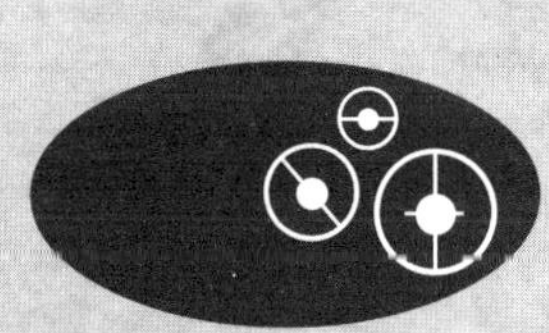

THE OUTER COLONIES
When Las Hermanitas were first colonized, disagreements erupted over what "flag" they should adopt. This solution, proposed by a child during a colonies-wide contest, suggested showing the three sisters sparkling in the sky beyond the dead skies of the "Casa de Los Muertos."

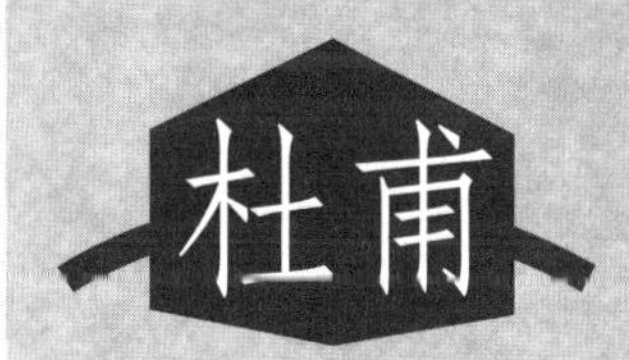

DU FU
Reflecting the combined Human/Nyx governance of Du Fu, the symbol of the planet contains the poet's name within the Nyxian symbol for "soundness" and "surety." Cargo ships from Du Fu will be making daily port calls.

THE PRAHSITT
On a bright green background, the Prahsitt flag features The Great Hunting Bird – the form their god assumed – superimposed on their symbol for luck and well-being. Rarely will you see Prahsitt in port.

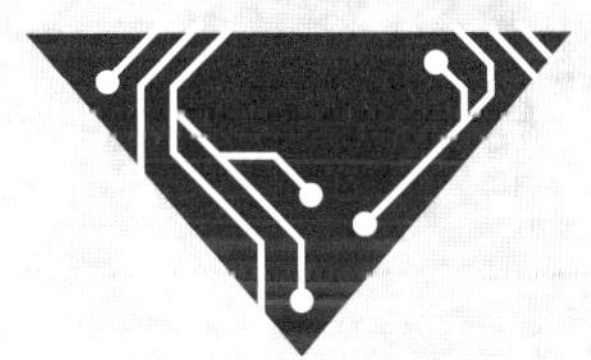

THE SILL
As the mark that once stood astride the galaxy, the symbol of the Sill is now rarely encountered. You will see it regularly, however, as there are currently two Sill archaelogical digs here on Lorca.

DRIVE CORPS REFERENCE SHEET FOR THE "PLEASURE PLANET" LORCA

Members of la Familia who earned truly terrible marks in Drive Corps were often assigned work as a Harbor Master. Since it was not a position that could be given to civilians (as it required access to military, Jinyiwei, and "secret" Familia transports), the tasks of the Harbor Master could only be taken on by Familia. But, as their grades were so terrible, these Harbor Masters were usually idiots. Hence, basic reference sheets like this.

HOW GO THINGS IN THE SALT MINE, COUSIN?
I THINK SHE'S READY TO FLY!
WANNA CHECK OUT A NEARBY STAR?

REALLY? YOU GOT THE DRIVE WORKING?
WON'T KNOW 'TIL WE TRY IT. HOP IN!
WAIT. WHAT HAPPENS IF YOU DID IT WRONG? WILL IT JUST...NOT WORK? OR WILL IT NOT WORK IN A WAY ***THAT SPLAYS US OUT AS A FINE MIST?***
I CAN THINK OF A FEW DOZEN MIST SCENARIOS, YES. LOTTA POSSIBLE MISTING, OR MIST-ADJACENT RESULTS.
HA! OK! LIVE FAST, DIE YOUNG!
WE AIN'T YOUNG.
BUT WE'LL DIE FAST!

HEY HEY HEY! LOOKIT US! GOIN' FASTER THAN LIIIIIIIGHT!
PINCHIN' SPACE LIKE A GRANDMA PINCHES CHEEKS!
BREAKIN' MONOPOLIES LIKE...
...A THING THAT BREAKS MONOPOLIES.
LOOK ON MY WORKS, YE MIGHTY, AND DESPAIR!
HA HA! ...SO WHERE TO, FRIENDS?
WHEREVER THE GOOD LADY WISHES TO DRIVE.
HOW ABOUT THE MOONS OF PARLAY? I HEAR THEY'RE LOVELY THIS TIME OF YEAR.
PARLAY IT IS! AND THE FIRST GLASS OF WINE IS ON ME!
WAIT. WAIT. HOLD UP. WHAT'S THIS PINGING RED LIGHT?
HMM. THAT'S MY WEAPONS-SENSOR SYSTEM. IT MIGHT BE SENSING A SHIP'S M-WAVES POWERING UP? ...MAYBE?
"MAYBE" ???
LOOK, CAP'N CRITIC: I SPENT WEEKS BUILDING A FLIPPIN' DRIVE, AND LIKE FIVE HOT SECONDS BUILDING THAT DUMB CONSOLE.
SO LET'S BRING IT DOWN A NOTCH.

STATUS REPORT!
IT WAS A MIST OF MICRO-PARTICLES. CRASHED US OUT OF PINCHED SPACE.
RING IS GONE. ENGINESSES IS GONE. COMMUNICATIONS IS GONE.
OOF.

ANYONE HURT?
NO. BUT WE HAVE MULTIPLE HULL BREACHES.
FWUMP
AND PROBABLY NO EVA SUITS TO FIX IT. THE ENTIRE CARGO BAY HAS VENTED INTO SPACE.
GASP! THAT MEANS FOOD'S GONE, TOO!
EXTRA GASP!

AND OUR AIR IS GOING, TOO. THE OXY SYSTEM IS DEAD.
BUT. BUT. BREATHING AND EATING ARE FAVORITE THINGS!
FAVORITE THINGS, 'NANDO.
WOBBIDA WOBBIDA WOBBIDA

I'M SO SORRY, EVERYONE. THIS IS MY FAULT WE'RE GONNA DIE.
STOP THAT TALK, 'NANDO. YOU WERE FLYING UNDER MY ORDERS.
TRIED TO TELL YOU, 'NANDO.
SKITTER!
TRIED TO TELL YOU WE WERE GONNA CRASH. COULDN'T GET MY LIPS TO MOVE.
YOU'RE ALIVE! YOU'RE HERE!
WAS...
HARD TO MOVE. HARD TO LEAVE.
THE QUIET.
HOW'D YOU DO IT? HOW'D YOU BREAK OUT?
TOOK ME A WHILE TO REMEMBER...ME? REMEMBER I WAS A SINGLE UNIT. THEN IT TOOK ME EVEN LONGER TO MOVE ME. HARD TO EXPLAIN.
BUT YOU DID IT! SO PROUD OF YOU!
SO SO SO PROUD!
NOSH, I'M NOT FIVE.
LOOK AT YOU! MOVING YOUR BIG-BOY ARMS!
IN YOUR BIG-BOY SHIRT!

OK: I GOT OUR SURROUNDING BREACHES SEALED UP. BUT WITH COMMUNICATIONS AND SENSORS OUT, ALL WE CAN DO IS...SIT.
HMM.
ACTUALLY. WITH APOLOGIES FOR NOT ASKING, I RIGGED UP AN OLD FILLIPOD TRICK.
I TORE UP A BIT OF CEILING, AND SOLDERED A SMALL MAGNETRON ONTO THE HULL.
THE SPIDER STEEL SHOULD AMPLIFY THE MAGNETRON TO CREATE A SORT OF PINGING THAT OTHER SHIPS MIGHT PICK UP.
A BEACON? OH, HOORAY! CUDDOW SAVES THE DAYS! NOW WE MUST ALL GET HIM BESTEST CHRISTMAS GIFTS! WITH BESTEST BOWS!
NO, NO: IT'S STILL A WEAK PLAY, NOSH. PASSING SHIPS WILL EITHER HAVE TO FLY VERY CLOSE, OR HAVE AMAZING SENSORS TO NOTICE THAT PINGING.
BUT IT'S SOMETHING. THANK YOU, CUDDOW.
INDEED. WHY DON'T WE ALL GO HAVE A LOOK: SEE IF WE CAN'T FIND A WAY TO BOOST THAT SIGNAL.
YES! IT'S OVER HERE!
OH GOSH HEY WOW MAYBE OUR CHIEF SCIENCE OFFICER COULD JOIN US?
SORRY! WAS STILL ON CHRISTMAS.

CAPTAIN, YOU NEED TO KNOW WE'RE NOT ALONE.
SKITTER, GO BACK TO BED.
BUT WE'RE NOT ALONE.
I KNOW, SON. YOU FELT A ONE-NESS IN THE GREAT QUIET. "NONE OF US ARE ALONE." I GET IT.
NO, CAPTAIN: WE-ARE-NOT-ALONE. THERE IS A SHIP HERE. RIGHT NOW.
SOMEONE'S OUTSIDE?
IN FACT, THEY'RE ABOUT TO
KLANG
...DOCK.
ORLA! IT'S A SHIP! ENGAGE OUR AIRLOCK!
PORT SIDE.
THE PORT SIDE!
KEEEEEOOORUNNCH
PLEASE TO HAVE FOOD. PLEASE TO HAVE FOOD. PLEASE TO HAVE FOOD.
KANG!
ting ting ting

KLANG
KEENG
KLANG

KRANK KRANK KRANK KRANK

NOT A QUICK PROCESS WHEN OUR AUTO-SYSTEMS ARE DOWN, IS IT?

TIKKA
TIKKA
TIKKA

TSSSSSSSSSSSSSSS

OH. HEY! Y'ALL ARE HUMANS.
...THEY HUMANS, AHMIS!
I AIN'T NEVER MET HUMANS A'FORE!
BUT I SEEN Y'ALL IN MAH PITCHER-BOOKS.

STOP WHERE YOU ARE.
DO. NOT. MOVE.
ORLA, WHAT ARE YOU DOING? IT'S A PRAHSITT! THEY'RE PEACEFUL!
FERNANDO, GET OUT OF THE WAY. THAT'S AN ORDER.
ORLA, PLEASE DO NOT SHOOT THE NICE PRAHSITTS WHO'VE COME TO SAVE US WITH THEIR WORKING ENGINES ON THEIR WORKING SHIP.
I'M NOT WORRIED ABOUT PRAHSITTS, MA'AM.
I'M WORRIED ABOUT THE MAKER BEHIND THEM.
KNOW
I BRING
YOUR DEATH?

KEEP THAT WEAPON UP, ORLA.
SEE AHMIS? THEY'S HUMANS!
OH, HUMANS ARE FROM TIERRA. WHOLE EMPIRE OF 'EM. THEY COME TO WIRUM ONCE-IN-A-WHILE.
OH YEAH SURE I CAN DO THAT.
HI, HELLO, HEY!
MY NAME'S KIK, AND THIS HERE'S AHMIS.
HE WANTS TO KNOW IF Y'ALL GOT A BIG OL' HONKIN' DICTIONARY HE COULD SPEED-READ REAL QUICK. TAKES HIM LIKE TWO MINUTES.
PRAHSITT, YOU ARE TRAVELING WITH A MAKER OF THE CONTINUUM, WHO WE ARE CURRENTLY AT WAR WITH.
WAIT, WHAT??? AHMIS AIN'T NEVER TALKED ABOUT A WAR.
HE SAYS NO THAT AIN'T TRUE BUT THEN AGAIN IT MIGHT BE TRUE ON ACCOUNTA HE'S BEEN SLEEPIN' FOR A WHILE.
OOOF. THAT IS HECKOF NAP
OH HOT DANG MISTER YOU GOT NO IDEA IT'S A WHOLE DEAL OF A STORY AND ALSO I LIKE YER COOL BUMP SITUATION.

KIK, WHILE THE MAKER READS UP, CAN I ASK YOU SOMETHING?
SURE, YEAH.
I'VE TRIED LEARNING THE LANGUAGE OF THESE MAKERS. IT'S A SURPRISINGLY BEAUTIFUL TONGUE!
WOULD IT OFFEND IF I TRIED TO GREET HIM IN THE WAY OF HIS PEOPLE?
OH, I BETCHA HE'D LOVE IT. HE'S REAL BIG ON LEARNIN'
...ALWAYS HAVING FOLKS STAY UP LATE TO READ ENGINEERING MANUALS, THEN QUIZZING THEM ENDLESSLY ON IT.
AH, WONDERFUL!
???
HE WANTS TO KNOW WHY YOU'RE PROPOSITIONING HIS HAT.

I BELIEVE I UNDERSTAND THE NUANCES OF YOUR HYBRID.
!SALUDOS A TODOS!
I AM AHMIS, AND THIS IS MY APPRENTICE, KIK.
I SAID THAT BIT.
WE NOTICED YOUR SIGNAL, AND DETERMINED YOU WERE DERELICT IN SPACE. WE HAVE COME TO OFFER ASSISTANCE, IF NEEDED.
YOU HAVE?
IF NEEDED.
YOU'LL FORGIVE US IF WE DON'T...QUITE... BELIEVE YOU?
GIVEN MY PEOPLE'S TROUBLE WITH YOUR PEOPLE, CAN I ASK.... WHY?
THE CONTINUUM'S QUARRELS ARE NOT MINE.
AND KIK HAS AN INTERESTING THESIS THAT THERE IS GOOD IN THE UNIVERSE. ...AND WE SHOULD AID IT WHERE POSSIBLE.
THAT SEEMS A SPIRITED LINE OF INQUIRY TO PURSUE.
...BEATS YELLIN'

WE CIRCLED YOUR SHIP A NUMBER OF TIMES BEFORE DOCKING.
AND I'M SORRY TO SAY
YER SHIP DONE ***BROKE.***
HA! WE KEEP BUILDING SHIP, THEN BREAKING SHIP, THEN BUILDING SHIP. LIFE IS GREAT BIG LOOPTY-LOOPTY-LOOP, NO?
IT, AH, IS INDEED.
..."LOOPTY?"
DON'T LOOKIT ME.
PERHAPS YOU CAN DISPLAY YOUR SHIP'S PLANS, SO WE CAN BEGIN TRIAGE?
SURE.
Va-Boop!
AH.
HMM.
HMM.
HMM.
HMMMMM.
HEE! YOU TWO MAKE A LOT OF SILLY HMMMS.

THERE IS ERROR, HERE.
ERROR...?
THIS IS NOT OF YOU. THIS IS NOT OF YOUR PEOPLE.
WHAT?? THAT'S *ABSOLUTELY* OF MY PEOPLE. THAT'S A CLASS-4 CRUZ DRIVE ...INVENTED BY MY FAMILY.
THIS IS A SPIRIT OF *SHEN*.
??? WHAT IS A "SHENS"?
"ON THE FIRST DAY, THERE WAS THE FIRST MAKER, AND HE WALKED THE LAND ALONE..."
"...AND IT WAS HE WHO BIRTHED THIS SPIRIT... AND WHO SAW IT WAS GOOD. AND IT WAS HE WHO NAMED IT."
SHEN, THE FIRST MAKER. AND THIS, THE FIRST SPIRIT. *THE CIRCLE'S END.*

AND YOU HAVE *STOLEN* IT.
THE MOST SACRED SPIRIT OF MY PEOPLE.

WHOA, NOW: LET'S TAKE A BREATH BEFORE WE START THROWING ACCUSATIONS AROUND.
YOUR DICTIONARY WAS VERY REVEALING ABOUT YOUR SPECIES. THERE WERE 8,000 WORDS EXPRESSING ***HOW MUCH YOUR OFFSPRING MEAN TO YOU.*** THEY ARE, IN SOME SENSE, ***YOUR DEFINING CREATION.***
WHEREAS ***MY*** PEOPLE PROCREATE BY PARTHENOGENESIS. *OUR OFFSPRING MEAN VERY LITTLE TO US.* WE DEFINE OURSELVES ***BY WHAT WE CREATE.***
OUR SPIRITS ARE THE ULTIMATE EXPRESSION OF WHO WE ARE. THEY ARE OUR LASTING MARK ON THE UNIVERSE. THEY ARE, ***IN YOUR TERMS,*** OUR CHILDREN.
SO WHEN A SPIRIT IS STOLEN, THINK OF IT AS STEALING WHAT MAKES YOU, *YOU.* SNATCHING A SOUL. ***KIDNAPPING A CHILD.***
FEE-DEET DEET
HEY! ***HOW'D YOU OPEN THAT FILE?!?*** THAT'S GOT FAMILIA LOCKS ON IT!
AND IN THIS CASE, KIDNAPPING A ***VERY SPECIFIC CHILD.***

I DON'T GET IT. HOW CAN YOU GET MAD IF WE *INDEPENDENTLY* INVENTED THE SAME TECH AS YOUR ANCESTOR?
IN THE MAIN, BECAUSE I **KNOW** YOU DID NOT INVENT IT.
EVEN IF THAT WERE SO: WHO DOES THAT OFFEND? THE RING WASN'T **YOUR** SPIRIT... OR ANY MAKER **CURRENTLY ALIVE.**
WE ARE NOT A GROUP OF MAKERS, OR A FEDERATION OF MAKERS. WE ARE A **CONTINUUM OF MAKERS.**
WHAT SPIRITS ARE BIRTHED ARE BIRTHED *FOR THE CONTINUUM.* TO ILLUMINATE THE UNIVERSE.
THERE! THEN IT **HAS!** ILLUMINATED THE UNIVERSE!
FOR THE CONTINUUM! FOR THE CLOSED LOOP OF THE CONTINUUM! WE STAND ALONE, AS A CIRCLE UNTO OURSELVES!
BUT AHMIS... Y'ALL AIN'T IN THE CONTINUUM NO MORE.

NO, KIK!
...NONE OF THEM UNDERSTAND THE OFFENSE THEY HAVE GIVEN!
A SPIRIT IS MORE THAN JUST SOME...FIXED CREATION. IT IS A SONG! ONE THAT JOINS A TIMELESS CHORUS!
...EACH MAKER ADDING THEIR VOICE TO THE CHOIR WITH THE SPIRITS THEY BIRTH!
THE GREAT SPIRITS OF OLD ARE A FOUNDATIONAL RHYTHM.
BUT THE CHOIR CHANGES AS IT MARCHES THROUGH TIME.
WITH EACH ADDED SPIRIT, EACH FRESH ITERATION, THE CHOIR BLENDS IN NEW NOTES TO THE TIMELESS MUSIC.
IT IS CONTINUANCE, BUT ALSO BEAUTIFUL EVOLUTION.
AHMIS...NONE OF US KNOW WHAT YER TALKING 'BOUT WITH THIS SONG STUFF.
THEY STOLE MY SPIRIT, KIK!
MINE!
THIS SHIP HAS MY ITERATION OF THE CIRCLE'S END!

WE ARE LEAVING, KIK.
NO.
YOU ARE NOT.

YES.

...WE ARE.

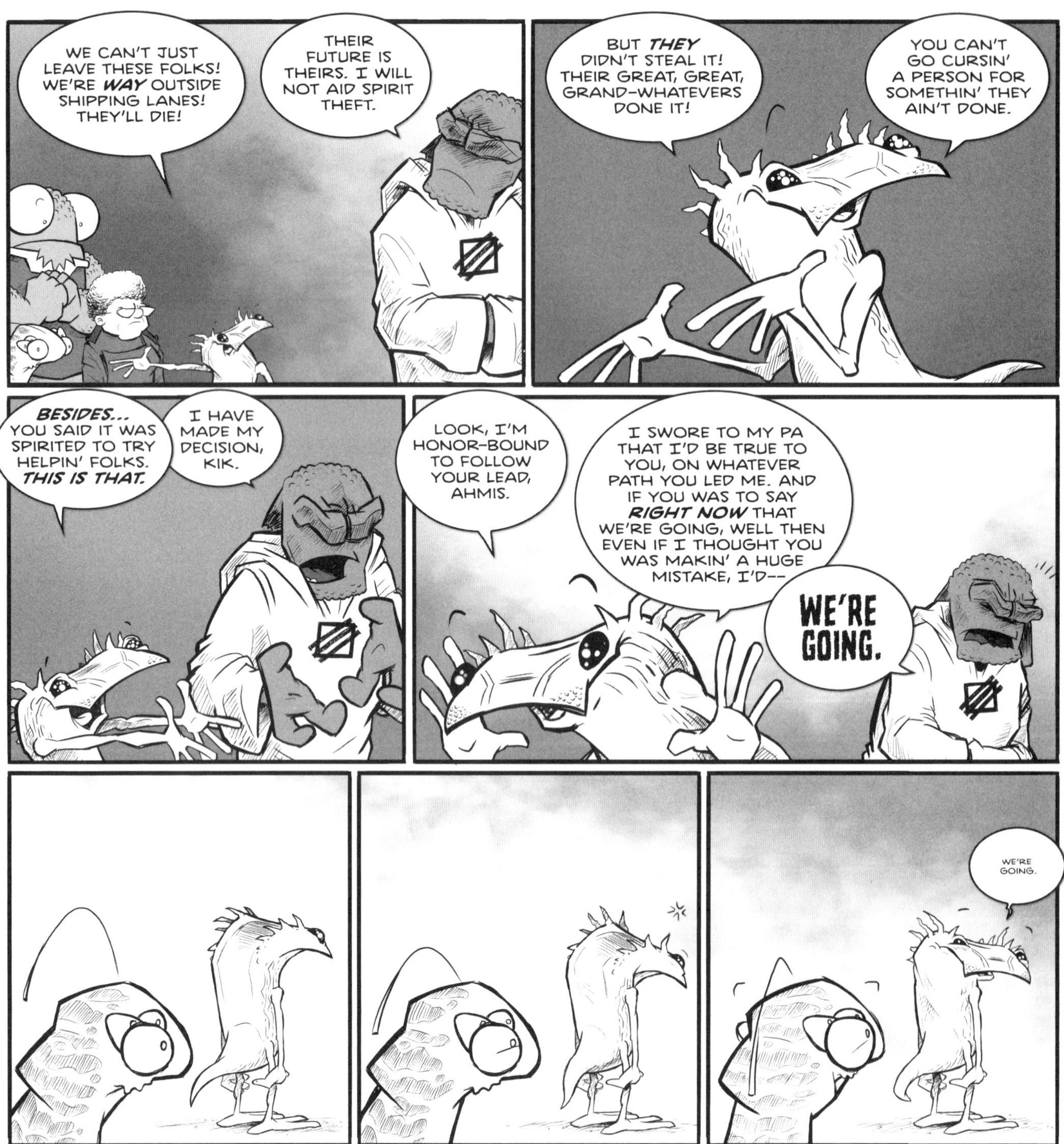
WE CAN'T JUST LEAVE THESE FOLKS! WE'RE *WAY* OUTSIDE SHIPPING LANES! THEY'LL DIE!
THEIR FUTURE IS THEIRS. I WILL NOT AID SPIRIT THEFT.
BUT *THEY* DIDN'T STEAL IT! THEIR GREAT, GREAT, GRAND-WHATEVERS DONE IT!
YOU CAN'T GO CURSIN' A PERSON FOR SOMETHIN' THEY AIN'T DONE.
BESIDES... YOU SAID IT WAS SPIRITED TO TRY HELPIN' FOLKS. *THIS IS THAT.*
I HAVE MADE MY DECISION, KIK.
LOOK, I'M HONOR-BOUND TO FOLLOW YOUR LEAD, AHMIS.
I SWORE TO MY PA THAT I'D BE TRUE TO YOU, ON WHATEVER PATH YOU LED ME. AND IF YOU WAS TO SAY *RIGHT NOW* THAT WE'RE GOING, WELL THEN EVEN IF I THOUGHT YOU WAS MAKIN' A HUGE MISTAKE, I'D--
WE'RE GOING.
WE'RE GOING.

CAPTAIN. DON'T REACT.
MAAAH.
I SAID DON'T REACT.
JUST SORTA... THINK FOR ME.
SKITTER?
YES, MA'AM.
I HOPE YOU DON'T MIND, BUT I SLIPPED AWAY WHEN THINGS WENT SOUTH. I GOT AN IDEA.
...SLIPPED AWAY?? WHERE ARE YOU?
WELL, SEE, THAT'S MY IDEA.
I SNUCK ONTO THE MAKER'S SHIP. ADVANTAGE OF BEIN' SMALL.
DEET DEET DEET
IMMA STEAL US A DRIVE SHIP.

IT'S WEIRD.
THE SYMBOLS ARE TOTALLY FOREIGN... BUT THE INTERFACE IS ODDLY SIMILAR TO OUR SYSTEMS... SO I CAN KINDA FIGURE IT OUT.
BE CAREFUL, MAKER SYSTEMS COULD BE RIDDLED WITH TRIPWIRES.
NO, NO: I...THINK I GOT THIS.
ZAAAAAAP

SKITTER? SKITTER?
WHAT JUST HAPPENED?!?

CAUSALITY, CAPTAIN. HE'LL BE FINE IN A FEW MINUTES. WE'LL RETURN HIM AS WE UNDOCK.
NO! YOU CAN NOT UNDOCK, MAKER-PERSON! I...I WILL STOP YOU!

HNNG.

HNNNGH.

HENNNGH
ZZT

...YOU OK THERE, COOL-BUMPS?
DANGITS.

MAKER, PLEASE: YOU HAVE MY WORD THAT--
CAPTAIN, STOP. THE TRANSGRESSION HERE IS REAL, AND WE ARE *LEAVING.*
BEFORE YOU GO, MAKER, MAY I SHARE SOMETHING ABOUT MY PEOPLE?
WHEN WE FINALLY CAME TO UNDERSTAND IT, *MY PEOPLE SAW YOUR SPIRIT AS A GIFT.* ...A GLIMPSE OF GREATER POETRY IN THE COSMOS.
YOU PULLED BACK A CURTAIN FOR US... AND WE SAW NEW PATHS WE MIGHT WALK DOWN. OUR MATHEMATICS BEGAN TO SING IN WAYS WE NEVER DREAMED.
SO.. I'D LIKE TO OFFER YOU A GIFT.
ONE FREELY GIVEN...FOR THE GIFT YOU'VE GIVEN US.
IT'S SOMETHING I AND MY PEOPLE HAVE BEEN WORKING ON WITH GREAT JOY. ...A SPIRIT OF OUR OWN MAKING.
BOOP
THIS...
WHAT IS...
...
IF THIS DOES WHAT I THINK IT DOES...
WHAT *ILLUMINATION* YOU HAVE BIRTHED.

...AND YOU BIRTHED THIS?
I HELPED!
THEY HAD AMAAAAZING DONUTS AT THE MEETINGS.
VERY ON-BRAND FOR MY PEOPLE
BOY. THIS TOOK A TURN.
INDEED. WHO KNEW CUDDOW WOULD BE OUR BEST PLAY!
LOOK HOW HAPPY HE IS!
GETTING PRAISE FOR BEING CREATIVE IS, LIKE, PEAK FILLIPOD.
BUT SHOULDN'T WE JOIN THE CHAT?
NO NO NO. LET THEM BUILD RAPPORT. NO ONE INTERRUPT.
UH-OH. CUDDOW IS GETTING OUT POEM.
INTERRUPT! INTERRUPT!

I AM HONORED BY THIS GIFT.
OH GOSH!
I'LL TRANSFER THE FILES TO YOUR SHIP BEFORE YOU LEAVE.
ABOUT THAT.
I...DON'T THINK I CAN ANSWER SO SPIRITED A GESTURE WITH ...AN EXIT.
TELL ME ABOUT YOURSELF. YOU SPEAK MY TONGUE... YOU GIFT AWAY EXTRAORDINARY SPIRITS... WHO ARE YOU?
WELL, MY NAME IS CUDDOW.
I AM FROM A PLANET CALLED TESSKIL.
AND IN MY DAYS I HAVE BEEN A SCIENTIST, AN ENGINEER, A LECTURER...
AND NOW, I AM THE SENIOR ADVISOR OF MY PEOPLE.
BUT DEEP DOWN, I'M JUST A BOY...
STANDING IN FRONT OF A UNIVERSE...
AND ASKING IT TO LOVE ME.
OK SLOW IT DOWN OVER THERE

AS YOU ALL KNOW, THE ATTACK BEGAN A WEEK AGO AT OUR SCIENCE COLONY ON *MIND'S EYE.*
AT MARD'S ORDER, WE ATTEMPTED TO RUSH *THE SHATTERED SENSE* INTO SERVICE....
THE REPLACEMENT FOR *THE SPARK OF THOUGHT?* IT'S READY?
DESPITE GALLANT ATTEMPTS, *NO.* A SPIRIT OF THAT SIZE CAN NOT BE RUSHED.

SO WHAT OF THE COLONY, THEN??
WE DECIDED TO SEND THE FULL *COGITATION* FLEET.
BUT IT ARRIVED... TOO LATE.
NO! ...HOW MANY MAKERS ON MIND'S EYE?
TWO HUNDRED. ...AND FOUR MILLION SEEKERS.

THEN MARD'S WARNINGS WERE CORRECT.
INDEED. IT WOULD SEEM THAT WE ARE NOW, QUITE FIRMLY....
...AT WAR WITH THESE VINN.
AND HOW UNFORTUNATE FOR THEM.

REPORT THEN, SEEKER. WHAT DID YOU SEE?
C-COLEGIUM! MAY YOUR SPIRITS ILLUMINATE THE UNIVERSE! I AM HONORED TO BE IN YOUR--
YES, YES. BUT WHAT DID YOU SEE ON MIND'S EYE?
I... I WORKED UNDER THE TUTELAGE OF THE MAKER THONN. HE OF THE GREAT HYDROPONIC SPIRITS USED IN TH
GET ON WITH IT!
ON THE THIRD DAY OF THE ATTACK, THE VINN DISCOVERED OUR REMOTE LAB IN THE MOUNTAINS.
THERE WERE ONLY FIVE OF US, SO THONN ORDERED US TO EVACUATE.
AS I PREPPED THE CIRCLE'S END, I LOOKED OUT THE COCKPIT TO SEE THE VINN OVERWHELM OUR LAB.
THEY...DIDN'T KILL. THEY HELD US DOWN, AND A...A SPRAY WAS ADMINISTERED FROM SOME TYPE OF GLAND. AIRBORNE, BUT CLOSE-QUARTERED.
I...FROZE TO SEE IT. AND STAYED FROZEN, UNTIL I HEARD A VOICE FROM UNDER THAT MOUNTAIN OF BODIES.
IT WAS THONN... YELLING AT ME, SCREAMING AT ME, TO FLEE. SO...I DID.
BUT AS I RAISED THE SHIP, I LOOKED BACK TO SEE THE VINN LETTING THONN GO.
HE STOOD UP SLOWLY, CAREFULLY, AND LOOKED DIRECTLY AT ME...
...
AND??
...AND WAS VINN.

HOW DO WE FIGHT THIS POWER, MARD? THESE ARE NOT THE GRASSKANS, OR THE HUMANS, OR THE
WE FACE A DIRE NEW CALCULUS, HERE.
WE CAN NOT ORDER NEW SEEKERS BIRTHED AT THE FRONT, AS WE NORMALLY WOULD.
...THAT WOULD JUST MULTIPLY OUR ENEMY'S RANKS, ULTIMATELY.
BUT EVEN STILL, THEY WILL COME AT US WITH ***UNENDING WAVES OF BODIES.***

THEN WHAT SAVES US?
WHAT HAS ALWAYS SAVED US...***SPIRITS.*** SOME NEW ILLUMINATION TO GUIDE OUR WAY.

I WILL ORDER THE HALL OF SPIRITS TO REVIEW ALL RECORDS, THEN!
GOOD. BUT I FEAR OUR SALVIFIC SPARK ***HAS YET TO BE BIRTHED.***

I WANT ALL MAKERS TO TURN THEIR MINDS TO A WAR FOOTING. STOP ALL OTHER PROJECTS...AND FOCUS ON THESE VINN.

THE ***ENTIRE*** CONTINUUM? MARD...ARE YOU THAT AFRAID?

I WOULD NEVER SAY THIS TO THE COLEGIUM...
BUT I CAN'T FORESEE HOW WE WIN THIS FIGHT.

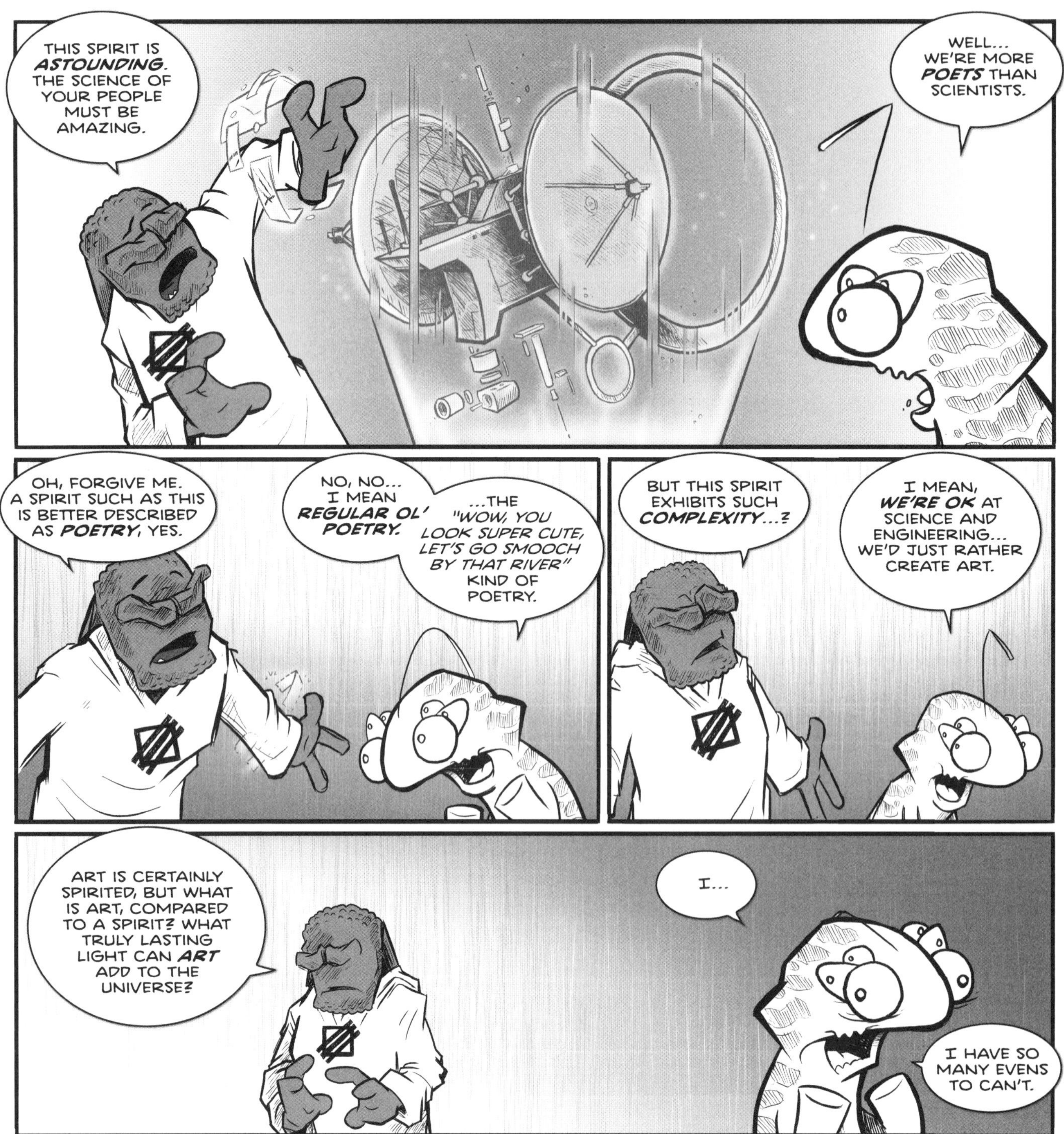
THIS SPIRIT IS ASTOUNDING. THE SCIENCE OF YOUR PEOPLE MUST BE AMAZING.
WELL... WE'RE MORE POETS THAN SCIENTISTS.
OH, FORGIVE ME. A SPIRIT SUCH AS THIS IS BETTER DESCRIBED AS POETRY, YES.
NO, NO... I MEAN REGULAR OL' POETRY.
...THE "WOW, YOU LOOK SUPER CUTE, LET'S GO SMOOCH BY THAT RIVER" KIND OF POETRY.
BUT THIS SPIRIT EXHIBITS SUCH COMPLEXITY...?
I MEAN, WE'RE OK AT SCIENCE AND ENGINEERING... WE'D JUST RATHER CREATE ART.
ART IS CERTAINLY SPIRITED, BUT WHAT IS ART, COMPARED TO A SPIRIT? WHAT TRULY LASTING LIGHT CAN ART ADD TO THE UNIVERSE?
I...
I HAVE SO MANY EVENS TO CAN'T.

YOURS IS A CURIOUS MIND. I WOULD LIKE TO KNOW YOU, IN TIME.
OOO! I HAVE A 'ZINE I SEND OUT MONTHLY-- I'LL ADD YOU!
THERE IS AN ODD JOY ABOUT YOU. ARE ALL YOUR PEOPLE LIKE THIS?
WELL, NOSH'S PEOPLE ARE ACTUALLY THE JOY-FILLED ONES...
IS FUN TO BE FUN!
...BUT WE DO OUR BEST
MY PEOPLE HAD A HARD ROAD, SO WE LOOK FOR JOY WHEREVER WE CAN FIND IT.
EVEN OUR MOST FAMOUS ANTI-VINN WEAPON WAS A SILLY, ACCIDENTAL INVENTION.
WE WERE ACTUALLY LOOKING FOR NEW WAYS TO POP POPCORN!
WAIT. ANTI-WHAT?

cuddow's fas

this month's poem:
"froyo"

A twist. A turn. A curl.
A light little hint of a twirl.
I jaunt, I sashay,
to friends I say "hey"--
and you hand me a froyo that's swirl'd.

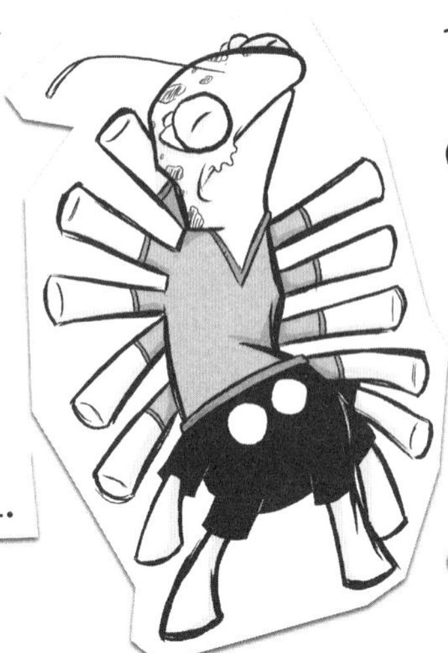

This month's constellation:
CHINCHILLA MAJOR

best YOU ever!

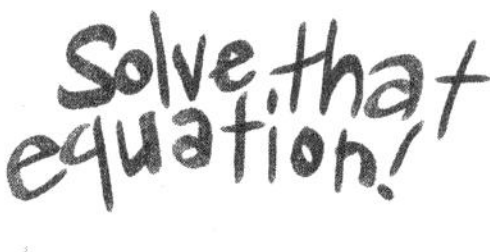
Solve that equation!

Just a smile!

Quantum Mechanics

love that!

get inspired!
Proton Party Style

Be yourself!

NAKED SINGULARITIES

Hel-looooooo!

hion zine!

big on fashion...
big on friendship!

this month's recipe:
papa cuddow's mint brownies!

1.) Make a normal pan of **BROWNIES!** (Human-style, not Veetan-style!)
2.) Then, let's make **MINTY BUTTERCREAMY GOODNESS!**
Using a mixer:
- Mix 2 sticks of butter until super fluffy. Fluffy like a super fat cat.
- Add in 4 cups powdered sugar and keep mixin'!
- Add 1 teaspoon mint extract and keep on mixin'!
- Add 4 tablespoons of heavy cream and–*you guessed it*–keep mixin'!
- Spread this Mint Buttercream over your brownies & pop in the fridge!
- (Lick the mixing whisk before putting in sink! THAT'S GOOD STUFF!)

3.) Then let's make our **CHOCO-TACO-TOPPER!**
- Heat 3/4 cup of heavy cream in a small saucepan, medium heat. When li'l baby bubbles start to break the surface, oh man quick turn off the heat!
- Put 1 1/2 to 2 cups of chocolate chips in metal mixing bowl.
- Pour that heated cream over those chocolate chips. Then just let the cream melt them chips! Go ahead and relax, maybe check the Vid-Pak Network for five minutes!
- After five minutes get to mixin' again! Whisk & whisk until that choco-cream is soupy-smooth and ready to go!
- Pull your brownies out of the fridge and pour your new Choco-Taco-Topper mix over the Mint Buttercream.
- Pop it back in the fridge for 30 minutes, or overnight!
- Serve it to Veetans as a form of payment, or to yourself, as your new **DELISH-DISH-WISH!**

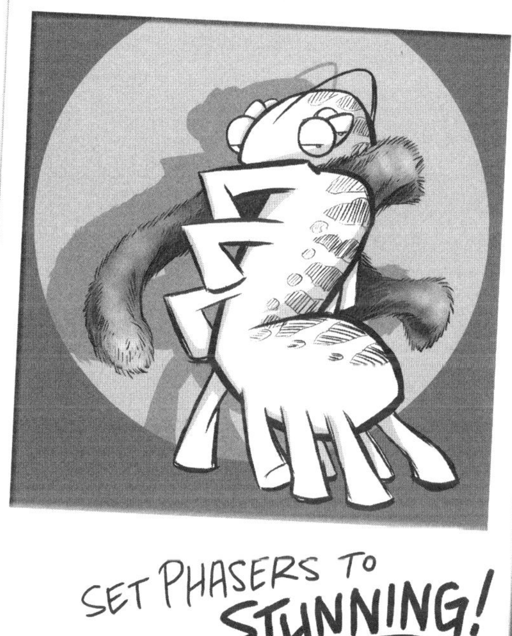

SET PHASERS TO **STUNNING!**

Reach for the STARS!

Let your
friendship
be a salad
with unlimited
croutons

LIVE
LAUGH
Love
Lanthanides

sent every third thursday to **325** subscribers on the vid-pack network
...tired of joy? holo-swipe **here** to unsubscribe

HOW DO YOU KNOW OF THE VINN?
THE VINN? THEY TRIED TO CONQUER HUMANS BACK IN THE DAY.
...BACK BEFORE THE HUMANS CONQUERED MY PEOPLE.
YOU WERE CONQUERED BY THESE HUMANS?
WELL, THEY REALLY CONQUERED THE TESSKANS, WHO HAD CONQUERED US EARLIER.
SO...IT WAS OUR CONQUERORS' CONQUERORS WHO FOUGHT THE VINN.
AND THEY... WON?
ALAS, NO. WHILE OUR CONQUERORS' CONQUERORS WENT UNCONQUERED
...THE VINN WERE UNCONQUERABLY STRONG.
AIN'T LANGUAGE JUST A COOL GLASS OF SPA WATER?

KIK AND I HAVE *HEARD* OF THESE VINN. THEY ARE A MASSIVE FORCE, CONQUERING EMPIRES EVEN *MY* PEOPLE COULDN'T SUBDUE.
humblebrag.
WE'VE SEEN IT, YES! THEY CONQUERED SLAUGHTER WHILE WE WERE THERE.
NO!
SLAUGHTER'S NEAR *MY* WORLD! AND...THEY FELL?
BUT THEY'RE SO *TOUGH*.
AS WERE THE GRASSKANS.
IT APPEARS THE GALAXY HAS FOUND ITS UNSTOPPABLE FORCE.
AND YET YOUR HUMAN EMPIRE STOPPED THEM?
ONLY TO A DRAW, ABOUT 70 YEARS AGO.
THEY'VE SINCE FOUND NEW DEFENSES TO OUR WEAPONS. WE LOST THOUSANDS JUST TRYING TO SAVE THESE TWO FROM VINN INFECTION.
WHAT?? *THESE TWO* WERE INFECTED, BUT ARE BACK TO NORMAL ???

CAPTAIN... THAT WAS NOT INFORMATION WE WERE AUTHORIZED TO SHARE.
WHA?
WE
SHOULD STOP TALKING
ABOUT THAT TOPIC.
OH FOR GOD'S SAKE...
THE JINYIWEI HAVE BEEN WORKING VERY HARD WITH THE FILLIPOD SCIENCE AUTHORITY TO KEEP THAT QUIET.
OH WELL LET'S JUST CALL EL PUÑO, SHALL WE?
OH WAIT
sigh
I KNOW...
IT'S JUST...
WE JUST...
I RESPECT WHAT YOU'RE SAYING, BUT THIS ISN'T A MOMENT TO GO "SECRET POLICE".
SECRET POLICE?
FORGIVE ME. THIS IS ORLA O'MALLEY. SHE'S A MEMBER OF OUR TOTALITARIAN SECRET POLICE FORCE, THE JINYIWEI.
WE'RE FUN AT PARTIES.

KEEP WHAT SECRETS YOU WILL. I AM NOT ONE TO STEAL KNOWLEDGE.
WE ACTUALLY KNOW VERY LITTLE! THE HUMANS *BARELY* BEAT BACK THE VINN WITH THEIR NEUTRON WEAPON ...AND MY PEOPLE HAVE BEEN STRUGGLING TO IMPROVE ON THE TECHNOLOGY.
WITH MY PEOPLE.
WITH HER POLICE, YES.
BUT THANKFULLY, THAT TECH LET US SAVE NOSH AND VINNIE, HERE. BOTH ARE VIRUS-FREE! ALBEIT... *CHANGED.*
MY I.B.S. IS GONE!
AND WHAT OF YOU? YOU'VE YET TO SPEAK SINCE I CAME ABOARD.
HE CAN'T YET.
HIS BRAIN IS STILL REWIRING BASIC PATHWAYS. HE *UNDERSTANDS,* BUT DOESN'T REALLY RESPOND.
EXCEPT IN DRAWINGS!
HE DREW US THIS PICTURE OF THE VINN'S *MISSING GODS!*

...IS
ROCKET
SHIP.
THIS
IS NO
ROCKET.

END
drive
the scifi comic
ACT TWO

CONTINUE READING ACT 3 ONLINE!
http://www.drivecomic.com/archive/180913.html

OR AIM YOUR PHONE'S CAMERA AT THIS QR CODE TO JUMP THERE IMMEDIATELY!

SHORT STORIES SET IN
THE DRIVE UNIVERSE

RARE EARTHS

written and drawn by
JON ROSENBERG

amultiverse.com

VREEEEREEEEEREEEEEREEEEREEEEREEEE CHUNK.
....AND THAT'S THE SOUND OF THE TERTIARY POWER SUPPLY FAILING.

HOW ABOUT THE DRIVE?
IT'S MOSTLY USELESS WITHOUT NAVIGATION.

AND EVEN IF WE DIDN'T STUMBLE INTO A BLACK HOLE WE'D BE DEAD LONG BEFORE WE REACHED THE NEXT INHABITED WORLD.
ON BRIGHT SIDE, LOCAL STAR SYSTEM IS LOVELY THIS TIME OF YEAR!

SCANS OF THE SINGLE ORBITING PLANET SHOW AN ATMOSPHERE THAT DISSOLVES HUMAN FLESH AND A STRANGE SPACE-TIME SIGNATURE THAT INDICATES SEVERE EXISTENCE INSTABILITY THROUGHOUT THE REGION.

SO IS UNIQUE FIXER-UPPER OPPORTUNITY!

THAT MIGHT EXPLAIN THE POWER FAILURES, CAPTAIN.
THE RARE EARTH ELEMENTS IN THE POWER CELLS DON'T PLAY NICE WITH EXISTENCE INSTABILITIES.

LIKE SCHRODINGER'S KITTEN.
HOW LONG DO WE HAVE?
MAYBE AN HOUR, IF **NOSH** DOESN'T HOG ALL THE OXYGEN.

MY LUNGS HAVE SPECIAL NEEDS!
AND I WAS ONLY TWO DECADES TO RETIREMENT.
IF **ONLY** WE HAD ACCESS TO A NEARBY SOURCE OF RARE EARTH ELEMENTS!
THEN WE COULD REBUILD THE POWER CELLS.

HEY GUYS! MY WEIRD MOHAWK THING IS DETECTING A WHOLE **MESS** OF LANTHANIDES DOWN ON THE PLANET.
ARE THOSE ANYTHING LIKE RARE EARTH ELE-MENTS?

THOSE ARE **EXACTLY** LIKE RARE EARTH ELEMENTS!
IT'S TOO BAD THE PLANET'S ATMOSPHERE DISSOLVES HUMAN FLESH OR WE COULD GO GET THEM.

YES. TOO BAD.
TERRIBLE.

...AND SEE IF YOU CAN FIND SOME CHEEZ PUFFS.
LANTHANIDES AND CHEEZ PUFFS.
NOT THE CRUNCHY KIND.

THE CRUNCHY KIND CREEPS ME OUT.

THIS METER THING SAYS WE WALK IN STRAIGHT LINE THROUGH ROCK WALL.
I COULD FLY OVER

AND WHAT AM I, SPACE LIVER?
IT'S NOT MY FAULT YOU DON'T HAVE A NEAT LITTLE HOVERCAR LIKE I DO.

YOU AND HOVERCAR. **SO** SUPERIOR.
THIS IS WHY NO ONE EVER BRING YOU HOME TO MEET PARENTS.
THE ONLY OTHER OPTION IS TO BURROW THROUGH THE WALL.

WHAT, YOU THINK I CANNOT BURROW?
I AM **WORLD-CLASS BURROWER** THREE YEARS RUNNING.
SO BURROW ALREADY.

DINK

THERE! I BURROW OPENING TO PERFECTLY SERVICEABLE NETWORK OF CAVES.
TECHNICALLY THAT WASN'T BURROWING.
THE CAVES WERE ALREADY THERE.

OPENING DID NOT BURROW ITSELF.
I'D SAY THAT WAS MORE OF AN **UNCOVERING.**

I THINK NOSH SAID SOMETHING ABOUT GRAVES?
LOVELY. CAN YOU BOOST RECEPTION?

UNLIKELY. THEY'VE GONE UNDERGROUND INTO A NETWORK OF CAVES.
CAVES! I BET HE WAS SAYING CAVES.
KEEP TRYING TO RAISE THEM. I WANT TO KNOW THE MINUTE THEY SCREW UP.

I THINK WE SCREW UP.
THIS CAVE LOOKS FAMILIAR, LIKE WE ARE GOING IN CIRCLES.
THE LANTHANIDE DEPOSITS ARE JUST AHEAD. I CAN FEEL THEM.

DO LANTHANIDES HAVE LEGS AND EYES?
NOT USUALLY. WHY DO YOU ASK?

NO REASON.

WHAT ARE THOSE THINGS?
THEY ARE LIKE LITTLE GLOW CHICKENS! THAT EAT ROCKS.

THOSE AREN'T JUST ROCKS.
THOSE ARE THE LANTHANIDES WE CAME DOWN HERE TO FIND!
WHO KNEW ROCKS COULD BE DELICIOUS.

ARE YOU DELICIOUS, GLOW CHICKEN?
NO. YOU ARE ADORABLE!
I CALL YOU BLOBBERS AND FEED YOU YUMMY ROCKS AND SING RUSSIAN NURSERY SONG TO SCARE YOU TO SLEEP.

DON'T GET TOO ATTACHED TO BLOBBERS THERE, NOSH.
WHY HOWCOME?

WE'RE ABOUT TO TAKE THEIR FOOD SUPPLY AWAY.
OH.

BUT IS NO FAIR!
TAKING AWAY OF GLOW CHICKENS' FOODS IS *WRONG* AND *BAD.*
IMAGINE POOR BLOBBERS STARVING IN CAVE, CRYING.
IT'S US OR THEM, NOSH.
AND I'M A BIG FAN OF *US.*
LET US BE ASKING CAPTAIN FOR SOLOMONESQUE WISDOM.
CAPTAIN TANEEL, IS NOSH. CAN YOU BE ROGERING?
NOBODY ANSWER.
GREAT. NOW WE ARE HAVING TO USE OWN BRAINS FOR WISDOM.
THESE CREATURES ARE GIVING OFF *TREMENDOUS* AMOUNTS OF ENERGY.
THAT MUST BE INTERFERING WITH COMMUNICATIONS.
WHY YOU ARE SO RACIST AGAINST ADORABLE GLOW CHICKENS?
DID GLOW CHICKEN POOP IN CORNFLAKES PERHAPS?

LOOK. YOU ARE SMART. I ALSO AM SMART.
TOGETHER WE HAVE TWO SMARTS WORTH.
SURELY WE CAN COME UP WITH SOLUTION FOR EVERYONE TO LIVE.
UNDER NORMAL CONDITIONS, SURE.
BUT WE'RE RUNNING OUT OF TIME.
ALSO, I DON'T KNOW IF YOU'VE NOTICED, BUT NONE OF THIS MAKES ANY SENSE.
HOW SO?
WE'RE IN A REGION OF SPACE THAT DRAINS RARE EARTH ELEMENTS DRY BUT HERE'S A WHOLE MESS OF 'EM IN PERFECT CONDITION.
ISN'T THAT A BIT TOO CONVENIENT?
CONVENIENCE IS NOT BAD THING.
THEY HAVE WHOLE STORES NOW JUST FOR CONVENIENCE.
THEY SELL CONVENIENCE?
THEY SELL CHIPS OF POTATO.
BUT IN MANY VARIETIES!

LOOK AT DISCOVERY I MADE!
TWO GLOW CHICKENS TOGETHER MAKE GLOW DOG.
YOU'RE NOT HELPING.
HERE BLOBBERS, I FEED YOU RARE EARTH, IS RARE EARTH DELICACY.
WHERE IS THE FOOD COMING FROM?
IT COMES FROM ROCKS, YES?
THEY'D RUN OUT.
THERE'S BARELY ENOUGH TO FUEL OUR SHIP.
AND IT'S NOT LIKE CHARGED LANTHANIDES ARE JUST GOING TO GROW DOWN HERE. UNLESS...
UNLESS YOU FARMED 'EM.

AAH! I SWEAR THAT THING JUST SPOKE TO ME.
I HEAR NOTHING.

ONLY YOU CAN HEAR ME.
YOUR FRIEND LACKS THE RECEPTORS FOR TELEPATHIC COMMUNI-CATION.

THAT'S NOT THE ONLY THING HE LACKS.
WE DO DETECT A NEURAL DEFECIT, BUT WERE TOO TIMID TO MENTION IT.

LAST WEEK I FOUND HIM TRYING TO TEACH THE CAT TO PLAY GUITAR.
HE DOESN'T KNOW HOW TO PLAY GUITAR.
AND WE DON'T HAVE A CAT.
OY.

STOP THAT!
I CAN SMELL YOU THINKING ABOUT ME.
WE WERE TALKING ABOUT SPORTS... BALL.
SPORTSBALL.
NICE SAVE.

SO YOU TALK TO BLOBBERS USING MOHAWK AND MIND WAVES?
I THINK SO.
THAT IS SOME HANDY MOHAWK THERE.
IT'S GROWING ON ME.
I HAVE DETECTED A CONCERN FOR LANTHANIDES IN YOUR THOUGHTSTREAM.
WE TOO SHARE THIS CONCERN.
YES! OUR POWER CELLS WERE MYSTERIOUSLY DEPLETED AS WE APPROACHED YOUR PLANET.
IT IS NO MYSTERY. THAT'S ON US.
SORRY DUDE.
WE USE OUR VAST MENTAL POWERS TO FUNNEL ENERGY FROM RARE EARTHS IN THE NEARBY ASTEROID BELT TO CHARGE DEPLETED DEPOSITS DOWN HERE IN OUR CAVES.
IT'S CHEAPER THAN EATING OUT.
...SO YOU CAN RECHARGE DEPLETED LANTHANIDES?
HELLO? STANDING RIGHT HERE.
SO AM CHOPPED SPACE LIVER AFTER ALL.
MOM WAS RIGHT.

...PKSHHHTTT FSHHHTTT KEEP TRYING, BECAUSE IT'S OBVIOUSLY A WASTE OF TIME.
SKITTER, NOSH, IF YOU'RE OUT THERE, PLEASE RESPOND. MESSAGE ENDS.

NO! NO ENDS! PLEASE BE BEGINNING AGAIN, WE ARE HERE!
NOSH? IT'S FERNANDO! WHAT HAPPENED TO YOU GUYS?

IS VERY COMPLICATED, THERE WERE CAVES AND LAVA BEASTS AND SKELETONS.
I WAS VERY BRAVE FIGHTING WITH FISTS.
THEN WE FIND GLOW CHICKENS!

THEY'VE AGREED TO HELP US!
I THINK GLOW CHICKENS FIX SPACESHIP MAYBE?

THE ATMOSPHERE OUT THERE MAY BE MILDLY HALLUCINOGENIC, CAPTAIN.
TELL THEM TO BRING BACK SAMPLES.

SEE? GLOW CHICKEN MAKE EVERYTHING BETTER.
UNBELIEVABLE. THEY'RE RE-ENERGIZING THE POWER CELLS.

LIFE SUPPORT IS BACK ONLINE!
DESSERT SERVICES SHOULD BE UP WITHIN THE HOUR.

GOOD WORK, SKITTER! WE ALL GET TO LIVE.
AND GOOD WORK NOSH TOO.

YES, NOSH. YOU TOO.
WOULD YOU LIKE A MEDAL?
IF IS NOT TOO MUCH TROUBLE.

ALL I GOT IS SOAP ON A ROPE.
IS ONE MORE SOAP AND ROPE THAN HAVE PREVIOUSLY.
SOAP

IS HARD, BEING HERO? YES.
BUT SOMEONE MUST WEAR SOAP.
SOAP

MEMORIES
IN FUTURE TENSE

written and drawn by
MIKE NORTON

---

battlepug.com

JUST A LITTLE LONGER, BUDDY. THESE TESTS WILL TAKE LESS TIME THAN THE LAST ROUND.
DON'T WORRY, LITTLE VINNY IS STRONG AS HE IS ADORABLE.
IT'S TOO BAD WE CAN'T REALLY TALK TO THE LITTLE GUY.

AH, BUT CUTENESS IS A UNIVERSAL LANGUAGE, NANDO!

I'M POSITIVE WE'LL BE ABLE TO ESTABLISH MORE COMMUNICATION BEYOND CRAYON DRAWINGS THE LONGER HE'S ON BOARD.
BAH, WHO NEEDS COMMUNICATION WHEN YOU HAVE...

POPSICLES!
!

STILL... IT'D BE NICE, Y'KNOW...
"...TO KNOW WHAT'S GOING ON IN THAT LITTLE HEAD."

"Dearest Binko,
I hope you weren't in the science temple when it fell in the first assault. I'm told the losses were great, but I still cling to hope"
"We've been charged with pushing back the Vinn invasion forces. Our only luck has been that they've been forced to ground since their ships cannot travel here."
"But still... They are many, and we are fewer and fewer each day."
"We were on the verge of giving up hope."
"Until the SARGE arrived."

"He ran through the Vinn as if they weren't even there."
"He screamed as he mowed through their ranks."
GET UP, YOU NURSLINGS!
GET UP AND FIGHT!
"And so we did. At great cost."
"But through his leadership we made it to the main citadel."
"And we found ourselves alive another day."

"We've made camp in the citadel center. We have no doubt the Vinn will return and in much greater numbers."
"Until then, I write to you and listen to the other survivors."

"Irole and Deerole made the evening portion with what rations we could assemble. They are snipers from the lower regions.
Irole makes fun of Deerole's eyesight."

"Tikko is the last survivor of his cadre. He hasn't stopped pacing since the attack. Muttering formula or something."

"Sinc is the medic. He is quiet and good-natured. He shared a story of his travels on the planet's surface."

"Ziff, our munitions officer, told what may be the filthiest joke in the universe. I'm serious. It made Sinc vomit."

"Rather than engage with them, I just listen..."

"...And think of you."

"And then, there's SARGE..."

LISTEN UP, NURSLINGS!
YOU KNOW WHAT'S ABOUT TO HAPPEN, SO I'LL SPARE NICETIES.
THE SCIENCE TEMPLE HAS FALLEN. THERE'S A GOOD CHANCE EVERYTHING YOU KNOW AND LOVE IS *GONE*.
BUT WE AREN'T!
AS LONG AS ONE OF US STILL CONTROLS OUR OWN MIND, THIS IS STILL *OUR* ROCK.
I WILL *NOT* BE SUBJUGATED!
AND I PROMISE THAT WHILE I DRAW LIFE FUME, I WILL NOT LET ANY OF YOU BE, EITHER. *WHATEVER* THE COST!
PKOW!

"And like that, our last hopes were dashed."
"Dearest Binko, I am not strong."
"I want to be strong like the others."
"They fight as I hide."
"But what can I do?"
"I am a lowly comms officer, and they..."
"... Are you."

"You see, Binko... I am not just writing this message in the hopes that it will one day find your eyes."

"I write this for *both* of us."

"I write so that when the time may come, we both will discover it."

"And remember."

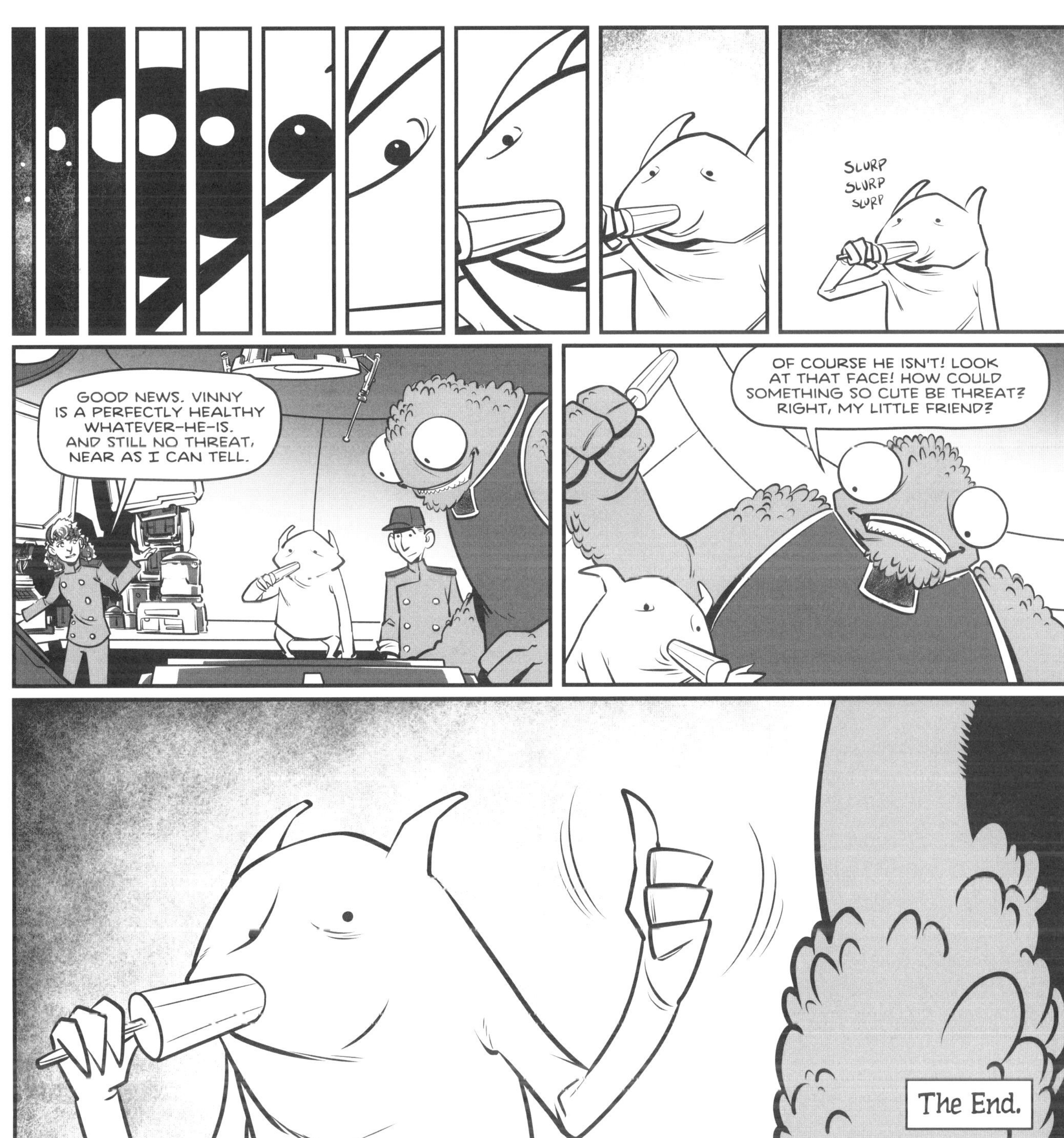
SLURP
SLURP
SLURP
GOOD NEWS. VINNY IS A PERFECTLY HEALTHY WHATEVER-HE-IS. AND STILL NO THREAT, NEAR AS I CAN TELL.
OF COURSE HE ISN'T! LOOK AT THAT FACE! HOW COULD SOMETHING SO CUTE BE THREAT? RIGHT, MY LITTLE FRIEND?
The End.

written and drawn by

CHRISTOPHER BALDWIN

---

spacetrawler.com

FREEDO, THIS IS MY CAPTAIN, TO WHOM I OWE SEVERAL LIFE DEBTS.
CAPTAIN TANEEL, IT'S A PLEASURE.
FOR ME TOO. A FRIEND OF NOSH'S IS A FRIEND OF MINE.
AFTER WE LEFT VEETAN SCHOOL FOR OUR WALKABOUTS I'D HOPED HE'D CREW WITH SOMEONE LIKE YOU.
HE'S MORE THAN CREW TO ME, HE'S FAMILY.
BUT WITHOUT TEENAGE REBELLION PHASE, I DO DISHES AND LAUNDRY WILLINGLY.
WELL, IT WAS NICE OF YOU TO STOP BY, BUT I'M AFRAID YOU BOTH SHOULD LEAVE.
OOH, YOU HAVE GIRL VEETAN COMING HERE FOR PLAY TWISTER WHILE DRINKING GRAPE SODA?
IS THAT TRUE? NO DOUBT YOU SERVE CUPCAKES AND MILKSHAKES TOO.
CUPCAKES AND MILKSHAKES? ONLY AS A PROPOSAL FOR MARRIAGE.
BUT, NO. IT'S BECAUSE THERE ARE A GROUP OF KLOCKNOCKS WHO WANT TO KILL ME.
YOU TRY AND ASK POLITELY THEM NOT TO?
THEY USUALLY HAVE A HARD TIME HEARING ME OVER THEIR GUNFIRE.
I FEEL SAME WHEN CAPTAIN FALLS ASLEEP ON DECK. LOUDER THAN SAWING LOGS ON LOOSE ALUMINUM ROOF.
I. DO. NOT. SNORE.

WHAT HAPPENED TO GET THESE KLOCKNOCKS TO WANT TO KILL YOU?
I ACCIDENTALLY KILLED ONE OF THEM.
BUT... BUT... YOU WOULD NOT HURT FLY. EVEN IF FLY OLD AND APPLIED FOR LEGAL EUTHANASIA.
WELL, MORE LIKE I DIDN'T ACTIVELY SAVE HIS LIFE, WHICH IS PRACTICALLY THE SAME THING ACCORDING TO HIS GOONS. DID I MENTION HE WAS A MAFIA BOSS?
IF YOU HAVE TROUBLE, FREEDO, WE STAY AND HELP YOU LIKE LIONS.
BIG ONES. NOT LION CUBS, THEY TOO CUTE AND FUNNY. I SEND PICTURES TO YOUR INBOX, YOU WILL SEE.
I COULDN'T ASK YOU TO STAY AND PUT YOURSELVES IN DANGER LIKE THAT.
NO NEED TO ASK, WE ARE LIKE FORK ALREADY THRUST IN CAKE.
IF IT WEREN'T FOR THE MASSIVE FREE TABLE OF JELLY DONUTS DOWN IN THE LOBBY, I'D ACCEPT YOUR OFFER.
NO DOUBT YOU CAN CALL IF THINGS GET HAIRY.
IF YOU HAD STAYED, I WOULD HAVE BEEN VERY WORRIED ABOUT YOU.

I FEEL SO LOST, I FEEL LIKE ALL GOOD IN UNIVERSE IS GONE.
IS SOMETHING THE MATTER, NOSH?
WE CANNOT FIND DONUT TABLE ANYWHERE. MAYBE FREEDO JUST TRY TO GET RID OF US!
YA THINK?
MAYBE HE NOT WANT US IN CROSSFIRE AND HAS SLIPPERY ESCAPE PLANNED LIKE OILED BABUSHKA.
OH LOOK, WHY ARE THOSE PEOPLE POINTING THOSE GUNS AT US?
DOWN!
WHY YOU DO THAT?
I WAS TRYING TO KNOCK YOU OUT OF THE WAY BEFORE THEY BLASTED A HOLE THROUGH YOU.
OH, I SEE. I WAS SUPPOSED TO FALL LIKE THIS.
POOM!
BLAST!
NEXT TIME IF YOU TELL ME FIRST, I CAN FALL DOWN WHEN YOU BUMP INTO ME.
MAYBE I'LL WRITE IT TO YOU ON A POSTCARD.
OOOH! LOVE POST-CARDS. MAYBE YOU WRITE "WISH YOU WERE NOT THERE WHERE YOU ARE STANDING." IT BE REAL FUNNY.

THAT'S FREEDO! GET HIM!
"FREEDO"? OH, DID HE COME DOWN LOOK FOR DONUT TABLE TOO?
NO, NOSH, THEY THINK YOU'RE FREEDO.
OH, ONLY MISUNDER-STANDING THEN. MAYBE THEY'RE NOT SUCH BAD KLOCKNOCKS.
HELLO! I'M NOSH! NOT FREEDO!
BLAST!
YES. THEY ONLY BIT CONFUSED. POOR LITTLE GUYS. YOU RIGHT. THEY THINK I AM FREEDO.
MAYBE WE SHOULD TRY RUNNING WHILE THEY CLEAR THEIR HEADS.
YOU DON'T THINK WE SHOULD STAND GROUND TO HELP CLARIFY? I HAVE CHEESECAKE FACTORY DISCOUNT CARD FOR I.D.
NO.
REALLY?
REALLY.
IS THIS PART OF BIT WHY YOU ARE CAPTAIN AND I'M SCIENCE ADVISOR?
A WEE BIT MAYBE.

BLAST!
YOU HOOLIGANS, YOU'RE SHOOTING AT **NOSH**, NOT **FREEDO!**
I THOUGHT YOU SAID TRYING TO CLARIFY WAS A WASTE OF BREATH.
BLAST!
NO, I SAID, "**STANDING STILL** WOULD RESULT IN OUR **FINAL** BREATH."
SKREEEEEEEEEEE
MANUEL JOSE CRUZ 2339-2401
CATCH
THANK YOU, NOSH. AT LEAST THAT'S **ONE** LESS LIFE DEBT YOU OWE ME.
WHAT? NO, I JUST TRY TO SAVE STATUE. SOME CRAFTSMAN PROBABLY SPEND WHOLE LIFE ON THIS. A MASTERPIECE.
HAMMERING AWAY, SACRIFICING YEARS OF PARTYING AND DATING, ALL FOR FINAL WORK OF BEAUTY. AND SO I REST IT DOWN GENTLY, WITH CARE AND REVERENCE.
JOI INNY, WE NEED ANOTHER STATUE OF THE EMPERADOR.
3-D STATUE CORP
PRESS
MAKE-A-STATUE ENGINE
EMPEROR CONRADO I
EMPEROR PABLO
SPORTS
FSTFD

WHAT WAY WE GO? LEFT OR RIGHT?
WHAT DO YOU THINK, NOSH?
BLAST!
BLAST!
I THINK I TRUST CAPTAIN.
FINE, LET'S GO LEFT.
BLAST!
BLAST!
THAT'S VERY INTERESTING.
SEE, THIS MUST BE ALSO WHY I'M NOT CAPTAIN.
"INTERESTING"? WHY, PRAY TELL?
BLAST!
BLAST!
BECAUSE THE GARAGE EXIT IS ON THE OTHER SIDE.
WHAT!
BLAST!
BLAST!
WHY NOT WE GO THAT WAY TO ESCAPE?
BLAST!
UM...
BLAST!
IS NOW NOT TIME FOR US TO KEEP MOVING BEFORE THEY GET US?
GIVE ME ONE MORE MINUTE TO REMIND MYSELF THAT I LIKE YOU, BEFORE I RASHLY DECIDE TO USE YOU AS A MEAT SHIELD.
BLAST!

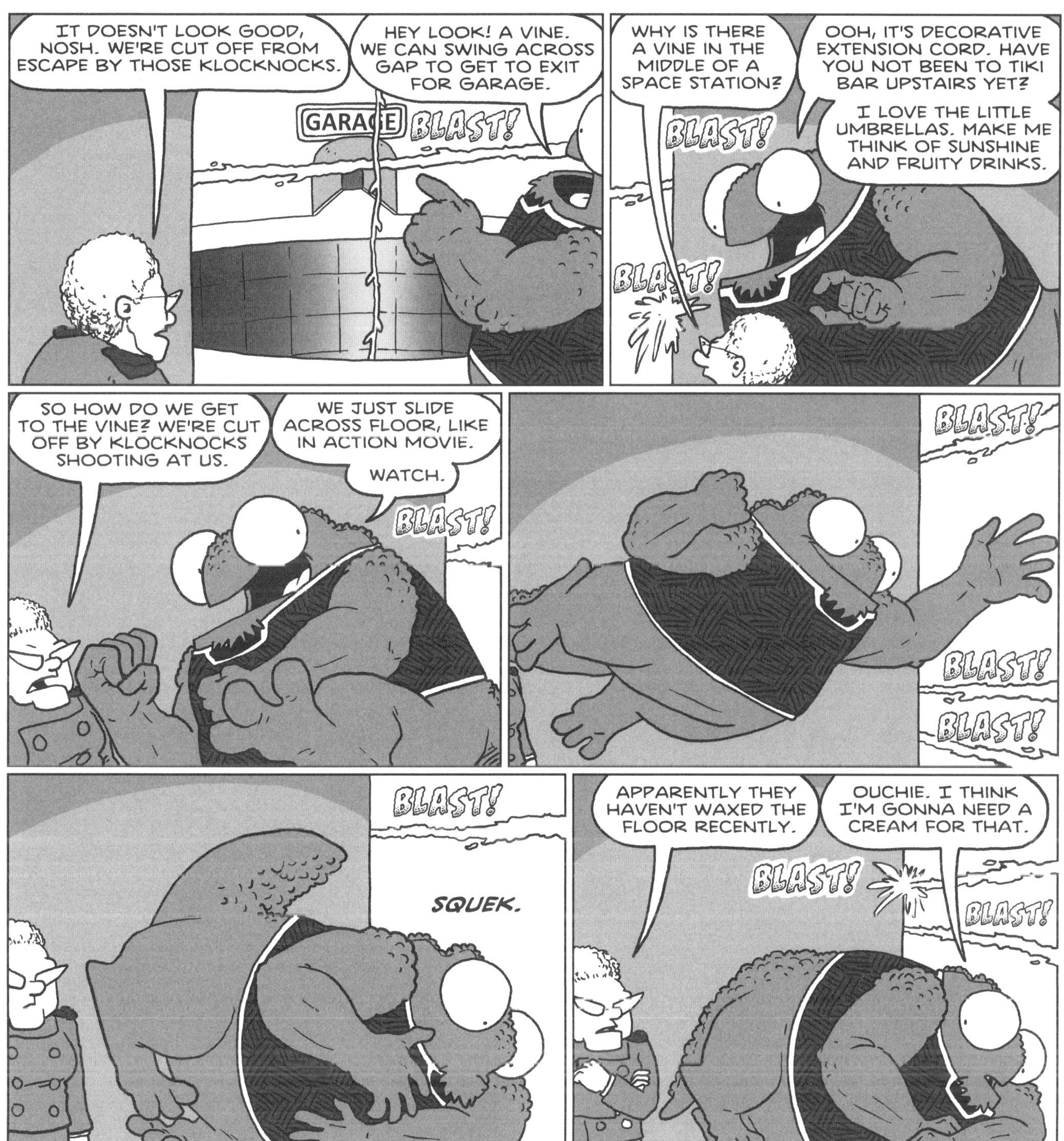
IT DOESN'T LOOK GOOD, NOSH. WE'RE CUT OFF FROM ESCAPE BY THOSE KLOCKNOCKS.
HEY LOOK! A VINE. WE CAN SWING ACROSS GAP TO GET TO EXIT FOR GARAGE.
GARAGE
BLAST!
WHY IS THERE A VINE IN THE MIDDLE OF A SPACE STATION?
OOH, IT'S DECORATIVE EXTENSION CORD. HAVE YOU NOT BEEN TO TIKI BAR UPSTAIRS YET?
I LOVE THE LITTLE UMBRELLAS. MAKE ME THINK OF SUNSHINE AND FRUITY DRINKS.
BLAST!
BLAST!
SO HOW DO WE GET TO THE VINE? WE'RE CUT OFF BY KLOCKNOCKS SHOOTING AT US.
WE JUST SLIDE ACROSS FLOOR, LIKE IN ACTION MOVIE.
WATCH.
BLAST!
BLAST!
BLAST!
BLAST!
BLAST!
SQUEK.
APPARENTLY THEY HAVEN'T WAXED THE FLOOR RECENTLY.
OUCHIE. I THINK I'M GONNA NEED A CREAM FOR THAT.
BLAST!
BLAST!

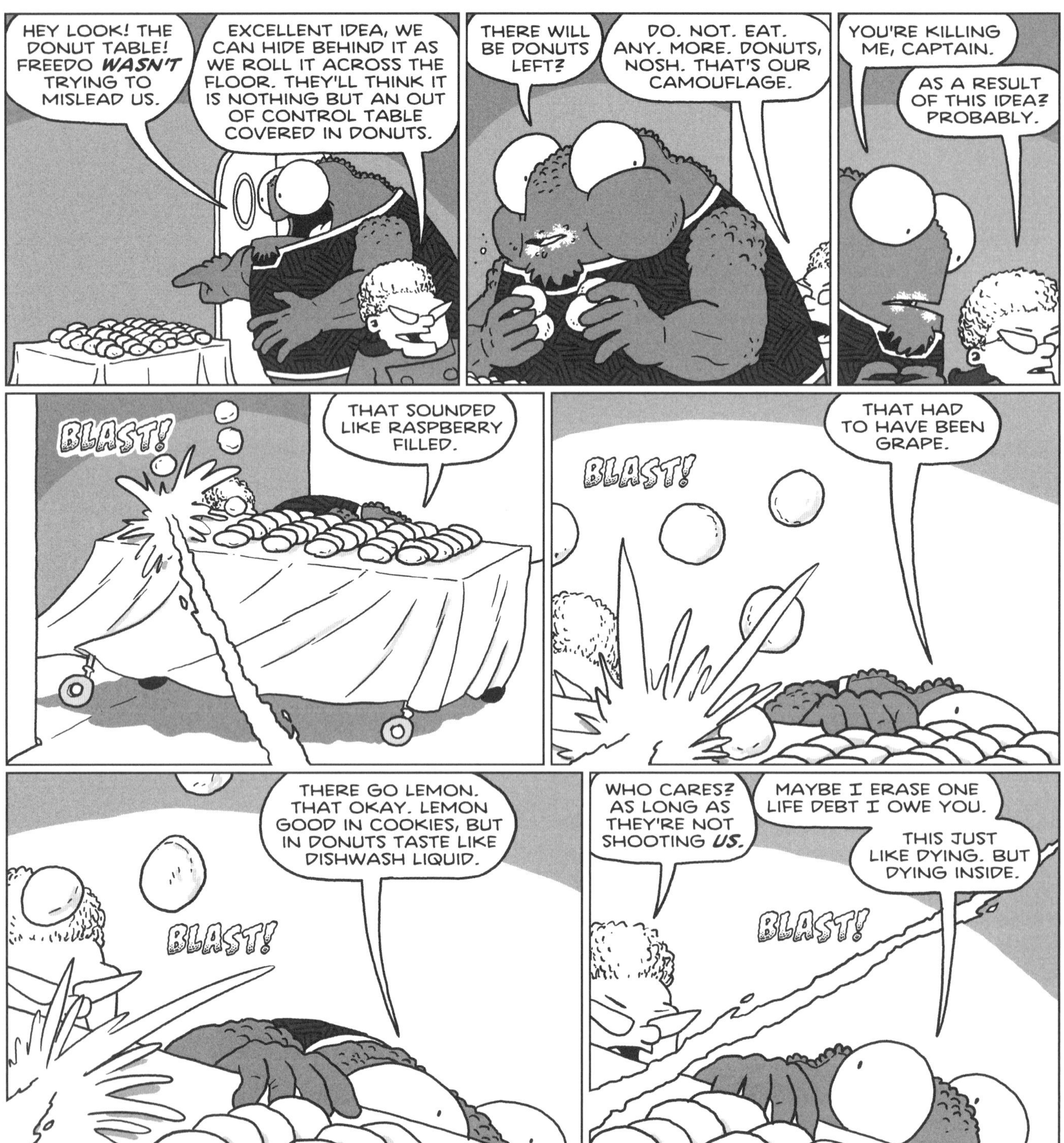
HEY LOOK! THE DONUT TABLE! FREEDO **WASN'T** TRYING TO MISLEAD US.
EXCELLENT IDEA, WE CAN HIDE BEHIND IT AS WE ROLL IT ACROSS THE FLOOR. THEY'LL THINK IT IS NOTHING BUT AN OUT OF CONTROL TABLE COVERED IN DONUTS.
THERE WILL BE DONUTS LEFT?
DO. NOT. EAT. ANY. MORE. DONUTS, NOSH. THAT'S OUR CAMOUFLAGE.
YOU'RE KILLING ME, CAPTAIN.
AS A RESULT OF THIS IDEA? PROBABLY.
BLAST!
THAT SOUNDED LIKE RASPBERRY FILLED.
BLAST!
THAT HAD TO HAVE BEEN GRAPE.
THERE GO LEMON. THAT OKAY. LEMON GOOD IN COOKIES, BUT IN DONUTS TASTE LIKE DISHWASH LIQUID.
BLAST!
WHO CARES? AS LONG AS THEY'RE NOT SHOOTING **US.**
MAYBE I ERASE ONE LIFE DEBT I OWE YOU.
THIS JUST LIKE DYING. BUT DYING INSIDE.
BLAST!

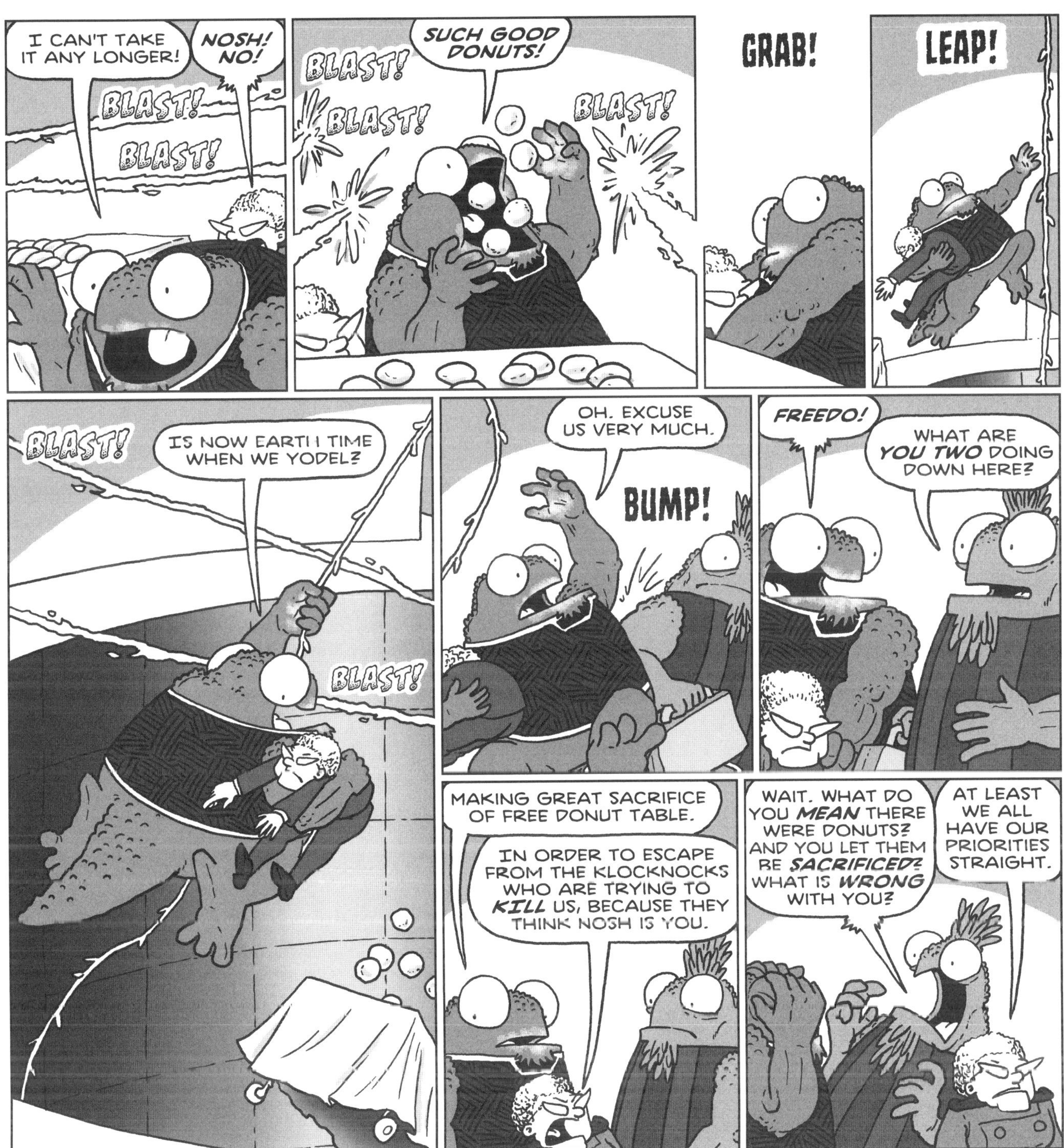
I CAN'T TAKE IT ANY LONGER!
NOSH! NO!
BLAST!
BLAST!
SUCH GOOD DONUTS!
BLAST!
BLAST!
BLAST!
GRAB!
LEAP!
BLAST!
IS NOW EARTH TIME WHEN WE YODEL?
BLAST!
OH. EXCUSE US VERY MUCH.
BUMP!
FREEDO!
WHAT ARE YOU TWO DOING DOWN HERE?
MAKING GREAT SACRIFICE OF FREE DONUT TABLE.
IN ORDER TO ESCAPE FROM THE KLOCKNOCKS WHO ARE TRYING TO KILL US, BECAUSE THEY THINK NOSH IS YOU.
WAIT. WHAT DO YOU MEAN THERE WERE DONUTS? AND YOU LET THEM BE SACRIFICED? WHAT IS WRONG WITH YOU?
AT LEAST WE ALL HAVE OUR PRIORITIES STRAIGHT.

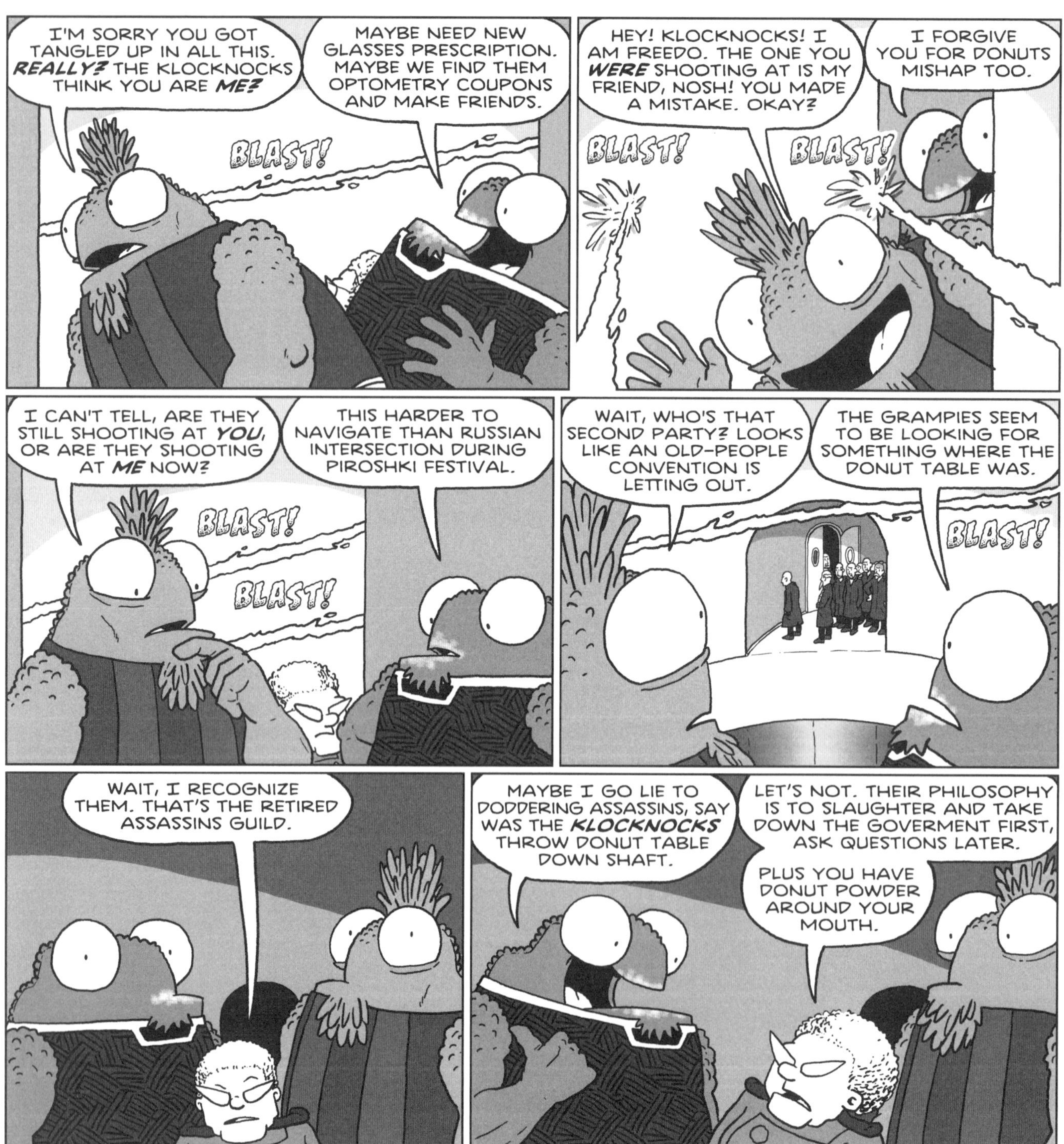

I'M SORRY YOU GOT TANGLED UP IN ALL THIS. REALLY? THE KLOCKNOCKS THINK YOU ARE ME?
MAYBE NEED NEW GLASSES PRESCRIPTION. MAYBE WE FIND THEM OPTOMETRY COUPONS AND MAKE FRIENDS.
BLAST!
HEY! KLOCKNOCKS! I AM FREEDO. THE ONE YOU WERE SHOOTING AT IS MY FRIEND, NOSH! YOU MADE A MISTAKE. OKAY?
I FORGIVE YOU FOR DONUTS MISHAP TOO.
BLAST!
BLAST!
I CAN'T TELL, ARE THEY STILL SHOOTING AT YOU, OR ARE THEY SHOOTING AT ME NOW?
THIS HARDER TO NAVIGATE THAN RUSSIAN INTERSECTION DURING PIROSHKI FESTIVAL.
BLAST!
BLAST!
WAIT, WHO'S THAT SECOND PARTY? LOOKS LIKE AN OLD-PEOPLE CONVENTION IS LETTING OUT.
THE GRAMPIES SEEM TO BE LOOKING FOR SOMETHING WHERE THE DONUT TABLE WAS.
BLAST!
WAIT, I RECOGNIZE THEM. THAT'S THE RETIRED ASSASSINS GUILD.
MAYBE I GO LIE TO DODDERING ASSASSINS, SAY WAS THE KLOCKNOCKS THROW DONUT TABLE DOWN SHAFT.
LET'S NOT. THEIR PHILOSOPHY IS TO SLAUGHTER AND TAKE DOWN THE GOVERMENT FIRST, ASK QUESTIONS LATER.
PLUS YOU HAVE DONUT POWDER AROUND YOUR MOUTH.

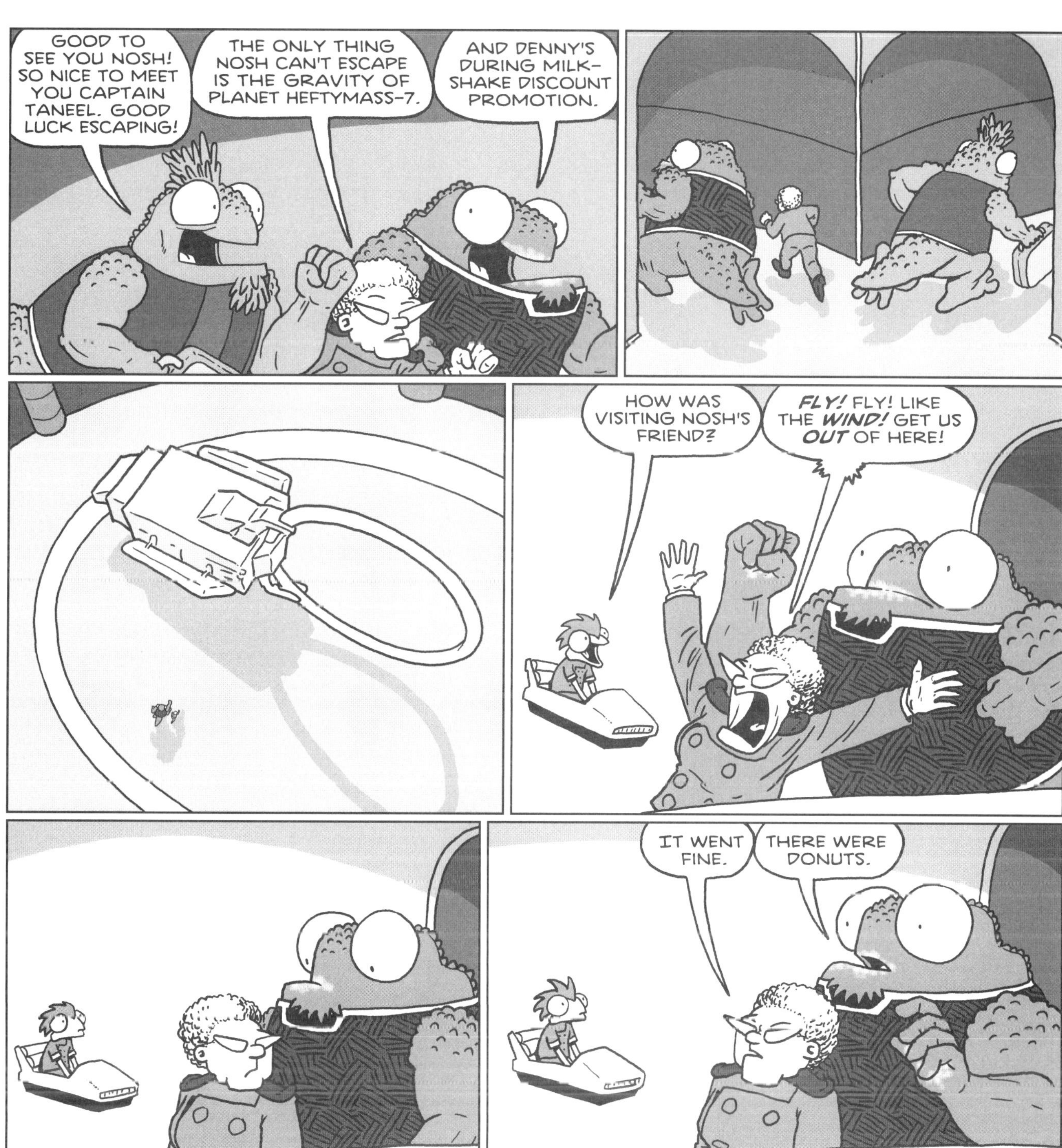
GOOD TO SEE YOU NOSH! SO NICE TO MEET YOU CAPTAIN TANEEL. GOOD LUCK ESCAPING!
THE ONLY THING NOSH CAN'T ESCAPE IS THE GRAVITY OF PLANET HEFTYMASS-7.
AND DENNY'S DURING MILK-SHAKE DISCOUNT PROMOTION.
HOW WAS VISITING NOSH'S FRIEND?
FLY! FLY! LIKE THE WIND! GET US OUT OF HERE!
IT WENT FINE.
THERE WERE DONUTS.

PEBBLE

written and drawn by

ADAM KOFORD

adamkoford.com

SPACE. IT'S FULL OF STUFF. HARD TO SEE, BUT IT'S ALL OUT THERE. JUST FLOATIN' AROUND.

BIG STUFF
COIN FOR SCALE

SMALL STUFF
COIN FOR SCALE
LIBERTY

ASSORTED GARBAGE
HEY!
AND SO MUCH MORE

TROUBLE IS, THERE'S SO MUCH, WELL, SPACE BETWEEN ALL THE STUFF THAT HARDLY ANY OF IT EVER INTERACTS.

AT LEAST NOT AS MUCH AS YOU'D THINK. EVERY ONCE IN A WHILE THOUGH, THINGS MEET AND SOMETHING SPECIAL HAPPENS. MORE ON THAT IN JUST A MOMENT. FIRST, THERE'S SOMEONE I'D LIKE YOU TO MEET.
THIS IS PEBBLE.

THIS IS THE STORY OF ONE TINY, INSIGNIFICANT PEBBLE OF IRON.

(ACTUAL SIZE)

BORN LONG BEFORE YOU AND I, IN FIRE AND CATACLYSM, OUR TINY FRIEND WAS SOON ACTED UPON BY FORCES BEYOND ITS CONTROL AND SET IN MOTION. LAWS BEING WHAT THEY ARE, IT THEREAFTER STAYED IN MOTION.

KA-BLOOIE!

FOR A LONG, LONG TIME ANYWAY.

BY THE WAY, I, YOUR INVISIBLE NARRATOR, HAVE TAKEN THE LIBERTY OF ASSIGNING A CUTE LITTLE FACE AND PERSONALITY TO THIS PEBBLE OF IRON, JUST TO KEEP THINGS INTERESTING.
WEEEEEEE!

AND SO OUR LITTLE FRIEND SAILED ALONG, UTTERLY ALONE.

EONS PASSED.

HOW DID IT SPEND THE TIME? HOW WOULD YOU SPEND THE TIME?
I WISH I'D BROUGHT A BOOK OR A MAGAZINE OR SOMETHIN'.

EVERY ONCE IN A WHILE, THE GRAVITATIONAL PULL OF NEARBY FIERY GIANTS AND PLANETS WOULD ATTEMPT TO SEDUCE OUR PAL TO A SUDDEN ENDING, BUT USUALLY THEY'D MERELY SUCCEED IN GENTLY ARCING ITS PATH.
DO YOU MIND?
hello

EVERY ONCE IN A WHILE, OUR FRIEND WAS WITNESS TO SIGNIFICANT HISTORICAL MOMENTS.

THE BIRTH OF A STAR
AWWW! WHAT A CUTIE!

THE DEATH OF A STAR
R.I.P.

HANNIBAL CROSSING THE ALPS
BEHOLD! AN OMEN!

THE FATEFUL MOMENT THE TESSKANS CONQUERED THE FILLIPODS.
OH, YA HATE TO SEE THAT.

SOMETIMES, UNFAMILIAR FORCES WOULD HAVE AN EFFECT. LIKE THE SPACE BENDING POWER OF DRIVE ENGINES PASSING NEARBY.
EEP!

AND THEN, ON A DAY LIKE ANY OTHER (IF YOU CAN EVEN CALL THE UNITS OF TIME SPENT DRIFTING IN SPACE "DAYS")
HO-HUM. SAILING ALONG AT A CONSTANT VELOCITY.

A SHIP APPEARED IN THE DISTANCE.
BWIP!

SMALL AT FIRST
WELL, WHAT DO YA KNOW?

BUT UNMISTAKABLY IN THE WAY.

EXCITED AT FIRST, PEBBLE SOON GREW CONCERNED.

UH, YOU'RE GONNA WANNA GET OUT OF THE WAY!

ALTHOUGH I DO SUPPOSE I'M TOO TINY TO BE SEEN BY ANY OF YOUR SENSORS.

THIS IS GONNA BE BAD.

CRACK!
BASH!
SORRYSORRY SORRYSORRYSORRY SORRYSORRY
KLONG!

EEEEWWWW! BRAIN GOO!

THAT WAS SO GROSS! OH! THAT POOR GUY! HE NEVER KNEW WHAT HIT HIM!

TIME PASSED, AND THAT ONE SMALL MOMENT OF UNAVOIDABLENESS WAS FORGOTTEN AS PEBBLE SPED ONWARD.
NOW... WHERE WAS I GOING AGAIN?

AND ONWARD
...OH YEAH. NOWHERE!

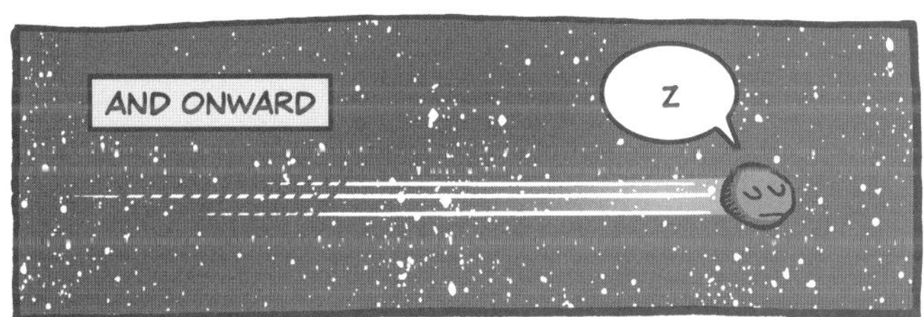
AND ONWARD
Z

AND ONWARD
HEY, THAT'S MY JOB.
SORRY.

UNTIL A PLANETARY SYSTEM SLOWLY APPEARED ON THE HORIZON. OR WHATEVER THE EQUIVALENT OF A "HORIZON" WOULD BE IN THE VAST EMPTINESS OF SPACE.
OOOH! THIS LOOKS COZY!

AND NOW, A PEEK AT AN IMPORTANT MOMENT WHICH OCCURRED ON THE VEETAN HOMEWORLD.
MY DECISION IS MADE. I'M SETTING OUT TO SEE THE WORLDS.
THAT'S NICE.

THERE'S NO WAY ANYONE WOULD HAVE NOTICED AT THE TIME. THERE WAS A LOT GOING ON.
PICK UP SOME MILK WHILE YOU'RE OUT.

BUT THERE! LOOK! IT'S PEBBLE.
NO, YOU ARE NOT UNDERSTANDING ME! I'M HEADING OUT FOR A LIFE OF EXPLORATION AND ADVENTURE IN THE GREAT EXPANSE OF THE UNIVERSE!

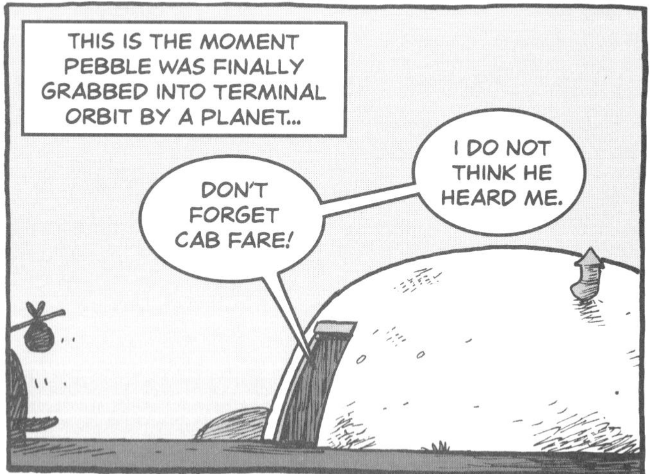
THIS IS THE MOMENT PEBBLE WAS FINALLY GRABBED INTO TERMINAL ORBIT BY A PLANET...
DON'T FORGET CAB FARE!
I DO NOT THINK HE HEARD ME.

BUT IT WASN'T AN INSTANT CRASH LANDING, AS BEFALLS SO MUCH ERRANT DEEP SPACE DETRITUS. IT WAS SOMETHING MORE GRACEFUL AND SUBTLE.
COME TO MAMA!
UH... OKAY.
REMEMBER: FACES AND PERSONALITIES HAVE BEEN ASSIGNED TO CELESTIAL BODIES TO KEEP THINGS INTERESTING.

AND SO OUR LITTLE PEBBLE FRIEND WAS SUCKED INTO ORBIT.

BUT ORBITS, AS YOU LIKELY KNOW, OFTEN DECAY IF UNSTABLE.
OH, HEY. PLANET DEAD AHEAD, AND SHE'S A BIGGIE.

AND AS PEBBLE ENTERED THE ATMOSPHERE OF THE VEETAN HOMEWORLD, THE UNDENIABLE EFFECTS OF FRICTION BEGAN TO WORK THEIR MAGIC.
SNIFF SNIFF IS THAT ME?

AH, BACK HOME AFTER ELEVEN YEARS. AND LOOK! IS SHOOTING STAR HERE TO GREET MY RETURN.
MUST BE AN OMEN.

AND SO THINGS FINALLY GOT EXCITING FOR OUR FRIEND THE ANCIENT YET INSIGNIFICANT LITTLE IRON PEBBLE.

YOU COULD SAY THINGS REALLY STARTED TO HEAT UP.
NOW IS NOT THE TIME FOR JOKES, PAL!

THIS CAN'T BE GOOD!

NOW IS PROBABLY A GOOD TIME TO REMIND YOU THAT THE CUTE LITTLE FACE AND PERSONALITY ARE MERELY ARTISTIC LICENSE ON MY PART. LET'S GO AHEAD AND ABANDON THAT NOTION NOW, BECAUSE WHAT FOLLOWS IS NOT PRETTY.
SMACK

AND SO THE PEBBLE WAS UTTERLY OBLITERATED. THE IMPACT WITH THE PLANET SHATTERED IT IRREPARABLY INTO BILLIONS OF MICROSCOPIC PIECES.

PIECES THAT SOON SETTLED INTO THE DUST.

DUST THAT WAS KICKED UP

AND MADE ITS WAY SOME-WHERE IT COULD HAVE ONE LAST IMPACTFUL EFFECT ON THE UNIVERSE.

WAH-CHOO
EW! COVER YOUR FACE WHEN YOU DO THAT!

BEFORE TIME & WITHOUT END

written and drawn by

ROSEMARY VALERO-O'CONNELL

hirosemary.com

You live,
but are not yet alive.
You see,
but are not yet seen.
You move,
but have not yet made.

A solitary atom of hydrogen at the heart of a single, trembling star.

A bead of water in the belly of a great sea.

One moment in time whose
only possible significance
can be the creation
on the moments
yet to come.

Nothing so vital, so fundamental, as the act of altering some small portion
of the void that cradles you, of adding something to it
that has not existed before, will not exist again,
could not have existed without you.

Proof, undeniable, galvanizing,
that though incalculable lives
have preceded yours and
incalculable ones will follow,
your atom,
your drop of water,
is forever folded into the
lifespan of the universe

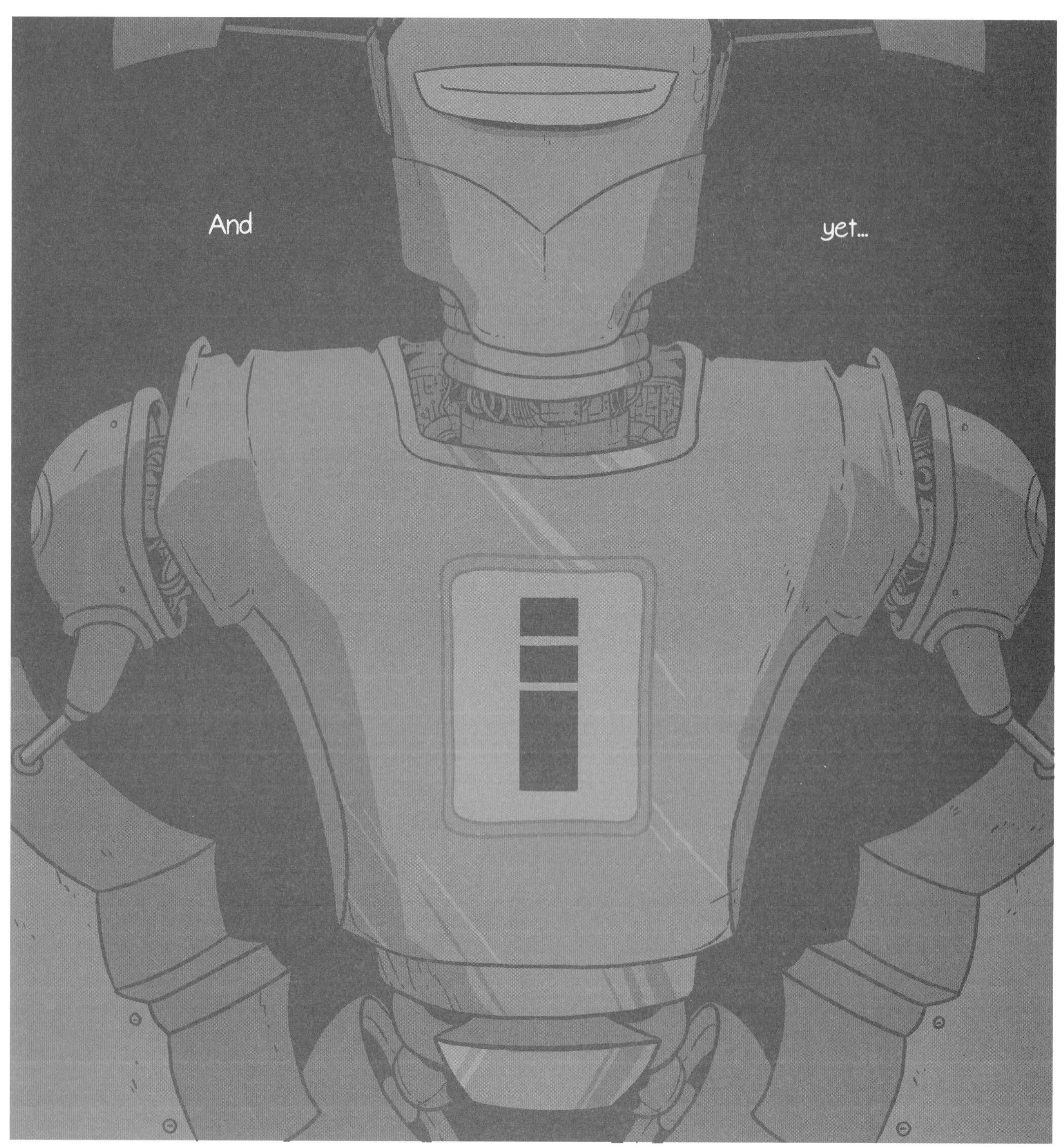
And
yet...

What bitter hypocrisy, to be cast out,
repudiated, for making something
that not only defined you but
could define itself.

No bauble, no useless trinket,
a spirit that embodies the essence
of everything you, all of you,
are.

The act
of creation made
physical, self-perpetuating,
never-ending.

Would that you had had more time
to learn it, for it to learn you.
To bask in possibility together, each learning
from the existence of the other.

Ah, well.
The line stretches on.

written by
Dylan Meconis
dylanmeconis.com

---

drawn by
Carissa Powell
simkaye.com

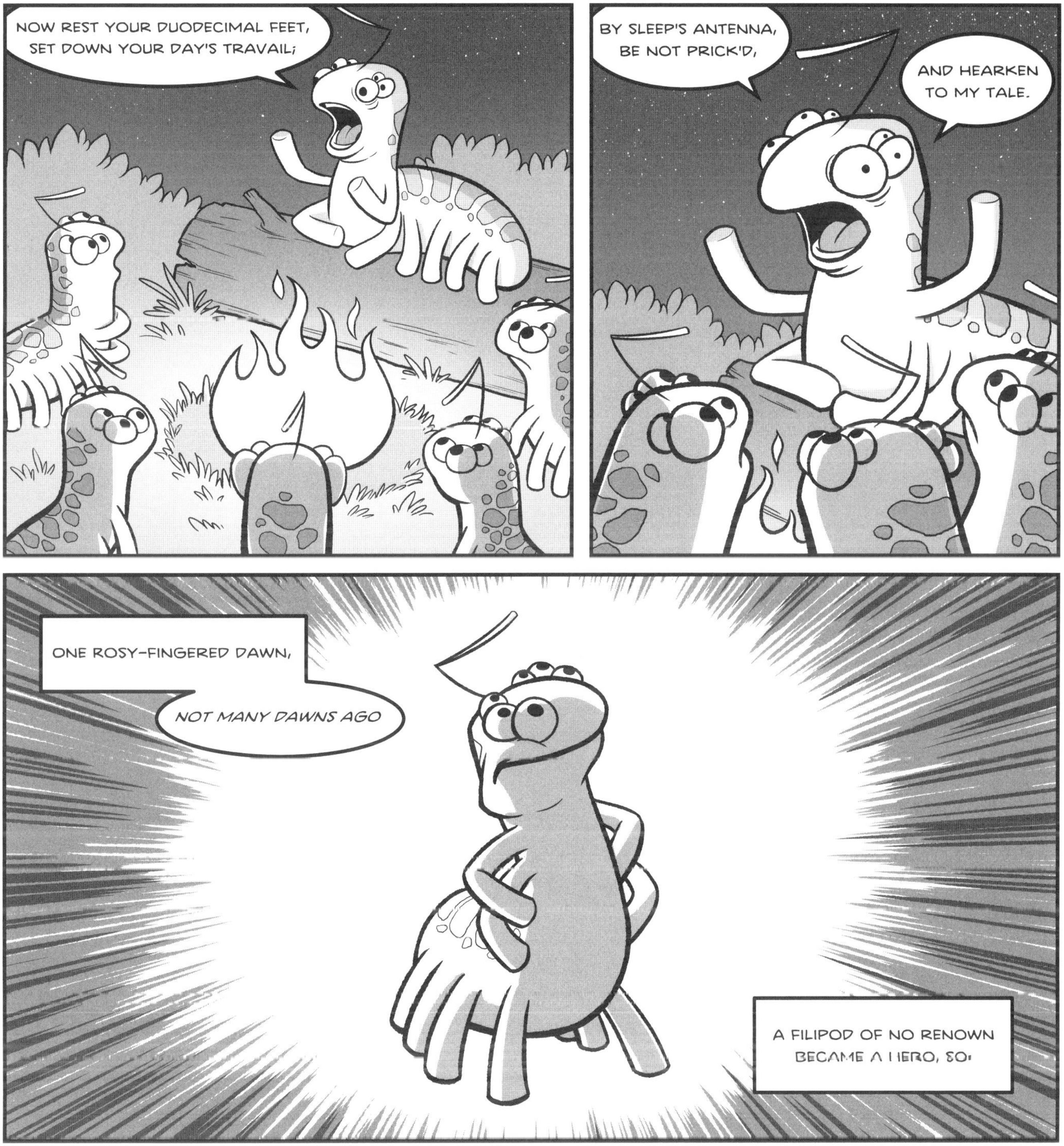

ONE ROSY-FINGERED DAWN,

NOT MANY DAWNS AGO

A FILIPOD OF NO RENOWN
BECAME A HERO, SO:

A CLERK, BY NAME
"FINTRESSLANOPE!"

A MINOR FUNCTIONARY,
WAS CALLED INTO
HIS BOSS'S SUITE--

SO WENT HE, THOUGH QUITE WARY.

"FINTRESSLANOPE,"
SAID HIM HIS BOSS,
"I HAVE A TASK FOR THEE;"

"BUT WHETHER YOU
CAN PULL IT OFF,"
"WELL, THAT
WE'VE YET TO SEE."

"FULL FATHOM FIVE, THE CLERKS WE'VE LOST,
UNTO THIS DREADFUL QUEST."

"NOW YOU MUST TAKE THIS BURDEN ON,
AT MANAGEMENT'S BEHEST."

"SEEK OUT THE LURKING MONGERBEAST,
WITHIN HER STINKING CAVE;
HER CLAWS ARE SHARP, HER EYES ARE RED –
TO *THINK* OF HER IS BRAVE!"

AND UNTO HIM WAS GIVEN THEN
A VORPAL BLADE AND SHIELD,
AND SENT UNTO THE DARKEST SPHERE,
TO SEE WHAT IT MIGHT YIELD.

HE ASKED HIM EVERY LOCAL BEAST
OF MINOR SENTIENCE PROVEN –
?

BUT NONE WOULD HELP HIM TO HIS GOAL,

NOT FEATHERED,

FINNED,

NOR HOOVEN.

AT LAST HE GREW BOTH HOT AND DIRE
AND LAID HIM BY A POND,

AND THERE HE MET A COMELY LASS
WHO FROM THAT GOO HAD SPAWN'D.

"I SEEK THE LURKING MONGERBEAST,
WITHIN HER STINKING CAVE;
HER CLAWS ARE SHARP, HER EYES ARE RED –
TO *THINK* OF HER IS BRAVE!"

"I KNOW OF WHOM YOU SPEAK, MY LORD,"
SAID SHE,
THIS GOOISH MAIDEN;

"AND KNOW HOW YOU COULD BE RELIEVED OF THAT WHICH YOU ARE LADEN."

WITHIN A BLINK, SHE HOVE HIM TO,

THEN DRAGGED HIM FURTHER DOWN,
INTO THE GLUEY, INKISH DEPTHS

'TIL HE WAS LIKE TO DROWN.

YET AS HIS THOUGHTS WERE BLINKING OUT, REMEMBERED HE ONE THING: ANTENNAE, AND THEIR DREAMSY DRUGS!

SHE FELT THE SLEEPY STING.

ONE LIGHT HE SAW, ONE WAY HE SWAM –

AND AIR WAS HIS REWARD!

THEN SAW HE WHERE HE HAD EMERGED –

THE MONGERBEAST, HER HOARD!

HER TEETH ALL SHARP! HER EYES ALL RED!
THOUGH CLOSED, AS IF TO NAP.
ALL QUIETLY DID HE APPROACH,

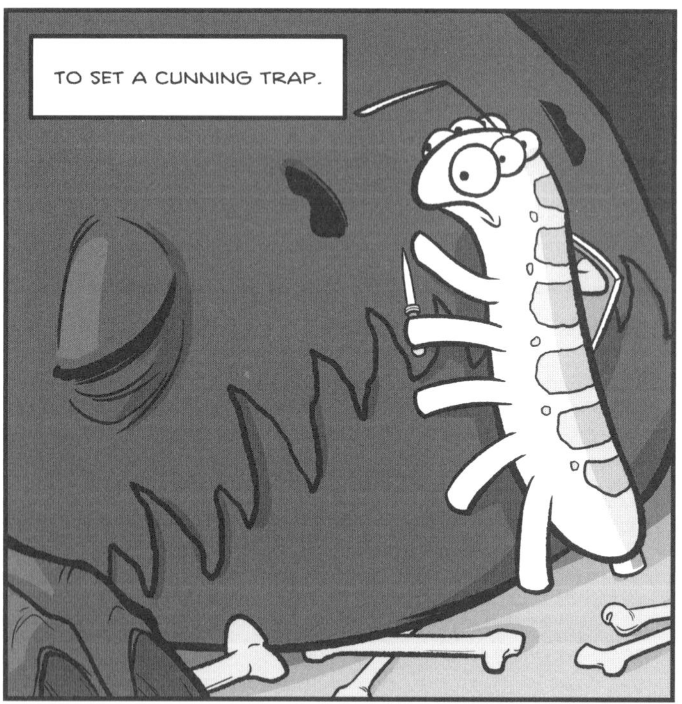

WHEN WOKE THE BEAST, FULL STARTLED BY
HIS FILLIPODIAN SCENT,
HE THRUST BEFORE VERMILION EYES
THE REASON HE'D BEEN SENT!

"WHAT HO, WHEREFORE,"
THE BEAST LASS ROARED,

"ART THOU WITHIN MY KEEP?
FULL FATHOM FIVE OF CLERKS HAVE DIED
FOR HINDERING MY SLEEP!"

"AH, MADAME, HEED MY URGENT WORD –
THIS MESSAGE YOU'VE BEEN SENT –
YOU'VE JURY DUTY IN SIX WEEKS.
THE STATE WILL NOT RELENT."

"NOW PLEASE SIGN HERE, TO INDICATE
OF THIS YOU'VE HAD RECEIPT."

"WITH THAT, I LEAVE YOU TO YOUR REST.
MY MISSION IS COMPLETE."

SO SWAM HE BACK TO LIGHT AND LIFE,
AND OFFICE WALLS SECURE.
OUR HERO, BRAVE FINTRESSLANOPE,
FULL MUCH DID HE ENDURE!

HE FOUND THE LURKING MONGERBEAST,
WITHIN HER STINKING CAVE;
HER CLAWS WERE SHARP, HER EYES WERE RED –
TO *THINK* OF HER WAS BRAVE!

THE DRIVE GALLERY

SELECT CONCEPT SKETCHES, CHARACTER DESIGNS, AND PROCESS SHOTS THAT WENT INTO THE CREATION OF DRIVE: ACT TWO

THE MAIN CAST

Drawn in a Japanese chibi-esque look, this is a cute-i-fied version of the main cast. The Captain is basically "two heads high" here, which is delightful.

CAST SKETCHES

The support DRIVE gets from Patreon.com/Drive genuinely makes the comic possible. So I began giving away weekly cast sketches on Patreon as a “thank you”. Here’s a selection from the past 18 months.

NOSH.
PUT DOWN the CAKE.
who wants a poem?!?

LOOKIT THIS CUTIE MUFFIN!

WATERCOLORS 'N WASHES

DRIVE is drawn digitally, in Photoshop, on a Wacom Cintiq: Basically, a very large, very accurate iPad. Working digitally has a billion advantages, but it's fun to bring DRIVE into traditional media just for fun. **LEFT** An ink-and-wash sketch of the Emperor and Nosh. **ABOVE** A watercolor of the Machito exiting orbit.

FILLIPOD (NAKED)

TESSKAN (NAKED)

TURNAROUNDS

It's been so fun to invite artists to play in the DRIVE universe. The stories and styles that result are a continuing delight. But before these artists start in on their "Tales," I try to give them as many resources as I can to help tell their story. So, for example, I'll put together quick turnarounds of aliens, like these Fillipods and Tesskans, here.

TEAM FILLIPOD

There are few aliens in DRIVE that are as joy-filled to draw as Fillipods. Stretchy, bendy, twisty and squishy, their little frames make for infinite joy. And when you remember that they're as soft as a microfiber, memory foam, full-body pillow, it makes them even more delightful. **ON THIS PAGE** There's clearly a preference for drawing them dancing – as it's a fun, goofy challenge just to decide where to "put" all those dang limbs. But it also speaks to the childlike joy they get from any and all art: These are the folks who would absolutely invite you to their gallery opening as soon as you were introduced. (Don't go, by the way: Just send a nice note.)

JUST DO IT.

YES. JUST DO IT.

...JUST EAT THIRD SLEEVE OF OREOS.

IS OK

NOSH

There are a billion absent-minded sketches of Nosh around my house. Any time I'm on the phone, or have a spare, day-dreaming moment, my hand begins to draw him. He's round and plump and joy-filled and loving, and there's never a bad time to sketch that.

THE MACHITO

The Machito, which means "little tough guy" in Spanish, is super fun to draw. Like all human ships, there's not a lot of "art" to the "artifice": They are clunky and boxy and utilitarian ships, when compared to crafts from the Continuum or the Grasskans. The Machito was originally a pleasure craft: A yacht built for rich Familia members who wanted a "Foodies" vacation on Veeta, or to go tech shopping on Du Fu, or to sit on the endless beaches of Lorca. It would've been expensive to build, expensive to maintain, and expensive to fly...but for a certain class of Familia members, that was never a concern. Which is probably why, when it was converted from a pleasure craft to a military scout ship, it took two weeks just to remove the 14-person hot tub from the upper deck

THE VINN

Coming up with new Vinn is endlessly fun, and endlessly necessary. Because they are constantly conquering new species, and those species only "last" a generation, we need a constantly rotating lineup of Vinn. And the fun part is, they can be large or small, menacing or sweet, furry or scaly. Anything you want, basically. Which is an artist's delight.

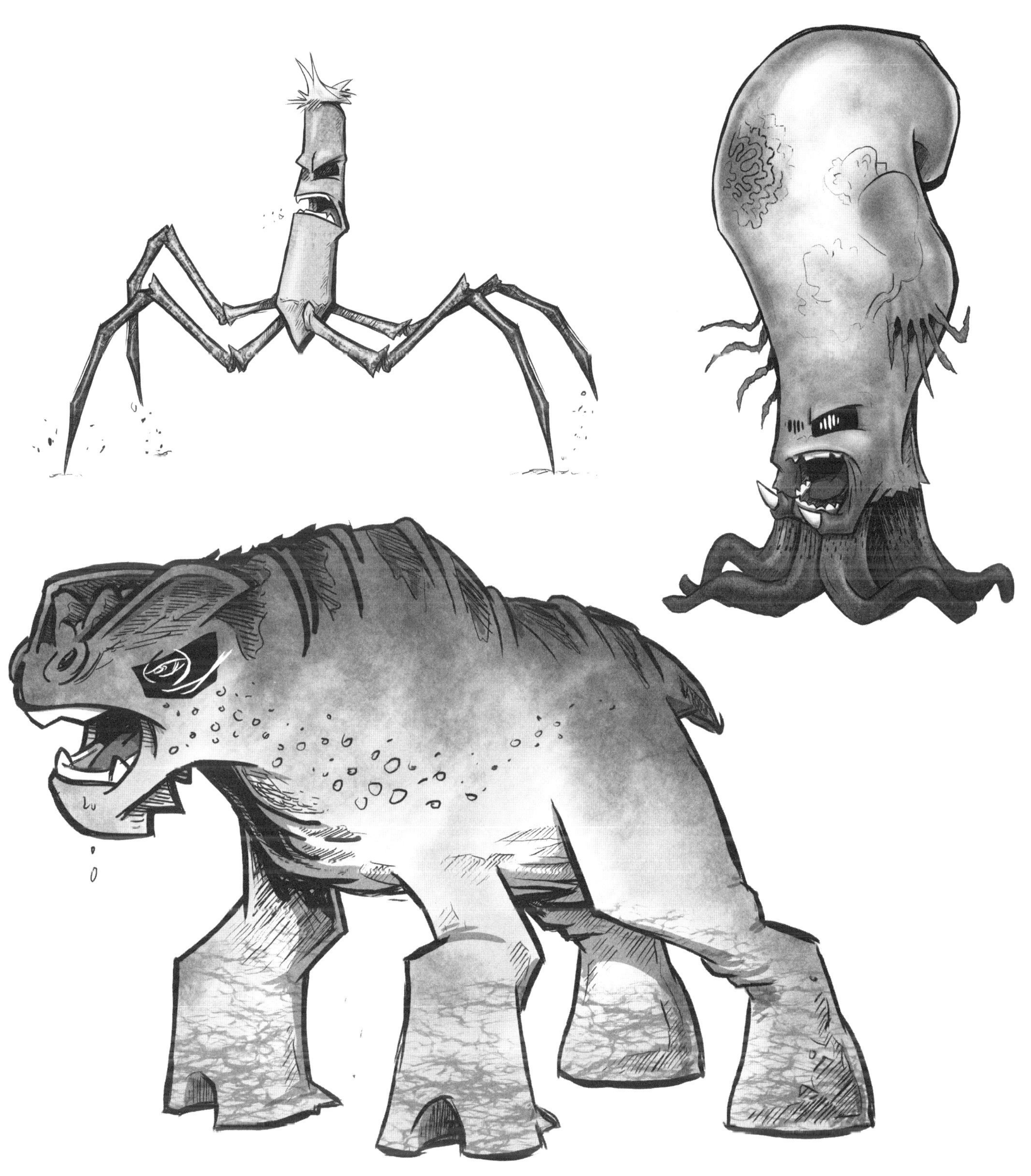

THE SILL

The Great Hold of the Sill – the empire that once stretched one-third the width of the galaxy, was a memory by the time humans began taking to the stars. But for Veetans, Fillipods, Grasskans and others, the Sill loom large in their early development – as they were their "first contact". In our story, we've only seen the Sill as the wrapped-in-rags, remnant Monks of Yanta – but they once towered above all other species, figuratively and literally. Their caste system featured soldiers and statesmen, artisans and archaeologists, builders and bankers. Each proudly displayed gold or platinum inlays carved into their massive shells, or markings to indicate how many planets they've conquered for The Hold. **ABOVE** A pen-and-wash sketch of a monk. **OPPOSITE PAGE** A goofy interaction between a monk and Nosh (who is rarely the "shorter" one in a conversation), and some very early concept sketches of what the Sill might look like walking on all fours, or as a decorated soldier of The Hold.

PILGRIM!
YOU COME SEEKING THE GREAT QUIET?

oh no no no AM HAPPY TO CHAT! AM PEOPLE PERSONS!

THE COVER

LEFT First sketches for the Act Two book cover started about two years before the book was finished. The initial idea was to feature a strongly parallel design, like those Soviet space-race posters that would say, "Progress! Onward and Upward!" So the initial sketch had the Captain's arm, the Nuevo Chile explosion, the Continuum ships, and Cuddow's arm and mouth...all moving in parallel. **BELOW** This design was brought all the way to a finished, painted, high resolution, cover-ready piece. Which did turn out quite nice, frankly! But when placed next to the first book, the design didn't feel like it worked as a companion piece. And both Nosh and the Captain felt kind of... squished in there. So, it was back to the drawing board...

LEFT To test out a new cover, elements from the first cover were brought into a quick, copy-n-paste job in Photoshop, where dozens of iterations could be easily tried out. (In Photoshop, all the elements can be handled like paper dolls: Characters can be picked up and moved around, and surrounding elements can be introduced or discarded.)

ABOVE The final cover sits very nicely next to the Act One cover, and feels like a continuation of that design. There's a nice converging flow from the Captain, the Monk, the Continuum ships, and Nosh's eyes, that then flows through Skitter and on to the Machito. And the full crew was featured on the cover, which felt better.

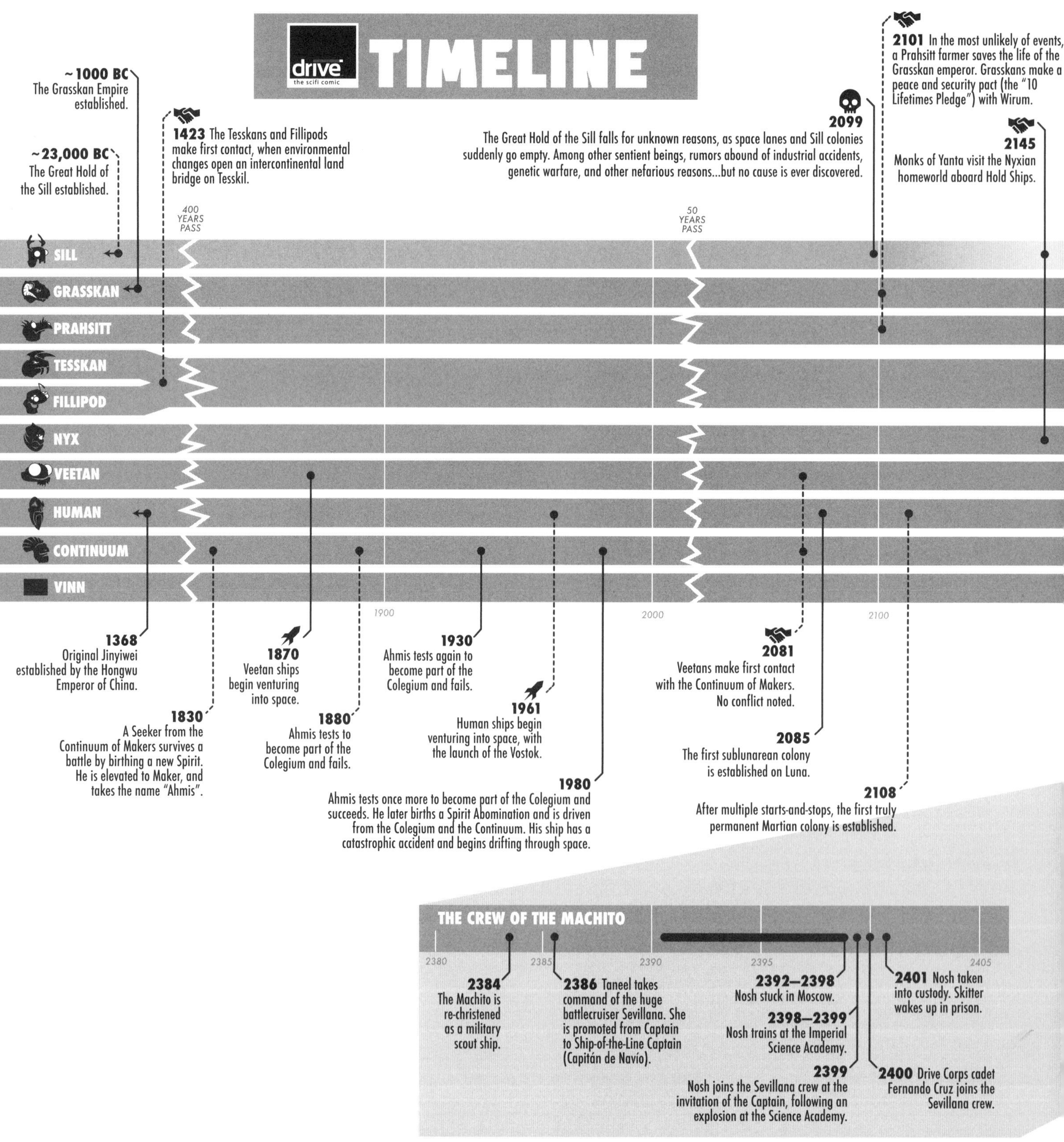

drive
the scifi comic
TIMELINE
~1000 BC The Grasskan Empire established.
~23,000 BC The Great Hold of the Sill established.
1423 The Tesskans and Fillipods make first contact, when environmental changes open an intercontinental land bridge on Tesskil.
2099 The Great Hold of the Sill falls for unknown reasons, as space lanes and Sill colonies suddenly go empty. Among other sentient beings, rumors abound of industrial accidents, genetic warfare, and other nefarious reasons...but no cause is ever discovered.
2101 In the most unlikely of events, a Prahsitt farmer saves the life of the Grasskan emperor. Grasskans make a peace and security pact (the "10 Lifetimes Pledge") with Wirum.
2145 Monks of Yanta visit the Nyxian homeworld aboard Hold Ships.
400 YEARS PASS
50 YEARS PASS
SILL
GRASSKAN
PRAHSITT
TESSKAN
FILLIPOD
NYX
VEETAN
HUMAN
CONTINUUM
VINN
1900
2000
2100
1368 Original Jinyiwei established by the Hongwu Emperor of China.
1870 Veetan ships begin venturing into space.
1930 Ahmis tests again to become part of the Colegium and fails.
2081 Veetans make first contact with the Continuum of Makers. No conflict noted.
1830 A Seeker from the Continuum of Makers survives a battle by birthing a new Spirit. He is elevated to Maker, and takes the name "Ahmis".
1880 Ahmis tests to become part of the Colegium and fails.
1961 Human ships begin venturing into space, with the launch of the Vostok.
2085 The first sublunarean colony is established on Luna.
1980 Ahmis tests once more to become part of the Colegium and succeeds. He later births a Spirit Abomination and is driven from the Colegium and the Continuum. His ship has a catastrophic accident and begins drifting through space.
2108 After multiple starts-and-stops, the first truly permanent Martian colony is established.
THE CREW OF THE MACHITO
2380
2385
2390
2395
2405
2384 The Machito is re-christened as a military scout ship.
2386 Taneel takes command of the huge battlecruiser Sevillana. She is promoted from Captain to Ship-of-the-Line Captain (Capitán de Navío).
2392—2398 Nosh stuck in Moscow.
2398—2399 Nosh trains at the Imperial Science Academy.
2399 Nosh joins the Sevillana crew at the invitation of the Captain, following an explosion at the Science Academy.
2401 Nosh taken into custody. Skitter wakes up in prison.
2400 Drive Corps cadet Fernando Cruz joins the Sevillana crew.

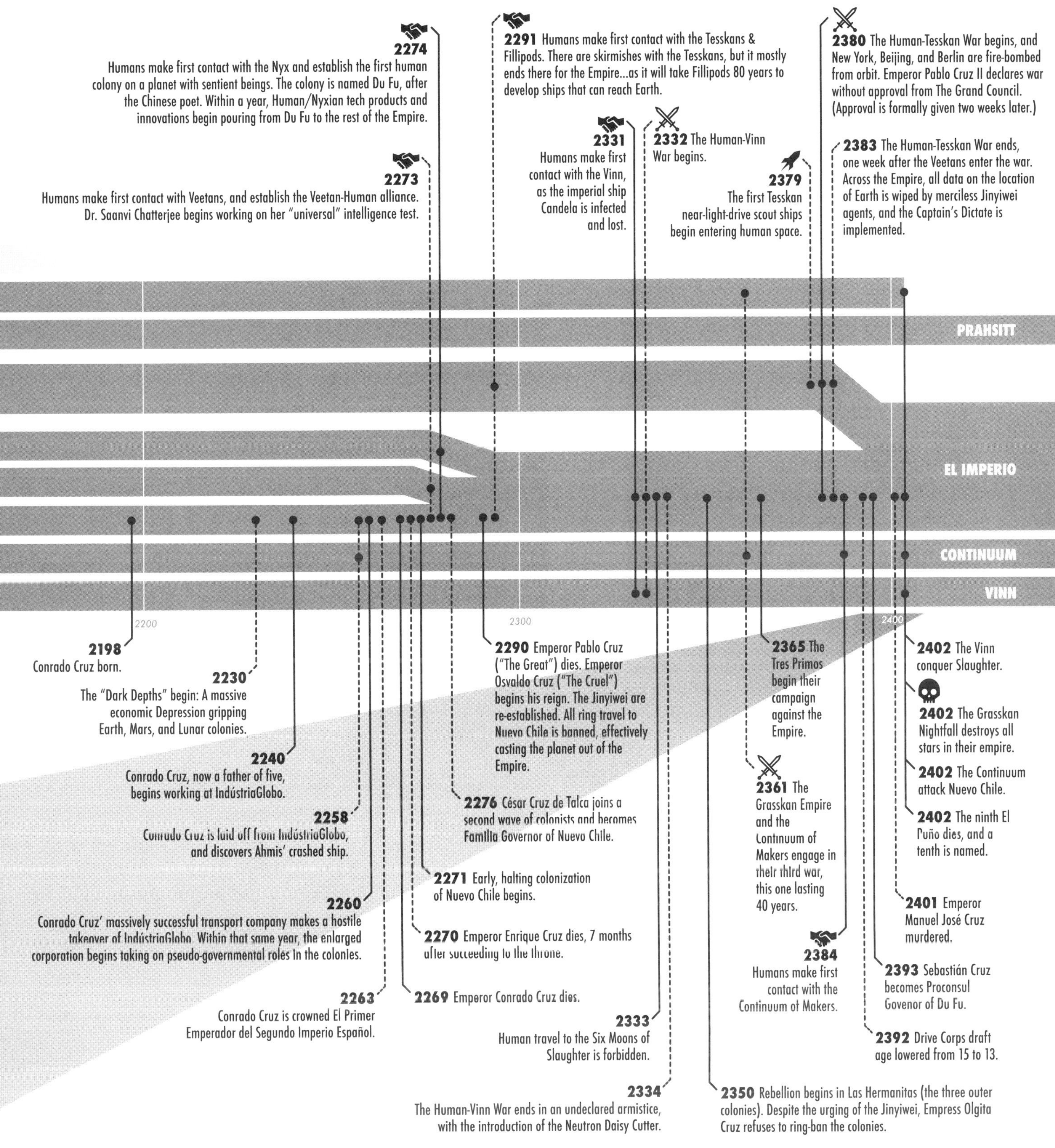
2274
Humans make first contact with the Nyx and establish the first human colony on a planet with sentient beings. The colony is named Du Fu, after the Chinese poet. Within a year, Human/Nyxian tech products and innovations begin pouring from Du Fu to the rest of the Empire.
2273
Humans make first contact with Veetans, and establish the Veetan-Human alliance. Dr. Saanvi Chatterjee begins working on her "universal" intelligence test.
2291 Humans make first contact with the Tesskans & Fillipods. There are skirmishes with the Tesskans, but it mostly ends there for the Empire...as it will take Fillipods 80 years to develop ships that can reach Earth.
2331
Humans make first contact with the Vinn, as the imperial ship Candela is infected and lost.
2332 The Human-Vinn War begins.
2379
The first Tesskan near-light-drive scout ships begin entering human space.
2380 The Human-Tesskan War begins, and New York, Beijing, and Berlin are fire-bombed from orbit. Emperor Pablo Cruz II declares war without approval from The Grand Council. (Approval is formally given two weeks later.)
2383 The Human-Tesskan War ends, one week after the Veetans enter the war. Across the Empire, all data on the location of Earth is wiped by merciless Jinyiwei agents, and the Captain's Dictate is implemented.
PRAHSITT
EL IMPERIO
CONTINUUM
VINN
2200
2300
2400
2198
Conrado Cruz born.
2230
The "Dark Depths" begin: A massive economic Depression gripping Earth, Mars, and Lunar colonies.
2240
Conrado Cruz, now a father of five, begins working at IndústriaGlobo.
2258
Conrado Cruz is laid off from IndústriaGlobo, and discovers Ahmis' crashed ship.
2260
Conrado Cruz' massively successful transport company makes a hostile takeover of IndústriaGlobo. Within that same year, the enlarged corporation begins taking on pseudo-governmental roles in the colonies.
2263
Conrado Cruz is crowned El Primer Emperador del Segundo Imperio Español.
2269 Emperor Conrado Cruz dies.
2270 Emperor Enrique Cruz dies, 7 months after succeeding to the throne.
2271 Early, halting colonization of Nuevo Chile begins.
2276 César Cruz de Talca joins a second wave of colonists and becomes Familia Governor of Nuevo Chile.
2290 Emperor Pablo Cruz ("The Great") dies. Emperor Osvaldo Cruz ("The Cruel") begins his reign. The Jinyiwei are re-established. All ring travel to Nuevo Chile is banned, effectively casting the planet out of the Empire.
2333
Human travel to the Six Moons of Slaughter is forbidden.
2334
The Human-Vinn War ends in an undeclared armistice, with the introduction of the Neutron Daisy Cutter.
2350 Rebellion begins in Las Hermanitas (the three outer colonies). Despite the urging of the Jinyiwei, Empress Olgita Cruz refuses to ring-ban the colonies.
2365 The Tres Primos begin their campaign against the Empire.
2361 The Grasskan Empire and the Continuum of Makers engage in their third war, this one lasting 40 years.
2384
Humans make first contact with the Continuum of Makers.
2402 The Vinn conquer Slaughter.
2402 The Grasskan Nightfall destroys all stars in their empire.
2402 The Continuum attack Nuevo Chile.
2402 The ninth El Puño dies, and a tenth is named.
2401 Emperor Manuel José Cruz murdered.
2393 Sebastián Cruz becomes Proconsul Govenor of Du Fu.
2392 Drive Corps draft age lowered from 15 to 13.

INDEX

ENCICLOPEDIA ENTRIES:

IMPERIAL MISCELLANY:

TALES OF THE DRIVE:

GENERAL MISCELLANY:

ABOUT THE AUTHOR:

Dave Kellett is a Southern California native who grew up loving all sorts of comics and sci-fi and nerdery. So no, not a big dater in high school, if that's what you're asking. But! It made him a prime candidate to draw sci-fi comics later in life, SO IT WORKED OUT OK.

The author is currently 5'8" tall.

COMICS & MORE BY DAVE KELLETT:

DRIVE
DriveComic.com

SHELDON
SheldonComics.com

STRIPPED
StrippedFilm.com

COMICLAB PODCAST
Patreon.com/ComicLab

GET MORE BOOKS FROM DAVE KELLETT...

If you love the humor of DRIVE, check out Dave's other comic strip, SHELDON! This all-ages comic centers on a 10-year old boy living with his grandfather and their house full of pets. Doofus pugs, adventurous lizards, egotistical ducks: It's a weird, wonderful menagerie...and a two-time Eisner nominee for "Best Humor," and the winner of the Silver Reuben Award!

GET 30% OFF
YOUR SHELDON ORDER!
POINT YOUR PHONE CAMERA AT THIS QR CODE TO JUMP TO THE STORE NOW!

Read Sheldon free at **SheldonComics.com**
Get the books at **SheldonStore.com**

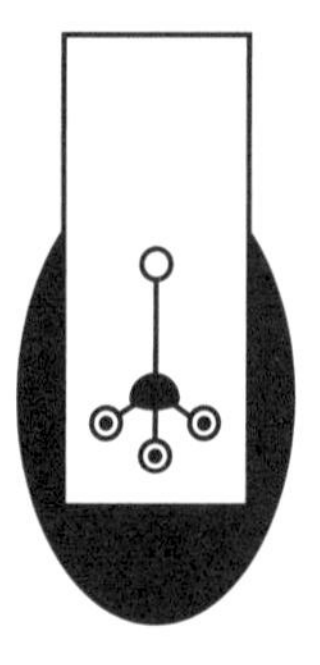

MAPA DE LAS GRANDES POTENCIAS GALÁCTICAS

HOLD THUMB, AIM OCSCAN TO ACTIVATE MAP'S HOLOGRAPHIC DISPLAY

SAGITTARIUS A

PERSEUS ARM

CONTINUU

FINNTESK

LEAGUE OF KAYN

KAYNTANNA

THE SHAPING

THE COLD ARMISTICE

CALLES NU

MCBRIDE'S WORMHOLE

PARLAY

THE MONASTERIES OF YANTA

THE CONSORT OF DUVELL

THE V

TESSKIL

THE TESSKAN ANNEXATION

VEETA

LAS HERMANITAS

CASA DE LOS MUERTOS

THE OUTER COLONIES

SOL

EL IMPE

NUEVO CHILE

CONTINUUM OCCUPIED SPACE

DAWN

DU FU